STANDARD ATOMIC WEIGHTS OF THE ELEMENTS 2010 Based on relative atomic mass of $^{12}C = 12$, where ^{12}C is a neutral atom in its nuclear and electronic ground state.[†]

Name	Symbol	Atomic Number	Atomic Weight	Name	Symbol	Atomic Number	Atomic Weight
Actinium*	Ac	89	(227)	Molybdenum	Mo	42	95.96(2)
Aluminum	Al	13	26.9815386(8)	Neodymium	Nd	60	144.22(3)
Americium*	Am	95	(243)	Neon	Ne	10	20.1797(6)
Antimony	Sb	51	121.760(1)	Neptunium*	Np	93	(237)
Argon	Ar	18	39.948(1)	Nickel	Ni	28	58.6934(4)
Arsenic	As	33	74.92160(2)	Niobium	Nb	41	92.90638(2)
Astatine*	At	85	(210)	Nitrogen	N	7	14.0067(2)
Barium	Ba	56	137.327(7)	Nobelium*	No	102	(259)
Berkelium*	Bk	97	(247)	Osmium	Os	76	190.23(3)
Beryllium	Be	4	9.012182(3)	Oxygen	O	8	15.9994(3)
Bismuth	Bi	83	208.98040(1)	Palladium	Pd	46	106.42(1)
Bohrium	Bh	107	(264)	Phosphorus	P	15	30.973762(2)
Boron	B	5	10.811(7)	Platinum	Pt	78	195.084(9)
Bromine	Br	35	79.904(1)	Plutonium*	Pu	94	(244)
Cadmium	Cd	48	112.411(8)	Polonium*	Po	84	(209)
Cesium	Cs	55	132.9054519(2)	Potassium	K	19	39.0983(1)
Calcium	Ca	20	40.078(4)	Praseodymium	Pr	59	140.90765(2)
Californium*	Cf	98	(251)	Promethium*	Pm	61	(145)
Carbon	C	6	12.0107(8)	Protactinium*	Pa	91	231.03588(2)
Cerium	Ce	58	140.116(1)	Radium*	Ra	88	(226)
Chlorine	Cl	17	35.453(2)	Radon*	Rn	86	(222)
Chromium	Cr	24	51.9961(6)	Rhenium	Re	75	186.207(1)
Cobalt	Co	27	58.933195(5)	Rhodium	Rh	45	102.90550(2)
Copernicium*	Cn	112	(285)	Roentgenium	Rg	111	(272)
Copper	Cu	29	63.546(3)	Rubidium	Rb	37	85.4678(3)
Curium*	Cm	96	(247)	Ruthenium	Ru	44	101.07(2)
Darmstadtium	Ds	110	(271)	Rutherfordium	Rf	104	(261)
Dubnium	Db	105	(262)	Samarium	Sm	62	150.36(2)
Dysprosium	Dy	66	162.500(1)	Scandium	Sc	21	44.955912(6)
Einsteinium*	Es	99	(252)	Seaborgium	Sg	106	(266)
Erbium	Er	68	167.259(3)	Selenium	Se	34	78.96(3)
Europium	Eu	63	151.964(1)	Silicon	Si	14	28.0855(3)
Fermium*	Fm	100	(257)	Silver	Ag	47	107.8682(2)
Fluorine	F	9	18.9984032(5)	Sodium	Na	11	22.9896928(2)
Francium*	Fr	87	(223)	Strontium	Sr	38	87.62(1)
Gadolinium	Gd	64	157.25(3)	Sulfur	S	16	32.065(5)
Gallium	Ga	31	69.723(1)	Tantalum	Ta	73	180.9488(2)
Germanium	Ge	32	72.64(1)	Technetium*	Tc	43	(98)
Gold	Au	79	196.966569(4)	Tellurium	Te	52	127.60(3)
Hafnium	Hf	72	178.49(2)	Terbium	Tb	65	158.92535(2)
Hassium	Hs	108	(277)	Thallium	Tl	81	204.3833(2)
Helium	He	2	4.002602(2)	Thorium*	Th	90	232.03806(2)
Holmium	Ho	67	164.93032(2)	Thulium	Tm	69	168.93421(2)
Hydrogen	H	1	1.00794(7)	Tin	Sn	50	118.710(7)
Indium	In	49	114.818(3)	Titanium	Ti	22	47.867(1)
Iodine	I	53	126.90447(3)	Tungsten	W	74	183.84(1)
Iridium	Ir	77	192.217(3)	Ununhexium	Uuh	116	(292)
Iron	Fe	26	55.845(2)	Ununoctium	Uuo	118	(294)
Krypton	Kr	36	83.798(2)	Ununpentium	Uup	115	(228)
Lanthanum	La	57	138.90547(7)	Ununquadium	Uuq	114	(289)
Lawrencium*	Lr	103	(262)	Ununseptium	Uus	117	(292)
Lead	Pb	82	207.2(1)	Ununtrium	Uut	113	(284)
Lithium	Li	3	6.941(2)	Uranium*	U	92	238.02891(3)
Lutetium	Lu	71	174.9668(1)	Vanadium	V	23	50.9415(1)
Magnesium	Mg	12	24.3050(6)	Xenon	Xe	54	131.293(6)
Manganese	Mn	25	54.938045(5)	Ytterbium	Yb	70	173.54(5)
Meitnerium	Mt	109	(268)	Yttrium	Y	39	88.90585(2)
Mendelevium*	Md	101	(258)	Zinc	Zn	30	65.38(2)
Mercury	Hg	80	200.59(2)	Zirconium	Zr	40	91.224(2)

[†]The atomic weights of many elements can vary depending on the origin and treatment of the sample. This is particularly true for Li; commercially available lithium-containing materials have Li atomic weights in the range of 6.939 and 6.996. The uncertainties in atomic weight values are given in parentheses following the last significant figure to which they are attributed.

*Elements with no stable nuclide; the value given in parentheses is the atomic mass number of the isotope of longest known half-life. However, three such elements (Th, Pa, and U) have a characteristic terrestial isotopic composition, and the atomic weight is tabulated for these. **http://www .chem.qmw.ac.uk/iupac/AtWt/**

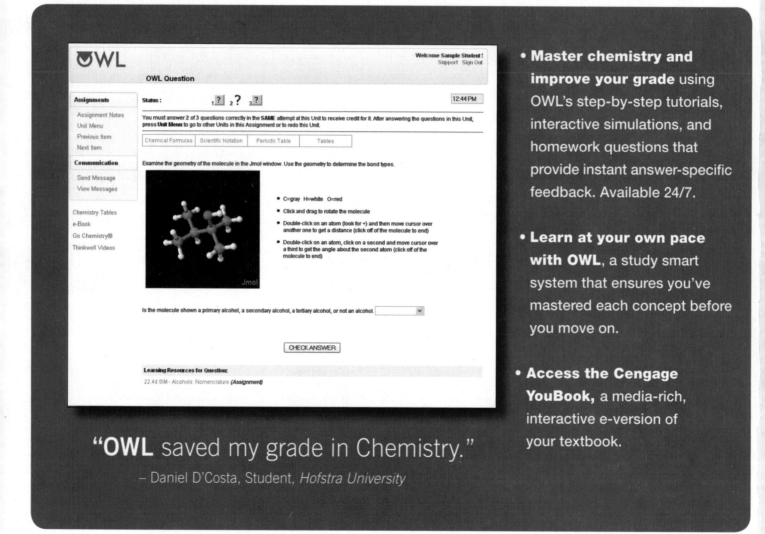

INTRODUCTION TO
General, Organic, and Biochemistry

TENTH EDITION

Frederick A. Bettelheim

William H. Brown
Beloit College

Mary K. Campbell
Mount Holyoke College

Shawn O. Farrell
Olympic Training Center

Omar J. Torres
College of the Canyons

BROOKS/COLE
CENGAGE Learning™

Australia • Brazil • Japan • Korea • Mexico • Singapore • Spain • United Kingdom • United States

BROOKS/COLE
CENGAGE Learning™

Introduction to General, Organic, and Biochemistry, Tenth Edition
Frederick A. Bettelheim, William H. Brown, Mary K. Campbell, Shawn O. Farrell, Omar J. Torres

Publisher: Mary Finch

Developmental Editor: Sandra Kiselica

Assistant Editor: Elizabeth Woods

Editorial Assistant: Krista Mastroianni

Senior Media Editor: Lisa Weber

Media Editor: Stephanie VanCamp

Marketing Manager: Nicole Hamm

Marketing Assistant: Julie Stefani

Marketing Communications Manager: Linda Yip

Content Project Manager: Teresa L. Trego

Design Director: Rob Hugel

Art Director: Maria Epes

Manufacturing Planner: Karen Hunt

Rights Acquisitions Specialist: Dean Dauphinais

Production Service: PreMediaGlobal

Text Designer: Bill Reuter Design

Photo Researcher: Chris Altof/Bill Smith Group

Text Researcher: Sue C. Howard

Copy Editor: PreMediaGlobal

Illustrator: PreMediaGlobal, 2064 Design

OWL Producers: Stephen Battisti, Cindy Stein, David Hart (Center for Educational Software Development, University of Massachusetts, Amherst)

Cover Designer: Bill Reuter Design

Cover Images: Main image ©Dave Reede/Getty Images, from left to right ©Ottmar Diez/Getty Images, ©Chris Hill/Getty Images, ©David Norton Photography/Alamy, © PhotoResearchers, Inc.

Compositor: PreMediaGlobal

For product information and technology assistance, contact us at **Cengage Learning Customer & Sales Support, 1-800-354-9706**

For permission to use material from this text or product, submit all requests online at **cengage.com/permissions**
Further permissions questions can be emailed to **permissionrequest@cengage.com**

Library of Congress Control Number: 2011934948

ISBN-13: 978-1-133-10508-4

ISBN-10: 1-133-10508-4

Brooks/Cole
20 Davis Drive
Belmont, CA 94002-3098
USA

Cengage Learning is a leading provider of customized learning solutions with office locations around the globe, including Singapore, the United Kingdom, Australia, Mexico, Brazil, and Japan. Locate your local office at: **www.cengage.com/global**

Cengage Learning products are represented in Canada by Nelson Education, Ltd.

For your course and learning solutions, visit **academic.cengage.com**

Purchase any of our products at your local college store or at our preferred online store **www.cengagebrain.com**

Printed in the United States of America
2 3 4 5 15 14 13 12

To Carolyn, with whom life is a joy. —WB

To my family and friends – thank you for all your support.

I couldn't have done it without you. — MC

To my lovely wife, Courtney — Between textbook revisions,

a full-time job, and school, I have been little more than

a ghost around the house, hiding in my study writing.

Courtney held the family together, taking care of our children

and our home while maintaining her own writing schedule.

None of this would have been possible without her love,

support, and tireless effort. —SF

To my loving family and friends who have supported me through

this journey: Mom, Dad, Lisa, Abuela, René, Ryan, Deanna,

and Dianne. I could not have made it without your urging

and support. I am truly blessed to have each of

you in my life. — OT

Contents in Brief

Biochemistry

(Find this chapter on this book's companion web site. To access, enter ISBN 1-133-10508-4 at www.cengagebrain.com)

Contents

       Receptors?   846
31.6   How Is the Immune Response Controlled?   848
31.7   Immunization   849
31.8   How Does the Body Distinguish
       "Self" from "Nonself"?   853
31.9   How Does the Human Immunodeficiency Virus
       Cause AIDS?   857

       Summary   866
       Problems   867
       Chemical Connections

       31A Monoclonal Antibodies Wage War on Breast
            Cancer   845
       31B Antibiotics: A Double-Edged Sword   855
       31C Why Are Stem Cells Special?   861
       31D A Little Swine Goes a Long Way   864
       31E Immunologists Take on the Flu Virus   865

Chapter 32 Body Fluids

(Find this chapter on this book's companion
web site. To access, enter ISBN 1-133-10508-4 at
www.cengagebrain.com)

Preface

Perceiving order in nature is a deep-seated human need. It is our primary aim to convey the relationship among facts and thereby present a totality of the scientific edifice built over the centuries. In this process, we marvel at the unity of laws that govern everything in the ever-exploding dimensions: from photons to protons, from hydrogen to water, from carbon to DNA, from genome to intelligence, from our planet to the galaxy and to the known Universe. Unity in all diversity.

As we prepare the tenth edition of our textbook, we cannot help but be struck by the changes that have taken place in the last 40 years. From the slogan of the '70s, "Better living through chemistry" to today's saying "Life by chemistry," one is able to sample the change in the focus. Chemistry helps to provide not just the amenities of a good life, but it is at the core of our concept and pre-occupation with life itself. This shift in emphasis demands that our textbook designed primarily for the education of future practitioners of Health Sciences should attempt to provide both the basics as well as the scope of the horizon within which chemistry touches our life.

The increasing use of our textbook made this new edition possible and we wish to thank our colleagues who adopted the previous editions for their courses. Testimony from colleagues and students indicates that we managed to convey our enthusiasm for the subject to students, who find this book to be a great help in studying difficult concepts.

Therefore, in the new edition we strive further to present an easily readable and understandable text along with more application problems related to the Health Sciences. At the same time, we emphasize the inclusion of new relevant concepts and examples in this fast growing discipline especially in the biochemistry chapters. We maintain an integrated view of chemistry. From the very beginning of the book, we include organic compounds and biochemical substances to illustrate the principles. The progress is ascension from the simple to the complex. We urge our colleagues to advance to the chapters of biochemistry as fast as possible, because there lies most of the material that is relevant to the future professions of our students.

Dealing with such a giant field in one course, and possibly the only course in which our students get an exposure to chemistry, makes the selection of the material an overarching enterprise. We are aware that even though we tried to keep the book to a manageable size and proportion, we included more topics than could be covered in the course. Our aim was to give enough material from which the instructor can select the topics he or she deems important. The wealth of problems, both drill and challenging, provides students with numerous ways to test their knowledge from a variety of angles.

Audience and Unified Approach

This book is intended for non-chemistry majors, mainly those entering health sciences and related fields, such as nursing, medical technology, physical therapy, and nutrition. In its entirety, it can be used for a one-year (two-semester or three-quarter) course in chemistry, or parts of the book can be used in a one-term chemistry course.

We assume that the students using this book have little or no background in chemistry. Therefore, we introduce the basic concepts slowly at the beginning and increase the tempo and the level of sophistication as we go on. We progress from the basic tenets of general chemistry to organic and then to biochemistry. Throughout we integrate the parts by keeping a unified view of chemistry. For example, we frequently use organic and biological substances to illustrate general principles.

While teaching the chemistry of the human body is our ultimate goal, we try to show that each subsection of chemistry is important in its own right, besides being required for future understanding.

Chemical Connections (Medical and Other Applications of Chemical Principles)

The Chemical Connections boxes contain applications of the principles discussed in the text. Comments from users of the earlier editions indicate that these boxes have been especially well received, and provide a much-requested relevance to the text. For example, in Chapter 1, students can see how cold compresses relate to waterbeds and to lake temperatures (Chemical Connections 1C). New up-to-date topics include coverage of omega-3 fatty acids and heart disease (Chemical Connections 21H), and the search for treatments for cystic fibrosis (Chemical Connections 26G).

The presence of Chemical Connections allows a considerably degree of flexibility. If an instructor wants to assign only the main text, the Chemical Connections do not interrupt continuity, and the essential material will be covered. However, because they enhance core material, most instructors will probably wish to assign at least some of the Chemical Connections. In our experience, students are eager to read the relevant Chemical Connections, without assignments and they do with discrimination. From such a large number of boxes, an instructor can select those that best fit the particular needs of the course. So that students can test their knowledge, we provide problems at the end of each chapter for all of the Chemical Connections; these problems are now identified within the boxes.

Metabolism: Color Code

The biological functions of chemical compounds are explained in each of the biochemistry chapters and in many of the organic chapters. Emphasis is placed on chemistry rather than physiology. Positive feedback about the organization the metabolism chapters has encouraged us to maintain the order (Chapters 26-28).

First, we introduce the common metabolic pathway through which all food will be utilized (the citric acid cycle, and oxidative phosphorylation), and only after that do we discuss the specific pathways leading to the common pathway. We find this a useful pedagogic device, and it enables us to sum the caloric values of each type of food because its utilization through the common pathway has already been learned. Finally, we separate the catabolic pathways from the anabolic pathways by treating them in different chapters, emphasizing the different ways the body breaks down and builds up different molecules.

The topic of metabolism is a difficult one for most students, and we have tried to explain it as clearly as possible. We enhance the clarity of presentation by the use of a color code for the most important biological compounds. Each type of compound is screened in a specific color, which remains the same throughout the three chapters. These colors are as follows:

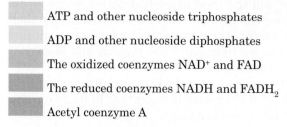

ATP and other nucleoside triphosphates

ADP and other nucleoside diphosphates

The oxidized coenzymes NAD+ and FAD

The reduced coenzymes NADH and FADH$_2$

Acetyl coenzyme A

In figures showing metabolic pathways, we display the numbers of the various steps in yellow. In addition to this main use of a color code, other figures in various parts of the book are color coded so that the same color is used for the same entity throughout. For example, in all figures that show enzyme-substrate interactions, enzymes are always shown in blue and substrates in orange.

Features

- **Problem-Solving Strategies** The in-text examples include a description of the strategy used to arrive at a solution. This will help students organize the information in order to solve the problem.

- **[New] Interactive Examples in OWL** allow students to work some of in-text examples multiple times in slightly different versions to encourage thinking their way through the example instead of passively reading through to the solution.

- **[UPDATED] Visual Impact** We have introduced illustrations with heightened pedagogical impact. These include ones that show the microscopic and macroscopic aspects of a topic under discussion, such as Figures 6.4 (Henry's Law) and 6.11 (electrolytic conductance). The Chemical Connections essays have been enhanced further with more photos that illustrate each topic.

- **Key Questions** We use a Key Questions framework to emphasize key chemical concepts. This focused approach guides students through each chapter by using section head questions.

- **[UPDATED] Chemical Connections** Over 150 essays describe applications of chemical concepts presented in the text, linking the chemistry to their real uses. Many new application boxes on diverse topics were added such as, Electrolyte Solutions in Body and Intravenous Fluids (Chapter 6), DDT, A Boon and a Curse (Chapter 13), Carbohydrates and Obesity (Chapter 20), and Depression and Nutrient Deficiency (Chapter 30).

- **Summary of Key Reactions** In each organic chemistry chapter (10–19) there is an annotated summary of all the new reactions introduced. Keyed to sections in which they are introduced, there is also an example of each reaction.

- **[UPDATED] Chapter Summaries** Summaries reflect the Key Questions framework. At the end of each chapter, the Key Questions are restated and the summary paragraphs that follow are designed to highlight the concepts associated with the questions.

- **[NEW] GOB OWL Problems** The number of end-of-chapter problems assignable in GOB OWL, the web-based homework system that accompanies this book, has doubled in this edition.

- **[UPDATED] Looking Ahead Problems** At the end of most chapters, the challenge problems are designed to show the application of principles in the chapter to material in the following chapters.

- **[UPDATED] Tying-It-Together and Challenge Problems** At the end of most chapters, these problems build on past material to test students' knowledge of the concepts. In the Challenge Problems, associated chapter references are given.

- **[UPDATED] How To Boxes** These boxes emphasize the skills students need to master the material. They include topics such as, "How to Determine the Number of Significant Figures in a Number" (Chapter 1) and "How to Draw Enantiomers" (Chapter 15).

- **Molecular Models** Ball-and-stick models, space-filling models, and electron-density maps are used throughout the text as appropriate aids to visualizing molecular properties and interactions.

- **Margin Definitions** Many terms are also defined in the margin to help students learn terminology. By skimming the chapter for these definitions students will have a quick summary of its contents.

- **Margin Notes** Additional bits of information, such as historical notes and reminders, complement nearby text.

- **Answers to all in-text and odd-numbered end-of-chapter problems** Answers to selected problems are provided at the end of the book. Detailed worked-out solutions to these same problems are provided in the Student Solutions Manual.

- **Glossary** The glossary at the back of the book gives a definition of each new term along with the number of the section in which the term is introduced.

Organization and Updates

General Chemistry (Chapters 1–9)

- **Chapter 1, Matter, Energy, and Measurement,** serves as a general introduction to the text and introduces the pedagogical elements that are new to this edition. A new example on unit conversions was added, as well as various application problems related to a clinical setting. Thirteen new problems were added.

- In **Chapter 2, Atoms,** we introduce four of the five ways we use to represent molecules throughout the text: we show water as a molecular formula, a structural formula, a ball-and-stick model, and a space-filling model. Nine new problems were added.

- **Chapter 3, Chemical Bonds,** begins with a discussion of ionic compounds, followed by a discussion of molecular compounds. New problems dealing with biomolecules were added.

- **Chapter 4, Chemical Reactions,** was reorganized in order for students to better understand the various intricacies in writing and balancing chemical reactions before stoichiometry is introduced. This chapter includes the How To box, *How to Balance a Chemical Equation,* which illustrates a step-by-step method for balancing an equation. Several challenge problems were added.

- In **Chapter 5, Gases, Liquids, and Solids,** we present intermolecular forces of attraction in order of increasing energy, namely London dispersion forces, dipole-dipole interactions, and hydrogen bonding. Section 5.9 was updated to include a more descriptive overview of the various types of crystalline solids, and increased coverage of allotropes of carbon. Nine new problems were added.

- **Chapter 6, Solutions and Colloids,** opens with a listing of the most common types of solutions, followed by a discussion of the factors that affect solubility, the most common units for concentration, and closes with an enhanced discussion of colligative properties. A new Chemical Connections box on electrolytes and eight new problems were added.

- **Chapter 7, Reaction Rates and Chemical Equilibrium,** shows how these two important topics are related to one another. A How To box shows how to *Interpret the Value of the Equilibrium Constant, K.* In addition, nine new problems were added.

- **Chapter 8, Acids and Bases,** introduces the use of curved arrows to show the flow of electrons in organic reactions. Specifically, we use them here to show the flow of electrons in proton-transfer reactions. The major theme in this chapter is the discussion of acid-base buffers and the Henderson-Hasselbalch equation. Information on the activity series and seven new problems were added.

- **Chapter 9, Nuclear Chemistry,** highlights nuclear applications to medicine.

Organic Chemistry (Chapters 10–19)

- **Chapter 10, Organic Chemistry,** is an introduction to the characteristics of organic compounds and to the most important organic functional groups. Seven new problems were added.

- In **Chapter 11, Alkanes and Cycloalkanes,** we introduce the concept of a line-angle formula and continue using them throughout the organic chapters. They are easier to draw than the usual condensed structural formulas and are easier to visualize. A new box on *How To… Draw Alternative Chair Conformations of Cyclohexane* was added, along with nine new problems.

- In **Chapter 12, Alkenes and Alkynes,** we introduce a new, simple way of looking at reaction mechanisms: add a proton, take a proton away, break a bond and make a bond. The purpose of this introduction to reaction mechanisms is to demonstrate to students that chemists are interested not only in what happens in a chemical reaction, but also in how it happens. Eight new problems were added to this chapter.

- **Chapter 13, Benzene and Its Derivatives,** includes a discussion of phenols and antioxidants. A new Chemical Connections box on *DDT* and 15 new problems were added.

- **Chapter 14, Alcohols, Ethers, and Thiols,** discusses the structures, names, and properties of alcohols first, and then gives a similar treatment to ethers, and finally thiols. Twelve new problems were added.

- In **Chapter 15, Chirality: The Handedness of Molecules,** the concept of a stereocenter and enantiomerism is slowly introduced, using 2-butanol as a prototype. We then treat molecules with two or more stereocenters and show how to predict the number of stereoisomers possible for a particular molecule. We also explain *R,S* convention for assigning absolute configuration to a tetrahedral stereocenter. Many new problems deal with drug development.

- In **Chapter 16, Amines,** we trace the development of new asthma medications from epinephrine, which can be viewed as a lead drug to albuterol (Proventil).

- **Chapter 17, Aldehydes and Ketones,** has a discussion of $NaBH_4$ as a carbonyl-reducing agent with emphasis on it as a hydride transfer reagent. We then make the parallel to NADH as a carbonyl reducing agent and hydride transfer agent. A new Chemical Connections box on *Warfarin* and eight new problems round out this chapter.

- **Chapter 18, Carboxylic Acids,** focuses on the chemistry and physical properties of carboxylic acids. There is a brief discussion of *trans* fatty acids, omega-3 fatty acids, and the significance of their presence in our diets. Ten new problems, many on pharmacology, complete this chapter.

- **Chapter 19, Anhydrides, Esters, and Amides,** describes the chemistry of these three important functional groups with emphasis on their acid-catalyzed and base-promoted hydrolysis, and reactions with amines and alcohols. New summary tables for reactions and ten new problems were added to this chapter.

Biochemistry (Chapters 20–31)

- **Chapter 20, Carbohydrates,** begins with the structure and nomenclature of monosaccharides, their oxidation, reduction, and formation of glycosides and concludes with a discussion of the structure of disaccharides, polysaccharides, and acidic polysaccharides. A new Chemical Connections box on *Carbohydrates and Obesity* and six new problems were added.

- **Chapter 21, Lipids,** covers the most important features of lipid biochemistry, including membrane structure, and the structures and functions of steroids. New Chemical Connections boxes on *Ceramides and Oxygen Deprivation* and on *Omega-3 Fatty Acids and Heart Disease* were included.

- **Chapter 22, Proteins,** covers the many facets of protein structure and function. It gives an overview of how proteins are organized, beginning with the nature of individual amino acids and how this organization leads to their many functions. This supplies the student with the basics needed to lead into the sections on enzymes and metabolism. A new section on *Transition Metals and their Effect on the Structure of Proteins* was added.

- **Chapter 23, Enzymes,** covers the important topic of enzyme catalysis and regulation. The focus is on how the structure of an enzyme leads to the vast increases in reaction rates with enzyme-catalyzed reactions. Specific medical applications of enzyme inhibition are included, as well as an introduction to the fascinating topic of transition state analogs and their use as potent inhibitors. A new Chemical Connections on *The Role of Enzymes* was added.

- In **Chapter 24, Chemical Communications,** we see the biochemistry of hormones and neurotransmitters. The health-related implications of how these substances act in the body is the main focus of this chapter. New information on the possible causes of Alzheimer's disease is explored. A new Chemical Connections on *Zebrafish, Nerve Synapses, and Sleep* was added.

- In **Chapter 25, Nucleic Acids, Nucleotides and Heredity,** introduces DNA and the processes surrounding its replication and repair. How nucleotides are linked together and the flow of genetic information that occurs due to the unique properties of these molecules is emphasized. The sections on the types of RNA have been greatly expanded as our knowledge increases daily about these important nucleic acids. The uniqueness of an individual's DNA is described with a chemical connections box that introduces DNA fingerprinting and how forensic science relies on DNA for positive identification. Three new Chemical Connections boxes were added: *Who Owns Your Genes, Synthetic Genome Created,* and *Did the Neandertals Go Extinct?*

- **Chapter 26, Gene Expression and Protein Synthesis,** shows how the information contained in the DNA blueprint of the cell is used to produce RNA and eventually protein. The focus is on how organisms control the expression of genes through transcription and translation. The chapter ends with the timely and important topic of gene therapy, which is the attempt to cure genetic diseases by giving an individual a gene he or she was missing. A new Chemical Connections on *Cystic Fibrosis* research was added.

- **Chapter 27, Bioenergetics,** is an introduction to metabolism that focuses strongly on the central pathways, namely the citric acid cycle, electron transport, and oxidative phosphorylation.

- In **Chapter 28, Specific Catabolic Pathways,** we address the details of carbohydrate, lipid, and protein breakdown, concentrating on the energy yield. A new Chemical Connections box on *How the Body Selects Proteins for Degradation* was included.

- **Chapter 29, Biosynthetic Pathways,** starts with a general consideration of anabolism and proceeds to carbohydrate biosynthesis in both plants and animals. Lipid biosynthesis is linked to the production of membranes, and the chapter concludes with an account of amino acid biosynthesis. New information on glucose consumption and metabolism was added.

- In **Chapter 30, Nutrition,** we take a biochemical approach to understanding nutrition concepts. Along the way, we look at a revised version of the Food Guide Pyramid, and debunk some of the myths about carbohydrates and fats. A new Chemical Connections box on *Depression and Nutrition* was added.

- **Chapter 31, Immunochemistry,** covers the basics of our immune system and how we protect ourselves from foreign invading organisms. Considerable time is spent on the acquired immunity system. No chapter on immunology would be complete without a description of the Human Immunodeficiency Virus. The chapter includes a new section on Immunization, new Chemical Connections boxes on *Influenza* and the *Flu Vaccine,* and eight new problems.

- **Chapter 32, Body Fluids,** can be found on the companion web site, which is accessible from **www.cengagebrain.com**. Search for this textbook's ISBN: 1-133-10508-4 to find this resource.

Alternate Version

General, Organic and Biochemistry, 10e, Hybrid Version with OWL ISBN-10: 1-133-10982-9; ISBN-13: 978-1-133-10982-2
This briefer paperback version of General, Organic and Biochemistry does not contain the end-of-chapter problems, which can be assigned in OWL. Access to OWL and the Cengage YouBook is packaged with the hybrid version. The Cengage YouBook is the full version of the text, with all end-of-chapter questions and problem sets.

Supporting Materials

OWL for General, Organic and Biochemistry/Allied Health
Instant Access OWL with Cengage YouBook (6 months)
ISBN: 978-1-133-22999-5
Instant Access OWL with Cengage YouBook (24 months)
ISBN: 978-1-133-35129-1

OWL
Online Web
Learning

By Roberta Day, Beatrice Botch and David Gross of the University of Massachusetts, Amherst, William Vining of the State University of New York at Oneonta, and Susan Young of Hartwick College. **OWL** Online Web Learning offers more assignable, gradable content (including end-of chapter questions specific to this textbook), and more reliability and flexibility than any other system. OWL's powerful course management tools allow instructors to control due dates, number of attempts, and whether students see answers or receive feedback on how to solve problems. OWL includes the **Cengage YouBook**, an interactive and customizable Flash-based eBook. Instructors can publish web links, modify the textbook narrative as needed with the text edit tool, quickly re-order entire sections and chapters, and hide any content they don't teach to create an eBook that perfectly matches their syllabus. The Cengage YouBook includes animated figures, video clips, highlighting, notes, and more.

Developed by chemistry instructors for teaching chemistry, OWL is the only system specifically designed to support **mastery learning**, where students work as long as they need to master each chemical concept and skill. OWL has already helped hundreds of thousands of students master chemistry through a wide range of assignment types, including tutorials, interactive simulations, and algorithmically generated homework questions that provide instant, answer-specific feedback.

OWL is continually enhanced with online learning tools to address the various learning styles of today's students such as:

- **Quick Prep** review courses that help students learn essential skills to succeed in General and Organic Chemistry.

- **Jmol** molecular visualization program for rotating molecules and measuring bond distances and angles.

- **Go Chemistry®** mini video lectures on key concepts that students can play on their computers or download to their video iPods, smart phones, or personal video players.

In addition, when you become an OWL user, you can expect service that goes far beyond the ordinary. To learn more or to see a demo, please contact your Cengage Learning representative or visit us at **www.cengage.com/owl**.

For Instructors

Power Lecture Instructor's CD/DVD Package with JoinIn® and ExamView®
ISBN: 978-1-133-10540-4
This digital library and presentation tool includes:

PowerLecture™

- **PowerPoint® lecture slides** written for this text by William H. Brown that instructors can customize by importing their own lecture slides or other materials.

- **Image libraries** that contain digital files for figures, photographs, and numbered tables from the text, as well as multimedia animations in a variety of digital formats. Use these files to print transparencies, create your own PowerPoint slides, and supplement your lectures.

- Digital files of the complete **Instructor's Manual** by Mark Erickson and Andrew Piefer (Hartwick College) and **ExamView Test Bank** by Bette Kruez (University of Michigan, Dearborn).
- Sample chapters from the **Student Solutions Manual and Study Guide**.
- **ExamView testing software** that enables you to create, deliver, and customize tests using the more than 1,500 test bank questions written specifically for this text.
- **JoinIn student response (clicker) questions** written for this book for use with the classroom response system of the instructor's choice.

Chemistry CourseMate
Instant Access (two semesters) ISBN: 978-1-4282-7175-3
Chemistry CourseMate is available to complement this text. Chemistry CourseMate includes an interactive eBook, interactive teaching and learning tools such as quizzes, flashcards, videos, and more. Chemistry CourseMate also includes the Engagement Tracker, a first-of-its-kind tool that monitors student engagement in the course. Go to **login.cengage.com** to access these resources. Look for the CourseMate icon, which denotes a resource available within CourseMate.

Instructor Companion Site Supporting materials are available to qualified adopters. Please consult your local Cengage Learning sales representative for details. Go to **login.cengage.com**, find this textbook, and choose Instructor Companion Site to see samples of these materials, request a desk copy, locate your sales representative, and download the WebCT or Blackboard versions of the Test Bank.

Instructor's Manual for *Laboratory Experiments for General, Organic & Biochemistry*, **Eighth Edition,** by Frederick A. Bettelheim and Joseph M. Landesberg. This manual will help instructors in grading the answers to questions and in assessing the range of experimental results obtained by students. The Instructor's Manual also contains important notes for students and details on how to handle the disposal of waste chemicals. Available on the Instructor Companion Site.

For Students

Visit cengagebrain.com
To access these and additional course materials, please visit **www.cengagebrain.com**. At the cengagebrain.com home page, search for this textbook's ISBN (from the back cover of your book). This will take you to the product page where these resources can be found. (Instructors can log in at **login.cengage.com**.)

Instant Access Go Chemistry® for General Chemistry
Pressed for time? Miss a lecture? Need more review? Go Chemistry® for General Chemistry is a set of 27 downloadable mini video lectures. Developed by award-winning chemists, Go Chemistry helps you quickly review essential topics—whenever and wherever you want! Each video contains animations and problems and can be downloaded to your computer desktop or portable video player (like iPod or iPhone) for convenient self-study and exam review. Selected Go Chemistry videos have e-flashcards to briefly introduce a key concept and then test student understanding of the basics with a series of questions. OWL includes five Go Chemistry videos. Students can enter ISBN: 978-0-495-38228-7 at **www.cengagebrain.com** to download two free videos or to purchase instant access to the 27-video set or to individual videos.

Chemistry CourseMate

Instant Access (two semesters) ISBN: 978-1-4282-7175-3

Chemistry CourseMate is available with this book to help you make the grade. Chemistry CourseMate includes an interactive eBook with highlighting, note taking and search capabilities, as well as interactive learning tools such as quizzes, flashcards, videos, and more. Go to **login.cengage.com** to access these resources, and look for the CourseMate icon to find resources related to your text in Chemistry CourseMate.

Student Study Guide by William Scovell of Bowling Green State University includes reviews of chapter objectives, important terms and comparisons, focused reviews of concepts, and self-tests. ISBN: 978-1-133-10541-1.

Student Solutions Manual by Mark Erickson and Andrew Piefer (Hartwick College). This ancillary manual contains complete worked-out solutions to all in-text and odd-numbered end-of-chapter problems. ISBN: 978-1-133-10910-5.

***Laboratory Experiments for General, Organic & Biochemistry*, Eighth Edition,** by Frederick A. Bettelheim and Joseph M. Landesberg (Adelphi University). Forty-eight experiments illustrate important concepts and principles in general, organic and biochemistry. Includes eleven organic chemistry experiments, seventeen biochemistry experiments, and twenty general chemistry experiments. Many experiments have been revised, and a new addition is an experiment on the properties of enzymes. All experiments have new Pre- and Post-lab Questions. The large number of experiments allows sufficient flexibility for the instructor. ISBN: 978-1-133-10602-9.

Survival Guide for General, Organic, and Biochemistry by Richard Morrison, Charles H. Atwood, and Joel Caughran (University of Georgia). Available free in a package with any Cengage Chemistry text or available for separate purchase at **www.cengagebrain.com.** Modeled after Atwood's widely popular *General Chemistry Survival Guide*, this straightforward, thorough guide helps students make the most of their study time for optimal exam results. The *Survival Guide* is packed with examples and exercises to help students master concepts and improve essential problem-solving skills through detailed step-by-step problem-solving sequences. This reader-friendly guide gives students the competency—and confidence—they need to survive and thrive in the GOB course. ISBN: 978-0-495-55469-1.

Acknowledgments

The publication of a book such as this requires the efforts of many more people than merely the authors. We would like to thank the following professors who offered many valuable suggestions for this new edition:

We are especially grateful to Bette Kruez, University of Michigan, Dearborn and Robert Keil, Moorpark College, who read page proofs with eyes for accuracy.

We give special thanks to Sandi Kiselica, our Senior Development Editor, who has been a rock of support through the entire revision process. We appreciate her constant encouragement as we worked to meet deadlines; she has also been a valuable resource person. We appreciate the help of our other colleagues at Brooks/Cole: executive editor Mary Finch, production manager Teresa Trego, assistant editor Krista Mastroianni, media editors Lisa Weber and Stephanie vanCamp, and senior project manager Patrick Franzen of PreMediaGlobal.

We so appreciate the time and expertise of our reviewers who have read our manuscript and given us helpful comments. They include:

Reviewers of the 10th Edition:

Julian Davis, *University of the Incarnate Word*

Robert Keil, *Moorpark College*

Margaret Kimble, *Indiana University, Purdue University, Fort Wayne*

Bette Kruez, *University of Michigan, Dearborn*

Timothy Marshall, *Pima Community College*

Donald Mitchell, *Delaware Technical and Community College*

Paul Root, *Henry Ford Community College*

Ahmed Sheikh, *West Virginia University*

Steven Socol, *McHenry County College*

Susan Thomas, *University of Texas, San Antonio*

Holly Thompson, *University of Montana*

Janice Webster, *Ivy Tech Community College*

Reviewers of the 9th Edition:

Allison J. Dobson, *Georgia Southern University*

Sara M. Hein, *Winona State University*

Peter Jurs, *The Pennsylvania State University*

Delores B. Lamb, *Greenville Technical College*

James W. Long, *University of Oregon*

Richard L. Nafshun, *Oregon State University*

David Reinhold, *Western Michigan University*

Paul Sampson, *Kent State University*

Garon C. Smith, *University of Montana*

Steven M. Socol, *McHenry County College*

Health-Related Topics

Matter, Energy, and Measurement

1

James Balog/Getty Images

A woman climbing a frozen waterfall in British Columbia.

1.1 Why Do We Call Chemistry the Study of Matter?

The world around us is made of chemicals. Our food, our clothing, the buildings in which we live are all made of chemicals. Our bodies are made of chemicals, too. To understand the human body, its diseases, and its cures, we must know all we can about those chemicals. There was a time—only a few hundred years ago—when physicians were powerless to treat many diseases. Cancer, tuberculosis, smallpox, typhus, plague, and many other sicknesses struck people seemingly at random. Doctors, who had no idea what caused any of these diseases, could do little or nothing about them. Doctors treated them with magic as well as by such measures as bleeding, laxatives, hot plasters, and pills made from powdered staghorn, saffron, or gold. None of these treatments was effective, and the doctors, because they came into direct contact with highly contagious diseases, died at a much higher rate than the general public.

OWL

Sign in to OWL at **www.cengage.com/owl** to view tutorials and simulations, develop problem-solving skills, and complete online homework assigned by your professor.

Medical practice over time.
(a) A woman being bled by a leech on her left forearm; a bottle of leeches is on the table. From a 1639 woodcut.
(b) Modern surgery in a well-equipped operating room.

(a)

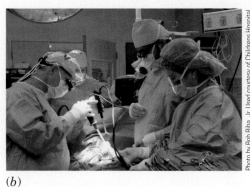

(b)

Medicine has made great strides since those times. We live much longer, and many once-feared diseases have been essentially eliminated or are curable. Smallpox has been eradicated, and polio, typhus, bubonic plague, diphtheria, and other diseases that once killed millions no longer pose a serious problem, at least not in the developed countries.

How has this medical progress come about? The answer is that diseases could not be cured until they were understood, and this understanding has emerged through greater knowledge of how the body functions. It is progress in our understanding of the principles of biology, chemistry, and physics that has led to these advances in medicine. Because so much of modern medicine depends on chemistry, it is essential that students who intend to enter the health professions have some understanding of basic chemistry. This book was written to help you achieve that goal. Even if you choose a different profession, you will find that the chemistry you learn in this course will greatly enrich your life.

The universe consists of matter, energy, and empty space. **Matter** is anything that has mass and takes up space. **Chemistry** is the science that deals with matter: the structure and properties of matter and the transformations from one form of matter to another. We will discuss energy in Section 1.8.

It has long been known that matter can change, or be made to change, from one form to another. In a **chemical change,** more commonly called a **chemical reaction,** substances are used up (disappear) and others are formed to take their places. An example is the burning of the mixture of hydrocarbons usually called "bottled gas." In this mixture of hydrocarbons, the main component is propane. When this chemical change takes place, propane and oxygen from the air are converted to carbon dioxide and water. Figure 1.1 shows another chemical change.

Matter also undergoes other kinds of changes, called **physical changes.** These changes differ from chemical reactions in that the identities of the substances do not change. Most physical changes involve changes of state—for example, the melting of solids and the boiling of liquids. Water remains water whether it is in the liquid state or in the form of ice or steam. The conversion from one state to another is a physical—not a chemical—change. Another important type of physical change involves making or separating mixtures. Dissolving sugar in water is a physical change.

When we talk about the **chemical properties** of a substance, we mean the chemical reactions that it undergoes. **Physical properties** are all properties that do not involve chemical reactions. For example, density, color, melting point, and physical state (liquid, solid, gas) are all physical properties.

(a)   (b)   (c)

FIGURE 1.1 A chemical reaction. (*a*) Bromine, an orange-brown liquid, and aluminum metal. (*b*) These two substances react so vigorously that the aluminum becomes molten and glows white hot at the bottom of the beaker. The yellow vapor consists of vaporized bromine and some of the product of the reaction, white aluminum bromide. (*c*) Once the reaction is complete, the beaker is coated with aluminum bromide and the products of its reaction with atmospheric moisture. (*Note:* This reaction is dangerous! Under no circumstances should it be done except under properly supervised conditions.)

1.2 What Is the Scientific Method?

Scientists learn by using a tool called the **scientific method.** The heart of the scientific method is the testing of theories. It was not always so, however. Before about 1600, philosophers often believed statements just because they sounded right. For example, the great philosopher Aristotle (384–322 BCE) believed that if you took the gold out of a mine it would grow back. He believed this idea because it fitted in with a more general picture that he had about the workings of nature. In ancient times, most thinkers behaved in this way. If a statement sounded right, they believed it without testing it.

About 1600 CE, the scientific method came into use. Let us look at an example to see how the scientific method operates. The Greek physician Galen (200–130 BCE) recognized that the blood on the left side of the heart somehow gets to the right side. This is a fact. A **fact** is a statement based on direct experience. It is a consistent and reproducible observation. Having observed this fact, Galen then proposed a hypothesis to explain it. A **hypothesis** is a statement that is proposed, without actual proof, to explain the facts and their relationship. Because Galen could not actually see how the blood got from the left side to the right side of the heart, he came up with the hypothesis that tiny holes must be present in the muscular wall that separates the two halves.

Up to this point, a modern scientist and an ancient philosopher would behave the same way. Each would offer a hypothesis to explain the facts. From this point on, however, their methods would differ. To Galen, his explanation sounded right and that was enough to make him believe it, even though he couldn't see any holes. His hypothesis was, in fact, believed by virtually all physicians for more than 1000 years. When we use the scientific method, however, we do not believe a hypothesis just because it sounds right. We test it, using the most rigorous testing we can imagine.

William Harvey (1578–1657) tested Galen's hypothesis by dissecting human and animal hearts and blood vessels. He discovered that one-way

Galen did not do experiments to test his hypothesis.

A PET scanner is an example of how modern scientists do experiments to test a hypothesis.

Hypothesis A statement that is proposed, without actual proof, to explain a set of facts and their relationship

valves separate the upper chambers of the heart from the lower chambers. He also discovered that the heart is a pump that, by contracting and expanding, pushes the blood out. Harvey's teacher, Fabricius (1537–1619), had previously observed that one-way valves exist in the veins, so that blood in the veins can travel only toward the heart and not the other way.

Harvey put these facts together to come up with a new hypothesis: Blood is pumped by the heart and circulates throughout the body. This was a better hypothesis than Galen's because it fitted the facts more closely. Even so, it was still a hypothesis and, according to the scientific method, had to be tested further. One important test took place in 1661, four years after Harvey died. Harvey had predicted that because there had to be a way for the blood to get from the arteries to the veins, tiny blood vessels must connect them. In 1661, the Italian anatomist Malpighi (1628–1694), using the newly invented microscope, found these tiny vessels, which are now called capillaries.

Malpighi's discovery supported the blood circulation hypothesis by fulfilling Harvey's prediction. When a hypothesis passes the tests, we have more confidence in it and call it a theory. A **theory** is the formulation of an apparent relationship among certain observed phenomena, which has been verified to some extent. In this sense, a theory is the same as a hypothesis except that we have a stronger belief in it because more evidence supports it. No matter how much confidence we have in a theory, however, if we discover new facts that conflict with it or if it does not pass newly devised tests, the theory must be altered or rejected. In the history of science, many firmly established theories have eventually been thrown out because they could not pass new tests.

One of the most important ways to test a hypothesis is by a controlled experiment. It is not enough to say that making a change causes an effect, we must also see that the lack of that change does not produce the observed effect. If, for example, a researcher proposes that adding a vitamin mixture to the diet of children improves growth, the first question is whether children in a control group who do not receive the vitamin mixture do not grow as quickly. Comparison of an experiment with a control is essential to the scientific method.

The scientific method is thus very simple. We don't accept a hypothesis or a theory just because it sounds right. We devise tests, and only if the hypothesis or theory passes the tests do we accept it. The enormous progress made since 1600 in chemistry, biology, and the other sciences is a testimony to the value of the scientific method.

You may get the impression from the preceding discussion that science progresses in one direction: facts first, hypothesis second, theory last. Real life is not so simple, however. Hypotheses and theories call the attention of scientists to discover new facts. An example of this scenario is the discovery of the element germanium. In 1871, Mendeleev's Periodic Table—a graphic description of elements organized by properties—predicted the existence of a new element whose properties would be similar to those of silicon. Mendeleev called this element eka-silicon. In 1886, it was discovered in Germany (hence the name), and its properties were truly similar to those predicted by theory.

On the other hand, many scientific discoveries result from **serendipity,** or chance observation. An example of serendipity occurred in 1926, when James Sumner of Cornell University left an enzyme preparation of jack bean urease in a refrigerator over the weekend. Upon his return, he found that his solution contained crystals that turned out to be a protein. This chance discovery led to the hypothesis that all enzymes are proteins. Of course, serendipity is not enough to move science forward. Scientists must have the creativity and insight to recognize the significance of their observations. Sumner fought for more than 15 years for his hypothesis to gain

Theory The formulation of an apparent relationship among certain observed phenomena, which has been verified. A theory explains many interrelated facts and can be used to make predictions about natural phenomena. Examples are Newton's theory of gravitation and the kinetic molecular theory of gases, which we will encounter in Section 6.6. This type of theory is also subject to testing and will be discarded or modified if it is contradicted by new facts.

acceptance because people believed that only small molecules can form crystals. Eventually his view won out, and he was awarded a Nobel Prize in chemistry in 1946.

1.3 How Do Scientists Report Numbers?

Scientists often have to deal with numbers that are very large or very small. For example, an ordinary copper penny (dating from before 1982, when pennies in the United States were still made of copper) contains approximately

$$29,500,000,000,000,000,000,000 \text{ atoms of copper}$$

and a single copper atom weighs

$$0.0000000000000000000000023 \text{ pound}$$

which is equal to

$$0.0000000000000000000000104 \text{ gram}$$

Many years ago, an easy way to handle such large and small numbers was devised. This method, which is called **exponential notation,** is based on powers of 10. In exponential notation, the number of copper atoms in a penny is written

$$2.95 \times 10^{22}$$

and the weight of a single copper atom is written

$$2.3 \times 10^{-25} \text{ pound}$$

which is equal to

$$1.04 \times 10^{-22} \text{ gram}$$

The origin of this shorthand form can be seen in the following examples:

$$100 = 1 \times 10 \times 10 = 1 \times 10^2$$
$$1000 = 1 \times 10 \times 10 \times 10 = 1 \times 10^3$$

What we have just said in the form of an equation is "100 is a one with two zeros after the one, and 1000 is a one with three zeros after the one." We can also write

$$1/100 = 1/10 \times 1/10 = 1 \times 10^{-2}$$
$$1/1000 = 1/10 \times 1/10 \times 1/10 = 1 \times 10^{-3}$$

where negative exponents denote numbers less than 1. The exponent in a very large or very small number lets us keep track of the number of zeros. That number can become unwieldy with very large or very small quantities, and it is easy to lose track of a zero. Exponential notation helps us deal with this possible source of determinant error.

When it comes to measurements, not all the numbers you can generate in your calculator or computer are of equal importance. Only the number of digits that are known with certainty are significant. Suppose that you measured the weight of an object as 3.4 g on a balance that you can read to the nearest 0.1 g. You can report the weight as 3.4 g but not as 3.40 or 3.400 g because you do not know the added zeros with certainty. This becomes even more important when you do calculations using a calculator. For example, you might measure a cube with a ruler and find that each side is 2.9 cm. If you are asked to calculate the volume, you multiply 2.9 cm $\times$ 2.9 cm $\times$ 2.9 cm. The calculator will then give you an answer that is 24.389 cm^3. However, your initial measurements were good to only one decimal place, so your final

Photos showing different orders of magnitude.
1. Group picnic in stadium parking lot (~10 meters)
2. Football field (~100 meters)
3. Vicinity of stadium (~1000 meters).

answer cannot be good to three decimal places. As a scientist, it is important to report data that have the correct number of **significant figures.** A detailed account of using significant figures is presented in Appendix II. The following How To box describes the way to determine the number of significant figures in a number. You will find boxes like this at places in the text where detailed explanations of concepts are useful.

How To . . .

Determine the Number of Significant Figures in a Number

1. **Nonzero digits are always significant.**
   For example, 233.1 m has four significant figures; 2.3 g has two significant figures.
2. **Zeros at the beginning of a number are never significant.**
   For example, 0.0055 L has two significant figures; 0.3456 g has four significant figures.
3. **Zeros between nonzero digits are always significant.**
   For example, 2.045 kcal has four significant figures; 8.0506 g has five significant figures.
4. **Zeros at the end of a number that contains a decimal point are always significant.**
   For example, 3.00 L has three significant figures; 0.0450 mm has three significant figures.
5. **Zeros at the end of a number that contains no decimal point may or may not be significant.**

We cannot tell whether they are significant without knowing something about the number. This is the ambiguous case. If you know that a certain small business made a profit of $36,000 last year, you can be sure that the 3 and 6 are significant, but what about the rest? The profit might have been $36,126 or $35,786.53, or maybe even exactly $36,000. We just don't know because it is customary to round off such numbers. On the other hand, if the profit were reported as $36,000.00, then all seven digits would be significant.

In science, to get around the ambiguous case, we use exponential notation. Suppose a measurement comes out to be 2500 g. If we made the measurement, then we know whether the two zeros are significant, but we need to tell others. If these digits are *not* significant, we write our number as 2.5×10^3. If one zero is significant, we write 2.50×10^3. If both zeros are significant, we write 2.500×10^3. Because we now have a decimal point, all the digits shown are significant. We are going to use decimal points throughout this text to indicate the number of significant figures.

Example 1.1 Exponential Notation and Significant Figures

Multiply:
(a) $(4.73 \times 10^5)(1.37 \times 10^2)$ (b) $(2.7 \times 10^{-4})(5.9 \times 10^8)$

Divide:
(c) $\dfrac{7.08 \times 10^{-8}}{300.}$ (d) $\dfrac{5.8 \times 10^{-6}}{6.6 \times 10^{-8}}$ (e) $\dfrac{7.05 \times 10^{-3}}{4.51 \times 10^5}$

Use your calculator for this example.

Strategy and Solution

The way to do calculations of this sort is to use a button on scientific calculators that automatically uses exponential notation. The button is usually labeled "E." (On some calculators, it is labeled "EE." In some cases, it is accessed by using the second function key.)

(a) Enter 4.73E5, press the multiplication key, enter 1.37E2, and press the "=" key. The answer is 6.48×10^7. The calculator will display this number as 6.48E7. This answer makes sense. We add exponents when we multiply, and the sum of these two exponents is correct $(5 + 2 = 7)$. We also multiply the numbers, 4.73×1.37. This is approximately $4 \times 1.5 = 6$, so 6.48 is also reasonable.

(b) Here we have to deal with a negative exponent, so we use the "+/−" key. Enter 2.7E+/−4, press the multiplication key, enter 5.9E8, and press the "=" key. The calculator will display the answer as 1.593E5. To have the correct number of significant figures, we should report our answer as 1.6E5. This answer makes sense because 2.7 is a little less than 3 and 5.9 is a little less than 6, so we predict a number slightly less than 18; also, the algebraic sum of the exponents $(-4 + 8)$ is equal to 4. This gives 16×10^4. In exponential notation, we normally prefer to report numbers between 1 and 10, so we rewrite our answer as 1.6×10^5. We made the first number 10 times smaller, so we increased the exponent by 1 to reflect that change.

(c) Enter 7.08E+/−8, press the division key, enter 300., and press the "=" key. The answer is 2.36×10^{-10}. The calculator will display this number as 2.36E − 10. We subtract exponents when we divide, and we can also write 300. as 3.00×10^2.

(d) Enter 5.8E+/−6, press the division key, enter 6.6E+/−8, and press the "=" key. The calculator will display the answer as 87.878787878788. We report this answer as 88 to get the right number of significant figures. This answer makes sense. When we divide 5.8 by 6.6, we get a number slightly less than 1. When we subtract the exponents algebraically $(-6 - [-8])$, we get 2. This means that the answer is slightly less than 1×10^2, or slightly less than 100.

(e) Enter 7.05E+/−3, press the division key, enter 4.51E5, and press the "=" key. The calculator displays the answer as 1.5632E−8, which, to the correct number of significant figures, is 1.56×10^{-8}. The algebraic subtraction of exponents is $-3 - 5 = -8$.

Problem 1.1

Multiply:

(a) $(6.49 \times 10^7)(7.22 \times 10^{-3})$ (b) $(3.4 \times 10^{-5})(8.2 \times 10^{-11})$

Divide:

(a) $\dfrac{6.02 \times 10^{23}}{3.10 \times 10^5}$ (b) $\dfrac{3.14}{2.30 \times 10^{-5}}$

1.4 How Do We Make Measurements?

In our daily lives, we are constantly making measurements. We measure ingredients for recipes, driving distances, gallons of gasoline, weights of fruits and vegetables, and the timing of TV programs. Doctors and nurses measure pulse rates, blood pressures, temperatures, and drug dosages. Chemistry, like other sciences, is based on measurements.

The label on this bottle of water shows the metric size (one liter) and the equivalent in quarts.

Metric system A system of units of measurement in which the divisions to subunits are made by a power of 10

Table 1.1 Base Units in the Metric System

Length	meter (m)
Volume	liter (L)
Mass	gram (g)
Time	second (s)
Temperature	Kelvin (K)
Energy	joule (J)
Amount of substance	mole (mol)

A measurement consists of two parts: a number and a unit. A number without a unit is usually meaningless. If you were told that a person's weight is 57, the information would be of very little use. Is it 57 pounds, which would indicate that the person is very likely a child or a midget, or 57 kilograms, which is the weight of an average woman or a small man? Or is it perhaps some other unit? Because so many units exist, a number by itself is not enough; the unit must also be stated.

In the United States, most measurements are made with the English system of units: pounds, miles, gallons, and so on. In most other parts of the world, however, few people could tell you what a pound or an inch is. Most countries use the **metric system,** a system that originated in France about 1800 and that has since spread throughout the world. Even in the United States, metric measurements are slowly being introduced (Figure 1.2). For example, many soft drinks and most alcoholic beverages now come in metric sizes. Scientists in the United States have been using metric units all along.

Around 1960, international scientific organizations adopted another system, called the **International System of Units** (abbreviated **SI**). The SI is based on the metric system and uses some of the metric units. The main difference is that the SI is more restrictive: It discourages the use of certain metric units and favors others. Although the SI has advantages over the older metric system, it also has significant disadvantages. For this reason, U.S. chemists have been very slow to adopt it. At this time, approximately 40 years after its introduction, not many U.S. chemists use the entire SI, although some of its preferred units are gaining ground.

In this book, we will use the metric system (Table 1.1). Occasionally we will mention the preferred SI unit.

A. Length

The key to the metric system (and the SI) is that there is one base unit for each kind of measurement and that other units are related to the base unit only by powers of 10. As an example, let us look at measurements of length. In the English system, we have the inch, the foot, the yard, and the mile (not to mention such older units as the league, furlong, ell, and rod). If you want to convert one unit to another unit, you must memorize or look up these conversion factors:

$$5280 \text{ feet} = 1 \text{ mile}$$

$$1760 \text{ yards} = 1 \text{ mile}$$

$$3 \text{ feet} = 1 \text{ yard}$$

$$12 \text{ inches} = 1 \text{ foot}$$

FIGURE 1.2 Road sign in Massachusetts showing metric equivalents of mileage.

Table 1.2 The Most Common Metric Prefixes

Prefix	Symbol	Value
giga	G	10^9 = 1,000,000,000 (one billion)
mega	M	10^6 = 1,000,000 (one million)
kilo	k	10^3 = 1000 (one thousand)
deci	d	10^{-1} = 0.1 (one-tenth)
centi	c	10^{-2} = 0.01 (one-hundredth)
milli	m	10^{-3} = 0.001 (one-thousandth)
micro	μ	10^{-6} = 0.000001 (one-millionth)
nano	n	10^{-9} = 0.000000001 (one-billionth)
pico	p	10^{-12} = 0.000000000001 (one-trillionth)

Exponential notation for quantities with multiple zeros is shown in parentheses.

All this is unnecessary in the metric system (and the SI). In both systems the base unit of length is the **meter (m).** To convert to larger or smaller units, we do not use arbitrary numbers like 12, 3, and 1760, but only 10, 100, 1/100, 1/10, or other powers of 10. This means that *to convert from one metric or SI unit to another, we only have to move the decimal point.* Furthermore, the other units are named by putting prefixes in front of "meter," and *these prefixes are the same throughout the metric system and the SI.* Table 1.2 lists the most important of these prefixes. If we put some of these prefixes in front of "meter," we have

Conversion factors are defined. We can use them to have as many significant figures as needed without limit. This point will not be the case with measured numbers.

$$1 \text{ kilometer (km)} = 1000 \text{ meters (m)}$$

$$1 \text{ centimeter (cm)} = 0.01 \text{ meter}$$

$$1 \text{ nanometer (nm)} = 10^{-9} \text{ meter}$$

For people who have grown up using English units, it is helpful to have some idea of the size of metric units. Table 1.3 shows some conversion factors.

Table 1.3 Some Conversion Factors Between the English and Metric Systems

Length	Mass	Volume
1 in. = 2.54 cm	1 oz = 28.35 g	1 qt = 0.946 L
1 m = 39.37 in.	1 lb = 453.6 g	1 gal = 3.785 L
1 mile = 1.609 km	1 kg = 2.205 lb	1 L = 33.81 fl oz
	1 g = 15.43 grains	1 fl oz = 29.57 mL
		1 L = 1.057 qt

Some of these conversions are difficult enough that you will probably not remember them and must, therefore, look them up when you need them. Some are easier. For example, a meter is about the same as a yard. A kilogram is a little over two pounds. There are almost four liters in a gallon. These conversions may be important to you someday. For example, if you rent a car in Europe, the price of gas listed on the sign at the gas station will be in Euros per liter. When you realize that you are spending two dollars per liter and you know that there are almost four liters to a gallon, you will realize why so many people take the bus or a train instead.

B. Volume

Volume is space. The volume of a liquid, solid, or gas is the space occupied by that substance. The base unit of volume in the metric system is the **liter (L).**

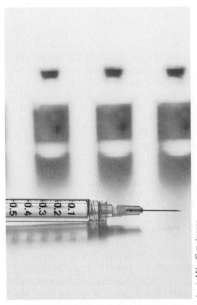

Hypodermic syringe. Note that the volumes are indicated in milliliters.

This unit is a little larger than a quart (Table 1.3). The only other common metric unit for volume is the milliliter (mL), which is equal to 10^{-3} L.

$$1 \text{ mL} = 0.001 \text{ L } (1 \times 10^{-3} \text{ L})$$

$$(1 \times 10^3 \text{ mL}) \, 1000 \text{ mL} = 1 \text{ L}$$

One milliliter is exactly equal to one cubic centimeter (cc or cm^3):

$$1 \text{ mL} = 1 \text{ cc}$$

Thus, there are 1000 (1×10^3) cc in 1 L.

C. Mass

Mass is the quantity of matter in an object. The base unit of mass in the metric system is the **gram (g).** As always in the metric system, larger and smaller units are indicated by prefixes. The ones in common use are

$$1 \text{ kilogram (kg)} = 1000 \text{ g}$$

$$1 \text{ milligram (mg)} = 0.001 \text{ g}$$

The gram is a small unit; there are 453.6 g in one pound (Table 1.3).

We use a device called a balance to measure mass. Figure 1.3 shows two types of laboratory balances.

There is a fundamental difference between mass and weight. Mass is independent of location. The mass of a stone, for example, is the same whether we measure it at sea level, on top of a mountain, or in the depths of a mine. In contrast, weight is not independent of location. **Weight** is the force a mass experiences under the pull of gravity. This point was dramatically demonstrated when the astronauts walked on the surface of the Moon. The Moon, being a smaller body than the Earth, exerts a weaker gravitational pull. Consequently, even though the astronauts wore space suits and equipment that would be heavy on Earth, they felt lighter on the Moon and could execute great leaps and bounces during their walks.

Although mass and weight are different concepts, they are related to each other by the force of gravity. We frequently use the words interchangeably

FIGURE 1.3 Two laboratory balances for measuring mass.

Chemical Connections 1A

Drug Dosage and Body Mass

In many cases, drug dosages are prescribed on the basis of body mass. For example, the recommended dosage of a drug may be 3 mg of drug for each kilogram of body weight. In this case, a 50 kg (110 lb) woman would receive 150 mg and an 82 kg (180 lb) man would get 246 mg. This adjustment is especially important for children, because a dose suitable for an adult will generally be too much for a child, who has much less body mass. For this reason, manufacturers package and sell smaller doses of certain drugs, such as aspirin, for children.

Drug dosage may also vary with age. Occasionally, when an elderly patient has an impaired kidney or liver function, the clearance of a drug from the body is delayed, and the drug may stay in the body longer than is normal. This persistence can cause dizziness, vertigo, and migraine-like headaches, resulting in falls and broken bones. Such delayed clearance must be monitored and the drug dosage adjusted accordingly.

This package of Advil has a chart showing the proper doses for children of a given weight.

Test your knowledge with Problems 1.69 and 1.70.

---

because we weigh objects by comparing their masses to standard reference masses (weights) on a balance, and the gravitational pull is the same on the unknown object and on the standard masses. Because the force of gravity is essentially constant, mass is always directly proportional to weight.

D. Time

Time is the one quantity for which the units are the same in all systems: English, metric, and SI. The base unit is the **second (s):**

$$60 \text{ s} = 1 \text{ min}$$

$$60 \text{ min} = 1 \text{ h}$$

E. Temperature

Most people in the United States are familiar with the Fahrenheit scale of temperature. The metric system uses the centigrade, or **Celsius,** scale. In this scale, the boiling point of water is set at 100°C and the freezing point at 0°C. We can convert from one scale to the other by using the following formulas:

$$°F = \frac{9}{5} °C + 32$$

$$°C = \frac{5}{9} (°F - 32)$$

The 32 in these equations is a defined number and is, therefore, treated as if it had an infinite number of zeros following the decimal point. (See Appendix II.)

Example 1.2 Temperature Conversion

Normal body temperature is 98.6°F. Convert this temperature to Celsius.

Strategy

We use the conversion formula that takes into account the fact that the freezing point of water, 0°C is equal to 32°F.

Solution

$$°C = \frac{5}{9} (98.6 - 32) = \frac{5}{9} (66.6) = 37.0°C$$

Problem 1.2

Convert:
(a) 64.0°C to Fahrenheit   (b) 47°F to Celsius

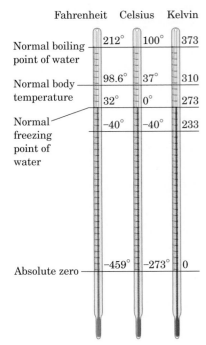

            Fahrenheit   Celsius   Kelvin

Normal boiling —   212°   100°   373
point of water

Normal body —   98.6°   37°   310
temperature   32°   0°   273

Normal   −40°   −40°   233
freezing
point of
water

Absolute zero —   −459°   −273°   0

FIGURE 1.4 Three temperature scales.

Figure 1.4 shows the relationship between the Fahrenheit and Celsius scales.

A third temperature scale is the **Kelvin (K)** scale, also called the absolute scale. The size of a Kelvin degree is the same as that of a Celsius degree; the only difference is the zero point. The temperature −273°C is taken as the zero point on the Kelvin scale. This makes conversions between Kelvin and Celsius very easy. To go from Celsius to Kelvin, just *add* 273; to go from Kelvin to Celsius, *subtract* 273:

$$K = °C + 273$$

$$°C = K - 273$$

Figure 1.4 also shows the relationship between the Kelvin and Celsius scales. Note that we don't use the degree symbol in the Kelvin scale: 100°C equals 373 K, not 373°K.

Why was −273°C chosen as the zero point on the Kelvin scale? The reason is that *−273°C, or 0 K, is the lowest possible temperature.* Because of this, 0 K is called **absolute zero.** Temperature reflects how fast molecules move. The more slowly they move, the colder it gets. At absolute zero, molecules stop moving altogether. Therefore, the temperature cannot get any lower. For some purposes, it is convenient to have a scale that begins at the lowest possible temperature; the Kelvin scale fulfills this need. The Kelvin is the SI unit.

It is very important to have a "gut feeling" about the relative sizes of the units in the metric system. Often, while doing calculations, the only thing that might offer a clue that you have made an error is your understanding of the sizes of the units. For example, if you are calculating the amount of a chemical that is dissolved in water and you come up with an answer of 254 kg/mL, does your answer make sense? If you have no intuitive feeling about the size of a kilogram or a milliliter, you will not know. If you realize that a milliliter is about the volume of a thimble and that a standard bag of sugar might weigh 2 kg, then you will realize that there is no way to pack 254 kg into a thimble of water, and you will know that you made a mistake.

1.5   What Is a Handy Way to Convert from One Unit to Another?

Factor-label method A procedure in which the equations are set up so that all the unwanted units cancel and only the desired units remain

We frequently need to convert a measurement from one unit to another. The best and most foolproof way to do this is the **factor-label method.** In this method, we follow the rule that *when multiplying numbers, we also multiply units and when dividing numbers, we also divide units.*

For conversions between one unit and another, it is always possible to set up two fractions, called **conversion factors.** Suppose we wish to convert the weight of an object from 381 grams to pounds. We are converting the units, but we are not changing the object itself. We want a ratio that reflects the change in units. In Table 1.3, we see that there are 453.6 grams in 1 pound. That is, the amount of matter in 453.6 grams is the same as the amount in 1 pound. In that sense, it is a one-to-one ratio, even though the units are not numerically the same. The conversion factors between grams and pounds therefore are

$$\frac{1 \text{ lb}}{453.6 \text{ g}} \quad \text{and} \quad \frac{453.6 \text{ g}}{1 \text{ lb}}$$

Conversion factor A ratio of two different units

To convert 381 grams to pounds, we must multiply by the proper conversion factor—but which one? Let us try both and see what happens.

First, let us multiply by 1 lb/453.6 g:

$$381 \text{ g} \times \frac{1 \text{ lb}}{453.6 \text{ g}} = 0.840 \text{ lb}$$

Following the procedure of multiplying and dividing units when we multiply and divide numbers, we find that dividing grams by grams cancels out the grams. We are left with pounds, which is the answer we want. Thus, 1 lb/453.6 g is the correct conversion factor because it converts grams to pounds.

Suppose we had done it the other way, multiplying by 453.6 g/1 lb:

$$381 \text{ g} \times \frac{453.6 \text{ g}}{1 \text{ lb}} = 173,000 \frac{\text{g}^2}{\text{lb}} \left(1.73 \times 10^5 \frac{\text{g}^2}{\text{lb}} \right)$$

When we multiply grams by grams, we get g^2 (grams squared). Dividing by pounds gives g^2/lb. This is not the unit we want, so we used the incorrect conversion factor.

In these conversions, we are dealing with measured numbers. Ambiguity can arise about the number of significant figures. The number 173,000 does not have six significant figures. We write 1.73×10^5 to show that it has three significant figures. We will use decimal points throughout this book to indicate significant figures in numbers with trailing zeros.

How To . . .
Do Unit Conversions by the Factor-Label Method

One of the most useful ways of approaching conversions is to ask three questions:

- What information am I given? This is the starting point.
- What do I want to know? This is the answer that you want to find.
- What is the connection between the first two? This is the conversion factor. Of course, more than one conversion factor may be needed for some problems.

Let's look at how to apply these principles to a conversion from pounds to kilograms. Suppose we want to know the weight in kilograms of a woman who weighs 125 lb. We see in Table 1.3 that there are 2.205 lb in 1 kg. Note that we are starting out with pounds and we want an answer in kilograms.

$$125 \text{ lb} \times \frac{1 \text{ kg}}{2.205 \text{ lb}} = 56.7 \text{ kg}$$

- The mass in pounds is the starting point. We were given that information.

- We wanted to know the mass in kilograms. That was the desired answer, and we found the number of kilograms.
- The connection between the two is the conversion factor in which the unit of the desired answer is in the numerator of the fraction, rather than the denominator. It is not simply a mechanical procedure to set up the equation so that units cancel; it is a first step to understanding the underlying reasoning behind the factor-label method. If you set up the equation to give the desired unit as the answer, you have made the connection properly.

If you apply this kind of reasoning, you can always pick the right conversion factor. Given the choice between

$$\frac{2.205 \text{ lb}}{1 \text{ kg}} \quad \text{and} \quad \frac{1 \text{ kg}}{2.205 \text{ lb}}$$

you know that the second conversion factor will give an answer in kilograms, so you use it. When you check the answer, you see that it is reasonable. You expect a number that is about one half of 125, which is 62.5. The actual answer, 56.7, is close to that value. The number of pounds and the number of kilograms are not the same, but they represent the same mass. That fact makes the use of conversion factors logically valid; the factor-label method uses the connection to obtain a numerical answer.

The advantage of the factor-label method is that it lets us know when we have made an incorrect calculation. *If the units of the answer are not the ones we are looking for, the calculation must be wrong.* Incidentally, this principle works not only in unit conversions but in all problems where we make calculations using measured numbers. Keeping track of units is a sure-fire way of doing conversions. It is impossible to overemphasize the importance of this way of checking on calculations.

The factor-label method gives the correct mathematical solution for a problem. However, it is a mechanical technique and does not require you to think through the problem. Thus, it may not provide a deeper understanding. For this reason and also to check your answer (because it is easy to make mistakes in arithmetic—for example, by punching the wrong numbers into a calculator), you should always ask yourself if the answer you have obtained is reasonable. For example, the question might ask the mass of a single oxygen atom. If your answer comes out 8.5×10^6 g, it is not reasonable. A single atom cannot weigh more than you do! In such a case, you have obviously made a mistake and should take another look to see where you went wrong. Of course, everyone makes mistakes at times, but if you check, you can at least determine whether your answer is reasonable. If it is not, you will immediately know that you have made a mistake and can then correct it.

Checking whether an answer is reasonable gives you a deeper understanding of the problem because it forces you to think through the relationship between the question and the answer. The concepts and the mathematical relationships in these problems go hand in hand. Mastery of the mathematical skills makes the concepts clearer, and insight into the concepts suggests ways to approach the mathematics. We will now give a few examples of unit conversions and then test the answers to see whether they are reasonable. To save space, we will practice this technique mostly in this chapter, but you should use a similar approach in all later chapters.

In unit conversion problems, you should always check two things. First, the numeric factor by which you multiply tells you whether the answer will be larger or smaller than the number being converted. Second, the factor tells you how much greater or smaller your answer should be when compared to your starting number. For example, if 100 kg is converted to pounds and there are 2.205 lb in 1 kg, then an answer of about 200 is reasonable— but an answer of 0.2 or 2000 (2.00×10^3) is not.

This is a good time to recall Thomas Edison's definition of genius: 99% perspiration and 1% inspiration.

⚇WL Interactive Example 1.3 Unit Conversion: Volume

The label on a container of olive oil says 1.844 gal. How many milliliters does the container hold?

Strategy

Here we use two conversion factors, rather than a single one. We still need to keep track of units.

Solution

Table 1.3 shows no factor for converting gallons to milliliters, but it does show that 1 gal = 3.785 L. Because we know that 1000 mL = 1 L, we can solve this problem by multiplying by two conversion factors, making certain that all units cancel except milliliters:

$$1.844 \ \cancel{\text{gal}} \times \frac{3.785 \ \cancel{\text{L}}}{1 \ \cancel{\text{gal}}} \times \frac{1000 \ \text{mL}}{1 \ \cancel{\text{L}}} = 6980. \ \text{mL}$$

Is this answer reasonable? The conversion factor in Table 1.3 tells us that there are more liters in a given volume than gallons. How much more? Approximately four times more. We also know that any volume in milliliters is 1000 times larger than the same volume in liters. Thus, we expect that the volume expressed in milliliters will be 4 × 1000, or 4000, times more than the volume given in gallons. The estimated volume in milliliters will be approximately 1.8 × 4000, or 7000 mL. But we also expect that the actual answer should be somewhat less than the estimated figure because we overestimated the conversion factor (4 rather than 3.785). Thus, the answer, 6980. mL, is quite reasonable. Note that the answer is given to four significant figures. The decimal point after the zero makes that point clear. We do not need a period after 1000 in the defined conversion factor; that is an exact number.

Problem 1.3

Calculate the number of kilometers in 8.55 miles. Check your answer to see whether it is reasonable.

⚇WL Interactive Example 1.4 Unit Conversion: Multiple Units

The maximum speed limit on many roads in the United States is 65 mi/h. How many meters per second (m/s) is this speed?

Strategy

We use four conversion factors in succession. It is more important than ever to keep track of units.

Solution

Here, we have essentially a double conversion problem: We must convert miles to meters and hours to seconds. We use as many conversion factors as necessary, always making sure that we use them in such a way that the proper units cancel:

$$65 \, \frac{\text{mi}}{\text{h}} \times \frac{1.609 \text{ km}}{1 \text{ mi}} \times \frac{1000 \text{ m}}{1 \text{ km}} \times \frac{1 \text{ h}}{60 \text{ min}} \times \frac{1 \text{ min}}{60 \text{ s}} = 29 \, \frac{\text{m}}{\text{s}}$$

Is this answer reasonable? To estimate the 65 mi/h speed in meters per second, we must first establish the relationship between miles and meters. As there are approximately 1.5 km in 1 mi, there must be approximately 1500 times more meters. We also know that in one hour, there are $60 \times 60 = 3600$ seconds. The ratio of meters to seconds will be approximately 1500/3600, which is about one half. Therefore, we estimate that the speed in meters per second will be about one half of that in miles per hour, or 32 m/s. Once again, the actual answer, 29 m/s, is not far from the estimate of 32 m/s, so the answer is reasonable.

Problem 1.4

Convert the speed of sound, 332 m/s to mi/h. Check your answer to see whether it is reasonable.

> Estimating the answer is a good thing to do when working any mathematical problem, not just unit conversions. We are not using decimal points after trailing zeros in the approximation.

Example 1.5 Unit Conversion: Multiple Units and Health Care

A physician recommends adding 100. mg of morphine to 500. cc of IV fluid and administering it at a rate of 20. cc/h to alleviate a patient's pain. Determine how many grams per second (g/s) the patient is receiving.

Strategy

Here, we use four conversion factors, rather than a single one. It is important to keep track of the desired units in the calculation and set up the other units in a way that allows us to cancel them out. It is also important to note that certain conversion factors do not need to be looked up in a table. Instead, they can be found in the problem (100. mg = 500. cc and 20. cc = 1 h).

Solution

We must convert the numerator from milligrams to grams and the denominator from cubic centimeters to seconds using the provided information. Because we know that 1000 mg = 1 g, 60 min = 1 h, and 60 s = 1 min, we can solve this problem by multiplying these conversion factors and making sure the units provided in the problem cancel out:

$$\frac{100. \text{ mg}}{500. \text{ cc}} \times \frac{20. \text{ cc}}{1 \text{ h}} \times \frac{1 \text{ g}}{1000 \text{ mg}} \times \frac{1 \text{ h}}{60 \text{ min}} \times \frac{1 \text{ min}}{60 \text{ s}} = 1.1 \times 10^{-6} \, \frac{\text{g}}{\text{s}}$$

Is this answer reasonable? Because this problem involves manipulating various conversion units, one way to estimate the final answer is to examine the ratio of the numerator to denominator. In this case, we know from our setup that the answer has to be less than one since it is obtained by dividing by a larger quantity than itself.

As shown in these examples, when canceling units, we do not cancel the numbers. The numbers are multiplied and divided in the ordinary way.

Problem 1.5

An intensive care patient is receiving an antibiotic IV at the rate of 50. mL/h. The IV solution contains 1.5 g of the antibiotic in 1000. mL. Calculate the mg/min of the drip. Check your answer to see if it is reasonable.

1.6 What Are the States of Matter?

Matter can exist in three states: gas, liquid, and solid. **Gases** have no definite shape or volume. They expand to fill whatever container they are put into. On the other hand, they are highly compressible and can be forced into small containers. Liquids also have no definite shape, but they do have a definite volume that remains the same when they are poured from one container to another. **Liquids** are only slightly compressible. **Solids** have definite shapes and definite volumes. They are essentially incompressible.

Whether a substance is a gas, a liquid, or a solid depends on its temperature and pressure. On a cold winter day, a puddle of liquid water turns to ice; it becomes a solid. If we heat water in an open pot at sea level, the liquid boils at 100°C; it becomes a gas—we call it steam. If we heated the same pot of water on the top of Mount Everest, it would boil at about 70°C due to the reduced atmospheric pressure. Most substances can exist in the three states: they are gases at high temperature, liquids at a lower temperature, and solids when their temperature becomes low enough. Figure 1.5 shows a single substance in the three different states.

The chemical identity of a substance does not change when it is converted from one state to another. Water is still water whether it is in the form of ice, steam, or liquid water. We discuss the three states of matter, and the changes between one state and another, at greater length in Chapter 5.

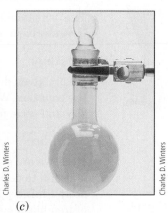

(a) *(b)* *(c)*

FIGURE 1.5 The three states of matter for bromine: (*a*) bromine as a solid, (*b*) bromine as a liquid, and (*c*) bromine as a gas.

The *Deepwater Horizon* oil spill (also referred to as the BP oil spill) in April 2010 flowed for three months in the Gulf of Mexico, releasing about 200 million barrels of crude oil. It is the largest accidental marine oil spill in the history of the petroleum industry. The spill continues to cause extensive damage to marine and wildlife habitats, as well as the Gulf's fishing and tourism industries.

1.7 What Are Density and Specific Gravity?

A. Density

One of the many pollution problems that the world faces is the spillage of petroleum into the oceans from oil tankers or from offshore drilling. When oil spills into the ocean, it floats on top of the water. The oil doesn't sink because it is not soluble in water and because water has a higher density than oil. When two liquids are mixed (assuming that one does not dissolve in the other), the one of lower density floats on top (Figure 1.6).

The **density** of any substance is defined as its *mass per unit volume*. Not only do all liquids have a density, but so do all solids and gases. Density is calculated by dividing the mass of a substance by its volume:

$$d = \frac{m}{V} \quad d = \text{density}, \ m = \text{mass}, \ V = \text{volume}$$

Example 1.6 Density Calculations

If 73.2 mL of a liquid has a mass of 61.5 g, what is its density in g/mL?

Strategy

We use the formula for density and substitute the values we are given for mass and volume.

Solution

$$d = \frac{m}{V} = \frac{61.5 \ \text{g}}{73.2 \ \text{mL}} = 0.840 \ \frac{\text{g}}{\text{mL}}$$

Problem 1.6

The density of titanium is 4.54 g/mL. What is the mass, in grams, of 17.3 mL of titanium? Check your answer to see whether it is reasonable.

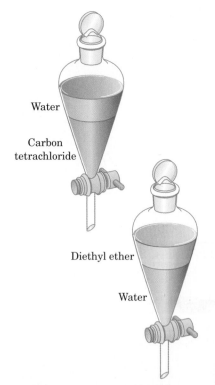

Example 1.7 Using Density to Find Volume

The density of iron is 7.86 g/cm³. What is the volume in milliliters of an irregularly shaped piece of iron that has a mass of 524 g?

Strategy

We are given density and mass. The volume is the unknown quantity in the equation. We substitute the known quantities in the formula for density and solve for volume.

Solution

Here, we are given the mass and the density. In this type of problem, it is useful to derive a conversion factor from the density. Since 1 cm³ is exactly 1 mL, we know that the density is 7.86 g/mL. This means that 1 mL of iron has a mass of 7.86 g. From this, we can get two conversion factors:

$$\frac{1 \ \text{mL}}{7.86 \ \text{g}} \quad \text{and} \quad \frac{7.86 \ \text{g}}{1 \ \text{mL}}$$

FIGURE 1.6 Two separatory funnels containing water and another liquid. The density of carbon tetrachloride is 1.589 g/mL, that of water is 1.00 g/mL, and that of diethyl ether is 0.713 g/mL. In each case, the liquid with the lower density is on top.

As usual, we multiply the mass by whichever conversion factor results in the cancellation of all but the correct unit:

$$524 \text{ g} \times \frac{1 \text{ mL}}{7.86 \text{ g}} = 66.7 \text{ mL}$$

Is this answer reasonable? The density of 7.86 g/mL tells us that the volume in milliliters of any piece of iron is always less than its mass in grams. How much less? Approximately eight times less. Thus, we expect the volume to be approximately 500/8 = 63 mL. As the actual answer is 66.7 mL, it is reasonable.

Problem 1.7

An unknown substance has a mass of 56.8 g and occupies a volume of 23.4 mL. What is its density in g/mL? Check your answer to see whether it is reasonable.

The density of any liquid or solid is a physical property that is constant, which means that it always has the same value at a given temperature. We use physical properties to help identify a substance. For example, the density of chloroform (a liquid formerly used as an inhalation anesthetic) is 1.483 g/mL at 20°C. If we want to find out if an unknown liquid is chloroform, one thing we might do is measure its density at 20°C. If the density is, say, 1.355 g/mL, we know the liquid isn't chloroform. If the density is 1.483 g/mL, we cannot be sure the liquid is chloroform, because other liquids might also have this density, but we can then measure other physical properties (the boiling point, for example). If all the physical properties we measure match those of chloroform, we can be reasonably sure the liquid is chloroform.

We have said that the density of a pure liquid or solid is a constant at a given temperature. Density does change when the temperature changes. Almost always, density decreases with increasing temperature. This is true because mass does not change when a substance is heated, but volume almost always increases because atoms and molecules tend to get farther apart as the temperature increases. Since $d = m/V$, if m stays the same and V gets larger, d must get smaller.

The most common liquid, water, provides a partial exception to this rule. As the temperature increases from 4°C to 100°C, the density of water does decrease, but from 0°C to 4°C, the density increases. That is, water has its maximum density at 4°C. This anomaly and its consequences are due to the unique structure of water and will be discussed in Chemical Connections 5E.

B. Specific Gravity

Because density is equal to mass divided by volume, it always has units, most commonly g/mL or g/cc or (g/L for gases). **Specific gravity** is numerically the same as density, but it has no units (it is dimensionless). The reason why there are no units is because specific gravity is defined as a comparison of the density of a substance with the density of water, which is taken as a standard. For example, the density of copper at 20°C is 8.92 g/mL. The density of water at the same temperature is 1.00 g/mL. Therefore, copper is 8.92 times as dense as water, and its specific gravity at 20°C is 8.92. Because water is taken as the standard and because the density of water is 1.00 g/mL at 20°C, the specific gravity of any substance is always numerically equal to its density, provided that the density is measured in g/mL or g/cc.

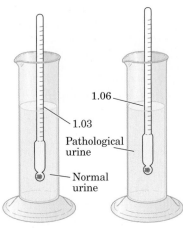

FIGURE 1.7 Urinometer.

1.06

1.03

Pathological urine

Normal urine

Specific gravity is often measured by a hydrometer. This simple device consists of a weighted glass bulb that is inserted into a liquid and allowed to float. The stem of the hydrometer has calibration marks, and the specific gravity is read where the meniscus (the curved surface of the liquid) hits the marking. The specific gravity of the acid in your car battery and that of a urine sample in a clinical laboratory are measured by hydrometers. A hydrometer measuring a urine sample is also called urinometer (Figure 1.7). Normal urine can vary in specific gravity from about 1.010 to 1.030. Patients with diabetes mellitus have an abnormally high specific gravity of their urine samples, while those with some forms of kidney disease have an abnormally low specific gravity.

Example 1.8 Specific Gravity

The density of ethanol at 20°C is 0.789 g/mL. What is its specific gravity?

Strategy

We use the definition of specific gravity.

Solution

$$\text{Specific gravity} = \frac{0.789 \text{ g/mL}}{1.00 \text{ g/mL}} = 0.789$$

Problem 1.8

The specific gravity of a urine sample at 20°C is 1.016. What is its density, in g/mL?

1.8 How Do We Describe the Various Forms of Energy?

Energy is defined as the capacity to do work. It can be described as being either kinetic energy or potential energy.

Kinetic energy (KE) is the energy of motion. Any object that is moving possesses kinetic energy. We can calculate how much kinetic energy by the formula $KE = \frac{1}{2}mv^2$, where m is the mass of the object and v is its velocity. This means that kinetic energy increases (1) when an object moves faster and (2) when a heavier object is moving. When a truck and a bicycle are moving at the same velocity, the truck has more kinetic energy.

Potential energy is stored energy. The potential energy possessed by an object arises from its capacity to move or to cause motion. For example, body weight in the up position on a seesaw contains potential energy—it is capable of doing work. If given a slight push, it will move down. The potential energy of the body in the up position is converted to kinetic energy as the body moves down on the seesaw. Work is done by gravity in the process. Figure 1.8 shows another way in which potential energy is converted to kinetic energy.

An important principle in nature is that things have a tendency to seek their lowest possible potential energy. We all know that water always flows downhill and not uphill.

Several forms of energy exist. The most important are (1) mechanical energy, light, heat, and electrical energy, which are examples of kinetic energy possessed by all moving objects, whether elephants or molecules or electrons, and (2) chemical energy and nuclear energy, which are examples of potential energy or stored energy. In chemistry, the more common form of potential energy is chemical energy—the energy stored within chemical substances

Potential energy is stored in this drawn bow and becomes kinetic energy in the arrow when released.

FIGURE 1.8 The water held back by the dam possesses potential energy, which is converted to kinetic energy when the water is released.

and given off when they take part in a chemical reaction. For example, a log possesses chemical energy. When the log is ignited in a fireplace, the chemical energy (potential) of the wood is turned into energy in the form of heat and light. Specifically, the potential energy has been transformed into thermal energy (heat makes molecules move faster) and the radiant energy of light.

The various forms of energy can be converted from one to another. In fact, we make such conversions all the time. A power plant operates either on the chemical energy derived from burning fuel or on nuclear energy. This energy is converted to heat, which is converted to the electricity that is sent over transmission wires into houses and factories. Here, we convert the electricity to light, heat (in an electrical heater, for example), or mechanical energy (in the motors of refrigerators, vacuum cleaners, and other devices).

Although one form of energy can be converted to another, the *total amount* of energy in any system does not change. *Energy can be neither created nor destroyed.* This statement is called the **law of conservation of energy.***

An example of energy conversion. Light energy from the sun is converted to electrical energy by solar cells. The electricity runs a refrigerator on the back of the camel, keeping the vaccines cool so that they can be delivered to remote locations.

1.9 How Do We Describe Heat and the Ways in Which It Is Transferred?

A. Heat and Temperature

One form of energy that is particularly important in chemistry is **heat.** This is the form of energy that most frequently accompanies chemical reactions. Heat is not the same as temperature, however. Heat is a form of energy, but temperature is not.

The difference between heat and temperature can be seen in the following example. If we have two beakers, one containing 100 mL of water and the other containing 1 L of water at the same temperature, the heat content of the water in the larger beaker is ten times that of the water in the smaller beaker, even though the temperature is the same in both. If you were to dip your hand accidentally into a liter of boiling water, you would be much more severely burned than if only one drop fell on your hand. Even though the water is at the same temperature in both cases, the liter of boiling water has much more heat.

As we saw in Section 1.4, temperature is measured in degrees. Heat can be measured in various units, the most common of which is the **calorie,** which is defined as the amount of heat necessary to raise the temperature

---

*This statement is not completely true. As discussed in Sections 9.8 and 9.9, it is possible to convert matter to energy, and vice versa. Therefore, a more correct statement would be *matter-energy can be neither created nor destroyed*. However, the law of conservation of energy is valid for most purposes and is highly useful.

Hypothermia and Hyperthermia

The human body cannot tolerate temperatures that are too low. A person outside in very cold weather (say, −20°F [−29°C]) who is not protected by heavy clothing will eventually freeze to death because the body loses heat. Normal body temperature is 37°C. When the outside temperature is lower than that, heat flows out of the body. When the air temperature is moderate (10°C to 25°C), this poses no problem and is, in fact, necessary because the body produces more heat than it needs and must lose some. At extremely low temperatures, however, too much heat is lost and body temperature drops, a condition called **hypothermia.** A drop in body temperature of 1 or 2°C causes shivering, which is the body's attempt to increase its temperature by the heat generated through muscular action. An even greater drop results in unconsciousness and eventually death.

The opposite condition is **hyperthermia.** It can be caused either by high outside temperatures or by the body itself when an individual develops a high fever. A sustained body temperature as high as 41.7°C (107°F) is usually fatal.

The label on this sleeping bag indicates the temperature range in which it can be used safely.

Test your knowledge with Problems 1.71 and 1.72.

of 1 g of liquid water by 1°C. This is a small unit, and chemists more often use the kilocalorie (kcal):

$$1 \text{ kcal} = 1000 \text{ cal}$$

Nutritionists use the word "Calorie" (with a capital "C") to mean the same thing as "kilocalorie"; that is, 1 Cal = 1000 cal = 1 kcal. The calorie is not part of the SI. The official SI unit for heat is the **joule (J),** which is about one-fourth as big as the calorie:

$$1 \text{ cal} = 4.184 \text{ J}$$

B. Specific Heat

As we noted, it takes 1 cal to raise the temperature of 1 g of liquid water by 1°C. **Specific heat (SH)** is the amount of heat necessary to raise the temperature of 1 g of any substance by 1°C. Each substance has its own specific heat, which is a physical property of that substance, like density or melting point. Table 1.4

Table 1.4 Specific Heats for Some Common Substances

Substance	Specific Heat (cal/g · °C)	Substance	Specific Heat (cal/g · °C)
Water	1.00	Wood (typical)	0.42
Ice	0.48	Glass (typical)	0.22
Steam	0.48	Rock (typical)	0.20
Iron	0.11	Ethanol	0.59
Aluminum	0.22	Methanol	0.61
Copper	0.092	Ether	0.56
Lead	0.031	Carbon tetrachloride	0.21

Cold Compresses, Waterbeds, and Lakes

The high specific heat of water is useful in cold compresses and makes them last a long time. For example, consider two patients with cold compresses: one compress made by soaking a towel in water and the other made by soaking a towel in ethanol. Both are at 0°C. Each gram of water in the water compress requires 25 cal to make the temperature of the compress rise to 25°C (after which it must be changed). Because the specific heat of ethanol is 0.59 cal/g · °C (see Table 1.4), each gram of ethanol requires only 15 cal to reach 25°C. If the two patients give off heat at the same rate, the ethanol compress is less effective because it will reach 25°C a good deal sooner than the water compress and will need to be changed sooner.

The ice on this lake will take days, or even weeks, to melt in the spring.

The high specific heat of water also means that it takes a great deal of heat to increase its temperature. That is why it takes a long time to get a pot of water to boil. Anyone who has a waterbed (300 gallons) knows that it takes days for the heater to bring the bed up to the desired temperature. It is particularly annoying when an overnight guest tries to adjust the temperature of your waterbed because the guest will probably have left before the change is noticed, but then you will have to set it back to your favorite temperature. This same effect in reverse explains why the outside temperature can be below zero (°C) for weeks before a lake will freeze. Large bodies of water do not change temperature very quickly.

Test your knowledge with Problem 1.73.

lists specific heats for a few common substances. For example, the specific heat of iron is 0.11 cal/g · °C. Therefore, if we had 1 g of iron at 20°C, it would require only 0.11 cal to increase the temperature to 21°C. Under the same conditions, aluminum would require twice as much heat. Thus, cooking in an aluminum pan of the same weight as an iron pan would require more heat than cooking in the iron pan. Note from Table 1.4 that ice and steam do not have the same specific heat as liquid water.

It is easy to make calculations involving specific heats. The equation is

Amount of heat = specific heat × mass × change in temperature

Amount of heat = SH × m × ΔT

where ΔT is the change in temperature.

We can also write this equation as

Amount of heat = SH × m × $(T_2 - T_1)$

where T_2 is the final temperature and T_1 is the initial temperature in °C.

Example 1.9 Specific Heat

How many calories are required to heat 352 g of water from 23°C to 95°C?

Strategy

We use the equation for the amount of heat and substitute the values given for the mass of water and the temperature change. We have already seen the value for the specific heat of water.

Solution

$$\text{Amount of heat} = \text{SH} \times m \times \Delta T$$

$$\text{Amount of heat} = \text{SH} \times m \times (T_2 - T_1)$$

$$= \frac{1.00 \text{ cal}}{g \cdot °C} \times 352 \text{ g} \times (95 - 23)°C$$

$$= 2.5 \times 10^4 \text{ cal}$$

Is this answer reasonable? Each gram of water requires one calorie to raise its temperature by one degree. We have approximately 350 g of water. To raise its temperature by one degree would therefore require approximately 350 calories. But we are raising the temperature not by one degree but by approximately 70 degrees (from 23 to 95). Thus, the total number of calories will be approximately $70 \times 350 = 24{,}500$ cal, which is close to the calculated answer. (Even though we were asked for the answer in calories, we should note that it will be more convenient to convert the answer to 25 kcal. We are going to see that conversion from time to time.)

Problem 1.9

How many calories are required to heat 731 g of water from 8°C to 74°C? Check your answer to see whether it is reasonable.

Example 1.10 Specific Heat and Temperature Change

If we add 450. cal of heat to 37 g of ethanol at 20.°C, what is the final temperature?

Strategy

The equation we have has a term for temperature change. We use the information we are given to calculate that change. We then use the value we are given for the initial temperature and the change to find the final temperature.

Solution

The specific heat of ethanol is 0.59 cal/g · °C (see Table 1.4).

$$\text{Amount of heat} = \text{SH} \times m \times \Delta T$$

$$\text{Amount of heat} = \text{SH} \times m \times (T_2 - T_1)$$

$$450. \text{ cal} = 0.59 \text{ cal/g} \cdot °C \times 37 \text{ g} \times (T_2 - T_1)$$

We can show the units in fraction form by rewriting this equation.

We write 450. cal with a decimal point to indicate three significant figures, also 20.°C to indicate two significant figures, but "40 times" and "20 calories" without decimal points to show quick approximations. Likewise, 230. cal has three significant figures.

$$450. \text{ cal} = 0.59 \frac{\text{cal}}{g \cdot °C} \times 37 \text{ g} \times (T_2 - T_1)$$

$$(T_2 - T_1) = \frac{\text{amount of heat}}{\text{SH} \times m}$$

$$(T_2 - T_1) = \frac{450. \text{ cal}}{\left[\dfrac{0.59 \text{ cal} \times 37 \text{ g}}{g \cdot °C}\right]} = \frac{21}{1/°C} = 21°C$$

(Note that we have the reciprocal of temperature in the denominator, which gives us temperature in the numerator. The answer has units of degrees Celsius). Because the starting temperature is 20°C, the final temperature is 41°C.

Is this answer reasonable? The specific heat of ethanol is 0.59 cal/g · °C. This value is close to 0.5, meaning that about half a calorie will raise the temperature of 1 g by 1°C. However, 37 g of ethanol need approximately 40 times as many calories for a rise, and $40 \times \frac{1}{2} = 20$ calories. We are adding 450. calories, which is about 20 times as much. Thus, we expect the temperature to rise by about 20°C, from 20°C to 40°C. The actual answer, 41°C, is quite reasonable.

Problem 1.10

A 100 g piece of iron at 25°C is heated by adding 230. cal. What will be the final temperature? Check your answer to see whether it is reasonable.

Example 1.11 Calculating Specific Heat

We heat 50.0 g of an unknown substance by adding 205 cal, and its temperature rises by 7.0°C. What is its specific heat? Using Table 1.4, identify the substance.

Strategy

We solve the equation for specific heat by substituting the values for mass, amount of heat, and temperature change. We compare the number we obtain with the values in Table 1.4 to identify the substance.

Solution

$$\text{SH} = \frac{\text{Amount of heat}}{m \times (\Delta T)}$$

$$\text{SH} = \frac{\text{Amount of heat}}{m \times (T_2 - T_1)}$$

$$\text{SH} = \frac{205 \text{ cal}}{50.0 \text{ g} \times 7.0°C} = 0.59 \text{ cal/g} \cdot °C$$

The substance in Table 1.4 having a specific heat of 0.59 cal/g · °C is ethanol.

Is this answer reasonable? If we had water instead of an unknown substance with SH = 1 cal/g · °C, raising the temperature of 50.0 g by 7.0°C would require $50 \times 7.0 = 350$ cal. But we added only approximately 200 cal. Therefore, the SH of the unknown substance must be less than 1.0. How much less? Approximately 200/350 = 0.6. The actual answer, 0.59 cal/g · °C, is quite reasonable.

Problem 1.11

It required 88.2 cal to heat 13.4 g of an unknown substance from 23°C to 176°C. What is the specific heat of the unknown substance? Check your answer to see whether it is reasonable.

Summary

OWL Sign in at **www.cengage.com/owl** to develop problem-solving skills and complete online homework assigned by your professor.

Section 1.1 Why Do We Call Chemistry the Study of Matter?

- **Chemistry** is the science that deals with the structure of matter and the changes it can undergo. In a **chemical change,** or **chemical reaction,** substances are used up and others are formed.
- Chemistry is also the study of energy changes during chemical reactions. In **physical changes,** substances do not change their identity.

Section 1.2 What Is the Scientific Method?

- The **scientific method** is a tool used in science and medicine. The heart of the scientific method is the testing of **hypotheses** and **theories** by collecting facts.

Section 1.3 How Do Scientists Report Numbers?

- Because we frequently use very large or very small numbers, we use powers of 10 to express these numbers more conveniently, a method called **exponential notation.**
- With exponential notation, we no longer have to keep track of so many zeros, and we have the added convenience of being able to see which digits convey information **(significant figures)** and which merely indicate the position of the decimal point.

Section 1.4 How Do We Make Measurements?

- In chemistry, we use the **metric system** for measurements.
- The base units are the meter for length, the liter for volume, the gram for mass, the second for time, and the joule for heat. Other units are indicated by prefixes that represent powers of 10. Temperature is measured in degrees Celsius or in Kelvins.

Section 1.5 What Is a Handy Way to Convert from One Unit to Another?

- Conversions from one unit to another are best done by the **factor-label method,** in which units are multiplied and divided to yield the units requested in the answer.

Section 1.6 What Are the States of Matter?

- There are three states of matter: **solid, liquid,** and **gas.**

Section 1.7 What Are Density and Specific Gravity?

- **Density** is mass per unit volume. **Specific gravity** is density relative to water and thus has no units. Density usually decreases with increasing temperature.

Section 1.8 How Do We Describe the Various Forms of Energy?

- **Kinetic energy** is energy of motion; **potential energy** is stored energy. Energy can be neither created nor destroyed, but it can be converted from one form to another.
- Examples of kinetic energy are: mechanical energy, light, heat, and electrical energy. Examples of potential energy are chemical energy and nuclear energy.

Section 1.9 How Do We Describe Heat and the Ways in Which It Is Transferred?

- **Heat** is a form of energy and is measured in calories. A calorie is the amount of heat necessary to raise the temperature of 1 g of liquid water by 1°C.
- Every substance has a **specific heat,** which is a physical constant. The specific heat is the number of calories required to raise the temperature of 1 g of a substance by 1°C.

Problems

OWL Interactive versions of these problems may be assigned in OWL.

Orange-numbered problems are applied.

Section 1.1 Why Do We Call Chemistry the Study of Matter?

1.12 The life expectancy of a citizen in the United States is 76 years. Eighty years ago it was 56 years. In your opinion, what was the major contributor to this spectacular increase in life expectancy? Explain your answer.

1.13 Define the following terms:
(a) Matter   (b) Chemistry

Section 1.2 What Is the Scientific Method?

1.14 In Table 1.4, you find four metals (iron, aluminum, copper, and lead) and three organic compounds (ethanol, methanol, and ether). What kind of hypothesis would you suggest about the specific heats of these chemicals?

1.15 In a newspaper, you read that Dr. X claimed that he has found a new remedy to cure diabetes.

The remedy is an extract of carrots. How would you classify this claim: (a) fact, (b) theory, (c) hypothesis, or (d) hoax? Explain your choice of answer.

1.16 Classify each of the following as a chemical or physical change:

(a) Burning gasoline

(b) Making ice cubes

(c) Boiling oil

(d) Melting lead

(e) Rusting iron

(f) Making ammonia from nitrogen and hydrogen

(g) Digesting food

Section 1.3 How Do Scientists Report Numbers?

Exponential Notation

1.17 Write in exponential notation:
(a) 0.351   (b) 602.1   (c) 0.000128   (d) 628122

1.18 Write out in full:
(a) 4.03×10^5 (b) 3.2×10^3
(c) 7.13×10^{-5} (d) 5.55×10^{-10}

1.19 Multiply:
(a) $(2.16 \times 10^5)(3.08 \times 10^{12})$
(b) $(1.6 \times 10^{-8})(7.2 \times 10^8)$
(c) $(5.87 \times 10^{10})(6.6 \times 10^{-27})$
(d) $(5.2 \times 10^{-9})(6.8 \times 10^{-15})$

1.20 Divide:
(a) $\dfrac{6.02 \times 10^{23}}{2.87 \times 10^{10}}$ (b) $\dfrac{3.14}{2.93 \times 10^{-4}}$
(c) $\dfrac{5.86 \times 10^{-9}}{2.00 \times 10^3}$ (d) $\dfrac{7.8 \times 10^{-12}}{9.3 \times 10^{-14}}$
(e) $\dfrac{6.83 \times 10^{-12}}{5.02 \times 10^{14}}$

1.21 Add:
(a) $(7.9 \times 10^4) + (5.2 \times 10^4)$
(b) $(8.73 \times 10^4) + (6.7 \times 10^3)$
(c) $(3.63 \times 10^{-4}) + (4.776 \times 10^{-3})$

1.22 Subtract:
(a) $(8.50 \times 10^3) - (7.61 \times 10^2)$
(b) $(9.120 \times 10^{-2}) - (3.12 \times 10^{-3})$
(c) $(1.3045 \times 10^2) - (2.3 \times 10^{-1})$

1.23 Solve:
$$\frac{(3.14 \times 10^3) \times (7.80 \times 10^5)}{(5.50 \times 10^2)}$$

1.24 Solve:
$$\frac{(9.52 \times 10^4) \times (2.77 \times 10^{-5})}{(1.39 \times 10^7) \times (5.83 \times 10^2)}$$

Significant Figures

1.25 How many significant figures are in the following?
(a) 0.012   (b) 0.10203
(c) 36.042   (d) 8401.0
(e) 32100   (f) 0.0402
(g) 0.000012

1.26 How many significant figures are in the following?
(a) 5.71×10^{13} (b) 4.4×10^5
(c) 3×10^{-6} (d) 4.000×10^{-11}
(e) 5.5550×10^{-3}

1.27 Round off to two significant figures:
(a) 91.621   (b) 7.329
(c) 0.677   (d) 0.003249
(e) 5.88

1.28 Multiply these numbers, using the correct number of significant figures in your answer:
(a) 3630.15×6.8
(b) 512×0.0081
(c) $5.79 \times 1.85825 \times 1.4381$

1.29 Divide these numbers, using the correct number of significant figures in your answer:
(a) $\dfrac{3.185}{2.08}$ (b) $\dfrac{6.5}{3.0012}$ (c) $\dfrac{0.0035}{7.348}$

1.30 Add these groups of measured numbers using the correct number of significant figures in your answer:
(a)   37.4083
       5.404
  10916.3
      3.94
      0.0006
(b)   84
    8.215
    0.01
  151.7
(c)   51.51
  100.27
   16.878
    3.6817

Section 1.4 How Do We Make Measurements?

1.31 In the SI system, the second is the base unit of time. We talk about atomic events that occur in picoseconds (10^{-12} s) or even in femtoseconds (10^{-15} s). But we don't talk about megaseconds or kiloseconds; the old standards of minutes, hours, and days prevail. How many minutes and hours are 20 kiloseconds?

1.32 How many grams are in the following?
(a) 1 kg   (b) 1 mg

1.33 Estimate without actually calculating which one is the shorter distance:
(a) 20 mm or 0.3 m
(b) 1 inch or 30 mm
(c) 2000 m or 1 mile

1.34 For each of these, tell which figure is closest to the correct answer:
(a) A baseball bat has a length of 100 mm or 100 cm or 100 m
(b) A glass of milk holds 23 cc or 230 mL or 23 L
(c) A man weighs 75 mg or 75 g or 75 kg
(d) A tablespoon contains 15 mL or 150 mL or 1.5 L

(e) A paper clip weighs 50 mg or 50 g or 50 kg

(f) Your hand has a width of 100 mm or 100 cm or 100 m

(g) An audiocassette weighs 40 mg or 40 g or 40 kg

1.35 You are taken for a helicopter ride in Hawaii from Kona (sea level) to the top of the volcano Mauna Kea. Which property of your body would change during the helicopter ride?

(a) height   (b) weight   (c) volume   (d) mass

1.36 Convert to Celsius and to Kelvin:

(a) 320°F   (b) 212°F   (c) 0°F   (d) −250°F

1.37 Convert to Fahrenheit and to Kelvin:

(a) 25°C   (b) 40°C   (c) 250°C   (d) −273°C

Section 1.5 What Is a Handy Way to Convert from One Unit to Another?

1.38 Make the following conversions (conversion factors are given in Table 1.3):

(a) 42.6 kg to lb   (b) 1.62 lb to g

(c) 34 in. to cm   (d) 37.2 km to mi

(e) 2.73 gal to L   (f) 62 g to oz

(g) 33.61 qt to L   (h) 43.7 L to gal

(i) 1.1 mi to km   (j) 34.9 mL to fl oz

1.39 Make the following metric conversions:

(a) 96.4 mL to L   (b) 275 mm to cm

(c) 45.7 kg to g   (d) 475 cm to m

(e) 21.64 cc to mL   (f) 3.29 L to cc

(g) 0.044 L to mL   (h) 711 g to kg

(i) 63.7 mL to cc   (j) 0.073 kg to mg

(k) 83.4 m to mm   (l) 361 mg to g

1.40 There are 2 bottles of cough syrup available on the shelf at the pharmacy. One contains 9.5 oz and the other has 300. cc. Which one has the larger volume?

1.41 A humidifier located at a nursing station holds 4.00 gallons of water. How many fluid ounces of water will completely fill the reservoir?

1.42 You drive in Canada where the distances are marked in kilometers. The sign says you are 80 km from Ottawa. You are traveling at a speed of 75 mi/h. Would you reach Ottawa within one hour, after one hour, or later than that?

1.43 The speed limit in some European cities is 80 km/h. How many miles per hour is this?

1.44 Your car gets 25.00 miles on a gallon of gas. What would be your car's fuel efficiency in km/L?

1.45 Children's Chewable Tylenol contains 80. mg of acetaminophen per tablet. If the recommended dosage is 10. mg/kg, how many tablets are needed for a 70.-lb child?

1.46 A patient weighs 186 lbs. She must receive an IV medication based on body weight. The order reads, "Give 2.0 mg per kilogram." The label reads "10. mg per cc." How many mL of medication would you give?

1.47 The doctor orders administration of a drug at 120. mg per 1000. mL at 400. mL/24 h. How many mg of drug will the patient receive every 8.0 hours?

1.48 The recommended pediatric dosage of Velosef is 20. mg/kg/day. What is the daily dose in mg for a child weighing 36 pounds? If the stock vial of Velosef

is labeled 208 mg/mL, how many mL would be given in a daily dose?

1.49 A critical care physician prescribes an IV of heparin to be administered at a rate of 1100 units per hour. The IV contains 26,000 units of heparin per liter. Determine the rate of the IV in cc/h.

1.50 If an IV is mixed so that each 150 mL contains 500. mg of the drug lidocaine, how many minutes will it take for 750 mg of lidocaine to be administered if the rate is set at 5 mL/min?

1.51 A nurse practitioner orders isotonic sodium lactate 50. mL/kg body mass to be administered intravenously for a 139-lb patient with severe acidosis. The rate of flow is 150 gtts/min, and the IV administration set delivers 20. gtts/mL, where the unit "gtts" stands for drops of liquid. What is the running time in minutes?

1.52 An order for a patient reads "Give 40. mg of pantoprazole IV and 5 g of $MgSO_4$ IV." The pantoprazole should be administered at a concentration of 0.4 mg/mL and the $MgSO_4$ should be administered at a concentration of 0.02 g/mL in separate IV infusion bags. What is the total fluid volume the patient has received from both IV infusions?

Section 1.6 What Are the States of Matter?

1.53 Which states of matter have a definite volume?

1.54 Will most substances be solids, liquids, or gases at low temperatures?

1.55 Does the chemical nature of a substance change when it melts from a solid to a liquid?

Section 1.7 What Are Density and Specific Gravity?

1.56 The volume of a rock weighing 1.075 kg is 334.5 mL. What is the density of the rock in g/mL? Express it to three significant figures.

1.57 The density of manganese is 7.21 g/mL, that of calcium chloride is 2.15 g/mL, and that of sodium acetate is 1.528 g/mL. You place these three solids in a liquid, in which they are not soluble. The liquid has a density of 2.15 g/mL. Which will sink to the bottom, which will stay on the top, and which will float in the middle of the liquid?

1.58 The density of titanium is 4.54 g/mL. What is the volume, in milliliters, of 163 g of titanium?

1.59 An injection of 4 mg of Valium has been prescribed for a patient suffering from muscle spasms. A sample of Valium labeled 5 mg/mL is on hand. How many mL should be injected?

1.60 The density of methanol at 20°C is 0.791 g/mL. What is the mass, in grams, of a 280 mL sample?

1.61 The density of dichloromethane, a liquid insoluble in water, is 1.33 g/cc. If dichloromethane and water are placed in a separatory funnel, which will be the upper layer?

1.62 A sample of 10.00 g of oxygen has a volume of 6702 mL. The same weight of carbon dioxide occupies 5058 mL.

(a) What is the density of each gas in g/L?

(b) Carbon dioxide is used as a fire extinguisher to cut off the fire's supply of oxygen. Do the densities of these two gases explain the fire-extinguishing ability of carbon dioxide?

1.63 Crystals of a material are suspended in the middle of a cup of water at 2°C. This means that the densities of the crystal and of the water are the same. How might you enable the crystals to rise to the surface of the water so that you can harvest them?

Section 1.8 How Do We Describe the Various Forms of Energy?

1.64 On many country roads, you see telephones powered by a solar panel. What principle is at work in these devices?

1.65 While you drive your car, your battery is being charged. How would you describe this process in terms of kinetic and potential energy?

Section 1.9 How Do We Describe Heat and the Ways in Which It Is Transferred?

1.66 How many calories are required to heat the following (specific heats are given in Table 1.4)?
(a) 52.7 g of aluminum from 100°C to 285°C
(b) 93.6 g of methanol from −35°C to 55°C
(c) 3.4 kg of lead from −33°C to 730°C
(d) 71.4 g of ice from −77°C to −5°C

1.67 If 168 g of an unknown liquid requires 2750 cal of heat to raise its temperature from 26°C to 74°C, what is the specific heat of the liquid?

1.68 The specific heat of steam is 0.48 cal/g · °C. How many kilocalories are needed to raise the temperature of 10.5 kg of steam from 120°C to 150°C?

Chemical Connections

1.69 (Chemical Connections 1A) If the recommended dose of a drug is 445 mg for a 180 lb man, what would be a suitable dose for a 135 lb man?

1.70 (Chemical Connections 1A) The average lethal dose of heroin is 1.52 mg/kg of body weight. Estimate how many grams of heroin would be lethal for a 200 lb man.

1.71 (Chemical Connections 1B) How does the body react to hypothermia?

1.72 (Chemical Connections 1B) Low temperatures often cause people to shiver. What is the function of this involuntary body action?

1.73 (Chemical Connections 1C) Which would make a more efficient cold compress, ethanol or methanol? (Refer to Table 1.4.)

Additional Problems

1.74 The meter is a measure of length. Tell what each of the following units measures:
(a) cm^3 (b) mL   (c) kg   (d) cal
(e) g/cc   (f) joule   (g) °C   (h) cm/s

1.75 A brain weighing 1.0 lb occupies a volume of 620 mL. What is the specific gravity of the brain?

1.76 If the density of air is 1.25×10^{-3} g/cc, what is the mass in kilograms of the air in a room that is 5.3 m long, 4.2 m wide, and 2.0 m high?

1.77 Classify these as kinetic or potential energy:
(a) Water held by a dam
(b) A speeding train
(c) A book on its edge before falling

(d) A falling book
(e) Electric current in a lightbulb

1.78 The kinetic energy possessed by an object with a mass of 1 g moving with a velocity of 1 cm/s is called 1 erg. What is the kinetic energy, in ergs, of an athlete with a mass of 127 lb running at a velocity of 14.7 mi/h?

1.79 A European car advertises an efficiency of 22 km/L, while an American car claims an economy of 30 mi/gal. Which car is more efficient?

1.80 In Potsdam, New York, you can buy gas for US$3.93/ gal. In Montreal, Canada, you pay US$1.22/L. (Currency conversions are outside the scope of this text, so you are not asked to do them here.) Which is the better buy? Is your calculation reasonable?

1.81 Shivering is the body's response to increase the body temperature. What kind of energy is generated by shivering?

1.82 When the astronauts walked on the Moon, they could make giant leaps in spite of their heavy gear.
(a) Why were their weights on the Moon so small?
(b) Were their masses different on the Moon than on the Earth?

1.83 Which of the following is the largest mass and which is the smallest?
(a) 41 g   (b) 3×10^3 mg
(c) 8.2×10^6 μg   (d) 4.1310×10^{-8} kg

1.84 Which quantity is bigger in each of the following pairs?
(a) 1 gigaton : 10. megaton
(b) 10. micrometer : 1 millimeter
(c) 10. centigram : 200. milligram

1.85 In Japan, high-speed "bullet trains" move with an average speed of 220. km/h. If Dallas and Los Angeles were connected by such a train, how long would it take to travel nonstop between these cities (a distance of 1490. miles)?

1.86 The specific heats of some elements at 25°C are as follows: aluminum = 0.215 cal/g · °C; carbon (graphite) = 0.170 cal/g · °C; iron = 0.107 cal/g · °C; mercury = 0.0331 cal/g · °C.
(a) Which element would require the smallest amount of heat to raise the temperature of 100 g of the element by 10°C?
(b) If the same amount of heat needed to raise the temperature of 1 g of aluminum by 25°C were applied to 1 g of mercury, by how many degrees would its temperature be raised?
(c) If a certain amount of heat is used to raise the temperature of 1.6 g of iron by 10°C, the temperature of 1 g of which element would also be raised by 10°C, using the same amount of heat?

1.87 Water that contains deuterium rather than ordinary hydrogen (see Section 2.4D) is called heavy water. The specific heat of heavy water at 25°C is 4.217 J/g · °C. Which requires more energy to raise the temperature of 10.0 g by 10°C, water or heavy water?

1.88 One quart of milk costs 80 cents and one liter costs 86 cents. Which is the better buy?

1.89 Consider butter, density 0.860 g/mL, and sand, density 2.28 g/mL.
   (a) If 1.00 mL of butter is thoroughly mixed with 1.00 mL of sand, what is the density of the mixture?
   (b) What would be the density of the mixture if 1.00 g of the same butter were mixed with 1.00 g of the same sand?

1.90 Which speed is the fastest?
   (a) 70 mi/h   (b) 140 km/h
   (c) 4.5 km/s   (d) 48 mi/min

1.91 In calculating the specific heat of a substance, the following data are used: mass = 92.15 g; heat = 3.200 kcal; rise in temperature = 45°C. How many significant figures should you report in calculating the specific heat?

1.92 A solar cell generates 500. kilojoules of energy per hour. To keep a refrigerator at 4°C, one needs 250. kcal/h. Can the solar cell supply sufficient energy per hour to maintain the temperature of the refrigerator?

1.93 The specific heat of urea is 1.339 J/g · °C . If one adds 60.0 J of heat to 10.0 g of urea at 20°C, what would be the final temperature?

1.94 You are waiting in line in a coffee shop. As you look at the selections, you see that the decaffeinated coffee is labeled "chemical-free." Comment on this label in light of the material in Section 1.1.

1.95 Which number has more significant figures?
   (a) 0.0000001   (b) 4.38

1.96 You are on vacation in Europe. You have bought a loaf of bread for a picnic lunch, and you want to buy some cheese to go with it. Do you buy 200. mg, 200. g, or 200. kg?

1.97 You have just left Tucson, Arizona, on I-19 to go on a trip to Mexico. Distances on this highway are shown in kilometers. A sign says that the border crossing is 95 km away. You estimate that is about 150 miles to go. When you get to the border, you find that you have traveled less than 60 miles. What went wrong in your calculation?

1.98 The antifreeze-coolant compound used in cars does not have the same density as water. Would a hydrometer be useful for measuring the amount of antifreeze in the cooling system?

1.99 In photosynthesis, light energy from the sun is used to produce sugars. How does this process represent a conversion of energy from one form to another?

1.100 What is the difference between aspirin tablets that contain 81 mg of aspirin and tablets that contain 325 mg?

1.101 In Canada, a sign indicates that the current temperature is 30°C. Are you most likely to be wearing a down parka and wool slacks, jeans and a long-sleeved shirt, or shorts and a T-shirt? What is the reason for your answer?

1.102 In very cold weather, ice fishing enthusiasts build small structures on the ice, drill holes, and put their fishing lines through the holes in the ice. How can fish survive under these conditions?

1.103 Most solids have a higher density than the corresponding liquid. Ice is less dense than water, with

corresponding expansion on freezing. How can this property be used to disrupt living cells by cycles of freezing and thawing?

1.104 A scientist claims to have found a treatment for ear infections in children. All the patients given this treatment showed improvement within three days. What comments do you have on this report?

Special Categories

Three special categories of problems—Tying It Together, Looking Ahead, and Challenge Problems—will appear from time to time at the ends of chapters. Not every chapter will have these problems, but they will appear to make specific points.

Tying It Together

1.105 Heats of reaction are frequently measured by monitoring the change in temperature of a water bath in which the reaction mixture is immersed. A water bath used for this purpose contains 2.000 L of water. In the course of the reaction, the temperature of the water rose 4.85°C. How many calories were liberated by the reaction? (You will need to use what you know about unit conversions and apply that information to what you know about energy and heat.)

1.106 You have samples of urea (a solid at room temperature) and pure ethanol (a liquid at room temperature). Which technique or techniques would you use to measure the amount of each substance?

Looking Ahead

1.107 You have a sample of material used in folk medicine. Suggest the approach you would use to determine whether this material contains an effective substance for treating disease. If you do find a new and effective substance, can you think of a way to determine the amount present in your sample? (Pharmaceutical companies have used this approach to produce many common medications.)

1.108 Many substances that are involved in chemical reactions in the human body (and in all organisms) contain carbon, hydrogen, oxygen, and nitrogen arranged in specific patterns. Would you expect new medications to have features in common with these substances, or would you expect them to be drastically different? What are the reasons for your answer?

Challenge Problems

1.109 If 2 kg of a given reactant is consumed in the reaction described in Problem 1.105, how many calories are liberated for each kilogram?

1.110 You have a water sample that contains a contaminant you want to remove. You know that the contaminant is much more soluble in diethyl ether than it is in water. You have a separatory funnel available. Propose a way to remove the contaminant.

1.111 In the hospital, your doctor orders 100. mg of medication per hour. The label on the IV bag reads 5.0 g/1000 mL. The IV administration set delivers 15. gtts/mL, where the unit gtts denotes drops of liquid as explained in Problem 1.51.
   (a) How many mL should infuse each hour?
   (b) The current drip rate is set to 10. gtts/min. Is this correct? If not, what is the correct drip rate?

Atoms

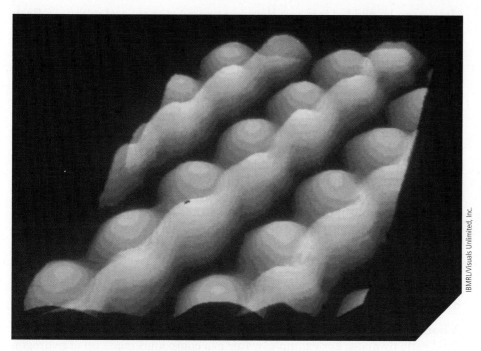

Image of atoms by SEM (scanning electron microscope).

IBMRL/Visuals Unlimited, Inc.

2.1 What Is Matter Made Of?

This question was discussed for thousands of years, long before humans had any reasonable way of getting an answer. In ancient Greece, two schools of thought tried to answer this question. One group, led by a scholar named Democritus (about 460–370 BCE), believed that all matter is made of very small particles—much too small to see. Democritus called these particles atoms (Greek *atomos*, meaning "not to cut"). Some of his followers developed the idea that there were different kinds of atoms, with different properties, and that the properties of the atoms caused ordinary matter to have the properties we all know.

Not all ancient thinkers, however, accepted this idea. A second group, led by Zeno of Elea (born about 450 BCE), did not believe in atoms at all. They insisted that matter is infinitely divisible. If you took any object, such as a piece of wood or a crystal of table salt, you could cut it or otherwise divide it into two parts, divide each of these parts into two more parts, and continue the process forever. According to Zeno and his followers, you would never reach a particle of matter that could no longer be divided.

Today we know that Democritus was right and Zeno was wrong. Atoms are the basic units of matter. Of course, there is a great difference in the way we now look at this question. Today our ideas are based on evidence. Democritus had no evidence to prove that matter cannot be divided an infinite number of times, just as Zeno had no evidence to support his claim that matter can be divided infinitely. Both claims were based not on evidence, but on visionary belief: one in unity, the other in diversity. In Section 2.3 we will discuss the evidence for the existence of atoms, but first we need to look at the diverse forms of matter.

2.2 How Do We Classify Matter?

Matter can be divided into two classes: pure substances and mixtures. Each class is then subdivided as shown in Figure 2.1.

A. Elements

An **element** is a substance (for example, carbon, hydrogen, and iron) that consists of identical atoms. At this time, 116 elements are known. Of these, 88 occur in nature; chemists and physicists have made the others in the laboratory. A list of the known elements appears on the inside front cover of this book, along with their symbols. Their symbols consist of one or two letters. Many symbols correspond directly to the name in English (for example, C for carbon, H for hydrogen, and Li for lithium), but a few are derived from the Latin or German names. Others are named for people who played significant roles in the development of science—in particular, atomic science (see Problem 2.12). Still other elements are named for geographic locations (See Problem 2.13).

B. Compounds

A **compound** is a pure substance made up of two or more elements in a fixed ratio by mass. For example, water is a compound made up of hydrogen and oxygen and table salt is a compound made up of sodium and chlorine. There are an estimated 20 million known compounds, only a few of which we will introduce in this book.

FIGURE 2.1 Classification of matter. Matter is divided into pure substances and mixtures. A pure substance may be either an element or a compound. A mixture may be either homogeneous or heterogeneous.

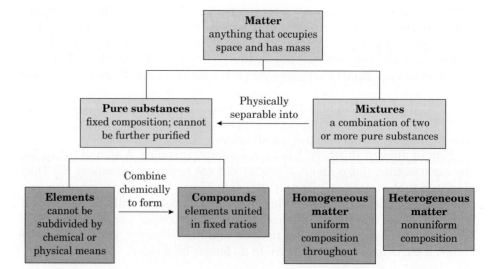

Chemical Connections 2A

Elements Necessary for Human Life

To the best of our knowledge, 20 of the 116 known elements are necessary for human life. The six most important of these—carbon, hydrogen, nitrogen, oxygen, phosphorus, and sulfur—are the subjects of organic chemistry and biochemistry (Chapters 10–31). Carbon, hydrogen, nitrogen, and oxygen are the big four in the human body. Seven other elements are also quite important, and our bodies use at least nine additional ones (trace elements) in very small quantities. The table lists these 20 major elements and their functions in the human body. Many of these elements are more fully discussed later in the book. For the average daily requirements for these elements, their sources in foods, and symptoms of their deficiencies, see Chapter 30.

Table 2A Elements and Their Functions in the Human Body

Element	Function	Element	Function
The Big Four		**The Trace Elements**	
Carbon (C)	The subject of Chapters 10–19	Chromium (Cr)	Increases effectiveness of insulin
Hydrogen (H)	(organic chemistry) and 20–31	Cobalt (Co)	Part of vitamin B_{12}
Nitrogen (N)	(biochemistry)	Copper (Cu)	Strengthens bones; assists in enzyme
Oxygen (O)			activity
The Next Seven		Fluorine (F)	Reduces the incidence of dental
			cavities
Calcium (Ca)	Strengthens bones and teeth; aids	Iodine (I)	An essential part of thyroid hormones
	blood clotting	Iron (Fe)	An essential part of some proteins,
Chlorine (Cl)	Necessary for normal growth and		such as hemoglobin, myoglobin,
	development		cytochromes, and FeS proteins
Magnesium (Mg)	Helps nerve and muscle action;	Manganese (Mn)	Present in bone-forming enzymes; aids
	present in bones		in fat and carbohydrate metabolism
Phosphorus (P)	Present as phosphates in bone, in	Molybdenum (Mo)	Helps regulate electrical balance in
	nucleic acids (DNA and RNA),		body fluids
	and involved in energy storage and	Zinc (Zn)	Necessary for the action of certain
	transfer		enzymes
Potassium (K)	Helps regulate electrical balance		
	in body fluids; essential for nerve		
	conduction		
Sulfur (S)	An essential component of proteins		
Sodium (Na)	Helps regulate electrical balance in		
	body fluids		

Test your knowledge with Problem 2.69.

A compound is characterized by its formula. The formula gives us the ratios of the compound's constituent elements and identifies each element by its atomic symbol. For example, in table salt, the ratio of sodium atoms to chlorine atoms is 1:1. Given that Na is the symbol for sodium and Cl is the symbol for chlorine, the formula of table salt is NaCl. In water, the combining ratio is two hydrogen atoms to one oxygen atom. The symbol for hydrogen is H, that for oxygen is O, and the formula of water is H_2O. The subscripts following the atomic symbols indicate the ratio of the combining elements. The number 1 in these ratios is omitted from the subscript. It is understood that NaCl means a ratio of 1:1 and that H_2O represents a ratio of 2:1. You will find out more about the nature of combining elements in a compound and their names and formulas in Chapter 3.

Figure 2.2 shows four representations for a water molecule. We will have more to say about molecular models as we move through this book.

FIGURE 2.2 Four representations of a water molecule.

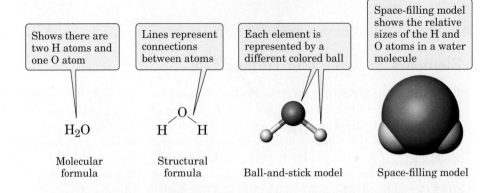

Shows there are two H atoms and one O atom

Lines represent connections between atoms

Each element is represented by a different colored ball

Space-filling model shows the relative sizes of the H and O atoms in a water molecule

H_2O H   H

Molecular formula   Structural formula   Ball-and-stick model   Space-filling model

Example 2.1 Formula of a Compound

(a) In the compound magnesium fluoride, magnesium (atomic symbol Mg) and fluorine (atomic symbol F) combine in a ratio of 1:2. What is the formula of magnesium fluoride?
(b) The formula of perchloric acid is $HClO_4$. What are the combining ratios of the elements in perchloric acid?

Strategy

The formula gives the atomic symbol of each element combined in the compound, and subscripts give the ratio of its constituent elements.

Solution

(a) The formula is MgF_2. We do not write a subscript of 1 after Mg.
(b) Both H and Cl have no subscripts, which means that hydrogen and chlorine have a combining ratio of 1:1. The subscript on oxygen is 4. Therefore, the combining ratios in $HClO_4$ are 1:1:4.

Problem 2.1

Write the formulas of compounds in which the combining ratios are as follows:
(a) Sodium: chlorine: oxygen, 1:1:3
(b) Aluminum (atomic symbol Al): fluorine (atomic symbol F), 1:3

C. Mixtures

A **mixture** is a combination of two or more pure substances. Most of the matter we encounter in our daily lives (including our own bodies) consists of mixtures rather than pure substances. For example, blood, butter, gasoline, soap, the metal in a ring, the air we breathe, and the Earth we walk on are all mixtures of pure substances. An important difference between a compound and a mixture is that the ratios by mass of the elements in a compound are fixed, whereas in a mixture, the pure substances can be present in any mass ratio.

For some mixtures—blood, for example (Figure 2.3)—the texture of the mixture is even throughout. If you examine a mixture under magnification, however, you can see that it is composed of different substances.

Other mixtures are homogeneous throughout, and no amount of magnification will reveal the presence of different substances. The air we breathe, for example, is a mixture of gases, primarily nitrogen (78 percent) and oxygen (21 percent).

An important characteristic of a mixture is that it consists of two or more pure substances, each having different physical properties. If we know the physical properties of the individual substances, we can use appropriate physical means to separate the mixture into its component parts. Figure 2.4 shows one example of how a mixture can be separated.

(a)   (b)   (c)

FIGURE 2.3 Mixtures. (a) A cup of noodle soup is a heterogeneous mixture. (b) A sample of blood may look homogeneous, but examination with an optical microscope shows that it is, in fact, a heterogeneous mixture of liquid and suspended particles (blood cells). (c) A homogeneous solution of salt, NaCl, in water. The models show that the salt solution contains Na^+ and Cl^- ions as separate particles in water, with each ion being surrounded by a sphere of six or more water molecules. The particles in this solution cannot be seen with an optical microscope because they are too small.

Repeated stirrings eventually leave a bright yellow sample of sulfur that cannot be purified further by this technique.

Iron and sulfur can be separated by stirring with a magnet.

The first time that the magnet is removed, much of the iron is removed with it.

The sulfur still looks dirty because a small quantity of iron remains.

FIGURE 2.4 Separating a mixture of iron and sulfur. (a) The iron–sulfur mixture is stirred with a magnet, which attracts the iron filings. (b) Much of the iron is removed after the first stirring. (c) Stirring continues until no more iron filings can be removed.

2.3 What Are the Postulates of Dalton's Atomic Theory?

In 1808, the English chemist John Dalton (1766–1844) put forth a model of matter that underlies modern scientific atomic theory. The major difference between Dalton's theory and that of Democritus (Section 2.1) is that Dalton based his theory on evidence rather than on a belief. First, let us state his theory. We will then see what kind of evidence supported it.

1. All matter is made up of very tiny, indivisible particles, which Dalton called **atoms.**

2. All atoms of a given element have the same chemical properties. Conversely, atoms of different elements have different chemical properties.

3. In ordinary chemical reactions, no atom of any element disappears or is changed into an atom of another element.

Atom The smallest particle of an element that retains the chemical properties of the element. The interaction among atoms accounts for the properties of matter.

4. Compounds are formed by the chemical combination of two or more different kinds of atoms. In a given compound, the relative numbers of atoms of each kind of element are constant and are most commonly expressed as integers.

5. A **molecule** is a tightly bound combination of two or more atoms that acts as a single unit.

A. Evidence for Dalton's Atomic Theory

The Law of Conservation of Mass

The great French chemist Antoine Laurent Lavoisier (1743–1794) discovered the **law of conservation of mass,** which states that matter can neither be created nor destroyed. In other words, there is no detectable change in mass in an ordinary chemical reaction. Lavoisier proved this law by conducting many experiments in which he showed that the total mass of matter at the end of the experiment was exactly the same as that at the beginning. Dalton's theory explained this fact in the following way: If all matter consists of indestructible atoms (postulate 1) and if no atoms of any element disappear or are changed into an atom of a different element (postulate 3), then any chemical reaction simply changes the attachments between atoms but does not destroy the atoms themselves. Thus, mass is conserved in a chemical reaction.

In the following illustration, a carbon monoxide molecule reacts with a lead oxide molecule to give a carbon dioxide molecule and a lead atom. All of the original atoms are still present at the end; they have merely changed partners. Thus, the total mass after this chemical change is the same as the mass that existed before the reaction took place.

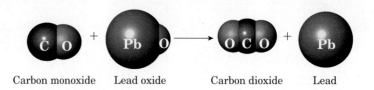

Carbon monoxide   Lead oxide   Carbon dioxide   Lead

The Law of Constant Composition

Another French chemist, Joseph Proust (1754–1826), demonstrated the **law of constant composition,** which states that any compound is always made up of elements in the same proportion by mass. For example, if you decompose water, you will always get 8.0 g of oxygen for each 1.0 g of hydrogen. The mass ratio of oxygen to hydrogen in pure water is always 8.0 to 1.0, whether the water comes from the Atlantic Ocean or the Missouri River or is collected as rain, squeezed out of a watermelon, or distilled from urine.

This fact was also evidence for Dalton's theory. If a water molecule consists of one atom of oxygen and two atoms of hydrogen and if an oxygen atom has a mass 16 times that of a hydrogen atom, then the mass ratio of these two elements in water must always be 8.0 to 1.0. The two elements can never be found in water in any other mass ratio.

Thus, if the atom ratio of the elements in a compound is fixed (postulate 4), then their proportions by mass must also be fixed.

B. Monatomic, Diatomic, and Polyatomic Elements

Some elements—for example, helium and neon—consist of single atoms that are not connected to each other—that is, they are **monatomic elements.** In contrast, oxygen, in its most common form, contains two atoms in each

Chemical Connections 2B

Abundance of Elements Present in the Human Body and in the Earth's Crust

Table 2B shows the abundance of the elements present in the human body. As you can see, oxygen is the most abundant element by mass, followed by carbon, hydrogen, and nitrogen. If we go by number of atoms, however, hydrogen is even more abundant in the human body than oxygen.

The table also shows the abundance of elements in the Earth's crust. Although 88 elements are found in the Earth's crust (we know very little about the interior of the Earth because we have not been able to penetrate into it very far), they are not present in anything close to equal amounts. In the Earth's crust as well as the human body, the most abundant element by mass is oxygen. But there the similarity ends. Silicon, aluminum, and iron, which are the second, third, and fourth most abundant elements in the Earth's crust, respectively, are not major elements in the body, whereas carbon, the second most abundant element by mass in the human body, is present to the extent of only 0.08 percent in the Earth's crust.

Table 2B The Relative Abundance of Elements Present in the Human Body and in the Earth's Crust, Including the Atmosphere and Oceans

	Percentage in Human Body		Percentage in Earth's Crust by Mass
Element	By Number of Atoms	By Mass	
H	63.0	10.0	0.9
O	25.4	64.8	49.3
C	9.4	18.0	0.08
N	1.4	3.1	0.03
Ca	0.31	1.8	3.4
P	0.22	1.4	0.12
K	0.06	0.4	2.4
S	0.05	0.3	0.06
Cl	0.03	0.2	0.2
Na	0.03	0.1	2.7
Mg	0.01	0.04	1.9
Si	—	—	25.8
Al	—	—	7.6
Fe	—	—	4.7
Others	0.01	—	—

Test your knowledge with Problem 2.70.

molecule, connected to each other by a chemical bond. We write the formula for an oxygen molecule as O_2, with the subscript showing the number of atoms in the molecule. Six other elements also occur as diatomic molecules (that is, they contain two atoms of the same element per molecule): hydrogen (H_2), nitrogen (N_2), fluorine (F_2), chlorine (Cl_2), bromine (Br_2), and iodine (I_2). It is important to understand that under normal conditions, free atoms of O, H, N, F, Cl, Br, and I do not exist. Rather, these seven elements occur only as **diatomic elements** (Figure 2.5).

Some elements have even more atoms in each molecule. Ozone, O_3, has three oxygen atoms in each molecule. In one form of phosphorus, P_4, each molecule has four atoms. One form of sulfur, S_8, has eight atoms per molecule. Some elements have molecules that are much larger. For example, diamond has millions of carbon atoms all bonded together in a gigantic cluster. Diamond and S_8 are referred to as **polyatomic elements.**

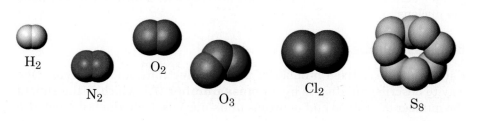

FIGURE 2.5 Some diatomic, triatomic, and polyatomic elements. Hydrogen, nitrogen, oxygen, and chlorine are diatomic elements. Ozone, O_3, is a triatomic element. One form of sulfur, S_8, is a polyatomic element.

2.4　What Are Atoms Made Of?

A. Three Subatomic Particles

There are many other subatomic particles, but we will not deal with them in this book.

Today, we know that matter is more complex than Dalton believed. A wealth of experimental evidence obtained over the last 100 years or so has convinced us that atoms are not indivisible, but rather, consist of even smaller particles called subatomic particles. Three subatomic particles make up all atoms: protons, electrons, and neutrons. Table 2.1 shows the charge, mass, and location of these particles in an atom.

Table 2.1 Properties and Location within Atoms of Protons, Neutrons, and Electrons

Subatomic Particle	Charge	Mass (g)	Mass (amu)	Mass (amu); Rounded to One Significant Figure	Location in an Atom
Proton	+1	1.6726×10^{-24}	1.0073	1	In the nucleus
Electron	−1	9.1094×10^{-28}	5.4858×10^{-4}	0.0005	Outside the nucleus
Neutron	0	1.6749×10^{-24}	1.0087	1	In the nucleus

Proton A subatomic particle with a charge of +1 and a mass of approximately 1 amu; it is found in a nucleus

Atomic mass unit (amu) A unit of the scale of relative masses of atoms: 1 amu = 1.6605×10^{-24} g. By definition, 1 amu is 1/12 the mass of a carbon atom containing 6 protons and 6 neutrons.

Electron A subatomic particle with a charge of −1 and a mass of approximately 0.0005 amu. It is found in the space surrounding a nucleus.

Neutron A subatomic particle with a mass of approximately 1 amu and a charge of zero; it is found in the nucleus

A **proton** has a positive charge. By convention we say that the magnitude of the charge is +1. Thus, one proton has a charge of +1, two protons have a charge of +2, and so forth. The mass of a proton is 1.6726×10^{-24} g, but this number is so small that it is more convenient to use another unit, called the **atomic mass unit (amu),** to describe its mass.

$$1 \text{ amu} = 1.6605 \times 10^{-24} \text{ g}$$

Thus, a proton has a mass of 1.0073 amu. For most purposes in this book, it is sufficient to round this number to one significant figure, and therefore, we say that the mass of a proton is 1 amu.

An **electron** has a charge of −1, equal in magnitude to the charge on a proton, but opposite in sign. The mass of an electron is approximately 5.4858×10^{-4} amu or 1/1837 that of the proton. It takes approximately 1837 electrons to equal the mass of one proton.

Like charges repel, and unlike charges attract. Two protons repel each other, just as two electrons also repel each other. A proton and an electron, however, attract each other.

A **neutron** has no charge. Therefore, neutrons neither attract nor repel each other or any other particle. The mass of a neutron is slightly greater than that of a proton: 1.6749×10^{-24} g or 1.0087 amu. Again, for our purposes, we round this number to 1 amu.

These three particles make up atoms, but where are they found? Protons and neutrons are found in a tight cluster in the center of an atom (Figure 2.6), which is called the **nucleus.** We will discuss the nucleus in greater detail in Chapter 9. Electrons are found as a diffuse cloud outside the nucleus.

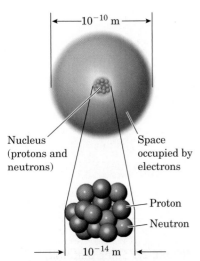

FIGURE 2.6 Relative sizes of the atomic nucleus and an atom (not to scale). The diameter of the region occupied by the electrons is approximately 10,000 times the diameter of the nucleus.

B. Mass Number

Each atom has a fixed number of protons, electrons, and neutrons. One way to describe an atom is by its **mass number** (A), which is the sum of the number of protons and neutrons in its nucleus. Note that an atom also contains electrons, but because the mass of an electron is so small compared to that of protons and neutrons (Table 2.1), electrons are not counted in determining mass number.

Mass number (A) = the number of protons + neutrons in the nucleus of an atom

For example, an atom with 5 protons, 5 electrons, and 6 neutrons has a mass number of 11.

OWL Interactive Example 2.2 Mass Number

What is the mass number of an atom containing:
(a) 58 protons, 58 electrons, and 78 neutrons?
(b) 17 protons, 17 electrons, and 20 neutrons?

Strategy

The mass number of an atom is the sum of the number of protons and neutrons in its nucleus.

Solution

(a) The mass number is 58 + 78 = 136.
(b) The mass number is 17 + 20 = 37.

Problem 2.2

What is the mass number of an atom containing:
(a) 15 protons, 15 electrons, and 16 neutrons?
(b) 86 protons, 86 electrons, and 136 neutrons?

C. Atomic Number

The **atomic number** (Z) of an element is the number of protons in its nucleus.

Atomic number (Z) = number of protons in the nucleus of an atom

Note that in a neutral atom, the number of electrons is equal to the number protons.

At the present time, 116 elements are known. These elements have atomic numbers from 1 to 116. The smallest atomic number belongs to the element hydrogen, which has only one proton and the largest (so far), to the as-yet-unnamed heaviest known element, which contains 116 protons.

If you know the atomic number and the mass number of an element, you can properly identify it. For example, the element with 6 protons, 6 electrons, and 6 neutrons has an atomic number of 6 and a mass number of 12. The element with atomic number 6 is carbon, C. Because its mass number is 12, we call this atomic nucleus carbon-12. Alternatively, we can write the symbol for this atomic nucleus as $^{12}_{6}C$. In this symbol, the mass number of the element is always written in the upper-left corner (as a superscript) of the symbol of the element and the atomic number in the lower-left corner (as a subscript).

Atomic numbers for all the known elements are given in the atomic weight table on the inside front cover. They are also given in the Periodic Table on the inside front cover.

If you know the name of the element, you can look up its atomic number and atomic weight from the atomic weight table on the inside front cover. Conversely, if you know the atomic number of the element, you can look up its symbol from the Periodic Table on the inside front cover.

Mass number (number of protons + neutrons) $^{12}_{6}C$ ←— Symbol of the element
Atomic number (number of protons)

Example 2.3 Atomic Number

Name the elements given in Example 2.2 and write the symbols for their atomic nuclei.

Strategy

Determine the atomic number (the number of protons in the nucleus) and then locate the element in the Periodic Table on the inside front cover.

Solution

(a) This element has 58 protons. We find in the Periodic Table that the element with atomic number 58 is cerium, and its symbol is Ce. An atom of this element has 58 protons and 78 neutrons, and therefore, its mass number is 136. We call it cerium-136. Its symbol is $^{136}_{58}$Ce.

(b) This atom has 17 protons, making it a chlorine (Cl) atom. Because its mass number is 37, we call it chlorine-37. Its symbol is $^{37}_{17}$Cl.

Problem 2.3

Name the elements given in Problem 2.2. Write the symbols of their atomic nuclei.

Example 2.4 Atomic Nuclei

A number of elements have an equal number of protons and neutrons in their nuclei. Among these are oxygen, nitrogen and, neon. What are the atomic numbers of these elements? How many protons and neutrons does an atom of each have? Write the name and the symbol of each of these atomic nuclei.

Strategy

Look at the Periodic Table to determine the atomic number of each element. Mass number is the number of protons plus the number of neutrons.

Solution

Atomic numbers for these elements are found in the list of elements on the inside back cover. This table shows that oxygen (O) has atomic number 8, nitrogen (N) has atomic number 7, and neon (Ne) has atomic number 10. This means that oxygen has 8 protons and 8 neutrons. Its name is oxygen-16, and its symbol is $^{16}_{8}$O. Nitrogen has 7 protons and 7 neutrons, its name is nitrogen-14, and its symbol is $^{14}_{7}$N. Neon has 10 protons and 10 neutrons, its name is neon-20, and its symbol is $^{20}_{10}$Ne.

Problem 2.4

(a) What are the atomic numbers of mercury (Hg) and lead (Pb)?
(b) How many protons does an atom of each have?
(c) If both Hg and Pb have 120 neutrons in their nuclei, what is the mass number of each isotope?
(d) Write the name and the symbol of each.

D. Isotopes

The fact that isotopes exist means that the second statement of Dalton's atomic theory (Section 2.3) is not correct.

Although we can say that an atom of carbon always has 6 protons and 6 electrons, we cannot say that an atom of carbon must have any particular number of neutrons. Some of the carbon atoms found in nature have 6 neutrons; the mass number of these atoms is 12, they are written as carbon-12, and their symbol is $^{12}_{6}C$. Other carbon atoms have 6 protons and 7 neutrons and, therefore, a mass number of 13; they are written as carbon-13, and their symbol is $^{13}_{6}C$. Still other carbon atoms have 6 protons and 8 neutrons; they are written as carbon-14 or $^{14}_{6}C$. Atoms with the same number of protons but different numbers of neutrons are called **isotopes.** All isotopes of carbon contain 6 protons and 6 electrons (or they wouldn't be carbon atoms). Each isotope, however, contains a different number of neutrons and, therefore, has a different mass number.

The properties of isotopes of the same element are almost identical, and for most purposes, we regard them as identical. They differ, however, in radioactive properties, which we discuss in Chapter 9.

Example 2.5 Isotopes

How many neutrons are in each isotope of oxygen? Write the symbol of each isotope.

(a) Oxygen-16   (b) Oxygen-17   (c) Oxygen-18

Strategy

Each oxygen atom has 8 protons. The difference between the mass number and the number of protons gives the number of neutrons.

Solution

(a) Oxygen-16 has $16 - 8 = 8$ neutrons. Its symbol is $^{16}_{8}O$.
(b) Oxygen-17 has $17 - 8 = 9$ neutrons. Its symbol is $^{17}_{8}O$.
(c) Oxygen-18 has $18 - 8 = 10$ neutrons. Its symbol is $^{18}_{8}O$.

Problem 2.5

Two iodine isotopes are used in medical treatments: iodine-125 and iodine-131. How many neutrons are in each isotope? Write the symbol for each isotope.

Most elements are found on Earth as mixtures of isotopes, in a more or less constant ratio. For example, all naturally occurring samples of the element chlorine contain 75.77% chlorine-35 (18 neutrons) and 24.23% chlorine-37 (20 neutrons). Silicon exists in nature in a fixed ratio of three isotopes, with 14, 15, and 16 neutrons, respectively. For some elements, the ratio of isotopes may vary slightly from place to place, but for most purposes, we can ignore these slight variations. The atomic masses and isotopic abundances are determined using an instrument called a mass spectrometer.

E. Atomic Weight

The **atomic weight** of an element given in the Periodic Table is a weighted average of the masses (in amu) of its isotopes found on the Earth. As an example of the calculation of atomic weight, let us examine chlorine. As we have just seen, two isotopes of chlorine exist in nature, chlorine-35 and chlorine-37. The mass of a chlorine-35 atom is 34.97 amu, and the mass of a chlorine-37 atom is 36.97 amu. Note that the atomic weight of each chlorine isotope (its mass in amu) is very close to its mass number (the number of protons and neutrons in its nucleus). This statement holds true for the isotopes of chlorine and those of all elements, because protons and neutrons have a mass of approximately (but not exactly) 1 amu.

Atomic weight The weighted average of the masses of the naturally occurring isotopes of the element. The units of atomic weight are atomic mass units (amu).

The atomic weight of chlorine is a weighted average of the masses of the two naturally occurring chlorine isotopes:

Chlorine-35   Chlorine-37

$$\left(\frac{75.77}{100} \times 34.97 \text{ amu}\right) + \left(\frac{24.23}{100} \times 36.97 \text{ amu}\right) = 35.45 \text{ amu}$$

| 17 |
| Cl |
| 35.4527 |

Atomic weight in the Periodic Table is given to four decimal places using more precise data than given here

Some elements—for example, gold, fluorine, and aluminum—occur naturally as only one isotope. The atomic weights of these elements are close to whole numbers (gold, 196.97 amu; fluorine, 18.998 amu; aluminum, 26.98 amu). A table of atomic weights is found facing the inside front cover of this book.

Example 2.6 Atomic Weight

The natural abundances of the three stable isotopes of magnesium are 78.99% magnesium-24 (23.98504 amu), 10.00% magnesium-25 (24.9858 amu), and 11.01% magnesium-26 (25.9829 amu). Calculate the atomic weight of magnesium and compare your value with that given in the Periodic Table.

Strategy

To calculate the weighted average of the masses of the isotopes, multiply each atomic mass by its abundance and then add.

Solution

Magnesium-24   Magnesium-25   Magnesium-26

$$\left(\frac{78.99}{100} \times 23.99 \text{ amu}\right) + \left(\frac{10.00}{100} \times 24.99 \text{ amu}\right) + \left(\frac{11.01}{100} \times 25.98 \text{ amu}\right) =$$

$$18.95 \quad + \quad 2.499 \quad + \quad 2.860 \quad = 24.31 \text{ amu}$$

The atomic weight of magnesium given in the Periodic Table to four decimal places is 24.3050.

Problem 2.6

The atomic weight of lithium is 6.941 amu. Lithium has only two naturally occurring isotopes: lithium-6 and lithium-7. Estimate which isotope of lithium is in greater natural abundance.

F. The Mass and Size of an Atom

A typical heavy atom (although not the heaviest) is lead-208, a lead atom with 82 protons, 82 electrons, and $208 - 82 = 126$ neutrons. It has a mass of 3.5×10^{-22} g. You would need 1.3×10^{24} atoms (a very large number) of lead-208 to make 1 lb. of lead. There are approximately 7 billion people on Earth right now. If you divided 1 lb. of these atoms among all the people on Earth, each person would get about 2.2×10^{14} atoms.

An atom of lead-208 has a diameter of about 3.1×10^{-10} m. If you could line them up with the atoms just touching, it would take 82 million lead atoms to make a line 1 inch long. Despite their tiny size, we can actually see atoms, in certain cases, by using a special instrument called a scanning tunneling microscope (Figure 2.7).

Virtually all of the mass of an atom is concentrated in its nucleus (because the nucleus contains the protons and neutrons). The nucleus of a lead-208 atom, for example, has a diameter of about 1.6×10^{-14} m. When you compare this with the diameter of a lead-208 atom, which is about 3.1×10^{-10} m, you see that the nucleus occupies only a tiny fraction of the total volume of the atom. If the nucleus of a lead-208 atom were the size of a baseball, then the entire atom would be much larger than a baseball stadium. In fact, it would be a sphere about one mile in diameter. Because a nucleus has such a relatively large mass concentrated in such a relatively small volume, a nucleus has a very high density. The density of a lead-208 nucleus, for example, is 1.6×10^{14} g/cm^3. Nothing in our daily life has a density anywhere near as high. If a paper clip had this density, it would weigh about 10 million (10^7) tons.

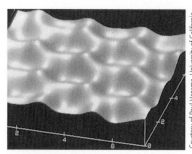

FIGURE 2.7 The surface of graphite is revealed with a scanning tunneling microscope. The contours represent the arrangement of individual carbon atoms on a crystal surface.

2.5 What Is the Periodic Table?

A. Origin of the Periodic Table

In the 1860s, the Russian scientist Dmitri Mendeleyev (1834–1907), then professor of chemistry at the University of St. Petersburg, produced one of the first Periodic Tables, the form of which we still use today. Mendeleyev started by arranging the known elements in order of increasing atomic weight beginning with hydrogen. He soon discovered that when the elements are arranged in the order of increasing atomic weight, certain sets of properties recur periodically. Mendeleyev then arranged those elements with recurring properties into **periods** (horizontal rows) by starting a new row each time he came to an element with properties similar to hydrogen. In this way, he discovered that lithium, sodium, potassium, and so forth, each start new rows. All are metallic solids at room temperature, all form ions with a charge of $+1$ (Li^+, Na^+, K^+, and so on), and all react with water to form metal hydroxides (LiOH, NaOH, KOH, and so on). Mendeleyev also discovered that elements in other vertical columns (families) have similar properties.

For example, the elements fluorine (atomic number 9), chlorine (17), bromine (35), and iodine (53) all fall in the same column of the table. These elements, which are called **halogens,** are all colored substances, with the color deepening as we go down the table (Figure 2.8). All form compounds with sodium that have the general formula NaX (for example, NaCl and NaBr), but not NaX_2, Na_2X, Na_3X, or anything else. Only the elements in this column share this property.

At this point, we must say a word about the numbering of the columns (families or groups) of the Periodic Table. Mendeleyev gave them numerals and added the letter A for some columns and B for others. This numbering pattern remains in common use in the United States today. In 1985, an alternative pattern was recommended by the International Union of Pure and Applied Chemistry (IUPAC). In this system, the groups are numbered 1 to 18, without added letters, beginning on the left. Thus, in Mendeleyev's numbering system, the halogens are in Group 7A; in the new international numbering system, they are in Group 17. Although this book uses the Mendeleyev numbering system, both patterns are shown on the Periodic Table on the inside front cover. The A group elements (Groups 1A and 2A on the left side of the table and Groups 3A through 8A at the right) are known collectively as **main-group elements.**

The elements in the B columns (Groups 3 to 12 in the new numbering system) are called **transition elements.** Notice that elements 58 to 71 and

Dmitri Mendeleyev.

Period of the Periodic Table A horizontal row of the Periodic Table

Family in the Periodic Table The elements in a vertical column of the Periodic Table

"X" is a commonly used symbol for a halogen.

Main-group element An element in the A groups (Groups 1A, 2A, and 3A–8A) of the Periodic Table

FIGURE 2.8 Four halogens. Fluorine and chlorine are gases, bromine is a liquid, and iodine is a solid.

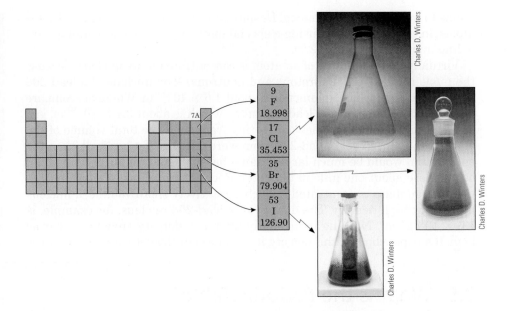

90 to 103 are not included in the main body of the table but rather are shown separately at the bottom. These sets of elements, called **inner transition elements,** actually belong in the main body of the Periodic Table, between columns 3 and 4 (between La and Hf and Ac and Rf). As is customary, we put them outside the main body solely to make a more compact presentation. If you like, you may mentally take a pair of scissors, cut through the heavy line between columns 3B and 4B, move them apart, and insert the inner transition elements. You will now have a table with 32 columns.

B. Classification of the Elements

There are three classes of elements: metals, nonmetals, and metalloids. The majority of elements are **metals**—only 24 are not. Metals are solids at room temperature (except for mercury, which is a liquid), shiny, conductors of electricity, ductile (they can be drawn into wires), and malleable (they can be hammered and rolled into sheets). In their reactions, metals tend to give up electrons. They also form alloys, which are solutions of one or more metals dissolved in another metal. Brass, for example, is an alloy of copper and zinc. Bronze is an alloy of copper and tin, and pewter is an alloy of tin, antimony, and lead. In their chemical reactions, metals tend to give up electrons (Section 3.2). Figure 2.9 shows a form of the Periodic Table in which the elements are classified by type.

Metal An element that is a solid at room temperature (except for mercury, which is a liquid), shiny, conducts electricity, is ductile and malleable, and forms alloys. In their reactions, metals tend to give up electrons.

FIGURE 2.9 Classification of the elements.

1A																	8A
H	2A		☐ Metals		☐ Metalloids		☐ Nonmetals					3A	4A	5A	6A	7A	He
Li	Be											B	C	N	O	F	Ne
Na	Mg	3B	4B	5B	6B	7B	8B	8B	8B	1B	2B	Al	Si	P	S	Cl	Ar
K	Ca	Sc	Ti	V	Cr	Mn	Fe	Co	Ni	Cu	Zn	Ga	Ge	As	Se	Br	Kr
Rb	Sr	Y	Zr	Nb	Mo	Te	Ru	Rh	Pd	Ag	Cd	In	Sn	Sb	Te	I	Xe
Cs	Ba	La	Hf	Ta	W	Re	Os	Ir	Pt	Au	Hg	Tl	Pb	Bi	Po	At	Rn
Fr	Ra	Ac	Rf	Db	Sg	Bh	Hs	Mt	Ds	Rg	Cn	≠	≠	≠	≠		

≠ Not yet named

Chemical Connections 2C

Strontium-90

Elements in the same column of the Periodic Table show similar properties. One important example is the similarity of strontium (Sr) and calcium (strontium is just below calcium in Group 2A). Calcium is an important element for humans because our bones and teeth consist largely of calcium compounds. We need some of this mineral in our diet every day, and we get it mostly from milk, cheese, and other dairy products.

One of the products released by test nuclear explosions in the 1950s and 1960s was the isotope strontium-90. This isotope is radioactive, with a half-life of 28.1 years. (Half-life is discussed in Section 9.4.) Strontium-90 was present in the fallout from aboveground nuclear test explosions. It was carried all over the Earth by winds and slowly settled to the ground, where it was eaten by cows and other animals. Strontium-90 got into milk and eventually into human bodies as well. If it were not so similar to calcium, our bodies

would eliminate it within a few days. Because it is similar, however, some of the strontium-90 became deposited in bones and teeth (especially in children), subjecting all of us to a small amount of radioactivity for long periods of time.

In 1958, pathologist Walter Bauer helped start the St. Louis Baby Tooth Survey to study the effects of nuclear fallout on children. The study helped establish an early '60s ban on aboveground A-bomb testing and led to similar surveys across the United States and the rest of the world. By 1970, the team had collected 300,000 shed primary teeth, which they discovered had absorbed nuclear waste from the milk of cows that were fed contaminated grass.

A 1963 treaty between the United States and the former Soviet Union banned aboveground nuclear testing. Although a few other countries still conducted occasional aboveground tests, there is reason to hope that such testing will be completely halted in the future.

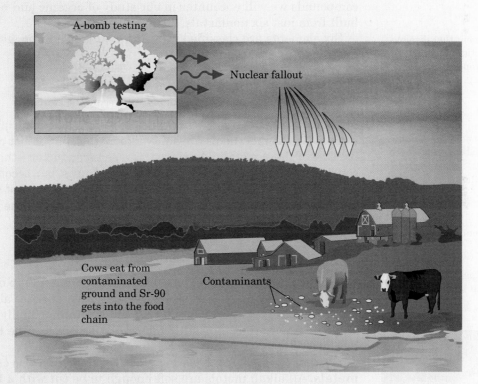

Test your knowledge with Problem 2.71.

Nonmetals are the second class of elements. With the exception of hydrogen, the 18 nonmetals appear to the right side of the Periodic Table. With the exception of graphite, which is one form of carbon, nonmetals do not conduct electricity. At room temperature, nonmetals such as phosphorus and iodine are solids. Bromine is a liquid, and the elements of Group 8A (the noble gases)—helium through radon—are gases. In their chemical reactions, nonmetals tend to accept electrons (Section 3.2). Virtually all of the

Nonmetal An element that does not have the characteristic properties of a metal and, in its reactions, tends to accept electrons. Eighteen elements are classified as nonmetals.

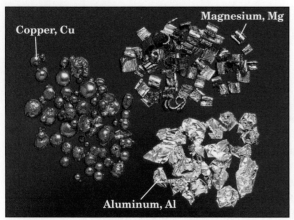

Copper, Cu

Magnesium, Mg

Aluminum, Al

Bromine, Br$_2$ Iodine, I$_2$

Forms of silicon

Charles D. Winters

(*a*) Metals   (*b*) Nonmetals   (*c*) Metalloids

FIGURE 2.10 Representative elements. (a) Magnesium, aluminum, and copper are metals. All can be drawn into wires and conduct electricity. (b) Only 18 or so elements are classified as nonmetals. Shown here are liquid bromine and solid iodine. (c) Only six elements are generally classified as metalloids. This photograph is of solid silicon in various forms, including a wafer on which electronic circuits are printed.

compounds we will encounter in our study of organic and biochemistry are built from just six nonmetals: H, C, N, O, P, and S.

Six elements are classified **metalloids:** boron, silicon, germanium, arsenic, antimony, and tellurium.

B	Si	Ge	As	Sb	Te
Boron	Silicon	Germanium	Arsenic	Antimony	Tellurium

These elements have some properties of metals and some of nonmetals. For example, some metalloids are shiny like metals, but do not conduct electricity. One of these metalloids, silicon, is a semiconductor—that is, it does not conduct electricity under certain applied voltages, but becomes a conductor at higher applied voltages. This semiconductor property of silicon makes it a vital element for Silicon Valley–based companies and the entire electronics industry (Figure 2.10).

Metalloid An element that displays some of the properties of metals and some of the properties of nonmetals. Six elements are classified as metalloids.

Although hydrogen (H) appears in Group 1A, it is not an alkali metal; it is a nonmetal. Hydrogen is placed in Group 1A because of its electron configuration (Section 2.7).

Halogen An element in Group 7A of the Periodic Table

Alkali metal An element, except hydrogen, in Group 1A of the Periodic Table

C. Examples of Periodicity in the Periodic Table

Not only do the elements in any particular column (group or family) of the Periodic Table share similar properties, but the properties also vary in some fairly regular ways as we go up or down a column (family). For instance, Table 2.2 shows that the melting and boiling points of the **halogens** regularly increase as we go down a column.

Another example involves the Group 1A elements, also called the **alkali metals.** All alkali metals are soft enough to be cut with a knife, and their softness increases going down the column. They have relatively low melting and boiling points, which decrease going down the columns (Table 2.3).

All alkali metals react with water to form hydrogen gas, H$_2$, and a metal hydroxide with the formula MOH, where "M" stands for the alkali metal. The violence of their reaction with water increases in going down the column.

$$2Na + 2H_2O \longrightarrow 2NaOH + H_2$$

Sodium   Water   Sodium hydroxide   Hydrogen gas

Charles D. Winters

Sodium metal can be cut with a knife.

Chemical Connections 2D

The Use of Metals as Historical Landmarks

The malleability of metals played an important role in the development of human society. In the Stone Age, tools were made from stone, which has no malleability. Then, about 11,000 BCE, it was discovered that the pure copper found on the surface of the Earth could be hammered into sheets, which made it suitable for use in vessels, utensils, and religious and artistic objects. This period became known as the Copper Age. Pure copper on the surface of the Earth, however, is scarce. Around 5000 BCE, humans found that copper could be obtained by putting malachite, $Cu_2CO_3(OH)_2$, a green copper-containing stone, into a fire. Malachite yielded pure copper at the relatively low temperature of 200°C.

Copper is a soft metal made of layers of large copper crystals. It can easily be drawn into wires because the layers of crystals can slip past one another. When hammered, the large crystals break into smaller ones with rough edges and the layers can no longer slide past one another. Therefore, hammered copper sheets are harder than drawn copper. Using this knowledge, the ancient profession of coppersmith was born, and beautiful plates, pots, and ornaments were produced.

Around 4000 BCE, it was discovered that an even greater hardness could be achieved by mixing molten copper with tin. The resulting alloy is called bronze. The Bronze Age was born somewhere in the Middle East and quickly spread to China and all over the world. Because hammered bronze takes an edge, knives and swords could be manufactured using it.

An even harder metal was soon to come. The first raw iron was found in meteorites. (The ancient Sumerian name of iron is "metal from heaven.") Around 2500 BCE, it was discovered that iron could be recovered from its ore by smelting, the process of recovering a metal from its ore by

Werner Forman/Art Resource, NY

Bronze Age artifact.

heating the ore. Thus began the Iron Age. More advanced technology was needed for smelting iron ores because iron melts only at a high temperature (about 1500°C). For this reason, it took a longer time to perfect the smelting process and to learn how to manufacture steel, which is about 90–95% iron and 5–10% carbon. Steel objects appeared first in India around 100 BCE.

Modern anthropologists and historians look back at ancient cultures and use the discovery of a new metal as a landmark for that age.

Test your knowledge with Problems 2.72 and 2.73.

Table 2.2 Melting and Boiling Points of the Halogens (Group 7A Elements)

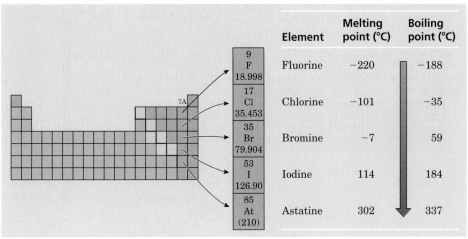

Element	Melting point (°C)	Boiling point (°C)
Fluorine	−220	−188
Chlorine	−101	−35
Bromine	−7	59
Iodine	114	184
Astatine	302	337

Table 2.3 Melting and Boiling Points of the Alkali Metals (Group 1A Elements)

Element	Melting point (°C)	Boiling point (°C)
Lithium	180	1342
Sodium	98	883
Potassium	63	760
Rubidium	39	686
Cesium	28	669

They also form compounds with the halogens with the formula MX, where "X" stands for the halogen.

$$2Na \; + \; Cl_2 \longrightarrow 2NaCl$$

Sodium　Chlorine　　Sodium chloride

The elements in Group 8A, often called the **noble gases,** provide yet another example of how the properties of elements change gradually within a column. Group 8A elements are gases under normal temperature and pressure, and they form either no compounds or very few compounds. Notice how close the melting and boiling points of the elements in this series are to one another (Table 2.4).

The Periodic Table is so useful that it hangs in nearly every chemistry classroom and chemical laboratory throughout the world. What makes it so useful is that it correlates a vast amount of data about the elements and their compounds and allows us to make many predictions about both chemical and physical properties. For example, if you were told that the boiling point of germane (GeH_4) is $-88°C$ and that of methane (CH_4) is $-164°C$,

Table 2.4 Melting and Boiling Points of the Noble Gases (Group 8A Elements)

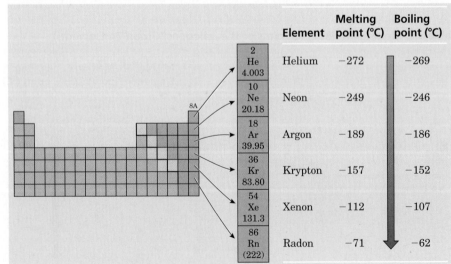

Element	Melting point (°C)	Boiling point (°C)
Helium	−272	−269
Neon	−249	−246
Argon	−189	−186
Krypton	−157	−152
Xenon	−112	−107
Radon	−71	−62

could you predict the boiling point of silane (SiH$_4$)? The position of silicon in the table, between germanium and carbon, might lead you to a prediction of about $-125°C$. The actual boiling point of silane is $-112°C$, not far from this prediction.

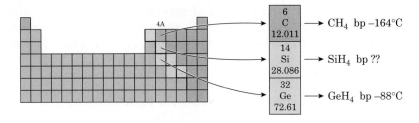

2.6 How Are the Electrons in an Atom Arranged?

We have seen that the protons and neutrons of an atom are concentrated in the atom's very small nucleus and that the electrons of an atom are located in the considerably larger space outside the nucleus. We can now ask how the electrons of an atom are arranged in this extranuclear space. Are they arranged randomly like seeds in a watermelon, or are they organized into layers like the layers of an onion?

Let us begin with hydrogen because it has only one electron and is the simplest atom. Before we do so, however, it is necessary to describe a discovery made in 1913 by the Danish physicist Niels Bohr (1885–1962). At the time, it was known that an electron is always moving around the nucleus and so possesses kinetic energy. Bohr discovered that only certain values are possible for this energy. This was a very surprising discovery. If you were told that you could drive your car at 23.4 mi/h or 28.9 mi/h or 34.2 mi/h, but never at any speed in between these values, you wouldn't believe it. Yet, that is just what Bohr discovered about electrons in atoms. The lowest possible energy level is the **ground state.**

If an electron is to have more energy than it has in the ground state, only certain values are allowed; values in between are not permitted. Bohr was unable to explain why these energy levels of electrons exist in atoms, but the accumulated evidence forced him to the conclusion that they do. We say that the energy of electrons in atoms is quantized. We can liken quantization to walking up a ramp compared with walking up a flight of stairs (Figure 2.11). You can put your foot on any stair step, but you cannot stand any place between two steps. You can stand only on steps.

Niels Bohr was awarded the 1922 Nobel Prize for physics. In addition, element 107 was named Bohrium in his honor.

Ground-state electron configuration The electron configuration of the lowest energy state of an atom

FIGURE 2.11 An energy stairway. A ramp, foreground (not quantized), and stair steps, background (quantized).

Charles D. Winters

A. Electrons Are Distributed in Shells, Subshells, and Orbitals

Principal energy level An energy level containing orbitals of the same number (1, 2, 3, 4, and so forth)

Shell All orbitals of a principal energy level of an atom

One conclusion reached by Bohr is that electrons in atoms do not move freely in the space around the nucleus, but rather remain confined to specific regions of space called **principal energy levels,** or more simply, **shells.** These shells are numbered 1, 2, 3, and 4, and so on, from the inside out. Table 2.5 gives the number of electrons that each of the first four shells can hold.

Table 2.5 Distribution of Electrons in Shells

Shell	Number of Electrons Shell Can Hold	Relative Energies of Electrons in Each Shell
4	32	Higher
3	18	
2	8	
1	2	Lower

Electrons in the first shell are closest to the positively charged nucleus and are held most strongly by it; these electrons are said to be the lowest in energy (hardest to remove). Electrons in higher-numbered shells are farther from the nucleus and are held less strongly to it; these electrons are said to be higher in energy (easier to remove).

Subshell All of the orbitals of an atom having the same principal energy level and the same letter designation (either s, p, d, or f)

Orbital A region of space around a nucleus that can hold a maximum of two electrons

Shells are divided into **subshells** designated by the letters s, p, d, and f. Within these subshells, electrons are grouped in **orbitals.** An orbital is a region of space and can hold two electrons (Table 2.6). The first shell contains a single s orbital and can hold two electrons. The second shell contains one s orbital and three p orbitals. All p orbitals come in sets of three and can hold six electrons. The third shell contains one s orbital, three p orbitals, and five d orbitals. All d orbitals come in sets of five and can hold ten electrons. The fourth shell also contains a set of f orbitals. All f orbitals come in sets of seven and can hold 14 electrons.

Table 2.6 Distribution of Orbitals within Shells

Shell	Orbitals Contained in Each Shell	Maximum Number of Electrons Shell Can Hold
4	One 4s, three 4p, five 4d, and seven 4f orbitals	2 + 6 + 10 + 14 = 32
3	One 3s, three 3p, and five 3d orbitals	2 + 6 + 10 = 18
2	One 2s and three 2p orbitals	2 + 6 = 8
1	One 1s orbital	2

B. Orbitals Have Definite Shapes and Orientations in Space

The d and f orbitals are less important to us, so we will not discuss their shapes.

All s orbitals have the shape of a sphere with the nucleus at the center of the sphere. Figure 2.12 shows the shapes of the 1s and 2s orbitals. Of the s orbitals, the 1s is the smallest sphere, the 2s is a larger sphere, and the 3s (not shown) is a still larger sphere. Figure 2.12 also shows the three-dimensional shapes of the three 2p orbitals. Each 2p orbital has the shape of a dumbbell with the nucleus at the midpoint of the dumbbell. The three 2p orbitals are at right angles to each other, with one orbital on the x-axis, the second on

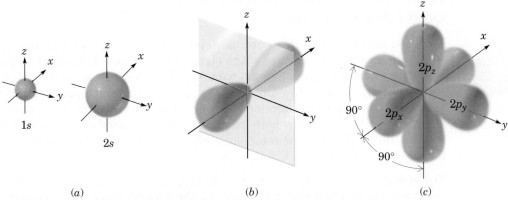

FIGURE 2.12 The 1s, 2s, and 2p orbitals. (a) A 1s orbital has the shape of a sphere, with the nucleus at the center of the sphere. A 2s orbital is a larger sphere than a 1s orbital, and a 3s orbital (not shown) is larger still. (b) A 2p orbital has the shape of a dumbbell, with the nucleus at the midpoint of the dumbbell. (c) Each 2p orbital is perpendicular to the other two. The 3p orbitals are similar in shape but larger. To make it easier for you to see the two lobes of each 2p orbital, one lobe is colored red and the other is colored blue.

the y-axis, and the third on the z-axis. The shapes of 3p orbitals are similar, but larger.

Because the vast majority of organic compounds and biomolecules consist of the elements H, C, N, O, P, and S, which use only 1s, 2s, 2p, 3s, and 3p orbitals for bonding, we will concentrate on just these and other elements of the first, second, and third periods of the Periodic Table.

C. Electron Configurations of Atoms Are Governed by Three Rules

The **electron configuration** of an atom is a description of the orbitals that its electrons occupy. The orbitals available to all atoms are the same—namely, 1s, 2s, 2p, 3s, 3p, and so on. In the ground state of an atom, only the lowest-energy orbitals are occupied; all other orbitals are empty. We determine the ground-state electron configuration of an atom using the following rules:

Rule 1. Orbitals fill in the order of increasing energy from lowest to highest.

> **Example:** In this book, we are concerned primarily with elements of the first, second, and third periods of the Periodic Table. Orbitals in these elements fill in the order 1s, 2s, 2p, 3s, and 3p. Figure 2.13 shows the order of filling through the third period.

Rule 2: Each orbital can hold up to two electrons with spins paired.

> **Example:** With four electrons, the 1s and 2s orbitals are filled and we write them as $1s^2 2s^2$. With an additional six electrons, the three 2p orbitals are filled and we write them either in the expanded form of $2p_x^2 \, 2p_y^2 \, 2p_z^2$, or in the condensed form $2p^6$. Spin pairing means that the electrons spin in opposite directions (Figure 2.14).

Rule 3: When there is a set of orbitals of equal energy, each orbital becomes half-filled before any of them becomes completely filled.

> **Example:** After the 1s and 2s orbitals are filled, a fifth electron is put into the $2p_x$ orbital, a sixth into the $2p_y$ orbital, and a seventh into the $2p_z$ orbital. Only after each 2p orbital has one electron is a second added to any 2p orbital.

Electron configuration A description of the orbitals of an atom or ion occupied by electrons

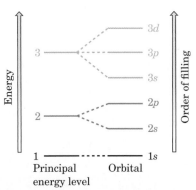

FIGURE 2.13 Energy levels for orbitals through the third shell.

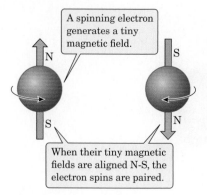

A spinning electron generates a tiny magnetic field.

When their tiny magnetic fields are aligned N-S, the electron spins are paired.

FIGURE 2.14 The pairing of electron spins.

D. Showing Electron Configurations: Orbital Box Diagrams

To illustrate how these rules are used, let us write the ground-state electron configurations for several of the elements in periods 1, 2, and 3. In the following **orbital box diagrams,** we use a box to represent an orbital, an arrow with its head up to represent a single electron, and a pair of arrows with heads in opposite directions to represent two electrons with paired spins. In addition, we show both expanded and condensed electron configurations. Table 2.7 gives the complete condensed ground-state electron configurations for elements 1 through 18.

Hydrogen (H)   The atomic number of hydrogen is 1, which means that its neutral atoms have a single electron. In the ground state, this electron is placed in the 1s orbital. Shown first is its orbital box diagram and then its electron configuration. A hydrogen atom has one unpaired electron.

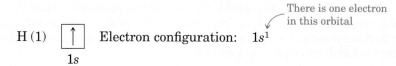

There is one electron in this orbital

H (1)   ↑   Electron configuration: $1s^1$
       1s

Helium (He)   The atomic number of helium is 2, which means that its neutral atoms have two electrons. In the ground state, both electrons are placed in the 1s orbital with paired spins, which fill the 1s orbital. All electrons in helium are paired.

This orbital is now filled with two electrons

He (2)   ⇅   Electron configuration: $1s^2$
        1s

Lithium (Li)   Lithium has atomic number 3, which means that its neutral atoms have three electrons. In the ground state, two electrons are placed in the 1s orbital with paired spins and the third electron is placed in the 2s orbital. A lithium atom has one unpaired electron.

Li has one unpaired electron

Li (3)   ⇅ ↑   Electron configuration: $1s^2 2s^1$
        1s 2s

Carbon (C)   Carbon, atomic number 6, has six electrons in its neutral atoms. Two electrons are placed in the 1s orbital with paired spins and two are placed in the 2s orbital with paired spins. The fifth and sixth electrons are placed one each in the $2p_x$ and $2p_y$ orbitals. The ground state of a carbon atom has two unpaired electrons.

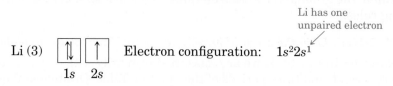

All orbitals of equal energy have at least one electron before any of them is filled

In a condensed electron configuration, orbitals of equal energy are grouped together

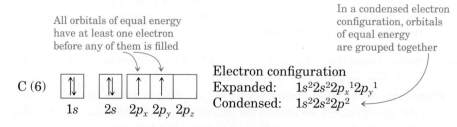

C (6)   ⇅ ⇅ ↑ ↑ ␣   Electron configuration
       1s 2s $2p_x$ $2p_y$ $2p_z$

Expanded: $1s^2 2s^2 2p_x^{1} 2p_y^{1}$
Condensed: $1s^2 2s^2 2p^2$

Table 2.7 Ground-State Electron Configurations of the First 18 Elements

	Orbital Box Diagram	Electron Configuration (Condensed)	Noble Gas Notation
	1s 2s 2px 2py 2pz 3s 3px 3py 3pz		
H (1)	↑	$1s^1$	
He (2)	↑↓	$1s^2$	
Li (3)	↑↓ ↑	$1s^2\,2s^1$	[He] $2s^1$
Be (4)	↑↓ ↑↓	$1s^2\,2s^2$	[He] $2s^2$
B (5)	↑↓ ↑↓ ↑	$1s^2\,2s^2\,2p^1$	[He] $2s^2\,2p^1$
C (6)	↑↓ ↑↓ ↑ ↑	$1s^2\,2s^2\,2p^2$	[He] $2s^2\,2p^2$
N (7)	↑↓ ↑↓ ↑ ↑ ↑	$1s^2\,2s^2\,2p^3$	[He] $2s^2\,2p^3$
O (8)	↑↓ ↑↓ ↑↓ ↑ ↑	$1s^2\,2s^2\,2p^4$	[He] $2s^2\,2p^4$
F (9)	↑↓ ↑↓ ↑↓ ↑↓ ↑	$1s^2\,2s^2\,2p^5$	[He] $2s^2\,2p^5$
Ne (10)	↑↓ ↑↓ ↑↓ ↑↓ ↑↓	$1s^2\,2s^2\,2p^6$	[He] $2s^2\,2p^6$
Na (11)	↑↓ ↑↓ ↑↓ ↑↓ ↑↓ ↑	$1s^2\,2s^2\,2p^6\,3s^1$	[Ne] $3s^1$
Mg (12)	↑↓ ↑↓ ↑↓ ↑↓ ↑↓ ↑↓	$1s^2\,2s^2\,2p^6\,3s^2$	[Ne] $3s^2$
Al (13)	↑↓ ↑↓ ↑↓ ↑↓ ↑↓ ↑↓ ↑	$1s^2\,2s^2\,2p^6\,3s^2\,3p^1$	[Ne] $3s^2\,3p^1$
Si (14)	↑↓ ↑↓ ↑↓ ↑↓ ↑↓ ↑↓ ↑ ↑	$1s^2\,2s^2\,2p^6\,3s^2\,3p^2$	[Ne] $3s^2\,3p^2$
P (15)	↑↓ ↑↓ ↑↓ ↑↓ ↑↓ ↑↓ ↑ ↑ ↑	$1s^2\,2s^2\,2p^6\,3s^2\,3p^3$	[Ne] $3s^2\,3p^3$
S (16)	↑↓ ↑↓ ↑↓ ↑↓ ↑↓ ↑↓ ↑↓ ↑ ↑	$1s^2\,2s^2\,2p^6\,3s^2\,3p^4$	[Ne] $3s^2\,3p^4$
Cl (17)	↑↓ ↑↓ ↑↓ ↑↓ ↑↓ ↑↓ ↑↓ ↑↓ ↑	$1s^2\,2s^2\,2p^6\,3s^2\,3p^5$	[Ne] $3s^2\,3p^5$
Ar (18)	↑↓ ↑↓ ↑↓ ↑↓ ↑↓ ↑↓ ↑↓ ↑↓ ↑↓	$1s^2\,2s^2\,2p^6\,3s^2\,3p^6$	[Ne] $3s^2\,3p^6$

Oxygen (O) Oxygen, atomic number 8, has eight electrons in its neutral atoms. The first four electrons fill the $1s$ and $2s$ orbitals. The next three electrons are placed in the $2p_x$, $2p_y$, and $2p_z$ orbitals so that each $2p$ orbital has one electron. The remaining electron now fills the $2p_x$ orbital. The ground state of an oxygen atom has two unpaired electrons.

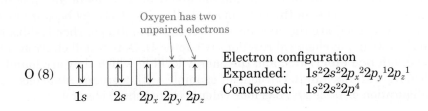

Oxygen has two unpaired electrons

Electron configuration
Expanded: $1s^2 2s^2 2p_x^2 2p_y^1 2p_z^1$
Condensed: $1s^2 2s^2 2p^4$

Neon (Ne) Neon, atomic number 10, has ten electrons in its neutral atoms, which completely fill all orbitals of the first and second shells. The ground state of a neon atom has no unpaired electrons.

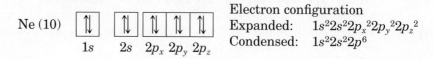

Ne (10)

1s 2s 2p_x 2p_y 2p_z

Electron configuration
Expanded: $1s^2 2s^2 2p_x{}^2 2p_y{}^2 2p_z{}^2$
Condensed: $1s^2 2s^2 2p^6$

Sodium (Na) Sodium, atomic number 11, has 11 electrons in its neutral atoms. The first 10 fill the 1s, 2s, and 2p orbitals. The 11th electron is placed in the 3s orbital. The ground state of a sodium atom has one unpaired electron.

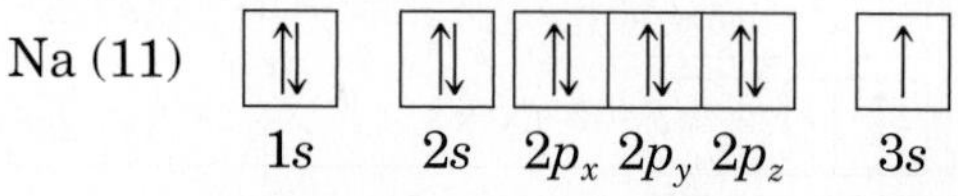

Na (11)

1s 2s 2p_x 2p_y 2p_z 3s

Electron configuration
Expanded: $1s^2 2s^2 2p_x{}^2 2p_y{}^2 2p_z{}^2 3s^1$
Condensed: $1s^2 2s^2 2p^6 3s^1$

Phosphorus (P) Phosphorus, atomic number 15, has 15 electrons in its neutral atoms. The first 12 fill the 1s, 2s, 2p, and 3s orbitals. Electrons 13, 14, and 15 are placed one each in the $3p_x$, $3p_y$, and $3p_z$ orbitals. The ground state of a phosphorus atom has three unpaired electrons.

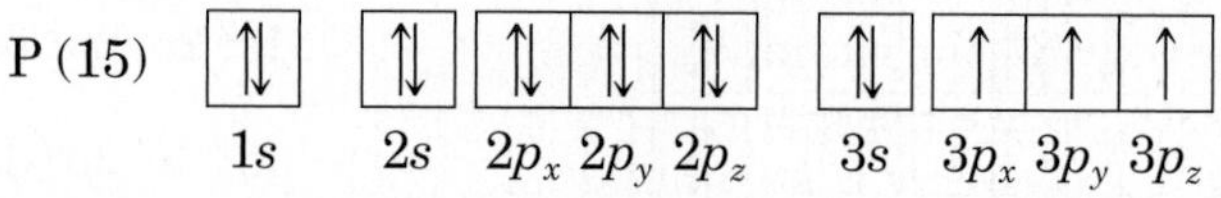

P (15)

1s 2s 2p_x 2p_y 2p_z 3s 3p_x 3p_y 3p_z

Electron configuration
Expanded: $1s^2 2s^2 2p_x{}^2 2p_y{}^2 2p_z{}^2 3s^2 3p_x{}^1 3p_y{}^1 3p_z{}^1$
Condensed: $1s^2 2s^2 2p^6 3s^2 3p^3$

E. Showing Electron Configurations: Noble Gas Notations

In an alternative way of writing ground-state electron configurations, we use the symbol of the noble gas immediately preceding the particular atom to indicate the electron configuration of all filled shells. The first shell of lithium, for example, is abbreviated [He] and the single electron in its 2s shell is indicated by $2s^1$. Thus, the electron configuration of a lithium atom is [He]$2s^1$ (right column of Table 2.7).

F. Showing Electron Configurations: Lewis Dot Structures

Valence electron An electron in the outermost occupied (valence) shell of an atom

Valence shell The outermost occupied shell of an atom

When discussing the physical and chemical properties of an element, chemists often focus on the outermost shell of its electrons because electrons in this shell are the ones involved in the formation of chemical bonds (Chapter 3) and in chemical reactions (Chapter 4). Outer-shell electrons are called **valence electrons,** and the energy level in which they are found is called the **valence shell.** Carbon, for example, with a ground-state electron configuration of $1s^2 2s^2 2p^2$, has four valence (outer-shell) electrons.

To show the outermost electrons of an atom, we commonly use a representation called a **Lewis dot structure,** named after the American chemist Gilbert N. Lewis (1875–1946), who devised this notation. A Lewis structure shows the symbol of the element surrounded by a number of dots equal to the number of electrons in the outer (valence) shell of an atom of that element. In a Lewis structure, the atomic symbol represents the nucleus and all filled inner shells. Table 2.8 shows Lewis structures for the first 18 elements of the Periodic Table.

Lewis dot structure The symbol of the element surrounded by a number of dots equal to the number of electrons in the valence shell of an atom of that element

Table 2.8 Lewis Dot Structures for Elements 1–18 of the Periodic Table

1A	2A	3A	4A	5A	6A	7A	8A
H·							He:
Li·	Be:	B:	·C:	·N:	:O:	:F:	:Ne:
Na·	Mg:	Al:	·Si:	·P:	:S:	:Cl·	:Ar:

Each dot represents one valence electron.

The noble gases helium and neon have filled valence shells. The valence shell of helium is filled with two electrons ($1s^2$); that of neon is filled with eight electrons ($2s^2 2p^6$). Neon and argon have in common an electron configuration in which the s and p orbitals of their valence shells are filled with eight electrons. The valence shells of all other elements shown in Table 2.8 contain fewer than eight electrons.

At this point, let us compare the Lewis structures given in Table 2.8 with the ground-state electron configurations given in Table 2.7. The Lewis structure of boron (B), for example, is shown in Table 2.8 with three valence electrons; these are the paired $2s$ electrons and the single $2p_x$ electron shown in Table 2.7. The Lewis structure of carbon (C) is shown in Table 2.8 with four valence electrons; these are the two paired $2s$ electrons and the unpaired $2p_x$ and $2p_y$ electrons shown in Table 2.7.

Example 2.7 Electron Configuration

The Lewis dot structure for nitrogen shows five valence electrons. Write the expanded electron configuration for nitrogen and show to which orbitals its five valence electrons are assigned.

Strategy

Locate nitrogen in the Periodic Table and determine its atomic number. In an electrically neutral atom, the number of negatively charged extranuclear electrons is the same as the number of positively charged protons in its nucleus. The order of filling of orbitals is $1s$ $2s$ $2p_x$ $2p_y$ $2p_z$ $3s$, etc.

Solution

Nitrogen, atomic number 7, has the following ground-state electron configuration:

$$1s^2 \, 2s^2 \, 2p_x^1 \, 2p_y^1 \, 2p_z^1$$

The five valence electrons of the Lewis dot structure are the two paired electrons in the $2s$ orbital and the three unpaired electrons in the $2p_x$, $2p_y$, and $2p_z$ orbitals.

Problem 2.7

Write the Lewis dot structure for the element that has the following ground-state electron configuration. What is the name of this element?

$$1s^2\, 2s^2\, 2p_x^2\, 2p_y^2\, 2p_z^2\, 3s^2\, 3p_x^1$$

2.7   How Are Electron Configuration and Position in the Periodic Table Related?

When Mendeleyev published his first Periodic Table in 1869, he could not explain why it worked—that is, why elements with similar properties became aligned in the same column. Indeed, no one had a good explanation for this phenomenon. It was not until the discovery of electron configurations that chemists finally understood why the Periodic Table works. The answer, they discovered, is very simple: Elements in the same column have the same ground-state electron configuration in their outer shells. Figure 2.15 shows the relationship between shells (principal energy levels) and orbitals being filled.

All main-group elements (those in A columns) have in common the fact that either their s or p orbitals are being filled. Notice that the $1s$ shell is filled with two electrons; there are only two elements in the first period. The $2s$ and $2p$ orbitals are filled with eight electrons; there are eight elements in period 2. Similarly, the $3s$ and $3p$ orbitals are filled with eight electrons; there are eight elements in period 3.

To create the elements of period 4, one $4s$, three $4p$, and five $3d$ orbitals are available. These orbitals can hold a total of 18 electrons; there are 18 elements in period 4. Similarly, there are 18 elements in period 5. Inner transition elements are created by filling f orbitals, which come in sets of seven and can hold a total of 14 electrons; there are 14 inner transition elements in the lanthanide series and 14 in the actinide series.

To see the similarities in electron configurations within the Periodic Table, let us look at the elements in column 1A. We already know the configurations for lithium, sodium, and potassium (Table 2.7). To this list we can add rubidium and cesium. All elements in column 1A have one electron in their valence shell (Table 2.9).

All Group 1A elements are metals, with the exception of hydrogen, which is a nonmetal. The properties of elements largely depend on the electron configuration of their outer shell. As a consequence, it is not surprising that

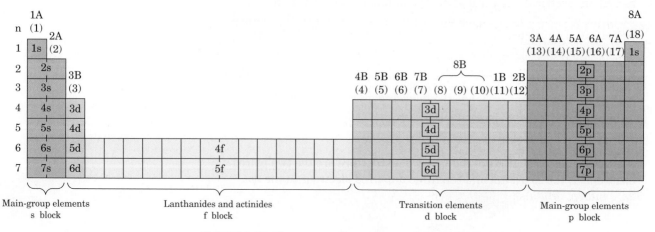

FIGURE 2.15 Electron configuration and the Periodic Table.

Table 2.9 Noble Gas Notation and Lewis Dot Structures for the Alkali Metals (Group 1A Elements)

Noble Gas Notation	Lewis Dot Structure
[He]$2s^1$	Li•
[Ne]$3s^1$	Na•
[Ar]$4s^1$	K•
[Kr]$5s^1$	Rb•
[Xe]$6s^1$	Cs•

Group 1A elements, all of which have the same outer-shell configuration, are metals (except for hydrogen) and have such similar physical and chemical properties.

2.8 What Is a Periodic Property?

As we have now seen, the Periodic Table originally was constructed on the basis of trends (periodicity) in physical and chemical properties. With an understanding of electron configurations, chemists realized that the periodicity in chemical properties could be understood in terms of the periodicity in ground-state electron configuration. As we noted in the opening of Section 2.7, The Periodic Table works because "elements in the same column have the same ground-state electron configuration in their outer shells." Thus, chemists could now explain why certain chemical and physical properties of elements changed in predictable ways in going down a column or going across a row of the Periodic Table, In this section, we will examine the periodicity of one physical property (atomic size) and one chemical property (ionization energy) to illustrate how periodicity is related to position in the Periodic Table.

A. Atomic Size

The size of an atom is determined by the size of its outermost occupied orbital. The size of a sodium atom, for example, is the size of its singly occupied $3s$ orbital. The size of a chlorine atom is determined by the size of its three $3p$ orbitals ($3s^2 3p^5$). The simplest way to determine the size of an atom is to determine the distance between bonded nuclei in a sample of the element. A chlorine molecule, for example, has a diameter of 198 pm (pm = picometer; 1 pm = 10^{-12} meter). The radius of a chlorine atom is thus 99 pm, which is one-half of the distance between two bonded chlorine nuclei in Cl_2.

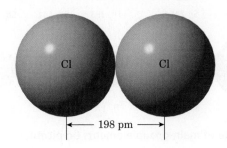

Similarly, the distance between bonded carbon nuclei in diamond is 154 pm, and so the radius of a carbon atom is 77 pm.

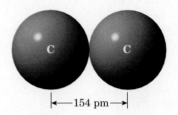

From measurements such as these, we can assemble a set of atomic radii (Figure 2.16).

From the information in this figure, we can see that for main group elements, (1) atomic radii increase going down a group and (2) decrease going from left to right across a period. Let us examine the correlation between each of these trends and ground-state electron configuration.

1. The size of an atom is determined by the size of its outermost electrons. In going down a column, the outermost electrons are assigned to higher and higher principal energy levels. The electrons of lower principal energy levels (those lying below the valence shell) must occupy some space, so the outer-shell electrons must be farther and farther from the nucleus, which rationalizes the increase in size in going down a column.

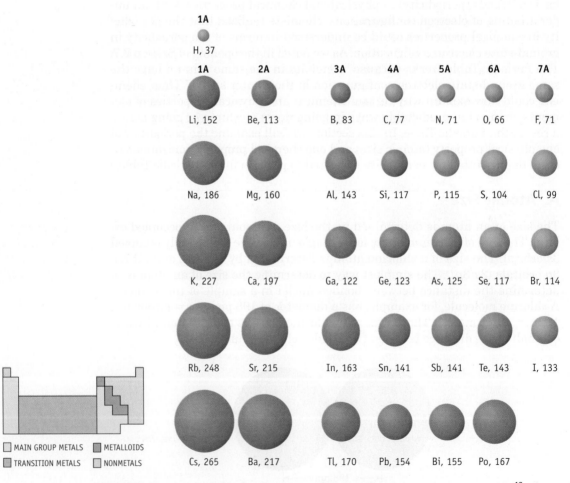

FIGURE 2.16 Atomic radii of main-group elements (in picometers, 1 pm = 10^{-12} m).

2. For elements in the same period, the principal energy level remains the same (for example, the valence electrons of all second-period elements occupy the second principal energy level). But in going from one element to the next across a period, one more proton is added to the nucleus, thus increasing the nuclear charge by one unit for each step from left to right. The result is that the nucleus exerts an increasingly stronger pull on the valence electrons and atomic radius decreases.

B. Ionization energy

Atoms are electrically neutral—the number of electrons outside the nucleus of an atom is equal to the number of protons inside the nucleus. Atoms do not normally lose or gain protons or neutrons, but they can lose or gain electrons. When a lithium atom, for example, loses one electron, it becomes a lithium **ion.** A lithium atom has three protons in its nucleus and three electrons outside the nucleus. When a lithium atom loses one of these electrons, it still has three protons in its nucleus (and, therefore, is still lithium), but now it has only two electrons outside the nucleus. The two remaining electrons cancel the charge of two of the protons, but there is no third electron to cancel the charge of the third proton. Therefore, a lithium ion has a charge of $+1$ and we write it as Li^+. The ionization energy for a lithium ion in the gas phase is 0.52 kJ/mol.

Ion An atom with an unequal number of protons and electrons

$$Li \ + \ energy \longrightarrow Li^+ \ + \ e^-$$

Lithium   Ionization   Lithium   Electron

atom   energy   ion

Ionization energy is a measure of how difficult it is to remove the most loosely held electron from an atom in the gaseous state. The more difficult it is to remove the electron, the higher the ionization energy. Ionization energies are always positive because energy must be supplied to overcome the attractive force between the electron and the positively charged nucleus. Figure 2.17 shows the ionization energies for the atoms of main-group elements 1 through 37 (hydrogen through rubidium).

Ionization energy The energy required to remove the most loosely held electron from an atom in the gas phase

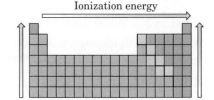

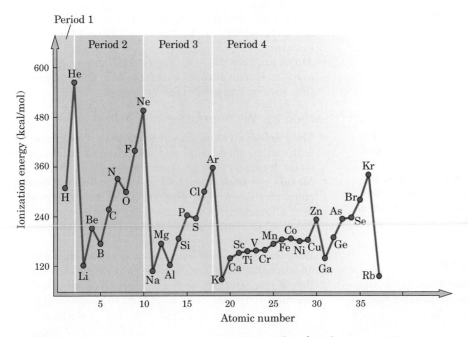

FIGURE 2.17 Ionization energy versus atomic number for elements 1–37.

As we see in Figure 2.17, ionization energy generally increases as we go up a column of the Periodic Table and, with a few exceptions, generally increases as we go from left to right across a row. For example, within the Group 1A metals, rubidium gives up its $5s$ electron most easily and lithium gives up its $2s$ electron least easily.

We explain this trend by saying that the $5s$ electron of rubidium is farther from the positively charged nucleus than is the $4s$ electron in potassium, which in turn is farther from the positively charged nucleus than is the $3s$ electron of sodium, and so forth. Furthermore, the $5s$ electron of rubidium is more "shielded" by inner-shell electrons from the attractive force of the positive nucleus than is the $4s$ electron of potassium, and so forth. The greater the shielding, the lower the ionization energy. Thus, going down a column of the Periodic Table, the shielding of an atom's outermost electrons increases and the element's ionization energies decrease.

We explain the increase in ionization energy across a row by the fact that the valence electrons across a row are in the same shell (principal energy level). Because the number of protons in the nucleus increases regularly across a row, the valence electrons experience an increasingly stronger pull by the nucleus, which makes them more difficult to remove. Thus, ionization energy increases from left to right across a row of the Periodic Table.

Summary

OWL Sign in at **www.cengage.com/owl** to develop problem-solving skills and complete online homework assigned by your professor.

Section 2.1 What Is Matter Made Of?

- The Greek philosopher Democritus (circa 460–370 BCE) was the first person to propose an atomic theory of matter. He stated that all matter is made of very tiny particles, which he called atoms.

Section 2.2 How Do We Classify Matter?

- We classify matter as **elements, compounds, or mixtures.**

Section 2.3 What Are the Postulates of Dalton's Atomic Theory?

- (1) All matter is made up of atoms; (2) all atoms of a given element are identical, and the atoms of any one element are different from those of any other element; (3) compounds are formed by the chemical combination of atoms; and (4) a molecule is a cluster of two or more atoms that acts as a single unit.
- Dalton's theory is based on the **law of conservation of mass** (matter can be neither created nor destroyed) and the **law of constant composition** (any compound is always made up of elements in the same proportion by mass).

Section 2.4 What Are Atoms Made Of?

- Atoms consist of protons and neutrons found inside the nucleus and electrons located outside it. An **electron** has a mass of approximately 0.0005 amu and a charge of -1. A **proton** has a mass of approximately 1 amu and a charge of $+1$. A **neutron** has a mass of approximately 1 amu and no charge.
- The **mass number** of an atom is the sum of the number of its protons and neutrons.
- The **atomic number** of an element is the number of protons in the nucleus of an atom of that element.
- **Isotopes** are atoms with the same atomic number but different mass numbers; that is, they have the same number of protons but different numbers of neutrons in their nuclei.
- The **atomic weight** of an element is a weighted average of the masses (in amu) of its isotopes as they occur in nature.
- Atoms are very tiny, with a very small mass, almost all of which is concentrated in the nucleus. The nucleus is tiny, with an extremely high density.

Section 2.5 What Is the Periodic Table?

- The **Periodic Table** is an arrangement of elements with similar chemical properties into columns; the properties gradually change as we move down a column.
- **Metals** are solids (except for mercury, which is a liquid), shiny, conductors of electricity, ductile, malleable, and form alloys, which are solutions of one or more metals dissolved in another metal. In their chemical reactions, metals tend to give up electrons.
- With the exception of hydrogen, the **nonmetals** appear on the right side of the Periodic Table. With the exception of graphite, they do not conduct electricity. In their chemical reactions, nonmetals tend to accept electrons.

- Six elements are classified as **metalloids:** boron, silicon, germanium, arsenic, antimony, and tellurium. These elements have some properties of metals and some properties of nonmetals.

Section 2.6 How Are the Electrons in an Atom Arranged?

- Electrons in atoms exist in **principal energy levels** or **shells.**
- All principal energy levels except the first are divided into **subshells** designated by the letters *s, p, d,* and *f.* Within each subshell, electrons are grouped into **orbitals.** An orbital is a region of space that can hold two electrons with paired spins. All *s* orbitals are spherical and can hold two electrons. All *p* orbitals come in sets of three, and each is shaped like a dumbbell, with the nucleus at the center of the dumbbell. A set of three *p* orbitals can hold six electrons. A set of five *d* orbitals can hold ten electrons, and a set of 7 *f* orbitals can hold fourteen electrons.
- Electrons are arranged in orbitals according to the following rules.
  (1) Orbitals fill in order of increasing energy; (2) each orbital can hold a maximum of two electrons with paired spins; (3) when filling orbitals of equivalent energy, each orbital adds one electron before any orbital adds a second electron.
- The electron configuration of an atom may be shown by an orbital notation, an orbital box diagram, or a noble gas notation.
- Electrons in the outermost or **valence shell** of an atom are called **valence electrons.** In a **Lewis dot** structure of an atom, the symbol of the element is surrounded by a number of dots equal to the number of its valence electrons.

Section 2.7 How Are Electron Configuration and Position in the Periodic Table Related?

- The Periodic Table works because elements in the same column have the same outer-shell electron configuration.

Section 2.8 What Is a Periodic Property?

- **Ionization energy** is the energy necessary to remove the most loosely held electron from an atom in the gas phase to form an **ion.** Ionization energy increases from bottom to top within a column of the Periodic Table because the valence shell of the atom becomes closer to the positively charged nucleus. It increases from left to right within a row because the positive charge on the nucleus increases in this direction.
- The **size of an atom (atomic radius)** is determined by the size of its outermost occupied orbital. Atomic size is a periodic property. For main-group elements, atomic size increases going down a group and decreases going from left to right across a period. In going down a column, the outermost electrons are assigned to higher and higher principal energy levels. For elements in the same period, the principal energy level remains the same from one element to the next, but the nuclear charge increases by one unit (by one proton). As a result of this increase in nuclear charge across a period, the nucleus exerts an increasingly stronger pull on the valence electrons and atomic size decreases.

Problems

⌾WL Interactive versions of these problems may be assigned in OWL.

Orange-numbered problems are applied.

Section 2.1 What Are Atoms Made Of?

2.8 In what way(s) was Democritus's atomic theory similar to that of Dalton's atomic theory?

Section 2.2 How Do We Classify Matter?

2.9 Answer true or false.
  (a) Matter is divided into elements and pure substances.
  (b) Matter is anything that has mass and volume (occupies space).
  (c) A mixture is composed of two or more pure substances.
  (d) An element is a pure substance.
  (e) A heterogeneous mixture can be separated into pure substances, but a homogeneous mixture cannot.
  (f) A compound consists of elements combined in a fixed ratio.
  (g) A compound is a pure substance.
  (h) All matter has mass.
  (i) All of the 118 known elements occur naturally on Earth.
  (j) The first six elements in the Periodic Table are the most important for human life.
  (k) The combining ratio of a compound tells you how many atoms of each element are combined in the compound.
  (l) The combining ratio of 1:2 in the compound CO_2 tells you that this compound is formed by the combination of one gram of carbon is with two grams of oxygen.

2.10 Classify each of the following as an element, a compound, or a mixture:
  (a) Oxygen   (b) Table salt
  (c) Sea water   (d) Wine
  (e) Air   (f) Silver
  (g) Diamond   (h) A pebble
  (i) Gasoline   (j) Milk
  (k) Carbon dioxide (l) Bronze

2.11 Name these elements (try not to look at a Periodic Table):

   (a) O   (b) Pb   (c) Ca   (d) Na

   (e) C   (f) Ti   (g) S   (h) Fe

   (i) H   (j) K   (k) Ag   (l) Au

2.12 The elements game, Part I. Name and give the symbol of the element that is named for each person.

   (a) Niels Bohr (1885–1962), Nobel Prize for physics in 1922

   (b) Pierre and Marie Curie, Nobel Prize for chemistry in 1903

   (c) Albert Einstein (1879–1955), Nobel Prize for physics in 1921

   (d) Enrico Fermi (1901–1954), Nobel Prize for physics in 1938

   (e) Ernest Lawrence (1901–1958), Nobel Prize for physics in 1939

   (f) Lise Meitner (1868–1968), codiscoverer of nuclear fission

   (g) Dmitri Mendeleyev (1834–1907), first person to formulate a workable Periodic Table

   (h) Alfred Nobel (1833–1896), discoverer of dynamite

   (i) Ernest Rutherford (1871–1937), Nobel Prize for chemistry in 1908

   (j) Glen Seaborg (1912–1999), Nobel Prize for chemistry in 1951

2.13 The elements game, Part II. Name and give the symbol of the element that is named for each geographic location.

   (a) The Americas

   (b) Berkeley, California

   (c) The state and University of California

   (d) Dubna, location in Russia of the Joint Institute of Nuclear Research

   (e) Europe

   (f) France

   (g) Gallia, the Latin name for ancient France

   (h) Germany

   (i) Hafnia, the Latin name for ancient Copenhagen

   (j) Hesse, a German state

   (k) Holmia, the Latin name for ancient Stockholm

   (l) Lutetia, the Latin name for ancient Paris

  (m) Magnesia, a district in Thessaly

   (n) Poland, the native country of Marie Curie

   (o) Rhenus, the Latin name for the river Rhine

   (p) Ruthenia, the Latin name for ancient Russia

   (q) Scandia, the Latin name for ancient Scandinavia

   (r) Strontian, a town in Scotland

   (s) Ytterby, a village in Sweden (three elements)

   (t) Thule, the earliest name for Scandinavia

2.14 The elements game, Part III. Give the names and symbols for the two elements named for planets. Note that the element plutonium was named for Pluto, which is no longer classified as a planet.

2.15 Write the formulas of compounds in which the combining ratios are as follows:

   (a) Potassium: oxygen, 2:1

   (b) Sodium: phosphorus: oxygen, 3:1:4

   (c) Lithium: nitrogen: oxygen, 1:1:3

2.16 Write the formulas of compounds in which the combining ratios are as follows:

   (a) Sodium: hydrogen: carbon: oxygen, 1:1:1:3

   (b) Carbon: hydrogen: oxygen, 2:6:1

   (c) Potassium: manganese: oxygen, 1:1:4

Section 2.3 What Are Postulates of Dalton's Atomic Theory?

2.17 How does Dalton's atomic theory explain:

   (a) the law of conservation of mass?

   (b) the law of constant composition?

2.18 When 2.16 g of mercuric oxide is heated, it decomposes to yield 2.00 g of mercury and 0.16 g of oxygen. Which law is supported by this experiment?

2.19 The compound carbon monoxide contains 42.9% carbon and 57.1% oxygen. The compound carbon dioxide contains 27.3% carbon and 72.7% oxygen. Does this disprove Proust's law of constant composition?

2.20 Calculate the percentage of hydrogen and oxygen in water, H_2O, and hydrogen peroxide, H_2O_2.

Section 2.4 What Are Atoms Made Of?

2.21 Answer true or false.

   (a) A proton and an electron have the same mass but opposite charges.

   (b) The mass of an electron is considerably smaller than that of a neutron.

   (c) An atomic mass unit (amu) is a unit of mass.

   (d) One amu is equal to 1 gram.

   (e) The protons and neutrons of an atom are found in the nucleus.

   (f) The electrons of an atom are found in the space surrounding the nucleus.

   (g) All atoms of the same element have the same number of protons.

   (h) All atoms of the same element have the same number of electrons.

   (i) Electrons and protons repel each other.

   (j) The size of an atom is approximately the size of its nucleus.

   (k) The mass number of an atom is the sum of the numbers of protons and neutrons in the nucleus of that atom.

   (l) For most atoms, their mass number is the same as their atomic number.

  (m) The three isotopes of hydrogen (hydrogen-1, hydrogen-2, and hydrogen-3) differ only in the number of neutrons in the nucleus.

   (n) Hydrogen-1 has one neutron in its nucleus, hydrogen-2 has two neutrons in its nucleus, and hydrogen-3 has three neutrons.

(o) All isotopes of an element have the same number of electrons.

(p) Most elements found on Earth are mixtures of isotopes.

(q) The atomic weight of an element given in the Periodic Table is the weighted average of the masses of its isotopes found on Earth.

(r) The atomic weights of most elements are whole numbers.

(s) Most of the mass of an atom is found in its nucleus.

(t) The density of a nucleus is its mass number expressed in grams.

2.22 Where in an atom are these subatomic particles located?

(a) Protons   (b) Electrons   (c) Neutrons

2.23 It has been said, "The number of protons determines the identity of the element." Do you agree or disagree with this statement? Explain.

2.24 What is the mass number of an atom with:

(a) 22 protons, 22 electrons, and 26 neutrons?

(b) 76 protons, 76 electrons, and 114 neutrons?

(c) 34 protons, 34 electrons, and 45 neutrons?

(d) 94 protons, 94 electrons, and 150 neutrons?

2.25 Name and give the symbol for each element in Problem 2.24.

2.26 Given these mass numbers and number of neutrons, what is the name and symbol of each element?

(a) Mass number 45; 24 neutrons

(b) Mass number 48; 26 neutrons

(c) Mass number 107; 60 neutrons

(d) Mass number 246; 156 neutrons

(e) Mass number 36; 18 neutrons

2.27 If each atom in Problem 2.26 acquired two more neutrons, what element would each then be?

2.28 How many neutrons are in:

(a) a carbon atom of mass number 13?

(b) a germanium atom of mass number 73?

(c) an osmium atom of mass number 188?

(d) a platinum atom of mass number 195?

2.29 How many protons and how many neutrons does each of these isotopes of radon contain?

(a) Rn-210   (b) Rn-218   (c) Rn-222

2.30 How many neutrons and protons are in each isotope?

(a) ^{22}Ne   (b) ^{104}Pd

(c) ^{35}Cl   (d) Tellurium-128

(e) Lithium-7   (f) Uranium-238

2.31 Tin-118 is one of the isotopes of tin. Name the isotopes of tin that contain two, three, and six more neutrons than tin-118.

2.32 What is the difference between atomic number and mass number?

2.33 Define:

(a) Ion   (b) Isotope

2.34 There are only two naturally occurring isotopes of antimony: ^{121}Sb (120.90 amu) and ^{123}Sb (122.90 amu). The atomic weight of antimony given in the Periodic Table is 121.75. Which of the two isotopes has the greater natural abundance?

2.35 The two most abundant naturally occurring isotopes of carbon are carbon-12 (98.90%, 12.000 amu) and carbon-13 (1.10%, 13.003 amu). From these abundances, calculate the atomic weight of carbon and compare your calculated value with that given in the Periodic Table.

2.36 Another isotope of carbon, carbon-14, occurs in nature but in such small amounts relative to carbon-12 and carbon-13 that it does not contribute to the atomic weight of carbon as recorded in the Periodic Table. Carbon-14 is invaluable in the science of radiocarbon dating (see Chemical Connections 9A). Give the number of protons, neutrons, and electrons in an atom of carbon-14.

2.37 The isotope carbon-11 does not occur in nature but has been made in the laboratory. This isotope is used in a medical imaging technique called positron emission tomography (PET, see Section 9.7A). Give the number of protons, neutrons, and electrons in an atom of carbon-11.

2.38 Other isotopes used in PET imaging are fluorine-18, nitrogen-13, and oxygen-15. None of these isotopes occurs in nature; all must be produced in the laboratory. Give the number of protons, neutrons, and electrons in an atom of each of these artificial isotopes.

2.39 Americium-241 is used in household smoke detectors. This element has 11 known isotopes, none of which occurs in nature, but must be made in the laboratory. Give the number of protons, neutrons, and electrons in an atom of americium-241.

2.40 In dating geological samples, scientists compare the ratio of rubidium-87 to strontium-87. Give the number of protons, neutrons, and electrons in an atom of each element.

Section 2.5 What Is the Periodic Table?

2.41 Answer true or false.

(a) Mendeleyev discovered that when elements are arranged in order of increasing atomic weight, certain sets of properties recur periodically.

(b) Main-group elements are those in the columns 3A to 8A of the Periodic Table.

(c) Nonmetals are found at the top of the Periodic Table, metalloids in the middle, and metals at the bottom.

(d) Among the 118 known elements, there are approximately equal numbers of metals and nonmetals.

(e) A horizontal row in the Periodic Table is called a group.

(f) The Group 1A elements are called the "alkali metals."

(g) The alkali metals react with water to give hydrogen gas and a metal hydroxide, MOH, where "M" is the metal.

(h) The halogens are Group 7A elements.

(i) The boiling points of noble gases (Group 8A elements) increase going from top to bottom of the column.

2.42 How many metals, metalloids, and nonmetals are there in the third period of the Periodic Table?

2.43 Which group(s) of the Periodic Table contain(s):

(a) Only metals?   (b) Only metalloids?

(c) Only nonmetals?

2.44 Which period(s) in the Periodic Table contain(s) more nonmetals than metals? Which contain(s) more metals than nonmetals?

2.45 Group the following elements according to similar properties (look at the Periodic Table): As, I, Ne, F, Mg, K, Ca, Ba, Li, He, N, P.

2.46 Which are transition elements?

(a) Pd   (b) K   (c) Co

(d) Ce   (e) Br   (f) Cr

2.47 Which element in each pair is more metallic?

(a) Silicon or aluminum   (b) Arsenic or phosphorus

(c) Gallium or germanium   (d) Gallium or aluminum

2.48 Classify these elements as metals, nonmetals, or metalloids:

(a) Argon   (b) Boron   (c) Lead

(d) Arsenic   (e) Potassium   (f) Silicon

(g) Iodine   (h) Antimony   (i) Vanadium

(j) Sulfur   (k) Nitrogen

Section 2.6 How Are the Electrons in an Atom Arranged?

2.49 Answer true or false.

(a) To say that "energy is quantized" means that only certain energy values are allowed.

(b) Bohr discovered that the energy of an electron in an atom is quantized.

(c) Electrons in atoms are confined to regions of space called "principal energy levels."

(d) Each principal energy level can hold a maximum of two electrons.

(e) An electron in a $1s$ orbital is held closer to the nucleus than an electron in a $2s$ orbital.

(f) An electron in a $2s$ orbital is harder to remove from an atom than an electron in a $1s$ orbital.

(g) An s orbital has the shape of a sphere, with the nucleus at the center of the sphere.

(h) Each $2p$ orbital has the shape of a dumbbell, with the nucleus at the midpoint of the dumbbell.

(i) The three $2p$ orbitals in an atom are aligned parallel to each other.

(j) An orbital is a region of space that can hold two electrons.

(k) The second shell contains one s orbital and three p orbitals.

(l) In the ground-state electron configuration of an atom, only the lowest-energy orbitals are occupied.

(m) A spinning electron behaves as a tiny bar magnet, with a North Pole and a South Pole.

(n) An orbital can hold a maximum of two electrons with their spins paired.

(o) Paired electron spins means that the two electrons are aligned with their spins North Pole to North Pole and South Pole to South Pole.

(p) An orbital box diagram puts all of the electrons of an atom in one box with their spins aligned.

(q) An orbital box diagram of a carbon atom shows two unpaired electrons.

(r) A Lewis dot structure shows only the electrons in the valence shell of an atom of the element.

(s) A characteristic of Group 1A elements is that each has one unpaired electron in its outermost occupied (valence) shell.

(t) A characteristic of Group 6A elements is that each has six unpaired electrons in its valence shell.

2.50 How many periods of the Periodic Table have two elements? How many have eight elements? How many have 18 elements? How many have 32 elements?

2.51 What is the correlation between the group number of the main-group elements (those in the A columns of the Mendeleyev system) and the number of valence electrons in an element in the group?

2.52 Given your answer to Problem 2.51, write the Lewis dot structure for each of the following elements using no information other than the number of the group in the Periodic Table to which the element belongs.

(a) Carbon (4A)   (b) Silicon (4A)

(c) Oxygen (6A)   (d) Sulfur (6A)

(e) Aluminum (3A)   (f) Bromine (7A)

2.53 Write the condensed ground-state electron configuration for each of the following elements. The element's atomic number is given in parentheses.

(a) Li (3)   (b) Ne (10)   (c) Be (4)

(d) C (6)   (e) Mg (12)

2.54 Write the Lewis dot structure for each element in Problem 2.53.

2.55 Write the condensed ground-state electron configuration for each of the following elements. The element's atomic number is given in parentheses.

(a) He (2)   (b) Na (11)   (c) Cl (17)

(d) P (15)   (e) H (1)

2.56 Write the Lewis dot structure for each element in Problem 2.55.

2.57 What is the same and what is different in the electron configurations of:

(a) Na and Cs?   (b) O and Te?   (c) C and Ge?

2.58 Silicon, atomic number 14, is in Group 4A. How many orbitals are occupied by the valence electrons of Si in its ground state?

2.59 You are presented with a Lewis dot structure of element X as X:. To which two groups in the Periodic Table might this element belong?

2.60 The electron configurations for the elements with atomic numbers higher than 36 follow the same rules as given in the text for the first 36 elements. In fact, you can arrive at the correct order of filling of orbitals from Figure 2.15 by starting with H and reading the orbitals from left to right across the first row, then the second row, and so on. Write the condensed ground-state electron configuration for:

(a) Rb   (b) Sr   (c) Br

Section 2.7 How Are Electron Configuration and Position in the Periodic Table Related?

2.61 Answer true or false.

(a) Elements in the same column of the Periodic Table have the same outer-shell electron configuration.

(b) All Group 1A elements have one electron in their valence shell.

(c) All Group 6A elements have six electrons in their valence shell.

(d) All Group 8A elements have eight electrons in their valence shell.

(e) Period 1 of the Periodic Table has one element, period 2 has two elements, period 3 has three elements, and so forth.

(f) Period 2 results from filling the 2s and 2p orbitals, and therefore, there are eight elements in period 2.

(g) Period 3 results from filling the 3s, 3p, and 3d orbitals, and therefore, there are nine elements in period 3.

(h) The main-group elements are s block and p block elements.

2.62 Why do the elements in column 1A of the Periodic Table (the alkali metals) have similar but not identical properties?

Section 2.8 What Is a Periodic Property?

2.63 Answer true or false.

(a) Ionization energy is the energy required to remove the most loosely held electron from an atom in the gas phase.

(b) When an atom loses an electron, it becomes a positively charged ion.

(c) Ionization energy is a periodic property because ground-state electron configuration is a periodic property.

(d) Ionization energy generally increases going from left to right across a period of the Periodic Table.

(e) Ionization energy generally increases in going from top to bottom within a column in the Periodic Table.

(f) The sign of an ionization energy is always positive; the process is always endothermic.

2.64 Consider the elements B, C, and N. Using only the Periodic Table, predict which of these three elements has:

(a) the largest atomic radius.

(b) the smallest atomic radius.

(c) the largest ionization energy.

(d) the smallest ionization energy.

2.65 Account for the following observations.

(a) The atomic radius of an anion is always larger than that of the atom from which it is derived. Examples: Cl 99 pm and Cl⁻ 181 pm; O 73 pm and O²⁻ 140 pm.

(b) The atomic radius of a cation is always smaller than that of the atom from which it is derived. Examples: Li 152 pm and Li⁺ 76 pm; Na 156 pm and Na⁺ 98 pm.

2.66 Using only the Periodic Table, arrange the elements in each set in order of increasing ionization energy:

(a) Li, Na, K   (b) C, N, Ne

(c) O, C, F   (d) Br, Cl, F

2.67 Account for the fact that the first ionization energy of oxygen is less than that of nitrogen.

2.68 Every atom except hydrogen has a series of ionization energies (IE) because they have more than one electron that can be removed. Following are the first three ionization energies for magnesium:

$$Mg(g) \longrightarrow Mg^+(g) + e^-(g) \quad IE_1 = 738 \text{ kJ/mol}$$
$$Mg^+(g) \longrightarrow Mg^{2+}(g) + e^-(g) \quad IE_2 = 1450 \text{ kJ/mol}$$
$$Mg^{2+}(g) \longrightarrow Mg^{3+}(g) + e^-(g) \quad IE_3 = 7734 \text{ kJ/mol}$$

(a) Write the ground-state electron configuration for Mg, Mg⁺, Mg²⁺, and Mg³⁺.

(b) Account for the large increase in ionization energy for the removal of the third electron compared with the ionization energies for removal of the first and second electrons.

Chemical Connections

2.69 (Chemical Connections 2A) Why does the body need sulfur, calcium, and iron?

2.70 (Chemical Connections 2B) Which are the two most abundant elements, by weight, in:

(a) the Earth's crust?   (b) the human body?

2.71 (Chemical Connections 2C) Why is strontium-90 more dangerous to humans than most other radioactive isotopes that were present in the Chernobyl fallout?

2.72 (Chemical Connections 2D) Bronze is an alloy of which two metals?

2.73 (Chemical Connections 2D) Copper is a soft metal. How can it be made harder?

Additional Problems

2.74 Give the designations of all subshells in the:

(a) 1 shell   (b) 2 shell

(c) 3 shell   (d) 4 shell

2.75 Tell whether metals or nonmetals are more likely to have each of the following characteristics:

(a) Conduct electricity and heat

(b) Accept electrons

(c) Be malleable

(d) Be a gas at room temperature

(e) Be a transition element

(f) Lose electrons

2.76 Explain why:

(a) atomic radius decreases going across a period in the Periodic Table.

(b) energy is required to remove an electron from an atom.

2.77 Name and give the symbol of the element with the given characteristic.

(a) Largest atomic radius in Group 2A.

(b) Smallest atomic radius in Group 2A.

(c) Largest atomic radius in the second period.

(d) Smallest atomic radius in the second period.

(e) Largest ionization energy in Group 7A.

(f) Lowest ionization energy in Group 7A.

2.78 What is the outer-shell electron configuration of the elements in:

(a) Group 3A?   (b) Group 7A?

(c) Group 5A?

2.79 Determine the number of protons, electrons, and neutrons present in:

(a) ^{32}P (b) ^{98}Mo (c) ^{44}Ca

(d) ^{3}H (e) ^{158}Gd (f) ^{212}Bi

2.80 What percentage of the mass of each element do neutrons contribute?

(a) Carbon-12   (b) Calcium-40

(c) Iron-55   (d) Bromine-79

(e) Platinum-195   (f) Uranium-238

2.81 Do isotopes of the heavy elements (for example, those from atomic number 37 to 53) contain more, the same, or fewer neutrons than protons?

2.82 What is the symbol for each of the following elements? (Try not to look at a Periodic Table.)

(a) Phosphorus   (b) Potassium

(c) Sodium   (d) Nitrogen

(e) Bromine   (f) Silver

(g) Calcium   (h) Carbon

(i) Tin   (j) Zinc

2.83 The natural abundance of boron isotopes is as follows: 19.9% boron-10 (10.013 amu) and 80.1% boron-11 (11.009 amu). Calculate the atomic weight of boron (watch the significant figures) and compare your calculated value with that given in the Periodic Table.

2.84 How many electrons are in the outer shell of each of the following elements?

(a) Si   (b) Br

(c) P   (d) K

(e) He   (f) Ca

(g) Kr   (h) Pb

(i) Se   (j) O

2.85 The mass of a proton is 1.67×10^{-24} g. The mass of a grain of salt is 1.0×10^{-2} g. How many protons would it take to have the same mass as a grain of salt?

2.86 (a) What are the charges of an electron, a proton, and a neutron?

(b) What are the masses (in amu, to one significant figure) of an electron, a proton, and a neutron?

2.87 What is the name of this element, and how many protons and neutrons does this isotope have in its nucleus: $^{131}_{54}X$?

2.88 Based on the data presented in Figure 2.16, which atom would have the highest ionization energy: I, Cs, Sn, or Xe?

2.89 Assume that a new element has been discovered with atomic number 117. Its chemical properties should be similar to those of astatine (At). Predict whether the new element's ionization energy will be greater than, the same as, or smaller than that of:

(a) At   (b) Ra

2.90 Explain why the sizes of atoms change when proceeding across a period of the Periodic Table.

2.91 These are the first two ionization energy for lithium:

$Li(g) \longrightarrow Li^+(g) + e^-(g)$

Ionization energy = 523 kJ/mol

$Li^+(g) \longrightarrow Li^{2+}(g) + e^-(g)$

Ionization energy = 7298 kJ/mol

(a) Explain the large increase in ionization energy that occurs for the removal of the second electron.

(b) The radius of Li^+ is 78 pm (1 pm = 10^{-12} m) while that of a lithium atom, Li, is 152 pm. Explain why the radius of Li^+ is so much smaller than the radius of Li.

2.92 Which has the largest radius: O^{2-}, F^- or F? Explain your reasoning.

2.93 Arrange the following elements in order of increasing size: Al, B, C, and Na. Try doing it without looking at Figure 2.16 and then check yourself by looking at the figure.

2.94 Using your knowledge of trends in element sizes in going across a period of the Periodic Table, explain why the density of the elements increases from potassium through vanadium. (Recall from Section 1.7

that specific gravity is numerically the same as density but has no units.)

Element	Specific Gravity
K	0.862
Ca	1.55
Se	2.99
Ti	4.54
V	6.11

2.95 Name the elements in Group 3A. What does the group designation tell you about the electron configuration of these elements?

2.96 Using the orbital box diagrams and the noble gas notation, write the electron configuration for each atom and ion.

(a) Ti   (b) Ti^{2+} (c) Ti^{4+}

2.97 Explain why the Ca^{3+} ion is not found in chemical compounds.

2.98 Explain how the ionization energy of atoms changes when proceeding down a group of the Periodic Table and explain why this change occurs.

Looking Ahead

2.99 Suppose that you face a problem similar to Mendeleyev: You must predict the properties of an element not yet discovered. What will element 118 be like if and when enough of it is made for chemists to study its physical and chemical properties?

2.100 Compare the neutron to proton ratio for the heavier and lighter elements. Does the value of this ratio generally increase, decrease, or remain the same as atomic number increases?

3

Chemical Bonds

Sign in to OWL at **www.cengage.com/owl** to view tutorials and simulations, develop problem-solving skills, and complete online homework assigned by your professor.

Sodium chloride crystal

Charles D. Winters

3.1 What Do We Need to Know Before We Begin?

In Chapter 2, we stated that compounds are tightly bound groups of atoms. In this chapter, we will see that the atoms in compounds are held together by powerful forces of attraction called chemical bonds. There are two main types: ionic bonds and covalent bonds. We begin by examining ionic bonds. To talk about ionic bonds, however, we must first discuss why atoms form the ions they do.

3.2 What Is the Octet Rule?

In 1916, Gilbert N. Lewis (Section 2.6) devised a beautifully simple model that unified many of the observations about chemical bonding and chemical reactions. He pointed out that the lack of chemical reactivity of the noble gases (Group 8A) indicates a high degree of stability of their electron configurations: helium with a filled valence shell of two electrons ($1s^2$), neon with a filled valence shell of eight electrons ($2s^2 2p^6$), argon with a valence shell of eight electrons ($3s^2 3p^6$), and so forth.

The tendency of atoms to react in ways that achieve an outer shell of eight valence electrons is particularly common among Group 1A–7A elements and is given the special name of the **octet rule**. An atom with almost eight valence electrons tends to gain the needed electrons to have eight electrons in its valence shell and an electron configuration like that of the noble gas nearest to it in atomic number. In gaining electrons, the atom becomes a negatively charged ion called an **anion**. An atom with only one or two valence electrons tends to lose the number of electrons required to have an electron configuration like the noble gas nearest it in atomic number. In losing electrons, the atom becomes a positively charged ion called a **cation**. When an ion forms, the number of protons and neutrons in the nucleus of the atom remains unchanged; only the number of electrons in the valence shell of the atom changes.

Noble Gas	Electron Configuration
He	$1s^2$
Ne	$[He]2s^2 2p^6$
Ar	$[Ne]3s^2 3p^6$
Kr	$[Ar]4s^2 4p^6 3d^{10}$
Xe	$[Kr]5s^2 5p^6 4d^{10}$

Octet rule When undergoing chemical reaction, atoms of Group 1A–7A elements tend to gain, lose, or share sufficient electrons to achieve an election configuration having eight valence electrons

Anion An ion with a negative electric charge

Cation An ion with a positive electric charge

Example 3.1 The Octet Rule

Show how the following chemical changes obey the octet rule:

(a) A sodium atom loses an electron to form a sodium ion, Na^+.

$$Na \longrightarrow Na^+ + e^-$$

A sodium atom   A sodium ion   An electron

(b) A chorine atom gains an electron to form a chloride ion, Cl^-.

$$Cl + e^- \longrightarrow Cl^-$$

A chlorine atom   An electron   A chloride ion

Strategy

To see how each chemical change follows the octet rule, first write the condensed ground-state electron configuration (Section 2.6C) of the atom involved in the chemical change and of the ion it forms and then compare them.

Solution

(a) The condensed ground-state electron configurations for Na and Na^+ are:

$$Na \text{ (11 electrons): } 1s^2 2s^2 2p^6 3s^1$$

$$Na^+ \text{ (10 electrons): } 1s^2 2s^2 2p^6$$

A Na atom has one electron ($3s^1$) in its valence shell. The loss of this one valence electron changes the Na atom to a Na ion, Na^+, which has a complete octet of electrons in its valence shell ($2s^2 2p^6$) and the same electron configuration as Ne, the noble gas nearest to it in atomic number. We can write this chemical change using Lewis dot structures (Section 2.6F):

$$Na\cdot \longrightarrow Na^+ + e^-$$

A sodium atom   A sodium ion   An electron

(a) Sodium chloride

(b) Sodium

(c) Chlorine

(a) The chemical compound sodium chloride (table salt) is composed of the elements sodium (b) and chlorine (c) in chemical combination. Salt is very different from the elements that constitute it.

(b) The condensed ground-state electron configurations for Cl and Cl⁻ are:

$$\text{Cl (17 electrons): } 1s^2 2s^2 2p^6 3s^2 3p^5$$

$$\text{Cl}^- \text{ (18 electrons): } 1s^2 2s^2 2p^6 3s^2 3p^6$$

A Cl atom has seven electrons in its valence shell ($3s^2 3p^5$). The gain of one electron changes the Cl atom to a Cl ion, Cl⁻, which has a complete octet of electrons in its valence shell ($3s^2 3p^6$) and the same electron configuration as Ar, the noble gas nearest to it in atomic number. We can write this chemical change using Lewis dot structures:

$$:\ddot{\text{Cl}}\cdot \;+\; e^- \;\longrightarrow\; :\ddot{\ddot{\text{Cl}}}:^-$$

A chlorine atom   An electron   A chloride ion

Problem 3.1

Show how the following chemical changes obey the octet rule:
(a) A magnesium atom forms a magnesium ion, Mg^{2+}.
(b) A sulfur atom forms a sulfide ion, S^{2-}.

The octet rule gives us a good way to understand why Group 1A–7A elements form the ions that they do. It is not perfect, however, for two reasons:

1. Ions of period 1A and 2A elements with charges greater than +2 are unstable. Boron, for example, has three valence electrons. If it lost these three electrons, it would become B^{3+} and have a complete outer shell like that of helium. It seems, however, that this is far too large a charge for an ion of this second-period element; consequently, this ion is not found in stable ionic compounds. By the same reasoning, carbon does not lose its four valence electrons to become C^{4+}, nor does it gain four valence electrons to become C^{4-}. Either of these changes would place too great a charge on this period 2 element.

2. The octet rule does not apply to Group 1B–7B elements (the transition elements), most of which form ions with two or more different positive charges. Copper, for example, can lose one valence electron to form Cu^+; alternatively, it can lose two valence electrons to form Cu^{2+}.

It is important to understand that there are enormous differences between the properties of an atom and those of its ion(s). Atoms and their ions are completely different chemical species and have completely different chemical and physical properties. Consider, for example, sodium and chlorine. Sodium, a soft metal made of sodium atoms, reacts violently with water. Chlorine atoms are very unstable and even more reactive than sodium atoms. Both sodium and chlorine are poisonous. NaCl, common table salt, is made up of sodium ions and chloride ions. These two ions are quite stable and unreactive. Neither sodium ions nor chloride ions react with water at all.

Because atoms and their ions are different chemical species, we must be careful to distinguish one from the other. Consider the drug commonly known as "lithium," which is used to treat bipolar disorder (also known as manic depression). The element lithium, like sodium, is a soft metal that reacts violently with water. The drug used to treat bipolar disorder is not composed of lithium atoms, Li, but rather lithium ions, Li^+, usually administered in the form of lithium carbonate, Li_2CO_3. Another example comes from the fluoridation of drinking water and of toothpastes and dental gels. The element fluorine, F_2, is an extremely poisonous and corrosive gas: it is not what is used for fluoridation. Instead, this process uses fluoride ions, F^-, in the form of sodium fluoride, NaF, a compound that is unreactive and nonpoisonous in the concentrations used.

3.3 How Do We Name Anions and Cations?

We form names for anions and cations using a system developed by the International Union of Pure and Applied Chemistry. We will refer to these names as "systematic" names. Many ions have "common" names that were in use long before chemists undertook the effort to name them systematically. In this and the following chapters, we will make every effort to use systematic names for ions, but where a long-standing common name remains in use, we will give it as well.

A. Naming Monatomic Cations

A monatomic (containing only one atom) cation forms when a metal loses one or more valence electrons. Elements of Groups 1A, 2A, and 3A form only one type of cation. For ions of these metals, the name of the cation is the name of the metal followed by the word "ion" (Table 3.1). There is no need to specify the charge on these cations, because only one charge is possible. For example, Na^+ is sodium ion and Ca^{2+} is calcium ion.

Table 3.1 Names of Cations from Some Metals That Form Only One Positive Ion

Group 1A		Group 2A		Group 3A	
Ion	Name	Ion	Name	Ion	Name
H^+	Hydrogen ion	Mg^{2+}	Magnesium ion	Al^{3+}	Aluminum ion
Li^+	Lithium ion	Ca^{2+}	Calcium ion		
Na^+	Sodium ion	Sr^{2+}	Strontium ion		
K^+	Potassium ion	Ba^{2+}	Barium ion		

Most transition and inner transition elements form more than one type of cation, and therefore, the name of the cation must show its charge. To show the charge in a systematic name, we write a Roman numeral (enclosed in parentheses), immediately following (with no space) the name of the metal (Table 3.2). For example, Cu^+ is copper(I) ion and Cu^{2+} is copper(II) ion. Note that even though silver is a transition metal, it forms only Ag^+; therefore, there is no need to use a Roman numeral to show this ion's charge.

In the older common system for naming metal cations with two different charges, the suffix -ous is used to show the smaller charge and -ic is used to show the larger charge (Table 3.2). These suffixes are often added to the stem part of the Latin name for the element.

Copper(I) oxide and copper(II) oxide. The different copper ion charges result in different colors.

Table 3.2 Names of Cations from Four Metals That Form Two Different Positive Ions

Ion	Systematic Name	Common Name	Origin of the Symbol of the Element or the Common Name of the Ion
Cu^+	Copper(I) ion	Cuprous ion	*Cupr-* from *cuprum*, the Latin name for copper
Cu^{2+}	Copper(II) ion	Cupric ion	
Fe^{2+}	Iron(II) ion	Ferrous ion	*Ferr-* from *ferrum*, the Latin name for iron
Fe^{3+}	Iron(III) ion	Ferric ion	
Hg^+	Mercury(I) ion	Mercurous ion	*Hg* from *hydrargyrum*, the Latin name for mercury
Hg^{2+}	Mercury(II) ion	Mercuric ion	
Sn^{2+}	Tin(II) ion	Stannous ion	*Sn* from *stannum*, the Latin name for tin
Sn^{4+}	Tin(IV) ion	Stannic ion	

Table 3.3 Names of the Most Common Monatomic Anions

Anion	Stem Name	Anion Name
F^-	*fluor*	Fluoride
Cl^-	*chlor*	Chloride
Br^-	*brom*	Bromide
I^-	*iod*	Iodide
O^{2-}	*ox*	Oxide
S^{2-}	*sulf*	Sulfide

B. Naming Monatomic Anions

A monatomic anion is named by adding *-ide* to the stem part of the name. Table 3.3 gives the names of the monatomic anions we deal with most often.

C. Naming Polyatomic Ions

A **polyatomic ion** contains more than one atom. Examples are the hydroxide ion, OH^-, and the phosphate ion, PO_4^{3-}. We will not be concerned with how these ions are formed, only that they exist and are present in the materials around us. Table 3.4 lists several important polyatomic ions.

The preferred system for naming polyatomic ions that differ in the number of hydrogen atoms is to use the prefixes di-, tri-, and so forth, to show the presence of more than one hydrogen. For example, HPO_4^{2-} is the hydrogen phosphate ion and $H_2PO_4^-$ is the dihydrogen phosphate ion. Because several hydrogen-containing polyatomic anions have common names that are still widely used, you should memorize them as well. In these common names, the prefix *bi-* is used to show the presence of one hydrogen.

Table 3.4 Names of Common Polyatomic Ions (Common names, where still widely used, are given in parentheses.)

Polyatomic Ion	Name	Polyatomic Ion	Name
NH_4^+	Ammonium	HCO_3^-	Hydrogen carbonate (bicarbonate)
OH^-	Hydroxide	SO_3^{2-}	Sulfite
NO_2^-	Nitrite	HSO_3^-	Hydrogen sulfite (bisulfite)
NO_3^-	Nitrate	SO_4^{2-}	Sulfate
CH_3COO^-	Acetate	HSO_4^-	Hydrogen sulfate (bisulfate)
CN^-	Cyanide	PO_4^{3-}	Phosphate
MnO_4^-	Permanganate	HPO_4^{2-}	Hydrogen phosphate
CrO_4^{2-}	Chromate	$H_2PO_4^-$	Dihydrogen phosphate
$Cr_2O_7^{2-}$	Dichromate		

Chemical Connections 3A

Coral Chemistry and Broken Bones

Bone is a highly structured matrix consisting of both inorganic and organic materials. The inorganic material is chiefly hydroxyapatite, $Ca_5(PO_4)_3OH$, which makes up about 70% of bone by dry weight. By comparison, the enamel of teeth consists almost entirely of hydroxyapatite. Chief among the organic components of bone are collagen fibers (proteins, see Chapter 22), which thread their way through the inorganic matrix, providing extra strength and allowing bone to flex under stress. Also weaving through the hydroxyapatite-collagen framework are blood vessels that supply nutrients.

A problem faced by orthopedic surgeons is how to repair bone damage. For a minor fracture, usually a few weeks in a cast suffices for the normal process of bone growth to repair the damaged area. For severe fractures, especially those

involving bone loss, a bone graft may be needed. An alternative to a bone graft is an implant of synthetic bone material. One such material, called Pro Osteon®, is derived by heating coral (calcium carbonate) with ammonium hydrogen phosphate to form a hydroxyapatite similar to that of bone. Throughout the heating process, the porous structure of the coral, which resembles that of bone, is retained.

$$5CaCO_3 + 3(NH_4)_2HPO_4 \xrightarrow[\text{24--60 hours}]{200^\circ C}$$

Coral

$$Ca_5(PO_4)_3OH + 3(NH_4)_2CO_3 + 2H_2CO_3$$

Hydroxyapatite

The surgeon can shape a piece of this material to match the bone void, implant it, stabilize the area by inserting metal plates and/or screws, and let new bone tissue grow into the pores of the implant.

In an alternative process, a dry mixture of calcium dihydrogen phosphate monohydrate, $Ca(H_2PO_4)_2 \cdot H_2O$; calcium phosphate, $Ca_3(PO_4)_2$; and calcium carbonate, $CaCO_3$, is prepared. Just before the surgical implant occurs, these chemicals are mixed with a solution of sodium phosphate to form a paste that is then injected into the bony area to be repaired. In this way, the fractured bony area is held in the desired position by the synthetic material while the natural process of bone rebuilding replaces the implant with living bone tissue.

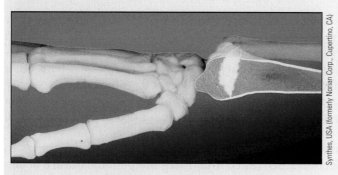

A wrist fracture repaired with bone cement (white area).

Synthes, USA (formerly Norian Corp., Cupertino, CA)

Test your knowledge with Problems 3.92 and 3.96.

3.4 What Are the Two Major Types of Chemical Bonds?

A. Ionic and Covalent Bonds

According to the Lewis model of chemical bonding, atoms bond together in such a way that each atom participating in a bond acquires a valence-shell electron configuration the same as that of the noble gas nearest to it in atomic number. Atoms acquire completed valence shells in two ways:

1. An atom may lose or gain enough electrons to acquire a filled valence shell, becoming an ion as it does so (Section 3.2). An **ionic bond** results from the force of electrostatic attraction between a cation and an anion.

2. An atom may share electrons with one or more other atoms to acquire a filled valence shell. A **covalent bond** results from the force of attraction between two atoms that share one or more pairs of electrons. A molecule or polyatomic ion is formed.

Ionic bond A chemical bond resulting from the attraction between positive and negative ions

Covalent bond A chemical bond resulting from the sharing of electrons between two atoms

We can now ask how to determine whether two atoms in a compound are bonded by an ionic bond or a covalent bond. One way to do so is to consider the relative positions of the two atoms in the Periodic Table. Ionic bonds usually form between a metal and a nonmetal. An example of an ionic bond is that formed between the metal sodium and the nonmetal chlorine in the compound sodium chloride, Na^+Cl^-. When two nonmetals or a metalloid and a nonmetal combine, the bond between them is usually covalent. Examples of compounds containing covalent bonds between nonmetals include Cl_2, H_2O, CH_4, and NH_3. Examples of compounds containing covalent bonds between a metalloid and a nonmetal include BF_3, $SiCl_4$, and AsH_3.

Another way to determine the bond type is to compare the electronegativities of the atoms involved, which is the subject of the next subsection.

B. Electronegativity and Chemical Bonds

Electronegativity is a measure of an atom's attraction for the electrons it shares in a chemical bond with another atom. The most widely used scale of electronegativities (Table 3.5) was devised in the 1930s by Linus Pauling. On the Pauling scale, fluorine, the most electronegative element, is assigned an electronegativity of 4.0 and all other elements are assigned values relative to fluorine.

As you study the electronegativity values in Table 3.5, note that they generally increase from left to right across a row of the Periodic Table and from bottom to top within a column. Values increase from left to right because of the increasing positive charge on the nucleus, which leads to a stronger attraction for electrons in the valence shell. Values increase going up a column because the decreasing distance of the valence electrons from the nucleus leads to stronger attraction between a nucleus and its valence electrons.

You might compare these trends in electronegativity with the trends in ionization energy (Section 2.8B). Each illustrates the periodic nature of elements within the Periodic Table. Ionization energy measures the amount of energy necessary to remove an electron from an atom. Electronegativity measures how tightly an atom holds the electrons that it shares with another atom. Notice that both electronegativity and ionization potential generally increase from left to right across a row of the Periodic Table from columns 1A to 7A. In addition, both electronegativity and ionization energy increases going up a column.

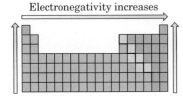

Electronegativity increases

Table 3.5 Electronegativity Values of the Elements (Pauling Scale)

1A	2A	3B	4B	5B	6B	7B	8B			1B	2B	3A	4A	5A	6A	7A
H 2.1																
Li 1.0	Be 1.5											B 2.0	C 2.5	N 3.0	O 3.5	F 4.0
Na 0.9	Mg 1.2											Al 1.5	Si 1.8	P 2.1	S 2.5	Cl 3.0
K 0.8	Ca 1.0	Sc 1.3	Ti 1.5	V 1.6	Cr 1.6	Mn 1.5	Fe 1.8	Co 1.8	Ni 1.8	Cu 1.9	Zn 1.6	Ga 1.6	Ge 1.8	As 2.0	Se 2.4	Br 2.8
Rb 0.8	Sr 1.0	Y 1.2	Zr 1.4	Nb 1.6	Mo 1.8	Tc 1.9	Ru 2.2	Rh 2.2	Pd 2.2	Ag 1.9	Cd 1.7	In 1.7	Sn 1.8	Sb 1.9	Te 2.1	I 2.5
Cs 0.7	Ba 0.9	La 1.1	Hf 1.3	Ta 1.5	W 1.7	Re 1.9	Os 2.2	Ir 2.2	Pt 2.2	Au 2.4	Hg 1.9	Tl 1.8	Pb 1.8	Bi 1.9	Po 2.0	At 2.2

⊎WL **Interactive Example 3.2** Electronegativity

Judging from their relative positions in the Periodic Table, which element in each pair has the larger electronegativity?

(a) Lithium or carbon   (b) Nitrogen or oxygen   (c) Carbon or oxygen

Strategy

The elements in each pair are in the second period of the Periodic Table. Within a period, electronegativity increases from left to right across the period.

Solution

(a) C > Li   (b) O > N   (c) O > C

Problem 3.2

Judging from their relative positions in the Periodic Table, which element in each pair has the larger electronegativity?

(a) Lithium or potassium   (b) Nitrogen or phosphorus
(c) Carbon or silicon

3.5 What Is an Ionic Bond?

A. Forming Ionic Bonds

According to the Lewis model of bonding, an ionic bond forms by the transfer of one or more valence-shell electrons from an atom of lower electronegativity to the valence shell of an atom of higher electronegativity. The more electronegative atom gains one or more valence electrons and becomes an anion; the less electronegative atom loses one or more valence electrons and becomes a cation. The compound formed by the electrostatic attraction of positive and negative ions is called an **ionic compound**.

As a guideline, we say that this type of electron transfer to form an ionic compound is most likely to occur if the difference in electronegativity between two atoms is approximately 1.9 or greater. A bond is more likely to be covalent if this difference is less than 1.9. You should be aware that the value of 1.9 for the formation of an ionic bond is somewhat arbitrary. Some chemists prefer a slightly larger value, others a slightly smaller value. The essential point is that the value of 1.9 gives us a guidepost against which to decide if a bond is more likely to be ionic or more likely to be covalent. Section 3.7 discusses covalent bonding.

An example of an ionic compound is that formed between the metal sodium (electronegativity 0.9) and the nonmetal chlorine (electronegativity 3.0). The difference in electronegativity between these two elements is 2.1. In forming the ionic compound NaCl, the single $3s$ valence electron of a sodium atom is transferred to the partially filled valence shell of a chlorine atom.

$$\text{Na } (1s^22s^22p^63s^1) + \text{Cl } (1s^22s^22p^63s^23p^5) \longrightarrow \text{Na}^+ (1s^22s^22p^6) + \text{Cl}^- (1s^22s^22p^63s^23p^6)$$

Sodium atom   Chlorine atom   Sodium ion   Chloride ion

In the following equation, we use a single-headed curved arrow to show the transfer of one electron from sodium to chlorine.

$$\text{Na} \cdot + \cdot \ddot{\text{Cl}} \colon \longrightarrow \text{Na}^+ \; \colon\!\ddot{\text{Cl}}\colon^-$$

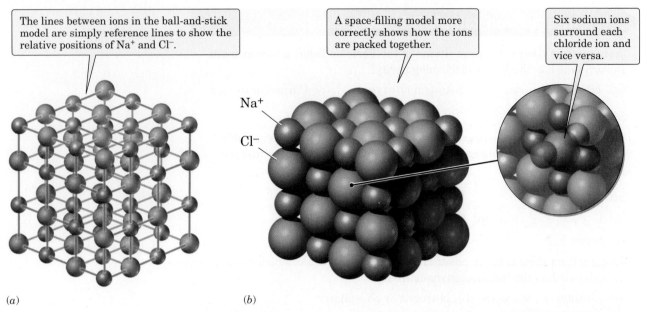

The lines between ions in the ball-and-stick model are simply reference lines to show the relative positions of Na$^+$ and Cl$^-$.

A space-filling model more correctly shows how the ions are packed together.

Six sodium ions surround each chloride ion and vice versa.

Na$^+$

Cl$^-$

(a)

(b)

FIGURE 3.1 The structure of a sodium chloride crystal. (*a*) Ball-and-stick models show the relative positions of the ions. (*b*) Space-filling models show the relative sizes of the ions.

The ionic bond in solid sodium chloride results from the force of electrostatic attraction between positive sodium ions and negative chloride ions. In its solid (crystalline) form, sodium chloride consists of a three-dimensional array of Na$^+$ and Cl$^-$ ions arranged as shown in Figure 3.1.

Although ionic compounds do not consist of molecules, they do have a definite ratio of one kind of ion to another; their formulas give this ratio. For example, NaCl represents the simplest ratio of sodium ions to chloride ions—namely, 1:1.

B. Predicting Formulas of Ionic Compounds

Ions are charged particles, but the matter we see all around us and deal with every day is electrically neutral (uncharged). If ions are present in any sample of matter, the total number of positive charges must equal the total number of negative charges. Therefore, we cannot have a sample containing only Na$^+$ ions. Any sample that contains Na$^+$ ions must also contain negative ions, such as Cl$^-$, Br$^-$, or S^{2-}, and the sum of the positive charges must equal the sum of the negative charges.

Example 3.3 Formulas of Ionic Compounds

Write the formulas for the ionic compounds formed from the following ions:

(a) Lithium ion and bromide ion   (b) Barium ion and iodide ion
(c) Aluminum ion and sulfide ion

Strategy

The formula of an ionic compound shows the simplest whole-number ratio between cations and anions. In an ionic compound, the total number of positive charges of the cations and the total number of negative charges of the anions must be equal. Therefore, to predict the formula of an ionic compound, you must know the charges of the ions involved.

Solution

(a) Table 3.1 shows that the charge on a lithium ion is $+1$, and Table 3.3 shows that the charge on a bromide ion is -1. Therefore, the formula for lithium bromide is LiBr.
(b) The charge on a barium ion is $+2$ (Table 3.1) and the charge on an iodide ion is -1 (Table 3.3). Two I^- ions are required to balance the charge of one Ba^{2+} ion. Therefore, the formula for barium iodide is BaI_2.
(c) The charge on an aluminum ion is $+3$ (Table 3.1) and the charge on a sulfide ion is -2 (Table 3.3). For the compound to have an overall charge of zero, the ions must combine in the ratio of two aluminum ions to three sulfur ions. The formula of aluminum sulfide is Al_2S_3.

Problem 3.3

Write the formulas for the ionic compounds formed from the following ions:

(a) Potassium ion and chloride ion   (b) Calcium ion and fluoride ion
(c) Iron(III) ion and oxide ion

Remember that the subscripts in the formulas for ionic compounds represent the ratio of the ions. Thus, a crystal of BaI_2 has twice as many iodide ions as barium ions. For ionic compounds, when both charges are 2, as in the compound formed from Ba^{2+} and O^{2-}, we must "reduce to lowest terms." That is, barium oxide is BaO, not Ba_2O_2. The reason is that we are looking at ratios only, and the ratio of ions in barium oxide is 1:1.

3.6 How Do We Name Ionic Compounds?

To name an ionic compound, we give the name of the cation first, followed by the name of the anion.

A. Binary Ionic Compounds of Metals That Form Only One Positive Ion

A **binary compound** contains only two elements. In a **binary ionic compound**, both of the elements are present as ions. The name of the compound consists of the name of the metal from which the cation (positive ion) was formed, followed by the name of the anion (negative ion). We generally ignore subscripts in naming binary ionic compounds. For example, $AlCl_3$ is named aluminum chloride. We know this compound contains three chloride ions because the positive and negative charges in the compound must be equal—that is, one Al^{3+} ion must combine with three Cl^- ions to balance the charges.

⬛WL Interactive Example 3.4 Binary Ionic Compounds

Name these binary ionic compounds:
(a) LiBr   (b) Ag_2S (c) NaBr

Strategy

The name of an ionic compound consists of two words: name of the cation followed by the name of the anion.

Solution

(a) Lithium bromide　　　(b) Silver sulfide　　　(c) Sodium bromide

Problem 3.4

Name these binary ionic compounds:

(a) MgO　　　　　(b) BaI_2　　　　　(c) KCl

⦿WL Interactive Example 3.5　Binary Ionic Compounds

Write the formulas for these binary ionic compounds:

(a) Barium hydride　　(b) Sodium fluoride　　(c) Calcium oxide

Strategy

Write the formula of the positive ion and then the formula of the negative ion. Remember that the number of positive and negative charges must be equal. Show the ratio of each ion in the formula of the compound by subscripts. Where only one of either ion is present, do not show a subscript.

Solution

(a) BaH_2　　　　　(b) NaF　　　　　(c) CaO

Problem 3.5

Write the formulas for these binary ionic compounds:

(a) Magnesium chloride　(b) Aluminum oxide　(c) Lithium iodide

B. Binary Ionic Compounds of Metals That Form More Than One Positive Ion

Table 3.2 shows that many transition metals form more than one positive ion. For example, copper forms both Cu^+ and Cu^{2+} ions. For systematic names, we use Roman numerals in the name to show the charge. For common names, we use the *-ous, -ic* system.

Example 3.6　Binary Ionic Compounds

Give each binary ionic compound a systematic name and a common name.

(a) CuO　　　　　　(b) Cu_2O

Strategy

The name of a binary ionic compound consists of two words. First is the name of the cation followed by the name of the anion. Because transition metals typically form more than one cation, the charge on the cation must be indicated by a Roman numeral in parentheses following the name of the transition metal or by using the suffix *-ic* to show the higher of the two possible cation charges or the suffix *-ous* to show the lower of the two possible cation charges.

Solution

(a) Systematic name: copper(II) oxide. Common name: cupric oxide.
(b) Systematic name: copper(I) oxide. Common name: cuprous oxide.

Remember in answering part (b) that we ignore subscripts in naming binary ionic compounds. Therefore, the 2 in Cu_2O is not indicated in the name. You know that two copper(I) ions are present because two positive charges are needed to balance the two negative charges on an O^{2-} ion.

Problem 3.6

Give each binary compound a systematic name and a common name.

(a) FeO   (b) Fe_2O_3

C. Ionic Compounds That Contain Polyatomic Ions

To name ionic compounds containing polyatomic ions, name the positive ion first and then the negative ion, each as a separate word.

⚲WL Interactive Example 3.7 Polyatomic Ions

Name these ionic compounds, each of which contains a polyatomic ion:

(a) $NaNO_3$ (b) $CaCO_3$ (c) $(NH_4)_2SO_3$ (d) NaH_2PO_4

Strategy

To name ionic compounds containing polyatomic ions (Table 3.4), name the positive ion first and then the negative ion, each as a separate word.

Solution

(a) Sodium nitrate   (b) Calcium carbonate
(c) Ammonium sulfite   (d) Sodium dihydrogen phosphate

Problem 3.7

Name these ionic compounds, each of which contains a polyatomic ion:

(a) K_2HPO_4 (b) $Al_2(SO_4)_3$ (c) $FeCO_3$

3.7 What Is a Covalent Bond?

A. Formation of a Covalent Bond

A covalent bond forms when electron pairs are shared between two atoms whose difference in electronegativity is less than 1.9. As we have already mentioned, the most common covalent bonds occur between two nonmetals or between a nonmetal and a metalloid.

According to the Lewis model, a pair of electrons in a covalent bond functions in two ways simultaneously: The two atoms share it, and it fills the valence shell of each atom. The simplest example of a covalent bond is that in a hydrogen molecule, H_2. When two hydrogen atoms bond, the single electrons from each atom combine to form an electron pair. A bond formed by sharing a pair of electrons is called a **single bond** and is represented by a single line between the two atoms. The electron pair shared between the two hydrogen atoms in H_2 completes the valence shell of each hydrogen. Thus, in H_2, each hydrogen has, in effect, two electrons in its valence shell

Chemical Connections 3B

Ionic Compounds in Medicine

Many ionic compounds have medical uses, some of which are shown in the table.

Formula	Name	Medical Use
$AgNO_3$	Silver nitrate	Antibiotic
$BaSO_4$	Barium sulfate	Radiopaque medium for X-ray work
$CaSO_4$	Calcium sulfate	Plaster of Paris casts
$FeSO_4$	Iron(II) sulfate	Treatment of iron deficiency
$KMnO_4$	Potassium permanganate	Anti-infective (external)
KNO_3	Potassium nitrate (saltpeter)	Diuretic
Li_2CO_3	Lithium carbonate	Treatment of bipolar disorder
$MgSO_4$	Magnesium sulfate (Epsom salts)	Cathartic
$NaHCO_3$	Sodium bicarbonate (baking soda)	Antacid
NaI	Sodium iodide	Iodine for thyroid hormones
NH_4Cl	Ammonium chloride	Acidification of the digestive system
$(NH_4)_2CO_3$	Ammonium carbonate	Expectorant
SnF_2	Tin(II) fluoride	To strengthen teeth (external)
ZnO	Zinc oxide	Astringent (external)

Drinking a "barium cocktail," which contains barium sulfate, makes the intestinal tract visible on an X-ray.

Test your knowledge with Problems 3.93, 3.94, and 3.95.

and an electron configuration like that of helium, the noble gas nearest to it in atomic number.

The single line represents a shared pair of electrons

$$H \cdot + \cdot H \longrightarrow H - H$$

B. Nonpolar and Polar Covalent Bonds

Although all covalent bonds involve the sharing of electrons, they differ widely in the degree of sharing. We classify covalent bonds into two categories, **nonpolar covalent** and **polar covalent**, depending on the difference in electronegativity between the bonded atoms. In a nonpolar covalent bond, electrons are shared equally. In a polar covalent bond, they are shared unequally. It is important to realize that no sharp line divides these two categories, nor, for that matter, does a sharp line divide polar covalent bonds and ionic bonds. Nonetheless, the rule-of-thumb guidelines given in Table 3.6 will help you decide whether a given bond is more likely to be nonpolar covalent, polar covalent, or ionic.

An example of a polar covalent bond is that in H—Cl, in which the difference in electronegativity between the bonded atoms is $3.0 - 2.1 = 0.9$. A covalent bond between carbon and hydrogen, for example, is classified as nonpolar covalent because the difference in electronegativity between these

Nonpolar covalent bond A covalent bond between two atoms whose difference in electronegativity is less than 0.5

Polar covalent bond A covalent bond between two atoms whose difference in electronegativity is between 0.5 and 1.9

Table 3.6 Classification of Chemical Bonds

Electronegativity Difference Between Bonded Atoms	Type of Bond	Most Likely Formed Between
Less than 0.5	Nonpolar covalent ⎫	Two nonmetals or a non-
0.5 to 1.9	Polar covalent ⎬	metal and a metalloid
Greater than 1.9	Ionic ⎭	A metal and a nonmetal

two atoms is only 2.5 − 2.1 = 0.4. You should be aware, however, that there is some slight polarity to a C—H bond, but because it is quite small, we arbitrarily say that a C—H bond is nonpolar. Increasing difference in electronegativity are related to increasing bond polarity.

Example 3.8 Classification of Chemical Bonds

Classify each bond as nonpolar covalent, polar covalent, or ionic.

(a) O—H   (b) N—H   (c) Na—F   (d) C—Mg   (e) C—S

Strategy

Using Table 3.5, determine the difference in electronegativity between bonded atoms. Then, use the values given in Table 3.6 to classify the type of bond formed.

Solution

Bond	Difference in Electronegativity	Type of Bond
(a) O—H	3.5 − 2.1 = 1.4	Polar covalent
(b) N—H	3.0 − 2.1 = 0.9	Polar covalent
(c) Na—F	4.0 − 0.9 = 3.1	Ionic
(d) C—Mg	2.5 − 1.2 = 1.3	Polar covalent
(e) C—S	2.5 − 2.5 = 0.0	Nonpolar covalent

Problem 3.8

Classify each bond as nonpolar covalent, polar covalent, or ionic.

(a) S—H   (b) P—H   (c) C—F   (d) C—Cl

An important consequence of the unequal sharing of electrons in a polar covalent bond is that the more electronegative atom gains a greater fraction of the shared electrons and acquires a partial negative charge, indicated by the symbol $\delta-$ (read "delta minus"). The less electronegative atom has a lesser fraction of the shared electrons and acquires a partial positive charge, indicated by the symbol $\delta+$ (read "delta plus"). This separation of charge produces a **dipole** (two poles). We commonly show the presence of a bond dipole by an arrow, with the head of the arrow near the negative end of the dipole and a cross on the tail of the arrow near the positive end (Figure 3.2).

We can also show the polarity of a covalent bond by an electron density map. In this type of molecular model, a blue color shows the presence of a $\delta+$ charge and a red color shows the presence of a $\delta-$ charge. Figure 3.2 also shows an electron density map of HCl. The ball-and-stick model in the center of the electron density map shows the orientation of atoms in space.

Dipole A chemical species in which there is a separation of charge; there is a positive pole in one part of the species and a negative pole in another part

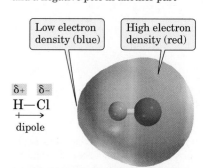

FIGURE 3.2 HCl is a polar covalent molecule. In the electron density map of HCl, red indicates a region of high electron density and blue indicates a region of low electron density.

The transparent surface surrounding the ball-and-stick model shows the relative sizes of the atoms (equivalent to the size shown by a space-filling model). Colors on the surface show the distribution of electron density. We see by the blue color that hydrogen bears a $\delta+$ charge and by the red color that chlorine bears a $\delta-$ charge.

Example 3.9 Polarity of A Covalent Bond

Using the symbols $\delta-$ and $\delta+$, indicate the polarity in each polar covalent bond.
(a) C—O　(b) N—H　(c) C—Mg

Strategy

The more electronegative atom of a covalent bond bears a partial negative charge, and the less electronegative atom bears a partial positive charge.

Solution

For (a), C and O are both in period 2 of the Periodic Table. Because O is farther to the right than C, it is more electronegative than C. For (c), Mg is a metal located to the far left in the Periodic Table and C is a nonmetal located to the right. All nonmetals, including H, have a greater electronegativity than do the metals in columns 1A and 2A. The electronegativity of each element is given below the symbol of the element.

$$
\begin{array}{lll}
\overset{\delta+\quad\delta-}{(a)\ \text{C—O}} & \overset{\delta-\quad\delta+}{(b)\ \text{N—H}} & \overset{\delta-\quad\delta+}{(c)\ \text{C—Mg}} \\
\quad\ 2.5\ \ 3.5 & \quad\ 3.0\ \ 2.1 & \quad\ 2.5\ \ 1.2
\end{array}
$$

Probem 3.9

Using the symbols $\delta-$ and $\delta+$, indicate the polarity in each polar covalent bond.
(a) C—N　(b) N—O　(c) C—Cl

C. Drawing Lewis Structures of Covalent Compounds

The ability to draw Lewis structures for covalent molecules is a fundamental skill for the study of chemistry. The following How To box will help you with this task.

How To . . .

Draw Lewis Structures

1. **Determine the number of valence electrons in the molecule.**
   Add up the number of valence electrons contributed by each atom. To determine the number of valence electrons, you only need to know the number of each kind of atom in the molecule. For each unit of negative charge on the ion, add one electron. For each unit of positive charge, subtract one electron.
   Example: The Lewis structure for formaldehyde, CH_2O, must show 12 valence electrons:

   4 (from C) + 2 (from the two H) + 6 (from O) = 12

2. **Determine the connectivity of the atoms (which atoms are bonded to each other) and connect bonded atoms by single bonds.**
Determining the connectivity of the atoms is often the most challenging part of drawing a Lewis structure. For some molecules, we ask you to propose connectivity. For most, however, we give you the experimentally determined connectivity and ask you to complete the Lewis structure.
Example: The atoms in formaldehyde are bonded in the following order. Note that we do not attempt at this point to show bond angles or the three-dimensional shape of the molecule; we just show what is bonded to what.

$$\begin{array}{c} O \\ | \\ H-C-H \end{array}$$

This partial structure shows six valence electrons in the three single bonds. In it, we have accounted for six of the 12 valence electrons.
3. **Arrange the remaining electrons so that each atom has a complete outer shell.**
Each hydrogen atom must be surrounded by two electrons. Each carbon, nitrogen, oxygen, and halogen atom must be surrounded by eight valence electrons. The remaining valence electrons may be shared between atoms in bonds or may be unshared pairs on a single atom. A pair of electrons involved in a covalent bond (**bonding electrons**) is shown as a single line; an unshared pair of electrons (**nonbonding electrons**) is shown as a pair of Lewis dots.

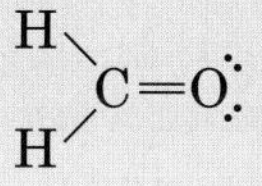

Bonding electrons Valence electrons involved in forming a covalent bond; that is, shared electrons

Nonbonding electrons Valence electrons not involved in forming covalent bonds; that is, unshared electrons

Single bond A bond formed by sharing one pair of electrons and represented by a single line between two atoms

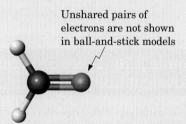

Unshared pairs of electrons are not shown in ball-and-stick models

Lewis structure

Ball-and-stick model of formaldehyde

Lewis structure A formula for a molecule or ion showing all pairs of bonding electrons as single, double, or triple bonds and all nonbonded electrons as pairs of Lewis dots

Structural formula A formula showing how atoms in a molecule or ion are bonded to each other; similar to a Lewis structure except that a structural formula typically shows only bonding pairs of electrons

By placing two pairs of bonding electrons between C and O, we give carbon a complete octet. By placing the remaining four electrons on oxygen as two Lewis dot pairs, we give oxygen eight valence electrons and a complete octet (octet rule). Note that we placed the two pairs of bonding electrons between C and O before we assigned the unshared pairs of electrons on the oxygen.
As a check on this structure, verify (1) that each atom has a complete valence shell (which each does) and (2) that the Lewis structure has the correct number of valence electrons (12, which it does).
4. In a **double bond**, two atoms share two pairs of electrons; we represent a double bond by two lines between the bonded atoms. Double bonds are most common between atoms of C, N, O, and S. In the organic and biochemistry chapters in particular, we shall see many examples of C=C, C=N, and C=O double bonds.
5. In a **triple bond**, two atoms share three pairs of electrons; we show a triple bond by three lines between the bonded atoms. Triple bonds are most common between atoms of C and N, as for example —C≡C— and —C≡N: triple bonds.

Double bond A bond formed by sharing two pairs of electrons and represented by two lines between the two bonded atoms

Triple bond A bond formed by sharing three pairs of electrons and represented by three lines between the two bonded atoms

Table 3.7 Lewis Structures for Several Small Molecules

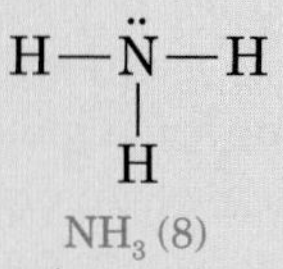

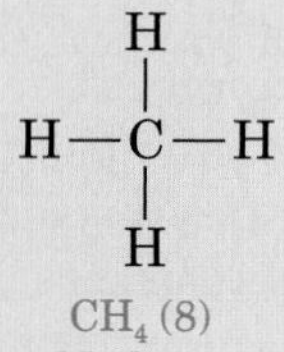

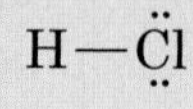

H_2O (8) NH_3 (8) CH_4 (8) HCl (8)
Water   Ammonia   Methane   Hydrogen chloride

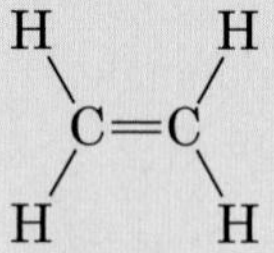

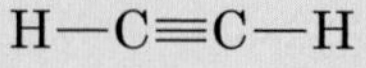

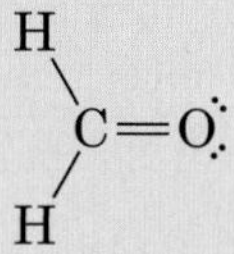

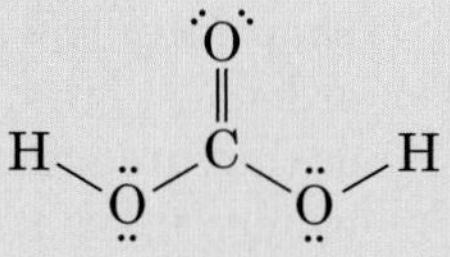

C_2H_4 (12) C_2H_2 (10) CH_2O (12) H_2CO_3 (24)
Ethylene   Acetylene   Formaldehyde   Carbonic acid

(The number of valence electrons in each molecule is given in parentheses after the molecular formula of the compound.)

Table 3.7 gives Lewis structures and names for several small molecules. Notice that each hydrogen is surrounded by two valence electrons and that each carbon, nitrogen, oxygen, and chlorine is surrounded by eight valence electrons. Furthermore, each carbon has four bonds, each nitrogen has three bonds and one unshared pair of electrons, each oxygen has two bonds and two unshared pairs of electrons, and chlorine (as well as the other halogens) has one bond and three unshared pairs of electrons.

Example 3.10 Lewis Structures of Covalent Compounds

State the number of valence electrons in each molecule and draw a Lewis structure for each:

(a) Hydrogen peroxide, H_2O_2 (b) Methanol, CH_3OH
(c) Acetic acid, CH_3COOH

Strategy

To determine the number of valence electrons in a molecule, add the number of valence electrons contributed by each kind of atom in the molecule. To draw a Lewis structure, determine the connectivity of the atoms and connect bonded atoms by single bonds. Then arrange the remaining valence electrons so that each atom has a complete outer shell.

Solution

(a) A Lewis structure for hydrogen peroxide, H_2O_2, must show the 14 valence electrons—six from each oxygen and the one from each hydrogen, for a total of $12 + 2 = 14$ valence electrons. We know that hydrogen forms only one covalent bond, so the connectivity of atoms must be as follows:

$$H—O—O—H$$

The three single bonds account for six valence electrons. The remaining eight valence electrons must be placed on the oxygen atoms to give each a complete octet:

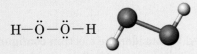

H—Ö—Ö—H

Ball-and-stick models show only nuclei, covalent bonds, and the shape of the molecule; they do not show unshared pairs of electrons

(b) A Lewis structure for methanol, CH_3OH, must show the four valence electrons from carbon, one from each hydrogen, and the six from oxygen for a total of $4 + 4 + 6 = 14$ valence electrons. The connectivity of atoms in methanol is given on the left, below. The five single bonds in this partial structure account for ten valence electrons. The remaining four valence electrons must be placed on oxygen as two Lewis dot pairs to give it a complete octet.

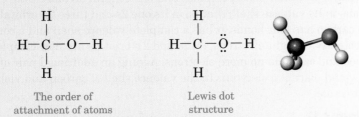

The order of attachment of atoms

Lewis dot structure

(c) A molecule of acetic acid, CH_3COOH, must contain the four valence electrons from each carbon, the six from each oxygen, and the one from each hydrogen for a total of $8 + 12 + 4 = 24$ valence electrons. The connectivity of atoms, shown on the left below, contains seven single bonds, which account for 14 valence electrons. The remaining ten electrons must be added in such a way that each carbon and oxygen atom has a complete outer shell of eight electrons. This can be done in only one way, which creates a double bond between carbon and one of the oxygens.

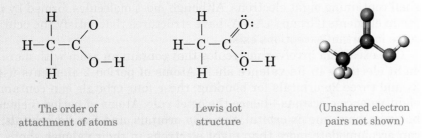

The order of attachment of atoms

Lewis dot structure

(Unshared electron pairs not shown)

In this Lewis structure, each carbon has four bonds: One carbon has four single bonds, and the other carbon has two single bonds and one double bond. Each oxygen has two bonds and two unshared pairs of electrons: One oxygen has one double bond and two unshared pairs of electrons, and the other oxygen has two single bonds and two unshared pairs of electrons.

Problem 3.10

Draw a Lewis structure for each molecule. Each has only one possible order of attachment of its atoms, which is left for you to determine.

(a) Ethane, C_2H_6 (b) Chloromethane, CH_3Cl
(c) Hydrogen cyanide, HCN

Example 3.11 Covalent Bonding of Carbon

Why does carbon have four bonds and no unshared pairs of electrons in some covalent compounds?

Strategy

In answering this question, you need to consider the electron configuration of carbon, the number of electrons its valence shell can hold, and the orbitals available to it for sharing electrons to form covalent bonds.

Solution

In forming covalent compounds, carbon reacts in such a way as to gain a filled valence shell; that is, a complete octet in its valence shell and an electron configuration resembling that of neon, the noble gas nearest it in atomic number.

Carbon is a second-period element and can contain no more than eight electrons in its valence shell; that is, in its one $2s$ and three $2p$ orbitals. When carbon has four bonds, it has a complete valence shell and a complete octet. With eight electrons, its $2s$ and $2p$ orbitals are now completely occupied and can hold no more electrons. Adding an additional pair of electrons would place ten electrons in the valence shell of carbon and violate the octet rule.

Problem 3.11

Draw a Lewis structure of a covalent compound in which carbon has:

(a) Four single bonds   (b) Two single bonds and one double bond
(c) Two double bonds   (d) One single bond and one triple bond

D. Exceptions to the Octet Rule

The Lewis model of covalent bonding focuses on valence electrons and the necessity for each atom other than hydrogen to have a completed valence shell containing eight electrons. Although most molecules formed by main-group elements (Groups 1A–7A) have structures that satisfy the octet rule, some important exceptions exist.

One exception involves molecules that contain an atom with more than eight electrons in its valence shell. Atoms of period 2 elements use one $2s$ and three $2p$ orbitals for bonding; these four orbitals can contain only eight valence electrons—hence the octet rule. Atoms of period 3 elements, however, have one $3s$ orbital, three $3p$ orbitals, and five $3d$ orbitals; they can accommodate more than eight electrons in their valence shells (Section 2.6A). In phosphine, PH_3, phosphorus has eight electrons in its valence shell and obeys the octet rule. The phosphorus atoms in phosphorus pentachloride, PCl_5, and phosphoric acid, H_3PO_4, have ten electrons in their valence shells and, therefore, are exceptions to the octet rule.

8 electrons in the valence shell of P	10 electrons in the valence shell of P	10 electrons in the valence shell of P
Phosphine	Phosphorus pentachloride	Phosphoric acid

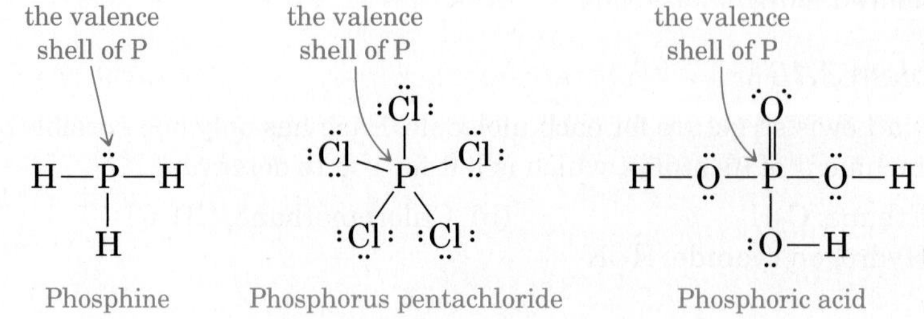

Sulfur, another period 3 element, forms compounds in which it has 8, 10, and even 12 electrons in its valence shell. The sulfur atom in H_2S has 8 electrons in its valence shell and obeys the octet rule. The sulfur atoms in SO_2 and H_2SO_4 have 10 and 12 electrons, respectively, in their valence shells and are exceptions to the octet rule.

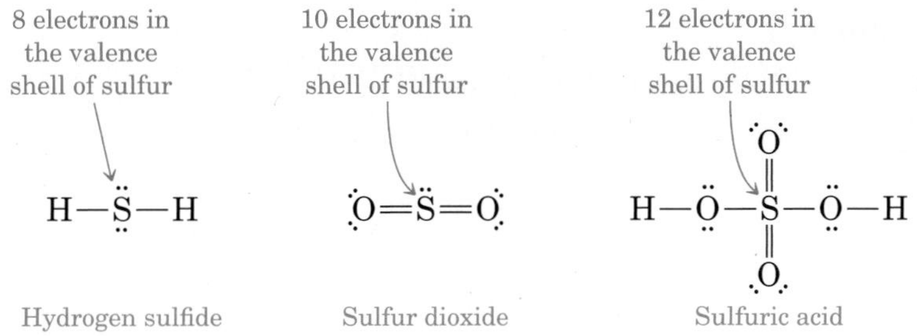

8 electrons in the valence shell of sulfur

10 electrons in the valence shell of sulfur

12 electrons in the valence shell of sulfur

Hydrogen sulfide   Sulfur dioxide   Sulfuric acid

3.8 How Do We Name Binary Covalent Compounds?

A **binary covalent compound** is a binary (two-element) compound in which all bonds are covalent. In naming a binary covalent compound:

1. First name the less electronegative element (see Table 3.5). Note that the less electronegative element is also generally written first in the formula.

2. Then name the more electronegative element. To name it, add *-ide* to the stem name of the element. Chlorine, for example, becomes chloride and oxygen becomes oxide (Table 3.3).

3. Use the prefixes *di-, tri-, tetra-,* and so on, to show the number of atoms of each element. The prefix *mono-* is omitted when it refers to the first atom named, and it is rarely used with the second atom. An exception to this rule is CO, which is named carbon monoxide.

The name is then written as two words.

Name of the first element in the formula; use prefixes *di-* and so forth if necessary	Name of the second element; use prefixes *mono-* and so forth if necessary

ᗢWL Interactive Example 3.12 Binary Covalent Compounds

Name these binary covalent compounds:

(a) NO   (b) SF_2 (c) N_2O

Strategy

The systematic name of a binary covalent compound consists of two words. The first word gives the name of the element that appears first in the formula. A prefix (*di-, tri-, tetra-,* and so forth) is used to show the number of atoms of that element in the formula. The second word consists of (1) a suffix designating the number of atoms of the second element, (2) the stem name of the second element, and (3) the suffix *-ide*.

Solution

(a) Nitrogen oxide (more commonly called nitric oxide)
(b) Sulfur difluoride
(c) Dinitrogen oxide (more commonly called nitrous oxide or laughing gas)

Problem 3.12

Name these binary covalent compounds:

(a) NO_2 (b) PBr_3 (c) SCl_2 (d) BF_3

Chemical Connections 3C

Nitric Oxide: Air Pollutant and Biological Messenger

Nitric oxide, NO, is a colorless gas whose importance in the environment has been known for several decades but whose biological importance is only now being fully recognized. This molecule has 11 valence electrons. Because its number of electrons is odd, it is not possible to draw a structure for NO that obeys the octet rule; there must be one unpaired electron, here shown on the less electronegative nitrogen atom.

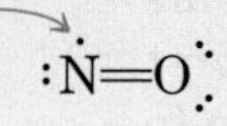

An unpaired electron → $:\!N\!=\!O\!:$

Nitric oxide

The importance of NO in the environment arises from the fact that it forms as a by-product during the combustion of fossil fuels. Under the temperature conditions of internal combustion engines and other combustion sources, nitrogen and oxygen of the air react to form small quantities of NO:

$$N_2 + O_2 \xrightarrow{\text{heat}} 2NO$$

Nitric oxide

When inhaled, NO passes from the lungs into the bloodstream. There it interacts with the iron in hemoglobin, decreasing its ability to carry oxygen. What makes nitric oxide so hazardous in the environment is that it reacts almost immediately with oxygen to form NO_2. When dissolved in water, NO_2 reacts with water to form nitric acid and nitrous acid, which are major acidifying components of acid rain.

$$2NO + O_2 \longrightarrow 2NO_2$$

Nitric oxide   Nitrogen dioxide

$$NO_2 + H_2O \longrightarrow HNO_3$$

Nitrogen dioxide   Nitric acid

Imagine the surprise when it was discovered within the last two decades that this highly reactive, seemingly hazardous compound is synthesized in humans and plays a vital role as a signaling molecule in the cardiovascular system (Chemical Connections 24F).

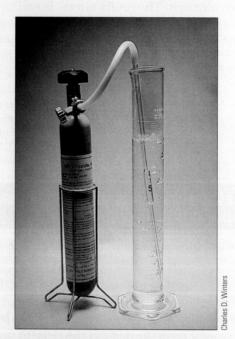

Charles D. Winters

Colorless nitric oxide, NO, coming from the tank, bubbles through the water. When it reaches the air, it is oxidized to brown nitrogen dioxide, NO_2.

Test your knowledge with Problem 3.97.

3.9 What Is Resonance?

As chemists developed a deeper understanding of covalent bonding in organic and inorganic compounds, it became obvious that for a great many molecules and ions, no single Lewis structure provides a truly accurate representation. For example, Figure 3.3 shows three Lewis structures for the carbonate ion, $CO_3{}^{2-}$. In each structure, carbon is bonded to three oxygen atoms by a combination of one double bond and two single bonds. Each Lewis structure implies that one carbon-oxygen bond is different from the other two. However, this is not the case. It has been determined experimentally that all three carbon-oxygen bonds are identical.

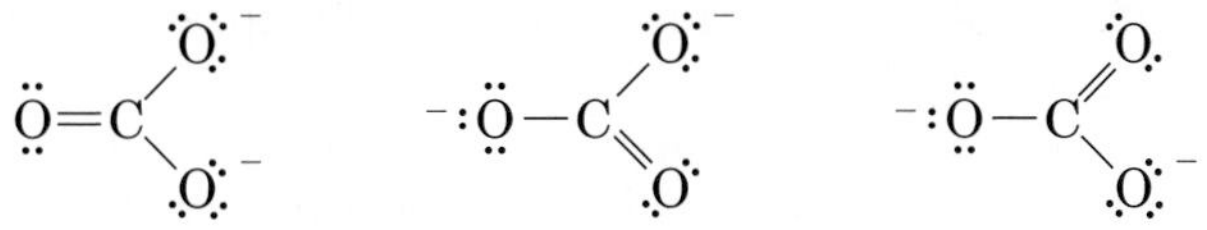

(a)   (b)   (c)

FIGURE 3.3 Three Lewis structures for the carbonate ion.

The problem for chemists, then, is how to describe the structure of molecules and ions for which no single Lewis structure is adequate and yet still retain Lewis structures. As an answer to this problem, Linus Pauling proposed the theory of resonance.

A. Theory of Resonance

According to the theory of **resonance**, many molecules and ions are best described by writing two or more Lewis structures and considering the real molecule or ion to be a hybrid of these structures. An individual Lewis structure is called a **contributing structure**. They are also sometimes referred to as **resonance structures** or **resonance contributors**. We show that the real molecule or ion is a **resonance hybrid** of the various contributing structures by interconnecting them with **double-headed arrows**. Do not confuse the double-headed arrow with the double arrow used to show chemical equilibrium. As we explain shortly, resonance structures are not in equilibrium with each other.

Figure 3.4 shows three contributing structures for the carbonate ion. These contributing structures are said to be equivalent. All three have identical patterns of covalent bonding.

The use of the term "resonance" for this theory of covalent bonding appears to suggest that bonds and electron pairs are constantly changing back and forth from one position to another over time. This notion is not at all correct. The carbonate ion, for example, has one—and only one—real structure. The problem is ours. How do we represent that one real structure? The resonance method offers a way to represent the real structure while simultaneously retaining Lewis structures with electron-pair bonds and showing all nonbonding pairs of electrons. Thus, although we realize that the carbonate ion is not accurately represented by any one contributing structure shown

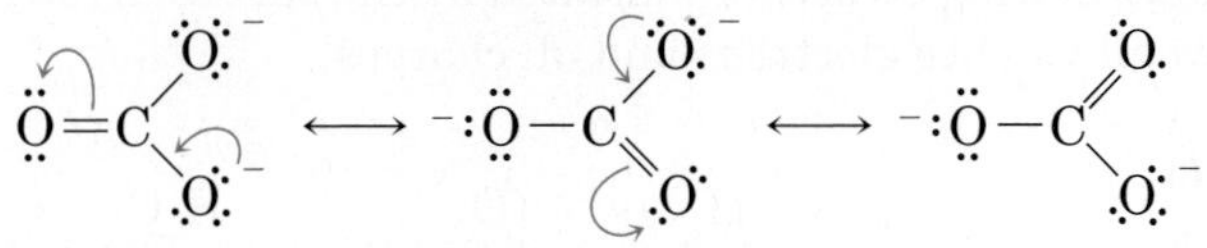

Resonance A theory that many molecules and ions are best described as a hybrid of two or more Lewis contributing structures

Contributing structure Representations of a molecule or ion that differ only in the distribution of valence electrons

Resonance hybrid A molecule or ion described as a composite or hybrid of a number of contributing structures

Double-headed arrow A symbol used to show that the structures on either side of it are resonance-contributing structures

(a)   (b)   (c)

FIGURE 3.4 The carbonate ion represented as a hybrid of three equivalent contributing structures. Curved arrows (in red) show how electron pairs are redistributed from one contributing structure to the next.

How To . . .

Draw Curved Arrows and Push Electrons

Notice in Figure 3.4 that the only difference among contributing structures (a), (b), and (c) is the position of the valence electrons. To generate one resonance structure from another, chemists use a **curved arrow**. The arrow indicates where a pair of electrons originates (the tail of the arrow) and where it is repositioned in an alternative contributing structure (the head of the arrow).

A curved arrow is nothing more than a bookkeeping symbol for keeping track of electron pairs or, as some call it, **electron pushing**. Do not be misled by its simplicity. Electron pushing will help you see the relationship among contributing structures.

Following are contributing structures for the nitrite and acetate ions. Curved arrows show how the contributing structures are interconverted. For each ion, the contributing structures are equivalent. They have the same bonding patterns.

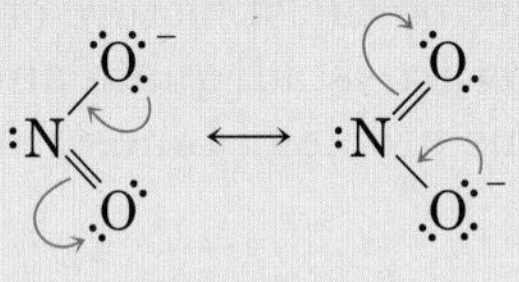

Nitrite ion
(equivalent contributing structures)

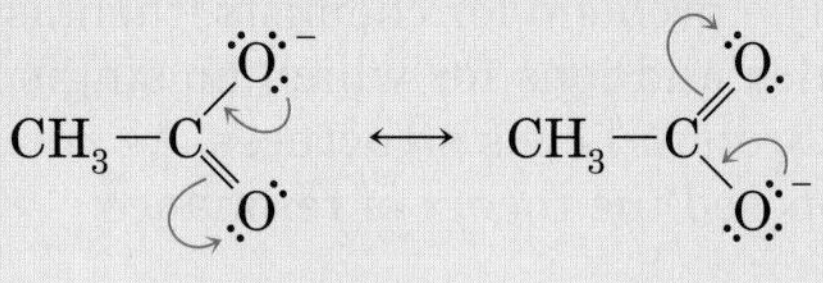

Acetate ion
(equivalent contributing structures)

A common mistake is to use curved arrows to indicate the movement of atoms or positive charges. This is never correct. Curved arrows are used only to show the repositioning of electron pairs when a new contributing structure is generated.

in Figure 3.4, we continue to represent it by one of them for convenience. We understand, of course, that we are referring to the resonance hybrid.

Resonance, when it exists, is a stabilizing factor—that is, a resonance hybrid is more stable than any one of its hypothetical contributing structures. We will see three particularly striking illustrations of the stability of resonance hybrids when we consider the unusual chemical properties of benzene and aromatic hydrocarbons in Chapter 13, the acidity of carboxylic acids in Chapter 18, and the geometry of the amide bonds in proteins in Chapter 19.

Example 3.13 Resonance

Draw the contributing structure indicated by the curved arrows. Be certain to show all valence electrons and all charges.

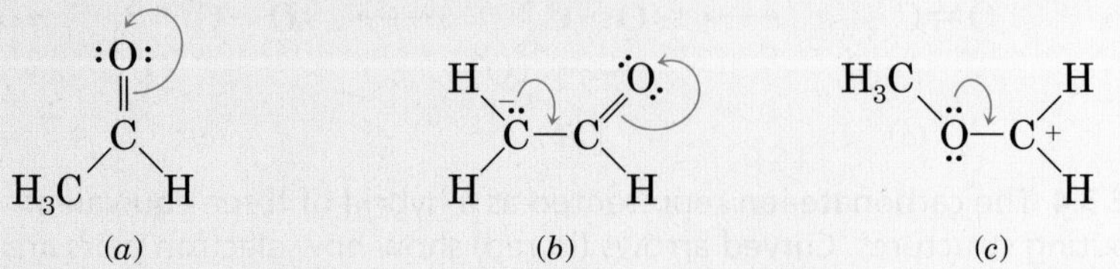

(a)   (b)   (c)

Strategy

Curved arrows show the repositioning of a pair of electrons either from a bond to an adjacent atom as in parts (a) and (b) or from an atom to an adjacent bond as in parts (b) and (c).

Solution

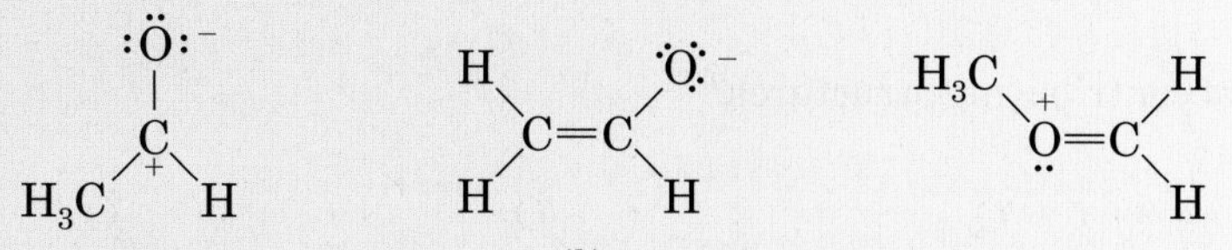

(a)   (b)   (c)

Problem 3.13

Draw the contributing structure indicated by the curved arrows. Be certain to show all valence electrons and all charges.

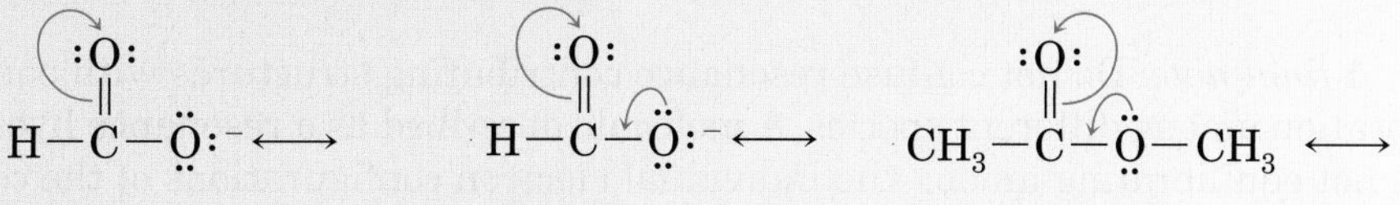

(a)   (b)   (c)

B. Writing Acceptable Contributing Structures

Certain rules must be followed to write acceptable contributing structures:

1. All contributing structures must have the same number of valence electrons.

2. All contributing structures must obey the rules of covalent bonding. In particular, no contributing structure may have more than two electrons in the valence shell of hydrogen or more than eight electrons in the valence shell of a second-period element. Third-period elements, such as phosphorus and sulfur, may have more than eight electrons in their valence shells.

3. The positions of all atomic nuclei must be the same in all resonance structures; that is, contributing structures differ only in the distribution of valence electrons.

Example 3.14 Resonance Contributing Structures

Which sets are valid pairs of contributing structures?

(a)   (b)

Strategy

The guideline tested in this example is that contributing structures involve only the redistribution of valence electrons. The position of all atoms remains the same.

Solution

(a) These are valid contributing structures. They differ only in the distribution (location) of valence electrons.

(b) These are not valid contributing structures. They differ in the arrangement of their atoms.

Problem 3.14

Which sets are valid pairs of contributing structures?

(a)   (b)

A final note: Do not confuse resonance contributing structures with equilibration among different species. A molecule described as a resonance hybrid is not equilibrating among the individual electron configurations of the contributing structures. Rather, the molecule has only one structure, which is best described as a hybrid of its various contributing structures. The colors on the color wheel provide a good analogy. Purple is not a primary color; the primary colors blue and red are mixed to make purple. You can think of molecules represented by resonance hybrids as being purple. Purple is not sometimes blue and sometimes red: Purple is purple. In an analogous way, a molecule described as a resonance hybrid is not sometimes one contributing structure and sometimes another: It is a single structure all the time.

3.10 How Do We Predict Bond Angles in Covalent Molecules?

In Section 3.7, we used a shared pair of electrons as the fundamental unit of covalent bonds and drew Lewis structures for several small molecules containing various combinations of single, double, and triple bonds (see, for example, Table 3.7). We can predict **bond angles** in these and other molecules by using the **valence-shell electron-pair repulsion (VSEPR) model**.

Bond angle The angle between two bonded atoms and a central atom

According to this model, the valence electrons of an atom may be involved in the formation of single, double, or triple bonds or they may be unshared. Each combination creates a negatively charged region of electron density around a nucleus. Because like charges repel each other, the various regions of electron density around a nucleus spread out so that each is as far away as possible from the others.

You can demonstrate the bond angles predicted by this model in a very simple way. Imagine that an inflated balloon represents a region of electron density. Two inflated balloons tied together by their ends assume the shape shown in Figure 3.5(a). The point where they are tied together represents the atom about which you want to predict a bond angle, and the balloons represent regions of electron density about that atom.

We use the VSEPR model and the balloon model analogy in the following way to predict the shape of a molecule of methane, CH_4. The Lewis structure for CH_4 shows a carbon atom surrounded by four regions of electron density. Each region contains a pair of electrons forming a single covalent bond to a hydrogen atom. According to the VSEPR model, the four regions point away from carbon so that they are as far away from one another as possible.

(a) Linear   (b) Trigonal planar   (c) Tetrahedral

FIGURE 3.5 Inflated balloon models to predict bond angles. (*a*) Two balloons assume a linear shape with a bond angle of 180° about the tie point. (*b*) Three balloons assume a trigonal planar shape with bond angles of 120° about the tie point. (*c*) Four balloons assume a tetrahedral shape with bond angles of 109.5° about the tie point.

The maximum separation occurs when the angle between any two regions of electron density is 109.5°. Therefore, we predict all H—C—H bond angles to be 109.5° and the shape of the molecule to be **tetrahedral** (Figure 3.5c and 3.6). The H—C—H bond angles in methane have been measured experimentally and found to be 109.5°. Thus, the bond angles and shape of methane predicted by the VSEPR model are identical to those observed experimentally.

We can predict the shape of an ammonia molecule, NH_3, in the same way. The Lewis structure of NH_3 shows nitrogen surrounded by four regions of electron density. Three regions contain single pairs of electrons that form covalent bonds with hydrogen atoms. The fourth region contains an unshared pair of electrons [Figure 3.7(*a*)]. Using the VSEPR model, we predict that the four regions are arranged in a tetrahedral manner and that the three H—N—H bond angles in this molecule are 109.5°. The observed bond angles are 107.5°. We can explain this small difference between the predicted and the observed angles by proposing that the unshared pair of electrons on nitrogen repels adjacent bonding electron pairs more strongly than the bonding pairs repel one another.

The geometry of an ammonia molecule is described as **pyramidal**; that is, the molecule is shaped like a triangular-based pyramid with the three hydrogens located at the base and the single nitrogen located at the apex.

Ammonia gas is drilled into the soil of a farm field. Most of the ammonia manufactured in the world is used as fertilizer because ammonia supplies the nitrogen needed by green plants.

(a)

H
|
H—C—H
|
H

(a)

H—N̈—H
|
H

(b)

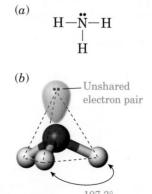

109.5°

109.5°

(b)

Unshared
electron pair

107.3°

FIGURE 3.6 The shape of a methane molecule, CH_4. (*a*) Lewis structure and (*b*) ball-and-stick model. The hydrogens occupy the four corners of a regular tetrahedron, and all H—C—H bond angles are 109.5°.

FIGURE 3.7 The shape of an ammonia molecule, NH_3. (*a*) Lewis structure and (*b*) ball-and-stick model. The H—N—H bond angles are 107.3°, slightly smaller than the H—C—H bond angles of methane.

Geometry is determined by the bonded atoms and not by the lone pair or lone pairs.

(a)

$$H-\overset{..}{\underset{..}{O}}-H$$

(b)

Unshared electron pairs

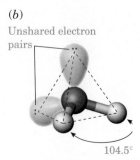

104.5°

FIGURE 3.8 The shape of a water molecule, H₂O, is tetrahedral. (a) Lewis structure and (b) ball-and-stick model.

Figure 3.8 shows a Lewis structure and a ball-and-stick model of a water molecule. In H_2O, oxygen is surrounded by four regions of electron density. Two of these regions contain pairs of electrons used to form single covalent bonds to hydrogens; the remaining two regions contain unshared electron pairs. Using the VSEPR model, we predict that the four regions of electron density around oxygen are arranged in a tetrahedral manner and that the H—O—H bond angle is 109.5°. Experimental measurements show that the actual H—O—H bond angle in a water molecule is 104.5°, a value smaller than that predicted. We can explain this difference between the predicted and the observed bond angle by proposing, as we did for NH_3, that unshared pairs of electrons repel adjacent pairs more strongly than bonding pairs do. Note that the distortion from 109.5° is greater in H_2O, which has two unshared pairs of electrons, than it is in NH_3, which has only one unshared pair.

A general prediction emerges from this discussion. If a Lewis structure shows four regions of electron density around an atom, the VSEPR model predicts a tetrahedral distribution of electron density and bond angles of approximately 109.5°.

In many of the molecules we will encounter, three regions of electron density surround an atom. Figure 3.9 shows Lewis structures and ball-and-stick models for molecules of formaldehyde, CH_2O, and ethylene, C_2H_4.

In the VSEPR model, we treat a double bond as a single region of electron density. In formaldehyde, three regions of electron density surround carbon. Two regions contain single pairs of electrons, each of which forms a single bond to a hydrogen; the third region contains two pairs of electrons, which form a double bond to oxygen. In ethylene, three regions of electron density also surround each carbon atom; two contain single pairs of electrons, and the third contains two pairs of electrons.

Three regions of electron density about an atom are farthest apart when they lie in a plane and make angles of 120° with one another. Thus, the predicted H—C—H and H—C—O bond angles in formaldehyde and the H—C—H and H—C—C bond angles in ethylene are all 120°. Furthermore, all atoms in each molecule lie in a plane. Thus, both formaldehyde and ethylene are planar molecules. The geometry about an atom surrounded by three regions of electron density, as in formaldehyde and ethylene, is described as **trigonal planar**.

In still other types of molecules, two regions of electron density surround a central atom. Figure 3.10 shows Lewis structures and ball-and-stick models of molecules of carbon dioxide, CO_2, and acetylene, C_2H_2.

In carbon dioxide, two regions of electron density surround carbon; each contains two pairs of electrons and forms a double bond to an oxygen atom.

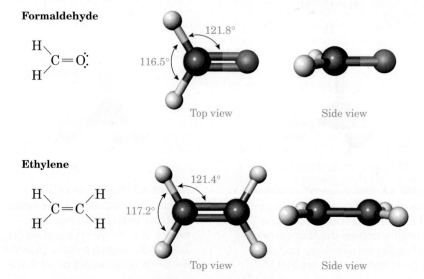

Formaldehyde

$$\underset{H}{\overset{H}{>}}C=\overset{..}{\underset{..}{O}}:$$

116.5°   121.8°

Top view   Side view

Ethylene

$$\underset{H}{\overset{H}{>}}C=C\underset{H}{\overset{H}{<}}$$

117.2°   121.4°

Top view   Side view

FIGURE 3.9 The shapes of formaldehyde, CH_2O, and ethylene, C_2H_4, are trigonal planar.

Carbon dioxide

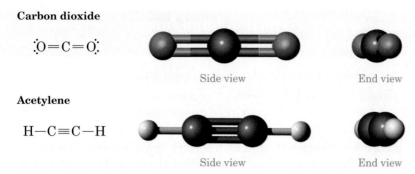

FIGURE 3.10 The shapes of carbon dioxide, CO_2, and acetylene, C_2H_2, are linear.

In acetylene, two regions of electron density also surround each carbon; one contains a single pair of electrons and forms a single bond to a hydrogen atom, and the other contains three pairs of electrons and forms a triple bond to a carbon atom. In each case, the two regions of electron density are farthest apart if they form a straight line through the central atom and create an angle of 180°. Both carbon dioxide and acetylene are linear molecules.

Table 3.8 summarizes the predictions of the VSEPR model. In this table, three-dimensional shapes are shown using a solid wedge to represent a bond coming toward you, out of the plane of the paper. A broken wedge represents a bond going away from you, behind the plane of the paper. A solid line represents a bond in the plane of the paper.

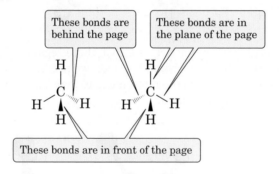

Table 3.8 Predicted Molecular Shapes (VSEPR model)

Regions of Electron Density Around Central Atom	Predicted Distribution of Electron Density	Predicted Bond Angles	Examples (Shape of the Molecule)
4	Tetrahedral	109.5°	Methane (tetrahedral) Ammonia (pyramidal) Water (bent)
3	Trigonal planar	120°	Ethylene (planar) Formaldehyde (planar)
2	Linear	180°	Carbon dioxide (linear) Acetylene (linear)

Example 3.15 Predicting Bond Angles In Covalent Compounds

Predict all bond angles and the shape of each molecule:

(a) CH_3Cl (b) $CH_2=CHCl$

Strategy

To predict bond angles, first draw a correct Lewis structure for the compound. Be certain to show all unpaired electrons. Then determine the number of regions of electron density (either 2, 3, or 4) around each atom and use that number to predict bond angles (either 109.5°, 120°, or 180°).

Solution

(a) The Lewis structure for CH_3Cl shows that four regions of electron density surround carbon. Therefore, we predict that the distribution of electron pairs about carbon is tetrahedral, all bond angles are 109.5°, and the shape of CH_3Cl is tetrahedral.

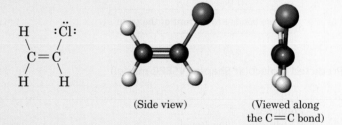

(b) In the Lewis structure for $CH_2 = CHCl$, three regions of electron density surround each carbon. Therefore, we predict that all bond angles are 120° and that the molecule is planar. The bonding about each carbon is trigonal planar.

(Side view)   (Viewed along the C=C bond)

Problem 3.15

Predict all bond angles for these molecules:

(a) CH_3OH (b) CH_2Cl_2 (c) H_2CO_3 (carbonic acid)

3.11 How Do We Determine If a Molecule Is Polar?

In Section 3.7B, we used the terms "polar" and "dipole" to describe a covalent bond in which one atom bears a partial positive charge and the other bears a partial negative charge. We also saw that we can use the difference in electronegativity between bonded atoms to determine the polarity of a covalent bond and the direction of its dipole. We can now combine our understanding of bond polarity and molecular geometry (Section 3.10) to predict the polarity of molecules. To discuss the physical and chemical properties of a molecule, it is essential to have an understanding of polarity. Many chemical reactions, for example, are driven by the interaction of the positive part of one molecule with the negative part of another molecule.

A molecule will be polar if (1) it has polar bonds and (2) its centers of partial positive charge and partial negative charge lie at different places within the molecule. Consider first carbon dioxide, CO_2, a molecule with two polar carbon-oxygen double bonds. The oxygen on the left pulls electrons of the $O\!=\!C$ bond toward it, giving it a partial negative charge. Similarly, the oxygen on the right pulls electrons of the $C\!=\!O$ bond toward it by the same amount, giving it the same partial negative charge as the oxygen on the left. Carbon bears a partial positive charge. We can show the polarity of these bonds by using the symbols $\delta+$ and $\delta-$. Alternatively, we can show that each carbon-oxygen bond has a dipole by using an arrow, where the head of the arrow points to the negative end of the dipole and the crossed tail is positioned at the positive end of the dipole. Because carbon dioxide is a linear molecule, its centers of negative and positive partial charge coincide. Therefore, CO_2 is a nonpolar molecule; that is, it has no dipole.

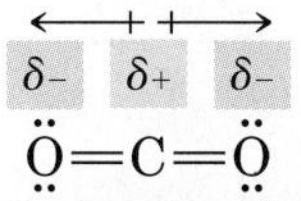

$$\overset{\displaystyle \longleftarrow\;+\;+\longrightarrow}{\underset{\displaystyle \underset{\ddot{}}{\overset{\ddot{}}{O}}=C=\underset{\ddot{}}{\overset{\ddot{}}{O}}}{\underset{\delta-\quad\delta+\quad\delta-}{}}$$

Carbon dioxide
(a nonpolar molecule)

In a water molecule, each O—H bond is polar. Oxygen, the more electronegative atom, bears a partial negative charge, and each hydrogen bears a partial positive charge. The center of partial positive charge in a water molecule is located halfway between the two hydrogen atoms, and the center of partial negative charge is on the oxygen atom. Thus, a water molecule has polar bonds and because of its geometry is a polar molecule.

Water
(a polar molecule)

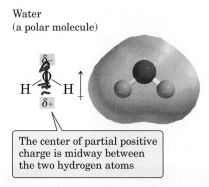

The center of partial positive charge is midway between the two hydrogen atoms

Ammonia has three polar N—H bonds. Because of its geometry, the centers of partial positive and partial negative charges are found at different places within the molecule. Thus, ammonia has polar bonds and because of its geometry is a polar molecule.

Ammonia
(a polar molecule)

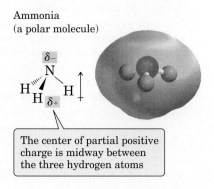

The center of partial positive charge is midway between the three hydrogen atoms

Example 3.16 Polarity of Covalent Molecules

Which of these molecules are polar? Show the direction of the molecular dipole by using an arrow with a crossed tail.

(a) CH_2Cl_2 (b) CH_2O (c) C_2H_2

Strategy

To determine whether a molecule is polar, first determine if it has polar bonds, and if it does, determine if the centers of positive and negative charge lie at the same or different places within the molecule. If they lie at the same place, the molecule is nonpolar; if they lie at different places, the molecule is polar.

Solution

Both dichloromethane, CH_2Cl_2, and formaldehyde, CH_2O, have polar bonds and because of their geometry are polar molecules. Because acetylene, C_2H_2, contains no polar bonds, it is a nonpolar molecule.

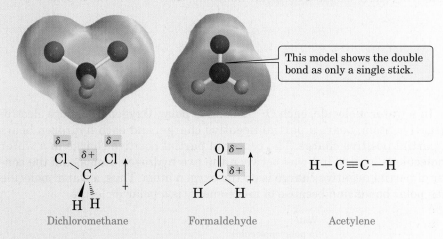

This model shows the double bond as only a single stick.

Dichloromethane   Formaldehyde   Acetylene

Problem 3.16

Which of these molecules are polar? Show the direction of the molecular dipole by using an arrow with a crossed tail.

(a) H_2S (b) HCN   (c) C_2H_6

Summary

OWL Sign in at **www.cengage.com/owl** to develop problem-solving skills and complete online homework assigned by your professor.

Section 3.2 What Is the Octet Rule?

- The **octet rule** states that elements of Groups 1A–7A tend to gain or lose electrons so as to achieve an outer shell containing eight valence electrons and the same electron configuration as that of the noble gas nearest it in atomic number.
- An atom with almost eight valence electrons tends to gain the needed electrons to have eight electrons in its valence shell—that is, to achieve the same electron configuration as the noble gas nearest to it in atomic number. In gaining electrons, the atom becomes a negatively charged ion called an **anion**.
- An atom with only one or two valence electrons tends to lose the number of electrons required to have eight valence electrons in its next lower shell—that is, to have the same electron configuration as the noble gas nearest it in atomic number. In losing electrons, the atom becomes a positively charged ion called a **cation**.

Section 3.3 How Do We Name Anions and Cations?

- For metals that form only one type of cation, the name of the cation is the name of the metal followed by the word "ion."
- For metals that form more than one type of cation, we show the charge on the ion by placing a Roman numeral enclosed in parentheses immediately following the name of the metal. Alternatively, for some elements, we use the suffix -*ous* to show the lower positive charge and -*ic* to show the higher positive charge.
- A **monatomic anion** is named by adding -*ide* to the stem part of the name.
- A **polyatomic ion** contains more than one type of atom.

Section 3.4 What Are the Two Major Types of Chemical Bonds?

- The two major types of chemical bonds are ionic bonds and covalent bonds.
- According to the Lewis model of chemical bonding, atoms bond together in such a way that each atom participating in the bond acquires a valence-shell electron configuration matching that of the noble gas nearest to it in atomic number.
- **Electronegativity** is a measure of the force of attraction that an atom exerts on electrons it shares in a chemical bond. It increases from left to right across a row and from bottom to top in a column of the Periodic Table.
- An **ionic bond** forms between two atoms if the difference in electronegativity between them is greater than 1.9.
- A **covalent bond** forms if the difference in electronegativity between the bonded atoms is 1.9 or less.

Section 3.5 What Is an Ionic Bond?

- An **ionic bond** forms by the transfer of valence-shell electrons from an atom of lower electronegativity to the valence shell of an atom of higher electronegativity.
- In an ionic compound, the total number of positive charges must equal the total number of negative charges.

Section 3.6 How Do We Name Ionic Compounds?

- For a **binary ionic compound**, name the cation first, followed by the name of the anion. Where a metal ion may form different cations, use a Roman numeral to show its positive charge. To name an ionic compound that contains one or more polyatomic ions, name the cation first, followed by the name of the anion.

Section 3.7 What Is a Covalent Bond?

- According to the Lewis model, a **covalent bond** forms when pairs of electrons are shared between two atoms whose difference in electronegativity is 1.9 or less.
- A pair of electrons in a covalent bond is shared by two atoms and at the same time fills the valence shell of each atom.

- A **nonpolar covalent bond** is a covalent bond in which the difference in electronegativity between bonded atoms is less than 0.5. A **polar covalent bond** is a covalent bond in which the difference in electronegativity between bonded atoms is between 0.5 and 1.9. In a polar covalent bond, the more electronegative atom bears a partial negative charge ($\delta-$) and the less electronegative atom bears a partial positive charge ($\delta+$). This separation of charge produces a **dipole**.
- A **Lewis structure** for a covalent compound must show (1) the correct arrangement of atoms, (2) the correct number of valence electrons, (3) no more than two electrons in the outer shell of hydrogen, and (4) no more than eight electrons in the outer shell of any second-period element.
- Exceptions to the octet rule include compounds of third-period elements, such as phosphorus and sulfur, which may have as many as 10 and 12 electrons, respectively, in their valence shells.

Section 3.8 How Do We Name Binary Covalent Compounds?

- To name a **binary covalent compound**, name the less electronegative element first, followed by the name of the more electronegative element. The name of the more electronegative element is derived by adding -*ide* to the stem name of the element. Use the prefixes *di-*, *tri-*, *tetra-*, and so on, to show the presence of two or more atoms of the same kind.

Section 3.9 What Is Resonance?

- According to the theory of resonance, a molecule or ion for which no single Lewis structure is adequate is best described by writing two or more **resonance contributing structures** and considering the real molecule or ion to be a hybrid of these contributing structures. To show how pairs of valence electrons are redistributed from one contributing structure to the next, we use curved arrows. A curved arrow extends from where a pair of electrons is initially shown (on an atom or in a covalent bond) to its new location (on an adjacent atom or an adjacent covalent bond).

Section 3.10 How Do We Predict Bond Angles in Covalent Molecules?

- The **valence-shell electron-pair repulsion (VSEPR) model** predicts bond angles of 109.5° about atoms surrounded by four regions of electron density, angles of 120° about atoms surrounded by three regions of electron density, and angles of 180° about atoms surrounded by two regions of electron density.

Section 3.11 How Do We Determine If a Molecule Is Polar?

- A molecule is polar (has a dipole) if it has polar bonds and the centers of its partial positive and partial negative charges do not coincide.
- If a molecule has polar bonds but the centers of its partial positive and negative charges coincide, the molecule is nonpolar (it has no dipole).

Problems

Orange-numbered problems are applied.

Section 3.2 What Is the Octet Rule?

3.17 Answer true or false.

(a) The octet rule refers to the chemical bonding patterns of the first eight elements of the Periodic Table.

(b) The octet rule refers to the tendency of certain elements to react in such a way that they achieve an outer shell of eight valence electrons.

(c) In gaining electrons, an atom becomes a positively charged ion called a cation.

(d) When an atom forms an ion, only the number of valence electrons changes; the number of protons and neutrons in the nucleus does not change.

(e) In forming ions, Group 2A elements typically lose two electrons to become cations with a charge of +2.

(f) In forming an ion, a sodium atom $(1s^2 2s^2 2p^6 3s^1)$ completes its valence shell by adding one electron to fill its 3s shell $(1s^2 2s^2 2p^6 3s^2)$.

(g) The elements of Group 6A typically react by accepting two electrons to become anions with a charge of -2.

(h) With the exception of hydrogen, the octet rule applies to all elements in periods 1,2, and 3.

(i) Atoms and the ions derived from them have very similar physical and chemical properties.

3.18 How many electrons must each atom gain or lose to acquire an electron configuration identical to the noble gas nearest to it in atomic number?

(a) Li   (b) Cl   (c) P   (d) Al
(e) Sr   (f) S   (g) Si   (h) O

3.19 Show how each chemical change obeys the octet rule.

(a) Lithium forms Li^+ (b) Oxygen forms O^{2-}

3.20 Show how each chemical change obeys the octet rule.

(a) Hydrogen forms H^- (hydride ion)

(b) Aluminum forms Al^{3+}

3.21 Write the formula for the most stable ion formed by each element.

(a) Mg   (b) F   (c) Al
(d) S   (e) K   (f) Br

3.22 Why is Li^- not a stable ion?

3.23 Predict which ions are stable:

(a) I^- (b) Se^{2+} (c) Na^+ (d) S^{2-} (e) Li^{2+} (f) Ba^{3+}

3.24 Predict which ions are stable:

(a) Br^{2-} (b) C^{4-} (c) Ca^+
(d) Ar^+ (e) Na^+ (f) Cs^+

3.25 Why are carbon and silicon reluctant to form ionic bonds?

3.26 Table 3.2 shows the following ions of copper: Cu^+ and Cu^{2+}. Do these violate the octet rule? Explain.

Section 3.3 How Do We Name Anions and Cations?

3.27 Answer true or false.

(a) For Group 1A and Group 2A elements, the name of the ion each forms is simply the name of the element followed by the word ion; for example, Mg^{2+} is named magnesium ion.

(b) H^+ is named hydronium ion, and H^- is named hydride ion.

(c) The nucleus of H^+ consists of one proton and one neutron.

(d) Many transition and inner transition elements form more than one positively charged ion.

(e) In naming metal cations with two different charges, the suffix -ous refers to the ion with a charge of +1 and -ic refers to the ion with a charge of +2.

(f) Fe^{3+} may be named either iron(III) ion or ferric ion.

(g) The anion derived from a bromine atom is named bromine ion.

(h) The anion derived from an oxygen atom is named oxide ion.

(i) HCO_3^- is named hydrogen carbonate ion.

(j) The prefix bi- in the name "bicarbonate" ion indicates that this ion has a charge of -2.

(k) The hydrogen phosphate ion has a charge of +1, and the dihydrogen phosphate ion has a charge of +2.

(l) The phosphate ion is PO_3^{4-}.

(m) The nitrite ion is NO_2^-, and the nitrate ion is NO_3^-.

(n) The carbonate ion is CO_3^{2-}, and the hydrogen carbonate ion is HCO_3^-.

3.28 Name each polyatomic ion.

(a) HCO_3^- (b) NO_2^- (c) SO_4^{2-}
(d) HSO_4^- (e) $H_2PO_4^-$

Section 3.4 What Are the Two Major Types of Chemical Bonds?

3.29 Answer true or false.

(a) According to the Lewis model of bonding, atoms bond together in such a way that each atom participating in the bond acquires an outer-shell electron configuration matching that of the noble gas nearest to it in atomic number.

(b) Atoms that lose electrons to achieve a filled valence shell become cations and form ionic bonds with anions.

(c) Atoms that gain electrons to achieve filled valence shells become anions and form ionic bonds with cations.

(d) Atoms that share electrons to achieve filled valence shells form covalent bonds.

(e) Ionic bonds tend to form between elements on the left side of the Periodic Table, and covalent bonds tend to form between elements on the right side of the Periodic Table.

(f) Ionic bonds tend to form between a metal and a nonmetal.

(g) When two nonmetals combine, the bond between them is usually covalent.

(h) Electronegativity is a measure of an atom's attraction for the electrons it shares in a chemical bond with another atom.

(i) Electronegativity generally increases with atomic number.

(j) Electronegativity generally increases with atomic weight.

(k) Electronegativity is a periodic property.

(l) Fluorine, in the upper-right corner of the Periodic Table, is the most electronegative element; hydrogen, in the upper-left corner, is the least electronegative element.

(m) Electronegativity depends on both the nuclear charge and the distance of the valence electrons from the nucleus.

(n) Electronegativity generally increases from left to right across a period of the Periodic Table.

(o) Electronegativity generally increases from top to bottom in a column of the Periodic Table.

3.30 Why does electronegativity generally increase going up a column (group) of the Periodic Table?

3.31 Why does electronegativity generally increase going from left to right across a row of the Periodic Table?

3.32 Judging from their relative positions in the Periodic Table, which element in each pair has the larger electronegativity?
(a) F or Cl   (b) O or S   (c) C or N   (d) C or F

3.33 Toward which atom are the bonding electrons shifted in a covalent bond between each of the following pairs:
(a) H and Cl   (b) N and O
(c) C and O   (d) Cl and Br
(e) C and S   (f) P and S   (g) H and O

3.34 Which of these bonds is the most polar? The least polar?
(a) C—N   (b) C—C   (c) C—O

3.35 Classify each bond as nonpolar covalent, polar covalent, or ionic.
(a) C—Cl   (b) C—Li   (c) C—N

3.36 Classify each bond as nonpolar covalent, polar covalent, or ionic.
(a) C—Br   (b) S—Cl   (c) C—P

Section 3.5 What Is an Ionic Bond?

3.37 Answer true or false.
(a) An ionic bond is formed by the combination of positive and negative ions.

(b) An ionic bond between two atoms forms by the transfer of one or more valence electrons from the atom of higher electronegativity to the atom of lower electronegativity.

(c) As a rough guideline, we say that an ionic bond will form if the difference in electronegativity between two atoms is approximately 1.9 or greater.

(d) In forming NaCl from sodium and chlorine atoms, one electron is transferred from the valence shell of sodium to the valence shell of chlorine.

(e) The formula of sodium sulfide is Na_2S.

(f) The formula of calcium hydroxide is CaOH.

(g) The formula of aluminum sulfide is AlS.

(h) The formula of iron(III) oxide is Fe_3O_2.

(i) Barium ion is Ba^{2+}, and oxide ion is O^{2-}; therefore, the formula of barium oxide is Ba_2O_2.

3.38 Complete the chart by writing formulas for the compounds formed:

	Br^-	MnO_4^-	O^{2-}	NO_3^-	SO_4^{2-}	PO_4^{3-}	OH^-
Li^+							
Ca^{2+}							
Co^{3+}							
K^+							
Cu^{2+}							

3.39 Write a formula for the ionic compound formed from each pair of elements.
(a) Sodium and bromine   (b) Sodium and oxygen
(c) Aluminum and chlorine   (d) Barium and chlorine
(e) Magnesium and oxygen

3.40 Although not a transition metal, lead can form Pb^{2+} and Pb^{4+} ions. Write the formula for the compound formed between each of these lead ions and the following anions:
(a) Chloride ion   (b) Hydroxide ion
(c) Oxide ion

3.41 Describe the structure of sodium chloride in the solid state.

3.42 What is the charge on each ion in these compounds?
(a) CaS   (b) MgF_2 (c) Cs_2O
(d) $ScCl_3$ (e) Al_2S_3

3.43 Write the formula for the compound formed from the following pairs of ions:
(a) Iron(III) ion and hydroxide ion
(b) Barium ion and chloride ion
(c) Calcium ion and phosphate ion
(d) Sodium ion and permanganate ion

3.44 Write the formula for the ionic compound formed from the following pairs of ions:
  (a) Iron(II) ion and chloride ion
  (b) Calcium ion and hydroxide ion
  (c) Ammonium ion and phosphate ion
  (d) Tin(II) ion and fluoride ion

3.45 Which formulas are not correct? For each that is not correct, write the correct formula.
  (a) Ammonium phosphate; $(NH_4)_2PO_4$
  (b) Barium carbonate; Ba_2CO_3
  (c) Aluminum sulfide; Al_2S_3
  (d) Magnesium sulfide; MgS

3.46 Which formulas are not correct? For each that is not correct, write the correct formula.
  (a) Calcium oxide; CaO_2
  (b) Lithium oxide; LiO
  (c) Sodium hydrogen phosphate; $NaHPO_4$
  (d) Ammonium nitrate; NH_4NO_3

Section 3.6 How Do We Name Ionic Compounds?

3.47 Answer true or false.
  (a) The name of a binary ionic compound consists of the name of the positive ion followed by the name of the negative ion.
  (b) In naming binary ionic compounds, it is necessary to state the number of each ion present in the compound.
  (c) The formula of aluminum oxide is Al_2O_3.
  (d) Both copper(II) oxide and cupric oxide are acceptable names for CuO.
  (e) The systematic name for Fe_2O_3 is iron(II) oxide.
  (f) The systematic name for $FeCO_3$ is iron carbonate.
  (g) The systematic name for NaH_2PO_4 is sodium dihydrogen phosphate.
  (h) The systematic name for K_2HPO_4 is dipotassium hydrogen phosphate.
  (i) The systematic name for Na_2O is sodium oxide.
  (j) The systematic name for PCl_3 is potassium chloride.
  (k) The formula of ammonium carbonate is NH_4CO_3.

3.48 Potassium chloride and potassium bicarbonate are used as potassium dietary supplements. Write the formula of each compound.

3.49 Potassium nitrite has been used as a vasodilator and as an antidote for cyanide poisoning. Write the formula of this compound.

3.50 Name the polyatomic ion(s) in each compound.
  (a) Na_2SO_3 (b) KNO_3 (c) Cs_2CO_3
  (d) NH_4OH (e) K_2HPO_4

3.51 Write the formulas for the ions present in each compound.
  (a) $NaBr$ (b) $FeSO_3$ (c) $Mg_3(PO_4)_2$
  (d) KH_2PO_4 (e) $NaHCO_3$ (f) $Ba(NO_3)_2$

3.52 Name these ionic compounds:
  (a) NaF (b) MgS (c) Al_2O_3
  (d) $BaCl_2$ (e) $Ca(HSO_3)_2$ (f) KI
  (g) $Sr_3(PO_4)_2$ (h) $Fe(OH)_2$ (i) NaH_2PO_4
  (j) $Pb(CH_3COO)_2$ (k) BaH_2 (l) $(NH_4)_2HPO_4$

3.53 Write formulas for the following ionic compounds:
  (a) Potassium bromide   (b) Calcium oxide
  (c) Mercury(II) oxide   (d) Copper(II) phosphate
  (e) Lithium sulfate   (f) Iron(III) sulfide

3.54 Write formulas for the following ionic compounds:
  (a) Ammonium hydrogen sulfite
  (b) Magnesium acetate
  (c) Strontium dihydrogen phosphate
  (d) Silver carbonate
  (e) Strontium chloride
  (f) Barium permanganate

Section 3.7 What Is a Covalent Bond?

3.55 Answer true or false.
  (a) A covalent bond is formed between two atoms whose difference in electronegativity is less than 1.9.
  (b) If the difference in electronegativity between two atoms is zero (they have identical electronegativities), then the two atoms will not form a covalent bond.
  (c) A covalent bond formed by sharing two electrons is called a double bond.
  (d) In the hydrogen molecule (H_2), the shared pair of electrons completes the valence shell of each hydrogen.
  (e) In the molecule CH_4, each hydrogen has an electron configuration like that of helium and carbon has an electron configuration like that of neon.
  (f) In a polar covalent bond, the more electronegative atom has a partial negative charge ($\delta-$) and the less electronegative atom has a partial positive charge ($\delta+$).
  (g) These bonds are arranged in order of *increasing* polarity $C—H < N—H < O—H$.
  (h) These bonds are arranged in order of *increasing* polarity $H—F < H—Cl < H—Br$.
  (i) A polar bond has a dipole with the negative end located at the more electronegative atom.
  (j) In a single bond, two atoms share one pair of electrons; in a double bond, they share two pairs of electrons; and in a triple bond, they share three pairs of electrons.
  (k) The Lewis structure for ethane, C_2H_6, must show eight valence electrons.
  (l) The Lewis structure for formaldehyde, CH_2O, must show 12 valence electrons.
  (m) The Lewis structure for the ammonium ion, NH_4^+, must show nine valence electrons.
  (n) Atoms of third-period elements can hold more than eight electrons in their valence shells.

3.56 How many covalent bonds are normally formed by each element?
(a) N (b) F (c) C (d) Br (e) O

3.57 What is:
(a) A single bond? (b) A double bond?
(c) A triple bond?

3.58 In Section 2.3B, we saw that there are seven diatomic elements.
(a) Draw Lewis structures for each of these diatomic elements.
(b) Which diatomic elements are gases at room temperature? Which are liquids? Which are solids?

3.59 Draw a Lewis structure for each covalent compound.
(a) CH_4 (b) C_2H_2 (c) C_2H_4
(d) BF_3 (e) CH_2O (f) C_2Cl_6

3.60 What is the difference between a molecular formula, a structural formula, and a Lewis structure?

3.61 State the total number of valence electrons in each molecule.
(a) NH_3 (b) C_3H_6 (c) $C_2H_4O_2$ (d) C_2H_6O
(e) CCl_4 (f) HNO_2 (g) CCl_2F_2 (h) O_2

3.62 Draw a Lewis structure for each of the following molecules and ions. In each case, the atoms can be connected in only one way.
(a) Br_2 (b) H_2S (c) N_2H_4 (d) N_2H_2
(e) CN^- (f) NH_4^+ (g) N_2 (h) O_2

3.63 What is the difference between (a) a bromine atom, (b) a bromine molecule, and (c) a bromide ion? Draw the Lewis structure for each.

3.64 Acetylene (C_2H_2), hydrogen cyanide (HCN), and nitrogen (N_2) each contain a triple bond. Draw a Lewis structure for each molecule. Which of these are polar molecules, and which are nonpolar molecules?

3.65 Why can't hydrogen have more than two electrons in its valence shell?

3.66 Why can't second-row elements have more than eight electrons in their valence shells? That is, why does the octet rule work for second-row elements?

3.67 Why does nitrogen have three bonds and one unshared pair of electrons in covalent compounds?

3.68 Draw a Lewis structure of a covalent compound in which nitrogen has:
(a) Three single bonds and one unshared pair of electrons
(b) One single bond, one double bond, and one unshared pair of electrons
(c) One triple bond and one unshared pair of electrons

3.69 Why does oxygen have two bonds and two unshared pairs of electrons in covalent compounds?

3.70 Draw a Lewis structure of a covalent compound in which oxygen has:
(a) Two single bonds and two unshared pairs of electrons

(b) One double bond and two unshared pairs of electrons

3.71 The ion O^{6+} has a complete outer shell. Why is this ion not stable?

3.72 Draw a Lewis structure for a molecule in which a carbon atom is bonded by a double bond to (a) another carbon atom, (b) an oxygen atom, and (c) a nitrogen atom.

3.73 Which of the following molecules have an atom that does not obey the octet rule (not all of these are stable molecules)?
(a) BF_3 (b) CF_2 (c) BeF_2 (d) C_2H_4
(e) CH_3 (f) N_2 (g) NO

Section 3.8 How Do We Name Binary Covalent Compounds?

3.74 Answer true or false.
(a) A binary covalent compound contains two kinds of atoms.
(b) The two types of atoms in a binary covalent compound are named in this order: first the more electronegative element and then the less electronegative element.
(c) The name for SF_2 is sulfur difluoride.
(d) The name for CO_2 is carbon dioxide.
(e) The name for CO is carbon oxide.
(f) The name for HBr is hydrogen bromide.
(g) The name for CCl_4 is carbon tetrachloride.

3.75 Name these binary covalent compounds.
(a) SO_2 (b) SO_3 (c) PCl_3 (d) CS_2

Section 3.9 What Is Resonance?

3.76 Write two acceptable contributing structures for the bicarbonate ion, HCO_3^- and show by the use of curved arrows how the first contributing structure is converted to the second.

3.77 Ozone, O_3, is an unstable blue gas with a characteristic pungent odor. In an ozone molecule, the connectivity of the atoms is O—O—O and both O—O bonds are equivalent.
(a) How many valence electrons must be present in an acceptable Lewis structure for an ozone molecule?
(b) Write two equivalent resonance contributing structures for ozone. Be certain to show any positive or negative charges that may be present in your contributing structures. By equivalent contributing structures, we mean that each has the same pattern of bonding.
(c) Show by the use of curved arrows how the first of your contributing structures may be converted to the second.
(d) Based on your contributing structures, predict the O—O—O bond angle in ozone.
(e) Explain why the following is not an acceptable contributing structure for an ozone molecule:
$$\ddot{O}=\ddot{O}=\ddot{O}$$

3.78 Nitrous oxide, N_2O, laughing gas, is a colorless, nontoxic, tasteless, and odorless gas. It is used as an inhalation anesthetic in dental and other surgeries. Because nitrous oxide is soluble in vegetable oils (fats), it is used commercially as a propellant in whipped toppings.

Charles D. Winters

Nitrous oxide dissolves in fats. The gas is added under pressure to cans of whipped topping. When the valve is opened, the gas expands, thus expanding (whipping) the topping and forcing it out of the can.

(a) How many valence electrons are present in a molecule of N_2O?

(b) Write two equivalent contributing structures for this molecule. The connectivity in nitrous oxide is N—N—O.

(c) Explain why the following is not an acceptable contributing structure:

$$:N{=}\!\!\!=\!\!\!=N{=}\ddot{\ddot{O}}$$

Section 3.10 How Do We Predict Bond Angles in Covalent Molecules?

3.79 Answer true or false.

(a) The letters VSEPR stand for valence-shell electron-pair repulsion.

(b) In predicting bond angles about a central atom in a covalent molecule, the VSEPR model considers only shared electron pairs (electron pairs involved in forming covalent bonds).

(c) The VSEPR model treats the two electron pairs of a double bond as one region of electron density and the three electron pairs of a triple bond as one region of electron density.

(d) In carbon dioxide, O═C═O, carbon is surrounded by four pairs of electrons and the VSEPR model predicts 109.5° for the O—C—O bond angle.

(e) For a central atom surrounded by three regions of electron density, the VSEPR model predicts bond angles of 120°.

(f) The geometry about a carbon atom surrounded by three regions of electron density is described as trigonal planar.

(g) For a central atom surrounded by four regions of electron density, the VSEPR model predicts bond angles of 360°/4 = 90°.

(h) For the ammonia molecule, NH_3, the VSEPR model predicts H—N—H bond angles of 109.5°.

(i) For the ammonium ion, NH_4^+, the VSEPR model predicts H—N—H bond angles of 109.5°.

(j) The VSEPR model applies equally well to covalent compounds of carbon, nitrogen, and oxygen.

(k) In water, H—O—H, the oxygen atom forms covalent bonds to two other atoms, and therefore, the VSEPR model predicts an H—O—H bond angle of 180°.

(l) If you fail to consider unshared pairs of valence electrons when you use the VSEPR model, you will arrive at an incorrect prediction.

(m) Given the assumptions of the VSEPR model, the only bond angles it predicts for compounds of carbon, nitrogen, and oxygen are 109.5°, 120°, and 180°.

3.80 State the shape of a molecule whose central atom is surrounded by:

(a) Two regions of electron density

(b) Three regions of electron density

(c) Four regions of electron density

3.81 Hydrogen and oxygen combine in different ratios to form H_2O (water) and H_2O_2 (hydrogen peroxide).

(a) How many valence electrons are found in H_2O? In H_2O_2?

(b) Draw Lewis structures for each molecule in part (a). Be certain to show all valence electrons.

(c) Using the VSEPR model, predict the bond angles about the oxygen atom in water and about each oxygen atom in hydrogen peroxide.

3.82 Hydrogen and nitrogen combine in different ratios to form three compounds: NH_3 (ammonia), N_2H_4 (hydrazine), and N_2H_2 (diimide).

(a) How many valence electrons must the Lewis structure of each molecule show?

(b) Draw a Lewis structure for each molecule.

(c) Predict the bond angles about the nitrogen atom(s) in each molecule.

3.83 Predict the shape of each molecule.

(a) CH_4　　(b) PH_3　　(c) CHF_3　　(d) SO_2

(e) SO_3　　(f) CCl_2F_2　　(g) NH_3　　(h) PCl_3

3.84 Predict the shape of each ion.

(a) NO_2^-　　(b) NH_4^+　　(c) CO_3^{2-}

Section 3.11 How Do We Determine If a Molecule Is Polar?

3.85 Answer true or false.

(a) To predict whether a covalent molecule is polar or nonpolar, you must know both the polarity of each bond and the geometry (shape) of the molecule.

(b) A molecule may have two or more polar bonds and still be nonpolar.

(c) All molecules with polar bonds are polar.

(d) If water were a linear molecule with an H—O—H bond angle of 180°, water would be a nonpolar molecule.

(e) H_2O and NH_3 are polar molecules, but CH_4 is nonpolar.

(f) In methanol, CH_3OH, the O—H bond is more polar than the C—O bond.

(g) Dichloromethane, CH_2Cl_2, is polar, but tetrachloromethane, CCl_4, is nonpolar.

(h) Ethanol, CH_3CH_2OH, the alcohol of alcoholic beverages, has polar bonds, has a net dipole, and is a polar molecule.

3.86 Both CO_2 and SO_2 have polar bonds. Account for the fact that CO_2 is nonpolar and SO_2 is polar.

3.87 Consider the molecule boron trifluoride, BF_3.

(a) Write a Lewis structure for BF_3.

(b) Predict the F—B—F bond angles using the VSEPR model.

(c) Does BF_3 have polar bonds? Is it a polar molecule?

3.88 Is it possible for a molecule to have polar bonds and yet have no dipole? Explain.

3.89 Is it possible for a molecule to have no polar bonds and yet have a dipole? Explain.

3.90 In each case, tell whether the bond is ionic, polar covalent, or nonpolar covalent.

(a) Br_2 (b) BrCl   (c) HCl   (d) SrF_2

(e) SiH_4 (f) CO   (g) N_2 (h) CsCl

3.91 Account for the fact that chloromethane, CH_3Cl, which has only one polar C—Cl bond, is a polar molecule, but carbon tetrachloride, CCl_4, which has four polar C—Cl bonds, is a nonpolar molecule.

Chemical Connections

3.92 (Chemical Connections 3A) What are the three main inorganic components of one dry mixture currently used to create synthetic bone?

3.93 (Chemical Connections 3B) Why is sodium iodide often present in the table salt we buy at the grocery store?

3.94 (Chemical Connections 3B) What is a medical use of barium sulfate?

3.95 (Chemical Connections 3B) What is a medical use of potassium permanganate?

3.96 (Chemical Connections 3A) What is the most prevalent metal ion in bone and tooth enamel?

3.97 (Chemical Connections 3C) In what way does the gas nitric oxide, NO, contribute to the acidity of acid rain?

Additional Problems

3.98 Explain why argon does not form either (a) ionic bonds or (b) covalent bonds.

3.99 Knowing what you do about covalent bonding in compounds of carbon, nitrogen, and oxygen and given the fact that silicon is just below carbon in the Periodic Table, phosphorus is just below nitrogen, and sulfur is just below oxygen, predict the molecular formula for the compound formed by (a) silicon and chlorine, (b) phosphorus and hydrogen, and (c) sulfur and hydrogen.

3.100 Use the valence-shell electron-pair repulsion model to predict the shape of a molecule in which a central atom is surrounded by five regions of electron density—as, for example, in phosphorus pentafluoride, PF_5. (Hint: Use molecular models or if you do not have a set handy, use marshmallows or gumdrops and toothpicks.)

3.101 Use the valence-shell electron-pair repulsion model to predict the shape of a molecule in which a central atom is surrounded by six regions of electron density, as, for example, in sulfur hexa-fluoride, SF_6.

3.102 Chlorine dioxide, ClO_2, is a yellow to reddish yellow gas at room temperature. This strong oxidizing agent is used for bleaching cellulose, paper pulp, and textiles and for water purification. It was the gas used to kill anthrax spores in the anthrax-contaminated Hart Senate Office Building.

(a) How many valence electrons are present in ClO_2?

(b) Draw a Lewis structure for this molecule. (Hint: The order of attachment of atoms in this molecule is O—Cl—O. Chlorine is a third-period element, and its valence shell may contain more than eight electrons.)

3.103 Using the information in Figure 2.16, estimate the H—O and H—S distances (the atom—atom distances) in H_2O and H_2S, respectively.

3.104 Arrange the single covalent bonds within each set in order of increasing polarity.

(a) C—H, O—H, N—H   (b) C—H, C—Cl, C—I

(c) C—C, C—O, C—N

3.105 Consider the structure of Vitamin E shown below, which is found most abundantly in wheat germ oil, sunflower, and safflower oils:

Vitamin E

(a) Identify the various types of geometries present in each central atom using VSEPR theory.

(b) Determine the various relative bond angles associated with each central atom using VSEPR theory.

(c) Which is the most polar bond in Vitamin E?

(d) Would you predict Vitamin E to be polar or nonpolar?

3.106 Consider the structure of Penicillin G shown below, an antibiotic used to treat bacterial infections caused by gram-positive organisms, derived from *Penicillium* fungi:

Penicillin G

(a) Identify the various types of geometries present in each central atom using VSEPR theory.

(b) Determine the various relative bond angles associated with each central atom using VSEPR theory.

(c) Which is the most polar bond in Penicillin G?

(d) Would you predict Penicillin G to be polar or nonpolar?

3.107 Ephedrine, a molecule at one time found in the dietary supplement ephedra, has been linked to adverse health reactions, such as heart attacks, strokes, and heart palpitations. The use of ephedra in dietary supplements is now banned by the FDA.

Ephedrine

(a) Which is the most polar bond in ephedra?

(b) Would you predict ephedra to be polar or nonpolar?

3.108 Allene, C_3H_4, has the structural formula $CH_2=C=CH_2$.

(a) Describe the shape of this molecule.

(b) Is allene polar or nonpolar?

3.109 Until several years ago, the two chlorofluorocarbons (CFCs) most widely used as heat transfer media in refrigeration systems were Freon-11 (trichlorofluoromethane, CCl_3F) and Freon-12 (dichlorodifluoromethane, CCl_2F_2). Draw a three-dimensional representation of each molecule and indicate the direction of its polarity.

Reading Labels

3.110 Name and write the formula for the fluorine-containing compound present in fluoridated toothpastes and dental gels.

3.111 If you read the labels of sun-blocking lotions, you will find that a common UV-blocking agent is a compound containing zinc. Name and write the formula of this zinc-containing compound.

3.112 On packaged table salt, it is common to see a label stating that the salt "supplies iodide, a necessary nutrient." Name and write the formula of the iodine-containing nutrient compound found in iodized salt.

3.113 We are constantly warned about the dangers of "lead-based" paints. Name and write the formula for a lead-containing compound found in lead-based paints.

3.114 If you read the labels of several liquid and tablet antacid preparations, you will find that in many of them, the active ingredients are compounds containing hydroxide ions. Name and write formulas for these hydroxide ion–containing compounds.

3.115 Iron forms Fe^{2+} and Fe^{3+} ions. Which ion is found in over-the-counter preparations intended to treat "iron-poor blood"?

3.116 Read the labels of several multivitamin/multimineral formulations. Among their components, you will find a number of so-called trace minerals—minerals required in the diet of a healthy adult in amounts less than 100 mg per day or present in the body in amounts less than 0.01% of total body weight. Following are 18 trace minerals. Name at least one form of each trace mineral present in multivitamin formulations.

(a) Phosphorus   (b) Magnesium
(c) Potassium   (d) Iron
(e) Calcium   (f) Zinc
(g) Manganese   (h) Titanium
(i) Silicon   (j) Copper
(k) Boron   (l) Molybdenum
(m) Chromium   (n) Iodine
(o) Selenium   (p) Vanadium
(q) Nickel   (r) Tin

3.117 Write formulas for these compounds.

(a) Calcium sulfite, which is used in preserving cider and other fruit juices

(b) Calcium hydrogen sulfite, which is used in dilute aqueous solutions for washing casks in brewing to prevent souring and cloudiness of beer and to prevent secondary fermentation

(c) Calcium hydroxide, which is used in mortar, plaster, cement, and other building and paving materials

(d) Calcium hydrogen phosphate, which is used in animal feeds and as a mineral supplement in cereals and other foods

3.118 Many paint pigments contain transition metal compounds. Name the compounds in these pigments using a Roman numeral to show the charge on a transition metal ion.

(a) Yellow, CdS   (b) Green, Cr_2O_3
(c) White, TiO_2 (d) Purple, $Mn_3(PO_4)_2$
(e) Blue, Co_2O_3 (f) Ochre, Fe_2O_3

Looking Ahead

3.119 Perchloroethylene, which is a liquid at room temperature, is one of the most widely used solvents for commercial dry cleaning. It is sold for this purpose under several trade names, including Perclene®. Does this molecule have polar bonds? Is it a polar molecule? Does it have a dipole?

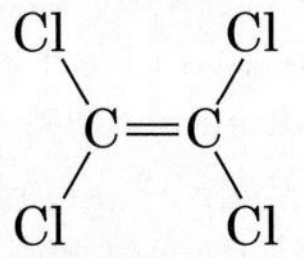

Perchloroethylene

3.120 Vinyl chloride is the starting material for the production of poly(vinyl chloride), abbreviated PVC. Its recycling code is "V". The major use of PVC is for tubing in residential and commercial construction (Section 12.7).

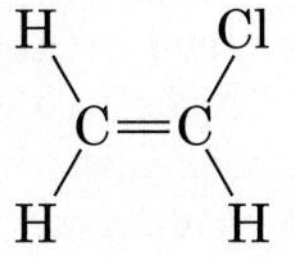

Vinyl chloride

(a) Complete the Lewis structure for vinyl chloride by showing all unshared pairs of electrons.
(b) Predict the H—C—H, H—C—C, and Cl—C—H bond angles in this molecule.
(c) Does vinyl chloride have polar bonds? Is it a polar molecule? Does it have a dipole?

3.121 Tetrafluoroethylene is the starting material for the production of poly(tetrafluoroethylene), PTFE, a polymer that is widely used for the preparation of nonstick coatings on kitchenware (Section 12.7). The most widely known trade name for this product is Teflon®.

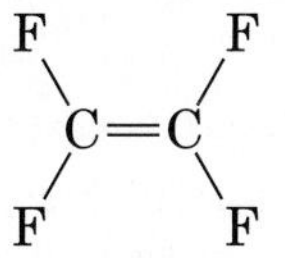

Tetrafluoroethylene

(a) Complete the Lewis structure for tetrafluoroethylene by showing all unshared pairs of electrons.
(b) Predict the F—C—F and F—C—C bond angles in this molecule.

(c) Does tetrafluoroethylene have polar bonds? Is it a polar molecule? Does it have a dipole?

3.122 Some of the following structural formulas are incorrect because they contain one or more atoms that do not have their normal number of covalent bonds. Which structural formulas are incorrect, and which atom or atoms in each have the incorrect number of bonds?

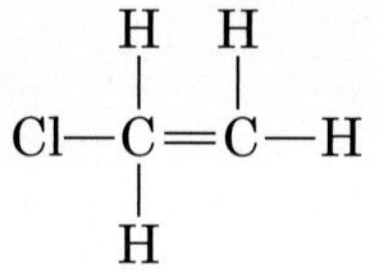

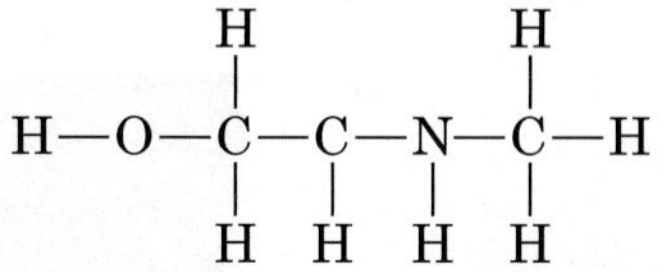

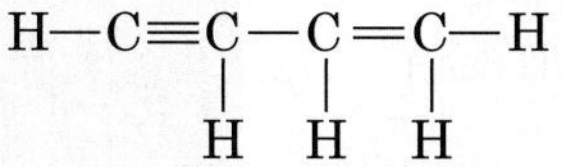

3.123 Sodium borhydride, $NaBH_4$, has found wide use as a reducing agent in organic chemistry. It is an ionic compound composed of one sodium ion, Na^+, and one borohydride ion, BH_4^-.

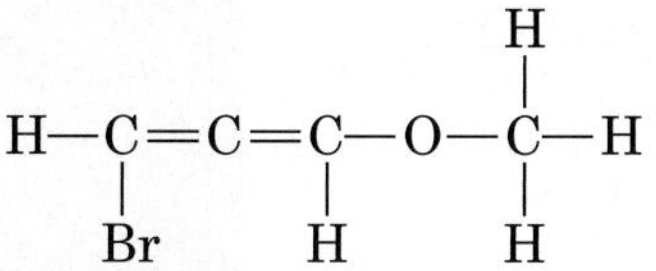

(a) How many valence electrons are present in the borohydride ion?
(b) Draw a Lewis structure for the borohydride ion.
(c) Predict the H—B—H bond angles in the borohydride ion.

3.124 Given your answer to Problem 3.123 and knowing that aluminum is immediately below boron in column 3A of the Periodic Table, propose a structure for lithium aluminum hydride, another widely used reducing agent in organic chemistry.

4

Chemical Reactions

Joseph Nettis/Photo Researchers, Inc.

Fireworks are spectacular displays of chemical reactions.

4.1 What Is a Chemical Reaction?

In Chapter 1, we learned that chemistry is mainly concerned with two things: the structure of matter and the transformations from one form of matter to another. In Chapters 2 and 3, we discussed the first of these topics, and now we are ready to turn our attention to the second. In a chemical change, also called a **chemical reaction**, one or more **reactants** (starting materials) are converted into one or more **products**. Chemical reactions occur all around us. They fuel and keep alive the cells of living tissues; they occur when we light a match, cook a dinner, start a car, listen to a CD player, or watch television. Most of the world's manufacturing processes involve chemical reactions; they include petroleum refining and food processing as well as the manufacture of drugs, plastics, synthetic fibers, fertilizers, explosives, and many other materials.

In this chapter, we discuss four aspects of chemical reactions: (1) how to write and balance chemical equations, (2) types of chemical reactions, (3) mass relationships in chemical reactions, and (4) heat gains and losses.

4.2 How Do We Balance Chemical Equations?

When propane, which is the major component in bottled gas or LPG (liquefied petroleum gas), burns in air, it reacts with the oxygen in the air. These two reactants are converted to the products carbon dioxide and water in a chemical reaction called **combustion**. We can write this chemical reaction in the form of a **chemical equation**, using chemical formulas for the reactants and products and an arrow to indicate the direction in which the reaction proceeds. In addition, it is important to show the state of each reactant and product; that is, whether it is a gas, liquid, or solid. We use the symbol (g) for gas, (l) for liquid, (s) for solid, and (aq) for a substance dissolved in water (aqueous). We place the appropriate symbol immediately following each reactant and product. In our combustion equation, propane, oxygen, and carbon dioxide are gases, and the flame produced when propane burns is hot enough so that the water that forms is a gas (steam).

Combustion Burning in air

Chemical equation A representation using chemical formulas of the process that occurs when reactants are converted to products

$$C_3H_8(g) + O_2(g) \longrightarrow CO_2(g) + H_2O(g)$$

Propane   Oxygen   Carbon   Water
                             dioxide

The equation we have written is incomplete, however. While it tells us the formulas of the starting materials and products (which every chemical equation must do) and the physical state of each reactant and product, it does not give the amounts correctly. It is not balanced, which means that the number of atoms on the left side of the equation is not the same as the number of atoms on the right side. From the law of conservation of mass (Section 2.3A), we know that atoms are neither destroyed nor created in chemical reactions; they merely shift from one substance to another. Thus, all of the atoms present at the start of the reaction (on the left side of the equation) must still be present at the end (on the right side of the equation). In the equation we have just written, three carbon atoms are on the left but only one is on the right.

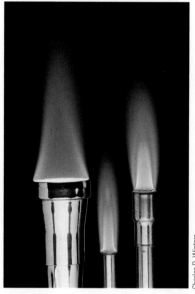

Propane burning in air

Charles D. Winters

How To . . .

Balance a Chemical Equation

To balance an equation, we place numbers in front of the formulas until the number of each kind of atom in the products is the same as that in the starting materials. These numbers are called **coefficients.** As an example, let us balance our propane equation:

$$C_3H_8(g) + O_2(g) \longrightarrow CO_2(g) + H_2O(g)$$

Propane   Oxygen   Carbon   Water
                             dioxide

To balance an equation:

1. Begin with atoms that appear in only one compound on the left and only one compound on the right. In the equation for the reaction of propane and oxygen, begin with either carbon or hydrogen.
2. If an atom occurs as a free element—as, for example O_2, in the reaction of propane with oxygen—balance this element last.
3. You can change only coefficients in balancing an equation; you cannot change chemical formulas. For example, if you have H_2O on the left side of an equation but need two oxygens, you can add the coefficient "2" to read $2H_2O$. You cannot, however, get two oxygens by changing

the formula to H_2O_2. Doing so would change the chemical composition of the expected product from water, H_2O, to hydrogen peroxide, H_2O_2.

In the equation for the combustion (burning) of propane with oxygen, we can begin with carbon. Three carbon atoms appear on the left and one on the right. If we put a 3 in front of the CO_2 (indicating that three CO_2 molecules are formed), three carbons will appear on each side and the carbons will be balanced:

$$\overbrace{\quad}^{\text{Three C on each side}}$$

$$C_3H_8(g) + O_2(g) \longrightarrow 3CO_2(g) + H_2O(g)$$

Next, we look at the hydrogens. There are eight on the left and two on the right. If we put a 4 in front of the H_2O, there will be eight hydrogens on each side and the hydrogens will be balanced:

$$\overbrace{\quad}^{\text{Eight H on each side}}$$

$$C_3H_8(g) + O_2(g) \longrightarrow 3CO_2(g) + 4H_2O(g)$$

The only atom still unbalanced is oxygen. Notice that we saved this reactant for last (rule 2). There are two oxygen atoms on the left and ten on the right. If we put a 5 in front of the O_2 on the left, we both balance the oxygen atoms and arrive at the balanced equation:

$$\overbrace{\quad}^{\text{Ten O on each side}}$$

$$C_3H_8(g) + 5O_2(g) \longrightarrow 3CO_2(g) + 4H_2O(g)$$

At this point, the equation ought to be balanced, but we should always check, just to make sure. In a balanced equation, there must be the same number of atoms of each element on both sides. A check of our work shows three C, ten O, and eight H atoms on each side. The equation is indeed balanced.

⬤WL Interactive Example 4.1 Balancing a Chemical Equation

Balance this equation:

$$\underset{\substack{\text{Calcium} \\ \text{hydroxide}}}{Ca(OH)_2(s)} + \underset{\substack{\text{Hydrogen} \\ \text{chloride}}}{HCl(g)} \longrightarrow \underset{\substack{\text{Calcium} \\ \text{chloride}}}{CaCl_2(s)} + H_2O(\ell)$$

Strategy

To balance an equation, we place numbers in front of the formulas until there are identical numbers of atoms on each side of the equation. Begin with atoms that appear in only one compound on the left and only one compound on the right.

Solution

The calcium is already balanced—there is one Ca on each side. There is one Cl on the left and two on the right. To balance chlorine, we add the coefficient 2 in front of HCl:

$$Ca(OH)_2(s) + 2HCl(g) \longrightarrow CaCl_2(s) + H_2O(\ell)$$

Looking at hydrogens, we see that there are four hydrogens on the left but only two on the right. Placing the coefficient 2 in front of H_2O balances the hydrogens. It also balances the oxygens and completes the balancing of the equation:

$$Ca(OH)_2(s) + 2HCl(g) \longrightarrow CaCl_2(s) + 2H_2O(\ell)$$

Problem 4.1

Following is an unbalanced equation for photosynthesis, the process by which green plants convert carbon dioxide and water to glucose and oxygen. Balance this equation:

$$CO_2(g) + H_2O(\ell) \xrightarrow{\text{Photosynthesis}} C_6H_{12}O_6(aq) + O_2(g)$$
$$\text{Glucose}$$

Example 4.2 Balancing a Chemical Equation

Balance this equation for the combustion of butane, the fluid most commonly used in pocket lighters:

$$C_4H_{10}(g) + O_2(g) \longrightarrow CO_2(g) + H_2O(g)$$
$$\text{Butane}$$

Strategy

The equation for the combustion of butane is very similar to the one we examined at the beginning of this section for the combustion of propane. To balance an equation, we place numbers in front of the formulas until there are identical numbers of atoms on each side of the equation.

Solution

To balance carbons, put a 4 in front of the CO_2 (because there are four carbons on the left). Then to balance hydrogens, place a 5 in front of the H_2O to give ten hydrogens on each side of the equation.

$$C_4H_{10}(g) + O_2(g) \longrightarrow 4CO_2(g) + 5H_2O(g)$$

When we count the oxygens, we find 2 on the left and 13 on the right. We can balance their numbers by putting 13/2 in front of the O_2.

$$C_4H_{10}(g) + \frac{13}{2} O_2(g) \longrightarrow 4CO_2(g) + 5H_2O(g)$$

Although chemists sometimes have good reason to write equations with fractional coefficients, it is common practice to use only whole-number coefficients. We accomplish that by multiplying everything by 2, which gives the balanced equation:

$$2C_4H_{10}(g) + 13O_2(g) \longrightarrow 8CO_2(g) + 10H_2O(g)$$

Problem 4.2

Balance this equation:

$$C_6H_{14}(g) + O_2(g) \longrightarrow CO_2(g) + H_2O(g)$$

A pocket lighter contains butane in both the liquid and gaseous state.

Example 4.3 Balancing a Chemical Equation

Balance this equation:

$$Na_2SO_3(aq) + H_3PO_4(aq) \longrightarrow H_2SO_3(aq) + Na_3PO_4(aq)$$

Sodium   Phosphoric   Sulfurous   Sodium
sulfite   acid   acid   phosphate

Strategy

The key to balancing equations like this one is to realize that polyatomic ions such as SO_3^{2-} and PO_4^{3-} remain intact on both sides of the equation.

Solution

We can begin by balancing the Na^+ ions. We put a 3 in front of Na_2SO_3 and a 2 in front of Na_3PO_4, giving us six Na^+ ions on each side:

Six Na on each side

$$3Na_2SO_3(aq) + H_3PO_4(aq) \longrightarrow H_2SO_3(aq) + 2Na_3PO_4(aq)$$

There are now three SO_3^{2-} units on the left and only one on the right, so we put a 3 in front of H_2SO_3:

Three SO_3^{2-} units on each side

$$3Na_2SO_3(aq) + H_3PO_4(aq) \longrightarrow 3H_2SO_3(aq) + 2Na_3PO_4(aq)$$

Now let's look at the PO_4^{3-} units. There are two PO_4^{3-} units on the right but only one on the left. To balance them, we put a 2 in front of H_3PO_4. In so doing, we balance not only the PO_4^{3-} units but also the hydrogens and arrive at the balanced equation:

Two PO_4^{3-} units on each side

$$3Na_2SO_3(aq) + 2H_3PO_4(aq) \longrightarrow 3H_2SO_3(aq) + 2Na_3PO_4(aq)$$

Problem 4.3

Balance this equation:

$$K_2C_2O_4(aq) + Ca_3(AsO_4)_2(s) \longrightarrow K_3AsO_4(aq) + CaC_2O_4(s)$$

Potassium   Calcium   Potassium   Calcium
oxalate   arsenate   arsenate   oxalate

One final point about balancing chemical equations. The following equation for the combustion of propane is correctly balanced.

$$C_3H_8(g) + 5O_2(g) \longrightarrow 3CO_2(g) + 4H_2O(g)$$
Propane

Would it be correct if we doubled all the coefficients?

$$2C_3H_8(g) + 10O_2(g) \longrightarrow 6CO_2(g) + 8H_2O(g)$$
Propane

Yes, this revised equation is mathematically and scientifically correct, but chemists do not normally write equations with coefficients that are all divisible by a common number. A correctly balanced equation is almost always written with the coefficients expressed in the lowest set of whole numbers.

4.3 How Can We Predict If Ions in Aqueous Solution Will React with Each Other?

Many ionic compounds are soluble in water. As we saw in Section 3.5, ionic compounds always consist of both positive and negative ions. When they dissolve in water, the positive and negative ions separate from each other. We call such separation **dissociation**. For example,

$$NaCl(s) \xrightarrow{H_2O} Na^+(aq) + Cl^-(aq)$$

What happens when we mix **aqueous solutions** of two different ionic compounds? Does a reaction take place between the ions? The answer depends on the ions. For example, if any of the negative and positive ions come together to form a water-insoluble compound, then a reaction takes place and a precipitate forms.

Suppose we prepare one solution by dissolving sodium chloride, NaCl, in water and a second solution by dissolving silver nitrate, $AgNO_3$, in water.

$$\text{Solution 1} \quad NaCl(s) \xrightarrow{H_2O} Na^+(aq) + Cl^-(aq)$$
$$\text{Solution 2} \quad AgNO_3(s) \xrightarrow{H_2O} Ag^+(aq) + NO_3^-(aq)$$

If we now mix the two solutions, four ions are present in the solution: Ag^+, Na^+, Cl^-, and NO_3^-. Two of these ions, Ag^+ and Cl^-, react to form the compound AgCl (silver chloride), which is insoluble in water. A reaction therefore takes place, forming a white precipitate of AgCl that slowly sinks to the bottom of the container (Figure 4.1). We write this reaction as follows:

$$\underset{\substack{\text{Silver} \\ \text{ion}}}{Ag^+(aq)} + \underset{\substack{\text{Nitrate} \\ \text{ion}}}{NO_3^-(aq)} + \underset{\substack{\text{Sodium} \\ \text{ion}}}{Na^+(aq)} + \underset{\substack{\text{Chloride} \\ \text{ion}}}{Cl^-(aq)} \longrightarrow \underset{\substack{\text{Silver} \\ \text{chloride}}}{AgCl(s)} + Na^+(aq) + NO_3^-(aq)$$

Notice that the Na^+ and NO_3^- ions do not participate in a reaction, but merely remain dissolved in the water. Ions that do not participate in a reaction are called **spectator ions**, certainly an appropriate name.

We can simplify the equation for the formation of silver chloride by omitting all spectator ions:

$$\underset{\substack{\text{Net ionic} \\ \text{equation:}}}{} \quad \underset{\substack{\text{Silver} \\ \text{ion}}}{Ag^+(aq)} + \underset{\substack{\text{Chloride} \\ \text{ion}}}{Cl^-(aq)} \longrightarrow \underset{\substack{\text{Silver} \\ \text{chloride}}}{AgCl(s)}$$

This kind of equation that we write for ions in solution is called a **net ionic equation**. Like all other chemical equations, net ionic equations must be balanced. We balance them in the same way we do other equations, except now we must make sure that charges balance as well as atoms.

Net ionic equations show only the ions that react—no spectator ions are shown. For example, consider the net ionic equation for the precipitation of arsenic(III) sulfide from aqueous solution:

$$\text{Net ionic equation: } 2As^{3+}(aq) + 3S^{2-}(aq) \longrightarrow As_2S_3(s)$$

Not only are there two arsenic and three sulfur atoms on each side, but the total charge on the left side is the same as the total charge on the right side; they are both zero.

In general, ions in solution react with each other only when one of these four things can happen:

1. Two ions form a solid that is insoluble in water. AgCl is one example, as shown in Figure 4.1.

H_2O above the arrow means that the reaction takes place in water.

Aqueous solution A solution in which the solvent is water

FIGURE 4.1 Adding Cl^- ions to a solution of Ag^+ ions produces a white precipitate of silver chloride, AgCl.

Charles Steele

Spectator ion An ion that appears unchanged on both sides of a chemical equation

Net ionic equation A chemical equation that does not contain spectator ions

FIGURE 4.2 When aqueous solutions of $NaHCO_3$ and HCl are mixed, a reaction between HCO_3^- and H_3O^+ ions produces CO_2 gas, which can be seen as bubbles.

FIGURE 4.3 (a) The beaker on the left contains a solution of potassium sulfate (colorless), and the beaker on the right contains a solution of copper(II) nitrate (blue). (b) When the two solutions are mixed, the blue color becomes lighter because the copper(II) nitrate is less concentrated, but no other chemical reaction occurs.

2. Two ions form a gas that escapes from the reaction mixture as bubbles. An example is the reaction of sodium bicarbonate, $NaHCO_3$, with HCl to form the gas carbon dioxide, CO_2 (Figure 4.2). The net ionic equation for this reaction, where the expected reactant ion $H^+(aq)$ from HCl is written as a hydrated ion or $H_3O^+(aq)$, is written as:

Net ionic equation:
$$\underset{\text{Bicarbonate ion}}{HCO_3^-(aq)} + H_3O^+(aq) \longrightarrow \underset{\text{Carbon dioxide}}{CO_2(g)} + 2H_2O(\ell)$$

3. An acid neutralizes a base to form water. Acid-base reactions are so important that we devote Chapter 8 to them.

4. One of the ions can oxidize another. We discuss this type of reaction in Section 4.4.

In many cases, no reaction takes place when we mix solutions of ionic compounds because none of these situations holds. For example, if we mix solutions of copper(II) nitrate, $Cu(NO_3)_2$, and potassium sulfate, K_2SO_4, we merely have a mixture containing Cu^{2+}, K^+, NO_3^-, and SO_4^{2-} ions dissolved in water. None of these ions react with each other; therefore, we see nothing happening (Figure 4.3).

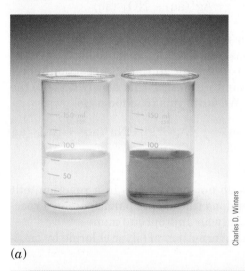

(a)　　　　(b)

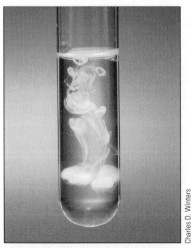

The mixing of solutions of barium chloride, $BaCl_2$, and sodium sulfate, Na_2SO_4, forms a white precipitate of barium sulfate, $BaSO_4$.

Example 4.4 Net Ionic Equation

When a solution of barium chloride, $BaCl_2$, is added to a solution of sodium sulfate, Na_2SO_4, a white precipitate of barium sulfate, $BaSO_4$, forms. Write the net ionic equation for this reaction.

Strategy

The net ionic equation shows only those ions that combine to form a precipitate.

Solution

Because both barium chloride and sodium sulfate are ionic compounds, each exists in water as its dissociated ions:

$$Ba^{2+}(aq) + 2Cl^-(aq) + 2Na^+(aq) + SO_4^{2-}(aq)$$

We are told that a precipitate of barium sulfate forms:

$$Ba^{2+}(aq) + 2Cl^-(aq) + 2Na^+(aq) + SO_4^{2-}(aq) \longrightarrow$$
$$\underset{\text{Barium sulfate}}{BaSO_4(s)} + 2Na^+(aq) + 2Cl^-(aq)$$

Because Na^+ and Cl^- ions appear on both sides of the equation (they are spectator ions), we cancel them and are left with the following net ionic equation:

Net ionic equation: $Ba^{2+}(aq) + SO_4^{2-}(aq) \longrightarrow BaSO_4(s)$

Problem 4.4

When a solution of copper(II) chloride, $CuCl_2$, is added to a solution of potassium sulfide, K_2S, a black precipitate of copper(II) sulfide, CuS, forms. Write the net ionic equation for the reaction.

Of the four ways for ions to react in water, one of the most common is the formation of an insoluble compound. We can predict when this result will happen if we know the solubilities of the ionic compounds. Some useful guidelines for the solubility of ionic compounds in water are given in Table 4.1.

Table 4.1 Solubility Rules for Ionic Compounds

Usually Soluble	
Li^+, Na^+, K^+, Rb^+, Cs^+, NH_4^+	All Group 1A (alkali metal) and ammonium salts are soluble.
Nitrates, NO_3^-	All nitrates are soluble.
Chlorides, bromides, iodides, Cl^-, Br^-, I^-	All common chlorides, bromides, and iodides are soluble except $AgCl$, Hg_2Cl_2, $PbCl_2$, $AgBr$, Hg_2Br_2, $PbBr_2$, AgI, Hg_2I_2, PbI_2
Sulfates, SO_4^{2-}	Most sulfates are soluble except $CaSO_4$, $SrSO_4$, $BaSO_4$, $PbSO_4$
Acetates, CH_3COO^-	All acetates are soluble.

Usually Insoluble	
Phosphates, PO_4^{3-}	All phosphates are insoluble except those of NH_4^+ and Group 1A (the alkali metal) cations.
Carbonates, CO_3^{2-}	All carbonates are insoluble except those of NH_4^+ and Group 1A (the alkali metal) cations.
Hydroxides, OH^-	All hydroxides are insoluble except those of NH_4^+ and Group 1A (the alkali metal) cations. $Sr(OH)_2$, $Ba(OH)_2$, and $Ca(OH)_2$ are only slightly soluble.
Sulfides, S^{2-}	All sulfides are insoluble except those of NH_4^+ and Group 1A (the alkali metal) and Group 2A cations. MgS, CaS, and BaS are only slightly soluble.

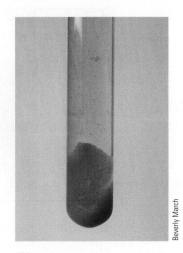

(a)   (b)

Both (a) $Fe(OH)_3$, iron(III) hydroxide, and (b) $CuCO_3$, copper(II) carbonate, are insoluble in water.

Sea animals of the mollusk family often use insoluble calcium carbonate, $CaCO_3$ to construct their shells.

Charles D. Winters

Beverly March

Beverly March

Solubility and Tooth Decay

The outermost protective layer of a tooth is the enamel, which is composed of approximately 95% hydroxyapatite, $Ca_5(PO_4)_3(OH)$, and 5% collagen (Figure 22.13). Like most other phosphates and hydroxides, hydroxyapatite is insoluble in water. In acidic media, however, it dissolves to a slight extent, yielding Ca^{2+}, PO_4^{3-}, and OH^- ions. This loss of enamel creates pits and cavities in the tooth.

Acidity in the mouth is produced by bacterial fermentation of remnants of food, especially carbohydrates.

Once pits and cavities form in the enamel, bacteria can hide there and cause further damage in the underlying softer material called dentin. The fluoridation of water brings F^- ions to the hydroxyapatite. There, F^- ions take the place of OH^- ions, forming the considerably less acid-soluble fluoroapatite, $Ca_5(PO_4)_3(OH)$. Fluoride-containing toothpastes enhance this exchange process and provide protection against tooth decay.

Test your knowledge with Problems 4.78 and 4.79.

4.4 What Are Oxidation and Reduction?

Oxidation The loss of electrons; the gain of oxygen atoms and/or the loss of hydrogen atoms

Reduction The gain of electrons; the loss of oxygen atoms and/or the gain of hydrogen atoms

Redox reaction An oxidation-reduction reaction

Oxidation-reduction is one of the most important and common types of chemical reactions. **Oxidation** is the loss of electrons. **Reduction** is the gain of electrons. An **oxidation-reduction reaction** (often called a **redox reaction**) involves the transfer of electrons from one species to another. An example is the oxidation of zinc by copper ions, the net ionic equation for which is:

$$Zn(s) + Cu^{2+}(aq) \longrightarrow Zn^{2+}(aq) + Cu(s)$$

When we put a piece of zinc metal into a beaker containing copper(II) ions in aqueous solution, three things happen (Figure 4.4):

1. Some of the zinc metal dissolves and goes into solution as Zn^{2+}.
2. Copper metal deposits on the surface of the zinc metal.
3. The blue color of the Cu^{2+} ions gradually disappears.

FIGURE 4.4 When a piece of zinc is added to a solution containing Cu^{2+} ions, Zn is oxidized by Cu^{2+} ions and Cu^{2+} ions are reduced by the Zn.

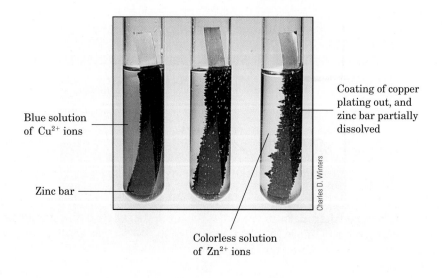

Blue solution of Cu^{2+} ions

Zinc bar

Coating of copper plating out, and zinc bar partially dissolved

Colorless solution of Zn^{2+} ions

Charles D. Winters

Zinc atoms lose electrons to copper ions and become zinc ions:

$$Zn(s) \longrightarrow Zn^{2+}(aq) + 2e^- \qquad \text{Zn is oxidized}$$

At the same time, Cu^{2+} ions gain electrons from the zinc atoms. The copper ions are reduced:

$$Cu^{2+}(aq) + 2e^- \longrightarrow Cu(s) \qquad \text{Cu^{2+} is reduced}$$

Oxidation and reduction are not independent reactions. That is, a species cannot gain electrons from nowhere, nor can a species lose electrons to nothing. In other words, no oxidation can occur without an accompanying reduction, and vice versa. In the preceding reaction, Cu^{2+} oxidizes Zn. We call Cu^{2+} an **oxidizing agent**. Similarly, Zn reduces Cu^{2+}, and we call Zn a **reducing agent**.

We summarize these oxidation-reduction relationships for the reaction of zinc metal with Cu^{2+} ion in the following way:

Negative ions such as Cl^- or NO_3^- are present to balance charges, but we do not show them because they are spectator ions.

Oxidizing agent An entity that accepts electrons in an oxidation-reduction reaction

Reducing agent An entity that donates electrons in an oxidation-reduction reaction

$$
\begin{array}{c}
\overset{\displaystyle 2e^-}{\overbrace{\hspace{3cm}}} \\[-2pt]
\underset{\substack{\text{Zinc is} \\ \text{oxidized}}}{Zn(s)} \quad + \quad \underset{\substack{\text{Copper} \\ \text{is reduced}}}{Cu^{2+}(aq)} \longrightarrow Zn^{2+}(aq) + Cu(s)
\end{array}
$$

Zinc is the reducing agent $\quad\times\quad$ Copper is the oxidizing agent

Note here that a curved arrow from $Zn(s)$ to Cu^{2+} shows the transfer of two electrons from zinc to copper ion. Refer to Section 8.6B for additional examples of redox reactions which feature active metals with strong acids.

Although the definitions we have given for oxidation (loss of electrons) and reduction (gain of electrons) are easy to apply in many redox reactions, they are not so easy to apply in other cases. For example, another redox reaction is the combustion (burning) of methane, CH_4, in which CH_4 is oxidized to CO_2 while O_2 is reduced to H_2O.

$$\underset{\text{Methane}}{CH_4(g)} + 2O_2(g) \longrightarrow CO_2(g) + 2H_2O(g)$$

It is not easy to see the electron loss and gain in such a reaction, so chemists developed another definition of oxidation and reduction, one that is easier to apply in many cases, especially where organic (carbon-containing) compounds are involved:

Oxidation: The gain of oxygen atoms and/or the loss of hydrogen atoms

Reduction: The loss of oxygen atoms and/or the gain of hydrogen atoms

Applying these alternative definitions to the reaction of methane with oxygen, we find the following:

$$
\underset{\substack{\text{Gains O and loses} \\ \text{H; is oxidized}}}{CH_4(g)} \quad + \quad \underset{\substack{\text{Gains H;} \\ \text{is reduced}}}{2O_2(g)} \longrightarrow CO_2(g) + 2H_2O(g)
$$

Is the reducing agent $\qquad$ Is the oxidizing agent

In fact, this second definition is much older than the one involving electron transfer; it is the definition given by Lavoisier when he first discovered oxidation and reduction more than 200 years ago. Note that we could not apply this definition to our zinc-copper example.

The rusting of iron and steel can be a serious problem in an industrial society. In rusting, iron is oxidized.

Example 4.5 Oxidation-Reduction

In each equation, identify the substance that is oxidized, the substance that is reduced, the oxidizing agent, and the reducing agent.

(a) $Al(s) + Fe^{3+}(aq) \longrightarrow Al^{3+}(aq) + Fe(s)$

(b) $CH_3OH(g) + O_2(g) \longrightarrow HCOOH(g) + H_2O(g)$
        Methanol   Formic acid

Strategy

The substance that is oxidized loses electrons and is a reducing agent. The substance that gains electrons is the oxidizing agent and is reduced. For organic compounds, oxidation involves the gain of oxygen atoms and/or the loss of hydrogen atoms. Reduction involves the gain of hydrogen atoms and/or the loss of oxygen atoms.

Solution

(a) $Al(s)$ loses three electrons and becomes Al^{3+}; therefore, aluminum is oxidized. In the process of being oxidized, $Al(s)$ gives its electrons to Fe^{3+}, and so $Al(s)$ is the reducing agent. Fe^{3+} gains three electrons and becomes $Fe(s)$ and is reduced. In the process of being reduced, Fe^{3+} accepts three electrons from $Al(s)$, and so Fe^{3+} is the oxidizing agent. To summarize:

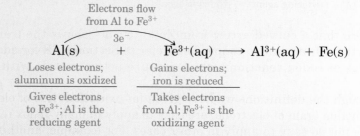

Electrons flow
from Al to Fe^{3+}

$3e^-$

$Al(s)$ + $Fe^{3+}(aq) \longrightarrow Al^{3+}(aq) + Fe(s)$

Loses electrons;   Gains electrons;
aluminum is oxidized   iron is reduced

Gives electrons   Takes electrons
to Fe^{3+}; Al is the   from Al; Fe^{3+} is the
reducing agent   oxidizing agent

(b) Because it is not easy to see the loss or gain of electrons in this example, we apply the second set of definitions. In converting CH_3OH to $HCOOH$, CH_3OH both gains oxygen atoms and loses hydrogen atoms; it is oxidized. In being converted to H_2O, O_2 gains hydrogen atoms; it is reduced. The compound oxidized is the reducing agent; CH_3OH is the reducing agent. The compound reduced is the oxidizing agent; O_2 is the oxidizing agent. To summarize:

$CH_3OH(g)$ + $O_2(g) \longrightarrow HCOOH(g) + H_2O(g)$

Is oxidized;   Is reduced;
methanol is   oxygen is
the reducing   the oxidizing
agent   agent

Problem 4.5

In each equation, identify the substance that is oxidized, the substance that is reduced, the oxidizing agent, and the reducing agent:

(a) $Ni^{2+}(aq) + Cr(s) \longrightarrow Ni(s) + Cr^{2+}(aq)$

(b) $CH_2O(g) + H_2(g) \longrightarrow CH_3OH(g)$
        Formaldehyde   Methanol

Air pollution is caused by incomplete fuel combustion.

We have said that redox reactions are extremely common. Here are some important categories:

1. **Combustion** All combustion (burning) reactions are redox reactions in which the compounds or mixtures that are burned are oxidized by oxygen, O_2. They include the burning of gasoline, diesel oil, fuel oil, natural gas, coal, wood, and paper. All these materials contain carbon, and all

Chemical Connections 4B

Voltaic Cells

In Figure 4.4, we see that when a piece of zinc metal is put in a solution containing Cu^{2+} ions, zinc atoms give electrons to Cu^{2+} ions. We can change the experiment by putting the zinc metal in one beaker and the Cu^{2+} ions in another and then connecting the two beakers by a length of wire and a salt bridge (see the accompanying figure). A reaction still takes place; that is, zinc atoms still give electrons to Cu^{2+} ions, but now the electrons must flow through the wire to get from the Zn to the Cu^{2+}. This flow of electrons produces an electric current, and the electrons keep flowing until either the Zn or the Cu^{2+} is used up. In this way, the apparatus generates an electric current by using a redox reaction. We call this device a **voltaic cell** or, more commonly, a battery.

The electrons produced at the zinc end carry negative charges. This end of the battery is a negative electrode (called the **anode**). The electrons released at the anode as zinc is oxidized go through an outside circuit and, in doing so, produce the battery's electric current. At the other end of the battery via the positively charged electrode (called the **cathode**), electrons are consumed as Cu^{2+} ions are reduced to copper metal.

To see why a salt bridge is necessary, we must look at the Cu^{2+} solution. Because we cannot have positive charges in any place without an equivalent number of negative charges, negative ions must be in the beaker as well—perhaps sulfate, nitrate, or some other anion. When electrons come over the wire, the Cu^{2+} is converted to Cu:

$$Cu^{2+}(aq) + 2e^- \longrightarrow Cu(s)$$

This reaction diminishes the number of Cu^{2+} ions, but the number of negative ions remains unchanged. The salt bridge is necessary to carry some of these negative ions to the other beaker, where they are needed to balance the Zn^{2+} ions being produced by the following reaction:

$$Zn(s) \longrightarrow Zn^{2+}(aq) + 2e^-$$

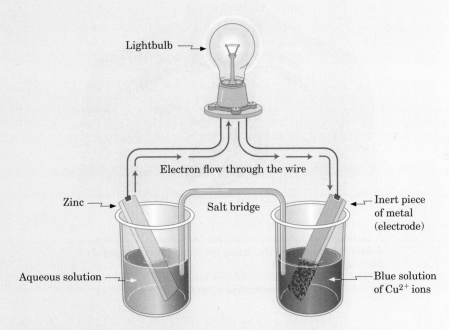

Voltaic cell. The electron flow over the wire from Zn to Cu^{2+} is an electric current that causes the lightbulb to glow.

Test your knowledge with Problem 4.80.

except coal also contain hydrogen. If the combustion is complete, carbon is oxidized to CO_2 and oxygen is reduced to H_2O. In an incomplete combustion, these elements are converted to other compounds, many of which cause air pollution.

Unfortunately, much of today's combustion that takes place in gasoline and diesel engines and in furnaces is incomplete and so contributes

Chemical Connections 4C

Artificial Pacemakers and Redox

An artificial pacemaker is a small electrical device that uses electrical impulses, delivered by electrodes contacting the heart muscles, to regulate the beating of the heart. The primary purpose of a pacemaker is to maintain an adequate heart rate, either because the heart's native pacemaker does not beat fast enough, or perhaps there is a blockage in the heart's electrical conduction system. When a pacemaker detects that the heart is beating too slowly, it sends an electrical signal to the heart, generated via a redox reaction, so that the heart muscle beats faster. Modern pacemakers are externally programmable and allow a cardiologist to select the optimum pacing modes for individual patients.

Early pacemakers generated an electrical impulse via the following redox reaction:

$$Zn + Hg^{2+} \longrightarrow Zn^{2+} + Hg$$

The zinc atom is oxidized to Zn^{2+}, and Hg^{2+} is reduced to Hg. Many contemporary artificial pacemakers contain a lithium-iodine battery, which has a longer battery life (10 years or more). Consider the unbalanced redox reaction for the lithium-iodine battery:

$$Li + I_2 \longrightarrow LiI$$

The lithium atom is oxidized to Li^+, and the I_2 molecule is reduced to I^-. When the pacemaker fails to sense a heartbeat within a normal beat-to-beat time period, an electrical signal produced from these reactions is initiated, stimulating the ventricle of the heart. This sensing and stimulating activity continues on a beat-by-beat basis. More complex systems include the ability to stimulate both the atrial and ventricular chambers.

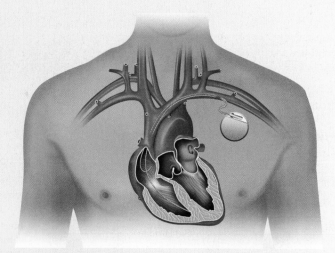

A pacemaker is a medical device that uses electrical impulses, delivered by electrodes contacting the heart muscles, to regulate the beating of the heart.

Test your knowledge with Problem 4.81.

to air pollution. In the incomplete combustion of methane, for example, carbon is oxidized to carbon monoxide, CO, because there is not a sufficient supply of oxygen to complete its oxidation to CO_2:

$$2CH_4(g) + 3O_2(g) \longrightarrow 2CO(g) + 4H_2O(g)$$
Methane

2. **Respiration** Humans and animals get their energy by respiration. The oxygen in the air we breathe oxidizes carbon-containing compounds in our cells to produce CO_2 and H_2O. Note that respiration is equivalent to combustion, except that it takes place more slowly and at a much lower temperature. We discuss respiration more fully in Chapter 27.

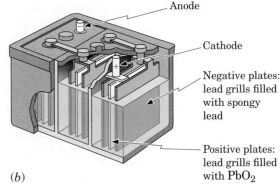

Anode

Cathode

Negative plates: lead grills filled with spongy lead

Positive plates: lead grills filled with PbO_2

Charles D. Winters

(a)   (b)

FIGURE 4.5 (a) Dry cell batteries. (b) A lead storage battery.

The important product of respiration is not CO_2 (which the body eliminates) or H_2O, but energy.

3. **Rusting** We all know that when iron or steel objects are left out in the open air, they eventually rust (steel is mostly iron but contains certain other elements as well). In rusting, iron is oxidized to a mixture of iron oxides. We can represent the main reaction by the following equation:

$$4Fe(s) + 3O_2(g) \longrightarrow 2Fe_2O_3(s)$$

4. **Bleaching** Most bleaching involves oxidation, and common bleaches are oxidizing agents. The colored compounds being bleached are usually organic compounds; oxidation converts them to colorless compounds.

5. **Batteries** A voltaic cell (Chemical Connections 4B) is a device in which electricity is generated from a chemical reaction. Such cells are often called batteries (Figure 4.5). We are all familiar with batteries in our cars and in such portable devices as radios, flashlights, cell phones, and computers. In all cases, the reaction that takes place in the battery is a redox reaction.

These household bleaches are oxidizing agents.

Charles D. Winters

4.5 What Are Formula Weights and Molecular Weights?

We begin our study of mass relationships with a discussion of formula weight. The **formula weight (FW)** of a compound is the sum of the atomic weights in atomic mass units (amu) of all the atoms in the compound's formula. The term "formula weight" can be used for both ionic and molecular compounds and tells nothing about whether the compound is ionic or molecular.

Another term, **molecular weight (MW)**, is strictly correct only when used for covalent compounds. In this book, we use "formula weight" for both ionic and covalent compounds and "molecular weight" only for covalent compounds.

A table of atomic weights is given on the inside front cover. Atomic weights can also be found in the Periodic Table on the inside front cover.

Molecular weight (MW) The sum of the atomic weights of all atoms in a molecular compound expressed in atomic mass units (amu)

Example 4.6 Molecular Weight

What is the molecular weight of (a) glucose, $C_6H_{12}O_6$, and (b) urea, $(NH_2)_2CO$?

Strategy

Molecular weight is the sum of the atomic weights of all atoms in the molecular formula expressed in atomic mass units (amu).

Solution

(a) Glucose, $C_6H_{12}O_6$

C	$6 \times 12.0 = 72.0$
H	$12 \times 1.0 = 12.0$
O	$6 \times 16.0 = 96.0$
	$C_6H_{12}O_6 = 180.0$ amu

(b) Urea, $(NH_2)_2CO$

N	$2 \times 14.0 = 28.0$
H	$4 \times 1.0 = 4.0$
C	$1 \times 12.0 = 12.0$
O	$1 \times 16.0 = 16.0$
	$(NH_2)_2CO = 60.0$ amu

Problem 4.6

What is (a) the molecular weight of ibuprofen, $C_{13}H_{18}O_2$, and (b) the formula weight of barium phosphate, $Ba_3(PO_4)_2$?

4.6 What Is a Mole and How Do We Use It to Calculate Mass Relationships?

Atoms and molecules are so tiny (Section 2.4F) that chemists are seldom able to deal with them one at a time. When we weigh even a very small quantity of a compound, huge numbers of formula units (perhaps 10^{19}) are present. The formula unit may be atoms, molecules, or ions. To overcome this problem, chemists long ago defined a unit called the **mole (mol)**. A mole is the amount of substance that contains as many atoms, molecules, or ions as there are atoms in exactly 12 g of carbon-12. The important point here is that whether we are dealing with a mole of iron atoms, a mole of methane molecules, or a mole of sodium ions, a mole always contains the same number of formula units. We are accustomed to scale-up factors in situations where there are large numbers of units involved in counting. We count eggs by the dozen and pencils by the gross. Just as the dozen (12 units) is a useful scale-up factor for eggs and the gross (144 units) a useful scale-up factor for pencils, the mole is a useful scale-up factor for atoms and molecules. We are soon going to see that the number of units is much larger for a mole than for a dozen or a gross.

The number of formula units in a mole is called **Avogadro's number** after the Italian physicist Amadeo Avogadro (1776–1856), who first proposed the concept of a mole but was not able to experimentally determine the number of units it represented. Note that Avogadro's number is not a defined value, but rather a value that must be determined experimentally. Its value is now known to nine significant figures.

$$\text{Avogadro's number} = 6.02214199 \times 10^{23} \text{ formula units per mole}$$

For most calculations in this text, we round this number to three significant figures to 6.02×10^{23} formula units per mole.

A mole of hydrogen atoms is 6.02×10^{23} hydrogen atoms, a mole of sucrose (table sugar) molecules is 6.02×10^{23} sugar molecules, a mole of apples is 6.02×10^{23} apples, and a mole of sodium ions is 6.02×10^{23} sodium ions. Just as we call 12 of anything a dozen, 20 a score, and 144 a gross, we call 6.02×10^{23} of anything a mole.

The **molar mass** of any substance (the mass of one mole of the substance) is the formula weight of the substance expressed in grams per mole. For instance, the formula weight of glucose, $C_6H_{12}O_6$ (Example 4.6), is 180.0 amu; therefore, 180.0 g of glucose is one mole of glucose. Likewise, the formula weight of urea, $(NH_2)_2CO$, is 60.0 amu and, therefore, one mole of urea is 60.0 grams of urea. For atoms, one mole is the atomic weight expressed in grams: 12.0 g of carbon is one mole of carbon atoms; 32.1 g of sulfur is one

Mole (mol) The formula weight of a substance expressed in grams

Avogadro's number 6.02×10^{23}, is the number of formula units per mole

One mole of pennies placed side by side would stretch for more than one million light–years, a distance far outside our solar system and even outside our own galaxy. Six moles of this text would weigh as much as the Earth.

Molar mass The mass of one mole of a substance expressed in grams; the formula weight of a compound expressed in grams

(a)   (b)

FIGURE 4.6 One-mole quantities of (a) six metals and (b) four compounds. (a) Top row (left to right): Cu beads (63.5 g), Al foil (27.0 g), and Pb shot (207.2 g). Bottom row (left to right): S powder (32.1 g), Cr chunks (52.0 g), and Mg shavings (24.4 g). (b) H_2O (18.0 g); small beaker, NaCl (58.4 g); large left beaker, aspirin, $C_9H_8O_4$ (180.2 g); and large right beaker, green $NiCl_2 \cdot 6H_2O$ (237.7 g).

mole of sulfur atoms; and so on. As you see, the important point here is that to talk about the mass of a mole, we need to know the chemical formula of the substance we are considering. Figure 4.6 shows one-mole quantities of several compounds.

Now that we know the relationship between moles and molar mass (g/mol), we can use molar mass as a conversion factor to convert from grams to moles and from moles to grams. For this calculation, we use molar mass as a conversion factor.

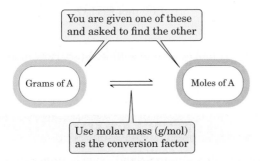

Suppose we want to know the number of moles of water in a graduated cylinder that contains 36.0 g of water. We know that the molar mass of water is 18.0 g/mol. If 18.0 g of water is one mole of water, then 36.0 g must be two moles of water.

$$36.0 \text{ g } H_2O \times \frac{1 \text{ mol } H_2O}{18.0 \text{ g } H_2O} = 2.00 \text{ mol } H_2O$$

Molar mass can also be used to convert from moles to grams. Suppose you have a beaker that contains 0.753 mol of sodium chloride and you want to calculate the number of grams of sodium chloride in the beaker. As a conversion factor, use the fact that the molar mass of NaCl is 58.5 g/mol.

$$0.753 \text{ mol NaCl} \times \frac{58.5 \text{ g NaCl}}{1 \text{ mol NaCl}} = 44.1 \text{ g NaCl}$$

Note that this type of calculation can be performed for ionic compounds such as NaF as well as for molecular compounds such as CO_2 and urea.

OWL Interactive Example 4.7 Moles

We have 27.5 g of sodium fluoride, NaF, the form of fluoride ions most commonly used in fluoride toothpastes and dental gels. How many moles of NaF is this?

Strategy

The formula weight of NaF = 23.0 + 19.0 = 42.0 amu. Thus, each mole of NaF has a mass of 42.0 g, allowing us to use the conversion factor 1 mol NaF = 42.0 g NaF.

Solution

$$27.5 \text{ g NaF} \times \frac{1 \text{ mol NaF}}{42.0 \text{ g NaF}} = 0.655 \text{ mol NaF}$$

Problem 4.7

A person drinks 1500. g of water per day. How many moles is this?

OWL Interactive Example 4.8 Moles

We wish to weigh 3.41 mol of ethanol, C_2H_6O. How many grams is this?

Strategy

The formula weight of C_2H_6O is 2(12.0) + 6(1.0) + 16.0 = 46.0 amu, so the conversion factor is 1 mol C_2H_6O = 46.0 g C_2H_6O.

Solution

$$3.41 \text{ mol } C_2H_6O \times \frac{46.0 \text{ g } C_2H_6O}{1.00 \text{ mol } C_2H_6O} = 157 \text{ g } C_2H_6O$$

Problem 4.8

We wish to weigh 2.84 mol of sodium sulfide, Na_2S. How many grams is this?

OWL Interactive Example 4.9 Moles

How many moles of nitrogen atoms and oxygen atoms are in 21.4 mol of the explosive trinitrotoluene (TNT), $C_7H_5N_3O_6$?

Strategy

The molecular formula $C_7H_5N_3O_6$ tells us that each molecule of TNT contains three nitrogen atoms and six oxygen atoms. It also tells us that each mole of TNT contains three moles of N atoms and six moles of O atoms. Therefore, we have the following conversion factors: 1 mol TNT = 3 mol N atoms, and 1 mol TNT = 6 mol O atoms.

Solution

The number of moles of N atoms in 21.4 moles of TNT is:

$$21.4 \text{ mol TNT} \times \frac{3 \text{ mol N atoms}}{1 \text{ mol TNT}} = 64.2 \text{ mol N atoms}$$

The number of moles of O atoms in 21.4 moles of TNT is:

$$21.4 \text{ mol TNT} \times \frac{6 \text{ mol O atoms}}{1 \text{ mol TNT}} = 128 \text{ mol O atoms}$$

Note that we give the answer to three significant figures because we were given the number of moles to three significant figures. The ratio of moles of O atoms to moles of TNT is an exact number.

Problem 4.9

How many moles of C atoms, H atoms, and O atoms are in 2.5 mol of glucose, $C_6H_{12}O_6$?

ⓌL Interactive Example 4.10 Moles

How many moles of sodium ions, Na^+, are in 5.63 g of sodium sulfate, Na_2SO_4?

Strategy

The formula weight of Na_2SO_4 is $2(23.0) + 32.1 + 4(16.0) = 142.1$ amu. In the conversion of grams Na_2SO_4 to moles, we use the conversion factors 1 mol $Na_2SO_4 = 142.1$ g Na_2SO_4 and 1 mole $Na_2SO_4 = 2$ moles Na^+.

Solution

First, we need to find out how many moles of Na_2SO_4 are in the sample.

$$5.63 \text{ g Na}_2\text{SO}_4 \times \frac{1 \text{ mol Na}_2\text{SO}_4}{142.1 \text{ g Na}_2\text{SO}_4} = 0.0396 \text{ mol Na}_2\text{SO}_4$$

The number of moles of Na^+ ions in 0.0396 mol of Na_2SO_4 is:

$$0.0396 \text{ mol Na}_2\text{SO}_4 \times \frac{2 \text{ mol Na}^+}{1 \text{ mol Na}_2\text{SO}_4} = 0.0792 \text{ mol Na}^+$$

Problem 4.10

How many moles of copper(I) ions, Cu^+, are there in 0.062 g of copper(I) nitrate, $CuNO_3$?

Example 4.11 Molecules per Gram

An aspirin tablet, $C_9H_8O_4$, contains 0.360 g of aspirin. (The rest of the tablet is starch or other fillers.) How many molecules of aspirin are present in this tablet?

Strategy

The formula weight of aspirin is $9(12.0) + 8(1.0) + 4(16.0) = 180.0$ amu, which gives us the conversion factor 1 mol aspirin = 180.0 g aspirin. To convert moles of aspirin to molecules of aspirin, we use the conversion factor 1 mole aspirin = 6.02×10^{23} molecules aspirin.

Solution

First, we need to find out how many moles of aspirin are in 0.360 g:

$$0.360 \text{ g aspirin} \times \frac{1 \text{ mol aspirin}}{180.0 \text{ g aspirin}} = 0.00200 \text{ mol aspirin}$$

The number of molecules of aspirin in a tablet is:

$$0.00200 \ \cancel{mol} \times 6.02 \times 10^{23} \ \frac{molecules}{\cancel{mol}} = 1.20 \times 10^{21} \ molecules$$

Problem 4.11

How many molecules of water, H_2O, are in a glass of water (235 g)?

4.7 How Do We Calculate Mass Relationships in Chemical Reactions?

A. Stoichiometry

As we saw in Section 4.2, a balanced chemical equation tells us not only which substances react and which are formed, but also the molar ratios in which they react. For example, using the molar ratios in a balanced chemical equation, we can calculate the mass of starting materials we need to produce a particular mass of a product. The quantitative relationship between the amounts of reactants consumed and products formed in chemical reactions as expressed by a balanced chemical equation is called **stoichiometry**.

Let us look once again at the balanced equation for the combustion of propane:

$$C_3H_8(g) + 5O_2(g) \longrightarrow 3CO_2(g) + 4H_2O(g)$$
Propane

This equation tells us that propane and oxygen are converted to carbon dioxide and water and that 1 mol of propane combines with 5 mol of oxygen to produce 3 mol of carbon dioxide and 4 mol of water; that is, we know the molar ratios involved. The same is true for any other balanced equation. This fact allows us to answer questions such as the following:

1. How many moles of any particular product are formed if we start with a given mass of a starting material?

2. How many grams (or moles) of one starting material are necessary to react completely with a given number of grams (or moles) of another starting material?

3. How many grams (or moles) of starting material are needed if we want to form a certain number of grams (or moles) of a certain product?

4. How many grams (or moles) of a product are obtained when a certain amount of another product is produced?

It might seem as if we have four different types of problems here. In fact, we can solve them all by the same very simple procedure summarized in the following diagram:

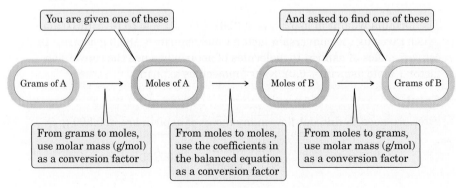

Stoichiometry The quantitative relationship between reactants and products in a chemical reaction as expressed by a balanced chemical equation

"Stoichiometry" comes from the Greek *stoicheion*, element, and *metron*, measure.

In Section 4.2, we saw that the coefficients in an equation represent numbers of molecules. Because moles are proportional to molecules (Section 4.6), the coefficients in an equation also represent numbers of moles.

You will always need a conversion factor that relates moles to moles. You will also need the conversion factors for grams to moles and from moles to grams according to the way the problem is asked; you may need one or both in some problems and not in others. It is easy to weigh a given number of grams, but the molar ratio determines the amount of substance involved in a reaction.

⦿WL Interactive Example 4.12 Stoichiometry

Ammonia is produced on an industrial scale by the reaction of nitrogen gas with hydrogen gas (the Haber process) according to this balanced equation:

$$N_2(g) + 3H_2(g) \longrightarrow 2NH_3(g)$$
$$\text{Ammonia}$$

How many grams of N_2 are necessary to produce 7.50 g of NH_3?

Strategy

The coefficients in an equation refer to the relative numbers of moles, not grams. Therefore, we must first find out how many moles of NH_3 are in 7.50 g of NH_3. To convert grams of NH_3 to moles of NH_3, we use the conversion factor 17.0 g NH_3 = 1 mol NH_3. We see from the balanced chemical equation that 2 mol NH_3 are produced from 1 mol of N_2, which gives us the conversion factor 2 mol NH_3 = 1.0 mol N_2. Finally, we convert moles of N_2 to grams of N_2, using the conversion factor 1 mol N_2 = 28.0 g N_2. Thus, solving this example requires three steps and three conversion factors.

Solution

Step 1: Convert 7.50 grams of NH_3 to moles of NH_3.

$$7.50 \text{ g } NH_3 \times \frac{1 \text{ mol } NH_3}{17.0 \text{ g } NH_3} = 0.441 \text{ mol } NH_3$$

Step 2: Convert moles of NH_3 to moles of N_2.

$$0.441 \text{ mol } NH_3 \times \frac{1 \text{ mol } N_2}{2 \text{ mol } NH_3} = 0.221 \text{ mol } N_2$$

Step 3: Convert moles N_2 to grams of N_2.

$$0.221 \text{ mol } N_2 \times \frac{28.0 \text{ g } N_2}{1 \text{ mol } N_2} = 6.18 \text{ g } N_2$$

Alternatively, one could perform the calculations in one continuous step.

$$7.50 \text{ g } \cancel{NH_3} \times \frac{1 \text{ mol } \cancel{NH_3}}{17.0 \text{ g } \cancel{NH_3}} \times \frac{1 \text{ mol } \cancel{N_2}}{2 \text{ mol } \cancel{NH_3}} \times \frac{28.0 \text{ g } N_2}{1 \text{ mol } \cancel{N_2}} = 6.18 \text{ g } N_2$$

In all such problems, we are given a mass (or number of moles) of one compound and asked to find the mass (or number of moles) of another compound. The two compounds can be on the same side of the equation or on opposite sides. We can do all such problems by the three steps we just used. The remaining problems in this chapter involving multiple steps will be alternatively solved using the one continuous step method shown directly above.

methane to → CO_2 H_2O combust

Problem 4.12

Pure aluminum is prepared by the electrolysis of aluminum oxide according to this equation:

$$Al_2O_3(s) \xrightarrow{\text{Electrolysis}} Al(s) + O_2(g)$$

Aluminum
oxide

(a) Balance this equation.
(b) What mass of aluminum oxide is required to prepare 27 g (1 mol) of aluminum?

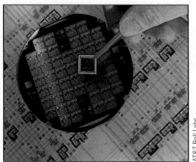

As microprocessor chips become increasingly smaller, the purity of the silicon becomes more important, because impurities can prevent the circuit from working properly

AT&T Bell Labs

⦿WL Interactive Example 4.13 Stoichiometry

Silicon to be used in computer chips is manufactured in a process represented by the following reaction:

$$SiCl_4(s) + 2Mg(s) \longrightarrow Si(s) + 2MgCl_2(s)$$

Silicon   Magnesium
tetrachloride   chloride

A sample of 225 g of silicon tetrachloride, $SiCl_4$, is reacted with an excess (more than necessary) of Mg. How many moles of Si are produced?

Strategy

To solve this example, we first convert grams of $SiCl_4$ to moles of $SiCl_4$, then moles of $SiCl_4$ to moles of Si.

Solution

Step 1: First, we convert grams of $SiCl_4$ to moles of $SiCl_4$. For this calculation, we use the conversion factor 1 mol $SiCl_4$ = 170 g $SiCl_4$:

$$225 \text{ g } \cancel{SiCl_4} \times \frac{1 \text{ mol } SiCl_4}{170 \text{ g } \cancel{SiCl_4}} = 1.32 \text{ mol } SiCl_4$$

Step 2: To convert moles of $SiCl_4$ to moles of Si, use the conversion factor 1 mol $SiCl_4$ = 1 mol Si, which we obtain from the balanced chemical equation. Now we do the arithmetic and obtain an answer of 1.32 mol Si:

$$225 \text{ g } SiCl_4 \times \frac{1 \text{ mol } \cancel{SiCl_4}}{170 \text{ g } \cancel{SiCl_4}} \times \frac{1 \text{ mol Si}}{1 \text{ mol } \cancel{SiCl_4}} = 1.32 \text{ mol Si}$$

Problem 4.13

In the industrial synthesis of acetic acid, methanol is reacted with carbon monoxide. How many moles of CO are required to produce 16.6 mol of acetic acid?

$$CH_3OH(g) + CO(g) \longrightarrow CH_3COOH(\ell)$$

Methanol   Carbon   Acetic acid
               monoxide

Example 4.14 Stoichiometry

When urea, $(NH_2)_2CO$, is acted on by the enzyme urease in the presence of water, ammonia and carbon dioxide are produced. Urease, the catalyst, is placed over the reaction arrow.

$$(NH_2)_2CO(aq) + H_2O(\ell) \xrightarrow{\text{Urease}} 2NH_3(aq) + CO_2(g)$$

Urea   Ammonia

If excess water is present (more than necessary for the reaction), how many grams each of CO_2 and NH_3 are produced from 0.83 mol of urea?

Strategy

We are given moles of urea and asked for grams of CO_2. First, we use the conversion factor 1 mol urea = 1 mol CO_2 to find the number of moles of CO_2 that will be produced and then convert moles of CO_2 to grams of CO_2. We use the same strategy to find the number of grams of NH_3 produced.

Solution

For grams of CO_2:

Step 1: We first convert moles of urea to moles of carbon dioxide using the conversion factor derived from the balanced chemical equation, 1 mol urea = 1 mol carbon dioxide.

Step 2: Use the conversion factor 1 mol CO_2 = 44 g CO_2 and then do the math to give the answer:

$$0.83 \text{ mol urea} \times \frac{1 \text{ mol CO}_2}{1 \text{ mol urea}} \times \frac{44 \text{ g CO}_2}{1 \text{mol CO}_2} = 37\text{g CO}_2$$

For grams of NH_3:

Steps 1 and 2 combine into one equation, where we follow the same procedure as for CO_2 but use different conversion factors:

$$0.83 \text{ mol urea} \times \frac{2 \text{ mol NH}_3}{1 \text{ mol urea}} \times \frac{17 \text{ g NH}_3}{1 \text{ mol NH}_3} = 28 \text{ g NH}_3$$

Problem 4.14

Ethanol is produced industrially by the reaction of ethylene with water in the presence of an acid catalyst. How many grams of ethanol are produced from 7.24 mol of ethylene? Assume that excess water is present.

$$C_2H_4(g) + H_2O(\ell) \xrightarrow[\text{catalyst}]{\text{Acid}} C_2H_6O(\ell)$$

Ethylene   Ethanol

B. Limiting Reagents

Frequently, reactants are mixed in molar proportions that differ from those that appear in a balanced equation. It often happens that one reactant is completely used up but one or more other reactants are not all used up. At times, we deliberately choose to have an excess of one reagent over another. As an example, consider an experiment in which NO is prepared by mixing five moles of N_2 with one mole of O_2. Only one mole of N_2 will react, consuming the one mole of O_2. The oxygen is used up completely, and four moles of nitrogen remain. These mole relationships are summarized under the balanced equation:

$$N_2(g) + O_2(g) \longrightarrow 2NO(g)$$

	N_2	O_2	$2NO$
Before reaction (moles)	5.0	1.0	0
After reaction (moles)	4.0	0	2.0

The **limiting reagent** is the reactant that is used up first. In this example, O_2 is the limiting reagent, because it governs how much NO can form. The other reagent, N_2, is in excess.

Limiting reagent The reactant that is consumed, leaving an excess of another reagent or reagents unreacted

Example 4.15 Limiting Reagent

Suppose 12 g of C is mixed with 64 g of O_2 and the following reaction takes place:

$$C(s) + O_2(g) \longrightarrow CO_2(g)$$

(a) Which reactant is the limiting reagent, and which reactant is in excess?
(b) How many grams of CO_2 will be formed?

Strategy

Determine how many moles of each reactant are present initially. Because C and O_2 react in a 1:1 molar ratio, the reactant present in the smaller molar amount is the limiting reagent and determines how many moles and, therefore, how many grams of CO_2 can be formed.

Solution

(a) We use the molar mass of each reactant to calculate the number of moles of each compound present before reaction.

$$12 \text{ g C} \times \frac{1 \text{ mol C}}{12 \text{ g C}} = 1.0 \text{ mol C}$$

$$64 \text{ g O}_2 \times \frac{1 \text{ mol O}_2}{32 \text{ g O}_2} = 2.0 \text{ mol O}_2$$

According to the balanced equation, reaction of one mole of C requires one mole of O_2. But two moles of O_2 are present at the start of the reaction. Therefore, C is the limiting reagent and O_2 is in excess.

(b) To calculate the number of grams of CO_2 formed, we use the conversion factor 1 mol CO_2 = 44 g CO_2.

$$12 \text{ g C} \times \frac{1 \text{ mol C}}{12 \text{ g C}} \times \frac{1 \text{ mol CO}_2}{1 \text{ mol C}} \times \frac{44 \text{ g CO}_2}{1 \text{ mol CO}_2} = 44 \text{ g CO}_2$$

We can summarize these numbers in the following table. Note that as required by the law of conservation of mass, the sum of the masses of the material present after reaction is the same as the amount present before any reaction took place, namely 76 g of material.

	C	+	O_2	$\longrightarrow$	CO_2
Before reaction	12 g		64 g		0
Before reaction	1.0 mol		2.0 mol		0
After reaction	0		1.0 mol		1.0 mol
After reaction	0		32.0 g		44.0 g

Problem 4.15

Assume that 6.0 g of C and 2.1 g of H_2 are mixed and react to form methane according to the following balanced equation:

$$C(s) + 2H_2(g) \longrightarrow CH_4(g)$$

Methane

(a) Which is the limiting reagent, and which reactant is in excess?
(b) How many grams of CH_4 are produced in the reaction?

C. Percent Yield

When carrying out a chemical reaction, we often get less of a product than we might expect from the type of calculation we discussed in Section 4.7. For example, suppose we react 32.0 g (1 mol) of CH_3OH with excess CO to form acetic acid:

$$CH_3OH \ + \ CO \longrightarrow CH_3COOH$$

Methanol   Carbon   Acetic acid
           monoxide

If we calculate the expected yield based on the stoichiometry of the balanced equation, we find that we should get 1 mol (60.0 g) of acetic acid. Suppose we get only 57.8 g of acetic acid. Does this result mean that the law of conservation of mass is being violated? No, it does not. We get less than 60.0 g of acetic acid because some of the CH_3OH does not react or because some of it reacts in another way or perhaps because our laboratory technique is not perfect and we lose a little in transferring it from one container to another.

At this point, we need to define three terms, all of which relate to yield of product in a chemical reaction:

Actual yield The mass of product actually formed or isolated in a chemical reaction.

Theoretical yield The mass of product that should form in a chemical reaction according to the stoichiometry of the balanced equation.

Percent yield The actual yield divided by the theoretical yield times 100.

$$\text{Percent yield} = \frac{\text{actual yield}}{\text{theoretical yield}} \times 100\%$$

We summarize the data for the preceding preparation of acetic acid in the following table:

	CH_3OH	+	CO	$\longrightarrow$	CH_3COOH
Before reaction	32.0 g		Excess		0
Before reaction	1.00 mol		Excess		0
Theoretical yield					1.00 mol
Theoretical yield					60.0 g
Actual yield					57.8 g

We calculate the percent yield in this experiment as follows:

$$\text{Percent yield} = \frac{57.8 \text{ acetic acid}}{60.0 \text{ g acetic acid}} \times 100\% = 96.3\%$$

A reaction that does not give the main product is called a side reaction.

Occasionally, the percent yield is larger than 100%. For example, if a chemist fails to dry a product completely before weighing it, the product weighs more than it should because it also contains water. In such cases, the actual yield may be larger than expected and the percent yield may be greater than 100%.

Example 4.16 Percent Yield

In an experiment forming ethanol, the theoretical yield is 50.5 g. The actual yield is 46.8 g. What is the percent yield?

Strategy

Percent yield is the actual yield divided by the theoretical yield times 100.

Solution

$$\% \text{ Yield} = \frac{46.8 \text{ g}}{50.5 \text{ g}} \times 100\% = 92.7\%$$

Problem 4.16

In an experiment to prepare aspirin, the theoretical yield is 153.7 g. If the actual yield is 124.3 g, what is the percent yield?

Why is it important to know the percent yield of a chemical reaction or a series of reactions? The most important reason often relates to cost. If the yield of commercial product is, say, only 10%, the chemists will most probably be sent back to the lab to vary experimental conditions in an attempt to improve the yield. As an example, consider a reaction in which starting material A is converted first to compound B, then to compound C, and finally to compound D.

$$A \longrightarrow B \longrightarrow C \longrightarrow D$$

Suppose the yield is 50% at each step. In this case, the yield of compound D is 13% based on the mass of compound A. If, however, the yield at each step is 90%, the yield of compound D increases to 73%; and if the yield at each step is 99%, the yield of compound D is 97%. These numbers are summarized in the following table.

If the Percent Yield per Step Is	The Percent Yield of Compound D Is
50%	$0.50 \times 0.50 \times 0.50 \times 100 = 13\%$
90%	$0.90 \times 0.90 \times 0.90 \times 100 = 73\%$
99%	$0.99 \times 0.99 \times 0.99 \times 100 = 97\%$

4.8 What Is Heat of Reaction?

In almost all chemical reactions, not only are starting materials converted to products, but heat is also either given off or absorbed. For example, when one mole of carbon is oxidized by oxygen to produce one mole of CO_2, 94.0 kilocalories of heat is given off per mole of carbon:

$$C(s) + O_2(g) \longrightarrow CO_2(g) + 94.0 \text{ kcal}$$

Heat of reaction The heat given off or absorbed in a chemical reaction

Exothermic reaction A chemical reaction that gives off heat

Endothermic reaction A chemical reaction that absorbs heat

The heat given off or gained in a reaction is called the **heat of reaction**. A reaction that gives off heat is **exothermic**; a reaction that absorbs heat is **endothermic**. The amount of heat given off or absorbed is proportional to the amount of material. For example, when 2 mol of carbon is oxidized by oxygen to give carbon dioxide, $2 \times 94.0 = 188$ kcal of heat is given off.

The energy changes accompanying a chemical reaction are not limited to heat. In some reactions, such as in voltaic cells (Chemical Connections 4B), the energy given off takes the form of electricity. In other reactions, such as photosynthesis (the reaction whereby plants convert water and carbon dioxide to carbohydrates and oxygen), the energy absorbed is in the form of light.

An example of an endothermic reaction is the decomposition of mercury(II) oxide:

$$2HgO(s) + 43.4 \text{ kcal} \longrightarrow 2Hg(\ell) + O_2(g)$$
Mercury(II) oxide
(Mercuric oxide)

This equation tells us that if we want to decompose 2 mol of mercury(II) oxide into the elements $Hg(l)$ and $O_2(g)$, we must add 43.4 kcal of energy to HgO. Incidentally, the law of conservation of energy tells us that the reverse reaction, the oxidation of mercury, must give off exactly the same amount of heat:

$$2Hg(\ell) + O_2(g) \longrightarrow 2HgO(s) + 43.4 \text{ kcal}$$

Especially important are the heats of reaction for combustion reactions. As we saw in Section 4.4, combustion reactions are the most important heat-producing reactions, because most of the energy required for modern society to function is derived from them. All combustions are exothermic. The heat given off in a combustion reaction is called the **heat of combustion**.

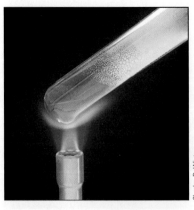

Charles D. Winters

Mercury (II) oxide, a red compound, decomposes into two elements when heated: mercury (a metal) and oxygen (a nonmetal). Mercury vapor condenses on the cooler upper portion of the test tube.

Summary

Section 4.2 How Do We Balance Chemical Equations?

- A **chemical equation** is an expression showing which reactants are converted to which products. A balanced chemical equation shows how many moles of each starting material are converted to how many moles of each product according to the law of conservation of mass.

Section 4.3 How Can We Predict If Ions in Aqueous Solution Will React with Each Other?

- When ions are mixed in aqueous solution, they react with one another only if (1) a precipitate forms, (2) a gas forms, (3) an acid neutralizes a base, or (4) an oxidation-reduction takes place.
- Ions that do not react are called **spectator ions**.
- A **net ionic equation** shows only those ions that react. In a net ionic equation, both the charges and the number (mass) of atoms must be balanced.

Section 4.4 What Are Oxidation and Reduction?

- **Oxidation** is the loss of electrons; **reduction** is the gain of electrons. These two processes must take place together; you cannot have one without the other. The joint process is often called a redox reaction.
- Oxidation can also be defined as the gain of oxygens and/or the loss of hydrogens; reduction can also be defined as the loss of oxygens and/or the gain of hydrogens.

Section 4.5 What Are Formula Weights and Molecular Weights?

- The **formula weight (FW)** of a compound is the sum of the atomic weights of all atoms in the compound expressed in atomic mass units (amu). Formula weight applies to both ionic and molecular compounds.
- The term **molecular weight**, also expressed in amu, applies to only molecular compounds.

Section 4.6 What Is a Mole and How Do We Use It to Calculate Mass Relationships?

- A **mole (mol)** of any substance is defined as Avogadro's number (6.02×10^{23}) of formula units of the substance.
- The **molar mass** of a substance is its formula weight expressed in grams.

Section 4.7 How Do We Calculate Mass Relationships in Chemical Reactions?

- **Stoichiometry** is the study of the mass relationships in chemical reactions.
- The reagent that is used up first in a reaction is called the **limiting reagent**.
- The **percent yield** for a reaction equals the **actual yield** divided by the **theoretical yield** multiplied by 100.

Section 4.8 What Is Heat of Reaction?

- Almost all chemical reactions are accompanied by either a gain or a loss of heat. This heat is called the heat of reaction.
- Reactions that give off heat are exothermic; those that absorb heat are endothermic.
- The heat given off in a combustion reaction is called the **heat of combustion**.

Problems

Orange-numbered problems are applied.

Section 4.2 How Do We Balance Chemical Equations?

4.17 Balance each equation.
- (a) $HI + NaOH \longrightarrow NaI + H_2O$
- (b) $Ba(NO_3)_2 + H_2S \longrightarrow BaS + HNO_3$
- (c) $CH_4 + O_2 \longrightarrow CO_2 + H_2O$
- (d) $C_4H_{10} + O_2 \longrightarrow CO_2 + H_2O$
- (e) $Fe + CO_2 \longrightarrow Fe_2O_3 + CO$

4.18 Balance each equation.
- (a) $H_2 + I_2 \longrightarrow HI$
- (b) $Al + O_2 \longrightarrow Al_2O_3$
- (c) $Na + Cl_2 \longrightarrow NaCl$
- (d) $Al + HBr \longrightarrow AlBr_3 + H_2$
- (e) $P + O_2 \longrightarrow P_2O_5$

4.19 If you blow carbon dioxide gas into a solution of calcium hydroxide, a milky-white precipitate of calcium carbonate forms. Write a balanced equation for the formation of calcium carbonate in this reaction.

4.20 Calcium oxide is prepared by heating limestone (calcium carbonate, $CaCO_3$) to a high temperature, at which point it decomposes to calcium oxide and carbon dioxide. Write a balanced equation for this preparation of calcium oxide.

4.21 The brilliant white light in some firework displays is produced by burning magnesium in air. The magnesium reacts with oxygen in the air to form magnesium oxide. Write a balanced equation for this reaction.

4.22   The rusting of iron is a chemical reaction of iron with oxygen in the air to form iron(III) oxide. Write a balanced equation for this reaction.

4.23   When solid carbon burns in a limited supply of oxygen gas, the gas carbon monoxide, CO, forms. This gas is deadly to humans because it combines with hemoglobin in the blood, making it impossible for the blood to transport oxygen. Write a balanced equation for the formation of carbon monoxide.

4.24   Solid ammonium carbonate, $(NH_4)_2CO_3$, decomposes at room temperature to form gaseous ammonia, carbon dioxide, and water. Because of the ease of decomposition and the penetrating odor of ammonia, ammonium carbonate can be used as smelling salts. Write a balanced equation for this decomposition.

4.25   In the chemical test for arsenic, the gas arsine, AsH_3, is prepared. When arsine is decomposed by heating, arsenic metal deposits as a mirror-like coating on the surface of a glass container and hydrogen gas, H_2, is given off. Write a balanced equation for the decomposition of arsine.

4.26   When a piece of aluminum metal is dropped into hydrochloric acid, HCl, hydrogen is released as a gas and a solution of aluminum chloride forms. Write a balanced equation for the reaction.

4.27   In the industrial chemical preparation of chlorine, Cl_2, electric current is passed through an aqueous solution of sodium chloride to give $Cl_2(g)$ and $H_2(g)$. The other product of this reaction is sodium hydroxide. Write a balanced equation for this reaction.

Section 4.3   How Can We Predict If Ions in Aqueous Solution Will React with Each Other?

4.28   Answer true or false.
   (a) A net ionic equation shows only those ions that undergo chemical reaction.
   (b) In a net ionic equation, the number of moles of starting material must equal the number of moles of product.
   (c) A net ionic equation must be balanced by both mass and charge.
   (d) As a generalization, all lithium, sodium, and potassium salts are soluble in water.
   (e) As a generalization, all nitrate (NO_3^-) salts are soluble in water.
   (f) As a generalization, most carbonate (CO_3^{2-}) salts are insoluble in water.
   (g) Sodium carbonate, Na_2CO_3, is insoluble in water.
   (h) Ammonium carbonate, $(NH_4)_2CO_3$, is insoluble in water.
   (i) Calcium carbonate, $CaCO_3$, is insoluble in water.
   (j) Sodium dihydrogen phosphate, NaH_2PO_4, is insoluble in water.
   (k) Sodium hydroxide, NaOH, is soluble in water.
   (l) Barium hydroxide, $Ba(OH)_2$, is soluble in water.

4.29   Balance these net ionic equations.
   (a) $Ag^+(aq) + Br^-(aq) \longrightarrow AgBr(s)$
   (b) $Cd^{2+}(aq) + S^{2-}(aq) \longrightarrow CdS(s)$

   (c) $Sc^{3+}(aq) + SO_4^{2-}(aq) \longrightarrow Sc_2(SO_4)_3(s)$
   (d) $Sn^{2+}(aq) + Fe^{2+}(aq) \longrightarrow Sn(s) + Fe^{3+}(aq)$
   (e) $K(s) + H_2O(\ell) \longrightarrow K^+(aq) + OH^-(aq) + H_2(g)$

4.30   In the equation

$$2Na^+(aq) + CO_3^{2-}(aq) + Sr^{2+}(aq) + 2Cl^-(aq) \longrightarrow$$
$$SrCO_3(s) + 2Na^+(aq) + 2Cl^-(aq)$$

   (a) Identify the spectator ions.
   (b) Write the balanced net ionic equation.

4.31   Predict whether a precipitate will form when aqueous solutions of the following compounds are mixed. If a precipitate will form, write its formula and write a net ionic equation for its formation. To make your predictions, use the solubility generalizations in Section 4.3.
   (a) $CaCl_2(aq) + K_3PO_4(aq) \longrightarrow$
   (b) $KCl(aq) + Na_2SO_4(aq) \longrightarrow$
   (c) $(NH_4)_2CO_3(aq) + Ba(NO_3)_2(aq) \longrightarrow$
   (d) $FeCl_2(aq) + KOH(aq) \longrightarrow$
   (e) $Ba(NO_3)_2(aq) + NaOH(aq) \longrightarrow$
   (f) $Na_2S(aq) + SbCl_3(aq) \longrightarrow$
   (g) $Pb(NO_3)_2(aq) + K_2SO_4(aq) \longrightarrow$

4.32   When a solution of ammonium chloride is added to a solution of lead(II) nitrate, $Pb(NO_3)_2$, a white precipitate, lead(II) chloride, forms. Write a balanced net ionic equation for this reaction. Both ammonium chloride and lead nitrate exist as dissociated ions in aqueous solution.

4.33   When a solution of hydrochloric acid, HCl, is added to a solution of sodium sulfite, Na_2SO_3, sulfur dioxide gas is released from the solution. Write a net ionic equation for this reaction. An aqueous solution of HCl contains H^+ and Cl^- ions, and Na_2SO_3 exists as dissociated ions in aqueous solution.

4.34   When a solution of sodium hydroxide is added to a solution of ammonium carbonate, H_2O is formed and ammonia gas, NH_3, is released when the solution is heated. Write a net ionic equation for this reaction. Both NaOH and $(NH_4)_2CO_3$ exist as dissociated ions in aqueous solution.

4.35   Using the solubility generalizations given in Section 4.3, predict which of these ionic compounds are soluble in water.
   (a) KCl   (b) NaOH   (c) $BaSO_4$
   (d) Na_2SO_4 (e) Na_2CO_3 (f) $Fe(OH)_2$

4.36   Using the solubility generalizations given in Section 4.3, predict which of these ionic compounds are soluble in water.
   (a) $MgCl_2$ (b) $CaCO_3$ (c) Na_2SO_3
   (d) NH_4NO_3 (e) $Pb(OH)_2$

Section 4.4   What Are Oxidation and Reduction?

4.37   Answer true or false.
   (a) When a substance is oxidized, it loses electrons.
   (b) When a substance gains electrons, it is reduced.
   (c) In a redox reaction, the oxidizing agent becomes reduced.

(d) In a redox reaction, the reducing reagent becomes oxidized.

(e) When Zn is converted to Zn^{2+} ion, zinc is oxidized.

(f) Oxidation can also be defined as the loss of oxygen atoms and/or the gain of hydrogen atoms.

(g) Reduction can also be defined as the gain of oxygen atoms and/or the loss of hydrogen atoms.

(h) When oxygen, O_2, is converted to hydrogen peroxide, H_2O_2, we say that O_2 is reduced.

(i) Hydrogen peroxide, H_2O_2, is an oxidizing agent.

(j) All combustion reactions are redox reactions.

(k) The products of complete combustion (oxidation) of hydrocarbon fuels are carbon dioxide, water, and heat.

(l) In the combustion of hydrocarbon fuels, oxygen is the oxidizing agent and the hydrocarbon fuel is the reducing agent.

(m) Incomplete combustion of hydrocarbon fuels can produce significant amounts of carbon monoxide.

(n) Most common bleaches are oxidizing agents.

4.38 In the reaction

$$Pb(s) + 2Ag^+(aq) \longrightarrow Pb^{2+}(aq) + 2Ag(s)$$

(a) Which species is oxidized and which is reduced?

(b) Which species is the oxidizing agent and which is the reducing agent?

4.39 In the reaction

$$C_7H_{12}(\ell) + 10O_2(g) \longrightarrow 7CO_2(g) + 6H_2O(\ell)$$

(a) Which species is oxidized and which is reduced?

(b) Which species is the oxidizing agent and which is the reducing agent?

4.40 When a piece of sodium metal is added to water, hydrogen is evolved as a gas and a solution of sodium hydroxide is formed.

(a) Write a balanced equation for this reaction.

(b) What is oxidized in this reaction? What is reduced?

Section 4.5 What Are Formula Weights and Molecular Weights?

4.41 Answer true or false.

(a) Formula weight is the mass of a compound expressed in grams.

(b) 1 atomic mass unit (amu) is equal to 1 gram (g).

(c) The formula weight of H_2O is 18 amu.

(d) The molecular weight of H_2O is 18 amu.

(e) The molecular weight of a covalent compound is the same as its formula weight.

4.42 Calculate the formula weight of:

(a) KCl   (b) Na_3PO_4 (c) $Fe(OH)_2$

(d) $NaAl(SO_3)_2$ (e) $Al_2(SO_4)_3$ (f) $(NH_4)_2CO_3$

4.43 Calculate the molecular weight of:

(a) Sucrose, $C_{12}H_{22}O_{11}$ (b) Glycine, $C_2H_5NO_2$

(c) DDT, $C_{14}H_9Cl_5$

Section 4.6 What Is a Mole and How Do We Use It to Calculate Mass Relationships?

4.44 Answer true or false.

(a) The mole is a counting unit, just as a dozen is a counting unit.

(b) Avogadro's number is the number of formula units in one mole.

(c) Avogadro's number, to three significant figures, is 6.02×10^{23} formula units per mole.

(d) 1 mol of H_2O contains $3 \times 6.02 \times 10^{23}$ formula units.

(e) 1 mol of H_2O has the same number of molecules as 1 mol of H_2O_2.

(f) The molar mass of a compound is its formula weight expressed in amu.

(g) The molar mass of H_2O is 18 g/mol.

(h) 1 mol of H_2O has the same molar mass as 1 mol of H_2O_2.

(i) 1 mol of ibuprofen, $C_{13}H_{18}O_2$, contains 33 mol of atoms.

(j) To convert moles to grams, multiply by Avogadro's number.

(k) To convert grams to moles, divide by molar mass.

(l) 1 mol of H_2O contains 1 mol of hydrogen atoms and one mol of oxygen atoms.

(m) 1 mol of H_2O contains 2 g of hydrogen atoms and 1 g of oxygen atoms.

(n) 1 mole of H_2O contains 18.06×10^{23} atoms.

4.45 Calculate the number of moles in:

(a) 32 g of methane, CH_4

(b) 345.6 g of nitric oxide, NO

(c) 184.4 g of chlorine dioxide, ClO_2

(d) 720. g of glycerin, $C_3H_8O_3$

4.46 Calculate the number of grams in:

(a) 1.77 mol of nitrogen dioxide, NO_2

(b) 0.84 mol of 2-propanol, C_3H_8O (rubbing alcohol)

(c) 3.69 mol of uranium hexafluoride, UF_6

(d) 0.348 mol of galactose, $C_6H_{12}O_6$

(e) 4.9×10^{-2} mol of vitamin C, $C_6H_8O_6$

4.47 Calculate the number of moles of:

(a) O atoms in 18.1 mol of formaldehyde, CH_2O

(b) Br atoms in 0.41 mol of bromoform, $CHBr_3$

(c) O atoms in 3.5×10^3 mol of $Al_2(SO_4)_3$

(d) Hg atoms in 87 g of HgO

4.48 Calculate the number of moles of:

(a) S^{2-} ions in 6.56 mol of Na_2S

(b) Mg^{2+} ions in 8.320 mol of $Mg_3(PO_4)_2$

(c) acetate ions, CH_3COO^-, in 0.43 mol of $Ca(CH_3COO)_2$

4.49 Calculate the number of:

(a) nitrogen atoms in 25.0 g of TNT, $C_7H_5N_3O_6$

(b) carbon atoms in 40.0 g of ethanol, C_2H_6O

(c) oxygen atoms in 500. mg of aspirin, $C_9H_8O_4$

(d) sodium atoms in 2.40 g of sodium dihydrogen phosphate, NaH_2PO_4

4.50 How many molecules are in each of the following?

(a) 2.9 mol of TNT, $C_7H_5N_3O_6$

(b) one drop (0.0500 g) of water

(c) 3.1×10^{-1} g of aspirin, $C_9H_8O_4$

4.51 What is the mass in grams of each number of molecules of formaldehyde, CH_2O?

(a) 100. molecules   (b) 3000. molecules

(c) 5.0×10^6 molecules   (d) 2.0×10^{24} molecules

4.52 The molecular weight of hemoglobin is about 68,000 amu. What is the mass in grams of a single molecule of hemoglobin?

4.53 A typical deposit of cholesterol, $C_{27}H_{46}O$, in an artery might have a mass of 3.9 mg. How many molecules of cholesterol are in this mass?

Section 4.7 How Do We Calculate Mass Relationships in Chemical Reactions?

4.54 Answer true or false.

(a) Stoichiometry is the study of mass relationships in chemical reactions.

(b) To determine mass relationships in a chemical reaction, you first need to know the balanced chemical equation for the reaction.

(c) To convert from grams to moles and vice versa, use Avogadro's number as a conversion factor.

(d) To convert from grams to moles and vice versa, use molar mass as a conversion factor.

(e) A limiting reagent is the reagent that is used up first.

(f) Suppose a chemical reaction between A and B requires 1 mol of A and 2 mol of B. If 1 mol of each is present, then B is the limiting reagent.

(g) Theoretical yield is the yield of product that should be obtained according to the balanced chemical equation.

(h) Theoretical yield is the yield of product that should be obtained if all limiting reagent is converted to product.

(i) Percent yield is the number of grams of product divided by the number of grams of the limiting reagent times 100.

(j) To calculate percent yield, divide the mass of product formed by the theoretical yield and multiply by 100.

4.55 For the reaction:

$$2N_2(g) + 3O_2(g) \longrightarrow 2N_2O_3(g)$$

(a) How many moles of N_2 are required to react completely with 1 mole of O_2?

(b) How many moles of N_2O_3 are produced from the complete reaction of 1 mole of O_2?

(c) How many moles of O_2 are required to produce 8 moles of N_2O_3?

4.56 Magnesium reacts with sulfuric acid according to the following equation. How many moles of H_2 are produced by the complete reaction of 230. mg of Mg with sulfuric acid?

$$Mg(s) + H_2SO_4(aq) \longrightarrow MgSO_4(aq) + H_2(g)$$

4.57 Chloroform, $CHCl_3$, is prepared industrially by the reaction of methane with chlorine. How many grams of Cl_2 are needed to produce 1.50 moles of chloroform?

$$\underset{\text{Methane}}{CH_4(g)} + 3Cl_2(g) \longrightarrow \underset{\text{Chloroform}}{CHCl_3(\ell)} + 3HCl(g)$$

4.58 At one time, acetaldehyde was prepared industrially by the reaction of ethylene with air in the presence of a copper catalyst. How many grams of acetaldehyde can be prepared from 81.7 g of ethylene?

$$\underset{\text{Ethylene}}{2C_2H_4(g)} + O_2(g) \xrightarrow{\text{Catalyst}} \underset{\text{Acetaldehyde}}{2C_2H_4O(g)}$$

4.59 Chlorine dioxide, ClO_2, is used for bleaching paper. It is also the gas used to kill the anthrax spores that contaminated the Hart Senate Office Building in the fall of 2001. Chlorine dioxide is prepared by treating sodium chlorite with chlorine gas.

$$\underset{\substack{\text{Sodium} \\ \text{chlorite}}}{NaClO_2(aq)} + Cl_2(g) \longrightarrow \underset{\substack{\text{Chlorine} \\ \text{dioxide}}}{ClO_2(g)} + NaCl(aq)$$

(a) Balance the equation for the preparation of chlorine dioxide.

(b) Calculate the weight of chlorine dioxide that can be prepared from 5.50 kg of sodium chlorite.

4.60 Ethanol, C_2H_6O, is added to gasoline to produce "gasohol," a fuel for automobile engines. How many grams of O_2 are required for complete combustion of 421 g of ethanol?

$$\underset{\text{Ethanol}}{C_2H_5OH(\ell)} + 3O_2(g) \longrightarrow 2CO_2(g) + 3H_2O$$

4.61 In photosynthesis, green plants convert CO_2 and H_2O to glucose, $C_6H_{12}O_6$. How many grams of CO_2 are required to produce 5.1 g of glucose?

$$6CO_2(g) + 6H_2O(\ell) \xrightarrow{\text{Photosynthesis}} \underset{\text{Glucose}}{C_6H_{12}O_6(aq)} + 6O_2(g)$$

4.62 Iron ore is converted to iron by heating it with coal (carbon), and oxygen according to the following equation:

$$2Fe_2O_3(s) + 6C(s) + 3O_2(g) \longrightarrow 4Fe(s) + 6CO_2(g)$$

If the process is run until 3940. g of Fe is produced, how many grams of CO_2 will also be produced?

4.63 Given the reaction in Problem 4.62, how many grams of C are necessary to react completely with 0.58 g of Fe_2O_3?

4.64 Aspirin is made by the reaction of salicylic acid with acetic anhydride. How many grams of aspirin are produced if 85.0 g of salicylic acid is treated with excess acetic anhydride?

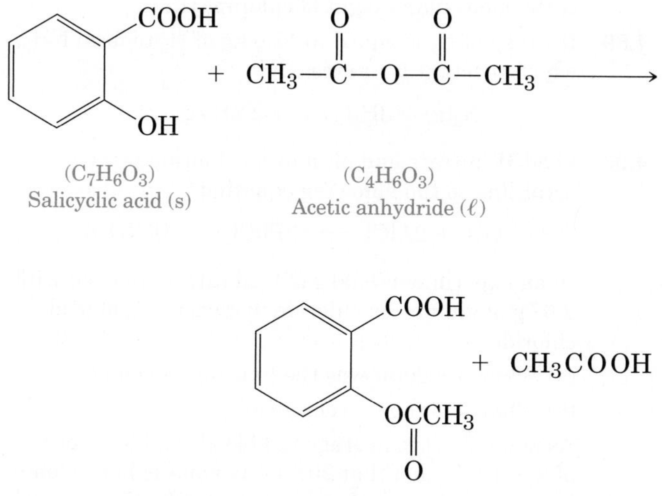

$(C_7H_6O_3)$
Salicyclic acid (s)

$(C_4H_6O_3)$
Acetic anhydride (ℓ)

+ CH_3COOH

$(C_9H_8O_4)$
Aspirin (s)

$(C_2H_4O_2)$
Acetic acid (ℓ)

4.65 Suppose the preparation of aspirin from salicylic acid and acetic anhydride (Problem 4.64) gives a yield of 75.0% of aspirin. How many grams of salicylic acid must be used to prepare 50.0 g of aspirin?

4.66 Benzene reacts with bromine to produce bromobenzene according to the following equation:

$$C_6H_6(\ell) + Br_2(\ell) \longrightarrow C_6H_5Br(\ell) + HBr(g)$$
Benzene   Bromine   Bromobenzene   Hydrogen bromide

If 60.0 g of benzene is mixed with 135 g of bromine,
(a) Which is the limiting reagent?
(b) How many grams of bromobenzene are formed in the reaction?

4.67 Ethyl chloride is prepared by the reaction of chlorine with ethane according to the following balanced equation.

$$C_2H_6(g) + Cl_2(g) \longrightarrow C_2H_5Cl(\ell) + HCl(g)$$
Ethane   Ethyl chloride

When 5.6 g of ethane is reacted with excess chlorine, 8.2 g of ethyl chloride forms. Calculate the percent yield of ethyl chloride.

4.68 Diethyl ether is made from ethanol according to the following reaction:

$$2C_2H_5OH(\ell) \longrightarrow (C_2H_5)_2O(\ell) + H_2O(\ell)$$
Ethanol   Diethyl ether

In an experiment, 517 g of ethanol gave 391 g of diethyl ether. What was the percent yield in this experiment?

Section 4.8 What Is Heat of Reaction?
4.69 Answer true or false.
(a) Heat of reaction is the heat given off or absorbed by a chemical reaction.

(b) An endothermic reaction is one that gives off heat.
(c) If a chemical reaction is endothermic, the reverse reaction is exothermic.
(d) All combustion reactions are exothermic.
(e) If the reaction of glucose ($C_6H_{12}O_6$) and O_2 in the body to give CO_2 and H_2O is an exothermic reaction, then photosynthesis in green plants (the reaction of CO_2 and H_2O to give glucose and O_2) is an endothermic process.
(f) The energy required to drive photosynthesis comes from the sun in the form of electromagnetic radiation.

4.70 What is the difference between exothermic and endothermic reactions?

4.71 Which of these reactions are exothermic, and which are endothermic?
(a) $2NH_3(g) + 22.0$ kcal $\longrightarrow N_2(g) + 3H_2(g)$
(b) $H_2(g) + F_2(g) \longrightarrow 2HF(g) + 124$ kcal
(c) $C(s) + O_2(g) \longrightarrow CO_2(g) + 94.0$ kcal
(d) $H_2(g) + CO_2(g) + 9.80$ kcal $\longrightarrow H_2O(g) + CO(g)$
(e) $C_3H_8(g) + 5O_2(g) \longrightarrow 3CO_2(g) +$
$$4H_2O(g) + 531 \text{ kcal}$$

4.72 In the following reaction, 9.80 kcal is absorbed per mole of CO_2 undergoing reaction. How much heat is given off if two moles of water are reacted with two moles of carbon monoxide?

$$H_2(g) + CO_2(g) + 9.80 \text{ kcal} \longrightarrow H_2O(g) + CO(g)$$

4.73 Following is the equation for the combustion of acetone:

$$2C_3H_6O(\ell) + 8O_2(g) \longrightarrow 6CO_2(g) +$$
Acetone
$$6H_2O(g) + 853.6 \text{ kcal}$$

How much heat will be given off if 0.37 mol of acetone is burned completely?

4.74 The oxidation of glucose, $C_6H_{12}O_6$, to carbon dioxide and water is exothermic. The heat liberated is the same whether glucose is metabolized in the body or burned in air.

$$C_6H_{12}O_6 + 6O_2 \longrightarrow 6CO_2 + 6H_2O + 670 \text{ kcal/mol}$$
Glucose

Calculate the heat liberated when 15.0 g of glucose is metabolized to carbon dioxide and water in the body.

4.75 The heat of combustion of glucose, $C_6H_{12}O_6$, is 670 kcal/mol. The heat of combustion of ethanol, C_2H_6O, is 327 kcal/mol. The heat liberated by oxidation of each compound is the same whether it is burned in air or metabolized in the body. On a kcal/g basis, metabolism of which compound liberates more heat?

4.76 A plant requires approximately 4178 kcal for the production of 1.00 kg of starch (Chapter 20) from carbon dioxide and water.
(a) Is the production of starch in a plant an exothermic process or an endothermic process?
(b) Calculate the energy in kilocalories required by a plant for the production of 6.32 g of starch.

4.77 To convert 1 mol of iron(III) oxide to its elements requires 196.5 kcal:

$$Fe_2O_3(s) + 196.5 \text{ kcal} \longrightarrow 2Fe(s) + \frac{3}{2}O_2(g)$$

How many grams of iron can be produced if 156.0 kcal of heat is absorbed by a large-enough sample of iron(III) oxide?

Chemical Connections

4.78 (Chemical Connections 4A) How does fluoride ion protect the tooth enamel against decay?

4.79 (Chemical Connections 4A) What ions are present in hydroxyapatite?

4.80 (Chemical Connections 4B) A voltaic cell is represented by the following equation:

$$Fe(s) + Zn^{2+}(aq) \longrightarrow Fe^{2+}(aq) + Zn(s)$$

Which electrode is the anode, and which is the cathode?

4.81 (Chemical Connections 4C) Balance the lithium-iodine battery redox reaction described in this section and identify the oxidizing and reducing agents present.

Additional Problems

4.82 When gaseous dinitrogen pentoxide, N_2O_5, is bubbled into water, nitric acid, HNO_3, forms. Write a balanced equation for this reaction.

4.83 In a certain reaction, Cu^+ is converted to Cu^{2+}. Is Cu^+ ion oxidized or reduced in this reaction? Is Cu^+ ion an oxidizing agent or a reducing agent in this reaction?

4.84 Using the equation:

$$Fe_2O_3(s) + 3CO(g) \longrightarrow 2Fe(s) + 3CO_2(g)$$

(a) Show that this is a redox reaction. Which species is oxidized, and which is reduced?

(b) How many moles of Fe_2O_3 are required to produce 38.4 mol of Fe?

(c) How many grams of CO are required to produce 38.4 mol of Fe?

4.85 Methyl tertiary butyl ether (or MTBE), a chemical compound with molecular formula $C_5H_{12}O$, is an additive used as an oxygenate to raise the octane number of gas, although its use has declined in the last few years in response to environmental and health concerns. Write the balanced molecular equation for the reaction involving the complete burning of liquid MTBE in air.

4.86 When an aqueous solution of Na_3PO_4 is added to an aqueous solution of $Cd(NO_3)_2$, a precipitate forms. Write a net ionic equation for this reaction and identify the spectator ions.

4.87 The active ingredient in an analgesic tablet is 488 mg of aspirin, $C_9H_8O_4$. How many moles of aspirin does the tablet contain?

4.88 Chlorophyll, the compound responsible for the green color of leaves and grasses, contains one atom of magnesium in each molecule. If the percentage by weight of magnesium in chlorophyll is 2.72%, what is the molecular weight of chlorophyll?

4.89 If 7.0 kg of N_2 is added to 11.0 kg of H_2 to form NH_3, which reactant is in excess?

$$N_2(g) + 3H_2(g) \longrightarrow 2NH_3(g)$$

4.90 Lead(II) nitrate and aluminum chloride react according to the following equation:

$$3Pb(NO_3)_2 + 2AlCl_3 \longrightarrow 3PbCl_2 + 2Al(NO_3)_3$$

In an experiment, 8.00 g of lead nitrate reacted with 2.67 g of aluminum chloride to give 5.55 g of lead chloride.

(a) Which reactant was the limiting reagent?

(b) What was the percent yield?

4.91 Assume that the average red blood cell has a mass of 2×10^{-8} g and that 20% of its mass is hemoglobin (a protein whose molar mass is 68,000). How many molecules of hemoglobin are present in one red blood cell?

4.92 Reaction of pentane, C_5H_{12}, with oxygen, O_2, gives carbon dioxide and water.

(a) Write a balanced equation for this reaction.

(b) In this reaction, what is oxidized and what is reduced?

(c) What is the oxidizing agent, and what is the reducing agent?

4.93 Ammonia is prepared industrially by the reaction of nitrogen and hydrogen according to the following equation:

$$N_2(g) + 3H_2(g) \longrightarrow 2NH_3(g)$$
Ammonia

If 29.7 kg of N_2 is added to 3.31 kg of H_2,

(a) Which reactant is the limiting reagent?

(b) How many grams of the other reactant are left over?

(c) How many grams of NH_3 are formed if the reaction goes to completion?

Tying It Together

4.94 2,3,7,8-Tetrachlorodibenzo-*p*-dioxin (TCDD) is a potent poison with the chemical formula $C_{12}H_4Cl_4O_2$. The average lethal dose in humans is approximately 2.9×10^{-2} mg per kg of body weight. How many molecules of TCDD constitute a lethal dose for an 82-kg individual?

4.95 Furan, an organic compound used in the synthesis of nylon and referenced in Section 20.2, has the molecular formula C_4H_4O.

(a) Determine the number of moles of furan in a 441 mg sample.

(b) If the density of furan is known to be 0.936 g/mL, how many carbon atoms are present in 0.060 L of furan?

(c) Calculate the mass in grams of 9.86×10^{25} molecules of furan.

4.96 A sample of gold consisting of 8.68×10^{23} atoms with a density of 19.3 g/mL is hammered into a sheet that covers an area of 1.00×10^2 ft^2. Determine the thickness of the sheet in centimeters.

4.97 Consider the production of KClO$_4$(*aq*) via the three balanced sequential reactions below, where the percentage yield of each reaction is written above the reaction arrows:

$$Cl_2(g) + 2KOH(aq) \xrightarrow{92.1\%}$$
$$KCl(aq) + KClO(aq) + H_2O(\ell)$$

$$3KClO(aq) \xrightarrow{86.7\%} 2KCl(aq) + KClO_3(aq)$$

$$4KClO_3(aq) \xrightarrow{75.3\%} 3KClO_4(aq) + KCl(aq)$$

Determine the mass in grams of KClO$_4$(*aq*) produced at the end of the three-step reaction sequence if a student begins with 966 kg of Cl$_2$(g).

4.98 Elemental chlorine is commonly used to kill microorganisms in drinking water supplies as well as to remove sulfides. For example, noxious-smelling hydrogen sulfide gas is removed from water via the following unbalanced chemical equation:

$$H_2S(aq) + Cl_2(aq) \longrightarrow HCl(aq) + S_8(s)$$

(a) Write a balanced equation for this reaction.

(b) Determine the mass in grams of elemental sulfur, S$_8$, which is produced when 50.0 L of water containing 1.5×10^{-5} g of H$_2$S per liter is treated with 1.0 g of Cl$_2$.

(c) Calculate the percent yield of the reaction if 5.8×10^{-4} g of S$_8$ is generated.

Looking Ahead

4.99 The two major sources of energy in our diets are fats and carbohydrates. Palmitic acid, one of the major components of both animal fats and vegetable oils, belongs to a group of compounds called fatty acids.

The metabolism of fatty acids is responsible for the energy from fats. The major carbohydrates in our diets are sucrose (table sugar; Section 20.4A) and starch (Section 20.5A). Both starch and sucrose are first converted in the body to glucose, and then glucose is metabolized to produce energy. The heat of combustion of palmitic acid is 2385 kcal/mol, and that of glucose is 670. kcal/mol. Below are unbalanced equations for the metabolism of each body fuel:

$$C_{16}H_{32}O_2(aq) + O_2(g) \longrightarrow$$
Palmitic acid
(256 g/mol)
$$CO_2(g) + H_2O(\ell) + 2385 \text{ kcal/mol}$$

$$C_6H_{12}O_6(aq) + O_2(g) \longrightarrow$$
Glucose
(180. g/mol)
$$CO_2(g) + H_2O(\ell) + 670 \text{ kcal/mol}$$

(a) Balance the equation for the metabolism of each fuel.

(b) Calculate the heat of combustion of each in kcal/g.

(c) In terms of kcal/mol, which of the two is the better source of energy for the body?

(d) In terms of kcal/g, which of the two is the better source of energy for the body?

4.100 The heat of combustion of methane, CH$_4$, the major component of natural gas, is 213 kcal/mol. The heat of combustion of propane, C$_3$H$_8$, the major component of LPG, or bottled gas, is 530. kcal/mol.

(a) Write a balanced equation for the complete combustion of each to CO$_2$ and H$_2$O.

(b) On a kcal/mol basis, which of these two fuels is the better source of heat energy?

(c) On a kcal/g basis, which of these two fuels is the better source of heat energy?

5 Gases, Liquids, and Solids

Sign in to OWL at **www.cengage.com/owl** to view tutorials and simulations, develop problem-solving skills, and complete online homework assigned by your professor.

Hot-air balloon, Utah.

© Vince Streano/CORBIS

5.1   What Are the Three States of Matter?

Various forces hold matter together causing it to take different forms. In an atomic nucleus, very strong forces of attraction keep the protons and neutrons together (Chapter 2). In an atom itself, there are attractions between the positive nucleus and the negative electrons that surround it. Within molecules, atoms are attracted to each other by covalent bonds, the arrangement of which causes the molecules to assume a particular shape. Within an ionic crystal, three-dimensional shapes arise because of electrostatic attractions between ions.

In addition to these forces, there are attractive forces between molecules. These forces, which are the subject of this chapter, are weaker than any of the forces already mentioned; nevertheless, they help to determine whether a particular compound is a solid, a liquid, or a gas at any given temperature.

These attractive forces hold matter together; in effect, they counteract another form of energy—kinetic energy—that tends to lead to a number of different ways for molecules to arrange themselves. Intermolecular attractive forces counteract the kinetic energy that molecules possess, which keeps them constantly moving in random, disorganized ways. Kinetic energy increases with increasing temperature. Therefore, the higher the temperature, the greater the tendency of particles to have more possible arrangements. The total energy remains the same, but it is more widely dispersed. This dispersal of energy will have some important consequences, as we will see shortly.

The physical state of matter thus depends on a balance between the kinetic energy of particles, which tends to keep them apart, and the attractive forces between them, which tend to bring them together (Figure 5.1).

At high temperatures, molecules possess a high kinetic energy and move so fast that the attractive forces between them are too weak to hold them together. This situation is called the **gaseous state**. At lower temperatures, molecules move more slowly, to the point where the forces of attraction between them become important. When the temperature is low enough, a gas condenses to form a **liquid state**. Molecules in the liquid state still move past each other, but they travel much more slowly than they do in the gaseous state. When the temperature is even lower, molecules no longer have enough energy to move past each other. In the **solid state**, each molecule has a certain number of nearest neighbors, and these neighbors do not change.

The attractive forces between molecules are the same in all three states. The difference is that in the gaseous state (and to a lesser degree in the liquid state), the kinetic energy of the molecules is great enough to overcome the attractive forces between them.

Most substances can exist in any of the three states. Typically a solid, when heated to a sufficiently high temperature, melts and becomes a liquid. The temperature at which this change takes place is called the melting point. Further heating causes the temperature to rise to the point at which the liquid boils and becomes a gas. This temperature is called the boiling point. Not all substances, however, can exist in all three states. For example, wood and paper cannot be melted. Upon heating, they either decompose or burn (depending on whether air is present), but they do not melt. Another example is sugar, which does not melt when heated but rather forms a dark substance called caramel.

5.2 What Is Gas Pressure and How Do We Measure It?

On the Earth, we live under a blanket of air that presses down on us and on everything else around us. As we know from weather reports, the **pressure** of the atmosphere varies from day to day.

A gas consists of molecules in rapid, random motion. The pressure a gas exerts on a surface, such as the walls of a container, results from the continual bombardment on the walls of the container by the rapidly moving gas molecules. We use an instrument called a **barometer** (Figure 5.2) to measure atmospheric pressure. One type of barometer consists of a long glass tube that is completely filled with mercury and then inverted into a pool of mercury in a dish. Because there is no air at the top of the mercury column inside the tube (there is no way air could get in), no gas pressure is exerted on the mercury column. The entire atmosphere, however, exerts its pressure on the mercury in the open dish. The difference in the heights of the two mercury levels is a measure of the atmospheric pressure.

Pressure is most commonly measured in **millimeters of mercury (mm Hg)**. Pressure is also measured in **torr**, a unit named in honor of the Italian physicist and mathematician Evangelista Torricelli (1608–1647), who invented the barometer. At sea level, the average pressure of the

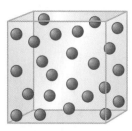

Gas

- Molecules far apart and disordered
- Negligible interactions between molecules

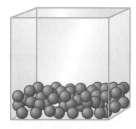

Liquid

- Intermediate situation

Solid

- Molecules close together and ordered
- Strong interactions between molecules

FIGURE 5.1 The three states of matter. A gas has no definite shape, and its volume is the volume of the container. A liquid has a definite volume but no definite shape. A solid has a definite shape and a definite volume.

Pressure The force per unit area exerted against a surface

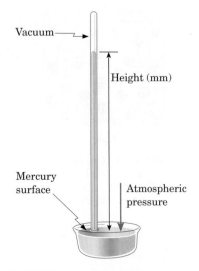

FIGURE 5.2 A mercury barometer.

$$1 \text{ atm} = 760 \text{ mm Hg}$$
$$= 760 \text{ torr}$$
$$= 101{,}325 \text{ Pa}$$
$$= 29.92 \text{ in. Hg}$$

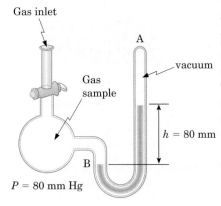

$P = 80 \text{ mm Hg}$

FIGURE 5.3 A mercury manometer.

These gas laws we describe hold not only for pure gases but also for mixtures of gases.

atmosphere is 760 mm Hg. We use this number to define still another unit of pressure, the **atmosphere (atm)**.

There are several other units with which to measure pressure. The SI unit is the pascal, and meteorologists report pressure in inches of mercury. In this book, we use only mm Hg and atm.

A barometer is adequate for measuring the pressure of the atmosphere, but to measure the pressure of a gas in a container, we use a simpler instrument called a **manometer**. One type of manometer consists of a U-shaped tube containing mercury (Figure 5.3). Arm A has been evacuated and sealed and has zero pressure. Arm B is connected to the container in which the gas sample is enclosed. The pressure of the gas depresses the level of the mercury in arm B. The difference between the two mercury levels gives the pressure directly in mm Hg. If more gas is added to the sample container, the mercury level in B will be pushed down and that in A will rise as the pressure in the bulb increases.

5.3 What Are the Laws That Govern the Behavior of Gases?

By observing the behavior of gases under different sets of temperatures and pressures, scientists have established a number of relationships. In this section, we study three of the most important of these.

A. Boyle's Law and the Pressure–Volume Relationship

Boyle's law states that for a fixed mass of an ideal gas at a constant temperature, the volume of the gas is inversely proportional to the applied pressure. If the pressure doubles, for example, the volume decreases by one-half. This law can be stated mathematically in the following equation, where P_1 and V_1 are the initial pressure and volume and P_2 and V_2 are the final pressure and volume:

$$PV = \text{constant} \quad \text{or} \quad P_1V_1 = P_2V_2$$

This relationship between pressure and volume is illustrated in Figure 5.4.

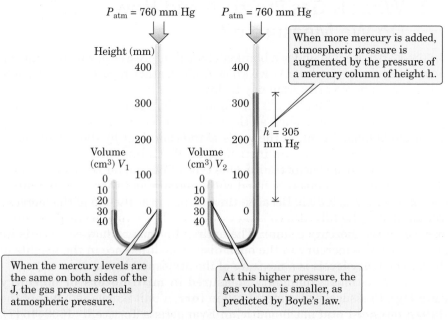

FIGURE 5.4 Boyle's law. Boyle's law experiment showing the compressibility of gases.

Chemical Connections 5A

Breathing and Boyle's Law

Under normal resting conditions, we breathe about 12 times per minute, each time inhaling and exhaling about 500. mL of air. When we inhale, we lower the diaphragm or raise the rib cage, either of which increases the volume of the chest cavity. In accord with Boyle's law, as the volume of the chest cavity increases, the pressure within it decreases and becomes lower than the outside pressure. As a result, air flows from the higher-pressure area outside the body into the lungs. While the difference in these two pressures is only about 3 mm Hg, it is enough to cause air to flow into the lungs. In exhaling, we reverse the process: We raise the diaphragm or lower

the rib cage. The resulting decrease in volume increases the pressure inside the chest cavity, causing air to flow out of the lungs.

In certain diseases, the chest becomes paralyzed and the affected person cannot move either the diaphragm or the rib cage. In such a case, a respirator is used to help the person breathe. The respirator first pushes down on the chest cavity and forces air out of the lungs. The pressure of the respirator is then lowered below atmospheric pressure, causing the rib cage to expand and draw air into the lungs.

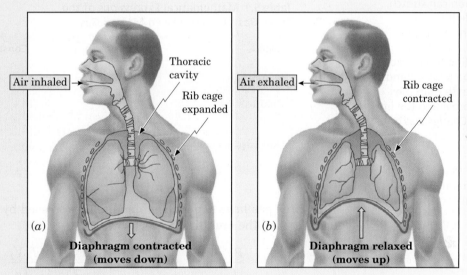

Schematic drawing of the chest cavity. (*a*) The lungs fill with air. (*b*) Air empties from the lungs.

Test your knowledge with Problem 5.87.

B. Charles's Law and the Temperature–Volume Relationship

Charles's law states that the volume of a fixed mass of an ideal gas at a constant pressure is directly proportional to the temperature in kelvins (K). In other words, as long as the pressure on a gas remains constant, increasing the temperature of the gas causes an increase in the volume occupied by the gas. Charles's law can be stated mathematically this way:

When using the gas laws, temperature must be expressed in kelvins (K). The zero in this scale is the lowest possible temperature.

$$\frac{V}{T} = \text{a constant} \quad \text{or} \quad \frac{V_1}{T_1} = \frac{V_2}{T_2}$$

This relationship between volume and temperature is the basis of the hot-air balloon operation (Figure 5.5).

FIGURE 5.5 Charles's law illustrated in a hot-air balloon. Because the balloon can stretch, the pressure inside it remains constant. When the air in the balloon is heated, its volume increases, expanding the balloon. As the air in the balloon expands, it becomes less dense than the surrounding air, providing the lift for the balloon. (Charles was one of the first balloonists.)

Combined gas law The pressure, volume, and temperature in kelvins of two samples of the same gas are related by the equation $P_1V_1/T_1 = P_2V_2/T_2$

C. Gay-Lussac's Law and the Temperature–Pressure Relationship

Gay-Lussac's law states that, for a fixed mass of a gas at constant volume, the pressure is directly proportional to the temperature in kelvins (K):

$$\frac{P}{T} = \text{a constant} \quad \text{or} \quad \frac{P_1}{T_1} = \frac{P_2}{T_2}$$

As the temperature of the gas increases, the pressure increases proportionately. Consider, for example, what happens inside an autoclave. Steam generated inside an autoclave at 1 atm pressure has a temperature of 100°C. As the steam is heated further, the pressure within the autoclave increases. A valve controls the pressure inside the autoclave; if the pressure exceeds the designated maximum, the valve opens, releasing the steam. At maximum pressure, the temperature may reach 120°C to 150°C. All microorganisms in the autoclave are destroyed at such high temperatures. Table 5.1 shows mathematical expressions of these three gas laws.

Table 5.1 Mathematical Expressions of the Three Gas Laws for a Fixed Mass of Gas

Name	Expression	Constant
Boyle's law	$P_1V_1 = P_2V_2$	T
Charles's law	$\dfrac{V_1}{T_1} = \dfrac{V_2}{T_2}$	P
Gay-Lussac's law	$\dfrac{P_1}{T_1} = \dfrac{P_2}{T_2}$	V

The three gas laws can be combined and expressed by a mathematical equation called the **combined gas law**:

$$\frac{PV}{T} = \text{a constant} \quad \text{or} \quad \frac{P_1V_1}{T_1} = \frac{P_2V_2}{T_2}$$

Example 5.1 The Combined Gas Law

A gas occupies 3.00 L at 2.00 atm pressure. Calculate its volume when we increase the pressure to 10.15 atm at the same temperature.

Strategy

First, we identify the known quantities. Because T_1 and T_2 are the same in this example and consequently cancel each other, we don't need to know the temperature. We use the relationship $P_1V_1 = P_2V_2$ and solve the combined gas law for V_2.

Solution

Initial: $P_1 = 2.00$ atm $V_1 = 3.00$ L

Final: $P_2 = 10.15$ atm $V_2 = ?$

$$V_2 = \frac{P_1V_1P_2}{P_1P_2} = \frac{(2.00 \ \text{atm})(3.00 \ \text{L})}{10.15 \ \text{atm}} = 0.591 \ \text{L}$$

Problem 5.1

A gas occupies 3.8 L at 0.70 atm pressure. If we expand the volume at constant temperature to 6.5 L, what is the final pressure?

Example 5.2 The Combined Gas Law

In an autoclave, steam at 100°C is generated at 1.00 atm. After the autoclave is closed, the steam is heated at constant volume until the pressure gauge indicates 1.13 atm. What is the final temperature in the autoclave?

Strategy

All temperatures in gas law calculations must be in kelvins; therefore, we must first convert the Celsius temperature to kelvins. Then we identify the known quantities. Because V_1 and V_2 are the same in this example and consequently cancel each other, we don't need to know the volume of the autoclave.

Solution

Step 1: Convert from degrees C to K.

$$100°C = 100 + 273 = 373 \text{ K}$$

Step 2: Identify the known quantities.

Initial:	$P_1 = 1.00$ atm	$T_1 = 373$ K
Final:	$P_2 = 1.13$ atm	$T_2 = ?$

Step 3: Solve the combined gas law equation for T_2, the new temperature.

$$T_2 = \frac{P_2 V_2 T_1}{P_1 V_1} = \frac{(1.13 \text{ atm})(373 \text{ K})}{1.00 \text{ atm}} = 421 \text{ K}$$

The final temperature is 421K, or $421 - 273 = 148°C$.

Problem 5.2

A constant volume of oxygen gas, O_2, is heated from 120.°C to 212.°C. The final pressure is 20.3 atm. What was the initial pressure?

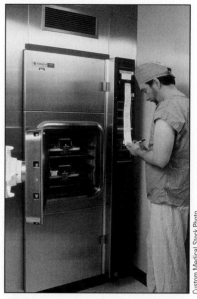

An autoclave used to sterilize hospital equipment.

⬢WL Interactive Example 5.3 The Combined Gas Law

A gas in a flexible container has a volume of 0.50 L and a pressure of 1.0 atm at 393 K. When the gas is heated to 500. K, its volume expands to 3.0 L. What is the new pressure of the gas in the flexible container?

Strategy

We identify the known quantities and then solve the combined gas law for the new pressure.

Solution

Step 1: The known quantities are:

Initial:	$P_1 = 1.0$ atm	$V_1 = 0.50$ L	$T_1 = 393$ K
Final:	$P_2 = ?$	$V_2 = 3.0$ L	$T_2 = 500.$ K

Step 2: Solving the combined gas law for P_2, we find:

$$P_2 = \frac{P_1 V_1 T_2}{T_1 V_2} = \frac{(1.0 \text{ atm})(0.50 \text{ L})(500. \text{ K})}{(3.0 \text{ L})(393 \text{ K})} = 0.21 \text{ atm}$$

Problem 5.3

A gas is expanded from an initial volume of 20.5 L at 0.92 atm at room temperature (23.0°C) to a final volume of 340.6 L. During the expansion, the gas cools to 12.0°C. What is the new pressure?

5.4 What Are Avogadro's Law and the Ideal Gas Law?

The relationship between the mass of gas present and its volume is described by **Avogadro's law**, which states that equal volumes of gases at the same temperature and pressure contain equal numbers of molecules. Thus, if the temperature, pressure, and volumes of two gases are the same, then the two gases contain the same number of molecules, regardless of their identity (Figure 5.6). Avogadro's law is valid for all gases, no matter what they are.

The actual temperature and pressure at which we compare two or more gases do not matter. It is convenient, however, to select one temperature and one pressure as standard, and chemists have chosen 1 atm as the standard pressure and 0°C (273 K) as the standard temperature. These conditions are called **standard temperature and pressure (STP)**.

All gases at STP or at any other combination of temperature and pressure contain the same number of molecules in any given volume. But how many molecules is that? In Chapter 4, we saw that one mole contains 6.02×10^{23} formula units. What volume of a gas at STP contains one mole of molecules? This quantity has been measured experimentally and found to be 22.4 L. Thus, one mole of any gas at STP occupies a volume of 22.4 L.

Avogadro's law allows us to write a gas law that is valid not only for any pressure, volume, and temperature, but also for any quantity of gas. This law, called **the ideal gas law**, is:

$$PV = nRT$$

where $P =$ pressure of the gas in atmospheres (atm)

$\quad\quad\ V =$ volume of the gas in liters (L)

$\quad\quad\ n =$ amount of the gas in moles (mol)

$\quad\quad\ T =$ temperature of the gas in kelvins (K)

$\quad\quad\ R =$ a constant for all gases, called the **ideal gas constant**

We can find the value of R by using the fact that one mole of any gas at STP occupies a volume of 22.4 L:

$$R = \frac{PV}{nT} = \frac{(1.00 \text{ atm})(22.4 \text{ L})}{(1.00 \text{ mol})(273 \text{ K})} = 0.0821 \frac{\text{L} \cdot \text{atm}}{\text{mol} \cdot \text{K}}$$

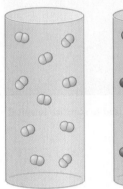
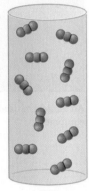

H₂   CO₂

T, P, and V
are equal in
both containers

FIGURE 5.6 Avogadro's law. Two tanks of gas of equal volume at the same temperature and pressure contain the same number of molecules.

Avogadro's law Equal volumes of gases at the same temperature and pressure contain the same number of molecules

Standard temperature and pressure (STP) 0°C (273 K) and one atmosphere pressure

Ideal gas law $PV = nRT$

Ideal gas A gas whose physical properties are described accurately by the ideal gas law

Ideal gas constant (R) 0.0821 L · atm · mol^{-1} · K^{-1}

The ideal gas law holds for all ideal gases at any temperature, pressure, and volume. But the only gases we have around us in the real world are real gases. How valid is it to apply the ideal gas law to real gases? The answer is that under most experimental conditions, real gases behave sufficiently like ideal gases that we can use the ideal gas law for them with little trouble. Thus, using $PV = nRT$, we can calculate any one quantity—P, V, T, or n—if we know the other three quantities.

Molar volume. The cube has a volume of 22.4 liters, which is the volume of one mole of gas at STP (standard temperature and pressure).

Example 5.4 Ideal Gas Law

1.00 mole of CH_4 gas occupies 20.0 L at 1.00 atm pressure. What is the temperature of the gas in kelvins?

Strategy

Solve the ideal gas law for T and plug in the given values:

Solution

$$T = \frac{PV}{nR} = \frac{PV}{n} \times \frac{1}{R} = \frac{(1.00\ \text{atm})(20.0\ \text{L})}{(1.00\ \text{mol})} \times \frac{\text{mol} \cdot \text{K}}{0.0821\ \text{L} \cdot \text{atm}} = 244\ \text{K}$$

Note that we calculated the temperature for 1.00 mol of CH_4 gas under these conditions. The answer would be the same for 1.00 mol of CO_2, N_2, NH_3, or any other gas under these conditions. Note also that we have shown the gas constant separately to make it clear what is happening with the units attached to all quantities. We are going to do this throughout.

Problem 5.4

If 2.00 mol of NO gas occupies 10.0 L at 295 K, what is the pressure of the gas in atmospheres?

Example 5.5 Ideal Gas Law

If there is 5.0 g of CO_2 gas in a 10. L cylinder at 25°C, what is the gas pressure within the cylinder?

Real gases behave most like ideal gases at low pressures (1 atm or less) and high temperatures (300 K or higher).

Strategy

We are given the quantity of CO_2 in grams, but to use the ideal gas law, we must express the quantity in moles. Therefore, we must first convert grams of CO_2 to moles CO_2 and then use this value in the ideal gas law. To convert from grams to moles, we use the conversion factor 1.00 mol CO_2 = 44 g CO_2.

Solution

Step 1: Convert grams of CO_2 to moles of CO_2.

$$5.0\ \text{g}\ CO_2 \times \frac{1\ \text{mol}\ CO_2}{44\ \text{g}\ CO_2} = 0.11\ \text{mol}\ CO_2$$

Step 2: We now use this value in the ideal gas equation to solve for the pressure of the gas. Note that temperature must be expressed in kelvins.

$$P = \frac{nRT}{V}$$

$$= \frac{nT}{V} \times R = \frac{(0.11\ \text{mol}\ CO_2)(298\ \text{K})}{10.\ \text{L}} \times \frac{0.0821\ \text{L} \cdot \text{atm}}{\text{mol} \cdot \text{K}} = 0.27\ \text{atm}$$

Problem 5.5

A certain quantity of neon gas is under 1.05 atm pressure at 303 K in a 10.0 L vessel. How many moles of neon are present?

Example 5.6 Ideal Gas Law

If 3.3 g of a gas at 40°C and 1.15 atm pressure occupies a volume of 1.00 L, what is the mass of one mole of the gas?

Strategy

This problem is more complicated than previous ones. We are given grams of gas and P, T, and V values and asked to calculate the mass of one mole of the gas (g/mol). We can solve this problem in two steps: (1) Use the ideal gas law to calculate the number of moles of gas present in the sample. (2) We are given the mass of gas (3.3 grams) and use the ratio grams/mole to determine the mass of one mole of the gas.

Solution

Step 1: Use the P, V, and T measurements and the ideal gas law to calculate the number of moles of gas present in the sample. To use the ideal gas law, we must first convert 40°C to kelvins: 40 + 273 = 313 K.

$$n = \frac{PV}{RT} = \frac{PV}{T} \times \frac{1}{R} = \frac{(1.15\ \text{atm})(1.00\ \text{L})}{313\ \text{K}} \times \frac{\text{mol} \cdot \text{K}}{0.0821\ \text{L} \cdot \text{atm}}$$

$$= 0.0448\ \text{mol}$$

Step 2: Calculate the mass of one mole of the gas by dividing grams by moles.

$$\text{Mass of one mole} = \frac{3.3\ \text{g}}{0.0448\ \text{mol}} = 74\ \text{g} \cdot \text{mol}^{-1}$$

Problem 5.6

An unknown amount of He gas occupies 30.5 L at 2.00 atm pressure and 300. K. What is the weight of the gas in the container?

5.5 What Is Dalton's Law of Partial Pressures?

In a mixture of gases, each molecule acts independently of all the others, provided that the gases behave as ideal gases and do not interact with each other in any way. For this reason, the ideal gas law works for mixtures of gases as well as for pure gases. **Dalton's law of partial pressures** states that the total pressure, P_T, of a mixture of gases is the sum of the partial pressures of each individual gas:

$$P_T = P_1 + P_2 + P_3 + \cdots$$

Partial pressure The pressure that a gas in a mixture of gases would exert if it were alone in the container

A corollary to Dalton's law is that the **partial pressure** of a gas in a mixture is the pressure that the gas would exert if it were alone in the

Chemical Connections 5B

Hyperbaric Medicine

Ordinary air contains 21% oxygen. Under certain conditions, the cells of tissues can become starved for oxygen (hypoxia), and quick oxygen delivery is needed. Increasing the percentage of oxygen in the air supplied to a patient is one way to remedy this situation, but sometimes even breathing pure (100%) oxygen may not be enough. For example, in carbon monoxide poisoning, hemoglobin, which normally carries most of the O_2 from the lungs to the tissues, binds CO and cannot take up any O_2 in the lungs. Without any help, tissues would soon become starved for oxygen and the patient would die. When oxygen is administered under a pressure of 2 to 3 atm, it dissolves in the plasma to such a degree that the tissues receive enough of it to recover without the help of the poisoned hemoglobin molecules. Other conditions for which hyperbaric medicine is used are treatment of gas gangrene, smoke inhalation, cyanide poisoning, skin grafts, thermal burns, and diabetic lesions.

Breathing pure oxygen for prolonged periods, however, is toxic. For example, if O_2 is administered at 2 atm for more than 6 hours, it may damage both lung tissue

Hyperbaric oxygen chamber at Medical City Dallas Hospital.

and the central nervous system. In addition, this treatment may cause nuclear cataract formation, necessitating postrecovery eye surgery. Therefore, recommended exposures to O_2 are 2 hours at 2 atm and 90 minutes at 3 atm. The benefits of hyperbaric medicine must be carefully weighed against these and other contraindications.

Test your knowledge with Problem 5.88.

container. The equation holds separately for each gas in the mixture as well as for the mixture as a whole.

Consider a mixture of nitrogen and oxygen illustrated in Figure 5.7. At constant volume and temperature, the total pressure of the mixture is equal to the pressure that the nitrogen alone plus the oxygen alone would exert. The pressure of one gas in a mixture of gases is called the partial pressure of that gas.

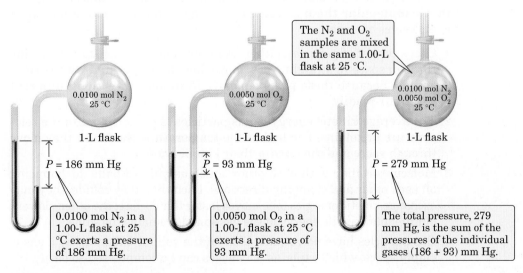

FIGURE 5.7 Dalton's law of partial pressures.

Example 5.7 Dalton's Law of Partial Pressures

To a tank containing N_2 at 2.0 atm and O_2 at 1.0 atm, we add an unknown quantity of CO_2 until the total pressure within the tank is 4.6 atm. What is the partial pressure of the CO_2?

Strategy

Dalton's law tells us that the addition of CO_2 does not affect the partial pressures of the N_2 or O_2 already present in the tank. The partial pressures of N_2 and O_2 remain at 2.0 atm and 1.0 atm, respectively, and their sum is 3.0 atm. The final total pressure within the tank, which is 4.6 atm, must be due to the partial pressure of the added CO_2.

Solution

If the final pressure is 4.6 atm, the partial pressure of the added CO_2 must be 1.6 atm. Thus, when the final pressure is 4.6 atm, the partial pressures are

$$4.6 \text{ atm} = 2.0 \text{ atm} + 1.0 \text{ atm} + 1.6 \text{ atm}$$

Total pressure	Partial pressure of N_2	Partial pressure of O_2	Partial pressure of CO_2

Problem 5.7

A vessel under 2.015 atm pressure contains nitrogen, N_2, and water vapor, H_2O. The partial pressure of N_2 is 1.908 atm. What is the partial pressure of the water vapor?

5.6   What Is the Kinetic Molecular Theory?

To this point, we have studied the macroscopic properties of gases—namely, the various laws dealing with the relationships among temperature, pressure, volume, and number of moles of gas in a sample. Now let us examine the behavior of gases at the molecular level and see how we can explain their macroscopic behavior in terms of molecules and the interactions between them.

The relationship between the observed behavior of gases and the behavior of individual gas molecules within the gas can be explained by the **kinetic molecular theory**, which makes the following assumptions about the molecules of a gas:

1. Gases consist of particles, either atoms or molecules, constantly moving through space in straight lines, in random directions, and with various speeds. Because these particles move in random directions, different gases mix readily.

2. The average kinetic energy of gas particles is proportional to the temperature in kelvins. The higher the temperature, the faster they move through space and the greater their kinetic energy.

3. Molecules collide with each other, much as billiard balls do, bouncing off each other and changing directions. Each time they collide, they may exchange kinetic energies (one moves faster than before; the other, slower), but the total kinetic energy of the gas sample remains the same.

4. Gas particles have no volume. Most of the volume taken up by a gas is empty space, which explains why gases can be compressed so easily.

5. There are no attractive forces between gas particles. They do not stick together after a collision occurs.

6. Molecules collide with the walls of the container, and these collisions constitute the pressure of the gas (Figure 5.8). The greater the number of collisions per unit time, the greater the pressure. The greater the average kinetic energy of the gas molecules, the greater the pressure.

These six assumptions of the kinetic molecular theory give us an idealized picture of the molecules of a gas and their interactions with one another (Figure 5.8). In real gases, however, forces of attraction between molecules do exist and molecules do occupy some volume. Because of these factors, a gas described by these six assumptions of the kinetic molecular theory is called an **ideal gas**. In reality, there is no ideal gas; all gases are real. At STP, however, most real gases behave in much the same way that an ideal gas would, so we can safely use these assumptions.

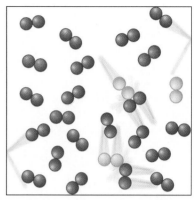

FIGURE 5.8 The kinetic molecular model of a gas. Molecules of nitrogen (blue) and oxygen (red) are in constant motion and collide with each other and with the walls of the container. Collisions of gas molecules with the walls of the container cause gas pressure. In air at STP, 6.02×10^{23} molecules undergo approximately 10 billion collisions per second.

5.7 What Types of Attractive Forces Exist Between Molecules?

As noted in Section 5.1, the strength of the intermolecular forces (forces between molecules) in any sample of matter determines whether the sample is a gas, a liquid, or a solid under given conditions of temperature and pressure. In general, the closer the molecules are to each other, the greater the effect of the intermolecular forces. When the temperature of a gas is high (room temperature or higher) and the pressure is low (1 atm or less), molecules of the gas are so far apart that we can effectively ignore attractions between them and treat the gas as ideal. When the temperature decreases, the pressure increases, or both, the distances between molecules decrease so that we can no longer ignore intermolecular forces. In fact, these forces become so important that they cause **condensation** (change from a gas to a liquid) and **solidification** (change from a liquid to a solid). Therefore, before discussing the structures and properties of liquids and solids, we must look at the nature of these intermolecular forces of attraction.

In this section, we discuss three types of intermolecular forces: London dispersion forces, dipole–dipole interactions, and hydrogen bonding. Table 5.2 shows the strengths of these three forces. Also shown for comparison are the strengths of ionic and covalent bonds, both of which are considerably stronger than the other three types of intermolecular forces. Although intermolecular forces are relatively weak compared to the strength of ionic and covalent bonds, it is the former that determines many of the physical properties of molecules, such as melting point, boiling point, and viscosity. As we will see in Chapters 21–31, these forces are also extremely important in influencing the three-dimensional shapes of biomolecules such as proteins and nucleic acids and in affecting how these types of biomolecules recognize and interact with one another.

Condensation The change of a substance from the vapor or gaseous state to the liquid state

Solidification The change of a substance from the liquid state to the solid state

A. London Dispersion Forces

Attractive forces exist between all molecules, whether they are polar or nonpolar. If the temperature falls far enough, even nonpolar molecules such as He, Ne, H_2, and CH_4 can be liquefied. Neon, for example, is a gas at room temperature and atmospheric pressure. It can be liquefied if cooled to $-246°C$. The fact that these and other nonpolar gases can be

Table 5.2 Forces of Attraction Between Molecules and Ions

Attractive Force	Example	Typical Energy (k cal/mol)
Ionic bonds	$Na^+ \text{IIIIIII} Cl^-$, $Mg^{2+} \text{IIIIIII} O^{2-}$	170–970
Single, double, and triple covalent bonds	$C-C$ $C=C$ $C\equiv C$ $O-H$	80–95 175 230 90–120
Hydrogen bonding	$\overset{\displaystyle H}{\underset{\displaystyle H}{\diagdown}} O \overset{\delta^-}{\text{IIIIIII}} \overset{\delta^+}{H} - O \diagdown H$	2–10
Dipole-dipole interaction	$\overset{\displaystyle H_3C}{\underset{\displaystyle H_3C}{\diagdown}} C = O \overset{\delta^-}{\text{IIIIIIIIIII}} \overset{\displaystyle H_3C}{\underset{\displaystyle H_3C}{\diagdown}} \overset{\delta^+}{C} = O$	1–6
London dispersion forces	$Ne \text{ IIIIII } Ne$	0.01–2.0

London dispersion forces Extremely weak attractive forces between atoms or molecules caused by the electrostatic attraction between temporary induced dipoles.

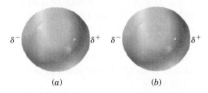

(a)   (b)

FIGURE 5.9 London dispersion forces. A temporary polarization of electron density in one neon atom creates positive and negative charges, which in turn induce temporary positive and negative charges in an adjacent atom. The intermolecular attractions between the temporary induced positive end of one dipole and the negative end of another temporary dipole are called London dispersion forces.

Dipole–dipole attraction The attraction between the positive end of a dipole of one molecule and the negative end of another dipole in the same or different molecule

liquefied means that some sort of interactions must occur between them to make them stick together in the liquid state. These weak attractive forces are called **London dispersion forces**, after the American chemist Fritz London (1900–1954), who was the first to explain them.

London dispersion forces have their origin in electrostatic interactions. To visualize the origin of these forces, it is necessary to think in terms of instantaneous distributions of electrons within an atom or a molecule. Consider, for example, a sample of neon atoms, which can be liquefied if cooled to −246°C. Over time, the distribution of electron density in a neon atom is symmetrical, and a neon atom has no permanent dipole; that is, there is no separation of positive and negative charges. However, at any given instant, the electron density in a neon atom may be shifted more toward one part of the atom than another, thus creating a temporary dipole (Figure 5.9). This temporary dipole, which lasts for only tiny fractions of a second, induces temporary dipoles in adjacent neon atoms. The attractions between these temporary induced dipoles are called London dispersion forces, and they make nonpolar molecules stick together to form the liquid state.

London dispersion forces exist between all molecules, but they are the only forces of attraction between nonpolar molecules. They range in strength from 0.01 to 2.0 kcal/mol depending on the mass, size, and shape of the interacting molecules. In general, their strength increases as the mass and number of electrons in a molecule increase. Even though London dispersion forces are very weak, they contribute significantly to the attractive forces between large molecules because they act over large surface areas.

B. Dipole–Dipole Interactions

As mentioned in Section 3.7B, many molecules are polar. The attraction between the positive end of one dipole and the negative end of another dipole is called a **dipole–dipole interaction**. These interactions can exist between two identical polar molecules or between two different polar molecules. To see the importance of dipole–dipole interactions, we can look at the differences in boiling points between nonpolar and polar molecules of comparable molecular weight. Butane, C_4H_{10}, with a molecular weight of 58 amu, is a nonpolar molecule with a boiling point of 0.5°C. Acetone, C_3H_6O, with the same molecular weight, has a boiling point of 58°C. Acetone is a polar molecule, and its molecules are held together in the liquid state by dipole–dipole attractions between the negative end of the C=O dipole of one acetone molecule and the positive end of the C=O dipole of another acetone molecule. Because it requires more energy to overcome the dipole–dipole interactions between acetone molecules than it does to overcome the considerably weaker London dispersion forces between butane molecules, acetone has a higher boiling point than butane.

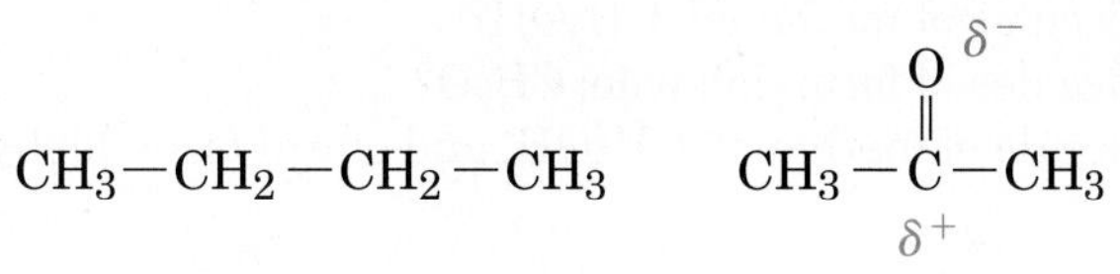

$$CH_3-CH_2-CH_2-CH_3 \qquad CH_3-\overset{\overset{\displaystyle O}{\parallel}}{\underset{\delta+}{C}}-CH_3 \quad \delta-$$

Butane
(bp 0.5°C)

Acetone
(bp 58°C)

C. Hydrogen Bonding

As we have just seen, the attraction between the positive end of one dipole and the negative end of another results in dipole–dipole attraction. When the positive end of one dipole is a hydrogen atom bonded to an O, N, or F (atoms of high electronegativity; see Table 3.5) and the negative end of the other dipole is an O, N, or F atom, the attractive interaction between dipoles is particularly strong and is given a special name: **hydrogen bonding**.

An example is the hydrogen bonding that occurs between molecules of water in both the liquid and solid states (Figure 5.10).

The strength of hydrogen bonding ranges from 2 to 10 kcal/mol. The strength in liquid water, for example, is approximately 5 kcal/mol. By comparison, the strength of the O—H covalent bond in water is approximately 119 kcal/mol. As can be seen by comparing these numbers, an O—H hydrogen bond is considerably weaker than an O—H covalent bond. Nonetheless, the presence of hydrogen bonds in liquid water has an important effect on the physical properties of water. Because of hydrogen bonding, extra energy is required to separate each water molecule from its neighbors—hence the relatively high boiling point of water. As we will see in later chapters, hydrogen bonds play an especially important role in biological molecules.

Hydrogen bonds are not restricted to water, however. They form between two molecules whenever one molecule has a hydrogen atom covalently

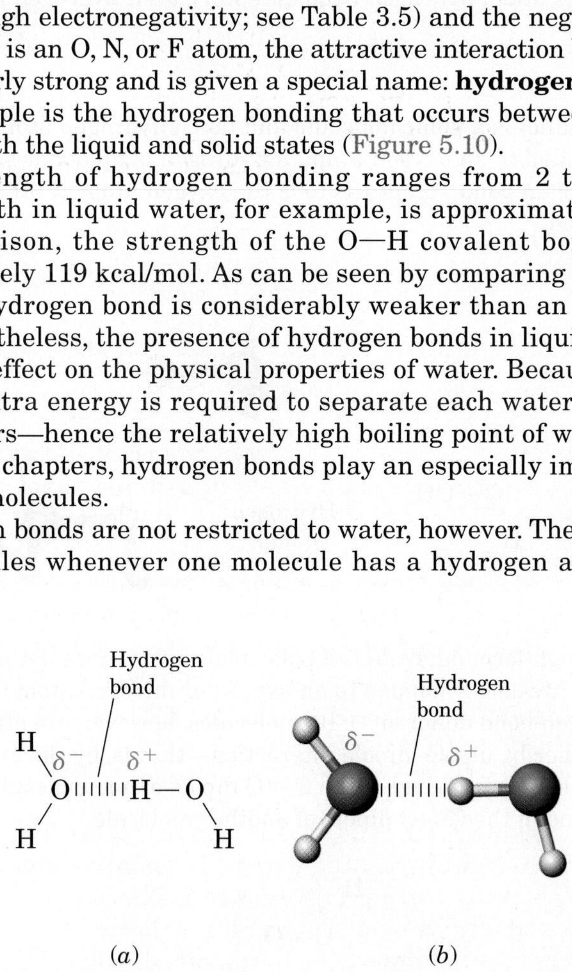

Hydrogen bond An intermolecular force of attraction between the partial positive charge on a hydrogen atom bonded to an atom of high electronegativity, most commonly oxygen or nitrogen, and the partial negative charge on a nearby oxygen or nitrogen

(a) (b)

FIGURE 5.10 Two water molecules joined by a hydrogen bond. (*a*) Structural formulas and (*b*) ball-and-stick models.

bonded to O, N, or F and the other molecule has an O, N, or F atom bearing a partial negative charge.

Because oxygen and nitrogen atoms are more commonly encountered in biochemical systems involving hydrogen bonding, we will focus our discussions on these two atoms.

Example 5.8 Hydrogen Bonding

Can a hydrogen bond form between:
(a) Two molecules of methanol, CH_3OH?
(b) Two molecules of formaldehyde, CH_2O?
(c) One molecule of methanol, CH_3OH, and one of formaldehyde, CH_2O?

Strategy

Examine the Lewis structure of each molecule and determine if there is a hydrogen atom bonded to either a nitrogen or oxygen atom. That is, determine if there is a polar O—H or N—H bond in one molecule in which hydrogen bears a partial positive charge. In other words, is there a hydrogen bond donor? Then examine the Lewis structure of the other molecule and determine if there is a polar bond in which either oxygen or nitrogen bears a partial negative charge. In other words, is there a potential hydrogen bond acceptor? If both features are present (a hydrogen bond donor and a hydrogen bond acceptor), then hydrogen bonding is possible.

Solution

(a) Yes. Methanol is a polar molecule and has a hydrogen atom covalently bonded to an oxygen atom (a hydrogen bond donor site). The hydrogen bond acceptor site is the oxygen atom of the polar O—H bond.

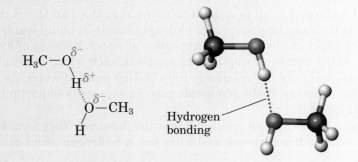

(b) No. Although formaldehyde is a polar molecule, it does not have a hydrogen covalently bonded to an oxygen or nitrogen atom (it has no hydrogen bond donor site). Its molecules, however, are attracted to each other by dipole–dipole interaction—that is, by the attraction between the negative end of the C=O dipole of one molecule and the positive end of the C=O dipole of another molecule.

(c) Yes. Methanol has a hydrogen atom bonded to an oxygen atom (hydrogen bond donor site) and formaldehyde has an oxygen atom bearing a partial negative charge (a hydrogen bond acceptor site).

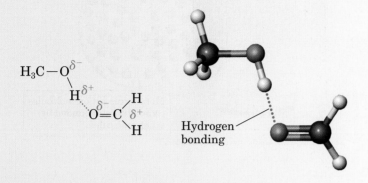

Problem 5.8

Will the molecules in each set form a hydrogen bond between them?
(a) A molecule of water and a molecule of methanol, CH_3OH
(b) Two molecules of methane, CH_4

5.8 How Do We Describe the Behavior of Liquids at the Molecular Level?

We have seen that we can describe the behavior of gases under most circumstances by the ideal gas law, which assumes that there are no attractive forces between molecules. As pressure increases in a real gas, however, the molecules of the gas become squeezed into a smaller space, with the result that attractions between molecules become increasingly more effective in causing molecules to stick together.

If the distances between molecules decrease so that they touch or almost touch each other, the gas condenses to a liquid. Unlike gases, liquids do not fill all the available space, but they do have a definite volume, irrespective of the container. Because gases have a lot of empty space between molecules, it is easy to compress them into a smaller volume. In contrast, there is very little empty space in liquids; consequently, liquids are difficult to compress. A great increase in pressure is needed to cause even a very small decrease in the volume of a liquid. Thus, liquids, for all practical purposes, are incompressible. In addition, the density of liquids is much greater than that of gases because the same mass occupies a much smaller volume in liquid form than it does in gaseous form.

The brake system in a car is based on hydraulics. The force you exert with the brake pedal is transmitted to the brake via cylinders filled with liquid. This system works very well until an air leak occurs. Once air gets into the brake line, pushing the brake pedal compresses the air and greatly reduces the ability of the fluid to transfer force into pressure.

The positions of molecules in the liquid state are random, and some irregular empty space is available into which molecules can slide. Molecules in the liquid state are, therefore, constantly changing their positions with respect to neighboring molecules. This property causes liquids to be fluid and explains why liquids have a constant volume but not a constant shape.

FIGURE 5.11 Surface tension. Molecules in the interior of a liquid have equal intermolecular attractions in every direction. Molecules at the surface (the liquid–gas interface), however, experience greater attractions toward the interior of the liquid than toward the gaseous state above it. Therefore, molecules on the surface are preferentially pulled toward the center of the liquid. This pull crowds the molecules on the surface, thereby creating a layer, like an elastic skin, that is tough to penetrate.

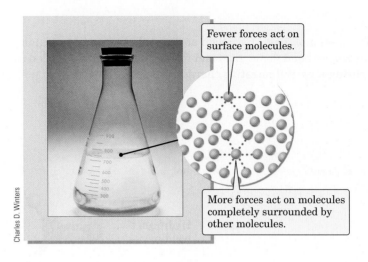

Fewer forces act on surface molecules.

More forces act on molecules completely surrounded by other molecules.

Charles D. Winters

Hermann Eisenbeiss/Photo Researchers, Inc.

A water-strider standing on water. The surface tension of water supports it.

A. Surface Tension

Unlike gases, liquids have surface properties, one of which is **surface tension** (Figure 5.11). The surface tension of a liquid is directly related to the strength of the intermolecular attraction between its molecules. It is defined as the energy required to increase the surface area of a liquid. Water has a high surface tension because of strong hydrogen bonding among water molecules. As a result, a steel needle can easily be made to float on the surface of water. If, however, the same needle is pushed below the elastic skin into the interior of the liquid, it sinks to the bottom. Similarly, water bugs gliding on the surface of a pond appear to be walking on an elastic skin of water.

B. Vapor Pressure

An important property of liquids is their tendency to evaporate. A few hours after a heavy rain, for example, most of the puddles have dried up; the water has evaporated and gone into the air. The same thing occurs if we leave a container of water or any other liquid out in the open. Let us explore how this change occurs.

In any liquid, there is a distribution of velocities among its molecules. Some of the molecules have high kinetic energy and move rapidly. Others have low kinetic energy and move slowly. Whether fast- or slow-moving, molecules in the interior of the liquid cannot go very far before they hit another molecule and have their speed and direction changed by the collision. Molecules at the surface, however, are in a different situation (Figure 5.12). If they are moving slowly (have a low kinetic energy), they cannot escape from the liquid because of the attractions of their neighboring molecules. If they are moving rapidly (have a high kinetic energy) and upward, however, they can escape from the liquid and enter the gaseous space above it.

In an open container, this process continues until all molecules have escaped. If the liquid is in a closed container, as in Figure 5.13, the molecules in the gaseous state cannot diffuse away (as they would do if the container were open). Instead, they remain in the air space above the liquid, where they move rapidly in straight lines until they strike something. Some of these vapor molecules move downward, strike the surface of the liquid, and are recaptured by it.

At this point, we have reached **equilibrium**. As long as the temperature does not change, the number of vapor molecules reentering the liquid

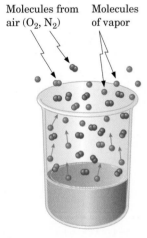

Molecules from air (O_2, N_2)   Molecules of vapor

FIGURE 5.12 Evaporation. Some molecules at the surface of the liquid are moving fast enough to escape into the gaseous space.

Vapor A gas

Equilibrium A condition in which two opposing processes occur at an equal rate

Chemical Connections 5C

Blood Pressure Measurement

Liquids, like gases, exert a pressure on the walls of their containers. Blood pressure, for example, results from pulsating blood pushing against the walls of the blood vessels. When the heart ventricles contract and push blood out into the arteries, the blood pressure is high (systolic pressure); when the ventricles relax, the blood pressure is lower (diastolic pressure). Blood pressure is usually expressed as a fraction showing systolic over diastolic pressure—for instance, 120/80. The normal range in young adults is 100 to 120 mm Hg systolic and 60 to 80 mm Hg diastolic. In older adults, the corresponding normal ranges are 115 to 135 and 75 to 85 mm Hg, respectively.

A sphygmomanometer—the instrument used to measure blood pressure—consists of a bulb, a cuff, a manometer, and a stethoscope. The cuff is wrapped around the upper arm and inflated by squeezing the bulb (Figure, part *a*). The inflated cuff exerts a pressure on the arm, which is read on the manometer. When the cuff is sufficiently inflated, its pressure collapses the brachial artery, preventing pulsating blood from flowing to the lower arm (Figure, part *b*). At

this pressure, no sound is heard in the stethoscope because the applied pressure in the cuff is greater than the blood pressure. Next, the cuff is slowly deflated, which decreases the pressure on the arm. The first faint tapping sound is heard when the pressure in the cuff just matches the systolic pressure as the ventricle contracts—that is, when the pressure in the cuff is low enough to allow pulsating blood to begin flowing into the lower arm. As the cuff pressure continues to decrease, the tapping first becomes louder and then begins to fade. At the point when the last faint tapping sound is heard, the cuff pressure matches the diastolic pressure when the ventricle is relaxed, thus allowing continuous blood flow into the lower arm (Figure, part *c*).

Digital blood pressure monitors are now available for home or office use. In these instruments, the stethoscope and the manometer are combined in a sensory device that records the systolic and diastolic blood pressures together with the pulse rate. The cuff and the inflation bulb are used the same way as in traditional sphygmomanometers.

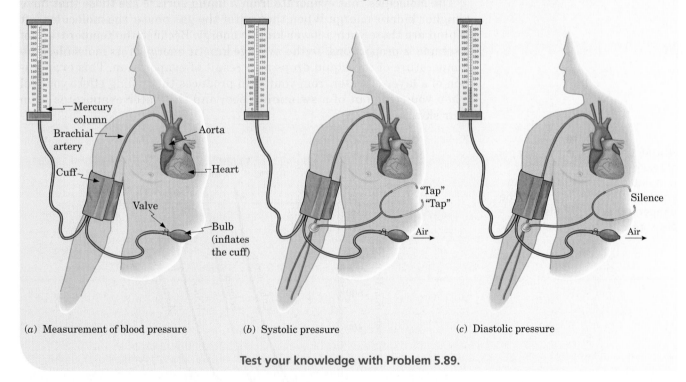

(*a*) Measurement of blood pressure (*b*) Systolic pressure (*c*) Diastolic pressure

Test your knowledge with Problem 5.89.

equals the number escaping from it. At equilibrium, the space above the liquid shown in Figure 5.13 contains both air and vapor molecules, and we can measure the partial pressure of the vapor, called **vapor pressure** of the liquid. Note that we measure the partial pressure of a gas but call it the vapor pressure of the liquid.

Vapor pressure The partial pressure of a gas in equilibrium with its liquid form in a closed container

At equilibrium, the rate of
vaporization equals the rate of
liquefaction.

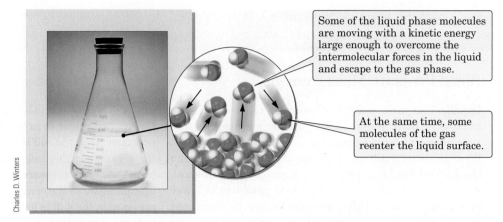

Some of the liquid phase molecules
are moving with a kinetic energy
large enough to overcome the
intermolecular forces in the liquid
and escape to the gas phase.

At the same time, some
molecules of the gas
reenter the liquid surface.

Charles D. Winters

FIGURE 5.13 Evaporation and condensation. In a closed container, molecules of
the liquid escape into the vapor phase and the liquid recaptures vapor molecules.

The vapor pressure of a liquid is a physical property of the liquid and a
function of temperature (Figure 5.14). As the temperature of a liquid increases, the average kinetic energy of its molecules increases and it becomes
easier for molecules to escape from the liquid state to the gaseous state. As
the temperature of the liquid increases, its vapor pressure continues to increase until it equals the atmospheric pressure. At this point, bubbles of
vapor form under the surface of the liquid, and then force their way upward
through the surface of the liquid, and the liquid boils.

The molecules that evaporate from a liquid surface are those that have
a higher kinetic energy. When they enter the gas phase, the molecules left
behind are those with a lower kinetic energy. Because the temperature of
a sample is proportional to the average kinetic energy of its molecules, the
temperature of the liquid drops as a result of evaporation. This evaporation of a layer of water from your skin produces the cooling effect you feel
when you come out of a swimming pool and the water evaporates from
your skin.

FIGURE 5.14 The change in vapor
pressure with temperature for
four liquids. The normal boiling
point of a liquid is defined as the
temperature at which its vapor
pressure equals 760 mm Hg.

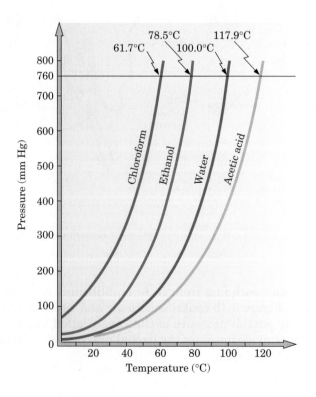

Because water has an appreciable vapor pressure at normal outdoor temperatures, water vapor is present in the atmosphere at all times. The vapor pressure of water in the atmosphere is expressed as **relative humidity**, which is the ratio of the actual partial pressure of the water vapor in the air, P_{H_2O}, to the equilibrium vapor pressure of water at the relevant temperature, $P°_{H_2O}$. The factor of 100 changes the fraction to a percentage.

$$\text{Relative humidity} = \frac{P_{H_2O}}{P°_{H_2O}} \times 100\%$$

For example, consider a typical warm day with an outdoor temperature of 25°C. The equilibrium vapor pressure of water at this temperature is 23.8 mm Hg. If the actual partial pressure of water vapor were 17.8 mm Hg, then the relative humidity would be 75%.

$$\text{Relative humidity} = \frac{17.8}{23.8} \times 100\% = 75\%$$

C. Boiling Point

The **boiling point** of a liquid is the temperature at which its vapor pressure is equal to the pressure of the atmosphere in contact with its surface. The boiling point when the atmospheric pressure is 1 atm is called the **normal boiling point**. For example, 100°C is the normal boiling point of water because that is the temperature at which water boils at 1 atm pressure (Figure 5.15).

The use of a pressure cooker is an example of boiling water at higher temperatures. In this type of pot, food is cooked at, say, 2 atm, at which pressure the boiling point of water is 121°C. Because the food has been raised to a higher temperature, it cooks faster than it would in an open pot, in which boiling water cannot get hotter than 100°C. Conversely, at low pressures, water boils at lower temperatures. For example, in Salt Lake City, Utah, where the average barometric pressure is about 650 mm Hg, the boiling point of water is about 95°C.

Boiling point The temperature at which the vapor pressure of a liquid is equal to the atmospheric pressure

Normal boiling point The temperature at which a liquid boils under a pressure of 1 atm

D. Factors That Affect Boiling Point

As Figure 5.14 shows, different liquids have different normal boiling points. Table 5.3 gives molecular formulas, molecular weights, and normal boiling points for five liquids.

FIGURE 5.15 Boiling point.

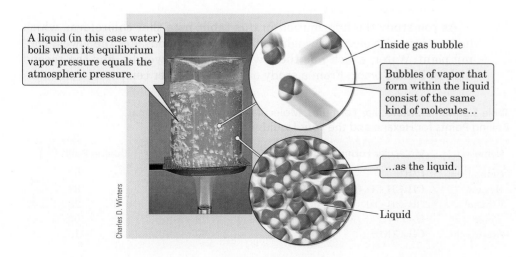

A liquid (in this case water) boils when its equilibrium vapor pressure equals the atmospheric pressure.

Inside gas bubble

Bubbles of vapor that form within the liquid consist of the same kind of molecules...

...as the liquid.

Liquid

Charles D. Winters

The Densities of Ice and Water

The hydrogen-bonded superstructure of ice contains empty spaces in the middle of each hexagon because the H_2O molecules in ice are not as closely packed as those in liquid water. For this reason, ice has a lower density (0.917 g/cm^3) than does liquid water (1.00 g/cm^3). As ice melts, some of the hydrogen bonds are broken and the hexagonal superstructure of ice collapses into the more densely packed organization of water. This change explains why ice floats on top of water instead of sinking to the bottom. Such behavior is highly unusual—most substances are denser in the solid state than they are in the liquid state. The lower density of ice keeps fish and microorganisms alive in rivers and lakes that would freeze solid each winter if the ice sank to the bottom. The presence of ice on top insulates the remaining water and keeps it from freezing.

The fact that ice has a lower density than liquid water means that a given mass of ice takes up more space than the same mass of liquid water. This factor explains the damage done to biological tissues by freezing. When parts of the body (usually fingers, toes, nose, and ears) are subjected to extreme cold, they develop a condition called frostbite. Water in cells freezes despite the blood's attempt to keep the temperature at 37°C. As liquid water freezes, it expands and in doing so ruptures the walls of cells containing it, causing damage. In some cases, frostbitten fingers or toes must be amputated.

Cold weather can damage plants in a similar way. Many plants are killed when the air temperature drops below the freezing point of water for several hours. Trees can survive cold winters because they have a low water content inside their trunks and branches.

Slow freezing is often more damaging to plant and animal tissues than quick freezing. In slow freezing, only a few crystals form, and these can grow to large sizes, rupturing cells. In quick freezing, such as that which can be achieved by cooling in liquid nitrogen (at a temperature of −196°C), many tiny crystals form. Because they do not grow much, tissue damage may be minimal.

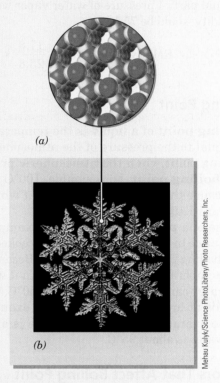

(a) In the structure of ice, each water molecule occupies a fixed position in a regular array or lattice. (b) The form of a snowflake reflects the hexagonal arrangement of water molecules within the ice crystal lattice.

Test your knowledge with Problems 5.90 and 5.91.

As you study the information in this table, note that chloroform, which has the largest molecular weight of the five compounds, has the lowest boiling point. Water, which has the lowest molecular weight, has the second highest boiling point. From a study of these and other compounds, chemists

Table 5.3 Names, Molecular Formulas, Molecular Weights, and Normal Boiling Points for Hexane and the Four Liquids in Figure 5.14

Name	Molecular Formula	Molecular Weight (amu)	Boiling Point (°C)
Chloroform	$CHCl_3$	120	62
Hexane	$CH_3CH_2CH_2CH_2CH_2CH_3$	86	69
Ethanol	CH_3CH_2OH	46	78
Water	H_2O	18	100
Acetic acid	CH_3COOH	60	118

have determined that the boiling point of covalent compounds depends primarily on three factors:

1. **Intermolecular forces** Water (H_2O, MW 18) and methane (CH_4, MW 16) have about the same molecular weight. The normal boiling point of water is 100°C, while that of methane is -164°C. The difference in boiling points reflects the fact that CH_4 molecules in the liquid state must overcome only the weak London dispersion forces to escape to the vapor state (low boiling point). In contrast, water molecules, being hydrogen-bonded to each other, need more kinetic energy (and a higher boiling temperature) to escape into the vapor phase. Thus the difference in boiling points between these two compounds is due to the greater strength of hydrogen bonding compared with the much weaker London dispersion forces.

2. **Number of sites for intermolecular interaction (surface area)** Consider the boiling points of methane, CH_4, and hexane, C_6H_{14}. Both are nonpolar compounds with no possibility for hydrogen bonding or dipole–dipole interactions between their molecules. The only force of attraction between molecules of either compound is London dispersion forces. The normal boiling point of hexane is 69°C, and that of methane is -164°C. The difference in their boiling points reflects the fact that hexane has more electrons and a larger surface area than methane. Because of its larger surface area, there are more sites for London dispersion forces to arise between hexane molecules than between methane molecules and, therefore, hexane has the higher boiling point.

3. **Molecular shape** When molecules are similar in every way except arrangement of the atoms, the strengths of London dispersion forces determine their relative boiling points. Consider pentane, bp 36.1°C, and 2,2-dimethylpropane, bp 9.5°C (Figure 5.16).

Both compounds have the same molecular formula, C_5H_{12}, and the same molecular weight, but the boiling point of pentane is approximately 26° higher than that of 2,2-dimethylpropane. This difference in boiling points is related to the arrangement of the atoms in the following way. The only forces of attraction between these nonpolar molecules are London dispersion forces. Pentane is a roughly straight chain molecule, whereas 2,2-dimethylpropane has a branched arrangement and a smaller surface area than pentane. As surface area decreases, contact between adjacent molecules, the strength of London dispersion forces, and boiling points all decrease. Consequently, London dispersion forces between molecules of 2,2-dimethylpropane are weaker than those between molecules of pentane and, therefore, 2,2-dimethylpropane has a lower boiling point.

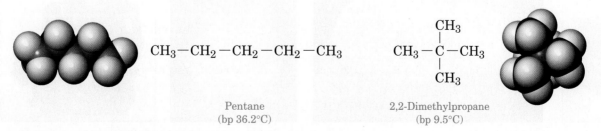

$$CH_3-CH_2-CH_2-CH_2-CH_3$$

Pentane
(bp 36.2°C)

$$CH_3-\overset{\overset{\textstyle CH_3}{|}}{\underset{\underset{\textstyle CH_3}{|}}{C}}-CH_3$$

2,2-Dimethylpropane
(bp 9.5°C)

FIGURE 5.16 Pentane and 2,2-dimethylpropane have the same molecular formula, C_5H_{12}, but quite different shapes.

5.9 What Are the Characteristics of the Various Types of Solids?

When liquids are cooled, their molecules come closer together and attractive forces between them become so strong that random motion stops, and a solid forms. Formation of a solid from a liquid is called solidification or, alternatively, **crystallization**.

Crystallization The formation of a solid from a liquid.

Even in the solid state, molecules and ions do not stop moving completely. They vibrate around fixed points.

All solids have a regular shape that, in many cases, is obvious to the eye (Figure 5.17). This regular shape often reflects the arrangement of the particles within the crystal. In table salt, for example, the Na^+ and Cl^- ions are arranged in a cubic system (Figure 3.1). Metals such as solid gold also consist of particles arranged in a regular crystal lattice (generally not cubic), but here the particles are atoms rather than ions. Because the particles in a solid are almost always closer together than they are in the corresponding liquid, solids almost always have a higher density than liquids. A notable exception to this generality is ice (see Chemical Connections 5D).

A fundamental distinction between kinds of solids is that some are crystalline and others are amorphous. **Crystalline solids** are those whose atoms, ions, or molecules have an ordered arrangement extending over a long range. These can be further categorized as ionic, molecular, polymeric, network covalent, or metallic as summarized in Table 5.4. By contrast, **amorphous solids** consist of randomly arranged particles that have no ordered long-range structure.

Ionic solids consist of orderly arrays of ions held together by ionic bonds in a crystal lattice. The strength of an ionic bond depends on the charges of the ions, their relative sizes, and directly impacts the physical properties of the solid. For example, NaCl, in which the ions have charges of 1+ and 1−, has a melting point of 801°C; MgO, in which the charges are 2+ and 2−, melts at 2852°C. Molecular solids consist of atoms or molecules held together by intermolecular forces (London dispersion forces, dipole-dipole interactions, and hydrogen boding). Because molecules are held only by intermolecular forces, which are much weaker than ionic bonds, molecular solids generally have lower melting points than ionic solids.

Other types of solids exist as well. For example, some are extremely large molecules, with each molecule having as many as 10^{23} atoms, all connected by covalent bonds. In such a case, the entire crystal is one big molecule. We call such molecules network covalent solids. A classic example is diamond [Figure 5.18(a)]. When you hold a diamond in your hand, you are holding a gigantic assembly of bonded atoms. Like ionic crystals, network covalent solids have very high melting points—if they can be melted at all. In many cases, they cannot be.

(a) Garnet

(b) Sulfur

(c) Quartz

(d) Pyrite

Beverly March

FIGURE 5.17 Some crystals.

Metallic solids consist of metal atoms surrounded by valence electrons. The bonding in metallic solids is too strong to be due to London dispersion forces, and yet there are not sufficient valence electrons for ordinary covalent bonds to be formed between atoms. Instead, the resulting metallic bond is due to the overlap of valence electrons that are scattered throughout the entire solid. The strength of the bonding increases as the number of electrons available for bonding increases. Table 5.4 further describes the various types of solids.

Table 5.4 Types of Solids

Type	Made Up of	Characteristics	Examples
Ionic	Ions in a crystal lattice	High melting point	$NaCl$, K_2SO_4
Molecular	Molecules in a crystal lattice	Low melting point	Ice, aspirin
Polymeric	Giant molecules; can be crystalline, semicrystalline, or amorphous	Low melting point or cannot be melted; soft or hard	Rubber, plastics, proteins
Network	A very large number of atoms connected by covalent bonds	Very hard; very high melting point or cannot be melted	Diamond, quartz
Amorphous	Randomly arranged atoms or molecules	Mostly soft, can be made to flow, but no melting point	Soot, tar, glass
Metallic	Metal atoms surrounded by a cloud of electrons	Soft to very hard; low to very high melting point	Au, Fe, Cu

Some elements exist in different forms in the same physical state. These forms have different chemical and physical properties and are known as **allotropes**. The best-known example is the element carbon, which exists in more than 40 known structural forms, five of which are crystalline (Figure 5.18) but most of which are amorphous. Diamond occurs when solidification takes place under very high pressure (thousands of atmospheres). Another form of carbon is the graphite in a pencil. Carbon atoms are packed differently in high-density hard diamonds than they are in low-density soft graphite.

Richard E. Smalley (1943–2005), Robert F. Curl, Jr. (1933–), and Harold Kroto (1939–) were awarded the 1996 Nobel Prize in chemistry for the discovery of these compounds.

FIGURE 5.18 Solid forms of carbon: *(a) diamond, (b) graphite, (c) "buckyball," (d) nanotube, and (e) soot.*

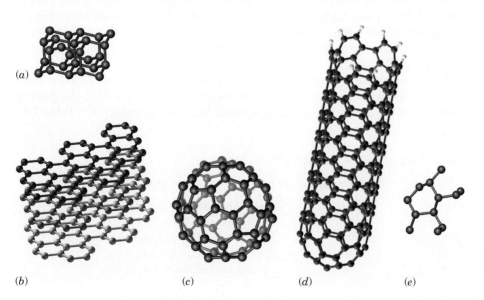

(a)

(b)　　　　(c)　　　　(d)　　　　(e)

In a third form of carbon, each molecule contains 60 carbon atoms arranged in a structure having 12 pentagons and 20 hexagons as faces, resembling a soccer ball [Figure 5.18(c)]. Because the famous architect Buckminster Fuller (1895–1983) invented domes of a similar structure (he called them geodesic domes), the C-60 substance was named buckminsterfullerene, or "buckyball" for short. The discovery of buckyballs has generated a whole new area of carbon chemistry. Similar cage-like structures containing 72, 80, and even larger numbers of carbon have been synthesized. As a group, they are called fullerenes.

New variations on the fullerenes are nanotubes [Figure 5.18(d)]. The *nano-* part of the name comes from the fact that the cross section of each tube is only nanometers in size (1 nm = 10^{-9} m). Nanotubes come in a variety of forms. Single-walled carbon nanotubes can vary in diameter from 1 to 3 nm and are about 20 mm long. These structures have generated great industrial interest because of their optical and electronic properties. They may play a role in the miniaturization of instruments, giving rise to a new generation of nanoscale devices.

Soot is the fifth form of solid carbon. This substance solidifies directly out of carbon vapor and is an amorphous solid [Figure 5.18(e)]. Another example of an amorphous solid is glass. In essence, glass is an immobilized liquid.

5.10 What Is a Phase Change and What Energies Are Involved?

A. The Heating Curve for H₂O(s) to H₂O(g)

Imagine the following experiment: We heat a piece of ice that is initially at −20°C. At first, we don't see any difference in its physical state. The temperature of the ice increases, but its appearance does not change. At 0°C, the ice begins to melt and liquid water appears. As we continue heating, more and more of the ice melts, but the temperature stays constant at 0°C until all the ice has melted and only liquid water remains. After all the ice has melted, the temperature of the water again increases as heat is added. At 100°C, the water boils. We continue heating as it continues to evaporate, but the temperature of the remaining liquid water does not change. Only after all the liquid water has changed to gaseous water (steam) does the temperature of the sample rise above 100°C.

These changes in state are called **phase changes**. A **phase** is any part of a system that looks uniform (homogeneous) throughout. Solid water (ice) is one phase, liquid water is another, and gaseous water is still another phase. Table 5.5 summarizes the energies for each step in the conversion of 1.0 g of ice to 1.0 g of steam.

Let us calculate the heat required to raise the temperature of 1.0 g of ice at −20°C to water vapor at 120°C and compare our results with the data in Table 5.5. We begin with ice, whose specific heat is 0.48 cal/g · °C (Table 1.4).

Phase change A change from one physical state (gas, liquid, or solid) to another.

The criterion of uniformity is the way it appears to our eyes and not as it is on the molecular level.

Table 5.5 Energy Required to Heat 1.0 g of Solid Water at −20°C to 120°C

Physical Change	Energy (cal)	Basis for Calculation of Energy Required
Warming ice from −20°C to 0°C	9.6	Specific heat of ice = 0.48 cal/g · °C
Melting ice; temperature = 0°C	80	Heat of fusion of ice = 80. cal/g
Warming water from 0°C to 100°C	100	Specific heat of liquid water = 1.00 cal/g · °C
Boiling water; temperature = 100°C	540	Heat of vaporization = 540 cal/g
Warming steam from 100°C to 120°C	9.6	Specific heat of steam = 0.48 cal/g · °C

Recall from Section 1.9B that it requires $0.48 \times 20 = 9.6$ cal to raise the temperature of 1.0 g of ice from -20 to $0°C$.

$$0.48 \; \frac{\text{cal}}{g \cdot °\!\!\!\!\!\diagup C} \times 1.0 \; \cancel{g} \times 20°\!\!\!\!\!\diagup C = 9.6 \; \text{cal}$$

After the ice reaches $0°C$, additional heat causes a phase change: Solid water melts and becomes liquid water. The heat necessary to melt 1.0 g of any solid is called its **heat of fusion**. The heat of fusion of ice is 80. cal/g. Thus it requires 80 cal to melt 1.0 g of ice—that is, to change 1.0 g of ice at $0°C$ to liquid water at $0°C$.

Only after the ice has completely melted does the temperature of the water rise again. The **specific heat** of liquid water is 1.00 cal/g · °C (Table 1.4). Thus it requires 100 cal to raise the temperature of 1.0 g of liquid water from $0°C$ to $100°C$. Contrast this with the 80 cal required to melt 1.0 g of ice.

When liquid water reaches $100°C$, the normal boiling point of water, the temperature of the sample remains constant while another phase change takes place: Liquid water vaporizes to gaseous water. The amount of heat necessary to vaporize 1.0 g of a liquid at its normal boiling point is called its **heat of vaporization**. For water, this value is 540 cal/g. Once all of the liquid water has been vaporized, the temperature again rises as the water vapor (steam) is heated. The specific heat of steam is 0.48 cal/g (Table 1.4). Thus it requires 9.6 cal to heat 1.0 g of steam from $100°C$ to $120°C$. The data for heating 1.0 g of water from $-20°C$ to $120°C$ can be shown in a graph called a **heating curve** (Figure 5.19). The total energy required for this conversion involves 739 cal of heat.

An important aspect of these phase changes is that each one of them is reversible. If we start with liquid water at room temperature and cool it by immersing the container in a dry ice bath ($-78°C$), the reverse process is observed. The temperature drops until it reaches $0°C$, and then ice begins to crystallize. During this phase change, the temperature of the sample stays constant but heat is given off. The amount of heat given off when 1.0 g of liquid water at $0°C$ freezes is exactly the same as the amount of heat absorbed when the 1.0 g of ice at $0°C$ melts.

A transition from the solid state directly into the vapor state without going through the liquid state is called **sublimation**. Solids usually sublime only at reduced pressures (less than 1 atm). At high altitudes, where the atmospheric pressure is low, snow sublimes. Solid CO_2 (dry ice) sublimes at $-78.5°C$ under 1 atm pressure. At 1 atm pressure, CO_2 can exist only as a solid or as a gas, never as a liquid.

Charles D. Winters

These freeze-dried coffee crystals were prepared by subliming water from frozen coffee.

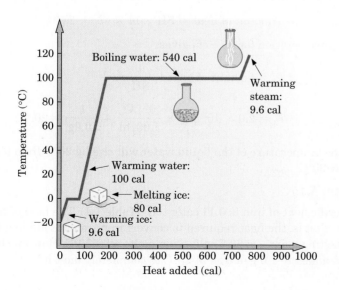

FIGURE 5.19 The heating curve of ice. The graph shows the effect of adding heat to 1.0 g of ice initially at $-20°C$ and raising its temperature to $120°C$.

Notice that the heat of vaporization is larger than the heat of fusion because all intermolecular forces must be overcome before vaporization can occur.

Example 5.9 Heat of Fusion

The heat of fusion of ice is 80. cal/g. How many calories are required to melt 1.0 mol of ice?

Strategy

We first convert moles of ice to grams of ice using the conversion factor 1 mol of ice = 18 g of ice. We then use the heat of fusion of ice (80. cal/g) to calculate the number of calories required to melt the given quantity of ice.

Solution

1.0 mole of H_2O has a mass of 18 g. We use the factor-label method to calculate the heat required to melt 1.0 mole of ice at 0°.

$$\frac{80.\text{cal}}{\text{g-ice}} \times 18 \text{ g-ice} = 1.4 \times 10^3 \text{ cal}$$

Problem 5.9

What mass of water at 100°C can be vaporized by the addition of 45.0 kcal of heat?

Example 5.10 Heat of Fusion and Phase Change

What will be the final temperature if we add 1000. cal of heat to 10.0 g of ice at 0°C?

Strategy

The first thing the added heat does is to melt the ice. So we must first determine if 1000 cal is sufficient to melt the ice completely. If less than 1000. cal is required to melt the ice to liquid water, then the remaining heat will serve to raise the temperature of the liquid water. The specific heat (SH; Section 1.9) of liquid water is 1.00 cal/g · °C (Table 1.4).

Solution

Step 1: This phase change will use 10.0 g × 80. cal/g = 8.0×10^2 cal, which leaves 2.0×10^2 cal to raise the temperature of the liquid water.

Step 2: The temperature of the liquid water is now raised by the remaining heat. The relationship between specific heat, mass, and temperature change is given by the following equation (Section 1.9):

$$\text{Amount of heat} = \text{SH} \times m \times (T_2 - T_1)$$

Solving this equation for $T_2 - T_1$ gives:

$$T_2 - T_1 = \text{amount of heat} \times \frac{1}{\text{SH}} \times \frac{1}{m}$$

$$T_2 - T_1 = 2.0 \times 10^2 \text{ cal} \times \frac{\text{g} \cdot \text{°C}}{1.00 \text{ cal}} \times \frac{1}{10.0 \text{ g}} = 20.\text{°C}$$

Thus, the temperature of the liquid water will rise by 20°C from 0°C, and it is now 20°C.

Problem 5.10

The specific heat of iron is 0.11 cal/g · °C (Table 1.4). The heat of fusion of iron—that is, the heat required to convert iron from a solid to a liquid at its melting point—is 63.7 cal/g. Iron melts at 1530°C. How much heat must be added to 1.0 g of iron at 25°C to completely melt it?

Chemical Connections 5E

Supercritical Carbon Dioxide

We are conditioned to think that a compound may exist in three phases: solid, liquid, or gas. Under certain pressures and temperatures, however, more phases may exist. A case in point is the plentiful nonpolar substance carbon dioxide. At room temperature and 1 atm pressure, CO_2 is a gas. Even when it is cooled to $-78°C$, it does not become a liquid but rather goes directly from a gas to a solid, which we call dry ice. At room temperature, a pressure of 60 atm is necessary to force molecules of CO_2 close enough together that they condense to a liquid.

Much more esoteric is the form of carbon dioxide called supercritical CO_2, which has some of the properties of a gas and some of the properties of a liquid. It has the den-

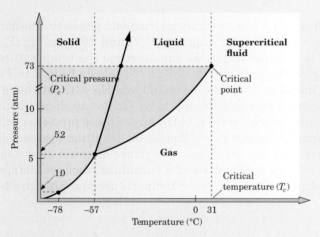

The phase diagram for carbon dioxide.

sity of a liquid but maintains its gas-like property of being able to flow with little viscosity or surface tension. What makes supercritical CO_2 particularly useful is that it is an excellent solvent for many organic materials. For example, supercritical CO_2 can extract caffeine from ground coffee beans, and after extraction when the pressure is released, it simply evaporates, leaving no traces behind. Similar processes can be performed with organic solvents, but traces of solvent may be left behind in the decaffeinated coffee and may alter its taste.

To understand the supercritical state, it is necessary to think about the interactions of molecules in the gas and liquid states. In the gaseous state, molecules are far apart, there is very little interaction between them, and most of the volume occupied by the gas is empty space. In the liquid state, molecules are held close together by the attractive forces between their molecules, and there is very little empty space between them. The supercritical state is something between these two states. Molecules are close enough together to give the sample some of the properties of a liquid but at the same time far enough apart to give it some of the properties of a gas.

The critical temperature and pressure for carbon dioxide are $31°C$ and 73 atm. When supercritical CO_2 is cooled below the critical temperature and/or compressed, there is a phase transition and the gas and liquid coexist. At a critical temperature and pressure, the two phases merge. Above critical conditions, the supercritical fluid exists, which exhibits characteristics that are intermediate between those of a gas and those of a liquid.

Test your knowledge with Problem 5.92.

We can show all phase changes for any substance on a **phase diagram**. Figure 5.20 is a phase diagram for water. Temperature is plotted on the *x*-axis and pressure on the *y*-axis. The three areas with different colors are labeled solid, liquid, and vapor. Within these areas water exists either as ice or liquid water or water vapor. The line (A–B) separating the solid phase from the liquid phase contains all the freezing (melting) points of water—for example, 0°C at 1 atm and 0.005°C at 400 mm Hg.

At the melting point, the solid and liquid phases coexist. The line separating the liquid phase from the gas phase (A–C) contains all the boiling points of water—for example, 100°C at 760 mm Hg and 84°C at 400 mm Hg. At the boiling points, the liquid and gas phases coexist.

Finally, the line separating the solid phase from the gas phase (A–D) contains all the sublimation points. At the sublimation points, the solid and gas phases coexist.

At a unique point (A) on the phase diagram, called the **triple point**, all three phases coexist. The triple point for water occurs at 0.01°C and 4.58 mm Hg pressure.

In this photo, vapors of gaseous CO_2 are cold enough to cause the moisture in the air above it to condense. The mixture of CO_2 vapors and condensed water vapor is heavier than air and slowly glides along the table or other surface on which the dry ice sits.

Charles D. Winters

FIGURE 5.20 Phase diagram of water. Temperature and pressure scales are greatly reduced.

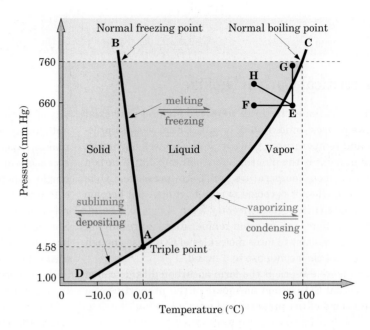

A phase diagram illustrates how one may go from one phase to another. For example, suppose we have water vapor at 95°C and 660 mm Hg (E). We want to condense it to liquid water. We can decrease the temperature to 70°C without changing the pressure (moving horizontally from E to F). Alternatively, we can increase the pressure to 760 mm Hg without changing the temperature (moving vertically from E to G). Or we can change both temperature and pressure (moving from E to H). Any of these processes will condense the water vapor to liquid water, although the resulting liquids will be at different pressures and temperatures. The phase diagram allows us to visualize what will happen to the phase of a substance when we change the experimental conditions from one set of temperature and pressure to another set.

Example 5.11 Phase Diagram

What will happen to ice at 0°C if the pressure decreases from 1 atm to 0.001 atm?

Strategy

Consult the location on the phase diagram of water (Figure 5.20) and find the point that corresponds to the given temperature and pressure conditions.

Solution

According to Figure 5.20, when the pressure decreases while the temperature remains constant, we move vertically from 1 atm (760 mm Hg) to 0.001 atm (0.76 mm Hg). During this process, we cross the boundary separating the solid phase from the vapor phase. Thus, when the pressure drops to 0.001 atm, ice sublimes and becomes vapor.

Problem 5.11

What will happen to water vapor if it is cooled from 100°C to −30°C while the pressure stays at 1 atm?

Summary

OWL Sign in at **www.cengage.com/owl** to develop problem-solving skills and complete online homework assigned by your professor.

Section 5.1 What Are the Three States of Matter?

- Matter can exist in three different states: gas, liquid, and solid.
- Attractive forces between molecules tend to hold matter together, whereas the kinetic energy of the molecules tends to disorganize matter.

Section 5.2 What Is Gas Pressure and How Do We Measure It?

- Gas pressure results from the bombardment of gas particles on the walls of its container.
- The pressure of the atmosphere is measured with a barometer. Three common units of pressure are millimeters of mercury, torr, and atmospheres. 1 mm Hg = 1 torr and 760 mm Hg = 1 atm.

Section 5.3 What Are the Laws That Govern the Behavior of Gases?

- **Boyle's law** states that for a fixed mass of gas at constant temperature, the volume of the gas is inversely proportional to the pressure.
- **Charles's law** states that the volume of a fixed mass of gas at constant pressure is directly proportional to the temperature in kelvins.
- **Gay-Lussac's law** states that for a fixed mass of gas at constant volume, the pressure is directly proportional to the temperature in kelvins.
- These laws are combined and expressed as the **combined gas law**:

$$\frac{P_1 V_1}{T_1} = \frac{P_2 V_2}{T_2}$$

Section 5.4 What Are Avogadro's Law and the Ideal Gas Law?

- **Avogadro's law** states that equal volumes of gases at the same temperature and pressure contain the same number of molecules.
- The ideal gas law, $PV = nRT$, incorporates Avogadro's law into the combined gas law.
- In summary, in problems involving gases, the two most important equations are:
  The **ideal gas law,** useful when three of the variables P, V, T, and n are given and you are asked to calculate the fourth variable:

$$PV = nRT$$

Section 5.5 What Is Dalton's Law of Partial Pressures?

- **Dalton's law of partial pressures** states that the total pressure of a mixture of gases is the sum of the partial pressures of each individual gas.

Section 5.6 What Is the Kinetic Molecular Theory?

- The **kinetic molecular theory** explains the behavior of gases. Molecules in the gaseous state move rapidly and randomly, allowing a gas to fill all the available space of its container. Gas molecules have no volume and there are no forces of attraction between them. In their random motion, gas molecules collide with the walls of the container and thereby exert pressure.

Section 5.7 What Types of Attractive Forces Exist Between Molecules?

- **Intermolecular forces of attraction** are responsible for the condensation of gases into the liquid state and for the solidification of liquids to the solid state. In order of increasing strength, the intermolecular forces of attraction are **London dispersion forces**, **dipole–dipole attractions**, and **hydrogen bonds**.

Section 5.8 How Do We Describe the Behavior of Liquids at the Molecular Level?

- **Surface tension** is the energy required to increase the surface area of a liquid.
- **Vapor pressure** is the pressure of a vapor (gas) above its liquid in a closed container. The vapor pressure of a liquid increases with increasing temperature.
- The **boiling point** of a liquid is the temperature at which its vapor pressure equals the atmospheric pressure. The boiling point of a liquid is determined by (1) the nature and strength of the intermolecular forces between its molecules, (2) the number of sites for intermolecular interaction, and (3) molecular shape.

Section 5.9 What Are the Characteristics of the Various Types of Solids?

- Solids crystallize in well-formed geometrical shapes that include crystalline (ionic, molecular, polymeric, network covalent, or metallic) and amorphous solids.
- The **melting point** is the temperature at which a substance changes from the solid state to the liquid state.
- **Crystallization** is the formation of a solid from a liquid.
- **Allotropes** are elements that exist in different forms while in the same physical state with different chemical and physical properties.

Section 5.10 What Is a Phase Change and What Energies Are Involved?

- A **phase** is any part of a system that looks uniform throughout. A **phase change** involves a change of matter from one physical state to another—that is, from a solid, liquid, or gaseous state to any one of the other two states.
- **Sublimation** is a change from a solid state directly to a gaseous state.
- **Heat of fusion** is the heat necessary to convert 1.0 g of any solid to a liquid.
- **Heat of vaporization** is the heat necessary to convert 1.0 g of any liquid to the gaseous state.

- A **phase diagram** allows the visualization of what happens to the phase of a substance when the temperature or pressure is changed.

- A phase diagram contains all the melting points, boiling points, and sublimation points where two phases coexist.
- A phase diagram also contains a unique triple point where all three phases coexist.

Problems

⊙WL Interactive versions of these problems may be assigned in OWL.

Orange-numbered problems are applied.

Section 5.2 What Is Gas Pressure and How Do We Measure It?

5.12 A weather report says that the barometric pressure is 29.5 inches of mercury. What is this pressure in atmospheres?

5.13 Use the kinetic molecular theory to explain why, at constant temperature, the pressure of a gas increases as its volume is decreased.

5.14 Use the kinetic molecular theory to explain why the pressure of a gas in a fixed-volume container increases as its temperature is increased.

5.15 Name three ways by which the volume of a gas can be decreased.

Section 5.3 What Are the Laws That Govern the Behavior of Gases?

5.16 Answer true or false.

(a) For a sample of gas at constant temperature, its pressure multiplied by its volume is a constant.

(b) For a sample of gas at constant temperature, increasing the pressure increases the volume.

(c) For a sample of gas at constant temperature, $P_1/V_1 = P_2/V_2$.

(d) As a gas expands at constant temperature, its volume increases.

(e) The volume of a sample of gas at constant pressure is directly proportional to its temperature—the higher its temperature, the greater its volume.

(f) A hot-air balloon rises because hot air is less dense than cooler air.

(g) For a gas sample in a container of fixed volume, an increase in temperature results in an increase in pressure.

(h) For a gas sample in a container of fixed volume, $P \times T$ is a constant.

(i) When steam at 100°C in an autoclave is heated to 120°C, the pressure within the autoclave increases.

(j) When a gas sample in a flexible container at constant pressure at 25°C is heated to 50°C, its volume doubles.

(k) Lowering the diaphragm causes the chest cavity to increase in volume and the pressure of air in the lungs to decrease.

(l) Raising the diaphragm decreases the volume of the chest cavity and forces air out of the lungs.

5.17 A sample of gas has a volume of 6.20 L at 20°C at a pressure of 1.10 atm. What is its volume at the same temperature and at a pressure of 0.925 atm?

5.18 Methane gas is compressed from 20. L to 2.5 L at a constant temperature. The final pressure is 12.2 atm. What was the original pressure?

5.19 A gas syringe at 20°C contains 20.0 mL of CO_2 gas. The pressure of the gas in the syringe is 1.0 atm. What is the pressure in the syringe at 20°C if the plunger is depressed to 10.0 mL?

5.20 Suppose that the pressure in an automobile tire is 2.30 atm at a temperature of 20.0°C. What will the pressure in the tire be if after 10 miles of driving the temperature of the tire increases to 47.0°C?

5.21 A sample of 23.0 L of NH_3 gas at 10.0°C is heated at constant pressure until it fills a volume of 50.0 L. What is the new temperature in °C?

5.22 If a sample of 4.17 L of ethane gas, C_2H_6, at 725°C is cooled to 175°C at constant pressure, what is the new volume?

5.23 A sample of SO_2 gas has a volume of 5.2 L. It is heated at constant pressure from 30. to 90.°C. What is its new volume?

5.24 A sample of B_2H_6 gas in a 35-mL container is at a pressure of 450. mm Hg and a temperature of 625°C. If the gas is allowed to cool at constant volume until the pressure is 375 mm Hg, what is the new temperature in °C?

5.25 A gas in a bulb as in Figure 5.3 registers a pressure of 833 mm Hg in the manometer in which the reference arm of the U-shaped tube (A) is sealed and evacuated. What will the difference in the mercury levels be if the reference arm of the U-shaped tube is open to atmospheric pressure (760 mm Hg)?

5.26 In an autoclave, a constant amount of steam is generated at a constant volume. Under 1.00 atm pressure, the steam temperature is 100.°C. What pressure setting should be used to obtain a 165°C steam temperature for the sterilization of surgical instruments?

5.27 A sample of the inhalation anesthetic gas Halothane, $C_2HBrClF_3$, in a 500-mL cylinder has a pressure of 2.3 atm at 0°C. What will be the pressure of the gas if its temperature is warmed to 37°C (body temperature)?

5.28 Complete this table:

V_1	T_1	P_1	V_2	T_2	P_2
546 L	43°C	6.5 atm	_____	65°C	1.9 atm
43 mL	−56°C	865 torr	_____	43°C	1.5 atm
4.2 L	234 K	0.87 atm	3.2 L	29°C	_____
1.3 L	25°C	740 mm Hg	_____	0°C	1.0 atm

5.29 Complete this table:

V_1	T_1	P_1	V_2	T_2	P_2
6.35 L	10°C	0.75 atm	_____	0°C	1.0 atm
75.6 L	0°C	1.0 atm	_____	35°C	735 torr
1.06 L	75°C	0.55 atm	3.2 L	0°C	_____

5.30 A balloon filled with 1.2 L of helium at 25°C and 0.98 atm pressure is submerged in liquid nitrogen at −196°C. Calculate the final volume of the helium in the balloon.

5.31 A balloon used for atmospheric research has a volume of 1×10^6 L. Assume that the balloon is filled with helium gas at STP and then allowed to ascend to an altitude of 10 km, where the pressure of the atmosphere is 243 mm Hg and the temperature is −33°C. What will the volume of the balloon be under these atmospheric conditions?

5.32 A gas occupies 56.44 L at 2.00 atm and 310. K. If the gas is compressed to 23.52 L and the temperature is lowered to 281 K, what is the new pressure?

5.33 A certain quantity of helium gas is at a temperature of 27°C and a pressure of 1.00 atm. What will the new temperature be if its volume is doubled at the same time that its pressure is decreased to one-half its original value?

5.34 A sample of 30.0 mL of krypton gas, Kr, is at 756 mm Hg and 25.0°C. What is the new volume if the pressure is decreased to 325 mm Hg and the temperature is decreased to −12.5°C?

5.35 A 26.4-mL sample of ethylene gas, C_2H_4, has a pressure of 2.50 atm at 2.5°C. If the volume is increased to 36.2 mL and the temperature is raised to 10°C, what is the new pressure?

Section 5.4 What Are Avogadro's Law and the Ideal Gas Law?

5.36 Answer true or false.

  (a) Avogadro's law states that equal volumes of gas at the same temperature and pressure contain equal numbers of molecules.

  (b) At STP, one mole of uranium hexafluoride (UF_6, MW 352 amu), the gas used in uranium enrichment programs, occupies a volume of 352 L.

  (c) If two gas samples have the same temperature, volume, and pressure, then both contain the same number of molecules.

  (d) The value of Avogadro's number is 6.02×10^{23} g/mol.

  (e) Avogadro's number is valid only for gases at STP.

  (f) The ideal gas law is $PV = nRT$.

  (g) When using the ideal gas law for calculations, temperature must be in degrees Celsius.

  (h) If one mole of ethane (CH_3CH_3) gas occupies 20.0 L at 1.00 atm, the temperature of the gas is 244 K.

  (i) One mole of helium (MW 4.0 amu) gas at STP occupies twice the volume of one mole of hydrogen (MW 2.0 amu).

5.37 A sample of a gas at 77°C and 1.33 atm occupies a volume of 50.3 L.

  (a) How many moles of the gas are present?

  (b) Does your answer depend on knowing what gas it is?

5.38 What is the volume in liters occupied by 1.21 g of Freon-12 gas, CCl_2F_2, at 0.980 atm and 35°C?

5.39 An 8.00-g sample of a gas occupies 22.4 L at 2.00 atm and 273 K. What is the molar mass of the gas?

5.40 What volume is occupied by 5.8 g of propane gas, C_3H_8, at 23°C and 1.15 atm pressure?

5.41 Does the density of a gas increase, decrease, or stay the same as the pressure increases at constant temperature? As the temperature increases at constant pressure?

5.42 What volume in milliliters does 0.275 g of uranium hexafluoride gas, UF_6, occupy at its boiling point of 56°C at 365 torr?

5.43 A hyperbaric chamber has a volume of 200. L.

  (a) How many moles of oxygen are needed to fill the chamber at room temperature (23°C) and 3.00 atm pressure?

  (b) How many grams of oxygen are needed?

5.44 One gulp of air has a volume of 2 L at STP. If air contains 20.9% oxygen, how many molecules of oxygen are in one gulp?

5.45 An average pair of lungs has a volume of 5.5 L. If the air they contain is 21% oxygen, how many molecules of O_2 do the lungs contain at 1.1 atm and 37°C?

5.46 Calculate the molar mass of a gas if 3.30 g of the gas occupies 660. mL at 735 mm Hg and 27°C.

5.47 The three main components of dry air and the percentage of each are N_2 (78.08%), O_2 (20.95%), and Ar (0.93%).

  (a) Calculate the mass of one mole of air.

  (b) Given the mass of one mole of air, calculate the density of air in g/L at STP.

5.48 The density of Freon-12, CCl_2F_2 at STP is 4.99 g/L, which means that it is approximately four times more dense than air. Show how the kinetic molecular theory of gases accounts for the fact that although Freon-12 is more dense than air, it nevertheless finds its way to the stratosphere, where it is implicated in the destruction of Earth's protective ozone layer.

5.49   Calculate the density in g/L of each of these gases at STP. Which gases are denser than air as calculated in Problem 5.47(b)? Which are less dense than air?

(a) SO_2 (b) CH_4 (c) H_2

(d) He   (e) CO_2

5.50   How many molecules of CO are in 100. L of CO at STP?

5.51   The density of liquid octane, C_8H_{18}, is 0.7025 g/mL. If 1.00 mL of liquid octane is vaporized at 100°C and 725 torr, what volume does the vapor occupy?

5.52   The density of acetylene gas, C_2H_2, in a 4-L container at 0°C and 2 atm pressure is 0.02 g/mL. What would be the density of the gas under identical temperature and pressure if the container were partitioned into two 2-L compartments?

5.53   Sodium metal reacts explosively with hydrochloric acid, HCl(aq), as shown in the following chemical equation:

$$2Na(s) + 2HCl(aq) \longrightarrow 2NaCl(aq) + H_2(g)$$

What volume of $H_2(g)$ is produced when 3.50 g of Na(s) is reacted with an excess of hydrochloric acid at a temperature of 18°C and a pressure of 0.995 atm?

5.54   Automobile air bags are inflated by nitrogen gas. When a significant collision occurs, an electronic sensor triggers the decomposition of sodium azide to form nitrogen gas and sodium metal. The nitrogen gas then inflates nylon bags, which protect the driver and front-seat passenger from impact with the dashboard and windshield.

$$2NaN_3(s) \longrightarrow 2Na(s) + 3N_2(g)$$

Sodium azide

What volume of nitrogen gas measured at 1 atm and 27°C is formed by the decomposition of 100. g of sodium azide?

Section 5.5 What Is Dalton's Law of Partial Pressures?

5.55   Answer true or false.

(a) Partial pressure is the pressure that a gas in a container would exert if it were alone in the container.

(b) The units of partial pressure are grams per liter.

(c) Dalton's law of partial pressures states that the total pressure of a mixture of gases is the sum of the partial pressures of each gas.

(d) If 1 mole of CH_4 gas at STP is added to 22.4 L of N_2 at STP, the final pressure in the 22.4 L container will be 1.00 atm.

5.56   The three main components of dry air and the percentage of each are nitrogen (78.08%), oxygen (20.95%), and argon (0.93%).

(a) Calculate the partial pressure of each gas in a sample of dry air at 760 mm Hg.

(b) Calculate the total pressure exerted by these three gases combined.

5.57   Air in the trachea contains oxygen (19.4%), carbon dioxide (0.4%), water vapor (6.2%), and nitrogen (74.0%). If the pressure in the trachea is assumed to be 1.0 atm, what are the partial pressures of these gases in this part of the body?

5.58   The partial pressures of a mixture of gases were as follows: oxygen, 210 mm Hg; nitrogen, 560 mm Hg; and carbon dioxide, 15 mm Hg. The total pressure of the gas mixture was 790 mm Hg. Was there another gas present in the mixture?

Section 5.6 What Is the Kinetic Molecular Theory?

5.59   Answer true or false.

(a) According to the kinetic molecular theory, gas particles have mass but no volume.

(b) According to the kinetic molecular theory, the average kinetic energy of gas particles is proportional to the temperature in degrees Celsius.

(c) According to the kinetic molecular theory, when gas particles collide, they bounce off each other with no change in total kinetic energy.

(d) According to the kinetic molecular theory, there are only weak intramolecular forces of attraction between gas particles.

(e) According to the kinetic molecular theory, the pressure of a gas in a container is the result of collisions of gas particles on the walls of the container.

(f) Warming a gas results in an increase in the average kinetic energy of its particles.

(g) When a gas is compressed, the increase in its pressure is the result of an increase in the number of collisions of its particles on the walls of the container.

(h) The kinetic molecular theory describes the behavior of ideal gases, of which there are only a few.

(i) As the temperature and volume of a gas increase, the behavior of the gas becomes more like the behavior predicted by the ideal gas law.

(j) If the assumptions of the kinetic molecular theory of gases are correct, then there is no combination of temperature and pressure at which a gas would become liquid.

5.60   Compare and contrast Dalton's atomic theory and the kinetic molecular theory.

Section 5.7 What Types of Attractive Forces Exist Between Molecules?

5.61   Answer true or false.

(a) Of the forces of attraction between particles, London dispersion forces are the weakest and covalent bonds are the strongest.

(b) All covalent bonds have approximately the same energy.

(c) London dispersion forces arise because of the attraction of temporary induced dipoles.

(d) In general, London dispersion forces increase as molecular size increases.

(e) London dispersion forces occur only between polar molecules—they do not occur between nonpolar atoms or molecules.

(f) The existence of London dispersion forces accounts for the fact that even small, nonpolar particles such as Ne, He, and H_2 can be liquefied if the temperature is low enough and the pressure is high enough.

(g) For nonpolar gases at STP, the average kinetic energy of its particles is greater than the force of attraction between gas particles.

(h) Dipole–dipole interaction is the attraction between the positive end of one dipole and the negative end of another dipole.

(i) Dipole–dipole interactions exist between CO molecules but not between CO_2 molecules.

(j) If two polar molecules have approximately the same molecular weight, the strength of the dipole–dipole interactions between the molecules of each will be approximately the same.

(k) Hydrogen bonding refers to the single covalent bond between the two hydrogen atoms in H—H.

(l) The strength of hydrogen bonding in liquid water is approximately the same as that of an O—H covalent bond in water.

(m) Hydrogen bonding, dipole–dipole interactions, and London dispersion forces have in common that the forces of attraction between particles are all electrostatic (positive for negative and negative for positive).

(n) Water (H_2O, bp 100°C) has a higher boiling point than hydrogen sulfide (H_2S, bp −61°C) because the hydrogen bonding between H_2O molecules is stronger than that between H_2S molecules.

(o) The hydrogen bonding among molecules containing N—H groups is stronger than that among molecules containing O—H groups.

5.62 Which forces are stronger, intramolecular covalent bonds or intermolecular hydrogen bonds?

5.63 Under which condition does water vapor behave most ideally?
(a) 0.5 atm, 400 K (b) 4 atm, 500 K
(c) 0.01 atm, 500 K

5.64 Can water and dimethyl sulfoxide, $(CH_3)_2S{=}O$, molecules form hydrogen bonds between them?

5.65 What kind of intermolecular interactions take place in (a) liquid CCl_4 and (b) liquid CO? Which will have the highest surface tension?

5.66 Ethanol, C_2H_5OH, and carbon dioxide, CO_2, have approximately the same molecular weight, yet carbon dioxide is a gas at STP and ethanol is a liquid. How do you account for this difference in physical property?

5.67 Can dipole–dipole interactions ever be weaker than London dispersion forces? Explain.

5.68 Which compound has a higher boiling point: butane, C_4H_{10}, or hexane, C_6H_{14}?

Section 5.8 How Do We Describe the Behavior of Liquids at the Molecular Level?

5.69 Answer true or false.
(a) The ideal gas law assumes that there are no attractive forces between molecules. If this were true, then there would be no liquids.

(b) Unlike a gas, whose molecules move freely in any direction, molecules in a liquid are locked into fixed positions, giving the liquid a constant shape.

(c) Surface tension is the force that prevents a liquid from being stretched.

(d) Surface tension creates an elastic-like layer on the surface of a liquid.

(e) Water has a high surface tension because H_2O is a small molecule.

(f) Vapor pressure is proportional to temperature—as the temperature of a liquid sample increases, its vapor pressure also increases.

(g) When molecules evaporate from a liquid, the temperature of the liquid drops.

(h) Evaporation is a cooling process because it leaves fewer molecules with high kinetic energy in the liquid state.

(i) The boiling point of a liquid is the temperature at which its vapor pressure equals the atmospheric pressure.

(j) As the atmospheric pressure increases, the boiling point of a liquid increases.

(k) The temperature of boiling water is related to how vigorously it is boiling—the more vigorous the boiling, the higher the temperature of the water.

(l) The most important factor determining the relative boiling points of liquids is molecular weight—the greater the molecular weight, the higher the boiling point.

(m) Ethanol (CH_3CH_2OH, bp 78.5°C) has a greater vapor pressure at 25°C than water (H_2O, bp 100°C).

(n) Hexane ($CH_3CH_2CH_2CH_2CH_2CH_3$, bp 69°C) has a higher boiling point than methane (CH_4, bp −164°C) because hexane has more sites for hydrogen bonding between its molecules than does methane.

(o) A water molecule can participate in hydrogen bonding through each of its hydrogen atoms and through its oxygen atom.

(p) For nonpolar molecules of comparable molecular weight, the more compact the shape of the molecule, the higher its boiling point.

5.70 The melting point of chloroethane, CH_3CH_2Cl, is −136°C and its boiling point is 12°C. Is chloroethane a gas, a liquid, or a solid at STP?

Section 5.9 What Are the Characteristics of the Various Types of Solids?

5.71 Answer true or false.
(a) Formation of a liquid from a solid is called melting; formation of a solid from a liquid is called crystallization.

(b) Most solids have a higher density than their liquid forms.

(c) Molecules in a solid are locked into fixed positions.

(d) Each element has one and only one solid (crystalline) form.

(e) Diamond and graphite are both crystalline forms of carbon.

(f) Diamond consists of hexagonal crystals of carbon arranged in a repeating pattern.

(g) The *nano* in nanotube refers to the structure dimensions, which are in the nanometer (10^{-9} m) range.

(h) Nanotubes have lengths up to 1 nm.

(i) A buckyball (C_{60}) has a diameter of 1 nm.

(j) All solids, if heated to a high enough temperature, can be melted.

(k) Glass is an amorphous solid.

5.72 Identify the type of crystalline solid (i.e., ionic, molecular, metallic, network covalent, polymeric) formed by each of the following:

(a) glucose, $C_6H_{12}O_6$

(b) silver, Ag

(c) silicon carbide, SiC

(d) bottle containing Arrowhead® brand mountain spring water

(e) potassium iodide, KI

(f) elemental sulfur powder, S_8

Section 5.10 What Is a Phase Change and What Energies Are Involved?

5.73 Answer true or false.

(a) A phase change from solid to liquid is called melting.

(b) A phase change from liquid to gas is called boiling.

(c) If heat is added slowly to a mixture of ice and liquid water, the temperature of the sample gradually increases until all of the ice is melted.

(d) Heat of fusion is the heat required to melt 1 g of a solid.

(e) Heat of vaporization is the heat required to evaporate 1 g of liquid at the normal boiling point of the liquid.

(f) Steam burns are more damaging to the skin than hot-water burns because the specific heat of steam is so much higher than the specific heat of hot water.

(g) The heat of vaporization of water is approximately the same as its heat of fusion.

(h) The specific heat of water is the heat required to raise the temperature of 1 g of water from 0° to 100°C.

(i) Melting a solid is an exothermic process; crystallization of a liquid is an endothermic process.

(j) Melting a solid is a reversible process; the solid can be converted to a liquid and the liquid back to a solid with no change in composition of the sample.

(k) Sublimation is a phase change from solid directly to gas.

5.74 Calculate the specific heat (Section 1.9) of gaseous Freon-12, CCl_2F_2, if it requires 170. cal to change the temperature of 36.6 g of Freon-12 from 30.°C to 50.°C.

5.75 The heat of vaporization of liquid Freon-12, CCl_2F_2, is 4.71 kcal/mol. Calculate the energy required to vaporize 39.2 g of this compound. The molecular weight of Freon-12 is 120.9 amu.

5.76 The specific heat (Section 1.9) of mercury is 0.0332 cal/g·°C. Calculate the energy necessary to raise the temperature of one mole of liquid mercury by 36°C.

5.77 Using Figure 5.14, estimate the vapor pressure of ethanol at (a) 30°C, (b) 40°C, and (c) 60°C.

5.78 CH_4 and H_2O have about the same molecular weight. Which has the higher vapor pressure at room temperature? Explain.

5.79 The normal boiling point of a substance depends on both the mass of the molecule and the attractive forces between molecules. Arrange the compounds in each set in order of increasing boiling point and explain your answer:

(a) HCl, HBr, HI   (b) O_2, HCl, H_2O_2

5.80 Refer to Figure 5.19. How many calories are required to bring one mole of ice at 0°C to a liquid state at room temperature (23°C)?

5.81 Compare the number of calories absorbed when 100. g of ice at 0°C is changed to liquid water at 37°C with the number of calories absorbed when 100. g of liquid water is warmed from 0°C to 37°C.

5.82 (a) How much energy is released when 10. g of steam at 100°C is condensed and cooled to body temperature (37°C)?

(b) How much energy is released when 100. g of liquid water at 100°C is cooled to body temperature (37°C)?

(c) Why are steam burns more painful than hot-water burns?

5.83 When iodine vapor hits a cold surface, iodine crystals form. Name the phase change that is the reverse of this condensation.

5.84 If a 156-g block of dry ice, CO_2, is sublimed at 25°C and 740 mm Hg, what volume does the gas occupy?

5.85 Trichlorofluoromethane (Freon-11, CCl_3F) as a spray is used to temporarily numb the skin around minor scrapes and bruises. It accomplishes this by reducing the temperature of the treated area, thereby numbing the nerve endings that perceive pain. Calculate the heat in kilocalories that can be removed from the skin by 1.00 mL of Freon-11. The density of Freon-11 is 1.49 g/mL, and its heat of vaporization is 6.42 kcal/mol.

5.86 Using the phase diagram of water (Figure 5.20), describe the process by which you can sublime 1 g of ice at −10°C and at 1 atm pressure to water vapor at the same temperature.

Chemical Connections

5.87 (Chemical Connections 5A) What happens when a person lowers the diaphragm in his or her chest cavity?

5.88 (Chemical Connections 5B) In carbon monoxide poisoning, the hemoglobin is incapable of transporting oxygen to the tissues. How does the oxygen get delivered to the cells when a patient is put into a hyperbaric chamber?

5.89 (Chemical Connections 5C) In a sphygmomanometer one listens to the first tapping sound as the constrictive pressure of the arm cuff is slowly released. What is the significance of this tapping sound?

5.90 (Chemical Connections 5D) Why is the damage by severe frostbite irreversible?

5.91 (Chemical Connections 5D) If you fill a glass bottle with water, cap it, and cool to −10°, the bottle will crack. Explain.

5.92 (Chemical Connections 5E) In what way does supercritical CO_2 have some of the properties of a gas and some of the properties of a liquid?

Additional Problems

5.93 Why is it difficult to compress a liquid or a solid?

5.94 Explain in terms of the kinetic molecular theory what causes (a) the pressure of a gas and (b) the temperature of a gas.

5.95 The unit of pressure most commonly used for checking the inflation of automobile and bicycle tires is pounds per square inch (lb/in^2), abbreviated psi. The conversion factor between atm and psi is 1.00 atm = 14.7 psi. Suppose an automobile tire is filled to a pressure of 34 psi. What is the pressure in atm in the tire?

5.96 The gas in an aerosol can is at a pressure of 3.0 atm at 23°C. What will the pressure of the gas in the can be if the temperature is raised to 400°C?

5.97 Why do aerosol cans carry the warning "Do not incinerate"?

5.98 Under certain weather conditions (just before rain), the air becomes less dense. How does this change affect the barometric pressure reading?

5.99 An ideal gas occupies 387 mL at 275 mm Hg and 75°C. If the pressure changes to 1.36 atm and the temperature increases to 105°C, what is the new volume?

5.100 Arrange the following solids in order of increasing expected melting points: $CO_2(s)$, $Xe(s)$, $CaO(s)$, $H_2O(s)$, $LiCl(s)$, and $HCl(s)$.

5.101 On the basis of what you have learned about intermolecular forces, predict which liquid has the highest boiling point:
(a) Pentane, C_5H_{12}
(b) Chloroform, $CHCl_3$
(c) Water, H_2O

5.102 A 10-L gas cylinder is filled with N_2 to a pressure of 35 in. Hg. How many moles of N_2 do you have to add to your container to raise the pressure to 60 in. Hg? Assume a constant temperature of 27°C.

5.103 When filled, a typical tank for an outdoor grill contains 20. lb of LP (liquefied petroleum) gas, the major component of which is propane, C_3H_8. For this problem, assume that propane is the only substance present.
(a) How do you account for the fact that when propane is put under pressure, it can be liquefied?

(b) How many kilograms of propane does a full tank contain?
(c) How many moles of propane does a full tank contain?
(d) If the propane in a full tank was released into a flexible container, what volume would it occupy at STP?

5.104 Explain why many gases are transparent.

5.105 The density of a gas is 0.00300 g/cm^3 at 100.°C and 1.00 atm. What is the mass of one mole of the gas?

5.106 The normal boiling point of hexane, C_6H_{14}, is 69°C, and that of pentane, C_5H_{12}, is 36°C. Predict which of these compounds has a higher vapor pressure at 20°C.

5.107 If 60.0 g of NH_3 occupies 35.1 L under a pressure of 77.2 in. Hg, what is the temperature of the gas, in °C?

5.108 Water is a liquid at STP. Hydrogen sulfide, H_2S, a heavier molecule, is a gas under the same conditions. Explain.

5.109 Why does the temperature of a liquid drop as a result of evaporation?

5.110 What volume of air (21% oxygen) measured at 25°C and 0.975 atm is required to completely oxidize 3.42 g of aluminum to aluminum oxide, Al_2O_3?

Tying It Together

5.111 Diving, particularly SCUBA (Self-Contained Underwater Breathing Apparatus) diving, subjects the body to increased pressure. Each 10. m (approximately 33 ft) of water exerts an additional pressure of 1 atm on the body.
(a) What is the pressure on the body at a depth of 100. ft?
(b) The partial pressure of nitrogen gas in air at 1 atm is 593 mm Hg. Assuming a SCUBA diver breathes compressed air, what is the partial pressure of nitrogen entering the lungs from a breathing tank at a depth of 100. ft?
(c) The partial pressure of oxygen gas in the air at 2 atm is 158 mm Hg. What is the partial pressure of oxygen in the air in the lungs at a depth of 100. ft?
(d) Why is it absolutely essential to exhale vigorously in a rapid ascent from a depth of 100. ft?

5.112 Consider the mixing of 3.5 L of $CO_2(g)$ and 1.8 L of $H_2O(g)$ at 35°C and 740 mm Hg. Determine the mass of $O_2(g)$ that can be produced from the unbalanced reaction:

$$CO_2(g) + H_2O(g) \longrightarrow C_4H_{10}(l) + O_2(g)$$

5.113 Ammonia and gaseous hydrogen chloride react to form ammonium chloride according to the following equation:

$$NH_3(g) + HCl(g) \longrightarrow NH_4Cl(s)$$

If 4.21 L of $NH_3(g)$ at 27°C and 1.02 atm is combined with 5.35 L of $HCl(g)$ at 26°C and 0.998 atm, what mass of $NH_4Cl(s)$ will be generated?

5.114 Carbon dioxide gas, saturated with water vapor, can be produced by the addition of aqueous acid to calcium carbonate based on the following balanced net ionic equation:

$$CaCO_3(s) + 2H^+(aq) \longrightarrow Ca^{2+}(aq) + H_2O(l) + CO_2(g)$$

(a) How many moles of wet $CO_2(g)$, collected at 60.°C and 774 torr total pressure, are produced by the complete reaction of 10.0 g of $CaCO_3$ with excess acid?

(b) What volume does this wet CO_2 occupy?

(c) What volume would the CO_2 occupy at 774 torr if a desiccant (a chemical drying agent) were added to remove the water? The vapor pressure of water at 60.°C is 149.4 mm Hg.

5.115 Ammonium nitrite decomposes upon heating to form nitrogen gas and water vapor according to the following unbalanced chemical reaction:

$$NH_4NO_2(s) \longrightarrow N_2(g) + H_2O(g)$$

When a sample is decomposed in a test tube, 511 mL of wet $N_2(g)$ is collected over water at 26°C and 745 torr total pressure. How many grams of dry $NH_4NO_2(s)$ were initially decomposed? The vapor pressure of water at 26°C is 25.2 torr.

5.116 How much total heat in calories is required to raise the temperature of 3.50 g of ice at $-10.0°C$ to water vapor at 115°C? Refer to Table 5.5 for relevant data.

5.117 Determine the total amount of heat lost in calories when 5.75 g of water vapor at 120.°C is cooled to $-20.°C$. Refer to Table 5.5 for relevant data.

Solutions and Colloids

6

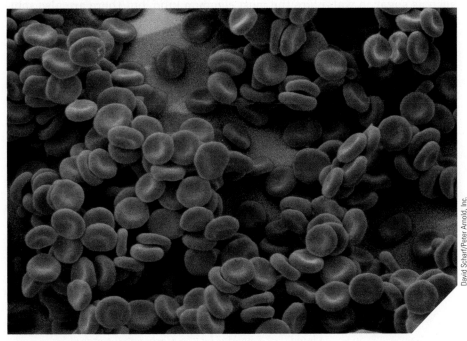

Human blood cells in an isotonic solution.

Key Questions

6.1 What Do We Need to Know as We Begin?

6.2 What Are the Most Common Types of Solutions?

6.3 What Are the Distinguishing Characteristics of Solutions?

6.4 What Factors Affect Solubility?

6.5 What Are the Most Common Units for Concentration?

6.6 Why Is Water Such a Good Solvent?

6.7 What Are Colloids?

6.8 What Is a Colligative Property?

6.1 What Do We Need to Know as We Begin?

In Chapter 2, we discussed pure substances—compounds made of two or more elements in a fixed ratio. Such systems are the easiest to study, so it was convenient to begin with them. In our daily lives, however, we more frequently encounter mixtures—systems consisting of more than one component. Air, smoke, seawater, milk, blood, and rocks, for example, are mixtures (Section 2.2C).

If a mixture is uniform throughout at the molecular level, we call it a homogeneous mixture or, more commonly, a solution. Filtered air and seawater, for example, are both solutions. They are clear and transparent. In contrast, in most rocks, we can see distinct regions separated from each other by well-defined boundaries. Such rocks are heterogeneous mixtures. Another example is a mixture of sand and sugar. We can easily distinguish between the two components; the mixing does not occur at the molecular level (Figure 2.3). Thus, mixtures are classified on the basis of how they look to the unaided eye.

OWL

Sign in to OWL at **www.cengage.com/owl** to view tutorials and simulations, develop problem-solving skills, and complete online homework assigned by your professor.

Making a homogeneous solution. A green solid, nickel nitrate, is stirred into water, where it dissolves to form a homogeneous solution.

Beer is a solution in which a liquid (alcohol), a solid (malt), and a gas (CO_2) are dissolved in the solvent, water.

Mixtures can be homogeneous, as with brass, which is a solid solution of copper and zinc. Alternatively, they can be heterogeneous, as with granite, which contains discrete regions of different minerals (feldspar, mica, and quartz).

Alloy A homogeneous mixture of two or more metals

Many alloys are solid solutions. One example is stainless steel, which is mostly iron but also contains carbon, chromium, and other elements (See also Chemical Connections 2E).

We normally do not use the terms "solute" and "solvent" when talking about solutions of gases in gases or solids in solids.

Some systems, however, fall between homogeneous and heterogeneous mixtures. Cigarette smoke, milk, and blood plasma may look homogeneous, but they do not have the transparency of air or seawater. These mixtures are classified as suspensions. We will deal with such systems in Section 6.7.

Although mixtures can contain many components, we will generally restrict our discussion to two-component systems, with the understanding that everything we say can be extended to multicomponent systems.

6.2 What Are the Most Common Types of Solutions?

When we think of a solution, we normally think of a liquid. Liquid solutions, such as sugar in water, are the most common kind, but there are also solutions that are gases or solids. In fact, all mixtures of gases are solutions. Because gas molecules are far apart from each other and much empty space separates them, two or more gases can mix with each other in any proportions. Because the mixing takes place at the molecular level, a true solution always forms; that is, there are no heterogeneous mixtures of gases.

With solids, we are at the other extreme. Whenever we mix solids, we almost always get a heterogeneous mixture. Because even microscopic pieces of solid still contain many billions of particles (molecules, ions, or atoms), there is no way to achieve mixing at the molecular level. Homogeneous mixtures of solids (or **alloys**), such as brass, do exist, but we make them by melting the solids, mixing the molten components, and allowing the mixture to solidify.

Table 6.1 lists the five most common types of solutions. Examples of other types (gas in solid, liquid in gas, and so on) are also known but are much less important.

Table 6.1 The Most Common Types of Solutions

Solute		Solvent	Appearance of Solution	Example
Gas	in	Liquid	Liquid	Carbonated water
Liquid	in	Liquid	Liquid	Wine
Solid	in	Liquid	Liquid	Salt water (saline solution)
Gas	in	Gas	Gas	Air
Solid	in	Solid	Solid	14-Carat gold

When a solution consists of a solid or a gas dissolved in a liquid, the liquid is called the **solvent** and the solid or gas is called the **solute.** A solvent may have several solutes dissolved in it, even of different types. A common example is spring water, in which gases (carbon dioxide and oxygen) and solids (salts) are dissolved in the solvent, water.

When one liquid is dissolved in another, a question may arise regarding which is the solvent and which is the solute. The one present in the greater amount is usually called the solvent.

6.3 What Are the Distinguishing Characteristics of Solutions?

The following are some properties of solutions:

1. **The distribution of particles in a solution is uniform.**
   Every part of the solution has exactly the same composition and properties as every other part. That, in fact, is the definition of "homogeneous." As a consequence, we cannot usually tell a solution from a pure solvent

Chemical Connections 6A

Acid Rain

The water vapor evaporated by the sun from oceans, lakes, and rivers condenses and forms clouds of water vapor that eventually fall as rain. The raindrops contain small amounts of CO_2, O_2, and N_2. The table shows that of these gases, CO_2 is the most soluble in water. When CO_2 dissolves in water, it reacts with a water molecule to give carbonic acid, H_2CO_3.

$$CO_2(g) + H_2O(\ell) \longrightarrow H_2CO_3(aq)$$
Carbonic acid

The acidity caused by the CO_2 is not harmful; however, contaminants that result from industrial pollution may create a serious acid rain problem. Burning coal or oil that contains sulfur generates sulfur dioxide, SO_2, which has a high solubility in water. Sulfur dioxide in the air is oxidized to sulfur trioxide, SO_3. The reaction of sulfur dioxide with water gives sulfurous acid, and the reaction of sulfur trioxide with water gives sulfuric acid.

$$SO_2 + H_2O \longrightarrow H_2SO_3$$
Sulfur dioxide   Sulfurous acid

$$SO_3 + H_2O \longrightarrow H_2SO_4$$
Sulfur trioxide   Sulfuric acid

Smelting, which involves melting or fusing an ore as part of the refining (or separation) process, produces other soluble gases as well. In many parts of the world, especially those located downwind from heavily industrialized areas, the result is acid rain that pours down on forests and lakes. It damages vegetation and kills fish. Acid rain has been observed with increasing frequency in the eastern United States, in North Carolina, in the Adirondack Mountains of New York State, and in parts of New England, as well as in eastern Canada.

Trees killed by acid rain at Mt. Mitchell in North Carolina.

Will McIntyre/Photo Researchers, Inc.

Table 6A **The Solubility of Some Gases in Water**

Gas	Solubility (g/kg H_2O at 20°C and 1 atm)
O_2	0.0434
N_2	0.0190
CO_2	1.688
H_2S	3.846
SO_2	112.80
NO_2	0.0617

Test your knowledge with Problems 6.78 and 6.79.

simply by looking at it. A glass of pure water looks the same as a glass of water containing dissolved salt or sugar. In some cases, we can tell by looking—for example, if the solution is colored and we know that the solvent is colorless.

2. **The components of a solution do not separate on standing.**
A solution of vinegar (acetic acid in water), for example, will never separate.

3. **A solution cannot be separated into its components by filtration.**
Both the solvent and the solute pass through a filter paper.

4. **For any given solute and solvent, it is possible to make solutions of many different compositions.**
For example, we can easily make a solution of 1 g of glucose in 100. g of water, or 2 g, or 6 g, or 8.7 g, or any other amount of glucose up to the solubility limit (Section 6.4).

We use the word "clear" to mean transparent. A solution of copper sulfate in water is blue, but clear.

5. **Solutions are almost always transparent.**
   They may be colorless or colored, but we can usually see through them. Solid solutions are exceptions.
6. **Solutions can be separated into pure components.**
   Common separation methods include distillation and chromatography, which we may learn about in the laboratory portion of this course. The separation of a solution into its components is a physical change, not a chemical one.

6.4 What Factors Affect Solubility?

The **solubility** of a solid in a liquid is the maximum amount of the solid that will dissolve in a given amount of a particular solvent at a given temperature. Suppose we wish to make a solution of table salt (NaCl) in water. We take some water, add a few grams of salt, and stir. At first, we see the particles of salt suspended in the water. Soon, however, all the salt dissolves. Now let us add more salt and continue to stir. Again, the salt dissolves. Can we repeat this indefinitely? The answer is no—there is a limit. The solubility of table salt is 36.2 g per 100. g of water at 30°C. If we add more salt than that amount, the excess solid does not dissolve but rather remains suspended as long as we keep stirring; it will sink to the bottom after we stop stirring.

Solubility is a physical constant, like melting point or boiling point. Each solid has a different solubility in every liquid. Some solids have a very low solubility in a particular solvent; we often call these solids *insoluble*. Others have a much higher solubility; we call these *soluble*. Even for soluble solids, however, there is always a solubility limit (see Section 4.3 for some useful solubility generalizations). The same is true for gases dissolved in liquids. Different gases have different solubilities in a solvent (see Chemical Connections 6A). Some liquids are essentially insoluble in other liquids (gasoline in water), whereas others are soluble to a limit. For example, 100. g of water dissolves about 6 g of diethyl ether (another liquid). If we add more ether than that amount, we see two layers (Figure 6.1).

Some liquids, however, are completely soluble in other liquids, no matter how much is present. An example is ethanol, C_2H_6O, and water, which form a solution no matter what quantities of each are mixed. We say that water and ethanol are **miscible** in all proportions.

When a solvent contains all the solute it can hold at a given temperature, we call the solution **saturated.** Any solution containing a lesser amount of solute is **unsaturated.** If we add more solute to a saturated solution at constant temperature, it looks as if none of the additional solid dissolves, because the solution already holds all the solute that it can. Actually, an equilibrium similar to the one discussed in Section 5.8B is at work in this situation. Some particles of the additional solute dissolve, but an equal quantity of dissolved solute comes out of solution. Thus, even though the concentration of dissolved solute does not change, the solute particles themselves are constantly going into and out of solution.

A **supersaturated** solution contains more solute in the solvent than it can normally hold at a given temperature under equilibrium conditions. A supersaturated solution is not stable; when disturbed in any way, such as by stirring or shaking, the excess solute precipitates—thus, the solution returns to equilibrium and becomes merely saturated.

Whether a particular solute dissolves in a particular solvent depends on several factors, as discussed next.

A. Nature of the Solvent and the Solute

The more similar two compounds are, the more likely that one will be soluble in the other. Here the rule is "like dissolves like." This is not an absolute rule, but it does apply in a great many cases.

FIGURE 6.1 Diethyl ether and water form two layers. A separatory funnel permits the bottom layer to be drawn off.

Diethyl ether

Water

We use the word "miscible" to refer to a liquid dissolving in a liquid.

Supersaturated solution
A solution that contains more than the equilibrium amount of solute at a given temperature and pressure

When we say "like," we mostly mean similar in terms of polarity. In other words, polar compounds dissolve in polar solvents and nonpolar compounds dissolve in nonpolar solvents. For example, the liquids benzene (C_6H_6) and carbon tetrachloride (CCl_4) are nonpolar compounds. They dissolve in each other, and other nonpolar materials, such as gasoline, dissolve in them. In contrast, ionic compounds such as sodium chloride (NaCl) and polar compounds such as table sugar ($C_{12}H_{22}O_{11}$) are insoluble in these solvents.

The most important polar solvent is water. We have already seen that most ionic compounds are soluble in water, as are small covalent compounds that can form hydrogen bonds with water. It is worth noting that even polar molecules are usually insoluble in water if they are unable either to react with water or to form hydrogen bonds with water molecules. Water as a solvent is discussed in Section 6.6.

B. Temperature

For most solids and liquids that dissolve in liquids, the general rule is that solubility increases with increasing temperature. Sometimes the increase in solubility is great, while at other times it is only moderate. For a few substances, solubility even decreases with increasing temperature (Figure 6.2).

For example, the solubility of glycine, H_2N—CH_2—$COOH$, a white crystalline solid and a polar building block of proteins, is 52.8 g in 100. g of water at 80°C but only 33.2 g at 30°C. If, for instance, we prepare a saturated solution of glycine in 100. g of water at 80°C, it will hold 52.8 g of glycine. If we then allow the solution to cool to 30°C where the solubility is 33.2 g, we might expect the excess glycine, 19.6 g, to precipitate from solution as crystals. It often does, but on many occasions, it does not. The latter case is an example of a **supersaturated solution**. Even though the solution contains more glycine than the water can normally hold at 30°C, the excess glycine stays in solution because the molecules need a seed—a surface on which to begin crystallizing. If no such surface is available, no precipitate will form.

Supersaturated solutions are not indefinitely stable, however. If we shake or stir the solution, we may find that the excess solid precipitates at once (Figure 6.3). Another way to crystallize the excess solute is to add a crystal of the solute, a process called **seeding.** The seed crystal provides the surface onto which the solute molecules can converge.

For gases, solubility in liquids almost always decreases with increasing temperature. The effect of temperature on the solubility of gases in water can have important consequences for fish, for example. Oxygen is only slightly soluble in water, and fish need that oxygen to live. When the temperature of a body of water increases, perhaps because of the output from a nuclear power plant, the solubility of oxygen decreases and may become so low that fish die. This situation is called thermal pollution.

C. Pressure

Pressure has little effect on the solubility of liquids or solids. For gases, however, **Henry's law** applies (Figure 6.4): the higher the pressure, the greater the solubility of a gas in a liquid. This concept is the basis of the hyperbaric medicine discussed in Chemical Connections 5B. When the pressure increases, more O_2 dissolves in the blood plasma and reaches tissues at higher-than-normal pressures (2 to 3 atm).

Henry's law also explains why a bottle of beer or an other carbonated beverage foams when it is opened. The bottle is sealed under greater than 1 atm of pressure. When opened at 1 atm, the solubility of CO_2 in the liquid decreases. The excess CO_2 is released, forming bubbles, and the gas pushes out some of the liquid.

Polar compounds dissolve in polar compounds because the positive end of the dipole of one molecule attracts the negative end of the dipole of the other molecule.

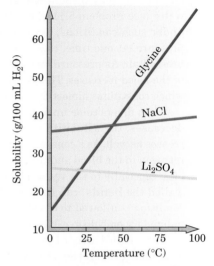

FIGURE 6.2 The solubilities of some solids in water as a function of temperature. The solubility of glycine increases rapidly, that of NaCl barely increases, and that of Li_2SO_4 decreases with increasing temperature.

FIGURE 6.3 When a supersaturated aqueous solution of sodium acetate ($CH_3COO^-Na^+$) is disturbed, the excess salt crystallizes rapidly.

Henry's law The solubility of a gas in a liquid is directly proportional to the pressure

The Bends

Deep-sea divers encounter high pressures while under water (see Problem 5.111). For them to breathe properly under such conditions, oxygen must be supplied under pressure. At one time, this goal was achieved with compressed air. As pressure increases, the solubility of gases in the blood increases. This is especially true for nitrogen, which constitutes almost 80% of our air.

When divers come up and the pressure on their bodies decreases, the solubility of nitrogen in their blood decreases as well. As a consequence, the previously dissolved nitrogen in the blood and tissues starts to form small bubbles, especially in the veins. The formation of gas bubbles (called the **bends**) can hamper blood circulation. If this condition is allowed to develop uncontrolled, a resulting pulmonary embolism can prove fatal.

If the diver's ascent is gradual, regular exhalation and diffusion through the skin remove the dissolved gases. Divers use decompression chambers, where the high pressure is gradually reduced to normal pressure.

If decompression disease develops after a dive, patients are put into a hyperbaric chamber (see Chemical Connections 5B), where they breathe pure oxygen at 2.8 atm pressure. In the standard form of treatment, the pressure is reduced to 1 atm over a period of 6 hours.

Nitrogen also has a narcotic effect on divers when they breathe compressed air at depths greater than 40 m. This effect, called "rapture of the deep," is similar to alcohol-induced intoxication.

Ascending too rapidly will cause dissolved nitrogen bubbles to be released and form bubbles in the blood.

Because of the problem caused by nitrogen, divers' tanks often are charged with a helium–oxygen mixture instead of with air. The solubility of helium in blood is affected less by pressure than is the solubility of nitrogen.

Sudden decompression and ensuing bends are important not only in deep-sea diving but also in high-altitude flight, especially orbital flight.

Test your knowledge with Problems 6.80 and 6.81.

Application of Henry's law. The greater the partial pressure of CO_2 over the soft drink in the bottle, the greater the concentration of dissolved CO_2. When the bottle is opened, the partial pressure of CO_2 drops and CO_2 bubbles out of the solution.

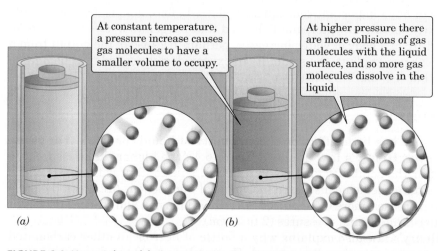

At constant temperature, a pressure increase causes gas molecules to have a smaller volume to occupy.

At higher pressure there are more collisions of gas molecules with the liquid surface, and so more gas molecules dissolve in the liquid.

FIGURE 6.4 Henry's law. (*a*) A gas sample in a liquid under pressure in a closed container. (*b*) The pressure is increased at constant temperature, causing more gas to dissolve.

6.5 What Are the Most Common Units for Concentration?

We can express the amount of a solute dissolved in a given quantity of solvent—that is, the **concentration** of the solution—in a number of ways. Some concentration units are better suited for some purposes than others are. Sometimes qualitative terms are good enough. For example, we may say that a solution is dilute or concentrated. These terms give us little specific information about the concentration, but we know that a concentrated solution contains more solute than a dilute solution does.

For most purposes, however, we need quantitative concentrations. For example, a nurse must know precisely how much glucose to give to a patient. Many methods of expressing concentration exist, but in this chapter, we deal with just the three most important: percent concentration, molarity, and parts per million (ppm).

A. Percent Concentration

Chemists represent **percent concentration** in three ways. The most common is mass of solute per volume of solution (w/v):

Percent concentration (% w/v) The number of grams of solute in 100. mL of solution

$$\text{Weight/volume (w/v)}\% = \frac{\text{mass of solute}}{\text{volume of solution}} \times 100$$

If we dissolve 10. g of sucrose (table sugar) in enough water so that the total volume is 100. mL, the concentration is 10.% w/v. Note that here we need to know the total volume of the solution, not the volume of the solvent.

Example 6.1 Percent Concentration

The label on a bottle of vinegar says it contains 5.0% w/v acetic acid, CH_3COOH. The bottle contains 240 mL of vinegar. How many grams of acetic acid are in the bottle?

Strategy

We are given the volume of the solution and its weight/volume concentration. To calculate the number of grams of CH_3COOH present in this solution, we use the conversion factor 5.0 g of acetic acid in 100. mL of solution.

Solution

$$240 \text{ mL solution} \times \frac{5.0 \text{ g } CH_3COOH}{100. \text{ mL solution}} = 12 \text{ g } CH_3COOH$$

Problem 6.1

How would we prepare 250 mL of a 4.4% w/v KBr solution in water? Assume that a 250-mL volumetric flask is available.

A second way to represent percent concentration is weight of solute per weight of solution (w/w):

$$\text{Weight/weight (w/w)}\% = \frac{\text{weight solute}}{\text{weight of solution}} \times 100$$

Example 6.2 Weight/Volume Percent

If 6.0 g of NaCl is dissolved in enough water to make 300. mL of solution, what is the w/v percent of NaCl?

Strategy

To calculate the w/v percent, we divide the weight of the solute by the volume of the solution and multiply by 100:

Solution

$$\frac{6.0 \text{ g NaCl}}{300. \text{ mL solution}} \times 100 = 2.0\% \text{ w/v}$$

Problem 6.2

If 6.7 g of lithium iodide, LiI, is dissolved in enough water to make 400. mL of solution, what is the w/v percent of LiI?

Calculations of w/w percent are essentially the same as w/v percent calculations, except that we use the weight of the solution instead of its volume. A volumetric flask is not used for these solutions. (Why not?)

Finally, we can represent percent concentration as volume of solute per volume of solution (v/v) percent:

$$\text{Volume/volume (v/v)}\% = \frac{\text{volume solute}}{\text{volume of solution}} \times 100$$

A 40.% v/v solution of ethanol in water is 80 proof. Proof is twice the percent concentration (v/v%) of ethanol in water.

The unit v/v percent is used only for solutions of liquids in liquids—most notably, alcoholic beverages. For example, 40.% v/v ethanol in water means that 40. mL of ethanol has been added to enough water to make 100. mL of solution. This solution might also be called 80 proof, where proof of an alcoholic beverage is twice the v/v percent concentration.

B. Molarity

For many purposes, it is easiest to express concentration by using the weight or volume percentage methods just discussed. When we want to focus on the number of molecules present, however, we need another concentration unit. For example, a 5% solution of glucose in water does not contain the same number of solute molecules as a 5% solution of ethanol in water. That is why chemists often use molarity. **Molarity (M)** is defined as the number of moles of solute dissolved in 1 L of solution. The units of molarity are moles per liter.

$$\text{Molarity } (M) = \frac{\text{moles solute } (n)}{\text{volume of solution (L)}}$$

Thus, in the same volume of solution, a 0.2 M solution of glucose, $C_6H_{12}O_6$, in water contains the same number of molecules of solute as a 0.2 M solution of ethanol, C_2H_6O, in water. In fact, this relationship holds true for equal volumes of any solution, as long as the molarities are the same.

We can prepare a solution of a given molarity in essentially the same way that we prepare a solution of given w/v concentration, except that we use moles instead of grams in our calculations. We can always find out how

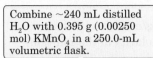

Combine ~240 mL distilled H_2O with 0.395 g (0.00250 mol) $KMnO_4$ in a 250.0-mL volumetric flask.

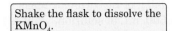

Shake the flask to dissolve the $KMnO_4$.

After the solid dissolves, add sufficient water to fill the flask to the mark etched in the neck, indicating a volume of 250.0 mL, and shake the flask again to thoroughly mix its contents.

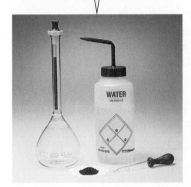

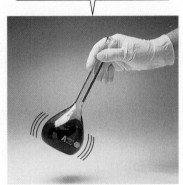

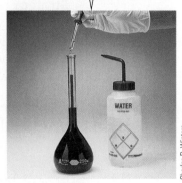

Charles D. Winters

FIGURE 6.5 Solution preparation from a solid solute, making 250.0 mL of 0.0100 M aqueous solution of $KMnO_4$.

many moles of solute are in any volume of a solution of known molarity by using the following relationship:

$$\text{Molarity} \times \text{volume in liters} = \text{number of moles}$$

$$\frac{\text{moles}}{\text{liters}} \times \text{liters} = \text{moles}$$

The solution is then prepared as shown in Figure 6.5.

Example 6.3 Molarity

How do we prepare 2.0 L of a 0.15 M aqueous solution of sodium hydroxide, NaOH?

Strategy

We are given solid NaOH and want 2.0 L of a 0.15 M solution. First, we find out how many moles of NaOH are present in 2.0 liters of this solution; then we convert this number moles to grams.

Solution

Step 1: Determine the number of moles of NaOH in 2.0 liters of this solution. For this calculation, we use molarity as a conversion factor.

$$\frac{0.15 \text{ mol NaOH}}{1.0 \text{ L}} \times 2.0 \text{ L} = 0.30 \text{ mol NaOH}$$

Step 2: To convert 0.30 mol of NaOH to grams of NaOH, we use the molar mass of NaOH (40.0 g/mol) as a conversion factor:

$$0.30 \text{ mol NaOH} \times \frac{40.0 \text{ g NaOH}}{1 \text{ mol NaOH}} = 12 \text{ g NaOH}$$

Step 3: To prepare this solution, we place 12 g of NaOH in a 2-L volumetric flask, add some water, swirl until the solid dissolves, and then fill the flask with water to the 2-L mark.

Problem 6.3

How would we prepare 2.0 L of a 1.06 M aqueous solution of KCl?

⬤WL Interactive Example 6.4 Molarity

If we dissolve 18.0 g of Li_2O (molar mass = 29.9 g/mol) in sufficient water to make 500. mL of solution, what is the molarity of the solution?

Strategy

We are given 18.0 g Li_2O in 500. mL of water and want the molarity of the solution. We first calculate the number of moles of Li_2O in 18.0 g of Li_2O and then convert from moles per 500. mL to moles per liter.

Solution

To calculate the number of moles of Li_2O in a liter of solution, we use two conversion factors: molar mass of Li_2O = 29.9 g and 1000 mL = 1 L.

$$\frac{18.0 \text{ g } \cancel{Li_2O}}{500. \text{ } \cancel{mL}} \times \frac{1 \text{ mol } Li_2O}{29.9 \text{ g } \cancel{Li_2O}} \times \frac{1000 \text{ } \cancel{mL}}{1 \text{ L}} = 1.20 \text{ } M$$

Problem 6.4

If we dissolve 0.440 g of KSCN in enough water to make 340. mL of solution, what is the molarity of the resulting solution?

Example 6.5 Molarity

The concentration of sodium chloride in blood serum is approximately 0.14 M. What volume of blood serum contains 2.0 g of NaCl?

Strategy

We are given the concentration in moles per liter and asked to calculate the volume of blood that contains 2.0 g NaCl. To find the volume of blood, we use two conversion factors: the molar mass of NaCl is 58.4 g and the concentration of NaCl in blood is 0.14 M.

Solution

$$2.0 \text{ g } \cancel{NaCl} \times \frac{1 \text{ } \cancel{mol} \text{ } \cancel{NaCl}}{58.4 \text{ g } \cancel{NaCl}} \times \frac{1 \text{ L}}{0.14 \text{ } \cancel{mol} \text{ } \cancel{NaCl}} = 0.24 \text{ L} = 2.4 \times 10^2 \text{ mL}$$

Note that the answer in mL must be expressed to no more than two significant figures because the mass of NaCl (2.0 g) is given to only two significant figures. To write the answer as 240. mL would be expressing it to three significant figures. We solve the problem of significant figures by expressing the answer in scientific notation.

Problem 6.5

If a 0.300 M glucose solution is available for intravenous infusion, how many milliliters of this solution are needed to deliver 10.0 g of glucose?

Example 6.6 Molarity

How many grams of HCl are in 225 mL of 6.00 M HCl?

Strategy

We are given 225 mL of 6.00 M HCl and asked to find grams of HCl present. We use two conversion factors—the molar mass of HCl = 36.5 g and 1000 mL = 1 L.

Blood serum is the liquid part of the blood that remains after the removal of cellular particulates and fibrinogen.

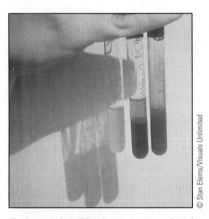

© Stan Elems/Visuals Unlimited

(Left to right): Blood serum, coagulated blood, and whole blood.

Solution

$$225 \text{ mL} \times \frac{1 \text{ L}}{1000 \text{ mL}} \times \frac{6.00 \text{ mol HCl}}{1 \text{ L}} \times \frac{36.5 \text{ g HCl}}{1 \text{ mol HCl}} = 49.3 \text{ g HCl}$$

Problem 6.6

A certain wine contains 0.010 M NaHSO$_3$ (sodium bisulfite) as a pre-servative. How many grams of sodium bisulfite must be added to a 100. gallon barrel of wine to reach this concentration? Assume no change in volume of wine upon addition of the sodium bisulfite.

C. Dilution

We frequently prepare solutions by diluting concentrated solutions rather than by weighing out pure solute (Figure 6.6). Because we add only solvent during dilution, the number of moles of solute remains unchanged. Before dilution, the equation that applies is:

$$M_1V_1 = \text{moles}$$

After dilution, the volume and molarity have both changed and we have:

$$M_2V_2 = \text{moles}$$

Because the number of moles of solute is the same both before and after dilution, we can say that:

$$M_1V_1 = M_2V_2$$

We can use this handy equation (the units of which are moles = moles) for all dilution problems.

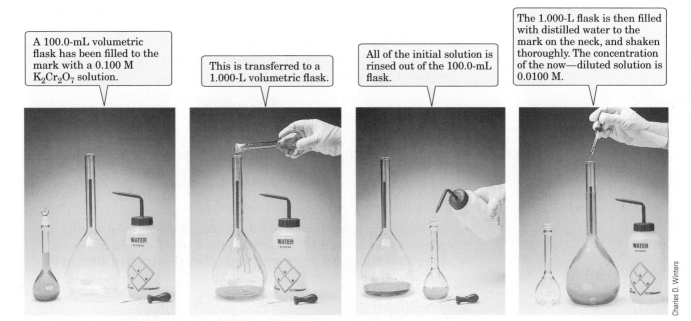

A 100.0-mL volumetric flask has been filled to the mark with a 0.100 M K$_2$Cr$_2$O$_7$ solution.

This is transferred to a 1.000-L volumetric flask.

All of the initial solution is rinsed out of the 100.0-mL flask.

The 1.000-L flask is then filled with distilled water to the mark on the neck, and shaken thoroughly. The concentration of the now—diluted solution is 0.0100 M.

FIGURE 6.6 Solution preparation by dilution. Here 100 mL of 0.100 M potassium dichromate, K$_2$Cr$_2$O$_7$ is diluted to 1.000 L. The result is dilution by a factor of 10.

Charles D. Winters

Example 6.7 Dilution

Suppose we have a bottle of concentrated acetic acid (6.0 M). How would we prepare 200. mL of a 3.5 M solution of acetic acid?

Strategy

We are given $M_1 = 6.0\ M$ and asked to calculate V_1. We are also given $M_2 = 3.5\ M$ and $V_2 = 200.$ mL, that is $V_2 = 0.200$ L.

Solution

$$M_1V_1 = M_2V_2$$

$$\frac{6.0\ \text{mol}}{1.0\ \text{L}} \times V_1 = \frac{3.5\ \text{mol}}{1.0\ \text{L}} \times 0.200\ \text{L}$$

Solving this equation for V_1 gives:

$$V_1 = \frac{3.5\ \text{mol} \times 0.200\ \text{L}}{6.0\ \text{mol}} = 0.12\ \text{L}$$

To prepare this solution, we place 0.12 L, or 120. mL, of concentrated acetic acid in a 200. mL volumetric flask, add some water and mix, and then fill to the calibration mark with water.

Problem 6.7

We are given a solution of 12.0 M HCl and want to prepare 300. mL of a 0.600 M solution. How would we prepare it?

A similar equation can be used for dilution problems involving percent concentrations:

$$\%_1V_1 = \%_2V_2$$

Example 6.8 Dilution

Suppose we have a solution of 50.% w/v NaOH on hand. How would we prepare 500. mL of a 0.50% w/v solution of NaOH?

Strategy

We are given 50.% w/v NaOH and asked to prepare 500. mL (V_2) of 0.50% solution V_1. We use the relationship:

$$\%_1V_1 = \%_2V_2$$

Solution

$$(50.\%) \times V_1 = (0.50\%) \times 500.\ \text{mL}$$

$$V_1 = \frac{0.50\% \times 500.\ \text{mL}}{50.\%} = 5.0\ \text{mL}$$

To prepare this solution, we add 5.0 mL of the 50.% w/v solution (the concentrated solution) to a 500.-mL volumetric flask, then some water and mix, and then fill to the mark with water. Note that this is a dilution by a factor of 100.

Problem 6.8

A concentrated solution of 15% w/v KOH solution is available. How would we prepare 20.0 mL of a 0.10% w/v KOH solution?

D. Parts per Million

Sometimes we need to deal with very dilute solutions—for example, 0.0001%. In such cases, it is more convenient to use the unit **parts per million (ppm)** to express concentration. For example, if drinking water is polluted with lead ions to the extent of 1 ppm, it means that there is 1 mg of lead ions in 1 kg (1 L) of water. When reporting concentration in ppm, the units must be the same for both solute and solution—for example, mg of solute per 10^6 mg of solution, or g of solute per g of solution. Some solutions are so dilute that we use **parts per billion (ppb)** to express their concentrations.

$$\text{ppm} = \frac{\text{g solute}}{\text{g solution}} \times 10^6$$

$$\text{ppb} = \frac{\text{g solute}}{\text{g solution}} \times 10^9$$

Example 6.9 Parts per Million (ppm)

Verify that 1 mg of lead in 1 kg of drinking water is equivalent to 1 ppm lead.

Strategy

The units we are given are milligrams and kilograms. To report ppm, we must convert them to a common unit, say grams. For this calculation, we use two conversion factors: 1000 mg = 1 g and 1 kg solution = 1000 g solution.

Solution

Step 1: First we find the mass (grams) of lead:

$$1 \text{ mg lead} \times \frac{1 \text{ g lead}}{1000 \text{ mg lead}} = 1 \times 10^{-3} \text{ g lead}$$

Step 2: Next, we find the mass (grams) of the solution:

$$1 \text{ kg solution} \times \frac{1000 \text{ g solution}}{1 \text{ kg solution}} = 1 \times 10^3 \text{ g solution}$$

Step 3: Finally use these values to calculate the concentration of lead in ppm:

$$\text{ppm} = \frac{1 \times 10^{-3} \text{ g lead}}{1 \times 10^3 \text{ g solution}} \times 10^6 = 1 \text{ ppm}$$

Problem 6.9

Sodium hydrogen sulfate, $NaHSO_4$, which dissolves in water to release H^+ ion, is used to adjust the pH of the water in swimming pools. Suppose we add 560. g of $NaHSO_4$ to a swimming pool that contains 4.5×10^5 L of water at 25°C. What is the Na^+ ion concentration in ppm?

Modern methods of analysis allow us to detect such minuscule concentrations. Some substances are harmful even at concentrations measured in ppb. One such substance is dioxin, an impurity in the 2,4,5-T herbicide sprayed by the United States as a defoliant in Vietnam.

6.6 Why Is Water Such a Good Solvent?

Water covers about 75% of the Earth's surface in the form of oceans, ice caps, glaciers, lakes, and rivers. Water vapor is always present in the atmosphere. Life evolved in water and without it life as we know it could not exist. The human body is about 60% water. This water is found both inside the cells of the body (intracellular) and outside the cells (extracellular).

Chemical Connections 6C

Electrolyte Solutions in Body and Intravenous Fluids

Body fluids typically contain a mixture of several electrolytes (see Section 6.6C) such as Na^+, Ca^{2+}, Cl^-, HCO_3^-, and HPO_4^{2-}. The ions present generally originate from more than one source. We measure each individual ion present in terms of an equivalent (Eq), which is the molar amount of an ion equal to one mole of positive or negative electrical charge. For example, 1 mole of Na^+ ion and HCO_3^- ion are each one equivalent because they supply one mole of electrical charge. Ions with a 2+ or 2− charge, such as Ca^{2+} and HPO_4^{2-}, are each two equivalents per one mole of ion.

The concentrations of electrolytes present in body fluids and in intravenous fluids given to a patient are often expressed in milliequivalents per liter (mEq/L) of solution. For example, lactated Ringer's solution is often used for fluid resuscitation after a patient suffers blood loss due to trauma, surgery, or a brain injury. It consists of three cations (130. mEq/L Na^+, 4 mEq/L K^+, and 3 mEq/L Ca^{2+}) and two anions (109 mEq/L Cl^- and 28 mEq/L $C_3H_5O_3^-$, lactate). Notice that the charge balance of the solution is maintained, and the total number of positive charges is equal to the total number of negative charges. The various electrolyte concentrations present in lactated

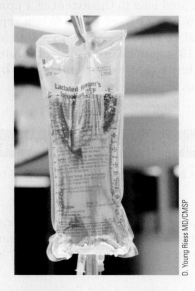

D. Young Riess MD/CMSP

Ringer's solution are one of many possible intravenous replacement solutions used in a clinical setting. The use of specific intravenous solutions depends on the fluid, electrolytic, and nutritional needs of an individual patient.

Test your knowledge with Problems 6.82 and 6.83.

Most of the important chemical reactions in living tissue occur in aqueous solution; water serves as a solvent to transport reactants and products from one place in the body to another. Water is also itself a reactant or product in many biochemical reactions. The properties that make water such a good solvent are its polarity and its hydrogen-bonding capacity (Section 5.7C).

A. How Does Water Dissolve Ionic Compounds?

We learned in Section 3.5 that ionic compounds in the solid state are composed of a regular array of ions in a crystal lattice. The crystal is held together by ionic bonds, which are electrostatic attractions between positive and negative ions. Water, of course, is a polar molecule. When a solid ionic compound is added to water, water molecules surround the ions at the surface of the crystal. The negative ions (anions) attract the positive poles of water molecules, and the positive ions (cations) attract the negative poles of water molecules (Figure 6.7). Each ion attracts multiple water molecules. When the combined force of attraction to water molecules is greater than the force of attraction of the ionic bonds that keeps the ions in the crystal, the ions will be completely dislodged. Water molecules now surround the ion removed from the crystal (Figure 6.8). Such ions are said to be **hydrated.** A more general term, covering all solvents, is **solvated.** The solvation layer—that is, the surrounding shell of solvent molecules—acts as a cushion. It prevents a solvated anion from colliding directly with a solvated cation, thereby keeping the solvated ions in solution.

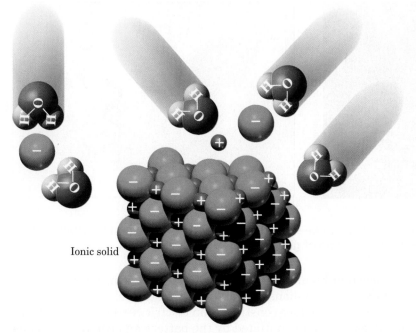

Ionic solid

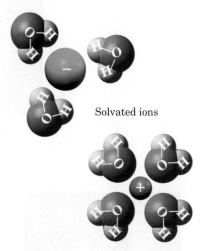

Solvated ions

FIGURE 6.7 (*a*) When water dissolves an ionic compound, water molecules remove anions and cations from the surface of the solid and water molecules surround the ions.

FIGURE 6.8 Anions and cations solvated by water.

Not all ionic solids are soluble in water. Some rules for predicting solubilities were given in Section 4.3.

B. Solid Hydrates

The attraction between ions and water molecules is so strong in some cases that the water molecules are an integral part of the crystal structure of the solids. Water molecules in a crystal are called **water of hydration.** The substances that contain water in their crystals are themselves called **hydrates.** For example, both gypsum and plaster of Paris are hydrates of calcium sulfate: gypsum is calcium sulfate dihydrate, $CaSO_4 \cdot 2H_2O$, and plaster of Paris is calcium sulfate monohydrate, $(CaSO_4)_2 \cdot H_2O$. Some hydrates hold on to their water molecules tenaciously. To remove them, the crystals must be heated for some time at a high temperature. The crystal without its water of hydration is called **anhydrous.** In many cases, anhydrous crystals attract water so strongly that they absorb from the water vapor in the air. That is, some anhydrous crystals become hydrated upon standing in air. Crystals that do so are called **hygroscopic.**

Hydrated crystals often look different from the anhydrous forms. For example, copper(II) sulfate pentahydrate, $CuSO_4 \cdot 5H_2O$, is blue but the anhydrous form, $CuSO_4$, is white (Figure 6.9).

The difference between hydrated and anhydrous crystals can sometimes have an effect in the body. For example, the compound sodium urate exists in the anhydrous form as spherical crystals, but in the monohydrate form as needle-shaped crystals (Figure 6.10). The deposition of sodium urate monohydrate in the joints (mostly in the big toe) causes gout.

C. Electrolytes

Ions in water migrate from one place to another, maintaining their charge as they migrate. As a consequence, solutions of ions conduct electricity. They can do so because ions in the solution migrate independently of one

The dot in the formula $CaSO_4 \cdot 2H_2O$ indicates that H_2O is present in the crystal, but it is not covalently bonded to the Ca^{2+} or SO_4^{2-} ions.

Hygroscopic substance A substance able to absorb water vapor from the air

If we want a hygroscopic compound to remain anhydrous, we must place it in a sealed container that contains no water vapor.

FIGURE 6.9 When blue hydrated copper(II) sulfate, $CuSO_4 \cdot 5H_2O$, is heated and the compound releases its water of hydration, it changes to white anhydrous copper(II) sulfate, $CuSO_4$.

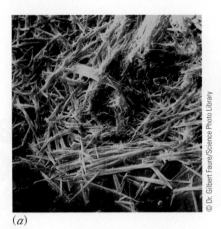

(a) *(b)*

FIGURE 6.10 (*a*) The needle-shaped sodium urate monohydrate crystals that cause gout. (*b*) The pain of gout as depicted by a cartoonist.

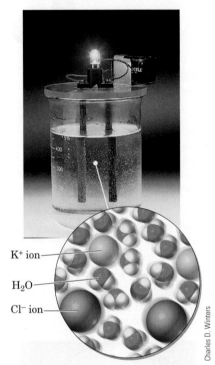

K$^+$ ion

H$_2$O

Cl$^-$ ion

FIGURE 6.11 Conductance by an electrolyte. When an electrolyte, such as KCl, is dissolved in water and provides ions that move about, their migration completes an electrical circuit and the lightbulb in the circuit glows. The ions of every KCl unit have dissociated to K$^+$ and Cl$^-$. The Cl$^-$ ions move toward the positive electrode and the K$^+$ ions move toward the negative electrode, thereby transporting electrical charge through the solution.

another. As shown in Figure 6.11, cations migrate to the negative electrode, called the **cathode,** and anions migrate to the positive electrode, called the **anode.** The movement of ions constitutes an electric current. The migration of ions completes the circuit initiated by the battery and can cause an electric bulb to light up (see also Chemical Connections 4B).

A substance, such as potassium chloride, that conducts an electric current when dissolved in water or when in the molten state is called an **electrolyte.** Hydrated K$^+$ ions carry positive charges, and hydrated Cl$^-$ ions carry negative charges; as a result, the bulb in Figure 6.11 lights brightly if these ions are present. A substance that does not conduct electricity is called a **nonelectrolyte.** Distilled water, for example, is a nonelectrolyte. The light bulb shown in Figure 6.11 does not light up if only distilled water is placed in the beaker. However, with tap water in the beaker, the bulb lights dimly. Tap water contains enough ions to carry electricity, but their concentration is so low that the solution conducts only a small amount of electricity.

As we see, electric conductance depends on the concentration of ions. The higher the ion concentration, the greater the electric conductance of the solution. Nevertheless, differences in electrolytes exist. If we take a 0.1 *M* aqueous NaCl and compare it with a 0.1 *M* aqueous acetic acid (CH$_3$COOH), we find that the NaCl solution lights a bulb brightly, but the acetic acid solution lights it only dimly. We might have expected the two solutions to behave similarly, because each has the same concentration, 0.1 *M*, and each compound provides two ions, a cation and an anion (Na$^+$ and Cl$^-$, H$^+$ and CH$_3$COO$^-$). The reason they behave differently is that, whereas NaCl dissociates completely to two ions (each hydrated and each moving independently), in the case of CH$_3$COOH, only a few of its molecules dissociate into ions. Most of the acetic acid molecules do not dissociate, and undissociated molecules do not conduct electricity. Compounds that dissociate completely are called **strong electrolytes,** and those that dissociate only partially into ions are called **weak electrolytes.**

Electrolytes are important components of the body because they help to maintain the acid–base balance and the water balance. The most important cations in tissues of the human body are Na$^+$, K$^+$, Ca^{2+}, and Mg^{2+}. The most important anions in the body are HCO$_3^-$, Cl$^-$, HPO$_4^{2-}$, and H$_2$PO$_4^-$.

D. How Does Water Dissolve Covalent Compounds?

Water is a good solvent not only for ionic compounds but also for many covalent compounds. In a few cases, the covalent compounds dissolve because they react with water. An example of a covalent compound that dissolves in

Chemical Connections 6D

Hydrates and Air Pollution: The Decay of Buildings and Monuments

Many buildings and monuments in urban areas throughout the world are decaying, ruined by air pollution. The main culprit in this process is acid rain, an end product of air pollution. The stones most commonly used for buildings and monuments are limestone and marble, both of which are largely calcium carbonate. In the absence of polluted air, these stones can last for thousands of years. Thus, many statues and buildings from ancient times (Babylonian, Egyptian, Greek, and others) survived until recently with little change. Indeed, they remain intact in many rural areas.

In urban areas, however, the air is polluted with SO_2 and SO_3, which come mostly from the combustion of coal and petroleum products containing small amounts of sulfur compounds as impurities (see Chemical Connections 6A). They react with the calcium carbonate at the surface of the stones to form calcium sulfate. When calcium sulfate interacts with rainwater, it forms the dihydrate gypsum.

$$SO_3(g) + H_2O(g) \longrightarrow H_2SO_4(\ell)$$

Sulfur trioxide   Sulfuric acid

$$CaCO_3(s) + H_2SO_4(\ell) \longrightarrow CaSO_4(s) + H_2O(g) + CO_2(g)$$

Calcium carbonate   Calcium sulfate
(marble, limestone)

$$CaSO_4(s) + 2H_2O(g) \longrightarrow CaSO_4(s) \cdot 2H_2O(s)$$

Calcium sulfate   Calcium sulfate dihydrate
(gypsum)

The problem is that gypsum has a larger volume than the original marble or limestone, and its presence causes the surface of the stone to expand. This activity, in turn, results in flaking. Eventually, statues such as those in the Parthenon (in Athens, Greece) become noseless and later faceless.

Acid rain damage to stonework on the walls of York Minster, York, England.

Test your knowledge with Problems 6.84 and 6.85.

water is HCl. HCl is a gas (with a penetrating, choking odor) that attacks the mucous membranes of the eyes, nose, and throat. When dissolved in water, HCl molecules react with water to give ions:

$$HCl(g) + H_2O(\ell) \longrightarrow Cl^-(aq) + H_3O^+(aq)$$

Hydrogen   Hydronium ion
chloride

Another example is the gas sulfur trioxide, which reacts with water as follows:

$$SO_3(g) + 2H_2O(\ell) \longrightarrow H_3O^+(aq) + HSO_4^-(aq)$$

Sulfur   Hydronium
trioxide   ion

Because HCl and SO_3 are completely converted to ions in dilute aqueous solution, these solutions are ionic solutions and behave just as other electrolytes do (they conduct a current). Nevertheless, HCl and SO_3 are themselves covalent compounds, unlike salts such as NaCl.

Sports drinks help to maintain the body's electrolyte balance.

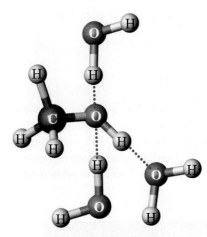

FIGURE 6.12 Solvation of a polar covalent compound by water. The dotted lines represent hydrogen bonds.

H^+ does not exist in aqueous solution; it combines with a water molecule and forms a hydronium ion, H_3O^+.

Most covalent compounds that dissolve in water do not, in fact, react with water. They dissolve because water molecules surround the entire covalent molecule and solvate it. For example, when methanol, CH_3OH, dissolves in water, the methanol molecules are solvated by the water molecules (Figure 6.12).

There is a simple way to predict which covalent compounds will dissolve in water and which will not. Covalent compounds will dissolve in water if they can form hydrogen bonds with water, provided that the solute molecules are fairly small. Hydrogen bonding is possible between two molecules if one of them contains an O, N, or F atom (a hydrogen bond acceptor) and the other contains an O—H, N—H, or F—H bond (a hydrogen bond donor). Every water molecule contains an O atom and O—H bonds. Therefore, water can form hydrogen bonds with any molecule that also contains an O, N, or F atom or an O—H, N—H, or F—H bond. If these molecules are small enough, they will be soluble in water. How small? In general, they should have no more than three C atoms for each O or N atom.

For example, acetic acid, CH_3COOH, is soluble in water, but benzoic acid, C_6H_5COOH, is not significantly soluble. Similarly, ethanol, C_2H_6O, is soluble in water, but dipropyl ether, $C_6H_{14}O$, is not. Table sugar, $C_{12}H_{22}O_{11}$ (Section 20.4A), is very soluble in water. Although each molecule of sucrose contains a large number (12) of carbon atoms, it has so many oxygen atoms (11) that it forms many hydrogen bonds with water molecules; thus, a sucrose molecule in aqueous solution is very well solvated.

As a generalization, covalent molecules that do not contain O or N atoms are almost always insoluble in water. For example, methanol, CH_3OH, is infinitely soluble in water, but chloromethane, CH_3Cl, is not. The exception to this generalization is the rare case where a covalent compound reacts with water—for instance, HCl.

E. Water in the Body

Water is important in the body not only because it dissolves ionic substances as well as some covalent compounds, but also because it hydrates all polar molecules in the body. In this way, water serves as a vehicle to transport most of the organic compounds, nutrients, and fuels used by the body, as well as waste material. Blood and urine are two examples of aqueous body fluids.

In addition, the hydration of macromolecules such as proteins, nucleic acids, and polysaccharides allows the proper motions within these molecules, which are necessary for such functions as enzyme activity (see Chapter 23).

6.7 What Are Colloids?

Up to now, we have discussed only solutions. The maximum diameter of the solute particles in a true solution is about 1 nm. If the diameter of the solute particles exceeds this size, then we no longer have a true solution—we have a **colloid**. In a colloid (also called a colloidal dispersion or colloidal system), the diameter of the solute particles ranges from about 1 to 1000. nm. The term *colloid* has acquired a new name recently. In Section 5.9, we encountered the term *nanotube*. The "nano" part refers to dimensions in the nanometer range (1 nm = 10^{-9} m), which is the size range of colloids. Thus, when we encounter terms such as "nanoparticle" or "nanoscience," they are equivalent to "colloidal particle" or "colloid science," although the former terms refer mostly to particles with a well-defined geometrical shape (such as tubes), while the latter terms are more general.

Colloidal particles usually have a very large surface area, which accounts for the two basic characteristics of colloidal systems:

1. They scatter light and therefore appear turbid, cloudy, or milky.
2. Although colloidal particles are large, they form stable dispersions—they do not form separate phases that settle out. As with true solutions, colloids can exist in a variety of phases: gas, liquid, or solid (Table 6.2).

All colloids exhibit the following characteristic effect. When we shine light through a colloid and look at the system from a 90° angle, we see the pathway of the light without seeing the colloidal particles themselves (they are too small to see). Rather, we see flashes of the light scattered by the particles in the colloid (Figure 6.13).

The **Tyndall effect** is due to light scattering by colloidal particles. Smoke, serum, and fog, to name a few examples, all exhibit the Tyndall effect. We are all familiar with the sunbeams that can be seen when sunlight passes through dusty air. This, too, is an example of the Tyndall effect. Again, we do not see the particles in dusty air, but only the light scattered by them.

Colloidal systems are stable. Mayonnaise, for example, stays emulsified and does not separate into oil and water. When the size of colloidal particles is larger than about 1000. nm, however, the system is unstable and separates into phases. Such systems are called **suspensions.**

For example, if we take a lump of soil and disperse it in water, we get a muddy suspension. The soil particles are anywhere from 10^3 to 10^9 nm in diameter. The muddy mixture scatters light and, therefore, appears turbid. It is not a stable system, however. If left alone, the soil particles soon settle, with clear water found above the sediment. Therefore, soil in water is a suspension, not a colloidal system.

Table 6.3 summarizes the properties of solutions, colloids, and suspensions.

What makes a colloidal dispersion stable? To answer this question, we must first realize that colloidal particles are in constant motion. Just look at the dust particles dancing in a ray of sunlight that enters a room. Actually, you do not see the dust particles themselves; they are too small. Rather, you see flashes of scattered light. The motion of the dust particles dispersed in air is a random, chaotic motion. This motion of any colloidal particle suspended in a solvent is called **Brownian motion** (Figure 6.14).

The constant buffeting and collisions by solvent molecules cause the colloidal particles to move in random Brownian motion. (In the case of the dust particles, the solvent is air.) This ongoing motion creates favorable conditions for collisions between particles. When such large particles collide, they stick together, combine to give larger particles, and finally settle out of the solution. That is what happens in a suspension.

Table 6.2 Types of Colloidal Systems

Type	Example
Gas in gas	None
Gas in liquid	Whipped cream
Gas in solid	Marshmallows
Liquid in gas	Clouds, fog
Liquid in liquid	Milk, mayonnaise
Liquid in solid	Cheese, butter
Solid in gas	Smoke
Solid in liquid	Jelly
Solid in solid	Dried paint

Tyndall effect Light passing through and scattered by a colloid viewed at a right angle

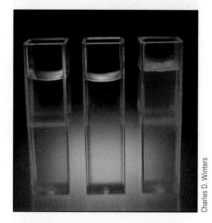

Charles D. Winters

FIGURE 6.13 The Tyndall effect. A narrow beam of light from a laser is passed through a colloidal mixture (left), then a NaCl solution, and finally a colloidal mixture of gelatin and water (right). This illustrates the light-scattering ability of the colloid-sized particles.

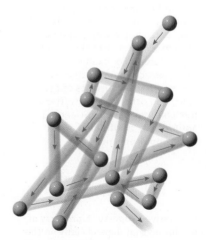

FIGURE 6.14 Brownian motion.

Table 6.3 Properties of Three Types of Mixtures

Property	Solutions	Colloids	Suspensions
Particle size (nm)	0.1–1.0	1–1000	>1000
Filterable with ordinary paper	No	No	Yes
Homogeneous	Yes	Borderline	No
Settles on standing	No	No	Yes
Behavior to light	Transparent	Tyndall effect	Translucent or opaque

Chemical Connections 6E

Emulsions and Emulsifying Agents

Oil and water do not mix. Even when we stir them vigorously and the oil droplets become dispersed in the water, the two phases separate as soon as we stop stirring. There are, however, a number of stable colloidal systems made of oil and water, known as **emulsions.** For example, the oil droplets in milk are dispersed in an aqueous solution. This is possible because milk contains a protective colloid—the milk protein called casein. Casein molecules surround the oil droplets, and because they are polar and carry a charge, they protect and stabilize the oil droplets. Casein is thus an emulsifying agent.

Another emulsifying agent is egg yolk. This ingredient in mayonnaise coats the oil droplets and prevents them from separating.

Test your knowledge with Problem 6.86.

Freshly made wines are often cloudy because of colloidal particles (left). Removing the particles clarifies the wine (right).

Emulsion A system, such as fat in milk, consisting of a liquid with or without an emulsifying agent in an immiscible liquid, usually as droplets of larger than colloidal size.

Nanotubes, nanowires, and nanopores in composite coatings have unusual electronic and optical properties because of their enormous surface areas. For example, titanium oxide particles smaller than 20 nm are used to coat surfaces of plastics, glass, and other materials. These thin coatings have self-cleaning, antifogging, antifouling, and sterilizing properties.

Colligative property A property of a solution that depends only on the number of solute particles and not on the chemical identity of the solute

So why do colloidal particles remain in solution despite all the collisions due to their Brownian motion? Two reasons explain this phenomenon:

1. Most colloidal particles carry a large solvation layer. If the solvent is water, as in the case of protein molecules in the blood, the colloidal particles are surrounded by a large number of water molecules, which move together with the colloidal particles and cushion them. When two colloidal particles collide as a result of Brownian motion, they do not actually touch each other; instead, only their solvent layers collide. As a consequence, the particles do not stick together and precipitate. Instead, they stay in solution.

2. The large surface area of colloidal particles acquires charges from the solution. All colloids in a particular solution acquire the same kind of charge—for example, a negative charge. This development leaves a net negative charge in the solvent. When a charged colloidal particle encounters another charged colloidal particle, the two repel each other because of their like charges.

Thus, the combined effects of the solvation layer and the surface charge keep colloidal particles in a stable dispersion. By taking advantage of these effects, chemists can either increase or decrease the stability of a colloidal system. If we want to get rid of a colloidal dispersion, we can remove the solvation layer, the surface charge, or both. For example, proteins in the blood form a colloidal dispersion. If we want to isolate a protein from blood, we may want to precipitate it. We can accomplish this task in two ways: by removing the hydration layer or by removing the surface charges. If we add a solvent such as ethanol or acetone, each of which has great affinity for water, water is removed from the solvation layer of the protein, and when unprotected protein molecules collide, they stick together and form sediment. Similarly, by adding an electrolyte such as NaCl to the solution, we can remove the charges from the surface of the proteins (by a mechanism too complicated to discuss here). Without their protective charges, two protein molecules will no longer repel each other. Instead, when they collide, they stick together and precipitate from the solution.

6.8 What Is a Colligative Property?

A **colligative property** is any property of a solution that depends only on the number of solute particles dissolved in the solvent and not on the nature of the solute particles. Several colligative properties exist, including freezing-point depression, boiling-point elevation, and osmotic pressure. Of these three, osmotic pressure is of paramount importance in biological systems.

A. Freezing-Point Depression

One mole of any particle, whether it is a molecule or ion, dissolved in 1000. g of water lowers the freezing point of the water by 1.86°C. The nature of the solute does not matter, only the number of particles.

$$\Delta T_f = \frac{-1.86°C}{mol} \times \text{mol of particles}$$

This principle is used in a number of practical ways. In winter, we use salts (sodium chloride and calcium chloride) to melt snow and ice on our streets. The salts dissolve in the melting snow and ice, which lowers the freezing point of the water. Another application is the use of antifreeze in automobile radiators. Because water expands upon freezing (see Chemical Connections 5D), the ice formed in a car's cooling system when the outside temperature falls below 0°C can crack the engine block. The addition of antifreeze prevents this problem, because it makes the water freeze at a much lower temperature. The most common automotive antifreeze is ethylene glycol, $C_2H_6O_2$.

Freezing-point depression The decrease in the freezing point of a liquid caused by adding a solute.

Note that in preparing a solution for this purpose, we do not use molarity. That is, we do not need to measure the total volume of the solution.

Example 6.10 Freezing-Point Depression

If we add 275 g of ethylene glycol, $C_2H_6O_2$, a nondissociating molecular compound, per 1000. g of water in a car radiator, what will be the freezing point of this solution?

Strategy

We are given 275 g of ethylene glycol (molar mass 62.1 g) per 1000. g water and asked to calculate the freezing point of the solution. We first calculate the moles of ethylene glycol present in the solution and then the freezing-point depression caused by that number of moles.

Solution

$$\Delta T = 275 \text{ g } C_2H_6O_2 \times \frac{1 \text{ mol } C_2H_6O_2}{62.1 \text{ g } C_2H_6O_2} \times \frac{1.86°C}{1 \text{ mol } C_2H_6O_2} = 8.26°C$$

The freezing point of the water will be lowered from 0°C to −8.26°C, and the radiator will not crack if the outside temperature remains above −8.26°C (17.13°F).

Problem 6.10

If we add 215 g of methanol, CH_3OH, to 1000. g of water, what will be the freezing point of the solution?

Salting lowers the freezing point of ice.

Charles D. Winters

If a solute is ionic, then each mole of solute dissociates to more than one mole of particles. For example, if we dissolve one mole (58.5 g) of NaCl in 1000. g of water, the solution contains two moles of solute particles: one mole each of Na^+ and Cl^-. The freezing point of water will be lowered by twice 1.86°C, that is, by 3.72°C per mole of NaCl.

Example 6.11 Freezing-Point Depression

What will be the freezing point of the resulting solution if we dissolve one mole of potassium sulfate, K_2SO_4, in 1000. g of water?

Strategy and Solution

One mole of K_2SO_4 dissociates to produce three moles of ions: two moles of K^+ and one mole of SO_4^{2-}. The freezing point will be lowered by $3 \times 1.86°C = 5.58°C$, and the solution will freeze at $-5.58°C$.

Problem 6.11

Which aqueous solution would have the lowest freezing point?

(a) 6.2 *M* NaCl   (b) 2.1 *M* $Al(NO_3)_3$ (c) 4.3 *M* K_2SO_3

B. Boiling-Point Elevation

The boiling point of a substance is the temperature at which the vapor pressure of the substance equals atmospheric pressure. A solution containing a nonvolatile solute has a lower vapor pressure than the pure solvent and must be at a higher temperature before its vapor pressure equals atmospheric pressure and it boils. Thus, the boiling point of a solution containing a nonvolatile solute is higher than that of the pure solvent.

One mole of any molecule or ion dissolved in 1000. g of water raises the boiling point of the water by 0.512°C. The nature of the solute does not matter, only the number of particles.

$$\Delta T_b = \frac{0.512°C}{mol} \times \text{mol of particles}$$

Example 6.12 Boiling-Point Elevation

Calculate the boiling point of a solution prepared by dissolving 275 g of ethylene glycol ($C_2H_6O_2$) in 1000. mL of water.

Strategy

To calculate the boiling point elevation, we must determine the number of moles of ethylene glycol dissolved in 1000. mL of water. We use the conversion factor 1.00 mole of ethylene glycol = 62.1 g of ethylene glycol.

Solution

Step 1: Calculate the number of moles of ethylene glycol (Egly) in the solution.

$$275 \text{ g Egly} \times \frac{1 \text{ mol Egly}}{62.1 \text{ g Egly}} = 4.43 \text{ mol Egly}$$

Step 2: The boiling point elevation is 2.20°C:

$$0.512 \times 4.43 = 2.27°C$$

The boiling point is raised by 2.27°C. Therefore, the solution boils at 102.3°C.

Problem 6.12

Calculate the boiling point of a solution prepared by dissolving 310. g of ethanol, CH_3CH_2OH, in 1000. mL of water.

Ethylene glycol, boiling point 199°C, is a nonvolatile alcohol widely used in automobile radiators. An aqueous solution of ethylene glycol raises the boiling point of the coolant mixture and prevents engine overheating in the summer. It lowers the freezing point of the coolant mixture and prevents it from freezing in the winter.

C. Osmotic Pressure

To understand osmotic pressure, let us consider the experimental setup shown in Figure 6.15. Suspended in the beaker is a bag containing a 5% solution of sugar in water. The bag is made of a **semipermeable membrane** that contains very tiny pores, far too small for us to see but large enough to allow solvent (water) molecules to pass through them but not the larger solvated sugar molecules.

When the bag is submerged in pure water, Figure 6.15(a), water flows into the bag by osmosis and raises the liquid level in the tube attached to the bag, Figure 6.15(b). Although sugar molecules are too big to pass through the membrane, water molecules easily move back and forth across it. However, this process cannot continue indefinitely because gravity prevents the difference in levels from becoming too great. Eventually a dynamic equilibrium is achieved. The height of the liquid in the tube remains unchanged and is a measure of osmotic pressure.

The liquid level in the glass tube and the breaker can be made equal again if we apply an external pressure through the glass tube. The amount of external pressure required to equalize the levels is called the **osmotic pressure** (Π).

Although this discussion assumes that one compartment contains pure solvent and the other a solution, the same principle applies if both compartments contain solutions, as long as their concentrations are different. The solution of higher concentration always has a higher osmotic pressure than the one of lower concentration, which means that the flow of solvent molecules always occurs from the more dilute solution into the more concentrated solution. Of course, the number of particles is the most important consideration. We must remember that in ionic solutions, each mole of solute gives rise to more than one mole of particles. For convenience in calculation, we define a new term, **osmolarity**, which is molarity (M) of the

Osmotic Pressure (Π) The amount of external pressure that must be applied to the more concentrated solution to stop the passage of solvent molecules across a semipermeable membrane

A semipermeable membrane is a thin slice of some material, such as cellophane, that contains very tiny holes that allows only solvent molecules to pass through them. Solvated solute particles are much larger than solvent particles and cannot pass through the membrane.

Some ions are small but still do not go through the membrane because they are solvated by a shell of water molecules (see Figure 6.8).

Osmosis The passage of solvent molecules from a less concentrated solution across a semipermeable membrane into a more concentrated solution

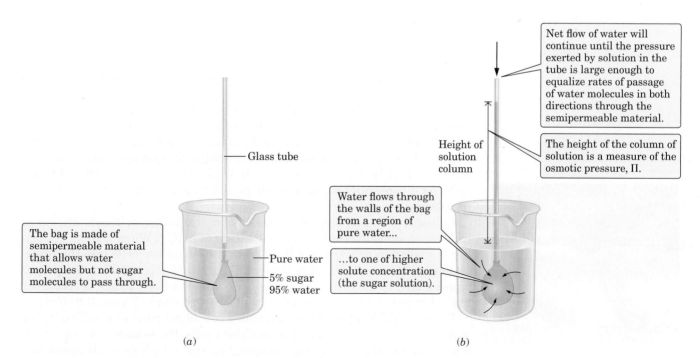

FIGURE 6.15 Demonstration of osmotic pressure.

solution multiplied by the number of particles (i) produced by each formula unit of solute.

$$\text{Osmolarity} = M \times i$$

A 5.5% glucose solution is also isotonic and is used in intravenous feeding.

Example 6.13 Osmolarity

A 0.89 percent w/v NaCl aqueous solution is referred to as a physiological or isotonic saline solution because it has the same concentration of salts as normal human blood. Although blood contains several salts, saline solution has only NaCl. What is the osmolarity of this solution?

Strategy

We are given a 0.89% solution—that is, a solution that contains 0.89 g NaCl per 100. mL of solution. Because osmolarity is based on grams of solute per 1000. grams of solution, we calculate that this solution contains 8.9 g of NaCl per 1000. g of solution. Given this concentration, we can then calculate the molarity of the solution.

Solution

$$\frac{0.89 \text{ g NaCl}}{100. \text{ mL}} \times \frac{1000 \text{ mL}}{1 \text{ L}} \times \frac{1 \text{ mol NaCl}}{58.4 \text{ g NaCl}} = \frac{0.15 \text{ mol NaCl}}{1 \text{ L}} = 0.15 \, M$$

Each formula unit of NaCl dissociates into two particles, namely Na^+ and Cl^-; therefore, the osmolarity is two times the molarity.

$$\text{Osmolarity} = 0.15 \times 2 = 0.30 \text{ osmol}$$

Problem 6.13

What is the osmolarity of a 3.3% w/v Na_3PO_4 solution?

As noted earlier, osmotic pressure is a colligative property. The osmotic pressure generated by a solution across a semipermeable membrane—the difference between the heights of the two columns in Figure 6.15(b)—depends on the osmolarity of the solution. If the osmolarity increases by a factor of 2, the osmotic pressure will also increase by a factor of 2. Osmotic pressure is very important in biological organisms because cell membranes are semipermeable. For example, red blood cells in the body are suspended in a medium called plasma, which must have the same osmolarity as the red blood cells. Two solutions with the same osmolarity are called **isotonic,** so plasma is said to be isotonic with red blood cells. As a consequence, no osmotic pressure is generated across the cell membrane.

Cell-shriveling by osmosis occurs when vegetables or meats are cured in brine (a concentrated aqueous solution of NaCl). When a fresh cucumber is soaked in brine, water flows from the cucumber cells into the brine, leaving behind a shriveled cucumber, Figure 6.16 (*right*). With the proper spices added to the brine, the cucumber becomes a tasty pickle. A cucumber soaked in pure water is affected very little, as shown in Figure 6.16 (*left*).

What would happen if we suspended red blood cells in distilled water instead of in plasma? Inside the red blood cells, the osmolarity is approximately the same as in a physiological saline solution—0.30 osmol (**an isotonic solution**). Distilled water has zero osmolarity. As a consequence, water flows into the red blood cells. The volume of the cells increases, and

FIGURE 6.16 Osmosis and vegetables.

Charles D. Winters

Chemical Connections 6F

Reverse Osmosis and Desalinization

In osmosis, the solvent flows spontaneously from the dilute solution compartment into the concentrated solution compartment. In reverse osmosis, the opposite happens. When we apply pressures greater than the osmotic pressure to the more concentrated solution, solvent flows from it to the more dilute solution by a process we call **reverse osmosis** (Figure 6.17).

Reverse osmosis is used to make drinkable water from seawater or brackish water. In large plants in the Persian Gulf countries, for example, more than 100. atm pressure is applied to seawater containing 35,000. ppm salt. The water that passes through the semipermeable membrane under this pressure contains only 400. ppm salt—well within the limits set by the World Health Organization for drinkable water.

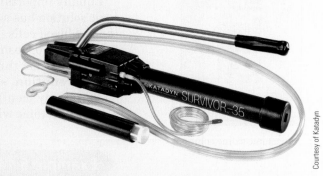

An emergency hand-operated water desalinator that works by reverse osmosis. It can produce 4.5 L of pure water per hour from seawater, which can save someone adrift at sea.

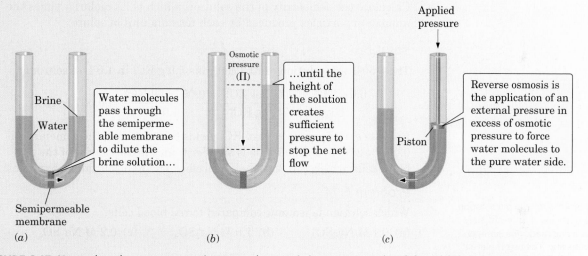

FIGURE 6.17 Normal and reverse osmosis. Normal osmosis is represented in (*a*) and (*b*). Reverse osmosis is represented in (*c*).

Test your knowledge with Problems 6.87 and 6.88.

they swell, as shown in Figure 6.18(b). The membrane cannot resist the osmotic pressure, and the red blood cells eventually burst, spilling their contents into the water. We call this process **hemolysis.**

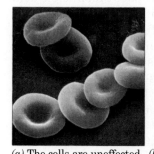

FIGURE 6.18 Red blood cells in solutions of different osmolarity or *tonicity*. Red blood cells in (a) isotonic, (b) hypotonic, and (c) hypertonic solutions.

(*a*) The cells are uneffected. (*b*) The cells swell by hemolysis. (*c*) The cells shrink by crenation.

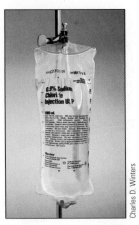

An isotonic saline solution.

Solutions in which the osmolarity (and hence osmotic pressure) is lower than that of suspended cells are called **hypotonic solutions.** Obviously, it is very important that we always use isotonic solutions and never hypotonic solutions in intravenous feeding and blood transfusion. Hypotonic solutions would simply kill the red blood cells by hemolysis.

Equally important, we should not use **hypertonic solutions.** A hypertonic solution has a greater osmolarity (and greater osmotic pressure) than the red blood cells. If red blood cells are placed in a hypertonic solution—for example, 0.5 osmol glucose solution—water flows from the cells into the glucose solution through the semipermeable cell membrane. This process, called **crenation,** shrivels the cells, as shown in Figure 6.18(c).

As already mentioned in Example 6.13, 0.89 w/v% NaCl (physiological saline) is isotonic with red blood cells and is used in intravenous injections.

Example 6.14 Toxicity

Is a 0.50% w/v aqueous solution of KCl (a) hypertonic, (b) hypotonic, or (c) isotonic compared to red blood cells?

Strategy

Calculate the osmolarity of the solution, which is its molarity times the number of particles produced by each formula unit of solute.

Solution

The 0.50% w/v solution of KCl contains 5.0 g KCl in 1.0 L of solution:

$$\frac{5.0 \text{ g KCl}}{1.0 \text{ L}} \times \frac{1.0 \text{ mol KCl}}{74.6 \text{ g KCl}} = \frac{0.067 \text{ mol KCl}}{1.0 \text{ L}} = 0.067 \ M \text{ KCl}$$

Because each formula unit of KCl yields two particles, the osmolarity is $0.067 \times 2 = 0.13$ osmol; this is smaller than the osmolarity of the red blood cells, which is 0.30 osmol. Therefore, the KCl solution is hypotonic.

Problem 6.14

Which solution is isotonic compared to red blood cells?

(a) 0.1 M Na$_2$SO$_4$ (b) 1.0 M Na$_2$SO$_4$ (c) 0.2 M Na$_2$SO$_4$

Dialysis A process in which a solution containing particles of different sizes is placed in a bag made of a semipermeable membrane. The bag is placed into a solvent or solution containing only small molecules. The solution in the bag reaches equilibrium with the solvent outside, allowing the small molecules to diffuse across the membrane but retaining the large molecules.

D. Dialysis

An osmotic semipermeable membrane allows only solvent and not solute molecules to pass. If, however, the openings in the membrane are somewhat larger, then small solute molecules can also pass through, but large solute molecules, such as macromolecular and colloidal particles, cannot. This process is called **dialysis.**

For example, ribonucleic acids are important biological molecules that we will study in Chapter 25. When biochemists prepare ribonucleic acid solutions, they must remove small particles, such as NaCl, from the solution to obtain a pure nucleic acid preparation. To do so, they place the nucleic acid solution in a dialysis bag (made of cellophane) of sufficient pore size to allow all the small particles to diffuse and retain only the large nucleic acid molecules. If the dialysis bag is suspended in flowing distilled water, all NaCl and small particles will leave the bag. After a certain amount of time, the bag will contain only the pure nucleic acids dissolved in water.

Our kidneys work in much the same way. The millions of nephrons, or kidney cells, have very large surface areas in which the capillaries of the

A portable dialysis unit.

Hemodialysis

The kidneys' main function is to remove toxic waste products from the blood. When the kidneys are not functioning properly, these waste products accumulate and may threaten life. **Hemodialysis** is a process that performs the same filtration function (see the figure).

In hemodialysis, the patient's blood circulates through a long tube of cellophane membrane suspended in an isotonic solution and then returns to the patient's vein. The cellophane membrane retains the large particles (for example, proteins) but allows the small ones, including the toxic wastes, to pass through. In this way, dialysis removes wastes from the blood.

If the cellophane tube were suspended in distilled water, other small molecules, such as glucose, and ions, such as Na^+ and Cl^-, would also be removed from the blood. That is something we don't want to happen. The isotonic solution used in hemodialysis consists of 0.6% NaCl, 0.04% KCl, 0.2% $NaHCO_3$, and 0.72% glucose (all w/v). It ensures that no glucose or Na^+ is lost from the blood.

A patient usually remains on an artificial kidney machine for four to seven hours. During this time, the isotonic bath is changed every two hours. Kidney machines allow people with kidney failure to lead a normal life, although they must take these hemodialysis treatments regularly.

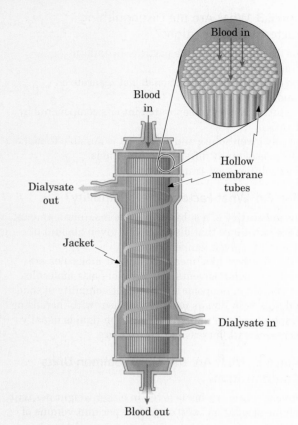

A schematic diagram of the hollow-fiber (or capillary) dialyzer, the most commonly used artificial kidney. During dialysis, blood flows through small tubes constructed of a semipermeable membrane; the tubes themselves are bathed in the dialyzing solution.

Test your knowledge with Problems 6.89 and 6.90.

blood vessels come in contact with the nephrons. The kidneys serve as a gigantic filtering machine. The waste products of the blood dialyse out through semipermeable membranes in the glomeruli and enter collecting tubes that carry the urine to the ureter. Meanwhile, large protein molecules and cells are retained in the blood.

The glomeruli of the kidneys are fine capillary blood vessels in which the body's waste products are removed from the blood.

Summary

OWL Sign in at **www.cengage.com/owl** to develop problem-solving skills and complete online homework assigned by your professor.

Section 6.1 What Do We Need to Know as We Begin?

- Systems containing more than one component are **mixtures.**
- **Homogeneous mixtures** are uniform throughout.

- **Heterogeneous mixtures** exhibit well-defined boundaries between phases.

Section 6.2 What Are the Most Common Types of Solutions?

- The most common types of solutions are gas in liquid, liquid in liquid, solid in liquid, gas in gas, and solid in solid.
- When a solution consists of a solid or gas dissolved in a liquid, the liquid acts as the **solvent,** and the solid

or gas is the **solute.** When one liquid is dissolved in another, the liquid present in greater amount is considered to be the solvent.

Section 6.3 What Are the Distinguishing Characteristics of Solutions?

- The distribution of solute particles is uniform throughout.
- The components of a solution do not separate on standing.
- A solution cannot be separated into its components by filtration.
- For any given solute and solvent, it is possible to make solutions of many different compositions.
- Most solutions are transparent.

Section 6.4 What Factors Affect Solubility?

- The **solubility** of a substance is the maximum amount of the substance that dissolves in a given amount of solvent at a given temperature.
- "Like dissolves like" means that polar molecules are soluble in polar solvents and that nonpolar molecules are soluble in nonpolar solvents. The solubility of solids and liquids in liquids usually increases with increasing temperature; the solubility of gases in liquids usually decreases with increasing temperature.

Section 6.5 What Are the Most Common Units for Concentration?

- Percent concentration is given in either weight per unit volume of solution (w/v) or volume per unit volume of solution (v/v).
- Percent weight/volume (w/v%) is the weight of solute per unit volume of solvent multiplied by 100.
- Percent volume/volume (v/v%) is the volume of solute per unit volume of solution multiplied by 100.
- **Molarity (M)** is the number of moles of solute per liter of solution.

Section 6.6 Why Is Water Such a Good Solvent?

- Water is the most important solvent, because it dissolves polar compounds and ions through hydrogen bonding and dipole–dipole interactions. Hydrated ions are surrounded by water molecules (as a solvation layer) that move together with the ion, cushioning it from collisions with other ions. Aqueous solutions of ions and molten salts are **electrolytes** and conduct electricity.

Section 6.7 What Are Colloids?

- Colloids exhibit a chaotic random motion, called **Brownian motion.** Colloids are stable mixtures despite the relatively large size of the colloidal particles (1 to 1000 nm). The stability results from the solvation layer that cushions the colloidal particles from direct collisions and from the electric charge on the surface of colloidal particles.

Section 6.8 What Is a Colligative Property?

- A **colligative property** is a property of a solution that depends only on the number of solute particles present.
- **Freezing-point depression, boiling-point elevation,** and **osmotic pressure** are examples of colligative properties.
- Osmotic pressure operates across an osmotic semipermeable membrane that allows only solvent molecules to pass but screens out all larger particles. In osmotic pressure calculations, concentration is measured in **osmolarity,** which is the molarity of the solution multiplied by the number of particles produced by dissociation of the solute.
- Red blood cells in a **hypotonic solution** swell and burst, a process called **hemolysis.**
- Red blood cells in a **hypertonic solution** shrink, a process called **crenation.** Some semipermeable membranes allow small solute particles to pass through along with solvent molecules.
- In **dialysis,** such membranes are used to separate larger particles from smaller ones.

Problems

⚙**WL** Interactive versions of these problems may be assigned in OWL.

Orange-numbered problems are applied.

References to previous chapters are given in parentheses.

Section 6.2 What Are the Most Common Types of Solutions?

6.15 Answer true or false.
  (a) A solute is the substance dissolved in a solvent to form a solution.
  (b) A solvent is the medium in which a solute is dissolved to form a solution.
  (c) Some solutions can be separated into their components by filtration.
  (d) Acid rain is a solution.

6.16 Answer true or false.
  (a) Solubility is a physical property like melting point and boiling point.
  (b) All solutions are transparent—that is, you can see through them.
  (c) Most solutions can be separated into their components by physical methods such as distillation and chromatography.

6.17 Vinegar is a homogeneous aqueous solution containing 6% acetic acid. Which is the solvent?

6.18 Suppose you prepare a solution by dissolving glucose in water. Which is the solvent, and which is the solute?

6.19 In each of the following, tell whether the solutes and solvents are gases, liquids, or solids.

(a) Bronze (see Chemical Connections 2E)

(b) Cup of coffee

(c) Car exhaust

(d) Champagne

6.20 Give a familiar example of solutions of each of these types:

(a) Liquid in liquid

(b) Solid in liquid

(c) Gas in liquid

(d) Gas in gas

6.21 Are mixtures of gases true solutions or heterogeneous mixtures? Explain.

Section 6.4 What Factors Affect Solubility?

6.22 Answer true or false.

(a) Water is a good solvent for ionic compounds because water is a polar liquid.

(b) Small covalent compounds dissolve in water if they can form hydrogen bonds with water molecules.

(c) The solubility of ionic compounds in water generally increases as temperature increases.

(d) The solubility of gases in liquids generally increases as temperature increases.

(e) Pressure has little effect on the solubility of liquids in liquids.

(f) Pressure has a major effect on the solubility of gases in liquids.

(g) In general, the greater the pressure of a gas over water, the greater the solubility of the gas in water.

(h) Oxygen, O_2, is insoluble in water.

6.23 We dissolved 0.32 g of aspartic acid in 115.0 mL of water and obtained a clear solution. After it stands for two days at room temperature, we notice a white powder at the bottom of the beaker. What may have happened?

6.24 The solubility of a compound is 2.5 g in 100. mL of aqueous solution at 25°C. If we put 1.12 g of the compound in a 50.-mL volumetric flask at 25°C and add sufficient water to fill it to the 50.-mL mark, what kind of solution do we get—saturated or unsaturated? Explain.

6.25 To a separatory funnel with two layers—the nonpolar diethyl ether and the polar water—is added a small amount of solid. After shaking the separatory funnel, in which layer will we find each of the following solids?

(a) NaCl   (b) Camphor ($C_{10}H_{16}O$)   (c) KOH

6.26 On the basis of polarity and hydrogen bonding, which solute would be the most soluble in benzene, C_6H_6?

(a) CH_3OH (b) H_2O (c) $CH_3CH_2CH_2CH_3$ (d) H_2SO_4

6.27 Suppose that you discover a stain on an oil painting and want to remove it without damaging the painting. The stain is not water-insoluble. Knowing the polarities of the following solvents, which one would you try first and why?

(a) Benzene, C_6H_6

(b) Isopropyl (rubbing) alcohol, C_3H_7OH

(c) Hexane, C_6H_{14}

6.28 Which pairs of liquids are likely to be miscible?

(a) H_2O and CH_3OH (b) H_2O and C_6H_6

(c) C_6H_{14} and CCl_4 (d) CCl_4 and CH_3OH

6.29 The solubility of aspartic acid in water is 0.500 g in 100. mL at 25°C. If we dissolve 0.251 g of aspartic acid in 50.0 mL of water at 50°C and let the solution cool to 25°C without stirring, shaking, or otherwise disturbing the solution, would the resulting solution be a saturated, unsaturated, or supersaturated solution? Explain.

6.30 Near a power plant, warm water is discharged into a river. Sometimes dead fish are observed in the area. Why do fish die in the warm water?

6.31 If a bottle of beer is allowed to stand for several hours after being opened, it becomes "flat" (it loses CO_2). Explain.

6.32 Would you expect the solubility of ammonia gas in water at 2 atm pressure to be:

(a) greater than,   (b) the same as, or

(c) smaller than at 0.5 atm pressure?

Section 6.5 What Are the Most Common Units for Concentration?

6.33 Verify the following statements.

(a) One part per million corresponds to one minute in two years, or a single penny in $10,000.

(b) One part per billion corresponds to one minute in 2000 years, or a single penny in $10 million.

6.34 Describe how we would make the following solutions:

(a) 500.0 mL of a 5.32% w/w H_2S solution in water

(b) 342.0 mL of a 0.443% w/w benzene solution in toluene

(c) 12.5 mL of a 34.2% w/w dimethyl sulfoxide solution in acetone

6.35 Describe how we would prepare the following solutions:

(a) 280. mL of a 27% v/v solution of ethanol, C_2H_6O, in water

(b) 435 mL of a 1.8% v/v solution of ethyl acetate, $C_4H_8O_2$, in water

(c) 1.65 L of an 8.00% v/v solution of benzene, C_6H_6, in chloroform, $CHCl_3$

6.36 Describe how we would prepare the following solutions:

(a) 250 mL of a 3.6% w/v solution of NaCl in water

(b) 625 mL of a 4.9% w/v solution of glycine, $C_2H_5NO_2$, in water

(c) 43.5 mL of a 13.7% w/v solution of Na_2SO_4 in water

(d) 518 mL of a 2.1% w/v solution of acetone, C_3H_6O, in water

6.37 Calculate the w/v percentage of each of these solutes:

(a) 623 mg of casein in 15.0 mL of milk

(b) 74 mg of vitamin C in 250 mL of orange juice

(c) 3.25 g of sucrose in 186 mL of coffee

6.38 Describe how we would prepare 250 mL of 0.10 M NaOH from solid NaOH and water.

6.39 Assuming that the appropriate volumetric flasks are available, describe how we would make these solutions:

(a) 175 mL of a 1.14 M solution of NH_4Br in water

(b) 1.35 L of a 0.825 M solution of NaI in water

(c) 330 mL of a 0.16 M solution of ethanol, C_2H_6O, in water

6.40 What is the molarity of each solution?

(a) 47 g of KCl dissolved in enough water to give 375 mL of solution

(b) 82.6 g of sucrose, $C_{12}H_{22}O_{11}$, dissolved in enough water to give 725 mL of solution

(c) 9.3 g of ammonium sulfate, $(NH_4)_2SO_4$, dissolved in enough water to give 2.35 L of solution

6.41 A teardrop with a volume of 0.5 mL contains 5.0 mg NaCl. What is the molarity of the NaCl in the teardrop?

6.42 The concentration of stomach acid, HCl, is approximately 0.10 M. What volume of stomach acid contains 0.25 mg of HCl?

6.43 The label on a sparkling cider says it contains 22.0 g glucose ($C_6H_{12}O_6$), 190. mg K^+, and 4.00 mg Na^+ per serving of 240. mL of cider. Calculate the molarities of these ingredients in the sparkling cider.

6.44 If 3.18 g $BaCl_2$ is dissolved in enough solvent to make 500.0 mL of solution, what is the molarity of this solution?

6.45 The label on a jar of jam says it contains 13 g of sucrose, $C_{12}H_{22}O_{11}$ per tablespoon (15 mL). What is the molarity of sucrose in the jam?

6.46 A particular toothpaste contains 0.17 g NaF in 75 mL toothpaste. What are the percent w/v and the molarity of NaF in the toothpaste?

6.47 A student has a bottle labeled 0.750% albumin solution. The bottle contains exactly 5.00 mL. How much water must the student add to make the concentration of albumin become 0.125%?

6.48 How many grams of solute are present in each of the following aqueous solutions?

(a) 575 mL of a 2.00 M solution of nitric acid, HNO_3

(b) 1.65 L of a 0.286 M solution of alanine, $C_3H_7NO_2$

(c) 320 mL of a 0.0081 M solution of calcium sulfate, $CaSO_4$

6.49 A student has a stock solution of 30.0% w/v H_2O_2 (hydrogen peroxide). Describe how the student should prepare 250 mL of a 0.25% w/v H_2O_2 solution.

6.50 To make 5.0 L of a fruit punch that contains 10% v/v ethanol, how much 95% v/v ethanol must be mixed with how much fruit juice?

6.51 A pill weighing 325 mg contains the following. What is the concentration of each in ppm?

(a) 12.5 mg Captopril, a medication for high blood pressure

(b) 22 mg Mg^{2+}

(c) 0.27 mg Ca^{2+}

6.52 One slice of enriched bread weighing 80. g contains 70. μg of folic acid. What is the concentration of folic acid in ppm and ppb?

6.53 Dioxin is considered to be poisonous in concentrations above 2 ppb. If a lake containing 1×10^7 L has been contaminated by 0.1 g of dioxin, did the concentration reach a dangerous level?

6.54 An industrial wastewater contains 3.60 ppb cadmium, Cd^{2+}. How many mg of Cd^{2+} could be recovered from a ton (1016 kg) of this wastewater?

6.55 According to the label on a piece of cheese, one serving of 28 g provides the following daily values: 2% of Fe, 6% of Ca, and 6% of vitamin A. The recommended daily allowance (RDA) of each of these nutrients are as follows: 15 mg Fe, 1200 mg Ca, and 0.800 mg vitamin A. Calculate the concentrations of each of these nutrients in the cheese in ppm.

Section 6.6 Why Is Water Such a Good Solvent?

6.56 Answer true or false.

(a) The properties that make water a good solvent are its polarity and its capacity for hydrogen bonding.

(b) When ionic compounds dissolve in water, their ions become solvated by water molecules.

(c) The term "water of hydration" refers to the number of water molecules that surround an ion in aqueous solution.

(d) The term "anhydrous" means "without water."

(e) An electrolyte is a substance that dissolves in water to give a solution that conducts electricity.

(f) In a solution that conducts electricity, cations migrate toward the cathode and anions migrate toward the anode.

(g) Ions must be present in a solution for the solution to conduct electricity.

(h) Distilled water is a nonelectrolyte.

(i) A strong electrolyte is a substance that dissociates completely into ions in aqueous solution.

(j) All compounds that dissolve in water are electrolytes.

6.57 Considering polarities, electronegativities, and similar concepts learned in Chapter 3, classify each of the following as a strong electrolyte, a weak electrolyte, or a nonelectrolyte.

(a) KCl   (b) C_2H_6O (ethanol)   (c) NaOH

(d) HCl   (e) $C_6H_{12}O_6$ (glucose)

6.58 Which of the following would produce the brightest light in the conductance apparatus shown in Figure 6.11?

(a) 0.1 M KCl   (b) 0.1 M $(NH_4)_3PO_4$

(c) 0.5 M sucrose

6.59 Ethanol is very soluble in water. Describe how water dissolves ethanol.

6.60 Predict which of these covalent compounds is soluble in water.

(a) C_2H_6 (b) CH_3OH (c) HF

(d) NH_3 (e) CCl_4

Section 6.7 What Are Colloids?

6.61 Answer true or false.

(a) A colloid is a state of matter intermediate between a solution and a suspension, in which particles are large enough to scatter light but too small to settle out from solution.

(b) Colloidal solutions appear cloudy because the colloidal particles are large enough to scatter visible light.

6.62 A type of car tire is made of synthetic rubber in which carbon black particles of the size of 200–500 nm are randomly dispersed. Because carbon black absorbs light, we do not see any turbidity (that is, a Tyndall effect). Do we consider a tire to be a colloidal system, and if so, what kind? Explain.

6.63 On the basis of Tables 6.1 and 6.2, classify the following systems as homogeneous, heterogeneous, or colloidal mixtures.

(a) Physiological saline solution   (b) Orange juice

(c) A cloud   (d) Wet sand   (e) Soap suds   (f) Milk

6.64 Table 6.2 shows no examples of a gas-in-gas colloidal system. Considering the definition of a colloid, explain why.

6.65 A solution of protein is transparent at room temperature. When it is cooled to 10°C, it becomes turbid. What causes this change in appearance?

6.66 What gives nanotubes their unique optical and electrical properties?

Section 6.8 What Is a Colligative Property?

6.67 Calculate the freezing points of solutions made by dissolving 1.00 mole of each of the following ionic solutes in 1000. g of H_2O.

(a) NaCl   (b) $MgCl_2$

(c) $(NH_4)_2CO_3$ (d) $Al(HCO_3)_3$

6.68 If we add 175 g of ethylene glycol, $C_2H_6O_2$, per 1000. g of water to a car radiator, what will be the freezing point of the solution?

6.69 Methanol, CH_3OH, is used as an antifreeze. How many grams of methanol would you need per 1000. g of water for an aqueous solution to stay liquid at −20.°C?

6.70 In winter, after a snowstorm, salt (NaCl) is spread to melt the ice on roads. How many grams of salt per 1000. g of ice is needed to make it liquid at −5°C?

6.71 A 4 M acetic acid (CH_3COOH) solution lowers the freezing point by −8°C; a 4 M KF solution yields a −15°C freezing-point depression. What can account for this difference?

Osmosis

6.72 In an apparatus using a semipermeable membrane, a 0.005 M glucose (a small molecule) solution yielded an osmotic pressure of 10 mm Hg. What kind of osmotic pressure change would you expect if instead of a semipermeable membrane you used a dialysis membrane?

6.73 In each case, tell which side (if either) rises and why. The solvent is water.

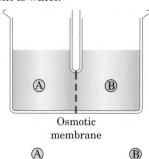

Osmotic membrane

Ⓐ	Ⓑ
(a) 1% glucose	5% glucose
(b) 0.1 M glucose	0.5 M glucose
(c) 1 M NaCl	1 M glucose
(d) 1 M NaCl	1 M K_2SO_4
(e) 3% NaCl	3% KCl
(f) 1 M NaBr	1 M KCl

6.74 An osmotic semipermeable membrane that allows only water to pass separates two compartments, A and B. Compartment A contains 0.9% NaCl, and compartment B contains 3% glycerol, $C_3H_8O_3$.

(a) In which compartment will the level of solution rise?

(b) Which compartment (if either) has the higher osmotic pressure?

6.75 Calculate the osmolarity of each of the following solutions.

(a) 0.39 M Na_2CO_3 (b) 0.62 M $Al(NO_3)_3$

(c) 4.2 M LiBr   (d) 0.009 M K_3PO_4

6.76 Two compartments are separated by a semipermeable osmotic membrane through which only water molecules can pass. Compartment A contains a 0.3 M KCl solution, and compartment B contains a 0.2 M Na_3PO_4 solution. Predict from which compartment the water will flow to the other compartment.

6.77 A 0.9% NaCl solution is isotonic with blood plasma. Which solution would crenate red blood cells?

(a) 0.3% NaCl   (b) 0.9 M glucose (MW 180)

(c) 0.9% glucose

Chemical Connections

6.78 (Chemical Connections 6A) Oxides of nitrogen (NO, NO_2, N_2O_3) are also responsible for acid rain. Which acids can be formed from these nitrogen oxides?

6.79 (Chemical Connections 6A) What makes normal rainwater slightly acidic?

6.80 (Chemical Connections 6B) Why do deep-sea divers use a helium–oxygen mixture in their tanks instead of air?

6.81 (Chemical Connections 6B) What is nitrogen narcosis?

6.82 (Chemical Connections 6C) A solution contains 54 mEq/L of Cl^- and 12 mEq/L of HCO_3^-. If Na^+ is the only cation present in the solution, what is the Na^+ concentration in milliequivalents per liter?

6.83 (Chemical Connections 6C) The concentration of Ca^{2+} ion present in a blood sample is found to be 4.6 mEq/L. How many milligrams of Ca^{2+} ion are present in 250.0 mL of the blood?

6.84 (Chemical Connections 6D) What is the chemical formula for the main component of limestone and marble?

6.85 (Chemical Connections 6D) Write balanced equations (two steps) for the conversion of marble to gypsum dihydrate.

6.86 (Chemical Connections 6E) What is the protective colloid in milk?

6.87 (Chemical Connections 6F) What is the minimum pressure on seawater that will force water to flow from the concentrated solution into the dilute solution?

6.88 (Chemical Connections 6F) The osmotic pressure generated across a semipermeable membrane by a solution is directly proportional to its osmolarity. Given the data in Chemical Connections 6F on the purification of seawater, estimate what pressure you would need to apply to purify brackish water containing 5000. ppm salt by reverse osmosis.

6.89 (Chemical Connections 6G) A manufacturing error occurred in the isotonic solution used in hemodialysis. Instead of 0.2% $NaHCO_3$, 0.2% of $KHCO_3$ was added. Did this error change the labeled tonicity of the solution? If so, is the resulting solution hypotonic or hypertonic? Would such an error create an electrolyte imbalance in the patient's blood? Explain.

6.90 (Chemical Connections 6G) The artificial kidney machine uses a solution containing 0.6% w/v NaCl, 0.04% w/v KCl, 0.2% w/v $NaHCO_3$, and 0.72% w/v glucose. Show that this is an isotonic solution.

Additional Problems

6.91 When a cucumber is put into a saline solution to pickle it, the cucumber shrinks; when a prune is put into the same solution, the prune swells. Explain what happens in each case.

6.92 A solution of As_2O_3 has a molarity of $2 \times 10^{-5}\ M$. What is this concentration in ppm? (Assume that the density of the solution is 1.00 g/mL.)

6.93 Two bottles of water are carbonated, with CO_2 gas being added, under 2 atm pressure and then capped. One bottle is stored at room temperature; the other is stored in the refrigerator. When the bottle stored at room temperature is opened, large bubbles escape, along with a third of the water. The bottle stored in the refrigerator is opened without frothing or bubbles escaping. Explain.

6.94 How many grams of ethylene glycol must be added to 1000. g of water to create an automobile radiator coolant mixture that will not freeze at $-15°C$?

6.95 Both methanol, CH_3OH, and ethylene glycol, $C_2H_6O_2$, are used as antifreeze. Which is more efficient—that is, which produces a lower freezing point if equal weights of each are added to the same weight of water?

6.96 We know that a 0.89% saline (NaCl) solution is isotonic with blood. In a real-life emergency, you run out of physiological saline solution and have only KCl as a salt and distilled water. Would it be acceptable to make a 0.89% aqueous KCl solution and use it for intravenous infusion? Explain.

6.97 Carbon dioxide and sulfur dioxide are soluble in water because they react with water. Write possible equations for these reactions.

6.98 A reagent label shows that the reagent contains 0.05 ppm lead as a contaminant. How many grams of lead are present in 5.0 g of the reagent?

6.99 A concentrated nitric acid solution contains 35% HNO_3. How would we prepare 300. mL of 4.5% solution?

6.100 Which will have greater osmotic pressure?

    (a) A 0.9% w/v NaCl solution

    (b) A 25% w/v solution of a nondissociating dextran with a molecular weight of 15,000.

6.101 Government regulations permit a 6 ppb concentration of a certain pollutant. How many grams of pollutant are allowed in 1 ton (1016 kg) of water?

6.102 The average osmolarity of seawater is 1.18. How much pure water would have to be added to 1.0 mL of seawater for it to achieve the osmolarity of blood (0.30 osmol)?

6.103 A swimming pool containing 20,000. L of water is chlorinated to have a final Cl_2 concentration of 0.00500 M. What is the Cl_2 concentration in ppm? How many kilograms of Cl_2 were added to the swimming pool to reach this concentration?

6.104 The density of a solution that is 20.0% $HClO_4$ is 1.138 g/mL. Calculate the molarity of the solution.

6.105 A 10.0% H_2SO_4 solution has a density of 1.07 g/mL. How many milliliters of solution contain 8.37 g of H_2SO_4?

Looking Ahead

6.106 Synovial fluid that exists in joints is a colloidal solution of hyaluronic acid (Section 20.6A) in water. To isolate hyaluronic acid from synovial fluid, a biochemist adds ethanol, C_2H_6O, to bring the solution to 65% ethanol. The hyaluronic acid precipitates upon standing. What makes the hyaluronic acid solution unstable and causes it to precipitate?

Challenge Problems

6.107 A solution is made by dissolving 25.0 g of magnesium chloride crystals in 1000. g of water.

    (a) What will be the freezing point of the new solution assuming complete dissociation of the $MgCl_2$ salt?

    (b) Determine the boiling point of the new solution assuming complete dissociation of the $MgCl_2$ salt.

6.108 Explain why saltwater fish do not survive when they are suddenly transferred to a freshwater aquarium.

6.109 Consider the reaction of 1.46 g Ca(s) with 115 mL of 0.325 M HBr(aq) according to the following unbalanced chemical equation:

$$Ca(s) + HBr(aq) \longrightarrow CaBr_2(aq) + H_2(g)$$

The hydrogen produced was collected by displacement of water at 22°C with a total pressure of 754 torr.

(a) Which reactant is the limiting reagent? (Chapter 4)

(b) Determine the volume (in L) of hydrogen gas produced if the vapor pressure of water at 22°C is 21 torr. (Chapter 5)

(c) How many grams of the other reactant are left over? (Chapter 4)

6.110 Vitamin B$_2$, riboflavin, is a nondissociating molecular compound soluble in water. If 370.3 g of riboflavin is dissolved in 1000.0 g of water, the resulting solution has a freezing point of −1.83°C.

(a) What is the molar mass of riboflavin? (Chapter 4)

(b) Consider the skeletal structure of riboflavin, where all the bonded atoms are shown but double bonds, triple bonds, and/or lone pairs are missing. Complete the structure as shown below. (Chapter 3)

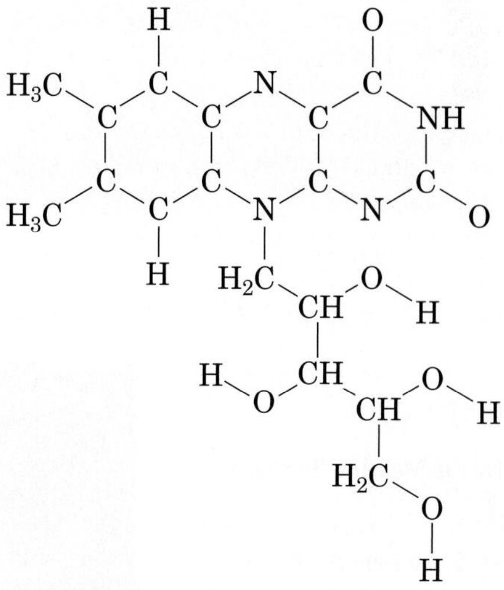

Riboflavin

7

Reaction Rates and Chemical Equilibrium

Sign in to OWL at **www.cengage.com/owl** to view tutorials and simulations, develop problem-solving skills, and complete online homework assigned by your professor.

In the course of several years, a few molecules of glucose and O_2 will react, but not enough for us to detect within a laboratory period.

Charles D. Winters

When magnesium burns in air, it creates a brilliant white light.

7.1 How Do We Measure Reaction Rates?

In this chapter, we are going to look at two closely related topics—reaction rates and chemical equilibrium. Knowing whether a reaction takes place quickly or slowly can give important information about the process in question. If the process has health implications, the information can be especially crucial. Sooner or later, many reactions will appear to stop, but that simply means that two reactions that are the reverse of each other are proceeding at the same rate. When this is the case, the reaction is said to be at equilibrium. The study of chemical equilibrium gives information about how to control reactions, including those that play key roles in life processes.

Some chemical reactions take place rapidly; others are very slow. For example, glucose and oxygen gas react with each other to form water and carbon dioxide:

$$C_6H_{12}O_6(s) + 6O_2(g) \longrightarrow 6CO_2(g) + 6H_2O(\ell)$$
Glucose

This reaction is extremely slow, however. A sample of glucose exposed to O_2 in the air shows no measurable change even after many years.

In contrast, consider what happens when you take one or two aspirin tablets for a slight headache. Very often, the pain disappears in half an hour or so. Thus, the aspirin must have reacted with compounds in the body within that time.

Many reactions occur even faster. For example, if we add a solution of silver nitrate to a solution of sodium chloride (NaCl), a precipitate of silver chloride (AgCl) forms almost instantaneously.

$$\text{Net ionic equation: } Ag^+(aq) + Cl^-(aq) \longrightarrow AgCl(s)$$

The precipitation of AgCl is essentially complete in considerably less than 1 s.

The study of reaction rates is called **chemical kinetics.** The **rate of a reaction** is the change in concentration of a reactant (or product) per unit time. Every reaction has its own rate, which must be measured in the laboratory.

Chemical kinetics The study of the rates of chemical reactions

Consider the following reaction carried out in the solvent acetone:

$$CH_3\text{—}Cl + I^- \xrightarrow{\text{Acetone}} CH_3\text{—}I + Cl^-$$
$$\text{Chloromethane} \qquad\qquad \text{Iodomethane}$$

To determine the reaction rate, we can measure the concentration of the product, iodomethane, in the acetone at periodic time intervals—say, every 10 min. For example, the concentration might increase from 0 to 0.12 mol/L over a period of 30 min. The rate of the reaction is the change in the concentration of iodomethane divided by the time interval:

$$\frac{(0.12 \text{ mol } CH_3I/L) - (0 \text{ mol } CH_3I/L)}{30 \text{ min}} = \frac{0.0040 \text{ mol } CH_3I/L}{\text{min}}$$

The rate could also be determined by following the decrease in concentration of CH_3Cl or of I^-, if that is more convenient.

This unit is read "0.0040 mole per liter per minute." During each minute of the reaction, an average of 0.0040 mol of chloromethane is converted to iodomethane for each liter of solution.

The rate of a reaction is not constant over a long period of time. At the beginning, in most reactions, the change in concentration is directly proportional to time. This period is shown as the linear portion of the graph in Figure 7.1. The rate calculated during this period, called the **initial rate,** is constant during this time interval. Later, as the reactant is used up, the rate of reaction decreases. Figure 7.1 shows a rate determined at a later time as well as the initial rate. The rate determined later is less than the initial rate.

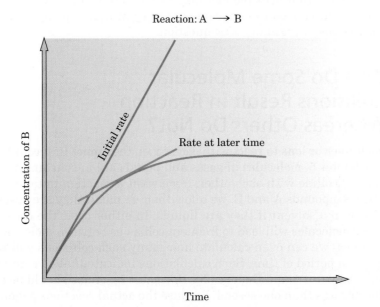

Reaction: A $\longrightarrow$ B

Initial rate

Rate at later time

Concentration of B

Time

FIGURE 7.1 Changes in the concentration of B in the A → B system with respect to time. The rate (the change in concentration of B per unit time) is largest at the beginning of the reaction and gradually decreases until it reaches zero at the completion of the reaction.

◉WL **Interactive Example 7.1** Reaction Rate

Another way to determine the rate of the reaction of chloromethane with iodide ion is to measure the disappearance of I^- from the solution. Suppose that the concentration of I^- was 0.24 mol I^-/L at the start of the reaction. At the end of 20 min, the concentration dropped to 0.16 mol I^-/L. This difference is equal to a change in concentration of 0.08 mol I^-/L. What is the rate of reaction?

Strategy

We use the definition of rate as the change in concentration in a unit of time. We can determine the change in concentration by subtraction. The time interval is given.

Solution

The rate of the reaction is:

$$\frac{(0.16 \text{ mol } I^-/L) - (0.24 \text{ mol } I^-/L)}{20 \text{ min}} = \frac{-0.0040 \text{ mol } I^-/L}{\text{min}}$$

Because the stoichiometry of the components is 1 : 1 in this reaction, we get the same numerical answer for the rate whether we monitor a reactant or a product. Note, however, that when we measure the concentration of a reactant that disappears with time, the rate of reaction is a negative number.

Problem 7.1

In the reaction

$$2HgO(s) \longrightarrow 2Hg(\ell) + O_2(g)$$

we measure the evolution of oxygen gas to determine the rate of reaction. At the beginning of the reaction (at 0 min), 0.020 L of O_2 is present. After 15 min, the volume of O_2 gas is 0.35 L. What is the rate of reaction?

Don't be confused by the use of the negative sign on the reaction rate. It implies that the reactant is being used up.

The rates of chemical reactions—both the ones that we carry out in the laboratory and the ones that take place inside our bodies—are very important. A reaction that goes more slowly than we need may be useless, whereas a reaction that goes too fast may be dangerous. Ideally, we would like to know what causes the enormous variety in reaction rates. In the next three sections, we examine this question.

7.2 Why Do Some Molecular Collisions Result in Reaction Whereas Others Do Not?

For two molecules or ions to react with each other, they must first collide. As we saw in Chapter 5, molecules in gases and liquids are in constant motion and frequently collide with each other. If we want a reaction to take place between two compounds A and B, we allow them to mix if they are gases or dissolve them in a solvent if they are liquids. In either case, the constant motion of the molecules will lead to frequent collisions between molecules of A and B. In fact, we can even calculate how many such collisions will take place in a given period of time. Such calculations indicate that so many collisions occur between A and B molecules that most reactions should be over in considerably less than one second. Because the actual reactions generally proceed much more slowly, we must conclude that most collisions do not result in a reaction. Typically, when a molecule of A collides with a molecule of B, the two simply bounce apart without reacting. Every once in awhile, though, molecules of A and B collide and react to form a new compound. A collision that results in a reaction is called an **effective collision.**

Why are some collisions effective whereas others are not? There are three main reasons:

1. In most cases, for a reaction to take place between A and B, one or more covalent bonds must be broken in A or B or both, and energy is required for this to happen. The energy comes from the collision between A and B. If the energy of the collision is large enough, bonds will break and a reaction will take place. If the collision energy is too low, the molecules will bounce apart without reacting. The minimum energy necessary for a reaction to occur is called the **activation energy.**

   The energy of any collision depends on the relative speeds (that is, on the relative kinetic energies) of the colliding objects and on their angle of approach. Much greater damage is done in a head-on collision of two cars both going 40 mi/h than in a collision in which a car going 20 mi/h sideswipes one going 10 mi/h. The same consideration applies with molecules, as Figure 7.2 shows.

Effective collision A collision between two molecules or ions that results in a chemical reaction

Activation energy The minimum energy necessary to cause a chemical reaction to occur

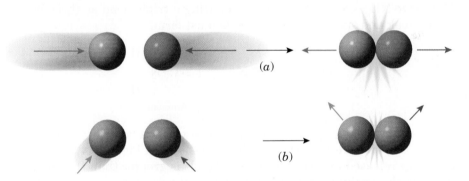
(a)
(b)

FIGURE 7.2 The energy of molecular collisions varies. (*a*) Two fast-moving molecules colliding head-on have a higher collision energy than (*b*) two slower-moving molecules colliding at an angle.

2. Even if two molecules collide with an energy greater than the activation energy, a reaction may not take place if the molecules are not oriented properly when they collide. Consider, for example, the reaction between H_2O and HCl:

$$H_2O(\ell) + HCl(g) \longrightarrow H_3O^+(aq) + Cl^-(aq)$$

For this reaction to take place, the molecules must collide in such a way that the H of the HCl hits the O of the water, as shown in Figure 7.3(a). A collision in which the Cl hits the O, as shown in Figure 7.3(b), cannot lead to a reaction, even if sufficient energy is available.

3. The frequency of collisions is another important factor. If more collisions take place, the chances are that more of them will have sufficient energy and the proper orientation of molecules for a reaction to take place.

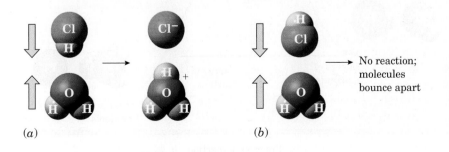
(a)
(b)
No reaction; molecules bounce apart

FIGURE 7.3 Molecules must be properly oriented for a reaction to take place. (*a*) HCl and H_2O molecules are oriented so that the H of HCl collides with the O of H_2O, and a reaction takes place. (*b*) No reaction takes place because Cl, and not H, collides with the O of H_2O. The colored arrows show the path of the molecules.

Returning to the example given at the beginning of this chapter, we can now see why the reaction between glucose and O_2 is so slow. The O_2 molecules are constantly colliding with glucose molecules, but the percentage of effective collisions is extremely tiny at room temperature.

7.3 What Is the Relationship Between Activation Energy and Reaction Rate?

Figure 7.4 shows a typical energy diagram for an exothermic reaction. The products have a lower energy than the reactants; we might, therefore, expect the reaction to take place rapidly. As the curve shows, however, the reactants cannot be converted to products without the necessary activation energy. The activation energy is like a hill. If we are in a mountainous region, we may find that the only way to go from one point to another is to climb over a hill. It is the same in a chemical reaction. Even though the products may have a lower energy than the reactants, the products cannot form unless the reactants "go over the hill" or over a high pass—that is, they must possess the necessary activation energy.

Let us look into this issue more closely. In a typical reaction, existing bonds are broken and new bonds form. For example, when H_2 reacts with N_2 to give NH_3, six covalent bonds (counting a triple bond as three bonds) must break and six new covalent bonds must form.

$$3H-H + N\equiv N \longrightarrow 2H-N\overset{\displaystyle H}{\underset{\displaystyle H}{\big\langle}}$$

Ammonia

Breaking a bond requires an input of energy, but a bond forming releases energy. In a "downhill" reaction of the type shown in Figure 7.4, the amount of energy released in creating the new bonds is greater than that required to break the original bonds. In other words, the reaction is exothermic. Yet it may well have a substantial activation energy, or energy barrier, because in most cases, at least one bond must break before any new bonds can form. Thus, energy must be put into the system before we get any back. This is

FIGURE 7.4 Energy diagram for the exothermic reaction.

$H_2O(l) + HCl(g) \longrightarrow$
$\qquad\qquad H_3O^+(aq) + Cl^-(aq)$

The energy of the reactants is greater than the energy of the products. The diagram shows the positions of all atoms before, at, and after the transition state.

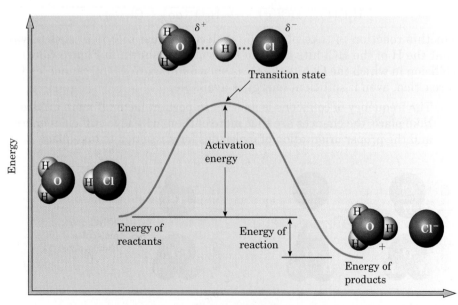

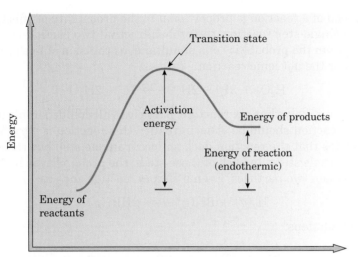

FIGURE 7.5 Energy diagram for an endothermic reaction. The energy of the products is greater than that of the reactants.

analogous to the following situation: Somebody offers to let you buy into a business from which, for an investment of $10,000, you could get an income of $40,000 per year, beginning in one year. In the long run, you would do very well. First, however, you need to put up the initial $10,000 (the activation energy) to start the business.

"Uphill" reactions are endothermic.

Notice that we just used an analogy for discussing energy by comparing energy changes to dollar amounts. Analogies can be useful to a point, but, at times, they are not enough. This point is especially true when we need exact information. Being precise in terminology is highly useful when we talk about energy changes, especially in view of the fact that scientists have developed a number of ways for describing transformations of energy under different conditions.

Every reaction has a different energy diagram. Sometimes, the energy of the products is higher than that of the reactants (Figure 7.5); that is, the reaction is "uphill". For almost all reactions, however, there is an energy "hill"—the activation energy. The activation energy is inversely related to the rate of the reaction. The lower the activation energy, the faster the reaction; the higher the activation energy, the slower the reaction.

The top of the hill on an energy diagram is called the **transition state.** When the reacting molecules reach this point, one or more original bonds are partially broken and one or more new bonds may be in the process of formation. The transition state for the reaction of iodide ion with chloromethane occurs as an iodide ion collides with a molecule of chloromethane in such a way that the iodide ion approaches the carbon atom (Figure 7.6).

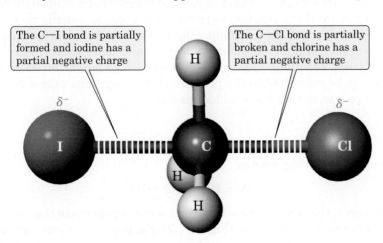

The C—I bond is partially formed and iodine has a partial negative charge

The C—Cl bond is partially broken and chlorine has a partial negative charge

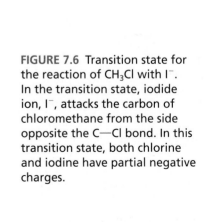

FIGURE 7.6 Transition state for the reaction of CH_3Cl with I^-. In the transition state, iodide ion, I^-, attacks the carbon of chloromethane from the side opposite the C—Cl bond. In this transition state, both chlorine and iodine have partial negative charges.

The speed of a reaction is proportional to the probability of effective collisions. In a single-step reaction, the probability that two particles will collide is greater than the probability of a simultaneous collision of five particles. If you consider the net ionic reaction

$$H_2O_2 + 3I^- + 2H^+ \longrightarrow I_3^- + 2H_2O$$

it is highly unlikely that six reactant particles will collide simultaneously; thus, this reaction should be slow. In reality, this reaction is very fast. This fact indicates that the reaction does not occur in one step but rather takes place in multiple steps. In each of those steps, the probability is high for collisions between two particles. Even a simple reaction such as

$$H_2(g) + Br_2(g) \longrightarrow 2HBr(g)$$

occurs in four steps:

$$\text{Step 1:} \quad Br_2 \xrightarrow{\text{slow}} 2Br\cdot$$

$$\text{Step 2:} \quad Br\cdot + H_2 \xrightarrow{\text{fast}} HBr + H\cdot$$

$$\text{Step 3:} \quad H\cdot + Br_2 \xrightarrow{\text{fast}} HBr + Br\cdot$$

$$\text{Step 4:} \quad Br\cdot + Br\cdot \xrightarrow{\text{fast}} Br_2$$

The dot ($\cdot$) indicates the single unpaired electron in the atom. The overall rate of the reaction will be controlled by the slowest of the four steps, just as the slowest-moving car controls the flow of traffic on a street. In the preceding reaction, step 1 is the slowest, because it has the highest activation energy.

7.4 How Can We Change the Rate of a Chemical Reaction?

In Section 7.2, we saw that reactions occur as a result of collisions between fast-moving molecules possessing a certain minimum energy (the activation energy). In this section, we examine some of the factors that affect activation energies and reaction rates.

A. Nature of the Reactants

In general, reactions that take place between ions in aqueous solution (Section 4.3) are extremely rapid, occurring almost instantaneously. Activation energies for these reactions are very low because usually no covalent bonds must be broken. As we might expect, reactions between covalent molecules, whether in aqueous solution or not, take place much more slowly. Many of these reactions require 15 min to 24 h or longer for most of the reactants to be converted to the products. Some reactions take a good deal longer, of course, but they are seldom useful.

B. Concentration

Consider the following reaction:

$$A + B \longrightarrow C + D$$

For reactions in the gas phase, an increase in pressure usually increases the rate.

In most cases, the reaction rate increases when we increase the concentration of either or both reactants (Figure 7.7). For many reactions—though by

Charles D. Winters

Leon Lewandowski

(a)

(b)

FIGURE 7.7 The reaction of steel wool with oxygen. (*a*) When heated in air, steel wool glows but does not burn rapidly because the concentration of O_2 in the air is only about 20%. (*b*) When the glowing steel wool is put into 100% O_2, it burns vigorously.

no means all—a direct relationship exists between concentration and reaction rate; that is, when the concentration of a reactant is doubled, the reaction rate also doubles. This outcome is easily understandable on the basis of the collision theory. If we double the concentration of A, there are twice as many molecules of A in the same volume, so the molecules of B in that volume now collide with twice as many A molecules per second than before. Given that the reaction rate depends on the number of effective collisions per second, the rate doubles.

We can express the relationship between rate and concentration mathematically. For example, for the reaction

$$2H_2O_2(\ell) \longrightarrow 2H_2O(\ell) + O_2(g)$$

the rate was determined to be -0.01 mol H_2O_2/L/min at a constant temperature when the initial concentration of H_2O_2 was 1 mol/L. In other words, every minute 0.01 mol/L of hydrogen peroxide was used up. Researchers also found that every time the concentration of H_2O_2 was doubled, the rate also doubled. Thus, the rate is directly proportional to the concentration of H_2O_2. We can write this relationship as

$$\text{Rate} = k[H_2O_2]$$

where k is a constant, called the **rate constant.** Rate constants are usually calculated from the **initial rates of reaction** and corresponding initial concentrations (Figure 7.1).

In the case where one of the reactants is a solid, the rate is affected by the surface area of the solid. For this reason, a substance in powder form reacts faster than the same substance in the form of large chunks.

The brackets [ ] stand for the molar concentration of the chemical species whose formula is between the brackets.

Rate constant A proportionality constant, k, between the molar concentration of reactants and the rate of reaction; rate = k [compound]

Example 7.2 Rate Constants

Calculate the rate constant, k, for the reaction

$$2H_2O_2(\ell) \longrightarrow 2H_2O(\ell) + O_2(g)$$

using the rate and the initial concentration mentioned in the preceding discussion:

$$\frac{-0.01 \text{ mol } H_2O_2}{L \cdot \text{min}} \qquad [H_2O_2] = \frac{1 \text{ mol}}{L}$$

Strategy and Solution

We start with the rate equation, solve it for k, and then insert the appropriate experimental values.

$$\text{Rate} = k[\text{H}_2\text{O}_2]$$

$$k = \frac{\text{Rate}}{[\text{H}_2\text{O}_2]}$$

$$= \frac{-0.01 \text{ mol } \text{H}_2\text{O}_2}{\text{L} \cdot \text{min}} \times \frac{\text{L}}{1 \text{ mol } \text{H}_2\text{O}_2}$$

$$= \frac{-0.01}{\text{min}}$$

Note that all the concentration units cancel and that the rate constant has units that indicate some event in a given time, which makes sense. The answer is also a reasonable number.

Problem 7.2

Calculate the rate for the reaction in Example 7.2 when the initial concentration of H_2O_2 is 0.36 mol/L.

C. Temperature

In virtually all cases, reaction rates increase with increasing temperature. A rule of thumb for many reactions is that every time the temperature goes up by 10°C, the rate of reaction doubles. This rule is far from exact, but it is not far from the truth in many cases. As you can see, this effect can be quite large. It says, for example, that if we run a reaction at 90°C instead of at room temperature (20°C), the reaction will go about 128 times faster. There are seven 10° increments between 20°C and 90°C, and $2^7 = 128$. Put another way, if it takes 20 h to convert 100 g of reactant A to product C at 20°C, then it would take only 10 min at 90°C. Temperature, therefore, is a powerful tool that lets us increase the rates of reactions that are inconveniently slow. It also lets us decrease the rates of reactions that are inconveniently fast. For example, we might choose to run reactions at low temperatures because explosions might result or the reactions would otherwise be out of control at room temperature.

What causes reaction rates to increase with increasing temperature? Once again, we turn to collision theory. Here temperature has two effects:

1. In Section 5.6, we learned that temperature is related to the average kinetic energy of molecules. When the temperature increases, molecules move more rapidly, which means that they collide more frequently. More frequent collisions mean higher reaction rates. However, this factor is much less important than the second factor.

2. Recall from Section 7.2 that a reaction between two molecules takes place only if an effective collision occurs—a collision with an energy equal to or greater than the activation energy. When the temperature increases, not only is the average speed (kinetic energy) of the molecules greater, but there is also a different distribution of speeds. The number of very fast molecules increases much more than the number with the average speed (Figure 7.8). As a consequence, the number of effective collisions rises even more than the total number of collisions. Not only do more collisions take place, but the percentage of collisions

In this dish, chloride ion, Cl^-, acts as a catalyst for the decomposition of NH_4NO_3.

Charles D. Winters

Chemical Connections 7A

Why High Fever Is Dangerous

Chemical Connections 1B points out that a sustained body temperature of 41.7°C (107°F) is invariably fatal. We can now see why a high fever is dangerous. Normal body temperature is 37°C (98.6°F), and all the many reactions in the body—including respiration, digestion, and the synthesis of various compounds—take place at that temperature. If an increase of 10°C causes the rates of most reactions to approximately double, then an increase of even 1°C makes them go significantly faster than normal.

Fever is a protective mechanism, and a small increase in temperature allows the body to kill germs faster by mobilizing the immune defense mechanism. This increase must be small, however: A rise of 1°C brings the temperature to 38°C (100.4°F); a rise of 3°C brings it to 40°C (104°F). A temperature higher than 104°F increases reaction rates to the danger point.

One can easily detect the increase in reaction rates when a patient has a high fever. The pulse rate increases and breathing becomes faster as the body attempts to supply increased amounts of oxygen for the accelerated

An overheated runner is at risk of serious health problems.

reactions. A marathon runner, for example, may become overheated on a hot and humid day. After a time, perspiration can no longer cool his or her body effectively, and the runner may suffer hyperthermia or heat stroke, which, if not treated properly, can cause brain damage.

Test your knowledge with Problems 7.42 and 7.43.

that have an energy greater than the activation energy also rises. This factor is mainly responsible for the sharp increase in reaction rates with increasing temperature.

D. Presence of a Catalyst

Any substance that increases the rate of a reaction without itself being used up is called a **catalyst.** Many catalysts are known—some that increase the rate of only one reaction and others that can affect several reactions. Although we have seen that we can speed up reactions by increasing the temperature, in some cases they remain too slow even at the highest

Catalyst A substance that increases the rate of a chemical reaction by providing an alternative pathway with a lower activation energy

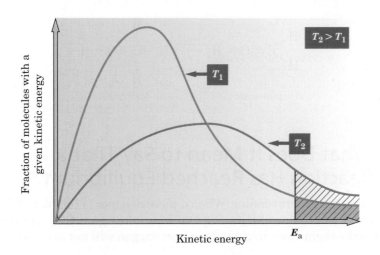

FIGURE 7.8 Distribution of kinetic energies (molecular velocities) at two temperatures. The kinetic energy on the *x*-axis designated E_a indicates the energy (molecular velocity) necessary to pass through the activation energy barrier. The shaded areas represent the fraction of molecules that have kinetic energies (molecular velocities) greater than the activation energy.

FIGURE 7.9 Energy diagram for a catalyzed reaction. The dashed line shows the energy curve for the uncatalyzed process. The catalyst provides an alternative pathway whose activation energy is lower.

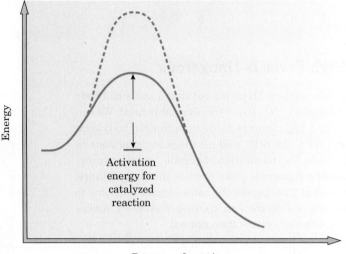

Progress of reaction

Heterogeneous catalyst A catalyst in a separate phase from the reactants—for example, the solid platinum, $Pt(s)$, in the reaction between $CH_2O(g)$ and $H_2(g)$

Homogeneous catalyst A catalyst in the same phase as the reactants—for example, enzymes in body tissues

temperatures we can conveniently reach. In other cases, it is not feasible to increase the temperature—perhaps because other unwanted reactions would be speeded up, too. In such cases, a catalyst, if we can find the right one for a given reaction, can prove very valuable. Many important industrial processes rely on catalysts (see Chemical Connections 7E), and virtually all reactions that take place in living organisms are catalyzed by enzymes (Chapter 22).

Catalysts work by allowing the reaction to take a different pathway, one with a lower activation energy. Without the catalyst, the reactants would have to get over the higher energy hill shown in Figure 7.9. The catalyst provides a lower hill. As we have seen, a lower activation energy means a faster reaction rate.

Each catalyst has its own way of providing an alternative pathway. Many catalysts provide a surface on which the reactants can meet. For example, the reaction between formaldehyde (HCHO) and hydrogen (H_2) to give methanol (CH_3OH) goes so slowly without a catalyst that it is not practical, even if we increase the temperature to a reasonable level. If the mixture of gases is shaken with finely divided platinum metal, however, the reaction takes place at a convenient rate (Section 12.6D). The formaldehyde and hydrogen molecules meet each other on the surface of the platinum, where the proper bonds can be broken and new bonds form and the reaction can proceed:

We often write the catalyst over or under the arrow.

$$\underset{\text{Formaldehyde}}{\overset{H}{\underset{H}{>}}C{=}O + H_2} \xrightarrow{\text{Pt}} \underset{\text{Methanol}}{H{-}\overset{\overset{\displaystyle H}{|}}{\underset{\underset{\displaystyle H}{|}}{C}}{-}O{-}H}$$

7.5 What Does It Mean to Say That a Reaction Has Reached Equilibrium?

Many reactions are irreversible. When a piece of paper is completely burned, the products are CO_2 and H_2O. Anyone who takes pure CO_2 and H_2O and tries to make them react to give paper and oxygen will not succeed.

Chemical Connections 7B

The Effects of Lowering Body Temperature

Like a significant increase in body temperature, a substantial decrease in body temperature below 37°C (98.6°F) can prove harmful because reaction rates are abnormally low. It is sometimes possible to take advantage of this effect. In some heart operations, for example, it is necessary to stop the flow of oxygen to the brain for a considerable time. At 37°C (98.6°F), the brain cannot survive without oxygen for longer than about 5 min without suffering permanent damage. When the patient's body temperature is deliberately lowered to about 28 to 30°C (82.4 to 86°F), however, the oxygen flow can be stopped for a considerable time without causing damage because reaction rates slow down. At 25.6°C (78°F), the body's oxygen consumption is reduced by 50%.

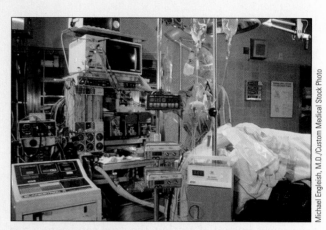

Michael Engleish, M.D./Custom Medical Stock Photo

An operating table unit monitors a patient packed in ice.

Test your knowledge with Problems 7.43 and 7.44.

A tree, of course, turns CO_2 and H_2O into wood and oxygen, and we, in sophisticated factories, make paper from the wood. These activities are not the same as directly combining CO_2, H_2O, and energy in a single process to get paper and oxygen, however. Therefore, we can certainly consider the burning of paper to be an irreversible reaction.

Other reactions are reversible. A **reversible reaction** can be made to go in either direction. For example, if we mix carbon monoxide with water in the gas phase at a high temperature, carbon dioxide and hydrogen are produced:

$$CO(g) + H_2O(g) \longrightarrow CO_2(g) + H_2(g)$$

If we desire, we can also make this reaction take place the other way. That is, we can mix carbon dioxide and hydrogen to get carbon monoxide and water vapor:

$$CO_2(g) + H_2(g) \longrightarrow CO(g) + H_2O(g)$$

Let us see what happens when we run a reversible reaction. We will add some carbon monoxide to water vapor in the gas phase. The two compounds begin to react at a certain rate (the forward reaction):

$$CO(g) + H_2O(g) \longrightarrow CO_2(g) + H_2(g)$$

As the reaction proceeds, the concentrations of CO and H_2O gradually decrease because both reactants are being used up. In turn, the rate of the reaction gradually decreases because it depends on the concentrations of the reactants (Section 7.4B).

But what is happening in the other direction? Before we added the carbon monoxide, no carbon dioxide or hydrogen was present. As soon as the forward reaction began, it produced small amounts of these substances, and we now have some CO_2 and H_2. These two compounds will now, of course, begin reacting with each other (the reverse reaction):

$$CO_2(g) + H_2(g) \longrightarrow CO(g) + H_2O(g)$$

Timed-Release Medication

It is often desirable that a particular medicine act slowly and maintain its action evenly in the body for 24 h. We know that a solid in powder form reacts faster than the same weight in pill form because the powder has a greater surface area at which the reaction can take place. To slow the reaction and to have the drug be delivered evenly to the tissues, pharmaceutical companies coat beads of some of their drugs. The coating prevents the drug from reacting for a time. The thicker the coating, the longer it takes the drug to react. A drug with a smaller bead size has more surface area than a drug with a larger bead size; hence, drugs packaged in a smaller bead size will react more rapidly. By combining the proper bead size with the proper amount of coating, the drug can be designed to deliver its effect over a 24-h period. In this way, the patient needs to take only one pill per day.

Coating can also prevent problems related to stomach irritation. For example, aspirin can cause stomach ulceration or bleeding in some people. Enteric (from the Greek

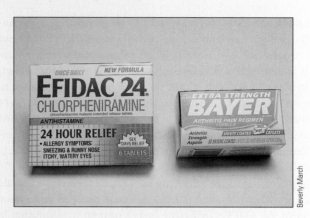

Two packages of timed-release medications.

enteron, which means affecting the intestines) coated aspirin tablets have a polymeric coat that is acid-resistant. Such a drug does not dissolve until it reaches the intestines, where it causes no harm.

Test your knowledge with Problem 7.45.

At first, the reverse reaction is very slow. As the concentrations of H_2 and CO_2 (produced by the forward reaction) gradually increase, the rate of the reverse reaction also gradually increases.

We have a situation, then, in which the rate of the forward reaction gradually decreases, while the rate of the reverse reaction (which began at zero) gradually increases. Eventually the two rates become equal. At this point, the process is in **dynamic equilibrium** (or just **equilibrium**).

Dynamic equilibrium A state in which the rate of the forward reaction equals the rate of the reverse reaction

$$CO_2(g) + H_2(g) \underset{\text{reverse}}{\overset{\text{forward}}{\rightleftharpoons}} CO(g) + H_2(g)$$

We use a double arrow to indicate that a reaction is reversible.

What happens in the reaction container once we reach equilibrium? If we measure the concentrations of the substances in the container, we find that no change in concentration takes place after equilibrium is reached (Figure 7.10). Whatever the concentrations of all the substances are at equilibrium, they remain the same forever unless something happens to disturb the equilibrium (as discussed in Section 7.7). This does not mean that all the concentrations must be the same—all of them can, in fact, be different and usually are—but it does mean that, whatever they are, they no longer change once equilibrium has been reached, no matter how long we wait.

Another way to look at this situation is to say that the concentration of carbon monoxide (and of the other three compounds) does not change at equilibrium, because it is being used up as fast as it is being formed.

Given that the concentrations of all the reactants and products no longer change, can we say that nothing is happening? No, we know that both reactions are occurring; all the molecules are constantly reacting—the CO and H_2O are being changed to CO_2 and H_2, and the CO_2 and H_2 are being changed to CO and H_2O. Because the rates of the forward and reverse reactions are the same, however, none of the concentrations change.

In the example just discussed, we approached equilibrium by adding carbon monoxide to water vapor. Alternatively, we could have added carbon

Reaction A + B $\rightleftharpoons$ C + D

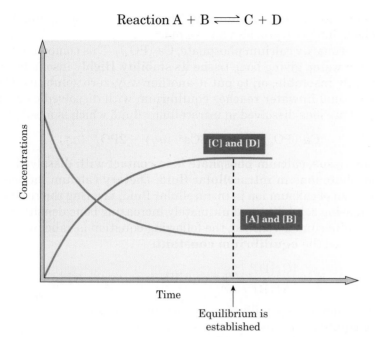

FIGURE 7.10 Changes in the concentrations of reactants (A and B) and products (C and D) as a system approaches equilibrium. Only A and B are present at the beginning of the reaction.

dioxide to hydrogen. In either case, we eventually get an equilibrium mixture containing the same four compounds (Figure 7.11).

It is not necessary to begin with equal amounts. We could, for example, take 10 moles of carbon monoxide and 0.2 mole of water vapor. We would still arrive at an equilibrium mixture of all four compounds.

7.6 What Is an Equilibrium Constant and How Do We Use It?

Chemical equilibria can be treated by a simple mathematical expression. First, let us write the following reaction as the general equation for all reversible reactions:

$$aA + bB \rightleftharpoons cC + dD$$

In this equation, the capital letters stand for substances—CO_2, H_2O, CO, and H_2, for instance—and the lowercase letters are the coefficients of the balanced equation. The double arrow shows that the reaction is reversible. In general, any number of substances can be present on either side.

In the laboratory, we study equilibrium reactions such as the one discussed in the preceding paragraph under carefully controlled conditions. Living things are a far cry from these laboratory conditions. The concept of equilibrium, however, can give useful insight into processes that take place

Be careful not to use a double-headed single arrow ($\longleftrightarrow$) to denote an equilibrium reaction. This symbol is used to show resonance (Section 3.9).

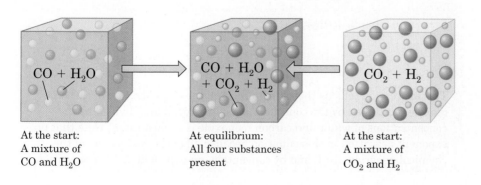

FIGURE 7.11 An equilibrium can be approached from either direction.

At the start: A mixture of CO and H_2O

At equilibrium: All four substances present

At the start: A mixture of CO_2 and H_2

in living organisms, such as humans. The importance of calcium in maintaining bone integrity provides an example.

Bone is primarily calcium phosphate, $Ca_3(PO_4)_2$. This compound is highly insoluble in water, giving bone tissue its stability. Highly insoluble does not mean totally insoluble, or, to put it another way, zero solubility. Calcium phosphate solid in water reaches equilibrium with dissolved calcium ions and phosphate ions dissolved in intracellular fluid, which is mostly water.

$$Ca_3(PO_4)_2(s) \rightleftharpoons 3Ca^{2+}(aq) + 2PO_4^{3-}(aq)$$

In bone tissue, calcium phosphate is in contact with dissolved calcium and phosphate ions in intracellular fluid. Dietary calcium increases the concentration of calcium ion in intracellular fluid, favoring the reverse reaction, decreasing solubility, and ultimately increasing bone density.

Once equilibrium is reached, the following equation is valid, where K is a constant called the **equilibrium constant:**

Equilibrium constant The ratio of product concentrations to reactant concentrations (with exponents that depend on the coefficients of the balanced equation)

$$K = \frac{[C]^c[D]^d}{[A]^a[B]^b}$$ **The equilibrium expression**

Let us examine the equilibrium expression. It tells us that when we multiply the equilibrium concentrations of the substances on the right side of the chemical equation and divide this product by the equilibrium concentrations of the substances on the left side (after raising each number to the appropriate power), we get the equilibrium constant, a number that does not change as long as temperature remains constant. In the example involving solid calcium phosphate above, one might expect that the equilibrium constant is:

$$K = \frac{[Ca^{2+}]^3[PO_4^{3-}]^2}{[Ca_3(PO_4)_2]}$$

However, as a general rule, pure solids and pure liquids are not included when writing an equilibrium expression (for reasons that extend beyond the scope of this book). The concentrations of gases and solutes in solution are included because only those concentrations can be varied, and therefore, it is important to state what they are. Therefore, the equilibrium expression actually becomes:

$$K = [Ca^{2+}]^3[PO_4^{3-}]^2$$

Let us look at several examples of how to set up equilibrium expressions.

Example 7.3 Equilibrium Expressions

Write the equilibrium expression for the reaction:

$$CO(g) + H_2O(g) \rightleftharpoons CO_2(g) + H_2(g)$$

Strategy and Solution

$$K = \frac{[CO_2][H_2]}{[CO][H_2O]}$$

It is understood that the concentration of a species within brackets is always expressed in moles per liter.

This expression tells us that at equilibrium, the concentration of carbon dioxide multiplied by the concentration of hydrogen and divided by the concentrations of water and carbon monoxide is a constant, K. Note that no exponent is written in this equation because all of the coefficients of the chemical equation are 1, and by convention, an exponent of 1 is not written.

Problem 7.3

Write the equilibrium expression for the reaction:

$$SO_3(g) + H_2O(\ell) \rightleftharpoons H_2SO_4(aq)$$

This reaction takes place in the atmosphere when water droplets react with the sulfur oxides formed in the combustion of fuels that contain sulfur. The resulting sulfuric acid is a component of acid rain.

Example 7.4 Equilibrium Expressions

Write the equilibrium expression for the reaction:

$$C_6H_{12}O_6(s) + 6O_2(g) \rightleftharpoons 6CO_2(g) + 6H_2O(\ell)$$

Strategy and Solution

$$K = \frac{[CO_2]^6}{[O_2]^6}$$

In this case, the chemical equation has coefficients other than unity, so the equilibrium expression contains exponents. Notice that the solid and liquid are excluded from the expression.

Problem 7.4

Write the equilibrium expression for the reaction:

$$2NH_3(g) \rightleftharpoons N_2(g) + 3H_2(g)$$

Now let us see how K is calculated.

Example 7.5 Equilibrium Constants

Some H_2 is added to I_2 at 427°C and the following reaction is allowed to come to equilibrium:

$$H_2(g) + I_2(g) \rightleftharpoons 2HI(g)$$

When equilibrium is reached, the concentrations are $[I_2] = 0.42$ mol/L, $[H_2] = 0.025$ mol/L, and $[HI] = 0.76$ mol/L. Calculate K at 427°C.

Strategy

Write the expression for the equilibrium constant; then substitute the values for the concentrations.
The equilibrium expression is:

$$K = \frac{[HI]^2}{[I_2][H_2]}$$

Solution

Substituting the concentrations, we get:

$$K = \frac{[0.76\ \textit{M}]^2}{[0.42\ \textit{M}][0.025\ \textit{M}]} = 55$$

Equilibrium constants are usually written without units. This is current practice among chemists.

Problem 7.5

What is the equilibrium constant for the following reaction? Equilibrium concentrations are given under the formula of each component.

$$PCl_3 \ + \ Cl_2 \ \rightleftharpoons \ PCl_5$$
$$\text{1.66 } M \qquad \text{1.66 } M \qquad \text{1.66 } M$$

Example 7.5 shows us that the reaction between I_2 and H_2 to give HI has an equilibrium constant of 55. What does this value mean? At constant temperature, equilibrium constants remain the same no matter what concentrations we have. That is, at 427°C, if we begin by adding, say, 5 moles of H_2 to 5 moles of I_2, the forward and backward reactions will take place, and equilibrium will eventually be reached. At that point, the value of K will equal 55. If we begin at 427°C with different numbers of moles of H_2 and I_2, perhaps 7 moles of H_2 and 2 moles of I_2, once equilibrium is reached, the value of $[HI]^2/[I_2][H_2]$ will again be 55. It makes no difference what the initial concentrations of the three substances are. At 427°C, as long as all three substances are present and equilibrium has been reached, the concentrations of the three substances will adjust themselves so that the value of the equilibrium constant equals 55.

The equilibrium constant is different for every reaction. Some reactions have a large K; others have a small K. A reaction with a very large K proceeds almost to completion (to the right). For example, K for the following reaction is about 100,000,000, or 10^8 at 25°C:

$$N_2(g) \ + \ 3H_2(g) \rightleftharpoons 2NH_3(g)$$

This value of 10^8 for K means that at equilibrium, $[NH_3]$ must be very large and $[N_2]$ and $[H_2]$ must be very small so that $[NH_3]^2/[N_2][H_2]^3 = 10^8$. Thus, if we add N_2 to H_2, we can be certain that when equilibrium is reached, an essentially complete reaction has taken place.

On the other hand, a reaction such as the following, which has a very small K, about 10^{-8} at 25°C, hardly goes forward at all:

$$AgCl(s) \rightleftharpoons Ag^+(aq) + Cl^-(aq)$$

Recall that pure solids (and pure liquids) are generally not included in the equilibrium expression.

This value of 10^{-8} for K means that at equilibrium, $[Ag^+]$ and $[Cl^-]$ must be very small so that $[Ag^+][Cl^-] = 10^{-8}$.

Equilibrium effects are most obvious in reactions with K values between 10^3 and 10^{-3}. In such cases, the reaction goes part of the way and significant concentrations of all substances are present at equilibrium. An example is the reaction between carbon monoxide and water discussed in Section 7.5, for which K is equal to 10 at 600°C.

How To . . .

Interpret the Value of the Equilibrium Constant, K

Position of Equilibrium

The first question about the value of an equilibrium constant is whether the number is larger than one or smaller than one. If the number is larger than one, it means the ratio of product concentrations to reactant concentrations favors products. In other words, *the equilibrium lies to the right*. If the number is smaller than one, it means the ratio of product concentrations to reactant concentrations favors reactants. In other words, *the equilibrium lies to the left*.

Numerical Value of *K*

The next question focuses on the numerical value of the equilibrium constant. As we saw in Section 1.3, we frequently write numbers with exponents, with positive exponents for very large numbers and negative exponents for very small numbers. The sign and the numerical value of the exponent for a given equilibrium constant conveys information about whether the equilibrium lies strongly to the right (reaction goes to completion), strongly to the left (very little product formed), or at some intermediate point with significant amounts of both reactants and products present when the reaction reaches equilibrium.

Very Large Values of *K* (above 10^3)

The conversion of NO gas to NO_2 in the presence of atmospheric oxygen is a reaction of environmental importance. Both these gases are pollutants and play a large role in the formation of smog and of acid rain.

$$2NO(g) + O_2(g) \rightleftharpoons 2NO_2(g)$$

The equilibrium constant for this reaction is 4.2×10^{12} at room temperature. If we start with 10.0 *M* NO, we find that only $2.2 \times 10^{-6} M$ is found at equilibrium and that the concentration of NO_2 is 10.0 *M* to within experimental error. Only a negligible amount of NO remains, and we say that the reaction has gone to completion.

Intermediate Values of *K* (less than 10^3 but more than 10^{-3})

Great care is taken in transporting and handling chlorine gas, especially with regard to fire prevention. Chlorine can react with carbon monoxide (also produced in fires) to produce phosgene ($COCl_2$), one of the poison gases used in World War I.

$$CO(g) + Cl_2(g) \rightleftharpoons COCl_2(g)$$
$$0.50\ M \quad 1.10\ M \qquad 0.10\ M$$

The equilibrium constant for this reaction is 0.20 (2.0×10^{-1}) at 600°C. The equilibrium concentrations are given below the formula of each component. They are similar in terms of order of magnitude, but the lower concentration for phosgene is consistent with the equilibrium constant less than one.

Very Small Values of *K* (less than 10^{-3})

Barium sulfate is a compound of low solubility widely used to coat the gastrointestinal tract in preparation for X-rays. The solid is in equilibrium with dissolved barium and sulfate ions.

$$BaSO_4(s) \rightleftharpoons Ba^{2+}(aq) + SO_4^{2-}(aq)$$

The equilibrium constant for this reaction is 1.10×10^{-10} at room temperature. The concentrations of barium ion and sulfate ion are each $1.05 \times 10^{-5} M$. This low number implies that very little of the solid has dissolved.

The equilibrium constant for a given reaction remains the same no matter what happens to the concentrations, but the same is not true for changes in temperature. The value of K, however, does change when the temperature changes.

As pointed out earlier in this section, the equilibrium expression is valid only after equilibrium has been reached. Before that point, there is no equilibrium, and the equilibrium expression is not valid. But how long does it take for a reaction to reach equilibrium? There is no easy answer to this question. Some reactions, if the reactants are well mixed, reach equilibrium in less than one second; others will not get there even after millions of years.

There is no relationship between the rate of a reaction (how long it takes to reach equilibrium) and the value of K. It is possible to have a large K and a slow rate, as in the reaction between glucose and O_2 to give CO_2 and H_2O, which does not reach equilibrium for many years (Section 7.1), or a small K and a fast rate. In other reactions, the rate and K are both large or both small.

7.7 What Is Le Chatelier's Principle?

When a reaction reaches equilibrium, the forward and reverse reactions take place at the same rate, and the equilibrium concentration of the reaction mixture does not change as long as we don't do anything to the system. But what happens if we do? In 1888, Henri Le Chatelier (1850–1936) put forth the statement known as **Le Chatelier's principle:** If an external stress is applied to a system in equilibrium, the system reacts in such a way as to partially relieve that stress. Let us look at five types of stress that can be put on chemical equilibria: adding a reactant or product, removing a reactant or product, and changing the temperature.

Le Chatelier's principle A principle stating that when a stress is applied to a system in chemical equilibrium, the position of the equilibrium shifts in the direction that will relieve the applied stress

Charles D. Winters/Cengage Learning

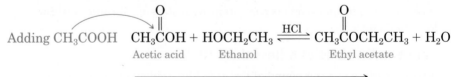

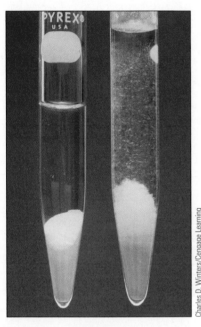

The tube on the left contains a saturated solution of silver acetate (Ag^+ ions and CH_3COO^- ions) in equilibrium with solid silver acetate. When more silver ions are added in the form of silver nitrate solution, the equilibrium shifts to the right, producing more silver acetate, as can be seen in the tube on the right.
$Ag^+(aq) + CH_3COO^-(aq) \rightleftharpoons CH_3COOAg(s)$

A. Addition of a Reaction Component

Suppose that the reaction between acetic acid and ethanol has reached equilibrium:

$$\underset{\text{Acetic acid}}{CH_3COOH(\ell)} + \underset{\text{Ethanol}}{C_2H_5OH(\ell)} \overset{HCl}{\rightleftharpoons} \underset{\text{Ethyl acetate}}{CH_3COOC_2H_5(\ell)} + H_2O(\ell)$$

This means that the reaction flask contains all four substances (plus the catalyst) and that their concentrations no longer change.

We now disturb the system by adding some acetic acid.

$$\text{Adding } CH_3COOH \quad \underset{\text{Acetic acid}}{CH_3\overset{\overset{O}{\|}}{C}OH} + \underset{\text{Ethanol}}{HOCH_2CH_3} \overset{HCl}{\rightleftharpoons} \underset{\text{Ethyl acetate}}{CH_3\overset{\overset{O}{\|}}{C}OCH_2CH_3} + H_2O$$

Equilibrium shifts to formation of more products →

The result is that the concentration of acetic acid suddenly increases, which increases the rate of the forward reaction. As a consequence, the concentrations of the products (ethyl acetate and water) begin to increase. At the same time, the concentrations of reactants decrease. Now, an increase in the concentrations of the products causes the rate of the reverse reaction to increase, but the rate of the forward reaction is decreasing, so eventually the two rates will be equal again and a new equilibrium will be established.

When that happens, the concentrations are once again constant, but they are not the same as they were before the addition of the acetic acid. The concentrations of ethyl acetate and water are higher now, and the concentration of ethanol is lower. The concentration of acetic acid is higher because we added some, but it is less than it was immediately after we made the addition.

When we add more of any component to a system in equilibrium, that addition constitutes a stress. The system relieves this stress by increasing the concentrations of the components on the other side of the equilibrium equation. We say that the equilibrium shifts in the opposite direction. The addition of acetic acid, on the left side of the equation, causes the rate of the forward reaction to increase and the reaction to move toward the right: More ethyl acetate and water form, and some of the acetic acid and ethanol are used up. The same thing happens if we add ethanol.

On the other hand, if we add water or ethyl acetate, the rate of the reverse reaction increases and the reaction shifts to the left:

$$\underset{\text{Acetic acid}}{CH_3COOH} + \underset{\text{Ethanol}}{C_2H_5OH} \xrightleftharpoons{\text{HCl}} \underset{\text{Ethyl acetate}}{CH_3COOC_2H_5} + H_2O \qquad \text{Adding ethyl acetate}$$

Equilibrium shifts toward formation of reactants

We can summarize by saying that the addition of any component causes the equilibrium to shift to the opposite side.

Example 7.6 Le Chatelier's Principle–Effect of Concentration

When dinitrogen tetroxide, a colorless gas, is enclosed in a vessel, a color indicating the formation of brown nitrogen dioxide soon appears (see Figure 7.12 later in this chapter). The intensity of the brown color indicates the amount of nitrogen dioxide formed. The equilibrium reaction is:

$$\underset{\substack{\text{Dinitrogen} \\ \text{tetroxide} \\ \text{(colorless)}}}{N_2O_4(g)} \rightleftharpoons \underset{\substack{\text{Nitrogen} \\ \text{dioxide} \\ \text{(brown)}}}{2NO_2(g)}$$

When more N_2O_4 is added to the equilibrium mixture, the brownish color becomes darker. Explain what happened.

Strategy and Solution

The darker color indicates that more nitrogen dioxide is formed. This happens because the addition of the reactant shifts the equilibrium to the right, forming more product.

Problem 7.6

What happens to the following equilibrium reaction when Br_2 gas is added to the equilibrium mixture?

$$2NOBr(g) \rightleftharpoons 2NO(g) + Br_2(g)$$

B. Removal of a Reaction Component

It is not always as easy to remove a component from a reaction mixture as it is to add one, but there are often ways to do it. The removal of a component, or even a decrease in its concentration, lowers the corresponding reaction rate and changes the position of the equilibrium. If we remove a reactant, the reaction shifts to the left, toward the side from which the reactant was removed. If we remove a product, the reaction shifts to the right, toward the side from which the product was removed.

In the case of the acetic acid–ethanol equilibrium, ethyl acetate has the lowest boiling point of the four components and can be removed by distillation. The equilibrium then shifts to that side so that more ethyl acetate is produced to compensate for the removal. The concentrations of acetic acid and ethanol decrease, and the concentration of water increases. The effect of removing a component is thus the opposite of adding one. The removal of a component causes the equilibrium to shift to the side from which the component was removed.

$$CH_3COOH + C_2H_5OH \xrightleftharpoons{HCl} H_2O + CH_3COOC_2H_5 \quad \text{Removing ethyl acetate}$$

Acetic acid   Ethanol   Ethyl acetate

Equilibrium shifts toward ———————————————————→
formation of more products

No matter what happens to the individual concentrations, the value of the equilibrium constant remains unchanged.

Example 7.7 Le Chatelier's Principle–Removal of a Reaction Component

The beautiful stone we know as marble is mostly calcium carbonate. When acid rain containing sulfuric acid attacks marble, the following equilibrium reaction can be written:

$$CaCO_3(s) + H_2SO_4(aq) \rightleftharpoons CaSO_4(s) + CO_2(g) + H_2O(\ell)$$

Calcium   Sulfuric   Calcium   Carbon
carbonate   acid   sulfate   dioxide

How does the fact that carbon dioxide is a gas influence the equilibrium?

Strategy and Solution

The gaseous CO_2 diffuses away from the reaction site, meaning that this product is removed from the equilibrium mixture. The equilibrium shifts to the right so that the statue continues eroding.

Problem 7.7

Consider the following equilibrium reaction for the decomposition of an aqueous solution of hydrogen peroxide:

$$2H_2O_2(aq) \rightleftharpoons 2H_2O(\ell) + O_2(g)$$

Hydrogen
peroxide

Oxygen has limited solubility in water (see the table in Chemical Connections 6A). What happens to the equilibrium after the solution becomes saturated with oxygen?

C. Change in Temperature

The effect of a change in temperature on a reaction that has reached equilibrium depends on whether the reaction is exothermic (gives off heat) or endothermic (requires heat). Let us look first at an exothermic reaction:

$$2H_2(g) + O_2(g) \rightleftharpoons 2H_2O(\ell) + 137,000 \text{ cal per mol } H_2O$$

This headstone has been damaged by acid rain.

John D. Cunningham/Visuals Unlimited

Chemical Connections 7D

Sunglasses and Le Chatelier's Principle

Heat is not the only form of energy that affects equilibria. The statements made in the text regarding endothermic and exothermic reactions can be generalized to reactions involving other forms of energy. A practical illustration of this generalization is the use of sunglasses with adjustable shading. The compound silver chloride, AgCl, is incorporated in the glasses. This compound, upon exposure to sunlight, produces metallic silver, Ag, and chlorine, Cl_2:

$$Light + 2Ag^+ + 2Cl^- \rightleftharpoons 2Ag(s) + Cl_2$$

The more silver metal produced, the darker the glasses. At night or when the wearer goes indoors, the reaction is reversed according to Le Chatelier's principle. In this case, the addition of energy in the form of sunlight drives the equilibrium to the right; its removal drives the equilibrium to the left.

Image courtesy of Transitions Optical, Inc.

Test your knowledge with Problems 7.46 and 7.47.

If we consider heat to be a product of this reaction, then we can use Le Chatelier's principle and the same type of reasoning as we did before. An increase in temperature means that we are adding heat. Because heat is a product, its addition pushes the equilibrium to the opposite side. We can therefore say that if this exothermic reaction is at equilibrium and we increase the temperature, the reaction goes to the left—the concentrations of H_2 and O_2 increase and that of H_2O decreases. This is true of all exothermic reactions.

- An increase in temperature drives an exothermic reaction toward the reactants (to the left).
- A decrease in temperature drives an exothermic reaction toward the products (to the right).

For an endothermic reaction, of course, the opposite is true.

- An increase in temperature drives an endothermic reaction toward the products (to the right).
- A decrease in temperature drives an endothermic reaction toward the reactants (to the left).

Recall from Section 7.4 that a change in temperature changes not only the position of equilibrium but also the value of K, the equilibrium constant.

Example 7.8 Le Chatelier's Principle–Effect of Temperature

The conversion of nitrogen dioxide to dinitrogen tetroxide is an exothermic reaction:

$$2NO_2(g) \rightleftharpoons N_2O_4(g) + 13,700 \text{ cal}$$

Nitrogen dioxide   Dinitrogen tetroxide
(brown)   (colorless)

FIGURE 7.12 Effect of temperature on the N_2O_4—NO_2 system at equilibrium. (*top*) At 50°C, the deep brown color indicates the predominance of NO_2. (*bottom*) At 0°C, N_2O_4, which is colorless, predominates.

In Figure 7.12 we see that the brown color is darker at 50°C than it is at 0°C. Explain.

Strategy and Solution

To go from 0°C to 50°C, heat must be added. But heat is a product of this equilibrium reaction, as it is written in the question. The addition of heat, therefore, shifts the equilibrium to the left. This shift produces more $NO_2(g)$, leading to the darker brown color.

Problem 7.8

Consider the following equilibrium reaction:

$$A \rightleftharpoons B$$

Increasing the temperature results in an increase in the equilibrium concentration of B. Is the conversion of A to B an exothermic reaction or an endothermic reaction? Explain.

D. Change in Pressure

A change in pressure influences the equilibrium only if one or more components of the reaction mixture are gases. Consider the following equilibrium reaction:

$$N_2O_4(g) \rightleftharpoons 2NO_2(g)$$

Dinitrogen   Nitrogen
tetroxide   dioxide
(colorless)   (brown)

In this equilibrium, we have one mole of gas as a reactant and two moles of gas as products. According to Le Chatelier's principle, an increase in pressure shifts the equilibrium in the direction that will decrease the moles in the gas phase and thus decrease the internal pressure. In the preceding reaction, the equilibrium will shift to the left.

• An increase in pressure shifts the reaction toward the side with fewer moles of gas.

• A decrease in pressure shifts the reaction toward the side with more moles of gas.

• When the moles of gas are the same in a balanced reaction, an increase or decrease in pressure results in no shift of the reaction.

Le Châtelier's principle does not apply to pressure increases caused by the addition of a nonreactive (inert) gas to the reaction mixture. This addition has no effect on the equilibrium position.

Example 7.9 Le Chatelier's Principle–Effect of Gas Pressure

In the production of ammonia, both reactants and products are gases:

$$N_2(g) + 3H_2(g) \rightleftharpoons 2NH_3(g)$$

What kind of pressure change would increase the yield of ammonia?

Strategy and Solution

There are four moles of gases on the left side and two moles on the right side. To increase the yield of ammonia, we must shift the equilibrium to the right. An increase in pressure shifts the equilibrium toward the side with fewer moles—that is, to the right. Thus, an increase in pressure will increase the yield of ammonia.

Problem 7.9

What happens to the following equilibrium reaction when the pressure is increased?

$$O_2(g) + 4ClO_2(g) \rightleftharpoons 2Cl_2O_5(g)$$

E. The Effects of a Catalyst

As we saw in Section 7.4D, a catalyst increases the rate of a reaction without itself being changed. For a reversible reaction, catalysts always increase the rates of both the forward and reverse reactions to the same extent. Therefore, the addition of a catalyst has no effect on the position of equilibrium. However, adding a catalyst to a system not yet at equilibrium causes it to reach equilibrium faster than it would without the catalyst.

Chemical Connections 7E

The Haber Process

Both humans and other animals need proteins and other nitrogen-containing compounds to live. Ultimately, the nitrogen in these compounds comes from the plants that we eat. Although the atmosphere contains plenty of N_2, nature converts it to compounds usable by biological organisms in only one way: Certain bacteria have the ability to "fix" atmospheric nitrogen—that is, convert it to ammonia. Most of these bacteria live in the roots of certain plants such as clover, alfalfa, peas, and beans. However, the amount of nitrogen fixed by such bacteria each year is far less than the amount necessary to feed all the humans and animals in the world.

The world today can support its population only by using fertilizers made by artificial fixing, primarily the **Haber process,** which converts N_2 to NH_3.

$$N_2(g) + 3H_2(g) \rightleftharpoons 2NH_3(g) + 22 \text{ kcal}$$

Early workers who focused on the problem of fixing nitrogen were troubled by a conflict between equilibrium and rate. Because the synthesis of ammonia is an exothermic reaction, an increase in temperature drives the equilibrium to the left; so the best results (largest possible yield) should be obtained at low temperatures. At low temperatures, however, the rate is too slow to produce any meaningful amounts of NH_3. In 1908, Fritz Haber (1868–1934) solved this problem when he discovered a catalyst that permits the reaction to take place at a convenient rate at 500°C.

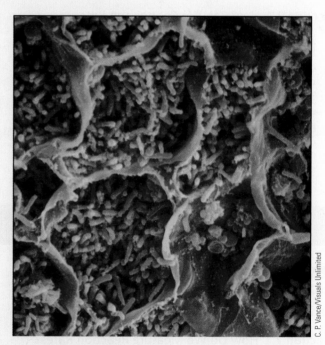

C. P. Vance/Visuals Unlimited

Ammonia is produced by bacteria in these root nodules.

The NH_3 produced by the Haber process is converted to fertilizers, which are used all over the world. Without these fertilizers, food production would diminish so much that widespread starvation would result.

Test your knowledge with Problem 7.48.

Summary

Section 7.1 How Do We Measure Reaction Rates?

- The **rate of a reaction** is the change in concentration of a reactant or a product per unit time. Some reactions are fast; others are slow.

Section 7.2 Why Do Some Molecular Collisions Result in Reaction Whereas Others Do Not?

- The rate of a reaction depends on the number of **effective collisions**—that is, collisions that lead to a reaction.
- The energy necessary for a reaction to take place is the **activation energy.** Effective collisions have (1) more than the activation energy required for the reaction to proceed forward and (2) the proper orientation in space of the colliding particles.

Section 7.3 What Is the Relationship Between Activation Energy and Reaction Rate?

- The lower the activation energy, the faster the reaction.
- An energy diagram shows the progress of a reaction.
- The position at the top of the curve in an energy diagram is called the **transition state.**

Section 7.4 How Can We Change the Rate of a Chemical Reaction?

- Reaction rates generally increase with increasing concentration and temperature; they also depend on the nature of the reactants.
- The rates of some reactions can be increased by adding a **catalyst,** a substance that provides an alternative pathway with a lower activation energy.
- A rate constant gives the relationship between the rate of the reaction and the concentrations of the reactants at a constant temperature.

Section 7.5 What Does It Mean to Say That a Reaction Has Reached Equilibrium?

- Many reactions are reversible and eventually reach equilibrium.
- At **equilibrium,** the forward and reverse reactions take place at equal rates and concentrations do not change.

Section 7.6 What Is an Equilibrium Constant and How Do We Use It?

- Every equilibrium has an **equilibrium expression** and an **equilibrium constant,** K, which does not change when concentrations change but does change when temperature changes.
- There is no relationship between the value of the equilibrium constant, K, and the rate at which equilibrium is reached.

Section 7.7 What Is Le Chatelier's Principle?

- **Le Chatelier's principle** tells us what happens when we put stress on a system in equilibrium.
- The addition of a component causes the equilibrium to shift to the opposite side.
- The removal of a component causes the equilibrium to shift to the side from which the component is removed.
- Increasing the temperature drives an exothermic equilibrium to the side of the reactants; increasing the temperature drives an endothermic equilibrium to the side of the products.
- Increasing the pressure of a mixture shifts the equilibrium in the direction that decreases the moles in the gas phase; decreasing the pressure of a mixture shifts the equilibrium in the direction that increases the moles in the gas phase.
- Addition of a catalyst has no effect on the position of equilibrium.

Problems

Orange-numbered problems are applied.

Section 7.1 How Do We Measure Reaction Rates?

7.10   The rate of disappearance of HCl was measured for the following reaction:

$$CH_3OH + HCl \longrightarrow CH_3Cl + H_2O$$

The initial concentration of HCl is 1.85 M. Its concentration decreases to 1.58 M in 54.0 min. What is the rate of reaction?

7.11   Consider the following reaction:

$$\underset{\text{Chloromethane}}{CH_3-Cl} + I^- \longrightarrow \underset{\text{Iodomethane}}{CH_3-I} + Cl^-$$

Suppose we start the reaction with an initial iodomethane concentration of 0.260 M. This concentration increases to 0.840 M over a period of 1 h 20 min. What is the rate of reaction?

Section 7.2 Why Do Some Molecular Collisions Result in Reaction Whereas Others Do Not?

7.12   Two kinds of gas molecules are reacted at a set temperature. The gases are blown into the reaction vessel from two tubes. In setup A, the two tubes are aligned parallel to each other; in setup B, they are 90° to each other; and in setup C, they are aligned directly opposite each other. Which setup would yield the most effective collisions?

7.13 Why are reactions between ions in aqueous solution generally much faster than reactions between covalent molecules?

Section 7.3 What Is the Relationship Between Activation Energy and Reaction Rate?

7.14 What is the likelihood that the following reaction occurs in a single step? Explain.

$$O_2(g) + 4ClO_2(g) \rightleftharpoons 2Cl_2O_5(g)$$

7.15 A certain reaction is exothermic by 9 kcal/mol and has an activation energy of 14 kcal/mol. Draw an energy diagram for this reaction and label the transition state.

Section 7.4 How Can We Change the Rate of a Chemical Reaction?

7.16 A quart of milk quickly spoils if left at room temperature but keeps for several days in a refrigerator. Explain.

7.17 If a certain reaction takes 16 h to go to completion at 10°C, what temperature should we run it if we want it to go to completion in 1 h?

7.18 In most cases, when we run a reaction by mixing a fixed quantity of substance A with a fixed quantity of substance B, the rate of the reaction begins at a maximum and then decreases as time goes by. Explain.

7.19 If you were running a reaction and wanted it to go faster, what three things might you try to accomplish this goal?

7.20 What factors determine whether a reaction run at a given temperature will be fast or slow?

7.21 Explain how a catalyst increases the rate of a reaction.

7.22 If you add a piece of marble, $CaCO_3$, to a 6 M HCl solution at room temperature, you will see some bubbles form around the marble as gas slowly rises. If you crush another piece of marble and add it to the same solution at the same temperature, you will see vigorous gas formation, so much so that the solution appears to be boiling. Explain.

Section 7.5 What Does It Mean to Say That a Reaction Has Reached Equilibrium?

7.23 Burning a piece of paper is an irreversible reaction. Give some other examples of irreversible reactions.

7.24 Suppose the following reaction is at equilibrium:

$$PCl_3 + Cl_2 \rightleftharpoons PCl_5$$

(a) Are the equilibrium concentrations of PCl_3, Cl_2, and PCl_5 necessarily equal? Explain.

(b) Is the equilibrium concentration of PCl_3 necessarily equal to that of Cl_2? Explain.

Section 7.6 What Is an Equilibrium Constant and How Do We Use It?

7.25 Write equilibrium expressions for these reactions.

(a) $2H_2O_2(g) \rightleftharpoons 2H_2O(g) + O_2(g)$

(b) $2N_2O_5(g) \rightleftharpoons 2N_2O_4(g) + O_2(g)$

(c) $6H_2O(g) + 6CO_2(g) \rightleftharpoons C_6H_{12}O_6(s) + 6O_2(g)$

7.26 Write the chemical equations corresponding to the following equilibrium expressions.

(a) $K = \dfrac{[H_2CO_3]}{[CO_2][H_2O]}$

(b) $K = \dfrac{[P_4][O_2]^5}{[P_4O_{10}]}$

(c) $K = \dfrac{[F_2]^3[PH_3]}{[HF]^3[PF_3]}$

7.27 Consider the following equilibrium reaction. Under each species is its equilibrium concentration. Calculate the equilibrium constant for the reaction.

$$CO(g) + H_2O(g) \rightleftharpoons CO_2(g) + H_2(g)$$

$\quad$ 0.933 M $\quad$ 0.720 M $\qquad$ 0.133 M $\quad$ 3.37 M

7.28 When the following reaction reached equilibrium at 325 K, the equilibrium constant was found to be 172. When a sample was taken of the equilibrium mixture, it was found to contain 0.0714 M NO_2. What was the equilibrium concentration of N_2O_4?

$$2NO_2(g) \rightleftharpoons N_2O_4(g)$$

7.29 The following reaction was allowed to reach equilibrium at 25°C. Under each component is its equilibrium concentration. Calculate the equilibrium constant, K, for this reaction.

$$2NOCl(g) \rightleftharpoons 2NO(g) + Cl_2(g)$$

$\quad$ 2.6 M $\qquad\qquad$ 1.4 M $\quad$ 0.34 M

7.30 Write the equilibrium expression for this reaction:

$$HNO_3(aq) + H_2O(\ell) \rightleftharpoons H_3O^+(aq) + NO_3^-(aq)$$

7.31 Here are equilibrium constants for several reactions. Which of them favor the formation of products, and which favor the formation of reactants?

(a) 4.5×10^{-8} $\qquad$ (b) 32

(c) 4.5 $\qquad\qquad\qquad$ (d) 3.0×10^{-7}

(e) 0.0032

7.32 A particular reaction has an equilibrium constant of 1.13 under one set of conditions and an equilibrium constant of 1.72 under a different set of conditions. Which conditions would be more advantageous in an industrial process that sought to obtain the maximum amount of products? Explain.

7.33 If a reaction is very exothermic—that is, if the products have a much lower energy than the reactants—can we be reasonably certain that it will take place rapidly?

7.34 If a reaction is very endothermic—that is, if the products have a much higher energy than the reactants—can we be reasonably certain that it will take place extremely slowly or not at all?

7.35 A reaction has a high rate constant but a small equilibrium constant. What does this mean in terms of producing an industrial product?

Section 7.7 What Is Le Chatelier's Principle?

7.36 Complete the following table showing the effects of changing reaction conditions on the equilibrium and value of the equilibrium constant, K.

Change in Condition	How the Reacting System Changes to Achieve a New Equilibrium	Does the Value of K Increase or Decrease?
Addition of a reactant	Shift to product formation	Neither
Removal of a reactant		
Addition of a product		
Removal of a product		
Increasing pressure		

7.37 Assume that the following exothermic reaction is at equilibrium:

$$H_2(g) + I_2(g) \rightleftharpoons 2HI(g)$$

Tell whether the position of equilibrium will shift to the right or the left if we:

(a) Remove some HI
(b) Add some I_2
(c) Remove some I_2
(d) Increase the temperature
(e) Add a catalyst

7.38 The following reaction is endothermic:

$$3O_2(g) \rightleftharpoons 2O_3(g)$$

If the reaction is at equilibrium, tell whether the equilibrium will shift to the right or the left if we:

(a) Remove some O_3
(b) Remove some O_2
(c) Add some O_3
(d) Decrease the temperature
(e) Add a catalyst
(f) Increase the pressure

7.39 The following reaction is exothermic: After it reaches equilibrium, we add a few drops of Br_2.

$$2NO(g) + Br_2(g) \rightleftharpoons 2NOBr(g)$$

(a) What will happen to the equilibrium?
(b) What will happen to the equilibrium constant?

7.40 Is there any change in conditions that change the equilibrium constant, K, of a given reaction?

7.41 The equilibrium constant at 1127°C for the following endothermic reaction is 571:

$$2H_2S(g) \rightleftharpoons 2H_2(g) + S_2(g)$$

If the mixture is at equilibrium, what happens to K if we:

(a) Add some H_2S?
(b) Add some H_2?
(c) Lower the temperature to 1000°C?

Chemical Connections

7.42 (Chemical Connections 7A) In a bacterial infection, body temperature may rise to 101°F. Does this body defense kill the bacteria directly by heat or by another mechanism? If so, by which mechanism?

7.43 (Chemical Connections 7A and 7B) Why is a high fever dangerous? Why is a low body temperature dangerous?

7.44 (Chemical Connections 7B) Why do surgeons sometimes lower body temperatures during heart operations?

7.45 (Chemical Connections 7C) A painkiller—for example, Tylenol—can be purchased in two forms, each containing the same amount of drug. One form is a solid coated pill, and the other is a capsule that contains tiny beads and has the same coat. Which medication will act faster? Explain.

7.46 (Chemical Connections 7D) What reaction takes place when sunlight hits the compound silver chloride?

7.47 (Chemical Connections 7D) You have a recipe to manufacture sunglasses: 3.5 g AgCl/kg glass. A new order comes in to manufacture sunglasses to be used in deserts like the Sahara. How would you change the recipe?

7.48 (Chemical Connections 7E) If the equilibrium for the Haber process is unfavorable at high temperatures, why do factories nevertheless use high temperatures?

Additional Problems

7.49 In the reaction between H_2 and Cl_2 to give HCl, a 10°C increase in temperature doubles the rate of reaction. If the rate of reaction at 15°C is 2.8 moles of HCl per liter per second, what are the rates at −5°C and at 45°C?

7.50 Draw an energy diagram for an exothermic reaction that yields 75 kcal/mol. The activation energy is 30 kcal/mol.

7.51 Draw a diagram similar to Figure 7.4. Draw a second line of the energy profile starting and ending at the same level as the first but having a smaller peak than the first line. Label them 1 and 2. What may have occurred to change the energy profile of a reaction from 1 to 2?

7.52 For the reaction

$$2NOBr(g) \rightleftharpoons 2NO(g) + Br_2(g)$$

the rate of the reaction was −2.3 mol NOBr/L/h when the initial NOBr concentration was 6.2 mol NOBr/L. What is the rate constant of the reaction?

7.53 The equilibrium constant for the following reaction is 25:

$$2NOBr(g) \rightleftharpoons 2NO(g) + Br_2(g)$$

A measurement made on the equilibrium mixture found that the concentrations of NO and Br_2 were each 0.80 M. What is the concentration of NOBr at equilibrium?

7.54 In the following reaction, the concentration of N_2O_4 in mol/L was measured at the end of the times shown. What is the initial rate of the reaction?

$$N_2O_4(g) \rightleftharpoons 2NO_2(g)$$

Time (s)	[N_2O_4]
0	0.200
10	0.180
20	0.162
30	0.146

7.55 How could you increase the rate of a gaseous reaction without adding more reactants or a catalyst and without changing the temperature?

7.56 In an endothermic reaction, the activation energy is 10.0 kcal/mol. Is the activation energy of the reverse reaction also 10.0 kcal/mol, or would it be more or less? Explain with the aid of a diagram.

7.57 Write the reaction to which the following equilibrium expression applies:

$$K = \frac{[NO_2]^4[H_2O]^6}{[NH_3]^4[O_2]^7}$$

7.58 The rate for the following reaction at 300 K was found to be 0.22 M NO_2/min. What would be the approximate rate at 320 K?

$$N_2O_4(g) \rightleftharpoons 2NO_2(g)$$

7.59 Assume that two different reactions are taking place at the same temperature. In reaction A, two different spherical molecules collide to yield a product. In reaction B, the shape of the colliding molecules is rodlike. Each reaction has the same number of collisions per second and the same activation energy. Which reaction goes faster?

7.60 Is it possible for an endothermic reaction to have zero activation energy?

7.61 In the following reaction, the rate of appearance of I_2 is measured at the times shown. What is the initial rate of the reaction?

$$2HI(g) \rightleftharpoons H_2(g) + I_2(g)$$

Time (s)	[I_2]
0	0
10	0.30
20	0.57
30	0.81

7.62 A reaction occurs in three steps with the following rate constants:

$$A \xrightarrow[\text{Step 1}]{k_1 = 0.3\,M} B \xrightarrow[\text{Step 2}]{k_2 = 0.05\,M} C \xrightarrow[\text{Step 3}]{k_3 = 4.5\,M} D$$

(a) Which step is the slow step?
(b) Which step has the lowest activation energy?

Looking Ahead

7.63 As we shall see in Chapter 8, weak acids such as acetic acid only partially dissociate in solution, as shown in the following simplified net ionic equilibrium reaction.

$$CH_3COOH \rightleftharpoons H^+ + CH_3COO^-$$
Acetic acid

(a) Suppose that initially, only 0.10 M acetic acid is present. Analysis of the equilibrium mixture shows that the concentration of acetic acid reduces to 0.098 M. Determine the equilibrium concentrations of H^+ and CH_3COO^-.
(b) What is the expected equilibrium constant, K, for this reaction?

7.64 As we shall see in Chapter 20, there are two forms of glucose, designated alpha (α) and beta (β), which are in equilibrium in aqueous solution. The equilibrium constant for the reaction is 1.5 at 30°C.

$$\alpha\text{-D-glucose}(aq) \rightleftharpoons \beta\text{-D-glucose}(aq)\ K = 1.5$$

(a) If you begin with a fresh 1.0 M solution of α-D-glucose in water, what will be its concentration when equilibrium is reached?
(b) Calculate the percentage of α-glucose and of β-glucose present at equilibrium in aqueous solution at 30°C.

7.65. Consider the reaction A $\longrightarrow$ B, for which you wish to determine the rate. You do not have any convenient method for determining the amount of B formed. You do, however, have a method for determining the amount of A left as the reaction proceeds. Does it make any difference if you determine the rate in terms of disappearance of A rather than appearance of B? Why or why not?

7.66 You have a choice of two methods for determining the rate of a reaction. In the first method, you have to extract part of the reaction mixture to test for the amount of product formed. In the second method, you can do continuous monitoring of the amount of product formed. Which method is preferable and why?

7.67 You want to measure reaction rates for some very fast reactions. What sort of technical difficulties do you expect to arise?

7.68 You make five measurements of the rate of a reaction, and for each measurement, you determine the rate constant. The values of four of the rate constants are close to each other (within experimental error). The other is quite different. Is your result likely to represent a different rate or an error in calculation? Why?

Tying It Together

7.69 Pure carbon exists is several forms, two of which are diamond and graphite. The conversion of the diamond form to the graphite form is exothermic to a very slight extent. How is it that jewelers can advertise "Diamonds are forever"?

7.70 You have decided to change the temperature at which you run a certain reaction in hopes of obtaining more product more quickly. You find that you actually get less of the desired product, although you get to the equilibrium state more quickly. What happened?

Challenge Problems

7.71 You have a beaker that contains solid silver chloride (AgCl) and a saturated solution of Ag^+ and Cl^- ions in equilibrium with the solid.

$$AgCl(s) \rightleftharpoons Ag^+(aq) + Cl^-(aq)$$

You add several drops of a sodium chloride solution. What happens to the concentration of silver ions?

7.72 What would happen to the reaction that produces ammonia if water is present in the reaction mixture?

$$N_2(g) + 3H_2(g) \rightleftharpoons 2NH_3(g)$$

Hint: Ammonia is very soluble in water.

7.73 The equilibrium constant, K, is 2.4×10^{-3} for the following reaction at a certain temperature.

$$H_2(g) + F_2(g) \rightleftharpoons 2HF(g)$$

If the concentrations of both $H_2(g)$ and $F_2(g)$ are found to be $0.0021\ M$ at equilibrium, what is the concentration of $HF(g)$ under these conditions?

7.74 It can be shown that a mathematical relationship exists between rate constants and equilibrium constants. For example, consider the following generic reaction, where the rate constant k refers to the rate of the forward reaction and the rate constant k' refers to the rate of the reverse reaction.

$$A + B \underset{k'}{\overset{k}{\rightleftharpoons}} C + D$$

Verify that the equilibrium constant for a reaction is equal to the ratio of the rate constants for the forward and reverse reactions.

$$K = \frac{k}{k'}$$

7.75 The following exothermic reaction is at equilibrium.

$$Zn(s) + 4H^+(aq) + 2NO_3^-(aq) \rightleftharpoons$$
$$Zn^{2+}(aq) + 2NO_2(g) + 2H_2O(\ell)$$

Consider each of the following changes separately and state the effect (increase, decrease, or no change) that the change from the first column has on the equilibrium value of the quantity listed in the second column.

Change	Quantity	Effect
Increase the pressure	Concentration of NO_3^-	
Add some Zn	Concentration of NO_2	
Decrease the H^+	Concentration of Zn^{2+}	
Add Pt catalyst	Equilibrium constant, K	
Add some Ar gas	Concentration of H^+	
Decrease the Zn^{2+}	Equilibrium constant, K	
Increase the temperature	Concentration of Zn	

7.76 When a $0.10\ M$ solution of glucose-1-phosphate is incubated with a catalytic amount of phosphoglucomutase, the glucose-1-phosphate is transformed into glucose-6-phosphate until equilibrium is established at 25°C.

$$\text{glucose-1-phosphate} \rightleftharpoons \text{glucose-6-phosphate}$$

If the equilibrium concentration of glucose-6-phosphate is found to be $9.6 \times 10^{-2}\ M$, determine the equilibrium constant, K, for this reaction at 25°C.

7.77 A certain endothermic reaction (see Figure 7.5) at equilibrium has an activation energy of 40. kJ. If the energy of reaction is found to be 30. kJ, what is the activation energy for the reverse reaction?

7.78 Consider the equilibrium of phosphorus pentachloride, PCl_5, with its decomposition products, where $K = 6.3 \times 10^{-4}$ at a certain temperature.

$$PCl_5(g) \rightleftharpoons PCl_3(g) + Cl_2(g)$$

At equilibrium, it is found that the concentration of PCl_5 is three times the concentration of PCl_3. Determine the concentration of Cl_2 under these conditions.

7.79 Consider the reaction shown below at a certain temperature.

$$2H_2O(g) \rightleftharpoons 2H_2(g) + O_2(g)$$

The equilibrium constant, K, is equal to 8.7×10^3 at a certain temperature. At equilibrium, it is found that $[H_2] = 1.9 \times 10^{-2}\ M$ and $[O_2] = 8.0 \times 10^{-2}\ M$. What is the concentration of H_2O at equilibrium?

Acids and Bases

8

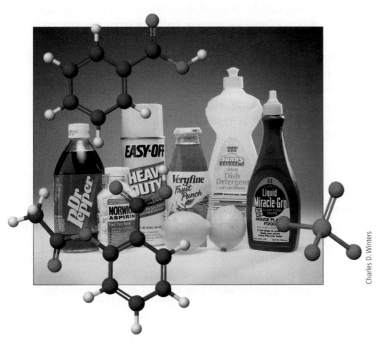

Charles D. Winters

Some foods and household products are very acidic, while others are basic. From your prior experiences, can you tell which ones belong to which category?

8.1 What Are Acids and Bases?

We frequently encounter acids and bases in our daily lives. Oranges, lemons, and vinegar are examples of acidic foods, and sulfuric acid is in our automobile batteries. As for bases, we take antacid tablets for heartburn and use household ammonia as a cleaning agent. What do these substances have in common? Why are acids and bases usually discussed together?

In 1884, a young Swedish chemist named Svante Arrhenius (1859–1927) answered the first question by proposing what was then a new definition of acids and bases. According to the Arrhenius definition, an **acid** is a substance that produces H_3O^+ ions in aqueous solution, and a **base** is a substance that produces OH^- ions in aqueous solution.

This definition of acid is a slight modification of the original Arrhenius definition, which stated that an acid produces hydrogen ions, H^+. Today we know that H^+ ions cannot exist in water. An H^+ ion is a bare proton, and a charge of $+1$ is too concentrated to exist on such a tiny particle (Section 3.2). Therefore, an H^+ ion in water immediately combines with an H_2O molecule to give a **hydronium ion**, H_3O^+.

Hydronium ion The H_3O^+ ion

$$H^+(aq) + H_2O(\ell) \longrightarrow H_3O^+(aq)$$
Hydronium ion

Apart from this modification, the Arrhenius definitions of acid and base are still valid and useful today, as long as we are talking about aqueous solutions. Although we know that acidic aqueous solutions do not contain H^+ ions, we frequently use the terms "H^+" and "proton" when we really mean "H_3O^+". The three terms are generally used interchangeably.

When an acid dissolves in water, it reacts with the water to produce H_3O^+. For example, hydrogen chloride, HCl, in its pure state is a poisonous gas. When HCl dissolves in water, it reacts with a water molecule to give hydronium ion and chloride ion:

$$H_2O(\ell) + HCl(aq) \longrightarrow H_3O^+(aq) + Cl^-(aq)$$

Thus, a bottle labeled aqueous "HCl" is actually not HCl at all, but rather an aqueous solution of H_3O^+ and Cl^- ions in water.

We can show the transfer of a proton from an acid to a base by using a curved arrow. First, we write the Lewis structure (Section 2.6F) of each reactant and product. Then we use curved arrows to show the change in position of electron pairs during the reaction. The tail of the curved arrow is located at the electron pair. The head of the curved arrow shows the new position of the electron pair.

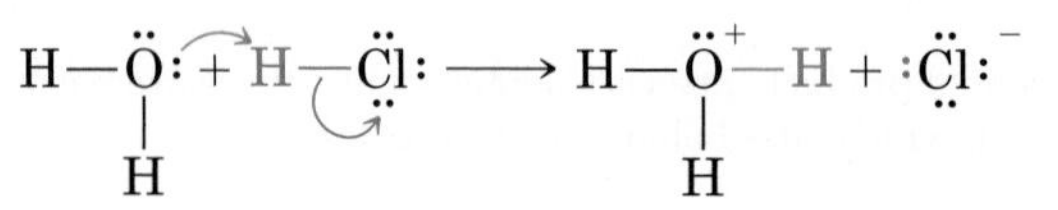

$$\text{H}-\ddot{\text{O}}: + \text{H}-\ddot{\text{Cl}}: \longrightarrow \text{H}-\overset{+}{\underset{|}{\ddot{\text{O}}}}-\text{H} + :\ddot{\text{Cl}}:^-$$
$$\underset{\text{H}}{|} \qquad\qquad\qquad \underset{\text{H}}{}$$

In this equation, the curved arrow on the left shows that an unshared pair of electrons on oxygen forms a new covalent bond with hydrogen. The curved arrow on the right shows that the pair of electrons of the H—Cl bond is given entirely to chlorine to form a chloride ion. Thus, in the reaction of HCl with H_2O, a proton is transferred from HCl to H_2O, and in the process, an O—H bond forms and an H—Cl bond is broken.

With bases, the situation is slightly different. Many bases are metal hydroxides, such as KOH, NaOH, $Mg(OH)_2$, and $Ca(OH)_2$. When these ionic solids dissolve in water, their ions merely separate, and each ion is solvated by water molecules (Section 6.6A). For example,

$$NaOH(s) \xrightarrow{H_2O} Na^+(aq) + OH^-(aq)$$

Other bases are not hydroxides. Instead, they produce OH^- ions in water by reacting with water molecules. The most important example of this kind of base is ammonia, NH_3, a poisonous gas. When ammonia dissolves in water, it reacts with water to produce ammonium ions and hydroxide ions.

$$NH_3(aq) + H_2O(\ell) \rightleftharpoons NH_4^+(aq) + OH^-(aq)$$

As we will see in Section 8.2, ammonia is a weak base, and the position of the equilibrium for its reaction with water lies considerably toward the left. In a 1.0 M solution of NH_3 in water, for example, only about 4 molecules of NH_3 out of every 1000 react with water to form NH_4^+ and OH^-. Thus, when ammonia is dissolved in water, it exists primarily as hydrated NH_3 molecules. Nevertheless, some OH^- ions are produced, and therefore, NH_3 is a base.

Bottles of NH_3 in water are sometimes labeled "ammonium hydroxide" or "NH_4OH," but this gives a false impression of what is really in the bottle. Most of the NH_3 molecules have not reacted with the water, so the bottle contains mostly NH_3 and H_2O and only a little NH_4^+ and OH^-.

We indicate how the reaction of ammonia with water takes place by using curved arrows to show the transfer of a proton from a water molecule to an ammonia molecule. Here, the curved arrow on the left shows that the unshared pair of electrons on nitrogen forms a new covalent bond with a hydrogen of a water molecule. At the same time as the new N—H bond forms, an O—H bond of a water molecule breaks and the pair of electrons forming the H—O bond moves entirely to oxygen, forming OH^-.

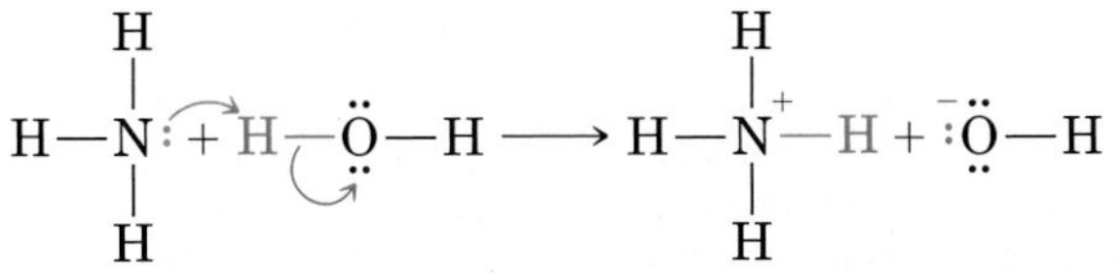

Thus, ammonia produces an OH^- ion by taking H^+ from a water molecule and leaving OH^- behind.

8.2 How Do We Define the Strength of Acids and Bases?

All acids are not equally strong. According to the Arrhenius definition, a **strong acid** is one that reacts completely or almost completely with water to form H_3O^+ ions. Table 8.1 gives the names and molecular formulas for six of the most common strong acids. They are strong acids because when they dissolve in water, they dissociate completely to give H_3O^+ ions.

Strong acid An acid that ionizes completely in aqueous solution

Table 8.1 Strong Acids and Bases

Acid Formula	Name	Base Formula	Name
HCl	Hydrochloric acid	LiOH	Lithium hydroxide
HBr	Hydrobromic acid	NaOH	Sodium hydroxide
HI	Hydroiodic acid	KOH	Potassium hydroxide
HNO_3	Nitric acid	$Ba(OH)_2$	Barium hydroxide
H_2SO_4	Sulfuric acid	$Ca(OH)_2$	Calcium hydroxide
$HClO_4$	Perchloric acid	$Sr(OH)_2$	Strontium hydroxide

Weak acids produce a much smaller concentration of H_3O^+ ions. Acetic acid, for example, is a weak acid. In water it exists primarily as acetic acid molecules; only a few acetic acid molecules (4 out of every 1000) are converted to acetate ions.

Weak acid An acid that is only partially ionized in aqueous solution

$$CH_3COOH(aq) + H_2O(\ell) \rightleftharpoons CH_3COO^-(aq) + H_3O^+(aq)$$

Acetic acid   Acetate ion

There are six common **strong bases** (Table 8.1), all of which are metal hydroxides. They are strong bases because, when they dissolve in water, they ionize completely to give OH^- ions. Another base, $Mg(OH)_2$, dissociates almost completely once dissolved, but it is also very insoluble in water to begin with.

Strong base A base that ionizes completely in aqueous solution

Chemical Connections 8A

Some Important Acids and Bases

STRONG ACIDS Sulfuric acid, H_2SO_4, is used in many industrial processes, such as manufacturing fertilizer, dyes and pigments, and rayon. In fact, sulfuric acid is one of the most widely produced single chemicals in the United States.

Hydrochloric acid, HCl, is an important acid in chemistry laboratories. Pure HCl is a gas, and the HCl in laboratories is an aqueous solution. HCl is the acid in the gastric fluid in your stomach, where it is secreted at a strength of about 5% w/v.

Nitric acid, HNO_3, is a strong oxidizing agent. A drop of it causes the skin to turn yellow because the acid reacts with skin proteins. A yellow color upon contact with nitric acid has long been a test for proteins.

WEAK ACIDS Acetic acid, CH_3COOH, is present in vinegar (about 5%). Pure acetic acid is called glacial acetic acid because of its melting point of 17°C, which means that it freezes on a moderately cold day.

Boric acid, H_3BO_3, is a solid. Solutions of boric acid in water were once used as antiseptics, especially for eyes. Boric acid is toxic when swallowed.

Phosphoric acid, H_3PO_4, is one of the strongest of the weak acids. The ions produced from it—$H_2PO_4^-$, HPO_4^{2-}, and PO_4^{3-}—are important in biochemistry (see Section 27.3).

STRONG BASES Sodium hydroxide, NaOH, also called lye, is the most important of the strong bases. It is a solid

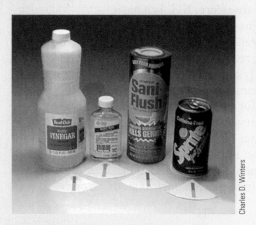

Weak acids are found in many common materials. In the foreground are strips of litmus paper that have been dipped into solutions of these materials. Acids turn litmus paper red.

Charles D. Winters

Weak bases are also common in many household products. These cleaning agents all contain weak bases. Bases turn litmus paper blue.

Charles D. Winters

whose aqueous solutions are used in many industrial processes, including the manufacture of glass and soap. Potassium hydroxide, KOH, also a solid, is used for many of the same purposes as NaOH.

WEAK BASES Ammonia, NH_3, the most important weak base, is a gas with many industrial uses. One of its chief uses is for fertilizers. A 5% solution is sold in supermarkets as a cleaning agent, and weaker solutions are used as "spirits of ammonia" to revive people who have fainted.

Magnesium hydroxide, $Mg(OH)_2$, is a solid that is insoluble in water. A suspension of about 8% $Mg(OH)_2$ in water is called milk of magnesia and is used as a laxative. $Mg(OH)_2$ is also used to treat wastewater in metal-processing plants and as a flame retardant in plastics.

Test your knowledge with Problem 8.73.

Weak base A base that is only partially ionized in aqueous solution

We classify it as a **weak base**. As we saw in Section 8.1, ammonia is a weak base because the equilibrium for its reaction with water lies far to the left.

It is important to understand that the strength of an acid or a base is not related to its concentration. HCl is a strong acid, whether it is concentrated or dilute, because it dissociates completely in water to chloride ions and hydronium ions. Acetic acid is a weak acid, whether it is concentrated or dilute, because the equilibrium for its reaction with water lies far to the left. When acetic acid dissolves in water, most of it is present as undissociated CH_3COOH molecules.

$$HCl(aq) + H_2O(\ell) \longrightarrow Cl^-(aq) + H_3O^+(aq)$$

$$\underset{\text{Acetic acid}}{CH_3COOH(aq)} + H_2O(\ell) \rightleftharpoons \underset{\text{Acetate ion}}{CH_3COO^-(aq)} + H_3O^+(aq)$$

In Section 6.6C, we saw that electrolytes (substances that produce ions in aqueous solution) can be strong or weak. The strong acids and bases in Table 8.1 are strong electrolytes. Almost all other acids and bases are weak electrolytes.

8.3 What Are Conjugate Acid–Base Pairs?

The Arrhenius definitions of acid and base are very useful in aqueous solutions. But what if water is not involved? In 1923, the Danish chemist Johannes Brønsted and the English chemist Thomas Lowry independently proposed the following definitions: An **acid** is a proton donor, a **base** is a proton acceptor, and an **acid–base reaction** is a proton-transfer reaction. Furthermore, according to the Brønsted-Lowry definitions, any pair of molecules or ions that can be interconverted by transfer of a proton is called a **conjugate acid–base pair**. When an acid transfers a proton to a base, the acid is converted to its **conjugate base.** When a base accepts a proton, it is converted to its **conjugate acid**.

We can illustrate these relationships by examining the reaction between acetic acid and ammonia:

Conjugate acid–base pair A pair of molecules or ions that are related to one another by the gain or loss of a proton

Conjugate base In the Brønsted-Lowry theory, a substance formed when an acid donates a proton to another molecule or ion

Conjugate acid In the Brønsted-Lowry theory, a substance formed when a base accepts a proton

$$CH_3COOH + NH_3 \rightleftharpoons CH_3COO^- + NH_4^+$$

Acetic acid   Ammonia   Acetate   Ammonium
                           ion   ion

(Acid)   (Base)   (Conjugate   (Conjugate
                                 base of   acid of
                              acetic acid)   ammonia)

We can use curved arrows to show how this reaction takes place. The curved arrow on the right shows that the unshared pair of electrons on nitrogen becomes shared to form a new H—N bond. At the same time that the H—N bond forms, the O—H bond breaks and the electron pair of the O—H bond moves entirely to oxygen to form —O⁻ of the acetate ion. The result of these two electron-pair shifts is the transfer of a proton from an acetic acid molecule to an ammonia molecule:

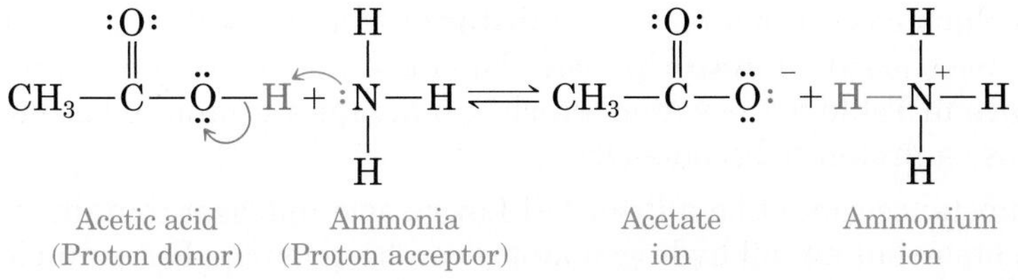

Acetic acid   Ammonia   Acetate   Ammonium
(Proton donor)   (Proton acceptor)   ion   ion

Table 8.2 gives examples of common acids and their conjugate bases. As you study the examples of conjugate acid–base pairs in Table 8.2, note the following points:

1. An acid can be positively charged, neutral, or negatively charged. Examples of these charge types are H_3O^+, H_2CO_3, and $H_2PO_4^-$, respectively.

2. A base can be negatively charged or neutral. Examples of these charge types are PO_4^{3-} and NH_3, respectively.

3. Acids are classified as monoprotic, diprotic, or triprotic depending on the number of protons each may give up. Examples of **monoprotic acids** include HCl, HNO_3, and CH_3COOH. Examples of **diprotic acids**

Monoprotic acid An acid that can give up only one proton

Diprotic acid An acid that can give up two protons

Triprotic acid An acid that can give up three protons

Table 8.2 Some Acids and Their Conjugate Bases

	Acid	Name		Conjugate Base	Name	
Strongly reducing	HI	Hydroiodic acid		I^-	Iodide ion	**Weak Bases**
	HCl	Hydrochloric acid		Cl^-	Chloride ion	
	H_2SO_4	Sulfuric acid		HSO_4^-	Hydrogen sulfate ion	
	HNO_3	Nitric acid		NO_3^-	Nitrate ion	
	H_3O^+	Hydronium ion		H_2O	Water	
	HSO_4^-	Hydrogen sulfate ion		SO_4^{2-}	Sulfate ion	
	H_3PO_4	Phosphoric acid		$H_2PO_4^-$	Dihydrogen phosphate ion	
	CH_3COOH	Acetic acid		CH_3COO^-	Acetate ion	
	H_2CO_3	Carbonic acid		HCO_3^-	Bicarbonate ion	
	H_2S	Hydrogen sulfide		HS^-	Hydrogen sulfide ion	
	$H_2PO_4^-$	Dihydrogen phosphate ion		HPO_4^{2-}	Hydrogen phosphate ion	
	NH_4^+	Ammonium ion		NH_3	Ammonia	
	HCN	Hydrocyanic acid		CN^-	Cyanide ion	
	C_6H_5OH	Phenol		$C_6H_5O^-$	Phenoxide ion	
	HCO_3^-	Bicarbonate ion		CO_3^{2-}	Carbonate ion	
	HPO_4^{2-}	Hydrogen phosphate ion		PO_4^{3-}	Phosphate ion	
Weakly reducing	H_2O	Water		OH^-	Hydroxide ion	**Strong Bases**
	C_2H_5OH	Ethanol		$C_2H_5O^-$	Ethoxide ion	

A box of Arm & Hammer baking soda (sodium bicarbonate). Sodium bicarbonate is composed of Na^+ and HCO_3^-, the amphiprotic bicarbonate ion.

Charles D. Winters

Amphiprotic A substance that can act as either an acid or a base

include H_2SO_4 and H_2CO_3. An example of a **triprotic acid** is H_3PO_4. Carbonic acid, for example, loses one proton to become bicarbonate ion and then a second proton to become carbonate ion.

$$H_2CO_3 + H_2O \rightleftharpoons HCO_3^- + H_3O^+$$
Carbonic   Bicarbonate
acid   ion

$$HCO_3^- + H_2O \rightleftharpoons CO_3^{2-} + H_3O^+$$
Bicarbonate   Carbonate
ion   ion

4. Several molecules and ions appear in both the acid and conjugate base columns; that is, each can function as either an acid or a base. The bicarbonate ion, HCO_3^-, for example, can give up a proton to become CO_3^{2-} (in which case it is an acid) or it can accept a proton to become H_2CO_3 (in which case it is a base). A substance that can act as either an acid or a base is called **amphiprotic**. The most important amphiprotic substance in Table 8.2 is water, which can accept a proton to become H_3O^+ or lose a proton to become OH^-.

5. A substance cannot be a Brønsted-Lowry acid unless it contains a hydrogen atom, but not all hydrogen atoms can be given up. For example, acetic acid, CH_3COOH, has four hydrogens but is monoprotic; it gives up only one of them. Similarly, phenol, C_6H_5OH, gives up only one of its six hydrogens:

$$C_6H_5OH + H_2O \rightleftharpoons C_6H_5O^- + H_3O^+$$
Phenol   Phenoxide
ion

This is because a hydrogen must be bonded to a strongly electronegative atom, such as oxygen or a halogen, to be acidic.

6. There is an inverse relationship between the strength of an acid and the strength of its conjugate base: The stronger the acid, the weaker its conjugate base. HI, for example, is the strongest acid listed in Table 8.2 and I^-,

its conjugate base, is the weakest base. As another example, CH_3COOH (acetic acid) is a stronger acid than H_2CO_3 (carbonic acid); conversely, CH_3COO^- (acetate ion) is a weaker base than HCO_3^- (bicarbonate ion).

Example 8.1 Diprotic Acids

Show how the amphiprotic ion hydrogen sulfate, HSO_4^-, can react as both an acid and a base.

Strategy

For a molecule to act as both an acid and a base, it must be able to both give up a hydrogen ion and accept a hydrogen ion. Therefore, we write two equations, one donating a hydrogen ion and the other accepting one.

Solution

Hydrogen sulfate reacts as an acid in the equation shown below:

$$HSO_4^- + H_2O \rightleftharpoons H_3O^+ + SO_4^{2-}$$

It can react as a base in the equation shown below:

$$HSO_4^- + H_3O^+ \rightleftharpoons H_2O + H_2SO_4$$

Problem 8.1

Draw the acid and base reactions for the amphiprotic ion HPO_4^{2-}.

How To . . .

Name Common Acids

The names of common acids are derived from the name of the anion that they produce when they dissociate. There are three common endings for these ions: –*ide*, –*ate, and* –*ite.*

Acids that dissociate into ions with the suffix –*ide* are named *hydro* _____ *ic acid*			
Cl^-	Chlor*ide* ion	HCl	*hydro*chlor*ic acid*
F^-	Fluor*ide* ion	HF	*hydro*fluor*ic acid*
CN^-	Cyan*ide* ion	HCN	*hydro*cyan*ic acid*

Acids that dissociate into ions with the suffix –*ate* are named _____ *ic acid*			
SO_4^{2-}	Sulf*ate* ion	H_2SO_4	Sulfur*ic acid*
PO_4^{3-}	Phosph*ate* ion	H_3PO_4	Phosphor*ic acid*
NO_3^-	Nitr*ate* ion	HNO_3	Nitr*ic acid*

Acids that dissociate into ions with the suffix –*ite* are named _____ *ous acid*			
SO_3^{2-}	Sulf*ite* ion	H_2SO_3	Sulfur*ous acid*
NO_2^-	Nitr*ite* ion	HNO_2	Nitr*ous acid*

8.4 How Can We Tell the Position of Equilibrium in an Acid–Base Reaction?

We know that HCl reacts with H_2O according to the following equilibrium:

$$HCl + H_2O \rightleftharpoons Cl^- + H_3O^+$$

We also know that HCl is a strong acid, which means the position of this equilibrium lies very far to the right. In fact, this equilibrium lies so far to the right that out of every 10,000 HCl molecules dissolved in water, all but one react with water molecules to give Cl^- and H_3O^+.

For this reason, we usually write the acid reaction of HCl with a unidirectional arrow, as follows:

$$HCl + H_2O \longrightarrow Cl^- + H_3O^+$$

As we have also seen, acetic acid reacts with H_2O according to the following equilibrium:

$$\underset{\text{Acetic acid}}{CH_3COOH} + H_2O \rightleftharpoons \underset{\text{Acetate ion}}{CH_3COO^-} + H_3O^+$$

Acetic acid is a weak acid. Only a few acetic acid molecules react with water to give acetate ions and hydronium ions, and the major species present in equilibrium in aqueous solution are CH_3COOH and H_2O. The position of this equilibrium, therefore, lies very far to the left.

In these two acid–base reactions, water is the base. But what if we have a base other than water as the proton acceptor? How can we determine which are the major species present at equilibrium? That is, how can we determine if the position of equilibrium lies toward the left or toward the right?

As an example, let us examine the acid–base reaction between acetic acid and ammonia to form acetate ion and ammonium ion. As indicated by the question mark over the equilibrium arrow, we want to determine whether the position of this equilibrium lies toward the left or toward the right.

$$\underset{\substack{\text{Acetic acid} \\ \text{(Acid)}}}{CH_3COOH} + \underset{\substack{\text{Ammonia} \\ \text{(Base)}}}{NH_3} \overset{?}{\rightleftharpoons} \underset{\substack{\text{Acetate ion} \\ \text{(Conjugate base} \\ \text{of } CH_3COOH)}}{CH_3COO^-} + \underset{\substack{\text{Ammonium ion} \\ \text{(Conjugate acid} \\ \text{of } NH_3)}}{NH_4^+}$$

In this equilibrium, there are two acids present: acetic acid and ammonium ion. There are also two bases present: ammonia and acetate ion. One way to analyze this equilibrium is to view it as a competition of the two bases, ammonia and acetate ion, for a proton. Which is the stronger base? The information we need to answer this question is found in Table 8.2. We first determine which conjugate acid is the stronger acid and then use this information along with the fact that the stronger the acid, the weaker its conjugate base. From Table 8.2, we see that CH_3COOH is the stronger acid, which means that CH_3COO^- is the weaker base. Conversely, NH_4^+ is the weaker acid, which means that NH_3 is the stronger base. We can now label the relative strengths of each acid and base in this equilibrium:

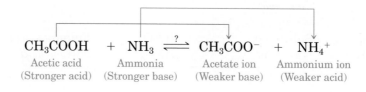

In an acid–base reaction, the equilibrium position always favors reaction of the stronger acid and stronger base to form the weaker acid and weaker base. Thus, at equilibrium, the major species present are the weaker acid and the weaker base. In the reaction between acetic acid and ammonia, therefore, the equilibrium lies to the right and the major species present are acetate ion and ammonium ion:

$$CH_3COOH \ + \ NH_3 \ \rightleftharpoons \ CH_3COO^- \ + \ NH_4^+$$

Acetic acid	Ammonia	Acetate ion	Ammonium ion
(Stronger acid)	(Stronger base)	(Weaker base)	(Weaker acid)

To summarize, we use the following four steps to determine the position of an acid–base equilibrium:

1. Identify the two acids in the equilibrium; one is on the left side of the equilibrium, and the other is on the right side.
2. Using the information in Table 8.2, determine which acid is the stronger acid and which acid is the weaker acid.
3. Identify the stronger base and the weaker base. Remember that the stronger acid gives the weaker conjugate base and the weaker acid gives the stronger conjugate base.
4. The stronger acid and stronger base react to give the weaker acid and weaker base. The position of equilibrium, therefore, lies on the side of the weaker acid and weaker base.

Example 8.2 Acid/Base Pairs

For each acid–base equilibrium, label the stronger acid, the stronger base, the weaker acid, and the weaker base. Then predict whether the position of equilibrium lies toward the right or toward the left.

(a) $H_2CO_3 \ + \ OH^- \rightleftharpoons HCO_3^- \ + \ H_2O$
(b) $HPO_4^{2-} \ + \ NH_3 \rightleftharpoons PO_4^{3-} \ + \ NH_4^+$

Strategy

Use Table 8.2 to identify the stronger acid from the weaker acid and the stronger base from the weaker base. Once you have done that, determine in which direction the equilibrium lies. It always lies in the direction of the stronger components moving towards the weaker components.

Solution

Arrows connect the conjugate acid–base pairs, with the red arrows showing the stronger acid. The position of equilibrium in (a) lies toward the right. In (b) it lies toward the left.

(a)

$$H_2CO_3 \ + \ OH^- \ \rightleftharpoons \ HCO_3^- \ + \ H_2O$$

Stronger acid	Stronger base	Weaker base	Weaker acid

(b)

$$HPO_4^{2-} \ + \ NH_3 \ \rightleftharpoons \ PO_4^{3-} \ + \ NH_4^+$$

Weaker acid	Weaker base	Stronger base	Stronger acid

Problem 8.2

For each acid–base equilibrium, label the stronger acid, the stronger base, the weaker acid, and the weaker base. Then predict whether the position of equilibrium lies toward the right or the left.

(a) $H_3O^+ + I^- \rightleftharpoons H_2O + HI$

(b) $CH_3COO^- + H_2S \rightleftharpoons CH_3COOH + HS^-$

8.5 How Do We Use Acid Ionization Constants?

In Section 8.2, we learned that acids vary in the extent to which they produce H_3O^+ when added to water. Because the ionizations of weak acids in water are all equilibria, we can use equilibrium constants (Section 7.6) to tell us quantitatively just how strong any weak acid is. The reaction that takes place when a weak acid, HA, is added to water is:

$$HA + H_2O \rightleftharpoons A^- + H_3O^+$$

The equilibrium constant expression for this ionization is:

$$K_a = \frac{[A^-][H_3O^+]}{[HA]}$$

Acid ionization constant (K_a)
An equilibrium constant for the ionization of an acid in aqueous solution to H_3O^+ and its conjugate base; also called an acid dissociation constant

The subscript a on K_a shows that it is an equilibrium constant for the ionization of an acid, so K_a is called the **acid ionization constant**.

The value of the acid ionization constant for acetic acid, for example, is 1.8×10^{-5}. Because acid ionization constants for weak acids are numbers with negative exponents, we often use an algebraic trick to turn them into numbers that are easier to use. To do so, we take the negative logarithm of the number. Acid strengths are therefore expressed as $-\log K_a$, which we call the pK_a. The "p" of anything is just the negative logarithm of that given item. The pK_a of acetic acid is $-\log (1.8 \times 10^{-5})$ which is equal to 4.75. Table 8.3 gives names, molecular formulas, and values of K_a and pK_a for some weak acids. As you study the entries in this table, note the inverse relationship between the values of K_a and pK_a. The weaker the acid, the smaller its K_a, but the larger its pK_a. Also note that for the common logarithm of each measured K_a value, the number of digits after the decimal point equals the number of significant figures in the original number. For example, the K_a of phenol is 1.3×10^{-10}. Because there are two significant figures (Section 1.3) in this number, the $pK_a = -\log (1.3 \times 10^{-10}) = 9.89$

Table 8.3 K_a and pK_a Values for Some Weak Acids

Formula	Name	K_a	pK_a
H_3PO_4	Phosphoric acid	7.5×10^{-3}	2.12
HCOOH	Formic acid	1.8×10^{-4}	3.75
$CH_3CH(OH)COOH$	Lactic acid	1.4×10^{-4}	3.86
CH_3COOH	Acetic acid	1.8×10^{-5}	4.75
H_2CO_3	Carbonic acid	4.3×10^{-7}	6.37
$H_2PO_4^-$	Dihydrogen phosphate ion	6.2×10^{-8}	7.21
H_3BO_3	Boric acid	7.3×10^{-10}	9.14
NH_4^+	Ammonium ion	5.6×10^{-10}	9.25
HCN	Hydrocyanic acid	4.9×10^{-10}	9.31
C_6H_5OH	Phenol	1.3×10^{-10}	9.89
HCO_3^-	Bicarbonate ion	5.6×10^{-11}	10.25
HPO_4^{2-}	Hydrogen phosphate ion	2.2×10^{-13}	12.66

Increasing acid strength

(two significant figures after the decimal point). We will conform to this rule of significant figures when performing calculations involving logarithms.

One reason for the importance of K_a is that it immediately tells us how strong an acid is. For example, Table 8.3 shows us that although acetic acid, formic acid, and phenol are all weak acids, their strengths as acids are not the same. Formic acid, with a K_a of 1.8×10^{-4}, is stronger than acetic acid, whereas phenol, with a K_a of 1.3×10^{-10}, is much weaker than acetic acid. Phosphoric acid is the strongest of the weak acids. We can tell that an acid is classified as a weak acid by the fact that we list a pK_a for it, and the pK_a is a positive number. If we tried to take the negative logarithm of the K_a for a strong acid, we would get a negative number.

Example 8.3 pK_a

K_a for benzoic acid is 6.5×10^{-5}. What is the pK_a of this acid?

Strategy

The pK_a is $-\log K_a$. Thus, use your calculator to find the log of the K_a and then take the negative of it.

Solution

Take the logarithm of 6.5×10^{-5} on your scientific calculator. The answer is -4.19. Because pK_a is equal to $-\log K_a$, you must multiply this value by -1 to get pK_a. The pK_a of benzoic acid is 4.19.

Problem 8.3

K_a for hydrocyanic acid, HCN, is 4.9×10^{-10}. What is its pK_a?

Example 8.4 Acid Strength

Which is the stronger acid?
(a) Benzoic acid with a K_a of 6.5×10^{-5} or hydrocyanic acid with a K_a of 4.9×10^{-10}?
(b) Boric acid with a pK_a of 9.14 or carbonic acid with a pK_a of 6.37?

Strategy

Relative acid strength is determined by comparing the K_a values or the pK_a values. If using K_a values, the stronger acid has the larger K_a. If using pK_a values, the stronger acid has the smaller pK_a.

Solution

(a) Benzoic acid is the stronger acid; it has the larger K_a value.
(b) Carbonic acid is the stronger acid; it has the smaller pK_a.

Problem 8.4

Which is the stronger acid?
(a) Carbonic acid, $pK_a = 6.37$, or ascorbic acid (vitamin C), $pK_a = 4.10$?
(b) Aspirin, $pK_a = 3.49$, or acetic acid, $pK_a = 4.75$?

All of these fruits and fruit drinks contain organic acids.

Charles D. Winters

8.6 What Are the Properties of Acids and Bases?

Today's chemists do not taste the substances they work with, but 200 years ago they routinely did so. That is how we know that acids taste sour and

FIGURE 8.1 Acids react with metals. A ribbon of magnesium metal reacts with aqueous HCl to give H_2 gas and aqueous $MgCl_2$.

bases taste bitter. The sour taste of lemons, vinegar, and many other foods, for example, is due to the acids they contain.

A. Neutralization

The most important reaction of acids and bases is that they react with each other in a process called neutralization. This name is appropriate because, when a strong corrosive acid such as hydrochloric acid reacts with a strong caustic base such as sodium hydroxide, the product (a solution of ordinary table salt in water) has neither acidic nor basic properties. We call such a solution neutral. Section 8.9 discusses neutralization reactions in detail.

B. Reaction with Metals

Strong acids react with certain metals (called active metals) to produce hydrogen gas, H_2, and a salt. Hydrochloric acid, for example, reacts with magnesium metal to give the salt magnesium chloride and hydrogen gas (Figure 8.1).

$$Mg(s) + 2HCl(aq) \longrightarrow MgCl_2(aq) + H_2(g)$$

| Magnesium | Hydrochloric acid | | Magnesium chloride | Hydrogen |

The reaction of an acid with an active metal to give a salt and hydrogen gas is a redox reaction (Section 4.4). The metal is oxidized to a metal ion and H^+ is reduced to H_2 as shown in the following net ionic equation:

$$Mg(s) + 2H_3O^+(aq) \longrightarrow Mg^{2+}(aq) + H_2(g) + 2H_2O(\ell)$$

Recall in Section 8.1, we learned that H_3O^+ is commonly written as H^+, although we know that acidic aqueous solutions do not contain H^+ ions. Therefore, the reaction can also be written as:

$$Mg(s) + 2H^+(aq) \longrightarrow Mg^{2+}(aq) + H_2(g)$$

Whether or not a reaction occurs between a metal and an acid depends on how easily each substance is reduced or oxidized. By noting the experimental results obtained from multiple reactions, we construct an **activity series**, which ranks the elements in order of their reducing abilities in aqueous solution. As noted in Table 8.4, the metals located above H_2 give up electrons and are stronger

Table 8.4 Activity Series of Certain Elements

	Oxidation Reaction	
Strongly reducing ↑	$Li \longrightarrow Li^+ + e^-$	
	$K \longrightarrow K^+ + e^-$	
	$Ca \longrightarrow Ca^{2+} + 2e^-$	
	$Na \longrightarrow Na^+ + e^-$	
	$Mg \longrightarrow Mg^{2+} + 2e^-$	
	$Al \longrightarrow Al^{3+} + 3e^-$	
	$Mn \longrightarrow Mn^{2+} + 2e^-$	These metals react rapidly with aqueous H_3O^+ ions (or acid) and release H_2 gas.
	$Zn \longrightarrow Zn^{2+} + 2e^-$	
	$Cr \longrightarrow Cr^{3+} + 3e^-$	
	$Fe \longrightarrow Fe^{2+} + 2e^-$	
	$Cd \longrightarrow Cd^{2+} + 2e^-$	
	$Ni \longrightarrow Ni^{2+} + 2e^-$	
	$Sn \longrightarrow Sn^{2+} + 2e^-$	
	$Pb \longrightarrow Pb^{2+} + 2e^-$	
	$H_2 \longrightarrow 2H^+ + 2e^-$	
Weakly reducing	$Cu \longrightarrow Cu^{2+} + 2e^-$	These metals do not react with aqueous H_3O^+ ions (or acid) and do not release H_2 gas.
	$Ag \longrightarrow Ag^+ + e^-$	
	$Au \longrightarrow Au^+ + e^-$	

reducing agents, resulting in a reaction between a given metal and an acid. In the preceding example, a ribbon of magnesium metal reacts with aqueous hydrochloric acid because Mg is ranked higher than H_2 on the activity series.

In contrast, metals located below H_2 do not give up electrons as readily and are weaker reducing agents, resulting in no reaction between a given metal and an acid. For example, silver metal will not react with aqueous nitric acid because H_2 is ranked higher than Ag on the activity series.

C. Reaction with Metal Hydroxides

Acids react with metal hydroxides to give a salt and water.

$$HCl(aq) \ + \ KOH(aq) \ \longrightarrow \ H_2O(\ell) \ + \ KCl(aq)$$

Hydrochloric   Potassium   Water   Potassium
acid   hydroxide   chloride

Both the acid and the metal hydroxide are ionized in aqueous solution. Furthermore, the salt formed is an ionic compound that is present in aqueous solution as anions and cations. Therefore, the actual equation for the reaction of HCl and KOH could be written showing all of the ions present (Section 4.3):

$$H_3O^+ + Cl^- + K^+ + OH^- \longrightarrow 2H_2O + Cl^- + K^+$$

We usually simplify this equation by omitting the spectator ions (Section 4.3), which gives the following equation for the net ionic reaction of any strong acid and strong base to give a soluble salt and water:

$$H_3O^+ + OH^- \longrightarrow 2H_2O$$

D. Reaction with Metal Oxides

Strong acids react with metal oxides to give water and a soluble salt, as shown in the following net ionic equation:

$$2H_3O^+(aq) + CaO(s) \longrightarrow 3H_2O(\ell) + Ca^{2+}(aq)$$

Calcium
oxide

E. Reaction with Carbonates and Bicarbonates

When a strong acid is added to a carbonate such as sodium carbonate, bubbles of carbon dioxide gas are rapidly given off. The overall reaction is a summation of two reactions. In the first reaction, carbonate ion reacts with H_3O^+ to give carbonic acid. Almost immediately, in the second reaction, carbonic acid decomposes to carbon dioxide and water. The following equations show the individual reactions and then the overall reaction:

$$2H_3O^+(aq) + CO_3{}^{2-}(aq) \longrightarrow H_2CO_3(aq) + 2H_2O(\ell)$$
$$H_2CO_3(aq) \longrightarrow CO_2(g) + H_2O(\ell)$$
$$\overline{2H_3O^+(aq) + CO_3{}^{2-}(aq) \longrightarrow CO_2(g) + 3H_2O(\ell)}$$

Strong acids also react with bicarbonates such as potassium bicarbonate to give carbon dioxide and water:

$$H_3O^+(aq) + HCO_3{}^-(aq) \longrightarrow H_2CO_3(aq) + H_2O(\ell)$$
$$H_2CO_3(aq) \longrightarrow CO_2(g) + H_2O(\ell)$$
$$\overline{H_3O^+(aq) + HCO_3{}^-(aq) \longrightarrow CO_2(g) + 2H_2O(\ell)}$$

To generalize, any acid stronger than carbonic acid will react with carbonate or bicarbonate ion to give CO_2 gas.

The production of CO_2 is what makes bread doughs and cake batters rise. The earliest method used to generate CO_2 for this purpose involved the addition of yeast, which catalyzes the fermentation of carbohydrates to produce carbon dioxide and ethanol (Chapter 28):

$$C_6H_{12}O_6 \xrightarrow{\text{Yeast}} 2CO_2 + 2C_2H_5OH$$

Glucose   Ethanol

The production of CO_2 by fermentation, however, is slow. Sometimes it is desirable to have its production take place more rapidly, in which case bakers use the reaction of $NaHCO_3$ (sodium bicarbonate, also called **baking soda**) and a weak acid. But which weak acid? Vinegar (a 5% solution of acetic acid in water) would work, but it has a potential disadvantage—it imparts a particular flavor to foods. For a weak acid that imparts little or no flavor, bakers use either sodium dihydrogen phosphate, NaH_2PO_4, or potassium dihydrogen phosphate, KH_2PO_4. The two salts do not react when they are dry, but when mixed with water in a dough or batter, they react quite rapidly to produce CO_2. The production of CO_2 is even more rapid in an oven!

$$H_2PO_4^-(aq) + H_2O(\ell) \rightleftharpoons HPO_4^{2-}(aq) + H_3O^+(aq)$$
$$HCO_3^-(aq) + H_3O^+(aq) \longrightarrow CO_2(g) + 2H_2O(\ell)$$
$$\overline{H_2PO_4^-(aq) + HCO_3^-(aq) \longrightarrow HPO_4^{2-}(aq) + CO_2(g) + H_2O(\ell)}$$

Baking powder contains a weak acid, either sodium or potassium dihydrogen phosphate, and a weak base, sodium or potassium bicarbonate. When they are mixed with water, they react to produce the bubbles of CO_2 seen in this picture.

Charles D. Winters

F. Reaction with Ammonia and Amines

Any acid stronger than NH_4^+ (Table 8.2) is strong enough to react with NH_3 to form a salt. In the following reaction, the salt formed is ammonium chloride, NH_4Cl, which is shown as it would be ionized in aqueous solution:

$$HCl(aq) + NH_3(aq) \longrightarrow NH_4^+(aq) + Cl^-(aq)$$

In Chapter 16, we will meet a family of compounds called amines, which are similar to ammonia except that one or more of the three hydrogen atoms of ammonia are replaced by carbon groups. A typical amine is methylamine, CH_3NH_2. The base strength of most amines is similar to that of NH_3, which means that amines also react with acids to form salts. The salt formed in the reaction of methylamine with HCl is methylammonium chloride, shown here as it would be ionized in aqueous solution:

$$HCl(aq) + CH_3NH_2(aq) \longrightarrow CH_3NH_3^+(aq) + Cl^-(aq)$$

Methylamine   Methylammonium
                        ion

The reaction of ammonia and amines with acids to form salts is very important in the chemistry of the body, as we will see in later chapters.

8.7 What Are the Acidic and Basic Properties of Pure Water?

We have seen that an acid produces H_3O^+ ions in water and that a base produces OH^- ions. Suppose that we have absolutely pure water, with no added acid or base. Surprisingly enough, even pure water contains a very small

Chemical Connections 8B

Drugstore Antacids

Stomach fluid is normally quite acidic because of its HCl content. At some time, you probably have gotten "heartburn" caused by excess stomach acidity. To relieve your discomfort, you may have taken an antacid, which, as the name implies, is a substance that neutralizes acids—in other words, a base.

The word "antacid" is a medical term, not one used by chemists. It is, however, found on the labels of many medications available in drugstores and supermarkets. Almost all of them use bases such as $CaCO_3$, $Mg(OH)_2$, $Al(OH)_3$, and $NaHCO_3$ to decrease the acidity of the stomach.

Also in drugstores and supermarkets are nonprescription drugs labeled "acid reducers." Among these brands are Zantac, Tagamet, Pepcid, and Axid. Instead of neutralizing acidity, these compounds reduce the secretion of acid into the stomach. In larger doses (sold only with a prescription), some of these drugs are used in the treatment of stomach ulcers.

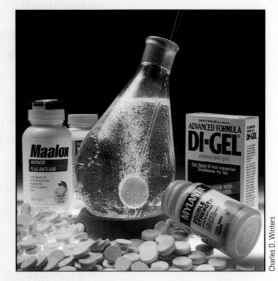

Commercial remedies for excess stomach acid.

Charles D. Winters

Test your knowledge with Problem 8.74.

number of H_3O^+ and OH^- ions. They are formed by the transfer of a proton from one molecule of water (the proton donor) to another (the proton acceptor).

$$H_2O \ + \ H_2O \ \rightleftharpoons \ OH^- \ + \ H_3O^+$$
$$\text{Acid} \qquad \text{Base} \qquad \text{Conjugate} \qquad \text{Conjugate}$$
$$\text{base of } H_2O \quad \text{acid of } H_2O$$

What is the extent of this reaction? We know from the information in Table 8.2 that in this equilibrium, H_3O^+ is the stronger acid and OH^- is the stronger base. Therefore, as shown by the arrows, the equilibrium for this reaction lies far to the left.

The equilibrium constant for the ionization of water, K_w, is called the **ion product of water**. In pure water at room temperature, K_w has a value of 1.0×10^{-14}.

K_w is the ion product of water, also called the water constant, and is equal to 1.0×10^{-14}.

$$K_w = [H_3O^+][OH^-]$$

$$K_w = 1.0 \times 10^{-14}$$

In pure water, H_3O^+ and OH^- form in equal amounts, so their concentrations must be equal. That is, in pure water:

$$[H_3O^+] = 1.0 \times 10^{-7} \text{ mol/L}$$
$$[OH^-] = 1.0 \times 10^{-7} \text{ mol/L}$$
$$\Big\} \text{ In pure water}$$

These are very small concentrations, not enough to make pure water a conductor of electricity. Pure water is not an electrolyte (Section 6.6C).

The equation for the ionization of water is important because it applies not only to pure water but also to any water solution. The product of $[H_3O^+]$ and $[OH^-]$ in any aqueous solution is equal to 1.0×10^{-14}. If, for example, we add 0.010 mol of HCl to 1 L of pure water, it reacts completely to give H_3O^+ ions and Cl^- ions. The concentration of H_3O^+ will be 0.010 M, or 1.0×10^{-2} M. This means that $[OH^-]$ must be $1.0 \times 10^{-14}/1.0 \times 10^{-2} = 1.0 \times 10^{-12}$ M.

OWL Interactive Example 8.5 Water Equation

The $[OH^-]$ of an aqueous solution is 1.0×10^{-4} M. What is its $[H_3O^+]$?

Strategy

To determine the hydrogen ion concentration when you know the hydroxide ion concentration, you simply divide the $[OH^-]$ into 10^{-14}.

Solution

We substitute into the equation:

$$[H_3O^+][OH^-] = 1.0 \times 10^{-14}$$

$$[H_3O^+] = \frac{1.0 \times 10^{-14}}{1.0 \times 10^{-4}} = 1.0 \times 10^{-10}\ M$$

Problem 8.5

The $[OH^-]$ of an aqueous solution is 1.0×10^{-12} M. What is its $[H_3O^+]$?

Aqueous solutions can have a very high $[H_3O^+]$, but the $[OH^-]$ must then be very low, and vice versa. Any solution with a $[H_3O^+]$ greater than 1.0×10^{-7} M is acidic. In such solutions, of necessity $[OH^-]$ must be less than 1.0×10^{-7} M. The higher the $[H_3O^+]$, the more acidic the solution. Similarly, any solution with an $[OH^-]$ greater than 1.0×10^{-7} M is basic. Pure water, in which $[H_3O^+]$ and $[OH^-]$ are equal (they are both 1.0×10^{-7} M), is neutral—that is, neither acidic nor basic.

How To . . .
Use Logs and Antilogs

When dealing with acids, bases, and buffers, we often have to use common or base 10 logarithms (logs). To most people, a logarithm is just a button they push on a calculator. Here we describe briefly how to handle logs and antilogs.

1. What is a logarithm and how is it calculated?
A common logarithm is the power to which you raise 10 to get another number. For example, the log of 100 is 2, because you must raise 10 to the second power to get 100.

$$\log 100 = 2 \quad \text{since} \quad 10^2 = 100$$

Other examples are:

$$\log 1000 = 3 \quad \text{since} \quad 10^3 = 1000$$
$$\log 10 = 1 \quad \text{since} \quad 10^1 = 10$$
$$\log 1 = 0 \quad \text{since} \quad 10^0 = 1$$
$$\log 0.1 = -1 \quad \text{since} \quad 10^{-1} = 0.1$$

The common logarithm of a number other than a simple power is usually obtained from a calculator by entering the number and then pressing log. For example,

$$\log 52 = 1.7$$

$$\log 4.5 = 0.65$$

$$\log 0.25 = -0.60$$

Try it now. Enter 100 and then press log. Did you get 2? If so, you did it right. Try again with 52. Enter 52 and press log. Did you get 1.72 (rounded to two decimal places)? Some calculators may have you press log first and then the number. Try it both ways to make sure you know how your calculator works.

2. What are antilogarithms (antilogs)?

An antilog is the reverse of a log. It is also called the inverse log. If you take 10 and raise it to a power, you are taking an antilog. For example,

$$\text{antilog } 5 = 100{,}000$$

because taking the antilog of 5 means raising 10 to the power of 5 or

$$10^5 = 100{,}000$$

Try it now on your calculator. What is the antilog of 3? Enter 3 on your calculator. Press INV (inverse) or 2nd (second function), and then press log. The answer should be 1000. Your calculator may be different, but the INV or 2nd function keys are the most common.

3. What is the difference between antilog and −log?

There is a huge and very important difference. Antilog 3 means that we take 10 and raise it to the power of 3, so we get 1000. In contrast, $-\log 3$ means that we take the log of 3, which equals 0.477, and take the negative of it. Thus, $-\log 3$ equals -0.48. For example,

$$\text{antilog } 2 = 100$$

$$-\log 2 = -0.3$$

In Section 8.8, we will use negative logs to calculate pH. The pH equals $-\log[\text{H}^+]$. Thus, if we know that $[\text{H}^+]$ is 0.01 *M*, to find the pH, we enter 0.01 into our calculator and press log. That gives an answer of -2. Then we take the negative of that value to give a pH of 2.

In the last example, what answer would we have gotten if we had taken the antilog instead of the negative log? We would have gotten what we started with: 0.01. Why? Because all we calculated was antilog log 0.01. If we take the antilog of the log, we have not done anything at all.

8.8 What Are pH and pOH?

Because hydronium ion concentrations for most solutions are numbers with negative exponents, these concentrations are more conveniently expressed as pH, where

$$\text{pH} = -\log[\text{H}_3\text{O}^+]$$

similarly to how we expressed pK_a values in Section 8.5.

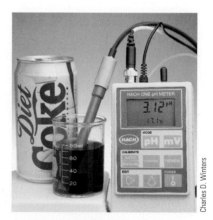

The pH of this soft drink is 3.12. Soft drinks are often quite acidic.

In Section 8.7, we saw that a solution is acidic if its $[H_3O^+]$ is greater than $1.0 \times 10^{-7} M$ and that it is basic if its $[H_3O^+]$ is less than $1.0 \times 10^{-7} M$. We can now state the definitions of acidic and basic solutions in terms of pH.

> A solution is acidic if its pH is less than 7.0
>
> A solution is basic if its pH is greater than 7.0
>
> A solution is neutral if its pH is equal to 7.0

Example 8.6 Calculating pH

(a) The $[H_3O^+]$ of a certain liquid detergent is $1.4 \times 10^{-9} M$. What is its pH? Is this solution acidic, basic, or neutral?
(b) The pH of black coffee is 5.3. What is its $[H_3O^+]$? Is it acidic, basic, or neutral?

Strategy

To determine the pH when given the concentration of a hydrogen ion, just take the negative of the log. If it is less than 7, the solution is acidic. If it is greater than 7, it is basic.

If given the pH, you can immediately determine if it is acidic, basic, or neutral according to how the number relates to 7. To convert the pH to the $[H_3O^+]$, take the inverse log of $-$pH.

Solution

(a) On your calculator, take the log of 1.4×10^{-9}. The answer is -8.85. Multiply this value by -1 to give the pH of 8.85. This solution is basic.
(b) Enter 5.3 into your calculator and then press the $+/-$ key to change the sign to minus and give -5.3. Then take the antilog of this number. The $[H_3O^+]$ of black coffee is 5×10^{-6}. This solution is acidic.

Problem 8.6

(a) The $[H_3O^+]$ of an acidic solution is $3.5 \times 10^{-3} M$. What is its pH?
(b) The pH of tomato juice is 4.1. What is its $[H_3O^+]$? Is this solution acidic, basic, or neutral?

Just as pH is a convenient way to designate the concentration of H_3O^+, pOH is a convenient way to designate the concentration of OH^-.

$$pOH = -\log [OH^-]$$

As we saw in the previous section, in aqueous solutions, the ion product of water, K_w, is 1×10^{-14}, which is equal to the product of the concentration of H^+ and OH^-:

$$K_w = 1 \times 10^{-14} = [H^+][OH^-]$$

By taking the logarithm of both sides, and the fact that $-\log (1 \times 10^{-14}) = 14$, we can rewrite this equation as shown below:

$$14 = pH + pOH$$

Thus, once we know the pH of a solution, we can easily calculate the pOH.

Example 8.7 Calculating pOH

The $[OH^-]$ of a strongly basic solution is 1.0×10^{-2}. What are the pOH and pH of this solution?

Strategy

When given the $[OH^-]$, determine the pOH by taking the negative logarithm. To calculate the pH, subtract the pOH from 14.

Solution

The pOH is $-\log 1.0 \times 10^{-2}$ or 2.00, and the pH is $14.00 - 2.00 = 12.00$.

Problem 8.7

The $[OH^-]$ of a solution is $1.0 \times 10^{-4}\ M$. What are the pOH and pH of this solution?

All fluids in the human body are aqueous; that is, the only solvent present is water. Consequently, all body fluids have a pH value. Some of them have a narrow pH range; others have a wide pH range. The pH of blood, for example, must be between 7.35 and 7.45 (slightly basic). If it goes outside these limits, illness and even death may result (Chemical Connections 8C). In contrast, the pH of urine can vary from 5.5 to 7.5. Table 8.5 gives pH values for some common materials.

The pH of three household substances. The colors of the acid–base indicators in the flasks show that vinegar is more acidic than club soda and the cleaner is basic.

Table 8.5 pH Values of Some Common Materials

Material	pH	Material	pH
Battery acid	0.5	Saliva	6.5–7.5
Gastric juice	1.0–3.0	Pure water	7.0
Lemon juice	2.2–2.4	Blood	7.35–7.45
Vinegar	2.4–3.4	Bile	6.8–7.0
Tomato juice	4.0–4.4	Pancreatic fluid	7.8–8.0
Carbonated beverages	4.0–5.0	Seawater	8.0–9.0
Black coffee	5.0–5.1	Soap	8.0–10.0
Urine	5.5–7.5	Milk of magnesia	10.5
Rain (unpolluted)	6.2	Household ammonia	11.7
Milk	6.3–6.6	Lye (1.0 M NaOH)	14.0

Strips of paper impregnated with indicator are used to find an approximate pH.

One thing you must remember when you see a pH value is that because pH is a logarithmic scale, an increase (or decrease) of one pH unit means a tenfold decrease (or increase) in the $[H_3O^+]$. For example, a pH of 3 does not sound very different from a pH of 4. The first, however, means a $[H_3O^+]$ of $10^{-3}\ M$, whereas the second means a $[H_3O^+]$ of $10^{-4}\ M$. The $[H_3O^+]$ of the pH 3 solution is ten times the $[H_3O^+]$ of the pH 4 solution.

There are two ways to measure the pH of an aqueous solution. One way is to use pH paper, which is made by soaking plain paper with a mixture of pH indicators. A pH **indicator** is a substance that changes color at a certain pH. When we place a drop of solution on this paper, the paper turns a certain color. To determine the pH, we compare the color of the paper with the colors on a chart supplied with the paper.

One example of an acid–base indicator is the compound methyl orange. When a drop of methyl orange is added to an aqueous solution with a pH of 3.2 or lower, this indicator turns red and the entire solution becomes red. When added to an aqueous solution with a pH of 4.4 or higher, this indicator turns yellow. These particular limits and colors apply only to methyl orange. Other indicators have other limits and colors (Figure 8.2). With pH

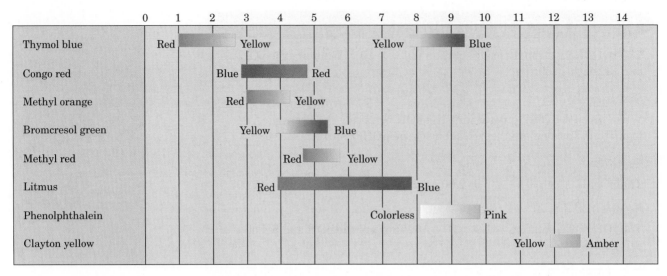

FIGURE 8.2 Some acid–base indicators. Note that some indicators have two color changes.

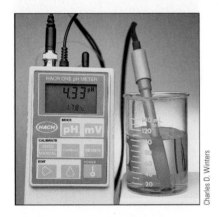

FIGURE 8.3 A pH meter can rapidly and accurately measure the pH of an aqueous solution.

Charles D. Winters

Titration An analytical procedure whereby we react a known volume of a solution of known concentration with a known volume of a solution of unknown concentration

Equivalence point The point at which there is a stoichiometrically equal number of moles of each reactant present

indicators, the chemical form of the indicator determines its color. The lower pH color is due to the acid form of the indicator, while the higher pH color is associated with the conjugate base form of the indicator.

The second way of determining pH is more accurate and more precise. In this method, we use a pH meter (Figure 8.3). We dip the electrode of the pH meter into the solution whose pH is to be measured and then read the pH on a display. The most commonly used pH meters read pH to the nearest hundredth of a unit. It should be mentioned that the accuracy of a pH meter, like that of any instrument, depends on correct calibration.

8.9 How Do We Use Titrations to Calculate Concentration?

Laboratories, whether medical, academic, or industrial, are frequently asked to determine the exact concentration of a particular substance in solution, such as the concentration of acetic acid in a given sample of vinegar, or the concentrations of iron, calcium, and magnesium ions in a sample of "hard" water. Determinations of solution concentrations can be made using an analytical technique called a **titration**.

In a titration, we react a known volume of a solution of known concentration with a known volume of a solution of unknown concentration. The solution of unknown concentration may contain an acid (such as stomach acid), a base (such as ammonia), an ion (such as Fe^{2+} ion), or any other substance whose concentration we are asked to determine. If we know the titration volumes and the mole ratio in which the solutes react, we can then calculate the concentration of the second solution.

Titrations must meet several requirements:

1. We must know the equation for the reaction so that we can determine the stoichiometric ratio of reactants to use in our calculations.

2. The reaction must be rapid and complete.

3. When the reactants have combined exactly, there must be a clear-cut change in some measurable property of the reaction mixture. We call the point at which the stoichiometrically correct number of moles of reactants combine exactly the **equivalence point** of the titration.

4. We must have accurate measurements of the amount of each reactant.

Let us apply these requirements to the titration of a solution of sulfuric acid of known concentration with a solution of sodium hydroxide of unknown concentration. We know the balanced equation for this acid–base reaction, so requirement 1 is met.

$$2NaOH(aq) + H_2SO_4(aq) \longrightarrow Na_2SO_4(aq) + 2H_2O(\ell)$$

(Concentration   (Concentration
not known)   known)

Sodium hydroxide ionizes in water to form sodium ions and hydroxide ions; sulfuric acid ionizes to form hydronium ions and sulfate ions. The reaction between hydroxide and hydronium ions is rapid and complete, so requirement 2 is met.

To meet requirement 3, we must be able to observe a clear-cut change in some measurable property of the reaction mixture at the equivalence point. For acid–base titrations, we use the sudden pH change that occurs at this point. Suppose we add the sodium hydroxide solution slowly. As it is added, it reacts with hydronium ions to form water. As long as any unreacted hydronium ions are present, the solution is acidic. When the number of hydroxide ions added exactly equals the original number of hydronium ions, the solution becomes neutral. Then, as soon as any extra hydroxide ions are added, the solution becomes basic. We can observe this sudden change in pH by reading a pH meter.

Another way to observe the change in pH at the equivalence point is to use an acid–base indicator (Section 8.8). Such an indicator changes color when the solution changes pH. Phenolphthalein, for example, is colorless in acid solution and pink in basic solution. If this indicator is added to the original sulfuric acid solution, the solution remains colorless as long as excess hydronium ions are present. After enough sodium hydroxide solution has been added to react with all of the hydronium ions, the next drop of base provides excess hydroxide ions and the solution turns pink (Figure 8.4). Thus, we have a clear-cut indication of the equivalence point. The point at

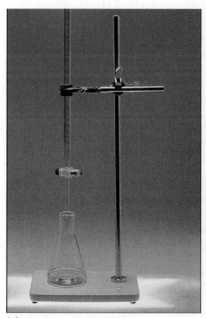 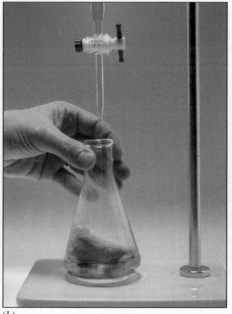 

(a)   (b)   (c)

Charles D. Winters

FIGURE 8.4 An acid–base titration. (*a*) An acid of known concentration is in the Erlenmeyer flask. (*b*) When a base is added from the buret, the acid is neutralized. (*c*) The end point is reached when the color of the indicator changes from colorless to pink.

The end point and equivalence point of a titration are not exactly the same, although an indicator is chosen that will change as close to the equivalent point as possible.

which an indicator changes color is called the **end point** of the titration. It is convenient if the end point and the equivalence point are the same, but there are many pH indicators whose end points are not at pH 7.

To meet requirement 4, which is that the volume of each solution used must be known, we use volumetric glassware such as volumetric flasks, burets, and pipets.

Data for a typical acid–base titration are given in Example 8.8. Note that the experiment is run in triplicate, a standard procedure for checking the precision of a titration.

Example 8.8 Titrations

Following are data for the titration of 0.108 M H_2SO_4 with a solution of NaOH of unknown concentration. What is the concentration of the NaOH solution?

	Volume of 0.108 M H_2SO_4	Volume of NaOH
Trial I	25.0 mL	33.48 mL
Trial II	25.0 mL	33.46 mL
Trial III	25.0 mL	33.50 mL

Strategy

Use the volume of the acid and its concentration to calculate how many moles of hydrogen ions are available to be titrated. At the equivalence point, the moles of base used will equal the moles of H^+ available. Divide the moles of H^+ by the volume of base used in liters to calculate the concentration of the base.

Solution

From the balanced equation for this acid–base reaction, we know the stoichiometry: Two moles of NaOH react with one mole of H_2SO_4. From the three trials, we calculate that the average volume of the NaOH required for complete reaction is 33.48 mL. Because the units of molarity are moles/liter, we must convert volumes of reactants from milliliters to liters. We can then use the factor-label method (Section 1.5) to calculate the molarity of the NaOH solution. What we wish to calculate is the number of moles of NaOH per liter of NaOH.

$$\frac{\text{mol NaOH}}{\text{L NaOH}} = \frac{0.108 \text{ mol } H_2SO_4}{1 \text{ L } H_2SO_4} \times \frac{0.0250 \text{ L } H_2SO_4}{0.03348 \text{ L NaOH}} \times \frac{2 \text{ mol NaOH}}{1 \text{ mol } H_2SO_4}$$

$$= \frac{0.161 \text{ mol NaOH}}{\text{L NaOH}} = 0.161 \ M$$

Problem 8.8

Calculate the concentration of an acetic acid solution using the following data. Three 25.0-mL samples of acetic acid were titrated to a phenolphthalein end point with 0.121 M NaOH. The volumes of NaOH were 19.96 mL, 19.73 mL, and 19.79 mL.

It is important to understand that a titration is not a method for determining the acidity (or basicity) of a solution. If we want to do that, we must measure the sample's pH, which is the only measurement of solution acidity or basicity. Rather, titration is a method for determining the total acid or base concentration of a solution, which is not the same as the acidity. For example, a 0.1 M solution of HCl in water has a pH of 1.0, but a 0.1 M solution of acetic acid has a pH of 2.9. These two solutions have the same concentration of acid and each neutralizes the same volume of NaOH solution, but they have very different acidities.

8.10 What Are Buffers?

As noted earlier, the body must keep the pH of blood between 7.35 and 7.45. Yet we frequently eat acidic foods such as oranges, lemons, sauerkraut, and tomatoes, and doing so eventually adds considerable quantities of H_3O^+ to the blood. Despite these additions of acidic or basic substances, the body manages to keep the pH of blood remarkably constant. The body manages this feat by using buffers. A **buffer** is a solution whose pH changes very little when small amounts of H_3O^+ or OH^- ions are added to it. In a sense, a pH buffer is an acid or base "shock absorber."

Buffer A solution that resists change in pH when limited amounts of an acid or a base are added to it; the most common example is an aqueous solution containing a weak acid and its conjugate base

The most common buffers consist of approximately equal molar amounts of a weak acid and a salt of the weak acid (or alternatively, a weak base and a salt of the weak base, which we will not consider here). For example, if we dissolve 1.0 mol of acetic acid (a weak acid) and 1.0 mol of its conjugate base (in the form of CH_3COONa, sodium acetate) in 1.0 L of water, we have a good buffer solution. The equilibrium present in this buffer solution is:

$$\overset{\substack{\text{Added as} \\ CH_3COOH}}{CH_3COOH} + H_2O \rightleftharpoons \overset{\substack{\text{Added as} \\ CH_3COO^-Na^+}}{CH_3COO^-} + H_3O^+$$

Acetic acid (A weak acid)   Acetate ion (Conjugate base of a weak acid)

A. How Do Buffers Work?

A buffer resists a drastic change in pH upon the addition of small quantities of acid or base. To see how, we will use an acetic acid–sodium acetate buffer as an example. If a strong acid such as HCl is added to this buffer solution, the added H_3O^+ ions react with CH_3COO^- ions and are removed from solution.

$$CH_3COO^- + H_3O^+ \longrightarrow CH_3COOH + H_2O$$

Acetate ion (Conjugate base of a weak acid)   Acetic acid (A weak acid)

There is a slight increase in the concentration of CH_3COOH as well as a slight decrease in the concentration of CH_3COO^-, but there is no appreciable change in pH. We say that this solution is buffered because it resists a change in pH upon the addition of small quantities of a strong acid.

If NaOH or another strong base is added to the buffer solution, the added OH^- ions react with CH_3COOH molecules and are removed from solution:

$$CH_3COOH + OH^- \longrightarrow CH_3COO^- + H_2O$$

Acetic acid (A weak acid)   Acetate ion (Conjugate base of a weak acid)

Here there is a slight decrease in the concentration of CH_3COOH as well as a slight increase in the concentration of CH_3COO^-, but, again, there is no appreciable change in pH.

The important point about this or any other buffer solution is that when the conjugate base of the weak acid removes H_3O^+, it is converted to the undissociated weak acid. Because a substantial amount of weak acid is already present, there is no appreciable change in its concentration, and because H_3O^+ ions are removed from solution, there is no appreciable change in pH. By the same token, when the weak acid removes OH^- ions from solution, it is converted to its conjugate base. Because OH^- ions are removed from solution, there is no appreciable change in pH.

The effect of a buffer can be quite powerful. Addition of either dilute HCl or NaOH to pure water, for example, causes a dramatic change in pH (Figure 8.5).

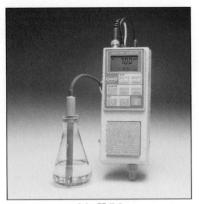
(a) pH 7.0

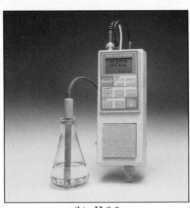
(b) pH 2.0

(c) pH 12.0

FIGURE 8.5 The addition of HCl and NaOH to pure water. (a) The pH of pure water is 7.0. (b) The addition of 0.01 mol of HCl to 1 L of pure water causes the pH to decrease to 2.0. (c) The addition of 0.01 mol of NaOH to 1 L of pure water causes the pH to increase to 12.0.

When HCl or NaOH is added to a phosphate buffer, the results are quite different. Suppose we have a phosphate buffer solution of pH 7.21 prepared by dissolving 0.10 mol NaH_2PO_4 (a weak acid) and 0.10 mol Na_2HPO_4 (its conjugate base) in enough water to make 1.00 L of solution. If we add 0.010 mol of HCl to 1.0 L of this solution, the pH decreases to only 7.12. If we add 0.01 mol of NaOH, the pH increases to only 7.30.

Phosphate buffer (pH 7.21) + 0.010 mol HCl   pH 7.21 $\longrightarrow$ 7.12

Phosphate buffer (pH 7.21) + 0.010 mol NaOH   pH 7.21 $\longrightarrow$ 7.30

Had the same amount of acid or base been added to 1 liter of pure water, the resulting pH values would have been 2 and 12, respectively.

Figure 8.6 shows the effect of adding acid to a buffer solution.

B. Buffer pH

In the previous example, the pH of the buffer containing equal molar amounts of $H_2PO_4^-$ and HPO_4^{2-} is 7.21. From Table 8.3, we see that 7.21 is the pK_a of the acid $H_2PO_4^-$. This is not a coincidence. If we make a buffer solution by mixing equimolar concentrations of any weak acid and its conjugate base, the pH of the solution will equal the pK_a of the weak acid.

This fact allows us to prepare buffer solutions to maintain almost any pH. For example, if we want to maintain a pH of 9.14, we could make a buffer solution from boric acid, H_3BO_3, and sodium dihydrogen borate, NaH_2BO_3, the sodium salt of its conjugate base (see Table 8.3).

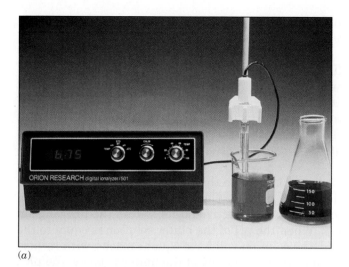

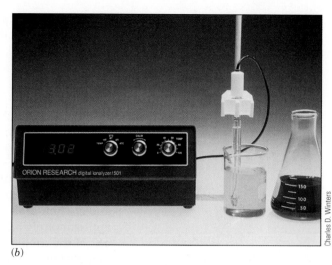

(a)   (b)

FIGURE 8.6 Buffer solutions. The solution in the Erlenmeyer flask on the right in both (a) and (b) is a buffer of pH 7.40, the same pH as human blood. The buffer solution also contains bromcresol green, an acid–base indicator that is blue at pH 7.40 (see Figure 8.2). (a) The beaker contains some of the pH 7.40 buffer and the bromcresol green indicator to which has been added 5.0 mL of 0.10 M HCl. After the addition of the HCl, the pH of the buffer solution drops only 0.65 unit to 6.75. (b) The beaker contains pure water and bromcresol green indicator to which has been added 5.0 mL of 0.10 M HCl. After the addition of the HCl, the pH of the unbuffered solution drops to 3.02.

Example 8.9 Buffers

What is the pH of a buffer solution containing equimolar quantities of:

(a) H_3PO_4 and NaH_2PO_4?   (b) H_2CO_3 and $NaHCO_3$?

Strategy

When there are equimolar quantities of a weak acid and its conjugate base in a buffer solution, the pH is always the same as the pK_a of the weak acid. Look up the pK_a of the weak acid in Table 8.3.

Solution

Because we are adding equimolar quantities of a weak acid and its conjugate base, the pH is equal to the pK_a of the weak acid, which we find in Table 8.3:

(a) pH = 2.12   (b) pH = 6.37

Problem 8.9

What is the pH of a buffer solution containing equimolar quantities of:

(a) NH_4Cl and NH_3?   (b) CH_3COOH and CH_3COONa?

C. Buffer Capacity

Buffer capacity is the amount of hydronium or hydroxide ions that a buffer can absorb without a significant change in its pH. We have already mentioned that a pH buffer is an acid–base "shock absorber." We now ask what makes one solution a better acid–base shock absorber than another

Buffer capacity The extent to which a buffer solution can prevent a significant change in pH of a solution upon addition of a strong acid or a strong base solution. The nature of the buffer capacity of a pH buffer depends on both its pH relative to its pK_a and its concentration.

pH:	The closer the pH of the buffer is to the pK_a of the weak acid, the more symmetric the buffer capacity, meaning the buffer can resist a pH change with added acid or added base.
Concentration:	The greater the concentration of the weak acid and its conjugate base, the greater the buffer capacity.

An effective buffer has a pH equal to the pK_a of the weak acid ± 1. For acetic acid, for example, the pK_a is 4.75. Therefore, a solution of acetic acid and sodium acetate functions as an effective buffer within the pH range of approximately 3.75–5.75. When the pH of the buffer solution is equal to the pK_a of the conjugate acid, the solution will have equal capacity with respect to additions of either acid or base. If the pH of the buffer is below the pK_a, the capacity will be greater toward addition of base. When the pH is above the pK_a, the acid buffer capacity will be greater than the base buffer capacity.

Buffer capacity also depends on concentration. The greater the concentration of the weak acid and its conjugate base, the greater the buffer capacity. We could make a buffer solution by dissolving 1.0 mol each of CH_3COONa and CH_3COOH in 1 L of H_2O, or we could use only 0.10 mol of each. Both solutions have the same pH of 4.75. However, the former has a buffer capacity ten times that of the latter. If we add 0.2 mol of HCl to the former solution, it performs the way we expect—the pH drops to 4.57. If we add 0.2 mol of HCl to the latter solution, however, the pH drops to 1.0 because the buffer has been used up. That is, the amount of H_3O^+ added has exceeded the buffer capacity. The first 0.10 mol of HCl completely neutralizes essentially all the CH_3COO^- present. After that, the solution contains only CH_3COOH and is no longer a buffer, so the second 0.10 mol of HCl decreases the pH to 1.0.

D. Blood Buffers

The average pH of human blood is 7.4. Any change larger than 0.10 pH unit in either direction may cause illness. If the pH goes below 6.8 or above 7.8, death may result. To hold the pH of the blood close to 7.4, the body uses three buffer systems: carbonate, phosphate, and proteins (proteins are discussed in Chapter 22).

The most important of these systems is the carbonate buffer. The weak acid of this buffer is carbonic acid, H_2CO_3; the conjugate base is the bicarbonate ion, HCO_3^-. The pK_a of H_2CO_3 is 6.37 (from Table 8.3). Because the pH of an equal mixture of a weak acid and its salt is equal to the pK_a of the weak acid, a buffer with equal concentrations of H_2CO_3 and HCO_3^- has a pH of 6.37.

Blood, however, has a pH of 7.4. The carbonate buffer can maintain this pH only if $[H_2CO_3]$ and $[HCO_3^-]$ are not equal. In fact, the necessary $[HCO_3^-]/[H_2CO_3]$ ratio is about 10 : 1. The normal concentrations of these species in blood are about 0.025 M HCO_3^- and 0.0025 M H_2CO_3. This buffer works because any added H_3O^+ is neutralized by the HCO_3^- and any added OH^- is neutralized by the H_2CO_3.

The fact that the $[HCO_3^-]/[H_2CO_3]$ ratio is 10 : 1 means that this system is a better buffer for acids, which lower the ratio and thus improve buffer efficiency, than for bases, which raise the ratio and decrease buffer capacity. This is in harmony with the actual functioning of the body because under normal conditions, larger amounts of acidic than basic substances enter the blood. The 10 : 1 ratio is easily maintained under normal conditions, because the body can very quickly increase or decrease the amount of CO_2 entering the blood.

The second most important buffering system of the blood is a phosphate buffer made up of hydrogen phosphate ion, HPO_4^{2-}, and dihydrogen phosphate ion, $H_2PO_4^-$. In this case, a $1.6:1$ $[HPO_4^{2-}]/[H_2PO_4^-]$ ratio is necessary to maintain a pH of 7.4. This ratio is well within the limits of good buffering action.

8.11 How Do We Calculate the pH of a Buffer?

Suppose we want to make a phosphate buffer solution of pH 7.00. The weak acid with a pK_a closest to this desired pH is $H_2PO_4^-$; it has a pK_a of 7.21. If we use equal concentrations of NaH_2PO_4 and Na_2HPO_4, however, we will have a buffer of pH 7.21. We want a phosphate buffer that is slightly more acidic than 7.21, so it would seem reasonable to use more of the weak acid, $H_2PO_4^-$, and less of its conjugate base, HPO_4^{2-}. But what proportions of these two salts do we use? Fortunately, we can calculate these proportions using the **Henderson-Hasselbalch** equation.

The Henderson-Hasselbalch equation is a mathematical relationship between pH, the pK_a of a weak acid, and the concentrations of the weak acid and its conjugate base. The equation is derived in the following way. Assume that we are dealing with weak acid, HA, and its conjugate base, A^-.

$$HA + H_2O \rightleftharpoons A^- + H_3O^+$$

$$K_a = \frac{[A^-][H_3O^+]}{[HA]}$$

Taking the logarithm of this equation gives:

$$\log K_a = \log [H_3O^+] + \log \frac{[A^-]}{[HA]}$$

Rearranging terms gives us a new expression, in which $-\log K_a$ is, by definition, pK_a and $-\log [H_3O^+]$ is, by definition, pH. Making these substitutions gives the Henderson-Hasselbalch equation.

$$-\log [H_3O^+] = -\log K_a + \log \frac{[A^-]}{[HA]}$$

$$pH = pK_a + \log \frac{[A^-]}{[HA]} \quad \text{Henderson-Hasselbalch equation}$$

The Henderson-Hasselbalch equation gives us a convenient way to calculate the pH of a buffer when the concentrations of the weak acid and its conjugate base are not equal.

Example 8.10 Buffer pH Calculation

What is the pH of a phosphate buffer solution containing 1.0 mol/L of sodium dihydrogen phosphate, NaH_2PO_4, and 0.50 mol/L of sodium hydrogen phosphate, Na_2HPO_4?

Strategy

Use the Henderson-Hasselbalch equation to determine the pH. You must know either the number of moles of both the conjugate acid and base or the concentrations of the conjugate acid or base. Divide the conjugate base by the conjugate acid, take the log of that ratio, and add it to the pK_a of the conjugate acid.

Solution

The weak acid in this problem is $H_2PO_4^-$; its ionization produces HPO_4^{2-}. The pK_a of this acid is 7.21 (from Table 8.3). Under the weak acid and its conjugate base are shown their concentrations.

$$H_2PO_4^- + H_2O \rightleftharpoons HPO_4^{2-} + H_3O^+ \quad pK_a = 7.21$$

1.0 mol/L   0.50 mol/L

Substituting these values in the Henderson-Hasselbalch equation gives a pH of 6.91.

$$pH = 7.21 + \log \frac{0.50}{1.0}$$

$$= 7.21 - 0.30 = 6.91$$

Problem 8.10

What is the pH of a boric acid buffer solution containing 0.25 mol/L of boric acid, H_3BO_3, and 0.50 mol/L of its conjugate base? See Table 8.3 for the pK_a of boric acid.

Returning to the problem posed at the beginning of this section, how do we calculate the proportions of NaH_2PO_4 and Na_2HPO_4 needed to make up a phosphate buffer of pH 7.00? We know that the pK_a of $H_2PO_4^-$ is 7.21 and that the buffer we wish to prepare has a pH of 7.00. We can substitute these two values in the Henderson-Hasselbalch equation as follows:

$$7.00 = 7.21 + \log \frac{[HPO_4^{2-}]}{[H_2PO_4^-]}$$

Rearranging and solving gives:

$$\log \frac{[HPO_4^{2-}]}{[H_2PO_4^-]} = 7.00 - 7.21 = -0.21$$

$$\frac{[HPO_4^{2-}]}{[H_2PO_4^-]} = 10^{-0.21} = \frac{0.62}{1.0}$$

Thus, to prepare a phosphate buffer of pH 7.00, we can use 0.62 mol of Na_2HPO_4 and 1.0 mol of NaH_2PO_4. Alternatively, we can use any other amounts of these two salts as long as their mole ratio is 0.62:1.0.

8.12 What Are TRIS, HEPES, and These Buffers with the Strange Names?

The original buffers used in the lab were made from simple weak acids and bases, such as acetic acid, phosphoric acid, and citric acid. It was eventually discovered that many of these buffers had limitations. For example, they often changed their pH too much if the solution was diluted or if the temperature changed. They often permeated cells in solution, thereby changing the chemistry of the interior of the cell. To overcome these shortcomings, a scientist named N. E. Good developed a series of buffers that consist of zwitterions, molecules with both positive and negative charges. Zwitterions

Respiratory and Metabolic Acidosis

The pH of blood is normally between 7.35 and 7.45. If the pH goes lower than that level, the condition is called **acidosis.** Acidosis leads to depression of the nervous system. Mild acidosis can result in dizziness, disorientation, or fainting; a more severe case can cause coma. If the acidosis persists for a sufficient period of time or if the pH gets too far away from 7.35 to 7.45, death may result.

Acidosis has several causes. One type, called **respiratory acidosis,** results from difficulty in breathing (hypoventilation). An obstruction in the windpipe or diseases such as pneumonia, emphysema, asthma, or congestive heart failure may diminish the amount of oxygen that reaches the tissues and the amount of CO_2 that leaves the body through the lungs. You can even produce mild acidosis by holding your breath. If you ever tried to see how long you could swim underwater in a pool without surfacing, you will have noticed a deep burning sensation in all your muscles when you finally came up for air. The pH of the blood decreases because the CO_2, unable to escape fast enough, remains in the blood, where it lowers the $[HCO_3^-]/[H_2CO_3]$ ratio. Rapid breathing as a result of physical exertion is more about getting rid of CO_2 than it is about breathing in O_2.

Acidosis caused by other factors is called **metabolic acidosis.** Two causes of this condition are starvation (or fasting) and heavy exercise. When the body doesn't get enough food, it burns its own fat, and the products of this reaction are acidic compounds that enter the blood. This problem sometimes happens to people on fad diets. Heavy exercise causes the muscles to produce excessive amounts of lactic acid, which makes muscles feel tired and sore. The lowering of the blood pH due to lactic acid is also what leads to the rapid breathing, dizziness, and nausea that athletes feel at the end of a sprint. In addition, metabolic acidosis is caused by a number of metabolic irregularities.

These runners just competed for the gold medal in the 4 × 400 m relay race at the 1996 Olympic Games. The buildup of lactic acid and lowered blood pH caused severe muscle pain and breathlessness.

For example, the disease diabetes mellitus produces acidic compounds called ketone bodies (Section 28.6).

Both types of acidosis can be related. When cells are deprived of oxygen, respiratory acidosis results. These cells are unable to produce the energy they need through aerobic (*oxygen-requiring*) pathways that we will learn about in Chapters 27 and 28. To survive, the cells must use the anaerobic (*without oxygen*) pathway called glycolysis. This pathway has lactic acid as an end product, leading to metabolic acidosis. The lactic acid is the body's way of buying time and keeping the cells alive and functioning a little longer. Eventually the lack of oxygen, called an oxygen debt, must be repaid, and the lactic acid must be cleared out. In extreme cases, the oxygen debt is too great, and the individual can die. This was the case of a famous cyclist, Tom Simpson, who died on the slopes of Mont Ventoux during the 1967 Tour de France. Under the influence of amphetamines, he rode so hard that he built up a fatal oxygen debt.

Test your knowledge with Problem 8.75.

do not readily permeate cell membranes. Zwitterionic buffers also are more resistant to concentration and temperature changes.

Most of the common synthetic buffers used today have complicated formulas, such as 3-[N-morpholino]propanesulfonic acid, which we abbreviate as MOPS. Table 8.5 gives a few examples.

The important thing to remember is that you don't really need to know the structure of these odd-sounding buffers to use them correctly. The important considerations are the pK_a of the buffer and the concentration you want to have. The Henderson-Hasselbalch equation works just fine whether or not you know the structure of the compound in question.

Example 8.11 Buffer pH Calculation

What is the pH of a solution if you mix 100. mL of 0.20 M HEPES in the acid form with 200. mL of 0.20 M HEPES in the basic form?

Strategy

To use the Henderson-Hasselbalch equation, you need the ratio of the conjugate base to weak acid forms of the buffer. Since the HEPES solutions have equal concentrations, the ratio of the volumes will give you the ratio of the moles used. Divide the volume of the conjugate base form by the volume of the weak acid form. Take the log of the ratio and add it to the pK_a for HEPES.

Solution

First, we must find the pK_a, which we see from Table 8.6 is 7.55. Then we must calculate the ratio of the conjugate base to the acid. The formula calls for the concentration, but in this situation, the ratio of the concentrations will be the same as the ratio of the moles, which will be the same as the ratio of the volumes, because both solutions had the same starting concentration of 0.20 M. Thus, we can see that the ratio of base to acid is 2:1 because we added twice the volume of base.

$$pH = pK_a + \log([A^-]/[HA]) = 7.55 + \log(2) = 7.85$$

Notice that we did not have to know anything about the structure of HEPES to work out this example.

Problem 8.11

What is the pH of a solution made by mixing 0.2 mol of TRIS acid and 0.05 mol of TRIS base in 500 mL of water?

Table 8.6 Acid and Base Forms of Some Useful Biochemical Buffers

Acid Form		Base Form	pKa
TRIS—H$^+$ (protonated form) $(HOCH_2)_3CNH_3^+$	N—*tris*[hydroxymethyl]aminomethane (TRIS) $\rightleftharpoons$	TRIS* (free amine) $(HOCH_2)_3CNH_2$	8.3
$^-$TES—H$^+$ (zwitterionic form) $(HOCH_2)_3CNH_2CH_2CH_2SO_3^-$	N—*tris*[hydroxymethyl]methyl-2-aminoethane sulfonate (TES) $\rightleftharpoons$	$^-$TES (anionic form) $(HOCH_2)_3CNHCH_2CH_2SO_3^-$	7.55
$^-$HEPES—H$^+$ (zwitterionic form) $HOCH_2CH_2N^+\,NCH_2CH_2SO_3^-$ H	N—2—hydroxyethylpiperazine-N'-2-ethane sulfonate (HEPES) $\rightleftharpoons$	$^-$HEPES (anionic form) $HOCH_2CH_2N\,NCH_2CH_2SO_3^-$	7.55
$^-$MOPS—H$^+$ (zwitterionic form) $O\,^+NCH_2CH_2CH_2SO_3^-$ H	3—[N—morpholino]propane-sulfonic acid (MOPS) $\rightleftharpoons$	$^-$MOPS (anionic form) $O\,NCH_2CH_2CH_2SO_3^-$	7.2
$^{2-}$PIPES—H$^+$ (protonated dianion) $^-O_3SCH_2CH_2N\,^+NCH_2CH_2SO_3^-$ H	Piperazine—N,N'-*bis*[2-ethanesulfonic acid] (PIPES) $\rightleftharpoons$	$^{2-}$PIPES (dianion) $^-O_3SCH_2CH_2N\,NCH_2CH_2SO_3^-$	6.8

*Note that TRIS is not a zwitterion.

Chemical Connections 8D

Alkalosis and the Sprinter's Trick

Reduced pH is not the only irregularity that can occur in the blood. The pH may also be elevated, a condition called **alkalosis** (blood pH higher than 7.45). It leads to overstimulation of the nervous system, muscle cramps, dizziness, and convulsions. It arises from rapid or heavy breathing, called hyperventilation, which may be caused by fever, infection, the action of certain drugs, or even hysteria. In this case, the excessive loss of CO_2 raises both the ratio of $[HCO_3^-]/[H_2CO_3]$ and the pH.

Athletes who compete in short-distance races that take about a minute to finish have learned how to use hyperventilation to their advantage. By hyperventilating right before the start, they force extra CO_2 out of their lungs. This causes more H_2CO_3 to dissociate into CO_2 and H_2O to replace the lost CO_2. In turn, the loss of the HA form of the bicarbonate blood buffer raises the pH of the blood. When an athlete starts an event with a slightly higher blood pH, he or she can absorb more lactic acid before the blood pH drops to the point where performance is impaired. Of course, the timing of this hyperventilation

© AP Photo/Ricardo Mazalan

Athletes often hyperventilate before the start of a short distance event. This raises the pH of the blood, allowing it to absorb more H^+ before their performance declines.

must be perfect. If the athlete artificially raises blood pH and then the race does not start quickly, the side effect of dizziness will occur.

Test your knowledge with Problems 8.76 and 8.77.

Summary

OWL Sign in at **www.cengage.com/owl** to develop problem-solving skills and complete online homework assigned by your professor.

Section 8.1 What Are Acids and Bases?

- By the **Arrhenius definitions**, acids are substances that produce H_3O^+ ions in aqueous solution.
- Bases are substances that produce OH^- ions in aqueous solution.

Section 8.2 How Do We Define the Strength of Acids and Bases?

- A strong acid reacts completely or almost completely with water to form H_3O^+ ions.
- A strong base reacts completely or almost completely with water to form OH^- ions.

Section 8.3 What Are Conjugate Acid–Base Pairs?

- The **Brønsted-Lowry definitions** expand the definitions of acid and base beyond water.
- An acid is a proton donor; a base is a proton acceptor.
- Every acid has a **conjugate base**, and every base has a **conjugate acid**. The stronger the acid, the weaker its conjugate base. Conversely, the stronger the base, the weaker its conjugate acid.

- An **amphiprotic substance**, such as water, can act as either an acid or a base.

Section 8.4 How Can We Tell the Position of Equilibrium in an Acid–Base Reaction?

- In an acid–base reaction, the position of equilibrium favors the reaction of the stronger acid and the stronger base to form the weaker acid and the weaker base.

Section 8.5 How Do We Use Acid Ionization Constants?

- The strength of a weak acid is expressed by its **ionization constant**, K_a.
- The larger the value of K_a, the stronger the acid.
- $pK_a = -\log [K_a]$.

Section 8.6 What Are the Properties of Acids and Bases?

- Acids react with metals, metal hydroxides, and metal oxides to give **salts**, which are ionic compounds made up of cations from the base and anions from the acid.
- Acids also react with carbonates, bicarbonates, ammonia, and amines to give salts.

Section 8.7 What Are the Acidic and Basic Properties of Pure Water?

- In pure water, a small percentage of molecules undergo ionization:

$$H_2O + H_2O \rightleftharpoons H_3O^+ + OH^-$$

- As a result, pure water has a concentration of $10^{-7}\ M$ for H_3O^+ and $10^{-7}\ M$ for OH^-.
- The **ion product of water**, K_w, is equal to 1.0×10^{-14}. $pK_w = 14.00$.

Section 8.8 What Are pH and pOH?

- Hydronium ion concentrations are generally expressed in **pH** units, with $pH = -\log[H_3O^+]$.
- **pOH** $= -\log[OH^-]$.
- Solutions with pH less than 7 are acidic; those with pH greater than 7 are basic. A **neutral solution** has a pH of 7.
- The pH of an aqueous solution is measured with an acid–base indicator or with a pH meter.

Section 8.9 How Do We Use Titrations to Calculate Concentration?

- We can measure the concentration of aqueous solutions of acids and bases using titration. In an acid–base titration, a base of known concentration is added to an acid of unknown concentration (or vice versa) until an equivalence point is reached, at which point the acid or base being titrated is completely neutralized.

Section 8.10 What Are Buffers?

- A **buffer** does not significantly change its pH very much when small amounts of either hydronium ions or hydroxide ions are added to it.

- Buffer solutions consist of approximately equal concentrations of a weak acid and its conjugate base.
- The **buffer capacity** depends on both its pH relative to its pK_a and its concentration. The most effective buffer solutions have a pH equal to the pK_a of the weak acid. The greater the concentration of the weak acid and its conjugate base, the greater the buffer capacity.
- The most important buffers for blood are bicarbonate and phosphate.

Section 8.11 How Do We Calculate the pH of a Buffer?

- The **Henderson-Hasselbalch** equation is a mathematical relationship between pH, the pK_a of a weak acid, and the concentrations of the weak acid and its conjugate base:

$$pH = pK_a + \log\frac{[A^-]}{[HA]}$$

Section 8.12 What Are TRIS, HEPES, and These Buffers with the Strange Names?

- Many modern buffers have been designed, and their names are often abbreviated.
- These buffers have qualities useful to scientists, such as not crossing membranes and resisting pH change with dilution or temperature change.
- You do not have to understand the structure of these buffers to use them. The important things to know are the molar mass and the pK_a of the weak acid form of the buffer.

Problems

⊙WL Interactive versions of these problems may be assigned in OWL.

Orange-numbered problems are applied.

References to previous chapters are given in parentheses.

Section 8.1 What Are Acids and Bases?

8.12 Define (a) an Arrhenius acid and (b) an Arrhenius base.

8.13 Write an equation for the reaction that takes place when each acid is added to water. For a diprotic or triprotic acid, consider only its first ionization.
   (a) HNO_3 (b) HBr (c) H_2SO_3
   (d) H_2SO_4 (e) HCO_3^- (f) NH_4^+

8.14 Write an equation for the reaction that takes place when each base is added to water.
   (a) $LiOH$ (b) $(CH_3)_2NH$

Section 8.2 How Do We Define the Strength of Acids and Bases?

8.15 For each of the following, tell whether the acid is strong or weak.
   (a) Acetic acid   (b) HCl
   (c) H_3PO_4 (d) H_2SO_4
   (e) HCN (f) H_2CO_3

8.16 For each of the following, tell whether the base is strong or weak.
   (a) $NaOH$ (b) Sodium acetate
   (c) KOH (d) Ammonia
   (e) Water

8.17 Answer True or False
   (a) If an acid has a pK_a of 2.1, it is a strong acid.
   (b) The pH of $0.1\ M\ HCl$ is the same as the pH of $0.1\ M$ acetic acid.

(c) HCl and HNO_3 are both strong acids.

(d) The concentration of $[H^+]$ is always higher in a solution of strong acid than weak acid.

(e) If two monoprotic acids have the same concentration, the hydrogen ion concentration will be higher in the stronger acid.

(f) If two strong acids have the same concentration, the hydrogen ion will be higher in a polyprotic acid than a monoprotic one.

(g) Ammonia is a strong base.

(h) Carbonic acid is a strong acid.

Section 8.3 What Are Conjugate Acid–Base Pairs?

8.18 Which of these acids are monoprotic, which are diprotic, and which are triprotic? Which are amphiprotic?

(a) $H_2PO_4^-$ (b) HBO_3^{2-} (c) $HClO_4$ (d) C_2H_5OH

(e) HSO_3^- (f) HS^- (g) H_2CO_3

8.19 Define (a) a Brønsted-Lowry acid and (b) a Brønsted-Lowry base.

8.20 Write the formula for the conjugate base of each acid.

(a) H_2SO_4 (b) H_3BO_3 (c) HI

(d) H_3O^+ (e) NH_4^+ (f) HPO_4^{2-}

8.21 Write the formula for the conjugate base of each acid.

(a) $H_2PO_4^-$ (b) H_2S

(c) HCO_3^- (d) CH_3CH_2OH

(e) H_2O

8.22 Write the formula for the conjugate acid of each base.

(a) OH^- (b) HS^- (c) NH_3

(d) $C_6H_5O^-$ (e) CO_3^{2-} (f) HCO_3^-

8.23 Write the formula for the conjugate acid of each base.

(a) H_2O (b) HPO_4^{2-} (c) CH_3NH_2

(d) PO_4^{3-} (e) NH_3

Section 8.4 How Can We Tell the Position of Equilibrium in an Acid–Base Reaction?

8.24 For each equilibrium, label the stronger acid, stronger base, weaker acid, and weaker base. For which reaction(s) does the position of equilibrium lie toward the right? For which does it lie toward the left?

(a) $H_3PO_4 + OH^- \rightleftharpoons H_2PO_4^- + H_2O$

(b) $H_2O + Cl^- \rightleftharpoons HCl + OH^-$

(c) $HCO_3^- + OH^- \rightleftharpoons CO_3^{2-} + H_2O$

8.25 For each equilibrium, label the stronger acid, stronger base, weaker acid, and weaker base. For which reaction(s) does the position of equilibrium lie toward the right? For which does it lie toward the left?

(a) $C_6H_5OH + C_2H_5O^- \rightleftharpoons C_6H_5O^- + C_2H_5OH$

(b) $HCO_3^- + H_2O \rightleftharpoons H_2CO_3 + OH^-$

(c) $CH_3COOH + H_2PO_4^- \rightleftharpoons CH_3COO^- + H_3PO_4$

8.26 Will carbon dioxide be evolved as a gas when sodium bicarbonate is added to an aqueous solution of each compound? Explain.

(a) Sulfuric acid

(b) Ethanol, C_2H_5OH

(c) Ammonium chloride, NH_4Cl

Section 8.5 How Do We Use Acid Ionization Constants?

8.27 Which has the larger numerical value?

(a) The pK_a of a strong acid or the pK_a of a weak acid

(b) The K_a of a strong acid or the K_a of a weak acid

8.28 In each pair, select the stronger acid.

(a) Pyruvic acid ($pK_a = 2.49$) or lactic acid ($pK_a = 3.08$)

(b) Citric acid ($pK_a = 3.08$) or phosphoric acid ($pK_a = 2.10$)

(c) Benzoic acid ($K_a = 6.5 \times 10^{-5}$) or lactic acid ($K_a = 8.4 \times 10^{-4}$)

(d) Carbonic acid ($K_a = 4.3 \times 10^{-7}$) or boric acid ($K_a = 7.3 \times 10^{-10}$)

8.29 Which solution will be more acidic; that is, which will have a lower pH?

(a) 0.10 M CH_3COOH or 0.10 M HCl

(b) 0.10 M CH_3COOH or 0.10 M H_3PO_4

(c) 0.010 M H_2CO_3 or 0.010 M $NaHCO_3$

(d) 0.10 M NaH_2PO_4 or 0.10 M Na_2HPO_4

(e) 0.10 M aspirin ($pK_a = 3.47$) or 0.10 M acetic acid

8.30 Which solution will be more acidic; that is, which will have a lower pH?

(a) 0.10 M C_6H_5OH (phenol) or 0.10 M C_2H_5OH (ethanol)

(b) 0.10 M NH_3 or 0.10 M NH_4Cl

(c) 0.10 M NaCl or 0.10 M NH_4Cl

(d) 0.10 M $CH_3CH(OH)COOH$ (lactic acid) or 0.10 M CH_3COOH

(e) 0.10 M ascorbic acid (vitamin C, $pK_a = 4.1$) or 0.10 M acetic acid

Section 8.6 What Are the Properties of Acids and Bases?

8.31 Write an equation for the reaction of HCl with each compound. Which are acid–base reactions? Which are redox reactions?

(a) Na_2CO_3 (b) Mg   (c) NaOH   (d) Fe_2O_3

(e) NH_3 (f) CH_3NH_2 (g) $NaHCO_3$ (h) Al

8.32 When a solution of sodium hydroxide is added to a solution of ammonium carbonate and is heated, ammonia gas, NH_3, is released. Write a net ionic equation for this reaction. Both NaOH and $(NH_4)_2CO_3$ exist as dissociated ions in aqueous solution.

Section 8.7 What Are the Acidic and Basic Properties of Pure Water?

8.33 Given the following values of $[H_3O^+]$, calculate the corresponding value of $[OH^-]$ for each solution.

(a) 10^{-11} M (b) 10^{-4} M (c) 10^{-7} M (d) 10 M

8.34 Given the following values of $[OH^-]$, calculate the corresponding value of $[H_3O^+]$ for each solution.
(a) $10^{-10}\ M$ (b) $10^{-2}\ M$ (c) $10^{-7}\ M$ (d) $10\ M$

Section 8.8 What Are pH and pOH?

8.35 What is the pH of each solution given the following values of $[H_3O^+]$? Which solutions are acidic, which are basic, and which are neutral?
(a) $10^{-8}\ M$ (b) $10^{-10}\ M$ (c) $10^{-2}\ M$
(d) $10^0\ M$ (e) $10^{-7}\ M$

8.36 What is the pH and pOH of each solution given the following values of $[OH^-]$? Which solutions are acidic, which are basic, and which are neutral?
(a) $10^{-3}\ M$ (b) $10^{-1}\ M$ (c) $10^{-5}\ M$ (d) $10^{-7}\ M$

8.37 What is the pH of each solution given the following values of $[H_3O^+]$? Which solutions are acidic, which are basic, and which are neutral?
(a) $3.0 \times 10^{-9}\ M$ (b) $6.0 \times 10^{-2}\ M$
(c) $8.0 \times 10^{-12}\ M$ (d) $5.0 \times 10^{-7}\ M$

8.38 Which is more acidic, a beer with $[H_3O^+] = 3.16 \times 10^{-5}$ or a wine with $[H_3O^+] = 5.01 \times 10^{-4}$?

8.39 What is the $[OH^-]$ and pOH of each solution?
(a) $0.10\ M$ KOH, pH = 13.0
(b) $0.10\ M$ Na_2CO_3, pH = 11.6
(c) $0.10\ M$ Na_3PO_4, pH = 12.0
(d) $0.10\ M$ $NaHCO_3$, pH = 8.4

Section 8.9 How Do We Use Titrations to Calculate Concentration?

8.40 What is the purpose of an acid–base titration?

8.41 What is the molarity of a solution made by dissolving 12.7 g of HCl in enough water to make 1.00 L of solution?

8.42 What is the molarity of a solution made by dissolving 3.4 g of $Ba(OH)_2$ in enough water to make 450 mL of solution? Assume that $Ba(OH)_2$ ionizes completely in water to Ba^{2+} and OH^- ions. What is the pH of the solution?

8.43 Describe how you would prepare each of the following solutions (in each case, assume that the base is a solid).
(a) 400.0 mL of $0.75\ M$ NaOH
(b) 1.0 L of $0.071\ M$ $Ba(OH)_2$
(c) 500.0 mL of $0.1\ M$ KOH
(d) 2.0 L of $0.3\ M$ sodium acetate

8.44 If 25.0 mL of an aqueous solution of H_2SO_4 requires 19.7 mL of $0.72\ M$ NaOH to reach the end point, what is the molarity of the H_2SO_4 solution?

8.45 A sample of 27.0 mL of $0.310\ M$ NaOH is titrated with $0.740\ M$ H_2SO_4. How many milliliters of the H_2SO_4 solution are required to reach the end point?

8.46 A $0.300\ M$ solution of H_2SO_4 was used to titrate 10.00 mL of NaOH; 15.00 mL of acid was required to neutralize the basic solution. What was the molarity of the base?

8.47 A solution of NaOH base was titrated with $0.150\ M$ HCl, and 22.0 mL of acid was needed to reach the end point of the titration. How many moles of the unknown base were in the solution?

8.48 The usual concentration of HCO_3^- ions in blood plasma is approximately 24 millimoles per liter (mmol/L). How would you make up 1.00 L of a solution containing this concentration of HCO_3^- ions?

8.49 What is the end point of a titration?

8.50 Why does a titration not tell us the acidity or basicity of a solution?

Section 8.10 What Are Buffers?

8.51 Write equations to show what happens when, to a buffer solution containing equimolar amounts of CH_3COOH and CH_3COO^-, we add:
(a) H_3O^+ (b) OH^-

8.52 Write equations to show what happens when, to a buffer solution containing equimolar amounts of HPO_4^{2-} and $H_2PO_4^-$, we add
(a) H_3O^+ (b) OH^-

8.53 We commonly refer to a buffer as consisting of approximately equal molar amounts of a weak acid and its conjugate base—for example, CH_3COOH and CH_3COO^-. Is it also possible to have a buffer consisting of approximately equal molar amounts of a weak base and its conjugate acid? Explain.

8.54 What is meant by buffer capacity?

8.55 How can you change the pH of a buffer? How can you change the capacity of a buffer?

8.56 What is the connection between buffer action and Le Chatelier's principle?

8.57 Give two examples of a situation where you would want a buffer to have unequal amounts of the conjugate acid and the conjugate base.

8.58 How is the buffer capacity affected by the ratio of the conjugate base to the conjugate acid?

8.59 Can 100 mL of $0.1\ M$ phosphate buffer at pH 7.2 act as an effective buffer against 20 mL of $1\ M$ NaOH?

Section 8.11 How Do We Calculate the pH of a Buffer?

8.60 What is the pH of a buffer solution made by dissolving 0.10 mol of formic acid, HCOOH, and 0.10 mol of sodium formate, HCOONa, in 1 L of water?

8.61 The pH of a solution made by dissolving 1.0 mol of propanoic acid and 1.0 mol of sodium propanoate in 1.0 L of water is 4.85.
(a) What would the pH be if we used 0.10 mol of each (in 1 L of water) instead of 1.0 mol?
(b) With respect to buffer capacity, how would the two solutions differ?

8.62 Show that when the concentration of the weak acid, [HA], in an acid–base buffer equals that of the conjugate base of the weak acid, $[A^-]$, the pH of the buffer solution is equal to the pK_a of the weak acid.

8.63 Show that the pH of a buffer is 1 unit higher than its pK_a when the ratio of A^- to HA is 10 to 1.

8.64 Calculate the pH of an aqueous solution containing the following:
(a) $0.80\ M$ lactic acid and $0.40\ M$ lactate ion
(b) $0.30\ M$ NH_3 and $1.50\ M$ NH_4^+

8.65 The pH of 0.10 M HCl is 1.0. When 0.10 mol of sodium acetate, CH_3COONa, is added to this solution, its pH changes to 2.9. Explain why the pH changes and why it changes to this particular value.

8.66 If you have 100 mL of a 0.1 M buffer made of NaH_2PO_4 and Na_2HPO_4 that is at pH 6.8 and you add 10 mL of 1 M HCl, will you still have a usable buffer? Why or why not?

Section 8.12 What Are TRIS, HEPES, and These Buffers with the Strange Names?

8.67 Write an equation showing the reaction of TRIS in the acid form with sodium hydroxide (do not write out the chemical formula for TRIS).

8.68 What is the pH of a solution that is 0.1 M in TRIS in the acid form and 0.05 M in TRIS in the basic form?

8.69 Explain why you do not need to know the chemical formula of a buffer compound to use it.

8.70 If you have a HEPES buffer at pH 4.75, will it be a usable buffer? Why or why not?

8.71 Which of the compounds listed in Table 8.6 would be the most effective for making a buffer at pH 8.15? Why?

8.72 Which of the compounds listed in Table 8.6 would be the most effective for making a buffer at pH 7.0?

Chemical Connections

8.73 (Chemical Connections 8A) Which weak base is used as a flame retardant in plastics?

8.74 (Chemical Connections 8B) Name the most common bases used in over-the-counter antacids.

8.75 (Chemical Connections 8C) What causes (a) respiratory acidosis and (b) metabolic acidosis?

8.76 (Chemical Connections 8D) Explain how the sprinter's trick works. Why would an athlete want to raise the pH of his or her blood?

8.77 (Chemical Connections 8D) Another form of the sprinter's trick is to drink a sodium bicarbonate shake before the event. What would be the purpose of doing so? Give the relevant equations.

Additional Problems

8.78 4-Methylphenol, $CH_3C_6H_4OH$ ($pK_a = 10.26$), is only slightly soluble in water, but its sodium salt, $CH_3C_6H_4O^-Na^+$, is quite soluble in water. In which of the following solutions will 4-methylphenol dissolve more readily than in pure water?
(a) Aqueous NaOH   (b) Aqueous $NaHCO_3$
(c) Aqueous NH_3

8.79 Benzoic acid, C_6H_5COOH ($pK_a = 4.19$), is only slightly soluble in water, but its sodium salt, $C_6H_5COO^-Na^+$, is quite soluble in water. In which of the following solutions will benzoic acid dissolve more readily than in pure water?
(a) Aqueous NaOH   (b) Aqueous $NaHCO_3$
(c) Aqueous Na_2CO_3

8.80 Assume that you have a dilute solution of HCl (0.10 M) and a concentrated solution of acetic acid (5.0 M). Which solution is more acidic? Explain.

8.81 Which of the two solutions from Problem 8.80 would take a greater amount of NaOH to hit a phenolphthalein end point assuming you had equal volumes of the two? Explain.

8.82 If the $[OH^-]$ of a solution is 1×10^{-14},
(a) What is the pH of the solution?
(b) What is the $[H_3O^+]$?

8.83 What is the molarity of a solution made by dissolving 0.583 g of the diprotic acid oxalic acid, $H_2C_2O_4$, in enough water to make 1.75 L of solution?

8.84 Following are three organic acids and the pK_a of each: butanoic acid, 4.82; barbituric acid, 5.00; and lactic acid, 3.85.
(a) What is the K_a of each acid?
(b) Which of the three is the strongest acid, and which is the weakest?
(c) What information do you need to predict which of the three acids would require the most NaOH to reach a phenolphthalein end point?

8.85 The pK_a value of barbituric acid is 5.0. If the H_3O^+ and barbiturate ion concentrations are each 0.0030 M, what is the concentration of the undissociated barbituric acid?

8.86 If pure water self-ionizes to give H_3O^+ and OH^- ions, why doesn't pure water conduct an electric current?

8.87 Can an aqueous solution have a pH of zero? Explain your answer using aqueous HCl as your example.

8.88 If an acid, HA, dissolves in water such that the K_a is 1000, what is the pK_a of that acid? Is this scenario possible?

8.89 A scale of K_b values for bases could be set up in a manner similar to that for the K_a scale for acids. However, this setup is generally considered unnecessary. Explain.

8.90 Do a 1.0 M CH_3COOH solution and a 1.0 M HCl solution have the same pH? Explain.

8.91 Do a 1.0 M CH_3COOH solution and a 1.0 M HCl solution require the same amount of 1.0 M NaOH to hit a titration end point? Explain.

8.92 Suppose you wish to make a buffer whose pH is 8.21. You have available 1 L of 0.100 M NaH_2PO_4 and solid Na_2HPO_4. How many grams of the solid Na_2HPO_4 must be added to the stock solution to accomplish this task? (Assume that the volume remains 1 L.)

8.93 In the past, boric acid was used to rinse an inflamed eye. What is the $H_3BO_3/H_2BO_3^-$ ratio in a borate buffer solution that has a pH of 8.40?

8.94 Suppose you want to make a CH_3COOH/CH_3COO^- buffer solution with a pH of 5.60. The acetic acid concentration is to be 0.10 M. What should the acetate ion concentration be?

8.95 For an acid–base reaction, one way to determine the position of equilibrium is to say that the larger of the equilibrium arrow pair points to the acid with the higher value of pK_a. For example,

$$CH_3COOH + HCO_3^- \rightleftharpoons CH_3COO^- + H_2CO_3$$
$$pK_a = 4.75 \qquad\qquad\qquad\qquad pK_a = 6.37$$

Explain why this rule works.

8.96 When a solution prepared by dissolving 4.00 g of an unknown monoprotic acid in 1.00 L of water is titrated with 0.600 M NaOH, 38.7 mL of the NaOH solution is needed to neutralize the acid. Determine the molarity of the acid solution. What is the molar mass of the unknown acid?

8.97 Write equations to show what happens when, to a buffer solution containing equal amounts of HCOOH and HCOO$^-$, we add:

(a) H$_3$O$^+$ (b) OH$^-$

8.98 If we add 0.10 mol of NH$_3$ to 0.50 mol of HCl dissolved in enough water to make 1.0 L of solution, what happens to the NH$_3$? Will any NH$_3$ remain? Explain.

8.99 Suppose you have an aqueous solution prepared by dissolving 0.050 mol of NaH$_2$PO$_4$ in 1 L of water. This solution is not a buffer, but suppose you want to make it into one. How many moles of solid Na$_2$HPO$_4$ must you add to this aqueous solution to make it into:

(a) A buffer of pH 7.21

(b) A buffer of pH 6.21

(c) A buffer of pH 8.21

8.100 The pH of a 0.10 M solution of acetic acid is 2.93. When 0.10 mol of sodium acetate, CH$_3$COONa, is added to this solution, its pH changes to 4.74. Explain why the pH changes and why it changes to this particular value.

8.101 Suppose you have a phosphate buffer (H$_2$PO$_4^-$/HPO$_4^{2-}$) of pH 7.21. If you add more solid NaH$_2$PO$_4$ to this buffer, would you expect the pH of the buffer to increase decrease, or remain unchanged? Explain.

8.102 Suppose you have a bicarbonate buffer containing carbonic acid, H$_2$CO$_3$, and sodium bicarbonate, NaHCO$_3$, and the pH of the buffer is 6.37. If you add more solid NaHCO$_3$ to this buffer solution, would you expect its pH to increase, decrease, or remain unchanged? Explain.

8.103 A student pulls a bottle of TRIS off a shelf and notes that the bottle says, "TRIS (basic form), pK_a = 8.3." The student tells you that if you add 0.1 mol of this compound to 100 mL of water, the pH will be 8.3. Is the student correct? Explain.

Looking Ahead

8.104 Unless under pressure, carbonic acid in aqueous solution breaks down into carbon dioxide and water, and carbon dioxide is evolved as bubbles of gas. Write an equation for the conversion of carbonic acid to carbon dioxide and water.

8.105 Following are pH ranges for several human biological materials. From the pH at the midpoint of each range, calculate the corresponding [H$_3$O$^+$]. Which materials are acidic, which are basic, and which are neutral?

(a) Milk, pH 6.6–7.6

(b) Gastric contents, pH 1.0–3.0

(c) Spinal fluid, pH 7.3–7.5

(d) Saliva, pH 6.5–7.5

(e) Urine, pH 4.8–8.4

(f) Blood plasma, pH 7.35–7.45

(g) Feces, pH 4.6–8.4

(h) Bile, pH 6.8–7.0

8.106 What is the ratio of HPO$_4^{2-}$/H$_2$PO$_4^-$ in a phosphate buffer of pH 7.40 (the average pH of human blood plasma)?

8.107 What is the ratio of HPO$_4^{2-}$/H$_2$PO$_4^-$ in a phosphate buffer of pH 7.9 (the pH of human pancreatic fluid)?

Challenge Problems

8.108 A concentrated hydrochloric acid solution contains 36.0% HCl (density = 1.18 g/mL). How many liters are required to produce 10.0 L of a solution that has a pH of 2.05?

8.109 The volume of an adult's stomach ranges from 50 mL when empty to 1 L when full. On a certain day, its volume is 600. mL and its contents have a pH of 2.00.

(a) Determine the number of moles of H$^+$ present. (Chapter 4)

(b) Assuming that all the H$^+$ is due to HCl(aq), how many grams of sodium hydrogen carbonate, NaHCO$_3$, will completely neutralize the stomach acid? (Chapter 4)

8.110 Consider an initial 0.040 M hypobromous acid (HOBr) solution at a certain temperature.

$$HOBr(aq) \rightleftharpoons H^+(aq) + OBr^-(aq)$$

At equilibrium after partial dissociation, its pH is found to be 5.05. What is the acid ionization constant, K_a, for hypobromous acid at this temperature?

8.111 A 1.00 L sample of HF gas at 20.0°C and 0.601 atm was dissolved in enough water to make 50.0 mL of hydrofluoric acid solution, HF(aq).

(a) What is the molarity of this solution?

(b) The solution above is allowed to come to equilibrium, and its pH is found to be 1.88. Calculate the acid ionization constant, K_a, for hydrofluoric acid.

8.112 A laboratory student is given an alloy or solid mixture that contains Ag and Pb. The student is directed to separate the two components from one another and decides to treat the mixture with excess concentrated hydrochloric acid. Explain whether this separation will be successful and write any relevant balanced net ionic equations.

8.113 When a solution prepared by dissolving 0.125 g of an unknown diprotic acid in 25.0 mL of water is titrated with 0.200 M NaOH, 30.0 mL of the NaOH solution is needed to neutralize the acid. Determine the molarity of the acid solution. What is the molar mass of the unknown diprotic acid?

8.114 A railroad tank car derails and spills 26 tons of concentrated sulfuric acid (1 ton = 907.185 kg). The acid is 98.0% H$_2$SO$_4$ with a density of 1.836 g/mL.

(a) What is the molarity of the acid?

(b) Sodium carbonate, Na$_2$CO$_3$, is used to neutralize the acid spill. Determine the kilograms of sodium carbonate required to completely neutralize the acid. (Chapter 4)

(c) How many liters of carbon dioxide at 18°C and 745 mm Hg are produced by this reaction? (Chapter 5)

Nuclear Chemistry

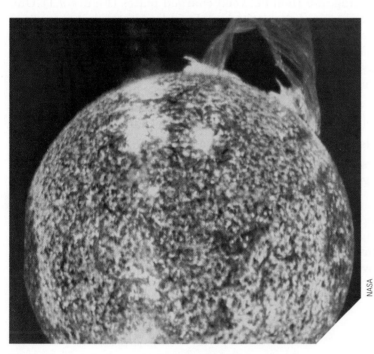

NASA

The sun's energy is the result of nuclear fusion.

9.1 How Was Radioactivity Discovered?

Every so often, a scientist makes the kind of discovery that changes the future of the world in some significant way. In 1896, a French physicist, Henri Becquerel (1852–1908), made one of these discoveries. At the time, Becquerel was engaged in a study of phosphorescent materials. In his experiments, he exposed certain salts, among them uranium salts, to sunlight for several hours, whereupon they phosphoresced. He then placed the glowing salts on a photographic plate that had been wrapped in opaque paper. Becquerel observed that by placing a coin or a metal cutout between the phosphorescing salts and the covered plate, he could create photographic images of the coin or metal cutout. He concluded that besides emitting visible light, the phosphorescent materials must have been emitting something akin to X-rays, which William Röntgen had discovered just the previous year. What was even more surprising to Becquerel was that his uranium salts continued to emit this same type of penetrating radiation long after their phosphorescence had ceased. What he had discovered was a type of radiation that Marie Curie was to call radioactivity. For this discovery, Becquerel shared the 1903 Nobel Prize for physics with Pierre and Marie Curie.

In this chapter, we will study the major types of radioactivity, their origin in the nucleus, the uses of radioactivity in the health and biological sciences, and its use as a source of power and energy.

9.2 What Is Radioactivity?

Early experiments identified three kinds of radiation, which were named alpha (α), beta (β), and gamma (γ) rays after the first three letters of the Greek alphabet. Each type of radiation behaves differently when passed between electrically charged plates. When a radioactive material is placed in a lead container that has a small opening, the emitted radiation passes through the opening and then between charged plates (Figure 9.1). One ray (β) is deflected toward the positive plate, indicating that it consists of negatively charged particles. A second ray (α) is deflected toward the negative plate, indicating that it consists of positively charged particles, and a third ray (γ) passes between the charged places without deflection, indicating that it has no charge.

Alpha particles are helium nuclei. Each contains two protons and two neutrons; each has an atomic number of 2 and a charge of +2.

Beta particles are electrons. Each has a charge of −1.

Gamma rays are high-energy electromagnetic radiation. They have no mass or charge.

Gamma rays are only one form of electromagnetic radiation. There are many others, including visible light, radio waves, and cosmic rays. All consist of waves (Figure 9.2).

The only difference between one form of electromagnetic radiation and another is the **wavelength** (λ, Greek letter lambda), which is the distance

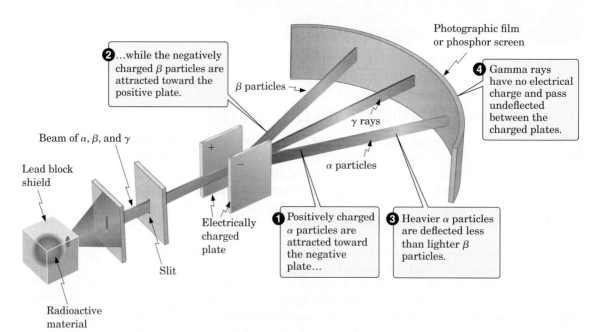

❷ ...while the negatively charged β particles are attracted toward the positive plate.

β particles

γ rays

Photographic film or phosphor screen

❹ Gamma rays have no electrical charge and pass undeflected between the charged plates.

Beam of α, β, and γ

Lead block shield

α particles

Electrically charged plate

Slit

Radioactive material

❶ Positively charged α particles are attracted toward the negative plate...

❸ Heavier α particles are deflected less than lighter β particles.

FIGURE 9.1 Electricity and radioactivity. Positively charged alpha (α) particles are attracted to the negative plate and negatively charged beta (β) particles are attracted to the positive plate. Gamma (γ) rays have no charge and are not deflected as they pass between the charged plates. Note that beta particles are deflected more than alpha particles.

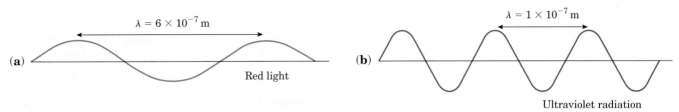

FIGURE 9.2 Two electromagnetic waves with different wavelengths.

from one wave crest to the next. The **frequency** (ν, Greek letter nu) of a radiation is the number of crests that pass a given point in one second. Mathematically, wavelength and frequency are related by the following equation, where c is the speed of light (3.0×10^8 m/s):

$$\lambda = \frac{c}{\nu}$$

As you can see from this relationship, the lower the frequency (ν), the longer the wavelength (λ); or conversely, the higher the frequency, the shorter the wavelength.

A relationship also exists between the frequency (ν) of electromagnetic radiation and its energy; the higher the frequency, the higher its energy. Electromagnetic radiation comes in packets; the smallest units are called **photons**.

Figure 9.3 shows the wavelengths of various types of radiation of the electromagnetic spectrum. Gamma rays are electromagnetic radiation of very high frequency (and high energy). Humans cannot see them because our eyes are not sensitive to waves of this frequency, but instruments (Section 9.5) can detect them. Another kind of radiation, called X-rays, can have higher energies than visible light but less than that of some gamma rays.

Materials that emit radiation (alpha, beta, or gamma) are called **radioactive**. Radioactivity comes from the atomic nucleus and not from the electron cloud that surrounds the nucleus. Table 9.1 summarizes the properties of the particles and rays that come out of radioactive nuclei, along with the properties of some other particles and rays. Note that X-rays are not considered to be a form of radioactivity, because they do not come out of the nucleus but are generated in other ways.

We have said that humans cannot see gamma rays. We cannot see alpha or beta particles either. Likewise, we cannot hear them, smell them, or feel

The only radiation known to have an even higher frequency (and energy) than gamma rays are cosmic rays.

UV stands for ultraviolet and IR for infrared.

The electron volt (eV) is a non-SI energy unit used frequently in nuclear chemistry.

$1 \text{ eV} = 1.602 \times 10^{-19} \text{ J}$
$= 3.829 \times 10^{-14} \text{ cal}$

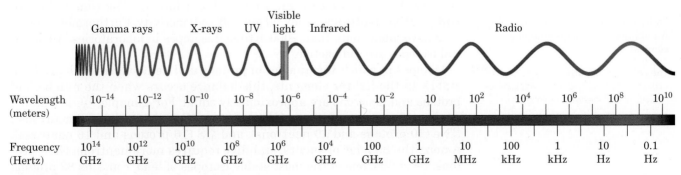

FIGURE 9.3 The electromagnetic spectrum.

Table 9.1 Particles and Rays Frequently Encountered in Radiation

Particle or Ray	Common Name of Radiation	Symbol	Charge	Atomic Mass Units	Penetrating Power[a]	Energy Range[b]
Proton	Proton beam	^1_1H	+1	1	1–3 cm	60 MeV
Electron	Beta particle	$^0_{-1}\text{e}$ or β^-	−1	$0.00055\left(\dfrac{1}{1835}\right)$	0–4 mm	1–3 MeV
Neutron	Neutron beam	^1_0n	0	1	—	—
Positron	—	$^0_{+1}\text{e}$ or β^+	+1	0.000555	—	—
Helium nucleus	Alpha particle	^4_2He or α	+2	4	0.02–0.04 mm	3–9 MeV
Energetic radiation	$\begin{cases}\text{Gamma ray} \\ \text{X-ray}\end{cases}$	γ	0 0	0 0	1–20 cm 0.01–1 cm	0.1–10 MeV 0.1–10 MeV

[a]Distance at which half of the radiation has been stopped.

[b]MeV = 1.602×10^{-13} J = 3.829×10^{-14} cal.

them. They are undetectable by our senses. We can detect radioactivity only by instruments, as discussed in Section 9.5.

9.3 What Happens When a Nucleus Emits Radioactivity?

As mentioned in Section 2.4D, different nuclei consist of different numbers of protons and neutrons. It is customary to indicate these numbers with subscripts and superscripts placed to the left of the atomic symbol. The atomic number (the number of protons in the nucleus) of an element is shown as a subscript and the mass number (the number of protons and neutrons in the nucleus) as a superscript. Following, for example, are symbols and names for the three known isotopes of hydrogen.

^1_1H	hydrogen-1	hydrogen	(not radioactive)
^2_1H	hydrogen-2	deuterium	(not radioactive)
^3_1H	hydrogen-3	tritium	(radioactive)

A. Radioactive and Stable Nuclei

Some isotopes are radioactive, whereas others are stable. Scientists have identified more than 300 naturally occurring isotopes. Of these, 264 are stable, meaning that the nuclei of these isotopes never give off any radioactivity. As far as we can tell, they will last forever. The remainder are **radioactive isotopes**—they do give off radioactivity. Furthermore, scientists have made more than 1000 artificial isotopes in laboratories. All artificial isotopes are radioactive.

Isotopes in which the number of protons and neutrons are balanced are stable. In the lighter elements, this balance occurs when the numbers of protons and neutrons are approximately equal. For example, $^{12}_6\text{C}$ is a stable nucleus (6 protons and 6 neutrons) as are $^{16}_8\text{O}$ (8 protons and 8 neutrons), $^{20}_{10}\text{Ne}$ (10 protons and 10 neutrons), and $^{32}_{16}\text{S}$ (16 protons and 16 neutrons). Among the heavier elements, stability requires more neutrons than protons. Lead-206, one of the most stable isotopes of lead, contains 82 protons and 124 neutrons.

Radioactive isotope (radioisotope) A radiation-emitting isotope of an element

The role of neutrons seems to be to provide binding energy to overcome the repulsion between protons.

If there is a serious imbalance in the proton-to-neutron ratio, either too few or too many neutrons, a nucleus will undergo a **nuclear reaction** to make the ratio more favorable and the nucleus more stable.

Nuclear reaction A reaction that changes the nucleus of an element (usually to the nucleus of another element)

B. Beta Emission

If a nucleus has more neutrons than it needs for stability, it can stabilize itself by converting a neutron to a proton and an electron.

$$\underset{\text{Neutron}}{^1_0\text{n}} \longrightarrow \underset{\text{Proton}}{^1_1\text{H}} + \underset{\text{Electron}}{^{\ 0}_{-1}\text{e}}$$

The proton remains in the nucleus and the electron is emitted from it. The emitted electron is called a **beta particle**, and the process is called **beta emission**. Phosphorus-32, for example, is a beta emitter:

$$^{32}_{15}\text{P} \longrightarrow ^{32}_{16}\text{S} + ^{\ 0}_{-1}\text{e}$$

A phosphorus-32 nucleus has 15 protons and 17 neutrons. The nucleus remaining after an electron has been emitted now has 16 protons and 16 neutrons; its atomic number is increased by 1, but its mass number is unchanged. The new nucleus is, therefore, sulfur-32. Thus, when the unstable phosphorus-32 (15 protons and 17 neutrons) is converted to sulfur-32 (16 protons and 16 neutrons), nuclear stability is achieved.

The changing of one element into another is called **transmutation**. It happens naturally every time an element gives off a beta particle. Every time a nucleus emits a beta particle, it is transformed into another nucleus with the same mass number but an atomic number one unit greater.

How To . . .
Balance a Nuclear Equation

In writing nuclear equations, we consider only the nucleus and disregard the surrounding electrons. There are two simple rules for balancing nuclear equations:

1. The sum of the mass numbers (superscripts) on both sides of the equation must be equal.
2. The sum of the atomic numbers (subscripts) on both sides of the equation must be equal. For the purposes of determining atomic numbers in a nuclear equation, an electron emitted from the nucleus has an atomic number of -1.

To see how to apply these rules, let us look at the decay of phosphorus-32, a beta emitter.

$$^{32}_{15}\text{P} \longrightarrow ^{32}_{16}\text{S} + ^{\ 0}_{-1}\text{e}$$

1. Mass number balance: the total mass number on each side of the equation is 32.
2. Atomic number balance: the atomic number on the left is 15. The sum of the atomic numbers on the right is $16 - 1 = 15$.

Thus, we see that in the phosphorus-32 decay equation, mass numbers are balanced (32 and 32) and atomic numbers are balanced (15 and 15); therefore, the nuclear equation is balanced.

Example 9.1 Beta Emission

Carbon-14, $^{14}_{6}C$, is a beta emitter. Write an equation for this nuclear reaction and identify the product formed.

$$^{14}_{6}C \longrightarrow ? + ^{0}_{-1}e$$

Strategy

In beta decay, a neutron is converted to a proton and an electron. The proton remains in the nucleus, and the electron is emitted as a beta particle.

Solution

The $^{14}_{6}C$ nucleus has six protons and eight neutrons. After beta decay, the nucleus has seven protons and seven neutrons:

$$^{14}_{6}C \longrightarrow ^{14}_{7}? + ^{0}_{-1}e$$

The sum of the mass numbers on each side of the equation is 14, and the sum of the atomic numbers on each side is 6. We now look in the Periodic Table to determine what element has atomic number 7 and see that it is nitrogen. The product of this nuclear reaction is therefore nitrogen-14, and we can now write a complete equation.

$$^{14}_{6}C \longrightarrow ^{14}_{7}N + ^{0}_{-1}e$$

Problem 9.1

Iodine-139 is a beta emitter. Write an equation for this nuclear reaction and identify the product formed.

C. Alpha Emission

For heavy elements, the loss of alpha (α) particles is an especially important stabilization process. For example:

$$^{238}_{92}U \longrightarrow ^{234}_{90}Th + ^{4}_{2}He$$

$$^{210}_{84}Po \longrightarrow ^{206}_{82}Pb + ^{4}_{2}He + \gamma$$

Note that the radioactive decay of polonium-210 emits both α particles and gamma rays.

A general rule for alpha emission is this: The new nucleus always has a mass number four units lower and an atomic number two units lower than the original.

Example 9.2 Alpha Emission

Polonium-218 is an alpha emitter. Write an equation for this nuclear reaction and identify the product formed.

Strategy

An alpha particle has a mass of 4 amu and a charge of +2, so that after alpha emission, the remaining nucleus has an atomic mass that is four units lower and an atomic number that is two units lower.

Solution

The atomic number of polonium is 84, so the partial equation is:

$$^{218}_{84}\text{Po} \longrightarrow ? + {}^{4}_{2}\text{He}$$

The mass number of the new isotope is $218 - 4 = 214$. The atomic number of the new isotope is $84 - 2 = 82$. We can now write:

$$^{218}_{84}\text{Po} \longrightarrow {}^{214}_{82}? + {}^{4}_{2}\text{He}$$

In the Periodic Table, we find that the element with an atomic number of 82 is lead, Pb. Therefore, the product is $^{214}_{82}\text{Pb}$, and we can now write the complete equation:

$$^{218}_{84}\text{Po} \longrightarrow {}^{214}_{82}\text{Pb} + {}^{4}_{2}\text{He}$$

Problem 9.2

Thorium-223 is an alpha emitter. Write an equation for this nuclear reaction and identify the product formed.

D. Positron Emission

A positron is a particle that has the same mass as an electron but a charge of +1 rather than −1. Its symbol is β^+ or $^{0}_{+1}e$. Positron emission is much rarer than alpha or beta emission. Because a positron has no appreciable mass, the nucleus is transmuted into another nucleus with the same mass number but an atomic number that is one unit less. Carbon-11, for example, is a positron emitter:

$$^{11}_{6}\text{C} \longrightarrow {}^{11}_{5}\text{B} + {}^{0}_{+1}e$$

In this balanced nuclear equation, the mass numbers on the left and right are 11. The atomic number on the left is 6; that on the right is also 6 $(5 + 1 = 6)$.

Example 9.3 Positron Emission

Nitrogen-13 is a positron emitter. Write an equation for this nuclear reaction and identify the product.

Strategy

A positron has a mass of 0 amu and a charge of +1.

Solution

We begin by writing the following partial equation:

$$^{13}_{7}\text{N} \longrightarrow ? + {}^{0}_{+1}e$$

Because a positron has no appreciable mass, the mass number of the new isotope is still 13. The sum of the atomic numbers on each side must be 7, which means that the atomic number of the new isotope must be 6. We find in the Periodic Table that the element with atomic number 6 is carbon. Therefore, the new isotope formed in this nuclear reaction is carbon-13 and the balanced nuclear equation is:

$$^{13}_{7}\text{N} \longrightarrow {}^{13}_{6}\text{C} + {}^{0}_{+1}e$$

Problem 9.3

Arsenic-74 is a positron emitter used in locating brain tumors. Write an equation for this nuclear reaction and identify the product.

E. Gamma Emission

Both alpha and beta emissions can be either "pure" or mixed with gamma rays.

Although rare, some nuclei are pure gamma emitters:

$$^{11}_{5}\text{B}^{*} \longrightarrow {}^{11}_{5}\text{B} + \gamma$$

Gamma emission often accompanies α and β emissions.

In this equation, $^{11}_{5}\text{B}^{*}$ symbolizes a boron nucleus in a high-energy (excited) state. In this case, no transmutation takes place. The element is still boron, but its nucleus is in a lower-energy (more stable) state after the emission of excess energy in the form of gamma rays. When all excess energy has been emitted, the nucleus returns to its most stable, lowest-energy state.

F. Electron Capture

In electron capture, an extranuclear electron is captured by the nucleus and there reacts with a proton to form a neutron. Thus, electron capture reduces the atomic number of the element, but the mass number is unchanged. Beryllium-7, for example, decays by electron capture to give lithium-7.

$$^{7}_{4}\text{Be} + {}^{0}_{-1}\text{e} \longrightarrow {}^{7}_{3}\text{Li}$$

Example 9.4 Electron Capture

Chromium-51, which is used to image the size and shape of the spleen, decays by electron capture and gamma emission. Write an equation for this nuclear decay and identify the product.

$$^{51}_{24}\text{Cr} + {}^{0}_{-1}\text{e} \longrightarrow ? + \gamma$$

Strategy and Solution

Because electron capture results in the conversion of one proton to a neutron and because there is no change in mass number upon gamma emission, the new nucleus has a mass number of 51. The new nucleus, however, has only 23 protons, one less than chromium-51. We find from the Periodic Table that the element with atomic number 23 is vanadium; therefore, the new element formed is vanadium-51. We can now write the complete equation for this nuclear decay.

$$^{51}_{24}\text{Cr} + {}^{0}_{-1}\text{e} \longrightarrow {}^{51}_{23}\text{V} + \gamma$$

Problem 9.4

Thallium-201, a radioisotope used to evaluate heart function in exercise stress tests, decays by electron capture and gamma emission. Write an equation for this nuclear decay and identify the product.

9.4 What Is Nuclear Half-Life?

Suppose we have 40 g of a radioactive isotope—say $^{90}_{38}\text{Sr}$. Strontium-90 nuclei are unstable and decay by beta emission to yttrium-39:

$$^{90}_{38}\text{Sr} \longrightarrow {}^{90}_{39}\text{Y} + {}^{0}_{-1}\beta$$

Our 40-gram sample of strontium-90 contains about 2.7×10^{23} atoms. We know that these nuclei decay, but at what rate do they decay? Do all of the nuclei decay at the same time, or do they decay over time? The answer is that they decay over time at a fixed rate. For strontium-90, the decay rate is such that one-half of our original sample (about 1.35×10^{23} atoms) will have decayed in 28.1 years. The time it takes for one-half of any sample of radioactive material to decay is called the **half-life**, $t_{1/2}$.

When a nucleus gives off radiation, it is said to *decay*.

It does not matter how big or small a sample is. For example, in the case of our 40 g of strontium-90, 20 g will be left at the end of 28.1 years (the rest has been converted to yttrium-90). It will then take another 28.1 years for half of the remainder to decay, so that after 56.2 years, we will have 10 g of strontium-90. If we wait for a third span of 28.1 years, then 5 g will be left. If we had begun with 100 g, then 50 g would be left after the first 28.1-year period.

Figure 9.4 shows the radioactive decay curve of iodine-131. Inspection of this graph shows that at the end of 8 days, half of the original has disappeared. Thus, the half-life of iodine-131 is 8 days. It would take a total of 16 days, or two half-lives, for three-fourths of the original amount of iodine-131 to decay.

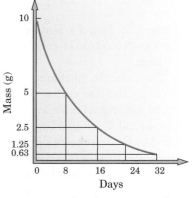

FIGURE 9.4 The decay curve of iodine-131.

Example 9.5 Nuclear Half-Life

If 10.0 mg of $^{131}_{53}I$ is administered to a patient, how much is left in the body after 32 days?

Strategy and Solution

We know from Figure 9.4 that $t_{1/2}$ of iodine-131 is eight days. The time span of 32 days corresponds to four half-lives. If we start with 10.0 mg, 5.00 mg remains after one half-life, 2.50 mg after two half-lives; 1.25 mg after three half-lives, and, 0.625 mg after four half-lives.

$$\overbrace{10.0 \text{ mg} \times \frac{1}{2} \times \frac{1}{2} \times \frac{1}{2} \times \frac{1}{2}}^{\text{32 days (4 half-lives)}} = 0.625 \text{ mg}$$

Problem 9.5

Barium-122 has a half-life of 2 minutes. Suppose you obtained a sample weighing 10.0 g and it takes 10 minutes to set up an experiment in which the barium-122 is to be used. How many grams of barium-122 will remain at the point when you begin the experiment?

It must be noted that in theory, it would take an infinite time period for all of a radioactive sample to decay. In reality, most of the radioactivity decays after five half-lives, by which time, only 3.1% of the original radioisotope remains.

$$\overbrace{\frac{1}{2} \times \frac{1}{2} \times \frac{1}{2} \times \frac{1}{2} \times \frac{1}{2}}^{\text{5 half-lives}} \times 100\% = 3.1\%$$

Radioactive Dating

Carbon-14, with a half-life of 5730 years, can be used to date archeological objects as old as 60,000 years. This dating technique relies on the principle that the carbon-12/carbon-14 ratio of an organism—whether plant or animal—remains constant during the lifetime of the organism. When the organism dies, the carbon-12 level remains constant (carbon-12 is not radioactive), but any carbon-14 present decays by beta emission to nitrogen-14.

$$^{14}_{6}\text{C} \longrightarrow {}^{14}_{7}\text{N} + {}^{0}_{-1}\text{e}$$

Using this fact, a scientist can calculate the changed carbon-12/carbon-14 ratio to determine the date of an artifact.

For example, in charcoal made from a tree that has recently died, the carbon-14 gives a radioactive count of 13.70 disintegrations/min per gram of carbon. In a piece of charcoal found in a cave in France near some ancient Cro-Magnon cave paintings, the carbon-14 count was 1.71 disintegrations/min for each gram of carbon. From this information, the cave paintings can be dated. After one half-life, the number of disintegrations/minute per gram is 6.85; after two half-lives, it is 3.42; and after three half-lives, it is 1.71. Therefore, three half-lives have passed since the paintings were created. Given that carbon-14 has a half-life of 5730 years, the paintings are approximately $3 \times 5730 = 17,190$ years old.

The famous Shroud of Turin, a piece of linen cloth with the image of a man's head on it, was believed by many to be the original cloth that was wrapped around the body of Jesus Christ after his death. However, radioactive dating showed with 95% certainty that the plants from which the linen was obtained were alive sometime between AD 1260 and 1380, proving that the cloth could not have been the shroud of Christ. Note that it was not necessary to destroy the shroud to perform the tests. In fact, scientists in different laboratories used only a few square centimeters of cloth from its edge.

Rock samples can be dated on the basis of their lead-206 and uranium-238 content. The underlying assumption is that lead-206 comes from the decay of uranium-238, which has a half-life of 4.5 billion years. One of the oldest rocks found on Earth is a granite outcrop in Greenland, dated at 3.7×10^9 years old. On the basis of dating of meteorites, the estimated age of the solar system is 4.6×10^9 years.

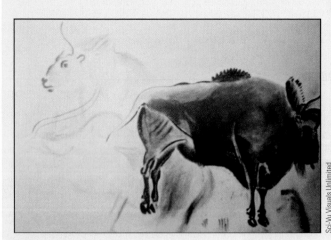

Cro-Magnon cave painting.

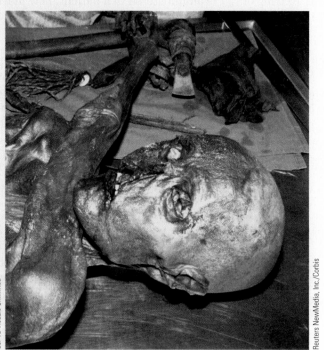

The Ice Man. This human mummy was found in 1991 in glacial ice high in the Alps. Carbon-14 dating determined that he lived about 5300 years ago. The mummy is exhibited at the South Tyrol Archeological Museum in Bolzano, Italy.

Test your knowledge with Problems 9.61, 9.62, 9.63, and 9.64.

The half-life of an isotope is independent of temperature and pressure—and, indeed, of all other physical and chemical conditions—and is a property of the particular isotope only. It does not depend on what other kind of atoms surround the particular nucleus (that is, what kind of molecule the nucleus is part of). We do not know any way to speed up radioactive decay or to slow it down.

Table 9.2 gives some half-lives. Even this brief sampling indicates that there are tremendous differences among half-lives. Some isotopes, such as technetium-99m, decay and disappear in a day; others, such as uranium-238, remain radioactive for billions of years. Very short-lived isotopes, especially the artificial heavy elements (Section 9.9) with atomic numbers greater than 100, have half-lives of the order of seconds.

Table 9.2 Half-Lives of Some Radioactive Nuclei

Name	Symbol	Half-Life	Radiation
Hydrogen-3 (tritium)	$^{3}_{1}\text{H}$	12.26 years	Beta
Carbon-14	$^{14}_{6}\text{C}$	5730 years	Beta
Phosphorus-28	$^{28}_{15}\text{P}$	0.28 second	Positrons
Phosphorus-32	$^{32}_{15}\text{P}$	14.3 days	Beta
Potassium-40	$^{40}_{19}\text{K}$	1.28×10^{9} years	Beta+gamma
Scandium-42	$^{42}_{21}\text{Sc}$	0.68 second	Positrons
Cobalt-60	$^{60}_{27}\text{Co}$	5.2 years	Gamma
Strontium-90	$^{90}_{38}\text{Sr}$	28.1 years	Beta
Technetium-99m	$^{99m}_{43}\text{Tc}$	6.0 hours	Gamma
Indium-116	$^{116}_{49}\text{In}$	14 seconds	Beta
Iodine-131	$^{131}_{53}\text{I}$	8 days	Beta+gamma
Mercury-197	$^{197}_{80}\text{Hg}$	65 hours	Gamma
Polonium-210	$^{210}_{84}\text{Po}$	138 days	Alpha
Radon-205	$^{205}_{86}\text{Rn}$	2.8 minutes	Alpha
Radon-222	$^{222}_{86}\text{Rn}$	3.8 days	Alpha
Uranium-238	$^{238}_{92}\text{U}$	4×10^{9} years	Alpha

The usefulness or inherent danger in radioactive isotopes is related to their half-lives. In assessing the long-range health effects of atomic-bomb damage or of nuclear power plant accidents like those at Three Mile Island, Pennsylvania in 1979, Chernobyl (in the former Soviet Union) in 1986 (Chemical Connections 9D), and Fukushima Daiichi (in Japan) in 2011, we can see that radioactive isotopes with long half-lives, such as $^{85}_{36}\text{Kr}$ ($t_{1/2} = 10$ years) or $^{60}_{27}\text{Co}$ ($t_{1/2} = 5.2$ years), are more important than short-lived ones. On the other hand, when a radioactive isotope is used in medical imaging or therapy, short-lived isotopes are more useful because they disappear faster from the body—for example, $^{99m}_{43}\text{Tc}$, $^{32}_{15}\text{P}$, $^{131}_{53}\text{I}$, and $^{197}_{80}\text{Hg}$.

9.5 How Do We Detect and Measure Nuclear Radiation?

As already noted, radioactivity is not detectable by our senses. We cannot see it, hear it, feel it, or smell it. How, then, do we know it is there? Alpha, beta, gamma, positron, and X-rays all have a property we can use to detect them. When these types of radiation interact with matter, they knock electrons out of the electron cloud surrounding an atomic nucleus, thereby creating positively charged ions from neutral atoms. For this reason, we call all of these rays **ionizing radiation**.

Ionizing radiation is characterized by two physical measurements: (1) its **intensity** (energy flux), which is the number of particles or photons emerging per unit time, and (2) the **energy** of each particle or photon emitted.

A. Intensity

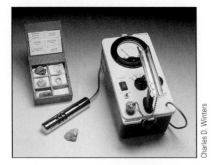

A Geiger-Müller Counter.

Charles D. Winters

A television picture tube works on a similar principle.

To measure intensity, we take advantage of the ionizing property of radiation. Instruments such as the **Geiger-Müller counter** (Figure 9.5) and the **proportional counter** contain a gas such as helium or argon. When a radioactive nucleus emits alpha or beta particles or gamma rays, these radiations ionize the gas, and the instrument registers this fact by indicating that an electric current has passed between two electrodes. In this way, the instrument counts radiation particle after particle.

Other measuring devices, such as **scintillation counters**, have a material called a phosphor that emits a unit of light for each alpha or beta particle or gamma ray that strikes it. Once again, the particles are counted one by one. The quantitative measure of radiation intensity can be reported in counts/minute or counts/second.

A common unit of radiation intensity is the **curie** (Ci), named in honor of Marie Curie, whose lifelong work with radioactive materials greatly helped our understanding of nuclear phenomena. One curie is defined as 3.7×10^{10} disintegrations per second (dps). This is radiation of very high intensity, the amount a person would get from exposure to 1.0 g of pure $^{286}_{88}\text{Ra}$. This intensity is too high for regular medical use, and the most common units used in the health sciences are small fractions of it. Another, albeit much smaller, unit of radiation activity (intensity) is the **becquerel** (Bq), which is the SI unit. One becquerel is one disintegration per second (dps).

$$1 \text{ becquerel (Bq)} = 1.0 \text{ dps}$$

$$1 \text{ curie (Ci)} = 3.7 \times 10^{10} \text{ dps}$$

$$1 \text{ millicurie (mCi)} = 3.7 \times 10^{7} \text{ dps}$$

$$1 \text{ microcurie } (\mu\text{Ci}) = 3.7 \times 10^{4} \text{ dps}$$

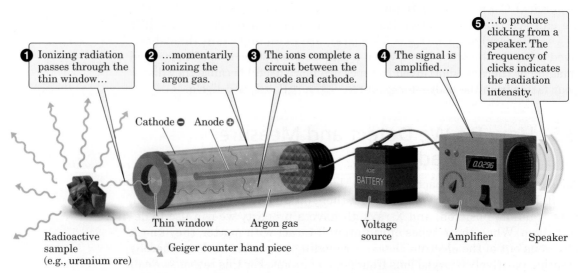

❶ Ionizing radiation passes through the thin window…

❷ …momentarily ionizing the argon gas.

❸ The ions complete a circuit between the anode and cathode.

❹ The signal is amplified…

❺ …to produce clicking from a speaker. The frequency of clicks indicates the radiation intensity.

Cathode ⊖ Anode ⊕

Thin window   Argon gas

Radioactive sample (e.g., uranium ore)

Geiger counter hand piece

Voltage source

Amplifier   Speaker

FIGURE 9.5 A schematic drawing of a Geiger-Müller counter.

ⓌWL Interactive Example 9.6 Intensity of Nuclear Radiation

A radioactive isotope with an intensity (activity) of 100. mCi per vial is delivered to a hospital. The vial contains 10. mL of liquid. The instruction is to administer 2.5 mCi intravenously. How many mL of the liquid should be administered?

Strategy and Solution

The intensity (activity) of a sample is directly proportional to the amount present, so:

$$2.5 \text{ mCi} \times \frac{10. \text{ mL}}{100. \text{ mCi}} = 0.25 \text{ mL}$$

Problem 9.6

A radioactive isotope in a 9.0-mL vial has an intensity of 300. mCi. A patient is required to take 50. mCi intravenously. How much liquid should be used for the injection?

The intensity of any radiation decreases with the square of the distance. If, for example, the distance (d) from a radiation source doubles, then the intensity (I) of the received radiation decreases by a factor of four.

$$\frac{I_1}{I_2} = \frac{d_2{}^2}{d_1{}^2}$$

Example 9.7 Intensity of Nuclear Radiation

If the intensity of radiation is 28 mCi at a distance of 1.0 m, what is the intensity at a distance of 2.0 m?

Strategy

As already noted, the intensity of any radiation decreases with the square of the distance.

Solution

From the preceding equation, we have:

$$\frac{28 \text{ mCi}}{I_2} = \frac{2.0^2}{1.0^2}$$

$$I_2 = \frac{28 \text{ mCi}}{4.0} = 7.0 \text{ mCi}$$

Thus, if the distance from a radioactive source increases by a factor of two, the intensity of the radiation at that distance is decreased by a factor of four.

Problem 9.7

If the intensity of radiation 1.0 cm from a source is 300. mCi, what is the intensity at 3.0 m?

B. Energy

The energies of different particles or photons vary. As shown in Table 9.1, each particle has a certain range of energy. For example, beta particles have an energy range of 1 to 3 MeV (megaelectron volts). This range may overlap

FIGURE 9.6 Penetration of radioactive emissions. Alpha particles, with a charge of +2 and a mass of 4 amu, interact strongly with matter but penetrate the least. They are stopped by several sheets of paper. Beta particles, with less mass and a lower charge than alpha particles, interact less strongly with matter. They easily penetrate paper but are stopped by a 0.5-cm sheet of lead. Gamma rays, with neither mass nor charge, have the greatest penetrating power. It takes 10 cm of lead to stop them.

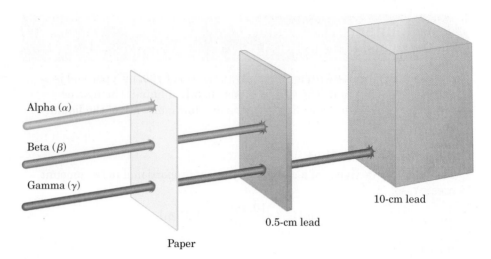

with the energy range of some other type of radiation—for example, gamma rays. The penetrating power of a radiation depends on its energy as well on the mass of its particles. Alpha particles are the most massive and the most highly charged and, therefore, the least penetrating; they are stopped by several sheets of ordinary paper, by ordinary clothing, and by the skin. Beta particles have less mass and lower charge than alpha particles and, consequently, have greater penetrating power. They can penetrate several millimeters of bone or tissue. Gamma radiation, which has neither mass nor charge, is the most penetrating of the three types of radiation. Gamma rays can pass completely through the body. Several centimeters of lead or one meter of concrete is required to stop gamma rays (Figure 9.6).

One easy way to protect against ionizing radiation is to wear lead aprons, covering sensitive organs. This practice is followed routinely when diagnostic X-rays are taken. Another way to lessen the damage from ionizing radiation is to move farther away from the source.

9.6 How Is Radiation Dosimetry Related to Human Health?

The term activity of a radiation *is the same as* intensity of a radiation.

In studying the effect of radiation on the body, neither the energy of the radiation (in kcal/mol) nor its intensity (in Ci) alone or in combination is of particular importance. Rather, the critical question is what kind of effects such radiation produces in the body. Three different units are used to describe the effects of radiation on the body: roentgens, rads, and rems.

Roentgens (R) Roentgens measure the energy delivered by a radiation source and are, therefore, a measure of exposure to a particular form of radiation. One roentgen is the amount of radiation that produces ions having 2.58×10^{-4} coulomb per kilogram (a coulomb is a unit of electrical charge).

Rads The rad, which stands for *r*adiation *a*bsorbed *d*ose, is a measure of the radiation absorbed from a radiation source. The SI unit is the gray (Gy), where 1 Gy = 100 rad. Roentgens (delivered energy) do not take into account the effect of radiation on tissue and the fact that different tissues absorb different amounts of delivered radiation. Radiation damages body tissue by causing ionization, and for ionization to occur, the tissue must absorb the delivered energy. The relationship between the delivered dose in roentgens and the absorbed dose in rads can be illustrated as follows: Exposure to

1 roentgen yields 0.97 rad of absorbed radiation in water, 0.96 rad in muscle, and 0.93 rad in bone. This relationship holds for high-energy photons. For lower-energy photons, such as "soft" X-rays, each roentgen yields 3 rads of absorbed dose in bone. This principle underlies diagnostic X-rays, wherein soft tissue lets the radiation through to strike a photographic plate but bone absorbs the radiation and casts a shadow on the plate.

Rems   The rem, which stands for *r*oentgen *e*quivalent for *m*an, is a measure of the effect of the radiation when a person absorbs 1 roentgen. Other units are the **millirem** (mrem; 1 mrem = 1×10^{-3} rem) and the **sievert** (Sv; 1 Sv = 100 rem). The sievert is the SI unit. The reason for the rem is that tissue damage from 1 rad of absorbed energy depends on the type of radiation. One rad from alpha rays, for example, causes ten times more damage than 1 rad from X-rays or gamma rays. Table 9.3 summarizes the various radiation units and what each measures.

Table 9.3 Radiation Dosimetry

Unit	What the Unit Measures	The SI Unit	Other Units
Roentgen	The amount of radiation delivered from a radiation source	Roentgen (R)	
Rad	The ratio between radiation absorbed by a tissue and that delivered to the tissue	Gray (Gy)	1 rad = 0.01 Gy
Rem	The ratio between the tissue damage caused by a rad of radiation and the type of radiation	Sievert (Sv)	1 rem = 0.01 Sv

Although alpha particles cause more damage than X-rays or gamma rays, they have a very low penetrating power (Table 9.1) and cannot pass through the skin. Consequently, they are not harmful to humans or animals as long as they do not enter the body. If they do get in, however, they can prove quite harmful. They can get inside, for example, if a person swallows or inhales a small particle of a substance that emits alpha particles. Beta particles are less damaging to tissue than alpha particles but penetrate farther and so are generally more harmful. Gamma rays, which can completely penetrate the skin, are by far the most dangerous and harmful form of radiation. Remember, of course, that once alpha particles such as those from radon-222 get in the body, they are very damaging. Therefore, for comparative purposes and for determining exposure from all kinds of sources, the equivalent dose is an important measure. If an organ receives radiation from different sources, the total effect can be summed up in rems (or mrem or Sv). For example, 10 mrem of alpha particles and 15 mrem of gamma radiation give a total of 25 mrem absorbed equivalent dose. Table 9.4 shows the amount of radiation exposure that an average person obtains yearly from both natural and artificial sources.

The naturally occurring background radiation varies with the geological location. For example, a level that is tenfold higher than the average radiation has been detected in some phosphate mines. People who work in nuclear medicine are, of course, exposed to greater amounts. To ensure that exposures do not get too high, they wear radiation badges. A single whole-body irradiation of 25 rem causes a noticeable reduction of white blood cells, and 100 rem causes the typical symptoms of radiation sickness, which include nausea, vomiting, a decrease in the white blood cell count, and loss of hair. A dose of 400 rem causes death within one month in 50% of exposed persons, and 600 rem is almost invariably lethal within a short time.

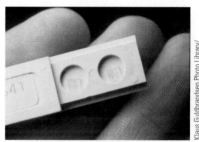

A radiation badge.

Klaus Guldbrandsen Photo Library/ Photo Researchers, Inc.

Cosmic rays High-energy particles, mainly protons, from outer space bombarding the Earth

Table 9.4 Average Exposure to Radiation from Common Sources

Source	Dose (mrem/year)
Naturally Occurring Radiation	
Cosmic rays	27
Terrestrial radiation (rocks, buildings)	28
Inside the human body (K-40 and Ra-226 in the bones)	39
Radon in the air	200
Total	294
Artificial Radiation	
Medical X-rays[a]	39
Nuclear medicine	14
Consumer products	10
Nuclear power plants	0.5
All others	1.5
Total	65
Grand total	359[b]

[a]Individual medical precedures may expose certain parts of the body to much higher levels. For instance, one chest X-ray gives 27 mrem and a diagnostic GI series gives 1970 mrem.

[b]The federal safety standard for allowable occupational exposure is about 5000 mrem/year. It has been suggested that this level be lowered to 4000 mrem/year or even lower to reduce the risk of cancer stemming from low levels of radiation.

SOURCE: *National Council on Radiation Protection and Measurements,* NCRP Report No. 93 (1993).

It should be noted that as much as 50,000 rem is needed to kill bacteria and as much as 10^6 rem to inactivate viruses.

Fortunately, most of us never get a single dose of more than a few rem and so never suffer from any form of radiation sickness. This does not mean, however, that small doses are totally harmless. The harm may arise in two ways:

1. Small doses of radioactivity over a period of years can cause cancer, especially blood cancers such as leukemia. Repeated exposure to sunlight also carries a risk of tissue damage. Most of the high-energy UV radiation of the sun is absorbed by the Earth's protective ozone layer in the stratosphere. In tanning, however, the repeated overexposure to UV radiation can cause skin cancer (see Chemical Connections 18D). No one knows how many cancers have resulted from this practice, because the doses are so small and continue for so many years that they cannot be measured accurately. Also, because so many other causes of cancer exist, it is difficult or impossible to decide if any particular case is caused by radiation.

2. If any form of radiation strikes an egg or sperm cell, it can cause a change in the genes (see Chemical Connections 25E). Such changes are called mutations. If an affected egg or sperm cell mates, grows, and becomes a new individual, that individual may have mutated characteristics, which are usually harmful and frequently lethal.

Because radiation carries so much potential for harm, it would be nice if we could totally escape it. But can we? Table 9.4 shows that this is impossible. Naturally occurring radiation, called **background radiation**, is present everywhere on Earth. As Table 9.4 shows, this background radiation vastly outstrips the average radiation level from artificial sources (mostly diagnostic X-rays). If we eliminated all forms of artificial radiation, including medical uses, we would still be exposed to the background radiation.

The Indoor Radon Problem

Most of our exposure to ionizing radiation comes from natural sources (Table 9.4), with radon gas being the main cause. Radon has more than 20 isotopes, all of which are radioactive. The most important is radon-222, an alpha emitter. Radon-222 is a natural decay product of uranium-238, which is widely distributed in the Earth's crust.

Radon poses a particular health hazard among radioactive elements because it is a gas at normal temperatures and pressures. As a consequence, it can enter our lungs with the air we breathe and become trapped in the mucous lining of the lungs. Radon-222 has a half-life of 3.8 days. It decays naturally and produces, among other isotopes, two harmful alpha emitters: polonium-218 and polonium-214. These polonium isotopes are solids and do not leave the lungs with exhalation. In the long run, they can cause lung cancer.

The U.S. Environmental Protection Agency has set a standard of 4 pCi/L (one picocurie, pCi, is 10^{-12} Ci) as a safe exposure level. A survey of single-family homes in the United States showed that 7% exceeded this level. Most radon seeps into dwellings through cracks in cement foundations and around pipes, then accumulates in basements. The remedy is to ventilate both basements and houses enough to reduce the radiation levels. In a notorious case, a group of houses in Grand Junction, Colorado, was built from bricks made from uranium tailings. Obviously, the

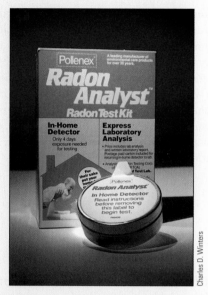

Charles D. Winters

Testing devices are available to determine whether radon is building up in a home.

radiation levels in these buildings were unacceptably high. Because they could not be controlled, the buildings had to be destroyed. In our modern radiation-conscious age, more and more homebuyers choose to request a certification of radon levels before buying a house.

Test your knowledge with Problem 9.65.

9.7 What Is Nuclear Medicine?

When we think of nuclear chemistry, our first thoughts may well be of nuclear power, atomic bombs, and weapons of mass destruction. True as this may be, it is also true that nuclear chemistry and the use of radioactive elements have become invaluable tools in all areas of science. Nowhere is this more important than in nuclear medicine; that is, in the use of radioactive isotopes as tools for both the diagnosis and treatment of diseases. To describe the full range of medical uses of nuclear chemistry would take far more space that we have in this text. What we have done, instead, is to choose several examples of each use to illustrate the range of applications of nuclear chemistry to the health sciences.

A. Medical Imaging

Medical imaging is the most widely used aspect of nuclear medicine. The goal of medical imaging is to create a picture of a target tissue. To create a useful image requires three things:

- A radioactive element administered in pure form or in a compound that becomes concentrated in the tissue to be imaged.

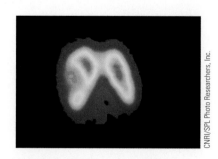

CNRI/SPL, Photo Researchers, Inc.

A scan of radiation released by radioactive iodine concentrated in thyroid tissue gives an image of the thyroid gland.

Chemical Connections 9C

How Radiation Damages Tissues: Free Radicals

As mentioned earlier, high-energy radiation damages tissue by causing ionization. That is, the radiation knocks electrons out of the molecules that make up the tissue (generally one electron per molecule), thereby forming unstable ions. For example, the interaction of high-energy radiation with water forms H_2O^+, an unstable cation. The positive charge on this cation means that one of the electrons normally present in the water molecule, either one from a covalent bond or from an unshared pair, is missing in this cation; it has been knocked out.

The unpaired electron
is on oxygen

$$[H-\overset{.}{\underset{..}{O}}-H]^+$$

Once formed, the H_2O^+ cation is unstable and decomposes to H^+ and a hydroxyl radical:

$$Energy + H_2O \longrightarrow H_2O^+ + e^-$$

$$H_2O^+ \longrightarrow H^+ + \cdot OH$$
Hydroxyl
radical

Whereas the oxygen atom in the hydroxide ion, OH^-, has a complete octet—it is surrounded by three unshared pairs of electrons and one shared pair—the oxygen atom in the hydroxyl radical is surrounded by only seven valence electrons—two unshared pairs, one shared pair, and one unpaired electron. Compounds that have unpaired electrons are called **free radicals**, or more simply, **radicals**.

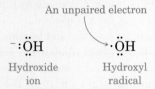

An unpaired electron

$^-:\overset{..}{\underset{..}{O}}H \qquad \cdot\overset{..}{\underset{..}{O}}H$

Hydroxide   Hydroxyl
ion   radical

The fact that the oxygen atom of the ·OH radical has an incomplete octet makes this radical extremely reactive. It rapidly interacts with other molecules, causing chemical reactions that damage tissues. These reactions have especially serious consequences if they occur inside cell nuclei and damage genetic material. In addition, they affect rapidly dividing cells more than they do stationary cells. Thus, the damage is greater to embryonic cells, cells of the bone marrow, the intestines, and cells in the lymph. Symptoms of radiation sickness include nausea, vomiting, a decrease in white blood cell count, and loss of hair.

- A method of detecting radiation from the radioactive source and recording its intensity and location.

- A computer to process the intensity–location data and transform it into a useful image.

Chemically and metabolically, a radioactive isotope in the body behaves in exactly the same way as do the nonradioactive isotopes of the same element. In the simplest form of imaging, a radioactive isotope is injected intravenously and a technician uses a detector to monitor how the radiation is distributed in the body of the patient. Table 9.5 lists some of the most important radioisotopes used in imaging and diagnosis.

The use of iodine-131, a beta and gamma emitter ($t_{1/2}$ = 8.04 days), to image and diagnose a malfunctioning thyroid gland is a good example. The thyroid gland in the neck produces a hormone, thyroxine, which controls the overall rate of metabolism (use of food) in the body. One molecule of thyroxine contains four iodine atoms. When radioactive iodine-131 is injected into the bloodstream, the thyroid gland takes it up and incorporates it into thyroxine (Chemical Connections 13C). A normally functioning thyroid absorbs about 12% of the administered iodine within a few hours. An overactive thyroid (hyperthyroidism) absorbs and localizes iodine-131 in the gland faster, and an underactive thyroid (hypothyroidism) does so much more slowly. By counting the gamma radiation emitted from the neck, one can determine the rate of uptake of iodine-131 into the thyroid gland and diagnose hyperthyroidism or hypothyroidism.

Table 9.5 Some Radioactive Isotopes Useful in Medical Imaging

	Isotope	Mode of Decay	Half-Life	Use in Medical Imaging
$^{11}_{6}C$	Carbon-11	β^+, γ	20.3 m	Brain scan to trace glucose metabolism
$^{18}_{9}F$	Fluorine-18	β^+, γ	109 m	Brain scan to trace glucose metabolism
$^{32}_{15}P$	Phosphorus-32	β	14.3 d	Detect eye tumors
$^{51}_{24}Cr$	Chromium-51	E.C., γ	27.7 d	Diagnose albinism; image the spleen and gastrointestinal tract
$^{59}_{26}Fe$	Iron-59	β, γ	44.5 d	Bone marrow function; diagnose anemias
$^{67}_{31}Ga$	Gallium-67	E.C., γ	78.3 h	Whole-body scan for tumors
$^{75}_{34}Se$	Selenium-75	E.C., γ	118 d	Pancreas scan
$^{81m}_{36}Kr$	Krypton-81m	γ	13.3 s	Lung ventilation scan
$^{81}_{38}Sr$	Strontium-81	β	22.2 m	Scan for bone diseases, including cancer
$^{99m}_{43}Tc$	Technetium-99m	γ	6.01 h	Brain, liver, kidney, bone scans; diagnosis of damaged heart muscle
$^{131}_{53}I$	Iodine-131	β, γ	8.04 d	Diagnosis of thyroid malfunction
$^{197}_{80}Hg$	Mercury-197	E,C., γ	64.1 h	Kidney scan
$^{201}_{81}Tl$	Thallium-201	E,C., γ	3.05 d	Heart scan and exercise stress test

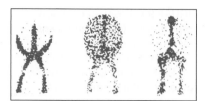
Normal

Meningioma (brain tumor)

"Brain death"

Scalp tumor

From CRC Handbook in Clinical Laboratory Science, Vol. 1. Nuclear Medicine. CRC Press, Inc.

FIGURE 9.7 A comparison of dynamic scan patterns for normal and pathological brains. The studies were performed by injecting technetium-99m into blood vessels.

Most organ scans are similarly based on the preferential uptake of some radioactive isotopes by a particular organ (Figure 9.7).

Another important type of medical imaging is called positron emission tomography (PET). This method is based on the property that certain isotopes (such as carbon-11 and fluorine-18) emit positrons (Section 9.3D). Fluorine-18 decays by positron emission to oxygen-18:

$$^{18}_{9}F \longrightarrow {}^{18}_{8}O + {}^{0}_{+1}e$$

Positrons have very short lives. When a positron and an electron collide, they annihilate each other, resulting in the emission of two gamma rays.

$$\underset{\text{Positron}}{{}^{0}_{+1}e} + \underset{\text{Electron}}{{}^{0}_{-1}e} \longrightarrow 2\,\gamma$$

Because electrons are present in every atom, there are always lots of them around, so positrons generated in the body do not last very long.

A favorite tagged molecule for following the uptake and metabolism of glucose, $C_6H_{12}O_6$, is 18-fluorodeoxyglucose (FDG), a molecule of glucose in which one of glucose's six oxygen atoms is replaced by fluorine-18. When FDG is administered intravenously, the tagged glucose soon enters the blood and from there moves to the brain. Gamma-ray detectors can pick up the signals that come from the areas where the tagged glucose accumulates. In this way, one can see which areas of the brain are involved when we process, for example, visual information (Figure 9.8). Whole body PET scans can be used to diagnose lung, colorectal, head and neck, and esophageal cancers as well as early stages of epilepsy and other diseases that involve abnormal glucose metabolism, such as schizophrenia.

Because tumors have high metabolic rates, PET scans using FDG have become the diagnostic choice for their detection and localization. FDG/PET has been used in the diagnosis of malignant melanoma and malignant lymphoma among other conditions.

Another important use of radioactive isotopes is to learn what happens to an ingested material. The foods and drugs swallowed or otherwise taken

FIGURE 9.8 Positron emission tomography (PET) brain scans. The upper scans show that 18-fluorodeoxyglucose can cross the blood–brain barrier. The lower scans show that visual stimulation increases blood flow and glucose concentration in certain areas of the brain. These areas are shown in red.

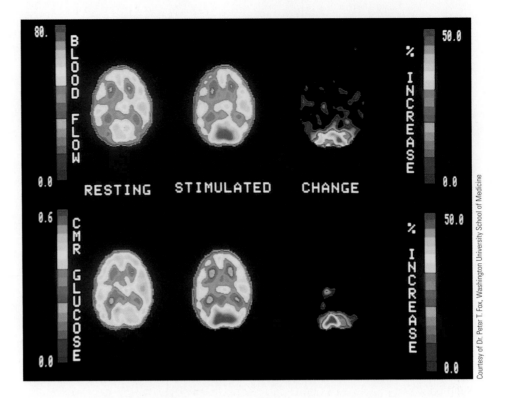

in by the body are transformed, decomposed, and excreted. To understand the pharmacology of a drug, it is important to know how and in what part of the body these processes occur. For example, a certain drug may be effective in treating certain bacterial infections. Before beginning a clinical trial of the drug, its manufacturer must prove that the drug is not harmful to humans. In a typical case, the drug is first tested in animal studies. It is synthesized, and some radioactive isotope, such as hydrogen-3, carbon-14, or phosphorus-32, is incorporated into its structure. The drug is administered to the test animals, and after a certain period, the animals are sacrificed. The fate of the drug is then determined by isolating from the body any radioactive compounds formed.

One typical pharmacological experiment studied the effects of tetracycline. This powerful antibiotic tends to accumulate in bones and is not given to pregnant women because it is transferred to the bones of the fetus. A particular tetracycline was tagged with the radioisotope tritium (hydrogen-3), and its uptake in rat bones was monitored in the presence and absence of a sulfa drug. With the aid of a scintillation counter, researchers measured the radiation intensity of maternal and fetal bones. They found that the sulfa drug helped to minimize the accumulation of tetracycline in the fetal bones.

The metabolic fate of essential chemicals in the body can also be followed with radioactive tracers. The use of radioactive isotopes has illuminated a number of normal and pathological body functions as well.

B. Radiation Therapy

The main use of radioactive isotopes in therapy is the selective destruction of pathological cells and tissues. Recall that radiation, whether from gamma rays, X-rays, or other sources, is detrimental to cells. Ionizing radiation damages cells, especially those that divide rapidly. This damage may be great enough to destroy diseased cells or to sufficiently alter the genes in them so that multiplication of the cells slows down.

In therapy applications, cancerous cells are the main targets for ionizing radiation. Radiation is typically used when a cancer is well localized; it may also be employed when the cancerous cells spread and are in a metastatic state. In addition, it is used for preventive purposes, namely to eliminate any possible remaining cancerous cells after surgery has been performed. The idea, of course, is to kill cancerous cells but not normal ones. Therefore, radiation such as high-energy X-rays or gamma rays from a cobalt-60 source is focused at a small part of the body where cancerous cells are suspected to reside. Besides X-rays and gamma rays from cobalt-60, other ionizing radiation is used to treat inoperable tumors. Proton beams from cyclotrons, for instance, have been used to treat ocular melanoma and tumors of the skull base and spine.

Despite this pinpointing technique, the radiation inevitably kills normal cells along with the cancerous cells. Because the radiation is most effective against rapidly dividing cancer cells rather than normal cells and because the radiation is aimed at a specific location, the damage to healthy tissues is minimized.

Another way to localize radiation damage in therapy is to use specific radioactive isotopes. In the case of thyroid cancer, large doses of iodine-131 are administered, which are taken up by the thyroid gland. The isotope, which has high radioactivity, kills all the cells of the gland (cancerous as well as normal ones), but does not appreciably damage other organs.

Another radioisotope, iodine-125, is used in the treatment of prostate cancer. Seeds of iodine-125, a gamma emitter, are implanted in the cancerous area of the prostate gland while being imaged with ultrasound. The seeds deliver 160 Gy (16,000 rad) over their lifetime.

A newer form of prostate cancer treatment with great potential relies on actinium-225, an alpha emitter. As discussed in Section 9.6, alpha particles cause more damage to the tissues than any other form of radiation, but they have low penetrating power. Researchers have developed a very clever way to deliver actinium-225 to the cancerous region of the prostate gland without damaging healthy tissues. The prostate tumor has a high concentration of prostate-specific antigen (PSA) on its surface. A monoclonal antibody (Section 30.4) homes in on the PSA and interacts with it. A single actinium-225 atom attached to such a monoclonal antibody can deliver the desired radiation, thereby destroying the cancer. Actinium-225 is especially effective because it has a half-life of ten days and it decays to three nuclides, themselves alpha emitters. In clinical trials, a single injection of antibody with an intensity in the kBq range (nanocuries) provided tumor regression without toxicity.

A metastatic state exists when the cancerous cells break off from their primary site(s) and begin moving to other parts of the body.

9.8 What Is Nuclear Fusion?

An estimated 98% of all matter in the universe is made up of hydrogen and helium. The "big bang" theory of the formation of the universe postulates that our universe started with an explosion (big bang) in which matter was formed out of energy and that at the beginning, only the lightest element, hydrogen, was in existence. Later, as the universe expanded, stars were born when hydrogen clouds collapsed under gravitational forces. In the cores of these stars, hydrogen nuclei fused to form helium.

The fusion of two hydrogen nuclei into a helium nucleus liberates a very large amount of energy in the form of photons, largely by the following reaction:

$$^2_1\text{H} + {}^3_1\text{H} \longrightarrow {}^4_2\text{He} + {}^1_0\text{n} + 5.3 \times 10^8 \text{ kcal/mol He}$$

Hydrogen-2   Hydrogen-3
(Deuterium)   (Tritium)

This process, called **fusion**, is how the sun makes its energy. Uncontrolled fusion is employed in the "hydrogen bomb." If we can ever achieve a controlled

Nuclear fusion The joining together of atomic nuclei to form a new nucleus heavier than either starting nuclei

The reactions occurring in the sun are essentially the same as those in hydrogen bombs.

version of this fusion reaction (which is unlikely to happen in the near term), we should be able to solve our energy problems.

As we have just seen, the fusion of deuterium and tritium nuclei to a helium nucleus gives off a very large amount of energy. What is the source of this energy? When we compare the mass of the reactants and products, we see that there is a loss of $5.0301 - 5.0113 = 0.0189$ g for each mole of helium formed:

$$\underset{2.01410 \text{ g}}{^{2}_{1}\text{H}} + \underset{3.0161 \text{ g}}{^{3}_{1}\text{H}} \longrightarrow \underset{4.0026 \text{ g}}{^{4}_{2}\text{He}} + \underset{1.0087 \text{ g}}{^{1}_{0}\text{n}}$$

$$\underset{5.0302 \text{ g}}{\qquad} \qquad \underset{5.0113 \text{ g}}{\qquad}$$

When the deuterium and tritium nuclei are converted to helium and a neutron, the extra mass has to go somewhere. Where does it go? The answer is that the missing mass is converted to energy. We even know, from the equation developed by Albert Einstein (1879–1955), how much energy we can get from the conversion of any amount of mass:

$$E = mc^2$$

This equation says that the mass (m), in kilograms, that is lost multiplied by the square of the velocity of light (c^2, where $c = 3.0 \times 10^8$ m/s), in meters squared per second squared (m^2/s^2), is equal to the amount of energy created (E), in joules. For example, 1 g of matter completely converted to energy would produce 8.8×10^{13} J, which is enough energy to boil 34,000,000 L of water initially at 20°C. This is equivalent to the amount of water in an Olympic-size swimming pool. As you can see, we get a tremendous amount of energy from a little bit of mass.

All of the **transuranium elements** (elements with atomic numbers greater than 92) are artificial and have been prepared by a fusion process in which heavy nuclei are bombarded with light ones. Many, as their names indicate, were first prepared at the Lawrence Laboratory of the University of California, Berkeley, by Glenn Seaborg (1912–1999; Nobel laureate in chemistry, 1951) and his colleagues:

$$^{244}_{96}\text{Cm} + {}^{4}_{2}\text{He} \longrightarrow {}^{245}_{97}\text{Bk} + {}^{1}_{1}\text{H} + 2\,{}^{1}_{0}\text{n}$$

$$^{238}_{92}\text{U} + {}^{12}_{6}\text{C} \longrightarrow {}^{246}_{98}\text{Cf} + 4\,{}^{1}_{0}\text{n}$$

$$^{252}_{98}\text{Cf} + {}^{10}_{5}\text{B} \longrightarrow {}^{257}_{103}\text{Lr} + 5\,{}^{1}_{0}\text{n}$$

These transuranium elements are unstable, and most have very short half-lives. For example, the half-life of Lawrencium-257 is 0.65 second. Many of the new superheavy elements have been obtained by bombarding lead isotopes with calcium-48 or nickel-64. So far, the creation of elements 110, 111, and 112–116 has been reported, even though their detection was based on the observation of the decay of a single atom.

A pioneer in developing radioisotopes for medical use, Glenn Seaborg was the first to produce iodine-131, used subsequently to treat his mother's abnormal thyroid condition. As a result of Seaborg's further research, it became possible to predict accurately the properties of many of the as-yet-undiscovered transuranium elements. In a remarkable 21-year span (1940–1961), Seaborg and his colleagues synthesized ten new transuranium elements (plutonium to lawrencium). He received the Nobel Prize in 1951 for his creation of new elements. In the 1990s, Seaborg was honored by having element 106 named for him.

Glenn Seaborg (1912–1999).

9.9 What Is Nuclear Fission and How Is It Related to Atomic Energy?

In the 1930s, Enrico Fermi (1901–1954) and his colleagues in Rome and Otto Hahn (1879–1968), Lise Meitner (1878–1968), and Fritz Strassman (1902–1980) in Germany tried to produce new transuranium elements by bombarding uranium-235 with neutrons. To their surprise, they found that, rather than fusion, they obtained **nuclear fission** (fragmentation of large nuclei into smaller pieces):

$$^{235}_{92}\text{U} + {^1_0}\text{n} \longrightarrow {^{141}_{56}}\text{Ba} + {^{92}_{36}}\text{Kr} + 3\,{^1_0}\text{n} + \gamma + \text{energy}$$

In this reaction, a uranium-235 nucleus first absorbs a neutron to become uranium-236 and then breaks into two smaller nuclei. The most important product of this nuclear decay is energy, which is produced because the products have less mass than the starting materials. This form of energy, called **atomic energy**, has been used for both war (in the atomic bomb) and peace.

With uranium-235, each fission produces three neutrons, which in turn can generate more fissions by colliding with other uranium-235 nuclei. If even one of these neutrons produces a new fission, the process becomes a self-propagating **chain reaction** (Figure 9.9) that continues at a constant rate. If all three neutrons are allowed to produce new fission, the rate of the reaction increases constantly and eventually culminates in a nuclear explosion. In nuclear power plants, the rate of reaction is controlled by inserting boron control rods into the reactor to absorb neutrons and thereby dampen the rate of fission.

In nuclear power plants, the energy produced by fission is sent to heat exchangers and used to generate steam, which drives a turbine to produce

Nuclear power plant in Salem, New Jersey.

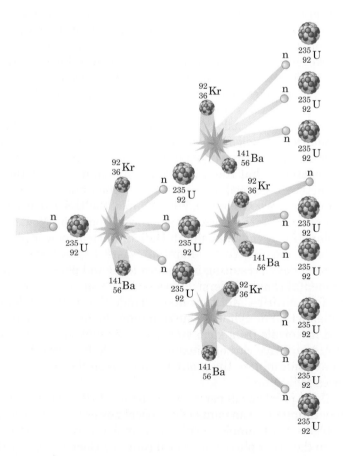

FIGURE 9.9 A chain reaction begins when a neutron collides with a nucleus of uranium-235.

electricity (Figure 9.10). Today, such plants supply more than 15% of the electrical energy in the United States. The opposition to nuclear plants is based on safety considerations and on the unsolved problems of waste disposal. Although nuclear plants in general have good safety records, accidents such as those at Fukushima Daiichi, Chernobyl (Chemical Connections 9D), and Three Mile Island have caused concern.

FIGURE 9.10 Schematic diagram of a nuclear power plant.

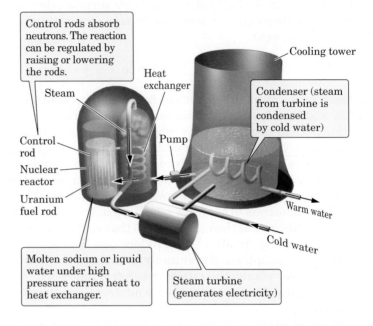

Control rods absorb neutrons. The reaction can be regulated by raising or lowering the rods.

Heat exchanger

Steam

Cooling tower

Condenser (steam from turbine is condensed by cold water)

Control rod

Pump

Nuclear reactor

Uranium fuel rod

Warm water

Cold water

Molten sodium or liquid water under high pressure carries heat to heat exchanger.

Steam turbine (generates electricity)

Storage of nuclear wastes in a storage room carved out of an underground salt mine.

U.S. Department of Energy

Nuclear waste disposal is a long-range problem. The fission products of nuclear reactors are highly radioactive themselves, with long half-lives. Spent fuel contains these high-level fission products as nuclear wastes, together with uranium and plutonium that can be recovered and reused as mixed oxide (MOX) fuel. Reprocessing is costly: although done routinely in Europe and Russia, it is not practiced by nuclear plants in the United States for economic reasons. However, this situation may change because cleaner extraction processes have been developed that use supercritical carbon dioxide (Chemical Connections 5E), thereby eliminating the need to dispose of solvent.

The United States has about 50,000 metric tons of spent fuel, stored under water and in dry casks at power plants. The Department of Energy stores the additional nuclear wastes from the nuclear weapons program, research reactors, and other sources in three major sites. After 40 years, the level of radioactivity that the wastes had immediately after their removal from the reactor is reduced a thousandfold. Such nuclear waste is a good candidate for underground burial. For example, the U.S. federal government gave its approval to store nuclear waste Yucca Mountain, Nevada. However, in light of Japan's recent Fukushima Daiichi nuclear power plant disaster, opponents to the ongoing Yucca Mountain waste disposal are determined to shut the project down, resulting in modern legal and political disputes.

Environmental concerns persist, however. The site cannot be guaranteed to stay dry for centuries. Moisture may corrode the steel cylinders and even the inner glass/ceramic cylinders surrounding the nuclear waste. Some fear that leaked materials from such storage tanks may escape if carbon-14 is oxidized to radioactive carbon dioxide or, less likely, that other radioactive nuclides may contaminate the groundwater, which lies far below the desert rock of Yucca Mountain.

To keep these problems in perspective, one must remember that most other ways of generating large amounts of electrical power have their own environmental problems. For example, burning coal or oil contributes to the accumulation of CO_2 in the atmosphere and to acid rain (see Chemical Connections 6A).

Chemical Connections 9D

Radioactive Fallout from Nuclear Accidents

On April 26, 1986, an accident occurred at the nuclear reactor in the town of Chernobyl in the former Soviet Union. It was a clear reminder of the dangers involved in this industry and of the far-reaching contamination that such accidents can produce. In Sweden, more than 500 miles away from the accident, the radioactive cloud increased the background radiation from 4 to 15 times the normal level. The radioactive cloud reached England, about 1300 miles away, one week later. There, it increased the natural background radiation by 15%. The radioactivity from iodine-131 was measured at 400 Bq/L in milk and 200 Bq kg in leafy vegetables. Even some 4000 miles away in Spokane, Washington, an iodine-131 activity of 242 Bq/L was found in rainwater; smaller activities—1.03 Bq/L of ruthenium-103 and 0.66 Bq/L of cesium-137—were recorded as well. These levels are not harmful.

Closer to the source of the nuclear accident, in neighboring Poland, potassium iodide pills were given to children. This step was taken to prevent radioactive iodine-131 (which might come from

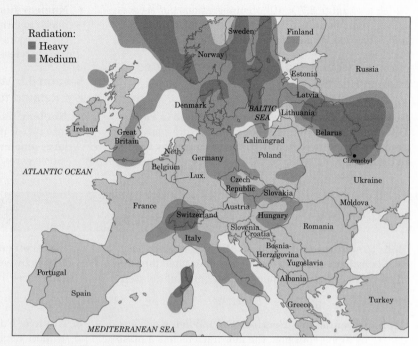

Map showing those areas most affected by the Chernobyl accident.

contaminated food) from concentrating in their thyroid glands, which could lead to cancer. In the wake of the September 11, 2001 terrorist attacks, Massachusetts became the first state to authorize the storage of KI pills in case of nuclear-related terrorist activity.

Test your knowledge with Problem 9.66.

Summary of Key Questions

OWL Sign in at **www.cengage.com/owl** to develop problem-solving skills and complete online homework assigned by your professor.

Section 9.1 How Was Radioactivity Discovered?

- Henri Becquerel discovered radioactivity in 1896.

Section 9.2 What Is Radioactivity?

- The four major types of radioactivity are **alpha particles** (helium nuclei), **beta particles** (electrons), **gamma rays** (high-energy photons), and **positrons** (positively charged electrons).

Section 9.3 What Happens When a Nucleus Emits Radioactivity?

- When a nucleus emits a **beta particle**, the new element has the same mass number but an atomic number one unit greater.

- When a nucleus emits an **alpha particle**, the new element has an atomic number two units lower and a mass number four units smaller.
- When a nucleus emits a **positron** (positive electron), the new element has the same mass number but an atomic number one unit smaller.
- In **gamma emission**, no transmutation takes place; only the energy of the nucleus is lowered.
- In **electron capture**, the new element has the same mass number but an atomic number one unit smaller.

Section 9.4 What Is Nuclear Half-Life?

- Each radioactive isotope decays at a fixed rate described by its **half-life**, which is the time required for half of the sample to decay.

Section 9.5 How Do We Detect and Measure Nuclear Radiation?

- Radiation is detected and counted by devices such as **Geiger-Müller counters**.
- The main unit of intensity of radiation is the **curie (Ci)**, which is equal to 3.7×10^{10} disintegrations per second. Other common units are the millicurie (mCi), the microcurie (μCi), and the becquerel (Bq).

Section 9.6 How Is Radiation Dosimetry Related to Human Health?

- For medical purposes and to measure potential radiation damage, the absorbed dose is measured in **rads**. Different particles damage body tissues differently; the **rem** is a measure of relative damage caused by the type of radiation.

Section 9.7 What Is Nuclear Medicine?

- Nuclear medicine is the use of radionuclei for diagnostic imaging and therapy.

Section 9.8 What Is Nuclear Fusion?

- **Nuclear fusion** is the combining (fusing) of two lighter nuclei to form a heavier nucleus. Helium is synthesized in the interiors of stars by the fusion of hydrogen nuclei. The energy released in this process is the energy of our sun.

Section 9.9 What Is Nuclear Fission and How Is It Related to Atomic Energy?

- **Nuclear fission** is the splitting of a heavier nucleus into two or more smaller nuclei. Nuclear fission releases large amounts of energy, which can be either controlled (nuclear reactors) or uncontrolled (nuclear weapons).

Summary of Key Reactions

1. **Beta (β) emission (Section 9.3B)** When a nucleus decays by beta emission, the new element has the same mass number but an atomic number one unit greater.

$$^{32}_{15}\text{P} \longrightarrow ^{32}_{16}\text{S} + ^{0}_{-1}\text{e}$$

2. **Alpha (α) emission (Section 9.3C)** When a nucleus decays by alpha emission, the new nucleus has a mass number four units smaller and an atomic number two units smaller.

$$^{238}_{92}\text{U} \longrightarrow ^{234}_{90}\text{Th} + ^{4}_{2}\text{He}$$

3. **Positron ($\beta+$) emission (Section 9.3D)** When a nucleus decays by positron emission, the new element has the same mass number but an atomic number one unit smaller.

$$^{11}_{6}\text{C} \longrightarrow ^{11}_{5}\text{B} + ^{0}_{+1}\text{e}$$

4. **Gamma (γ) emission (Section 9.3E)** When a nucleus emits gamma radiation, there is no change in either mass number or atomic number of the nucleus.

$$^{11}_{5}\text{B}^* \longrightarrow ^{11}_{5}\text{B} + \gamma$$

5. **Electron capture (Section 9.3F)** When a nucleus decays by electron capture, the product nucleus has the same mass number but an atomic number one unit smaller.

$$^{7}_{4}\text{Be} + ^{0}_{-1}\text{e} \longrightarrow ^{7}_{3}\text{Li}$$

6. **Nuclear Fusion (Section 9.8)** In nuclear fusion, two or more nuclei react to form a larger nucleus. In the process, there is a slight decrease in mass; the sum of the masses of the fusion products is less than the sum of the masses of the starting nuclei. The lost mass appears as energy.

$$^{2}_{1}\text{H} + ^{3}_{1}\text{H} \longrightarrow ^{4}_{2}\text{He} + ^{1}_{0}\text{n} + 5.3 \times 10^{8} \text{ kcal/mol He}$$

7. **Nuclear Fission (Section 9.9)** In nuclear fission, a nucleus captures a neutron to form a nucleus with a mass number increased by one unit. The new nucleus then splits into two smaller nuclei.

$$^{235}_{92}\text{U} + ^{1}_{0}\text{n} \longrightarrow ^{141}_{56}\text{Ba} + ^{92}_{36}\text{Kr} + 3\,^{1}_{0}\text{n} + \gamma + \text{energy}$$

Problems

OWL Interactive versions of these problems may be assigned in OWL.

Orange-numbered problems are applied.

Section 9.2 What Is Radioactivity?

9.8 What is the relationship between frequency and wavelength? frequency and energy? wavelength and energy?

9.9 What is the difference between an alpha particle and a proton?

9.10 Microwaves are a form of electromagnetic radiation that is used for the rapid heating of foods. What is the frequency of a microwave with a wavelength of 5.8 cm?

9.11 In each case, given the frequency, give the wavelength in centimeters or nanometers and tell what kind of radiation it is.
(a) 7.5×10^{14}/s
(b) 1.0×10^{10}/s
(c) 1.1×10^{15}/s
(d) 1.5×10^{18}/s

9.12 Red light has a wavelength of 650 nm. What is its frequency?

9.13 Which has the longest wavelength: (a) infrared, (b) ultraviolet, or (c) X-rays? Which has the highest energy?

9.14 Write the symbol for a nucleus with the following components:

(a) 9 protons and 10 neutrons

(b) 15 protons and 17 neutrons

(c) 37 protons and 50 neutrons

9.15 In each pair, tell which isotope is more likely to be radioactive:

(a) Nitrogen-14 and nitrogen-13

(b) Phosphorus-31 and phosphorus-33

(c) Lithium-7 and lithium-9

(d) Calcium-39 and calcium-40

9.16 Which isotope of boron is the most stable: boron-8, boron-10, or boron-12?

9.17 Which isotope of oxygen is the most stable: oxygen-14, oxygen-16, or oxygen-18?

Section 9.3 What Happens When a Nucleus Emits Radioactivity?

9.18 Answer true or false.

(a) The majority (greater than 50%) of the more than 300 naturally occurring isotopes are stable.

(b) More artificial isotopes have been created in the laboratory than there are naturally occurring stable isotopes.

(c) All artificial isotopes created in the laboratory are radioactive.

(d) The terms "beta particle," "beta emission," and "beta ray" all refer to the same type of radiation.

(e) When balancing a nuclear equation, the sum of the mass numbers and the sum of the atomic numbers on each side of the equation must be the same.

(f) The symbol of a beta particle is $_{-1}^{0}\beta$.

(g) When a nucleus emits a beta particle, the new nucleus has the same mass number but an atomic number one unit higher.

(h) When iron-59 ($_{26}^{59}$Fe) emits a beta particle, it is converted to cobalt-59 ($_{27}^{59}$Co).

(i) When a nucleus emits a beta particle, it first captures an electron from outside the nucleus and then emits it.

(j) For the purposes of determining atomic numbers in a nuclear equation, an electron is assumed to have a mass number of zero and an atomic number of -1.

(k) The symbol for an alpha particle is $_{2}^{4}$He.

(l) When a nucleus emits an alpha particle, the new nucleus has an atomic number two units higher and a mass number four units higher.

(m) When uranium-238 ($_{92}^{238}$U) undergoes alpha emission, the new nucleus is thorium-234 ($_{90}^{234}$Th).

(n) The symbol of a positron is $_{+1}^{0}\beta$.

(o) A positron is sometimes referred to as a positive electron.

(p) When a nucleus emits a positron, the new nucleus has the same mass number but an atomic number one unit lower.

(q) When carbon-11 ($_{6}^{11}$C) emits a positron, the new nucleus formed is boron-11 ($_{5}^{11}$B).

(r) Alpha emission and positron emission both result in the formation of a new nucleus with a lower atomic number.

(s) The symbol for gamma radiation is γ.

(t) When a nucleus emits gamma radiation, the new nucleus formed has the same mass number and the same atomic number.

(u) When a nucleus captures an extranuclear electron, the new nucleus formed has the same atomic number but a mass number one unit lower.

(v) When gallium-67 ($_{31}^{67}$Ga) undergoes electron capture, the new nucleus formed is germanium-67 ($_{32}^{67}$Ge).

9.19 Samarium-151 is a beta emitter. Write an equation for this nuclear reaction and identify the product nucleus.

9.20 The following nuclei turn into new nuclei by emitting beta particles. Write an equation for each nuclear reaction and identify the product nucleus.

(a) $_{63}^{159}$Eu (b) $_{56}^{141}$Ba (c) $_{95}^{242}$Am

9.21 Chromium-51 is used in diagnosing the pathology of the spleen. The nucleus of this isotope captures an electron according to the following equation. What is the transmutation product?

$$_{24}^{51}\text{Cr} + {}_{-1}^{0}\text{e} \longrightarrow ?$$

9.22 The following nuclei decay by emitting alpha particles. Write an equation for each nuclear reaction and identify the product nucleus.

(a) $_{83}^{210}$Bi (b) $_{94}^{238}$Pu (c) $_{72}^{174}$Hf

9.23 Curium-248 was bombarded, yielding antimony-116 and cesium-160. What was the bombarding nucleus?

9.24 Phosphorus-29 is a positron emitter. Write an equation for this nuclear reaction and identify the product nucleus.

9.25 For each of the following, write a balanced nuclear equation and identify the radiation emitted.

(a) Beryllium-10 changes to boron-10

(b) Europium-151^{m} changes to europium-151

(c) Thallium-195 changes to mercury-195

(d) Plutonium-239 changes to uranium-235

9.26 In the first three steps in the decay of uranium-238, the following isotopic species appear: uranium-238 decays to thorium, which then decays to protactinium-234, which then decays to uranium-234. What kind of emission occurs in each step?

9.27 What kind of emission does *not* result in transmutation?

9.28 Complete the following nuclear reactions.

(a) $_{8}^{16}\text{O} + {}_{8}^{16}\text{O} \longrightarrow ? + {}_{2}^{4}\text{He}$

(b) $_{92}^{235}\text{U} + {}_{0}^{1}\text{n} \longrightarrow {}_{38}^{90}\text{Sr} + ? + 3\,{}_{0}^{1}\text{n}$

(c) $_{6}^{13}\text{C} + {}_{2}^{4}\text{He} \longrightarrow {}_{8}^{16}\text{O} + ?$

(d) $_{83}^{210}\text{Bi} \longrightarrow ? + {}_{-1}^{0}\text{e}$

(e) $_{6}^{12}\text{C} + {}_{1}^{1}\text{H} \longrightarrow ? + \gamma$

9.29 Americium-240 is made by bombarding plutonium-239 with α particles. In addition to americium-240, a proton and two neutrons are also formed. Write a balanced equation for this nuclear reaction.

Section 9.4 What Is Nuclear Half-Life?

9.30 Answer true or false.
   (a) Half-life is the time its takes one-half of a radioactive sample to decay.
   (b) The concept of half-life refers to nuclei undergoing alpha, beta, and positron emission; it does not apply to nuclei undergoing gamma emission.
   (c) At the end of two half-lives, one-half of the original radioactive sample remains; at the end of three half-lives, one-third of the original sample remains.
   (d) If the half-life of a particular radioactive sample is 12 minutes, a time of 36 minutes represents three half-lives.
   (e) At the end of three half-lives, only 12.5% of an original radioactive sample remains.

9.31 Iodine-125 emits gamma rays and has a half-life of 60 days. If a 20-mg pellet of iodine-125 is implanted into a prostate gland, how much iodine-125 remains there after one year?

9.32 Polonium-218, a decay product of radon-222 (see Chemical Connections 9B), has a half-life of 3 min. What percentage of the polonium-218 formed will remain in the lung 9 min after inhalation?

9.33 A rock containing 1 mg of plutonium-239 per kg of rock is found in a glacier. The half-life of plutonium-239 is 25,000 years. If this rock was deposited 100,000 years ago during an ice age, how much plutonium-239 per kilogram of rock was in the rock at that time?

9.34 The element radium is extremely radioactive. If you converted a piece of radium metal to radium chloride (with the weight of the radium remaining the same), would it become less radioactive?

9.35 In what ways can we increase the rate of radioactive decay? Decrease it?

9.36 Suppose 50.0 mg of potassium-45, a beta emitter, was isolated in pure form. After one hour, only 3.1 mg of the radioactive material was left. What is the half-life of potassium-45?

9.37 A patient receives 200 mCi of iodine-131, which has a half-life of eight days.
   (a) If 12% of this amount is taken up by the thyroid gland after two hours, what will be the activity of the thyroid after two hours, in millicuries and in counts per minute?
   (b) After 24 days, how much activity will remain in the thyroid gland?

Section 9.5 How Do We Detect and Measure Nuclear Radiation?

9.38 Answer true or false.
   (a) Ionizing radiation refers to any radiation that interacts with neutral atoms or molecules to create positive ions.
   (b) Ionizing radiation creates positive ions by striking a nucleus and knocking one or more electrons from the nucleus.
   (c) Ionizing radiation creates positive ions by knocking one or more extranuclear electrons from a neutral atom or molecule.
   (d) The curie (Ci) and becquerel (Bq) are both units by which we report radiation intensity.
   (e) The units of a curie (Ci) are disintegrations per second (dps).
   (f) A microcurie (μCi) is a smaller unit than a curie (Ci).
   (g) The intensity of radiation is inversely related to the square of the distance from the radiation source; for example, the intensity at three meters from the source is 1/9 of what it is at the source.
   (h) Alpha particles are the most massive and highly charged type of nuclear radiation and, therefore, are the most penetrating type of nuclear radiation.
   (i) Beta particles have both a smaller mass and a smaller charge than alpha particles and, therefore, are more penetrating than alpha particles.
   (j) Gamma rays, with neither mass nor charge, are the least penetrating type of nuclear radiation.
   (k) After one half-life, the mass of a radioactive sample remaining is approximately 50% of the original mass.

9.39 If you work in a lab containing radioisotopes emitting all kinds of radiation, from which emission should you seek the most protection?

9.40 What do Geiger-Müller counters measure: (a) the intensity or (b) the energy of radiation?

9.41 It is known that radioactivity is being emitted with an intensity of 175 mCi at a distance of 1.0 m from the source. How far in meters from the source should you stand if you wish to be subjected to no more than 0.20 mCi?

Section 9.6 How Is Radiation Dosimetry Related to Human Health?

9.42 Briefly contrast the three different units used to describe the effects of radiation on the body.

9.43 Does a curie (Ci) measure radiation intensity or energy?

9.44 What property is measured with each of the following terms?
   (a) Rad   (b) Rem   (c) Roentgen
   (d) Curie   (e) Gray   (f) Becquerel
   (g) Sievert

9.45 A radioactive isotope with an activity (intensity) of 80.0 mCi per vial is delivered to a hospital. The vial contains 7.0 cc of liquid. The instruction is to administer 7.2 mCi intravenously. How many cubic centimeters of liquid should be used for one injection?

9.46 Why does exposure of a hand to alpha rays not cause serious damage to the person, whereas entry of an alpha emitter into the lung as an aerosol produces very serious damage to the person's health?

9.47 A certain radioisotope has an intensity of 10^6 Bq at 1-cm distance from the source. What would be the intensity at 20 cm? Give your answer in both Bq and μCi units.

9.48 Assuming the same amount of absorbed radiation, in rads from three sources, which would be the most damaging to the tissues: alpha particles, beta particles, or gamma rays?

9.49 In an accident involving radioactive exposure, person A received 3.0 Sv while person B received 0.50 mrem exposure. Who was hurt more seriously?

Section 9.7 What Is Nuclear Medicine?

9.50 Answer true or false.

(a) Of the radioisotopes listed in Table 9.5, the majority decay by beta emission.

(b) Isotopes that decay by alpha emission are rarely if ever used in nuclear imaging because alpha emitters are rare.

(c) Gamma emitters are so widely used in medical imaging because gamma radiation is penetrating and, therefore, can easily be measured by radiation detectors outside the body.

(d) When selenium-75 ($^{75}_{34}$Se) decays by electron capture and gamma emission, the new nucleus formed is arsenic-75 ($^{75}_{34}$As).

(e) When iodine-131 ($^{131}_{53}$I) decays by beta and gamma emission, the new nucleus formed is xenon-131 ($^{131}_{54}$Xe).

(f) In positron emission tomography (a PET scan), the detector counts the number of positrons emitted by a tagged material and the location within the body where the tagged material accumulates.

(g) The use of 18-fluorodeoxyglucose (FDG) in PET scans of the brain depends on the fact that FDG behaves in the body as does glucose.

(h) A goal of radiation therapy is to destroy pathological cells and tissues without at the same time damaging normal cells and tissues.

(i) In external beam radiation, radiation from an external source is directed at a tissue either on the surface of the body or within the body.

(j) In internal beam radiation, a radioactive material is implanted in a target tissue to destroy cells in the target tissue without doing appreciable damage to surrounding normal tissues.

9.51 In 1986, the nuclear reactor in Chernobyl had an accident and spewed radioactive nuclei that were carried by the winds for hundreds of miles. Today, among the child survivors of the event, the most common damage is thyroid cancer. What radioactive nucleus do you expect to be responsible for these cancers?

9.52 Cobalt-60, with a half-life of 5.26 years, is used in cancer therapy. The energy of the radiation from cobalt-62 is even higher (half-life = 14 minutes). Why isn't cobalt-62 also used for cancer therapy?

9.53 Match the radioactive isotope with its proper use:

_____ (a) Cobalt-60   1. Heart scan during exercise

_____ (b) Thallium-201   2. Measure water content of body

_____ (c) Tritium   3. Kidney scan

_____ (d) Mercury-197   4. Cancer therapy

Section 9.8 What Is Nuclear Fusion?

9.54 Answer true or false.

(a) In nuclear fusion, two nuclei combine to form a new nucleus.

(b) The energy of the sun is derived from the fusion of two hydrogen-1 ($^{1}_{1}$H) nuclei to form a helium-4 ($^{4}_{2}$He) nucleus.

(c) The energy of the sun occurs because once two hydrogen nuclei fuse, the two positive charges no longer repel each other.

(d) Fusion of hydrogen nuclei in the sun results in a small decrease in mass, which appears as an equivalent amount of energy.

(e) Einstein's famous $E = mc^2$ equation refers to the energy released when two particles of the same mass collide with the speed of light.

(f) Nuclear fusion occurs only in the sun.

(g) Nuclear fusion can be carried out in the laboratory.

9.55 What are the products of the fusion of hydrogen-2 and hydrogen-3 nuclei?

9.56 Assuming that one proton and two neutrons will be produced in an alpha-bombardment fusion reaction, what target nucleus would you use to obtain berkelium-249?

9.57 Element 109 was first prepared in 1982. A single atom of this element ($^{266}_{109}$Mt), with a mass number of 266, was made by bombarding a bismuth-209 nucleus with an iron-58 nucleus. What other products, if any, must have been formed besides $^{266}_{109}$Mt?

9.58 A new element was formed when lead-208 was bombarded by krypton-86. One could detect four neutrons as the product of the fusion. Identify the new element.

9.59 Boron-10 is used in control rods for nuclear reactors. This nucleus absorbs a neutron and then emits an alpha particle. Write an equation for each nuclear reaction and identify each product nucleus.

9.60 The most abundant isotope of uranium, ^{238}U, does not undergo fission. Instead, it captures a neutron and emits two β particles to make a fissionable isotope of plutonium, which can then be used as fuel in a nuclear reactor. Write an equation for the nuclear reaction and identify the product nucleus.

Chemical Connections

9.61 (Chemical Connections 9A) Why is it accurate to assume that the carbon-14 to carbon-12 ratio in a living plant is constant over the lifetime of the plant?

9.62 (Chemical Connections 9A) In a recent archeological dig in the Amazon region of Brazil, charcoal paintings were found in a cave. The carbon-14 content of the charcoal was one-fourth of what is found in charcoal prepared from that year's tree harvest. How long ago was the cave settled?

9.63 (Chemical Connections 9A) Carbon-14 dating of the Shroud of Turin indicated that the plant from which the shroud was made was alive around AD 1350. To how many half-lives does this time period correspond?

9.64 (Chemical Connections 9A) The half-life of carbon-14 is 5730 years. The wrapping of an Egyptian mummy gave off 7.5 counts per minute per gram of carbon. A piece of linen purchased today would give an activity of 15 counts per minute per gram of carbon. How old is the mummy?

9.65 (Chemical Connections 9B) How does radon-222 produce polonium-218?

9.66 (Chemical Connections 9D) In a nuclear accident, one of the radioactive nuclei that concerns people is iodine-131. Iodine is easily vaporized and can be carried by the winds to locations that are hundreds—even thousands—of miles away. Why is iodine-131 especially harmful?

Additional Problems

9.67 Phosphorus-32 ($t_{1/2} = 14.3$ h) is used in the medical imaging and diagnosis of eye tumors. Suppose a patient is given 0.010 mg of this isotope. Prepare a graph showing the mass in milligrams remaining in the patient's body after one week. (Assume that none is excreted from the body.)

9.68 During the bombardment of argon-40 with protons, one neutron is emitted for each proton absorbed. What new element is formed?

9.69 Neon-19 and sodium-20 are positron emitters. What products result in each case?

9.70 The half-life of nitrogen-16 is 7 seconds. How long does it take for 100 mg of nitrogen-16 to be reduced to 6.25 mg?

9.71 Do the curie and the becquerel measure the same or different properties of radiation?

9.72 Selenium-75 has a half-life of 120.4 days, so it would take 602 days (five half-lives) to diminish to 3% of the original quantity. Yet this isotope is used for pancreatic scans without any fear that the radioactivity will cause undue harm to the patient. Suggest a possible explanation.

9.73 Use Table 9.4 to determine the percentage of annual radiation we receive from the following sources:
(a) Naturally occurring sources
(b) Diagnostic medical sources
(c) Nuclear power plants

9.74 $^{225}_{89}$Ac is an alpha emitter. In its decay process, it produces three more alpha emitters in succession. Identify each of the decay products.

9.75 Which radiation will cause more ionization, X-rays or radar?

9.76 You have an old wristwatch that still has radium paint on its dial. Measurement of the radioactivity of the watch shows a beta-ray count of 0.50 count/s. If 1.0 microcurie of this sort of radiation produces 1000 mrem/year, how much radiation in mrem do you expect from the wristwatch if you wear it for one year?

9.77 Americium-241, which is used in some smoke detectors, has a half-life of 432 years and is an alpha emitter. What is the decay product of americium-241, and approximately what percentage of the original americium-241 will be still around after 1000 years?

9.78 On rare occasions, a nucleus will capture a beta particle instead of emitting one. Berkelium-246 is such a nucleus. What is the product of this nuclear transmutation?

9.79 A patient is reported to have been irradiated by a dose of 1 sievert in a nuclear accident. Is he in mortal danger?

9.80 What is the ground state of a nucleus?

9.81 Explain the following:
(a) It is impossible to have a completely pure sample of any radioactive isotope.
(b) Beta emission of a radioactive isotope creates a new isotope with an atomic number one unit higher than that of the radioactive isotope.

9.82 Yttrium-90, which emits beta particles, is used in radiotherapy. What is the decay product of yttrium-90?

9.83 The half-lives of some oxygen isotopes are as follows:

Oxygen-14 = 71 s   Oxygen-15 = 124 s

Oxygen-19 = 29 s   Oxygen-20 = 14 s

Oxygen-16 is the stable, nonradioactive isotope. Do the half-lives indicate anything about the stability of the other oxygen isotopes?

9.84 $^{225}_{89}$Ac is effective in prostate cancer therapy when administered at kBq levels. If an antibody tagged with $^{225}_{89}$Ac has an intensity of 2 million Bq/mg and if a solution contains 5 mg/L tagged antibody, how many milliliters of the solution should you use for an injection to administer 1 kBq intensity?

9.85 When $^{208}_{82}$Pb is bombarded with $^{64}_{28}$Ni, a new element and six neutrons are produced. Identify the new element.

9.86 Americium-241, the isotope used in smoke detectors, has a half-life of 432 years, which is sufficiently long to allow for handling it in large quantities. This isotope is prepared in the laboratory by bombarding plutonium-239 with α particles. In this reaction, plutonium-239 absorbs two neutrons and then decays by emission of a β particle. Write an equation for this nuclear reaction and identify the isotope formed as an intermediate between plutonium-239 and americium-241.

9.87 Boron-10, an effective absorber of neutrons, is used in control rods of uranium-235 fission reactors (see Figure 9.10) to absorb neutrons and thereby control the rate of reaction. Boron-10 absorbs a neutron and then emits an α particle. Write a balanced equation for this nuclear reaction and identify the nucleus formed as an intermediate between boron-10 and the final nuclear product.

9.88 Tritium, $^3_1$H, is a beta emitter widely used as a radioactive tracer in chemical and biochemical research. Tritium is prepared by the bombardment of lithium-6 with neutrons. Complete the following nuclear equation:

$$^6_3\text{Li} + {}^1_0\text{n} \longrightarrow {}^3_1\text{H} + ?$$

9.89 Radon-222 decays to a stable nucleus by a series of three alpha and two beta decays. Determine the stable nucleus that is formed.

9.90 Neptunium-237 decays by a series of steps to bismuth-209. How many alpha and beta particles are produced by this overall decay process?

9.91 Thorium-232 decays by a 10-step process, ultimately yielding lead-208. How many alpha particles and how many beta particles are emitted?

Organic Chemistry

Key Questions

10.1 What Is Organic Chemistry?

10.2 Where Do We Obtain Organic Compounds?

10.3 How Do We Write Structural Formulas of Organic Compounds?

10.4 What Is a Functional Group?

Tom and Pat Leeson/Photo Researchers, Inc.

The bark of the Pacific yew contains paclitaxel, a substance that has proven effective in treating certain types of ovarian and breast cancer (see Chemical Connections 10A).

10.1 What Is Organic Chemistry?

Organic chemistry is the chemistry of the compounds of carbon. As you study Chapters 10–19 (organic chemistry) and 20–31 (biochemistry), you will see that organic compounds are everywhere around us. They are in our foods, flavors, and fragrances; in our medicines, toiletries, and cosmetics; in our plastics, films, fibers, and resins; in our paints, varnishes, and glues; and, of course, in our bodies and the bodies of all other living organisms.

Perhaps the most remarkable feature of organic compounds is that they involve the chemistry of carbon and only a few other elements—chiefly, hydrogen, oxygen, and nitrogen. While the majority of organic compounds contain carbon and just these three elements, many also contain sulfur, a halogen (fluorine, chlorine, bromine, or iodine), and phosphorus.

As of the writing of this text, there are 118 known elements. Organic chemistry concentrates on carbon, just one of the 118. The chemistry of the other 117 elements comes under the field of inorganic chemistry. As we see in Figure 10.1, carbon is far from being among the most abundant elements

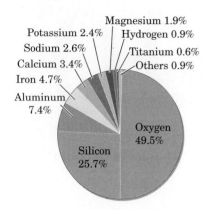

FIGURE 10.1 Abundance of the elements in the Earth's crust.

in the Earth's crust. In terms of elemental abundance, approximately 75% of the Earth's crust is composed of just two elements: oxygen and silicon. These two elements are the components of silicate minerals, clays, and sand. In fact, carbon is not even among the ten most abundant elements. Instead, it is merely one of the elements making up the remaining 0.9% of the Earth's crust. Why, then, do we pay such special attention to just one element from among 117?

The first reason is largely historical. In the early days of chemistry, scientists thought organic compounds were those produced by living organisms and that inorganic compounds were those found in rocks and other nonliving matter. At that time, they believed that a "vital force," possessed only by living organisms, was necessary to produce organic compounds. In other words, chemists believed that they could not synthesize any organic compound by starting with only inorganic compounds. This theory was very easy to disprove if, indeed, it was wrong. It required only one experiment in which an organic compound was made from inorganic compounds. In 1828, Friedrich Wöhler (1800–1882) carried out just such an experiment. He heated an aqueous solution of ammonium chloride and silver cyanate, both inorganic compounds and—to his surprise—obtained urea, an "organic" compound found in urine.

$$NH_4Cl + AgNCO \xrightarrow{\text{heat}} H_2N-\overset{\overset{\displaystyle O}{\|}}{C}-NH_2 + AgCl$$

Ammonium   Silver   Urea   Silver
chloride   cyanate   chloride

Urea

Although this single experiment of Wöhler's was sufficient to disprove the "doctrine of vital force," it took several years and a number of additional experiments for the entire scientific community to accept the fact that organic compounds could be synthesized in the laboratory. This discovery meant that the terms "organic" and "inorganic" no longer had their original meanings because, as Wöhler demonstrated, organic compounds could be obtained from inorganic materials. A few years later, August Kekulé (1829–1896) put forth a new definition—organic compounds are those containing carbon—and his definition has been accepted ever since.

A second reason for the study of carbon compounds as a separate discipline is the sheer number of organic compounds. Chemists have discovered or synthesized more than 10 million of them, and an estimated 10,000 new ones are reported each year. By comparison, chemists have discovered or synthesized an estimated 1.7 million inorganic compounds. Thus, approximately 85% of all known compounds are organic compounds.

A third reason—and one particularly important for those of you going on to study biochemistry—is that biochemicals, including carbohydrates, lipids, proteins, enzymes, nucleic acids (DNA and RNA), hormones, vitamins, and almost all other important chemicals in living systems are organic compounds. Furthermore, their reactions are often strikingly similar to those occurring in test tubes. For this reason, knowledge of organic chemistry is essential for an understanding of biochemistry.

One final point about organic compounds. They generally differ from inorganic compounds in many of their properties, some of which are shown in Table 10.1. Most of these differences stem from the fact that the bonding in organic compounds is almost entirely covalent, while most inorganic compounds have ionic bonds.

Table 10.1 A Comparison of Properties of Organic and Inorganic Compounds

Organic Compounds	Inorganic Compounds
Bonding is almost entirely covalent.	Most have ionic bonds.
Many are gases, liquids, or solids with low melting points (less than 360°C).	Most are solids with high melting points.
Most are insoluble in water.	Many are soluble in water.
Most are soluble in organic solvents such as diethyl ether, toluene, and dichloromethane.	Almost all are insoluble in organic solvents.
Aqueous solutions do not conduct electricity.	Aqueous solutions form ions that conduct electricity.
Almost all burn and decompose.	Very few burn.
Reactions are usually slow.	Reactions are often very fast.

Organic and inorganic compounds differ in their properties because they differ in their structure and composition—not because they obey different natural laws. One set of natural laws applies to all compounds.

Of course, the entries in Table 10.1 are generalizations, but they are largely true for the vast majority of compounds of both types.

10.2 Where Do We Obtain Organic Compounds?

Chemists obtain organic compounds in two principal ways: isolation from nature and synthesis in the laboratory.

A. Isolation from Nature

Living organisms are "chemical factories." Each terrestrial, marine, and freshwater plant (flora) and animal (fauna)—even microorganisms such as bacteria—make thousands of organic compounds by a process called biosynthesis. One way, then, to get organic compounds is to extract, isolate, and purify them from biological sources. In this book, we will encounter many compounds that are or have been isolated in this way. Some important examples include vitamin E, the penicillins, table sugar, insulin, quinine, and the anticancer drug paclitaxel (Taxol, see Chemical Connections 10A). Nature also supplies us with three other important sources of organic compounds: natural gas, petroleum, and coal. We will discuss them in Section 11.4.

B. Synthesis in the Laboratory

Ever since Wöhler synthesized urea, organic chemists have sought to develop more ways to synthesize the same compounds or design derivatives of those found in nature. In recent years, the methods for doing so have become so sophisticated that there are few natural organic compounds, no matter how complicated, that chemists cannot synthesize in the laboratory.

Compounds made in the laboratory are identical in both chemical and physical properties to those found in nature—assuming, of course, that each is 100% pure. There is no way that anyone can tell whether a sample of any particular compound was made by chemists or obtained from nature. As a consequence, pure ethanol made by chemists has exactly the same physical and chemical properties as pure ethanol prepared by distilling wine. The same is true for ascorbic acid (vitamin C). There is no advantage, therefore,

Taxol: A Story of Search and Discovery

In the early 1960s, the National Cancer Institute undertook a program to analyze samples of native plant materials in the hope of discovering substances that would prove effective in the fight against cancer. Among the materials tested was an extract of the bark of the Pacific yew, *Taxus brevifolia*, a slow-growing tree found in the old-growth forests of the Pacific Northwest. This biologically active extract proved to be remarkably effective in treating certain types of ovarian and breast cancer, even in cases where other forms of chemotherapy failed. The structure of the cancer-fighting component of yew bark was determined in 1962, and the compound was named paclitaxel (Taxol).

Paclitaxel
(Taxol)

Pacific yew bark being stripped for Taxol extraction

Pete K. Ziminski/Visuals Unlimited

Unfortunately, the bark of a single 100-year-old yew tree yields only about 1 g of Taxol, not enough for effective treatment of even one cancer patient. Furthermore, isolating Taxol means stripping the bark from trees, which kills them. In 1994, chemists succeeded in synthesizing Taxol in the laboratory, but the cost of the synthetic drug was far too high to be economical. Fortunately, an alternative natural source of the drug was found. Researchers in France discovered that the needles of a related plant, *Taxus baccata*, contain a compound that can be converted in the laboratory to Taxol. Because the needles can be gathered without harming the plant, it is not necessary to kill trees to obtain the drug.

Taxol inhibits cell division by acting on microtubules, a key component of the scaffolding of cells. Before cell division can take place, the cell must disassemble these microtubule units, and Taxol prevents this disassembly. Because cancer cells divide faster than normal cells, the drug effectively controls their spread.

The remarkable success of Taxol in the treatment of breast and ovarian cancer has stimulated research efforts to isolate and/or synthesize other substances that will act upon the human body in the same way and that may be even more effective anticancer agents than Taxol.

Test your knowledge with Problems 10.39 and 10.40.

in paying more money for vitamin C obtained from a natural source than for synthetic vitamin C, because the two are identical in every way.

Organic chemists, however, have not been satisfied with merely duplicating nature's compounds. They have also designed and synthesized compounds not found in nature. In fact, the majority of the more than 10 million known organic compounds are purely synthetic and do not exist in living organisms. For example, many modern drugs—Valium, albuterol, Prozac,

Zantac, Zoloft, Lasix, Viagra, and Enovid—are synthetic organic compounds not found in nature. Even the over-the-counter drugs aspirin and ibuprofen are synthetic organic compounds not found in nature.

10.3 How Do We Write Structural Formulas of Organic Compounds?

A structural formula shows all the atoms present in a molecule as well as the bonds that connect the atoms to each other. The structural formula for ethanol, whose molecular formula is C_2H_6O, for example, shows all nine atoms and the eight bonds that connect them:

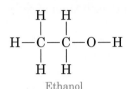

Ethanol

The vitamin C in an orange is identical to its synthetic tablet form.

The Lewis model of bonding (Section 3.7C) enables us to see how carbon forms four covalent bonds that may be various combinations of single, double, and triple bonds. Furthermore, the valence-shell electron-pair repulsion (VSEPR) model (Section 3.10) tells us that the most common bond angles about carbon atoms in covalent compounds are approximately 109.5°, 120°, and 180°, for tetrahedral, trigonal planar, and linear geometries, respectively.

Table 10.2 shows several covalent compounds containing carbon bonded to hydrogen, oxygen, nitrogen, and chlorine. From these examples, we see the following:

- Carbon normally forms four covalent bonds and has no unshared pairs of electrons.
- Nitrogen normally forms three covalent bonds and has one unshared pair of electrons.
- Oxygen normally forms two covalent bonds and has two unshared pairs of electrons.
- Hydrogen forms one covalent bond and has no unshared pairs of electrons.
- A halogen (fluorine, chlorine, bromine, and iodine) normally forms one covalent bond and has three unshared pairs of electrons.

Table 10.2 Single, Double, and Triple Bonds in Compounds of Carbon. Bond angles and geometries for carbon are predicted using the VSEPR model.

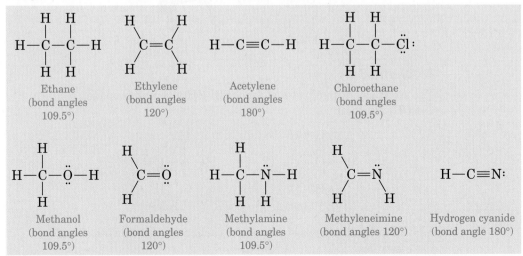

Example 10.1 Writing Structural Formulas

Following are structural formulas for acetic acid, CH_3COOH, and ethyl-amine, $CH_3CH_2NH_2$.

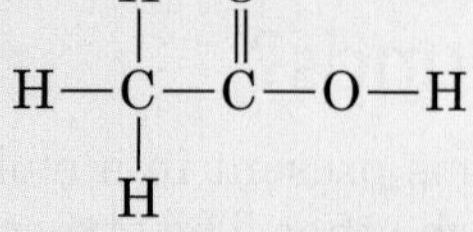

$$
\begin{array}{cc}
\underset{\text{Acetic acid}}{
\begin{array}{c}
\text{H} \quad \text{O} \\
| \qquad \| \\
\text{H}-\text{C}-\text{C}-\text{O}-\text{H} \\
| \\
\text{H}
\end{array}}
&
\underset{\text{Ethylamine}}{
\begin{array}{c}
\text{H} \quad \text{H} \\
| \qquad | \\
\text{H}-\text{C}-\text{C}-\text{N}-\text{H} \\
| \qquad | \quad | \\
\text{H} \quad \text{H} \quad \text{H}
\end{array}}
\end{array}
$$

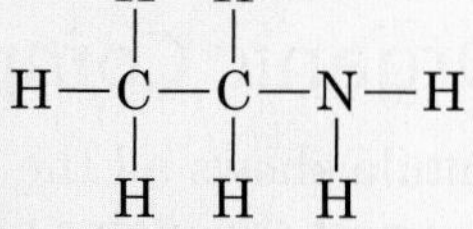

(a) Complete the Lewis structure for each molecule by adding unshared pairs of electrons so that each atom of carbon, oxygen, and nitrogen has a complete octet.

(b) Using the VSEPR model (Section 3.10), predict all bond angles in each molecule.

Strategy and Solution

(a) Each carbon atom must be surrounded by eight valence electrons to have a complete octet. Each oxygen must have two bonds and two unshared pairs of electrons to have a complete octet. Each nitrogen must have three bonds and one unshared pair of electrons to have a complete octet.

(b) To predict bond angles about a carbon, nitrogen, or oxygen atom, count the number of regions of electron density (lone pairs and bonding pairs of electrons about it). If four regions of electron density surround the atom, the predicted bond angles are 109.5°. If three regions surround it, the predicted bond angles are 120°. If two regions surround it, the predicted bond angle is 180°.

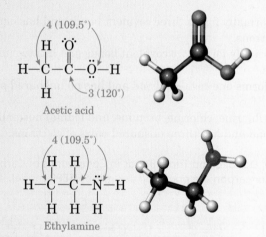

Acetic acid

Ethylamine

Problem 10.1

The structural formulas for ethanol, CH_3CH_2OH, and propene, $CH_3CH=CH_2$, are:

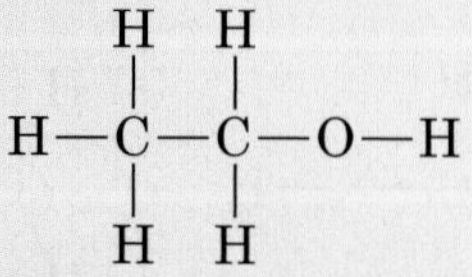

$$
\begin{array}{cc}
\underset{\text{Ethanol}}{
\begin{array}{c}
\text{H} \quad \text{H} \\
| \qquad | \\
\text{H}-\text{C}-\text{C}-\text{O}-\text{H} \\
| \qquad | \\
\text{H} \quad \text{H}
\end{array}}
&
\underset{\text{Propene}}{
\begin{array}{c}
\qquad \text{H} \\
\qquad | \\
\text{H}-\text{C}-\text{C}=\text{C}-\text{H} \\
| \qquad | \quad | \\
\text{H} \quad \text{H} \quad \text{H}
\end{array}}
\end{array}
$$

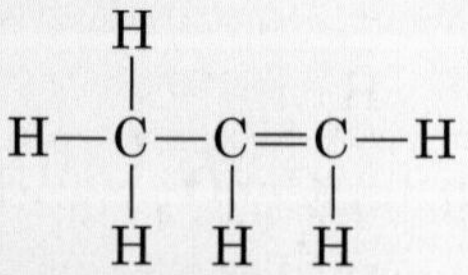

(a) Complete the Lewis structure for each molecule showing all valence electrons.

(b) Using the VSEPR model, predict all bond angles in each molecule.

10.4 What Is a Functional Group?

As noted earlier in this chapter, more than 10 million organic compounds have been discovered and synthesized by organic chemists. It might seem an almost impossible task to learn the physical and chemical properties of so many compounds. Fortunately, the study of organic compounds is not as formidable a task as you might think. While organic compounds can undergo a wide variety of chemical reactions, only certain portions of their structures undergo chemical transformations. We call the atoms or groups of atoms of an organic molecule that undergo predictable chemical reactions a **functional group**. As we will see, the same functional group, in whatever organic molecule it occurs, undergoes the same types of chemical reactions. Therefore, we do not have to study the chemical reactions of even a fraction of the 10 million known organic compounds. Instead, we need to identify only a few characteristic functional groups and then study the chemical reactions that each undergoes.

Functional group An atom or group of atoms within a molecule that shows a characteristic set of predictable physical and chemical behaviors

Functional groups are also important because they are the units by which we divide organic compounds into families of compounds. For example, we group those compounds that contain an —OH (hydroxyl) group bonded to a tetrahedral carbon into a family called alcohols; compounds containing a —COOH (carboxyl group) belong to a family called carboxylic acids. Table 10.3 introduces six of the most common functional groups. A complete list of all functional groups that we will study appears on the inside back cover of the text.

Table 10.3 Six Common Functional Groups

Family	Functional Group	Example	Name
Alcohol	—OH	CH_3CH_2OH	Ethanol
Amine	—NH_2	$CH_3CH_2NH_2$	Ethylamine
Aldehyde	$-\overset{\overset{\displaystyle O}{\|\|}}{C}-H$	$CH_3\overset{\overset{\displaystyle O}{\|\|}}{C}H$	Acetaldehyde
Ketone	$-\overset{\overset{\displaystyle O}{\|\|}}{C}-$	$CH_3\overset{\overset{\displaystyle O}{\|\|}}{C}CH_3$	Acetone
Carboxylic acid	$-\overset{\overset{\displaystyle O}{\|\|}}{C}-OH$	$CH_3\overset{\overset{\displaystyle O}{\|\|}}{C}OH$	Acetic acid
Ester	$-\overset{\overset{\displaystyle O}{\|\|}}{C}-OR$	$CH_3\overset{\overset{\displaystyle O}{\|\|}}{C}OCH_2CH_3$	Ethyl acetate

At this point, our concern is simply pattern recognition—that is, how to recognize and identify one of these six common functional groups when you see it and how to draw structural formulas of molecules containing it. We will have more to say about the physical and chemical properties of these and several other functional groups in Chapters 11–19.

Functional groups also serve as the basis for naming organic compounds. Ideally, each of the 10 million or more organic compounds must have a unique name different from the name of every other organic compound. We will show how these names are derived in Chapters 11–19 as we study individual functional groups in detail.

To summarize, functional groups:

- Are sites of predictable chemical behavior—a particular functional group, in whatever compound it is found, undergoes the same types of chemical reactions.

- Determine in large measure the physical properties of a compound.
- Serve as the units by which we classify organic compounds into families.
- Serve as a basis for naming organic compounds.

A. Alcohols

Alcohol A compound containing an —OH (hydroxyl) group bonded to a tetrahedral carbon atom

Hydroxyl group An —OH group bonded to a tetrahedral carbon atom

As previously mentioned, the functional group of an **alcohol** is an **—OH (hydroxyl) group** bonded to a tetrahedral carbon atom (a carbon having bonds to four atoms). In the general formula of an alcohol (shown below on the left), the symbol R— indicates either a hydrogen or another carbon group. The important point of the general structure is the —OH group bonded to a tetrahedral carbon atom.

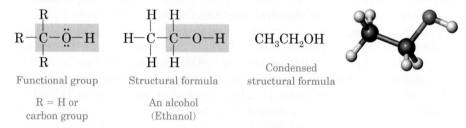

Functional group   Structural formula   Condensed structural formula

R = H or carbon group   An alcohol (Ethanol)

Here we represent the alcohol as a **condensed structural formula**, CH_3CH_2OH. In a condensed structural formula, CH_3 indicates a carbon bonded to three hydrogens, CH_2 indicates a carbon bonded to two hydrogens, and CH indicates a carbon bonded to one hydrogen. Unshared pairs of electrons are generally not shown in condensed structural formulas.

Alcohols are classified as **primary (1°), secondary (2°), or tertiary (3°)**, depending on the number of carbon atoms bonded to the carbon bearing the —OH group.

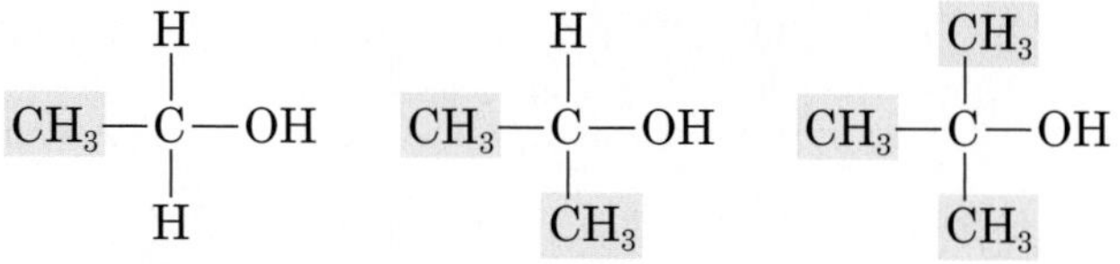

H
|
CH_3—C—OH
|
H
A 1° alcohol

H
|
CH_3—C—OH
|
CH_3
A 2° alcohol

CH_3
|
CH_3—C—OH
|
CH_3
A 3° alcohol

Example 10.2 Drawing Structural Formulas of Alcohols

Draw Lewis structures and condensed structural formulas for the two alcohols with molecular formula C_3H_8O. Classify each as primary, secondary, or tertiary.

Strategy and Solution

Begin by drawing the three carbon atoms in a chain. The oxygen atom of the hydroxyl group may be bonded to the carbon chain at two different positions on the chain: either to an end carbon or to the middle carbon.

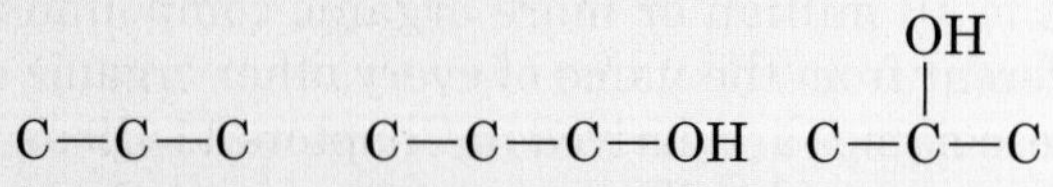

OH
|
C—C—C   C—C—C—OH   C—C—C
Carbon chain   The two locations for the —OH group

Finally, add seven more hydrogens, giving a total of eight as shown in the molecular formula. Show unshared electron pairs on the Lewis structures but not on the condensed structural formulas.

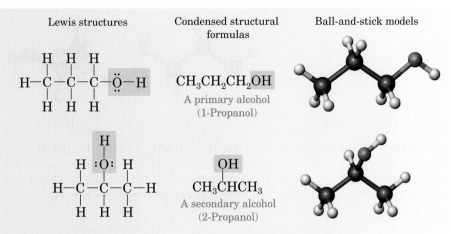

Lewis structures | Condensed structural formulas | Ball-and-stick models

CH$_3$CH$_2$CH$_2$OH
A primary alcohol
(1-Propanol)

CH$_3$CHCH$_3$
A secondary alcohol
(2-Propanol)

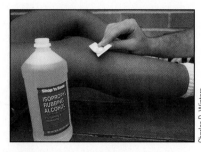

2-Propanol (isopropyl alcohol) is used to disinfect cuts and scrapes.

The secondary alcohol 2-propanol, whose common name is isopropyl alcohol, is the cooling, soothing component in rubbing alcohol.

Problem 10.2

Draw Lewis structures and condensed structural formulas for the four alcohols with the molecular formula C$_4$H$_{10}$O. Classify each alcohol as primary, secondary, or tertiary. (*Hint*: First consider the connectivity of the four carbon atoms; they can be bonded either four in a chain or three in a chain with the fourth carbon as a branch on the middle carbon. Then consider the points at which the —OH group can be bonded to each carbon chain.)

B. Amines

The functional group of an amine is an **amino group**—a nitrogen atom bonded to one, two, or three carbon atoms. In a **primary (1°) amine**, nitrogen is bonded to two hydrogens and one carbon group. In a **secondary (2°) amine**, it is bonded to one hydrogen and two carbon groups. In a **tertiary (3°) amine**, it is bonded to three carbon groups. The second and third structural formulas can be written in a more abbreviated form by collecting the CH$_3$ groups and writing them as (CH$_3$)$_2$NH and (CH$_3$)$_3$N, respectively. The latter are known as condensed structural formulas.

Amine An organic compound in which one, two, or three hydrogens of ammonia are replaced by carbon groups; RNH$_2$, R$_2$NH, or R$_3$N

Amino group A —NH$_2$, RNH$_2$, R$_2$NH, or R$_3$N group

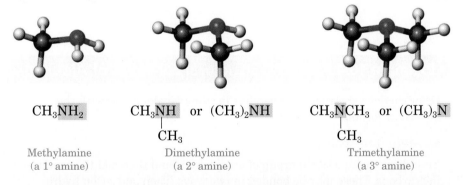

CH$_3$NH$_2$

Methylamine
(a 1° amine)

CH$_3$NH or (CH$_3$)$_2$NH
|
CH$_3$

Dimethylamine
(a 2° amine)

CH$_3$NCH$_3$ or (CH$_3$)$_3$N
|
CH$_3$

Trimethylamine
(a 3° amine)

Example 10.3 Drawing Structural Formulas of Amines

Draw condensed structural formulas for the two primary amines with the molecular formula C$_3$H$_9$N.

Strategy and Solution

For a primary amine, draw a nitrogen atom bonded to two hydrogens and one carbon.

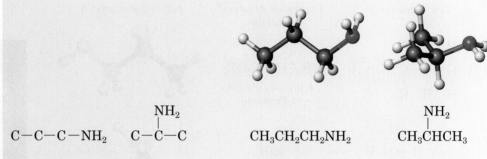

$$C-C-C-NH_2 \qquad C-\overset{\overset{\displaystyle NH_2}{|}}{C}-C$$

The three carbons may be
bonded to nitrogen in two ways

$$CH_3CH_2CH_2NH_2 \qquad CH_3\overset{\overset{\displaystyle NH_2}{|}}{C}HCH_3$$

Add seven hydrogens to give each carbon four
bonds and give the correct molecular formula

Problem 10.3

Draw structural formulas for the three secondary amines with the molecular formula $C_4H_{11}N$.

C. Aldehydes and Ketones

Carbonyl group A C=O group

Aldehyde A compound containing a carbonyl group bonded to a hydrogen; a —CHO group

Ketone A compound containing a carbonyl group bonded to two carbon groups

Both aldehydes and ketones contain a **C=O (carbonyl) group**. The **aldehyde** functional group contains a carbonyl group bonded to a hydrogen. Formaldehyde, CH_2O, the simplest aldehyde, has two hydrogens bonded to its carbonyl carbon. In a condensed structural formula, the aldehyde group may be written showing the carbon-oxygen double bond as CH=O or, alternatively, it may be written —CHO. The functional group of a **ketone** is a carbonyl group bonded to two carbon atoms. In the general structural formula of each functional group, we use the symbol R to represent other groups bonded to carbon to complete the tetravalence of carbon.

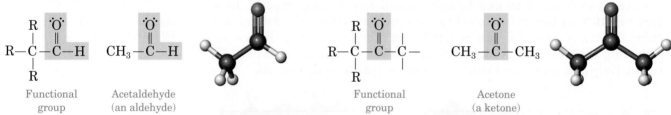

Functional group   Acetaldehyde (an aldehyde)   Functional group   Acetone (a ketone)

Example 10.4 Drawing Structural Formulas of Aldehydes

Draw condensed structural formulas for the two aldehydes with the molecular formula C_4H_8O.

Strategy and Solution

First draw the functional group of an aldehyde and then add the remaining carbons. These may be bonded in two ways. Then add seven hydrogens to complete the tetravalence of each carbon.

$$CH_3\overset{\overset{\displaystyle O}{\|}}{\underset{\underset{\displaystyle CH_3}{|}}{C}}HCH \qquad or \qquad CH_3\underset{\underset{\displaystyle CH_3}{|}}{C}HCHO$$

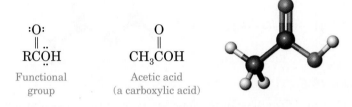

$$CH_3CH_2CH_2\overset{\overset{\textstyle O}{\|}}{CH} \quad \text{or} \quad CH_3CH_2CH_2CHO$$

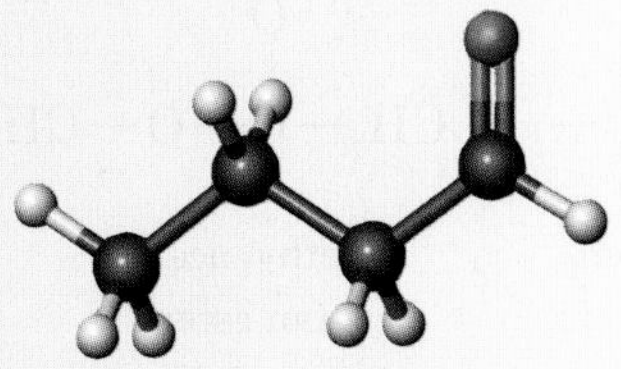

Problem 10.4

Draw condensed structural formulas for the three ketones with the molecular formula $C_5H_{10}O$.

D. Carboxylic Acids

The functional group of a carboxylic acid is a —**COOH** (**carboxyl**: <u>carb</u>onyl + hyd<u>roxyl</u>) **group**. In a condensed structural formula, a carboxyl group may also be written —CO_2H.

Carboxyl group A —COOH group

Carboxylic acid A compound containing a —COOH group

$$:\overset{\overset{\textstyle :O:}{\|}}{R\overset{..}{C}\overset{..}{O}H}$$

Functional group

$$CH_3\overset{\overset{\textstyle O}{\|}}{C}OH$$

Acetic acid (a carboxylic acid)

Example 10.5 Drawing Structural Formulas of Carboxylic Acids

Draw a condensed structural formula for the single carboxylic acid with the molecular formula $C_3H_6O_2$.

Srategy and Solution

The only way the carbon atoms can be written is three in a chain, and the —COOH group must be on an end carbon of the chain.

$$CH_3CH_2\overset{\overset{\textstyle O}{\|}}{C}OH \quad \text{or} \quad CH_3CH_2CO_2H$$

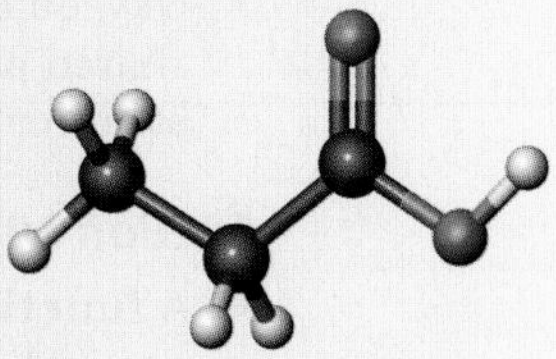

Problem 10.5

Draw condensed structural formulas for the two carboxylic acids with the molecular formula $C_4H_8O_2$.

E. Carboxylic Esters

A **carboxylic ester**, commonly referred to as simply an **ester**, is a derivative of a carboxylic acid in which a carbon group replaces the hydrogen of the carboxyl group. The ester group is written —COOR or —CO_2R in this text.

Carboxylic ester A derivative of a carboxylic acid in which the H of the carboxyl group is replaced by a carbon group

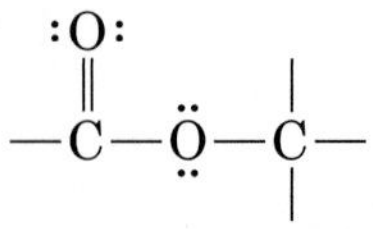

$$
\begin{matrix}
:O: \\
\| \\
-C-\ddot{O}-C- \\
| \\
\end{matrix}
\qquad
\begin{matrix}
O \\
\| \\
CH_3-C-O-CH_3
\end{matrix}
\quad or \quad
CH_3COOCH_3
$$

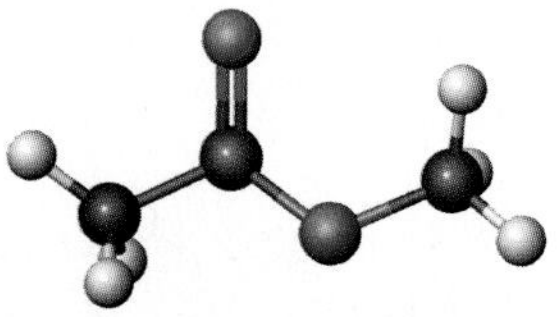

Functional group   Methyl acetate
                            (an ester)

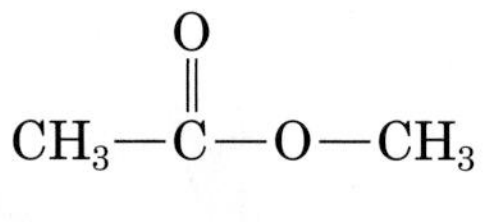

Example 10.6 Drawing Structural Formulas of Esters

The molecular formula of methyl acetate is $C_3H_6O_2$. Draw the structural formula of another ester with the same molecular formula.

Strategy and Solution

There is only one other ester with this molecular formula. Its structural formula is:

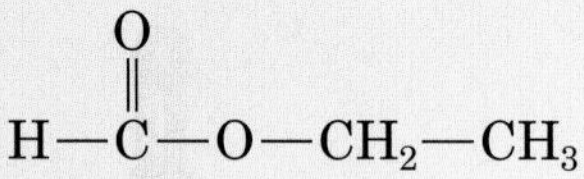

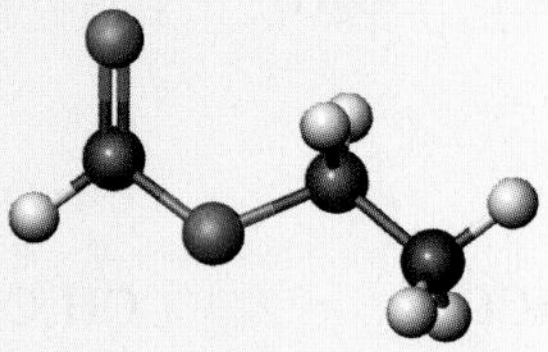

$$
\begin{matrix}
O \\
\| \\
H-C-O-CH_2-CH_3
\end{matrix}
$$

Ethyl formate

Problem 10.6

Draw structural formulas for the four esters with the molecular formula $C_4H_8O_2$.

Summary

Section 10.1 What Is Organic Chemistry?

- Organic chemistry is the study of compounds containing carbon.

Section 10.2 Where Do We Obtain Organic Compounds?

- Chemists obtain organic compounds either by isolation from plant and animal sources or by synthesis in the laboratory.

Section 10.3 How Do We Write Structural Formulas of Organic Compounds?

- Carbon normally forms four bonds and has no unshared pairs of electrons. Its four bonds may be four single bonds, two single bonds and one double bond, or one single bond and one triple bond.

- Nitrogen normally forms three bonds and has one unshared pair of electrons. Its bonds may be three single bonds, one single bond and one double bond, or one triple bond.
- Oxygen normally forms two bonds and has two unshared pairs of electrons. Its bonds may be two single bonds or one double bond.

Section 10.4 What Is a Functional Group?

- A **functional group** is a site of chemical reactivity; a particular functional group, in whatever compound it is found, always undergoes the same types of chemical reactions.
- In addition, functional groups are the characteristic structural units by which we both classify and name organic compounds. Important functional groups include the **hydroxyl group** of 1°, 2°, and 3° alcohols; the **amino group** of 1°, 2°, and 3° amines; the **carbonyl group** of aldehydes and ketones; the **carboxyl group** of carboxylic acids; and the **ester group**.

OWL Interactive versions of these problems may be assigned in OWL.

Orange-numbered problems are applied.

References to previous chapters are given in parentheses.

Section 10.1 What Is Organic Chemistry?

10.7 Answer true or false.

(a) All organic compounds contain one or more atoms of carbon.

(b) The majority of organic compounds are built from carbon, hydrogen, oxygen, and nitrogen.

(c) By number of atoms, carbon is the most abundant element in the Earth's crust.

(d) Most organic compounds are soluble in water.

Section 10.2 Where Do We Obtain Organic Compounds?

10.8 Answer true or false.

(a) Organic compounds can only be synthesized in living organisms.

(b) Organic compounds synthesized in the laboratory have the same chemical and physical properties as those synthesized in living organisms.

(c) Chemists have synthesized many organic compounds that are not found in nature.

10.9 Is there any difference between vanillin made synthetically and vanillin extracted from vanilla beans?

10.10 Suppose that you are told that only organic substances are produced by living organisms. How would you rebut this assertion?

10.11 What important experiment did Wöhler carry out in 1828?

Section 10.3 How Do We Write Structural Formulas of Organic Compounds?

10.12 Answer true or false.

(a) In organic compounds, carbon normally has four bonds and no unshared pairs of electrons.

(b) When found in organic compounds, nitrogen normally has three bonds and one unshared pair of electrons.

(c) The most common bond angles about carbon in organic compounds are approximately 109.5° and 180°.

10.13 List the four principal elements that make up organic compounds and give the number of bonds each typically forms.

10.14 Think about the types of substances in your immediate environment and make a list of those that are organic—for example, textile fibers. We will ask you to return to this list later in the course and to refine, correct, and possibly expand it.

10.15 How many electrons are in the valence shell of each of the following atoms? Write a Lewis structure for

an atom of each element. (*Hint*: Use the Periodic Table.)

(a) Carbon   (b) Oxygen

(c) Nitrogen   (d) Fluorine

10.16 What is the relationship between the number of electrons in the valence shell of each of the following atoms and the number of covalent bonds it forms?

(a) Carbon   (b) Oxygen

(c) Nitrogen   (d) Hydrogen

10.17 Write Lewis structures for these compounds. Show all valence electrons. None of them contains a ring of atoms. (*Hint*: Remember that carbon has four bonds, nitrogen has three bonds and one unshared pair of electrons, oxygen has two bonds and two unshared pairs of electrons, and each halogen has one bond and three unshared pairs of electrons.)

(a) H_2O_2 (b) N_2H_4
    Hydrogen peroxide   Hydrazine

(c) CH_3OH (d) CH_3SH
    Methanol   Methanethiol

(e) CH_3NH_2 (f) CH_3Cl
    Methylamine   Chloromethane

10.18 Write Lewis structures for these compounds. Show all valence electrons. None of them contains a ring of atoms.

(a) CH_3OCH_3 (b) C_2H_6
    Dimethyl ether   Ethane

(c) C_2H_4 (d) C_2H_2
    Ethylene   Acetylene

(e) CO_2 (f) CH_2O
    Carbon dioxide   Formaldehyde

(g) H_2CO_3 (h) CH_3COOH
    Carbonic acid   Acetic acid

10.19 Write Lewis structures for these ions.

(a) HCO_3^- (b) CO_3^{2-}
    Bicarbonate ion   Carbonate ion

(c) CH_3COO^- (d) Cl^-
    Acetate ion   Chloride ion

10.20 Why are the following molecular formulas impossible?

(a) CH_5 (b) C_2H_7

Section 10.3 Review of the VSEPR Model

10.21 Explain how to use the valence-shell electron-pair repulsion (VSEPR) model to predict bond angles and geometry about atoms of carbon, oxygen, and nitrogen.

10.22 Suppose you forget to take into account the presence of the unshared pair of electrons on nitrogen in the molecule NH_3. What would you then predict for the H—N—H bond angles and the geometry (bond angles and shape) of ammonia?

10.23 Suppose you forget to take into account the presence of the two unshared pairs of electrons on the oxygen atom of ethanol, CH_3CH_2OH. What would you then predict for the C—O—H bond angle and the geometry of ethanol?

10.24 Use the VSEPR model to predict the bond angles and geometry about each highlighted atom. (*Hint*: Remember to take into account the presence of unshared pairs of electrons.)

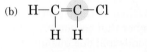

(a)

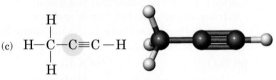

(b) H—C=C—Cl (with H, H below)

(c)

10.25 Use the VSEPR model to predict the bond angles about each highlighted atom.

(a) H—C̈—Ö—H   (b) H—C—N̈—H (with H, H)

(c) H—Ö—N̈=Ö

Section 10.4 What Is a Functional Group?

10.26 Answer true or false.
   (a) A functional group is a group of atoms in an organic molecule that undergoes a predictable set of chemical reactions.
   (b) The functional group of an alcohol, an aldehyde, and a ketone have in common the fact that each contains a single oxygen atom.
   (c) A primary alcohol has one —OH group, a secondary alcohol has two —OH groups, and a tertiary alcohol has three —OH groups.
   (d) There are two alcohols with the molecular formula C_3H_8O.
   (e) There are three amines with the molecular formula C_3H_9N.
   (f) Aldehydes, ketones, carboxylic acids, and esters all contain a carbonyl group.
   (g) A compound with the molecular formula of C_3H_6O may be either an aldehyde, a ketone, or a carboxylic acid.
   (h) Bond angles about the carbonyl carbon of an aldehyde, a ketone, a carboxylic acid, and an ester are all approximately 109.5°.
   (i) The molecular formula of the smallest aldehyde is C_3H_6O, and that of the smallest ketone is also C_3H_6O.

(j) The molecular formula of the smallest carboxylic acid is $C_2H_4O_2$.

10.27 What is meant by the term *functional group*?

10.28 List three reasons why functional groups are important in organic chemistry.

10.29 Draw Lewis structures for each of the following functional groups. Show all valence electrons in each functional group.
   (a) A carbonyl group
   (b) A carboxyl group
   (c) A hydroxyl group
   (d) A primary amino group
   (e) An ester group

10.30 Complete the following structural formulas by adding enough hydrogens to complete the tetravalence of each carbon. Then write the molecular formula of each compound.

(a) C—C=C—C—C (with C above third carbon)

(b) C—C—C—C—OH (with O double bond on fourth carbon)

(c) C—C—C—C (with O double bond on third carbon)

(d) C—C—C—H (with O double bond on second carbon, C below first carbon)

(e) C—C—C—C—NH₂ (with C above and C below second carbon)

(f) C—C—C—OH (with O double bond on third carbon, NH₂ below second carbon)

(g) C—C—C—C—C (with OH above second carbon)

(h) C—C—C—C—OH (with OH above second carbon, O double bond on fourth carbon)

(i) C=C—C—OH

10.31 What is the meaning of the term *tertiary* (3°) when it is used to classify alcohols?

10.32 Draw a structural formula for the one tertiary (3°) alcohol with the molecular formula $C_4H_{10}O$.

10.33 What is the meaning of the term *tertiary* (3°) when it is used to classify amines?

10.34 Draw condensed structural formulas for all compounds with the molecular formula C_4H_8O that contain a carbonyl group (there are two aldehydes and one ketone).

10.35 Draw structural formulas for each of the following:
(a) The four primary (1°) alcohols with the molecular formula $C_5H_{12}O$.
(b) The three secondary (2°) alcohols with the molecular formula $C_5H_{12}O$.
(c) The one tertiary (3°) alcohol with the molecular formula $C_5H_{12}O$.

10.36 Draw structural formulas for the six ketones with the molecular formula $C_6H_{12}O$.

10.37 Draw structural formulas for the eight carboxylic acids with the molecular formula $C_6H_{12}O_2$.

10.38 Draw structural formulas for each of the following:
(a) The four primary (1°) amines with the molecular formula $C_4H_{11}N$.
(b) The three secondary (2°) amines with the molecular formula $C_4H_{11}N$.
(c) The one tertiary (3°) amine with the molecular formula $C_4H_{11}N$.

Chemical Connections

10.39 (Chemical Connections 10A) How was Taxol discovered?

10.40 (Chemical Connections 10A) In what way does Taxol interfere with cell division?

Additional Problems

10.41 Use the VSEPR model to predict the bond angles about each atom of carbon, nitrogen, and oxygen in these molecules. (*Hint*: First, add unshared pairs of electrons as necessary to complete the valence shell of each atom and then predict the bond angles.)

(a) $CH_3CH_2CH_2OH$

(b) $CH_3CH_2\overset{\overset{\displaystyle O}{\|}}{C}H$

(c) $CH_3CH{=}CH_2$

(d) $CH_3C{\equiv}CCH_3$

(e) $CH_3\overset{\overset{\displaystyle O}{\|}}{C}OCH_3$

(f) $CH_3\overset{\overset{\displaystyle CH_3}{|}}{N}CH_3$

10.42 Silicon is immediately below carbon in Group 4A of the Periodic Table. Predict the C—Si—C bond angles in tetramethylsilane, $(CH_3)_4Si$.

10.43 Phosphorus is immediately below nitrogen in Group 5A of the Periodic Table. Predict the C—P—C bond angles in trimethylphosphine, $(CH_3)_3P$.

10.44 Draw the structure for a compound with the molecular formula:
(a) C_2H_6O that is an alcohol
(b) C_3H_6O that is an aldehyde
(c) C_3H_6O that is a ketone
(d) $C_3H_6O_2$ that is a carboxylic acid

10.45 Draw structural formulas for the eight aldehydes with the molecular formula $C_6H_{12}O$.

10.46 Draw structural formulas for the three tertiary (3°) amines with the molecular formula $C_5H_{13}N$.

10.47 Which of these covalent bonds are polar, and which are nonpolar? (*Hint*: Review Section 3.7B.)
(a) C—C   (b) C═C   (c) C—H   (d) C—O
(e) O—H   (f) C—N   (g) N—H   (h) N—O

10.48 Of the bonds in Problem 10.47, which is the most polar? Which is the least polar?

10.49 Using the symbol $\delta+$ to indicate a partial positive charge and $\delta-$ to indicate a partial negative charge, indicate the polarity of the most polar bond (or bonds if two or more have the same polarity) in each of the following molecules.

(a) CH_3OH

(b) CH_3NH_2

(c) $HSCH_2CH_2NH_2$

(d) $CH_3\overset{\overset{\displaystyle O}{\|}}{C}CH_3$

(e) $H\overset{\overset{\displaystyle O}{\|}}{C}H$

(f) $CH_3\overset{\overset{\displaystyle O}{\|}}{C}OH$

Looking Ahead

10.50 Following is a structural formula and a ball-and-stick model of benzene, C_6H_6.

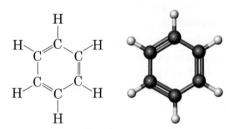

(a) Predict each H—C—C and C—C—C bond angle in benzene.
(b) Predict the shape of a benzene molecule.

10.51 Following is a structural formula for naphthalene. It was first obtained by heating coal to a high temperature in the absence of air (oxygen). At one time, it was used in "mothballs."

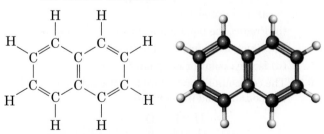

Naphthalene

(a) Predict the shape of naphthalene.
(b) Is a naphthalene molecule polar or nonpolar?

10.52 Identify the functional group(s) in each compound.

(a) $CH_3CH_2\overset{\overset{\displaystyle O}{\|}}{C}CH_3$

2-Butanone
(a solvent for
paints and lacquers)

(b) $\overset{O}{\overset{\|}{HOCCH_2CH_2CH_2CH_2COH}}$

Hexanedioic acid
(the second component
of nylon-66)

(c) $H_2NCH_2CH_2CH_2CH_2\overset{O}{\overset{\|}{CHCOH}}$
$\underset{NH_2}{|}$

Lysine
(one of the 20 amino acid
building blocks of proteins)

(d) $\overset{O}{\overset{\|}{HOCH_2CCH_2OH}}$

Dihydroxyacetone
(a component of several
artificial tanning lotions)

10.53 Consider molecules with the molecular formula $C_4H_8O_2$. Write the structural formula for a molecule with this molecular formula that contains:

(a) A carboxyl group

(b) An ester group

(c) A ketone group and a 2° alcohol

(d) An aldehyde and a 3° alcohol

10.54 Urea, $(NH_2)_2CO$, is used in plastics and in fertilizers. It is also the primary nitrogen-containing substance excreted by humans.

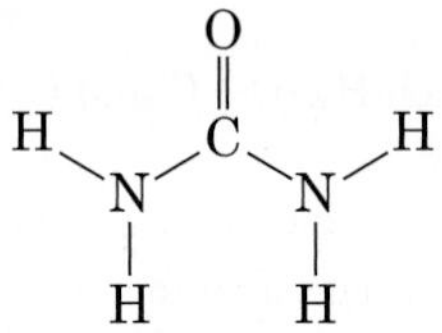

Urea

(a) Complete the Lewis structure of urea, showing all valence electrons.

(b) Predict the bond angle about each C and N.

(c) Which is the most polar bond in the molecule?

(d) Is urea polar or nonpolar?

10.55 The compound drawn here is lactic acid, a natural compound found in sour milk.

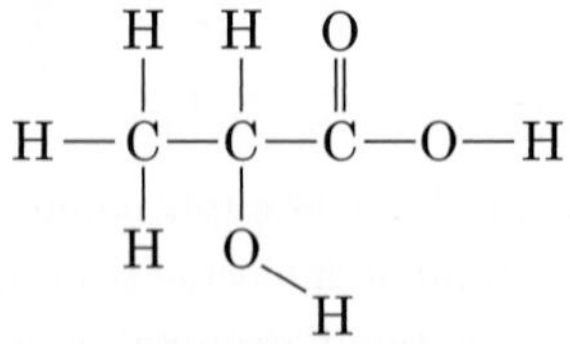

Lactic acid

(a) What is the molecular formula of lactic acid?

(b) Name the two functional groups in lactic acid.

(c) Predict the bond angles about each carbon atom.

(d) Which bonds are polar, and which are nonpolar?

(e) Would you predict that lactic acid is polar or nonpolar?

Tying It Together

10.56 Following is the structural formula of acetylsalicylic acid, better known by its common name aspirin.

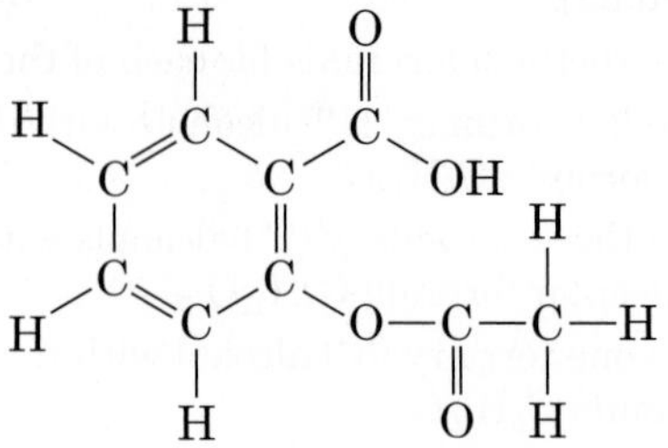

Acetylsalicylic acid
(Aspirin)

(a) Name the two oxygen-containing functional groups in aspirin.

(b) What is the molecular formula of aspirin? (Chapter 4)

10.57 Aspirin is prepared by the reaction of salicylic acid with acetic anhydride as shown in the following equation. The stoichiometry of the reaction is given in the equation. Acetic acid is a by-product of the reaction and must be separated and removed so that aspirin can then be sold as a pure product. How many grams of aspirin can be prepared from 120 grams of salicylic acid? Assume that there is an excess of acetic anhydride. (Chapter 4)

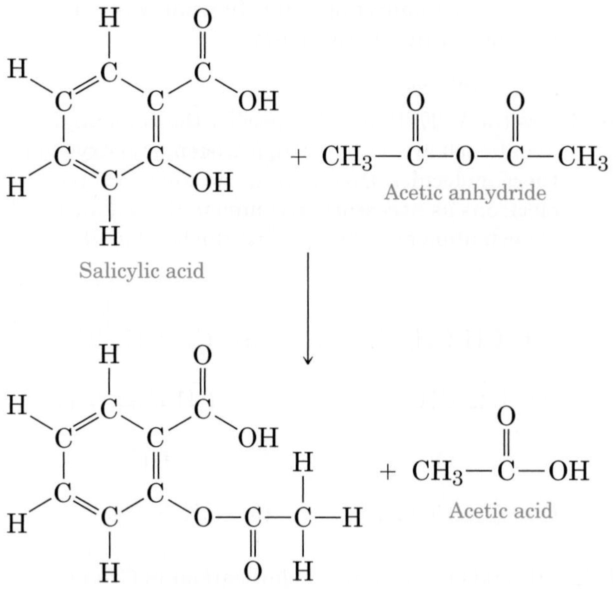

Salicylic acid   Acetic anhydride

Acetylsalicylic acid
(Aspirin)   Acetic acid

10.58 Following is the structural formula of acetamide, a molecule that contains a functional group called an **amide**. In an amide, the —OH of a carboxyl group is replaced by an amino group.

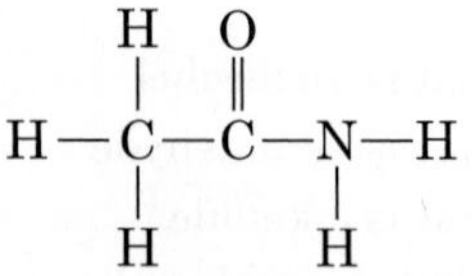

Acetamide
(an amide)

(a) Complete the Lewis structure for acetamide, showing all valence electrons.

(b) Use the valence-shell electron-pair repulsion (VSEPR) model (Section 3.10) to predict all bond angles in acetamide.

(c) Which is the most polar bond in acetamide?

10.59 The amide group, in acetamide as well as in all other amides, is best represented as a resonance hybrid (Section 3.9). Following are two contributing structures for the hybrid.

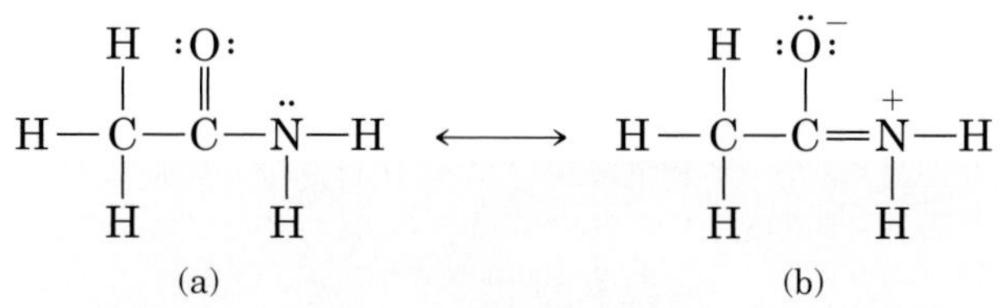

(a) Show by the use of curved arrows how contributing structure (a) is converted into contributing structure (b).

(b) Notice that structure (b) contains an oxygen atom with one bond and three unshared pairs of electrons and that the oxygen bears a negative charge. Compare the Lewis structure of this oxygen with the oxygen atom in the hydroxide ion.

(c) Notice that the nitrogen atom of structure (b) has four bonds and bears a positive charge. Compare the Lewis structure and bonding of this nitrogen with the nitrogen in the ammonium ion, NH_4^+.

(d) If the acetamide hybrid is best represented by contributing structure (a), predict the H—N—H bond angle.

(e) If, on the other hand, the acetamide hybrid is best represented by contributing structure (b), predict the H—N—H bond angle.

(f) Proteins are molecules that can be described as polyamides (Chapter 22). Linus Pauling, in his pioneering studies on the structure of proteins, discovered that the actual H—N—H bond angle in each amide bond of a protein is 120°. What does this fact tell you about the relative importance of contributing structures (a) and (b) in the resonance hybrid?

11

Alkanes

Sign in to OWL at **www.cengage.com/owl** to view tutorials and simulations, develop problem-solving skills, and complete online homework assigned by your professor.

© L. Lefkawitz/Taxi/Getty

Bunsen burners burn natural gas, which is primarily methane, with small amounts of ethane, propane, and butane.

11.1 What Are Alkanes?

In this chapter, we examine the physical and chemical properties of **alkanes**, the simplest type of organic compounds. Actually, alkanes are members of a larger class of organic compounds called hydrocarbons. A **hydrocarbon** is a compound composed of only carbon and hydrogen. Figure 11.1 shows the four classes of hydrocarbons, along with the characteristic type of bonding between carbon atoms in each class. Alkanes are **saturated hydrocarbons**; that is, they contain only carbon–carbon single bonds. Saturated in this context means that each carbon in the hydrocarbon has the maximum number of hydrogens bonded to it. A hydrocarbon that contains one or more carbon–carbon double bonds, triple bonds, or benzene rings is classified as an **unsaturated hydrocarbon**. We study alkanes (saturated hydrocarbons) in this chapter and alkenes, alkynes, and arenes (unsaturated hydrocarbons) in Chapters 12 and 13.

We often refer to alkanes as **aliphatic hydrocarbons** because the physical properties of the higher members of this class resemble those of the long carbon-chain molecules we find in animal fats and plant oils (Greek: *aleiphar*, fat or oil).

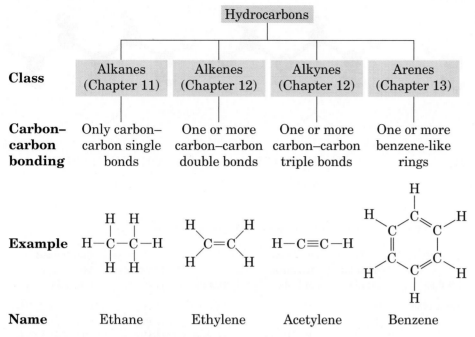

FIGURE 11.1 The four classes of hydrocarbons.

Alkane A saturated hydrocarbon whose carbon atoms are arranged in a chain

Hydrocarbon A compound that contains only carbon and hydrogen atoms

Saturated hydrocarbon A hydrocarbon that contains only carbon–carbon single bonds

Aliphatic hydrocarbon An alkane

11.2 How Do We Write Structural Formulas of Alkanes?

Methane, CH_4, and ethane, C_2H_6, are the first two members of the alkane family. Lewis structures and ball-and-stick models for these molecules are shown below. The shape of methane is tetrahedral, and all H—C—H bond angles are 109.5°. Each carbon atom in ethane is also tetrahedral, and the bond angles in it are all approximately 109.5° as well. Although the three-dimensional shapes of larger alkanes are more complex than those of methane and ethane, the four bonds about each carbon atom are still arranged in a tetrahedral manner, and all bond angles are still approximately 109.5°.

Methane

Ethane

The next members of the alkane family are propane, butane, and pentane. In the following representations, these hydrocarbons are drawn as condensed structural formulas, which show all carbons and hydrogens. They can also be drawn in a more abbreviated form called a **line-angle formula**. In this type of representation, a line represents a carbon–carbon bond and a vertex represents a carbon atom. A line ending in space represents a —CH_3 group. To count hydrogens from a line-angle formula, you simply add enough hydrogens in your mind to give each carbon its required four bonds. Chemists use line-angle formulas because they are easier and faster to draw than condensed structural formulas.

Structural formulas for alkanes can be written in yet another condensed form. For example, the structural formula of pentane contains three CH_2 (**methylene**) groups in the middle of the chain. We can group them together and write the structural formula $CH_3(CH_2)_3CH_3$. Table 11.1 gives names and molecular formulas for the first ten alkanes with unbranched chains. Note that the names of all these alkanes end in "-ane." We will have more to say about naming alkanes in Section 11.3.

Line-angle formula An abbreviated way to draw structural formulas in which each vertex and line terminus represents a carbon atom and each line represents a bond

Ball-and-stick model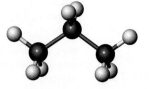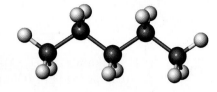

Line-angle formula

Condensed structural
formula

$CH_3CH_2CH_3$ $CH_3CH_2CH_2CH_3$ $CH_3CH_2CH_2CH_2CH_3$

Propane   Butane   Pentane

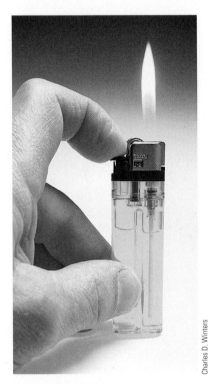

Butane, $CH_3CH_2CH_2CH_3$, is the fuel
in this lighter. Butane molecules are
present in both the liquid and gaseous
states in the lighter.

Charles D. Winters

Table 11.1 The First Ten Alkanes with Unbranched Chains

Name	Molecular Formula	Condensed Structural Formula	Name	Molecular Formula	Condensed Structural Formula
methane	CH_4	CH_4	hexane	C_6H_{14}	$CH_3(CH_2)_4CH_3$
ethane	C_2H_6	CH_3CH_3	heptane	C_7H_{16}	$CH_3(CH_2)_5CH_3$
propane	C_3H_8	$CH_3CH_2CH_3$	octane	C_8H_{18}	$CH_3(CH_2)_6CH_3$
butane	C_4H_{10}	$CH_3(CH_2)_2CH_3$	nonane	C_9H_{20}	$CH_3(CH_2)_7CH_3$
pentane	C_5H_{12}	$CH_3(CH_2)_3CH_3$	decane	$C_{10}H_{22}$	$CH_3(CH_2)_8CH_3$

Example 11.1 Drawing Line-Angle Formulas

Table 11.1 gives the condensed structural formula for hexane. Draw a
line-angle formula for this alkane and number the carbons on the chain
beginning at one end and proceeding to the other end.

Strategy and Solution

Hexane contains six carbons in a chain. Its line-angle formula is:

Problem 11.1

Following is a line-angle formula for an alkane. What are the name and
the molecular formula of this alkane?

11.3 What Are Constitutional Isomers?

Constitutional isomers Compounds
with the same molecular formula
but a different connectivity of their
atoms

Constitutional isomers have also
been called structural isomers, an
older term that is still in use.

Constitutional isomers are compounds that have the same molecular formula but different structural formulas. By "different structural formulas,"
we mean that they differ in the kinds of bonds (single, double, or triple) and/
or in the connectivity of their atoms. For the molecular formulas CH_4, C_2H_6,
and C_3H_8, only one connectivity of their atoms is possible, so there are no
constitutional isomers for these molecular formulas. For the molecular formula C_4H_{10}, however, two structural formulas are possible: In butane, the
four carbons are bonded in a chain; in 2-methylpropane, three carbons are
bonded in a chain and the fourth carbon is a branch on the chain. The two
constitutional isomers with the molecular formula C_4H_{10} are drawn here

both as condensed structural formulas and as line-angle formulas. Also shown are ball-and-stick models of each.

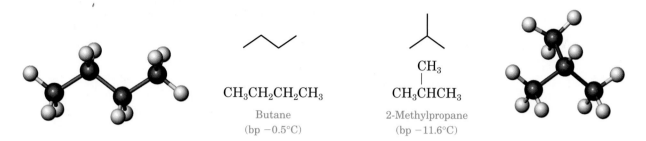

$$CH_3CH_2CH_2CH_3$$

Butane
(bp −0.5°C)

$$\underset{\text{2-Methylpropane}}{\overset{CH_3}{\underset{|}{CH_3CHCH_3}}}$$

2-Methylpropane
(bp −11.6°C)

Butane and 2-methylpropane are different compounds and have different physical and chemical properties. Their boiling points, for example, differ by approximately 11°C.

In Section 10.4, we encountered several examples of constitutional isomers, although we did not call them that at the time. We saw, for example, that there are two alcohols with the molecular formula C_3H_8O, two primary amines with the molecular formula C_3H_9N, two aldehydes with the molecular formula C_4H_8O, and two carboxylic acids with the molecular formula $C_4H_8O_2$.

To determine whether two or more structural formulas represent constitutional isomers, write the molecular formula of each and then compare them. All compounds that have the same molecular formula but different structural formulas are constitutional isomers.

Example 11.2 Constitutional Isomerism

Do the structural formulas in each of the following sets represent the same compound or constitutional isomers? (*Hint*: You will find it helpful to redraw each molecule as a line-angle formula, which will make it easier for you to see similarities and differences in molecular structure.)

(a) $CH_3CH_2CH_2CH_2CH_2CH_3$ and $\underset{\qquad CH_2CH_2CH_3}{CH_3CH_2CH_2}$ (Each is C_6H_{14})

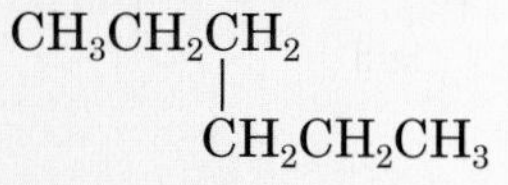

(b) 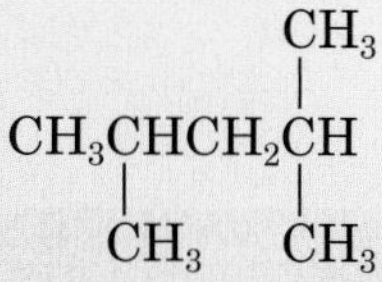 $\underset{\overset{|}{CH_3}}{\overset{\overset{CH_3}{|}}{CH_3CHCH_2CH}}$ and $\underset{\overset{|}{CH_3}}{\overset{\overset{CH_3}{|}}{CH_3CH_2CHCHCH_3}}$ (Each is C_7H_{16})

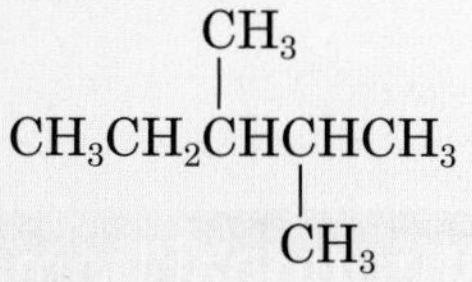

Strategy

First, find the longest chain of carbon atoms. It makes no difference whether the chain is drawn as straight or bent; as structural formulas are drawn in this problem, there is no attempt to show three-dimensional shapes. Second, number the longest chain from the end nearest the first branch. Third, compare the lengths of the two chains and the sizes and locations of any branches. Structural formulas that have the same molecular formula and the same connectivity of their atoms represent the same compound; those that have the same molecular formula but different connectivities of their atoms represent constitutional isomers.

Solution

(a) Each structural formula has an unbranched chain of six carbons; they are identical and represent the same compound.

$$CH_3CH_2CH_2CH_2CH_2CH_3 \quad \text{and} \quad CH_3CH_2CH_2$$
$$CH_2CH_2CH_3$$

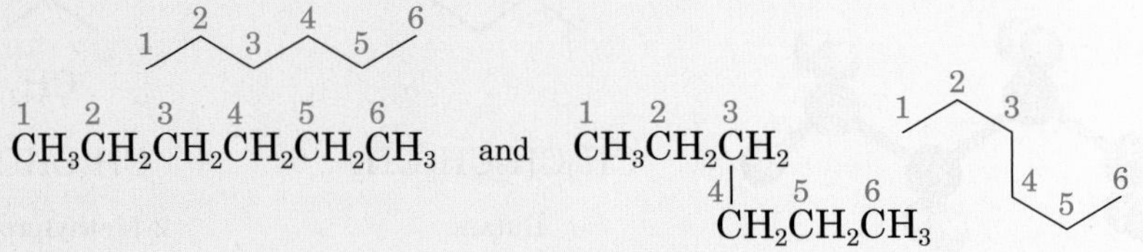

(b) Each structural formula has the same molecular formula, C_7H_{16}. In addition, each has a chain of five carbons with two CH_3 branches. Although the branches are identical, they are at different locations on the chains. Therefore, these structural formulas represent constitutional isomers.

$$CH_3CHCH_2CH \quad \text{and} \quad CH_3CH_2CHCHCH_3$$

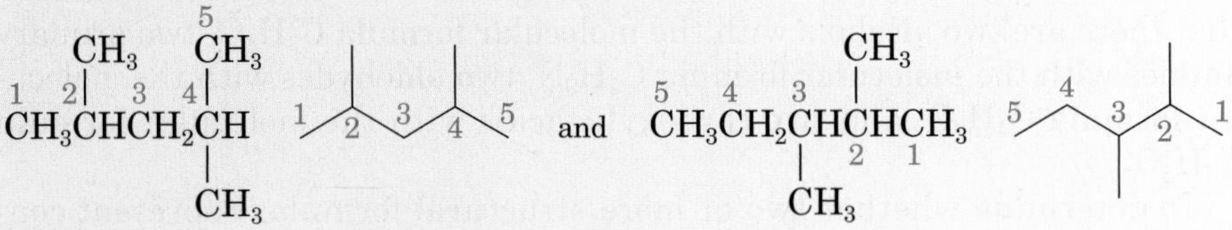

Problem 11.2

Do the line-angle formulas in each of the following sets represent the same compound or constitutional isomers?

(a) 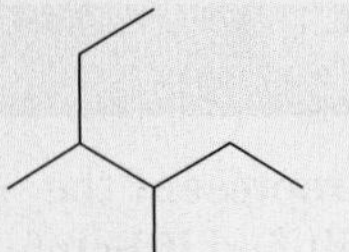 and

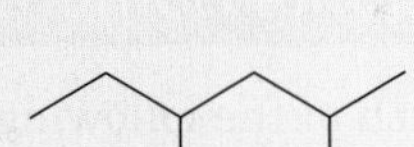

(b) 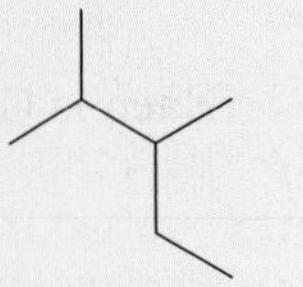 and

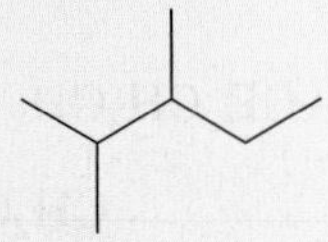

Example 11.3 Constitutional Isomerism

Draw line-angle formulas for the five constitutional isomers with the molecular formula C_6H_{14}.

Strategy

In solving problems of this type, you should devise a strategy and then follow it. Here is one possible strategy. First, draw a line-angle formula for the constitutional isomer with all six carbons in an unbranched chain. Then, draw line-angle formulas for all constitutional isomers with five carbons in a chain and one carbon as a branch on the chain. Finally, draw line-angle formulas for all constitutional isomers with four carbons in a chain and two carbons as branches.

Solution

Here are line-angle formulas for all constitutional isomers with six, five, and four carbons in the longest chain. No constitutional isomers for C_6H_{14} having only three carbons in the longest chain are possible.

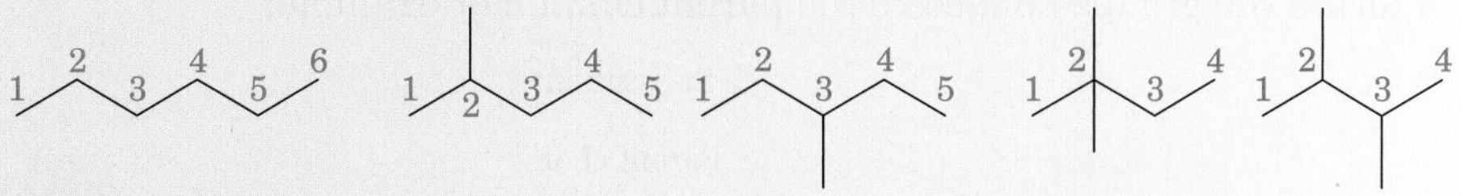

Six carbons in an unbranched chain

Five carbons in a chain; one carbon as a branch

Four carbons in a chain; two carbons as branches

Problem 11.3

Draw structural formulas for the three constitutional isomers with the molecular formula C_5H_{12}.

The ability of carbon atoms to form strong, stable bonds with other carbon atoms results in a staggering number of constitutional isomers, as the table to the right shows.

Thus, for even a small number of carbon and hydrogen atoms, a very large number of constitutional isomers is possible. In fact, the potential for structural and functional group individuality among organic molecules made from just the basic building blocks of carbon, hydrogen, nitrogen, and oxygen is practically limitless.

Molecular Formula	Number of Constitutional Isomers
CH_4	1
C_5H_{12}	3
$C_{10}H_{22}$	75
$C_{15}H_{32}$	4347
$C_{25}H_{52}$	36,797,588
$C_{30}H_{62}$	4,111,846,763

11.4 How Do We Name Alkanes?

A. The IUPAC System

Ideally, every organic compound should have a name from which its structural formula can be drawn. For this purpose, chemists have adopted a set of rules established by the **International Union of Pure and Applied Chemistry (IUPAC)**.

The IUPAC name for an alkane with an unbranched chain of carbon atoms consists of two parts: (1) a prefix that shows the number of carbon atoms in the chain and (2) the suffix **-ane**, which shows that the compound is a saturated hydrocarbon. Table 11.2 gives the prefixes used to show the presence of 1 to 20 carbon atoms.

The IUPAC chose the first four prefixes listed in Table 11.2 because they were well established long before the nomenclature was systematized. For example, the prefix *but-* appears in the name butyric acid, a compound of four carbon atoms formed by the air oxidation of butterfat (Latin: *butyrum*,

Table 11.2 Prefixes Used in the IUPAC System to Show the Presence of 1 to 20 Carbons in an Unbranched Chain

Prefix	Number of Carbon Atoms	Prefix	Number of Carbon Atoms	Prefix	Number of Carbon Atoms	Prefix	Number of Carbon Atoms
meth-	1	hex-	6	undec-	11	hexadec-	16
eth-	2	hept-	7	dodec-	12	heptadec-	17
prop-	3	oct-	8	tridec-	13	octadec-	18
but-	4	non-	9	tetradec-	14	nonadec-	19
pent-	5	dec-	10	pentadec-	15	eicos-	20

butter). The prefixes that denote five or more carbons are derived from Latin numbers. Table 11.1 gives the names, molecular formulas, and condensed structural formulas for the first ten alkanes with unbranched chains.

IUPAC names of alkanes with branched chains consist of a parent name that shows the longest chain of carbon atoms and substituent names that indicate the groups bonded to the parent chain. For example:

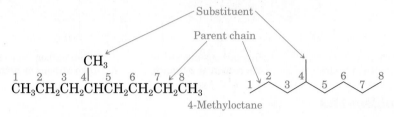

4-Methyloctane

Alkyl group A group derived by removing a hydrogen from an alkane; given the symbol R—

R— A symbol used to represent an alkyl group

A substituent group derived from an alkane by removal of a hydrogen atom is called an **alkyl group** and is commonly represented by the symbol **R—**. Alkyl groups are named by dropping the *-ane* from the name of the parent alkane and adding the suffix *-yl*. Table 11.3 gives the names and condensed structural formulas for eight of the most common alkyl groups. The prefix *sec-* is an abbreviation for "secondary," meaning a carbon bonded to two other carbons. The prefix *tert-* is an abbreviation for "tertiary," meaning a carbon bonded to three other carbons.

Table 11.3 Names of the Eight Most Common Alkyl Groups

Name	Condensed Structural Formula	Name	Condensed Structural Formula
methyl	$-CH_3$	butyl	$-CH_2CH_2CH_2CH_3$
ethyl	$-CH_2CH_3$	isobutyl	$-CH_2CHCH_3$ $\quad\ \ \ \ CH_3$
propyl	$-CH_2CH_2CH_3$	*sec*-butyl	$-CHCH_2CH_3$ $\quad\ CH_3$
isopropyl	$-CHCH_3$ $\quad\ CH_3$	*tert*-butyl	CH_3 $-CCH_3$ $\quad\ CH_3$

The rules of the IUPAC system for naming alkanes are as follows:

1. The name for an alkane with an unbranched chain of carbon atoms consists of a prefix showing the number of carbon atoms in the parent chain and the suffix *-ane*.

2. For branched-chain alkanes, take the longest chain of carbon atoms as the parent chain and its name becomes the root name.

3. Give each substituent on the parent chain a name and a number. The number shows the carbon atom of the parent chain to which the substituent is bonded. Use a hyphen to connect the number to the name.

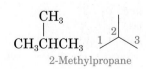

2-Methylpropane

4. If there is one substituent, number the parent chain from the end that gives the substituent the lower number.

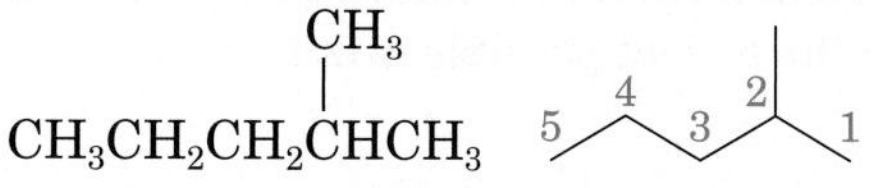

$$CH_3$$
$$|$$
$$CH_3CH_2CH_2CHCH_3$$

2-Methylpentane
(not 4-methylpentane)

5. If the same substituent occurs more than once, number the parent chain from the end that gives the lower number to the substituent encountered first. Indicate the number of times the substituent occurs by a prefix *di-, tri-, tetra-, penta-, hexa-*, and so on. Use a comma to separate position numbers.

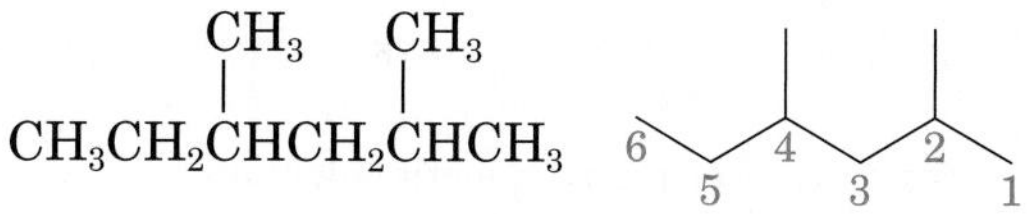

$$CH_3 \quad CH_3$$
$$| \qquad |$$
$$CH_3CH_2CHCH_2CHCH_3$$

2,4-Dimethylhexane
(not 3,5-dimethylhexane)

6. If there are two or more different substituents, list them in alphabetical order and number the chain from the end that gives the lower number to the substituent encountered first. If there are different substituents in equivalent positions on opposite ends of the parent chain, give the substituent of lower alphabetical order the lower number.

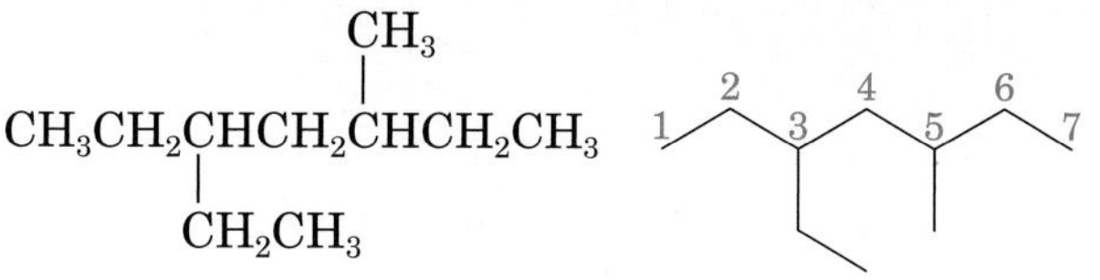

$$CH_3$$
$$|$$
$$CH_3CH_2CHCH_2CHCH_2CH_3$$
$$|$$
$$CH_2CH_3$$

3-Ethyl-5-methylheptane
(not 3-methyl-5-ethylheptane)

7. Do not include the prefixes *di-, tri-, tetra-*, and so, on or the hyphenated prefixes *sec-* and *tert-* in alphabetizing. Alphabetize the names of substituents first and then insert these prefixes. In the following example, the alphabetizing parts are **ethyl** and **methyl**, not *ethyl* and *dimethyl*.

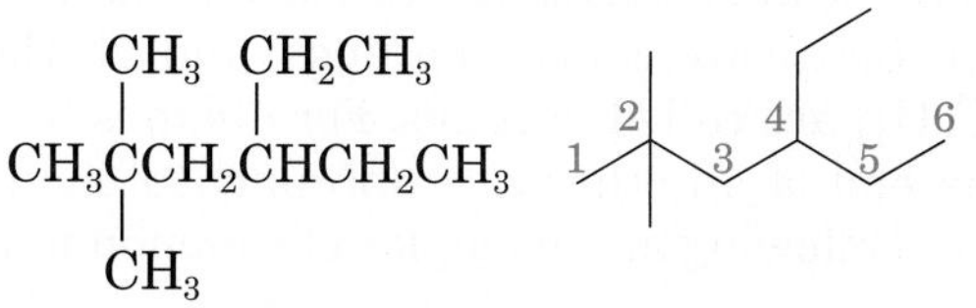

$$CH_3 \quad CH_2CH_3$$
$$| \qquad |$$
$$CH_3CCH_2CHCH_2CH_3$$
$$|$$
$$CH_3$$

4-Ethyl-2,2-dimethylhexane
(not 2,2-dimethyl-4-ethylhexane)

⬢WL **Interactive Example 11.4** IUPAC Names of Alkanes

Write the molecular formula and IUPAC name for each alkane.

(a) 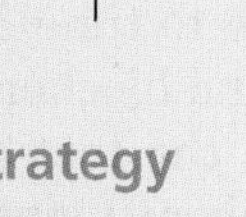   (b)

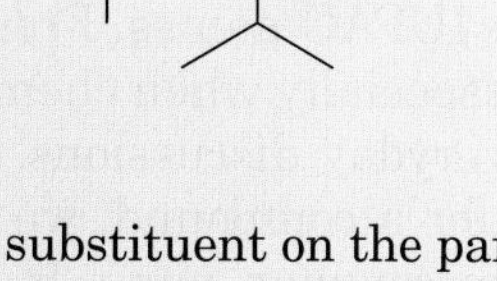

Strategy

If there is only one substituent on the parent chain, as in (a), number the parent chain from the end that gives the substituent the lowest possible number. If there are two or more substituents on the parent chain, as in (b), number the parent chain from the end that gives the substituent of lowest alphabetical order the lowest possible number.

Solution

The molecular formula of (a) is C_5H_{12}, and that of (b) is $C_{11}H_{24}$. In (a), number the longest chain from the end that gives the methyl substituent the lower number (rule 4). In (b), list isopropyl and methyl substituents in alphabetical order (rule 6).

(a) 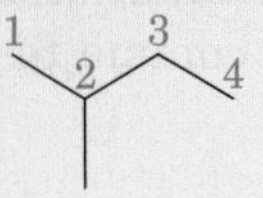   2-Methylbutane

(b) 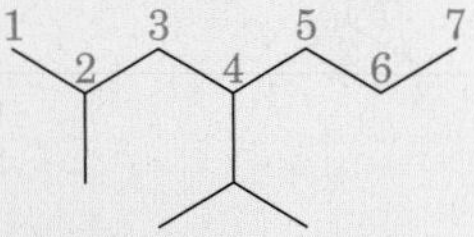   4-Isopropyl-2-methylheptane

Problem 11.4

Write the molecular formula and IUPAC name for each alkane.

(a) 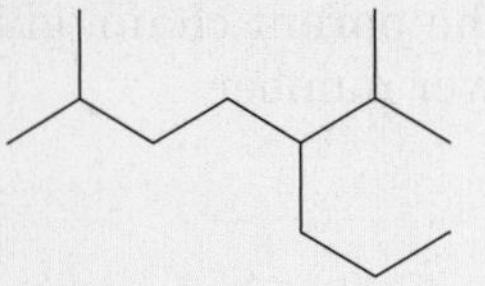   (b)

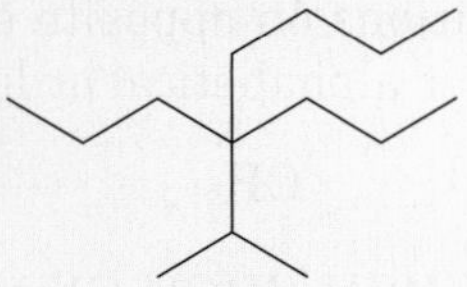

B. Common Names

In the older system of **common nomenclature**, the total number of carbon atoms in an alkane, regardless of their arrangement, determines the name.

The first three alkanes are methane, ethane, and propane. All alkanes with the molecular formula C_4H_{10} are called butanes, all those with the molecular formula C_5H_{12} are called pentanes, and all those with the molecular formula C_6H_{14} are called hexanes. For alkanes beyond propane, **iso** indicates that one end of an otherwise unbranched chain terminates in a $(CH_3)_2CH—$group. Following are examples of common names:

$$CH_3CH_2CH_2CH_3 \qquad CH_3\underset{\displaystyle |}{\overset{\displaystyle CH_3}{CH}}CH_3 \qquad CH_3CH_2CH_2CH_2CH_3 \qquad CH_3CH_2\underset{\displaystyle |}{\overset{\displaystyle CH_3}{CH}}CH_3$$

Butane   Isobutane   Pentane   Isopentane

This system of common names has no way of handling other branching patterns; therefore, for more complex alkanes, we must use the more flexible IUPAC system.

In this book, we concentrate on IUPAC names. From time to time, however, we also use common names, especially when chemists and biochemists use them almost exclusively in everyday discussions. When the text gives both IUPAC and common names for a compound, we will always give the IUPAC name first, followed by the common name in parentheses. In this way, you should have no doubt about which name is which.

11.5 Where Do We Obtain Alkanes?

The two major sources of alkanes are natural gas and petroleum. **Natural gas** consists of approximately 90 to 95% methane, 5 to 10% ethane, and a mixture of other relatively low-boiling alkanes—chiefly propane, butane, and 2-methylpropane.

Petroleum is a thick, viscous liquid mixture of thousands of compounds, most of them hydrocarbons, formed from the decomposition of marine plants and animals. Petroleum and petroleum-derived products fuel automobiles, aircraft, and trains. They provide most of the greases and lubricants required for the machinery utilized by our highly industrialized society. Furthermore, petroleum, along with natural gas, provides nearly 90% of the organic raw materials for the synthesis and manufacture of synthetic fibers, plastics, detergents, drugs, dyes, and a multitude of other products.

The fundamental separation process in refining petroleum is fractional distillation (Figure 11.2). Practically all crude petroleum that enters a refinery goes to distillation units, where it is heated to temperatures as high as 370 to 425°C and separated into fractions. Each fraction contains a mixture of hydrocarbons that boils within a particular range.

11.6 What Are Cycloalkanes?

A hydrocarbon that contains carbon atoms joined to form a ring is called a **cyclic hydrocarbon**. When all carbons of the ring are saturated (only carbon–carbon single bonds are present), the hydrocarbon is called a **cycloalkane**. Cycloalkanes of ring sizes ranging from 3 to more than 30 carbon atoms are found in nature, and in principle there is no limit to ring size. Five-membered (cyclopentane) and six-membered (cyclohexane) rings are especially abundant in nature; for this reason, we concentrate on them in this book.

A petroleum fractional distillation tower.

Cycloalkane A saturated hydrocarbon that contains carbon atoms bonded to form a ring

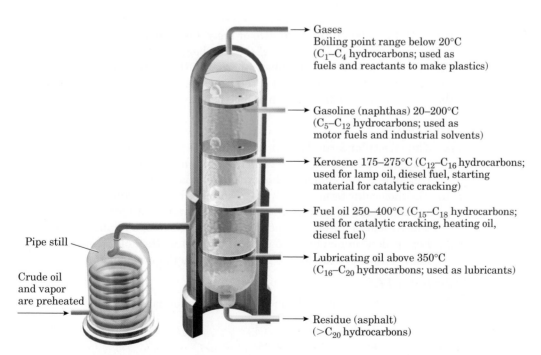

Gases
Boiling point range below 20°C
(C_1–C_4 hydrocarbons; used as fuels and reactants to make plastics)

Gasoline (naphthas) 20–200°C
(C_5–C_{12} hydrocarbons; used as motor fuels and industrial solvents)

Kerosene 175–275°C (C_{12}–C_{16} hydrocarbons; used for lamp oil, diesel fuel, starting material for catalytic cracking)

Fuel oil 250–400°C (C_{15}–C_{18} hydrocarbons; used for catalytic cracking, heating oil, diesel fuel)

Lubricating oil above 350°C (C_{16}–C_{20} hydrocarbons; used as lubricants)

Residue (asphalt) (>C_{20} hydrocarbons)

Pipe still

Crude oil and vapor are preheated

FIGURE 11.2 Fractional distillation of petroleum. The lighter, more volatile fractions are removed from higher up the column; the heavier, less volatile fractions are removed from lower down.

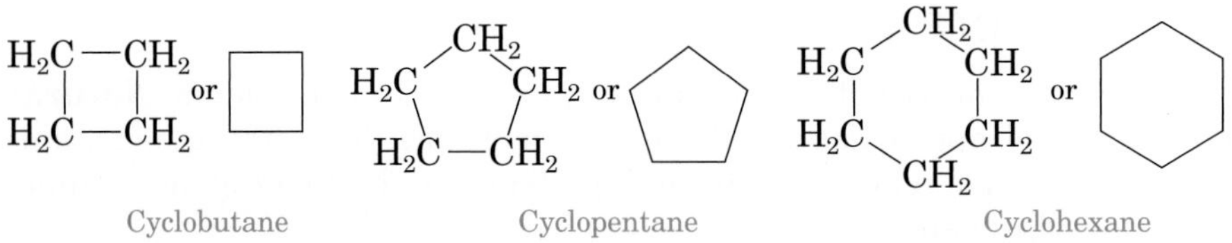

FIGURE 11.3 Examples of cycloalkanes.

Cyclobutane   Cyclopentane   Cyclohexane

Organic chemists rarely show all carbons and hydrogens when writing structural formulas for cycloalkanes. Rather, we use line-angle formulas to represent cycloalkane rings and represent each ring by a regular polygon having the same number of sides as there are carbon atoms in the ring. For example, we represent cyclobutane by a square, cyclopentane by a pentagon, and cyclohexane by a hexagon (Figure 11.3).

To name a cycloalkane, prefix the name of the corresponding open-chain alkane with *cyclo-* and name each substituent on the ring. If there is only one substituent on the ring, there is no need to give it a location number. If there are two or more substituents, number the ring beginning with the substituent of lower alphabetical order.

Example 11.5 IUPAC Names of Cycloalkanes

Write the molecular formula and IUPAC name for each cycloalkane.

(a)

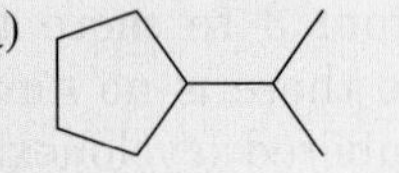

(b)

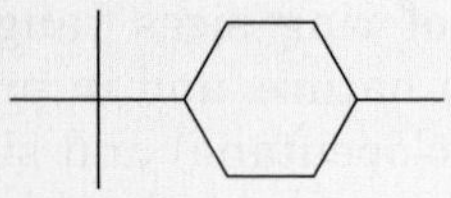

Strategy

For cycloalkanes, the parent name of the ring is the prefix *cyclo-*, plus the name of the alkane with the same number of carbon atoms as are in the ring. If there is only one substituent on the ring, as in (a), there is no need to give it a number. If there are two or more substituents on the ring, as in (b), number the carbon atoms of the ring beginning at the carbon with the substituent of lowest alphabetical order. If there are three or more substituents, number the atoms of the ring so as to give the substituents the lowest set of numbers and then list them in alphabetical order.

Solution

(a) The molecular formula of this compound is C_8H_{16}. Because there is only one substituent, there is no need to number the atoms of the ring. The IUPAC name of this cycloalkane is isopropylcyclopentane.

(b) The molecular formula is $C_{11}H_{22}$. To name this compound, first number the atoms of the cyclohexane ring beginning with *tert*-butyl, the substituent of lower alphabetical order (remember, alphabetical order here is determined by the *b* of butyl and not by the *t* of *tert*-). The name of this cycloalkane is 1-*tert*-butyl-4-methylcyclohexane.

Problem 11.5

Write the molecular formula and IUPAC name for each cycloalkane.

(a)

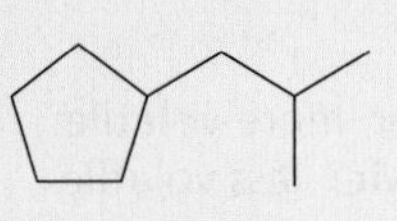

(b)

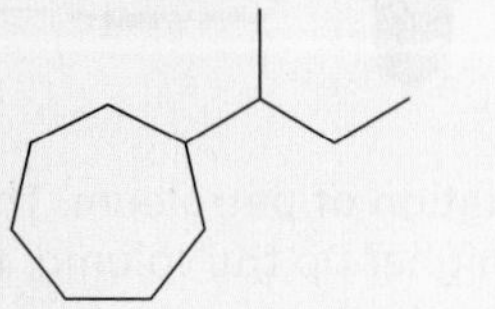

(c)

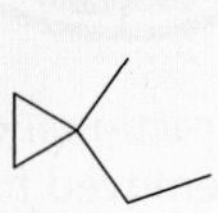

11.7 What Are the Shapes of Alkanes and Cycloalkanes?

In this section, we concentrate on ways to visualize molecules as three-dimensional objects and to visualize bond angles and relative distances between various atoms and functional groups within a molecule. We urge you to build molecular models of these compounds and to study and manipulate those models. Organic molecules are three-dimensional objects, and it is essential that you become comfortable in dealing with them as such.

At this point, you should review Section 3.10 and use of the valence-shell electron-pair repulsion (VSEPR) model to predict bond angles and shapes of molecules.

A. Alkanes

Although the VSEPR model gives us a way to predict the geometry about each carbon atom in an alkane, it provides us with no information about the three-dimensional shape of an entire molecule. There is, in fact, free rotation about each carbon–carbon bond in an alkane. As a result, even a molecule as simple as ethane has an infinite number of possible three-dimensional shapes, or **conformations**.

Figure 11.4 shows three conformations for a butane molecule. Conformation (a) is the most stable because the methyl groups at the ends of the four-carbon chain are farthest apart. Conformation (b) is formed by a rotation of 120° about the single bond joining carbons 2 and 3. In this conformation, some crowding of groups occurs because the two methyl groups are closer together than they are in conformation (a). Rotation about the C_2—C_3 single bond by another 60° gives conformation (c), which is the most crowded because the two methyl groups face each other.

Figure 11.4 shows only three of the possible conformations for a butane molecule. In fact, there are an infinite number of possible conformations that differ only in the angles of rotation about the various C—C bonds within the molecule. In an actual sample of butane, the conformation of each molecule constantly changes as a result of the molecule's collisions with other butane molecules and with the walls of the container. Even so, at any given time, a majority of butane molecules are in the most stable, fully extended conformation. There are the fewest butane molecules in the most crowded conformation.

To summarize, for any alkane (except, of course, for methane), there are an infinite number of conformations. The majority of molecules in any sample will be in the least crowded conformation; the fewest will be in the most crowded conformation.

Conformation Any three-dimensional arrangement of atoms in a molecule that results from rotation about a single bond

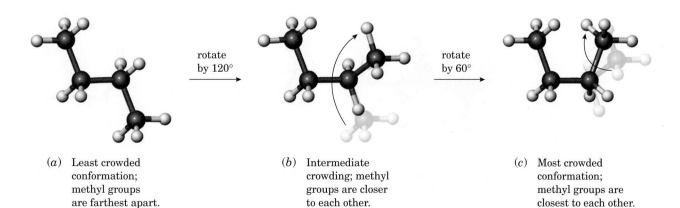

(a) Least crowded conformation; methyl groups are farthest apart.

rotate by 120°

(b) Intermediate crowding; methyl groups are closer to each other.

rotate by 60°

(c) Most crowded conformation; methyl groups are closest to each other.

FIGURE 11.4 Three conformations of a butane molecule.

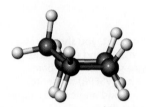

FIGURE 11.5 The most stable conformation of cyclopentane.

B. Cycloalkanes

We limit our discussion to the conformations of cyclopentanes and cyclo-hexanes because they are the carbon rings most commonly found in nature. Nonplanar or puckered conformations are favored in all cycloalkanes larger than cyclopropane.

Cyclopentane

The most stable conformation of cyclopentane is the **envelope conforma-tion** shown in Figure 11.5. In it, four carbon atoms are in a plane and the fifth carbon atom is bent out of the plane, like an envelope with its flap bent upward. All bond angles in cyclopentane are approximately 109.5°.

Cyclohexane

The most stable conformation of cyclohexane is the **chair conformation**, in which all bond angles are approximately 109.5°.

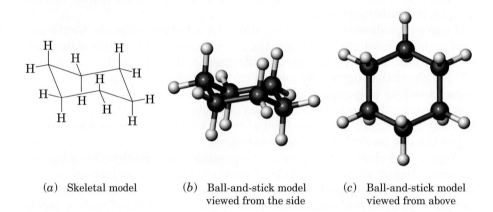

(a) Skeletal model

(b) Ball-and-stick model
     viewed from the side

(c) Ball-and-stick model
     viewed from above

Equatorial position A position on a chair conformation of a cyclohexane ring that extends from the ring roughly perpendicular to the imaginary axis of the ring

Axial position A position on a chair conformation of a cyclohexane ring that extends from the ring parallel to the imaginary axis of the ring

In a chair conformation, the 12 C—H bonds are arranged in two different orientations. Six of them are **axial bonds**, and the other six are **equatorial bonds**. One way to visualize the difference between these two types of bonds is to imagine an axis running through the center of the chair (Figure 11.6). Axial bonds are oriented parallel to this axis. Three of the axial bonds point up; the other three point down. Notice also that axial bonds alternate, first up and then down, as you move from one carbon to the next.

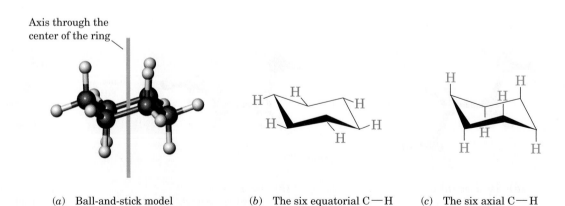

(a) Ball-and-stick model
     showing all 12 hydrogens

(b) The six equatorial C—H
     bonds shown in red

(c) The six axial C—H
     bonds shown in blue

FIGURE 11.6 Chair conformation of cyclohexane showing equatorial and axial C—H bonds.

Equatorial bonds are oriented approximately perpendicular to the imaginary axis of the ring and also alternate first slightly up and then slightly down as you move from one carbon to the next. Notice also that if the axial bond on a carbon points upward, the equatorial bond on that carbon points slightly downward. Conversely, if the axial bond on a particular carbon points downward, the equatorial bond on that carbon points slightly upward.

Finally, notice that each equatorial bond is oriented parallel to two ring bonds on opposite sides of the ring. A different pair of parallel C—H bonds is shown in each of the following structural formulas, along with the two ring bonds to which each pair is parallel.

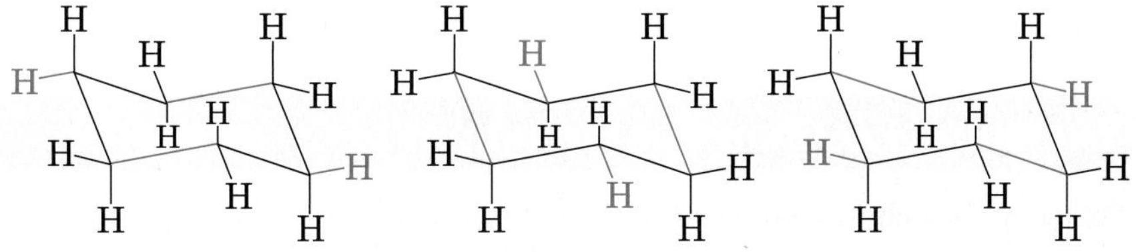

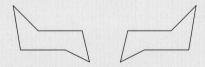

How To . . .

Draw Alternative Chair Conformations of Cyclohexane

You will be asked frequently to draw three-dimensional representations of chair conformations of cyclohexane and to show spatial relationships among atoms and groups of atoms bonded to the ring. Here are four steps that will help you to draw them. With a little practice, you will find it easy to draw them.

Step 1: Draw two sets of parallel lines, one line in each set offset from the other in the set as shown.

Step 2: Complete the chair by drawing the head and foot pieces, one up and the other down.

Step 3: Draw the equatorial bonds using ring bonds as a guide. Remember that each equatorial bond is parallel to two ring bonds and that equatorial bonds on opposite carbons of the ring are parallel to one another. Sets of parallel equatorial bonds are shown here in color.

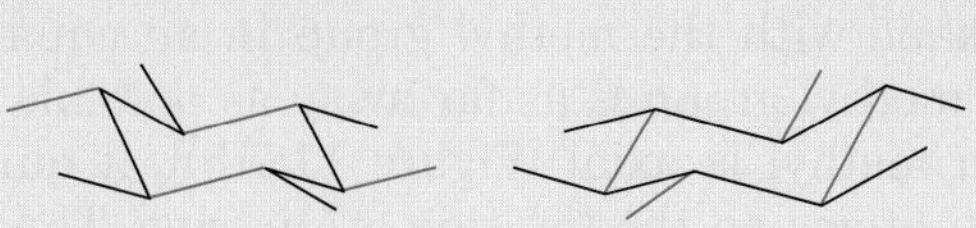

Equatorial bonds

Step 4: Draw the six axial bonds as vertical lines. Remember that all axial bonds are parallel to each other. Sets of parallel axial bonds are shown in color.

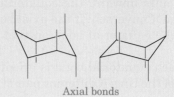

Axial bonds

Example 11.6 Chair Conformations of Cyclohexanes

Following is a chair conformation of methylcyclohexane showing a methyl group and one hydrogen. Indicate by a label whether each is equatorial or axial.

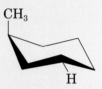

Strategy

Equatorial bonds are approximately perpendicular to the imaginary axis of the ring and form an equator about the ring. Axial bonds are parallel to the imaginary axis of the ring.

Solution

The methyl group is axial, and the hydrogen is equatorial.

Problem 11.6

Following is a chair conformation of cyclohexane with carbon atoms numbered 1 through 6. Draw methyl groups that are equatorial on carbons 1, 2, and 4.

    Suppose that —CH_3 or another group on a cyclohexane ring occupies either an equatorial or an axial position. Chemists have discovered that a six-membered ring is more stable when the maximum number of substituent groups are equatorial. Perhaps the simplest way to confirm this relationship is to examine molecular models. Figure 11.7(a) shows a space-filling model of methylcyclohexane with the methyl group in an equatorial position. In this position, the methyl group is as far away as possible from other atoms of the ring. When methyl is axial [Figure 11.7(b)], it quite literally bangs into two hydrogen atoms on the top side of the ring. Thus, the more stable conformation of a substituted cyclohexane ring has the substituent group(s) as equatorial.

The Poisonous Puffer Fish

Nature is by no means limited to carbon in six-membered rings. Tetrodotoxin, one of the most potent toxins known, is composed of a set of interconnected six-membered rings, each in a chair conformation. All but one of these rings contains atoms other than carbon.

Tetrodotoxin is produced in the liver and ovaries of many species of *Tetraodontidae*, one of which is the puffer fish, so called because it inflates itself to an almost spherical spiny ball when alarmed. It is evidently a species highly preoccupied with defense, but the Japanese are not put off by its prickly appearance. They regard the puffer, called *fugu* in Japanese, as a delicacy. To serve it in a public restaurant, a chef must be registered as sufficiently skilled in removing the toxic organs so as to make the flesh safe to eat.

Tetrodotoxin blocks the sodium ion channels, which are essential for neurotransmission (Section 24.3). This blockage prevents communication between neurons and muscle cells and results in weakness, paralysis, and eventual death.

A puffer fish with its body inflated.

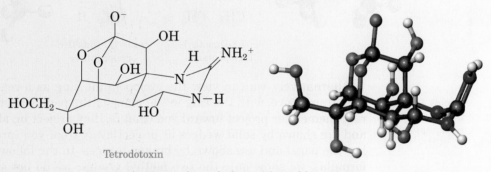

Tetrodotoxin

Test your knowledge with Problem 11.59.

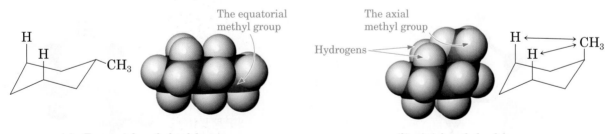

(a) Equatorial methylcyclohexane

(b) Axial methylcyclohexane

FIGURE 11.7 Methylcyclohexane. The three hydrogens of the methyl group are shown in green to make them stand out more clearly.

11.8 What Is *Cis-Trans* Isomerism in Cycloalkanes?

Cycloalkanes with substituents on two or more carbons of the ring show a type of isomerism called ***cis-trans* isomerism**. Cycloalkane *cis-trans* isomers have (1) the same molecular formula and (2) the same connectivity of their atoms, but (3) a different arrangement of their atoms in space because of restricted rotation around the carbon–carbon single bonds of the ring.

Cis-trans isomers Isomers that have the same connectivity of their atoms but a different arrangement of their atoms in space due to the presence of either a ring or a carbon–carbon double bond

Planar representations of five- and six-membered rings are not spatially accurate because these rings normally exist as envelope and chair conformations. Planar representations are, however, adequate for showing *cis-trans* isomerism.

We study *cis-trans* isomerism in cycloalkanes in this chapter and that of alkenes in Chapter 12.

We can illustrate *cis-trans* isomerism in cycloalkanes by using 1,2-dimethylcyclopentane as an example. In the following structural formulas, the cyclopentane ring is drawn as a planar pentagon viewed through the plane of the ring. (In determining the number of *cis-trans* isomers in a substituted cycloalkane, it is adequate to draw the cycloalkane ring as a planar polygon.) Carbon–carbon bonds of the ring projecting toward you are shown as heavy lines. When viewed from this perspective, substituents bonded to the cyclopentane ring project above and below the plane of the ring. In one isomer of 1,2-dimethylcyclopentane, the methyl groups are on the same side of the ring (either both above or both below the plane of the ring); in the other isomer, they are on opposite sides of the ring (one above and one below the plane of the ring).

The prefix *cis* (Latin: on the same side) indicates that the substituents are on the same side of the ring; the prefix *trans* (Latin: across) indicates that they are on opposite sides of the ring.

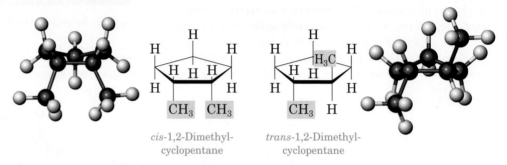

cis-1,2-Dimethyl-cyclopentane   *trans*-1,2-Dimethyl-cyclopentane

Occasionally hydrogen atoms are written before the carbon, H_3C- to avoid crowding or to emphasize the C—C bond, as in H_3C-CH_3.

Alternatively, we can view the cyclopentane ring as a regular pentagon seen from above, with the ring in the plane of the page. Substituents on the ring then either project toward you (that is, they project up above the page) and are shown by solid wedges or project away from you (project down below the page) and are shown by broken wedges. In the following structural formulas, we show only the two methyl groups; we do not show hydrogen atoms of the ring.

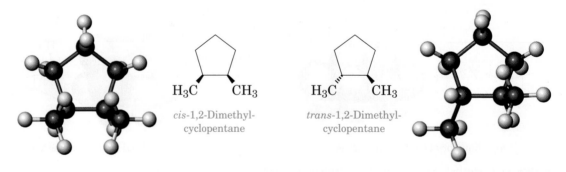

cis-1,2-Dimethyl-cyclopentane   *trans*-1,2-Dimethyl-cyclopentane

Stereocenter A tetrahedral atom, most commonly carbon, at which exchange of two groups produces a stereoisomer

Configuration Refers to the arrangement of atoms about a stereocenter; that is, to the relative arrangement of parts of a molecule in space

We say that 1,2-dimethylcyclopentane has two stereocenters. A **stereocenter** is a tetrahedral atom, most commonly carbon, at which exchange of two groups produces a stereoisomer. Both carbons 1 and 2 of 1,2-dimethylcyclopentane, for example, are stereocenters; in this molecule, exchange of H and CH_3 groups at either stereocenter converts a *trans* isomer to a *cis* isomer, or vice versa.

Alternatively, we can refer to the stereoisomers of 1,2-dimethylcyclobutane as having either a *cis* or a *trans* configuration. **Configuration** refers to the arrangement of atoms about a stereocenter. We say, for example, that exchange of groups at either stereocenter in the *cis* configuration gives the isomer with the *trans* configuration.

Cis and *trans* isomers are also possible for 1,2-dimethylcyclohexane. We can draw a cyclohexane ring as a planar hexagon and view it through the plane of the ring. Alternatively, we can view it as a regular hexagon viewed from above with substituent groups pointing toward us, shown by solid wedges, or pointing away from us, shown by broken wedges.

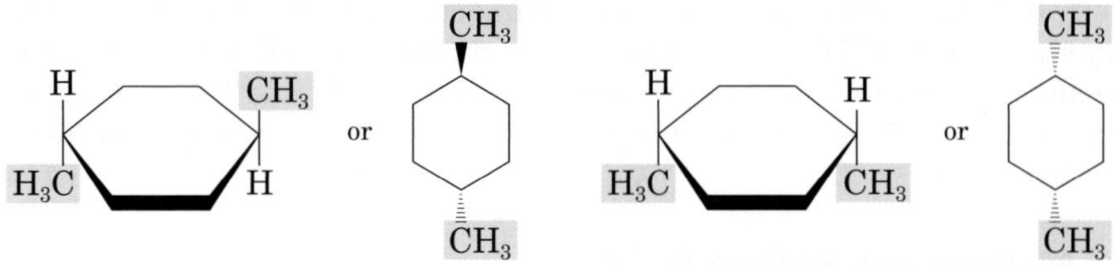

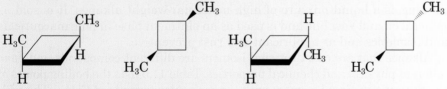

trans-1,4-Dimethylcyclohexane   *cis*-1,4-Dimethylcyclohexane

Because *cis-trans* isomers differ in the orientation of their atoms in space, they are **stereoisomers**. *Cis-trans* isomerism is one type of stereoisomerism. We will study another type of stereoisomerism, called enantiomerism, in Chapter 15.

Stereoisomers Isomers that have the same connectivity of their atoms but a different orientation of their atoms in space

Example 11.7 *Cis-Trans* Isomerism in Cycloalkanes

Which of the following cycloalkanes show *cis-trans* isomerism? For each that does, draw both isomers.

(a) Methylcyclopentane   (b) 1,1-Dimethylcyclobutane

(c) 1,3-Dimethylcyclobutane

Strategy

For a cycloalkane to show *cis-trans* isomerism, it must have at least two substituents, each on a different carbon of the ring.

Solution

(a) Methylcyclopentane does not show *cis-trans* isomerism; it has only one substituent on the ring.

(b) 1,1-Dimethylcyclobutane does not show *cis-trans* isomerism because only one arrangement is possible for the two methyl groups. Because both methyl groups are bonded to the same carbon, they must be *trans* to each other—one above the ring, the other below it.

(c) 1,3-Dimethylcyclobutane shows *cis-trans* isomerism. The two methyl groups may be *cis*, or they may be *trans*.

cis-1,3-Dimethylcyclobutane   *trans*-1,3-Dimethylcyclobutane

Problem 11.7

Which of the following cycloalkanes show *cis-trans* isomerism? For each that does, draw both isomers.

(a) 1,3-Dimethylcyclopentane

(b) Ethylcyclopentane

(c) 1,3-Dimethylcyclohexane

11.9　What Are the Physical Properties of Alkanes and Cycloalkanes?

The most important physical property of alkanes and cycloalkanes is their almost complete lack of polarity. We saw in Section 3.4B that the electronegativity difference between carbon and hydrogen is $2.5 - 2.1 = 0.4$ on the Pauling scale. Given this small difference, we classify a C—H bond as nonpolar covalent. Therefore, alkanes are nonpolar compounds and the only interactions between their molecules are the very weak London dispersion forces (Section 5.7A).

A. Melting and Boiling Points

The boiling points of alkanes are lower than those of almost any other type of compound with the same molecular weight. In general, both boiling and melting points of alkanes increase with increasing molecular weight (Table 11.4).

Table 11.4　Physical Properties of Some Unbranched Alkanes

Name	Condensed Structural Formula	Molecular Weight (amu)	Melting Point (°C)	Boiling Point (°C)	Density of Liquid (g/mL at 0°C)*
methane	CH_4	16.0	−182	−164	(a gas)
ethane	CH_3CH_3	30.1	−183	−88	(a gas)
propane	$CH_3CH_2CH_3$	44.1	−190	−42	(a gas)
butane	$CH_3(CH_2)_2CH_3$	58.1	−138	0	(a gas)
pentane	$CH_3(CH_2)_3CH_3$	72.2	−130	36	0.626
hexane	$CH_3(CH_2)_4CH_3$	86.2	−95	69	0.659
heptane	$CH_3(CH_2)_5CH_3$	100.2	−90	98	0.684
octane	$CH_3(CH_2)_6CH_3$	114.2	−57	126	0.703
nonane	$CH_3(CH_2)_7CH_3$	128.3	−51	151	0.718
decane	$CH_3(CH_2)_8CH_3$	142.3	−30	174	0.730

*For comparison, the density of H_2O is 1.000 g/mL at 4°C.

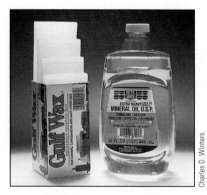

Paraffin wax and mineral oil are mixtures of alkanes.

Charles D. Winters

Alkanes containing 1 to 4 carbons are gases at room temperature. Alkanes containing 5 to 17 carbons are colorless liquids. High-molecular-weight alkanes (those containing 18 or more carbons) are white, waxy solids. Several plant waxes are high-molecular-weight alkanes. The wax found in apple skins, for example, is an unbranched alkane with the molecular formula $C_{27}H_{56}$. Paraffin wax, a mixture of high-molecular-weight alkanes, is used for wax candles, in lubricants, and to seal home-canned jams, jellies, and other preserves. Petrolatum, so named because it is derived from petroleum refining, is a liquid mixture of high-molecular-weight alkanes. It is sold as mineral oil and Vaseline and is used as an ointment base in pharmaceuticals and cosmetics and as a lubricant and rust preventive.

Alkanes that are constitutional isomers are different compounds and have different physical and chemical properties. Table 11.5 lists the boiling points of the five constitutional isomers with the molecular formula of C_6H_{14}. The boiling point of each branched-chain isomer is lower than that of hexane itself; the more branching, the lower the boiling point. These differences in boiling points are related to molecular shape in the following way. As branching increases, the alkane molecule becomes more compact and its surface area decreases. As we learned in Section 5.7A, as surface area decreases, London dispersion forces act over a smaller surface area. Hence, the attraction between molecules decreases and boiling point decreases. Thus, for any group of alkane constitutional isomers, the least-branched isomer generally has the highest boiling point and the most-branched isomer generally has the lowest boiling point.

Table 11.5 Boiling Points of the Five Isomeric Alkanes with the Molecular Formula C_6H_{14}

Name	bp (°C)
hexane	68.7
3-methylpentane	63.3
2-methylpentane	60.3
2,3-dimethylbutane	58.0
2,2-dimethylbutane	49.7

Hexane
(bp 68.7°)

Larger surface area, an increase in London dispersion forces, and a higher boiling point

2,2-Dimethylbutane
(bp 49.7°)

Smaller surface area, a decrease in London dispersion forces, and a lower boiling point

B. Solubility: A Case of "Like Dissolves Like"

Because alkanes are nonpolar compounds, they are not soluble in water, which dissolves only ionic and polar compounds. Recall that water is a polar substance and that its molecules associate with one another through hydrogen bonding (Section 6.6D). Alkanes do not dissolve in water because they cannot form hydrogen bonds with water. Alkanes, however, are soluble in each other, an example of "like dissolves like" (Section 6.4A). Alkanes are also soluble in other nonpolar organic compounds, such as toluene and diethyl ether.

C. Density

The average density of the liquid alkanes listed in Table 11.4 is about 0.7 g/mL; that of higher-molecular-weight alkanes is about 0.8 g/mL. All liquid and solid alkanes are less dense than water (1.000 g/mL), and because they are insoluble in water, they float on water.

Example 11.8 Physical Properties of Alkanes

Arrange the alkanes in each set in order of increasing boiling point.
(a) Butane, decane, and hexane
(b) 2-Methylheptane, octane, and 2,2,4-trimethylpentane

Strategy

The compounds in each set are alkanes, and the only forces of attraction between alkane molecules are very weak London dispersion forces. As the number of carbons in a hydrocarbon chain increases, London dispersion forces between chains increase; therefore, boiling point also increases (Section 5.7A). For alkanes that are constitutional isomers, the strength of London dispersion forces between molecules depends on shape. The more compact the shape, the weaker the intermolecular forces of attraction and the lower the boiling point.

Solution

(a) All three compounds are unbranched alkanes. Decane has the longest carbon chain, the strongest London forces between its molecules, and the highest boiling point. Butane has the shortest carbon chain and the lowest boiling point.

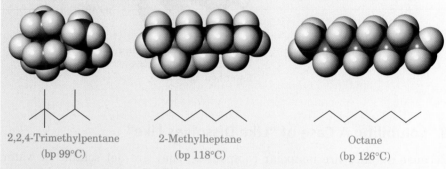

Butane   Hexane   Decane
bp −0.5°C   bp 69°C   bp 174°C

(b) These three alkanes are constitutional isomers with the molecular formula C_8H_{18}. 2,2,4-Trimethylpentane is the most highly branched isomer and therefore has the smallest surface area and the lowest boiling point. Octane, the unbranched isomer, has the largest surface area and the highest boiling point.

2,2,4-Trimethylpentane   2-Methylheptane   Octane
(bp 99°C)   (bp 118°C)   (bp 126°C)

Problem 11.8

Arrange the alkanes in each set in order of increasing boiling point.
(a) 2-Methylbutane, pentane, and 2,2-dimethylpropane
(b) 3,3-Dimethylheptane, nonane, and 2,2,4-trimethylhexane

11.10 What Are the Characteristic Reactions of Alkanes?

The most important chemical property of alkanes and cycloalkanes is their inertness. They are quite unreactive under the normal ionic reaction conditions we studied in Chapters 5 and 8. Under certain conditions, however, alkanes react with oxygen, O_2. By far, their most important reaction with oxygen is oxidation (combustion) to form carbon dioxide and water. They also react with bromine and chlorine to form halogenated hydrocarbons.

A. Reaction with Oxygen: Combustion

Oxidation of hydrocarbons, including alkanes and cycloalkanes, is the basis for their use as energy sources for heat [natural gas, liquefied petroleum gas (LPG), and fuel oil] and power (gasoline, diesel, and aviation fuel). Following are balanced equations for the complete combustion of methane, the major component of natural gas, and propane, the major component of LPG or bottled gas. The heat liberated when an alkane is oxidized to carbon dioxide and water is called its heat of combustion (Section 4.8).

$$CH_4 \;+\; 2O_2 \longrightarrow CO_2 + 2H_2O \;+\; 212 \text{ kcal/mol}$$
Methane

$$CH_3CH_2CH_3 + 5O_2 \longrightarrow 3CO_2 + 4H_2O \;+\; 530 \text{ kcal/mol}$$
Propane

Chemical Connections 11B

Octane Rating: What Those Numbers at the Pump Mean

Gasoline is a complex mixture of C_6 to C_{12} hydrocarbons. The quality of gasoline as a fuel for internal combustion engines is expressed in terms of an octane rating. Engine knocking occurs when a portion of the air–fuel mixture explodes prior to the piston reaching the top of its stroke (usually as a result of heat developed during the compression) and independent of ignition by the spark plug. The resulting shockwave of the piston against the cylinder wall reverberates, creating a characteristic metallic "pinging" sound.

Two compounds were selected as reference fuels for rating gasoline quality. One of these, 2,2,4-trimethylpentane (isooctane) has very good antiknock properties and was assigned an octane rating of 100. Heptane, the other reference compound, has poor antiknock properties and was assigned an octane rating of 0.

The **octane rating** of a particular gasoline is the percent of 2,2,4-trimethylpentane in a mixture of 2,2,4-trimethylpentane and heptane that has antiknock properties equivalent to that of the test gasoline. For example, the antiknock properties of 2-methylhexane are equivalent to those of a mixture of 42% 2,2,4-trimethylpentane and 58% heptane; therefore, the octane rating of 2-methylhexane is 42. Ethanol, which is added to gasoline to produce gasohol, has an octane rating of 105. Octane itself has an octane rating of −20.

Typical octane ratings of commonly available gasolines.

2,2,4-Trimethylpentane
(octane rating 100)

Heptane
(octane rating 0)

Test your knowledge with Problems 11.60–11.62.

B. Reaction with Halogens: Halogenation

If we mix methane with chlorine or bromine in the dark at room temperature, nothing happens. If, however, we heat the mixture to 100°C or higher or expose it to light, a reaction begins at once. The products of the reaction between methane and chlorine are chloromethane and hydrogen chloride. What occurs is a substitution reaction—in this case, the substitution of chlorine for hydrogen in methane.

$$CH_4 + Cl_2 \xrightarrow{\text{heat or light}} CH_3Cl + HCl$$

Methane — Chloromethane (Methyl chloride)

If chloromethane is allowed to react with more chlorine, further chlorination produces a mixture of dichloromethane, trichloromethane, and tetrachloromethane.

$$CH_3Cl + Cl_2 \xrightarrow{\text{heat}} CH_2Cl_2 + HCl$$

Chloromethane (Methyl chloride) — Dichloromethane (Methylene chloride)

$$CH_2Cl_2 \xrightarrow[\text{heat}]{Cl_2} CHCl_3 \xrightarrow[\text{heat}]{Cl_2} CCl_4$$

Dichloromethane (Methylene chloride) — Trichloromethane (Chloroform) — Tetrachloromethane (Carbon tetrachloride)

In the last equation, the reagent Cl_2 is placed over the reaction arrow and the equivalent amount of HCl formed is not shown. Placing reagents over reaction arrows and omitting by-products is commonly done to save space.

We derive IUPAC names of haloalkanes by naming the halogen atom as a substituent (*fluoro-, chloro-, bromo-,* and *iodo-*) and alphabetizing it along with other substituents. Common names consist of the common name of the alkyl group followed by the name of the halogen (chloride, bromide, and so forth) as a separate word. Dichloromethane (methylene chloride) is a widely used solvent for organic compounds.

Example 11.9 Halogenation of Alkanes

Write a balanced equation for the reaction of ethane with chlorine to form chloroethane, C_2H_5Cl.

Strategy

The reaction of ethane with chlorine results in the substitution of one of the hydrogen atoms of ethane with a chlorine atom.

Solution

$$CH_3CH_3 + Cl_2 \xrightarrow{\text{heat or light}} CH_3CH_2Cl + HCl$$

Ethane · · · · · · · · · · · · · · · · · · · Chloroethane
(Ethyl chloride)

Problem 11.9

Reaction of propane with chlorine gives two products, each with the molecular formula C_3H_7Cl. Draw structural formulas for these two compounds and give each an IUPAC name and a common name.

11.11 What Are Some Important Haloalkanes?

One of the major uses of haloalkanes is as intermediates in the synthesis of other organic compounds. Just as we can replace a hydrogen atom of an alkane, we can, in turn, replace the halogen atom with a number of other functional groups. In this way, we can construct more complex molecules. In contrast, alkanes that contain several halogens are often quite unreactive, a fact that has proved especially useful in the design of several classes of consumer products.

A. Chlorofluorocarbons

Of all the haloalkanes, the **chlorofluorocarbons (CFCs)** manufactured under the trade name Freons are the most widely known. CFCs are non-toxic, nonflammable, odorless, and noncorrosive. Originally, they seemed to be ideal replacements for hazardous compounds such as ammonia and sulfur dioxide formerly used as heat-transfer agents in refrigeration systems. Among the CFCs most widely used for this purpose were trichlorofluoro-methane (CCl_3F, Freon-11) and dichlorodifluoromethane (CCl_2F_2, Freon-12). The CFCs also found wide use as industrial cleaning solvents to prepare surfaces for coatings, to remove cutting oils and waxes from millings, and to remove protective coatings. In addition, they were employed as propellants for aerosol sprays.

B. Solvents

Several low-molecular-weight haloalkanes are excellent solvents in which to carry out organic reactions and to use as cleaners and degreasers. Carbon

Chemical Connections 11C

The Environmental Impact of Freons

Concern about the environmental impact of CFCs arose in the 1970s, when researchers found that more than 4.5×10^5 kg/yr of these compounds were being emitted into the atmosphere. In 1974, Sherwood Rowland and Mario Molina, both of the United States, announced their theory, which has since been amply confirmed, that these compounds destroy the stratospheric ozone layer. When released into the air, CFCs escape to the lower atmosphere. Because of their inertness, however, they do not decompose there. Slowly they find their way to the stratosphere, where they absorb ultraviolet radiation from the Sun and then decompose. As they do so, they set up chemical reactions that lead to the destruction of the stratospheric ozone layer, which shields the Earth against short-wavelength ultraviolet radiation from the Sun. An increase in short-wavelength ultraviolet radiation reaching the Earth is believed to promote the destruction of certain crops and agricultural species and to increase the incidence of skin cancer in light-skinned individuals.

This concern prompted two conventions, one in Vienna in 1985 and one in Montreal in 1987, held by the United Nations Environmental Program. The 1987 meeting produced the Montreal Protocol, which set limits on the production and use of ozone-depleting CFCs and urged the complete phase-out of their production by 1996. This phase-out resulted in enormous costs for manufacturers and is not yet complete in developing countries.

Rowland, Molina, and Paul Crutzen (a Dutch chemist at the Max Planck Institute for Chemistry in Germany) were awarded the 1995 Nobel Prize for chemistry. As noted in the award citation by the Royal Swedish Academy of Sciences, "By explaining the chemical mechanisms that affect the thickness of the ozone layer, these three researchers have contributed to our salvation from a global environmental problem that could have catastrophic consequences."

The chemical industry has responded to this crisis by developing replacement refrigerants that have much lower ozone-depleting potential. The most prominent of these replacements are the hydrofluorocarbons (HFCs) and hydrochlorofluorocarbons (HCFCs).

HFC-134a   HCFC-141b

These compounds are much more chemically reactive in the atmosphere than the Freons and are destroyed before they reach the stratosphere. However, they are not compatible in the air conditioners of cars manufactured before 1995.

Test your knowledge with Problems 11.63–11.65.

tetrachloride ("carbon tet") was the first of these compounds to find wide application, but its use for this purpose has since been discontinued because it is now known that carbon tet is both toxic and a carcinogen. Today, the most widely used haloalkane solvent is dichloromethane, CH_2Cl_2.

Summary of Key Questions

⊙WL Sign in at **www.cengage.com/owl** to develop problem-solving skills and complete online homework assigned by your professor.

Section 11.1 What Are Alkanes?

- A **hydrocarbon** is a compound composed of only carbon and hydrogen.
- A **saturated hydrocarbon** contains only carbon–carbon single bonds.
- An **unsaturated hydrocarbon** is a hydrocarbon that contains one or more carbon–carbon double or triple bonds or benzene rings.

- **Alkanes** are saturated hydrocarbons whose carbon atoms are arranged in an open chain.

Section 11.2 How Do We Write Structural Formulas of Alkanes?

- In a **line-angle formula**, a line represents a carbon–carbon bond and a vertex represents a carbon atom. To count hydrogens from a line-angle formula, add enough hydrogens to give each carbon its required four bonds.

Section 11.3 What Are Constitutional Isomers?

- **Constitutional isomers** have the same molecular formula but a different connectivity of their atoms.

Section 11.4 How Do We Name Alkanes?

- Alkanes are named according to a set of rules developed by the **International Union of Pure and Applied Chemistry (IUPAC)**.
- The IUPAC name of an alkane consists of two parts: a prefix that tells the number of carbon atoms in the parent chain and the ending **-ane**. Substituents derived from alkanes by removal of a hydrogen atom are called **alkyl groups** and are denoted by the symbol **R—**.

Section 11.5 Where Do We Obtain Alkanes?

- **Natural gas** consists of 90 to 95% methane with lesser amounts of ethane and other lower-molecular-weight hydrocarbons.
- **Petroleum** is a liquid mixture of thousands of different hydrocarbons.

Section 11.6 What Are Cycloalkanes?

- A **cycloalkane** is an alkane that contains carbon atoms bonded to form a ring.
- To name a cycloalkane, prefix the name of the open-chain alkane with cyclo-.

Section 11.7 What Are the Shapes of Alkanes and Cycloalkanes?

- A **conformation** is any three-dimensional arrangement of the atoms of a molecule that results from rotation about a single bond.
- The lowest-energy conformation of cyclopentane is an **envelope conformation**.
- The lowest-energy conformation of cyclohexane is a **chair conformation**. In a chair conformation, six C—H bonds are **axial** and six C—H bonds are **equatorial**. A substituent on a six-membered ring is more stable when it is equatorial than when it is axial.

Section 11.8 What Is *Cis-Trans* Isomerism in Cycloalkanes?

- ***Cis-trans* isomers** of cycloalkanes have (1) the same molecular formula and (2) the same connectivity of their atoms, but (3) a different orientation of their atoms in space because of the restricted rotation around the C—C bonds of the ring.
- For *cis-trans* isomers of cycloalkanes, *cis* means that substituents are on the same side of the ring; *trans* means that they are on opposite sides of the ring.

Section 11.9 What Are the Physical Properties of Alkanes and Cycloalkanes?

- Alkanes are nonpolar compounds, and the only forces of attraction between their molecules are London dispersion forces.
- At room temperature, low-molecular-weight alkanes are gases, higher-molecular-weight alkanes are liquids, and very-high-molecular-weight alkanes are waxy solids.
- For any group of alkane constitutional isomers, the least-branched isomer generally has the highest boiling point and the most-branched isomer generally has the lowest boiling point.
- Alkanes are insoluble in water but soluble in each other and in other nonpolar organic solvents such as toluene. All liquid and solid alkanes are less dense than water.

Section 11.10 What Are the Characteristic Reactions of Alkanes?

- The most important chemical property of alkanes and cycloalkanes is their inertness. They are quite unreactive under normal ionic reaction conditions.
- By far, the most important reaction of alkanes and cycloalkanes is their reaction with oxygen (**combustion**) in which they are converted to carbon dioxide and water.
- Combustion of alkanes and cycloalkanes is the basis for their use as energy sources and power.
- Alkanes and cycloalkanes also react with chlorine, Cl_2, and bromine, Br_2, by **substitution**, in which an atom of chlorine or bromine is substituted for a hydrogen of the alkane or cycloalkane. The products of this type of substitution reaction are haloalkanes and halocycloalkanes.

Section 11.11 What Are Some Important Haloalkanes?

- Haloalkanes are often used as intermediates in the synthesis of other organic compounds. Just as we can replace a hydrogen atom with a halogen atom, we can in turn replace the halogen atom with other functional groups.
- Low-molecular-weight haloalkanes such as dichloromethane, CH_2Cl_2, are excellent solvents in which to carry out organic reactions and to use as degreasers and solvents for cleaning.

Summary of Key Reactions

1. **Oxidation of Alkanes (Section 11.10A)** Oxidation of alkanes to carbon dioxide and water, an exothermic reaction, is the basis for our use of them as sources of heat and power.

$$CH_3CH_2CH_3 + 5O_2 \longrightarrow 3CO_2 + 4H_2O + 530 \text{ kcal/mol}$$
Propane

2. **Halogenation of Alkanes (Section 11.10B)** Reaction of an alkane with chlorine or bromine results in the substitution of a halogen atom for a hydrogen.

$$CH_3CH_3 + Cl_2 \xrightarrow{\text{heat or light}} CH_3CH_2Cl + HCl$$
Ethane   Chloroethane
(Ethyl chloride)

Problems

Section 11.2 How Do We Write Structural Formulas of Alkanes?

11.10 Answer true or false.

(a) A hydrocarbon is composed only of the elements carbon and hydrogen.

(b) Alkanes are saturated hydrocarbons.

(c) The general formula of an alkane is C_nH_{2n+2}, where n is the number of carbons in the alkane.

(d) Alkenes and alkynes are unsaturated hydrocarbons.

11.11 Define:

(a) Hydrocarbon

(b) Alkane

(c) Saturated hydrocarbon

11.12 Why is it not accurate to describe an unbranched alkane as a "straight-chain" hydrocarbon?

11.13 What is meant by the term *line-angle formula* as applied to alkanes and cycloalkanes?

11.14 For each condensed structural formula, write a line-angle formula.

$$\text{(a) } CH_3CH_2\overset{\overset{\displaystyle CH_2CH_3}{|}}{C}H CH \overset{\overset{\displaystyle CH_3}{|}}{C}H_2 CH CH_3$$
$$\underset{\underset{\displaystyle CH(CH_3)_2}{|}}{}$$

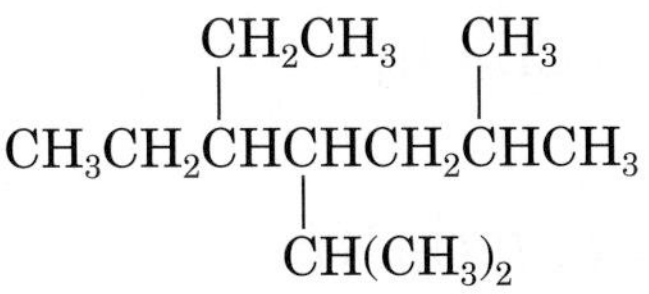

$$\text{(b) } CH_3 \overset{\overset{\displaystyle CH_3}{|}}{\underset{\underset{\displaystyle CH_3}{|}}{C}} CH_3$$

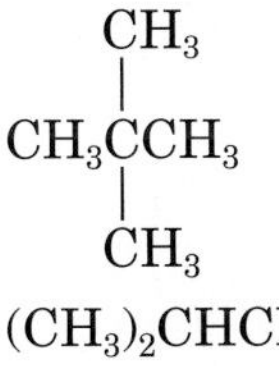

(c) $(CH_3)_2CHCH(CH_3)_2$

$$\text{(d) } CH_3CH_2 \overset{\overset{\displaystyle CH_2CH_3}{|}}{\underset{\underset{\displaystyle CH_2CH_3}{|}}{C}} CH_2CH_3$$

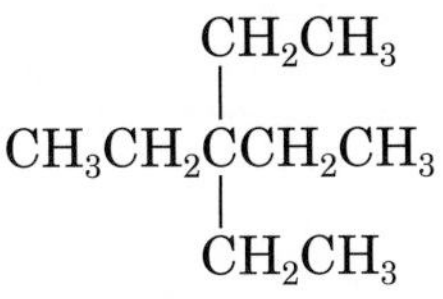

(e) $(CH_3)_3CH$

(f) $CH_3(CH_2)_3CH(CH_3)_2$

11.15 Write the molecular formula for each alkane.

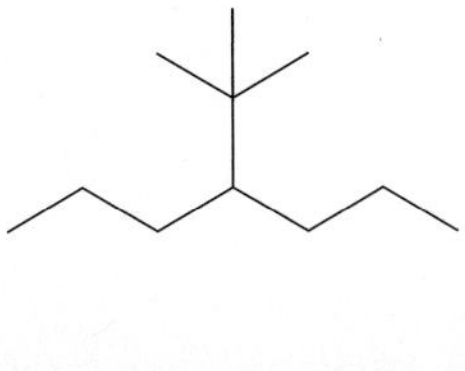

Section 11.3 What Are Constitutional Isomers?

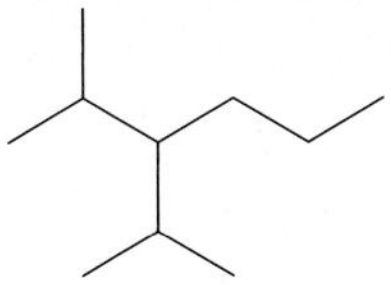

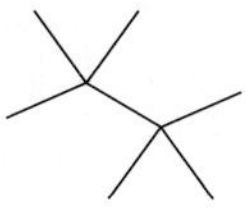

11.16 Answer true or false.

(a) Constitutional isomers have the same molecular formulas and the same connectivity of their atoms.

(b) There are two constitutional isomers with the molecular formula C_3H_8.

(c) There are four constitutional isomers with the molecular formula C_4H_{10}.

(d) There are five constitutional isomers with the molecular formula C_5H_{12}.

11.17 Which statements are true about constitutional isomers?

(a) They have the same molecular formula.

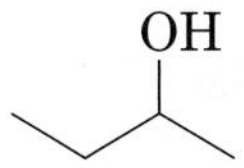

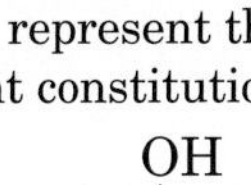

(b) They have the same molecular weight.

(c) They have the same connectivity of their atoms.

(d) They have the same physical properties.

11.18 Each member of the following set of compounds is an alcohol; that is, each contains an —OH (hydroxyl group; see Section 10.4A). Which structural formulas represent the same compound, and which represent constitutional isomers?

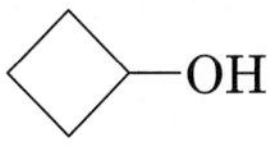

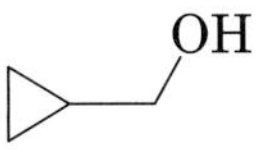

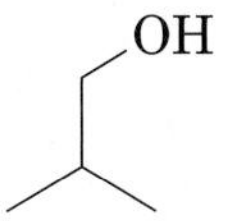

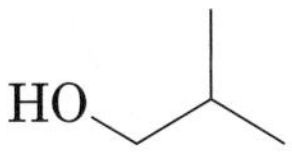

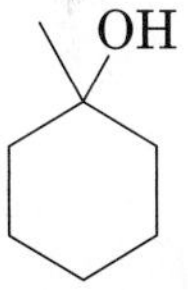

11.19 Each member of the following set of compounds is an amine; that is, each contains a nitrogen bonded to one, two, or three carbon groups (Section 10.4B). Which structural formulas represent the same compound, and which represent constitutional isomers?

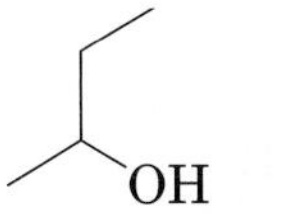

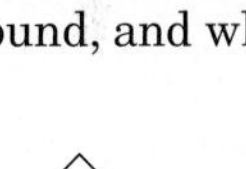

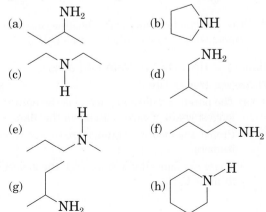

11.20 Each member of the following set of compounds is either an aldehyde or a ketone (Section 10.4C). Which structural formulas represent the same compound, and which represent constitutional isomers?

(a)

(b)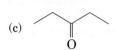

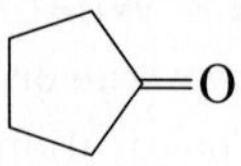

(c)

(d)

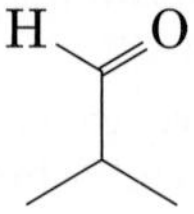

(e)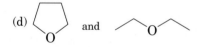

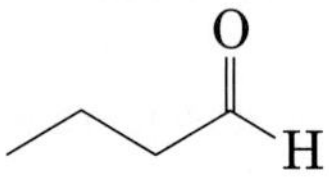

(f)

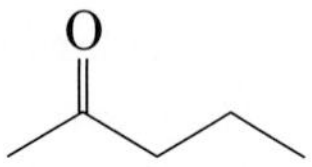

(g)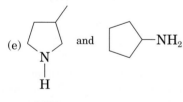

(h)

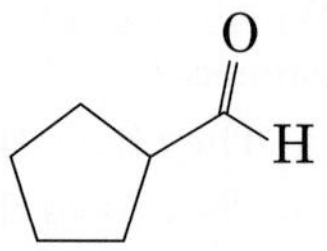

11.21 In the six following sets, which pairs of structural formulas represent constitutional isomers?

(a) ☐   and

(b) 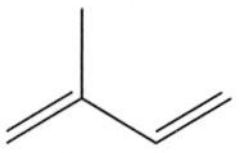   and

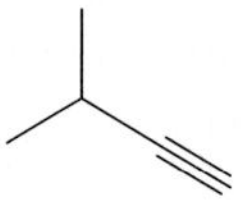

(c)   and

(d)   and

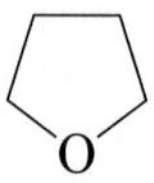

(e)   and   —NH₂

(f)   and   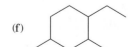OH

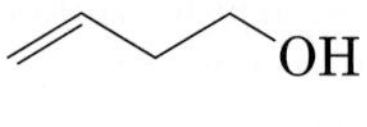

11.22 Draw line-angle formulas for the nine constitutional isomers with the molecular formula C_7H_{16}.

Section 11.4 How Do We Name Alkanes?

11.23 Answer true or false.

(a) The parent name of an alkane is the name of the longest chain of carbon atoms in the alkane.

(b) Propyl and isopropyl groups are constitutional isomers.

(c) There are four alkyl groups with the molecular formula C_4H_9.

11.24 Name these alkyl groups:

(a) CH_3CH_2—

(b) CH_3CH— with CH_3 above

(c) CH_3CHCH_2— with CH_3 below

(d) CH_3C— with CH_3 above and CH_3 below

11.25 Write the IUPAC names for isobutane and isopentane.

Section 11.5 Where Do We Obtain Alkanes?

11.26 Answer true or false.

(a) The two major sources of alkanes the world over are petroleum and natural gas.

(b) The octane number of a particular gasoline is the number of grams of octane per liter of the fuel.

(c) Octane and 2,2,4-trimethylpentane are constitutional isomers and have the same octane number.

Section 11.6 What Are Cycloalkanes?

11.27 Answer true or false.

(a) Cycloalkanes are saturated hydrocarbons.

(b) Hexane and cyclohexane are constitutional isomers.

(c) The parent name of a cycloalkane is the name of the unbranched alkane with the same number of carbon atoms as are in the cycloalkane ring.

11.28 Write the IUPAC names for these alkanes and cycloalkanes.

(a) $CH_3CHCH_2CH_2CH_3$ with CH_3 below

(b) $CH_3CHCH_2CH_2CHCH_3$ with CH_3 and CH_3 below

(c) $CH_3(CH_2)_4CHCH_2CH_3$ with CH_2CH_3 below

(d)

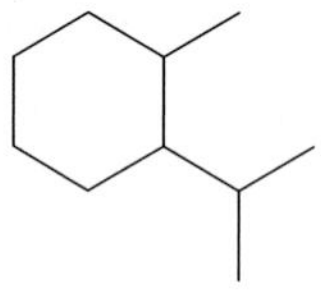

(e)

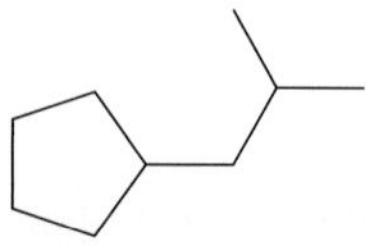

(f)

11.29 Write line-angle formulas for these alkanes and cycloalkanes.

 (a) 2,2,4-Trimethylhexane

 (b) 2,2-Dimethylpropane

 (c) 3-Ethyl-2,4,5-trimethyloctane

 (d) 5-Butyl-2,2-dimethylnonane

 (e) 4-Isopropyloctane

 (f) 3,3-Dimethylpentane

 (g) *trans*-1,3-Dimethylcyclopentane

 (h) *cis*-1,2-Diethylcyclobutane

Section 11.7 What Are the Shapes of Alkanes and Cycloalkanes?

11.30 Answer true or false.

 (a) Conformations have the same molecular formula and the same connectivity but differ in the three-dimensional arrangement of their atoms in space.

 (b) In all conformations of ethane, propane, butane, and higher alkanes, all C—C—C and H—C—H bond angles are approximately 109.5°.

 (c) In a cyclohexane ring, if an axial bond is above the plane of the ring on a particular carbon, axial bonds on the two adjacent carbons are below the plane of the ring.

 (d) In a cyclohexane ring, if an equatorial bond is above the plane of the ring on a particular carbon, equatorial bonds on the two adjacent carbons are below the plane of the ring.

 (e) The more stable chair conformation of a cyclohexane ring has more substituent groups in equatorial positions.

11.31 The condensed structural formula of butane is $CH_3CH_2CH_2CH_3$. Explain why this formula does not show the geometry of the real molecule.

11.32 Draw a conformation of ethane in which hydrogen atoms on adjacent carbons are as far apart as possible. Also draw a conformation in which they are as close together as possible. In a sample of ethane molecules at room temperature, which conformation is the more likely?

11.33 Calculate the actual C—C—C bond angles in planar (a) cyclopropane and (b) cyclopentane and compare them with optimal bond angles.

Section 11.8 What Is *Cis-Trans* Isomerism in Cycloalkanes?

11.34 Answer true or false.

 (a) *Cis-* and *trans*-cycloalkanes have the same molecular formula but a different connectivity of their atoms.

 (b) A *cis* isomer of a cycloalkane can be converted to its *trans* isomer by rotation about an appropriate carbon–carbon single bond.

 (c) A *cis* isomer of a cycloalkane can be converted to its *trans* isomer by exchange of two groups at a stereocenter in the *cis*-cycloalkane.

 (d) Configuration refers to the arrangement in space of the atoms or groups of atoms at a stereocenter.

 (e) *Cis*-1,4-dimethylcyclohexane and *trans*-1,4-dimethylcyclohexane are classified as conformations.

11.35 What structural feature of cycloalkanes makes *cis-trans* isomerism in them possible?

11.36 Is *cis-trans* isomerism possible in alkanes?

11.37 Name and draw structural formulas for the *cis* and *trans* isomers of 1,2-dimethylcyclopropane.

11.38 Name and draw structural formulas for the six cycloalkanes with the molecular formula C_5H_{10}. Include *cis-trans* isomers as well as constitutional isomers.

11.39 Why is equatorial methylcyclohexane more stable than axial methylcyclohexane?

11.40 Following is a structural formula and a ball-and-stick model of cholestanol, a close relative of cholesterol.

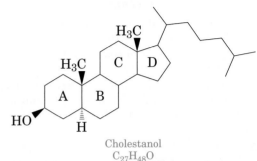

Cholestanol
$C_{27}H_{48}O$

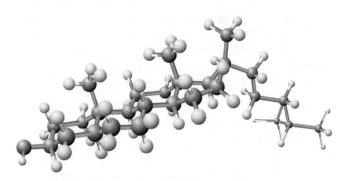

 (a) Describe the conformation of each six-membered ring and the one five-membered ring.

 (b) Is the —OH group on ring **A** in an axial or equatorial position?

 (c) Is the —CH_3 group between rings **A** and **B** in an axial or an equatorial position?

11.41 Consider a cyclohexane ring substituted with one methyl group and one hydroxyl group. Draw the structural formula for a compound with this composition that:

 (a) Does not show cis/trans isomerism.

 (b) Does show cis/trans isomerism.

Section 11.9 What Are the Physical Properties of Alkanes and Cycloalkanes?

11.42 Answer true or false.

 (a) Boiling points among alkanes with unbranched chains increase as the number of carbons in the chain increases.

 (b) Alkanes that are liquid at room temperature are more dense than water.

 (c) *Cis* and *trans* isomers have the same molecular formula, the same connectivity, and the same physical properties.

(d) Among alkane constitutional isomers, the least branched isomer generally has the lowest boiling point.

(e) Alkanes and cycloalkanes are insoluble in water.

(f) Liquid alkanes are soluble in each other.

11.43 In Problem 11.22, you drew structural formulas for the nine constitutional isomers with molecular formula C_7H_{16}. Predict which isomer has the lowest boiling point and which has the highest boiling point.

11.44 Which unbranched alkane (Table 11.4) has about the same boiling point as water? Calculate the molecular weight of this alkane and compare it with the molecular weight of water.

11.45 What generalizations can you make about the densities of alkanes relative to the density of water?

11.46 What generalization can you make about the solubility of alkanes in water?

11.47 Suppose that you have samples of hexane and octane. Could you tell the difference by looking at them? What color would each be? How could you tell which is which?

11.48 As you can see from Table 11.4, each CH_2 group added to the carbon chain of an alkane increases its boiling point. This increase is greater going from CH_4 to C_2H_6 and from C_2H_6 to C_3H_8 than it is going from C_8H_{18} to C_9H_{20} or from C_9H_{20} to $C_{10}H_{22}$. What do you think is the reason for this difference?

11.49 How are the boiling points of hydrocarbons during petroleum refining related to their molecular weight?

Section 11.10 What Are the Characteristic Reactions of Alkanes?

11.50 Answer true or false.

(a) Combustion of alkanes is an endothermic reaction.

(b) The products of complete combustion of an alkane are carbon dioxide and water.

(c) Halogenation of an alkane converts it to a haloalkane.

11.51 Write balanced equations for the combustion of each of the following hydrocarbons. Assume that each is converted completely to carbon dioxide and water.

(a) Hexane

(b) Cyclohexane

(c) 2-Methylpentane

11.52 The heat of combustion of methane, a component of natural gas, is 212 kcal/mol. That of propane, a component of LP gas, is 530 kcal/mol. On a gram-for-gram basis, which hydrocarbon is the better source of heat energy?

11.53 Draw structural formulas for these haloalkanes.

(a) Bromomethane

(b) Chlorocyclohexane

(c) 1,2-Dibromoethane

(d) 2-Chloro-2-methylpropane

(e) Dichlorodifluoromethane (Freon-12)

11.54 The reaction of chlorine with pentane gives a mixture of three chloroalkanes, each with the molecular formula $C_5H_{11}Cl$. Write a line-angle formula and the IUPAC name for each chloroalkane.

11.55 Complete and balance the equation for the complete combustion of each hydrocarbon.

(a) Hexane   (b) Cyclohexane

(c 2-Methylpentane

11.56 Name and draw structural formulas for all possible monochlorination products that might be formed in each reaction.

(a) + Cl_2 $\xrightarrow{\text{heat}}$

(b) + Cl_2 $\xrightarrow{\text{heat}}$

(c) + Cl_2 $\xrightarrow{\text{heat}}$

11.57 There are three constitutional isomers with the molecular formula C_5H_{12}. When heated with chlorine at 300°C, isomer A gives a mixture of four monochlorination products. Under the same experimental conditions, isomer B gives a mixture of three monochlorination products and isomer C gives only one monochlorination product. From this information, deduce the structural formulas of isomers A, B, and C.

Section 11.11 What Are Some Important Haloalkanes?

11.58 Answer true or false.

(a) The Freons are members of a class of organic compounds called chlorofluorocarbons (CFCs).

(b) An advantage of Freons as heat-transfer agents in refrigeration systems, propellants in aerosol sprays, and solvents for industrial cleaning is that they are nontoxic, nonflammable, odorless, and noncorrosive.

(c) Freons in the stratosphere interact with ultraviolet radiation and thereby set up chemical reactions that lead to the destruction of the stratospheric ozone layer.

(d) Alternative names for the important laboratory and industrial solvent CH_2Cl_2 are dichloromethane, methylene chloride, and chloroform.

Chemical Connections

11.59 (Chemical Connections 11A) How many rings in tetrodotoxin contain only carbon atoms? How many contain nitrogen atoms? How many contain two oxygen atoms?

11.60 (Chemical Connections 11B) What is an "octane rating"? What two reference hydrocarbons are used for setting the scale of octane ratings?

11.61 (Chemical Connections 11B) Octane has an octane rating of −20. Will it produce more or less engine knocking than heptane does?

11.62 (Chemical Connections 11B) Ethanol is added to gasoline to produce E-15 and E-85. It promotes more complete combustion of the gasoline and is an octane booster. Compare the heats of combustion of 2,2,4-trimethylpentane (1304 kcal/mol) and ethanol (327 kcal/mol). Which has the higher heat of combustion in kcal/mol? In kcal/g?

11.63 (Chemical Connections 11C) What are Freons? Why were they considered ideal compounds to use as heat-transfer agents in refrigeration systems? Give structural formulas of two Freons used for this purpose.

11.64 (Chemical Connections 11C) In what way do Freons negatively affect the environment?

11.65 (Chemical Connections 11C) What are HFCs and HCFCs? How does their use in refrigeration systems prevent the environmental problems associated with the use of Freons?

Additional Problems

11.66 Tell whether the compounds in each set are constitutional isomers.

(a) CH_3CH_2OH and CH_3OCH_3

(b) $CH_3\overset{\displaystyle O}{\overset{\displaystyle \|}{C}}CH_3$ and $CH_3CH_2\overset{\displaystyle O}{\overset{\displaystyle \|}{C}}H$

(c) $CH_3\overset{\displaystyle O}{\overset{\displaystyle \|}{C}}OCH_3$ and $CH_3CH_2\overset{\displaystyle O}{\overset{\displaystyle \|}{C}}OH$

(d) $CH_3\overset{\displaystyle OH}{\overset{\displaystyle |}{C}}HCH_2CH_3$ and $CH_3\overset{\displaystyle O}{\overset{\displaystyle \|}{C}}CH_2CH_3$

(e) and $CH_3CH_2CH_2CH_2CH_2CH_3$

(f) and $CH_2{=}CHCH_2CH_2CH_3$

11.67 Explain why each of the following is an incorrect IUPAC name. Write the correct IUPAC name for the compound.

(a) 1,3-Dimethylbutane
(b) 4-Methylpentane
(c) 2,2-Diethylbutane
(d) 2-Ethyl-3-methylpentane
(e) 2-Propylpentane
(f) 2,2-Diethylheptane
(g) 2,2-Dimethylcyclopropane
(h) 1-Ethyl-5-methylcyclohexane

11.68 Which of the following compounds can exist as *cis-trans* isomers? For each that can, draw both isomers using solid and dashed wedges to show the orientation in space of the —OH and —CH₃ groups.

(a)

(b)

(c)

11.69 Tetradecane, $C_{14}H_{30}$, is an unbranched alkane with a melting point of 5.9°C and a boiling point of 254°C. Is tetradecane a solid, liquid, or gas at room temperature?

11.70 Dodecane, $C_{12}H_{26}$, is an unbranched alkane. Predict the following:

(a) Will it dissolve in water?
(b) Will it dissolve in hexane?
(c) Will it burn when ignited?
(d) Is it a liquid, solid, or gas at room temperature and atmospheric pressure?
(e) Is it more or less dense than water?

Looking Ahead

11.71 Following is a structural formula for 2-isopropyl-5-methylcyclohexanol:

2-Isopropyl-5-methylcyclohexanol

Using a planar hexagon representation for the cyclohexane ring, draw a structural formula for the *cis-trans* isomer with isopropyl *trans* to —OH and methyl *cis* to —OH. If you answered this part correctly, you have drawn the isomer found in nature and given the name menthol.

11.72 On the left is a representation of the glucose molecule. Convert this representation to the alternative representations using the rings on the right. (We discuss the structure and chemistry of glucose in Chapter 20.)

Planar hexagon representation   Chair conformation

11.73 On the left is a representation for 2-deoxy-D-ribose. This molecule is the "D" of DNA. Convert this representation to the alternative representation using the ring on the right. (We discuss the structure and chemistry of this compound in more detail in Chapter 20.)

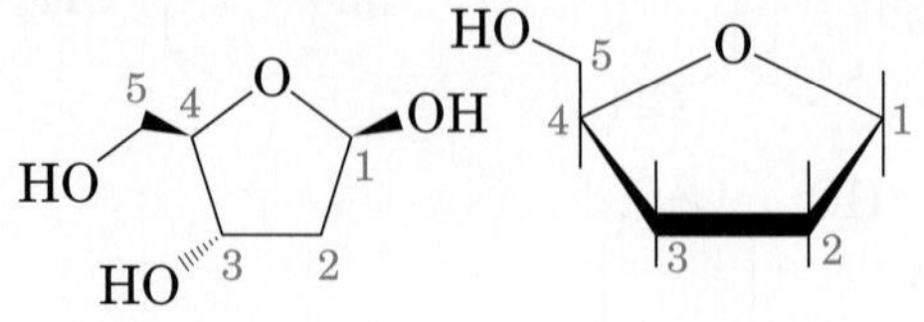

2-Deoxy-D-ribose

11.74 As stated in Section 11.9, the wax found in apple skins is an unbranched alkane with the molecular formula $C_{27}H_{56}$. Explain how the presence of this alkane in apple skins prevents the loss of moisture from within the apple.

11.75 Following are structural formulas for dimethyl ether and ethanol. These compounds are constitutional isomers with the molecular formula C_2H_6O.

$$CH_3OCH_3 \qquad CH_3CH_2OH$$

Dimethyl ether   Ethanol

One of these compounds has a boiling point of 78°C; the other has a boiling point of −24°C.

(a) Which compound has which boiling point? Explain your reasoning. (Chapter 5)

(b) One compound is soluble in water in all proportions, that is, it is infinitely soluble. The other has a water solubility of only 7.8 g/100 mL. Which compound has which water solubility? How do you account for this difference in water solubility? We discuss the physical properties of alcohols and ethers in detail in Chapter 14. (Chapter 6)

Alkenes and Alkynes

Carotene is a naturally occurring polyene in carrots and tomatoes (Problems 12.61 and 12.62).

Key Questions

12.1 What Are Alkenes and Alkynes?

12.2 What Are the Structures of Alkenes and Alkynes?

12.3 How Do We Name Alkenes and Alkynes?

12.4 What Are the Physical Properties of Alkenes and Alkynes?

12.5 What Are the Characteristic Reactions of Alkenes?

12.6 What Are the Important Polymerization Reactions of Ethylene and Substituted Ethylenes?

12.1 What Are Alkenes and Alkynes?

In this chapter, we begin our study of unsaturated hydrocarbons. Recall from Section 11.1 that these unsaturated compounds contain one or more carbon–carbon double bonds, triple bonds, or benzene-like rings. In this chapter, we study **alkenes** and **alkynes**. **Alkenes** are unsaturated hydrocarbons that contain one or more carbon–carbon double bonds. The simplest alkene is ethylene.

Alkene An unsaturated hydrocarbon that contains a carbon–carbon double bond

Alkene An unsaturated hydrocarbon that contains a carbon–carbon triple bond

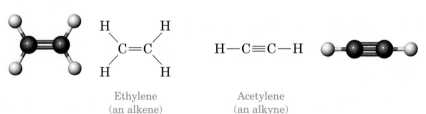

Ethylene
(an alkene)

Acetylene
(an alkyne)

Alkynes are unsaturated hydrocarbons that contain one or more carbon–carbon triple bonds. The simplest alkyne is acetylene. Because alkynes are not widespread in nature and have little importance in biochemistry, we will not study their chemistry in depth.

OWL

Sign in to OWL at **www.cengage.com/owl** to view tutorials and simulations, develop problem-solving skills, and complete online homework assigned by your professor.

Ethylene: A Plant Growth Regulator

As we noted below, ethylene occurs only in trace amounts in nature. True enough, but scientists have discovered that this small molecule is a natural ripening agent for fruits. Thanks to this knowledge, fruit growers can now pick fruit while it is still green and less susceptible to bruising. Then, when the growers are ready to pack the fruit for shipment, they can treat it with ethylene gas. Alternatively, the fruit can be treated with ethephon (Ethrel), which slowly releases ethylene and initiates fruit ripening.

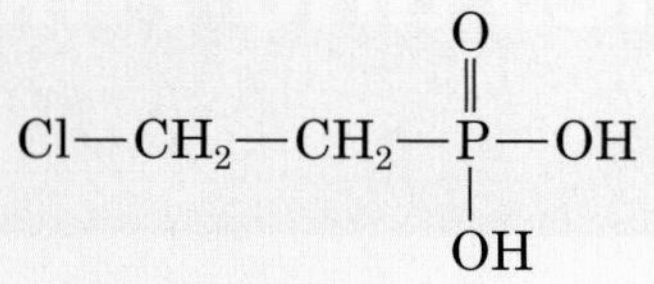

$$Cl-CH_2-CH_2-\overset{\displaystyle O}{\underset{\displaystyle OH}{\overset{\displaystyle \|}{\underset{|}{P}}}}-OH$$

Ethephon

The next time you see ripe bananas in the market, you might wonder when they were picked and whether their ripening was artificially induced.

Text your knowledge with Problem 12.58.

Compounds containing carbon–carbon double bonds are especially widespread in nature. Furthermore, several low-molecular-weight alkenes, including ethylene and propene, have enormous commercial importance in our modern industrialized society. The organic chemical industry worldwide produces more pounds of ethylene than any other organic chemical. Annual production in the United States alone exceeds 55 billion pounds.

What is unusual about ethylene is that it occurs only in trace amounts in nature. The enormous amounts of it required to meet the needs of the chemical industry are derived the world over by thermal cracking of hydrocarbons. In the United States and other areas of the world with vast reserves of natural gas, the major process for the production of ethylene is thermal cracking of the small quantities of ethane extracted from natural gas. In **thermal cracking**, a saturated hydrocarbon is converted to an unsaturated hydrocarbon plus H_2. Ethane is thermally cracked by heating it in a furnace to 800–900°C for a fraction of a second.

$$CH_3CH_3 \xrightarrow[\text{(thermal cracking)}]{800-900°C} CH_2{=}CH_2 + H_2$$

Ethane   Ethylene

Europe, Japan, and other parts of the world with limited supplies of natural gas depend almost entirely on the thermal cracking of petroleum for their ethylene.

From the perspective of the chemical industry, the single most important reaction of ethylene and other low-molecular-weight alkenes is polymerization, which we discuss in Section 12.6. The crucial point to recognize is that ethylene and all of the commercial and industrial products synthesized from it are derived from either natural gas or petroleum—both nonrenewable natural resources!

12.2 What Are the Structures of Alkenes and Alkynes?

A. Alkenes

Using the VSEPR model (Section 3.10), we predict bond angles of 120° about each carbon in a double bond. The observed H—C—C bond angle in ethylene, for example, is 121.7°, close to the predicted value. In other alkenes, deviations from the predicted angle of 120° may be somewhat larger because of interactions between alkyl groups bonded to the doubly bonded carbons. The C—C—C bond angle in propene, for example, is 124.7°.

Cis Double Bonds in Fatty Acids

Fatty acids consist of a carboxylic acid group on one end and a fatty chain on the other. The most common have between 12 and 20 carbons in the unbranched hydrocarbon chain. They are the major building blocks of all fats and oils, including animal fats and vegetable oils. Those that have no carbon–carbon double bond in their hydrocarbon chain are called saturated fatty acids. Those with a single carbon–carbon double bond are called monounsaturated fatty acids, and those with two or more carbon–carbon double bonds in their hydrocarbon chain are called polyunsaturated fatty acids. Following are structural formulas for three C_{18} fatty acids. Notice that all carbon–carbon double bonds in oleic and linolenic acids (both unsaturated fatty acids) have the *cis* configuration. In addition, note that the melting points decrease as the number of double bonds increases. We will describe the physical state of fatty acids in more detail in Chapter 18.

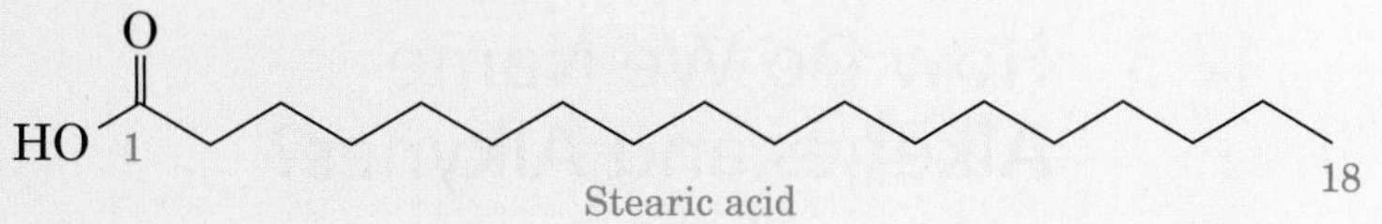

Stearic acid
mp 70°C
(a C_{18} saturated fatty acid)

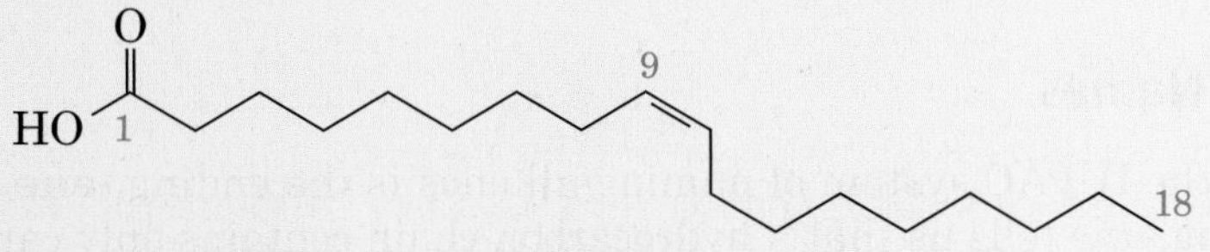

Oleic acid
mp 13°C
(a C_{18} monounsaturated fatty acid)

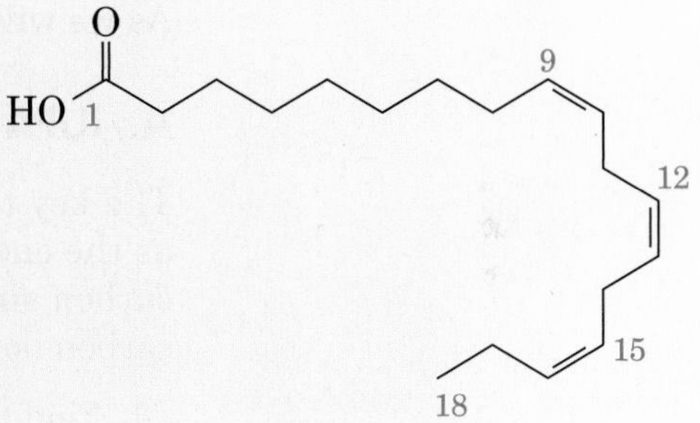

Linolenic acid
mp −17°C
(a C_{18} polyunsaturated fatty acid)

Test your knowledge with Problems 12.59, 12.60, and 12.61.

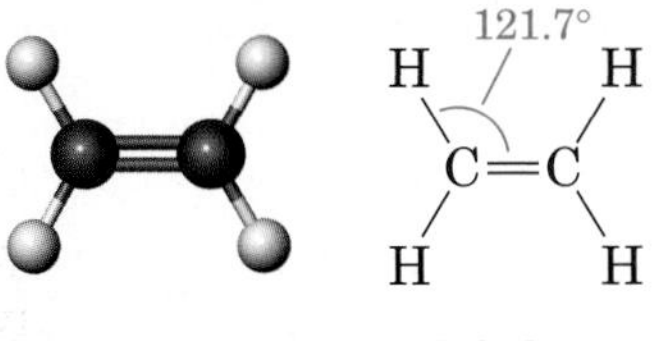

Ethylene

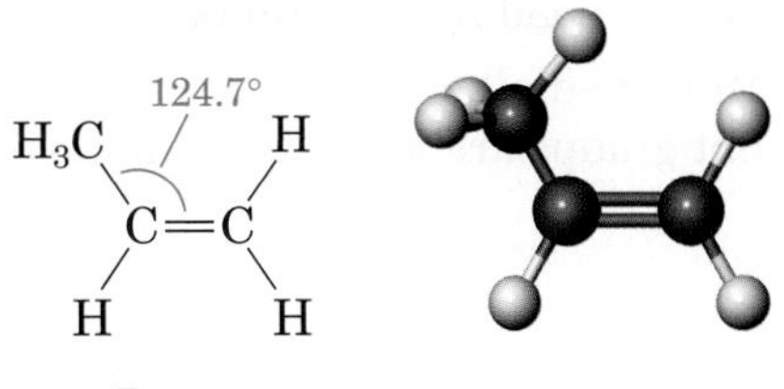

Propene

If we look at a molecular model of ethylene, we see that the two carbons of the double bond and the four hydrogens bonded to them all lie in the same plane—that is, ethylene is a flat or planar molecule. Furthermore, chemists have discovered that under normal conditions, no rotation is possible about the carbon–carbon double bond of ethylene or, for that matter, of any other alkene. Whereas free rotation occurs about each carbon–carbon single bond in an alkane (Section 11.6A), rotation about the carbon–carbon double bond in an alkene does not normally take place because the bond is so rigid. For an important exception to this generalization about carbon–carbon double bonds, see Chemical Connections 12D on *cis-trans* isomerism in vision.

B. *Cis-Trans* Stereoisomerism in Alkenes

Because of the restricted rotation about a carbon–carbon double bond, an alkene in which each carbon of the double bond has two different groups bonded to it shows *cis-trans* isomerism (a type of stereoisomerism). For

Cis-trans isomers Isomers that have the same connectivity of their atoms but a different arrangement of their atoms in space. Specifically, *cis* and *trans* stereoisomers result from the presence of either a ring or a carbon–carbon double bond.

example, 2-butene has two *cis-trans* isomers. In *cis*-2-butene, the two methyl groups are located on the same side of the double bond and the two hydrogens are on the other side. In *trans*-2-butene, the two methyl groups are located on opposite sides of the double bond. *Cis*-2-butene and *trans*-2-butene are different compounds and have different physical and chemical properties.

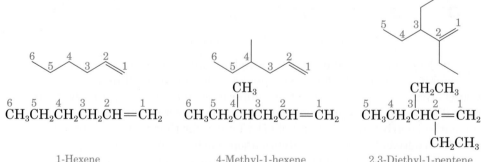

cis-2-Butene
mp −139°C, bp 4°C

trans-2-Butene
mp −106°C, bp 1°C

12.3 How Do We Name Alkenes and Alkynes?

Alkenes and alkynes are named using the IUPAC system of nomenclature. As we will see, some are still referred to by their common names.

A. IUPAC Names

The key to the IUPAC system of naming alkenes is the ending **-ene**. Just as the ending *-ane* tells us that a hydrocarbon chain contains only carbon–carbon single bonds, so the ending *-ene* tells us that it contains a carbon–carbon double bond. To name an alkene:

1. Find the longest carbon chain that includes the double bond. Indicate the length of the parent chain by using a prefix that tells the number of carbon atoms in it (see Table 11.2) and the suffix *-ene* to show that it is an alkene.

2. Number the chain from the end that gives the lower set of numbers to the carbon atoms of the double bond. Designate the position of the double bond by the number of its first carbon.

3. Branched alkenes are named in a manner similar to alkanes; substituent groups are located and named.

$$\underset{\text{1-Hexene}}{\overset{6\quad5\quad4\quad3\quad2\quad1}{CH_3CH_2CH_2CH_2CH=CH_2}}$$

$$\underset{\text{4-Methyl-1-hexene}}{\overset{6\quad5\quad4\quad3\quad2\quad1}{CH_3CH_2CHCH_2CH=CH_2}}$$
$$\underset{4}{\overset{}{\underset{|}{CH_3}}}$$

$$\underset{\text{2,3-Diethyl-1-pentene}}{\overset{5\quad4\quad3\quad2\quad1}{CH_3CH_2CHC=CH_2}}$$
$$\underset{}{\overset{}{\underset{|}{CH_2CH_3}}}$$

Note that although 2,3-diethyl-1-pentene has a six-carbon chain, the longest chain that contains the double bond has only five carbons. The parent alkene is, therefore, a pentene rather than a hexene, and the molecule is named as a disubstituted 1-pentene.

The key to the IUPAC name of an alkyne is the ending **-yne**, which shows the presence of a carbon–carbon triple bond. Thus, HC≡CH is ethyne (or acetylene) and $CH_3C≡CH$ is propyne. In higher alkynes, number the longest carbon chain that contains the triple bond from the end that gives the lower set of numbers to the triply bonded carbons. Indicate the location of the triple bond by the number of its first carbon atom.

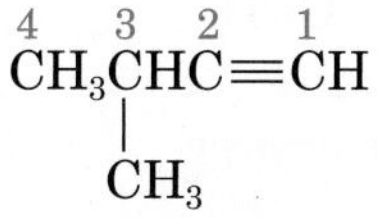

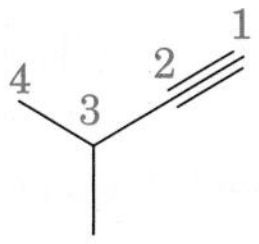

$$\overset{4}{C}H_3\overset{3}{C}H\overset{2}{C}\!\!\equiv\!\!\overset{1}{C}H$$
$$\underset{CH_3}{|}$$

3-Methyl-1-butyne

$$\overset{1}{C}H_3\overset{2}{C}H_2\overset{3}{C}\!\!\equiv\!\!\overset{4}{C}\overset{5}{C}H_2\overset{6}{C}CH_3$$

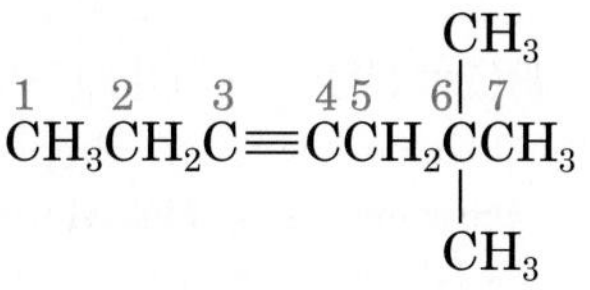

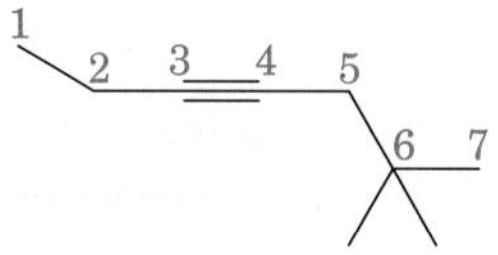

6,6-Dimethyl-3-heptyne

Example 12.1 IUPAC Names of Alkenes and Alkynes

Write the IUPAC name of each unsaturated hydrocarbon.

(a) $CH_2\!\!=\!\!CH(CH_2)_5CH_3$

(b) 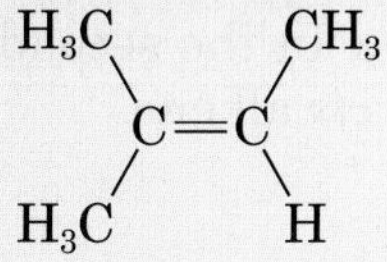
$$\begin{array}{cc} H_3C & CH_3 \\ & C\!\!=\!\!C \\ H_3C & H \end{array}$$

(c) $CH_3(CH_2)_2C\!\!\equiv\!\!CCH_3$

Strategy

Step 1: Locate the parent chain—the longest chain of carbon atoms that contains the carbon–carbon double or triple bond.

Step 2: Number the parent chain from the direction that gives the carbons of the double or triple bond the lower set of numbers. Show the presence of the multiple bond by the suffix -*ene* (for a double bond) or -*yne* (for a triple bond). Indicate the presence of the multiple bond by its first number.

Step 3: Name and locate all substituents on the parent chain. List them in alphabetical order.

Solution

(a) The parent chain contains eight carbons; thus, the parent alkene is octene. To show the presence of the carbon–carbon double bond, use the suffix -*ene*. Number the chain beginning with the first carbon of the double bond. This alkene is 1-octene.

(b) Because there are four carbon atoms in the chain containing the carbon–carbon double bond, the parent alkene is butene. The double bond is between carbons 2 and 3 of the chain, and there is a methyl group on carbon 2. This alkene is 2-methyl-2-butene.

(c) There are six carbons in the parent chain, with the triple bond between carbons 2 and 3. This alkyne is 2-hexyne.

Problem 12.1

Write the IUPAC name of each unsaturated hydrocarbon.

(a)  (b) 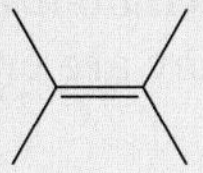 (c)

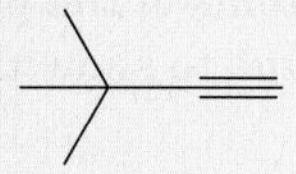

B. Common Names

Despite the precision and universal acceptance of IUPAC nomenclature, some alkenes and alkynes—particularly those of low molecular weight—are known almost exclusively by their common names. Three examples follow:

	$CH_2\!\!=\!\!CH_2$	$CH_3CH\!\!=\!\!CH_2$	$CH_3\overset{\textstyle CH_3}{\overset{\textstyle |}{C}}\!\!=\!\!CH_2$
IUPAC name:	Ethene	Propene	2-Methylpropene
Common name:	Ethylene	Propylene	Isobutylene

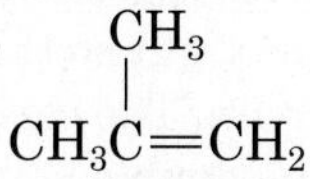

We derive common names for alkynes by prefixing the names of the substituents on the carbon–carbon triple bond to the name *acetylene*:

	$HC{\equiv}CH$	$CH_3C{\equiv}CH$	$CH_3C{\equiv}CCH_3$
IUPAC name:	Ethyne	Propyne	2-Butyne
Common name:	Acetylene	Methylacetylene	Dimethylacetylene

C. *Cis* and *Trans* Configurations of Alkenes

The orientation of the carbon atoms of the parent chain determines whether an alkene is *cis* or *trans*. If the carbons of the parent chain are on the same side of the double bond, the alkene is *cis*; if they are on opposite sides, it is a *trans* alkene. In the first example below, they are on opposite sides and the compound is a *trans* alkene. In the second example, they are on the same side and the compound is a *cis* alkene.

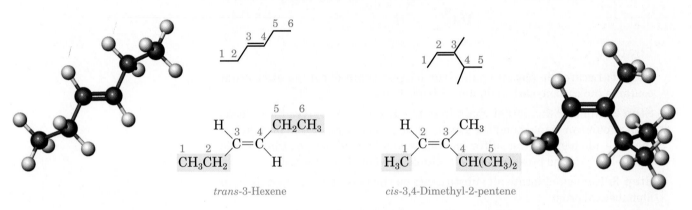

trans-3-Hexene   *cis*-3,4-Dimethyl-2-pentene

Example 12.2 Naming Alkene *Cis* and *Trans* Isomers

Name each alkene and specify its configuration by indicating *cis* or *trans* where appropriate.

(a) 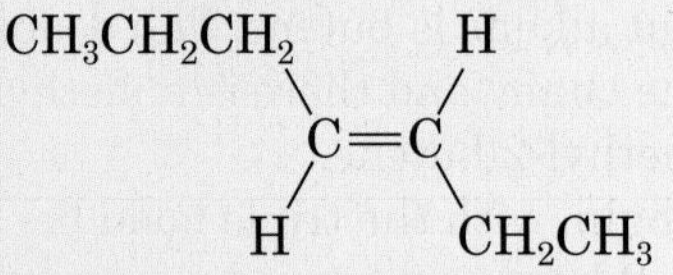
$$CH_3CH_2CH_2 \quad\quad H$$
$$C{=}C$$
$$H \quad\quad CH_2CH_3$$

(b) 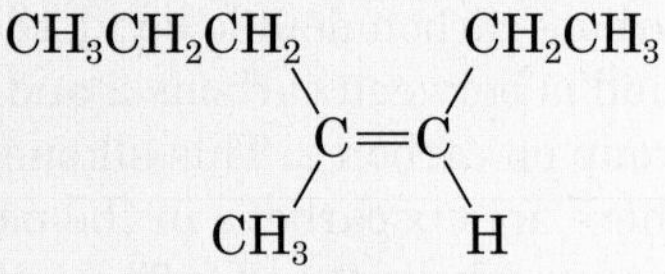
$$CH_3CH_2CH_2 \quad\quad CH_2CH_3$$
$$C{=}C$$
$$CH_3 \quad\quad H$$

Strategy

For alkenes that show *cis-trans* isomerism, use the designator *cis* to show that the carbon atoms of the parent chain are on the same side of the double bond and *trans* to show that they are on opposite sides of the double bond.

Solution

(a) The chain contains seven carbon atoms and is numbered from the right to give the lower number to the first carbon of the double bond. The carbon atoms of the parent chain are on opposite sides of the double bond. This alkene is *trans*-3-heptene.

(b) The longest chain contains seven carbon atoms and is numbered from the right so that the first carbon of the double bond is carbon 3 of the chain. The carbon atoms of the parent chain are on the same side of the double bond. This alkene is *cis*-4-methyl-3-heptene.

Problem 12.2

Name each alkene and specify its configuration.

(a)

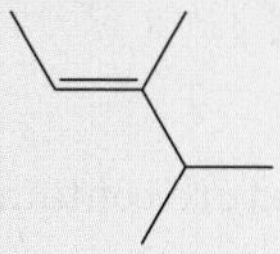

(b)

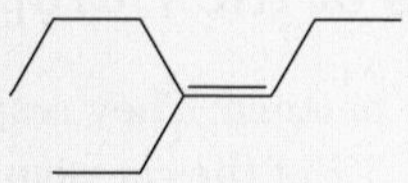

D. Cycloalkenes

In naming **cycloalkenes**, number the carbon atoms of the ring double bond 1 and 2 in the direction that gives the substituent encountered first the lower number. It is not necessary to use a location number for the carbons of the double bond because according to the IUPAC system of nomenclature, they will always be 1 and 2. Number substituents and list them in alphabetical order.

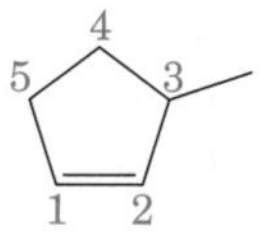

3-Methylcyclopentene
(not 5-methylcyclopentene)

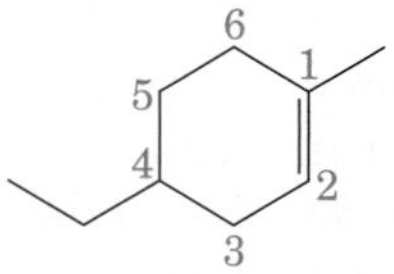

4-Ethyl-1-methylcyclohexene
(not 5-ethyl-2-methylcyclohexene)

ⓞWL Interactive Example 12.3 Naming Cycloalkenes

Write the IUPAC name for each cycloalkene.

(a) 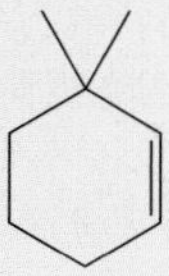   (b) 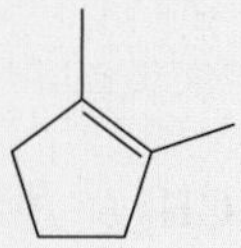   (c)

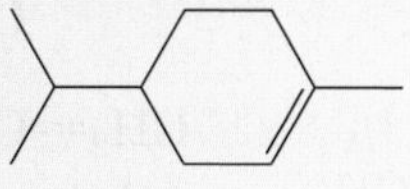

Strategy

In naming cycloalkenes, the carbon atoms of the double bond are always numbered 1 and 2 in the direction that gives the substituent encountered first the lowest possible number. If there are multiple substituents, list them in alphabetical order.

Solution

(a) 3,3-Dimethylcyclohexene
(b) 1,2-Dimethylcyclopentene
(c) 4-Isopropyl-1-methylcyclohexene

Problem 12.3

Write the IUPAC name for each cycloalkene.

(a) 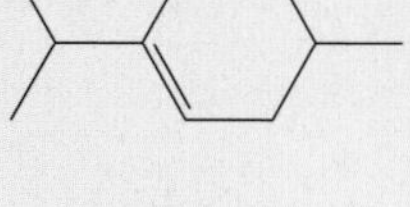   (b) 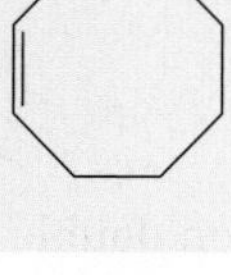   (c)

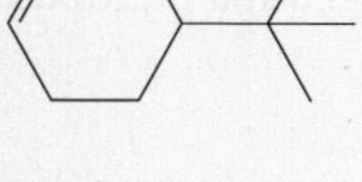

The Case of the Iowa and New York Strains of the European Corn Borer

Although humans communicate largely by sight and sound, the vast majority of other species in the animal world communicate by chemical signals. Often, communication within a species is specific for one of two or more stereoisomers. For example, a member of a given species might respond to the *cis* isomer of a chemical but not to the *trans* isomer. Alternatively, it might respond to a precise ratio of *cis* and *trans* isomers but not to other ratios of these same stereoisomers.

Several groups of scientists have studied the components of the sex **pheromones** of both the Iowa and New York strains of the European corn borer. Females of these closely related species secrete the sex attractant 11-tetradecenyl acetate. Males of the Iowa strain show their maximum response to a mixture containing 96% of the *cis* isomer and 4% of the *trans* isomer. When the pure *cis* isomer is used alone, males are only weakly attracted. Males of the New York strain show an entirely different response

pattern: They respond maximally to a mixture containing 3% of the *cis* isomer and 97% of the *trans* isomer.

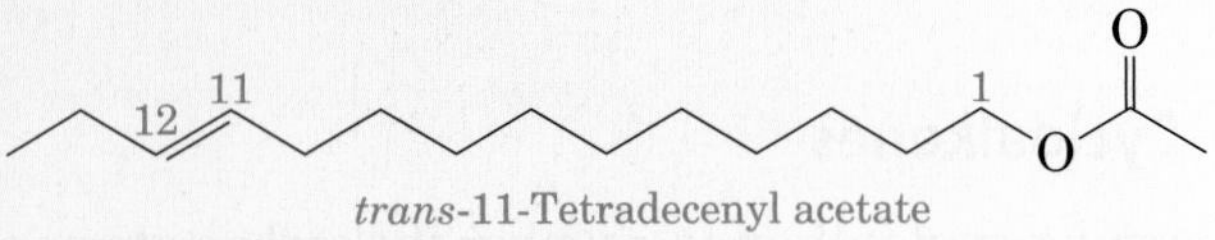

trans-11-Tetradecenyl acetate

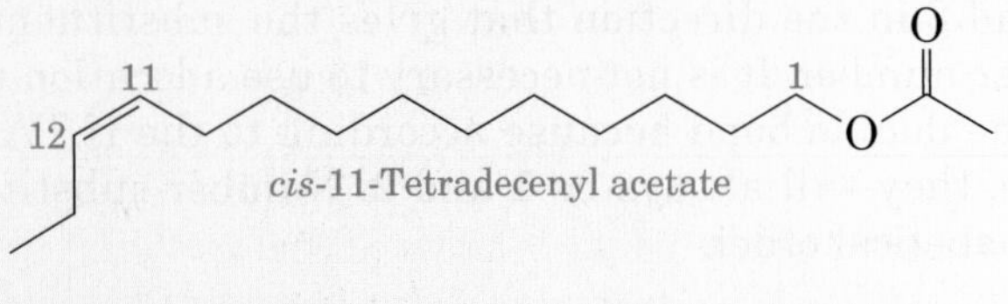

cis-11-Tetradecenyl acetate

Evidence suggests that an optimal response to a narrow range of stereoisomers, as we see here, is widespread in nature and that most insects maintain species isolation for mating and reproduction by the stereochemical mixtures of their pheromones.

Test your knowledge with Problems 12.62 and 12.63.

E. Dienes, Trienes, and Polyenes

We name alkenes that contain more than one double bond as alkadienes, alkatrienes, and so on. We often refer to those that contain several double bonds more generally as polyenes (Greek: *poly*, many). Following are three dienes:

CH_2=$CHCH_2CH$=CH_2
1,4-Pentadiene

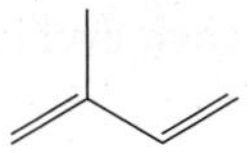

CH_2=CCH=CH_2
2-Methyl-1,3-butadiene
(Isoprene)

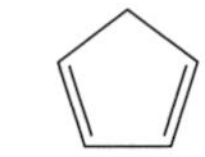

1,3-Cyclopentadiene

We saw earlier that for an alkene with one carbon–carbon double bond that can show *cis-trans* isomerism, two stereoisomers are possible. For an alkene with n carbon–carbon double bonds, each of which can show *cis-trans* isomerism, 2^n stereoisomers are possible.

Example 12.4 *Cis-Trans* Isomerism

How many stereoisomers are possible for 2,4-heptadiene?

$$CH_3—CH=CH—CH=CH—CH_2—CH_3$$
2,4-Heptadiene

Strategy

To show *cis-trans* isomerism, each carbon of the double bond must have two different groups bonded to it.

Solution

This molecule has two carbon–carbon double bonds, each of which shows *cis-trans* isomerism. As shown in the following table, $2^2 = 4$

stereoisomers are possible. Line-angle formulas for two of these dienes are drawn here.

	Double-Bond	
	C_2—C_3	C_4—C_5
(1)	*trans*	*trans*
(2)	*trans*	*cis*
(3)	*cis*	*trans*
(4)	*cis*	*cis*

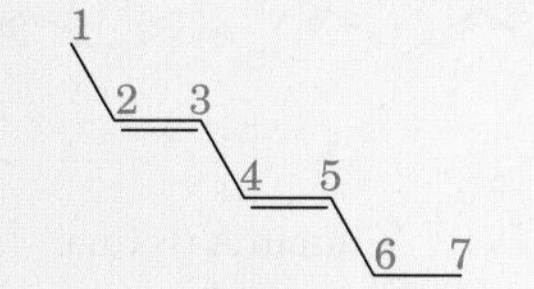

trans,trans-2,4-Heptadiene 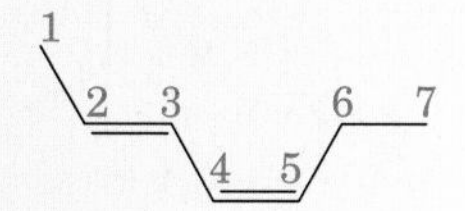   *trans,cis*-2,4-Heptadiene

Problem 12.4

Draw structural formulas for the other two stereoisomers of 2,4-heptadiene.

Example 12.5 Drawing Alkene *Cis-Trans* Isomers

Draw all stereoisomers that are possible for the following unsaturated alcohol.

$$CH_3C{=}CHCH_2CH_2C{=}CHCH_2OH$$

with CH_3 groups

Strategy

To show *cis-trans* isomerism, each carbon of the double bond must have two different groups bonded to it. If a molecule has *n* double bonds about which *cis-trans* isomerism is possible, then 2^n isomers are possible, where *n* is the number of double bonds that show *cis-trans* isomerism.

Solution

Cis-trans isomerism is possible only about the double bond between carbons 2 and 3 of the chain. It is not possible for the other double bond because carbon 7 has two identical groups on it (review Section 12.2B). Thus, $2^1 = 2$ stereoisomers (one *cis-trans* pair) are possible. The *trans* isomer of this alcohol, named geraniol, is a major component of the oils of rose, citronella, and lemon grass.

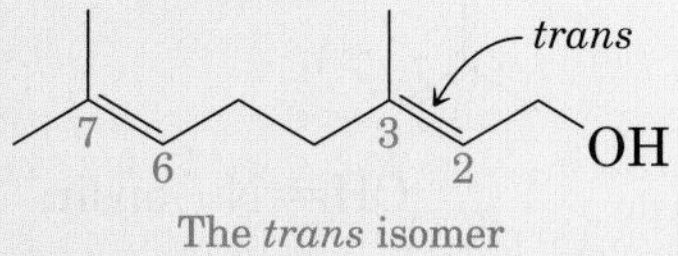

trans

The *trans* isomer

cis

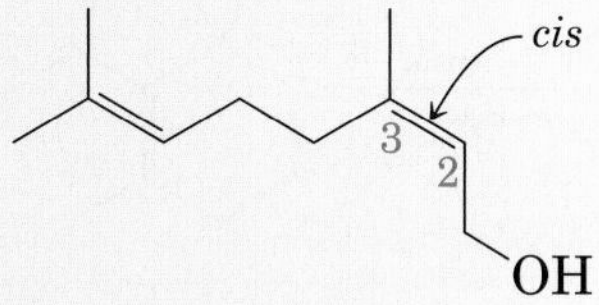

The *cis* isomer

David Sieren/Visuals Unlimited

Lemon grass from Shelton Herb Farm, North Carolina.

Problem 12.5

How many stereoisomers are possible for the following unsaturated alcohol?

$$CH_3C{=}CHCH_2CH_2C{=}CHCH_2CH_2C{=}CHCH_2OH$$

with CH_3 groups

An example of a biologically important polyunsaturated alcohol for which a number of *cis-trans* stereoisomers are possible is vitamin A. Each of the four carbon–carbon double bonds in the chain of carbon atoms bonded to the substituted cyclohexene ring has the potential for *cis-trans* isomerism. There are, therefore, $2^4 = 16$ stereoisomers possible for this structural formula. Vitamin A, the stereoisomer shown here, is the all-*trans* isomer.

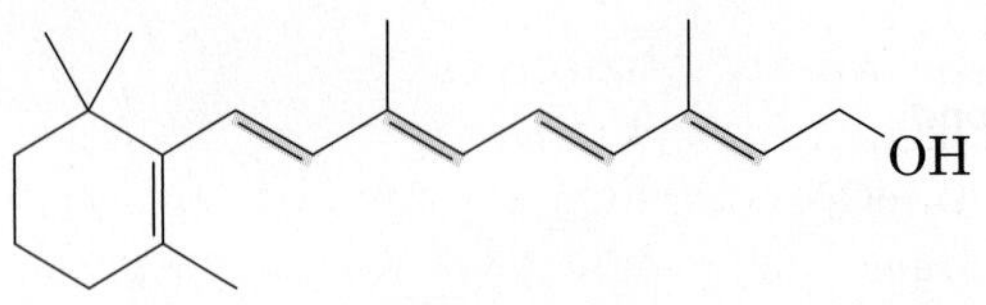

Vitamin A (retinol)

Chemical Connections 12D

Cis-Trans Isomerism in Vision

The retina, the light-detecting layer in the back of our eyes, contains reddish compounds called visual pigments. Their name, rhodopsin, is derived from the Greek word meaning "rose-colored." Each rhodopsin molecule is a combination of one molecule of a protein called opsin and one molecule of 11-*cis*-retinal, a derivative of vitamin A in which the CH_2OH group of carbon 15 is converted to an aldehyde group, $-CH=O$.

When rhodopsin absorbs light energy, the less stable 11-*cis* double bond is converted to the more stable 11-*trans* double bond. This isomerization changes the shape of the rhodopsin molecule, which in turn causes

the neurons of the optic nerve to fire and produce a visual image.

The retinas of vertebrates contain two kinds of rhodopsin-containing cells: rods and cones. Cones function in bright light and are used for color vision; they are concentrated in the central portion of the retina, called the macula, and are responsible for the greatest visual acuity. The remaining area of the retina consists mostly of rods, which are used for peripheral and night vision. 11-*cis*-retinal is present in both cones and rods. Rods have one kind of opsin, whereas cones have three kinds: one for blue, one for green, and one for red color vision.

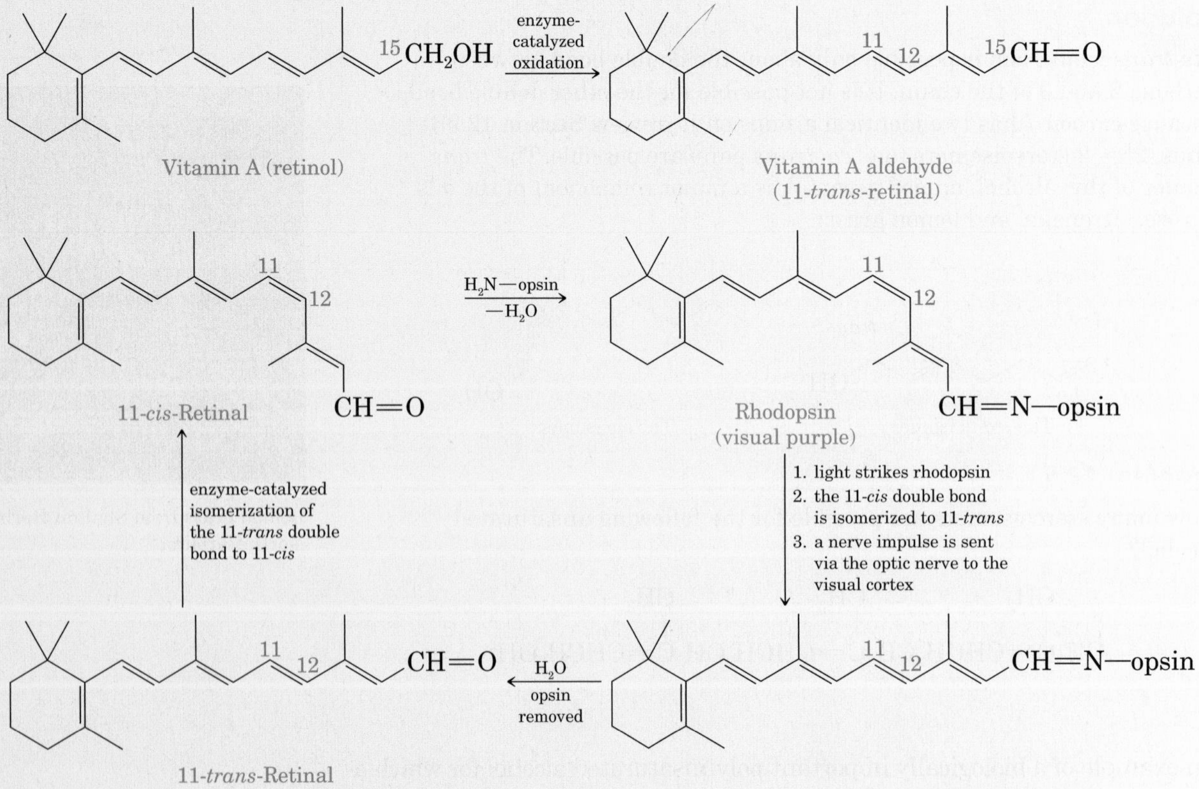

Vitamin A (retinol)

Vitamin A aldehyde
(11-*trans*-retinal)

11-*cis*-Retinal

Rhodopsin
(visual purple)

1. light strikes rhodopsin
2. the 11-*cis* double bond is isomerized to 11-*trans*
3. a nerve impulse is sent via the optic nerve to the visual cortex

enzyme-catalyzed isomerization of the 11-*trans* double bond to 11-*cis*

11-*trans*-Retinal

Test your knowledge with Problems 12.64 and 12.65.

12.4 What Are the Physical Properties of Alkenes and Alkynes?

Alkenes and alkynes are nonpolar compounds, and the only attractive forces between their molecules are very weak London dispersion forces (Section 5.7A). Their physical properties, therefore, are similar to those of alkanes with the same carbon skeletons. Alkenes and alkynes that are liquid at room temperature have densities less than 1.0 g/mL (they float on water). They are insoluble in water but soluble in one another and in other nonpolar organic liquids.

12.5 What Are the Characteristic Reactions of Alkenes?

The most characteristic reaction of alkenes is an addition to their carbon–carbon double bond: The double bond is broken, and in its place, single bonds form between the carbons and two new atoms or groups of atoms. Table 12.1 shows several examples of alkene addition reactions along with the descriptive name(s) associated with each reaction.

At this point, you might ask why a carbon–carbon double bond is a site of chemical reactivity, whereas carbon–carbon single bonds are quite unreactive under most experimental conditions. One way to answer this question is to focus on the changes in bonding that occur as a result of an alkene addition reaction. Consider, for example, the addition of hydrogen (H_2) to ethylene. As a result of this addition, one double bond and one single bond (the H—H bond of H_2) are replaced by three single bonds, giving a net conversion of one bond of the double bond to two single bonds.

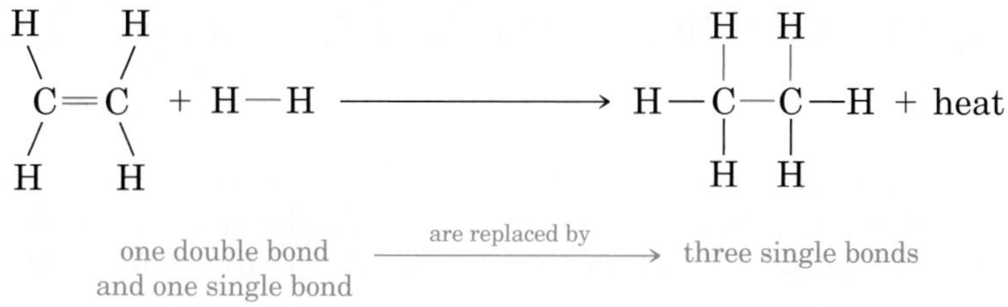

Table 12.1 Characteristic Addition Reactions of Alkenes

Reaction	Descriptive Name(s)
$\rangle C{=}C\langle + HCl \longrightarrow -\overset{H}{\underset{\vert}{C}}-\overset{Cl}{\underset{\vert}{C}}-$	hydrochlorination
$\rangle C{=}C\langle + H_2O \longrightarrow -\overset{H}{\underset{\vert}{C}}-\overset{OH}{\underset{\vert}{C}}-$	hydration
$\rangle C{=}C\langle + Br_2 \longrightarrow -\overset{Br}{\underset{\vert}{C}}-\overset{Br}{\underset{\vert}{C}}-$	bromination
$\rangle C{=}C\langle + H_2 \longrightarrow -\overset{H}{\underset{\vert}{C}}-\overset{H}{\underset{\vert}{C}}-$	hydrogenation (reduction)

This and almost all other addition reactions of alkenes are exothermic, which means that the products are more stable (have lower energy) than the reactants. Just because an alkene addition reaction is exothermic, however, doesn't mean that it occurs rapidly. The rate of a chemical reaction depends on its activation energy, not on how exothermic or endothermic it is (Section 7.3). In fact, the addition of H_2 to an alkene is immeasurably slow at room temperature but, as we will see soon, proceeds quite rapidly in the presence of a suitable transition metal catalyst.

A. Addition of Hydrogen Halides (Hydrohalogenation)

Haloalkanes (alkyl halides) are formed when a hydrogen halide (HCl, HBr, and HI) is added to an alkene. Addition of HCl to ethylene, for example, gives chloroethane (ethyl chloride):

$$CH_2{=}CH_2 + HCl \longrightarrow \overset{\displaystyle H \quad\ Cl}{CH_2{-}CH_2}$$

Ethylene   Chloroethane
(Ethyl chloride)

Addition of HCl to propene gives 2-chloropropane (isopropyl chloride); hydrogen adds to carbon 1 of propene and chlorine adds to carbon 2. If the orientation of addition were reversed, 1-chloropropane (propyl chloride) would form. The observed result is that almost no 1-chloropropane forms. Because 2-chloropropane is the observed product, we say that addition of HCl to propene is **regioselective**.

Regioselective reaction A reaction in which one direction of bond forming or bond breaking occurs in preference to all other directions

$$\overset{2}{C}H_3\overset{1}{C}H{=}CH_2 + HCl \longrightarrow \overset{\displaystyle Cl\quad\ H}{CH_3CH{-}CH_2} \qquad \overset{\displaystyle H\quad\ Cl}{CH_3CH{-}CH_2}$$

Propene   2-Chloropropane   1-Chloropropane
(not formed)

This regioselectivity was noted by Vladimir Markovnikov (1838–1904), who made the following generalization, known as **Markovnikov's rule**: In the addition of HX (where X = halogen) to an alkene, hydrogen adds to the doubly bonded carbon that already has the greater number of hydrogens bonded to it; halogen adds to the other carbon.

Markovnikov's rule In the addition of HX to an alkene, hydrogen adds to the carbon of the double bond having the greater number of hydrogens

Markovnikov's rule is often paraphrased as "the rich get richer."

Example 12.6 Addition of HX to an Alkene

Draw a structural formula for the product of each alkene addition reaction.

(a) $\overset{\displaystyle CH_3}{CH_3C{=}CH_2} + HI \longrightarrow$
         (b) 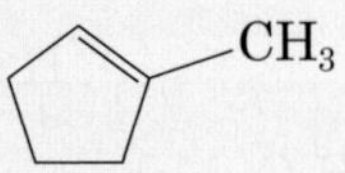 (cyclopentene with CH₃) + HCl $\longrightarrow$

Strategy

Use Markovnikov's rule to predict the structural formula for the product of each reaction. The H from the HI and HCl will add to the carbon of the double bond that already has the greater number of H atoms bonded to it.

Solution

(a) Markovnikov's rule predicts that the hydrogen of HI adds to carbon 1 and iodine adds to carbon 2 to give 2-iodo-2-methylpropane.

(b) H adds to carbon 2 of the ring and Cl adds to carbon 1 to give 1-chloro-1-methylcyclopentane.

(a)

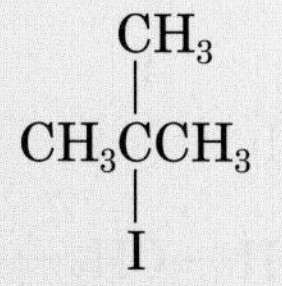

$$CH_3CCH_3$$

with CH_3 above and I below the central carbon

2-Iodo-2-methylpropane

(b)

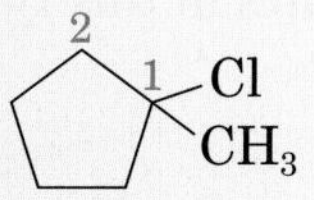

1-Chloro-1-methylcyclopentane

(cyclopentane ring with Cl and CH_3 on carbon 1, carbon 2 labeled)

Problem 12.6

Draw a structural formula for the product of each alkene addition reaction.

(a) $CH_3CH{=}CH_2 + HBr \longrightarrow$

(b) 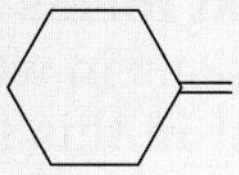 (cyclohexane ring)$={=}CH_2 + HBr \longrightarrow$

Markovnikov's rule tells us what happens when we add HCl, HBr, or HI to a carbon–carbon double bond. We know that in the addition of HCl or another halogen acid, one bond of the double bond and the H—Cl bond are broken and that new C—H and C—Cl bonds form. But chemists also want to know how this conversion happens. Are the C=C and H—X bonds broken and both new covalent bonds formed all at the same time? Or does this reaction take place in a series of steps? If the latter, what are these steps and in what order do they take place?

Before we tackle specific mechanisms, let us take a moment for an overview of the most common steps we will encounter and see how these steps build on the chemistry already presented.

Pattern 1: Add a proton. In Section 8.1, we learned that "an acid is a proton donor, a base is a proton acceptor, and an acid–base reaction is a proton-transfer reaction." We also saw that we can use curved arrows to show how a proton-transfer reaction takes place as, for example, in the acid–base reaction between acetic acid and ammonia to form acetate ion and ammonium ion.

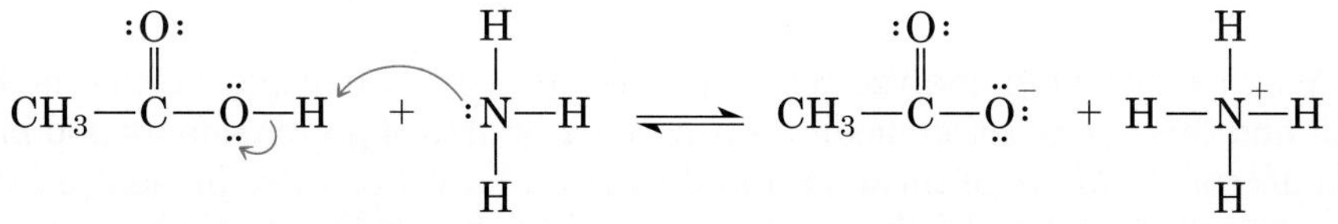

| Acetic acid (proton donor) | Ammonia (proton acceptor) | Acetate ion | Ammonium ion |

$$CH_3{-}\overset{\overset{\displaystyle :O:}{\|}}{C}{-}\overset{..}{O}{-}H \;+\; :N{-}H \rightleftharpoons CH_3{-}\overset{\overset{\displaystyle :O:}{\|}}{C}{-}\overset{..}{\overset{..}{O}}{:}^{-} \;+\; H{-}\overset{\overset{\displaystyle H}{|}}{\underset{\displaystyle H}{N}}{}^{+}{-}H$$

Pattern 2: Take a proton away. If we run the above reaction in reverse, then it corresponds to taking a proton away from the ammonium ion and transferring it to the acetate ion. We can also use curved arrows to show the flow of electron pairs in this type of reaction.

Pattern 3: Reaction of an electrophile and a nucleophile to form a new covalent bond. Another characteristic pattern is the reaction between an **electrophile** (an electron-poor species that can accept a pair of electrons to form a new covalent bond) and a **nucleophile** (an electron-rich species that can donate a pair of electrons to form a new covalent bond). An example of this type of reaction is that between a carbocation and a halide ion.

The driving force behind this reaction is the strong attraction between the positive and negative charges of the reacting species and the energy released when the covalent bond forms. The following equation shows the flow of electron pairs in this type of reaction.

$$\underset{\substack{\text{A carbocation} \\ \text{(an electrophile)}}}{\overset{+}{CH_2}CH-CH_2-CH_3} \; + \; \underset{\substack{\text{Chloride ion} \\ \text{(a nucleophile)}}}{:\overset{..}{\underset{..}{Cl}}:^-} \quad \longrightarrow \quad \underset{}{\overset{\displaystyle :\overset{..}{\underset{..}{Cl}}: \atop |}{CH_3CH}-CH_2-CH_3}$$

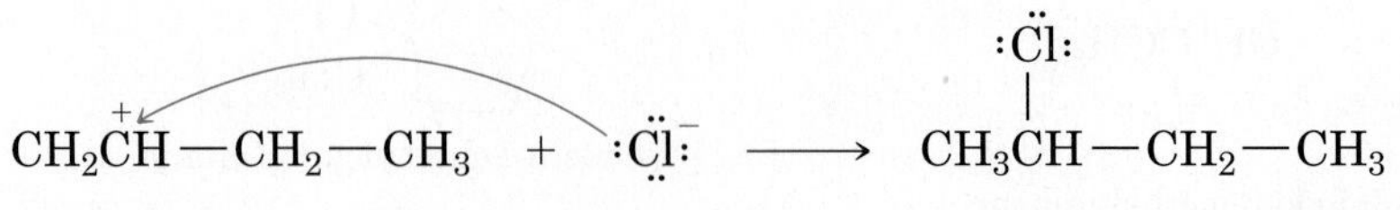

Pattern 4: Reaction of a proton donor with a carbon–carbon double bond to form a new covalent bond. The double bond provides the pair of electrons that forms a new covalent bond. This pattern is typical in all alkene reactions in which the reaction is catalyzed by an acid. As you study it, note that in this step, the carbon–carbon double bond serves as the nucleophile that provides the electron pair to form the new covalent bond. Remember that in a carbon–carbon double bond, two pairs of electrons are shared between the two carbons. An acid–base reaction in which a double bond provides the pair of electrons for the hydrogen transfer creates a carbocation. And remember that, as shown in Section 8.1, a proton, H^+, does not exist as such in aqueous solution. Instead, it immediately combines with a water molecule to form the hydronium ion, H_3O^+. The following reaction shows the flow of electrons in this type of reaction.

$$\underset{\substack{\text{An alkene} \\ \text{(proton acceptor)}}}{CH_3-CH=CH-CH_3} + \underset{\substack{\text{Hydronium ion} \\ \text{(proton donor)}}}{H-\overset{+}{\underset{\underset{H}{|}}{\overset{..}{O}}}-H} \longrightarrow \underset{\substack{\text{A carbocation} \\ \text{(an electrophile)}}}{\overset{+}{CH_2}CH-CH_2-CH_3} + H-\overset{..}{\underset{..}{O}}-H$$

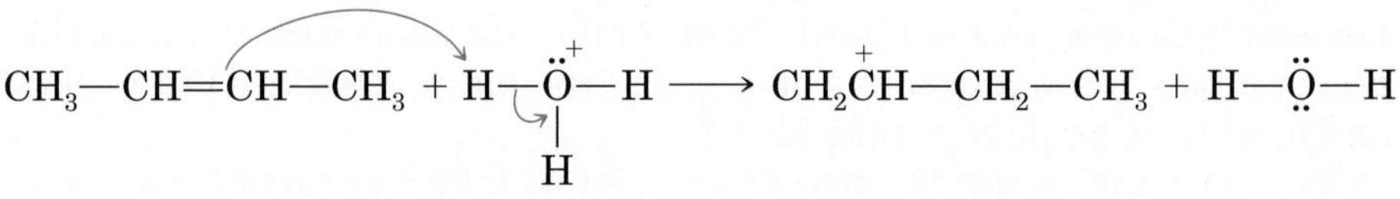

While the above equation is the most accurate way to show the proton transfer in an aqueous solution, we will simplify the equation to show just the proton and the formation of the new covalent bond.

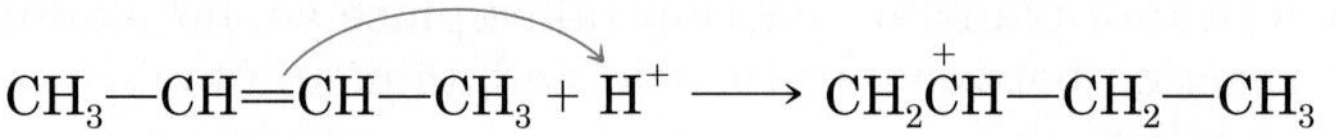

$$\underset{\substack{\text{An alkene} \\ \text{(proton acceptor)}}}{CH_3-CH=CH-CH_3} + \underset{\substack{\text{A proton}}}{H^+} \longrightarrow \underset{\substack{\text{A carbocation} \\ \text{(an electrophile)}}}{\overset{+}{CH_2}CH-CH_2-CH_3}$$

When we come to analyze reactions taking place in biological systems, we will find that the reaction mediums are rich mixtures of proton donors and proton acceptors. Many of these proton donors and acceptors are present on the enzymes that catalyze biochemical reactions. In fact, some biochemists even talk about wraparound enzymes, which have within their three-dimensional structures both proton acceptors and proton donors. It is the presence of these proton donors and acceptors that gives enzymes their remarkable catalytic ability.

Reaction mechanism A step-by-step description of how a chemical reaction occurs

Chemists account for the addition of HX to an alkene by defining a two-step **reaction mechanism**, which we illustrate for the reaction of 2-butene with hydrogen chloride to give 2-chlorobutane. Step 1 is the addition of H^+ to 2-butene. To show this addition, we use a **curved arrow** that shows the repositioning of an electron pair from its origin (the tail of the arrow) to its new location (the head of the arrow). Recall that we used curved arrows in Section 8.1 to show bond breaking and bond formation in proton-transfer reactions. We now use curved arrows in the same way to show bond breaking and bond formation in a reaction mechanism.

Carbocation A species containing a carbon atom with only three bonds to it and bearing a positive charge

Step 1 results in the formation of an organic cation. One carbon atom in this cation has only six electrons in its valence shell, so it carries a charge of $+1$. A species containing a positively charged carbon atom is called a **carbocation** (*carbon* + *cation*). Carbocations are classified as primary (1°), secondary (2°), or tertiary (3°) depending on the number of carbon groups bonded to the carbon bearing the positive charge.

Mechanism: Addition of HCl to 2-Butene

Step 1: Add a proton. Reaction of the carbon–carbon double bond of the alkene with H^+ forms a 2° carbocation intermediate. In forming this intermediate, one bond of the double bond breaks and its pair of electrons is used to form a new covalent bond with H^+. One carbon of the double bond is then left with only six electrons in its valence shell and therefore has a positive charge.

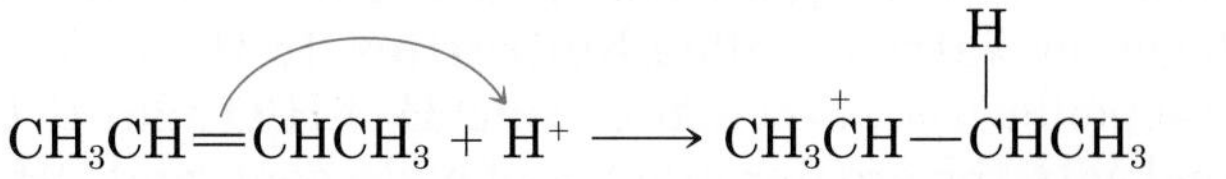

$$CH_3CH=CHCH_3 + H^+ \longrightarrow CH_3\overset{+}{C}H-\overset{\overset{H}{|}}{C}HCH_3$$

A 2° carbocation intermediate

Step 2: Reaction of an electrophile and a nucleophile to form a new covalent bond. Reaction of the 2° carbocation intermediate with a chloride ion completes the valence shell of carbon and gives 2-chlorobutane.

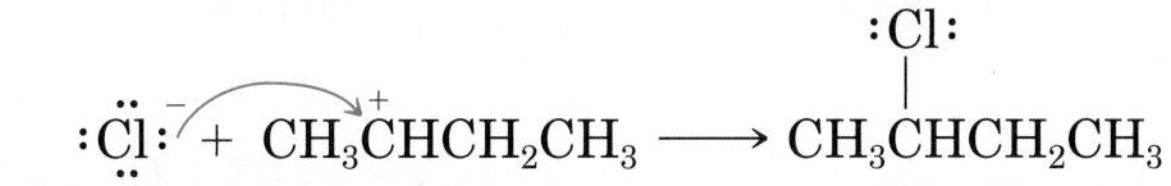

$$:\!\overset{..}{\underset{..}{Cl}}\!:^- + CH_3\overset{+}{C}HCH_2CH_3 \longrightarrow CH_3\overset{\overset{:\overset{..}{Cl}:}{|}}{C}HCH_2CH_3$$

Chloride ion · A 2° carbocation intermediate · 2-Chlorobutane (*sec*-Butyl chloride)

Example 12.7 Mechanism of Addition of HX to an Alkene

Propose a two-step mechanism for the addition of HI to methylenecyclohexane to give 1-iodo-1-methylcyclohexane.

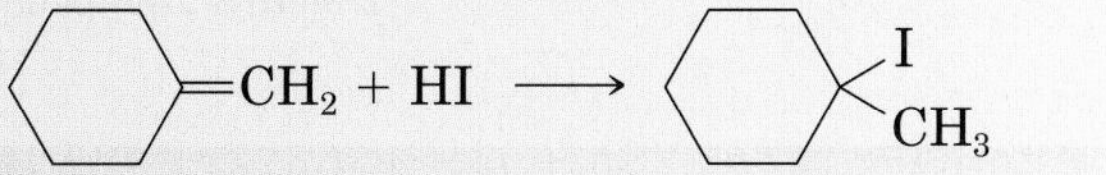

Methylenecyclohexane · 1-Iodo-1-methylcyclohexane

Strategy

The mechanism for the addition of HI to an alkene is similar to the two-step mechanism proposed for the addition of HCl to 2-butene.

Solution

Step 1: Add a proton. In this step, the carbon–carbon double bond is a nucleophile and H^+ is the electrophile. Reaction of H^+ with the carbon–carbon double bond forms a new C—H bond with the carbon bearing the greater number of hydrogens and gives a 3° carbocation intermediate.

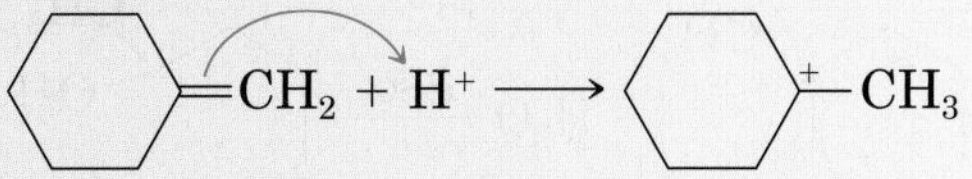

A 3° carbocation intermediate

Step 2: Reaction of an electrophile and a nucleophile to form a new covalent bond. Reaction of the 3° carbocation intermediate with iodide ion completes the valence shell of carbon and gives the product.

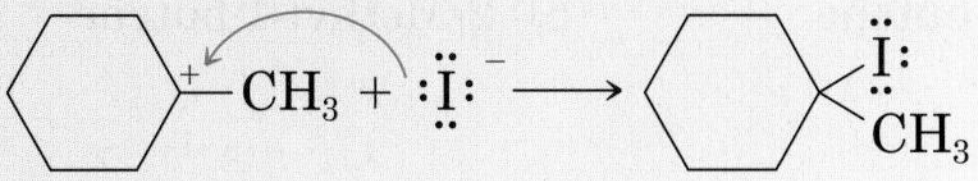

Problem 12.7

Propose a two-step mechanism for the addition of HBr to 1-methylcyclohexene to give 1-bromo-1-methylcyclohexane.

B. Addition of Water: Acid-Catalyzed Hydration

Hydration Addition of water

Most industrial ethanol is made by the acid-catalyzed hydration of ethylene.

In the presence of an acid catalyst, most commonly concentrated sulfuric acid, water adds to the carbon–carbon double bond of an alkene to give an alcohol. Addition of water is called **hydration**. In the case of simple alkenes, hydration follows Markovnikov's rule: H of H_2O adds to the carbon of the double bond with the greater number of hydrogens, and OH of H_2O adds to the carbon with the smaller number of hydrogens.

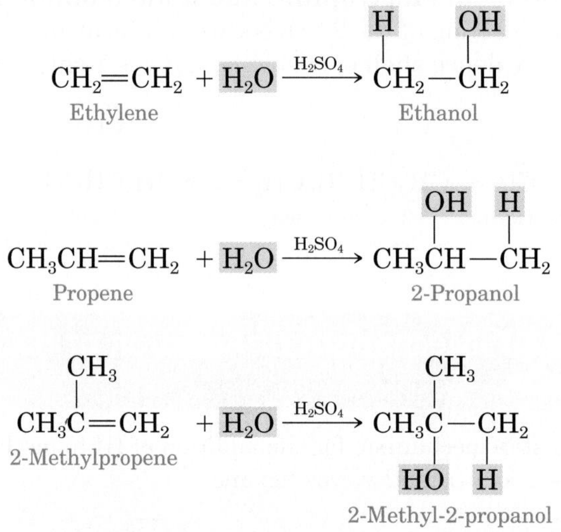

$$CH_2{=}CH_2 + H_2O \xrightarrow{H_2SO_4} \overset{\displaystyle H \quad\ OH}{CH_2{-}CH_2}$$

Ethylene   Ethanol

$$CH_3CH{=}CH_2 + H_2O \xrightarrow{H_2SO_4} \overset{\displaystyle OH \quad H}{CH_3CH{-}CH_2}$$

Propene   2-Propanol

$$\overset{\displaystyle CH_3}{CH_3C{=}CH_2} + H_2O \xrightarrow{H_2SO_4} \overset{\displaystyle CH_3}{\underset{\displaystyle HO \quad H}{CH_3C{-}CH_2}}$$

2-Methylpropene   2-Methyl-2-propanol

Example 12.8 Acid-Catalyzed Hydration of an Alkene

Draw a structural formula for the alcohol formed by the acid-catalyzed hydration of 1-methylcyclohexene.

Strategy

Markovnikov's rule predicts that H adds to the carbon with the greater number of hydrogens.

Solution

H adds to carbon 2 of the cyclohexene ring, and OH then adds to carbon 1.

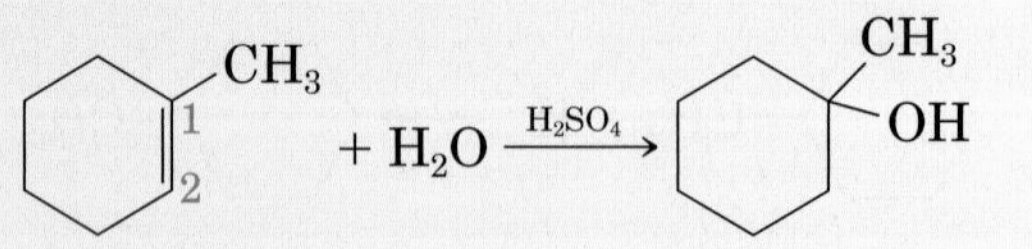

1-Methylcyclohexene $+ H_2O \xrightarrow{H_2SO_4}$ 1-Methylcyclohexanol

Problem 12.8

Draw a structural formula for the alcohol formed by acid-catalyzed hydration of each alkene:
(a) 2-Methyl-2-butene   (b) 2-Methyl-1-butene

The mechanism for the acid-catalyzed hydration of an alkene is similar to what we proposed for the addition of HCl, HBr, and HI to an alkene and is illustrated by the hydration of propene. This mechanism is consistent with the fact that acid is a catalyst. One H^+ is consumed in Step 1, but another is generated in Step 3.

Mechanism: Acid-Catalyzed Hydration of Propene

Step 1: Add a proton. Addition of H^+ to the carbon of the double bond with the greater number of hydrogens gives a 2° carbocation intermediate.

$$CH_3CH{=}CH_2 + H^+ \longrightarrow CH_3\overset{+}{C}HCH_2$$

A 2° carbocation
intermediate

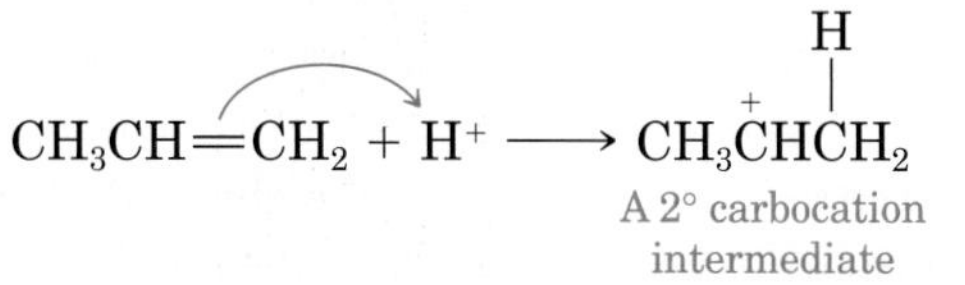

Step 2: Reaction of an electrophile and a nucleophile to form a new covalent bond. The carbocation intermediate completes its valence shell by forming a new covalent bond with an unshared pair of electrons of the oxygen atom of H_2O to give an **oxonium ion**.

$$CH_3\overset{+}{C}HCH_3 + {:}\overset{..}{O}{-}H \longrightarrow CH_3CHCH_3$$

An oxonium ion

Oxonium ion An ion in which oxygen is bonded to three other atoms and bears a positive charge

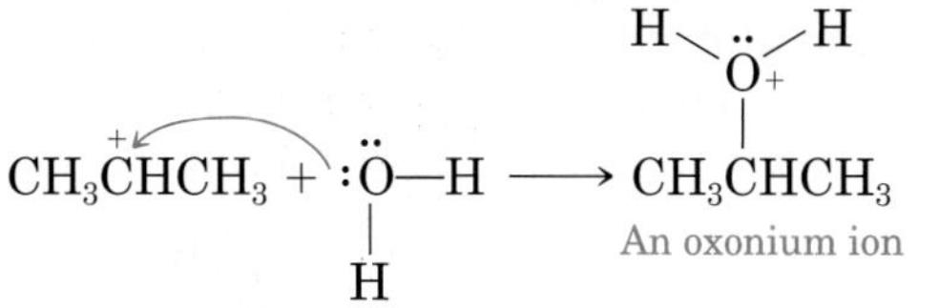

Step 3: Take a proton away. Loss of H^+ from the oxonium ion gives the alcohol and generates a new H^+ catalyst.

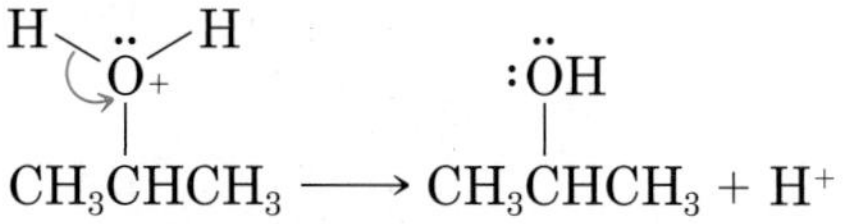

$$CH_3CHCH_3 \longrightarrow CH_3CHCH_3 + H^+$$

Example 12.9 Acid-Catalyzed Hydration of an Alkene

Propose a three-step reaction mechanism for the acid-catalyzed hydration of methylenecyclohexane to give 1-methylcyclohexanol.

Strategy

The reaction mechanism for the acid-catalyzed hydration of methylenecyclohexane is similar to the three-step mechanism proposed for the acid-catalyzed hydration of propene.

Solution

Step 1: Add a proton. Reaction of the carbon–carbon double bond with H^+ gives a 3° carbocation intermediate.

$$\overset{\textstyle\bigcirc}{}{=}CH_2 + H^+ \longrightarrow \overset{\textstyle\bigcirc}{}{-}CH_3$$

Methylenecyclohexane   A 3° carbocation
intermediate

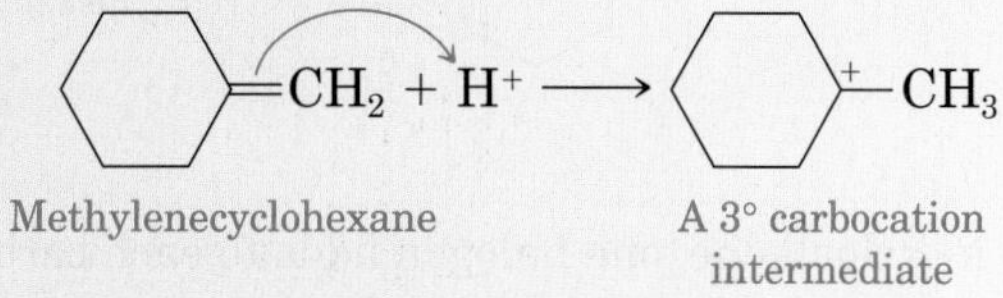

Step 2: Reaction of the carbocation intermediate with water completes the valence shell of carbon and gives an oxonium ion.

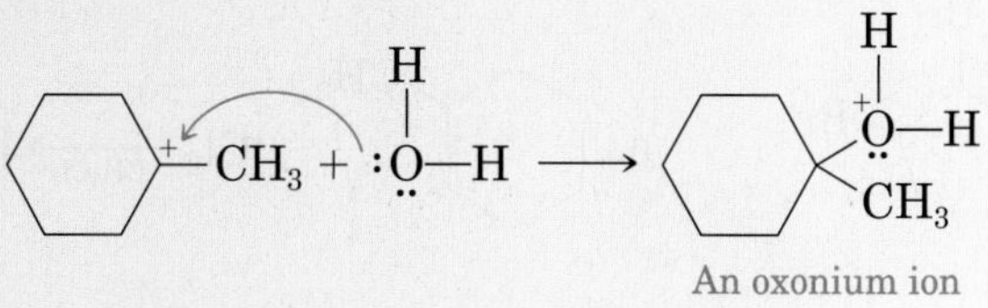

$$\overset{\textstyle\bigcirc}{}{-}CH_3 + {:}\overset{..}{O}{-}H \longrightarrow$$

An oxonium ion

Step 3: Take a proton away. Loss of H^+ from the oxonium ion completes the reaction and generates a new H^+ catalyst.

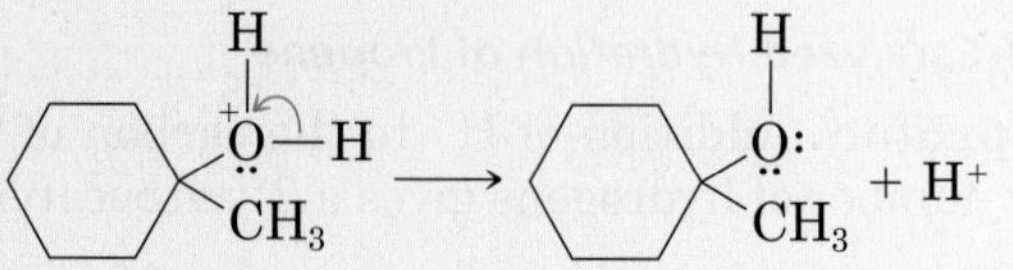

Problem 12.9

Propose a three-step reaction mechanism for the acid-catalyzed hydration of 1-methylcyclohexene to give 1-methylcyclohexanol.

C. Addition of Bromine and Chlorine (Halogenation)

Chlorine, Cl_2, and bromine, Br_2, react with alkenes at room temperature by addition of halogen atoms to the carbon atoms of the double bond. This reaction is generally carried out by mixing the pure reagents together or by mixing them in an inert solvent such as dichloromethane, CH_2Cl_2.

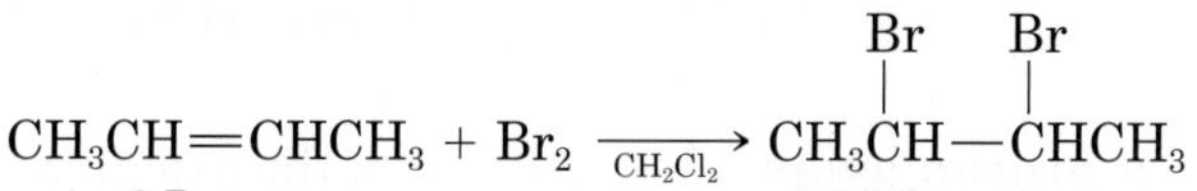

2-Butene 2,3-Dibromobutane

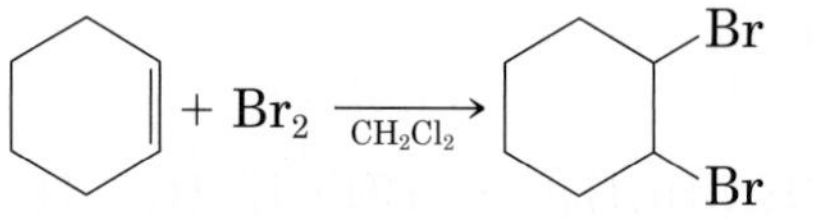

Cyclohexene 1,2-Dibromocyclohexane

Addition of bromine is a useful qualitative test for the presence of an alkene. If we dissolve bromine in carbon tetrachloride, the solution is red. In contrast, alkenes and dibromoalkanes are colorless. If we mix a few drops of the red bromine solution with an unknown sample suspected of being an alkene, disappearance of the red color as bromine adds to the double bond tells us that an alkene is, indeed, present.

Example 12.10 Addition of Halogens to an Alkene

Complete these reactions.

(a) 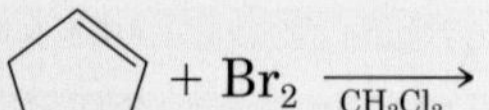

(b) 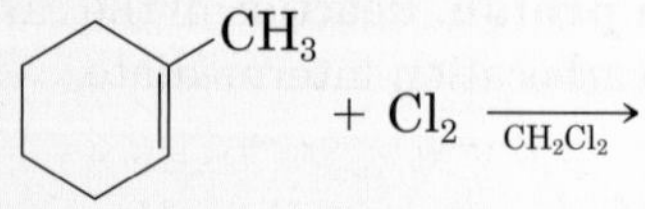

Strategy

When Br_2 or Cl_2 are added to a cycloalkene, one halogen adds to each carbon of the double bond.

Solution

(a) 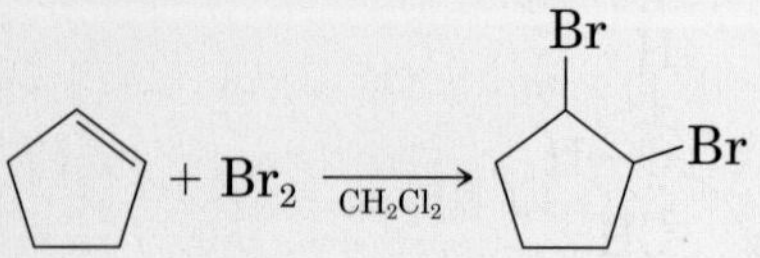

(b) 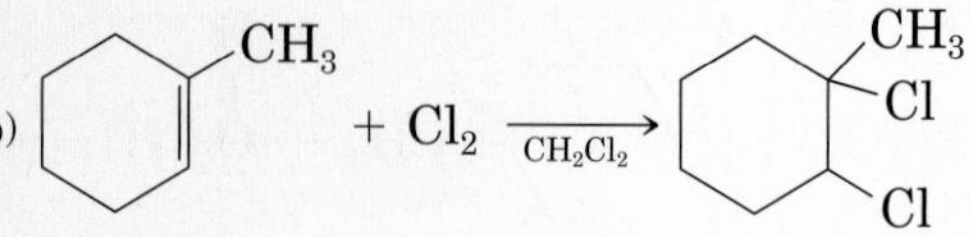

Problem 12.10

Complete these reactions.

(a) $CH_3CCH = CH_2 + Br_2 \xrightarrow{CH_2Cl_2}$
with CH_3 substituents above and below the central carbon

(b) cyclohexane ring with $=CH_2$ group $+ Cl_2 \xrightarrow{CH_2Cl_2}$

D. Addition of Hydrogen: Reduction (Hydrogenation)

Virtually all alkenes react quantitatively with molecular hydrogen, H_2, in the presence of a transition metal catalyst to give alkanes. Commonly used transition metal catalysts include platinum, palladium, ruthenium, and nickel. Because the conversion of an alkene to an alkane involves reduction by hydrogen in the presence of a catalyst, the process is called **catalytic reduction** or, alternatively, **catalytic hydrogenation**.

In Section 21.3, we will see how catalytic hydrogenation is used to solidify liquid vegetable oils so that margarines and semisolid cooking fats are formed.

$$\underset{\text{trans-2-Butene}}{\overset{H_3C}{\underset{H}{>}}C=C\overset{H}{\underset{CH_3}{<}}} + H_2 \xrightarrow[25°C,\ 3\ atm]{Pd} \underset{\text{Butane}}{CH_3CH_2CH_2CH_3} \xleftarrow[25°C,\ 3\ atm]{Pd} H_2 + \underset{\text{cis-2-Butene}}{\overset{H}{\underset{CH_3}{>}}C=C\overset{H}{\underset{CH_3}{<}}}$$

$$\underset{\text{Cyclohexene}}{\text{cyclohexene}} + H_2 \xrightarrow[25°C,\ 3\ atm]{Pd} \underset{\text{Cyclohexane}}{\text{cyclohexane}}$$

The metal catalyst is used in the form of a finely powdered solid. The reaction is carried out by dissolving the alkene in ethanol or another nonreacting organic solvent, adding the solid catalyst, and exposing the mixture to hydrogen gas at pressures ranging from 1 to 150 atm.

Mechanism: Catalytic Reduction

The transition metal catalysts used in catalytic hydrogenation are able to absorb large quantities of hydrogen onto their surfaces, probably by forming metal–hydrogen bonds. Similarly, alkenes are absorbed on metal surfaces with the formation of carbon–metal bonds. It is noted here that carbon–metal bonds are formed, but this is not really addressed in Figure 12.1. Addition of hydrogen atoms to an alkene occurs in two steps (Figure 12.1).

FIGURE 12.1 The addition of hydrogen to an alkene involving a transition metal catalyst. (*a*) Hydrogen and the alkene are absorbed on the metal surface, and (*b*) one hydrogen atom is transferred to the alkene, forming one new C—H bond. The other carbon remains absorbed on the metal surface. (*c*) A second C—H bond is formed, and the alkene is released.

12.6 What Are the Important Polymerization Reactions of Ethylene and Substituted Ethylenes?

A. Structure of Polyethylenes

Polymer From the Greek *poly*, many, and *meros*, part; any long-chain molecule synthesized by bonding together many single parts called monomers

Monomer From the Greek *mono*, single, and *meros*, part; the simplest nonredundant unit from which a polymer is synthesized

From the perspective of the chemical industry, the single most important reaction of alkenes is the formation of **chain-growth polymers** (Greek: *poly*, many, and *meros*, part). In the presence of certain compounds called initiators, many alkenes form polymers made by the stepwise addition of **monomers** (Greek: *mono*, one, and *meros*, part) to a growing polymer chain, as illustrated by the formation of polyethylene from ethylene. In alkene polymers of industrial and commercial importance, n is a large number, typically several thousand.

$$n\,CH_2\!=\!CH_2 \xrightarrow[\text{(polymerization)}]{\text{initiator}} \left(CH_2CH_2\right)_n$$

Ethylene Polyethylene

To show the structure of a polymer, we place parentheses around the repeating monomer unit. The structure of an entire polymer chain can be reproduced by repeating this enclosed structure in both directions. A subscript n is placed outside the parentheses to indicate that this unit is repeated n times, as illustrated for the conversion of propylene to polypropylene.

(a)

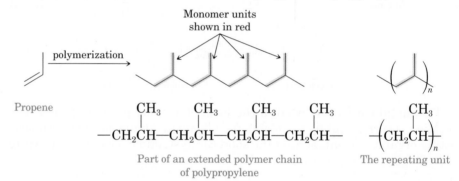

Monomer units
shown in red

Propene

Part of an extended polymer chain
of polypropylene

The repeating unit

The most common method of naming a polymer is to attach the prefix **poly-** to the name of the monomer from which the polymer is synthesized—for example, polyethylene and polystyrene. When the name of the monomer consists of two words (for example, the monomer vinyl chloride), its name is enclosed in parentheses.

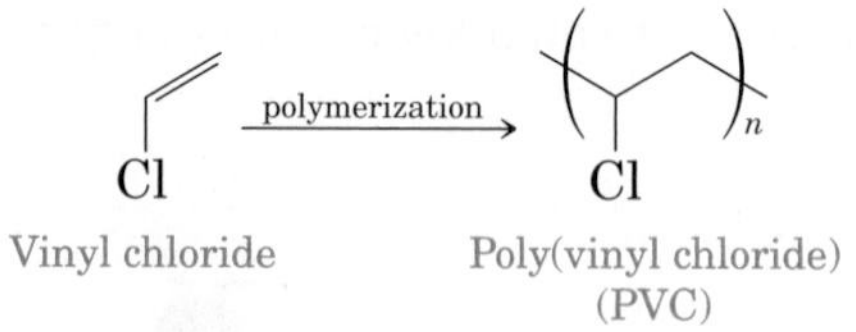

(b)

Some articles made from chain-growth polymers. (*a*) Saran Wrap, a copolymer of vinyl chloride and 1,1-dichloroethylene. (*b*) Articles made from polystyrene.

Vinyl chloride Poly(vinyl chloride) (PVC)

Table 12.2 lists several important polymers derived from ethylene and substituted ethylene, along with their common names and most important uses.

B. Low-Density Polyethylene (LDPE)

Peroxide Any compound that contains an —O—O— bond as, for example, hydrogen peroxide, H—O—O—H

The first commercial process for ethylene polymerization used **peroxide** initiators at 500°C and 1000 atm and yielded a tough, transparent polymer known as **low-density polyethylene (LDPE)**. At the molecular level,

Chemical Connections 12E

Recycling Plastics

Plastics are polymers that can be molded when hot and that retain their shape when cooled. Because they are durable and lightweight, plastics are probably the most versatile synthetic materials in existence. In fact, the current production of plastics in the United States exceeds the U.S. production of steel. Plastics have come under criticism, however, for their role in the solid waste crisis. They

and other lightweight materials. After any remaining labels and adhesives are eliminated with a detergent wash, the PET chips are dried. The PET produced by this method is 99.9% free of contaminants and sells for about half the price of the virgin material. The biggest market for recycled PET in 2005 was fibers. The carpet maker Mohawk Industries, for example, starts with about 250 million lb of recycled

Code	Polymer	Common Uses
1 PET	poly(ethylene terephthalate)	soft drink bottles, household chemical bottles, films, textile fibers
2 HDPE	high-density polyethylene	milk and water jugs, grocery bags, squeezable bottles
3 V	poly(vinyl chloride), PVC	shampoo bottles, pipes, shower curtains, vinyl siding, wire insulation, floor tiles
4 LDPE	low-density polyethylene	shrink wrap, trash and grocery bags, sandwich bags, squeeze bottles
5 PP	polypropylene	plastic lids, clothing fibers, bottle caps, toys, diaper linings
6 PS	polystyrene	Styrofoam cups, egg cartons, disposable utensils, packaging materials, appliances
7	all other plastics	various

account for approximately 21% of the volume and 8% of the weight of solid wastes, with most plastic waste consisting of disposable packaging and wrapping.

Six types of plastics are commonly used for packaging applications. In 1988, manufacturers adopted recycling code letters developed by the Society of the Plastics Industry as a means of identifying them.

Currently, only poly(ethylene terephthalate) (PET) and high-density polyethylene (HDPE) are recycled in large quantities. In fact, bottles made of these plastics account for more than 99% of the plastics recycled in the United States.

The synthesis and structure of PET, a polyester, is described in Section 19.6B.

The process for recycling most plastics is simple, with separation of the plastic from other contaminants being the most labor-intensive step. For example, PET soft drink bottles usually have a paper label and adhesive that must be removed before the PET can be reused. Recycling begins with hand or machine sorting, after which the bottles are chopped into small chips. Any ferrous metals are removed by magnets. Any nonferrous metal contaminants are removed by electric eddy currents that cause them to jump like fleas into a bin as they move down a conveyor belt during the soarting process. An air cyclone then removes paper

PET bottles per year and ends up with 80 to 100 million sq yards of carpet. The largest domestic use of recycled HDPE resins in 2005 was bottles.

These students are wearing jackets made from recycled PET soda bottles.

Polyethylene films are produced by extruding molten plastic through a ring-like gap and inflating the film into a balloon.

The Stock Market

Table 12.2 Polymers Derived from Ethylene and Substituted Ethylenes, Along with their Common Names and Most Important Uses.

Monomer Formula	Common Name	Polymer Name(s) and Common Uses
$CH_2{=}CH_2$	ethylene	polyethylene, Polythene; break-resistant containers and packaging materials
$CH_2{=}CHCH_3$	propylene	polypropylene, Herculon; textile and carpet fibers
$CH_2{=}CHCl$	vinyl chloride	poly(vinyl chloride), PVC; construction tubing
$CH_2{=}CCl_2$	1,1-dichloroethylene	poly(1,1-dichloroethylene); Saran Wrap is a copolymer with vinyl chloride
$CH_2{=}CHCN$	acrylonitrile	polyacrylonitrile, Orlon; acrylics and acrylates
$CF_2{=}CF_2$	tetrafluoroethylene	polytetrafluoroethylene, PTFE; Teflon, nonstick coatings
$CH_2{=}CHC_6H_5$	styrene	polystyrene, Styrofoam; insulating materials
$CH_2{=}CHCOOCH_2CH_3$	ethyl acrylate	poly(ethyl acrylate), latex paint
$CH_2{=}\underset{\underset{CH_3}{\mid}}{C}COOCH_3$	methyl methacrylate	poly(methyl methacrylate), Lucite; Plexiglas; glass substitutes

LDPE chains are highly branched, with the result that they do not pack well together and the London dispersion forces (Section 5.7A) between them are weak. LDPE softens and melts at about 115°C, which means that it cannot be used in products that will be exposed to boiling water.

Today, approximately 65% of all LDPE is used for the manufacture of films by the blow-molding technique illustrated in Figure 12.2. LDPE film is inexpensive, which makes it ideal for packaging such consumer items as baked goods and vegetables and for the manufacture of trash bags.

C. High-Density Polyethylene (HDPE)

In the 1950s, Karl Ziegler of Germany and Giulio Natta of Italy developed an alternative method for the polymerization of alkenes, which does not rely on peroxide initiators. Polyethylene from Ziegler-Natta systems, termed **high-density polyethylene (HDPE)**, has little chain branching. Consequently, its chains pack together more closely than those of LDPE, with the result that the London dispersion forces between chains of HDPE are stronger than those in LDPE.

Linear polyethylene
(high density)

HDPE has a higher melting point than LDPE and is three to ten times stronger.

Approximately 45% of all HDPE products are made by the blow-molding process shown in Figure 12.3. HDPE is used for consumer items such as milk and water jugs, grocery bags, and squeezable bottles.

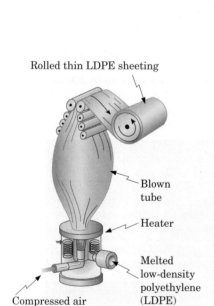

Rolled thin LDPE sheeting

Blown tube

Heater

Melted low-density polyethylene (LDPE)

Compressed air

FIGURE 12.2 Fabrication of LDPE film. A tube of melted LDPE along with a jet of compressed air is forced through an opening and blown into a gigantic, thin-walled bubble. The film is then cooled and taken up onto a roller. This double-walled film can be slit down the side to give LDPE film or sealed at points along its length to make LDPE bags.

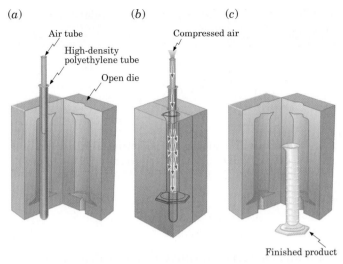

FIGURE 12.3 Blow molding an HDPE container. (*a*) A short length of HDPE tubing is placed in an open die and the die is closed, sealing the bottom of the tube. (*b*) Compressed air is forced into the hot polyethylene/die assembly, and the tubing is literally blown up to take the shape of the mold. (*c*) After the assembly cools, the die is opened, and there is the container!

Summary of Key Questions

OWL Sign in at **www.cengage.com/owl** to develop problem-solving skills and complete online homework assigned by your professor.

Section 12.1 What Are Alkenes and Alkynes?

- An **alkene** is an unsaturated hydrocarbon that contains a carbon–carbon double bond.
- An **alkyne** is an unsaturated hydrocarbon that contains a carbon–carbon triple bond.

Section 12.2 What Are the Structures of Alkenes and Alkynes?

- The structural feature that makes ***cis-trans* stereo-isomerism** possible in alkenes is restricted rotation about the two carbons of the double bond. The *cis* or *trans* configuration of an alkene is determined by the orientation of the atoms of the parent chain about the double bond. If atoms of the parent chain are located on the same side of the double bond, the configuration of the alkene is *cis*; if they are located on opposite sides, the configuration is *trans*.

Section 12.3 How Do We Name Alkenes and Alkynes?

- In IUPAC names, the presence of a carbon–carbon double bond is indicated by a prefix showing the number of carbons in the parent chain and the ending **-ene.** Substituents are numbered and named in alphabetical order.
- The presence of a carbon–carbon triple bond is indicated by a prefix that shows the number of carbons in the parent chain and the ending **-yne.**

- The carbon atoms of the double bond of a cycloalkene are numbered 1 and 2 in the direction that gives the smaller number to the first substituent.
- Compounds containing two double bonds are called **dienes**, those with three double bonds are called **trienes**, and those containing four or more double bonds are called **polyenes**.

Section 12.4 What Are the Physical Properties of Alkenes and Alkynes?

- Because alkenes and alkynes are nonpolar compounds and the only interactions between their molecules are London dispersion forces, their physical properties are similar to those of alkanes with similar carbon skeletons.

Section 12.5 What Are the Characteristic Reactions of Alkenes?

- A characteristic reaction of alkenes is addition to the double bond.
- In addition, the double bond breaks and bonds to two new atoms or groups of atoms form in its place.
- A **reaction mechanism** is a step-by-step description of how a chemical reaction occurs, including the role of the catalyst (if one is present).
- A **carbocation** contains a carbon with only six electrons in its valence shell and bears a positive charge.

Section 12.6 What Are the Important Polymerization Reactions of Ethylene and Substituted Ethylenes?

- **Polymerization** is the process of bonding together many small **monomers** into large, high-molecular-weight **polymers**.

Summary of Key Reactions

1. **Addition of HX (Hydrohalogenation) (Section 12.6A)** Addition of HX to the carbon–carbon double bond of an alkene follows Markovnikov's rule. The reaction occurs in two steps and involves formation of a carbocation intermediate.

$$\text{(cyclopentene with } CH_3) + HCl \longrightarrow \text{(cyclopentane with } Cl, CH_3)$$

2. **Acid-Catalyzed Hydration (Section 12.6B)** Addition of H_2O to the carbon–carbon double bond of an alkene follows Markovnikov's rule. Reaction occurs in three steps and involves formation of carbocation and oxonium ion intermediates.

$$\underset{\overset{|}{CH_3}}{CH_3C}=CH_2 + H_2O \xrightarrow{H_2SO_4} CH_3\underset{\overset{|}{OH}}{\overset{\overset{|}{CH_3}}{C}}CH_3$$

3. **Addition of Bromine and Chlorine (Halogenation) (Section 12.6C)** Addition to a cycloalkene gives a 1,2-dihalocycloalkane.

$$\text{(cyclohexene)} + Br_2 \xrightarrow{CH_2Cl_2} \text{(1,2-dibromocyclohexane)}$$

Cyclohexene   1,2-Dibromocyclohexane

4. **Reduction: Formation of Alkanes (Hydrogenation) (Section 12.6D)** Catalytic reduction involves addition of hydrogen to form two new C—H bonds.

$$\text{(cyclohexene)} + H_2 \xrightarrow[\text{metal catalyst}]{\text{transition}} \text{(cyclohexane)}$$

5. **Polymerization of Ethylene and Substituted Ethylenes (Section 12.7A)** In polymerization of alkenes, monomer units bond together without the loss of any atoms.

$$n\,CH_2{=}CH_2 \xrightarrow{\text{initiator}} \left(CH_2CH_2\right)_n$$

Problems

⏻**WL** Interactive versions of these problems may be assigned in OWL.

Orange-numbered problems are applied.

Section 12.1 What Are Alkenes and Alkynes?

12.11 Answer true or false.

(a) There are two classes of unsaturated hydrocarbons—alkenes and alkynes.

(b) The bulk of the ethylene used by the chemical industry worldwide is obtained from renewable resources.

(c) Ethylene and acetylene are constitutional isomers.

(d) Cyclohexane and 1-hexene are constitutional isomers.

Section 12.2 What Are the Structures of Alkenes and Alkynes?

12.12 Answer true or false.

(a) Both ethylene and acetylene are planar molecules.

(b) An alkene in which each carbon of the double bond has two different groups bonded to it will show *cis-trans* isomerism.

(c) *Cis-trans* isomers have the same molecular formula but a different connectivity of their atoms.

(d) *Cis*-2-butene and *trans*-2-butene can be interconverted by rotation about the carbon–carbon double bond.

(e) *Cis-trans* isomerism is possible only among appropriately substituted alkenes.

(f) Both 2-hexene and 3-hexene can exist as pairs of *cis-trans* isomers.

(g) Cyclohexene can exist as a pair of *cis-trans* isomers.

(h) 1-Chloropropene can exist as a pair of *cis-trans* isomers.

12.13 What is the difference in structure between a saturated hydrocarbon and an unsaturated hydrocarbon?

12.14 There are three compounds with the molecular formula $C_2H_2Br_2$. Two of these are polar and have a dipole, and one has no dipole (it is not polar). Draw structural formulas for these three compounds and explain why two are polar and the third is not.

12.15 Name and draw structural formulas for all alkenes with the molecular formula C_5H_{10}. As you draw these alkenes, remember that *cis* and *trans* isomers are different compounds and must be counted separately.

12.16 Name and draw structural formulas for alkenes with the molecular formula C_6H_{12} that have the following carbon skeletons (remember *cis* and *trans* isomers).

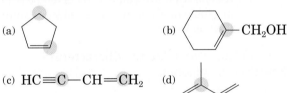

(a)

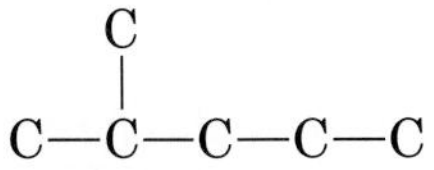

(b)

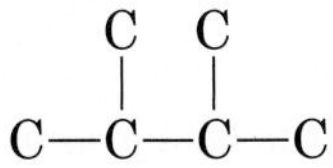

(c)

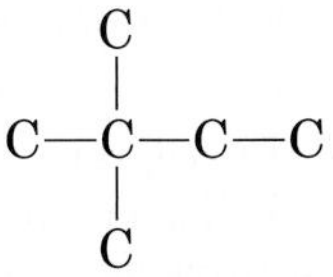

12.17 Draw a structural formula for at least one bromo-alkene with the molecular formula C_5H_9Br that (a) shows *cis/trans* isomerism and (b) does not show *cis/trans* isomerism.

12.18 Each carbon atom in ethane and in ethylene is surrounded by eight valence electrons and has four bonds to it. Explain how the VSEPR model (Section 3.10) predicts a bond angle of 109.5° about each carbon in ethane but an angle of 120° about each carbon in ethylene.

12.19 Predict all bond angles about each highlighted carbon atom.

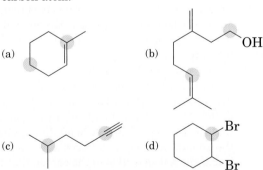

(a)   (b) —CH$_2$OH

(c) HC≡C—CH=CH$_2$ (d)

12.20 Predict all bond angles about each highlighted carbon atom.

(a)   (b) 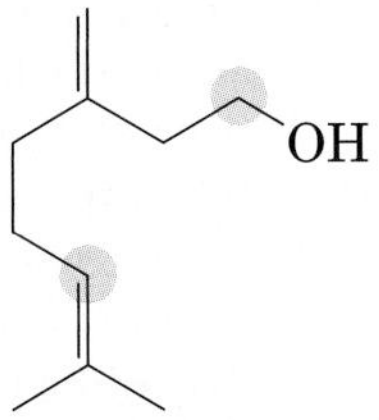   OH

(c)   (d)   Br
                                Br

Section 12.3 How Do We Name Alkenes and Alkynes?

12.21 Answer true or false.

(a) The IUPAC name of an alkene is derived from the name of the longest carbon chain that contains the carbon–carbon double bond.

(b) The IUPAC name of $CH_3CH{=}CHCH_3$ is 1,2-dimethylethene.

(c) 2-Methyl-2-butene shows *cis-trans* isomerism.

(d) 1,2-dimethylcyclohexene shows *cis-trans* isomerism.

(e) The IUPAC name of $CH_2{=}CHCH{=}CHCH_3$ is 1,3-pentadiene.

(f) 1,3-Butadiene has two carbon–carbon double bonds and $2^2 = 4$ stereoisomers are possible for it.

12.22 Draw a structural formula for each compound.

(a) *trans*-2-Methyl-3-hexene

(b) 2-Methyl-3-hexyne

(c) 2-Methyl-1-butene

(d) 3-Ethyl-3-methyl-1-pentyne

(e) 2,3-Dimethyl-2-pentene

12.23 Draw a structural formula for each compound.

(a) 3-Chloropropene

(b) 3-Methylcyclohexene

(c) 1,2-Dimethylcyclohexene

(d) *trans*-3,4-Dimethyl-3-heptene

(e) Cyclopropene

(f) 3-Hexyne

12.24 Write the IUPAC name for each unsaturated hydrocarbon.

(a) $CH_2{=}CH(CH_2)_4CH_3$

(b)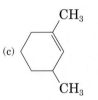

(c)

(d) $(CH_3)_2CHCH{=}C(CH_3)_2$

(e) $CH_3(CH_2)_5C{\equiv}CH$

(f) $CH_3CH_2C{\equiv}CC(CH_3)_3$

(g)

12.25 Write the IUPAC name for each unsaturated hydrocarbon.

(a) 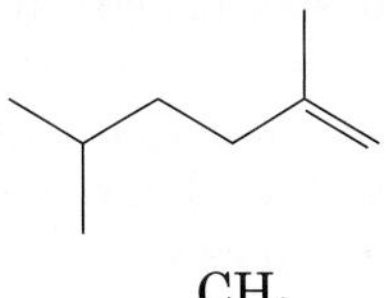   (b)

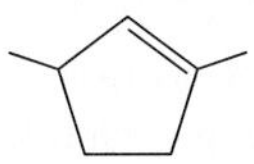

(c) 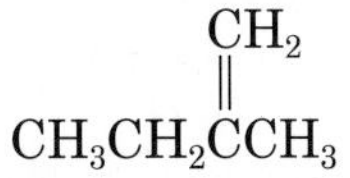    (d)

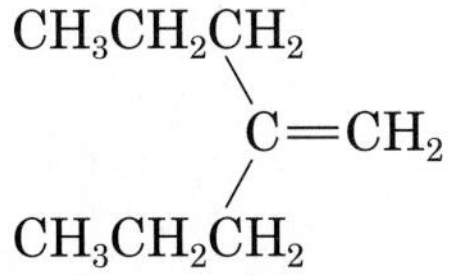

12.26 Explain why each name is incorrect and then write a correct name.

(a) 1-Methylpropene   (b) 3-Pentene

(c) 2-Methylcyclohexene   (d) 3,3-Dimethylpentene

(e) 4-Hexyne   (f) 2-Isopropyl-2-butene

12.27 Explain why each name is incorrect and then write a correct name.

(a) 2-Ethyl-1-propene

(b) 5-Isopropylcyclohexene

(c) 4-Methyl-4-hexene

(d) 2-*sec*-Butyl-1-butene

(e) 6,6-Dimethylcyclohexene

(f) 2-Ethyl-2-hexene

12.28 What structural feature in alkenes makes *cis-trans* isomerism in them possible? What structural feature in cycloalkanes makes *cis-trans* isomerism in them possible? What do these two structural features have in common?

12.29 Which of these alkenes show *cis-trans* isomerism? For each that does, draw structural formulas for both isomers.

(a) 1-Hexene (b) 2-Hexene
(c) 3-Hexene (d) 2-Methyl-2-hexene
(e) 3-Methyl-2-hexene (f) 2,3-Dimethyl-2-hexene

12.30 Which of these alkenes shows *cis-trans* isomerism? For each that does, draw structural formulas for both isomers.

(a) 1-Pentene
(b) 2-Pentene
(c) 3-Ethyl-2-pentene
(d) 2,3-Dimethyl-2-pentene
(e) 2-Methyl-2-pentene
(f) 2,4-Dimethyl-2-pentene

12.31 Cyclodecene exists as both *cis* and *trans* isomers. Draw line-angle formulas for each isomer, showing the configuration of the double bond in each.

12.32 Arachidonic acid is a naturally occurring polyunsaturated fatty acid. Draw a line-angle formula for arachidonic acid showing the *cis* configuration about each double bond.

$$CH_3(CH_2)_4(CH=CHCH_2)_4CH_2CH_2COOH$$
Arachidonic acid

12.33 Following is the structural formula of a naturally occurring unsaturated fatty acid.

$$CH_3(CH_2)_7CH=CH(CH_2)_7COOH$$

The *cis* stereoisomer is named oleic acid, and the *trans* isomer is named elaidic acid. Draw a line-angle formula of each acid, showing clearly the configuration of the carbon–carbon double bond in each.

12.34 If you examine the structural formulas for the following cycloalkenes, you will see that the configuration of the double bond is *cis* in each.

Cyclopentene   Cyclohexene   Cycloheptene

All attempts to synthesize these cycloalkenes in which the double bond has a *trans* configuration have failed. Apparently, it is impossible to have a *trans* configuration in these cycloalkenes. Offer an explanation for why this is so.

12.35 For each molecule that shows *cis-trans* isomerism, draw the *cis* isomer.

(a) (b)

(c) (d)

12.36 Name and draw structural formulas for all compounds with the molecular formula C_5H_{10} that are:

(a) Alkenes that do not show *cis-trans* isomerism
(b) Alkenes that show *cis-trans* isomerism
(c) Cycloalkanes that do not show *cis-trans* isomerism
(d) Cycloalkanes that show *cis-trans* isomerism

12.37 β-Ocimene, a triene found in the fragrance of cotton blossoms and several essential oils, has the IUPAC name *cis*-3,7-dimethyl-1,3,6-octatriene. (*Cis* refers to the configuration of the double bond between carbons 3 and 4, the only double bond in this molecule about which *cis-trans* isomerism is possible.) Draw a structural formula for β-ocimene.

Section 12.4 What Are the Physical Properties of Alkenes and Alkynes?

12.38 Answer true or false.

(a) Alkenes and alkynes are nonpolar molecules.
(b) The physical properties of alkenes are similar to those of alkanes of the same carbon skeletons.
(c) Alkenes that are liquid at room temperature are insoluble in water and when added to water, will float on water.

Section 12.5 What Are the Characteristic Reactions of Alkenes?

12.39 Answer true or false.

(a) Complete combustion of an alkene gives carbon dioxide and water.
(b) Addition reactions of alkenes involve breaking one of the bonds of the carbon–carbon double bond and formation of two new single bonds in its place.
(c) Markovnikov's rule refers to the regioselectivity of addition reactions of carbon–carbon double bonds.
(d) According to Markovnikov's rule, in the addition of HCl, HBr, or HI to an alkene, hydrogen adds to the carbon of the double bond that already has the greater number of hydrogen atoms bonded to it and the halogen adds to the carbon that has the lesser number of hydrogens bonded to it.
(e) A carbocation is a carbon atom with four bonds that bears a positive charge.
(f) The carbocation derived from ethylene is $CH_3CH_2^+$.
(g) The reaction mechanism for the addition of a halogen acid (HX) to an alkene is divided into two steps, (1) formation of a carbocation and (2) reaction of the carbocation with halide ion, which complete the reaction.
(h) Acid-catalyzed addition of H_2O to an alkene is called *hydration*.
(i) If a compound fails to react with Br_2, it is unlikely that the compound contains a carbon–carbon double bond.

(j) Addition of H_2 to a double bond is a reduction reaction.

(k) Catalytic reduction of cyclohexene gives hexane.

(l) According to the mechanism presented in the text for acid-catalyzed hydration of an alkene, the H and —OH groups added to the carbon–carbon double bond both arise from the same molecule of H_2O.

(m) The conversion of ethylene, $CH_2{=}CH_2$, to ethanol, CH_3CH_2OH, is an oxidation reaction.

(n) Acid-catalyzed hydration of 1-butene gives 1-butanol. Acid-catalyzed hydration of 2-butene gives 2-butanol.

12.40 Define *alkene addition reaction*. Write an equation for an addition reaction of propene.

12.41 What reagent and/or catalysts are necessary to bring about each conversion?

(a) $CH_3CH{=}CHCH_3 \longrightarrow CH_3CH_2\overset{\overset{\displaystyle Br}{|}}{C}HCH_3$

(b) $CH_3\overset{\overset{\displaystyle CH_3}{|}}{C}{=}CH_2 \longrightarrow CH_3\overset{\overset{\displaystyle CH_3}{|}}{\underset{\underset{\displaystyle OH}{|}}{C}}CH_3$

(c) $\longrightarrow$

(d) $CH_3\overset{\overset{\displaystyle CH_3}{|}}{C}{=}CH_2 \longrightarrow CH_3\overset{\overset{\displaystyle CH_3}{|}}{\underset{\underset{\displaystyle Br}{|}}{C}}{-}\underset{\underset{\displaystyle Br}{|}}{C}H_2$

12.42 Complete these equations.

(a) $-CH_2CH_3 + HCl \longrightarrow$

(b) $-CH_2CH_3 + H_2O \xrightarrow{H_2SO_4}$

(c) $CH_3(CH_2)_5CH{=}CH_2 + HI \longrightarrow$

(d) $+ HCl \longrightarrow$

(e) $CH_3CH{=}CHCH_2CH_3 + H_2O \xrightarrow{H_2SO_4}$

(f) $CH_2{=}CHCH_2CH_2CH_3 + H_2O \xrightarrow{H_2SO_4}$

12.43 Draw structural formulas for all possible carbocations formed by the reaction of each alkene with HCl. Label each carbocation as primary, secondary, or tertiary.

(a) $CH_3CH_2\overset{\overset{\displaystyle CH_3}{|}}{C}{=}CHCH_3$

(b) $CH_3CH_2CH{=}CHCH_3$

(c)

(d) $={=}CH_2$

12.44 Draw a structural formula for the product formed by treatment of 2-methyl-2-pentene with each reagent.

(a) HCl (b) H_2O in the presence of H_2SO_4

12.45 Draw a structural formula for the product of each reaction.

(a) 1-Methylcyclohexene $+ Br_2$

(b) 1,2-Dimethylcyclopentene $+ Cl_2$

12.46 Draw a structural formula for an alkene with the indicated molecular formula that gives the compound shown as the major product. Note that more than one alkene may give the same compound as the major product.

(a) $C_5H_{10} + H_2O \xrightarrow{H_2SO_4} CH_3\overset{\overset{\displaystyle CH_3}{|}}{\underset{\underset{\displaystyle OH}{|}}{C}}CH_2CH_3$

(b) $C_5H_{10} + Br_2 \longrightarrow CH_3\overset{\overset{\displaystyle CH_3}{|}}{C}HCH\underset{\underset{\displaystyle Br}{|}}{C}\underset{\underset{\displaystyle Br}{|}}{}H_2$

(c) $C_7H_{12} + HCl \longrightarrow$

12.47 Draw a structural formula for an alkene with the molecular formula C_5H_{10} that reacts with Br_2 to give each product.

(a) $CH_3\overset{\overset{\displaystyle CH_3}{|}}{\underset{\underset{\displaystyle Br}{|}}{C}}{-}\underset{\underset{\displaystyle Br}{|}}{C}HCH_3$

(b) $CH_2\overset{\overset{\displaystyle CH_3}{|}}{\underset{\underset{\displaystyle Br}{|}}{C}}CH_2CH_3$ (with Br on CH_2)

(c) $\underset{\underset{\displaystyle Br}{|}}{C}H_2\underset{\underset{\displaystyle Br}{|}}{C}HCH_2CH_2CH_3$

12.48 Draw a structural formula for an alkene with the molecular formula C_5H_{10} that reacts with HCl to give the indicated chloroalkane as the major product. More than one alkene may give the same compound as the major product.

(a) $CH_3\overset{\overset{\displaystyle CH_3}{|}}{\underset{\underset{\displaystyle Cl}{|}}{C}}CH_2CH_3$

(b) $CH_3\overset{\overset{\displaystyle CH_3}{|}}{C}H\underset{\underset{\displaystyle Cl}{|}}{C}HCH_3$

(c) $CH_3\underset{\underset{\displaystyle Cl}{|}}{C}HCH_2CH_2CH_3$

12.49 With the notable exception of ethanol, the acid-catalyzed hydration of alkenes cannot be used to prepare primary alcohols. It can only be used to prepare secondary and tertiary alcohols from alkenes in good yield. Explain why this is so and illustrate your reasoning with specific examples.

12.50 Draw the structural formula of an alkene that undergoes acid-catalyzed hydration to give the indicated alcohol as the major product. More than one alkene may give each alcohol as the major product.

(a) 3-Hexanol

(b) 1-Methylcyclobutanol

(c) 2-Methyl-2-butanol

(d) 2-Propanol

12.51 Acid-catalyzed hydration of 2-pentene gives a mixture of two alcohols, each with the molecular formula $C_5H_{12}O$. Draw the structural formula for both alcohols. Similar treatment of 3-hexene gives only one alcohol with the molecular formula $C_6H_{14}O$. Draw the structural formula for this alcohol.

12.52 Terpin, $C_{10}H_{20}O_2$, is prepared commercially by the acid-catalyzed hydration of limonene.

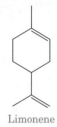

Limonene

(a) Propose a structural formula for terpin.

(b) How many *cis-trans* isomers are possible for the structural formula you propose?

(c) Terpin hydrate, the isomer of terpin in which the methyl and isopropyl groups are *trans* to each other, is used as an expectorant in cough medicines. Draw a structural formula for terpin hydrate showing the *trans* orientation of these groups.

12.53 Draw the product formed by treatment of each alkene with H_2/Ni.

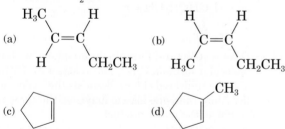

12.54 Hydrocarbon A, C_5H_8, reacts with 2 moles of Br_2 to give 1,2,3,4-tetrabromo-2-methylbutane. What is the structure of hydrocarbon A?

12.55 Show how to convert ethylene to these compounds.

(a) Ethane

(b) Ethanol

(c) Bromoethane

(d) 1,2-Dibromoethane

(e) Chloroethane

12.56 Show how to convert 1-butene to these compounds.

(a) Butane

(b) 2-Butanol

(c) 2-Bromobutane

(d) 1,2-Dibromobutane

Section 12.6 What Are the Important Polymerization Reactions of Ethylene and Substituted Ethylenes?

12.57 Answer true or false.

(a) Ethylene contains one carbon–carbon double bond, and polyethylene contains many carbon–carbon double bonds.

(b) All C—C—C bond angles in both LDPE and HDPE are approximately 120°.

(c) Low-density polyethylene (LDPE) is a highly branched polymer.

(d) High-density polyethylene (HDPE) consists of carbon chains with little branching.

(e) The density of polyethylene polymers is directly related to the degree of chain branching; the greater the branching, the lower the density of the polymer.

(f) PS and PVC are currently recycled.

Chemical Connections

12.58 (Chemical Connections 12A) What is one function of ethylene as a plant growth regulator?

12.59 (Chemical Connections 12B) What is the meaning of the term *pheromone*?

12.60 (Chemical Connections 12B) What is the molecular formula of 11-tetradecenyl acetate? What is its molecular weight?

12.61 (Chemical Connections 12B) Assume that 1×10^{-12} g of 11-tetradecenyl acetate is secreted by a single corn borer. How many molecules is this?

12.62 (Chemical Connections 12C) What different functions are performed by the rods and cones in the eye?

12.63 (Chemical Connections 12C) In which isomer of retinal is the end-to-end distance longer, the all-*trans* isomer or the 11-*cis* isomer?

12.64 (Chemical Connections 12D) What types of consumer products are made of high-density polyethylene? What types of products are made of low-density polyethylene? One type of polyethylene is currently recyclable, and the other is not. Which is which?

12.65 (Chemical Connections 12D) In recycling codes, what do these abbreviations stand for?

(a) V   (b) PP   (c) PS

Additional Problems

12.66 Write line-angle formulas for all compounds with the molecular formula C_4H_8. Which are sets of constitutional isomers? Which are sets of *cis-trans* isomers?

12.67 Below is the structural formula of lycopene, $C_{40}H_{56}$, a deep-red compound that is partially responsible for the red color of ripe fruits, especially tomatoes. Approximately 20 mg of lycopene can be isolated from 1 kg of fresh ripe tomatoes. How many of the carbon–carbon double bonds in lycopene have the possibility for *cis-trans* isomerism? Lycopene is the all-*trans* isomer.

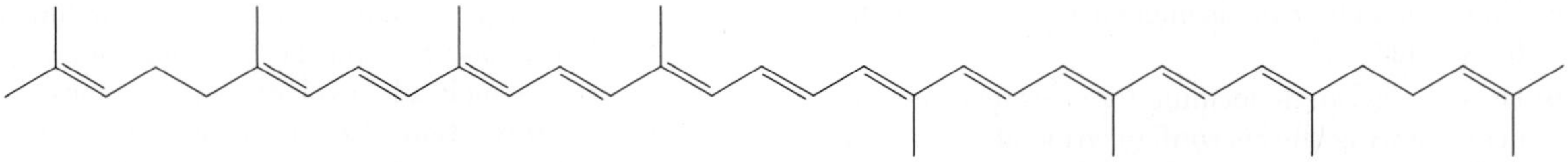

Lycopene

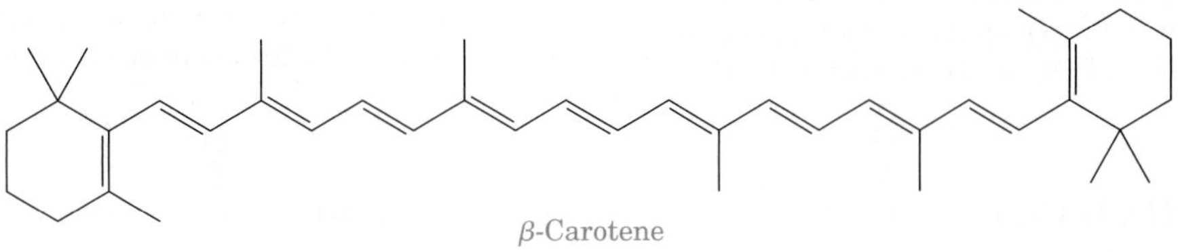

β-Carotene

12.68 As you might suspect, β-carotene, $C_{40}H_{56}$, a precursor to vitamin A, was first isolated from carrots. Dilute solutions of β-carotene are yellow—hence its use as a food coloring. In plants, this compound is almost always present in combination with chlorophyll to assist in the harvesting of the energy of sunlight. As tree leaves die in the fall, the green of their chlorophyll molecules is replaced by the yellows and reds of carotene and carotene-related molecules (see β-carotene skeleton above).

Compare the carbon skeletons of β-carotene and lycopene. What are the similarities? What are the differences?

12.69 Draw the structural formula for a cycloalkene with the molecular formula C_6H_{10} that reacts with Cl_2 to give each compound.

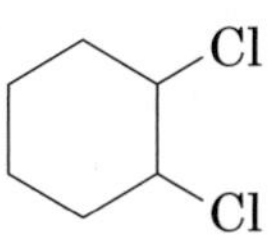

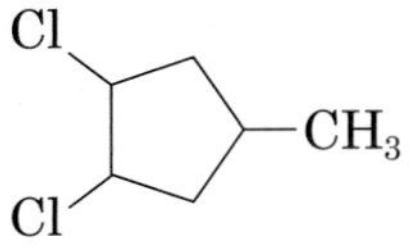

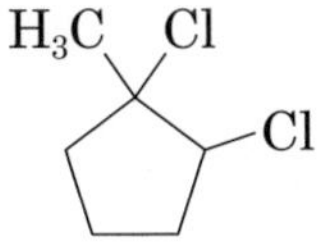

12.70 Propose a structural formula for the product(s) when each of the following alkenes is treated with H_2O/H_2SO_4. Why are two products formed in part (b), but only one in parts (a) and (c)?

(a) 1-Hexene gives one alcohol with a molecular formula of $C_6H_{14}O$.

(b) 2-Hexene gives two alcohols, each with a molecular formula of $C_6H_{14}O$.

(c) 3-Hexene gives one alcohol with a molecular formula of $C_6H_{14}O$.

12.71 *cis*-3-Hexene and *trans*-3-hexene are different compounds and have different physical and chemical properties. Yet, when treated with H_2O/H_2SO_4, each gives the same alcohol. What is this alcohol, and how do you account for the fact that each alkene gives the same one?

12.72 Draw the structural formula of an alkene that undergoes acid-catalyzed hydration to give each of the following alcohols as the major product. More than one alkene may give each compound as the major product.

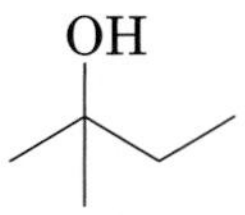

12.73 Show how to convert cyclopentene into these compounds.

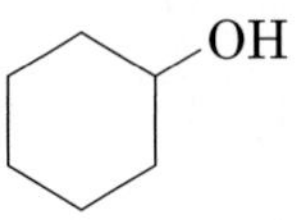

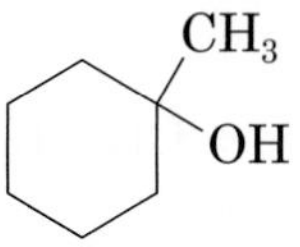

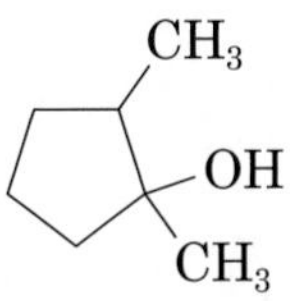

(a) 1,2-Dibromocyclopentane

(b) Cyclopentanol

(c) Iodocyclopentane

(d) Cyclopentane

Looking Ahead

12.74 In Chapter 21 on the biochemistry of lipids, we will study the three long-chain unsaturated carboxylic acids shown below. Each has 18 carbons and is a component of animal fats, vegetable oils, and biological membranes. Because they consist of a carboxylic acid group on one end and a fatty hydrocarbon chain on the other, they are called fatty acids.

Oleic acid $CH_3(CH_2)_7CH=CH(CH_2)_7COOH$
Linoleic acid $CH_3(CH_2)_4(CH=CHCH_2)_2(CH_2)_6COOH$
Linolenic acid $CH_3CH_2(CH=CHCH_2)_3(CH_2)_6COOH$

12.75 The fatty acids in Problem 12.74 occur in animal fats, vegetable oils, and biological membranes almost exclusively as the all-*cis* isomers. Draw line-angle formulas for each fatty acid showing the *cis* configuration about each carbon–carbon double bond.

12.76 In omega-3 fatty acids, the last carbon of the last double bond of the hydrocarbon chain ends three carbons from the methyl terminal end of the chain. The last carbon of the chain is called the omega carbon, hence, the designation omega-3. Eicosapentaenoic acid is a common omega-3 fatty acid found in cold water fatty fish and health food supplements.

$$CH_3(CH_2CH=CH)_5CH_2CH_2CH_2CO_2H$$

Eicosapentaenoic acid
a C_{20} polyunsaturated fatty acid

(a) How many *cis-trans* isomers are possible for this fatty acid?

(b) Draw a line-angle formula for eicosapentaenoic acid, showing the *cis* configuration of all carbon–carbon double bonds in the hydrocarbon chain.

12.77 The following series of three reactions occurs in the metabolic pathway known as the β-oxidation of fatty acids (Section 28.5). Fatty acids are metabolized by this pathway to produce energy. In the following structural formulas, the symbol CoA stands for coenzyme A, an organic molecule involved as a cofactor in many biological reactions. Note that coenzyme A is bound to the fatty acid chain as a thioester, that is, an ester in which oxygen is replaced by a sulfur atom. Name the type of reaction that occurs in each step and suggest a reason why the sequence is called β-oxidation.

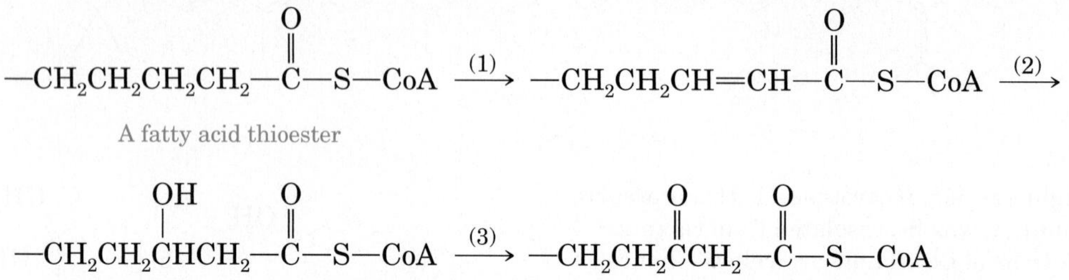

$$-CH_2CH_2CH_2CH_2-\overset{\overset{\displaystyle O}{\|}}{C}-S-CoA \xrightarrow{(1)} -CH_2CH_2CH=CH-\overset{\overset{\displaystyle O}{\|}}{C}-S-CoA \xrightarrow{(2)}$$

A fatty acid thioester

$$-CH_2CH_2\overset{\overset{\displaystyle OH}{|}}{C}HCH_2-\overset{\overset{\displaystyle O}{\|}}{C}-S-CoA \xrightarrow{(3)} -CH_2CH_2\overset{\overset{\displaystyle O}{\|}}{C}CH_2-\overset{\overset{\displaystyle O}{\|}}{C}-S-CoA$$

Benzene and Its Derivatives

Peppers of the Capsicum family (see Chemical Connections 13F).

13.1 What Is the Structure of Benzene?

So far we have described three classes of hydrocarbons—alkanes, alkenes, and alkynes—called aliphatic hydrocarbons. More than 150 years ago, organic chemists realized that yet another class of hydrocarbons exists, one whose properties are quite different from those of aliphatic hydrocarbons. Because some of these new hydrocarbons have pleasant odors, they were called **aromatic compounds**. Today we know that not all aromatic compounds share this characteristic. Some do have pleasant odors, but some have no odor at all, and others have downright unpleasant odors. A more appropriate definition of an aromatic compound is any compound that has one or more benzene-like rings.

We use the term **arene** to describe aromatic hydrocarbons. Just as a group derived by removal of an H from an alkane is called an alkyl group and given the symbol R—, a group derived by removal of an H from an arene is called an **aryl group** and given the symbol **Ar**—.

Benzene, the simplest aromatic hydrocarbon, was discovered in 1825 by Michael Faraday (1791–1867). Its structure presented an immediate problem to chemists of the day. Benzene has the molecular formula C_6H_6, and a compound with so few hydrogens for its six carbons (compare hexane,

Aromatic compound Benzene or one of its derivatives

Arene A compound containing one or more benzene-like rings

Aryl group A group derived from an arene by removal of a H atom from an arene and given the symbol Ar—

Ar— The symbol used for an aryl group

© Eric Delmar/iStockphoto.com

Benzene is an important compound in both the chemical industry and the laboratory, but it must be handled carefully. Not only is it poisonous if ingested in liquid form, but its vapor is also toxic and can be absorbed either by breathing or through the skin. Long-term inhalation can cause liver damage and cancer.

C_6H_{14}, and cyclohexane C_6H_{12}), chemists argued, should be unsaturated. But benzene does not behave like an alkene (the only class of unsaturated hydrocarbons known at that time). Whereas 1-hexene, for example, reacts instantly with Br_2 (Section 12.6C), benzene does not react at all with this reagent. Nor does benzene react with HBr, H_2O/H_2SO_4, or H_2/Pd—all reagents that normally add to carbon–carbon double bonds.

A. Kekulé's Structure of Benzene

The first structure for benzene was proposed by Friedrich August Kekulé in 1872 and consisted of a six-membered ring with alternating single and double bonds, with one hydrogen bonded to each carbon.

A Kekulé structure showing all atoms

A Kekulé structure as a line-angle drawing

Although Kekulé's proposal was consistent with many of the chemical properties of benzene, it was contested for years. The major objection was its failure to account for the unusual chemical behavior of benzene. If benzene contains three double bonds, Kekulé's critics asked, why doesn't it undergo reactions typical of alkenes?

B. Resonance Structure of Benzene

Resonance hybrid A molecule best described as a composite of two or more Lewis structures

The concept of resonance, developed by Linus Pauling in the 1930s, provided the first adequate description of the structure of benzene. According to the theory of resonance, certain molecules and ions are best described by writing two or more Lewis structures and considering the real molecule or ion to be a **resonance hybrid** of these structures. Each individual Lewis structure is called a **contributing structure**. To show that the real molecule is a resonance hybrid of the two Lewis structures, we position a double-headed arrow between them.

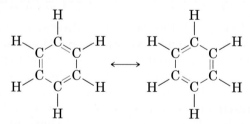

Alternative Lewis contributing structures for benzene

The two contributing structures for benzene are often called Kekulé structures.

A note about resonance hybrids. Do not confuse resonance contributing structures with equilibration among different chemical species. A molecule described as a resonance hybrid will not equilibrate among the electron configurations of the various contributing structures. Rather, the molecule has only one structure, which is best described as a hybrid of its various contributing structures. A molecule described as a resonance hybrid is not sometimes one contributing structure and sometimes another; it is a single structure all the time.

The resonance hybrid has some of the characteristics of each Lewis contributing structure. For example, the carbon–carbon bonds are neither single nor double but rather something intermediate between the two extremes. It has been determined experimentally that the length of the carbon–carbon bond in benzene is not as long as a carbon–carbon single bond nor as short as a carbon–carbon double bond, but rather is midway between the two. The closed loop of six electrons (two from the second bond of each double bond) characteristic of a benzene ring is sometimes called an **aromatic sextet**.

Wherever we find resonance, we find stability. The real structure is generally more stable than any of the hypothetical Lewis contributing structures. The benzene ring is greatly stabilized by resonance, which explains why it does not undergo the addition reactions typical of alkenes.

13.2 How Do We Name Aromatic Compounds?

A. One Substituent

Monosubstituted alkylbenzenes are named as derivatives of benzene—for example, ethylbenzene. The IUPAC system retains certain common names for several of the simpler monosubstituted alkylbenzenes, including **toluene** and **styrene**.

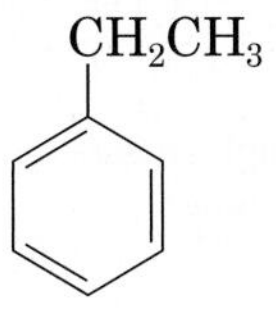

CH_2CH_3 CH_3 $CH=CH_2$

Ethylbenzene 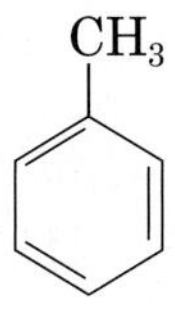   Toluene 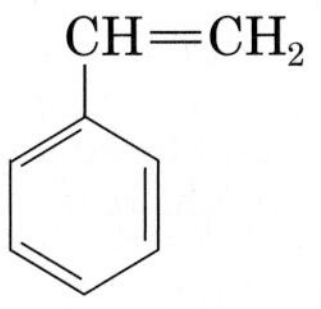   Styrene

The IUPAC system also retains common names for the following compounds:

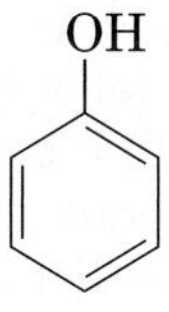

OH OCH_3 NH_2 $\overset{O}{\overset{\|}{C}}-H$ $\overset{O}{\overset{\|}{C}}-OH$

Phenol 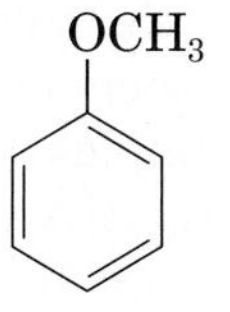   Anisole 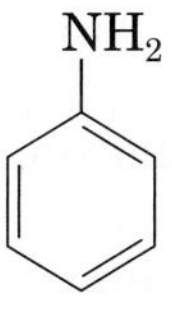   Aniline 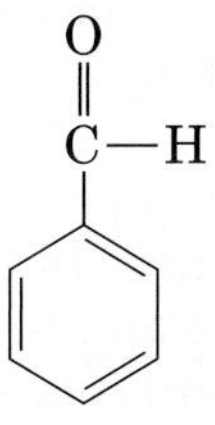   Benzaldehyde 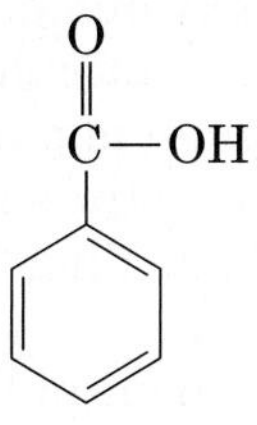   Benzoic acid

The substituent group derived by loss of an H from benzene is called a **phenyl group**, C_6H_5—, the common symbol for which is **Ph**—. In molecules containing other functional groups, phenyl groups are often named as substituents.

Phenyl group C_6H_5— the aryl group derived by removing a hydrogen atom from benzene. The name is derived from *phene*, an earlier name for benzene.

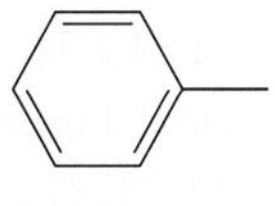

Phenyl group   1-Phenylcyclohexene 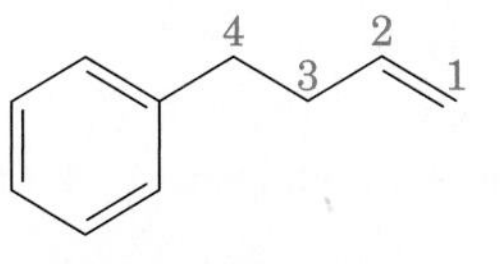   4-Phenyl-1-butene
(C_6H_5—; Ph—)

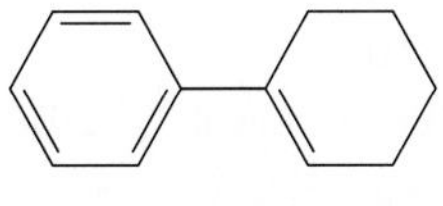

B. Two Substituents

When two substituents occur on a benzene ring, three isomers are possible. We locate the substituents either by numbering the atoms of the ring or by using the locators *ortho (o), meta (m)*, and *para (p)*. The numbers 1,2- are equivalent to *ortho* (Greek: straight); 1,3- to *meta* (Greek: after); and 1,4- to *para* (Greek: beyond).

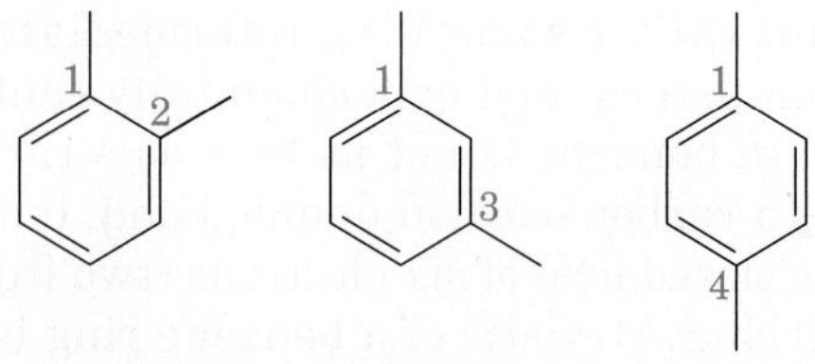

1,2- or *ortho-*   1,3- or *meta*   1,4- or *para*

When one of the two substituents on the ring imparts a special name to the compound (for example, —CH₃, —OH, —NH₂, or —COOH), we name the compound as a derivative of that parent molecule and assume that the substituent occupies ring position number 1. The IUPAC system retains the common name **xylene** for the three isomeric dimethylbenzenes. Where neither substituent imparts a special name, we locate the two substituents and list them in alphabetical order before the ending "benzene." The carbon of the benzene ring with the substituent of lower alphabetical ranking is numbered C—1.

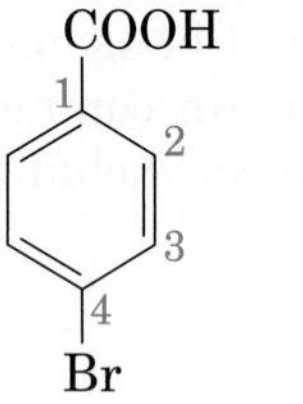

4-Bromobenzoic acid   3-Chloroaniline   1,3-Dimethylbenzene   1-Chloro-4-ethylbenzene
(*p*-Bromobenzoic acid)   (*m*-Chloroaniline)   (*m*-Xylene)   (*p*-Chloroethylbenzene)

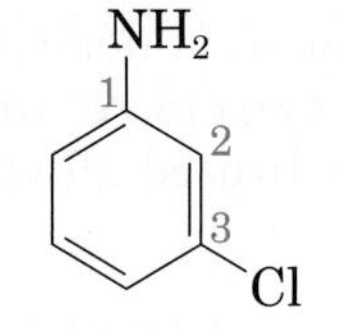

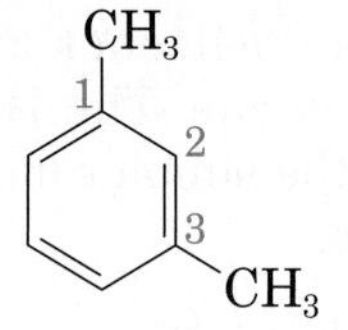

C. Three or More Substituents

p-Xylene is a starting material for the synthesis of poly(ethylene terephthalate). Consumer products derived from this polymer include Dacron polyester fibers and Mylar films (Section 19.6B).

When three or more substituents are present on a benzene ring, specify their locations by numbers. If one of the substituents imparts a special name, then name the molecule as a derivative of that parent molecule. If none of the substituents imparts a special name, then locate the substituents, number them to give the smallest set of numbers, and list them in alphabetical order before the ending "benzene." In the following examples, the first compound is a derivative of toluene and the second is a derivative of phenol. Because no substituent in the third compound imparts a special name, list its three substituents in alphabetical order followed by the word "benzene."

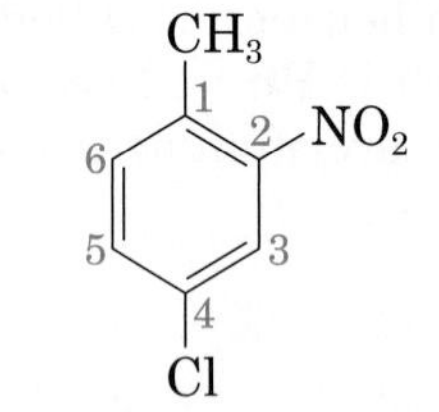

4-Chloro-2-nitrotoluene   2,4,6-Tribromophenol   2-Bromo-1-ethyl-4-nitrobenzene

⊙WL Interactive Example 13.1 Naming Aromatic Compounds

Write names for these compounds.

(a)

(b)

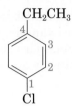
(c)

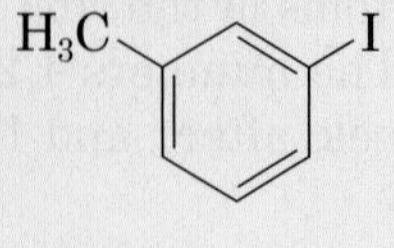

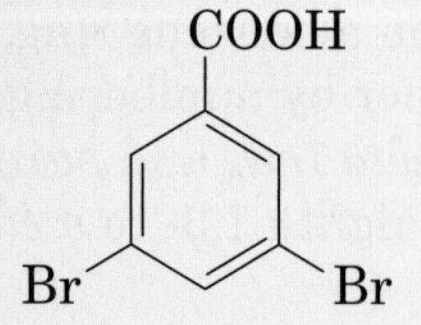

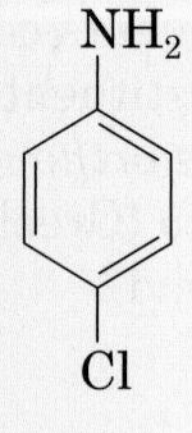

Strategy

First check to see if one of the substituents on the benzene ring imparts a special name. If one of them does, then name the compound as a derivative of that parent molecule.

Solution

(a) The parent is toluene, and the compound is 3-iodotoluene or *m*-iodotoluene.
(b) The parent is benzoic acid, and the compound is 3,5-dibromobenzoic acid.
(c) The parent is aniline, and the compound is 4-chloroaniline or *p*-chloroaniline.

Problem 13.1

Write names for these compounds.

(a)

OH

(b)

NH$_2$

Cl

Cl

(c)

COOH

NO$_2$

Chemical Connections 13A

DDT: A Boon and a Curse

Probably the best-known insecticide worldwide is *di*chlo-ro*di*phenyl*tri*chloroethane (not an IUPAC name), commonly abbreviated DDT.

Cl—⬡—CH—⬡—Cl
          |
        CCl$_3$

Dichlorodiphenyltrichloroethane
(DDT)

This compound was first prepared in 1874, but it was not until the late 1930s that its potential as an insecticide was recognized. First used for this purpose in 1939, it proved extremely effective in ridding large areas of the world of the insect hosts that transmit malaria and typhus. In addition, it was so effective in killing crop-destroying insect pests that crop yields in many areas of the world increased dramatically.

Widespread use of DDT, however, has proven to be a double-edged sword. Despite DDT's well-known benefits, it has an enormous disadvantage. Because it resists biodegradation, it remains in the soil for years—and this persistence in the environment creates the problem. Scientists estimate that the tissues of adult humans contain, on average, five to ten parts per million of DDT.

Rachel Carson dramatically portrayed the dangers associated with the persistence of DDT in the environment in

DDT was sprayed on crops in the United States until its use was banned in the 1970s.

Courtesy of the U.S. Department of Agriculture

her 1962 book, *Silent Spring*, which documented serious declines in the populations of eagles and other raptors as well as many other kinds of birds and songbirds. Scientists discovered soon thereafter that DDT inhibits the mechanism by which these birds incorporate calcium into their eggshells. As a result, the shells become so thin and weak that they break during incubation, killing the embryo inside.

Because of these problems, almost all nations have now banned the use of DDT for agricultural purposes. It is still used, however, in some areas to control the population of disease-spreading insects.

Test your knowledge with Problems 13.34–13.38.

Carcinogenic Polynuclear Aromatics and Smoking

A **carcinogen** is a compound that causes cancer. The first carcinogens to be identified were a group of polynuclear aromatic hydrocarbons, all of which have at least four aromatic rings. Among them is benzo[a]pyrene, one of the most carcinogenic of the aromatic hydrocarbons. It forms whenever incomplete combustion of organic compounds occurs. Benzo[a]pyrene is found, for example, in cigarette smoke, automobile exhaust, and charcoal-broiled meats.

Benzo[a]pyrene causes cancer in the following way. Once it is absorbed or ingested, the body attempts to convert it into a more water-soluble compound that can be excreted easily. By a series of enzyme-catalyzed reactions, benzo[a]pyrene is transformed into a **diol** (two —OH groups) **epoxide** (a three-membered ring, one atom of which is oxygen). This compound can bind to DNA by reacting with one of its amino groups, thereby altering the structure of DNA and producing a cancer-causing mutation.

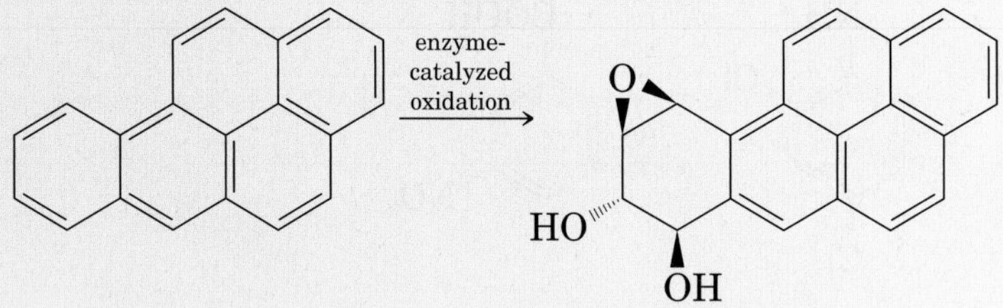

Benzo[a]pyrene   A diol epoxide

Test your knowledge with Problem 13.39.

D. Polynuclear Aromatic Hydrocarbons

Polynuclear aromatic hydrocarbon
A hydrocarbon containing two or more benzene rings, each of which shares two carbon atoms with another benzene ring

Polynuclear aromatic hydrocarbons (PAHs) contain two or more benzene rings, with each pair of rings sharing two adjacent carbon atoms. Naphthalene, anthracene, and phenanthrene, the most common PAHs, and substances derived from them are found in coal tar and high-boiling petroleum residues.

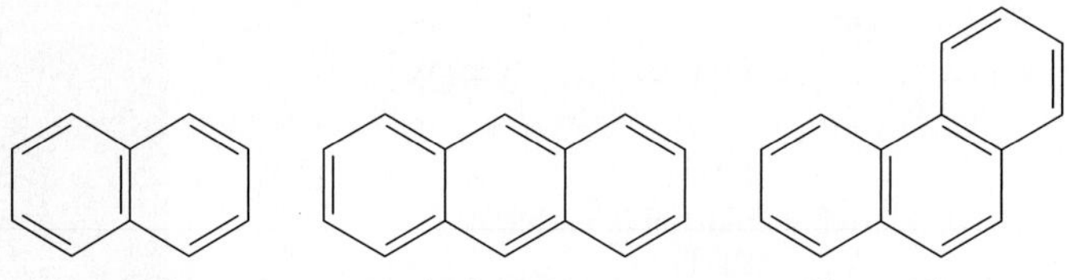

Naphthalene   Anthracene   Phenanthrene

At one time, naphthalene was used in mothballs and as an insecticide to preserve woolens and furs, but its use decreased after the introduction of chlorinated hydrocarbons such as *p*-dichlorobenzene.

13.3 What Are the Characteristic Reactions of Benzene and Its Derivatives?

By far the most characteristic reaction of aromatic compounds is substitution at a ring carbon, which we give the name **aromatic substitution**. Groups we can introduce directly on the ring include the halogens, the nitro (—NO$_2$) group, and the sulfonic acid (—SO$_3$H) group.

A. Halogenation

As noted in Section 13.1, chlorine and bromine do not react with benzene, in contrast to their instantaneous reaction with cyclohexene and other alkenes

Chemical Connections 13C

Iodide Ion and Goiter

One hundred years ago, goiter, an enlargement of the thyroid gland caused by iodine deficiency, was common in the central United States and central Canada. This disease results from underproduction of thyroxine, a hormone synthesized in the thyroid gland. Young mammals require this hormone for normal growth and development. A deficiency of thyroxine during fetal development results in mental retardation. Low levels of thyroxine in adults result in hypothyroidism, commonly called goiter, the symptoms of which are lethargy, obesity, and dry skin.

Iodine is an element that comes primarily from the sea. Rich sources of it, therefore, are fish and other seafoods. The iodine in our diets that doesn't come from the sea most commonly is derived from food additives. Most of the iodide ion in the North American diet comes from table salt fortified with sodium iodide, commonly referred to as iodized salt. Another source is dairy products, which accumulate iodide because of the iodine-containing additives used in cattle feeds and the iodine-containing disinfectants used on milking machines and milk storage tanks.

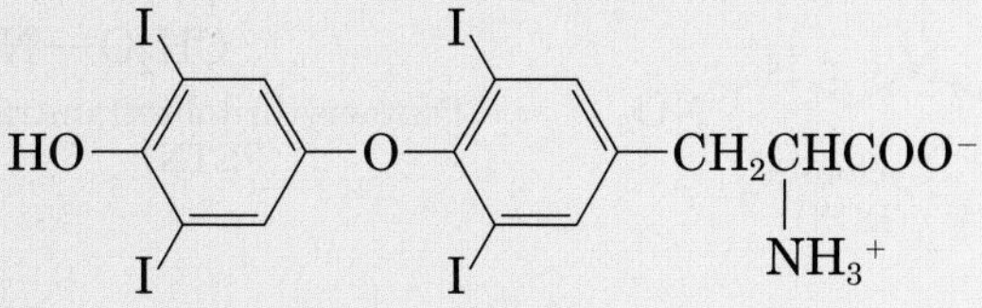

Thyroxine

Test your knowledge with Problem 13.40.

(Section 12.6C). In the presence of an iron catalyst, however, chlorine reacts rapidly with benzene to give chlorobenzene and HCl:

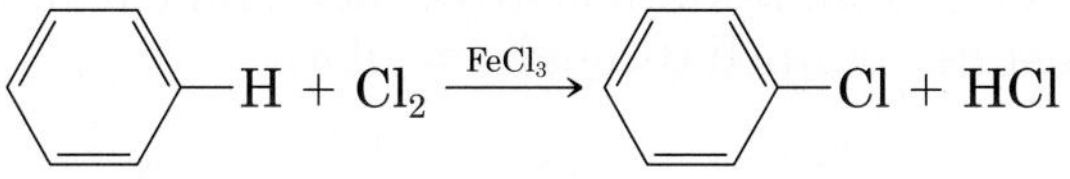

Benzene   Chlorobenzene

Treatment of benzene with bromine in the presence of $FeCl_3$ results in the formation of bromobenzene and HBr.

B. Nitration

When we heat benzene or one of its derivatives with a mixture of concentrated nitric and sulfuric acids, a nitro ($-NO_2$) group replaces one of the hydrogen atoms bonded to the ring.

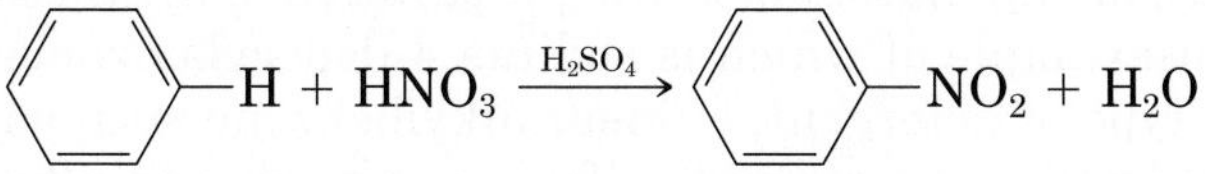

Nitrobenzene

A particular value of nitration is that we can reduce the resulting $-NO_2$ group to a primary amino group, $-NH_2$, by catalytic reduction using hydrogen in the presence of a transition-metal catalyst. In the following example, neither the benzene ring nor the carboxyl group is affected by these experimental conditions:

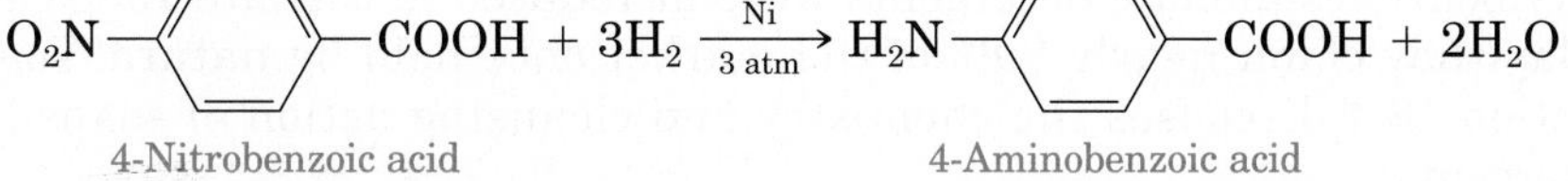

4-Nitrobenzoic acid   4-Aminobenzoic acid
(p-Nitrobenzoic acid)   (p-Aminobenzoic acid, PABA)

Chemical Connections 13D

The Nitro Group in Explosives

Treatment of toluene with three moles of nitric acid in the presence of sulfuric acid as a catalyst results in nitration of toluene three times to form the explosive 2, 4, 6-trinitrotoluene, TNT. The presence of these three nitro groups confers explosive properties to TNT. Similarly, the presence of three nitro groups leads to the explosive properties of nitroglycerin.

In recent years, several new explosives have been discovered, all of which contain multiple nitro groups. Among them are RDX and PETN. The plastic explosive Semtex, for example, is a mixture of RDX and PETN. It was used in the destruction of Pan Am flight 103 over Lockerbie, Scotland, in December 1988.

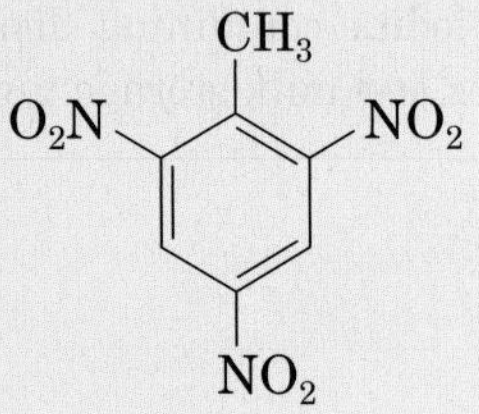

2,4,6-Trinitrotoluene
(TNT)

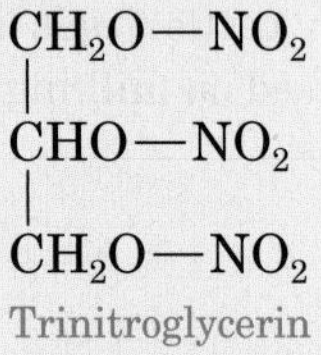

CH_2O-NO_2
$CHO-NO_2$
CH_2O-NO_2
Trinitroglycerin
(Nitroglycerin)

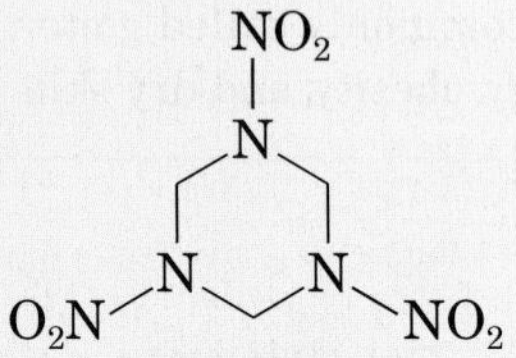

Cyclonite
(RDX)

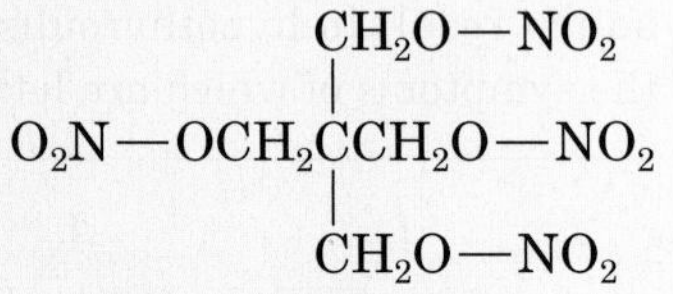

CH_2O-NO_2
$O_2N-OCH_2CCH_2O-NO_2$
CH_2O-NO_2
Pentaerythritol tetranitrate
(PETN)

Test your knowledge with Problem 13.41.

Bacteria require p-aminobenzoic acid for the synthesis of folic acid (Section 30.4), which is in turn required for the synthesis of the heterocyclic aromatic amine bases of nucleic acids (Section 25.2). Whereas bacteria can synthesize folic acid from p-aminobenzoic acid, folic acid is a vitamin for humans and must be obtained through the diet.

C. Sulfonation

Heating an aromatic compound with concentrated sulfuric acid results in formation of an arenesulfonic acid, all of which are strong acids, comparable in strength to sulfuric acid.

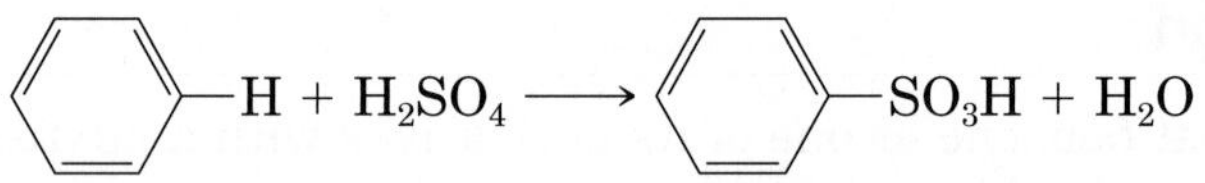

$$\text{C}_6\text{H}_5{-}\text{H} + \text{H}_2\text{SO}_4 \longrightarrow \text{C}_6\text{H}_5{-}\text{SO}_3\text{H} + \text{H}_2\text{O}$$

Benzenesulfonic acid

A major use of sulfonation is in the preparation of synthetic detergents, an important example of which is sodium 4-dodecybenzenesulfonate. To prepare this type of detergent, a linear alkylbenzene such as dodecylbenzene is treated with concentrated sulfuric acid to give an alkylbenzenesulfonic acid. The sulfonic acid is then neutralized with sodium hydroxide.

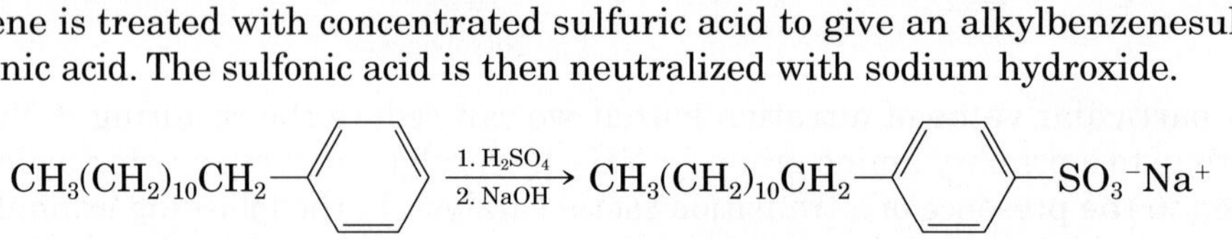

$$CH_3(CH_2)_{10}CH_2{-}\text{C}_6\text{H}_4 \xrightarrow[\text{2. NaOH}]{\text{1. H}_2\text{SO}_4} CH_3(CH_2)_{10}CH_2{-}\text{C}_6\text{H}_4{-}SO_3^-Na^+$$

Dodecylbenzene

Sodium 4-dodecylbenzenesulfonate, SDS
(an anionic detergent)

Alkylbenzenesulfonate detergents were introduced in the late 1950s, and today they claim nearly 90% of the market once held by natural soaps. Section 18.4 discusses the chemistry and cleansing action of soaps and detergents.

13.4 What Are Phenols?

A. Structure and Nomenclature

The functional group of a **phenol** is a hydroxyl group bonded to a benzene ring. Substituted phenols are named either as derivatives of phenol or by common names.

Phenol A compound that contains an —OH group bonded to a benzene ring

Phenol   3-Methylphenol (*m*-Cresol)   1,2-Benzenediol (Catechol)   1,3-Benzenediol (Resorcinol)   1,4-Benzenediol (Hydroquinone)

Phenol in crystalline form.

Phenols are widely distributed in nature. Phenol itself and the isomeric cresols (*o*-, *m*-, and *p*-cresol) are found in coal tar. Thymol and vanillin are important constituents of thyme and vanilla beans, respectively. Urushiol is the main component of the irritating oil of poison ivy. It can cause severe contact dermatitis in sensitive individuals.

2-Isopropyl-5-methylphenol (Thymol)   4-Hydroxy-3-methoxy-benzaldehyde (Vanillin)   Urushiol

B. Acidity of Phenols

Phenols are weak acids, with pK_a values of approximately 10 (Table 8.3). Most phenols are insoluble in water, but they react with strong bases such as NaOH and KOH to form water-soluble salts.

$$\text{Phenol} + \text{NaOH} \longrightarrow \text{—O}^-\text{Na}^+ + \text{H}_2\text{O}$$

Phenol
$pK_a = 9.95$
(stronger acid)

Sodium hydroxide
(stronger base)

Sodium phenoxide
(weaker base)

Water
$pK_a = 15.7$
(weaker acid)

Poison ivy.

Most phenols are such weak acids that they do not react with weak bases such as sodium bicarbonate; that is, they do not dissolve in aqueous sodium bicarbonate.

C. Phenols as Antioxidants

An important reaction for living systems, foods, and other materials that contain carbon–carbon double bonds is **autoxidation**—that is, oxidation requiring oxygen and no other reactant. If you open a bottle of cooking oil that has stood for a long time, you may notice a hiss of air entering the bottle. This sound occurs because the consumption of oxygen by autoxidation of the oil creates a negative pressure inside the bottle.

Cooking oils contain esters of polyunsaturated fatty acids. We will discuss the structure and chemistry of esters in Chapter 19. The important

FD & C No. 6 (a.k.a. Sunset Yellow)

Did you ever wonder what gives gelatin desserts their red, green, orange, or yellow color? Or what gives margarines their yellow color? Or what gives maraschino cherries their red color? If you read the content labels, you will see code names such as FD & C Yellow No. 6 and FD & C Red No. 40.

At one time, the only colorants for foods were compounds obtained from plant or animal materials. Beginning as early as the 1890s, however, chemists discovered a series of synthetic food dyes that offered several advantages over natural dyes, such as greater brightness, better stability, and lower cost. Opinion remains divided on the safety of their use. No synthetic food colorings are allowed, for example, in Norway and Sweden. In the United States, the Food and Drug Administration (FDA) has certified seven synthetic dyes for use in foods, drugs, and cosmetics

(FD & C)—two yellows, two reds, two blues, and one green. When these dyes are used alone or in combinations, they can approximate the color of almost any natural food.

Following are structural formulas for Allura Red (Red No. 40) and Sunset Yellow (Yellow No. 6). These and the other five food dyes certified in the United States have in common three or more benzene rings and two or more ionic groups, either the sodium salt of a carboxylic acid group, —COO⁻Na⁺, or the sodium salt of sulfonic acid group, —SO₃⁻Na⁺. These ionic groups make the dyes soluble in water.

To return to our original questions, maraschino cherries are colored with FD & C Red No. 40 and margarines are colored with FD & C Yellow No 6. Gelatin desserts use either one or a combination of the seven certified food dyes to create their color.

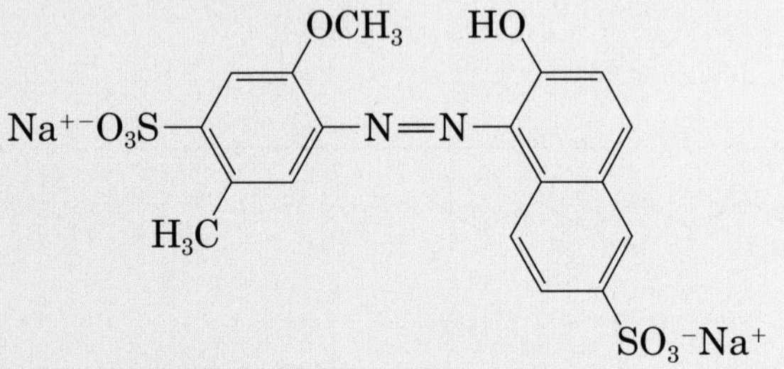

Allura Red
(FD & C Red No. 40)

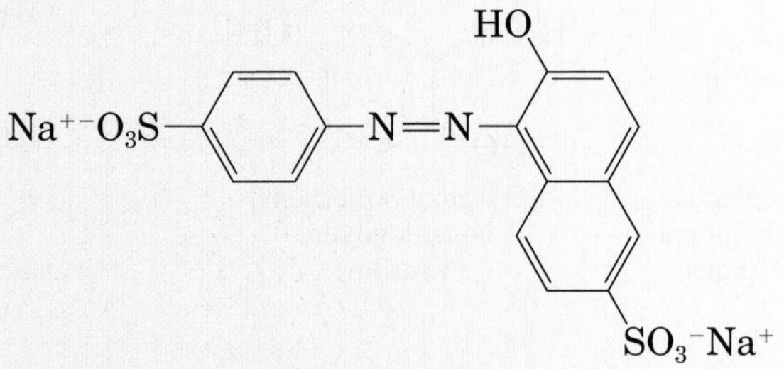

Sunset Yellow
(FD & C Yellow No. 6)

Test your knowledge with Problems 13.43–13.45.

point here is that all vegetable oils contain fatty acids with long hydrocarbon chains, many of which have one or more carbon–carbon double bonds (see Problems 12.29 and 12.72 for the structures of several of these fatty acids). Autoxidation takes place adjacent to one or more of their double bonds.

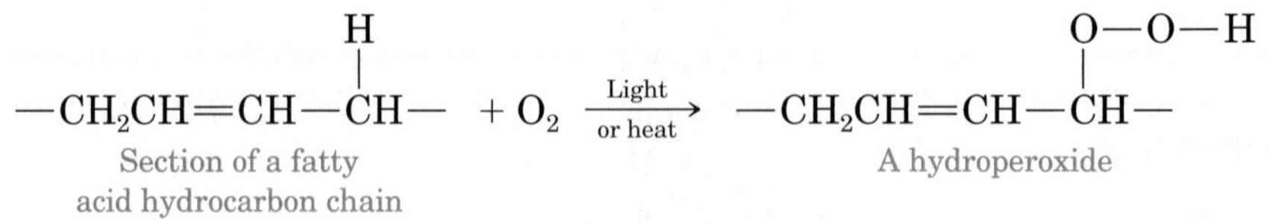

$$-\text{CH}_2\text{CH}=\text{CH}-\overset{\overset{\text{H}}{|}}{\text{CH}}- \ + \ \text{O}_2 \ \xrightarrow[\text{or heat}]{\text{Light}} \ -\text{CH}_2\text{CH}=\text{CH}-\overset{\overset{\text{O}-\text{O}-\text{H}}{|}}{\text{CH}}-$$

Section of a fatty
acid hydrocarbon chain

A hydroperoxide

Autoxidation is a radical-chain process that converts an R—H group to an R—O—O—H group, called a hydroperoxide. This process begins when a hydrogen atom with one of its electrons (H·) is removed from a carbon adjacent to one of the double bonds in a hydrocarbon chain. The carbon losing the H· has only seven electrons in its valence shell, one of which is unpaired. An atom or a molecule with an unpaired electron is called a **radical**.

Mechanism of Autoxidation

Step 1: Chain Initiation—Formation of a Radical from a Nonradical Compound
Removal of a hydrogen atom (H·) may be initiated by light or heat. The

Chemical Connections 13F

Capsaicin, for Those Who Like It Hot

Capsaicin, the pungent principle from the fruit of various species of peppers (*Capsicum* and *Solanaceae*), was isolated in 1876, and its structure was determined in 1919. Capsaicin contains both a phenol and a phenol ether.

Capsaicin
(from various types of peppers)

The inflammatory properties of capsaicin are well known; the human tongue can detect as little as one drop in 5 L of water. We all know of the burning sensation in the mouth and sudden tearing in the eyes caused by a good dose of hot chili peppers. For this reason, capsaicin-containing extracts from these flaming foods are used in sprays to ward off dogs or other animals that might nip at your heels while you are running or cycling.

Paradoxically, capsaicin is able to both cause and relieve pain. Currently, two capsaicin-containing creams, Mioton and Zostrix, are prescribed to treat the burning pain associated with postherpetic neuralgia, a complication of the disease known as shingles. They are also prescribed for diabetics to relieve persistent foot and leg pain.

Red chilies being dried in the sun.

Test your knowledge with Problems 13.46–13.49.

product formed is a carbon radical; that is, it contains a carbon atom with one unpaired electron.

$$-CH_2CH=CH-\overset{H}{\underset{|}{CH}}- \xrightarrow[\text{or heat}]{\text{light}} -CH_2CH=CH-\dot{C}H-$$

Section of a fatty acid hydrocarbon chain   A carbon radical

Step 2a: Chain Propagation—Reaction of a Radical to Form a New Radical The carbon radical reacts with oxygen, itself a diradical, to form a hydroperoxy radical. The new covalent bond of the hydroperoxy radical forms by the combination of one electron from the carbon radical and one electron from the oxygen diradical.

These two unpaired electrons combine to form a C—O single bond

$$-CH_2CH=CH-\dot{C}H- + \cdot O-O\cdot \longrightarrow -CH_2CH=CH-\overset{O-O\cdot}{\underset{|}{CH}}-$$

Oxygen is a diradical   A hydroperoxy radical

Step 2b: Chain Propagation—Reaction of a Radical and a Molecule to Form a New Radical   The hydroperoxy radical removes a hydrogen atom (H·) from a new fatty acid hydrocarbon chain to complete the formation of a hydroperoxide and at the same time produce a new carbon radical.

$$
\begin{array}{c}
\overset{\displaystyle O-O\cdot}{\underset{\displaystyle |}{}} \\
-CH_2CH{=}CH-CH-
\end{array}
\;+\;
\begin{array}{c}
\overset{\displaystyle H}{\underset{\displaystyle |}{}} \\
-CH_2CH{=}CH-CH-
\end{array}
\;\longrightarrow
$$

Section of a new fatty
acid hydrocarbon chain

$$
\begin{array}{c}
\overset{\displaystyle O-O-H}{\underset{\displaystyle |}{}} \\
-CH_2CH{=}CH-CH-
\end{array}
\;+\;
-CH_2CH{=}CH-\overset{\displaystyle \cdot}{C}H-
$$

A hydroperoxide   A new carbon radical

The most important point about the pair of chain propagation steps (Steps 2a and 2b) is that they form a continuous cycle of reactions, in the following way. The new radical formed in Step 2b reacts with another molecule of O_2 by Step 2a to give a new hydroperoxy radical. This new hydroperoxy radical then reacts with a new hydrocarbon chain to repeat Step 2b, and so forth. This cycle of propagation steps repeats over and over in a chain reaction. Thus, once a radical is generated in Step 1, the cycle of propagation steps repeats many thousands of times and, in so doing, generates thousands of hydroperoxide molecules. The number of times the cycle of chain propagation steps repeats is called the **chain length**.

Hydroperoxides themselves are unstable and, under biological conditions, degrade to short-chain aldehydes and carboxylic acids with unpleasant "rancid" smells. These odors may be familiar to you if you have ever smelled old cooking oil or aged foods that contain polyunsaturated fats or oils. Similar formation of hydroperoxides in the low-density lipoproteins (Section 27.4) deposited on the walls of arteries leads to cardiovascular disease in humans. In addition, many effects of aging are thought to result from the formation and subsequent degradation of hydroperoxides.

Fortunately, nature has developed a series of defenses against the formation of these and other destructive hydroperoxides, including the phenol vitamin E (Section 30.6). This compound is one of "nature's scavengers." It inserts itself into either Step 2a or 2b, donates an H· from its —OH group to the carbon radical, and converts the carbon radical back to its original hydrocarbon chain. Because the vitamin E radical is stable, it breaks the cycle of chain propagation steps, thereby preventing further formation of destructive hydroperoxides. While some hydroperoxides may form, their numbers are very small and they are easily decomposed to harmless materials by one of several possible enzyme-catalyzed reactions.

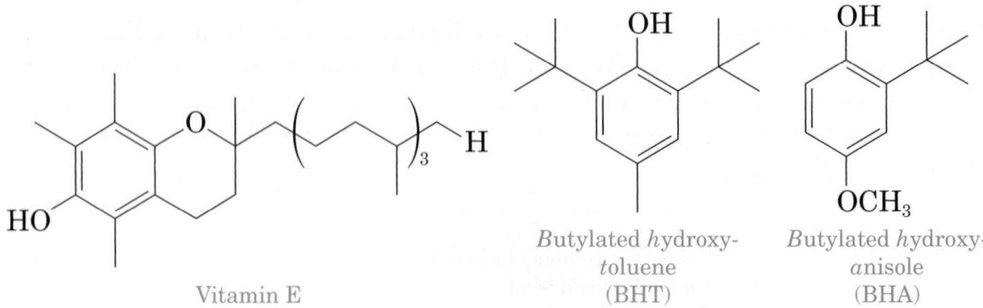

Vitamin E

Butylated hydroxy-
toluene
(BHT)

Butylated hydroxy-
anisole
(BHA)

Unfortunately, vitamin E is removed in the processing of many foods and food products. To make up for this loss, phenols such as BHT and BHA are added to foods to "retard spoilage" (as they say on the packages) by autoxidation. Likewise, similar compounds are added to other materials, such as plastics and rubber, to protect them against autoxidation.

Summary of Key Questions

OWL Sign in at **www.cengage.com/owl** to develop problem-solving skills and complete online homework assigned by your professor.

Section 13.1 What Is the Structure of Benzene?

- **Benzene** and its alkyl derivatives are classified as **aromatic hydrocarbons**, or **arenes**.
- The first structure for benzene was proposed by Friederich August Kekulé in 1872.
- The theory of **resonance**, developed by Linus Pauling in the 1930s, provided the first adequate structure for benzene.

Section 13.2 How Do We Name Aromatic Compounds?

- Aromatic compounds are named according to the IUPAC system.
- The C_6H_5— group is named **phenyl**.
- Two substituents on a benzene ring may be located by numbering the atoms of the ring or by using the locators *ortho* (*o*), *meta* (*m*), and *para* (*p*).

- **Polynuclear aromatic hydrocarbons** contain two or more benzene rings, each sharing two adjacent carbon atoms with another ring.

Section 13.3 What Are the Characteristic Reactions of Benzene and Its Derivatives?

- A characteristic reaction of aromatic compounds is **aromatic substitution**, in which another atom or group of atoms is substituted for a hydrogen atom of the aromatic ring.
- Typical aromatic substitution reactions are halogenation, nitration, and sulfonation.

Section 13.4 What Are Phenols?

- The functional group of a phenol is an —OH group bonded to a benzene ring.
- Phenol and its derivatives are weak acids, with pK_a equal to approximately 10.0.
- Vitamin E, a phenolic compound, is a natural antioxidant.
- Phenolic compounds such as BHT and BHA are synthetic antioxidants.

Summary of Key Reactions

1. **Halogenation (Section 13.3A)** Treatment of an aromatic compound with Cl_2 or Br_2 in the presence of an $FeCl_3$ catalyst substitutes a halogen for an H.

$$\text{benzene} + Cl_2 \xrightarrow{FeCl_3} \text{C}_6\text{H}_5\text{—Cl} + HCl$$

Chlorobenzene

2. **Nitration (Section 13.3B)** Heating an aromatic compound with a mixture of concentrated nitric and sulfuric acids substitutes a nitro group for an H.

$$\text{benzene} + HNO_3 \xrightarrow[\text{heat}]{H_2SO_4} \text{C}_6\text{H}_5\text{—NO}_2 + H_2O$$

Nitrobenzene

3. **Sulfonation (Section 13.3C)** Heating an aromatic compound with concentrated sulfuric acid substitutes a sulfonic acid group for an H.

$$\text{benzene} + H_2SO_4 \xrightarrow{\text{heat}} \text{C}_6\text{H}_5\text{—SO}_3\text{H} + H_2O$$

Benzenesulfonic acid

4. **Reaction of Phenols with Strong Bases (Section 13.4B)** Phenols are weak acids that react with strong bases to form water-soluble salts.

$$\text{C}_6\text{H}_5\text{—OH} + NaOH \longrightarrow \text{C}_6\text{H}_5\text{—O}^-\text{Na}^+ + H_2O$$

Problems

OWL Interactive versions of these problems may be assigned in OWL.

Orange-numbered problems are applied.

Section 13.1 What Is the Structure of Benzene?

13.2 Answer true or false.

    (a) Alkenes, alkynes, and arenes are unsaturated hydrocarbons.

    (b) Aromatic compounds were so named because many of them have pleasant odors.

    (c) According to the resonance model of bonding, benzene is best described as a hybrid of two equivalent contributing structures.

    (d) Benzene is a planar molecule.

13.3 What is the difference in structure between a saturated and an unsaturated compound?

13.4 Define *aromatic compound*.

13.5 Why are alkenes, alkynes, and aromatic compounds said to be unsaturated?

13.6 Do aromatic rings have double bonds? Are they unsaturated? Explain.

13.7 Can an aromatic compound be a saturated compound?

13.8 Draw at least two structural formulas for each of the following. (Several constitutional isomers are possible for each part.)

(a) An alkene of six carbons

(b) A cycloalkene of six carbons

(c) An alkyne of six carbons

(d) An aromatic hydrocarbon of eight carbons

13.9 Write a structural formula and the name for the simplest (a) alkane, (b) alkene, (c) alkyne, and (d) aromatic hydrocarbon.

13.10 Account for the fact that the six-membered ring in benzene is planar but the six-membered ring in cyclohexane is not.

13.11 The compound 1,4-dichlorobenzene (*p*-dichlorobenzene) has a rigid geometry that does not allow free rotation. Yet no *cis-trans* isomers exist for this structure. Explain why it does not show *cis-trans* isomerism.

13.12 One analogy often used to explain the concept of a resonance hybrid is to relate a rhinoceros to a unicorn and a dragon. Explain the reasoning in this analogy and how it might relate to a resonance hybrid.

Section 13.2 How Do We Name Aromatic Compounds?

13.13 Answer true or false.

(a) A phenyl group has the molecular formula C_6H_5— and is represented by the symbol Ph—.

(b) Para substituents occupy adjacent carbons on a benzene ring.

(c) 4-Bromobenzoic acid can be separated into *cis* and *trans* isomers.

(d) Naphthalene is a planar molecule.

(e) Benzene, naphthalene, and anthracene are polynuclear aromatic hydrocarbons (PAHs).

(f) Benzo[a]pyrene causes cancer by binding to DNA and producing a cancer-causing mutation.

13.14 Name these compounds.

(a) 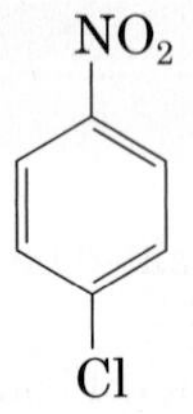

(b) 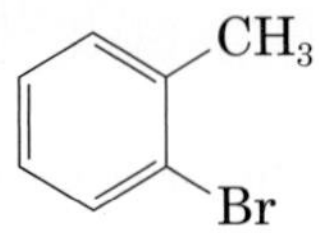

(c) $C_6H_5CH_2CH_2CH_2Cl$

(d) 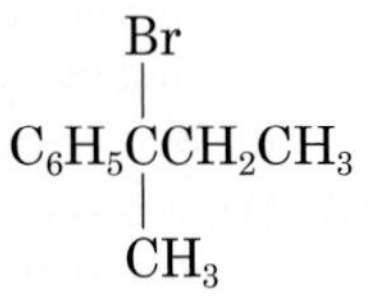

(e) 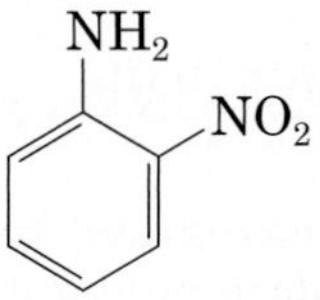

(f) 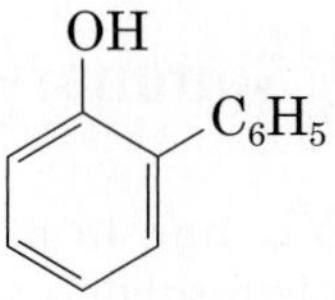

(g) 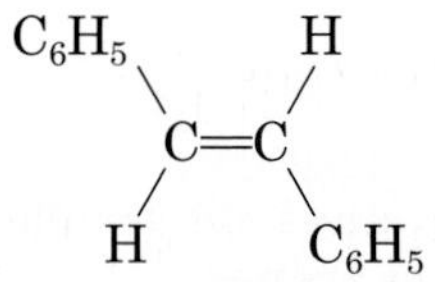

(h) 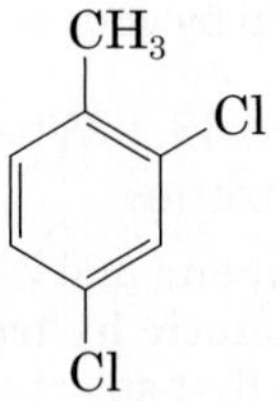

13.15 Draw structural formulas for these compounds.

(a) 1-Bromo-2-chloro-4-ethylbenzene

(b) 4-Bromo-1,2-dimethylbenzene

(c) 2,4,6-Trinitrotoluene

(d) 4-Phenyl-1-pentene

(e) *p*-Cresol

(f) 2,4-Dichlorophenol

13.16 We say that naphthalene, anthracene, phenanthrene, and benzo[a]pyrene are polynuclear aromatic hydrocarbons. In this context, what does "polynuclear" mean? What does "aromatic" mean? What does "hydrocarbon" mean?

13.17 Following is the structural formula of styrene, the monomer from which the polymer polystyrene is prepared.

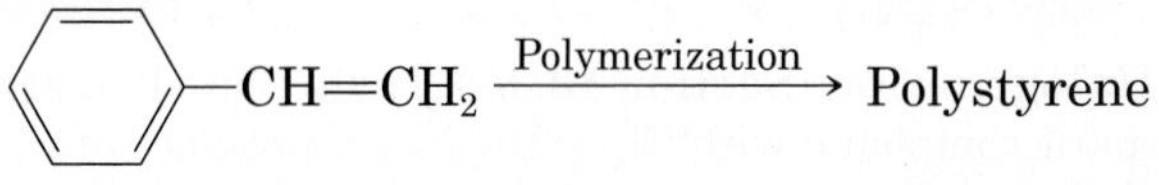

Styrene

Draw a section of a polystyrene chain showing at least three monomer units.

Section 13.3 What Are the Characteristic Reactions of Benzene and Its Derivatives?

13.18 Answer true or false.

(a) Benzene does not undergo the addition reactions that are characteristic of alkenes.

(b) A defining feature of aromatic compounds is that they are highly unsaturated but do not undergo characteristic alkene addition reactions.

(c) Nitration of benzene adds a $—NO_2$ group to one of the carbons of the aromatic ring.

(d) Halogenation of an alkene is an addition reaction; halogenation of an arene is a substitution reaction.

13.19 Suppose you have unlabeled bottles of benzene and cyclohexene. What chemical reaction could you use to tell which bottle contains which chemical? Explain what you would do, what you would expect to see, and how you would explain your observations. Write an equation for a positive test.

13.20 Three products with the molecular formula C_6H_4BrCl form when bromobenzene is treated with chlorine, Cl_2, in the presence of $FeCl_3$ as a catalyst. Name and draw a structural formula for each product.

13.21 The reaction of bromine with toluene in the presence of FeCl$_3$ gives a mixture of three products, all with the molecular formula C$_7$H$_7$Br. Name and draw a structural formula for each product.

13.22 What reagents and/or catalysts are necessary to carry out each conversion?

(a) Benzene to nitrobenzene

(b) 1,4-dichlorobenzene to 2-bromo-1,4-dichlorobenzene

(c) Benzene to aniline

13.23 What reagents and/or catalysts are necessary to carry out each conversion? Each conversion requires two steps.

(a) Benzene to 3-nitrobenzenesulfonic acid

(b) Benzene to 1-bromo-4-chlorobenzene

13.24 Aromatic substitution can be done on naphthalene. Treatment of naphthalene with concentrated H$_2$SO$_4$ gives two (and only two) different sulfonic acids. Draw a structural formula for each.

Section 13.4 What Are Phenols?

13.25 Answer true or false.

(a) Phenols and alcohols have in common the presence of an —OH group.

(b) Phenols are weak acids and react with strong bases to give water-soluble salts.

(c) The pK_a of phenol is smaller than that of acetic acid.

(d) Autoxidation converts an R—H group to an R—OH (hydroxyl) group.

(e) A carbon radical has only seven electrons in the valence shell of one of its carbons, and this carbon bears a positive charge.

(f) A characteristic of a chain initiation step is conversion of a nonradical to a radical.

(g) Autoxidation is a radical-chain process.

(h) A characteristic of the chain propagation step is reaction of a radical and a molecule to form a new radical and a new molecule.

(i) Vitamin E and other natural antioxidants function by interrupting the cycle of chain propagation steps that occurs in autoxidation.

13.26 Both phenol and cyclohexanol are only slightly soluble in water. Account for the fact that phenol dissolves in aqueous sodium hydroxide but cyclohexanol does not.

13.27 Define autoxidation.

13.28 Autoxidation is described as a *radical-chain reaction*. What is meant by the term "radical" in this context? By the term "chain"? By the term "chain length"?

13.29 Show that if you add Steps 2a and 2b of the radical-chain mechanism for the autoxidation of a fatty acid hydrocarbon chain, you arrive at the following net equation:

$$\underset{\substack{\text{Section of a fatty acid}\\\text{hydrocarbon chain}}}{-CH_2CH{=}CH-\overset{\text{H}}{\underset{|}{CH}}-} + \underset{\text{Oxygen}}{\cdot O{-}O\cdot} \longrightarrow$$

$$\underset{\text{A hydroperoxide}}{-CH_2CH{=}CH-\overset{\overset{\text{O—O—H}}{|}}{CH}-}$$

13.30 How does vitamin E function as an antioxidant?

13.31 What structural features are common to vitamin E, BHT, and BHA (the three antioxidants presented in Section 13.4C)?

13.32 Black-and-white photography is a commercial process that involves a phenol. Black-and-white film is coated with an emulsion containing silver bromide or silver iodide crystals that become activated by exposure to light. The activated silver ions then react with hydroquinone in the developing stage as shown in the following balanced equation.

$$+ 2Ag^+ \longrightarrow \qquad + 2Ag^0 + 2H^+$$

1,-Benzenediol (Hydroquinone)   1,4-Benzoquinone (p-Quinone)

All silver halide not activated by light is removed in the fixing process, and the result is a black image (a negative) left by the deposited metallic silver where the film had been struck by light. In this redox reaction state:

(a) What is reduced and what is the reducing agent?

(b) What is oxidized and what is the oxidizing agent?

13.33 Following is the structural formula of albuterol (Proventil), one of the most widely used inhalation bronchodilators.

Albuterol

(a) Name the functional groups present in albuterol.

(b) Draw a structural formula for the product formed when albuterol is treated with one equivalent of aqueous sodium hydroxide.

(c) Draw a structural formula for the product formed when albuterol is treated with one equivalent of hydrochloric acid.

Chemical Connections

13.34 (Chemical Connections 13A) From what parts of its common name are the letters DDT derived?

13.35 (Chemical Connections 13A) What are the advantages and disadvantages of using DDT as an insecticide?

13.36 (Chemical Connections 13A) Would you expect DDT to be soluble or insoluble in water? Explain.

13.37 (Chemical Connections 13A) One of the degradation products of DDT is dichlorodiphenyldichloroethylene (DDE), which is formed by the loss of HCl from adjacent carbons of DDT. DDE inhibits the enzyme responsible for the incorporation of calcium ion into bird eggshells. Draw a structural formula for DDE.

13.38 (Chemical Connections 13A) What is meant by the term *biodegradable*?

13.39 (Chemical Connections 13B) What is a carcinogen? What kind of carcinogen is found in cigarette smoke?

13.40 (Chemical Connections 13C) In the absence of iodine in the diet, goiter develops. Explain why goiter is a regional disease.

13.41 (Chemical Connections 13D) Calculate the molecular weight of each explosive in this Chemical Connection. In which explosive do the nitro groups contribute the largest percentage of molecular weight?

13.42 (Chemical Connections 13E) What are the differences in structure between Allura Red and Sunset Yellow?

13.43 (Chemical Connections 13F) Which features of Allura Red and Sunset Yellow make them water-soluble?

13.44 (Chemical Connections 13G) What color would you get if you mixed Allura Red and Sunset Yellow? (*Hint*: Remember the color wheel.)

13.45 (Chemical Connections 13E)

(a) Review the structural formulas for lycopene (Problem 12.63) and β-carotene (Problem 16.40), two natural food-coloring agents, and compare their structural formulas with those of Allura Red and Sunset Yellow.

(b) What structural feature or features do these four compounds have in common that might account for the fact that each is colored?

(c) Why are these two food and cosmetic dyes soluble in water while lycopene and β-carotene are not?

13.46 (Chemical Connections 13F) From what types of plants is capsaicin isolated?

13.47 (Chemical Connections 13F) Find the phenol group in capsaicin.

13.48 (Chemical Connections 13F) How many *cis-trans* isomers are possible for capsaicin? Is the structural formula shown in this Chemical Connection the *cis* isomer or the *trans* isomer?

13.49 (Chemical Connections 13F) In what ways is capsaicin used in medicine?

Additional Problems

13.50 The structure for naphthalene given in Section 13.2D is only one of three possible resonance structures. Draw the other two.

13.51 Draw structural formulas for these compounds.
(a) 1-Phenylcyclopropanol (b) Styrene
(c) *m*-Bromophenol   (d) 4-Nitrobenzoic acid
(e) Isobutylbenzene   (f) *m*-Xylene

13.52 2,6-Di-*tert*-butyl-4-methylphenol (BHT, Section 13.4C) is an antioxidant added to processed foods to "retard spoilage." How does BHT accomplish this goal?

13.53 Write the structural formula for the product of each reaction.

(a) benzene $+ HNO_3 \xrightarrow{H_2SO_4}$

(b) 1,4-dimethylbenzene $+ Br_2 \xrightarrow{FeCl_3}$

(c) 1,4-dibromobenzene $+ H_2SO_4 \longrightarrow$

13.54 Styrene reacts with bromine to give a compound with the molecular formula $C_8H_8Br_2$. Draw a structural formula for this compound.

13.55 When toluene is treated with Br_2 in the presence of an $AlBr_3$ catalyst, a mixture of three compounds is formed, all with the molecular formula C_7H_7Br. Draw structural formulas and name each of these products.

Looking Ahead

13.56 Benzene, as we have seen in this chapter, is the simplest aromatic compound. Pyridine is an analog of benzene in which a CH group is replaced with a nitrogen atom. Pyrimidine is an analog of benzene in which two CH groups are replaced with nitrogen atoms. Each nitrogen-containing compound shows the characteristic reactions of benzene and its derivatives—it is highly unsaturated but does not undergo the characteristic addition reactions of alkenes.

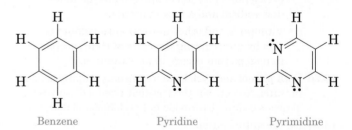

Benzene   Pyridine   Pyrimidine

(a) Show by the use of curved arrows that benzene, pyridine, and pyrimidine can be represented as hybrids of two contributing structures.

(b) Show that each aromatic compound has an aromatic sextet—that is, a loop of six electrons within a cyclic system.

(c) Predict the bond angles in pyridine and pyrimidine and the shape of each molecule.

Alcohols, Ethers, and Thiols

14

Fermentation vats of wine grapes at the Beaulieu Vineyards, California.

In this chapter, we study the physical and chemical properties of alcohols and ethers, two classes of oxygen-containing organic compounds. We also study thiols, a class of sulfur-containing organic compounds. Thiols are like alcohols in structure, except that they contain an —SH group rather than an —OH group.

$$CH_3CH_2OH \qquad CH_3CH_2OCH_2CH_3 \qquad CH_3CH_2SH$$

Ethanol   Diethyl ether   Ethanethiol
(An alcohol)   (An ether)   (A thiol)

These three compounds are certainly familiar to you. Ethanol is the fuel additive in E85 and E15, the alcohol in alcoholic beverages, and an important laboratory and industrial solvent. Diethyl ether was the first inhalation anesthetic used in general surgery. It is also an important laboratory and industrial solvent. Ethanethiol, like other low-molecular-weight thiols, has a stench. Traces of ethanethiol are added to natural gas so that gas leaks can be detected by the smell of the thiol.

Solutions in which ethanol is the solvent are called tinctures.

OWL

Sign in to OWL at **www.cengage.com/owl** to view tutorials and simulations, develop problem-solving skills, and complete online homework assigned by your professor.

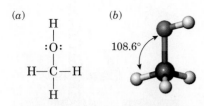

FIGURE 14.1 Methanol, CH_3OH. (a) Lewis structure and (b) ball-and-stick model. The H—C—O bond angle is 108.6°, very close to the tetrahedral angle of 109.5°.

14.1 What Are the Structures, Names, and Physical Properties of Alcohols?

A. Structure of Alcohols

The functional group of an **alcohol** is an **—OH (hydroxyl) group** bonded to a tetrahedral carbon atom (Section 10.4A). Figure 14.1 shows a Lewis structure and a ball-and-stick model of methanol, CH_3OH, the simplest alcohol.

B. Nomenclature

IUPAC names of alcohols are derived in the same manner as those for alkenes and alkynes, with the exception that the ending of the parent alkane is changed from -e to -ol. The ending -ol tells us that the compound is an alcohol.

1. Select as the parent alkane the longest carbon chain that contains the —OH group and number it from the end that gives —OH the lower number. In numbering the parent chain, the location of the —OH group takes precedence over alkyl groups, aryl groups, and halogens.

2. Change the ending of the parent alkane from -e to -ol and use a number to show the location of the —OH group. For cyclic alcohols, numbering begins at the carbon bearing the —OH group; this carbon is automatically carbon 1.

3. Name and number substituents and list them in alphabetical order.

To derive common names for alcohols, name the alkyl group bonded to —OH and then add the word "alcohol." Following are IUPAC names and, in parentheses, common names for eight low-molecular-weight alcohols:

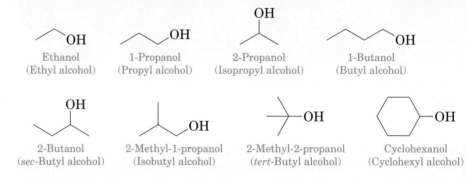

Ethanol
(Ethyl alcohol)

1-Propanol
(Propyl alcohol)

2-Propanol
(Isopropyl alcohol)

1-Butanol
(Butyl alcohol)

2-Butanol
(*sec*-Butyl alcohol)

2-Methyl-1-propanol
(Isobutyl alcohol)

2-Methyl-2-propanol
(*tert*-Butyl alcohol)

Cyclohexanol
(Cyclohexyl alcohol)

⊙WL Interactive Example 14.1 Systematic Names of Alcohols

Write the IUPAC name of each alcohol.

(a) [structure with OH]   (b) [cyclohexane structure with OH]

Strategy

Follow the steps outlined above.

Step 1: Identify the parent chain.

Step 2: Change the ending of the parent alkane from -e to -ol and use a number to show the location of the —OH group.

Step 3: Name and number substituents and list them in alphabetical order.

Step 4: Specify configuration if *cis-trans* isomerism exists.

Solution

(a) The parent alkane is pentane. Number the parent chain from the direction that gives the lower number to the carbon bearing the —OH group. This alcohol is 4-methyl-2-pentanol.

(b) The parent cycloalkane is cyclohexane. Number the atoms of the ring beginning with the carbon bearing the —OH group as carbon 1 and specify that the methyl and hydroxyl groups are *trans* to each other. This alcohol is *trans*-2-methylcyclohexanol.

Problem 14.1

Write the IUPAC name of each alcohol.

(a) 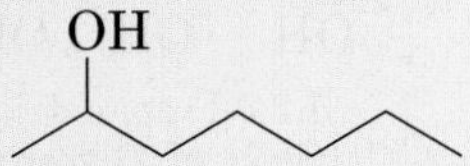   (b) 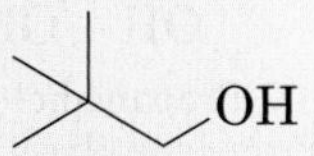   (c)

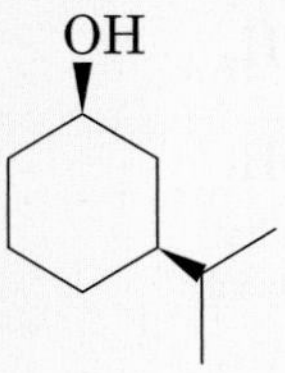

We classify alcohols as **primary (1°), secondary (2°),** or **tertiary (3°),** depending on the number of carbon groups bonded to the carbon bearing the —OH group (Section 10.4A).

Example 14.2 Classification of Alcohols

Classify each alcohol as primary, secondary, or tertiary.

(a) 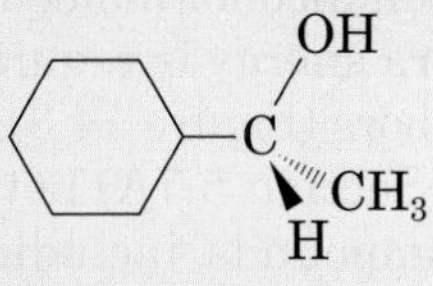   (b) 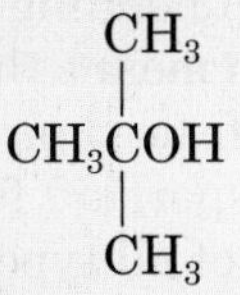 $CH_3\overset{\underset{\displaystyle CH_3}{|}}{\underset{\displaystyle CH_3}{\overset{\displaystyle CH_3}{|}}}COH$ (c)

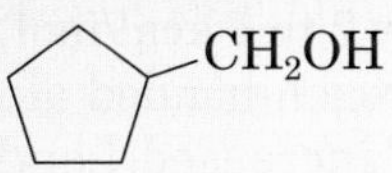

Strategy

Locate the carbon bearing the OH group and count the number of carbon groups bonded to that carbon.

Solution

(a) Secondary (2°)   (b) Tertiary (3°)   (c) Primary (1°)

Problem 14.2

Classify each alcohol as primary, secondary, or tertiary.

(a) 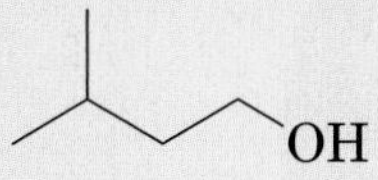   (b)

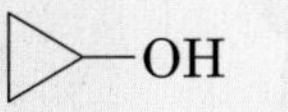

(c) 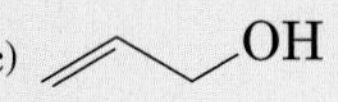   (d)

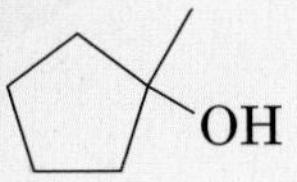

Diol A compound with —OH (hydroxyl) groups on adjacent carbons

Glycol A compound with hydroxyl (—OH) groups on adjacent carbons

Ethylene glycol is colorless; the color of most antifreezes comes from additives. Review Section 6.8A for a discussion of freezing-point depression.

In the IUPAC system, a compound containing two hydroxyl groups is named as a **diol**, one containing three hydroxyl groups as a **triol**, and so on. In IUPAC names for diols, triols, and so on, the final *-e* in the name of the parent alkane is retained—for example, 1,2-ethanediol.

As with many other organic compounds, common names for certain diols and triols have persisted. Compounds containing hydroxyl groups on adjacent carbons are often referred to as **glycols**. Ethylene glycol and propylene glycol are synthesized from ethylene and propylene, respectively—hence their common names.

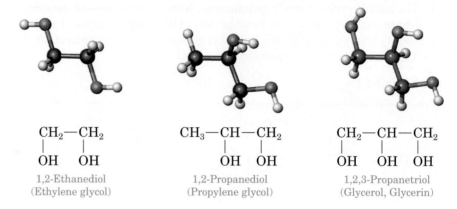

$$CH_2 - CH_2$$
$$\;\;|\qquad\;\;|$$
$$OH\quad OH$$

1,2-Ethanediol
(Ethylene glycol)

$$CH_3 - CH - CH_2$$
$$\qquad\quad|\qquad\;\;|$$
$$\qquad\quad OH\quad OH$$

1,2-Propanediol
(Propylene glycol)

$$CH_2 - CH - CH_2$$
$$\;\;|\qquad\;\;|\qquad\;\;|$$
$$OH\quad OH\quad OH$$

1,2,3-Propanetriol
(Glycerol, Glycerin)

Ethylene glycol is a polar molecule that dissolves readily in the polar solvent water.

Charles D. Winters

C. Physical Properties of Alcohols

The most important physical property of alcohols is the polarity of their —OH groups. Because of the large difference in electronegativity (Table 3.5) between oxygen and carbon ($3.5 - 2.5 = 1.0$) and between oxygen and hydrogen ($3.5 - 2.1 = 1.4$), both the C—O and O—H bonds of alcohols are polar-covalent and alcohols are polar molecules, as illustrated in Figure 14.2 for methanol.

Alcohols have higher boiling points than do alkanes, alkenes, and alkynes of similar molecular weight (Table 14.1), because alcohol molecules associate with one another in the liquid state by **hydrogen bonding** (Section 5.7C). The strength of hydrogen bonding between alcohol molecules is approximately 2 to 5 kcal/mol, which means that extra energy is required to separate hydrogen-bonded alcohols from their neighbors (Figure 14.3).

Because of increased London dispersion forces (Section 5.7A) between larger molecules, the boiling points of all types of compounds, including alcohols, increase with increasing molecular weight.

FIGURE 14.2 Polarity of the C—O—H bonds in methanol. (*a*) There are partial positive charges on carbon and hydrogen and a partial negative charge on oxygen. (*b*) An electron density map showing the partial negative charge (in red) around oxygen and a partial positive charge (in blue) around the hydrogen of the OH group.

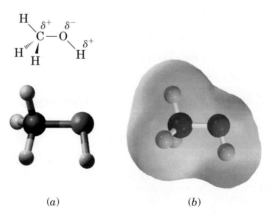

(*a*)   (*b*)

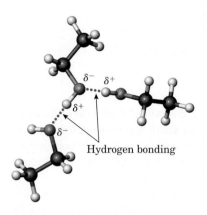

FIGURE 14.3 The association of
ethanol molecules in the liquid
state. Each O—H can participate
in up to three hydrogen bonds
(one through hydrogen and two
through oxygen). Only two of
these three possible hydrogen
bonds per molecule are shown.

Alcohols are much more soluble in water than are hydrocarbons of similar molecular weight (Table 14.1), because alcohol molecules interact by hydrogen bonding with water molecules. Methanol, ethanol, and 1-propanol are soluble in water in all proportions. As molecular weight increases, the water solubility of alcohols becomes more like that of hydrocarbons of similar molecular weight. Higher-molecular-weight alcohols are much less soluble in water because the size of the hydrocarbon portion of their molecules (which decreases water solubility) becomes so large relative to the size of the —OH group (which increases water solubility).

Table 14.1 Boiling Points and Solubilities in Water of Four
Sets of Alcohols and Alkanes of Similar Molecular Weight

Structural Formula	Name	Molecular Weight (amu)	Boiling Point (°C)	Solubility in Water
CH_3OH	methanol	32	65	infinite
CH_3CH_3	ethane	30	−89	insoluble
CH_3CH_2OH	ethanol	46	78	infinite
$CH_3CH_2CH_3$	propane	44	−42	insoluble
$CH_3CH_2CH_2OH$	1-propanol	60	97	infinite
$CH_3CH_2CH_2CH_3$	butane	58	0	insoluble
$CH_3CH_2CH_2CH_2OH$	1-butanol	74	117	8 g/100 g
$CH_3CH_2CH_2CH_2CH_3$	pentane	72	36	insoluble

14.2 What Are the Characteristic Reactions of Alcohols?

In this section, we study the acidity of alcohols, their dehydration to alkenes, and their oxidation to aldehydes, ketones, and carboxylic acids.

A. Acidity of Alcohols

Alcohols have about the same pK_a values as water (Table 8.3), which means that aqueous solutions of alcohols have approximately the same pH as that of pure water. In Section 13.4B, we studied the acidity of phenols, another class of compounds that contains an —OH group. Phenols are weak acids that react with aqueous sodium hydroxide to form water-soluble salts.

$$\text{Phenol} + NaOH \xrightarrow{H_2O} \text{Sodium phenoxide} + H_2O$$

Phenol   Sodium phenoxide
(a water-soluble salt)

Chemical Connections 14A

Nitroglycerin: An Explosive and a Drug

In 1847, Ascanio Sobrero (1812–1888) discovered that 1,2,3-propanetriol, more commonly called glycerin, reacts with nitric acid in the presence of sulfuric acid to give a pale yellow, oily liquid called nitroglycerin. Sobrero also discovered the explosive properties of this compound; when he heated a small quantity of it, it exploded!

$$
\begin{array}{c}
CH_2-OH \\
| \\
CH-OH \\
| \\
CH_2-OH
\end{array}
+ 3HNO_3 \xrightarrow{H_2SO_4}
\begin{array}{c}
CH_2-ONO_2 \\
| \\
CH-ONO_2 \\
| \\
CH_2-ONO_2
\end{array}
+ 3H_2O
$$

1,2,3-Propanetriol   1,2,3-Propanetriol trinitrate
(Glycerol, Glycerin)   (Nitroglycerin)

Nitroglycerin very soon became widely used for blasting in the construction of canals, tunnels, roads, and mines and, of course, for warfare.

One problem with the use of nitroglycerin was soon recognized: It was difficult to handle safely, and accidental explosions occurred all too frequently. This problem was solved by the Swedish chemist Alfred Nobel (1833–1896), who discovered that a clay-like substance called diatomaceous earth absorbs nitroglycerin so that it will not explode without a fuse. Nobel gave the name *dynamite* to this mixture of nitroglycerin, diatomaceous earth, and sodium carbonate.

Surprising as it may seem, nitroglycerin is used in medicine to treat angina pectoris, the symptoms of which are sharp chest pains caused by reduced flow of blood in the coronary artery. Nitroglycerin, which is available in liquid form (diluted with alcohol to render it nonexplosive), tablet form, and paste form, relaxes the smooth muscles of blood vessels, causing dilation of the coronary artery. This dilation, in turn, allows more blood to reach the heart.

When Nobel became ill with heart disease, his physicians advised him to take nitroglycerin to relieve his chest pains. He refused, saying he could not understand how the explosive could relieve chest pains. It took science more than 100 years to find the answer. We now know that nitric oxide, NO, derived from the nitro groups of nitroglycerin, relieves the pain (see Chemical Connections 24E).

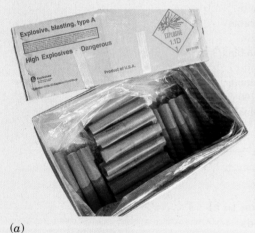

(a)

(b)

(a) Charles D. Winters (b) Bettmann Archives/Corbis

(*a*) Nitroglycerin is more stable when absorbed onto an inert solid, a combination called dynamite. (*b*) The fortune of Alfred Nobel (1833–1896), built on the manufacture of dynamite, now funds the Nobel Prizes.

Test your knowledge with Problems 14.50–14.52.

Alcohols are considerably weaker acids than phenols and do not react in this manner.

B. Acid-Catalyzed Dehydration of Alcohols

Dehydration Elimination of a molecule of water from an alcohol. In the dehydration of an alcohol, OH is removed from one carbon and an H is removed from an adjacent carbon.

We can convert an alcohol to an alkene by eliminating a molecule of water from adjacent carbon atoms in a reaction called **dehydration**. In the laboratory, dehydration of an alcohol is most often brought about by heating it with either 85% phosphoric acid or concentrated sulfuric acid. Primary alcohols—the most difficult to dehydrate—generally require heating in concentrated sulfuric acid at temperatures as high as 180°C. Secondary alcohols undergo acid-catalyzed dehydration at somewhat lower temperatures. Tertiary alcohols generally undergo acid-catalyzed dehydration at temperatures only slightly above room temperature.

$$CH_3CH_2OH \xrightarrow[180°C]{H_2SO_4} CH_2{=}CH_2 + H_2O$$

Ethanol   Ethylene

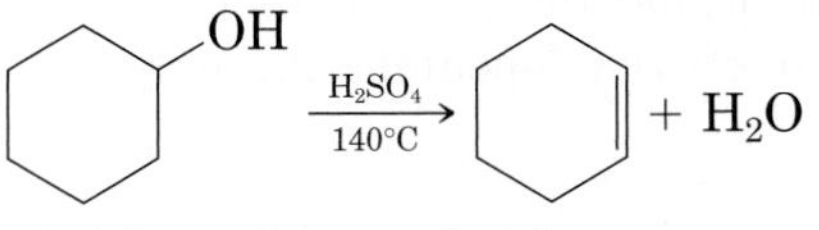

Cyclohexanol   Cyclohexene

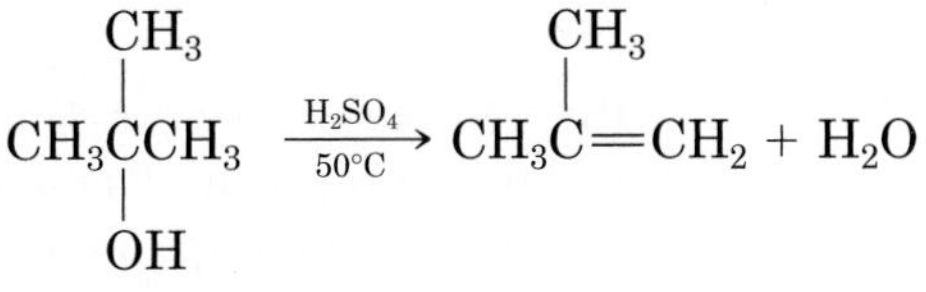

$$\underset{\substack{| \\ OH}}{\overset{\substack{CH_3 \\ |}}{CH_3CCH_3}} \xrightarrow[50°C]{H_2SO_4} \overset{\substack{CH_3 \\ |}}{CH_3C}{=}CH_2 + H_2O$$

2-Methyl-2-propanol   2-Methylpropene
(*tert*-Butyl alcohol)   (Isobutylene)

Thus, the ease of acid-catalyzed dehydration of alcohols follows this order:

1° alcohols   2° alcohols   3° alcohols

Ease of dehydration of alcohols ⟶

When the acid-catalyzed dehydration of an alcohol yields isomeric alkenes, the alkene having the greater number of alkyl groups on the double bond generally predominates. In the acid-catalyzed dehydration of 2-butanol, for example, the major product is 2-butene, which has two alkyl groups (two methyl groups) on its double bond. The minor product is 1-butene, which has only one alkyl group (an ethyl group) on its double bond.

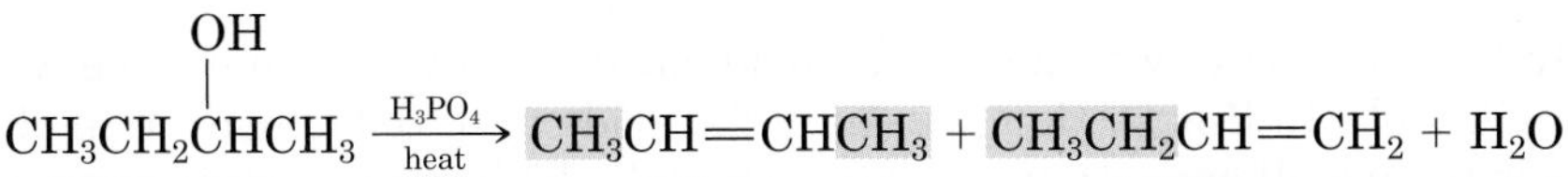

$$\underset{\text{2-Butanol}}{\overset{\substack{OH \\ |}}{CH_3CH_2CHCH_3}} \xrightarrow[\text{heat}]{H_3PO_4} \underset{\substack{\text{2-Butene} \\ (80\%)}}{CH_3CH{=}CHCH_3} + \underset{\substack{\text{1-Butene} \\ (20\%)}}{CH_3CH_2CH{=}CH_2} + H_2O$$

Example 14.3 Acid-Catalyzed Dehydration of Alcohols

Draw structural formulas for the alkenes formed by the acid-catalyzed dehydration of each alcohol. For each part, predict which alkene will be the major product.

(a) 3-Methyl-2-butanol   (b) 2-Methylcyclopentanol

Strategy

In the acid-catalyzed dehydration of an alcohol, H and OH are removed from adjacent carbon atoms. When dehydration yields isomeric alkenes, the alkene with the greater number of alkyl groups on the carbon atoms of the double bond generally predominates.

Solution

(a) Elimination of H_2O from carbons 2–3 gives 2-methyl-2-butene; elimination of H_2O from carbons 1–2 gives 3-methyl-1-butene. 2-Methyl-2-butene has three alkyl groups (three methyl groups) on its double bond and is the major product. 3-Methyl-1-butene has only one alkyl group (an isopropyl group) on its double bond and is the minor product.

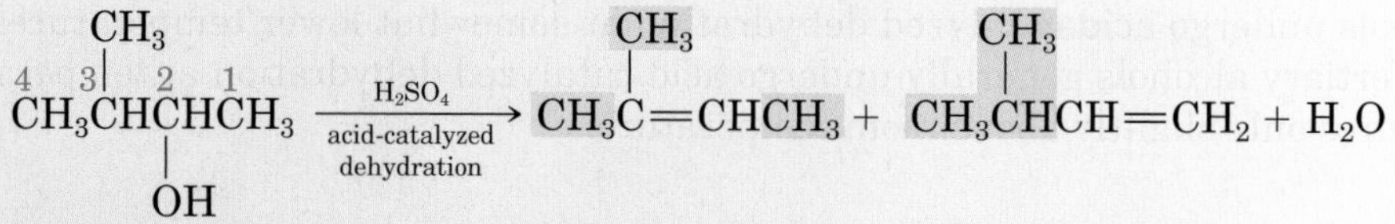

3-Methyl-2-butanol | 2-Methyl-2-butene (major product) | 3-Methyl-1-butene

(b) The major product, 1-methylcyclopentene, has three alkyl groups on its double bond. The minor product, 3-methylcyclopentene, has only two alkyl groups on its double bond.

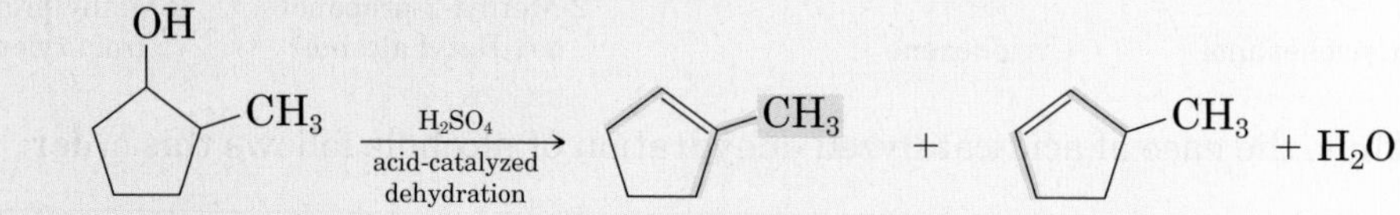

2-Methylcyclopentanol | 1-Methylcyclopentene (major product) | 3-Methylcyclopentene

Problem 14.3

Draw structural formulas for the alkenes formed by the acid-catalyzed dehydration of each alcohol. For each part, predict which alkene will be the major product.

(a) 2-Methyl-2-butanol (b) 1-Methylcyclopentanol

In Section 12.6B, we studied the acid-catalyzed hydration of alkenes to give alcohols. In this section, we study the acid-catalyzed dehydration of alcohols to give alkenes. In fact, hydration–dehydration reactions are reversible. Alkene hydration and alcohol dehydration are competing reactions, and the following equilibrium exists:

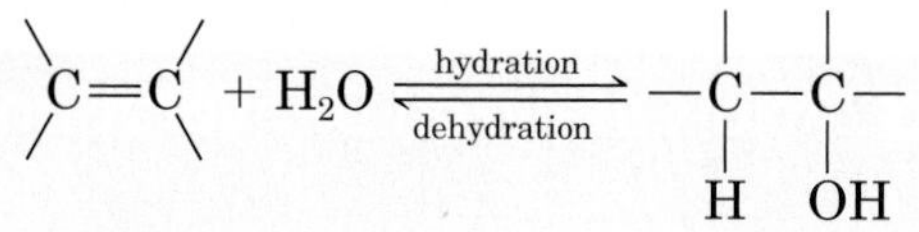

An alkene | An alcohol

In accordance with Le Chatelier's principle (Section 7.7), large amounts of water (in other words, using dilute aqueous acid) favor alcohol formation, whereas scarcity of water (using concentrated acid) or experimental conditions where water is removed (heating the reaction mixture above 100°C) favor alkene formation. Thus, depending on the experimental conditions, we can use the hydration–dehydration equilibrium to prepare both alcohols and alkenes, each in high yields.

Example 14.4 Acid-Catalyzed Dehydration of Alcohols and Hydration of Alkenes

In part (a), acid-catalyzed dehydration of 2-methyl-3-pentanol gives predominantly Compound A. Treatment of Compound A with water in the presence of sulfuric acid in part (b) gives Compound B. Propose structural formulas for Compounds A and B.

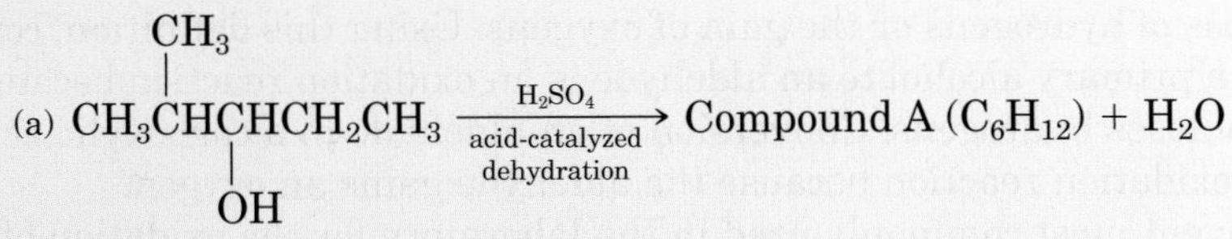

$$\text{(a) } \underset{\underset{\text{OH}}{|}}{\overset{\overset{\text{CH}_3}{|}}{\text{CH}_3\text{CHCHCH}_2\text{CH}_3}} \xrightarrow[\text{dehydration}]{\overset{\text{H}_2\text{SO}_4}{\text{acid-catalyzed}}} \text{Compound A } (\text{C}_6\text{H}_{12}) + \text{H}_2\text{O}$$

$$\text{(b) Compound A } (\text{C}_6\text{H}_{12}) + \text{H}_2\text{O} \xrightarrow{\text{H}_2\text{SO}_4} \text{Compound B } (\text{C}_6\text{H}_{14}\text{O})$$

Strategy

The key to part (a) is that when acid-catalyzed dehydration of an alcohol can yield isomeric alkenes, the alkene with the greater number of alkyl groups on the carbon atoms of the double bond generally predominates. After the structural formula of A is determined, use Markovnikov's Rule to predict the structural formula of compound B.

Solution

(a) Acid-catalyzed dehydration of 2-methyl-3-pentanol gives predominantly 2-methyl-2-pentene, an alkene with three substituents on its double bond: two methyl groups and one ethyl group.

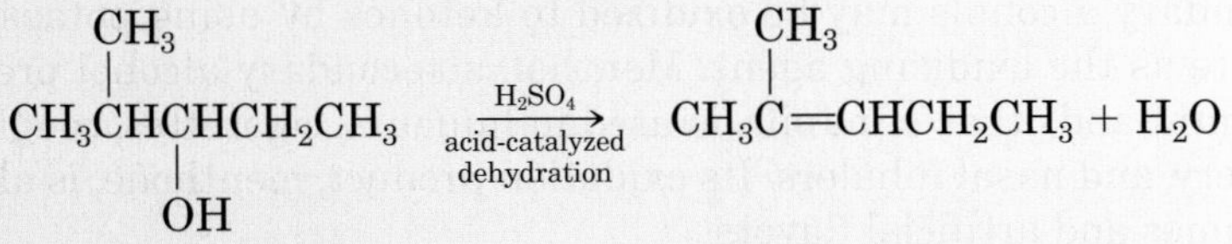

$$\underset{\underset{\text{OH}}{|}}{\overset{\overset{\text{CH}_3}{|}}{\text{CH}_3\text{CHCHCH}_2\text{CH}_3}} \xrightarrow[\text{dehydration}]{\overset{\text{H}_2\text{SO}_4}{\text{acid-catalyzed}}} \overset{\overset{\text{CH}_3}{|}}{\text{CH}_3\text{C}}{=}\text{CHCH}_2\text{CH}_3 + \text{H}_2\text{O}$$

2-Methyl-3-butanol   2-Methyl-2-pentene (major product)

(b) Acid-catalyzed addition of water to this alkene gives 2-methyl-2-pentanol in accord with Markovnikov's rule (Section 12.6B).

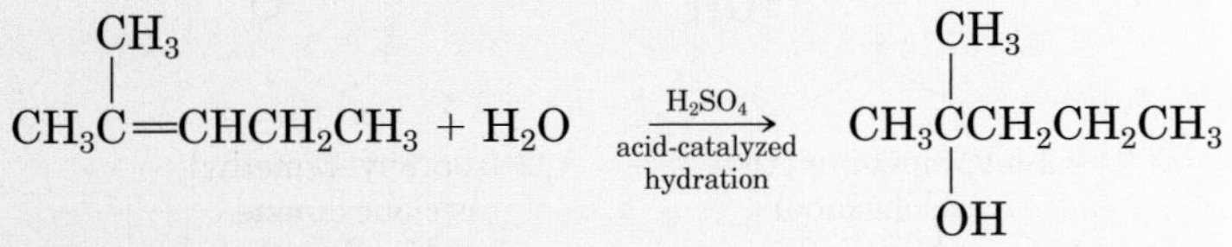

$$\overset{\overset{\text{CH}_3}{|}}{\text{CH}_3\text{C}}{=}\text{CHCH}_2\text{CH}_3 + \text{H}_2\text{O} \xrightarrow[\text{hydration}]{\overset{\text{H}_2\text{SO}_4}{\text{acid-catalyzed}}} \underset{\underset{\text{OH}}{|}}{\overset{\overset{\text{CH}_3}{|}}{\text{CH}_3\text{CCH}_2\text{CH}_2\text{CH}_3}}$$

Compound A (C_6H_{12})   Compound B ($\text{C}_6\text{H}_{14}\text{O}$)

Problem 14.4

Acid-catalyzed dehydration of 2-methylcyclohexanol gives predominantly Compound C (C_7H_{12}). Treatment of Compound C with water in the presence of sulfuric acid gives Compound D ($\text{C}_7\text{H}_{14}\text{O}$). Propose structural formulas for Compounds C and D.

C. Oxidation of Primary and Secondary Alcohols

A primary alcohol can be oxidized to an aldehyde or to a carboxylic acid, depending on the experimental conditions. Following is a series of transformations in which a primary alcohol is oxidized first to an aldehyde and then to a carboxylic acid. The letter O in brackets over the reaction arrow indicates that each transformation involves oxidation.

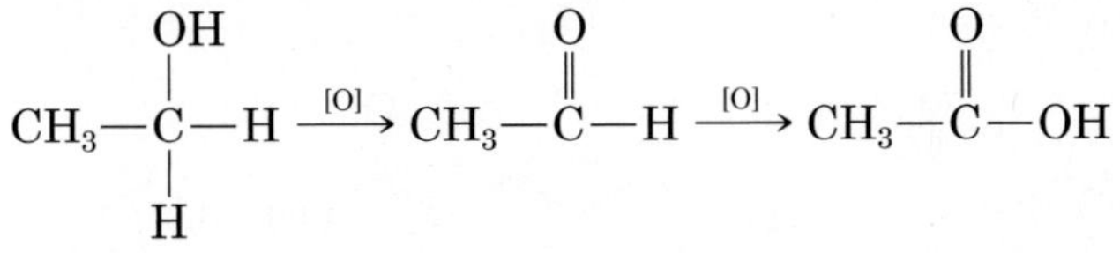

$$\underset{\underset{\text{H}}{|}}{\overset{\overset{\text{OH}}{|}}{\text{CH}_3{-}\text{C}{-}\text{H}}} \xrightarrow{[\text{O}]} \overset{\overset{\text{O}}{\|}}{\text{CH}_3{-}\text{C}{-}\text{H}} \xrightarrow{[\text{O}]} \overset{\overset{\text{O}}{\|}}{\text{CH}_3{-}\text{C}{-}\text{OH}}$$

A primary alcohol   An aldehyde   A carboxylic acid

Recall from Section 4.7 that according to one definition, oxidation is either the loss of hydrogens or the gain of oxygens. Using this definition, conversion of a primary alcohol to an aldehyde is an oxidation reaction because the alcohol loses hydrogens. Conversion of an aldehyde to a carboxylic acid is also an oxidation reaction because the aldehyde gains an oxygen.

The reagent most commonly used in the laboratory for the oxidation of a primary alcohol to a carboxylic acid is potassium dichromate, $K_2Cr_2O_7$, dissolved in aqueous sulfuric acid. Using this reagent, oxidation of 1-octanol, for example, gives octanoic acid. This experimental condition is more than sufficient to oxidize the intermediate aldehyde to a carboxylic acid.

$$CH_3(CH_2)_6CH_2OH \xrightarrow[H_2SO_4]{K_2Cr_2O_7} CH_3(CH_2)_6\overset{O}{\overset{\|}{C}}H \xrightarrow[H_2SO_4]{K_2Cr_2O_7} CH_3(CH_2)_6\overset{O}{\overset{\|}{C}}OH$$

1-Octanol   Octanal   Octanoic acid

Although the usual product of oxidation of a primary alcohol is a carboxylic acid, it is often possible to stop the oxidation at the aldehyde stage by distilling the mixture. That is, the aldehyde, which usually has a lower boiling point than either the primary alcohol or the carboxylic acid, is removed from the reaction mixture before it can be oxidized further.

Secondary alcohols may be oxidized to ketones by using potassium dichromate as the oxidizing agent. Menthol, a secondary alcohol present in peppermint and other mint oils, is used in liqueurs, cigarettes, cough drops, perfumery, and nasal inhalers. Its oxidation product, menthone, is also used in perfumes and artificial flavors.

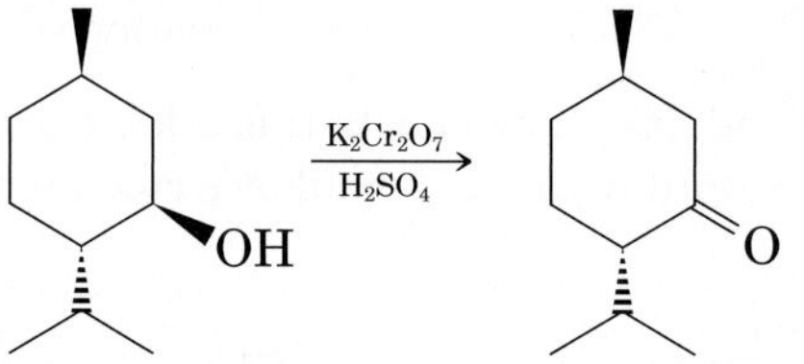

2-Isopropyl-5-methyl-
cyclohexanol
(Menthol)

2-Isopropyl-5-methyl-
cyclohexanone
(Menthone)

Tertiary alcohols resist oxidation because the carbon bearing the —OH is bonded to three carbon atoms and, therefore, cannot form a carbon–oxygen double bond.

Example 14.5 Oxidation of Alcohols

Draw a structural formula for the product formed by oxidation of each of the following alcohols with potassium dichromate.

(a) 1-Hexanol   (b) 2-Hexanol

Strategy

Oxidation of a 1-hexanol, a primary alcohol, gives either an aldehyde or a carboxylic acid, depending on the experimental conditions. Oxidation of 2-hexanol, a secondary alcohol, gives a ketone.

Solution

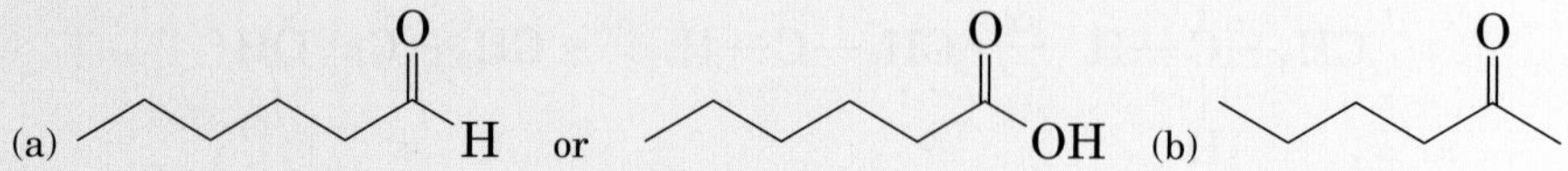

(a)   Hexanal   or   Hexanoic acid   (b)   2-Hexanone

Problem 14.5

Draw the product formed by oxidation of each of the following alcohols with potassium dichromate.

(a) Cyclohexanol

(b) 2-Pentanol

Chemical Connections 14B

Breath-Alcohol Screening

Potassium dichromate oxidation of ethanol to acetic acid is the basis for the original breath-alcohol screening test used by law enforcement agencies to determine a person's blood alcohol content (BAC). The test is based on the difference in color between the dichromate ion (reddish orange) in the reagent and the chromium(III) ion (green) in the product.

$$CH_3CH_2OH \ + \ Cr_2O_7^{2-}$$

Ethanol   Dichromate ion
            (reddish orange)

$$\xrightarrow[\text{H}_2\text{O}]{\text{H}_2\text{SO}_4} \ CH_3\overset{\displaystyle O}{\overset{\displaystyle \|}{C}}OH \ + \ Cr^{3+}$$

           Acetic acid   Chromium(III) ion
                            (green)

In its simplest form, breath-alcohol screening uses a sealed glass tube that contains a potassium dichromate–sulfuric acid reagent impregnated on silica gel. To administer the test, the ends of the tube are broken off, a mouthpiece is fitted to one end, and the other end is inserted into the neck of a plastic bag. The person being tested then blows into the mouthpiece to inflate the plastic bag.

As breath containing ethanol vapor passes through the tube, reddish orange dichromate ion is reduced to green chromium(III) ion. To estimate the concentration of ethanol in the breath, one measures how far the green color extends along the length of the tube. When it reaches beyond the halfway point, the person is judged as having a sufficiently high blood alcohol content to warrant further, more precise testing.

The test described here measures the alcohol content of the breath. The legal definition of being under the influence of alcohol, however, is based on alcohol content in the blood, not in the breath. The correlation between these two measurements is based on the fact that air deep within the lungs is in equilibrium with blood passing through the pulmonary arteries, and thus an equilibrium is established between blood alcohol and breath alcohol. Based on tests in persons drinking alcohol, researchers have determined that 2100 mL of breath contains the same amount of ethanol as 1.00 mL of blood.

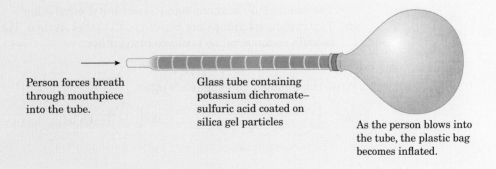

Person forces breath through mouthpiece into the tube.

Glass tube containing potassium dichromate–sulfuric acid coated on silica gel particles

As the person blows into the tube, the plastic bag becomes inflated.

Test your knowledge with Problems 14.53 and 14.54.

14.3 What Are the Structures, Names, and Properties of Ethers?

A. Structure

The functional group of an **ether** is an atom of oxygen bonded to two carbon atoms. Figure 14.4 shows a Lewis structure and a ball-and-stick model of dimethyl ether, CH_3OCH_3, the simplest ether.

Ether A compound containing an oxygen atom bonded to two carbon atoms

B. Nomenclature

Although the IUPAC system can be used to name ethers, chemists almost invariably use common names for low-molecular-weight ethers. Common names are derived by listing the alkyl groups bonded to oxygen in alphabetical order and adding the word *ether*. Alternatively, one of the groups on oxygen is named as an alkoxy group. The —OCH_3 group, for example, is named "methoxy" to indicate a <u>methy</u>l group bonded to <u>oxygen</u>.

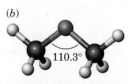

(a)

(b) 110.3°

FIGURE 14.4 Dimethyl ether, CH_3OCH_3. (a) Lewis structure and (b) ball-and-stick model. The C—O—C bond angle is 110.3°, close to the tetrahedral angle of 109.5°.

$$CH_3CH_2OCH_2CH_3$$

Diethyl ether

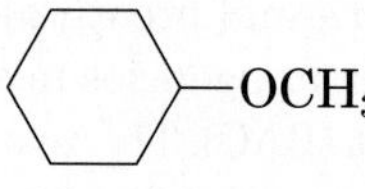

Cyclohexyl methyl ether
(Methoxycyclohexane)

—OCH_3

Example 14.6 Common Names for Ethers

Write the common name for each ether.

$$CH_3$$
(a) 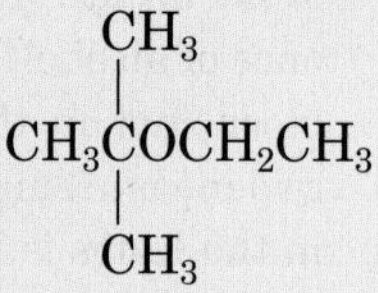 $CH_3\overset{\overset{\displaystyle CH_3}{|}}{\underset{\underset{\displaystyle CH_3}{|}}{C}}OCH_2CH_3$

(b)

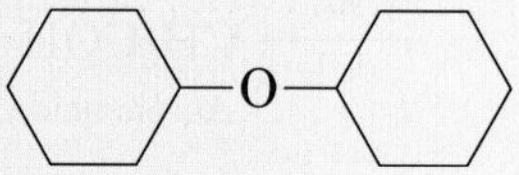

Strategy

To derive the common name of an ether, list the groups bonded to oxygen in alphabetical order.

Solution

(a) The groups bonded to the ether oxygen are *tert*-butyl and ethyl. The compound's common name is *tert*-butyl ethyl ether.

(b) Two cyclohexyl groups are bonded to the ether oxygen. The compound's common name is dicyclohexyl ether.

Problem 14.6

Write the common name for each ether.

(a)

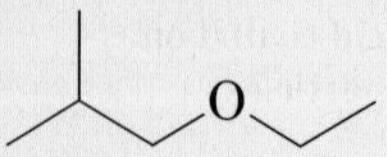

(b) 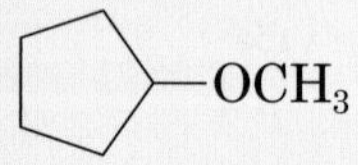 —OCH_3

Cyclic ether An ether in which the ether oxygen is one of the atoms of a ring

In **cyclic ethers**, one of the atoms in a ring is oxygen. These ethers are also known by their common names. Ethylene oxide is an important building block for the organic chemical industry (Section 14.5). Tetrahydrofuran is a useful laboratory and industrial solvent.

Chemical Connections 14C

Ethylene Oxide: A Chemical Sterilant

Ethylene oxide is a colorless, flammable gas with a boiling point of 11°C. Because it is such a highly strained molecule (the normal tetrahedral bond angles of both C and O are compressed from the normal tetrahedral angle of 109.5° to approximately 60°), ethylene oxide reacts with the amino (—NH$_2$) and sulfhydryl (—SH) groups present in biological materials.

At sufficiently high concentrations, it reacts with enough molecules in cells to cause the death of microorganisms. This toxic property is the basis for ethylene oxide's use as a fumigant in foodstuffs and textiles and its use in hospitals to sterilize surgical instruments.

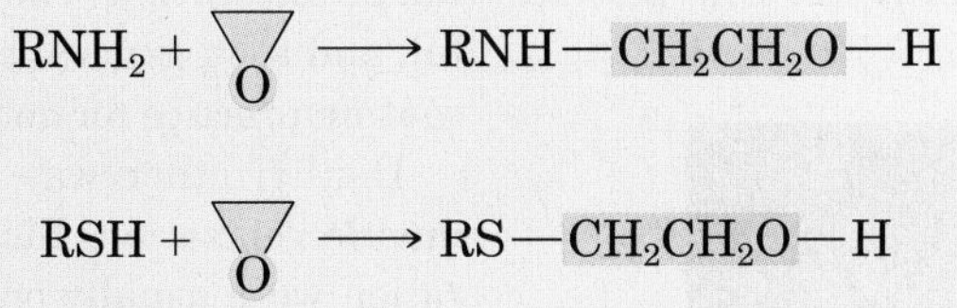

$$\text{RNH}_2 + \underset{\text{O}}{\bigtriangledown} \longrightarrow \text{RNH—CH}_2\text{CH}_2\text{O—H}$$

$$\text{RSH} + \underset{\text{O}}{\bigtriangledown} \longrightarrow \text{RS—CH}_2\text{CH}_2\text{O—H}$$

Test your knowledge with Problem 14.55.

Ethylene oxide   Tetrahydrofuran (THF)

C. Physical Properties

Ethers are polar compounds in which oxygen bears a partial negative charge and each attached carbon bears a partial positive charge (Figure 14.5). However, only weak forces of attraction exist between ether molecules in the pure liquid form. Consequently, the boiling points of ethers are close to those of hydrocarbons of similar molecular weight.

The effect of hydrogen bonding on physical properties is illustrated dramatically by comparing the boiling points of ethanol (78°C) and its constitutional isomer dimethyl ether (−24°C). The difference in boiling point between these two compounds is due to the presence in ethanol of a polar

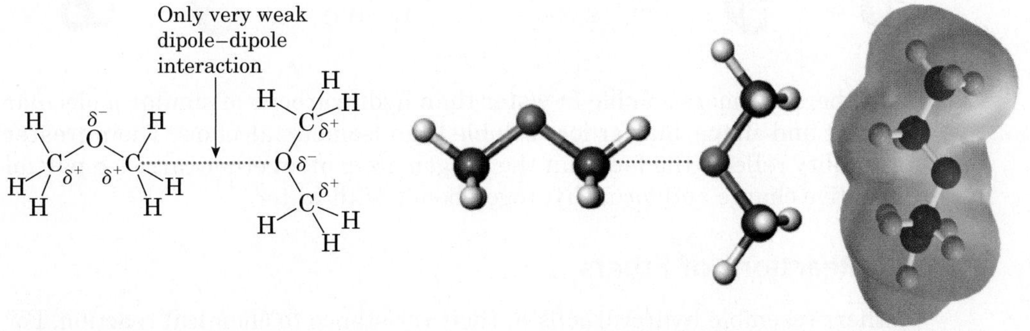

FIGURE 14.5 Ethers are polar molecules, but only weak attractive interactions exist between ether molecules in the liquid state. Shown on the right is an electron density map of diethyl ether.

Chemical Connections 14D

Ethers and Anesthesia

Before the mid-1800s, surgery was performed only when absolutely necessary, because no truly effective general anesthetic was available. More often than not, patients were drugged, hypnotized, or simply tied down.

In 1772, Joseph Priestley isolated nitrous oxide, N_2O, a colorless gas. In 1799, Sir Humphry Davy demonstrated this compound's anesthetic effect, naming it "laughing gas." In 1844, an American dentist, Horace Wells, introduced nitrous oxide into general dental practice. One patient

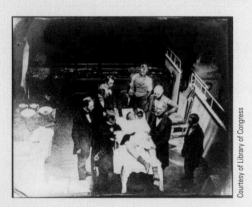

Courtesy of Library of Congress

This photo shows the first use of ether as an anesthetic in 1848. Dr. Robert John Collins was removing a tumor from the patient's neck and dentist W. T. G. Morton—who discovered the anesthetic properties—administered the ether.

awakened prematurely, however, screaming with pain; another died during the procedure. Wells was forced to withdraw from practice, became embittered and depressed, and committed suicide at age 33. In the same period, a Boston chemist, Charles Jackson, anesthetized himself with diethyl ether; he also persuaded a dentist, William Morton, to use it. Subsequently, they persuaded a surgeon, John Warren, to give a public demonstration of surgery under anesthesia. The operation was completely successful, and soon general anesthesia by diethyl ether became routine practice for general surgery.

Diethyl ether was easy to use and caused excellent muscle relaxation. Blood pressure, pulse rate, and respiration were usually only slightly affected. Diethyl ether's chief drawbacks are its irritating effect on the respiratory passages and its aftereffect of nausea.

Among the inhalation anesthetics used today are several halogenated ethers, the most important of which are enflurane and isoflurane.

Enflurane
(Ethrane)

Isoflurane
(Forane)

Test your knowledge with Problems 14.56–14.58.

O—H group, which is capable of forming hydrogen bonds. This hydrogen bonding increases intermolecular associations, thereby giving ethanol a higher boiling point than dimethyl ether.

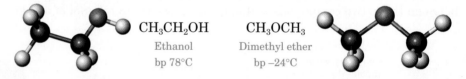

CH_3CH_2OH
Ethanol
bp 78°C

CH_3OCH_3
Dimethyl ether
bp −24°C

Ethers are more soluble in water than hydrocarbons of similar molecular weight and shape, but far less soluble than isomeric alcohols. Their greater solubility reflects the fact that the oxygen atom of an ether carries a partial negative charge and forms hydrogen bonds with water.

D. Reactions of Ethers

Ethers resemble hydrocarbons in their resistance to chemical reaction. For example, they do not react with oxidizing agents, such as potassium dichromate. Likewise, they do not react with reducing agents such as H_2 in the presence of a metal catalyst (Section 12.6D). Furthermore, most acids and bases at moderate temperatures do not affect them. Because of their general inertness to chemical reaction and their good solvent properties, ethers are excellent solvents in which to carry out many organic reactions. The most important ether solvents are diethyl ether and tetrahydrofuran.

14.4 What Are the Structures, Names, and Properties of Thiols?

A. Structure

The functional group of a **thiol** is an —SH **(sulfhydryl) group** bonded to a tetrahedral carbon atom. Figure 14.6 shows a Lewis structure and a ball-and-stick model of methanethiol, CH_3SH, the simplest thiol.

B. Nomenclature

The sulfur analog of an alcohol is called a thiol (*thi-* from the Greek: *theion*, sulfur) or, in the older literature, a **mercaptan**, which literally means "mercury capturing." Thiols react with Hg^{2+} in aqueous solution to give sulfide salts as insoluble precipitates. Thiophenol, C_6H_5SH, for example, gives $(C_6H_5S)_2Hg$.

In the IUPAC system, thiols are named by selecting the longest carbon chain that contains the —SH group as the parent alkane. To show that the compound is a thiol, we add the suffix *-thiol* to the name of the parent alkane. The parent chain is numbered in the direction that gives the —SH group the lower number.

Common names for simple thiols are derived by naming the alkyl group bonded to —SH and adding the word *mercaptan*.

$$CH_3CH_2SH \qquad \overset{\overset{\displaystyle CH_3}{|}}{CH_3CHCH_2SH}$$

Ethanethiol   2-Methyl-1-propanethiol
(Ethyl mercaptan) (Isobutyl mercaptan)

Thiol A compound that contains an —SH (sulfhydryl) group bonded to a tetrahedral carbon atom

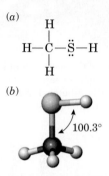

(a)

(b)

100.3°

FIGURE 14.6 Methanethiol, CH_3SH. (*a*) Lewis structure and (*b*) ball-and-stick model. The H—S—C bond angle is 100.3°, somewhat smaller than the tetrahedral angle of 109.5°.

Mercaptan A common name for any molecule containing an —SH group

---

Example 14.7 Systematic Names of Thiols

Write the IUPAC name of each thiol.

(a) [skeletal structure with SH] (b) [skeletal structure with SH]

Strategy

To derive the IUPAC name of a thiol, select as the parent alkane the longest carbon chain that contains the —SH group. Show that the compound is a thiol by adding the suffix *-thiol* to the name of the parent alkane. Number the parent chain in the direction that gives the —SH group the lower number.

Solution

(a) The parent alkane is pentane. Show the presence of the —SH group by adding "thiol" to the name of the parent alkane. The IUPAC name of this thiol is 1-pentanethiol. Its common name is pentyl mercaptan.
(b) The parent alkane is butane. The IUPAC name of this thiol is 2-butanethiol. Its common name is *sec*-butyl mercaptan.

Problem 14.7

Write the IUPAC name of each thiol.

The most notable property of low-molecular-weight **thiols** is their stench. They are responsible for unpleasant odors such as those from skunks, rotten eggs, and sewage. The scent of skunks is due primarily to two thiols:

$$CH_3CH=CHCH_2SH \qquad CH_3\overset{\overset{\displaystyle CH_3}{|}}{C}HCH_2CH_2SH$$

2-Butene-1-thiol   3-Methyl-1-butanethiol

C. Physical Properties

Because of the small difference in electronegativity between sulfur and hydrogen $(2.5 - 2.1 = 0.4)$, we classify the S—H bond as nonpolar covalent. Because of this lack of polarity, thiols show little association by hydrogen bonding. Consequently, they have lower boiling points and are less soluble in water and other polar solvents than are alcohols of similar molecular weight. Table 14.2 gives boiling points for three low-molecular-weight thiols. Shown for comparison are the boiling points of alcohols with the same number of carbon atoms.

Table 14.2 Boiling Points of Three Thiols and Alcohols with the Same Number of Carbon Atoms

Thiol	Boiling Point (°C)	Alcohol	Boiling Point (°C)
methanethiol	6	methanol	65
ethanethiol	35	ethanol	78
1-butanethiol	98	1-butanol	117

Earlier we illustrated the importance of hydrogen bonding in alcohols by comparing the boiling points of ethanol (78°C) and its constitutional isomer dimethyl ether (−24°C). By contrast, the boiling point of ethanethiol is 35°C and that of its constitutional isomer dimethyl sulfide is 37°C. Because the boiling points of these constitutional isomers are almost identical, we know that little or no association by hydrogen bonding occurs between thiol molecules.

$$CH_3CH_2SH \qquad CH_3SCH_3$$

Ethanethiol   Dimethyl sulfide
bp 35°C   bp 37°C

D. Reactions of Thiols

Thiols are weak acids ($pK_a = 10$) that are comparable in strength to phenols (Section 13.4B). Thiols react with strong bases such as NaOH to form thiolate salts.

$$CH_3CH_2SH + NaOH \xrightarrow{H_2O} CH_3CH_2S^-Na^+ + H_2O$$

Ethanethiol   Sodium
(pK_a 10)   ethanethiolate

The scent of the spotted skunk, native to the Sonoran Desert, is a mixture of two thiols, 2-buten-1-thiol and 3-methyl-1-butanethiol.

Steven J. Krasemann/Photo Researchers, Inc.

The most common reaction of thiols in biological systems is their oxidation to disulfides, the functional group of which is a **disulfide** (—S—S—) **bond**. Thiols are readily oxidized to disulfides by molecular oxygen. In fact, they are so susceptible to oxidation that they must be protected from contact with air during their storage. Disulfides, in turn, are easily reduced to thiols by several reducing agents. This easy interconversion between thiols and disulfides is very important in protein chemistry, as we will see in Chapters 22 and 23.

Disulfide A compound containing an (—S—S—) group

$$2HOCH_2CH_2SH \underset{\text{reduction}}{\overset{\text{oxidation}}{\rightleftharpoons}} HOCH_2CH_2S-SCH_2CH_2OH$$

A thiol   A disulfide

To derive the common name of a disulfide, list the names of the groups bonded to sulfur and add the word *disulfide*.

14.5 What Are the Most Commercially Important Alcohols?

As you study the alcohols described in this section, you should pay particular attention to two key points. First, they are derived almost entirely from petroleum, natural gas, or coal—all nonrenewable resources. Second, many are themselves starting materials for the synthesis of valuable commercial products, without which our modern industrial society could not exist.

At one time, **methanol** was derived by heating hard woods in a limited supply of air—hence the name "wood alcohol." Today methanol is obtained entirely from the catalytic reduction of carbon monoxide. Methanol, in turn, is the starting material for the preparation of several important industrial and commercial chemicals, including acetic acid and formaldehyde. Treatment of methanol with carbon monoxide in the presence of a rhodium catalyst gives acetic acid. Partial oxidation of methanol gives formaldehyde. An important use of this one-carbon aldehyde is in the preparation of phenol-formaldehyde and urea-formaldehyde glues and resins, which are used as molding materials and as adhesives for plywood and particle board for the construction industry.

$$\text{Coal or methane} \xrightarrow{[O]} \underset{\substack{\text{Carbon} \\ \text{monoxide}}}{CO} \xrightarrow{2H_2} \underset{\text{Methanol}}{CH_3OH} \begin{cases} \xrightarrow[\text{catalyst}]{CO} \underset{\text{Acetic acid}}{CH_3COOH} \\ \xrightarrow[\text{oxidation}]{O_2} \underset{\text{Formaldehyde}}{CH_2O} \end{cases}$$

The bulk of the **ethanol** produced worldwide is prepared by acid-catalyzed hydration of ethylene, itself derived from the cracking of the ethane separated from natural gas (Section 11.4). Ethanol is also produced by fermentation of the carbohydrates in plant materials, particularly corn and molasses. The majority of the fermentation-derived ethanol is used as an "oxygenate" additive to produce E85, which is a blend of up to 85% ethanol in gasoline. Combustion of E85 produces less air pollution than combustion of gasoline itself.

$$\underset{\text{Ethylene}}{CH_2{=}CH_2} \begin{cases} \xrightarrow{H_2O, H_2SO_4} \underset{\text{Ethanol}}{CH_3CH_2OH} \xrightarrow[180°C]{H_2SO_4} \underset{\text{Diethyl ether}}{CH_3CH_2OCH_2CH_3} + H_2O \\ \xrightarrow[\text{catalyst}]{O_2} \underset{\substack{\text{Ethylene} \\ \text{oxide}}}{H_2C\overset{\displaystyle O}{\frown}CH_2} \xrightarrow{H_2O, H_2SO_4} \underset{\text{Ethylene glycol}}{HOCH_2CH_2OH} \end{cases}$$

Acid-catalyzed dehydration of ethanol gives **diethyl ether**, an important laboratory and industrial solvent. Ethylene is also the starting material for the preparation of **ethylene oxide**. This compound itself has few direct uses. Rather, ethylene oxide's importance derives from its role as an intermediate in the production of **ethylene glycol**, a major component of automobile antifreeze. Ethylene glycol freezes at −12°C and boils at 199°C, which makes it ideal for this purpose. In addition, reaction of ethylene glycol with the methyl ester of terephthalic acid gives the polymer poly(ethylene terephthalate), abbreviated PET or PETE (Section 19.6B). Ethylene glycol is also used as a solvent in the paint and plastics industry and in the formulation of printer's inks, inkpads, and inks for ballpoint pens.

Isopropyl alcohol, the alcohol in rubbing alcohol, is made by acid-catalyzed hydration of propene. It is also used in hand lotions, aftershave lotions, and similar cosmetics. A multistep process also converts propene to epichlorohydrin, one of the key components in the production of epoxy glues and resins.

Glycerin, also known as glycerol, is a by-product of the manufacture of soaps by saponification of animal fats and tropical oils (Section 21.3). The bulk of the glycerin used for industrial and commercial purposes, however, is prepared from propene. Perhaps the best-known use of glycerin is for the manufacture of nitroglycerin. Glycerin is also used as an emollient in skin care products and cosmetics, in liquid soaps, and in printing inks.

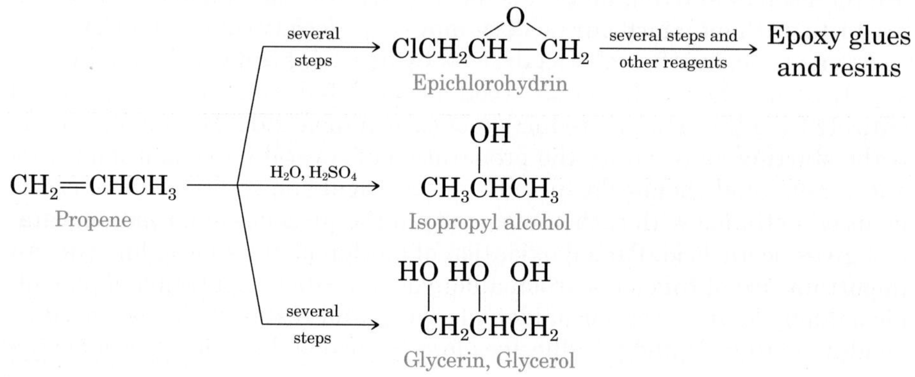

Summary of Key Questions

OWL Sign in at www.cengage.com/owl to develop problem-solving skills and complete online homework assigned by your professor.

Section 14.1 What Are the Structures, Names, and Physical Properties of Alcohols?

- The functional group of an **alcohol** is an **—OH (hydroxyl)** group bonded to a tetrahedral carbon atom.
- The functional group of an **ether** is an atom of oxygen bonded to two carbon atoms.
- The IUPAC name of an alcohol is derived by changing the *-e* of the parent alkane to *-ol*. The parent chain is numbered from the end that gives the carbon bearing the —OH group the lower number.
- The common name for an alcohol is derived by naming the alkyl group bonded to the —OH group and adding the word "alcohol."
- Alcohols are classified as **1°, 2°,** or **3°,** depending on the number of carbon groups bonded to the carbon bearing the —OH group.
- Compounds containing hydroxyl groups on adjacent carbons are called **glycols**.
- Alcohols are polar compounds in which oxygen bears a partial negative charge and both the carbon and hydrogen bonded to it bear partial positive charges. Alcohols associate in the liquid state by **hydrogen bonding**. As a consequence, their boiling points are higher than those of hydrocarbons of similar molecular weight.
- Because of increased London dispersion forces, the boiling points of alcohols increase with their increasing molecular weight.
- Alcohols interact with water by hydrogen bonding and are more soluble in water than are hydrocarbons of similar molecular weight.

- Alcohols have about the same pK_a values as pure water. For this reason, aqueous solutions of alcohols have the same pH as that of pure water.

Section 14.3 What Are the Structures, Names, and Properties of Ethers?

- Common names for ethers are derived by naming the two groups bonded to oxygen followed by the word "ether."
- In a **cyclic ether**, oxygen is one of the atoms in a ring.
- Ethers are weakly polar compounds. Their boiling points are close to those of hydrocarbons of similar molecular weight.
- Because ethers form hydrogen bonds with water, they are more soluble in water than are hydrocarbons of similar molecular weight.

Section 14.4 What Are the Structures, Names, and Properties of Thiols?

- A **thiol** contains an —SH (**sulfhydryl**) group.
- Thiols are named in the same manner as alcohols, but the suffix -e of the parent alkane is retained and -thiol is added.
- Common names for thiols are derived by naming the alkyl group bonded to —SH and adding the word **"mercaptan."**
- The S—H bond is nonpolar, and the physical properties of thiols resemble those of hydrocarbons of similar molecular weight.

Summary of Key Reactions

1. **Acid-Catalyzed Dehydration of an Alcohol (Section 14.2B)**
   When isomeric alkenes are possible, the major product is generally the more substituted alkene.

$$CH_3CH_2CHCH_3 \overset{OH}{|}$$

$$\xrightarrow[\text{heat}]{H_3PO_4} CH_3CH{=}CHCH_3 + CH_3CH_2CH{=}CH_2 + H_2O$$
Major product

2. **Oxidation of a Primary Alcohol (Section 14.2C)**
   Oxidation of a primary alcohol by potassium dichromate gives either an aldehyde or a carboxylic acid, depending on the experimental conditions.

$$CH_3(CH_2)_6CH_2OH \xrightarrow[H_2SO_4]{K_2Cr_2O_7} CH_3(CH_2)_6\overset{O}{\overset{||}{C}}H$$

$$\xrightarrow[H_2SO_4]{K_2Cr_2O_7} CH_3(CH_2)_6\overset{O}{\overset{||}{C}}OH$$

3. **Oxidation of a Secondary Alcohol (Section 14.2C)**
   Oxidization of a secondary alcohol by potassium dichromate gives a ketone.

$$CH_3(CH_2)_4\overset{OH}{\overset{|}{C}}HCH_3 \xrightarrow[H_2SO_4]{K_2Cr_2O_7} CH_3(CH_2)_4\overset{O}{\overset{||}{C}}CH_3$$

4. **Acidity of Thiols (Section 14.4D)** Thiols are weak acids, with pK_a values of approximately 10. They react with strong bases to form water-soluble thiolate salts.

$$CH_3CH_2SH + NaOH \xrightarrow{H_2O} CH_3CH_2S^-Na^+ + H_2O$$
Ethanethiol   Sodium
(pK_a 10)   ethanethiolate

5. **Oxidation of a Thiol to a Disulfide (Section 14.4D)**
   Oxidation of a thiol gives a disulfide. Reduction of a disulfide gives two thiols.

$$2HOCH_2CH_2SH \underset{\text{reduction}}{\overset{\text{oxidation}}{\rightleftharpoons}} HOCH_2CH_2S{-}SCH_2CH_2OH$$
A thiol   A disulfide

Problems

OWL Interactive versions of these problems may be assigned in OWL.

Orange-numbered problems are applied.

References to previous chapters are given in parentheses.

Section 14.1 What Are the Structures, Names, and Physical Properties of Alcohols?

14.8 Answer true or false.

(a) The functional group of an alcohol is the —OH (hydroxyl) group.

(b) The parent name of an alcohol is the name of the longest carbon chain that contains the —OH group.

(c) A primary alcohol contains one —OH group, and a tertiary alcohol contains three —OH groups.

(d) In the IUPAC system, the presence of three —OH groups is shown by the ending –triol.

(e) A glycol is a compound that contains two —OH groups. The simplest glycol is ethylene glycol, $HOCH_2CH_2$—OH.

(f) Because of the presence of an —OH group, all alcohols are polar compounds.

(g) The boiling points of alcohols increase with increasing molecular weight.

(h) The solubility of alcohols in water increases with increasing molecular weight.

14.9 What is the difference in structure between a primary, a secondary, and a tertiary alcohol?

14.10 Which of the following are secondary alcohols?

(a) [structure: cyclohexane ring with CH₃ and OH]

(b) $(CH_3)_3COH$

(c) [structure with OH]

(d) [cyclopentane ring with OH]

14.11 Which of the alcohols in Problem 14.10 are primary? Which are tertiary?

14.12 Write the IUPAC name of each compound.

(a) [chain with OH]

(b) HO [chain] OH

(c) [structure with OH and OH]

(d) HO [branched structure]

(e) [cyclohexane with OH and OH, wedge/dash]

(f) [cyclohexane with OH]

14.13 Draw a structural formula for each alcohol.

(a) Isopropyl alcohol
(b) Propylene glycol
(c) 5-Methyl-2-hexanol
(d) 2-Methyl-2-propyl-1,3-propanediol
(e) 1-Octanol
(f) 3,3-Dimethylcyclohexanol

14.14 Draw a structural formula for each of the following alcohols.

(a) Isobutyl alcohol
(b) 1,4-Butanediol
(c) 5-Methyl-1-hexanol
(d) 1,3-Pentanediol
(e) *trans*-1,4-Cyclohexanediol
(f) 1-Chloro-2-propanol

14.15 Both alcohols and phenols contain an —OH group. What structural feature distinguishes these two classes of compounds? Illustrate your answer by drawing the structural formulas of a phenol with six carbon atoms and an alcohol with six carbon atoms.

14.16 Name the functional groups in each compound.

(a) [structure of Prednisone]

Prednisone
(a synthetic anti-inflammatory steroid)

(b) [structure of Estradiol]

Estradiol
(a female sex hormone;
Section 21.10)

14.17 Explain in terms of noncovalent interactions why the low-molecular-weight alcohols are soluble in water but the low-molecular-weight alkanes and alkynes are not.

14.18 Explain in terms of noncovalent interactions why the low-molecular-weight alcohols are more soluble in water than the low-molecular-weight ethers.

14.19 Why does the water solubility of low-molecular-weight alcohols decrease as molecular weight increases?

14.20 Show hydrogen bonding between methanol and water in the following ways.

(a) Between the oxygen of methanol and a hydrogen of water
(b) Between the hydrogen of methanol's OH group and the oxygen of water

14.21 Show hydrogen bonding between the oxygen of diethyl ether and a hydrogen of water.

14.22 Arrange these compounds in order of increasing boiling point. Values in °C are −42, 78, 117, and 198.

(a) $CH_3CH_2CH_2CH_2OH$
(b) CH_3CH_2OH
(c) $HOCH_2CH_2OH$
(d) $CH_3CH_2CH_3$

14.23 Arrange these compounds in order of increasing boiling point. Values in °C are 0, 35, and 97.

(a) $CH_3CH_2CH_2OH$
(b) $CH_3CH_2OCH_2CH_3$
(c) $CH_3CH_2CH_2CH_3$

14.24 2-Propanol (isopropyl alcohol) is commonly used as rubbing alcohol to cool the skin. 2-Hexanol, also a liquid, is not suitable for this purpose. Why?

14.25 Explain why glycerol is much thicker (more viscous) than ethylene glycol, which in turn is much thicker than ethanol.

14.26 From each pair, select the compound that is more soluble in water.

(a) CH_3OH or CH_3OCH_3

(b) $CH_3\overset{\text{OH}}{\underset{|}{C}}HCH_3$ or $CH_3\overset{\text{CH}_2}{\underset{\|}{C}}CH_3$

(c) $CH_3CH_2CH_2SH$ or $CH_3CH_2CH_2OH$

14.27 Arrange the compounds in each set in order of decreasing solubility in water.

(a) Ethanol, butane, and diethyl ether
(b) 1-Hexanol, 1,2-hexanediol, and hexane

Synthesis of Alcohols (Review Chapter 12)

14.28 Give the structural formula of an alkene or alkenes from which each alcohol can be prepared.

(a) 2-Butanol

(b) 1-Methylcyclohexanol

(c) 3-Hexanol

(d) 2-Methyl-2-pentanol

(e) Cyclopentanol

Section 14.2 What Are the Characteristic Reactions of Alcohols?

14.29 Answer true or false.

(a) The two most important reactions of alcohols are their acid-catalyzed dehydration to give alkenes and their oxidation to aldehydes, ketones, and carboxylic acids.

(b) The acidity of alcohols is comparable to that of water.

(c) Water-insoluble alcohols and water-insoluble phenols react with strong bases to give water-soluble salts.

(d) Acid-catalyzed dehydration of cyclohexanol gives cyclohexane.

(e) When the acid-catalyzed dehydration of an alkene can yield isomeric alkenes, the alkene with the greater number of hydrogens on the carbons of the double bond generally predominates.

(f) The acid-catalyzed dehydration of 2-butanol gives predominantly 1-butene.

(g) The oxidation of a primary alcohol gives either an aldehyde or a carboxylic acid depending on experimental conditions.

(h) The oxidation of a secondary alcohol gives a carboxylic acid.

(i) Acetic acid, CH_3COOH, can be prepared from ethylene, $CH_2{=}CH_2$, by treatment of ethylene with H_2O/H_2SO_4, followed by treatment with $K_2Cr_2O_7/H_2SO_4$.

(j) Treatment of propene, $CH_3CH{=}CH_2$, with H_2O/H_2SO_4 followed by treatment with $K_2Cr_2O_7/H_2SO_4$ gives propanoic acid, CH_3CH_2COOH.

14.30 Show how to distinguish between cyclohexanol and cyclohexene by a simple chemical test. Tell what you would do, what you would expect to see, and how you would interpret your observation.

14.31 Compare the acidity of alcohols and phenols, which are both classes of organic compounds that contain an —OH group.

14.32 Both 2,6-diisopropylcyclohexanol and the intravenous anesthetic Propofol are insoluble in water. Show how these two compounds can be distinguished by their reaction with aqueous sodium hydroxide.

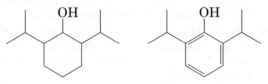

2,6-Diisopropylcyclohexanol   2,6-Diisopropylphenol
                                    (Propofol)

14.33 Write equations for the reaction of 1-butanol, a primary alcohol, with these reagents.

(a) H_2SO_4, heat

(b) $K_2Cr_2O_7$, H_2SO_4

14.34 Write equations for the reaction of 2-butanol with these reagents.

(a) H_2SO_4, heat

(b) $K_2Cr_2O_7$, H_2SO_4

14.35 Write equations for the reaction of each of the following compounds with $K_2Cr_2O_7/H_2SO_4$.

(a) 1-Octanol

(b) 1,4-Butanediol

14.36 Show how to convert cyclohexanol to these compounds.

(a) Cyclohexene

(b) Cyclohexane

(c) Cyclohexanone

(d) Bromocyclohexane

14.37 Show reagents and experimental conditions to synthesize each compound from 1-propanol.

14.38 Show how to convert this alcohol to compounds (a) and (b).

14.39 Name two important alcohols derived from ethylene and give two important uses of each.

14.40 Name two important alcohols derived from propene and give two important uses of each.

Section 14.3 What Are the Structures, Names, and Properties of Ethers?

14.41 Answer true or false.

(a) Ethanol and dimethyl ether are constitutional isomers.

(b) The solubility of low-molecular-weight ethers in water is comparable to that of low-molecular-weight alcohols in water.

(c) Ethers undergo many of the same reactions that alcohols do.

14.42 Write the common name for each ether.

(a)   (b) $[CH_3(CH_2)_4]_2O$

(c)
$$\begin{array}{c} CH_3 \ \ CH_3 \\ | \qquad | \\ CH_3CHOCHCH_3 \end{array}$$

14.43 Write the common name for each of the following ethers.

(a)

(b)

(c)

Section 14.4 What Are the Structures, Names, and Properties of Thiols?

14.44 Answer true or false.

(a) The functional group of a thiol is the —SH (sulfhydryl) group.

(b) The parent name of a thiol is the name of the longest carbon chain that contains the —SH group.

(c) The S—H bond is nonpolar covalent.

(d) The acidity of ethanethiol is comparable to that of phenol.

(e) Both phenols and thiols are classified as weak acids.

(f) The most common biological reaction of thiols is their oxidation to disulfides.

(g) The functional group of a disulfide is the —S—S— group.

(h) Conversion of a thiol to a disulfide is a reduction reaction.

14.45 Write the IUPAC name of each thiol.

(a)
$$\begin{array}{c} SH \\ | \\ CH_3CH_2CHCH_3 \end{array}$$
    (b) $CH_3CH_2CH_2CH_2SH$

(c)

14.46 Write the common name for each thiol in Problem 14.45.

14.47 Following are structural formulas for 1-butanol and 1-butanethiol. One of these compounds has a boiling point of 98°C and the other has a boiling point of 117°C. Which compound has which boiling point?

$$CH_3CH_2CH_2CH_2OH \qquad CH_3CH_2CH_2CH_2SH$$
1-Butanol   1-Butanethiol

14.48 Explain why methanethiol, CH_3SH, has a lower boiling point (6°C) than methanol, CH_3OH (65°C), even though methanethiol has a higher molecular weight.

Section 14.5 What Are the Most Commercially Important Alcohols?

14.49 Answer true or false.

(a) Today, the major carbon sources for the synthesis of methanol are coal and methane (natural gas), both nonrenewable resources.

(b) Today the major carbon sources for the synthesis of ethanol are petroleum and natural gas, both nonrenewable resources.

(c) Intermolecular acid-catalyzed dehydration of ethanol gives diethyl ether.

(d) Conversion of ethylene to ethylene glycol involves oxidation to ethylene oxide, followed by acid-catalyzed hydration (addition of water) to ethylene oxide.

(e) Ethylene glycol is soluble in water in all proportions.

(f) A major use of ethylene glycol is as automobile antifreeze.

Chemical Connections

14.50 (Chemical Connections 14A) When was nitroglycerin discovered? Is this substance a solid, a liquid, or a gas?

14.51 (Chemical Connections 14A) What was Alfred Nobel's discovery that made nitroglycerin safer to handle?

14.52 (Chemical Connections 14A) What is the relationship between the medical use of nitroglycerin to relieve the sharp chest pains (angina) associated with heart disease and the gas nitric oxide, NO?

14.53 (Chemical Connections 14B) What is the color of dichromate ion, $Cr_2O_7{}^{2-}$? What is the color of chromium(III) ion, Cr^{3+}? Explain how the conversion of one to the other is used in breath-alcohol screening.

14.54 (Chemical Connections 14B) The legal definition of being under the influence of alcohol is based on blood alcohol content. What is the relationship between breath alcohol content and blood alcohol content?

14.55 (Chemical Connections 14C) What does it mean to say that ethylene oxide is a highly strained molecule?

14.56 (Chemical Connections 14D) What are the advantages and disadvantages of using diethyl ether as an anesthetic?

14.57 (Chemical Connections 14D) Show that enflurane and isoflurane are constitutional isomers.

14.58 (Chemical Connections 14D) Would you expect enflurane and isoflurane to be soluble in water? Would you expect them to be soluble in organic solvents such as hexane?

Additional Problems

14.59 Write a balanced equation for the complete combustion of ethanol, the alcohol blended with gasoline to produce E85.

14.60 Knowing what you do about electronegativity, the polarity of covalent bonds, and hydrogen bonding, would you expect an N—H---N hydrogen bond to be stronger than, the same strength as, or weaker than an O—H---O hydrogen bond?

14.61 Draw structural formulas and write IUPAC names for the eight isomeric alcohols with the molecular formula $C_5H_{12}O$.

14.62 Draw structural formulas and write common names for the six isomeric ethers with the molecular formula $C_5H_{12}O$.

14.63 Explain why the boiling point of ethylene glycol (198°C) is so much higher than that of 1-propanol (97°C), even though their molecular weights are about the same.

14.64 Following is a structural formula for Erythromycin A, a macrolactone (macrolide) antibiotic.

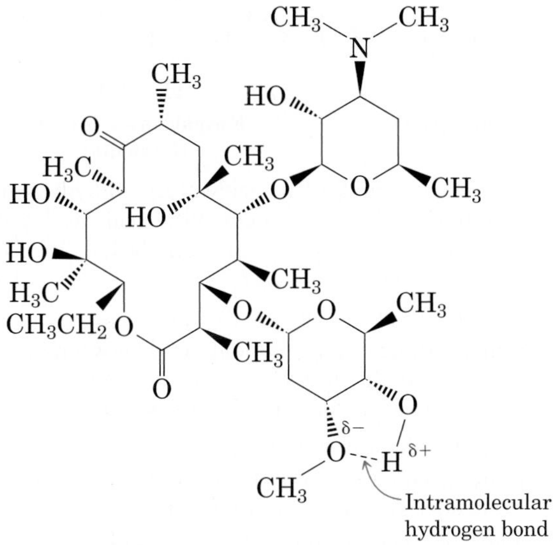

Erythromycin A
$C_{31}H_{57}NO_9$

(a) How many hydroxyl groups are present? Classify each as primary (1°), secondary (2°), or tertiary (3°). (Chapter 10)

(b) How many amine groups are present? Classify each as primary (1°), secondary (2°), or tertiary (3°). (Chapter 10)

(c) Locate the ester group within the large 15-member ring.

(d) Four of the hydroxyl groups within Erythromycin A are involved in intramolecular (internal) hydrogen bonds. One of these is pointed out on the structural formula. Note that this hydrogen bond creates a five-membered ring. Locate the other three intramolecular hydrogen bonds and specify the size of the ring created by each. (Chapter 5)

14.65 1,4-Butanediol, hexane, and 1-pentanol have similar molecular weights. Their boiling points, arranged from lowest to highest, are 69°C, 138°C, and 230°C. Which compound has which boiling point?

14.66 Of the three compounds given in Problem 14.65, one is insoluble in water, another has a solubility of 2.3 g/100 g water, and one is infinitely soluble in water. Which compound has which solubility?

14.67 Each of the following compounds is a common organic solvent. From each pair of compounds, select the solvent with the greater solubility in water.
(a) CH_2Cl_2 or CH_3CH_2OH
(b) $CH_3CH_2OCH_2CH_3$ or CH_3CH_2OH

14.68 Show how to prepare each compound from 2-methyl-1-propanol.
(a) 2-Methylpropene
(b) 2-Methyl-2-propanol
(c) 2-Methylpropanoic acid, $(CH_3)_2CHCOOH$

14.69 Show how to prepare each compound from 2-methylcyclohexanol.

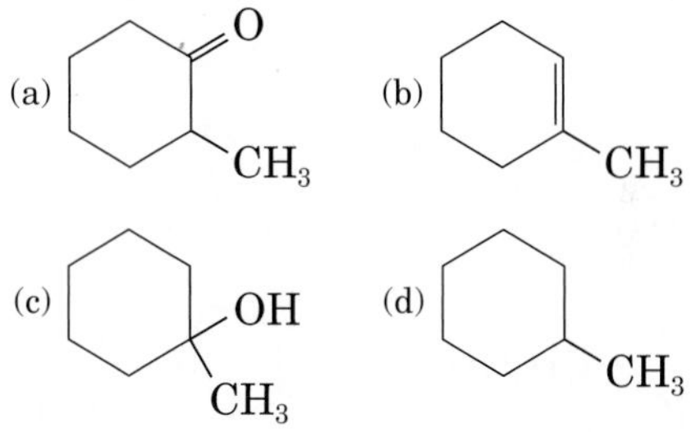

14.70 The mechanism of the acid-catalyzed dehydration of an alcohol to an alkene is the reverse of the acid-catalyzed hydration of an alkene. The dehydration mechanism occurs by the following three steps.

Step 1: Add a proton.

Step 2: Break a bond to form stable molecules or ions.

Step 3: Take away a proton.

These three steps are illustrated here by the dehydration of 2-butanol to give 2-butene.

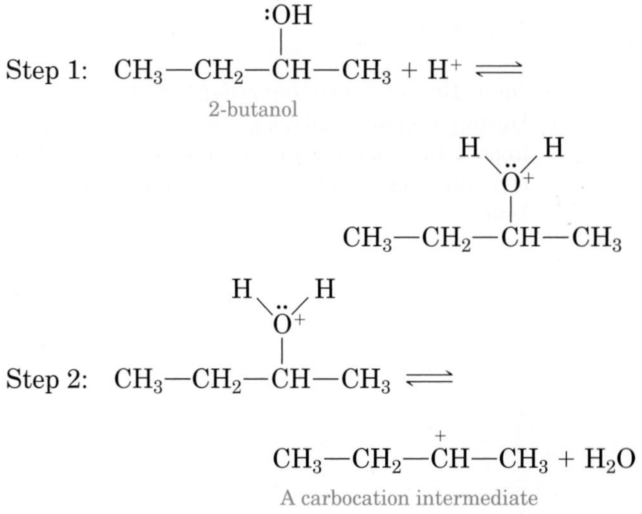

$$\text{Step 3: } \overset{\overset{\displaystyle H}{|}}{CH_3-CH-\overset{+}{CH}-CH_3} \rightleftharpoons$$

$$\underset{\text{2-butene}}{CH_3-CH=CH-CH_3} + H^+$$

Use curved arrows to show the flow of electrons in each step; that is, show how each bond-making or bond-breaking step occurs.

14.71 Use the reactions we learned in Chapter 12 and this chapter to show how to bring about the following chemical transformations. Some transformations will require only one step. Others will require two or more steps.

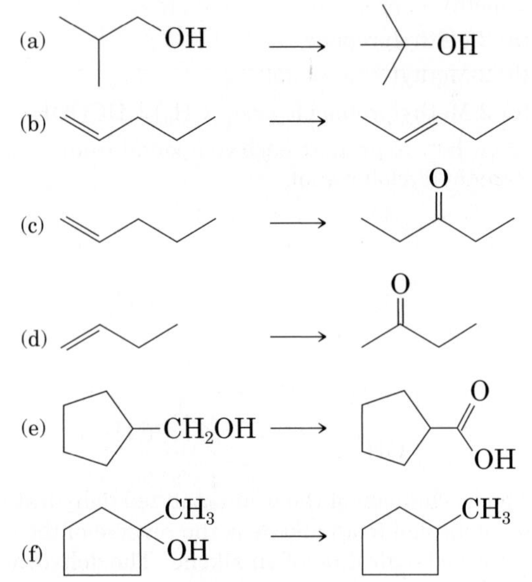

(a) [structure] OH ⟶ [structure] OH

(b) [structure] ⟶ [structure]

(c) [structure] ⟶ [structure with C=O]

(d) [structure] ⟶ [structure with C=O]

(e) [cyclopentane]—CH$_2$OH ⟶ [cyclopentane] with C(=O)OH

(f) [cyclopentane] with CH$_3$ and OH ⟶ [cyclopentane] with CH$_3$

Looking Ahead

14.72 Lipoic acid is a growth factor for many bacteria and protozoa and an essential component of several enzymes involved in human metabolism.

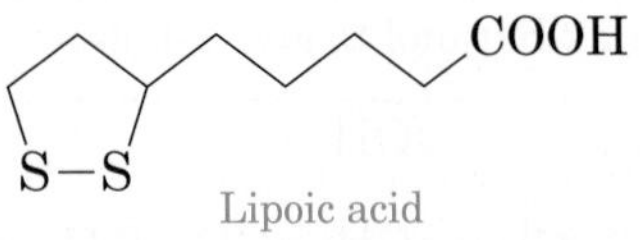

[structure] COOH

S—S
Lipoic acid

(a) Name the two functional groups in lipoic acid.

(b) During human metabolism, the disulfide bond of lipoic acid is reduced to two thiol groups. Draw a structural formula for this reduced form of lipoic acid.

14.73 Following is a structural formula for the amino acid cysteine:

$$HS-CH_2-\underset{\underset{\displaystyle NH_2}{|}}{CH}-\overset{\overset{\displaystyle O}{||}}{C}-OH$$

(a) Name the three functional groups in cysteine. (Chapter 10)

(b) In the human body, cysteine is oxidized to cystine, a disulfide. Draw a structural formula for cystine.

14.74 As we will see in Chapter 16, the carbonyl group of aldehydes and ketones reacts with water to form a compound called a hydrate. The hydration of a carbonyl group is catalyzed by acid. Formaldehyde, for example, reacts as shown below. When dissolved in water, formaldehyde exists almost entirely as its hydrate. This aqueous solution is given the name formalin.

$$\underset{\text{Formaldehyde}}{\overset{\overset{\displaystyle H}{\diagdown}}{\underset{\underset{\displaystyle H}{\diagup}}{C}}=O + H_2O} \overset{H_3O^+}{\rightleftharpoons} \underset{\substack{\text{Formaldehyde hydrate} \\ \text{(Formalin)}}}{\overset{H \quad OH}{\underset{H \quad OH}{C}}}$$

The mechanism for this acid-catalyzed hydration is similar to the mechanism of the acid-catalyzed hydration of an alkene to give an alcohol and occurs by the following three steps.

Step 1: Add a proton.

Step 2: Make a new covalent bond between an electrophile (an electron-seeking species) and a nucleophile (an electron-donating species).

Step 3: Take a proton away.

Write an equation for each step and show by the use of curved arrows how each of these steps occurs. Show all bond-forming steps and bond-breaking steps.

14.75 Draw the structural formula for the hydrate formed by the reaction of acetone with water.

$$\underset{\text{Acetone}}{CH_3-\overset{\overset{\displaystyle O}{||}}{C}-CH_3} + H_2O \overset{H_3O^+}{\rightleftharpoons}$$

15

Chiralty: The Handedness of Molecules

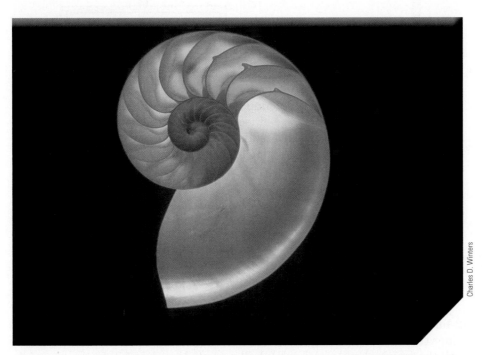

Charles D. Winters

Median cross section through the shell of a chambered nautilus found in the deep waters of the Pacific Ocean. The shell shows handedness; this cross section is a left-handed spiral.

Key Questions

15.1 What Is Enantiomerism?

How To . . . Draw Enantiomers

15.2 How Do We Specify the Configuration of a Stereocenter?

15.3 How Many Stereoisomers Are Possible for Molecules with Two or More Stereocenters?

15.4 What Is Optical Activity, and How Is Chirality Detected in the Laboratory?

15.5 What Is the Significance of Chirality in the Biological World?

15.1 What Is Enantiomerism?

In Chapters 11 through 14, we studied two types of stereoisomers, namely the *cis-trans* isomers of certain disubstituted cycloalkanes and appropriately substituted alkenes. Recall that stereoisomers have the same connectivity of their atoms but a different orientation of their atoms in space.

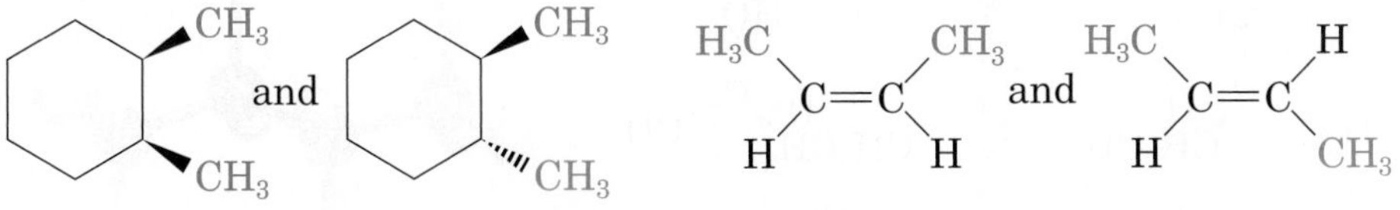

cis-1,2-Dimethylcyclohexane and *trans*-1,2-Dimethylcyclohexane; *cis*-2-Butene and *trans*-2-Butene

In this chapter, we study the relationship between objects and their **mirror images**; that is, we study stereoisomers called enantiomers and diastereomers. Figure 15.1 summarizes the relationship among these isomers and those we studied in Chapters 11 through 14.

OWL

Sign in to OWL at **www.cengage.com/owl** to view tutorials and simulations, develop problem-solving skills, and complete online homework assigned by your professor.

FIGURE 15.1 Relationships among isomers. In this chapter, we study enantiomers and diastereomers.

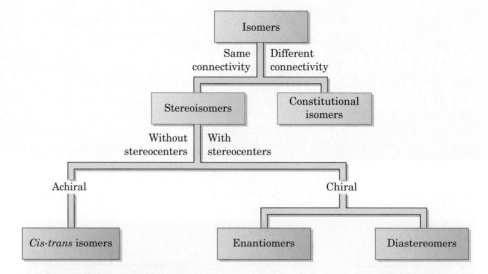

The significance of enantiomers is that, except for inorganic compounds and a few simple organic compounds, the vast majority of molecules in the biological world show this type of isomerism. Furthermore, approximately one half of all medications used to treat humans display this property.

As an example of enantiomerism, let us consider 2-butanol. As we discuss this molecule, we will focus on carbon 2, the carbon bearing the —OH group. What makes this carbon interesting is the fact that it has four different groups bonded to it: CH_3, H, OH, and CH_2CH_3.

$$\overset{\displaystyle OH}{\underset{}{\overset{|}{CH_3CHCH_2CH_3}}}$$

2-Butanol

This structural formula does not show the three-dimensional shape of 2-butanol or the orientation of its atoms in space. To do so, we must consider the molecule as a three-dimensional object. On the left is what we will call the "original molecule" and a ball-and-stick model of it. In this drawing, the —OH and —CH_3 groups are in the plane of the paper, —H is behind the plane (shown as a broken wedge), and —CH_2CH_3 is in front of it (shown as a solid wedge). In the middle is a mirror. On the right is a **mirror image** of the original molecule along with a ball-and-stick model of the mirror image. Every molecule—and, in fact, every object in the world around us—has a mirror image.

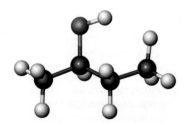

$$\underset{\text{Original molecule}}{H_3C \overset{\displaystyle OH}{\underset{\displaystyle CH_2CH_3}{\overset{|}{\underset{}{C}}}} H}$$

$$\underset{\text{Mirror image}}{CH_3CH_2 \overset{\displaystyle HO}{\underset{\displaystyle CH_3}{\overset{|}{\underset{}{C}}}} H}$$

The question we now need to ask is "What is the relationship between the original molecule of 2-butanol and its mirror image?" To answer this question, we need to imagine that we can pick up the mirror image and move it in 3-D space in any way we wish. If we can move the mirror image in space and find that it fits over the original molecule so that every bond, atom, and detail of the mirror image matches exactly the bonds, atoms, and details of the original, then the two are **superposable**. In other words, the mirror image and the original represent the same molecule; they are merely oriented differently in space. If, however, no matter how we turn the mirror image in space, it will not fit exactly on the original with every detail matching, then the two are **nonsuperposable**; they are different molecules.

One way to see that the mirror image of 2-butanol is not superposable on the original molecule is illustrated in the following drawings. Imagine that we hold the mirror image by the C—OH bond and rotate the bottom part of the molecule by 180° about this bond. The —OH group retains its position in space, but the —CH₃ group, which was to the right and in the plane of the paper, remains in the plane of the paper but is now to the left. Similarly, the —CH₂CH₃ group, which was in front of the plane of the paper and to the left, is now behind the plane and to the right.

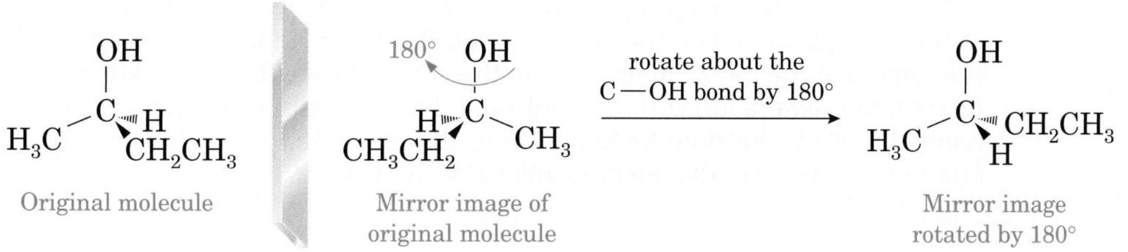

The terms "superposable" and "superimposable" mean the same thing and are both used currently.

Original molecule | Mirror image of original molecule | Mirror image rotated by 180°

Now move the mirror image in space and try to fit it on the original molecule so that all bonds and atoms match.

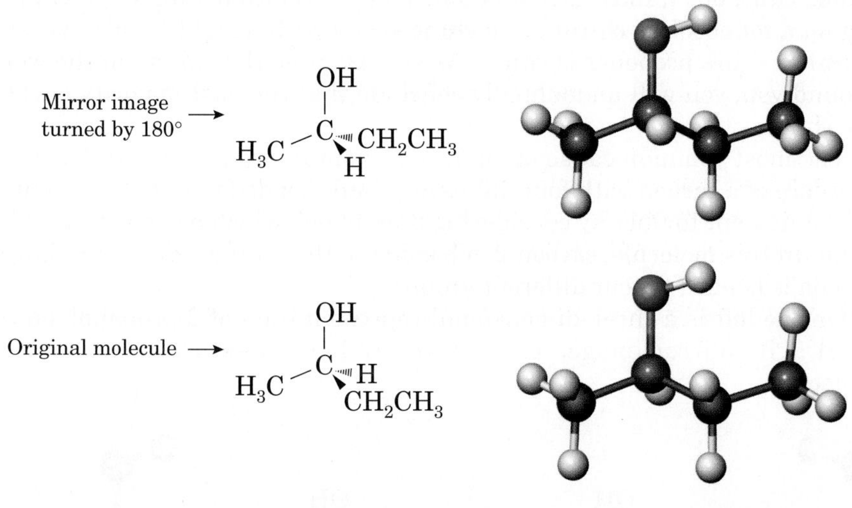

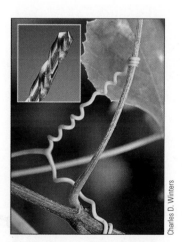

Charles D. Winters

Mirror image turned by 180° →

Original molecule →

By turning the mirror image as we did, its —OH and —CH₃ groups now fit exactly on top of the —OH and —CH₃ groups of the original molecule. However, the —H and the —CH₂CH₃ groups of the two do not match. The —H is away from us in the original but toward us in the mirror image; the —CH₂CH₃ group is toward us in the original but away from us in the mirror image. We conclude that the original molecule of 2-butanol and its mirror image are nonsuperposable and therefore are different compounds.

To summarize, we can turn and rotate the mirror image of 2-butanol in any direction in space, but as long as no bonds are broken and rearranged, we can make only two of the four groups bonded to carbon 2 of the mirror image coincide with those on its original molecule. Because 2-butanol and its mirror image are not superposable, they are isomers. Isomers such as these are called **enantiomers**. Enantiomers, like gloves, always occur in pairs.

The threads of a drill or screw twist along the axis of the helix, and some plants climb by sending out tendrils that twist helically. The drill bit shown here has a left-handed twist, and the tendril has a right-handed twist.

Enantiomers Stereoisomers that are nonsuperposable mirror images; refers to a relationship between pairs of objects

(a)

(b)

Charles D. Winters

FIGURE 15.2 Mirror images. (a) Two woodcarvings. The mirror images cannot be superposed on the actual model. The man's right arm rests on the camera in the mirror image, but in the actual statue, the man's left arm rests on the camera. (b) Left- and right-handed sea shells. If you cup a right-handed shell in your right hand with your thumb pointing from the narrow end to the wide end, the opening will be on your right.

Chiral From the Greek *cheir*, "hand"; an object that is not superposable on its mirror image

Objects that are not superposable on their mirror images are said to be **chiral** (pronounced "ki-ral," rhymes with spiral; from the Greek: *cheir*, "hand"); that is, they show handedness. We encounter chirality in three-dimensional objects of all sorts. Our left hand is chiral, and so is our right hand. Thus, our hands have an enantiomeric relationship. A spiral binding on a notebook is chiral. A machine screw with a right-handed twist is chiral. A ship's propeller is chiral. As you examine the objects in the world around you, you will undoubtedly conclude that the vast majority of them are chiral.

The most common cause of enantiomerism in organic molecules is the presence of a carbon with four different groups bonded to it. Let us examine this statement further by considering 2-propanol, which has no such carbon atom. In this molecule, carbon 2 is bonded to three different groups, but no carbon is bonded to four different groups.

On the left is a three-dimensional representation of 2-propanol; on the right is its mirror image. Also shown are ball-and-stick models of each molecule.

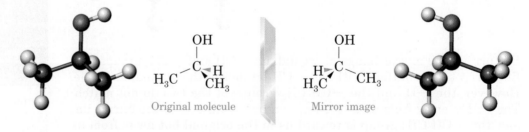

Original molecule

Mirror image

The question we now ask is "What is the relationship of the mirror image to the original?" This time, let us rotate the mirror image by 120° about the C—OH bond and then compare it to the original. After performing this rotation, we see that all atoms and bonds of the mirror image fit exactly on the original. Thus, the structures we first drew for the original molecule and its mirror image are, in fact, the same molecule—just viewed from different perspectives (Figure 15.3).

If an object and its mirror image are superposable, then they are identical and enantiomerism is not possible. We say that such an object is **achiral**

Achiral An object that lacks chirality; an object that is superposable on its mirror image

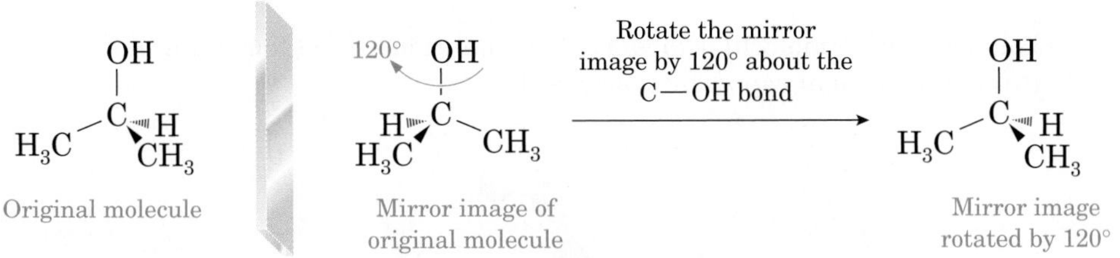

FIGURE 15.3 Rotation of the mirror image about the C—OH bond by 120° does not change the configuration of the stereocenter, but makes it easier to see that the mirror image is superposable on the original molecule.

(without chirality); that is, it has no handedness. Examples of achiral objects include an undecorated cup, an unmarked baseball bat, a regular tetrahedron, a cube, and a sphere.

To repeat, the most common cause of chirality in organic molecules is a tetrahedral carbon atom with four different groups bonded to it. We call such a chiral carbon atom a **stereocenter**. 2-Butanol has one stereocenter; 2-propanol has none. As another example of a molecule with a stereocenter, let us consider 2-hydroxypropanoic acid, more commonly named lactic acid. Lactic acid is a product of anaerobic glycolysis. (See Section 28.2 and Chemical Connections 28A.) It is also what gives sour cream its sour taste.

Figure 15.4 shows three-dimensional representations of lactic acid and its mirror image. In these representations, all bond angles about the central carbon atom are approximately 109.5° and the four bonds from it are directed toward the corners of a regular tetrahedron. Lactic acid displays enantiomerism or chirality; that is, the original molecule and its mirror image are not superposable but rather are different compounds.

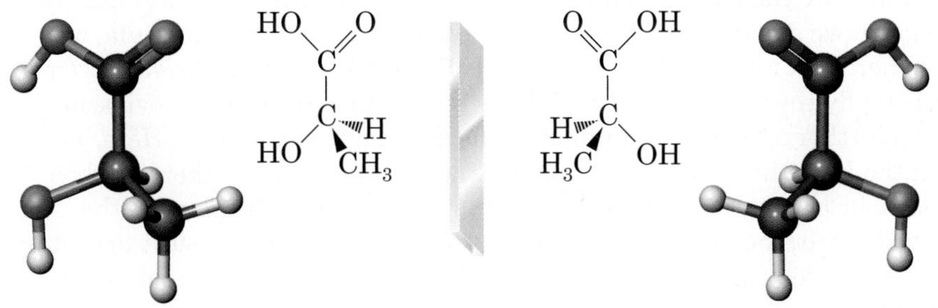

Stereocenter A tetrahedral carbon atom that has four different groups bonded to it

FIGURE 15.4 Three-dimensional representations of lactic acid and its mirror image.

An equimolar mixture of two enantiomers is called a **racemic mixture**, a term derived from the name racemic acid (Latin: *racemus*, "a cluster of grapes"). Racemic acid is the name originally given to the equimolar mixture of the enantiomers of tartaric acid that forms as a by-product during the fermentation of grape juice to produce wine.

Racemic mixture A mixture of equal amounts of two enantiomers

How To . . .

Draw Enantiomers

Now that we know what enantiomers are, we can think about how to represent their three-dimensional structures on a two-dimensional surface. Let us take one of the enantiomers of 2-butanol as an example.

Following are a molecular model of one enantiomer and four different three-dimensional representations of it:

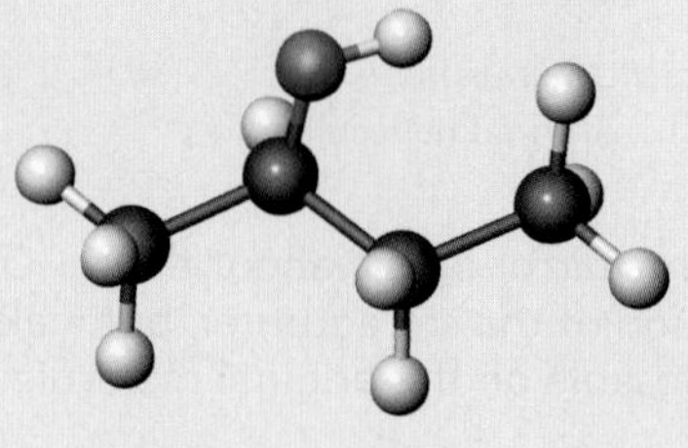

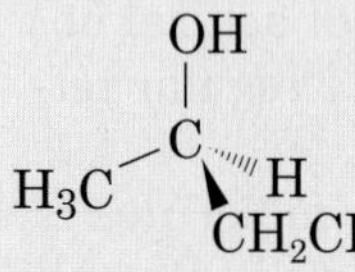

In our initial discussions of 2-butanol, we used representation (1) to show the tetrahedral geometry of the stereocenter. In this representation, two groups (OH and CH_3) are in the plane of the paper, one (CH_2CH_3) is coming out of the plane toward us, and one (H) is behind the plane and going away from us. We can turn representation (1) slightly in space and tip it a bit to place the carbon framework in the plane of the paper. Doing so gives us representation (2), in which there are still two groups in the plane of the paper, one coming toward us and one going away from us.

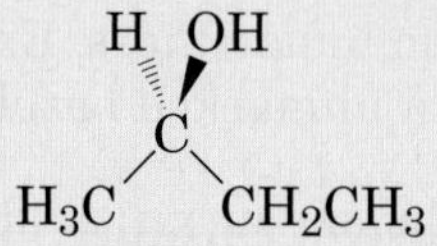

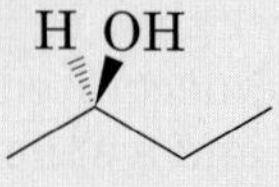

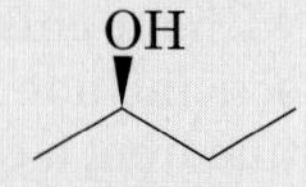

For an even more abbreviated representation of this enantiomer of 2-butanol, we can change representation (2) into the line-angle formula (3). Although we do not normally show hydrogens in a line-angle formula, we do so here just to remind ourselves that the fourth group on the stereocenter is really there and that it is H. Finally, we can carry the abbreviation a step further and write 2-butanol as the line-angle formula (4). Here, we omit the H on the stereocenter, but we know that it must be there (carbon needs four bonds), and we know that it must be behind the plane of the paper. Clearly, the abbreviated formulas (3) and (4) are the easiest to write, and we will rely on this type of representation throughout the remainder of the text.

When you have to write three-dimensional representations of stereocenters, try to keep the carbon framework in the plane of the paper and the other two atoms or groups of atoms on the stereocenter toward and away from you, respectively. Using representation (4) as a model, we get the following alternative representations of its mirror image:

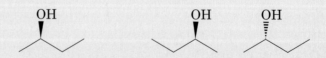

| One enantiomer of 2-butanol | Alternative representations of its mirror image |

Example 15.1 Drawing Mirror Images

Each of the following molecules has one stereocenter marked by an asterisk. Draw three-dimensional representations for the enantiomers of each molecule.

(a)  $CH_3\overset{*}{C}HCH_2CH_3$ with Cl on the stereocenter

(b) 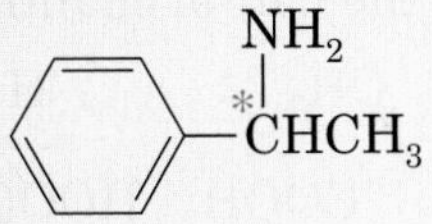 phenyl–$\overset{*}{C}HCH_3$ with NH_2

Strategy

First, draw the carbon stereocenter showing the tetrahedral orientation of its four bonds. One way to do this is to draw two bonds in the plane of the paper, a third bond toward you in front of the plane, and the fourth bond away from you behind the plane. Next, place the four groups bonded to the stereocenter on these positions. This completes the stereodrawing of one enantiomer. To draw the other enantiomer, interchange any two of the groups on the original stereodrawing.

Solution

To draw an original of (a), for example, place the CH_3 and CH_2CH_3 groups in the plane of the paper. Place H away from you and the Cl toward you; this orientation gives the enantiomer of (a) on the left. Its mirror image is on the right.

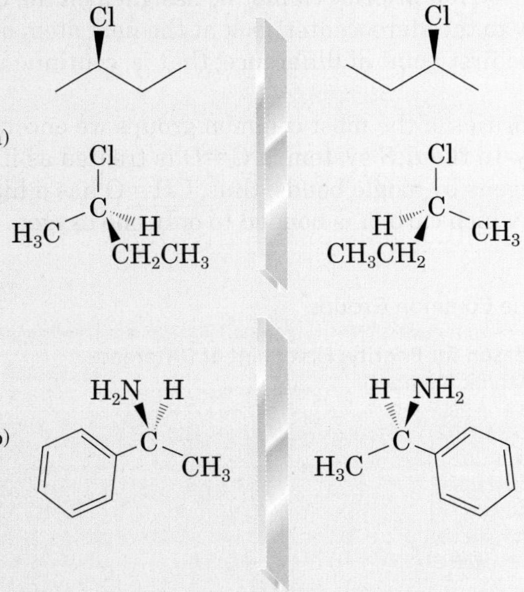

Problem 15.1

Each of the following molecules has one stereocenter marked by an asterisk. Draw three-dimensional representations for the enantiomers of each molecule.

(a) 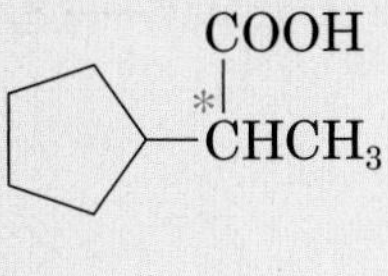 cyclopentyl–$\overset{*}{C}HCH_3$ with COOH

(b) 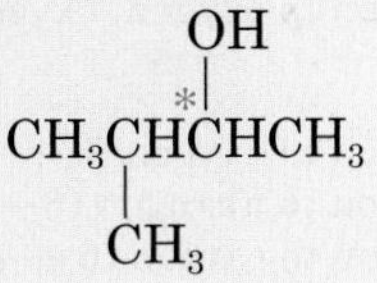 $CH_3\overset{*}{C}HCHCH_3$ with OH and CH_3

15.2 How Do We Specify the Configuration of a Stereocenter?

Because enantiomers are different compounds, each must have a different name. The over-the-counter drug ibuprofen, for example, displays enantiomerism and can exist as the pair of enantiomers shown here:

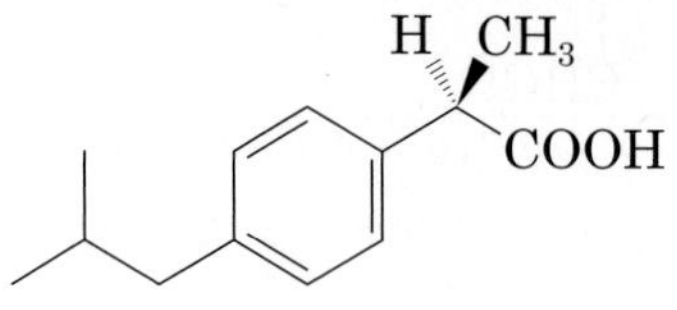

The inactive enantiomer of ibuprofen

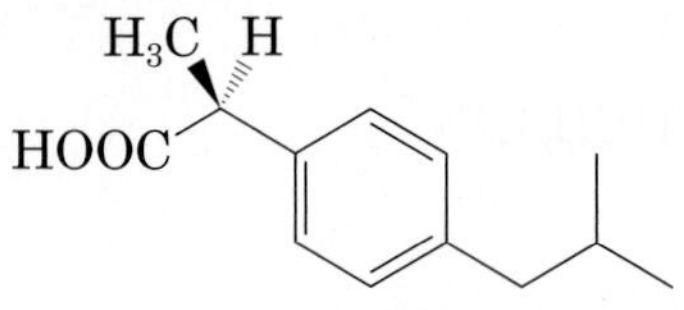

The active enantiomer of ibuprofen

Only one enantiomer of ibuprofen is biologically active. It reaches therapeutic concentrations in the human body in approximately 12 minutes, whereas the racemic mixture takes approximately 30 minutes to achieve this feat. However, in this case, the inactive enantiomer is not wasted. The body converts it to the active enantiomer, but this process takes time.

What we need is a system to designate which enantiomer of ibuprofen (or one of any other pair of enantiomers, for that matter) is which without having to draw and point to one or the other of the enantiomers. To do so, chemists have developed the ***R,S* system**. The first step in assigning an *R* or *S* configuration to a stereocenter is to arrange the groups bonded to it in order of priority. Priority is based on atomic number: The higher the atomic number, the higher the priority. If a priority cannot be assigned on the basis of the atoms bonded directly to the stereocenter, look at the next atom or set of atoms and continue to the first point of difference; that is, continue until you can assign a priority.

Table 15.1 shows the priorities of the most common groups we encounter in organic and biochemistry. In the *R,S* system, a C=O is treated as if carbon were bonded to two oxygens by single bonds; thus, CH=O has a higher priority than —CH_2OH, in which carbon is bonded to only one oxygen.

R,S system A set of rules for specifying the configuration about a stereocenter

Table 15.1 *R,S* Priorities of Some Common Groups

Atom or Group	Reason for Priority: First Point of Difference (Atomic Number)
—I	iodine (53)
—Br	bromine (35)
—Cl	chlorine (17)
—SH	sulfur (16)
—OH	oxygen (8)
—NH_2	nitrogen (7)
$\overset{O}{\underset{\|}{—COH}}$	carbon to oxygen, oxygen, then oxygen (6 ⟶ 8, 8, 8)
$\overset{O}{\underset{\|}{—CNH_2}}$	carbon to oxygen, oxygen, then nitrogen (6 ⟶ 8, 8, 7)
$\overset{O}{\underset{\|}{—CH}}$	carbon to oxygen, oxygen, then hydrogen (6 ⟶ 8, 8, 1)
—CH_2OH	carbon to oxygen (6 ⟶ 8)
—CH_2NH_2	carbon to nitrogen (6 ⟶ 7)
—CH_2CH_3	carbon to carbon (6 ⟶ 6)
—CH_2H	carbon to hydrogen (6 ⟶ 1)
—H	hydrogen (1)

Increasing priority →

Example 15.2 Using the *R,S* System

Assign priorities to the groups in each set.

(a) —CH₂OH and —CH₂CH₂OH
(b) —CH₂CH₂OH and —CH₂NH₂

Strategy and Solution

(a) The first point of difference is O of the —OH group compared to C of the —CH₂OH group.

First point of difference

—CH₂OH   —CH₂CH₂OH
Higher priority   Lower priority

(b) The first point of difference is C of the CH₂OH group compared to N of the NH₂ group.

First point of difference

—CH₂CH₂OH   —CH₂NH₂
Lower priority   **Higher priority**

Problem 15.2

Assign priorities to the groups in each set.

$$
\text{(a) } -CH_2OH \quad \text{and} \quad -CH_2CH_2\overset{\displaystyle O}{\overset{\displaystyle \|}{C}}OH
$$

$$
\text{(b) } -CH_2NH_2 \quad \text{and} \quad -CH_2\overset{\displaystyle O}{\overset{\displaystyle \|}{C}}OH
$$

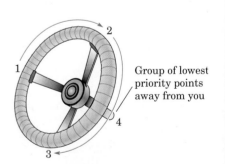

To assign an *R* or *S* configuration to a stereocenter:

1. Assign a priority from 1 (highest) to 4 (lowest) to each group bonded to the molecule's stereocenter.

2. Orient the molecule in space so that the lowest-priority group (4) is directed away from you, as would be, for instance, the steering column of a car. The three higher-priority groups (1–3) then project toward you, as would the spokes of a steering wheel.

3. Read the three groups projecting toward you in order from highest (1) to lowest (4) priority.

4. If reading the groups 1-2-3 proceeds in a clockwise direction (to the right), the configuration is designated as **R** (Latin: *rectus*, "straight"); if reading the groups 1-2-3 proceeds in a counterclockwise direction (to the left), the configuration is **S** (Latin: *sinister*, "left"). You can also visualize this system as follows: Turning the steering wheel to the right equals *R*, and turning it to the left equals *S*.

R Used in the *R,S* system to show that when the lowest-priority group is away from you, the order of priority of groups on a stereocenter is clockwise

S Used in the *R,S* system to show that when the lowest-priority group is away from you, the order of priority of groups on a stereocenter is counterclockwise

Example 15.3 Assigning an *R* or *S* Configuration

Assign an *R* or *S* configuration to each stereocenter.

(a)

$$\begin{array}{c} OH \\ | \\ H_3C \underset{CH_2CH_3}{\overset{C\cdots H}{\diagup \diagdown}} \end{array}$$

2-Butanol

(b)

$$\begin{array}{c} H_2N \quad H \\ \diagdown \diagup \\ C \\ \diagup \diagdown \\ H_3C \quad COOH \end{array}$$

Alanine

Strategy and Solution

View each molecule through the stereocenter and along the bond from the stereocenter to the group of lowest priority.

(a) The order of decreasing priority about the stereocenter in this enantiomer of 2-butanol is —OH > —CH₂CH₃ > —CH₃ > —H. Therefore, view the molecule along the C—H bond with the H pointing away from you. Reading the other three groups in the order 1-2-3 follows in the clockwise direction. Therefore, the configuration is *R* and this enantiomer is (*R*)-2-butanol.

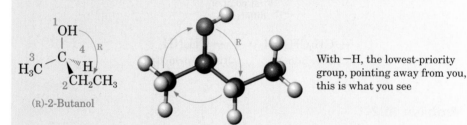

(*R*)-2-Butanol

With —H, the lowest-priority group, pointing away from you, this is what you see

(b) The order of decreasing priority in this enantiomer of alanine is —NH₂ > —COOH > —CH₃ > —H. View the molecule along the C—H bond with H pointing away from you. Reading the groups in the order 1-2-3 follows in a clockwise direction; therefore, the configuration is *R* and this enantiomer is (*R*)-alanine.

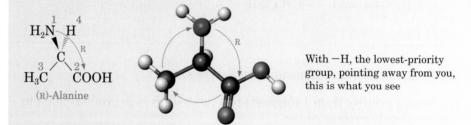

(*R*)-Alanine

With —H, the lowest-priority group, pointing away from you, this is what you see

Problem 15.3

Assign an *R* or *S* configuration to the single stereocenter in glyceraldehyde, the simplest carbohydrate (Chapter 20).

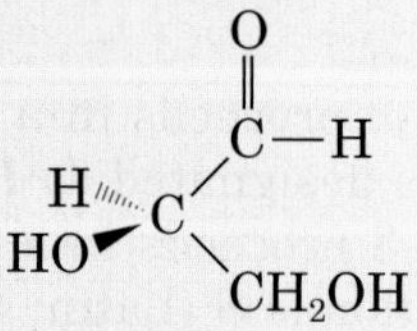

$$\begin{array}{c} O \\ \| \\ C-H \\ | \\ H\cdots C \\ HO \diagdown \quad \diagdown CH_2OH \end{array}$$

Glyceraldehyde

Now let us return to our three-dimensional drawing of the enantiomers of ibuprofen and assign each an R or S configuration. In order of decreasing priority, the groups bonded to the stereocenter are $—COOH$ (1) $>$ $—C_6H_5$ (2) $>$ $—CH_3$ (3) $>$ H (4). In the enantiomer on the left, reading the groups on the stereocenter in order of priority is clockwise, and therefore, this enantiomer is (R)-ibuprofen. Its mirror image is (S)-ibuprofen.

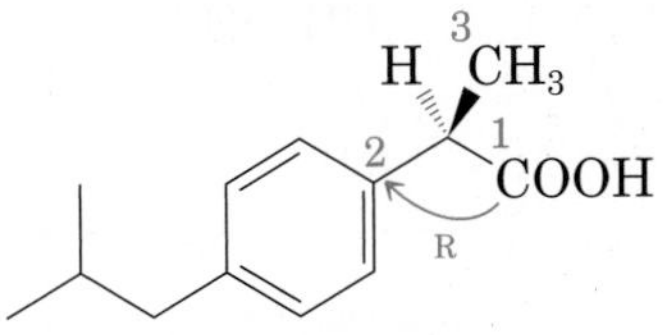

(R)-Ibuprofen
(the inactive enantiomer)

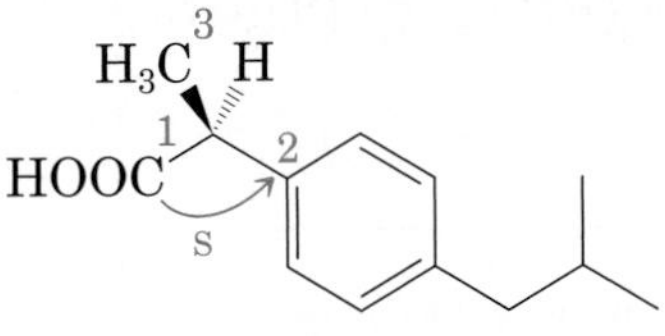

(S)-Ibuprofen
(the active enantiomer)

The R,S system can be used to specify the configuration of any stereocenter in any molecule. It is not, however, the only system used for this purpose. There is also a D,L system, which is used primarily to specify the configuration of carbohydrates (Chapter 20) and amino acids (Chapter 22).

In closing, note that the purpose of Section 15.2 is to show you how chemists assign a configuration to a stereocenter that specifies the relative orientation of the four groups on the stereocenter. What is important is that when you see a name such as (S)-Naproxen or (R)-Plavix, you realize that the compound is chiral and that the compound is not a racemic mixture. Rather, it is a pure enantiomer. We use the symbol (R,S) to show that a compound is a racemic mixture, as for example, (R,S)-Naproxen.

15.3 How Many Stereoisomers Are Possible for Molecules with Two or More Stereocenters?

For a molecule with n stereocenters, the maximum number of stereoisomers possible is 2^n. We have already verified that for a molecule with one stereocenter, $2^1 = 2$ stereoisomers (one pair of enantiomers) are possible. For a molecule with two stereocenters, a maximum of $2^2 = 4$ stereoisomers (two pairs of enantiomers) is possible; for a molecule with three stereocenters, a maximum of $2^3 = 8$ stereoisomers (four pairs of enantiomers) is possible; and so forth.

A. Molecules with Two Stereocenters

We begin our study of molecules with two stereocenters by considering 2,3,4-trihydroxybutanal, a molecule with two stereocenters.

$$CHO$$
$$|$$
$${}^*CHOH$$
$$|$$
$${}^*CHOH$$
$$|$$
$$CH_2OH$$

2,3,4-Trihydroxybutanal
(2 stereocenters;
4 stereoisomers
are possible)

The maximum number of stereoisomers possible for this molecule is $2^2 = 4$, each of which is drawn in Figure 15.5.

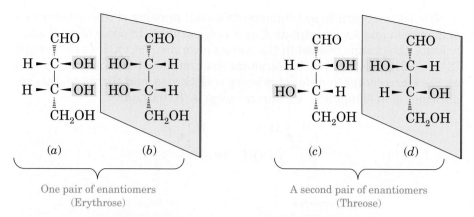

FIGURE 15.5 The four stereoisomers of 2,3,4-trihydroxybutanal.

Stereoisomers (a) and (b) are nonsuperposable mirror images and are, therefore, a pair of enantiomers. Stereoisomers (c) and (d) are also nonsuperposable mirror images and are a second pair of enantiomers. We describe the four stereoisomers of 2,3,4-trihydroxybutanal by saying that they consist of two pairs of enantiomers. Enantiomers (a) and (b) are named **erythrose**. Erythrose is synthesized in erythrocytes (red blood cells); hence the derivation of its name. Enantiomers (c) and (d) are named **threose**. Erythrose and threose belong to the class of compounds called carbohydrates, which we will discuss in Chapter 20.

We have specified the relationship between (a) and (b) and that between (c) and (d). What is the relationship between (a) and (c), between (a) and (d), between (b) and (c), and between (b) and (d)? The answer is that they are **diastereomers**—stereoisomers that are not mirror images.

Diastereomers Stereoisomers that are not mirror images

> ## Example 15.4 Enantiomers and Diastereomers

1,2,3-Butanetriol has two stereocenters (carbons 2 and 3); thus, $2^2 = 4$ stereoisomers are possible for it. Following are three-dimensional representations for each.

```
     CH2OH           CH2OH           CH2OH           CH2OH
  H—C—OH          H—C—OH         HO—C—H          HO—C—H
 HO—C—H           H—C—OH         HO—C—H           H—C—OH
     CH3             CH3             CH3             CH3
     (1)             (2)             (3)             (4)
```

(a) Which stereoisomers are pairs of enantiomers?
(b) Which stereoisomers are diastereomers?

Strategy

First, identify those structures that are mirror images. These, then, are the pairs of enantiomers. All other pairs of structures are diastereomers.

Solution

(a) Enantiomers are stereoisomers that are nonsuperposable mirror images. Compounds (1) and (4) are one pair of enantiomers and compounds (2) and (3) are a second pair of enantiomers.

(b) Diastereomers are stereoisomers that are not mirror images. Compounds (1) and (2), (1) and (3), (2) and (4), and (3) and (4) are diastereomers.

The diagram shows the relationship among these four stereoisomers.

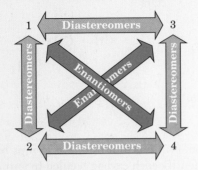

Problem 15.4

3-Amino-2-butanol has two stereocenters (carbons 2 and 3); thus, $2^2 = 4$ stereoisomers are possible for it.

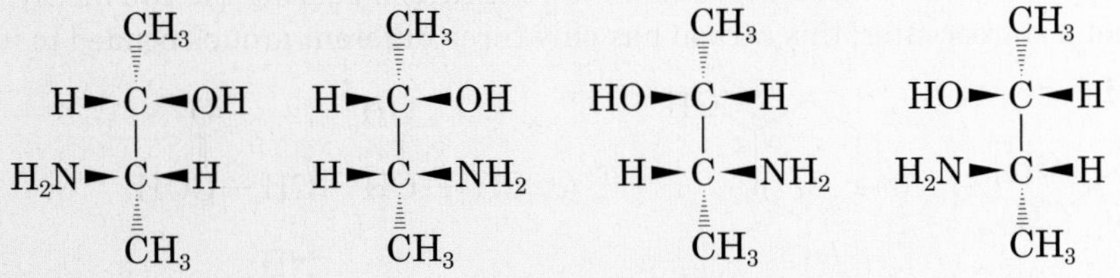

$$\begin{array}{cccc}
CH_3 & CH_3 & CH_3 & CH_3 \\
H\!-\!C\!-\!OH & H\!-\!C\!-\!OH & HO\!-\!C\!-\!H & HO\!-\!C\!-\!H \\
H_2N\!-\!C\!-\!H & H\!-\!C\!-\!NH_2 & H\!-\!C\!-\!NH_2 & H_2N\!-\!C\!-\!H \\
CH_3 & CH_3 & CH_3 & CH_3 \\
(1) & (2) & (3) & (4)
\end{array}$$

(a) Which stereoisomers are pairs of enantiomers?
(b) Which sets of stereoisomers are diastereomers?

We can analyze chirality in cyclic molecules with two stereocenters in the same way we analyzed it in acyclic compounds.

Example 15.5 Enentriomerism in Cyclic Compounds

How many stereoisomers are possible for 3-methylcyclopentanol?

Strategy and Solution

Carbons 1 and 3 of this compound are stereocenters. Therefore, $2^2 = 4$ stereoisomers are possible for this molecule. The *cis* isomer exists as one pair of enantiomers, the *trans* isomer exists as a second pair of enantiomers.

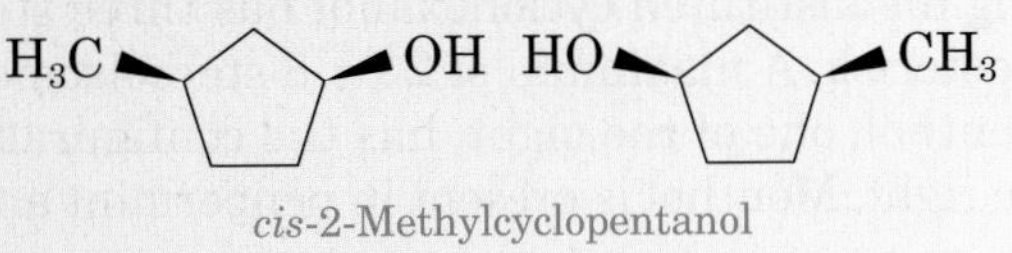

cis-2-Methylcyclopentanol
(a pair of enantiomers)

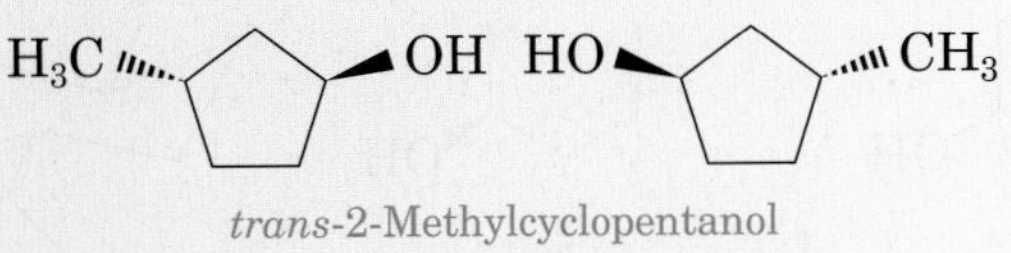

trans-2-Methylcyclopentanol
(a second pair of enantiomers)

Problem 15.5

How many stereoisomers are possible for 3-methylcyclohexanol?

Example 15.6 Locating Stereocenters

Mark the stereocenters in each compound with an asterisk. How many stereoisomers are possible for each?

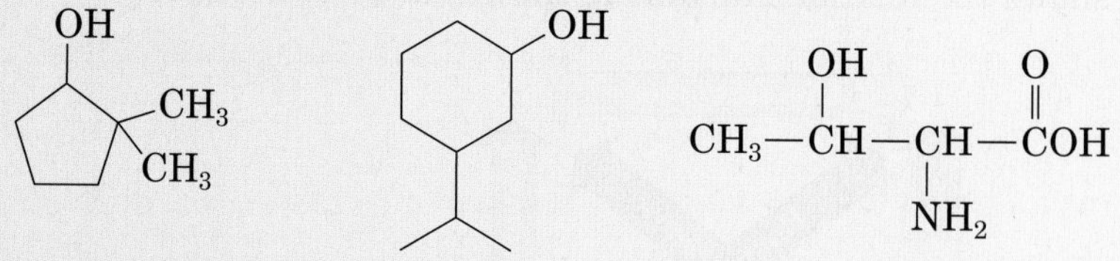

(a)   (b)   (c)

Strategy

A stereocenter is a carbon atom that has four different groups bonded to it. Therefore, you are being asked to identify each carbon bonded to four different groups.

Solution

Each stereocenter is marked with an asterisk, and the number of stereoisomers possible for it appears under each compound. In (a), the carbon bearing the two methyl groups is not a stereocenter; this carbon has only three different groups bonded to it.

(a)   (b)   (c) $CH_3-CH-CH-COH$
 NH_2

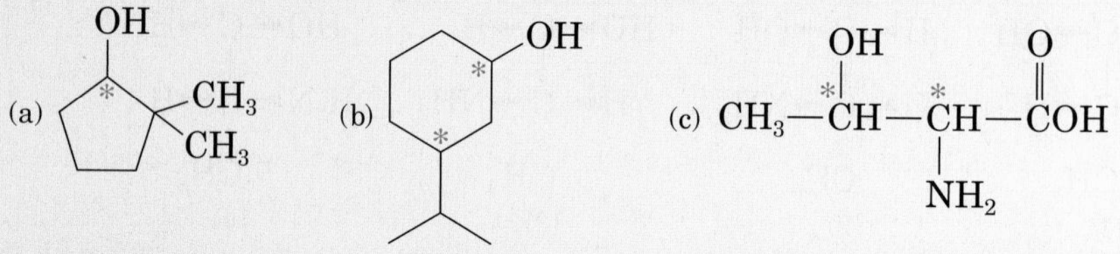

$2^1 = 2$ $2^2 = 4$ $2^2 = 4$

Problem 15.6

Mark all stereocenters in each compound with an asterisk. How many stereoisomers are possible for each?

(a)   (b) $CH_2=CHCHCH_2CH_3$ (c)

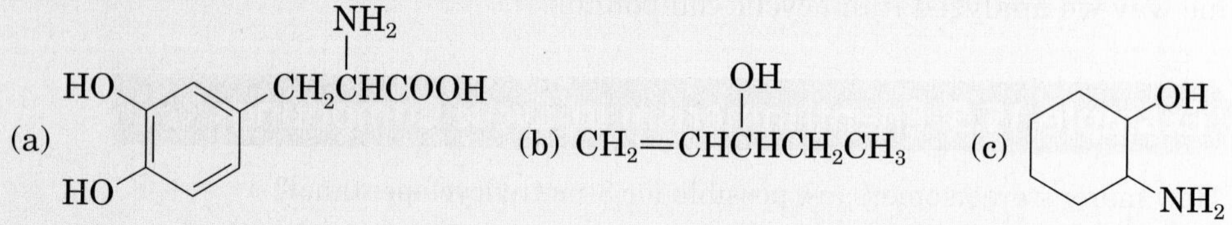

B. Molecules with Three or More Stereocenters

The 2^n rule applies equally well to molecules with three or more stereocenters. The following disubstituted cyclohexanol has three stereocenters, each marked with an asterisk. A maximum of $2^3 = 8$ stereoisomers is possible for this molecule. Menthol, one of the eight, has the configuration shown in the middle and on the right. Menthol is present in peppermint and other mint oils.

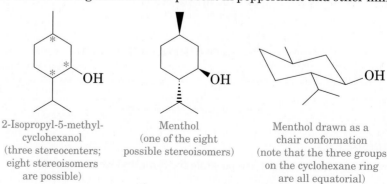

2-Isopropyl-5-methyl-
cyclohexanol
(three stereocenters;
eight stereoisomers
are possible)

Menthol
(one of the eight
possible stereoisomers)

Menthol drawn as a
chair conformation
(note that the three groups
on the cyclohexane ring
are all equatorial)

Chiral Drugs

Some common drugs used in human medicine—for example, aspirin—are achiral. Others, such as the penicillin and erythromycin classes of antibiotics and the drug Captopril, are chiral and are sold as single enantiomers. Captopril is very effective for the treatment of high blood pressure and congestive heart failure. It is manufactured and sold as the (S,S)-stereoisomer.

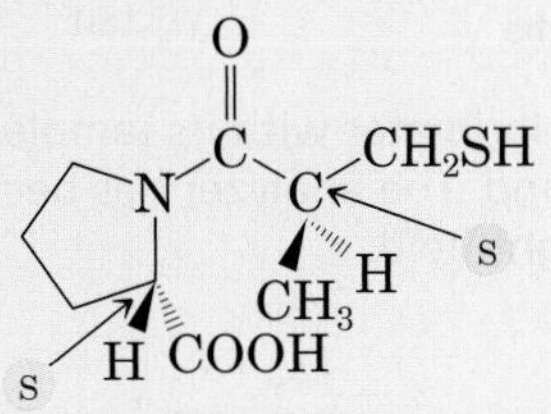

Captopril

A large number of chiral drugs, however, are sold as racemic mixtures. The popular analgesic ibuprofen (the active ingredient in Motrin, Advil, and many other non-aspirin analgesics) is an example.

Recently, the U.S. Food and Drug Administration established new guidelines for the testing and marketing of chiral drugs. After reviewing these guidelines, many drug companies have decided to develop only single enantiomers of new chiral drugs.

In addition to regulatory pressure, pharmaceutical developers must deal with patent considerations. If a company has a patent on a racemic mixture of a drug, a new patent can often be taken out on one of its enantiomers.

Test your knowledge with Problem 15.30.

Cholesterol, a more complicated molecule, has eight stereocenters. To identify them, remember to add an appropriate number of hydrogens to complete the tetravalence of each carbon you think might be a stereocenter.

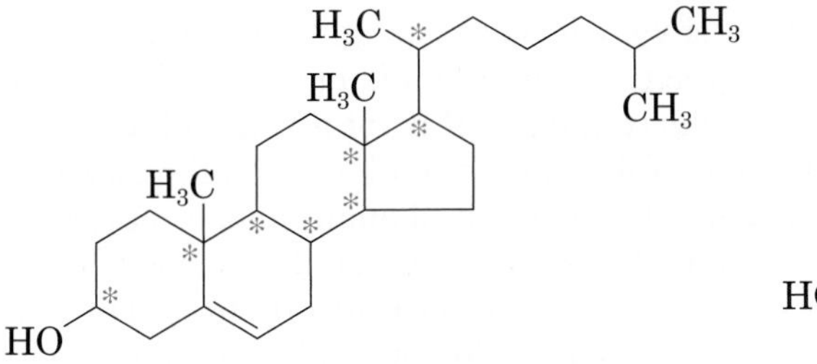

Cholesterol has eight stereocenters;
256 stereoisomers are possible

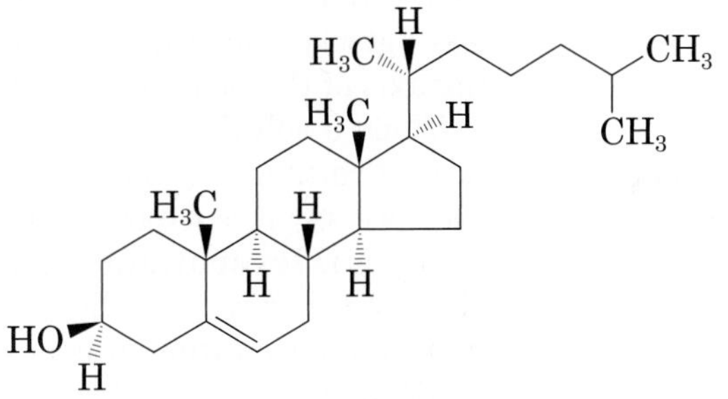

This is the stereoisomer
found in human metabolism

15.4 What Is Optical Activity, and How Is Chirality Detected in the Laboratory?

A. Plane-Polarized Light

As we have already established, the two members of a pair of enantiomers are different compounds, and we must expect, therefore, that some of their properties differ. One such property relates to their effect on the plane of polarized light. Each member of a pair of enantiomers rotates the plane of polarized light; for this reason, each enantiomer is said to be **optically active**. To understand how optical activity is detected in the laboratory, we must first understand what plane-polarized light is and how a polarimeter, the instrument used to detect optical activity, works.

Ordinary light consists of waves vibrating in all planes perpendicular to its direction of propagation. Certain materials, such as a Polaroid sheet (a plastic film like that used in polarized sunglasses), selectively transmit

Optically active Showing that a compound rotates the plane of polarized light

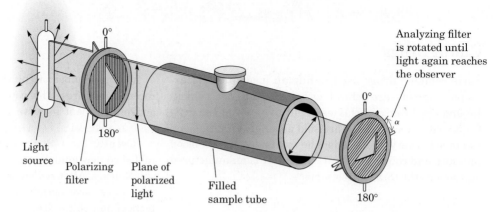

FIGURE 15.6 Schematic diagram of a polarimeter with its sample tube containing a solution of an optically active compound. The analyzer has been turned clockwise by α degrees to restore the light field.

light waves vibrating only in parallel planes. Electromagnetic radiation vibrating in only parallel planes is said to be **plane polarized**.

B. A Polarimeter

A **polarimeter** consists of a light source emitting unpolarized light, a polarizer, an analyzer, and a sample tube (Figure 15.6). If the sample tube is empty, the intensity of light reaching the detector (in this case, your eye) is at its maximum when the axes of the polarizer and analyzer are parallel to each other. If the analyzer is turned either clockwise or counterclockwise, less light is transmitted. When the axis of the analyzer is at right angles to the axis of the polarizer, the field of view is dark (no light passes through).

When a solution of an optically active compound is placed in the sample tube, it rotates the plane of the polarized light. If it rotates the plane clockwise, we say it is **dextrorotatory**; if it rotates the plane counterclockwise, we say it is **levorotatory**. Each member of a pair of enantiomers rotates the plane of polarized light by the same number of degrees, but in opposite directions. If one enantiomer is dextrorotatory, the other is levorotatory. Thus, racemic mixtures (as well as achiral compounds) do not display optical activity.

The number of degrees by which an optically active compound rotates the plane of polarized light is called its **specific rotation** and is given the symbol $[\alpha]$. Specific rotation is defined as the observed rotation of an optically active substance at a concentration of 1 g/mL in a sample tube that is 10 cm long. A dextrorotatory compound is indicated by a plus sign in parentheses, (+), and a levorotatory compound is indicated by a minus sign in parentheses, (−). It is common practice to report the temperature (in °C) at which the measurement is made and the wavelength of light used. The most common wavelength of light used in polarimetry is the sodium D line, the same wavelength responsible for the yellow color of sodium-vapor lamps.

Following are specific rotations for the enantiomers of lactic acid measured at 21°C and using the D line of a sodium-vapor lamp as the light source:

Plane-polarized light Light with waves vibrating in only parallel planes

Dextrorotatory The clockwise (to the right) rotation of the plane of polarized light in a polarimeter

Levorotatory The counterclockwise (to the left) rotation of the plane of polarized light in a polarimeter

The (+) enantiomer of lactic acid is produced by muscle tissue in humans. The (−) enantiomer is found in sour cream and sour milk.

COOH
|
H₃C—C‴H
       OH

(S)-(+)-Lactic acid
$[\alpha]_{\mathrm{D}}^{21} = +2.6°$

COOH
|
H‴C—CH₃
HO

(R)-(−)-Lactic acid
$[\alpha]_{\mathrm{D}}^{21} = -2.6°$

15.5 What Is the Significance of Chirality in the Biological World?

Except for inorganic salts and a few low-molecular-weight organic substances, the majority of molecules in living systems—both plant and animal—are chiral. Although these molecules can exist as a number of stereoisomers, almost invariably, only one stereoisomer is found in nature. Of course, instances do occur in which more than one stereoisomer is found, but these isomers rarely exist together in the same biological system.

A. Chirality in Biomolecules

Perhaps the most conspicuous examples of chirality among biological molecules are the enzymes, all of which have many stereocenters. Consider chymotrypsin, an enzyme in the intestines of animals that catalyzes the digestion of proteins (Chapter 23). Chymotrypsin has 251 stereocenters. The maximum number of stereoisomers possible is 2^{251}—a staggeringly large number, almost beyond comprehension. Fortunately, nature does not squander its precious energy and resources unnecessarily; any given organism produces and uses only one of these stereoisomers.

B. How an Enzyme Distinguishes Between a Molecule and Its Enantiomer

An enzyme catalyzes a biological reaction of a molecule by first positioning the molecule at a **binding site** on the enzyme surface. An enzyme with specific binding sites for three of the four groups on a stereocenter can distinguish between a chiral molecule and its enantiomer or one of its diastereomers. Assume, for example, that an enzyme involved in catalyzing a reaction of glyceraldehyde has three binding sites: one specific for —H, a second specific for —OH, and a third specific for —CHO. Assume further that the three sites are arranged on the enzyme surface as shown in Figure 15.7. The enzyme can distinguish (R)-glyceraldehyde (the natural or biologically active form) from its enantiomer because the natural enantiomer can be adsorbed with three groups interacting with their appropriate binding sites; for the S enantiomer, at best only two groups can interact with these three binding sites.

Because interactions between molecules in living systems take place in a chiral environment, it should come as no surprise that a molecule and its enantiomer will elicit a different physiological response when compared to the same molecule's diastereomer. As we have already seen, (S)-ibuprofen is

The horns of this African Gazelle show chirality; one horn is the mirror image of the other.

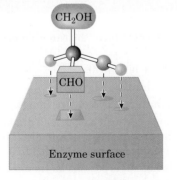

(R)-Glyceraldehyde
fits the three binding
sites on surface

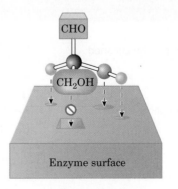

(S)-Glyceraldehyde
fits only two of the
three binding sites

FIGURE 15.7 A schematic diagram of an enzyme surface that can interact with (R)-glyceraldehyde at three binding sites, but with (S)-glyceraldehyde at only two of these sites.

active as a pain and fever reliever, while its *R* enantiomer is inactive. The *S* enantiomer of the closely related analgesic naproxen is also the active pain reliever of this compound, but its *R* enantiomer is a liver toxin!

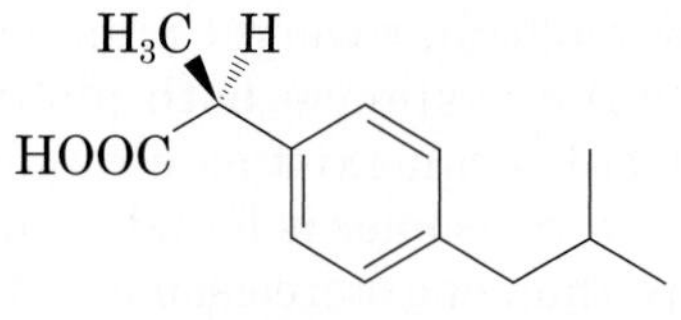

(*S*)-Ibuprofen

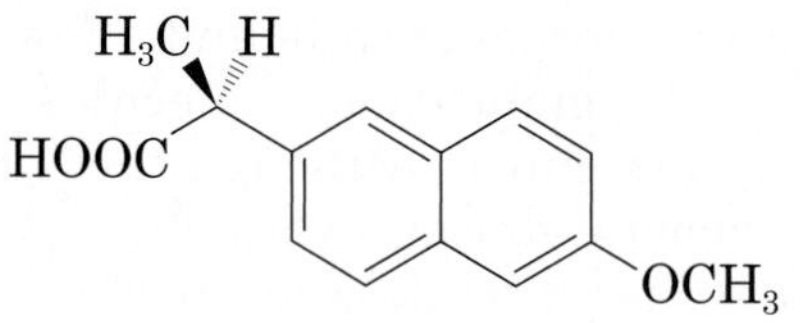

(*S*)-Naproxen

Summary

OWL Sign in at www.cengage.com/owl to develop problem-solving skills and complete online homework assigned by your professor.

Section 15.1 What Is Enantiomerism?

- A **mirror image** is the reflection of an object in a mirror.
- **Enantiomers** are a pair of stereoisomers that are nonsuperposable mirror images.
- A **racemic mixture** contains equal amounts of two enantiomers and does not rotate the plane of polarized light.
- **Diastereomers** are stereoisomers that are not mirror images.
- An object that is not superposable on its mirror image is said to be **chiral**; it has handedness. An **achiral** object lacks chirality (handedness); that is, it has a superposable mirror image.
- The most common cause of chirality in organic molecules is the presence of a tetrahedral carbon atom with four different groups bonded to it. Such a carbon is called a **stereocenter**.

Section 15.2 How Do We Specify the Configuration of a Stereocenter?

- We use the ***R,S* system** to specify the configuration of a stereocenter.

Section 15.3 How Many Stereoisomers Are Possible for Molecules with Two or More Stereocenters?

- For a molecule with n stereocenters, the maximum number of stereoisomers possible is 2^n.

Section 15.4 What Is Optical Activity, and How Is Chirality Detected in the Laboratory?

- Light with waves that vibrate in only parallel planes is said to be **plane polarized**.
- We use a **polarimeter** to measure optical activity. A compound is said to be **optically active** if it rotates the plane of polarized light.
- If a compound rotates the plane clockwise, it is **dextrorotatory**; if it rotates the plane counterclockwise, it is **levorotatory**.
- Each member of a pair of enantiomers rotates the plane of polarized light an equal number of degrees, but in opposite directions.

Section 15.5 What Is the Significance of Chirality in the Biological World?

- An enzyme catalyzes biological reactions of molecules by first positioning them at binding sites on its surface. An enzyme with binding sites specific for three of the four groups on a stereocenter can distinguish between a molecule and its enantiomer or one of its diastereomers.

Problems

OWL Interactive versions of these problems may be assigned in OWL.

Orange-numbered problems are applied.

References to previous chapters are given in parentheses.

Section 15.1 What Is Enantiomerism?

15.7 Answer true or false.

(a) The *cis* and *trans* stereoisomers of 2-butene are achiral.

(b) The carbonyl carbon of an aldehyde, ketone, carboxylic acid, or ester cannot be a stereocenter.

(c) Stereoisomers have the same connectivity of their atoms.

(d) Constitutional isomers have the same connectivity of their atoms.

(e) An unmarked cube is achiral.

(f) A human foot is chiral.

(g) Every object in nature has a mirror image.

(h) The most common cause of chirality in organic molecules is the presence of a tetrahedral carbon atom with four different groups bonded to it.

(i) If a molecule is not superposable on its mirror image, the molecule is chiral.

15.8 What does the term "chiral" mean? Give an example of a chiral molecule.

15.9 What does the term "achiral" mean? Give an example of an achiral molecule.

15.10 Define the term "stereoisomer." Name three types of stereoisomers.

15.11 In what way are constitutional isomers different from stereoisomers? In what way are they the same?

15.12 Which of the following objects are chiral (assume that there is no label or other identifying mark)?
  (a) Pair of scissors (b) Tennis ball
  (c) Paper clip (d) Beaker
  (e) The swirl created in water as it drains out of a sink or bathtub

15.13 2-Pentanol is chiral, but 3-pentanol is not. Explain.

15.14 2-Butene exists as a pair of *cis-trans* isomers. Is *cis*-2-butene chiral? Is *trans*-2-butene chiral? Explain.

15.15 Explain why the carbon of a carbonyl group cannot be a stereocenter.

15.16 Which of the following compounds contain stereocenters?
  (a) 2-Chloropentane (b) 3-Chloropentane
  (c) 3-Chloro-1-butene (d) 1,2-Dichloropropane

15.17 Which of the following compounds contain stereocenters?
  (a) Cyclopentanol
  (b) 1-Chloro-2-propanol
  (c) 2-Methylcyclopentanol
  (d) 1-Phenyl-1-propanol

15.18 Using only C, H, and O, write structural formulas for the lowest-molecular-weight chiral molecule of each class.
  (a) Alkane (b) Alkene
  (c) Alcohol (d) Aldehyde
  (e) Ketone (f) Carboxylic acid

15.19 Draw the mirror image for each molecule:

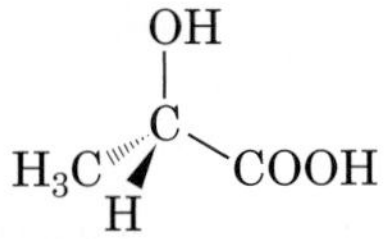

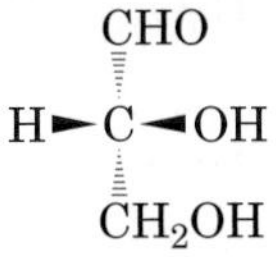

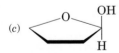

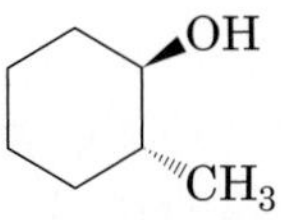

15.20 Draw the mirror image for each molecule:

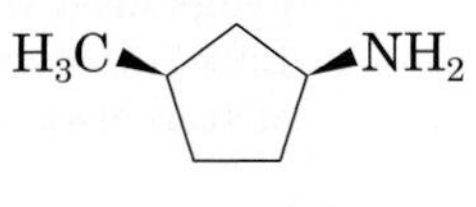

Section 15.3 How Many Stereoisomers Are Possible for Molecules with Two or More Stereocenters?

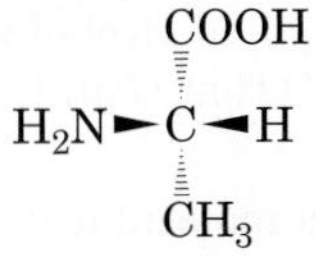

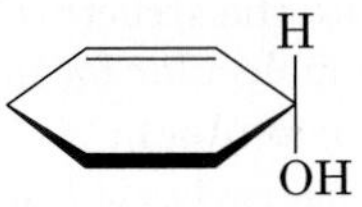

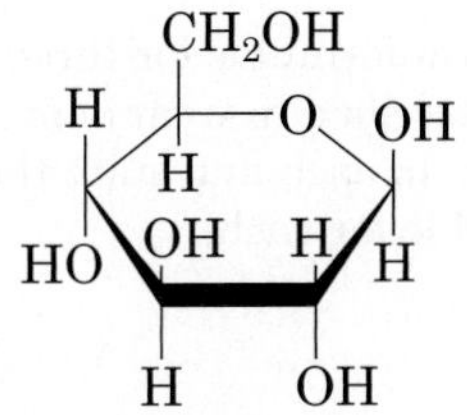

15.21 Answer true or false.
  (a) For a molecule with two stereocenters, $2^2 = 4$ stereoisomers are possible.
  (b) For a molecule with three stereocenters, $3^2 = 9$ stereoisomers are possible.
  (c) Enantiomers, like gloves, occur in pairs.
  (d) 2-Pentanol and 3-pentanol are both chiral and show enantiomerism.
  (e) 1-Methylcyclohexanol is achiral and does not show enantiomerism.
  (f) Diastereomers are stereoisomers that are not mirror images.

15.22 Mark each stereocenter in these molecules with an asterisk. Note that not all contain stereocenters.

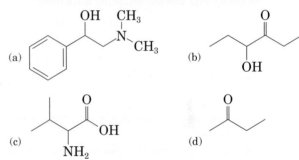

15.23 Mark each stereocenter in these molecules with an asterisk. Note that not all contain stereocenters.

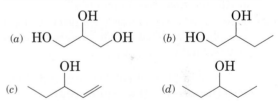

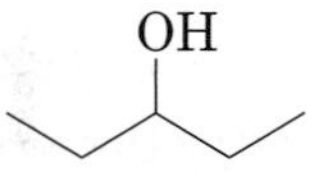

15.24 Label all stereocenters in each molecule with an asterisk. How many stereoisomers are possible for each molecule?

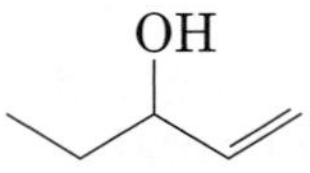

(a) $CH_3CHCHCOOH$ (b) ...

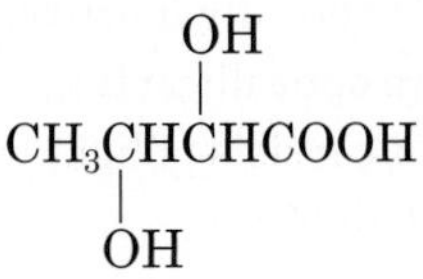

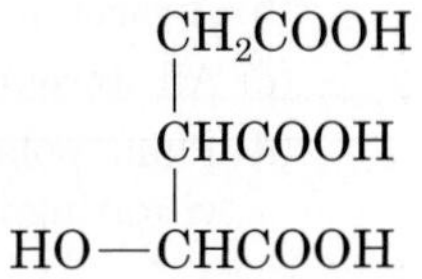

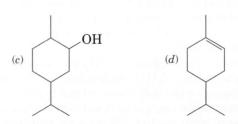

15.25 Label all stereocenters in each molecule with an asterisk. How many stereoisomers are possible for each molecule?

(a)

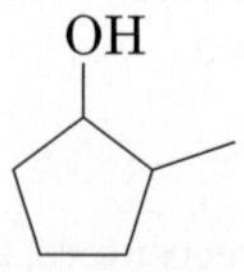

(b)

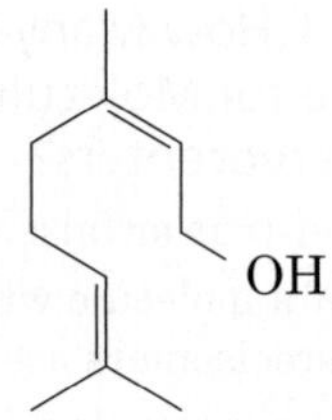

(c)

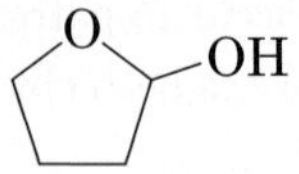

(d)

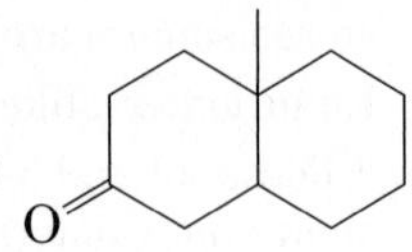

15.26 For centuries, Chinese herbal medicine has used extracts of *Ephedra sinica* to treat asthma. The asthma-relieving component of this plant is ephedrine, a very potent dilator of the air passages of the lungs. The naturally occurring stereoisomer is levorotatory and has the following structure.

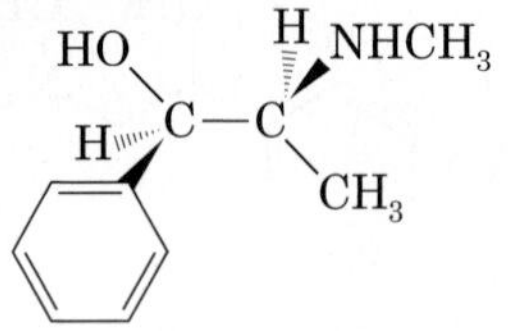

Ephedrine $[\alpha]_D^{21} = -41°$

(a) Mark each stereocenter in ephedrine with an asterisk.

(b) How many stereoisomers are possible for this compound?

15.27 The specific rotation of naturally occurring ephedrine, shown in Problem 15.26, is −41°. What is the specific rotation of its enantiomer?

15.28 What is a racemic mixture? Is a racemic mixture optically active? That is, will it rotate the plane of polarized light?

Section 15.4 What Is Optical Activity, and How Is Chirality Detected in the Laboratory?

15.29 Answer true or false.

(a) If a chiral compound is dextrorotatory, its enantiomer is levorotatory by the same number of degrees.

(b) A racemic mixture is optically inactive.

(c) All stereoisomers are optically active.

(d) Plane-polarized light consists of light waves vibrating in parallel planes.

Chemical Connections

15.30 (Chemical Connections 15A) What does it mean to say that a drug is *chiral*? If a drug is chiral, will it be optically active? That is, will it rotate the plane of polarized light?

15.31 (Chemical Connections 15A) When a successful drug is discovered, there often follows an effort to synthesize compounds closely related in structure to the original, in hopes of discovering new drugs that are even more effective. Following are structural

formulas for Captopril and three related angiotensin converting enzyme (ACE) inhibitors, which are used to treat hypertension. Which of these three are chiral, and how many stereoisomers are possible for each? Compare these three structures with that of Captopril and determine the similarities in structure among the four drugs.

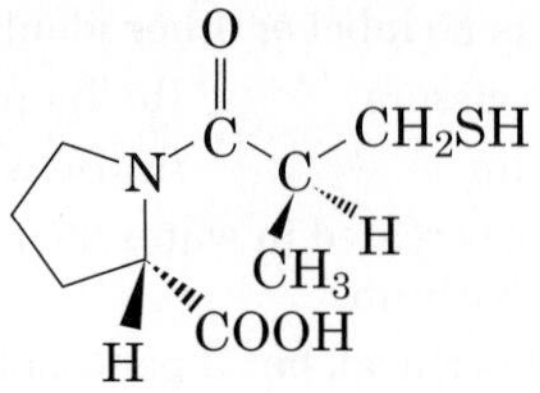

Captopril

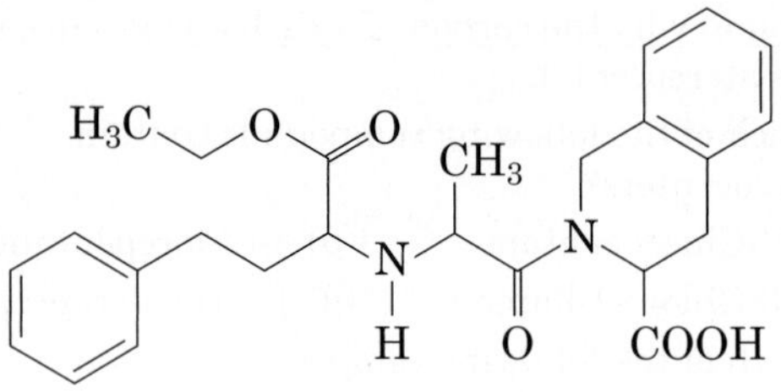

Quinapril (Accupril)

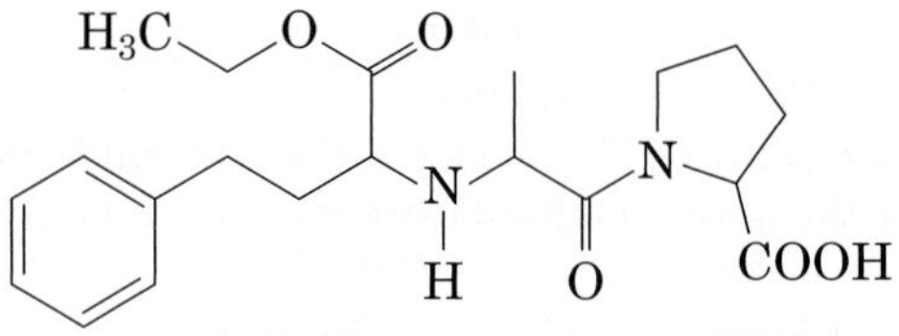

Enalopril (Altace)

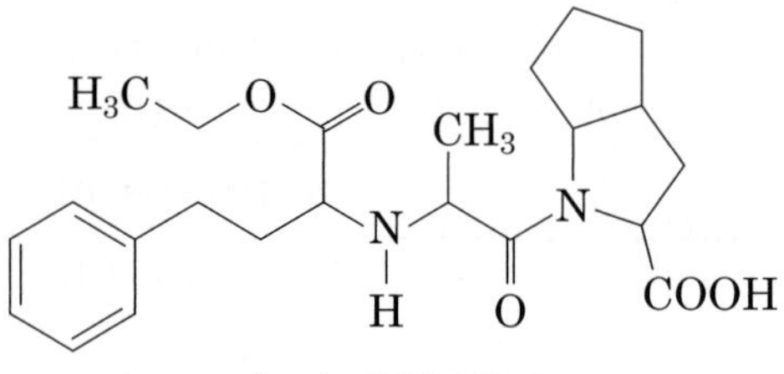

Ramipril (Vasotec)

Additional Problems

15.32 Which of the eight alcohols with a molecular formula of $C_5H_{12}O$ are chiral?

15.33 Write the structural formula of an alcohol with the molecular formula $C_6H_{14}O$ that contains two stereocenters.

15.34 Which carboxylic acids with a molecular formula of $C_6H_{12}O_2$ are chiral?

15.35 Following are structural formulas for three of the drugs most widely prescribed to treat depression. Label all stereocenters in each and state the number of stereoisomers possible for each.

(a)

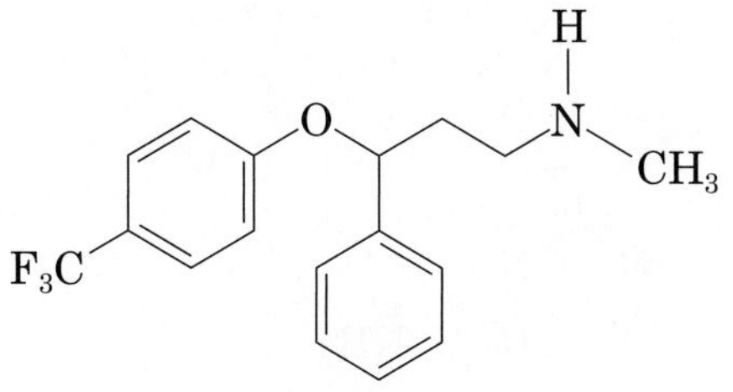

Fluoxetine
(Prozac)

(b)

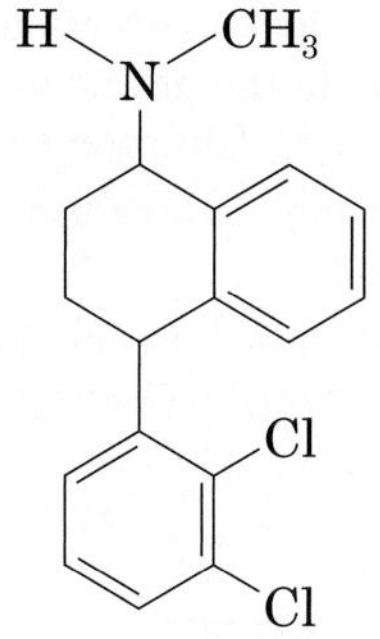

Sertraline
(Zoloft)

(c)

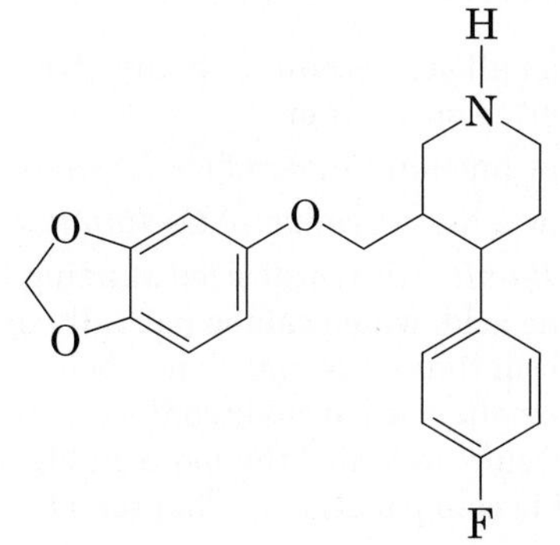

Paroxetine
(Paxil)

15.36 Label the four stereocenters in amoxicillin, which belongs to the family of semisynthetic penicillins.

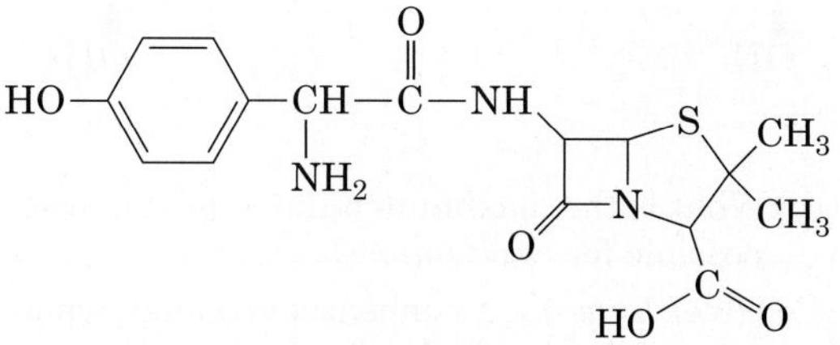

Amoxicillin

15.37 Consider a cyclohexane ring substituted with one hydroxyl group and one methyl group. Draw a structural formula for a compound of this composition that:

(a) Does not show *cis-trans* isomerism and has no stereocenters.

(b) Shows *cis-trans* isomerism but has no stereocenters.

(c) Shows *cis-trans* isomerism and has two stereocenters.

15.38 The next time you have the opportunity to view a collection of seashells that have a helical twist, study the chirality (handedness) of their twists. For each kind of shell, do you find an equal number of left-handed and right-handed twists or, for example, do they all have the same handedness?

15.39 The next time you have an opportunity to examine any of the seemingly endless varieties of blond spiral pasta (rotini, fusilli, radiatori, tortiglione, and so forth), examine their twists. Do the twists of any one kind all have a right-handed twist or a left-handed twist, or are they a racemic mixture?

15.40 Think about the helical coil of a telephone cord or the spiral binding on a notebook. Suppose that you view the spiral from one end and find that it has a left-handed twist. If you view the same spiral from the other end, does it have a left-handed twist from that end as well or does it have a right-handed twist?

15.41 Compound $A(C_5H_8)$ is not optically active and cannot be separated into enantiomers. It reacts with bromine in carbon tetrachloride to discharge the purple color of bromine and form Compound $B(C_5H_8Br_2)$. When Compound A is treated with H_2 in the presence of a transition metal catalyst, it is converted to compound $C(C_5H_{10})$. When treated with HCl, compound A is converted to compound $D(C_5H_9Cl)$. Given this information, propose structural formulas for compounds A, B, C, and D. *Hint:* There are at least three possibilities for Compound A and, in turn, three possibilities for Compounds B, C, and D.

Looking Ahead

15.42 Following is a chair conformation of glucose, the most prevalent carbohydrate in the biological world (Chapter 20).

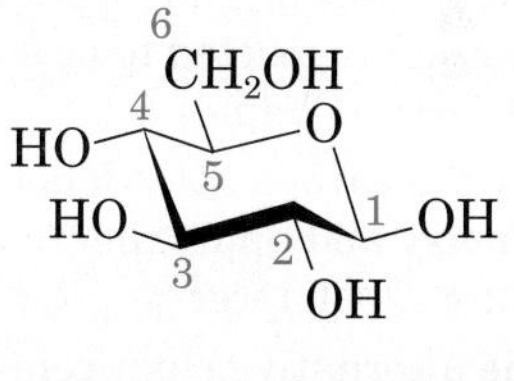

(a) Identify the five stereocenters in this molecule.

(b) How many stereoisomers are possible?

(c) How many pairs of enantiomers are possible?

15.43 Triamcinolone acetonide, the active ingredient in Azmacort Inhalation Aerosol, is a steroid used to treat bronchial asthma.

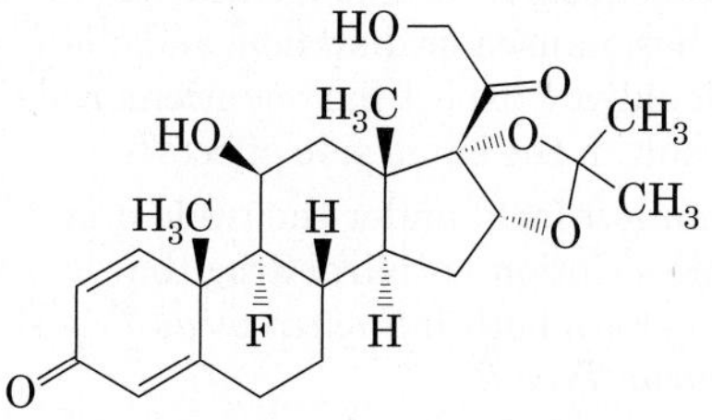

Triamcinolone acetonide

(a) Label the eight stereocenters in this molecule.

(b) How many stereoisomers are possible for it? (Of these, the stereoisomer with the configuration shown here is the active ingredient in Azmacort.)

Challenge Problems

15.44 Consider the structure of the immunosuppressant *FK-506*, a molecule shown to disrupt calcineurin-mediated signal transduction in *T*-lymphocytes.

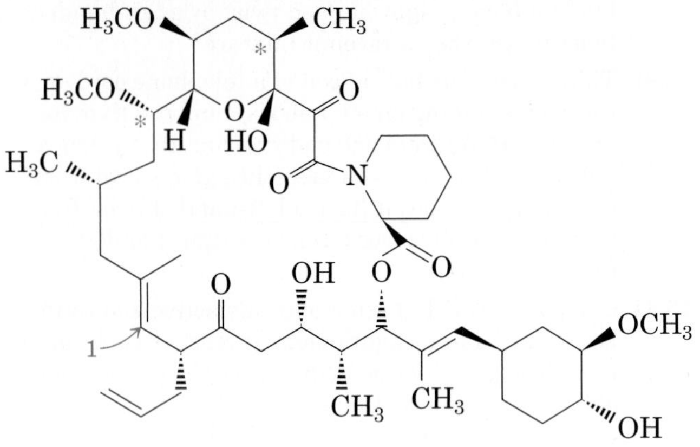

(a) What is the molecular formula of this immunosuppressant? (Chapter 10)

(b) How many stereocenters are present in *FK-506*? Determine the maximum number of stereoisomers possible.

(c) Identify and label the various functional groups present. (Chapter 10)

(d) Consider the two carbons in this structure labeled with asterisks (*). Determine the absolute configuration of each stereocenter.

(e) *FK-506* has been shown to exhibit moderate solubility in various organic solvents. Is this immunosuppressant expected to be soluble in ethanol (CH_3CH_2OH)? (Chapter 6)

(f) Consider the carbon atom labeled "**1**." What is the geometry and approximate bond angle of this carbon atom? (Chapter 3)

(g) Draw the alternative chair conformations of the cyclohexane ring at the lower right of *FK-506* and label the more stable conformation. (Chapter 11)

(h) Are there any aromatic components present in *FK-506*? (Chapter 13)

(i) Patients taking *FK-506* have reported several side effects from this medication, including headaches, nausea or diarrhea, and slight shaking. Would you expect the enantiomer of this drug to result in the same side effects?

15.45 *Oseltamivir* (sold under the trade name *Tamiflu*®) is a prescription antiviral drug that is used in the treatment of both *Influenza virus Type A* and *Influenzavirus Type B*.

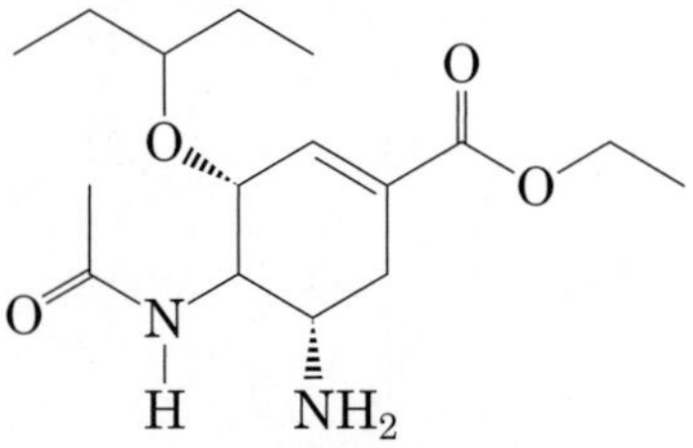

Oseltamivir
(Tamiflu)

(a) Identify the various functional groups present in *Oseltamivir*. Is the amine group primary, secondary, or tertiary? (Chapter 10)

(b) What is the molecular formula of *Oseltamivir*? (Chapter 10)

(c) Place a check mark in the box next to all the words that describe *Oseltamivir*:

☐ Chiral

☐ Achiral

☐ Optically active

☐ Optically inactive

☐ Racemic

(d) Label all stereocenters (if any) present in *Oseltamivir* as either *R* or *S*.

(e) Draw an enantiomer of *Oseltamivir*.

(f) Draw a diastereomer of *Oseltamivir*.

(g) *Oseltamivir* is synthesized starting from shikimic acid, which can be partially hydrogenated to form the compound shown below on the right. Draw alternative chair conformations of this molecule and label the more stable and less stable conformations. (Chapter 11)

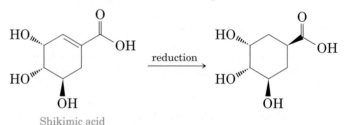

reduction

Shikimic acid

(h) What is the maximum number of stereoisomers possible for *Oseltamivir*?

15.46 Consider *Lunesta*, a nonbenzodiazepine hypnotic agent (i.e., sleep-inducing drug) that is frequently advertised on TV commercials. Answer the following questions with respect to the given structure:

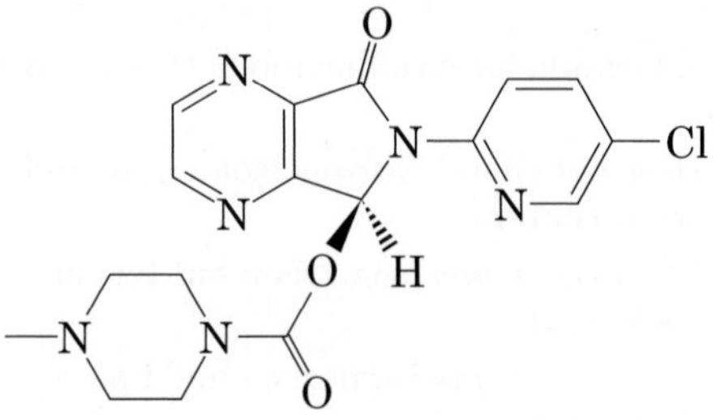

Lunesta

(a) Determine the molecular formula for *Lunesta*. (Chapter 10)

(b) Identify the functional groups present in *Lunesta*. (Chapter 10)

(c) How many of *Lunesta's* rings are aromatic? (Chapter 13)

(d) Fill in the blanks as shown: *Lunesta* has _____ stereocenter(s) and therefore _____ possible stereoisomer(s). Of the possible stereocenter(s), _____ is/are *R* and _____ is/are *S*.

(e) Does *Lunesta* have an enantiomer? Does it have a diastereomer?

(f) Which of the following is true about an enantiomer of *Lunesta*? Identify all that apply:

(1) The enantiomer rotates plane-polarized light in the opposite direction as *Lunesta*.

(2) The enantiomer is a mirror image of *Lunesta*.

(3) The enantiomer has the opposite biological effects as *Lunesta* (i.e., it keeps you awake).

(4) *Lunesta* does not have an enantiomer.

(g) Draw an enantiomer of *Lunesta*.

(h) Examine the derivative of the representation of the six-membered ring found in *Lunesta*. Draw the alternative chair conformations of this ring and label the more stable chair conformation. (Chapter 11)

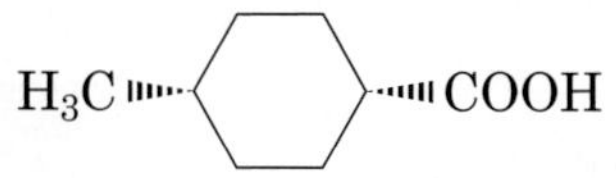

H_3C ⅢⅢ⋯ ⟨ ⟩ ⋯ⅢⅢ $COOH$

16

Amines

This inhaler delivers a puff of albuterol (Proventil), a potent synthetic bronchodilator whose structure is patterned after that of epinephrine. See Chemical Connections 16E.

Aliphatic amine An amine in which nitrogen is bonded only to alkyl groups or hydrogens

OWL

Sign in to OWL at **www.cengage.com/owl** to view tutorials and simulations, develop problem-solving skills, and complete online homework assigned by your professor.

16.1 What Are Amines?

Carbon, hydrogen, and oxygen are the three most common elements in organic compounds. Because of the wide distribution of amines in the biological world, nitrogen is the fourth most common element of organic compounds. The most important chemical property of amines is their basicity.

Amines (Section 10.4B) are classified as **primary (1°), secondary (2°)**, or **tertiary (3°)**, depending on the number of carbon groups bonded to nitrogen.

$$CH_3-NH_2 \qquad CH_3-\overset{\overset{\displaystyle H}{|}}{N}-CH_3 \qquad CH_3-\overset{\overset{\displaystyle CH_3}{|}}{N}-CH_3$$

Methylamine (a 1° amine)   Dimethylamine (a 2° amine)   Trimethylamine (a 3° amine)

Amines are further classified as aliphatic or aromatic. An **aliphatic amine** is one in which all the carbons bonded to nitrogen are derived from

Chemical Connections 16A

Amphetamines (Pep Pills)

Amphetamine, methamphetamine, and phentermine—all synthetic amines—are powerful stimulants of the central nervous system. Like most other amines, they are stored and administered as their salts. The sulfate salt of amphetamine is named Benzedrine, the hydrochloride salt of the S enantiomer of methamphetamine is named Methedrine, and the hydrochloride salt of phentermine is named Fastin.

These three amines have similar physiological effects and are referred to by the general name **amphetamines**. Structurally, they have in common a benzene ring with a three-carbon side chain and an amine nitrogen on the second carbon of the side chain. Physiologically, they share an ability to reduce fatigue and diminish hunger

by raising the glucose level of the blood. Because of these properties, amphetamines are widely prescribed to counter mild depression, reduce hyperactivity in children, and suppress appetite in people who are trying to lose weight. These drugs are also used illegally to reduce fatigue and elevate mood.

Abuse of amphetamines can have severe effects on both body and mind. They are addictive, concentrate in the brain and nervous system, and can lead to long periods of sleeplessness, loss of weight, and paranoia.

The action of amphetamines is similar to that of epinephrine (Chemical Connections 16E), the hydrochloride salt of which is named adrenaline.

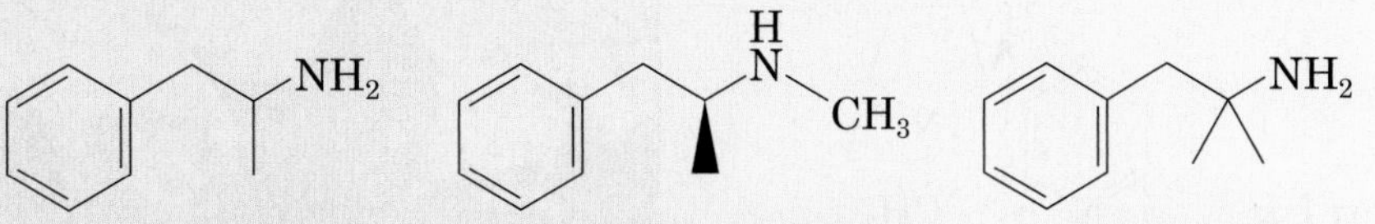

Amphetamine
(Benzedrine)

(s)-Methamphetamine
(Methedrine)

Phentermine
(Fastin)

Test your knowledge with Problems 16.33 and 16.34.

alkyl groups. An **aromatic amine** is one in which one or more of the groups bonded to nitrogen are aryl groups.

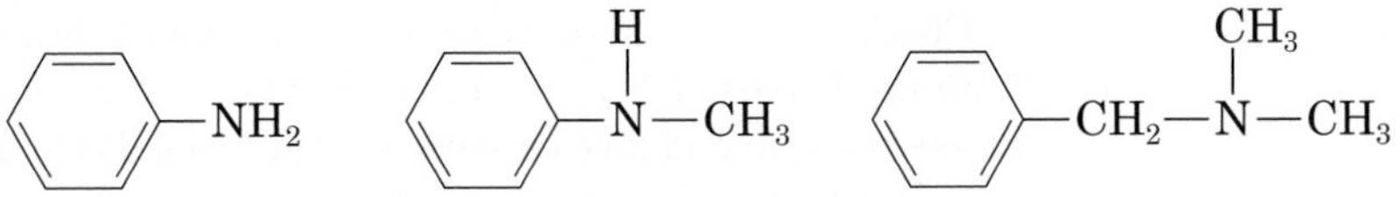

Aniline
(a 1° aromatic amine)

N-Methylaniline
(a 2° aromatic amine)

Benzyldimethylamine
(a 3° aliphatic amine)

Aromatic amine An amine in which nitrogen is bonded to one or more aromatic rings

An amine in which the nitrogen atom is part of a ring is classified as a **heterocyclic amine**. When the ring is saturated, the amine is classified as a **heterocyclic aliphatic amine**. When the nitrogen is part of an aromatic ring (Section 13.1), the amine is classified as a **heterocyclic aromatic amine**. Two of the most important heterocyclic aromatic amines are pyridine and pyrimidine, in which nitrogen atoms replace first one and then two CH groups of a benzene ring. Pyrimidine and purine serve as the building blocks for the amine bases of DNA and RNA (Chapter 25).

Heterocyclic amine An amine in which nitrogen is one of the atoms of a ring

Heterocyclic aromatic amine An amine in which nitrogen is one of the atoms of an aromatic ring

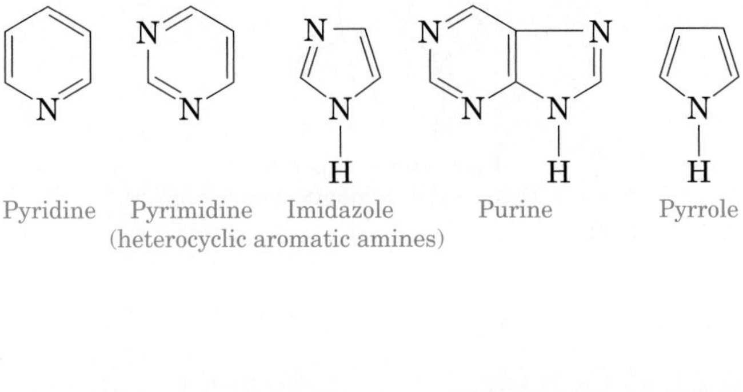

Pyridine Pyrimidine Imidazole   Purine   Pyrrole
(heterocyclic aromatic amines)

Chemical Connections 16B

Alkaloids

Alkaloids are basic nitrogen-containing compounds found in the roots, bark, leaves, berries, or fruits of plants. In almost all alkaloids, the nitrogen atom is part of a ring. The name "alkaloid" was chosen because these compounds are alkali-like (*alkali* is an older term for a basic substance) and react with strong acids to give water-soluble salts. Thousands of different alkaloids, many of which are used in modern medicine, have been extracted from plant sources.

When administered to animals, including humans, alkaloids have pronounced physiological effects. Whatever their individual effects, most alkaloids are toxic in large enough doses. For some, the toxic dose is very small!

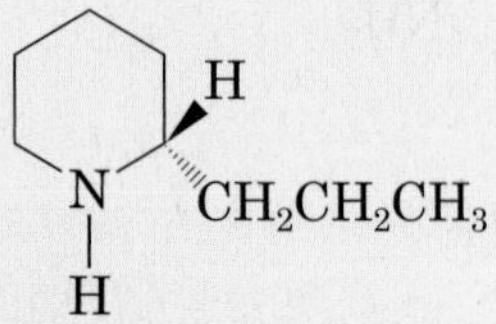

(*S*)-Coniine is the toxic principle of water hemlock (a member of the carrot family). Its ingestion can cause weakness, labored respiration, paralysis, and eventually death. It was the toxic substance in the "poison hemlock" used in the death of Socrates. Water hemlock is easily confused with Queen Anne's lace, a type of wild carrot—a mistake that has killed numerous people.

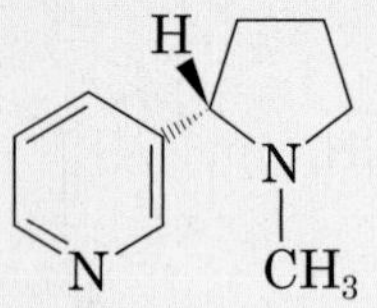

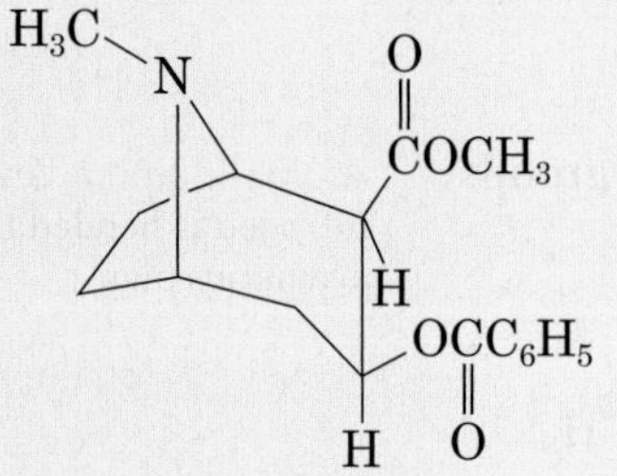

(*S*)-Nicotine occurs in the tobacco plant. In small doses, it is an addictive stimulant. In larger doses, this substance causes depression, nausea, and vomiting. In still larger doses, it is a deadly poison. Solutions of nicotine in water are used as insecticides.

(*s*)-Coniine   (*s*)-Nicotine

Cocaine

Tobacco plants.

Cocaine is a central nervous system stimulant obtained from the leaves of the coca plant. In small doses, it decreases fatigue and gives a sense of well-being. Prolonged use of cocaine leads to physical addiction and depression.

Test your knowledge with Problems 16.35–16.38.

⬤WL Interactive Example 16.1 Structure of Amines

How many hydrogen atoms does piperidine have? How many hydrogen atoms does pyridine have? Write the molecular formula of each amine.

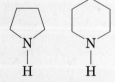

Pyrrolidine   Piperidine
(heterocyclic aliphatic amines)

Strategy

Remember that hydrogen atoms bonded to carbon are not shown in line-angle formulas. To determine the number of hydrogens present, add a sufficient number to give four bonds to each carbon and three bonds to each nitrogen.

Solution

Piperidine has 11 hydrogen atoms, and its molecular formula is $C_5H_{11}N$. Pyridine has 5 hydrogen atoms, and its molecular formula is C_5H_5N.

Problem 16.1

How many hydrogen atoms does pyrrolidine have? How many does purine have? Write the molecular formula of each amine.

16.2 How Do We Name Amines?

A. IUPAC Names

IUPAC names for aliphatic amines are derived just as they are for alcohols. The final -e of the parent alkane is dropped and replaced with -**amine**. Indicate the location of the amino group on the parent chain by a number.

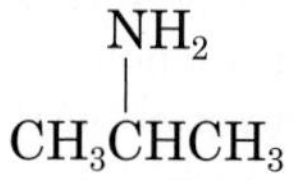

CH$_3$CHCH$_3$ (NH$_2$)
2-Propanamine 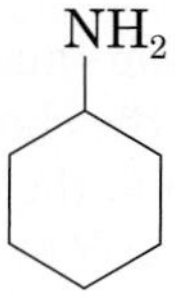   Cyclohexanamine   1,6-Hexanediamine

IUPAC nomenclature retains the common name **aniline** for $C_6H_5NH_2$, the simplest aromatic amine. Its simple derivatives are named using numbers to locate substituents or, alternatively, using the locators *ortho* (*o*), *meta* (*m*), and *para* (*p*). Several derivatives of aniline have common names that remain in use. Among them is **toluidine** for a methyl-substituted aniline.

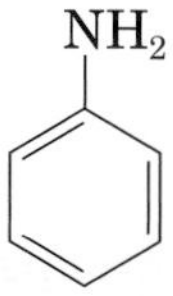

Aniline 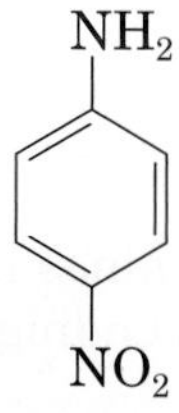   4-Nitroaniline (*p*-Nitroaniline) 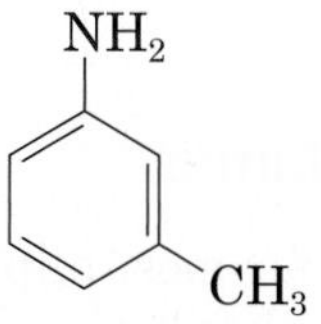   3-Methylaniline (*m*-Toluidine)

Unsymmetrical secondary and tertiary amines are commonly named as *N*-substituted primary amines. The largest group bonded to nitrogen is taken as the parent amine; the smaller groups bonded to nitrogen are named, and their locations are indicated by the prefix *N* (indicating that they are bonded to nitrogen).

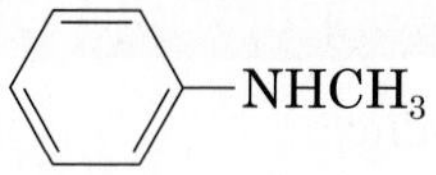

N-Methylaniline 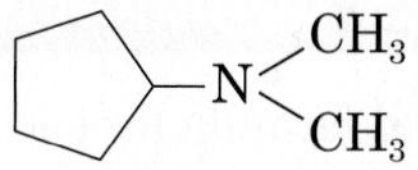   *N,N*-Dimethyl-cyclopentanamine

Example 16.2 IUPAC Names of Amines

Write the IUPAC name for each amine. Try to specify the configuration of the stereocenter in (c).

(a) [structure: NH₂ on 2-butanamine]   (b) H₂N(CH₂)₅NH₂   (c) [structure: phenyl group with CH₂CH(NH₂)CH₃, wedge/dash at stereocenter, H NH₂]

Strategy

The parent chain is the longest chain that contains the amino group. Number the parent chain from the end that gives the amino group the lowest possible number.

Solution

(a) The parent alkane has four carbon atoms and is butane. The amino group is on carbon 2, giving the IUPAC name 2-butanamine.
(b) The parent chain has five carbon atoms and is pentane. There are amino groups on carbons 1 and 5, giving the IUPAC name 1,5-pentanediamine. The common name of this diamine is cadaverine, which should give you a hint of where it occurs in nature and its odor. Cadaverine, one of the end products of decaying flesh, is quite poisonous.
(c) The parent chain has three carbon atoms and is propane. To have the lowest numbers possible, number the chain from the end that places the phenyl group on carbon 1 and the amino group on carbon 2. The priorities for determining R or S configuration are $NH_2 > C_6H_5CH_2 > CH_3 > H$. This amine's systematic name is (R)-1-phenyl-2-propanamine. It is the (R)-enantiomer of the stimulant amphetamine.

Problem 16.2

Write a structural formula for each amine.

(a) 2-Methyl-1-propanamine   (b) Cyclopentanamine
(c) 1,4-Butanediamine

B. Common Names

Common names for most aliphatic amines list the groups bonded to nitrogen in alphabetical order in one word ending in the suffix -**amine**.

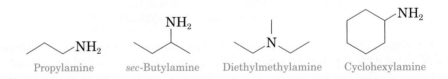

Propylamine   *sec*-Butylamine   Diethylmethylamine   Cyclohexylamine

Example 16.3 Common Names of Amines

Write a structural formula for each amine.
(a) Isopropylamine   (b) Cyclohexylmethylamine
(c) Triethylamine

Strategy and Solution

In these common names, the names of the groups bonded to carbon are listed in alphabetical order followed by the suffix -**amine**.

(a) $(CH_3)_2CHNH_2$ (b) 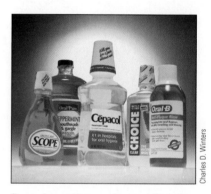—$NHCH_3$ (c) $(CH_3CH_2)_3N$

Problem 16.3

Write a structural formula for each amine.

(a) 2-Aminoethanol   (b) Diphenylamine

(c) Diisopropylamine

When four atoms or groups of atoms are bonded to a nitrogen atom—as, for example, in NH_4^+ and $CH_3NH_3^+$—nitrogen bears a positive charge and is associated with an anion as a salt. The compound is named as a salt of the corresponding amine. The ending -**amine** (or aniline, pyridine, or the like) is replaced with -**ammonium** (or *anilinium*, *pyridinium*, or the like), and the name of the anion (chloride, acetate, and so on) is added.

$$(CH_3CH_2)_3NH^+Cl^-$$

Triethylammonium chloride

16.3   What Are the Physical Properties of Amines?

Like ammonia, low-molecular-weight amines have very sharp, penetrating odors. Trimethylamine, for example, is the pungent principle in the smell of rotting fish. Two other particularly pungent amines are 1,4-butanediamine (putrescine) and 1,5-pentanediamine (cadaverine).

Amines are polar compounds because of the difference in electronegativity between nitrogen and hydrogen $(3.0 - 2.1 = 0.9)$. Both primary and secondary amines have N—H bonds and can form hydrogen bonds with one another (Figure 16.1). Tertiary amines do not have a hydrogen bonded to nitrogen and, therefore, do not form hydrogen bonds with one another.

An N—H----N hydrogen bond is weaker than an O—H----O hydrogen bond, because the difference in electronegativity between nitrogen and hydrogen $(3.0 - 2.1 = 0.9)$ is less than that between oxygen and hydrogen $(3.5 - 2.1 = 1.4)$. To see the effect of hydrogen bonding between alcohols and amines of comparable molecular weight, compare the boiling points of ethane, methanamine, and methanol. Ethane is a nonpolar hydrocarbon, and the only attractive forces between its molecules are weak London dispersion forces (Section 5.7A). Both methanamine and methanol have polar molecules that interact in the pure liquid by hydrogen bonding. Methanol has the highest boiling point of the three compounds, because the hydrogen bonding between its molecules is stronger than that between methanamine molecules.

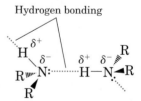

Several over-the-counter mouthwashes contain an *N*-alkylpyridinium chloride as an antibacterial agent.

Hydrogen bonding

FIGURE 16.1 Hydrogen bonding between two molecules of a secondary amine.

	CH$_3$CH$_3$	CH$_3$NH$_2$	CH$_3$OH
Molecular weight (amu)	30.1	31.1	32.0
Boiling point (°C)	−88.6	−6.3	65.0

Tranquilizers

Most people face anxiety and stress at some time in their lives, and each person develops various ways to cope with these factors. Perhaps this strategy involves meditation, exercise, psychotherapy, or drugs. One modern coping technique is to use tranquilizers, drugs that provide relief from the symptoms of anxiety or tension.

The first modern tranquilizers were derivatives of a compound called benzodiazepine. The first of these compounds, chlorodiazepoxide, better known as Librium, was introduced in 1960 and was soon followed by more than two dozen related compounds. Diazepam, better known as Valium, became one of the most widely used of these drugs.

Librium, Valium, and other benzodiazepines are central nervous system sedatives/hypnotics. As sedatives, they diminish activity and excitement, thereby exerting a calming effect. As hypnotics, they produce drowsiness and sleep.

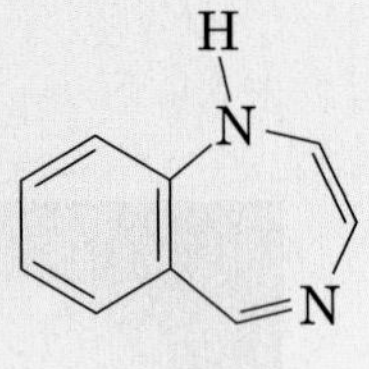

Benzodiazepine

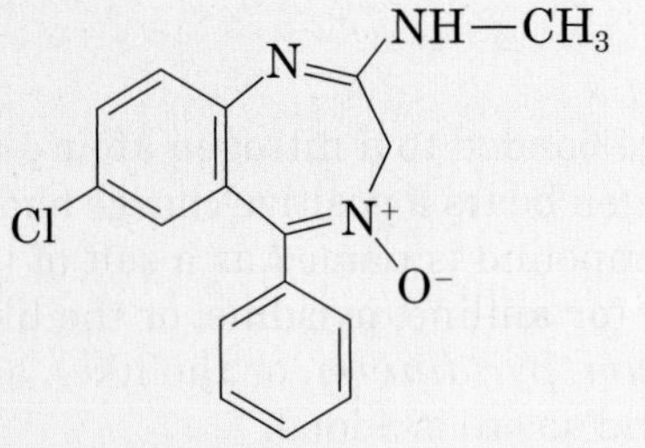

Chlorodiazepoxide
(Librium)

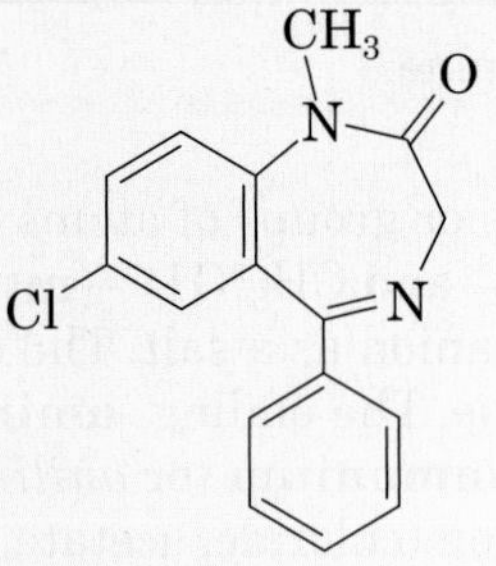

Diazepam
(Valium)

Test your knowledge with Problems 16.39–16.41.

All classes of amines form hydrogen bonds with water and are more soluble in water than are hydrocarbons of comparable molecular weight. Most low-molecular-weight amines are completely soluble in water, but higher-molecular-weight amines are only moderately soluble in water or are insoluble.

16.4 How Do We Describe the Basicity of Amines?

Like ammonia, amines are weak bases, and aqueous solutions of amines are basic. The following acid–base reaction between an amine and water is written using curved arrows to emphasize that in this proton-transfer reaction (Section 8.1), the unshared pair of electrons on nitrogen forms a new covalent bond with hydrogen and displaces a hydroxide ion.

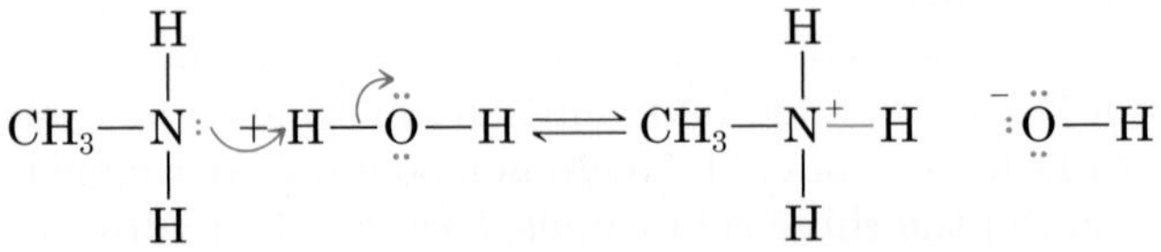

Methylamine
(a base)

Methylammonium
hydroxide

The base dissociation constant, K_b, for the reaction of an amine with water, has the following form, illustrated here for the reaction of methylamine with water to give methylammonium hydroxide. pK_b is defined as the negative logarithm of K_b.

$$K_b = \frac{[CH_3NH_3^+][OH^-]}{[CH_3NH_2]} = 4.37 \times 10^{-4}$$

$$pK_b = -\log 4.37 \times 10^{-4} = 3.360$$

All aliphatic amines have approximately the same base strength, pK_b 3.0 − 4.0, and are slightly stronger bases than ammonia (Table 16.1). Aromatic amines and heterocyclic aromatic amines (pK_b 8.5 − 9.5) are considerably weaker bases than aliphatic amines. One additional point about the basicities of amines: While aliphatic amines are weak bases by comparison with inorganic bases such as NaOH, they are strong bases among organic compounds.

Table 16.1 Approximate Base Strengths of Amines

Class	pK_b	Example	Name	
Aliphatic	3.0–4.0	$CH_3CH_2NH_2$	Ethanamine	Stronger base
Ammonia	4.74			↑
Aromatic	8.5–9.5	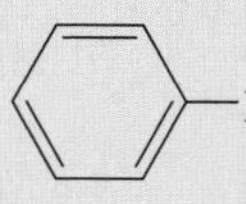 ⬡—NH_2	Aniline	Weaker base

Given the basicities of amines, we can determine which form of an amine exists in body fluids—say, blood. In a normal healthy person, the pH of blood is approximately 7.40, which is slightly basic. If an aliphatic amine is dissolved in blood, it is present predominantly as its protonated or conjugate acid form.

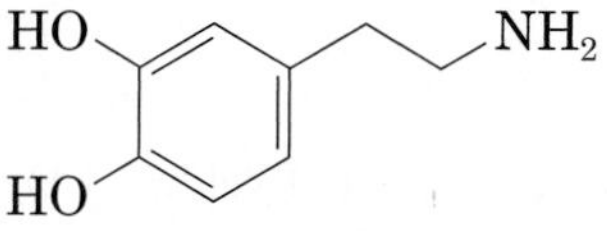

Dopamine 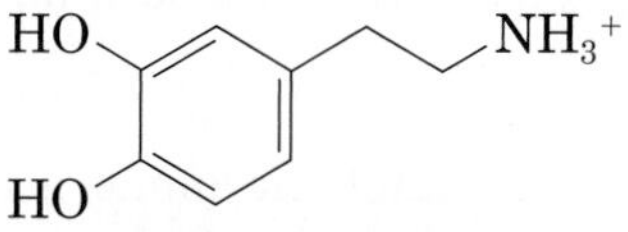   Conjugate acid of dopamine
                           (the major form present
                            in blood plasma)

We can show that an aliphatic amine such as dopamine dissolved in blood is present largely as its protonated or conjugate acid form in the following way. Assume that the amine RNH_2 has a pK_b of 3.50 and that it is dissolved in blood, pH 7.40. We first write the base dissociation constant for the amine and then solve for the ratio of RNH_3^+ to RNH_2.

$$RNH_2 + H_2O \rightleftharpoons RNH_3^+ + OH^-$$

$$K_b = \frac{[RNH_3^+][OH^-]}{[RNH_2]}$$

$$\frac{K_b}{[OH^-]} = \frac{[RNH_3^+]}{[RNH_2]}$$

We now substitute the appropriate values for K_b and $[OH^-]$ in this equation. Taking the antilog of 3.50 gives a K_b of 3.2×10^{-4}. Calculating the concentration of hydroxide requires two steps. First, recall from Section 8.8 that pH + pOH = 14. If the pH of blood is 7.40, then its pOH is 6.60 and its

$[OH^-]$ is 2.5×10^{-7}. Substituting these values in the appropriate equation gives a ratio of 1300 parts RNH_3^+ to 1 part RNH_2.

$$\frac{3.2 \times 10^{-4}}{2.5 \times 10^{-7}} = \frac{[RNH_3^+]}{[RNH_2]} = 1300$$

As this calculation demonstrates, an aliphatic amine present in blood is more than 99.9% in the protonated form. Thus, even though we may write the structural formula for dopamine as the free amine, it is present in blood as the protonated form. It is important to realize, however, that the amine and ammonium ion forms are always in equilibrium, so some of the unprotonated form is nevertheless present in solution.

Aromatic amines, by contrast, are considerably weaker bases than aliphatic amines and are present in blood largely in the unprotonated form. Performing the same type of calculation for an aromatic amine, $ArNH_2$, with a pK_b of approximately 10, we find that the aromatic amine is more than 99.0% in its unprotonated ($ArNH_2$) form.

Example 16.4 Basicity of Amines

Select the stronger base in each pair of amines.

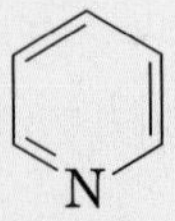

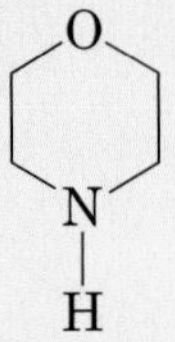

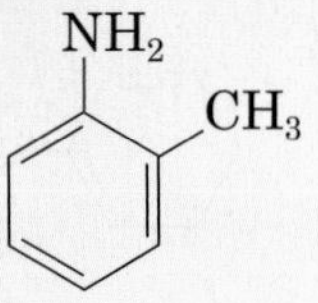

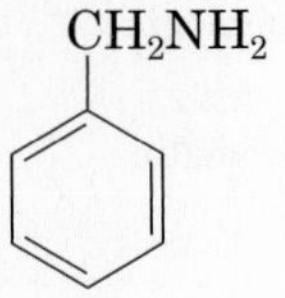

Strategy

Determine whether the amine is an aromatic or an aliphatic amine. Aliphatic amines are stronger bases than aromatic amines.

Solution

(a) Morpholine (B), a 2° aliphatic amine, is the stronger base. Pyridine (A), a heterocyclic aromatic amine, is the weaker base.

(b) Benzylamine (D), a 1° aliphatic amine, is the stronger base. Even though it contains an aromatic ring, it is not an aromatic amine because the amine nitrogen is not bonded to the aromatic ring. o-Toluidine (C), a 1° aromatic amine, is the weaker base.

Problem 16.4

Select the stronger base from each pair of amines.

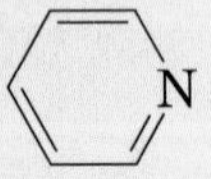

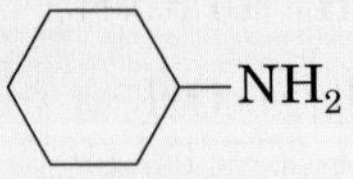

16.5 What Are the Characteristic Reactions of Amines?

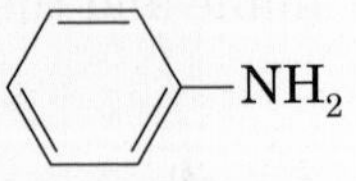

The most important chemical property of amines is their basicity. Amines, whether soluble or insoluble in water, react quantitatively with strong acids to form water-soluble salts, as illustrated by the reaction of (R)-norepinephrine (noradrenaline) with aqueous HCl to form a hydrochloride salt.

The Solubility of Drugs in Body Fluids

Many drugs have "•HCl" or some other acid as part of their chemical formula and occasionally as part of their generic name. Invariably, these drugs are amines that are insoluble in aqueous body fluids such as blood plasma and cerebrospinal fluid. For the administered drug to be absorbed and carried by body fluids, it must be treated with an acid to form a water-soluble ammonium salt. Methadone, a narcotic analgesic, is marketed as its water-soluble hydrochloride salt. Novocain, one of the first local anesthetics, is the hydrochloride salt of procaine.

Methadone ·HCl

Procaine ·HCl
(Novocain, a local anesthetic)

These two drugs are amino salts and are labeled as hydrochlorides.

There is another reason besides increased water solubility for preparing these and other amine drugs as salts. Amines are very susceptible to oxidation and decomposition by atmospheric oxygen, with a corresponding loss of biological activity. By comparison, their amine salts are far less susceptible to oxidation; they retain their effectiveness for a much longer time.

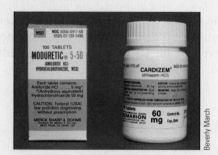

(R)-Norepinephrine
(only slightly soluble in water)

(R)-Norepinephrine hydrochloride
(a water-soluble salt)

$+ \ HCl \xrightarrow{\text{H}_2\text{O}}$

Example 16.5 Basicity of Amines

Complete the equation for each acid–base reaction and name the salt formed.

(a) $(CH_3CH_2)_2NH + HCl \longrightarrow$ (b) ⬡N $+ \ CH_3COOH \longrightarrow$

Strategy

Each acid–base reaction involves a proton transfer from the acid to the amino group (a base). The product is named as an ammonium salt.

Solution

(a) (CH$_3$CH$_2$)$_2$NH$_2$$^+Cl^-$
     Diethylammonium chloride

(b) 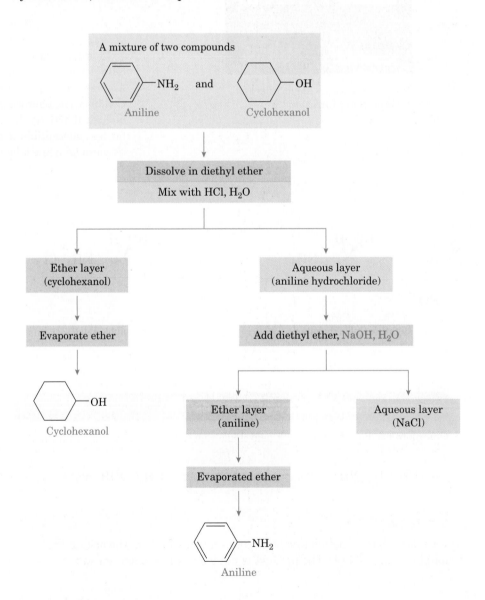 CH$_3$COO$^-$

Pyridinium acetate

Problem 16.5

Complete the equation for each acid–base reaction and name the salt formed.

(a) (CH$_3$CH$_2$)$_3$N + HCl $\longrightarrow$

(b) ⬡NH + CH$_3$COOH $\longrightarrow$

The basicity of amines and the solubility of amine salts in water gives us a way to separate water-insoluble amines from water-insoluble nonbasic compounds. Figure 16.2 is a flowchart for the separation of aniline from cyclohexanol, a neutral compound.

FIGURE 16.2 Separation and purification of an amine and a neutral compound.

Epinephrine: A Prototype for the Development of New Bronchodilators

Epinephrine was first isolated in pure form in 1897 and its structure determined in 1901. It occurs in the adrenal gland (hence the common name adrenalin) as a single enantiomer with the *R* configuration at its stereocenter. Epinephrine is commonly referred to as a catecholamine: the common name of 1,2-dihydroxybenzene is catechol (Section 13.4A), and amines containing a benzene ring with *ortho*-hydroxyl groups are called catecholamines.

Early on, it was recognized that epinephrine is a vasoconstrictor, a bronchodilator, and a cardiac stimulant. The fact that it has these three major effects stimulated much research, one line of which sought to develop compounds that are even more effective bronchodilators than epinephrine but, at the same time, are free from epinephrine's cardiac-stimulating and vasoconstricting effects.

Soon after epinephrine became commercially available, it emerged as an important treatment of asthma and hay fever. It has been marketed for the relief of bronchospasms under several trade names, including Bronkaid Mist and Primatine Mist.

enzyme-catalyzed reaction that converts one of the two —OH groups on the catechol unit to an OCH_3 group. A strategy to circumvent this enzyme-catalyzed inactivation was to replace the catechol unit with one that would allow the drug to bind to the catecholamine receptors in the bronchi but would not be inactivated by this enzyme.

In terbutaline (Brethaire), inactivation is prevented by placing the —OH groups *meta* to each other on the aromatic ring. In addition, the isopropyl group of isoproterenol is replaced with a *tert*-butyl group. In albuterol (Proventil), the commercially most successful of the antiasthma medications, one —OH group of the catechol unit is replaced with a —CH_2OH group and the isopropyl group is replaced with a *tert*-butyl group. When terbutaline and albuterol were introduced into clinical medicine in the 1960s, they almost immediately replaced isoproterenol as the drugs of choice for the treatment of asthmatic attacks. The *R* enantiomer of albuterol is 68 times more effective in the treatment of asthma than the *S* enantiomer.

Epinephrine

Terbutaline

(*R*)-Isoproterenol

(*R*)-Albuterol

One of the most important of the first synthetic catecholamines was isoproterenol, the levorotatory enantiomer of which retains the bronchodilating effects of epinephrine but is free from its cardiac-stimulating effects. (*R*)-Isoproterenol was introduced into clinical medicine in 1951; for the next two decades, it was the drug of choice for the treatment of asthmatic attacks. Interestingly, the hydrochloride salt of (*s*)-isoproterenol is a nasal decongestant and was marketed under several trade names, including Sudafed.

A problem with the first synthetic catecholamines (and with epinephrine itself) is that they are inactivated by an

In their search for a longer-acting bronchodilator, scientists reasoned that extending the side chain on nitrogen might strengthen the binding of the drug to the adrenoreceptors in the lungs, thereby increasing the duration of the drug's action. This line of reasoning led to the synthesis and introduction of salmeterol (Serevent), a bronchodilator that is approximately ten times more potent than albuterol and much longer acting.

Salmeterol

Test your knowledge with Problem 16.44.

Summary of Key Questions

Sign in at **www.cengage.com/owl** to develop problem-solving skills and complete online homework assigned by your professor.

Section 16.1 What Are Amines?

- Amines are classified as **primary**, **secondary**, or **tertiary**, depending on the number of carbon atoms bonded to nitrogen.
- In an **aliphatic amine**, all carbon atoms bonded to nitrogen are derived from alkyl groups.
- In an **aromatic amine**, one or more of the groups bonded to nitrogen are aryl groups.
- In a **heterocyclic amine**, the nitrogen atom is part of a ring.

Section 16.2 How Do We Name Amines?

- In IUPAC nomenclature, aliphatic amines are named by changing the final -**e** of the parent alkane to -**amine** and using a number to locate the amino group on the parent chain.
- In the common system of nomenclature, aliphatic amines are named by listing the carbon groups bonded to nitrogen in alphabetical order in one word ending in the suffix -**amine**.

Section 16.3 What Are the Physical Properties of Amines?

- Amines are polar compounds, and primary and secondary amines associate by intermolecular hydrogen bonding.
- All classes of amines form hydrogen bonds with water and are more soluble in water than are hydrocarbons of comparable molecular weight.

Section 16.4 How Do We Describe the Basicity of Amines?

- Amines are weak bases, and aqueous solutions of amines are basic.
- The base ionization constant for an amine in water is denoted by the symbol K_b.
- Aliphatic amines are stronger bases than aromatic amines.

Section 16.5 What Are the Characteristic Reactions of Amines?

- All amines, whether soluble or insoluble in water, react with strong acids to form water-soluble salts.
- We can use this property to separate water-insoluble amines from water-insoluble nonbasic compounds.

Summary of Key Reactions

1. **Basicity of Aliphatic Amines (Section 16.4)** Most aliphatic amines have about the same basicity (pK_b 3.0 – 4.0) and are slightly stronger bases than ammonia (pK_b 4.74).

$$CH_3NH_2 + H_2O \rightleftharpoons CH_3NH_3^+ + OH^- \quad pK_b = 3.36$$

2. **Basicity of Aromatic Amines (Section 16.4)** Most aromatic amines (pK_b 9.0 – 10.0) are considerably weaker bases than ammonia and aliphatic amines.

3. **Reaction with Acids (Section 16.5)** All amines, whether water-soluble or water-insoluble, react quantitatively with strong acids to form water-soluble salts.

Problems

Interactive versions of these problems may be assigned in OWL.

Orange-numbered problems are applied.

References to previous chapters are given in parentheses.

Section 16.1 What Are Amines?

16.6 Answer true or false.
 (a) *tert*-Butylamine is a 3° amine.
 (b) In an aromatic amine, one or more of the groups bonded to nitrogen is an aromatic ring.
 (c) In a heterocyclic amine, the amine nitrogen is one of the atoms of a ring.
 (d) The Lewis structures of both NH_4^+ and CH_4 show the same number (eight) of valence electrons, and the VSEPR model predicts tetrahedral geometry for each.
 (e) There are four constitutional isomers with the molecular formula C_3H_9N.

16.7 We use the terms *primary*, *secondary*, and *tertiary* to classify both alcohols and amines. Cyclohexanol, for example, is classified as a secondary alcohol, and cyclohexanamine is classified as a primary amine. In each compound, the functional group is bonded to a carbon of the cyclohexane ring. Explain why one compound is classified as secondary, whereas the other is classified as primary.

16.8 What is the difference in structure between an aliphatic amine and an aromatic amine?

16.9 In what way are pyridine and pyrimidine related to benzene?

Section 16.2 How Do We Name Amines?

16.10 Answer true or false.
(a) In the IUPAC system, primary aliphatic amines are named as alkanamines.
(b) The IUPAC name of $CH_3CH_2CH_2CH_2CH_2NH_2$ is 1-pentylamine.
(c) 2-Butanamine is chiral and shows enantiomerism.
(d) *N,N*-Dimethylaniline is a 3° aromatic amine.

16.11 Draw a structural formula for each amine.
(a) 2-Butanamine
(b) 1-Octanamine
(c) 2,2-Dimethyl-1-propanamine
(d) 1,5-Pentanediamine
(e) 2-Bromoaniline
(f) Tributylamine

16.12 Draw a structural formula for each amine.
(a) 4-Methyl-2-pentanamine
(b) *trans*-2-Aminocyclohexanol
(c) *N,N*-Dimethylaniline
(d) Dicyclohexylamine
(e) *sec*-Butylamine
(f) 2,4-Dimethylaniline

16.13 Classify each amino group as primary, secondary, or tertiary and as aliphatic or aromatic.

(a)

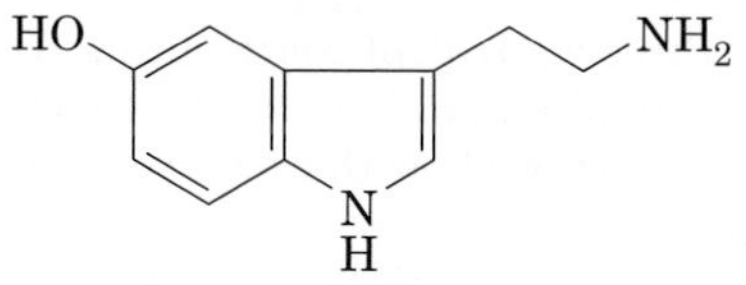

Serotonin
(a neurotransmitter)

(b)

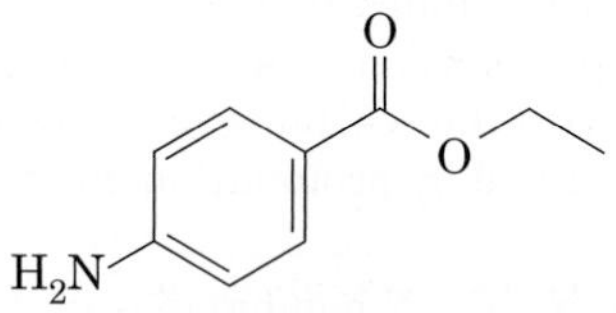

Benzocaine
(a topical anesthetic)

(c)

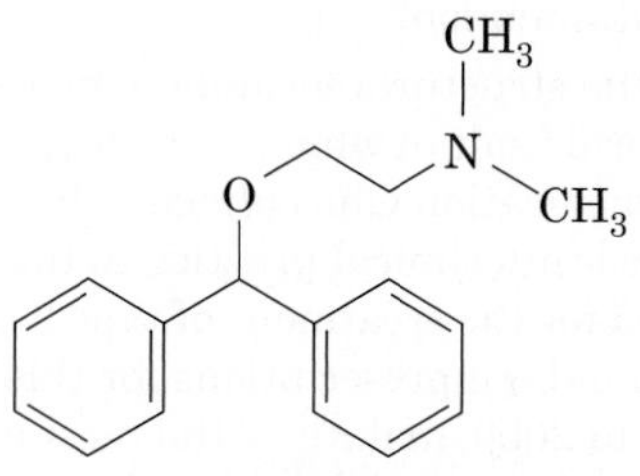

Diphenhydramine
(the hydrochloride salt is the antihistamine Benadryl)

(d)

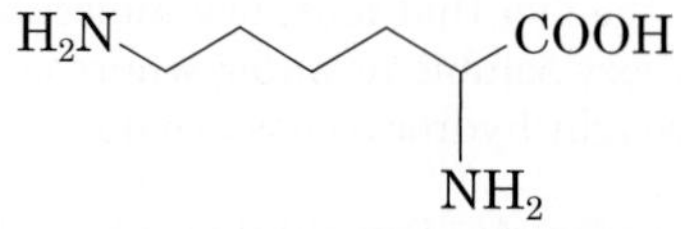

Lysine
(an amino acid)

(e)

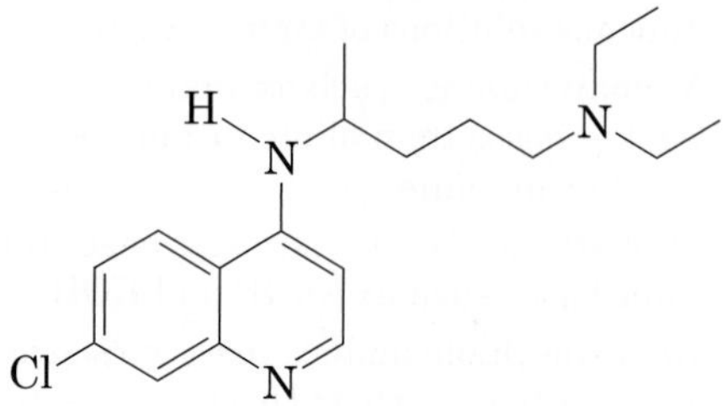

Chloroquine
(an antimalaria drug)

(f) H_2N⌐COOH

4-Aminobutanoic acid
(a neurotransmitter)

16.14 There are eight constitutional isomers with the molecular formula $C_4H_{11}N$.
(a) Name and draw a structural formula for each amine.
(b) Classify each amine as primary, secondary, or tertiary.
(c) Which are chiral?

16.15 There are eight primary amines with the molecular formula $C_5H_{13}N$.
(a) Name and draw a structural formula for each amine.
(b) Which are chiral?

Section 16.3 What Are the Physical Properties of Amines?

16.16 Answer true or false.
(a) Hydrogen bonding between 2° amines is stronger than that between 2° alcohols.
(b) Primary and secondary amines generally have higher boiling points than hydrocarbons with comparable carbon skeletons.
(c) The boiling points of amines increase as the molecular weight of the amine increases.

16.17 Propylamine (bp 48°C), ethylmethylamine (bp 37°C), and trimethylamine (bp 3°C) are constitutional isomers with the molecular formula C_3H_9N. Account for the fact that trimethylamine has the lowest boiling point of the three and propylamine has the highest boiling point.

16.18 Account for the fact that 1-butanamine (bp 78°C) has a lower boiling point than 1-butanol (bp 117°C).

16.19 2-Methylpropane (bp −12°C), 2-propanol (bp 82°C), and 2-propanamine (bp 32°C) all have approximately the same molecular weight, yet their boiling points are quite different. Explain the reason for these differences.

16.20 Account for the fact that most low-molecular-weight amines are very soluble in water, whereas low-molecular-weight hydrocarbons are not.

Section 16.4 How Do We Describe the Basicity of Amines?

16.21 Answer true or false.

(a) Aqueous solutions of amines are basic.

(b) Aromatic amines such as aniline in general are weaker bases than aliphatic amines such as cyclohexanamine.

(c) Aliphatic amines are stronger bases than inorganic bases such as NaOH and KOH.

(d) Water-insoluble amines react with strong aqueous acids such as HCl to form water-soluble salts.

(e) If the pH of an aqueous solution of a 1° aliphatic amine, RNH_2, is adjusted to pH 2.0 by the addition of concentrated HCl, the amine will be present in solution almost entirely as its conjugate acid, RNH_3^+.

(f) If the pH of an aqueous solution of a 1° aliphatic amine, RNH_2, is adjusted to pH 10.0 by the addition of NaOH, the amine will be present in solution almost entirely as the free base, RNH_2.

(g) For a 1° aliphatic amine, the concentrations of RNH_3^+ and RNH_2 will be equal when the pH of the solution is equal to the pK_b of the amine.

16.22 Compare the base strengths of amines with those of alcohols.

16.23 Write a structural formula for each amine salt.

(a) Ethyltrimethylammonium hydroxide

(b) Dimethylammonium iodide

(c) Tetramethylammonium chloride

(d) Anilinium bromide

16.24 Name these amine salts.

(a) $CH_3CH_2NH_3^+Cl^-$

(b) $(CH_3CH_2)_2NH_2^+Cl^-$

(c) $-NH_3^+HSO_4^-$

16.25 From each pair of compounds, select the stronger base.

(a) or

(b) or

(c) or

16.26 The pK_b of amphetamine is approximately 3.2.

Amphetamine

(a) Which form of amphetamine (the base or its conjugate acid) would you expect to be present at pH 1.0, the pH of stomach acid?

(b) Which form of amphetamine would you expect to be present at pH 7.40, the pH of blood plasma?

16.27 Guanidine, pK_a 13.6, is a very strong base, almost as basic as hydroxide ion.

$$pK_b = 13.6$$

(a) Complete the Lewis structure for guanidine, showing all valence electrons.

(b) The remarkable basicity of guanidine is attributed to the fact that the positive charge on the guanidinium ion is delocalized by resonance over the three nitrogen atoms. This delocalization increases the stability of the guanidinium ion relative to the ammonium ion or substituted ammonium ions.

Draw three equivalent contributing structures for the guanidinium ion and show by the use of curved arrows how these three contributors are related.

(c) Propose an explanation for the fact that protonation occurs on the C=NH nitrogen rather than on one of the —NH_2 nitrogens. (*Hint:* Consider the resonance stabilization of the structure formed by protonation on a —NH_2 nitrogen compared with the resonance stabilization of the structure formed by protonation on the =NH nitrogen.)

(d) Predict the N—C—N bond angles in the hybrid.

(e) Which is the stronger acid, the ammonium ion or the guanidinium ion?

16.28 Following is the structural formula of metformin, the hydrochloride salt of which is marketed as the antidiabetic medication Glucophage. Metformin was introduced into clinical practice in the United States in 1995 for the treatment of type 2 diabetes. More than 25 million prescriptions for this drug were written in 2000, making it the most commonly prescribed brand-name diabetes medication in the nation.

Metformin

(a) Complete the Lewis structure for metformin, showing all valence electrons.

(b) Which nitrogen is the most likely site of protonation?

(c) Draw the structural formula of Glucophage.

Section 16.5 What Are the Characteristic Reactions of Amines?

16.29 Suppose you have two test tubes, one containing 2-methylcyclohexanol and the other containing 2-methylcyclohexanamine (both of which are insoluble in water) and that you do not know which test tube contains which compound. Describe a simple chemical test by which you could tell which compound is the alcohol and which is the amine.

16.30 Complete the equations for the following acid–base reactions.

(a) CH_3COH +

Acetic acid   Pyridine

(b)   + HCl ⟶

1-Phenyl-2-propanamine
(Amphetamine)

(c)   + H_2SO_4 ⟶

Methamphetamine

16.31 Pyridoxamine is one form of vitamin B_6.

Pyridoxamine
(Vitamin B_6)

(a) Which nitrogen atom of pyridoxamine is the stronger base?

(b) Draw a structural formula for the salt formed when pyridoxamine is treated with one mole of HCl.

16.32 Many tumors of the breast are correlated with estrogen levels in the body. Drugs that interfere with estrogen binding have antitumor activity and may even help prevent tumor occurrence. A widely used antiestrogen drug is tamoxifen.

Tamoxifen

(a) Name the functional groups in tamoxifen.

(b) Classify the amino group in tamoxifen as primary, secondary, or tertiary.

(c) How many stereoisomers are possible for tamoxifen?

(d) Would you expect tamoxifen to be soluble or insoluble in water? In blood?

Chemical Connections

16.33 (Chemical Connections 16A) What are the differences in structure between the natural hormone epinephrine (Chemical Connections 16E) and the synthetic pep pill amphetamine? Between amphetamine and methamphetamine?

16.34 (Chemical Connections 16A) What are the possible negative effects of illegal use of amphetamines such as methamphetamine?

16.35 (Chemical Connections 16B) What is an alkaloid? Are all alkaloids basic to litmus?

16.36 (Chemical Connections 16B) Identify all stereocenters in coniine and nicotine. How many stereoisomers are possible for each?

16.37 (Chemical Connections 16B) Which of the two nitrogen atoms in nicotine is converted to its salt by reaction with one mole of HCl? Draw a structural formula for this salt.

16.38 (Chemical Connections 16B) Cocaine has four stereocenters. Identify each. Draw a structural formula for the salt formed by treatment of cocaine with one mole of HCl.

16.39 (Chemical Connections 16C) What structural feature is common to all benzodiazepines?

16.40 (Chemical Connections 16C) Is Librium chiral? Is Valium chiral?

16.41 (Chemical Connections 16C) Benzodiazepines affect neural pathways in the central nervous system that are mediated by GABA, whose IUPAC name is 4-aminobutanoic acid. Draw a structural formula for GABA.

16.42 (Chemical Connections 16D) Suppose you saw this label on a decongestant: phenylephrine · HCl. Should you worry about being exposed to a strong acid such as HCl? Explain.

16.43 (Chemical Connections 16D) Give two reasons why amine-containing drugs are most commonly administered as their salts.

16.44 (Chemical Connections 16E) Classify each amino group in epinephrine and albuterol as primary, secondary, or tertiary. In addition, list the similarities and differences between the structural formulas of these two compounds.

Additional Problems

16.45 Draw a structural formula for a compound with the given molecular formula that is:

(a) A 2° aromatic amine, C_7H_9N

(b) A 3° aromatic amine, $C_8H_{11}N$

(c) A 1° aliphatic amine, C_7H_9N

(d) A chiral 1° amine, $C_4H_{11}N$

(e) A 3° heterocyclic amine, $C_5H_{11}N$

(f) A trisubstituted 1° aromatic amine, $C_9H_{13}N$

(g) A chiral quaternary ammonium salt, $C_9H_{22}NCl$

16.46 Arrange these three compounds in order of decreasing ability to form intermolecular hydrogen bonds: CH_3OH, CH_3SH, and $(CH_3)_2NH$.

16.47 Consider these three compounds: CH_3OH, CH_3SH, and $(CH_3)_2NH$.

(a) Which is the strongest acid?

(b) Which is the strongest base?

(c) Which has the highest boiling point?

(d) Which forms the strongest intermolecular hydrogen bonds in the pure state?

16.48 Arrange these compounds in order of increasing boiling point: $CH_3CH_2CH_2CH_3$, $CH_3CH_2CH_2OH$, and $CH_3CH_2CH_2NH_2$. Boiling point values from lowest to highest are $-0.5°C$, $7.2°C$, and $77.8°C$.

16.49 Account for the fact that amines have about the same solubility in water as alcohols of similar molecular weight.

16.50 If you dissolve $CH_3CH_2CH_2OH$ and $CH_3CH_2CH_2NH_2$ in the same container of water and lower the pH of the solution to 2.0 by adding HCl, would anything happen to the structures of these compounds? Write the formula of the species present in solution at pH 2.0.

16.51 The compound phenylpropanolamine hydrochloride is used as both a decongestant and an anorexic. The IUPAC name of this compound is 1-phenyl-2-amino-1-propanol.

(a) Draw a structural formula for 1-phenyl-2-amino-1-propanol.

(b) How many stereocenters are present in this molecule? How many stereoisomers are possible for it?

16.52 Procaine was one of the first local anesthetics. Its hydrochloride salt is marketed as Novocain.

Procaine

(a) Is procaine chiral? Does it contain a stereocenter? (Chapter 15)

(b) Which nitrogen atom of procaine is the stronger base?

(c) Draw a structural formula for the salt formed by treating procaine with one mole of HCl, showing which nitrogen is protonated and bears the positive charge.

16.53 Following are two structural formulas for gabapentin, a drug used to treat epilepsy. Is gabapentin better represented by structure (A) or (B)? Explain.

$-NH_3^+$ or $-NH_2$

$-COO^-$ $-COOH$

(A)   (B)

16.54 Several poisonous plants, including *Atropa belladonna*, contain the alkaloid atropine. The name "belladonna" (which means "beautiful lady") probably comes from the fact that Roman women used extracts from this plant to make themselves more attractive. Atropine is widely used by ophthalmologists and optometrists to dilate the pupils for eye examination.

Atropine

(a) Classify the amino group in atropine as primary, secondary, or tertiary. (Chapter 10)

(b) Locate all stereocenters in atropine. (Chapter 15)

(c) Account for the fact that atropine is almost insoluble in water (1 g in 455 mL of cold water) but atropine hydrogen sulfate is very soluble (1 g in 5 mL of cold water). (Chapter 6)

(d) Account for the fact that a dilute aqueous solution of atropine is basic (pH approximately 10.0). (Chapter 8)

16.55 Epibatadine, a colorless oil isolated from the skin of the Equadorian poison arrow frog *Epipedobates tricolor*, has several times the analgesic potency of morphine. It is the first chlorine-containing, non-opioid (nonmorphine-like in structure) analgesic ever isolated from a natural source.

(a) Which of the two nitrogen atoms in epibatadine is the stronger base?

(b) Mark the three stereocenters in this molecule. (Chapter 15)

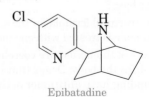

Epibatadine

16.56 Following are two structural formulas for 4-amino-butanoic acid, a neurotransmitter. Is this compound better represented by structural formula (A) or (B)? Explain.

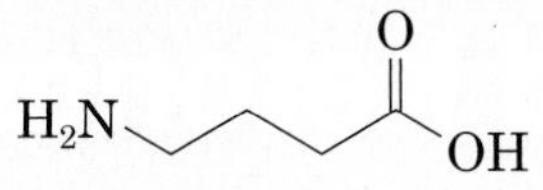

(A)   or   (B)

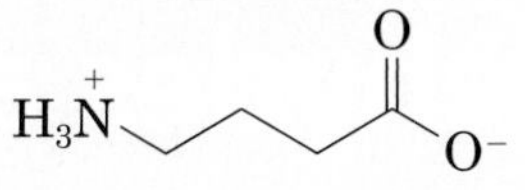

16.57 Alanine, $C_3H_7O_2N$, is one of the 20 amino acid building blocks of proteins (Chapter 22). Alanine contains a primary amino group (—NH_2) and a carboxyl group (—COOH) and has one stereocenter. Given this information, draw a structural formula for alanine.

Challenge Problems

16.58 Following is a structural formula of desosamine, a sugar component of several macrolide antibiotics, including the erythromycins. The configuration shown here is that of the natural product. Erythromycin is produced by a strain of *Streptomyces erythreus* originally found in a soil sample from the Philippine Archipelago.

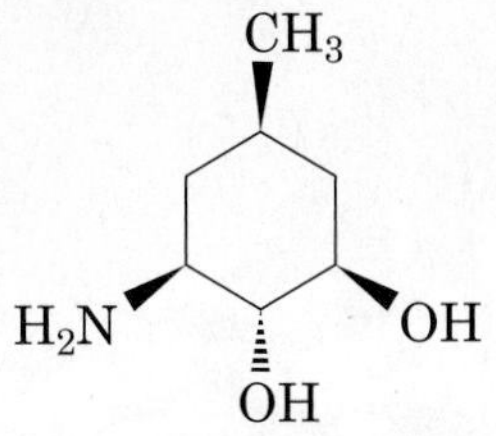

Desosamine

(a) Name all the functional groups in desosamine. (Chapter 10)

(b) How many chiral centers are present in desosamine? How many stereoisomers are possible for it? How many pairs of enantiomers are possible for it? (Chapter 15)

(c) Draw the alternative chair conformations for desosamine and label which groups are equatorial and which are axial. (Chapter 11)

(d) Which of the alternative chair conformations for desosamine is more stable? (Chapter 11)

16.59 Morphine and its O-methylated derivative, codeine, are among the most effective painkillers known. However, they possess two serious drawbacks: they are addictive, and repeated use induces a tolerance to the drug.

R = H; Morphine
R = CH_3; Codeine

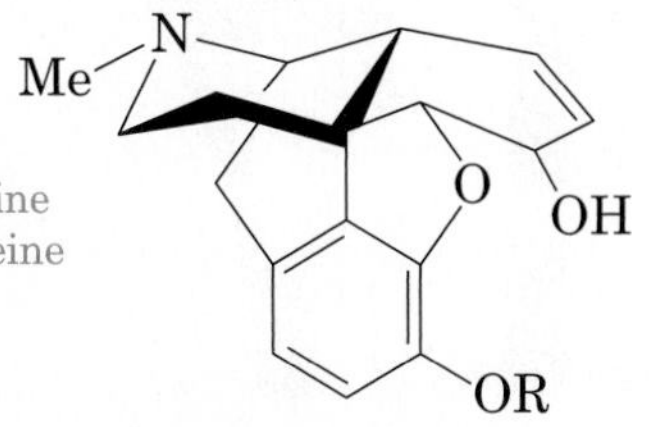

Many morphine analogs have been prepared in an effort to find drugs that are equally effective as painkillers but that have less risk of physical dependence and potential abuse. Following are three of these.

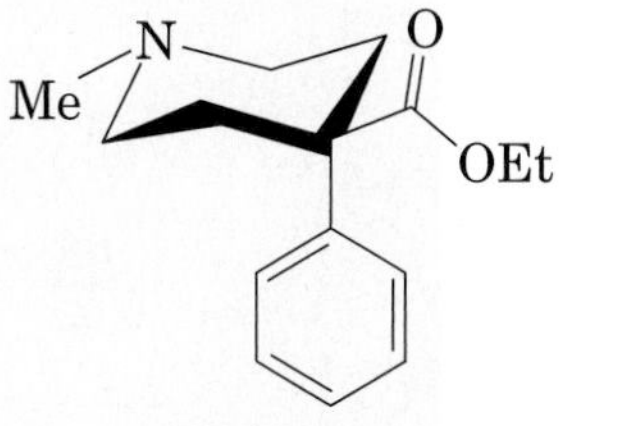

Meperidine
(Demerol)

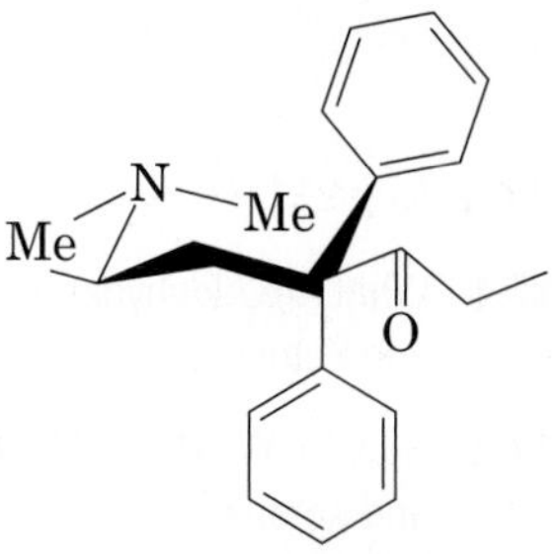

Methadone

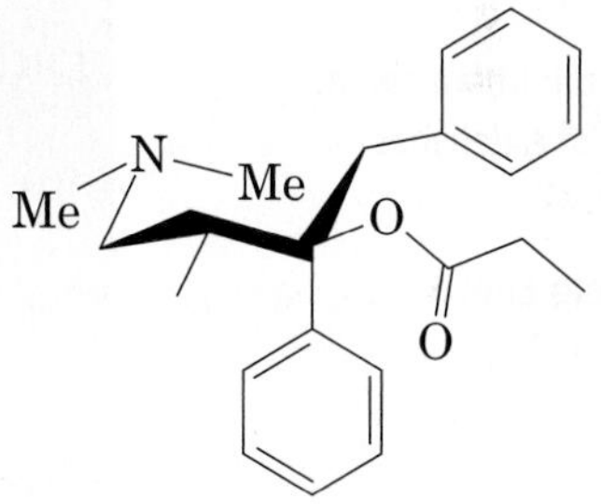

Propoxyphene
(Darvon)

(a) List the structural features common to each of these molecules. (Chapters 10–11, 13–15)

(b) The Beckett-Casey rules are a set of empirical rules to predict the structure of molecules that will bind to morphine receptors and act as analgesics. According to these rules, to provide an effective morphine-like analgesic, a molecule must have (1) an aromatic ring bonded to (2) a quaternary carbon and (3) a nitrogen at a distance equal to two carbon–carbon single bond lengths from the quaternary center. Show that these structural requirements are present in the four molecules shown in this problem. (Chapters 10, 13)

17

Aldehydes and Ketones

Charles D. Winters

Benzaldehyde is found in the kernels of bitter almonds, and cinnamaldehyde is found in Ceylonese and Chinese cinnamon oils.

OWL

Sign in to OWL at www.cengage.com/owl to view tutorials and simulations, develop problem-solving skills, and complete online homework assigned by your professor.

17.1 What Are Aldehydes and Ketones?

In this and the three following chapters, we study the physical and chemical properties of compounds containing the **carbonyl group**, C=O. Because the carbonyl group is present in aldehydes, ketones, and carboxylic acids and their derivatives as well as in carbohydrates, it is one of the most important functional groups in organic chemistry. Its chemical properties are straightforward, and an understanding of its characteristic reaction patterns leads very quickly to an understanding of a wide variety of organic and biochemical reactions.

The functional group of an **aldehyde** is a carbonyl group bonded to a hydrogen atom (Section 10.4C). In methanal, the simplest aldehyde, the carbonyl group is bonded to two hydrogen atoms. In other aldehydes, it is bonded to one hydrogen atom and one carbon atom. The functional group of a **ketone** is a carbonyl group bonded to two carbon atoms (Section 10.4C). Acetone is the simplest ketone.

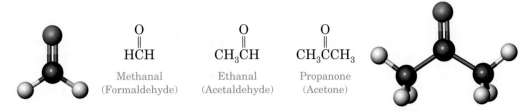

$$HCH$$
Methanal
(Formaldehyde)

$$CH_3CH$$
Ethanal
(Acetaldehyde)

$$CH_3CCH_3$$
Propanone
(Acetone)

Because aldehydes always contain at least one hydrogen bonded to the C=O group, they are often written RCH=O or RCHO. Similarly, ketones are often written RCOR′.

17.2 How Do We Name Aldehydes and Ketones?

A. IUPAC Names

The IUPAC names for aldehydes and ketones follow the familiar pattern of selecting as the parent alkane the longest chain of carbon atoms that contains the functional group (Section 11.3A). To name an aldehyde, we change the suffix *-e* of the parent alkane to *-al*. Because the carbonyl group of an aldehyde can appear only at the end of a parent chain and numbering must start with it as carbon 1, there is no need to use a number to locate the aldehyde group.

For **unsaturated aldehydes**, we show the presence of the carbon–carbon double bond and the aldehyde by changing the ending of the parent alkane from *-ane* to *-enal*: "-en-" to show the carbon–carbon double bond and "-al" to show the aldehyde. We show the location of the carbon–carbon double bond by the number of its first carbon.

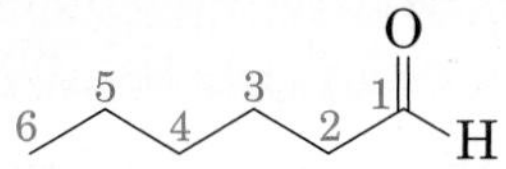

Hexanal 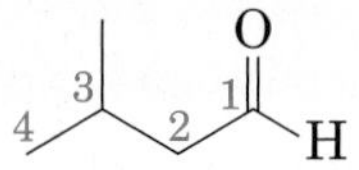   3-Methylbutanal 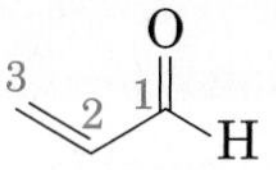   2-Propenal
                                           (Acrolein)

In the IUPAC system, we name ketones by selecting as the parent alkane the longest chain that contains the carbonyl group and then indicating the presence of this group by changing the *-e* of the parent alkane to *-one*. The parent chain is numbered from the direction that gives the smaller number to the carbonyl carbon. While the systematic name of the simplest ketone is 2-propanone, the IUPAC retains its common name, acetone.

Acetone   5-Methyl-3-hexanone   2-Methylcyclohexanone

⊘WL Interactive Example 17.1 IUPAC Names for Aldehydes and Ketones

Write the IUPAC name for each compound:

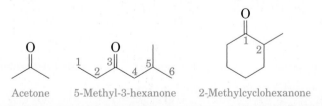

(a) 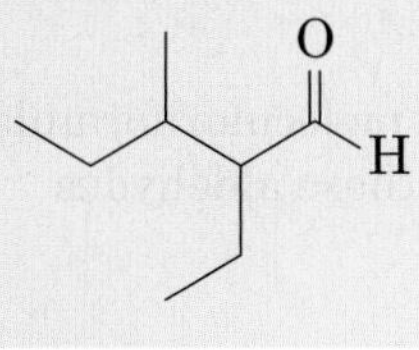   (b) 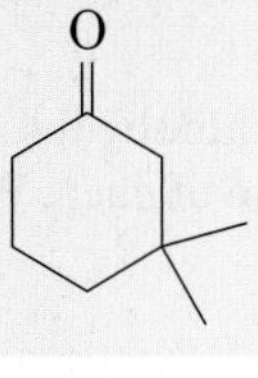   (c)

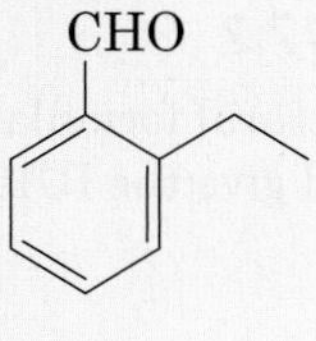

Strategy and Solution

(a) The longest chain has six carbons, but the longest chain that contains the carbonyl carbon has only five carbons. Its IUPAC name is 2-ethyl-3-methylpentanal.

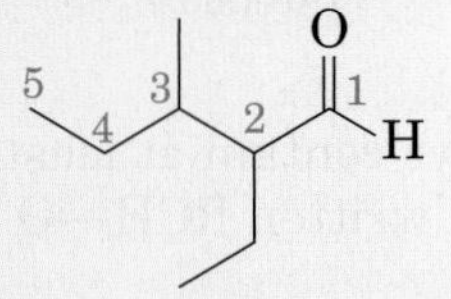

2-Ethyl-3-methylpentanal

(b) Number the six-membered ring beginning with the carbonyl carbon. Its IUPAC name is 3,3-dimethylcyclohexanone.

(c) This molecule is derived from benzaldehyde. Its IUPAC name is 2-ethylbenzaldehyde.

Problem 17.1

Write the IUPAC name for each compound.

(a) 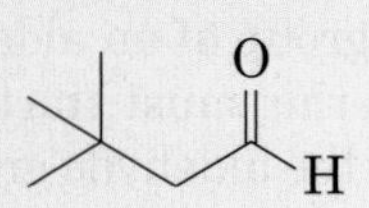   (b) 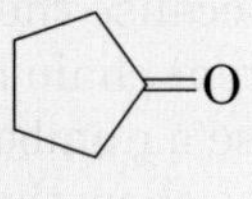   (c)

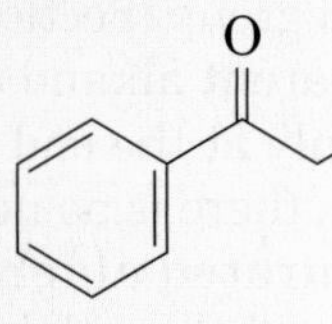

Example 17.2 Structural Formulas for Ketones

Write structural formulas for all ketones with the molecular formula $C_6H_{12}O$ and give the IUPAC name of each. Which of these ketones are chiral?

Strategy and Solution

There are six ketones with this molecular formula: two with a six-carbon chain, three with a five-carbon chain and a methyl branch, and one with a four-carbon chain and two methyl branches. Only 3-methyl-2-pentanone has a stereocenter and is chiral.

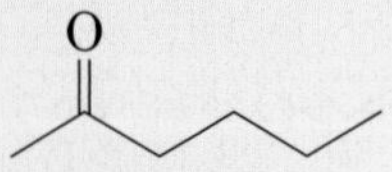

2-Hexanone 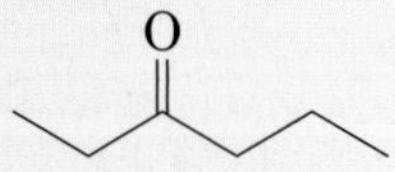   3-Hexanone 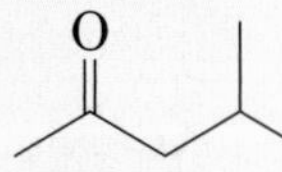   4-Methyl-2-pentanone

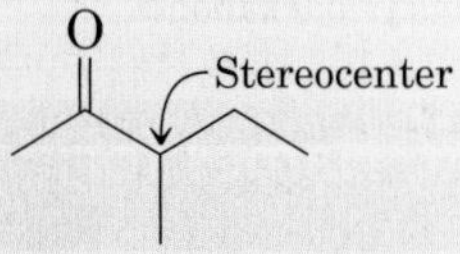

3-Methyl-2-pentanone 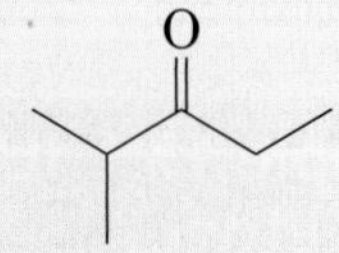   2-Methyl-3-pentanone 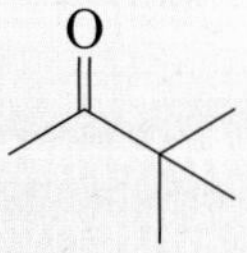   3,3-Dimethyl-2-butanone

Problem 17.2

Write structural formulas for all aldehydes with the molecular formula $C_6H_{12}O$ and give the IUPAC name of each. Which of these aldehydes are chiral?

In naming aldehydes or ketones that also contain an —OH or —NH$_2$ group elsewhere in the molecule, the parent chain is numbered to give the carbonyl group the lower number. An —OH substituent is indicated by *hydroxy*, and an —NH$_2$ substituent is indicated by *amino-*. Hydroxy and amino substituents are numbered and alphabetized along with any other substituents that might be present.

Example 17.3 Naming Difunctional Aldehydes and Ketones

Write the IUPAC name for each compound.

(a) 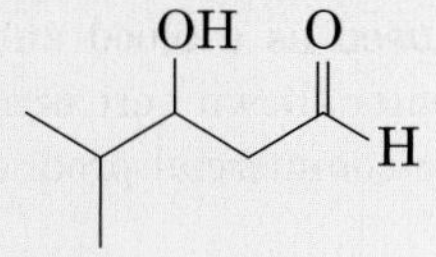 (b)

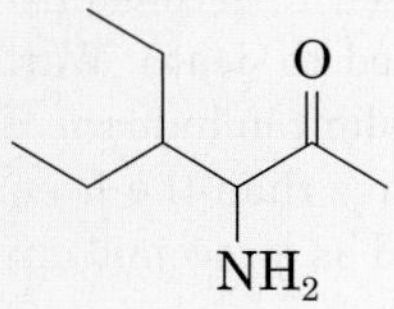

Strategy and Solution

(a) We number the parent chain beginning with CHO as carbon 1. There is a hydroxyl group on carbon 3 and a methyl group on carbon 4. The IUPAC name of this compound is 3-hydroxy-4-methylpentanal. Note that this hydroxyaldehyde is chiral and can exist as a pair of enantiomers.

(b) The longest chain that contains the carbonyl is six carbons; the carbonyl group is on carbon 2, and the amino group is on carbon 3. The IUPAC name of this compound is 3-amino-4-ethyl-2-hexanone. Note that this ketoamine is also chiral and can exist as a pair of enantiomers.

Problem 17.3

Write the IUPAC name for each compound.

(a) (b) 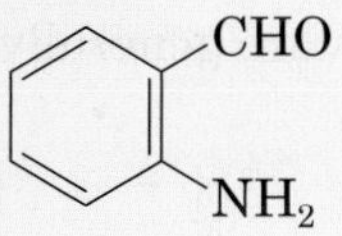 (c)

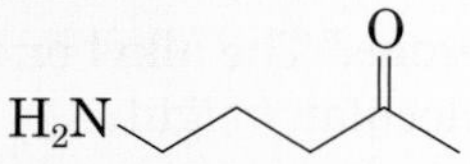

B. Common Names

We derive the common name for an aldehyde from the common or IUPAC names of the corresponding carboxylic acid. The word "acid" is dropped and the suffix *-ic* or *-oic* is changed to *-aldehyde*. Because we have not yet studied names for carboxylic acids, we are not in a position to discuss common names for aldehydes. We can, however, illustrate how they are derived by reference to two common names with which you are familiar. The name formaldehyde is derived from formic acid, and the name acetaldehyde is derived from acetic acid.

| Formaldehyde | Formic acid | Acetaldehyde | Acetic acid |

We derive common names for ketones by naming each alkyl or aryl group bonded to the carbonyl group as a separate word, followed by the word

Chemical Connections 17A

From Moldy Clover to a Blood Thinner

In 1933, a disgruntled farmer delivered a pail of unclotted blood to the laboratory of Dr. Karl Link at the University of Wisconsin, along with tales of cows bleeding to death from minor cuts. Over the next couple of years, Link and his collaborators discovered that when cows are fed moldy clover, their blood clotting is inhibited and they bleed to death from minor cuts and scratches. From the moldy clover, they isolated the anticoagulant dicoumarol, a substance that delays or prevents blood clotting. Dicoumarol exerts its anticoagulation effect by interfering with vitamin K activity. Within a few years after its discovery, dicoumarol became widely used to treat victims of heart attack and others at risk for developing blood clots.

Dicoumarol is a derivative of coumarin, a lactone that gives sweet clover its pleasant smell. Coumarin, which does not interfere with blood clotting, is converted to dicoumarol as sweet clover becomes moldy.

In a search for even more potent anticoagulants, Link developed warfarin (named for the Wisconsin Alumni Research Foundation), now used primarily as a rat poison. When rats consume it, their blood fails to clot and they bleed to death. Warfarin is also used as a blood anticoagulant in humans. The *S* enantiomer shown here is more active than the *R* enantiomer. The commercial product is sold as a racemic mixture.

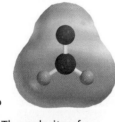

Coumarin
(from sweet clover)

as sweet clover
becomes moldy

Dicoumarol
(an anticoagulant)

(*S*)-Warfarin
(a synthetic anticoagulant)

Test your knowledge with Problem 17.46.

"ketone." The alkyl or aryl groups are generally listed in order of increasing molecular weight.

Ethyl isopropyl ketone   Methyl ethyl ketone   Dicyclohexyl ketone

2-Butanone, more commonly called methyl ethyl ketone (MEK), is used as a solvent for paints and varnishes.

17.3 What Are the Physical Properties of Aldehydes and Ketones?

Oxygen is more electronegative than carbon (3.5 compared with 2.5; see Table 3.5). Therefore, a carbon–oxygen double bond is polar, with oxygen bearing a partial negative charge and carbon bearing a partial positive charge (Figure 17.1).

In liquid aldehydes and ketones, intermolecular attractions occur between the partial positive charge on the carbonyl carbon of one molecule and the partial negative charge on the carbonyl oxygen of another molecule. There is no possibility for hydrogen bonding between aldehyde or ketone molecules, which explains why these compounds have lower boiling points than alcohols (Section 14.1C) and carboxylic acids (Section 18.3D), compounds in which hydrogen bonding between molecules does occur.

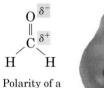

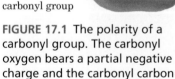

Polarity of a
carbonyl group

FIGURE 17.1 The polarity of a carbonyl group. The carbonyl oxygen bears a partial negative charge and the carbonyl carbon bears a partial positive charge.

Table 17.1 Boiling Points of Six Compounds of Comparable Molecular Weight

Name	Structural Formula	Molecular Weight	Boiling Point (°C)
diethyl ether	$CH_3CH_2OCH_2CH_3$	74	34
pentane	$CH_3CH_2CH_2CH_2CH_3$	72	36
butanal	$CH_3CH_2CH_2CHO$	72	76
2-butanone	$CH_3CH_2COCH_3$	72	80
1-butanol	$CH_3CH_2CH_2CH_2OH$	74	117
propanoic acid	CH_3CH_2COOH	74	141

Table 17.1 lists structural formulas and boiling points of six compounds of similar molecular weight. Of the six, pentane and diethyl ether have the lowest boiling points. The boiling point of 1-butanol, which can associate by intermolecular hydrogen bonding, is higher than that of either butanal or 2-butanone. Propanoic acid, in which intermolecular association by hydrogen bonding is the strongest, has the highest boiling point.

Because the oxygen atom of each carbonyl group is a hydrogen bond acceptor, the low-molecular-weight aldehydes and ketones are more soluble in water than are nonpolar compounds of comparable molecular weight. Formaldehyde, acetaldehyde, and acetone are infinitely soluble in water. As the hydrocarbon portion of the molecule increases in size, aldehydes and ketones become less soluble in water.

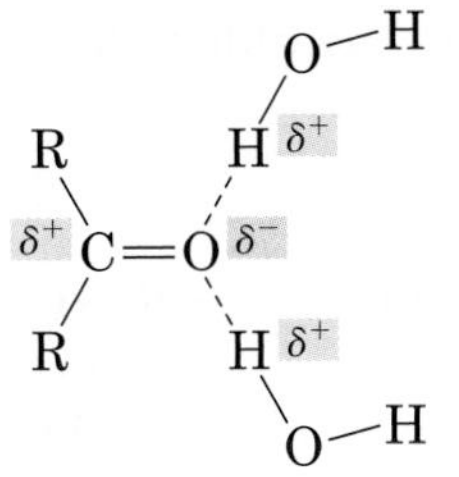

Most aldehydes and ketones have strong odors. The odors of ketones are generally pleasant, and many are used in perfumes and as flavoring agents. The odors of aldehydes vary. You may be familiar with the smell of formaldehyde; if so, you know that it is not pleasant. Many higher aldehydes, however, have pleasant odors and are used in perfumes.

17.4 What Are the Characteristic Reactions of Aldehydes and Ketones?

A. Oxidation

Aldehydes are oxidized to carboxylic acids by a variety of oxidizing agents, including potassium dichromate (Section 14.2C).

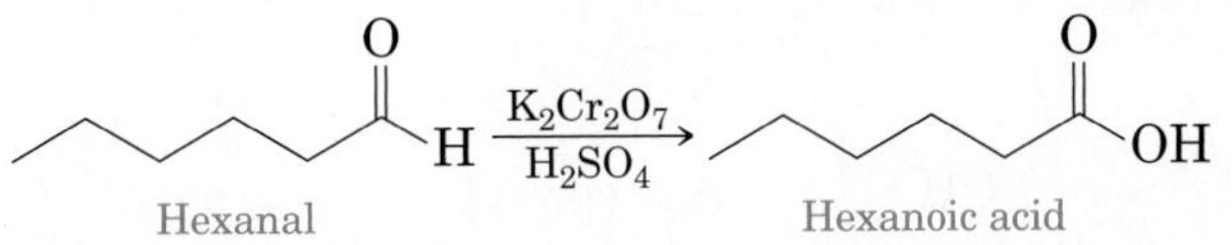

Hexanal — Hexanoic acid

The body uses nicotinamide adenine dinucleotide, NAD⁺, for this type of oxidation (Section 27.3).

Aldehydes are also oxidized to carboxylic acids by the oxygen in the air. In fact, aldehydes that are liquid at room temperature are so sensitive to oxidation that they must be protected from contact with air during storage. Often this is done by sealing the aldehyde in a container under an atmosphere of nitrogen.

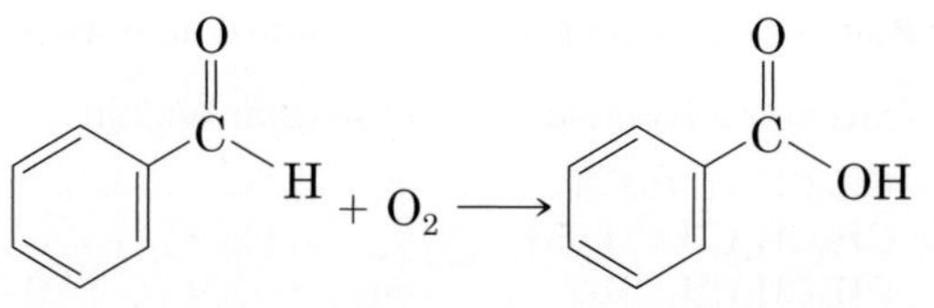

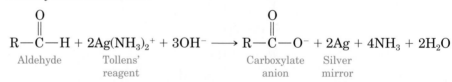

Benzaldehyde + $O_2 \longrightarrow$ Benzoic acid

Ketones, in contrast, resist oxidation by most oxidizing agents, including potassium dichromate and molecular oxygen.

The fact that aldehydes are so easy to oxidize and ketones are not allows us to use simple chemical tests to distinguish between these types of compounds. Suppose that we have a compound we know is either an aldehyde or a ketone. To determine which it is, we can treat the compound with a mild oxidizing agent. If it can be oxidized, it is an aldehyde; otherwise, it is a ketone. One reagent that has been used for this purpose is Tollens' reagent.

Tollens' reagent is prepared from silver nitrate and ammonia in water. When these two compounds are mixed, silver ion combines with NH_3 to form the complex ion $Ag(NH_3)_2^+$. When this solution is added to an aldehyde, the aldehyde acts as a reducing agent and reduces the complexed silver ion to silver metal. If this reaction is carried out properly, the silver metal precipitates as a smooth, mirror-like deposit on the inner surface of the reaction vessel, leading to the name **silver-mirror test**. If the remaining solution is then acidified with HCl, the carboxylic anion $RCOO^-$ formed during the aldehyde's oxidation is converted to the carboxylic acid RCOOH.

A silver mirror has been deposited on the inside of this flask by the reaction between an aldehyde and Tollens' reagent.

$$R-\overset{\overset{\displaystyle O}{\|}}{C}-H + 2Ag(NH_3)_2^+ + 3OH^- \longrightarrow R-\overset{\overset{\displaystyle O}{\|}}{C}-O^- + 2Ag + 4NH_3 + 2H_2O$$

Aldehyde   Tollens' reagent   Carboxylate anion   Silver mirror

Today, silver(I) is rarely used for the oxidation of aldehydes because of its high cost and because of the availability of other more convenient methods for this oxidation. This reaction, however, is still used for making (silvering) mirrors.

Example 17.4 Oxidation of Aldehydes and Ketones

Draw a structural formula for the product formed by treating each compound with Tollens' reagent followed by acidification with aqueous HCl.

(a) Pentanal   (b) 4-Hydroxybenzaldehyde

Strategy and Solution

The aldehyde group in each compound is oxidized to a carboxylate anion, $-COO^-$. Acidification with HCl converts the carboxylate anion to a carboxylic acid, $-COOH$.

(a)

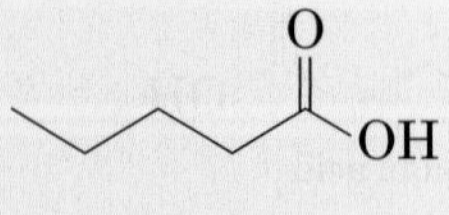

Pentanoic acid

(b) 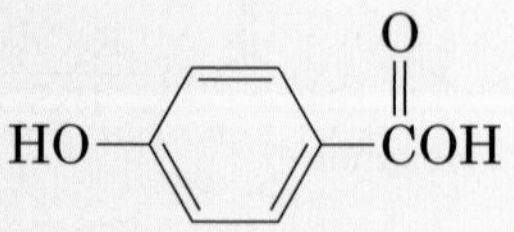 HO— —COH

4-Hydroxybenzoic acid

Problem 17.4

Complete equations for these oxidations.
(a) Hexanedial + $O_2 \longrightarrow$
(b) 3-Phenylpropanal + $Ag(NH_3)_2^+ \longrightarrow$

B. Reduction

In Section 12.6D, we saw that the C=C double bond of an alkene can be reduced by hydrogen in the presence of a transition metal catalyst to a C—C
single bond. The same is true of the C=O double bond of an aldehyde or
ketone. Aldehydes are reduced to primary alcohols and ketones are reduced
to secondary alcohols.

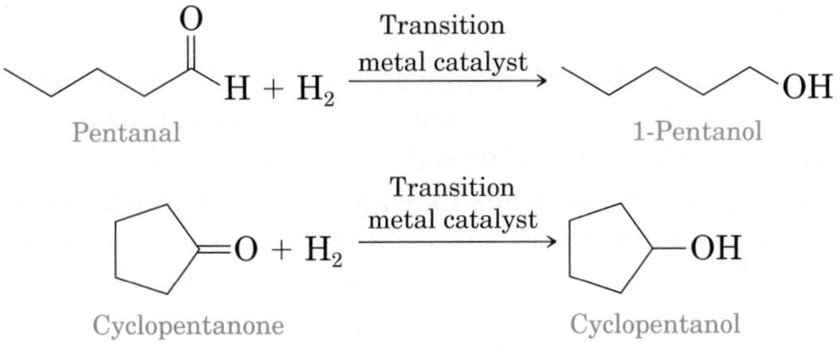

Pentanal   1-Pentanol

Cyclopentanone   Cyclopentanol

The reduction of a C=O double bond under these conditions is slower than
the reduction of a C=C double bond. Thus, if the same molecule contains
both C=O and C=C double bonds, the C=C double bond is reduced first.

The reagent most commonly used in the laboratory for the reduction of an
aldehyde or a ketone is sodium borohydride, $NaBH_4$. This reagent behaves
as a source of hydride ions, $H:^-$. In the hydride ion, hydrogen has two valence
electrons and bears a negative charge. In a reduction by sodium borohydride,
hydride ion is attracted to and then adds to the partially positive carbonyl
carbon, which leaves a negative charge on the carbonyl oxygen. Reaction of
this alkoxide intermediate with aqueous acid gives the alcohol.

Hydride   Alkoxide
ion   ion

Of the two hydrogens added to the carbonyl group in this reduction, one
comes from the reducing agent and the other comes from aqueous acid. Reduction of cyclohexanone, for example, with this reagent gives cyclohexanol:

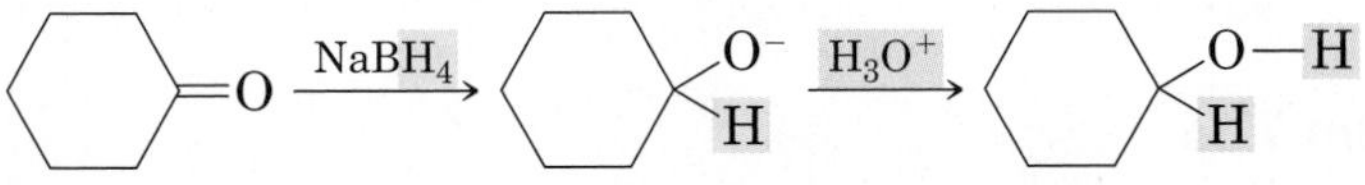

An advantage of using $NaBH_4$ over the H_2/metal reduction is that $NaBH_4$
does not reduce carbon–carbon double bonds. The reason for this selectivity
is quite straightforward. There is no polarity (no partial positive or negative
charges) on a carbon–carbon double bond. Therefore, a C=C double bond
has no partially positive site to attract the negatively charged hydride ion.
In the following example, $NaBH_4$ selectively reduces the aldehyde to a primary alcohol:

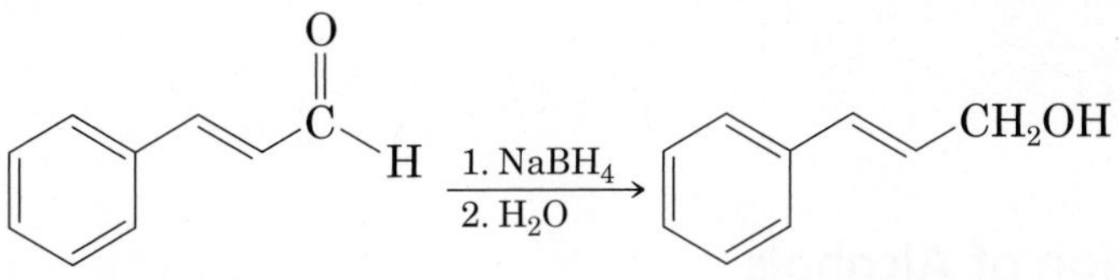

Cinnamaldehyde   Cinnamyl alcohol

In biological systems, the agent for the reduction of aldehydes and ketones is the reduced form of the coenzyme nicotinamide adenine dinucleotide, abbreviated NADH (Section 27.3). This reducing agent, like $NaBH_4$,

delivers a hydride ion to the carbonyl carbon of the aldehyde or ketone. Reduction of pyruvate, for example, by NADH gives lactate:

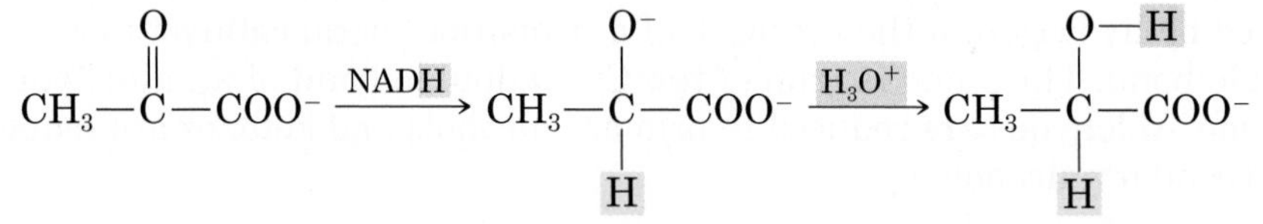

$$CH_3-\overset{\overset{O}{\|}}{C}-COO^- \xrightarrow{\text{NADH}} CH_3-\overset{\overset{O^-}{|}}{\underset{\underset{H}{|}}{C}}-COO^- \xrightarrow{H_3O^+} CH_3-\overset{\overset{O-H}{|}}{\underset{\underset{H}{|}}{C}}-COO^-$$

Pyruvate   Lactate

Pyruvate is the end product of glycolysis, a series of enzyme-catalyzed reactions that converts glucose to two molecules of this ketoacid (Section 28.1). Under anaerobic conditions, NADH reduces pyruvate to lactate. The build-up of lactate in the bloodstream leads to acidosis, and in muscle tissue, it is associated with muscle fatigue. When blood lactate reaches a concentration of about 0.4 mg/100 mL, muscle tissue becomes almost completely exhausted.

Example 17.5 Reduction of Aldehydes and Ketones

Complete the equations for these reductions.

(a) 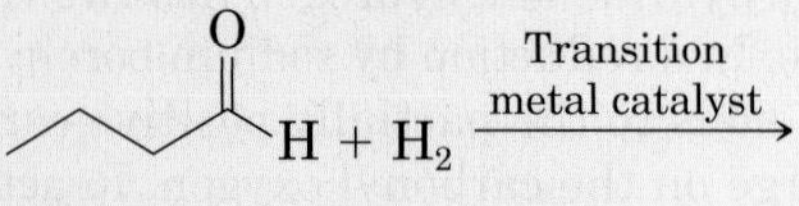 (structure) $H + H_2 \xrightarrow[\text{metal catalyst}]{\text{Transition}}$ (b) 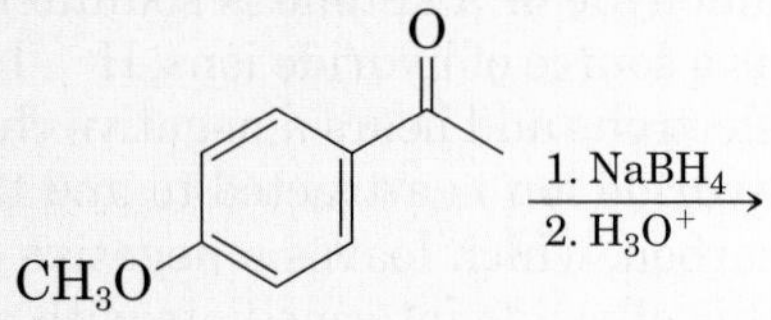 (structure with CH_3O) $\xrightarrow[\text{2. H}_3O^+]{\text{1. NaBH}_4}$

Strategy and Solution

The carbonyl group of the aldehyde in (a) is reduced to a primary alcohol, and that of the ketone in (b) is reduced to a secondary alcohol.

(a) 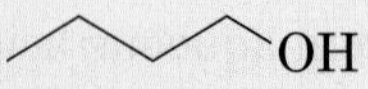 (structure) OH   (b) 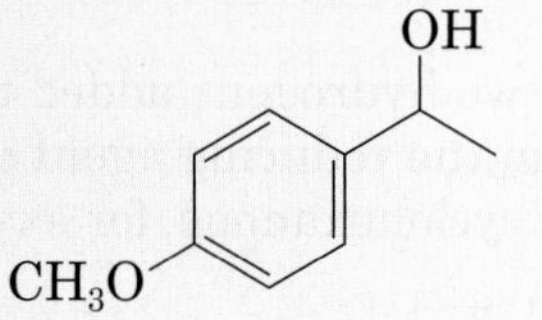 (structure with CH_3O and OH)

Problem 17.5

Which aldehyde or ketone gives these alcohols upon reduction with H_2/metal catalyst?

(a) 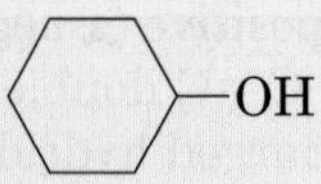 (cyclohexyl) —OH   (b) 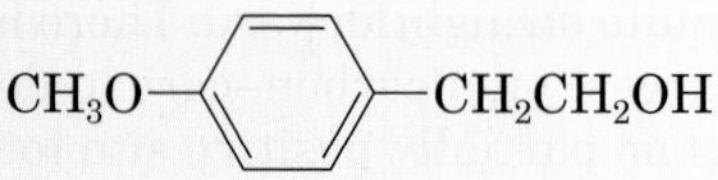 CH_3O— (benzene) —CH_2CH_2OH

(c) 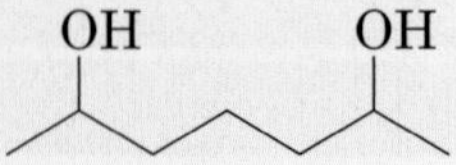 OH   OH (structure)

C. Addition of Alcohols

Hemiacetal A molecule containing a carbon bonded to one —OH group and one —OR group; the product of adding one molecule of alcohol to the carbonyl group of an aldehyde or ketone

Addition of a molecule of alcohol to the carbonyl group of an aldehyde or a ketone forms a **hemiacetal** (a half-acetal). The functional group of a hemiacetal is a carbon bonded to one —OH group and one —OR group. In forming a hemiacetal, the H of the alcohol adds to the carbonyl oxygen and the OR group of the alcohol adds to the carbonyl carbon. Shown here are the hemiacetals formed by addition of one molecule of ethanol to benzaldehyde and to cyclohexanone:

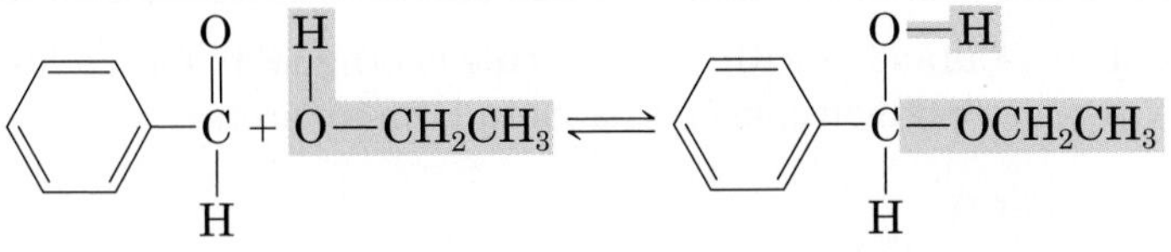

Benzaldehyde   Ethanol   A hemiacetal

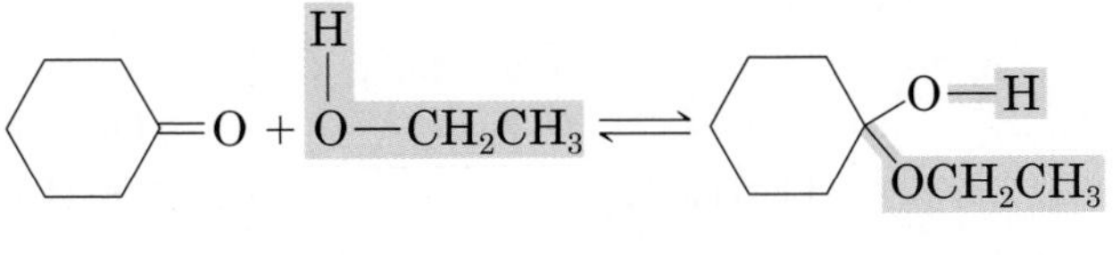

Cyclohexanone   Ethanol   A hemiacetal

Hemiacetals are generally unstable and are only minor components of an equilibrium mixture, except in one very important type of molecule. When a hydroxyl group is part of the same molecule that contains the carbonyl group and a five- or six-membered ring can form, the compound exists almost entirely in a cyclic hemiacetal form. In this case, the —OH group adds to the C=O group of the same molecule. We will have much more to say about cyclic hemiacetals when we consider the chemistry of carbohydrates in Chapter 20.

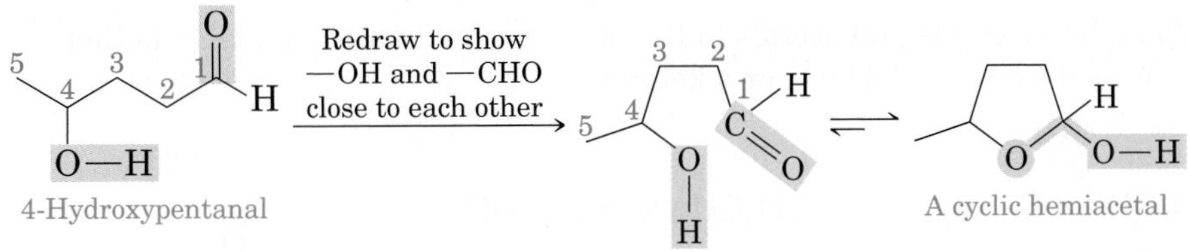

4-Hydroxypentanal   A cyclic hemiacetal

Hemiacetals can react further with alcohols to form **acetals** plus water. This reaction is acid-catalyzed. The functional group of an acetal is a carbon bonded to two —OR groups.

Acetal A molecule containing two —OR groups bonded to the same carbon

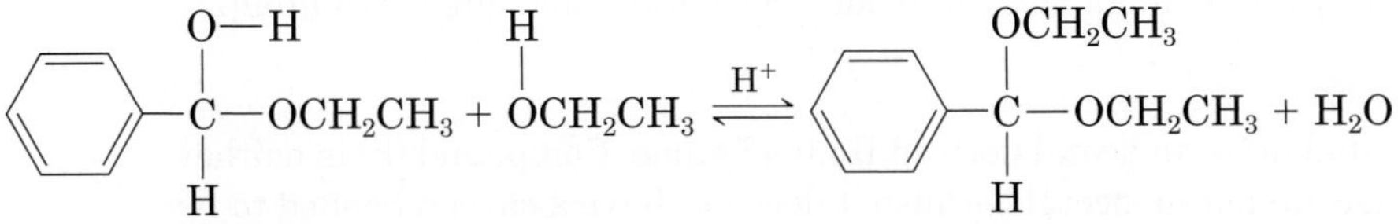

A hemiacetal (from benzaldehyde)   Ethanol   An acetal

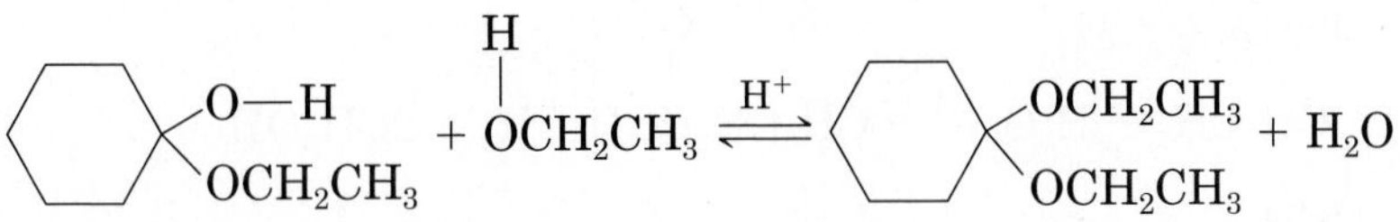

A hemiacetal (from cyclohexanone)   Ethanol   An acetal

All steps in hemiacetal and acetal formation are reversible. As with any other equilibrium, we can make this one go in either direction by using Le Chatelier's principle (Section 7.7). If we want to drive it to the right (formation of the acetal), we either use a large excess of alcohol or remove water from the equilibrium mixture. If we want to drive it to the left (hydrolysis of the acetal to the original aldehyde or ketone and alcohol), we use a large excess of water.

Example 17.6 Formation of Hemiacetals and Acetals

Show the reaction of 2-butanone with one molecule of ethanol to form a hemiacetal and then with a second molecule of ethanol to form an acetal.

Strategy and Solution

Given are structural formulas for the hemiacetal and then the acetal.

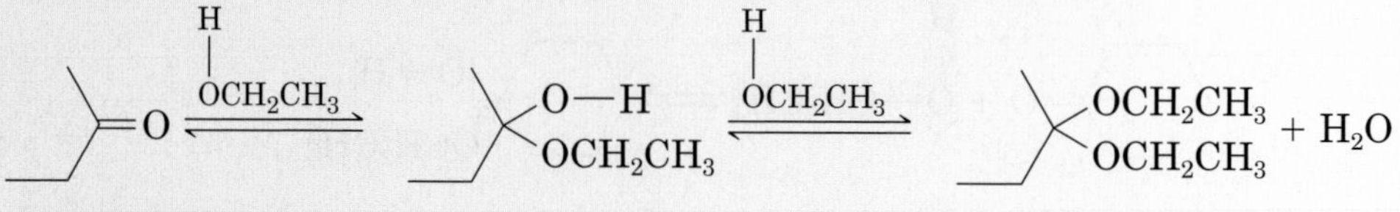

2-Butanone   A hemiacetal   An acetal

Problem 17.6

Show the reaction of benzaldehyde with one molecule of methanol to form a hemiacetal and then with a second molecule of methanol to form an acetal.

Example 17.7 Recognizing the Presence of a Hemiacetal and an Acetal

Identify all hemiacetals and acetals in the following structures and tell whether each is formed from an aldehyde or a ketone.

(a) $CH_3CH_2\underset{\underset{OCH_3}{|}}{\overset{\overset{OCH_3}{|}}{C}}CH_2CH_3$ (b) $CH_3CH_2OCH_2CH_2OH$ (c)

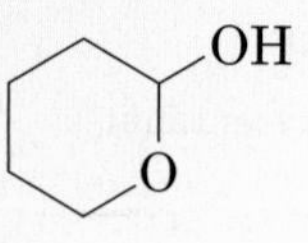

Strategy

An acetal contains a carbon atom bonded to two OR groups; a hemiacetal contains a carbon atom bonded to one —OH group and one —OR group.

Solution

Compound (a) is an acetal derived from a ketone. Compound (b) is neither a hemiacetal nor an acetal because it does not have a carbon bonded to two oxygens; its functional groups are an ether and a primary alcohol. Compound (c) is a cyclic hemiacetal derived from an aldehyde.

(a) $CH_3CH_2\underset{\underset{OCH_3}{|}}{\overset{\overset{OCH_3}{|}}{C}}CH_2CH_3 + H_2O \xrightarrow{H^+} CH_3CH_2\overset{\overset{O}{||}}{C}CH_2CH_3 + 2CH_3OH$

An acetal   2-Pentanone

(c)   A hemiacetal   5-Hydroxypentanal

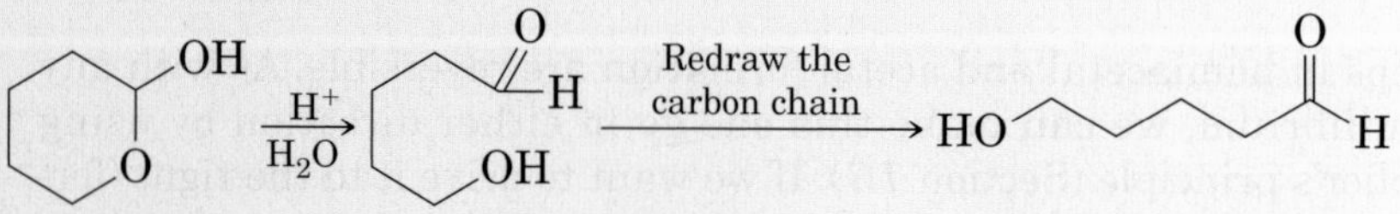

Problem 17.7

Identify all hemiacetals and acetals in the following structures and tell whether each is formed from an aldehyde or a ketone.

(a) $CH_3CH_2\overset{\overset{\displaystyle OH}{|}}{\underset{\underset{\displaystyle OCH_2CH_3}{|}}{C}}CH_2CH_3$

(b) $CH_3OCH_2CH_2OCH_3$

(c)

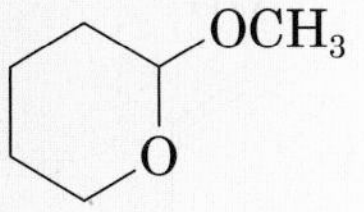

17.5 What Is Keto-Enol Tautomerism?

A carbon atom adjacent to a carbonyl group is called an α-carbon, and a hydrogen atom bonded to it is called an α-hydrogen.

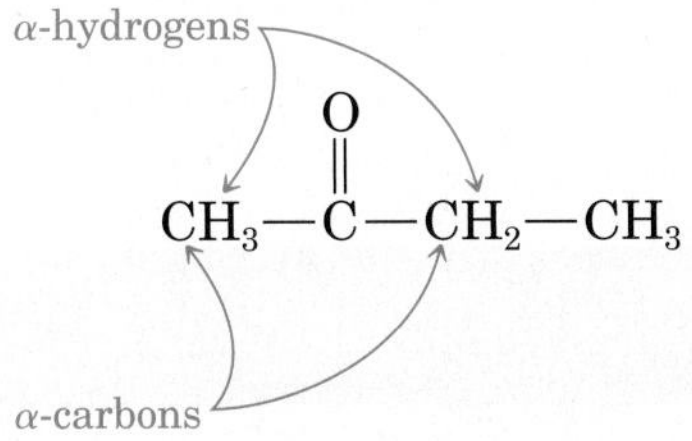

α-hydrogens

$$CH_3-\overset{\overset{\displaystyle O}{||}}{C}-CH_2-CH_3$$

α-carbons

A carbonyl compound that has a hydrogen on an α-carbon is in equilibrium with a constitutional isomer called an **enol**. The name "enol" is derived from the IUPAC designation of it as both an alkene (*-en-*) and an alcohol (*-ol*).

Enol A molecule containing an —OH group bonded to a carbon of a carbon–carbon double bond

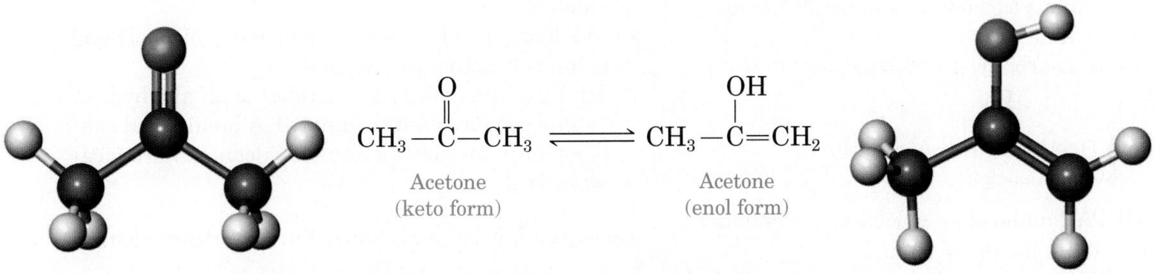

$$CH_3-\overset{\overset{\displaystyle O}{||}}{C}-CH_3 \rightleftharpoons CH_3-\overset{\overset{\displaystyle OH}{|}}{C}=CH_2$$

Acetone (keto form)   Acetone (enol form)

Keto and enol forms are examples of **tautomers**, constitutional isomers in equilibrium with each other that differ in the location of a hydrogen atom and a double bond. This type of isomerism is called **keto-enol tautomerism**. For any pair of keto-enol tautomers, the keto form generally predominates at equilibrium.

Tautomers Constitutional isomers that differ in the location of a hydrogen atom and a double bond

Example 17.8 Keto-Enol Tautomerism

Draw structural formulas for the two enol forms for each ketone.

(a)

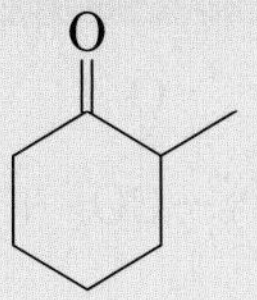

(b)

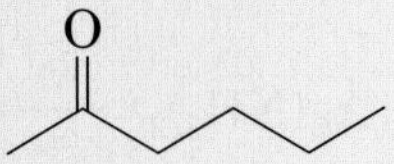

Strategy and Solution

Any aldehyde or ketone with one hydrogen on its α-carbon can show keto-enol tautomerism.

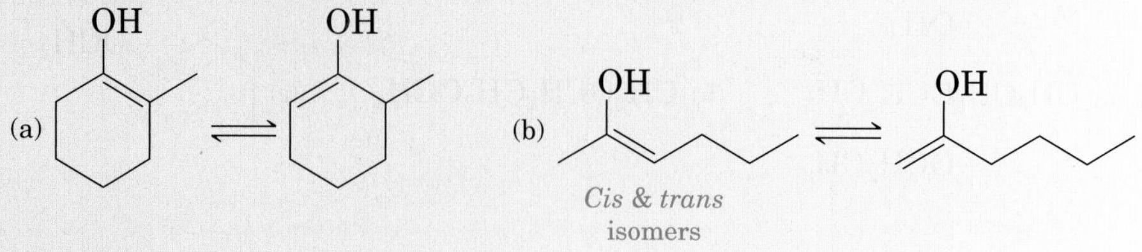

(a) [structure: 2-methylcyclohexen-1-ol] ⇌ [structure]

(b) [structure] ⇌ [structure]

Cis & trans isomers

Problem 17.8

Draw a structural formula for the keto form of each enol.

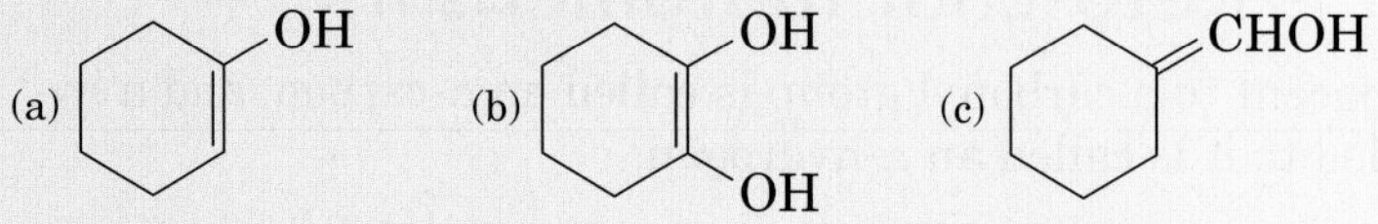

(a) [cyclohexene with OH]   (b) [cyclohexene with two OH]   (c) [cyclohexane with =CHOH]

Summary of Key Questions

Section 17.1 What Are Aldehydes and Ketones?

- An **aldehyde** contains a carbonyl group bonded to at least one hydrogen atom.
- A **ketone** contains a carbonyl group bonded to two carbon atoms.

Section 17.2 How Do We Name Aldehydes and Ketones?

- We derive the IUPAC name of an aldehyde by changing the -*e* of the parent alkane to -*al*.
- We derive the IUPAC name of a ketone by changing the -*e* of the parent alkane to -*one* and using a number to locate the carbonyl carbon.

Section 17.3 What Are the Physical Properties of Aldehydes and Ketones?

- Aldehydes and ketones are polar compounds. They have higher boiling points and are more soluble in water than nonpolar compounds of comparable molecular weight.

Section 17.4 What Are the Characteristic Reactions of Aldehydes and Ketones?

- Aldehydes are oxidized to carboxylic acids, but ketones resist oxidation.
- **Tollens' reagent** is used to test for the presence of aldehydes.
- Aldehydes can be reduced to primary alcohols and ketones to secondary alcohols.
- Addition of a molecule of alcohol to an aldehyde or a ketone produces a **hemiacetal**. A hemiacetal can react with another molecule of alcohol to produce an **acetal**.

Section 17.5 What Is Keto-Enol Tautomerism?

- A molecule containing an —OH group bonded to a carbon of a carbon–carbon double bond is called an **enol**.
- Constitutional isomers that differ in the location of a hydrogen atom and a double bond are called **tautomers**.

Summary of Key Reactions

1. **Oxidation of an Aldehyde to a Carboxylic Acid (Section 17.4A)**
   The aldehyde group is among the most easily oxidized organic functional groups. Oxidizing agents include $K_2Cr_2O_7$, Tollens' reagent, and O_2.

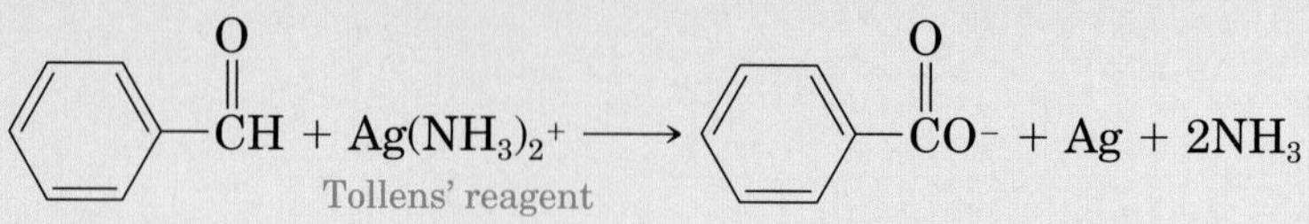

[reaction structure] $+ Ag(NH_3)_2^+ \longrightarrow$ [product structure] $+ Ag + 2NH_3$

Tollens' reagent

2. Reduction (Section 17.4B)

Aldehydes are reduced to primary alcohols and ketones to secondary alcohols by H_2 in the presence of a transition metal catalyst such as Pt or Ni. They are also reduced to alcohols by sodium borohydride, $NaBH_4$, followed by protonation.

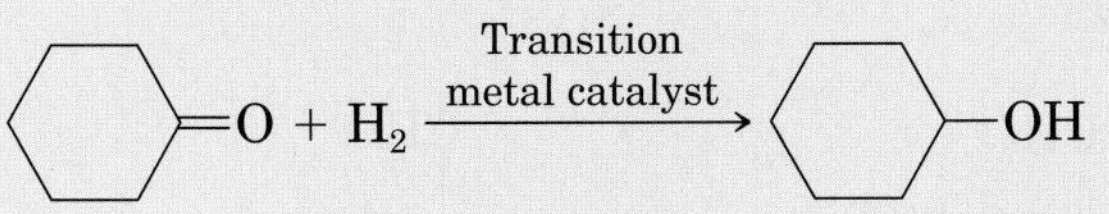

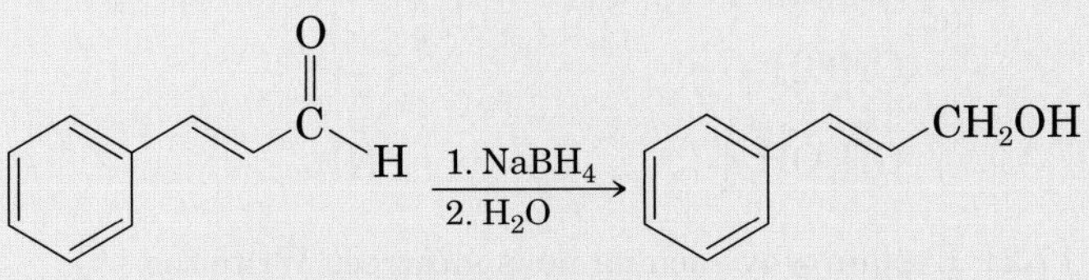

3. Addition of Alcohols to Form Hemiacetals (Section 17.4C)

Hemiacetals are only minor components of an equilibrium mixture of an aldehyde or a ketone and an alcohol, except where the —OH and C=O groups are parts of the same molecule and a five- or six-membered ring can form.

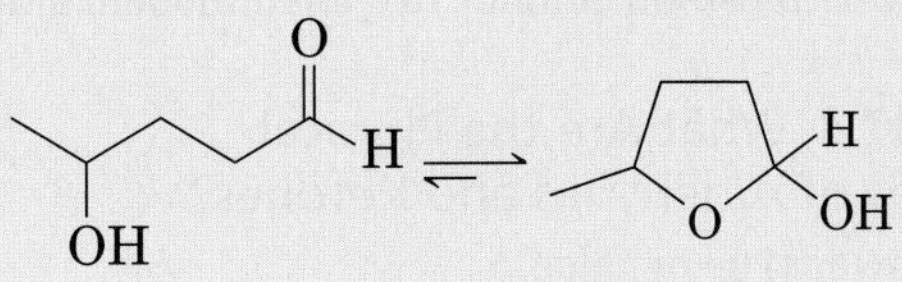

4. Addition of Alcohols to Form Acetals (Section 17.4C)

Formation of acetals is catalyzed by acid. Acetals are hydrolyzed in aqueous acid to an aldehyde or a ketone and two molecules of an alcohol.

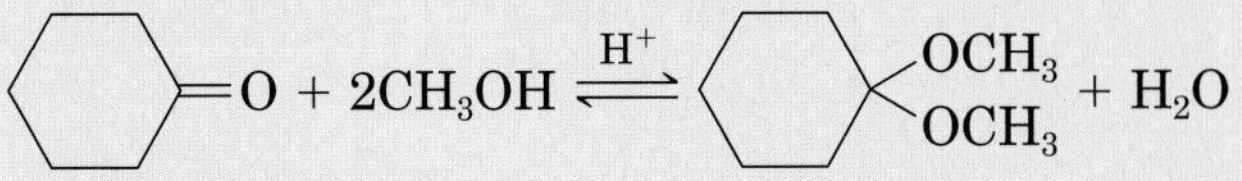

Cyclohexanone   Methanol   An acetal

5. Keto-Enol Tautomerism (Section 17.5)

The keto form generally predominates at equilibrium.

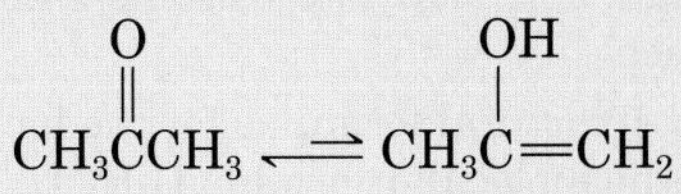

$$CH_3\overset{O}{\overset{\|}{C}}CH_3 \rightleftharpoons CH_3\overset{OH}{\overset{|}{C}}{=}CH_2$$

Keto form   Enol form
(approximately 99.9%)

Problems

Interactive versions of these problems may be assigned in OWL.

Orange-numbered problems are applied.

References to previous chapters are given in parentheses.

Section 17.1 What Are Aldehydes and Ketones?

17.9 Answer true or false.

(a) The one aldehyde and the one ketone with a molecular formula of C_3H_6O are constitutional isomers.

(b) Aldehydes and ketones both contain a carbonyl group.

(c) The VSEPR model predicts bond angles of 120° about the carbonyl carbon of aldehydes and ketones.

(d) The carbonyl carbon of a ketone is a stereocenter.

17.10 What is the difference in structure between an aldehyde and a ketone?

17.11 What is the difference in structure between an aromatic aldehyde and an aliphatic aldehyde?

17.12 Is it possible for the carbon atom of a carbonyl group to be a stereocenter? Explain.

17.13 Which compounds contain carbonyl groups?

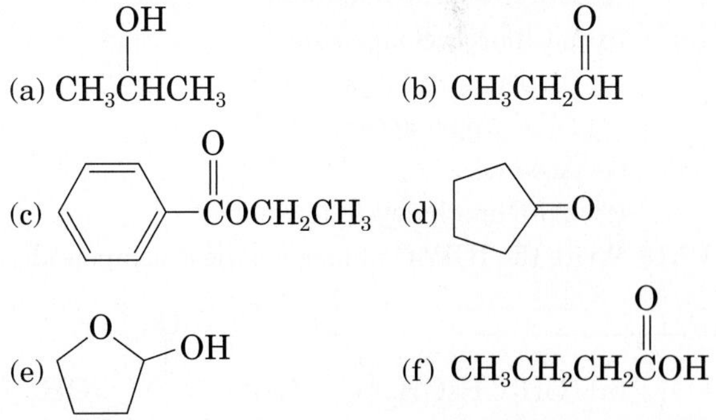

(a) $CH_3\overset{OH}{\overset{|}{C}}HCH_3$

(b) $CH_3CH_2\overset{O}{\overset{\|}{C}}H$

(c)

(d)

(e)

(f) $CH_3CH_2CH_2\overset{O}{\overset{\|}{C}}OH$

17.14 Following are structural formulas for two steroid hormones.

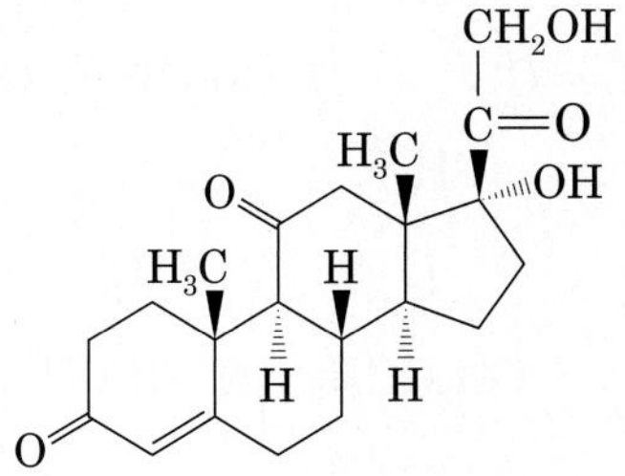

Cortisone

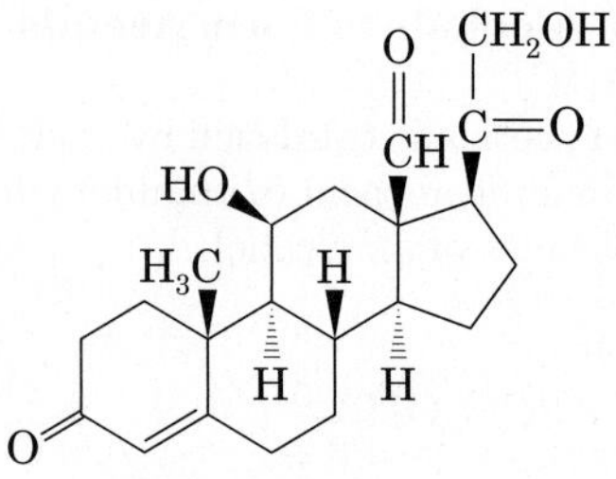

Aldosterone

(a) Name the functional groups in each.

(b) Mark all stereocenters in each hormone and state how many stereoisomers are possible for each.

17.15 Draw structural formulas for the four aldehydes with the molecular formula $C_5H_{10}O$. Which of these aldehydes are chiral?

Section 17.2 How Do We Name Aldehydes and Ketones?

17.16 Answer true or false.

(a) An aldehyde is named as an alkanal, and a ketone is named as an alkanone.

(b) The names for aldehydes and ketones are derived from the name of the longest carbon chain that contains the carbonyl group.

(c) In an aromatic aldehyde, the carbonyl carbon is bonded to an aromatic ring.

17.17 Draw structural formulas for these aldehydes.

(a) Formaldehyde   (b) Propanal

(c) 3,7-Dimethyloctanal   (d) Decanal

(e) 4-Hydroxybenzaldehyde

(f) 2,3-Dihydroxypropanal

17.18 Draw structural formulas for these ketones.

(a) Ethyl isopropyl ketone

(b) 2-Chlorocyclohexanone

(c) 2,4-Dimethyl-3-pentanone

(d) Diisopropyl ketone

(e) Acetone

(f) 2,5-Dimethylcyclohexanone

17.19 Write the IUPAC names for these compounds.

(a) $(CH_3CH_2CH_2)_2C{=}O$ (b)

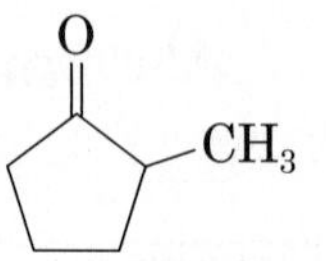

(c)

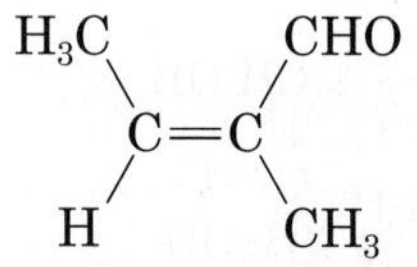

(d) 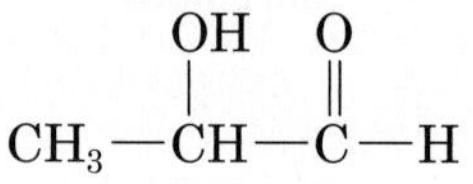 $CH_3{-}\overset{OH}{\underset{|}{CH}}{-}\overset{O}{\overset{\|}{C}}{-}H$

(e)

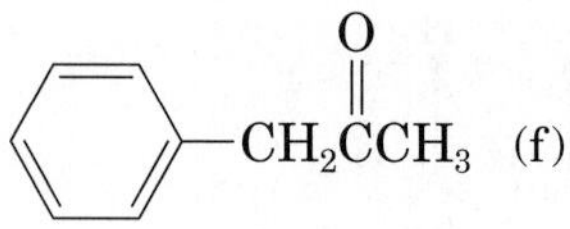

(f) 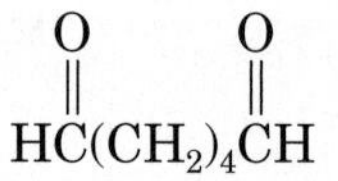 $HC(CH_2)_4CH$ (with carbonyl O on each end)

17.20 Write the IUPAC names for these compounds.

(a)

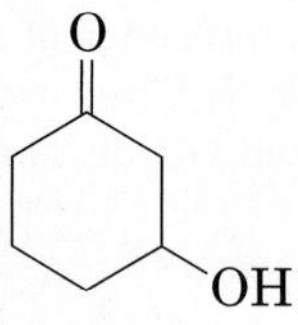

(b)

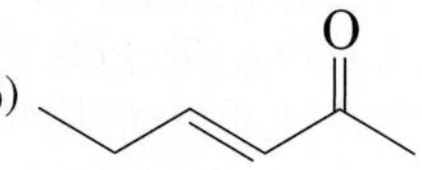

(c) 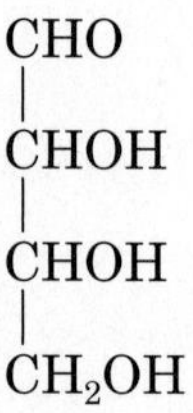 $\begin{array}{c} CHO \\ | \\ CHOH \\ | \\ CHOH \\ | \\ CH_2OH \end{array}$

(d)

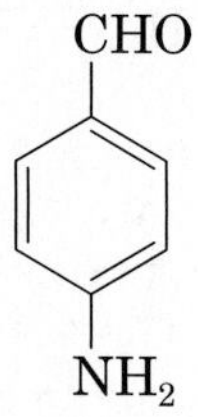

17.21 Explain why each name is incorrect. Write the correct IUPAC name for the intended compound.

(a) 3-Butanone   (b) 1-Butanone

(c) 4-Methylbutanal   (d) 2,2-Dimethyl-3-butanone

17.22 Explain why each name is incorrect. Write the correct IUPAC name for the intended compound.

(a) 2-Pentanal   (b) Cyclopentanal

(c) 3-Ethyl-2-butanone   (d) 5-Aminobenzaldehyde

Section 17.3 What Are the Physical Properties of Aldehydes and Ketones?

17.23 Answer true or false.

(a) Aldehydes and ketones are polar compounds.

(b) Aldehydes have lower boiling points than alcohols with comparable carbon skeletons.

(c) Low-molecular-weight aldehydes and ketones are very soluble in water.

(d) There is no possibility for hydrogen bonding between molecules of aldehydes and ketones.

17.24 In each pair of compounds, select the one with the higher boiling point.

(a) Acetaldehyde or ethanol

(b) Acetone or 3-pentanone

(c) Butanal or butane

(d) Butanone or 2-butanol

17.25 Acetone is completely soluble in water, but 4-heptanone is completely insoluble in water. Explain.

17.26 Account for the fact that acetone has a higher boiling point (56°C) than ethyl methyl ether (11°C) even though their molecular weights are almost the same.

17.27 Pentane, 1-butanol, and butanal all have approximately the same molecular weights but different boiling points. Arrange them in order of increasing boiling point. Explain the basis for your ranking.

17.28 Show how acetaldehyde can form hydrogen bonds with water.

17.29 Why can't two molecules of acetone form a hydrogen bond with each other?

Section 17.4 What Are the Characteristic Reactions of Aldehydes and Ketones?

17.30 Answer true or false.

(a) The reduction of an aldehyde always gives a primary alcohol.

(b) The reduction of a ketone always gives a secondary alcohol.

(c) The oxidation of an aldehyde gives a carboxylic acid.

(d) The oxidation of a primary alcohol gives a ketone.

(e) Tollens' reagent can be used to distinguish between an aldehyde and a ketone.

(f) Sodium borohydride, $NaBH_4$, reduces an aldehyde to a primary alcohol.

(g) The addition of one molecule of alcohol to the carbonyl group of a ketone gives a hemiacetal.

(h) The reaction of an aldehyde with two molecules of alcohol gives an acetal, plus a molecule of water.

(i) The formation of hemiacetals and acetals is reversible.

(j) The cyclic hemiacetal formed from 4-hydroxypentanal has two stereocenters and can exist as a mixture of $2^2 = 4$ stereoisomers.

17.31 Draw a structural formula for the principal organic product formed when each compound is treated with $K_2Cr_2O_7/H_2SO_4$. If there is no reaction, say so.

(a) Butanal   (b) Benzaldehyde

(c) Cyclohexanone   (d) Cyclohexanol

17.32 Draw a structural formula for the principal organic product formed when each compound in Problem 17.31 is treated with Tollens' reagent. If there is no reaction, say so.

17.33 What simple chemical test could you use to distinguish between the members of each pair of compounds? Tell what you would do, what you would expect to observe, and how you would interpret your experimental observation.

(a) Pentanal and 2-pentanone

(b) 2-Pentanone and 2-pentanol

17.34 Explain why liquid aldehydes are often stored under an atmosphere of nitrogen rather than in air.

17.35 Suppose that you take a bottle of benzaldehyde (a liquid, bp 179°C) from a shelf and find a white solid in the bottom of the bottle. The solid turns litmus red; that is, it is acidic. Yet aldehydes are neutral compounds. How can you explain these observations?

17.36 Explain why the reduction of an aldehyde always gives a primary alcohol and the reduction of a ketone always gives a secondary alcohol.

17.37 Write a structural formula for the principal organic product formed by treating each compound with H_2/transition metal catalyst. Which products are chiral?

(a) $CH_3CCH_2CH_3$ (with O double bond)   (b) $CH_3(CH_2)_4CH$ (with O double bond)

(c) [structure: cyclopentanone with CH₃ substituent]

(d) [structure: benzene ring with CH (O double bond) and OH]

17.38 Write a structural formula for the principal organic product formed by treating each compound in Problem 17.37 with $NaBH_4$ followed by H_2O.

17.39 1,3-Dihydroxy-2-propanone, more commonly known as dihydroxyacetone, is the active ingredient in artificial tanning agents such as Man-Tan and Magic Tan.

(a) Write a structural formula for this compound.

(b) Would you expect it to be soluble or insoluble in water?

(c) Write a structural formula for the product formed by its reduction with $NaBH_4$.

17.40 Draw a structural formula for the product formed by treatment of butanal with each set of reagents.

(a) H_2/metal catalyst

(b) $NaBH_4$, then H_2O

(c) $Ag(NH_3)_2^+$ (Tollens' reagent)

(d) $K_2Cr_2O_7/H_2SO_4$

17.41 Draw a structural formula for the product formed by treatment of acetophenone, $C_6H_5COCH_3$, with each set of reagents given in Problem 17.40.

Section 17.5 What Is Keto-Enol Tautomerism?

17.42 Mark each statement true or false.

(a) Keto and enol tautomers are constitutional isomers.

(b) For a pair of keto-enol tautomers, the keto form generally predominates.

17.43 Which of these compounds undergo keto-enol tautomerism?

(a) CH_3CH (with O double bond)

(b) CH_3CCH_3 (with O double bond)

(c) [benzene ring]—CH (with O double bond)

(d) [benzene ring]—CCH₃ (with O double bond)

(e) [benzene ring]—C—[benzene ring] (with O double bond)

(f) [cyclohexanone structure]

17.44 Draw all enol forms of each aldehyde and ketone.

(a) CH_3CH_2CH (with O double bond)

(b) $CH_3CCH_2CH_3$ (with O double bond)

(c) [cyclopentanone with CH₃ substituent]

17.45 Draw a structural formula for the keto form of each enol.

(a) [cyclopentene with OH]

(b) $CH_3C=CHCH_2CH_2CH_3$ (with OH)

(c) [benzene ring]—CH=CCH₃ (with OH)

Chemical Connections

17.46 Warfarin is widely used in medicine as a blood anticoagulant.

(a) Identify all the functional groups in warfarin.

(b) The S enantiomer is more active than the R enantiomer. Draw a stereorepresentation of the R enantiomer.

(c) Warfarin, as drawn in the Chemical Connections box, contains an enol group. Draw the structural formula of the keto form of warfarin.

(d) Draw the product formed by treating warfarin with $NaBH_4$.

(e) Warfarin exerts it anticoagulation property in many individuals at a dose of 4 mg/day. How many molecules of warfarin are present in a 4 mg tablet?

Addition of Alcohols

17.47 What is the characteristic structural feature of a hemiacetal? Of an acetal?

17.48 Which compounds are hemiacetals, which are acetals, and which are neither?

(a) [structure with OCH₃, –CHOCH₃]   (b) $CH_3CH_2CHOCH_3$ (with OH)

(c) $CH_3OCH_2OCH_3$ (d) [structure with CH₂CH₃]

(e) [structure with OCH₂CH₃]   (f) [structure with OH]

17.49 Which compounds are hemiacetals, which are acetals, and which are neither?

(a) [HO OCH₃ structure]   (b) [CH₃O OCH₃ structure]

(c) [OCH₃ OCH₃ structure]   (d) [OCH₃ OH structure]

(e) [O OCH₃ structure]   (f) [OCH₃ OCH₃ structure]

17.50 Draw the hemiacetal and then the acetal formed in each reaction. In each case, assume an excess of the alcohol.

(a) Propanal + methanol →

(b) Cyclopentanone + methanol →

17.51 Draw the structures of the aldehydes or ketones and alcohols formed when these acetals are treated with aqueous acid and hydrolyzed.

(a) [O O structure]   (b) [benzene with OCH₃, OCH₃, OCH₃]

(c) [structure with OCH₃, OCH₃]   (d) [structure with O, OCH₃]

17.52 The following compound is a component of the fragrance of jasmine:

From which carbonyl-containing compound and alcohol is this compound derived?

17.53 Following is the structure of immunosuppressant FK-506, a molecule shown to disrupt calcineurin-mediated signal transduction in T-lymphocytes.

Immunosuppressant FK-506

(a) There are three carbon–carbon double bonds present in this molecule. Which of the three has the potential for *cis/trans* isomerism? Assign a *cis* or *trans* configuration to each carbon–carbon double bond that has this possibility.

(b) How many stereocenters are present in this molecule? How many stereoisomers are possible for it?

(c) Are there any aromatic components in this molecule?

(d) Consider the two carbon atoms marked with asterisks. Determine the absolute (R or S) configuration of each stereocenter.

(e) Because of the presence of a 21-member ring, this molecule is described as a macrocycle. This ring is fashioned by three types of bonds, several carbon–carbon bonds, one ester, one hemiacetal, and one amide. Locate the ester and the hemiacetal.

(f) Draw the structural formula of the long chain compound that would result if the hemiacetal were to cleave to an alcohol and a carbonyl compound.

17.54 What is the difference in meaning between the terms "hydration" and "hydrolysis"? Give an example of each.

17.55 What is the difference in meaning between the terms "hydration" and "dehydration"? Give an example of each.

17.56 List the reagents and experimental conditions to convert cyclohexanone to each of the following compounds.

17.57 Draw a structural formula for an aldehyde or a ketone that can be reduced to produce each alcohol. If none exists, say so.

(a) CH₃CHCH₃ with OH

(b) benzene ring—CH₂OH

(c) CH₃OH

(d) cyclohexane ring with OH and CH₃

17.58 Draw a structural formula for an aldehyde or a ketone that can be reduced to produce each alcohol. If none exists, say so.

(a) cyclopentane—OH

(b) cyclohexane—CH₂OH

(c) CH₃COH with two CH₃ groups

(d) HO⌢⌢⌢OH

17.59 1-Propanol can be prepared by the reduction of an aldehyde, but it cannot be prepared by the acid-catalyzed hydration of an alkene. Explain why it cannot be prepared from an alkene.

17.60 Show how to bring about these conversions. In addition to the given starting material, use any other organic or inorganic reagents as necessary.

(a) $C_6H_5CCH_2CH_3 \longrightarrow C_6H_5CHCH_2CH_3 \longrightarrow C_6H_5CH=CHCH_3$

(b) cyclopentanone=O → cyclopentane—OH → cyclopentene → cyclopentane—Cl

17.61 Show how to bring about these conversions. In addition to the given starting material, use any other organic or inorganic reagents as necessary.

(a) 1-Pentene to 2-pentanone

(b) Cyclohexene to cyclohexanone

17.62 Describe a simple chemical test by which you could distinguish between the members of each pair of compounds.

(a) Cyclohexanone and aniline

(b) Cyclohexene and cyclohexanol

(c) Benzaldehyde and cinnamaldehyde

Additional Problems

17.63 Indicate the aldehyde or ketone group in these compounds.

(a) $HCCH_2CH_2CH_2CCH_3$ (b) cyclohexanone with —CH group

(c) $HOCH_2CHCH$ with HO and O groups

(d) bicyclic structure with O

(e) benzene ring—CCH_2CH_3 with O

(f) CH_3O and HO substituted benzene ring—CH with O

17.64 Draw a structural formula for the product formed by treating each compound in Problem 17.63 with sodium borohydride, $NaBH_4$.

17.65 Draw structural formulas for the (a) one ketone and (b) two aldehydes with the molecular formula C_4H_8O.

17.66 Draw structural formulas for these compounds.

(a) 1-Chloro-2-propanone

(b) 3-Hydroxybutanal

(c) 4-Hydroxy-4-methyl-2-pentanone

(d) 3-Methyl-3-phenylbutanal

(e) 1,3-Cyclohexanedione

(f) 5-Hydroxyhexanal

17.67 Why does acetone have a lower boiling point (56°C) than 2-propanol (82°C) even though their molecular weights are almost the same?

17.68 Propanal (bp 49°C) and 1-propanol (bp 97°C) have about the same molecular weight, yet their boiling points differ by almost 50°C. Explain this fact.

17.69 What simple chemical test could you use to distinguish between the members of each pair of compounds? Tell what you would do, what you would expect to observe, and how you would interpret your experimental observation.

(a) Benzaldehyde and cyclohexanone

(b) Acetaldehyde and acetone

17.70 5-Hydroxyhexanal forms a six-membered cyclic hemiacetal, which predominates at equilibrium in aqueous solution.

(a) Draw a structural formula for this cyclic hemiacetal.

(b) How many stereoisomers are possible for 5-hydroxyhexanal? (Chapter 15)

(c) How many stereoisomers are possible for this cyclic hemiacetal? (Chapter 15)

17.71 The following molecule is an enediol; each carbon of the double bond carries an —OH group. Draw structural formulas for the α-hydroxyketone and the α-hydroxyaldehyde with which this enediol is in equilibrium.

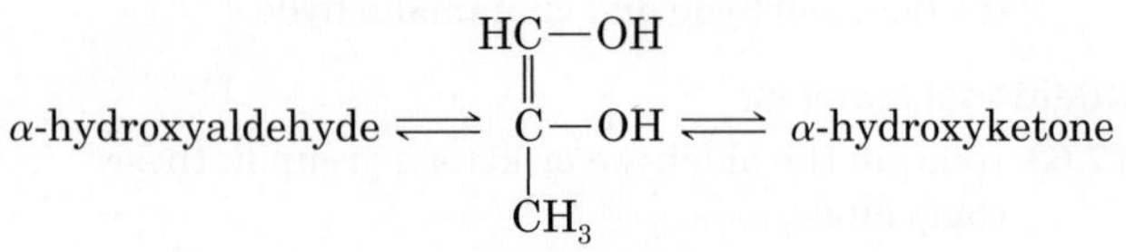

$$\alpha\text{-hydroxyaldehyde} \rightleftharpoons \overset{\displaystyle HC-OH}{\underset{\displaystyle CH_3}{\overset{\displaystyle \|}{C-OH}}} \rightleftharpoons \alpha\text{-hydroxyketone}$$

An enediol

17.72 Alcohols can be prepared by the acid-catalyzed hydration of alkenes (Section 12.6B) and by the reduction of aldehydes and ketones (Section 17.4B). Show how you might prepare each of the following alcohols by (1) acid-catalyzed hydration of an alkene and (2) reduction of an aldehyde or a ketone.

(a) Ethanol   (b) Cyclohexanol

(c) 2-Propanol   (d) 1-Phenylethanol

Looking Ahead

17.73 Glucose, $C_6H_{12}O_6$, contains an aldehyde group but exists predominantly in the form of the cyclic hemiacetal shown here. We will discuss this cyclic form of glucose in Chapter 20.

β-D-Glucose

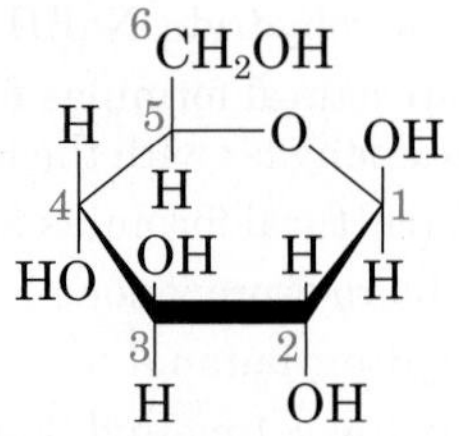

A cyclic hemiacetal is formed when the —OH group of one carbon bonds to the carbonyl group of another carbon. Which carbon in glucose provides the —OH group and which provides the CHO group?

17.74 Ribose, $C_5H_{10}O_5$, contains an aldehyde group but exists predominantly in the form of the cyclic hemiacetal shown here. We will discuss this cyclic form of ribose in Chapter 20.

β-D-Ribose

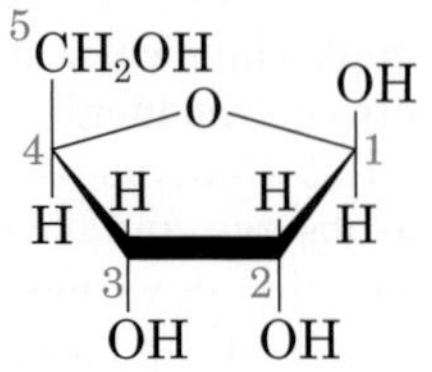

Which carbon of ribose provides the —OH group and which provides the CHO group for formation of this cyclic hemiacetal?

17.75 Name the functional groups in each steroid hormone. Methandrostenolone is a synthetic anabolic steroid; that is, it promotes tissue and muscle growth and development.

(a) What are the differences in the structural formula between testosterone and methandrostenolone? (Chapters 10, 14–15)

(b) Mark all stereocenters in each molecule and state the number of stereoisomers possible for each. The stereochemistry shown on the structural formulas is that of the biologically active stereoisomer. (Chapter 15)

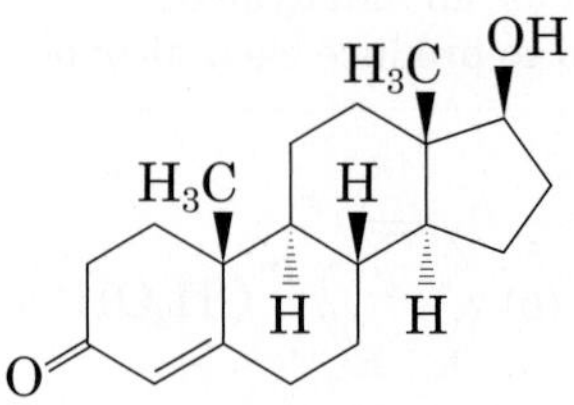

Testosterone   Methandrostenolone
(a)   (b)

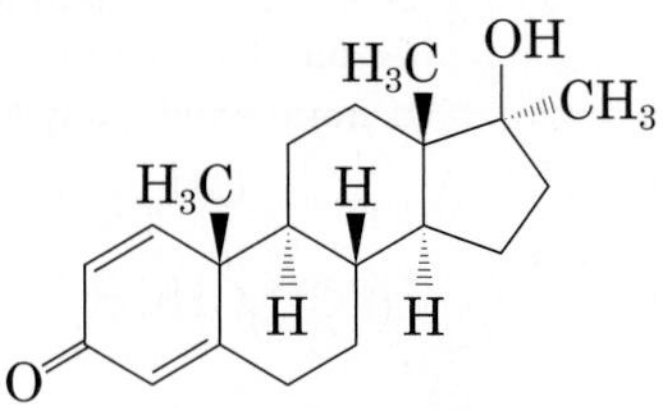

17.76 Sodium borohydride is a laboratory reducing agent. NADH is a biological reducing agent. In what way is the chemistry by which they reduce aldehydes and ketones similar?

Tying It Together

17.77 Complete the following equations for these reactions.

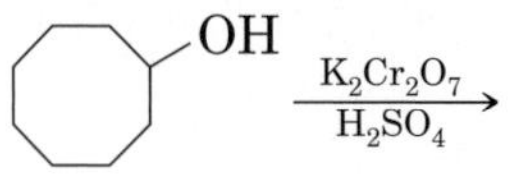

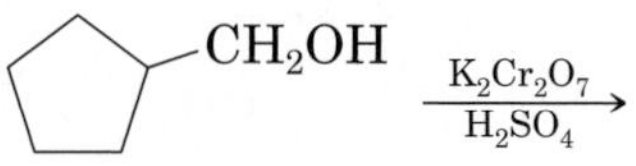

17.78 Write an equation for each conversion.

(a) 1-Pentanol to pentanal

(b) 1-Pentanol to pentanoic acid

(c) 2-Pentanol to 2-pentanone

(d) 2-Propanol to acetone

(e) Cyclohexanol to cyclohexanone

Carboxylic Acids

© Igor Dutina/iStockphoto.com

Citrus fruits are sources of citric acid, a tricarboxylic acid.

18.1 What Are Carboxylic Acids?

In this chapter, we study carboxylic acids, another class of organic compounds containing the carbonyl group. The functional group of a **carboxylic acid** is a **carboxyl group** (Section 10.4D), which can be represented in any one of three ways:

$$\underset{\underset{\displaystyle \overset{\|}{C}}{\|}}{\overset{O}{-}}-OH \qquad -COOH \qquad -CO_2H$$

18.2 How Do We Name Carboxylic Acids?

A. IUPAC Names

We derive the IUPAC name of an acyclic carboxylic acid from the name of the longest carbon chain that contains the carboxyl group. Drop the final *-e* from the name of the parent alkane and replace it with *-oic acid*. Number

OWL

Sign in to OWL at **www.cengage.com/owl** to view tutorials and simulations, develop problem-solving skills, and complete online homework assigned by your professor.

the chain beginning with the carbon of the carboxyl group. Because the carboxyl carbon is understood to be carbon 1, there is no need to give it a number. In the following examples, the common name is given in parentheses.

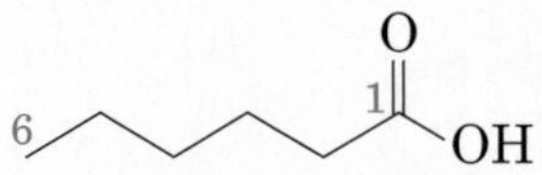

Hexanoic acid
(Caproic acid)

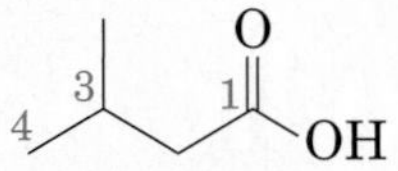

3-Methylbutanoic acid
(Isovaleric acid)

When a carboxylic acid also contains an —OH (hydroxyl) group, we indicate its presence by adding the prefix *hydroxy-*. When it contains a primary (1°) amine, we indicate the presence of the —NH₂ group with *amino-*.

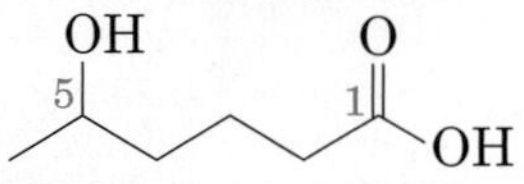

5-Hydroxyhexanoic acid

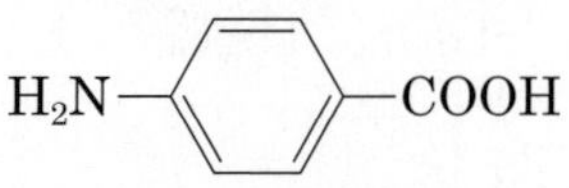

4-Aminobenzoic acid
(*p*-Aminobenzoic acid)

To name dicarboxylic acids, we add the suffix *-dioic acid* to the name of the parent alkane that contains both carboxyl groups. The numbers of the carboxyl carbons are not indicated because they can be only at the ends of the parent chain.

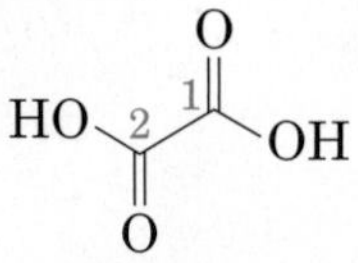

Ethanedioic acid
(Oxalic acid)

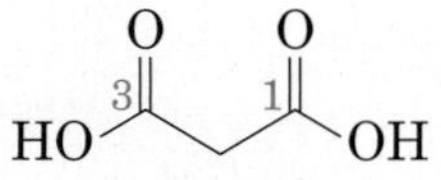

Propanedioic acid
(Malonic acid)

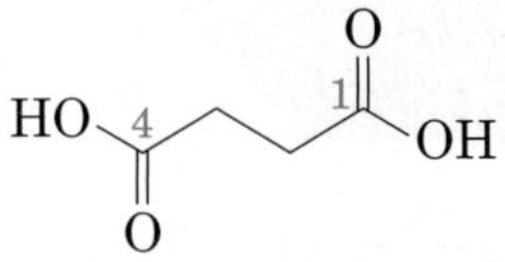

Butanedioic acid
(Succinic acid)

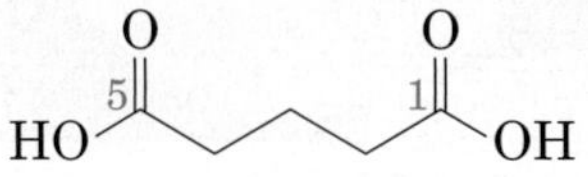

Pentanedioic acid
(Glutaric acid)

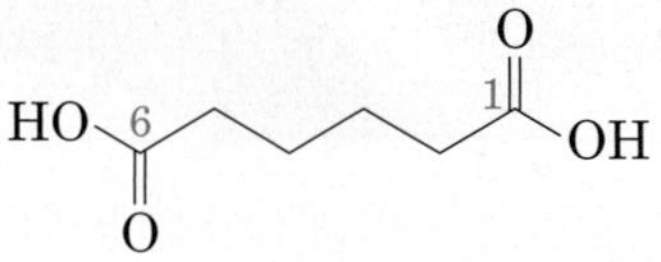

Hexanedioic acid
(Adipic acid)

The name *oxalic acid* is derived from one of its sources in the biological world—plants of the genus *Oxalis*, one of which is rhubarb. Oxalic acid also occurs in human and animal urine, and calcium oxalate is a major component of kidney stones. Succinic acid is an intermediate in the citric acid cycle (Section 27.4). Adipic acid is one of the two monomers required for the synthesis of the polymer nylon-66 (Section 19.6B).

B. Common Names

Common names for aliphatic carboxylic acids, many of which were known long before the development of IUPAC nomenclature, are often derived from the name of a natural source from which the acid could be isolated. Table 18.1 lists several of the unbranched aliphatic carboxylic acids found in the biological world along with the common name of each. Those with 16, 18, and 20 carbon atoms are particularly abundant in animal fats and vegetable oils (Section 21.2) and the phospholipid components of biological membranes (Section 21.5).

Formic acid was first obtained in 1670 from the destructive distillation of ants, whose Latin genus is *Formica*. It is one of the components of the venom injected by stinging ants.

Table 18.1 Several Aliphatic Carboxylic Acids and Their Common Names

Structure	IUPAC Name	Common Name	Derivation
HCOOH	methanoic acid	formic acid	Latin: *formica,* ant
CH$_3$COOH	ethanoic acid	acetic acid	Latin: *acetum,* vinegar
CH$_3$CH$_2$COOH	propanoic acid	propionic acid	Greek: *propion,* first fat
CH$_3$(CH$_2$)$_2$COOH	butanoic acid	butyric acid	Latin: *butyrum,* butter
CH$_3$(CH$_2$)$_3$COOH	pentanoic acid	valeric acid	Latin: *valere,* to be strong
CH$_3$(CH$_2$)$_4$COOH	hexanoic acid	caproic acid	Latin: *caper,* goat
CH$_3$(CH$_2$)$_6$COOH	octanoic acid	caprylic acid	Latin: *caper,* goat
CH$_3$(CH$_2$)$_8$COOH	decanoic acid	capric acid	Latin: *caper,* goat
CH$_3$(CH$_2$)$_{10}$COOH	dodecanoic acid	lauric acid	Latin: *laurus,* laurel
CH$_3$(CH$_2$)$_{12}$COOH	tetradecanoic acid	myristic acid	Greek: *myristikos,* fragrant
CH$_3$(CH$_2$)$_{14}$COOH	hexadecanoic acid	palmitic acid	Latin: *palma,* palm tree
CH$_3$(CH$_2$)$_{16}$COOH	octadecanoic acid	stearic acid	Greek: *stear,* solid fat
CH$_3$(CH$_2$)$_{18}$COOH	eicosanoic acid	arachidic acid	Greek: *arachis,* peanut

The unbranched carboxylic acids having between 12 and 20 carbon atoms are known as fatty acids. We study them further in Chapter 21.

When common names are used, the Greek letters alpha (α), beta (β), gamma (γ), and so forth, are often added as a prefix to locate substituents.

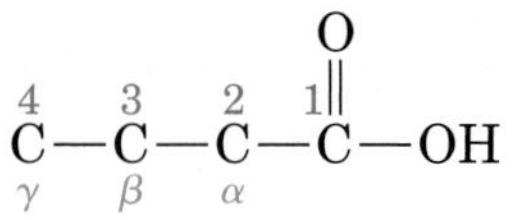

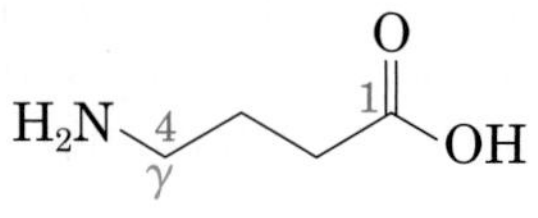

4-Aminobutanoic acid
(γ-Aminobutyric acid; GABA)

GABA is a neurotransmitter in the central nervous system.

OWL Interactive Example 18.1 IUPAC Names for Carboxylic Acids

Write the IUPAC name for each carboxylic acid:

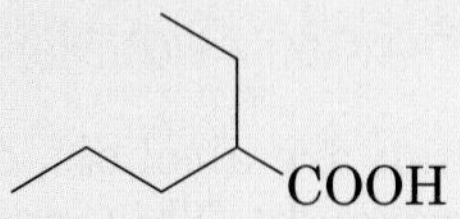

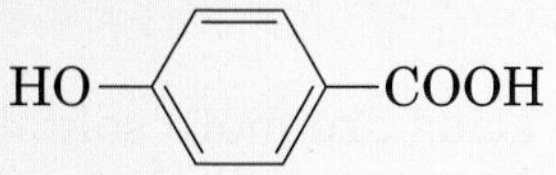

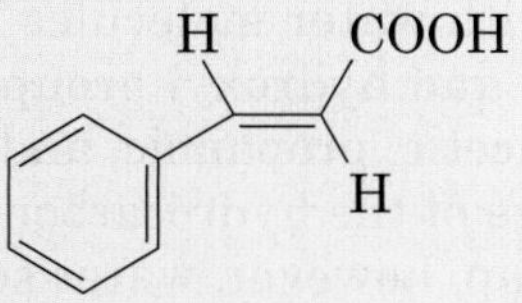

Strategy and Solution

(a) The longest carbon chain that contains the carboxyl group has five carbons; therefore, the parent alkane is pentane. The IUPAC name is 2-ethylpentanoic acid.

(b) 4-Hydroxybenzoic acid

(c) *trans*-3-Phenyl-2-propenoic acid (cinnamic acid)

Problem 18.1

Each of the following compounds has a well-recognized and widely used common name. A derivative of glyceric acid is an intermediate in glycolysis (Section 28.2). β-Alanine is a building block of pantothenic acid

(Section 27.5). Mevalonic acid is an intermediate in the biosynthesis of steroids (Section 27.4). Write the IUPAC name for each compound.

(a)

    COOH
    |
    CHOH
    |
    CH₂OH

    Glyceric acid

(b) $H_2NCH_2CH_2COOH$

    β-Alanine

(c)

    HO CH₃
              |
    HO   COOH

    Mevalonic acid

18.3 What Are the Physical Properties of Carboxylic Acids?

A major feature of carboxylic acids is the polarity of the carboxyl group (Figure 18.1). This group contains three polar covalent bonds: C=O, C—O, and O—H. The polarity of these bonds determines the major physical properties of carboxylic acids.

Carboxylic acids have significantly higher boiling points than other types of organic compounds of comparable molecular weight (Table 18.2). Their higher boiling points result from their polarity and the fact that hydrogen bonding between two carboxyl groups creates a dimer that behaves as a higher-molecular-weight compound.

Hydrogen bonding between two molecules

$$H_3C-C \overset{\displaystyle O \cdots H-O}{\underset{\displaystyle O-H\cdots O}{}} C-CH_3$$

A hydrogen-bonded dimer of acetic acid

Carboxylic acids are more soluble in water than are alcohols, ethers, aldehydes, and ketones of comparable molecular weight. This increased solubility is due to their strong association with water molecules by hydrogen bonding through both their carbonyl and hydroxyl groups. The first four aliphatic carboxylic acids (formic, acetic, propanoic, and butanoic) are infinitely soluble in water. As the size of the hydrocarbon chain increases relative to that of the carboxyl group, however, water solubility decreases. The solubility of hexanoic acid (six carbons) in water is 1.0 g/100 mL water.

FIGURE 18.1 Polarity of a carboxyl group.

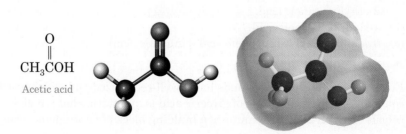

    O
    ‖
    CH₃COH

    Acetic acid

Table 18.2 Boiling Points and Solubilities in Water of Two Groups of Compounds of Comparable Molecular Weight

Structure	Name	Molecular Weight	Boiling Point (°C)	Solubility (g/100 mL H₂O)
CH_3COOH	acetic acid	60.1	118	infinite
$CH_3CH_2CH_2OH$	1-propanol	60.1	97	infinite
CH_3CH_2CHO	propanal	58.1	48	16
$CH_3(CH_2)_2COOH$	butanoic acid	88.1	163	infinite
$CH_3(CH_2)_3CH_2OH$	1-pentanol	88.1	137	2.3
$CH_3(CH_2)_3CHO$	pentanal	86.1	103	slight

We must mention two other properties of carboxylic acids. First, the liquid carboxylic acids from propanoic acid to decanoic acid have sharp, often disagreeable odors. Butanoic acid is found in stale perspiration and is a major component of "locker room odor." Pentanoic acid smells even worse, and goats, which secrete C_6, C_8, and C_{10} carboxylic acids (Table 18.1), are not famous for their pleasant odors. Second, carboxylic acids have a characteristic sour taste. The sour taste of pickles and sauerkraut, for example, is due to the presence of lactic acid. The sour tastes of limes (pH 1.9), lemons (pH 2.3), and grapefruit (pH 3.2) are due to the presence of citric and other acids.

18.4 What Are Soaps and Detergents?

A. Fatty Acids

More than 500 different **fatty acids** have been isolated from various cells and tissues. Given in Table 18.3 are common names and structural formulas for the most abundant fatty acids. The number of carbons in a fatty acid and the number of carbon–carbon double bonds in its hydrocarbon chain are shown by two numbers separated by a colon. In this notation, linoleic

Fatty acids Long, unbranched-chain carboxylic acids, most commonly consisting of 12 to 20 carbons. They are derived from the hydrolysis of animal fats, vegetable oils, and the phospholipids of biological membranes (Chapter 21).

Table 18.3 The Most Abundant Fatty Acids in Animal Fats, Vegetable Oils, and Biological Membranes

Carbon Atoms: Double Bonds*	Structure	Common Name	Melting Point (°C)
Saturated Fatty Acids			
12:0	$CH_3(CH_2)_{10}COOH$	lauric acid	44
14:0	$CH_3(CH_2)_{12}COOH$	myristic acid	58
16:0	$CH_3(CH_2)_{14}COOH$	palmitic acid	63
18:0	$CH_3(CH_2)_{16}COOH$	stearic acid	70
20:0	$CH_3(CH_2)_{18}COOH$	arachidic acid	77
Unsaturated Fatty Acids			
16:1	$CH_3(CH_2)_5CH{=}CH(CH_2)_7COOH$	palmitoleic acid	1
18:1	$CH_3(CH_2)_7CH{=}CH(CH_2)_7COOH$	oleic acid	16
18:2	$CH_3(CH_2)_4(CH{=}CHCH_2)_2(CH_2)_6COOH$	linoleic acid	−5
18:3	$CH_3CH_2(CH{=}CHCH_2)_3(CH_2)_6COOH$	linolenic acid	−11
20:4	$CH_3(CH_2)_4(CH{=}CHCH_2)_4(CH_2)_2COOH$	arachidonic acid	−49

*The first number is the number of carbons in the fatty acid; the second number is the number of carbon–carbon double bonds in its hydrocarbon chain.

acid, for example, is designated as an 18:2 fatty acid; its 18-carbon chain contains two carbon–carbon double bonds.

Following are several characteristics of the most abundant fatty acids in higher plants and animals:

1. Nearly all fatty acids have an even number of carbon atoms, most between 12 and 20, in an unbranched chain.

2. The three most abundant fatty acids in nature are palmitic acid (16:0), stearic acid (18:0), and oleic acid (18:1).

3. In most unsaturated fatty acids, the *cis* isomer predominates; the *trans* isomer is rare.

4. Unsaturated fatty acids have lower melting points than their saturated counterparts. The greater the degree of unsaturation, the lower the melting point. Compare, for example, the melting points of the following 18-carbon fatty acids: Linolenic acid, with three carbon–carbon double bonds, has the lowest melting point of these four fatty acids.

COOH   Stearic acid (18:0)
(mp 70°C)

COOH   Oleic acid (18:1)
(mp 16°C)

COOH   Linoleic acid (18:2)
(mp –5°C)

COOH   Linolenic acid (18:3)
(mp –11°C)

Fatty acids can be divided into two groups: saturated and unsaturated. Saturated fatty acids have only carbon–carbon single bonds in their hydrocarbon chains. Unsaturated fatty acids have at least one C=C double bond in the chain. All unsaturated fatty acids listed in Table 18.3 are the *cis* isomer.

Saturated fatty acids are solids at room temperature, because the regular nature of their hydrocarbon chains allows their molecules to pack together in close parallel alignment. When packed in this manner, the attractive interactions between adjacent hydrocarbon chains (London dispersion forces, Section 5.7A) are maximized. Although London dispersion forces are weak interactions, the regular packing of hydrocarbon chains allows these forces to operate over a large portion of their chains, ensuring that a considerable amount of energy is needed to separate and melt them.

Chemical Connections 18A

Trans Fatty Acids: What Are They and How Do You Avoid Them?

Animal fats are rich in saturated fatty acids, whereas plant oils (for example, corn, soybean, canola, olive, and palm oils) are rich in unsaturated fatty acids. Fats are added to processed foods to provide a desirable firmness along with a moist texture and pleasant taste. To supply the demand for dietary fats of the appropriate consistency, the *cis* double bonds of vegetable oils are partially hydrogenated. The greater the extent of hydrogenation, the higher the melting point of the triglyceride. The extent of hydrogenation is carefully controlled, usually by employing a Ni catalyst and a calculated amount of H_2 as a limiting reagent. Under these conditions, the H_2 is used up before all double bonds are reduced, so that only partial hydrogenation and the desired overall consistency is achieved. For example, by controlling the degree of hydrogenation, an oil with a melting point below room temperature can be converted to a semisolid or even a solid product.

The mechanism of catalytic hydrogenation of alkenes was discussed in Section 12.6D. Recall that a key step in this mechanism involves interaction of the carbon–carbon double bond of the alkene with the metal catalyst to form a carbon–metal bond. Because the interaction of a carbon–carbon double bond with the Ni catalyst is reversible, many of the double bonds remaining in the oil may be isomerized from the less stable *cis* configuration to the more stable *trans* configuration. Thus, the equilibration between the *cis* and *trans* configurations may occur when H_2 is the limiting reagent. For example, elaidic acid is the C_{18} *trans* fatty acid analog of oleic acid, a common C_{18} *cis* fatty acid.

The oils used for frying in fast-food restaurants are usually partially hydrogenated plant oils, and therefore, they contain substantial amounts of *trans* fatty acids that are transferred to the foods cooked in them. Other major sources of *trans* fatty acids in the diet include stick margarine, certain commercial bakery products, creme-filled cookies, potato and corn chips, frozen breakfast foods, and cake mixes.

Recent studies have shown that consuming significant amounts of *trans* fatty acids can lead to serious health problems related to serum cholesterol levels. Low overall serum cholesterol and a decreased ratio of low-density lipoprotein (LDL) cholesterol to high-density lipoprotein (HDL) cholesterol are associated with good cardiovascular health. High serum cholesterol levels and an elevated ratio of LDL cholesterol to HDL cholesterol are linked to a high incidence of cardiovascular disease, especially atherosclerosis. Research has indicated that consuming a diet high in either saturated fatty acids or *trans* fatty acids substantially increases the risk of cardiovascular disease.

The FDA has announced that processed foods must list the amount of *trans* fatty acids they contain, so that consumers can make better choices about the foods they eat. A diet low in saturated and *trans* fatty acids is recommended, along with consumption of more fish, whole grains, fruits, and vegetables, and especially daily exercise, which is tremendously beneficial regardless of diet.

Monounsaturated and polyunsaturated fatty acids have not produced similar health risks in most studies, although too much fat of any kind in the diet can lead to obesity, a major health problem that is associated with several diseases, one of which is diabetes. Some polyunsaturated (*cis*) fatty acids, such as those found in certain types of fish, have even been shown to have beneficial effects in some studies. These are the so-called omega-3 fatty acids.

In omega-3 fatty acids, the last carbon of the last double bond of the hydrocarbon chain ends three carbons in from the methyl terminal end of the chain. The last carbon of the hydrocarbon chain is called the omega (the last letter of the Greek alphabet) carbon—hence the designation of omega-3. The two most commonly found in health food supplements are eicosapentaenoic acid and docosahexaenoic acid.

Eicosapentaenoic acid, $C_{20}H_{30}O_2$, is an important fatty acid in the marine food chain and serves as a precursor in humans of several members of the prostacyclin and thromboxane families (Chapter 21). Note how the name of this fatty acid is derived. *Eicosa-* is the prefix indicating 20 carbons in the chain, *-pentaene-* indicates five carbon–carbon double bonds, and *–oic acid* shows the carboxyl functional group.

Docosahexaenoic acid, $C_{22}H_{32}O_2$, is found in fish oils and many phospholipids. It is a major structural component of excitable membranes in the retina and brain and is synthesized in the liver from linoleic acid.

Elaidic acid
mp 46°C
(a *trans* C_{18} fatty acid)

Trans Fatty Acids: What Are They and How Do You Avoid Them? (*continued*)

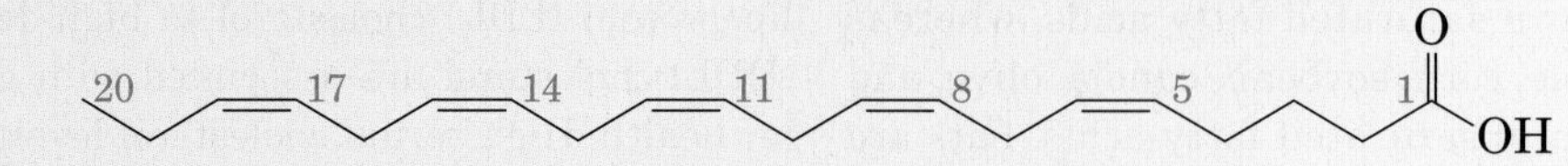

Eicosapentaenoic acid

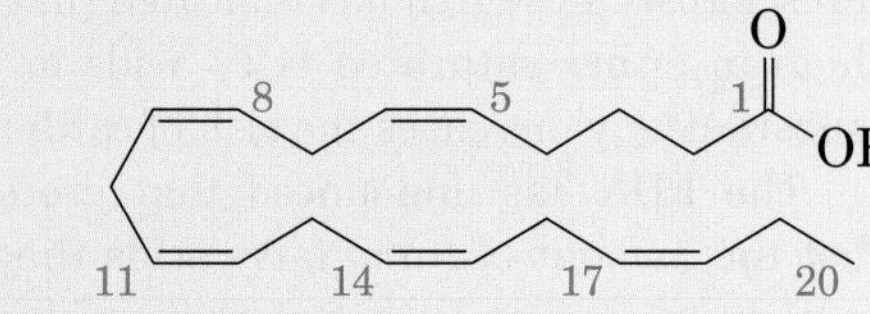

Eicosapentaenoic acid, drawn in a more compact form

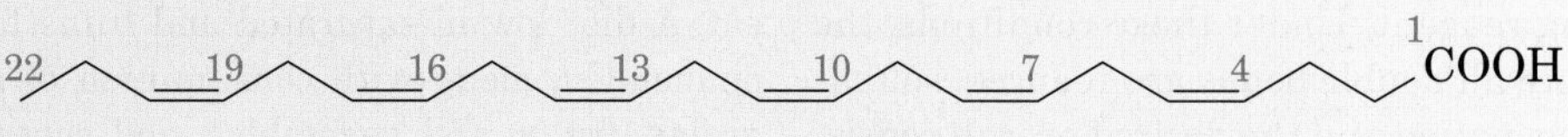

Docosahexaenoic acid

All common *cis* unsaturated fatty acids are liquids at room temperature because the *cis* double bonds interrupt the regular packing of the chains and the London dispersion forces act over only shorter segments of the chains, so that less energy is required to melt them. The greater the degree of unsaturation, the lower the melting point, because each double bond interrupts the regular packing of the fatty acid molecules.

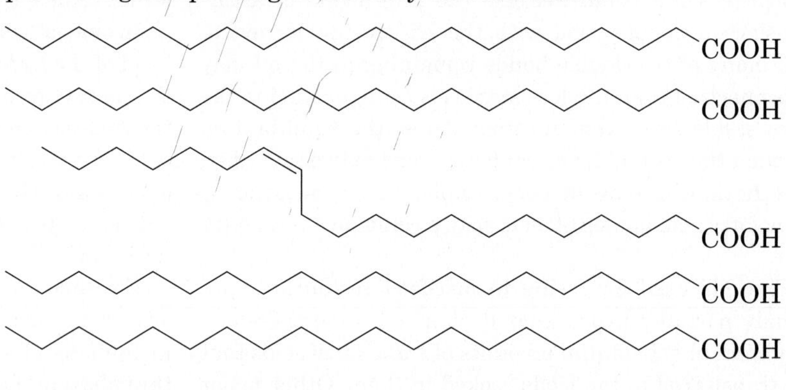

B. Structure and Preparation of Soaps

Natural **soaps** are most commonly prepared from a blend of tallow and coconut oils. In the preparation from tallow, the solid fats of cattle are melted with steam, and the tallow layer that forms on the top is removed. The preparation of soaps begins by boiling these triglycerides with sodium hydroxide. The reaction that takes place is called **saponification** (Latin: *saponem*, "soap"):

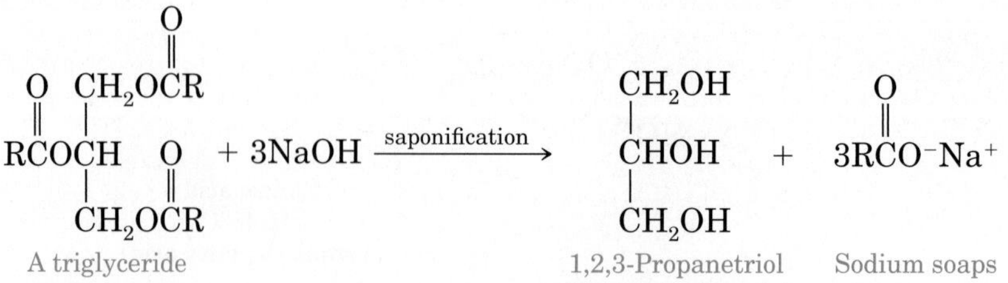

Saponification The hydrolysis of an ester in aqueous NaOH or KOH to an alcohol and the sodium or potassium salt of a carboxylic acid (Section 19.4A)

A triglyceride + 3NaOH → saponification → 1,2,3-Propanetriol (Glycerol; Glycerin) + 3RCO⁻Na⁺ Sodium soaps

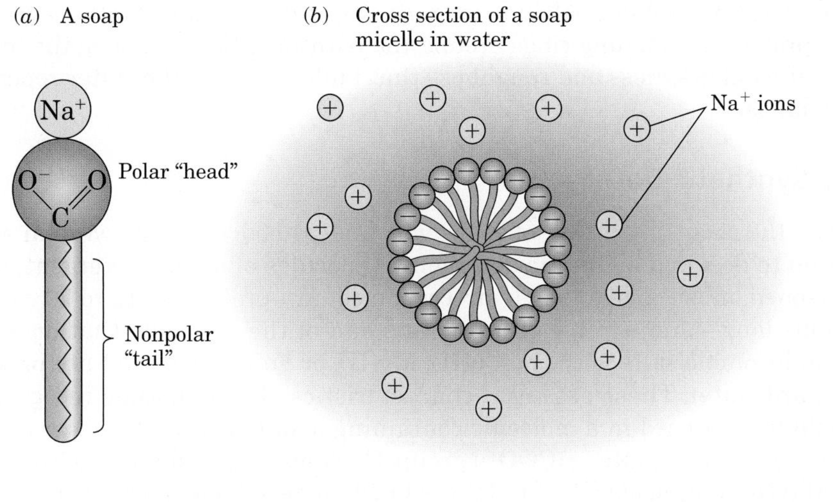

(a) A soap

(b) Cross section of a soap micelle in water

Na⁺

Polar "head"

O⁻   O
   C

Nonpolar "tail"

Na⁺ ions

FIGURE 18.2 Soap micelles. Nonpolar (hydrophobic) hydrocarbon chains cluster in the interior of the micelle, and polar (hydrophilic) carboxylate groups are on the surface of the micelle. Soap micelles repel each other because of their negative surface charges.

At the molecular level, saponification corresponds to base-promoted hydrolysis (Section 19.4A) of the ester groups in triglycerides. A triglyceride is a triester of glycerol. The resulting soaps contain mainly the sodium salts of palmitic, stearic, and oleic acids from tallow and the sodium salts of lauric and myristic acids from coconut oil.

After hydrolysis is complete, sodium chloride is added to precipitate the sodium salts as thick curds of soap. The water layer is then drawn off, and glycerol is recovered by vacuum distillation. The crude soap contains sodium chloride, sodium hydroxide, and other impurities that are removed by boiling the curd in water and reprecipitating with more sodium chloride. After several purifications, the soap can be used as an inexpensive industrial soap without further processing. Other treatments transform the crude soap into pH-controlled cosmetic soaps, medicated soaps, and the like.

C. How Soap Cleans

Soap owes its remarkable cleansing properties to its ability to act as an emulsifying agent. Because the long hydrocarbon chains of natural soaps are insoluble in water, soap molecules tend to cluster in such a way as to minimize contact of their hydrocarbon chains with surrounding water molecules. The polar carboxylate groups, by contrast, tend to remain in contact with the surrounding water molecules. Thus, in water, soap molecules spontaneously cluster into **micelles** (Figure 18.2).

Many of the things we commonly think of as dirt (such as grease, oil, and fat stains) are nonpolar and insoluble in water. When soap and this type of dirt are mixed together, as in a washing machine, the nonpolar hydrocarbon inner parts of the soap micelles "dissolve" the nonpolar substances. In effect, new soap micelles form, with the nonpolar dirt molecules in the center (Figure 18.3). In this way, nonpolar organic grease, oil, and fat are "dissolved" and washed away in the polar wash water.

Soaps, however, have their disadvantages, foremost among which is the fact that they form water-insoluble salts when used in water containing Ca(II), Mg(II), or Fe(III) ions (hard water):

Micelle A spherical arrangement of molecules in aqueous solution such that their hydrophobic (water-hating) parts are shielded from the aqueous environment and their hydrophilic (water-loving) parts are on the surface of the sphere and in contact with the aqueous environment.

Soap micelle with "dissolved" grease

Grease

Soap

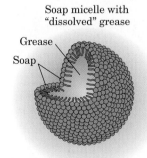

FIGURE 18.3 A soap micelle with a "dissolved" oil or grease droplet.

$$2CH_3(CH_2)_{14}COO^-Na^+ + Ca^{2+} \longrightarrow [CH_3(CH_2)_{14}COO^-]_2Ca^{2+} + 2Na^+$$

A sodium soap
(soluble in water as micelles)

Calcium salt of a fatty acid
(insoluble in water)

These water-insoluble calcium, magnesium, and iron salts of fatty acids create problems, including rings around the bathtub, films that spoil the luster of hair, and grayness and roughness that build up on textiles after repeated washings.

D. Synthetic Detergents

After the cleansing action of soaps was understood, chemists were in a position to design a synthetic detergent. Molecules of a good detergent, they reasoned, must have a long hydrocarbon chain—preferably 12 to 20 carbon atoms long—and a polar group at one end of the molecule that does not form insoluble salts with the Ca(II), Mg(II), or Fe(III) ions that are present in hard water. These essential characteristics of a soap, they recognized, could be produced in a molecule containing a sulfonate ($-SO_3^-$) group instead of a carboxylate ($-COO^-$) group. Calcium, magnesium, and iron salts of alkylsulfonic acids ($R-SO_3H$) are much more soluble in water than comparable salts of fatty acids.

The most widely used synthetic detergents today are the linear alkylbenzenesulfonates (LAS). One of the most common of these is sodium 4-dodecylbenzenesulfonate. To prepare this type of detergent, a linear alkylbenzene is treated with sulfuric acid to form an alkylbenzenesulfonic acid (Section 13.3C), followed by neutralization of the sulfonic acid with sodium hydroxide:

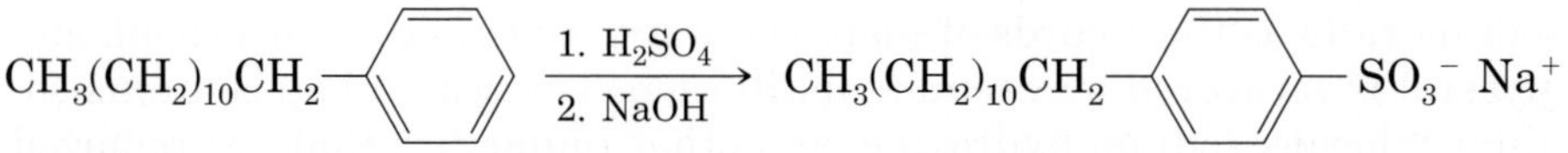

$$CH_3(CH_2)_{10}CH_2 - \bigcirc \xrightarrow[\text{2. NaOH}]{\text{1. } H_2SO_4} CH_3(CH_2)_{10}CH_2 - \bigcirc - SO_3^- \ Na^+$$

Dodecylbenzene   Sodium 4-dodecylbenzenesulfonate
               (an anionic detergent)

The product is mixed with builders and then spray-dried to give a smooth-flowing powder. The most common builder is sodium silicate. Alkylbenzenesulfonate detergents were introduced in the late 1950s, and today they account for close to 90% of the market that was once held by natural soaps.

Among the most common additives to detergent preparations are foam stabilizers, bleaches, and optical brighteners. A common foam stabilizer added to liquid soaps, but not to laundry detergents (for obvious reasons: think of a top-loading washing machine with foam spewing out of the lid!), is the amide prepared from dodecanoic acid (lauric acid) and 2-aminoethanol (ethanolamine). The most common bleach is sodium perborate tetrahydrate, which decomposes at temperatures higher than 50°C to give hydrogen peroxide, the actual bleaching agent.

$$CH_3(CH_2)_{10} \overset{\overset{\displaystyle O}{\|}}{C} NHCH_2CH_2OH \qquad\qquad O{=}B{-}O{-}O^- \ Na^+ \bullet \ 4H_2O$$

N-(2-Hydroxyethyl)dodecanamide   Sodium perborate tetrahydrate
(a foam stabilizer)   (a bleach)

Also added to laundry detergents are optical brighteners (optical bleaches). These substances are absorbed into fabrics and, after absorbing ambient light, fluoresce with a blue color, offsetting the yellow color that develops in fabric as it ages. Optical brighteners produce a "whiter-than-white" appearance. You most certainly have observed their effects if you have seen the glow of white T-shirts or blouses when they are exposed to black light (UV radiation).

18.5 What Are the Characteristic Reactions of Carboxylic Acids?

A. Acidity

Carboxylic acids are weak acids. Values of K_a for most unsubstituted aliphatic and aromatic carboxylic acids fall within the range of 10^{-4} to 10^{-5} (pK_a = 4.0–5.0). The value of K_a for acetic acid, for example, is 1.74×10^{-5}; its pK_a is 4.76 (Section 8.5).

$$CH_3\overset{\overset{\displaystyle O}{\|}}{C}OH + H_2O \rightleftharpoons CH_3\overset{\overset{\displaystyle O}{\|}}{C}O^- + H_3O^+ \qquad K_a = \frac{[CH_3COO^-][H_3O^+]}{[CH_3COOH]} = 1.74 \times 10^{-5}$$

$$pK_a = 4.76$$

Substituents of high electronegativity (especially —OH, —Cl, and —NH$_3^+$) near the carboxyl group increase the acidity of carboxylic acids, often by several orders of magnitude. Compare, for example, the acidities of acetic acid and the chlorine-substituted acetic acids. Both dichloroacetic acid and trichloroacetic acid are stronger acids than acetic acid (pK_a 4.75) and H_3PO_4 (pK_a 2.12).

Dichloroacetic acid is used as a topical astringent and as a treatment for genital warts in males.

Dentists use a 50% aqueous solution of trichloroacetic acid to cauterize gums. This strong acid stops the bleeding, kills diseased tissue, and allows the growth of healthy gum tissue.

Formula:	CH_3COOH	$ClCH_2COOH$	$Cl_2CHCOOH$	Cl_3CCOOH
Name:	Acetic acid	Chloroacetic acid	Dichloroacetic acid	Trichloroacetic acid
pK_a:	4.76	2.86	1.48	0.70

Increasing acid strength →

Electronegative atoms on the carbon adjacent to a carboxyl group increase acidity because they pull electron density away from the O—H bond, thereby facilitating ionization of the carboxyl group and making it a stronger acid.

One final point about carboxylic acids: When a carboxylic acid is dissolved in an aqueous solution, the form of the carboxylic acid present depends on the pH of the solution in which it is dissolved. Consider typical carboxylic acids, which have pK_a values in the range of 4.0 to 5.0. When the pH of the solution is equal to the pK_a of the carboxylic acid (that is, when the pH of the solution is in the range 4.0–5.0), the acid, RCOOH, and its conjugate base, RCOO$^-$, are present in equal concentrations, which we can demonstrate by using the Henderson-Hasselbalch equation (Section 8.11).

$$pH = pK_a + \log\frac{[A^-]}{[HA]} \qquad \text{Henderson-Hasselbalch Equation}$$

Consider the ionization of a weak acid, HA, in aqueous solution. When the pH of the solution is equal to the pK_a of the carboxylic acid, the Henderson-Hasselbalch equation reduces to:

$$\log\frac{[A^-]}{[HA]} = 0$$

Taking the antilog gives us the ratio of $[A^-]$ to $[HA]$ and tells us that the concentrations of the two are equal.

$$\frac{[A^-]}{[HA]} = 1$$

If the pH of the solution is adjusted to 2.0 or lower by the addition of a strong acid, the carboxylic acid then is present in solution almost entirely as RCOOH. If the pH of the solution is adjusted to 7.0 or higher, the carboxylic acid is present almost entirely as its anion. Thus, even in a neutral solution (pH 7.0), a carboxylic acid is present predominantly as its anion.

$$\underset{\substack{\text{Predominant}\\\text{species when}\\\text{pH of the}\\\text{solution is} \leq 2.0}}{\text{R}-\overset{\displaystyle \overset{O}{\|}}{\text{C}}-\text{OH}} \underset{\text{H}^+}{\overset{\text{OH}^-}{\rightleftharpoons}} \underset{\substack{\text{Present in equal}\\\text{concentrations when}\\\text{pH of the solution} =\\\text{p}K_a \text{ of the acid}}}{\text{R}-\overset{\displaystyle \overset{O}{\|}}{\text{C}}-\text{OH} \; + \; \text{R}-\overset{\displaystyle \overset{O}{\|}}{\text{C}}-\text{O}^-} \underset{\text{H}^+}{\overset{\text{OH}^-}{\rightleftharpoons}} \underset{\substack{\text{Predominant}\\\text{species when}\\\text{pH of the}\\\text{solution is} \geq 7.0}}{\text{R}-\overset{\displaystyle \overset{O}{\|}}{\text{C}}-\text{O}^-}$$

B. Reaction with Bases

All carboxylic acids, whether soluble or insoluble in water, react with NaOH, KOH, and other strong bases to form water-soluble salts.

$$\underset{\substack{\text{Benzoic acid}\\\text{(slightly soluble in water)}}}{\text{C}_6\text{H}_5-\text{COOH}} \; + \; \text{NaOH} \xrightarrow{\text{H}_2\text{O}} \underset{\substack{\text{Sodium benzoate}\\\text{(60 g/100 mL water)}}}{\text{C}_6\text{H}_5-\text{COO}^-\text{Na}^+} + \text{H}_2\text{O}$$

Sodium benzoate and calcium propanoate are fungal growth inhibitors that are added to baked goods "to retard spoilage."

Sodium benzoate, a fungal growth inhibitor, is often added to baked goods "to retard spoilage." Calcium propanoate is used for the same purpose. Carboxylic acids also form water-soluble salts with ammonia and amines.

$$\underset{\substack{\text{Benzoic acid}\\\text{(slightly soluble in water)}}}{\text{C}_6\text{H}_5-\text{COOH}} \; + \; \text{NH}_3 \xrightarrow{\text{H}_2\text{O}} \underset{\substack{\text{Ammonium benzoate}\\\text{(20 g/100 mL water)}}}{\text{C}_6\text{H}_5-\text{COO}^-\text{NH}_4^+}$$

Carboxylic acids react with sodium bicarbonate and sodium carbonate to form water-soluble sodium salts and carbonic acid (a weaker acid). Carbonic acid, in turn, decomposes to give water and carbon dioxide, which evolves as a gas (Section 8.6E).

$$\text{CH}_3\text{COOH(aq)} + \text{NaHCO}_3(\text{aq}) \longrightarrow \text{CH}_3\text{COO}^-\text{Na}^+(\text{aq}) + \text{CO}_2(\text{g}) + \text{H}_2\text{O(l)}$$

Salts of carboxylic acids are named in the same manner as the salts of inorganic acids: The cation is named first and then the anion. The name of the anion is derived from the name of the carboxylic acid by dropping the suffix *-ic acid* and adding the suffix *-ate*.

Example 18.2 Acidity of Carboxylic Acids

Complete each acid–base reaction and name the carboxylate salt formed.

(a) CH₃CH₂CH₂COOH + NaOH ⟶

(b) (with OH) COOH + NaHCO₃ ⟶
     (s)-Lactic acid

Strategy and Solution

Each carboxylic acid is converted to its sodium salt. In (b), carbonic acid is formed and decomposes to carbon dioxide and water.

(a) $\diagup\!\!\diagdown\!\!\diagup$ COOH + NaOH $\longrightarrow$ $\diagup\!\!\diagdown\!\!\diagup$ COO$^-$Na$^+$ + H$_2$O

    Butanoic acid   Sodium butanoate

         OH   OH

(b) $\diagup\!\!\diagdown$ COOH + NaHCO$_3$ $\longrightarrow$ $\diagup\!\!\diagdown$ COO$^-$Na$^+$ + H$_2$O + CO$_2$

    (s)-Lactic acid   Sodium (s)-lactate

Problem 18.2

Write equations for the reaction of each acid in Example 18.2 with ammonia and name the carboxylate salt formed.

A consequence of the water solubility of carboxylic acid salts is that water-insoluble carboxylic acids can be converted to water-soluble ammonium or alkali metal salts and extracted into aqueous solution. The salt, in turn, can be transformed back to the free carboxylic acid by treatment with HCl, H$_2$SO$_4$, or another strong acid. These reactions allow an easy separation of water-insoluble carboxylic acids from water-insoluble nonacidic compounds.

Shown in Figure 18.4 is a flowchart for the separation of benzoic acid, a water-insoluble carboxylic acid, from benzyl alcohol, a nonacidic

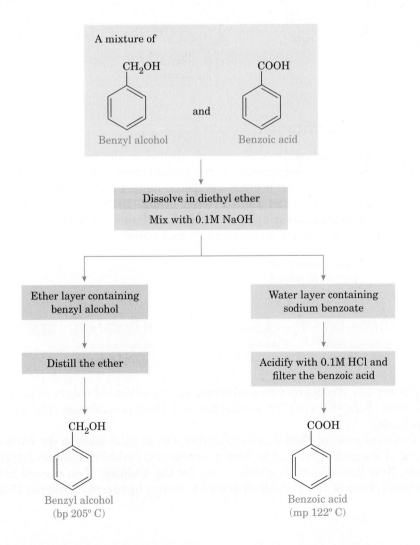

FIGURE 18.4 Flowchart for separation of benzoic acid from benzyl alcohol.

compound. First, the mixture of benzoic acid and benzyl alcohol is dissolved in diethyl ether. When the ether solution is shaken with aqueous NaOH or another strong base, benzoic acid is converted to its water-soluble sodium salt. Then the ether and aqueous phases are separated. The ether solution is distilled, yielding first diethyl ether (bp 35°C) and then benzyl alcohol (bp 205°C). The aqueous solution is acidified with HCl, and benzoic acid precipitates as a white crystalline solid (mp 122°C), which is recovered by filtration.

C. Reduction

The carboxyl group is one of the organic functional groups that is most resistant to reduction. It is not affected by catalytic reduction under conditions that readily reduce alkenes to alkanes (Section 12.6D) or by sodium borohydride, ($NaBH_4$), which readily reduces aldehydes to 1° alcohols and ketones to 2° alcohols (Section 17.4B).

The most common reagent for the reduction of a carboxylic acid to a 1° alcohol is the very powerful reducing agent lithium aluminum hydride, $LiAlH_4$. Reduction of a carboxyl group with this reagent is commonly carried out in diethyl ether. The initial product is an aluminum alkoxide, which is then treated with water to give the primary alcohol and lithium and aluminum hydroxides. These two hydroxides are insoluble in diethyl ether and are removed by filtration. Evaporation of the ether solvent yields the primary alcohol.

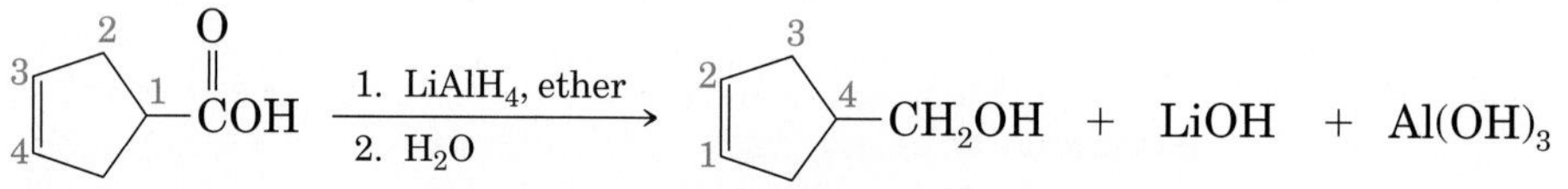

3-Cyclopentene-carboxylic acid → 4-Hydroxymethyl-cyclopentene + LiOH + $Al(OH)_3$

D. Fischer Esterification

Ester A compound in which the —OH of a carboxyl group, RCOOH, is replaced with an alkoxy (—OR) or aryloxy (—OAr) group

Fischer esterification The process of forming an ester by refluxing a carboxylic acid and an alcohol in the presence of an acid catalyst, commonly sulfuric acid

Treatment of a carboxylic acid with an alcohol in the presence of an acid catalyst—most commonly, concentrated sulfuric acid—gives an **ester**. This method of forming an ester is given the special name **Fischer esterification**, after the German chemist Emil Fischer (1852–1919). As an example of Fischer esterification, treating acetic acid with ethanol in the presence of concentrated sulfuric acid gives ethyl acetate and water:

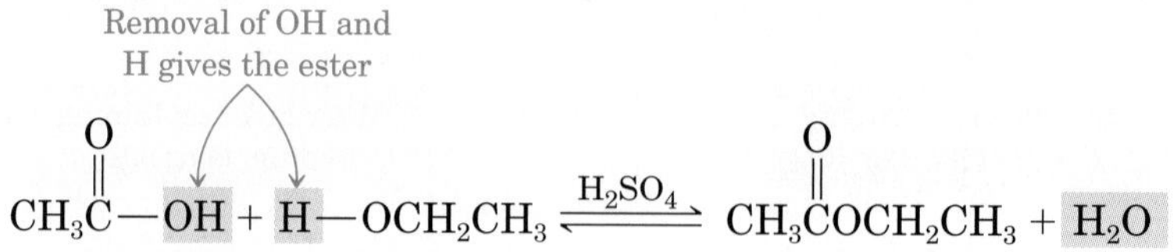

Removal of OH and H gives the ester

CH_3C—OH + H—OCH_2CH_3 $\xrightarrow{H_2SO_4}$ $CH_3COCH_2CH_3$ + H_2O

Ethanoic acid (Acetic acid) — Ethanol (Ethyl alcohol) — Ethyl ethanoate (Ethyl acetate)

We study the structure, nomenclature, and reactions of esters in detail in Chapter 19. In this chapter, we discuss only their preparation from carboxylic acids.

In the process of Fischer esterification, the alcohol adds to the carbonyl group of the carboxylic acid to form a tetrahedral carbonyl addition intermediate. Note how closely this step resembles the addition of an alcohol to the carbonyl group of an aldehyde or ketone to form a hemiacetal (Section 17.4C).

In the case of Fischer esterification, the intermediate collapses by loss of a molecule of water to give the ester.

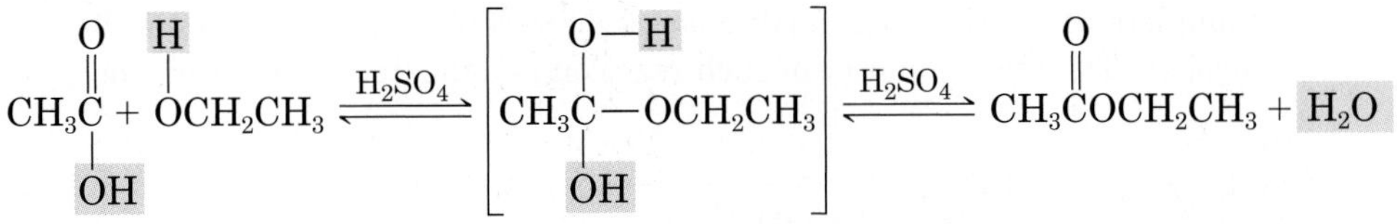

A tetrahedral carbonyl addition intermediate

Acid-catalyzed esterification is reversible, and at equilibrium, the quantities of remaining carboxylic acid and alcohol are generally appreciable. By controlling the experimental conditions, however, we can use Fischer esterification to prepare esters in high yields. If the alcohol is inexpensive compared with the carboxylic acid, we can use a large excess of the alcohol (one of the starting reagents) to drive the equilibrium to the right and achieve a high conversion of carboxylic acid to its ester. Alternatively, we can remove water (one of the products of the reaction) as it forms and drive the equilibrium to the right (review Le Chatelier's principle, Section 7.7).

Chemical Connections 18B

Esters as Flavoring Agents

Flavoring agents are the largest class of food additives. At present, more than 1000 synthetic and natural flavors are available. The majority of these are concentrates or extracts from the material whose flavor is desired; they are often complex mixtures of tens to hundreds of compounds. A number of ester flavoring agents are synthesized industrially. Many have flavors very close to the target flavor, and so adding only one or a few of them is sufficient to make ice cream, soft drinks, or candies taste natural. The table shows the structures of a few of the esters used as flavoring agents.

Structure	Name	Flavor
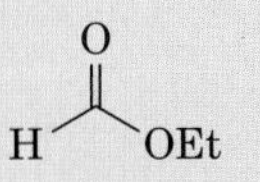	Ethyl formate	Rum
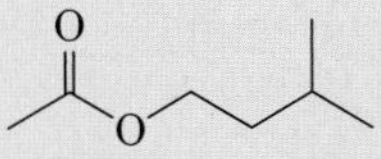	Isopentyl acetate	Banana
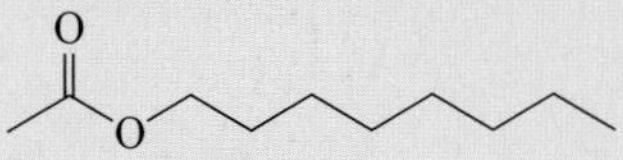	Octyl acetate	Orange
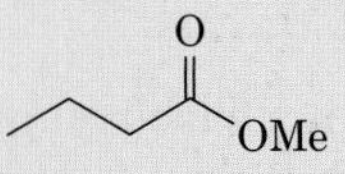	Methyl butanoate	Apple
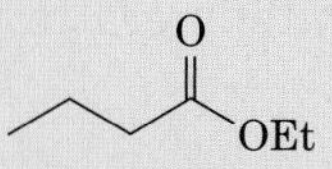	Ethyl butanoate	Pineapple
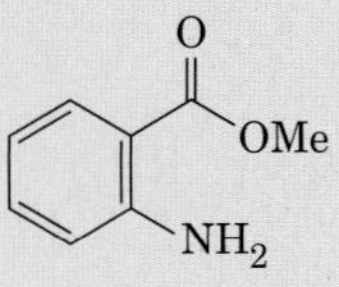	Methyl 2-aminobenzoate (Methyl anthranilate)	Grape

Example 18.3 Fischer Esterification

Complete these Fischer esterification reactions (assume an excess of the alcohol). The stoichiometry of each reaction is indicated in the problem.

(a)

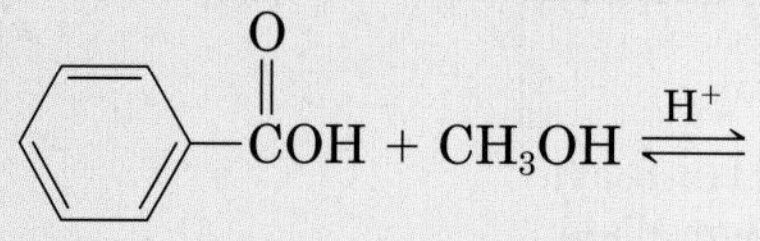

Benzoic acid

$$\text{+ CH}_3\text{OH} \xrightleftharpoons{\text{H}^+}$$

(b)

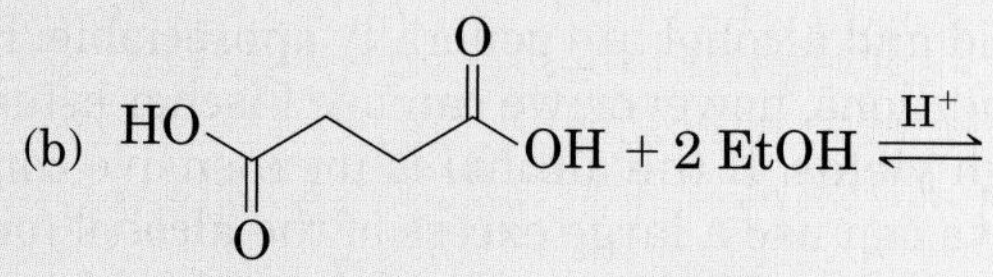

Butanedioic acid   (excess)
(Succinic acid)

$$\text{+ 2 EtOH} \xrightleftharpoons{\text{H}^+}$$

Strategy and Solution

Substitution of the —OR group of the alcohol for the —OH group of the carboxyl group gives an ester. Here are the structural formulas and names for the ester produced in each reaction.

(a)

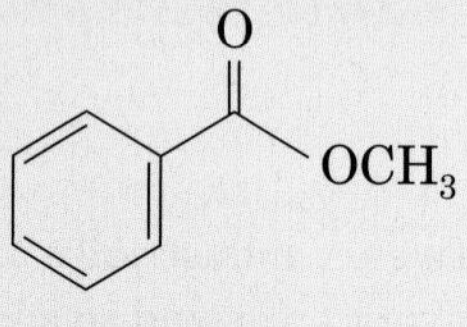

Methyl benzoate

(b)

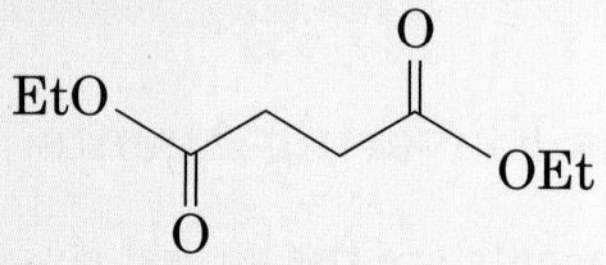

Diethyl butanedioate
(Diethyl succinate)

Problem 18.3

Complete these Fischer esterification reactions:

(a)

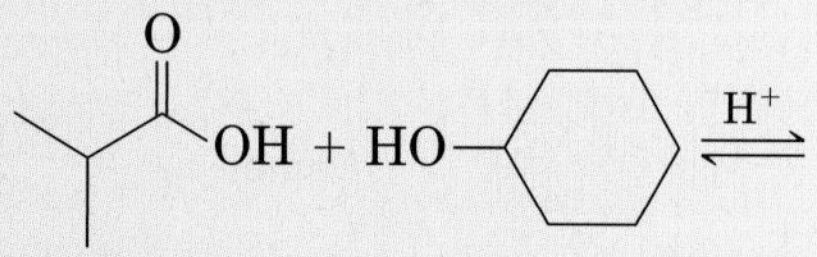

$$\xrightleftharpoons{\text{H}^+}$$

(b)

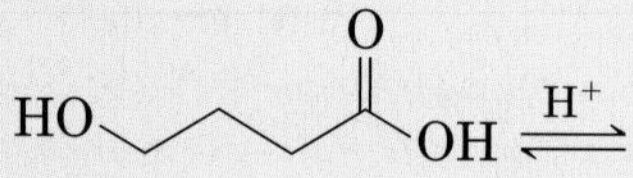

$$\xrightleftharpoons{\text{H}^+} \text{(a cyclic ester)}$$

E. Decarboxylation

Decarboxylation is the loss of CO_2 from a carboxyl group. Almost any carboxylic acid, when heated to a very high temperature, undergoes thermal decarboxylation:

$$\underset{\text{RCOH}}{\overset{\text{O}}{\|}} \xrightarrow[\text{high temperature}]{\text{decarboxylation}} \text{RH} + CO_2$$

Most carboxylic acids, however, are quite resistant to moderate heat and melt or even boil without decarboxylation. Exceptions are carboxylic acids that have a carbonyl group β to the carboxyl group. This type of carboxylic acid undergoes decarboxylation quite readily on mild heating. For example, when 3-oxobutanoic acid (acetoacetic acid) is heated moderately, it undergoes decarboxylation to give acetone and carbon dioxide:

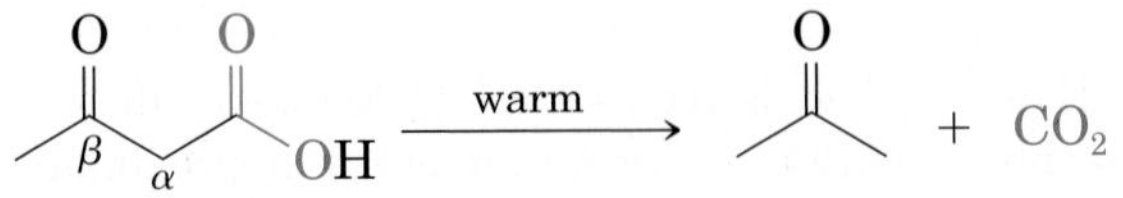

3-Oxobutanoic acid
(Acetoacetic acid)
(an α-ketoacid)

Acetone

Decarboxylation on moderate heating is a unique property of β-ketoacids and is not observed with other classes of ketoacids.

Mechanism: Decarboxylation of a β-Ketocarboxylic Acid

Step 1: Redistribution of six electrons in a cyclic six-membered transition state gives carbon dioxide and an enol:

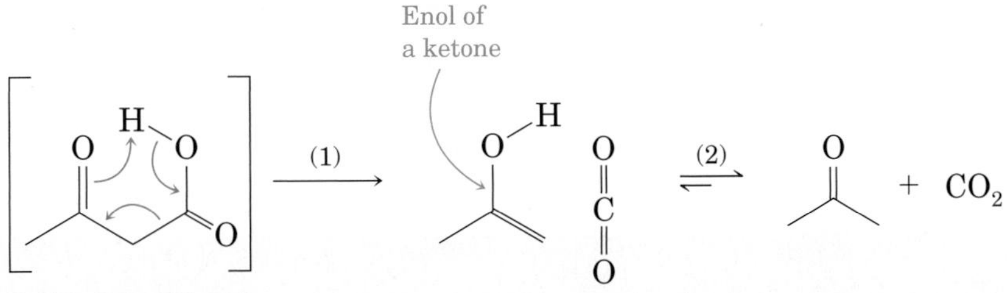

Enol of a ketone

Cyclic six-membered transition state

Step 2: Keto-enol tautomerism (Section 17.5) of the enol gives the more stable keto form of the product:

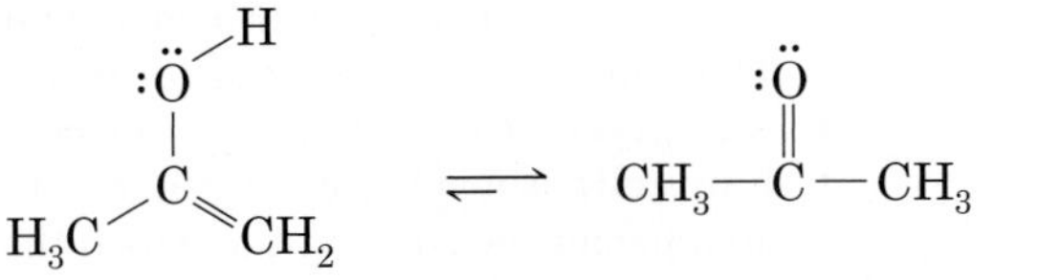

An important example of decarboxylation of a β-ketoacid in the biological world occurs during the oxidation of foodstuffs in the tricarboxylic acid (TCA) cycle (Chapter 28). Oxalosuccinic acid, one of the intermediates in this cycle, undergoes spontaneous decarboxylation to produce α-ketoglutaric acid. Only one of the three carboxyl groups of oxalosuccinic acid has a carbonyl group in the position β to it, and this carboxyl group is lost as CO_2:

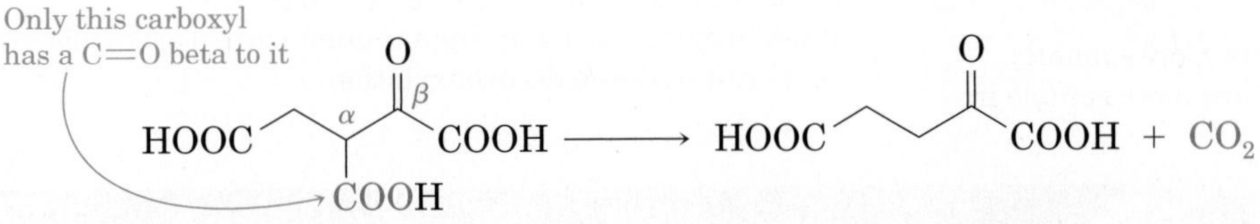

Oxalosuccinic acid

α-Ketoglutaric acid

Note that thermal decarboxylation is a reaction unique to β-ketoacids—it does not occur with α-ketoacids. In the biochemistry chapters that follow, however, we will see examples of decarboxylation of α-ketoacids—as for example, the decarboxylation of α-ketoglutarate. Because the decarboxylation of α-ketoacids requires an oxidizing agent (NAD^+), this reaction is called oxidative decarboxylation.

Chemical Connections 18C

Ketone Bodies and Diabetes

3-Oxobutanoic acid (acetoacetic acid) and its reduction product, 3-hydroxybutanoic acid, are synthesized in the liver from acetyl-CoA (Section 28.5), a product of the metabolism of fatty acids and certain amino acids.

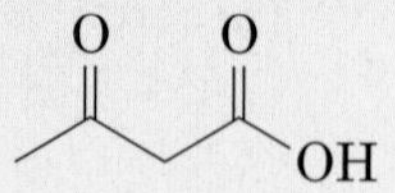

3-Oxobutanoic acid
(Acetoacetic acid)

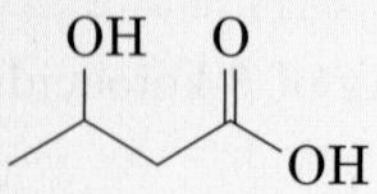

3-Hydroxybutanoic acid
(β-Hydroxybutyric acid)

3-Oxobutanoic acid and 3-hydroxybutanoic acid are known collectively as ketone bodies.

The concentration of ketone bodies in the blood of healthy, well-fed humans is approximately 0.01 mM/L. However, in persons suffering from starvation or diabetes mellitus, the concentration of ketone bodies may increase to as much as 500 times the normal level. Under these conditions, the concentration of acetoacetic acid increases to the point where it undergoes spontaneous decarboxylation to form acetone and carbon dioxide. Acetone is not metabolized by humans and is excreted through the kidneys and the lungs. The odor of acetone is responsible for the characteristic "sweet smell" on the breath of severely diabetic patients.

Summary of Key Questions

OWL Sign in at **www.cengage.com/owl** to develop problem-solving skills and complete online homework assigned by your professor.

Section 18.1 What Are Carboxylic Acids?

• The functional group of a **carboxylic acid** is the **carboxyl group**, —COOH.

Section 18.2 How Do We Name Carboxylic Acids?

• IUPAC names of carboxylic acids are derived from the name of the parent alkane by dropping the suffix -e and adding -oic acid.
• Dicarboxylic acids are named as -dioic acids.
• Common names for many carboxylic and dicarboxylic acids are still widely used.

Section 18.3 What Are the Physical Properties of Carboxylic Acids?

• Carboxylic acids are polar compounds. Consequently, they have higher boiling points and are more soluble in water than alcohols, aldehydes, ketones, and ethers of comparable molecular weight.

Section 18.4 What Are Soaps and Detergents?

• Fatty acids are long, unbranched-chain carboxylic acids. They can be saturated or unsaturated.
• A **triglyceride** is a triester of glycerol.
• A **micelle** is a spherical arrangement of molecules in an aqueous environment in which the hydrocarbon parts are on the inside and the hydrophilic parts are on the surface.

Section 18.5 What Are the Characteristic Reactions of Carboxylic Acids?

• Carboxylic acids are weak acids that react with strong bases to form water-soluble salts.
• Treatment of a carboxylic acid with an alcohol in the presence of an acid catalyst gives an ester.
• When exposed to a very high temperature, carboxylic acids can undergo decarboxylation.

Summary of Key Reactions

1. **Acidity of Carboxylic Acids (Section 18.5A)** Values of pK_a for most unsubstituted carboxylic acids are within the range of 4 to 5.

$$CH_3COH + H_2O \rightleftharpoons CH_3CO^- + H_3O^+$$

$$K_a = \frac{[CH_3COO^-][H_3O^+]}{[CH_3COOH]} = 1.74 \times 10^{-5}$$

$$pK_a = 4.76$$

2. Reaction of Carboxylic Acids with Bases

(Section 18.5B) Carboxylic acids, whether water-soluble or insoluble, react with alkali metal hydroxides, carbonates and bicarbonates, and ammonia and amines to form water-soluble salts.

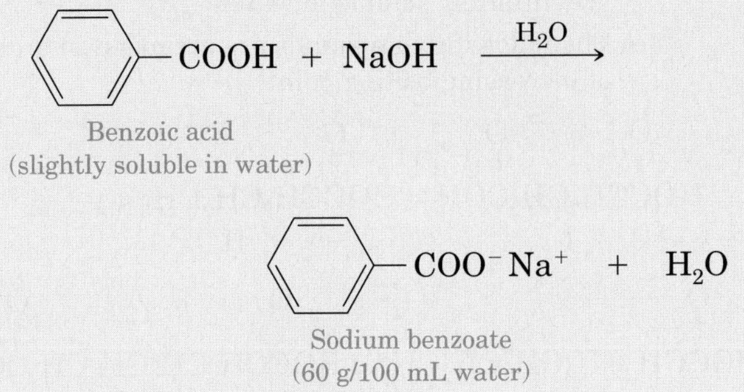

$$\text{Benzoic acid} - \text{COOH} + \text{NaOH} \xrightarrow{\text{H}_2\text{O}}$$

Benzoic acid
(slightly soluble in water)

$$- \text{COO}^- \text{Na}^+ + \text{H}_2\text{O}$$

Sodium benzoate
(60 g/100 mL water)

$$\text{CH}_3\text{COOH} + \text{NaHCO}_3 \longrightarrow$$

$$\text{CH}_3\text{COO}^-\text{Na}^+ + \text{CO}_2 + \text{H}_2\text{O}$$

3. Reduction by Lithium Aluminum Hydride

(Section 18.5C) Lithium aluminum hydride reduces a carboxyl group to a primary alcohol. This reagent does not normally reduce carbon–carbon double bonds, but it does reduce aldehydes to 1° alcohols and ketones to 2° alcohols.

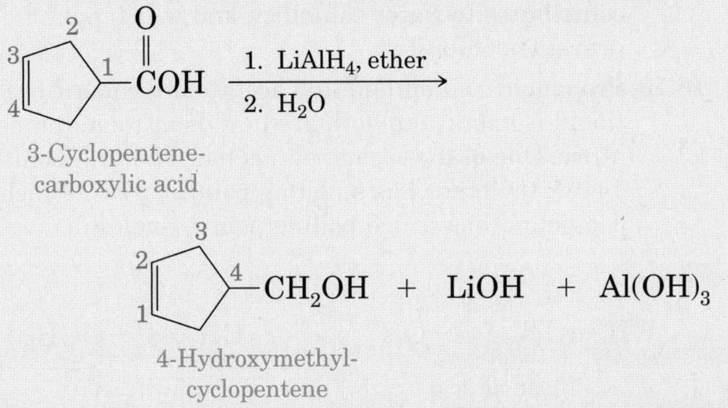

$$\xrightarrow[\text{2. H}_2\text{O}]{\text{1. LiAlH}_4, \text{ether}}$$

3-Cyclopentene-carboxylic acid

$$-\text{CH}_2\text{OH} + \text{LiOH} + \text{Al(OH)}_3$$

4-Hydroxymethyl-cyclopentene

4. Fischer Esterification (Section 18.5D) Fischer esterification is reversible.

$$\text{CH}_3\text{COH} + \text{CH}_3\text{CH}_2\text{OH} \underset{}{\overset{\text{H}_2\text{SO}_4}{\rightleftharpoons}}$$

Ethanoic acid
(Acetic acid)

Ethanol
(Ethyl alcohol)

$$\text{CH}_3\text{COCH}_2\text{CH}_3 + \text{H}_2\text{O}$$

Ethyl ethanoate
(Ethyl acetate)

One way to force the equilibrium to the right is to use an excess of the alcohol. Alternatively, water can be removed from the reaction mixture as it is formed.

5. Decarboxylation (Section 18.5E) Thermal decarboxylation is a unique property of β-ketoacids. The immediate products of thermal decarboxylation of β-ketoacids are carbon dioxide and an enol. Loss of CO_2 is followed immediately by keto-enol tautomerism.

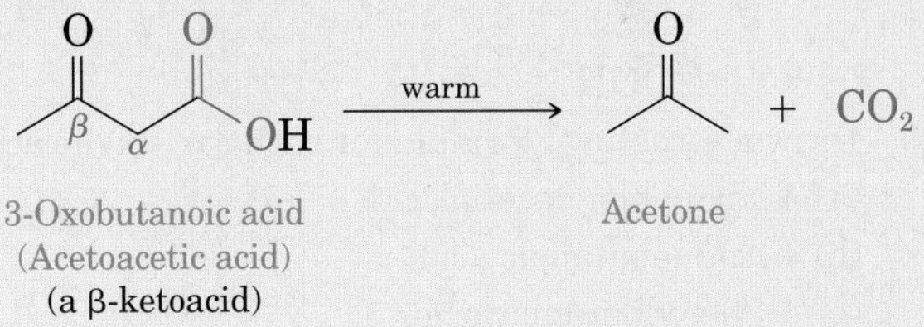

$$\xrightarrow{\text{warm}} \quad + \text{CO}_2$$

3-Oxobutanoic acid
(Acetoacetic acid)
(a β-ketoacid)

Acetone

Problems

Orange-numbered problems are applied.

References to previous chapters are given in parentheses.

For preparation of carboxylic acids, review Chapters 14, Alcohols, Esters, and Thiols, and 17, Aldehydes and Ketones

Section 18.1 What Are Carboxylic Acids?

18.4 Answer true or false.

(a) The functional groups of a carboxylic acid are a carbonyl group bonded to a hydroxyl group.

(b) The VSEPR model predicts bond angles of 180° about the carbonyl carbon of a carboxyl group.

(c) The VSEPR model predicts bond angles of 109.5° about the oxygen of the OH group of a carboxyl group.

(d) The carbonyl carbon of a carboxyl group can be a stereocenter, depending on its location within a molecule.

(e) Carboxylic acids can be prepared by chromic acid oxidation of primary alcohols and of aldehydes.

(f) The product of chromic acid oxidation of hexanoic acid is 1-hexanol.

Section 18.2 How Do We Name Carboxylic Acids?

18.5 Answer true or false.

(a) The general name of an aliphatic carboxylic acid is alkanoic acid.

(b) A molecule containing two COOH groups is called a dicarboxylic acid.

(c) Ethanedioic acid (oxalic acid) is the simplest dicarboxylic acid.

(d) 3-Methylbutanoic acid is chiral.

(e) The simplest carboxylic acid is methanoic acid (common name: formic acid), HCO_2H.

(f) Benzoic acid is an aromatic carboxylic acid.

(g) Formic acid, which is the common name for HCO_2H, is derived from the word *formica*, which is the Latin name for ants.

(h) (*S*)-Lactic acid, CH_3—CHOH—COOH, contains two functional groups: a 2° alcohol and a carboxyl group.

18.6 Name and draw structural formulas for the four carboxylic acids with the molecular formula $C_5H_{10}O_2$. Which of these carboxylic acids are chiral?

18.7 Write the IUPAC name for each carboxylic acid.

(a) [structural formula] (b) [structural formula]

(c) [structural formula]

18.8 Write the IUPAC name for each carboxylic acid.

(a) [structural formula] (b) [structural formula]

(c) CCl_3COOH

18.9 Draw a structural formula for each carboxylic acid.

(a) 4-Nitrophenylacetic acid

(b) 4-Aminobutanoic acid

(c) 4-Phenylbutanoic acid

(d) *cis*-3-Hexenedioic acid

18.10 Draw a structural formula for each carboxylic acid.

(a) 2-Aminopropanoic acid

(b) 3,5-Dinitrobenzoic acid

(c) Dichloroacetic acid

(d) *o*-Aminobenzoic acid

18.11 Draw a structural formula for each salt.

(a) Sodium benzoate   (b) Lithium acetate

(c) Ammonium acetate   (d) Disodium adipate

(e) Sodium salicylate   (f) Calcium butanoate

18.12 Calcium oxalate is a major component of kidney stones. Draw a structural formula for this compound.

18.13 The monopotassium salt of oxalic acid is present in certain leafy vegetables, including rhubarb. Both oxalic acid and its salts are poisonous in high concentrations. Draw a structural formula for monopotassium oxalate.

Section 18.3 What Are the Physical Properties of Carboxylic Acids?

18.14 Answer true or false.

(a) Carboxylic acids are polar compounds.

(b) The most polar bond of a carboxyl group is the C—O single bond.

(c) Carboxylic acids have significantly higher boiling points than aldehydes, ketones, and alcohols of comparable molecular weight.

(d) The low-molecular-weight carboxylic acids (formic, acetic, propanoic, and butanoic acids) are infinitely soluble in water.

(e) The following compounds are arranged in order of increasing boiling point:

$$\underset{I}{HOCCH_2CH_2COH} < \underset{II}{HOCCH_2CH_2CH_3} <$$

$$\underset{III}{HOCCH_2CH_2CH_2CH_2CH_3} < \underset{IV}{HOCCH_2CH_2CH_2CH_2COH}$$

18.15 Draw a structural formula for the dimer formed when two molecules of formic acid interact by hydrogen bonding.

18.16 Propanedioic (malonic) acid forms an intramolecular hydrogen bond in which the H of one COOH group forms a hydrogen bond with an O of the other COOH group. Draw a structural formula to show this internal hydrogen bonding. (There are two possible answers.)

18.17 Hexanoic (caproic) acid has a solubility in water of about 1 g/100 mL water. Which part of the molecule contributes to water solubility, and which part prevents solubility?

18.18 Propanoic acid and methyl acetate are constitutional isomers, and both are liquids at room temperature. One of these compounds has a boiling point of 141°C; the other has a boiling point of 57°C. Which compound has which boiling point? Explain.

$$\underset{\text{Propanoic acid}}{CH_3-CH_2-\overset{O}{\overset{\|}{C}}-OH} \qquad \underset{\text{Methyl acetate}}{CH_3-\overset{O}{\overset{\|}{C}}-OCH_3}$$

18.19 The following compounds have approximately the same molecular weight: hexanoic acid, heptanal, and 1-heptanol. Arrange them in order of increasing boiling point.

18.20 The following compounds have approximately the same molecular weight: propanoic acid, 1-butanol, and diethyl ether. Arrange them in order of increasing boiling point.

18.21 Arrange these compounds in order of increasing solubility in water: acetic acid, pentanoic acid, decanoic acid.

Section 18.4 What Are Soaps and Detergents?

18.22 Answer true or false.

(a) Fatty acids are long-chain carboxylic acids, with most consisting of between 12 to 20 carbons in an unbranched chain.

(b) An unsaturated fatty acid contains one or more carbon–carbon double bonds in its hydrocarbon chain.

(c) In most unsaturated fatty acids found in animal fats, vegetable oils, and biological membranes, the *cis* isomer predominates.

(d) In general, unsaturated fatty acids have lower melting points than saturated fatty acids with the same number of carbon atoms.

(e) Natural soaps are sodium or potassium salts of fatty acids.

(f) Soaps remove grease, oil, and fat stains by incorporating these substances into the nonpolar interior of soap micelles.

(g) "Hard water," by definition, is water that contains Ca^{2+}, Mg^{2+}, or Fe^{3+} ions, all of which react with soap molecules to form water-insoluble salts.

(h) The structure of synthetic detergents is patterned after that of natural soaps.

(i) The most widely used synthetic detergents are the linear alkylbenzenesulfonates (LAS).

(j) Present-day synthetic detergents do not form water-insoluble salts with hard water.

(k) Most detergent preparations contain foam stabilizers, a bleach, and optical brighteners (optical bleaches).

18.23 Characterize the structural features necessary to make a good synthetic detergent.

18.24 The detergents illustrated in this chapter are classified as anionic detergents. Following are structural formulas for two other classifications of synthetic detergents: cationic detergents and neutral detergents.

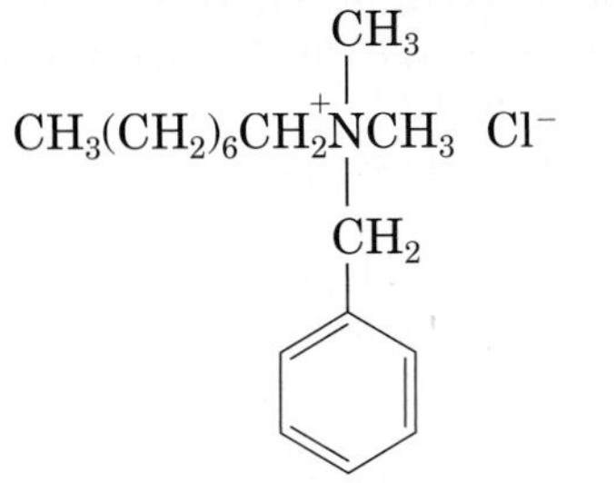

$$\underset{}{CH_3(CH_2)_6CH_2\overset{+}{N}CH_3} \quad Cl^-$$

Benzyldimethyloctylammonium chloride
(a cationic detergent)

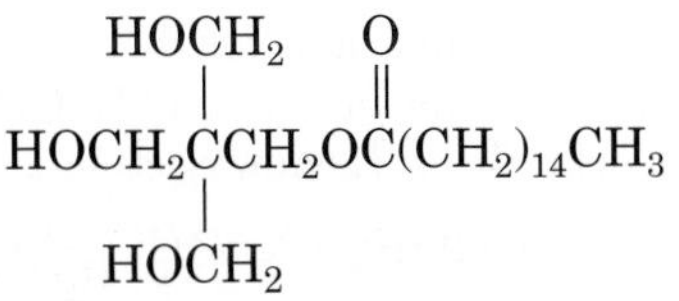

$$HOCH_2CCH_2OC(CH_2)_{14}CH_3$$

Pentaerythrityl palmitate
(a neutral detergent)

Explain how each compound is able to function as a detergent.

18.25 Following are structural formulas for two more cationic detergents. Each is a mild surface-acting detergent and fungicide and is used as a topical antiseptic and disinfectant. They are examples of quaternary ammonium detergents, commonly called "quats." Account for the detergent properties of each.

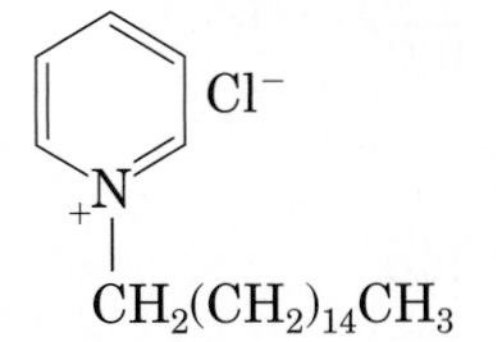

Cetylpyridinium chloride 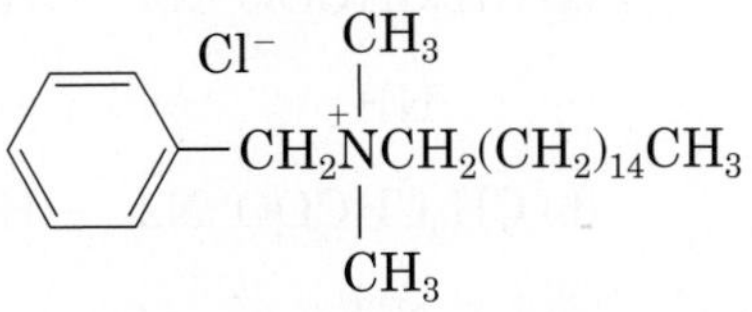   Benzylcetyldimethylammonium chloride

Section 18.5 What Are the Characteristic Reactions of Carboxylic Acids?

18.26 Answer true or false.

(a) Carboxylic acids are weak acids compared to mineral acids such as HCl, H_2SO_4, and HNO_3.

(b) Phenols, alcohols, and carboxylic acids have in common the presence of an —OH group.

(c) Carboxylic acids are stronger acids than alcohols but weaker acids than phenols.

(d) The order of acidity of the following carboxylic acids is:

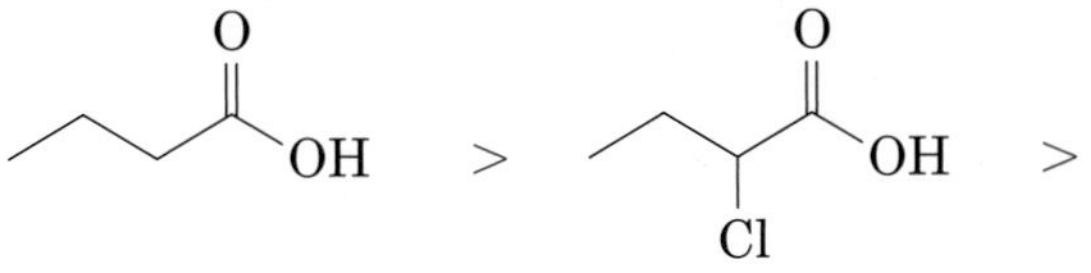

A   >   B 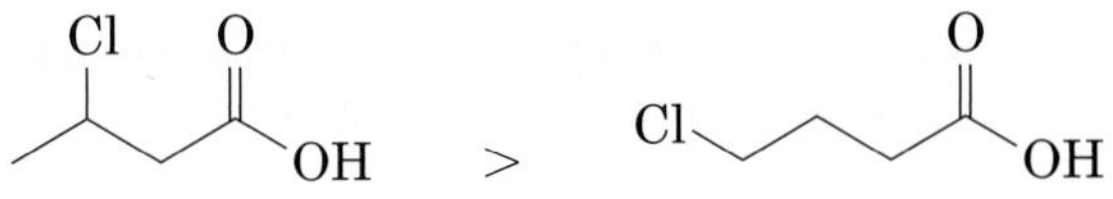   >   C   >   D

(e) The order of acidity of the following carboxylic acids is:

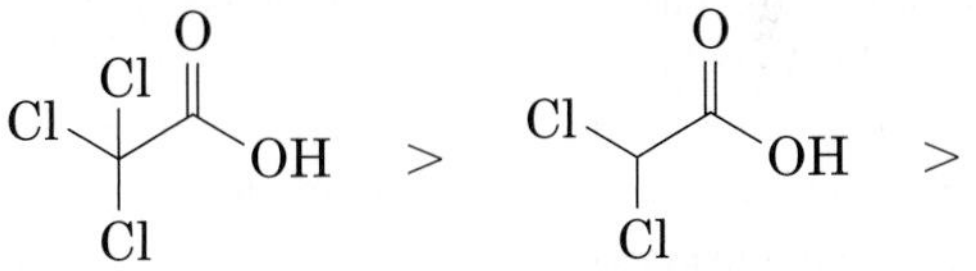

A   >   B 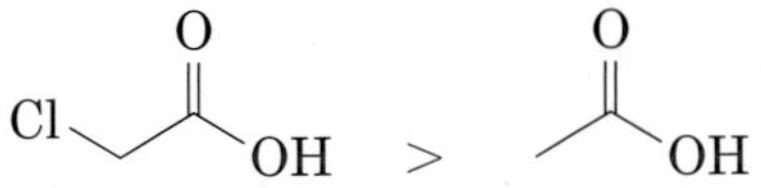   >   C   >   D

(f) The reaction of benzoic acid with aqueous sodium hydroxide gives sodium benzoate.

(g) A mixture of the following compounds is extracted in order with (1) 1 M HCl, (2) 1 M NaOH, and (3) diethyl ether. Only compound II is extracted into the basic layer.

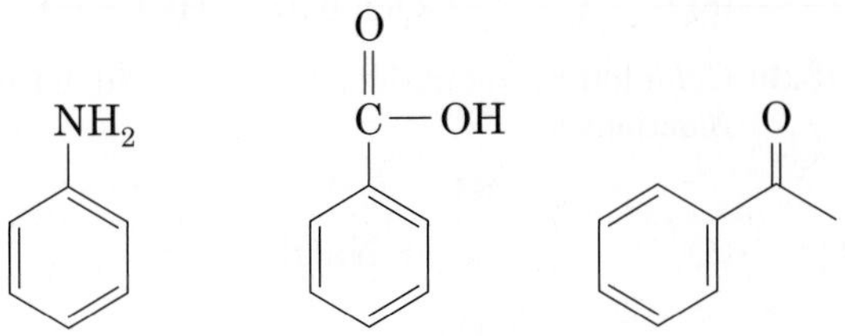

I   II   III

(h) Conversion of compound I to compound II is best accomplished by reduction with $NaBH_4$.

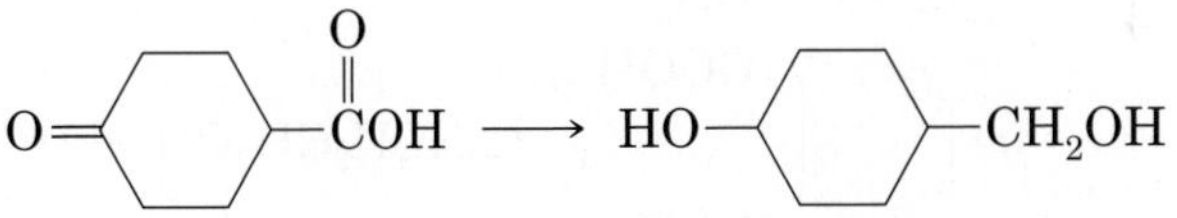

I   →   II

(i) The following ester can be prepared by treating benzoic acid with 1-butanol in the presence of a catalytic amount of H_2SO_4:

$$\text{(structure: 2-hydroxybenzoate ester)} \quad OCH_2CH_2CH_2CH_3$$

OH

(j) Thermal decarboxylation of this β-ketoacid gives benzoic acid and carbon dioxide:

$$\text{(structure: 3-oxo-3-phenylpropanoic acid)} \quad OH$$

(k) Thermal decarboxylation of this β-ketoacid gives 2-pentanone and carbon dioxide.

$$CH_3CH_2CH_2\overset{O}{\underset{\|}{C}}CH_2\overset{O}{\underset{\|}{C}}OH$$

18.27 Alcohols, phenols, and carboxylic acids all contain an —OH group. Which are the strongest acids? Which are the weakest acids?

18.28 Arrange these compounds in order of increasing acidity: benzoic acid, benzyl alcohol, phenol.

18.29 Complete the equations for these acid–base reactions.

(a) (benzene ring)—$CH_2COOH + NaOH \longrightarrow$

(b) (structure)—$COOH + NaHCO_3 \longrightarrow$

(c) $\text{(benzene ring with } COOH \text{ and } OCH_3) + NaHCO_3 \longrightarrow$

(d) $CH_3\underset{\underset{OH}{|}}{C}HCOOH + H_2NCH_2CH_2OH \longrightarrow$

(e) (structure)—$COO^-Na^+ + HCl \longrightarrow$

18.30 Complete the equations for these acid–base reactions.

(a) $\text{(benzene ring with } OH \text{ and } CH_3) + NaOH \longrightarrow$

(b) $\text{(benzene ring with } COO^-Na^+ \text{ and } OH) + HCl \longrightarrow$

(c) $\text{(benzene ring with } COOH \text{ and } OCH_3) + H_2NCH_2CH_2OH \longrightarrow$

(d) (cyclohexyl)—$COOH + NaHCO_3 \longrightarrow$

18.31 Formic acid is one of the components responsible for the sting of biting ants and is injected under the skin by bee and wasp stings. The pain can be relieved by rubbing the area of the sting with a paste of baking soda ($NaHCO_3$) and water, which neutralizes the acid. Write an equation for this reaction.

18.32 Starting with the definition of K_a of a weak acid, HA, as

$$HA + H_2O \rightleftharpoons A^- + H_3O^+ \qquad K_a = \frac{[A^-][H_3O^+]}{[HA]}$$

show that:

$$\frac{[A^-]}{[HA]} = \frac{K_a}{[H_3O^+]}$$

18.33 Using the equation from Problem 18.32 that shows the relationship between K_a, $[H_3O^+]$, $[A^-]$, and $[HA]$, calculate the ratio of $[A^-]$ to $[HA]$ in a solution whose pH is:

(a) 2.0   (b) 5.0

(c) 7.0   (d) 9.0

(e) 11.0

Assume that the pK_a of the weak acid is 5.0.

18.34 The pK_a of acetic acid is 4.76. What form(s) of acetic acid are present at pH 2.0 or lower? At pH 4.76? At pH 8.0 or higher?

18.35 The normal pH range for blood plasma is 7.35 to 7.45. Under these conditions, would you expect the carboxyl group of lactic acid (pK_a 4.07) to exist primarily as a carboxyl group or as a carboxylate anion? Explain.

18.36 The pK_a of ascorbic acid (Vitamin C, see Chemical Connections 20B) is 4.10. Would you expect ascorbic acid dissolved in blood plasma, pH 7.35–7.45, to exist primarily as ascorbic acid or as ascorbate anion? Explain.

18.37 Complete the equations for the following acid–base reactions. Assume one mole of NaOH per mole of amino acid. (*Hint*: Review Section 8.4.)

(a) $CH_3\underset{\underset{NH_3^+}{|}}{C}HCOOH + NaOH \xrightarrow{H_2O}$

(b) $CH_3\underset{\underset{NH_3^+}{|}}{C}HCOO^-Na^+ + NaOH \xrightarrow{H_2O}$

18.38 Which is the stronger base: $CH_3CH_2NH_2$ or $CH_3CH_2COO^-$? Explain.

18.39 Complete the equations for the following acid–base reactions. Assume one mole of HCl per mole of amino acid.

(a) $CH_3\underset{\underset{NH_2}{|}}{C}HCOO^-Na^+ + HCl \xrightarrow{H_2O}$

(b) $CH_3\underset{\underset{NH_3^+}{|}}{C}HCOO^-Na^+ + HCl \xrightarrow{H_2O}$

Section 18.5D Fischer Esterification

18.40 Define and give an example of Fischer esterification.

18.41 Complete these examples of Fischer esterification. In each case, assume an excess of the alcohol.

(a)

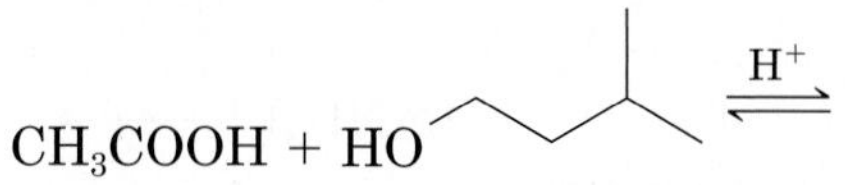

$$CH_3COOH + HO \underset{H^+}{\overset{H^+}{\rightleftharpoons}}$$

(b) $CH_3COOH + HO \overset{H^+}{\rightleftharpoons}$

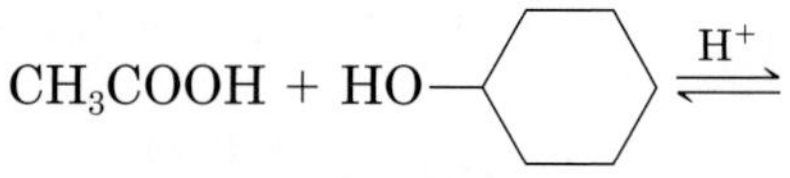

(c)

COOH
$+ CH_3CH_2OH \overset{H^+}{\rightleftharpoons}$
COOH

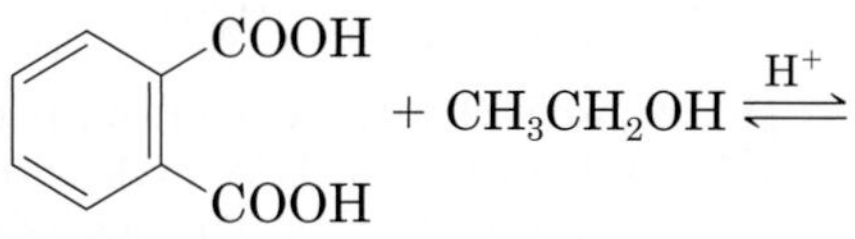

18.42 From what carboxylic acid and alcohol is each ester derived?

(a) $CH_3CO- -OCCH_3$

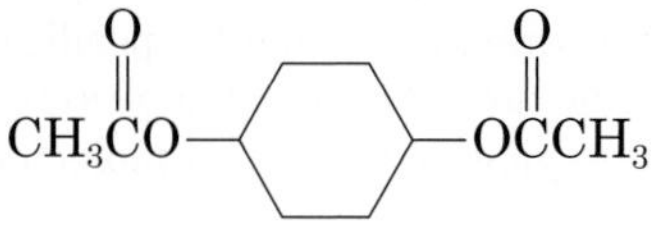

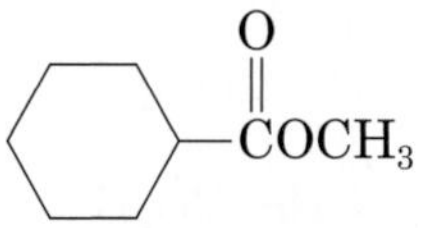

(b) $-COCH_3$

(c) $CH_3OCCH_2CH_2COCH_3$

(d)

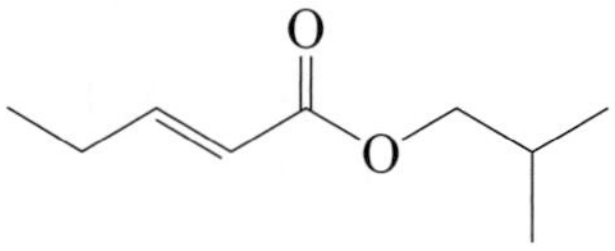

18.43 Methyl 2-hydroxybenzoate (methyl salicylate) has the odor of oil of wintergreen. This compound is prepared by Fischer esterification of 2-hydroxybenzoic acid (salicylic acid) with methanol. Draw a structural formula for methyl 2-hydroxybenzoate.

18.44 Show how you could convert cinnamic acid to each compound.

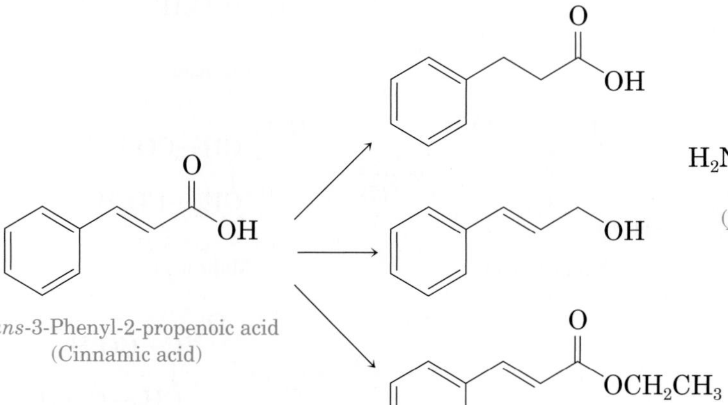

trans-3-Phenyl-2-propenoic acid
(Cinnamic acid)

OCH_2CH_3

Additional Problems

18.45 Give the expected organic product formed when phenylacetic acid, $C_6H_5CH_2COOH$, is treated with each of the following reagents:

(a) $NaHCO_3$, H_2O (b) $NaOH$, H_2O
(c) NH_3, H_2O (d) $LiAlH_4$ then H_2O
(e) $NaBH_4$ then H_2O (f) $CH_3OH + H_2SO_4$ (catalyst)
(g) H_2/Ni

18.46 Procaine (its hydrochloride salt is marketed as Novocaine) was one of the first local anesthetics developed for infiltration and regional anesthesia. It is synthesized by the following Fischer esterification:

H_2N- OH
4-Aminobenzoic acid

$+ HO-N$
2-Diethylaminoethanol

$\xrightarrow{\text{Fischer esterification}}$ Procaine

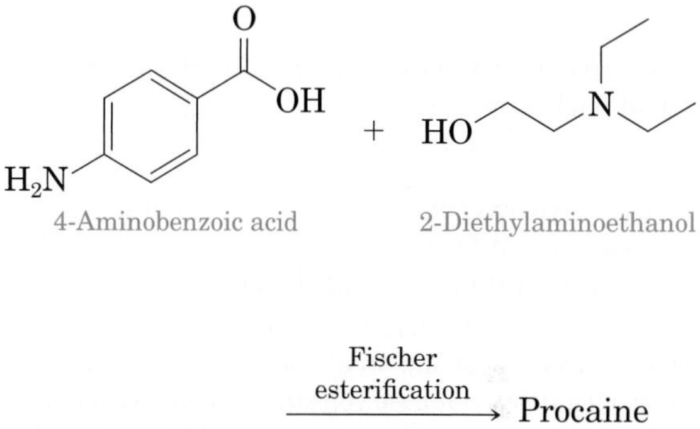

(a) Draw the structural formula of Procaine.
(b) Which of the two nitrogen atoms in Procaine is the stronger base? (Chapter 16)
(c) Draw the structural formula of Novocaine (the hydrochloride salt of Procaine). (Chapter 16)
(d) Would you predict Procaine or Novocaine to be more soluble in blood? (Chapters 5, 8)

18.47 Methylparaben and propylparaben are used as preservatives in foods, beverages, and cosmetics.

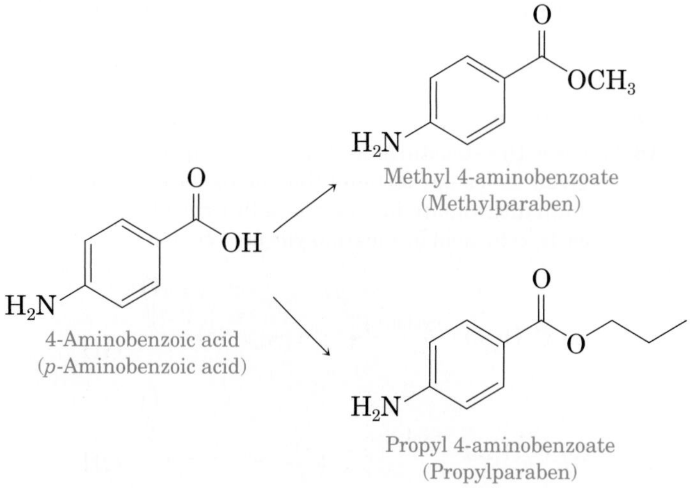

H_2N- OH
4-Aminobenzoic acid
(*p*-Aminobenzoic acid)

H_2N- OCH_3
Methyl 4-aminobenzoate
(Methylparaben)

H_2N- O
Propyl 4-aminobenzoate
(Propylparaben)

Show how each of these preservatives can be prepared from 4-aminobenzoic acid.

18.48 4-Aminobenzoic acid is prepared from benzoic acid by the following two steps.

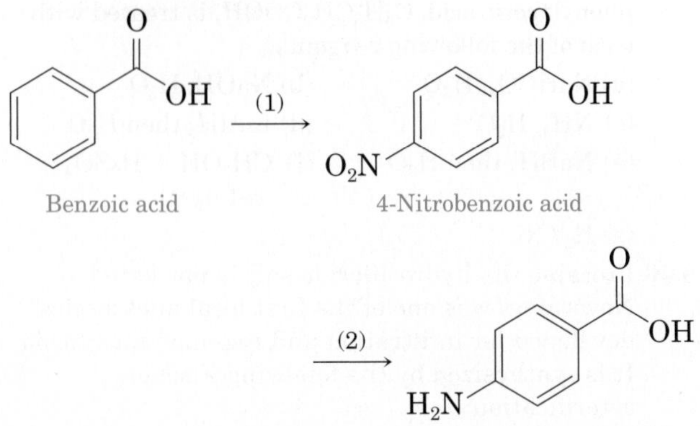

Benzoic acid   4-Nitrobenzoic acid

4-Aminobenzoic acid

Show reagents and experimental conditions to bring about each step.

Looking Ahead

18.49 When 5-hydroxypentanoic acid is treated with an acid catalyst, it forms a lactone (a cyclic ester). Draw a structural formula for this lactone.

18.50 We have seen that esters can be prepared by treatment of a carboxylic acid and an alcohol in the presence of an acid catalyst. Suppose you start instead with a dicarboxylic acid such as 1,6-hexanedioic acid (adipic acid) and a diol such as 1,2-ethanediol (ethylene glycol).

$$HO\overset{O}{\overset{\|}{C}}CH_2CH_2CH_2CH_2\overset{O}{\overset{\|}{C}}OH +$$

1,6-Hexanedioic acid
(Adipic acid)

$$HOCH_2CH_2OH \longrightarrow A\ polyester$$

1,2-Ethanediol
(Ethylene glycol)

Show how Fischer esterification in this case can produce a polymer (a macromolecule with a molecular weight several thousand times that of the starting materials).

Tying It Together

18.51 Draw the structural formula of a compound of the given molecular formula that, on oxidation by potassium dichromate in aqueous sulfuric acid, gives the carboxylic acid or dicarboxylic acid shown.

(a) $C_6H_{14}O \xrightarrow{\text{oxidation}}$

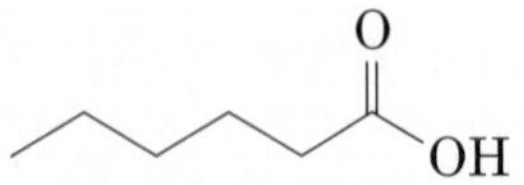

(b) $C_6H_{12}O \xrightarrow{\text{oxidation}}$

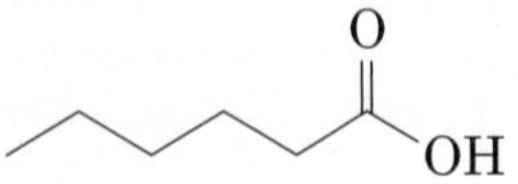

(c) $C_6H_{14}O_2 \xrightarrow{\text{oxidation}}$

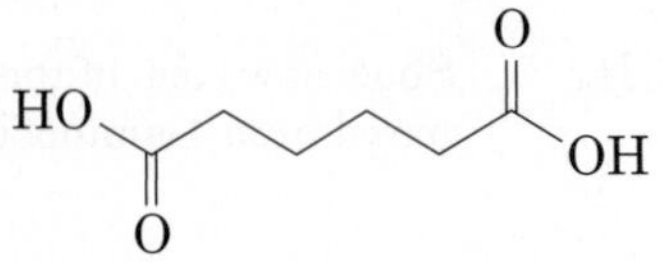

18.52 Complete the equations for these oxidations:

(a) $CH_3(CH_2)_4CH_2OH \xrightarrow[H_2SO_4]{K_2Cr_2O_7}$

(b) 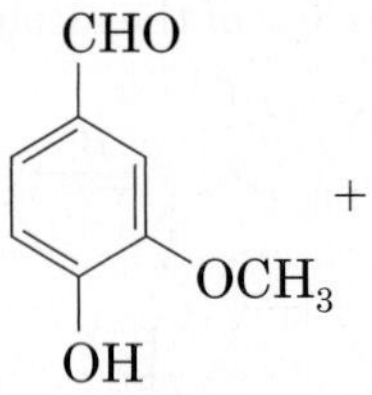 $+ Ag(NH_3)_2^+ \longrightarrow$

(c) $HO\!-\!\!\bigcirc\!\!-\!CH_2OH \xrightarrow[H_2SO_4]{K_2Cr_2O_7}$

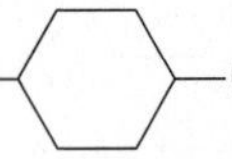

18.53 In Chapter 27, we will study a metabolic pathway called the tricarboxylic acid (TCA) cycle, also known as the citric acid or Krebs cycle. We have already seen examples of most of the reactions in this pathway. Following is an outline of the pathway beginning with the molecule for which the pathway is named. A particular enzyme that is highly specific catalyzes each of these reactions. Each enzyme-catalyzed reaction gives a high yield of the target molecule.

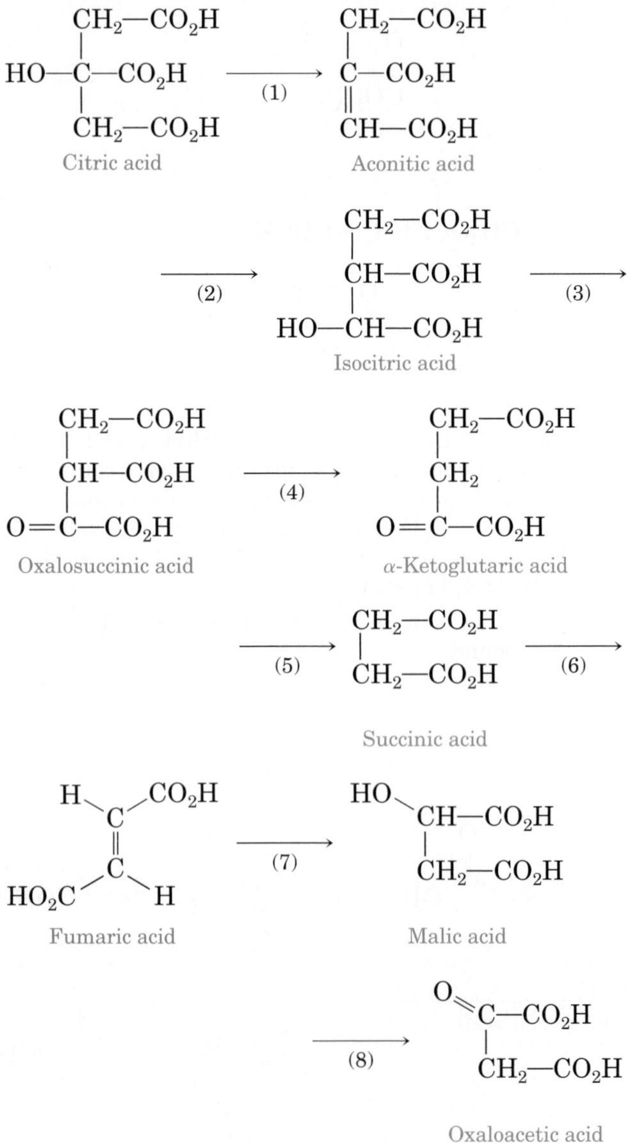

Citric acid → (1) → Aconitic acid

→ (2) → Isocitric acid → (3) →

Oxalosuccinic acid → (4) → α-Ketoglutaric acid

→ (5) → Succinic acid → (6) →

Fumaric acid → (7) → Malic acid

→ (8) → Oxaloacetic acid

(a) Which of these TCA-cycle intermediates are chiral? Which intermediate has the greatest number of chiral centers? Which intermediates show *cis/trans* isomerism? (Chapter 15)

(b) Name the type of reaction that takes place in Steps 1–3. (Chapters 12, 14)

(c) Notice that the hydration of aconitic acid to give isocitric acid does not follow Markovnikov's rule. If the hydration of aconitic acid were to follow Markovnikov's rule, what product would be formed? Offer an explanation for the formation of this non-Markovnikov product. (Chapter 12)

(d) What type of reactions take place in Steps 4–8? (Chapter 14)

(e) The only reaction we have not studied is Step 5. Because it involves a loss of CO_2, it is classified as a decarboxylation. Show that it also involves oxidation. Because it involves both oxidation and decarboxylation, it is classified as an oxidative decarboxylation.

(f) Reaction 4 is also classified as a decarboxylation. Does this decarboxylation also involve an oxidation?

18.54 Azithromycin is a broad-spectrum antibiotic derived from erythromycin. It is one of a large number of what are called macrocyclic antibiotics, so named because they contain a large ring as part of their structure. Like many of the macrocyclic antibiotics, azithromycin has an incredibly complex structure, and it was an enormous challenge for chemists to determine its structural formula.

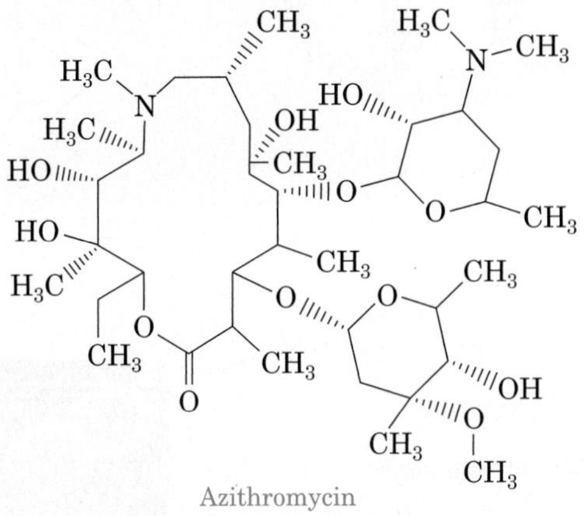

Azithromycin
$C_{38}H_{72}N_2O_{12}$

(a) What is the size of the large ring in this antibiotic? (Chapter 10)

(b) Name the functional groups present in azithromycin. (Chapter 10)

(c) Is azithromycin chiral? If so, how many stereoisomers are possible for this structural formula? (Chapter 15)

19

Carboxylic Anhydrides, Esters, and Amides

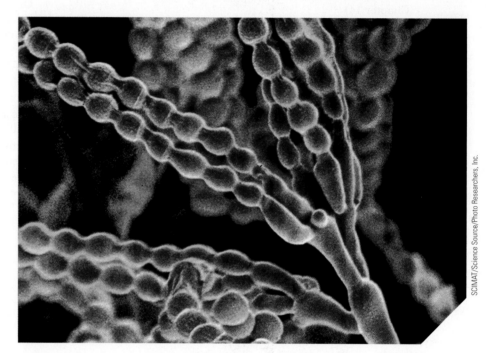

Colored scanning electron micrograph of *Penicillium* fungus. The stalk-like objects are condiophores to which are attached numerous round condia. The condia are the fruiting bodies of the fungus. See Chemical Connections 19B.

SCIMAT/Science Source/Photo Researchers, Inc.

ØWL

Sign in to OWL at **www.cengage.com/owl** to view tutorials and simulations, develop problem-solving skills, and complete online homework assigned by your professor.

19.1 What Are Carboxylic Anhydrides, Esters, and Amides?

In Chapter 18, we studied the structure and preparation of esters, a class of organic compounds derived from carboxylic acids. In this chapter, we study anhydrides and amides, two more classes of carboxylic derivatives. Following, under the general formula for each carboxylic acid derivative, is a drawing to help you see how the functional group of each derivative is formally related to a carboxyl group. The loss of —OH from a carboxyl group and —H from an alcohol, for example, gives an ester. Loss of —OH from a carboxyl group and —H from ammonia or an amine gives an amide.

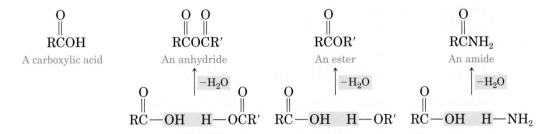

Of these three carboxylic derivatives, anhydrides are so reactive that they are rarely found in nature. Esters and amides, however, are widespread in the biological world.

A. Anhydrides

The functional group of an **anhydride** consists of two carbonyl groups bonded to an oxygen atom. The anhydride may be symmetrical (from two identical **acyl groups**) or mixed (from two different acyl groups). To name anhydrides, we drop the word *acid* from the name of the carboxylic acid from which the anhydride is derived and add the word *anhydride*.

$$CH_3\overset{O}{\overset{\|}{C}}-O-\overset{O}{\overset{\|}{C}}CH_3$$
Acetic anhydride

B. Esters

The functional group of an ester is a carbonyl group bonded to an —OR group. Both IUPAC and common names of esters are derived from the names of the parent carboxylic acids (Chapter 18). The alkyl group bonded to oxygen is named first, followed by the name of the acid in which the suffix -**ic acid** is replaced with the suffix -**ate**.

$$CH_3\overset{O}{\overset{\|}{C}}OCH_2CH_3$$
Ethyl ethanoate
(Ethyl acetate)

Diethyl pentanedioate
(Diethyl glutarate)

Recall that cyclic esters are called **lactones**.

C. Amides

The functional group of an **amide** is a carbonyl group bonded to a nitrogen atom. We name amides by dropping the suffix -**oic acid** from the IUPAC name of the parent acid, or -**ic acid** from its common name, and adding -*amide*. If the nitrogen atom of the amide is bonded to an alkyl or aryl group, the group is named and its location on nitrogen is indicated by *N*-. Two alkyl groups are indicated by *N,N*-di-.

$$CH_3\overset{O}{\overset{\|}{C}}NH_2$$
Acetamide
(a 1° amide)

$$CH_3\overset{O}{\overset{\|}{C}}NHCH_3$$
N-Methylacetamide
(a 2° amide)

$$H\overset{O}{\overset{\|}{C}}N(CH_3)_2$$
N,N-Dimethylformamide
(a 3° amide)

The Pyrethrins—Natural Insecticides of Plant Origin

Pyrethrym is a natural insecticide obtained from the powdered flower heads of several species of *Chrysanthemum*, particularly *C. cinerariaefolium*. The active substances in pyrethrum, principally pyrethrins I and II, are contact poisons for insects and cold-blooded vertebrates. Because their concentrations in the pyrethrum powder used in *Chrysanthemum*-based insecticides are nontoxic to plants and higher animals, pyrethrum powder has found wide application in household and livestock sprays, as well as in dusts for edible plants. Natural pyrethrins are esters of chrysanthemic acid.

While pyrethrum powders are effective insecticides, the active substances in them are destroyed rapidly in the environment. In an effort to develop synthetic compounds that are as effective as these natural insecticides but offer greater biostability, chemists have prepared a series of esters related in structure to chrysanthemic acid. Permethrin is one of the most commonly used synthetic pyrethrin-like compounds in household and agricultural products.

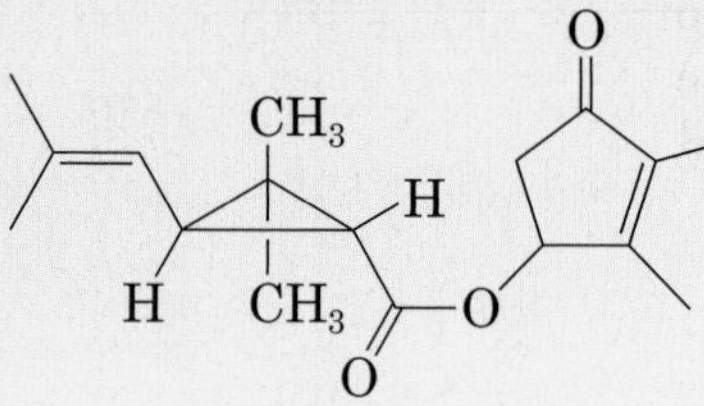

Pyrethrin I

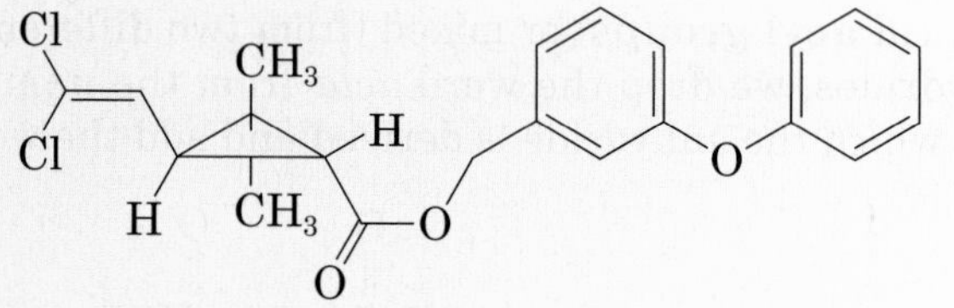

Permethrin

Test your knowledge with Problems 19.22–19.24.

Cyclic amides are called **lactams**. Following are structural formulas for a four-membered and a seven-membered lactam. A four-membered lactam is essential to the functioning of the penicillin and cephalosporin antibiotics (Chemical Connections 19B).

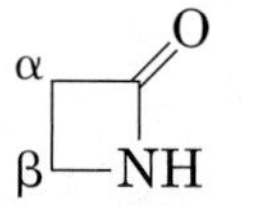

A four-membered
lactam
(α β-lactam)

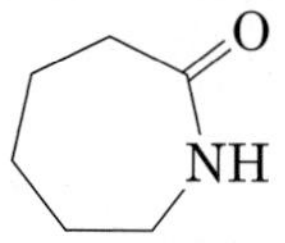

A seven-membered
lactam

WL Interactive Example 19.1 IUPAC Names for Amides

Write the IUPAC name for each amide.

(a) $CH_3CH_2CH_2CNH_2$ (b) 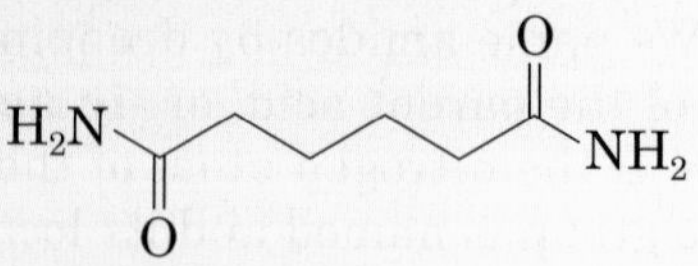 H_2N ... NH_2

Strategy and Solution

To name an amide, start with the systematic name of the corresponding carboxylic acid, drop the suffix **-oic acid** and replace it with **-amide.** Given is the IUPAC name and then, in parentheses, the common name.

(a) Butanamide (butyramide, from butyric acid)
(b) Hexanediamide (adipamide, from adipic acid)

Problem 19.1

Draw a structural formula for each amide.

(a) *N*-Cyclohexylacetamide   (b) Benzamide

The Penicillins and Cephalosporins: β-Lactam Antibiotics

The **penicillins** were discovered in 1928 by the Scottish bacteriologist Sir Alexander Fleming. Thanks to the brilliant experimental work of Sir Howard Florey, an Australian pathologist, and Ernst Chain, a German chemist who fled Nazi Germany, penicillin was introduced into the practice of medicine in 1943. For their pioneering work in developing one of the most effective antibiotics of all time, in 1945, Fleming, Florey, and Chain were awarded the Nobel Prize in Physiology or Medicine.

The mold from which Fleming discovered penicillin was *Penicillium notatum,* a strain that gives a relatively low yield of penicillin. It was replaced in commercial production of the antibiotic by *P. chrysogenum,* a strain cultured from a mold found growing on a grapefruit in a market in Peoria, Illinois. The structural feature common to all penicillins is the four-membered β-lactam ring, bonded to a five-membered sulfur-containing ring. The penicillins owe their antibacterial activity to a common mechanism that inhibits the biosynthesis of a vital part of bacterial cell walls.

Penicillin G

Amoxicillin

Soon after the penicillins were introduced into medical practice, however, penicillin-resistant strains of bacteria began to appear. They have since proliferated dramatically. One approach to combating resistant strains is to synthesize newer, more effective penicillins such as ampicillin, methicillin, and amoxicillin.

Another approach is to search for newer, more effective β-lactam antibiotics. The most effective of these agents discovered so far are the **cephalosporins**, first of which was isolated from the fungus *Cephalosporium acremonium.* This class of β-lactam antibiotics has an even broader spectrum of antibacterial activity than the penicillins and is effective against many penicillin-resistant bacterial strains. Cephalexin (Keflex) is currently one of the most widely prescribed of the cephalosporin antibiotics.

Cephalexin
(a β-lactam antibiotic)

The commonly prescribed formulation Augmentin is a combination of amoxicillin trihydrate, a penicillin, and clavulanic acid, a β-lactamase inhibitor that is isolated from *Streptomyces clavuligerus.*

Clavulanic acid

Clavulanic acid, which also contains a β-lactam ring, reacts with and inhibits the β-lactamase enzyme before the enzyme can catalyze the inactivation of the penicillin. Augmentin is used as a second line of defense against childhood ear infections when penicillin resistance is suspected. Most children know it as a white liquid with a banana taste.

Test your knowledge with Problem 19.25.

19.2 How Do We Prepare Esters?

The most common method for the preparation of esters is Fischer esterification (Section 18.5D). As an example of Fischer esterification, treating acetic acid with ethanol in the presence of concentrated sulfuric acid gives ethyl acetate and water:

$$\underset{\substack{\text{Ethanoic acid}\\ \text{(Acetic acid)}}}{CH_3\overset{\displaystyle O}{\overset{\|}{C}}OH} + \underset{\substack{\text{Ethanol}\\ \text{(Ethyl alcohol)}}}{CH_3CH_2OH} \underset{}{\overset{H_2SO_4}{\rightleftharpoons}} \underset{\substack{\text{Ethyl ethanoate}\\ \text{(Ethyl acetate)}}}{CH_3\overset{\displaystyle O}{\overset{\|}{C}}OCH_2CH_3} + H_2O$$

From Willow Bark to Aspirin and Beyond

The story of the development of this modern pain reliever goes back more than 2000 years. In 400 BCE, the Greek physician Hippocrates recommended chewing the bark of the willow tree to alleviate the pain of childbirth and to treat eye infections.

The active component of willow bark was found to be salicin, a compound composed of salicyl alcohol bonded to a unit of β-D-glucose (Section 20.4A). Hydrolysis of salicin in aqueous acid followed by oxidation gave salicylic acid. Salicylic acid proved to be an even more effective reliever of pain, fever, and inflammation than salicin, and without the latter's extremely bitter taste. Unfortunately, patients quickly recognized salicylic acid's major side effect: It causes severe irritation of the mucous membrane lining of the stomach.

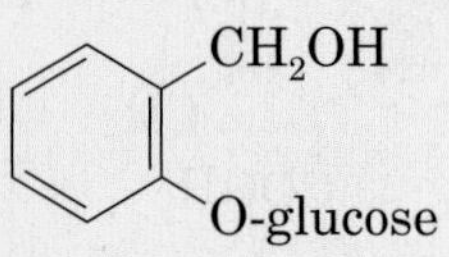

Salicin 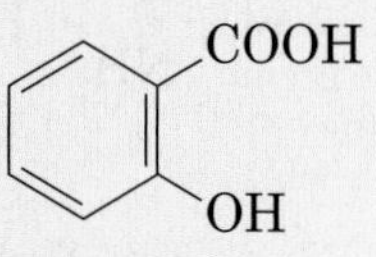   Salicylic acid

In the search for less irritating but still effective derivatives of salicylic acid, chemists at the Bayer division of I. G. Farben in Germany in 1883 treated salicylic acid with acetic anhydride and prepared acetylsalicylic acid. They gave this new compound the name aspirin.

Acetylsalicylic acid (Aspirin)

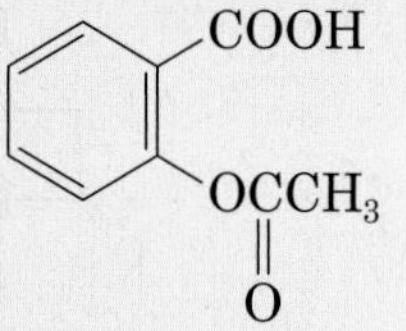

Aspirin proved to be less irritating to the stomach than salicylic acid as well as more effective in relieving the

pain and inflammation of rheumatoid arthritis. Aspirin, however, remains irritating to the stomach, and frequent use of it can cause duodenal ulcers in susceptible persons.

In the 1960s, in a search for even more effective and less irritating analgesics and nonsteroidal anti-inflammatory drugs (NSAIDs), chemists at the Boots Pure Drug Company in England, who were studying compounds structurally related to salicylic acid, discovered an even more potent compound, which they named ibuprofen. Soon thereafter, Syntex Corporation in the United States developed naproxen, the active ingredient in Aleve. Both ibuprofen and naproxen have one stereocenter and can exist as a pair of enantiomers. For each drug, the active form is the S enantiomer. Naproxen is administered as its water-soluble sodium salt.

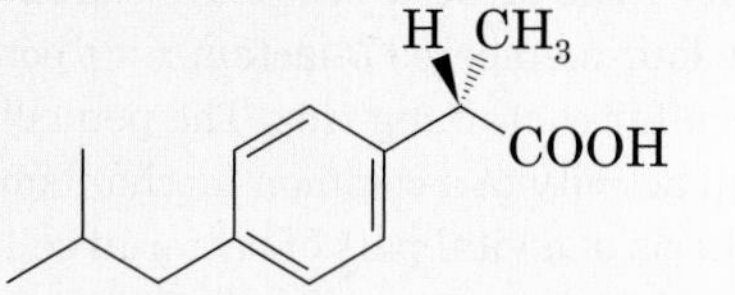

(S)-Ibuprofen

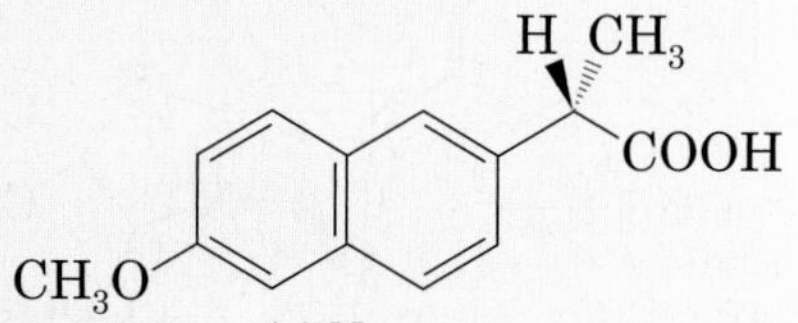

(S)-Naproxen

In the 1960s, researchers discovered that aspirin acts by inhibiting cyclooxygenase (COX), a key enzyme in the conversion of arachidonic acid to prostaglandins (Chemical Connections 21H). With this discovery, it became clear why only one enantiomer of ibuprofen and naproxen is active: Only the S enantiomer of each has the correct handedness to bind to COX and inhibit its activity.

Test your knowledge with Problems 19.26–19.29.

19.3 How Do We Prepare Amides?

In principle, we can form an amide by treating a carboxylic acid with an amine and removing —OH from the acid and an —H from the amine. In practice, mixing these two leads to an acid–base reaction that forms an ammonium salt. If this salt is heated to a high enough temperature, water splits out and an amide forms.

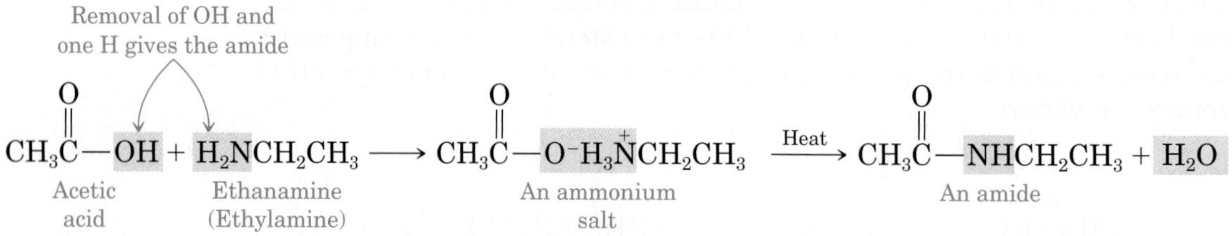

It is much more common, however, to prepare amides by treating an anhydride with an amine (Section 19.4C).

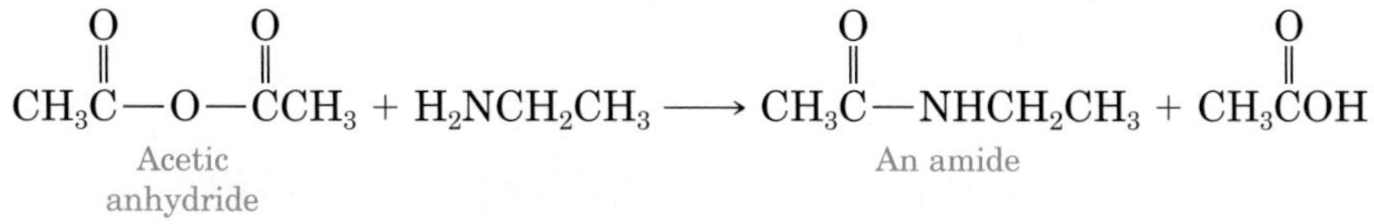

$$CH_3\overset{O}{\underset{}{C}}-O-\overset{O}{\underset{}{C}}CH_3 + H_2NCH_2CH_3 \longrightarrow CH_3\overset{O}{\underset{}{C}}-NHCH_2CH_3 + CH_3\overset{O}{\underset{}{C}}OH$$

Acetic anhydride   An amide

19.4 What Are the Characteristic Reactions of Anhydrides, Esters, and Amides?

The most common reaction of each of these three functional groups is with compounds that contain either an —OH group, as in water (H—OH) or an alcohol (H—OR), or an H—N group, as in ammonia (H—NH$_2$), or in a primary or secondary amine (H—NR$_2$ or H—NHR). These reactions have in common the addition of the oxygen or nitrogen atom to the carboxyl carbon and the hydrogen atom to the carbonyl oxygen to give a tetrahedral carbonyl addition intermediate. This intermediate then collapses to regenerate the carbonyl group and give either a new carboxyl derivative or a carboxylic acid itself. We illustrate here with the reaction of an ester with water:

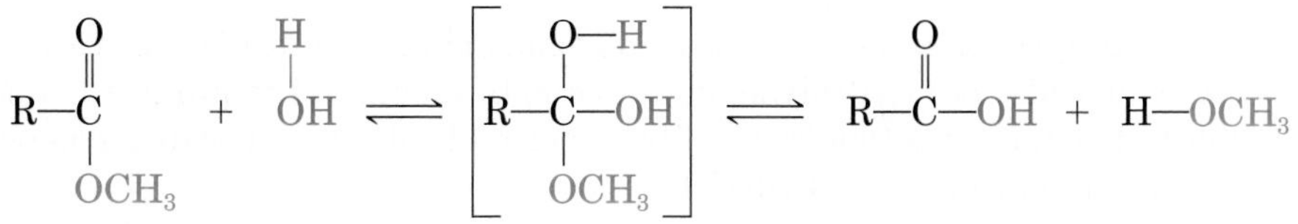

$$R-\overset{O}{\underset{OCH_3}{C}} + OH \rightleftharpoons \left[R-\overset{O-H}{\underset{OCH_3}{C}}-OH\right] \rightleftharpoons R-\overset{O}{\underset{}{C}}-OH + H-OCH_3$$

Compare the formation of this tetrahedral carbonyl addition intermediate with that formed by the addition of an alcohol to the carbonyl group of an aldehyde or ketone and formation of a hemiacetal (Section 17.4C) and that formed by the addition of an alcohol to the carbonyl group of a carboxylic acid during Fischer esterification (Section 18.5D).

A. Reaction with Water: Hydrolysis

Hydrolysis is a chemical decomposition involving breaking a bond and the addition of the elements of water.

Anhydrides

Carboxylic anhydrides, particularly the low-molecular-weight ones, react readily with water to give two carboxylic acids. In the hydrolysis of an anhydride, one of the C—O bonds breaks and OH is added to carbon and H is added to oxygen of what was the C—O bond. Hydrolysis of acetic anhydride gives two molecules of acetic acid.

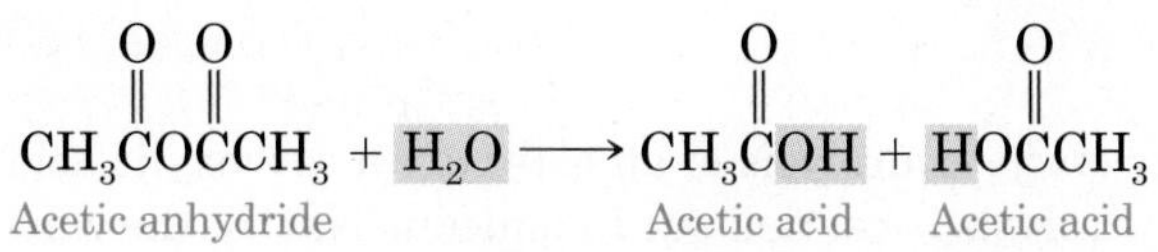

$$CH_3\overset{O}{\underset{}{C}}OC\overset{O}{\underset{}{C}}CH_3 + H_2O \longrightarrow CH_3\overset{O}{\underset{}{C}}OH + HO\overset{O}{\underset{}{C}}CH_3$$

Acetic anhydride   Acetic acid   Acetic acid

Esters

All esters are hydrolyzed very slowly, even when in boiling water. Hydrolysis becomes considerably more rapid, however, when the ester is heated in aqueous acid or base. When we discussed acid-catalyzed Fischer esterification in Section 18.5D, we pointed out that it is an equilibrium reaction.

Hydrolysis of esters in aqueous acid, also an equilibrium reaction, is the reverse of Fischer esterification. A large excess of water drives the equilibrium to the right to form the carboxylic acid and alcohol (Le Chatelier's principle, Section 7.7).

$$CH_3\overset{O}{\overset{\|}{C}}OCH_2CH_3 + H_2O \underset{}{\overset{H^+}{\rightleftharpoons}} CH_3\overset{O}{\overset{\|}{C}}OH + CH_3CH_2OH$$

Ethyl acetate   Acetic acid   Ethanol

Hydrolysis of an ester can also be carried out using a hot aqueous base, such as aqueous NaOH. This reaction is often called **saponification**, a reference to its use in the manufacture of soaps (Section 18.4B). The carboxylic acid formed in the hydrolysis reacts with hydroxide ion to form a carboxylate anion. Thus, each mole of ester hydrolyzed requires one mole of base, as shown in the following balanced equation:

$$CH_3\overset{O}{\overset{\|}{C}}OCH_2CH_3 + NaOH \xrightarrow{H_2O} CH_3\overset{O}{\overset{\|}{C}}O^-Na^+ + CH_3CH_2OH$$

Ethyl acetate   Sodium   Sodium   Ethanol
                     hydroxide   acetate

There are two major differences between hydrolysis of esters in aqueous acid and aqueous base.

1. For hydrolysis of an ester in aqueous acid, acid is required in only catalytic amounts. For hydrolysis in aqueous base, base is required in stoichiometric amounts (one mole of base per mole of ester), because base is a reactant, not merely a catalyst.

2. Hydrolysis of an ester in aqueous acid is reversible. Hydrolysis in aqueous base is irreversible because the carboxylate anion does not react with water or hydroxide ion.

Example 19.2 Hydrolysis of an Ester

Complete the equation for each hydrolysis reaction. Show the products as they are ionized under the given experimental conditions.

(a) 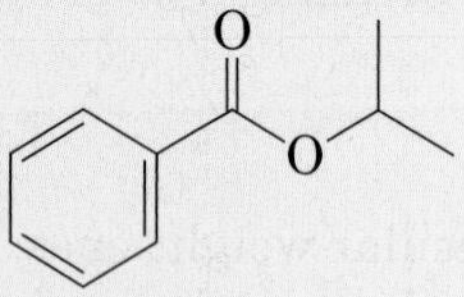 + NaOH $\xrightarrow{H_2O}$

(b) $CH_3\overset{O}{\overset{\|}{C}}OCH_2CH_2O\overset{O}{\overset{\|}{C}}CH_3 + 2NaOH \xrightarrow{H_2O}$

Strategy

The products of the hydrolysis of an ester are a carboxylic acid and an alcohol. If hydrolysis is carried out in aqueous NaOH, the carboxylic acid is converted to its sodium salt.

Solution

The products of hydrolysis of compound (a) are benzoic acid and 2-propanol. In aqueous NaOH, benzoic acid is converted to its sodium salt. In this reaction, one mole of NaOH is required for the hydrolysis of each mole of this ester. Compound (b) is a diester of ethylene glycol that requires two moles of NaOH for its complete hydrolysis.

(a)

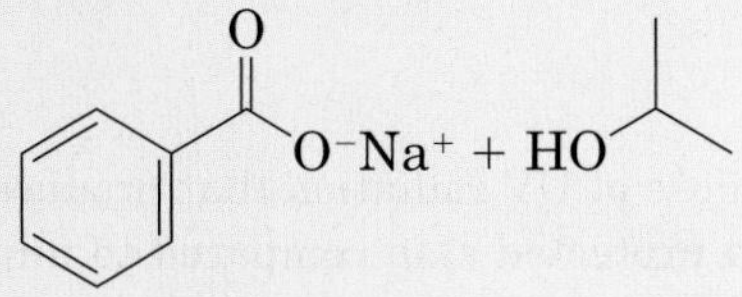

$$O^-Na^+ + HO$$

Sodium
benzoate

2-Propanol
(Isopropyl alcohol)

(b) $2CH_3CO^-Na^+ + HOCH_2CH_2OH$

Sodium acetate   1,2-Ethanediol
(Ethylene glycol)

Problem 19.2

Complete the equation for each hydrolysis reaction. Show all products as they are ionized under these experimental conditions.

(a)

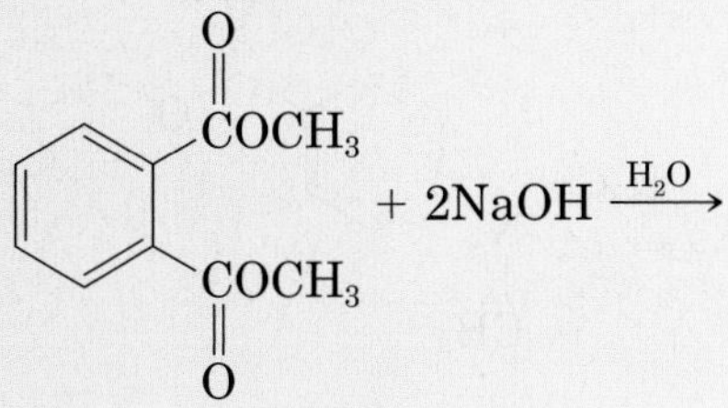

$$COCH_3$$
$$COCH_3$$
$$+ 2NaOH \xrightarrow{H_2O}$$

(b)

$$+ H_2O \xrightarrow{HCl}$$

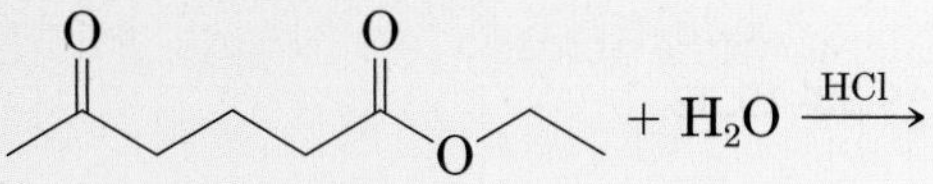

Amides

Amides require more vigorous conditions for hydrolysis in both acid and base than do esters. Hydrolysis in hot aqueous acid gives a carboxylic acid and an ammonium ion. This reaction is driven to completion by the acid–base reaction between ammonia or the amine and the acid to form an ammonium ion. Complete hydrolysis requires one mole of acid per mole of amide.

$$CH_3CH_2CH_2CNH_2 + H_2O + HCl \xrightarrow[\text{heat}]{H_2O} CH_3CH_2CH_2COH + NH_4^+ Cl^-$$

Butanamide   Butanoic acid

The products of amide hydrolysis in aqueous base are a carboxylic acid salt and ammonia or an amine. The acid–base reaction between the carboxylic acid and base to form a carboxylate salt drives this hydrolysis to completion. Thus, complete hydrolysis of an amide requires one mole of base per mole of amide.

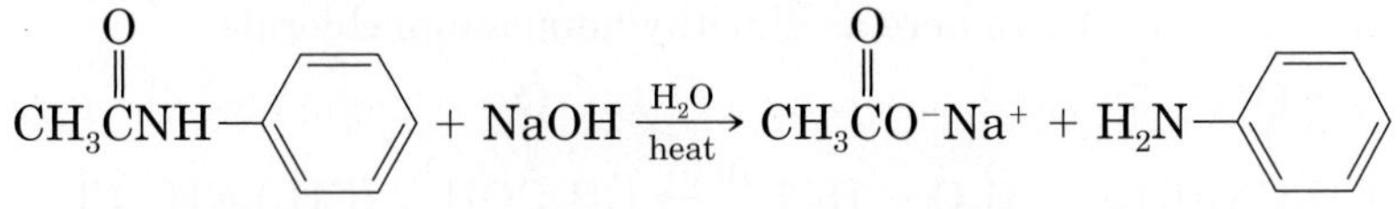

$$CH_3CNH- + NaOH \xrightarrow[\text{heat}]{H_2O} CH_3CO^-Na^+ + H_2N-$$

Acetanilide   Sodium acetate   Aniline

Ultraviolet Sunscreens and Sunblocks

Ultraviolet (UV) radiation (Section 9.2) penetrating the Earth's ozone layer is arbitrarily divided by wavelength into two regions: UVB (290–320 nm) and UVA (320–400 nm). UVB, a more energetic form of radiation than UVA, creates more radicals and hence does more oxidative damage to tissue (Section 25.7). UVB radiation interacts directly with biomolecules of the skin and eyes, causing skin cancer, skin aging, eye damage leading to cataracts, and delayed sunburn that appears 12–24 hours after exposure. UVA radiation, by contrast, causes tanning. It also damages skin, albeit much less efficiently than UVB. Its role in promoting skin cancer is less well understood.

Commercial sunscreen products are rated according to their sun protection factor (SPF), which is defined as the minimum effective dose of UV radiation that produces a delayed sunburn on protected skin compared to unprotected skin. Two types of active ingredients are found in commercial sunblocks and sunscreens. The most common sunblock agent is zinc oxide, ZnO, which reflects and scatters UV radiation. Sunscreens, the second type of active ingredient, absorb UV radiation and then reradiate it as heat. These compounds are most effective in screening UVB radiation, but they do not screen UVA radiation. Thus, they allow tanning but prevent the UVB-associated damage. Given here are structural formulas for three common esters used as UVB-screening agents, along with the name by which each is most commonly listed in the Active Ingredients labels on commercial products:

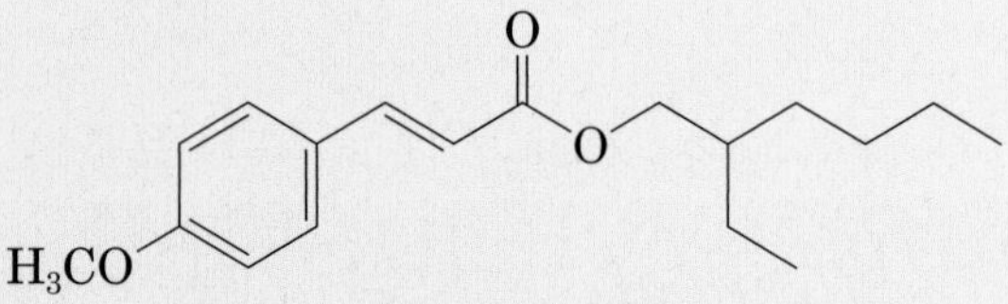

Octyl *p*-methoxycinnamate 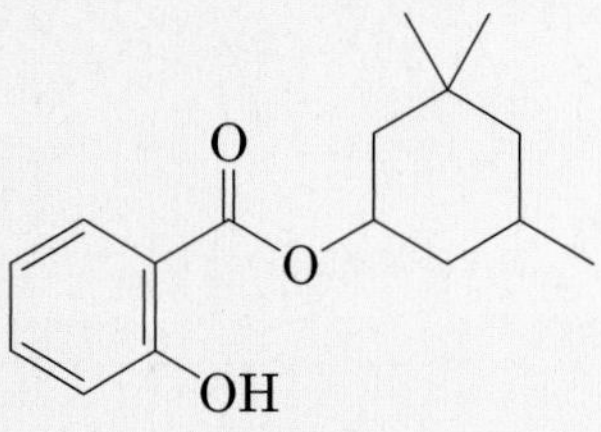   Homosalate 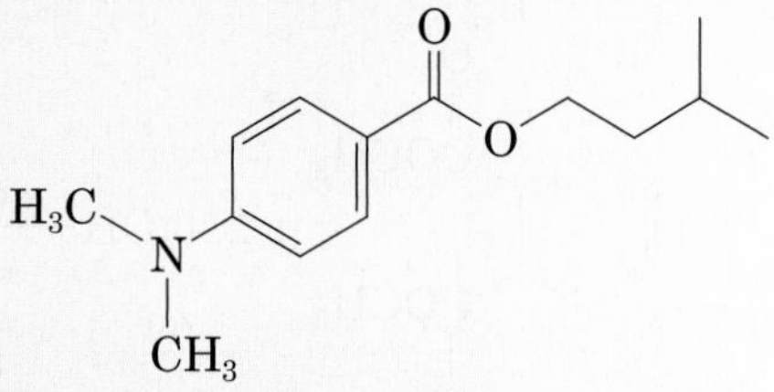   Padimate A

Test your knowledge with Problems 19.30–19.32.

Example 19.3 Hydrolysis of an Amide

Write a balanced equation for the hydrolysis of each amide in concentrated aqueous HCl. Show all products as they exist in aqueous HCl.

(a) $CH_3CN(CH_3)_2$ (b)

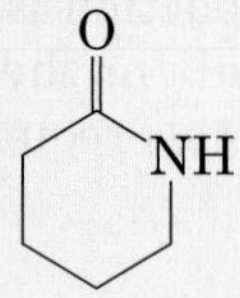

Strategy

Hydrolysis of an amide gives a carboxylic acid and an amine. If the hydrolysis is carried out in aqueous acid, the amine is converted to its ammonium salt. Hydrolysis of an amide in aqueous acid requires one mole of acid per mole of amide.

Solution

(a) Hydrolysis of *N,N*-dimethylacetamide gives acetic acid and dimethylamine. Dimethylamine, a base, reacts with HCl to form dimethylammonium ion, shown here as dimethylammonium chloride.

$$CH_3CN(CH_3)_2 + H_2O + HCl \xrightarrow{\text{Heat}} CH_3COH + (CH_3)_2NH_2{}^+Cl^-$$

(b) Hydrolysis of this lactam gives the protonated form of 5-aminopentanoic acid.

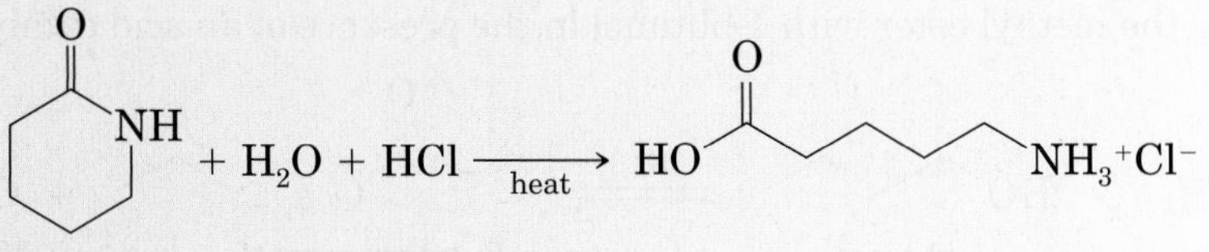

$$\text{(lactam)} + H_2O + HCl \xrightarrow{\text{heat}} \text{HO} \overset{O}{\underset{}{\parallel}} \text{C—(CH}_2)_3\text{—NH}_3{}^+Cl^-$$

Problem 19.3

Write a balanced equation for the hydrolysis of each amide in Example 19.3 in concentrated aqueous NaOH. Show all products as they exist in aqueous NaOH.

Table 19.1 Summary of the Reactions of Anhydrides, Esters, and Amides with Water

$$R-\overset{O}{\underset{\parallel}{C}}-O-\overset{O}{\underset{\parallel}{C}}-R + H_2O \longrightarrow R-\overset{O}{\underset{\parallel}{C}}-OH + HO-\overset{O}{\underset{\parallel}{C}}-R$$

An anhydride

$$R-\overset{O}{\underset{\parallel}{C}}-OR' + H_2O \begin{cases} \xrightarrow{H_2SO_4} R-\overset{O}{\underset{\parallel}{C}}-OH + R'OH \\ \xrightarrow{NaOH} R-\overset{O}{\underset{\parallel}{C}}-O^-Na^+ + R'OH \end{cases}$$

An ester

$$R-\overset{O}{\underset{\parallel}{C}}-NH_2 + H_2O \begin{cases} \xrightarrow{H_2SO_4} R-\overset{O}{\underset{\parallel}{C}}-OH + NH_4{}^+HSO_4{}^- \\ \xrightarrow{NaOH} R-\overset{O}{\underset{\parallel}{C}}-O^-Na^+ + NH_3 \end{cases}$$

An amide

B. Reaction with Alcohols

Anhydrides

Anhydrides react with alcohols and phenols to give one mole of ester and one mole of a carboxylic acid.

$$\underset{\text{Acetic anhydride}}{\overset{O\ \ O}{CH_3\overset{\parallel}{C}O\overset{\parallel}{C}CH_3}} + \underset{\text{Ethanol}}{HOCH_2CH_3} \longrightarrow \underset{\text{Ethyl acetate}}{CH_3\overset{O}{\overset{\parallel}{C}}OCH_2CH_3} + \underset{\text{Acetic acid}}{HO\overset{O}{\overset{\parallel}{C}}CH_3}$$

Thus, the reaction of an alcohol with an anhydride is a useful method for the synthesis of esters. Aspirin (Chemical Connections 19C) is synthesized on an industrial scale by the reaction of acetic anhydride with salicylic acid.

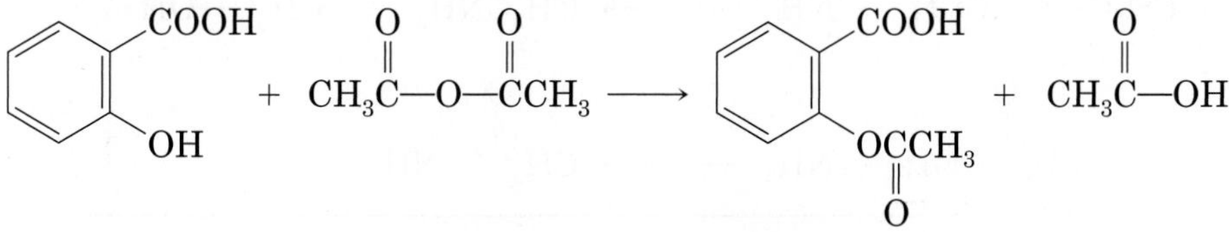

$$\underset{\text{Salicylic acid}}{\text{(COOH, OH benzene)}} + \underset{\text{Acetic anhydride}}{CH_3\overset{O}{\overset{\parallel}{C}}-O-\overset{O}{\overset{\parallel}{C}}CH_3} \longrightarrow \underset{\substack{\text{Acetylsalicylic acid} \\ \text{(Aspirin)}}}{\text{(COOH, OCCH}_3 \text{ benzene)}} + \underset{\text{Acetic acid}}{CH_3\overset{O}{\overset{\parallel}{C}}-OH}$$

Esters

Esters react with alcohols in an acid-catalyzed reaction called **transesterification**. For example, it is possible to convert a methyl ester to a butyl ester by heating the methyl ester with 1-butanol in the presence of an acid catalyst.

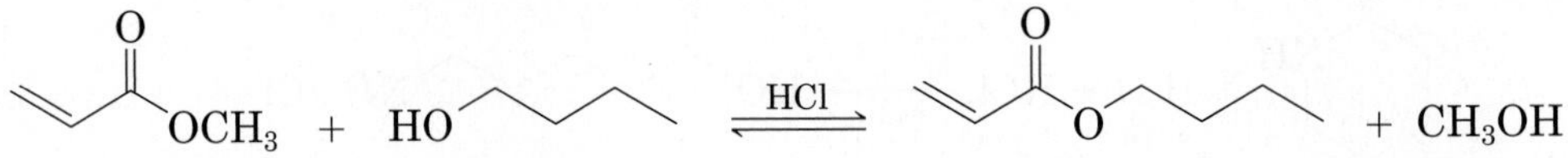

Methyl propenoate
Methyl acrylate
(bp 81°C)

1-Butanol
(bp 117°C)

Butyl propenoate
Butyl acrylate
(bp 147°C)

Methanol
(65°C)

Transesterification is an equilibrium reaction that can be driven in either direction by control of the experimental conditions. For example, in the reaction of a methyl ester, transesterification is carried out at slightly above the boiling point of methanol (the lowest boiling component in the mixture). Methanol distills from the reaction mixture, thus shifting the position of equilibrium in favor of butyl ester. Conversely, reaction of butyl acrylate with a large excess of methanol shifts the equilibrium to favor the formation of methyl acrylate.

Amides

Amides do not react with alcohols.

Table 19.2 Summary of the Reactions of Anhydrides, Esters, and Amides with Alcohols

$$\underset{\text{An anhydride}}{\text{R}-\overset{\overset{\displaystyle \text{O}}{\|}}{\text{C}}-\text{O}-\overset{\overset{\displaystyle \text{O}}{\|}}{\text{C}}-\text{R}} + \text{R'OH} \longrightarrow \text{R}-\overset{\overset{\displaystyle \text{O}}{\|}}{\text{C}}-\text{OR'} + \text{HO}-\overset{\overset{\displaystyle \text{O}}{\|}}{\text{C}}-\text{R}$$

$$\underset{\text{An ester}}{\text{R}-\overset{\overset{\displaystyle \text{O}}{\|}}{\text{C}}-\text{OR'}} + \text{R''OH} \overset{\text{H}_2\text{SO}_4}{\rightleftharpoons} \text{R}-\overset{\overset{\displaystyle \text{O}}{\|}}{\text{C}}-\text{OR''} + \text{R'OH}$$

$$\underset{\text{An amide}}{\text{R}-\overset{\overset{\displaystyle \text{O}}{\|}}{\text{C}}-\text{NH}_2} + \text{ROH} \quad \text{No reaction}$$

C. Reaction with Ammonia and Amines

Anhydrides

Anhydrides react with ammonia and with 1° and 2° amines to form amides. Two moles of amine are required: one to form the amide and one to neutralize the carboxylic acid by-product. We show this reaction here in two steps: (1) formation of the amide and the carboxylic acid by-product and (2) an acid-base reaction of the carboxylic acid by-product with the second mole of ammonia to give an ammonium salt.

$$\text{CH}_3\overset{\overset{\displaystyle \text{O}}{\|}}{\text{C}}-\text{O}-\overset{\overset{\displaystyle \text{O}}{\|}}{\text{C}}\text{CH}_3 + \text{NH}_3 \longrightarrow \text{CH}_3\overset{\overset{\displaystyle \text{O}}{\|}}{\text{C}}\text{NH}_2 + \text{CH}_3\overset{\overset{\displaystyle \text{O}}{\|}}{\text{C}}-\text{OH}$$

$$\text{CH}_3\overset{\overset{\displaystyle \text{O}}{\|}}{\text{C}}-\text{OH} + \text{NH}_3 \longrightarrow \text{CH}_3\overset{\overset{\displaystyle \text{O}}{\|}}{\text{C}}\text{O}^-\text{NH}_4^+$$

$$\underset{\text{Acetic anhydride}}{\text{CH}_3\overset{\overset{\displaystyle \text{O}}{\|}}{\text{C}}-\text{O}-\overset{\overset{\displaystyle \text{O}}{\|}}{\text{C}}\text{CH}_3} + 2\text{NH}_3 \longrightarrow \underset{\text{Acetamide}}{\text{CH}_3\overset{\overset{\displaystyle \text{O}}{\|}}{\text{C}}\text{NH}_2} + \underset{\text{Ammonium acetate}}{\text{CH}_3\overset{\overset{\displaystyle \text{O}}{\|}}{\text{C}}\text{O}^-\text{NH}_4^+}$$

Chemical Connections 19E

Barbiturates

In 1864, Adolph von Baeyer (1835–1917) discovered that heating the diethyl ester of malonic acid with urea in the presence of sodium ethoxide (like sodium hydroxide, a strong base) gives a cyclic compound that he named barbituric acid. Some say that Baeyer named it after a friend of his named Barbara. Others claim that he named it after St. Barbara, the patron saint of artillerymen.

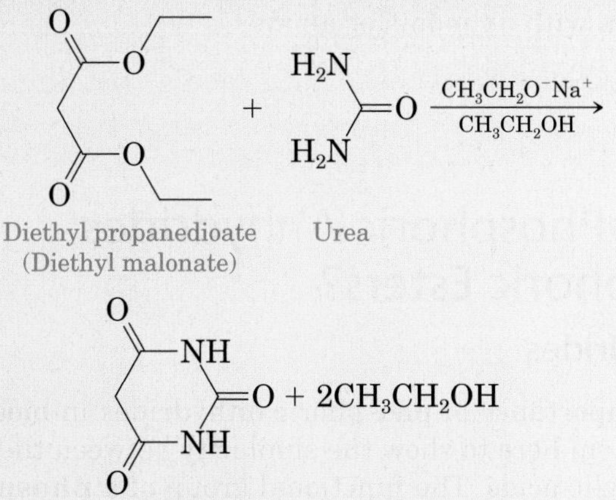

Diethyl propanedioate   Urea
(Diethyl malonate)

Barbituric acid

A number of derivatives of barbituric acid have powerful sedative and hypnotic effects. One such derivative is pentobarbital. Like other derivatives of barbituric acid, pentobarbital is quite insoluble in water and body fluids. To increase its solubility in these fluids, pentobarbital is converted to its sodium salt, which is given the name Nembutal. Phenobarbital, also administered as its sodium salt, is an anticonvulsant, sedative, and hypnotic.

Technically speaking, only the sodium salts of these compounds should be called barbiturates. In practice, however, all derivatives of barbituric acid are called barbiturates

whether they are the un-ionized form or the ionized, water-soluble salt form.

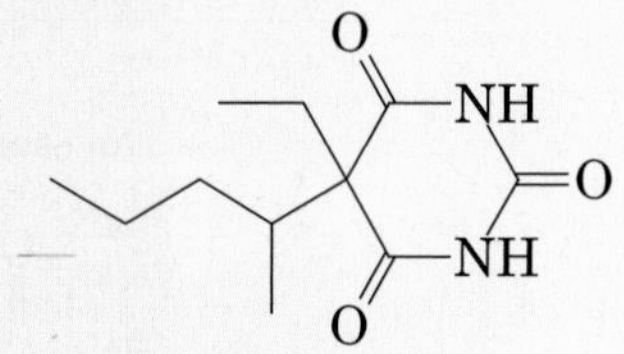

Pentobarbital

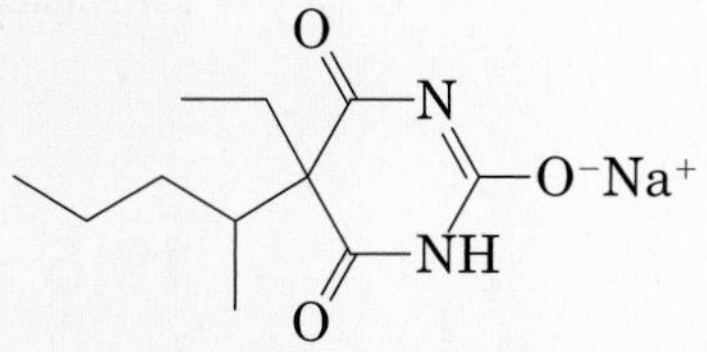

Sodium pentobarbital
(Nembutal)

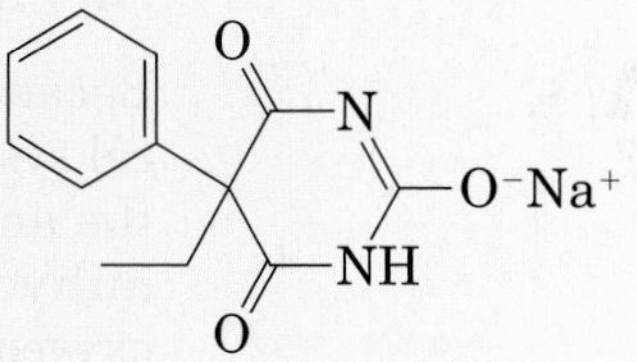

Phenobarbital

Barbiturates have two principal effects. In small doses, they are sedatives (tranquilizers); in larger doses, they induce sleep. Barbituric acid, in contrast, has neither of these effects. Barbiturates are dangerous because they are addictive, which means that a regular user will suffer withdrawal symptoms when their use is stopped. They are especially dangerous when taken with alcohol because the combined effect (called a synergistic effect) is usually greater than the sum of the effects of either drug taken separately.

Test your knowledge with Problem 19.33.

Esters

Esters react with ammonia and with 1° and 2° amines to form amides.

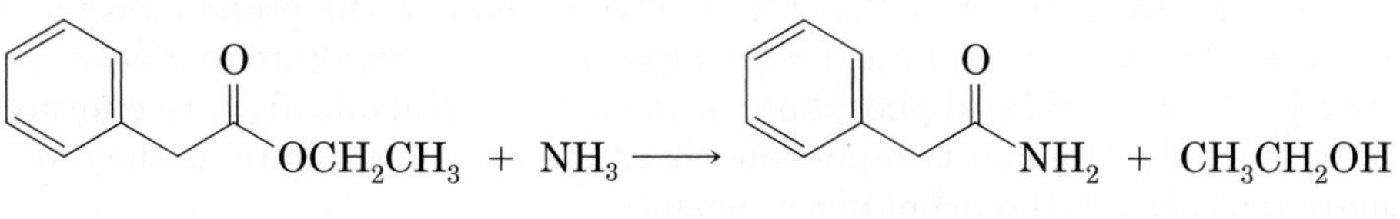

Ethyl 2-phenyl acetate   2-Phenylacetamide

Thus, as seen in this section, amides can be prepared readily from esters. Because carboxylic acids are easily converted to esters by Fischer esterification, we have a good way to convert a carboxylic acid to an amide. This method of amide formation is, in fact, much more useful and applicable than converting a carboxylic acid to an ammonium salt and then heating the salt to form an amide.

Amides

Amides do not react with ammonia or with primary or secondary amines.

Table 19.3 Summary of the Reactions of Anhydrides, Esters, and Amides with Ammonia and Amines

$$\underset{\text{An anhydride}}{R-\overset{\displaystyle O}{\overset{\|}{C}}-O-\overset{\displaystyle O}{\overset{\|}{C}}-R} + NH_3 \longrightarrow R-\overset{\displaystyle O}{\overset{\|}{C}}-NH_2 + R-\overset{\displaystyle O}{\overset{\|}{C}}-O^-NH_4{}^+$$

$$\underset{\text{An ester}}{R-\overset{\displaystyle O}{\overset{\|}{C}}-OR'} + NH_3 \longrightarrow R-\overset{\displaystyle O}{\overset{\|}{C}}-NH_2 + R'OH$$

$$\underset{\text{An amide}}{R-\overset{\displaystyle O}{\overset{\|}{C}}-NH_2} \quad \text{No reaction with ammonia or amines}$$

19.5 What Are Phosphoric Anhydrides and Phosphoric Esters?

A. Phosphoric Anhydrides

Because of the special importance of phosphoric anhydrides in biochemical systems, we discuss them here to show the similarity between them and the anhydrides of carboxylic acids. The functional group of a **phosphoric anhydride** consists of two phosphoryl ($P=O$) groups bonded to the same oxygen atom. Shown here are structural formulas for two anhydrides of phosphoric acid and the ions derived by ionization of the acidic hydrogens of each:

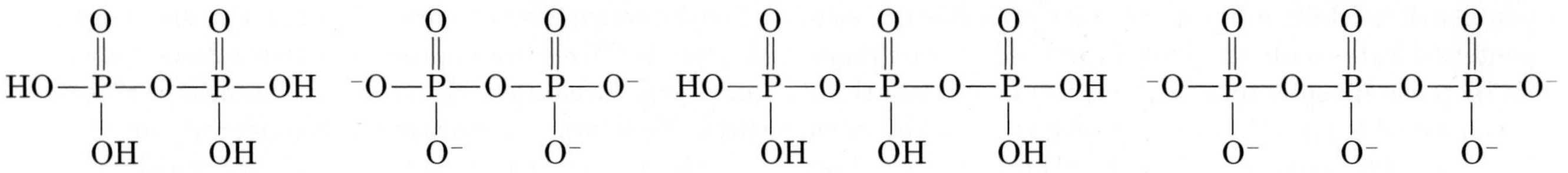

Diphosphoric acid (Pyrophosphoric acid)   Diphosphate ion (Pyrophosphate ion)   Triphosphoric acid   Triphosphate ion

B. Phosphoric Esters

Phosphoric acid has three —OH groups and forms mono-, di-, and triphosphoric esters, which we name by giving the name(s) of the alkyl group(s) bonded to oxygen followed by the word "phosphate" (for example, dimethyl phosphate). In more complex **phosphoric esters**, it is common practice to name the organic molecule and then indicate the presence of the phosphoric ester by including either the word *phosphate* or the prefix *phospho-*. Dihydroxyacetone phosphate, for example, is an intermediate in glycolysis (Section 28.2). Pyridoxal phosphate is one of the metabolically active forms of vitamin B_6. The last two phosphoric esters are shown here as they are ionized at pH 7.4, the pH of blood plasma.

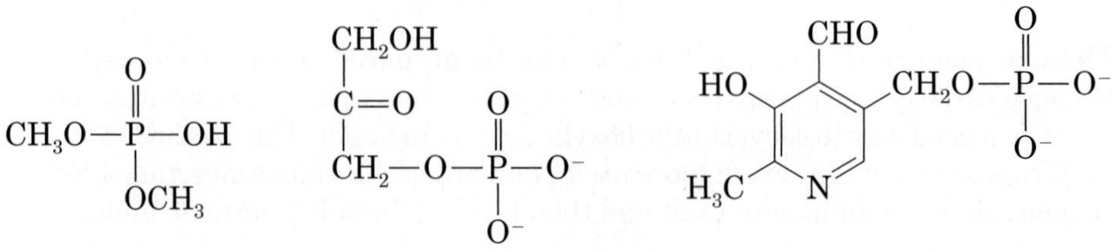

Dimethyl phosphate   Dihydroxyacetone phosphate   Pyridoxal phosphate

19.6 What Is Step-Growth Polymerization?

Step-growth polymers form from the reaction of molecules containing two functional groups, with each new bond being created in a separate step. In this section, we discuss three types of step-growth polymers: polyamides, polyesters, and polycarbonates.

A. Polyamides

In the early 1930s, chemists at E. I. DuPont de Nemours & Company began fundamental research into the reactions between dicarboxylic acids and di-amines to form **polyamides**. In 1934, they synthesized nylon-66, the first purely synthetic fiber. Nylon-66 is so named because it is synthesized from two different monomers, each containing six carbon atoms.

In the synthesis of nylon-66, hexanedioic acid and 1,6-hexanediamine are dissolved in aqueous ethanol and then heated in an autoclave to 250°C and an internal pressure of 15 atm. Under these conditions, —COOH and —NH$_2$ groups react by loss of H$_2$O to form a polyamide, similar to the formation of amides described in Section 19.3.

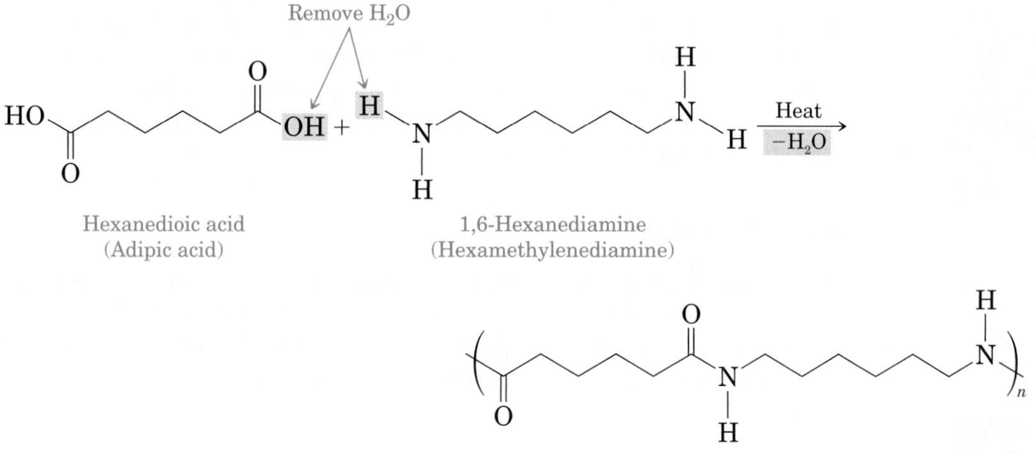

Hexanedioic acid
(Adipic acid)

1,6-Hexanediamine
(Hexamethylenediamine)

Nylon-66
(a polyamide)

Based on extensive research into the relationships between molecular structure and bulk physical properties, scientists at DuPont reasoned that a polyamide containing benzene rings would be even stronger than nylon-66. This line of reasoning eventually produced a polyamide that DuPont named Kevlar.

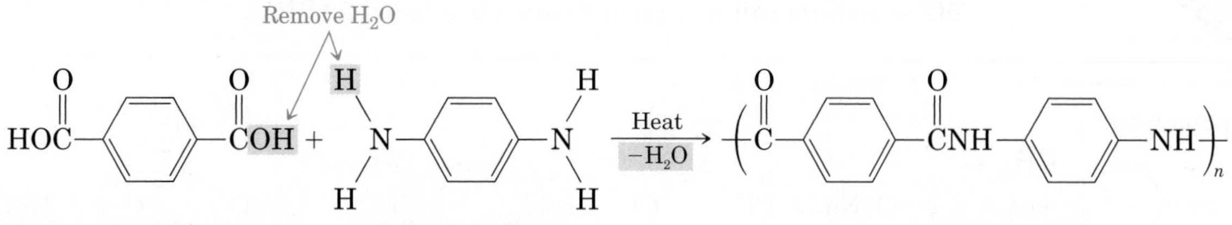

1,4-Benzenedicarboxylic acid
(Terephthalic acid)

1,4-Benzenediamine
(*p*-Phenylenediamine)

Kevlar
(a polyaromatic amide)

One remarkable feature of Kevlar is that it weighs less than other materials of similar strength. For example, a cable woven of Kevlar has a strength equal to that of a similarly woven steel cable. Yet, the Kevlar cable has only 20% of the weight of the steel cable! Kevlar is now used in such articles as anchor cables for offshore drilling rigs and reinforcement fibers for automobile tires. It is also woven into a fabric that is so tough that it can be used for bulletproof vests, jackets, and raincoats.

B. Polyesters

The first **polyester**, developed in the 1940s, involved polymerization of benzene 1,4-dicarboxylic acid with 1,2-ethanediol to give poly(ethylene terephthalate), abbreviated PET. Virtually all PET is now made from the dimethyl ester of terephthalic acid by the following reaction:

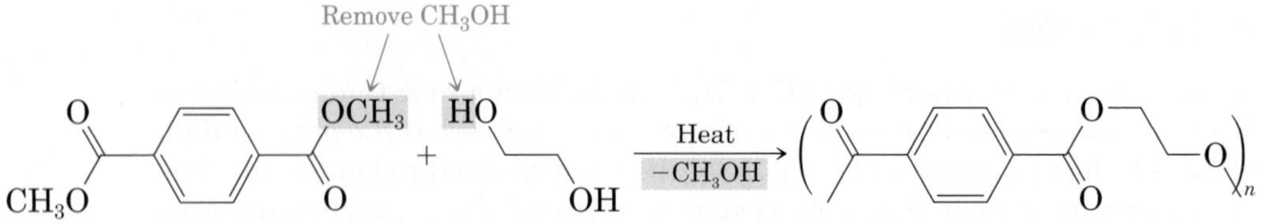

Remove CH_3OH

Dimethyl terephthalate   +   1,2-Ethanediol (Ethylene glycol)   →[Heat, $-CH_3OH$]   Poly(ethylene terephthalate) (Dacron, Mylar)

Mylar can be made into extremely strong films. Because the film has a very small pore size, it is used for balloons that can be inflated with helium; the helium atoms diffuse extremely slowly through the pores of the film.

The crude polyester can be melted, extruded, and then drawn to form the textile fiber called Dacron polyester. Dacron's outstanding features include its stiffness (about four times that of nylon-66), very high tensile strength, and remarkable resistance to creasing and wrinkling. Because the early Dacron polyester fibers were harsh to the touch due to their stiffness, they were usually blended with cotton or wool to make acceptable textile fibers. Newly developed fabrication techniques now produce less harsh Dacron polyester textile fibers. PET is also fabricated into Mylar films and recyclable plastic beverage containers.

C. Polycarbonates

A **polycarbonate**, the most familiar of which is Lexan, forms from the reaction between the disodium salt of bisphenol A and phosgene. Phosgene is a derivative of carbonic acid, H_2CO_3, in which both —OH groups have been replaced with chlorine atoms. An ester of carbonic acid is called a carbonate.

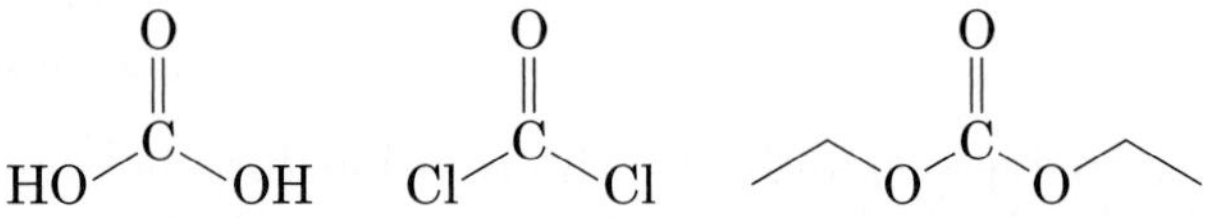

A polycarbonate hockey mask.

Carbonic acid (H_2CO_3)   Phosgene   Diethyl carbonate (a carbonate ester)

In forming a polycarbonate, each mole of phosgene reacts with two moles of the sodium salt of a phenol called bisphenol A (BPA).

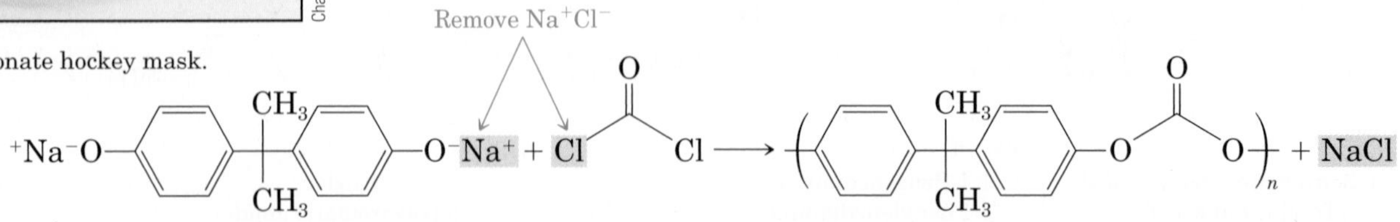

Remove Na^+Cl^-

Disodium salt of Bisphenol A   +   Phosgene   →   Lexan (a polycarbonate)   +   NaCl

Lexan is a tough, transparent polymer that has high impact and tensile strength that retains its properties over a wide temperature range. It is used in sporting equipment (helmets and face masks); to make light, impact-resistant housings for household appliances; and in the manufacture of safety glass and unbreakable windows.

Chemical Connections 19F

Stitches That Dissolve

As the technological capabilities of medicine have expanded, the demand for synthetic materials that can be used inside the body has increased as well. Polymers already have many of the characteristics of an ideal biomaterial: They are lightweight and strong, are inert or biodegradable depending on their chemical structure, and have physical properties (softness, rigidity, elasticity) that are easily tailored to match those of natural tissues.

Even though most medical uses of polymeric materials require biostability, some applications require them to be biodegradable. An example is the polyester of glycolic acid

and lactic acid used in absorbable sutures, which are marketed under the trade name of Lactomer.

A health care specialist must remove traditional suture materials such as catgut after they have served their purpose. Stitches of Lactomer, however, are hydrolyzed slowly over a period of approximately two weeks. By the time the torn tissues have healed, the stitches have hydrolyzed, and no suture removal is necessary. The body metabolizes and excretes the glycolic and lactic acids formed during this hydrolysis.

Glycolic acid   Lactic acid   A polymer of
                                                    glycolic acid and lactic acid

Test your knowledge with Problem 19.34.

Summary of Key Questions

OWL Sign in at **www.cengage.com/owl** to develop problem-solving skills and complete online homework assigned by your professor.

Section 19.1 What Are Carboxylic Anhydrides, Esters, and Amides?

- A **carboxylic anhydride** contains two carbonyl groups bonded to the same oxygen.
- A carboxylic **ester** contains a carbonyl group bonded to an —OR group derived from an alcohol or a phenol.
- A carboxylic **amide** contains a carbonyl group bonded to a nitrogen atom derived from an amine.

Section 19.2 How Do We Prepare Esters?

- The most common laboratory method for the preparation of esters is Fischer esterification (Section 18.5D).

Section 19.3 How Do We Prepare Amides?

- Amides can be prepared by the reaction of an amine with a carboxylic anhydride.

Section 19.4 What Are the Characteristic Reactions of Carboxylic Anhydrides, Esters, and Amides?

- Hydrolysis is a chemical process in which a bond is split and the elements of H_2O are added.

- Hydrolysis of a carboxylic anhydride gives two molecules of carboxylic acid.
- Hydrolysis of a carboxylic ester requires the presence of either concentrated aqueous acid or base. Acid is a catalyst, and the reaction is the reverse of **Fischer esterification**. Base is a reactant that is required in stoichiometric amounts.
- Hydrolysis of a carboxylic amide requires the presence of either aqueous acid or base. Both acid and base are reactants that are required in stoichiometric amounts.

Section 19.5 What Are Phosphoric Anhydrides and Phosphoric Esters?

- Phosphoric anhydrides consist of two phosphoryl groups (P=O) bonded to the same oxygen atom.

Section 19.6 What Is Step-Growth Polymerization?

- Step-growth polymerization involves the stepwise reaction of difunctional monomers. Important commercial polymers synthesized through step-growth processes include polyamides, polyesters, and polycarbonates.

Summary of Key Reactions

1. Fischer Esterification (Section 19.2) Fischer esterification is reversible. To achieve high yields of ester, it is necessary to force the equilibrium to the right. One way to maximize the yield of ester is to use an excess of the alcohol. Another way is to remove the water as it is formed.

$$CH_3COH + CH_3CH_2CH_2OH$$

$$\overset{H_2SO_4}{\rightleftharpoons} CH_3COCH_2CH_2CH_3 + H_2O$$

2. Preparation of an Amide (Section 19.3) Reaction of an anhydride with ammonia or a 1° or 2° amine gives an amide.

$$CH_3C-O-CCH_3 + H_2NCH_2CH_3$$

$$\longrightarrow CH_3C-NHCH_2CH_3 + CH_3COH$$

3. Hydrolysis of an Anhydride (Section 19.4A) Anhydrides, particularly low-molecular-weight ones, react readily with water to give two carboxylic acids.

$$CH_3COCCH_3 + H_2O$$

$$\longrightarrow CH_3COH + HOCCH_3$$

4. Hydrolysis of an Ester (Section 19.4A) Esters are hydrolyzed rapidly only in the presence of acid or base. Acid-catalyzed hydrolysis is the reverse of Fischer esterification. Acid is a catalyst. Base is a reactant and therefore is required in an equimolar amount.

$$CH_3COCH_2CH_3 + H_2O$$

$$\overset{H^+}{\rightleftharpoons} CH_3COH + HOCH_2CH_3$$

$$CH_3COCH_2CH_3 + NaOH$$

$$\overset{H_2O}{\longrightarrow} CH_3CO^-Na^+ + CH_3CH_2OH$$

5. Hydrolysis of an Amide (Section 19.4A) Amides require more vigorous conditions for hydrolysis than do esters. Either acid or base is required in an amount equivalent to that of the amide: acid to convert the resulting amine to an ammonium salt and base to convert the resulting carboxylic acid to a carboxylate salt.

$$CH_3CH_2CH_2CNH_2 + H_2O + HCl$$

$$\overset{H_2O}{\underset{heat}{\longrightarrow}} CH_3CH_2CH_2COH + NH_4^+Cl^-$$

$$CH_3CH_2CH_2CNH_2 + NaOH$$

$$\overset{H_2O}{\underset{heat}{\longrightarrow}} CH_3CH_2CH_2CO^-Na^+ + NH_3$$

6. Reaction of Anhydrides with Alcohols (Section 19.4B) Anhydrides react with alcohols to give one mole of ester and one mole of a carboxylic acid.

$$CH_3COCCH_3 + HOCH_2CH_3$$

$$\longrightarrow CH_3COCH_2CH_3 + HOCCH_3$$

7. Reaction of Anhydrides with Ammonia and Amines (Section 19.4C) Anhydrides react with ammonia and with 1° and 2° amines to give amides. Two moles of amine are required: one mole to give the amide and one mole to neutralize the carboxylic acid by-product.

$$CH_3C-O-CCH_3 + 2NH_3$$

$$\longrightarrow CH_3CNH_2 + CH_3CO^-NH_4^+$$

8. Reaction of Esters with Ammonia and with 1° and 2° Amines (Section 19.4C) Esters react with ammonia and with 1° and 2° amines to give an amide and an alcohol.

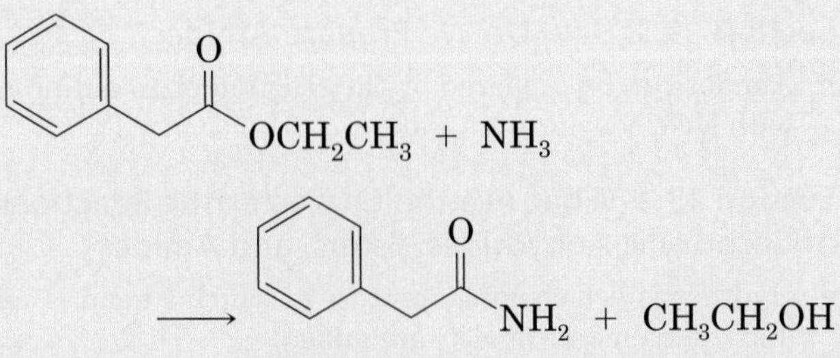

Problems

Interactive versions of these problems may be assigned in OWL.

Orange-numbered problems are applied.

References to previous chapters are given in parentheses.

Section 19.1 What Are Carboxylic Anhydrides, Esters, and Amides?

19.4 Draw a structural formula for each compound.
  (a) Dimethyl carbonate
  (b) *p*-Nitrobenzamide
  (c) Ethyl 3-hydroxybutanoate
  (d) Diethyl oxalate
  (e) Ethyl *trans*-2-pentenoate
  (f) Butanoic anhydride

19.5 Write the IUPAC name for each compound.

(a)

(b) $CH_3(CH_2)_8 \overset{\displaystyle O}{\overset{\displaystyle \|}{C}} OCH_3$

(c) $CH_3(CH_2)_4 \overset{\displaystyle O}{\overset{\displaystyle \|}{C}} NHCH_3$

(d)

(e) $CH_3 \overset{\displaystyle O}{\overset{\displaystyle \|}{C}} O$—

(f) $CH_3 \overset{\displaystyle OH}{C} H CH_2 \overset{\displaystyle O}{\overset{\displaystyle \|}{C}} OCH_2CH_3$

19.6 When oil from the head of the sperm whale is cooled, spermaceti, a translucent wax with a white, pearly luster, crystallizes from the mixture. Spermaceti, which makes up 11% of sperm whale oil, is composed of mainly hexadecyl hexadecanoate (cetyl palmitate). At one time, spermaceti was widely used in the making of cosmetics, fragrant soaps, and candles. Draw the structural formula of spermaceti. (*Hint:* Hexadecane, the parent hydrocarbon, is $CH_3(CH_2)_{14}CH_3$.)

Section 19.4 What Are the Characteristic Reactions of Anhydrides, Esters, and Amides?

19.7 What product forms when ethyl benzoate is treated with each of the following sets?
  (a) H_2O, NaOH, heat (b) H_2O, HCl, heat

19.8 What product forms when benzamide, $C_6H_5CONH_2$, is treated with each of the following sets?
  (a) H_2O, NaOH, heat (b) H_2O, HCl, heat

19.9 Which of these types of compounds will produce bubbles of CO_2 when added to an aqueous solution of sodium bicarbonate?
  (a) A carboxylic acid (b) A carboxylic ester
  (c) The sodium salt of a carboxylic acid

19.10 Complete the equations for these reactions.

(a) CH_3O——$NH_2 + CH_3 \overset{\displaystyle O}{\overset{\displaystyle \|}{C}} O \overset{\displaystyle O}{\overset{\displaystyle \|}{C}} CH_3 \longrightarrow$

(b) $NH + CH_3 \overset{\displaystyle O}{\overset{\displaystyle \|}{C}} O \overset{\displaystyle O}{\overset{\displaystyle \|}{C}} CH_3 \longrightarrow$

19.11 The analgesic phenacetin is synthesized by treating 4-ethoxyaniline with acetic anhydride. Draw a structural formula for phenacetin.

CH_3CH_2O——NH_2

4-Ethoxyaniline

19.12 Phenobarbital is a long-acting sedative, hypnotic, and anticonvulsant.
  (a) Name all functional groups in this compound. (Chapter 10)
  (b) Draw structural formulas for the products from complete hydrolysis of all amide groups in aqueous NaOH.

Phenobarbital

19.13 Following is a structural formula for aspartame, an artificial sweetener about 180 times as sweet as sucrose (table sugar).

Aspartame

  (a) Is aspartame chiral? If so, how many stereoisomers are possible for it? (Chapter 15)
  (b) Name each functional group in aspartame. (Chapter 10)
  (c) Estimate the net charge on an aspartame molecule in aqueous solution at pH 7.0.

(d) Would you expect aspartame to be soluble in water? Explain. (Chapter 5)

(e) Draw structural formulas for the products of complete hydrolysis of aspartame in aqueous HCl. Show each product as it would be ionized in this solution.

(f) Draw structural formulas for the products of complete hydrolysis of aspartame in aqueous NaOH. Show each product as it would be ionized in this solution.

19.14 Why are nylon-66 and Kevlar referred to as polyamides?

19.15 Draw short sections of two parallel chains of nylon-66 (each chain running in the same direction) and show how it is possible to align them such that there is hydrogen bonding between the N—H groups of one chain and the C=O groups of the parallel chain.

19.16 Why are Dacron and Mylar referred to as polyesters?

Section 19.5 What Are Phosphoric Anhydrides and Phosphoric Esters?

19.17 What type of structural feature do the anhydrides of phosphoric acid have in common with carboxylic acids?

19.18 Draw structural formulas for the mono-, di-, and triethyl esters of phosphoric acid.

19.19 1,3-Dihydroxy-2-propanone (dihydroxyacetone) and phosphoric acid form a monoester called dihydroxyacetone phosphate, which is an intermediate in glycolysis (Section 28.2). Draw a structural formula for this monophosphate ester.

19.20 Show how triphosphoric acid can form from three molecules of phosphoric acid. How many moles of H_2O are produced per mole of triphosphoric acid?

19.21 Write an equation for the hydrolysis of trimethyl phosphate to dimethyl phosphate and methanol in aqueous base. Show each product as it would be ionized in this solution.

Chemical Connections

19.22 (Chemical Connections 19A) Locate the ester group in pyrethrin I and draw a structural formula for chrysanthemic acid, the carboxylic acid from which this ester is derived.

19.23 (Chemical Connections 19A) What structural features do pyrethrin I (a natural insecticide) and permethrin (a synthetic pyrethrenoid) have in common?

19.24 (Chemical Connections 19A) A commercial Clothing & Gear Insect Repellant gives the following information about permethrin, its active ingredient:

Cis/trans ratio: Minimum 35% (+/−) *cis* and maximum 65% (+/−) *trans*

(a) To what does the *cis/trans* ratio refer?

(b) To what does the designation "(+/−)" refer?

19.25 (Chemical Connections 19B)

(a) Identify the β-lactam portion of amoxicillin and cephalexin.

(b) Penicillin-resistant organisms are able to synthesize β-lactamase, an enzyme that catalyzes the hydrolysis of the β-lactam ring of penicillin. The integrity of the β-lactam ring is essential for antibacterial activity. Draw the structural formula

of the molecule formed by the enzyme-catalyzed hydrolysis of the β-lactam ring of penicillin G.

19.26 (Chemical Connections 19C) What is the compound in willow bark that is responsible for its ability to relieve pain? How is this compound related to salicylic acid?

19.27 (Chemical Connections 19C) Name the two functional groups in aspirin.

19.28 (Chemical Connections 19C) Once it has been opened, and particularly if it has been left open to the air, a bottle of aspirin may develop a vinegar-like odor. Explain how this might happen.

19.29 (Chemical Connections 19C) What is the structural relationship between aspirin and ibuprofen? Between aspirin and naproxen?

19.30 (Chemical Connections 19D) What is the difference in meaning between *sunblock* and *sunscreen*?

19.31 (Chemical Connections 19D) How do sunscreens prevent UV radiation from reaching the skin?

19.32 (Chemical Connections 19D) What structural features do the three sunscreens given in this Chemical Connection have in common?

19.33 (Chemical Connections 19E) Barbiturates are derived from urea. Identify the portion of the structure of pentobarbital and phenobarbital that is derived from urea.

19.34 (Chemical Connections 19F) Why do Lactomer stitches dissolve within 2 to 3 weeks following surgery?

Additional Problems

19.35 Benzocaine, a topical anesthetic, is prepared by treating 4-aminobenzoic acid with ethanol in the presence of an acid catalyst, followed by neutralization. Draw a structural formula for benzocaine.

19.36 The analgesic acetaminophen is synthesized by treating 4-aminophenol with one equivalent of acetic anhydride. Write an equation for the formation of acetaminophen. (*Hint:* The —NH₂ group is more reactive with acetic anhydride than the —OH group.)

19.37 1,3-Diphosphoglycerate, an intermediate in glycolysis (Section 28.2), contains a mixed anhydride (an anhydride of a carboxylic acid and phosphoric acid) and a phosphoric ester. Draw structural formulas for the products formed by hydrolysis of the anhydride and ester bonds in this molecule. Show each product as it would exist in solution at pH 7.4.

1,3-Diphosphoglycerate

Looking Ahead

19.38 In Chapter 22, we will discuss a class of compounds called amino acids, so named because they contain both an amino group and a carboxyl group. Following is a structural formula for the amino acid alanine.

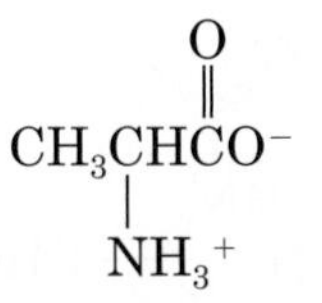

$$CH_3CHCO^-$$

O
‖

|
NH₃⁺

Alanine

What would you expect to be the major form of alanine present in aqueous solution (a) at pH 2.0, (b) at pH 5–6, and (c) at pH 11.0? Explain.

19.39 We have seen that an amide can be formed from a carboxylic acid and an amine. Suppose that you start instead with an amino acid such as alanine. Show how amide formation in this case can lead to a macromolecule of molecular weight several thousands of times that of the starting materials. We will study these polyamides in Chapter 22 (proteins).

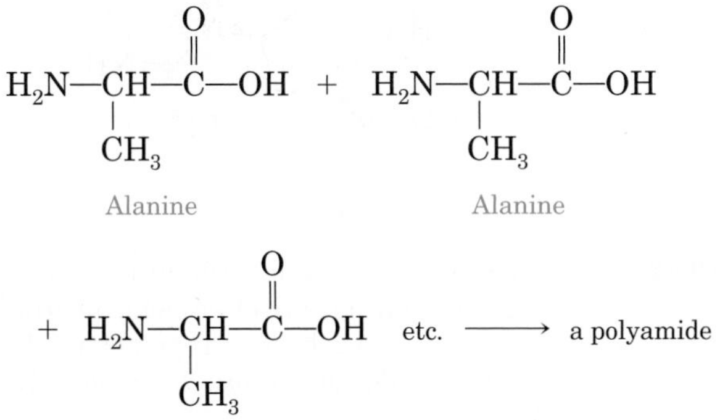

O   O
‖   ‖
H₂N—CH—C—OH + H₂N—CH—C—OH
| |
CH₃   CH₃

Alanine   Alanine

O
‖
+ H₂N—CH—C—OH   etc. ⟶ a polyamide
|
CH₃

Alanine

19.40 We will encounter the following molecule in our discussion of glycolysis, the biochemical pathway that converts glucose to pyruvic acid (Section 28.2).

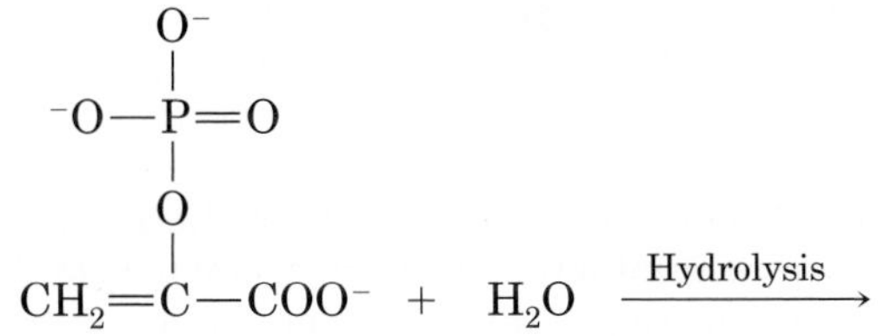

O⁻
|
⁻O—P=O
|
O
|
CH₂=C—COO⁻ + H₂O   —Hydrolysis→

Phosphoenolpyruvate

(a) Draw structural formulas for the products of hydrolysis of the ester bond in phosphoenolpyruvate.

(b) Why are the letters *enol* a part of the name of this compound?

Tying It Together

19.41 Starting with ethylene as the only source of carbon atoms, show how you could synthesize ethyl acetate.

19.42 From what carboxylic acid and amine or ammonia can each amide be synthesized?

O
‖
(a) 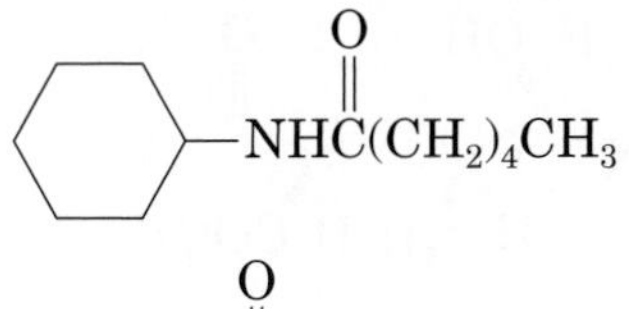 ⬡—NHC(CH₂)₄CH₃

O
‖
(b) 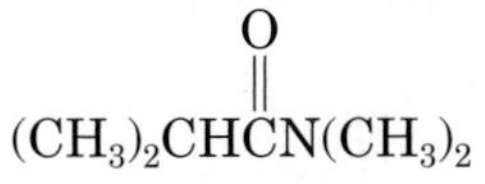 (CH₃)₂CHCN(CH₃)₂

O   O
‖   ‖
(c) 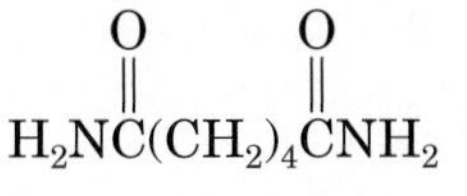 H₂NC(CH₂)₄CNH₂

19.43 *N,N*-Diethyl *m*-toluamide (DEET) is the active ingredient in several common insect repellents. From what acid and amine can DEET by synthesized?

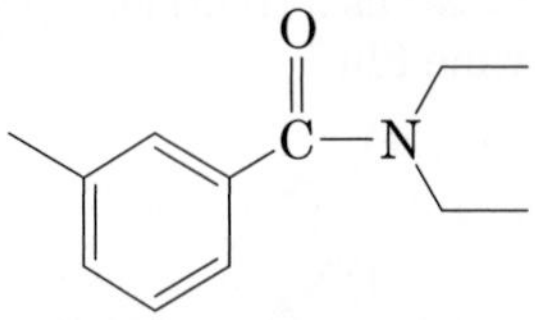

N,N-Diethyl *m*-toluamide
(DEET)

19.44 Following are structural formulas for two local anesthetics used in dentistry. Lidocaine was introduced in 1948 and is now the most widely used local anesthetic for infiltration and regional anesthesia. Its hydrochloride salt is marketed under the name Xylocaine. Mepivacaine is faster-acting and somewhat longer in duration than lidocaine. Its hydrochloride salt is marketed under the name Carbocaine.

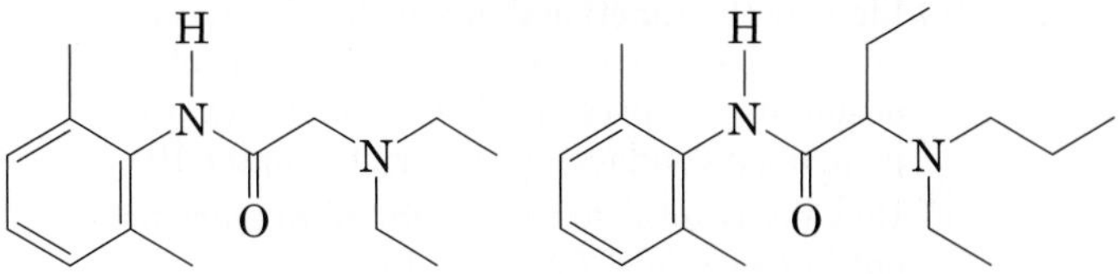

Lidocaine   Etidocaine
(Xylocaine)   (Duranest)

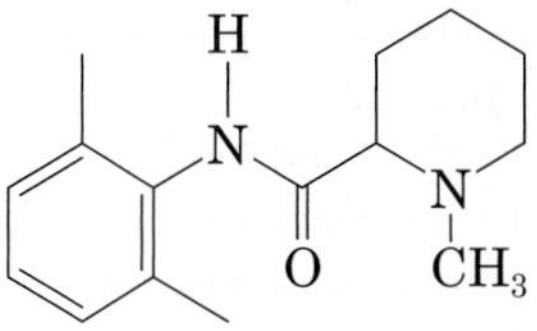

Mepivacaine
(Carbocaine)

(a) Name the functional groups in each anesthetic. (Chapter 10)

(b) Which of these anesthetics are chiral? (Chapter 15)

(c) Which nitrogen atom in each is the more basic? (Chapter 16)

(d) What similarities in structure do you find between these compounds?

19.45 Barbiturates are prepared by treating diethyl malonate or a derivative of diethyl malonate with urea in the presence of sodium ethoxide as a catalyst. Following is an equation for the preparation of barbital from a substituted diethyl malonate and urea. Barbital, a long-duration hypnotic and sedative, is prescribed under a dozen or more names. Draw the structural formula of barbital.

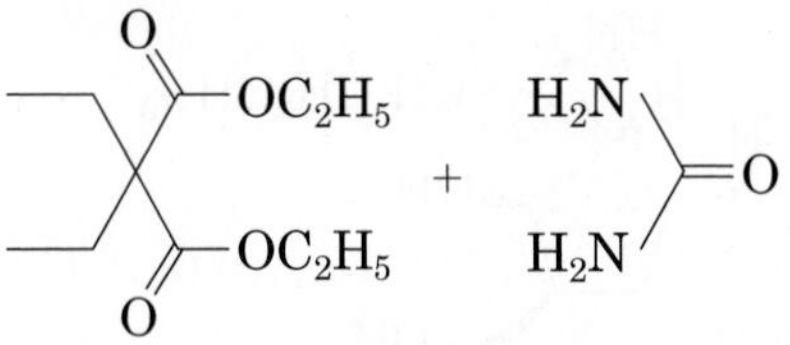

Diethyl   Urea
2,2-diethylmalonate

1. CH₃CH₂O⁻Na⁺
————————————→ Barbital + 2CH₃CH₂OH
2. H₂O

19.46 Consider the experimental antiviral drug *Peramivir*, shown below, a neuraminidase inhibitor recently supported by the U.S. Department of Health and Human Services as a treatment for Influenza A (H1N1), "Swine Flu."

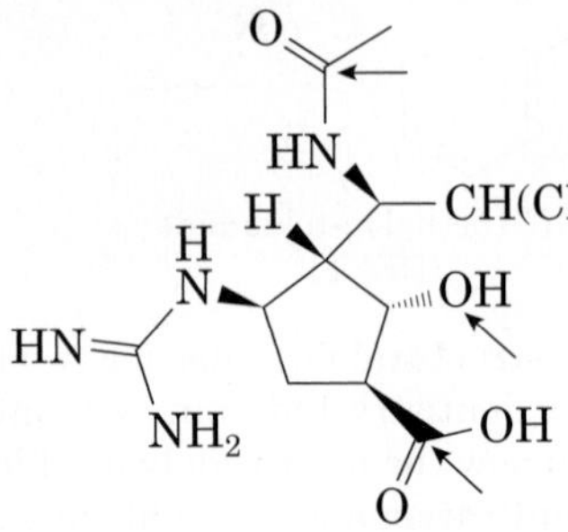

Peramivir

(a) What is the molecular formula of *Peramivir*? (Chapter 10)

(b) Identify the functional groups labeled with arrows in the structure above. If any alcohols or amines are present, determine if they are primary, secondary, or tertiary. (Chapter 10)

(c) How many total stereocenters (if any) are present in *Peramivir*? (Chapter 15)

(d) Determine the maximum number of possible stereoisomers for this molecule. (Chapter 15)

(e) Fill in the blanks: There is/are _____ enantiomer(s) and _____ diastereomer(s) of *Peramivir*. (Chapter 15)

(f) One of the stereocenters in *Peramivir* is known to have the *R*-configuration. Label this stereocenter in the structure. (Chapter 15)

(g) Compare each molecule below with *Peramivir* and identify it as "identical," "enantiomer," "diastereomer," or "none of these." (Chapter 15)

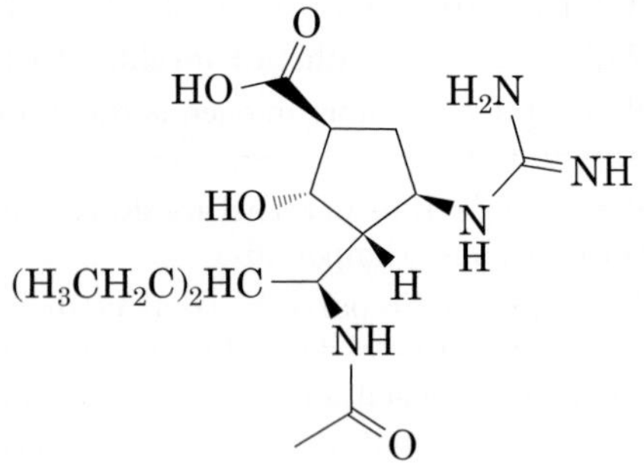

A

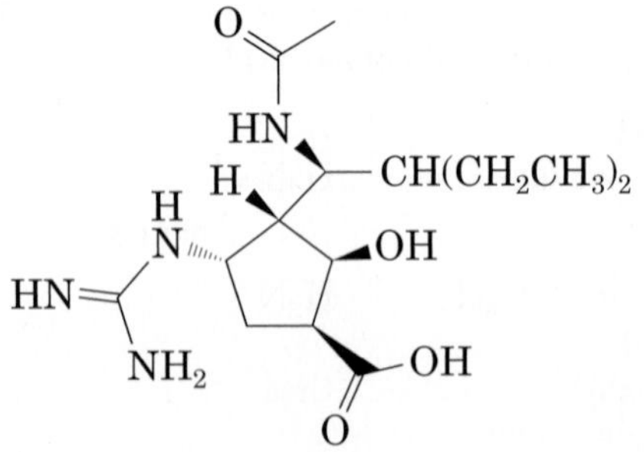

B

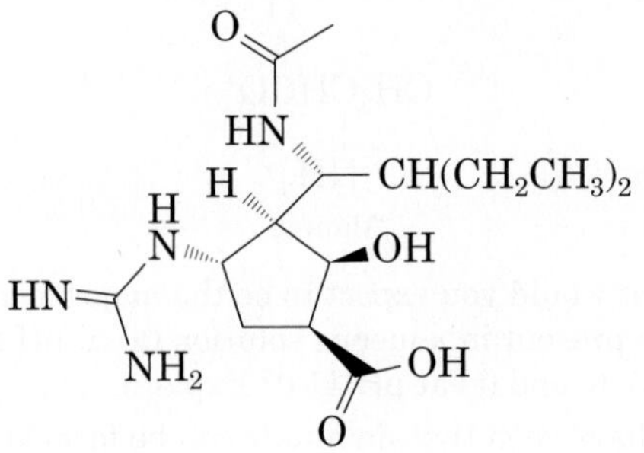

C

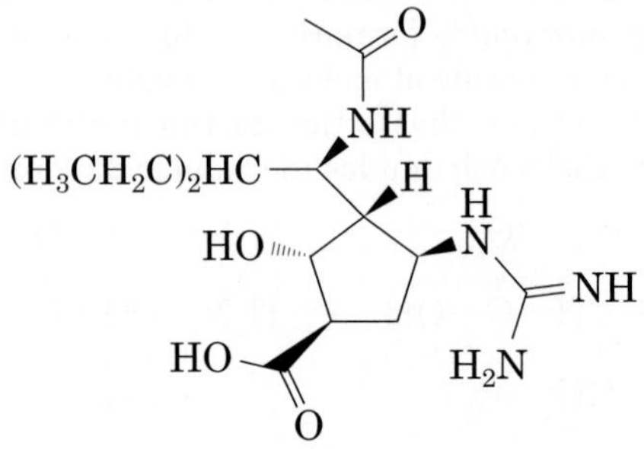

D

19.47 In pharmaceutical research, structural analogs (or derivatives) of drugs are often made in order to see if they have similar biological properties. Below is a structural analog of *Peramivir* (Problem 19.46) called **A**.

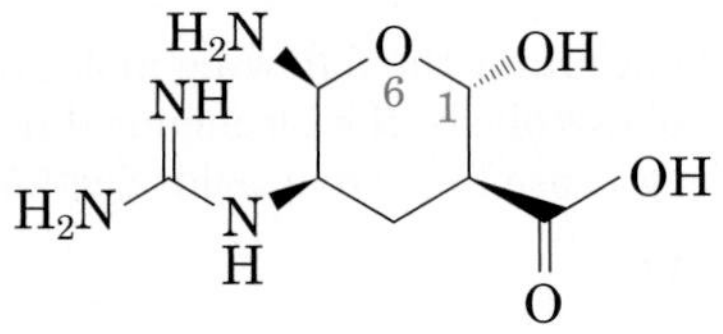

A

(a) Complete the drawings of the alternative chair conformations of **A**, making sure to note the numbering of the ring atoms. Which of the alternative chair conformations is more stable? (Chapter 11)

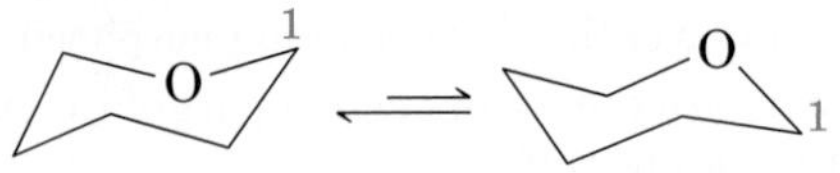

(b) **A** is formed from the cyclization of **B**. Which two functional groups in **B** react to form **A**? What is the name of the new functional group formed when these two groups react to close the ring? (*Hint:* We studied this reaction in Chapter 17.)

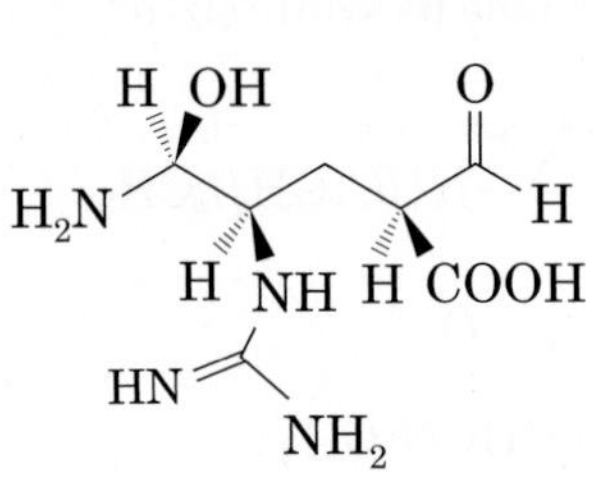

B

Carbohydrates

Charles D. Winters

Breads, grains, and pasta are sources of carbohydrates.

20.1 Carbohydrates: What Are Monosaccharides?

Carbohydrates are the most abundant organic compounds in the plant world. They act as storehouses of chemical energy (glucose, starch, glycogen); are components of supportive structures in plants (cellulose), crustacean shells (chitin), and connective tissues in animals (acidic polysaccharides); and are essential components of nucleic acids (D-ribose and 2-deoxy-D-ribose). Carbohydrates account for approximately three-fourths of the dry weight of plants. Animals (including humans) get their carbohydrates by eating plants, but they do not store much of what they consume. In fact, less than 1% of the body weight of animals is made up of carbohydrates.

The word *carbohydrate* means "hydrate of carbon" and derives from the formula $C_n(H_2O)_m$. Two examples of carbohydrates with this general molecular formula that can be written alternatively as hydrates of carbon are:

- Glucose (blood sugar): $C_6H_{12}O_6$, which can be written as $C_6(H_2O)_6$
- Sucrose (table sugar): $C_{12}H_{22}O_{11}$, which can be written as $C_{12}(H_2O)_{11}$

⊙WL

Sign in to OWL at **www.cengage.com/owl** to view tutorials and simulations, develop problem-solving skills, and complete online homework assigned by your professor.

Not all carbohydrates, however, have this general formula. Some contain too few oxygen atoms to fit it; some contain too many oxygens. Some also contain nitrogen. The term *carbohydrate* has become so firmly rooted in the chemical nomenclature that, although not completely accurate, it persists as the name for this class of compounds.

At the molecular level, most **carbohydrates** are polyhydroxyaldehydes, polyhydroxyketones, or compounds that yield them after hydrolysis. The simpler members of the carbohydrate family are often referred to as **saccharides** because of their sweet taste (Latin: *saccharum,* "sugar"). Carbohydrates are classified as monosaccharides, oligosaccharides, or polysaccharides depending on the number of simple sugars they contain.

Carbohydrate A polyhydroxyaldehyde or polyhydroxyketone or a substance that gives these compounds on hydrolysis

A. Structure and Nomenclature

Monosaccharide A carbohydrate that cannot be hydrolyzed to a simpler compound

Monosaccharides have the general formula $C_nH_{2n}O_n$, with one of the carbons being the carbonyl group of either an aldehyde or a ketone. The most common monosaccharides have three to nine carbon atoms. The suffix **-ose** indicates that a molecule is a carbohydrate, and the prefixes **tri-**, **tetr-**, **pent-**, and so forth, indicate the number of carbon atoms in the chain. Monosaccharides containing an aldehyde group are classified as **aldoses**; those containing a ketone group are classified as **ketoses**.

Aldose A monosaccharide containing an aldehyde group

Ketose A monosaccharide containing a ketone group

There are only two trioses: the aldotriose glyceraldehyde and the ketotriose dihydroxyacetone.

$$
\begin{array}{cc}
\text{CHO} & \text{CH}_2\text{OH} \\
| & | \\
\text{CHOH} & \text{C}=\text{O} \\
| & | \\
\text{CH}_2\text{OH} & \text{CH}_2\text{OH} \\
\text{Glyceraldehyde} & \text{Dihydroxyacetone} \\
\text{(an aldotriose)} & \text{(a ketotriose)}
\end{array}
$$

Often the designations *aldo-* and *keto-* are omitted, and these molecules are referred to simply as trioses, tetroses, and the like.

B. Fischer Projection Formulas

Glyceraldehyde contains a stereocenter and therefore exists as a pair of enantiomers (Figure 20.1).

Fischer projection A two-dimensional representation showing the configuration of a stereocenter; horizontal lines represent bonds projecting forward from the stereocenter, and vertical lines represent bonds projecting toward the rear

Chemists commonly use two-dimensional representations called **Fischer projections** to show the configuration of carbohydrates. To draw a Fischer projection, draw a three-dimensional representation of the molecule oriented so that the vertical bonds from the stereocenter are directed away from you and the horizontal bonds from it are directed toward you (none of the bonds to the stereocenter are in the plane of the paper). Then write the molecule as a cross, with the stereocenter indicated by the point at which the bonds cross.

Emil Fischer, who in 1902 became the second Nobel Prize winner in chemistry, made many fundamental discoveries in the chemistry of carbohydrates, proteins, and other areas of organic and biochemistry.

$$
\begin{array}{ccc}
\text{CHO} & & \text{CHO} \\
| & \xrightarrow[\text{Fischer projection}]{\text{convert to a}} & | \\
\text{H}-\text{C}-\text{OH} & & \text{H}-\!\!\!-\!\!\!-\text{OH} \\
| & & | \\
\text{CH}_2\text{OH} & & \text{CH}_2\text{OH} \\
\text{(R)-Glyceraldehyde} & & \text{(R)-Glyceraldehyde} \\
\text{(three-dimensional} & & \text{(Fischer projection)} \\
\text{representation)} & &
\end{array}
$$

The horizontal segments of this Fischer projection represent bonds directed toward you, and the vertical segments represent bonds directed away from you. The only atom in the plane of the paper is the stereocenter.

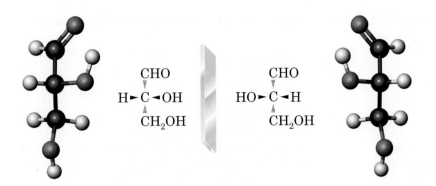

FIGURE 20.1 The enantiomers of glyceraldehyde.

C. D- and L-Monosaccharides

Even though the *R,S* system is widely accepted today as a standard for designating configuration, the configuration of carbohydrates is commonly designated using the D,L system proposed by Emil Fischer in 1891. At that time, it was known that one enantiomer of glyceraldehyde has a specific rotation (Section 15.4B) of +13.5°; the other has a specific rotation of −13.5°. Fischer proposed that these enantiomers be designated D and L, but he had no experimental way to determine which enantiomer had which specific rotation. Fischer, therefore, did the only possible thing—he made an arbitrary assignment. He assigned the dextrorotatory enantiomer the following configuration and named it D-glyceraldehyde. He named its enantiomer L-glyceraldehyde. Fischer could have been wrong, but by a stroke of good fortune, he wasn't. In 1952, scientists proved that his assignment of the D,L-configuration to the enantiomers of glyceraldehyde was correct.

$$
\begin{array}{cc}
\text{CHO} & \text{CHO} \\
\text{H}\!-\!\!\!-\!\!\!-\!\text{OH} & \text{HO}\!-\!\!\!-\!\!\!-\!\text{H} \\
\text{CH}_2\text{OH} & \text{CH}_2\text{OH}
\end{array}
$$

D-Glyceraldehyde   L-Glyceraldehyde
$[\alpha]_\mathrm{D}^{25} = +13.5°$ $[\alpha]_\mathrm{D}^{25} = -13.5°$

D-glyceraldehyde and L-glyceraldehyde serve as reference points for the assignment of relative configurations to all other aldoses and ketoses. The reference point is the penultimate carbon—that is, the next-to-the-last carbon on the chain. A **D-monosaccharide** has the same configuration at its penultimate carbon as D-glyceraldehyde (its —OH group is on the right) in a Fischer projection; an **L-monosaccharide** has the same configuration at its penultimate carbon as L-glyceraldehyde (its —OH group is on the left).

Tables 20.1 and 20.2 show names and Fischer projections for all D-aldo- and D-2-ketotetroses, pentoses, and hexoses. Each name consists of three parts. The D specifies the configuration at the stereocenter farthest from the carbonyl group. Prefixes such as *rib-*, *arabin-*, and *gluc-* specify the configuration of all other stereocenters in the monosaccharide relative to one another. The suffix *-ose* indicates that the compound is a carbohydrate.

The three most abundant hexoses in the biological world are D-glucose, D-galactose, and D-fructose. The first two are D-aldohexoses; the third is a D-2-ketohexose. Glucose, by far the most abundant of the three, is also known as dextrose because it is dextrorotatory. Other names for this monosaccharide include grape sugar and blood sugar. Human blood normally contains 65–110 mg of glucose/100 mL of blood.

D-Monosaccharide A monosaccharide that, when written as a Fischer projection, has the —OH group on its penultimate carbon to the right

L-Monosaccharide A monosaccharide that, when written as a Fischer projection, has the —OH group on its penultimate carbon to the left

Table 20.1 Configurational Relationships among the Isomeric D-Aldotetroses, D-Aldopentoses, and D-Aldohexoses*

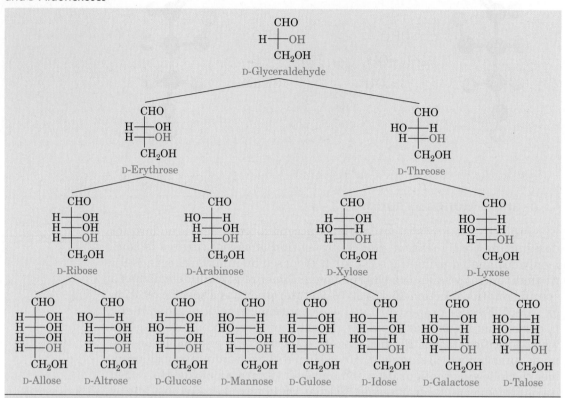

*The configuration of the reference —OH group on the penultimate carbon is shown in color.

Table 20.2 Configurational Relationships among the D-2-Ketopentoses and D-2-Ketohexoses

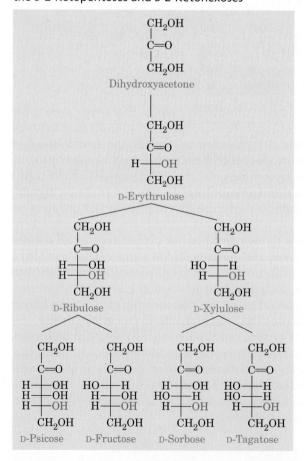

Example 20.1 Drawing Fischer Projections

Draw Fischer projections for the four aldotetroses. Which are D-monosaccharides, which are L-monosaccharides, and which are enantiomers? Refer to Table 20.1 and write the name of each aldotetrose.

Strategy

Start with the Fischer projections of the two aldotrioses, D-glyceraldehyde and L-glyceraldehyde. Draw structures with four carbons, adding the fourth carbon between the one that determines the D,L designation and the aldehyde carbon.

Solution

Following are Fischer projections for the four aldotetroses. The D- and L- refer to the configuration of the penultimate carbon, which, in the case of aldotetroses, is carbon 3. In the Fischer projection of a D-aldotetrose, the —OH group on carbon 3 is on the right; in an L-aldotetrose, it is on the left.

Gloved hand holding an intravenous (IV) drip bag containing 0.15% potassium chloride (saline) and 5% glucose.

Claire Paxton & Jacqui Farrow/Science Photo Library/Photo Researchers Inc.

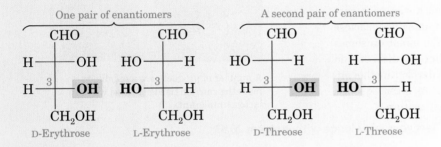

Problem 20.1

Draw Fischer projections for all 2-ketopentoses. Which are D-2-ketopentoses, which are L-2-ketopentoses, and which are enantiomers? Refer to Table 20.2 and write the name of each 2-ketopentose.

D. Amino Sugars

Amino sugar A monosaccharide in which an —OH group is replaced with an —NH$_2$ group

Amino sugars contain an —NH$_2$ group in place of an —OH group. Only three amino sugars are common in nature: D-glucosamine, D-mannosamine, and D-galactosamine.

CHO	CHO	CHO	CHO
H——NH$_2$	H$_2$N——²——H	H——NH$_2$	H——NHCCH$_3$
HO——H	HO——H	HO——H	HO——H
H——OH	H——OH	HO——⁴——H	H——OH
H——OH	H——OH	H——OH	H——OH
CH$_2$OH	CH$_2$OH	CH$_2$OH	CH$_2$OH
D-Glucosamine	D-Mannosamine (C-2 stereoisomer of D-glucosamine)	D-Galactosamine (C-4 stereoisomer of D-glucosamine)	N-Acetyl-D-glucosamine

N-Acetyl-D-glucosamine, a derivative of D-glucosamine, is a component of many polysaccharides, including connective tissue such as cartilage. It is

Galactosemia

One out of every 18,000 infants is born with a genetic defect that renders the child unable to utilize the monosaccharide galactose. Galactose is part of lactose (milk sugar, Section 20.4B). When the body cannot absorb galactose, it accumulates in the blood and in the urine. This buildup in the blood is harmful because it can lead to mental retardation; failure to grow; cataract formation in the eyes; and, in severe cases, death due to liver damage. When galactose accumulation results from a transient disorder in infants, known as galactosuria, it has only mild symptoms. When the enzyme galactose-1-phosphate uridinyltransferase is defective, however, the disorder is called galactosemia, and its symptoms are severe.

The deleterious effects of galactosemia can be avoided by giving the infant a milk formula in which sucrose is substituted for lactose. Because sucrose contains no galactose and cannot be converted to galactose, the infant consumes a galactose-free diet. A galactose-free diet is critical only in infancy. With maturation, most children develop another enzyme capable of metabolizing galactose. As a consequence, they are able to tolerate galactose as they mature.

A number of products are available to meet the calcium needs of those who are lactose intolerant.

Test your knowledge with Problem 20.53

also a component of chitin, the hard, shell-like exoskeleton of lobsters, crabs, shrimp, and other shellfish. Several other amino sugars are components of naturally-occurring antibiotics.

E. Physical Properties of Monosaccharides

Monosaccharides are colorless, crystalline solids. Because hydrogen bonding is possible between their polar —OH groups and water, all monosaccharides are very soluble in water. They are only slightly soluble in ethanol and are insoluble in nonpolar solvents such as diethyl ether, dichloromethane, and benzene.

20.2 What Are the Cyclic Structures of Monosaccharides?

In Section 17.4C, we saw that aldehydes and ketones react with alcohols to form **hemiacetals**. We also saw that cyclic hemiacetals form very readily when hydroxyl and carbonyl groups are part of the same molecule and that their interaction produces a ring. For example, 4-hydroxypentanal forms a five-membered cyclic hemiacetal. Note that 4-hydroxypentanal contains one stereocenter and that hemiacetal formation generates a second stereocenter at carbon 1.

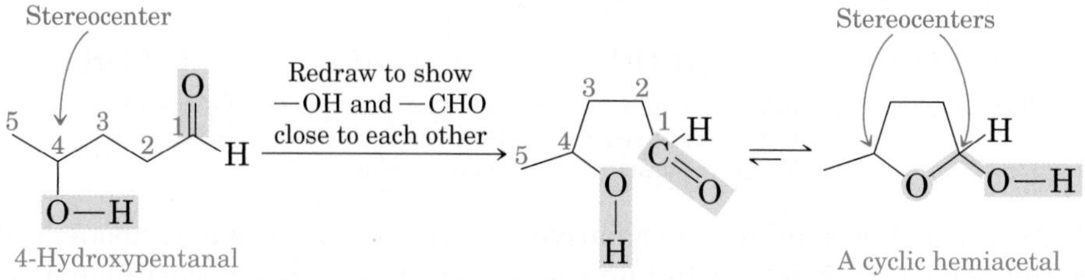

Stereocenter

Redraw to show
—OH and —CHO
close to each other

Stereocenters

4-Hydroxypentanal

A cyclic hemiacetal

Monosaccharides have hydroxyl and carbonyl groups in the same molecule. As a result, they exist almost exclusively as five- and six-membered cyclic hemiacetals.

A. Haworth Projections

A common way of representing the cyclic structure of monosaccharides is the **Haworth projection**, named after the English chemist Sir Walter N. Haworth (Nobel Prize for chemistry, 1937). In a Haworth projection, a five- or six-membered cyclic hemiacetal is represented as a planar pentagon or hexagon, respectively, lying roughly perpendicular to the plane of the paper. Groups bonded to the carbons of the ring then lie either above or below the plane of the ring. The new carbon stereocenter created in forming the cyclic structure is called an **anomeric carbon**. Stereoisomers that differ in configuration only at the anomeric carbon are called **anomers**. The anomeric carbon of an aldose is carbon 1; that of the most common ketoses is carbon 2.

Typically, Haworth projections are most commonly drawn with the anomeric carbon to the right and the hemiacetal oxygen to the back (Figure 20.2).

In the terminology of carbohydrate chemistry, the designation β means that the —OH on the anomeric carbon of the cyclic hemiacetal lies on the same side of the ring as the terminal —CH$_2$OH. Conversely, the designation α means that the —OH on the anomeric carbon of the cyclic hemiacetal lies on the side of the ring opposite the terminal —CH$_2$OH.

A six-membered hemiacetal ring is indicated by **-pyran-**, and a five-membered hemiacetal ring is indicated by **-furan-**. The terms **furanose** and **pyranose** are used because monosaccharide five- and six-membered rings correspond to the heterocyclic compounds furan and pyran.

Haworth projection A way to view furanose and pyranose forms of monosaccharides; the ring is drawn flat and viewed through its edge, with the anomeric carbon on the right and the oxygen atom to the rear

Anomeric carbon The hemiacetal carbon of the cyclic form of a monosaccharide

Anomers Monosaccharides that differ in configuration only at their anomeric carbons

Furanose A five-membered cyclic hemiacetal form of a monosaccharide

Pyranose A six-membered cyclic hemiacetal form of a monosaccharide

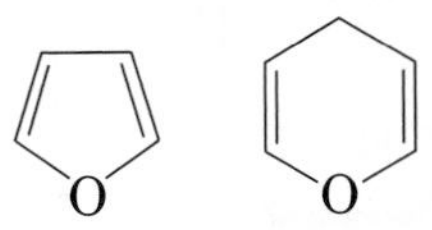

Furan   Pyran

Because the α and β forms of glucose are six-membered cyclic hemiacetals, they are named α-D-glucopyranose and β-D-glucopyranose, respectively. The

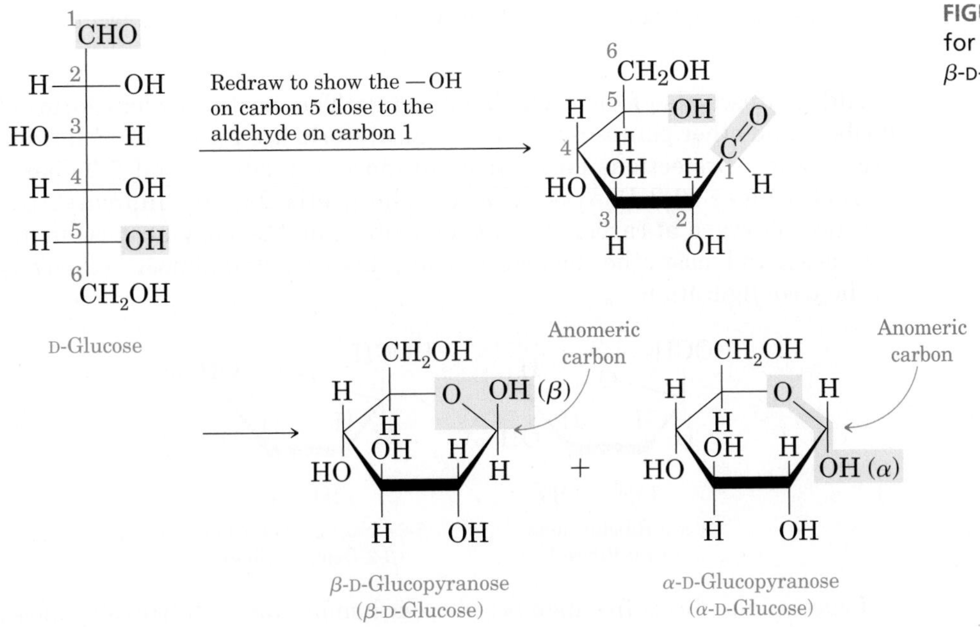

FIGURE 20.2 Haworth projections for α-D-glucopyranose and β-D-glucopyranose.

D-Glucose

Redraw to show the —OH on carbon 5 close to the aldehyde on carbon 1

β-D-Glucopyranose
(β-D-Glucose)

α-D-Glucopyranose
(α-D-Glucose)

designations *-furan-* and *-pyran-* are not always used in monosaccharide names, however. Thus, the glucopyranoses, for example, are often named simply α-D-glucose and β-D-glucose.

You would do well to remember the configurations of the groups on the Haworth projections of α-D-glucopyranose and β-D-glucopyranose as reference structures. Knowing how the open-chain configuration of any other aldohexose differs from that of D-glucose, you can construct Haworth projections for the aldohexose by referring to the Haworth projection of D-glucose.

Example 20.2 Drawing Haworth Projections

Draw Haworth projections for the α and β anomers of D-galactopyranose.

Strategy

A comparison of α and β anomers appears in Figure 20.2, with glucose as the example. The only modification needed is to change the structure of glucose to that of galactose.

Solution

One way to arrive at these projections is to use the α and β forms of D-glucopyranose as references and to remember (or discover by looking at Table 20.1) that D-galactose differs from D-glucose only in the configuration at carbon 4. Thus, you can begin with the Haworth projection shown in Figure 20.2 and then invert the configuration at carbon 4.

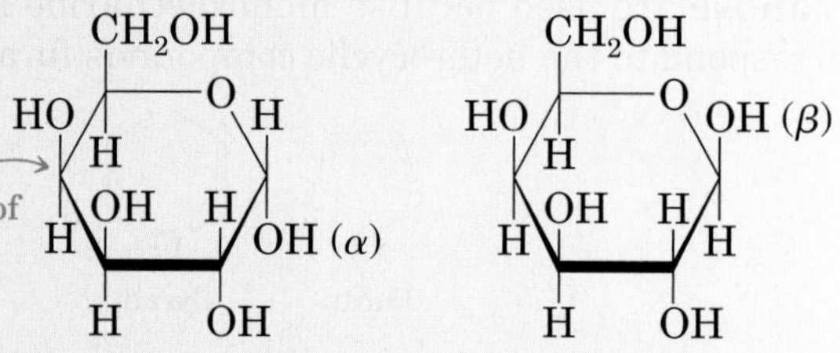

α-D-Galactopyranose
(α-D-Galactose)

β-D-Galactopyranose
(β-D-Galactose)

Problem 20.2

D-Mannose exists in aqueous solution as a mixture of α-D-mannopyranose and β-D-mannopyranose. Draw Haworth projections for these molecules.

Aldopentoses also form cyclic hemiacetals. The most prevalent forms of D-ribose and other pentoses in the biological world are furanoses. Following are Haworth projections for α-D-ribofuranose (α-D-ribose) and β-2-deoxy-D-ribofuranose (β-2-Deoxy-D-ribose). The prefix *2-deoxy* indicates the absence of oxygen at carbon 2. Units of D-ribose and 2-deoxy-D-ribose in nucleic acids and most other biological molecules are found almost exclusively in the β configuration.

α-D-Ribofuranose
(α-D-Ribose)

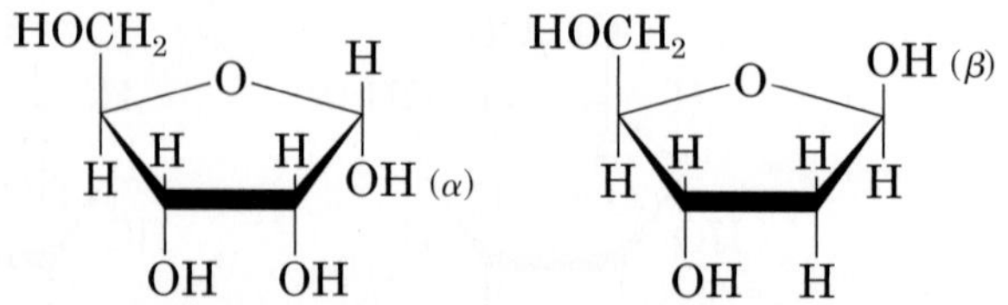

β-2-Deoxy-D-ribofuranose
(β-2-Deoxy-D-ribose)

Fructose also forms five-membered cyclic hemiacetals. β-D-fructofuranose, for example, is found in the disaccharide sucrose (Section 20.4A).

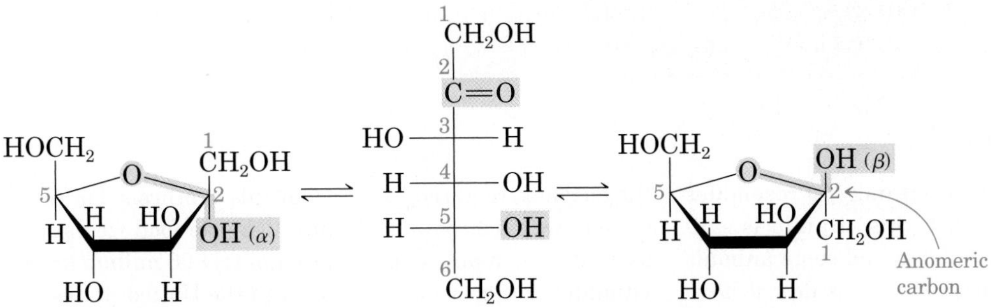

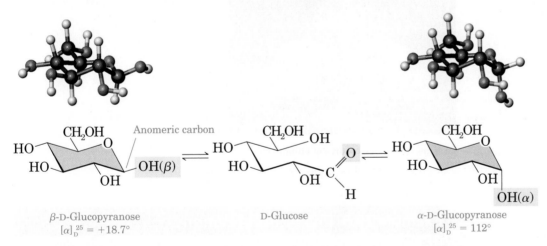

α-D-Fructofuranose
(α-D-Fructose)

D-Fructose

β-D-Fructofuranose
(β-D-Fructose)

Anomeric carbon

B. Conformation Representations

A five-membered furanose ring is so close to being planar that Haworth projections provide adequate representations of furanoses. For pyranoses, however, the six-membered ring is more accurately represented as a **chair conformation** (Section 11.6B). Following are structural formulas for α-D-glucopyranose and β-D-glucopyranose, both drawn as chair conformations. Also shown is the open-chain or free aldehyde form with which the cyclic hemiacetal forms are in equilibrium in aqueous solution. Notice that each group, including the anomeric —OH group, on the chair conformation of β-D-glucopyranose is equatorial. Notice also that the —OH group on the anomeric carbon is axial in α-D-glucopyranose. Because the —OH on the anomeric carbon of β-D-glucopyranose is in the more stable equatorial position (Section 11.6B), the β anomer predominates in aqueous solution.

We do not show hydrogen atoms bonded to the ring in chair conformations. We often show them, however, in Haworth projections.

Anomeric carbon

β-D-Glucopyranose
$[\alpha]_D^{25} = +18.7°$

D-Glucose

α-D-Glucopyranose
$[\alpha]_D^{25} = 112°$

At this point, let us compare the relative orientations of groups on the D-glucopyranose ring in the Haworth projection and the chair conformation. The orientations of groups on carbons 1 through 5 of β-D-glucopyranose, for example, are up, down, up, down, and up in both representations. Note that in β-D-glucopyranose, all groups other than hydrogen atoms are in the stable equatorial position.

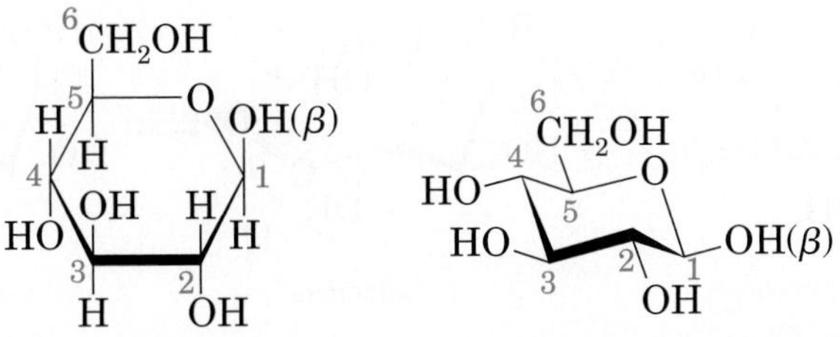

β-D-Glucopyranose
(Haworth projection)

β-D-Glucopyranose
(chair conformation)

Chemical Connections 20B

L-Ascorbic Acid (Vitamin C)

The structure of L-ascorbic acid (vitamin C) resembles that of a monosaccharide. In fact, this vitamin is synthesized both biochemically by plants and some animals and commercially from D-glucose. Humans do not have the enzymes required to carry out this synthesis. For this reason, we must obtain vitamin C in the food we eat or as a vitamin supplement. Approximately 66 million kg of vitamin C is synthesized annually in the United States.

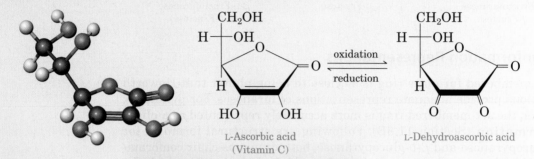

L-Ascorbic acid
(Vitamin C)

L-Dehydroascorbic acid

Oranges are a major source of vitamin C.

Test your knowledge with Problems 20.54 and 20.55

Example 20.3 Chair Conformations

Draw chair conformations for α-D-galactopyranose and β-D-galactopyranose. Label the anomeric carbon in each.

Strategy and Solution

The configuration of D-galactose differs from that of D-glucose only at carbon 4. Therefore, draw the α and β forms of D-glucopyranose and then interchange the positions of the —OH and —H groups on carbon 4.

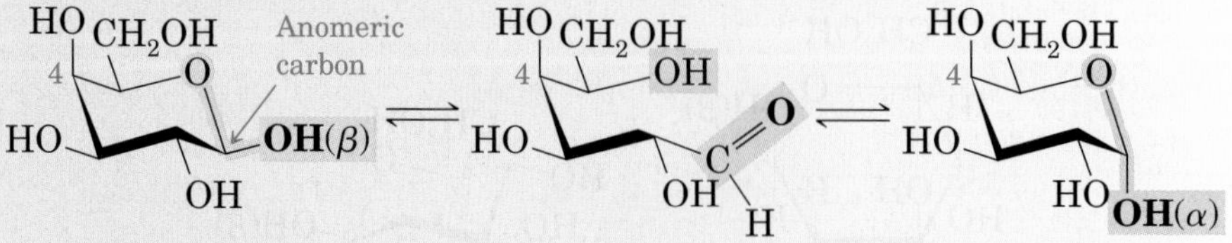

β-D-Galactopyranose
(β-D-Galactose)

D-Galactose

α-D-Galactopyranose
(α-D-Galactose)

Draw chair conformations for α-D-mannopyranose and β-D-mannopyranose. Label the anomeric carbon in each.

C. Mutarotation

Mutarotation is the change in specific rotation that accompanies the equilibration of α and β anomers in aqueous solution. For example, a solution prepared by dissolving crystalline α-D-glucopyranose in water has a specific rotation of $+112°$, which gradually decreases to an equilibrium value of $+52.7°$ as α-D-glucopyranose reaches equilibrium with β-D-glucopyranose. A solution of β-D-glucopyranose also undergoes mutarotation, during which the specific rotation changes from $+18.7°$ to the same equilibrium value of $+52.7°$. The equilibrium mixture consists of 64% β-D-glucopyranose and 36% α-D-glucopyranose, with only a trace (0.003%) of the open-chain form. Mutarotation is common to all carbohydrates that exist in hemiacetal forms.

Mutarotation The change in specific rotation that occurs when an α or β form of a carbohydrate is converted to an equilibrium mixture of the two forms

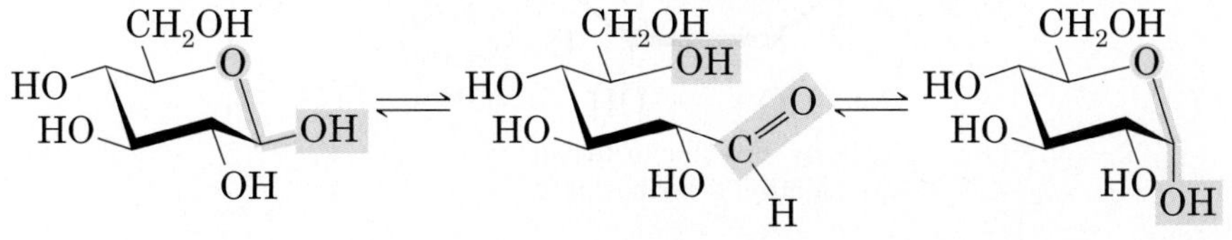

β-D-Glucopyranose
$[\alpha]_D^{25} = +18.7°$

Open-chain form

α-D-Glucopyranose
$[\alpha]_D^{25} = +112°$

20.3 What Are the Characteristic Reactions of Monosaccharides?

A. Formation of Glycosides (Acetals)

As we saw in Section 17C, treatment of an aldehyde or a ketone with one molecule of alcohol yields a hemiacetal and treatment of the hemiacetal with a molecule of alcohol yields an acetal. Treatment of a monosaccharide—all forms of which exist almost exclusively as cyclic hemiacetals—with an alcohol also yields an acetal, as illustrated by the reaction of β-D-glucopyranose with methanol.

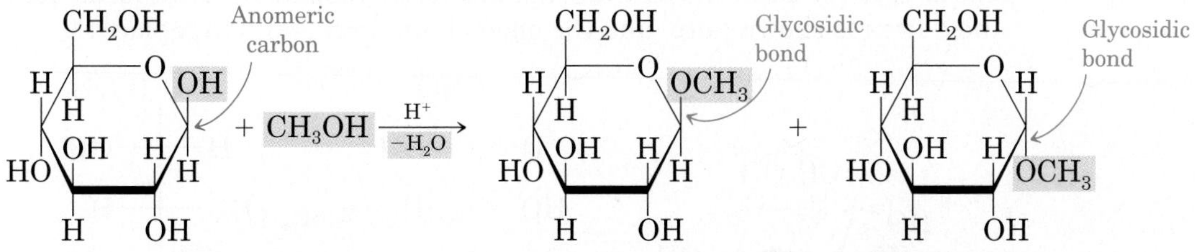

β-D-Glucopyranose
(β-D-Glucose)

Methyl β-D-glucopyranoside
(Methyl β-D-glucoside)

Methyl α-D-glucopyranoside
(Methyl α-D-glucoside)

A cyclic acetal derived from a monosaccharide is called a **glycoside**, and the bond from the anomeric carbon to the —OR group is called a **glycosidic bond**. Mutarotation is not possible in a glycoside because an acetal—unlike a hemiacetal—is no longer in equilibrium with the open-chain carbonyl-containing compound. Glycosides are stable in water and aqueous base; like other acetals (Section 17.4C), however, they are hydrolyzed in aqueous acid to an alcohol and a monosaccharide.

Glycoside A carbohydrate in which the —OH group on its anomeric carbon is replaced with an —OR group

Glycosidic bond The bond from the anomeric carbon of a glycoside to an —OR group

We name glycosides by listing the alkyl or aryl group bonded to oxygen, followed by the name of the carbohydrate in which the ending **-e** is replaced with **-ide**. For example, the methyl glycoside derived from β-D-glucopyranose is named methyl β-D-glucopyranoside; that derived from β-D-ribofuranose is named methyl β-D-ribofuranoside.

Example 20.4 Finding the Anomeric Carbon and Glycosidic Bond

Draw a structural formula for methyl β-D-ribofuranoside (methyl β-D-riboside). Label the anomeric carbon and the glycosidic bond.

Strategy

Furanosides are five-membered rings. The anomeric carbon is carbon one, and the glycosidic bond is formed at the anomeric carbon.

Solution

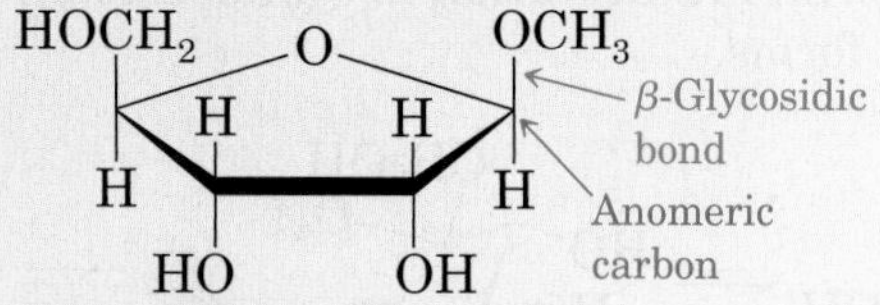

Methyl β-D-ribofuranoside
(Methyl β-D-riboside)

Problem 20.4

Draw a Haworth projection and a chair conformation for methyl α-D-mannopyranoside (methyl α-D-mannoside). Label the anomeric carbon and the glycosidic bond.

Oxidations and reductions of monosaccharides in nature are catalyzed by specific enzymes classified as oxidases—for example, glucose oxidase.

Alditol The product formed when the CHO group of a monosaccharide is reduced to a CH₂OH group

B. Reduction to Alditols

The carbonyl group of a monosaccharide can be reduced to a hydroxyl group by a variety of reducing agents, including hydrogen in the presence of a transition metal catalyst and sodium borohydride (Section 17.4C). The reduction products are known as **alditols**. Reduction of D-glucose gives D-glucitol, more commonly known as D-sorbitol. Here, D-glucose is shown in the open-chain form. Only a small amount of this form is present in solution, but as it is reduced, the equilibrium between cyclic hemiacetal forms (only the β form is shown here) and the open-chain form shifts to replace it.

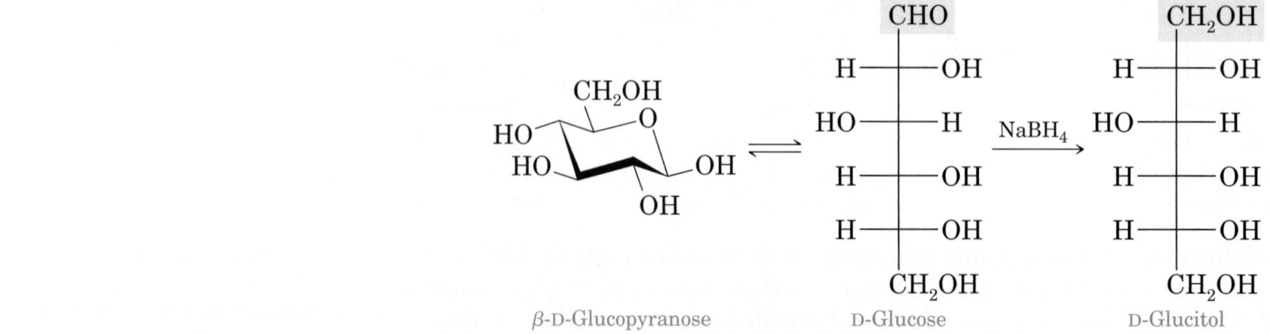

β-D-Glucopyranose   D-Glucose   D-Glucitol
                                               (D-Sorbitol)

We name alditols by dropping the **-ose** from the name of the monosaccharide and adding **-itol**. Sorbitol is found in the plant world in many berries and in cherries, plums, pears, apples, seaweed, and algae. It is about 60% as sweet as sucrose (table sugar) and is used in the manufacture of candies and as a sugar substitute for diabetics. Other alditols commonly found in the biological world include erythritol, D-mannitol, and xylitol. Xylitol is used as a sweetening agent in "sugarless" gum, candy, and sweet cereals.

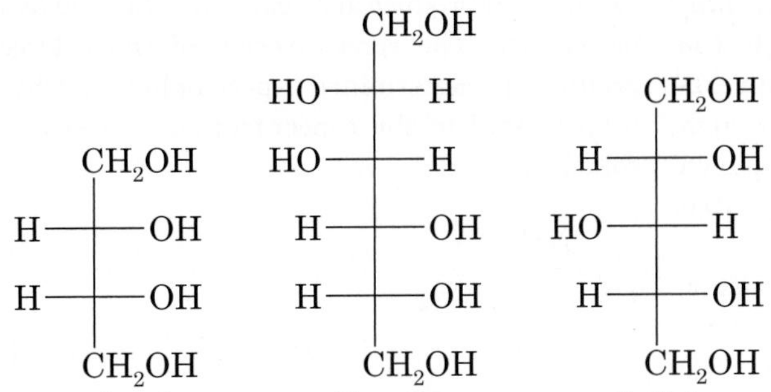

Many "sugar-free" products contain sugar alcohols, such as sorbitol and xylitol.

Erythritol · D-Mannitol · Xylitol

C. Oxidation to Aldonic Acids (Reducing Sugars)

As we saw in Section 17.4A, aldehydes (RCHO) are oxidized to carboxylic acids (RCOOH) by several agents, including oxygen, O_2. Similarly, the aldehyde group of an aldose can be oxidized, under basic conditions, to a carboxylate group. Under these conditions, the cyclic form of the aldose is in equilibrium with the open-chain form, which is then oxidized by the mild oxidizing agent. D-Glucose, for example, is oxidized to D-gluconate (the anion of D-gluconic acid).

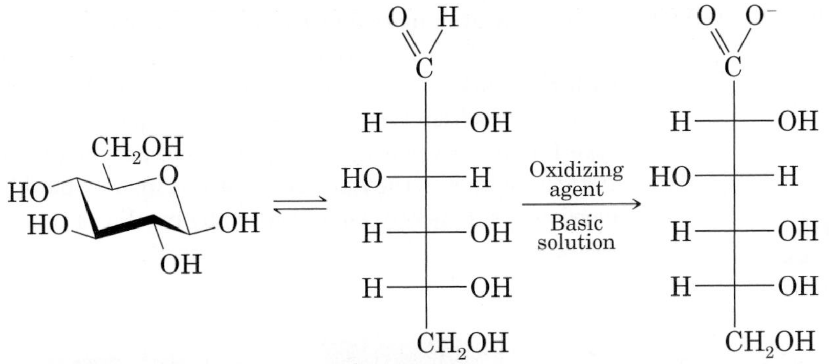

β-D-Glucopyranose
(β-D-Glucose)

D-Glucose

D-Gluconate
(an aldonic acid)

Any carbohydrate that reacts with an oxidizing agent to form an aldonic acid is classified as a **reducing sugar** (it reduces the oxidizing agent).

Surprisingly, 2-ketoses are also reducing sugars. Carbon 1 (a CH_2OH group) of a ketose is not oxidized directly. Instead, under the basic conditions of this oxidation, a 2-ketose exists in equilibrium with an aldose by way of an enediol intermediate. The aldose is then oxidized by the mild oxidizing agent.

Reducing sugar A carbohydrate that reacts with a mild oxidizing agent under basic conditions to give an aldonic acid; the carbohydrate reduces the oxidizing agent

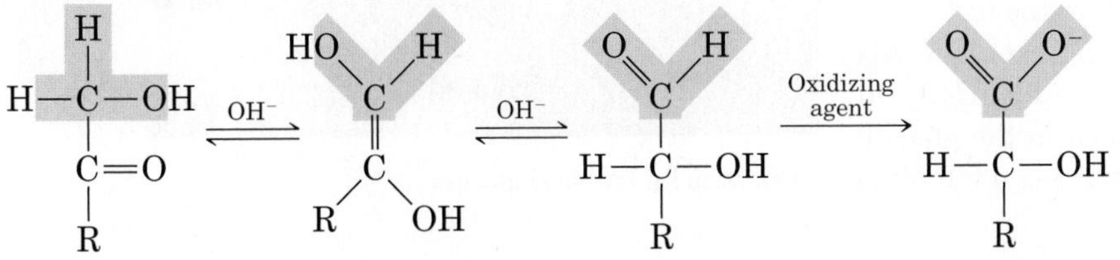

A 2-ketose · An enediol · An aldose · An aldonate

Chemical Connections 20C

Testing for Glucose

The analytical procedure most often performed in a clinical chemistry laboratory is the determination of glucose in blood, urine, or other biological fluids. The high frequency with which this test is performed reflects the high incidence of diabetes mellitus. Nearly 20 million known diabetics live in the United States, and it is estimated that millions more remain undiagnosed.

Diabetes mellitus (Chemical Connections 24F) is characterized by insufficient blood levels of the hormone insulin. If the blood concentration of insulin is too low, muscle and liver cells do not absorb glucose from the blood; this problem, in turn, leads to increased levels of blood glucose (hyperglycemia), impaired metabolism of fats and proteins, ketosis, and possible diabetic coma. A rapid test for blood glucose levels is critical for early diagnosis and effective management of this disease. In addition to giving results rapidly, a test must be specific for D-glucose; that is, it must give a positive test for glucose but not react with any other substances normally present in biological fluids.

Today, blood glucose levels are measured by an enzyme-based procedure using the enzyme glucose oxidase. This enzyme catalyzes the oxidation of β-D-glucose to D-gluconic acid.

β-D-Glucopyranose
(β-D-Glucose)

$$+ O_2 + H_2O$$

Glucose oxidase

COOH
H——OH
HO——H
H——OH
H——OH
CH₂OH

D-Gluconic acid

$$+ H_2O_2$$
Hydrogen peroxide

Glucose oxidase is specific for β-D-glucose. Therefore, complete oxidation of any sample containing both β-D-glucose and α-D-glucose requires conversion of the α form to the β form. Fortunately, this interconversion is rapid and complete in the short time required for the test.

Molecular oxygen, O_2, is the oxidizing agent in this reaction and is reduced to hydrogen peroxide, H_2O_2. In one procedure, hydrogen peroxide formed in the glucose oxidase–catalyzed reaction oxidizes colorless o-toluidine

to a colored product in a reaction catalyzed by the enzyme peroxidase. The concentration of the colored oxidation product is determined spectrophotometrically and is proportional to the concentration of glucose in the test solution.

2-Methylaniline
(o-Toluidine)

$$+ H_2O_2 \xrightarrow{\text{Peroxidase}} \text{Colored product}$$

Several commercially available test kits use the glucose oxidase reaction for the qualitative determination of glucose in urine. In addition, contact lenses that change color with fluctuations in the level of glucose in tears are in the development stage. This test depends on nanotechnology. Nanoparticles whose color depends on the glucose level are embedded in the lenses.

Because diabetes has become much more prevalent over the last few decades, monitoring of glucose levels has become even more important. In particular, long-term monitoring of glucose levels has become a topic of interest with the marked rise in the incidence of type 2 diabetes, which differs from type 1 diabetes in an important way. In type 1 diabetes, the body does not produce enough insulin. In type 2 diabetes, the issue is not lack of insulin, but high blood sugar (hyperglycemia) caused by insulin resistance.

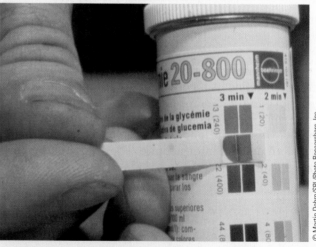

Chemstrip kit for blood glucose.

© Martin Dohrn/SPL/Photo Researchers, Inc.

Chemical Connections 20C

Testing for Glucose (*continued*)

This form was once called adult-onset diabetes, but that name as fallen into disuse as type 2 diabetes has started to occur more frequently in younger people, even among adolescents.

For long-term monitoring of glucose in the body, researchers have made use of the fact that sugars such as glucose can form covalent bonds with proteins. Red blood cells contain hemoglobin, the protein that carries oxygen from the lungs to the cells. Before the cells turn over, hemoglobin in them is glycated, that is to say, it forms covalent bonds with glucose. The modification takes place to a greater or lesser extent depending on the amount of glucose in the bloodstream. This modified form of glucose has several designations, with A1c being the one in most common use. Various tests exist for determining the amount of A1c in the blood. One of these tests is immunological and depends on the fact that the bound sugars can act as antigens in a way similar to the sugars on cell surfaces that determine A, B, AB, and O blood types (Chemical Connection 20D). These tests are offered in the pharmacy departments of large drugstore chains.

A certain amount of controversy exists in some of the clinical uses of the results of these tests. The dangers of hyperglycemia are well known, including damage to blood vessels, kidneys, and nerves. Questions arise, however, when blood sugar is reduced to very low levels (hypoglycemia). Studies have suggested that control of blood sugar to the point of hypoglycemia has no cardiovascular benefit. Besides no significant lowering of heart disease and stroke, some researchers questioned whether inducing very low blood sugar levels was causing more harm than good. Ongoing research is being done to determine optimum blood glucose levels.

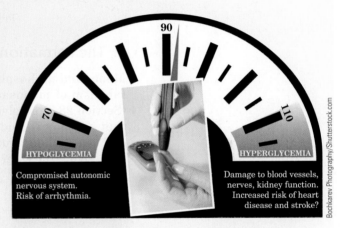

Health problems can arise if the level of blood glucose is too high or too low.

Test your knowledge with Problem 20.56.

D. Oxidation to Uronic Acids

Enzyme-catalyzed oxidation of the primary alcohol at carbon 6 of a hexose yields a uronic acid. Enzyme-catalyzed oxidation of D-glucose, for example, yields D-glucuronic acid, shown here in both its open-chain and cyclic hemiacetal forms:

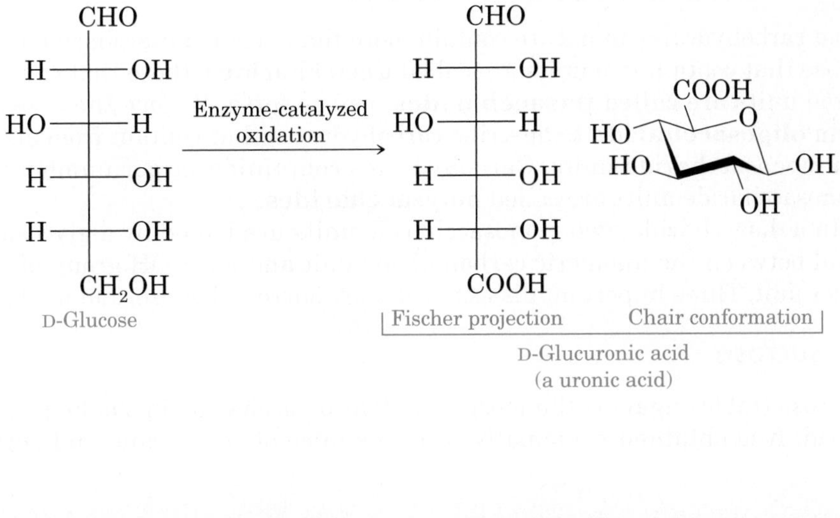

D-Glucose → (Enzyme-catalyzed oxidation) → D-Glucuronic acid (a uronic acid)

Fischer projection   Chair conformation

D-glucuronic acid is widely distributed in both the plant and animal worlds. In humans, it serves as an important component of the acidic polysaccharides of connective tissues (Section 20.6A). The body also uses D-glucuronic acid to detoxify foreign phenols and alcohols. In the liver, these compounds are converted to glycosides of glucuronic acid (glucuronides) and excreted in the urine. The intravenous anesthetic propofol, for example, is converted to the following glucuronide and then excreted in urine:

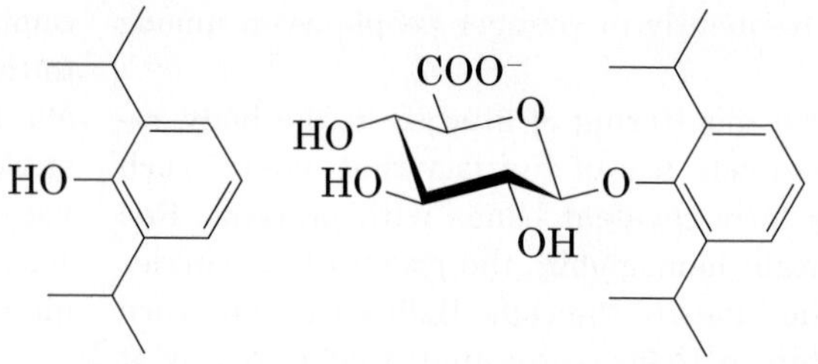

Propofol   A urine-soluble D-glucuronide

E. The Formation of Phosphoric Esters

Mono- and diphosphoric esters are important intermediates in the metabolism of monosaccharides. For example, the first step in glycolysis (Section 28.2) involves conversion of glucose to glucose 6-phosphate. Note that phosphoric acid is a strong enough acid so that at the pH of cellular and intercellular fluids, both acidic protons of a phosphoric ester are ionized, giving the ester a charge of -2.

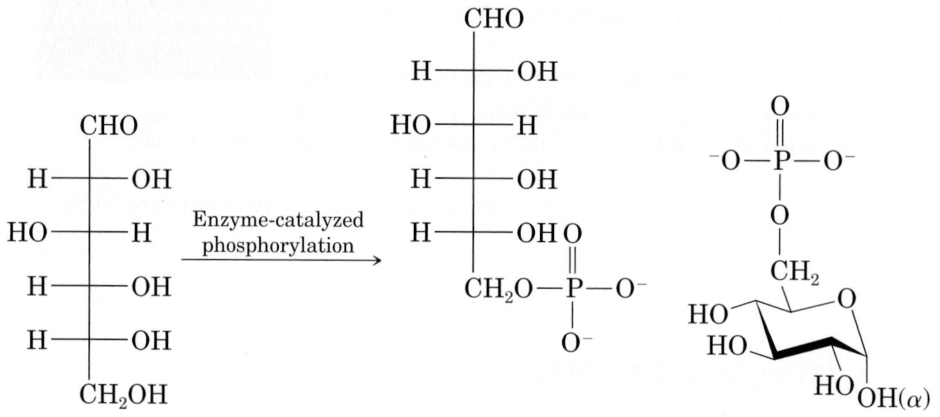

D-Glucose   D-Glucose 6-phosphate α-D-Glucose 6-phosphate

20.4 What Are Disaccharides and Oligosaccharides?

Disaccharide A carbohydrate containing two monosaccharide units joined by a glycosidic bond

Oligosaccharide A carbohydrate containing from six to ten monosaccharide units, each joined to the next by a glycosidic bond

Polysaccharide A carbohydrate containing a large number of monosaccharide units, each joined to the next by one or more glycosidic bonds

Most carbohydrates in nature contain more than one monosaccharide unit. Those that contain two units are called **disaccharides**, those that contain three units are called **trisaccharides**, and so forth. We use the general term **oligosaccharide** to describe carbohydrates that contain from six to ten monosaccharide units. Carbohydrates containing larger numbers of monosaccharide units are called **polysaccharides**.

In a disaccharide, two monosaccharide units are joined by a glycosidic bond between the anomeric carbon of one unit and an —OH group of the other unit. Three important disaccharides are sucrose, lactose, and maltose.

A. Sucrose

Sucrose (table sugar) is the most abundant disaccharide in the biological world. It is obtained principally from the juice of sugar cane and sugar

A, B, AB, and O Blood Types

Membranes of animal plasma cells have large numbers of relatively small carbohydrates bound to them. In fact, the outsides of most plasma cell membranes are literally "sugar-coated." These membrane-bound carbohydrates are part of the mechanism by which cell types recognize one another and, in effect, act as biochemical markers. Typically, they contain from 4 to 17 monosaccharide units consisting primarily of relatively few monosaccharides, the most common of which are D-galactose, D-mannose, L-fucose, N-acetyl-D-glucosamine, and N-acetyl-D-galactosamine. L-Fucose is a 6-deoxyaldohexose.

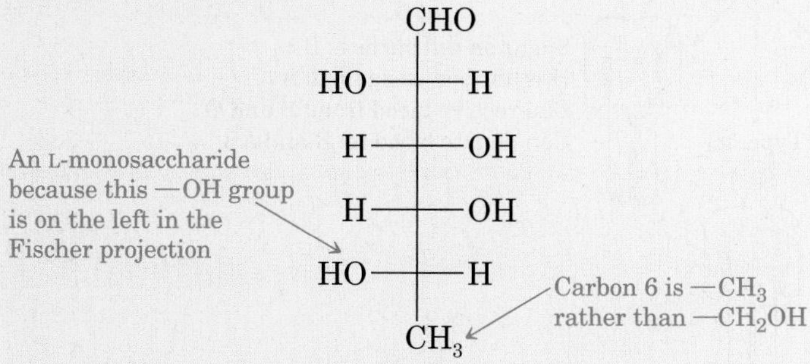

CHO
HO———H
H———OH
H———OH
HO———H
CH₃

An L-monosaccharide because this —OH group is on the left in the Fischer projection

Carbon 6 is —CH₃ rather than —CH₂OH

L-Fucose

To see the importance of these membrane-bound carbohydrates, consider the ABO blood group system, discovered in 1900 by Karl Landsteiner (1868–1943). Whether an individual belongs to type A, B, AB, or O is genetically determined and depends on the type of trisaccharide or tetrasaccharide bound to the surface of the red blood cells. These surface-bound carbohydrates, designated as A, B, and O, act as antigens. The type of glycosidic bond joining each monosaccharide is shown in the figure.

The blood carries antibodies against foreign substances. When a person receives a blood transfusion, the antibodies clump together (aggregate) the foreign blood cells. Type A blood, for example, has A antigens (N-acetyl-D-galactosamine) on the surfaces of its red blood cells and carries anti-B antibodies (against B antigen). B-type blood carries B antigen (D-galactose) and has anti-A antibodies (against A antigens). Transfusion of type A blood into a person with type B blood can be fatal, and vice versa. The relationships between blood type and donor–receiver interactions are summarized in the figure.

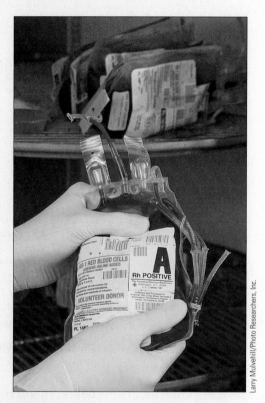

Bag of blood showing blood type.

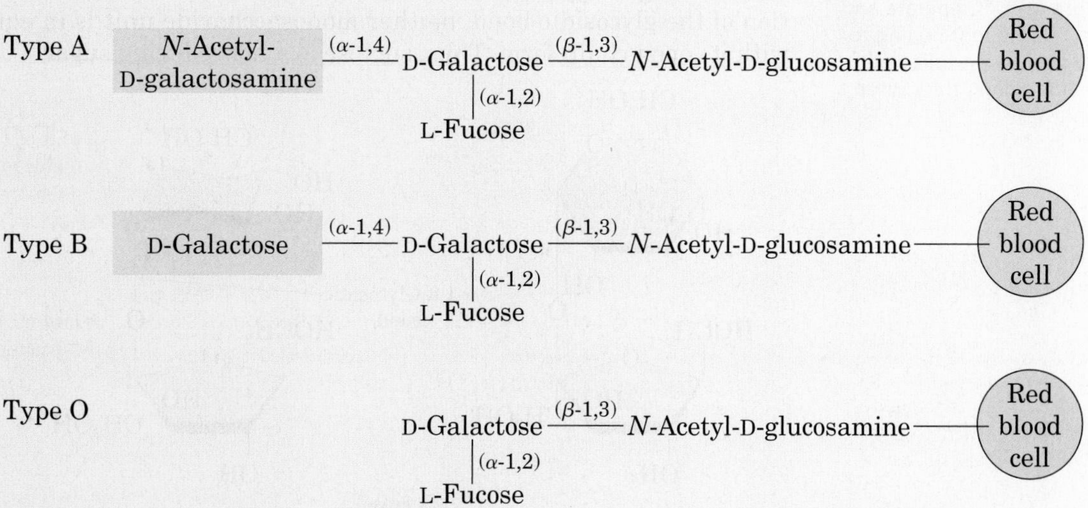

Larry Mulvehill/Photo Researchers, Inc.

Type A   N-Acetyl-D-galactosamine $\xrightarrow{(\alpha\text{-}1,4)}$ D-Galactose $\xrightarrow{(\beta\text{-}1,3)}$ N-Acetyl-D-glucosamine ——— Red blood cell
$\big|(\alpha\text{-}1,2)$
L-Fucose

Type B   D-Galactose $\xrightarrow{(\alpha\text{-}1,4)}$ D-Galactose $\xrightarrow{(\beta\text{-}1,3)}$ N-Acetyl-D-glucosamine ——— Red blood cell
$\big|(\alpha\text{-}1,2)$
L-Fucose

Type O   D-Galactose $\xrightarrow{(\beta\text{-}1,3)}$ N-Acetyl-D-glucosamine ——— Red blood cell
$\big|(\alpha\text{-}1,2)$
L-Fucose

A, B, AB, and O Blood Types (*continued*)

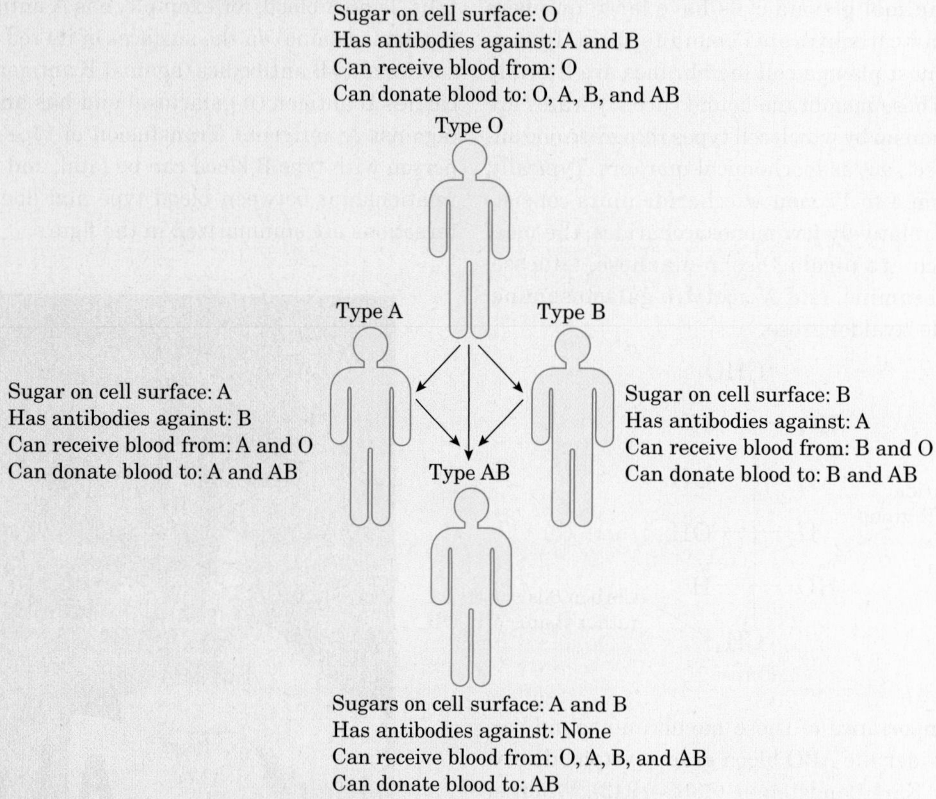

Sugar on cell surface: O
Has antibodies against: A and B
Can receive blood from: O
Can donate blood to: O, A, B, and AB

Type O

Type A

Type B

Sugar on cell surface: A
Has antibodies against: B
Can receive blood from: A and O
Can donate blood to: A and AB

Sugar on cell surface: B
Has antibodies against: A
Can receive blood from: B and O
Can donate blood to: B and AB

Type AB

Sugars on cell surface: A and B
Has antibodies against: None
Can receive blood from: O, A, B, and AB
Can donate blood to: AB

People with type O blood are universal donors, and those with type AB blood are universal acceptors. People with type A blood can accept blood from type A or type O donors only. Those with type B blood can accept blood from type B or type O donors only. Type O people can accept blood only from type O donors.

Test your knowledge with Problems 20.57–20.59.

In the production of sucrose, sugar cane or sugar beet is boiled with water and the resulting solution is cooled. Sucrose crystals separate and are collected. Subsequent boiling to concentrate the solution followed by cooling yields a dark, thick syrup known as molasses.

beets. In sucrose, carbon 1 of α-D-glucopyranose bonds to carbon 2 of D-fructofuranose by an α-1,2-glycosidic bond. Because the anomeric carbons of both the glucopyranose and fructofuranose units are involved in formation of the glycosidic bond, neither monosaccharide unit is in equilibrium with its open-chain form. Thus, sucrose is a nonreducing sugar.

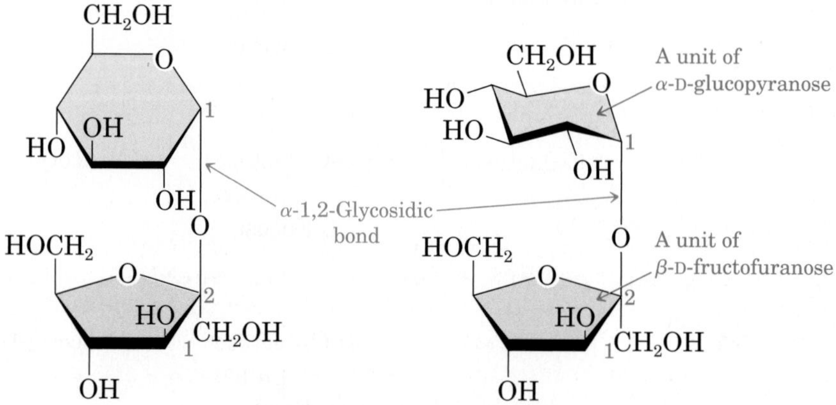

CH₂OH

α-1,2-Glycosidic bond

A unit of α-D-glucopyranose

A unit of β-D-fructofuranose

Sucrose

B. Lactose

Lactose is the principal sugar present in milk. It accounts for 5 to 8% of human milk and 4 to 6% of cow's milk. This disaccharide consists of D-galactopyranose bonded by a β-1,4-glycosidic bond to carbon 4 of D-glucopyranose. Lactose is a reducing sugar, because the cyclic hemiacetal of the D-glucopyranose unit is in equilibrium with its open-chain form and can be oxidized to a carboxyl group.

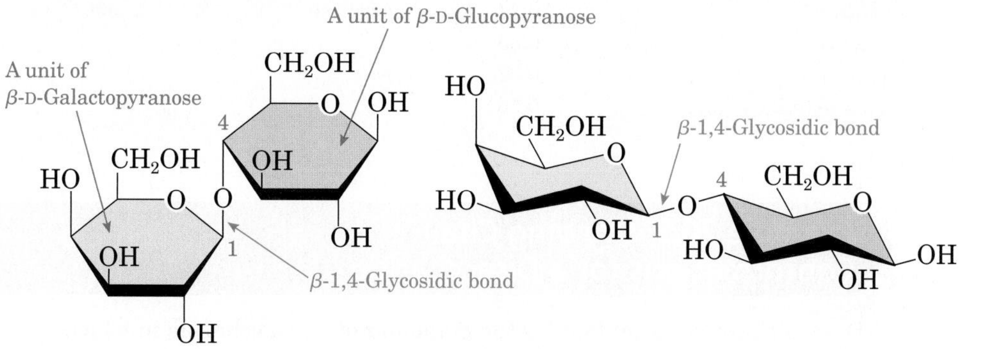

Lactose

C. Maltose

Maltose derives its name from its presence in malt, the juice from sprouted barley and other cereal grains. It consists of two units of D-glucopyranose joined by a glycosidic bond between carbon 1 (the anomeric carbon) of one unit and carbon 4 of the other unit. Because the oxygen atom on the anomeric carbon of the first glucopyranose unit is alpha, the bond joining the two units is an α-1,4-glycosidic bond. Following are a Haworth projection and a chair conformation for β-maltose, so named because the —OH groups on the anomeric carbon of the glucose unit on the right are beta.

Maltose is an ingredient in most syrups.

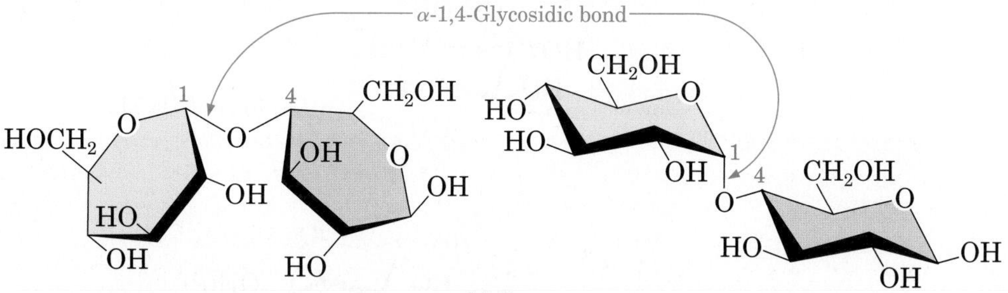

Maltose

Maltose is a reducing sugar; the hemiacetal group on the right unit of D-glucopyranose is in equilibrium with the free aldehyde and can be oxidized to a carboxylic acid.

D. Relative Sweetness

Among the common sweetening agents, D-fructose tastes the sweetest—even sweeter than sucrose (Table 20.3). The sweet taste of honey is due largely to D-fructose and D-glucose. Lactose has almost no sweetness and is sometimes added to foods as a filler. Some people cannot tolerate lactose well, however, and should avoid these foods.

We have no mechanical way to measure sweetness. Such testing is done by having a group of people taste and rate the sweetness of solutions of varying sweetening agents.

Table 20.3 Relative Sweetness of Some Carbohydrate and Artificial Sweetening Agents

Carbohydrate	Sweetness Relative to Sucrose	Artificial Sweetener	Sweetness Relative to Sucrose
fructose	1.74	saccharine	450
sucrose (table sugar)	1.00	acesulfame-K	200
honey	0.97	aspartame	180
glucose	0.74	sucralose	600
maltose	0.33		
galactose	0.32		
lactose (milk sugar)	0.16		

Example 20.5 Drawing Chair Conformations for a Disaccharide

Draw a chair conformation for the β anomer of a disaccharide in which two units of D-glucopyranose are joined by an α-1,6-glycosidic bond.

Strategy

Three points are needed here. The first is the chair conformation of α-D-glucopyranose. The second is the α-1,6-glycosidic bond between the two glucopyranose molecules. The third is the correct conformation of the anomeric carbon at the reducing end, β in this case.

Solution

First, draw a chair conformation of α-D-glucopyranose. Then connect the anomeric carbon of this monosaccharide to carbon 6 of a second D-glucopyranose unit by an α-1,6-glycosidic bond. The resulting molecule is either α or β depending on the orientation of the —OH group on the reducing end of the disaccharide. The disaccharide shown here is the β form.

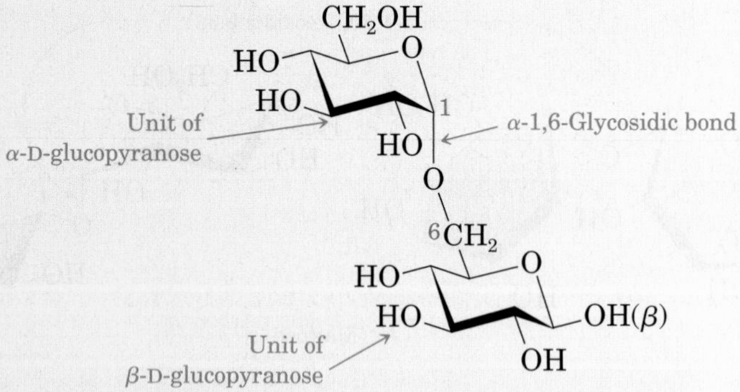

Problem 20.5

Draw a chair conformation for the α form of a disaccharide in which two units of D-glucopyranose are joined by a β-1,3-glycosidic bond.

20.5 What Are Polysaccharides?

Polysaccharides consist of large numbers of monosaccharide units bonded together by glycosidic bonds. Three important polysaccharides, all made up of glucose units, are starch, glycogen, and cellulose.

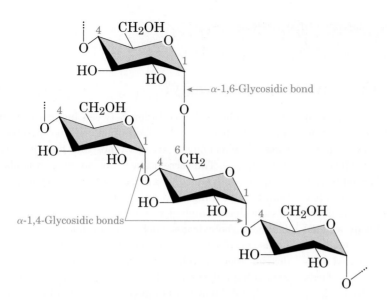

FIGURE 20.3 Amylopectin is a branched polymer of approximately 10,000 D-glucose units joined by α-1,4-glycosidic bonds. Branches consist of 24–30 D-glucose units that start with an α-1,6 glycosidic bond.

A. Starch: Amylose and Amylopectin

Starch is used for energy storage in plants. It is found in all plant seeds and tubers and is the form in which glucose is stored for later use. Starch can be separated into two principal polysaccharides: amylose and amylopectin. Although the starch from each plant is unique, most starches contain 20 to 25% amylose and 75 to 80% amylopectin.

Complete hydrolysis of both amylose and amylopectin yields only D-glucose. Amylose is composed of continuous, unbranched chains of as many as 4000 D-glucose units joined by α-1,4-glycosidic bonds. Amylopectin contains chains of as many as 10,000 D-glucose units also joined by α-1,4-glycosidic bonds. In addition, considerable branching from this linear network occurs. New chains of 24 to 30 units are started at branch points by α-1,6-glycosidic bonds (Figure 20.3).

B. Glycogen

Glycogen acts as the energy-reserve carbohydrate for animals. Like amylopectin, it is a branched polysaccharide containing approximately 10^6 glucose units joined by α-1,4- and α-1,6-glycosidic bonds. The total amount of glycogen in the body of a well-nourished adult human is about 350 g, divided almost equally between liver and muscle.

C. Cellulose

Cellulose, the most widely distributed plant skeletal polysaccharide, constitutes almost half of the cell-wall material of wood. Cotton is almost pure cellulose.

Cellulose is a linear polysaccharide of D-glucose units joined by β-1,4-glycosidic bonds (Figure 20.4). It has an average molecular weight of

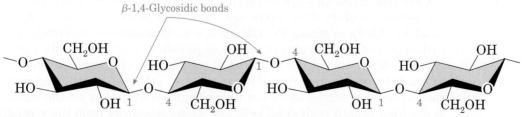

FIGURE 20.4 Cellulose is a linear polysaccharide containing as many as 3000 units of D-glucose joined by β-1,4-glycosidic bonds.

Is There a Connection Between Carbohydrates and Obesity?

What does consumption of carbohydrates have to do with obesity? The answer is "quite a lot." Clearly, consumption of too many nutritional calories from any source and a lack of physical activity will lead to weight gain. The number of overweight and obese people in the United States has increased markedly over the last several decades. It is estimated that one-third of the population is obese (overweight by more than 25% of their ideal body weight) and another third is overweight. This trend has been accompanied by an avoidance of fats and an increased preference for carbohydrates. Many high-carbohydrate foods have even been promoted as healthful on the basis of their low fat content.

The rationale behind the decreased intake of fats, especially saturated fats, has been to promote cardiovascular health and to lower the incidence of heart disease and strokes. The incidence of heart disease remains high, and the obesity rate has skyrocketed, as has that of type-2 diabetes. Obviously, there is more to the situation than the simple dictum that all fats are bad and all carbohydrates are good. The question arises as to whether the kind of carbohydrates and the kind of fats in the diet make a difference. Is there a difference between carbohydrates such as easily digested mono- and disaccharides on the one hand and complex carbohydrates on the other hand? Is there a difference between saturated fats and unsaturated fats? The answer appears to be that in both cases, there is a difference. We will have more to say about fats in the next chapter, but we can draw some conclusions about carbohydrates now.

A number of studies have been done on diets designed to promote weight loss. All these diets curtailed calories, but the amounts of carbohydrates and fats have varied. Weight loss was the common result in all diets, but other health benefits were noted with the low-carbohydrate regimens. Lowering of blood pressure took place to a greater extent with low-carbohydrate diets than with the others. Other studies indicate a correlation between increased diabetes and heart disease and consumption of carbohydrates that raise the blood sugar strongly (ones that have a high glycemic index). The correlation is not noted with foods that have a low glycemic index. Processed foods such as pastries tend to have a high glycemic index. In general, foods high in sugar have a high glycemic index, as do those made with refined carbohydrates, such as white flour. Most of the fibrous carbohydrates in vegetables have a low glycemic index, but root vegetables such as beets, carrots, and potatoes have a high glycemic index. Instead of counting calories, a more helpful way to plan meals is to take into account the glycemic index of foods, and there are pocket guides that list a wide variety of foods and how many grams of carbohydrates are contained in a serving. Research is continuing to determine criteria for healthful diets, but it is clear that sugary treats are harmful when eaten in excess.

Doughnuts are a high-carbohydrate food.

Test your knowledge with Problem 20.60.

400,000 g/mol, corresponding to approximately 2200 glucose units per molecule.

Cellulose molecules act much like stiff rods, a characteristic that enables them to align themselves side by side into well-organized, water-insoluble fibers in which the OH groups form numerous intermolecular hydrogen bonds. This arrangement of parallel chains in bundles gives cellulose fibers their high mechanical strength. It also explains why cellulose is insoluble in water. When a piece of cellulose-containing material is placed in water, there are not enough —OH groups on the surface of the fiber to pull individual cellulose molecules away from the strongly hydrogen-bonded fiber.

Humans and other animals cannot use cellulose as food because our digestive systems do not contain β-glucosidases, enzymes that catalyze the hydrolysis of β-glucosidic bonds. Instead, we have only α-glucosidases; hence, we use the polysaccharides starch and glycogen as sources of glucose. In contrast, many bacteria and microorganisms do contain β-glucosidases and so can digest cellulose. Termites (much to our regret) have such bacteria in their intestines and can use wood as their principal food. Ruminants (cud-chewing animals) and horses can also digest grasses and hay because β-glucosidase-containing microorganisms are present in their alimentary systems.

20.6 What Are Acidic Polysaccharides?

Acidic polysaccharides are a group of polysaccharides that contain carboxyl groups and/or sulfuric ester groups. Acidic polysaccharides play important roles in the structure and function of connective tissues. Because they contain amino sugars, a more current name for these substances is glycosaminoglycans. There is no single general type of connective tissue. Rather, a large number of highly specialized forms exist, such as cartilage, bone, synovial fluid, skin, tendons, blood vessels, intervertebral disks, and cornea. Most connective tissues consist of collagen, a structural protein, combined with a variety of acidic polysaccharides (glycosaminoglycans) that interact with collagen to form tight or loose networks.

In rheumatoid arthritis, inflammation of the synovial tissue results in swelling of the joints.

A. Hyaluronic Acid

Hyaluronic acid is the simplest acidic polysaccharide present in connective tissue. It has a molecular weight between 10^5 and 10^7 g/mol and contains from 300 to 100,000 repeating units, depending on the organ in which it occurs. It is most abundant in embryonic tissues and in specialized connective tissues such as synovial fluid, the lubricant of joints in the body, and the vitreous of the eye, where it provides a clear, elastic gel that holds the retina in its proper position. Hyaluronic acid is also a common ingredient in lotions, moisturizers, and cosmetics.

Hyaluronic acid is composed of D-glucuronic acid joined by a β-1,3-glycosidic bond to *N*-acetyl-D-glucosamine, which is in turn linked to D-glucuronic acid by a β-1,4-glycosidic bond.

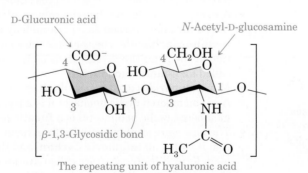

The repeating unit of hyaluronic acid

B. Heparin

Heparin is a heterogeneous mixture of variably sulfonated polysaccharide chains, ranging in molecular weight from 6000 to 30,000 g/mol. This acidic

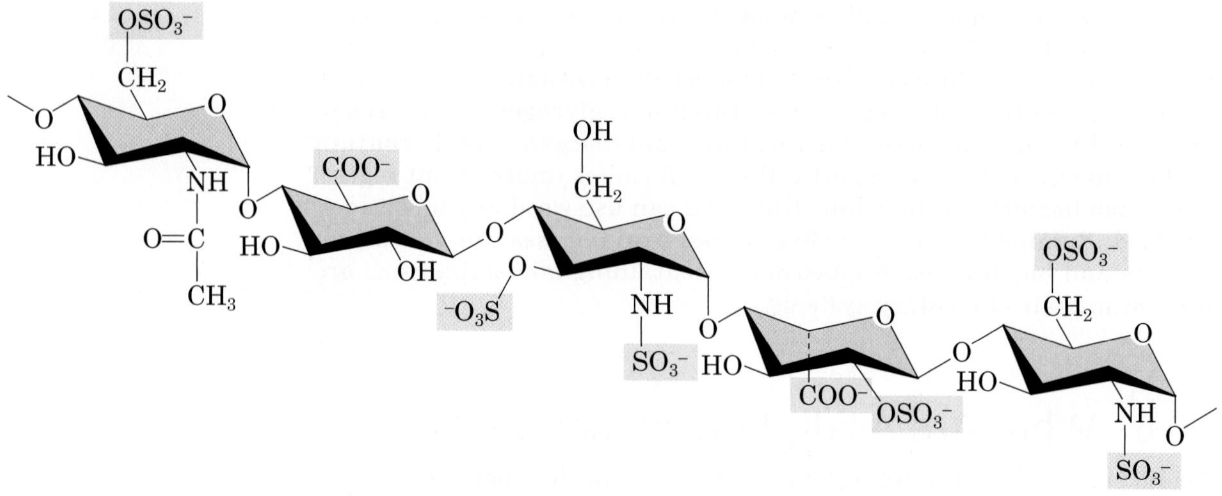

FIGURE 20.5 A repeating pentasaccharide unit of heparin.

polysaccharide is synthesized and stored in mast cells (cells that are part of the immune system and that occur in several types of tissues) of various tissues—particularly the liver, lungs, and gut. Heparin has many biological functions, the best known and fully understood of which is its anticoagulant activity. It binds strongly to antithrombin III, a plasma protein involved in terminating the clotting process. A heparin preparation with good antico-agulant activity contains a minimum of eight repeating units (Figure 20.5). The larger the molecule, the greater its anticoagulant activity. Because of this anticoagulant activity, it is widely used in medicine.

Summary of Key Questions

OWL Sign in at **www.cengage.com/owl** to develop problem-solving skills and complete online homework assigned by your professor.

Section 20.1 Carbohydrates: What Are Monosaccharides?

- **Monosaccharides** are polyhydroxyaldehydes or polyhydroxyketones.
- The most common have the general formula $C_nH_{2n}O_n$, where n varies from 3 to 8.
- Their names contain the suffix *-ose*, and the prefixes *tri-*, *tetr-*, and so on, indicate the number of carbon atoms in the chain. The prefix *aldo-* indicates an alde-hyde, and the prefix *keto-* indicates a ketone.

Section 20.2 What Are the Cyclic Structures of Monosaccharides?

- In a **Fischer projection** of a monosaccharide, we write the carbon chain vertically with the most highly oxi-dized carbon toward the top. Horizontal lines represent groups projecting above the plane of the page; vertical

lines represent groups projecting behind the plane of the page.
- The penultimate carbon of a monosaccharide is the next-to-last carbon of a Fischer projection.
- A monosaccharide that has the same configuration at the penultimate carbon as D-glyceraldehyde is called a **D-monosaccharide**; one that has the same configura-tion at the penultimate carbon as L-glyceraldehyde is called an **L-monosaccharide**.
- Monosaccharides exist primarily as cyclic hemiacetals.
- A six-membered cyclic hemiacetal is a **pyranose**; a five-membered cyclic hemiacetal is a **furanose**.
- The new stereocenter resulting from hemiacetal forma-tion is called an **anomeric carbon**, and the stereoiso-mers formed in this way are called **anomers**.
- The symbol **β-** indicates that the —OH group on the anomeric carbon lies on the same side of the ring as the terminal —CH₂OH.
- The symbol **α-** indicates that the —OH group on the anomeric carbon lies on the opposite side from the terminal —CH₂OH.

- Furanoses and pyranoses can be drawn as **Haworth projections**.
- Pyranoses can also be drawn as **chair conformations**.
- **Mutarotation** is the change in specific rotation that accompanies formation of an equilibrium mixture of α and β anomers in aqueous solution.

Section 20.3 What Are the Characteristic Reactions of Monosaccharides?

- A **glycoside** is a cyclic acetal derived from a monosaccharide.
- An **alditol** is a polyhydroxy compound formed when the carbonyl group of a monosaccharide is reduced to a hydroxyl group.
- An **aldonic acid** is a carboxylic acid formed when the aldehyde group of an aldose is oxidized to a carboxyl group.
- Any carbohydrate that reacts with an oxidizing agent to form an aldonic acid is classified as a **reducing sugar** (it reduces the oxidizing agent).

Section 20.4 What Are Disaccharides and Oligosaccharides?

- A **disaccharide** contains two monosaccharide units joined by a glycosidic bond.
- Terms applied to carbohydrates containing larger numbers of monosaccharides are **trisaccharide, tetrasaccharide, oligosaccharide,** and **polysaccharide**.

- **Sucrose** is a disaccharide consisting of D-glucose joined to D-fructose by an α-1,2-glycosidic bond.
- **Lactose** is a disaccharide consisting of D-galactose joined to D-glucose by a β-1,4-glycosidic bond.
- **Maltose** is a disaccharide of two molecules of D-glucose joined by an α-1,4-glycosidic bond.

Section 20.5 What Are Polysaccharides?

- **Starch** can be separated into two fractions: amylose and amylopectin.
- **Amylose** is a linear polysaccharide of as many as 4000 units of D-glucopyranose joined by α-1,4-glycosidic bonds.
- **Amylopectin** is a highly branched polysaccharide of D-glucose joined by α-1,4-glycosidic bonds and, at branch points, by α-1,6-glycosidic bonds.
- **Glycogen**, the reserve carbohydrate of animals, is a highly branched polysaccharide of D-glucopyranose joined by α-1,4-glycosidic bonds and, at branch points, by α-1,6-glycosidic bonds.
- **Cellulose**, the skeletal polysaccharide of plants, is a linear polysaccharide of D-glucopyranose joined by β-1,4 glycosidic bonds.

Section 20.6 What Are Acidic Polysaccharides?

- The carboxyl and sulfate groups of **acidic polysaccharides** are ionized to $-COO^-$ and $-SO_3^-$ at the pH of body fluids, which gives these polysaccharides net negative charges.

Summary of Key Reactions

1. **Formation of Cyclic Hemiacetals (Section 20.2)** A monosaccharide existing as a five-membered ring is a furanose; one existing as a six-membered ring is a pyranose. A pyranose is most commonly drawn as either a Haworth projection or a chair conformation.

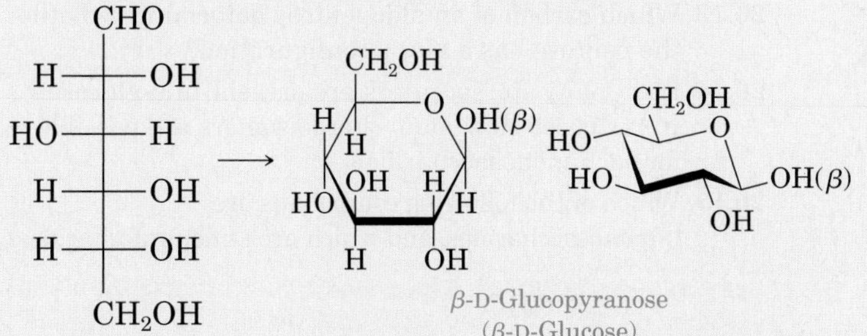

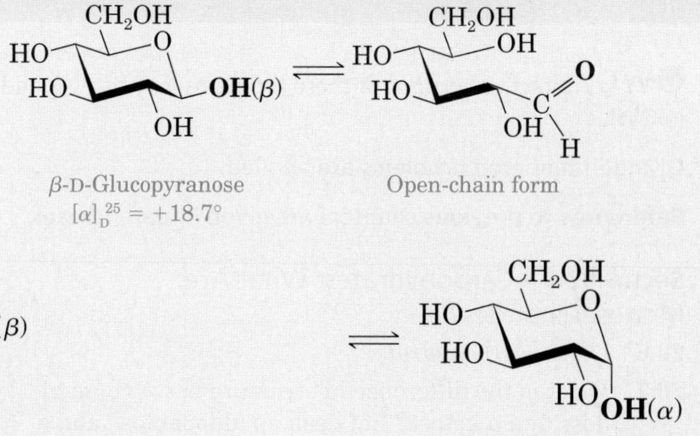

β-D-Glucopyranose
$[\alpha]_D^{25} = +18.7°$

Open-chain form

β-D-Glucopyranose
(β-D-Glucose)

D-Glucose

α-D-Glucopyranose
$[\alpha]_D^{25} = 112°$

2. **Mutarotation (Section 20.2C)** Anomeric forms of a monosaccharide are in equilibrium in aqueous solution. Mutarotation is the change in specific rotation that accompanies this equilibration.

3. **Formation of Glycosides (Section 20.3A)** Treatment of a monosaccharide with an alcohol in the presence of an acid catalyst forms a cyclic acetal called

a glycoside. The bond to the new —OR group is called a glycosidic bond.

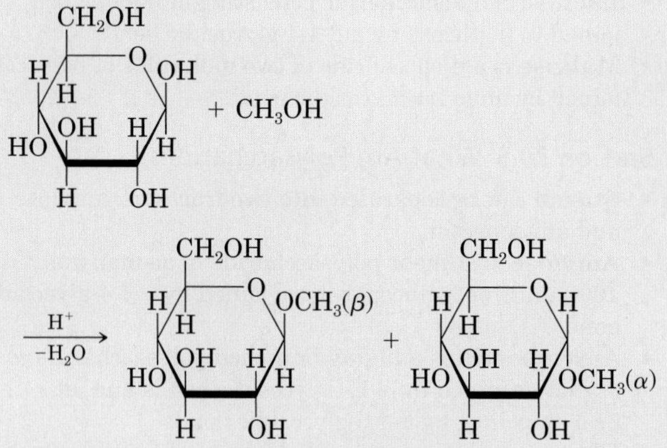

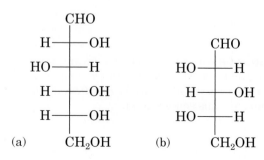

CH₂OH

H / H O OH
HO OH H / H
H   OH

+ CH₃OH

H⁺
−H₂O →

CH₂OH

H / H O OCH₃(β)
HO OH H / H
H   OH

+

CH₂OH

H / H O H
HO OH H / H
H   OH OCH₃(α)

glycosides

4. Reduction to Alditols (Section 20.3B) Reduction of the carbonyl group of an aldose or ketose to a hydroxyl group yields a polyhydroxy compound called an alditol.

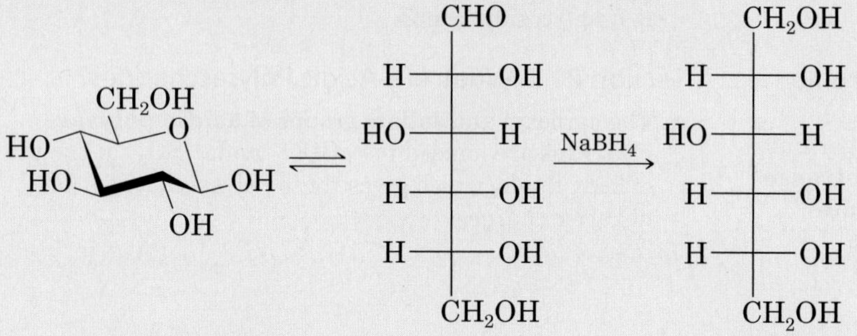

HO CH₂OH O
HO   OH
     OH

β-D-Glucopyranose

⇌

CHO
H——OH
HO——H
H——OH
H——OH
CH₂OH

D-Glucose

NaBH₄ →

CH₂OH
H——OH
HO——H
H——OH
H——OH
CH₂OH

D-Glucitol
(D-Sorbitol)

5. Oxidation to an Aldonic Acid (Section 20.3C) Oxidation of the aldehyde group of an aldose to a carboxyl group by a mild oxidizing agent gives a polyhydroxycarboxylic acid called an aldonic acid.

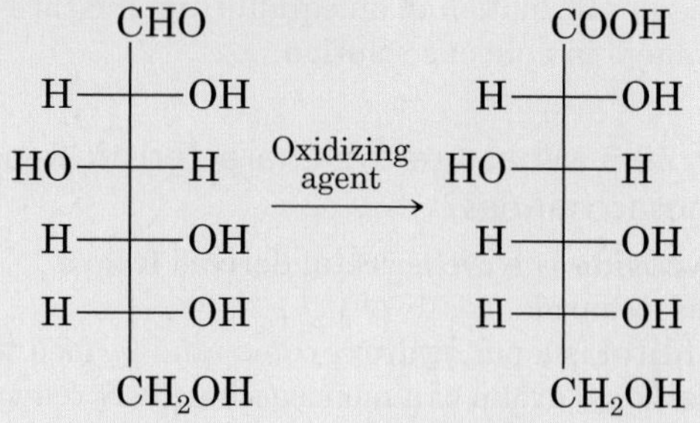

CHO
H——OH
HO——H
H——OH
H——OH
CH₂OH

D-Glucose

Oxidizing
agent →

COOH
H——OH
HO——H
H——OH
H——OH
CH₂OH

D-Gluconic acid

Problems

⏻WL Interactive versions of these problems may be assigned in OWL.

Orange-numbered problems are applied.

References to previous chapters are given in parentheses.

Section 20.1 Carbohydrates: What Are Monosaccharides?

20.6 Define *carbohydrate*.

20.7 What is the difference in structure between an aldose and a ketose? Between an aldopentose and a ketopentose?

20.8 Of the eight D-aldohexoses, which is the most abundant in the biological world?

20.9 Name the three most abundant hexoses in the biological world. Which are aldohexoses, and which are ketohexoses?

20.10 Which hexose is also known as "dextrose"?

20.11 What does it mean to say that D- and L-glyceraldehyde are enantiomers?

20.12 Explain the meaning of the designations D and L as used to specify the configuration of a monosaccharide.

20.13 Which carbon of an aldopentose determines whether the pentose has a D or L configuration?

20.14 How many stereocenters are present in D-glucose? In D-ribose? How many stereoisomers are possible for each monosaccharide?

20.15 Which of the following compounds are D-monosaccharides, and which are L-monosaccharides?

CHO
H——OH
HO——H
H——OH
H——OH
(a)   CH₂OH

CHO
HO——H
H——OH
HO——H
(b)   CH₂OH

CH₂OH
|
C=O
H———OH
H———OH
(c)   CH₂OH

20.16 Draw Fischer projections for L-ribose and L-arabinose.

20.17 Draw a Fischer projection for a D-2-ketoheptose.

20.18 Explain why all mono- and disaccharides are soluble in water.

20.19 What is an amino sugar? Name the three amino sugars most commonly found in nature.

Section 20.2 What Are the Cyclic Structures of Monosaccharides?

20.20 Define the term *anomeric carbon*. Which carbon is the anomeric carbon in glucose? In fructose?

20.21 Define (a) pyranose and (b) furanose.

20.22 Explain the conventions for using α and β to designate the configurations of the cyclic forms of monosaccharides.

20.23 Are α-D-glucose and β-D-glucose anomers? Explain. Are they enantiomers? Explain.

20.24 Are the hydroxyl groups on carbons 1, 2, 3, and 4 of α-D-glucose all in equatorial positions?

20.25 In what way are chair conformations a more accurate representation of the molecular shapes of hexopyranoses than Haworth projections?

20.26 Convert each of the following Haworth projections to an open-chain form and to a Fischer projection. Name the monosaccharides you have drawn.

(a) CH₂OH ... H H HO OH H H OH OH

(b) CH₂OH ... H H HO H OH HO OH H

20.27 Convert each of the following chair conformations to an open-chain form and to a Fischer projection. Name the monosaccharides you have drawn.

(a) HO CH₂OH O HO HO OH

(b) HO CH₂OH O HO OH OH HO

20.28 Explain the phenomenon of mutarotation. How is it detected?

20.29 The specific rotation of α-D-glucose is +112.2°. What is the specific rotation of α-L-glucose?

20.30 When α-D-glucose is dissolved in water, the specific rotation of the solution changes from +112.2° to +52.7°. Does the specific rotation of α-L-glucose also change when it is dissolved in water? If so, to what value does it change?

20.31 Define the terms *glycoside* and *glycosidic bond*.

20.32 What is the difference in meaning between the terms *glycosidic bond* and *glucosidic bond*?

20.33 Do glycosides undergo mutarotation?

Section 20.3 What Are the Characteristic Reactions of Monosaccharides?

20.34 Draw Fischer projections for the product formed by treating each of the following monosaccharides with sodium borohydride, NaBH₄, in water.

    (a) D-Galactose   (b) D-Ribose

20.35 Reduction of D-glucose by NaBH₄ gives D-sorbitol, a compound used in the manufacture of sugar-free gums and candies. Draw a structural formula for D-sorbitol.

20.36 Reduction of D-fructose by NaBH₄ gives two alditols, one of which is D-sorbitol. Name and draw a structural formula for the other alditol.

20.37 Ribitol and β-D-ribose 1-phosphate are derivatives of D-ribose. Draw a structural formula for each compound.

Section 20.4 What Are Disaccharides and Oligosaccharides?

20.38 Name three important disaccharides. From which monosaccharides is each derived?

20.39 What does it mean to describe a glycosidic bond as β-1,4-? To describe it as α-1,6-?

20.40 Both maltose and lactose are reducing sugars, but sucrose is a nonreducing sugar. Explain why.

20.41 Following is a structural formula for a disaccharide.

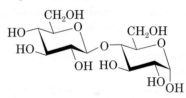

    (a) Name the two monosaccharide units in the disaccharide.

    (b) Describe the glycosidic bond.

    (c) Is this disaccharide a reducing sugar or a nonreducing sugar?

    (d) Will this disaccharide undergo mutarotation?

20.42 The disaccharide trehalose is found in young mushrooms and is the chief carbohydrate in the blood of certain insects.

CH₂OH
HO O
HO
HO 1
O O CH₂OH
Trehalose   HO OH
OH

    (a) Name the two monosaccharide units in trehalose.

    (b) Describe the glycosidic bond in trehalose.

    (c) Is trehalose a reducing sugar or a nonreducing sugar?

    (d) Will trehalose undergo mutarotation?

Section 20.5 What Are Polysaccharides?

20.43 What is the difference in structure between oligosaccharides and polysaccharides?

20.44 Name three polysaccharides that are composed of units of D-glucose. In which polysaccharide are the glucose units joined by α-glycosidic bonds? In which are they joined by β-glycosidic bonds?

20.45 Starch can be separated into two principal polysaccharides, amylose and amylopectin. What is the major difference in structure between the two?

20.46 Where is glycogen stored in the human body?

20.47 Why is cellulose insoluble in water?

20.48 How is it possible that cows can digest grass but humans cannot?

20.49 A Fischer projection of N-acetyl-D-glucosamine is given in Section 20.1D.

(a) Draw a Haworth projection and a chair conformation for the β-pyranose form of this monosaccharide.

(b) Draw a Haworth projection and a chair conformation for the disaccharide formed by joining two units of the pyranose form of N-acetyl-D-glucosamine with a β-1,4-glycosidic bond. If you draw them correctly, you will have a structural formula for the repeating dimer of chitin, the structural polysaccharide component of the shell of lobsters and other crustaceans.

20.50 Propose structural formulas for the repeating disaccharide unit in these polysaccharides.

(a) Alginic acid, isolated from seaweed, is used as a thickening agent in ice cream and other foods. Alginic acid is a polymer of D-mannuronic acid in the pyranose form joined by β-1,4-glycosidic bonds.

(b) Pectic acid is the main component of pectin, which is responsible for the formation of jellies from fruits and berries. Pectic acid is a polymer of D-galacturonic acid in the pyranose form joined by α-1,4-glycosidic bonds.

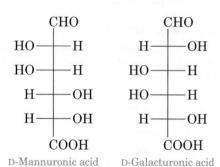

D-Mannuronic acid   D-Galacturonic acid

Section 20.6 What Are Acidic Polysaccharides?

20.51 Hyaluronic acid acts as a lubricant in the synovial fluid of joints. In rheumatoid arthritis, inflammation breaks hyaluronic acid down to smaller molecules. Under these conditions, what happens to the lubricating power of the synovial fluid?

20.52 The anticlotting property of heparin is partly due to the negative charges it carries.

(a) Identify the functional groups that provide the negative charges.

(b) Which type of heparin is a better anticoagulant, one with a high degree or a low degree of polymerization?

Chemical Connections

20.53 (Chemical Connections 20A) Why does congenital galactosemia appear only in infants? Why can galactosemia be relieved by feeding an affected infant a formula containing sucrose as the only carbohydrate?

20.54 (Chemical Connections 20B) What is the difference in structure between L-ascorbic acid and L-dehydroascorbic acid? What does the designation L indicate in these names?

20.55 (Chemical Connections 20B) When L-ascorbic acid participates in a redox reaction, it is converted to L-dehydroascorbic acid. In this reaction, is L-ascorbic acid oxidized or reduced? Is L-ascorbic acid a biological oxidizing agent or a biological reducing agent?

20.56 (Chemical Connections 20C) Why is the glucose assay one of the most common analytical tests performed in clinical chemistry laboratories?

20.57 (Chemical Connections 20D) What monosaccharides do types A, B, and O blood have in common? In which monosaccharides do they differ?

20.58 (Chemical Connections 20D) L-Fucose is a monosaccharide unit common to A, B, AB, and O blood types.

(a) Is this monosaccharide an aldose or a ketose? A hexose or a pentose?

(b) What is unusual about this monosaccharide?

(c) If L-fucose were to undergo a reaction in which its terminal —CH$_3$ group was converted to a —CH$_2$OH group, which monosaccharide would result?

20.59 (Chemical Connections 20D) Why can't a person with type A blood donate to a person with type B blood?

20.60 (Chemical Connections 20E) Low-fat and low-carbohydrate diets can both produce weight loss. Are there any health-related differences in the outcomes of the two kinds of diets? Explain.

Additional Problems

20.61 2,6-Dideoxy-D-altrose, also known as D-digitoxose, is a monosaccharide obtained from the hydrolysis of digitoxin, a natural product extracted from purple foxglove (*Digitalis purpurea*). Digitoxin has found wide use in cardiology because it reduces pulse rate, regularizes heart rhythm, and strengthens heartbeat. Draw a structural formula for the open-chain form of 2,6-dideoxy-D-altrose.

20.62 In making candy or sugar syrups, sucrose is boiled in water with a little acid, such as lemon juice. Why does the product mixture taste sweeter than the initial sucrose solution?

20.63 Hot-water extracts of ground willow bark are an effective pain reliever (Chemical Connections 19C). Unfortunately, the liquid is so bitter that most people refuse it. The pain reliever in these infusions is salicin. Name the monosaccharide unit in salicin.

Salicin

20.64 Show how D-sorbitol, used in "sugarless" chewing gum, is produced from D-glucose.

20.65 Carbohydrates in most foods have roughly the same molecular weight. True or false?

20.66 Ribose and fructose have an important similarity in structure in that they both have five-membered rings. How do they differ in structure?

20.67 In Section 20.4A, two structures are shown for sucrose. In one, both the glucose and fructose moieties are shown in the Haworth representation. In the other, the chair form of the glucose moiety is shown with a glycosidic bond to the fructose moiety shown in the Haworth form. Why does the Haworth form for the fructose moiety appear in both structures?

20.68 Blood samples for research or medical tests sometimes have heparin added. Why is this done?

20.69 What is the difference in the glycosidic bonds in starch and cellulose? How does this difference affect their biological function?

20.70 What are the structural differences between vitamin C and sugars? Do these structural differences play a role in the susceptibility of this vitamin to air oxidation?

20.71 A substance called laetrile is structurally related to carbohydrates. It has been suggested as a treatment for cancer, but it is not available in the United States. Its efficacy is doubtful, and it is inherently dangerous. Here is the structure. What is its relationship to carbohydrates? Do you think that the presence of the cyanide (—CN) group in the structure has a connection to the dangers associated with this compound? Explain.

Laetrile

20.72 Why are five- and six-membered rings encountered more frequently than any other possible ring size in the cyclic structures of sugars?

20.73 What is the structural difference between glucose and galactose? Is it possible that galactose could be converted to glucose in the body? If so, what kind of process takes place?

20.74 Concentrated sulfuric acid can be used as a highly effective dehydrating agent. When concentrated sulfuric acid is added to a beaker that contains sucrose, the white powder reacts, leaving behind a black substance that is mostly carbon. What sort of reaction has taken place?

20.75 Chitin is a polysaccharide found in shrimp and lobster shells. It is a polymer of N-acetyl-D-glucosamine, where the bonding between monomer units is a β-1,4-glycosidic bond. Propose a structure for chitin.

Tying It Together

20.76 In Section 20.2, we saw that in glucose, all atoms other than hydrogen are in the stable equatorial position. Suggest a correlation between the structure of glucose and its central role in the structure and function of carbohydrates.

Looking Ahead

20.77 One pathway for the metabolism of D-glucose-6-phosphate is its enzyme-catalyzed conversion to D-fructose-6-phosphate. Show that this transformation can be regarded as two enzyme-catalyzed keto-enol tautomerizations (Section 17.5).

D-Glucose 6-phosphate   D-Fructose 6-phosphate

20.78 One step in glycolysis, the pathway that converts glucose to pyruvate (Section 28.2), involves an enzyme-catalyzed conversion of dihydroxyacetone phosphate to D-glyceraldehyde 3-phosphate. Show that this transformation can be regarded as two enzyme-catalyzed keto-enol tautomerizations (Section 17.5).

Dihydroxyacetone   D-Glyceraldehyde
phosphate   3-phosphate

20.79 Following is a Haworth projection and a chair conformation for the repeating disaccharide unit in chondroitin 6-sulfate. This biopolymer acts as the flexible connecting matrix between the tough protein filaments in cartilage. It is available as a dietary supplement, often combined with D-glucosamine sulfate. Some believe this combination can strengthen and improve joint flexibility.

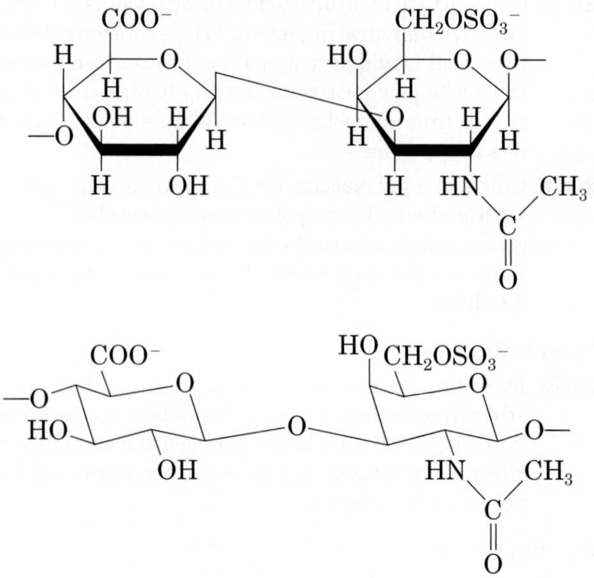

(a) From what two monosaccharide units is the repeating disaccharide unit of chondroitin 6-sulfate derived?

(b) Describe the glycosidic bond between the two units.

20.80 Below is the structural formula of coenzyme A, an important biomolecule.

(a) Is coenzyme A chiral? (Chapter 15)

(b) Name each functional group in coenzyme A. (Chapter 10)

(c) Would you expect coenzyme A to be soluble in water? Explain. (Chapter 5)

(d) Draw structural formulas for the products of complete hydrolysis of coenzyme A in aqueous HCl. Show each product as it would be ionized in this solution. (Chapter 19)

(e) Draw structural formulas for the products of complete hydrolysis of coenzyme A in aqueous NaOH. Show each product as it would be ionized in this solution. (Chapter 19)

Challenge Problem

20.81 Paper consists primarily of cellulose fibers with no specific orientation molded into a sheet. It is well known that paper loses most of its mechanical strength when it is wet with water. The loss of strength does not take place when the paper is wet with oil. Propose an explanation for this observation.

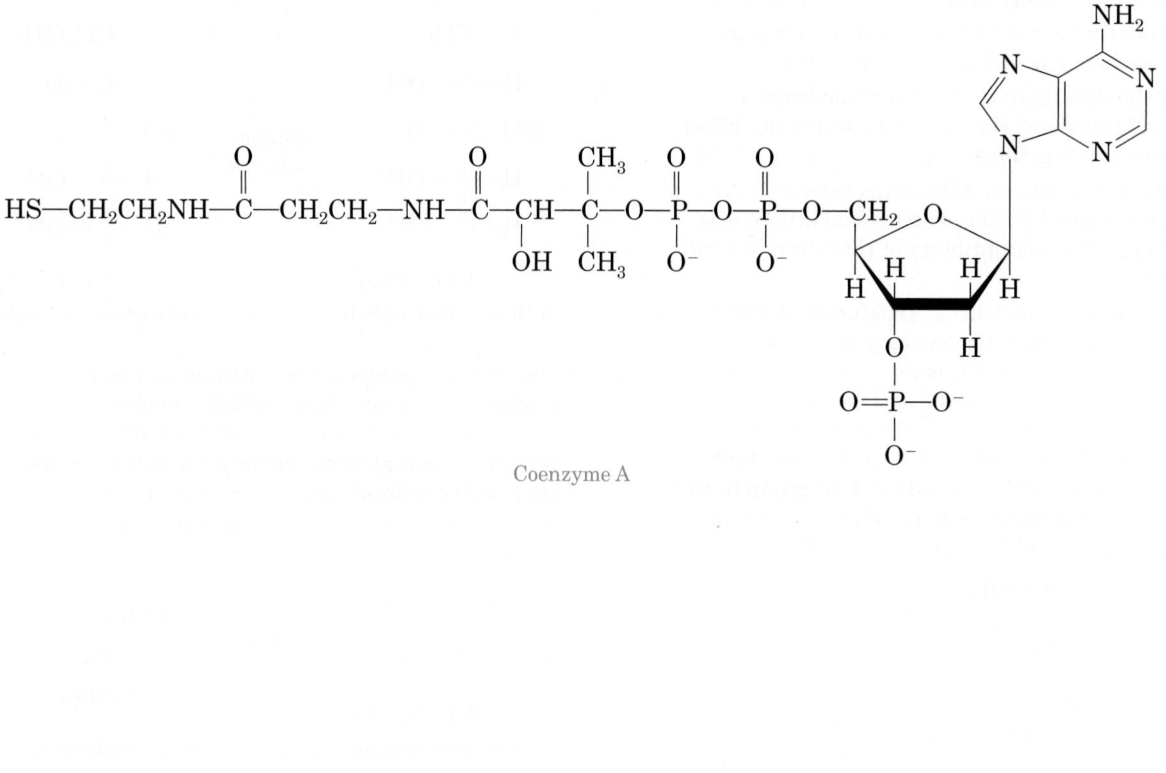

Coenzyme A

Lipids

Sea lions are marine mammals that require a heavy layer of fat to survive in cold waters.

21.1 What Are Lipids?

Found in living organisms, **lipids** are a family of substances that are insoluble in water but soluble in nonpolar solvents and solvents of low polarity, such as diethyl ether. Unlike carbohydrates, we define lipids in terms of a property and not in terms of their structure.

A. Classification by Function

Lipids play three major roles in human biochemistry: (1) They store energy within fat cells, (2) they are parts of membranes that separate compartments of aqueous solutions from each other, and (3) they serve as chemical messengers.

Storage

An important use for lipids, especially in animals, is the storage of energy. As we saw in Section 20.5, plants store energy in the form of starch. Animals (including humans) find it more economical to use fats instead. Although our bodies do store some carbohydrates in the form of glycogen for quick energy when we need it, energy stored in the form of fats has much greater importance for us. The reason is simple: The burning of fats produces more than twice as much energy (about 9 kcal/g) as the burning of an equal weight of carbohydrates (about 4 kcal/g).

Membrane Components

The lack of water solubility of lipids is an important property because our body chemistry is so heavily based on water. Most body constituents, including carbohydrates and proteins, are soluble in water. However, the body also needs insoluble compounds for the membranes that separate compartments containing aqueous solutions, whether they are cells or organelles within the cells. Lipids provide these membranes. Their water insolubility derives from the fact that the polar groups they contain are much smaller than their alkane-like (nonpolar) portions. These nonpolar portions provide the water-repellent, or *hydrophobic*, property.

Messengers

Lipids also serve as chemical messengers. Primary messengers such as steroid hormones deliver signals from one part of the body to another part. Secondary messengers such as prostaglandins and thromboxanes mediate the hormonal response.

B. Classification by Structure

For purposes of study, we can classify lipids into four groups: (1) simple lipids such as fats and waxes; (2) complex lipids; (3) steroids; and (4) prostaglandins, thromboxanes, and leukotrienes.

21.2   What Are the Structures of Triglycerides?

Animal fats and plant oils are triglycerides. **Triglycerides** are triesters of glycerol and long-chain carboxylic acids called fatty acids. In Section 19.1, we saw that esters are made up of an alcohol part and an acid part. As the name indicates, the alcohol of triglycerides is always glycerol.

$$CH_2\!-\!OH$$
$$|$$
$$CH\!-\!OH$$
$$|$$
$$CH_2\!-\!OH$$

Glycerol

In contrast to the alcohol part, the acid component of triglycerides may be any number of fatty acids (Section 18.4A). These fatty acids do, however, have certain things in common:

1. Fatty acids are practically all unbranched carboxylic acids.
2. They range in size from about 10 to 20 carbons.
3. They contain an even number of carbon atoms.
4. Apart from the —COOH group, they have no functional groups, except that some do have double bonds.
5. In most fatty acids that have double bonds, the *cis* isomers predominate.

Only even-numbered acids are found in triglycerides because the body builds these acids entirely from acetate units and therefore puts the carbons in two at a time (Section 29.2).

In **triglycerides** (also called **triacylglycerols**), all three hydroxyl groups of glycerol are esterified. Thus, a typical triglyceride molecule might be:

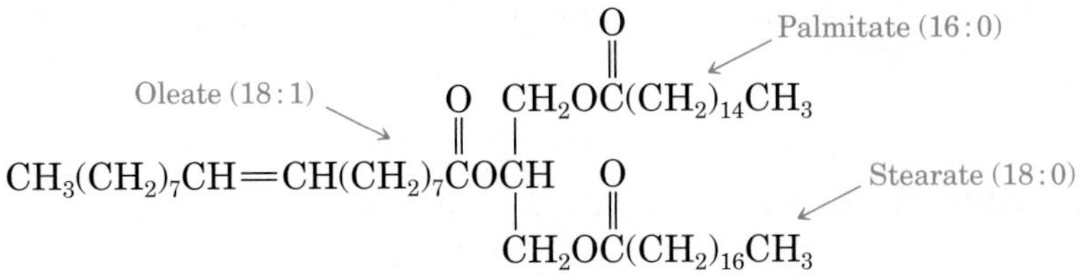

$$\text{Oleate (18:1)} \qquad \text{Palmitate (16:0)}$$

$$CH_3(CH_2)_7CH{=}CH(CH_2)_7COCH \qquad CH_2OC(CH_2)_{14}CH_3$$

$$\text{Stearate (18:0)}$$

$$CH_2OC(CH_2)_{16}CH_3$$

A triglyceride

Triglycerides are the most common lipid materials, although **mono-** and **diglycerides** are not infrequent. In the latter two types, only one or two —OH groups of the glycerol are esterified by fatty acids.

Triglycerides are complex mixtures. Although some of the molecules contain three identical fatty acids, in most cases, two or three different acids are present. The hydrophobic character of triglycerides is caused by the long hydrocarbon chains. The ester groups ($-\overset{\text{O}}{\overset{\|}{\text{C}}}-\text{O}-$), although polar themselves, are buried in a nonpolar environment, which makes the triglycerides insoluble in water.

21.3 What Are Some Properties of Triglycerides?

A. Physical State

With some exceptions, **fats** that come from animals are generally solids at room temperature and those from plants or fish are usually liquids. Liquid fats are often called **oils**, even though they are esters of glycerol just like solid fats and should not be confused with petroleum, which is mostly alkanes.

What is the structural difference between solid fats and liquid oils? In most cases, it is the degree of unsaturation. The physical properties of the fatty acids are carried over to the physical properties of the triglycerides. Solid animal fats contain mainly saturated fatty acids, whereas vegetable oils contain high amounts of unsaturated fatty acids. Table 21.1 shows the average fatty acid content of some common fats and oils. Note that even solid fats contain some unsaturated acids and that liquid fats contain some saturated acids. Some unsaturated fatty acids (linoleic and linolenic acids) are called *essential fatty acids* because the body cannot synthesize them from precursors; they must, therefore, be consumed as part of the diet.

Although most vegetable oils contain high amounts of unsaturated fatty acids, there are exceptions. Coconut oil, for example, has only a small amount of unsaturated fatty acids. This oil is a liquid not because it contains many double bonds, but because it is rich in low-molecular-weight fatty acids (chiefly lauric acid).

Oils with an average of more than one double bond per fatty acid chain are called *polyunsaturated*. Their role in the human diet is discussed in Section 30.4.

Pure fats and oils are colorless, odorless, and tasteless. This statement may seem surprising because we all know the tastes and colors of such fats and oils as butter and olive oil. The tastes, odors, and colors are caused by small amounts of other substances dissolved in the fat or oil.

Fat A mixture of triglycerides containing a high proportion of long-chain, saturated fatty acids

Oil A mixture of triglycerides containing a high proportion of long-chain, unsaturated fatty acids or short-chain, saturated fatty acids

Table 21.1 Average Percentage of Fatty Acids of Some Common Fats and Oils

	Saturated				Unsaturated			
	Lauric	Myristic	Palmitic	Stearic	Oleic	Linoleic	Linolenic	Other
Animal Fats								
Beef tallow	—	6.3	27.4	14.1	49.6	2.5	—	0.1
Butter	2.5	11.1	29.0	9.2	26.7	3.6	—	17.9
Human	—	2.7	24.0	8.4	46.9	10.2	—	7.8
Lard	—	1.3	28.3	11.9	47.5	6.0	—	5.0
Vegetable Oils								
Coconut	45.4	18.0	10.5	2.3	7.5	—	—	16.3
Corn	—	1.4	10.2	3.0	49.6	34.3	—	1.5
Cottonseed	—	1.4	23.4	1.1	22.9	47.8	—	3.4
Linseed	—	—	6.3	2.5	19.0	24.1	47.4	0.7
Olive	—	—	6.9	2.3	84.4	4.6	—	1.8
Peanut	—	—	8.3	3.1	56.0	26.0	—	6.6
Safflower	—	—	6.8	—	18.6	70.1	3.4	1.1
Soybean	0.2	0.1	9.8	2.4	28.9	52.3	3.6	2.7
Sunflower	—	—	6.1	2.6	25.1	66.2	—	—

B. Hydrogenation

In Section 12.6D, we learned that we can reduce carbon–carbon double bonds to single bonds by treating them with hydrogen and a catalyst. It is, therefore, not difficult to convert unsaturated liquid oils to solids. For example:

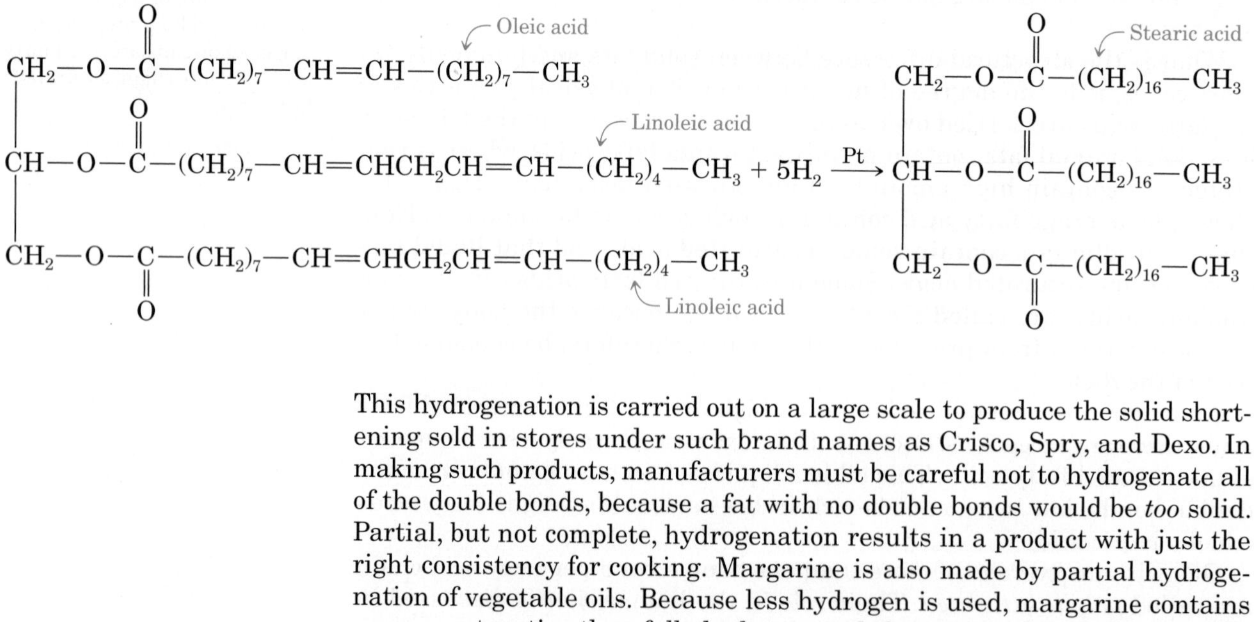

This hydrogenation is carried out on a large scale to produce the solid shortening sold in stores under such brand names as Crisco, Spry, and Dexo. In making such products, manufacturers must be careful not to hydrogenate all of the double bonds, because a fat with no double bonds would be *too* solid. Partial, but not complete, hydrogenation results in a product with just the right consistency for cooking. Margarine is also made by partial hydrogenation of vegetable oils. Because less hydrogen is used, margarine contains more unsaturation than fully hydrogenated shortenings. The hydrogenation process is the source of *trans* fatty acids, as we have already seen (Chemical Connections 18A). The food-processing industry is taking steps to address this issue. Many food labels specifically call attention to the fact that there are "no *trans* fats" in the product.

Waxes

Plant and animal waxes are simple esters. They are solids because of their high molecular weights. As in fats, the acid portions of the esters consist of a mixture of fatty acids; the alcohol portions are not glycerol, however, but rather simple long-chain alcohols. For example, a major component of beeswax is 1-triacontyl palmitate:

Palmitic acid portion   1-Triacontanol portion

$$CH_3(CH_2)_{13}CH_2\overset{\displaystyle O}{\overset{\|}{C}}-OCH_2(CH_2)_{28}CH_3$$

1-Triacontyl palmitate

Waxes generally have higher melting points than fats (60 to 100°C) and are harder. Animals and plants often

use them for protective coatings. For example, the leaves of most plants are coated with wax, which helps to prevent microorganisms from attacking them and allows the plants to conserve water. The feathers of birds and the fur of animals are also coated with wax.

Important waxes include carnauba wax (from a Brazilian palm tree), lanolin (from lamb's wool), beeswax, and spermaceti (from whales). These substances are used to make cosmetics, polishes, candles, and ointments. Paraffin waxes are not esters, but are mixtures of high-molecular-weight alkanes. Neither is earwax a simple ester. This gland secretion contains a mixture of fats (triglycerides), phospholipids, and esters of cholesterol.

Test your knowledge with Problem 21.51.

C. Saponification

Glycerides, being esters, are subject to hydrolysis, which can be carried out with either acids or bases. As we saw in Section 19.4, the use of bases is more practical. An example of the saponification of a typical fat is:

$$\underset{\text{A triglyceride}}{\begin{matrix} O & CH_2OCR \\ \| & | \\ RCOCH & O \\ | & \| \\ CH_2OCR \end{matrix}} + 3NaOH \xrightarrow{\text{saponification}} \underset{\substack{\text{1,2,3-Propanetriol} \\ \text{(Glycerol; Glycerin)}}}{\begin{matrix} CH_2OH \\ | \\ CHOH \\ | \\ CH_2OH \end{matrix}} + \underset{\text{Sodium soaps}}{3RCO^-Na^+}$$

Thus, saponification is the base-promoted hydrolysis of fats and oils, producing glycerol and a mixture of fatty acid salts called soaps. **Soap** has been used for thousands of years, and saponification is one of the oldest known chemical reactions.

Many common products contain hydrogenated vegetable oils.

Charles D. Winters

21.4 What Are the Structures of Complex Lipids?

The triglycerides discussed in the previous sections are significant components of fat storage cells. Other kinds of lipids, called complex lipids, are important in a different way. They constitute the main components of membranes (Section 21.5). Complex lipids can be classified into two groups: phospholipids and glycolipids.

Phospholipids contain an alcohol, two fatty acids, and a phosphate group. There are two types: **glycerophospholipids** and **sphingolipids**. In glycerophospholipids, the alcohol is glycerol (Section 21.6). In sphingolipids, the alcohol is sphingosine (Section 21.7).

Glycolipids are complex lipids that contain carbohydrates (Section 21.8). Figure 21.1 shows schematic structures for all of these lipids.

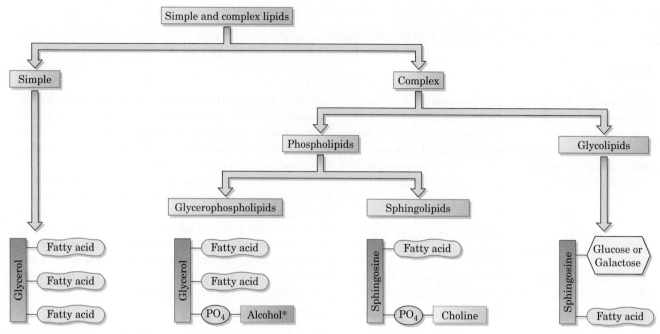

FIGURE 21.1 Schematic diagram of simple and complex lipids.*

*The alcohol can be choline, serine, ethanolamine, inositol, or certain others.

21.5 What Role Do Lipids Play in the Structure of Membranes?

The complex lipids mentioned in Section 21.4 form the **membranes** around body cells and around small structures inside the cells. (These small structures inside the cell are called *organelles*.) Unsaturated fatty acids are important components of these lipids. Most lipid molecules in the bilayer contain at least one unsaturated fatty acid. The cell membranes separate cells from the external environment and provide selective transport for nutrients and waste products into and out of cells.

These membranes are made of **lipid bilayers** (Figure 21.2). In a lipid bilayer, two rows (layers) of complex lipid molecules are arranged tail to tail. The hydrophobic tails point toward each other, which enables them to get as far away from the water as possible. This arrangement leaves the hydrophilic heads projecting to the inner and outer surfaces of the membrane. Cholesterol (Section 21.9), another membrane component, also positions its hydrophilic portion on the surface of the molecule on the surface of the membranes and the hydrophobic portion inside the bilayer.

The unsaturated fatty acids prevent the tight packing of the hydrophobic chains in the lipid bilayer, thereby providing a liquid-like character to the membranes. This effect is similar to the one that causes unsaturated fatty acids to have lower melting points than saturated fatty acids. This property of membrane fluidity is of extreme importance because many products of the body's biochemical processes must cross the membrane, and the liquid nature of the lipid bilayer allows such transport.

The lipid part of the membrane serves as a barrier against any movement of ions or polar compounds into and out of the cells. In the lipid bilayer, protein molecules are either suspended on the surface (peripheral proteins) or partly or fully embedded in the bilayer (integral proteins). These proteins stick out either on the inside or on the outside of the membrane. Others are thoroughly embedded, going through the bilayer and projecting from both sides. The model shown in Figure 21.2, called the **fluid mosaic model** of membranes, allows the passage of nonpolar compounds by diffusion, as

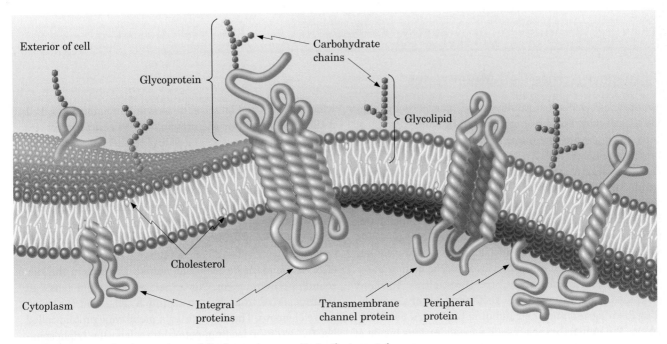

FIGURE 21.2 The fluid mosaic model of membranes. Note that proteins are embedded in the lipid matrix.

these compounds are soluble in the lipid membranes. The term *mosaic* refers to the topography of the bilayers: protein molecules dispersed in the lipid. The term *fluid* is used because the free lateral motion in the bilayers makes membranes liquid-like. In contrast, polar compounds are transported either via specific channels through the protein regions or by a mechanism called active transport (see Chemical Connections 21B). For any transport process, the membrane must behave like a nonrigid material that has liquid properties so that the proteins can move sideways within the membrane.

21.6 What Are Glycerophospholipids?

The structure of glycerophospholipids (also called phosphoglycerides) is very similar to that of fats. Glycerophospholipids are membrane components of cells throughout the body. The alcohol is glycerol. Two of the three hydroxyl groups are esterified by fatty acids. As with the simple fats, these fatty acids may be any long-chain carboxylic acids, with or without double bonds. In all glycerophospholipids, lecithins, cephalins, and phosphatidylinositols, the fatty acid on carbon 2 of glycerol is always unsaturated. The third group is esterified not by a fatty acid, but by a phosphate group, which is also esterified to another alcohol. If the other alcohol is choline, a quaternary ammonium compound, the glycerophospholipids are called **phosphatidylcholines** (common name **lecithin**):

Chemical Connections 21B

Transport Across Cell Membranes

Membranes are not just random assemblies of complex lipids that provide a nondescript barrier. In human red blood cells, for example, the outer part of the bilayer is made largely of phosphatidylcholine and sphingomyelin, while the inner part consists mostly of phosphatidylethanolamine and phosphatidylserine (Sections 21.6 and 21.7). In another example, in the membrane called the sarcoplasmic reticulum in the heart muscles, phosphatidylethanolamine is found in the outer part of the membrane, phosphatidylserine is found in the inner part, and phosphatidylcholine is equally distributed in the two layers of the membrane.

Membranes are not static structures, either. In many processes, they fuse with one another; in other processes, they disintegrate and their building blocks are used elsewhere. When membranes fuse in vacuole fusions inside of cells, for example, certain restrictions prevent incompatible membranes from intermixing.

The protein molecules are not dispersed randomly in the bilayer. Sometimes they cluster in patches; at other times, they appear in regular geometric patterns. An example of the latter are **gap junctions**, channels made of six proteins that create a central pore. These channels allow neighboring cells to communicate. Gap junctions are an example of **passive transport**. Small polar molecules—which include such essential nutrients as inorganic ions, sugars, amino acids, and nucleotides—can readily pass through gap junctions. Large molecules such as proteins, polysaccharides, and nucleic acids cannot.

In **facilitated transport**, a specific interaction takes place between the transporter and the transported molecule. Consider the **anion transporter** of the red blood cells, through which chloride and bicarbonate ions are exchanged in a 1 : 1 ratio. The transporter is a protein with 14 helical structures that span the membrane. One side of the helices contains the hydrophobic parts of the protein, which can interact with the lipid membrane. The other side of the helices forms a channel. This channel contains the hydrophilic portions of the protein, which can interact with the hydrated ions. In this manner, anions can pass through the erythrocyte membrane.

Active transport involves the passage of ions through a concentration gradient. For example, a higher concentration of K^+ is found inside cells than outside the cells in the surrounding environment. Nevertheless, potassium ions can be transported from the outside into a cell, albeit at the expense of energy. The transporter, a membrane protein called Na^+/K^+–ATPase (sodium-potassium adenosine triphosphatase), uses the energy from the hydrolysis of ATP molecules to change the conformation of the transporter, which brings in K^+ and exports Na^+. Detailed studies of K^+ ion channels have revealed that K^+ ions enter the channel in pairs. Each of the hydrated ions carries eight water molecules in its solvation layers, with the negative pole of the water molecules (the oxygen atoms) surrounding the positive ion. Deeper in the channel, K^+ ions encounter a constriction, called a selectivity filter. To pass through, the K^+ ions must shed their water molecules. The pairing of the naked K^+ ions generates enough electrostatic repulsion to keep the ions moving through the channel. The channel itself is lined with oxygen atoms to provide an attractive environment similar to that offered by the stable hydrated form before entry.

Polar compounds, in general, are transported through specific **transmembrane channels**.

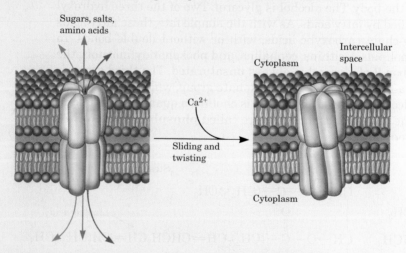

Gap junctions are made of six cylindrical protein subunits. They line up in two plasma membranes parallel to each other, forming a pore. The pores of gap junctions are closed by a sliding and twisting motion of the cylindrical subunits.

Test your knowledge with Problems 21.52–21.54.

This typical lecithin molecule has stearic acid on one end and linoleic acid in the middle. Other lecithin molecules contain other fatty acids, but the one on the end is always saturated and the one in the middle is always unsaturated. Lecithin is a major component of egg yolk. Because it includes both polar and nonpolar portions within one molecule, it is an excellent emulsifier (see Chemical Connections 6D) and is used in mayonnaise.

Note that lecithin has a negatively charged phosphate group and a positively charged quaternary nitrogen from the choline. These charged parts of the molecule provide a strongly hydrophilic head, whereas the rest of the molecule is hydrophobic. Thus, when a phospholipid such as lecithin is part of a lipid bilayer, the hydrophobic tail points toward the middle of the bilayer and the hydrophilic heads line both the inner and outer surfaces of the membranes (Figures 21.2 and 21.3).

Lecithins are just one example of glycerophospholipids. Another is the **cephalins**, which are similar to the lecithins in every way except that, instead of choline, they contain other alcohols such as ethanolamine or serine:

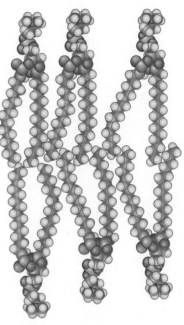

From *Biochemistry* by Lubert Stryer © 1975, 1981, 1988, 1995 by Lubert Stryer; © 2002 by W. H. Freeman and Company. Used with the permission of W. H. Freeman and Company.

FIGURE 21.3 Space-filling molecular models of complex lipids in a bilayer.

A phosphatidylethanolamine
(a cephalin)

A phosphatidylserine
(a cephalin)

R = hydrocarbon tail of fatty acid portion

Another important group of glycerophospholipids is the **phosphatidylinositols (PI)**. In PI, the alcohol inositol is bonded to the rest of the molecule by a phosphate ester bond. Such compounds not only are integral structural parts of the biological membranes, but also, in their higher phosphorylated form, such as **phosphatidylinositol 4,5-bisphosphates (PIP2)**, serve as signaling molecules in chemical communication (see Chapter 24).

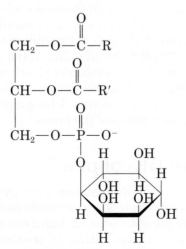

Phosphatidylinositols, PI

21.7 What Are Sphingolipids?

Myelin, the coating of nerve axons, contains a different kind of complex lipid: **sphingolipids**. In sphingolipids, the alcohol portion is sphingosine:

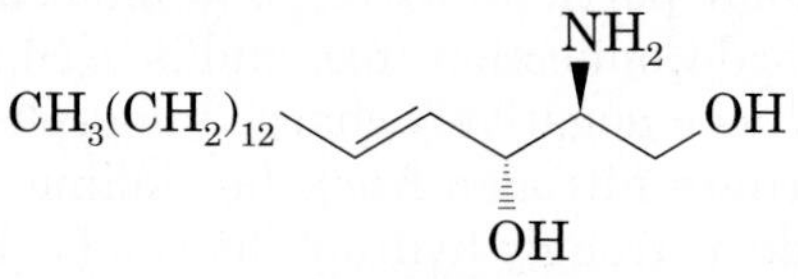

Sphingosine

A long-chain fatty acid is connected to the —NH₂ group by an amide bond, and the —OH group at the end of the chain is esterified by phosphorylcholine:

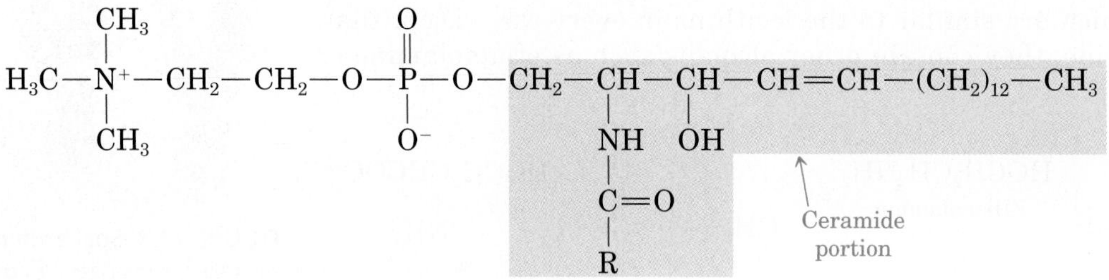

A sphingomyelin
(a sphingolipid)

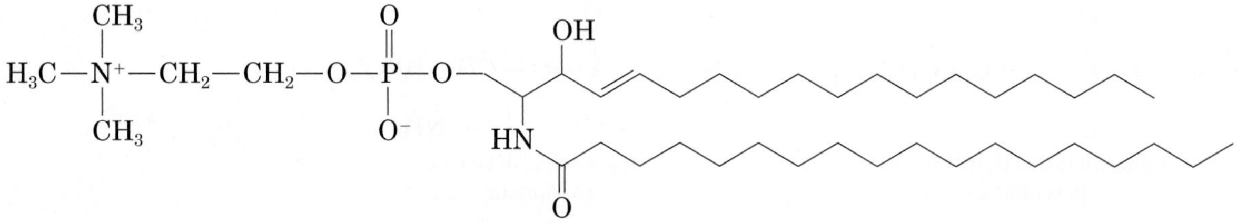

Sphingomyelin
(schematic diagram)

The combination of a fatty acid and sphingosine (shown in color) is called the **ceramide** portion of the molecule, because many of these compounds are also found in cerebrosides (Section 21.8). The ceramide part of complex lipids may contain different fatty acids. Stearic acid, for example, occurs mainly in sphingomyelin.

Sphingomyelins are the most important lipids in the myelin sheaths of nerve cells and are associated with diseases such as multiple sclerosis. Sphingolipids are not randomly distributed in membranes. In viral membranes, for example, most of the sphingomyelin appears on the inside of the membrane. Johann Thudichum, who discovered sphingolipids in 1874, named these brain lipids after a monster of Greek mythology, the sphinx. Part woman and part winged lion, the sphinx devoured all who could not provide the correct answer to her riddles. Sphingolipids appeared to Thudichum as part of a dangerous riddle of the brain.

21.8 What Are Glycolipids?

Glycolipids are complex lipids that contain carbohydrates and ceramides. One group, the **cerebrosides**, consists of ceramide mono- or oligosaccharides. Other groups, such as the **gangliosides**, contain a more complex carbohydrate structure (see Chemical Connections 21D). In cerebrosides, the fatty acid of the ceramide part may contain either 18-carbon or 24-carbon chains; the latter form is found only in these complex lipids. A glucose or galactose carbohydrate unit forms a beta-glycosidic bond with the ceramide portion of the molecule.

The cerebrosides occur primarily in the brain (accounting for 7% of the brain's dry weight) and at nerve synapses.

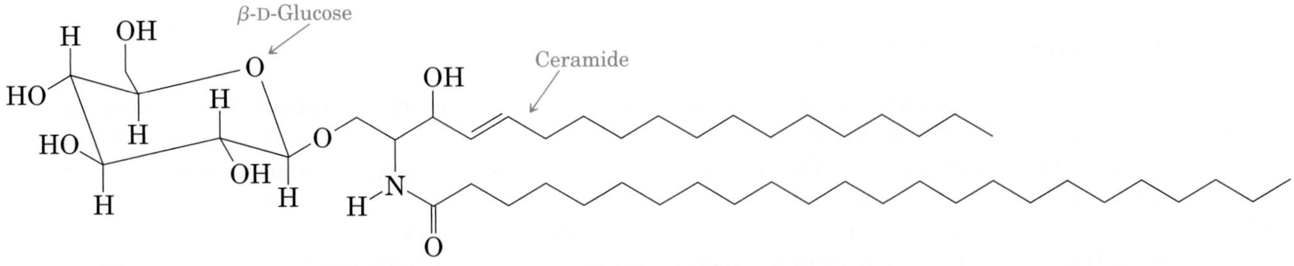

Glucocerebroside

Example 21.1 Lipid Structures

A lipid isolated from the membrane of red blood cells has the following structure:

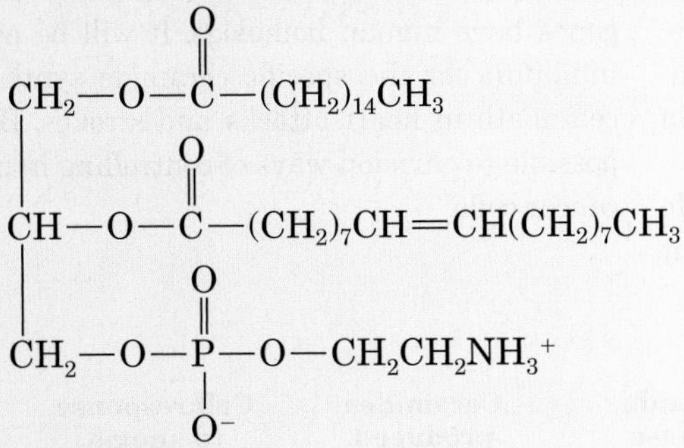

(a) To what group of complex lipids does this compound belong?
(b) What are the components?

Strategy

Part (b) of the question, about the component parts, is key to the whole answer. Once the parts are identified, they indicate the class of compound.

Solution

(a) The molecule is a triester of glycerol and contains a phosphate group; therefore, it is a glycerophospholipid.
(b) Besides glycerol and phosphate, it has palmitic acid and oleic acid components. The other alcohol is ethanolamine. Therefore, it belongs to the subgroup of cephalins.

Problem 21.1

A complex lipid has the following structure:

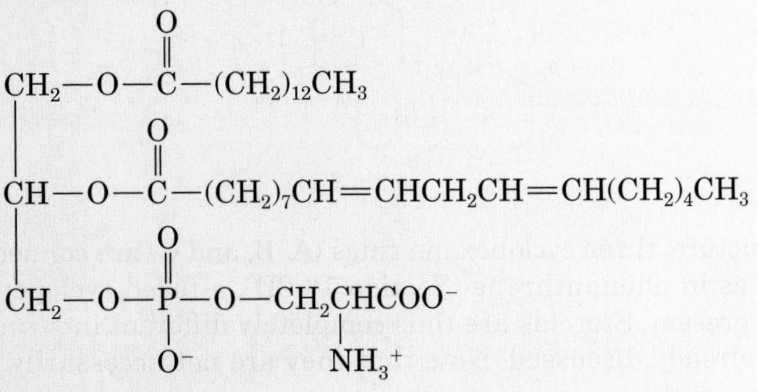

(a) To what group of complex lipids does this compound belong?
(b) What are the components?

Chemical Connections 21C

Ceramides, Oxygen, Cancer, and Strokes

A connection between carbon chain-length in ceramides and the response to oxygen deprivation (hypoxia) may have implications for serious issues in human health, including cancer and strokes. Cancer cells thrive under hypoxic conditions, and heart attacks and strokes cause damage due to oxygen deprivation of cells. Thus, sensitivity to hypoxia and mechanisms that control it are of interest to medical researchers. The connection in question came to light in studies on ceramide synthases in the roundworm *Caenorhabditis elegans*, specifically in their response to oxygen deprivation. Ceramides with different chain-lengths were produced in response to low oxygen levels, having different effects on survival. After observing the process in *C. elegans*, the question immediately arose as to whether human ceramide synthases play a role in response to the oxygen deprivation in cells that accompanies a heart attack or a stroke.

As shown in the figure, two different *C. elegans* ceramide synthases, hyl-1 and hyl-2, produce ceramides with different length carbon chains. Hyl-1 produces 24–26 carbon chains, while hyl-2 produces 20–22 carbon chains. Researchers used mutant flatworms lacking one of the two enzymes to study the effects of oxygen deprivation in the presence or absence of the different chain-length ceramides. Production of the longer chain-length ceramides by hyl-1 led to cell death, while production of shorter chain-length ceramides by hyl-2 led to cell survival. This survival raises a number of questions. The complete genome of *C. elegans* is known, making it possible to analyze mutations involved in the case that led to cell survival. This information can have implications in human health, since approximately 35% of *C. elegans* genes have human homologs. It will be possible to design inhibitors for the specific ceramide synthases involved in cell death in heart attacks and strokes. Beyond that, it is possible to envision ways of controlling hypoxia so as to kill cancer cells.

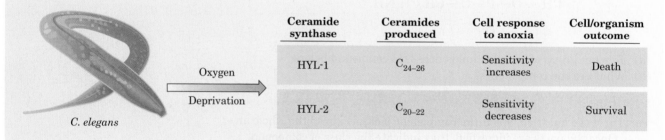

Ceramide synthase	Ceramides produced	Cell response to anoxia	Cell/organism outcome
HYL-1	C_{24-26}	Sensitivity increases	Death
HYL-2	C_{20-22}	Sensitivity decreases	Survival

Test your knowledge with Problem 21.55.

21.9 What Are Steroids?

The third major class of lipids is the **steroids**, which are compounds containing the following ring system:

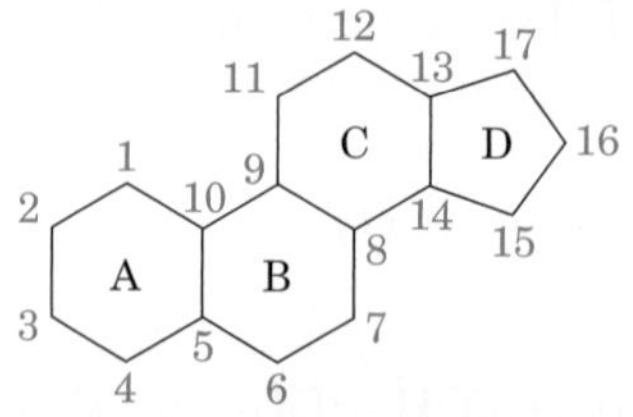

In this structure, three cyclohexane rings (A, B, and C) are connected in the same way as in phenanthrene (Section 13.2D); a fused cyclopentane ring (D) is also present. Steroids are thus completely different in structure from the lipids already discussed. Note that they are not necessarily esters, although some of them are.

Lipid Storage Diseases

Complex lipids are constantly being synthesized and decomposed in the body. In several genetic diseases classified as lipid storage diseases, some of the enzymes needed to decompose the complex lipids are defective or missing. As a consequence, the complex lipids accumulate and cause an enlarged liver and spleen, mental retardation, blindness, and, in certain cases, early death. Table 21D summarizes some of these diseases and indicates the missing enzyme and the accumulating complex lipid in each.

At present, no treatment is available for these diseases. The best way to prevent them is by genetic counseling. Some of the diseases can be diagnosed during fetal development. For example, Tay-Sachs disease, which affects about 1 in every 30 Jewish Americans (versus 1 in 300 in the non-Jewish population), can be diagnosed from amniotic fluid obtained by amniocentesis.

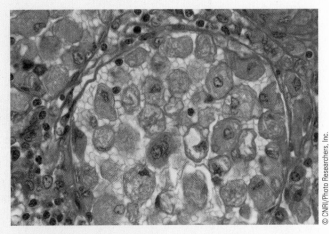

The accumulation of glucocerebrosides in the cell of a patient with Gaucher's disease. These Gaucher cells infiltrate the bone marrow.

Table 21D Lipid Storage Diseases

Name	Accumulating Lipid	Missing or Defective Enzyme Type
Gaucher's disease	Glucocerebroside	β-Glucosidase
Krabbe's leukodystrophy	Galactocerebroside	β-Galactosidase
Fabry's disease	Ceramide trihexoside	α-Galactosidase

Chemical Connections 21D

Lipid Storage Diseases (*continued*)

Table 21D Lipid Storage Diseases

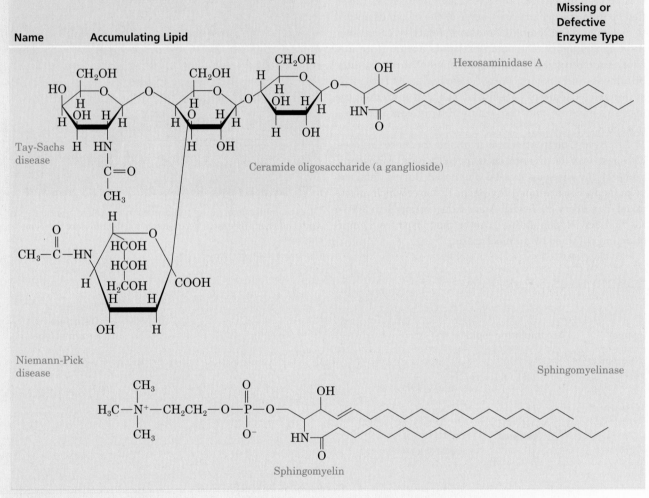

Name	Accumulating Lipid	Missing or Defective Enzyme Type

Ceramide oligosaccharide (a ganglioside) — Hexosaminidase A — Tay-Sachs disease

Niemann-Pick disease — Sphingomyelin — Sphingomyelinase

Test your knowledge with Problems 21.56 and 21.57.

A. Cholesterol

The most abundant steroid in the human body, and the most important, is **cholesterol**:

Cholesterol

Cholesterol serves as a plasma membrane component in all animal cells—for example, in red blood cells. Its second important function is to serve as a raw material for the synthesis of other steroids, such as the sex and adreno-corticoid hormones (Section 21.10) and bile salts (Section 21.11).

Cholesterol exists both in the free form and esterified with fatty acids. Gallstones contain free cholesterol (Figure 21.4).

Because the correlation between high serum cholesterol levels and such diseases as atherosclerosis has received so much publicity, many people are afraid of cholesterol and regard it as some kind of poison. Far from being poisonous, cholesterol is, in fact, necessary for human life. In essence, our livers manufacture cholesterol that satisfies our needs even without dietary intake. When the cholesterol level in the blood exceeds 150 mg/100 mL, cholesterol synthesis in the liver is reduced to half the normal rate of production. The amount of cholesterol is regulated, and it is an excess—rather than the presence—of cholesterol that is associated with disease.

A similar misconception is that eating too much cholesterol-containing food is what leads to high cholesterol levels. This is only partially true. A mere 10–15% of the cholesterol in our systems comes from cholesterol we ingest. The biggest victim of this misconception is the egg. Many people avoid eating egg yolks due to the cholesterol levels found in this part of the egg. However, most of the cholesterol in our bodies that come from the food we ingest comes from excess calories. Too much fat intake or even too much carbohydrate intake will lead to higher production of cholesterol. Thus, it would make little sense for a person to avoid egg yolks but then eat an otherwise fatty diet.

Cholesterol in the body is in a dynamic state. It constantly circulates in the blood. Cholesterol and esters of cholesterol, being hydrophobic, need a water-soluble carrier to circulate in the aqueous medium of blood.

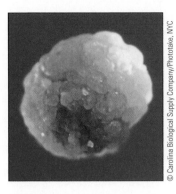

FIGURE 21.4 A human gallstone is almost pure cholesterol. This gallstone measures 5 mm in diameter.

B. Lipoproteins: Carriers of Cholesterol

Cholesterol, along with fat, is transported by **lipoproteins**. Most lipoproteins contain a core of hydrophobic lipid molecules surrounded by a shell of hydrophilic molecules such as proteins and phospholipids (Figure 21.5). As summarized in Table 21.2, there are four kinds of lipoproteins:

Lipoproteins Spherically shaped clusters containing both lipid molecules and protein molecules

- **High-density lipoprotein (HDL) ("good cholesterol")**, which consists of about 33% protein and about 30% cholesterol

- **Low-density lipoprotein (LDL) ("bad cholesterol")**, which contains only 25% protein but 50% cholesterol

- **Very-low-density lipoprotein (VLDL)**, which mostly carries triglycerides (fats) synthesized by the liver

- **Chylomicrons**, which carry dietary lipids synthesized in the intestines

C. Transport of Cholesterol in LDL

The transport of cholesterol from the liver starts out as a large VLDL particle (55 nanometers in diameter). The core of VLDL contains triglycerides and cholesteryl esters, mainly cholesteryl linoleate. It is surrounded by a polar coat of phospholipids and proteins (Figure 21.5). The VLDL is carried in the serum. When the capillaries reach muscle or fat tissues, the triglycerides and all proteins except a protein called apoB-100 are removed from the VLDL. At this point, the diameter of the lipoprotein shrinks to 22 nanometers and its core contains only cholesteryl esters. Because of the removal of fat, its density increases and it becomes LDL. Low-density lipoprotein stays in the plasma for about 2.5 days.

The LDL carries cholesterol to the cells, where specific LDL-receptor molecules line the cell surface in certain concentrated areas called **coated pits**. The apoB-100 protein on the surface of the LDL binds specifically to

FIGURE 21.5 Low-density lipoprotein.

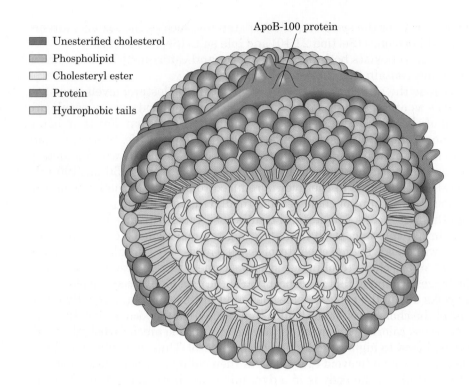

- ■ Unesterified cholesterol
- ▨ Phospholipid
- ▢ Cholesteryl ester
- ■ Protein
- ▨ Hydrophobic tails

ApoB-100 protein

Table 21.2 Compositions and Properties of Human Lipoproteins

Property	HDL	LDL	VLDL	Chylomicrons
Core				
Cholesterol and cholesteryl esters (%)	30	50	22	8
Triglycerides (%)	8	4	50	84
Surface				
Phospholipids (%)	29	21	18	7
Proteins (%)	33	25	10	1–2
Density (g/mL)	1.05–1.21	1.02–1.06	0.95–1.00	<0.95
Diameter (nm)	5–15	18–28	30–80	100–500

Percentages are given as % dry weight.

the LDL-receptor molecules in the coated pits. After such binding, the LDL is taken inside the cell (endocytosis), where enzymes break down the lipoprotein. In the process, they liberate free cholesterol from cholesteryl esters. In this manner, the cell can, for example, use cholesterol as a component of a membrane. This is the normal fate of LDL and the normal course of cholesterol transport. Michael Brown and Joseph Goldstein of the University of Texas shared the Nobel Prize in medicine in 1986 for the discovery of the LDL-receptor–mediated pathway. If the LDL receptors are not sufficient in number, cholesterol accumulates in the blood; this accumulation can happen even with low intake of dietary cholesterol. Both genetics and diet play a role in determining cholesterol levels in the blood.

D. Transport of Cholesterol in HDL

High-density lipoprotein transports cholesterol from peripheral tissues to the liver and transfers cholesterol to LDL. While in the serum, the free cholesterols in HDL are converted to cholesteryl esters. These esterified cholesterols are delivered to the liver for synthesis of bile acids and steroid hormones. The cholesterol uptake from HDL differs from that noted with LDL. This process does not involve endocytosis and degradation of the lipoprotein particle. Instead, in a selective lipid uptake, the HDL binds to the liver cell surface and transfers its cholesteryl ester to the cell. The HDL, depleted from its lipid content, then reenters the circulation. It is desirable to have a high level of HDL in the blood because of the way it removes cholesterol from the bloodstream.

E. Levels of LDL and HDL

Like all lipids, cholesterol is insoluble in water. If its level is elevated in the blood serum, plaque-like deposits may form on the inner surfaces of the arteries. The resulting decrease in the diameter of the blood vessels may, in turn, decrease the flow of blood. This **atherosclerosis**, along with accompanying high blood pressure, may lead to heart attack, stroke, or kidney dysfunction.

Atherosclerosis may exacerbate the blockage of some arteries by enabling a clot to form at the point where the arteries are constricted by plaque. Furthermore, blockage may deprive cells of oxygen, causing them to cease to function. The death of heart muscles due to lack of oxygen is called *myocardial infarction*.

Most cholesterol is transported by low-density lipoproteins. If a sufficient number of LDL receptors are found on the surface of the cells, LDL is effectively removed from the circulation and its concentration in the blood plasma drops. The number of LDL receptors on the surface of cells is controlled by a feedback mechanism (see Section 23.6). That is, when the concentration of cholesterol molecules inside the cells is high, the synthesis of the LDL receptor is suppressed. As a consequence, less LDL is taken into the cells from the plasma and the LDL concentration in the plasma rises. Conversely, when the cholesterol level inside the cells is low, the synthesis of the LDL receptor increases. As a consequence, the LDL is taken up more rapidly and its level in the plasma falls.

In certain cases, however, there are not enough LDL receptors. In the disease called *familial hypercholesterolemia*, the cholesterol level in the plasma may be as high as 680 mg/100 mL, compared to 175 mg/100 mL in normal subjects. These high levels of cholesterol can lead to premature atherosclerosis and heart attack. The high plasma cholesterol levels in affected patients occur because the body lacks enough functional LDL receptors, or if enough are produced, they are not concentrated in the coated pits.

In general, high LDL content means high cholesterol content in the plasma because LDL cannot enter the cells and be metabolized. Therefore, a high LDL level combined with a low HDL level is a symptom of faulty cholesterol transport and a warning for possible atherosclerosis.

The serum cholesterol level controls the amount of cholesterol synthesized by the liver. When serum cholesterol is high, synthesis is at a low level. Conversely, when the serum cholesterol level is low, synthesis of cholesterol increases.

Diets low in cholesterol and saturated fatty acids usually reduce the serum cholesterol level, and a number of drugs can inhibit the synthesis of cholesterol in the liver. Commonly used statin drugs such as atorvastatin (Lipitor) and simvastatin (Zocor) inhibit one of the key enzymes (biocatalysts) in cholesterol synthesis, HMG-CoA reductase (Section 29.4). In this way, they block the synthesis of cholesterol inside the cells and stimulate the synthesis of LDL-receptor proteins. More LDL then enters the cells, diminishing the amount of cholesterol that will be deposited on the inner walls of arteries.

It is generally considered desirable to have high levels of HDL and low levels of LDL in the bloodstream. High-density lipoproteins carry cholesterol from plaques deposited in the arteries to the liver, thereby reducing the risk of atherosclerosis. Premenopausal women have more HDL than men, which is why women have a lower risk of coronary heart disease. HDL levels can be increased by exercise and weight loss.

F. Membrane Cholesterol Functions

While decades of research have gone into establishing connections between excess cholesterol and health problems such as atherosclerosis, researchers are also discovering that cholesterol in the cell membrane affects other processes. Membrane cholesterol levels are controlled by the ratio of cholesterol synthesis and removal. Until recently, most attempts at regulating cholesterol levels have focused on inhibiting its synthesis, as described in the previous section. However, in 2010, researchers discovered a link between HDL, cholesterol, and the regulation of stem cell (Chapter 31) proliferation, and this link seems to be controlled by the removal of cholesterol from the cell membrane.

Many cell surface receptors are assembled into membrane "rafts" that have a high content of cholesterol and glycolipids. One example is the growth factor receptor for interleukin-3 (IL-3). When the receptor binds to the growth factor, IL-3, it promotes division and multiplication of many types of immune cells. The amount of cholesterol in the membrane near the IL-3 receptor influences the sensitivity of the receptor for IL-3. Cholesterol in the membranes is transported to HDL particles via ATP binding cassette (ABC) transporters, as shown in Figure 21.6, and two different versions of

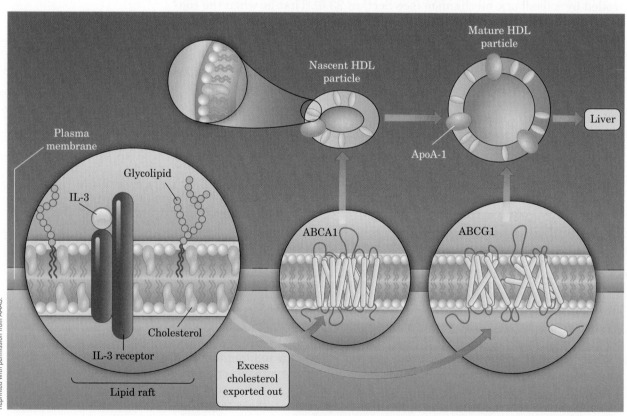

G. K. Hansson and M. Björkholm, "Tackling Two Diseases with HDL," Science, 328, p. 1641, (2010). Copyright © 2010 AAAS. Reprinted with permission from AAAS.

FIGURE 21.6 Interleukin 3 (IL-3) and its receptor are involved in cell-signaling pathways that lead to proliferation of many types of immune cells. The presence of cholesterol in the membrane near the IL-3 receptor affects sensitivity for IL-3. Two transporters, ABCA1 and ABCG1, transport cholesterol out of the membrane and into HDL particles. Lack of one or both of these transporters leads to proliferation of immune cells and can result in disease such as leukemia.

the transporters, ABCA1 and ABCG1, were discovered. Mutant mice lacking one or both of these transporters were found to have increased numbers of immune cells, such as neutrophils and monocytes, as well as increased hematopoietic stem cells, the common progenitor cells in bone marrow. The decreased removal of cholesterol from the membrane around the IL-3 receptor caused an oversensitivity of the response to IL-3, leading to an increased proliferation of the immune cells.

There are many known, often fatal, diseases, called myeloproliferative diseases, based on a patient having an overabundance of certain immune cell types, with the most well-known of these being leukemia. The results of these experiments were interesting for a couple of reasons. First, they showed the link between membrane cholesterol and cell receptor function. Second, the link between myeloproliferative diseases and cholesterol removal could offer doctors another weapon against these diseases. If cholesterol efflux can be enhanced, perhaps by stimulating the ABC proteins, then the cell proliferation could be slowed or stopped.

21.10 What Are Some of the Physiological Roles of Steroid Hormones?

Cholesterol is the starting material for the synthesis of steroid hormones. In this process, the aliphatic side chain on the D ring is shortened by the removal of a six-carbon unit and the secondary alcohol group on carbon 3 is oxidized to a ketone. The resulting molecule, *progesterone*, serves as the starting compound for both the sex hormones and the adrenocorticoid hormones (Figure 21.7).

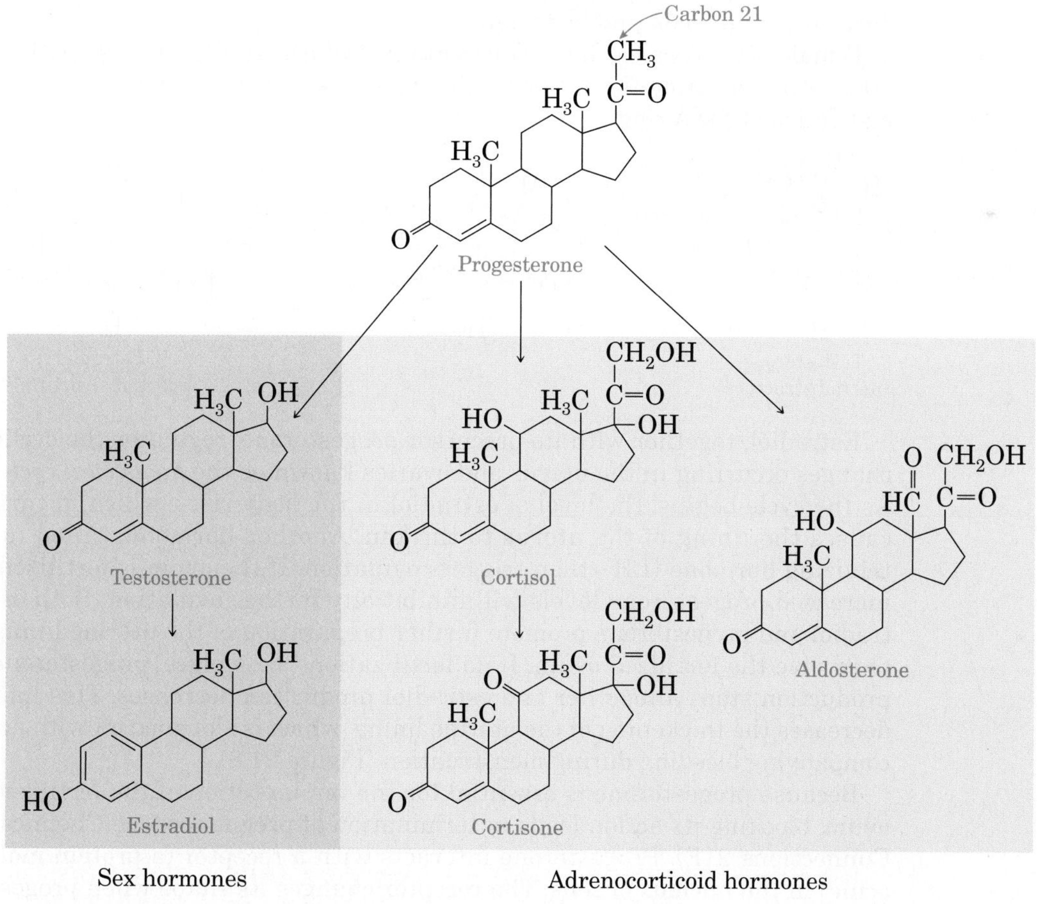

FIGURE 21.7 The biosynthesis of hormones from progesterone.

A. Adrenocorticoid Hormones

The adrenocorticoid hormones (Figure 21.7) are products of the adrenal glands. The term *adrenal* means "adjacent to the renal" (which refers to the kidney). We classify these hormones into two groups according to function: *Mineralocorticoids* regulate the concentrations of ions (mainly Na^+ and K^+), and *glucocorticoids* control carbohydrate metabolism. The term *corticoid* indicates that the site of the secretion is the cortex (outer part) of the gland.

Aldosterone is one of the most important mineralocorticoids. Increased secretion of aldosterone enhances the reabsorption of Na^+ and Cl^- ions in the kidney tubules and increases the loss of K^+. Because Na^+ concentration controls water retention in the tissues, aldosterone controls tissue swelling.

Cortisol is the major glucocorticoid. Its function is to increase the glucose and glycogen concentrations in the body. This accumulation occurs at the expense of other nutrients. Fatty acids from fat storage cells and amino acids from body proteins are transported to the liver, which, under the influence of cortisol, manufactures glucose and glycogen from these sources.

Cortisol and its ketone derivative *cortisone* have remarkable anti-inflammatory effects. These or similar synthetic derivatives, such as prednisolone, are used to treat inflammatory diseases of many organs, rheumatoid arthritis, and bronchial asthma.

B. Sex Hormones

The most important male sex hormone is testosterone (Figure 21.7). This hormone, which promotes the normal growth of the male genital organs, is synthesized in the testes from cholesterol. During puberty, increased testosterone production leads to such secondary male sexual characteristics as deep voice and facial and body hair.

Female sex hormones, the most important of which is estradiol (Figure 21.7), are synthesized from the corresponding male hormone (testosterone) by aromatization of the A ring:

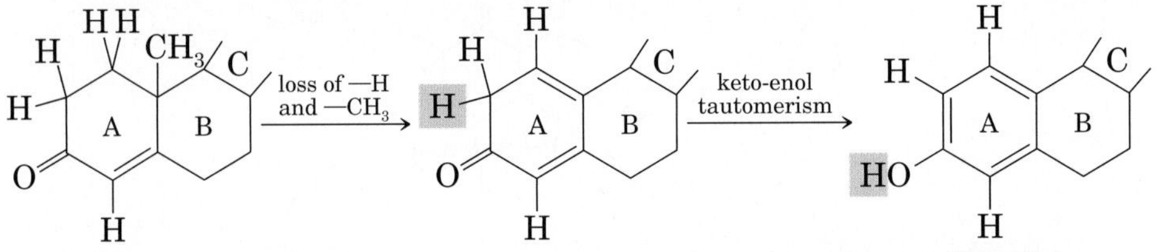

Testosterone
(partial structure)

Estradiol
(partial structure)

Estradiol, together with its precursor progesterone, regulates the cyclic changes occurring in the uterus and ovaries known as the *menstrual cycle*. As the cycle begins, the level of estradiol in the body rises, which in turn causes the lining of the uterus to thicken. Another hormone, called luteinizing hormone (LH), then triggers ovulation. If the ovum is fertilized, increased progesterone levels will inhibit any further ovulation. Both estradiol and progesterone promote further preparation of the uterine lining to receive the fertilized ovum. If no fertilization takes place, progesterone production stops altogether and estradiol production decreases. This halt decreases the thickening of the uterine lining, which is sloughed off with accompanying bleeding during menstruation (Figure 21.8).

Because progesterone is essential for the implantation of the fertilized ovum, blocking its action leads to termination of pregnancy (see Chemical Connections 21F). Progesterone interacts with a receptor (a protein molecule) in the nucleus of cells. The receptor changes its shape when progesterone binds to it (see Section 24.7).

Chemical Connections 21E

Anabolic Steroids

Testosterone, the principal male hormone, is responsible for the buildup of muscles in men. Recognizing this fact, many athletes have taken this drug in an effort to increase their muscular development. The practice is especially common among athletes in sports in which strength and muscle mass are important, including weightlifting, shot put, and hammer throw. Participants in other sports, such as running, swimming, and cycling, would also like larger and stronger muscles.

Although used by many athletes, testosterone has two disadvantages:

1. Besides its effect on muscles, it affects secondary sexual characteristics and too much of it can result in undesired side effects.
2. It is not very effective when taken orally and must be injected to achieve the best results.

For these reasons, a large number of other anabolic steroids, all of them synthetic, have been developed. Examples include the following compounds:

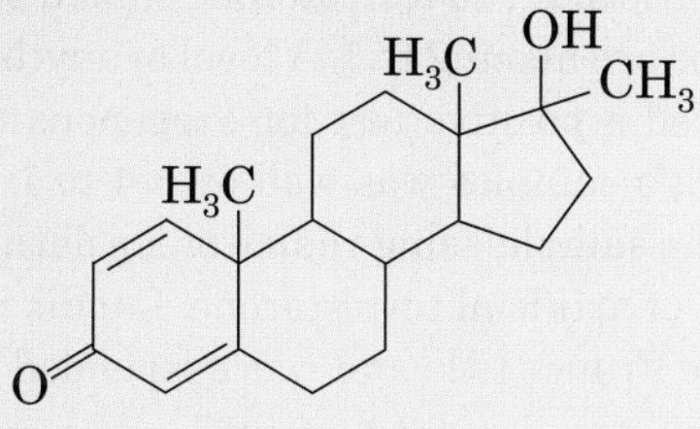

Methandienone

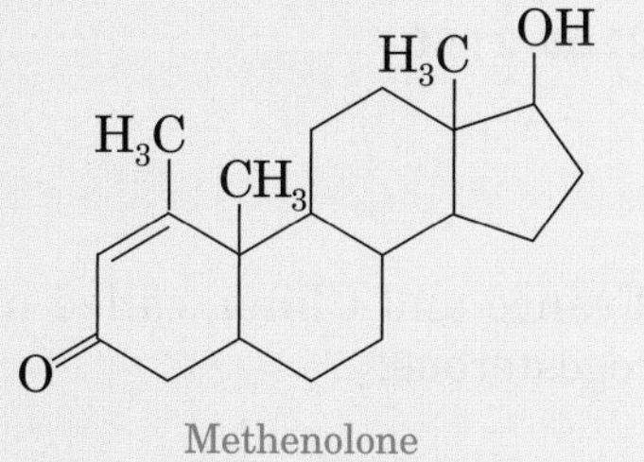

Methenolone

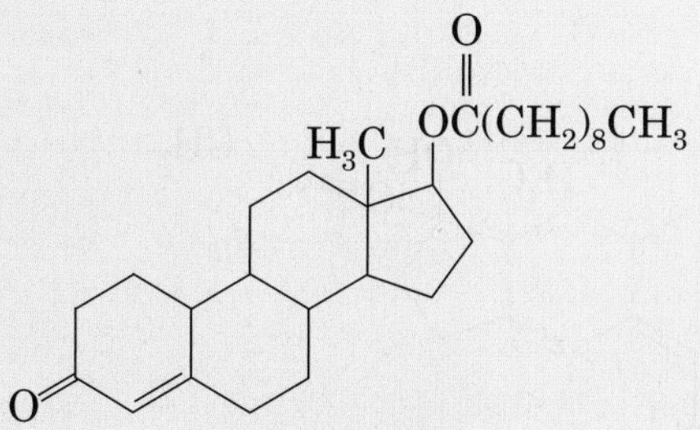

Nandrolone decanoate

Some women athletes use anabolic steroids, just as their male counterparts do. Because their bodies produce only small amounts of testosterone, women actually have much more to gain from anabolic steroids than men do.

Another way to increase testosterone concentration is to use prohormones, which the body converts to testosterone. One such prohormone is 4-androstenedione, or "andro." Some athletes have used it to enhance performance.

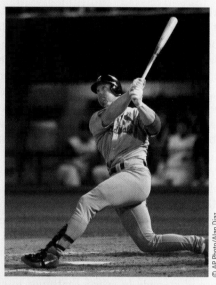

Former home run champion Mark McGwire used androstenedione, a muscle-building dietary supplement that is allowed in baseball but is banned in professional football, college athletics, and the Olympic Games.

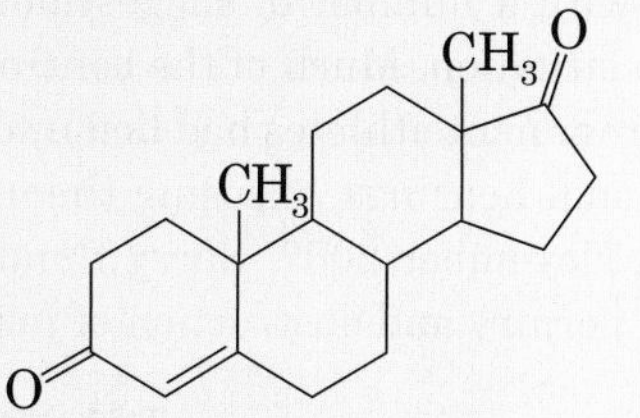

4-Androstene-3,17-dione

The use of anabolic steroids is forbidden in many sporting events, especially in international competition, largely for two reasons: (1) It gives competitors an unfair advantage, and (2) these drugs can have many unwanted and even dangerous side effects, ranging from acne to liver tumors. Side effects can be especially disadvantageous for women; they can include growth of facial hair, baldness, deepening of the voice, and menstrual irregularities.

Olympic sprinter Marion Jones went to prison for lying under oath about steriod use.

Chemical Connections 21E

Anabolic Steroids (*continued*)

All athletes participating in the Olympic Games are required to pass a urine test for anabolic steroids. A number of medal-winning athletes have had their victories taken away because they tested positive for steroid use. For example, the Canadian Ben Johnson, a world-class sprinter, was stripped of both his world record and his gold medal in the 1988 Olympiad. A positive test for andro resulted in the U.S. shot put champion Randy Barnes being banned from competition in 1998. Prohormones such as andro are not listed under the Anabolic Steroid Act of 1990; hence, their nonmedical use is not a federal offense, as is the case with anabolic steroids. Mark McGwire hit his record-breaking home runs in 1998 while taking andro, because baseball rules did not prohibit its usage. Even so, the International Olympic Committee has banned the use of both prohormones and anabolic steroids.

The use of steroids in sports continues to cause controversy. In early 2008, a commission led by former Senator George Mitchell announced that a number of baseball players had used steroids. Congressional hearings followed, along with a number of suggestions about how to deal with the situation. Much of the controversy centered on whether prominent athletes had lied under oath during the congressional hearings, exposing them to accusations of perjury. In November 2007, Barry Bonds was indicted on charges of perjury and obstruction of justice. As of May 2008, only one prominent athlete had been convicted of perjury in a high-profile case. In October 2007, former Olympic sprinter Marion Jones, winner of five medals at the 2000 games in Sydney, admitted to steroid use for a two-year period that included the Olympics. Before that, she had vehemently denied steroid use. She was sentenced to six months in prison for lying under oath and started her term in federal prison in March 2008.

While steroid use is usually associated with power sports, athletes in endurance sports also use steroids. In an endurance event like a marathon or long-distance cycling, use of steroids at low levels has the effect of aiding recovery. Floyd Landis was the third American to win the prestigious Tour de France, standing on the victor's podium in Paris in 2006. Unfortunately, a urine sample he gave after stage 17 of the race was found to have an abnormal ratio of epitestosterone (ET) to testosterone, the former of which is a precursor. This is a common initial test for someone taking testosterone. While testosterone levels can vary greatly in men, the ratio of ET to testosterone should be a constant. Normal ratios are about 2 to 1. A level of anything over 4 to 1 is considered a positive test for exogenous testosterone. One of Landis's samples was well over 4 to 1. Subsequent analysis of the sample using radioisotope dating confirmed the presence of artificial testosterone. Landis was stripped of his Tour de France title and was suspended from cycling for 2 years.

Test your knowledge with Problems 21.58–21.60.

A drug, now widely used in France and China, called mifepristone or RU486 acts as a competitor to progesterone:

Mifepristone
(RU486)

Mifepristone blocks the action of progesterone by binding to the same receptor sites. Because the progesterone molecule cannot reach the receptor molecule, the uterus is not prepared for the implantation of the fertilized ovum,

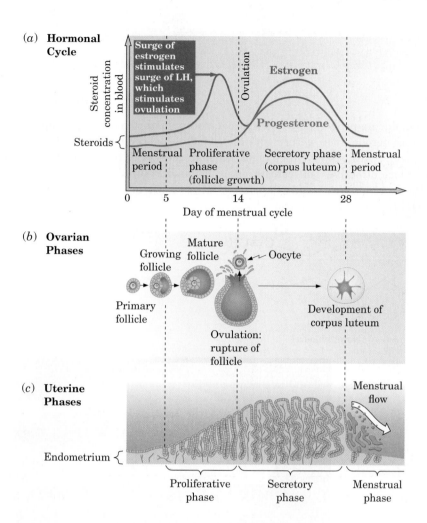

(a) **Hormonal Cycle**

Surge of estrogen stimulates surge of LH, which stimulates ovulation

Ovulation

Estrogen

Progesterone

Steroids {

Menstrual period | Proliferative phase (follicle growth) | Secretory phase (corpus luteum) | Menstrual period

0   5   14   28

Day of menstrual cycle

(b) **Ovarian Phases**

Growing follicle | Mature follicle | Oocyte

Primary follicle

Ovulation: rupture of follicle

Development of corpus luteum

(c) **Uterine Phases**

Menstrual flow

Endometrium {

Proliferative phase | Secretory phase | Menstrual phase

FIGURE 21.8 Events of the menstrual cycle. (a) Levels of sex hormones in the bloodstream during the phases of one menstrual cycle in which pregnancy does not occur. (b) Development of an ovarian follicle during the cycle. (c) Phases of development of the endometrium, the lining of the uterus. The endometrium thickens during the proliferative phase. In the secretory phase, which follows ovulation, the endometrium continues to thicken and the glands secrete a glycogen-rich nutritive material in preparation to receive an embryo. If no embryo is implanted, the new outer layers of the endometrium disintegrate and the blood vessels rupture, producing the menstrual flow.

and the ovum is aborted. Once pregnancy has been established, RU486 can be taken up through 49 days of gestation. This chemical form of abortion has been approved by the U.S. Food and Drug Administration (FDA) and in recent years has found clinical application as a supplement to surgical abortion. RU486 binds to the receptors of glucocorticoid hormones as well. Its use as an antiglucocorticoid is also recommended to alleviate a disease known as Cushing's syndrome, involving the overproduction of cortisone.

A completely different approach is the "morning after pill" (emergency contraceptive pill [ECP]), which can be taken orally up to 72 hours after unprotected intercourse. The "morning after pill" is not an abortion pill, because it acts before pregnancy takes place. Actually, the components of the pill are regular contraceptives. Two kinds are on the market as prescription drugs: a progesterone-like compound, called levonorgestrel, and a combination of levonorgestrel and ethinyl estradiol marketed as Preven.

Estradiol and progesterone also regulate secondary female sex characteristics, such as the growth of breasts. Thanks to this property, RU486, as an antiprogesterone, has been reported to be effective against certain types of breast cancer.

Testosterone and estradiol are not exclusive to either males or females. A small amount of estradiol production occurs in males, and a small amount of testosterone production is normal in females. Only when the proportion of these two hormones (hormonal balance) becomes upset can one observe symptoms of abnormal sexual differentiation.

Chemical Connections 21F

Oral Contraception

Because progesterone prevents ovulation during pregnancy, it occurred to investigators that progesterone-like compounds might be used for birth control. Synthetic analogs of progesterone proved to be more effective than the natural compound because they had a much longer half-life, with more potent effects. In "the Pill,"

a synthetic progesterone-like compound is supplied together with an estradiol-like compound (the latter prevents irregular menstrual flow). Triple-bond derivatives of testosterone, such as norethindrone, norethynodrel, and ethynodiol diacetate, are used most often in birth-control pills:

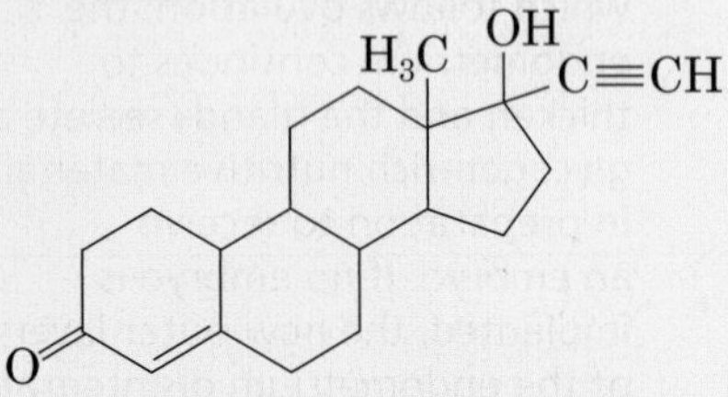

Norethindone

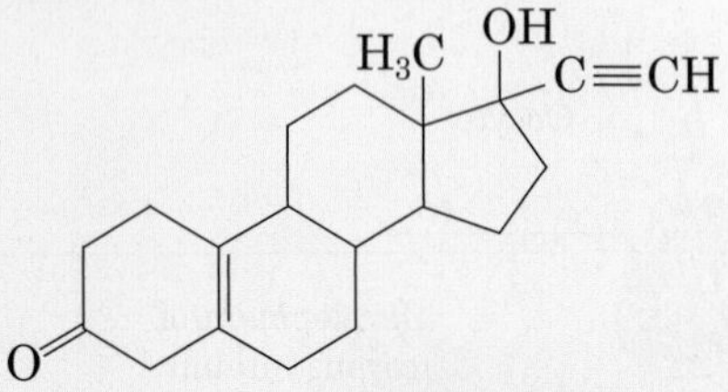

Norethynodrel

Ethynodiol diacetate

The packaging of oral contraceptives is a reminder of their daily use.

Test your knowledge with Problem 21.61.

21.11 What Are Bile Salts?

Bile salts are oxidation products of cholesterol. First, the cholesterol is oxidized to the trihydroxy derivative, and the end of the aliphatic chain is oxidized to the carboxylic acid. The latter, in turn, forms an amide bond with an amino acid, either glycine or taurine:

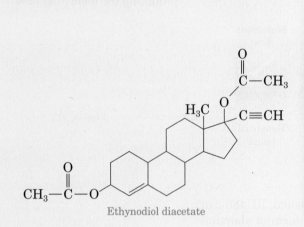

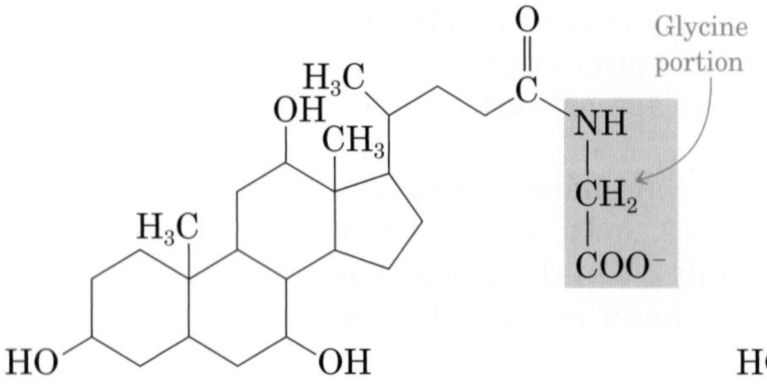

Glycocholate

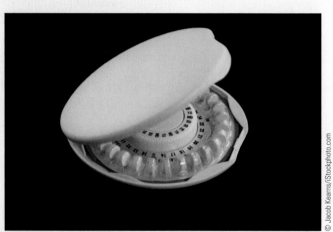

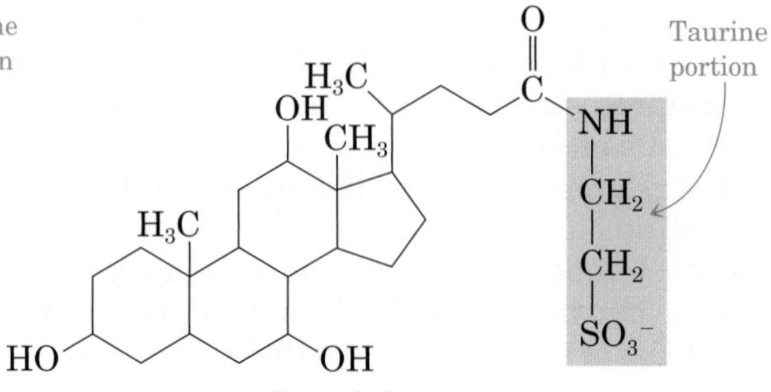

Taurocholate

Taurine has developed a certain amount of commercial importance in recent years as an ingredient in sports drinks. The drink marketed under the trade name Red Bull (*taurus* is the Latin word for "bull") contains various sugars (Chapter 20), caffeine, and B vitamins (Section 30.6) in addition to taurine.

Bile salts are powerful detergents. One end of the molecule is strongly hydrophilic because of the negative charge, and the rest of the molecule is largely hydrophobic. As a consequence, bile salts can disperse dietary lipids in the small intestine into fine emulsions, thereby facilitating digestion. The dispersion of dietary lipids by bile salts is similar to the action of soap on dirt.

Because they are eliminated in the feces, bile salts remove excess cholesterol in two ways: (1) They are themselves breakdown products of cholesterol (so cholesterol is eliminated via bile salts), and (2) they solubilize deposited cholesterol in the form of bile salt–cholesterol particles.

21.12 What Are Prostaglandins, Thromboxanes, and Leukotrienes?

Prostaglandins, a group of fatty-acid-like substances, were discovered by Kurzrok and Leib in the 1930s, when they demonstrated that seminal fluid caused a hysterectomized uterus to contract. Ulf von Euler of Sweden (winner of the Nobel Prize in physiology and medicine in 1970) isolated these compounds from human semen and, thinking that they had come from the prostate gland, named them **prostaglandin**. Even though the seminal gland secretes 0.1 mg of prostaglandin per day in mature males, small amounts of prostaglandins are present throughout the body in both sexes.

Prostaglandins are synthesized in the body from arachidonic acid by a ring closure at carbons 8 and 12. The enzyme catalyzing this reaction is called **cyclooxygenase** (COX for short). The product, known as PGG$_2$, is the common precursor of other prostaglandins, including PGE and PGF. The prostaglandin E group (PGE) has a carbonyl group at carbon 9; the subscript indicates the number of double bonds in the hydrocarbon chain. The prostaglandin F group (PGF) has two hydroxyl groups on the ring at carbons 9 and 11. Other prostaglandins (PGAs and PGBs) are derived from PGE.

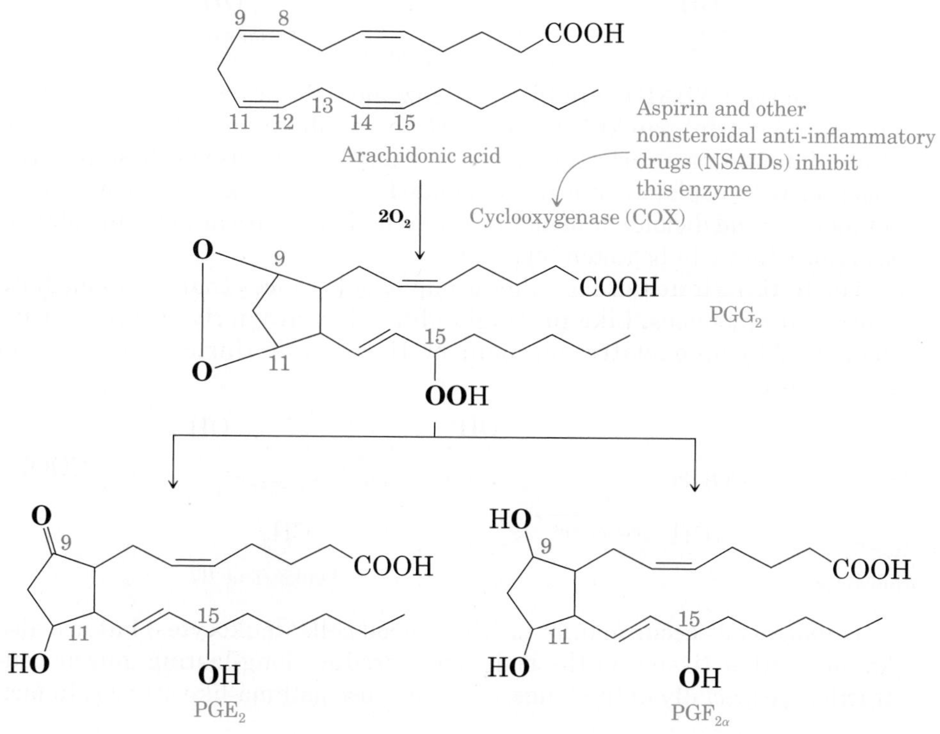

The COX enzyme comes in two forms in the body: COX-1 and COX-2. COX-1 catalyzes the normal physiological production of prostaglandins, which are always present in the body. For example, PGE_2 and $PGF_{2\alpha}$ stimulate uterine contractions and induce labor. PGE_2 lowers blood pressure by relaxing the smooth muscles cells inside blood vessels. In aerosol form, this prostaglandin is used to treat asthma; it opens up the bronchial tubes by relaxing the surrounding muscles. PGE_1 is used as a decongestant; it opens up nasal passages.

COX-2, by contrast, produces prostaglandins in response to inflammation. When a tissue is injured or damaged, special inflammatory cells invade the injured tissue and interact with resident cells—for example, smooth muscle cells. This interaction activates the COX-2 enzyme, and prostaglandins are synthesized. Such tissue injury may occur in a heart attack (myocardial infarction), rheumatic arthritis, and ulcerative colitis. Nonsteroidal anti-inflammatory drugs (NSAIDs) such as aspirin inhibit both COX enzymes (see Chemical Connections 21G).

Another class of arachidonic acid derivatives is the **thromboxanes**. Their synthesis also includes a ring closure. These substances are derived from PGH_2, but their ring is a cyclic acetal. Thromboxane is known to induce platelet aggregation. When a blood vessel is ruptured, the first line of defense is the platelets circulating in the blood, which form an incipient clot. Thromboxane A_2 causes other platelets to clump, thereby increasing the size of the blood clot. Aspirin and similar anti-inflammatory agents inhibit the COX enzyme. Consequently, PGH_2 and thromboxane synthesis is inhibited, and blood clotting is impaired. This effect has prompted many physicians to recommend a daily dose of 81 mg aspirin for people at risk for heart attack or stroke. It also explains why physicians forbid patients to use aspirin and other anti-inflammatory agents for a week before a planned surgery—aspirin and other NSAIDs may cause excessive bleeding.

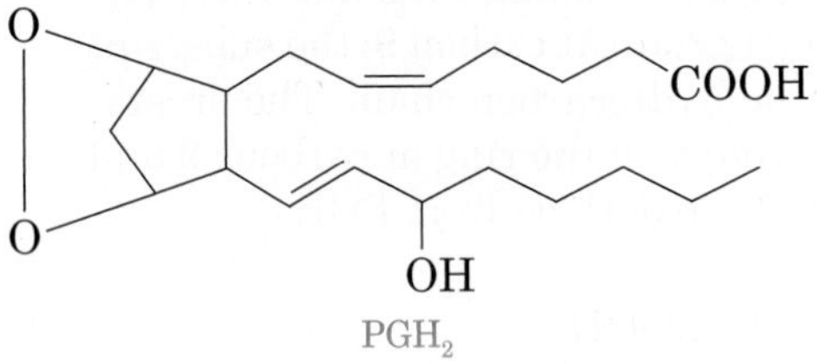

PGH$_2$ 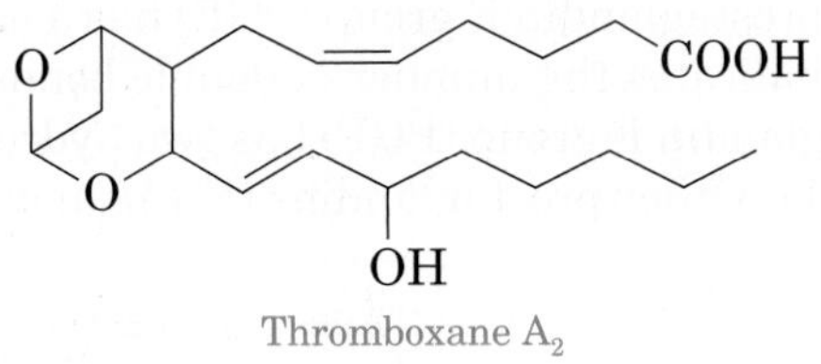   Thromboxane A$_2$

A variety of NSAIDs inhibit COX enzymes. Ibuprofen and indomethacin, both powerful painkillers, can block the inhibitory effect of aspirin and thus eliminate its anticlotting benefits. Therefore, the use of these NSAIDs together with aspirin is not recommended. Other painkillers such as acetaminophen and diclofenac do not interfere with aspirin's anticlotting ability; therefore, they can be taken together.

The **leukotrienes** are another group of substances that act to mediate hormonal responses. Like prostaglandins, they are derived from arachidonic acid by an oxidative mechanism. However, in this case, there is no ring closure.

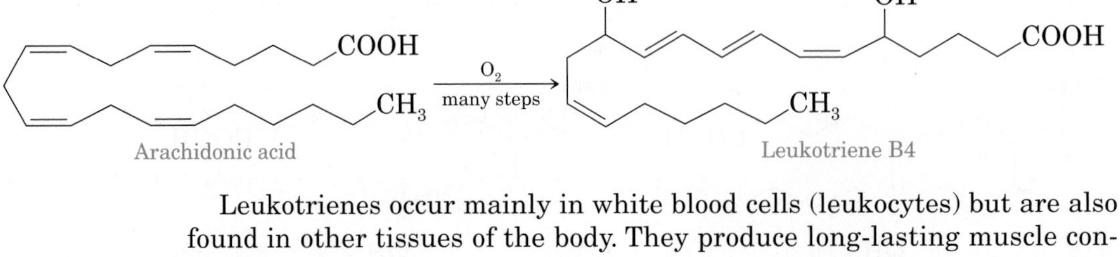

Arachidonic acid   Leukotriene B4

Leukotrienes occur mainly in white blood cells (leukocytes) but are also found in other tissues of the body. They produce long-lasting muscle contractions, especially in the lungs, and can cause asthma-like attacks. In fact,

Action of Anti-inflammatory Drugs

Anti-inflammatory steroids (such as cortisone; Section 21.10) exert their function by inhibiting phospholipase A_2, the enzyme that releases unsaturated fatty acids from complex lipids in the membranes. For example, arachidonic acid, one of the components of membranes, is made available to the cell through this process. Because arachidonic acid is the precursor of prostaglandins, thromboxanes, and leukotrienes, inhibiting its release stops the synthesis of these compounds and prevents inflammation.

Steroids such as cortisone are associated with many undesirable side effects (duodenal ulcer and cataract formation, among others). Therefore, their use must be controlled. A variety of nonsteroidal anti-inflammatory agents, including aspirin, ibuprofen, ketoprofen, and indomethacin, are available to serve this function.

Aspirin and other NSAIDs (see Chemical Connections 19C) inhibit the cyclooxygenase enzymes, which synthesize prostaglandins and thromboxanes. Aspirin (acetylsalicylic acid), for example, acetylates the enzymes, thereby blocking the entrance of arachidonic acid to the active site. This inhibition of both COX-1 and COX-2 explains why aspirin and the other anti-inflammatory agents have undesirable side effects. NSAIDs also interfere with the COX-1 isoform of the enzyme, which is needed for normal physiological function. Their side effects include stomach and duodenal ulceration and renal (kidney) toxicity.

Obviously, it would be desirable to have an anti-inflammatory agent without such side effects, and one that inhibits only the COX-2 isoform. To date, the FDA has approved two COX-2 inhibitor drugs: Celebrex, which quickly became the most frequently prescribed drug, and Vioxx, a more recent entrant. Despite their selective inhibition of COX-2, however, these drugs also have ulcer-causing side effects. Many other COX-2 inhibitors remain in the clinical trial stage of development.

The use of COX-2 inhibitors is not limited to rheumatoid arthritis and osteoarthritis. Celebrex has been approved by the FDA to treat a type of colon cancer called familial adenomateous polyposis, in an approach called chemoprevention. All anti-inflammatory agents reduce pain and relieve fever and swelling by reducing the prostaglandin production, but they do not affect the leukotriene production. As a consequence, asthmatic patients must beware of using these anti-inflammatory agents. Even though they inhibit the prostaglandin synthesis, these drugs may shift the available arachidonic acid to leukotriene production, which could precipitate a severe asthma reaction.

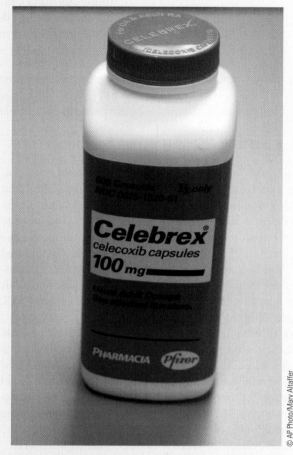

Packages of the NSAID Celebrex contain a warning label.

During the fall of 2004, studies indicated that high doses of Vioxx correlated with a higher incidence of heart attacks and strokes; concerns were also raised about other COX-2 inhibitors, particularly Celebrex. The inhibition of prostaglandin synthesis allows formation of other lipids, including those that build up in atherosclerotic plaque. Vioxx was taken off the U.S. market, and some physicians became hesitant to prescribe Celebrex. These events caused consternation among patients who had come to depend on these drugs, as well as among the pharmaceutical companies that produced them. In February 2005, a panel looked into the controversy for the FDA. This group concluded that COX-2 inhibitors should stay on the market, but their use should be strictly monitored. Warning labels must now appear on the packaging for these drugs.

Test your knowledge with Problems 21.62–21.65.

they are 100 times more potent than histamines. Both prostaglandins and leukotrienes cause inflammation and fever, so the inhibition of their production in the body is a major pharmacological concern. One way to counteract the effects of leukotrienes is to inhibit their uptake by leukotriene receptors (LTRs) in the body. A new antagonist of LTRs, zafirlukast (brand name Accolate), is used to treat and control chronic asthma. Another anti-asthmatic drug, zileuton (brand name Zyflo), inhibits 5-lipoxygenase, which is the initial enzyme in leukotriene biosynthesis from arachidonic acid.

Chemical Connections 21H

Why Should We Eat More Salmon?

Platelets are elements in the blood that initiate blood clotting and tissue repair by releasing clotting factors and platelet-derived growth factor (PDGF). Turbulence in the bloodstream may cause platelets to rupture. Fat deposits and bifurcations of arteries lead to such turbulence, so platelets and PDGF are implicated in blood clotting and the growth of atherosclerotic plaque. Furthermore, the anaerobic conditions that exist under a large plaque deposit may lead to weakness and dead cells in the arterial wall, aggravating the problem.

In cultures that depend on fish as a major food source, including some Eskimo tribes, very little heart disease is diagnosed even though people in these groups eat high-fat diets and have high levels of blood cholesterol. Analysis of their diet led to the discovery that certain highly unsaturated fatty acids are found in the oils of fish and diving mammals. One class of these fatty acids is called omega-3 ($\omega 3$), an example of which is eicosapentaenoic acid (EPA).

$$CH_3CH_2(CH=CHCH_2)_5(CH_2)_2COOH$$
Eicosapentaenoic acid (EPA)

Note the presence of a double bond at the third carbon atom from the end of the hydrocarbon tail. The omega system of nomenclature is based on numbering the double bonds from the last carbon in the fatty acid instead of the carbonyl group (the delta Δ system). Omega is the last letter in the Greek alphabet.

The omega-3 fatty acids inhibit the formation of certain prostaglandins and thromboxane A, which is similar in structure to prostaglandins. Thromboxane released by ruptured arteries causes other platelets to clump in the

immediate area and increases the size of the blood clot. Any disruption in thromboxane synthesis results in a lower tendency to form blood clots and thus in a lower potential for arterial damage.

Raw salmon is very popular in sushi restaurants.

Test your knowledge with Problems 21.66 and 21.67.

Summary

⬛WL Sign in at **www.cengage.com/owl** to develop problem-solving skills and complete online homework assigned by your professor.

Section 21.1 What Are Lipids?

- **Lipids** are water-insoluble substances.
- Lipids are classified into four groups: fats (triglycerides); complex lipids; steroids; and prostaglandins, thromboxanes, and leukotrienes.

Section 21.2 What Are the Structures of Triglycerides?

- **Fats** consist of fatty acids and glycerol. In saturated fatty acids, the hydrocarbon chains have only single bonds; unsaturated fatty acids have hydrocarbon chains with one or more double bonds, all in the *cis* configuration.

Section 21.3 What Are Some Properties of Triglycerides?

- Solid fats contain mostly saturated fatty acids, whereas **oils** contain substantial amounts of unsaturated fatty acids.
- The alkali salts of fatty acids are called **soaps**.

Section 21.4 What Are the Structures of Complex Lipids?

- **Complex lipids** can be classified into two groups: phospholipids and glycolipids.
- **Phospholipids** are made of a central alcohol (glycerol or sphingosine), fatty acids, and a nitrogen-containing phosphate ester such as phosphorylcholine or inositol phosphate.
- **Glycolipids** contain sphingosine and a fatty acid, collectively known as the ceramide portion of the molecule, and a carbohydrate portion.

Section 21.5 What Role Do Lipids Play in the Structure of Membranes?

- Many phospholipids and glycolipids are important components of cell **membranes.**
- Membranes are made of a **lipid bilayer** in which the hydrophobic parts of phospholipids (fatty acid residues) point toward the middle of the bilayer and the hydrophilic parts point toward the inner and outer surfaces of the membrane.

Section 21.6 What Are Glycerophospholipids?

- **Glycerophospholipids** are complex lipids that consist of a central glycerol moiety to which two fatty acids are esterified. The third alcohol group of the glycerol is esterified to a nitrogen-containing phosphate ester.

Section 21.7 What Are Sphingolipids?

- **Sphingolipids** are complex lipids that consist of the long-chain alcohol sphingosine esterified to a fatty acid (the ceramide moiety). Nitrogen-containing phosphate esters are also bonded to the sphingosine moiety.

Section 21.8 What Are Glycolipids?

- **Glycolipids** are complex lipids that consist of two parts: a ceramide portion and a carbohydrate portion.

Section 21.9 What Are Steroids?

- The third major group of lipids comprises the **steroids**. The characteristic feature of the steroid structure is a fused four-ring nucleus.
- The most common steroid, **cholesterol**, serves as a starting material for the synthesis of other steroids, such as bile salts and sex and other hormones. Cholesterol is also an integral part of membranes, occupying the hydrophobic region of the lipid bilayer. Because of their low solubility in water, cholesterol deposits are implicated in the formation of gallstones and the plaque-like deposits of atherosclerosis.
- Cholesterol is transported in the blood plasma mainly by two kinds of lipoproteins: **HDL** and **LDL**. LDL delivers cholesterol to the cells to be used mostly as a membrane component. HDL delivers cholesteryl esters mainly to the liver to be used in the synthesis of bile acids and steroid hormones.
- High levels of LDL and low levels of HDL are symptoms of faulty cholesterol transport, indicating greater risk of atherosclerosis.

Section 21.10 What Are Some of the Physiological Roles of Steroid Hormones?

- An oxidation product of cholesterol is progesterone, a **sex hormone**. It also gives rise to the synthesis of other sex hormones, such as testosterone and estradiol.
- Progesterone is also a precursor of the **adrenocorticoid hormones**. Within this group, cortisol and cortisone are best known for their anti-inflammatory action.

Section 21.11 What Are Bile Salts?

- **Bile salts** are oxidation products of cholesterol. They emulsify all kinds of lipids, including cholesterol, and are essential in the digestion of fats.

Section 21.12 What Are Prostaglandins, Thromboxanes, and Leukotrienes?

- **Prostaglandins, thromboxanes**, and **leukotrienes** are derived from arachidonic acid. They have a wide variety of effects on body chemistry. Among other things, they can lower or raise blood pressure, cause inflammation and blood clotting, and induce labor. In general, they mediate hormone action.

Problems

OWL Interactive versions of these problems may be assigned in OWL.

Orange-numbered problems are applied.

Section 21.1 What Are Lipids?

21.2 Why are fats a good source of energy for storage in the body?

21.3 Proteins, nucleic acids, and carbohydrates are grouped by common structural features found within their group. What is the basis for grouping substances as lipids?

Section 21.2 What Are the Structures of Triglycerides?

21.4 Draw the structural formula of a fat molecule (triglyceride) made of myristic acid, oleic acid, palmitic acid, and glycerol.

21.5 Oleic acid has a melting point of 16°C. If you converted the *cis* double bond into a *trans* double bond, what would happen to the melting point? Explain.

21.6 Draw schematic formulas for all possible 1,3-diglycerides made up of glycerol, oleic acid, or stearic acid. How many are there? Draw the structure of one of the diglycerides.

Section 21.3 What Are Some Properties of Triglycerides?

21.7 For the diglycerides in Problem 21.6, predict which two will have the highest melting points and which two will have the lowest melting points.

21.8 Predict which acid in each pair has the higher melting point and explain why.
(a) Palmitic acid or stearic acid
(b) Arachidonic acid or arachidic acid

21.9 Which has the higher melting point: (a) a triglyceride containing only lauric acid and glycerol or (b) a triglyceride containing only stearic acid and glycerol?

21.10 Explain why the melting points of the saturated fatty acids increase as we move from lauric acid to stearic acid.

21.11 Predict the order of the melting points of triglycerides containing fatty acids, as follows:
(a) Palmitic, palmitic, stearic
(b) Oleic, stearic, palmitic
(c) Oleic, linoleic, oleic

21.12 Look at Table 21.1. Which animal fat has the highest percentage of unsaturated fatty acids?

21.13 Rank the following in order of increasing solubility in water (assuming that all are made with the same fatty acids): (a) triglycerides, (b) diglycerides, and (c) monoglycerides. Explain your answer.

21.14 How many moles of H_2 are used up in the catalytic hydrogenation of one mole of a triglyceride containing glycerol, palmitic acid, oleic acid, and linoleic acid?

21.15 Name the products of the saponification of this triglyceride:

$$CH_2-O-\overset{\overset{O}{\|}}{C}-(CH_2)_{14}CH_3$$
$$CH-O-\overset{\overset{O}{\|}}{C}-(CH_2)_{16}CH_3$$
$$CH_2-O-\overset{\overset{O}{\|}}{C}-(CH_2)_7(CH=CHCH_2)_3CH_3$$

21.16 Using the equation in Section 21.3C as a guideline for stoichiometry, calculate the number of moles of NaOH needed to saponify 5 mol of (a) triglycerides, (b) diglycerides, and (c) monoglycerides.

Section 21.4 What Are the Structures of Complex Lipids?

21.17 What are the main types of complex lipids, and what are the main characteristics of their structures?

Section 21.5 What Role Do Lipids Play in the Structure of Membranes?

21.18 Which portion of the phosphatidylinositol molecule contributes to (a) the fluidity of the bilayer and (b) the surface polarity of the bilayer?

21.19 How do the unsaturated fatty acids of the complex lipids contribute to the fluidity of a membrane?

21.20 Which type of lipid molecule is most likely to be present in membranes?

21.21 What is the difference between an integral and a peripheral membrane protein?

Section 21.6 What Are Glycerophospholipids?

21.22 Which glycerophospholipid has the most polar groups capable of forming hydrogen bonds with water?

21.23 Draw the structure of a phosphatidylinositol that contains oleic acid and arachidonic acid.

21.24 Among the glycerophospholipids containing palmitic acid and linolenic acid, which will have the greatest solubility in water: (a) phosphatidylcholine, (b) phosphatidylethanolamine, or (c) phosphatidylserine? Explain.

Section 21.7 What Are Sphingolipids?

21.25 Name all the groups of complex lipids that contain ceramides.

21.26 Are the various phospholipids randomly distributed in membranes? Give an example.

Section 21.8 What Are Glycolipids?

21.27 Enumerate the functional groups that contribute to the hydrophilic character of (a) glucocerebroside and (b) sphingomyelin.

Section 21.9 What Are Steroids?

21.28 Cholesterol has a fused four-ring steroid nucleus and is a part of body membranes. The —OH group on carbon 3 is the polar head, and the rest of the molecule provides the hydrophobic tail that does not fit into the zig-zag packing of the hydrocarbon portion of the saturated fatty acids. Considering this structure, predict whether small amounts of cholesterol that are well dispersed in the membrane will contribute to the stiffening (rigidity) or the fluidity of a membrane. Explain.

21.29 Where can pure cholesterol crystals be found in the body?

21.30 (a) Find all of the carbon stereocenters in a cholesterol molecule.

    (b) How many total stereoisomers are possible?

    (c) How many of these stereoisomers do you think are found in nature?

21.31 Look at the structures of cholesterol and the hormones shown in Figure 21.6. Which ring of the steroid structure undergoes the most substitution?

21.32 What makes LDL soluble in blood plasma?

21.33 How does LDL deliver its cholesterol to the cells?

21.34 How does lovastatin reduce the severity of atherosclerosis?

21.35 How does VLDL become LDL?

21.36 How does HDL deliver its cholesteryl esters to liver cells?

21.37 How does the serum cholesterol level control both cholesterol synthesis in the liver and LDL uptake?

Section 21.10 What Are Some of the Physiological Roles of Steroid Hormones?

21.38 What physiological functions are associated with cortisol?

21.39 Estradiol in the body is synthesized starting from progesterone. What chemical modifications occur when estradiol is synthesized?

21.40 Describe the difference in structure between the male hormone testosterone and the female hormone estradiol.

21.41 Considering that RU486 can bind to the receptors of progesterone as well as to the receptors of cortisone and cortisol, what can you say regarding the importance of the functional group on carbon 11 of the steroid ring in drug and receptor binding?

21.42 (a) How does the structure of RU486 resemble that of progesterone?

    (b) How do the two structures differ?

21.43 What are the structural features common to oral contraceptive pills, including mifepristone?

Section 21.11 What Are Bile Salts?

21.44 List all of the functional groups that make taurocholate water-soluble.

21.45 Explain how the constant elimination of bile salts through the feces can reduce the danger of plaque formation in atherosclerosis.

Section 21.12 What Are Prostaglandins, Thromboxanes, and Leukotrienes?

21.46 What is the basic structural difference between:

    (a) Arachidonic acid and prostaglandin PGE_2?

    (b) PGE_2 and $PGF_{2\alpha}$?

21.47 Find and name all of the functional groups in (a) glycocholate, (b) cortisone, (c) prostaglandin PGE_2, and (d) leukotriene B4.

21.48 What are the chemical and physiological functions of the COX-2 enzyme?

21.49 How does aspirin, an anti-inflammatory drug, prevent strokes caused by blood clots in the brain?

Chemical Connections

21.50 (Chemical Connections 21A) What do plants and animals use waxes for?

21.51 (Chemical Connections 21A) What makes waxes harder and more difficult to melt than fats?

21.52 (Chemical Connections 21B) How do the gap junctions prevent the passage of proteins from cell to cell?

21.53 (Chemical Connections 21B) How does the anion transporter provide a suitable environment for the passage of hydrated chloride ions?

21.54 (Chemical Connections 21B) In what sense is the active transport of K^+ selective? How does K^+ pass through the transporter?

21.55 (Chemical Connections 21C) What is the proposed link between carbon chain-length in ceramides and the response to hypoxia? Why is this link important to human health?

21.56 (Chemical Connections 21D) Compare the complex lipid structures listed for the lipid storage diseases with missing or defective enzymes. Explain why the missing enzyme in Fabry's disease is α-galactosidase and not β-galactosidase.

21.57 (Chemical Connections 21D) Identify the monosaccharides in the accumulating glycolipid of Fabry's disease.

21.58 (Chemical Connections 21E) How does the oral anabolic steroid methenolone differ structurally from testosterone?

21.59 (Chemical Connections 21E) In what ways do athletes use steroids to enhance athletic performance?

21.60 (Chemical Connections 21E) Why were Mark McGwire and Floyd Landis not given the same penalties for taking steroids in their sports?

21.61 (Chemical Connections 21F) What is the role of progesterone and similar compounds in contraceptive pills?

21.62 (Chemical Connections 21G) How does cortisone prevent inflammation?

21.63 (Chemical Connections 21G) How does indomethacin act in the body to reduce inflammation?

21.64 (Chemical Connections 21G) What kind of prostaglandins are synthesized by COX-1 and COX-2 enzymes?

21.65 (Chemical Connections 21G) Steroids prevent asthma-causing leukotriene synthesis, as well as inflammation-causing prostaglandin synthesis. Nonsteroidal anti-inflammatory agents (NSAIDs) such as aspirin reduce only prostaglandin production. Why do NSAIDs not affect leukotriene production?

21.66 (Chemical Connections 21H) Define omega-3 fatty acid.

21.67 (Chemical Connections 21H) Why is very little heart disease found among people who eat a lot of fish?

Additional Problems

21.68 What is the role of taurine in lipid digestion?

21.69 Draw a schematic diagram of a lipid bilayer. Show how the bilayer prevents the passage by diffusion of a polar molecule such as glucose. Show why non-polar molecules such as CH_3CH_2—O—CH_2CH_3 can diffuse through the membrane.

21.70 How many different triglycerides can you create using three different fatty acids (A, B, and C) in each case?

21.71 Prostaglandins have a five-membered ring closure; thromboxanes have a six-membered ring closure. The synthesis of both groups of compounds is prevented by COX inhibitors; the COX enzymes catalyze ring closure. How can these facts be correlated?

21.72 Which lipoprotein is instrumental in removing the cholesterol deposited in the plaque on arteries?

21.73 What are coated pits? What is their function?

21.74 What are the constituents of sphingomyelin?

21.75 (Chemical Connections 21B) What is the difference between a facilitated transporter and an active transporter?

21.76 Which part of LDL interacts with the LDL receptor?

21.77 What is the major difference between aldosterone and the other hormones listed in Figure 21.6?

21.78 (Chemical Connections 21G) The anti-inflammatory drug Celebrex does not have the usual side effect of stomach upset or ulceration commonly observed with the other NSAIDs. Why?

21.79 How many grams of H_2 are needed to saturate 100.0 g of a triglyceride made of glycerol and one unit each of lauric, oleic, and linoleic acids?

21.80 Prednisolone is the synthetic glucocorticoid medicine most frequently prescribed to combat autoimmune diseases. Compare its structure to the natural glucocorticoid hormone cortisone. What are the similarities and differences in structure?

21.81 You have just isolated a pure lipid that contains only sphingosine and a fatty acid. To what class of lipid does it belong?

21.82 Suggest a reason why the same protein system moves both sodium and potassium ions into and out of the cell.

21.83 Do all proteins associated with membranes span the membrane from one side to the other? Explain.

21.84 In the preparation of sauces that involve mixing water and melted butter, egg yolks are added to prevent separation. How do the egg yolks prevent separation? (*Hint:* Egg yolks are rich in phosphatidylcholine [lecithin].)

21.85 Which of the following statements is (are) consistent with what is known about membranes?

(a) A membrane consists of a layer of proteins sandwiched between two layers of lipids.

(b) The compositions of the inner and outer lipid layers are the same in any individual membrane.

(c) Membranes contain glycolipids and glycoproteins.

(d) Lipid bilayers are an important component of membranes.

(e) Covalent bonding takes place between lipids and proteins in most membranes.

21.86 Suggest a reason why animals that live in cold climates tend to have higher proportions of polyunsaturated fatty acid residues in their lipids than do animals that live in warm climates.

21.87 Which statements are consistent with the fluid mosaic model of membranes?

(a) All membrane proteins are bound to the interior of the membrane.

(b) Both proteins and lipids undergo transverse (flip-flop) diffusion from the inside to the outside of the membrane.

(c) Some proteins and lipids undergo lateral diffusion along the inner or outer surface of the membrane.

21.88 Suggest a reason why the cell membranes of bacteria grown at 20°C tend to have a higher proportion of unsaturated fatty acids than the membranes of bacteria of the same species grown at 37°C. In other words, the bacteria grown at 37°C have a higher proportion of saturated fatty acids in their cell membranes.

Tying It Together

21.89 Lipids and carbohydrates are both vehicles for energy storage. How are they similar in terms of molecular structure, and how do they differ? What does the molecular structure of each class of substance imply about the polarity of typical molecules?

21.90 To what extent do lipids and carbohydrates play structural roles in living organisms? Do these roles differ in plants and in animals?

21.91 Which substances would you expect to consist primarily of carbohydrates and which primarily of lipids: olive oil, butter, cotton, cotton candy?

21.92 To what extent would you expect to find the following functional groups in lipids and in carbohydrates: aldehyde groups, carboxylic acid groups, ester bonds, hydroxyl groups?

Looking Ahead

21.93 Sports drinks tend to contain large amounts of sugars, and some contain taurine in small amounts. Would you expect more of the effect of these performance aids to come from dietary carbohydrates or from the role of taurine in breaking down fats?

21.94 Which of the following foods consist primarily of carbohydrates and which of fats: soft drinks (not diet drinks), salad dressing, canned fruit, cream cheese?

21.95 The ester bonds in lipids do not give rise to macromolecules, but the amide bonds in proteins do. Comment on the underlying reason for this difference.

21.96 Given the structural differences between steroids and other kinds of lipids, would you expect the synthesis of steroids in living organisms to differ from the synthesis of other lipids?

21.97 What are stem cells (*Hint:* See Chapter 31), and how are they related to myeloproliferative diseases?

21.98 What is the reported link between cholesterol, HDL, and myeloproliferative diseases?

Challenge Problems

21.99 Some of the lipid molecules that occur in membranes are bulkier than others. Are the bulkier molecules more or less likely to be found on the cytoplasmic side of the cell membrane or on the side facing the exterior of the cell?

21.100 What are the functions of a cell membrane? To what extent is a bilayer that consists entirely of lipids able to carry out these functions?

21.101 Glycerophospholipids tend to have both a positive charge and a negative charge in their hydrophilic portions. Does this fact help or hinder lipid packing in membranes? Explain.

21.102 Leukotrienes differ from prostaglandins and thromboxanes in that they lack a ring closure. They also differ from prostaglandins and thromboxanes (and from all other lipids) in another feature of their structure. What is that structural feature? (*Hint:* It has to do with the position of their double bonds.)

22

Proteins

OWL

Sign in to OWL at **www.cengage.com/owl** to view tutorials and simulations, develop problem-solving skills, and complete online homework assigned by your professor.

Spider silk is a fibrous protein that exhibits unmatched strength and toughness.

Hans Strand/Stone/Getty Images

22.1 What Are the Many Functions of Proteins?

Proteins are by far the most important of all biological compounds. The very word "protein" is derived from the Greek *proteios,* meaning "of first importance," and the scientists who named these compounds more than 100 years ago chose an appropriate term. Many types of proteins exist, and they perform a variety of functions, including the following roles:

1. **Structure** In Section 20.5, we saw that the main structural material for plants is cellulose. For animals, it is structural proteins, which are the chief constituents of skin, bones, hair, and nails. Two important structural proteins are collagen and keratin.

2. **Catalysis** Virtually all the reactions that take place in living organisms are catalyzed by proteins called enzymes. Without enzymes, the reactions would take place so slowly as to be useless. We will discuss enzymes in depth in Chapter 23.

3. **Movement** Every time we crook a finger, climb stairs, or blink an eye, we use our muscles. Muscle expansion and contraction are involved in every movement we make. Muscles are made up of protein molecules called myosin and actin.

4. **Transport** A large number of proteins perform transportation duties. For example, hemoglobin, a protein in the blood, carries oxygen from the lungs to the cells in which it is used and carbon dioxide from the cells to the lungs. Other proteins transport molecules across cell membranes.

5. **Hormones** Many hormones are proteins, including insulin, erythropoietin, and human growth hormone.

6. **Protection** When a protein from an outside source or some other foreign substance (called an antigen) enters the body, the body makes its own proteins (called antibodies) to counteract the foreign protein. This antibody production is one of the major mechanisms that the body uses to fight disease. Blood clotting is another protective function carried out by a protein, this one called fibrinogen. Without blood clotting, we would bleed to death from any small wound.

7. **Storage** Some proteins store materials in the way that starch and glycogen store energy. For example, casein in milk and ovalbumin in eggs store nutrients for newborn mammals and birds. Ferritin, a protein in the liver, stores iron.

8. **Regulation** Some proteins not only control the expression of genes, thereby regulating the kind of proteins synthesized in a particular cell, but also dictate when such manufacture takes place.

These are not the only functions of proteins, but they are among the most important. Clearly, any individual needs a great many proteins to carry out these varied functions. A typical cell contains about 9000 different proteins; the entire human body has about 100,000 different proteins.

We can classify proteins into two major types: **fibrous proteins**, which are insoluble in water and are used mainly for structural purposes, and **globular proteins**, which are more or less soluble in water and are used mainly for nonstructural purposes.

22.2 What Are Amino Acids?

Although a wide variety of proteins exist, they all have basically the same structure: They are chains of amino acids. As its name implies, an **amino acid** is an organic compound containing an amino group and a carboxyl group. Organic chemists can synthesize many thousands of amino acids, but nature is much more restrictive and uses 20 common amino acids to make up proteins. Furthermore, all but one of the 20 fit the formula:

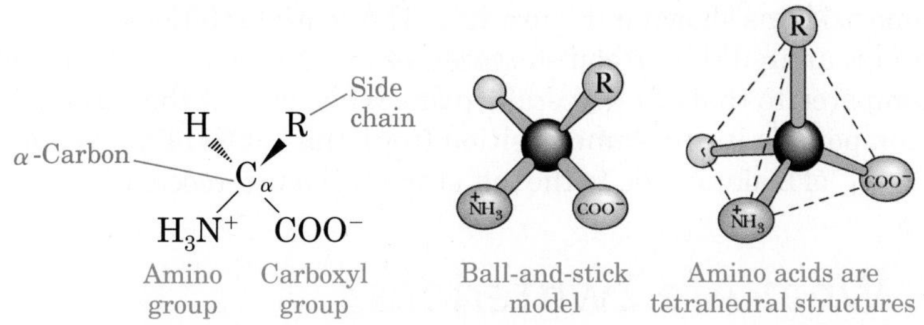

Even the one amino acid that doesn't fit this formula (proline) comes fairly close: It differs only in that it has a bond between the R and the N. The 20 amino acids commonly found in proteins are called **alpha amino acids**. They are listed in Table 22.1, which also shows the one- and three-letter abbreviations that chemists and biochemists use for them.

Protein A large biological molecule made of numerous amino acids linked by amide bonds

Alpha (α) amino acid An amino acid in which the amino group is linked to the carbon atom next to the —COOH carbon

Table 22.1 The 20 Amino Acids Commonly Found in Proteins

Name	3-Letter Abbreviation	1-Letter Abbreviation	Isoelectric Point
Alanine	Ala	A	6.01
Arginine	Arg	R	10.76
Asparagine	Asn	N	5.41
Aspartic acid	Asp	D	2.77
Cysteine	Cys	C	5.07
Glutamic acid	Glu	E	3.22
Glutamine	Gln	Q	5.65
Glycine	Gly	G	5.97
Histidine	His	H	7.59
Isoleucine	Ile	I	6.02
Leucine	Leu	L	5.98
Lysine	Lys	K	9.74
Methionine	Met	M	5.74
Phenylalanine	Phe	F	5.48
Proline	Pro	P	6.48
Serine	Ser	S	5.68
Threonine	Thr	T	5.87
Tryptophan	Trp	W	5.88
Tyrosine	Tyr	Y	5.66
Valine	Val	V	5.97

The most important aspect of the R groups is their polarity. On that basis, we can classify amino acids into four groups, as shown in Figure 22.1: non-polar, polar but neutral, acidic, and basic. Note that the nonpolar side chains are *hydrophobic* (they repel water), whereas polar but neutral, acidic, and basic side chains are *hydrophilic* (attracted to water). This aspect of the R groups is very important in determining both the structure and the function of each protein molecule.

When we look at the general formula for the 20 amino acids, we see at once that all of them (except glycine, in which R = H) are chiral with (carbon) stereocenters, since R, H, COOH, and NH_2 are four different groups. Thus, each of the amino acids with one stereocenter exists as two enantiomers. As is the case for most examples of this kind, nature makes only one of the two possible enantiomers for each amino acid, and it is virtually always the L-isomer. Except for glycine, which is achiral, all the amino acids in all the proteins in your body are the L-isomer. D amino acids are extremely rare in nature; some are found, for example, in the cell walls of a few types of bacteria.

In Section 20.1C, we learned about the systematic use of the D,L system. There we used glyceraldehyde as a reference point for the assignment of relative configuration. Here again, we can use glyceraldehyde as a reference point with amino acids, as shown in Figure 22.2. The spatial relationship of the functional groups around the carbon stereocenter in L-amino acids, as in L-alanine, can be compared to that of L-glyceraldehyde. When we put the carbonyl groups of both compounds in the same position (top), the —OH of L-glyceraldehyde and the NH_3^+ of L-alanine lie to the left of the carbon stereocenter.

22.3 What Are Zwitterions?

In Section 18.5B, we learned that carboxylic acids, RCOOH, cannot exist in the presence of a moderately weak base (such as NH_3). They donate a proton to become carboxylate ions, $RCOO^-$. Likewise, amines, RNH_2 (Section 16.5), cannot exist as such in the presence of a moderately weak acid (such as acetic acid). They gain a proton to become substituted ammonium ions, RNH_3^+.

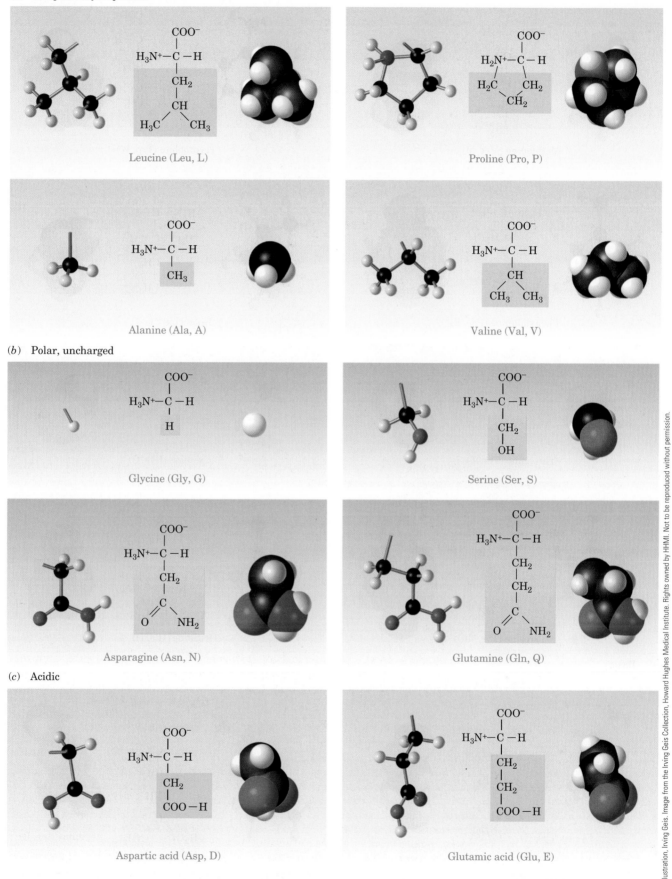

FIGURE 22.1 The 20 amino acids that are the building blocks of proteins can be classified as *(a)* nonpolar (hydrophobic), *(b)* polar but neutral, *(c)* acidic, or *(d)* basic. Also shown here are the one-letter and three-letter codes used to denote amino acids. For each amino acid, the ball-and-stick *(left)* and space-filling *(right)* models show only the side chain.

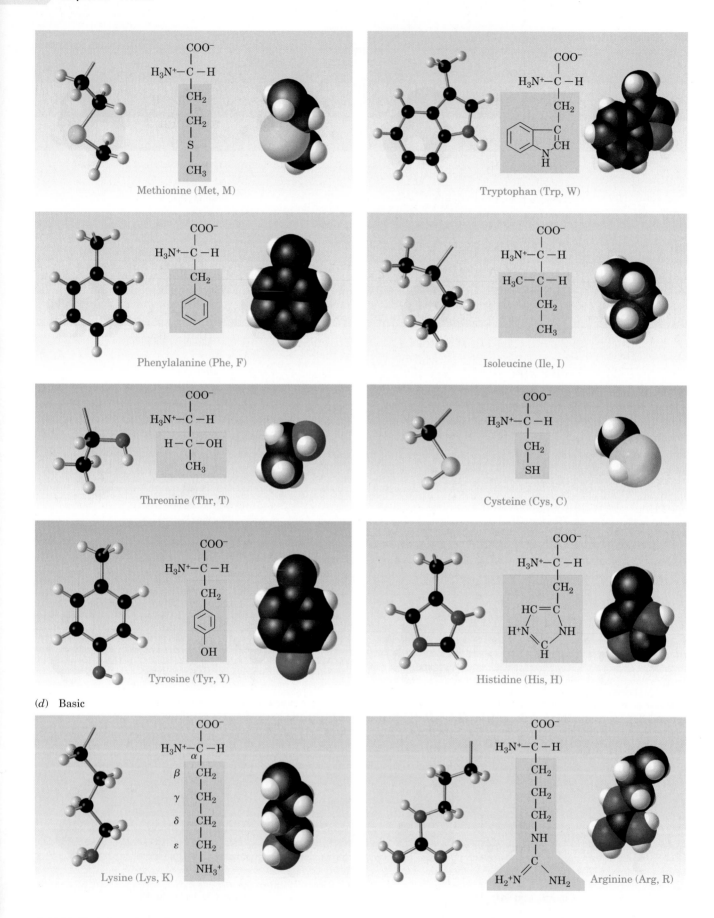

Methionine (Met, M)

Tryptophan (Trp, W)

Phenylalanine (Phe, F)

Isoleucine (Ile, I)

Threonine (Thr, T)

Cysteine (Cys, C)

Tyrosine (Tyr, Y)

Histidine (His, H)

(d)   Basic

Lysine (Lys, K)

Arginine (Arg, R)

FIGURE 22.1 (*continued*)

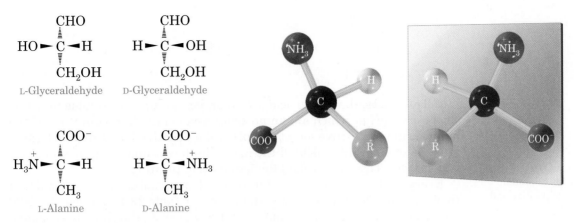

FIGURE 22.2 Stereochemistry of alanine and glyceraldehyde. The amino acids found in proteins have the same chirality as L-glyceraldehyde, which is opposite that of D-glyceraldehyde.

An amino acid has —COOH and —NH$_2$ groups in the same molecule. Therefore, in water solution, the —COOH donates a proton to the —NH$_2$ so that an amino acid actually has the structure:

$$\begin{array}{c} \text{H} \\ | \\ \text{R---C---COO}^- \\ | \\ \text{NH}_3{}^+ \end{array}$$

Compounds that have a positive charge on one atom and a negative charge on another are called **zwitterions**, from the German word *zwitter*, meaning "hybrid." Amino acids are zwitterions, not only in water solution but also in the solid state. They are therefore ionic compounds—that is, internal salts. *Un-ionized RCH(NH$_2$)COOH molecules do not actually exist, in any form.*

The fact that amino acids are zwitterions explains their physical properties. All of them are solids with high melting points (for example, glycine melts at 262°C), just as we would expect for ionic compounds. The 20 common amino acids are also fairly soluble in water, as ionic compounds generally are. If they had no charges, we would expect only the smaller ones to be soluble.

If we add an amino acid to water, it dissolves and then has the same zwitterionic structure that it has in the solid state. Let us see what happens if we change the pH of the solution, as we can easily do by adding a source of H$_3$O$^+$, such as HCl solution (to lower the pH), or a strong base, such as NaOH (to raise the pH). Because H$_3$O$^+$ is a stronger acid than a typical carboxylic acid (Section 18.1), it donates a proton to the —COO$^-$ group, turning the zwitterion into a positive ion. This happens to all amino acids if the pH is sufficiently lowered—say, to pH 0.

$$\begin{array}{c} \text{H} \\ | \\ \text{R---C---COO}^- \\ | \\ \text{NH}_3{}^+ \end{array} + \text{H}_3\text{O}^+ \longrightarrow \begin{array}{c} \text{H} \\ | \\ \text{R---C---COOH} \\ | \\ \text{NH}_3{}^+ \end{array} + \text{H}_2\text{O}$$

Addition of OH$^-$ to the zwitterion causes the —NH$_3{}^+$ to donate its proton to OH$^-$, turning the zwitterion into a negative ion. This happens to all amino acids if the pH is sufficiently raised—say, to pH 14.

$$\underset{\displaystyle NH_3^+}{R-\overset{\displaystyle H}{\underset{\displaystyle |}{\overset{\displaystyle |}{C}}}-COO^-} + OH^- \longrightarrow \underset{\displaystyle NH_2}{R-\overset{\displaystyle H}{\underset{\displaystyle |}{\overset{\displaystyle |}{C}}}-COO^-} + H_2O$$

Isoelectric point (pI) A pH at which a sample of amino acids or protein has an equal number of positive and negative charges

In both cases, the amino acid is still an ion, so it is still soluble in water. There is no pH at which an amino acid has no ionic character at all. If the amino acid is a positive ion at low pH and a negative ion at high pH, there must be some pH at which all the molecules have equal positive and negative charges. This pH is called the **isoelectric point (pI)**.

Every amino acid has a different isoelectric point, although most of them are not very far apart (see the values in Table 22.1). Fifteen of the 20 amino acids have isoelectric points near 6. However, the three basic amino acids have higher isoelectric points, and the two acidic amino acids have lower values.

At or near the isoelectric point, amino acids exist in aqueous solution largely or entirely as zwitterions. As we have seen, they react with either a strong acid, by taking a proton (the —COO$^-$ becomes —COOH), or a strong base, by giving a proton (the —NH$_3^+$ becomes —NH$_2$). To summarize:

$$\underset{NH_3^+}{R-\overset{H}{\underset{|}{\overset{|}{C}}}-COOH} \underset{H_3O^+}{\overset{OH^-}{\rightleftarrows}} \underset{NH_3^+}{R-\overset{H}{\underset{|}{\overset{|}{C}}}-COO^-} \underset{H_3O^+}{\overset{OH^-}{\rightleftarrows}} \underset{NH_2}{R-\overset{H}{\underset{|}{\overset{|}{C}}}-COO^-}$$

In Section 8.3, we learned that a compound that is both an acid and a base is called amphiprotic. We also learned in Section 8.10 that a solution that neutralizes both acid and base is a buffer solution. Amino acids are therefore *amphiprotic* compounds, and aqueous solutions of them are *buffers*.

22.4 What Determines the Characteristics of Amino Acids?

Since the side chains are the only differences between the amino acids, ultimately the functions of amino acids and their polymers, proteins, are determined by these side chains. For example, one of the 20 amino acids in Table 22.1 has a chemical property not shared by any of the others. This amino acid, cysteine, can easily be dimerized by many mild oxidizing agents:

A disulfide bond

$$2HS-CH_2-\underset{\underset{Cysteine}{NH_3^+}}{\overset{|}{CH}}-COO^- \underset{[H]}{\overset{[O]}{\rightleftarrows}} {}^-OOC-\underset{NH_3^+}{\overset{|}{CH}}-CH_2-S-S-CH_2-\underset{\underset{Cystine}{NH_3^+}}{\overset{|}{CH}}-COO^-$$

The dimer of cysteine, which is called **cystine**, can in turn be fairly easily reduced to give two molecules of cysteine. As we shall see, the presence of cystine has important consequences for the chemical structure and shape of the protein molecules of which it is part. The bond (shown in color) is also called a **disulfide bond** (Section 14.4D).

Several of the amino acids have acidic or basic properties. Two amino acids, glutamic acid and aspartic acid, have carboxyl groups in their side chains in addition to the one present in all amino acids. A carboxyl group can lose a proton, forming the corresponding carboxylate anion—glutamate and aspartate, respectively, in the case of these two amino acids. Because of the presence of the carboxylate, the side chains of these two amino acids are negatively charged at neutral pH.

Three amino acids—histidine, lysine, and arginine—have basic side chains. The side chains of lysine and arginine are positively charged at or near neutral pH. In lysine, the side-chain amino group is attached to an aliphatic hydrocarbon tail. In arginine, the side-chain basic group, the guanidino group, is more complex in structure than the amino group, but it is also bonded to an aliphatic hydrocarbon tail. In free histidine, the pK_a of the side-chain imidazole group is 6.0, which is not far from physiological pH. The pK_a values for amino acids depend on the environment and can change significantly within the confines of a protein. Histidine can be found in the protonated or unprotonated forms in proteins, and the properties of many proteins depend on whether individual histidine residues are charged or not. The charged amino acids are often found in the active sites of enzymes, which we will study in Chapter 23.

The amino acids phenylalanine, tryptophan, and tyrosine have aromatic rings in their side chains. They are important for a number of reasons. As a matter of practicality, these amino acids allow us to locate and measure proteins because the aromatic rings absorb strongly at 280 nm and can be detected using a spectrophotometer. These amino acids are also very important physiologically because they are both key precursors to neurotransmitters (substances involved in the transmission of nerve impulses). Tryptophan is converted to serotonin, more properly called 5-hydroxytryptamine, which has a calming effect. Very low levels of serotonin are associated with depression, whereas extremely high levels produce a manic state. Manic-depressive schizophrenia (also called bipolar disorder) can be managed by controlling the levels of serotonin and its further metabolites.

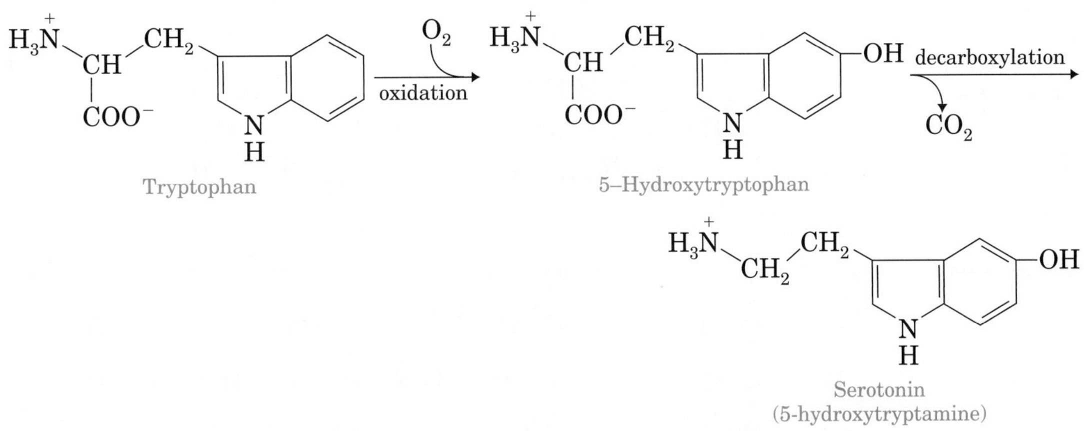

Tryptophan → (O₂, oxidation) → 5–Hydroxytryptophan → (decarboxylation, CO₂) → Serotonin (5-hydroxytryptamine)

Tyrosine, itself normally derived from phenylalanine, is converted to the neurotransmitter class called catecholamines, which includes epinephrine, commonly known by its proprietary name, adrenalin.

L-Dihydroxyphenylalanine (L-dopa) is an intermediate in the conversion of tyrosine. Lower-than-normal levels of L-dopa are involved in Parkinson's disease. Tyrosine or phenylalanine supplements might increase the levels of dopamine, although L-dopa, the immediate precursor, is usually prescribed because it passes into the brain quickly through the blood–brain barrier.

Tyrosine and phenylalanine are precursors to norepinephrine and epinephrine, both of which are stimulatory. Epinephrine is commonly known as the "flight-or-fight" hormone. It causes the release of glucose and other nutrients into the blood and stimulates brain function.

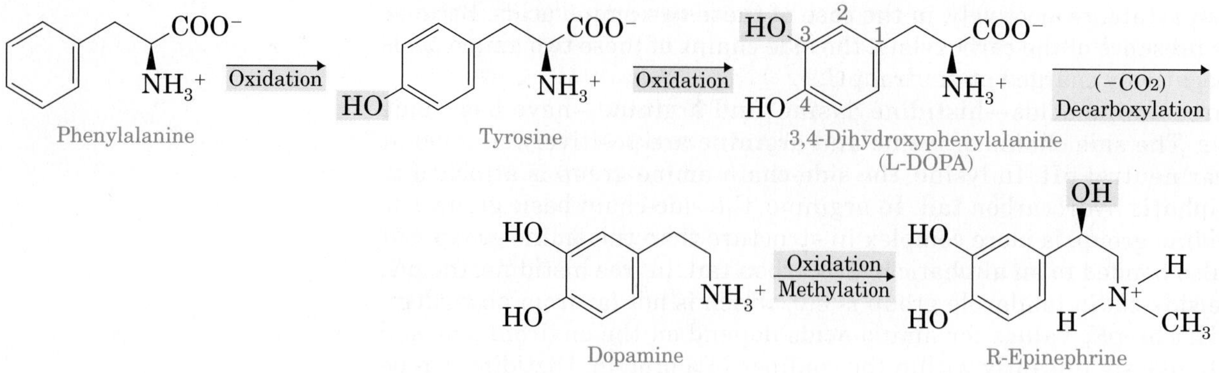

It has been suggested that tyrosine and phenylalanine may have un-expected effects in some people. For example, a growing body of evidence indicates that some people get headaches from the phenylalanine in aspartame, an artificial sweetener commonly found in diet soft drinks. Some people insist that supplements of tyrosine give them a morning life; others claim that tryptophan helps them sleep at night. Milk proteins have high levels of tryptophan; a glass of warm milk before bed is widely believed to be an aid in inducing sleep.

22.5 What Are Uncommon Amino Acids?

Many other amino acids in addition to the ones listed in Table 22.1 are known to exist. They occur in some, but by no means all, proteins. Figure 22.3 shows some examples of the many possibilities. These uncommon amino acids are derived from the common amino acids and are produced by modification of

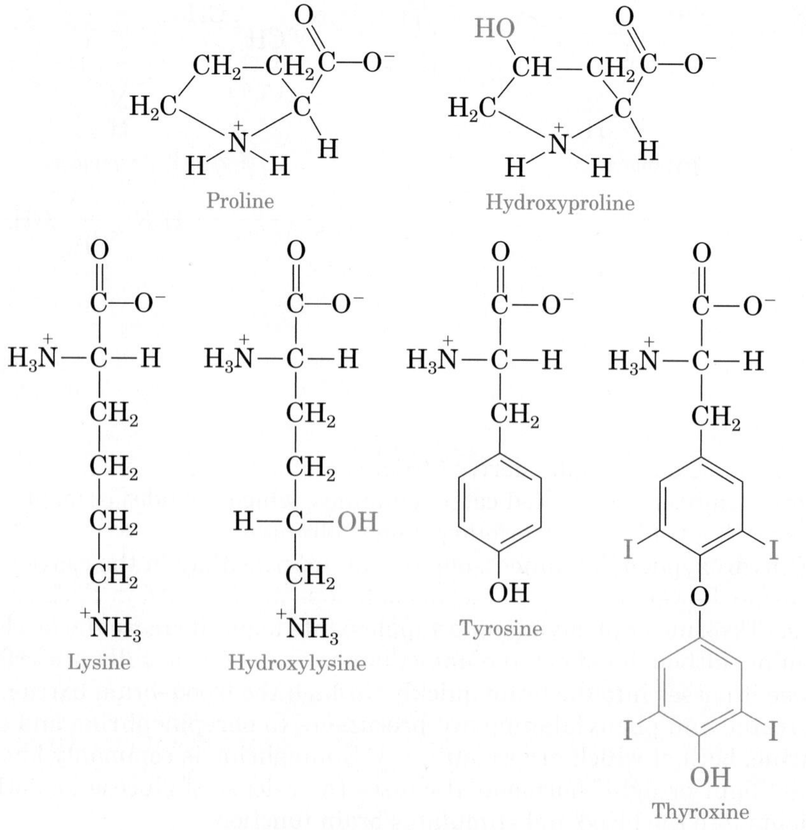

FIGURE 22.3 Structures of hydroxyproline, hydroxylysine, and thyroxine. The structures of the parent amino acids—proline for hydroxyproline, lysine for hydroxylysine, and tyrosine for thyroxine—are shown for comparison. All amino acids are shown in their predominant ionic forms at pH 7.

the parent amino acid after the protein is synthesized by the organism in a process called post-translational modification (Chapter 26). Hydroxyproline and hydroxylysine differ from their parent amino acids in that they have hydroxyl groups on their side chains; they are found only in a few connective tissue proteins, such as collagen. Thyroxine differs from tyrosine in that it has an extra iodine-containing aromatic group on the side chain; it is found only in the thyroid gland, where it is formed by post-translational modification of tyrosine residues in the protein thyroglobulin. Thyroxine is then released as a hormone by proteolysis of thyroglobulin. Both animals and humans that exhibit sluggishness and slow metabolism are often given thyroxine to help ramp up their metabolism.

22.6 How Do Amino Acids Combine to Form Proteins?

Each amino acid has a carboxyl group and an amino group. In Chapter 19, we saw that a carboxylic acid and an amine could be combined to form an amide:

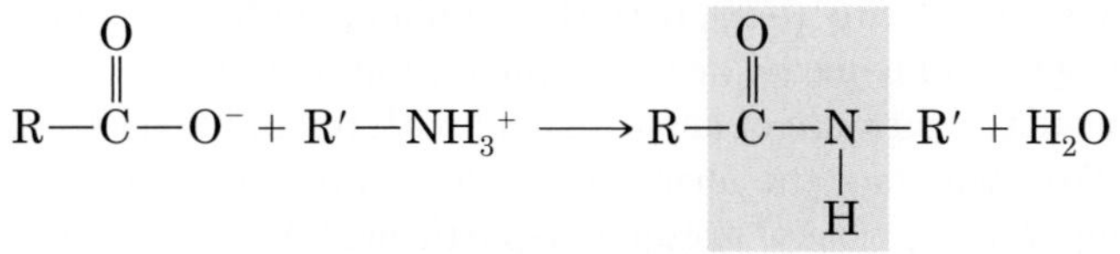

$$R-\overset{O}{\overset{\|}{C}}-O^- + R'-NH_3^+ \longrightarrow R-\overset{O}{\overset{\|}{C}}-\overset{}{\underset{H}{N}}-R' + H_2O$$

In the same way, the $-COO^-$ group of one amino acid molecule—say, glycine—can combine with the $-NH_3^+$ group of a second molecule—say, alanine:

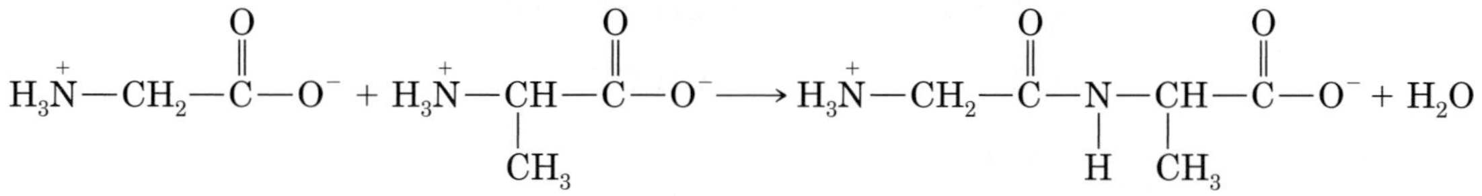

$$\overset{+}{H_3N}-CH_2-\overset{O}{\overset{\|}{C}}-O^- + \overset{+}{H_3N}-\underset{CH_3}{CH}-\overset{O}{\overset{\|}{C}}-O^- \longrightarrow \overset{+}{H_3N}-CH_2-\overset{O}{\overset{\|}{C}}-\underset{H}{N}-\underset{CH_3}{CH}-\overset{O}{\overset{\|}{C}}-O^- + H_2O$$

Glycine   Alanine   Glycylalanine (Gly—Ala)

This reaction takes place in the cells by a mechanism that we will examine in Section 26.5. The product is an amide. The two amino acids are joined together by a **peptide bond** (also called a **peptide linkage**). The product is a **dipeptide**.

Peptide bond An amide bond that links two amino acids

It is important to realize that glycine and alanine could also be linked the other way:

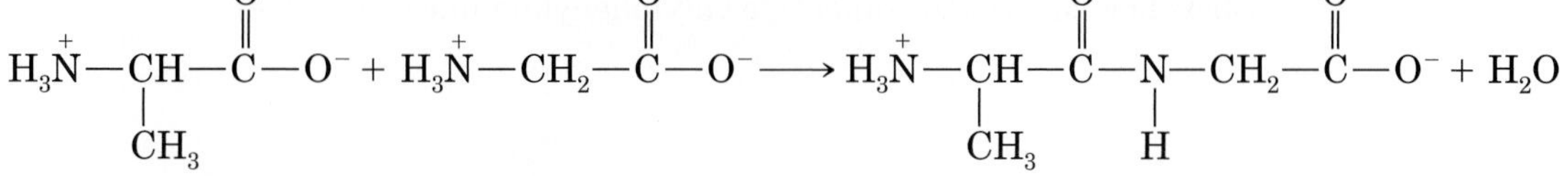

$$\overset{+}{H_3N}-\underset{CH_3}{CH}-\overset{O}{\overset{\|}{C}}-O^- + \overset{+}{H_3N}-CH_2-\overset{O}{\overset{\|}{C}}-O^- \longrightarrow \overset{+}{H_3N}-\underset{CH_3}{CH}-\overset{O}{\overset{\|}{C}}-\underset{H}{N}-CH_2-\overset{O}{\overset{\|}{C}}-O^- + H_2O$$

Alanine   Glycine   Alanylglycine (Ala—Gly)

In this case, we get a *different* dipeptide. The two dipeptides are constitutional isomers, of course; they are different compounds in all respects, with different properties. The phrase "do much, talk little" has the same words as "do little, talk much," but the meaning of the two are quite different. In the same way, the order of amino acids in a peptide or protein is critical to both the structure and function.

Example 22.1 Peptide Formation

Show how to form the dipeptide aspartylserine (Asp—Ser).

Strategy

Start by drawing the two amino acids. Orient them so that both read (from left to right): amino group, alpha carbon, carboxyl group. Then draw the reaction between the first amino acid's carboxyl group and the second amino acid's amino group to give the peptide bond.

Solution

The name implies that this dipeptide is made of two amino acids, aspartic acid (Asp) and serine (Ser), with the amide being formed between the α-carboxyl group of aspartic acid and the α-amino group of serine. Therefore, we write the formula of aspartic acid with its amino group on the left side. Next, we place the formula of serine to the right, with its amino group facing the α-carboxyl group of aspartic acid. Finally, we eliminate a water molecule between the —COO$^-$ and —NH$_3^+$ groups that are next to each other, forming the peptide bond:

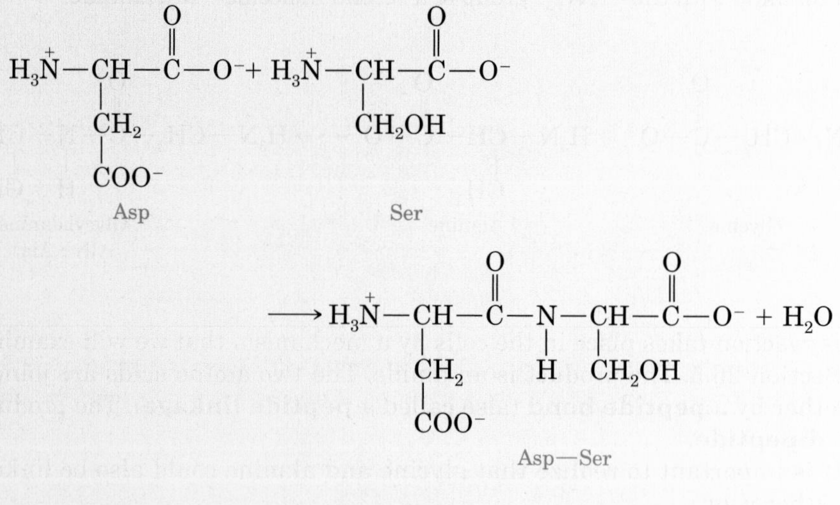

Problem 22.1

Show how to form the dipeptide valylphenylalanine (Val—Phe).

Any two amino acids, whether the same or different, can be linked together to form dipeptides in a similar manner. But the possibilities do not end there. Each dipeptide still contains a —COO$^-$ and an —NH$_3^+$ group. We can, therefore, add a third amino acid to alanylglycine—say, lysine: The product is a **tripeptide**. Because it also contains a —COO$^-$ and an —NH$_3^+$

Chemical Connections 22A

Aspartame, the Sweet Peptide

The dipeptide L-aspartyl-L-phenylalanine is of considerable commercial importance. The aspartyl residue has a free α-amino group, the N-terminal end of the molecule, and the phenylalanyl residue has a free carboxyl group, the C-terminal end. This dipeptide is about 200 times sweeter than sugar. A methyl ester derivative of this dipeptide is of even greater commercial importance than the dipeptide itself. The derivative has a methyl group at the C-terminal end in an ester linkage to the carboxyl group. The methyl ester derivative is called *aspartame* and is marketed as a sugar substitute under the trade name NutraSweet.

Common table sugar is consumed in the United States at about 100 pounds per person per year. Many people want to curtail their sugar intake in the interest of fighting obesity.

Others must limit their sugar intake because of diabetes. One of the most common ways of doing so is by drinking diet soft drinks. The soft drink industry is one of the largest markets for aspartame. The use of this sweetener was approved by the U.S. Food and Drug Administration in 1981 after extensive testing, although there is still considerable controversy about its safety. Diet soft drinks sweetened with aspartame carry warning labels about the presence of phenylalanine. This information is of vital importance to people who have phenylketonuria, a genetic disease of phenylalanine metabolism. Note that both amino acids have the L configuration. If a D-amino acid is substituted for either amino acid or for both of them, the resulting derivative is bitter rather than sweet.

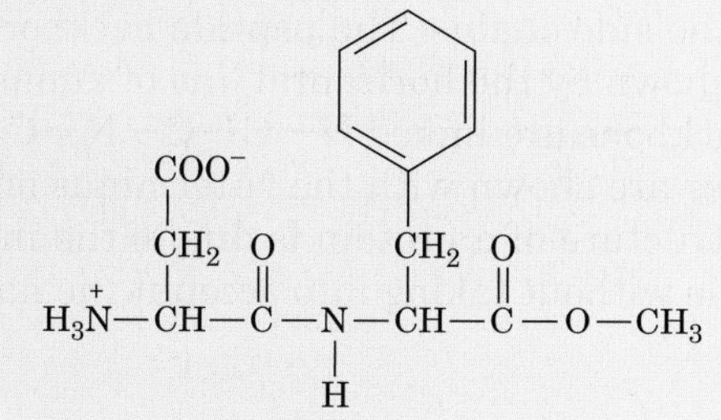

L-Aspartyl-L-phenylalanine (methyl ester)

Test your knowledge with Problem 22.69.

group, we can continue the process to get a tetrapeptide, a pentapeptide, and so on, until we have a chain containing hundreds or even thousands of amino acids. These chains of amino acids are the proteins that serve so many important functions in living organisms.

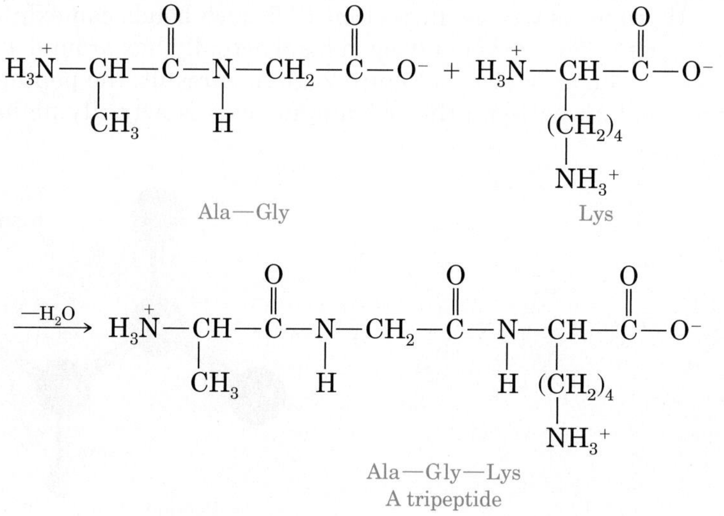

Ala—Gly   Lys

Ala—Gly—Lys
A tripeptide

A word must be said about the terms used to describe these compounds. The shortest chains are often simply called **peptides**, longer ones are **polypeptides**, and still longer ones are **proteins**, but chemists differ about where to draw the line. Many chemists use the terms "polypeptide" and "protein" almost interchangeably. In this book, we will consider a protein to be a polypeptide chain that contains a minimum of 30 to 50 amino acids. The amino acids in a chain are often called **residues**. It is customary to use either the one-letter or the three-letter abbreviations shown in Table 22.1 to represent peptides and proteins. For example, the tripeptide shown on the previous page, alanylglycyllysine, is AGK or Ala—Gly—Lys. The **C-terminal amino acid**, or **C-terminus**, is the residue with the free α-COO$^-$ group (lysine in Ala—Gly—Lys), and the **N-terminal amino acid**, or **N-terminus**, is the residue with the free α-NH$_3^+$ group (alanine in Ala—Gly—Lys). It is the universal custom to write peptide and protein chains with the N-terminal residue on the left. This decision is not as arbitrary as it might seem. We read left to right, and proteins are synthesized from N-terminus to C-terminus, as we will see in Chapter 26.

C-terminus The amino acid at the end of a peptide that has a free α-carboxyl group

N-terminus The amino acid at the end of a peptide that has a free α-amino group

22.7 What Are the Properties of Proteins?

The properties of proteins are based on properties of the peptide backbone and properties of the side chains. The peptide backbone consists of the repeating structure shown by the horizontal line of atoms in Figure 22.4. The atoms along the backbone are linked N—C—C—N—C—C— and so on. By convention, peptides are shown with the N-terminus on the left. As it turns out, much of the structure of a protein is due to the interactions of the atoms in the backbone without taking into account the nature of the R groups on the side chains.

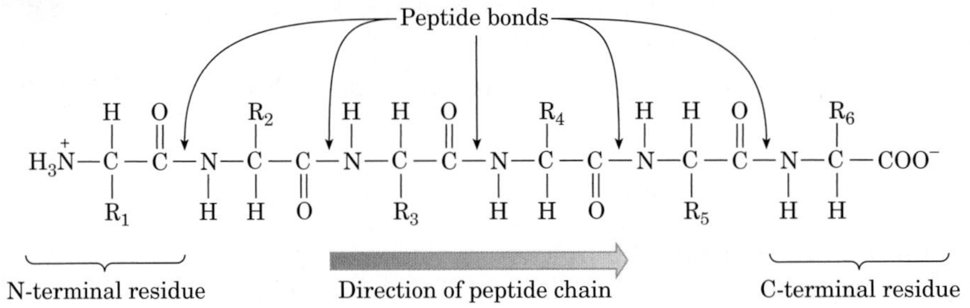

FIGURE 22.4 A small peptide showing the direction of the peptide chain (N-terminal to C-terminal).

Although the peptide bond is typically written as a carbonyl group bonded to an N—H group, as we saw in Section 17.5, such bonds can exhibit keto-enol tautomerism. The carbon–nitrogen bond actually has around 40% double bond character, as shown in Figure 22.5. As a result, the peptide group that forms the link between the two amino acids is actually planar. This

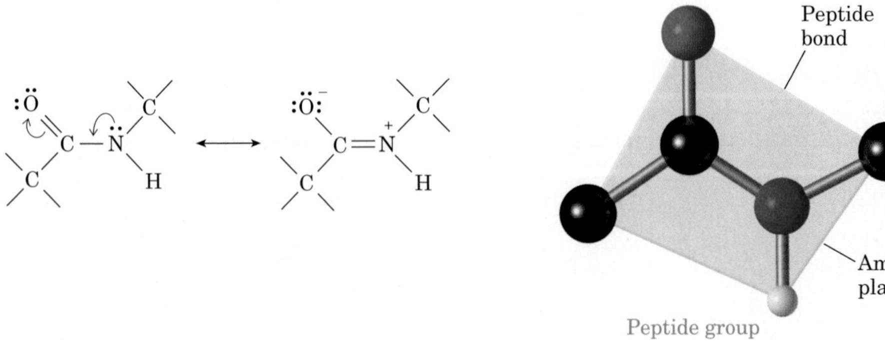

FIGURE 22.5 The resonance structures of the peptide bond lead to a planar group.

Chemical Connections 22B

AGE and Aging

A reaction can take place between a primary amine and an aldehyde or a ketone, linking the two molecules (shown here for an aldehyde):

$$R-\overset{\overset{\displaystyle O}{\|}}{C}-H + H_2N-R' \longrightarrow R-CH=N-R' + H_2O$$

An imine

Because proteins have NH_2 groups and carbohydrates have aldehyde or ketone groups, they can undergo this reaction, establishing a link between a sugar and a protein molecule. When this reaction is not catalyzed with the controlling action of an enzyme, it is a haphazard process that impairs the functioning of biomolecules and is called *glycation* of proteins. The process, however, does not stop there. When these linked products are heated in a test tube, high-molecular-weight water-insoluble brownish complexes form. These complexes are called **advanced glycation end-products (AGEs)**. In the body, they cannot be heated, but the same result happens over long periods of time.

The longer we live, and the higher the blood sugar concentration becomes, the more AGE products accumulate in the body. These AGEs can alter the function of proteins. Such AGE-dependent changes are thought to contribute to circulation, joint, and vision problems in people with diabetes. People with diabetes have high blood sugar due to a lack of transport of glucose out of the blood and into the cells. AGE products show up in all of the afflicted organs of diabetic patients: in the lens of the eye (cataracts), in the capillary blood vessels of the retina (diabetic retinopathy), and in the glomeruli of the kidneys (kidney failure). AGEs have been linked to atherosclerosis, as AGE-modified cells can bind to endothelial cells in blood vessels. AGE-modified collagen causes stiffening of arteries.

For people who do not have diabetes, these harmful protein modifications become disturbing only in an individual's advanced years. In a young person, metabolism functions properly and the AGE products decompose and are eliminated from the body. In an older person, metabolism slows and the AGE products accumulate. The AGE products themselves are thought to enhance oxidative damages.

Scientists are searching for ways to combat the harmful effects of AGEs. One approach is to use antioxidants, including the B vitamin thiamine. A few other anti-AGE drugs have been developed, including aminoguanidine and metformin. Both of these drugs have been studied in animal models but have not yet reached large-scale use in humans. Another approach that is being studied for several metabolic problems, including normal aging, is caloric restriction. A vast amount of evidence from animal models and human models alike indicates that lifespan can be extended by living a lean existence. This lifestyle has also been shown to reduce the level of AGEs.

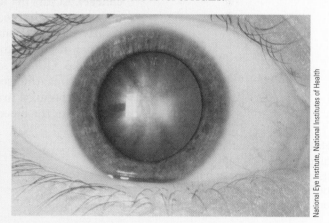

Cataracts in the eye are caused by advanced glycation end-products (AGEs).

National Eye Institute, National Institutes of Health

Test your knowledge with Problem 22.70.

grouping is called the amide plane, and it has a tremendous influence on protein structure. There is freedom of rotation about the two bonds from the alpha carbon, but there is no rotation of the carbon–nitrogen bonds. A chain of amino acids linked via peptide bonds can be thought of as a series of playing cards linked by a swivel at their corners, as shown in Figure 22.6. The rigidity of the amide plane limits the possible orientation of the peptide.

The 20 different amino acid side chains supply variety and determine the rest of the physical and chemical properties of proteins. Among these properties, acid–base behavior is one of the most important. Like amino acids (Section 22.3), proteins behave as zwitterions. The side chains of glutamic and aspartic acids provide acidic groups, whereas lysine and arginine provide basic groups (histidine does as well, but this side chain is less basic than the other two). (See the structures of these amino acids in Figure 22.1.)

FIGURE 22.6 Planar nature of peptide bond. The rigid planar peptide groups (called "playing cards" in the text) are shaded.

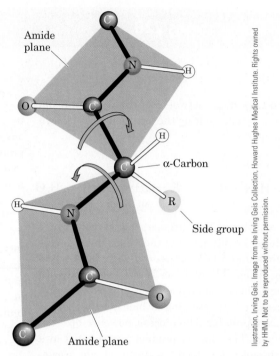

The isoelectric point of a protein occurs at the pH at which there are an equal number of positive and negative charges (the protein has no *net* charge). At any pH above the isoelectric point, the protein molecules have a net negative charge; at any pH below the isoelectric point, they have a net positive charge. Some proteins, such as hemoglobin, have an almost equal number of acidic and basic groups; the isoelectric point of hemoglobin is at pH 6.8. Others, such as serum albumin, have more acidic groups than basic groups; the isoelectric point of this protein is at pH 4.9. In each case, however, because proteins behave like zwitterions, they act as buffers—for example, in the blood (Figure 22.7).

FIGURE 22.7 Schematic diagram of a protein (*a*) at its isoelectric point and its buffering action when (*b*) H$^+$ or (*c*) OH$^-$ ions are added.

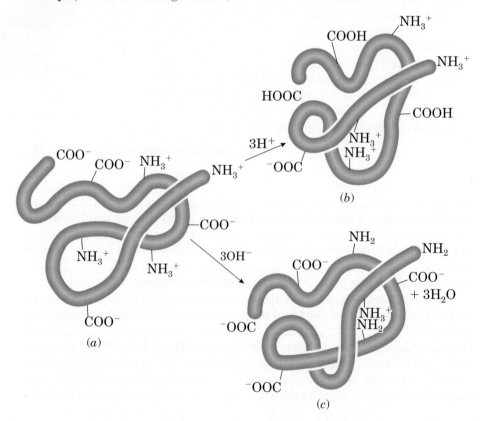

The water solubility of large molecules such as proteins often depends on the repulsive forces between like charges on their surfaces. When protein molecules are at a pH at which they have a net positive or negative charge, the presence of these like charges causes the protein molecules to repel one another. These repulsive forces are smallest at the isoelectric point, when the net charges are zero. When there are no repulsive forces, the protein molecules tend to clump together to form aggregates of two or more molecules, reducing their solubility. As a consequence, *proteins are least soluble in water at their isoelectric points and can be precipitated from their solutions.*

As we pointed out in Section 22.1, proteins have many functions. To understand these functions, we must look at four levels of organization in their structures. The *primary structure* describes the linear sequence of amino acids in the polypeptide chain. The *secondary structure* refers to certain repeating patterns, such as the α-helix conformation or the pleated sheet (Section 22.9) or the absence of a repeating pattern, as with the random coil (Section 22.9). The *tertiary structure* describes the overall 3D conformation of the polypeptide chain (Section 22.10). The *quaternary structure* (Section 22.11) applies mainly to proteins containing more than one polypeptide chain (subunit) and deals with how the different chains are spatially related to one another.

22.8 What Is the Primary Structure of a Protein?

Very simply, the **primary structure** of a protein consists of the sequence of amino acids that makes up the chain. Each of the very large number of peptide and protein molecules in biological organisms has a different sequence of amino acids—and that sequence allows the protein to carry out its function, whatever it may be.

Primary structure The sequence of amino acids in a protein

How can so many different proteins arise from different sequences of 20 amino acids? Let us look at a little arithmetic, starting with a dipeptide. How many different dipeptides can be made from 20 amino acids? There are 20 possibilities for the N-terminal amino acid, and for each of these 20, there are 20 possibilities for the C-terminal amino acid. This means that there are $20 \times 20 = 400$ different dipeptides possible from the 20 amino acids. What about tripeptides? We can form a tripeptide by taking any of the 400 dipeptides and adding any of the 20 amino acids. Thus, there are $20 \times 20 \times 20 = 8000$ tripeptides, all different. It is easy to see that we can calculate the total number of possible peptides or proteins for a chain of n amino acids simply by raising 20 to the nth power (20^n).

Taking a typical small protein to be one with 60 amino acid residues, the number of proteins that can be made from the 20 amino acids is $20^{60} = 10^{78}$. This is an enormous number, possibly greater than the total number of atoms in the universe. Clearly, only a tiny fraction of all possible protein molecules have ever been made by biological organisms.

Each peptide or protein in the body has its own unique sequence of amino acids. As with naming of peptides, the *assignment of positions of the amino acids in the sequence starts at the N-terminal end.* Thus, in Figure 22.8, glycine is in the number 1 position on the A chain and phenylalanine is number 1 on the B chain. We mentioned that proteins also have secondary, tertiary, and (in some cases) quaternary structures. We will deal with these in Sections 22.9, 22.10, and 22.11, but here we can say that *the primary structure of a protein determines to a large extent the native* (most frequently occurring) *secondary and tertiary structures.* That is, the particular sequence of amino acids on the chain enables the whole chain

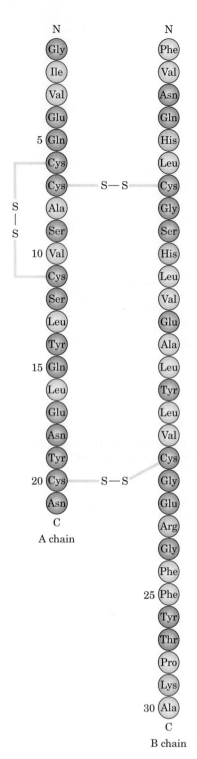

FIGURE 22.8 The hormone insulin consists of two polypeptide chains, A and B, held together by two disulfide cross-bridges (S—S). The sequence shown is for bovine insulin.

to fold and curl in such a way as to assume its final shape. As we will see in Section 22.12, without its particular three-dimensional shape, a protein cannot function.

Just how important is the exact amino acid sequence to the function of a protein? Can a protein perform the same function if its sequence is a little different? The answer to this question is that a change in amino acid sequence may or may not matter, depending on what kind of a change it is. Consider, as an example, cytochrome c, which is a protein of terrestrial vertebrates. Its chain consists of 104 amino acid residues. It performs the same function (electron transport) in humans, chimpanzees, sheep, and other animals. While humans and chimpanzees have exactly the same amino acid sequence of this protein, sheep cytochrome c differs in 10 positions out of the 104. (You can find more about biochemical evolution in Chemical Connections 26F.)

Another example is the hormone insulin. Human insulin consists of two chains having a total of 51 amino acids. The two chains are connected by disulfide bonds. Figure 22.8 shows the sequence of amino acids. Insulin is necessary for proper utilization of carbohydrates (Section 28.1), and people with severe diabetes (Chemical Connections 22C) must take insulin injections. The amount of human insulin available is far too small to meet the need for it, so bovine insulin (from cattle) or insulin from hogs or sheep is used instead. Insulin from these sources is similar but not identical to human insulin. The differences are entirely in the 8, 9, and 10 positions of the A chain and the C-terminal position (30) of the B chain, as shown in Table 22.2. The remainder of the molecule is the same in all four varieties of insulin. Despite the slight differences in structure, all of these insulins perform the same function and even can be used by humans. However, none of the other three is quite as effective in humans as human insulin. This is one of the reasons that recombinant DNA techniques are now used to produce human insulin from bacteria (Section 26.8 and Chemical Connections 22C).

Another factor showing the effect of substituting one amino acid for another is that sometimes patients become allergic to, say, bovine insulin but can switch to hog or sheep insulin without experiencing an allergic reaction.

In contrast to the previous examples, some small changes in amino acid sequence make a great deal of difference. Consider two peptide hormones, oxytocin and vasopressin (Figure 22.9). These nonapeptides have identical structures, including a disulfide bond, except for different amino acids in positions 2 and 7. Yet their biological functions are quite different. Vasopressin is an antidiuretic hormone. It increases the amount of water reabsorbed by the kidneys and raises blood pressure. Oxytocin has no effect on water in the kidneys and slightly lowers blood pressure. It affects contractions of the uterus in childbirth and the muscles in the breast that aid in the secretion of milk. Vasopressin also stimulates uterine contractions, albeit to a much lesser extent than oxytocin.

Table 22.2 Amino Acid Sequence Differences for Human, Bovine, Hog, and Sheep Insulin

	A Chain			B Chain
	8	9	10	30
Human	—Thr—	Ser—	Ile—	—Thr
Bovine	—Ala—	Ser—	Val—	—Ala
Hog	—Thr—	Ser—	Ile—	—Ala
Sheep	—Ala—	Gly—	Val—	—Ala

The Use of Human Insulin

Although human insulin, as manufactured by recombinant DNA techniques (see Section 26.8), is on the market, many diabetic patients continue to use hog or sheep insulin because it is less expensive. Changing from animal to human insulin creates an occasional problem for diabetics. All diabetics experience an insulin reaction (hypoglycemia) when the insulin level in the blood is too high relative to the blood sugar level. Hypoglycemia is preceded by symptoms of hunger, sweating, and poor coordination. These symptoms, called hypoglycemic awareness, signal the patient that hypoglycemia is coming and that it must be reversed, which the patient can do by eating sugar.

Some diabetics who changed from animal to human insulin reported that the hypoglycemic awareness from recombinant DNA human insulin is not as strong as that from animal insulin. This lack of recognition can create some hazards, and the effect is probably due to different rates of absorption in the body. The literature supplied with human insulin now incorporates a warning that hypoglycemic awareness may be altered.

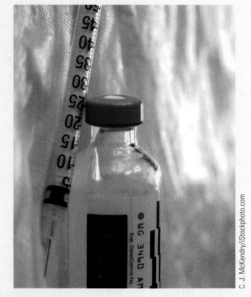

C. J. McKendry/iStockphoto.com

Insulin medication can be made from animal sources, such as hog or sheep, or created by recombinant DNA techniques to give a human protein.

Test your knowledge with Problem 22.71.

Another instance where a minor change makes a major difference is in the blood protein hemoglobin. A change in only one amino acid in a chain of 146 is enough to cause a fatal disease—sickle cell anemia (Chemical Connections 22D).

In some cases, slight changes in amino acid sequence make little or no difference to the functioning of peptides and proteins, but most times, the sequence is highly important. The sequences of tens of thousands of protein and peptide molecules have now been determined. The methods for doing so are complicated and will not be discussed in this book.

$$
\begin{array}{ccccccc}
9 & & & 4 & 3 & 2 & 1 \\
\text{Cys} - \text{S} - \text{S} - \text{Cys} - \text{Pro} - \text{Arg} - \text{Gly} - \text{NH}_2 \\
| & & & | \\
8\,\text{Tyr} & & & \text{Asn 5} \\
& \text{Phe} - \text{Gln} \\
& 7 \quad\;\; 6
\end{array}
$$

Vasopressin

$$
\begin{array}{ccccccc}
9 & & & 4 & 3 & 2 & 1 \\
\text{Cys} - \text{S} - \text{S} - \text{Cys} - \text{Pro} - \text{Leu} - \text{Gly} - \text{NH}_2 \\
| & & & | \\
8\,\text{Tyr} & & & \text{Asn 5} \\
& \text{Ile} - \text{Gln} \\
& 7 \quad\;\; 6
\end{array}
$$

Oxytocin

FIGURE 22.9 The structures of vasopressin and oxytocin. Differences are shown in color.

Chemical Connections 22D

Sickle Cell Anemia

Normal adult human hemoglobin has two alpha chains and two beta chains (see Figure 22.17). Some people, however, have a slightly different kind of hemoglobin in their blood. This hemoglobin (called HbS) differs from the normal type only in the beta chains and only in one position on these two chains: The glutamic acid in the sixth position of normal Hb is replaced with a valine residue in HbS.

	4	5	6	7	8	9
Normal Hb	—Thr—Pro—Glu—Glu—Lys—Ala—
Sickle cell Hb	—Thr—Pro—Val—Glu—Lys—Ala—

This change affects only two positions in a molecule containing 574 amino acid residues, yet it is enough to produce a very serious disease, **sickle cell anemia**.

Red blood cells carrying HbS behave normally when there is an ample oxygen supply. When the oxygen pressure decreases, however, the red blood cells become sickle-shaped, as shown in the figure. This malformation occurs in the capillaries. As a result of this change in shape, the cells may clog the capillaries. The body's defenses destroy the clogging cells, and the loss of the blood cells causes anemia.

This change at only a single position of a chain consisting of 146 amino acids is severe enough to cause a high death rate. A child who inherits two genes programmed to produce sickle cell hemoglobin (a homozygote) has an 80% smaller chance of surviving to adulthood than a child with only one such gene (a heterozygote) or a child with two normal genes. Despite the high mortality of homozygotes, the genetic trait survives. In central Africa, 40% of the population in malaria-ridden areas carry the sickle cell gene and 4% are homozygotes. It seems that the sickle cell genes help to acquire immunity against malaria in early childhood so that in malaria-ridden areas, the transmission of these genes is advantageous.

There is no known cure for sickle cell anemia. Recently, the U.S. Food and Drug Administration approved hydroxyurea

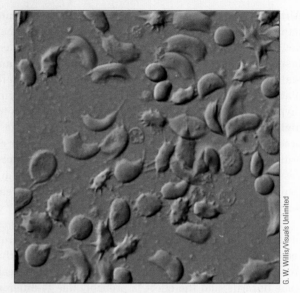

Blood cells from a patient with sickle cell anemia. Both normal cells (round) and sickle cells (shriveled) are visible.

G. W. Willis/Visuals Unlimited

(sold under the name Droxia) to treat and control the symptoms of the disease.

$$H_2NCN \quad \overset{O}{\underset{}{\|}} \quad \overset{H}{\underset{OH}{\diagdown}}$$

Hydroxyurea

Hydroxyurea prompts the bone marrow to manufacture fetal hemoglobin (HbF), which does not have beta chains where the mutation occurs. Thus, red blood cells containing HbF do not sickle and do not clog the capillaries. With hydroxyurea therapy, the bone marrow still manufactures mutated HbS, but the presence of cells with fetal hemoglobin dilutes the concentration of the sickling cells, thereby relieving the symptoms of the disease.

Test your knowledge with Problem 22.72.

22.9 What Is the Secondary Structure of a Protein?

Secondary structure A repetitive conformation of the protein backbone

Proteins can fold or align themselves in such a manner that certain patterns repeat themselves. These repeating patterns are referred to as **secondary structures**. The two most common secondary structures encountered in proteins are the α-helix and the β-pleated sheet (Figure 22.10), which were proposed by Linus Pauling and Robert Corey in the 1940s. In contrast, those protein conformations that do not exhibit a repeated pattern are called random coils (Figure 22.11).

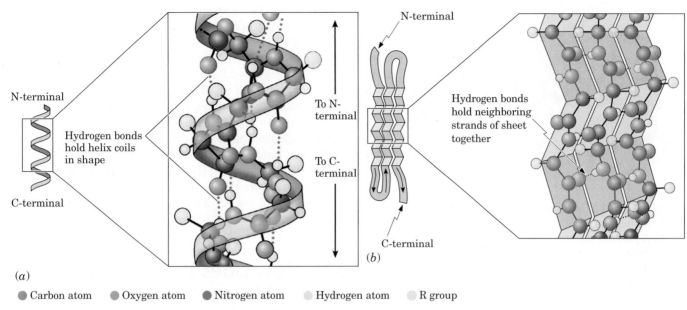

(a)

● Carbon atom ● Oxygen atom ● Nitrogen atom ● Hydrogen atom ● R group

FIGURE 22.10 (*a*) The α-helix. (*b*) The β-pleated sheet structure.

In the α-**helix** form, a single protein chain twists in such a manner that its shape resembles a right-handed coiled spring—that is, a helix. The shape of the helix is maintained by numerous **intramolecular hydrogen bonds** that exist between the backbone —C=O and H—N— groups. As shown in Figure 22.10, there is a hydrogen bond between the —C=O oxygen atom of each peptide linkage and the —N—H hydrogen atom of another peptide linkage four amino acid residues farther along the chain. These hydrogen bonds are in just the right position to cause the molecule (or a portion of it) to maintain a helical shape. Each —N—H points upward and each C=O points downward, roughly parallel to the axis of the helix. All the amino acid side chains point outward from the helix.

The other important orderly structure in proteins is the **β-pleated sheet**. In this case, the orderly alignment of protein chains is maintained by **intermolecular** or **intramolecular hydrogen bonds**. The β-sheet structure can occur between molecules when polypeptide chains run parallel (all N-terminal ends on one side) or antiparallel (neighboring N-terminal ends on opposite sides). β-Pleated sheets can also occur intramolecularly, when the polypeptide chain makes a U-turn, forming a hairpin structure, and the pleated sheet is antiparallel (Figure 22.10).

In all secondary structures, the hydrogen bonding is between backbone —C=O and H—N— groups, a characteristic that distinguishes between secondary and tertiary structures. In the latter, as we shall see, the hydrogen bonding can take place between R groups on the side chains.

Few proteins have predominantly α-helix or β-sheet structures. Most proteins, especially globular ones, have only certain portions of their molecules in these conformations. The rest of the molecule consists of **random coil**. Many globular proteins contain all three kinds of secondary structures in different parts of their molecules: α-helix, β-sheet, and random coil. Figure 22.12 shows a schematic representation of such a structure.

Keratin, a fibrous protein of hair, fingernails, horns, and wool, is one protein that does have a predominantly α-helix structure. Silk is made of fibroin, another fibrous protein, which exists mainly in the β-pleated sheet form. Silkworm silk and especially spider silk exhibit a combination of strength and toughness that is unmatched by high-performance synthetic fibers. In its primary structure, silk contains sections that consist of only

Alpha (α)-helix A secondary structure where the protein folds into a coil held together by hydrogen bonds parallel to the axis of the coil

Beta (β)-pleated sheet A secondary protein structure in which the backbone of two protein chains in the same or different molecules is held together by hydrogen bonds

FIGURE 22.11 A random coil.

FIGURE 22.12 Schematic structure of the enzyme carboxypeptidase. The β-pleated sheet portions are shown in blue, the green structures are the α-helix portions, and the orange strings are the random coil areas.

FIGURE 22.13 The triple helix of collagen.

Lanthanide series The 14 elements (58–71) immediately following lanthanum in period 6, in which the 4f shell is being filled

Actinide series The 14 elements (90–103) immediately following actinium in period 7, in which the 5f shell is being filled

alanine (25%) and glycine (42%). The formation of β-pleated sheets, largely by the alanine sections, allows microcrystals to orient themselves along the fiber axis, which accounts for the material's superior tensile strength.

Another repeating pattern classified as a secondary structure is the **extended helix** of collagen (Figure 22.13). It is quite different from the α-helix. Collagen is the structural protein of connective tissues (bone, cartilage, tendon, blood vessels, skin), where it provides strength and elasticity. The most abundant protein in humans, it makes up about 30% by weight of all the body's protein. The extended helix structure is made possible by the primary structure of collagen. Each strand of collagen consists of repetitive units that can be symbolized as Gly—X—Y; that is, every third amino acid in the chain is glycine. Glycine, of course, has the shortest side chain (—H) of all amino acids. About one-third of the X amino acid is proline, and the Y is often hydroxyproline.

22.10   Interlude: How Does the Presence of Transition Metals Affect the Structure of Proteins?

It is important to remember that the levels of structure in proteins are simply convenient tools for discussion. Each protein has its own unique structure, and the levels we assign are strictly ways of dividing up the topic to make it easier for us to understand. Protein structure is complex, so it can help our understanding to back off in the course of studying it, look at a related topic, and then come back to the main discussion.

We will look at an allied concept, the bonding of transition metals, because it directly relates to the tertiary structure of proteins. These bonding concepts are important in understanding the three-dimensional structure of many biologically important compounds, so it can help with learning about protein structure. The bonding of transition metals is very important because it involves *d* orbitals and relates most directly to the biochemical subjects in this book. *The key point is that bonding of transition metals produces definite geometries, such as those we have already seen in the bonding of carbon compounds. If a transition metal is bonded to several side chains of a given protein, those side chains are held in a definite position. The positions of the side chains, in turn, are a part of the overall three-dimensional structure of the protein.*

So far, our discussions of ionic and covalent bonding have focused on the chemical changes and bonding that involve electrons in *s* and *p* orbitals. The chemistry of second-period elements is essentially the chemistry involved in the sharing and transfer of electrons among 2*s* and 2*p* orbitals. Now let us take a cursory look at some of the chemistry of transition metals. Recall from Section 2.7 that a characteristic of the electron configuration of transition and inner transition elements is that they involve successive filling of sets of *d* and *f* orbitals. The transition elements lie in column 3B–2B of the Periodic Table (Figure 2.15). Alternatively, according to the newer IUPAC numbering system of the Periodic Table, they lie in columns 3–12. Because *d* orbitals occur in sets of five, *d* orbitals can hold 10 electrons, and thus, there are 10 transition elements in periods 4, 5, 6, and 7. We will not need to discuss bonding that involves electrons in *f* orbitals, but we will point out their existence for the sake of completeness. The inner transition elements are created by filling *f* orbitals, which occur in sets of seven, and therefore, the inner transition elements occur in groups of 14. There are 14 inner transition elements in the **lanthanide series** (elements 58–71) and 14 in the **actinide series** (elements 90–103).

How many valence electrons are present in a transition metal? The new IUPAC numbering system for the columns of the Periodic Table makes this easy to figure out. The number of the column gives the number of valence electrons directly. Iron, cobalt, and nickel, for example, have 8, 9, and 10 valence electrons, respectively.

In this section, we consider a group of transition metal compounds called **coordination compounds**. These compounds typically contain a transition metal as a component of either an ion or a neutral molecule, complexed with a cluster of other molecules or ions called **ligands**. For example, when ammonia is added to an aqueous solution of nickel(II) chloride, $NiCl_2$, a lilac-colored solution forms, from which another compound, $[Ni(NH_3)_6Cl_2]$ can be isolated. This and many other transition metal complexes are ionic compounds that consist of positive and negative ions. In the example just given, the respective ions are $Ni(NH_3)_6^{2+}$ ions and Cl^- ions. As another example, when ammonia is added to an aqueous solution of zinc chloride, $ZnCl_2$, a new compound can be isolated that is composed of $Zn(NH_3)_2^{2+}$ and Cl^- ions. As a final example, when finely divided iron metal, Fe, is treated with carbon monoxide, CO, a toxic liquid, $Fe(CO)_5$, forms. Because this coordination compound is prepared from a neutral iron atom and neutral carbon monoxide molecules, there are no ions present and the charge on this complex is zero.

The first synthetic transition metal complex to make its mark on modern medicine was the compound cisplatin, $[Pt(NH_3)_2Cl_2]$, discovered in 1960 by Dr. Barnet Rosenberg, head of the Barros Research Institute in East Lansing, Michigan. This compound is still used today as a treatment for many types of cancer and has been remarkably successful in treating testicular cancer.

⦻WL Interactive Example 22.2 Oxidation States in Transition Metal Complexes

Assuming no oxidation or reduction took place during the preparation of the complex $Fe(CO)_5$, what is the charge on the iron in this complex?

Strategy

Consider the overall charge on the compound and the charge on each of the ligands. Their sum gives the charge on the compound.

Solution

There is no net charge shown for the compound. Each carbon monoxide ligand is also uncharged. Therefore, there is no charge on the iron atom.

Problem 22.2

What is the oxidation number (the charge) on the platinum atom in cisplatin, $[Pt(NH_3)_2Cl_2]$?

Now that we have seen several examples of coordination compounds formed by transition metals, we then ask, "How do chemists explain the bonding between the metal and its ligands?" Chemists assume that (1) the bonds formed between the metal and its ligands are covalent and that (2) the driving force for complex formation is that the transition metal can become more stable if it has an electron configuration the same as that of the nearest noble gas.

The valence shell of a transition metal contains nine orbitals: one s orbital, three p orbitals, and five d orbitals. These nine orbitals can contain a maximum of 18 valence electrons. Thus, a transition metal can achieve a greater stability if it becomes surrounded by 18 valence electrons, hence the **18-electron rule**. Just as second-period elements are governed by the octet rule, the transition metals are governed by the 18-electron rule.

To see how this rule is applied, let us consider the coordination compound $Co(NH_3)_3Cl_3$. We start with the fact given in the Periodic Table that a neutral cobalt atom has 9 valence electrons. Cobalt in this complex is present as Co^{3+} (it has lost three electrons and has a charge of $+3$), leaving six valence electrons. Cobalt can achieve a complete valence shell and a greater stability by acquiring an additional 12 electrons in its valence shell. This is accomplished by having three ammonia molecules and three chloride ions each donate one unshared pair of electrons to cobalt, thus completing the valence shell of cobalt with 18 valence electrons.

The single covalent bond between a metal and a ligand is commonly called a **coordinate covalent bond**.

18-electron rule When transition metals form compounds, maximum stability is achieved with 18 electrons in bonds and unshared pairs, analogous to the octet rule with second row elements

Coordinate covalent bond A covalent bond between two elements, typically between a metal and a ligand, in which one of the bonded atoms provides both of the electrons for the bond

Example 22.3 Bonding in Coordination Compounds

Cobalt is essential for the synthesis and activity of vitamin B_{12}. The oxidation state of the cobalt changes as it performs its function. In one form, cobalt is present as Co^{3+} and is bound in a rigid complex formed by coordination with four nitrogen atoms of a large ring called corrin. In this complex, cobalt, Co^{3+}, forms coordinate covalent bonds with four nitrogen atoms of the corrin ring system, one to a nitrogen-containing ligand covalently bonded to the corrin ring, and one coordinate covalent bond with the $-CH_{2-}$ group of a molecule of adenosine. Thus, Co^{3+} is held rigidly in a framework created by six coordinate covalent bonds. Show how this coordination compound of cobalt and corrin satisfies the 18-electron rule.

Strategy

From the Periodic Table, determine the number of valence electrons in a cobalt atom and then in a Co^{3+} ion. Then determine how many electron pairs are supplied by ligands to satisfy the 18-electron rule.

Solution

From the Periodic Table, determine that cobalt has 9 valence electrons and that Co^{3+} has 6 valence electrons. To satisfy the 18-electron rule, cobalt must form a complex in which it shares an additional 12 electrons. These are supplied in vitamin B_{12} by one pair from each of the four nitrogen atoms of the corrin ring, one pair from the other nitrogen-containing ligand, and one pair from the $-CH_{2-}$ group of adenosine.

Problem 22.3

Describe the bonding in the complex ion $[Co(NH_3)_5Cl]^{2+}$.

Many biologically important compounds have iron as part of their structures in the form of coordination complexes, so we will pay particular attention to their bonding. The protein myoglobin is responsible for storage of oxygen in cells. The protein hemoglobin, which we will discuss in detail in Section 22.12, is responsible for carrying oxygen in the bloodstream. In both cases, the iron is the binding site for the oxygen. Myoglobin consists of a protein called globin and the iron-containing coordination complex called heme. Hemoglobin consists of four subunits, each of which is very similar to myoglobin. In myoglobin and in hemoglobin, the site of oxygen binding is similar. The standard images showing the detailed binding of oxygen were first determined in myoglobin, so we will use them here. The nitrogens of a large ring called porphyrin provide unshared pairs of electrons for four

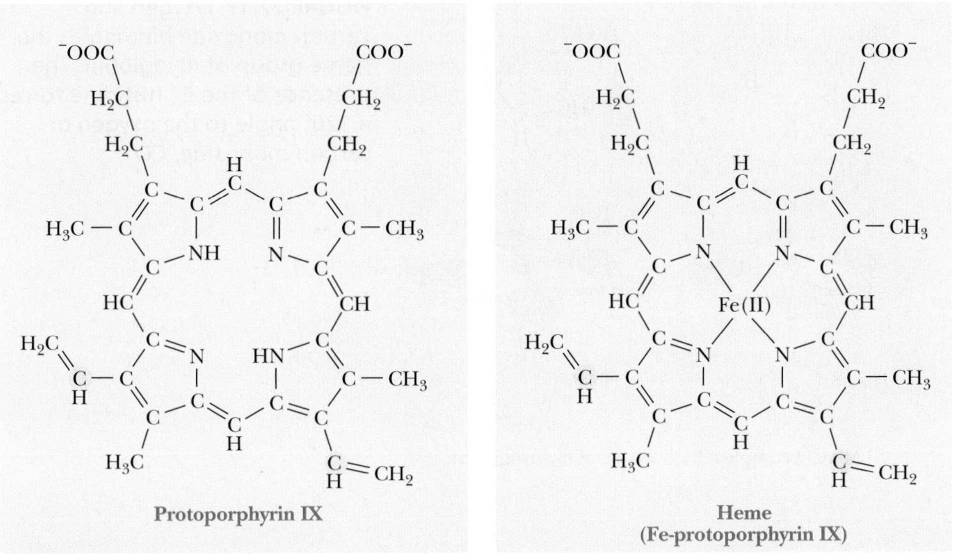

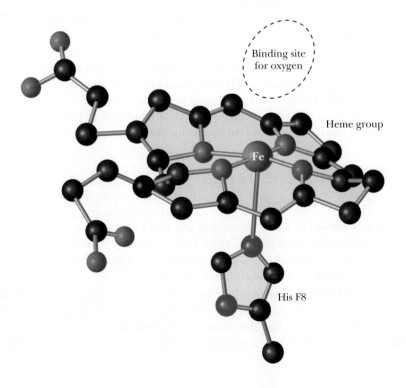

Protoporphyrin IX

Heme
(Fe-protoporphyrin IX)

FIGURE 22.14 The structure of the heme group. Several isomeric porphyrin rings are possible, depending on the nature and arrangement of the side chains. The porphyrin isomer found in heme is protoporphyrin IX. The addition of iron to protoporphyrin IX produces the heme group.

coordinate covalent bonds (Figure 22.14). Note that the ring is planar. A histidine side chain of globin is bonded to the iron as a fifth ligand. The bond is a coordinate covalent bond in which an unshared pair of electrons on nitrogen in the histidine side chain contributes both electrons. Oxygen is bound to the iron as the sixth ligand in the complex.

The iron coordination bonds in hemoglobin provide another example of the 18-electron rule. Iron is present as the Fe^{2+} ion. A glance at the Periodic Table indicates that this ion has 6 valence electrons. The 12 electrons needed come from the coordinate covalent bonds: four pairs from the nitrogens in the porphyrin ring, a fifth pair from the nitrogen in the histidine side chain, and the sixth from an unshared pair on oxygen (Figure 22.15).

FIGURE 22.15 The porphyrin ring occupies four of the six coordination sites of the Fe(II). Histidine occupies the fifth coordination site and oxygen is found as the sixth coordination site of the iron. (Leonard Lessin/Waldo Feng/Mt. Sinai (CORE).

Binding site for oxygen

Heme group

Fe

His F8

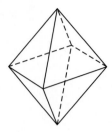

FIGURE 22.16 Octahedral binding geometry. Six ligands are positioned at the corners of an octahedron. Four of the ligands lie at 90° angles to each other in the same plane as the central atom. The other two ligands are positioned above and below the plane at 90° angles.

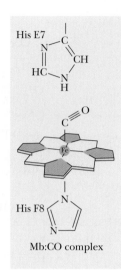

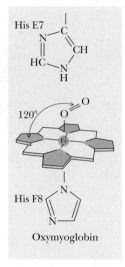

Mb:CO complex   Oxymyoglobin

FIGURE 22.17 Oxygen and carbon monoxide binding to the heme group of myoglobin. The presence of the E7 histidine forces a 120° angle to the oxygen or carbon monoxide, CO.

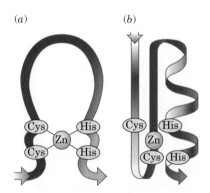

FIGURE 22.18 Cys$_2$His$_2$ zinc finger. (*a*) The coordination between zinc and cysteine and histidine residues. (*b*) The secondary structure.

The six bonds from the iron to its ligands are directed to the corners of an octahedron (Figure 22.16), with the four bonds to the porphyrin nitrogens in a plane at 90° to one another, the bond to the histidine perpendicular to the plane of the porphyrin, and the sixth bond to the oxygen also perpendicular to the porphyrin on the other side of the ring. This geometry is precisely what is predicted by VSEPR theory for six centers of electron density. This strictly prescribed spatial arrangement puts constraints on the three-dimensional structure of myoglobin and hemoglobin. Carbon monoxide binds to the same site as oxygen, with the carbon providing the electron pair for the coordinate covalent bond (see Problem 22.51). The Fe—C—O bond angle is approximately 120° because the approach of CO is blocked by another histidine side chain elsewhere in the protein (Figure 22.17). This situation keeps the CO from binding so strongly that it becomes nearly impossible to remove it from the binding site. The bond angle is the same for oxygen binding. It also allows myoglobin and hemoglobin to give up oxygen when it is needed by the body.

We can mention one more important example of metal-ion binding before we leave the topic. A number of proteins require zinc ion for activity. The Zn^{2+} ion is bound to a protein by coordinate covalent bonds. Because Zn^{2+} has 10 valence electrons, it needs eight more to satisfy the 18-electron rule. Four coordinate covalent bonds supply the needed electrons. VSEPR theory indicates that these bonds will take a tetrahedral arrangement, and that is what is observed. A particularly important class of zinc-binding proteins interacts with DNA sequences to control gene expression. In these proteins, called zinc fingers, the zinc forms bonds to the sulfurs of two cysteine side chains and to the nitrogens of two histidine side chains (Figure 22.18).

22.11 What Is the Tertiary Structure of a Protein?

The **tertiary structure** of a protein is the three-dimensional arrangement of every atom in the molecule. Unlike the secondary structure, it includes interactions of the side chains, and not just the peptide backbone. In general, tertiary structures are stabilized five ways:

1. **Covalent Bonds** The covalent bond most often involved in stabilization of the tertiary structure of proteins is the disulfide bond. In Section 22.4, we noted that the amino acid cysteine is easily converted to the dimer cystine. When a cysteine residue is in one chain and another

Tertiary structure The complete three-dimensional arrangement of the atoms in a protein

cysteine residue is in another chain (or in another part of the same chain), formation of a disulfide bond provides a covalent linkage that binds together the two chains or the two parts of the same chain:

$$\xi\!-\!SH \; HS\!-\!\xi \xrightarrow{\;[O]\;} \xi\!-\!S\!-\!S\!-\!\xi$$

Examples of both types are found in the structure of insulin (Figure 22.8).

2. **Hydrogen Bonding** In Section 22.9, we saw that secondary structures are stabilized by hydrogen bonding between backbone $-C=O$ and $-N-H$ groups. Tertiary structures are stabilized by hydrogen bonding between polar groups on side chains or between side chains and the peptide backbone [Figure 22.19(a)].

3. **Salt Bridges** Salt bridges, also called electrostatic attractions, occur between two amino acids with ionized side chains—that is, between an acidic amino acid ($-COO^-$) and a basic amino acid ($-NH_3^+$ or $=NH_2^+$) side chain. The two are held together by simple ion–ion attraction [Figure 22.19(b)].

4. **Hydrophobic Interactions** In aqueous solution, globular proteins usually turn their polar groups outward, toward the aqueous solvent, and their nonpolar groups inward, away from the water molecules. The nonpolar groups prefer to interact with each other, excluding water from these regions. The result is a series of hydrophobic interactions (see Section 21.1) [Figure 22.19(c)]. Although this type of interaction is weaker than hydrogen bonding or salt bridges, it usually acts over large surface areas, so that the interactions are collectively strong enough to stabilize a loop or some other tertiary structure formation.

5. **Metal Ion Coordination** Two side chains with the same charge would normally repel each other, but they can also be linked via a metal ion. For example, two glutamic acid side chains ($-COO^-$) would both be attracted to a magnesium ion (Mg^{2+}), forming a bridge. This is one reason the human body requires certain trace minerals—they are necessary components of proteins [Figure 22.19(d)].

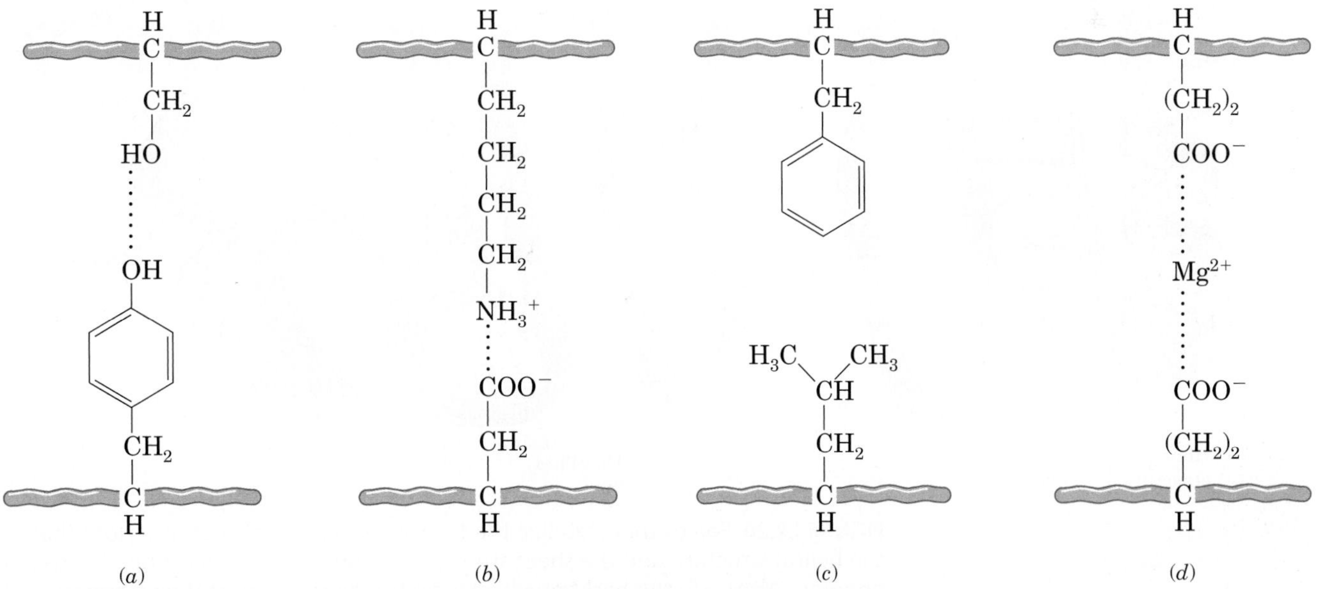

FIGURE 22.19 Noncovalent interactions that stabilize the tertiary and quaternary structures of proteins: (*a*) hydrogen bonding, (*b*) salt bridge (electrostatic interaction), (*c*) hydrophobic interaction, and (*d*) metal ion coordination.

Example 22.4 Amino Acid Interactions

What kind of noncovalent interaction occurs between the side chains of serine and glutamine?

Strategy

Analyze the types of functional groups in the side chains and then look for potential interactions.

Solution

The side chain of serine ends in an —OH group; that of glutamine ends in an amide, the CO—NH$_2$ group. The two groups can form hydrogen bonds.

Problem 22.4

What kind of noncovalent interaction occurs between the side chains of arginine and glutamic acid?

In Section 22.8, we pointed out that the primary structure of a protein largely determines its secondary and tertiary structures. We can now see the reason for this relationship. When the particular R groups are in the proper positions, all of the hydrogen bonds, salt bridges, disulfide linkages, and hydrophobic interactions that stabilize the three-dimensional structure of that molecule can form. Figure 22.20 illustrates the possible combinations of forces that lead to tertiary structure.

The side chains of some proteins allow them to fold (form a tertiary structure) in only one way; other proteins, especially those with long polypeptide chains, can fold in a number of possible ways. Certain proteins in living

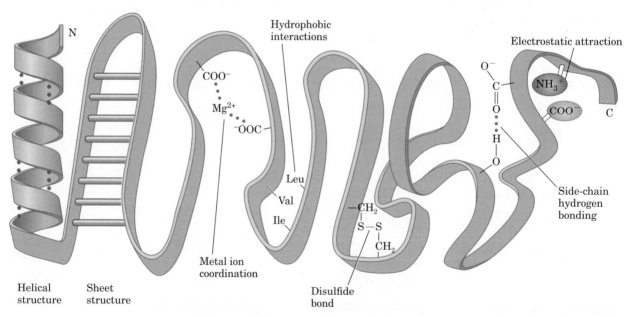

FIGURE 22.20 Forces that stabilize the tertiary structures of proteins. Note that the helical structure and the sheet structure are two kinds of backbone hydrogen bonding. Although the backbone hydrogen bonding is part of the secondary structure, the conformation of the backbone puts constraints on the possible arrangement of the side chains.

Protein/Peptide Conformation–Dependent Diseases

In a number of diseases, a normal protein or peptide becomes pathological when its conformation changes. A common feature of these proteins is the property to self-assemble into β-sheet–forming amyloid (starch-like) plaques. These amyloid structures appear in several diseases.

One example of this process involves the prion protein, the discovery of which brought Stanley Prusiner of the University of California, San Francisco, the Nobel Prize in 1997. Prions are small proteins found in nerve tissue, although their exact function remains a mystery. When prions undergo conformational change, they can cause diseases such as mad cow disease and scrapie in sheep. During the conformational change, the α-helical content of the normal prion protein unfolds and reassembles in the β-sheet form. This new form has the potential to cause more normal prion proteins to undergo conformational change. In humans, it causes spongiform encephalitis; Creutzfeld-Jakob disease is one variant that mainly afflicts elderly people. Although the transmission of this infection from diseased cows to humans is rare, fear of it caused the wholesale slaughter of British

cattle in 1998 and for a while, an embargo was placed on the importation of such meat in most of Europe and America. β-Amyloid plaques also appear in the brains of patients with Alzheimer's disease (see Chemical Connections 24C).

The modus operandi of prion diseases stumped scientists for many years. On the one hand, human spongiform encephalopathies behave like inheritable diseases in that they can be traced through families. On the other hand, they behave like infectious diseases, which can be acquired from someone else. It is now believed that the mechanism of spread is a combination of the two. There is a genetic component in that a person could have a 100% wild-type prion protein that would not adopt the alternate form. Several mutations that lead to the abnormal prion form have been identified. However, there appears to be a need for a triggering event as well. This characteristic was seen in studies of sheep in New Zealand, where isolated groups were found to have the right mutations to get a prion disease, but none of the sheep did, generation after generation, because they were never infected with a mutant prion.

Schematic representation of a possible mechanism of amyloid fibril formation. After synthesis, the protein is assumed to fold in native (N) secondary structure aided by chaperones. Under certain conditions, the native structure can partially unfold (I) and form sheets of amyloid fibrils or even completely unfold (U) as a random coil.

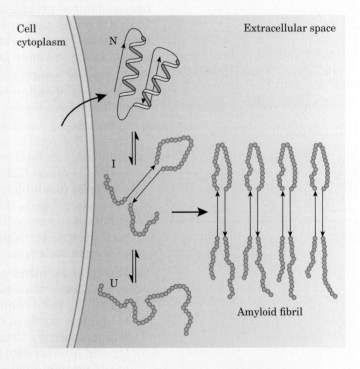

Test your knowledge with Problem 22.73.

Chaperone A protein that helps other proteins to fold into the biologically active conformation and enables partially denatured proteins to regain their biologically active conformation

Quaternary structure The spatial relationship and interactions between subunits in a protein that has more than one polypeptide chain

The designations α and β with respect to hemoglobin have nothing to do with the same designations for the α-helix and β-pleated sheet.

cells, called **chaperones**, help a newly synthesized polypeptide chain assume the proper secondary and tertiary structures that are necessary for the functioning of that molecule and prevent foldings that would yield biologically inactive molecules.

22.12 What Is the Quaternary Structure of a Protein?

The highest level of protein organization is the **quaternary structure**, which applies to proteins with more than one polypeptide chain. Figure 22.21 summarizes schematically the four levels of protein structure. Quaternary structure determines how the different subunits of the protein fit into an organized whole. The subunits are packed and held together by hydrogen bonds, salt bridges, and hydrophobic interactions—the same forces that operate within tertiary structures.

1. **Hemoglobin** Hemoglobin in adult humans is made of four chains (called globins): two identical α chains of 141 amino acid residues each and two identical β chains of 146 residues each. Figure 22.22 shows how the four chains fit together.

   In hemoglobin, each globin chain surrounds an iron-containing heme unit, the structure of which is shown in Figure 22.23. Proteins that contain non-amino acid portions are called **conjugated proteins**. The non-amino acid portion of a conjugated protein is called a **prosthetic group**. In hemoglobin, the globins are the amino acid portions and the heme units are the prosthetic groups.

   Hemoglobin containing two alpha and two beta chains is not the only kind existing in the human body. In the early developmental stage of the fetus, the hemoglobin contains two alpha and two gamma chains. The fetal hemoglobin has greater affinity for oxygen than does the adult hemoglobin. In this way, the mother's red blood cells carrying oxygen can pass it to the fetus for its own use. Fetal hemoglobin also alleviates some of the symptoms of sickle cell anemia (see Chemical Connections 22D).

2. **Collagen** Another example of quaternary structure and higher organizations of subunits can be seen in collagen. The triple helix units, called *tropocollagen,* constitute the soluble form of collagen; they are stabilized by hydrogen bonding between the backbones of the three chains. Collagen consists of many tropocollagen units. Tropocollagen is found only in fetal or young connective tissues. With aging, the triple helixes (Figure 22.13) that organize themselves into fibrils cross-link and form insoluble collagen. In collagen, the **cross-linking** consists of covalent bonds that link together two lysine residues on adjacent chains of the helix. This cross-linking of collagen is an example of the tertiary structures that stabilize the three-dimensional conformations of protein molecules.

3. **Integral Membrane Proteins** Integral membrane proteins partly or completely traverse a membrane bilayer. (See Figure 21.2.) An estimated one-third of all proteins are integral membrane proteins. To keep the protein stable in the nonpolar environment of a lipid bilayer, it must form quaternary structures in which the outer surface is largely nonpolar and interacts with the lipid bilayer. Thus, most of the polar groups of the protein must turn inward. Two such quaternary structures exist in integral membrane proteins: (1) often 6 to 10 α-helices crossing the membrane and (2) β-barrels made of 8, 12, 16, or 18 antiparallel β-sheets (Figures 22.24 and 22.25).

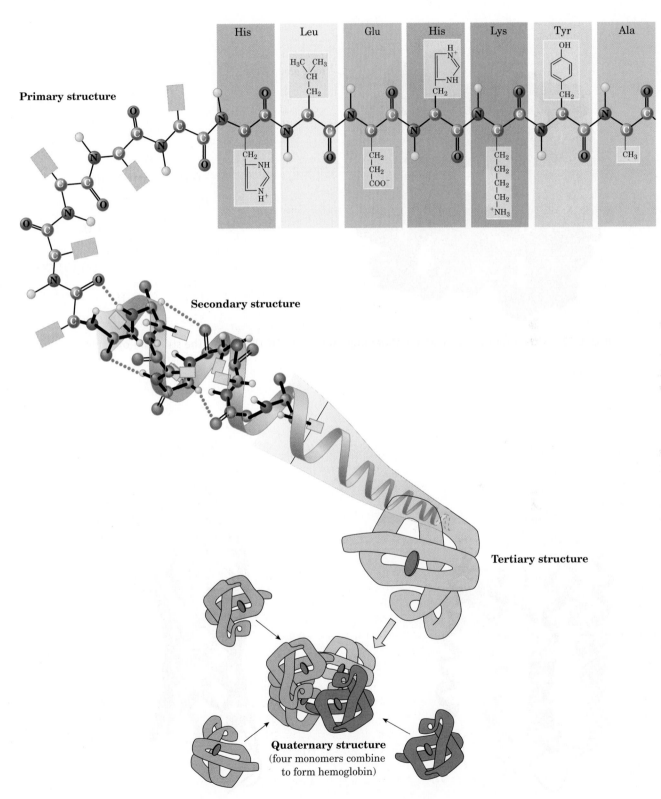

FIGURE 22.21 Primary, secondary, tertiary, and quaternary structures of a protein.

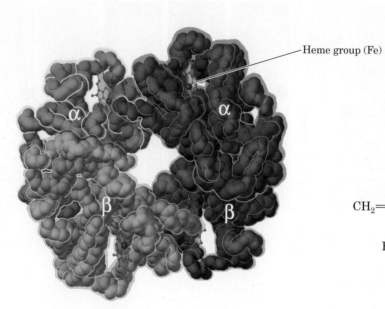

FIGURE 22.22 The quaternary structure of hemoglobin.

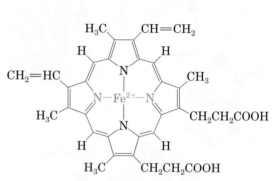

FIGURE 22.23 The structure of heme.

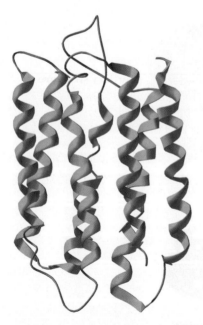

FIGURE 22.24 Integral membrane protein of rhodopsin, made of α-helices.

FIGURE 22.25 An integral membrane protein from the outer mitochondrial membrane forming a β-barrel from eight β-pleated sheets.

Chemical Connections 22F

Proteomics, Ahoy!

Proteins in the body are in a state of dynamic flux. Their multiple functions necessitate that they change constantly: Some are rapidly synthesized; others have their synthesis inhibited. Some are degraded; others are modified. The complement of proteins expressed by a genome is called its **proteome**. Today, a concerted effort is being made to catalogue all the proteins in their various forms in a particular cell or a tissue. The name of this venture is *proteomics,* a term coined as an analogy to *genomics* (see Chemical Connections 25D), in which all the genes of an organism and their locations in the chromosomes are determined. The approximately 30,000 genes defined by the Human Genome Project translate into 300,000 to 1 million proteins when alternate splicing and post-translational modifications are considered (Chapter 26).

While a genome remains unchanged to a large extent, the proteins in any particular cell change dramatically as genes are turned on and off in response to its environment. In proteomics, all the proteins and peptides of a cell or tissue are separated and then studied by a variety of procedures, including some very new technologies. The first procedure is necessary to separate proteins from one another. The main way this separation has been achieved is two-dimensional polyacrylamide gel electrophoresis (2-D PAGE; see Chapter 25 for more on electrophoresis). 2-D PAGE can achieve the separation of several thousand different proteins in one gel. High-resolution 2-D PAGE can resolve as many as 10,000 protein spots per gel. In one dimension, the proteins are separated by charge (isoelectric point; Section 22.3); in the second dimension, they are separated by mass. In isoelectric focusing, the proteins migrate in a pH gradient to the pH at which they have no net charge (the isoelectric point). Most commonly, proteins are separated by size in the vertical direction and by isoelectic point in the horizontal direction.

Mass spectrometry is used for mass determination and can be adapted for protein identification. A mass spectrometer separates proteins according to their mass-to-charge (m/Z) ratio. The molecule is ionized by one of several techniques, and the ion is propelled into a mass analyzer by an electric field that resolves each ion according to its m/Z ratio. The detector passes the information to the computer for analysis.

A new technology that holds much promise for the future of protein analysis is **protein microarrays**. They can be used for protein purification, expression profiling, or protein interaction profiling. Many types of substances may be bound to protein arrays, including antibodies,

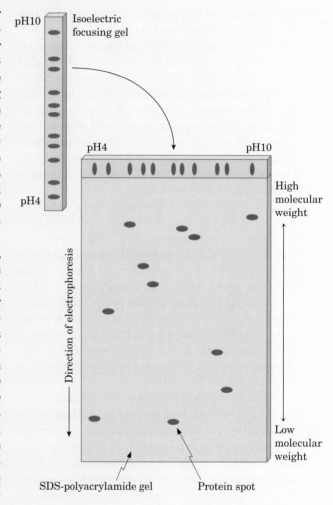

receptors, ligands, nucleic acids, carbohydrates, or chromatographic (cationic, anionic, hydrophobic, hydrophilic) surfaces. Some surfaces have broad specificity and bind entire classes of proteins; others are highly specific and bind only a few proteins from a complex sample. Some protein arrays contain antibodies (Chapter 31) that are covalently immobilized onto the array surface and that capture corresponding antigens from a complex mixture. Many analyses can follow from this binding. Other proteins of interest can also be immobilized on the array. Bound receptors can reveal ligands, and binding domains for protein–protein interactions can be detected.

The goal of such techniques is to obtain information about the dynamic states of a large number of proteins and the status of a cell or tissue, thereby ascertaining whether it is healthy or pathological.

Test your knowledge with Problem 22.74.

Quaternary Structure and Allosteric Proteins

Quaternary structure is a property of proteins that consist of more than one polypeptide chain. Each chain is called a subunit. The number of chains can range from two to more than a dozen, and the chains may be identical or different. The chains interact with one another noncovalently via electrostatic attractions, hydrogen bonds, and hydrophobic interactions. As a result of these noncovalent interactions, subtle changes in structure at one site on a protein molecule may cause drastic changes in properties at a distant site. Proteins that exhibit this property are called **allosteric proteins**. Not all multisubunit proteins exhibit allosteric effects, but many do.

A classic illustration of the quaternary structure of proteins and its effect on properties is a comparison of hemoglobin, an allosteric protein, with myoglobin, which consists of a single polypeptide chain. Both hemoglobin and myoglobin bind to oxygen via a heme group (Figure 22.18). As we have seen, hemoglobin is a **tetramer**, a molecule consisting of four polypeptide chains: two α chains and two β chains. The two α chains of hemoglobin are identical, as are the two β chains. The overall structure of hemoglobin is $\alpha_2\beta_2$ in Greek-letter notation. Both the α and β chains of hemoglobin are very similar to the myoglobin chain.

The α chain is 141 residues long, and the β chain is 146 residues long. For comparison, the myoglobin chain is 153 residues long. Many of the amino acids of the α chain, the β chain, and myoglobin are *homologous;* that is, the same amino acid residues occupy the same positions.

The heme group is the same in both myoglobin and hemoglobin. One molecule of myoglobin binds to one oxygen

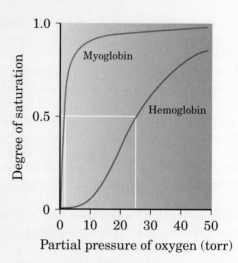

A comparison of the oxygen-binding behavior of myoglobin and hemoglobin. The oxygen-binding curve of myoglobin is hyperbolic, whereas that of hemoglobin is sigmoidal. Myoglobin is 50% saturated with oxygen at 1 torr partial pressure; hemoglobin does not reach 50% saturation until the partial pressure of oxygen reaches 26 torr.

molecule. Four molecules of oxygen can bind to one hemoglobin molecule. Both hemoglobin and myoglobin bind oxygen reversibly, but the binding of oxygen to hemoglobin exhibits **positive cooperativity**, whereas oxygen binding to myoglobin does not. Positive cooperativity means that when one oxygen molecule is bound, it becomes easier for the next molecule to bind. A graph of the oxygen-binding properties of hemoglobin and myoglobin is one of the best ways to illustrate this point.

22.13 How Are Proteins Denatured?

Denaturation The loss of the secondary, tertiary, and quaternary structures of a protein by a chemical or physical agent that leaves the primary structure intact

Protein conformations are stabilized in their native states by secondary and tertiary structures and through the aggregation of subunits in quaternary structure. Any physical or chemical agent that destroys these stabilizing structures changes the conformation of the protein (Table 22.3). We call this process **denaturation**.

Table 22.3 Modes of Protein Denaturation (Destruction of Secondary and Higher Structures)

Denaturing Agent	Affected Regions
Heat	H bonds
6 M urea	H bonds
Detergents	Hydrophobic regions
Acids, bases	Salt bridges, H bonds
Salts	Salt bridges
Reducing agents	Disulfide bonds
Heavy metals	Disulfide bonds
Alcohol	Hydration layers

Quaternary Structure and Allosteric Proteins (*continued*)

When the degree of saturation of myoglobin with oxygen is plotted against oxygen pressure, a steady rise is observed until complete saturation is approached and the curve levels off. The oxygen-binding curve of myoglobin is thus said to be **hyperbolic**. In contrast, the oxygen-binding curve for hemoglobin is **sigmoidal**. This shape indicates that the binding of the first oxygen molecule facilitates the binding of the second oxygen, which facilitates the binding of the third oxygen, which in turn facilitates the binding of the fourth oxygen. This is precisely what is meant by the term "cooperative binding."

The two types of behavior are related to the functions of these proteins. Myoglobin has the function of oxygen *storage* in muscle. It must bind strongly to oxygen at very low pressures, and it is 50% saturated at a partial pressure of oxygen of 1 torr (Section 6.2). The function of hemoglobin is oxygen *transport,* and it must be able to bind strongly to oxygen and to release oxygen easily, depending upon conditions. In the alveoli of lungs (where hemoglobin must bind oxygen for transport to the tissues), the oxygen pressure is 100 torr. At this pressure, hemoglobin is 100% saturated with oxygen. In the capillaries running through active muscles, the pressure of oxygen is 20 torr, corresponding to less than 50% saturation of hemoglobin, which occurs at 26 torr. In other words, hemoglobin gives up oxygen easily in capillaries, where the need for oxygen is great.

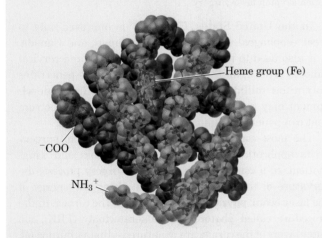

The structure of myoglobin.

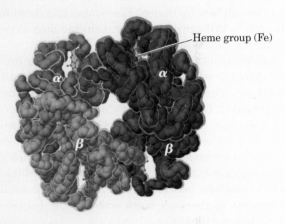

The structure of hemoglobin.

Test your knowledge with Problem 22.75.

For example, heat cleaves hydrogen bonds, so boiling a protein solution destroys the α-helical and β-pleated sheet structure. In collagen, the triple helixes disappear upon boiling and the molecules have a largely random-coil conformation in the denatured state, which is gelatin. In other proteins, especially globular proteins, heat causes the unfolding of the polypeptide chains; because of subsequent intermolecular protein–protein interactions, precipitation or coagulation then takes place. That is what happens when we boil an egg.

Similar conformational changes can be brought about by the addition of denaturing chemicals. Solutions such as 6 *M* aqueous urea, $H_2N—CO—NH_2$, break hydrogen bonds and cause the unfolding of globular proteins. Surface-active agents (detergents) change protein conformation by opening up the hydrophobic regions, whereas acids, bases, and salts affect both salt bridges and hydrogen bonds.

Reducing agents such as 2-mercaptoethanol ($HOCH_2CH_2SH$) can break the —S—S— disulfide bonds, reducing them to —SH groups. The processes of permanent waving and straightening of curly hair are examples of the latter effect (Figure 22.26). The protein keratin, which makes up human hair, contains a high percentage of disulfide bonds. These bonds are

FIGURE 22.26 A permanent wave alters the shape of hair through reduction and oxidation of disulfides and thiols.

Laser Surgery and Protein Denaturation

Proteins can be denatured by physical means, most notably by heat. For instance, bacteria are killed and surgical instruments are sterilized by heat. A special method of heat denaturation that is seeing increasing use in medicine relies on lasers. A laser beam (a highly coherent light beam of a single wavelength) is absorbed by tissues, and its energy is converted to heat energy. This process can be used to cauterize incisions so that a minimal amount of blood is lost during the operation.

Laser beams can be delivered by an instrument called a **fiberscope**. The laser beam is guided through tiny fibers, thousands of which are fitted into a tube only 1 mm in diameter. In this way, the laser delivers the energy for denaturation only where it is needed. It can, for example, seal wounds or join blood vessels without the necessity of cutting through healthy tissues. Fiberscopes have been used successfully to diagnose and treat many bleeding ulcers in the stomach, intestines, and colon.

A novel use of the laser fiberscope is in treating tumors that cannot be reached for surgical removal. A drug called Photofrin®, which is activated by light, is given to patients intravenously. The drug in this form is inactive and harmless. The patient then waits 24 to 48 hours, during which time the drug accumulates in the tumor but is removed and excreted from healthy tissues. A laser fiberscope that emits red light at 630 nm is then directed toward the tumor. An exposure between 10 and 30 minutes is applied. The energy of the laser beam activates the Photofrin®, which destroys the tumor.

This technique does not offer a complete cure, because the tumor may grow back or it may have spread before the treatment. The treatment has only one side effect: The patient remains sensitive to exposure to strong light for approximately 30 days (so sunlight must be avoided). Of course, this inconvenience is minor compared to the pain, nausea, hair loss, and other side effects that accompany traditional radiation or chemotherapy of tumors.

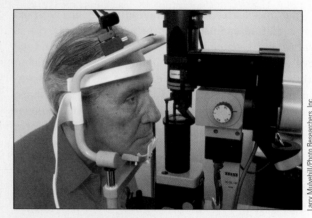

Argon krypton laser surgery.

In the United States, Photofrin® is approved only to treat esophageal cancer. In Europe, Japan, and Canada, it is also used to treat lung, bladder, gastric, and cervical cancers. The light that activates Photofrin® penetrates only a few millimeters, but the new drugs under development may use radiation in the near-infrared spectrum that can penetrate tumors up to a few centimeters.

The most common use of laser technology in surgery is its application to correct near-sightedness and astigmatism. In a computer-assisted laser surgery process, the curvature of the cornea is changed. Using the energy of the laser beam, physicians remove part of the cornea. In the procedure called photorefractive keratectomy (PRK), the outer layers of the cornea are denatured—that is, burned off. In the LASIK (laser-assisted in situ keratomileusis) procedure, the surgeon creates a flap or a hinge of the outer layers of the cornea and then with the laser beam, burns off a computer-programmed amount under the flap to change the shape of the cornea. After the 5- to 10-minute procedure is complete, the flap is put back, and it heals without stitches. In successful surgeries, patients regain good vision one day after the surgery and no longer need prescription lenses.

Test your knowledge with Problem 22.76.

primarily responsible for the shape of the hair, whether straight or curly. In either permanent waving or straightening, the hair is first treated with a reducing agent that cleaves some of the —S—S— bonds. This treatment allows the molecules to lose their rigid orientations and become more flexible. The hair is then set into the desired shape, using curlers or rollers, and an oxidizing agent is applied. The oxidizing agent reverses the preceding reaction, forming new disulfide bonds, which now hold the molecules together in the desired positions.

Heavy metal ions (for example, Pb^{2+}, Hg^{2+}, and Cd^{2+}) also denature protein by attacking the —SH groups. They form salt bridges, as in —$S^- Hg^{2+} S$—. This very feature is taken advantage of in the antidote for oral heavy metal poisoning: raw egg whites and milk. The egg and milk proteins are denatured by the metal ions, forming insoluble precipitates in the stomach. These must be pumped out or removed by inducing vomiting. In this way, the poisonous metal ions are removed from the body. If the antidote is not pumped out of the stomach, the digestive enzymes would degrade the proteins and release the poisonous heavy metal ions, which would then be absorbed into the bloodstream.

Other chemical agents such as alcohol also denature proteins, coagulating them. This process is used in sterilizing the skin before injections. At a concentration of 70%, ethanol penetrates bacteria and kills them by coagulating their proteins, whereas 95% alcohol denatures only surface proteins.

Denaturation changes secondary, tertiary, and quaternary structures. It does not affect primary structures (that is, the sequence of amino acids that make up the chain). If these changes occur to a small extent, denaturation can be reversed. For example, when we remove a denatured protein from a urea solution and put it back into water, it often reassumes its secondary and tertiary structures. This process is called reversible denaturation. In living cells, some denaturation caused by heat can be reversed by chaperones. These proteins help a partially heat-denatured protein to regain its native secondary, tertiary, and quaternary structures. Some denaturation, however, is irreversible. We cannot unboil a hard-boiled egg, for example.

Raw egg whites are an antidote to heavy metal poisoning.

Summary

OWL Sign in at www.cengage.com/owl to develop problem-solving skills and complete online homework assigned by your professor.

Section 22.1 What Are the Many Functions of Proteins?

- **Proteins** are giant molecules made of amino acids linked together by **peptide bonds**.
- Proteins have many functions: structural (collagen), enzymatic, carrier (hemoglobin), storage (casein), protective (immunoglobulin), and hormonal (insulin).

Section 22.2 What Are Amino Acids?

- **Amino acids** are organic compounds containing an amino and a carboxyl group.
- The 20 common amino acids found in proteins are classified by their side chains: nonpolar, polar but neutral, acidic, and basic.
- All amino acids in human tissues are L-amino acids.

Section 22.3 What Are Zwitterions?

- Amino acids in the solid state, as well as in water, carry both positive and negative charges; they are called **zwitterions**.
- The pH at which the number of positive charges equals the number of negative charges is the **isoelectric point** of an amino acid or protein.

Section 22.4 What Determines the Characteristics of Amino Acids?

- Amino acids are nearly identical in most ways except for their side chain (R—) groups.
- It is the unique nature of the side chain that gives an amino acid its particular properties.

- Some amino acids have charged side chains (Glu, Asp, Lys, Arg, His).
- Cysteine is a special amino acid because its side chain (—SH) can form disulfide bridges with another cysteine.
- The aromatic amino acids (Phe, Tyr, Trp) are important physiologically, because they are precursors of neurotransmitters. They also absorb ultraviolet light and allow us to easily measure and locate them.

Section 22.5 What Are Uncommon Amino Acids?

- Besides the 20 common amino acids found in proteins, other amino acids are known.
- These amino acids are normally produced after one of the standard amino acids has been incorporated into a protein.
- Examples include hydroxyproline (collagen), hydroxylysine, and thyroxine.

Section 22.6 How Do Amino Acids Combine to Form Proteins?

- When the amino group of one amino acid condenses with the carboxyl group of another amino acid, an amide (peptide) bond is formed, with the elimination of water.
- Two amino acids form a dipeptide. Three amino acids form a tripeptide.
- Many amino acids form a **polypeptide chain**. Proteins are made of one or more polypeptide chains.

Section 22.7 What Are the Properties of Proteins?

- The properties of proteins are based on properties of the peptide backbone and the properties of the side chains.

- Although the peptide bond is typically written as a carbonyl group bonded to an N—H group, such bonds can exhibit keto-enol tautomerism. As a result, the peptide bond that links two amino acids is planar.
- The planar nature of the peptide bond limits the possible orientations that peptides and proteins can take.
- The nature of the amino acid side chains determines most of the nature of a protein.
- Some amino acids have acidic or basic side chains. The isoelectric point of a protein is the pH where all the negative charges match all the positive charges and the net charge on the protein is zero.

Section 22.8 What Is the Primary Structure of a Protein?

- The linear sequence of amino acids is the **primary structure** of the protein.
- The primary structure is largely responsible for the eventual higher-order structures of proteins.

Section 22.9 What Is the Secondary Structure of a Protein?

- The repeating short-range conformations (**α-helix, β-pleated sheet, extended helix of collagen**, and **random coil**) are the **secondary structures** of proteins.
- Secondary structure refers to those repetitive structures that are held together via hydrogen bonds between groups on the peptide backbone only.

Section 22.10 Interlude: How Does the Presence of Transition Metals Affect the Structure of Proteins?

- Transition metals form biologically important compounds with strictly prescribed three-dimensional structures.
- The bonding in these compounds, which are called coordination complexes, depends on coordinate covalent

bonds in which both electrons are donated to the bond by the substance other than the transition metal.
- Bonding in transition-metal complexes is described by the 18-electron rule, which takes into account the contribution of d-electrons.

Section 22.11 What Is the Tertiary Structure of a Protein?

- The **tertiary structure** is the three-dimensional conformation of the protein molecule.
- Tertiary structures are maintained by covalent cross-links such as **disulfide bonds** and by **salt bridges, hydrogen bonds, metal ion coordination**, and **hydrophobic interactions** between the side chains.

Section 22.12 What Is the Quaternary Structure of a Protein?

- The precise fit of polypeptide subunits into an aggregated whole is called the **quaternary structure**.
- Not all proteins have a quaternary structure—only those proteins that have subunits have this structure.
- Hemoglobin is an example of a protein that exhibits a quaternary structure.

Section 22.13 How Are Proteins Denatured?

- Secondary and tertiary structures stabilize the native conformations of proteins.
- Physical and chemical agents, such as heat or urea, destroy these structures and **denature** proteins.
- Protein functions depend on native conformation; when a protein is denatured, it can no longer carry out its function.
- Some (but not all) denaturation is reversible; in some cases, **chaperone** molecules may reverse denaturation.

Problems

⟳WL Interactive versions of these problems may be assigned in OWL.

Orange-numbered problems are applied.

References to previous chapters are given in parentheses.

Section 22.1 What Are the Many Functions of Proteins?

22.5 What are the functions of (a) ovalbumin and (b) myosin?

22.6 The members of which class of proteins are insoluble in water and can serve as structural materials?

22.7 What is the function of an immunoglobulin?

22.8 What are the two basic types of proteins?

Section 22.2 What Are Amino Acids?

22.9 What is the difference in structure between tyrosine and phenylalanine?

22.10 Classify the following amino acids as nonpolar, polar but neutral, acidic, or basic.
  (a) Arginine   (b) Leucine
  (c) Glutamic acid   (d) Asparagine
  (e) Tyrosine   (f) Phenylalanine
  (g) Glycine

22.11 Which amino acid has the highest percentage of nitrogen (g N/100 g amino acid)?

22.12 Why does glycine have no D or L form?

22.13 Draw the structure of proline. To which class of heterocyclic compounds does this molecule belong? (See Section 16.1.)

22.14 Which amino acid is also a thiol?

22.15 Why is it necessary to have proteins in our diets?

22.16 Which amino acids in Table 22.1 have more than one stereocenter?

22.17 What are the similarities and differences in the structures of alanine and phenylalanine?

22.18 Draw the structures of L- and D-valine.

Section 22.3 What Are Zwitterions?

22.19 Why are all amino acids solids at room temperature?

22.20 Show how alanine, in solution at its isoelectric point, acts as a buffer (write equations to show why the pH does not change much if we add an acid or a base).

22.21 Explain why an amino acid cannot exist in an un-ionized form at any pH.

22.22 Draw the structure of valine at pH 1 and at pH 12.

22.23 Draw the most predominant form of aspartic acid at its isoelectric point.

22.24 Draw the most predominant form of histidine at its isoelectric point.

22.25 Draw the most predominant form of lysine at its isoelectric point.

22.26 Draw the sequential transition of glutamic acid as it passes from its fully protonated form to its fully deprotonated form as the pH rises.

Section 22.4 What Determines the Characteristics of Amino Acids?

22.27 Which of the three functional groups on histidine is the most unique?

22.28 How are aromatic amino acids related to neurotransmitters?

22.29 Why is histidine considered a basic amino acid when the pK_a of its side chain is 6.0?

22.30 Which are the acidic amino acids?

22.31 Which are the basic amino acids?

22.32 Why does proline not absorb light at 280 nm?

Section 22.5 What Are Uncommon Amino Acids?

22.33 Two of the 20 amino acids listed in Table 22.1 can be obtained by hydroxylation of other amino acids. What are those two, and what are their precursor amino acids?

22.34 When a protein contains hydroxyproline, at what point in the production of the protein is the proline hydroxylated?

22.35 What is the effect of thyroxine on metabolism?

22.36 How is thyroxine made?

Section 22.6 How Do Amino Acids Combine to Form Proteins?

22.37 Show by chemical equations how alanine and glutamine can be combined to give two different dipeptides.

22.38 A tetrapeptide is abbreviated as DPKH. Which amino acid is at the N-terminal, and which is at the C-terminal?

22.39 Draw the structure of a tripeptide made of threonine, arginine, and methionine.

22.40 (a) Use the three-letter abbreviations to write a representation of the following tripeptide:

(b) Which amino acid is at the C-terminal end, and which is at the N-terminal end?

22.41 A polypeptide chain is made of alternating valine and phenylalanine. Which part of the polypeptide is polar (hydrophilic)?

Section 22.7 What Are the Properties of Proteins?

22.42 (a) How many atoms of the peptide bond lie in the same plane?

(b) Which atoms are they?

22.43 (a) Draw the structural formula of the tripeptide met—ser—cys.

(b) Draw the different ionic structures of this tripeptide at pH 2.0, 7.0, and 10.0.

22.44 How can a protein act as a buffer?

22.45 Proteins are least soluble at their isoelectric points. What would happen to a protein precipitated at its isoelectric point if a few drops of dilute HCl were added?

Section 22.8 What Is the Primary Structure of a Protein?

22.46 How many different tripeptides can be made (a) using one, two, or three residues each of leucine, threonine, and valine and (b) using all 20 amino acids?

22.47 How many different tetrapeptides can be made (a) if the peptides contain the residues of asparagine, proline, serine, and methionine and (b) if all 20 amino acids can be used?

22.48 How many amino acid residues in the A chain of insulin are the same in insulin from humans, cattle (bovine), hogs, and sheep?

22.49 Based on your knowledge of the chemical properties of amino acid side chains, suggest a substitution for leucine in the primary structure of a protein that would probably not change the character of the protein very much.

Section 22.9 What Is the Secondary Structure of a Protein?

22.50 Is a random coil a (a) primary, (b) secondary, (c) tertiary, or (d) quaternary structure? Explain.

22.51 Decide whether the following structures that exist in collagen are primary, secondary, tertiary, or quaternary.

(a) Tropocollagen

(b) Collagen fibril

(c) Collagen fiber

(d) The proline—hydroxyproline—glycine repeating sequence

22.52 Proline is often called an α-helix terminator; that is, it is usually in the random-coil secondary structure following an α-helical portion of a protein chain. Why does proline not fit easily into an α-helix structure?

Section 22.10 Interlude: How Does the Presence of Transition Metals Affect the Structure of Proteins?

22.53 Consider the coordination compound Fe(CO)₅ and show that in forming this compound, the 18-electron rule is satisfied for iron.

22.54 Consider the molecule carbon monoxide, CO.

(a) Explain why the following is not an acceptable Lewis structure for CO. (Chapter 3)

$$:C=\ddot{O}$$

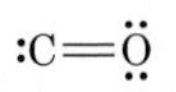

(b) Write an acceptable structure for CO. (Chapter 3)

(c) Propose which electron pair of CO is donated to iron to form the coordinate covalent bond Fe(CO)₅. (Chapter 3)

(c) Draw a Lewis structure for an oxygen molecule, O_2, and suggest how an oxygen molecule might form a coordinate covalent bond with an Fe^{2+} ion that is present in hemoglobin and myoglobin. (Chapter 3)

22.55 Consider the coordination compound $[Zn(NH_3)_2Cl_2]$, which contains $Zn(NH_3)_2^{2+}$ and Cl^- ions. Show that in forming this coordination compound, the 18-electron rule is satisfied for zinc.

22.56 Knowing what you do about VSEPR, predict the shape (geometry) of the cobalt binding site in vitamin B_{12}.

Section 22.11 What Is the Tertiary Structure of a Protein?

22.57 Polyglutamic acid (a polypeptide chain made only of glutamic acid residues) has an α-helix conformation below pH 6.0 and a random-coil conformation above pH 6.0. What is the reason for this conformational change?

22.58 Distinguish between intermolecular and intramolecular hydrogen bonding between backbone groups. Where in protein structures do you find one, and where do you find the other?

22.59 Identify the primary, secondary, and tertiary structures in the numbered boxes:

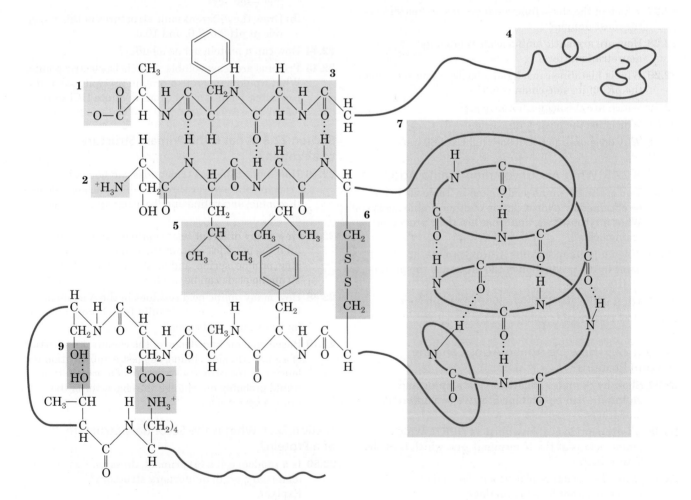

Section 22.12 What Is the Quaternary Structure of a Protein?

22.60 If both cysteine residues on the B chain of insulin were changed to alanine residues, how would it affect the quaternary structure of insulin?

22.61 (a) What is the difference in the quaternary structure between fetal hemoglobin and adult hemoglobin?

(b) Which can carry more oxygen?

(c) What would the oxygen saturation curve of fetal hemoglobin look like compared to that of mygoglobin and regular adult hemoglobin?

22.62 Where are the nonpolar side chains of proteins located in an integral membrane protein?

22.63 The cytochrome *c* protein is important in producing energy from food. It contains a heme surrounded by a polypeptide chain. What kind of structure do these two entities form? To which group of proteins does cytochrome *c* belong?

22.64 Hemoglobin is an important protein for many reasons and has interesting physical characteristics. How would you classify hemoglobin?

Section 22.13 How Are Proteins Denatured?

22.65 In a 6 *M* urea solution, a protein that contained mostly antiparallel β-sheets became a random coil. Which groups and bonds were affected by urea?

22.66 What kind of changes are necessary to transform a protein having a predominantly α-helical structure into one having a β-pleated sheet structure?

22.67 Which amino acid side chain is most frequently involved in denaturation by reduction?

22.68 What does the reducing agent do in straightening curly hair?

22.69 Silver nitrate is sometimes put into the eyes of newborn infants as a preventive measure against gonorrhea. Silver is a heavy metal. Explain how this treatment may work against bacteria.

22.70 Why do nurses and physicians use 70% alcohol to wipe the skin before giving injections?

Chemical Connections

22.71 (Chemical Connections 22A) Why must some people avoid drinking diet sodas with Nutrasweet?

22.72 (Chemical Connections 22B) AGE products become disturbing only in elderly people, even though they also form in younger people. Why don't they harm younger people?

22.73 (Chemical Connections 22C) Define *hypoglycemic awareness*.

22.74 (Chemical Connections 22D) How does hydroxyurea therapy alleviate the symptoms of sickle cell anemia?

22.75 (Chemical Connections 22E) What is the difference in the conformation between normal prion protein and the amyloid prion that causes mad cow disease?

22.76 (Chemical Connections 22F) What is the aim of proteomics?

22.77 (Chemical Connections 22G) Explain the difference in oxygen-binding behavior of myoglobin and hemoglobin.

22.78 (Chemical Connections 22H) How does the fiberscope help to heal bleeding ulcers?

Additional Problems

22.79 Which diseases are associated with amyloid plaques?

22.80 How many different dipeptides can be made (a) using only alanine, tryptophan, glutamic acid, and arginine and (b) using all 20 amino acids?

22.81 Denaturation is usually associated with transitions from helical structures to random coils. If an imaginary process were to transform the keratin in your hair from an α-helix to a β-pleated sheet structure, would you call the process denaturation? Explain.

22.82 Draw the structure of lysine (a) above, (b) below, and (c) at its isoelectric point.

22.83 In collagen, some of the chains of the triple helices in tropocollagen are cross-linked by covalent bonds between two lysine residues. What kind of structure is formed by these cross-links? Explain.

22.84 Considering the vast number of animal and plant species on Earth (including those now extinct) and the large variety of protein molecules in each organism, have all possible protein molecules been used already by some species or other? Explain.

22.85 What kind of noncovalent interaction occurs between the following amino acids?

(a) Valine and isoleucine

(b) Glutamic acid and lysine

(c) Tyrosine and threonine

(d) Alanine and alanine

22.86 How many different decapeptides (peptides containing 10 amino acids each) can be made from the 20 amino acids?

22.87 Which amino acid does not rotate the plane of polarized light?

22.88 Write the expected products of the acid hydrolysis of the following tetrapeptide:

22.89 What charges are on aspartic acid at pH 2.0?

22.90 How many ways can you link the two amino acids lysine and valine in a dipeptide? Which of these peptide bonds will you find in proteins?

Looking Ahead

22.91 Enzymes are biological catalysts and usually proteins. They catalyze common organic reactions. Why are amino acids such as histidine, aspartic acid,

632 Chapter 22 Proteins

and serine found more often near the reaction catalysis site than amino acids such as leucine and valine?

22.92 Hormones are molecules that are released from
one tissue but have their effect in another tissue.
Give an example of a hormone encountered in this
chapter that would be ineffective if taken orally.
Give an example of one that could be effective if
taken orally.

22.93 Using what you know about protein denaturation,
what is one reason you must maintain a body temperature in a strict range?

22.94 What is the difference between genomics and
proteomics?

22.95 Why does knowing the complete genome of an organism not necessarily tell you about the nature of
all the proteins in the organism?

22.96 Why is collagen not a very good source of dietary
protein?

22.97 A recent diet supplement advertised that it would
repair your muscles while allowing you to burn fat
because the product had collagen protein. Evaluate
this claim.

Enzymes

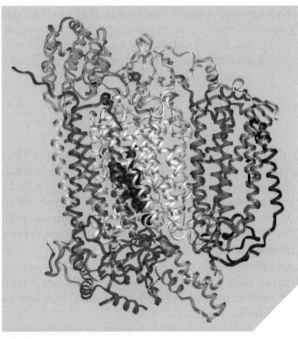

Ribbon diagram of cytochrome *c* oxidase, the enzyme that directly uses oxygen during respiration.

23.1 What Are Enzymes?

The cells in your body are chemical factories. Only a few of the thousands of compounds necessary for the operation of the human organism are obtained from the diet. Instead, most of these substances are synthesized within the cells, which means that hundreds of chemical reactions take place in your cells every second of your life.

Nearly all of these reactions are catalyzed by **enzymes**, which are large molecules that increase the rates of chemical reactions without themselves undergoing any change. Without enzymes to act as biological catalysts, life as we know it would not be possible.

The vast majority of all known enzymes are globular proteins, and we will devote most of our study to protein-based enzymes. However, proteins are not the only biological catalysts. **Ribozymes** are enzymes made of ribonucleic acids. They catalyze the self-cleavage of certain portions of their own molecules and have been implicated in the reaction that generates peptide bonds (Chapter 22). Many biochemists believe that during evolution, RNA catalysts emerged first, with protein enzymes arriving on the scene later. (We will learn more about RNA catalysts in Section 25.4.)

Like all catalysts, enzymes do not change the position of equilibrium. That is, enzymes cannot make a reaction take place that would not occur without them. Instead, they increase the reaction rate: They cause reactions to take place faster by lowering the activation energy (recall this term from Chapter 7). As catalysts, enzymes are remarkable in two respects:

1. They are extremely effective, increasing reaction rates by anywhere from 10^9 to 10^{20} times.

Substrate specificity The limitation of an enzyme to catalyze specific reactions with specific substrates

2. Most of them are extremely **specific**.

As an example of their effectiveness, consider the oxidation of glucose. A lump of glucose or even a glucose solution exposed to oxygen under sterile conditions would show no appreciable change for months. In the human body, however, the same glucose is oxidized within seconds.

Every organism has many enzymes—many more than 3000 in a single cell. Most enzymes are very specific, each of them speeding up only one particular reaction or class of reactions. For example, the enzyme urease catalyzes only the hydrolysis of urea and not that of other amides, even closely related ones.

$$(NH_2)_2C{=}O + H_2O \xrightarrow{\text{urease}} 2\,NH_3 + CO_2$$
$$\text{Urea}$$

Another type of specificity can be seen with trypsin, an enzyme that cleaves the peptide bonds of protein molecules—but not every peptide bond, only those on the carboxyl side of lysine and arginine residues (Figure 23.1).

The enzyme carboxypeptidase specifically catalyzes the hydrolysis of only the last amino acid on a protein chain—the one at the C-terminal end. Lipases are less specific: They catalyze the hydrolysis of any triglyceride, but they still do not affect carbohydrates or proteins.

The specificity of enzymes also extends to stereospecificity. The enzyme arginase hydrolyzes the amino acid L-arginine (the naturally occurring form) to a compound called L-ornithine and urea (Section 28.8) but has no effect on its mirror image, D-arginine.

Enzymes are distributed according to the body's need to catalyze specific reactions. A large number of protein-splitting enzymes are in the blood, ready to promote clotting. Digestive enzymes, which also catalyze the hydrolysis of proteins, are located in the secretions of the stomach and pancreas. Even within the cells themselves, some enzymes are localized according to the need for specific reactions. The enzymes that catalyze the oxidation of compounds that are part of the citric acid cycle (Section 27.4) are located in the mitochondria, for example, and special organelles such as lysosomes contain an enzyme (lysozyme) that catalyzes the dissolution of bacterial cell walls.

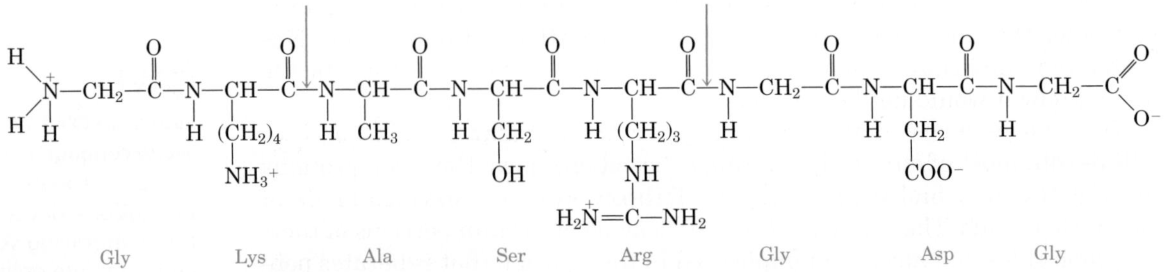

FIGURE 23.1 A typical amino acid sequence. The enzyme trypsin catalyzes the hydrolysis of this chain only at the points marked with an arrow (the carboxyl side of lysine and arginine).

Muscle Relaxants and Enzyme Specificity

In the body, nerves transmit signals to the muscles. Acetylcholine is a neurotransmitter (Section 24.1) that operates between the nerve endings and muscles. It attaches itself to a specific receptor in the muscle end plate. This attachment transmits a signal to the muscle to contract; shortly thereafter, the muscle relaxes. A specific enzyme, acetylcholinesterase, then catalyzes the hydrolysis of the acetylcholine, removing it from the receptor site and preparing it for the next signal transmission—that is, the next contraction.

Succinylcholine is sufficiently similar to acetylcholine so that it, too, binds to the receptor of the muscle end plate. However, acetylcholinesterase can hydrolyze succinylcholine only very slowly. While it remains bound to the receptor, no new signal can reach the muscle to allow it to contract again. Thus, the muscle stays relaxed for a long time.

This feature makes succinylcholine a good muscle relaxant during minor surgery, especially when a tube must be inserted into the bronchus (bronchoscopy). For example, after intravenous administration of 50 mg of succinylcholine, paralysis and respiratory arrest are observed within 30 seconds. While respiration is carried on artificially, the bronchoscopy can be performed within minutes.

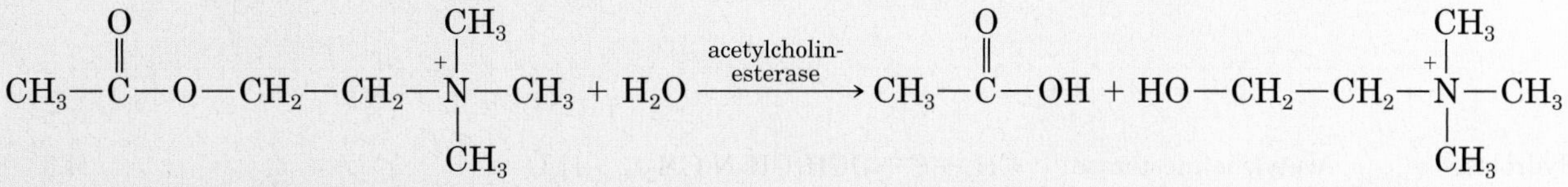

Acetylcholine → Acetic acid + Choline

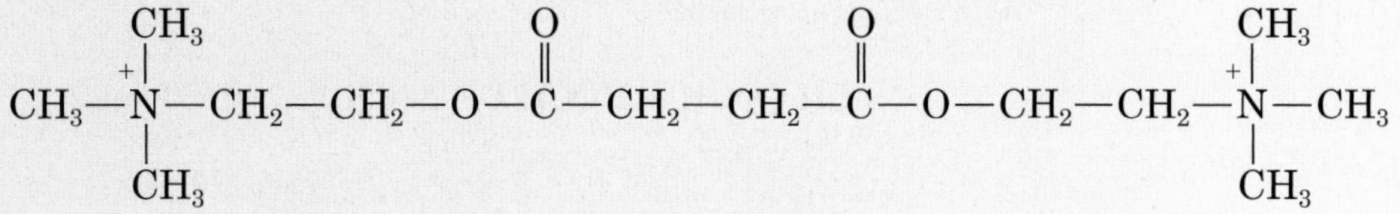

Succinylcholine

Test your knowledge with Problems 23.51 and 23.52.

23.2 How Are Enzymes Named and Classified?

Enzymes are commonly given names derived from the reaction that they catalyze and/or the compound or type of compound on which they act. For example, lactate dehydrogenase speeds up the removal of hydrogen from lactate (an oxidation reaction). Acid phosphatase catalyzes the hydrolysis of phosphate ester bonds under acidic conditions. As can be seen from these examples, the names of most enzymes end in "-ase." Some enzymes, however, have older names, which were assigned before their actions were clearly understood. Among these are pepsin, trypsin, and chymotrypsin—all enzymes of the digestive tract.

Enzymes can be classified into six major groups according to the type of reaction they catalyze (see also Table 23.1):

1. **Oxidoreductases** catalyze oxidations and reductions.
2. **Transferases** catalyze the transfer of a group of atoms, such as from one molecule to another.
3. **Hydrolases** catalyze hydrolysis reactions.
4. **Lyases** catalyze the addition of two groups to a double bond or the removal of two groups from adjacent atoms to create a double bond.
5. **Isomerases** catalyze isomerization reactions.
6. **Ligases**, or synthetases, catalyze the joining of two molecules.

Table 23.1 Classification of Enzymes

Class	Typical Example	Reaction Catalyzed	Section Number in This Book
1. Oxidoreductases	Lactate dehydrogenase	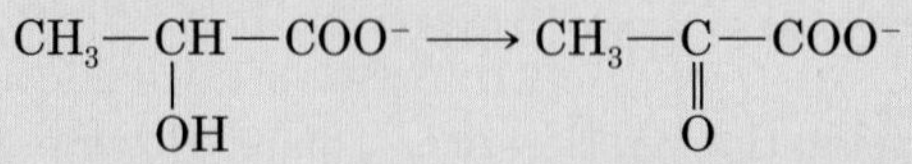	28.2
2. Transferases	Aspartate amino transferase or aspartate transaminase		28.8
3. Hydrolases	Acetylcholinesterase		24.3
4. Lyases	Aconitase		27.4
5. Isomerases	Phosphohexose isomerase		28.2
6. Ligases	Tyrosine-tRNA synthetase	$ATP + L\text{-tyrosine} + tRNA \longrightarrow L\text{-tyrosyltRNA} + AMP + PP_i$	26.6

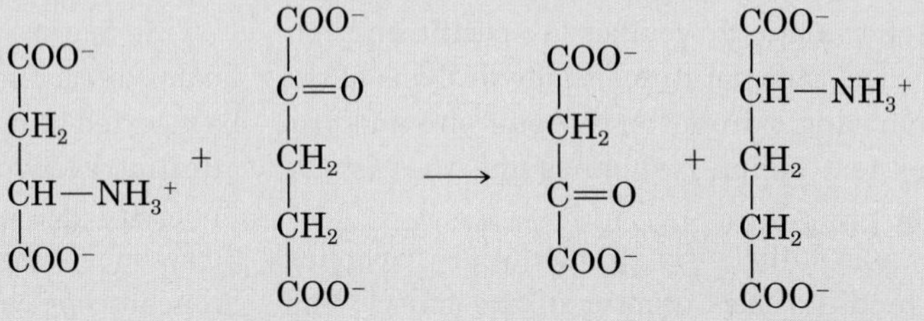

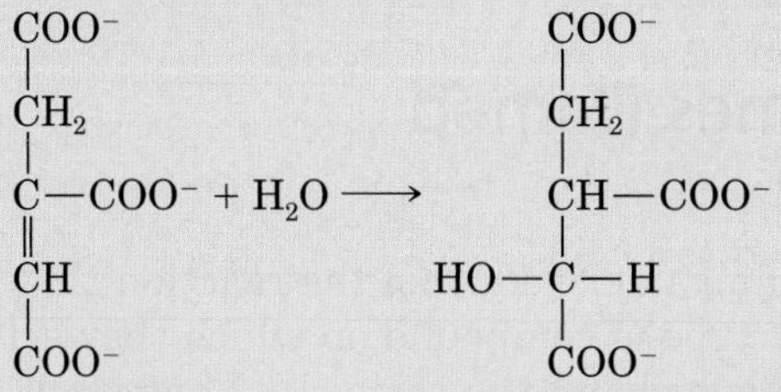

1. Oxidoreductases — Lactate dehydrogenase:

$$CH_3\text{—}\underset{\underset{OH}{|}}{CH}\text{—}COO^- \longrightarrow CH_3\text{—}\underset{\underset{O}{\|}}{C}\text{—}COO^-$$

L-(+)-Lactate → Pyruvate

2. Transferases — Aspartate aminotransferase:

Aspartate + α-Ketoglutarate → Oxaloacetate + Glutamate

3. Hydrolases — Acetylcholinesterase:

$$CH_3\text{—}\underset{\underset{O}{\|}}{C}\text{—}OCH_2CH_2\overset{+}{N}(CH_3)_3 + H_2O$$

Acetylcholine

$$\longrightarrow CH_3COOH + HOCH_2CH_2\overset{+}{N}(CH_3)_3$$

Acetic acid + Choline

4. Lyases — Aconitase:

cis-Aconitate + H₂O → Isocitrate

5. Isomerases — Phosphohexose isomerase:

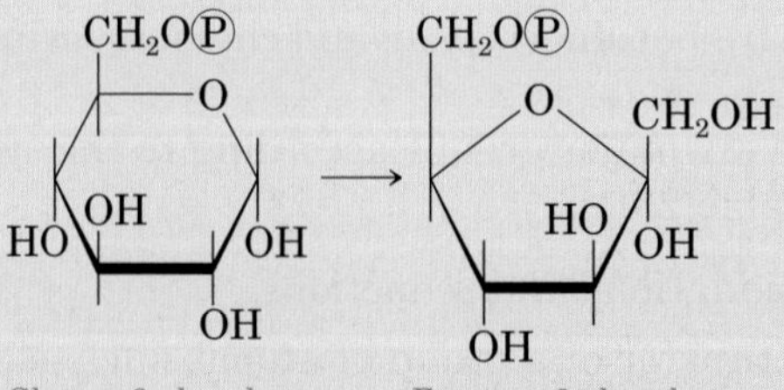

Glucose 6-phosphate → Fructose 6-phosphate

23.3 What Is the Terminology Used with Enzymes?

Some enzymes, such as pepsin and trypsin, consist of polypeptide chains only. Other enzymes contain nonprotein portions called **cofactors**. The protein (polypeptide) portion of the enzyme is called an **apoenzyme**.

The cofactors may be metallic ions, such as Zn^{2+} or Mg^{2+}, or organic compounds. Organic cofactors are called **coenzymes**. An important group of coenzymes are the B vitamins, which are essential to the activity of many enzymes (Section 27.3). Another important coenzyme is heme (Figure 22.16), which is part of several oxidoreductases as well as part of hemoglobin. In any case, an apoenzyme cannot catalyze a reaction without its cofactor, nor can the cofactor function without the apoenzyme. When a metal ion is a cofactor, it can bind directly to the protein or to the coenzyme, if the enzyme contains one.

The compound on which the enzyme works, and whose reaction it speeds up, is called the **substrate**. The substrate usually binds to the enzyme surface while it undergoes the reaction. The substrate binds to a specific portion of the enzyme during the reaction, called the **active site**. If the enzyme has coenzymes, they are located at the active site. Therefore, the substrate is simultaneously surrounded by parts of the apoenzyme, coenzyme, and metal ion cofactor (if any), as shown in Figure 23.2.

Activation is any process that initiates or increases the action of an enzyme. It can be the simple addition of a cofactor to an apoenzyme or the cleavage of a polypeptide chain of a proenzyme (Section 23.6B).

Inhibition is the opposite—any process that makes an active enzyme less active or inactive (Section 23.5). Inhibitors are compounds that accomplish this task, and there are many types of enzyme inhibition. **Competitive inhibitors** bind to the active site of the enzyme surface, thereby preventing the binding of substrate. **Noncompetitive inhibitors** bind to some other portion of the enzyme surface and sufficiently alter the tertiary structure of the enzyme so that its catalytic effectiveness is eliminated. That is, the enzyme cannot catalyze while the inhibitor is bound. Both competitive and noncompetitive inhibition are *reversible,* but some compounds alter the structure of the enzyme *permanently* and thus make it *irreversibly* inactive.

23.4 What Factors Influence Enzyme Activity?

Enzyme activity is a measure of how much reaction rates are increased. In this section, we examine the effects of concentration, temperature, and pH on enzyme activity.

A. Enzyme and Substrate Concentration

If we keep the concentration of substrate constant and increase the concentration of enzyme, the rate increases linearly (Figure 23.3). That is, if the enzyme concentration doubles, the rate doubles as well; if the enzyme concentration triples, the rate also triples. This is the case in practically all enzyme reactions, because the molar concentration of enzyme is almost always much lower than that of substrate (that is, many more molecules of substrate are typically present than molecules of enzyme).

Conversely, if we keep the concentration of enzyme constant and increase the concentration of substrate, we get an entirely different type of curve,

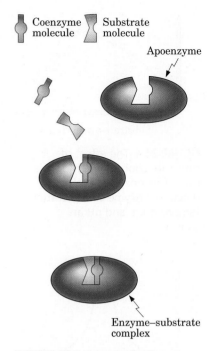

FIGURE 23.2 Schematic diagram of the active site of an enzyme and the participating components.

Cofactor The nonprotein part of an enzyme necessary for its catalytic function

Coenzyme A nonprotein organic molecule, frequently a B vitamin, that acts as a cofactor

Active site A three-dimensional cavity of the enzyme with specific chemical properties that enable it to accommodate the substrate

Inhibitor A compound that binds to an enzyme and lowers its activity

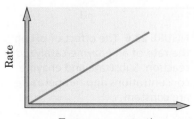

FIGURE 23.3 The effect of enzyme concentration on the rate of an enzyme-catalyzed reaction. Substrate concentration, temperature, and pH are constant.

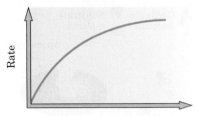

FIGURE 23.4 The effect of substrate concentration on the rate of an enzyme-catalyzed reaction. Enzyme concentration, temperature, and pH are constant.

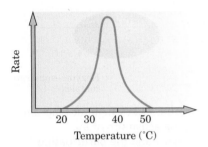

FIGURE 23.5 The effect of temperature on the rate of an enzyme-catalyzed reaction. Substrate and enzyme concentrations and pH are constant.

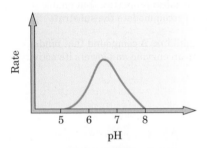

FIGURE 23.6 The effect of pH on the rate of an enzyme-catalyzed reaction. Substrate and enzyme concentrations and temperature are constant.

called a saturation curve (Figure 23.4). In this case, the rate does not increase continuously. Instead, a point is reached after which the rate stays the same even if we increase the substrate concentration further. This happens because at the saturation point, substrate molecules are bound to all available active sites of the enzymes. Because the reactions take place at the active sites, once they are all occupied, the reaction is proceeding at its maximum rate. Increasing the substrate concentration can no longer increase the rate because the excess substrate cannot find any active sites to which to bind.

B. Temperature

Temperature affects enzyme activity because it changes the conformation of the enzyme. In uncatalyzed reactions, the rate usually increases as the temperature increases (Section 8.4). Changing the temperature has a different effect on enzyme-catalyzed reactions. When we start at a low temperature (Figure 23.5), an increase in temperature first causes an increase in rate. However, protein conformations are very sensitive to temperature changes. Once the optimal temperature is reached, any further increase in temperature alters the enzyme conformation. The substrate may then not fit properly onto the changed enzyme surface, so the rate of reaction actually *decreases*.

After a *small* temperature increase above the optimum, the decreased rate could be increased again by lowering the temperature because over a narrow temperature range, changes in conformation are reversible. However, at some higher temperature above the optimum, we reach a point where the protein denatures (Section 22.12); the three-dimensional conformation is then altered irreversibly, and the polypeptide chain cannot refold to its native conformation. At this point, the enzyme is completely inactivated. The inactivation of enzymes at low temperatures is used in the preservation of food by refrigeration.

Most enzymes from bacteria and higher organisms have an optimal temperature around 37°C. However, the enzymes of organisms that live at the ocean floor at 2°C have an optimal temperature in that range. Other organisms live in ocean vents under extreme conditions, and their enzymes have optimal conditions at ranges from 90 to 105°C. The enzymes of these hyperthermophile organisms also have other extreme requirements, such as pressures up to 100 atm, and some of them have an optimal pH in the range of 1 to 4. Enzymes from these hyperthermophiles, especially polymerases that catalyze the polymerization of DNA (Section 25.6), have gained commercial importance.

C. pH

As the pH of its environment changes the conformation of a protein (Section 22.12), we would expect pH-related effects to resemble those observed when the temperature changes. Each enzyme operates best at a certain pH (Figure 23.6). Once again, within a narrow pH range, changes in enzyme activity are reversible. However, at extreme pH values (either acidic or basic), enzymes are denatured irreversibly and enzyme activity cannot be restored by changing back to the optimal pH.

23.5 What Are the Mechanisms of Enzyme Action?

We have seen that the action of enzymes is highly specific for a substrate. What kind of mechanism can account for such specificity? About 100 years ago, Arrhenius suggested that catalysts speed up reactions by combining with the substrate to form some kind of intermediate compound. In an enzyme-catalyzed reaction, this intermediate is called the **enzyme–substrate complex**.

A. Lock-and-Key Model

To account for the high substrate specificity of most enzyme-catalyzed reactions, a number of models have been proposed. The simplest and most frequently referenced is the **lock-and-key model** (Figure 23.7). This model assumes that the enzyme is a rigid three-dimensional body. The surface that contains the active site has a restricted opening into which only one kind of substrate can fit, just as only the proper key can fit exactly into a lock and turn it open.

According to the lock-and-key model, an enzyme molecule has its particular shape because that shape is necessary to maintain the active site in exactly the conformation required for that particular reaction. An enzyme molecule is very large (typically consisting of 100 to 200 amino acid residues), but the active site is usually composed of only two or a few amino acid residues, which may well be located at different places in the chain. The other amino acids—those that are not part of the active site—are located in the sequence in which we find them because that sequence causes the molecule as a whole to fold up in exactly the required way. This arrangement emphasizes that the shape and the functional groups on the surface of the active site are of utmost importance in recognizing a substrate.

The lock-and-key model was the first to explain the action of enzymes. For most enzymes, however, this model is too restrictive. Enzyme molecules are in a dynamic state, not a static one. There are constant motions within them, so that the active site has some flexibility. Also, while the lock-and-key model does a good job explaining why the enzyme binds to the substrate, if the fit is that perfect, there would be no reason for the reaction to occur, as the enzyme-substrate complex would be too stable.

B. Induced-Fit Model

From X-ray diffraction, we know that the size and shape of the active site cavity change when the substrate enters. To explain this phenomenon, an American biochemist, Daniel Koshland, introduced the **induced-fit model** (Figure 23.8), in which he compared the changes occurring in the shape of the cavity upon substrate binding to the changes in the shape of a glove when a hand is inserted. That is, the enzyme modifies the shape of the active site to accommodate the substrate. Recent experiments during actual catalysis have demonstrated that not only does the shape of the active site change with the binding of substrate, but even in the bound state, both the backbone and the side chains of the enzyme are in constant motion.

Both the lock-and-key and induced-fit models explain the phenomenon of competitive inhibition (Section 23.3). The inhibitor molecule fits into the active site cavity in the same way the substrate does (Figure 23.9), thereby preventing the substrate from entering. The result: Whatever reaction is supposed to take place on the substrate does not occur.

Many cases of noncompetitive inhibition can also be explained by the induced-fit model. In this case, the inhibitor does not bind to the active site, but rather, binds to a different part of the enzyme. Nevertheless, the binding causes a change in the three-dimensional shape of the enzyme molecule, which so alters the shape of the active site that once the substrate is bound, there can be no catalysis (Figure 23.10).

If we compare enzyme activity in the presence and the absence of an inhibitor, we can tell whether competitive or noncompetitive inhibition is taking place (Figure 23.11). The maximum reaction rate is the same without

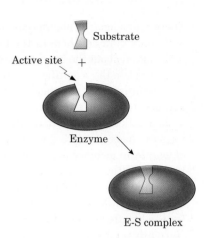

FIGURE 23.7 The lock-and-key model of the enzyme mechanism.

Lock-and-key model A model explaining the specificity of enzyme action by comparing the active site to a lock and the substrate to a key

Induced-fit model A model explaining the specificity of enzyme action by comparing the active site to a glove and the substrate to a hand

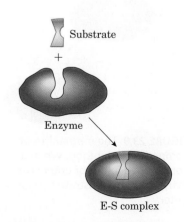

FIGURE 23.8 The induced-fit model of the enzyme mechanism.

Enzymes and Memory

There are thousands of different enzymes in a cell, and we will study many of them in the following chapters. New information about the importance of enzymes is published every week in the scientific literature. One class of enzyme important in many metabolic processes is the class called **kinases** (a type of transferase; see Table 23.1). One such kinase, called protein kinase Mζ (PKMζ) (ζ is the symbol for zeta), has recently been implicated in the maintenance of long-term memory. Scientists created a drug called ZIP that blocks this enzyme. They gave rats saccharine-laced water and then induced nausea shortly afterwards. Control rats then had an aversion to saccharine-laced water for weeks afterwards. Humans have the same response: Normally a person who throws up shortly after eating a specific type of food will remember the experience and not want to consume the same food. Researchers then injected the cerebral cortex of test rats and found that they lost their aversion to saccharine within two hours. Since blocking the PKMζ eliminated the memory, here was a

first indication that this specific enzyme is required for long-term memory retention, a novel finding. The next step will be to determine if the drug ZIP eliminates all learning past a certain point or whether it could be used selectively. Researchers have been looking for ways to selectively block memories, such as the painful memories of trauma survivors.

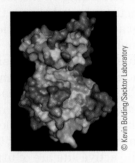

Memory molecule. PKMζ sustains long-term memory in the cerebral cortex of rats

Test your knowledge with Problems 23.53–23.55.

an inhibitor and in the presence of a competitive inhibitor. The only difference is that this maximum rate is achieved at a low substrate concentration with no inhibitor but at a high substrate concentration when an inhibitor is present. This is the true sign of competitive inhibition, because in this scenario the substrate and the inhibitor are competing for the same active site. If the substrate concentration is sufficiently increased, the inhibitor will be displaced from the active site by Le Chatelier's principle, thus allowing for increased binding of the substrate and an increased rate of reaction.

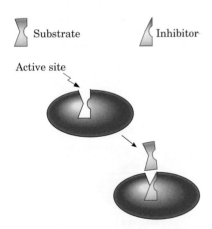

FIGURE 23.9 The mechanism of competitive inhibition. When a competitive inhibitor enters the active site, the substrate cannot enter.

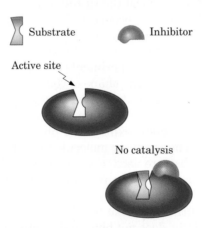

FIGURE 23.10 Mechanism of noncompetitive inhibition. The inhibitor binds itself to a site other than the active site (allosterism), thereby changing the conformation of the active site. The substrate still binds, but there is no catalysis.

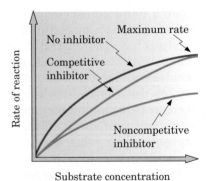

FIGURE 23.11 Enzyme kinetics in the presence and the absence of inhibitors.

Chemical Connections 23C

Active Sites

The perception of the active site as either a rigid cavity (lock-and-key model) or a partly flexible template (induced-fit model) is an oversimplification. Not only is the geometry of the active site important, but so are the specific interactions that take place between the enzyme surface and the substrate. To illustrate, we take a closer look at the active site of pyruvate kinase. This enzyme catalyzes the transfer of the phosphate group from phosphoenolpyruvate (PEP) to ADP, an important step in glycolysis (Section 28.2).

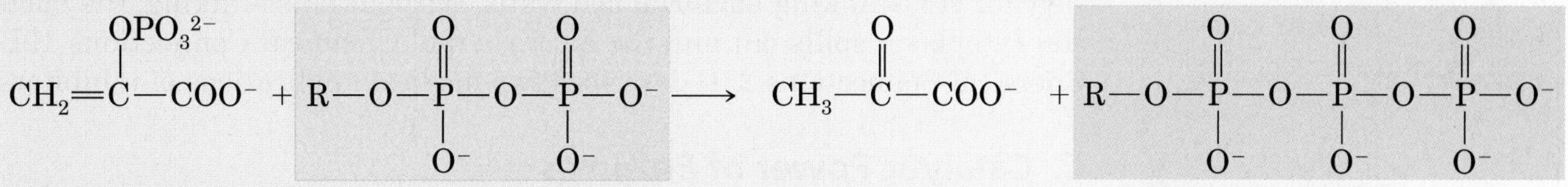

$$CH_2=\underset{\underset{\text{Phosphoenolpyruvate}}{\displaystyle |}}{\overset{\overset{OPO_3^{2-}}{\displaystyle |}}{C}}-COO^- + \underset{\text{ADP}}{R-O-\underset{\underset{O^-}{\displaystyle |}}{\overset{\overset{O}{\displaystyle ||}}{P}}-O-\underset{\underset{O^-}{\displaystyle |}}{\overset{\overset{O}{\displaystyle ||}}{P}}-O^-} \longrightarrow CH_3-\underset{\text{Pyruvate}}{\overset{\overset{O}{\displaystyle ||}}{C}}-COO^- + \underset{\text{ATP}}{R-O-\underset{\underset{O^-}{\displaystyle |}}{\overset{\overset{O}{\displaystyle ||}}{P}}-O-\underset{\underset{O^-}{\displaystyle |}}{\overset{\overset{O}{\displaystyle ||}}{P}}-O-\underset{\underset{O^-}{\displaystyle |}}{\overset{\overset{O}{\displaystyle ||}}{P}}-O^-}$$

The active site of the enzyme binds both substrates, PEP and ADP (see figure below left). The rabbit muscle pyruvate kinase has two cofactors, K^+ and either Mn^{2+} or Mg^{2+}. The divalent cation is coordinated to the carbonyl and carboxylate oxygen of the pyruvate substrate and to the glutamate 271 and aspartate 295 residues of the enzyme. (The numbers indicate the position of the amino acid in the sequence.) The nonpolar $=CH_2$ group lies in a hydrophobic pocket formed by an alanine 292, glycine 294, and threonine 327 residue. The K^+ on the other side of the active site is coordinated with the phosphate of the substrate and the serine 76 and asparagine 74 residue of the enzyme. Lysine 269 and arginine 72 are also part of the catalytic apparatus anchoring the ADP. This arrangement of the active site illustrates that specific folding into secondary and tertiary structures is required to bring important functional groups together. The residues of amino acids participating in the active sites are sometimes close in the sequence (asparagine 74 and serine 76), but mostly remain far apart (glutamate 271 and aspartate 295). The figure below right illustrates the secondary and tertiary structures providing such a stable active site.

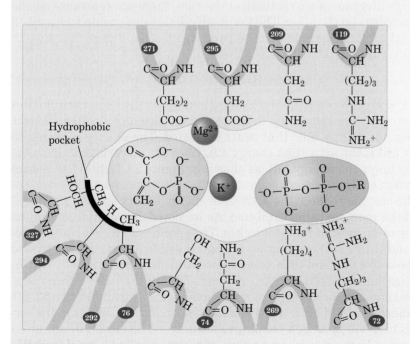

The active site and substrates of pyruvate kinase.

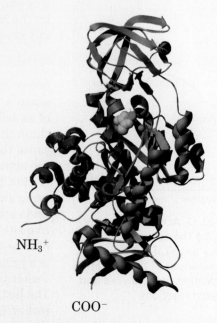

Ribbon cartoon of pyruvate kinase. Pyruvate, Mg^{2+}, and K^+ are depicted as space-filling models.

Test your knowledge with Problems 23.56 and 23.57.

If the inhibitor is noncompetitive, it cannot be displaced by addition of excess substrate because it is bound to a different site that is unaffected by the addition of excess substrate. In this case, the enzyme cannot be restored to its maximum activity and the maximum rate of the reaction is lower than it would be in the absence of the inhibitor. With a noncompetitive inhibitor, the net effect is always as if less enzyme were available. Competitive and noncompetitive inhibition are the two most common extremes of enzyme inhibition. Many other types of reversible inhibitors exist, but they are beyond the scope of this book.

Enzymes can also be inhibited irreversibly if a compound is bound covalently and permanently to or near the active site. Such inhibition occurs with penicillin, which inhibits the enzyme transpeptidase; this enzyme is necessary for cross-linking bacterial cell walls. Without cross-linking, the bacterial cytoplasm spills out and the bacteria die (Chemical Connections 19B). Chemical Connections 23D describes two medical applications of inhibitors.

C. Catalytic Power of Enzymes

Both the lock-and-key model and the induced-fit model emphasize the shape of the active site. However, the chemistry at the active site is actually the most important factor. A survey of known active sites of enzymes shows that five amino acids participate in the active sites in more than 65% of all cases. They are, in order of their dominance, His > Cys > Asp > Arg > Glu. A quick glance at Table 22.1 reveals that most of these amino acids have either acidic or basic side chains. Thus, acid–base chemistry often underlies the mode of catalysis. The example given in Chemical Connections 23C confirms this relationship. Out of the eleven amino acids in the catalytic site of pyruvate kinase, two are Arg, one is Glu, one is Asp, and two are the related Asn.

We have said that enzymes cannot change the thermodynamic relationships between the substrates and the products of a reaction; rather, they speed up the reaction. But how do they really accomplish this feat? If we look at an energy diagram of a hypothetical reaction, there are reactants on one side and products on the other. The thermodynamic relationship is described by the height difference between the two, as shown in Figure 23.12(a). In any reaction that can be written as follows:

$$A + B \rightleftharpoons C + D$$

before A and B can become C and D, they must pass through a **transition state** where they are something in between. This situation is often thought of as being an "energy hill" that must be scaled. The energy required to climb this hill is the activation energy. Enzymes are powerful catalysts because they lower the energy hill, as shown in Figure 23.12(b). They reduce the activation energy.

How the enzyme reduces the activation energy is very specific to the enzyme and the reaction being catalyzed. As we noted, however, a few amino acids show up in most of the active sites. The specific amino acids in the active site and their exact orientation make it possible for the substrate(s) to bind to the active site and then react to form products. For example, papain is a protease, an enzyme that cleaves peptide bonds as we saw with trypsin. Two critical amino acids are present in the active site of papain (Figure 23.13). The histidine (shown in blue) helps attract the peptide and hold it in the correct orientation via hydrogen bonding (shown as red dashes). The sulfur on the cysteine side chain performs a type of reaction called a **nucleophilic attack** on the carbonyl carbon of the peptide bond, and the C—N bond is broken. Such nucleophilic attacks appear in the vast majority of enzyme mechanisms, and they are possible because of the precise arrangement of the amino acid side chains that can participate in this type of organic reaction.

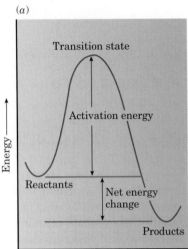

(a)

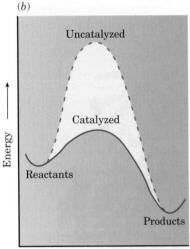

(b)

FIGURE 23.12 Activation energy profiles. (a) The activation energy profile for a typical reaction. (b) A comparison of the activation energy profiles for catalyzed and uncatalyzed reactions.

Nucleophilic attack A chemical reaction where an electron-rich atom such as oxygen or sulfur bonds to an electron-deficient atom such as carbonyl carbon

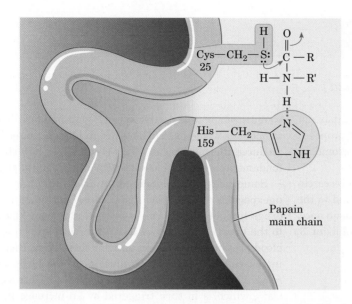

FIGURE 23.13 Papain is a cysteine protease. A critical cysteine residue is involved in the nucleophilic attack on the peptide bond it hydrolyzes.

Chemical Connections 23D

Medical Uses of Inhibitors

A key strategy in the treatment of acquired immunodeficiency syndrome (AIDS) has been to develop specific inhibitors that selectively block the actions of enzymes unique to the human immunodeficiency virus (HIV), which causes AIDS. Many laboratories are using this approach to develop therapeutic agents.

One of the most important targets is HIV protease, an enzyme essential to the production of new virus particles in infected cells. HIV protease is unique to this virus. It catalyzes the processing of viral proteins in an infected cell. Without these proteins, viable virus particles cannot be released to cause further infection. The structure of HIV protease, including its active site, was elucidated by X-ray crystallography. With this structure in mind, scientists then designed and synthesized competitive inhibitors to bind to the active site. Improvements were made in the drug design by obtaining structures of a series of inhibitors bound to the active site of HIV protease. These structures were also elucidated by X-ray crystallography. This process eventually led to several HIV protease inhibitors: saquinavir from Hoffmann-La Roche, ritonavir from Abbott, indinavir from Merck, viracept from Agouron Pharmaceuticals, and amprenavir from Vertex Pharmaceuticals. (These companies maintain highly informative home pages on the World Wide Web.)

Treatment of AIDS is most effective when a combination of drug therapies is used, and HIV protease inhibitors play an important role. Especially promising results (such as lowering of levels of the virus in the bloodstream) are obtained when HIV protease inhibitors are part of drug regimens for AIDS.

Another use for enzyme inhibitors in medicine centers on alleviating side effects from cancer chemotherapy.

A group of researchers in North Carolina worked on the drug camptothecin (CPT-11). This compound is used to treat colon cancer, but it causes severe diarrhea. CPT-11 is processed in the body by the enzyme β-glucoronidase found in intestinal bacteria. The product of the reaction is ultimately responsible for the diarrhea. The intestinal

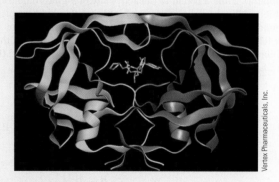

Active site of HIV-1 protease complexed with VX-478.

Structure of amprenavir (VX-478), an HIV protease inhibitor.

Chemical Connections 23D

Medical Uses of Inhibitors (*continued*)

bacteria do many useful things for the body, so it is not a good idea to kill them with antibiotics. It was proposed that selectively inhibiting the bacterial glucoronidase should lessen this side effect.

Using the crystal structure of the enzyme, the researchers synthesized compounds that were likely to bind to the active site. Four inhibitors showed promising results in mice. A combination of two, called Inhibitor 2 and Inhibitor 3,

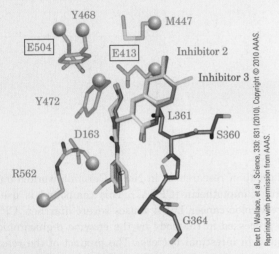

Bret D. Wallace, et al.. *Science*, 330: 831 (2010). Copyright © 2010 AAAS. Reprinted with permission from AAAS.

Potent β-glucuronidase inhibitors. Crystal structures of Inhibitors 2 and 3 bound to the active site of β-glucuronidase. Inhibitor 2 is shown in orange and Inhibitor 3 in light purple. Amino acid residues in the active site are shown in green to the left of the inhibitors and in red and blue to the right of the inhibitors. Amino acid abbreviations: D, Asp; E, Glu; F, Phe; G, Gly; L, Leu; M, Met; R, Arg; S, Ser; Y, Tyr.

bound strongly to the active site of β-glucoronidase. Now the compounds would need to be tested in humans to see if they produced the same results. This study shows how chemical fundamentals can lead to improvements in clinical practice.

Sometimes, the designing of a drug inhibitor leads to unexpected results. Scientists have long sought better drugs to fight *angina* (chest pains due to poor blood flow to the heart) and *hypertension* (otherwise known as high blood pressure), a common ailment these days. Blood flow increases when the smooth muscles in the blood vessels relax. This relaxation is due to a decrease in intracellular Ca^{2+}, which is in turn triggered by an increase in cyclic GMP (cGMP, Chapter 25). Cyclic GMP is degraded by enzymes called phosphodiesterases. Scientists thought that if they could design an inhibitor of these phosphodiesterases, the cGMP would last longer, the blood vessels would stay open longer, and blood pressure would decrease. Scientists developed a drug to mimic cGMP in the hopes of inhibiting phosphodiesterases. The structural name of the drug is sildenafil citrate, but a company called Pfizer marketed it under the name of Viagra.

Unfortunately, Viagra showed no significant benefits for reducing the pain of angina or decreasing blood pressure. However, some men in the clinical trials of the drug noted penile erections. Apparently, the drug did work to inhibit the phosphodiesterases in the penile vascular tissue, leading to smooth muscle relaxation and increased blood flow. Despite the fact that the drug did not accomplish what it was intended for, this competitive inhibitor became a very big seller for the companies that produce it.

Note the structural similarity between cGMP (left) and Viagra.

Test your knowledge with Problems 23.58–23.61.

23.6 How Are Enzymes Regulated?

A. Feedback Control

Enzymes are often regulated by environmental conditions. **Feedback control** is an enzyme regulation process in which formation of a product inhibits an earlier reaction in the sequence. The reaction product of one enzyme may control the activity of another, especially in a complex system in which enzymes work cooperatively. For example, in such a system, each step is catalyzed by a different enzyme:

$$A \xrightarrow{E_1} B \xrightarrow{E_2} C \xrightarrow{E_3} D$$

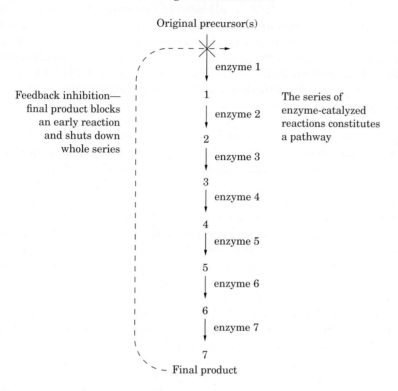

A schematic representation of a pathway
showing feedback inhibition

Original precursor(s)

enzyme 1

Feedback inhibition—
final product blocks
an early reaction
and shuts down
whole series

1

enzyme 2

The series of
enzyme-catalyzed
reactions constitutes
a pathway

2

enzyme 3

3

enzyme 4

4

enzyme 5

5

enzyme 6

6

enzyme 7

7

Final product

The final product in the chain may inhibit the activity of the first enzyme (by competitive, noncompetitive, or some other type of inhibition). When the concentration of the final product is low, all of the reactions proceed rapidly. As the concentration increases, however, the action of enzyme 1 becomes inhibited and eventually stops. In this manner, the accumulation of the final product serves as a message that tells enzyme 1 to shut down because the cell has enough final product for its present needs. Shutting down enzyme 1 stops the entire process.

B. Proenzymes

Some enzymes are manufactured by the body in an inactive form. To make them active, a small part of their polypeptide chain must be removed. These inactive forms of enzymes are called **proenzymes** or **zymogens**. After the excess polypeptide chain is removed, the enzyme becomes active.

For example, trypsin is manufactured in the pancreas as the inactive molecule trypsinogen (a zymogen). When a fragment containing six amino acid residues is removed from the N-terminal end, the molecule becomes a fully active trypsin molecule. Removal of the fragment not only shortens the chain, but also changes the three-dimensional structure (the tertiary structure), thereby allowing the molecule to achieve its active form.

Why does the body go through so much trouble? Why not just make the fully active trypsin to begin with? The reason is very simple. Trypsin is a protease—it catalyzes the hydrolysis of peptide bonds (Figure 23.1)—and is, therefore, an important catalyst for the digestion of the proteins we eat. But it would not be good if it cleaved the proteins of which our own bodies are

Proenzyme (zymogen) A protein that becomes an active enzyme after undergoing a chemical change

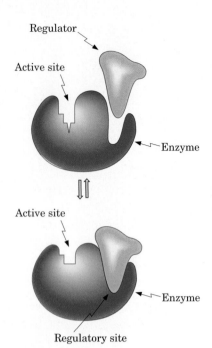

FIGURE 23.14 The allosteric effect. Binding of a regulator to a site other than the active site changes the shape of the active site.

Allosteric enzyme An enzyme in which the binding of a regulator on one site on the enzyme modifies the enzyme's ability to bind the substrate in the active site

made! Therefore, the pancreas makes trypsin in an inactive form; only after trypsin enters the digestive tract does it become active.

C. Allosterism

Sometimes regulation takes place by means of an event that occurs at a site other than the active site, but that eventually affects the active site. This type of interaction is called **allosterism**, and any enzyme regulated by this mechanism is called an **allosteric enzyme**. If a substance binds noncovalently and reversibly to a site *other than the active site,* it may affect the enzyme in either of two ways: It may inhibit enzyme action (**negative modulation**) or it may stimulate enzyme action (**positive modulation**).

The substance that binds to an allosteric enzyme is called a **regulator**, and the site to which it attaches is called a **regulatory site**. In most cases, allosteric enzymes contain more than one polypeptide chain (subunits); the regulatory site is on one polypeptide chain, and the active site is on another.

Specific regulators can bind reversibly to the regulatory sites. For example, the enzyme depicted in Figure 23.14 is an allosteric enzyme. In this case, the enzyme has only one polypeptide chain, so it carries both the active site and the regulatory site at different points in this chain. The regulator binds reversibly to the regulatory site. As long as the regulator remains bound to the regulatory site, the total enzyme–regulator complex will be inactive. When the regulator is removed from the regulatory site, the enzyme becomes active. In this way, the regulator controls the allosteric enzyme action.

Allosteric enzymes have two kinetic states, with a corresponding conformational change in each of these states that results in two different forms. One form is more likely to bind the substrate and produce the product than the other form. This more active form is referred to as the **R form**, where "R" stands for *relaxed*. The less active form is referred to as the **T form**, where "T" stands for *taut* (Figure 23.15). There is an equilibrium between the T form and the R form. When the enzyme is in the R form, it will bind substrate well and catalyze the reaction. Allosteric regulators are seen to function by binding to the enzyme and favoring one form versus the other.

D. Protein Modification

The activity of an enzyme may also be controlled by **protein modification**. The modification is usually a change in the primary structure, typically by addition of a functional group covalently bound to the apoenzyme. The

A dimeric protein that can exist in either of two states: R_0 or T_0. This protein can bind three ligands:

1. Substrate (S) ▉: Binds only to R

2. Activator (A) ◣: Binds only to R

3. Inhibitor (I) ◤: Binds only to T

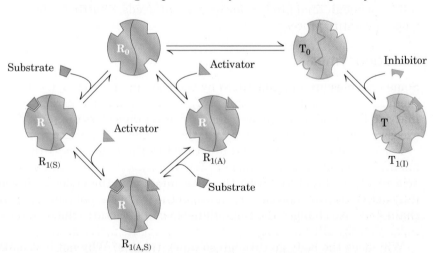

FIGURE 23.15 Effects of binding activators and inhibitors to allosteric enzymes. The enzyme has an equilibrium between the T form and the R form. An activator is anything that binds to the regulatory site and favors the R form. An inhibitor binds to the regulatory site and favors the T form.

best-known example of protein modification is the activation or inhibition of enzymes by phosphorylation. A phosphate group is often bonded to a serine or tyrosine residue. In some enzymes, such as glycogen phosphorylase (Section 29.1), the phosphorylated form is the active form of the enzyme. Without it, the enzyme is less active or inactive.

The opposite example is the enzyme pyruvate kinase (PK, discussed in Chemical Connections 23C). Pyruvate kinase from the liver is inactive when it is phosphorylated. Enzymes that catalyze such phosphorylation go by the common name of *kinases*. When the activity of PK is not needed, it is phosphorylated (to PKP) by a protein kinase using ATP as a substrate as well as a source of energy (Section 27.3). When the system wants to turn on PK activity, the phosphate group, Pi, is removed by another enzyme, phosphatase, which renders PK active.

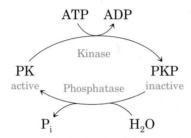

E. Isoenzymes

Another type of regulation of enzyme activity occurs when the same enzyme appears in different forms in different tissues. Lactate dehydrogenase (LDH) catalyzes the oxidation of lactate to pyruvate, and vice versa (Figure 28.3, step 11). The enzyme has four subunits (tetramer). Two kinds of subunits, called H and M, exist. The enzyme that dominates in the heart is an H_4 enzyme, meaning that all four subunits are of the H-type, although some M-type subunits are present as well. In the liver and skeletal muscles, the M-type dominates. Other types of tetramer combinations exist in different tissues: H_3M, H_2M_2, and HM_3. These different forms of the same enzyme are called **isozymes** or **isoenzymes**.

The different subunits confer subtle, yet important differences to the function of the enzyme in relation to the tissue. The heart is a purely aerobic organ, except perhaps during a heart attack. Under normal circumstances, LDH is used to convert lactate to pyruvate in the heart. The H_4 enzyme is allosterically inhibited by high levels of pyruvate (its product) and has a higher affinity for lactate (its substrate) than does the M_4 enzyme, which is optimized for the opposite reaction. The M_4 isozyme favors the production of lactate.

The distribution of LDH isozymes can be seen using the technique of electrophoresis, where samples are separated in a gel using an electric field. Besides their kinetic differences, the two subunits of LDH carry different charges. Therefore, each combination of subunits travels in the electric field at a different rate (Figure 23.16).

Isozymes (Isoenzymes) Enzymes that perform the same function but have different combinations of subunits and thus different quaternary structures

23.7 How Are Enzymes Used in Medicine?

Most enzymes are confined within the cells of the body. However, small amounts of them can also be found in body fluids such as blood, urine, and cerebrospinal fluid. The level of enzyme activity in these fluids can easily be monitored. This information can prove extremely useful: Abnormal activity (either high or low) of particular enzymes in various body fluids signals either the onset of certain diseases or their progression. Table 23.2 lists some enzymes used in medical diagnosis and their activities in normal body fluids.

FIGURE 23.16 The isozymes of lactate dehydrogenase (LDH). (*a*) The five combinations possible from mixing two types of subunits, H and M, in all permutations to make a tetramer. (*b*) An electrophoresis gel depiction of the relative isozyme types found in different tissues.

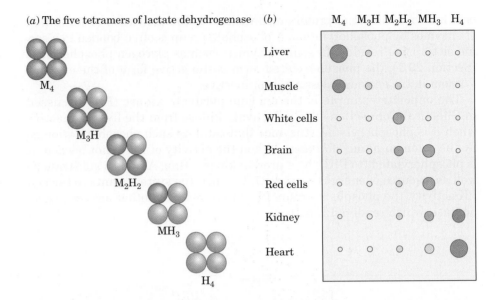

(*a*) The five tetramers of lactate dehydrogenase   (*b*)

For example, a number of enzymes are assayed (measured) during myocardial infarction to diagnose the severity of the heart attack. Dead heart muscle cells spill their enzyme contents into the serum. As a consequence, the level of creatine phosphokinase (CPK) in the serum rises rapidly, reaching a maximum within two days. This increase is followed by a rise in aspartate aminotransferase (AST; formerly called glutamate-oxaloacetate transaminase, or GOT). This second enzyme reaches a maximum two to three days after the heart attack. In addition to CPK and AST, lactate dehydrogenase (LDH) levels are monitored; they peak after five to six days. In infectious hepatitis, the alanine aminotransferase (ALT; formerly called glutamate-pyruvate transaminase, or GPT) level in the serum can rise to 10 times normal. There is also a concurrent increase in AST activity in the serum.

In some cases, administration of an enzyme is part of therapy. After duodenal or stomach ulcer operations, for instance, patients are advised to take tablets containing digestive enzymes that are in short supply in the stomach after surgery. Such enzyme preparations contain lipases, either alone or combined with proteolytic enzymes. They are sold under such names as pancreatin, Arco-lase, and Ku-zyme.

Table 23.2 Enzyme Assays Useful in Medical Diagnosis

Enzyme	Normal Activity	Body Fluid	Disease Diagnosed
Alanine aminotransferase (ALT)	3–17 U/L*	Serum	Hepatitis
Acid phosphatase	2.5–12 U/L	Serum	Prostate cancer
Alkaline phosphatase (ALP)	13–38 U/L	Serum	Liver or bone disease
Amylase	19–80 U/L	Serum	Pancreatic disease or mumps
Aspartate aminotransferase (AST)	7–19 U/L	Serum	Heart attack or hepatitis
	7–49 U/L	Cerebrospinal fluid	
Lactate dehydrogenase (LDH)	100–350 WU/mL	Serum	Heart attack
Creatine phosphokinase (CPK)	7–60 U/L	Serum	
Phosphohexose isomerase (PHI)	15–75 U/L	Serum	

*IU/L = International units per liter; WU/mL = Wrobleski units per milliliter.

Glycogen Phosphorylase: A Model of Enzyme Regulation

An excellent example of the subtle elegance of enzyme regulation can be seen in the enzyme glycogen phosphorylase, an enzyme that breaks down glycogen (Chapter 20) into glucose when the human body needs energy. Glycogen phosphorylase is a dimer controlled by modification (covalent control) and by allosterism.

Two forms of phosphorylase exist called phosphorylase *b* and phosphorylase *a,* as shown in the figure. Phosphorylase *a* has a phosphate attached to each subunit, which was put there by the enzyme phosphorylase kinase. Phosphorylase *b* does not have the phosphate. The kinase is activated by hormonal signals that indicate a need for quick energy or a need for more blood glucose, depending on the tissue.

Phosphorylase is also controlled allosterically by a variety of regulators. The *b* form is converted to a more active form in the presence of AMP. Glucose-6-phosphate, glucose, and caffeine convert it to the less active form. The *a* form is also converted to the less active form by glucose and caffeine. In general, the equilibrium lies more toward the active form of phosphorylase *a* than with phosphorylase *b.*

This combination of regulation is very beneficial, because it leads to both quick changes and long-lasting ones. When you need quick action for the "fight-or-flight" response, the first few muscle contractions will cause ATP to be broken down. AMP will increase, which will, in much less than a second, convert some of the phosphorylase *b* to the more active form (R state). At the same time, you will probably experience an adrenaline rush that will cause the activation of phosphorylase kinase and the subsequent phosphorylation of glycogen phosphorylase from the *b* form to the *a* form. This conversion will then cause even more phosphorylase to shift to the active R form (see right side of figure, arrow going down). The hormonal response is a

Glycogen phosphorylase activity is subject to allosteric control and covalent modification via phosphorylation. The phosphorylated form is more active. The enzyme that puts a phosphate group on phosphorylase is called phosphorylase kinase.

little slower, taking seconds to minutes to take effect, but it is more long-lasting because the equilibrium will stay shifted to the R form until other enzymes (phosphatases) remove the phosphates. Thus, the combination of allosteric and covalent control gives us the best of both worlds.

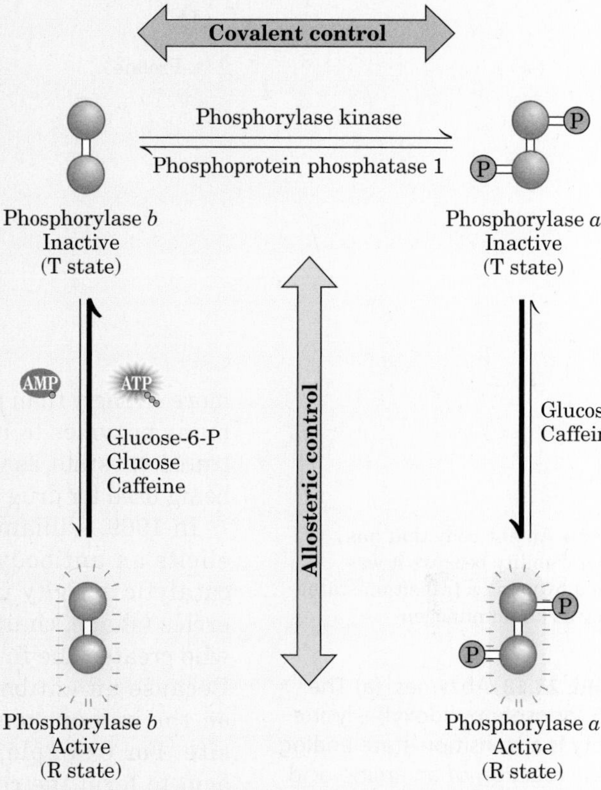

Test your knowledge with Problems 23.62 and 23.63.

23.8 What Are Transition-State Analogs and Designer Enzymes?

As we saw in Section 23.5C, an enzyme lowers the activation energy for a reaction, making the transition state more favorable. It does so by having an active site that actually fits best to the transition state conformation rather than to the substrates or the products. This has been documented by the use of **transition-state analogs**, molecules with a shape that mimics the transition state of the substrate.

Proline racemase, for example, catalyzes a reaction that converts L-proline to D-proline. During this reaction, the α-carbon must change from a tetrahedral arrangement to a planar form and then back to a tetrahedral form, but with the orientation of two bonds reversed (Figure 23.17). An inhibitor of the reaction is pyrrole-2-carboxylate, a chemical that is structurally similar to what proline would look like at its transition state because it is always planar at the equivalent carbon. This inhibitor binds to proline racemase 160 times

Transition-state analog A molecule that mimics the transition state of a chemical reaction and that is used as an inhibitor of an enzyme

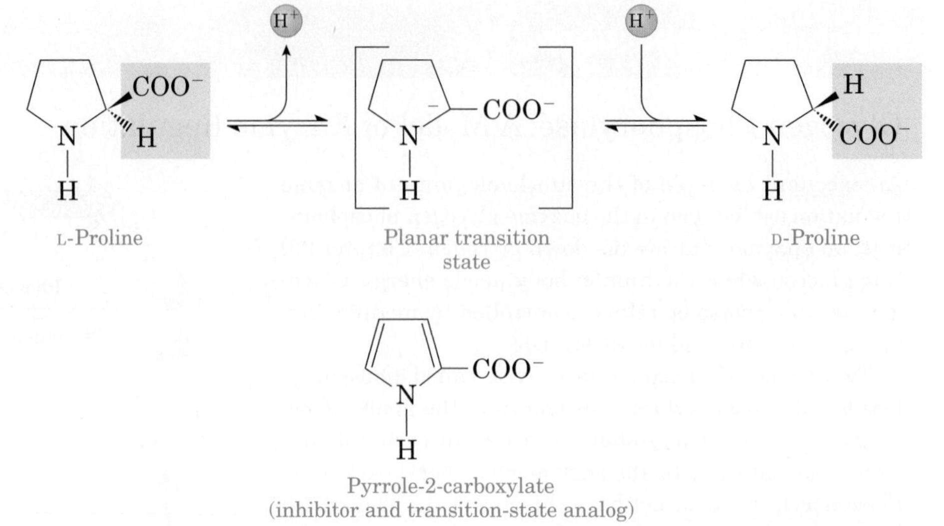

L-Proline

Planar transition state

D-Proline

Pyrrole-2-carboxylate
(inhibitor and transition-state analog)

Abzyme An antibody that has catalytic ability because it was created by using a transition-state analog as an immunogen

FIGURE 23.18 Abzymes. (a) The N^α-(5′-phosphopyridoxyl)-L-lysine moiety is a transition-state analog for the reaction of an amino acid with pyridoxal-5′-phosphate. When this moiety is attached to a protein and injected into a host, it acts like an antigen and the host produces antibodies that have catalytic activity (abzymes). (b) The abzyme is then used to catalyze the reaction.

more strongly than proline does. Transition-state analogs have been used with many enzymes to help verify a suspected mechanism and structure of the transition state, as well as to inhibit an enzyme selectively. They are currently being used for drug design when a specific enzyme is the cause of a disease.

In 1969, William Jencks proposed that an immunogen (a molecule that elicits an antibody response) would elicit antibodies (Chapter 31) with catalytic activity if the immunogen mimicked the transition state properties (shape, charge) of the reaction. Richard Lerner and Peter Schultz, who created the first catalytic antibodies, verified this hypothesis in 1986. Because an antibody is a protein designed to bind to specific molecules on the immunogen, the antibody will, in essence, serve as a fake active site. For example, the reaction of pyridoxal phosphate and an amino acid to form the corresponding α-ketoacid and pyridoxamine phosphate is a very important reaction in amino acid metabolism. The molecule N^α-(5′-phosphopyridoxyl)-L-lysine serves as a transition-state analog for this reaction. When this antigen molecule was used to elicit antibodies, these antibodies, or **abzymes**, had catalytic activity (Figure 23.18). Thus, in addition to helping us to verify the nature of the transition state or making an inhibitor, transition-state analogs now offer the possibility of creating designer enzymes to catalyze a wide variety of reactions.

(a)

N^α-(5′-Phosphopyridoxyl)-L-lysine moiety
(antigen)

(b)

D-Alanine

Pyridoxal 5′-P

Abzyme (antibody)

Pyruvate

Pyridoxamine 5′-P

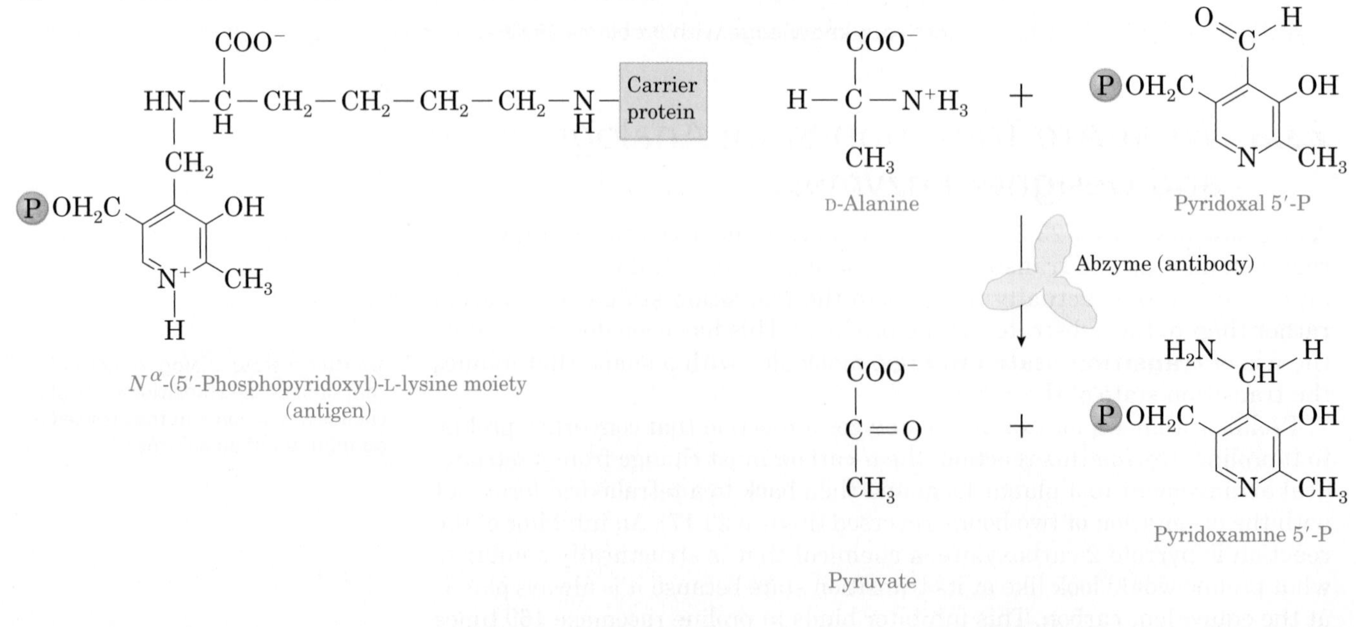

Enzymes Are First-Rate Organic Chemists

Enzymes are constantly carrying out reactions that would be a challenge to the most skilled organic chemists. In a sense, they do microsurgery on substrates, producing elegant and simple results in challenging situations.

In some cases, enzymes even synthesize the co-factors they need to carry out reactions. The co-factors thus produced can be electron acceptors, which are needed to react with the nucleophilic side chains of many proteins. (Recall the term "nucleophilic attack" from Section 23.5C. It refers to a reaction where an electron-rich atom, such as oxygen, bonds to an electron-deficient atom, such as a carbonyl carbon.) An enzyme called methylamine dehydrogenase (MADH) provides an example that draws on many concepts we have already seen. The co-factor is tryptophan tryptophyl quinone (TTQ), which is synthesized in place in the β-subunit of MADH. The reaction consists of two oxidation steps, with H_2O_2 as the oxidizing agent in each case from tryptophan residues 57 and 108. The electron transfer is facilitated by the protein MauG, which contains two iron-containing heme groups. (The heme groups are shown in the figure as Fe in a circle.) The "surgery" takes place by the transfer of electrons over a distance of about 400 nm, which is a large one in terms of molecular dimensions. Experimental evidence indicates that large conformational changes in the proteins do not play a role in the process. Note the unusual oxidation state of the iron in both heme groups. Protein structures can produce unique environments that cannot exist in smaller molecules.

Given the way that enzymes can carry out reactions, the next question is whether we can make use of their properties for reactions other than the ones they normally catalyze. We can do so within limits. It has been possible, for example, to engineer a protein that can perform a Diels-Alder reaction, in which new bonds are formed and a cyclic compound is the result. This is quite a feat, one that is likely to open new avenues for future research.

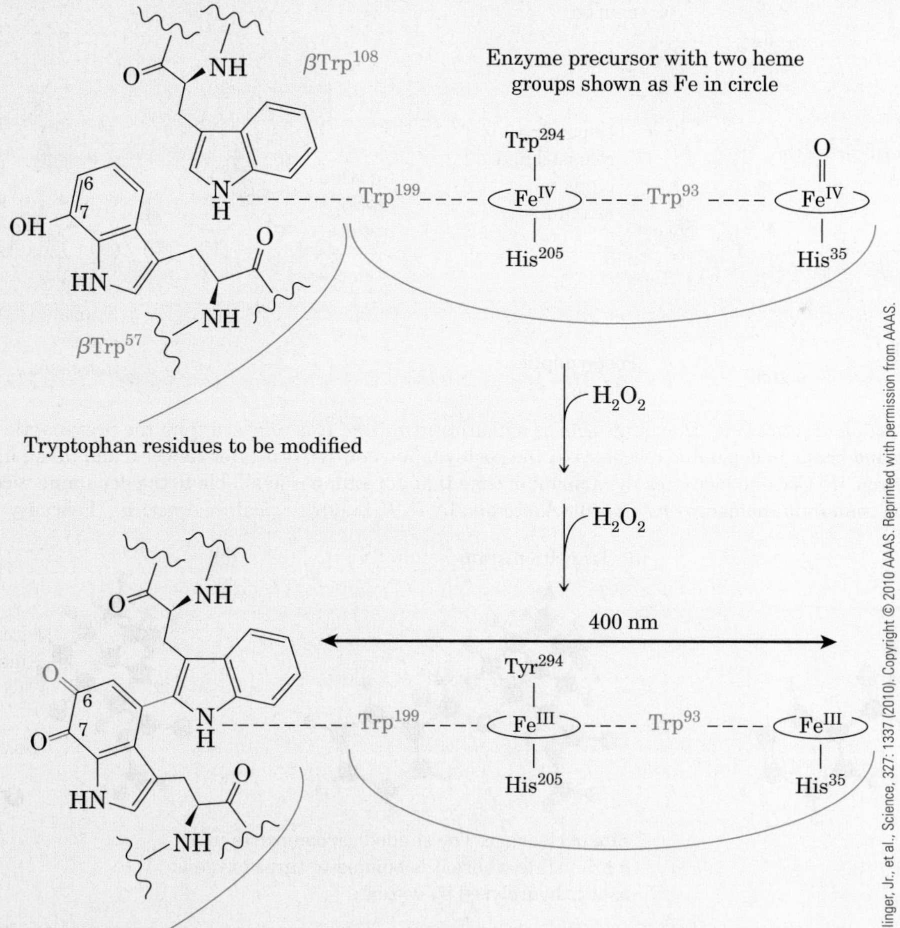

Schematic representation of the MADH·MauG complex and synthesis of the TTQ cofactor. Bonds of the cofactor installed by MauG are shown in red, and residues that are part of the surgeon's "scope" are in blue.

J. Martin Bollinger, Jr., et al., Science, 327: 1337 (2010). Copyright © 2010 AAAS. Reprinted with permission from AAAS.

Test your knowledge with Problems 23.64–23.66.

Catalytic Antibodies Against Cocaine

Many addictive drugs, including heroin, operate by binding to a particular receptor in the neurons, mimicking the action of a neurotransmitter. When a person is addicted to such a drug, a common way to attempt to treat the addiction is to use a compound that blocks the receptor and denies the drug access to it. Cocaine addiction has always been difficult to treat, due primarily to its unique *modus operandi*. As shown below, cocaine blocks the reuptake of the neurotransmitter dopamine. As a result, dopamine stays in the system longer, overstimulating neurons and leading to the reward signals in the brain that lead to addiction. There is promising research on the horizon that explores blocking certain brain receptors to help end cocaine addiction.

Cocaine (Section 16.2) can be degraded by a specific esterase, an enzyme that hydrolyzes an ester bond that is part of cocaine's structure. In the process of this hydrolysis, the cocaine must pass through a transition state that changes its shape. Catalytic antibodies to the transition state of the hydrolysis of cocaine have now been created. When administered to patients suffering from cocaine addiction, the antibodies successfully hydrolyze cocaine to two harmless degradation products—benzoic acid and ecgonine methyl ester. When degraded, the cocaine cannot block dopamine reuptake. No prolongation of the neuronal stimulus occurs, and the addictive effects of the drug vanish over time.

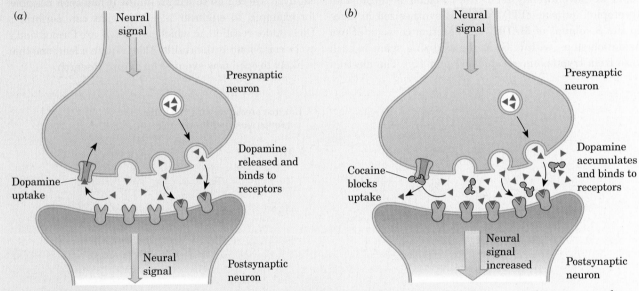

The mechanism of action of cocaine. (*a*) Dopamine acts as a neurotransmitter. It is released from the presynaptic neuron, travels across the synapse, and bonds to dopamine receptors on the postsynaptic neuron. It is later released and taken up into vesicles in the presynaptic neuron. (*b*) Cocaine increases the amount of time that dopamine is available to the dopamine receptors by blocking its uptake. (Adapted from *Immunotherapy for Cocaine Addiction,* by D. W. Landry, *Scientific American,* February 1997, pp 42–45.)

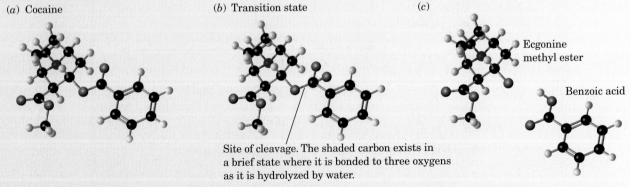

(*a*) Cocaine

(*b*) Transition state

Site of cleavage. The shaded carbon exists in a brief state where it is bonded to three oxygens as it is hydrolyzed by water.

(*c*) Ecgonine methyl ester

Benzoic acid

Degradation of cocaine by esterases or catalytic antibodies. Cocaine (*a*) passes through a transition state (*b*) on its way to being hydrolyzed to benzoic acid and ecgonine methyl ester (*c*). Transition-state analogs are used to generate catalytic antibodies for this reaction. (Adapted from *Immunotherapy for Cocaine Addiction,* by D. W. Landry, *Scientific American,* February 1997, pp 42–45.)

Test your knowledge with Problems 23.67–23.69.

Summary

Section 23.1 What Are Enzymes?

- **Enzymes** are macromolecules that catalyze chemical reactions in the body. Most enzymes are very specific—they catalyze only one particular reaction.
- The compound whose reaction is catalyzed by an enzyme is called the **substrate**.
- Most enzymes are proteins, although some are made of RNA.

Section 23.2 How Are Enzymes Named and Classified?

- Enzymes are classified into six major groups according to the type of reaction they catalyze.
- Enzymes are typically named after the substrate and the type of reaction they catalyze by adding the ending "-ase."

Section 23.3 What Is the Terminology Used with Enzymes?

- Some enzymes are made of polypeptide chains only. Others have, besides the polypeptide chain (the **apoenzyme**), nonprotein **cofactors**, which are either organic compounds (**coenzymes**) or inorganic ions.
- Only a small part of the enzyme surface, called the **active site**, participates in the actual catalysis of chemical reactions. Cofactors, if any, are part of the active site.
- Compounds that slow enzyme action are called **inhibitors**.
- A **competitive inhibitor** attaches itself to the active site. A **noncompetitive inhibitor** binds to other parts of the enzyme surface.

Section 23.4 What Factors Influence Enzyme Activity?

- The higher the enzyme and substrate concentrations, the higher the enzyme activity. At sufficiently high substrate concentrations, however, a saturation point is reached. After this point, increasing the substrate concentration no longer increases the reaction rate.
- Each enzyme has an optimal temperature and pH at which it has its greatest activity.

Section 23.5 What Are the Mechanisms of Enzyme Action?

- Two closely related models that seek to explain enzyme activity and specificity are the **lock-and-key model** and the **induced-fit model**.
- Enzymes lower the **activation energy** required for a biochemical reaction to occur.

Section 23.6 How Are Enzymes Regulated?

- Enzyme activity is regulated by five mechanisms.
- In **feedback control**, the concentration of products influences the rate of the reaction.
- Some enzymes, called **proenzymes** or **zymogens**, must be activated by removing a small portion of the polypeptide chain.
- In **allosterism**, an interaction takes place at a position other than the active site but affects the active site either positively or negatively.
- Enzymes can be activated or inhibited by **protein modification**.
- Enzyme activity is also regulated by **isozymes** (isoenzymes), which are different forms of the same enzyme.

Section 23.7 How Are Enzymes Used in Medicine?

- Abnormal enzyme activity can be used to diagnose certain diseases.

Section 23.8 What Are Transition-State Analogs and Designer Enzymes?

- The active site of an enzyme favors the production of the **transition state** conformation.
- Molecules having a shape that mimics the shape of the substrate molecule during the transition state are called **transition-state analogs**, and they make potent enzyme inhibitors.

Problems

Orange-numbered problems are applied.

Section 23.1 What Are Enzymes?

23.1 What is the difference between a *catalyst* and an *enzyme*?

23.2 What are ribozymes made of?

23.3 Would a lipase hydrolyze two triglycerides, one containing only oleic acid and the other containing only palmitic acid, with equal ease?

23.4 Compare the activation energy in uncatalyzed reactions and in enzyme-catalyzed reactions.

23.5 Why does the body need so many different enzymes?

23.6 Trypsin cleaves polypeptide chains at the carboxyl side of a lysine or arginine residue (Figure 23.1). Chymotrypsin cleaves polypeptide chains on the carboxyl side of an aromatic amino acid residue or any other nonpolar, bulky side chain. Which enzyme is more specific? Explain.

Section 23.2 How Are Enzymes Named and Classified?

23.7 Both lyases and hydrolases catalyze reactions involving water molecules. What is the difference in the types of reactions that these two enzymes catalyze?

23.8 Monoamine oxidases are important enzymes in brain chemistry. Judging from the name, which of the following would be a suitable substrate for this class of enzymes:

(a)
$$HO-\!\!\!\raisebox{0pt}{\text{(benzene ring)}}\!\!\!-\overset{\displaystyle OH}{\underset{\displaystyle |}{CH}}-CH_2NH_2$$

(b) $CH_3-\overset{\displaystyle O}{\overset{\displaystyle \|}{C}}-N(CH_3)_2$

(c) (benzene ring)$-NO_2$

23.9 On the basis of the classification given in Section 23.2, decide to which group each of the following enzymes belongs:

(a) Phosphoglyceromutase

$$^-OOC-\underset{\underset{\displaystyle OH}{|}}{CH}-CH_2-OPO_3{}^{2-}$$

3-Phosphoglycerate

$$\rightleftharpoons {}^-OOC-\underset{\underset{\displaystyle OPO_3{}^{2-}}{|}}{CH}-CH_2-OH$$

2-Phosphoglycerate

(b) Urease

$$H_2N-\overset{\displaystyle O}{\overset{\displaystyle \|}{C}}-NH_2 + H_2O \rightleftharpoons 2NH_3 + CO_2$$

Urea

(c) Succinate dehydrogenase

$$^-OOC-CH_2-CH_2-COO^- \quad + \quad FAD$$

Succinate $\qquad$ Coenzyme
(oxidized form)

$$\rightleftharpoons \underset{^-OOC}{\overset{H}{\diagup}}C=C\underset{H}{\overset{COO^-}{\diagup}} \quad + \quad FADH_2$$

Fumarate $\qquad$ Coenzyme
(reduced form)

(d) Aspartase

$$\underset{^-OOC}{\overset{H}{\diagup}}C=C\underset{H}{\overset{COO^-}{\diagup}} \quad + \quad NH_4{}^+$$

Fumarate

$$\rightleftharpoons {}^-OOC-CH_2-\underset{\underset{\displaystyle NH_3{}^+}{|}}{CH}-COO^-$$

L-Asparate

23.10 What kind of reaction does each of the following enzymes catalyze?

(a) Deaminases $\qquad$ (b) Hydrolases

(c) Dehydrogenases $\qquad$ (d) Isomerases

Section 23.3 What Is the Terminology Used with Enzymes?

23.11 What is the difference between a *coenzyme* and a *cofactor*?

23.12 In the citric acid cycle, an enzyme converts succinate to fumarate (see the reaction in Problem 23.9c). The enzyme consists of a protein portion and an organic molecule portion called FAD. What terms do we use to refer to (a) the protein portion and (b) the organic molecule portion?

23.13 What is the difference between reversible and irreversible noncompetitive inhibition?

Section 23.4 What Factors Influence Enzyme Activity?

23.14 In most enzyme-catalyzed reactions, the rate of reaction reaches a constant value with increasing substrate concentration. This relationship is described in a saturation curve diagram (Figure 23.4). If the enzyme concentration on a molar basis is twice the maximum substrate concentration, would you obtain a saturation curve?

23.15 At a very low concentration of a certain substrate, we find that when the substrate concentration doubles, the rate of the enzyme-catalyzed reaction also doubles. Would you expect the same finding at a very high substrate concentration? Explain.

23.16 If we wish to double the rate of an enzyme-catalyzed reaction, can we do so by increasing the temperature by 10°C? Explain.

23.17 A bacterial enzyme has the following temperature-dependent activity.

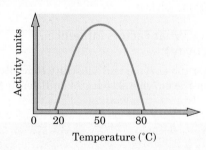

(a) Is this enzyme more or less active at normal body temperature than when a person has a fever?

(b) What happens to the enzyme activity if the patient's temperature is lowered to 35°C?

23.18 The optimal temperature for the action of lactate dehydrogenase is 36°C. It is irreversibly inactivated at 85°C, but a yeast containing this enzyme can survive for months at −10°C. Explain how this can happen.

23.19 The activity of pepsin was measured at various pH values. When the temperature and the concentrations of pepsin and substrate were held constant, the following activities were obtained:

pH	Activity
1.0	0.5
1.5	2.6
2.0	4.8
3.0	2.0
4.0	0.4
5.0	0.0

(a) Plot the pH dependence of pepsin activity.

(b) What is the optimal pH?

(c) Predict the activity of pepsin in the blood at pH 7.4.

23.20 How can the pH profile of an enzyme tell you something about the reaction mechanism if you know the amino acids at the active site?

Section 23.5 What Are the Mechanisms of Enzyme Action?

23.21 Urease can catalyze the hydrolysis of urea but not the hydrolysis of diethylurea. Explain why diethylurea is not hydrolyzed.

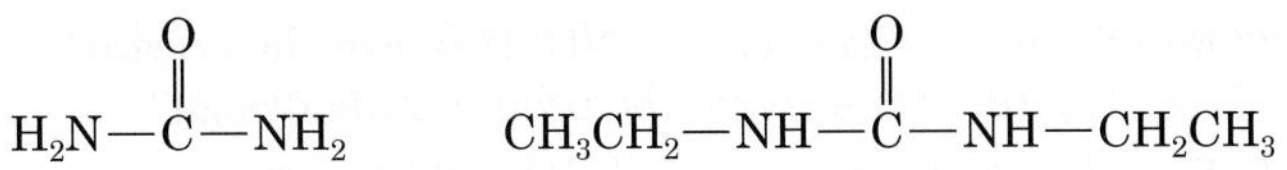

$$H_2N-\overset{\overset{\displaystyle O}{\|}}{C}-NH_2 \qquad CH_3CH_2-NH-\overset{\overset{\displaystyle O}{\|}}{C}-NH-CH_2CH_3$$

Urea   Diethylurea

23.22 The following reaction may be represented by the cartoon figures:

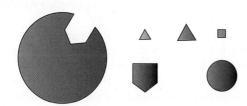

Glucose + ATP $\rightleftharpoons$ glucose 6-phosphate + ADP

In this enzyme-catalyzed reaction, Mg^{2+} is a cofactor, fluoroglucose is a competitive inhibitor, and Cd^{2+} is a noncompetitive inhibitor. Identify each component of the reaction by a cartoon figure and assemble them to show (a) the normal enzyme reaction, (b) a competitive inhibition, and (c) a noncompetitive inhibition.

23.23 Which amino acids appear most frequently in the active sites of enzymes?

23.24 What kind of chemical reaction occurs most frequently at the active site?

23.25 Which of the following is a correct statement describing the induced-fit model of enzyme action? Substrates fit into the active site:

(a) because both are exactly the same size and shape.

(b) by changing their size and shape to match those of the active site.

(c) by changing the size and shape of the active site upon binding.

23.26 What is the maximum rate that can be achieved in competitive inhibition compared with noncompetitive inhibition?

23.27 Enzymes are long protein chains, usually containing more than 100 amino acid residues. Yet the active site contains only a few amino acids. Explain why the other amino acids of the chain are present and what would happen to the enzyme activity if the enzyme's structure were changed significantly.

23.28 On some baking product labels, you might see an ingredient called "invert sugar." This is made by hydrolyzing sucrose (common table sugar) to glucose and fructose. The reaction is catalyzed by the enzyme invertase. Using the following data, determine whether the inhibition by 2 M urea is competitive or noncompetitive.

Sucrose Concentration (M)	Velocity (arbitrary units)	Velocity + Inhibitor
0.0292	0.182	0.083
0.0584	0.265	0.119
0.0876	0.311	0.154
0.117	0.330	0.167
0.175	0.372	0.192

Section 23.6 How Are Enzymes Regulated?

23.29 The hydrolysis of glycogen to yield glucose is catalyzed by the enzyme phosphorylase. Caffeine, which is not a carbohydrate and not a substrate for the enzyme, inhibits phosphorylase. What kind of regulatory mechanism is at work?

23.30 Can the product of a reaction that is part of a sequence act as an inhibitor for another reaction in the sequence? Explain.

23.31 What is the difference between a *zymogen* and a *proenzyme*?

23.32 The enzyme trypsin is synthesized by the body in the form of a long polypeptide chain containing 235 amino acids (trypsinogen), from which a piece must be cut before the trypsin can be active. Why does the body not synthesize trypsin directly?

23.33 Give the structure of a tyrosyl residue of an enzyme modified by a protein kinase.

23.34 What is an *isozyme*?

23.35 The enzyme glycogen phosphorylase initiates the phosphorolysis of glycogen to glucose 1-phosphate. It comes in two forms: Phosphorylase *b* is less active, and phosphorylase *a* is more active. The difference between the *b* and *a* forms is the modification of the apoenzyme. Phosphorylase *a* has two phosphate groups added to the polypeptide chain. In analogy with the pyruvate kinase discussed in the text, give a scheme indicating the transition between the *b* and *a* forms. Which enzymes and which cofactors control this reaction?

23.36 How can you tell if an enzyme is allosteric by plotting velocity versus substrate?

23.37 Explain the nature of the two types of control of glycogen phosphorylase. What is the advantage to having both control types?

23.38 Which type of regulation discussed in Section 23.6 is the least reversible? Explain.

23.39 The enzyme phosphofructokinase (PFK) (Chapter 28) has two types of subunits, M and L, for muscle and liver, respectively. These subunits combine to form a tetramer. How many isozymes of PFK exist? What are their designations?

23.40 If you separated PFK using electrophoresis, how would the isozymes migrate if the M subunit has a lower pI than the L subunit?

Section 23.7 How Are Enzymes Used in Medicine?

23.41 After a heart attack, the levels of certain enzymes rise in the serum. Which enzyme would you monitor within 24 hours following a suspected heart attack?

23.42 The enzyme formerly known as GPT (glutamate-pyruvate transaminase) has a new name: ALT (alanine aminotransferase). Looking at the equation in Section 28.9, which is catalyzed by this enzyme, what prompted this change of name?

23.43 If an examination of a patient indicated elevated levels of AST but normal levels of ALT, what would be your tentative diagnosis?

23.44 Which LDH isozyme is monitored in the case of a heart attack?

23.45 Chemists who have been exposed for years to organic vapors usually show higher-than-normal activity when given the alkaline phosphatase test. Which organ in the body do organic vapors affect?

23.46 Which enzyme preparation is given to patients after duodenal ulcer surgery?

23.47 Chymotrypsin is secreted by the pancreas and passed into the intestine. The optimal pH for this enzyme is 7.8. If a patient's pancreas cannot manufacture chymotrypsin, would it be possible to supply it orally? What happens to chymotrypsin's activity during its passage through the gastrointestinal tract?

Section 23.8 What Are Transition-State Analogs and Designer Enzymes?

23.48 Explain why transition-state analogs are potent inhibitors.

23.49 How do transition-state analogs relate to the idea of the induced-fit model of enzymes?

23.50 Explain the relationship between transition-state analogs and abzymes.

Chemical Connections

23.51 (Chemical Connections 23A) Acetylcholine causes muscles to contract. Succinylcholine, a close relative, is a muscle relaxant. Explain the different effects of these related compounds.

23.52 (Chemical Connections 23A) An operating team usually administers succinylcholine before bronchoscopy. What is achieved by this procedure?

23.53 (Chemical Connections 23B) PKMζ is a type of enzyme called a kinase. Kinases are very important in metabolism. Look through the metabolism chapters (Chapters 27 and 28) and find two examples of kinases. What reactions do kinases catalyze?

23.54 (Chemical Connections 23B) Explain how researchers used the drug ZIP to test its effect on long-term memory. How did they know that food aversion was a long-term memory phenomenon?

23.55 (Chemical Connections 23B) Why would researchers want to be able to selectively block long-term memory?

23.56 (Chemical Connections 23C) What role does Mn^{2+} play in anchoring the substrate in the active site of protein kinase?

23.57 (Chemical Connections 23C) Which amino acids of the active site interact with the $=CH_2$ group of the phosphoenol pyruvate? Do these amino acids provide the same surface environment? What is the nature of the interaction?

23.58 (Chemical Connections 23D) What is the strategy in drug design to fight AIDS?

23.59 (Chemical Connections 23D) Why did scientists want to create a drug to inhibit cGMP diesterases?

23.60 (Chemical Connections 23D) How can the crystal structures of enzymes be used in drug design?

23.61 (Chemical Connections 23D) Would reducing the population of intestinal bacteria be a useful way to alleviate the side effects of treatment for colon cancer?

23.62 (Chemical Connections 23E) Explain the difference between phosphorylase a and phosphorylase b. Which is more active and why?

23.63 (Chemical Connections 23E) What is the relationship between protein modification and allosteric control of glycogen phosphorylase?

23.64 (Chemical Connections 23F) Are the functional groups in the side chains of enzymes mostly nucleophiles, mostly electrophiles, or a mixture of both?

23.65 (Chemical Connections 23F) Can heme groups play a role in the action of enzymes?

23.66 (Chemical Connections 23F) Can enzymes be engineered to carry out reactions that are known in organic chemistry?

23.67 (Chemical Connections 23G) Explain how catalytic antibodies are produced to fight cocaine addiction.

23.68 (Chemical Connections 23G) Why can inhibitors not be used to block a cocaine receptor as is often done with other drug addictions?

23.69 (Chemical Connections 23G) How does cocaine work as a drug?

Additional Problems

23.70 Where can one find enzymes that are both stable and active at 90°C?

23.71 Food can be preserved by inactivation of enzymes that would cause spoilage—for example, by refrigeration. Give an example of food preservation in which the enzymes are inactivated (a) by heat and (b) by lowering the pH.

23.72 Enzyme therapy (administration of digestive enzymes) is suggested as a treatment for various medical conditions, including autism. How likely is it that this method will be effective?

23.73 Would you expect to find active digestive enzymes in a cooked hot dog? What is the reason for your answer?

23.74 Why is enzyme activity during myocardial infarction measured in patients' serum rather than in his or her urine?

23.75 What is the common characteristic of the amino acids of which the carboxyl groups of the peptide bonds can be hydrolyzed by trypsin?

23.76 Many enzymes are active only in the presence of Zn^{2+}. What common term is used for ions such as Zn^{2+} when discussing enzyme activity?

23.77 An enzyme has the following pH dependence:

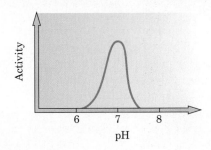

At what pH do you think this enzyme works best?

23.78 What enzyme is monitored in the diagnosis of infectious hepatitis?

23.79 The enzyme chymotrypsin catalyzes the following type of reaction:

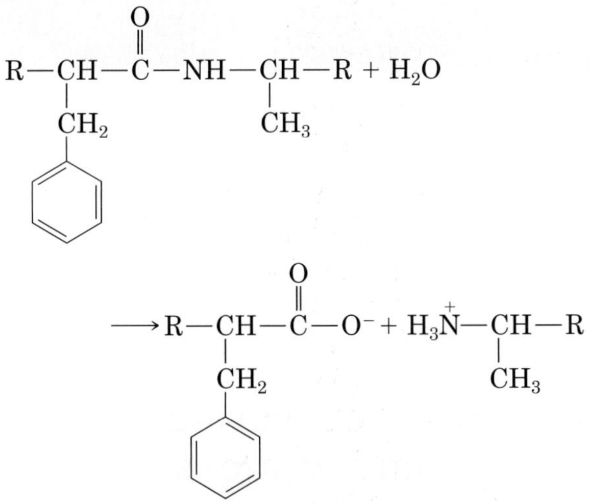

$$R—\overset{\displaystyle CH}{\underset{\displaystyle CH_2}{|}}—\overset{\displaystyle O}{\overset{\|}{C}}—NH—\overset{\displaystyle CH}{\underset{\displaystyle CH_3}{|}}—R + H_2O$$

$$\longrightarrow R—\overset{\displaystyle CH}{\underset{\displaystyle CH_2}{|}}—\overset{\displaystyle O}{\overset{\|}{C}}—O^- + \overset{+}{H_3}N—\overset{\displaystyle CH}{\underset{\displaystyle CH_3}{|}}—R$$

On the basis of the classification given in Section 23.2, to which group of enzymes does chymotrypsin belong?

23.80 Nerve gases operate by forming covalent bonds at the active site of cholinesterase. Is this an example of competitive inhibition? Can the nerve gas molecules

be removed by simply adding more substrate (acetylcholine) to the enzyme?

23.81 What would be the appropriate name for an enzyme that catalyzes each of the following reactions:

(a) $CH_3CH_2OH \longrightarrow CH_3\overset{\displaystyle O}{\overset{\|}{C}}—H$

(b) $CH_3\overset{\displaystyle O}{\overset{\|}{C}}—O—CH_2CH_3 + H_2O$

$\longrightarrow CH_3\overset{\displaystyle O}{\overset{\|}{C}}—OH + CH_3CH_2OH$

23.82 In Section 29.5, a reaction between pyruvate and glutamate to form alanine and α-ketoglutarate is given. How would you classify the enzyme that catalyzes this reaction?

23.83 A liver enzyme is made of four subunits: 2A and 2B. The same enzyme, when isolated from the brain, has the following subunits: 3A and 1B. What would you call these two enzymes?

23.84 What is the function of a ribozyme?

23.85 Can an enzyme catalyze the forward reaction but not the backward reaction for its substrate-product pair(s)? Explain.

23.86 Why was the discovery of ribozymes a remarkable event?

23.87 In some health food stores, it is possible to buy supplements that contain isolated enzymes. For example, the enzyme superoxide dismutase functions as an antioxidant in cells that are exposed to oxygen. Do you think that these supplements are likely to have significant value?

Looking Ahead

23.88 Caffeine is a stimulant that is taken by many people in the form of coffee, tea, chocolate, and cola beverages. It is also used by many athletes. Caffeine has many effects, including stimulating lipases. Given its effect on lipases and on glycogen phosphorylase, would you predict caffeine to be more effective as an aid to a runner in a 10K race or in a 1-mile race?

23.89 Caffeine is also a diuretic, which means it increases the movement of water through the kidneys and into the urine. Why would this potentially offset its value to a distance athlete?

23.90 Until the discovery of thermophilic bacteria that live in conditions of extreme heat and pressure, it was impossible to have an automated system of DNA synthesis. Explain why this was so, given that it takes temperatures of around 90°C to separate strands of DNA.

23.91 What characteristics of RNA make it likely to have catalytic ability? Why is DNA less likely to have catalytic activity?

Chemical Communications: Neurotransmitters and Hormones

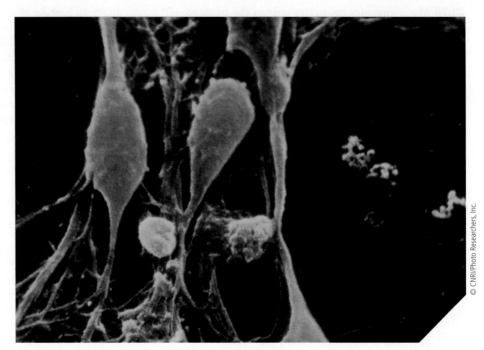

© CNRI/Photo Researchers, Inc.

Nerve cells. Colored scanning electron micrograph (SEM) of neurons (nerve cells) from the cerebral cortex (the outer, heavily folded gray matter of the brain). Neurons exist in varying sizes and shapes throughout the nervous system, but all have a similar basic structure; a large central cell body (colored gold, containing a nucleus) and processes of two types: a single axon (a nerve fiber) which is the effector part of the cell that terminates on other neurons (or organs) and one or more dendrites, smaller processes that act as sensory receptors. Similar types of neurons are arranged in layers in the cerebral cortex. Magnification: 470X at 6 × 7 cm size. Magnification: 235X at 35 mm size.

24.1 What Molecules Are Involved in Chemical Communications?

The importance of chemical communications in health becomes apparent from a quick glance at Table 24.1. This list contains only a small sample of drugs pertaining to this chapter, and it is of utmost interest to health care providers. In fact, a vast number of drugs in the pharmacopoeia act, in one way or another, to influence chemical communications.

Each cell in the body is an isolated entity enclosed in its own membrane. Furthermore, within each cell of higher organisms, organelles, such as the nucleus or the mitochondrion, are enclosed by membranes separating them from the rest of the cell. If cells could not communicate with one another, the thousands of reactions in each cell would be uncoordinated. The same is true for organelles within a cell. Such communication allows the activity of a cell in one part of the body to be coordinated with the activity of cells in a different part of the body. There are three principal types of molecules for communications:

- **Receptors** are protein molecules that bind to ligands and effect some type of change. They may be on the surface of cells, imbedded in the membrane of subcellular organelles, or free in solution. Most of the receptors we will study are membrane bound.
- **Chemical messengers**, also called ligands, interact with the receptors. (Chemical messengers fit into the receptor sites in a manner reminiscent of the lock-and-key model mentioned in Section 23.5.)
- **Secondary messengers** in many cases carry the message from the receptor to the inside of the cell and amplify the message.

If your house is on fire and the fire threatens your life, external signals—light, smoke, and heat—register alarm at specific receptors in your eyes, nose, and skin. From there, the signals are transmitted by specific compounds to nerve cells, or **neurons**. Nerve cells are present throughout the body and, together with the brain, constitute the nervous system. In the neurons, the signals travel as electric impulses along the axons (Figure 24.1). When they reach the end of the neuron, the signals are transmitted to adjacent neurons by specific compounds called **neurotransmitters**. Communication between the eyes and the brain, for example, is by neural transmission.

As soon as the danger signals are processed in the brain, other neurons carry messages to the muscles and to the endocrine glands. The message to the muscles is to run away or to take some other action in response to the fire (save the baby or run to get the fire extinguisher, for example). To do so, the muscles must be activated. Again, neurotransmitters carry the necessary messages from the neurons to the muscle cells and the endocrine glands. The endocrine glands are stimulated, and a different chemical signal, called a **hormone**, is secreted into your bloodstream. "The adrenaline begins to flow." Adrenaline is a hormone that binds to specific receptors in muscle and liver cells. Once bound, it triggers the production of a second messenger, cyclic AMP (cAMP). The second messenger leads to modification of several enzymes involved in carbohydrate metabolism. The immediate result is that the cells produce quick energy so that the muscles can fire rapidly and often, allowing the organism to use its strength and speed in the moment of crisis. We will revisit second messengers in Section 24.6.

Without these chemical communicators, the whole organism—you—would not survive because there is a constant need for coordinated efforts to face a complex outside world. The chemical communications between different cells and different organs play a role in the proper functioning of our bodies. Its significance is illustrated by the fact that *a large percentage of the drugs we encounter in medical practice try to influence this communication.* The scope of these drugs covers all fields—from prescriptions to treat hypertension, to heart disease, to antidepressants, to painkillers, just to mention a few. There are several ways these drugs act in the body. A drug may affect the messenger, the receptor, the secondary messenger, or any one of a host of enzymes that is activated or inhibited as part of a metabolic pathway (see Chapter 23).

1. An **antagonist** drug blocks the receptor and prevents its stimulation.

2. An **agonist** drug competes with the natural messenger for the receptor site. Once there, it stimulates the receptor.

3. Other drugs decrease the concentration of the messenger by controlling the release of messengers from their storage.

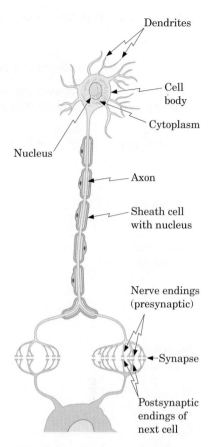

FIGURE 24.1 Neuron and synapse.

Neurotransmitter A chemical messenger between a neuron and another target cell: neuron, muscle cell, or cell of a gland

Hormone A chemical messenger released by an endocrine gland into the bloodstream and transported there to reach its target cell

Table 24.1 Drugs That Affect Nerve Transmission

Messenger	Drugs That Affect Receptor Sites		Drugs That Affect Available Concentration of the Neurotransmitter or Its Removal from Receptor Sites	
	Agonists (Activate Receptor Sites)	Antagonists (Block Receptor Sites)	Increase Concentration	Decrease Concentration
Acetylcholine (cholinergic)	Nicotine Succinylcholine	Curare Atropine	Malathion Nerve gases Succinylcholine Donepezil (Aricept)	*Clostridium botulinum* toxin
Calcium ion		Nifedipine (Adalat) Diltiazem (Cardizem)	Digoxin (Lanoxicaps)	
Epinephrine (α-adrenergic)	Terazosin (Hytrin)			
Norepinephrine (β-adrenergic)	Phenylephrine Epinephrine (Adrenalin)	Propranolol (Inderal)	Amphetamines	Reserpine, Methyldopa (Aldomet) Metyrosine (Demser)
Dopamine (adrenergic)		Clozapine (Clozaril)	Entacapone (Comtan)	
Serotonin (adrenergic)		Ondansetron (Zofran)	Antidepressant Fluoxetine (Prozac)	
Histamine (adrenergic)	2-Methylhistamine	Fexofenadine (Allegra) Diphenhydramine (Benadryl) Ranitidine (Zantac) Cimetidine (Tagamet)	Histidine	Hydrazinohistidine
Glutamic acid (amino acid)	*N*-Methyl-D-aspartate	Phencyclidine		
Enkephalin (peptidergic)	Opiate Morphine Heroin Meperidine (Demerol)		Naloxone (Narcan)	

4. Other drugs increase the concentration of the messenger by inhibiting its removal from the receptors.

5. Still others act to inhibit or activate specific enzymes inside the cells.

Table 24.1 presents selected drugs and their modes of action that affect neurotransmission. We will take a closer look at chemical communication as it relates to enzyme control later.

24.2 How Are Chemical Messengers Classified as Neurotransmitters and Hormones?

As mentioned earlier, neurotransmitters are compounds that communicate between two nerve cells or between a nerve cell and another cell (such as a muscle cell). A nerve cell (Figure 24.1) consists of a main cell body from which projects a long, fiber-like part called an **axon**. Coming off the other side of the main body are hair-like structures called **dendrites**.

Typically, neurons do not touch each other. Between the axon end of one neuron and the cell body or dendrite end of the next neuron is a space filled with an aqueous fluid, called a **synapse**. If the chemical signal travels, say, from axon to dendrite, we call the nerve ends on the axon the **presynaptic** site. The neurotransmitters are stored at the presynaptic site in **vesicles**, which are small, membrane-enclosed packages. Receptors are located on the **postsynaptic** site of the cell body or the dendrite.

Hormones are diverse compounds secreted by specific tissues (the endocrine glands), released into the bloodstream, and then adsorbed onto specific receptor sites, usually relatively far from their source. (This is the physiological definition of a hormone.) Table 24.2 lists some of the principal hormones. Figure 24.2 shows the target organs of hormones secreted by the pituitary gland.

The distinction between hormones and neurotransmitters is physiological, not chemical. Whether a certain compound is considered to be a neurotransmitter or a hormone depends on whether it acts over a short distance across a synapse (2×10^{-6} cm), in which case it is a neurotransmitter, or over a long distance (20 cm) from the secretory gland through the bloodstream to

Synapse An aqueous small space between the tip of a neuron and its target cell

Table 24.2 The Principal Hormones and Their Actions

Gland	Hormone	Action	Structures Shown in
Parathyroid	Parathyroid hormone	Increases blood calcium Excretion of phosphate by the kidneys	
Thyroid	Thyroxine (T_4)	Growth, maturation, and metabolic rate	Section 22.5
	Triiodothyronine (T_3)	Metamorphosis	
Pancreatic islets			
Beta cells	Insulin	Hypoglycemic factor Regulation of carbohydrates, fats, and proteins	Section 22.8 Chemical Connections 24G
Alpha cells	Glucagon	Liver glycogenolysis	
Adrenal medulla	Epinephrine Norepinephrine	Liver and muscle glycogenolysis	Section 24.5
Adrenal cortex	Cortisol	Carbohydrate metabolism	Section 21.10
	Aldosterone	Mineral metabolism	Section 21.10
	Adrenal androgens	Androgenic activity (especially females)	
Kidney	Renin	Hydrolysis of blood precursor protein to yield angiotensin	
Anterior pituitary	Luteinizing hormone	Causes ovulation	
	Interstitial cell-stimulating hormone	Formation of testosterone and progesterone in interstitial cells	
	Prolactin	Growth of mammary gland	
	Mammotropin	Lactation Corpus luteum function	
Posterior pituitary	Vasopressin	Contraction of blood vessels Kidney reabsorption of water	Section 22.8
	Oxytocin	Stimulates uterine contraction and milk ejection	Section 22.8
Ovaries	Estradiol	Estrous cycle	Section 21.10
	Progesterone	Female sex characteristics	Section 21.10
Testes	Testosterone	Male sex characteristics	Section 21.10
	Androgens	Spermatogenesis	

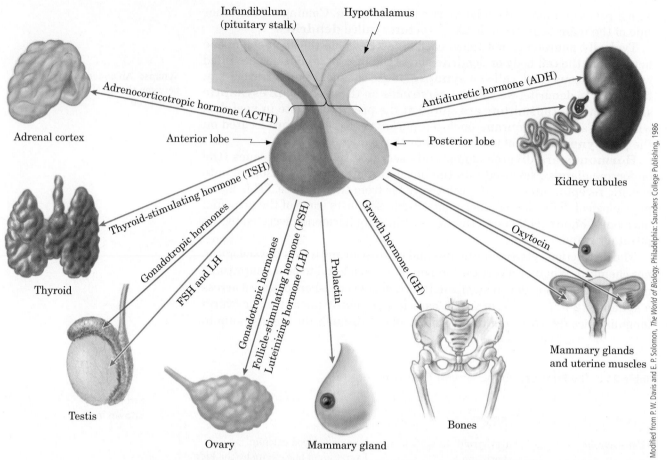

FIGURE 24.2 The pituitary gland is suspended from the hypothalamus by a stalk of neural tissue. The hormones secreted by the anterior and posterior lobes of the pituitary gland and the target tissues they act on are shown.

Modified from P. W. Davis and E. P. Solomon, *The World of Biology*. Philadelphia: Saunders College Publishing, 1986

the target cell, in which case it is a hormone. For example, epinephrine and norepinephrine are both neurotransmitters and hormones.

There are, broadly speaking, five classes of chemical messengers: *cholinergic, amino acid, adrenergic, peptidergic,* and *steroid* messengers. This classification is based on the chemical nature of the molecules in each group. Neurotransmitters can belong to all five classes, and hormones can belong to the last three classes.

Messengers can also be classified according to how they work. Some of them—epinephrine, for example—*activate enzymes.* Others affect the *synthesis of enzymes and proteins* by turning on the genes that produce them (Section 26.2). Steroid hormones (Section 21.10) work in this manner. Finally, some affect the *permeability of membranes;* acetylcholine and insulin belong to this class.

Yet another way of classifying messengers is according to their potential to *act directly* or through a *secondary messenger.* The steroid hormones act directly. They can penetrate the cell membrane and pass through the membrane of the nucleus. For example, estradiol stimulates uterine growth.

Other chemical messengers act through secondary messengers. For example, epinephrine, glucagon, luteinizing hormone, norepinephrine, and vasopressin use cAMP as a secondary messenger (details in Section 24.5C).

In the following sections, we will sample the mode of communication within each of the five chemical categories of messengers.

24.3 How Does Acetylcholine Act as a Messenger?

The main **cholinergic neurotransmitter** is acetylcholine:

$$CH_3-\overset{\displaystyle O}{\overset{\displaystyle \|}{C}}-O-CH_2-CH_2-\overset{\displaystyle CH_3}{\underset{\displaystyle CH_3}{\overset{\displaystyle |}{\underset{\displaystyle |}{N^+}}}}-CH_3$$

Acetylcholine

A. Cholinergic Receptors

There are two kinds of receptors for this messenger. We will look at one that exists on the motor end plates of skeletal muscles or in the sympathetic ganglia. The nerve cells that transmit messages contain stored acetylcholine in the vesicles in their axons. The receptor on the muscle cells or neurons is also known as the nicotinic receptor because nicotine (see Chemical Connections 16B) inhibits the neurotransmission of these nerves. The receptor itself is a *transmembrane protein* (Figure 21.2) made of five different subunits. The central core of the receptor is an ion channel through which, when open, Na^+ and K^+ ions can pass (Figure 24.3). When the ion channels are closed, the K^+ ion concentration is higher inside the cell than outside; the reverse is true for the Na^+ ion concentration.

B. Storage of Messengers

Events begin when a message is transmitted from one neuron to the next by neurotransmitters. The message is initiated by calcium ions. When the concentration in a neuron reaches a certain level (more than 0.1 μM), the vesicles containing acetylcholine fuse with the presynaptic membrane of the nerve cells. Then they empty the neurotransmitters into the synapse. The messenger molecules travel across the synapse and are adsorbed onto specific receptor sites.

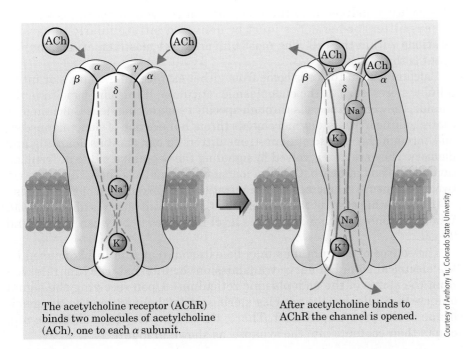

The acetylcholine receptor (AChR) binds two molecules of acetylcholine (ACh), one to each α subunit.

After acetylcholine binds to AChR the channel is opened.

Courtesy of Anthony Tu, Colorado State University

FIGURE 24.3 Acetylcholine in action. The receptor protein has five subunits. When two molecules of acetylcholine bind to the two α subunits, a channel opens to allow the passage of Na^+ and K^+ ions by facilitated transport (Chemical Connections 21C).

Chemical Connections 24A

Zebrafish, Synapses, and Sleep

Everyone understands that muscles need rest after strenuous exercise, but there are still many questions regarding what actually causes muscle fatigue. Even less is known about what causes "brain fatigue," but the chemistry of the neurons in the brain must be involved. Nerve impulses are chemical reactions requiring the combination of membrane depolarization, release of neurotransmitters, movement of the transmitters across the synaptic space, and binding to receptors. The brain is more constant than a muscle when it comes to the steady-state firing of its neurons, but even the brain needs rest.

Scientists have wondered for years why mammals and other animals need sleep. While examining the nature of the neurons and their chemistry, scientists could be divided into two camps. One believes that sleep is necessary because during sleep, the brain can actively reinforce firing patterns that lead to memory retention. Another camp believes it is just the opposite, that during sleep, some of the neurons shut down, allowing them to "rest" by not using unimportant synapses for a time. Thus, one theory would predict an increase in synapse activity for this "active strengthening" of memories, while the other would predict a decrease in synapse activity.

In 2010, researchers at Stanford University made progress in answering this question by using zebrafish to study

brain function. The larvae of zebrafish have two very useful characteristics. One is that like humans, they have a sleep cycle. The other is that they are transparent, allowing scientists to actually watch their brains work. The Stanford research team of Lior Appelbaum and Philippe Mourrain used a fluorescent protein to stain the synapses of zebrafish brains. The synapses that were active would appear green, while those that were inactive would be black. By watching the change in fluorescence over time, they were able to show that during the sleep cycle, fewer of the synapses were active, supporting the theory that the brain rests its synapses at night so they can be more functional during the day.

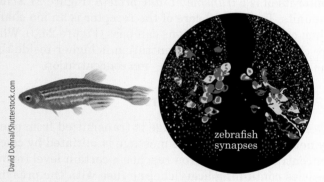

David Dohnal/Shutterstock.com

zebrafish synapses

Test your knowledge with Problems 24.47 and 24.48.

C. Calcium as a Signaling Agent (Secondary Messenger)

The message delivered to the receptors on the cell membranes by neurotransmitters or hormones must be delivered intracellularly to various locations within the cell. The most universal yet most versatile signaling agent is the cation Ca^{2+}.

Calcium ions in the cells come from either extracellular sources or intracellular stores, such as the endoplasmic reticulum. If the ions come from the outside, they enter the cells through specific calcium channels. Calcium ions control our heartbeats; our movements through the action of skeletal muscles; and through the release of neurotransmitters in our neurons, learning and memory. They are also involved in signaling the beginning of life at fertilization and its end at death. Calcium ion signaling controls these functions via two mechanisms: (1) increased concentration and (2) duration of the signals.

In the resting state of the neuron, the Ca^{2+} concentration is about 0.1 μM. When neurons are stimulated, this level increases to between 0.5 μM and 25 μM.

The source of calcium ions may be external (calcium influx caused by the electrical signal of nerve transmission) or internal (calcium released from the stores of the endoplasmic reticulum). Upon receiving the signal of increased calcium, the vesicles storing acetylcholine travel to the membrane of the presynaptic cleft. There, they fuse with the membrane and empty their contents into the synapse, as shown in Figure 24.4.

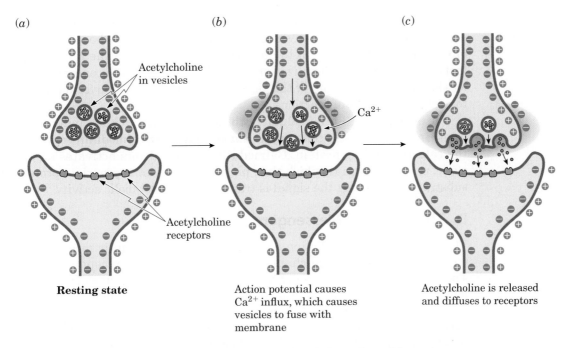

(*a*) (*b*) (*c*)

Acetylcholine in vesicles

Acetylcholine receptors

Ca^{2+}

Resting state

Action potential causes Ca^{2+} influx, which causes vesicles to fuse with membrane

Acetylcholine is released and diffuses to receptors

FIGURE 24.4 Calcium signaling results in acetylcholine being released from the vesicles of a neuron. (*a*) In the resting state, acetylcholine is sequestered inside vesicles in the presynaptic site. (*b*) A nerve transmission depolarizes the membrane and causes the concentration of Ca^{2+} to increase, which causes the vesicles to fuse with the membrane. (*c*) Acetylcholine is released into the synaptic space and binds to receptors on the postsynaptic site.

Chemical Connections 24B

Botulism and Acetylcholine Release

When meat or fish is improperly cooked or preserved, a deadly type of food poisoning, called botulism, may result. The culprit is the bacterium *Clostridium botulinum,* whose toxin prevents the release of acetylcholine from the presynaptic vesicles. Therefore, no neurotransmitter reaches the receptors on the surface of muscle cells, and the muscles no longer contract. Left untreated, the affected person may die.

Surprisingly, the botulin toxin has a valuable medical use. It is used in the treatment of involuntary muscle spasms—for example, in facial tics. These tics are caused by the uncontrolled release of acetylcholine. Controlled administration of the toxin, when applied locally to the facial muscles, stops the uncontrolled contractions and relieves the facial distortions.

"Facial distortion" has multiple meanings in the cosmetics industry. "Frown lines" and other wrinkles can also be removed by temporarily paralyzing facial muscles. The U.S. Food and Drug Administration has approved Botox (botulin toxin) for cosmetic use, meaning that its manufacturer can now advertise its rejuvenating appeal. Its use

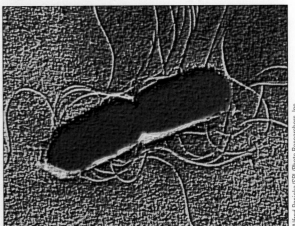

Clostridium botulinum is a food-poisoning bacterium.

is spreading fast. Botox has been used on an off-label basis for this application for a number of years, especially in Hollywood. Indeed, several movie directors have complained that some actors have used so much Botox that they can no longer show a variety of facial expressions.

Test your knowledge with Problems 24.49 and 24.50.

Calcium ions can also control signaling by controlling the duration of the signal. The signal in arterial smooth muscle lasts for 0.1 to 0.5 s. The wave of Ca^{2+} in the liver lasts 10 to 60 s. The calcium wave in the human egg lasts 1 to 35 min after fertilization. Thus, by combining the concentration, localization, and duration of the signal, calcium ions can deliver messages to perform a variety of functions.

The effects of Ca^{2+} are modulated through specific calcium-binding proteins. In all nonmuscle cells and in smooth muscles, calmodulin serves as the calcium-binding protein. Calmodulin-bound calcium activates an enzyme, protein kinase II, which then phosphorylates an appropriate protein substrate. In this way, the signal is translated into metabolic activity.

D. The Action of Messengers

The presence of acetylcholine molecules at the postsynaptic receptor site triggers a conformational change (Section 22.10) in the receptor protein. This opens the *ion channel* and allows ions to cross membranes freely. Na^+ ions have a higher concentration outside the neuron than do K^+ ions; thus, more Na^+ enters the cell than K^+ leaves. Because it involves ions, which carry electric charges, this process is translated into an electrical signal. After a few milliseconds, the channel closes again. The acetylcholine still occupies the receptor. For the channel to reopen and transmit a new signal, the acetylcholine must be removed and the neuron must be reactivated.

E. The Removal of Messengers

Acetylcholine is removed rapidly from the receptor site by the enzyme *acetylcholinesterase,* which hydrolyzes it.

$$CH_3-\overset{O}{\overset{\|}{C}}-O-CH_2-CH_2-\overset{CH_3}{\overset{|}{\underset{CH_3}{N^+}}}-CH_3 + H_2O \xrightarrow{\text{Acetylcholin-esterase}} CH_3-\overset{O}{\overset{\|}{C}}-O^- + HO-CH_2-CH_2-\overset{CH_3}{\overset{|}{\underset{CH_3}{N^+}}}-CH_3 + H^+$$

Acetylcholine   Acetate   Choline

This rapid removal enables the nerves to transmit more than 100 signals per second. By this means, with acetylcholine occupying a receptor, being removed, and then reactivation of the neuron with additional acetylcholine, the message moves from neuron to neuron until it is finally transmitted, again by acetylcholine molecules, to the muscles or endocrine glands that are the ultimate target of the message.

The action of the acetylcholinesterase enzyme is essential to the entire process. When this enzyme is inhibited, the removal of acetylcholine is incomplete and nerve transmission ceases.

Chemical Connections 24C

Alzheimer's Disease and Chemical Communication

Alzheimer's disease causes severe memory loss and other senile behavior that afflicts about 1.5 million people in the United States. People with Alzheimer's disease are forgetful, especially about recent events. As the disease advances, they become confused and, in severe cases, lose their ability to speak; at that point, they need total care. As yet, there is no cure for this disease. Postmortem identification of this disease focuses on three pathological hallmarks in the brain: (1) buildup of protein deposits known as β-amyloid plaques outside the nerve cells, (2) neurofibrillar tangles composed of tau proteins, and (3) brain shrinkage. Controversy exists as to which one is the primary cause of the neurodegeneration observed in Alzheimer's disease. Each has its advocates.

Chemical Connections 24C

Alzheimer's Disease and Chemical Communication (*continued*)

Besides understanding the nature of β-amyloid plaques and tau tangles, scientists are actively working to discover the timeline of the disease. The failure of several drug candidates has led to the conclusion that much of the damage happens years in advance of noticeable symptoms and that by the time these symptoms become apparent, it may be too late. Therefore, the current focus is the arrest of the disease before the notable loss of memory and other symptoms.

Much work is being done in Medellin, Colombia, where there is a group of 25 extended families with over 5000 members who develop Alzheimer's before the age of 50, which is 15–20 years earlier than most. Using the volunteer members of this group as test subjects, many new drugs, therapies, and study techniques have been tried.

Progress of the Disease

The earliest seen effect is the formation of aggregates of β-amyloid in the neurons in the centers of the brain that form new memories, as shown in the figure below: The β-amyloid plaques block the receptors for the neurotransmitters, reducing nerve transmission in the brain. This effect is seen 5–20 years before a person has noticeable symptoms.

One to five years before symptoms are noted, tau protein buildup can be seen. The tau proteins become phosphorylated and they detach from the microtubules where they belong, which leads to destruction of the microtubules and aggregates of tau proteins that disrupt nerve cell function, as shown in the second figure below:

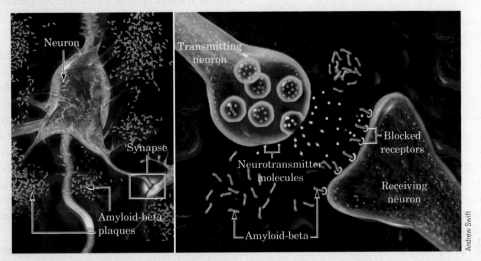

Amyloid-beta fragments are one of the problems seen in Alzheimer's disease. Left: amyloid-beta fragments (purple) congest the neuron cells. Right: pieces of amyloid-beta interfere with proper nerve cell functioning by blocking the receiving neuron receptors.

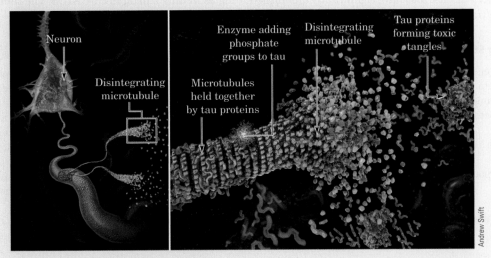

Tau tangles are another problem seen in Alzheimer's disease. Left: the overall problem is disintegration of microtubules. Right: tau proteins become phosphorylated and begin to dissociate from the microtubules, causing their disintegration.

Alzheimer's Disease and Chemical Communication (*continued*)

Finally, about one to five years before diagnosis, magnetic resonance imaging (MRI) will show that the brain is shrinking. Areas that involve memory, like the hippocampus, are hit particularly hard (see figure at right).

Usually, by the time the brain has shrunk, the patient or his or her family has noticed a loss of memory, although the patient may not yet have been formally diagnosed with Alzheimer's.

Knowing that there are telltale signs that precede formal diagnosis has given scientists tools for identifying Alzheimer's patients earlier. For example, an MRI can spot brain shrinkage. Testing cerebrospinal fluid can show altered levels of tau proteins (increased) or β-amyloid protein (decreased), giving an early warning. This then allows doctors to prescribe drugs while they might still have a chance.

While many researchers focus on the β-amyloid plaques and the tau proteins, it is not clear whether these are the real culprits that cause neuron death. Another chemical messenger, Ca^{2+}, may also be involved. Current research is leading to the conclusion that the calcium flux into neurons is disrupted in Alzheimer's patients. The β-amyloid proteins are believed to form channels in the neuron outer membrane, leading to higher-than-normal levels of intracellular calcium. Enzymes called presenilins may also play a role in the way calcium ion is released from intracellular stores, primarily from the endoplasmic reticulum (ER). Mutant presenilins from Alzheimer's patients are thought to provide a leak from the ER into the cytosol, as well as possibly affect the protein called SERCA that is supposed to clear calcium ion from the cytosol. While β-amyloid and tau proteins are the most notable and obvious characteristics of brain tissue from the disease, many believe that an overload of calcium ion actually causes the cell death. Patients with Alzheimer's disease also have significantly diminished acetylcholine transferase activity in their brains.

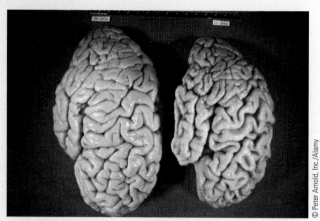

Pictures of a normal brain (left) and an Alzheimer's brain (right). The brain on the right is shrunken and gaps have formed between the folds.

The diminished concentration of acetylcholine can be partially compensated for by inhibiting the enzyme acetylcholinesterase, which decomposes acetylcholine. Certain drugs that act as acetylcholinesterase inhibitors have been shown to improve memory and other cognitive functions in some people with the disease. Drugs such as donepezil (Aricept), rivastigmine (Exelon), and galantamine (Razadyne) belong to this category; they all moderate the symptoms of Alzheimer's disease. The alkaloid huperzine A, an active ingredient of Chinese herb tea that has been used for centuries to improve memory, is also a potent inhibitor of acetylcholinesterase.

Unfortunately, there have been no huge successes to date in creating a super drug that will stop Alzheimer's. In fact, there have been some huge failures. Drug companies continue the search for such a drug, but it is as challenging as finding a drug to cure AIDS. The table below shows some of the drug types under study:

Drugs Under Study	What They Do
Inhibitors of enzymes that produce amyloid-beta	Such inhibitors block or modify the action of enzymes that cut a large protein (the amyloid precursor protein) in a way that releases the amyloid-beta peptides.
Vaccines or antibodies that clear amyloid-beta	Vaccines induce the body to produce antibodies that bind to amyloid and clear them from the brain. Unfortunately, in clinical trials, both vaccines and antibodies have induced side effects of varying severity in some patients.
Amyloid-beta aggregation blockers	Agents that prevent amyloid fragments from clumping could prevent damage to neurons.
Antitau compounds	These agents, although fewer in number than those that target the amyloid pathway, take various approaches, such as blocking production of the toxic form of the tau protein or impeding its aggregation into tangles.
Neuroprotective agents	Different strategies attempt to boost natural brain chemicals that enhance the health of neurons. In one, a gene is delivered into the brain to start production of a protective substance.

Gary Stix, "Alzheimer's: Forestalling the Darkness," *Scientific American*, June 2010, p. 56. Reproduced with permission. Copyright © 2010 Scientific American, a division of Nature America, Inc. All rights reserved.

Test your knowledge with Problems 24.51–24.58.

F. Control of Neurotransmission

Acetylcholinesterase is inhibited reversibly by succinylcholine (Chemical Connections 23A) and decamethonium bromide.

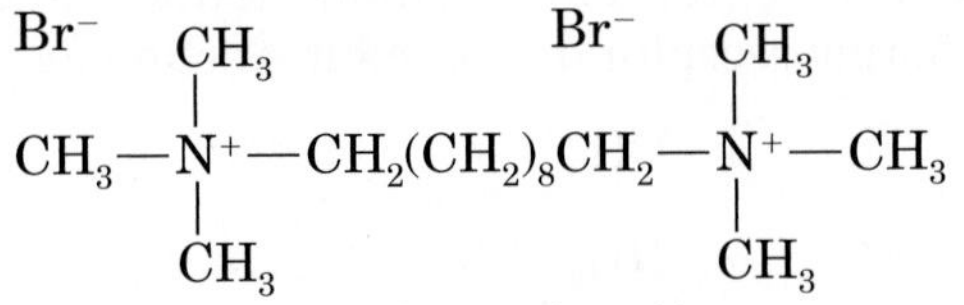

$$
\text{CH}_3-\overset{\overset{\displaystyle \text{CH}_3}{|}}{\underset{\underset{\displaystyle \text{CH}_3}{|}}{\text{N}^+}}-\text{CH}_2(\text{CH}_2)_8\text{CH}_2-\overset{\overset{\displaystyle \text{CH}_3}{|}}{\underset{\underset{\displaystyle \text{CH}_3}{|}}{\text{N}^+}}-\text{CH}_3
$$

$$\text{Br}^- \qquad\qquad\qquad \text{Br}^-$$

Decamethonium bromide

Succinylcholine and decamethonium bromide resemble the choline end of acetylcholine and therefore act as competitive inhibitors of acetylcholinesterase. In small doses, these reversible inhibitors relax the muscles temporarily, making them useful as muscle relaxants in surgery. In large doses, they are deadly.

The inhibition of acetylcholinesterase is but one way in which cholinergic neurotransmission is controlled. Another way is to modulate the action of the receptor. Because acetylcholine enables the ion channels to open and propagate signals, this mode of action is called *ligand-gated ion channels*. The attachment of the ligand to the receptor is critical in signaling. Nicotine given in low doses is a stimulant; it is an agonist because it prolongs the receptor's biochemical response. When given in large doses, however, nicotine becomes an antagonist and blocks the action on the receptor. As such, it may cause convulsions and respiratory paralysis. Succinylcholine, besides being a reversible inhibitor of acetylcholinesterase, also has this concentration-dependent agonist/antagonist effect on the receptor. A strong antagonist, which blocks the receptor completely, can interrupt the communication between neuron and muscle cell. The venom of a number of snakes, such as cobratoxin, exerts its deadly influence in this manner. The plant extract curare, which was used in poisoned arrows by the Amazon Indians, works in the same way. In small doses, curare is used as a muscle relaxant.

Finally, the supply of the acetylcholine messenger can influence the proper nerve transmission. If the acetylcholine messenger is not released from its storage, as in botulism (Chemical Connections 24B), or if its synthesis is impaired, as in Alzheimer's disease (Chemical Connections 24C), the concentration of acetylcholine is reduced and nerve transmission is impaired.

24.4 What Amino Acids Act as Neurotransmitters?

A. Messengers

Amino acids are distributed throughout the neurons individually or as parts of peptides and proteins. They can also act as neurotransmitters. Some of them, such as glutamic acid, aspartic acid, and cysteine, act as **excitatory neurotransmitters** similar to acetylcholine and norepinephrine. Others, such as glycine, β-alanine, taurine (Section 21.11), and mainly γ-aminobutyric acid (GABA), are **inhibitory neurotransmitters**; they reduce neurotransmission. Note that β-alanine and γ-aminobutyric acid (GABA) are not found in proteins.

$$^+\text{H}_3\text{NCH}_2\text{CH}_2\text{SO}_3^- \qquad ^+\text{H}_3\text{NCH}_2\text{CH}_2\text{COO}^- \qquad ^+\text{H}_3\text{NCH}_2\text{CH}_2\text{CH}_2\text{COO}^-$$

Taurine   β-Alanine   γ-Aminobutyric acid
(GABA)
(IUPAC name:
4-aminobutanoic acid)

B. Receptors

Each of these amino acids has its own receptor; in fact, glutamic acid has at least five subclasses of receptors. The best known is the *N*-methyl-D-aspartate (NMDA) receptor. This ligand-gated ion channel is similar to the nicotinic cholinergic receptor discussed in Section 24.3:

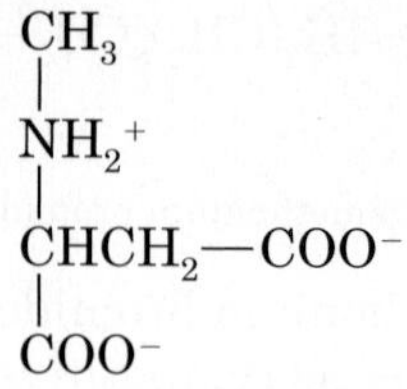

$$
\begin{array}{c}
CH_3 \\
| \\
NH_2{}^+ \\
| \\
CHCH_2\!\!-\!\!COO^- \\
| \\
COO^-
\end{array}
$$

N-Methyl-D-aspartate

When glutamic acid binds to this receptor, the ion channel opens, Na^+ and Ca^{2+} flow into the neuron, and K^+ flows out of the neuron. The same thing happens when NMDA, being an agonist, stimulates the receptor. The gate of this channel is closed by a Mg^{2+} ion.

Phencyclidine (PCP), an antagonist of this receptor, induces hallucination. PCP, known by the street name "angel dust," is a controlled substance; it causes bizarre psychotic behavior and long-term psychological problems.

C. Removal of Messengers

In contrast to acetylcholine, there is no enzyme that would degrade glutamic acid and thereby remove it from its receptor once the signaling has occurred. Glutamic acid is removed by **transporter** molecules, which bring it back through the presynaptic membrane into the neuron. This process is called **reuptake**.

Transporter A protein molecule that carries small molecules, such as glucose or glutamic acid, across a membrane

24.5 What Are Adrenergic Messengers?

A. Monoamine Messengers

The third class of neurotransmitters/hormones, the adrenergic messengers, includes such monoamines as epinephrine, serotonin, dopamine, and histamine. (Structures of these compounds can be found later in this section and in Chemical Connections 24D.) These monoamines transmit signals by a mechanism whose beginning is similar to the action of acetylcholine. That is, they are adsorbed on a receptor.

B. Signal Transduction

Once a hormone or neurotransmitter binds to a receptor, some mechanism must propagate the signal to the cell. The process by which the initial signal is spread and amplified throughout the cell is called signal transduction. The process involves intermediate compounds that pass the signal on to the ultimate targets. Eventually, enzymes are modified to alter their activity or membrane channels are opened or closed. The term "signal transduction" and much of the pioneering research on this topic came from the work of Martin Rodbell (1925–1998) of the National Institutes of Health, winner of the 1994 Nobel Prize in physiology and medicine.

The action of monoamine neurotransmitters is a prime example. Once the monoamine neurotransmitter/hormone (for example, norepinephrine) is adsorbed onto the receptor site, the signal will be amplified inside the cell. In the example shown in Figure 24.5, the receptor has an associated protein called G-protein. This protein is the key to the cascade that produces many signals

Signal transduction A cascade of events through which the signal of a neurotransmitter or hormone delivered to its receptor is carried inside the target cell and amplified into many signals that can cause protein modifications, enzyme activation, and the opening of membrane channels

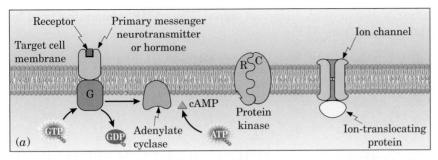

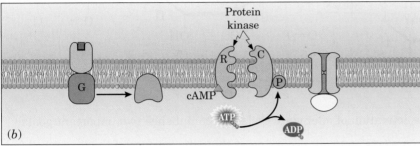

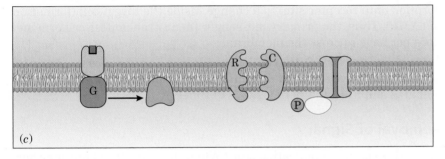

FIGURE 24.5 The sequence of events in the postsynaptic membrane when norepinephrine is absorbed onto the receptor site. (*a*) The active G-protein hydrolyzes GTP. The energy of hydrolysis of GTP to GDP activates adenylate cyclase. A molecule of cAMP forms when adenylate cyclase cleaves ATP into cAMP and pyrophosphate. (*b*) Cyclic AMP activates a protein kinase by dissociating the regulatory (R) unit from the catalytic unit (C). A second molecule of ATP, shown in (*b*), has phosphorylated the catalytic unit and been converted to ADP. (*c*) The catalytic unit phosphorylates the ion-translocating protein that blocked the channel for ion flow. The phosphorylated ion-translocating protein changes its shape and position and opens the ion gates.

inside the cell (amplification). The active G-protein has an associated nucleotide, guanosine triphosphate (GTP). It is an analog of adenosine triphosphate (ATP), in which the aromatic base adenine is substituted by guanine (Section 25.2). The G-protein becomes inactive when its associated nucleotide is hydrolyzed to guanosine diphosphate (GDP). Signal transduction starts with the active G-protein, which activates the enzyme adenylate cyclase.

G-protein also participates in another signal transduction cascade, which involves inositol-based compounds (Section 21.6) as signaling molecules. Phosphatidylinositol diphosphate (PIP$_2$) mediates the action of hormones and neurotransmitters. These messengers can stimulate the phosphorylation of enzymes, in a manner similar to the cAMP cascade (described next). They also play an important role in the release of calcium ions from their storage areas in the endoplasmic reticulum (ER) or sarcoplasmic reticulum (SR).

C. Secondary Messengers

Adenylate cyclase produces a secondary messenger inside the cell, cyclic AMP (cAMP). The manufacture of cAMP activates processes that result in the transmission of an electrical signal. The cAMP is manufactured by adenylate cyclase from ATP:

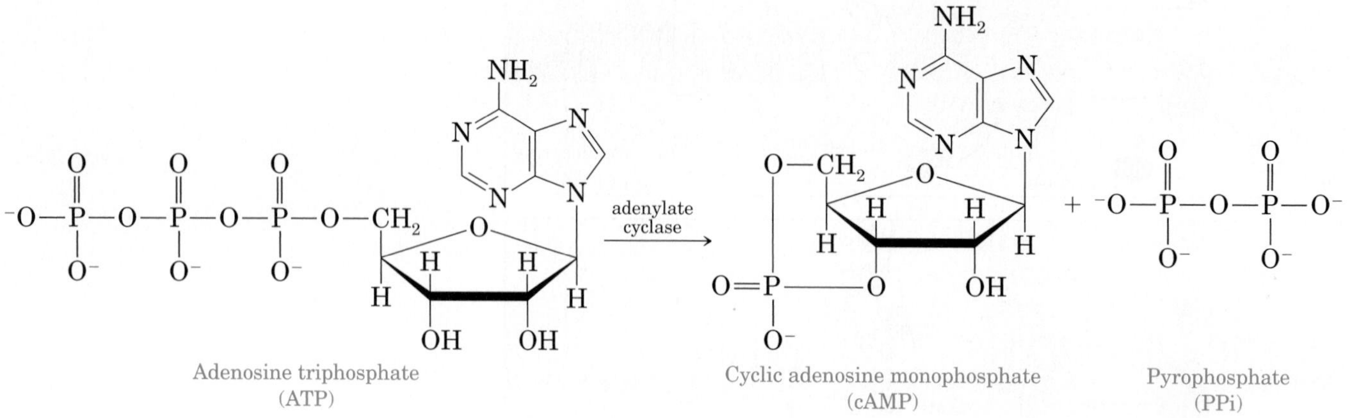

| Adenosine triphosphate (ATP) | Cyclic adenosine monophosphate (cAMP) | Pyrophosphate (PPi) |

The activation of adenylate cyclase accomplishes two important goals:

1. It converts an event occurring at the outer surface of the target cell (adsorption onto receptor site) to a change inside the target cell (formation of cAMP). Thus, the primary messenger (neurotransmitter or hormone) does not have to cross the membrane.

2. It amplifies the signal. One molecule adsorbed on the receptor triggers the adenylate cyclase to make many cAMP molecules. In this way, the signal is amplified many thousands of times.

D. Removal of Signal

How does this signal amplification stop? When the neurotransmitter or hormone dissociates from the receptor, the adenylate cyclase halts the manufacture of cAMP. The cAMP already produced is destroyed by the enzyme phosphodiesterase, which catalyzes the hydrolysis of the phosphoric ester bond, yielding AMP.

The amplification through the secondary messenger (cAMP) is a relatively slow process. It may take from 0.1 s to a few minutes. Therefore, in cases where the transmission of signals must be fast (milliseconds or seconds), a neurotransmitter such as acetylcholine acts on membrane permeability directly, without the mediation of a secondary messenger.

E. Control of Neurotransmission

The G-protein–adenylate cyclase cascade in transduction signaling is not limited to monoamine messengers. A wide variety of peptide hormones and neurotransmitters (Section 24.6) use this signaling pathway. Included among them are glucagon, vasopressin, luteinizing hormone, enkephalins, and P-protein. Neither is the opening of ion channels, depicted in Figure 24.4, the only target of this signaling. A number of enzymes can be phosphorylated by protein kinases, and the phosphorylation controls whether these enzymes will be active or inactive (Section 23.6).

The fine control of the G-protein–adenylate cyclase cascade is essential for health. Consider the toxin of the bacterium *Vibrio cholerae,* which permanently activates G-protein. The result is the symptoms of cholera—namely, severe dehydration as a result of diarrhea. This problem arises because the activated G-proteins overproduce cAMP. This excess, in turn, opens the ion channels, which leads to a large outflow of ions and accompanying water from the epithelial cells to the intestines. Therefore, the first measure taken in treating cholera victims is to replace the lost water and salt.

F. Removal of Neurotransmitters

The inactivation of the adrenergic neurotransmitters differs somewhat from the inactivation of the cholinergic transmitters. While acetylcholine is decomposed by acetylcholinesterase, most of the adrenergic neurotransmitters are inactivated in a different way. *The body inactivates monoamines by oxidizing them to aldehydes.* Enzymes that catalyze these reactions, called monoamine oxidases (MAOs), are very common in the body. For example, one MAO converts both epinephrine and norepinephrine to the corresponding aldehyde:

Epinephrine (salt form)   Norepinephrine (salt form)

MAO   MAO

Many drugs that are used as antidepressants or antihypertensive agents are MAO inhibitors—for example, Marplan and Nardil. They prevent MAOs from converting monoamines to aldehydes, thereby increasing the concentration of the active adrenergic neurotransmitters.

There is also an alternative way to remove adrenergic neurotransmitters. Shortly after adsorption onto the postsynaptic membrane, the neurotransmitter comes off the receptor site and is reabsorbed through the presynaptic membrane and stored again in the vesicles.

G. Histamines

The neurotransmitter histamine is present in mammalian brains. It is synthesized from the amino acid histidine by decarboxylation:

Histidine $\xrightarrow{H^+}$ Histamine $+ \; CO_2$

The action of histamine as a neurotransmitter is very similar to that of other monoamines. There are two kinds of receptors for histamine. One receptor, H_1, can be blocked by antihistamines such as dimenhydrinate (Dramamine) and diphenhydramine (Benadryl). The other receptor, H_2, can be blocked by ranitidine (Zantac) and cimetidine (Tagamet).

H_1 receptors are found in the respiratory tract. They affect the vascular, muscular, and secretory changes associated with hay fever and asthma.

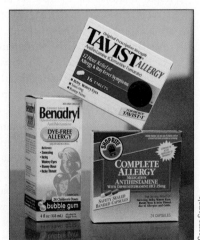

Antihistamines block the H_1 receptor for histamine.

Chemical Connections 24D

Parkinson's Disease: Depletion of Dopamine

Parkinson's disease is characterized by spastic motion of the eyelids as well as rhythmic tremors of the hands and other parts of the body, often when the patient is at rest. The posture of the patient changes to a forward, bent-over position; walking becomes slow, with shuffling footsteps. The cause of this degenerative nerve disease is unknown, but genetic factors and environmental effects (for example, exposure to pesticides or high concentrations of metals such as Mn^{2+} ion) have been implicated.

The neurons affected contain, under normal conditions, mostly dopamine as a neurotransmitter. People with Parkinson's disease have depleted amounts of dopamine in their brains, but the dopamine receptors are not affected. Thus, the first line of remedy is to *increase the concentration of dopamine*. Dopamine cannot be administered directly, because it cannot penetrate the blood–brain barrier and therefore does not reach the tissue where its action is needed. L-dopa, by contrast, is transported through the arterial wall and converted to dopamine in the brain:

(S)-3,4-Dihydroxyphenylalanine
(L-Dopa)

enzyme-catalyzed decarboxylation $\longrightarrow$

Dopamine

When L-dopa is administered, many patients with Parkinson's disease are able to synthesize dopamine and resume normal nerve transmission. In these individuals, L-dopa reverses the symptoms of their disease, although the respite is only temporary. In other patients, the L-dopa regimen provides little benefit.

Another way to increase dopamine concentration is to *prevent its metabolic elimination*. The drug entacapon (Comtan) inhibits an enzyme that is instrumental in clearing dopamine from the brain. The enzyme (catechol-O-methyl transferase, COMT) converts dopamine to 3-methoxy-4-hydroxy-L-phenylalanine, which is then eliminated. Entacapon is usually administered together with L-dopa. Another drug, (R)-selegiline (L-Deprenyl), is a monoamine oxidase (MAO) inhibitor. L-Deprenyl, which is also given in combination with L-dopa, can reduce the symptoms of Parkinson's disease and even increase the lifespan of patients. It increases the level of dopamine by *preventing its oxidation by MAOs*.

Other drugs may treat only the symptoms of Parkinson's disease: the spastic motions and the tremors. These drugs, such as benztropine (Cogentin), are similar to atropine and act on cholinergic receptors, thereby preventing muscle spasms.

The real cure for Parkinson's disease may lie in transplanting human embryonic dopamine neurons. In preliminary work, such grafts have been functionally integrated in patients' brains and have produced dopamine. In the most successful cases, patients have been able to resume a normal, independent life following the transplant.

Certain drugs designed to affect one neurotransmitter may also affect another. An example is the drug methylphenidate (Ritalin). In high doses, this drug enhances the dopamine concentration in the brain and acts as a stimulant. In small doses, it is prescribed to calm hyperactive children or to minimize ADD (attention deficit disorder). It seems that in smaller doses, Ritalin raises the concentration of serotonin. This neurotransmitter decreases hyperactivity without affecting the dopamine levels of the brain.

Serotonin

The close connection between two monoamine neurotransmitters, dopamine and serotonin, is also evident in their roles in controlling the nausea and vomiting that often follow general anesthesia and chemotherapy. Blockers of dopamine receptors in the brain, such as promethazine (Phenergan), can alleviate the symptoms after anesthesia. A blocker of serotonin receptors in the brain as well as on the terminals of the vagus nerve in the stomach, such as ondansetron (Zofran), is the drug of choice for preventing chemotherapy-induced vomiting.

Synthesis and degradation of dopamine are not the only way that the brain keeps its concentration at a steady state. The concentration is also controlled by specific proteins, called *transporters*, that ferry the used dopamine from the receptor back across the synapse into the original neuron for reuptake. Cocaine addiction works through such a transporter. Cocaine binds to the dopamine transporter, like a reversible inhibitor, thereby preventing the reuptake of dopamine. As a consequence, dopamine is not transported back to the original neuron and stays in the synapse, increasing the continuous firing of signals, which is the psychostimulatory effect associated with a cocaine "high."

Test your knowledge with Problems 24.59–24.62 and 24.89.

Chemical Connections 24E

Nitric Oxide as a Secondary Messenger

The toxic effects of the simple gaseous molecule NO have long been recognized (see Chemical Connections 3C). Therefore, it came as a surprise to find that this compound plays a major role in chemical communications. This simple molecule is synthesized in the cells when arginine is converted to citrulline. (These two compounds appear in the urea cycle; see Section 28.8.) Nitric oxide is a relatively nonpolar molecule, and shortly after it has been produced in the nerve cell, it quickly diffuses across the lipid bilayer membrane. During its short half-life (4–6 s), it can reach a neighboring cell. Because NO passes through membranes, it does not need extracellular receptors to deliver its message. NO is very unstable, so there is no need for a special mechanism to carry out its destruction.

NO acts as an intercellular messenger between the endothelial cells surrounding the blood vessels and the smooth muscles encompassing these cells. It relaxes the muscle cells, thereby dilating the blood vessels. The outcome is less-restricted blood flow and a drop in blood pressure. This reason also explains why nitroglycerin (Chemical Connections 13D) is effective against angina: It produces NO in the body.

Another role of NO in dilating blood vessels lies in remedying impotence. The impotence-relieving drug Viagra enhances the activity of NO by inhibiting an enzyme (phosphodiesterase) that otherwise would reduce NO's effect on smooth muscles. When the NO concentration is sufficiently high, the blood vessels dilate, allowing enough blood to flow to provide an erection. In most cases, this happens within an hour after taking the pill.

Sometimes the dilation of blood vessels is not so beneficial. Headaches are caused by dilated arteries in the head. NO-producing compounds in food—nitrites in smoked and cured meats and sodium glutamate in seasoning—can cause such headaches. Nitroglycerin itself often induces headaches.

Nitric oxide is toxic, as discussed in Chemical Connections 3C. This toxicity is used by our immune system (Section 31.2B) to fight infections caused by viruses.

Monosodium glutamate (MSG) is a flavor enhancer. However, as a NO-producing molecule, it can also cause headaches. Because many Chinese restaurants used to use MSG, people who got headaches after eating Chinese food were often said to have the "Chinese Restaurant Syndrome."

The toxic effect of NO is also evident in strokes. In a stroke, a blocked artery restricts the blood flow to certain parts of the brain; the oxygen-starved neurons then die. Next, neurons in the surrounding area, 10 times larger than the place of the initial attack, release glutamic acid, which stimulates other cells. They, in turn, release NO, which kills all cells in the area. Thus, the damage to the brain is spread tenfold. A concentrated effort is under way to find inhibitors of the NO-producing enzyme, nitric oxide synthase, that can be used as antistroke drugs. For the discovery of NO and its role in blood pressure control, three pharmacologists—Robert Furchgott, Louis Ignarro, and Ferid Murad—received the 1998 Nobel Prize in physiology.

Test your knowledge with Problems 24.63–24.65 and 24.83.

Therefore, antihistamines that block H_1 receptors relieve these symptoms. The H_2 receptors are found mainly in the stomach and affect the secretion of HCl. Cimetidine and ranitidine, both H_2 blockers, reduce acid secretion and, therefore, are effective drugs for ulcer patients. The main culprit in the formation of most ulcers, however, is the bacteria *Helicobacter pylori*. Sir James W. Black of the United Kingdom received the 1988 Nobel Prize in medicine for the invention of cimetidine and other such drugs as propranolol that kill the ulcer-causing bacteria (Table 24.1).

Example 24.1 Identifying Enzymes
in the Adrenergic Pathway

Three enzymes in the adrenergic neurotransmission pathway affect the transduction of the signals. Identify them and describe how they affect the neurotransmission.

Solution

Adenylate cyclase amplifies the signal by producing cAMP secondary messengers. Phosphatase terminates the signal by hydrolyzing cAMP. Monoamine oxidase (MAO) reduces the frequency of signals by oxidizing the monoamine neurotransmitters to the corresponding aldehydes.

Problem 24.1

What is the functional difference between G-protein and GTP?

24.6 What Is the Role of Peptides in Chemical Communication?

A. Messengers

Many of the most important hormones affecting metabolism belong to the peptidergic messengers group. Among them are insulin (Section 22.8 and Chemical Connections 24F) and glucagon, hormones of the pancreatic islets, and vasopressin and oxytocin (Section 22.8), products of the posterior pituitary gland.

In the last few years, scientists have isolated a number of brain peptides that have affinity for certain receptors and, therefore, act as if they were neurotransmitters. Some 25 or 30 such peptides are now known.

The first brain peptides isolated were the **enkephalins**. These pentapeptides are present in certain nerve cell terminals. They bind to specific pain receptors and seem to control pain perception. Because they bind to the receptor site that also binds the pain-killing alkaloid morphine, it is assumed that the N-terminal end of the pentapeptide fits the receptor (Figure 24.6).

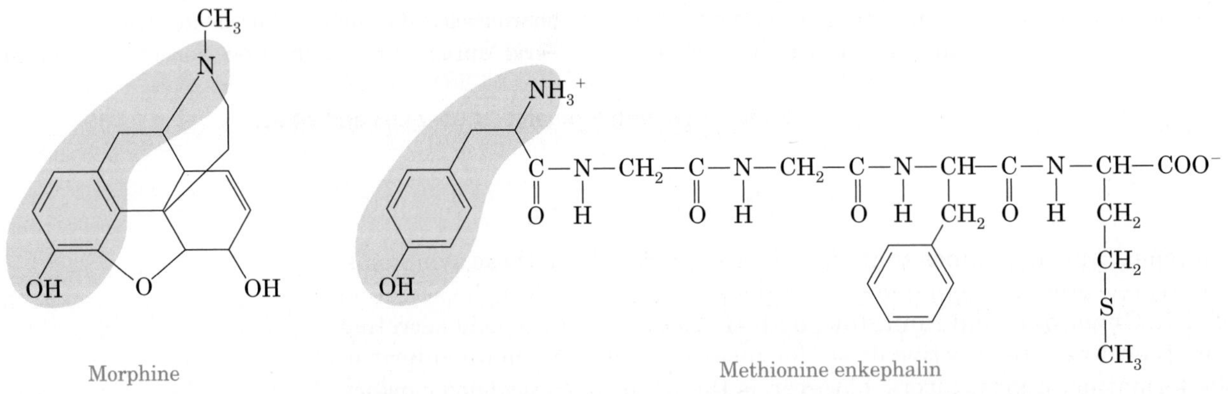

Morphine

Methionine enkephalin

FIGURE 24.6 Similarities between the structure of morphine and that of the brain's own pain regulators, the enkephalins.

Even though morphine remains the most effective agent in reducing pain, its clinical use is limited because of its side effects. These include respiratory depression; constipation; and, most significantly, addiction. The clinical use of enkephalins has yielded only modest relief. The challenge is to develop analgesic drugs that do not involve the opiate receptors in the brain.

Another brain peptide, **neuropeptide Y**, affects the hypothalamus, a region that integrates the body's hormonal and nervous systems. Neuropeptide Y is a potent orexic (appetite-stimulating) agent. When its receptors are blocked (for example, by leptin, the "thin" protein), appetite is suppressed. Leptin is an anorexic agent.

Yet another peptidergic neurotransmitter is **substance P** (*P* for "pain"). This 11-amino-acid peptide is involved in transmission of pain signals. In injury or inflammation, sensory nerve fibers transmit pain signals from the peripheral nervous system (where the injury occurred) to the spinal cord, which processes the pain. The peripheral neurons synthesize and release substance P, which bonds to receptors on the surface of the spinal cord. Substance P, in its turn, removes the magnesium block at the *N*-methyl-D-aspartate (NMDA) receptor. Glutamic acid, an excitatory amino acid, can then bind to this receptor. In doing so, it amplifies the pain signal going to the brain.

B. Secondary Messengers and Control of Metabolism

All peptidergic messengers, hormones, and neurotransmitters act through secondary messengers. Glucagon, luteinizing hormone, antidiuretic hormone, angiotensin, enkaphalin, and substance P use the G-protein–adenylate cyclase cascade that we saw in the previous section.

Glucagon is a peptide hormone that is critical for maintaining blood glucose levels. When the pancreas senses that blood glucose is dropping, it releases glucagon. When glucagon is released, it binds to receptors on liver cells and acts through a series of reactions to raise blood glucose. The method of action is far from simple, however. When glucagon binds to its receptor and activates the G-protein cascade, the second messenger, cAMP, activates protein kinase, an enzyme that phosphorylates many target enzymes. As shown in Figure 24.7, protein kinase phosphorylates two key enzymes in carbohydrate metabolism, fructose bisphosphatase 2 (FBP-2) and phosphfructokinase 2 (PFK-2). Phosphorylating these two enzymes has opposite effects. The kinase is inactivated, and the phosphatase is activated. This lowers the intracellular concentration of fructose 2,6 bisphosphate, a key metabolic regulator. The reduced level of the regulator increases the activity of the pathway called **gluconeogenesis** (Chapter 29) and reduces the activity of the pathway called **glycolysis** (Chapter 28). Gluconeogenesis produces glucose, and glycolysis uses it. Thus, by turning on gluconeogenesis and turning off glycolysis, the liver produces more glucose for the blood.

Insulin is another peptide hormone produced by the pancreas, but its overall effect is roughly the opposite of glucagon's. Insulin binds to insulin receptors on liver and muscle cells, as shown in Figure 24.8. The receptor is an example of a protein called a tyrosine kinase. A specific tyrosine residue becomes phosphorylated on the receptor, initiating its kinase activity. A target protein called IRS (insulin receptor substrate) is then phosphorylated by the active tyrosine kinase. The phosphorylated IRS acts as the second messenger. It causes the phosphorylation of many target enzymes in the cell. The effect is to reduce the level of glucose in the blood by increasing the rate of pathways that use glucose and slowing the rate of pathways that make glucose.

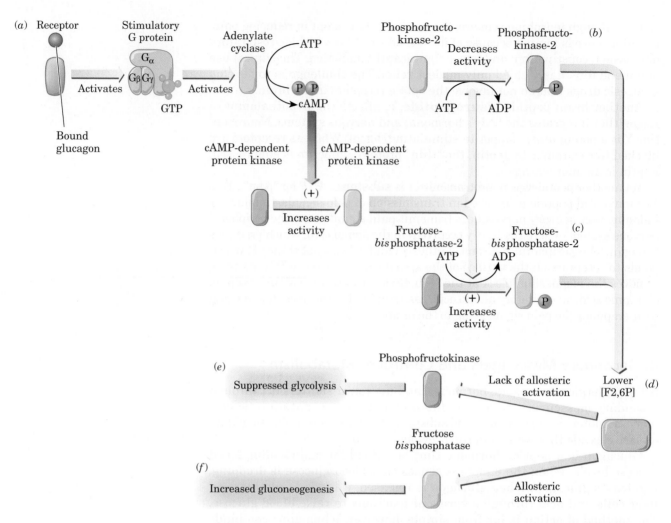

FIGURE 24.7 Glucagon action. (*a*) Binding of a glucagon to its receptor sets off a chain of events that leads to the activation of a cAMP-dependent protein kinase. The enzymes phosphorylated in this case are (*b*) phosphofructokinase-2, which is inactivated, and (*c*) fructose-*bis* phosphate-2, which is activated. The combined result of phosphorylating these two enzymes is to (*d*) lower the concentration of fructose-2, 6-*bis* phosphate (F2,6P). A lower concentration of F2,6P leads to lack of allosteric activation of phosphofructokinase-1 and (*e*) lowered glycolysis while also leading to (*f*) allosteric activation of fructose-*bis* phosphatase-1 and increased gluconeogenesis.

24.7 How Do Steroid Hormones Act as Messengers?

In Section 21.10, we saw that a large number of hormones possess steroid ring structures. These hormones, which include the sex hormones, are hydrophobic; therefore, they can cross the hydrophobic plasma membranes of the cell by passive diffusion.

There is no need for special receptors embedded in the membrane for these hormones. It has been shown, however, that **steroid hormones** interact inside the cell with protein receptors. Most of these receptors are localized in the nucleus of the cell, but small amounts also exist in the cytoplasm. When they interact with steroids, they facilitate their migration through the aqueous cytoplasm, since the protein receptors themselves are hydrophilic.

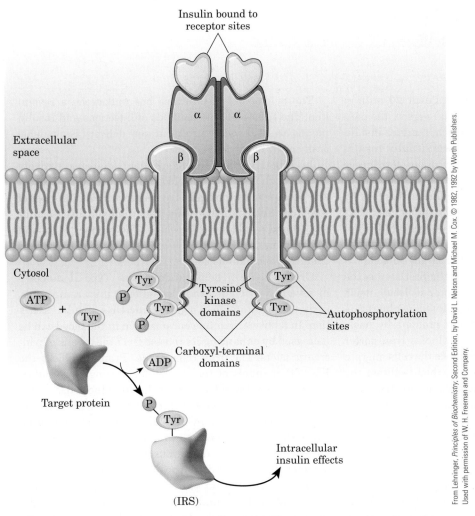

Insulin bound to
receptor sites

α　　α

Extracellular
space

β　　β

Cytosol

ATP

+

Tyr

P

Tyr

Tyr

P

Tyr

Tyr

Tyr

Tyrosine
kinase
domains

Autophosphorylation
sites

Carboxyl-terminal
domains

Target protein

ADP

P

Tyr

Intracellular
insulin effects

(IRS)

FIGURE 24.8 The insulin receptor has two types of subunits, α and β. The α-subunit is on the extracellular side of the membrane, and it binds to insulin. The β-subunit spans the membrane. When insulin binds to the α-subunit, the β-subunits autophosphorylate on tyrosine residues. These then phosphorylate target proteins called insulin receptor substrates (IRS). The IRSs act as the second messengers in the cells.

CH₃

CH₃　C=O

H₃C

O

Progesterone

Once inside the nucleus, the steroid–receptor complex can either bind directly to the DNA or combine with a **transcription factor**, a protein that binds to DNA and alters the expression of a gene (Section 26.2), influencing the synthesis of a certain key protein. Thyroid hormones also have large hydrophobic domains as well as protein receptors, which facilitate their transport through hydrophobic cell membraines.

Chemical Connections 24F

Diabetes

The disease diabetes mellitus affects over 20 million people in the United States. In a normal person, the pancreas, a large gland behind the stomach, secretes insulin and several other hormones. Diabetes usually results from low insulin secretion. Insulin is necessary for glucose molecules to penetrate such cells as brain, muscle, and fat cells, where they can be used. It accomplishes this task by being adsorbed onto the receptors in the target cells. This adsorption triggers the manufacture of cyclic GMP (not cAMP); this secondary messenger, in turn, increases the transport of glucose molecules into the target cells.

In the resulting cascade of events, the first step is the self-(auto)phosphorylation of the receptor molecule itself, on the cytoplasmic side. The phosphorylated insulin receptor activates enzymes and regulatory proteins by phosphorylating them. As a consequence, glucose transporter molecules (GLUT4) that are stored inside the cells migrate to the plasma membrane. Once there, they facilitate the movement of glucose across the membrane. This transport relieves the accumulation of glucose in blood serum and makes it available for metabolic activity inside the cells. The glucose can then be used as an energy source, stored as glycogen, or even diverted to enter fat and other molecular biosynthetic pathways.

In diabetic patients, the glucose level can rise to 600 mg/100 mL of blood or higher (normal is 80 to 100 mg/100 mL). Two kinds of diabetes exist. In insulin-dependent diabetes, patients do not manufacture enough of this hormone in the pancreas. So-called Type I disease develops early, before the age of 20, and must be treated with daily injections of insulin. Even with daily injections of insulin, the blood sugar level fluctuates, which may cause other disorders, such as cataracts, retinal dystrophy leading to blindness, kidney disease, heart attack, nervous disorders, and peripheral vascular disease (PVD).

One way to counteract these fluctuations is to monitor the blood sugar and, as the glucose level rises, to administer insulin. Such monitoring requires pricking the finger six times per day, an invasive regimen that few diabetics follow faithfully. Recently, noninvasive monitoring techniques have been developed. One of the most promising employs contact lenses. The blood sugar's fall and rise is reflected in the glucose content of tears. A fluorescent sensor in the contact lens monitors these glucose fluctuations; its data can be read via a hand-held fluorimeter. Thus, the patterns of glucose fluctuation can be obtained noninvasively, and if the level rises to the danger zone, it can be counteracted by insulin intake.

The delivery of insulin also has undergone a revolution. The tried-and-true methods of injections and insulin pumps are still widely used, but new delivery is available orally or by nasal delivery.

In Type II (non–insulin-dependent) diabetes, patients have enough insulin in the blood but cannot utilize it properly because the target cells have an insufficient number of receptors. Such patients typically develop the disease after age 40 and are likely to be obese. The number of insulin receptors in the adipose (fat) cells of overweight people is usually lower than normal.

Oral drugs can help patients with Type II diabetes in several ways. For example, sulfonyl urea compounds, such as tolbutamide (Orinase), increase insulin secretion. In addition, insulin concentration in the blood can be increased by enhancing its release from the β-cells of pancreatic islets. The drug repaglinide (Prandin) blocks the K^+-ATP channels of the β-cells, facilitating Ca^{2+} influx, which induces the release of insulin from the cells.

The oral drugs seem to control the symptoms of diabetes, but fluctuations in insulin levels may turn high blood sugar (hyperglycemia) into low blood sugar (hypoglycemia), which is just as dangerous. Other drugs for Type II diabetes that do not elicit hypoglycemia attempt to control the glucose level at its source. Miglitol (Glyset), an anti-a-glucosidase drug, inhibits the enzyme that converts glycogen or dietary starch into glucose. The drug metformin (Glucophage) decreases glucose production in the liver, carbohydrate absorption in the intestines, and glucose uptake by fat cells.

Tolbutamide
(Orinase)

Metformin
(Glucophage is the
hydrochloride of metformin)

Diabetes (*continued*)

There is a noticeable link between obesity and Type II diabetes, although it is not clear whether diabetes leads to obesity or vice versa. The GLUT4 transporter is one of many glucose transporters, and it is the one most affected by insulin levels. It is also a protein whose levels can be affected by physical training. Studies have shown that one of the major changes associated with physical activity is an increase in the amount of GLUT4 in the muscles.

In the trained state, a person transports more glucose into the cell than when untrained. Studies showed that only one week of moderate exercise (1–2 hours a day at 70% of maximal oxygen uptake) would double the GLUT4 protein content of the muscles of sedentary people.

By definition, loss of function of glucose transport is Type II diabetes. It takes only a few days without physical training for the activity of GLUT4 to decrease to half its normal level. Fortunately, the intensity of the training has less to do with the effect, at least in young to middle-aged people.

With the apparent link between Type II diabetes and obesity, one method of maintaining proper glucose transport appears to be staying light and fit.

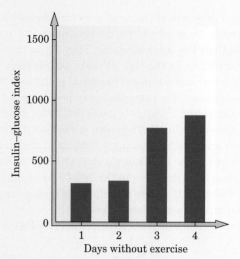

Insulin-glucose index versus detraining time. Moderately trained middle-aged men were tested for the effect of detraining on their muscles' ability to clear glucose out of the blood (measured as the insulin-glucose index or the amount of insulin it takes to clear glucose from the blood). On the third day without training, there is a pronounced increase in the amount of insulin required to clear glucose. (*Adapted from "Metabolic basis of the health benefits of exercise" by Adrianne Hardman, the Biochemist 20 [3], pp. 18–22 [1998].*)

Test your knowledge with Problems 24.66–24.73.

The steroid hormonal response through protein synthesis is not fast. In fact, it takes hours to occur. Steroids can also act at the cell membrane, influencing ligand-gated ion channels. Such a response would take only seconds. An example of such a fast response occurs in fertilization. The sperm head contains proteolytic enzymes, which act on the egg to facilitate its penetration. These enzymes are stored in acrosomes, organelles found on the sperm head. During fertilization, progesterone originating from the follicle cells surrounding the egg acts on the acrosome outer membrane, which disintegrates within seconds and releases the proteolytic enzymes.

The same steroid hormones depicted in Figure 21.7 act as neurotransmitters, too. These neurosteroids are synthesized in the brain cells in both neurons and glia, and they affect receptors—mainly the NMDA and GABA receptors (Section 24.4). For both sexes, progesterone and progesterone metabolites in brain cells can induce sleep, have analgesic and anticonvulsive effects, and can even serve as natural anesthetics.

Chemical Connections 24G

Hormones and Biological Pollutants

Hormones are some of the most powerful chemicals when we consider their effects on development and metabolism and their low concentrations: They act at the level of parts per billion. In the last 20 years, people have begun to worry about many types of biological pollutants that affect development. For example, many people prefer to eat organically raised chickens to avoid pesticides ingested by factory-farmed animals. They also prefer to eat hormone-free meat and poultry, fearing the effects of the hormones on their children. It is known that the age of onset of puberty has been dropping in the last 30 years, and many believe this drop is due to the effects of biological pollutants that mimic human hormones.

One such chemical is called bisphenol A (BPA).

BPA is an estrogen-like compound used to make polycarbonate bottles, such as baby bottles, as well as the linings of food cans and plastic storage containers. Small amounts of BPA can leach into food, and many people have detectable levels of the compound in their blood, although the levels found are below the maximum safe dose set by the Environmental Protection Agency. In 1997, a reproductive biologist found that very low levels of BPA fed to pregnant mice caused enlargement of the prostate of their male offspring. Other studies found increases in chromosomal abnormalities in the eggs of mice after BPA leached out of plastic mouse cages. Epidemiological studies have linked human health problems to BPA, such as breast cancer and early onset of puberty. The National Toxicology Program formed a twelve-member panel to study the effects of BPA, but so far the results have been inconclusive. While they concluded that the dangers of BPA were negligible, they

did have some concern about possible risk to fetuses and children. Reproductive biologist Dr. John Vandenbergh of North Carolina State University, a member of the NTP panel, was quoted as saying, "I think there is a human risk. What we are trying to do here is define what the risk is."

Depending on who you listen to, BPA is either perfectly harmless or a serious health threat. Even research scientists do not agree on the answer. An Internet search for bisphenol A will lead to many websites dedicated to discussing the possible dangers. One thing is clear, however. Hormones are critical to human physiology, and even tiny perturbations in their levels may affect our health.

Polycarbonate used for some plastic bowls and baby bottles contains bisphenol A (BPA), a substance of concern to the Environmental Protection Agency for its link to human health problems.

Test your knowledge with Problems 24.74–24.76.

Summary

Section 24.1 What Molecules Are Involved in Chemical Communications?

- Cell-to-cell communications are carried out by three kinds of molecules.

- **Receptors** are protein molecules embedded in the membranes of cells.
- **Chemical messengers,** or ligands, interact with receptors.
- **Secondary messengers** carry and amplify the signals from the receptor to inside the cell.

Section 24.2 How Are Chemical Messengers Classified as Neurotransmitters and Hormones?

- **Neurotransmitters** send chemical messages across a short distance—the **synapse** between two neurons or between a neuron and a muscle or an endocrine gland cell. This communication occurs in milliseconds.
- **Hormones** transmit their signals more slowly and over a longer distance, from the source of their secretion (endocrine gland), through the bloodstream, and into target cells.
- **Antagonists** block receptors; **agonists** stimulate receptors.
- Five kinds of chemical messengers exist: **cholinergic, amino acid, adrenergic, peptidergic,** and **steroid.** Neurotransmitters may belong to all five classes, hormones to the last three classes. Acetylcholine is cholinergic, glutamic acid is an amino acid, epinephrine (adrenaline) and norepinephrine are adrenergic, enkephalins are peptidergic, and progesterone is a steroid.

Section 24.3 How Does Acetylcholine Act as a Messenger?

- Nerve transmission starts with the neurotransmitters, such as acetylcholine packaged in **vesicles** in the **presynaptic end** of neurons.
- When neurotransmitters are released, they cross the membrane and the synapse and are adsorbed onto receptor sites on the **postsynaptic** membranes. This adsorption triggers an electrical response.
- Some neurotransmitters act directly, whereas others act through a secondary messenger, **cyclic AMP.**
- After the electrical signal is triggered, the neurotransmitter molecules must be removed from the postsynaptic ends. In the case of acetylcholine, this removal is done by an enzyme called acetylcholinesterase.

Section 24.4 What Amino Acids Act as Neurotransmitters?

- Amino acid neurotransmitters, many of which differ from the amino acids found in proteins, bind to their receptors, which are ligand-gated ion channels.
- Removal of amino acid messengers takes place by **reuptake** through the presynaptic membrane rather than by hydrolysis.

Section 24.5 What Are Adrenergic Messengers?

- The mode of action of monoamines such as epinephrine, serotonin, dopamine, and histamine is similar to that of acetylcholine in the sense that they start with binding to a receptor.
- Cyclic AMP is an important secondary messenger.
- The mode of removal of monoamines differs from the hydrolysis of acetylcholine. In the case of monoamines, enzymes (MAOs) oxidize them to aldehydes.

Section 24.6 What Is the Role of Peptides in Chemical Communication?

- Peptides and proteins bind to receptors on the target cell membrane and use secondary messengers to exert their influence.
- **Signal transduction** is a term that was discussed in Section 24.5.

Section 24.7 How Do Steroid Hormones Act as Messengers?

- Steroids penetrate the cell membrane, and their receptors are found in the cytoplasm. Together with their receptors, they penetrate the cell nucleus.
- Steroid hormones can act in three ways: (1) They activate enzymes, (2) they affect the gene transcription of an enzyme or protein, and (3) they change membrane permeability.
- The same steroids can also act as neurotransmitters when synthesized in neurons.

Problems

◑WL Interactive versions of these problems may be assigned in OWL.

Orange-numbered problems are applied.

Section 24.1 What Molecules Are Involved in Chemical Communications?

24.2 What kind of signal travels along the axon of a neuron?

24.3 What is the difference between a *chemical messenger* and a *secondary messenger?*

Section 24.2 How Are Chemical Messengers Classified as Neurotransmitters and Hormones?

24.4 Define the following:
  (a) Synapse   (b) Receptor
  (c) Presynaptic   (d) Postsynaptic
  (e) Vesicle

24.5 What is the role of Ca^{2+} in releasing neurotransmitters into the synapse?

24.6 Which signal takes longer: (a) a neurotransmitter or (b) a hormone? Explain.

24.7 Which gland controls lactation?

24.8 To which of the three groups of chemical messengers do these hormones belong?
  (a) Norepinephrine
  (b) Thyroxine
  (c) Oxytocin
  (d) Progesterone

Section 24.3 How Does Acetylcholine Act as a Messenger?

24.9 How does acetylcholine transmit an electrical signal from neuron to neuron?

24.10 Which end of the acetylcholine molecule fits into the receptor site?

24.11 Cobra venom and botulin are both deadly toxins, but they affect cholinergic neurotransmission differently. How does each cause paralysis?

24.12 Different ion concentrations across a membrane generate a potential (voltage). We call such a membrane *polarized*. What happens when acetylcholine is adsorbed on its receptor?

24.13 What is the difference between calcium sparks (puffs) and calcium waves?

24.14 What is the role of calmodulin in signaling by Ca^{2+} ions?

24.15 To enable a fusion between the synaptic vesicle and the plasma membrane, the calcium concentration is increased. How many-fold of an increase in Ca^{2+} concentration is needed?

Section 24.4 What Amino Acids Act as Neurotransmitters?

24.16 List two features by which taurine differs from the amino acids found in proteins.

24.17 How is glutamic acid removed from its receptor?

24.18 What is unique in the structure of GABA that distinguishes it from the amino acids that are present in proteins?

24.19 What is the structural difference between NMDA, an agonist of a glutamic acid receptor, and L-aspartic acid?

Section 24.5 What Are Adrenergic Messengers?

24.20 (a) Identify two monoamine neurotransmitters in Table 24.1.
(b) Explain how they act.
(c) Which medication controls the diseases caused by a lack of monoamine neurotransmitters?

24.21 What bond is hydrolyzed and what bond is formed in the synthesis of cAMP?

24.22 How is the catalytic unit of protein kinase activated in adrenergic neurotransmission?

24.23 The formation of cyclic AMP is described in Section 24.5. Show by analogy how cyclic GMP is formed from GTP.

24.24 Using the action of MAO on epinephrine as an analogy, write the structural formula of the product of the corresponding oxidation of dopamine.

24.25 The action of protein kinase is the next-to-last step in the signal transduction of the G-protein–adenylate cyclase cascade. What kind of effects can elicit the phosphorylation by this enzyme?

24.26 Explain how adrenergic neurotransmission is affected by (a) amphetamines and (b) reserpine. (See Table 24.1.)

24.27 Which step in the events depicted in Figure 24.3 provides an electrical signal?

24.28 What kind of product is the MAO-catalyzed oxidation of epinephrine?

24.29 How is histamine removed from the receptor site?

24.30 Cyclic AMP affects the permeability of membranes for ion flow.
(a) What blocks the ion channel?
(b) How is this blockage removed?
(c) What is the direct role of cAMP in this process?

24.31 Dramamine and cimetidine are both antihistamines. Would you expect Dramamine to cure ulcers and cimetidine to relieve the symptoms of asthma? Explain.

Section 24.6 What Is the Role of Peptides in Chemical Communication?

24.32 What is the chemical nature of enkephalins?

24.33 What is the mode of action of Demerol as a painkiller? (See Table 24.1.)

24.34 What enzyme catalyzes the formation of inositol-1,4,5-triphosphate from inositol-1,4-diphosphate? Give the structures of the reactant and the product.

24.35 How is the secondary messenger inositol-1,4,5-triphosphate inactivated?

24.36 What second messenger is formed in response to glucagon binding to its receptor?

24.37 Which organ produces glucagon and why?

24.38 What is the direct target of the second messenger produced when glucagon binds to its receptor?

24.39 In the course of the glucagon effect, what does protein kinase A do?

24.40 Why does glucagon lead to the activation of gluconeogenesis and the inhibition of glycolysis?

24.41 How is fructose-2,6-*bis*phosphate involved in glucose metabolism?

24.42 Describe the signaling pathway involving insulin.

24.43 Does insulin use a G-protein signaling pathway? What is the nature of the insulin receptor?

Section 24.7 How Do Steroid Hormones Act as Messengers?

24.44 Where are the receptors for steroid hormones located—on the cell surface or elsewhere?

24.45 Do steroid hormones affect protein synthesis? If so, does this effect have any implications for the time the hormonal response can take?

24.46 Can steroid hormones act as neurotransmitters?

Chemical Connections

24.47 (Chemical Connections 24A) What are the two prevalent theories about what happens in the brain during sleep?

24.48 (Chemical Connections 24A) How did fluorescent staining of zebrafish brain synapses help distinguish between the two theories in 24.47?

24.49 (Chemical Connections 24B) What is the mode of action of botulinum toxin?

24.50 (Chemical Connections 24B) How can a deadly botulinum toxin contribute to the facial beauty of Hollywood actors and actresses?

24.51 (Chemical Connections 24C) What are the neurofibrillar tangles in the brains of patients with Alzheimer's disease made of? How do they affect the cell structure?

24.52 (Chemical Connections 24C) What are the plaques in the brains of patients with Alzheimer's disease made of?

24.53 (Chemical Connections 24C) Alzheimer's disease causes loss of memory. What kind of drugs may provide some relief for—if not cure—this disease? How do they act?

24.54 (Chemical Connections 24C) How are the β-amyloid proteins and the presenilins involved with calcium flux in brain cells?

24.55 (Chemical Connections 24C) Why are scientists studying the extended families in Medellin, Colombia?

24.56 (Chemical Connections 24C) How are MRIs and spinal taps used to study Alzheimer's?

24.57 (Chemical Connections 24C) What is the progression of the physical effects on the brain of patients with Alzheimer's disease?

24.58 (Chemical Connections 24C) What are the purposes of the drugs that are used to fight Alzheimer's?

24.59 (Chemical Connections 24D) Why would a dopamine pill be ineffective in treating Parkinson's disease?

24.60 (Chemical Connections 24D) What is the mechanism by which cocaine stimulates the continuous firing of signals between neurons?

24.61 (Chemical Connections 24D) Parkinson's disease is due to a paucity of dopamine neurons, yet its symptoms are relieved by drugs that block cholinergic receptors. Explain.

24.62 (Chemical Connections 24D) In certain cases, embryonic dopamine neurons transplanted into the brains of patients with advanced Parkinson's disease resulted in complete remission. How was this result possible?

24.63 (Chemical Connections 24E) How does NO cause headaches?

24.64 (Chemical Connections 24E) How is NO synthesized in the cells?

24.65 (Chemical Connections 24E) How is the toxicity of NO detrimental in strokes?

24.66 (Chemical Connections 24F) Tolbutamide is called a sulfonyl urea compound. Identify the sulfonyl urea moiety in the structure of the drug.

24.67 (Chemical Connections 24F) What is the difference between insulin-dependent and non–insulin-dependent diabetes?

24.68 (Chemical Connections 24F) How does insulin facilitate the absorption of glucose from blood serum into adipocytes (fat cells)?

24.69 (Chemical Connections 24F) Diabetic patients must frequently monitor the fluctuation of glucose levels in their blood. What is the advantage of the latest technique for monitoring the glucose in tears over the older technique that required frequent blood samples?

24.70 (Chemical Connections 24D) Why can Type II diabetes be referred to as a lifestyle disease?

24.71 (Chemical Connections 24D) What is the effect of moderate training on the level of the GLUT4 transporter?

24.72 (Chemical Connections 24D) What is the insulin/glucose index, and why is it relevant to studying diabetes?

24.73 (Chemical Connections 24D) What is the effect of the intensity of the exercise on the GLUT4 transporter?

24.74 (Chemical Connections 24G) Why type of compound is bisphenol A, and where does it come from?

24.75 (Chemical Connections 24G) What are the possible biological effects of ingesting bisphenol A?

24.76 (Chemical Connections 24G) What experimental evidence led scientists to be concerned about bisphenol A?

24.77 (Chemical Connections 24E) List a number of effects that are caused when NO, acting as a secondary messenger, relaxes smooth muscles.

24.78 (Chemical Connections 24D) Ritalin is used to alleviate hyperactivity in childhood attention deficit disorder (ADD). How does this drug work?

Additional Problems

24.79 Considering its chemical nature, how does aldosterone (Section 21.10) affect mineral metabolism (Table 24.2)?

24.80 What is the function of the ion-translocating protein in adrenergic neurotransmission?

24.81 Decamethonium bromide acts as a muscle relaxant. If an overdose of decamethonium bromide occurs, can paralysis be prevented by administering large doses of acetyl choline? Explain.

24.82 Endorphin, a potent painkiller, is a peptide containing 22 amino acids; among them are the same five N-terminal amino acids found in the enkephalins. Does this explain endorphin's pain-killing action?

24.83 How do alanine and β-alanine differ in structure?

24.84 Where is a G-protein located in adrenergic neurotransmission?

24.85 (a) In terms of their action, what do the hormone vasopressin and the neurotransmitter dopamine have in common?

(b) What is the difference in their modes of action?

24.86 What is the difference in the modes of action between acetylcholinesterase and acetylcholine transferase?

24.87 How does cholera toxin exert its effect?

24.88 Give the formulas for the following reaction:

$$GTP + H_2O \rightleftharpoons GDP + P_i$$

24.89 Insulin is a hormone that, when it binds to a receptor, enables glucose molecules to enter the cell and be metabolized. If you have a drug that is an agonist, how would the glucose level in the serum change upon administering the drug?

24.90 In females, the pituitary gland releases luteinizing hormone (LH), which enhances the production of progesterone in the uterus. Classify these two messengers and discuss how each delivers its message.

Tying It Together

24.91 Why are receptors proteins rather than any other kind of molecule?

24.92 Why is it useful for organisms to have several different classes of neurotransmitters and hormones?

24.93 What relationship do adrenergic messengers have to amino acid messengers, and what does this relationship say about the biochemical origin of adrenergic messengers?

24.94 What functional groups are found in the structures of chemical messengers? What do these structural features imply about the active sites of the enzymes that process these messengers?

Looking Ahead

24.95 Why is insulin not administered orally in the treatment of insulin-dependent diabetes?

24.96 One of the challenges in treating cholera is that of preventing dehydration. What can make this a double challenge? (*Hint:* See Chapter 30; cholera is frequently a water-borne disease.)

24.97 Do any chemical messengers have a *direct* effect on the synthesis of nucleic acids?

24.98 Would you expect the role of chemical messengers to have any bearing on the body's requirements for energy?

Challenge Problems

24.99 Do all chemical messengers require the same time to elicit a response? If there are differences, how do the underlying response mechanisms differ?

24.100 A number of agricultural pesticides are acetylcholinesterase inhibitors. Why is their use carefully controlled?

24.101 What benefit is it to an organism to have two different enzymes for the synthesis and breakdown of acetylcholine—acetylcholine transferase and acetylcholinesterase, respectively?

24.102 Which would be a better form of therapy for cocaine addiction—an inhibitor for the dopamine transporter or a substance that degrades cocaine?

Nucleotides, Nucleic Acids, and Heredity

25

AP/Wide World Photos.

While these dogs might appear to be a normal mother and puppy, the latter is really the first cloned dog, Snuppy. The larger dog is a male Afghan hound whose DNA was used to create the clone.

Key Questions

25.1 What Are the Molecules of Heredity?

25.2 What Are Nucleic Acids Made Of?

25.3 What Is the Structure of DNA and RNA?

25.4 What Are the Different Classes of RNA?

25.5 What Are Genes?

25.6 How Is DNA Replicated?

25.7 How Is DNA Repaired?

25.8 How Do We Amplify DNA?

25.1 What Are the Molecules of Heredity?

Each cell of our bodies contains thousands of different protein molecules. Recall from Chapter 22 that all of these molecules are made up of the same 20 amino acids, just arranged in different sequences. That is, the hormone insulin has a different amino acid sequence from the globin of the red blood cell. Even the same protein—for example, insulin—has a different sequence in different species (Section 22.8). Within the same species, individuals may have some differences in their proteins, although the differences are much less dramatic than those seen between species. This variation is most obvious in cases where people have such conditions as hemophilia, albinism, or color-blindness because they lack certain proteins that normal people have or because the sequence of their amino acids is somewhat different (see Chemical Connections 22D).

After scientists became aware of the differences in amino acid sequences, their next quest was to determine how cells know which proteins to synthesize out of the extremely large number of possible amino acid sequences. The answer is that an individual gets the information from its parents

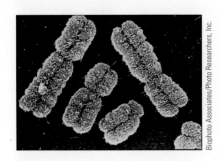

Human chromosomes magnified about 8000 times.

Biophoto Associates/Photo Researchers, Inc.

Gene The unit of heredity; a DNA segment that codes for one protein or one type of RNA

Bases Purines and pyrimidines, which are components of nucleotides, DNA, and RNA

through *heredity*. Heredity is the transfer of characteristics, anatomical as well as biochemical, from generation to generation. We all know that a pig gives birth to a pig and a mouse gives birth to a mouse.

It was easy to determine that the information is obtained from the parent or parents, but what form does this information take? During the last 60 years, revolutionary developments have enabled us to answer this question—the transmission of heredity occurs on the molecular level.

From about the end of the nineteenth century, biologists suspected that the transmission of hereditary information from one generation to another took place in the nucleus of the cell. More precisely, they believed that structures within the nucleus, called **chromosomes**, had something to do with heredity. Different species have different numbers of chromosomes in the nucleus. The information that determines external characteristics (red hair, blue eyes) and internal characteristics (blood group, hereditary diseases) was thought to reside in **genes** located inside the chromosomes.

Chemical analysis of nuclei showed that they are largely made up of special basic proteins called *histones* and a type of compound called *nucleic acids*. By 1940, it became clear through the work of Oswald Avery (1877–1955) that of all the material in the nucleus, only a nucleic acid called deoxyribonucleic acid (DNA) carries the hereditary information. That is, the genes are located in the DNA. Other work in the 1940s by George Beadle (1903–1989) and Edward Tatum (1909–1975) demonstrated that each gene controls the manufacture of one protein and that external and internal characteristics are expressed through this gene. Thus, the expression of the gene (DNA) in terms of an enzyme (protein) led to the study of protein synthesis and its control. *The information that tells the cell which proteins to manufacture is carried in the molecules of DNA.* We now know that not all genes lead to the production of protein, but all genes do lead to the production of another type of nucleic acid, called ribonucleic acid (RNA).

25.2 What Are Nucleic Acids Made Of?

Two kinds of nucleic acids are found in cells: **ribonucleic acid (RNA)** and **deoxyribonucleic acid (DNA)**. Each has its own role in the transmission of hereditary information. As we just saw, DNA is present in the chromosomes of the nuclei of eukaryotic cells. RNA is not found in the chromosomes, but rather, is located elsewhere in the nucleus and even outside the nucleus, in the cytoplasm. As we will see in Section 25.4, there are six types of RNA, all with specific structures and functions.

Both DNA and RNA are polymers. Just as proteins consist of chains of amino acids and polysaccharides consist of chains of monosaccharides, nucleic acids are also chains. The building blocks (monomers) of nucleic acid chains are *nucleotides*. Nucleotides themselves, however, are composed of three simpler units: a base, a monosaccharide, and a phosphate. We will look at each of these components in turn.

A. Bases

The **bases** found in DNA and RNA are chiefly those shown in Figure 25.1. All of them are basic because they are heterocyclic aromatic amines (Section 16.1). Two of these bases—adenine (A) and guanine (G)—are purines; the other three—cytosine (C), thymine (T), and uracil (U)—are pyrimidines. The two purines (A and G) and one of the pyrimidines (C) are found in both DNA and RNA, but uracil (U) is found only in RNA, and thymine (T) is found only in DNA. Note that thymine differs from uracil only in the methyl group at the 5 position. Thus, both DNA and RNA contain four bases: two pyrimidines and two purines. For DNA, the bases are A, G, C, and T; for RNA, the bases are A, G, C, and U.

Purines

Pyrimidines

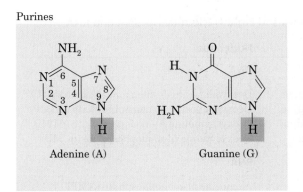

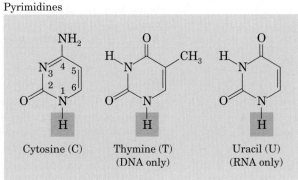

Adenine (A)   Guanine (G)   Cytosine (C)   Thymine (T) (DNA only)   Uracil (U) (RNA only)

FIGURE 25.1 The five principal bases of DNA and RNA. Note how the rings are numbered. The hydrogens shown in blue are lost when the bases bond to monosaccharides.

B. Sugars

The sugar component of RNA is D-ribose (Section 20.1C). In DNA, it is 2-deoxy-D-ribose (hence the name deoxyribonucleic acid).

β-D-Ribose   β-2-Deoxy-D-ribose

The full name of β-D-ribose is β-D ribofuranose, and that of β-2-deoxy-D-ribose is β-2-deoxy-D-ribofuranose (see Section 20.2A).

The combination of sugar and base is known as a **nucleoside**. The purine bases are linked to C-1 of the monosaccharide through N-9 (the nitrogen at position 9 of the five-membered ring) by a β-*N*-glycosidic bond:

Nucleoside A compound composed of ribose or deoxyribose and a base

Adenine

β-D-Ribose

+ H_2O

β-*N*-glycosidic bond

Adenosine

The nucleoside made of guanine and ribose is called **guanosine**. Table 25.1 gives the names of the other nucleosides.

The pyrimidine bases are linked to C-1 of the monosaccharide through their N-1 by a β-*N*-glycosidic bond.

Table 25.1 The Eight Nucleosides and Eight Nucleotides in DNA and RNA

Base	Nucleoside	Nucleotide
		DNA
Adenine (A)	Deoxyadenosine	Deoxyadenosine 5'-monophosphate (dAMP)*
Guanine (G)	Deoxyguanosine	Deoxyguanosine 5'-monophosphate (dGMP)*
Thymine (T)	Deoxythymidine	Deoxythymidine 5'-monophosphate (dTMP)*
Cytosine (C)	Deoxycytidine	Deoxycytidine 5'-monophosphate (dCMP)*
		RNA
Adenine (A)	Adenosine	Adenosine 5'-monophosphate (AMP)
Guanine (G)	Guanosine	Guanosine 5'-monophosphate (GMP)
Uracil (U)	Uridine	Uridine 5'-monophosphate (UMP)
Cytosine (C)	Cytidine	Cytidine 5'-monophosphate (CMP)

*The d indicates that the sugar is deoxyribose.

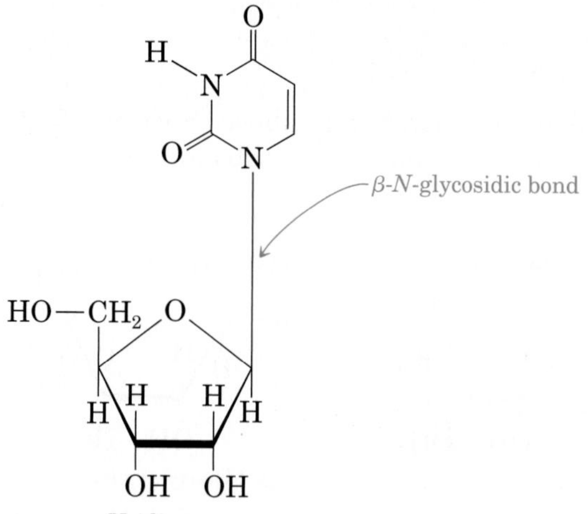

Uridine

C. Phosphate

The third component of nucleic acids is phosphoric acid. When this group forms a phosphate ester (Section 19.5) bond with a nucleoside, the result is a compound known as a **nucleotide**. For example, adenosine combines with phosphate to form the nucleotide adenosine 5'-monophosphate (AMP):

Nucleotide A nucleoside bonded to one, two, or three phosphate groups

The ' sign in adenosine 5'-monophosphate is used to distinguish which molecules the phosphate is bound to. Numbers without primes refer to positions on the purine or pyrimidine base. Numbers on the sugar are denoted with primes.

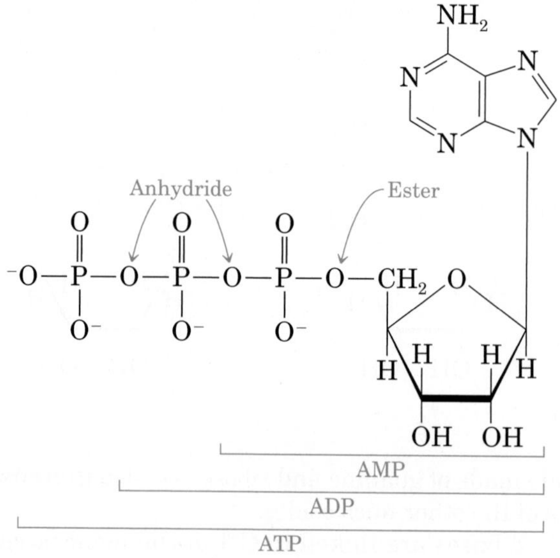

Table 25.1 gives the names of the other nucleotides. Some of these nucleotides play important roles in metabolism. They are part of the structure of key coenzymes, cofactors, and activators (Sections 27.3 and 29.2). Most notably, adenosine 5′-triphosphate (ATP) serves as a common currency into which the energy gained from food is converted and stored. In ATP, two more phosphate groups are joined to AMP with phosphate anhydride bonds (Section 19.5). In adenosine 5′-diphosphate (ADP), only one phosphate group is bonded to the AMP. All other nucleotides have important multiphosphorylated forms. For example, guanosine exists as GMP, GDP, and GTP.

In Section 25.3, we will see how DNA and RNA are chains of nucleotides. In summary:

A nucleoside = Base + Sugar

A nucleotide = Base + Sugar + Phosphate

A nucleic acid = A chain of nucleotides

Example 25.1 Nucleotide Structure

GTP is an important store of energy. Draw the structure of guanosine triphosphate.

Strategy

When drawing nucleotides, there are three things to consider. First, determine if the sugar is ribose or deoxyribose. Then attach the correct base to the C1 position of the sugar. Finally, put in the correct number of phosphates.

Solution

The base guanine is linked to a ribose unit by a β-N-glycosidic linkage. The triphosphate is linked to C-5′ of the ribose by an ester bond.

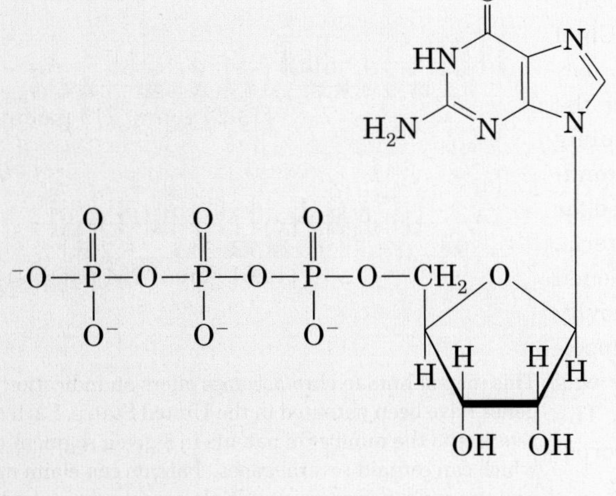

Problem 25.1

Draw the structure of UMP.

Chemical Connections 25A

Who Owns Your Genes

"There is a gene in your body's cells that plays a key role in early spinal cord development. It belongs to Harvard University. Incyte Corporation, based in Wilmington, Delaware, has patented the gene for a receptor for histamine, the compound released by cells during the hay fever season. About half of all the genes known to be involved in cancer are patented."[†] Following the explosion in information that came from the Human Genome Project, commercial firms, universities, and even government agencies began to look for patents on genes, which was the beginning of a long philosophical and legal battle that continues to this day. Human cells have about 24,000 genes that are the blueprint for the 100 trillion cells in our body. About 20 percent of the human genome has been patented. As of 2006, Incyte Corporation owned about 10 percent of all known human genes.

So the question that comes to mind is, "How can a company patent a biological entity?" Well, clearly it cannot actually patent you or your genes, at least not the ones you carry around. What can be patented is purified DNA containing the sequence of the gene and the techniques that allow the study of the genes. The idea of patenting information began with a landmark case in 1972 when Ananda M. Chakrabarty, a General Electric engineer, filed for a patent on a strain of *Pseudomonas* bacteria that could break down oil slicks more efficiently. He experimented with the bacteria, getting them to take up DNA from plasmids (rings of DNA, see Chapter 13) that conferred the clean-up ability. The patent office rejected the patent on the grounds that products of nature and live organisms cannot be patented. However, the battle was not over, and in 1980, the Supreme Court heard the appeal in the same year that the techniques of molecular biology and recombinant DNA technology really began to take off. Chief Justice Warren Burger declared arguments against patenting life as irrelevant by stating, "anything under the sun that is made by man" could be patented. The ruling was close, only 5–4 in favor of Chakrabarty, and the ramifications continue to this day. Patents have been issued for gene sequences, whole organisms such as specific bacteria, and for cell types such as stem cells. A patent on a cloned gene or the protein it produces gives the owner exclusivity in marketing the protein, such as insulin or erythropoietin. As of 2005, the largest scientific patent holder was the University of California, with over 1000 patents. The U.S. government was second with 926, and the first corporation, Sanofi Aventis, came in third at 587.

There are many issues stirring the controversy. Proponents for the patent system point out that it takes money to drive research. Companies will not want to invest hundreds of thousands to millions of dollars into research if they cannot get a tangible gain. Allowing them to patent a product means they can eventually recover their investment. Opponents believe a patent on what amounts to information stifles more research and even prevents the advancement of medicine. If a company holds the patent to a gene known to be involved in a disease, then others cannot study it effectively and perhaps come up with better

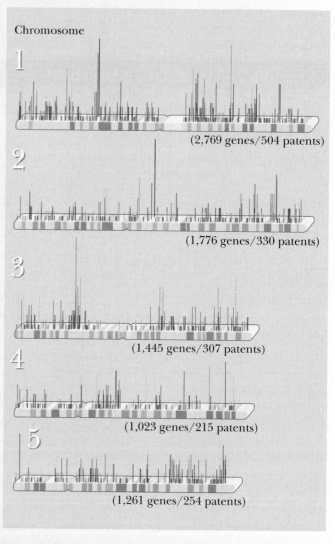

This map of human chromosomes offers an indication of how often genes have been patented in the United States. Each colored bar represents the number of patents in a given segment of a chromosome, which can contain several genes. Patents can claim multiple genes, and one gene may receive multiple patents. As a result, the number of patents indicated for each chromosome does not necessarily match the sum of the values represented by the colored bars.

[†]From page 78, "Owning the Stuff of Life," by Gary Stix, *Scientific American*, February 2006.

Who Owns Your Genes (*continued*)

or cheaper treatments. The latter has come under intense scrutiny recently because patents on diagnostic genes inhibit both research and clinical medicine. At the heart of the conflict are patents for two genes related to breast cancer, *BRCA1 and BRCA2*, both owned by Myriad Genetics Inc., of Salt Lake City. In 2009, a group of patients, doctors, and research professionals brought a lawsuit against the company to invalidate those patents. They argue that the two genes are "products of nature" and should never have been patented in the first place. The long-term effects of

such a lawsuit are important enough that the American Civil Liberties Union has decided to support the plaintiffs.

In April 2010, Judge Robert Sweet of the federal district court of New York City dropped a legal bombshell by invalidating a handful of gene patents, including BRCA1 and 2, claiming that they are products of nature, not human inventions. This decision is currently applicable only in New York and will likely be appealed, but this was the outcome hoped for by cancer patients and medical professionals who oppose gene patents.

Test your knowledge with Problems 25.76–25.79.

25.3 What Is the Structure of DNA and RNA?

In Chapter 22, we saw that proteins have primary, secondary, and higher-order structures. Nucleic acids, which are chains of monomers, also have primary, secondary, and higher-order structures.

Nucleic acid A polymer composed of nucleotides

A. Primary Structure

Nucleic acids are polymers of nucleotides, as shown schematically in Figure 25.2. Their primary structure is the sequence of nucleotides. Note that it can be divided into two parts: (1) the backbone of the molecule and (2) the bases that are the side-chain groups. The backbone in DNA consists of alternating deoxyribose and phosphate groups. Each phosphate group is linked to the 3′ carbon of one deoxyribose unit and simultaneously to the 5′ carbon of the next deoxyribose unit (Figure 25.3). Similarly, each monosaccharide unit forms a phosphate ester at the 3′ position and another at the 5′ position. The primary structure of RNA is the same except that each sugar is ribose (so an —OH group appears in the 2′ position) rather than deoxyribose, and U is present instead of T.

Thus, the backbone of the DNA and RNA chains has two ends: a 3′ —OH end and a 5′ —OH end. These two ends have roles similar to those of the C-terminal and N-terminal ends in proteins. The backbone provides structural stability for the DNA and RNA molecules.

As noted earlier, the bases that are linked, one to each sugar unit, are the side chains. They carry all of the information necessary for protein synthesis. Through analysis of the base composition of DNA molecules from many different species, Erwin Chargaff (1905–2002) showed that the quantity of adenine (in moles) is always approximately equal to the quantity of thymine and that the quantity of guanine is always approximately equal to the quantity of cytosine, although the adenine/guanine ratio varies widely from species to species (see Table 25.2). This important information helped to establish the secondary structure of DNA, as we will soon see.

Just as the order of the amino acid residues of protein side chains determines the primary structure of the protein (for example, —Ala—Gly—Glu—Met—), the order of the bases (for example, —ATTGAC—) provides the primary structure of DNA. As with proteins, we need a convention to tell us which

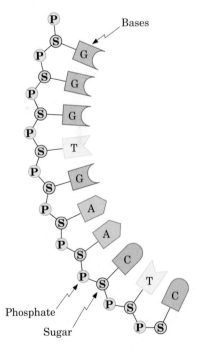

FIGURE 25.2 Schematic diagram of a nucleic acid molecule. The four bases of each nucleic acid are arranged in various specific sequences.

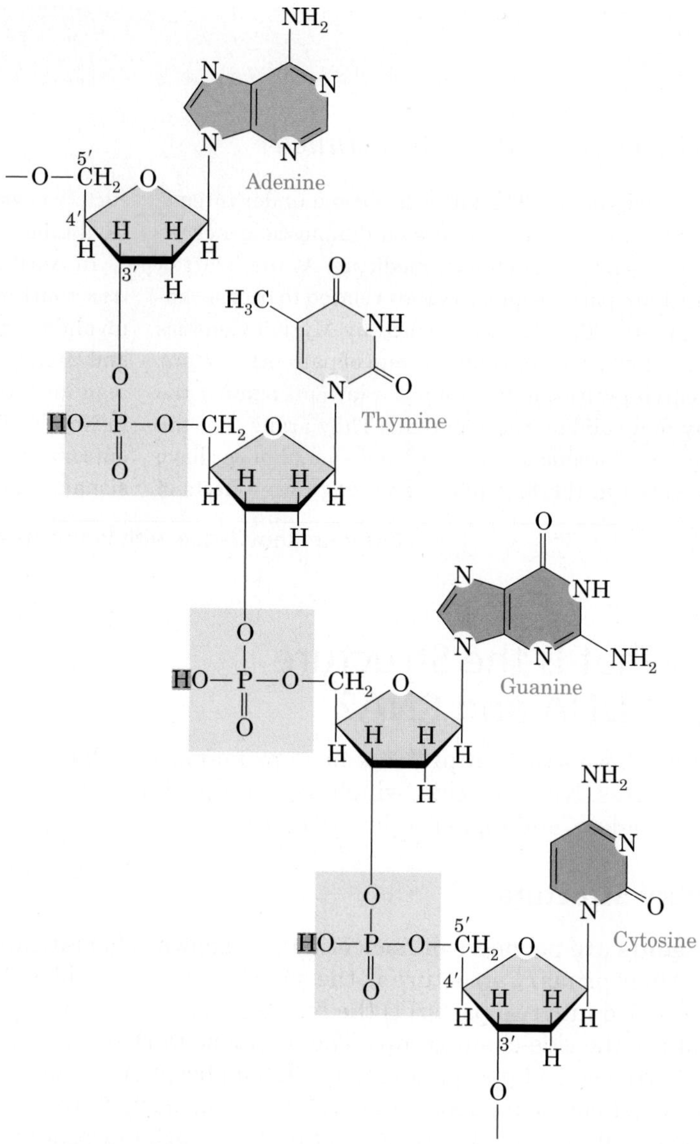

FIGURE 25.3 Primary structure of the DNA backbone. The hydrogens shown in blue cause the acidity of nucleic acids. In the body, at neutral pH, the phosphate groups carry a charge of −1 and the hydrogens are replaced by Na^+ and K^+.

end to start with when we write the sequence of bases. For nucleic acids, the convention is to begin the sequence with the nucleotide that has the free 5′ terminus. Thus, the sequence AGT means that adenine is the base at the 5′ terminus and thymine is the base at the 3′ terminus.

B. Secondary Structure of DNA

In 1953, James Watson (1928–) and Francis Crick (1916–2004) established the three-dimensional structure of DNA. Their work is a cornerstone in the history of biochemistry. The model of DNA developed by Watson and Crick was based on two important pieces of information obtained by other workers: (1) the Chargaff rule that (A and T) and (G and C) are present

Table 25.2 Base Composition and Base Ratio in Two Species

Organism	Base Composition (mol %)				Base Ratio	
	A	**G**	**C**	**T**	**A/T**	**G/C**
Human	30.9	19.9	19.8	29.4	1.05	1.01
Wheat germ	27.3	22.7	22.8	27.1	1.01	1.00

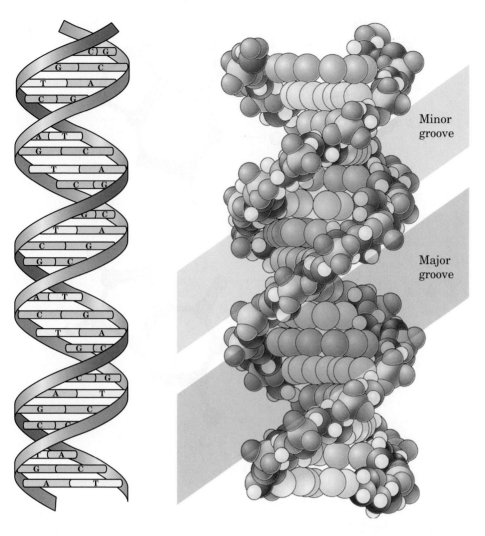

FIGURE 25.4 Three-dimensional structure of the DNA double helix.

Minor groove

Major groove

in equimolar quantities and (2) X-ray diffraction photographs of DNA obtained by Rosalind Franklin (1920–1958) and Maurice Wilkins (1916–2004). By the clever use of these facts, Watson and Crick concluded that DNA is composed of two strands entwined around each other in a **double helix**, as shown in Figure 25.4.

In the DNA double helix, the two polynucleotide chains run in opposite directions (which is called antiparallel). Thus, at each end of the double helix, there is one 5′ —OH and one 3′ —OH terminus. The sugar–phosphate backbone is on the outside, exposed to the aqueous environment, and the bases point inward. The bases are hydrophobic, so they try to avoid contact with water. Through their hydrophobic interactions, they stabilize the double helix. The bases are paired according to Chargaff's rule: For each adenine on one chain, a thymine is aligned opposite it on the other chain; each guanine on one chain has a cytosine aligned with it on the other chain. *The bases so paired form hydrogen bonds with each other, two for A—T and three for G—C, thereby stabilizing the double helix* (Figure 25.5). *A—T* and *G—C* are **complementary base pairs**.

The important thing, as Watson and Crick realized, is that only adenine could fit with thymine and only guanine could fit with cytosine. Let us consider the other possibilities. Can two purines (AA, GG, or AG) fit opposite each other? Figure 25.6 shows that they would overlap. How about two pyrimidines (TT, CC, or CT)? As shown in Figure 25.6, they would be too far apart. *There must be a pyrimidine opposite a purine.* But could A fit opposite

Double helix The arrangement in which two strands of DNA are coiled around each other in a screw-like fashion

Watson, Crick, and Wilkins were awarded the 1962 Nobel Prize in medicine for their discovery. Franklin died in 1958. The Nobel Committee does not award the Nobel Prize posthumously.

FIGURE 25.5 A and T pair up by forming two hydrogen bonds; G and C pair up by forming three hydrogen bonds.

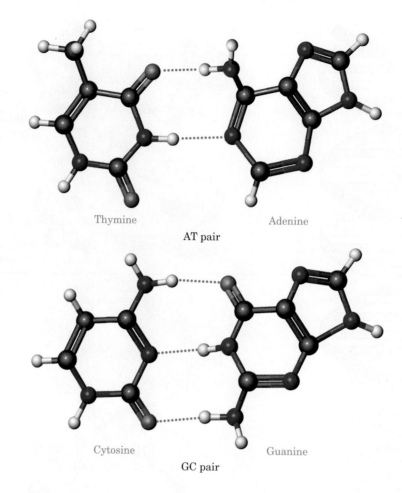

Thymine   Adenine

AT pair

Cytosine   Guanine

GC pair

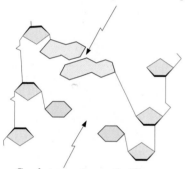

Two purines overlap

Gap between two pyrimidines

FIGURE 25.6 The bases of DNA cannot stack properly in the double helix if a purine is opposite a purine or if a pyrimidine is opposite a pyrimidine.

C, or G opposite T? Figure 25.7 shows that the hydrogen bonding would be much weaker.

The entire action of DNA—and of the heredity mechanism—depends on the fact that *wherever there is an adenine on one strand of the helix, there must be a thymine on the other strand because that is the only base that fits and forms strong hydrogen bonds with adenine, and similarly for G and C.* The entire heredity mechanism rests on these aligned hydrogen bonds (Figure 25.5), as we will see in Section 25.6.

Rosalind Franklin (1920–1958).

Watson and Crick with their model of the DNA molecule.

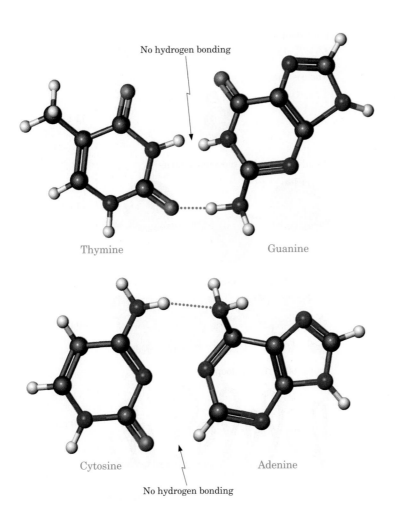

No hydrogen bonding

Thymine   Guanine

Cytosine   Adenine

No hydrogen bonding

FIGURE 25.7 Only one hydrogen bond is possible for TG or CA. These combinations are not found in DNA. Compare this figure with Figure 25.5.

The form of the DNA double helix shown in Figure 25.4 is called B-DNA. It is the most common and most stable form. Other forms become possible where the helix is wound more tightly or more loosely or is wound in the opposite direction. With B-DNA, a distinguishing feature is the presence of a **major groove** and a **minor groove**, which arise because the two strands are not equally spaced around the helix. Interactions of proteins and drugs with the major and minor grooves of DNA serve as an active area of research.

C. Higher-Order Structures of DNA

If a human DNA molecule were fully stretched out, its length would be perhaps 1 m. However, the DNA molecules in the nuclei are not stretched out, but rather, coiled around basic protein molecules called **histones**. The acidic DNA and the basic histones attract each other by electrostatic (ionic) forces, combining to form units called **nucleosomes**. In a nucleosome, eight histone molecules form a core, around which a 147-base-pair DNA double helix is wound. Nucleosomes are further condensed into **chromatin** when a 30-nm-wide fiber forms in which nucleosomes are wound in a **solenoid** fashion, with six nucleosomes forming a repeating unit (Figure 25.8). Chromatin fibers are organized still further into loops, and loops are arranged into bands to provide the superstructure of chromosomes.

The beauty of establishing the three-dimensional structure of the DNA molecule was that the knowledge of this structure immediately led to the explanation for the transmission of heredity—how the genes transmit traits from one generation to another. Before we look at the mechanism of

Chromatin The DNA complexed with histone and nonhistone proteins that exists in eukaryotic cells between cell divisions

Solenoid A coil wound in the form of a helix

FIGURE 25.8 Superstructure of chromosomes. In nucleosomes, the bandlike DNA double helix winds around cores consisting of eight histones. Solenoids of nucleosomes form a 30 nm filament. Loops and minibands are other substructures.

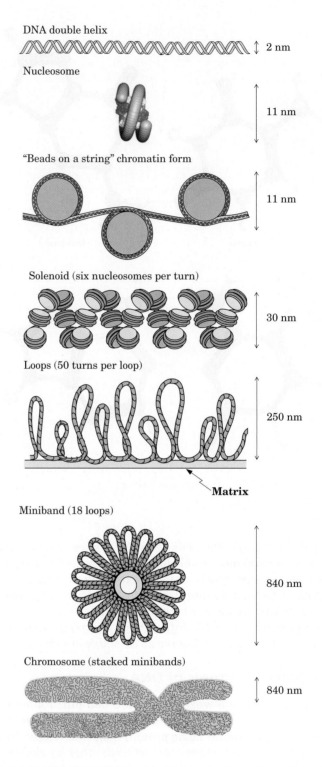

DNA double helix

2 nm

Nucleosome

11 nm

"Beads on a string" chromatin form

11 nm

Solenoid (six nucleosomes per turn)

30 nm

Loops (50 turns per loop)

250 nm

Matrix

Miniband (18 loops)

840 nm

Chromosome (stacked minibands)

840 nm

DNA replication (in Section 25.6), let us summarize the three differences in structure between DNA and RNA:

1. DNA has four bases: A, G, C, and T. RNA has three of these bases —A,G, and C— but its fourth base is U, not T.

2. In DNA, the sugar is 2-deoxy-D-ribose. In RNA, it is D-ribose.

3. DNA is almost always double-stranded, with the helical structure shown in Figure 25.4.

There are several kinds of RNA (as we will see in Section 25.4); none of them has a repetitive double-stranded structure like DNA, although base-pairing can occur within a chain (see, for example, Figure 25.10 on the next page). When it does, adenine pairs with uracil because thymine is not present. Other combinations of hydrogen-bonded bases are also possible outside the confines of a double helix, and Chargaff's rule does not apply.

25.4 What Are the Different Classes of RNA?

We previously noted that there are six types of RNA.

1. **Messenger RNA (mRNA)** mRNA molecules are produced in the process called **transcription**, and they carry the genetic information from the DNA in the nucleus directly to the cytoplasm, where most of the protein is synthesized. Messenger RNA consists of a chain of nucleotides whose sequence is exactly complementary to that of one of the strands of the DNA. This type of RNA is not long-lived, however. It is synthesized as needed and then degraded, so its concentration at any given time is rather low. The size of mRNA varies widely, with the average unit containing perhaps 750 nucleotides. Figure 25.9 shows the basic flow of genetic information and the major types of RNA.

Messenger RNA (mRNA) The RNA that carries genetic information from DNA to the ribosome and acts as a template for protein synthesis

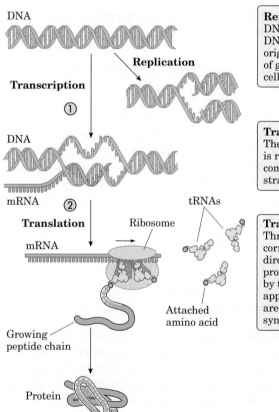

Replication
DNA replication yields two DNA molecules identical to the original one, ensuring transmission of genetic information to daughter cells with exceptional fidelity.

Transcription
The sequence of bases in DNA is recorded as a sequence of complementary bases in a single-stranded mRNA molecule.

Translation
Three-base codons on the mRNA corresponding to specific amino acids direct the sequence of building a protein. These codons are recognized by tRNAs (transfer RNAs) carrying the appropriate amino acids. Ribosomes are the "machinery" for protein synthesis.

FIGURE 25.9 The fundamental process of information transfer in cells. **(1)** Information encoded in the nucleotide sequence of DNA is transcribed through synthesis of an RNA molecule whose sequence is dictated by the DNA sequence. **(2)** As the sequence of this RNA is read (as groups of three consecutive nucleotides) by the protein synthesis machinery, it is translated into the sequence of amino acids in a protein. This information transfer system is encapsulated in what is known as the central dogma of molecular biology: DNA ⟶ RNA ⟶ protein.

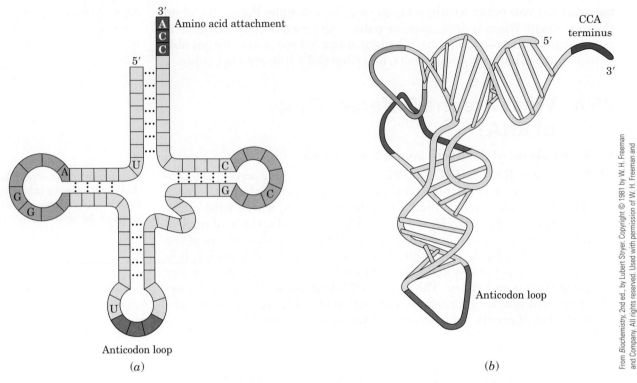

FIGURE 25.10 Structure of tRNA. (*a*) Two-dimensional simplified cloverleaf structure. (*b*) Three-dimensional structure.

Transfer RNA (tRNA) The RNA that transports amino acids to the site of protein synthesis in ribosomes

2. Transfer RNA (tRNA) Containing from 73 to 93 nucleotides per chain, tRNAs are relatively small molecules. There is at least one different tRNA molecule for each of the 20 amino acids from which the body makes its proteins. The three-dimensional tRNA molecules are L-shaped, but they are conventionally represented as a cloverleaf in two dimensions. Figure 25.10 shows a typical structure. Transfer RNA molecules contain not only cytosine, guanine, adenine, and uracil, but also several other modified nucleotides, such as 1-methylguanosine.

1-Methylguanosine

Ribosomal RNA (rRNA) The RNA complexed with proteins in ribosomes

Ribosome Small spherical bodies in the cell made of protein and RNA; the site of protein synthesis

3. Ribosomal RNA (rRNA) Ribosomes, which are small spherical bodies located in the cells but outside the nuclei, contain rRNA. They consist of about 35% protein and 65% ribosomal RNA (rRNA). These large molecules have molecular weights up to 1 million. As discussed in Section 25.5, protein synthesis takes place on the ribosomes.

Dissociation of ribosomes into their components has proved to be a useful way of studying their structure and properties. A particularly important endeavor has been to determine both the number and the kind of RNA and protein molecules that make up ribosomes. This approach has helped elucidate the role of ribosomes in protein synthesis. In both prokaryotes and eukaryotes, a ribosome consists of two subunits, one larger than the other. In turn, the smaller subunit consists of one large RNA molecule and about 20 different proteins; the larger subunit consists of two RNA molecules in prokaryotes (three in eukaryotes) and about 35 different proteins in prokaryotes (about 50 in eukaryotes) (Figure 25.11). The subunits are easily dissociated from one another in the laboratory by lowering the Mg^{2+} concentration of the medium. Raising the Mg^{2+} concentration to its original level reverses the process, and active ribosomes can be reconstituted by this method.

4. **Small Nuclear RNA (snRNA)** A recently discovered RNA molecule is snRNA, which is found, as the name implies, in the nucleus of eukaryotic cells. This type of RNA is small, about 100 to 200 nucleotides long, but it is neither a tRNA molecule nor a small subunit of rRNA. In the cell, it is complexed with proteins to form **small nuclear ribonucleoprotein particles**, **snRNPs**, pronounced "snurps." Their function is to help with the processing of the initial mRNA transcribed from DNA into a mature form that is ready for export out of the nucleus. This process is often referred to as **splicing**, and it is an active area of research. While studying splicing, scientists realized that part of the splicing reaction involved catalysis by the RNA portion of a snRNP and not the protein portion. This recognition led to the discovery of **ribozymes**, RNA-based enzymes, for which Thomas Cech received the Nobel Prize in 1989. Splicing will be discussed further in Chapter 26.

5. **Micro RNA (miRNA)** A very recent discovery is another type of small RNA, miRNA. These RNAs are only 20–22 nucleotides long but are important in the timing of an organism's development. They play important roles in cancer, stress responses, and viral infections. They inhibit translation of mRNA into protein and promote the degradation of mRNA. It was recently discovered, however, that these versatile RNAs can also stimulate protein production in cells when the cell cycle has been arrested.

6. **Small Interfering RNA (siRNA)** The process called RNA interference was heralded as the breakthrough of the year in 2002 by *Science* magazine. Short stretches of RNA (20–30 nucleotides long), called small interfering RNA, have been found to have an enormous control over gene expression. This process serves as a protective mechanism in many species, with the siRNAs being used to eliminate expression of an undesirable gene, such as one that causes uncontrolled cell growth or one that came from a virus. siRNAs degrade specific mRNA molecules to control gene activity. Scientists who wish to study gene expression are also using these small RNAs. In what has become an explosion of new biotechnology, many companies have been created to produce and market designer siRNAs to knock out hundreds of known genes. This technology also has medical applications, as siRNA has been used to protect the mouse liver from hepatitis and to help clear infected liver cells of the disease.

Table 25.3 summarizes the basic types of RNA.

Splicing The removal of an internal RNA segment and the joining of the remaining ends of the RNA molecule

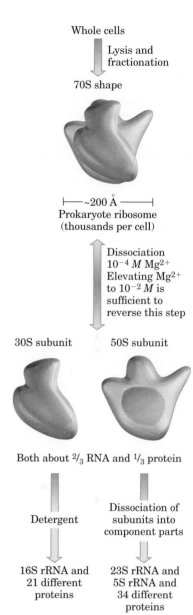

FIGURE 25.11 The structure of a typical prokaryotic ribosome. The individual components can be mixed, producing functional subunits. Reassociation of subunits gives rise to an intact ribosome. The designation S refers to Svedberg, a unit of relative size determined when molecules are separated by centrifugation.

Table 25.3 The Roles of Different Kinds of RNA

RNA Type	Size	Function
Transfer RNA (tRNA)	Small	Transports amino acids to site of protein synthesis
Ribosomal RNA (rRNA)	Several kinds— variable in size	Combines with proteins to form ribosomes, the site of protein synthesis
Messenger RNA (mRNA)	Variable	Directs amino acid sequence of proteins
Small nuclear RNA (snRNA)	Small	Processes initial mRNA to its mature form in eukaryotes
Micro RNA (miRNA)	Small	Affects gene expression; important in growth and development
Small interfering RNA (siRNA)	Small	Affects gene expression; used by scientists to knock out a gene being studied

25.5   What Are Genes?

A gene is a stretch of DNA, containing a few hundred nucleotides, that carries one particular message—for example, "make a globin molecule" or "make a tRNA molecule." One DNA molecule may have between 1 million and 100 million bases. Therefore, many genes are present in one DNA molecule. In bacteria, this message is continuous; in higher organisms, it is not. That is, stretches of DNA that spell out (encode) the amino acid sequence to be assembled are interrupted by long stretches that seemingly do not code for anything. The coding sequences are called **exons**, short for "expressed sequences," and the noncoding sequences are called **introns**, short for "intervening sequences."

For example, the globin gene has three exons broken up by two introns. Because DNA contains both exons and introns, the mRNA transcribed from it also contains both exons and introns. The introns are spliced out by ribozymes, and the exons are spliced together before the mRNA is used to synthesize a protein. In other words, the introns function as spacers and, in rare instances, as enzymes, catalyzing the splicing of exons into mature mRNA. Figure 25.12 shows the difference between prokaryotic and eukaryotic production of proteins.

In prokaryotes, the genes on a stretch of DNA are next to each other. These are turned into a sequence of mRNA, which is then translated by ribosomes to make proteins, all of which happens simultaneously. In eukaryotes, the genes are separated by introns and the processes take place in different compartments. The DNA is turned into RNA in the nucleus, but then the initial mRNA, containing introns, is transported to the cytosol where the introns are spliced out. The final mRNA is then translated to protein. The process of making RNA and protein is the subject of Chapter 26.

In humans, only 3% of the DNA codes for proteins or RNA with clear functions. Introns are not the only noncoding DNA sequences, however. **Satellites** are DNA molecules in which short nucleotide sequences are repeated hundreds or thousands of times. Large satellite stretches appear at the ends and centers of chromosomes and provide stability for the chromosomes.

Smaller repetitive sequences, called **mini-satellites** or **microsatellites**, are associated with cancer when they mutate.

Exon Nucleotide sequence in DNA or mRNA that codes for a protein

Intron A nucleotide sequence in DNA or mRNA that does not code for a protein

25.6   How Is DNA Replicated?

The DNA in the chromosomes carries out two functions: (1) It reproduces itself and (2) it supplies the information necessary to make all the RNA and proteins in the body, including enzymes. The second function is covered in Chapter 26. Here, we are concerned with the first, **replication**.

Each gene is a section of a DNA molecule that contains a specific sequence of the four bases A, G, T, and C, typically comprising about 1000 to 2000 nucleotides. The base sequence of the gene carries the information necessary to produce one protein molecule. If the sequence is changed (for example, if one A is replaced by a G or if an extra T is inserted), a different protein is produced, which might have an impaired function, as in sickle cell anemia (Chemical Connections 22D).

Consider the monumental task that must be accomplished by the organism. When an individual is conceived, the egg and sperm cells unite to form the zygote. This cell contains only a small amount of DNA, but it nevertheless provides all the genetic information that the individual will ever have.

Replication The process by which copies of DNA are made during cell division

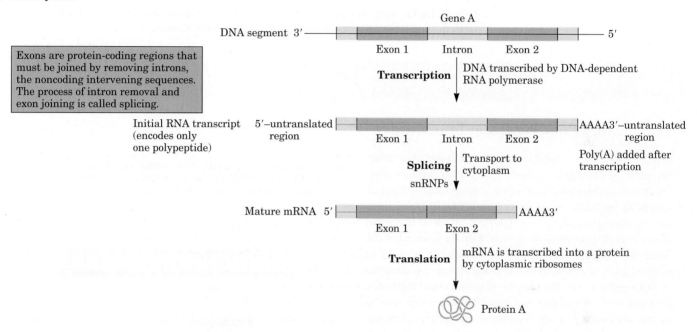

FIGURE 25.12 The properties of mRNA molecules in (*a*) prokaryotic versus (*b*) eukaryotic cells during transcription and translation.

In a human cell, some 3 billion base pairs must be duplicated at each cell cycle, and a fully grown human being may contain more than 1 trillion cells. Each cell contains the same amount of DNA as the original single cell. Furthermore, cells are constantly dying and being replaced. Thus, there must be a mechanism by which DNA molecules can be copied over and over again without error. In Section 25.7, we will see that such errors sometimes do occur and can have serious consequences. Here, however, we want to examine this remarkable mechanism that takes place every day in billions of organisms, from microbes to whales, and has been taking place for billions of years—with only a tiny percentage of errors.

Replication begins at a point in the DNA called an **origin of replication**. In human cells, the average chromosome has several hundred origins of replication where the copying occurs simultaneously. The DNA double helix has two strands running in opposite directions. The point on the DNA where replication is proceeding is called the **replication fork** (Figure 25.13).

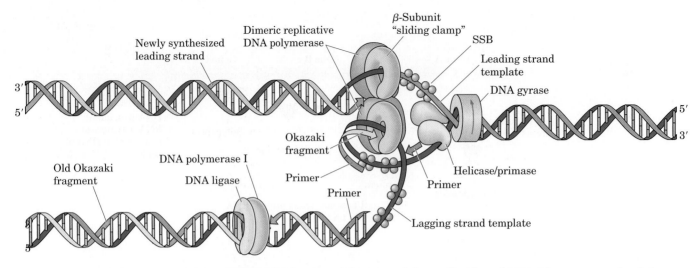

FIGURE 25.13 General features of the replication of DNA. The two strands of the DNA double helix are shown separating at the replication fork.

Chemical Connections 25B

Telomeres, Telomerase, and Immortality

Every person has a genetic makeup consisting of about 3 billion pairs of nucleotides, distributed over 46 chromosomes. Telomeres are specialized structures at the ends of chromosomes. In vertebrates, telomeres are TTAGGG sequences that are repeated hundreds to thousands of times. In normal **somatic cells** that divide in a cyclic fashion throughout the life of the organism (via mitosis), chromosomes lose about 50 to 200 nucleotides from their telomeres at each cell division.

DNA polymerase, the enzyme that links the fragments, does not work at the end of linear DNA. This fact results in the shortening of the telomeres at each replication. The telomere shortening acts as a clock by which the cells count the number of times they have divided. After a certain number of divisions, the cells stop dividing, having reached the limit of the aging process.

In contrast to somatic cells, all immortal cells (germ cells in proliferative stem cells, normal fetal cells, and cancer cells) possess an enzyme, telomerase, that can extend the shortened telomeres by synthesizing new chromosomal ends. Telomerase is a ribonucleoprotein; that is, it is made of RNA and protein. The activity of this enzyme seems to confer immortality to the cells.

(a) **Replication at the end of a linear template**

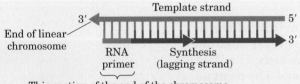

This portion of the end of the chromosome will be lost when the primer is removed.

(b) **A mechanism by which telomerase may work. (In this case, RNA of the telomerase acts as a template for reverse transcription)**

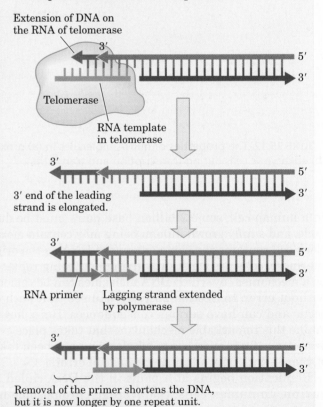

Removal of the primer shortens the DNA, but it is now longer by one repeat unit.

The telomerase extension cycle is repeated until there is an adequate number of DNA repeats for the end of the chromosome to survive.

Test your knowledge with Problems 25.80–25.83.

If the unwinding of the double helix begins in the middle, the synthesis of new DNA molecules on the old templates continues in both directions until the entire molecule is duplicated. Alternatively, the unwinding can start at one end and proceed in one direction until the entire double helix is unwound.

Replication is bidirectional and takes place at the same speed in both directions. An interesting detail of DNA replication is that the two daughter strands are synthesized in different ways. One of the syntheses is continuous along the 3′ to 5′ strand (see Figure 25.13). It is called the **leading strand**. Along the other strand that runs in the 5′ ⟶ 3′ direction, the synthesis is discontinuous. It is called the **lagging strand**.

The replication process is called **semiconservative** because each daughter molecule has one parental strand (conserved) and one newly synthesized one.

Replication always proceeds from the 5′ to the 3′ direction from the perspective of the chain that is being synthesized. The actual reaction occurring is a nucleophilic attack by the 3′ hydroxyl of the deoxyribose of one nucleotide against the first phosphate on the 5′ carbon of the incoming nucleoside triphosphate, as shown in Figure 25.14.

One of the more interesting aspects of DNA replication is that the basic reaction of synthesis always requires an existing chain with a nucleotide that has a free 3′-hydroxyl to do the nucleophilic attack. DNA replication cannot begin without this preexisting chain to latch onto. We call this chain a **primer**. In all known forms of replication, the primer is made out of RNA, not DNA.

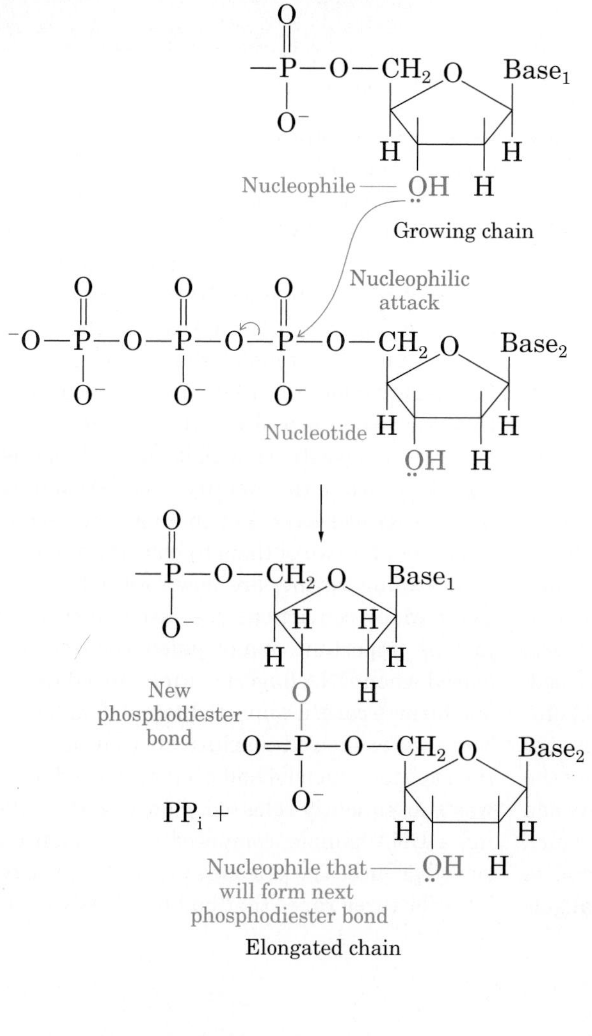

FIGURE 25.14 The addition of a nucleotide to a growing DNA chain. The 3′-hydoxyl group at the end of the growing DNA chain is a nucleophile. It attacks at the phosphorus adjacent to the sugar in the nucleotide, which will be added to the growing chain.

DNA Fingerprinting

The base sequence in the nucleus of every one of our billions of cells is identical. However, except for people who have an identical twin, the base sequence in the total DNA of one person is different from that of every other person. This uniqueness makes it possible to identify suspects in criminal cases from a bit of skin or a trace of blood left at the scene of the crime and to prove the identity of a child's father in paternity cases.

To do so, the nuclei of the cells of the criminal evidence are extracted. Their DNA is amplified by PCR techniques (Section 25.8). With the aid of restriction enzymes, the DNA molecules are cut at specific points. The resulting DNA fragments are put on a gel and subjected to **electrophoresis**. In this process, the DNA fragments move with different velocities; the smaller fragments move faster, and the larger fragments move slower. After a sufficient amount of time, the fragments separate. When they are made visible in the form of an autoradiogram, one can discern bands in a lane. This sequence is called a **DNA fingerprint**.

When the DNA fingerprint made from a sample taken from a suspect matches that from a sample obtained at the scene of the crime, the police have a positive identification. The accompanying figure shows DNA fingerprints derived by using one particular restriction enzyme. Here, a total of nine lanes can be seen. Three (numbers 1, 5, and 9) are control lanes. They contain the DNA fingerprint of a virus, using one particular restriction enzyme.

Three other lanes (2, 3, and 4) were used in a paternity suit: They contain the DNA fingerprints of the mother, the child, and the alleged father. The child's DNA fingerprint (lane 3) contains six bands. The mother's fingerprint (lane 4) has five bands, all of which match those of the child. The alleged father's DNA fingerprint (lane 2) also contains six bands, three of which match those of the child. This is a positive identification. In such cases, one cannot expect a perfect match even if the man is really the father because the child has inherited only half of his or her genes from the father. Each band in the child's DNA had to come from one of the parents. If the child has a band and the mother does not, then that band must be represented in the DNA of the alleged father; otherwise, he was not the child's father. In the case just described, the paternity suit was won on the basis of the DNA fingerprint matching. However, this gel by itself is not conclusive. Most of the DNA bands are matches between the mother and child, with only one unique match between the alleged father and child. Paternity testing is much more readily used to exclude a potential father than to prove a person is the father. Usually it takes several gels

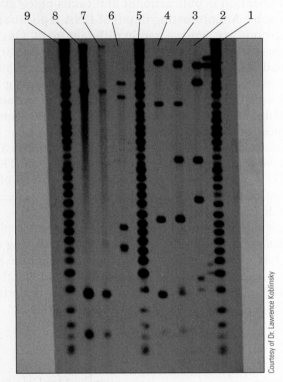

DNA fingerprint.

using several different enzymes to get enough data to conclude a positive match.

In the left area of the radiogram are three more lanes (6, 7, and 8). These DNA fingerprints were used in an attempt to identify a rapist. Lanes 7 and 8 show the DNA fingerprints of semen obtained from the rape victim. Lane 6 is the DNA fingerprint of the suspect. The DNA fingerprints of the semen do not match those of the suspect. This result is a negative identification and excluded the suspect from the case. When a positive identification occurs, the probability that a positive match is due to chance is 1 in 100 billion. Thus, while the identity is not absolutely proven, the law of averages says that there are not enough people on the planet for two of them to have the same DNA pattern.

DNA fingerprints are now routinely accepted in court cases. Many convictions are based on such evidence, and, just as important, many jailed suspects have been released when DNA fingerprinting proved them innocent. In one bizarre case, a convicted rapist demanded a DNA test. The results showed conclusively that he was not guilty of the rape for which he had been sentenced to prison, and he was subsequently released. However, the police, now having a DNA sample, compared it to evidence in other unsolved crimes. The released prisoner was arrested a week later for three rapes that had previously gone unsolved.

DNA Fingerprinting (*continued*)

In the United States, states maintain databases of DNA patterns from crime scenes and samples that have been collected from people. California has over a million samples from convicted felons. Since 2008, the state has allowed a new technique of forensic science called **familial DNA searches**. In this technique, police can use DNA samples already collected to not only pinpoint exact DNA matches, but also use the DNA database to match DNA left at a crime scene with family members, if sufficient similarity can be demonstrated between a criminal in the database and his or her blood relatives. In a recent high-profile case, police used DNA collected from a piece of pizza to solve a case that had been cold for 25 years. There was DNA evidence from a string of murders, but it did not match any person in the police database. So then a search was done to see if any similarities (not exact matches) could be found in the police database. A DNA sample had been collected from Christopher Franklin, the son of Lonnie Franklin, because Christopher had been convicted of a felony weapons charge. There was similarity between the crime scene DNA and Christopher Franklin. Even though he had a long criminal record, Lonnie's DNA was not in the database, but the familial similarity between the crime scene DNA and Christopher's DNA was noted. Undercover police officers collected evidence at a restaurant, including a pizza crust, where Lonnie ate. The pizza DNA from Lonnie matched the DNA on a murder victim. Lonnie Franklin was eventually convicted of being "The Grim Sleeper," a serial killer in Los Angeles who had murdered 10 women.

Test your knowledge with Problems 25.84–25.87.

Replication is a very complex process involving a number of enzymes and binding proteins. A growing body of evidence indicates that these enzymes assemble their products in "factories" through which the DNA moves. Such factories may be bound to membranes in bacteria. In higher organisms, the replication factories are not permanent structures. Instead, they may be disassembled and their parts reassembled in ever-larger factories. These assemblies of enzyme "factories" go by the name of **replisomes**, and they contain key enzymes such as polymerases, helicases, and primases (Table 25.4). The primases can shuttle in and out of the replisomes. Other proteins, such as clamp loaders and clamp proteins, through which the newly synthesized primer is threaded, are also parts of the replisomes.

The replication of DNA occurs in a number of distinct steps. A few of the salient features are enumerated here:

1. **Opening Up the Superstructure** During replication, the very condensed superstructure of chromosomes must be opened so that it becomes accessible to enzymes and other proteins. A complicated signal transduction mechanism accomplishes this feat. One notable step of the signal transduction is the acetylation and deacetylation of key lysine residues of histones. When histone acetylase, an enzyme, puts acetyl

Table 25.4 Components of Replisomes and Their Functions

Component	Function
Helicase	Unwinds the DNA double helix
Primase	Synthesizes short oligonucleotides (primers)
Clamp protein	Allows the leading strand to be threaded through
DNA polymerase	Joins the assembled nucleotides
Ligase	Joins Okazaki fragments in the lagging strand

Chemical Connections 25D

The Human Genome Project: Treasure or Pandora's Box?

The Human Genome Project (HGP) was a massive attempt to sequence the entire human genome, some 3.3 billion base pairs spread over 23 pairs of chromosomes. This project, started formally in 1990, was a worldwide effort driven forward by two groups. One is a private company called Celera Genomics, and its preliminary results were published in *Science* in February 2001. The other is a publicly funded group of researchers called the International Human Genome Sequencing Consortium. The Human Genome Project was completed in the year 2000. The preliminary results were published in *Nature* in February 2001. Researchers were surprised to find that there were only about 30,000 genes in the human genome. This figure has since dwindled to 25,000. This is similar to many other eukaryotes, including some as simple as the roundworm *Caenorhabditis elegans*.

What does one do with the information? From it, we will eventually be able to identify all human genes and to determine which sets of genes are likely to be involved in all human genetic traits, including diseases that have a genetic basis. There is always an elaborate interplay of genes, so it may never be possible to say that a defect in a given gene will ensure that the individual will develop a particular disease. Nevertheless, some forms of genetic screening will certainly become a routine part of medical testing in the future.

As technology has improved, the year 2007 saw the birth of a new industry—personal genomics. It is now possible for individuals to have their DNA completely sequenced, although only the most affluent can afford the $350,000 price tag. However, several companies offer a partial screen to scan for up to one million known DNA markers. The cost of this "recreational genomics" is much lower, $1000 to $2500, and requires only a saliva swab.

Many people are concerned that the availability of genetic information could lead to genetic discrimination. For that reason, the HGP is a rare example of scientific projects in which definite percentages of financial support and research effort have been devoted to the ethical, legal, and social implications (ELSI) of the research. The question is often posed in this form: Who has a right to know your genetic information? You? Your doctor? Your potential spouse or employer? An insurance company? These questions are not trivial, but they have not yet been answered definitively. The 1997 movie *GATTACA* depicted a society in which one's social and economic classes are established at birth on the basis of one's genome. Many citizens have expressed concern that genetic screening could lead to a new type of prejudice and bigotry aimed against "genetically challenged" people.

Many people have suggested that there is no point in screening for potentially disastrous genes if there is no meaningful therapy for the disease they may "cause." However, couples often want to know in advance if they are likely to pass on a potentially lethal disease to their children.

Two specific examples are pertinent:

1. There is no advantage in testing for the breast cancer gene if a woman is *not* in a family at high risk for the disease. The presence of a "normal" gene in such a low-risk individual tells nothing about whether a mutation might occur in the future.
2. The presence of a gene has not always predicted the development of the disease. Some individuals who have been shown to be carriers of the gene for Huntington's disease have lived to old age without developing the disease. On the other hand, some males who are functionally sterile have been found to have cystic fibrosis. They learn this when they go to a clinic to assess the nature of their fertility problem even though they may never have shown true symptoms of the disease as a child, other than perhaps a high occurrence of respiratory ailments.

Another major area for concern about the HGP is the possibility of gene therapy, which many people fear is akin to "playing God." Some people envision an era of so-called designer babies, with attempts made to create the "perfect" human. A more moderate view has been that gene therapy may be useful in correcting diseases that impair quality of life or are lethal. Tests with human subjects are already under way for cystic fibrosis, the "bubble boy" type of immune deficiency, and some other diseases. Current guidelines in the United States allow for gene therapy of somatic cells, but they do not allow for genetic modifications that would be passed on to the next generation.

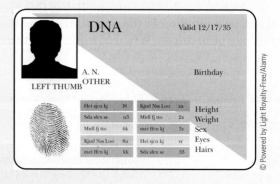

Your genome could appear on an ID card in the foreseeable future.

Test your knowledge with Problems 25.88–25.90.

groups on key lysine residues, some positive charges are eliminated and the strength of the DNA–histone interaction is weakened:

$$\text{Histone}-(CH_2)_4-NH_3^+ + CH_3-COO^- \underset{\text{deacetylation}}{\overset{\text{acetylation}}{\rightleftharpoons}} \text{Histone}-(CH_2)_4-\overset{\displaystyle H}{\underset{\displaystyle |}{N}}-\overset{\displaystyle O}{\underset{\displaystyle \|}{C}}-CH_3 + H_2O$$

This process allows the opening up of key regions on the DNA molecule. When another enzyme, histone deacetylase, removes these acetyl groups, the positive charges are reestablished. That, in turn, facilitates regaining the highly condensed structure of **chromatin**.

2. **Relaxation of Higher-Order Structures of DNA** Topoisomerases (also called gyrases) are enzymes that facilitate the relaxation of supercoiling in DNA. They do so during replication by temporarily introducing either single- or double-strand breaks in DNA. The transient break forms a phosphodiester linkage between a tyrosyl residue of the enzyme and either the 5′ or 3′ end of a phosphate on the DNA. Once the supercoiling is relaxed, the broken strands are joined together and the topoisomerase diffuses from the location of the replicating fork. Topoisomerases are also involved in the untangling of the replicated chromosomes, before cell division can occur.

3. **Unwinding the DNA Double Helix** The replication of DNA molecules starts with the unwinding of the double helix, which can occur at either end or in the middle. Special unwinding protein molecules, called **helicases**, attach themselves to one DNA strand (Figure 25.13) and cause the separation of the double helix. Helicases of eukaryotes are made of six different protein subunits. The subunits form a ring with a hollow core, where the single-stranded DNA sits. The helicases hydrolyze ATP as the DNA strand moves through. The energy of the hydrolysis promotes this movement.

4. **Primers/Primases** Primers are short—4 to 15 nucleotides long—RNA oligonucleotides synthesized from ribonucleoside triphosphates. They are needed to initiate the synthesis of both daughter strands. The enzyme catalyzing this synthesis is called primase. Primases form complexes with DNA polymerase in eukaryotes. Primers are placed about every 50 nucleotides in the lagging-strand synthesis.

5. **DNA Polymerase** The key enzymes in replication are the DNA polymerases. Once the two strands are separated at the replication fork, the DNA nucleotides must be lined up. All four kinds of free DNA nucleotide molecules are present in the vicinity of the replication fork. These nucleotides constantly move into the area and try to fit themselves into new chains. The key to the process is that, as we saw in Section 25.3, *only thymine can fit opposite adenine and only cytosine can fit opposite guanine.* Wherever a cytosine, for example, is present on one of the strands of an unwound portion of the helix, all four nucleotides may approach, but three of them will be turned away because they do not fit. Only the nucleotide of guanine fits.

In the absence of an enzyme, this alignment is extremely slow. The speed and specificity are provided by DNA polymerase. The active site of this enzyme is quite snug. It surrounds the end of the DNA template–primer complex, creating a specifically shaped pocket for the incoming nucleotide. With such a close contact, the activation energy is lowered and the polymerase enables complementary base pairing with high specificity at a rate of 100 times per second. While the bases of the newly arrived nucleotides are being hydrogen-bonded to their partners, polymerases join the nucleotide backbones.

Okazaki fragment A short
DNA segment made of about 200
nucleotides in higher organisms
(eukaryotes) and of 2000 nucleotides
in prokaryotes

Along the lagging strand $3' \longrightarrow 5'$, the enzymes can synthesize only short fragments because the only way they can work is from $5'$ to $3'$. These short fragments consist of about 200 nucleotides each, named **Okazaki fragments** after their discoverer.

6. **Ligation** The Okazaki fragments and any nicks remaining are eventually joined together by another enzyme, DNA ligase. At the end of the process, there are two double-stranded DNA molecules, each exactly the same as the original molecule because only thymine fits opposite adenine and only guanine fits against cytosine in the active site of the polymerase.

25.7   How Is DNA Repaired?

The viability of cells depends on DNA repair enzymes that can detect, recognize, and remove mutations from DNA. Such mutations may arise from external or internal sources. Externally, UV radiation or highly reactive oxidizing agents such as superoxide may damage a base. Errors in copying or internal chemical reactions—for example, deamination of a base—can create damage internally. Deamination of the base cytosine turns it into uracil (Figure 25.1), which creates a mismatch. The former C—G pair becomes a U—G mispair that must be removed.

The repair can be effected in a number of ways. One of the most common is called BER, *base excision repair* (Figure 25.15). This pathway contains two parts:

1. A specific DNA glycosylase recognizes the damaged base (1). It hydrolyzes the N—C' β-glycosidic bond between the uracil base and the deoxyribose, then releases the damaged base, completing the excision. The sugar–phosphate backbone is still intact. At the **AP site** (*ap*urinic or *ap*yrimidinic site) created in this way, the backbone is cleaved by a second enzyme, endonuclease (2). A third enzyme, exonuclease (3), liberates the sugar–phosphate unit of the damaged site.

2. In the synthesis step, the enzyme DNA polymerase (4) inserts the correct nucleotide, cytidine, and the enzyme DNA ligase seals (5) the backbone to complete the repair.

A second repair mechanism removes not a single mismatched base but rather whole nucleotides—as many as 25 to 32 residue oligonucleotides. Known as NER, for *nucleotide excision repair*, it similarly involves a number of repair enzymes.

Any defect in the repair mechanisms may lead to harmful or even lethal mutations. For example, individuals with inherited XP (xeroderma pigmentosa), a condition in which an enzyme in the NER repair pathway is missing or defective, have a 1000 times greater risk of developing skin cancer than do normal individuals.

25.8   How Do We Amplify DNA?

To study DNA for basic or applied scientific purposes, we must have enough of it to work with. There are several ways of amplifying DNA. One approach is to allow a rapidly growing organism, like bacteria, to replicate DNA for us. This process, which is usually referred to as **cloning**, will be discussed further in Chapter 26. Millions of copies of selected DNA fragments can also be made within a few hours with high precision by a technique called the **polymerase chain reaction (PCR)**, which was discovered by Kary B. Mullis (1944–), who shared the 1993 Nobel Prize in chemistry for this achievement.

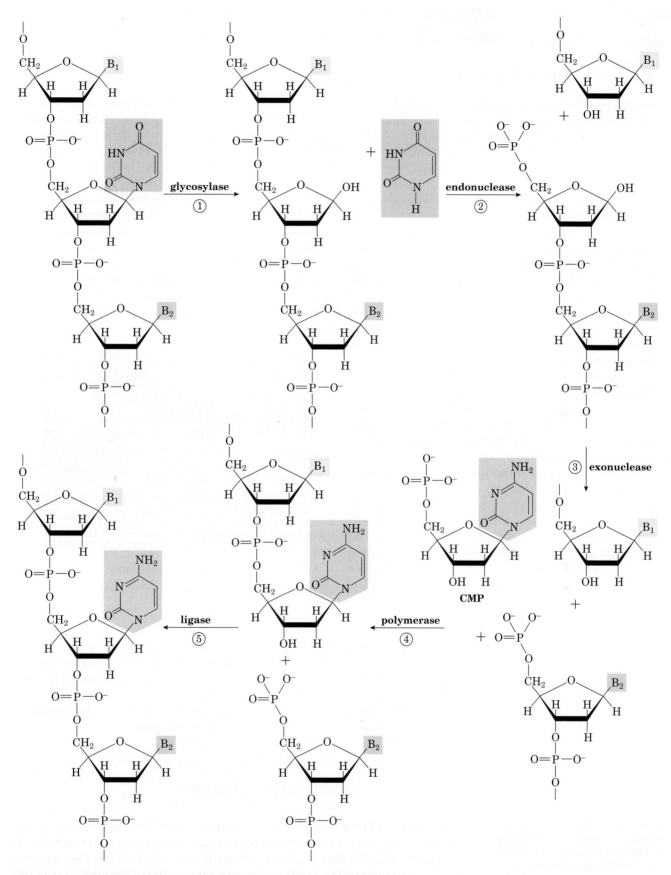

FIGURE 25.15 The base excision repair (BER) pathway. A uracil is replaced by a cytosine.

Chemical Connections 25E

Synthetic Genome Created

For those researchers interested in studying the nature of life and its relationship to DNA chemistry, the Holy Grail of experiments is to demonstrate that synthetic DNA can lead to life. In May 2010, science took a giant step towards that goal. The laboratory of Human Genome Project pioneer J. Craig Venter successfully designed synthetic DNA and used it to drive the reproduction of a bacterial species.

The experiment was completed in several stages, eventually taking 40 million dollars and a team of over 20 researchers a decade to complete. It began when Venter and two colleagues, Clyde Hutchinson and Hamilton Smith, demonstrated that they could transplant the DNA from one bacterial species into another. In 2008, they created an artificial chromosome of the bacterium *Mycoplasma genitalium*, which they chose because it has the smallest genome of any free-living organism, with only 600,000 bases. Besides the natural DNA sequence, their synthetic version contained constructed "watermark" sequences that allowed them to tell the synthetic version from the natural one. Unfortunately, the *M. genitalium* bacteria reproduced too slowly to be studied efficiently, so they switched species to the faster-growing *Mycoplasma mycoides*, which contains 1 million bases.

In 2009, they demonstrated that they could transplant natural *M. mycoides* DNA into a close cousin, *M. capricolum*. They culminated the experiment in 2010 by taking the synthetic version of the *M. mycoides* DNA and transplanting it into *M. capricolum* cells that had had their DNA removed. The watermark sequences cause the growing bacterial colonies to become blue in color, which

proves that they have the synthetic instead of the natural DNA (see figure). Some scientists have called this experiment "life re-created" because it did not quite demonstrate the creation of life from chemistry, since they transplanted DNA into cells that were previously alive. However, this experiment is a keystone demonstration of the importance of DNA to all life processes, since the chemically synthesized DNA was able to take over the evacuated *M. capricolum* cells and begin to grow colonies of *M. mycoides*.

While it will be years or decades before scientists can begin to create designer organisms, the potential to create microbes that can synthesize pharmaceuticals or fuels has molecular biologists excited and eagerly anticipating what organisms will come out of Venter's and other researchers' labs in the future.

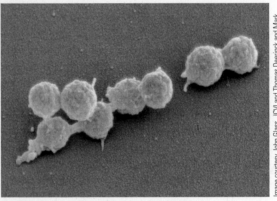

Image courtesy John Glass, JCVI and Thomas Deerinck and Mark Ellisman, NCMIR

Life re-created; SEM of a bacterial group.

Test your knowledge with Problems 25.91 and 25.92.

PCR techniques can be used if the sequence of a gene to be copied is known, or at least a sequence bordering the desired DNA is known. In such a case, one can synthesize two primers that are complementary to the ends of the gene or to the bordering DNA. The primers are oligonucleotides consisting of 12 to 16 nucleotides. When added to a target DNA segment, they hybridize with the end of each strand of the gene.

5′CATAGGACAGC—OH   Primer

3′TACGTATCCTGTCGTAGG—   Gene

Hybridization The process in which two strands of nucleic acids or segments of nucleic acid strands form a double-stranded structure through hydrogen bonding of complementary base pairs

In cycle 1 (Figure 25.16), the polymerase extends the primers in each direction as individual nucleotides are assembled and connected on the template DNA. In this way, two new copies are created. The two-step process is repeated (cycle 2) when the primers are **hybridized** with the new strands, and the primers are extended again. At that point, four new copies

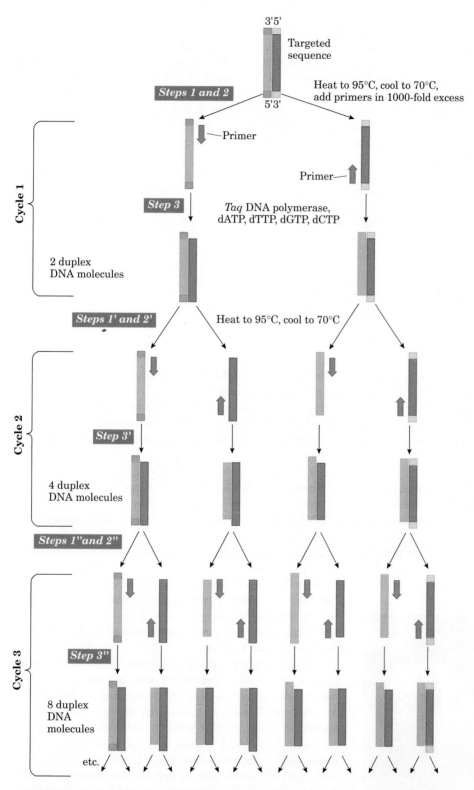

FIGURE 25.16 Polymerase chain reaction (PCR). Oligonucleotides complementary to a given DNA sequence prime the synthesis of only that sequence. Heat-stable *Taq* DNA polymerase survives many cycles of heating. Theoretically, the amount of the specific primed sequence is doubled in each cycle.

have been created. The process continues, and in 25 cycles, 2^{25}, or some 33 million, copies can be made. In practice, only a few million are produced, which is sufficient for the isolation of a gene.

This fast process is practical because of the discovery of heat-resistant polymerases isolated from bacteria that live in hot springs and in hot thermal vents on the sea floor (Section 23.4B). A temperature of 95°C is needed because the double helix must be unwound to hybridize the primer to the target DNA. Once single strands of DNA have been exposed, the mixture is cooled to 70°C. The primers are hybridized at this temperature, and subsequent extensions take place. The 95°C and 70°C cycles are repeated over and over. No new enzyme is required because the polymerase is stable at both temperatures.

PCR techniques are routinely used when a gene or a segment of DNA must be amplified from a few molecules. It is used in studying genomes, in obtaining evidence from a crime scene (Chemical Connections 25C), and even in obtaining the genes of long-extinct species found fossilized in amber.

Did the Neandertals Go Extinct?

Since modern man first discovered the bones of prehistoric humans in caves, archeologists have been fascinated by the common ancestors of humans. Neandertals appeared in the fossil record as early as 400,000 years ago and disappeared from the fossil record about 30,000 years ago. Modern humans lived in caves for tens of thousands of years, dating back more than 100,000 years ago, meaning that they must have coexisted with the Neandertals. Until recently, dogma held that although *Homo neandertalis* and *Homo sapiens* coexisted, they did not interbreed and eventually the Neandertal line went extinct. A typical evolutionary tree (see the figure on the next page) indicates that chimpanzees and humans diverged from a common ancestor around 5 million years ago. Then,

Mock photo of a Neanderthal and a modern human.

between 270,000 and 440,000 years ago, Neandertals diverged from modern humans.

However, after sequencing the Neandertal genome from three bones of three female Neandertals found in a cave in Croatia from 38,000 years ago, scientists found that their genomes are 99.84% identical to the genomes of modern humans. In effect, modern humans who originated in Asia or Europe have between 1 to 4% Neandertal DNA. That may not seem like much, but with over 3 billion base pairs of DNA, that amounts to quite a bit. Numbers aside, the conclusions were startling and "dogma-breaking" for evolutionary biologists, who are now convinced that there was interbreeding between modern humans and Neandertals. Equally interesting, and critical to the conclusions, is the fact that this incorporation of Neandertal DNA was not seen in humans who originated in Africa. In other words, the Neandertals are more closely related to Europeans and Asians than to Africans.

While much of the DNA is the same between the two species, there were significant and interesting differences, including genes that affect cognition, metabolism, and skeletal development. Researchers are focusing on a handful of genes that are different between the two species, such as *THADA*, a gene that varies in modern humans and is associated with Type 2 diabetes. Other genes, such as *NRG3*, when mutated in modern humans, can lead to schizophrenia. The figure on page 716 lists some of the genes currently under study that are different between modern humans and Neandertals.

Neanderthal Museum/H. Neumann

Chemical Connections 25F

Did the Neandertals Go Extinct? (*continued*)

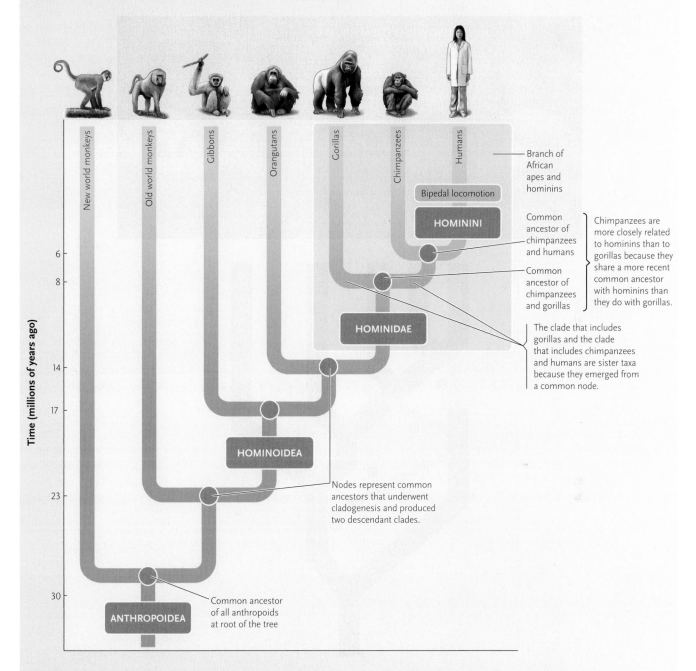

Branch of African apes and hominins

Bipedal locomotion

HOMININI

Common ancestor of chimpanzees and humans

Common ancestor of chimpanzees and gorillas

Chimpanzees are more closely related to hominins than to gorillas because they share a more recent common ancestor with hominins than they do with gorillas.

HOMINIDAE

The clade that includes gorillas and the clade that includes chimpanzees and humans are sister taxa because they emerged from a common node.

HOMINOIDEA

Nodes represent common ancestors that underwent cladogenesis and produced two descendant clades.

ANTHROPOIDEA

Common ancestor of all anthropoids at root of the tree

New world monkeys
Old world monkeys
Gibbons
Orangutans
Gorillas
Chimpanzees
Humans

Time (millions of years ago)

6
8
14
17
23
30

By studying the genotypes of modern humans, Neandertals, and more distant cousins such as chimpanzees, scientists will learn a lot about human evolution. This will include new information about population behavior and the chemical nature of the genes and how the changes in them caused us to become who we are today. One thing is certain—the old evolutionary tree will have to be redrawn, as it appears that the divergence of Neandertal from modern humans was not a one-way road without turns.

Chemical Connections 25F

Did the Neandertals Go Extinct? (*continued*)

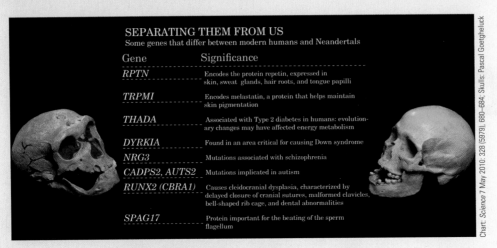

SEPARATING THEM FROM US
Some genes that differ between modern humans and Neandertals

Gene	Significance
RPTN	Encodes the protein repetin, expressed in skin, sweat glands, hair roots, and tongue papilli
TRPMI	Encodes melastatin, a protein that helps maintain skin pigmentation
THADA	Associated with Type 2 diabetes in humans: evolutionary changes may have affected energy metabolism
DYRKIA	Found in an area critical for causing Down syndrome
NRG3	Mutations associated with schizophrenia
CADPS2, AUTS2	Mutations implicated in autism
RUNX2 (CBRA1)	Causes cleidocranial dysplasia, characterized by delayed closure of cranial sutures, malformed clavicles, bell-shaped rib cage, and dental abnormalities
SPAG17	Protein important for the beating of the sperm flagellum

Chart: *Science* 7 May 2010: 328 (5979), 680–684; Skulls: Pascal Goetgheluck

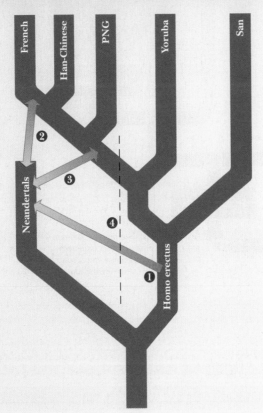

A possible modification of the DNA family tree to include the introduction of Neandertal genes into non-African DNA.

Test your knowledge with Problems 25.93–25.96.

Summary

⏺WL Sign in at **www.cengage.com/owl** to develop problem-solving skills and complete online homework assigned by your professor.

Section 25.1 What Are the Molecules of Heredity?

- Heredity is based on genes located in chromosomes.
- Genes are sections of DNA that encode specific RNA molecules.

Section 25.2 What Are Nucleic Acids Made Of?

- **Nucleic acids** are composed of sugars, phosphates, and organic bases.
- Two kinds of nucleic acids exist: **ribonucleic acid (RNA)** and **deoxyribonucleic acid (DNA)**.
- In DNA, the sugar is the monosaccharide 2-deoxy-D-ribose; in RNA, it is D-ribose.
- In DNA, the heterocyclic amine bases are adenine (A), guanine (G), cytosine (C), and thymine (T).
- In RNA, they are A, G, C, and uracil (U).
- Nucleic acids are giant molecules with backbones made of alternating units of sugar and phosphate. The bases are side chains joined by β-N-glycosidic bonds to the sugar units.

Section 25.3 What Is the Structure of DNA and RNA?

- DNA is made of two strands that form a double helix. The sugar–phosphate backbone runs on the outside of the double helix, and the hydrophobic bases point inward.
- **Complementary pairing** of the bases occurs in the double helix, such that each A on one strand is hydrogen-bonded to a T on the other strand and each G is hydrogen-bonded to a C. No other pairs fit.
- DNA is coiled around basic protein molecules called **histones**. Together they form **nucleosomes**, which are further condensed into chromatin.
- The DNA molecule carries, in the sequence of its bases, all the information necessary to maintain life. When cell division occurs and this information is passed from parent cell to daughter cells, the sequence of the parent DNA is copied.

Section 25.4 What Are the Different Classes of RNA?

- There are six kinds of RNA: **messenger RNA (mRNA)**, **transfer RNA (tRNA)**, **ribosomal RNA (rRNA)**, **small nuclear RNA (snRNA)**, **micro RNA (miRNA)**, and **small interfering RNA (siRNA)**.
- mRNA, tRNA, and rRNA are involved in all protein synthesis.
- Small nuclear RNA is involved in splicing reactions and has been found in some cases to have catalytic activity.
- RNA with catalytic activity is called a **ribozyme**.

Section 25.5 What Are Genes?

- A **gene** is a segment of a DNA molecule that carries the sequence of bases that directs the synthesis of one particular protein or RNA molecule.
- DNA in higher organisms contains sequences, called **introns**, that do not code for proteins.
- The sequences that do code for proteins are called **exons**.

Section 25.6 How Is DNA Replicated?

- DNA replication occurs in several distinct steps.
- The superstructures of chromosomes are initially loosened by the acetylation of histones. **Topoisomerases** relax the higher structures. **Helicases** at the replication fork separate the two strands of DNA.
- RNA primers and primases are needed to start the synthesis of daughter strands. The **leading strand** is synthesized continuously by **DNA polymerase**. The **lagging strand** is synthesized discontinuously as **Okazaki fragments**.
- **DNA ligase** seals the nicks and the Okazaki fragments.

Section 25.7 How Is DNA Repaired?

- An important **DNA repair** mechanism is BER, or single base excision repair.

Section 25.8 How Do We Amplify DNA?

- The **polymerase chain reaction (PCR)** technique can make millions of copies with high precision in a few hours.

Problems

⏺WL Interactive versions of these problems may be assigned in OWL.

Orange-numbered problems are applied.

Section 25.1 What Are the Molecules of Heredity?

25.2 What structures of the cell, visible in a microscope, contain hereditary information?

25.3 Name one hereditary disease.

25.4 What is the basic unit of heredity?

Section 25.2 What Are Nucleic Acids Made Of?

25.5 (a) Where in a cell is the DNA located?
(b) Where in a cell is the RNA located?

25.6 What are the components of (a) a nucleotide and (b) a nucleoside?

25.7 What are the differences between DNA and RNA?

25.8 Draw the structures of ADP and GDP. Are these structures parts of nucleic acids?

25.9 What is the difference in structure between thymine and uracil?

25.10 Which DNA and RNA bases contain a carbonyl group?

25.11 Draw the structures of (a) cytidine and (b) deoxycytidine.

25.12 Which DNA and RNA bases are primary amines?

25.13 What is the difference in structure between D-ribose and 2-deoxy-D-ribose?

25.14 What is the difference between a nucleoside and a nucleotide?

25.15 RNA and DNA refer to nucleic *acids*. Which part of the molecule is acidic?

25.16 What type of bond exists between the ribose and the phosphate in AMP?

25.17 What type of bond exists between the two phosphates in ADP?

25.18 What type of bond connects the base to the ribose in GTP?

Section 25.3 What Is the Structure of DNA and RNA?

25.19 In RNA, which carbons of the ribose are linked to the phosphate group and which are linked to the base?

25.20 What constitutes the backbone of DNA?

25.21 Draw the structures of (a) UDP and (b) dAMP.

25.22 In DNA, which carbon atoms of 2-deoxy-D-ribose are bonded to the phosphate groups?

25.23 The sequence of a short DNA segment is ATGGCAATAC.

(a) What name do we give to the two ends (terminals) of a DNA molecule?

(b) In this segment, which end is which?

(c) What would be the sequence of the complementary strand?

25.24 Chargaff showed that in samples of DNA taken from many different species, the molar quantity of A was always approximately equal to the molar quantity of T; the same is true for C and G. How did this information help to establish the structure of DNA?

25.25 How many hydrogen bonds can form between uracil and adenine?

25.26 How many histones are present in a nucleosome?

25.27 What is the nature of the interaction between histones and DNA in nucleosomes?

25.28 What are chromatin fibers made of?

25.29 What constitutes the superstructure of chromosomes?

25.30 What is the primary structure of DNA?

25.31 What is the secondary structure of DNA?

25.32 What is the major groove of a DNA helix?

25.33 What are the higher-order structures of DNA that eventually make up a chromosome?

Section 25.4 What Are the Different Classes of RNA?

25.34 Which type of RNA has enzyme activity? Where does it function mostly?

25.35 Which has the longest chains: tRNA, mRNA, or rRNA?

25.36 Which type of RNA contains modified nucleotides?

25.37 Which type of RNA has a sequence exactly complementary to that of DNA?

25.38 Where is rRNA located in the cell?

25.39 What kind of functions do ribozymes, in general, perform?

25.40 Which of the RNA types are always involved in protein synthesis?

25.41 What is the purpose of snRNA?

25.42 What is the purpose of siRNA?

25.43 What is the difference between miRNA and siRNA?

Section 25.5 What Are Genes?

25.44 Define:

(a) Intron   (b) Exon

25.45 Does mRNA also have introns and exons? Explain.

25.46 (a) What percentage of human DNA codes for proteins?

(b) What is the function of the rest of the DNA?

25.47 Do satellites code for a particular protein?

25.48 Do all genes code for a protein? If not, what do they code for?

Section 25.6 How Is DNA Replicated?

25.49 A DNA molecule normally replicates itself millions of times, with almost no errors. What single fact about the structure is most responsible for this fidelity of replication?

25.50 Which functional groups on the bases form hydrogen bonds in the DNA double helix?

25.51 Draw the structures of adenine and thymine and show with a diagram the two hydrogen bonds that stabilize A—T pairing in DNA.

25.52 Draw the structures of cytosine and guanine and show with a diagram the three hydrogen bonds that stabilize C—G pairing in nucleic acids.

25.53 How many different bases are present in a DNA double helix?

25.54 What is a replication fork? How many replication forks may exist simultaneously on an average human chromosome?

25.55 Why is replication called semiconservative?

25.56 How does the removal of some positive charges from histones enable the opening of the chromosomal superstructure?

25.57 Write the chemical reaction for the deacetylation of acetyl-histone.

25.58 What is the quaternary structure of helicases in eukaryotes?

25.59 What are helicases? What is their function?

25.60 Can dATP serve as a source for a primer?

25.61 What are the side products of the action of primase in forming primers?

25.62 What do we call the enzymes that join nucleotides into a DNA strand?

25.63 In which direction is the DNA molecule synthesized continuously?

25.64 What kind of bond formation do polymerases catalyze?

25.65 Which enzyme catalyzes the joining of Okazaki fragments?

25.66 What is the nature of the chemical reaction that joins nucleotides together?

25.67 From the perspective of the chain being synthesized, in which direction does DNA synthesis proceed?

Section 25.7 How Is DNA Repaired?

25.68 As a result of damage, a few of the guanine residues in a gene are methylated. What kind of mechanism could the cell use to repair the damage?

25.69 What is the function of endonuclease in the BER repair mechanism?

25.70 When cytosine is deaminated, uracil is formed. Uracil is a naturally occurring base. Why would the cell use base excision repair to remove it?

25.71 Which bonds are cleaved by glycosylase?

25.72 What are AP sites? Which enzyme creates them?

25.73 Why are patients with xeroderma pigmentosa 1000 times more likely to develop skin cancer than normal individuals are?

Section 25.8 How Do We Amplify DNA?

25.74 What is the advantage of using DNA polymerase from thermophilic bacteria that live in hot thermal vents in PCR?

25.75 What 12-nucleotide primer would you use in the PCR technique when you want to amplify a gene whose end is as follows: 3′TACCGTCATCCGGTG5′?

Chemical Connections

25.76 (Chemical Connections 25A) What are the two principal opposing views regarding the patenting of genes?

25.77 (Chemical Connections 25A) Describe the landmark case that set the stage for the biotech patent battles of today.

25.78 (Chemical Connections 25A) What two genes are at the heart of a current lawsuit, and what is their importance?

25.79 (Chemical Connections 25A) Do biotech firms actually own your genes?

25.80 (Chemical Connections 25B) What sequence of nucleotides is repeated many times in telomeres?

25.81 (Chemical Connections 25B) Why are as many as 200 nucleotides lost at each replication?

25.82 (Chemical Connections 25B) How does telomerase make a cancer cell immortal?

25.83 (Chemical Connections 25B) Why is DNA loss with replication not a problem for bacteria? (*Hint:* Bacteria have a circular genome.)

25.84 (Chemical Connections 25C) After having been cut by restriction enzymes, how are DNA fragments separated from each other?

25.85 (Chemical Connections 25C) How is DNA fingerprinting used in paternity suits?

25.86 (Chemical Connections 25C) Why is it easier to exclude someone via DNA fingerprinting than it is to prove that he or she is the person whose sample is being tested?

25.87 (Chemical Connections 25C) What is the principle behind paternity testing via DNA fingerprinting?

25.88 (Chemical Connection 25D) What would be some advantages to having your own genome sequenced?

25.89 (Chemical Connection 25D) How could the information from sequencing your genome be used against you theoretically if it fell into the wrong hands?

25.90 (Chemical Connection 25D) Why might a person make bad personal or lifestyle choices if she had knowledge of her genome?

25.91 (Chemical Connections 25E) What were the different phases of Venter's experiments on synthetic genomes?

25.92 (Chemical Connections 25E) Why did some scientists refer to the synthetic genome experiment as "life re-created?"

25.93 (Chemical Connections 25F) Why can it be argued that *Homo neandertalis* is not truly extinct?

25.94 (Chemical Connections 25F) Why do scientists now believe there was interbreeding between coexisting species of humans?

25.95 (Chemical Connections 25F) What genes associated with cognitive problems in modern humans are different in Neandertals?

25.96 (Chemical Connections 25F) What gene associated with skeletal structure is different in Neandertals?

Additional Problems

25.97 What is the active site of a ribozyme?

25.98 Why is it important that a DNA molecule be able to replicate itself millions of times without error?

25.99 Draw the structures of (a) uracil and (b) uridine.

25.100 How would you classify the functional groups that bond together the three different components of a nucleotide?

25.101 Which nucleic acid molecule is the largest?

25.102 What kind of bonds are broken during replication? Does the primary structure of DNA change during replication?

25.103 In sheep DNA, the mol % of adenine (A) was found to be 29.3. Based on Chargaff's rule, what would be the approximate mol % of G, C, and T?

Looking Ahead

25.104 DNA is the blueprint for the cell, but not all genes in DNA lead to protein. Gene expression is the study of how genes are used to make their particular product. What are some examples of gene products that do not lead to proteins?

25.105 In a process similar to DNA replication, RNA is produced via the process called transcription. The enzyme used is RNA polymerase. When RNA is synthesized, what is the direction of the synthesis reaction?

25.106 The Human Genome Project showed that human DNA is not considerably bigger than much simpler organisms, with about 30,000 total genes. However, humans make over 100, 000 different proteins. How is this possible? (*Hint:* Think about splicing.)

25.107 One of the biggest differences between DNA replication and transcription is that RNA polymerase does not require a primer. How does this fact relate to the theory that primordial life was based on RNA and not DNA?

25.108 How could life have evolved if DNA leads to RNA which leads to protein, but it takes many proteins to replicate DNA and to transcribe DNA into RNA?

25.109 When DNA is heated sufficiently, the strands separate. The energy that it takes to separate the DNA is related to the amount of guanine and cytosine bases. Why is this so?

25.110 If you wanted to amplify DNA using a technique similar to PCR but you had no source of a heat-stable DNA polymerase, what would you have to do to get the amplification?

25.111 Why do you think that DNA synthesis has evolved to have extensive proofreading and repair mechanisms while RNA synthesis has far fewer?

Gene Expression and Protein Synthesis

26

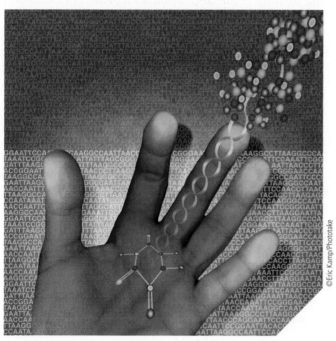

©Eric Kamp/Phototake

In transcription, the template strand of DNA is used to produce a complementary strand of RNA. Transcription is the most controlled and best understood part of gene regulation.

Key Questions

26.1 How Does DNA Lead to RNA and Protein?

26.2 How Is DNA Transcribed into RNA?

26.3 What Is the Role of RNA in Translation?

26.4 What Is the Genetic Code?

26.5 How Is Protein Synthesized?

26.6 How Are Genes Regulated?

26.7 What Are Mutations?

26.8 How and Why Do We Manipulate DNA?

26.9 What Is Gene Therapy?

26.1 How Does DNA Lead to RNA and Protein?

We have seen that the DNA molecule is a storehouse of information. We can compare it to a loose-leaf cookbook, each page of which contains one recipe. The pages are the genes. To prepare a meal, we use a number of recipes. Similarly, to provide a certain inheritable trait, a number of genes (Chapter 25)—segments of DNA—are needed.

Of course, the recipe itself is not the meal. The information in the recipe must be expressed in the proper combination of food ingredients. Similarly, the information stored in DNA must be expressed in the proper combination of amino acids representing a particular protein. The way this expression works is now so well established that it is called the **central dogma of molecular biology**. The dogma states that *the information contained in DNA molecules is transferred to RNA molecules, and then from the RNA molecules the information is expressed in the structure of proteins.*

OWL

Sign in to OWL at **www.cengage.com/owl** to view tutorials and simulations, develop problem-solving skills, and complete online homework assigned by your professor.

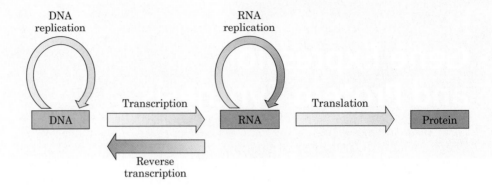

FIGURE 26.1 The central dogma of molecular biology. The yellow arrows represent the general cases, and the blue arrows represent special cases in RNA viruses.

Gene expression The activation of a gene to produce a specific protein; it involves both transcription and translation

Gene expression is the turning on, or activation, of a gene. Transmission of information occurs in two steps: transcription and translation.

Figure 26.1 shows the central dogma of gene expression. In some viruses (shown in blue), gene expression does not work this way. In some viruses with RNA genomes, replication proceeds from RNA to RNA. In retroviruses, RNA is reverse-transcribed to DNA.

A. Transcription

Transcription The process in which information encoded in a DNA molecule is copied into an mRNA molecule

Because the information (that is, the DNA) is in the nucleus of a eukaryotic cell and the amino acids are assembled outside the nucleus, the information must first be carried out of the nucleus. This step is analogous to copying a recipe from a cookbook. All the necessary information is copied, albeit in a slightly different format, as if we were converting the printed page into handwriting. On the molecular level, this task is accomplished by transcribing the information from the DNA molecule into a molecule of messenger RNA, so named because it carries the message from the nucleus to the site of protein synthesis. Other RNAs are similarly transcribed. rRNA is needed to form ribosomes, and tRNA is required to carry out the translation into protein language (Chapter 25). The transcribed information on the different RNA molecules is then carried out of the nucleus.

B. Translation

Translation The process in which information encoded in an mRNA molecule is used to assemble a specific protein

The mRNA serves as a template on which the amino acids are assembled in the proper sequence. To complete the assembly, the information that is written in the language of nucleotides must be translated into the language of amino acids. The translation is done by another type of RNA, transfer RNA (Section 25.4). An exact word-to-word translation occurs. Each amino acid in the protein language has a corresponding word in the RNA language. Each word in the RNA language is a sequence of three bases. This correspondence between three bases and one amino acid is called the genetic code (we will discuss the code in Section 26.4).

In higher organisms (eukaryotes), transcription and translation occur sequentially. The transcription takes place in the nucleus. After RNA leaves the nucleus and enters the cytoplasm, the translation takes place there. In lower organisms (prokaryotes), there is no nucleus, and thus, transcription and translation occur simultaneously in the cytoplasm. This extended form of the "central dogma" was challenged in 2001, when it was found that even in eukaryotes, about 15% of the proteins are produced in the nucleus itself. Clearly, some simultaneous transcription and translation do occur even in higher organisms.

We know more about bacterial transcription and translation because they are simpler than the processes operating in higher organisms and have been studied for a longer time. Nevertheless, we will concentrate on studying gene expression and protein synthesis in eukaryotic systems because they are more relevant to human health care.

26.2 How Is DNA Transcribed into RNA?

Transcription starts when the DNA double helix begins to unwind at a point near the gene that is to be transcribed (Figure 26.2). As we saw in Section 25.3C, nucleosomes form chromatin and higher condensed structures in the chromosomes. To make the DNA available for transcription, these superstructures change constantly. Specific **binding proteins** bind to the nucleosomes, making the DNA become less dense and more accessible. Only then can the enzyme **helicase**, a ring-shaped complex of six protein subunits, unwind the double helix.

Only one strand of the DNA molecule is transcribed. The strand that serves as a template for the formation of RNA has several names, including the **template strand**, the **(−) strand**, and the **antisense strand**. The other strand, while not used as a template, actually has a sequence that matches the RNA that will be produced. This strand is called the **coding strand**, the **(+) strand**, and the **sense strand**. Of these names, coding strand and template strand are the most commonly used.

Ribonucleotides assemble along the unwound DNA strand in the complementary sequence. Opposite each C on the DNA is a G on the growing mRNA; the other complementary bases follow the patterns G $\longrightarrow$ C, A $\longrightarrow$ U, and T $\longrightarrow$ A. The ribonucleotides, when aligned in this way, are linked to form the appropriate RNA.

In eukaryotes, three kinds of **polymerases** catalyze transcription. RNA polymerase I (pol I) catalyzes the formation of most of the rRNA; Pol II catalyzes mRNA formation; and Pol III catalyzes tRNA formation as well as one ribosomal subunit and other small regulatory RNA types, like snRNA. Each enzyme is a complex of 10 or more subunits. Some subunits are unique to each kind of polymerase, whereas other subunits appear in all three polymerases. Figure 26.3 shows the architecture of yeast RNA polymerase II.

The eukaryotic gene has two major parts: the **structural gene** itself, which is transcribed into RNA, and a **regulatory** portion that controls the transcription. The structural gene is made of exons and introns (Figure 26.4). The regulatory portion is not transcribed, but rather, has control elements.

One such control element is a **promoter**. On the DNA strand, there is always a sequence of bases that the polymerase recognizes as an **initiation signal**, saying, in essence, "Start here." The promoter is unique to each

Template strand The strand of DNA that serves as the template during RNA synthesis

Coding strand The strand of DNA with a sequence that matches the RNA produced during transcription

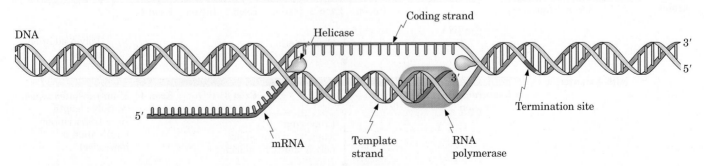

FIGURE 26.2 Transcription of a gene. The information in one DNA strand is transcribed to a strand of RNA. The termination site is the locus of termination of transcription.

FIGURE 26.3 Architecture of yeast RNA polymerase II. Transcription of DNA (helical structure) into RNA (red) is shown. The template strand of DNA is shown in blue and the coding strand in green. Transcription takes place in the clamp region of the active site, shown at center right. The jaws that hold DNA downstream of the active site are shown at lower left.

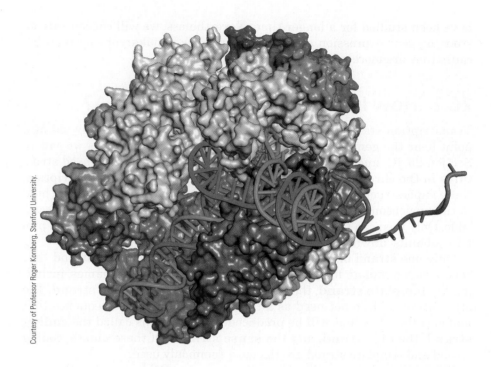

Courtesy of Professor Roger Kornberg, Stanford University.

gene. Besides unique nucleotide sequences, promoters contain **consensus sequences**, such as the TATA box, which gets its name from the sequence beginning TATAAT. A TATA box lies approximately 26 base pairs upstream—that is, before the beginning of the transcription process (see Figure 26.4). By convention, all sequences of DNA used to describe transcription are given from the perspective of the coding strand. TATA boxes are common to all eukaryotes. All three RNA polymerases interact with their promoter regions via **transcription factors** that are binding proteins.

Another type of control element is an enhancer, a DNA sequence that can be far removed from the promoter region. Enhancers also bind to transcription factors, enhancing transcription above the basal level that would be seen without such binding. Enhancers will be discussed in Section 26.6.

After initiation, the RNA polymerase zips up the complementary bases by forming a phosphate ester bond (Section 19.5) between each ribose and the next phosphate group. This process is called **elongation**.

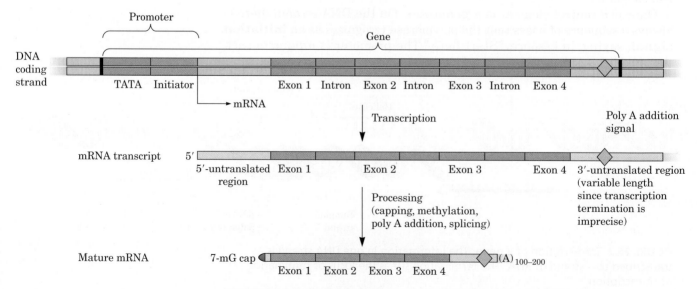

FIGURE 26.4 Organization and transcription of a split eukaryote gene.

At the end of the gene is a **termination sequence** that tells the enzyme, "Stop the transcription." The enzyme Pol II has two different forms. At the C-terminal domain, Pol II has serine and threonine repeats that can be phosphorylated. When Pol II starts the initiation, the enzyme is in its unphosphorylated form. Upon phosphorylation, it performs the elongation process. After termination of the transcription, Pol II is dephosphorylated by a phosphatase. In this manner, Pol II is constantly recycled between its initiation and elongation roles.

The enzyme synthesizes the mRNA molecule from the 5′ to the 3′ end (the zipper can move in only one direction). Because complementary nucleotide chains (RNA and DNA) run in opposite directions, however, the enzyme moves along the DNA template strand in the 3′ ⟶ 5′ direction (Figure 26.2). As the RNA is synthesized, it moves away from the DNA template, which then rewinds to the original double-helix form. Transfer RNA and ribosomal RNA are also synthesized on DNA templates in this manner.

The RNA products of transcription are not necessarily the functional RNAs. Previously, we have seen that in higher organisms, mRNA contains exons and introns (Section 25.5). To ensure that mRNA is functional, the transcribed product is capped at both ends. The 5′ end acquires a methylated guanine (7-mG cap), and the 3′ end has a poly-A tail that may contain 100 to 200 adenine residues. Once the two ends are capped, the introns are spliced out in a **post-transcription process** (Figure 26.4). Similarly, a transcribed tRNA must be trimmed and capped, and some of its nucleotides methylated, before it becomes functional tRNA. Functional rRNA also undergoes post-transcriptional methylation.

Example 26.1 DNA Polymerases

Polymerase II both initiates the transcription and performs the elongation. What are the two forms of the enzyme in these processes? What chemical bond formation occurs in the conversion between these two forms?

Solution

The phosphorylated form of Pol II performs the elongation, and the unphosphorylated form initiates the transcription. The chemical bond formed in the phosphorylation is the phosphoric ester between the —OH of serine and threonine residues of the enzyme and phosphoric acid.

Problem 26.1

DNA is highly condensed in the chromosomes. What is the sequence of events that enables the transcription of a gene to begin?

26.3 What Is the Role of RNA in Translation?

Translation is the process by which the genetic information preserved in the DNA and transcribed into the mRNA is converted to the language of proteins—that is, the amino acid sequence. Three types of RNA (mRNA, rRNA, and tRNA) participate in the process.

The synthesis of proteins takes place on the ribosomes (Section 25.4). These spheres dissociate into two parts—a larger and a smaller body. Each of these bodies contains rRNA and some polypeptide chains that act as

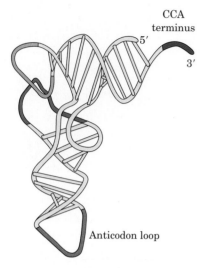

CCA
terminus

5'

3'

Anticodon loop

FIGURE 26.5 Three-dimensional structure of tRNA.

Codon The sequence of three nucleotides in messenger RNA that codes for a specific amino acid

Anticodon A sequence of three nucleotides on tRNA complementary to the codon in mRNA

Genetic code The sequence of triplets of nucleotides (codons) that determines the sequence of amino acids in a protein

enzymes, speeding up the synthesis. In higher organisms, including humans, the larger ribosomal body is called the 60S ribosome and the smaller one is called the 40S ribosome. The designation "S" refers to *Svedberg,* a measure of density used in centrifugation. In prokaryotes, the corresponding ribosomal subunits are called the 50S and the 30S, respectively. Messenger RNA is attached to the smaller ribosomal body and later joined by the larger body. Together, they form a unit on which the mRNA is stretched out. Triplets of bases on the mRNA are called **codons**. After the mRNA is attached to the ribosome in this way, the 20 amino acids are brought to the site, each carried by its own particular tRNA molecule.

The most important segments of the tRNA molecule are (1) the site to which enzymes attach the amino acids and (2) the recognition site. Figure 26.5 shows that the 3' terminus of the tRNA molecule is single-stranded; this end carries the amino acid.

As we have said, each tRNA is specific for one amino acid only. How does the body make sure that alanine, for example, attaches only to the one tRNA molecule that is specific for alanine? The answer is that each cell carries specific enzymes for this purpose. These **aminoacyl-tRNA synthetases** recognize specific tRNA molecules and amino acids. The enzyme then attaches the amino acid to the terminal group of the tRNA, forming an ester bond.

The second important segment of the tRNA molecule carries the **codon recognition site**, which is a sequence of three bases called an **anticodon** located at the opposite end of the molecule in the three-dimensional structure of tRNA (see Figure 26.5). This triplet of bases is complementary to the sequence of the codon and allows the tRNA to align with the mRNA. Thus, the mRNA and tRNA are antiparallel at the point of contact.

26.4 What Is the Genetic Code?

By 1961, it was apparent that the order of bases in a DNA molecule corresponds to the order of amino acids in a particular protein, but the code was unknown. Obviously, it could not be a one-to-one correspondence. There are only four bases, so if A coded for glycine, G for alanine, C for valine, and T for serine, there would be 16 amino acids that could not be coded.

In 1961, Marshall Nirenberg (1927–2010) and his co-workers attempted to break the code in a very ingenious way. They made a synthetic molecule of mRNA consisting of uracil bases only. They put this molecule into a cell-free system that synthesized proteins and then supplied the system with all 20 amino acids. The only polypeptide produced was a chain consisting solely of the amino acid phenylalanine. This experiment showed that the code for phenylalanine must be UUU or some other multiple of U.

A series of similar experiments by Nirenberg and other workers followed, and by 1967, the entire genetic code had been broken. *Each amino acid is coded for by a sequence of three bases,* called a *codon.* Table 26.1 shows the complete code.

The first important aspect of the **genetic code** is that it is almost universal. In virtually every organism, from a bacterium to an elephant to a human, the same sequence of three bases codes for the same amino acid. The universality of the genetic code implies that all living matter on Earth arose from the same primordial organisms. This finding is perhaps the strongest evidence supporting Darwin's theory of evolution.

Some exceptions to the genetic code in Table 26.1 occur in mitochondrial DNA. Because of that fact and other evidence, it is thought that the mitochondrion may have been an ancient free-living entity. During evolution,

Table 26.1 The Genetic Code

First Position (5'-end)	Second Position								Third Position (3'-end)
	U		C		A		G		
U	UUU	Phe	UCU	Ser	UAU	Tyr	UGU	Cys	U
	UUC	Phe	UCC	Ser	UAC	Tyr	UGC	Cys	C
	UUA	Leu	UCA	Ser	UAA	Stop	UGA	Stop	A
	UUG	Leu	UCG	Ser	UAG	Stop	UGG	Trp	G
C	CUU	Leu	CCU	Pro	CAU	His	CGU	Arg	U
	CUC	Leu	CCC	Pro	CAC	His	CGC	Arg	C
	CUA	Leu	CCA	Pro	CAA	Gln	CGA	Arg	A
	CUG	Leu	CCG	Pro	CAG	Gln	CGG	Arg	G
A	AUU	Ile	ACU	Thr	AAU	Asn	AGU	Ser	U
	AUC	Ile	ACC	Thr	AAC	Asn	AGC	Ser	C
	AUA	Ile	ACA	Thr	AAA	Lys	AGA	Arg	A
	AUG*	Met	ACG	Thr	AAG	Lys	AGG	Arg	G
G	GUU	Val	GCU	Ala	GAU	Asp	GGU	Gly	U
	GUC	Val	GCC	Ala	GAC	Asp	GGC	Gly	C
	GUA	Val	GCA	Ala	GAA	Glu	GGA	Gly	A
	GUG	Val	GCG	Ala	GAG	Glu	GGG	Gly	G

*AUG also serves as the principal initiation codon.

it developed a symbiotic relationship with eukaryotic cells. For example, some of the respiratory enzymes located on the cristae of the mitochondrion (see Section 27.2) are encoded in the mitochondrial DNA, and other members of the same respiratory chain are encoded in the nucleus of the eukaryotic cell.

There are 20 amino acids in proteins, but 64 possible combinations of four bases into triplets. All 64 codons (triplets) have been deciphered. Three of them—UAA, UAG, and UGA—are "stop signs." They terminate protein synthesis. The remaining 61 codons code for amino acids. Because there are only 20 amino acids, there must be more than one codon for each amino acid. Indeed, some amino acids have as many as six codons. Leucine, for example, is coded by UUA, UUG, CUU, CUC, CUA, and CUG.

Just as there are three stop signs in the code, there is also an initiation sign. The initiation sign is AUG, which is also the codon for the amino acid methionine. This means that in all protein synthesis, the first amino acid initially put into the protein is always methionine. Methionine can also be put into the middle of the chain.

Although all protein synthesis starts with methionine, most proteins in the body do not have a methionine residue at the N-terminus of the chain. In most cases, the initial methionine is removed by an enzyme before the polypeptide chain is completed. The code on the mRNA is always read in the 5' ⟶ 3' direction, and the first amino acid to be linked to the initial methionine is the N-terminal end of the translated polypeptide chain.

The genetic code is said to be continuous and unpunctuated. If the mRNA is AUGGGCCAA, then the AUG is one codon and specifies the first amino acid. The GGC is the second codon and specifies the second amino acid. The CCA is the third codon and specifies the third amino acid. There are no overlapping codons and no nucleotides interspersed.

> ## ⓌWL **Interactive Example 26.2** The Genetic Code
>
> Which amino acid is represented by the codon CGU? What is its anticodon?
>
> ### Solution
>
> Looking at Table 26.1, we find that CGU corresponds to arginine; the anticodon is GCA (read 3′ to 5′ to show how the codon and anticodon match up).
>
> ### Problem 26.2
>
> What are the codons for histidine? What are the anticodons?

26.5 How Is Protein Synthesized?

So far we have met the molecules that participate in protein synthesis (Section 26.3) and the dictionary of the translation, the genetic code. Now let us look at the actual mechanism by which the polypeptide chain is assembled.

There are four major stages in protein synthesis: activation, initiation, elongation, and termination. At each stage, a number of molecular entities participate in the process (Table 26.2). We will look specifically at prokaryotic translation because it has been studied longer, and we have more complete information on it. The details of eukaryotic translation are very similar, however.

Table 26.2 Molecular Components of Reactions at Four Stages of Protein Synthesis

Stage	Molecular Components
Activation	Amino acids, ATP, tRNAs, aminoacyl-tRNA synthetases
Initiation	fMet–tRNA^{fMet}, 30S ribosome, initiation factors, mRNA with Shine–Dalgarno sequence, 50S ribosome, GTP
Elongation	30S and 50S ribosomes, aminoacyl-tRNAs, elongation factors, mRNA, GTP
Termination	Release factors, GTP

A. Activation

Each amino acid is first activated by reacting with a molecule of ATP:

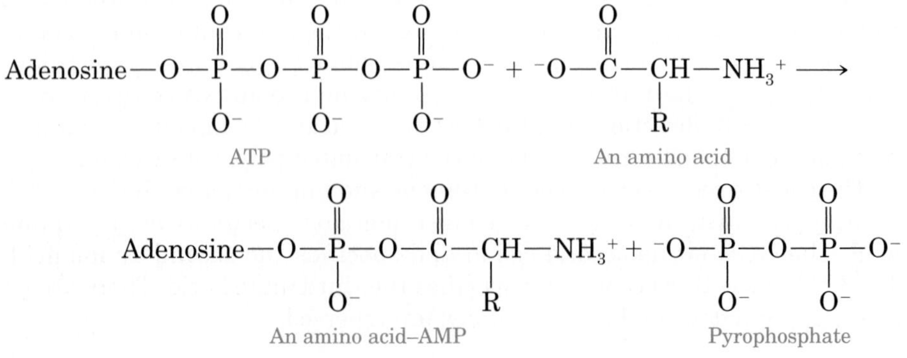

The activated amino acid is then bound to its own particular tRNA molecule with the aid of an enzyme (a synthetase) that is specific for that particular amino acid and that particular tRNA molecule:

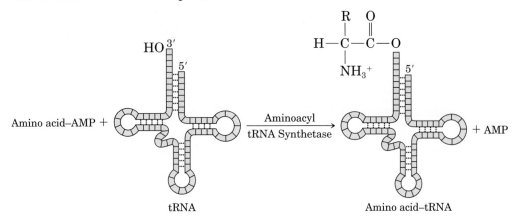

The different synthetases recognize their substrates by stretches of nucleotide sequences on the tRNA. The specific recognition by an enzyme, aminoacyl-tRNA synthetase, of its proper tRNA and amino acid is often referred to as the **second genetic code**. This step is very important because once the amino acid is on the tRNA, there is no other opportunity to check for the correct pairing. In other words, the anticodon of the tRNA will match up with its correct codon on the mRNA regardless of whether it is carrying the correct amino acid, so the aminoacyl-tRNA synthetases have to get it right.

B. Initiation

The initiation stage consists of three steps:

1. **Forming the pre-initiation complex** To initiate the protein synthesis, a unique tRNA is used, designated as **tRNA^{fMet}**. This tRNA carries a formylated methionine (fMet) residue, but it is used solely for the initiation step. It is attached to the 30S ribosomal body and forms the pre-initiation complex, along with GTP, [(Figure 26.6)(1)]. Just as in transcription, each step in translation is aided by a number of factors; these proteins are called **initiation factors**.

2. **Migration to mRNA** Next, the pre-initiation complex binds to the mRNA (2). The ribosome is aligned on the mRNA by recognizing a special RNA sequence called the **Shine-Dalgarno** sequence, which is complementary to a sequence on the 30S ribosomal subunit. The anticodon of the fMet–tRNA^{fMet}, UAC, lines up against the start codon, AUG.

3. **Forming the full ribosomal complex** The 50S ribosomal body joins the 30S ribosomal complex (3). The complete ribosome carries three sites. The one shown in the middle in Figure 26.6 is called the **P site**, because the growing peptide chain will bind there. The one next to it on the right is called the **A (acceptor) site**, because it accepts the incoming tRNA bringing the next amino acid. As the full initiation complex is completed, the initiation factors dissociate and the GTP is hydrolyzed to GDP.

C. Elongation

1. **Binding to the A site** At this point, the A site is vacant and each of the aminoacycl-tRNA molecules can try to fit itself in. However, only one of the tRNAs carries the right anticodon that corresponds to the next codon on the mRNA. This is an alanine tRNA in Figure 26.6.

FIGURE 26.6 The formation of an initiation complex. The 30S ribosomal subunit binds to mRNA and the fMet-tRNA^fMet in the presence of GTP and the three initiation factors, forming the 30S initiation complex (Step 1). The 50S ribosomal subunit is added, forming the full initiation complex (Step 2).

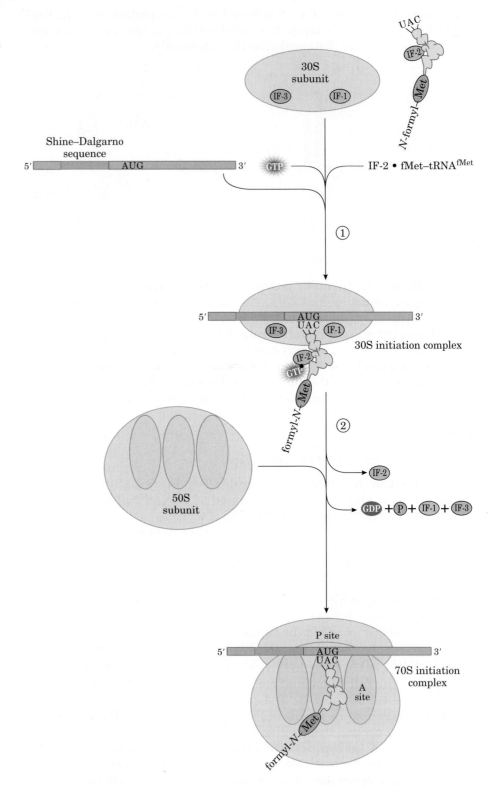

The binding of this tRNA to the A site takes place with the aid of proteins called **elongation factors** and GTP [Figure 26.7 (2)].

2. **Forming the first peptide bond** At the A site, the new amino acid, alanine (Ala), is linked to the fMet in a peptide bond by the enzyme **peptidyl transferase**. The empty tRNA remains on the P site [Figure 26.7 (3)].

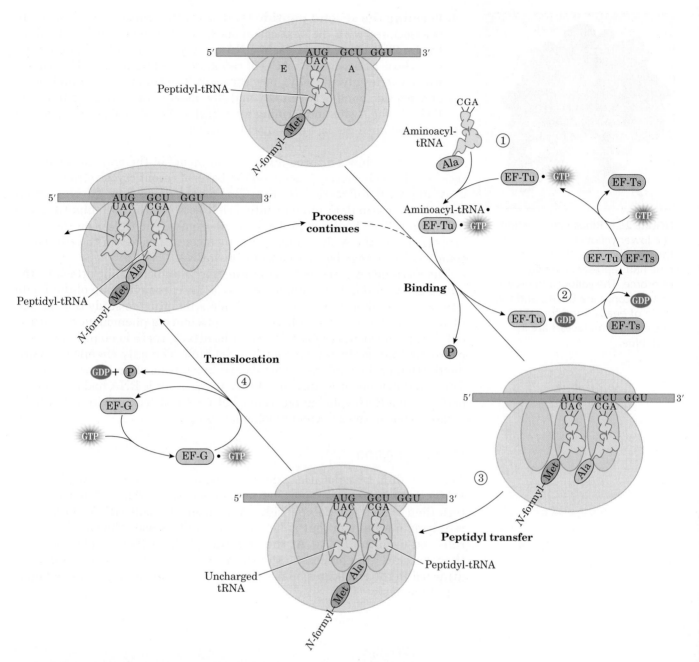

FIGURE 26.7 The steps in chain elongation. (1) An aminoacyl-tRNA is bound to the A site on the ribosome. Elongation factors and GTP are required. The P site on the ribosome is already occupied. (2) Elongation factors are recycled to prepare another incoming tRNA, and GTP is hydrolyzed. (3) The peptide bond is formed, leaving an uncharged tRNA at the P site. (4) In the translocation step, the uncharged tRNA is pushed into the E site and more GTP is hydrolyzed. The A site is now over the next codon on the mRNA.

3. Translocation In the next phase of elongation, the whole ribosome moves one codon along the mRNA. Simultaneously with this move, the dipeptide is **translocated** from the A site to the P site (4). The empty tRNA is moved to the E site. When this cycle occurs one more time, the empty tRNA will be ejected and go back to the pool of tRNA that is available for activation with an amino acid.

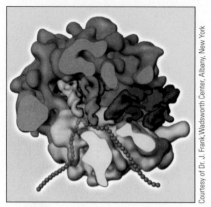

Courtesy of Dr. J. Frank, Wadsworth Center, Albany, New York

FIGURE 26.8 Ribosome in action. The lower yellow half represents the 30S ribosome; the upper blue half represents the 50S ribosome. The yellow and green twisted cones are tRNAs, and the chain of beads stand for mRNA. The elongation factors are in dark blue.

4. Forming the second peptide bond After the translocation, the A site is associated with the next codon on the mRNA, which is 5′ GGU 3′ in Figure 26.7. Once again, each tRNA can try to fit itself in, but only the one whose anticodon is 5′ACC 3′ can align itself with GGU. This tRNA, which carries glycine (Gly), now comes in. The transferase establishes a new peptide bond between Gly and Ala, moving the dipeptide from the P site to the A site and forming a tripeptide. These elongation steps are repeated until the last amino acid is attached.

Figure 26.8 shows a three-dimensional model of the translational process, which has been constructed on the basis of recent cryoelectron microscopy and X-ray diffraction studies. This model shows how the elongation factor proteins (in dark blue) fit into a cleft between the 50S (blue) and the 30S (pale yellow) bodies of prokaryotic ribosomes. The tRNAs on the P site (green) and on the A site (yellow) occupy a central cavity in the ribosomal complex. The orange beads represent the mRNA.

The mechanism of peptide bond formation is a nucleophilic attack by the amino group of the A site amino acid upon the carbonyl group of the P site amino acid, as shown in Figure 26.9. While attempting to study this mechanism in detail, researchers discovered a fascinating phenomenon. It turns out that in the vicinity of the nucleophilic attack, there is no protein in the ribosome that could catalyze such a reaction. The only chemical groups nearby that could catalyze a reaction are on a purine of the ribosomal RNA. Thus, the ribosome is a ribozyme. Previously, catalytic RNA had been found only in some RNA splicing reactions, but here is a scenario in which RNA catalyzes one of the principal reactions of life.

D. Termination

After the final translocation, the next codon reads "stop" (UAA, UGA, or UAG). At this point, no more amino acids can be added. Releasing factors then cleave the polypeptide chain from the last tRNA via a GTP-requiring mechanism that is not yet fully understood. The tRNA itself is released from the P site. At the end, the whole mRNA is released from the ribosome. This process is shown in Figure 26.10. While the mRNA is attached to the ribosomes, many polypeptide chains are synthesized on it simultaneously.

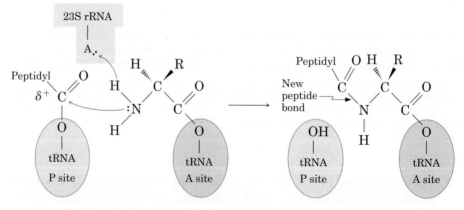

FIGURE 26.9 Peptide bond formation in protein synthesis. Nucleophilic attack by the amino group of the A site aminoacyl-tRNA on the carbonyl carbon of the P site peptidyl-tRNA is facilitated when a purine moiety of the rRNA abstracts a proton.

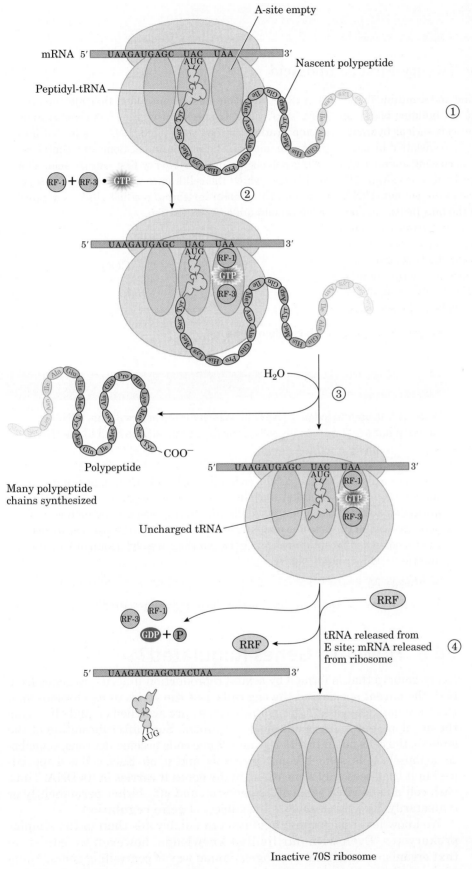

FIGURE 26.10 The events in peptide chain termination. As the ribosome moves along the mRNA, it encounters a stop codon, such as UAA (Step 1). Release factors and GTP bind to the A site (Step 2). The peptide is hydrolyzed from the tRNA (Step 3). Finally, the entire complex dissociates, and the ribosome, mRNA, and other factors can be recycled (Step 4).

Chemical Connections 26A

Breaking the Dogma: The Twenty-First Amino Acid

Many amino acids, such as citrulline and ornithine found in the urea cycle (Chapter 28), are not building blocks of proteins. Other nonstandard amino acids such as hydroxyproline (Chapter 22) are formed after translation by post-translational modification. When discussing amino acids and translation in the past, the magic number was always 20. That is, only 20 standard amino acids were put onto tRNA molecules for protein synthesis. In the late 1980s, another amino acid was found in proteins from eukaryotes and prokaryotes alike, including humans. It is selenocysteine, a cysteine residue that has the sulfur replaced by selenium.

Selenocysteine is formed by placing a serine onto a special tRNA molecule called tRNA^{sec}. Once bound, the oxygen in the serine side chain is replaced by selenium.

This tRNA molecule has an anticodon that matches the UGA "stop" codon. In special cases, the UGA is not read as a "stop"; rather, the selenocysteine-tRNA^{sec} is loaded into the A site and translation continues. Some are, therefore, calling selenocysteine the twenty-first amino acid. The methods by which the cell knows when to put selenocysteine into the protein instead of reading UGA as a "stop" codon remain under investigation.

$$H-Se-CH_2-\underset{\underset{NH_3^+}{|}}{\overset{\overset{H}{|}}{C}}-COO^-$$

Selenocysteine

Test your knowledge with Problem 26.58.

Example 26.3 Translation

A tRNA has an anticodon, 5′ AAG 3′. Which amino acid will this tRNA carry? What are the steps necessary for the amino acid to bind to the tRNA?

Solution

The amino acid is leucine, as the codon is 5′ CUU 3′. Remember that sequences are read from left to right as 5′ $\longrightarrow$ 3′, so you have to flip the anticodon around to see how it would bind to the codon. Leucine has to be activated by ATP. A specific enzyme, leucine-tRNA synthetase, catalyzes the carboxyl ester bond formation between the carboxyl group of leucine and the —OH group of tRNA.

Problem 26.3

What are the reactants in the reaction forming valine-tRNA?

26.6 How Are Genes Regulated?

Every embryo that is formed by sexual reproduction inherits its genes from both the parent sperm and the egg cells. But the genes in its chromosomal DNA are not active all the time. Rather, they are switched on and off during the development and growth of the organism. Soon after formation of the embryo, the cells begin to differentiate. Some cells become neurons, some become muscle cells, some become liver cells, and so on. Each cell is a specialized unit that uses only some of the many genes it carries in its DNA. Thus, each cell must switch some of its genes on and off—either permanently or temporarily. How this is done is the subject of **gene regulation**.

We know less about gene regulation in eukaryotes than in the simpler prokaryotes. Even with our limited knowledge, however, we can state that organisms do not have a single, unique way of controlling genes. Many gene regulations occur at the **transcriptional level** (DNA $\longrightarrow$ RNA). Others operate at the **translational level** (mRNA $\longrightarrow$ protein). A few of these processes are listed here as examples.

Gene regulation The control process by which the expression of a gene is turned on or off. Because RNA synthesis proceeds in one direction (5′ $\longrightarrow$ 3′), the gene (DNA) to be transcribed runs from 3′ to 5′. Thus, the control sites are in front of, or upstream of, the 3′ end of the structural gene.

Viruses

Nucleic acids are essential for life as we know it. No living thing can exist without them because they carry the information necessary to make protein molecules. The smallest forms of life, the viruses, consist only of a molecule of nucleic acid surrounded by a "coat" of protein molecules. In some viruses, the nucleic acid is DNA; in others, it is RNA. No virus has both. Whether viruses can be considered a true form of life became a topic of debate recently. In the summer of 2002, a group of scientists at the State University of New York, Stony Brook, reported that they had synthesized the polymyelitis virus in the laboratory from fragments of DNA. This new "synthetic" virus caused the same polio symptoms and death as the wild virus.

The shapes and sizes of viruses vary greatly, as shown in the figure. Because their structures are so simple, viruses cannot reproduce themselves in the absence of other organisms. They carry DNA or RNA but do not have the nucleotides, enzymes, amino acids, and other molecules necessary to replicate their nucleic acid (Section 25.6) or to synthesize proteins (Section 26.5). Instead, viruses invade the cells of other organisms and cause those cells (the hosts) to do these tasks for them. Typically, the protein coat of a virus remains outside the host cell, attached to the cell membrane, while the DNA or RNA is pushed inside. Once the viral nucleic acid is inside the cell, the cell stops replicating its own DNA and making its own proteins. Instead, it replicates the viral nucleic acid and synthesizes the viral protein according to the instructions on the viral nucleic acid. One host cell can make many copies of the virus.

In many cases, the cell bursts when a large number of new viruses have been synthesized, sending the new viruses out into the intercellular material, where they can infect other cells. This kind of process causes the host organism to get sick and perhaps to die. Among the many human diseases caused by viruses are measles, hepatitis, mumps, influenza, the common cold, rabies, and smallpox. There is no cure for most viral diseases. Antibiotics, which can kill bacteria, have no effect on viruses. So far, the best defense against these diseases has been immunization (Chemical Connections 31B), which under the proper circumstances can work spectacularly well. Smallpox, once one of the most dreaded diseases, has been eradicated from this planet by many years of vaccination, and comprehensive programs of vaccination against such diseases as polio and measles have greatly reduced the incidence of these diseases.

Lately, a number of antiviral agents have been developed. They completely stop the reproduction of viral nucleic acids (DNA or RNA) inside infected cells without preventing the DNA of normal cells from replicating. One such drug is called vidarabine, or Ara-A, and is sold under the trade name Vira-A.

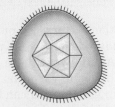

Herpes virus

Influenza virus

Poliomyelitis virus

The shape of three viruses. (Illustration by Irving Geis, from *Scientific American*, January 1963. Rights owned by Howard Hughes Medical Institute. Not to be reproduced without permission.)

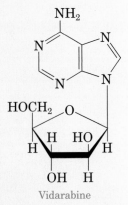

Vidarabine

Antiviral agents often act like anticancer drugs in that they have structures similar to one of the nucleotides necessary for the synthesis of nucleic acids. Vidarabine is the same as adenosine, except that the sugar is arabinose instead of ribose. Vidarabine is used to fight a life-threatening viral illness, herpes encephalitis. It is also effective in neonatal herpes infection and chickenpox. However, like many other anticancer and antiviral drugs, vidarabine is toxic, causing nausea and diarrhea. In some cases, it has caused chromosomal damage.

Test your knowledge with Problems 26.59 and 26.60.

A. Control at the Transcriptional Level

In eukaryotes, transcription is regulated by three entities: promoters, enhancers, and response elements.

1. Promoters of a gene are located adjacent to the transcription site. They are defined by an initiator and conserved sequences, such as a TATA box (see also Section 26.2 and Figure 26.4), or one or more copies of other sequences, such as the GGGCGG sequence called a GC box. In eukaryotes, the enzyme RNA polymerase has little affinity for binding to DNA. Instead, different transcription factors, or binding proteins, bind to the different modules of the promoter.

   There are two basic types of transcription factors. The first is called a **general transcription factor** (**GTF**). These proteins form a complex with RNA polymerase and the DNA and help to position the RNA polymerase correctly and stimulate the initiation of transcription. For the transcription of genes that will lead to mRNA (that is, Pol II transcription), there are six GTFs, all named as TFII and then a letter, for Pol II transcription factor. All of these transcription factors are necessary to establish the initiation of transcription. As can be seen in Figure 26.11, the events in the initiation of Pol II transcription are very complicated. Six transcription factors must bind to the DNA and RNA polymerase to initiate transcription. They first form the **pre-initiation complex**. The critical event in starting the transcription is the conversion to the **open complex**, which involves the phosphorylation of the C-terminal end of the RNA polymerase. Only when the open complex has formed can transcription take place. During elongation, three transcription factors (B, E, and H) are released. Transcription factor F remains bound to Pol II with D bound to the TATA box. Only factor F continues with the polymerase.

   With the aid of these transcription factors, the promoter functions to control the transcription on a steady, normal level. Transcription factors may allow the synthesis of mRNA (and from there the target protein) to vary by a factor of 1 million. This wide variation in eukaryotic cells is exemplified by the α-A-crystallin gene, which can be expressed in the lens of the eye at a rate a millionfold higher than that in the liver cell of the same organism.

2. Another group of transcription factors speeds up the transcription process by binding to DNA sequences that may be located several thousand nucleotides away from the transcription site. These sequences are known as **enhancers**. To stimulate transcription, an enhancer is brought to the vicinity of the promoter by the formation of a loop. Figure 26.12 shows how the transcription factor binds to the enhancer element and forms a bridge to the basal transcription unit. This complex then allows the RNA polymerase II to speed up the transcription when higher-than-normal production of proteins is needed.

   Other DNA sequences bind transcription factors but have the opposite effect—they slow down transcription. These are called **silencers**.

3. The third type of transcription control involves a type of enhancer called **response elements**. These enhancers are activated by their transcription factors in response to an outside stimulus. The stimulus may be heat shock, heavy metal toxicity, or simply a hormonal signal, such as the binding of a steroid hormone to its receptor. The response element of steroids is in front of, and 260 base pairs upstream from, the starting point of transcription. Only the receptor with the bound steroid hormone can interact with its response element, thereby initiating transcription.

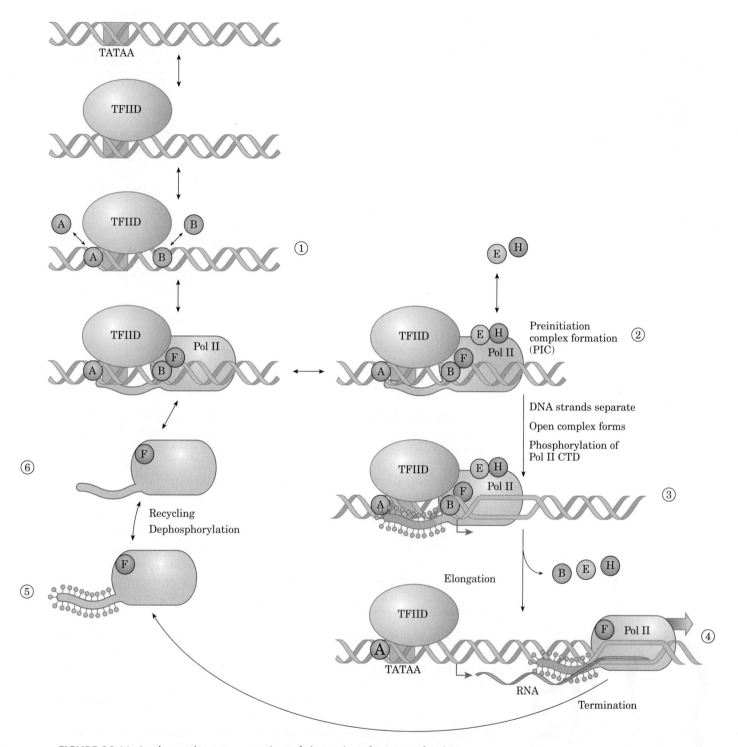

FIGURE 26.11 A schematic representation of the order of events of Pol II transcription. General Transcription Factor TFIID binds to the TATA box on the DNA and recruits TFIIA and TFIIB (Step 1). RNA Polymerase II carrying TFIIF binds to the DNA, followed by TFIIE and TFIIH to form the pre-initiation complex (PIC) (Step 2). The C-terminal domain of Pol II is phosphorylated and the DNA strands are separated to form the open complex (Step 3). TFIIB, TFIIE, and TFIIH are released as the polymerase synthesizes RNA in the process of elongation (Step 4). Transcription is terminated when the mRNA is complete and the Pol II is released (Step 5). The Pol II is dephosphorylated and is ready to be recycled for another round of transcription (Step 6).

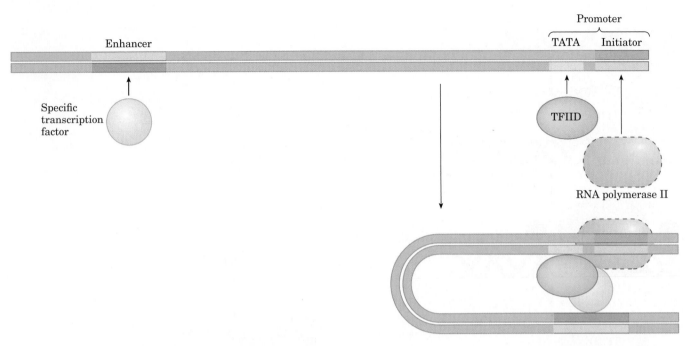

FIGURE 26.12 DNA looping brings enhancers in contact with transcription factors and RNA polymerase.

The difference between an enhancer and a response element is largely a matter of our own understanding of the system. We call something a response element when we understand the bigger picture of how controlling the gene is related to a pattern of metabolism. Many response elements may be controlling a particular process, and a given gene may be under the control of more than one response element.

4. Transcription does not occur at the same rate throughout the cell's entire life cycle. Instead, it is accelerated or slowed down as the need arises. The signal to speed up transcription may originate from outside the cell. One such signal, in the GTP–adenylate cyclase–cAMP pathway (Section 24.5B), produces **phosphorylated protein kinase**. This enzyme enters the nucleus, where it phosphorylates transcription factors, which aid in the transcription cascade.

How do these transcription factors find the specific gene control sequences into which they fit, and how do they bind to them? The interaction between the protein and DNA involves nonspecific electrostatic interactions (positive ions attracting negative ions and repelling other positive ions) as well as more specific hydrogen bonding. The transcription factors find their targeted sites by twisting their protein chains so that a certain amino acid sequence is present at the surface. One such conformational twist is provided by **metal-binding fingers** (Figure 26.13). These finger shapes are created by ions, which form covalent bonds with the amino acid side chains of the protein.

The zinc fingers interact with specific DNA (or sometimes RNA) sequences. The recognition comes by hydrogen bonding between a nucleotide (for example, guanine) and the side chain of a specific amino acid (for example, arginine) on the zinc finger. Zinc fingers allow the proteins to bind in the major groove of DNA, as shown in Figure 26.14.

Besides metal-binding fingers, at least two other prominent transcription factors exist: **helix-turn-helix** and **leucine zipper**.

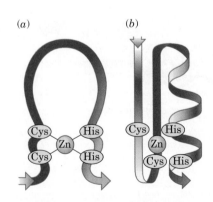

FIGURE 26.13 Cys$_2$His$_2$ zinc finger motifs. (*a*) The coordination between zinc and cysteine and histidine residues. (*b*) The secondary structure. (Adapted from Evans R. M. and Hollenberg, S. M., *Cell, 52:1, 1988,* Figure 1).

B. Control on the Post-transcriptional Level

In the spring of 2000, scientists were eagerly awaiting the results of the Human Genome Project and an accurate count of the number of genes in the human genome. The odds-on favorite would have been 100,000 to 150,000 genes. After all, it was known that humans produce 90,000 different proteins. The dogma stated that "one gene leads to one mRNA leads to one protein." The only exceptions to this rule were thought to occur in the production of antibodies and other immunoglobulin-based proteins. These proteins were known to undergo a type of post-transcriptional modification called alternative splicing, whereby the primary mRNA transcript can be spliced in different ways to give multiple mature mRNAs and therefore multiple proteins.

It came as a big shock, therefore, when the data revealed that humans have about 30,000 genes, a number close to that of the roundworm or corn. If 30,000 genes can lead to 90,000 proteins, there must be far more alternative splicing to account for it. Scientists now believe that splicing RNA in different ways is a very important process that leads to the differences in species that are otherwise similar. Chimpanzees and humans, for example, share 96% of their DNA. They also produce very similar protein complements. However, significant differences have been found in some tissues, most notably the brain, where certain human genes are more active and others generate different proteins by alternative splicing.

Figure 26.15 summarizes the various ways that alternative splicing can produce many different proteins. Exons may be included in all products, or they may be present on only some of them. Different splicing sites can appear either on the 5′ or 3′ side. In some cases, introns may even be retained in the final product.

Alternative splicing provides another powerful technique for controlling gene regulation. Within the same cell or the same organism, different genes can be spliced in different ways at different times, controlling what the eventual products of the gene are.

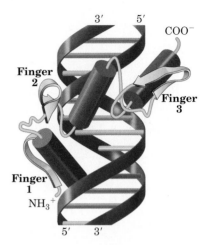

FIGURE 26.14 Zinc finger proteins follow the major groove of DNA. (Adapted with permission from Pavletich N. and Pabo C. O., *Science* 252: 809, 1991 Figure 2. Copyright © 1991 AAAS.)

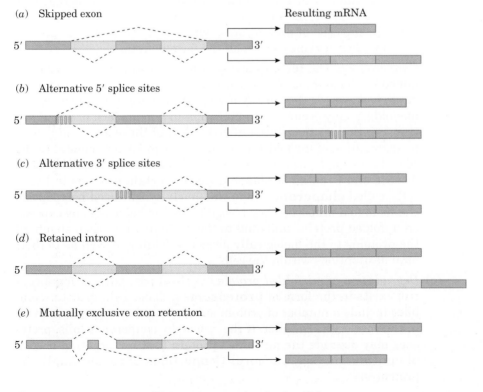

(a) Skipped exon

Resulting mRNA

(b) Alternative 5′ splice sites

(c) Alternative 3′ splice sites

(d) Retained intron

(e) Mutually exclusive exon retention

■ Exon always spliced in   ■ Exon alternatively spliced   ■ Intron

FIGURE 26.15 Alternative splicing. A gene's primary transcript can be edited in several different ways where splicing activity is indicated by dashed lines. An exon may be left out (*a*). Splicing machinery may recognize alternative 5′ splice sites for an intron (*b*) or alternative 3′ splice sites (*c*). An intron may be retained in the final mRNA transcript (*d*). Exons may be retained on a mutually exclusive basis (*e*). (*Scientific American*).

C. Control on the Translational Level

During translation, a number of mechanisms ensure quality control.

1. **The specificity of a tRNA for its unique amino acid** First, the attachment of the proper amino acid to the proper tRNA must be achieved. The enzyme that catalyzes this reaction, aminoacyl-tRNA synthetase (AARS), is specific for each amino acid. For those amino acids that have more than one type of tRNA, the same synthetase catalyzes the reaction for all of the tRNA types for that amino acid. The AARS enzymes recognize their tRNAs by specific nucleotide sequences. Furthermore, the active site of the enzyme has two **sieving portions**. For example, in isoleucyl-tRNA synthetases, the first sieve excludes any amino acids that are larger than isoleucine. If a similar amino acid such as valine, which is smaller than isoleucine, arrives at the active site, the second sieve eliminates it. The second sieving site is thus a "proofreading" site.

2. **Recognition of the stop codon** Another quality-control measure occurs at protein termination. The stop codons must be recognized by release factors, leading to the release of the polypeptide chain and allowing the recycling of the ribosomes. Otherwise, a longer polypeptide chain may be toxic. The release factor combines with GTP and binds to the ribosomal A site when that site is occupied by the termination codon. Both the GTP and the peptidyl-tRNA ester bond are hydrolyzed. This hydrolysis releases the polypeptide chain and the deacylated tRNA. Finally, the ribosome dissociates from the mRNA. As we saw in Chemical Connections 26A, sometimes the stop codon is used to continue translating by inserting a very uncommon amino acid, such as selenocysteine.

3. **Post-translational Controls**

   (a) *Removal of methionine.* In most proteins, the methionine residue at the N-terminus, which was added in the initiation step, is removed. A special enzyme, methionine aminopeptidase, cleaves the peptide bond. In the case of prokaryotes, when the N-terminus is methionine, another enzyme cleaves off the formyl group.

   (b) *Chaperoning.* The tertiary structure of a protein is largely determined by the amino acid sequence (primary structure). Proteins begin to fold even as they are synthesized on ribosomes. Nevertheless, misfolding may occur due to mutation in a gene, lack of fidelity in transcription, or translational errors. All of these errors may lead to aggregation of misfolded proteins that can be detrimental to the cell—for example, as seen in amyloid diseases such as Alzheimer's disease and Creutzfeldt-Jakob disease. Certain proteins in living cells, called **chaperones**, help the newly synthesized polypeptide chains to fold properly. They recognize hydrophobic regions exposed on unfolded proteins and bind to them. Chaperones then shepherd the proteins to the biologically desirable folding as well as to their local destinations within the cell.

   (c) *Degradation of misfolded proteins.* A third post-translational control exists in the form of **proteasomes**. These cylindrical assemblies include a number of protein subunits with proteolytic activity in the core of the cylinder. If the rescue by chaperones fails, proteases may degrade the misfolded protein first by targeting it with ubiquitination (see Chemical Connections 28E) and finally by proteolysis.

26.7 What Are Mutations?

In Section 25.6, we saw that the base-pairing mechanism provides an almost perfect way to copy a DNA molecule during replication. The key word here is "almost." No machine, not even the copying mechanism of DNA replication, is totally without error. It has been estimated that, on average, one error occurs for every 10^{10} bases (that is, one in 10 billion). An error in the copying of a sequence of bases is called a **mutation**. Mutations can occur during replication. Base errors can also occur during transcription (a noninheritable error).

These errors may have widely varying consequences. For example, the codon for valine in mRNA can be GUA, GUG, GUC, or GUU. In DNA, these codons correspond to GTA, GTG, GTC, and GTT, respectively. Assume that the original codon in the DNA is GTA. If a mistake is made during replication and GTA is spelled as GTG in the copy, there will be no harmful mutation. Instead, when a protein is synthesized, the GTG will appear on the mRNA as GUG, which also codes for valine. Therefore, although a mutation has occurred, the same protein is manufactured.

All sequences on DNA are given as coding strand sequences. Thus, the codon, which is on the mRNA, has the same sequence as the coding strand DNA, except that T is replaced with U.

Chemical Connections 26C

Mutations and Biochemical Evolution

We can trace the genetic relationship of different species through the variability of their amino acid sequences in different proteins. For example, the blood of all mammals contains hemoglobin, but the amino acid sequences of the hemoglobins are not identical. In the table below, we see that the first 10 amino acids in the β-globin of humans and gorillas are exactly the same. In fact, there is only one amino acid difference, at position 104, between us and apes. The β-globin of the pig differs from ours at 10 positions, of which 2 are in the N-terminal decapeptide. That of the horse differs from ours in 26 positions, of which 4 are in this decapeptide. β-Globin seems to have gone through many mutations during the evolutionary process, because only 26 of the 146 sites are invariant—that is, exactly the same in all species studied so far.

The relationship between different species can also be established by finding similarities in their mRNA primary structures. Because the mutations actually occurred on the original DNA molecule and then were perpetuated in the progeny by the mutant DNA, it is instructive to learn how a point mutation may occur in different species. Looking at position 4 of the β-globin molecule, we see

a change from serine to threonine. The code for serine is AGU or AGC, whereas that for threonine is ACU or ACC (Table 26.1). Thus, a change from G to C in the second position of the codon created the divergence between the β-globins of humans and horses. The genes of closely related species, such as humans and apes, have very similar primary structures, presumably because these two species diverged on the evolutionary tree only recently. In contrast, species far removed from each other diverged long ago and have undergone more mutations, which show up as differences in the primary structures of their DNA, mRNA, and consequently proteins.

The *number* of amino acid substitutions is significant in the evolutionary process caused by mutation, but the *kind* of substitution is even more important. If the substitution involves an amino acid with physicochemical properties similar to those of the amino acid in the ancestor protein, the mutation is most probably viable. For example, in human and gorilla β-globin, position 4 is occupied by threonine, but it is occupied by serine in the pig and horse. Both amino acids provide an —OH carrying side chain.

				Amino Acid Sequence of the N-Terminal Decapeptides of β-Globin in Different Species					
				Position					
Species	**1**	**2**	**3**	**4**	**5**	**6**	**7**	**8**	**9**	**10**
Human	Val	His	Leu	Thr	Pro	Glu	Glu	Lys	Ser	Ala
Gorilla	Val	His	Leu	Thr	Pro	Glu	Glu	Lys	Ser	Ala
Pig	Val	His	Leu	Ser	Ala	Glu	Glu	Lys	Ser	Ala
Horse	Val	Glu	Leu	Ser	Gly	Glu	Glu	Lys	Ala	Ala

Test your knowledge with Problem 26.61.

Chemical Connections 26D

Silent Mutations

A silent mutation is a mutation that changes the DNA but not the amino acid incorporated. For example, if the DNA coding strand has a TTC, the mRNA will be UUC and it will code for phenylalanine. If a mutation in the DNA changes the sequence to TTT, the DNA has undergone a silent mutation because in the resultant mRNA, UUC and UUU both code for the same amino acid. At least that is what scientists believed for decades. Recent evidence, however, has shown that this is not always true. Researchers at the National Cancer Institute were studying a gene called *MDR1*, which is named for its association with multiple drug resistance in tumor cells. They had sequences of this gene and knew that there were some common silent mutations. Interestingly, they discovered that there was a response to silent mutations of this gene that influenced

patients' response to certain drugs. A silent mutation leading to an observable change was striking, as a silent mutation should have no effect on the final product.

Apparently, not all codons are translated equally. Different codons may require alternate versions of the tRNA for a particular amino acid. Even though the amino acid incorporated is the same, the pace with which the ribosome is able to incorporate the amino acid differs depending on which codon it is. As shown below in the figure, translation kinetics can affect the form of the final protein. If the wild type codon is used, translation proceeds normally and produces the normal conformation of the protein. However, if a silent mutation changes the pace of the movement of the ribosome, folding differences will result in the creation of an abnormal protein conformation.

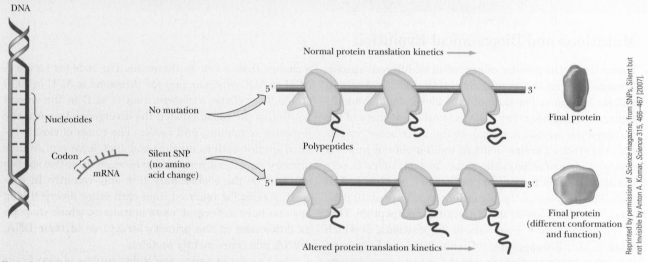

Reprinted by permission of *Science* magazine, from SNPs, Silent but not Invisible by Anton A. Komar, *Science* 315, 466–467 [2007].

Translation kinetics and protein folding. Unaffected translation kinetics results in a correctly folded protein. Abnormal kinetics, caused by the ribosome moving faster or slower through certain mRNA regions, can produce a different final protein conformation. Abnormal kinetics may arise from a silent single nucleotide polymorphism (SNP) in a gene that creates a codon synonymous to the wild type codon. However, this synonymous codon substitution may lead to different kinetics of mRNA translation, thus yielding a protein with a different final structure and function.

Test your knowledge with Problems 26.62–26.66.

Now assume that the original sequence in the gene's DNA is GAA, which will also be GAA in the mRNA and which codes for glutamic acid. If a mutation occurs during replication and GAA becomes TAA, a very serious mutation will have occurred. The TAA on the DNA coding strand would be UAA on the mRNA, which does not code for any amino acid; rather, it is a stop signal. Thus, instead of continuing to build a protein chain with glutamic acid, the synthesis will stop altogether. An important protein will not be manufactured, or at least will be manufactured incorrectly, and the organism may be sick or even die. As we saw in Chemical Connections 22D, sickle cell anemia is caused by a single base mutation that causes a glutamic acid to be replaced with valine.

Ionizing radiation (X-rays, ultraviolet light, gamma rays) can cause mutations. Furthermore, a large number of organic compounds can induce mutations by reacting with DNA. Such compounds are called **mutagens**. Many changes caused by radiation and mutagens do not become mutations because

Chemical Connections 26E

p53: A Central Tumor Suppressor Protein

There are some 36 known **tumor suppressor genes**, the products of which are proteins that control cell growth. None of them is more important than the protein with a molar mass of 53,000, simply named **p53**. This protein responds to a variety of cellular stresses, including DNA damage, lack of oxygen (hypoxia), and aberrantly activated oncogenes. In about 40% of all cancer cases, the tumor contains p53 that underwent mutation. Mutated p53 protein can be found in 55% of lung cancers; about half of all colon and rectal cancers; and some 40% of lymphomas, stomach cancers, and pancreatic cancers. In addition, in one-third of all soft tissue sarcomas, p53 is inactive, even though it did not undergo mutation.

These statistics indicate that the normal function of the p53 protein is to suppress tumor growth. When it is mutated or otherwise not present in sufficient or active form, it is unable to perform this protective function and cancer spreads. The p53 protein binds to specific sequences of double-stranded DNA. When X-rays or γ-rays damage DNA, an increase in p53 protein concentration is observed. The increased binding of p53 controls the cell cycle by holding it between cell division and DNA replication. The time gained in this arrested cell cycle allows the DNA to repair its damage. If that fails, the p53 protein triggers apoptosis, the programmed death of the injured cell.

Recently, it was reported that p53 performs a finely tuned function in cells. It suppresses tumor growth, but if p53 is overexpressed (that is, its concentration is high), it contributes to the premature aging of the organism.

In those conditions, p53 arrests the cell cycle not only of damaged cells, but also of stem cells. These cells normally differentiate into various types (muscle cells, nerve cells, and so forth) and replace those cells that die with aging. Excess p53 slows down this differentiation. In mice, an abundance of p53 protein made the animal cancer-free but at a cost: The mice lost weight and muscle, their bones became brittle, and their wounds took longer to heal. All in all, their average life span was 20% shorter than that of normal mice.

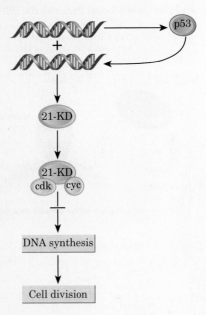

Test your knowledge with Problems 26.67 and 26.68.

the cell has repair mechanisms such as nucleotide excision repair (NER), which can prevent mutations by cutting out damaged areas and resynthesizing them (see Section 25.7 for a description of repair mechanisms). Despite this defense mechanism, certain errors in copying that result in mutations do slip by. Many compounds (both synthetic and natural) are mutagens, and some can cause cancer when introduced into the body. These substances are called **carcinogens** (Chemical Connections 13B). One of the main tasks of the U.S. Food and Drug Administration and the Environmental Protection Agency is to identify carcinogens and eliminate them from our food, drugs, and environment. Even though most carcinogens are mutagens, the reverse is not true.

Not all mutations are harmful. Some are beneficial because they enhance the survival rate of the species. For example, mutation is used to develop new strains of plants that can withstand pests.

If a mutation is harmful, it results in an inborn genetic disease. This condition may be carried as a recessive gene from generation to generation with no individual demonstrating the symptoms of the disease. When both parents carry recessive mutations, however, an offspring has a 25% chance of inheriting the disease. If the defective gene is dominant, on the other hand, every carrier will develop symptoms.

Chemical Connections 26F

Human Diversity and Transcription Factors

Research on the human genome showed that there was not enough diversity in the base genome of structural genes to explain the vast differences between species or even between individuals of a species. However, when we add in genes for transcription factors and the DNA sequences of enhancers and silencers, we have a much wider range of possible differences. One example can be seen in the human population. There is a particular protein usually found on the surface of red blood cells called Duffy. The DNA that encodes Duffy is regulated by a specific enhancer. It has been found that nearly 100% of West

Africans lack the Duffy proteins on their red blood cells. The lack of the Duffy protein has been traced to a single nucleotide mutation in the enhancer region of the Duffy gene. It turns out that the Duffy protein is a docking site for malaria-carrying parasites, and red blood cells lacking Duffy are resistant to malaria. This is an example of human evolution in progress. There would be significant evolutionary pressure in favor of the mutation that disrupts the synthesis of the Duffy protein in areas like West Africa, where malaria is present.

The human genome, like those of flies and fish, also displays evidence of evolution through changes to enhancer DNA. One example is the adaptive loss of a protein called Duffy on red blood cells in a West African population living in areas where malaria is endemic.

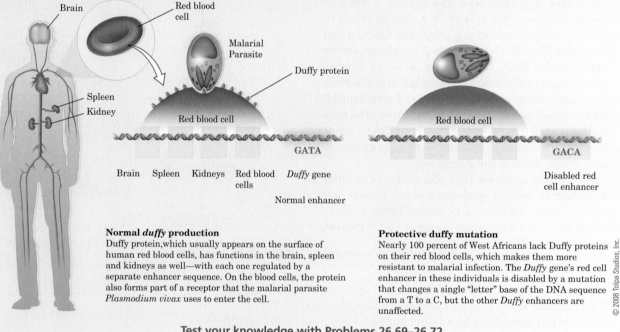

Normal *duffy* production
Duffy protein, which usually appears on the surface of human red blood cells, has functions in the brain, spleen and kidneys as well—with each one regulated by a separate enhancer sequence. On the blood cells, the protein also forms part of a receptor that the malarial parasite *Plasmodium vivax* uses to enter the cell.

Protective duffy mutation
Nearly 100 percent of West Africans lack Duffy proteins on their red blood cells, which makes them more resistant to malarial infection. The *Duffy* gene's red cell enhancer in these individuals is disabled by a mutation that changes a single "letter" base of the DNA sequence from a T to a C, but the other *Duffy* enhancers are unaffected.

© 2008 Tolpa Studios, Inc.

Test your knowledge with Problems 26.69–26.72.

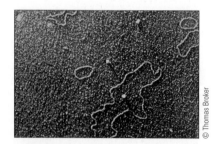

FIGURE 26.16 Plasmids from a bacterium used in the recombinant DNA technique.

© Thomas Broker

26.8 How and Why Do We Manipulate DNA?

There are no cures for the inborn genetic diseases discussed in Section 26.7. The best we can do is detect the carriers and, through genetic counseling of prospective parents, try not to perpetuate the defective genes. However, **recombinant DNA techniques** give us some hope for the future. At this time, these DNA techniques are being used mostly in bacteria, plants, and test animals (such as mice), but they are slowly being applied to humans as well, as will be described in Section 26.9.

One example of recombinant DNA techniques begins with certain circular DNA molecules found in the cells of the bacterium *Escherichia coli*. These molecules, called **plasmids** (Figure 26.16), consist of double-stranded

DNA arranged in a ring. Certain highly specific enzymes called **restriction endonucleases** cleave DNA molecules at specific locations (a different location for each enzyme). For example, one of these enzymes may split a double-stranded DNA as follows:

$$\text{B}\sim\!\!\!\sim\underset{\text{B}\sim\!\!\!\sim\text{CTTAAG}\sim\!\!\!\sim\text{B}}{\text{B}\sim\!\!\!\sim\text{GAATTC}\sim\!\!\!\sim\text{B}} \xrightarrow[\text{enzyme}]{\text{restriction}} \underset{\text{B}\sim\!\!\!\sim\text{CTTAA}}{\text{B}\sim\!\!\!\sim\text{G}} + \underset{\text{G}\sim\!\!\!\sim\text{B}}{\text{AATTC}\sim\!\!\!\sim\text{B}}$$

The enzyme is so programmed that whenever it finds this specific sequence of bases in a DNA molecule, it cleaves it as shown. Because a plasmid is circular, cleaving it in this way produces a double-stranded chain with two ends (Figure 26.17). These are called "sticky ends" because on one strand, each has several free bases that are ready to pair up with a complementary section if they can find one.

The next step is to give the strands such a section. This is done by adding a gene from some other species. The gene is a strip of double-stranded DNA that contains the necessary base sequence. For example, we can put in the human gene that manufactures insulin, which we can get in two ways:

1. It can be made in a laboratory by chemical synthesis; that is, chemists can combine the nucleotides in the proper sequence to make the gene.

2. We can cut a human chromosome in the desired place with the same restriction enzyme. Because it is the same enzyme, it cuts the human gene so as to leave the same sticky ends:

$$\underset{\text{H}-\text{CTTAAG}-\text{H}}{\text{H}-\text{GAATTC}-\text{H}} \xrightarrow[\text{enzyme}]{\text{restriction}} \underset{\text{H}-\text{CTTAA}}{\text{H}-\text{G}} + \underset{\text{G}-\text{H}}{\text{AATTC}-\text{H}}$$

The human gene must be cut at two places so that a piece of DNA that carries two sticky ends is freed. To splice the human gene into the plasmid, the two are mixed in the presence of DNA ligase and the sticky ends come together:

$$\underset{\text{H}-\text{CTTAA}}{\text{H}-\text{G}} + \underset{\text{G}-\text{B}}{\text{AATTC}-\text{B}} \xrightarrow[\text{ligase}]{\text{DNA}} \underset{\text{H}-\text{CTTAAG}-\text{B}}{\text{H}-\text{GAATTC}-\text{B}}$$

This reaction takes place at both ends of the human gene, turning the plasmid into a circle once again (Figure 26.17).

The modified plasmid is then put back into a bacterial cell, where it replicates naturally every time the cell divides. Bacteria multiply quickly, so soon we have a large number of bacteria, all containing the modified plasmid. All these cells now manufacture human insulin by transcription and translation. In this way, we can use bacteria as a factory to manufacture specific proteins. This new industry has tremendous potential for lowering the price of drugs that are currently manufactured by isolation from human or animal tissues (for example, human interferon, a molecule that fights infection).

> ### Example 26.4 Restriction Endonucleases
>
> Two different restriction endonucleases act on the following sequence of a double-stranded DNA:
>
> $$\sim\!\!\!\sim\text{AATGAATTCGAGGC}\sim\!\!\!\sim$$
> $$\sim\!\!\!\sim\text{TTACTTAAGCTCCG}\sim\!\!\!\sim$$

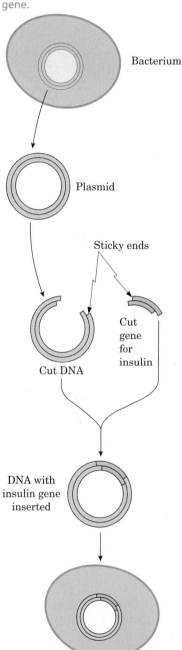

We use "B" to indicate the remaining DNA in a bacterial plasmid.

We use "H" to indicate a human gene.

Bacterium

Plasmid

Sticky ends

Cut DNA

Cut gene for insulin

DNA with insulin gene inserted

Bacterium with inserted recombinant DNA

FIGURE 26.17 The recombinant DNA technique can be used to turn a bacterium into an insulin "factory."

One endonuclease, EcoRI, recognizes the sequence GAATTC and cuts the sequence between G and A. The other endonuclease, TaqI, recognizes the sequence TCGA and cuts the sequence between T and C. What are the sticky ends that each of these endonucleases will create?

Solution

EcoRI   ∿∿∿AATG   AATTCGAGGC∿∿∿
            ∿∿∿TTACTTAA   GCTCCG∿∿∿

TaqI   ∿∿∿AATGAATT   CGAGGC∿∿∿
        ∿∿∿TTACTTAAGC   TCCG∿∿∿

Problem 26.4

Show the sticky end of the following double-stranded DNA sequence that is cut by TaqI:

∿∿∿CCTCGATTG∿∿∿
∿∿∿GGAGCTAAC∿∿∿

26.9 What Is Gene Therapy?

While viruses have traditionally been seen as problems for humans, there is one field where they are now being used for good. Viruses can be used to alter somatic cells, where a genetic disease is treated by the introduction of a gene for a missing protein. This process is called **gene therapy**.

The most successful form of gene therapy to date involves the gene for **adenosine deaminase** (**ADA**), an enzyme involved in purine catabolism (Section 25.8). If this enzyme is missing, dATP builds up in tissues, inhibiting the action of the enzyme ribonucleotide reductase. The result is a deficiency of the other three deoxyribonucleoside triphosphates (dNTPs). The dATP (in excess) and the other three dNTPs (deficient) are precursors for DNA synthesis. This imbalance particularly affects DNA synthesis in lymphocytes, on which much of the immune response depends (Chapter 31). Individuals who are homozygous for adenosine deaminase deficiency develop **severe combined immune deficiency** (**SCID**), the "bubble boy" syndrome. They are prone to infection because of their highly compromised immune systems. The ultimate goal of the planned gene therapy is to take bone marrow cells from affected individuals; introduce the gene for adenosine deaminase into the cells using a virus as a vector; and then reintroduce the bone marrow cells into the body, where they will produce the desired enzyme. The first clinical trials for a cure to ADA-SCID were simple enzyme replacement therapies begun in 1982. The patients in these trials were given injections of ADA. Later clinical trials sought to correct the gene in mature T cells. In 1990, transformed T cells were given to recipients via transfusions. In trials at the National Institutes of Health (NIH), two girls, aged 4 and 9 years at the start of treatment, showed improvement to the extent that they could attend regular public schools and had no more than the average number of infections. Administration of bone marrow stem cells in addition to T cells was the next step; clinical trials of this procedure were undertaken with two infants, aged 4 months and 8 months, in 2000. After 10 months, the children were healthy and had restored immune systems.

There are two types of delivery methods in human gene therapy. The first, called **ex vivo**, is the type used to combat SCID. *Ex vivo* means that somatic cells are removed from the patient, altered with the gene therapy, and then returned to the patient. The most common vector for this approach is the **Maloney murine leukemia virus** (**MMLV**). Figure 26.18 shows how the virus is used for gene therapy. Some of the MMLV is altered to remove certain genes, thereby rendering the virus unable to replicate. These genes are replaced with an **expression cassette**, which contains the gene being administered, such as the ADA gene, along with a suitable promoter. This mutated virus is used to infect a packaging cell line. Normal MMLV is also used to infect the packaging cell line. The normal MMLV will not replicate in

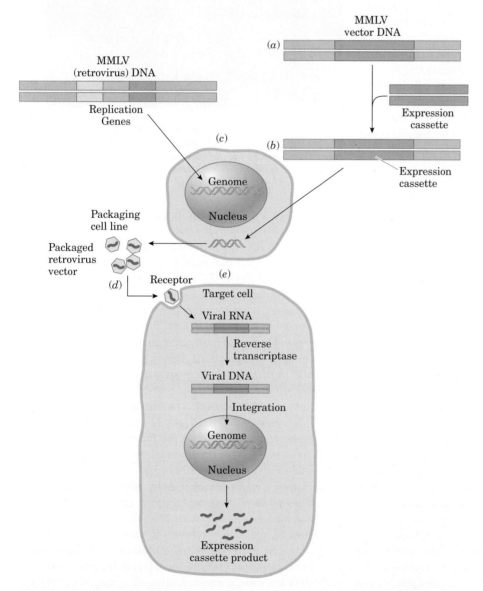

FIGURE 26.18 Gene therapy via retroviruses. The Maloney murine leukemia virus (MMLV) is used for ex vivo gene therapy. Replication genes are removed from the virus and replaced with an expression cassette containing the gene being replaced with gene therapy. The altered virus is grown in a packaging cell line that will allow replication. Viruses are collected and used to infect cultured target cells from the host needing the gene therapy. The altered virus produces RNA, which in turn produces DNA via reverse transcription. The DNA becomes integrated in the host cell's genome, and the host cells produce the desired protein. The cultured cells are returned to the host.

Twenty Years of Cystic Fibrosis Trials and Tribulations

There are many ways to fight a disease. Some of them can be classified as treating the symptoms, while others are considered to be fixing the problem. Scientists have been actively fighting the disease cystic fibrosis (CF) for decades, and have tried both of these techniques. Cystic fibrosis is an inherited disease characterized by sticky mucous secretions in the lungs and elsewhere, generally leading to a debilitated lifestyle and, eventually, the death of the patient. Twenty years ago, the mean life span for someone with CF was 20 years. Today, that figure is up to the mid- to late-thirties, but there is still no definitive cure.

Although the field of gene therapy was still in its infancy at that time, the isolation of the gene, the fact that it was a disease of the lungs, and the fact that scientists had recently begun to use adenovirus as a possible vector all led to the expectations that CF would be one of the first human diseases to be cured by gene therapy. It seemed like a slam dunk. Adenovirus is a common cold virus that can infect the lungs. If the correct CFTR gene could be infected into lungs of people with the disease, the correct gene could replace the defective one, and the patient would be cured.

Unfortunately for scientists in the field as well as CF patients, the reality and the promise never matched. While there were plenty of clinical trials, not a single gene therapy technique ever reached the market, and most of the trials were disappointing. In a test tube, scientists were able to replace lung cells carrying the CF gene and protein with the correct one, leading to the belief that they could do so *in vivo*. Other basic research demonstrated that the mutant proteins were misfolded, making CF one of the first protein-folding diseases known.

However, good theory aside, the gene therapy technique never really worked. Only about 1% of the cells in the lungs took up the virally-carried protein. There was also the problem of the lung tissue fighting off the virus. Eons of evolution have led to human physiology that will not just sit by idly and be invaded by viruses, even relatively benign ones. By the mid 1990s, many researchers were giving up their attempts at gene therapy, although basic research into CF continued. By 1998, scientists knew a lot more about the disease and the protein responsible. For example, there are hundreds of different mutations affecting the protein. To complicate matters further, the human response to the disease was variable. Two people with the exact same mutation could have very different responses, one requiring a lung transplant by the age of 12 and the other running marathons in his late twenties.

Steve Barrett

Cystic fibrosis is caused by a faulty transmembrane protein in the lungs that acts as a chloride ion channel. In 1989, researchers isolated the gene for the protein, called the cystic fibrosis transmembrane conductance regulator *(CFTR)*, which was heralded as the breakthrough of the year.

There are three different directions that research on CF is currently taking. One path is the search for a better animal model. The original mouse CF model proved frustrating, as it turned out that mice had a second chloride channel to rely on if the CFTR protein was defective. In 2008, scientists developed a pig CF model that is very promising. Pig physiology is very similar to humans, and if scientists can use pigs to study the disease, it will speed up the process tremendously. A second research direction is to develop drugs to fight the symptoms. This is a more traditional approach to medicine. The drug company Vertex took up the challenge and recently released two drugs, VX-809 and VX-770, that are very promising. The first one helps the mutated proteins make it to the cell membrane where they belong. The second one helps improve the function of the mutated CFTR protein. A phase II clinical trial showed that VX-770 improved lung function by 12%, which surpasses any other treatment's improvement. Finally, researchers have not completely given up on gene therapy. A group in the United Kingdom launched a study with 100 volunteers, where they are using fat particles as vectors to carry the therapeutic gene to the cells. With so many mutation possibilities and over 20 different functions for the CF protein, gene therapy still offers the most promise. A given drug can solve only a single problem with a single origin, but if it is possible to replace the defective gene, that solves the problem. Thus, many CF researchers still believe that making gene therapy work is the best approach.

Test your knowledge with Problems 26.73–26.78.

the packaging cell line, but it will restore the mutated virus's ability to replicate, albeit only in this cell line. These controls are necessary to keep mutant viruses from escaping to other tissues. The mutated virus particles are collected from the packaging cell line and used to infect the target cells—bone marrow cells in the case of SCID. MMLV is a retrovirus, so it infects the target cell and produces DNA from its RNA genome; this DNA can then become incorporated into the host genome, along with the promoter and ADA gene. In this way, the target cells that were collected are transformed, and they will produce ADA. These cells are then put back into the patient.

In the second delivery method, called **in vivo**, the virus is used to directly infect the patient's tissues. The most common vector for this delivery is the DNA virus **adenovirus**. A particular vector can be chosen based on specific receptors on the target tissue. Adenovirus has receptors in lung and liver cells, and it has been used in clinical trials for gene therapy of cystic fibrosis and ornithine transcarbamoylase deficiency. In mice, gene therapy has been successful in fighting diabetes.

The field of gene therapy is exciting and full of promise, but many obstacles to its success in humans remain. There are also many risks, such as a dangerous immunological response to the vector carrying the gene or the danger of a gene becoming incorporated in the host chromosome at a location that activates a cancer-causing gene. Both undesirable outcomes have happened in a limited number of human cases. See Chemical Connections 26G for a discussion of gene therapy and cystic fibrosis.

Gene therapy has been approved in humans only for the manipulation of somatic cells. It is illegal to tamper with human gametes in an attempt to create a heritable change in the human genome.

Summary

OWL Sign in at **www.cengage.com/owl** to develop problem-solving skills and complete online homework assigned by your professor.

Section 26.1 How Does DNA Lead to RNA and Protein?

- A **gene** is a segment of a DNA molecule that carries the sequence of bases that directs the synthesis of a specific RNA molecule. When the RNA is mRNA, it specifies the synthesis of a specific protein.
- The information stored in the DNA is transcribed onto RNA and then expressed in the synthesis of a protein molecule. This process involves two steps: **transcription** and **translation**.

Section 26.2 How Is DNA Transcribed into RNA?

- In transcription, the information is copied from DNA onto mRNA by complementary base pairing. There are also start and stop signals.
- The enzyme that synthesizes RNA is called RNA polymerase. In eukaryotes, three kinds of polymerase are used for the different types of RNA.

Section 26.3 What Is the Role of RNA in Translation?

- mRNA is strung out along the ribosomes.
- Transfer RNA carries individual amino acids, with each tRNA going to a specific site on the mRNA.

- A sequence of three bases (a triplet) on mRNA constitutes a **codon**. It spells out the particular amino acid that the tRNA brings to this site.
- Each tRNA has a recognition site, the **anticodon**, that pairs up with the codon.
- When two tRNA molecules are aligned at adjacent sites, the amino acids that they carry are linked by an enzyme, forming a peptide bond.
- The translation process continues until the entire protein is synthesized.

Section 26.4 What Is the Genetic Code?

- The **genetic code** provides the correspondence between a codon and an amino acid.
- In most cases, there is more than one codon for each amino acid, but the opposite is not true: A given codon will specify only one amino acid.

Section 26.5 How Is Protein Synthesized?

- Protein synthesis takes place in four stages: activation, initiation, elongation, and termination.
- Several of the steps of translation require an input of energy in the form of GTP.
- Ribosomes have three sites: the A site, the P site, and the E site.
- No protein is found in the area where the peptide synthesis is catalyzed. Thus, the ribosome is a ribozyme.

Section 26.6 How Are Genes Regulated?

- Most of human DNA (96 to 98%) does not code for proteins.
- A number of mechanisms for gene regulation exist on both the transcriptional level and the translational level.
- **Promoters** have initiator and conserved sequences.
- **Transcription factors** bind to the promoter, thereby regulating the rate of transcription.
- **Enhancers** are nucleotide sequences far removed from the transcription sites. They bind to transcription factors and increase the level of transcription.
- Some translational controls, such as **release factors**, act during translation; others, such as **chaperones**, act after translation is completed.

Section 26.7 What Are Mutations?

- A change in the sequence of bases is called a **mutation**.
- Mutations can be caused by an internal mistake or induced by chemicals or radiation. In fact, a change in just one base can cause a mutation.
- A mutation may be harmful or beneficial, or it may cause no change in the amino acid sequence. If a mutation is very harmful, the organism may die.
- Chemicals that cause mutations are called **mutagens**. Chemicals that cause cancer are called **carcinogens**. Many carcinogens are mutagens, but the reverse is not true.

Section 26.8 How and Why Do We Manipulate DNA?

- With the discovery of restriction enzymes that can cut DNA molecules at specific points, scientists have found ways to splice DNA segments together.
- A human gene (for example, the one that codes for insulin) can be spliced into a bacterial plasmid. The bacteria, when multiplied, can then transmit this new information to the daughter cells so that the ensuing generations of bacteria can manufacture human insulin. This powerful method is called the **recombinant DNA technique**.
- Genetic engineering is the process by which genes are inserted into cells.

Section 26.9 What Is Gene Therapy?

- Gene therapy is a technique whereby a missing gene is replaced using a viral vector.
- In **ex vivo** gene therapy, cells are removed from a patient and given the missing gene; then the cells are given back to the patient.
- In **in vivo** therapy, the patient is given the virus directly.

Problems

OWL Interactive versions of these problems may be assigned in OWL.

Orange-numbered problems are applied.

Section 26.1 How Does DNA Lead to RNA and Protein?

26.5 Does the term *gene expression* refer to:
(a) transcription, (b) translation, or (c) transcription plus translation?

26.6 In what part of the eukaryote cell does transcription occur?

26.7 Where does most of the translation occur in a eukaryote cell?

Section 26.2 How Is DNA Transcribed into RNA?

26.8 What is the function of RNA polymerase?

26.9 What is the role of helicase in transcription?

26.10 Where is an initiation signal located?

26.11 Which end of the DNA contains the termination signal?

26.12 What would happen to the transcription process if a drug added to a eukaryotic cell inhibited the phosphatase?

26.13 Where is the methyl group located in the guanine cap?

26.14 How are the adenine nucleotides linked together in the poly-A tails?

26.15 What is the difference in the requirement for a primer in RNA transcription compared to DNA replication (see section 25.6)?

26.16 What are the different names used for the two strands of DNA involved in transcription?

26.17 What is a consensus sequence in transcription?

26.18 What is a promoter sequence?

26.19 What is an intron?

26.20 What is an exon?

Section 26.3 What Is the Role of RNA in Translation?

26.21 Where are the codons located?

26.22 What are the two most important sites on tRNA molecules?

26.23 What are the ribosomal subunits for eukaryotic translation?

Section 26.4 What Is the Genetic Code?

26.24 (a) If a codon is GCU, what is the anticodon?
(b) For which amino acid does this codon code?

26.25 If a segment of DNA is 981 units long, how many amino acids appear in the protein encoded by this DNA segment? (Assume that the entire segment is used to code for the protein and that there is no methionine at the N-terminal end of the protein.)

26.26 In what sense does the universality of the genetic code support the theory of evolution?

26.27 Which amino acids have the most possible codons? Which have the fewest?

26.28 Using the first column of Table 26.1, explain how changing the second base of the codon is more detrimental to a protein than changing the first or the third base.

26.29 A genetic code in which two bases encode a single amino acid would not be adequate for protein synthesis. Give a reason why.

26.30 What is meant by the genetic code being continuous and unpunctuated?

Section 26.5 How Is Protein Synthesized?

26.31 To which end of the tRNA is the amino acid bonded? Where does the energy come from to form the tRNA–amino acid bond?

26.32 There are three sites on the ribosome, each participating in the translation. Identify them and describe what is happening at each site.

26.33 What is the main role of (a) the 40S ribosome and (b) the 60S ribosome?

26.34 What are the prokaryotic equivalents of the eukaryotic ribosomal subunits?

26.35 What is the function of elongation proteins?

26.36 What are the stages of protein synthesis?

26.37 Explain the nature of the tRNA used to initiate translation.

26.38 Explain what happens to the fMet initially put at the N-terminus.

26.39 Explain why scientists now refer to the ribosome as a ribozyme.

26.40 Why is amino acid activation called the second genetic code?

Section 26.6 How Are Genes Regulated?

26.41 Which molecules are involved in gene regulation at the transcriptional level?

26.42 Where are enhancers located? How do they work?

26.43 Where are the sieving portions of AARS enzymes located? What is their function?

26.44 What are the two types of transcription factors, and how do they work?

26.45 What is the difference between an enhancer and a response element?

26.46 How does alternative splicing lead to protein diversity?

26.47 What is the function of proteosomes in quality control?

26.48 What kind of interactions exist between metal-binding fingers and DNA?

Section 26.7 What Are Mutations?

26.49 Using Table 26.1, give an example of a mutation that: (a) does not change anything in a protein molecule and (b) might cause fatal changes in a protein.

26.50 How do cells repair mutations caused by X-rays?

26.51 Can a harmful mutation-causing genetic disease exist from generation to generation without exhibiting the symptoms of the disease? Explain.

26.52 Are all mutagens also carcinogens?

Section 26.8 How and Why Do We Manipulate DNA?

26.53 How do restriction endonucleases operate?

26.54 What are sticky ends?

26.55 A new genetically engineered corn has been approved by the Food and Drug Administration. This new corn shows increased resistance to a destructive insect called a corn borer. What is the difference, in principle, between this genetically engineered corn and one that developed insect resistance by mutation (natural selection)?

26.56 EcoRI restriction endonuclease recognizes the sequence GAATTC and cuts it between G and A. What will be the sticky ends of the following double-helical sequence when EcoRI acts on it?

CAAAGAATTCG
GTTTCTTAAGC

26.57 Why can it be argued that the discovery of restriction enzymes was the key to the beginning of modern molecular biology?

Chemical Connections

26.58 (Chemical Connections 26A) Why is selenocysteine called the twenty-first amino acid? Why were amino acids such as hydroxyproline and hydroxylysine not counted as additional amino acids?

26.59 (Chemical Connections 26B) What is a viral "coat"?

26.60 (Chemical Connections 26B) Where do the ingredients amino acids, enzymes, and so forth, necessary to synthesize the viral coat come from?

26.61 (Chemical Connections 26C) What is an invariant site?

26.62 (Chemical Connections 26D) What is a silent mutation?

26.63 (Chemical Connections 26D) If a mRNA codon has the sequence UCU, can there be a mutation in the third base that is not a silent mutation? Why or why not?

26.64 (Chemical Connections 26D) If a mRNA codon has the sequence UAU, which mutations of the third base would be the worst? Why?

26.65 (Chemical Connections 26D) Why can a silent mutation sometimes lead to a different protein product?

26.66 (Chemical Connections 26D) How was the gene *MDR1* involved in the discovery that silent mutations could lead to observable changes?

26.67 (Chemical Connections 26E) What is p53? Why is its mutated form associated with cancer?

26.68 (Chemical Connections 26E) How does p53 promote DNA repair?

26.69 (Chemical Connections 26F) What is the Duffy protein, and how is it important in the epidemiology of malaria?

26.70 (Chemical Connections 26F) What is the nature of the mutation by which West Africans do not produce the Duffy protein?

26.71 (Chemical Connections 26F) Consider a gene X that makes a protein Y. Give several examples of mutations that could affect the production of protein Y.

26.72 (Chemical Connections 26F) How is the Duffy protein thought to be related to human evolution?

26.73 (Chemical Connections 26G) Why was cystic fibrosis originally thought to be a perfect disease for the use of gene therapy?

26.74 (Chemical Connections 26G) What is the cause of cystic fibrosis?

26.75 (Chemical Connections 26G) Why was it a major milestone to isolate the CFTR gene?

26.76 (Chemical Connections 26G) Why did the first attempts at gene therapy with the CFTR gene fail?

26.77 (Chemical Connections 26G) How do the two drugs by Vertex help minimize the effects of CF?

26.78 (Chemical Connections 26G) Describe the most current attempt to revitalize the gene therapy approach to curing CF.

Additional Problems

26.79 In both the transcription and the translation steps of protein synthesis, a number of different molecules come together to act as a factor unit. What are these units of (a) transcription and (b) translation?

26.80 In the tRNA structure, there are stretches where complementary base pairing is necessary and other areas where it is absent. Describe two functionally critical areas (a) where base pairing is mandatory and (b) where it is absent.

26.81 Is there any way to prevent a hereditary disease? Explain.

26.82 How does the cell ensure that a specific amino acid (say, valine) attaches itself only to the one tRNA molecule that is specific for valine?

26.83 (a) What is a plasmid?

(b) How does it differ from a gene?

26.84 Why do we call the genetic code *degenerate?*

26.85 Glycine, alanine, and valine are classified as nonpolar amino acids. Compare their codons. What similarities do you find? What differences do you find?

26.86 Looking at the multiplicity (degeneracy) of the genetic code, you may get the impression that the third base of the codon is irrelevant. Point out where this is not the case. Out of the 16 possible combinations of the first and second bases, in how many cases is the third base irrelevant?

26.87 Which polypeptide is coded for by the mRNA sequence 5′-GCU-GAA-GUC-GAG-GUG-UGG-3′?

26.88 A new endonuclease is found. It cleaves double-stranded DNA at every location where C and G are paired on opposite strands. Could this enzyme be used in producing human insulin by the recombinant DNA technique? Explain.

Bioenergetics: How the Body Converts Food to Energy

27

A & L Sinibaldi/Stone/Getty Images

Wailua Falls, Hawaii, is a natural demonstration of two pathways ending in a common pool.

27.1 What Is Metabolism?

Living cells are in a dynamic state, which means that compounds are constantly being synthesized and then broken down into smaller fragments. Thousands of different reactions take place at the same time. **Metabolism** is the sum total of all the chemical reactions involved in maintaining the dynamic state of the cell.

In general, we can classify metabolic reactions into two broad groups: (1) those in which molecules are broken down to provide the energy needed by cells and (2) those that synthesize the compounds needed by cells—both simple and complex. **Catabolism** is the process of breaking down molecules to supply energy. The process of synthesizing (building up) molecules is **anabolism**. The same compounds may be synthesized in one part of a cell and broken down in a different part of the cell.

The topic of metabolism is so central to biochemistry that the journal *Science*, one of the leading scientific publications in the world, devoted a special section to it in its issue of 3 December 2010. The main outlines of metabolic pathways are well known and, in some cases, have been known for decades. Molecular biology is starting to add a new layer of understanding to the topic as we see how signaling mechanisms and genetic control play a large part in determining the physiological state of a cell. Cancer cell growth, circadian rhythms, and longevity affect and are affected by the metabolism of cells. Circadian rhythms are a good example because it is obvious that cells are in a more active metabolic state when an animal is awake than when it is asleep.

To focus on cancer, we know that genes that promote cancer (oncogenes) and mutations in genes that suppress cancer (tumor suppressor genes) can shift metabolic patterns to those characteristic of tumor cells from those found in normal cells. For example, it has been known for decades that when cancer cells metabolize sugars, the products do not enter the central metabolic pathway. Instead of entering that pathway, the intermediates are used in ways that aid the uncontrolled cell growth, which is a characteristic of cancer cells. We are now starting to understand the genetic control involved. Using this knowledge, we can try to alter the metabolism of cancer cells and thus find new treatment methods.

In spite of the large number of chemical reactions, a mere handful dominate cell metabolism. In this chapter and Chapter 28, we focus our attention on the catabolic pathways that yield energy. A **biochemical pathway** is a series of consecutive biochemical reactions. We will see the actual reactions that enable the chemical energy stored in our food to be converted to the energy we use every minute of our lives—to think; to breathe; and to use our muscles to walk, write, eat, and do everything else. In Chapter 29, we will look at some synthetic (anabolic) pathways.

The food we eat consists of many types of compounds, largely the ones we discussed in earlier chapters: carbohydrates, lipids, and proteins. All of them can serve as fuel, and we derive our energy from them. To convert those compounds to energy, the body uses a different pathway for each type of compound. *All of these diverse pathways converge to one* **common catabolic pathway**, which is illustrated in Figure 27.1. In the figure, the diverse pathways are shown as different food streams. The small molecules produced from the original large molecules in food drop into an imaginary collecting funnel that represents the common catabolic pathway. At the end of the funnel appears the energy carrier molecule adenosine triphosphate (ATP).

The purpose of catabolic pathways is to convert the chemical energy in foods to molecules of ATP. In the process, foods also yield metabolic intermediates, which the body can use for synthesis. In this chapter, we deal with the common catabolic pathway. In Chapter 28, we discuss the ways in which the different types of food (carbohydrates, lipids, and proteins) feed molecules into the common catabolic pathway.

Common catabolic pathway A series of chemical reactions through which foodstuffs are oxidized to yield energy in the form of ATP; the common catabolic pathway consists of (1) the citric acid cycle (Section 27.4) and (2) oxidative phosphorylation (Sections 27.5 and 27.6)

27.2 What Are Mitochondria and What Role Do They Play in Metabolism?

A typical animal cell has many components, as shown in Figure 27.2. Each serves a different function. For example, the replication of DNA (Section 25.6) takes place in the **nucleus, lysosomes** remove damaged cellular components and some unwanted foreign materials, and **Golgi bodies** package and process proteins for secretion and delivery to other cellular compartments. The specialized structures within cells are called **organelles**.

The **mitochondria**, which possess two membranes (Figure 27.3), are the organelles in which the common catabolic pathway takes place in higher organisms. The enzymes that catalyze the common pathway are all located in these organelles. Because these enzymes are synthesized in the cytosol,

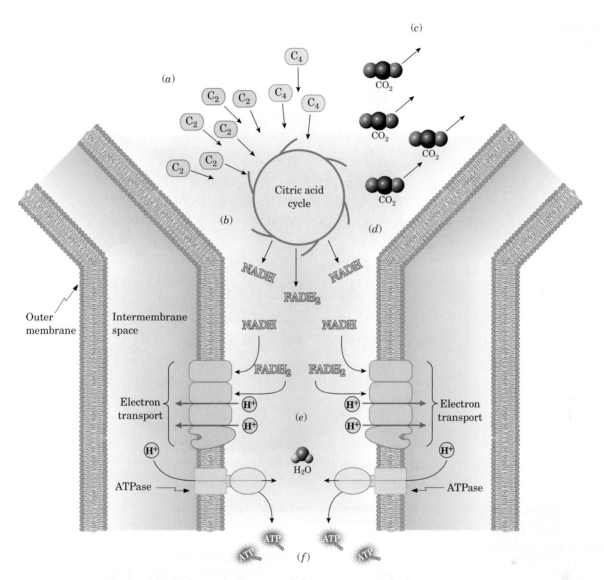

FIGURE 27.1 In this simplified schematic diagram of the common catabolic pathway, an imaginary funnel represents what happens in the cell. (*a*) The diverse catabolic pathways drop their products into the funnel of the common catabolic pathway, mostly in the form of C_2 fragments (Section 27.4). (The source of the C_4 fragments will be shown in Section 28.9.) (*b*) The spinning wheel of the citric acid cycle breaks these molecules down further. (*c*) The carbon atoms are released in the form of CO_2, and (*d*) the hydrogen atoms and electrons are picked up by special compounds such as NAD^+ and FAD. (*e*) Then the reduced NADH and $FADH_2$ cascade down into the stem of the funnel, where the electrons are transported inside the walls of the stem and the H^+ ions are expelled to the outside. (*f*) In their drive to get back, the H^+ ions form the energy carrier ATP. Once back inside, they combine with the oxygen that picked up the electrons and produce water.

they must be imported through the two membranes. They cross the outer membrane through translocator outer membrane (TOM) channels and are accepted in the intermembrane space by chaperone-like translocator inner membrane (TIM) complexes, which also insert them in the inner membrane.

Because the enzymes are located inside the inner membrane of mitochondria, the starting materials of the reactions in the common pathway must pass through the two membranes to enter the mitochondria. Products must leave the same way.

FIGURE 27.2 Diagram of a rat liver cell, a typical higher animal cell.

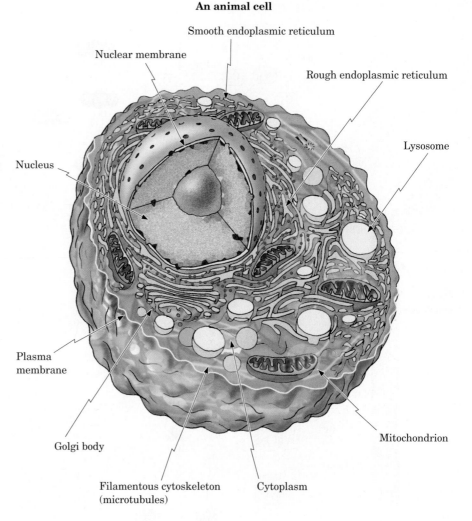

An animal cell

Smooth endoplasmic reticulum

Nuclear membrane

Rough endoplasmic reticulum

Lysosome

Nucleus

Plasma membrane

Golgi body

Filamentous cytoskeleton (microtubules)

Cytoplasm

Mitochondrion

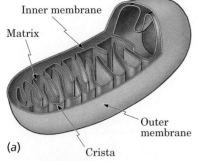

Inner membrane

Matrix

Outer membrane

(a) Crista

© Dennis Kunkel/Phototake

(b)

FIGURE 27.3 (a) Schematic of a mitochondrion cut to reveal the internal organization. (b) Colorized transmission electron micrograph of a mitochondrion in a heart muscle cell.

The inner membrane of a mitochondrion is quite resistant to the penetration of any ions and of most uncharged molecules. (*Mitochondria* is the plural form; *mitochondrion* is the singular form.) However, ions and molecules can still get through the membrane—they are transported across it by the numerous protein molecules embedded in it (Figure 21.2). The outer membrane, by contrast, is quite permeable to small molecules and ions and does not have transporting membrane proteins.

The **matrix** is the inner nonmembranous portion of a mitochondrion (Figure 27.3). The inner membrane is highly corrugated and folded. On the basis of electron microscopic studies, the Romanian-born American cell biologist George Palade (1912–2008) proposed his baffle model of the mitochondrion in 1952. The baffles, which are called **cristae**, project into the matrix like the bellows of an accordion. The enzymes of the oxidative phosphorylation cycle are localized on the cristae. The space between the inner and outer membranes is the **intermembrane space**. This classical baffle model of mitochondria underwent some changes in the late 1990s after three-dimensional pictures were obtained by the new technique called electron-microscope tomography. The 3-D images indicate that the cristae have narrow tubular connections to the inner membrane. These tubular connections may control the diffusion of metabolites from the inside to the intermembrane space. Furthermore, the space between the outer and the inner membrane varies during metabolism, possibly controlling the rate of reactions.

The enzymes of the citric acid cycle are located in the matrix. We will soon see in detail how the specific sequence of these enzymes causes the

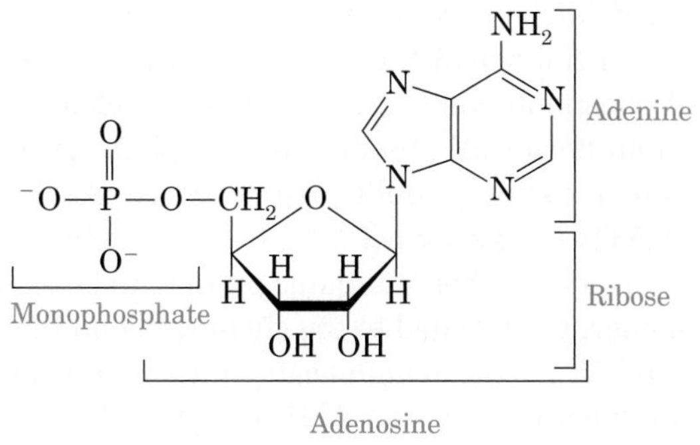

FIGURE 27.4 Adenosine 5'-monophosphate (AMP).

chain of events in the common catabolic pathway. Furthermore, we will discuss the ways in which nutrients and reaction products move in and out of the mitochondria.

27.3 What Are the Principal Compounds of the Common Metabolic Pathway?

The common catabolic pathway has two parts: the **citric acid cycle** (also called the tricarboxylic acid cycle or the Krebs cycle) and the **electron transport chain** and **phosphorylation**, together called the **oxidative phosphorylation pathway**. To understand what actually happens in these reactions, we must first introduce the principal compounds participating in the common catabolic pathway.

A. Agents for Storage of Energy and Transfer of Phosphate Groups

The most important of these agents are three rather complex compounds: **adenosine monophosphate (AMP)**, **adenosine diphosphate (ADP)**, and **adenosine triphosphate (ATP)** (Figures 27.4 and 27.5). All three of these molecules contain the heterocyclic amine adenine (Section 25.2) and the sugar D-ribose (Section 20.2) joined together by a β-N-glycosidic bond, forming adenosine (Section 25.2).

AMP, ADP, and ATP all contain adenosine connected to phosphate groups. The only difference between the three molecules is the number of phosphate groups. As you can see from Figure 27.5, each phosphate is attached to the next by an anhydride bond (Section 19.5A). ATP contains three phosphates—one phosphoric ester and two phosphoric anhydride bonds. In all three molecules, the first phosphate is attached to the ribose by a phosphoric ester bond (Section 19.5B).

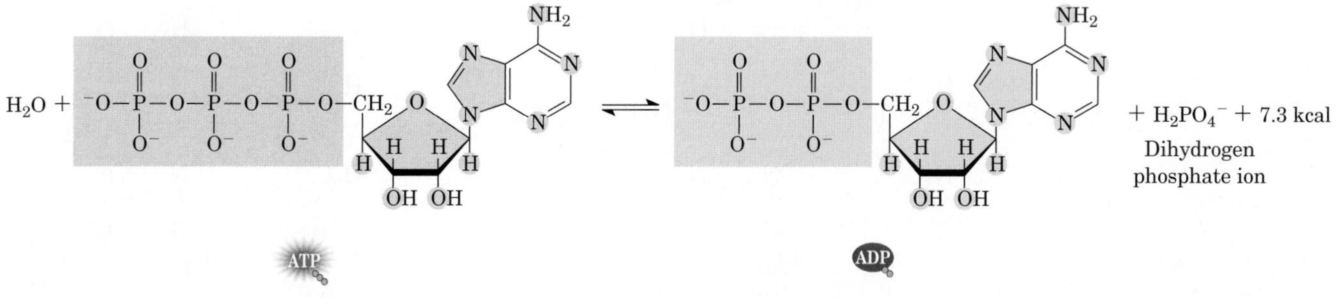

FIGURE 27.5 Hydrolysis of ATP produces ADP plus dihydrogen phosphate ion plus energy.

A phosphoric anhydride bond contains more chemical energy (7.3 kcal/mol) than a phosphoric ester linkage (3.4 kcal/ mol). Thus, when ATP and ADP are hydrolyzed to yield phosphate ion (Figure 27.5), they release more energy per phosphate group than does AMP. When one phosphate group is hydrolyzed from each, the following energy yields are obtained: AMP = 3.4 kcal/mol; ADP = 7.3 kcal/mol; ATP = 7.3 kcal/mol. (The PO_4^{3-} ion is generally called inorganic phosphate.) Conversely, when inorganic phosphate bonds to AMP or ADP, greater amounts of energy are added to the chemical bond than when it bonds to adenosine. ADP and ATP contain *high-energy* phosphoric anhydride bonds.

ATP releases the most energy and AMP releases the least energy when each gives up one phosphate group. This property makes ATP a very useful compound for energy storage and release. The energy gained in the oxidation of food is stored in the form of ATP, albeit only for a short while. ATP molecules in the cells normally do not last longer than about 1 min. They are hydrolyzed to ADP and inorganic phosphate to yield energy that drives other processes, such as muscle contraction, nerve signal conduction, and biosynthesis. As a consequence, ATP is constantly being formed and decomposed. Its turnover rate is very high. Estimates suggest that the human body manufactures and degrades as much as 40 kg (approximately 88 lb) of ATP every day. Even with these considerations, the body is able to extract only 40 to 60% of the total caloric content of food.

B. Agents for Transfer of Electrons in Biological Oxidation–Reduction Reactions

Two other actors in this drama are the coenzymes (Section 23.3) NAD^+ (nicotinamide adenine dinucleotide) and FAD (flavin adenine dinucleotide), both of which contain an ADP core (Figure 27.6). (The + in NAD^+ refers to the positive charge on the nitrogen.) In NAD^+, the operative part of the coenzyme is the nicotinamide part. In FAD, the operative part is the flavin portion. In both molecules, ADP is the handle by which the apoenzyme holds on to the coenzyme; the other end of the molecule carries out the actual chemical reaction. For example, when NAD^+ is reduced, the nicotinamide part of the molecule gets reduced:

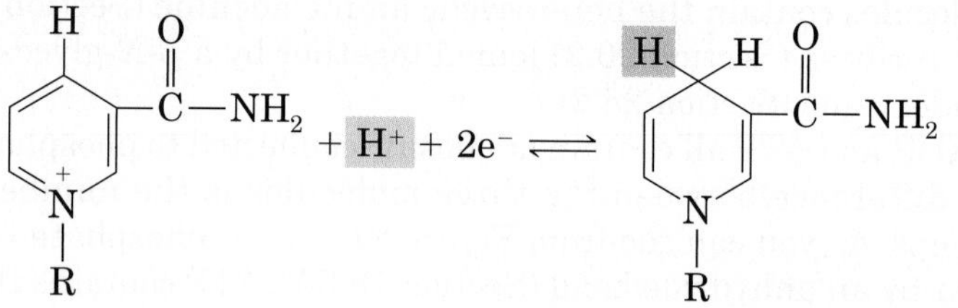

The reduced form of NAD^+ is called NADH. The same reduction happens on the two nitrogens of the flavin portion of FAD:

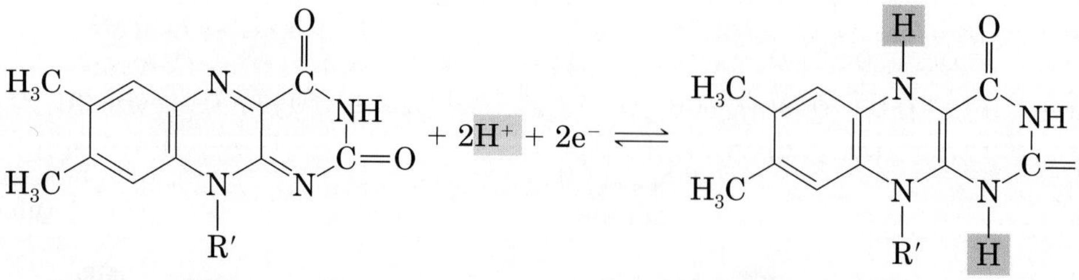

The reduced form of FAD is called $FADH_2$. We view NAD^+ and FAD coenzymes as the **hydrogen ion** and **electron-transporting molecules**.

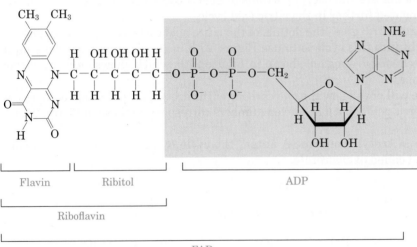

FIGURE 27.6 The structures of NAD⁺ and FAD.

C. Agent for Transfer of Acetyl Groups

The final principal compound in the common catabolic pathway is **coenzyme A** (CoA; Figure 27.7), which is the **acetyl ($CH_3CO—$)-transporting molecule**. Coenzyme A also contains ADP, but here the next structural unit is pantothenic acid, another B vitamin. Just as ATP can be viewed as an ADP molecule to which a $—PO_3^{2-}—$ group is attached by a high-energy bond, so **acetyl coenzyme A** can be considered a CoA molecule linked to an acetyl group by a high-energy thioester bond, for which the energy of hydrolysis is 7.51 kcal/mol. The active part of coenzyme A is the mercaptoethylamine. The acetyl group of acetyl coenzyme A is attached to the SH group:

Acetyl group The group $CH_3CO—$

$$CH_3 \overset{\overset{\displaystyle O}{\displaystyle \|}}{-C} -S-CoA$$

Acetyl coenzyme A

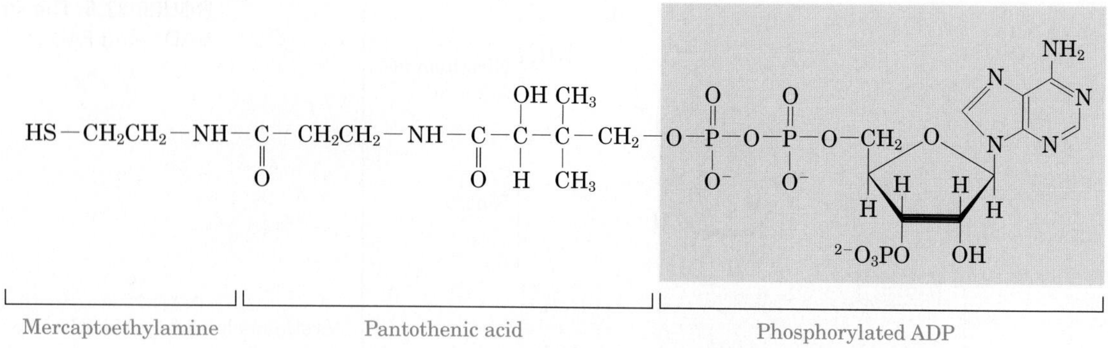

Mercaptoethylamine   Pantothenic acid   Phosphorylated ADP

FIGURE 27.7 The structure of coenzyme A.

27.4 What Role Does the Citric Acid Cycle Play in Metabolism?

The common catabolism of carbohydrates and lipids begins when they have been broken down into pieces of two carbon atoms each. The two-carbon fragments are the acetyl portions of acetyl coenzyme A. The acetyl is now fragmented further in the citric acid cycle.

Figure 27.8 gives the details of the citric acid cycle. A good way to gain an insight into this cycle is to use Figure 27.8 in conjunction with the simplified schematic diagram shown in Figure 27.9, which shows only the carbon balance.

We will now follow the two carbons of the acetyl group through each step in the citric acid cycle. The circled numbers correspond to those in Figure 27.8.

Step 1 Acetyl coenzyme A enters the cycle by combining with a C_4 compound called oxaloacetate:

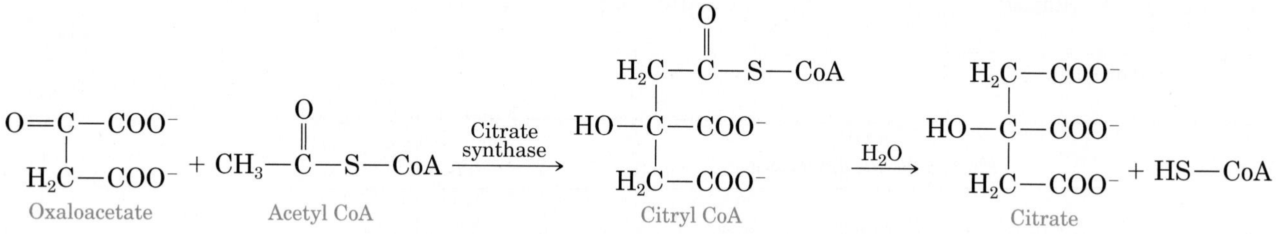

The first thing that happens is the addition of the $-CH_3$ group of the acetyl CoA to the $C=O$ of the oxaloacetate, catalyzed by the enzyme citrate synthase. This event is followed by hydrolysis of the thioester to produce the C_6 compound citrate ion and CoA. Therefore, Step ① is a building-up, rather than a breaking-down, process. In Step ⑧, we will see where the oxaloacetate comes from.

Step 2 The citrate ion is dehydrated to *cis*-aconitate, after which the *cis*-aconitate is hydrated, but this time to isocitrate instead of citrate:

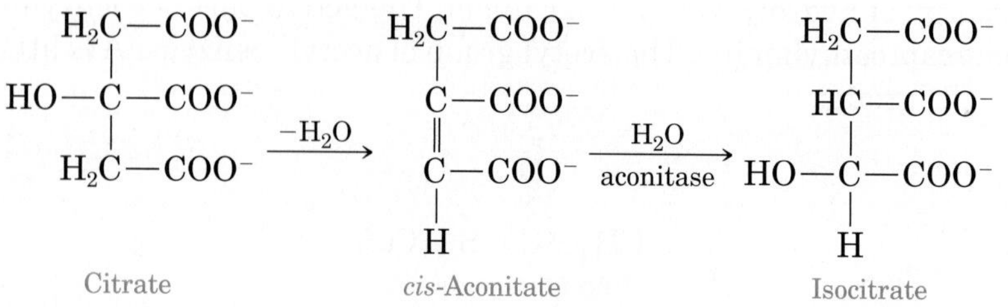

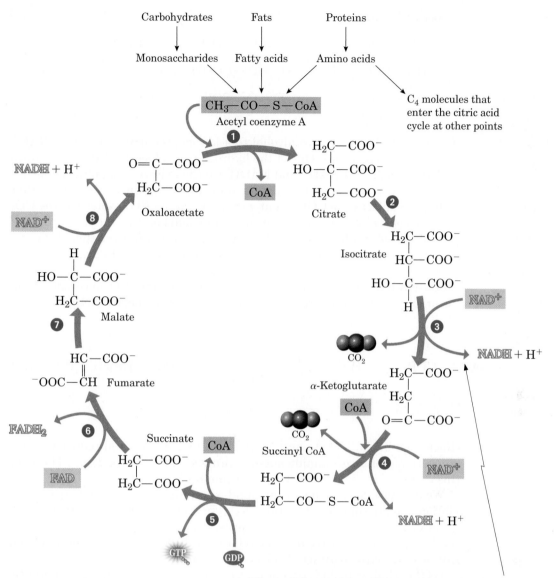

FIGURE 27.8 The citric acid (Krebs) cycle. The numbered steps are explained in detail in the text. [Hans Krebs (1900–1981), Nobel laureate in 1953, established the relationships among the different components of the cycle.]

The curved arrows are a shorthand way of showing the reactants and products. For example, in step ❸ NAD⁺ reacts with isocitrate to produce α-ketoglutarate, CO_2, NADH, and H⁺. The last two then leave the site of the reaction.

In citrate, the alcohol is a tertiary alcohol. We learned in Section 14.2 that tertiary alcohols cannot be oxidized. The alcohol in the isocitrate is a secondary alcohol, which upon oxidation yields a ketone.

Step 3 The isocitrate is now oxidized and **decarboxylated** at the same time:

Decarboxylation The process that leads to the loss of CO_2 from a —COOH group

Oxidation:

$$
\begin{array}{c}
\text{H}_2\text{C}-\text{COO}^- \\
|\\
\text{HC}-\text{COO}^- \\
|\\
\text{HO}-\text{C}-\text{COO}^- \\
|\\
\text{H}
\end{array}
+ \textbf{NAD}^+
\xrightarrow[\text{dehydrogenase}]{\text{Isocitrate}}
\begin{array}{c}
\text{H}_2\text{C}-\text{COO}^- \\
|\\
\text{HC}-\text{COO}^- \\
|\\
\text{O}=\text{C}-\text{COO}^-
\end{array}
+ \textbf{NADH} + \text{H}^+
$$

Isocitrate   Oxalosuccinate

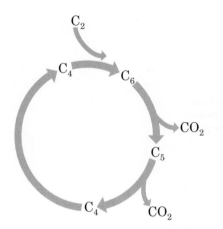

FIGURE 27.9 A simplified view of the citric acid cycle showing only the carbon balance.

Decarboxylation:

$$\underset{\text{Oxalosuccinate}}{\begin{matrix} H_2C-COO^- \\ | \\ HC-COO^- \\ | \\ O=C-COO^- \end{matrix}} + H^+ \longrightarrow \begin{matrix} H_2C-COO^- \\ | \\ HC-COOH \\ | \\ O=C-COO^- \end{matrix} \longrightarrow \underset{\alpha\text{-Ketoglutarate}}{\begin{matrix} H_2C-COO^- \\ | \\ CH_2 \\ | \\ O=C-COO^- \end{matrix}} + CO_2$$

In oxidizing the secondary alcohol to a ketone, the oxidizing agent NAD^+ removes two hydrogens. One of the hydrogens is added to NAD^+ to produce NADH. (Recall that NAD^+ and NADH are the oxidized and reduced forms, respectively, of nicotinamide adenine dinucleotide [Figure 27.6]). The other hydrogen replaces the COO^- that goes into making CO_2. Note that the CO_2 given off comes from the original oxaloacetate and not from the two carbons of the acetyl CoA. Both of these carbons are still present in the α-ketoglutarate. Also note that we are now down to a C_5 compound, α-ketoglutarate.

Steps 4 and 5 Next, a complex system removes another CO_2 once again from the original oxaloacetate portion rather than from the acetyl CoA portion:

$$\underset{\alpha\text{-Ketoglutarate}}{\begin{matrix} H_2C-COO^- \\ | \\ H_2C \\ | \\ O=C-COO^- \end{matrix}} + NAD^+ + GDP + P_i + H_2O \xrightarrow[\text{system}]{\text{Complex}} \underset{\text{Succinate}}{\begin{matrix} H_2C-COO^- \\ | \\ H_2C-COO^- \end{matrix}} + CO_2 + NADH + H^+ + GTP$$

(Recall again that NAD^+ and NADH are the oxidized and reduced forms, respectively, of nicotinamide adenine dinucleotide [Figure 27.6]). The P_i in this equation is the usual notation for inorganic phosphate.

We are now down to a C_4 compound, succinate. This oxidative decarboxylation is more complex than the first. It occurs in many steps and requires a number of cofactors. For our purpose, it is sufficient to know that during this second oxidative decarboxylation, a high-energy compound called **guanosine triphosphate** (**GTP**) is also formed.

GTP is similar to ATP, except that guanine replaces adenine. Otherwise, the linkages of the base to ribose and the phosphates are exactly the same as in ATP. The function of GTP is also similar to that of ATP—namely, it stores energy in the form of high-energy phosphoric anhydride bonds (chemical energy). The energy from the hydrolysis of GTP drives many important biochemical reactions—for example, the signal transduction in neurotransmission (Section 24.5).

As a final note on the decarboxylation steps, the CO_2 molecules given off in Steps ③ and ④ are the ones we exhale.

Step 6 In this step, succinate is oxidized by FAD, which removes two hydrogens to give fumarate (the double bond in this molecule is *trans*):

$$\underset{\text{Succinate}}{\begin{matrix} H_2C-COO^- \\ | \\ H_2C-COO^- \end{matrix}} + FAD \xrightarrow[\text{dehydrogenase}]{\text{Succinate}} \underset{\text{Fumarate}}{\begin{matrix} HC-COO^- \\ \| \\ {}^-OOC-CH \end{matrix}} + FADH_2$$

This reaction cannot be carried out in the laboratory, but with the aid of an enzyme catalyst, the body does it easily. (Recall that FAD and $FADH_2$ are the oxidized and reduced forms, respectively, of flavin adenine dinucleotide [Figure 27.6]).

Step 7 The fumarate is now hydrated to give the malate ion:

$$\underset{\text{Fumarate}}{\overset{\text{HC}-\text{COO}^-}{\underset{^-\text{OOC}-\text{CH}}{\|}}} + H_2O \xrightarrow{\text{Fumarase}} \underset{\text{Malate}}{\overset{\text{H}}{\underset{H_2\text{C}-\text{COO}^-}{\overset{|}{\underset{|}{\text{HO}-\text{C}-\text{COO}^-}}}}}$$

Step 8 In the final step of the cycle, malate is oxidized to give oxaloacetate:

$$\underset{\text{Malate}}{\overset{\text{H}}{\underset{H_2\text{C}-\text{COO}^-}{\overset{|}{\underset{|}{\text{HO}-\text{C}-\text{COO}^-}}}}} + \text{NAD}^+ \xrightarrow[\text{dehydrogenase}]{\text{Malate}} \underset{\text{Oxaloacetate}}{\overset{\text{O}=\text{C}-\text{COO}^-}{\underset{H_2\text{C}-\text{COO}^-}{|}}} + \text{NADH} + \text{H}^+$$

(Recall that NAD^+ and NADH are the oxidized and reduced forms, respectively, of nicotinamide adenine dinucleotide [Figure 27.6]). Thus, the final product of the Krebs cycle is oxaloacetate, which is the compound with which we started in Step ①.

In this process, the original two acetyl carbons of acetyl CoA were added to the C_4 oxaloacetate to produce a C_6 unit, which then lost two carbons in the form of CO_2 to produce, at the end of the process, the C_4 unit oxaloacetate. The net effect is that one two-carbon acetyl group enters the cycle and two carbon dioxides are given off.

How does the citric acid cycle produce energy? We have already learned that one step in the process produces a high-energy molecule of GTP. However, most of the energy is produced in the other steps that convert NAD^+ to NADH and FAD to $FADH_2$. These reduced coenzymes carry the H^+ and electrons that eventually provide the energy for the synthesis of ATP (discussed in detail in Sections 27.5 and 27.6).

This stepwise degradation and oxidation of acetate in the citric acid cycle results in the most efficient extraction of energy. Rather than being generated in one burst, the energy is released in small packets that are carried away, step-by-step in the form of NADH and $FADH_2$.

The cyclic nature of this acetate degradation has other advantages besides maximizing energy yield:

1. The citric acid cycle components provide raw materials for amino acid synthesis as the need arises (Chapter 29). For example, α-ketoglutaric acid is used to synthesize glutamic acid.

2. The many-component cycle provides an excellent method for regulating the speed of catabolic reactions.

The regulation can occur at many different parts of the cycle, so that feedback information can be used at many points to speed up or slow down the process as necessary.

The following equation represents the overall reactions in the citric acid cycle:

$$\text{GDP} + P_i + \text{CH}_3-\text{CO}-\text{S}-\text{CoA} + 2H_2O + 3\text{NAD}^+ + \text{FAD}$$
$$\longrightarrow \text{CoA} + \text{GTP} + 2\text{CO}_2 + 3\text{NADH} + \text{FADH}_2 + 3\text{H}^+ \qquad \text{(Eq.27.1)}$$

The citric acid cycle is controlled by feedback mechanisms. When the essential product of this cycle, NADH + H^+, and the end product of the common catabolic pathway, ATP, accumulate, they inhibit some of the enzymes in the cycle. Citrate synthase (Step ①), isocitrate dehydrogenase (Step ③), and α-ketoglutarate dehydrogenase (part of the complex enzyme system in

Step ④) are inhibited by ATP and/or by NADH + H⁺. This inhibition slows down or shuts off the cycle. Conversely, when the feed material, acetyl CoA, is in abundance, the cycle accelerates. The enzyme isocitrate dehydrogenase (Step ③) is stimulated by ADP and NAD⁺, which are the essential reactants from which the end products of the cycle are derived.

27.5 How Do Electron and H⁺ Transport Take Place?

The reduced coenzymes NADH and FADH$_2$ are end products of the citric acid cycle. They carry hydrogen ions and electrons and, therefore, have the potential to yield energy when these combine with oxygen to form water:

$$4H^+ + 4e^- + O_2 \longrightarrow 2H_2O + \text{energy}$$

This simple exothermic reaction is carried out in many steps. The oxygen in this reaction is the oxygen we breathe.

A number of enzymes are involved in this reaction, all of which are embedded in the inner membrane of the mitochondria. These enzymes are situated in a particular *sequence* in the membrane so that the product from one enzyme can be passed on to the next enzyme, in a kind of assembly line. The enzymes are arranged in order of increasing affinity for electrons, so electrons flow through the enzyme system (Figure 27.10).

The sequence of the electron-carrying enzyme systems starts with complex I. The largest complex, it contains some 40 subunits, among them a flavoprotein and several FeS clusters. **Coenzyme Q** (CoQ; also called ubiquinone) is associated with complex I, which oxidizes the NADH produced in the citric acid cycle and reduces the CoQ:

$$\boxed{NADH + H^+} + CoQ \rightarrow \boxed{NAD^+} + CoQH_2$$

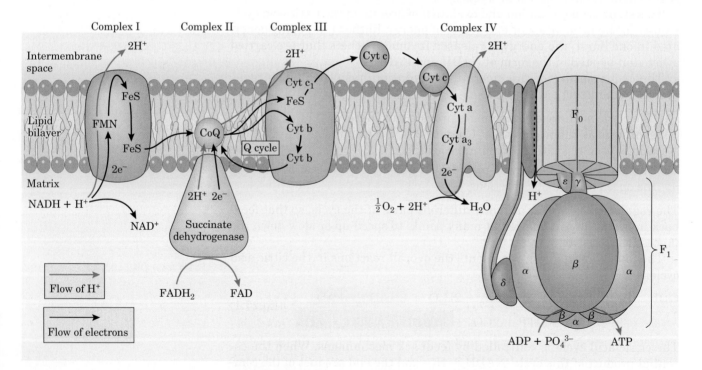

FIGURE 27.10 Schematic diagram of the electron and H⁺ transport chain and subsequent phosphorylation. The combined processes are also known as oxidative phosphorylation.

Chemical Connections 27A

Uncoupling and Obesity

The health concerns that surround the growing number of obese people in developed countries have led to research into the causes and alleviation of obesity. A number of weight-reducing drugs exist. Some of them operate as uncouplers of electron transport and oxidative phosphorylation.

The discovery of a role for uncouplers in weight reduction occurred more or less by accident. During World War I, many ammunition workers were exposed to 2,4-dinitrophenol (DNP), a compound used to prepare the explosive picric acid, which is structurally related to the well-known explosive trinitrotoluene (TNT). After it was observed that these workers lost weight, DNP was used as a weight-reducing drug during the 1920s. Unfortunately, DNP eliminated not only the fat but sometimes also the dieter, and its use as a diet pill was discontinued after 1929.

Today we know why DNP works as a weight-reducing drug: It is an effective protonophore—a compound that transports ions through cell membranes passively, without the expenditure of energy. As noted earlier, H⁺ ions accumulate in the intermembrane space of mitochondria and, under normal conditions, drive the synthesis of ATP while they are going back inside. This process is Mitchell's chemiosmotic principle in action. When DNP is ingested, it transfers the H⁺ back to the mitochondrion easily, and no ATP is manufactured. The energy of the electron separation is dissipated as heat and is not built in as chemical energy in ATP. The loss of this energy-storing compound makes the utilization of food much less efficient, resulting in weight loss.

The role of brown fat in hibernation may be related to obesity in humans.

A similar mechanism provides heat in hibernating bears. The bears have brown fat; its color is derived from the numerous mitochondria in the tissue. The brown fat also contains an uncoupling protein called thermogenin, a protonophore that allows the ions to stream back into the mitochondrial matrix without manufacturing ATP. The heat generated in this manner keeps the animal alive during cold winter days. In similar fashion, an uncoupling protein is known to be involved in obesity, but it is not known what relationship, if any, exists between this protein and hibernation. The question of human obesity and its prevention is important enough, however, to make uncoupling in brown fat a point of departure for obesity research.

DNP, TNT, Picric acid

Test your knowledge with Problems 27.54 and 27.55.

Some of the energy released in this reaction is used to move 2H⁺ across the membrane, from the matrix to the intermembrane space. The CoQ is soluble in lipids and can move laterally within the membrane. (The figure of 2H⁺ transported across the membrane is the minimum number that allows the overall oxidation process to take place. Some researchers would say that the number of protons transported by some of these respiratory complexes should be higher.)

Complex II also catalyzes the transfer of electrons to CoQ. The source of the electrons is the oxidation of succinate in the citric acid cycle, producing $FADH_2$. The final reaction is:

$$FADH_2 + CoQ \longrightarrow FAD + CoQH_2$$

The energy of this reaction is not sufficient to pump two protons across the membrane, however, nor is there an appropriate channel for such a transfer.

Complex III delivers the electrons from $CoQH_2$ to **cytochrome c**. This integral membrane complex contains 11 subunits, including cytochrome b, cytochrome c_1, and FeS clusters. (The letters used to designate the cytochromes were given in order of their discovery.) Each cytochrome has an iron-ion–containing heme center (Section 22.11) embedded in its own protein. Complex III has two channels through which two H^+ ions are pumped from the $CoQH_2$ into the intermembrane space. The process is very complicated. To simplify matters, we can imagine that it occurs in two distinct steps, as does the electron transfer. Because each cytochrome c can pick up only one electron, two cytochrome c units are needed:

$$CoQH_2 + 2 \text{ cytochrome c (oxidized)}$$
$$\longrightarrow CoQ + 2H^+ + 2 \text{ cytochrome c (reduced)}$$

Cytochrome c is also a mobile carrier of electrons—it can move laterally in the intermembrane space.

Complex IV, known as cytochrome oxidase, contains 13 subunits—most importantly, cytochrome a_3, a heme that has an associated copper center. Complex IV is an integral membrane protein complex. The electron movement flows from cytochrome c to cytochrome a to cytochrome a_3. There, the electrons are transferred to the oxygen molecule, and the O—O bond is cleaved. The oxidized form of the enzyme takes up two H^+ ions from the matrix for each oxygen atom. The water molecule formed in this way is released into the matrix:

$$\tfrac{1}{2}O_2 + 2H^+ + 2e^- \longrightarrow H_2O$$

During this process, two more H^+ ions are pumped out of the matrix into the intermembrane space. Although the mechanism of pumping out protons from the matrix is not known, the energy driving this process is derived from the energy of water formation. This final pumping into the intermembrane space makes a total of six H^+ ions per $NADH + H^+$ and four H^+ ions per $FADH_2$ molecule.

27.6 What Is the Role of the Chemiosmotic Pump in ATP Production?

Chemiosmotic theory Mitchell's proposal that electron transport is accompanied by an accumulation of protons in the intermembrane space of the mitochondrion, which, in turn, creates osmotic pressure; the protons driven back to the mitochondrion under this pressure generate ATP

How do the electron and H^+ transports produce the chemical energy of ATP? In 1961, Peter Mitchell (1920–1992), an English chemist, proposed the **chemiosmotic theory** to answer this question: The energy in the electron transfer chain creates a proton gradient. A **proton gradient** is a continuous variation in the H^+ concentration along a given region. In this case, there is a higher concentration of H^+ in the intermembrane space than inside the mitochondrion. The driving force, which is the result of the spontaneous flow of ions from a region of high concentration to a region of low concentration, propels the protons back to the mitochondrion through a complex known as **proton-translocating ATPase**. This compound is located on the inner membrane of the mitochondrion (Figure 27.10) and is the active enzyme that catalyzes the conversion of ADP and inorganic phosphate to ATP (the reverse of the reaction shown in Figure 27.5):

$$\text{ADP} + \text{P}_i \underset{}{\overset{\text{ATPase}}{\rightleftharpoons}} \text{ATP} + \text{H}_2\text{O}$$

Subsequent studies have confirmed this theory, and Mitchell received the Nobel Prize in chemistry in 1978.

The proton-translocating ATPase is a complex "rotor engine" made of 16 different proteins. The F_0 sector, which is embedded in the membrane, contains the **proton channel** (Figure 27.10). The 12 subunits that form this channel rotate every time a proton passes from the cytoplasmic side (intermembrane) to the matrix side of the mitochondrion. This rotation is transmitted to a "rotor" in the F_1 sector. F_1 contains five kinds of polypeptides. The rotor (γ and ε subunits) is surrounded by the catalytic unit (made of α and β subunits) that synthesizes the ATP. The catalytic unit converts the mechanical energy of the rotor into chemical energy of the ATP molecule. The last unit, the "stator," containing the δ subunit, stabilizes the whole complex. The proton-translocating ATPase can catalyze the reaction in both directions. When protons that have accumulated on the outer surface of the mitochondrion stream inward, the enzyme manufactures ATP and stores the electrical energy (due to the flow of charges) in the form of chemical energy. In the reverse reaction, the enzyme hydrolyzes ATP and, as a consequence, pumps out H^+ from the mitochondrion. Each pair of protons that is translocated gives rise to the formation of one ATP molecule. Only when the two parts of the proton-translocating ATPase F_1 and F_0 are linked is energy production possible. When the interaction between F_1 and F_0 is disrupted, the energy transduction is lost.

The protons that enter a mitochondrion combine with the electrons transported through the electron transport chain and with oxygen to form water. The net result of the two processes (electron/H^+ transport and ATP formation) is that each oxygen molecule we inhale combines with four H^+ ions and four electrons to give two water molecules. The four H^+ ions and four electrons come from the NADH and $FADH_2$ molecules produced in the citric acid cycle. The oxygen, therefore, has two functions:

- It oxidizes NADH to NAD^+ and $FADH_2$ to FAD so that these molecules can go back and participate in the citric acid cycle.

- It provides energy for the conversion of ADP to ATP.

The latter function is accomplished indirectly, not through the reduction of O_2 to H_2O. The entrance of the H^+ ions into the mitochondrion drives the ATP formation, but the H^+ ions enter the mitochondrion because the O_2 depleted the H^+ ion concentration when water was formed. This rather complex process involves the transport of electrons along a whole series of enzyme molecules (which catalyze all these reactions); however, the cell cannot utilize the O_2 molecules without the electron transport chain and eventually will die.

The electron and H^+ transport chain and the subsequent phosphorylation process are collectively known as **oxidative phosphorylation**. The following equations represent the overall reactions in oxidative phosphorylation:

$$\text{NADH} + 3\,\text{ADP} + \tfrac{1}{2}\text{O}_2 + 3\text{P}_i + \text{H}^+ \longrightarrow \text{NAD}^+ + 3\,\text{ATP} + \text{H}_2\text{O} \quad \text{(Eq.27.2)}$$

$$\text{FADH}_2 + 2\,\text{ADP} + \tfrac{1}{2}\text{O}_2 + 2\text{P}_i \longrightarrow \text{FAD} + 2\,\text{ATP} + \text{H}_2\text{O} \quad \text{(Eq.27.3)}$$

27.7 What Is the Energy Yield Resulting from Electron and H^+ Transport?

The energy released during electron transport is finally captured in the chemical energy of the ATP molecule. Therefore, it is instructive to look at the energy yield in the universal biochemical currency: the number of ATP molecules.

Each pair of protons entering a mitochondrion results in the production of one ATP molecule. For each NADH molecule, three pairs of protons are pumped into the intermembrane space in the electron transport process. Therefore, for each NADH molecule, we get three ATP molecules, as can be seen in Equation 27.2. For each $FADH_2$ molecule, only four protons are pumped out of the mitochondrion. Therefore, only two ATP molecules are produced for each, as seen in Equation 27.3. Note that ATP production is reported to the nearest whole number. The process is complex, and these numbers represent the least complicated way of looking at it.

Now we can produce the energy balance for the entire common catabolic pathway (citric acid cycle and oxidative phosphorylation combined). For each C_2 fragment entering the citric acid cycle, we obtain three NADH and one $FADH_2$ (Equation 27.1) plus one GTP, which is equivalent in energy to one ATP. Thus, the total number of ATP molecules produced per C_2 fragment is:

$$3 \text{ NADH} \times 3 \text{ ATP/NADH} = 9 \text{ ATP}$$
$$1 \text{ FADH}_2 \times 2 \text{ ATP/FADH}_2 = 2 \text{ ATP}$$
$$1 \text{ GTP} = \underline{1 \text{ ATP}}$$
$$= 12 \text{ ATP}$$

Each C_2 fragment that enters the cycle produces 12 ATP molecules and uses up two O_2 molecules. The total effect of the energy-production chain of reactions discussed in this chapter (the common catabolic pathway) is to oxidize one C_2 fragment with two molecules of O_2 to produce two molecules of CO_2 and 12 molecules of ATP:

$$C_2 + 2O_2 + 12ADP + 12P_i \longrightarrow 12ATP + 2CO_2$$

The important thing is not the waste product, CO_2, but rather, the 12 ATP molecules. These molecules now release their energy when they are converted to ADP.

27.8 How Is Chemical Energy Converted to Other Forms of Energy?

As mentioned in Section 27.3, the storage of chemical energy in the form of ATP lasts only a short time. Usually, within a minute, the ATP is hydrolyzed (an exothermic reaction) and releases its chemical energy. How does the body use this chemical energy? To answer this question, let us look at the different forms in which energy is needed in the body.

A. Conversion to Other Forms of Chemical Energy

The activity of many enzymes is controlled and regulated by phosphorylation. For example, the enzyme phosphorylase, which catalyzes the breakdown of glycogen (Chemical Connections 28B), occurs in an inactive form, phosphorylase *b*. When ATP transfers a phosphate group to a serine residue, the enzyme becomes active. Thus, the chemical energy of ATP is used in the form of chemical energy to activate phosphorylase *b* so that glycogen can be utilized. We will see several other examples of this energy conversion in Chapters 28 and 29.

B. Electrical Energy

The body maintains a high concentration of K^+ ions inside the cells despite the fact that the K^+ concentration is low outside the cells. The reverse is true for Na^+. So that K^+ does not diffuse out of the cells and Na^+ does not enter them, special transport proteins in the cell membranes constantly pump K^+ into and Na^+ out of the cells. This pumping requires energy, which is supplied by the hydrolysis of ATP to ADP. Because of this pumping, the charges inside and outside the cell are unequal, which generates an electric potential. Thus, the chemical energy of ATP is transformed into electrical energy, which operates in neurotransmission (Section 24.2).

C. Mechanical Energy

ATP is the immediate source of energy in muscle contraction. In essence, muscle contraction takes place when thick and thin filaments slide past each other (Figure 27.11). The thick filament is myosin, an ATPase enzyme (that is, one that hydrolyzes ATP). The thin filament, actin, binds strongly to myosin in the contracted state. However, when ATP binds to myosin, the actin–myosin complex dissociates, and the muscle relaxes. When myosin hydrolyzes ATP, it interacts with actin once more, and a new contraction occurs. In this way, the hydrolysis of ATP drives the alternating association and dissociation of actin and myosin and, consequently, the contraction and relaxation of the muscle.

D. Heat Energy

One molecule of ATP upon hydrolysis to ADP yields 7.3 kcal/mol. Some of this energy is released as heat and used to maintain body temperature. If we estimate that the specific heat of the body is about the same as that of water, a person weighing 60 kg would need to hydrolyze approximately 99 moles (approximately 50 kg) of ATP to raise the temperature of the body from room temperature, 25°C, to 37°C. Not all body heat is derived from ATP hydrolysis; some other exothermic reactions in the body also make heat contributions.

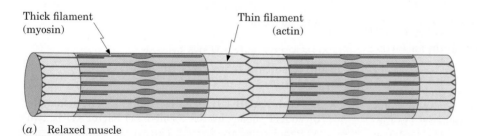

Thick filament (myosin)   Thin filament (actin)

(*a*) Relaxed muscle

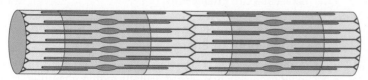

(*b*) Contracted muscle

FIGURE 27.11 Schematic diagram of muscle contraction.

Summary

Section 27.1 What Is Metabolism?

- The sum total of all the chemical reactions involved in maintaining the dynamic state of cells is called **metabolism**.
- The breaking down of molecules is **catabolism**; the building up of molecules is **anabolism**.

Section 27.2 What Are Mitochondria and What Role Do They Play in Metabolism?

- Many metabolic activities in cells take place in specialized structures called **organelles**.
- **Mitochondria** are the organelles in which the reactions of the **common catabolic pathway** take place.

Section 27.3 What Are the Principal Compounds of the Common Metabolic Pathway?

- The common metabolic pathway oxidizes a two-carbon C_2 fragment (acetyl) from different foods. The products of C_2 oxidation are water and carbon dioxide.
- The energy from oxidation is built into the high-chemical-energy-storing molecule **ATP**. As the C_2 fragments are oxidized, protons (H^+) and electrons are released and passed along to carriers.
- The principal carriers in the common catabolic pathway are as follows: ATP is a phosphate carrier, **CoA** is the C_2 fragment carrier, and **NAD$^+$** and **FAD** carry the hydrogen ions (protons) and electrons. The unit common to all of these carriers is **ADP**. The nonactive end of the carriers acts as a handle that fits into the active sites of the enzymes.

Section 27.4 What Role Does the Citric Acid Cycle Play in Metabolism?

- In the **citric acid cycle**, the C_2 fragment first combines with a C_4 fragment (oxaloacetate) to yield a C_6 fragment (citrate). An oxidative decarboxylation yields a C_5 fragment. One CO_2 is released, and one NADH + H^+ is passed to the **electron transport chain** for further oxidation.
- Another oxidative decarboxylation provides a C_4 fragment. Once again, a CO_2 is released and another NADH + H^+ is passed to the electron transport chain.
- The enzymes of the citric acid cycle are located in the mitochondrial matrix. Control of this cycle takes place by a feedback mechanism.

Section 27.5 How Do Electron and H^+ Transport Take Place?

- The electrons of NADH enter the electron transport chain at the complex I stage. The coenzyme Q (CoQ) of this complex picks up the electrons and the H^+ and becomes $CoQH_2$. The energy of this reduction reaction is used to expel two H^+ ions from the matrix into the intermembrane space.

- Complex II also has CoQ. Electrons and H^+ are passed to this complex, and complex II catalyzes the transfer of electrons from $FADH_2$. However, no H^+ ions are pumped into the intermembrane space at this point.
- Electrons are passed along by $CoQH_2$ to complex III of the electron transport chain. At complex III, the two H^+ ions from the $CoQH_2$ are expelled into the intermembrane space. Cytochrome c of complex III transfers electrons to complex IV through redox reactions.
- As the electrons are transported from cytochrome c to complex IV, two more H^+ ions are expelled from the matrix of the mitochondrion to the intermembrane space.
- For each NADH, six H^+ ions are expelled. For each $FADH_2$, four H^+ ions are expelled.
- The electrons passed to complex IV return to the matrix, where they combine with oxygen and H^+ to form water.

Section 27.6 What Is the Role of the Chemiosmotic Pump in ATP Production?

- Both the citric acid cycle and **oxidative phosphorylation** take place in the mitochondria. The enzymes of the citric acid cycle are found in the mitochondrial matrix, whereas the enzymes of the electron transport chain and oxidative phosphorylation are located on the inner mitochondrial membrane. Some of them project into the intermembrane space.
- When the H^+ ions expelled by electron transport stream back into the mitochondrion, they drive a complex enzyme called **proton-translocating ATPase**, which makes one ATP molecule for each two H^+ ions that enter the mitochondrion.
- The proton-translocating ATPase is a complex "rotor engine." The proton channel part (F_0) is embedded in the membrane, and the catalytic unit (F_1) converts mechanical energy to chemical energy of the ATP molecule.

Section 27.7 What Is the Energy Yield Resulting from Electron and H^+ Transport?

- For each NADH + H^+ coming from the citric acid cycle, three ATP molecules are formed. For each $FADH_2$, two ATP molecules are formed. The overall result: For each C_2 fragment that enters the citric acid cycle, 12 ATP molecules are produced.

Section 27.8 How Is Chemical Energy Converted to Other Forms of Energy?

- Chemical energy is stored in ATP only for a short time—ATP is quickly hydrolyzed, usually within a minute.
- This chemical energy is used to do chemical, mechanical, and electrical work in the body and to maintain body temperature.

Problems

WL Interactive versions of these problems may be assigned in OWL.

Orange-numbered problems are applied.

Section 27.1 What Is Metabolism?

27.1 To what end product is the energy of foods converted in the catabolic pathways?

27.2 (a) How many sequences are in the common catabolic pathway?

(b) Name these sequences.

Section 27.2 What Are Mitochondria and What Role Do They Play in Metabolism?

27.3 (a) How many membranes do mitochondria have?

(b) Which membrane is permeable to ions and small molecules?

27.4 How do the enzymes of the common pathway find their way into the mitochondria?

27.5 What are cristae, and how are they related to the inner membrane of mitochondria?

27.6 (a) Where are the enzymes of the citric acid cycle located?

(b) Where are the enzymes of oxidative phosphorylation located?

Section 27.3 What Are the Principal Compounds of the Common Metabolic Pathway?

27.7 How many high-energy phosphate bonds are in the ATP molecule?

27.8 What are the products of the following reaction? Complete the equation.

$$AMP + H_2O \xrightarrow{H^+}$$

27.9 Which yields more energy, (a) the hydrolysis of ATP to ADP or (b) the hydrolysis of ADP to AMP?

27.10 How much ATP is needed for normal daily activity in humans?

27.11 What kind of chemical bond exists between the ribitol and the phosphate group in FAD?

27.12 When NAD^+ is reduced, two electrons enter the molecule, together with one H^+ ion. Where in the product will the two electrons be located?

27.13 Which atoms in the flavin portion of FAD are reduced to yield $FADH_2$?

27.14 NAD^+ has two ribose units in its structure; FAD has a ribose and a ribitol. What is the relationship between these molecules?

27.15 In the common catabolic pathway, a number of important molecules act as carriers (transfer agents).

(a) Which is the carrier of phosphate groups?

(b) Which are the coenzymes transferring hydrogen ions and electrons?

(c) What kind of groups does coenzyme A carry?

27.16 The ribitol in FAD is bound to phosphate. What is the nature of this bond? On the basis of the energies of the different bonds in ATP, estimate how much energy (in kcal/mol) would be obtained from the hydrolysis of this bond.

27.17 What kind of chemical bond exists between the pantothenic acid and mercaptoethylamine in the structure of CoA?

27.18 Name the vitamin B molecules that are part of the structure of (a) NAD^+, (b) FAD, and (c) coenzyme A.

27.19 In both NAD^+ and FAD, the vitamin B portion of the molecule is the active part. Is this also true for CoA? Explain.

27.20 What type of compound is formed when coenzyme A reacts with acetate?

27.21 The fats and carbohydrates metabolized by our bodies are eventually converted to a single compound. What is it?

Section 27.4 What Role Does the Citric Acid Cycle Play in Metabolism?

27.22 The first step in the citric acid cycle is abbreviated as:

$$C_2 + C_4 = C_6$$

(a) What do these symbols stand for?

(b) What are the common names of the three compounds involved in this reaction?

27.23 What is the only C_5 compound in the citric acid cycle?

27.24 Identify by number those steps of the citric acid cycle that are not redox reactions.

27.25 Which substrate in the citric acid cycle is oxidized by FAD? What is the oxidation product?

27.26 In Steps ③ and ⑤ of the citric acid cycle, the compounds are shortened by one carbon each time. What is the form of this one-carbon compound? What happens to it in the body?

27.27 According to Table 23.1, to what class of enzymes does fumarase belong?

27.28 List all the enzymes or enzyme systems of the citric acid cycle that could be classified as oxidoreductases.

27.29 Is ATP directly produced during any step of the citric acid cycle? Explain.

27.30 There are four dicarboxylic acid compounds, each containing four carbons, in the citric acid cycle. Which is (a) the least oxidized and (b) the most oxidized?

27.31 Why is a many-step cyclic process more efficient in utilizing energy from food than a single-step combustion?

27.32 Did the two CO_2 molecules given off in one turn of the citric acid cycle originate from the entering acetyl group?

27.33 Which intermediates of the citric acid cycle contain $C{=}C$ double bonds?

27.34 The citric acid cycle can be regulated by the body; that is, it can be slowed down or speeded up. What mechanism controls this process?

27.35 Oxidation is defined as loss of electrons. When oxidative decarboxylation occurs, as in Step ④ of the citric acid cycle, where do the electrons of the α-ketoglutarate go?

Section 27.5 How Do Electron and H^+ Transport Take Place?

27.36 What is the main function of oxidative phosphorylation (the electron transport chain)?

27.37 What are the mobile electron carriers of oxidative phosphorylation?

27.38 In each complex of the electron transport system, the redox reaction occurs mostly around Fe ions.
   (a) Identify the compounds that contain such Fe centers.
   (b) Identify the compounds that contain ion centers other than iron.

27.39 What kind of motion is set up in the proton-translocating ATPase by the passage of H^+ from the intermembrane space into the matrix?

27.40 The following reaction is a reversible reaction:

$$NADH \rightleftharpoons NAD^+ + H^+ + 2e^-$$

   (a) Where does the forward reaction occur in the common catabolic pathway?
   (b) Where does the reverse reaction occur?

27.41 In oxidative phosphorylation, water is formed from H^+, e^-, and O_2. Where does this take place?

27.42 At what points in oxidative phosphorylation are the H^+ ions and the electrons separated from each other?

27.43 How many ATP molecules are generated (a) for each H^+ translocated through the ATPase complex and (b) for each C_2 fragment that goes through the complete common catabolic pathway?

27.44 When H^+ is pumped out into the intermembrane space, is the pH there increased, decreased, or unchanged compared with that in the matrix?

Section 27.6 What Is the Role of the Chemiosmotic Pump in ATP Production?

27.45 What is the channel through which ions reenter the matrix of mitochondria?

27.46 The proton gradient accumulated in the intermembrane area of a mitochondrion drives the ATP-manufacturing enzyme, ATPase. Why do you think Mitchell called this concept the "chemiosmotic theory"?

27.47 Which part of the proton-translocating ATPase machinery is the catalytic unit? What chemical reaction does it catalyze?

27.48 When the interaction between the two parts of proton-translocating ATPase, F_0 and F_1, are disrupted, no energy production is possible. Which subunits maintain connections between F_0 and F_1, and what names are designated for these subunits?

Section 27.7 What Is the Energy Yield Resulting from Electron and H^+ Transport?

27.49 If each mole of ATP yields 7.3 kcal of energy upon hydrolysis, how many kilocalories of energy would you get from 1 g of CH_3COO^- entering the citric acid cycle?

27.50 A hexose (C_6) enters the common metabolic pathway in the form of two C_2 fragments.
   (a) How many molecules of ATP are produced from one hexose molecule?
   (b) How many O_2 molecules are used up in the process?

Section 27.8 How Is Chemical Energy Converted to Other Forms of Energy?

27.51 (a) How do muscles contract?
   (b) Where does the energy used in muscle contraction come from?

27.52 Give an example of the conversion of the chemical energy of ATP to electrical energy.

27.53 How is the enzyme phosphorylase activated?

Chemical Connections

27.54 (Chemical Connections 27A) What is a protonophore?

27.55 (Chemical Connections 27A) Oligomycin is an antibiotic that allows electron transport to continue, but stops phosphorylation in both bacteria and humans. Would you use it as an antibacterial drug for people? Explain.

Additional Problems

27.56 (a) What is the difference in structure between ATP and GTP?
   (b) Compared with ATP, would you expect GTP to carry more, less, or about the same amount of energy?

27.57 How many grams of CH_3COOH (from acetyl CoA) molecules must be metabolized in the common metabolic pathway to yield 87.6 kcal of energy?

27.58 What is the basic difference in the functional groups between citrate and isocitrate?

27.59 The passage of ions from the cytoplasmic side into the matrix generates mechanical energy. Where is this energy of motion exhibited first?

27.60 What kind of reaction occurs in the citric acid cycle when a C_6 compound is converted to a C_5 compound?

27.61 What structural characteristics do citric acid and malic acid have in common?

27.62 Two ketoacids are important in the citric acid cycle. Identify them and tell how they are manufactured.

27.63 Which filament of muscles is an enzyme, catalyzing the reaction that converts ATP to ADP?

27.64 One of the end products of food metabolism is water. How many molecules of H_2O are formed from the entry of each molecule of (a) $NADH + H^+$ and (b) $FADH_2$? (*Hint*: Use Figure 27.10.)

27.65 How many stereocenters are in isocitrate?

27.66 Acetyl CoA is labeled with radioactive carbon as shown: $CH_3*CO—S—CoA$. This compound enters the citric acid cycle. If the cycle is allowed to progress to only the α-ketoglutarate level, will the CO_2 expelled by the cell be radioactive? Explain.

27.67 Where is the H^+ ion channel located in the proton-translocating ATPase complex?

27.68 Is the passage of H^+ ion through the channel converted directly into chemical energy?

27.69 Does all the energy used in ATP synthesis come from the mechanical energy of rotation?

27.70 (a) In the citric acid cycle, how many steps can be classified as decarboxylation reactions?

(b) In each case, what is the concurrent oxidizing agent? (*Hint:* See Table 23.1.)

27.71 What is the role of succinate dehydrogenase in the citric acid cycle?

27.72 How many stereocenters are in malate?

27.73 What is the source of carbon dioxide that we exhale?

27.74 Does oxygen combine directly with carbon-containing molecules to produce carbon dioxide?

27.75 Some soft drinks contain citric acid as flavoring. Is it a good nutrient?

27.76 Is mitochondrial ATPase an integral membrane protein? Explain your answer.

27.77 Do all complexes of the electron transport chain generate enough energy to produce ATP?

27.78 Why is mitochondrial ATPase considered a motor protein?

27.79 Does the metabolism of cancer cells differ from that of normal cells?

27.80 Do genes that play a role in the development of cancer also affect metabolism?

27.81 Is there a connection between circadian rhythms and metabolism?

27.82 Does the suppression of metabolism, especially catabolic pathways, have anything to do with obesity?

27.83 In this chapter, we have primarily discussed oxidative processes in metabolism that provide energy. Many body processes, such as the biosynthesis of proteins and nucleic acids, require energy. Do you expect that these metabolic pathways will be reduction reactions? What is the reason for your answer?

27.84 Do you expect that the citric acid cycle will release energy or require energy? What is the reason for your answer?

27.85 Mitochondrial processes are more important in athletic events of long duration than ones of short duration. In other words, active mitochondria play a more important role in a marathon than in a 100-meter sprint. Why is this so?

27.86 Why is it somewhat misleading to study biochemical pathways separately?

Tying It Together

27.87 Why does citrate isomerize to isocitrate before any oxidation steps take place in the citric acid cycle?

27.88 Why is the material in this chapter called the common catabolic pathway, rather than giving that designation to any other metabolic reactions?

27.89 What are two ways in which iron is part of the structure of the proteins of the electron transport pathway?

27.90 Why is it necessary for the proteins of the electron transport chain to be integral membrane proteins?

27.91 Why is it necessary to have mobile electron carriers as part of the electron transport chain?

27.92 Why does the loss of CO_2 make the citric acid cycle irreversible?

Looking Ahead

27.93 Why is the citric acid cycle central to biosynthetic pathways as well as to catabolism?

27.94 Is there a significant difference in the energy yield of the central catabolic pathway if FAD is used as an electron carrier rather than NAD^+?

27.95 Are biosynthetic pathways likely to involve oxidation, like the common catabolic pathway, or reduction? Why?

27.96 Are biosynthetic pathways likely to release energy, like the common catabolic pathway, or require energy? Why?

Challenge Problems

27.97 In a typical human, body weight fluctuates vary little during the course of a day. How can this statement be consistent with the estimate that the human body manufactures as much as 40 kg of ATP every day?

27.98 When the electron transport pathway was first studied, researchers used inhibitors to block the flow of electrons in their work. Why is it likely that such inhibitors could help establish the order of carriers?

27.99 Oxygen does not appear in any of the reactions of the citric acid cycle, but it is considered part of aerobic metabolism. Why?

27.100 Is it likely that some of the important molecules for the transfer of phosphate groups, electrons, and acetyl groups will appear in other metabolic pathways discussed in future chapters?

28

Specific Catabolic Pathways: Carbohydrate, Lipid, and Protein Metabolism

Sign in to OWL at **www.cengage.com/owl** to view tutorials and simulations, develop problem-solving skills, and complete online homework assigned by your professor.

© Dennis Degnan/Corbis

The ballet dancer derives energy from catabolism of nutrients.

28.1 What Is the General Outline of Catabolic Pathways?

The food we eat serves two main purposes: (1) It fulfills our energy needs, and (2) it provides the raw materials to build the compounds our bodies need. Before either of these processes can take place, food—carbohydrates, fats, and proteins—must be broken down into small molecules that can be absorbed through the intestinal walls. We will deal with most of the details of digestion in Chapter 30. In this chapter, along with the preceding and following chapters, we will keep to our main train of thought: the chemical aspects of metabolism.

A. Carbohydrates

Complex carbohydrates (di- and polysaccharides) in the diet are broken down by enzymes and stomach acid to produce monosaccharides, the most important of which is glucose (Section 30.3). Glucose also comes from the enzymatic breakdown of glycogen that is stored in the liver and muscles until needed. Once monosaccharides are produced, they can be used either to build new oligo- and polysaccharides or to provide energy. The specific pathway by which energy is extracted from monosaccharides is called glycolysis (Sections 28.2 and 28.3).

B. Lipids

Ingested fats are hydrolyzed by lipases to glycerol and fatty acids or to monoglycerides, which are absorbed through the intestine (Section 30.4). In a similar fashion, complex lipids are hydrolyzed to smaller units before their absorption. As with carbohydrates, these smaller molecules (fatty acids, glycerol, and so on) can be used to build the complex molecules needed in membranes, they can be oxidized to provide energy, or they can be stored in **fat storage depots** (Figure 28.1). The stored fats can later be hydrolyzed to glycerol and fatty acids whenever they are needed as fuel.

The specific pathway by which energy is extracted from glycerol involves the same glycolysis pathway as that used for carbohydrates (Section 28.4). The specific pathway used by the cells to obtain energy from fatty acids is called β-oxidation (Section 28.5).

C. Proteins

As you might expect from your knowledge of their structures, proteins are hydrolyzed by HCl in the stomach and by digestive enzymes in the stomach (pepsin) and intestines (trypsin, chymotrypsin, and carboxypeptidases) to produce their constituent amino acids. The amino acids absorbed through the intestinal wall enter the **amino acid pool**. They serve as building blocks for proteins as needed and, to a smaller extent (especially during starvation), as a fuel for energy. In the latter case, the nitrogen of the amino acids is catabolized through oxidative deamination and the urea cycle and is expelled from the body as urea in the urine (Section 28.8). The carbon skeletons of the amino acids enter the common catabolic pathway (Chapter 27) as either α-ketoacids (pyruvic, oxaloacetic, and α-ketoglutaric acids) or acetyl coenzyme A (Section 28.9).

In all cases, *the specific pathways of carbohydrate, triglyceride (fat), and protein catabolism converge to the common catabolic pathway* (Figure 28.2). In this way, the body needs fewer enzymes to get energy from diverse food materials. Efficiency is achieved because a minimal number of chemical steps are required and because the energy-producing factories of the body are localized in the mitochondria.

28.2 What Are the Reactions of Glycolysis?

Glycolysis is the specific pathway by which the body gets energy from monosaccharides. The detailed steps in glycolysis are shown in Figure 28.3, and the most important features are shown schematically in Figure 28.4.

A. Glycolysis of Glucose

In the first steps of glucose metabolism, energy is consumed rather than released. At the expense of two molecules of ATP (which are converted to ADP), glucose is phosphorylated. First, glucose 6-phosphate is formed in Step ①; then, after isomerization to fructose 6-phosphate in Step ②, a second phosphate group is bonded to yield fructose 1,6-bisphosphate in Step ③. We can consider these steps to be the activation process.

In the second stage, the C_6 compound, fructose 1,6-bisphosphate, is broken into two C_3 fragments in Step ④. The two C_3 fragments, glyceraldehyde 3-phosphate and dihydroxyacetone phosphate, are in equilibrium (they can

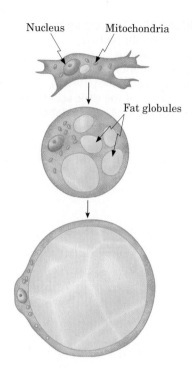

FIGURE 28.1 Storage of fat in a fat cell. As more and more fat droplets accumulate in the cytoplasm, they coalesce to form a very large globule of fat. Such a fat globule may occupy most of the cell, pushing the cytoplasm and the organelles to the periphery. (Modified from C. A. Villee, E. P. Solomon, and P. W. Davis, *Biology*, Philadelphia, Saunders College Publishing, 1985.)

Amino acid pool The free amino acids found both inside and outside cells throughout the body

Glycolysis The biochemical pathway that breaks down glucose to pyruvate, which yields chemical energy in the form of ATP and reduced coenzymes

FIGURE 28.2 The convergence of the specific pathways of carbohydrate, fat, and protein catabolism into the common catabolic pathway, which is made up of the citric acid cycle and oxidative phosphorylation.

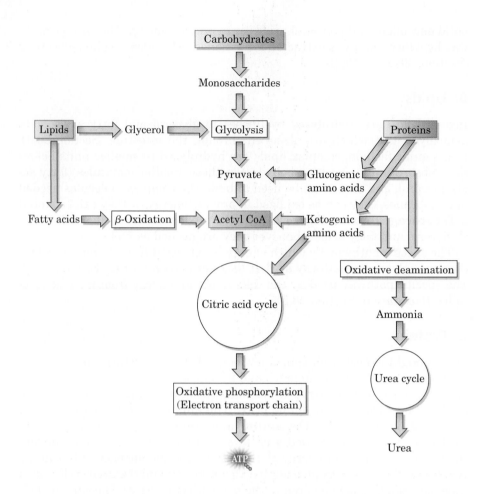

be converted to each other). Only glyceraldehyde 3-phosphate is oxidized in glycolysis, but as this species is removed from the equilibrium mixture, the equilibrium shifts (see the discussion of Le Chatelier's principle in Section 7.7) and dihydroxyacetone phosphate is converted to glyceraldehyde 3-phosphate.

In the third stage, glyceraldehyde 3-phosphate is oxidized to 1,3-bisphosphoglycerate in Step ⑤. The hydrogen of the aldehyde group is removed by the NAD^+ coenzyme. In Step ⑥, the phosphate from the carboxyl group is transferred to ADP, yielding ATP and 3-phosphoglycerate. The latter compound, after isomerization in Step ⑦ and dehydration in Step ⑧, is converted to phosphoenolpyruvate, which loses its remaining phosphate in Step ⑨ and yields pyruvate and another ATP molecule. [In Step ⑨, after hydrolysis of the phosphate, the resulting enol of pyruvic acid tautomerizes to the more stable keto form (Section 17.5).] Step ⑨ is also the "payoff" step, as the two ATP molecules produced here (one for each C_3 fragment) represent the net yield of ATPs in glycolysis. Step ⑨ is catalyzed by an enzyme, pyruvate kinase, whose active site was depicted in Chemical Connections 23C. This enzyme plays a key role in the regulation of glycolysis. For example, pyruvate kinase is inhibited by ATP and activated by AMP. Thus, when plenty of ATP is available, glycolysis is shut down; when ATP is scarce and AMP levels are high, the glycolytic pathway is speeded up.

All of these glycolysis reactions occur in the cytoplasm outside the mitochondria. Because they occur in the absence of O_2, they are also called reactions of the **anaerobic pathway**. As indicated in Figure 28.4, the end product of glycolysis, pyruvate, does not accumulate in the body. In certain bacteria and yeast, pyruvate undergoes decarboxylation in Step ⑩ to produce ethanol. In some bacteria, and in mammals in the absence of oxygen,

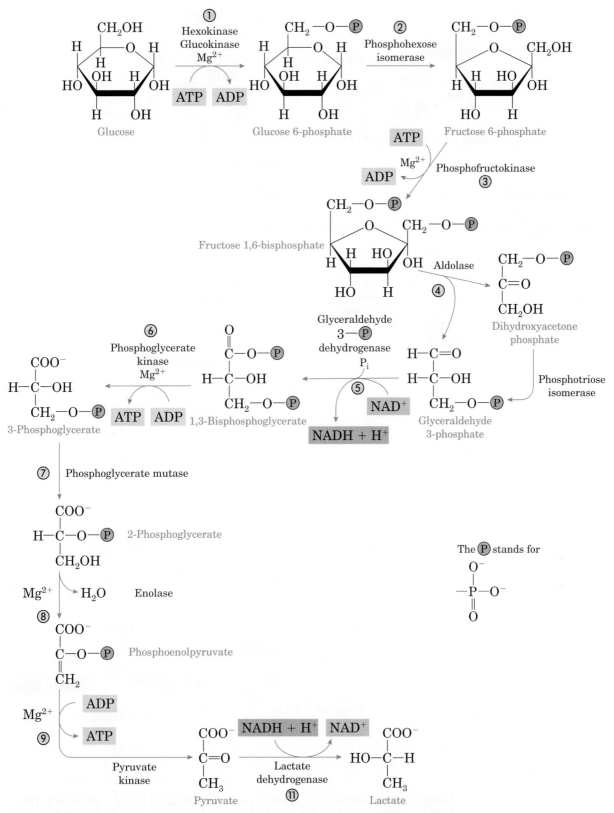

FIGURE 28.3 Glycolysis, the pathway of glucose metabolism. (Steps ⑩, ⑫, and ⑬ are shown in Figure 28.4.) Some of the steps are reversible, but equilibrium arrows are not shown (they appear in Figure 28.4).

pyruvate is reduced to lactate in Step ⑪. In Section 27.1, we mentioned that the products of sugar metabolism in cancer cells do not enter the central metabolic pathway by entering the citric acid cycle. Now we are in a position to look at the specific difference between cancer cells and normal cells. In cancer

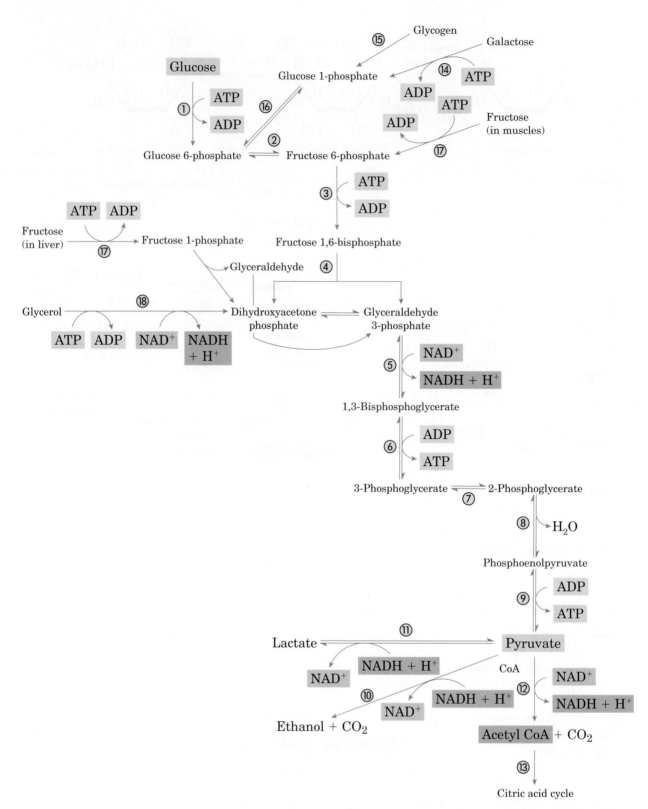

FIGURE 28.4 An overview of glycolysis and the entries to it and exits from it. The equilibrium arrows represent reversible steps. A change in conditions can, and does, affect the relative amounts of starting material metabolized by glycolysis, or the fate of the pyruvate produced.

cells, pyruvate is primarily converted to lactate. In normal cells that carry out aerobic metabolism, pyruvate enters the citric acid cycle. The reactions that produce ethanol in organisms capable of alcoholic fermentation operate in reverse when humans metabolize ethanol. Acetaldehyde (Section 17.2), which

Lactate Accumulation

Many athletes suffer muscle cramps when they engage in strenuous exercise (see Chapter 8). This problem results from a shift from normal glucose catabolism (glycolysis: citric acid cycle: oxidative phosphorylation) to that of lactate production (see Step ⑪ in Figure 28.4). During exercise, oxygen is used up rapidly, which slows down the rate of the common catabolic pathway. The demand for energy makes anaerobic glycolysis proceed at a high rate, but because the aerobic (oxygen-demanding) pathways are slowed down, not all the pyruvate produced in glycolysis can enter the citric acid cycle. The excess pyruvate ends up as lactate, which causes painful muscle contractions.

The same shift in catabolism occurs in heart muscle when coronary thrombosis leads to cardiac arrest. The blockage of the artery leading to the heart muscles cuts off the oxygen supply. The common catabolic pathway and its ATP production are consequently shut off. Glycolysis proceeds at an accelerated rate, causing lactate to accumulate. The heart muscle contracts, producing a cramp. Just as in skeletal muscle, massage of heart muscles can relieve the cramp and start the heart beating. Even if heartbeat is

restored within 3 minutes (the amount of time the brain can survive without being damaged), acidosis may develop as a result of the cardiac arrest. Therefore, at the same time that efforts are underway to start the heart beating by chemical, physical, or electrical means, an intravenous infusion of 8.4% bicarbonate solution is given to combat acidosis.

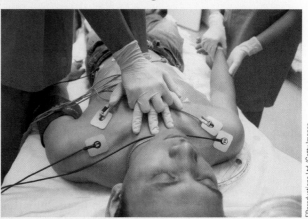

Knowledge of biochemistry is a big help in treating cardiac arrest.

Test your knowledge with Problems 28.43 and 28.44.

is a product of one of these reactions, is a toxic substance that's responsible for much of the damage in fetal alcohol syndrome. Transfer of nutrients and oxygen to the fetus is depressed, with tragic consequences.

B. Entrance to the Citric Acid Cycle

Pyruvate is not the end product of glucose metabolism. Most importantly, pyruvate goes through an oxidative decarboxylation in the presence of coenzyme A in Step ⑫ to produce acetyl CoA:

$$NAD^+ + CH_3 - \overset{\overset{O}{\|}}{C} - COO^- + CoA - SH \longrightarrow CH_3 - \overset{\overset{O}{\|}}{C} - S - CoA + CO_2 + NADH + H^+$$

Pyruvate   Acetyl coenzyme A

This reaction is catalyzed by the enzyme pyruvate dehydrogenase, which sits on the inner membrane of the mitochondrion. The reaction produces acetyl CoA, CO_2, and NADH + H$^+$. The acetyl CoA then enters the citric acid cycle in Step ⑬ and goes through the common catabolic pathway.

In summary, after converting complex carbohydrates to glucose, the body gets energy from glucose by converting it to acetyl CoA (by way of pyruvate) and then using the acetyl CoA as a starting material for the common catabolic pathway.

C. Pentose Phosphate Pathway

As we saw in Figure 28.4, glucose 6-phosphate plays a central role in several different entries into the glycolytic pathway. However, glucose 6-phosphate can

FIGURE 28.5 Simplified schematic representation of the pentose phosphate pathway, also called a shunt. Steps 19 and 21 in this figure represent multiple steps in the actual pathway.

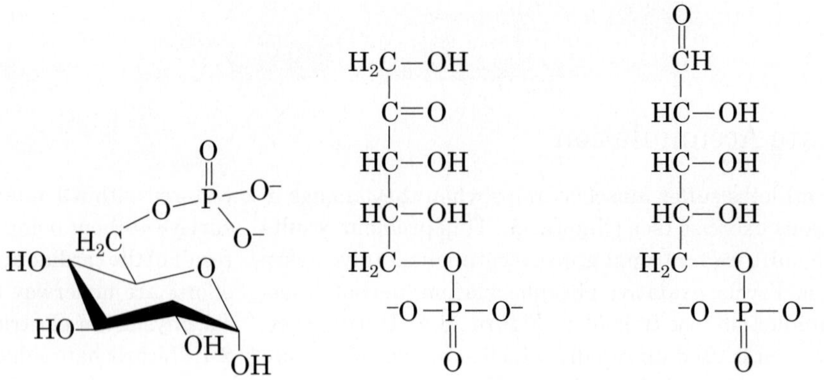

α-D-Glucose 6-phosphate

Ribulose 5-phosphate

Ribose 5-phosphate

Glucose 6-phosphate $+$ 2NADP$^+$ $\xrightarrow{\text{⑲}}$ Ribulose 5-phosphate $+$ 2NADPH $+$ CO$_2$

⑳

Ribose 5-phosphate

㉑

Glyceraldehyde 3-phosphate

Pentose phosphate pathway The biochemical pathway that produces ribose and NADPH from glucose 6-phosphate or, alternatively, releases energy

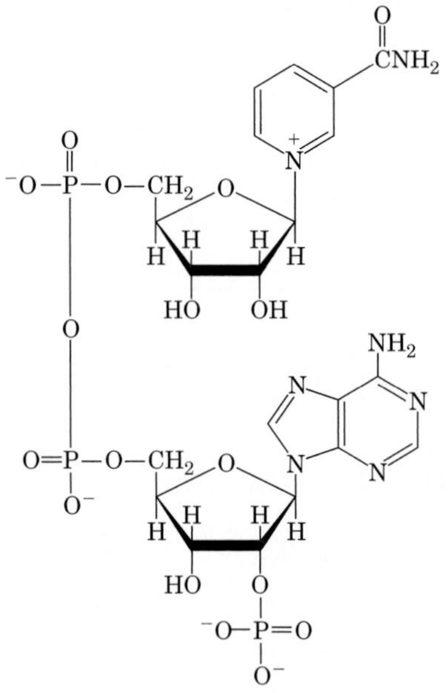

Nicotinamide adenine dinucleotide phosphate (NADP$^+$)

also be used by the body for other purposes, not just for the production of energy in the form of ATP. Most importantly, glucose 6-phosphate can be shunted to the **pentose phosphate pathway** in Step ⑲ (Figure 28.5). This pathway has the capacity to produce NADPH and ribose in Step ⑳ as well as energy.

NADPH is needed in many biosynthetic processes, including synthesis of unsaturated fatty acids (Section 29.3), cholesterol, and amino acids as well as photosynthesis (Chemical Connections 29A) and the reduction of ribose to deoxyribose for DNA. Ribose is needed for the synthesis of RNA (Section 25.3). Therefore, when the body needs these synthetic ingredients more than energy, glucose 6-phosphate is used in the pentose phosphate pathway. When energy is needed, glucose 6-phosphate remains in the glycolytic pathway—even ribose 5-phosphate can be channeled back to glycolysis through glyceraldehyde 3-phosphate. Through this reversible reaction, the cells can also obtain ribose directly from the glycolytic intermediates. In addition, NADPH is badly needed in red blood cells as a defense against oxidative damages. Glutathione is the main agent used to keep hemoglobin in its reduced form. It is regenerated by NADPH, so an insufficient supply of NADPH leads to the destruction of red blood cells, causing severe anemia.

28.3 What Is the Energy Yield from Glucose Catabolism?

Using Figure 28.4, let us sum up the energy derived from glucose catabolism in terms of ATP production. First, however, we must take into account the fact that glycolysis takes place in the cytoplasm, whereas oxidative phosphorylation occurs in the mitochondria. Therefore, the NADH + H$^+$ produced in glycolysis in the cytoplasm must be converted to NADH in the mitochondria before it can be used in oxidative phosphorylation.

NADH is too large to cross the mitochondrial membrane. Two routes are available to get the electrons into the mitochondria; they have different efficiencies. In glycerol 3-phosphate transport, which operates in muscle and nerve cells, only two ATP molecules are produced for each NADH + H$^+$.

Table 28.1 ATP Yield from Complete Glucose Metabolism

Step Numbers in Figure 28.4	Chemical Steps	Number of ATP Molecules Produced
① ② ③	Activation (glucose $\longrightarrow$ 1,6-fructose bisphosphate)	−2
⑤	Phosphorylation 2 (glyceraldehyde 3-phosphate $\longrightarrow$ 1,3-bisphosphoglycerate), producing 2 (NADH + H$^+$) in cytosol	4
⑥ ⑨	Dephosphorylation 2 (1,3-bisphosphoglycerate $\longrightarrow$ pyruvate)	4
⑫	Oxidative decarboxylation 2 (pyruvate $\longrightarrow$ acetyl CoA), producing 2 (NADH + H$^+$) in the mitochondrion	6
⑬	Oxidation of two C$_2$ fragments in the citric acid cycle and oxidative phosphorylation common pathways, producing 12 ATP for each C$_2$ fragment	24
	Total	36

In the other transport route, which operates in the heart and the liver, three ATP molecules are produced for each NADH + H$^+$ produced in the cytoplasm, as is the case in the mitochondria (Section 27.7). Because most energy production takes place in skeletal muscle cells, when we construct the energy balance sheet, we use two ATP molecules for each NADH + H$^+$ produced in the cytoplasm. (Muscles attached to bones are called skeletal muscles; cardiac muscle is in a different category.)

Armed with this knowledge, we are ready to calculate the energy yield of glucose in terms of ATP molecules produced in skeletal muscles. Table 28.1 shows this calculation. In the first stage of glycolysis (Steps ①, ②, and ③), two ATP molecules are used up, but this loss is more than compensated for by the production of 14 ATP molecules in Steps ⑤, ⑥, ⑨, and ⑫ and in the conversion of pyruvate to acetyl CoA. The net yield of these steps is 12 ATP molecules. As we saw in Section 27.7, the oxidation of one acetyl CoA molecule produces 12 ATP molecules and one glucose molecule provides two acetyl CoA molecules. Therefore, the total net yield from metabolism of one glucose molecule in skeletal muscle is 36 molecules of ATP, or 6 ATP molecules per carbon atom.

$$C_6H_{12}O_6 + 6O_2 \longrightarrow 6CO_2 + 6H_2O$$

If the same glucose is metabolized in the heart or liver, the electrons of the two NADH molecules produced in glycolysis are transported into the mitochondrion by the malate-aspartate shuttle. Through this shuttle, the two NADH molecules yield a total of 6 ATP molecules, so that in this case, 38 ATP molecules are produced for each glucose molecule. It is instructive to note that most of the energy (in the form of ATP) from glucose is produced in the common metabolic pathway. Recent investigations suggest that approximately 30 to 32 ATP molecules are actually produced per glucose molecule (2.5 ATP/NADH and 1.5 ATP/FADH$_2$). Further research into the complexity of the oxidative phosphorylation pathway is needed to verify these numbers. We alluded to this point briefly in Chapter 27 when we said that the ATP yields were reported to the nearest whole number. Here we see how the complexity of oxidative phosphorylation affects energy metabolism.

Glucose is not the only monosaccharide that can be used as an energy source. Other hexoses, such as galactose (Step ⑭) and fructose (Step ⑰), enter the glycolysis pathway at the stages indicated in Figure 28.4. They also yield 36 molecules of ATP per hexose molecule. Furthermore, glycogen stored in liver, muscle cells, and elsewhere can be converted by enzymatic breakdown and phosphorylation to glucose 1-phosphate (Step ⑮). This compound, in turn, isomerizes to glucose 6-phosphate, providing an entry into the glycolytic pathway (Step ⑯). The pathway in which glycogen breaks down to glucose is called **glycogenolysis**.

Glycogenolysis The biochemical pathway for the breakdown of glycogen to glucose

Now that we have seen the catabolic reactions of carbohydrates, we turn our attention to another major source of energy, the catabolism of lipids. Recall that for triglycerides, which are the main storage form of the chemical energy of lipids, we have to consider two parts, glycerol and fatty acids.

28.4 How Does Glycerol Catabolism Take Place?

The glycerol hydrolyzed from fats or complex lipids (Chapter 21) can also be a rich energy source. The first step in glycerol utilization is an activation step. The body uses one ATP molecule to form glycerol 1-phosphate, which is the same as glycerol 3-phosphate:

$$
\begin{array}{ccc}
\text{CH}_2\text{OH} & \text{CH}_2\text{O}-\textcircled{P} & \text{CH}_2\text{O}-\textcircled{P} \\
| & | & | \\
\text{CHOH} & \xrightarrow{\text{ATP} \quad \text{ADP}} \quad \text{CHOH} & \xrightarrow{\text{NAD}^+ \quad \text{NADH} + \text{H}^+} \quad \text{C}=\text{O} \\
| & | & | \\
\text{CH}_2\text{OH} & \text{CH}_2\text{OH} & \text{CH}_2\text{OH} \\
\text{Glycerol} & \text{Glycerol 1-phosphate} & \text{Dihydroxyacetone phosphate}
\end{array}
$$

The glycerol phosphate is oxidized by NAD$^+$ to dihydroxyacetone phosphate, yielding NADH + H$^+$ in the process. Dihydroxyacetone phosphate then enters the glycolysis pathway (Step ⑱ in Figure 28.4) and is isomerized to glyceraldehyde 3-phosphate. A net yield of 20 ATP molecules is produced from each glycerol molecule, or 6.7 ATP molecules per carbon atom.

28.5 What Are the Reactions of β-Oxidation of Fatty Acids?

As early as 1904, Franz Knoop, working in Germany, proposed that the body utilizes fatty acids as an energy source by breaking them down into fragments. Prior to fragmentation, the β-carbon (the second carbon atom from the COOH group) is oxidized:

$$
-\text{C}-\text{C}-\text{C}-\overset{\beta}{\text{C}}-\overset{\alpha}{\text{C}}-\text{COOH}
$$

β-oxidation The biochemical pathway that degrades fatty acids to acetyl CoA by removing two carbons at a time and yielding energy

The name **β-oxidation** has its origin in Knoop's prediction. It took about 50 years to establish the mechanism by which fatty acids are utilized as an energy source.

Figure 28.6 depicts the overall process of fatty acid metabolism with a saturated fatty acid. As is the case with the other foods we have seen, the first step involves activation. In the general case of lipid catabolism, this activation occurs in the cytosol, where the fat was previously hydrolyzed to glycerol and fatty acids. It converts ATP to AMP and inorganic phosphate (Step ①), which is equivalent to the cleavage of two high-energy phosphate bonds. The chemical energy derived from the hydrolysis of ATP is built into the compound acyl CoA, which forms when the fatty acid combines with coenzyme A. The fatty acid oxidation occurs inside the mitochondrion, so the acyl group of acyl CoA must pass through the mitochondrial membrane. Carnitine is the acyl group transporter. The enzyme system that catalyzes the process is carnitine acyltransferase.

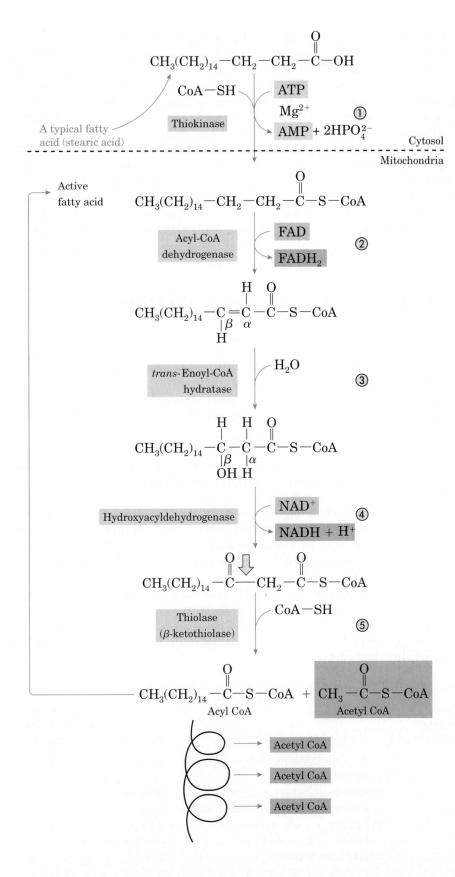

FIGURE 28.6 The β-oxidation spiral of fatty acids. Each loop in the spiral contains two dehydrogenations, one hydration, and one fragmentation. At the end of each loop, one acetyl CoA molecule is released.

Once the fatty acid in the form of acyl CoA is inside the mitochondrion, the β-oxidation starts. In the first oxidation (dehydrogenation; Step ②), two hydrogens are removed, creating a *trans* double bond between the alpha and beta carbons of the acyl chain. The hydrogens and electrons are picked up by FAD.

In Step ③, the double bond is hydrated. An enzyme specifically places the hydroxyl group on C-3, the beta carbon. The second oxidation (dehydrogenation; Step ④) requires NAD^+ as a coenzyme. The two hydrogens and electrons removed are transferred to the NAD^+ to form $NADH + H^+$. In the process, a secondary alcohol is oxidized to a ketone at the beta carbon. In Step ⑤, the enzyme thiolase cleaves the terminal C_2 fragment (an acetyl CoA) from the chain and the rest of the molecule is bonded to a new molecule of coenzyme A.

The cycle then starts again with the remaining acyl CoA, which is now two carbon atoms shorter. At each turn of the cycle, one acetyl CoA is produced. Most fatty acids contain an even number of carbon atoms. The cyclic spiral continues until it reaches the last four carbon atoms. When this fragment enters the cycle, two acetyl CoA molecules are produced in the fragmentation step.

The β-oxidation of unsaturated fatty acids proceeds in the same way. An extra step is involved, in which the *cis* double bond is isomerized to a *trans* bond, but otherwise the spiral is the same.

28.6 What Is the Energy Yield from Stearic Acid Catabolism?

To compare the energy yield from fatty acids with that of other foods, let us select a typical and quite abundant fatty acid—stearic acid, the C_{18} saturated fatty acid.

We start with the initial step, in which energy is used up rather than produced. The reaction breaks two high-energy phosphoric anhydride bonds:

$$\text{ATP} \longrightarrow \text{AMP} + 2P_i + \text{energy}$$

This reaction is equivalent to hydrolyzing two molecules of ATP to ADP. In each cycle of the spiral, we obtain one $FADH_2$, one $NADH + H^+$, and one acetyl CoA. Stearic acid (C_{18}) goes through seven cycles in the spiral before it reaches the final stage. In the last (eighth) cycle, one $FADH_2$, one $NADH + H^+$, and two acetyl CoA molecules are produced. Now we can add up the energy. Table 28.2 shows that for a C_{18} fatty acid, we obtain a total of 146 ATP molecules.

It is instructive to compare the energy yield from fats with that from carbohydrates, as both are important constituents of the diet. In Section 28.2, we saw that glucose produces 36 ATP molecules—that is, 6 ATP molecules for each carbon atom. For stearic acid, there are 146 ATP molecules and 18 carbons, or $146/18 = 8.1$ ATP molecules per carbon atom. The ATP produced from the glycerol portion of fats adds to the total. Fatty acids have a higher caloric value than carbohydrates.

Table 28.2 ATP Yield from Complete Stearic Acid Metabolism

Step Number in Figure 28.6	Chemical Steps	Happens	Number of ATP Molecules Produced
①	Activation (stearic acid ⟶ stearyl CoA	Once	−2
②	Dehydrogenation (acyl CoA ⟶ *trans*-enoyl CoA), producing $FADH_2$	8 times	16
④	Dehydrogenation (hydroxyacyl CoA ⟶ ketoacyl CoA), producing $NADH + H^+$	8 times	24
	C_2 fragment (acetyl CoA ⟶ common catabolic pathway), producing 12 ATP for each C_2 fragment	9 times	108
		Total	146

28.7 What Are Ketone Bodies?

In spite of the high caloric value of fats, the body preferentially uses glucose as an energy supply. When an animal is well fed (plenty of sugar intake), fatty acid oxidation is inhibited and fatty acids are stored in the form of neutral fat in fat depots. When physical exercise demands energy, when the glucose supply dwindles (as in fasting or starvation), or when glucose cannot be utilized (as in the case of diabetes), the β-oxidation pathway of fatty acid metabolism is mobilized. In some pathological conditions, glucose may not be available at all, giving added importance to this point.

Unfortunately, low glucose supply also slows down the citric acid cycle. This lag happens because some oxaloacetate is essential for the continuous operation of the citric acid cycle (Figure 27.8). Oxaloacetate is produced from malate, but it is also produced by the carboxylation of phosphoenolpyruvate (PEP):

$$CO_2 + \underset{\text{PEP}}{\begin{array}{c} O \\ \| \\ {}^{-}O-P-O \\ | \\ O^{-} \\ | \\ C-COO^{-} \\ \| \\ CH_2 \end{array}} \quad \xrightarrow[\text{GDP} \quad \text{GTP}]{} \quad \underset{\text{Oxaloacetate}}{\begin{array}{c} O \\ \| \\ C-COO^{-} \\ | \\ H_2C-COO^{-} \end{array}}$$

If there is no glucose, there will be no glycolysis, no PEP formation, and, therefore, greatly reduced oxaloacetate production.

Thus, even though the fatty acids are oxidized, not all of the resulting fragments (acetyl CoA) can enter the citric acid cycle because not enough oxaloacetate is present. As a result, acetyl CoA builds up in the body, with the following consequences.

The liver is able to condense two acetyl CoA molecules to produce acetoacetyl CoA:

$$\underset{\text{Acetyl CoA}}{2CH_3-\overset{\overset{\text{O}}{\|}}{C}-SCoA} \longrightarrow \underset{\text{Acetoacetyl CoA}}{CH_3-\overset{\overset{\text{O}}{\|}}{C}-CH_2-\overset{\overset{\text{O}}{\|}}{C}-SCoA} + CoASH$$

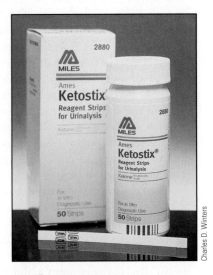

Test kit for the presence of ketone bodies in the urine.

Charles D. Winters

When the acetoacetyl CoA is hydrolyzed, it yields acetoacetate, which can be reduced to form β-hydroxybutyrate:

$$\underset{\text{Acetoacetyl CoA}}{CH_3-\overset{\overset{\text{O}}{\|}}{C}-CH_2-\overset{\overset{\text{O}}{\|}}{C}-SCoA} \xrightarrow{H_2O} \underset{\text{Acetoacetate}}{CH_3-\overset{\overset{\text{O}}{\|}}{C}-CH_2-\overset{\overset{\text{O}}{\|}}{C}-O^{-}} + CoASH + H^{+}$$

NADH + H⁺ → NAD⁺

H⁺ → CO₂

$$\underset{\beta\text{-Hydroxybutyrate}}{CH_3-\overset{\overset{\text{H}}{|}}{\underset{\overset{|}{OH}}{C}}-CH_2-\overset{\overset{\text{O}}{\|}}{C}-O^{-}}$$

$$\underset{\text{Acetone}}{CH_3-\overset{\overset{}{\underset{\overset{\|}{O}}{C}}}-CH_3}$$

These two compounds, together with smaller amounts of acetone, are collectively called **ketone bodies**. Under normal conditions, the liver sends these compounds into the bloodstream to be carried to tissues and utilized there as a source of energy via the common catabolic pathway. The brain,

Ketone bodies A collective name for acetone, acetoacetate, and β-hydroxybutyrate; compounds produced from acetyl CoA in the liver that are used as a fuel for energy production by muscle cells and neurons

Chemical Connections 28B

Ketoacidosis in Diabetes

In untreated diabetes, the glucose concentration in the blood is high because a lack of insulin prevents utilization of glucose by the cells. Regular injections of insulin can remedy this situation. However, in some stressful conditions, **ketoacidosis** can still develop.

A typical case is a diabetic patient who has been admitted to the hospital in a semi-comatose state. He showed signs of dehydration, his skin was inelastic and wrinkled, his urine showed high concentrations of glucose and ketone bodies, and his blood contained excess glucose and had a pH of 7.0, a drop of 0.4 pH units from normal, which is an indication of severe acidosis. The patient's urine also contained the bacterium *Escherichia coli*. This indication of a urinary tract infection explained why the normal doses of insulin were insufficient to prevent ketoacidosis.

The stress of infection can upset the normal control of diabetes by changing the balance between administered insulin and other hormones produced in the body. This imbalance happened during the patient's infection, and his body started to produce ketone bodies in large quantities. Both glucose and ketone bodies appear in the blood before they show up in the urine.

The acidic nature of ketone bodies (acetoacetic acid and β-hydroxybutyric acid) lowers the blood pH. A large drop in pH is prevented by the bicarbonate/carbonic acid buffer (Section 9.10), but even a drop of 0.3 to 0.5 pH units is sufficient to decrease the Na^+ concentration. Such a decrease of Na^+ ions in the interstitial fluids draws out K^+ ions from the cells. This, in turn, impairs brain function and leads to coma. During the secretion of ketone bodies and glucose in the urine, a lot of water is lost, the body becomes dehydrated, and the blood volume shrinks. As a consequence, the blood pressure drops, and the pulse rate increases to compensate for it. Smaller quantities of nutrients reach the brain cells, which can also cause coma.

The patient mentioned here was infused with physiological saline solution to remedy his dehydration. Extra doses of insulin restored his glucose level to normal, and antibiotics cured the urinary infection.

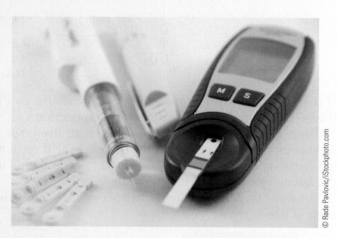

Devices that allow diabetics to monitor blood glucose levels are widely marketed.

Test your knowledge with Problems 28.45 and 28.46.

for example, normally uses glucose as an energy source. During periods of starvation, however, ketone bodies may serve as the major energy source for the brain. Normally, the concentration of ketone bodies in the blood is low. During starvation and in untreated diabetes mellitus, however, ketone bodies accumulate in the blood and can reach high concentrations. When this buildup occurs, the excess is secreted in the urine. A check of urine for ketone bodies is used in the diagnosis of diabetes.

Example 28.1 Counting ATPs

Ketone bodies are a source of energy, especially during dieting and starvation. If acetoacetate is metabolized through β-oxidation and the common pathway, how many ATP molecules would be produced?

Strategy

First determine from Section 28.7 that two acetyl CoA molecules are produced. Then consult Table 28.2, the final step.

Solution

In Step ①, activation of acetoacetate to acetoacetyl-CoA requires 2 ATPs. Step ⑤ yields two acetyl CoA molecules, which enter the common catabolic pathway, yielding 12 ATPs for each acetyl CoA, for a total of 24 ATPs. The net yield, therefore, is 22 ATP molecules.

Problem 28.1

Which fatty acid yields more ATP molecules per carbon atom: (a) stearic acid or (b) lauric acid?

28.8 How Is the Nitrogen of Amino Acids Processed in Catabolism?

The proteins of our foods are hydrolyzed to amino acids in digestion. These amino acids are primarily used to synthesize new proteins. Unlike carbohydrates and fats, however, they cannot be stored, so excess amino acids are catabolized for energy production. Section 28.9 explains what happens to the carbon skeletons of the amino acids. Here, we discuss the catabolic fate of the nitrogen. Figure 28.7 gives an overview of the entire process of protein catabolism.

In the tissues, amino groups ($-NH_2$) freely move from one amino acid to another. The enzymes that catalyze these reactions are the transaminases. In essence, nitrogen catabolism in the liver occurs in three stages: transamination, oxidative deamination, and the urea cycle.

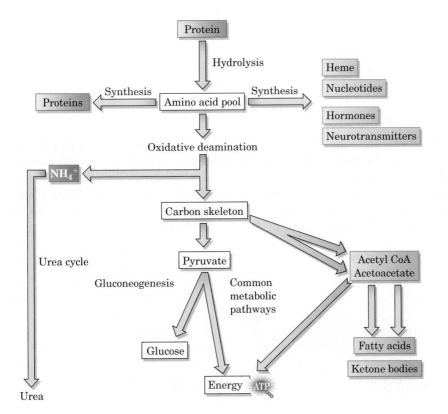

FIGURE 28.7 The overview of pathways in protein catabolism.

A. Transamination

Transamination The exchange of the amino group of an amino acid and a keto group of an α-ketoacid

In the first stage, **transamination**, amino acids transfer their amino groups to α-ketoglutarate:

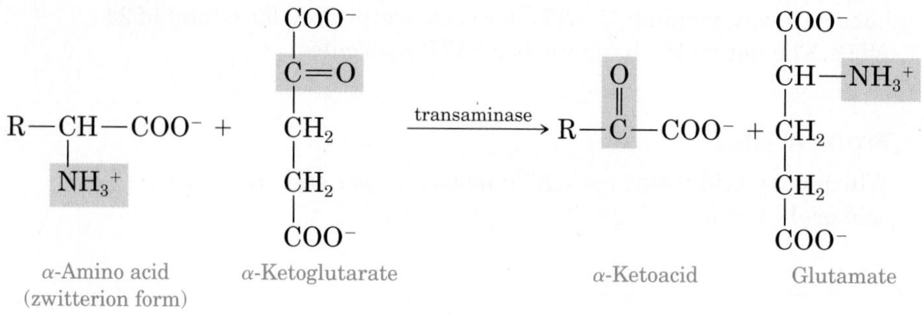

$$
\begin{array}{ccc}
& COO^- & \\
& | & \\
& C=O & \\
& | & \\
R-CH-COO^- + & CH_2 & \xrightarrow{\text{transaminase}} \\
| & | & \\
NH_3^+ & CH_2 & \\
& | & \\
& COO^- &
\end{array}
\qquad
\begin{array}{ccc}
& & COO^- \\
O & & | \\
\| & & CH-NH_3^+ \\
R-C-COO^- + & & CH_2 \\
& & | \\
& & CH_2 \\
& & | \\
& & COO^-
\end{array}
$$

α-Amino acid (zwitterion form) α-Ketoglutarate   α-Ketoacid   Glutamate

The carbon skeleton of the amino acid remains behind as an α-ketoacid, the catabolism of which is discussed in the next section.

B. Oxidative Deamination

Oxidative deamination The reaction in which the amino group of an amino acid is removed and an α-ketoacid is formed

The second stage of nitrogen catabolism is the **oxidative deamination** of glutamate, which occurs in the mitochondrion:

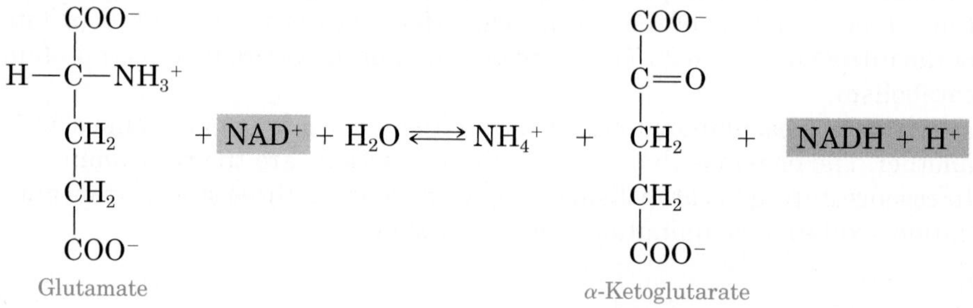

$$
\begin{array}{c}
COO^- \\
| \\
H-C-NH_3^+ \\
| \\
CH_2 \\
| \\
CH_2 \\
| \\
COO^-
\end{array}
\; + NAD^+ + H_2O \rightleftharpoons NH_4^+ +
\begin{array}{c}
COO^- \\
| \\
C=O \\
| \\
CH_2 \\
| \\
CH_2 \\
| \\
COO^-
\end{array}
+ \; NADH + H^+
$$

Glutamate   α-Ketoglutarate

The oxidative deamination yields NH_4^+ and regenerates α-ketoglutarate, which can again participate in the first stage (transamination). The $NADH + H^+$ produced in the second stage enters the oxidative phosphorylation pathway and eventually produces three ATP molecules. The body must get rid of NH_4^+ because both it and NH_3 are toxic.

C. Urea Cycle

Urea cycle A cyclic pathway that produces urea from ammonia and carbon dioxide

In the third stage, the NH_4^+ is converted to urea through the **urea cycle** (Figure 28.8). In Step ①, NH_4^+ is condensed with CO_2 in the mitochondrion to form an unstable compound, carbamoyl phosphate. This condensation occurs at the expense of two ATP molecules. In Step ②, carbamoyl phosphate is condensed with ornithine, a basic amino acid similar in structure to lysine, which does not occur in proteins, to produce citrulline.

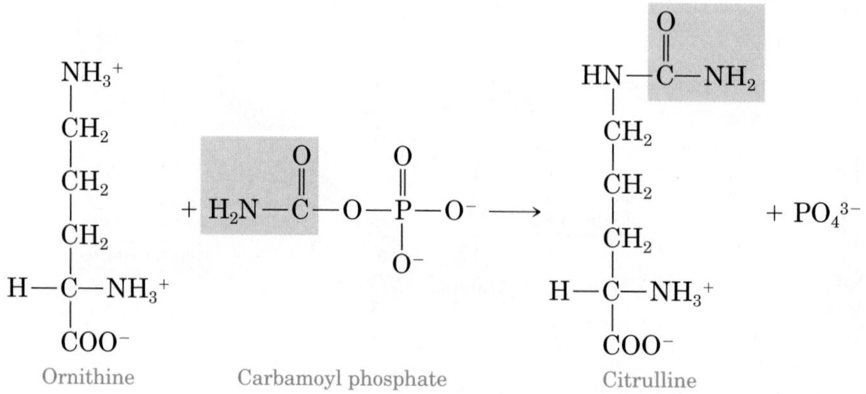

$$
\begin{array}{c}
NH_3^+ \\
| \\
CH_2 \\
| \\
CH_2 \\
| \\
CH_2 \\
| \\
H-C-NH_3^+ \\
| \\
COO^-
\end{array}
\; +
\begin{array}{c}
\;\;\;\;O\;\;\;\;\;\;O \\
\;\;\;\;\|\;\;\;\;\;\;\| \\
H_2N-C-O-P-O^- \\
\;\;\;\;\;\;\;\;\;\;\;| \\
\;\;\;\;\;\;\;\;\;\;\;O^-
\end{array}
\longrightarrow
\begin{array}{c}
\;\;\;\;\;\;\;\;O \\
\;\;\;\;\;\;\;\;\| \\
HN-C-NH_2 \\
| \\
CH_2 \\
| \\
CH_2 \\
| \\
CH_2 \\
| \\
H-C-NH_3^+ \\
| \\
COO^-
\end{array}
\; + PO_4^{3-}
$$

Ornithine   Carbamoyl phosphate   Citrulline

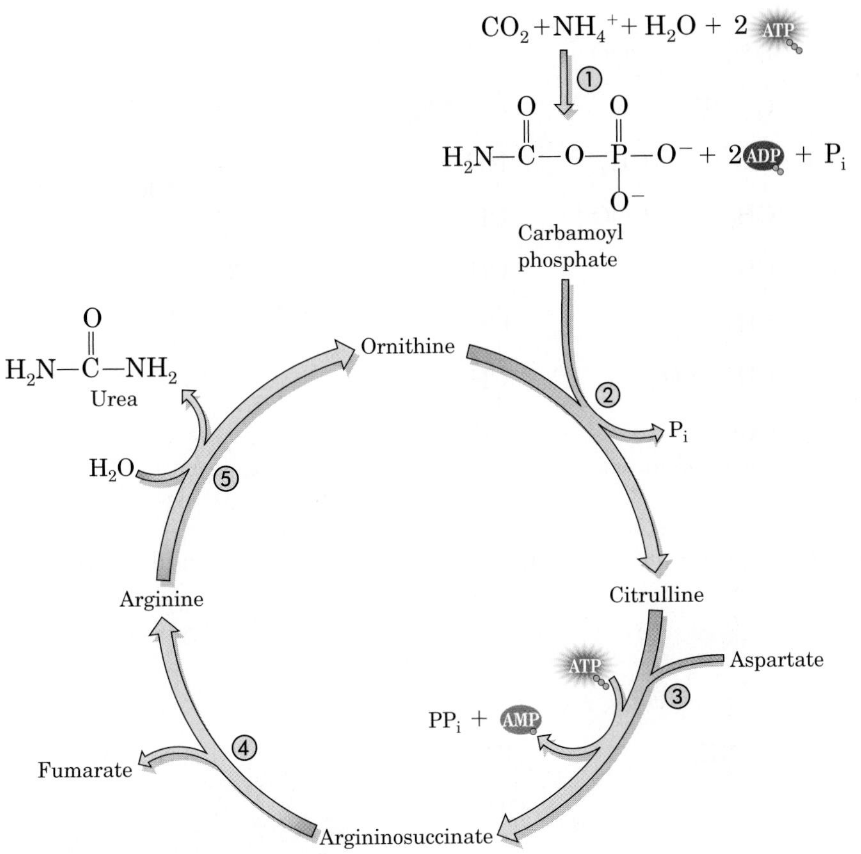

$$CO_2 + NH_4^+ + H_2O + 2 \text{ ATP}$$

(1)

$$H_2N - \overset{\overset{\displaystyle O}{\|}}{C} - O - \overset{\overset{\displaystyle O}{\|}}{\underset{\underset{\displaystyle O^-}{|}}{P}} - O^- + 2\text{ ADP} + P_i$$

Carbamoyl
phosphate

$$H_2N - \overset{\overset{\displaystyle O}{\|}}{C} - NH_2$$
Urea

Ornithine

(2)

P_i

H_2O

(5)

Arginine

Citrulline

ATP

(3)

Aspartate

PP_i + AMP

(4)

Fumarate

Argininosuccinate

FIGURE 28.8 The urea cycle.

The resulting citrulline diffuses out of the mitochondrion into the cytoplasm.

A second condensation reaction in the cytoplasm takes place between citrulline and aspartate, forming argininosuccinate (Step ③):

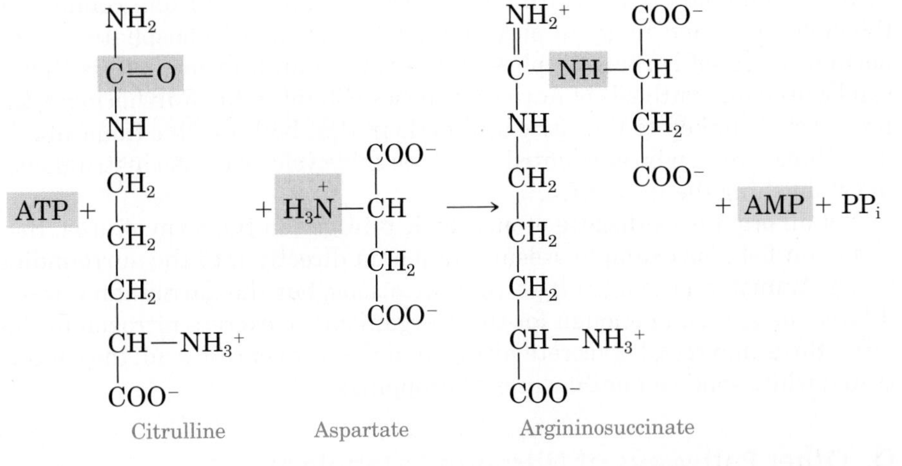

$$
\text{ATP} +
\begin{array}{c}
NH_2 \\
| \\
C{=}O \\
| \\
NH \\
| \\
CH_2 \\
| \\
CH_2 \\
| \\
CH_2 \\
| \\
CH{-}NH_3^+ \\
| \\
COO^-
\end{array}
+
\begin{array}{c}
COO^- \\
| \\
\overset{+}{H_3N}{-}CH \\
| \\
CH_2 \\
| \\
CH_2 \\
| \\
COO^-
\end{array}
\longrightarrow
\begin{array}{cc}
NH_2^+ & COO^- \\
\| & | \\
C{-}NH{-}CH \\
| & | \\
NH & CH_2 \\
| & | \\
CH_2 & COO^- \\
| \\
CH_2 \\
| \\
CH_2 \\
| \\
CH{-}NH_3^+ \\
| \\
COO^-
\end{array}
+ \text{AMP} + PP_i
$$

Citrulline   Aspartate   Argininosuccinate

The energy for this reaction comes from the hydrolysis of ATP to AMP and pyrophosphate (PP_i).

In Step ④, the argininosuccinate is split into arginine and fumarate:

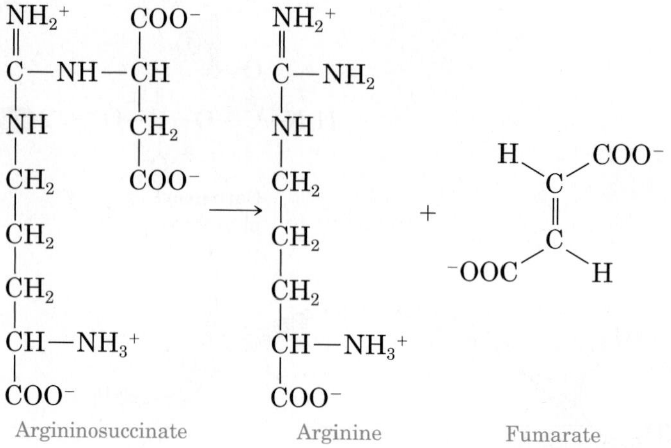

Argininosuccinate   Arginine   Fumarate

In Step ⑤, the final step, arginine is hydrolyzed to urea and ornithine:

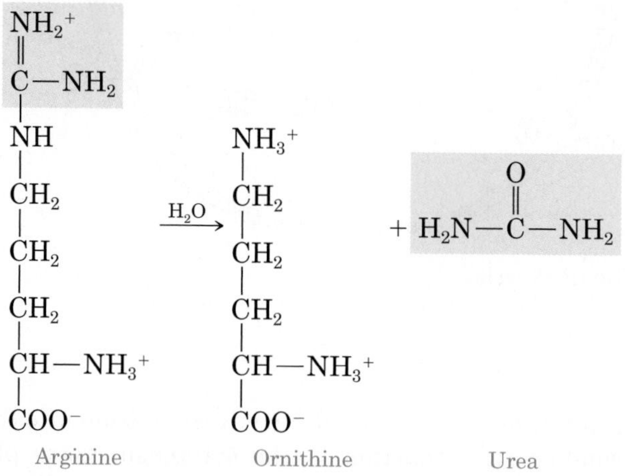

Arginine   Ornithine   Urea

The final product of the three stages is urea, which is excreted in the urine of mammals. The ornithine reenters the mitochondrion, completing the cycle. It is then ready to pick up another carbamoyl phosphate. An important aspect of carbamoyl phosphate's role as an intermediate is that it can be used for synthesis of nucleotide bases (Chapter 25). Furthermore, the urea cycle is linked to the citric acid cycle in that both involve fumarate. In fact, Hans Krebs, who elucidated the citric acid cycle, was also instrumental in establishing the urea cycle.

Not all organisms dispose of metabolic nitrogen in the form of urea. Bacteria and fish, for example, release ammonia directly into the surrounding water. Ammonia is toxic in high concentrations, but the surrounding water dilutes the ammonia enough for these organisms to excrete nitrogen in this form. Birds and reptiles secrete nitrogen in the form of uric acid, the concentrated white solid so familiar in bird droppings.

D. Other Pathways of Nitrogen Catabolism

The urea cycle is not the only way that the body can dispose of the toxic NH_4^+ ions. The oxidative deamination process, which produced the NH_4^+ in the first place, is reversible. Therefore, the buildup of glutamate from α-ketoglutarate and NH_4^+ is always possible. A third possibility for

How Does the Body Select Proteins for Degradation? Why Is It Important?

There is no storage form for proteins, as is the case with fats (lipid droplets in fat cells) and carbohydrates (glycogen in the liver). There is constant turnover with ongoing synthesis of new proteins. The process of selecting the right proteins for turnover is obviously an important one.

A number of mechanisms exist for selecting proteins for degradation. We will concentrate on one with important health implications. Some proteins that are most in need of degradation are those that are misfolded. They are of no use to the body, and in some cases are harmful. The misfolded proteins known as **prions** cause diseases in the nervous system, so it is particularly important to remove them from the body. Some of these diseases are Creutzfeldt-Jakob disease, Alzheimer's disease, Parkinson's disease, and Huntington's

disease. Mad cow disease and similar diseases in other species of animals such as sheep and goats are also prion diseases.

One of the most common ways of tagging proteins for degradation requires a polypeptide called **ubiquitin**. This polypeptide consists of 76 amino acid residues and is widely distributed in species of all kinds, with the exception of bacteria. (Ubiquitin molecules are shown in the figure as small beige objects.) It becomes linked to proteins to be degraded by a process called ubiquitination, in which several molecules are bonded to the targeted protein. The process uses ATP for energy. The tagged protein is degraded in a **proteasome**, which is a very large complex of many protein subunits. Ubiquitin is released for further rounds of protein degradation.

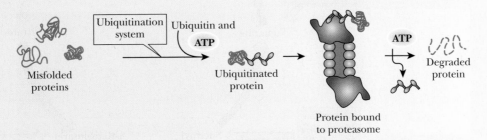

Misfolded proteins can be targeted for destruction by ubiquitination. (*From* Biochemistry, *by R. H. Garrett and C. M. Grisham, Brooks Cole, Cengage Learning, 2010, p. 1004.*)

Test your knowledge with Problems 28.47–28.50.

disposing of NH_4^+ is the ATP-dependent amidation of glutamate to yield glutamine:

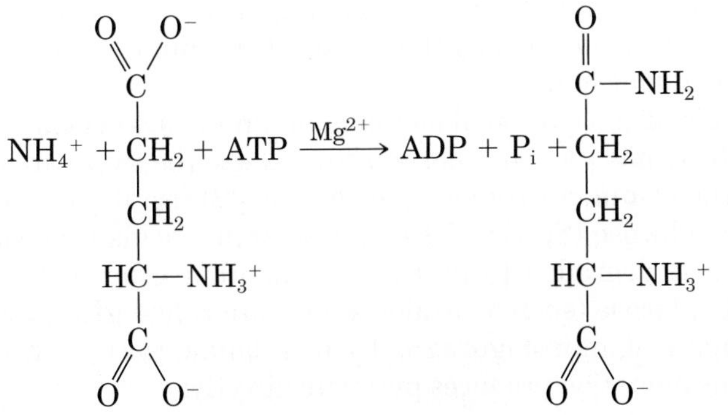

$$NH_4^+ + \underset{\text{Glutamate}}{\begin{matrix} O & O^- \\ \diagdown // \\ C \\ | \\ CH_2 \\ | \\ CH_2 \\ | \\ HC-NH_3^+ \\ | \\ C \\ // \diagdown \\ O & O^- \end{matrix}} + ATP \xrightarrow{Mg^{2+}} ADP + P_i + \underset{\text{Glutamine}}{\begin{matrix} O \\ \| \\ C-NH_2 \\ | \\ CH_2 \\ | \\ CH_2 \\ | \\ HC-NH_3^+ \\ | \\ C \\ // \diagdown \\ O & O^- \end{matrix}}$$

28.9 How Are the Carbon Skeletons of Amino Acids Processed in Catabolism?

After transamination of amino acids (Section 28.8A) to glutamate, the alpha amino group is removed from glutamate by oxidative deamination (Section 28.8B). The remaining carbon skeletons are used as an energy source (Figure 28.8). We will not study the pathways involved except to point

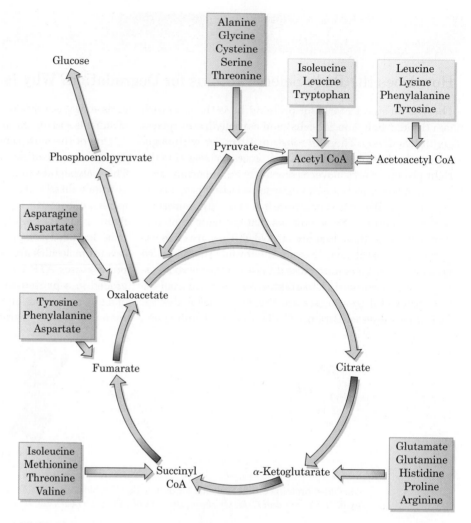

FIGURE 28.9 Catabolism of the carbon skeletons of amino acids. The glucogenic amino acids are in the purple boxes; the ketogenic ones in the gold boxes.

out the eventual fate of the skeletons. Not all of the carbon skeletons of amino acids are used as fuel. Some may be degraded up to a certain point, and the resulting intermediate may then be used as a building block to construct another needed molecule.

For example, if the carbon skeleton of an amino acid is catabolized to pyruvate, the body has two possible choices: (1) use the pyruvate as an energy supply via the common catabolic pathway or (2) use it as a building block to synthesize glucose (Section 29.1). Those amino acids that yield a carbon skeleton that is degraded to pyruvate or another intermediate capable of conversion to glucose (such as oxaloacetate) are called **glucogenic**.

One example is alanine (Figure 28.9). When alanine reacts with α-ketoglutaric acid, the transamination produces pyruvate directly:

Hereditary Defects in Amino Acid Catabolism: PKU

Many hereditary diseases involve missing or malfunctioning enzymes that catalyze the breakdown of amino acids. The oldest known of such diseases is cystinuria, which was described as early as 1810. In this disease, cystine shows up as flat hexagonal crystals in the urine. Stones form because of cystine's low solubility in water. This problem leads to blockage in the kidneys or the ureters and requires surgery to resolve it. One way to reduce the amount of cystine secreted is to remove as much methionine as possible from the diet. Beyond that, an increased fluid intake increases the volume of the urine, reducing the solubility problem. In addition, penicillamine can prevent cystinuria.

An even more important genetic defect is the absence of the enzyme phenylalanine hydroxylase, which causes a disease called phenylketonuria (PKU). In normal catabolism, this enzyme helps degrade phenylalanine by converting it to tyrosine. If the enzyme is defective, phenylalanine is converted to phenylpyruvate (see the discussion of the conversion of alanine to pyruvate in Section 28.9). Phenylpyruvate (an α-ketoacid) accumulates in the body and inhibits the conversion of pyruvate to acetyl CoA, thereby depriving the cells of energy via the common catabolic pathway. This effect is most important in the brain, which gets its energy from the utilization of glucose. PKU results in mental retardation.

This genetic defect can be detected early because phenylpyruvic acid appears in the urine and blood. A federal regulation requires that all infants be tested for this disease. When PKU is detected, mental retardation can be prevented by restricting the intake of phenylalanine in the diet. In particular, patients with PKU should avoid the artificial sweetener aspartame because it yields phenylalanine when hydrolyzed in the stomach.

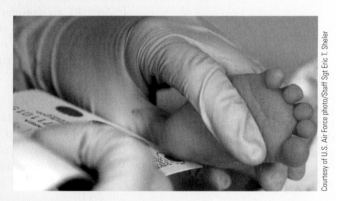

A newborn being tested for PKU.

Test your knowledge with Problem 28.51.

In contrast, many amino acids are degraded to acetyl CoA and acetoacetic acid. These compounds cannot form glucose but are capable of yielding ketone bodies; they are called **ketogenic**. Leucine is an example of a ketogenic amino acid. Some amino acids are both glucogenic and ketogenic—for example, phenylalanine.

Both glucogenic and ketogenic amino acids, when used as an energy supply, enter the citric acid cycle at some point (Figure 28.9) and are eventually oxidized to CO_2 and H_2O. The oxaloacetate (a C_4 compound) produced in this manner enters the citric acid cycle, adding to the oxaloacetate produced from PEP and in the cycle itself.

28.10 What Are the Reactions of Catabolism of Heme?

Carbohydrates, lipids, and proteins are the principal sources of energy in catabolism. Other cellular components contribute far less energy when they are catabolized. Their breakdown products can, however, affect the body. We will use the catabolism of heme as an example of an easily visible result of degradation.

Red blood cells are continuously manufactured in the bone marrow. Their life span is relatively short—about four months. Aged red blood cells are destroyed in the phagocytic cells. (Phagocytes are specialized blood cells that destroy foreign bodies.) When a red blood cell is destroyed, its hemoglobin is metabolized: The globin (Section 22.11) is hydrolyzed to amino acids, and the heme is first oxidized to biliverdin and finally reduced to bilirubin

Heme Biliverdin Bilirubin

FIGURE 28.10 Heme degradation from heme to biliverdin to bilirubin.

(Figure 28.10). The color change observed in bruises signals the redox reactions occurring in heme catabolism: Black and blue are due to the congealed blood, green to the biliverdin, and yellow to the bilirubin. The iron is preserved in ferritin, an iron-carrying protein, and reused. The bilirubin enters the liver via the blood and is then transferred to the gallbladder, where it is stored in the bile and finally excreted via the small intestine and colon. The color of feces is provided by urobilin, an oxidation product of bilirubin.

Postscript

It is useful to summarize the main points of catabolic pathways by showing how they are related. Figure 28.11 shows how all catabolic pathways lead to the citric acid cycle, producing ATP by the reoxidation of NADH and FADH$_2$. We saw the common metabolic pathway in Chapter 27, and here we see how it is related to all of catabolism.

FIGURE 28.11 A summary of catabolism showing the role of the common metabolic pathway. Note that the end products of the catabolism of carbohydrates, lipids, and amino acids all appear. (TA is transamination; → → → is a pathway with many steps.)

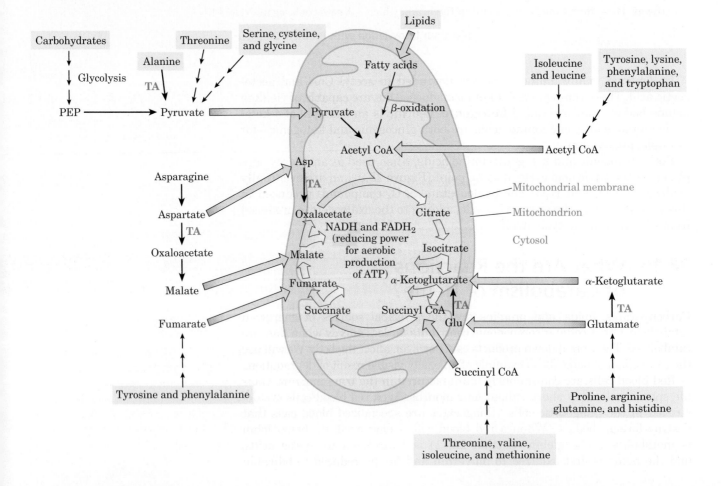

Summary

Section 28.1 What Is the General Outline of Catabolic Pathways?

- The foods we eat consist of carbohydrates, lipids, and proteins.
- There are specific breakdown pathways for each kind of nutrient.

Section 28.2 What Are the Reactions of Glycolysis?

- The specific pathway of carbohydrate catabolism is **glycolysis**.
- Hexose monosaccharides are activated by ATP and eventually converted to two C_3 fragments, dihydroxyacetone phosphate and glyceraldehyde phosphate.
- The glyceraldehyde phosphate is further oxidized and eventually ends up as pyruvate. All of these reactions occur in the cytosol.
- Pyruvate is converted to acetyl CoA, which is further catabolized in the common pathway.
- When the body needs intermediates for synthesis rather than energy, the glycolytic pathway can be shunted to the **pentose phosphate pathway**. NADPH, which is necessary for reduction and synthesis, is obtained in this way.
- The pentose phosphate pathway also yields ribose, which is necessary for the synthesis of RNA.

Section 28.3 What Is the Energy Yield from Glucose Catabolism?

- When completely metabolized, a hexose molecule yields the energy of 36 ATP molecules.

Section 28.4 How Does Glycerol Catabolism Take Place?

- Fats are broken down to glycerol and fatty acids.
- Glycerol is catabolized in the glycolysis pathway and yields 20 ATP molecules.

Section 28.5 What Are the Reactions of β-Oxidation of Fatty Acids?

- Fatty acids are broken down into fragments in the **β-oxidation** spiral.
- At each turn of the spiral, one acetyl CoA is released along with one $FADH_2$ and one $NADH + H^+$. These products go through the common catabolic pathway.

Section 28.6 What Is the Energy Yield from Stearic Acid Catabolism?

- Stearic acid, a C_{18} compound, yields 146 molecules of ATP.

Section 28.7 What Are Ketone Bodies?

- In starvation and under certain pathological conditions, not all of the acetyl CoA produced in the β-oxidation of fatty acids enters the common catabolic pathway.
- Some acetyl CoA forms acetoacetate, β-hydroxybutyrate, and acetone, commonly called **ketone bodies**.
- Excess ketone bodies in the blood are secreted in urine.

Section 28.8 How Is the Nitrogen of Amino Acids Processed in Catabolism?

- Proteins are broken down to amino acids. The nitrogen of the amino acids is first transferred to glutamate.
- Glutamate is **oxidatively deaminated** to yield ammonia.
- Mammals get rid of the toxic ammonia by converting it to urea in the **urea cycle**; urea is secreted in urine.

Section 28.9 How Are the Carbon Skeletons of Amino Acids Processed in Catabolism?

- The carbon skeletons of amino acids are catabolized via the citric acid cycle.
- Some amino acids, called **glucogenic amino acids**, enter as pyruvate or other intermediates of the citric acid cycle.
- Other amino acids are incorporated into acetyl CoA or ketone bodies and are called **ketogenic amino acids**.

Section 28.10 What Are the Reactions of Catabolism of Heme?

- Heme is catabolized to bilirubin, which is excreted in the feces.

Problems

Orange-numbered problems are applied.

Section 28.1 What Is the General Outline of Catabolic Pathways?

28.2 What are the products of lipase-catalyzed hydrolysis of fats?

28.3 What is the main use of amino acids in the body?

Section 28.2 What Are the Reactions of Glycolysis?

28.4 Although catabolism of a glucose molecule eventually produces a lot of energy, the first step uses up energy. Explain why this step is necessary.

28.5 In one step of the glycolysis pathway, a chain is broken into two fragments, only one of which can be further degraded in the glycolysis pathway. What happens to the other fragment?

28.6 Kinases are enzymes that catalyze the addition (or removal) of a phosphate group to (or from) a substance. ATP is also involved. How many kinases are in glycolysis? Name them.

28.7 (a) Which steps in glycolysis of glucose need ATP?

(b) Which steps in glycolysis yield ATP directly?

28.8 At which intermediate of the glycolytic pathway does oxidation—and hence energy production—begin? In what form is the energy produced?

28.9 At what point in glycolysis can ATP act as an inhibitor? What kind of enzyme regulation occurs in this inhibition?

28.10 The end product of glycolysis, pyruvate, cannot enter as such into the citric acid cycle. Which process converts this C_3 compound to a C_2 compound?

28.11 What essential compound is produced in the pentose phosphate pathway that is needed for synthesis as well as for defense against oxidative damages?

28.12 Which of the following steps yields energy and which consumes energy?

(a) Pyruvate $\longrightarrow$ lactate

(b) Pyruvate $\longrightarrow$ acetyl CoA + CO_2

28.13 How many moles of lactate are produced from 3 moles of glucose?

28.14 How many moles of net NADH + H^+ are produced from 1 mole of glucose going to

(a) Acetyl CoA? (b) Lactate?

Section 28.3 What Is the Energy Yield from Glucose Catabolism?

28.15 Of the 36 molecules of ATP produced by the complete metabolism of glucose, how many are produced directly in glycolysis alone—that is, before the common pathway?

28.16 How many net ATP molecules are produced in the skeletal muscles for each glucose molecule

(a) In glycolysis alone (up to pyruvate)?

(b) In converting pyruvate to acetyl CoA?

(c) In the total oxidation of glucose to CO_2 and H_2O?

28.17 (a) If fructose is metabolized in the liver, how many moles of net ATP are produced from each mole during glycolysis?

(b) How many moles are produced if the same thing occurs in a muscle cell?

28.18 In Figure 28.3, Step ⑤ yields one NADH. Yet in Table 28.1, the same step indicates a yield of 2 NADH + H^+. Is there a discrepancy between these two statements? Explain.

Section 28.4 How Does Glycerol Catabolism Take Place?

28.19 Based on the names of the enzymes participating in glycolysis, what would be the name of the enzyme catalyzing the activation of glycerol?

28.20 Which yields more energy upon hydrolysis, ATP or glycerol 1-phosphate? Why?

Section 28.5 What Are the Reactions of β-Oxidation of Fatty Acids?

28.21 Two enzymes participating in β-oxidation have the word "thio" in their names.

(a) Name the two enzymes.

(b) To which chemical group does this name refer?

(c) What is the common feature in the action of these two enzymes?

28.22 (a) Which part of the cells contains the enzymes needed for β-oxidation of fatty acids?

(b) How does the activated fatty acid get there?

28.23 Assume that lauric acid (C_{12}) is metabolized through β-oxidation. What are the products of the reaction after three turns of the spiral?

28.24 Is the β-oxidation of fatty acid (without the subsequent metabolism of C_2 fragments via the common metabolic pathway) more efficient with a short-chain fatty acid than with a long-chain fatty acid? Is more ATP produced per carbon atom in a short-chain fatty acid than in a long-chain fatty acid during β-oxidation?

Section 28.6 What Is the Energy Yield from Stearic Acid Catabolism?

28.25 Calculate the number of ATP molecules obtained in the β-oxidation of myristic acid, $CH_3(CH_2)_{12}COOH$.

28.26 Assume that the *cis-trans* isomerization in the β-oxidation of unsaturated fatty acids does not require energy. Which fatty acid yields the greater amount of energy, saturated (stearic acid) or mono-unsaturated (oleic acid)? Explain.

28.27 Assuming that both fats and carbohydrates are available, which does the body preferentially use as an energy source?

28.28 If equal weights of fats and carbohydrates are eaten, which will give more calories? Explain.

Section 28.7 What Are Ketone Bodies?

28.29 Acetoacetate is the common source of acetone and β-hydroxybutyrate. Name the type of reactions that yield these ketone bodies from acetoacetate.

28.30 Do ketone bodies have nutritional value?

28.31 What happens to the oxaloacetate produced from carboxylation of phosphoenolpyruvate?

Section 28.8 How Is the Nitrogen of Amino Acids Processed in Catabolism?

28.32 What kind of reaction is the following, and what is its function in the body?

$$\begin{array}{c} \text{H}_3\text{C} \\ \quad\diagdown \\ \quad\text{CH}-\!\!\underset{\underset{\text{NH}_3^+}{|}}{\overset{\overset{\text{H}}{|}}{\text{C}}}\!\!-\text{COO}^- \\ \quad\diagup \\ \text{H}_3\text{C} \end{array} \;+\; \begin{array}{c} \text{COO}^- \\ | \\ \text{C}=\!\text{O} \\ | \\ \text{CH}_2 \\ | \\ \text{CH}_2 \\ | \\ \text{COO}^- \end{array}$$

$$\longrightarrow \begin{array}{c} \text{H}_3\text{C} \\ \quad\diagdown \\ \quad\text{CH}-\!\!\underset{\underset{\text{O}}{\|}}{\text{C}}\!\!-\text{COO}^- \\ \quad\diagup \\ \text{H}_3\text{C} \end{array} \;+\; \begin{array}{c} \text{COO}^- \\ | \\ \text{H}-\text{C}-\text{NH}_3^+ \\ | \\ \text{CH}_2 \\ | \\ \text{CH}_2 \\ | \\ \text{COO}^- \end{array}$$

28.33 Write an equation for the oxidative deamination of alanine.

28.34 Ammonia, NH_3, and ammonium ion, NH_4^+, are both soluble in water and could easily be excreted in the urine. Why does the body convert them to urea rather than excreting them directly?

28.35 What are the sources of the nitrogen in urea?

28.36 What compound is common to both the urea and citric acid cycles?

28.37 (a) What is the toxic product of the oxidative deamination of glutamate?

(b) How does the body get rid of it?

28.38 If the urea cycle is inhibited, in what other ways can the body get rid of NH_4^+ ions?

Section 28.9 How Are the Carbon Skeletons of Amino Acids Processed in Catabolism?

28.39 The metabolism of the carbon skeleton of tyrosine yields pyruvate. Why is tyrosine a glucogenic amino acid?

Section 28.10 What Are the Reactions of Catabolism of Heme?

28.40 Why is a high bilirubin content in the blood an indication of liver disease?

28.41 When hemoglobin is fully metabolized, what happens to the iron in it?

28.42 Describe which groups on biliverdin (Figure 28.10) are the oxidation products and which are the reduction products in the degradation of heme.

Chemical Connections

28.43 (Chemical Connections 28A) What causes cramps of the muscles when a person is fatigued?

28.44 (Chemical Connections 28A) Why is lactic acid accumulation particularly dangerous in cardiac arrest?

28.45 (Chemical Connections 28B) What system counteracts the acidic effect of ketone bodies in the blood?

28.46 (Chemical Connections 28B) The patient whose condition is described in Chemical Connections 28B was transferred to a hospital in an ambulance. Could a nurse in the ambulance tentatively diagnose his diabetic condition without running blood and urine tests? Explain.

28.47 (Chemical Connections 28C) Is protein turnover an important process? Why or why not?

28.48 (Chemical Connections 28C) What role do misfolded proteins play in disease?

28.49 (Chemical Connections 28C) What are ubiquitins, and what role do they play in the processing of proteins?

28.50 (Chemical Connections 28C) What are proteasomes, and what is their role in the body?

28.51 (Chemical Connections 28D) Draw structural formulas for each reaction component and complete the following equation:

$$\text{Phenylalanine} \longrightarrow \text{Phenylpyruvate} \;+\; ?$$

Additional Problems

28.52 If you receive a laboratory report showing the presence of a high concentration of ketone bodies in the urine of a patient, which disease would you suspect?

28.53 Which compounds give the sequence in coloration of bruises from black and blue to green and yellow?

28.54 (a) At which step of the glycolysis pathway does NAD^+ participate (see Figures 28.3 and 28.4)?

(b) At which step does $NADH + H^+$ participate?

(c) As a result of the overall pathway, is there a net increase of NAD^+, of $NADH + H^+$, or of neither?

28.55 What is the net energy yield in moles of ATP produced when yeast converts one mole of glucose to ethanol?

28.56 Can the intake of alanine, glycine, and serine relieve hypoglycemia caused by starvation? Explain.

28.57 How can glucose be utilized to produce ribose for RNA synthesis?

28.58 Write the products of the transamination reaction between alanine and oxaloacetate:

$$\begin{array}{c} \text{COO}^- \\ | \\ \text{CH}-\text{NH}_3^+ \\ | \\ \text{CH}_3 \end{array} \;+\; \begin{array}{c} \text{COO}^- \\ | \\ \text{C}=\!\text{O} \\ | \\ \text{CH}_2 \\ | \\ \text{COO}^- \end{array} \longrightarrow$$

28.59 Phosphoenolpyruvate (PEP) has a high-energy phosphate bond that has more energy than the anhydride bonds in ATP. Which step in glycolysis suggests that this is so?

28.60 Suppose that a fatty acid labeled with radioactive carbon-14 is fed to an experimental animal. Where would you look for the radioactivity?

28.61 Which functional groups are present in carbamoyl phosphate?

28.62 Is the urea cycle an energy-producing or an energy-consuming pathway?

28.63 Which intermediate of the glycolytic pathway can replenish oxaloacetate in the citric acid cycle?

28.64 How many turns of the spiral are there in the β-oxidation of (a) lauric acid and (b) palmitic acid?

28.65 Why is glycolysis considered an oxidative pathway even though only one reaction is an oxidation reaction?

28.66 One of the enzymes needed for the metabolism of ethanol uses a derivative of vitamin B_1 as a cofactor. Why is it not surprising that alcoholics develop beri-beri, a disease caused by a vitamin B_1 deficiency?

28.67 How does the catabolism of glycerol provide a link between the pathways of fat catabolism and glycolysis?

28.68 What is the role of carnitine in fatty acid catabolism?

Tying It Together

28.69 The equations of glycolysis indicate that there is a net gain of two ATP molecules for each molecule of glucose processed. Why is it that we see the figure of 36 ATP molecules in Table 28.1?

28.70 What reactions can pyruvate undergo once it is formed? Are these reactions aerobic, anaerobic, or both?

28.71 Is lactate a dead-end product of metabolism, or does it play some role in generating (or regenerating) some useful compound?

28.72 Why do ketone bodies occur in the blood of people who are on severely restricted diets?

28.73 Can amino acids be catabolized to yield energy?

28.74 Suggest a reason why the carbon skeletons and nitrogen-containing portions of amino acids are catabolized separately.

Looking Ahead

28.75 Put the following words into two related groups: energy-yielding, oxidative, anabolism, reductive, energy-requiring, catabolism.

28.76 Would you expect the biosynthesis of a protein from constituent amino acids to require energy or to release energy? Explain.

28.77 In what ways can the production of glucose from CO_2 and H_2O in photosynthesis be considered the exact reversal of the complete aerobic catabolism of glucose? In what ways is it different?

28.78 Why is the citric acid cycle *the* central pathway in metabolism?

Challenge Problems

28.79 With their oxygen-containing functional groups, sugars are more oxidized than the hydrocarbon side chains of fatty acids. Does this fact have any bearing on the energy yield of carbohydrates compared to that of fats?

28.80 Many soft drinks contain citric acid to add flavor. Is it likely to be a good nutrient?

28.81 The intermediates of glycolysis have been phosphorylated and carry phosphate groups that are charged. The intermediates of the citric acid cycle are not phosphorylated. Suggest a reason for this difference. (*Hint:* In what parts of the cell do these pathways occur?)

28.82 One occasionally hears diet advice that proteins and carbohydrates should not be eaten at the same meal. Does this advice make sense to you in light of Figure 28.11?

28.83 The production of ATP is not shown explicitly in Figure 28.11. What point in this figure indicates that ATP production does indeed take place?

28.84 Many metabolic pathways, including those of catabolism, are long and complex. Suggest a reason for this observation.

Biosynthetic Pathways

Key Questions

29.1 What Is the General Outline of Biosynthetic Pathways?

29.2 How Does the Biosynthesis of Carbohydrates Take Place?

29.3 How Does the Biosynthesis of Fatty Acids Take Place?

29.4 How Does the Biosynthesis of Membrane Lipids Take Place?

29.5 How Does the Biosynthesis of Amino Acids Take Place?

Algae on mudflats.

29.1 What Is the General Outline of Biosynthetic Pathways?

In the human body, and in most other living tissues, the pathways by which a compound is synthesized (anabolism) are usually different from the pathways by which it is degraded (catabolism). (Anabolic pathways are also called biosynthetic pathways, and we will use these terms interchangeably.) There are several reasons why it is biologically advantageous for anabolic and catabolic pathways to be different. We will give two of them here:

1. **Flexibility** If the normal **biosynthetic pathway** is blocked, the body can often use the reverse of the degradation pathway (recall that most steps in degradation are reversible), thereby providing another way to make the necessary compounds.

2. **Overcoming the effect of Le Chatelier's principle** This point can be illustrated by the cleavage of a glucose unit from a glycogen molecule, an equilibrium process:

$$(\text{Glucose})_n + \text{P}_i \xrightleftharpoons{\text{phosphorylase}} (\text{Glucose})_{n-1} + \text{Glucose 1-phosphate} \quad (29.1)$$

Glycogen

Glycogen
(one unit smaller)

◯WL

Sign in to OWL at **www.cengage.com/owl** to view tutorials and simulations, develop problem-solving skills, and complete online homework assigned by your professor.

Phosphorylase catalyzes not only glycogen degradation (the forward reaction), but also glycogen synthesis (the reverse reaction). However, the body contains a large excess of inorganic phosphate, P_i. This excess would drive the reaction, on the basis of Le Chatelier's principle, to the right, which represents glycogen degradation. To provide a method for the synthesis of glycogen even in the presence of excess inorganic phosphate, a different pathway is needed in which P_i is not a reactant. Thus, the body uses the following synthetic pathway:

$$\text{(Glucose)}_{n-1} + \text{UDP-glucose} \longrightarrow \text{(Glucose)}_n + \text{UDP} \quad (29.2)$$

Glycogen   Glycogen
(one unit larger)

Not only do the synthetic pathways differ from the catabolic pathways, but the energy requirements are also different, as are the pathways' locations. Most catabolic reactions occur in the mitochondria, whereas anabolic reactions generally take place in the cytoplasm. We will not describe the energy balances of the biosynthetic processes in detail as we did for catabolism. However, keep in mind that while energy (in the form of ATP) is *obtained* in the degradative processes, biosynthetic processes *consume* energy.

29.2 How Does the Biosynthesis of Carbohydrates Take Place?

We discuss the biosynthesis of carbohydrates by looking at three examples:

- Conversion of atmospheric CO_2 to glucose in plants
- Synthesis of glucose in animals and humans
- Conversion of glucose to other carbohydrate molecules in animals and humans

A. Conversion of Atmospheric Carbon Dioxide to Glucose in Plants

Photosynthesis The process in which plants synthesize carbohydrates from CO_2 and H_2O with the help of sunlight and chlorophyll

The most important biosynthesis of carbohydrates takes place in plants, green algae, and cyanobacteria, with the last two representing an important part of the marine food web. In the process of **photosynthesis**, the energy of the sun is built into the chemical bonds of carbohydrates. The overall reaction is:

$$6H_2O + 6CO_2 \xrightarrow[\text{chlorophyll}]{\substack{\text{energy in} \\ \text{the form of} \\ \text{sunlight}}} C_6H_{12}O_6 + 6O_2 \quad (29.3)$$

Glucose

Although the primary product of photosynthesis is glucose, it is largely converted to other carbohydrates, mainly cellulose and starch. The very complicated process of glucose biosynthesis takes place in large protein–cofactor complexes (Chemical Connections 29A). We will not discuss it further here except to note that the carbohydrates of plants—starch, cellulose, and other mono- and polysaccharides—serve as the basic carbohydrate supply of all animals, including humans.

B. Synthesis of Glucose in Animals

In Chapter 28, we saw that when the body needs energy, carbohydrates are broken down via glycolysis. When energy is not needed, glucose can be synthesized from the intermediates of the glycolytic and citric acid pathways. This process is called **gluconeogenesis**. As shown in Figure 29.1, a large number of intermediates—pyruvate, lactate, oxaloacetate, malate, and several amino acids (the glucogenic amino acids we met in Section 28.9)—can serve as starting compounds. Gluconeogenesis proceeds in the reverse order from glycolysis, and many of the enzymes of glycolysis also catalyze gluconeogenesis. At four points, however, unique enzymes (marked in Figure 29.1) catalyze only gluconeogenesis and not the breakdown reactions. These four enzymes make *gluconeogenesis a pathway that is distinct from glycolysis*. Note that ATP is used up in gluconeogenesis and produced in glycolysis, another difference between the two pathways.

During periods of strenuous exercise, the body needs to replenish its carbohydrate supply. The Cori cycle makes use of lactate produced in glycolysis (Section 28.2) as the starting point for gluconeogenesis. Lactate produced in working muscle is then transported in the bloodstream to the liver, where

Gluconeogenesis The process by which glucose is synthesized in the body

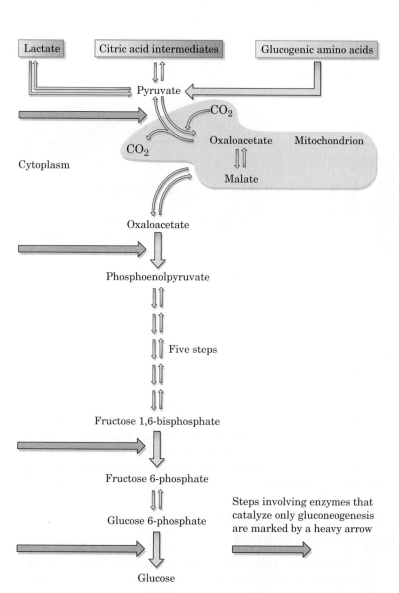

FIGURE 29.1 Gluconeogenesis. All reactions take place in the cytosol, except for those shown in the mitochondria.

Chemical Connections 29A

Photosynthesis

Photosynthesis requires sunlight, water, CO_2, and pigments found in plants, mainly chlorophyll. The overall reaction shown in Equation 29.3 actually occurs in two distinct steps. First, light interacts with the pigments that are located in highly membranous organelles of plants, called **chloroplasts**. Chloroplasts resemble mitochondria (Section 27.2) in many respects: They contain a whole chain of oxidation–reduction enzymes similar to the cytochrome and iron–sulfur complexes of mitochondrial membranes, and they contain a proton-translocating ATPase. In a manner similar to mitochondria, the proton gradient accumulated in the intermembrane region drives the synthesis of ATP in chloroplasts (see the discussion of the chemiosmotic pump in Section 27.6).

Chlorophyll is the central part of a complex machinery called photosystem I and II. The detailed structure of photosystem I was elucidated in 2001. The photosystem consists of three monomeric entities, which are designated as I, II, and III. Each monomer contains 12 different proteins; 96 chlorophyll molecules and 30 cofactors that include iron clusters, lipids, and Ca^{2+} ions. Its most important feature shows that a central Mg^{2+} is bound to the sulfur of a methionine residue of a surrounding protein. This Mg-S linkage makes this whole assembly a powerful oxidizing agent, meaning that it can readily accept electrons.

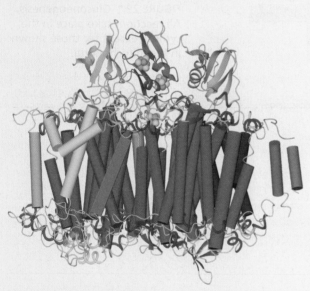

Side view of monomer III in photosystem I.

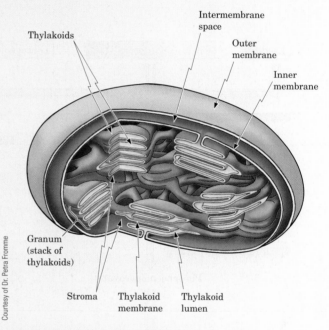

Courtesy of Dr. Petra Fromme

Inside of a chloroplast.

gluconeogenesis converts it to glucose (Figure 29.2). The newly formed glucose is then transported back to the muscle by the blood, where it can fuel further exercise. Note that the two different pathways, glycolysis and gluconeogenesis, take place in different organs. This division of labor ensures that both pathways are not simultaneously active in the same tissues, which is highly inefficient.

C. Conversion of Glucose to Other Carbohydrates in Animals

The third important biosynthetic pathway for carbohydrates is the conversion of glucose to other hexoses and hexose derivatives and the synthesis of di-, oligo-, and polysaccharides. The common step in all of these

Chemical Connections 29A

Photosynthesis (*continued*)

Chlorophyll itself, buried in a complex protein that traverses the chloroplast membranes, is a molecule similar to the heme we have already encountered in hemoglobin (Figure 22.18). In contrast to heme, however, chlorophyll contains Mg^{2+} instead of Fe^{2+}.

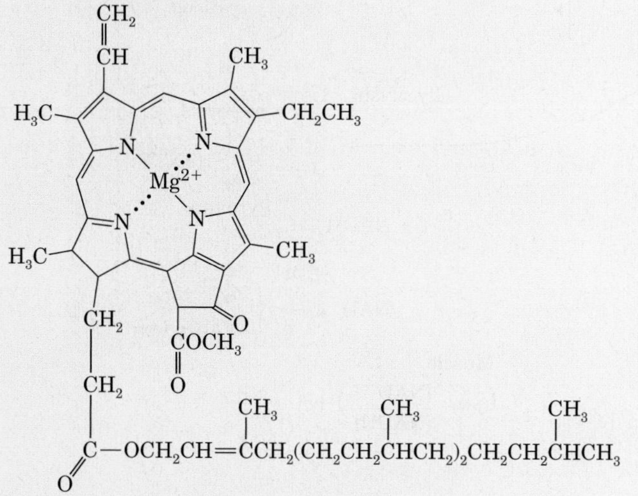

Chlorophyll a

The reactions in photosynthesis, collectively called the light reactions, are the ones in which chlorophyll captures the energy of sunlight and, with its aid, strips the electrons and protons from water to form oxygen, ATP, and $NADPH + H^+$ (see Section 28.2C):

$$H_2O + ADP + P_i + NADP^+ + \text{sunlight}$$
$$\longrightarrow \tfrac{1}{2}O_2 + ATP + NADPH + H^+$$

Another group of reactions, called the dark reactions because they do not need light, in essence converts CO_2 to carbohydrates:

$$6CO_2 + 12\ NADPH + 18\ ATP \longrightarrow$$
$$C_6H_{12}O_6 + 12\ NADP^+ + 18\ ADP + 18\ Pi$$
Carbohydrate

Energy, now in the form of ATP, is used to help $NADPH + H^+$ reduce carbon dioxide to carbohydrates. Thus, the protons and electrons stripped in the light reactions are added to the carbon dioxide in the dark reactions. This reduction takes place in a multistep cyclic process called the **Calvin cycle**, named after its discoverer, Melvin Calvin (1911–1997), who was awarded the 1961 Nobel Prize in chemistry for his work. In this cycle, CO_2 is first attached to a C_5 fragment, producing a C_6 molecule, which then breaks down to two C_3 fragments (triose phosphates). Through a complex series of steps, these fragments are converted to a C_6 compound and, eventually, to glucose.

$$CO_2 + C_5 = 2C_3 = C_6$$

The critical step in the dark reactions (Calvin cycle) is the attachment of CO_2 to ribulose 1,5-bisphosphate, a compound derived from ribulose (Table 20.2). The enzyme that catalyzes this reaction, ribulose-1,5-bisphosphate carboxylase-oxygenase, nicknamed RuBisCO, is one of the slowest in nature. As in traffic, the slowest-moving vehicle determines the overall flow, so RuBisCO is the main factor in the low efficiency of the Calvin cycle. Because of the low efficiency of this enzyme, most plants convert less than 1% of the absorbed radiant energy into carbohydrates. To overcome its inefficiency, plants must synthesize large quantities of this enzyme. More than half of the soluble proteins in plant leaves are RuBisCO enzymes, whose synthesis requires a large energy expenditure.

Test your knowledge with Problem 29.35.

processes is the activation of glucose by uridine triphosphate (UTP) to form UDP-glucose:

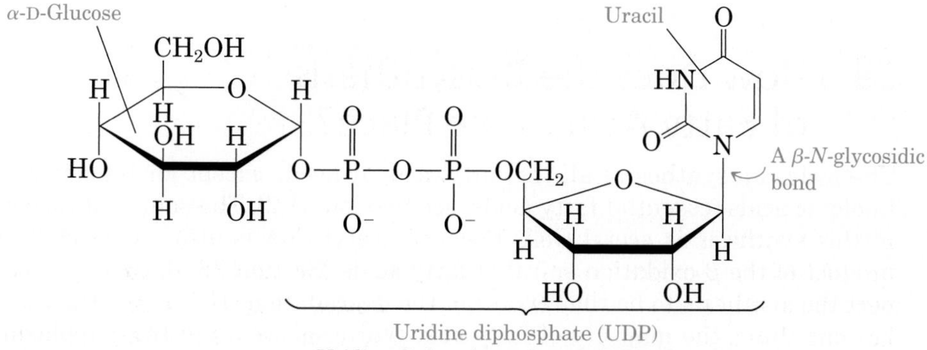

Uridine diphosphate (UDP)
Uridine diphosphate glucose (UDP-glucose)

FIGURE 29.2 The Cori cycle is named for its discovers, Gerty and Carl Cori. Lactate produced in muscles by glycolysis is transported by the blood to the liver. Gluconeogenesis in the liver converts the lactate back to glucose, which can be carried back to the muscles by the blood. (NTP stands for nucleoside triphosphate and LDH for lactate dehydrogenase.)

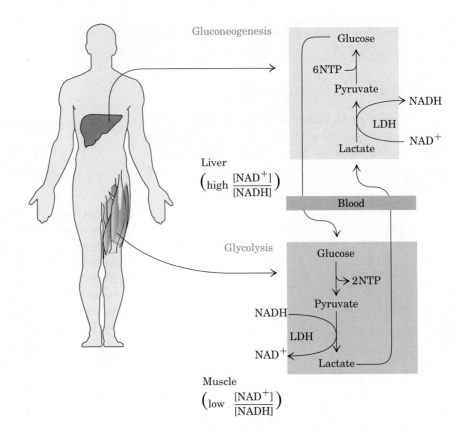

Glycogenesis The conversion of glucose to glycogen

UDP is similar to ADP except that the base is uracil instead of adenine. UTP, an analog of ATP, contains two high-energy phosphate anhydride bonds. For example, when the body has excess glucose and wants to store it as glycogen (a process called **glycogenesis**), the glucose is first converted to glucose 1-phosphate, but then a special enzyme catalyzes the reaction:

$$\text{glucose 1-phosphate} + \boxed{\text{UTP}} \longrightarrow \text{UDP-glucose} + \; ^-O-\underset{\underset{O^-}{|}}{\overset{\overset{O}{\|}}{P}}-O-\underset{\underset{O^-}{|}}{\overset{\overset{O}{\|}}{P}}-O^-$$

$$\underset{\text{Glycogen}}{\text{UDP-glucose} + (\text{glucose})_n} \longrightarrow \boxed{\text{UDP}} + \underset{\substack{\text{Glycogen} \\ \text{(one unit larger)}}}{(\text{glucose})_{n+1}}$$

The biosynthesis of many other di- and polysaccharides and their derivatives also uses the common activation step: forming the appropriate UDP compound.

29.3 How Does the Biosynthesis of Fatty Acids Take Place?

The body can synthesize all the fatty acids it needs except for linoleic and linolenic acids (essential fatty acids; see Section 21.2). The source of carbon in this synthesis is acetyl CoA. Because acetyl CoA is also a degradation product of the β-oxidation spiral of fatty acids (Section 28.5), we might expect the synthesis to be the reverse of the degradation. This is not the case. For one thing, the majority of fatty acid synthesis occurs in the cytoplasm, whereas degradation takes place in the mitochondria. Fatty acid synthesis is catalyzed by a multienzyme system.

However, one aspect of fatty acid synthesis is the same as in fatty acid degradation: Both processes involve acetyl CoA, so both proceed in units of two carbons. Fatty acids are built up two carbons at a time, just as they are broken down two carbons at a time (Section 28.5).

Most of the time, fatty acids are synthesized when excess food is available. That is, when we eat more food than we need for energy, our bodies turn the excess acetyl CoA (produced by catabolism of carbohydrates; see Section 28.2) into fatty acids and then to fats. The fats are stored in the fat depots, which are specialized fat-carrying cells (see Figure 28.1)

The key to fatty acid synthesis is an **acyl carrier protein** (**ACP**). It can be looked upon as a merry-go-round—a rotating protein molecule to which the growing chain of fatty acids is bonded. As the growing chain rotates with the ACP, it sweeps over the multienzyme complex; at each enzyme, one reaction of the chain is catalyzed (Figure 29.3).

At the beginning of this cycle, the ACP picks up an acetyl group from acetyl CoA and delivers it to the first enzyme, fatty acid synthase, here called synthase for short:

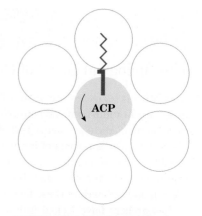

FIGURE 29.3 The biosynthesis of fatty acids. The ACP (central blue sphere) has a long side chain (—) that carries the growing fatty acid (⌇). The ACP rotates counterclockwise, and its side chain sweeps over a multienzyme system (empty spheres). As each cycle is completed, a C_2 fragment is added to the growing fatty acid chain.

$$\underset{\text{Acetyl CoA}}{CH_3\overset{\overset{\displaystyle O}{\|}}{C}-S-CoA} + HS-ACP \longrightarrow HS-CoA + \underset{\text{Acetyl ACP}}{CH_3\overset{\overset{\displaystyle O}{\|}}{C}-S-ACP}$$

$$CH_3\overset{\overset{\displaystyle O}{\|}}{C}-S-ACP + synthase-SH \longrightarrow CH_3\overset{\overset{\displaystyle O}{\|}}{C}-S-synthase + HS-ACP$$

The —SH group is the site of acyl binding as a thioester.

The C_2 fragment on the synthase is condensed with a C_3 fragment attached to the ACP in a process in which CO_2 is given off:

$$CH_3\overset{\overset{\displaystyle O}{\|}}{C}-S-synthase + \underset{\substack{| \\ COO^-}}{CH_2}\overset{\overset{\displaystyle O}{\|}}{-C}-S-ACP$$

$$\text{Malonyl-ACP}$$

$$\longrightarrow \underset{\text{Acetoacetyl-ACP}}{CH_3\overset{\overset{\displaystyle O}{\|}}{C}-CH_2\overset{\overset{\displaystyle O}{\|}}{-C}-S-ACP} + CO_2 + synthase-SH$$

The result is a C_4 fragment, which is reduced twice and dehydrated before it becomes a fully saturated C_4 group. This event marks the end of one cycle of the merry-go-round. These three steps are the reverse of what we saw in the β-oxidation of fatty acids in Section 28.5.

In the next cycle, the fragment is transferred to the synthase and another malonyl-ACP (C_3 fragment) is added. When CO_2 splits out, a C_6 fragment is obtained. The merry-go-round continues to turn. At each turn, another C_2 fragment is added to the growing chain. Chains up to C_{16} (palmitic acid) can be obtained in this process. If the body needs longer fatty acids—for example, stearic (C_{18})—another fragment is added to palmitic acid by a different enzyme system.

The Biological Basis of Obesity

It is difficult to overestimate the effects of obesity on public health in industrialized countries. Obesity plays a role in so many medical problems, including diabetes and heart disease, that it is a focus of intense research to find ways to address the issue. It is known that three factors play a role in obesity: genetics, metabolism, and brain function. We can address each of these factors in turn.

Researchers have looked into a connection between increased consumption of glucose and changes in gene expression that affect metabolism. (Recall that we mentioned this kind of connection in Section 27.1). In this scenario, increased glycolysis leads to higher levels of citrate in the mitochondria. The citrate thus produced is transported to the cell nucleus or to the cytosol. In both locations, the enzyme ATP-citrate lyase (ACL) converts citrate to acetyl-CoA. In the cytosol, acetyl-CoA serves as the starting point for the synthesis of lipids. In the nucleus, acetyl-CoA is the substrate for histone acetyltransferases (HATs), the enzymes that attach acetyl groups to histones. Acetylation and deacetylation of histones play a key role in gene expression (Section 25.6).

The second area of research on obesity concerns an enzyme that controls the level of a key intermediate in fatty acid biosynthesis, malonyl-CoA . This intermediate has two very important functions in metabolism. First, it is converted to fatty acids and not to any other compound in biosynthesis. Second, it strongly inhibits the enzyme that helps transfer fatty acids to the mitochondrion and therefore inhibits fatty acid oxidation. The level of malonyl-CoA in the cytosol can determine whether the cell will be oxidizing fats or storing fats. The enzyme that produces malonyl-CoA is acetyl-CoA carboxylase, or ACC. There are two forms of this enzyme, each encoded by separate genes. ACC1 is found in the liver and adipose tissue, while ACC2 is found in cardiac and skeletal muscle. High glucose concentrations and high insulin concentrations lead to stimulation of ACC2. Exercise has the opposite effect. During exercise, an AMP-dependent protein kinase phosphorylates ACC2 and inactivates it.

Some recent studies looked at the nature of weight gain and weight loss with respect to ACC2. The investigators created a strain of mice lacking the gene for ACC2. These mice ate more than their wild-type counterparts but had significantly lower stores of lipids (30%–40% less in skeletal muscle and 10% less in cardiac muscle). Even their adipose tissue, which still had ACC1, showed a reduction in stored triacylglycerols of up to 50%. The mice showed no other abnormalities. They grew and reproduced normally and had normal life spans. The investigators concluded that reduced pools of malonyl-CoA due to the lack of ACC2 led to two results: increased β-oxidation via removal of the block on transfer of fatty acids to the mitochondrion and a decrease in fatty acid synthesis. They speculate that ACC2 would be a good target for drugs used to combat obesity.

It is well established that parts of the nervous system, particularly the hypothalamus, play an important role in feelings of hunger and fullness. It is possible to make connections between environment and behavior, including overeating behavior. The most successful way to lose weight and to maintain a lower weight is by long-term changes in behavior. What works best is to eat less and to exercise more, and to do so consistently. *Fad diets do not work*. Programs are being formed to help with behavior modification, especially by use of support groups. By attacking the problem of obesity on several fronts, researchers hope to make progress in reversing what truly is an epidemic.

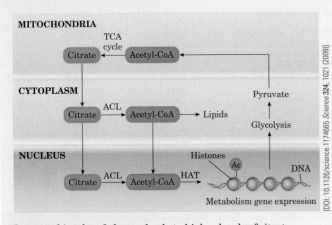

Increased intake of glucose leads to higher levels of citrate, affecting gene expression.

[DOI: 10.1126/science.1174665 *Science* 324, 1021 (2009)]

Test your knowledge with Problems 29.38–29.41.

Unsaturated fatty acids are obtained from saturated fatty acids by an oxidation step in which hydrogen is removed and combined with O_2 to form water;

$$R-CH_2-CH_2-(CH_2)_n COOH + O_2 + \boxed{NADPH} + H^+$$

$$\xrightarrow{\text{enzyme}} \underset{R}{\overset{H}{\underset{}{C}}} = \underset{(CH_2)_n COOH}{\overset{H}{\underset{}{C}}} + 2H_2O + \boxed{NADP^+}$$

An example of such elongation and unsaturation is docosahexaenoic acid, a 22-carbon fatty acid with 6 *cis* double bonds (22:6). Docosahexaenoic acid is part of the glycerophospholipids prevalent in the membranes of the eye where the visual pigment rhodopsin resides. Its presence is necessary to provide fluidity to the membrane so that it can process the light signals reaching our retina.

We have already seen that lipids are a highly efficient form of energy storage. They accumulate all too easily when we take in excess energy in the form of nutritional "calories." Obesity-related health problems are becoming all too common in developed countries, leading to research on ways to address the situation (Chemical Connections 29B).

29.4 How Does the Biosynthesis of Membrane Lipids Take Place?

The various membrane lipids (Sections 21.6 to 21.8) are assembled from their constituents. We just saw how fatty acids are synthesized in the body. These fatty acids are activated by CoA, forming acyl CoA. Glycerol 1-phosphate, which is obtained from the reduction of dihydroxyacetone phosphate (a C_3 fragment of glycolysis; see Figure 28.4), is the second building block of glycerophospholipids. This compound combines with two acyl CoA molecules, which can be the same or different:

$$
\begin{array}{c}
\text{CH}_2\text{—OH} \\
| \\
\text{CH—OH} \\
| \quad\quad\quad\;\; \text{O} \\
\quad\quad\quad\quad\quad\; \| \\
\text{CH}_2\text{—O—P—O}^- \\
\quad\quad\quad\quad\;\; | \\
\quad\quad\quad\quad\;\; \text{O}^-
\end{array}
\;+\;
\begin{array}{c}
\text{O} \\
\| \\
\text{R—C—S—CoA} \\
\text{O} \\
\| \\
\text{R}'\text{—C—S—CoA}
\end{array}
\;\longrightarrow\;
\begin{array}{c}
\quad\quad\quad\;\; \text{O} \\
\quad\quad\quad\;\; \| \\
\text{CH}_2\text{—O—C—R} \\
| \quad\quad\quad\;\; \text{O} \\
\quad\quad\quad\;\; \| \\
\text{CH—O—C—R}' \;+\; 2\text{CoA—SH} \\
| \quad\quad\quad\;\; \text{O} \\
\quad\quad\quad\;\; \| \\
\text{CH}_2\text{—O—P—O}^- \\
\quad\quad\quad\;\; | \\
\quad\quad\quad\;\; \text{O}^-
\end{array}
$$

Glycerol 1-phosphate   Acyl CoA   A phosphatidate

We encountered glycerol 1-phosphate as a vehicle for transporting electrons in and out of mitochondria (Section 28.3). To complete the molecule, an activated serine or choline or ethanolamine is added to the —OPO_3^{2-} group, forming phosphate esters (see the structures in Section 21.6; Figure 29.4 shows a model of phosphatidylcholine). Choline is activated by cytidine triphosphate (CTP), yielding CDP-choline. This process is similar to the activation of glucose by UTP (Section 29.2C) except that the base is cytosine rather than uracil (Section 25.2). Sphingolipids (Section 21.7) are similarly built up from smaller molecules. An activated phosphocholine is added to the sphingosine part of ceramide (Section 21.7) to make sphingomyelin.

The glycolipids are constructed in a similar fashion. Ceramide is assembled as described above, and the carbohydrate is added one unit at a time in the form of activated monosaccharides (UDP-glucose and so on).

Cholesterol, the molecule that controls the fluidity of membranes and is a precursor of all steroid hormones and bile salts, is also synthesized by the human body. It is assembled in the liver from fragments that come from the acetyl group of acetyl CoA. All of the carbon atoms of cholesterol come from the carbon atoms of acetyl CoA molecules (Figure 29.5). Cholesterol in the brain is synthesized in nerve cells themselves; its presence is necessary to form synapses. Cholesterol from our diet and cholesterol that is synthesized in the liver

FIGURE 29.4 A model of phosphatidylcholine, commonly called lecithin.

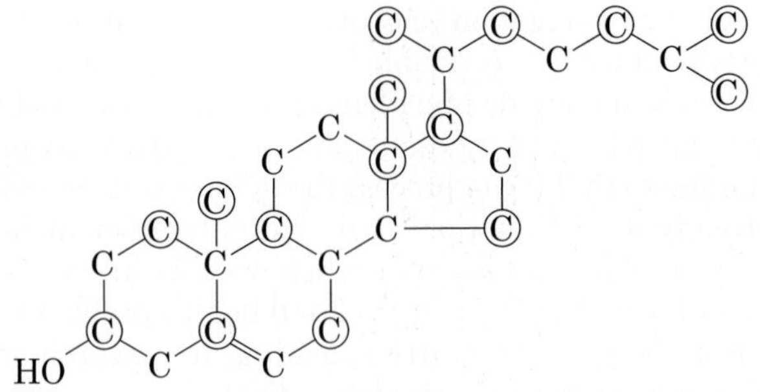

FIGURE 29.5 Biosynthesis of cholesterol. The circled carbon atoms come from the —CH_3 group, and the other carbon atoms come from the —CO— group of the acetyl group of acetyl CoA.

circulate in the plasma as LDL (see Section 21.9). However, LDL is not available for synapse formation because it cannot cross the blood–brain barrier.

Cholesterol synthesis starts with the sequential condensation of three acetyl CoA molecules to form a compound 3-hydroxy-3-methylglutaryl CoA (HMGCoA):

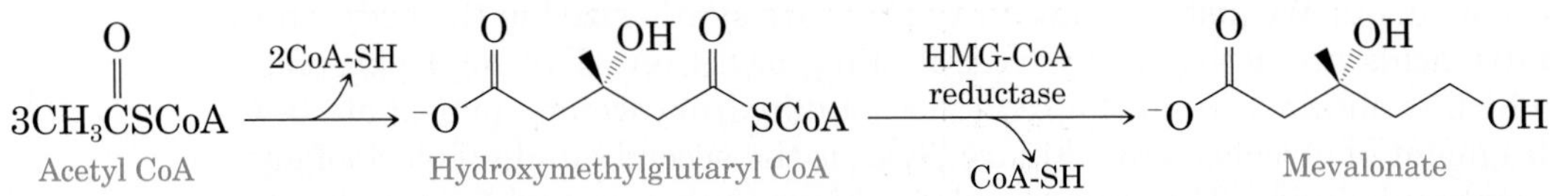

A key enzyme, HMG-CoA reductase, controls the rate of cholesterol synthesis. It reduces the thioester of HMGCoA to a primary alcohol, yielding CoA in the process. The resulting compound, mevalonate, undergoes phosphorylation and decarboxylation to yield a C_5 compound, isopentenyl pyrophosphate:

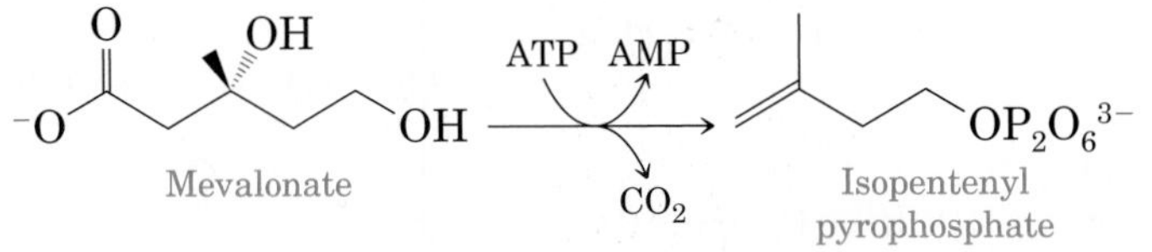

From this basic C_5 unit, other multiple-C_5 compounds are formed that eventually lead to cholesterol synthesis. These intermediates are geranyl, C_{10}, and farnesyl, C_{15}, pyrophosphates:

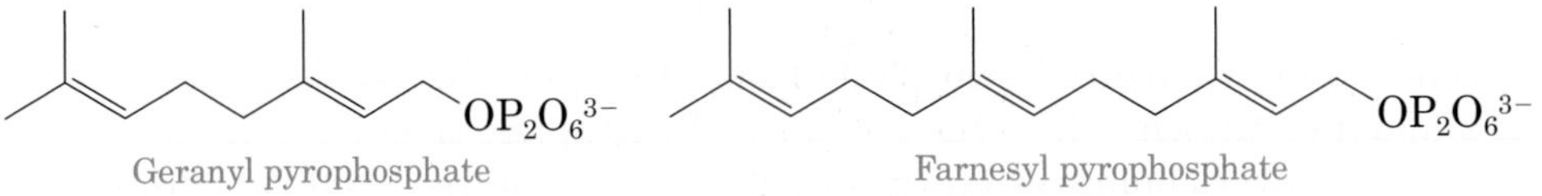

Finally, cholesterol is synthesized from the condensation of two farnesyl pyrophosphate molecules.

The statin drugs, such as lovastatin, competitively inhibit the key enzyme HMG-CoA reductase and, thereby, the biosynthesis of cholesterol. They are frequently prescribed to control the cholesterol level in the blood so as to prevent atherosclerosis (Section 21.9E).

Note that the intermediates in cholesterol synthesis, the geranyl and farnesyl pyrophosphates, are made of isoprene units; we discussed these C_5 compounds in Section 12.5. The C_{10} and C_{15} compounds are also used to enable protein molecules to be dispersed in the lipid bilayers of membranes. When these multiple-isoprene units are attached to a protein in a process called **prenylation**, the protein becomes more hydrophobic and is able to move laterally within the bilayer with greater ease. (The name *prenylation* originates from isoprene, the five-carbon unit from which the cholesterol intermediates C_{10}, C_{15}, and C_{30} are made.) Prenylation marks proteins to be associated with membranes and to perform other cellular functions, such as the signal transduction of G-protein (Section 24.5B).

29.5 How Does the Biosynthesis of Amino Acids Take Place?

The human body needs 20 different amino acids to make its protein chains—all 20 are found in a normal diet. Some of the amino acids can be synthesized from other compounds; these are the nonessential amino acids. Others cannot be synthesized by the human body and must be supplied in the diet; these are the **essential amino acids** (see Section 30.6). Most nonessential amino acids are synthesized from some intermediate of either glycolysis (Section 28.2) or the citric acid cycle (Section 27.4). Glutamate plays a central role in the synthesis of five nonessential amino acids. Glutamate itself is synthesized from α-ketoglutarate, one of the intermediates in the citric acid cycle:

$$\boxed{NADH} + H^+ + NH_4^+ + \begin{array}{c} COO^- \\ | \\ C{=}O \\ | \\ CH_2 \\ | \\ CH_2 \\ | \\ COO^- \end{array} \rightleftharpoons \begin{array}{c} COO^- \\ | \\ CH{-}NH_3^+ \\ | \\ CH_2 \\ | \\ CH_2 \\ | \\ COO^- \end{array} + \boxed{NAD^+} + H_2O$$

α-Ketoglutarate   Glutamate

The forward reaction is the synthesis, and the reverse reaction is the oxidative deamination (degradation) reaction we encountered in the catabolism

Chemical Connections 29C

Essential Amino Acids

The biosynthesis of proteins requires the presence of all of the protein's constituent amino acids. If any of the 20 amino acids is missing or in short supply, protein biosynthesis is inhibited.

Some organisms, including bacteria, can synthesize all the amino acids that they need. Other species, including humans, must obtain some amino acids from dietary sources. The essential amino acids in human nutrition are listed in Table 29.1. The body can synthesize some of these amino acids, but not in sufficient quantities for its needs, especially in the case of growing children (particularly children's requirement for arginine and histidine).

Amino acids are not stored (except in proteins), so dietary sources of essential amino acids are needed at regular intervals. Protein deficiency—especially a prolonged

deficiency of the essential amino acids—leads to the disease **kwashiorkor**. The problem in this disease, which is particularly severe in growing children, is not simply starvation, but the breakdown of the body's own proteins.

Charles D. Winters

The label on this supplement lists the amino acid content and points out which ones are essential amino acids.

Table 29.1 Amino Acid Requirements in Humans

Essential		Nonessential	
Arginine	Methionine	Alanine	Glutamine
Histidine	Phenylalanine	Asparagine	Glycine
Isoleucine	Threonine	Aspartate	Proline
Leucine	Tryptophan	Cysteine	Serine
Lysine	Valine	Glutamate	Tyrosine

Test your knowledge with Problem 29.42.

of amino acids (Section 28.8B). In this case, the synthetic and degradative pathways are exactly the reverse of each other.

Glutamate can serve as an intermediate in the synthesis of alanine, serine, aspartate, asparagine, and glutamine. For example, the transamination reaction we saw in Section 28.8A leads to alanine formation:

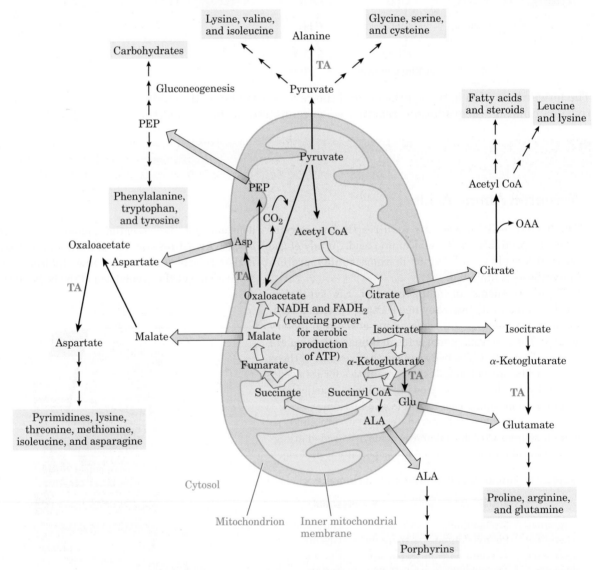

FIGURE 29.6 A summary of anabolism showing the role of the central metabolic pathway. Note that carbohydrates, lipids, and amino acids all appear as products. (OAA is oxaloacetate; ALA is a derivative of succinyl CoA; TA is transamination; ⟶ ⟶ ⟶ is a pathway with many steps.)

Besides being the building blocks of proteins, amino acids serve as intermediates for a large number of biological molecules. We have already seen that serine is needed in the synthesis of membrane lipids (Section 29.4). Certain amino acids are also intermediates in the synthesis of heme and of the purines and pyrimidines that are the raw materials for DNA and RNA (Chapter 25).

Postscript

It is useful to summarize the main points of anabolic pathways by considering how they are related. Figure 29.6 on the previous page shows how all anabolic pathways start at the citric acid cycle, using ATP and the reducing power of NADH and $FADH_2$. Chapter 28 introduced the central metabolic pathway, and here we see how it is related to all of anabolism.

Summary

Section 29.1 What Is the General Outline of Biosynthetic Pathways?

- For most biochemical compounds, the biosynthetic pathways are different from the degradation pathways.

Section 29.2 How Does the Biosynthesis of Carbohydrates Take Place?

- In **photosynthesis**, carbohydrates are synthesized in plants from CO_2 and H_2O, using sunlight as an energy source.
- Glucose can be synthesized by animals from the intermediates of glycolysis, from the intermediates of the citric acid cycle, and from glucogenic amino acids. This process is called **gluconeogenesis**.
- When glucose or other monosaccharides are built into di-, oligo-, and polysaccharides, each monosaccharide unit in its activated form is added to a growing chain.

Section 29.3 How Does the Biosynthesis of Fatty Acids Take Place?

- Fatty acid biosynthesis is accomplished by a multi-enzyme system.
- The key to fatty acid biosynthesis is the **acyl carrier protein (ACP)**, which acts as a merry-go-round transport system: It carries the growing fatty acid chain over

a number of enzymes, each of which catalyzes a specific reaction.
- With each complete turn of the merry-go-round, a C_2 fragment is added to the growing fatty acid chain.
- The source of the C_2 fragment is malonyl-ACP, a C_3 compound bonded to ACP. It becomes C_2 with the loss of CO_2.

Section 29.4 How Does the Biosynthesis of Membrane Lipids Take Place?

- Glycerophospholipids are synthesized from glycerol 1-phosphate, fatty acids that are activated by conversion to acyl CoA, and activated alcohols such as choline.
- Cholesterol is synthesized from acetyl CoA. Three C_2 fragments are condensed to form C_6, hydroxymethylglutaryl CoA.
- After reduction and decarboxylation, isoprene C_5 units are formed that condense to the C_{10} and C_{15} intermediates from which cholesterol is built.

Section 29.5 How Does the Biosynthesis of Amino Acids Take Place?

- Many nonessential amino acids are synthesized in the body from the intermediates of glycolysis or the intermediates of the citric acid cycle.
- In half of these cases, glutamate is the donor of the amino group in transamination.
- Amino acids serve as building blocks for proteins.

Problems

Orange-numbered problems are applied.

Section 29.1 What Is the General Outline of Biosynthetic Pathways?

29.1 Why are the pathways that the body uses for anabolism and catabolism mostly different?

29.2 How does the large excess of inorganic phosphate in a cell affect the amount of glycogen present?

29.3 Glycogen can be synthesized in the body by the same enzymes that degrade it. Why is this process utilized in glycogen synthesis only to a small extent, while most glycogen biosynthesis occurs via a different synthetic pathway?

29.4　Do most anabolic and catabolic reactions take place in the same location?

Section 29.2 How Does the Biosynthesis of Carbohydrates Take Place?

29.5　What is the difference in the overall chemical equations for photosynthesis and for respiration?

29.6　In photosynthesis, what are the sources of (a) carbon, (b) hydrogen, and (c) energy?

29.7　Name a compound that can serve as a raw material for gluconeogenesis and is (a) from the glycolytic pathway, (b) from the citric acid cycle, and (c) an amino acid.

29.8　How is glucose activated for glycogen synthesis?

29.9　Glucose is the only carbohydrate compound that the brain can use for energy. Which pathway is mobilized to supply the need of the brain during starvation: (a) glycolysis, (b) gluconeogenesis, or (c) glycogenesis? Explain.

29.10　Are the enzymes that combine two C_3 compounds into a C_6 compound in gluconeogenesis the same as or different from those that cleave the C_6 compound into two C_3 compounds in glycolysis?

29.11　Devise a scheme in which maltose is formed, starting with UDP-glucose.

29.12　Glycogen is written as $(glucose)_n$.
(a) What does n stand for?
(b) What is the approximate value of n?

29.13　What are the constituents of UTP?

Section 29.3 How Does the Biosynthesis of Fatty Acids Take Place?

29.14　What is the source of carbon in fatty acid synthesis?

29.15　(a) Where in the body does fatty acid synthesis occur?
(b) Does fatty acid degradation occur in the same location?

29.16　Is ACP an enzyme?

29.17　In fatty acid biosynthesis, which compound is added repeatedly to the synthase?

29.18　(a) What is the name of the first enzyme in fatty acid synthesis?
(b) What does it do?

29.19　From which compound is CO_2 released in fatty acid synthesis?

29.20　What are the common functional groups in CoA, ACP, and synthase?

29.21　In the synthesis of unsaturated fatty acids, $NADPH + H^+$ is converted to $NADP^+$, yet this is an oxidation step and not a reduction step. Explain.

29.22　Which of these fatty acids can be synthesized by the multienzyme fatty acid synthesis complex alone?
(a) Oleic　　　　　(b) Stearic
(c) Myristic　　　 (d) Arachidonic
(e) Lauric

29.23　Some enzymes can use NADH as well as NADPH as a coenzyme. Other enzymes use one or the other

exclusively. Which features would prevent NADPH from fitting into the active site of an enzyme that otherwise can accommodate NADH?

29.24　Are fatty acids for energy, in the form of fat, synthesized in the same way as fatty acids for the lipid bilayer of membrane?

29.25　Linoleic and linolenic acids cannot be synthesized in the human body. Does this mean that the human body cannot make an unsaturated fatty acid from a saturated one?

Section 29.4 How Does the Biosynthesis of Membrane Lipids Take Place?

29.26　When the body synthesizes the following membrane lipid, from which building blocks is it assembled?

$$
\begin{array}{l}
\text{CH}_2\!-\!\text{O}\!-\!\overset{\displaystyle\overset{\text{O}}{\|}}{\text{C}}\!-\!(\text{CH}_2)_{14}\text{CH}_3 \\[2mm]
\text{CH}\!-\!\text{O}\!-\!\overset{\displaystyle\overset{\text{O}}{\|}}{\text{C}}\!-\!(\text{CH}_2)_{10}\text{CH}_3 \\[2mm]
\text{CH}_2\!-\!\text{O}\!-\!\overset{\displaystyle\overset{\text{O}}{\|}}{\underset{\underset{\text{O}^-}{|}}{\text{P}}}\!-\!\text{O}\!-\!\text{CH}_2\!-\!\underset{\underset{\text{NH}_3^+}{|}}{\text{CH}}\!-\!\text{COO}^-
\end{array}
$$

29.27　Name the activated constituents necessary to form the glycolipid glucoceramide.

29.28　Why is HMG-CoA reductase a key enzyme in cholesterol synthesis?

29.29　Describe by carbon skeleton designation how a C_2 compound ends up as a C_5 compound.

Section 29.5 How Does the Biosynthesis of Amino Acids Take Place?

29.30　Which reaction is the reverse of the synthesis of glutamate from α-ketoglutarate, ammonia, and $NADH + H^+$?

29.31　Which amino acid will be synthesized by the following process?

$$
\begin{array}{l}
\text{COO}^- \\
| \\
\text{C}\!=\!\text{O} \\
| \qquad\quad + \text{ NADH} + \text{H}^+ + \text{NH}_4^+ \longrightarrow \\
\text{CH}_2 \\
| \\
\text{COO}^-
\end{array}
$$

29.32　Draw the structure of the compound needed to synthesize asparagine from glutamate by transamination.

29.33　Name the products of the following transamination reaction:

$$
(\text{CH}_3)_2\text{CH}\!-\!\overset{\displaystyle\overset{\text{O}}{\|}}{\text{C}}\!-\!\text{COO}^-
$$

$$
+ \ ^-\text{OOC}\!-\!\text{CH}_2\!-\!\text{CH}_2\!-\!\underset{\underset{\text{NH}_3^+}{|}}{\text{CH}}\!-\!\text{COO}^- \longrightarrow
$$

Chemical Connections

29.34 (Chemical Connections 29A) Photosystem I and II are complex factories of proteins, chlorophyll, and many cofactors. Where are these photosystems located in plants, and in which reaction of photosynthesis do they participate?

29.35 (Chemical Connections 29A) Which coenzyme reduces CO_2 in the Calvin cycle?

29.36 (Chemical Connections 29B) What is the metabolic importance of malonyl-CoA?

29.37 (Chemical Connections 29B) What enzyme provides a possible target for drugs to treat obesity?

29.38 (Chemical Connections 29B) Which parts of a cell are affected by increased consumption of glucose?

29.39 (Chemical Connections 29B) How can higher glucose consumption affect gene expression?

29.40 (Chemical Connections 29B) Is it necessary to avoid specific foods to lose weight?

29.41 (Chemical Connections 29B) Can behavior modification be used to increase the effectiveness of weight loss programs?

29.42 (Chemical Connections 29C) What is the result of eating only proteins that do not contain all 20 amino acids?

Additional Problems

29.43 In the structure of $NADP^+$, what are the bonds that connect nicotinamide and adenine to the ribose units?

29.44 Which C_3 fragment carried by ACP is used in fatty acid synthesis?

29.45 When glutamate transaminates phenylpyruvate, which amino acid is produced?

$$C_6H_5-CH_2-\overset{\overset{\displaystyle O}{\|}}{C}-COO^-$$

$$+ \ ^-OOC-CH_2-CH_2-\underset{\underset{\displaystyle NH_3^+}{|}}{CH}-COO^- \longrightarrow$$

29.46 Name three compounds based on isoprenoid units that play a role in cholesterol biosynthesis.

29.47 Each activation step in the synthesis of complex lipids occurs at the expense of one ATP molecule. How many ATP molecules are used in the synthesis of one molecule of lecithin?

29.48 Consider the fact that the deamination of glutamic acid and its synthesis from α-ketoglutaric acid are equilibrium reactions. Which way will the equilibrium shift when the human body is exposed to cold temperature?

29.49 Which compound reacts with glutamate in a transamination process to yield serine?

29.50 What are the names of the C_{10} and C_{15} intermediates in cholesterol biosynthesis?

29.51 Which is carbon 1 in HMG-CoA (3-hydroxy-3-methylglutaryl CoA)?

29.52 In most biosynthetic processes, the reactant is reduced to obtain the desired product. Verify whether this statement holds for the overall reaction of photosynthesis.

29.53 What is the major difference in structure between chlorophyll and heme?

29.54 Can the complex enzyme system participating in every fatty acid synthesis manufacture fatty acids of any length?

29.55 How does the biosynthesis of carbohydrates in plants by photosynthesis differ from gluconeogenesis in animals?

29.56 Is it possible for cholesterol levels in the body to be higher than the amount of cholesterol obtained from the diet?

Tying It Together

29.57 How does fatty acid biosynthesis differ from catabolism of fatty acids?

29.58 How does the energy source differ in carbohydrate biosynthesis in plants and in animals?

29.59 The enzyme that catalyzes carbon dioxide fixation in photosynthesis is one of the least efficient known. What does this fact imply about energy requirements in photosynthesis?

Looking Ahead

29.60 A vegan diet is one that excludes all animal products. Is it possible to get all essential nutrients from such a diet? Will it be more difficult or easier to achieve this goal with a diet that allows animal products?

29.61 Many key proteins in the immune system are glycoproteins (proteins that incorporate sugars in their structure). Would you expect the biosynthesis of such proteins to be affected by a lack of essential amino acids, a low-carbohydrate diet, or both? Explain.

Challenge Problems

29.62 The foods that we eat supply carbohydrates, fats, and proteins. Based on what you have learned in this chapter, which would you predict that we could do without? Explain.

29.63 In general, catabolic and biosynthetic processes do not take place in the same part of the cell. Why is this separation advantageous?

29.64 Would you expect feedback inhibition to play a role in long biosynthetic pathways? Give the reason for your answer.

29.65 If laboratory rats are fed all the amino acids except one of the essential ones and then fed the missing amino acid four hours later, what will be the effect on protein synthesis and why?

29.66 Do humans possess all the anabolic pathways shown in Figure 29.6? If any are missing in humans, what are they likely to be?

30

Nutrition

Foods high in fiber include whole grains, legumes, fruits, and vegetables.

30.1 How Do We Measure Nutrition?

In Chapters 27 and 28, we saw what happens to the food that we eat in its final stages—after the proteins, lipids, and carbohydrates have been broken down into their components. In this chapter, we discuss the earlier stages—nutrition and diet—and then the digestive processes that break down these large molecules into the small ones that undergo metabolism. Food provides energy and new molecules to replace those that the body uses. This synthesis of new molecules is especially important for the period during which a child becomes an adult.

The components of food and drink that provide for growth, replacement, and energy are called **nutrients**. Not all components of food are nutrients. Some components of food and drinks, such as those that provide flavor, color, or aroma, enhance our pleasure in the food but are not themselves nutrients.

Nutritionists classify nutrients into six groups:

1. Carbohydrates
2. Lipids
3. Proteins

4. Vitamins

5. Minerals

6. Water

For food to be used in our bodies, it must be absorbed through the intestinal walls into the bloodstream or lymph system. Some nutrients, such as vitamins, minerals, glucose, and amino acids, can be absorbed directly. Others, such as starch, fats, and proteins, must first be broken down into smaller components before they can be absorbed. This breakdown process is called **digestion**.

A healthy body needs the proper intake of all nutrients. However, nutrient requirements vary from one person to another. For example, more energy is needed to maintain the body temperature of an adult than that of a child. For this reason, nutritional requirements are usually given per kilogram of body weight. Furthermore, the energy requirements of a physically active individual are greater than those of a person in a sedentary occupation. Therefore, when average values are given, as in **Dietary Reference Intakes (DRI)** and in the former guidelines called **Recommended Daily Allowances (RDA)**, one should be aware of the wide range that these average values represent.

The public interest in nutrition and diet changes with time and geography. Seventy or eighty years ago, the main nutritional interest of most Americans was getting enough food to eat and avoiding diseases caused by vitamin deficiencies, such as scurvy or beriberi. These issues are still the main concern of the large majority of the world's population. In affluent societies such as the industrialized nations, however, today's nutritional message is no longer "eat more," but rather "eat less and discriminate more in your selection of food." Dieting to reduce body weight is a constant effort in a sizable percentage of the American population. Many people discriminate in their selection of food to avoid cholesterol (Section 21.9E) and saturated fatty acids to reduce the risk of heart attacks.

Along with such **discriminatory curtailment diets** came many faddish diets. **Diet faddism** is an exaggerated belief in the effects of nutrition upon health and disease. This phenomenon is not new; it has been prevalent for many years. Many times it is driven by visionary beliefs, but backed by little science. In the nineteenth century, Dr. Kellogg (of cornflakes fame) recommended a largely vegetarian diet based on his belief that meat produces sexual excess. Eventually his religious fervor withered and his brother made a commercial success of his inventions of grain-based food. Another fad is raw food, which bans any application of heat higher than 106°F to food. Heat, these faddists maintain, depletes the nutritional value of proteins and vitamins and concentrates pesticides in food. Obviously a raw food diet is vegetarian, as it excludes meat and meat products.

A recommended food is rarely as good and a condemned food is rarely as bad as faddists claim. Each food contains a large variety of nutrients. For example, a typical breakfast cereal lists the following items as its ingredients: milled corn sugar, salt, malt flavoring, and vitamins A, B, C, and D, plus flavorings and preservatives. U.S. consumer laws require that most packaged food be labeled in a uniform manner to show the nutritional values of the food. Figure 30.1 shows a typical label of the type found on almost every can, bottle, or box of food that we buy.

Such labels must list the percentages of **Daily Values** for four key vitamins and minerals: vitamins A and C, calcium, and iron. If other vitamins or minerals have been added or if the product makes a nutritional claim about other nutrients, their values must be shown as well. The percent daily values on the labels are based on a daily intake of 2000 Cal. For anyone who eats more than that amount, the actual percentage figures would be lower

Digestion The process in which the body breaks down large molecules into smaller ones that can be absorbed and metabolized

Discriminatory curtailment diet A diet that avoids certain food ingredients that are considered harmful to the health of an individual—for example, low-sodium diets for people with high blood pressure

Nutrition Facts

Serving size 1 Bar (28g)
Servings per Container 6

Amount Per Serving

Calories 120   Calories from Fat 35

	% Daily Value*
Total Fat 4g	6%
Saturated Fat 2g	10%
Cholesterol 0mg	0%
Sodium 45mg	2%
Potassium 100mg	3%
Total Carbohydrate 19g	6%
Dietary fiber 2g	8%
Sugars 13g	
Protein 2g	

Vitamin A 15%	•	Vitamin C 15%
Calcium 15%	•	Iron 15%
Vitamin D 15%	•	Vitamin E 15%
Thiamin 15%	•	Riboflavin 15%
Niacin 15%	•	Vitamin B$_6$ 15%
Folate 15%	•	Vitamin B$_{12}$ 15%
Biotin 10%	•	Pantothenic Acid 10%
Phosphorus 15%	•	Iodine 2%
Magnesium 4%	•	Zinc 4%

*Percent Daily Values are based on a 2,000 calorie diet. Your daily values may be higher or lower depending on your calorie needs.

		Calories:	2,000	2,500
Total Fat	Less than		65g	80g
Sat Fat	Less than		20g	25g
Cholesterol	Less than		300mg	300mg
Sodium	Less than		2,400mg	2,400mg
Potassium			3,500mg	3,500mg
Total Carbohydrate			300g	375g
Dietary Fiber			25g	30g

FIGURE 30.1 A food label for a peanut butter crunch bar. The portion at the bottom (following the asterisk) gives the same categories of information on all labels that carry it.

(and higher for those who eat less). Note that each label specifies the serving size; the percentages are based on that portion, not on the contents of the entire package. The section at the bottom of the label is exactly the same on all labels, no matter what the food; it shows the daily amounts of nutrients recommended by the government based on consumption of either 2000 or 2500 Cal. Some food packages are allowed to carry shorter labels, either because they have only a few nutrients or because the package has limited label space. The uniform labels make it much easier for consumers to know exactly what they are eating.

In 1992, the U.S. Department of Agriculture (USDA) issued a set of guidelines regarding what constitutes a healthy diet, depicted in the form of a pyramid (Figure 30.2). These guidelines considered the basis of a healthy diet to be the foods richest in starch (bread, rice, and so on), plus lots of

FIGURE 30.2 The Food Guide Pyramid developed by the U.S. Department of Agriculture as a general guide to a healthful diet.

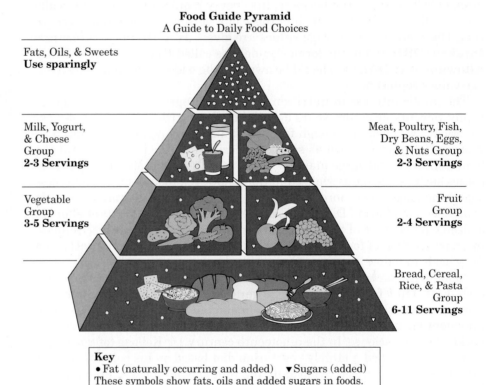

Food Guide Pyramid
A Guide to Daily Food Choices

Fats, Oils, & Sweets
Use sparingly

Milk, Yogurt, & Cheese Group
2-3 Servings

Meat, Poultry, Fish, Dry Beans, Eggs, & Nuts Group
2-3 Servings

Vegetable Group
3-5 Servings

Fruit Group
2-4 Servings

Bread, Cereal, Rice, & Pasta Group
6-11 Servings

Key
● Fat (naturally occurring and added)   ▼ Sugars (added)
These symbols show fats, oils and added sugars in foods.

USDA, 1992

FOOD GROUP SERVING SIZES

Bread, Cereal, Rice, and Pasta
1/2 cup cooked cereal, rice, pasta
1 ounce dry cereal
1 slice bread
2 cookies
1/2 medium doughnut

Vegetables
1/2 cup cooked or raw chopped vegetables
1 cup raw leafy vegetables
3/4 cup vegetable juice
10 french fries

Fruit
1 medium apple, banana, or orange
1/2 cup chopped, cooked, or canned fruit
3/4 cup fruit juice
1/4 cup dried fruit

Milk, Yogurt, and Cheese
1 cup milk or yogurt
$1\frac{1}{2}$ ounces natural cheese
2 ounces processed cheese
2 cups cottage cheese
$1\frac{1}{2}$ cups ice cream
1 cup frozen yogurt

Meat, Poultry, Fish, Dry Beans, Eggs, and Nuts
2–3 ounces cooked lean meat, fish, or poultry
2–3 eggs
4–6 tablespoons peanut butter
$1\frac{1}{2}$ cups cooked dry beans
1 cup nuts

Fats, Oils, and Sweets
Butter, mayonnaise, salad dressing, cream cheese, sour cream, jam, jelly

Chemical Connections 30A

The New Food Guide

Over time, scientists began questioning some aspects of the original food pyramid shown in Figure 30.2. For example, certain types of fat are known to be essential to health and actually reduce the risk of heart disease. Also, little evidence exists to back up the claim that a high intake of carbohydrates is beneficial, although for certain sports it is essential. The original pyramid glorified carbohydrates while casting all fats as the "bad guys." In fact, plenty of evidence does link the consumption of saturated fat with high cholesterol and risk of heart disease—but mono- and polyunsaturated fats have the opposite effect. While many scientists recognized the distinction between the various types of fats, they thought that the average person would not understand them. So the original pyramid was designed to just send a simple message: "Fat is bad." The implied corollary to "fat is bad" was that "carbohydrates are good." However, after years of study, no evidence proved that a diet in which 30% or fewer of the calories come from fat is healthier than one with a higher level of fat consumption.

In an attempt to reconcile the latest nutritional data while presenting them in a form understandable to the average person, the USDA initially created a website at www.mypyramid.gov. This interactive website allowed visitors to take a brief tutorial about a revised pyramid as well as to calculate their ideal amount of the various food types. However, the revised food pyramid did not last too long. In 2011, the government completely changed the presentation of basic nutritional information. Instead of a pyramid, the new presentation is a plate. The new website that contains the information is ChooseMyPlate.gov.

The USDA believed that a simpler message could be given using the plate presentation. It divides the plate into four different items—fruits, vegetables, protein, and grains. There is a side plate for dairy. On the site, you can click on any of the parts of a well-balanced diet and get specific information,

including lists of foods that fall into those groups, and tips for the consumer. For example, if you click on the dairy plate, there is a consumer tip suggesting that liquid dairy products should be non-fat. Also, calcium-fortified soy milk is included in the dairy section.

One of the biggest differences in this presentation is the change of focus away from nutrient types such as protein, fat, and carbohydrate toward food types such as grains and vegetables. The underlying message is that if you maintain the healthy proportion of the food groups on the plate, you will have a diet balanced in the three key nutrient types.

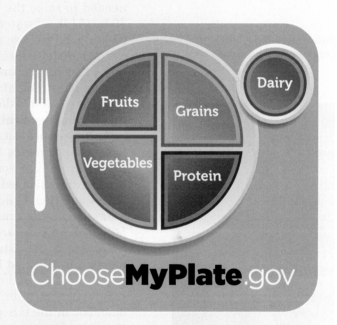

The newest presentation of nutritional recommendations is a plate instead of a pyramid. Basic nutritional recommendations as well as in-depth information can be found at ChooseMyPlate.gov.

Test your knowledge with Problems 30.43 and 30.44.

---

fruits and vegetables (which are rich in vitamins and minerals). Protein-rich foods (meat, fish, dairy products) were to be consumed more sparingly, and fats, oils, and sweets were not considered necessary at all. The pyramid shape demonstrated the relative importance of each type of food group, with the most important forming the base and the least important or unnecessary appearing at the top. This pictorial description has been used in many textbooks and taught in schools to children of all ages since its initial publication. However, the USDA has revised the information and appearance twice since its initial publication. Chemical Connections 30A discusses the most recent version and appearance of this information. The pyramid has been replaced with a new description, called MyPlate.

Fiber The cellulose-based non-nutrient component in our food

An important non-nutrient in some foods is **fiber**, which generally consists of the indigestible portion of vegetables and grains. Lettuce, cabbage, celery, whole wheat, brown rice, peas, and beans are all high in fiber. Chemically, fiber is made of cellulose, which, as we saw in Section 20.5C, cannot be digested by humans. Although we cannot digest it, fiber is necessary for proper operation of the digestive system; without it, constipation may result. In more serious cases, a diet lacking sufficient fiber may lead to colon cancer. The DRI recommendation is to ingest 35 g/day for men age 50 years and younger and 25 g/day for women of the same age.

30.2   Why Do We Count Calories?

The largest part of our food supply goes to provide energy for our bodies. As we saw in Chapters 27 and 28, this energy comes from the oxidation of carbohydrates, fats, and proteins. The energy derived from food is usually measured in calories. One nutritional calorie (Cal) equals 1000 cal, or 1 kcal. Thus, when we say that the average daily nutritional requirement for a young adult male is 3000 Cal, we mean the same amount of energy needed to raise the temperature of 3000 kg of water by 1°C or to raise 30 kg (64 lb) of water by 100°C (Section 1.9B). A young adult female needs 2100 Cal/day. These are peak requirements—children and older people, on average, require less energy. Keep in mind that these energy requirements apply to active people. For bodies completely at rest, the corresponding energy requirement for young adult males is 1800 Cal/day and that for females is 1300 Cal/day. The requirement for a resting body is called the **basal caloric requirement**.

Basal caloric requirement The caloric requirement for a resting body

An imbalance between the caloric requirement of the body and the caloric intake creates health problems. Chronic caloric starvation exists in many parts of the world where people simply do not have enough food to eat because of prolonged drought, the devastation of war, natural disasters, or overpopulation. Famine particularly affects infants and children. Chronic starvation, called **marasmus**, increases infant mortality by as much as 50%. It results in arrested growth, muscle wasting, anemia, and general weakness. Even if starvation is later alleviated, it leaves permanent brain damage, insufficient body growth, and lowered resistance to disease.

At the other end of the caloric spectrum is excessive caloric intake. It results in *obesity,* or the accumulation of body fat. Obesity is becoming an epidemic in the U.S. population, with important consequences: It increases the risk of hypertension, cardiovascular disease, and diabetes. Obesity is defined by the National Institutes of Health as applying to a person who has a body mass index (BMI) of 30 or greater. The BMI is a measure of body fat based on height and weight that applies to both adult men and women. For example, a person 70 inches tall is *normal* (BMI less than 25) if he or she weighs 174 lb or less. A person of the same height is *overweight* if he or she weighs more than 174 lb but less than 209 lb; an individual is *obese* if he or she weighs more than 209 lb. More than 200 million Americans are overweight or obese.

Different activity levels are associated with different caloric needs.

Reducing diets aim at decreasing caloric intake without sacrificing any essential nutrients. A combination of exercise and lower caloric intake can eliminate obesity, but usually these diets must achieve their goal over an extended period. Crash diets give the illusion of quick weight loss, but most of this decrease is due to loss of water, which can be regained very quickly. To reduce obesity, we must lose body fat, not water. Achieving this goal takes a lot of effort, because fats contain so much energy. A pound of body fat is equivalent to 3500 Cal. Thus, to lose 10 lb, it is necessary to

EDHAR/Shutterstock.com

consume 35,000 fewer Cal, which can be achieved if one reduces caloric intake by 350 Cal every day for 100 days (or by 700 Cal daily for 50 days) or uses up, through exercise, the same number of food calories. See Chemical Connections 29B for further information on the chemical nature of obesity.

Chemical Connections 30B

Why Is It So Hard to Lose Weight?

One of the great tragedies of being human is that it is far too easy to gain weight and far too difficult to lose it. If we had to analyze the specific chemical reactions involved in this reality, we would look very carefully at the citric acid cycle, especially the decarboxylation reactions. Of course, all foods consumed in excess can be stored as fat. This is true for carbohydrates proteins, and, of course, fats. In addition, these molecules can be interconverted, with the exception that fats cannot give a net yield of carbohydrates. Why can fats not yield carbohydrates? The only way that a fat molecule could make glucose would be to enter the citric acid cycle as acetyl-CoA and then be drawn off as oxaloacetate for gluconeogenesis (Section 29.2). Unfortunately, the two carbons that enter into the citric acid cycle are effectively lost by the decarboxylations (Section 27.4). This leads to an imbalance in the catabolic versus anabolic pathways.

All roads lead to fats, but fats cannot lead back to carbohydrates. Humans are very sensitive to glucose levels in the blood because so much of our metabolism is geared toward protecting our brain cells, which prefer glucose as a fuel. If we eat more carbohydrates than we need, the excess carbohydrates will turn to fats. As we know, it is very easy to put on fat, especially as we age.

What about the reverse? Why don't we just stop eating? Won't that reverse the process? Yes and no. When we start eating less, fat stores become mobilized for energy production. Fat is an excellent source of energy because it forms acetyl-CoA and gives a steady influx for the citric acid cycle. Thus, we can lose some weight by reducing caloric intake. Unfortunately, our blood sugar will also drop as soon as our glycogen stores run out. We have very little stored glycogen that could be tapped to maintain our blood glucose levels.

When the blood glucose drops, we become depressed, sluggish, and irritable. We start having negative thoughts like, "this dieting thing is really stupid. I should eat a pint of Oreo cookie ice cream." If we continue the diet, and given that we cannot turn fats into carbohydrates, where will the blood glucose come from? Only one source is left—proteins. Proteins will be degraded to amino acids and eventually be converted to pyruvate for gluconeogenesis. Thus, we will begin to lose muscle as well as fat.

There is a bright side to this process, however. Using our knowledge of biochemistry, we can see that there is a better way to lose weight than dieting—exercise! If you exercise correctly, you can train your body to use fats to supply acetyl-CoA for the citric acid cycle. If you consume a normal diet, you will maintain your blood glucose and not degrade proteins for that purpose; your ingested carbohydrates will be sufficient to maintain both blood glucose and carbohydrate stores. With the proper ratio of exercise to food intake, and the proper balance of the right types of nutrients, we can increase the breakdown of fats without sacrificing carbohydrate stores or proteins. In essence, it is easier and healthier to train off the weight than to diet off the weight. This fact has been known for a long time. Now we are in a position to see why it is so, biochemically.

Working out is far superior to dieting if you want to lose weight.

Test your knowledge with Problems 30.45–30.50.

30.3 How Does the Body Process Dietary Carbohydrates?

Carbohydrates are the major source of energy in the diet. They also furnish important compounds for the synthesis of cell components (Chapter 29). The main dietary carbohydrates are the polysaccharide starch, the disaccharides lactose and sucrose, and the monosaccharides glucose and fructose. Before the body can absorb carbohydrates, it must break down di-, oligo-, and polysaccharides into monosaccharides, because only monosaccharides can pass into the bloodstream.

The monosaccharide units are connected to each other by glycosidic bonds. Glycosidic bonds are cleaved by hydrolysis. In the body, this hydrolysis is catalyzed by acids and by enzymes. When a metabolic need arises, storage polysaccharides—amylose, amylopectin, and glycogen—are hydrolyzed to yield glucose and maltose.

This hydrolysis is aided by a number of enzymes:

- **α-Amylase** attacks all three storage polysaccharides at random, hydrolyzing the α-1,4-glycosidic bonds.

- **β-Amylase** also hydrolyzes the α-1,4-glycosidic bonds but in an orderly fashion, cutting disaccharidic maltose units one by one from the nonreducing end of a chain.

- The **debranching enzyme** hydrolyzes the α-1,6-glycosidic bonds (Figure 30.3).

In acid-catalyzed hydrolysis, storage polysaccharides are cut at random points. At body temperature, acid catalysis is slower than enzyme-catalyzed hydrolysis.

The digestion (hydrolysis) of starch and glycogen in our food supply starts in the mouth, where α-amylase is one of the main components of saliva. Hydrochloric acid in the stomach and other hydrolytic enzymes in the intestinal tract hydrolyze starch and glycogen to produce mono- and disaccharides (D-glucose and maltose).

D-glucose enters the bloodstream and is carried to the cells to be utilized (Section 28.2). For this reason, D-glucose is often called blood sugar. In healthy people, little or none of this sugar ends up in the urine except for short periods of time (binge eating). In diabetes mellitus, however, glucose is not completely metabolized and does appear in the urine. As a consequence, it is necessary to test the urine of diabetic patients for the presence of glucose (Chemical Connections 20C).

The latest DRI guideline, issued by the National Academy of Sciences in 2002, recommends a minimum carbohydrate intake of 130 g/day. Most people exceed this value. Artificial sweeteners (Chemical Connections 30C) can be used to reduce mono- and disaccharide intake.

Not everyone considers this carbohydrate recommendation to be the final word. Certain diets, such as the Atkins diet, recommend reducing carbohydrate intake to force the body to burn stored fat for energy. For example, during the introductory two-week period in the Atkins diet, a maximum of 20 g of carbohydrates per day are recommended in the form of salad and vegetables; no fruits or starchy vegetables are allowed. For the longer term, an additional 5 g of carbohydrates per day is added in the form of fruits. This restriction induces ketosis, the production of ketone bodies (Section 28.7) that may create muscle weakness and kidney problems.

Many fad diets exist today. The Atkins diet was preceded by the Zone Diet and the Sugar Buster's Diet, both of which attempted to limit carbohydrate intake. Another diet suggests that you match the foods you eat to your

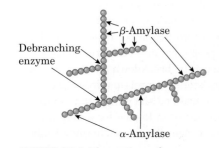

FIGURE 30.3 The action of different enzymes on glycogen and starch.

ABO blood type. To date, little scientific evidence supports any of these approaches, although some aspects of many diets have merit.

30.4   How Does the Body Process Dietary Fats?

Fats are the most concentrated source of energy. About 98% of the lipids in our diet are fats and oils (triglycerides); the remaining 2% consist of complex lipids and cholesterol.

The lipids in the food we eat must be hydrolyzed into smaller components before they can be absorbed into the blood or lymph system through the intestinal walls. The enzymes that promote this hydrolysis are located in the small intestine and are called *lipases*. However, because lipids are insoluble in the aqueous environment of the gastrointestinal tract, they must be dispersed into fine colloidal particles before the enzymes can act on them.

Bile salts (Section 21.11) perform this important function. Bile salts are manufactured in the liver from cholesterol and stored in the gallbladder. From there, they are secreted through the bile ducts into the intestine. Lipases act on the emulsion produced by bile salts and dietary fats, breaking the fats into glycerol and fatty acids and the complex lipids into fatty acids, alcohols (glycerol, choline, ethanolamine, sphingosine), and carbohydrates. These hydrolysis products are then absorbed through the intestinal walls.

Only two fatty acids are essential in higher animals, including humans: linolenic and linoleic acids (Section 21.3). Nutritionists occasionally list arachidonic acid as an **essential fatty acid**. In reality, our bodies can synthesize arachidonic acid from linoleic acid.

30.5   How Does the Body Process Dietary Protein?

Although the proteins in our diet can be used for energy (Section 28.9), their main use is to furnish amino acids from which the body synthesizes its own proteins (Section 26.5).

The digestion of dietary proteins begins with cooking, which denatures proteins. (Denatured proteins are hydrolyzed more easily by hydrochloric acid in the stomach and by digestive enzymes than are native proteins.) *Stomach acid* contains about 0.5% HCl. This HCl both denatures the proteins and hydrolyzes the peptide bonds randomly. *Pepsin,* the proteolytic enzyme of stomach juice, hydrolyzes peptide bonds on the amino side of the aromatic amino acids: tryptophan, phenylalanine, and tyrosine (see Figure 30.4).

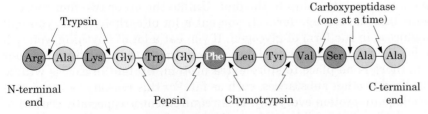

FIGURE 30.4 Different enzymes hydrolyze peptide chains in different but specific ways. Note that both chymotrypsin and pepsin would hydrolyze all of the same amino acids, but they are shown here hydrolyzing separate ones for comparison of the side they hydrolyze.

Most protein digestion occurs in the small intestine. There, the enzyme *chymotrypsin* hydrolyzes internal peptide bonds at the same amino acids as does pepsin, except it does so on the other side, leaving these amino acids as the carboxyl termini of their fragments. Another enzyme, *trypsin*, hydrolyzes them only on the carboxyl side of arginine and lysine. Other enzymes, such as *carboxypeptidase*, hydrolyze amino acids one by one from the C-terminal end of the protein. The amino acids and small peptides are then absorbed through the intestinal walls.

The human body is incapable of synthesizing ten of the amino acids in sufficient quanities needed to make proteins. These ten **essential amino acids** must be obtained from our food; they are shown in Table 29.1. The body hydrolyzes food proteins into their amino acid constituents and then puts the amino acids together again to make body proteins. For proper nutrition, the human diet should contain about 20% protein.

A dietary protein that contains all of the essential amino acids is called a **complete protein**. Casein, the protein of milk, is a complete protein, as are most other animal proteins—those found in meat, fish, and eggs. People who eat adequate quantities of meat, fish, eggs, and dairy products get all the amino acids they need to keep healthy. About 50 g/day of complete proteins constitutes an adequate quantity.

An important animal protein that is not complete is gelatin, which is made by denaturing collagen (Section 22.12). Gelatin lacks tryptophan and is low in several other amino acids, including isoleucine and methionine. Many people on quick-reducing diets consume "liquid protein." This substance is simply denatured and partially hydrolyzed collagen (gelatin). Therefore, if it is the only protein source in the diet, some essential amino acids will be lacking.

Most plant proteins are incomplete. For example, corn protein lacks lysine and tryptophan; rice protein lacks lysine and threonine; wheat protein lacks lysine; and legumes are low in methionine and cysteine. Even soy protein, one of the best plant proteins, is very low in methionine and is not a complete protein. Adequate amino acid nutrition is possible with a vegetarian diet, but only if a wide range of vegetables is eaten. **Protein complementation** is one such diet. In protein complementation, two or more foods complement the others' deficiencies. For example, grains and legumes complement each other, with grains being low in lysine but high in methionine. Over time, such protein complementation in vegetarian diets became the staple in many parts of the world—corn tortillas and beans in Central and South America, rice and lentils in India, and rice and tofu in China and Japan.

In many developing countries, protein deficiency diseases are widespread because the people get their protein mostly from plants. Among these diseases is **kwashiorkor**, whose symptoms include a swollen stomach, skin discoloration, and retarded growth.

Proteins are inherently different from carbohydrates and fats when it comes to their relationship to the diet. Unlike the other two fuel sources, proteins have no storage form. If you eat a lot of carbohydrate, you will store glucose in the form of glycogen. If you eat a lot of anything, you will store fat. However, if you eat a lot of protein (more than required for your needs), there is no place to store extra protein. Protein in excess will be metabolized to other substances, such as fat. For this reason, you must consume adequate protein every day. This requirement is especially critical in athletes and growing children. If an athlete works out intensely one day but eats incomplete protein on that day, he or she cannot repair the damaged muscles. The fact that the athlete may have eaten an excess of a complete protein the day before will not help.

Essential amino acid An amino acid that the body cannot synthesize in the required amounts and so must be obtained in the diet

30.6 What Is the Importance of Vitamins, Minerals, and Water?

Vitamins and **minerals** are essential for good nutrition. Animals maintained on diets that contain sufficient carbohydrates, fats, and proteins and provided with an ample water supply cannot survive on these alone; they also need the essential organic components called vitamins and the inorganic ions called minerals. Many vitamins, especially those in the B group, act as coenzymes, and inorganic ions act as cofactors in enzyme-catalyzed reactions (Table 30.1). Table 30.2 lists the structures, dietary sources, and functions of the vitamins and minerals. Deficiencies in vitamins and minerals lead to many nutritionally controllable diseases (one example is shown in Figure 30.5); these conditions are also listed in Table 30.2.

The recent trend in vitamin appreciation is connected to their general role rather than to any specific action they have against a particular disease. For example, today the role of vitamin C in the prevention of scurvy is barely mentioned, but it is hailed as an important antioxidant. Similarly, other antioxidant vitamins or vitamin precursors dominate the medical literature. As an example, it has been shown that consumption of carotenoids (other than β-carotene) and vitamins E and C contributes significantly to respiratory health. The most important of the three is vitamin E. Furthermore, the loss of vitamin C during hemodialysis contributes significantly to oxidative damage in patients, leading to accelerated atherosclerosis.

Some vitamins have unusual effects besides their normal participation as coenzymes in many metabolic pathways or action as antioxidants in the body. Among these are the well-known effect of riboflavin, vitamin B_2, as a photosensitizer that worsens damage caused by solar radiation. Niacinamide, the amide form of vitamin B (niacin), is used in megadoses (2 g/day) to treat an autoimmune disease called bullous pemphigoid, which causes blisters on the skin. Conversely, the same megadoses in healthy patients may cause harm.

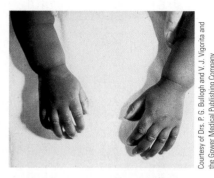

FIGURE 30.5 Symptoms of rickets, a vitamin D deficiency in children. The nonmineralization of the bones of the radius and the ulna results in prominence of the wrist.

Courtesy of Drs. P. G. Bullogh and V. J. Vigorita and the Gower Medical Publishing Company

Table 30.1 Vitamins and Trace Elements as Coenzymes and Cofactors

Vitamin/Trace Element	Form of Coenzyme	Representative Enzyme	Reference
B_1, thiamine	Thiamine pyrophosphate, TPP	Pyruvate dehydrogenase	Step 12, Section 28.2
B_2, riboflavin	Flavin adenine dinucleotide, FAD	Succinate dehydrogenase	Step 6, Section 27.4
Niacin	Nicotinamide adenine dinucleotide, NAD^+	D-Glyceraldehyde-3-phosphate dehydrogenase	Step 5, Section 28.2
Pantothenic acid	Coenzyme A, CoA	Fatty acid synthase	Step 1, Section 29.3
B_6, pyridoxal	Pyridoxal phosphate, PLP	Aspartate amino transferase	Class 2, Section 23.2
B_{12}, cobalamin		Ribose reductase	Step 1, Section 25.2
Biotin	N-carboxybiotin	Acetyl CoA carboxylase	Malonyl-CoA, Section 29.3
Folic acid		Purine biosynthesis	Section 25.2
Mg		Pyruvate kinase	Chemical Connections 23C
Fe		Cytochrome oxidase	Section 27.5
Cu		Cytochrome oxidase	Section 27.5
Zn		DNA polymerase	Section 25.6
Mn		Arginase	Step 5, Section 28.8
K		Pyruvate kinase	Chemical Connections 23C
Ni		Urease	Section 23.1
Mo		Nitrate reductase	
Se		Glutathione peroxidase	Chemical Connections 22A

Table 30.2 Vitamins and Minerals: Sources, Functions, Deficiency Diseases, and Daily Requirements

Name	Structure	Best Food Source	Function	Deficiency Symptoms and Diseases	Recommended Dietary Allowance[a]
Fat-Soluble Vitamins					
A	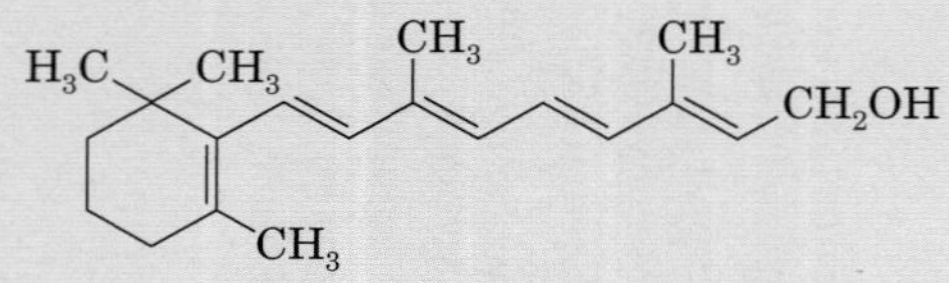 *(chemical structure: retinol)*	Liver, butter, egg yolk, carrots, spinach, sweet potatoes	Vision; to heal eye and skin injuries	Night blindness; blindness; keratinization of epithelium and cornea	800 μg (1500 μg)[b]
D	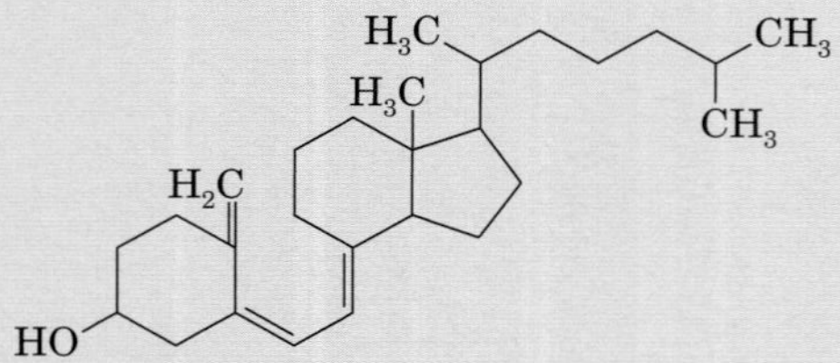 *(chemical structure: vitamin D)*	Salmon, sardines, cod liver oil, cheese, eggs, milk	Promotes calcium and phosphate absorption and mobilization	Rickets (in children): pliable bones; osteomalacia (in adults); fragile bones	5–10 μg; exposure to sunlight
E	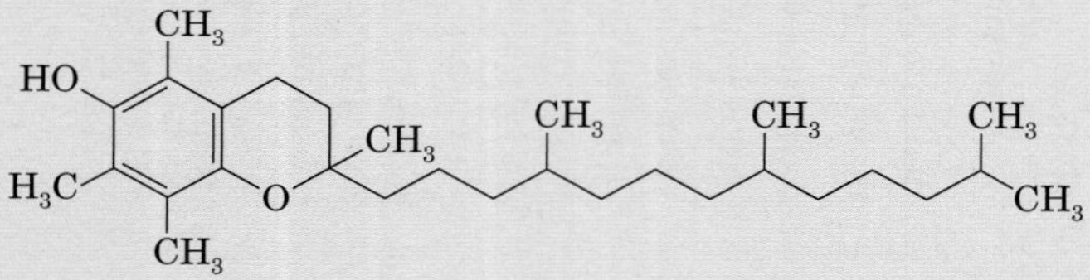 *(chemical structure: tocopherol)*	Vegetable oils, nuts, potato chips, spinach	Antioxidant	In cases of malabsorption such as in cystic fibrosis: anemia. In premature infants: anemia	8–10 mg
K	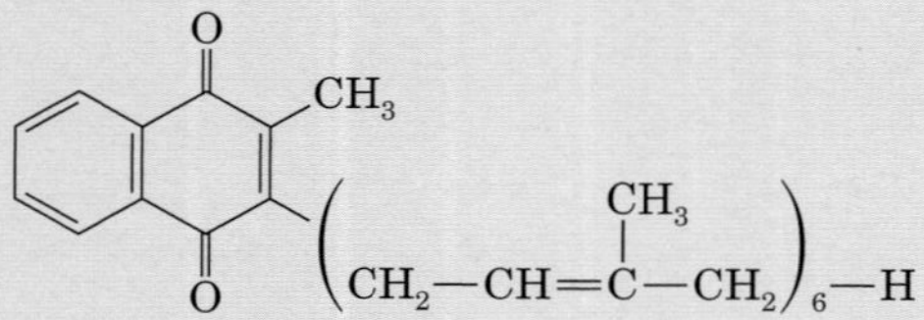 *(chemical structure: vitamin K)*	Spinach, potatoes, cauliflower, beef liver	Blood clotting	Uncontrolled bleeding (mostly in newborn infants)	65–80 μg
Water-Soluble Vitamins					
B$_1$ (thiamine)	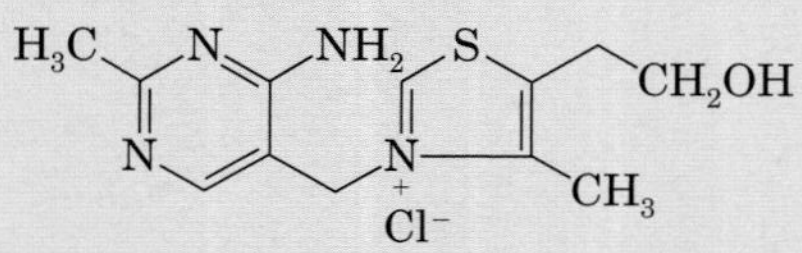 *(chemical structure: thiamine)*	Beans, soybeans, cereals, ham, liver	Coenzyme in oxidative decarboxylation and in pentose phosphate shunt	Beriberi. In alcoholics: heart failure; pulmonary congestion	1.1 mg
B$_2$ (riboflavin)	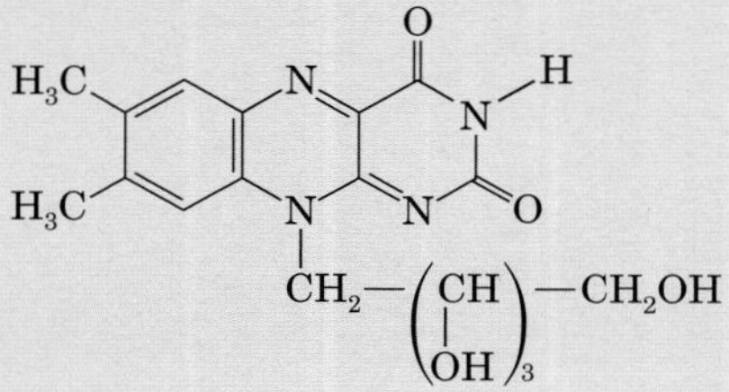 *(chemical structure: riboflavin)*	Kidney, liver, yeast, almonds, mushrooms, beans	Coenzyme of oxidative processes	Invasion of cornea by capillaries; cheilosis; dermatitis	1.4 mg

Vitamin	Sources	Function	Deficiency	RDA
Nicotinic acid (niacin)	Chickpeas, lentils, prunes, peaches, avocados, figs, fish, meat, mushrooms, peanuts, bread, rice, beans, berries	Coenzyme of oxidative processes	Pellagra	15–18 mg
B_6 (pyridoxal)	Meat, fish, nuts, oats, wheat germ, potato chips	Coenzyme in transamination; heme synthesis	Convulsions; chronic anemia; peripheral neuropathy	1.6–2.2 mg
Folic acid	Liver, kidney, eggs, spinach, beets, orange juice, avocados, cantaloupe	Coenzyme in methylation and in DNA synthesis	Anemia	400 μg
B_{12}	Oysters, salmon, liver, kidney	Part of methyl-removing enzyme in folate metabolism	Patchy demyelination; degradation of nerves, spinal cord, and brain	1–3 μg

Table 30.2 Vitamins and Minerals: Sources, Functions, Deficiency Diseases, and Daily Requirements (continued)

Name Structure	Best Food Source	Function	Deficiency Symptoms and Diseases	Recommended Dietary Allowance[a]
Water-Soluble Vitamins				
Pantothenic acid 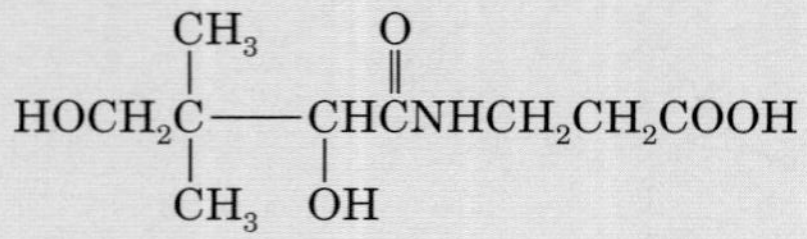	Peanuts, buckwheat, soybeans, broccoli, lima beans, liver kidney, brain, heart	Part of CoA; fat and carbohydrate metabolism	Gastrointestinal disturbances; depression	4–7 mg
Biotin 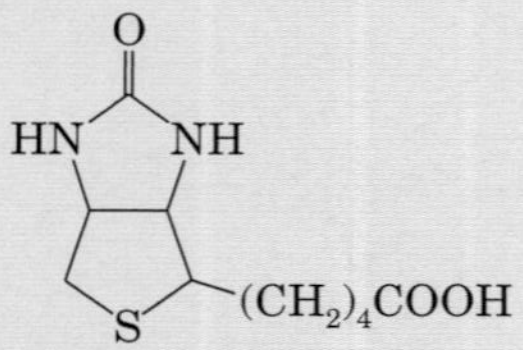	Yeast, liver, kidney, nuts, egg yolk	Synthesis of fatty acids	Dermatitis; nausea; depression	30–100 μg
C (ascorbic acid) 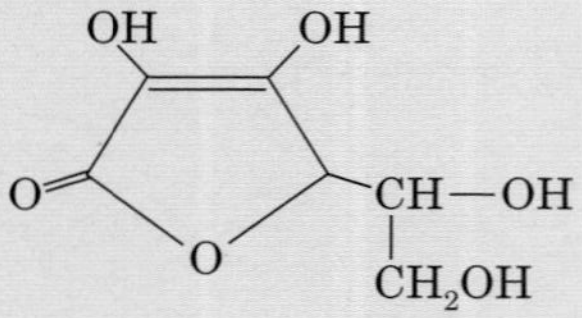	Citrus fruit, berries, broccoli, cabbage, peppers, tomatoes	Hydroxylation of collagen, wound healing; bond formation; antioxidant	Scurvy; capillary fragility	60 mg
Minerals				
Potassium	Apricots, bananas, dates, figs, nuts, raisins, beans, chickpeas, cress, lentils	Provides membrane potential	Muscle weakness	3500 mg
Sodium	Meat, cheese, cold cuts, smoked fish, table salt	Osmotic pressure	None	2000–2400 mg
Calcium	Milk, cheese, sardines, caviar	Bone formation; hormonal function; blood coagulation; muscle contraction	Muscle cramps; osteoporosis; fragile bones	800–1200 mg
Chloride	Meat, cheese, cold cuts, smoked fish, table salt	Osmotic pressure	None	1700–5100 mg

Mineral	Sources	Function	Deficiency/Excess	RDA[a]
Phosphorus	Lentils, nuts, oats, grain flours, cocoa, egg yolk, cheese, meat (brain, sweetbreads)	Balancing calcium in diet	Excess causes structural weakness in bones	800–1200 mg
Magnesium	Cheeses, cocoa, chocolate, nuts, soybeans, beans	Cofactor in enzymes	Hypocalcemia	280–350 mg
Iron	Raisins, beans, chickpeas, parsley, smoked fish, liver, kidney, spleen, heart, clams, oysters	Oxidative phosphorylation; hemoglobin	Anemia	15 mg
Zinc	Yeast, soybeans, nuts, corn, cheese, meat, poultry	Cofactor in enzymes, insulin	Retarded growth; enlarged liver	12–15 mg
Copper	Oysters, sardines, lamb, liver	Oxidative enzymes cofactor	Loss of hair pigmentation, anemia	1.5–3 mg
Manganese	Nuts, fruits, vegetables, whole-grain cereals	Bone formation	Low serum cholesterol levels; retarded growth of hair and nails	2.0–5.0 mg
Chromium	Meat, beer, whole wheat and rye flours	Glucose metabolism	Glucose not available to cells	0.05–0.2 mg
Molybdenum	Liver, kidney, spinach, beans, peas	Protein synthesis	Retarded growth	0.075–0.250 mg
Cobalt	Meat, dairy products	Component of vitamin B_{12}	Pernicious anemia	0.05 mg (20–30 mg)[b]
Selenium	Meat, seafood	Fat metabolism	Muscular disorders	0.05–0.07 mg (2.4–3.0 mg)[b]
Iodine	Seafood, vegetables, meat	Thyroid glands	Goiter	150–170 μg (1000 μg)[b]
Fluorine	Fluoridated water; fluoridated toothpaste	Enamel formation	Tooth decay	1.5–4.0 mg (8–20 mg)[b]

[a]The U.S. RDAs are set by the Food and Nutrition Board of the National Research Council. The numbers given here are based on the latest recommendations (*National Research Council Recommended Dietary Allowances*, 10th ed., 1989, National Academy Press, Washington, D.C.). The RDA varies with age, sex, and level of activity; the numbers given are average values for both sexes between the ages of 18 and 54.

[b]Toxic if doses above the level shown in parentheses are taken.

Iron: An Example of a Mineral Requirement

Iron, whether in the form Fe(II) or Fe(III), is usually found in the body in association with proteins. Little or no iron can be found "free" in the blood. Because iron-containing proteins of the body are ubiquitous, there is a dietary requirement for this mineral. Severe deficits can lead to iron-deficiency anemia.

Iron usually occurs as the Fe(III) form in food. This is also the form released from iron pots when food is cooked in them. However, iron must be in the Fe(II) state to be absorbed. Reduction from Fe(III) to Fe(II) can be accomplished by ascorbate (vitamin C) or by succinate. Factors that affect absorption include the solubility of a given compound of iron, the presence of antacids in the digestive tract, and the source of the iron. To give some examples, iron may form insoluble complexes with phosphate or oxalate, and the presence of antacids in the digestive tract may decrease iron absorption. Iron from meats is more easily absorbed than iron from plant sources.

Requirements for iron vary according to age and gender. Infants and adult men need 10 mg per day; infants are born with a three- to six-month supply. Children and women (ages 16 through 50) need 15 to 18 mg per day. Women lose 20 to 23 mg of iron during each menstrual period. Pregnant and lactating women need more than 18 mg per day. After a blood loss, anyone, regardless of age or gender, needs more than these amounts. Distance runners, particularly marathoners, are also at risk of becoming anemic through loss of blood cells in the feet caused by the pounding of the thousands of foot falls

that occur during a long run. People with iron deficiencies may experience a craving for nonfood items like clay, chalk, and ice.

Iron, in the form of ferrous sulfate, is a common over-the-counter mineral supplement.

Charles D. Winters

Test your knowledge with Problems 30.51–30.54.

Although the concept of the RDA has been used since 1940 and periodically updated as new knowledge dictated, a new concept is being developed in the field of nutrition. The DRI is designed to replace the RDA and is tailor-made to different ages and genders. It gives a set of two to four values for a particular nutrient in RDI:

- The estimated average requirement
- The recommended dietary allowance
- The adequate intake level
- The tolerable upper intake level

For example, the RDA for vitamin D is 5–10 μg, the adequate intake listed in the DRI for vitamin D for a person between age 9 and 50 years is 5 μg, and the tolerable upper intake level for the same age group is 50 μg.

A third set of standards appears on food labels in the United States, the Daily Values discussed in Section 30.2. Each gives a single value for each nutrient and reflects the need of an average healthy person eating 2000 to 2500 calories per day. The Daily Value for vitamin D, as it appears on vitamin bottles, is 400 International Units, which is 10 μg, the same as the RDA.

Food for Performance Enhancement

Athletes do whatever they can to enhance their performance. While the press focuses on those illegal methods employed by some athletes, such as using steroids or erythropoietin (EPO), many athletes continue to seek legal ways of enhancing their performance through diet and diet supplements. Any substance that aids performance is called an **ergogenic aid**.

After vigorous exercise lasting 30 minutes or more, performance typically declines because the glycogen stores in the muscle are depleted. After 90 minutes to 2 hours, liver glycogen stores also become depleted. There are two ways to combat this outcome. First, one can start the event with a full load of glycogen in the muscle and liver. This is why many athletes load up on carbohydrates in the form of pasta or other high-carbohydrate meals in the days before the event. Second, one can maintain the blood glucose levels during the event, so that liver glycogen does not have to be used for this purpose and ingested sugars can help support the athlete's energy needs, sparing some of the muscle and liver glycogen. This is why athletes often consume energy bars and sport drinks during an event.

The most often used ergogenic aid, although many athletes might not realize it, is caffeine. Caffeine works in two different ways. First, it acts as a general stimulant of the central nervous system, giving the athlete the sensation of having a lot of energy. Second, it stimulates the breakdown of fatty acids for fuel via its role as an activator of the lipases that break down triacylglycerols (Section 30.4). Caffeine is a "double-edged sword," however, because it can also cause dehydration and actually inhibits the breakdown of glycogen.

In recent years, a new performance-enhancing food has appeared on the market and quickly become a best seller: creatine. It is sold over the counter. Creatine is a naturally occurring amino acid in the muscles, which store energy in the form of high-energy phosphocreatine. During a short and strenuous bout of exercise, such as a 100-meter sprint, the muscles first use up the ATP obtainable from the reaction of phosphocreatine with ADP; only then do they rely on the glycogen stores.

Creatine is sold over the counter.

Both creatine and carbohydrates are natural food and body components and, therefore, cannot be considered equivalent to banned performance enhancers such as anabolic steroids or "andro" (Chemical Connections 21E). They are beneficial in improving performance in sports where short bursts of energy are needed, such as weight lifting, jumping, and sprinting. Lately, creatine has been used experimentally to preserve muscle neurons in degenerative diseases such as Parkinson's disease, Huntington's disease, and muscular dystrophy. Creatine also has few known hazards, even with long-term use. It is a highly nitrogenated compound, however, and overuse of creatine leads to the same problems as eating a diet characterized by excessive protein. The molecule must be hydrated, so water can be tied up with creatine that should be hydrating the body. The kidneys must also deal with excretion of extra nitrogen.

While athletes spend a lot of time and money on commercial ergogenic aids, the most important ergogenic aid remains water, the elixir of life that has been almost forgotten as it is replaced by its more expensive sports-drink cousins. A 1% level of dehydration during an event can adversely affect athletic performance. For a 150-pound athlete, this means losing 1.5 pounds of water as sweat. An athlete could easily lose much more than that in running a 10-kilometer race on all but the coldest days. To make things more difficult, performance is affected before thirst becomes noticeable. Only by drinking often during a long event can dehydration be prevented, and the athlete should begin drinking before noticing that she or he is thirsty.

$$
\begin{array}{c}
\text{O}^- \\
| \\
{}^-\text{O}-\text{P}=\text{O} \\
| \\
\text{N}-\text{H} \\
| \\
\text{C}=\text{NH}_2^+ + \text{ADP} \longrightarrow \\
| \\
\text{H}_3\text{C}-\text{N} \\
| \\
\text{CH}_2 \\
| \\
\text{COO}^-
\end{array}
\qquad
\begin{array}{c}
\text{NH}_2 \\
| \\
\text{C}=\text{NH}_2^+ \\
| \\
\text{H}_3\text{C}-\text{N} \quad + \text{ATP} \\
| \\
\text{CH}_2 \\
| \\
\text{COO}^-
\end{array}
$$

Phosphocreatine   Creatine

Test your knowledge with Problems 30.55–30.59.

Chemical Connections 30E

Organic Food—Hope or Hype?

The organic food industry is growing rapidly around the world. Organic markets like Whole Foods are flourishing along with associated organic restaurants. For a food to be labeled organic in the United States, it must be certified by the USDA as being grown in an environment without synthetic pesticides and fertilizers. Organic meat and poultry must be raised without hormones on feed grown organically. While it may seem intuitive to prefer such organically grown products, the price is often discouraging, between 10% and 100% more expensive than its nonorganic counterparts. The true benefits of organic food depend on many variables. For example, research shows fewer benefits to eating organic bananas, as most pesticides in nonorganic bananas are thrown away with the peel. Other fruits, such as peaches, strawberries, apples, and pears, are the opposite. A 2002 study by the USDA showed that 98% of these nonorganic fruits tested had measurable levels of pesticides.

Researchers agree that pregnant women and children have the highest risk from pesticides. Pesticides have been shown to cross the placenta during pregnancy, and one study found that there was a positive link between pesticides present in some New York apartments and impaired fetal growth. A University of Washington study found that preschoolers fed nonorganic diets had six times the levels of certain pesticides as children fed organic food. High doses of certain pesticides have been shown to cause neurological and reproductive disorders in children exposed to them. With young bodies less able to rid themselves of toxins, eating organic food would be most important for pregnant women and breast-feeding mothers. For those concerned about nonorganic meat and poultry, the issue is the use of antibiotics and growth hormones in raising the animals. Antibiotics are a worry as they can help produce resistant strains of bacteria. Some

studies have suggested a link between certain types of cancer and the use of growth hormones.

Besides the potential dangers of the pesticides comes the question of the quality of the nutrients in the food. This is a more difficult parameter to measure, and the results have been less conclusive. Recent studies on wheat have shown no significant differences in the quality of the wheat's nutrients when grown using synthetic pesticides compared to organic processes. Subjectively, however, many people prefer the taste of organically grown fruits, vegetables, and grains, despite the bitter taste the price tag might leave. If the rapid growth of the organic food industry is any indication, the general public believes the price tag is worth it.

USDS

The well-known symbol for government-certified organic food.

Test your knowledge with Problems 30.60 and 30.61.

While many people like to ingest "megadoses" of vitamins, in the belief that more is always better, care must be taken, especially with certain vitamins, like A, D, and E. These are fat-soluble vitamins, and overdoses must be avoided. They are stored in adipose tissue, and excesses can be toxic when large amounts of fat-soluble vitamins accumulate. Excess vitamin A is especially toxic. With water-soluble vitamins, the danger of overdose is lessened by the quick turnover of the vitamin in the body, and the ability of the body to easily eliminate them in the urine.

Water makes up 60% of our body weight. Most of the compounds in our body are dissolved in water, which also serves as a transporting medium to carry nutrients and waste materials. We must maintain a proper balance between water intake and water excretion via urine, feces, sweat, and exhalation of breath. A normal diet requires about 1200 to 1500 mL of water per day, in addition to the water consumed as part of our foods. Public drinking water systems in the United States are regulated by the Environmental Protection Agency (EPA), which sets minimum standards for protection of public health. Public water supplies are treated with disinfectants (typically chlorine) to kill microorganisms. Chlorinated water may have both an aftertaste and an odor. Private wells are not regulated by EPA standards.

Bottled water may come from springs, streams, or public water sources. Because bottled water is classified as a food, it comes under the FDA supervision, which requires it to meet the same standards for purity and sanitation as tap water. Most bottled water is disinfected with ozone, which leaves no flavor or odor in water.

Chemical Connections 30F

Depression in America—Don't Worry; Be Happy

Some believe it is caused by our rat-race lifestyles. Others think it is due to toxins. Others think it is due to poor nutrition. Whatever the cause, there is no doubt that millions of Americans suffer from depression. The pharmaceutical industry and the fields of psychiatry and psychology are heavily funded by revenues from patients fighting depression. But before you reach for a bottle of Prozac, there might be some safer and cheaper options to try first. In the last few years, a deficiency in many compounds has been found to correlate with depression.

Fatty Acids

In early 2011, researchers reported online in the journal *Nature Neuroscience* that deficiencies in omega-3 polyunsaturated fatty acids (Chemical Connections 21H) alters the functioning of the **endocannabinoid** system, a group of lipids and their receptors that are involved in mood, pain sensations, and other processes. They noted that mice subjected to a diet low in omega-3 polyunsaturated fatty acids had lower omega-3 levels in the brain. This was associated with an alteration in the functioning of the endocannabinoid system, specifically, a deficit in the signaling of the CB1 **cannabinoid receptor** in the prefrontal cortex of the brain. The cannabinoid receptors are a class of cell membrane receptors under the G-protein-coupled receptor superfamily (Section 24.5). Cannabinoid receptors are activated by three major groups of ligands—endocannabinoids (produced by the mammalian body), plant cannabinoids (such as THC, produced by the cannabis plant), and synthetic cannabinoids (such as HU-210). All of the endocannabinoids and plant cannabinoids are lipophilic (i.e., fat-soluble, compounds). The CB1 cannabinoid receptor has been linked to depressive disorders.

Other studies followed over 600 cases of depression and showed that patients who had diets higher in *trans* fatty acids (Section 21.3B) were almost 50% more likely to be depressed. While much research remains to be done, this could indicate that eating salmon and other sources of omega-3 fatty acids, as well as diets low in *trans* fatty acids, can help one's head as well as one's heart.

Vitamin D

A clinical trial is being planned to test the effect of vitamin D supplementation on insulin resistance and mood in diabetic women. Increased insulin resistance (type 2 diabetes) has been associated with depression. Higher vitamin D levels have been associated with a reduced risk of depression, diabetes, and other ailments. The study plans to administer 50,000 international units of vitamin D per week for 6 months to 80 stable type 2 diabetic women, aged 18 to 70 with signs of depression. Participants will be evaluated at three time points for serum vitamin D levels and other factors. There is evidence to suggest that vitamin D supplementation may decrease insulin resistance. Vitamin D supplementation has also been linked to reduction in several other chronic conditions, including osteoporosis, immune dysfunction, cardiovascular disease, and cancer. Type 2 diabetes is on the rise in America, and some evidence exists that several factors, including diet and vitamin deficiencies, can lead to this disease, as well as the associated depression.

Vitamin B_{12}

Vitamin B_{12} has the largest and most complex chemical structure of all the vitamins. It is unique among vitamins in that it contains a metal ion, cobalt (see Table 30.2). **Cobalamin** is the term used to refer to compounds having vitamin B_{12} activity. Methylcobalamin and 5-deoxyadenosyl cobalamin are the forms of vitamin B_{12} used in the human body. Most supplements contain cyanocobalamin, which is readily converted to 5-deoxyadenosyl and methylcobalamin in the body. In mammals, cobalamin is a cofactor for only two enzymes, methionine synthase and L-methylmalonyl-CoA mutase (2).

Methylcobalamin is required for the function of the folate-dependent enzyme methionine synthase. This enzyme is required for the synthesis of the amino acid methionine from homocysteine. Methionine, in turn, is required for the synthesis of S-adenosylmethionine, a methyl group donor used in many biological methylation reactions, including

Chemical Connections 30F

Depression in America—Don't Worry; Be Happy (continued)

the methylation of a number of sites within DNA and RNA. Methylation of DNA may be important in cancer prevention. Inadequate function of methionine synthase can lead to an accumulation of homocysteine, which has been associated with increased risk of cardiovascular diseases.

Besides being linked to cancer and Alzheimer's disease, Vitamin B_{12} deficiency is also believed to be involved in depression. Studies have found that up to 30% of patients hospitalized for depression are deficient in vitamin B_{12}. A study of several hundred physically disabled women over the age of 65 found that women deficient in vitamin B_{12} were more likely to be severely depressed than non-deficient women. A study of over three thousand elderly men and women with depression showed that those with vitamin B_{12} deficiency were almost 70% more

likely to experience depression than those with normal vitamin B_{12} levels. The relationship between vitamin B_{12} deficiency and depression is not clear, but may involve S-adenosylmethionine (SAMe). Vitamin B_{12} and folate are required for the synthesis of SAMe, a methyl group donor essential for the metabolism of neurotransmitters, whose bioavailability has been related to depression.

Just as the best way to lose weight involves exercise and reduced calories rather than diet pills, the best way to fight depression may also be found with a basic understanding of nutrition rather than synthetic chemicals. While antidepressants have helped millions of people, there would seem to be little downside to trying vitamin supplements and a correct diet prior to starting down that path.

Test your knowledge with Problems 30.62–30.66.

Summary

◑WL Sign in at **www.cengage.com/owl** to develop problem-solving skills and complete online homework assigned by your professor.

Section 30.1 How Do We Measure Nutrition?

- **Nutrients** are components of foods that provide for growth, replacement, and energy.
- Nutrients are classified into six groups: carbohydrates, lipids, proteins, vitamins, minerals, and water.
- Each food contains a variety of nutrients. The largest part of our food intake is used to provide energy for our bodies.

Section 30.2 Why Do We Count Calories?

- A typical young adult needs 3000 Cal (male) or 2100 Cal (female) as an average daily caloric intake.
- **Basal caloric requirements**, the energy needed when the body is completely at rest, are less than the normal requirements.
- An imbalance between energy needs and caloric intake may create health problems. For example, chronic starvation increases infant mortality, whereas obesity leads to hypertension, cardiovascular disease, and diabetes.

Section 30.3 How Does the Body Process Dietary Carbohydrates?

- Carbohydrates are the major source of energy in the human diet.
- Monosaccharides are directly absorbed in the intestines, while oligo- and polysaccharides, such as starch, are

digested with the aid of stomach acid, α- and β-amylases, and debranching enzymes.

Section 30.4 How Does the Body Process Dietary Fats?

- Fats are the most concentrated source of energy.
- Fats are emulsified by bile salts and digested by lipases before being absorbed as fatty acids and glycerol in the intestines.
- Essential fatty and amino acids are needed as building blocks because the human body cannot synthesize them.

Section 30.5 How Does the Body Process Dietary Protein?

- Proteins are hydrolyzed by stomach acid and further digested by enzymes such as pepsin and trypsin before being absorbed as amino acids.
- There is no storage form for proteins, so good protein sources must be consumed in the diet every day.

Section 30.6 What Is the Importance of Vitamins, Minerals, and Water?

- Vitamins and minerals are essential constituents of the diet that are needed in small quantities.
- The fat-soluble vitamins are A, D, E, and K.
- Vitamins C and the B group are water-soluble vitamins.
- Most of the B vitamins are essential coenzymes.
- The most important dietary minerals are Na^+, Cl^-, K^+, PO_4^{3-}, Ca^{2+}, Fe^{2+}, and Mg^{2+}, but trace minerals are also necessary.
- Water makes up 60% of body weight.

Problems

⊙WL Interactive versions of these problems may be assigned in OWL.

Orange-numbered problems are applied.

Section 30.1 How Do We Measure Nutrition?

30.1 Are nutrient requirements uniform for everyone?

30.2 Is banana flavoring, isopentyl acetate, a nutrient?

30.3 If sodium benzoate, a food preservative, is excreted as such and if calcium propionate, another food preservative, is metabolized to CO_2 and H_2O, would you consider either of these preservatives to be a nutrient? If so, why?

30.4 Is corn grown solely with organic fertilizers more nutritious than corn grown with chemical fertilizers?

30.5 Which part of the Nutrition Facts label found on food packages is the same for all labels carrying it?

30.6 Of which kinds of food does the U.S. government recommend that we have the most servings each day?

30.7 What is the importance of fiber in the diet?

30.8 Can a chemical that, in essence, goes through the body unchanged be an essential nutrient? Explain.

Section 30.2 Why Do We Count Calories?

30.9 A young adult female needs a caloric intake of 2100 Cal/day. Her basal caloric requirement is only 1300 Cal/day. Why is the extra 800 Cal needed?

30.10 What ill effects may obesity bring?

30.11 Assume that you want to lose 20 lb of body fat in 60 days. Your present dietary intake is 3000 Cal/day. What should your caloric intake be, in Cal/day, to achieve this goal, assuming no change in exercise habits?

30.12 What is marasmus?

30.13 Diuretics help to secrete water from the body. Would diuretic pills be a good way to reduce body weight?

Section 30.3 How Does the Body Process Dietary Carbohydrates?

30.14 Humans cannot digest wood; termites do so with the aid of bacteria in their digestive tract. Is there a basic difference in the digestive enzymes present in humans and termites?

30.15 What is the product of the reaction when α-amylase acts on amylose?

30.16 Does HCl in the stomach hydrolyze both the 1,4- and 1,6-glycosidic bonds?

30.17 Beer contains maltose. Can beer consumption be detected by analyzing the maltose content of a blood sample? Explain.

Section 30.4 How Does the Body Process Dietary Fats?

30.18 Which nutrient provides energy in its most concentrated form?

30.19 What is the precursor of arachidonic acid in the body?

30.20 How many (a) essential fatty acids and (b) essential amino acids do humans need in their diets?

30.21 Do lipases degrade (a) cholesterol or (b) fatty acids?

30.22 What is the function of bile salts in the digestion of fats?

Section 30.5 How Does the Body Process Dietary Protein?

30.23 Is it possible to get a sufficient supply of nutritionally adequate proteins by eating only vegetables?

30.24 Suggest a way to cure kwashiorkor.

30.25 What is the difference between protein digestion by trypsin and by HCl?

30.26 Which one will be digested faster: (a) a raw egg or (b) a hard-boiled egg? Explain.

30.27 What do we mean when we say there is no storage form for protein? How is this different from fats and carbohydrates?

30.28 Why would muscle not be considered a storage form of protein?

Section 30.6 What Is the Importance of Vitamins, Minerals, and Water?

30.29 In a prison camp during a war, the prisoners are fed plenty of rice and water but nothing else. What would be the result of such a diet in the long run?

30.30 (a) How many milliliters of water per day does a normal diet require?
(b) How many calories does this amount of water contribute?

30.31 Why did British sailing ships carry a supply of limes?

30.32 What are the symptoms of vitamin A deficiency?

30.33 What is the function of vitamin K?

30.34 (a) Which vitamin contains cobalt?
(b) What is the function of this vitamin?

30.35 Vitamin C is recommended in megadoses by some people for prevention of all kinds of diseases, ranging from colds to cancer. What disease has been scientifically proven to be prevented when sufficient daily doses of vitamin C are in the diet?

30.36 Why is the Recommended Dietary Allowance (RDA) being phased out in favor of the Daily Reference Intake (DRI)?

30.37 What are the nonspecific effects of vitamin E, C, and carotenoids?

30.38 What are the best dietary sources of calcium, phosphorus, and cobalt?

30.39 Which vitamins contain a sulfur atom?

30.40 What are the symptoms of vitamin B_{12} deficiency?

30.41 It has been suggested that limits be put on the dose of vitamin A sold in stores. Why might this limitation be a good idea?

30.42 Why would many athletes believe that taking large doses of B vitamins would be helpful?

Chemical Connections

30.43 (Chemical Connections 30A) What is the difference between the original Food Guide Pyramid published in 1992 and the revised presentation?

30.44 (Chemical Connections 30A) What is a recommendation from ChooseMyPlate.gov regarding dairy products?

30.45 (Chemical Connections 30B) Explain what is meant by this statement: "All nutrients in excess can turn into fat, but fat cannot be turned into carbohydrate."

30.46 (Chemical Connections 30B) What does blood glucose have to do with dieting?

30.47 (Chemical Connections 30B) What is the most effective weight loss method?

30.48 (Chemical Connections 30B) How can the difference between weight loss through dieting and through exercise be explained by biochemistry?

30.49 (Chemical Connections 30B) Plants have a pathway that humans lack, called the glyoxylate pathway. It allows acetyl-CoA to bypass the two decarboxylation steps of the citric acid cycle. How would dieting be different if humans had this pathway?

30.50 (Chemical Connections 30B) Besides glucose, what other fuel source can the brain use? (*Hint:* See Section 28.7)

30.51 (Chemical Connections 30C) Why is there a requirement for iron in the diet?

30.52 (Chemical Connections 30C) What is the form of iron found in the body?

30.53 (Chemical Connections 30C) What factors influence the absorption of iron from the digestive system?

30.54 (Chemical Connections 30C) What factors influence a person's requirement for iron?

30.55 (Chemical Connections 30D) Looking in Table 22.1, which lists the common amino acids found in proteins, which amino acid most resembles creatine?

30.56 (Chemical Connections 30D) Which single compound will have the greatest effect on athletic performance?

30.57 (Chemical Connections 30D) Identify two ways carbohydrates are used for athletic performance.

30.58 (Chemical Connections 30D) Why is creatine an effective ergogenic aid? For which types of competitions is it effective?

30.59 (Chemical Connections 30D) How is caffeine used as an ergogenic aid? What are possible downsides to caffeine use?

30.60 (Chemical Connections 30E) What is meant by "organic" as it relates to food?

30.61 (Chemical Connections 30E) What are the major considerations surrounding organic vs. nonorganic foods?

30.62 (Chemical Connections 30F) What is the endocannabinoid system?

30.63 (Chemical Connections 30F) What types of molecules stimulate endocannabinoid receptors?

30.64 (Chemical Connections 30F) How are omega-3 fatty acids believed to affect the endocannabinoid system?

30.65 (Chemical Connections 30F) How are *trans* fatty acids thought to play a role in depression?

30.66 (Chemical Connections 30F) How are omega-3 fatty acids thought to be related to depression?

30.67 (Chemical Connections 30F) What evidence indicates a relationship between vitamin B_{12} and depression?

Additional Problems

30.68 Which two chemicals are used most frequently to disinfect public water supplies?

30.69 Which vitamin is part of coenzyme A (CoA)? List a step (or the enzyme) that has CoA as a coenzyme in (a) glycolysis and (b) fatty acid synthesis.

30.70 Which vitamin is prescribed in megadoses to combat autoimmune blisters?

30.71 Why is it necessary to have protein in our diets?

30.72 Which chemical processes take place during digestion?

30.73 According to the U.S. government's Food Guide Pyramid, are there any foods that we can completely omit from our diets and still be healthy?

30.74 Does the debranching enzyme help in digesting amylose?

30.75 As an employee of a company that markets walnuts, you are asked to provide information for an ad that would stress the nutritional value of walnuts. What information would you provide?

30.76 In diabetes, insulin is administered intravenously. Explain why this hormone protein cannot be taken orally.

30.77 Egg yolk contains a lot of lecithin (a phosphoglyceride). After ingesting a hard-boiled egg, would you find an increase in the lecithin level of your blood? Explain.

30.78 What would you call a diet that scrupulously avoids phenylalanine-containing compounds? Could aspartame be used in such a diet?

30.79 What kind of supplemental enzyme would you recommend for a patient after a peptic ulcer operation?

30.80 In a trial, a woman was accused of poisoning her husband by adding arsenic to his meals. Her attorney stated that this supplementation was done to promote her husband's health, as arsenic is an essential nutrient. Would you accept this argument? Why or why not?

Immunochemistry

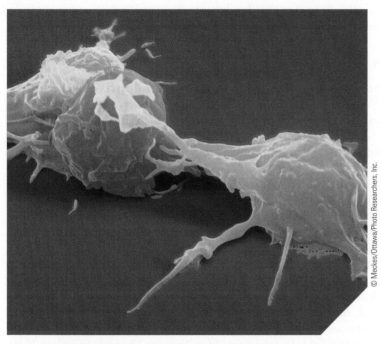

Two natural killer (NK) cells, shown in yellow-orange, attacking a leukemia cell, shown in red.

© Meckes/Ottawa/Photo Researchers, Inc.

31.1 How Does the Body Defend Itself from Invasion?

When you were in elementary school, you may have had chickenpox along with many other children. The viral disease passed from one person to the next and ran its course, but after the children recovered, they never had chickenpox again. Those who were infected became *immune* to this disease. Humans and other vertebrates possess a highly developed, complex immune system that defends the body against foreign invaders. The immune system is very involved with multiple layers of protection against invading organisms. In this section, we will briefly introduce the major parts of the immune system. These topics will then be expanded in the sections that follow.

A. Innate Immunity

When one considers the tremendous number of bacteria, viruses, parasites, and toxins that our bodies encounter, it is a wonder that we are not continually sick. Most students learn about antibodies in high school, and these

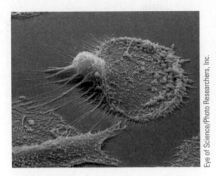

Eye of Science/Photo Researchers, Inc.

FIGURE 31.1 Dendritic cells get their name from their tentacle-like arms. The one shown here is from a human.

Innate immunity The natural non-specific resistance of the body against foreign invaders, which has no memory

days everyone learns about T cells due to their relationship to AIDS. When discussing immunity, however, many more weapons of defense exist other than T cells and antibodies. In reality, you only discover that you are sick once a pathogen has managed to beat the front-line defense, which is called **innate immunity**.

Innate immunity includes several components. One part, called **external innate immunity**, includes physical barriers such as skin, mucus, and tears. All of these barriers act to hinder penetration by pathogens and do not require specialized cells to fight a pathogen. If a pathogen—whether a bacterium, virus, or parasite—is able to breach this outer layer of defense, the cellular warriors of the innate system come into play. The cells of the innate immune system that we will discuss are **dendritic cells**, **macrophages**, and **natural killer** (**NK**) **cells**. Among the first and most important cells to join the fight are the dendritic cells, so called due to their long, tentacle-like projections (Figure 31.1).

B. Adaptive Immunity

Vertebrates have a second line of defense, called **adaptive** or **acquired immunity**. We refer to this type of immunity when we talk colloquially about the **immune system**. The key features of the immune system are *specificity* and *memory*. The key cellular components of acquired immunity are **T cells** and **B cells**. The immune system uses antibodies and cell receptors designed specifically for each type of invader. In a second encounter with the same invader, the response is more rapid, more vigorous, and more prolonged than it was in the first case, because the immune system remembers the nature of the invader from the first encounter.

The invaders may be bacteria, viruses, molds, or pollen grains. A body with no defense against such invaders could not survive. In a rare genetic disease, a person is born without a functioning immune system. Attempts have been made to bring up such children in an enclosure totally sealed from the environment. While in this environment, they can survive; when the environment is removed, however, such people always die within a short time. The severity of this disease, called severe combined immunodeficiency (SCID) explains why it was the first disease treated with gene therapy (Section 26.9). The disease AIDS (Section 31.9) slowly destroys the immune system, particularly a type of T cell, leaving its victims to die from some invading organism that a person without AIDS would easily be able to fight off.

As we shall see, the beauty of the body's immune system lies in its flexibility. The system is capable of making millions of potential defenders, so it can almost always find just the right one to counter the invader, even when it has never seen that particular organism before.

C. Components of the Acquired Immune System

Antigen A substance foreign to the body that triggers an immune response

Antibody A glycoprotein molecule that interacts with an antigen

Immunoglobulin superfamily Glycoproteins that are composed of constant and variable protein segments having significant homologies to suggest that they evolved from a common ancestry

Foreign substances that invade the body are called **antigens**. The immune system is made of both cells and molecules. Two types of **white blood cells**, called lymphocytes, fight against the invaders: (1) T cells kill the invader by contact and (2) B cells manufacture **antibodies**, which are soluble immunoglobulin molecules that immobilize antigens.

The basic molecules of the immune system belong to the **immunoglobulin superfamily**. All molecules of this class have a certain portion of the molecule that can interact with antigens, and all are glycoproteins. The polypeptide chains in this superfamily have two domains: a constant region and a variable region. The constant region has the same amino acid sequence in each of the same class of molecules. In contrast, the variable region is antigen-specific, which means that the amino acid sequence in this region is unique for each antigen. The variable regions are designed to recognize only one specific antigen.

There are three representatives of the immunoglobulin superfamily in the immune system:

1. Antibodies are soluble immunoglobulins secreted by plasma cells (see Section 31.2C).

2. Receptors on the surface of T cells (**TcR**) recognize and bind antigens presented to them.

3. Molecules that present antigens also belong to this superfamily. They reside inside the cells. These protein molecules are known as major histocompatibility complexes (MHC).

When a cell is infected by an antigen, MHC molecules interact with it and bring a characteristic portion of the antigen to the surface of the cell. Such a surface presentation then marks the diseased cell for destruction. This can happen in a cell that was infected by a virus, and it can happen in macrophages that engulf and digest bacteria and viruses.

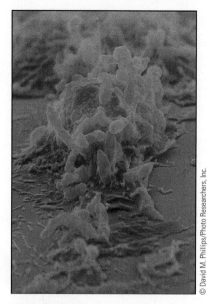

Macrophage ingesting bacteria (the rod-shaped structures). The bacteria will be pulled inside the cell within a membrane-bound vesicle and quickly killed.

31.2 What Organs and Cells Make Up the Immune System?

The blood plasma circulates in the body and comes in contact with the other body fluids through the semipermeable membranes of the blood vessels. Therefore, blood can exchange chemical compounds with other body fluids and, through them, with the cells and organs of the body (Figure 31.2).

A. Lymphoid Organs

The lymphatic capillary vessels drain the fluids that bathe the cells of the body. The fluid within these vessels is called **lymph**. Lymphatic vessels circulate throughout the body and enter certain organs, called **lymphoid organs**, such as the thymus, spleen, tonsils, and lymph nodes (Figure 31.3). The cells primarily responsible for the functioning of the immune system are the specialized white blood cells called **lymphocytes**. As their name implies, these cells are mostly found in the lymphoid organs. Lymphocytes may be either specific for a given antigen or nonspecific.

T cells are lymphocytes that originate in the bone marrow but mature in the thymus gland. B cells are lymphocytes that originate and mature in the bone marrow. Both B and T cells are found mostly in the lymph, where they circulate looking for invaders. Small numbers of lymphocytes are also

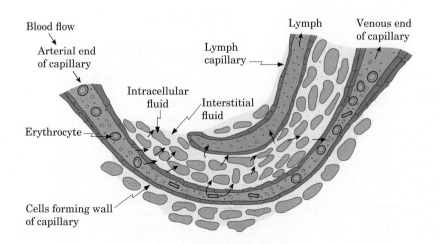

FIGURE 31.2 Exchange of compounds among three body fluids: blood, interstitial fluid, and lymph. (After Holum, J. R., *Fundamentals of General, Organic and Biological Chemistry*, New York: John Wiley & Sons, 1978, p. 569)

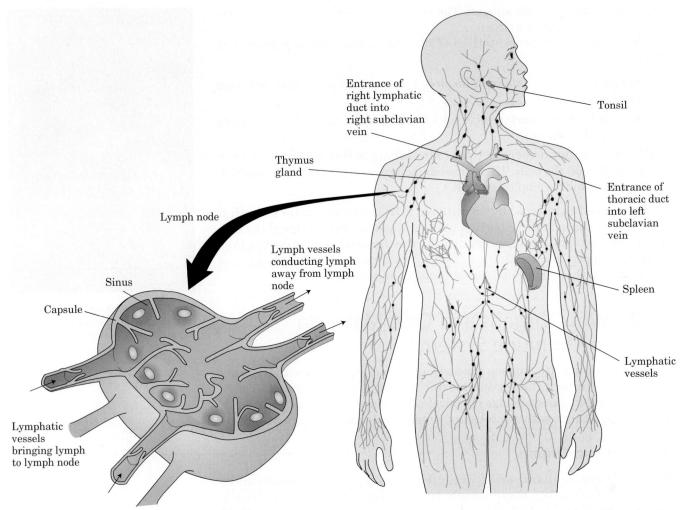

FIGURE 31.3 The lymphatic system is a web of lymphatic vessels containing a clear fluid called lymph and various lymphatic tissues and organs located throughout the body. Lymph nodes are masses of lymphatic tissue covered with a fibrous capsule. Lymph nodes filter the lymph. In addition, they are packed with macrophages and lymphocytes.

found in the blood. To get there, they must squeeze through tiny openings between the endothelial cells. This process is aided by signaling molecules called **cytokines** (Section 31.6). The sequence of the response of the body to a foreign invader is depicted schematically in Figure 31.4.

FIGURE 31.4 Interactions Among the Different Cells of the Immune System

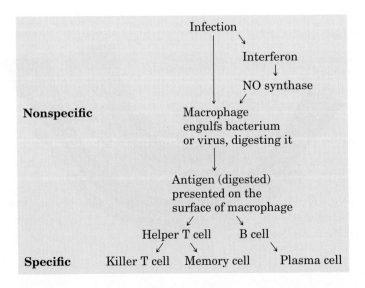

B. Cells of the Internal Innate Immunity

As mentioned in Section 31.1A, the major cells of innate immunity are dendritic cells, macrophages, and natural killer cells.

Dendritic cells are found in the skin, mucus membranes, the lungs, and the spleen. They are the first cells of the innate system that will have a crack at any virus or bacterium that wanders into their path. Using suction-cup like receptors, they grab onto invaders and then engulf them by endocytosis. These cells then chop up the devoured pathogens and bring parts of their proteins to the cellular surface. Here, the protein fragments are displayed on a protein called a **major histocompatibility complex** (**MHC**). The dendritic cells travel through the lymph to the spleen, where they present these antigens to other cells of the immune system, the **helper T cells** (T_H **cells**) as shown in Figure 31.5. Dendritic cells are members of a class of cells referred to as **antigen presenting cells** (**APCs**), and they are the starting point in most of the responses that are traditionally associated with the immune system.

Macrophages are the first cells in the blood that encounter an antigen. They belong to the internal innate immune system; inasmuch as they are nonspecific, macrophages attack virtually anything that is not recognized as part of the body, including pathogens, cancer cells, and damaged tissues. Macrophages engulf an invading bacterium or virus and kill it. The "magic bullet" in this case is the NO molecule, which is toxic (Chemical Connections 4C) and can act as a secondary messenger.

The NO molecule is short-lived and must be constantly manufactured anew. When an infection begins, the immune system manufactures the protein interferon. It, in turn, activates a gene that produces an enzyme, nitric oxide synthase. With the aid of this enzyme, the macrophages, endowed with NO, kill the invading organisms. Macrophages then digest the engulfed antigen and display a small portion of it on their surface.

Natural killer (**NK**) **cells** target abnormal cells. Once in physical contact with such cells, NK cells release proteins, aptly called perforins, that perforate the target cell membranes and create pores. The membrane of the target cell becomes leaky, allowing hypotonic (Section 6.8C) liquid from the surroundings to enter the cell, which swells and eventually bursts.

C. Cells of Adaptive Immunity: T and B Cells

T cells interact with the antigens presented by APCs and produce other T cells that are highly specific to the antigen. When these T cells differentiate, some of them become **killer T cells**, also called **cytotoxic T cells** (T_c **cells**), which kill the invading foreign cells by cell-to-cell contact. Killer T cells, like NK cells, act through perforins, which attach themselves to the target cell, in effect, punching holes in its membranes. Through these holes, water rushes into the target cell; it swells and eventually bursts.

Other T cells become **memory cells**. They remain in the bloodstream, so that if the same antigen enters the body again, even years after the primary infection, the body will not need to build up its defenses anew but is ready to kill the invader instantly.

A third type of T cell is the **helper T cell** (T_H **cell**). These cells do not kill other cells directly, but rather, are involved in recognizing antigens on APCs and recruiting other cells to help fight the infection.

The production of antibodies is the task of **plasma cells**. These cells are derived from B cells after the B cells have been exposed to an antigen.

The lymphatic vessels, in which most of the antigen attacking takes place, flow through many lymph nodes (Figure 31.3). These nodes are essentially filters. Most plasma cells reside in lymph nodes, so most antibodies are produced there. Each lymph node is also packed with millions of

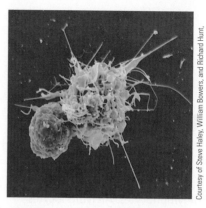

FIGURE 31.5 Dendritic cells and the other cells of the immune system. This figure shows a rat dendritic cell interacting with a T cell. Through these interactions, the dendritic cells teach the acquired immunity system what to attack.

Both natural killer cells and killer T cells act the same way, through perforin. The T_c cells attack specific targets; the NK cells attack all suspicious targets.

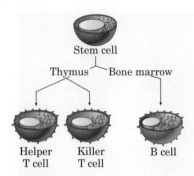

Stem cell

Thymus — Bone marrow

Helper T cell   Killer T cell   B cell

FIGURE 31.6 The development of lymphocytes. All lymphocytes are ultimately derived from the stem cells of the bone marrow. In the thymus, T cells develop into helper T cells and killer T cells. B cells develop in the bone marrow.

There are rare exceptions: In an autoimmune disease, the body mistakes its own protein as foreign.

Epitope The smallest number of amino acids on an antigen that elicits an immune response

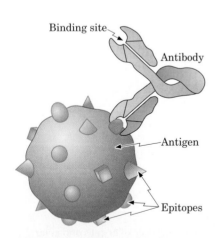

Binding site

Antibody

Antigen

Epitopes

FIGURE 31.7 Antibody binding to the epitope of an antigen.

other lymphocytes. More than 99% of all invading bacteria and foreign particles are filtered out in the lymph nodes. As a consequence, the outflowing lymph is almost free of invaders and is packed with antibodies produced by the plasma cells. All lymphocytes derive from stem cells in the bone marrow. Stem cells are undifferentiated cells that can become many different cell types. As shown in Figure 31.6, they can differentiate into T cells in the thymus or B cells in the bone marrow.

31.3 How Do Antigens Stimulate the Immune System?

A. Antigens

Antigens are foreign substances that elicit an immune response; for this reason, they are also called immunogens. Three features characterize an antigen. The first condition is foreignness—molecules of your own body should not elicit an immune response. The second condition is that the antigen must be of molecular weight greater than 6000. The third condition is that the molecule must have sufficient complexity. A polypeptide made of lysine only, for example, is not immunogenic.

Antigens can be proteins, polysaccharides, or nucleic acids, as all of these substances are large molecules. Antigens may be soluble in the cytoplasm, or they may be found at the surface of cells, either embedded in the membrane or just absorbed on it. An example of polysaccharidic antigenicity is ABO blood groups (Chemical Connections 20D).

In protein antigens, only part of the primary structure is needed to cause an immune response. Between 5 and 7 amino acids are needed to interact with an antibody, and between 10 and 15 amino acids are necessary to bind to a receptor on a T cell. The smallest unit of an antigen capable of binding with an antibody is called the **epitope**. The amino acids in an epitope do not have to be in sequence in the primary structure, as folding and secondary structures may bring amino acids that are not in sequence into each other's proximity. For example, the amino acids in positions 20 and 28 may form part of an epitope. *Antibodies can recognize all types of antigens, but T-cell receptors recognize only peptide antigens.*

As noted earlier, antigens may be in the interior of an infected cell or on the surface of a virus or bacterium that penetrated the cell. To elicit an immune reaction, the antigen or its epitope must be brought to the surface of the infected cell. Similarly, after a macrophage swallows up and partially digests an antigen, the macrophage must bring the epitope back to its surface to elicit an immune response from T cells (Figure 31.4).

Once the antigen is presented at the surface of a cell or if the antigen is already soluble, the immunoglobulin can bind to the epitope, as shown in Figure 31.7.

B. Major Histocompatibility Complexes

The task of bringing an antigen's epitope to the infected cell's surface is performed by the major histocompatibility complex (MHC). The name derives from the fact that its role in the immune response was first discovered in organ transplants. MHC molecules are transmembrane proteins belonging to the immunoglobulin superfamily. Two classes of MHC molecules exist (Figure 31.8), both of which have peptide-binding variable domains. Class I MHC is made of a single polypeptide chain, whereas class II MHC is a dimer. Class I MHC molecules seek out antigen molecules that have been

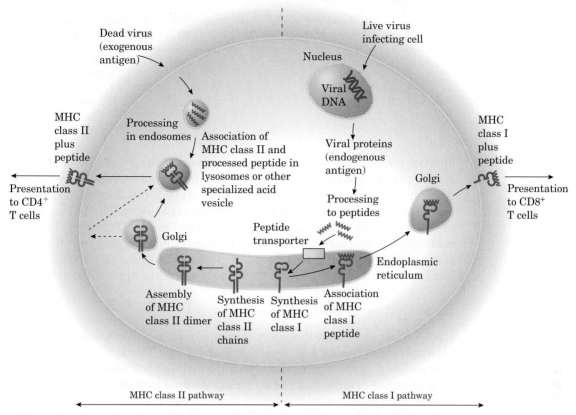

FIGURE 31.8 Differential processing of antigens in the MHC class II pathway (left) or MHC class I pathway (right). Cluster determinants (CD) are parts of the T-cell receptor complex (see Section 31.5B).

synthesized inside a virus-infected cell. Class II MHC molecules pick up exogenous *"dead" antigens.* In each case, the epitope attached to the MHC is brought to the cell surface to be presented to T cells.

For example, if a macrophage engulfed and digested a virus, the result would be dead antigens. The digestion occurs in several steps. First, the antigen is processed in lysosomes, special organelles of cells that contain proteolytic enzymes. An enzyme called GILT (gamma-interferon inducible lysosomal thiol reductase) breaks the disulfide bridges of the antigen by reduction. The reduced peptide antigen unfolds and is exposed to proteolytic enzymes that hydrolyze it to smaller peptides. These peptides serve as epitopes that are recognized by class II MHC. The difference between MHC I and MHC II becomes significant when we look at the functions of T cells. Antigens bound to MHC I will interact with killer T cells, while those bound to MHC II will interact with helper T cells.

31.4   What Are Immunoglobulins?

A. Classes of Immunoglobulins

Immunoglobulins are glycoproteins—that is, carbohydrate-carrying protein molecules. Not only do the different classes of immunoglobulins vary in molecular weight and carbohydrate content, but their concentration in the blood differs dramatically as well (Table 31.1). The IgG and IgM antibodies are the most important antibodies in the blood. They interact with antigens

Table 31.1 Immunoglobulin Classes

Class	Molecular Weight (MW)	Carbohydrate Content (%)	Concentration in Serum (mg/100 mL)
IgA	200,000–700,000	7–12	90–420
IgD	160,000	<1	1–40
IgE	190,000	10–12	0.01–0.1
IgG	150,000	2–3	600–1800
IgM	950,000	10–12	50–190

and trigger the swallowing up (phagocytosis) of these cells by phagocytes. Inside the phagocytes, the antigens are destroyed inside the lysosomes. Antigens bound to antibodies are also destroyed in the blood system by a complicated procedure called the complement pathway. The IgA molecules are found mostly in secretions: tears, milk, and mucus. Therefore, these immunoglobulins attack the invading material before it gets into the bloodstream. The IgE molecules play a part in such allergic reactions as asthma and hay fever and are involved in the body's defense against parasites.

B. Structure of Immunoglobulins

Each immunoglobulin molecule is made of four polypeptide chains: two identical light chains and two identical heavy chains. The four polypeptide chains are arranged symmetrically, forming a Y shape (Figure 31.9). Four disulfide bridges link the four chains into a unit. Both light and heavy chains have constant and variable regions. The constant regions have the same amino acid sequences in different antibodies, and the variable regions have different amino acid sequences in different antibodies.

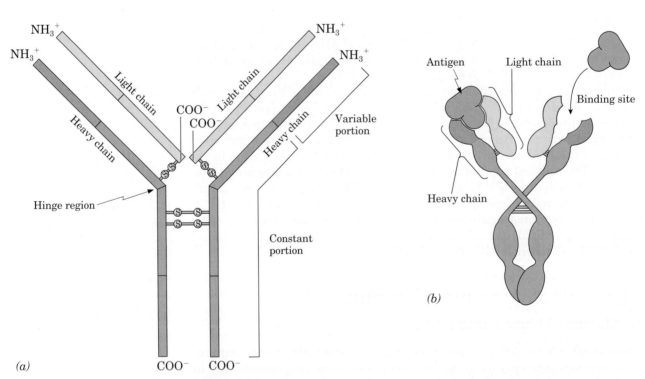

FIGURE 31.9 (a) Schematic diagram of an IgG-type antibody consisting of two heavy chains and two light chains connected by disulfide bonds. The amino terminal end of each chain has the variable portion. (b) Model showing how an antibody bonds to an antigen.

The variable regions of the antibody recognize the foreign substance (the antigen) and bind to it (Figure 31.10). Because each antibody contains two variable regions, it can bind two antigens, thereby forming a large aggregate, as shown in Figure 31.11.

The binding of the antigen to the variable region of the antibody occurs not by covalent bonds, but rather, by much weaker intermolecular forces such as London dispersion forces, dipole–dipole interactions, and hydrogen bonds (Section 5.7). This binding is similar to the way in which substrates bind to enzymes or hormones and neurotransmitters bind to a receptor site. That is, the antigen must fit into the antibody surface. Humans have more than 10,000 different antibodies circulating at measurable levels, which enables our bodies to fight a large number of foreign invaders. However, the potential number of antibodies that can be created by the available genes is in the millions.

C. B Cells and Antibodies

Each B cell synthesizes only one unique immunoglobulin antibody, and that antibody contains a unique antigen-binding site to one epitope. Before encountering an antigen, these antibodies are inserted in the plasma membrane of the B cells, where they serve as receptors. When an antigen interacts with its receptor, it stimulates the B cell to divide and differentiate into plasma cells. These daughter cells secrete soluble antibodies that have the same antigen-binding sites as the original antibody/receptor. The soluble secreted antibodies appear in the serum (the noncellular part of blood) and can react with the antigen. Thus, an immunoglobulin produced in B cells acts both as a receptor to be stimulated by the antigen and as a secreted messenger ready to neutralize and eventually destroy the antigen (Figure 31.12).

D. How Does the Body Acquire the Diversity Needed to React to Different Antigens?

From the moment of conception, an organism contains all of the DNA it will ever have, including that DNA that will lead to antibodies and T-cell receptors. Thus, the organism is born with a repertoire of genes necessary to fight infections. During B-cell development, the variable regions of the heavy chains are assembled by a process called V(J)D recombination. A number of exons are present in each of three different areas—V, J, and D—of the immunoglobin gene. Combining one exon from each area yields a *new V(J)D gene*. This process creates a great diversity because of the large number of ways that this combination can be performed (Figure 31.13). For one type of antibody light chain, called kappa, there are roughly 40 V genes and 5 J genes. That alone gives rise to 40 × 5, or 200, combinations of V and J. For another type of light chain, called lambda, about 120 combinations are possible. With the heavy chains, there is even greater diversity: about 50 V genes, 27 D genes, and 6 J genes. By the time one gets done calculating the possible combinations of all of the combinations of V, J, and D and the C regions of both the heavy and light chains, there are more than 2 million possible combinations.

However, that is merely the first step. A second level of diversity is created by mutation of the V(J)D genes in somatic cells. As cells proliferate in response to recognizing an antigen, such mutations can cause a 1000-fold increase in the binding affinity of an antigen to an antibody. This process is called **affinity maturation**.

There are three ways to create mutations. Two affect the variable regions, and one alters the constant region.

1. Somatic hypermutation (SHM) creates a point mutation (only one nucleotide). The resultant protein of the mutation is able to bind more strongly with the antigen.

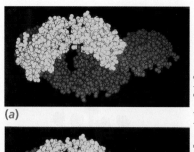

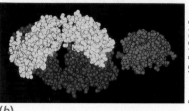

(a)

(b)

Courtesy of Dr. R. J. Pojak, Pasteur Institute, Paris, France

FIGURE 31.10 (*a*) An antigen–antibody complex. The antigen (shown in green) is lysozyme. The heavy chain of the antibody is shown in blue; the light chain in yellow. The most important amino acid residue (glutamine in the 121 position) on the antigen is the one that fits into the antibody groove (shown in red). (*b*) The antigen–antibody complex has been pulled apart. Note how they fit each other.

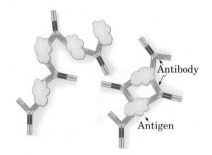

FIGURE 31.11 An antigen–antibody reaction forms a precipitate. An antigen, such as a bacterium or virus, typically has several binding sites for antibodies. Each variable region of an antibody (each prong of the Y) can bind to a different antigen. The aggregate thus formed precipitates and is attacked by phagocytes and the complement system.

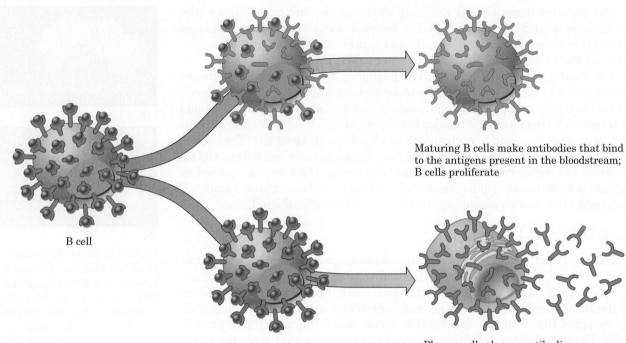

Maturing B cells make antibodies that bind to the antigens present in the bloodstream; B cells proliferate

B cell

Plasma cell releases antibodies into bloodstream

FIGURE 31.12 B cells have antibodies on their surfaces, which allow them to bind to antigens. The B cells with antibodies for the antigens present grow and develop. When B cells develop into plasma cells, they release circulating antibodies into the bloodstream. (Adapted from "How the Immune System Develops," by Irving L. Weissman and Max D. Cooper; illustrated by Jared Schneidman, Scientific American, September 1993.)

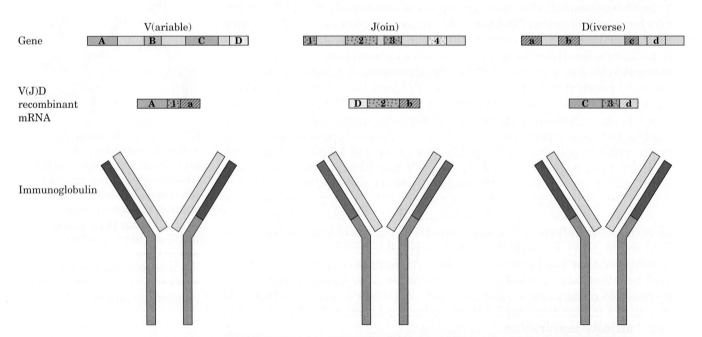

FIGURE 31.13 Diversification of immunoglobulins by V(J)D recombination. Exons of three genes—the V(ariable) (A, B, C, D); J(oin) (1, 2, 3, 4); and D(iverse) (a, b, c, d) genes—combine to form new V(J)D genes that are transcribed to the corresponding mRNAs. The expression of these new genes results in a wide variety of immunoglobulins having different variable regions on their heavy chains.

Chemical Connections 31A

Monoclonal Antibodies Wage War on Breast Cancer

Breast cancer is currently the second leading cause of cancer-related deaths in North America, but this status is likely to change in the near future. The survival rate for women diagnosed with breast cancer has been rising for the last ten years. Among the combination of contributing factors are increasing breast cancer awareness, leading to earlier detection, and the development of many new drugs and techniques to battle the disease.

Cancers result from a wide variety of complicated errors in metabolism. To combat cancer, scientists first identify specific differences between cancer cells and normal cells and then look for ways of stopping the change from normal cell to cancer cell or of attacking the cancer cell once it has formed. Many drugs used to combat breast cancer, as well as other cancers, work by directing monoclonal antibodies against specific cell surface proteins that have been identified as being active in cancer. One protein found in many breast cancers is Human Epidermal Growth Factor 2 (HER2), a member of a larger class of epidermal growth factors that are implicated in many cancer types. These proteins are receptors that bind to specific ligands, causing rapid cell growth. Studies reveal that many breast cancer types show increased levels of HER2.

In breast cancer, HER2 causes aggressive tumor growth, so any drug that can stop its action can be a potent anticancer agent. One such potent weapon is a monoclonal antibody called trastuzumab, approved for use in 1998; it significantly increases the life expectancy of patients in both early-stage and metastatic breast cancer. The success of trastuzumab led to the creation of newer drugs, such as pertuzumab, which attacks the HER2 protein at a different site and also keeps it from interacting with other receptors that have been linked to cancer.

Several strategies use monoclonal antibodies to combat breast cancer, as shown in the figure below. The antibody (shown in orange) may bind directly to the chemical growth factor before it binds to its receptor, as shown at the top of the figure. The antibody may also block the binding site on the receptor so the growth factor cannot bind. Many cellular effects are initiated by the dimerization of two cell receptors, and monoclonal antibodies can also stop that process. Some of the cell receptors that can lead to cancer are based on a tyrosine kinase, and monoclonal antibodies have been created that act as inhibitors of this activity (Chapter 23). Finally, new technologies are being developed that link a monoclonal antibody to a specific toxin. When the antibody binds to a critical cell receptor on a cancerous cell, the toxin is carried inside the cell and kills it.

Much progress is being made in the development of individualized therapy in which profiling of a patient's specific cells lets doctors know which cell proteins are the culprits. Once they identify a specific protein target, they can use the proper combination of drugs against it. This ability is already making a significant impact in the survival rates of breast cancer patients, and we can expect even more progress in the years to come.

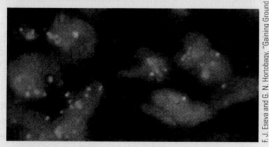

F. J. Eseva and G. N. Hortobagy, "Gaining Ground on Breast Cancer" *Scientific American*, June 2008, p. 63. Copyright © 2008 *Scientific American*. Used with permission.

Red fluorescent tags spot the HER2 receptor. In cancer cells, the protein is duplicated many times (bottom) compared to a control cell (top).

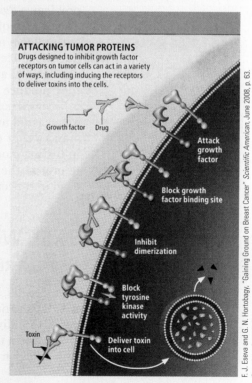

ATTACKING TUMOR PROTEINS
Drugs designed to inhibit growth factor receptors on tumor cells can act in a variety of ways, including inducing the receptors to deliver toxins into the cells.

Growth factor   Drug

Attack growth factor

Block growth factor binding site

Inhibit dimerization

Block tyrosine kinase activity

Toxin

Deliver toxin into cell

F. J. Eseva and G. N. Hortobagy, "Gaining Ground on Breast Cancer" *Scientific American*, June 2008, p. 63. Copyright © 2008 *Scientific American*. Used with permission.

Test your knowledge with Problems 31.68–31.73.

2. In a gene conversion (GC) mutation, stretches of nucleotide sequences are copied from a pseudogene V and entered into the V(J)D. This also enables the proteins synthesized from the GC mutation to make stronger contact with the antigen than without the mutation.

3. Mutations on the constant region of the chain can be achieved by class switch recombination (CSR). In this case, exons of the constant regions are swapped between highly repetitive regions.

The diversity of antibodies created by the V(J)D recombination is greatly amplified and finely tuned by the mutations on these genes. Because the response to the antigen occurs on the gene level, it is easily preserved and transmitted from one generation of cells to the next.

While the possible combination of genes leading to antibody diversity seems limitless, it is important to remember that the basis of antibody diversity is the genetic blueprint that the organism was given. Antibodies do not appear because they are needed; rather, they are selected and proliferated because they already existed in small quantities before they were stimulated by recognition of an antigen.

E. Monoclonal Antibodies

When an antigen is injected into an organism (for example, human lysozyme into a rabbit), the initial response is quite slow. It may take from one to two weeks before the anti-lysozyme antibody shows up in the rabbit's serum. Those antibodies, however, are not uniform. The antigen may have many epitopes, and the antisera contains a mixture of immunoglobulins with varying specificity for all the epitopes. Even antibodies to a single epitope usually have a variety of specificities.

Each B cell (and each progeny plasma cell) produces only one kind of antibody. In principle, each such cell should represent a potential source of a supply of homogeneous antibody by cloning. This is not possible in practice because lymphocytes do not grow continuously in culture. In the late 1970s, Georges Köhler and César Milstein developed a method to circumvent this problem, a feat for which they received the Nobel Prize in physiology in 1984. Their technique requires fusing lymphocytes that make the desired antibody with mouse myeloma cells. The resulting **hybridoma** (hybrid myeloma), like all cancer cells, can be cloned in culture (Figure 31.14) and produces the desired antibody. Because the clones are the progeny of a single cell, they produce homogeneous **monoclonal antibodies**. With this technique, it becomes possible to produce antibodies to almost any antigen in quantity. Monoclonal antibodies can, for instance, be used to assay for biological substances that can act as antigens. A striking example of their usefulness is in testing blood for the presence of HIV; this procedure has become routine to protect the public blood supply. Monoclonal antibodies are also commonly used in cancer treatment, as described in Chemical Connections 31A.

31.5 What Are T Cells and T-Cell Receptors?

A. T-Cell Receptors

Like B cells, T cells carry on their surface unique receptors that interact with antigens. We noted earlier that T cells respond only to protein antigens. An individual has millions of different T cells, each of which carries on its surface a unique T-cell receptor (TcR) that is specific for one antigen only. The TcR is a glycoprotein made of two subunits cross-linked by disulfide

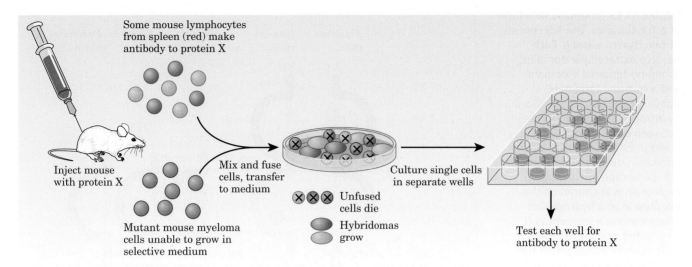

FIGURE 31.14 A procedure for producing monoclonal antibodies against a protein antigen X. A mouse is immunized against the antigen X, and some of its spleen lymphocytes produce antibody. The lymphocytes are fused with mutant myeloma cells that cannot grow in a given medium because they lack an enzyme found in the lymphocytes. Unfused cells die because lymphocytes cannot grow in culture and the mutant myeloma cells cannot survive in this medium. The individual cells are grown in culture in separate wells and tested for antibody to protein X.

bridges. Like immunoglobulins, TcRs have constant (C) and variable (V) regions. The antigen binding occurs on the variable region. The similarity in amino acid sequence between immunoglobulins (Ig) and TcR, as well as the organization of the polypeptide chains, makes TcR molecules members of the immunoglobulin superfamily.

There are, however, some fundamental differences between immunoglobulins and TcRs. For instance, immunoglobulins have four polypeptide chains, whereas TcRs contain only two subunits. Immunoglobulins can interact directly with antigens, but TcRs can interact with them only when the epitope of an antigen is presented by an MHC molecule. Lastly, immunoglobulins can undergo somatic mutation. This kind of mutation can occur in all body cells except the ones involved in sexual reproduction. Thus, Ig molecules can increase their diversity by somatic mutation; TcRs cannot.

B. T-Cell Receptor Complex

A TcR is anchored in the membrane by hydrophobic transmembrane segments (Figure 31.15). TcR alone, however, is not sufficient for antigen binding. Also needed are other protein molecules that act as coreceptors and/or signal transducers. These molecules go under the name of CD3, CD4, and CD8, where "CD" stands for **cluster determinant**. TcR and CD together form the **T-cell receptor complex**.

The CD3 molecule adheres to the TcR in the complex not through covalent bonds, but rather, by intermolecular forces (Section 5.7). It is a signal transducer because, upon antigen binding, CD3 becomes phosphorylated. This event sets off a signaling cascade inside the cell, which is carried out by different kinases. We saw a similar signaling cascade in neurotransmission (Section 24.5).

The CD4 and CD8 molecules act as **adhesion molecules** as well as signal transducers. A T cell has either a CD4 or a CD8 molecule to help

Adhesion molecules Various protein molecules that help to bind an antigen to the T-cell receptor and dock the T cell to another cell via an MHC

FIGURE 31.15 Schematic structure of a TcR complex. The TcR consists of two chains: α and β. Each has two extracellular domains: an amino-terminal V domain and a carboxyl-terminal C domain. Domains are stabilized by intrachain disulfide bonds between cysteine residues. The α and β chains are linked by an interchain disulfide bond near the cell membrane (hinge region). Each chain is anchored on the membrane by a hydrophobic transmembrane segment and ends in the cytoplasm with a carboxyl-terminal segment rich in cationic residues. Both chains are glycosylated (red spheres). The cluster determinant (CD) coreceptor consists of three chains: γ, δ, and ε. Each is anchored in the plasma membrane by a hydrophobic transmembrane segment. Each is also cross-linked by a disulfide bridge, and the carboxylic terminal is located in the cytoplasm.

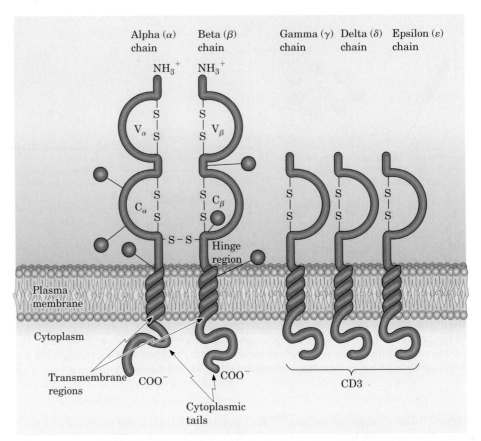

FIGURE 31.16 Interaction between helper T cells and antigen-presenting cells. Foreign peptides are displayed on the surface by MHC II proteins. These bind to the T-cell receptor of a helper T cell. A docking protein called CD4 helps link the two cells together.

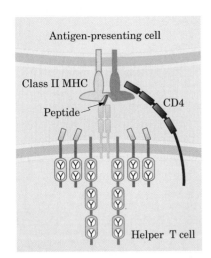

Cytokine A glycoprotein that traffics between cells and alters the function of a target cell

bind the antigen to the receptor and to dock the T cell to an APC or B cell (Figure 31.16). A unique characteristic of the CD4 molecule is that it binds strongly to a special glycoprotein that has a molecular weight of 120,000 (gp120). This glycoprotein exists on the surface of the human immunodeficiency virus (HIV). Through this binding to CD4, HIV can enter and infect helper T cells and cause AIDS. Helper T cells die as a result of HIV infection, which depletes the T cell population so drastically that the immune system can no longer function. As a consequence, the body succumbs to opportunistic pathogen infections (Section 31.9).

31.6 How Is the Immune Response Controlled?

A. Nature of Cytokines

Cytokines are glycoprotein molecules that are produced by one cell but alter the function of another cell. They have no antigen specificity. Cytokines transmit intercellular communications between different types of cells at diverse sites in the body. They are short-lived and are not stored in cells.

Cytokines facilitate a coordinated and appropriate inflammatory response by controlling many aspects of the immune reaction. They are released in bursts, in response to all manner of insult or injury (real or perceived). They travel and bind to specific cytokine receptors on the surface of macrophages and B and T cells to induce cell proliferation.

One set of cytokines are called **interleukins** (**ILs**) because they communicate between and coordinate the actions of leukocytes (all kinds of white blood cells). Macrophages secrete IL-1 upon a bacterial infection. The presence of IL-1 then induces shivering and fever. The elevated body temperature both reduces the bacterial growth and speeds up the mobilization of the immune system (Chemical Connections 7A). One leukocyte may make many different cytokines, and one cell may be the target of many cytokines.

B. Classes of Cytokines

Cytokines can be classified according to their mode of action, origin, or target. The best way to classify them is by their structure—namely, the secondary structure of their polypeptide chains.

1. One class of cytokines is made of four α-helical segments. A typical example is interleukin-2 (IL-2), which is a 15,000-MW polypeptide chain. A prominent source of IL-2 is T cells. IL-2 activates other B and T cells and macrophages. Its action is to enhance proliferation and differentiation of the target cells.

2. Another class of cytokines has only β-pleated sheets in its secondary structure. Tumor necrosis factor (TNF), for example, is produced mostly by T cells and macrophages. Its name comes from its ability to destroy susceptible tumor cells through lysis, after it binds to receptors on the tumor cell.

3. A third class of cytokines has both α-helix and β-sheet secondary structures. A representative of this class is epidermal growth factor (EGF), which is a cysteine-rich protein. As its name indicates, EGF stimulates the growth of epidermal cells; its main function is in wound healing.

4. A subgroup of cytokines, the chemotactic cytokines, are also called **chemokines**. Humans have some 40 chemokines, all low-molecular-weight proteins with distinct structural characteristics. They attract leukocytes to a site of infection or inflammation. All chemokines have four cysteine residues, which form two disulfide bridges: Cys1-Cys3 and Cys2-Cys4. Chemokines have a variety of names, such as interleukin-8 (IL-8) and monocyte chemotactic proteins (MCP-1 to MCP-4). Chemokines interact with specific receptors, which consist of seven helical segments coupled to GTP-binding proteins.

Chemokine A low-molecular-weight polypeptide that interacts with special receptors on target cells and alters their functions

C. Mode of Action of Cytokines

When a tissue is injured, leukocytes are rushed to the inflamed area. Chemokines help leukocytes migrate out of the blood vessels to the site of injury. There leukocytes, in all their forms—neutrophils, monocytes, lymphocytes—accumulate and attack the invaders by engulfing them (phagocytosis) and later killing them. Other phagocytic cells, the macrophages, which reside in the tissues and thus do not have to migrate, do the same. These activated phagocytic cells destroy their prey by releasing endotoxins that kill bacteria and/or by producing such highly reactive oxygen intermediates as superoxide, singlet oxygen, hydrogen peroxide, and hydroxyl radicals.

Chemokines are also major players in chronic inflammation, autoimmune diseases, asthma, and other forms of allergic inflammation and even in transplant rejection.

Singlet oxygen is a form of O_2 where the outer electrons are raised to a higher energy level state. There are two common forms of singlet oxygen, and both are reactive oxygen species.

31.7 Immunization

Smallpox was a scourge over many centuries, with each outbreak leaving many people dead and others maimed by deep pits on the face and body. A form of **immunization** was practiced in ancient China and the Middle East by intentionally exposing people to scabs and fluids from the lesions of smallpox victims. This practice was known as variolation in the Western

Edward Jenner developed the world's first vaccine in 1796. It was a safe and effective way to prevent smallpox and has led to eradication of this disease.

world, where the disease was called variola. Variolation was introduced to England and the American Colonies in 1721. Edward Jenner, an English physician, noted that milkmaids who had contracted cowpox from infected cows seemed to be immune to smallpox. Cowpox was a mild disease, whereas smallpox could be lethal. In 1796, Jenner performed a potentially deadly experiment: He dipped a needle into the pus of a cowpox-infected milkmaid and then scratched a boy's hand with the needle. Two months later, Jenner injected the boy with a lethal dose of smallpox-carrying agent. The boy survived and did not develop any symptoms of the disease. The word spread, and Jenner was soon established in the immunization business. When the news reached France, skeptics there coined a derogatory term, **vaccination**, which literally means "encowment." The derision did not last long, however, and the practice was soon adopted worldwide.

About a century later, in 1879, Louis Pasteur found that tissue infected with rabies had much weakened virus in it. When injected into patients, it elicited an immune response that protected against rabies. Pasteur named these protective antigens "vaccines" in honor of Jenner's work. Today, immunization and vaccination are synonymous. In the case of smallpox, the practice was so successful that the disease was officially declared to be eradicated in 1979.

A. How Are Vaccines Made?

A vaccine is something that will elicit an immune response in the host. Of course, the disease itself elicits the same response. So in one sense, catching a disease is the ultimate form of immunization, but the goal is to get the benefit without having to suffer from the disease. To accomplish this, a vaccine must be capable of eliciting the immune response without making the host sick.

There are three principal ways of creating a vaccine. The first, and the one used by Pasteur, is to use an **attenuated vaccine** (See Figure 31.17).

A live bacterium or virus is used, but in a weakened state. In its weakened state, it cannot reproduce very well, so it does not "out-reproduce" the immune response. Because it is alive, however, it stays in the system long enough to produce a powerful immune response. The host is actually infected with the disease but does not get the disease. The second way is to use a completely **inactivated vaccine** from a virus or bacterium. In this case, the disease agent is "killed" and is completely unable to reproduce, but its various antigens are still available to elicit the immune response in the host. The last way is to use a **subunit vaccine**. In this case, pieces or subunits of the pathogen are used to elicit the immune response. If it

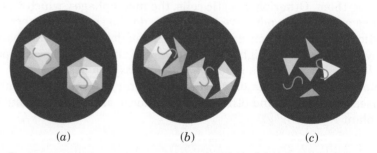

(a)　　　　(b)　　　　(c)

FIGURE 31.17 There are three common types of vaccines. (a) Attenuated vaccines are live but weakened whole viruses or bacteria. Their minimal reproduction extends immune cells' exposure to the antigen without causing disease. (b) Inactivated vaccines are whole viruses that are unable to reproduce or cause disease. (c) Subunit vaccines are fragments of the pathogen, such as genetic material or external proteins that provide an antigen for the immune cells to recognize.

works, this is considered the safest route since the host is never injected with the actual disease organism.

B. How Do Vaccines Work?

An ideal vaccine would accomplish several things. It would elicit a powerful immune response against the pathogen. It would confer lifelong immunity against the same pathogen, and it would do so without the need for multiple vaccinations. If we wanted to get even more demanding, the vaccine also would work against similar but non-identical pathogens, such as different strains of the flu. To accomplish any of these things, the vaccine must engage the whole host of cellular weaponry that we saw in Sections 31.2–31.5. To work effectively, a vaccine must mimic an infection. Figure 31.18 shows the basics of immunization with a viral vaccine.

The injected virus first encounters the cells of the innate immunity system, macrophages and dendritic cells. These cellular warriors work to rid the body of the virus particles. The macrophages engulf and ingest bacteria, and the dendritic cells activate lymphocytes and stimulate the secretion of cytokines. The dendritic cells mature and migrate to the lymph nodes, where they interact with immature T cells. These then mature into helper T cells and killer T cells. The helper T cells interact with B cells, triggering them to release antibodies. Some of the T cells remain as memory cells, as shown in Figure 31.19.

In this way, long-term immunity is conferred. If a second invasion occurs, these memory cells divide directly into antibody-secreting plasma cells and more memory cells. This time, the response is faster because it does not

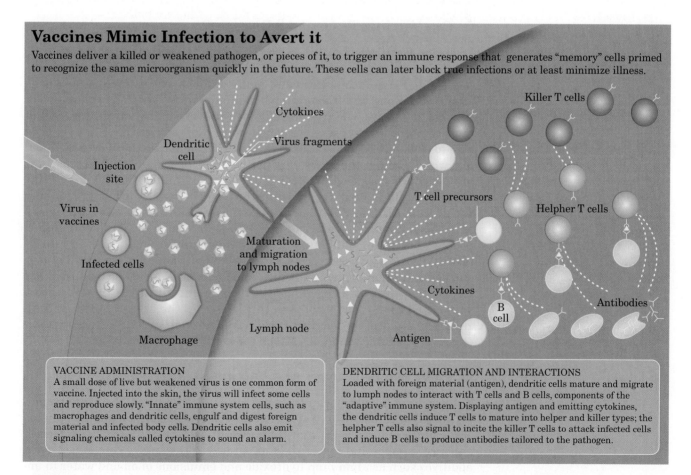

FIGURE 31.18. The basics of how vaccines work.

FIGURE 31.19. Vaccines lead to memory cells that can quickly fight off a subsequent infection of the same pathogen.

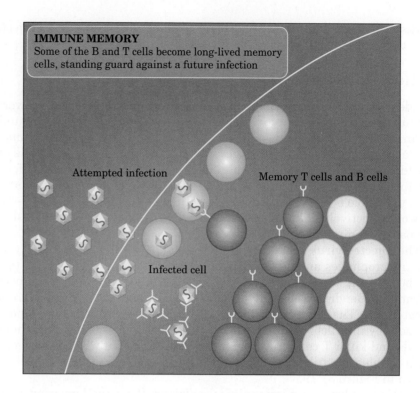

IMMUNE MEMORY
Some of the B and T cells become long-lived memory cells, standing guard against a future infection

Attempted infection

Memory T cells and B cells

Infected cell

have to go through the process of activation and differentiation into plasma cells, which usually takes two weeks.

C. Can Vaccines Prevent All Diseases?

Years ago, birth defects were caused by rubella. Children crippled with polio lived in iron lungs, and babies struggled to breathe when they had whooping cough. Fortunately these and many other diseases have largely been eradicated, along with the original smallpox disease studied by Jenner. For the last 200 years, vaccines have helped prevent many such diseases, but there are others that are more troublesome. Researchers have not yet been able to develop effective vaccines to fight HIV, hepatitis C, tuberculosis, and malaria, although these are very active areas of study. The common flu is also a major focus (see Chemical Connections 31D and 31E). While to many people, the flu is just a yearly nuisance, but in reality, it has killed more people than AIDS or the black plague and seems to have been with humans as long as humans have existed. In 1918, a flu pandemic killed 50 million people worldwide, making it the worst single disease outbreak in human history. In theory, vaccines could be used to fight a host of other afflictions, such as cancer, allergies, and Alzheimer's disease. Any disease state that can be identified by a specific antigen can theoretically be treated with a vaccine. Of course, the devil is in the details. In the case of cancer and Alzheimer's disease, the vaccine would have to be very selective in order to be effective, so that it would not attack some of the host's own cells.

D. How Can the Power of a Vaccine Be Increased?

The power of a vaccine can be increased by the addition of certain chemical additives called **adjuvants**, from the Latin *adjuvare*, meaning, "to help." Some of these have been known for over 100 years and have been used to enhance vaccines and cancer therapies. Early 20th-century researchers tried additives such as aluminum hydroxide and emulsions of oil and water to improve a vaccine's effect, as well as bacterial extracts of **lipopolysaccharides** (**LPS**), a component in the bacterial cell wall. Unfortunately, many of the

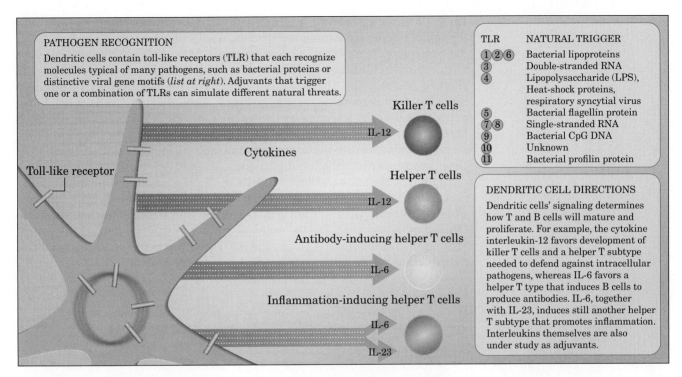

FIGURE 31.20 Toll-like receptors bind to the fragments of many pathogens and activate a strong immune response.

intended adjuvants had side effects, such as inflammation. For this reason, interest in adjuvant research faded until the late 1980s, when the search for an HIV vaccine that worked became desperate and researchers tried anything to improve a vaccine's effectiveness.

A major breakthrough in adjuvant research came in 1997 with the discovery that specific receptors on dendritic cells are specific for the recognition of parts of microorganisms, such as the protein flagellin, which is found in the tails of many different bacteria. These receptors generate the danger signal that induces the dendritic cell to release cytokines. Among these receptors is a class called **toll-like receptors** (**TLRs**) that recognize invaders that have breached physical barriers such as the skin or the intestinal tract mucosa. There are at least 10 TLRs that recognize specific motifs on viruses and bacteria. For example, TLR-4 recognizes bacterial lipopolysaccharide. This explained why LPS had been discovered to be beneficial in early adjuvant research. The LPS had boosted the vaccine's power by stimulating the immune system through activation of the dendritic cells, as shown in Figure 31.20.

Novel adjuvants are being created to combat many diseases, such as hepatitis A and B, human papillomavirus, influenza, malaria, and non-small-cell lung cancer (NSCLC).

31.8 How Does the Body Distinguish "Self" from "Nonself"?

One major problem facing body defenses is how to recognize the foreign body as "not self" and thereby avoid attacking the "self"—that is, the healthy cells of the organism.

A. Selection of T and B Cells

The members of the adaptive immunity system, B cells and T cells, are all specific and have memory, so they target only truly foreign invaders. The T cells mature in the thymus gland. During the maturation process, those T cells that fail to recognize and interact with MHC, and thus cannot respond to foreign antigens, are eliminated through a selection process. They essentially die by neglect. T cells that express receptors (TcR) that are prone to interact with normal self antigens are also eliminated through the selection process (Figure 31.21). Thus, the activated T cells leaving the thymus gland carry TcRs that can respond to foreign antigens. Even if some T cells prone to react with self antigen escape the selection detection, they can be deactivated through the signal transduction system that, among other functions, performs tyrosine kinase activation and phosphatase deactivation, similar to those processes seen in adrenergic neurotransmitter signaling (Section 24.5).

Similarly, the maturation of B cells in the bone marrow depends on the engagement of their receptors, BcR, with antigen. Those B cells that are prone to interact with self antigen are also eliminated before they leave the bone marrow. As with T cells, many signaling pathways control the proliferation of B cells. Among them, the activation by tyrosine kinase and the deactivation by phosphatase provide a secondary control.

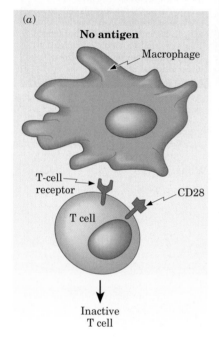
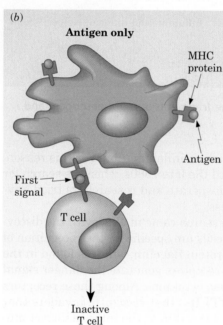
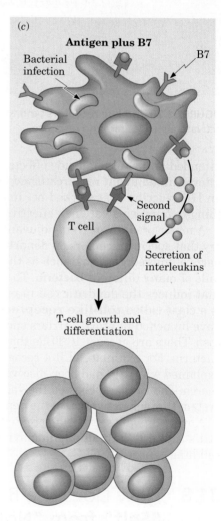

FIGURE 31.21 A two-stage process leads to the growth and differentiation of T cells. (a) In the absence of antigen, proliferation of T cells does not take place. Those T-cell lines die by neglect. (b) In the presence of antigen alone, the T-cell receptor binds to antigen presented on the surface of a macrophage cell by the MHC protein. There is still no proliferation of T cells because a second required signal is missing. In this way, the body can avoid an inappropriate response to its own antigens. This process occurs early in the development of T cells, effectively eliminating those cells that would otherwise be activated by the self antigens. (c) When an infection occurs, a B7 protein is produced in response to the infection. The B7 protein on the surface of the infected cell binds to a CD28 protein on the surface of the immature T cell, giving the second signal that allows it to grow and proliferate. (Adapted from "How the Immune System Recognizes Invaders," by Charles A. Janeway, Jr.; illustration by Ian Warpole, Scientific American, September 1993.)

Antibiotics: A Double-Edged Sword

Living in the modern world, you have undoubtedly taken advantage of antibiotics. Indeed, many of the diseases of the past have been all but eradicated by these drugs, which can stop a bacterial life cycle in its tracks. Common infections that may have proved fatal at the beginning of the 1900s are often treated successfully today with penicillin or another common antibiotic such as erythromycin or cephalosporin.

However, antibiotics can also cause several problems. Many people are allergic to penicillin and its derivatives, and these antibiotic allergies can be very potent. A person may take an antibiotic once and have no symptoms. A subsequent use of the same antibiotic may cause a severe skin rash or hives. A third exposure could be fatal. Indiscriminate use of antibiotics can also be harmful. Many diseases are caused by viruses, which do not respond to treatment with antibiotics. However, patients do not want to hear that there is nothing they can do except wait out the disease, so they often are given antibiotics. Antibiotics are also prescribed before the exact nature of the infection is known. This indiscriminate use of antibiotics is the major cause of the increasing incidence of drug-resistant microorganisms.

One disease that has flourished due to misuse of antibiotics is gonorrhea. One strain of *Neisseria gonorrhoeae* produces β-lactamase, an enzyme that degrades penicillin. These strains are referred to as PPNG, penicillinase-producing *N. gonorrhoeae*. Before 1976, almost no cases of PPNG were reported in the United States. Today, thousands of cases occur nationwide. The source of the problem has been traced to military bases in the Philippines, where soldiers acquired the disease from prostitutes. Among prostitutes, there is a common practice of using low doses of antibiotics in an attempt to prevent the spread of sexually transmitted diseases. In reality, constant use of antibiotics has just the opposite effect—it causes development of drug-resistant strains.

An antibiotic commonly given to young children for earaches is amoxicillin, a penicillin derivative. Severe and repeated earaches can cause hearing loss, so parents are often quick to put their children on antibiotics. There are two downsides, however, to overuse of this antibiotic. First, it has minimal efficiency because the bacteria causing the earaches are localized inside the ear, where the antibiotic has little access. Second, overuse of the antibiotic affects the inside of the growing teeth, causing them to remain soft and leading to future dental problems.

When a person with a bacterial infection uses antibiotics early in the infection, he or she never has the opportunity to elicit a true immune response. For this reason, the person will be susceptible to the same disease again and again. This issue is seen nowadays with diseases such as strep throat, which many people have almost yearly. Some doctors are trying to avoid prescribing antibiotics until their patients have had a chance to fight the disease on their own. Some patients are purposely avoiding using antibiotics for the same reason. While such a strategy is attractive and intuitive, if we take a closer look at strep throat, we can see yet another side to this story.

Rheumatic fever is a complication of untreated strep throat. It is characterized by fever and widespread inflammation of the joints and heart. These effects are produced by the body's immune response to the M protein of the group A streptococci. The M protein resembles a major protein in heart tissue. As a result, the antibodies attack the heart valves as well as the bacterial M protein. About 3% of individuals who do not get treatment with antibiotics when they have strep throat develop rheumatic fever. About 40% of patients with rheumatic fever develop heart valve damage, which may not become evident for 10 years or more. The best way to avoid this complication is early treatment of strep throat with antibiotics.

In summary, antibiotics are a very important weapon in our arsenal against disease, but they should not be used indiscriminately. If they are used, the entire course of treatment should be taken to completion. The last thing you would want to do is kill off most—but not all—of the bacteria you were infected with. This would leave behind a few "superbacteria," which would then be drug-resistant.

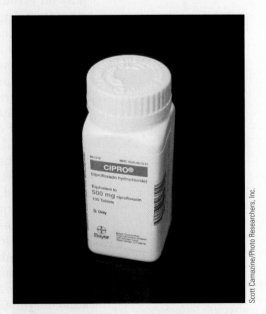

Cipro is a common antibiotic, often used to fight strep throat.

Test your knowledge with Problems 31.74–31.78.

B. Discrimination of the Cells of the Innate Immunity System

The first line of defense is the innate immunity system, in which cells such as natural killer cells or macrophages have no specific targets and no memory of which epitope signals danger. Nevertheless, these cells must somehow discriminate between normal and abnormal cells to identify targets. The mechanism by which this identification is accomplished has only recently been explored and is not yet fully understood. The main point is that the cells of innate immunity have two kinds of receptors on their surface: an **activating receptor** and an **inhibitory receptor**. When a healthy cell of the body encounters a macrophage or a natural killer cell, the inhibitory receptor on the latter's surface recognizes the epitope of the normal cell, binds to it, and prevents the activation of the killer cell or macrophage. Conversely, when a macrophage encounters a bacterium with a foreign antigen on its surface, the antigen binds to the activating receptor of the macrophage. This ligand binding prompts the macrophage to engulf the bacterium by phagocytosis. Such foreign antigens may be the lipopolysaccharides of gram-negative bacteria or the peptidoglycans of gram-positive bacteria.

When a cell is infected, damaged, or transformed into a malignant cell, the epitopes that signaled healthy cells diminish greatly, and unusual epitopes are presented on the surface of these cancer cells. The effect is that fewer inhibitory receptors of macrophages or natural killer cells can bind to the surface of the target cell and more activating receptors find inviting ligands. As a consequence, the balance shifts in favor of activation and the macrophages or killer cells will do their job.

C. Autoimmune Diseases

In spite of the safeguards in the body intended to prevent acting against "self," in the form of healthy cells, many diseases exist in which some part of a pathway in the immune system goes awry. The skin disease psoriasis is thought to be a T-cell–mediated disease in which cytokines and chemokines play an essential role. Other autoimmune diseases, such as myasthenia gravis, rheumatoid arthritis, multiple sclerosis, and insulin-dependent diabetes (Chemical Connections 24F), also involve cytokines and chemokines. Allergies are another example of malfunctioning of the immune system. Pollens and animal furs are allergens that can provoke asthma attacks. Some people are so sensitive to certain food allergens that even a knife once used in smearing peanut butter may prove fatal in a person known to be allergic to peanuts.

The major drug treatment for autoimmune diseases involves the glucocorticoids, the most important of which is cortisol (Section 21.10A). It is a standard therapy for rheumatoid arthritis, asthma, inflammatory bone diseases, psoriasis, and eczema. The beneficial effects of glucocorticoids are overshadowed, however, by their undesirable side effects, which include osteoporosis, skin atrophy, and diabetes. Glucocorticoids regulate the synthesis of cytokines either directly, by interacting with their genes, or indirectly, through transcription factors.

Macrolide drugs are used to suppress the immune system during tissue transplantation or in the case of certain autoimmune diseases. Drugs such as cyclosporin A and rapamycin bind to receptors in the cytosol and, through secondary messengers, inhibit the entrance of nuclear factors into the nucleus. Normally, those nuclear factors signal a need for transcription, so their absence prevents the transcription of cytokines—for example, interleukin-2.

Macrolide drugs are a class of drug, mostly antibiotics, that all have a large macrocyclic lactone ring system. Common examples are erythromycin and clarithromycin.

31.9 How Does the Human Immunodeficiency Virus Cause AIDS?

Human immunodeficiency virus (HIV) is the most infamous of the retroviruses, as it is the causative agent of acquired immunodeficiency syndrome (AIDS). This disease affects more than 40 million people worldwide and has continually thwarted attempts to eradicate it. The best medicine today can slow it down, but nothing has been able to stop AIDS.

The HIV genome is a single-stranded RNA that has a number of proteins packed around it, including virus-specific reverse transcriptase and protease. A protein coat surrounds the RNA–protein assemblage, giving the overall shape of a truncated cone. Finally, a membrane envelope encloses the protein coat. The envelope consists of a phospholipid bilayer formed from the plasma membrane of cells infected earlier in the life cycle of the virus, as well as some specific glycoproteins, such as gp41 and gp120, as shown in Figure 31.22.

HIV offers a classic example of the mode of operation of retroviruses. The HIV infection begins when the virus particle binds to receptors on the surface of a cell (Figure 31.23). The viral core is inserted into the cell and partially disintegrates. The reverse transcriptase catalyzes the production of DNA from the viral RNA. The viral DNA is integrated into the DNA of the host cell. The DNA, including the integrated viral DNA, is transcribed to RNA. Smaller RNAs are produced first, specifying the amino acid sequences of viral regulatory proteins. Larger RNAs, which specify the amino acid sequences of viral enzymes and coat proteins, are made next. The viral protease assumes particular importance in the budding of new virus particles. Both the viral RNA and viral proteins are included in the budding virus, as is some of the membrane of the infected cell.

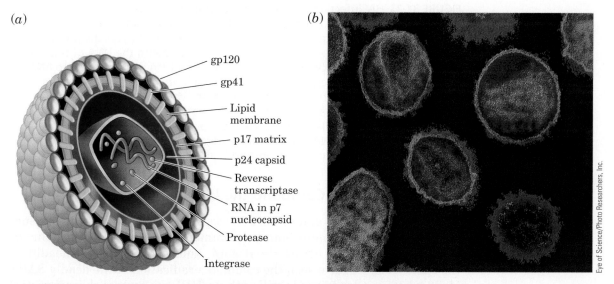

(a)

gp120
gp41
Lipid membrane
p17 matrix
p24 capsid
Reverse transcriptase
RNA in p7 nucleocapsid
Protease
Integrase

(b)

Eye of Science/Photo Researchers, Inc.

FIGURE 31.22 The architecture of HIV. (*a*) The RNA genome is surrounded by P7 nucleocapsid proteins and by several viral enzymes—namely, reverse transcriptase, integrase, and protease. The truncated cone consists of P24 capsid protein subunits. The P17 matrix (another layer of protein) lies inside the envelope, which consists of a lipid bilayer and glycoproteins such as gp41 and gp120. (*b*) An electron micrograph shows both mature virus particles, in which the core (the truncated cone) is visible, and immature virus particles, in which it is not.

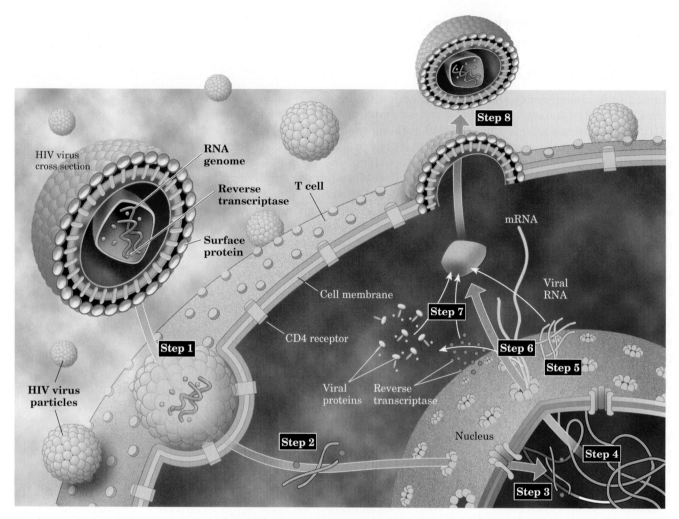

FIGURE 31.23 HIV infection begins when the virus particle binds to CD4 receptors on the surface of the cell (Step 1). The viral core is inserted into the cell and partially disintegrates (Step 2). The reverse transcriptase catalyzes the production of DNA from the viral RNA. The viral DNA is integrated into the DNA of the host cell (Step 3). The DNA, including the integrated viral DNA, is transcribed to RNA (Step 4). Smaller RNAs are produced first, specifying the amino acid sequence of viral regulatory proteins (Step 5). Larger RNAs, which specify the amino acid sequences of viral enzymes and coat proteins, are made next (Step 6). The viral protease assumes particular importance in the budding of new virus particles (Step 7). Both the viral RNA and viral proteins are included in the budding virus, as is some of the membrane of the infected cell (Step 8).

A. HIV's Ability to Confound the Immune System

Why is HIV so deadly and so hard to stop? Many viruses, such as adenovirus, cause nothing more than the common cold; others, such as the virus that causes severe accute respiratory syndrome (SARS), are deadly. At the same time, we have seen the complete eradication of the deadly SARS virus, whereas adenovirus is still with us. HIV has several characteristics that lead to its persistence and eventual lethality. Ultimately, it is deadly because of its targets, the helper T cells. The immune system is under constant attack by the virus, and millions of helper T cells and killer T cells are called up to fight billions of virus particles. Through degradation of the T-cell membrane via budding and the activation of enzymes that lead to cell death, the T-cell count diminishes to the point that the infected person is no longer able to mount a suitable immune response. As a result, the individual eventually succumbs to pneumonia or another opportunistic disease.

There are many reasons that the disease is so persistent. For example, it is slow acting. SARS was eradicated quickly because the virus was quick to act, making it easy to find infected people before they had a chance to spread the disease. In contrast, HIV-infected individuals can go years before they become aware that they have the disease. However, this is only a small part of what makes HIV so difficult to kill.

HIV is difficult to kill because it is difficult to find. For an immune system to fight a virus, it needs to locate specific macromolecules that can be bound to antibodies or T-cell receptors. The reverse transcriptase of HIV is very inaccurate in replication. The result is rapid mutation of HIV, a situation that presents a considerable challenge to those who want to devise treatments for AIDS. The virus mutates so rapidly that multiple strains of HIV may be present in a single individual.

Another trick the virus plays is a conformational change of the gp120 protein when it binds to the CD4 receptor on a T cell. The normal shape of the gp120 monomer may elicit an antibody response, but these altered antibodies are largely ineffective. The gp120 forms a complex with gp41 and changes shape when it binds to CD4. It also binds to a secondary site on the T cell that normally binds to a cytokine. This change exposes a part of gp120 that was previously hidden and, therefore, cannot elicit antibodies.

HIV is also adept at evading the innate immunity system. Natural killer cells attempt to attack the virus, but HIV binds a particular cell protein, called cyclophilin, to its capsid, which blocks the antiviral agent restriction factor-1. Another of HIV's proteins blocks the viral inhibitor called CEM-15, which normally disrupts the viral life cycle.

Finally, HIV hides from the immune system by cloaking its outer membrane in sugars that are very similar to the natural sugars found on most of its host's cells, rendering the immune system blind to it.

B. The Search for a Vaccine

The attempt to find a vaccine for HIV is akin to the search for the Holy Grail, and it has met with about as much success. One strategy for using a vaccine to stimulate the body's immunity to HIV is shown in Figure 31.24. DNA for a unique HIV gene, such as the *gag* gene, is injected into muscle. The *gag* gene leads to the Gag protein, which is taken up by antigen-presenting cells and displayed on their cell surfaces. This then elicits the cellular immune response, stimulating killer and helper T cells. It also stimulates the humoral immune response, spurring production of antibodies. Figure 31.24 also shows a second phase of the treatment, a booster shot of an altered adenovirus that carries the *gag* gene.

Unfortunately, most attempts at making antibodies have proved unsuccessful to date. The most thorough attempt was made by the VaxGen Company, which carried the research through the third stage of clinical trials, testing the vaccine on more than 1000 high-risk people and comparing them to 1000 individuals who did not receive the vaccine. In the study, 5.7% of the people who received the vaccine eventually became infected, compared to 5.8% of the placebo group. Many people analyzed the data, and despite attempts to show a better response in certain ethnic groups, the trials had to be declared a failure. The vaccine, called AIDSVAX, was based on gp120.

C. Antiviral Therapy

While the search for an effective vaccine continued with little to minor success, pharmaceutical companies flourished by designing drugs that would inhibit retroviruses. By 1996, there were 16 drugs used to inhibit either the HIV reverse transcriptase or the protease. Several others are in clinical

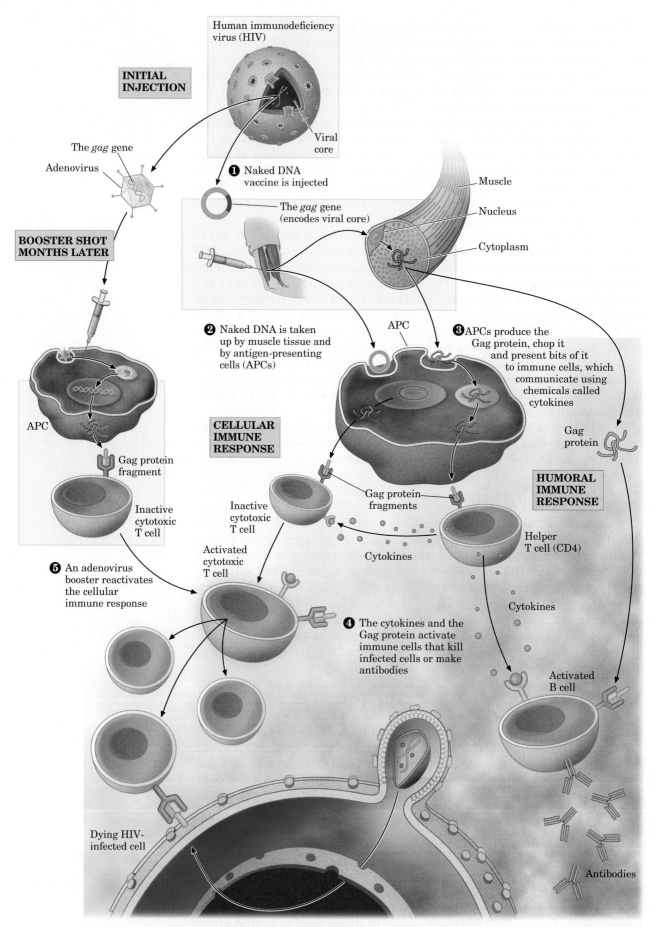

FIGURE 31.24 One strategy for an AIDS vaccine. (Reprinted by permission from Ezzell C., "Hope in a Vial," *Sci. Am.* 39–45 June 2002.)

Why Are Stem Cells Special?

Stem cells are the precursors of all other cell types, including T and B lymphocytes. These undifferentiated cells have the ability to form any cell type, as well as to replicate into more stem cells. Stem cells are often called **progenitor cells** due to their ability to differentiate into many cell types. A **pluripotent** stem cell is able to give rise to all cell types in an embryo or an adult. Some cells are called **multipotent** because they can differentiate into more than one cell type, but not into all cell types. The further from a zygote a cell is in the course of its development, the less potent the cell type. The use of stem cells, especially **embryonic stem cells**, has been an exciting field of research for several years.

History of Stem Cell Research

Stem cell research began in the 1970s with studies on teratocarcinoma cells, which are found in testicular cancers. These cells are bizarre blends of differentiated and undifferentiated cells. Such **embryonal carcinoma cells** (**EC**) were found to be pluripotent, which led to the idea of using them for therapy. However, this line of research was suspended because the cells came from tumors, which made their use dangerous, and they were **aneuploid**, which means they had the wrong number of chromosomes.

Early work with embryonic stem cells (ES) involved cells that were grown in culture after being taken from embryos. It was found that these stem cells could be maintained for long periods. In contrast, most differentiated cells will not grow for extended periods in culture. Stem cells are maintained in culture by addition of certain factors such as leukemia-inhibiting factor (LIF) or feeder cells (nonmitotic cells such as fibroblasts). Once released from these controls, ES cells will differentiate into all kinds of cells.

Stem Cells Offer Hope

Stem cells placed into a particular tissue, such as blood, will differentiate and grow into blood cells. Others, when placed into brain tissue, will grow brain cells. This discovery is very exciting because it had been believed that there was little hope for patients with spinal cord and other nerve damage, because these cells do not normally regenerate. In theory, neurons could be produced to treat neurodegenerative diseases such as Alzheimer's and Parkinson's disease. Muscle cells could be produced to treat muscular dystrophies and heart disease. In one study, mouse stem cells were injected into a mouse heart that had undergone a myocardial infarction. The cells spread from an unaffected region into the infarcted zone and began to grow new heart tissue. Human pluripotent stem cells have been used to regenerate nerve tissue in rats with nerve injuries and have been shown to improve motor and cognitive ability in rats that underwent a stroke. Results such as these have led some scientists to claim that stem cell technology will be the most important advancement in science since cloning.

Truly pluripotent stem cells have been harvested primarily from embryonic tissue, and these cells show the

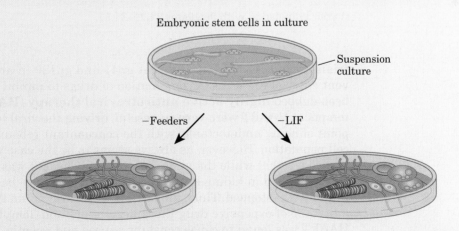

Embryonic stem cells in culture

Pluripotent embryonic stem cells can be grown in cell culture. They can be maintained in an undifferentiated state by growing them on certain feeder cells, such as fibroblasts, or by using leukemia inhibitory factor (LIF). When removed from the feeder cells or when the LIF is removed, they begin to differentiate into a wide variety of tissue types, which can then be harvested and grown for tissue therapy. (Taken from Donovan P. J., and Gearhart J., *Nature*, 414:92–97, 2001)

Why Are Stem Cells Special? (*continued*)

greatest ability to differentiate into various tissues and to reproduce in cell culture. Stem cells have also been taken from adult tissues, as some stem cells are always present in an organism, even at the adult stage. These cells are usually multipotent, as they can form several different cell types, but are not as versatile as ES cells. Many scientists believe that the ES cells represent a better source for tissue therapy than adult stem cells for this reason.

The acquisition and use of stem cells can be related to a technique called cell reprogramming, which is a necessary component of whole-mammal cloning, such as the process that created the world's most famous sheep, Dolly. Most somatic cells in an organism contain the same genes, but the cells develop as different tissues with wildly different patterns of gene expression. A mechanism that alters expression of genes without changing the actual DNA sequence is called an **epigenetic** mechanism. An epigenetic state of the DNA in a cell is a heritable trait that allows a "molecular memory" to exist in the cells. In essence, a liver cell remembers where it came from and will continue to divide and remain a liver cell. These epigenetic states involve methylation of cytosine–guanine dinucleotides and interactions with proteins of chromatin (Section 25.3C). Mammalian genes possess an additional level of epigenetic information called **imprinting**, which allows the DNA to retain a molecular memory of its germ-line origin. The paternal DNA is imprinted differently than the maternal DNA. In normal development, only DNA that came from both parents would be able to combine and lead to a viable offspring.

The epigenetic states of somatic cells are generally locked in a way that the differentiated tissues remain stable. The key to whole-organism cloning was the ability to erase the epigenetic state and return to the state of a fertilized egg, which has the potential to produce all cell types. If the nucleus of a somatic cell is injected into a recipient oocyte, the epigenetic state of the DNA can be reprogrammed, or at least "partially" reprogrammed. The molecular memory is erased, and the cell begins to behave like a true zygote. This development can be used to derive pluripotent stem cells or to transfer a blastocyst into a mother-carrier for growth and development. In November 2001, the first human cloned blastocyst was created in this way, with the aim of growing enough cells to harvest pluripotent stem cells for research.

Currently, debate continues to rage worldwide over the use of embryonic stem cells. The issue at hand is one of ethics and the definition of life. Embryonic stem cells come from many sources, including aborted fetuses, umbilical cords, and embryos from in vitro fertilization clinics. The report of the cloned human embryonic cells merely added to the controversy. The U.S. government has banned funding for stem cell research but allows research to continue on all existing embryonic cell lines. Some big questions that people will have to answer are the following: Do a few cells created by therapeutic cloning of your own somatic cells constitute life? If these cells do constitute life, do they have the same rights as a human being conceived naturally? If it were possible, should someone be allowed to grow his or her own therapeutic clone into an adult?

Test your knowledge with Problems 31.79–31.81.

trials, including drugs that target gp41 and gp120 in an attempt to prevent entry of the virus. A combination of drugs to inhibit retroviruses has been dubbed **highly active antiretroviral therapy** (**HAART**). Initial attempts at HAART were very successful, driving the viral load almost to the point of being undetectable, with the concomitant rebounding of the CD4 cell population. However, as always seems to be the case with HIV, it later turned out that while the virus was knocked down, it was not knocked out. HIV remained in hiding in the body and would bounce back as soon as the therapy was stopped. Thus, the best-case scenario for an AIDS patient was a lifetime of expensive drug therapies. In addition, long-term exposure to HAART was found to cause constant nausea and anemia, as well as diabetes symptoms, brittle bones, and heart disease.

D. A Second Chance for Antibodies

In the wake of the realization that patients could not stay on the HAART program indefinitely, several researchers attempted to combine HAART with vaccination. Even though most of the vaccines were not found to be effective alone, they proved more effective in combination with HAART. In addition, once on the vaccine, patients were able to take a rest from the other drugs, giving their bodies and minds time to recover from the side effects of the antiviral therapy.

E. The Future of Antibody Research

Early attempts at creating a vaccine appear to have failed because the vaccine elicited too many useless antibodies. What patients need is a **neutralizing antibody**, one capable of completely eliminating its target. Researchers discovered a patient who had HIV for six years but never developed AIDS. They then studied his blood and found a rare antibody, which they labeled **b12**. In laboratory trials, **b12** was found to stop most strains of HIV. What made b12 different from the other antibodies? Structural analysis revealed that this antibody has a different shape from a normal immunoglobulin. It has sections of long tendrils that fit into a fold in gp120. This fold in gp120 cannot mutate very much; otherwise, the protein will not be able to dock properly with the CD4 receptor.

Another antibody was found in a different patient who seemed resistant to HIV. This antibody was actually a dimer and had a shape more like an "I" than the traditional "Y". This antibody, called **2G12**, recognizes some of the sugars on HIV's outer membrane that are unique to HIV.

By identifying a few such antibodies, researchers have been able to search for a vaccine in the opposite direction from the normal way. In **retrovaccination**, researchers have the antibody and need to find a vaccine to elicit it, instead of injecting vaccines and noting which antibodies they produce.

Another active area in vaccine research is the search for antibodies that can bind to more than one antigen. This goes against the old immunology dogma of "one antigen—one antibody." But it would be a boon to research on HIV; influenza; hepatitis; or any other pathogen that mutates quickly, making a moving target. If an antibody can bind to more than one antigen, it would theoretically be able to attack multiple targets. A recent success was the design of a two-in-one antibody against two different antigens, vascular endothelial growth factor (VEGF) and human epidermal growth factor receptor 2 (HER2) (see Chemical Connections 31A). Figure 31.25 shows a schematic of the results of this study. The antibody produced has a binding site that has areas specific for HER2 (shown in red) and VEGF (shown in green), as well as one shared area (shown in purple). Successes such as this give hope to the idea of designing even better antibodies against the elusive HIV virus.

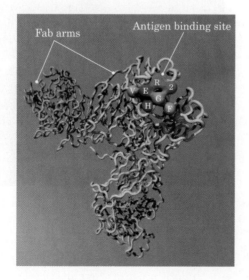

FIGURE 31.25 An antibody consists of four polypeptides, two heavy and two light chains, that form two "Fab (fragment antigen binding) arms." Each arm harbors an antigen-binding site, formed by loops from the heavy and light chains. The binding site in the two-in-one antibody shown can interact with HER2 (red) and VEGF (green) through mostly unique, but also some shared (blue), elements. An affinity-matured antibody has been generated in a process to create antibodies with increased binding affinities. When affinity-matured, the antibody inhibits both HER2 and VEGF activity *in vitro* and *in vivo*.

Chemical Connections 31D

A Little Swine Goes a Long Way

In the fall of 2009, a common phrase heard in the school-yard was, "he's got the swine," referring to the outbreak of swine flu that had begun in the spring of that year. Certainly anyone reading this book has had influenza—the flu—a disease that most people take for granted as an annoying fact of life. There are yearly epidemics around the world, with some being very serious. In 1918, there was a worldwide flu pandemic that led to the deaths of 50 million people, one of the worst epidemics in history, surpassing even the black plague of the Middle Ages. By comparison, there are only about 40 million people today living with the HIV virus, and it has taken 30 years to get to that point. The flu virus has been with us for thousands of years and has never been fully controlled by modern medicine.

A single particle of the influenza virus (a virion) contains a genome that is a single-stranded RNA template with a protein coat that protrudes through a lipid bilayer envelope. The figure shows the structural features of the influenza virus.

There are three major types of influenza, designated A, B, and C, depending on differences in the proteins. Influenza viruses cause infections of the upper respiratory tract that lead to fever, muscle pain, headaches, nasal congestion, sore throat, and coughing. One of the biggest problems is that people who catch the flu often get secondary infections, including pneumonia, which is what makes the flu potentially lethal.

We are going to talk about the influenza A virus because, of the three, it is responsible for most human illness. The most prominent features of the virus envelope are two spike proteins. One is called hemagglutinin (HA), which gets its name because it causes erythrocytes to clump together. The second is neuraminidase (NA), an enzyme that catalyzes the hydrolysis of a linkage of sialic acid to galactose or galactosamine (see Chapter 20). HA is believed to help the virus in recognizing target cells. NA is believed to help the virus get through mucous membranes and enter cells of the host. Sixteen subtypes of HA are known (designated H1–H16), and nine subtypes of neuraminidase (designated N1–N9) have been cataloged. H1, H2, H3, N1, and N2 occur in most of the known viruses that affect humans. Individual influenza A viruses are named by giving the subtypes of HA and NA—for example, H1N1 and H3N2. The virus that causes the avian influenza that has been in the news is H5N1. The presence of the H5 protein affects humans, but so far to a lesser extent than the other HA subtypes. It does, of course, affect birds, with many fatalities among chickens, ducks, and geese.

The nature of the virus subtype determines its effect on humans. The relevant factors to epidemiologists are the transmissibility and the mortality. For example, there have been only a few hundred cases of people contracting the avian flu worldwide, so its transmissibility is relatively low. However, of those who have gotten the virus, more than 60% have died, so it is still a big concern. In contrast, the recent swine flu is the H1N1 variety, and it is more transmissible but far less deadly to those who get it. In many cases, its symptoms are no worse than those of any common flu, and there have been few fatalities.

While the flu has been with us for millennia, it is always changing, and it is the possibility of such changes that worries agencies responsible for public health, such as the Centers for Disease Control (CDC) and the World

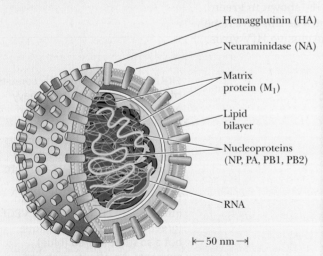

A cutaway diagram of the influenza virion. The HA and NA spikes are embedded in a lipid bilayer that forms the virion's outer envelope. A matrix protein, M1, coats the inside of this membrane. The virion core contains the eight single-stranded segments that constitute its genome in a complex with the proteins NP, PA, PB1, and PB2 to form helical structures called neocapsids. (Reprinted with permission from the Estate of Bunji Tagawa.)

Hemagglutinin (HA)
Neuraminidase (NA)
Matrix protein (M_1)
Lipid bilayer
Nucleoproteins (NP, PA, PB1, PB2)
RNA

← 50 nm →

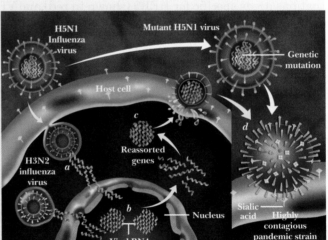

H5N1 Influenza virus
Mutant H5N1 virus
Genetic mutation
Host cell
c
Reassorted genes
H3N2 influenza virus
a
d
b
Viral RNA
Nucleus
Sialic acid
Highly contagious pandemic strain

Two possible strategies for mutating viruses. The H5N1 strain might undergo a mutation that would make it bind more easily to the cell of the host and therefore be more infective (pink path). The H5H1 and H3H2 strains might both bind to the same cell and then mix their RNA to form reassorted genes (yellow path).

A Little Swine Goes a Long Way (*continued*)

Health Organization (WHO). Mutations occur frequently with viruses, and the biggest worry is that a strain with a high mortality could mutate into one that is also very transmissible. The figure shows how the deadly avian strain could potentially change.

In one possibility, the virus (H5N1 in this example) mutates and changes its surface proteins, making it more able to bind to human cells and infect them (pink path). The other possibility is that two viruses might infect the same cell (H5N1 and H3N2 in this example, yellow path). The viral RNAs could get mixed and produce reassorted genes, leading to different capabilities in a mutated new strain. The deadly flu of 1918 was also an H1N1 swine flu.

In 1997, a flu that was mostly human in origin was found in North American pigs. A year later, researchers found another version that combined genes from human, avian, and swine sources, a triple reassortant. The 2009 swine flu is also a triple reassortant, which combines pieces from three different sources. Such combinations demonstrate that flu viruses do not stay contained in one species for long. This is the main reason that scientists worry about what the next jumbling of flu genes will do. It is also why the CDC and WHO take every case of flu seriously. A combination having the mortality of the avian flu with the transmissibility of the 2009 swine flu could lead to the next plague. Fortunately, it has not happened as of this writing.

Test your knowledge with Problems 31.82–31.85.

Immunologists Take on the Flu Virus

Immunologists are always looking for a better flu vaccine. Millions of people get flu vaccines every year, and the effectiveness has been variable. Because the flu viruses change quickly and recombine (see Chemical Connections 31D), it has been hard to find the perfect vaccine. Researchers are always one step behind in trying to identify the current assortment of epitopes in the "flu of the year" and then produce enough vaccine to fight it. The biggest

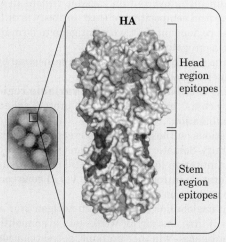

HA

Head region epitopes

Stem region epitopes

Most antibodies against the influenza A virus (inset shows the 2009 H1N1 strain) bind to the highly variable part of the hemagglutinin (HA) glycoprotein at the surface of the virus particle (head region). In the H1 subtype, these antibodies recognize four major sites (Sa, Sb, Ca, and Cb are shown in green, pink, cyan, and yellow, respectively). The HA structure of the 2009 H1N1 virus is shown (PDB [Protein Data Bank] code 3LZG).

problem is that many of the vaccines target the head portions of the HA molecule that are the most variable. Recently, some new vaccines have been developed that target portions of the stem of the HA molecule. These are more constant and may lead to a "universal flu vaccine."

The researchers who developed this vaccine determined that the key to its production was the methods used to generate it. In their case, they used a two-pronged approach. First, they used a DNA vaccine comprised of a plasmid that would express the HA protein of the seasonal flu vaccine. They then gave hosts a booster shot of the seasonal flu vaccine. While this approach does require more trips to the caregiver than a normal single-shot vaccine, the effect was the generation of a much better set of antibodies that recognize a constant region of HA, which were more effective at neutralizing a broader set of flu virus types.

Besides generating vaccines against the flu virus, scientists are also trying to generate monoclonal antibodies that will fight the disease. Monoclonal antibodies are expensive to make, and in many cases, the attempt is frustrated by the quickly changing nature of the virus. Therefore, researchers have been looking to create an antibody that will also attack the constant part (identical in all viruses of the same isotype) of the virus. In early 2009, two independent teams reported that they had created monoclonal antibodies that would react with a constant portion of the flu virus's hemagglutinin (HA) protein.

The teams identified 10 different antibodies that recognized the H5 subtype found in the avian flu and found that those antibodies would also block 8 of the 15 other HA types.

Chemical Connections 31E

Immunologists Take on the Flu Virus (*continued*)

They tested their antibodies in mice before and after they were dosed with lethal quantities of avian flu. Most of the rodents survived, indicating that these antibodies would successfully work to prevent or cure the disease.

The medical community is excited about the prospects of adding another weapon to the eternal fight against the flu. Such antibodies can be used to provide an immediate passive immunity to people who do not respond well to vaccines, such as the elderly or those whose immune systems are compromised. They will also allow a strong countermeasure to any impending pandemics. The usual downside to such new discoveries is the cost.

There are not enough cheap vaccines to meet the needs of people around the world, and many economically deprived countries have trouble getting them, especially an expensive monoclonal antibody. Still, many feel that if governments can afford them, it would be wise to stockpile some of these new antibodies as protection against the next pandemic.

Test your knowledge with Problems 31.86 and 31.87.

Summary

OWL Sign in at **www.cengage.com/owl** to develop problem-solving skills and complete online homework assigned by your professor.

Section 31.1 How Does the Body Defend Itself from Invasion?

- The human immune system protects us against foreign invaders. It consists of two parts: (1) the natural resistance of the body, called innate immunity, and (2) adaptive or acquired immunity.
- **Innate immunity** is nonspecific. **Macrophages** and **natural killer (NK) cells** are cells of innate immunity that function as police officers.
- **Adaptive** or **acquired immunity** is highly specific, being directed against one particular invader.
- Acquired immunity (known as the immune system) also has memory, unlike innate immunity.

Section 31.2 What Organs and Cells Make Up the Immune System?

- The principal cellular components of the immune system are white blood cells, or **leukocytes**. The specialized leukocytes in the lymph system are called **lymphocytes**. They circulate mostly in the **lymphoid organs**.
- The **lymph** system is a collection of vessels extending throughout the body and connected to the interstitial fluid on the one hand and to the blood vessels on the other hand.
- Lymphocytes that mature in the bone marrow and produce soluble immunoglobulins are **B cells**. Lymphocytes that mature in the thymus gland are **T cells**.

Section 31.3 How Do Antigens Stimulate the Immune System?

- **Antigens** are large, complex molecules of foreign origin. They can be proteins, polysaccharides, or nucleic acids and can originate from bacteria, viruses, fungus, yeast, parasites, pollen, or a toxin.
- An antigen may interact with antibodies, T-cell receptors (TcR), or major histocompatibility complex (MHC) molecules. All three types of molecules belong to the **immunoglobulin superfamily**.
- An **epitope** is the smallest part of an antigen that binds to antibodies, TcRs, and MHCs.

Section 31.4 What Are Immunoglobulins?

- Antibodies are **immunoglobulins**. These water-soluble glycoproteins are made of two heavy chains and two light chains. All four chains are linked together by disulfide bridges.
- Immunoglobulins contain variable regions in which the amino acid composition of each antibody is different. These regions interact with antigens to form insoluble large aggregates.
- A large variety of antibodies are synthesized by a number of processes in the body.
- During B-cell development, the **variable regions** of the heavy chains are assembled by a process called V(J)D recombination.
- Mutations on these new genes create even greater diversity. Somatic hypermutation (SHM) that creates a point mutation (only one nucleotide) is one way. Patches of mutation introduced into the V(J)D gene constitute gene conversion (GC) mutation.
- Immunoglobulins respond to an antigen over a long span of time, lasting for weeks and even months.
- All antigens—whether proteins, polysaccharides, or nucleic acids—interact with soluble immunoglobulins produced by B cells.

Section 31.5 What Are T Cells and T-Cell Receptors?

- Protein antigens interact with T cells. The binding of the epitope to the TcR is facilitated by MHC, which carries the epitope to the T-cell surface, where it is presented to the receptor.

- Upon epitope binding to the receptor, the T cell is stimulated. It proliferates and can differentiate into (1) killer T cells, (2) memory cells, or (3) helper T cells.

- The TcR has a number of helper molecules, such as CD4 or CD8, that enable it to bind the epitope tightly and to bind to other cells via MHC proteins.

- CD (**cluster determinant**) molecules also belong to the immunoglobulin superfamily.

- Antibodies can recognize all types of antigens, but TcRs recognize only peptide antigens.

Section 31.6 How Is the Immune Response Controlled?

- The control and coordination of the immune response are handled by **cytokines**, which are small protein molecules.

- Chemotactic cytokines, the **chemokines**, such as interleukin-8, facilitate the migration of leukocytes from blood vessels into the site of injury or inflammation. Other cytokines activate B and T cells and macrophages, enabling them to engulf the foreign body, digest it, or destroy it by releasing special toxins.

- Some cytokines, such as tumor necrosis factor (TNF), can lyse tumor cells.

Section 31.7 Immunization

- Edward Jenner was the first to attempt vaccination after realizing that people infected with cowpox never got smallpox. He used fluids from people infected with cowpox as the first vaccine.

- There are three common types of vaccines: attenuated, inactivated, and subunit.

- Vaccines work by mimicking an infection. They trigger responses by the innate immunity system, which then stimulates maturation of T cells and B cells.

- Vaccines are used for many diseases. The use of vaccines has eliminated smallpox and several other diseases. Vaccines have been only partially successful at preventing diseases such as HIV, malaria, and influenza.

- Adjuvants are chemicals that boost the power of a vaccine. They may be simple chemicals, such as oil and water, or purified extracts of bacteria, such as lipopolysaccharides.

- Some adjuvants have been shown to work due to their stimulation of toll-like receptors on dendritic cells.

Section 31.8 How Does the Body Distinguish "Self" from "Nonself"?

- A number of mechanisms in the body ensure that the body recognizes "self."

- In adaptive immunity, T and B cells that are prone to interact with "self" antigens are eliminated.

- In innate immunity, two kinds of receptors exist on the surfaces of T and B cells: an **activating receptor** and an **inhibitory receptor**. The inhibitory receptor recognizes the epitope of a normal cell, binds to it, and prevents the activation of the killer T cell or macrophage.

- Many autoimmune diseases are T-cell–mediated diseases in which cytokines and chemokines play an essential role.

- The standard drug treatment for autoimmune diseases involves glucocorticoids, which prevent the transcription and hence the synthesis of cytokines.

Section 31.9 How Does the Human Immunodeficiency Virus Cause AIDS?

- HIV is a retrovirus that enters helper T cells.

- The virus weakens the immune system by destroying these helper T cells through damage to their cell membranes and activation of enzymes that cause apoptosis.

- HIV has been studied for 25 years in an attempt to find a cure, but none has been discovered as yet. The virus hides from the host's immune system and mutates so frequently that no effective antibody response can be mounted.

- A combination of therapies combining enzyme inhibitors and antibodies has achieved the most success.

Problems

⏻WL Interactive versions of these problems may be assigned in OWL.

Orange-numbered problems are applied.

Section 31.1 How Does the Body Defend Itself from Invasion?

31.1 Give two examples of external innate immunity in humans.

31.2 Which form of immunity is characteristic of vertebrates only?

31.3 How does the skin fight bacterial invasion?

31.4 T-cell receptors and MHC molecules both interact with antigens. What is the difference in the mode of interaction between the two?

31.5 What differentiates innate immunity from adaptive (acquired) immunity?

Section 31.2 What Organs and Cells Make Up the Immune System?

31.6 Where in the body do you find the largest concentration of antibodies, as well as T cells?

31.7 Where do T and B cells mature and differentiate?

31.8 What are memory cells? What is their function?

31.9 What are the favorite targets of macrophages? How do they kill the target cells?

Section 31.3 How Do Antigens Stimulate the Immune System?

31.10 Would a foreign substance such as aspirin (MW 180) be considered an antigen by the body?

31.11 What kind of antigen does a T cell recognize?

31.12 What is the smallest unit of an antigen that is capable of binding to an antibody?

31.13 How does the body process antigens to be recognized by class II MHC?

31.14 What role do MHC molecules play in the immune response of the ABO blood groups?

31.15 To which class of compounds do MHCs belong? Where would you find them?

31.16 What is the difference in function between class I and class II MHC molecules?

Section 31.4 What are Immunoglobulins?

31.17 When a foreign substance is injected in a rabbit, how long does it take to find antibodies against the foreign substance in the rabbit serum?

31.18 Distinguish among the roles of the IgA, IgE, and IgG immunoglobulins.

31.19 (a) Which immunoglobulin has the highest carbohydrate content and the lowest concentration in the serum?

(b) What is its main function?

31.20 Chemical Connections 20D states that the antigen in the red blood cells of a person with B-type blood is a galactose unit. Show schematically how the antibody of a person with A-type blood would aggregate the red blood cells of a B-type person if such a transfusion were made by mistake.

31.21 In the immunoglobulin structure, the "hinge region" joins the stem of the Y to the arms. The hinge region can be cleaved by a specific enzyme to yield one F_c fragment (the stem of the Y) and two F_{ab} fragments (the two arms). Which of these two kinds of fragments can interact with an antigen? Explain.

31.22 How are the light and heavy chains of an antibody held together?

31.23 What do we mean by the term *immunoglobulin superfamily?*

31.24 If you could isolate two monoclonal antibodies from a certain population of lymphocytes, in what sense would they be similar to each other and in what sense would they differ?

31.25 What kind of interaction takes place between an antigen and an antibody?

31.26 How is a new protein created on the variable portion of a heavy chain by V(J)D recombination?

31.27 What accounts for antibody diversity?

Section 31.5 What Are T Cells and T-Cell Receptors?

31.28 T-cell receptor molecules are made of two polypeptide chains. Which part of the chain acts as a binding site, and what binds to it?

31.29 What is the difference between a T-cell receptor (TcR) and a TcR complex?

31.30 What kind of tertiary structure characterizes the TcR?

31.31 What are the components of the TcR complex?

31.32 By what chemical process does CD3 transduct signals inside the cell?

31.33 Which adhesion molecule in the TcR complex helps HIV infect a leukocyte?

31.34 Three kinds of molecules in the T cell belong to the immunoglobulin superfamily. List them and briefly indicate their functions.

31.35 What functions do CD4 and CD8 serve in the immune response?

Section 31.6 How Is the Immune Response Controlled?

31.36 What kind of molecules are cytokines?

31.37 With what do cytokines interact? Do they bind to antigens?

31.38 As in most biochemistry literature, one encounters a veritable alphabet soup when reading about cytokines. Identify these cytokines by their full names: (a) TNF, (b) IL, and (c) EGF.

31.39 What are chemokines? How do they deliver their message?

31.40 What is the characteristic chemical signature in the structure of chemokines?

31.41 What are the chemical characteristics of cytokines that allow their classification?

31.42 Which amino acid appears in all chemokines?

Section 31.7 Immunization

31.43 What made Edward Jenner "the father of immunization"? In your opinion, could one legally do such an experiment today?

31.44 What observation led Edward Jenner to attempt his experiment?

31.45 What is the derivation of the word "vaccination"?

31.46 What are the three main types of vaccines?

31.47 Of the vaccine types listed for the answer to 31.46, which one would be the safest?

31.48 Which cell types of the immune system are involved in the body's response to immunization?

31.49 What is the importance of dendritic cells to immunization?

31.50 What are some diseases that have been largely eradicated by the use of vaccination?

31.51 What are some diseases where the use of vaccines has been less successful to date?

31.52 What are adjuvants?

31.53 In the early 20th century, scientists discovered that bacterial lipopolysaccharide (LPS) acted as a booster for vaccines. Why did LPS work?

31.54 How are toll-like receptors involved in the immune response?

Section 31.8 How Does the Body Distinguish "Self" from "Nonself"?

31.55 How does the body prevent T cells from being active against a "self" antigen?

31.56 What makes a tumor cell different from a normal cell?

31.57 Name a signaling pathway that controls the maturation of B cells and prevents those with an affinity for self antigen from becoming active.

31.58 How does the inhibitory receptor on a macrophage prevent an attack on normal cells?

31.59 Which components of the immune system are principally involved in autoimmune diseases?

31.60 How do glucocorticoids make individuals with autoimmune disease feel more comfortable?

Section 31.9 How Does the Human Immunodeficiency Virus Cause AIDS?

31.61 Which cells are attacked by HIV?

31.62 How does HIV gain entry into the cells it attacks?

31.63 How does HIV confound the human immune system?

31.64 What types of therapy are used to fight AIDS?

31.65 Why have vaccines been relatively unsuccessful in stopping AIDS?

31.66 What are the structural features of the two types of neutralizing antibodies that have been the most successful at combating AIDS? What makes these antibodies more effective?

31.67 How is the development of "two-in-one" antibodies potentially significant to AIDS research?

Chemical Connections

31.68 (Chemical Connections 31A) What is contributing to the higher survival rate of women with breast cancer?

31.69 (Chemical Connections 31A) Why are monoclonal antibodies a good choice for a weapon against breast cancer?

31.70 (Chemical Connections 31A) Explain a situation in which a monoclonal antibody would be superior to a polyclonal antibody as a cancer drug.

31.71 (Chemical Connections 31A) What type of evidence suggests that the HER2 protein is important in many breast cancers?

31.72 (Chemical Connections 31A) How are monoclonal antibodies used to fight cancer?

31.73 (Chemical Connections 31A) What do tyrosine kinases have to do with cancer?

31.74 (Chemical Connections 31B) What happens when a person takes an antibiotic he or she is allergic to?

31.75 (Chemical Connections 31B) Why are allergies to antibiotics dangerous?

31.76 (Chemical Connections 31B) What do we mean by the term "indiscriminate use of antibiotics"?

31.77 (Chemical Connections 31B) Why has the sexually transmitted disease (STD) gonorrhea benefited from indiscriminate use of antibiotics?

31.78 (Chemical Connections 31B) What are the downsides to the use of amoxicillin to combat earaches in children?

31.79 (Chemical Connections 31B) Why can strep throat be a serious condition apart from the problems directly associated with the sore throat?

31.80 (Chemical Connections 31C) What are the different types of stem cells?

31.81 (Chemical Connections 31C) Why are stem cells special? Why do scientists think they offer so much hope?

31.82 (Chemical Connections 31C) What are epigenetic states? Why do scientists want to be able to manipulate the epigenetic state of stem cells?

31.83 (Chemical Connections 31D) Why can the flu be considered more dangerous than AIDS?

31.84 (Chemical Connections 31D) What are the two major proteins that we study on a flu virus?

31.85 (Chemical Connections 31D) Is it a correct statement that H1N1 is less dangerous than H5N1?

31.86 (Chemical Connections 31D) How can flu viruses change? What worries the CDC and WHO with respect to flu viruses?

31.87 (Chemical Connections 31E) What is special about the recently made antibodies to the hemagglutinin portion of the flu virus?

31.88 (Chemical Connections 31E) How did the studies on mice using antibodies against the flu virus demonstrate its efficacy?

Additional Problems

31.89 Which immunoglobulins form the first line of defense against invading bacteria?

31.90 Which cells of the innate immunity system are the first to interact with an invading pathogen?

31.91 Which compound or complex of compounds of the immune system is mostly responsible for the proliferation of leukocytes?

31.92 Name a process beside V(J)D recombination that can enhance immunoglobulin diversity in the variable region.

31.93 Name a tumor cell marker, a synthetic analog of which may be the first anticancer vaccine.

31.94 Is the light chain of an immunoglobulin the same as the V region?

31.95 Where are TNF receptors located?

31.96 The variable regions of immunoglobulins bind antigens. How many polypeptide chains carry variable regions in one immunoglobulin molecule?

Appendix I

Exponential Notation

Exponential notation is also called scientific notation.

For example, 10^6 means a one followed by six zeros, or 1,000,000, and 10^2 means 100.

The **exponential notation** system is based on powers of 10 (see table). For example, if we multiply $10 \times 10 \times 10 = 1000$, we express this as 10^3. The 3 in this expression is called the **exponent** or the **power**, and it indicates how many times we multiplied 10 by itself and how many zeros follow the 1.

There are also negative powers of 10. For example, 10^{-3} means 1 divided by 10^3:

$$10^{-3} = \frac{1}{10^3} = \frac{1}{1000} = 0.001$$

Numbers are frequently expressed like this: 6.4×10^3. In a number of this type, 6.4 is the **coefficient** and 3 is the exponent, or power of 10. This number means exactly what it says:

$$6.4 \times 10^3 = 6.4 \times 1000 = 6400$$

Similarly, we can have coefficients with negative exponents:

$$2.7 \times 10^{-5} = 2.7 \times \frac{1}{10^5} = 2.7 \times 0.00001 = 0.000027$$

For numbers greater than 10 in exponential notation, we proceed as follows: *Move the decimal point to the left,* to just after the first digit. The (positive) exponent is equal to the number of places we moved the decimal point.

APP. 1.1 Examples of Exponential Notation

$$10,000 = 10^4$$
$$1000 = 10^3$$
$$100 = 10^2$$
$$10 = 10^1$$
$$1 = 10^0$$
$$0.1 = 10^{-1}$$
$$0.01 = 10^{-2}$$
$$0.001 = 10^{-3}$$

Example

$$3\,7\,5\,0\,0 = 3.75 \times 10^4 \quad \text{4 because we went four places to the left}$$

Four places to the left   Coefficient

$$628 = 6.28 \times 10^2$$

Two places to the left   Coefficient

$$859{,}600{,}000{,}000 = 8.596 \times 10^{11}$$

Eleven places to the left   Coefficient

We don't really have to place the decimal point after the first digit, but by doing so we get a coefficient between 1 and 10, and that is the custom.

Using exponential notation, we can say that there are 2.95×10^{22} copper atoms in a copper penny. For large numbers, the exponent is always *positive*. Note that we do not usually write out the zeros at the end of the number.

For small numbers (less than 1), we move the decimal point *to the right,* to just after the first nonzero digit, and use a *negative exponent.*

Example

$$0.00346 = 3.46 \times 10^{-3}$$

Three places to
the right

$$0.000004213 = 4.213 \times 10^{-6}$$

Six places to
the right

In exponential notation, a copper atom weighs 2.3×10^{-25} pounds.

To convert exponential notation into fully written-out numbers, we do the same thing backward.

Example

Write out in full: (a) 8.16×10^{7} (b) 3.44×10^{-4}

Solution

(a) $8.16 \times 10^{7} = 81,600,000$

Seven places to the right
(add enough zeros)

(b) $3.44 \times 10^{-4} = 0.000344$

Four places to the left

When scientists add, subtract, multiply, and divide, they are always careful to express their answers with the proper number of digits, called significant figures. This method is described in Appendix II.

A. Adding and Subtracting Numbers in Exponential Notation

We are allowed to add or subtract numbers expressed in exponential notation *only if they have the same exponent.* All we do is add or subtract the coefficients and leave the exponent as it is.

Example

Add 3.6×10^{-3} and 9.1×10^{-3}.

Solution

$$\begin{array}{r} 3.6 \times 10^{-3} \\ + \ 9.1 \times 10^{-3} \\ \hline 12.7 \times 10^{-3} \end{array}$$

The answer could also be written in other, equally valid ways:

$$12.7 \times 10^{-3} = 0.0127 = 1.27 \times 10^{-2}$$

A calculator with exponential notation changes the exponent automatically.

When it is necessary to add or subtract two numbers that have different exponents, we first must change them so that the exponents are the same.

Example

Add 1.95×10^{-2} and 2.8×10^{-3}.

Solution

To add these two numbers, we make both exponents -2. Thus, $2.8 \times 10^{-3} = 0.28 \times 10^{-2}$. Now we can add:

$$
\begin{aligned}
1.95 &\times 10^{-2} \\
+\ 0.28 &\times 10^{-2} \\
\hline
2.23 &\times 10^{-2}
\end{aligned}
$$

B. Multiplying and Dividing Numbers in Exponential Notation

To multiply numbers in exponential notation, we first multiply the coefficients in the usual way and then algebraically *add* the exponents.

Example

Multiply 7.40×10^5 by 3.12×10^9.

Solution

$$7.40 \times 3.12 = 23.1$$

Add exponents:

$$10^5 \times 10^9 = 10^{5+9} = 10^{14}$$

Answer:

$$23.1 \times 10^{14} = 2.31 \times 10^{15}$$

Example

Multiply 4.6×10^{-7} by 9.2×10^4

Solution

$$4.6 \times 9.2 = 42$$

Add exponents:

$$10^{-7} \times 10^4 = 10^{-7+4} = 10^{-3}$$

Answer:

$$42 \times 10^{-3} = 4.2 \times 10^{-2}$$

To divide numbers expressed in exponential notation, the process is reversed. We first divide the coefficients and then algebraically *subtract* the exponents.

Example

Divide: $\dfrac{6.4 \times 10^8}{2.57 \times 10^{10}}$

Solution

$$6.4 \div 2.57 = 2.5$$

Subtract exponents:

$$10^8 \div 10^{10} = 10^{8-10} = 10^{-2}$$

Answer:

$$2.5 \times 10^{-2}$$

Example

Divide: $\dfrac{1.62 \times 10^{-4}}{7.94 \times 10^7}$

Solution

$$1.62 \div 7.94 = 0.204$$

Subtract exponents:

$$10^{-4} \div 10^7 = 10^{-4-7} = 10^{-11}$$

Answer:

$$0.204 \times 10^{-11} = 2.04 \times 10^{-12}$$

Scientific calculators do these calculations automatically. All that is necessary is to enter the first number, press $+$, $-$, $\times$, or $\div$, enter the second number, and press $=$. (The method for entering numbers of this form varies; consult the instructions that come with the calculator.) Many scientific calculators also have a key that will automatically convert a number such as 0.00047 to its scientific notation form (4.7×10^{-4}), and vice versa. For problems relating to exponential notation, see Chapter 1, Problems 1.17 through 1.24.

Significant Figures

If you measure the volume of a liquid in a graduated cylinder, you might find that it is 36 mL, to the nearest milliliter, but you cannot tell if it is 36.2, or 35.6, or 36.0 mL because this measuring instrument does not give the last digit with any certainty. A buret gives more digits, and if you use one you should be able to say, for instance, that the volume is 36.3 mL and not 36.4 mL. But even with a buret, you could not say whether the volume is 36.32 or 36.33 mL. For that, you would need an instrument that gives still more digits. This example should show you that *no measured number can ever be known exactly.* No matter how good the measuring instrument, there is always a limit to the number of digits it can measure with certainty.

We define the number of **significant figures** as the number of digits of a measured number that have uncertainty only in the last digit.

What do we mean by this definition? Assume that you are weighing a small object on a laboratory balance that can weigh to the nearest 0.1 g, and you find that the object weighs 16 g. Because the balance weighs to the nearest 0.1 g, you can be sure that the object does not weigh 16.1 g or 15.9 g. In this case, you would write the weight as 16.0 g. To a scientist, there is a difference between 16 g and 16.0 g. Writing 16 g says that you don't know the digit after the 6. Writing 16.0 g says that you do know it: It is 0. However, you don't know the digit after that. Several rules govern the use of significant figures in reporting measured numbers.

A. Determining the Number of Significant Figures

In Section 1.3, we saw how to determine the number of significant figures in a reported number. We summarize those guidelines here:

1. Nonzero digits are always significant.
2. Zeros at the beginning of a number are never significant.
3. Zeros between nonzero digits are always significant.
4. Zeros at the end of a number that contains a decimal point are always significant.
5. Zeros at the end of a number that contains no decimal point may or may not be significant.

We use periods as decimal points throughout this text to indicate the significant figures in numbers with trailing zeros. For example, 1000. mL has four significant figures; 20. m has two significant figures.

B. Multiplying and Dividing

The rule in multiplication and division is that the final answer should have the *same* number of significant figures as there are in the number with the *fewest* significant figures.

Example

Do the following multiplications and divisions:

(a) 3.6×4.27

(b) 0.004×217.38

(c) $\dfrac{42.1}{3.695}$

(d) $\dfrac{0.30652 \times 138}{2.1}$

Solution

(a) 15 (3.6 has two significant figures)

(b) 0.9 (0.004 has one significant figure)

(c) 11.4 (42.1 has three significant figures)

(d) 2.0×10^1 (2.1 has two significant figures)

C. Adding and Subtracting

In addition and subtraction, the rule is completely different. The number of significant figures in each number doesn't matter. The answer is given to the *same number of decimal places* as the term with the fewest decimal places.

Example

Add or subtract:

(a) 320.084
    80.47
  200.23
    20.0
  620.8

(b) 61.4532
  13.7
  22
   0.003
  97

(c) 14.26
  -1.05041
  13.21

Solution

In each case, we add or subtract in the normal way but then round off so that the only digits that appear in the answer are those in the columns in which every digit is significant.

D. Rounding Off

When we have too many significant figures in our answer, it is necessary to round off. In this book we have used the rule that if *the first digit dropped* is 5, 6, 7, 8, or 9, we raise *the last digit kept* to the next number; otherwise, we do not.

Example

In each case, drop the last two digits:

(a) 33.679   (b) 2.4715   (c) 1.1145   (d) 0.001309   (e) 3.52

Solution

(a) $33.679 = 33.7$

(b) $2.4715 = 2.47$

(c) $1.1145 = 1.11$

(d) $0.001309 = 0.0013$

(e) $3.52 = 4$

E. Counted or Defined Numbers

All of the preceding rules apply to *measured* numbers and **not** to any numbers that are *counted* or *defined*. Counted and defined numbers are known exactly. For example, a triangle is defined as having 3 sides, not 3.1 or 2.9. Here, we treat the number 3 as if it has an infinite number of zeros following the decimal point.

Example

Multiply 53.692 (a measured number) $\times$ 6 (a counted number).

Solution

$$322.15$$

Because 6 is a counted number, we know it exactly, and 53.692 is the number with the fewest significant figures. All we really are doing is adding 53.692 six times.

For problems relating to significant figures, see Chapter 1, Problems 1.25 to 1.30.

Answers

Chapter 1 Matter, Energy, and Measurement

1.1 multiplication (a) 4.69×10^5 (b) 2.8×10^{-15}; division (a) 2.00×10^{18} (b) 1.37×10^5

1.2 (a) 147°F (b) 8.3°C

1.3 13.8 km

1.4 743 mph

1.5 1.3 mg/min

1.6 78.5 g

1.7 2.43 g/mL

1.8 1.016 g/mL

1.9 4.8×10^3 cal = 48 kcal

1.10 46°C

1.11 0.0430 cal/g · deg

1.13 (a) Matter is anything that has mass and takes up space. (b) Chemistry is the science that studies matter.

1.15 Dr. X's claim that the extract cured diabetes would be classified as (c) a hypothesis. No evidence had been provided to prove or disprove the claim.

1.17 (a) 3.51×10^{-1} (b) 6.021×10^2 (c) 1.28×10^{-4} (d) 6.28122×10^5

1.19 (a) 6.65×10^{17} (b) 1.2×10^1 (c) 3.9×10^{-16} (d) 3.5×10^{-23}

1.21 (a) 1.3×10^5 (b) 9.40×10^4 (c) 5.139×10^{-3}

1.23 4.45×10^6

1.25 (a) 2 (b) 5 (c) 5 (d) 5 (e) ambiguous, better to write as 3.21×10^4 (three significant figures) or 32100. (five significant figures) (f) 3 (g) 2

1.27 (a) 92 (b) 7.3 (c) 0.68 (d) 0.0032 (e) 5.9

1.29 (a) 1.53 (b) 2.2 (c) 0.00048

1.31 330 min = 5.6 h

1.33 (a) 20 mm (b) 1 inch (c) 1 mile

1.35 Weight would change slightly. Mass is independent of location, but weight is a force exerted by a body influenced by gravity. The influence of the Earth's gravity decreases with increasing distance from sea level.

1.37 (a) 77°F, 298 K (b) 104°F, 313 K (c) 482°F, 523 K, (d) −459°F, 0 K

1.39 (a) 0.0964 L (b) 27.5 cm (c) 4.57×10^4 g (d) 4.75 m (e) 21.64 mL (f) 3.29×10^3 cc (g) 44 mL (h) 0.711 kg (i) 63.7 cc (j) 7.3×10^4 mg (k) 8.34×10^4 mm (l) 0.361 g

1.41 512 fl oz.

1.43 50 mi/h

1.45 4 tablets

1.47 16 mg

1.49 42 cc/h

1.51 420 min

1.53 solids and liquids

1.55 No, melting is a physical change.

1.57 bottom: manganese; top: sodium acetate; middle: calcium chloride

1.59 0.8 mL

1.61 water

1.63 One should raise the temperature of water to 4°C. During this temperature change, the density of the crystals decreases, while the density of water increases. This brings the less dense crystals to the surface of the more dense water.

1.65 The motion of the wheels of the car generates kinetic energy, which is stored in your battery as potential energy.

1.67 0.34 cal/g · C°

1.69 334 mg

1.71 The body shivers. Further temperature lowering results in unconsciousness and then death.

1.73 Methanol, because its higher specific heat allows it to retain the heat longer.

1.75 0.732

1.77 kinetic: (b), (d), (e); potential: (a), (c)

1.79 the European car

1.81 kinetic energy

1.83 The largest is 41 g. The smallest is 4.1310×10^{-8} kg.

1.85 10.9 h

1.87 The heavy water. When converting the specific heat given in J/g · °C to cal/g · °C, one finds that the specific heat of heavy water is 1.008 cal/g · °C, which is somewhat greater than that of ordinary water.

1.89 (a) 1.57 g/mL (b) 1.25 g/mL

1.91 two

1.93 60 J would raise the temperature by 4.5°C; thus, the final temperature will be 24.5°C.

1.95 Number (b), 4.38, has three significant figures. Number (a), 0.00000001, has only one significant figure. The zeros merely indicate the location of the decimal point.

1.97 To do this calculation, you need a conversion factor from kilometers to miles. Table 1.3 gives 1 mile = 1.609 km.

$$95 \text{ km} \times \frac{1 \text{ mile}}{1.609 \text{ km}} \sim 59 \text{ km}$$

If you use the other possible conversion factor:

$$95 \text{ km} \times \frac{1.609 \text{ km}}{\text{mi}} \sim \frac{153 \text{ km}^2}{\text{mi}}$$

Both the numbers and the units are incorrect.

1.99 In photosynthesis, the radiant energy of sunlight is converted to chemical energy in the sugars produced.

1.101 Converting 30°C from the Celsius to Fahrenheit temperature scales gives 86°F. You are most likely to be wearing a T-shirt and shorts.

1.103 Cells that have been exposed to several cycles of freezing and thawing will have expanded quite a bit. The expansion process tends to break open the cells to make their contents available for fractionation and further study.

1.105 We use the specific heat of water and the information that a liter of water weighs 1000. grams.

$$\text{Amount of heat} = \text{SH} \times \text{m} \times (T_2 - T_1)$$

$$\text{Amount of heat} = \frac{1.00 \text{ cal}}{\cancel{g}\cancel{C}} \times 2.000 \cancel{L} \times \frac{\cancel{1000. \text{ grams}}}{\cancel{L}} \times 4.85\cancel{C}$$

$$\text{Amount of heat} = 9.70 \times 10^3 \text{ calories}$$

1.107 Determining the amount of substance and its effectiveness can be done together. You separate the components of the original material and, in the process, determine its amount. One possible way is to weigh the amounts of recovered material. You would then test the substance to see whether the individual compound produces the predicted results.

1.109 4.85×10^3 calories

1.111 (a) 20. mL (b) No; 5 gtts/min

Chapter 2 Atoms

2.1 (a) $NaClO_3$ (b) AlF_3

2.2 (a) The mass number is $15 + 16 = 31$.

(b) The mass number is $86 + 136 = 222$.

2.3 (a) The element is phosphorus (P); its symbol is $^{31}_{15}\text{P}$.

(b) The element is radon (Rn); its symbol is $^{222}_{86}\text{Rn}$.

2.4 (a) The atomic number of mercury (Hg) is 80; that of lead (Pb) is 82.

(b) An atom of Hg has 80 protons; an atom of Pb has 82 protons.

(c) The mass number of this isotope of Hg is 200; the mass number of this isotope of Pb is 202.

(d) The symbols of these isotopes are $^{200}_{80}\text{Hg}$ and $^{202}_{82}\text{Pb}$.

2.5 The atomic number of iodine (I) is 53. The number of neutrons in each isotope is 72 for iodine-125 and 78 for iodine-131. The symbols for these two isotopes are $^{125}_{53}\text{I}$ and $^{131}_{53}\text{I}$, respectively.

2.6 Lithium-7 is the more abundant isotope (92.50%). The natural abundance of lithium-6 is 7.50%.

2.7 The element is aluminum (Al). Its Lewis dot structure is:

$$\dot{\text{Al}}:$$

2.9 (a) F (b) T (c) T (d) T (e) F (f) T
(g) T (h) T (i) F (j) F (k) T (l) F

(a) False: Matter is divided into pure substances and mixtures.

(e) False: Mixtures can be separated into their component pure substances.

(i) False: Technetium, promethium, and all of the elements beyond uranium are man-made.

(j) False: H, O, C, N, Ca, and P are the six most important elements in the human body.

(l) False: The combining ratio is based on the ratio of atoms, not the ratio of masses.

2.11 (a) Oxygen (b) Lead (c) Calcium
(d) Sodium (e) Carbon (f) Titanium (g) Sulfur
(h) Iron (i) Hydrogen (j) Potassium (k) Silver (l) Gold

2.13 (a) Americium (b) Berkelium (c) Californium
(d) Dubnium (e) Europium (f) Francium
(g) Gallium (h) Germanium (i) Hafnium
(j) Hassium (k) Holmium (l) Lutetium
(m) Magnesium (n) Polonium (o) Rhenium
(p) Ruthenium (q) Scandium (r) Strontium
(s) Ytterbium, Yttrium, Terbium (t) Thulium

2.15 (a) K_2O (b) Na_3PO_4 (c) $Li\,NO_3$

2.17 (a) The law of conservation of mass states that matter can be neither created nor destroyed. Dalton's theory explains

this because if all matter is made up of indestructible atoms, then any chemical reaction just changes the attachments between atoms and does not destroy the atoms themselves.

(b) The law of constant composition states that any compound is always made up of elements in the same proportion by mass. Dalton's theory explains this because molecules consist of tightly bound groups of atoms, each of which has a particular mass. Therefore, each element in a compound always constitutes a fixed proportion of the total mass.

2.19 No. CO and CO_2 are different compounds, and each obeys the law of constant composition for that compound.

2.21 (a) F (b) T (c) T (d) F (e) T (f) T (g) T (h) T
(i) F (j) F (k) T (l) F (m) T (n) F (o) T (p) T (q) T
(r) F (s) T (t) F

(a) Electrons and protons have equal but opposite charges. Electrons have a much lighter mass than protons.

(d) False: 1 amu has a mass of 1.6605×10^{-24} grams.

(i) False: Electrons and protons have opposite charges and attract each other.

(j) False: The size of an atom includes the space occupied by its electrons. The nucleus is a small fraction of the size of an atom.

(l) False: The mass number is the number of protons and neutrons.

(n) False: ^1H has no neutrons. ^2H has one neutron, and ^3H has two neutrons.

(r) False: Atomic weights are averages of the known isotopes.

(t) False: Density is mass/volume.

2.23 The statement is true in the sense that the number of protons (the atomic number) determines the identity of the element.

2.25 (a) The element with 22 protons is titanium (Ti).

(b) The element with 76 protons is osmium (Os).

(c) The element with 34 protons is selenium (Se).

(d) The element with 94 protons is plutonium (Pu).

2.27 Each would still be the same element, because the number of protons has not changed.

2.29 Radon (Rn) has an atomic number of 86, so each isotope has 86 protons. The number of neutrons is mass number − atomic number.

(a) Radon-210 has $210 - 86 = 124$ neutrons

(b) Radon-218 has $218 - 86 = 132$ neutrons

(c) Radon-222 has $222 - 86 = 136$ neutrons

2.31 Two more neutrons: tin-120

Three more neutrons: tin-121

Six more neutrons: tin-124

2.33 (a) An ion is an atom or a group of bonded atoms with an unequal number of protons and electrons.

(b) Isotopes are atoms of the same element with the same number of protons in their nuclei but a different number of neutrons.

2.35 Rounded to four significant figures, the calculated value is 12.01 amu; the value given in the Periodic Table is 12.011 amu.

$$\frac{98.90}{100} \times 12.00 + \frac{1.10}{100} \times 13.003 = 12.01 \text{ amu}$$

2.37 Carbon-11 has 6 protons, 6 electrons, and 5 neutrons.

2.39 Americium-241 (Am) has an atomic number of 95. This isotope has 95 protons, 95 electrons, and $241 - 95 = 146$ neutrons.

2.41 (a) T (b) F (c) F (d) F (e) F (f) T (g) T
(h) T (i) T

(b) False: The main group elements go from group 1A through 8A.

(c) False: Very roughly, the nonmetals exist in a diagonal starting in the lower right corner, moving up to the middle of the Periodic Table. Nonmetals exist above the diagonal and metals below it.

(d) False: There are more metals than nonmetals.

(e) False: Horizontal rows are called periods.

2.43 (a) Groups 2A, 3B, 4B, 5B, 6B, 7B, 8B, 1B, and 2B contain only metals. Note that Group 1A contains one nonmetal, hydrogen.

(b) No groups contain only metalloids.

(c) Only Groups 7A and 8A contain only nonmetals.

2.45 Elements in the same group of the Periodic Table should have similar properties.

N, P, and As   I and F   Ne and He   Mg, Ca, and Ba   K and Li

2.47 (a) Aluminum > silicon   (b) Arsenic > phosphorus

(c) Gallium > germanium   (d) Gallium > aluminum

2.49 (a) T (b) T (c) T (d) F (e) T (f) F (g) T

(h) T (i) F (j) T (k) T (l) T (m) T (n) T (o) F

(p) F (q) T (r) T (s) T (t) F

(d) False: Principal energy level 1 can contain a maximum of 2 electrons, principal energy level 2 can contain a maximum of 8 electrons, principal energy level 3 can contain a maximum of 18 electrons, and principal energy level 4 can contain a maximum of 32 electrons.

(f) False: A $2s$ electron is easier to remove than a $1s$ electron because it is further from the influence of the positively charged nucleus.

(i) False: The three $2p$ orbitals are at right angles to each other.

(o) False: Paired electrons have spins in opposite directions.

(p) False: Each box represents an orbital, and each orbital can accommodate two electrons. When an orbital is completely filled, one electron in the pair must spin in the opposite direction from the other.

(t) False: Group 6A elements have six electrons in their valence shells, only two of which are unpaired.

2.51 The group number tells the number of electrons in the valence shell of an element in the group.

2.53 (a) Li(3): $1s^2 2s^1$ (b) Ne(10): $1s^2 2s^2 2p^6$ (c) Be(4): $1s^2 2s^2$

(d) C(6): $1s^2 2s^2 2p^2$ (d) Mg(12): $1s^2 2s^2 2p^6 3s^2$

2.55 (a) He(2): $1s^2$ (b) Na(11): $1s^2 2s^2 2p^6 3s^1$

(c) Cl(17): $1s^2 2s^2 2p^6 3s^2 3p^5$ (d) P(15): $1s^2 2s^2 2p^6 3s^2 3p^3$

(e) H(1): $1s^1$

2.57 In (a), (b), and (c), the outer shell electron configurations are the same. The only difference is the number of the valence shell being filled.

2.59 The element might be in Group 2A, all of which have two valence electrons. It might also be helium, in Group 8A.

2.61 (a) T (b) T (c) T (d) F (e) F (f) T (g) F (h) T

(d) False: Helium, a group 8A element, has only two valence electrons.

(e) False: The period number has nothing to do with the number of valence electrons.

(g) False: Period 3 has eight elements.

2.63 (a) T (b) T (c) T (d) T (e) F (g) T

(e) False: Ionization energy decreases going from top to bottom within a column of the Periodic Table.

2.65 (a) Fact: The atomic radius of an anion is always larger than that of the atom from which it is derived. For anions, the nuclear charge is unchanged, but an electron added to the valence shell introduces new repulsions and the electron cloud swells because of the increased electron-to-electron repulsions.

(b) Fact: The atomic radius of a cation is always smaller than that of the atom from which it is derived. When an electron is removed from an atom, the nuclear charge remains the same but there are fewer electrons repelling each other. Consequently, the positive nucleus attracts the remaining elections more strongly, causing the electron cloud to contract.

2.67 Here are ground state electron configurations for each O, O^+, and N, N^+.

One of these
electrons is lost

$$O \; 1s^2 \, 2s^2 \, 2p_x^2 \, 2p_y^1 \, 2p_z^1 \longrightarrow O^+ \; 1s^2 \, 2s^2 \, 2p_x^1 \, 2p_y^1 \, 2p_z^1 + e^-$$

This electron is lost

$$N \; 1s^2 \, 2s^2 \, 2p_x^1 \, 2p_y^1 \, 2p_z^1 \longrightarrow O^+ \; 1s^2 \, 2s^2 \, 2p_x^1 \, 2p_y^1 \qquad + e^-$$

The electron removed from O is one of the paired electrons in the doubly occupied $2p_x$ orbital, whereas the electron removed from N is an electron from the singly occupied $2p_z$ orbital. There is some repulsion between the two paired electrons in the case of oxygen, which means that it is easier to remove an electron from O than it is to remove an electron from the singly occupied $2p_z$ orbital for nitrogen.

2.69 Sulfur and iron are essential components of proteins, and calcium is a major component of bones and teeth.

2.71 Calcium is an essential element in human bones and teeth. Because strontium behaves chemically much like calcium, strontium-90 gets into our bones and teeth and gives off radioactivity directly into our bodies for many years.

2.73 Copper can be hardened by hammering it.

2.75 (a) Metals (b) Nonmetals (c) Metals

(d) Nonmetals (e) Metals (f) Metals

2.77 (a) The largest atomic radius in Group 2A is radium (Ra).

(b) The smallest atomic radius in Group 2A is beryllium (Be).

(c) The largest atomic radius in the second Period is Neon (Ne).

(d) The smallest atomic radius in the second Period is Lithium (Li).

(e) The largest ionization energy in Group 7A is fluorine (F).

(f) The lowest ionization energy in Group 7A is astatine (At).

2.79

	Atomic Number	Protons	Neutrons	Electrons
(a) Phosphorus-32	15	15	17	15
(b) Molybdenum-98	42	42	56	42
(c) Calcium-44	20	20	24	20
(d) Hydrogen-3	1	1	2	1
(e) Gadolinium-158	64	64	94	64
(f) Bismuth 212	83	83	129	83

2.81 Isotopes of elements 37 to 53 contain more neutrons than protons. Here are some elements and the percent of their mass contributed by neutrons:

(a) Carbon-12 has 6 protons and 6 neutrons. Neutrons make up 50% of its mass.

(b) Calcium-40 has 20 protons and 20 neutrons. Neutrons make up 50% of its mass.

(c) Iron-55 has 26 protons and 29 neutrons. Neutrons make up 53% of its mass.

(d) Bromine-79 has 35 protons and 44 neutrons. Neutrons make up 56% of its mass.

(e) Platinum-195 has 78 protons and 117 neutrons. Neutrons make up 60% of its mass.

(f) Uranium-238 has 92 protons and 146 neutrons. Neutrons make up 61% of its mass.

2.83 Rounded to three significant figures, the calculated value from this information is 10.8 amu. The value given in the Periodic Table is 10.81.

$$\frac{19.9}{100} \times 10.013 + \frac{80.1}{100} \times 11.009 = 10.8 \text{ amu}$$

$$1.992 \qquad\qquad 8.818$$

2.85 The dimensions of the solution are protons per grain. Using the concept of dimensional analysis developed in Chapter 1, arrange the calculation so that these dimensions result.

$$1.0 \times 10^{-2}\frac{\text{grams}}{\text{grain}} \times \frac{1 \text{ proton}}{1.67 \times 10^{-24} \text{ grams}}$$

$$= 6.0 \times 10^{21} \text{ protons/grain}$$

2.87 The atomic number of this element is 54, which means that the element is xenon (Xe). This isotope of xenon has 54 protons, 54 electrons, and $131 - 54 = 77$ neutrons.

2.89 (a) Ionization energy generally decreases down a column in the Periodic Table, so the ionization energy of element 117 should be less than that of At (85).

(b) Ionization energy generally increases from left to right across a row in the Periodic Table, so the ionization energy of element 117 should be greater than that of radium Ra (88).

2.91 The key is to consider the electron configuration of a lithium atom and a lithium ion. Lithium has an atomic number of 3.

$$\text{Li}(3)\ 1s^2 2s^1 \longrightarrow \text{Li}^+ + e^-$$
$$\text{Li}^+\ 1s^2 \longrightarrow \text{Li}^{2+} + e^-$$

In forming lithium ion, Li^+, a lithium atom loses one electron from its valence shell, that is, from its $2s$ orbital. In forming Li^{2+} ion, a lithium ion now loses an electron from its $1s$ orbital. A $1s$ orbital is smaller than a $2s$ orbital, and electrons in a $1s$ orbital are held more tightly than those in a $2s$ orbital. $1s$ electrons are held more tightly because they are closer to the positively charged nucleus. Therefore, the second ionization energy is considerably larger than the first ionization energy.

2.93 Atomic size decreases in going across a period and increases in going down a column. In order of increasing atomic size, these elements are:

$$\text{C}(77) < \text{B}(83) < \text{Al}(143) < \text{Na}(186)$$

2.95 The group 3A elements are boron, aluminum, gallium, indium, and thallium. The characteristic that is common for all is that each has three valence electrons.

2.97 Ca^{3+}; a charge of +3 is too large for a small ion.

2.99 Element 118 will fall in Group 8A. Predict that it will have the chemical properties of the other elements in this family and that it will have a melting point higher than $-71°\text{C}$ and a boiling point above $-62°\text{C}$; in other words, it will be a gas at room temperature.

Chapter 3 Chemical Bonds

3.1 By losing two electrons, a Mg atom becomes Mg^{2+} and acquires a complete octet. By gaining two electrons, a sulfur atom becomes a sulfide ion, S^{2-}, with an eight-valence electron configuration the same as that of argon.

(a) Mg (12 electrons): $1s^2 2s^2 2p^6 3s^2 \longrightarrow \text{Mg}^{2+}$ (10 electrons): $1s^2 2s^2 2p^6 + 2e^-$

(b) S (16 electrons): $1s^2 2s^2 2p^6 3s^2 3p^4 + 2e^- \longrightarrow \text{S}^{2-}$ (18 electrons): $1s^2 2s^2 2p^6 3s^2 3p^6$

3.2 Each pair of elements is in the same column of the Periodic Table, and electronegativity increases from bottom to top within a column. Therefore:

(a) Li > K (b) N > P (c) C > Si

3.3 (a) KCl (b) CaF_2 (c) Fe_2O_3

3.4 (a) Magnesium oxide (b) Barium iodide
(c) Potassium chloride

3.5 (a) MgCl_2 (b) Al_2O_3 (c) LiI

3.6 (a) Iron(II) oxide, ferrous oxide (b) Iron(III) oxide, ferric oxide

3.7 (a) Potassium hydrogen phosphate (b) Aluminum sulfate
(c) Iron(II) carbonate, ferrous carbonate

3.8 (a) S—H $(2.5 - 2.1 = 0.4)$; nonpolar covalent
(b) P—H $(2.1 - 2.1 = 0.0)$; nonpolar covalent
(c) C—F $(4.0 - 2.5 = 1.5)$; polar covalent
(d) C—Cl $(3.0 - 2.5 = 0.5)$; polar covalent

3.9 (a) $\overset{\delta+}{\text{C}}-\overset{\delta-}{\text{N}}$ (b) $\overset{\delta+}{\text{N}}-\overset{\delta-}{\text{O}}$ (c) $\overset{\delta+}{\text{C}}-\overset{\delta-}{\text{Cl}}$

3.10 (a) (b) (c) H—C≡N:

3.11 (a) (b)

4 single bonds

2 single bonds and 1 double bond

(c) (d) H—C≡C—H

2 double bonds

1 single bond and 1 triple bond

3.12 (a) Nitrogen dioxide (b) Phosphorus tribromide
(c) Sulfur dichloride (d) Boron trifluoride

3.13 (a) (b)

(c)

3.14 (a) A valid pair of contributing structures.
(b) Not a valid pair. The contributing structure on the right has 10 electrons in the valence shell of carbon and thus violates the octet rule. The valence shell of carbon consists of one $2s$ orbital and three $2p$ orbitals, which can hold a maximum of 8 valence electrons, hence the octet rule.

3.15 Given are three-dimensional structures showing all bonds and unshared electron pairs.

(a) (b) (c)

3.16 (a) H_2S; the difference in electronegativity between H and S is $2.5 - 2.1 = 0.4$. Therefore, H—S bonds are nonpolar and the molecule is nonpolar.

H—S̈—H

Nonpolar

(b) HCN contains a polar C—N bond and is a polar molecule.

H—C≡N:

Nonpolar

(c) C_2H_6 contains no polar bonds and is not a polar molecule.

(structure of ethane)

Nonpolar

3.17 (a) F (b) T (c) F (d) T (e) T (f) F (g) T (h) F (i) F
(a) False: It helps us understand the bonding patterns of the Group 1A-7A elements.
(c) False: Atoms that gain electrons become anions.
(f) False: Sodium typically forms a positive ion, a cation, by losing its single $3s$ electron.
(h) False: Phosphorus, sulfur, and chlorine can expand their valence shells to accommodate more than eight electrons.
(i) False: They couldn't be more different, starting with the most obvious difference—that of their charges.

$$\text{Li}: 1s^2\,2s^1 \longrightarrow \text{Li}^+\,1s^2 + e^-$$

3.19 (a) A lithium (3) atom has the electron configuration $1s^2 2s^1$. When a Li atom loses its single $2s$ electron, it forms lithium ion, Li^+, which has the electron configuration $1s^2$. This configuration is the same as that of helium, the noble gas nearest Li in atomic number.

$$\text{Li}: 1s^2\,2s^1 \longrightarrow \text{Li}^+\,1s^2 + e^-$$

(b) An oxygen (8) atom has the electron configuration $1s^2 2s^2 2p^4$. When an O atom gains two electrons, it forms O^{2-}, which has the electron configuration $1s^2 2s^2 2p^6$. This configuration is the same as that of neon, the noble gas nearest oxygen in atomic number.

$$\text{O}: 1s^2\,2s^2 2p^4 + 2e^- \longrightarrow O^{2-}: 1s^2 2s^2 2p^6 \text{ (complete octet)}$$

3.21 (a) Mg^{2+} (b) F^- (c) Al^{3+} (d) S^{2-} (e) K^+ (f) Br^-
3.23 The stable ions are: (a) I^-, (c) Na^+, and (d) S^{2-}.
3.25 Being intermediate in electronegativity, carbon and silicon are reluctant to accept electrons from a metal or lose electrons to a halogen to form ionic bonds. Instead, carbon and silicon share electrons in nonpolar covalent and polar covalent bonds.
3.27 (a) T (b) F (c) F (d) T (e) F (f) T (g) F (h) T (i) T (j) F (k) F (l) F (m) T (n) T
(b) False: H^+ is named hydrogen ion, or more commonly, a proton, because it is a nucleus consisting of a single proton. The hydronium ion is H_3O^+.
(c) False: H^+ has one proton and no neutrons.
(e) False: –ous refers to the ion with the lower charge; –ic refers to the ion with the higher charge.
(g) False: The anion derived from a bromine atom is named bromide ion.

(j) False: The prefix "bi" indicates the presence of a single hydrogen in this polyatomic ion.
(k) False: The hydrogen phosphate ion has a charge of -2, and the dihydrogen phosphate ion has a charge of -1.
(l) False: The numbers in the superscripts and subscripts are reversed. The phosphate ion is $PO_4{}^{3-}$.
3.29 (a) T (b) T (c) T (d) T (e) F (f) T (g) T (h) T (i) F (j) F (k) T (l) F (m) T (n) T (o) F
(e) False: Ionic bonds usually form between elements on the far left and the far right of the Periodic Table.
(i) False: Electronegativity is a periodic property.
(j) False: Electronegativity increases going from left to right in a period and decreases going down a group in the Periodic Table.
(l) False: Fluorine is the most electronegative element, and francium is the least electronegative element
(o) False: The opposite is true.
3.31 Electronegativity generally increases going from left to right across a row of the Periodic Table because the number of positive charges in the nucleus of each successive element in the row increases going from left to right. The increasing nuclear charge exerts a stronger and stronger pull of the valence electrons.
3.33 Electrons are shifted toward the more electronegative atom.
(a) Cl (b) O (c) O (d) Cl (e) Negligible
(f) Negligible (g) O
3.35 (a) C—Cl polar covalent (b) C—Li polar covalent
(c) C—N polar covalent
3.37 (a) T (b) F (c) T (d) T (e) T (f) F (g) F (h) F (i) F
(b) False: Ionic bonds form by the transfer of one or more electrons from the atom of lower electronegativity to the atom of higher electronegativity.
(f) False: The formula of calcium hydroxide is $Ca(OH)_2$.
(g) False: The formula of aluminum sulfide is Al_2S_3.
(h) False: The formula of iron(III) oxide is Fe_2O_3.
(i) The formula of barium oxide is BaO.
3.39 (a) NaBr (b) Na_2O (c) $AlCl_3$ (d) $BaCl_2$ (e) MgO
3.41 Sodium chloride in the solid state forms a lattice in which each Na^+ ion is surrounded by six Cl^- ions, and each Cl^- ion is surrounded by six Na^+ ions.
3.43 (a) $Fe(OH)_3$ (b) $BaCl_2$ (c) $Ca_3(PO_4)_2$ (d) $NaMnO_4$
3.45 (a) The formula $(NH_4)_2PO_4$ is not correct. The correct formula is $(NH_4)_3PO_4$.
(b) The formula Ba_2CO_3 is not correct. The correct formula is $BaCO_3$.
(c) The formula Al_2S_3 is correct.
(d) The formula MgS is correct.
3.47 (a) T (b) F (c) T (d) T (e) F (f) F (g) T (h) F (i) T (j) F (k) F
(b) False: The name includes no indication of the number of ions present.
(e) False: The systematic name of Fe_2O is iron(III) oxide.
(f) False: The systematic name of $FeCO_3$ is iron(II) carbonate.
(h) False: The systematic name of K_2HPO_4 is potassium hydrogen phosphate.
(j) False: The name of PCl_3 is phosphorus trichloride.
(k) False: The correct formula is $(NH_4)_2CO_3$.
3.49 The formula for potassium nitrite is KNO_2.
3.51 (a) Na^+, Br^- (b) Fe^{2+}, $SO_3{}^{2-}$ (c) Mg^{2+}, $PO_4{}^{3-}$ (d) K^+, $H_2PO_4{}^-$ (e) Na^+, $HCO_3{}^-$ (f) Ba^{2+}, $NO_3{}^-$
3.53 (a) KBr (b) CaO (c) HgO (d) $Cu_3(PO_4)_2$ (e) Li_2SO_4 (f) Fe_2S_3

3.55 (a) T (b) F (c) F (d) T (e) T (f) T (g) T (h) F
(i) T (j) T (k) F (l) T (m) F (n) T
(b) False: They will form a nonpolar covalent bond.
(c) False: A bond formed by sharing two electrons is a single bond. A double bond is a bond formed by sharing two pairs of electrons.
(h) False: The order given here is reversed.
(k) Ethane, C_2H_6, must show 14 valence electrons.
(m) False: The Lewis structure for the ammonium ion, NH_4^+, must show eight valence electrons.
3.57 (a) A single bond results when one electron pair is shared between two atoms.
(b) A double bond results when two electron pairs are shared between two atoms.
(c) A triple bond results when three electron pairs are shared between two atoms.

3.59 (a) (b) H—C≡C—H

(c) (d)

(e) (f)

3.61 The total number of valence electrons for each compound is:
(a) NH_3 has 8 (b) C_3H_6 has 18 (c) $C_2H_4O_2$ has 24
(d) C_2H_6O has 20 (e) CCl_4 has 32 (f) HNO_2 has 18
(g) CCl_2F_2 has 32 (h) O_2 has 12
3.63 (a) A bromine atom has seven electrons in its valence shell. (b) A bromine molecule has two bromine atoms bonded by a single covalent bond. (c) A bromide ion is an anion; it has a complete octet of eight valence electrons and a charge of -1.

(a) :B̈r· (b) :B̈r—B̈r: (c) :B̈r:⁻

3.65 Hydrogen has the electron configuration $1s^1$. Hydrogen's valence shell has only a $1s$ orbital, which can hold only two electrons.
3.67 Nitrogen has five valence electrons. By sharing three more electrons with other atoms or another atom, nitrogen can achieve the outer shell electron configuration of neon, the noble gas nearest to it in atomic number. The three shared pairs of electrons may be in the form of three single bonds or one double bond and one single bond or one triple bond. With these combinations, there is one unshared pair of electrons on nitrogen.
3.69 Oxygen has six valence electrons. By sharing two electrons with another atom or atoms, oxygen can achieve the outer shell electron configuration of neon, the noble gas nearest it in atomic number. The two shared pairs of electrons may be in the form of one double bond or two single bonds. With either of these configurations, there are two unshared pairs of electrons on oxygen.
3.71 O^{6+} has a charge too concentrated and too large for a small ion.
3.73 (a) BF_3 does not obey the octet rule because in this compound, boron has only six electrons in its valence shell.
(b) CF_2 does not obey the octet rule because in this compound, carbon has only four electrons in its valence shell.

(c) BeF_2 does not obey the octet rule because in this compound, beryllium has only four electrons in its valence shell.
(d) C_2H_4, ethylene, obeys the octet rule. In this compound, each carbon has a double bond to the other carbon and single bonds to two hydrogen atoms, giving each carbon a complete octet.
(e) CH_3 does not obey the octet rule. In this compound, carbon has single bonds to three hydrogens, which gives carbon only six electrons in its valence shell.
(f) N_2 obeys the octet rule. Each nitrogen has one triple bond and one unshared pair of electrons and, therefore, eight electrons in its valence shell.
(g) NO does not obey the octet rule. This compound has 11 valence electrons, and any Lewis structure drawn for it will show either oxygen or nitrogen with only seven electrons in its valence shell.
3.75 (a) Sulfur dioxide (b) Sulfur trioxide
(c) Phosphorus trichloride (d) Carbon disulfide
3.77 (a) An acceptable structure for the ozone molecule must show 18 valence electrons.
(b) Two equivalent contributing structures for ozone are:

(c) The curved arrows in (b) show the redistribution of electron pairs between the two contributing structures.
(d) The two contributing structures for ozone are equivalent and, therefore, make equal contributors to the hybrid. In each contributing structure, three regions of electron density surround the central oxygen, and therefore, the O—O—O bond angles are predicted to be 120°.
(e) This structure is not acceptable because the central oxygen atom has 10 electrons in its valence shell, which violates the octet rule. The valence shell of oxygen has one $2s$ orbital and three $2p$ orbitals, which between them can hold no more than eight electrons, hence the octet rule.
3.79 (a) T (b) F (c) T (d) F (e) T (f) T (g) F (h) T
(i) T (j) T (k) F (l) T (m) T
(b) False: Predicting bond angles must also consider nonbonding pairs of electrons.
(d) False: In CO_2, carbon is surrounded by two regions of electron density, and the VSEPR model predicts a bond angle of 180°.
(g) False: Four regions of electron density around a central atom result in bond angles of approximately 109.5°.
(k) False: The oxygen atom in H_2O is surrounded by four regions of electron density, and therefore, the VSEPR model predicts the H—O—H bond angle of approximately 109.5°.
3.81 (a) H_2O has 8 valence electrons, and H_2O_2 has 14 valence electrons.
(b) Each Lewis structure must show two bonds to oxygen and two unshared pairs on each oxygen. Lewis structures are:

Water   Hydrogen peroxide

(c) Predict bond angles of 109.5° about each oxygen atom.

3.83 (a) (b) (c)

Tetrahedral   Pyramidal   Tetrahedral
(109.5°)   (109.5°)   (109.5°)

(d) $\overset{..}{\underset{..}{O}}=\overset{..}{\underset{..}{S}}=\overset{..}{\underset{..}{O}}$

Bent
(120°)

(e) $\overset{..}{\underset{..}{O}}=\overset{\overset{\displaystyle :O:}{\|}}{\underset{..}{S}}=\overset{..}{\underset{..}{O}}$

Trigonal planar
(120°)

(f) $F-\overset{\displaystyle Cl}{\underset{\displaystyle F}{C}}-Cl$

Tetrahedral
(109.5°)

(g) $H-\overset{..}{\underset{\displaystyle H}{N}}-H$

Pyramidal
(109.5°)

(h) $Cl-\overset{..}{\underset{\displaystyle Cl}{P}}-Cl$

Pyramidal
(109.5°)

3.85 (a) T (b) T (c) F (d) T (e) T (f) T (g) T (h) T
(c) False: If the dipole moments of polar bonds cancel each other by acting in equal but opposite directions, then the molecule will be nonpolar.

3.87 (a) The Lewis structure of BF_3 is:

$$:\overset{..}{\underset{..}{F}}-\overset{\overset{\textstyle}{}}{\underset{\overset{\textstyle}{:\overset{..}{\underset{..}{F}}:}}{B}}-\overset{..}{\underset{..}{F}}:$$

(b) The predicted F—B—F bond angles are 120°.
(c) BF_3 has three polar bonds, but the three bond dipole moments cancel each other by acting in opposite directions, and therefore, the molecule is nonpolar.

3.89 No, it is not possible to have a polar molecule with all nonpolar bonds.

3.91 The individual C—Cl bond dipoles in CCl_4 act in equal but opposite directions, canceling each other's effect on the molecular dipole.

3.93 Sodium iodide, NaI, and potassium iodide, KI, are used as iodide sources in table salt. Iodide is necessary for proper thyroid function and for the formation of thyroid hormones.

3.95 Potassium permanganate is used as an external antiseptic.

3.97 Nitric oxide, NO, quickly oxidizes in air to nitrogen dioxide, NO_2, which then dissolves in rainwater to form nitric acid, HNO_3.

3.99 (a) $SiCl_4$ (b) PH_3 (c) H_2S

3.101 The predicted shape is created by putting together the bases of two square-based pyramids. This shape is called octahedral, because it has eight faces.

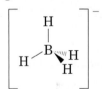

3.103 To arrive at the atom-atom distance in H_2O and H_2S, add the atomic radii:

H—O = 103 pm   H—S = 141 pm

3.105 (a) The following types of geometries are present in vitamin E: tetrahedral and trigonal planar.
(b) Bond angles about the carbon atom participating in four single bonds are 109.5°, and that about the single oxygen atom is also 109.5°.
(c) Vitamin E has only one polar bond, the —OH group, and large regions containing only nonpolar covalent bonds. Because the nonpolar regions are so large compared with the size of the one polar covalent region, predict that vitamin E is a nonpolar molecule.

3.107 (a) The most polar bond in ephedrine is the O—H bond.
(b) Predict that ephedrine is a polar molecule because it has polar covalent O—H, C—O, N—H, and C—N bonds.

3.109 Both are polar molecules, with the negative end of the dipole determined by the more electronegative fluorine atoms.

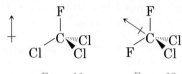

Freon-11   Freon-12

3.111 The compound is white zinc oxide, ZnO.

3.113 The lead-containing compound is primarily lead(IV) oxide, PbO_2.

3.115 Fe^{2+} is utilized in over-the-counter iron supplements.

3.117 (a) $CaSO_3$ (b) $Ca(HSO_3)_2$ (c) $Ca(OH)_2$
(d) $CaHPO_4$

3.119 Perchloroethylene has four polar C—Cl bonds, but given its geometry, the molecule is nonpolar.

3.121 (a) The Lewis structure for tetrafluoroethylene is:

$$\overset{\textstyle :\overset{..}{\underset{..}{F}}: \quad :\overset{..}{\underset{..}{F}}:}{\underset{\textstyle :\overset{..}{\underset{..}{F}}: \quad :\overset{..}{\underset{..}{F}}:}{C=C}}$$

(b) Predict 120° for each F—C—F bond angle.
(c) No, it does not have a dipole moment.

3.123 (a) The borohydride ion, BH_4^{-}, has $(3 + 4 + 1) = 8$ valence electrons.
(b) The Lewis structure of the borohydride ion shows boron surrounded by four regions of electron density.

(c) Predict each H—B—H bond angle to be 109.5°.

Chapter 4 Chemical Reactions

4.1 The balanced equation is

$$6CO_2(g) + 6H_2O(\ell) \xrightarrow{\text{photosynthesis}} C_6H_{12}O_6(aq) + 6O_2(g)$$

4.2 The balanced equation is

$$2C_6H_{14}(g) + 19O_2(g) \longrightarrow 12CO_2(g) + 14H_2O(g)$$

4.3 The balanced equation is

$$3K_2C_2O_4(aq) + Ca_3(AsO_4)_2(s) \longrightarrow$$
$$2K_3AsO_4(aq) + 3CaC_2O_4(s)$$

4.4 The net ionic equation is:

$$Cu^{2+}(aq) + S^{2-}(aq) \longrightarrow CuS(s).$$

4.5 (a) Ni^{2+} gained two electrons so is reduced. Cr lost two electrons so is oxidized. Ni^{2+} is the oxidizing agent, and Cr is the reducing agent.
(b) CH_2O gained hydrogens so is reduced. H_2 gains oxygens in being converted to CH_3OH and so is oxidized. CH_2O is the oxidizing agent, and H_2 is the reducing agent.

4.6 (a) ibuprofen, $C_{13}H_{18}O_2 = 206.1$ amu
(b) $Ba_3(PO_4)_2 = 601$ amu

4.7   1500. g H_2O is 83.3 mol H_2O.
4.8   2.84 mol Na_2S is 222 g Na_2S.
4.9   In 2.5 mol of glucose, there are 15 mol of C atoms, 30 mol of H atoms, and 15 mol of O atoms.
4.10   0.062 g $CuNO_3$ contains 4.9×10^{-4} mol Cu^+.
4.11   235 g H_2O contains 7.86×10^{24} molecules H_2O.
4.12   (a) The balanced equation is:

$$2Al_2O_3(s) \xrightarrow{\text{electrolysis}} 4Al(s) + 3O_2(g)$$

(b) It requires 51 g of alumina to prepare 27 g of aluminum.
4.13   From the balanced equation, we see that the molar ratio of CO required to produce CH_3COOH is 1:1. Therefore, it requires 16.6 moles of CO to produce 16.6 moles of CH_3COOH.
4.14   From the balanced equation, we see that the molar ratio of ethylene to ethanol is 1:1. Therefore, 7.24 mol of ethylene gives 7.24 mole of ethanol, which is 334 g of ethanol.
4.15   (a) H_2 (1.1 mole) is in excess, and C (0.50 mole) is the limiting reagent.
(b) 8.0 g CH_4 is produced.
4.16   The percent yield is 80.87%.
4.17   Following are the balanced equations.
(a) $HI + NaOH \longrightarrow NaI + H_2O$
(b) $Ba(NO_3)_2 + H_2S \longrightarrow BaS + 2HNO_3$
(c) $CH_4 + 2O_2 \longrightarrow CO_2 + 2H_2O$
(d) $2C_4H_{10} + 13O_2 \longrightarrow 8CO_2 + 10H_2O$
(e) $2Fe + 3CO_2 \longrightarrow Fe_2O_3 + 3CO$
4.19 $CO_2(g) + Ca(OH)_2(aq) \longrightarrow CaCO_3(s) + H_2O(\ell)$
4.21 $2Mg(s) + O_2(g) \longrightarrow 2MgO(s)$
4.23 $2C(s) + O_2(g) \longrightarrow 2CO(g)$
4.25 $2AsH_3(g) \xrightarrow{\text{heat}} 2As(s) + 3H_2(g)$
4.27 $2NaCl(aq) + 2H_2O(\ell) \xrightarrow{\text{electrolysis}}$

$$Cl_2(g) + 2NaOH(aq) + H_2(g)$$

4.29   The following chemical reactions are balanced net ionic equations.
(a) $Ag^+(aq) + Br^-(aq) \longrightarrow AgBr(s)$
(b) $Cd^{2+}(aq) + S^{2-}(aq) \longrightarrow CdS(s)$
(c) $2Sc^{3+}(aq) + 3SO_4^{2-}(aq) \longrightarrow Sc_2(SO_4)_3(s)$
(d) $Sn^{2+}(aq) + 3Fe^{2+}(aq) \longrightarrow Sn(s) + 2Fe^{3+}(aq)$
(e) $2K(s) + 2H_2O(\ell) \longrightarrow 2K^+(aq) + 2OH^-(aq) + H_2(g)$
4.31   (a) $Ca_3(PO_4)_2$ will precipitate.

$$3Ca^{2+}(aq) + 2PO_4^{3-}(aq) \longrightarrow Ca_3(PO_4)_2(s)$$

(b) No precipitate will form (Group 1 chlorides and sulfates are soluble).
(c) $BaCO_3$ will precipitate.

$$Ba^{2+}(aq) + CO_3^{2-}(aq) \longrightarrow BaCO_3(s)$$

(d) $Fe(OH)_2$ will precipitate.

$$Fe^{2+}(aq) + 2OH^-(aq) \longrightarrow Fe(OH)_2(s)$$

(e) $Ba(OH)_2$ will precipitate.

$$Ba^{2+}(aq) + 2OH^-(aq) \longrightarrow Ba(OH)_2(s)$$

(f) Sb_2S_3 will precipitate.

$$2Sb^{2+}(aq) + 3S^{2-}(aq) \longrightarrow Sb_2S_3(s)$$

(g) $PbSO_4$ will precipitate.

$$Pb^{2+}(aq) + SO_4^{2-}(aq) \longrightarrow PbSO_4(s)$$

4.33   The net ionic equation is

$$SO_3^{2-}(aq) + 2H^+(aq) \longrightarrow SO_2(g) + H_2O(\ell)$$

4.35   (a) KCl (soluble: all Group 1 chlorides are soluble).
(b) NaOH (soluble: all sodium salts are soluble).
(c) $BaSO_4$ (insoluble: most sulfates are insoluble).
(d) Na_2SO_4 (soluble: all sodium salts are soluble).
(e) Na_2CO_3 (soluble: all sodium salts are soluble).
(f) $Fe(OH)_2$ (insoluble: most hydroxides are insoluble).
4.37   (a) T (b) T (c) T (d) T (e) T (f) F (g) F (h) T (i) T (j) T (k) T (l) T (m) T (n) T
4.39   (a) C_7H_{12} is oxidized (the carbons gain oxygens in going to CO_2), and O_2 is reduced.
(b) O_2 is the oxidizing agent and C_7H_{12} is the reducing agent.
4.41   (a) F (b) F (c) T (d) T (e) T
4.43   (a) sucrose, $C_{12}H_{22}O_{11}$ 342.3 amu
(b) glycine, $C_2H_5NO_2$ 75.07 amu
(c) DDT, $C_{14}H_9Cl_5$ 354.5 amu
4.45   (a) 32 g CH_4 = 2.0 mol CH_4
(b) 345.6 g NO = 11.52 mol NO
(c) 184.4 g ClO_2 = 2.734 mol ClO_2
(d) 720. g glycerine = 7.82 mol glycerine
4.47   (a) 18.1 mol CH_2O = 18.1 mol O atoms
(b) 0.41 mol $CHBr_3$ = 1.2 mol Br atoms
(c) 3.5×10^3 mol $Al_2(SO_4)_3$ = 4.2×10^4 mol O atoms
(d) 87 g HgO = 0.40 mol Hg atoms
4.49   (a) 25.0 g TNT (MW = 227 g/mol) contains 1.99×10^{23} N atoms
(b) 40 g ethanol (MW = 46 g/mol) = 1.0×10^{24} mol C atoms
(c) 500. mg aspirin (MW 180.2 g/mol) = 6.68×10^{21} O atoms
(d) 2.40 g NaH_2PO_4 (MW 120 g/mol) = 1.20×10^{22} Na atoms
4.51   (a) 100. molecules CH_2O (MW 30 g/mol) = 4.98×10^{-21} gram CH_2O.
(b) 3000. molecules CH_2O (MW 30 g/mol) = 1.495×10^{-19} g CH_2O.
(c) 5.0×10^6 molecules CH_2O = 2.5×10^{16} grams CH_2O molecules.
(d) 2.0×10^{24} molecules CH_2O = 100 g CH_2O.
4.53   3.9 mg cholesterol (MW 386.7 g/mol) = 6.1×10^{18} molecules cholesterol.
4.55   (a) 1 mol O_2 requires 0.67 mol of N_2.
(b) 0.67 mol of N_2O_3 are produced from 1 mol of O_2.
(c) To produce 8 mol N_2O_3 requires 12 mol O_2.
4.57   1.50 mol $CHCl_3$ requires 319 g of Cl_2.
4.59   (a) $2NaClO_2(aq) + Cl_2(g) \longrightarrow 2ClO_2(g) + 2NaCl(aq)$
(b) 5.5 kg of $NaClO_2$ will yield 4.10 kg of ClO_2.
4.61   To produce 5.1 g of glucose requires 7.5 g of CO_2.
4.63   To completely react with 0.58 g of Fe_2O_3, we need 0.13 g C.
4.65   51.1 g of salicyclic acid.
4.67   The theoretical yield from 5.6 g of ethane is 12 g of chloroethane. The percentage yield is 68%.
4.69   (a) T (b) F (c) T (d) T (e) T (f) T
4.71   (a) endothermic (22.0 kcal appears as a reactant).
(b) exothermic (124 kcal appears as a product).
(c) exothermic (94.0 kcal appears as a product).
(d) endothermic (9.80 kcal appears as a reactant).
(e) exothermic (531 kcal appears as a product).
4.73 1.6×10^2 kcal of heat is evolved in burning 0.37 mol of acetone.
4.75   Ethanol has a greater heat of combustion per gram (7.09 kcal/g) than glucose 3.72 kcal/g).

4.77 156.0 kcal will produce 88.68 g of Fe metal.
4.79 Hydroxyapatite is composed of calcium ions, phosphate ions, and hydroxide ions.
4.81 Li is oxidized, and I_2 is reduced. I_2 is the oxidizing agent, and Li is the reducing agent.
4.83 Cu^+ is oxidized. The species that is oxidized during the course of the reaction gives up an electron and is the reducing agent. Therefore, Cu^+ is the reducing agent.
4.85 $2C_5H_{12}O(l) + 15O_2(g) \longrightarrow 10CO_2(g) + 12H_2O(g)$
4.87 488 mg of aspirin (MW 180.2 g/mol) is equal to 2.71×10^{-3} mol aspirin.
4.89 N_2 is the limiting reagent, and H_2 is in excess.
4.91 4×10^{10} molecules of hemoglobin are present in a red blood cell.
4.93 29.7 kg N_2 = 1061 mol N_2 and 3.31 kg H_2 = 1655 mol H_2
(a) From the balanced chemical equation, we see that the two gases react in the ratio $3H_2/N_2$. Complete reaction of 1061 mol N_2 requires 3183 mol H_2, but less than this number of moles of H_2 is present. Therefore, H_2 is the limiting reagent.
(b) Under the balanced equation are moles of each present before reaction, moles reacting, and moles present after complete reaction.

$$N_2 \ + \ 3H_2 \longrightarrow 2NH_3$$

	N_2	$3H_2$	$2NH_3$
Before rexn	1061	1655	0
Reacting	551	1655	0
After rexn	510	0	1102

551 mol N_2 = 14.3 kg of N_2 remains after the reaction.
(c) 1102 moles of NH_3 = 18.7 kg of NH_3 is formed.
4.95 (a) 441 mg of furan = 6.48×10^{-3} mol of furan
(b) 0.060 L of furan = 2.0×10^{24} atoms of C
(c) 9.86×10^{25} molecules of furan = 1.11×10^4 g of furan
4.97 2.84×10^5 g $KClO_4(aq)$
4.99 (a) Following are balanced equations for each oxidation.
$C_{16}H_{32}O_2(s) + 23O_2(g) \longrightarrow 16CO_2 + 16H_2O(\ell) + 238.5$ kcal/mol
$C_6H_{12}O_6(s) + 6O_2(g) \longrightarrow 6CO_2 + 6H_2O(\ell) + 670$ kcal/mol
(b) The heat of combustion of palmitic acid is 9.302 kcal/gram. The heat of combustion of glucose is 3.72 kcal/gram.
(c) Palmitic acid has the greater heat of combustion per mole.
(d) Palmitic acid also has the greater heat of combustion per gram.

Chapter 5 Gases, Liquids, and Solids

5.1 0.41 atm
5.2 16.4 atm
5.3 0.053 atm
5.4 4.84 atm
5.5 0.422 mol Ne
5.6 9.91 g He
5.7 0.107 atm of H_2O vapor
5.8 (a) Yes, there can be hydrogen bonding between water and methanol, because a hydrogen atom on each molecule is bonded to an electronegative oxygen atom. The O—H hydrogen can form a hydrogen bond to an oxygen lone pair on another molecule.
(b) No, there is no polarity to a C—H bond, and therefore, it cannot participate in a hydrogen bonding.
5.9 The heat of vaporization of water is 540 cal/g. 45.0 kcal is sufficient to vaporize 83.3g H_2O.
5.10 The heat required to heat 1.0 g of iron to melting = 2.3×10^2 cal.

Heat (up to melting) = 166 cal
Heat to melt = 63.7 cal
5.11 According to the phase diagram of water (Figure 5.20), the vapor will first condense to liquid water and then freeze to give ice.
5.13 As the volume of a gas decreases, the concentration of gas molecules per unit of volume increases and the number of gas molecules colliding with the walls of the container increases. Because gas pressure results from the collisions of gas molecules with the walls of the container, as volume decreases, pressure increases.
5.15 The volume of a gas can be decreased by (1) increasing the pressure on the gas or (2) lowering the temperature (cooling) of the gas. (3) The volume of the gas can be decreased by removing some of the gas,
5.17 7.37 L
5.19 2.0 atm of CO_2 gas
5.21 615 K
5.23 6.2 L of SO_2 gas upon heating
5.25 The pressure read by the manometer is the difference between the gas in the bulb and the atmospheric pressure: 833 mm Hg − 760 mm Hg = 73 mm Hg
5.27 2.6 atm of halothane
5.29

V_1	T_1	P_1	V_2	T_2	P_2
6.35 L	10°C	0.75 atm	**4.6 L**	0°C	1.0 atm
75.6 L	0°C	1.0 atm	**88 L**	35°C	735 torr
1.06 L	75°C	0.55 atm	3.2 L	0°C	**0.14 atm**

5.31 The volume of the balloon will be 3×10^6 L.
5.33 The new temperature is 300 K.
5.35 1.87 atm
5.37 (a) 2.33 mol of gas are present.
(b) No, the only information you need to know about the gas is that it is an ideal gas.
5.39 Using the ideal gas law $PV = nRT$ and n(moles) = mass/MW, the following equation can be derived and solved for the molecular weight of the gas.

$$MW = \frac{(\text{mass})RT}{PV} =$$
$$\frac{(8.00\text{g})(0.0821 \text{ L} \cdot \text{atm} \cdot \text{mol}^{-1} \cdot \text{K}^{-1})(273\text{K})}{(2.00 \text{ atm})(22.4\text{L})} = 4.00 \text{ g/mol}$$

5.41 At constant temperature, gas density increases as pressure increases. At constant pressure, gas density decreases as temperature increases.
5.43 (a) 24.7 mol O_2 are needed to fill the chamber.
(b) 790 g of O_2 are needed to fill the chamber.
5.45 5.5 L of air contains 1.16 L of O_2, which, under these conditions, is 0.050 mol of O_2.
0.050 mol of O_2 contains 13.0×10^{22} molecules of O_2.
5.47 (a) The mass of one mol of air is 28.95 grams.
(b) The density of air is 1.29 g/L.
5.49 The density of each gas is
(a) SO_2 = 2.86 g/L (b) CH_4 = 0.714 g/L (c) H_2 = 0.0892 g/L
(d) He = 0.179 g/L (e) CO_2 = 1.96 g/L
Gas comparison: SO_2 and CO_2 are more dense than air; He, H_2, and CH_4 are less dense than air.
5.51 The density of octane is 0.7025 g/mL.
The mass of 1.00 mL of octane is 0.07025g.
Using the ideal gas equation, calculate that this mass of octane occupies 0.197 L.
5.53 When 3.50 g of Na(s) is reacted at a temperature of 18°C and a pressure of 0.995 atm, 3.65 L of $H_2(g)$ is produced.

5.55   (a) T   (b) F   (c) T   (d) F

$P_T = P_{N_2} + P_{O_2} + P_{CO_2} + P_{H_2O}$

$P_{N_2} = (0.740)(1.0 \text{ atm}) = 0.740 \text{ atm} (562.4 \text{ mm Hg})$

$P_{O_2} = (0.194)(1.0 \text{ atm}) = 0.194 \text{ atm} (147.5 \text{ mm Hg})$

$P_{H_2O} = (0.062)(1.0 \text{ atm}) = 0.062 \text{ atm} (47.1 \text{ mm Hg})$

$P_{CO_2} = (0.004)(1.0 \text{ atm}) = 0.004 \text{ atm} (3.0 \text{ mm Hg})$

$P_T = 1.00 \text{ atm} (760.0 \text{ mm Hg})$

5.57   0.194 atm oxygen, 0.004 atm carbon dioxide, 0.062 atm water vapor, and 0.740 atm nitrogen

5.59   (a) T   (b) F   (c) T   (d) F   (e) T   (f) T   (g) T   (h) F
(i) T   (j) T

5.61   (a) F   (b) F   (c) T   (d) T   (e) F   (f) T   (g) T   (h) T
(i) F   (j) F   (k) F   (l) F   (m) T   (n) T   (o) F

5.63   Gases behave most ideally under low pressure and high temperature to minimize non-ideal intermolecular interactions. Therefore, choice (c) best suits these conditions.

5.65   (a) CCl_4 is nonpolar; London dispersion forces
(b) CO is polar; dipole−dipole interactions
The most polar molecule (CO) will have the largest surface tension.

5.67   Yes, London dispersion forces range from 0.001 to 0.2 kcal/mol, whereas the lower end of dipole−dipole attractive forces can be as low as 0.1 kcal/mol.

5.69   (a) T   (b) F   (c) F   (d) T   (e) F   (f) T   (g) T   (h) T
(i) T   (j) T   (k) F   (l) F   (m) T   (n) F   (o) T   (p) F

5.71   (a) T   (b) T   (c) T   (d) F   (e) T   (f) F   (g) T   (h) F
(i) F   (j) F   (k) T

5.73   (a) T   (b) T   (c) F   (d) T   (e) T   (f) F   (g) F   (h) F
(i) F   (j) T   (k) T

5.75   1.53 kcal is required to vaporize one mol of CF_2Cl_2.

5.77   The vapor pressures are approximately:
(a) 90 mm Hg   (b) 120 mm Hg   (c) 490 mm Hg

5.79   (a) HI > HBr > HCl   The increasing size in this series increases London dispersion forces.
(b) H_2O_2 > HCl > $O_2 \cdot O_2$ has only weak London dispersion intermolecular forces to overcome for boiling to occur, whereas HCl is a polar molecule with stronger dipole−dipole attractions to overcome for boiling to occur. H_2O_2 has the strongest intermolecular forces (hydrogen bonding) to overcome for boiling to occur.

5.81   The difference between heating water from 0°C to 37°C and heating ice from 0°C to 37°C is the heat of fusion.
The energy required to heat 100. g of ice from 0°C to 37°C is 11700. cal.
The energy required to heat 100. g of water from 0°C to 37°C is 3700 cal.

5.83   The name of the phase change is sublimation, which is the conversion of a solid to a gas, bypassing the liquid phase.

5.85   1.00 mL of Freon-11 is 1.08×10^{-2} mol of Freon-11. Vaporizing this volume of Freon-11 from the skin will remove 6.96×10^{-2} kcal.

5.87   When a person lowers the diaphragm, the volume of the chest cavity increases, thus lowering the pressure in the lungs relative to atmospheric pressure. Air at atmospheric pressure then is drawn into the lungs, beginning the breathing process.

5.89   The first tapping sound one hears is the systolic pressure, which occurs when the sphygmomanometer pressure matches the blood pressure and the ventricle contracts, pushing blood into the arm.

5.91   When water freezes, it expands (water is one of the few substances that expands on freezing) and will break the bottle when the expansion exceeds the volume of the bottle.

5.93   Compressing a liquid or a solid is difficult because their molecules or atoms are already close together and there is very little empty space between them.

5.95   34 psi = 2.3 atm

5.97   Aerosol cans already contain gases under pressure. Gay-Lussac's law predicts that the pressure inside the can will increase as it is heated, with the potential of explosive rupturing of the can causing injury.

5.99   112 mL

5.101   Water, which forms strong intermolecular hydrogen bonds, has the highest boiling point. Boiling points of these three compounds are:
(a) pentane, C_5H_{12} (36°C)   (b) chloroform, $CHCl_3$ (61°C)
(c) water, H_2O (100°C)

5.103   (a) As a gas is compressed under pressure, the molecules are forced closer together and the intermolecular forces pull molecules together, forming a liquid.
(b) 9.1 kg of propane   (c) 2.1×10^2 moles of propane
(d) 4.6×10^3 L of propane

5.105   The density of the gas is 3.00 g/L.
Using the ideal gas law, show that

$$MW = \frac{\text{mass}RT}{PV}$$

and then calculate that the molecular weight of the gas is 91.9 g/mol.

5.107   313K (40°C)

5.109   The temperature of a liquid drops during evaporation because as the molecules with higher kinetic energy leave the liquid and enter the gas phase, the average kinetic energy of molecules remaining in the liquid decreases. The temperature of the liquid is directly proportional to the average kinetic energy of molecules in the liquid phase and as the average kinetic energy decreases, the temperature decreases.

5.111   (a) The pressure on the body at 100 feet is 3.0 atm.
(b) At 1.00 atm, P_{N_2} = 593 mm Hg (0.780 atm) and thus makes up 78.0% of the gas mixture, which does not change at a depth of 100 feet. At this depth, the total pressure on the lungs, which is equalized by pressure of air delivered by the SCUBA tank, is 3.0 atm and the partial pressure of N_2 is 2.34 atm.
(c) At 2 atm, P_{O_2} = 158 mm Hg (0.208 atm) and thus makes up 20.8% of the gas mixture at 2 atm, which does not change at a depth of 100 feet. At this depth, the total pressure on the lungs, which is equalized by pressure of air delivered by the scuba tank, is 3.0 atm. Thus, at 100 feet, the partial pressure of O_2 = 0.63 atm.
(d) As a diver ascends from 100 ft, the external pressure on the lungs decreases, and therefore, the volume of gases in the lungs increases. If the diver does not exhale during a rapid ascent, the diver's lungs could overinflate due to the expansion of gases in the lungs, causing injury.

5.113   9.33 g NH_4Cl

5.115   1.26 g dry NH_4NO_2

5.117   4250 cal

Chapter 6   Solutions and Colloids

6.1   To 11 g of KBr, add a quantity of water sufficient to dissolve the KBr. Following dissolution of the KBr, add water to the 250 mL mark, stopper, and mix.

6.2   1.7 %w/v

6.3   First, calculate the number of moles and mass of KCl needed, which is 2.12 mol and 158 g of KCl. To prepare the solution, place 158 g of KCl in a 2 L volumetric flask, add some water until the solid has dissolved, and then fill the flask with water to the 2.0 L mark.

6.4 Because the units of molarity are moles of solute/L of solution, grams of KSCN must be converted to moles of KSCN and mL of solution must be converted to L of solution. When these conversions are completed, find that the concentration of the solution is 0.0133M.

6.5 First, convert grams of glucose into moles of glucose; then convert moles of glucose into mL of solution. 10.0 g of glucose is 0.0556 mol of glucose. This mass of glucose is contained in 185 mL of the given solution.

6.6 First, convert 100 gallons to liters of solution. 3.9×10^2 g $NaHSO_3$ must be added to the 100-gallon barrel.

6.7 Place 15.0 mL of 12.0 M HCl solution into a 300 mL volumetric flask, add some water, swirl to mix completely, and then fill the flask with water to the 300 mL mark.

6.8 Place 0.13 mL of the 15% KOH solution into a 20 mL volumetric flask, add some water, swirl until completely dissolved, and then fill the flask with water to the 20 mL mark.

6.9 The Na^+ concentration is 0.24 ppm Na^+.

6.10 215 g of CH_3OH (molecular weight 32.0 g/mol) is 6.72 mol of CH_3OH.
$\Delta T = (1.86°C/mol)(6.72 \text{ mol}) = 12.5°C$ The freezing point is lowered by 12.5°C. The new freezing point is $-12.5°C$.

6.11 Compare the number of moles of ions or molecules in each solution. The solution with the most ions or molecules in solution will have the lowest freezing point.

Solution	Particles in solution
(a) 6.2 M NaCl	$2 \times 6.2\ M = 12.4\ M$ ions
(b) 2.1 M Al(NO$_3$)$_3$	$4 \times 2.1\ M = 8.4\ M$ ions
(c) 4.3 M K$_2$SO$_3$	$3 \times 4.3\ M = 12.9\ M$ ions

Solution (c) has the highest concentration of solute particles (ions); therefore, it will have the lowest freezing point.

6.12 The boiling point is raised by 3.50°C. The new boiling point is 103.5°C.

6.13 The molarity of the solution prepared by dissolving 3.3 g Na$_3$PO$_4$ in 100 mL of water is 0.20M Na$_3$PO$_4$. Each formula unit of Na$_3$PO$_4$ dissolved in water gives 3 Na$^+$ ions and 1 PO$_4^{3-}$ ion, for a total of 4 particles. The osmolarity of the solution is (0.20 M)(4 ions) = 0.80 osmol.

6.14 The osmolarity of red blood cells is 0.30 osmol.

Solution	Mol particles/L
(a) 0.1 M Na$_2$SO$_4$	$3 \times 0.1\ M = 0.30$ osmol
(b) 1.0 M Na$_2$SO$_4$	$3 \times 1.0\ M = 3.0$ osmol
(c) 0.2 M Na$_2$SO$_4$	$3 \times 0.2\ M = 0.6$ osmol

Solution (a) has the same osmolarity as red blood cells and, therefore, is isotonic with red blood cells.

6.15 (a) T (b) T (c) T (d) T

6.17 The solvent is water.

6.19 (a) both tin and copper are solids
(b) solid solute (caffeine, flavorings) and liquid solvent (water).
(c) both CO_2 and H_2O (steam) are gases.
(d) gas (CO_2) and liquid (ethanol) solutes in a liquid solvent (water).

6.21 Mixtures of gases are true solutions because they mix in all proportions, molecules are distributed uniformly, and the component gases do not separate upon standing.

6.23 The prepared aspartic acid solution was unsaturated. Over the two days, some of the solvent (water) evaporated and the solution has become saturated. When water continued evaporating, the remaining water could not hold all the dissolved solute, so the excess aspartic acid precipitated as a white solid.

6.25 (a) NaCl is an ionic solid and will be dissolved in the water layer.
(b) Camphor is a nonpolar molecular compound and will dissolve in the nonpolar diethyl ether layer.
(c) KOH is an ionic solid and will be dissolved in the water layer.

6.27 Isopropyl alcohol would be a good first choice. The oil base in the paint is nonpolar. Both benzene and hexane are nonpolar solvents and may dissolve the oil-based paint, thus destroying the painting.

6.29 The solubility of aspartic acid in water at 25°C is 0.250 g in 50.0 mL of water. The cooled solution of 0.251 g of aspartic acid in 50.0 mL water will be supersaturated by 0.001 g of aspartic acid.

6.31 According to Henry's law, the solubility of a gas in a liquid is directly proportional to pressure. A closed bottle of a carbonated beverage is under pressure. After the bottle is opened, the pressure is released and the carbon dioxide becomes less soluble and escapes, leaving the contents "flat."

6.33 (a) $\dfrac{1 \text{ min}}{1.0 \times 10^6 \text{ min}} \times 10^6 = 1 \text{ ppm}$

$\dfrac{1 \text{ p}}{1.05 \times 10^6 \text{ p}} \times 10^6 = 1 \text{ ppm}$

(b) $\dfrac{1 \text{ min}}{1.05 \times 10^9 \text{ min}} \times 10^9 = 1 \text{ ppb}$

$\dfrac{1 \text{ p}}{1.05 \times 10^9 \text{ p}} \times 10^9 = 1 \text{ ppb}$

6.35 (a) Dissolve 76 mL of ethanol in 204 mL of water (to give 280 mL of solution).
(b) Dissolve 8.0 mL of ethyl acetate in 427 mL of water (to give 435 mL to solution).
(c) Dissolve 0.13 L of benzene in 1.52 L chloroform (to give 1.65 L of solution).

6.37 (a) 4.15% w/v casein   (b) 0.030% w/v vitamin C
(c) 1.75% w/v sucrose

6.39 (a) Place 19.5 g NH$_4$Br in a 175 mL volumetric flask, add some water, swirl until completely dissolved, and then fill the flask with water to the 175 mL mark.
(b) Place 167 g of NaI in a 1.35 L volumetric flask, add some water, swirl until completely dissolved, and then fill the flask with water to the 1.35 L mark.
(c) Place 2.4 g of ethanol in a 330 mL volumetric flask, add some water, swirl until completely dissolved, and then fill the flask with water to the 330 mL mark.

6.41 0.2 M NaCl

6.43 0.509 M glucose
0.0202 M K$^+$
7.25×10^{-4} M Na$^+$

6.45 2.5 M sucrose

6.47 The total volume of the dilution is 30.0 mL. Starting with 5.00 mL of the stock solution, add 25.0 mL of water to reach a final volume of 30.0 mL. Note that this is a dilution by a factor of 6.

6.49 Place 2.1 mL of the 30.0% H_2O_2 into a 250 mL volumetric flask, add some water, swirl until completely mixed, and then fill the flask with water to the 250 mL mark.

6.51 (a) 3.85×10^4 ppm Captopril   (b) 6.8×10^4 ppm Mg^{2+}
(c) 8.3×10^2 ppm Ca^{2+}

6.53 Assume the density of the lake water to be 1.00 g/mL. The dioxin concentration is 0.01 ppb dioxin.
No, the dioxin level in the lake did not reach a dangerous level.

6.55 (a) 10 ppm Fe or 1×10^1 ppm
(b) 3×10^3 ppm Ca
(c) 2 ppm vitamin A

6.57 (a) KCl An ionic compound, very soluble in water: a strong electrolyte
(b) Ethanol A covalent compound: a nonelectrolyte
(c) NaOH An ionic compound, very soluble in water: a strong electrolyte
(d) HCl; A strong acid that dissociates completely in water: a strong electrolyte
(e) Glucose A covalent compound, very soluble in water: a nonelectrolyte

6.59 Water dissolves ethanol by forming hydrogen bonds with it. The O—H group of ethanol is both a hydrogen bond acceptor and a hydrogen bond donor.

6.61 (a) T (b) T

6.63 (a) homogeneous (b) heterogeneous (c) colloidal (d) heterogeneous (e) colloidal (f) colloidal

6.65 As the temperature of the solution decreased, the protein molecules must have aggregated and formed a colloidal mixture. The turbid appearance is the result of the Tyndall effect.

6.67 (a) 1.0 mol NaCl, freezing point $-3.72°C$
(b) 1.0 mol $MgCl_2$ freezing point $-5.58°C$
(c) 1 mol $(NH_4)_2CO_3$ freezing point $-5.58°C$
(d) 1 mol $Al(HCO_3)_3$ freezing point $-7.44°C$

6.69 Methanol dissolves in water but does not dissociate; it is a nonelectrolyte. It would require 344 g of CH_3OH in 1000. g of water to lower the freezing point to $-20°C$.

6.71 Acetic acid, a weak acid, is only weakly dissociated in water. KF is a strong electrolyte, completely dissociating in water and nearly doubling the effect on freezing-point depression compared with that of acetic acid.

6.73 In each case, side with greater osmolarity rises.
(a) B (b) B (c) A (d) B (e) neither (f) neither

6.75 (a) 0.39 M Na_2CO_3 = 0.39 M × 3 particles/formula unit = 1.2 osmol
(b) 0.62 M $Al(NO_3)_3$ = 0.62 × 4 particles/formula unit = 2.5 osmol
(c) 4.2 M LiBr = 4.2 × 2 particles/formula unit = 8.4 osmol
(d) 0.009 M K_3PO_4 = 0.009M × 4 particles/formula unit = 0.04 osmol

6.77 Cells in hypertonic solutions undergo crenation (shrink).
(a) 0.3% NaCl = 0.3 osmol NaCl
(b) 0.9 M glucose = 0.9 osmol glucose
(c) 0.9% glucose = 0.05 osmol glucose
Solution (b) has a concentration greater than the isotonic solution, so it will crenate red blood cells.

6.79 Carbon dioxide (CO_2) dissolves in rainwater to form a dilute solution of carbonic acid (H_2CO_3), which is a weak acid.

6.81 Nitrogen narcosis is the intoxication caused by the increased solubility of nitrogen in the blood as a result of high pressures as divers descend.

6.83 23 mg Ca^{2+} ion

6.85

$$CaCO_3(s) + H_2SO_4(aq) \longrightarrow CaSO_4(s) + CO_2(g) + H_2O(\ell)$$

$$CaSO_4 + 2H_2O \longrightarrow CaSO_4 \cdot 2H_2O$$

Gypsum dihydrate

6.87 The minimum pressure required for reverse osmosis in the desalinization of seawater exceeds 100 atm (the osmotic pressure of seawater).

6.89 Yes, the change made a change in the tonicity. A 0.2% $NaHCO_3$ solution is 0.05 osmol. A 0.2% solution of $KHCO_3$ is 0.04 osmol. This difference arises because of the difference in formula weight of $NaHCO_3$ (84 g/mol) compared with that of $KHCO_3$ (100.1 g/mol). The error in replacing $NaHCO_3$ with $KHCO_3$ results in a hypotonic solution and an electrolyte imbalance by reducing the number of ions (osmolarity) in solution.

6.91 When a cucumber is placed in a saline solution, the osmolarity of the saline is greater than the water in the cucumber; so water moves from the cucumber to the saline solution. When a prune (a partially dehydrated plum) is placed in the same solution, it expands because the osmolarity in the prune is greater than the saline solution; so water moves from the saline solution to inside the prune.

6.93 The solubility of a gas is directly proportional to the pressure (Henry's law) and inversely proportional to the temperature. The dissolved carbon dioxide formed a saturated solution in water when bottled under 2 atm pressure. When the bottles are opened at atmospheric pressure, the gas becomes less soluble in water. The excess carbon dioxide escapes through bubbles and frothing. In the other bottle, the solution of carbon dioxide in water is unsaturated at lower temperature and does not lose carbon dioxide.

6.95 Methanol is more efficient at lowering the freezing point of water. A given mass of methanol (32 g/mol) contains a greater number of moles than the same mass of ethylene glycol (62 g/mol).

6.97 $$CO_2(g) + H_2O(\ell) \longrightarrow H_2CO_3(aq)$$
Carbonic acid

$$SO_2(g) + H_2O(\ell) \longrightarrow H_2SO_3(aq)$$
Sulfurous acid

6.99 Place 39 mL of 35% HNO_3 into a 300 mL volumetric flask, add some water, swirl until completely mixed, and then fill the flask with water to the 300 mL mark.

6.101 6×10^{-3} g of pollutant

6.103 Assume that the density of the pool water is 1.00 g/mL. The Cl_2 concentration in the pool is 355 ppm. 7.09 kg of Cl_2 must be added to reach this concentration.

6.105 78.2 mL H_2SO_4 solution

6.107 (a) One mole of $MgCl_2$ dissociates to produce three moles of ions: one mole of Mg^{2+} and two moles of Cl^-. The freezing point will be lowered by $3 \times \left(\dfrac{1.86°C}{mol}\right) \times 0.263$ mol = $1.47°C$, and the solution will freeze at $-1.47°C$.
(b) One mole of $MgCl_2$ dissociates to produce three moles of ions: one mole of Mg^{2+} and two moles of Cl^-. The boiling point will be raised by $3 \times \left(\dfrac{0.512°C}{mol}\right) \times 0.263$ mol = $0.404°C$, and the solution will boil at $100.404°C$.

6.109 (a) From the balanced chemical equation, $Ca(s) + 2HBr(aq) \longrightarrow CaBr_2(aq) + H_2(g)$, we see that the two reactants react in the ratio of 2HBr/Ca. The complete reaction of 0.0364 mol Ca requires 0.0728 mol of HBr, but less than this number of moles of HBr is present. Therefore, HBr is the limiting reagent, producing 0.0187 mol $H_2(g)$.
(b) The volume of dry H_2 produced is 0.469 L $H_2(g)$.
(c) 0.709 g of $Ca(s)$ remains after the reaction.

Chapter 7 Reaction Rates and Chemical Equilibrium

7.1 rate of O_2 formation = 0.022 L O_2/min

7.2 rate = 4×10^{-2} mol H_2O_2/L · min for disappearance of H_2O_2

7.3 $K = \dfrac{[H_2SO_4]}{[SO_3][H_2O]}$

7.4 $K = \dfrac{[N_2][H_2]^3}{[NH_3]^2}$

7.5 $K = 0.602 \; M^{-1}$

7.6 Le Chatelier's principle predicts that adding Br_2 (a product) will shift the equilibrium to the left—that is, toward the formation of more NOBr(g).

7.7 Because oxygen's solubility in water is exceeded, oxygen bubbles out of the solution, driving the equilibrium toward the right.

7.8 If the equilibrium shifts to the right with the addition of heat, heat must have been a reactant, and the reaction is endothermic.

7.9 The equilibrium in a reaction where there is an increase in pressure favors the side with fewer moles of gas. Therefore, this equilibrium shifts to the right.

7.11 rate of formation of $CH_3I = 7.3 \times 10^{-3} \; M$ CH_3I/min

7.13 Reactions involving ions in aqueous solution of ions are faster because they do not require bond breaking and have low activation energies. In addition, the attractive force between positive and negative ions provides energy to drive the reaction. Reactions between covalent compounds require the breaking of covalent bonds and have higher activation energies and, therefore, slower reaction rates.

7.15

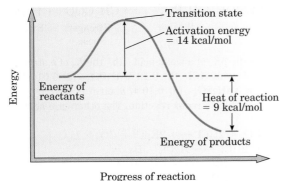

7.17 A general rule for the effect of temperature on the rate of reaction states that for every temperature increase of 10°C, the reaction rate doubles. In this case, a reaction temperature of 50°C would predict completion of the reaction in 1 h.

7.19 You might (a) increase the temperature, (b) increase the concentration of reactants, or (c) add a catalyst.

7.21 A catalyst increases the rate of a reaction by providing an alternative reaction pathway with lower activation energy.

7.23 Other examples of irreversible reactions include digesting a piece of candy, rusting of iron, exploding TNT, and the reaction of Na or K metal with water.

7.25 (a) $K = [H_2O]^2[O_2]/[H_2O_2]^2$

(b) $K = [N_2O_4]^2[O_2]/[N_2O_5]^2$

(c) $K = [O_2]^6/[H_2O]^6[CO_2]^6$

7.27 $K = 0.667$

7.29 $K = 0.099 \; M$

7.31 Products are favored in (b) and (c). Reactants are favored in (a), (d), and (e).

7.33 No, the rate of reaction is independent of the energy difference between products and reactants—that is, it is independent of the heat of reaction.

7.35 The reaction reaches equilibrium quickly, but the position of equilibrium favors the reactants. It would not be a very good industrial process unless products are constantly drawn off to shift the equilibrium to the right.

7.37 (a) right (b) right (c) left (d) left (e) no shift

7.39 (a) Adding Br_2 (a reactant) will shift the equilibrium to the right.

(b) The equilibrium constant will remain the same.

7.41 (a) no change (b) no change (c) smaller

7.43 As temperatures increase, the rates of most chemical processes increase. A high body temperature is dangerous because metabolic processes (including digestion, respiration, and biosynthesis of essential compounds) take place at a faster rate than is safe for the body. As temperatures decrease, so do the rates of most chemical reactions. As body temperature decreases below normal, the vital chemical reactions will slow to rates slower than is safe for the body.

7.45 The capsule with the tiny beads will act faster than the solid coated-pill form. The small bead size increases the drug's surface area, allowing the drug to react faster and deliver its therapeutic effects more quickly.

7.47 Assuming that there is an excess of AgCl from the previous recipe, the recipe does not need to be changed. The desert conditions add nothing that would affect the coating process.

7.49 At −5°C, the rate is 0.70 moles per liter per second. At 45°C, the rate is 22 moles per liter per second.

7.51

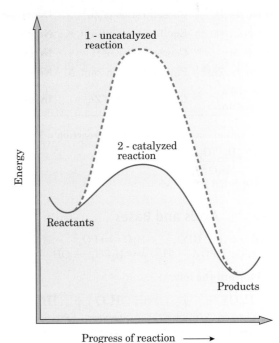

Profile 2 represents the addition of a catalyst.

7.53 0.14 M

7.55 The rate of a gaseous reaction could be increased by decreasing the volume of the container. This would increase the number of collisions between molecules.

7.57 $4NH_3 + 7O_2 \longrightarrow 4NO_2 + 6H_2O$

7.59 The reaction with spherical molecules will proceed more rapidly since in the case of the rod-like molecules, some collisions will be ineffective because the molecules will not interact with the proper orientations.

7.61 Initial rate = 0.030 moles I_2 per liter per second

7.63 (a) If the initial concentration of acetic acid is 0.10 M and the final concentration is 0.098 M, then a change of 0.0020 M has occurred. Therefore, the equilibrium concentration of each product is 0.0020 M.
(b) $K = 4.1 \times 10^{-5}$

7.65 Monitoring the disappearance of a reactant is a possible way to determine the rate of a reaction. It will do just as well as monitoring the formation of product because the stoichiometry of the reaction relates the concentrations of products and reactants to one another.

7.67 Some reactions are so fast that they are over before you can turn on a stopwatch or timer. You need specialized instruments with sophisticated electronics to follow the rates of very fast reactions.

7.69 The rate of conversion of diamond to graphite is so slow that it does not take place in any measurable length of time.

7.71 When you add sodium chloride, the presence of more chloride ions increases the concentration of one of the products of the reaction. The equilibrium shifts to the left, increasing the amount of solid silver chloride.

7.73 $[HF] = 1.0 \times 10^{-4} M$

7.75

Change	Quantity	Effect
Increase the pressure	Concentration of NO_3^-	**Increase**
Add some Zn	Concentration of NO_2	**No change**
Decrease the H^+	Concentration of Zn^{2+}	**Decrease**
Add Pt catalyst	Equilibrium constant, K	**No change**
Add some Ar	Concentration of H^+	**No change**
Decrease the Zn^{2+}	Equilibrium constant, K	**No change**
Increase the temperature	Concentration of Zn	**Increase**

7.77 Activation energy for the reverse reaction = 18 kJ.

7.79 $K = \dfrac{[H_2]^2[O_2]}{[H_2O]^2}$

$[H_2O] = 5.8 \times 10^{-5} M$

Chapter 8 Acids and Bases

8.1 acid reaction: $HPO_4^{2-} + H_2O \rightleftharpoons PO_4^{3-} + H_3O^+$;
base reaction: $HPO_4^{2-} + H_2O \rightleftharpoons H_2PO_4^- + OH^-$

8.2 (a) toward the left;

$$H_3O^+ + I^- \rightleftharpoons H_2O + HI$$
Weaker acid — Weaker base — Stronger base — Stronger acid

(b) toward the left;

$$CH_3COO^- + H_2S \rightleftharpoons CH_3COOH + HS^-$$
Weaker base — Weaker acid — Stronger acid — Stronger base

8.3 pK_a is 9.31

8.4 (a) ascorbic acid (b) aspirin

8.5 1.0×10^{-2}

8.6 (a) 2.46 (b) 7.9×10^{-5}, acidic

8.7 pOH = 4, pH = 10

8.8 0.0960 M

8.9 (a) 9.25 (b) 4.74

8.10 9.44

8.11 7.7

8.13 (a) $HNO_3(aq) + H_2O(\ell) \longrightarrow NO_3^-(aq) + H_3O^+(aq)$
(b) $HBr(aq) + H_2O(\ell) \longrightarrow Br^-(aq) + H_3O^+(aq)$
(c) $H_2SO_3(aq) + H_2O(\ell) \longrightarrow HSO_3^-(aq) + H_3O^+(aq)$
(d) $H_2SO_4(aq) + H_2O(\ell) \longrightarrow HSO_4^-(aq) + H_3O^+(aq)$
(e) $HCO_3^-(aq) + H_2O(\ell) \longrightarrow CO_3^{2-}(aq) + H_3O^+(aq)$
(f) $NH_4^+(aq) + H_2O(\ell) \longrightarrow NH_3(aq) + H_3O^+(aq)$

8.15 (a) weak (b) strong (c) weak (d) strong
(e) weak (f) weak

8.17 (a) false (b) false (c) true (d) false (e) true
(f) true (g) false (h) false

8.19 (a) A Brønsted-Lowry acid is a proton donor.
(b) A Brønsted-Lowry base is a proton acceptor.

8.21 (a) HPO_4^{2-} (b) HS^- (c) CO_3^{2-}
(d) $CH_3CH_2O^-$ (e) OH^-

8.23 (a) H_3O^+ (b) $H_2PO_4^-$ (c) $CH_3NH_3^+$
(d) HPO_4^{2-} (e) NH_4^+

8.25 The equilibrium favors the side with the weaker acid−weaker base combination. Equilibria (b) and (c) lie to the left; equilibriam (a) lies to the right.
(a)
$$C_6H_5OH + C_2H_5O^- \rightleftharpoons C_6H_5O^- + C_2H_5OH$$
Stronger acid — Stronger base — Weaker base — Weaker acid

(b)
$$HCO_3^- + H_2O \rightleftharpoons H_2CO_3 + OH^-$$
Weaker base — Weaker acid — Stronger acid — Stronger base

(c)
$$CH_3COOH + H_2PO_4^- \rightleftharpoons CH_3COO^- + H_3PO_4$$
Weaker acid — Weaker base — Stronger base — Stronger acid

8.27 (a) the pK_a of a weak acid (b) the K_a of a strong acid

8.29 (a) 0.10 M HCl (b) 0.10 M H_3PO_4 (c) 0.010 M H_2CO_3
(d) 0.10 M NaH_2PO_4 (e) 0.10 M aspirin

8.31 Only (b) is a redox reaction. The others are acid−base reactions.
(a) $Na_2CO_3 + 2HCl \longrightarrow 2NaCl + CO_2 + H_2O$
(b) $Mg + 2HCl \longrightarrow MgCl_2 + H_2$
(c) $NaOH + HCl \longrightarrow NaCl + H_2O$
(d) $Fe_2O_3 + 6HCl \longrightarrow 2FeCl_3 + 3H_2O$
(e) $NH_3 + HCl \longrightarrow NH_4Cl$
(f) $CH_3NH_2 + HCl \longrightarrow CH_3NH_3Cl$
(g) $NaHCO_3 + HCl \longrightarrow NaCl + H_2O + CO_2$
(h) $2Al + 6HCl \longrightarrow 2AlCl_3 + 3H_2$

8.33 (a) $10^{-3} M$ (b) $10^{-10} M$ (c) $10^{-7} M$ (d) $10^{-15} M$

8.35 (a) pH = 8 (basic) (b) pH = 10 (basic) (c) pH = 2 (acidic) (d) pH = 0 (acidic) (e) pH = 7 (neutral)

8.37 (a) pH = 8.5 (basic) (b) pH = 1.2 (acidic)
(c) pH = 11.1 (basic) (d) pH = 6.3 (acidic)

8.39 (a) pOH = 1.0, $[OH^-]$ = 0.10 M
(b) pOH = 2.4, $[OH^-]$ = $4.0 \times 10^{-3} M$
(c) pOH = 2.0, $[OH^-]$ = $1.0 \times 10^{-2} M$
(d) pOH = 5.6, $[OH^-]$ = $2.5 \times 10^{-6} M$

8.41 0.348 M

8.43 (a) 12 g of NaOH diluted to 400 mL of solution:

$$400 \text{ mL sol} \left(\frac{1 \text{ L sol}}{1000 \text{ mL sol}}\right)\left(\frac{0.75 \text{ mol NaOH}}{1 \text{ L sol}}\right) \times$$

$$\left(\frac{40.0 \text{ g NaOH}}{1 \text{ mol NaOH}}\right) = 12 \text{ g NaOH}$$

(b) 12 g of $Ba(OH)_2$ diluted to 1.0 L of solution:

$$\left(\frac{0.071 \; \text{mol Ba(OH)}_2}{1 \; \text{L sol}}\right)\left(\frac{171.4 \; Ba(OH)_2}{1 \; \text{mol Ba(OH)}_2}\right) = 12 \; \text{g Ba(OH)}_2$$

(c) 2.81 g KOH diluted to 500 mL

(d) 49.22 g sodium acetate diluted to 2 liters

8.45 5.66 mL

8.47 3.30×10^{-3} mol

8.49 The point at which the observed change occurs during a titration. It is usually so close to the equivalence point that the difference between the two becomes insignificant.

8.51 (a)
$H_3O^+ + CH_3COO^- \rightleftharpoons CH_3COOH + H_2O$ (removal of H_3O^+)
(b)
$HO^- + CH_3COOH \rightleftharpoons CH_3COO^- + H_2O$ (removal of OH^-)

8.53 Yes, the conjugate acid becomes the weak acid and the weak base becomes the conjugate base.

8.55 The pH of a buffer can be changed by altering the weak acid/conjugate base ratio, according to the Henderson-Hasselbalch equation. The buffer capacity can be changed without a change in pH by increasing or decreasing the amount of weak acid/conjugate base mixture while keeping the ratio of the two constant.

8.57 This would occur in a couple of cases. One is very common: You are using a buffer, such as Tris with a pK_a of 8.3, but you do not want the solution to have a pH of 8.3. If you wanted a pH of 8.0, for example, you would need unequal amounts of the conjugate acid and base, with there being more conjugate acid. Another case might be a situation where you are performing a reaction that you know will generate H^+ but you want the pH to be stable. In that situation, you might start with a buffer that was initially set to have more of the conjugate base so that it could absorb more of the H^+ that you know will be produced.

8.59 No, 100 mL of 0.1 M phosphate at pH 7.2 has a total of 0.01 mole of weak acid and conjugate base with equimolar amounts of each. 20 mL of 1 M NaOH has 0.02 mole of base, so there is more total base than there is buffer to neutralize it. This buffer would be ineffective.

8.61 (a) According to the Henderson-Hasselbalch equation, no change in pH will be observed as long as the weak acid/conjugate base ratio remains the same.
(b) The buffer capacity increases with increasing amounts of weak acid/conjugate base concentrations; therefore, 1.0 mol amounts of each diluted to 1 L would have a greater buffer capacity than 0.1 mol of each diluted to 1 L.

8.63 From the Henderson-Hasselbach equation,
$$pH = pK_a + \log(A^-/HA)$$
$A^-/HA = 10$, $\log(A^-/HA) = 1$ since $10^1 = 10$
$$pH = pK_a + 1$$

8.65 When 0.10 mol of sodium acetate is added to 0.10 M HCl, the sodium acetate completely neutralizes the HCl to acetic acid and sodium chloride. The pH of the solution is determined by the incomplete ionization of acetic acid.

$$K_a = \frac{[CH_3COO^-][H_3O^+]}{[CH_3COOH]} \qquad [H_3O^+]=[CH_3COO^-] = x$$

$$\sqrt{x^2} = \sqrt{K_a[CH_3COOH]} = \sqrt{(1.8 \times 10^{-5})(0.10)}$$

$x = [H_3O^+] = 1.34 \times 10^{-3} \; M$

$pH = -\log[H_3O^+] = 2.9$

8.67 $TRIS\text{-}H^+ + NaOH \longrightarrow TRIS + H_2O + Na^+$

8.69 The only parameter you need to know about a buffer is its pK_a. Choosing a buffer involves identifying the acid form that has a pK_a within one unit of the desired pH.

8.71 Choosing a buffer involves identifying the acid form that has a pK_a within one unit of the desired pH (a pH of 8.15). The TRIS buffer with a $pK_a = 8.3$ best fits this criteria.

8.73 $Mg(OH)_2$ is a weak base used in flame-retardant plastics.

8.75 (a) Respiratory acidosis is caused by hypoventilation, which occurs due to a variety of breathing difficulties such as a windpipe obstruction, asthma, or pneumonia. (b) Metabolic acidosis is caused by starvation or heavy exercise.

8.77 Sodium bicarbonate is the weak base form of one of the blood buffers. It tends to raise the pH of blood, which is the purpose of the sprinter's trick, so that the person can absorb more H^+ during the event. By putting $NaHCO_3$ into the system, the following reaction will occur:
$HCO_3^- + H^+ \rightleftharpoons H_2CO_3$. The loss of H^+ means that the blood pH will rise.

8.79 (a) Benzoic acid is soluble in aqueous NaOH.
$C_6H_5COOH + NaOH \rightleftharpoons C_6H_5COO^- + H_2O$
$pK_a = 4.19$ $\qquad\qquad\qquad\qquad pK_a = 15.56$
(b) Benzoic acid is soluble in aqueous $NaHCO_3$.
$C_6H_5COOH + NaHCO_3 \rightleftharpoons CH_3C_6H_4O^- + H_2CO_3$
$pK_a = 4.19$ $\qquad\qquad\qquad\qquad pK_a = 6.37$
(c) Benzoic acid is soluble in aqueous Na_2CO_3.
$C_6H_5COOH + CO_3^{2-} \rightleftharpoons CH_3C_6H_4O^- + HCO_3^-$
$pK_a = 4.19$ $\qquad\qquad\qquad\qquad pK_a = 10.25$

8.81 The strength of an acid is not important to the amount of NaOH that would be required to hit a phenolphthalein end point. Therefore, the more concentrated acid, the acetic acid, would require more NaOH.

8.83 $3.70 \times 10^{-3} \; M$

8.85 0.9 M

8.87 Yes, a pH of 0 is possible. A 1.0 M solution of HCl has $[H_3O^+] = 1.0 \; M$. $pH = -\log[H_3O^+] = -\log[1.0 \; M] = 0$

8.89 The qualitative relationship between acids and their conjugate bases states that the stronger the acid, the weaker its conjugate base. This can be quantified in the equation $K_b \times K_a = K_w$ or $K_b + 1.0 \times 10^{-14}/K_a$, where K_b is the base dissociation equilibrium constant for the conjugate base, K_a is the acid dissociation equilibruim constant for the acid, and K_w is the ionization equilibrium constant for water.

8.91 Yes, the strength of the acid is irrelevant. Both acetic acid and HCl have one H^+ to give up, so equal moles of either will require equal moles of NaOH to titrate to an end point.

8.93 You would need a ratio of 0.182 parts of the conjugate base to 1 part of the conjugate acid.

8.95 An equilibrium will favor the side of the weaker acid/weaker base. The larger the pK_a value, the weaker the acid.

8.97 (a) $HCOO^- + H_3O^+ \rightleftharpoons HCOOH + H_2O$
(b) $HCOOH + OH^- \rightleftharpoons HCOO^- + H_2O$

8.99 (a) 0.050 mol (b) 0.0050 mol (c) 0.50 mol

8.101 According to the Henderson-Hasselbalch equation,

$$pH = 7.21 + \log \frac{[HPO_4^{2-}]}{[H_2PO_4^-]}$$

As the concentration of $H_2PO_4^-$ increases, the $\log \frac{[HPO_4^{2-}]}{[H_2PO_4^-]}$ becomes negative, lowering the pH and becoming more acidic.

8.103 No, a buffer will have a pH equal to its pK_a only if equimolar amounts of the conjugate acid and base forms are present. If this is the basic form of Tris, then just putting any amount of it into water will give a pH much higher than the pK_a value.

8.105   (a) pH = 7.1, $[H_3O^+]$ = 7.9 × 10^{-8} M, basic
(b) pH = 2.0, $[H_3O^+]$ = 1.0 × 10^{-2} M, acidic
(c) pH = 7.4, $[H_3O^+]$ = 4.0 × 10^{-8} M, basic
(d) pH = 7.0, $[H_3O^+]$ = 1.0 × 10^{-7} M, neutral
(e) pH = 6.6, $[H_3O^+]$ = 2.5 × 10^{-7} M, acidic
(f) pH = 7.4, $[H_3O^+]$ = 4.0 × 10^{-8} M, basic
(g) pH = 6.5, $[H_3O^+]$ = 3.2 × 10^{-7} M, acidic
(h) pH = 6.9, $[H_3O^+]$ = 1.3 × 10^{-7} M, acidic
8.107   4.9:1, or 5:1 to one significant figure
8.109   (a) The concentration of H^+ is 1.0 × 10^{-2} M. Therefore, the number of moles of H^+ present is equal to 6.0 × 10^{-3} mol.
(b) From the balanced chemical equation, $HCl + NaHCO_3 \longrightarrow NaCl + CO_2 + H_2O$, the amount of sodium hydrogen carbonate needed to completely neutralize the stomach acid, is equal to 0.504 g $NaHCO_3$.
8.111   (a) 0.500 M HF(aq)
(b) K_a = 3.57 × 10^{-4}
8.113   From the balanced chemical equation, 2 moles of NaOH are needed for every 1 mole of the unknown diprotic acid. Therefore, the molarity of the unknown acid is 0.120 M, and its molar mass is 41.7 g/mol.

Chapter 9 Nuclear Chemistry

9.1 $^{139}_{53}I \longrightarrow ^{139}_{54}Xe + ^{0}_{-1}e$

9.2 $^{223}_{90}Th \longrightarrow ^{4}_{2}He + ^{219}_{88}Ra$

9.3 $^{74}_{33}As \longrightarrow ^{0}_{+1}e + ^{74}_{32}Ge$

9.4 $^{201}_{81}Tl + ^{0}_{-1}e \longrightarrow ^{201}_{80}Hg + \gamma$

9.5   Barium-122 has decayed through five half-lives, leaving 0.31 g. 10 g $\longrightarrow$ 5.0 g $\longrightarrow$ 2.5 g $\longrightarrow$ 1.25 g $\longrightarrow$ 0.625 g $\longrightarrow$ 0.31 g

9.6   The dose is 1.5 mL.

9.7   The intensity at 3.0 m is 3.3 × 10^{-3} mCi.

9.9   Alpha rays are He^{2+} ions ($^{4}_{2}He$) whereas protons are positively charged H^+ ions ($^{1}_{1}H$).

9.11   (a) 4.0 × 10^{-5} cm, which is visible light (blue).
(b) 3.0 cm (microwave radiation)
(c) 2.7 × 10^{-5} cm (ultraviolet light)
(d) 2.0 × 10^{-8} cm (X-ray)

9.13   (a) Infrared has the longest wavelength.
(b) X-rays have the highest energy.

9.15   (a) nitrogen-13   (b) phosphorus-33   (c) lithium-9   (d) calcium-39

9.17   oxygen-16

9.19 $^{151}_{62}Sm \longrightarrow ^{0}_{-1}e + ^{151}_{63}Eu$

9.21 $^{51}_{24}Cr + ^{0}_{-1}e \longrightarrow ^{51}_{23}V$

9.23 $^{248}_{96}Cm + ^{28}_{10}X \longrightarrow ^{116}_{51}Sb + ^{160}_{55}Cs$

The bombarding nucleus was neon $^{28}_{10}Ne$.

9.25   (a) beta emission   (b) gamma emission   (c) positron emission   (d) alpha emission

9.27   Gamma emission does not result in transmutation.

9.29 $^{239}_{94}Pu + ^{4}_{2}He \longrightarrow ^{240}_{95}Am + ^{1}_{1}H + 2^{1}_{0}n$

9.31   Iodine-125 decayed through approximately six half-lives, with 0.31 mg remaining: 20 mg $\longrightarrow$ 10 mg $\longrightarrow$ 5.0 mg $\longrightarrow$ 2.5 mg $\longrightarrow$ 1.25 mg $\longrightarrow$ 0.625 mg $\longrightarrow$ 0.31 mg

9.33   The plutonium underwent four half-lives since the glacier deposited it. There were 16 mg of plutonium/kg at the time of deposition.

16 mg $\longrightarrow$ 8 mg $\longrightarrow$ 4 mg $\longrightarrow$ 2 mg $\longrightarrow$ 1 mg

9.35   The rate of radioactive decay is independent of all conditions and is a property of each specific isotope. There is no way we can increase or decrease the rate.

9.37   (a) The iodine-131 remaining after two hours will be 8.88 × 10^8 counts/s.   (b) After 24 days, three half-lives have passed: 1/2 × 1/2 × 1/2 = 1/8, or 12.5% of the original amount remains. 24.0 mCi × 0.125 = 3.0 mCi.

9.39   Gamma radiation has the greatest penetrating power; therefore, protection from it requires the largest amount of shielding.

9.41   30 m

9.43   The curie (Ci) measures radiation intensity.

9.45   0.63 cc

9.47   At 20 cm, the intensity would be 3 × 10^3Bq (8 × 10^{-2}μCi).

9.49   Person A was exposed to the larger dose of radiation and injured more seriously.

9.51   Iodine-131 is concentrated in the thyroid and would be expected to induce the cancer.

9.53   (a) Cobalt-60 is used for (4) cancer therapy.
(b) Thallium-201 is used in (1) heart scans and exercise stress tests.   (c) Tritium is used for (2) measuring water content of the body.   (d) Mercury-197 is used for (3) kidney scans.

9.55   The product of fusion of hydrogen-2 and hydrogen-3 nuclei is helium-4 plus a neutron and energy.

9.57 $^{209}_{83}Bi + ^{58}_{26}Fe \longrightarrow ^{1}_{0}n + ^{266}_{109}Mt$

9.59 $^{10}_{5}B + ^{1}_{0}n \longrightarrow ^{11}_{5}B$

$^{11}_{5}B \longrightarrow ^{7}_{3}Li + ^{4}_{2}He$

9.61   The assumption of a constant carbon-14 to carbon-12 ratio rests on two assumptions: (1) that carbon-14 is continuously generated in the upper atmosphere by the production and decay of nitrogen-15 and (2) that carbon-14 is incorporated into carbon dioxide, CO_2, and other carbon compounds and then distributed worldwide as part of the carbon cycle. The continual formation of carbon-14; transfer of the isotope within the oceans, atmosphere, and biosphere; and decay of living matter keep the supply of carbon-14 constant.

9.63   2003 − 1350 = 653 years (if the experiment was run in 2003). 653 years/5730 years = 0.111 half-lives.

9.65   Radon-222 decays by alpha emission to polonium-218.

$$^{222}_{86}Rn \longrightarrow ^{218}_{84}Po + ^{4}_{2}He$$

9.67

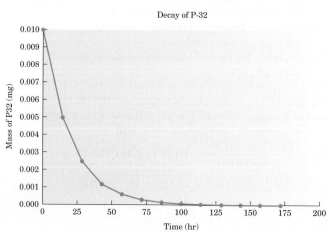

Decay of P-32

9.69 Neon-19 decays to fluorine-19, and sodium-20 decays to neon-20.

$$^{19}_{10}\text{Ne} \longrightarrow \,^{0}_{+1}\text{e} + \,^{19}_{9}\text{F}$$

$$^{20}_{11}\text{Na} \longrightarrow \,^{0}_{+1}\text{e} + \,^{20}_{10}\text{Ne}$$

9.71 Both the curie and the becquerel have units of disintegrations/second, a measure of radiation intensity.

9.73 (a) Natural sources = 82%

(b) Diagnostic medical sources = 11%

(c) Nuclear power plants = 0.1%

9.75 X-rays will cause more ionization than radar waves. X-rays have higher energy.

9.77 The decay product is neptunium-237. 1000/432 = 2.3 half-lives, so somewhat less that 25% of the original americium will remain after 1000 years.

9.79 One sievert is equal to 100 rem. This is sufficient to cause radiation sickness but not certain death.

9.81 (a) Radioactive elements are constantly decaying to other elements or isotopes, and these decay products are mixed with the original sample.

(b) Beta emission results from the decay of a neutron in the nucleus to a proton (the increase in atomic number) and an electron (the beta particle).

9.83 Oxygen-16 is stable because it has an equal number of protons and neutrons. The others are unstable because the numbers of protons and neutrons are unequal. In this case, the greater the difference in numbers of protons and neutrons, the faster the isotope decays.

9.85 The new element is darmstadtium-266.

$$^{208}_{82}\text{Pb} + \,^{64}_{28}\text{Ni} \longrightarrow \,^{266}_{110}\text{Ds} + 6\,^{1}_{0}\text{n}$$

9.87 The intermediate nucleus is boron-11.

$$^{10}_{5}\text{B} + \,^{1}_{0}\text{n} \longrightarrow \,^{11}_{5}\text{B}$$

$$^{11}_{5}\text{B} \longrightarrow \,^{7}_{3}\text{Li} + \,^{4}_{2}\text{He}$$

9.89 $^{206}_{80}\text{Hg}$

9.91 6 alpha and 4 beta particles are emitted.

Chapter 10 Organic Chemistry

10.1 Following are Lewis structures showing all valence electrons and all bond angles.

(a)

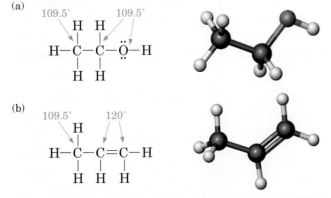

(b)

10.2 Of the four alcohols with the molecular formula $C_4H_{10}O$, two are 1°, one is 2°, and one is 3°. For the Lewis structures of the 3° alcohol and one of the 1° alcohols, some C—CH_3 bonds are drawn longer to avoid crowding in the formulas.

$CH_3CH_2CH_2CH_2OH$ — Primary (1°)

$CH_3CH_2CHCH_3$ (OH) — Secondary (2°)

CH_3CHCH_2OH (CH_3) — Primary (1°)

CH_3COH (CH_3, CH_3) — Tertiary (3°)

10.3 The three secondary (2°) amines with the molecular formula $C_4H_{11}N$ are:

$CH_3CH_2CH_2NHCH_3$ $CH_3CHNHCH_3$ (CH_3) $CH_3CH_2NHCH_2CH_3$

10.4 The three ketones with the molecular formula $C_5H_{10}O$ are:

$$CH_3CH_2CH_2\overset{\text{O}}{\overset{\|}{\text{C}}}CH_3 \quad CH_3CH_2\overset{\text{O}}{\overset{\|}{\text{C}}}CH_2CH_3 \quad CH_3\overset{\text{O}}{\overset{\|}{\text{C}}}CHCH_3$$
                                                        CH_3

10.5 The two carboxylic acids with the molecular formula $C_4H_8O_2$ are:

$$CH_3CH_2CH_2\overset{\text{O}}{\overset{\|}{\text{C}}}OH \quad \text{and} \quad CH_3\overset{\text{O}}{\overset{\|}{\text{C}}}HCOH$$
                                                  CH_3

10.6 The four esters with the molecular formula $C_4H_8O_2$ are:

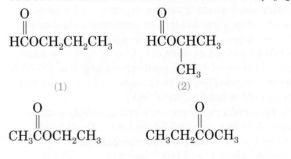

$$H\overset{\text{O}}{\overset{\|}{\text{C}}}OCH_2CH_2CH_3 \qquad H\overset{\text{O}}{\overset{\|}{\text{C}}}OCHCH_3$$
                                             CH_3
(1)   (2)

$$CH_3\overset{\text{O}}{\overset{\|}{\text{C}}}OCH_2CH_3 \qquad CH_3CH_2\overset{\text{O}}{\overset{\|}{\text{C}}}OCH_3$$
(3)   (4)

10.7 (a) T (b) T (c) F (d) F
(c) False: Carbon isn't even close. Silicon and oxygen are the two most abundant elements in the Earth's crust.
(d) False: Most organic compounds are insoluble in water.
10.9 Assuming that each is pure, there are no differences in their chemical or physical properties.
10.11 Wöhler heated ammonium chloride and silver cyanate, both inorganic compounds, and obtained urea, an organic compound
10.13 The four most principal elements that make up organic compounds and the number of bonds each typically forms are:
H forms one bond.
C forms four bonds.
O forms two bonds.
N forms three bonds.
10.15 Following are Lewis dot structures for each element:

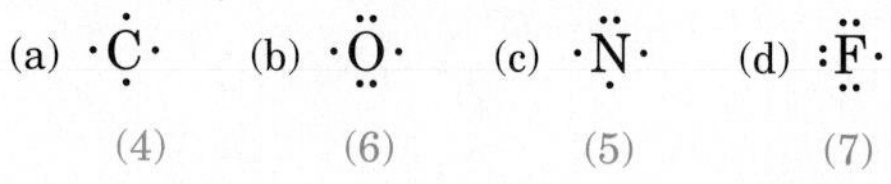

(a) $\cdot \overset{\cdot}{\underset{\cdot}{C}} \cdot$ (b) $\cdot \overset{\cdot \cdot}{\underset{\cdot}{O}} \cdot$ (c) $\cdot \overset{\cdot \cdot}{N} \cdot$ (d) $: \overset{\cdot \cdot}{\underset{\cdot \cdot}{F}} \cdot$

    (4)   (6)   (5)   (7)

10.17 (a) $H-\overset{\cdot\cdot}{\underset{\cdot\cdot}{O}}-\overset{\cdot\cdot}{\underset{\cdot\cdot}{O}}-H$

Hydrogen peroxide

(b) $H-\overset{\cdot\cdot}{N}-\overset{\cdot\cdot}{N}-H$ with H H below

Hydrazine

(c) $H-\overset{\overset{H}{|}}{\underset{\underset{H}{|}}{C}}-\overset{\cdot\cdot}{\underset{\cdot\cdot}{O}}-H$

Methanol

(d) $H-\overset{\overset{H}{|}}{\underset{\underset{H}{|}}{C}}-\overset{\cdot\cdot}{S}-H$

Methanethiol

(e) $H-\overset{\overset{H}{|}}{\underset{\underset{H}{|}}{C}}-\overset{\cdot\cdot}{N}-H$ (with H at bottom)

Methylamine

(f) $H-\overset{\overset{H}{|}}{\underset{\underset{H}{|}}{C}}-\overset{\cdot\cdot}{\underset{\cdot\cdot}{Cl}}:$

Chloromethane

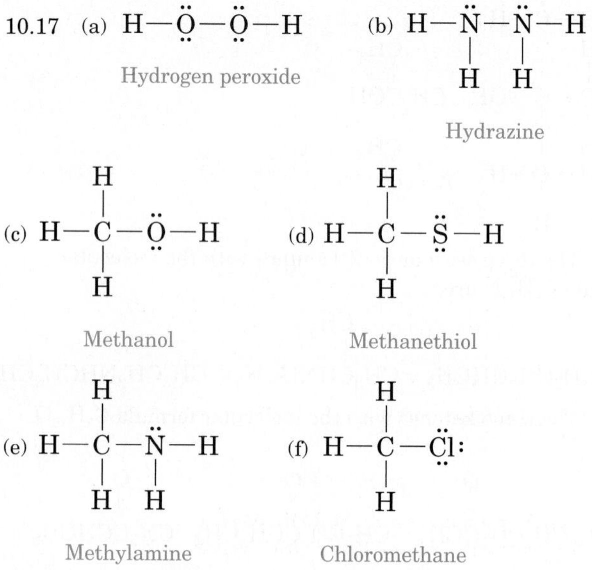

10.19 Following is a Lewis structure for each ion.

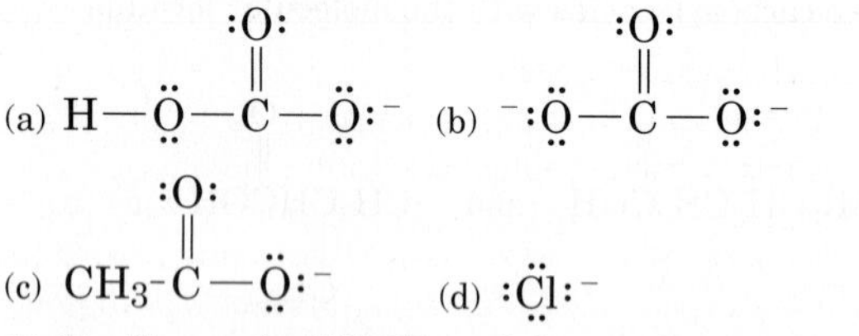

(a) $H-\overset{\cdot\cdot}{\underset{\cdot\cdot}{O}}-\overset{\overset{:O:}{\|}}{C}-\overset{\cdot\cdot}{\underset{\cdot\cdot}{O}}:^-$ (b) $^-:\overset{\cdot\cdot}{\underset{\cdot\cdot}{O}}-\overset{\overset{:O:}{\|}}{C}-\overset{\cdot\cdot}{\underset{\cdot\cdot}{O}}:^-$

(c) $CH_3-\overset{\overset{:O:}{\|}}{C}-\overset{\cdot\cdot}{\underset{\cdot\cdot}{O}}:^-$ (d) $:\overset{\cdot\cdot}{\underset{\cdot\cdot}{Cl}}:^-$

10.21 To use the VSEPR model to predict bond angles and the geometry about atoms of carbon, nitrogen, and oxygen:
(1) Write the Lewis structure for the target molecule showing all valence electrons. (2) Determine the number of regions of electron density around an atom of C, O, or N. (3) If you find four regions of electron density, predict bond angles of 109.5°. If you find three regions, predict bond angles of 120°. If you find two regions, predict bond angles of 180°.
10.23 You would find two regions of electron density around oxygen and, therefore, predict 180° for the C—O—H bond angle. The actual bond angle is approximately 109.5°.
10.25 (a) 120° about C and 109.5° about O
(b) 109.5° about N
(c) 120° about N

10.27 A functional group is a group of atoms that undergoes a predictable set of chemical reactions.

10.29 (a) $-\overset{\overset{:O:}{\|}}{C}-$ (b) $-\overset{\overset{:O:}{\|}}{C}-\overset{\cdot\cdot}{\underset{\cdot\cdot}{O}}-H$ (c) $-\overset{\cdot\cdot}{\underset{\cdot\cdot}{O}}-H$

(d) $-\overset{\cdot\cdot}{\underset{\underset{H}{|}}{N}}-H$ (e) $-\overset{\overset{:O:}{\|}}{C}-\overset{\cdot\cdot}{\underset{\cdot\cdot}{O}}-$

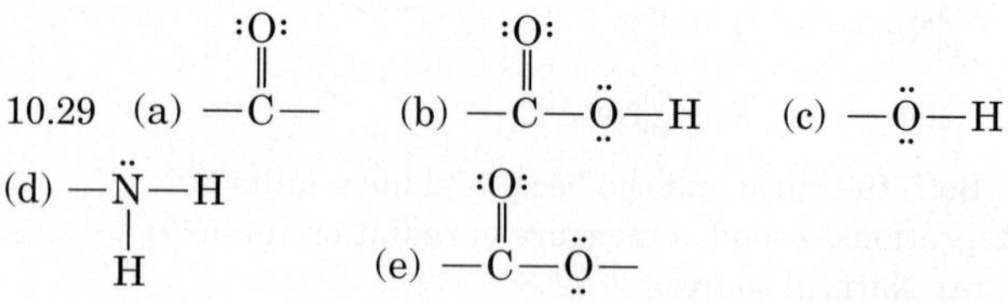

10.31 When applied to alcohols, tertiary (3°) means that the carbon bearing the –OH group is bonded to three other carbon atoms.
10.33 When applied to amines, tertiary (3°) means that the amine nitrogen is bonded to three other carbon groups.
10.35 (a) The four primary (1°) alcohols with the molecular formula $C_5H_{12}O$ are:

$CH_3CH_2CH_2CH_2CH_2OH$ $CH_3CH_2\underset{\underset{CH_3}{|}}{CH}CH_2OH$

$CH_3\overset{\overset{CH_3}{|}}{\underset{\underset{CH_3}{|}}{C}}CH_2OH$ $CH_3\underset{\underset{CH_3}{|}}{CH}CH_2CH_2OH$

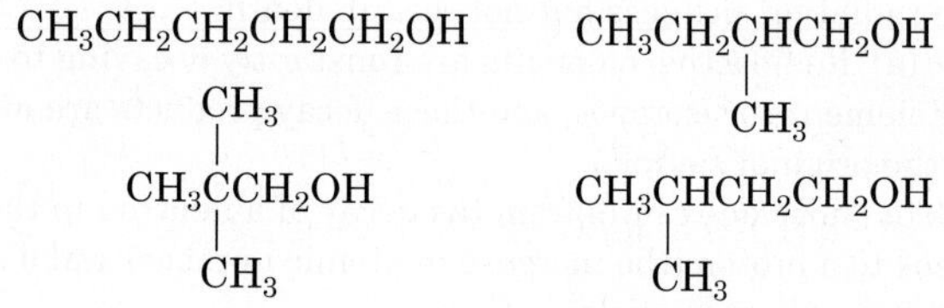

(b) The three secondary (2°) alcohols with the molecular formula $C_5H_{12}O$ are:

$CH_3\overset{\overset{OH}{|}}{CH}CH_2CH_2CH_3$ $CH_3CH_2\overset{\overset{OH}{|}}{CH}CH_2CH_3$ $CH_3\overset{\overset{OH}{|}}{CH}\underset{\underset{CH_3}{|}}{CH}CH_3$

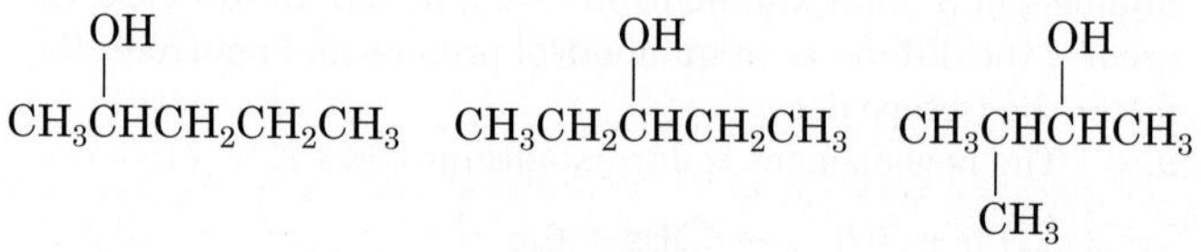

(c) The one tertiary (3°) alcohol with the molecular formula $C_5H_{12}O$ is:

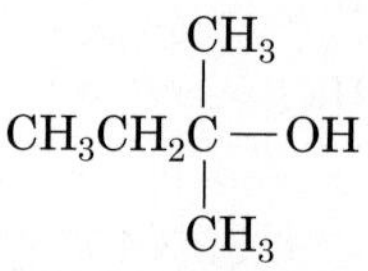

$CH_3CH_2\overset{\overset{CH_3}{|}}{\underset{\underset{CH_3}{|}}{C}}-OH$

10.37 The eight carboxylic acids with the molecular formula $C_6H_{12}O_2$ are:

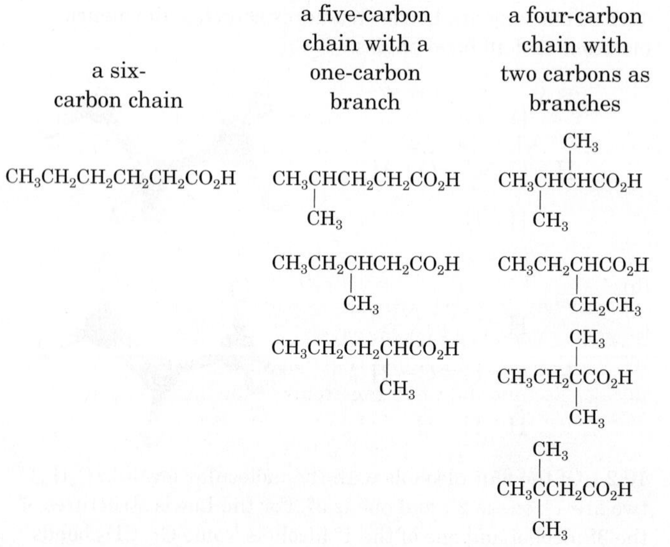

a six-carbon chain	a five-carbon chain with a one-carbon branch	a four-carbon chain with two carbons as branches

$CH_3CH_2CH_2CH_2CH_2CO_2H$ $CH_3\underset{\underset{CH_3}{|}}{CH}CH_2CH_2CO_2H$ $CH_3\underset{\underset{CH_3}{|}}{CH}\overset{\overset{CH_3}{|}}{CH}CO_2H$

$CH_3CH_2\underset{\underset{CH_3}{|}}{CH}CH_2CO_2H$ $CH_3CH_2\underset{\underset{CH_2CH_3}{|}}{CH}CO_2H$

$CH_3CH_2CH_2\underset{\underset{CH_3}{|}}{CH}CO_2H$ $CH_3CH_2\overset{\overset{CH_3}{|}}{\underset{\underset{CH_3}{|}}{C}}CO_2H$

$CH_3\overset{\overset{CH_3}{|}}{\underset{\underset{CH_3}{|}}{C}}CH_2CO_2H$

10.39 Taxol was discovered during a survey of indigenous plants for those containing phytochemicals that exhibited anti-tumor activity. It was sponsored by the National Cancer Institute with the goal of discovering new chemicals for fighting cancer.

10.41 The arrows point to atoms and show bond angles about each atom.

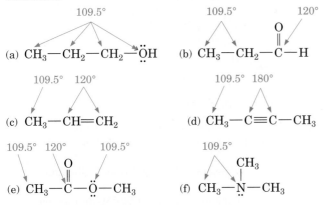

(a) $109.5°$ $CH_3-CH_2-CH_2-\ddot{O}H$

(b) $109.5°$ $120°$ $CH_3-CH_2-\overset{\overset{\textstyle O}{\|}}{C}-H$

(c) $109.5°$ $120°$ $CH_3-CH=CH_2$

(d) $109.5°$ $180°$ $CH_3-C\equiv C-CH_3$

(e) $109.5°$ $120°$ $109.5°$ $CH_3-\overset{\overset{\textstyle O}{\|}}{C}-\ddot{\ddot{O}}-CH_3$

(f) $109.5°$ $CH_3-\overset{\overset{\textstyle CH_3}{|}}{N}-CH_3$

10.43 Predict 109.5° for C—P—C bond angles.

$109.5°$ $CH_3-\overset{\overset{\textstyle ..}{}}{\underset{\underset{\textstyle CH_3}{|}}{P}}-CH_3$

10.45 The eight aldehydes with the molecular formula $C_6H_{12}O$ are below. The aldehyde functional group is written CHO.

a six-carbon chain	a five-carbon chain with a one-carbon branch	a four-carbon chain with two carbons as branches
$CH_3CH_2CH_2CH_2CH_2CHO$	$CH_3\overset{\underset{\textstyle CH_3}{\|}}{CH}CH_2CH_2CHO$	$CH_3\overset{\underset{\textstyle CH_3}{\|}}{CH}CHCHO$
	$CH_3CH_2\overset{\underset{\textstyle CH_3}{\|}}{CH}CH_2CHO$	$CH_3CH_2\overset{\underset{\textstyle CH_2CH_3}{\|}}{CH}CHO$
	$CH_3CH_2CH_2\overset{\underset{\textstyle CH_3}{\|}}{CH}CHO$	$CH_3CH_2\overset{\underset{\textstyle CH_3}{\|}}{C}CHO$ $\overset{\textstyle CH_3}{}$
		$CH_3\overset{\underset{\textstyle CH_3}{\|}}{C}CH_2CHO$ $\overset{\textstyle CH_3}{}$

10.47 (a) nonpolar covalent (b) nonpolar covalent
(c) nonpolar covalent (d) polar covalent (e) polar covalent
(f) polar covalent (g) polar covalent (h) polar covalent
10.49 Under each formula is given the difference in electronegativity between the atoms of the most polar bond.

(a) $H-\overset{\overset{\textstyle H}{|}}{\underset{\underset{\textstyle H}{|}}{C}}\overset{\delta-}{}\overset{\delta+}{}O-H$

(b) $H-\overset{\overset{\textstyle H}{|}}{\underset{\underset{\textstyle H}{|}}{C}}\overset{\delta-}{}\overset{\delta+}{}N-H$ $H^{\delta+}$

$O-H$ $(3.5-2.1 = 1.4)$ $N-H$ $(3.0-2.1 = 0.9)$

(c) $H-S-\overset{\overset{\textstyle H}{|}}{\underset{\underset{\textstyle H}{|}}{C}}-\overset{\overset{\textstyle H}{|}}{\underset{\underset{\textstyle H}{|}}{C}}\overset{\delta-}{}\overset{\delta+}{}\overset{}{\underset{\underset{\textstyle H^{\delta+}}{|}}{N}}-H$

(d) $H-\overset{\overset{\textstyle H}{|}}{\underset{\underset{\textstyle H}{|}}{C}}-\overset{\overset{\textstyle \delta^- O}{\|}}{\underset{}{C}}\overset{\delta+}{}-\overset{\overset{\textstyle H}{|}}{\underset{\underset{\textstyle H}{|}}{C}}-H$

$N-H$ $(3.0-2.1 = 0.9)$ $C=O$ $(3.5-2.5 = 1.0)$

(e) $\overset{\textstyle H}{\underset{\textstyle H}{}}\overset{\delta+}{}C=O_{\delta-}$

(f) $H-\overset{\overset{\textstyle H}{|}}{\underset{\underset{\textstyle H}{|}}{C}}-\overset{\overset{\textstyle O}{\|}}{C}\overset{\delta-}{}\overset{\delta+}{}O-H$

$C=O$ $(3.5-2.5 = 1.0)$ $O-H$ $(3.5-2.1 = 1.4)$

10.51 (a) All bond angles are approximately 120°, and the molecule is planar.
(b) Naphthalene is nonpolar.
10.53 The following all have the molecular formula $C_4H_8O_2$.

(a) Two carboxylic acids: $CH_3CH_2CH_2\overset{\overset{\textstyle O}{\|}}{C}-OH$ $CH_3\overset{\underset{\textstyle CH_3}{\|}}{C}H\overset{\overset{\textstyle O}{\|}}{C}-OH$

(b) Two esters: $CH_3CH_2\overset{\overset{\textstyle O}{\|}}{C}-OCH_3$ $CH_3\overset{\overset{\textstyle O}{\|}}{C}-OCH_2CH_3$

(c) One ketone + 2° alcohol: $CH_3\overset{\underset{\textstyle OH}{\|}}{C}H\overset{\overset{\textstyle O}{\|}}{C}CH_3$

(d) One aldehyde + 3° alcohol: $CH_3\overset{\overset{\textstyle HO}{|}}{\underset{\underset{\textstyle CH_3}{|}}{C}}-\overset{\overset{\textstyle O}{\|}}{C}-H$

10.55 (a) The molecular formula of lactic acid is $C_3H_6O_3$.
(b) The two functional groups are a 2° hydroxyl group and a carboxyl group.
(c) Predict bond angles of 120° about the carbonyl carbon and bond angles of 109.5° about the two other carbons and about the oxygen of each hydroxyl group.
(d) The C=O, O—H, and O—H bonds are polar covalent.
(e) Predict that lactic acid is a polar molecule.
10.57 Convert grams of salicylic acid (138 g/mol) to moles of salicylic acid. From the balanced equation, see that one mole of salicylic acid gives one mole of aspirin (180 g/mol). Finally, convert moles of aspirin to grams of aspirin. Doing this math gives 157 grams of aspirin.

$$120 \times \frac{180}{138} = 157 \text{ grams of aspirin}$$

10.59 (a) Curved arrows show the repositioning of two pairs of electrons.

$H-\overset{\overset{\textstyle H}{|}}{\underset{\underset{\textstyle H}{|}}{C}}-\overset{\overset{\textstyle :\ddot{O}:}{\|}}{C}\overset{}{\underset{\underset{\textstyle H}{|}}{N}}-H$ $\longleftrightarrow$ $H-\overset{\overset{\textstyle H}{|}}{\underset{\underset{\textstyle H}{|}}{C}}-\overset{\overset{\textstyle :\ddot{O}:^-}{|}}{C}=\overset{+}{\underset{\underset{\textstyle H}{|}}{N}}-H$

(a)   (b)

(b) Hydroxide ion shows 6 + 1 + 1 = 8 valence electrons. In both the hydroxide ion and the oxygen atom of contributing structure B, the oxygen in question has a complete octet. That is, each has a single bond and three nonbonding electron pairs and bears a negative charge.

$$H—\ddot{\underset{\cdot\cdot}{O}}:^{-}$$

(c) The ammonium ion contains 5 + 4 − 1 = 8 valence electrons. In both the ammonium ion and one nitrogen atom of contributing structure B, the nitrogen atom has a complete octet; it has four bonds and bears a positive charge.

$$H—\overset{\overset{\textstyle H}{|}}{\underset{\underset{\textstyle H}{|}}{N}}^{+}—H$$

(d) If the hybrid is best represented by contributing structure A, then predict the H—N—H bond angles to be approximately 109.5°.
(e) If the hybrid is best represented by contributing structure B, then predict the H—N—H bond angles to be 120°.
(f) The discovery that the actual H—N—H bond angles about each amide nitrogen atom in a protein are 120° suggests that contributing structure B makes a greater contribution to the hybrid than contributing structure A.

Chapter 11 Alkanes

11.1 This compound is octane, and its molecular formula is C_8H_{18}.
11.2 (a) Constitutional isomers (b) Same compound
11.3 Line-angle formulas for the three constitutional isomers with the molecular formula C_5H_{12} are:

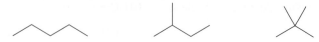

11.4 (a) 5-Isopropyl-2-methyloctane; its molecular formula is $C_{12}H_{26}$.
(b) 4-Isopropyl-4-propyloctane; its molecular formula is $C_{14}H_{30}$.
11.5 (a) Isobutylcyclopentane, C_9H_{18}
(b) *sec*-Butylcycloheptane, $C_{11}H_{22}$
(c) 1-Ethyl-1-methylcyclopropane, C_6H_{12}
11.6 The chair conformation with the three methyl groups equatorial is:

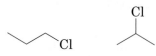

11.7 Cycloalkanes (a) and (c) show *cis-trans* isomerism:

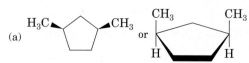

cis-1,3-Dimethylcyclopentane

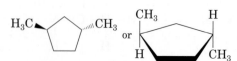

trans-1,3-Dimethylcyclopentane

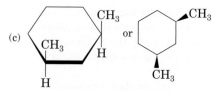

(c)

cis-1,3-Dimethylcyclohexane

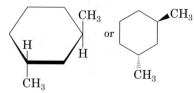

trans-1,3-Dimethylcyclohexane

11.8 In order of increasing boiling point, they are:
(a) 2,2-Dimethylpropane (9.5°C), 2-Methylbutane (27.8°C), Pentane (36.1°C)
(b) 2,2,4-Trimethylhexane (127°C), 3,3-Dimethylheptane (137°C), Nonane (151°C)
11.9 The two chloroalkanes with their IUPAC and common names are:

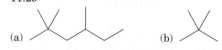

1-Chloropropane   2-Chloropropane
(Propyl chloride)   (Isopropyl chloride)

11.11 (a) A *hydrocarbon* is a compound that contains only hydrogen and carbon atoms.
(b) An *alkane* is a saturated hydrocarbon.
(c) A *saturated hydrocarbon* contains only carbon-carbon single bonds.
11.13 In a *line-angle formula*, each line terminus and vertex represents a carbon atom. Single, double, and triple carbon-carbon bonds are represented by one, two, and three lines, respectively.
11.15 (a) $C_{10}H_{22}$ (b) C_8H_{18} (c) $C_{11}H_{24}$
11.17 (a) T (b) T (c) F (d) F
(c) False: Constitutional isomers have the same molecular formula but a different connectivity of their atoms.
(d) False: Constitutional isomers are different compounds and, therefore, have different physical properties.
11.19 Structures (a) and (g) represent the same compound. Compounds (a,g), (c), (d), (e), and (f) each have the molecular formula $C_4H_{11}N$ and represent constitutional isomers.
11.21 Sets (b), (c), (e), and (f) represent pairs of constitutional isomers.
11.23 (a) T (b) T (c) T
11.25 2-Methylpropane and 2-methylbutane
11.27 (a) T (b) F (c) F
(b) False: They are not constitutional isomers because each has a different molecular formula: hexane is C_6H_{14}, and cyclohexane is C_6H_{12}.
(c) False: The parent name of a cycloalkane is the name of the unbranched alkane with the same number of carbon atoms that are in the ring with the prefix *cyclo-*.
11.29

(a)   (b)

(c)

(d)

(e)

(f)

(g)

(h)

11.31   A condensed structural formula shows only the order of bonding of the atoms in the compound. It does not show bond angles or the molecular shape.

11.33   The formula for calculating the internal angles in a regular polygon is:

$$\frac{(n-2)108°}{n}$$ where n = the number of sides

The actual bond angles in cyclopropane (a triangle) are approximately 60°, and the optimal bond angles are 109.5°. The actual bond angles in cyclopentane (a pentagon) are 108°, and the optimal bond angles are 109.5°.

11.35   The structural feature of cycloalkanes that makes *cis-trans* isomers possible in them is restricted rotation about the carbon-carbon bonds of the ring.

11.37   Following are the structural formulas and names for this pair of *cis* and *trans* isomers:

H_3C CH_3 H_3C CH_3

trans-1,2-Dimethyl-   *cis*-1,2-Dimethyl-
cyclopropane   cyclopropane

11.39   In equatorial methylcyclohexane, the methyl group is maximally separated from interaction with other atoms on the ring. In axial methylcyclohexane, the methyl group literally bangs into two axial hydrogen atoms on the same side of the ring.

11.41   (a) CH_3 OH (b) CH_3 OH or

CH_3 or CH_3

HO

OH

11.43   While the difference between the boiling points of some of the constitutional isomers with molecular formula C_7H_{16} is not large, you would be correct if you predicted that the isomer with the highest boiling point is heptane, the unbranched isomer, and the isomer with the lowest boiling point is 2,2-dimethylpentane, the most highly branched isomer.

Heptane   2,2-Dimethylpentane
C_7H_{16}, bp 98.4°C C_7H_{16}, bp 79.2°C

11.45   All liquid alkanes are less dense than water.
11.47   Both hexane and octane are colorless liquids, so you could not tell which is which simply by looking at them. To tell which is which, you would need to measure some physical property in which they differ, such as boiling point.

Hexane   Octane
C_6H_{12}, bp 69°C C_8H_{18}, bp 125.7°C

11.49   In general, boiling points increase as molecular weight increases.
11.51   (a) $2C_6H_{14} + 19O_2 \longrightarrow 12CO_2 + 14H_2O$
(b) $C_6H_{12} + 9O_2 \longrightarrow 6CO_2 + 6H_2O$
(c) $2C_6H_{12} + 19O_2 \longrightarrow 12CO_2 + 14H_2O$

11.53   (a) CH_3Br (b)   (c)

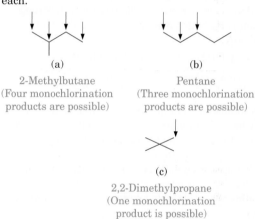

(d) Cl (e) CCl_2F_2

11.55   Hexane (a) and 2-methylpentane (c) have the same molecular formula, C_6H_{14}, and are constitutional isomers. Cyclohexane (b) has the molecular formula C_6H_{12}.
Hexane and 2-methylpentane
$$2C_6H_{14} + 19O_2 \longrightarrow 12CO_2 + 14H_2O$$
Cyclohexane
$$C_6H_{12} + 9O_2 \longrightarrow 6CO_2 + 6H_2O$$

11.57   Consider the possible structural formulas for C_5H_{12} and then the number of monochlorination products derived from each.

(a)   (b)

2-Methylbutane   Pentane
(Four monochlorination   (Three monochlorination
products are possible)   products are possible)

(c)

2,2-Dimethylpropane
(One monochlorination
product is possible)

11.59   Only one ring contains just carbon atoms. One ring contains two nitrogen atoms, and one ring contains two oxygen atoms.
11.61   Octane will produce more knocking than heptane.
11.63   The Freons are a class of chlorofluorocarbons. They were ideal for use as heat transfer agents in refrigeration systems because they are nontoxic, noncorrosive, and nonflammable. Two Freons used for this purpose were CCl_3F (Freon-11) and CCl_2F_2 (Freon-12).
11.65   They are hydrofluorocarbons and hydrochlorofluorocarbons. These compounds are much more chemically reactive in the atmosphere than the Freons and are destroyed before they reach the stratosphere.
11.67   (a) The longest chain is pentane. Its IUPAC name is 2-methylpentane.
(b) The pentane chain is numbered incorrectly. Its IUPAC name is 2-methylpentane.
(c) The longest chain is pentane. Its IUPAC name is 3-ethyl-3-methylpentane.
(d) The longest chain is hexane. Its IUPAC name is 3,4-dimethylhexane.

(e) The longest chain is heptane. Its IUPAC name is 4-methylheptane.

(f) The longest chain is octane. The IUPAC name is 3-ethyl-3-methyloctane.

(g) The ring is numbered incorrectly. Its IUPAC name is 1,1-dimethylcyclopropane.

(h) The ring is numbered incorrectly. The IUPAC name is 1-ethyl-3-methylcyclohexane.

11.69 Tetradecane is a liquid at room temperature.

11.71

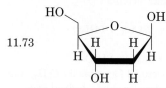

2-Isopropyl-5-methylcyclohexanol

In this chair conformation, all groups on the ring are in equatorial positions

11.73

HO, OH, H, H, H, H, O, OH, H

11.75 (a) Ethanol has the higher boiling point. In the pure liquid, ethanol molecules associate by relatively strong hydrogen bonding. Hydrogen bonding is not possible between molecules of diethyl ether.

CH_3OCH_3 CH_3CH_2OH

Dimethyl ether   Ethanol
(bp −24°C)   bp (78°C)

(c) Ethanol is soluble in all proportions in water. Dimethyl ether has only limited solubility. Ethanol associates with water molecules by hydrogen bonding through both the oxygen and hydrogen of its —OH group. Dimethyl ether associates with water molecules only through its partially negative oxygen atom. Intermolecular association with water molecules by hydrogen bonding is considerably greater for ethanol than for dimethyl ether.

Chapter 12 Alkenes and Alkynes

12.1 (a) 3,3-Dimethyl-1-pentene (b) 2,3-Dimethyl-2-butene
(c) 3,3-Dimethyl-1-butyne

12.2 (a) *trans*-3,4-Dimethyl-2-pentene
(b) *cis*-4-Ethyl-3-heptene

12.3 (a) 1-Isopropyl-4-methylcyclohexene (b) Cyclooctene
(c) 4-*tert*-Butylcyclohexene

12.4 Line-angle formulas for the other two heptadienes are:

cis,trans-2,4-Heptadiene   *cis,cis*-2,4-Heptadiene

12.5 Four stereoisomers are possible (two pair of *cis-trans* isomers).

12.6 (a) CH_3CHCH_3 with Br (b) cyclohexane with Br and CH_3

12.7 Propose a two-step mechanism similar to that which describes addition of HCl to propene.

Step 1: Add a proton. Reaction of H+ with carbon-2 of the carbon-carbon double bond gives a 3° carbocation intermediate.

cyclohexane–CH_3 + H+ ⟶ cyclohexane (+)–CH_3

A 3° carbocation intermediate

Step 2: Reaction of an electrophile and a nucleophile to form a new covalent bond. Reaction of the 3° carbocation intermediate with bromide ion completes the valence shell of carbon and gives the product.

cyclohexane(+)–CH_3 + :Br:⁻ ⟶ cyclohexane with Br: and CH_3

12.8 The product from each acid-catalyzed hydration is the same alcohol.

$$CH_3\overset{\displaystyle CH_3}{\underset{\displaystyle OH}{C}}CH_2CH_3$$

2-Methyl-2-butanol

12.9 Propose a three-step mechanism similar to that which describes the acid-catalyzed hydration of propene.

Step 1: Add a proton. Reaction of the carbon-carbon double bond with H+ gives a 3° carbocation intermediate.

cyclohexane–CH_3 + H+ ⟶ cyclohexane(+)–CH_3

A 3° carbocation intermediate

Step 2: Reaction of an electrophile and a nucleophile to form a new covalent bond. Reaction of the 3° carbocation intermediate with water completes the valence shell of carbon and gives an oxonium ion.

cyclohexane(+)–CH_3 + :O—H with H ⟶ cyclohexane with (+)O—H and CH_3

An oxonium ion

Step 3: Take a proton away. Loss of H+ from the oxonium ion completes the reaction and generates a new H+ catalyst.

cyclohexane with (+)O—H and CH_3 ⟶ cyclohexane with O: and CH_3 + H+

12.10 (a) $CH_3-\overset{\displaystyle CH_3}{\underset{\displaystyle H_3C}{C}}-\overset{\displaystyle}{\underset{\displaystyle Br}{CH}}-\overset{\displaystyle}{\underset{\displaystyle Br}{CH_2}}$ (b) cyclohexane with Cl and CH_2Cl

12.11 (a) F (b) F (c) F (d) T
(a) False: There are three classes of unsaturated hydrocarbons: alkenes, alkynes, and arenes. Arenes are discussed in Chapter 13.

(b) False: In the United States and other areas of the world that have vast supplies of natural gas, ethylene is derived from cracking the ethane that is separated from natural gas. In areas of the world that do not have vast reserves of natural gas, ethylene is derived from the thermal cracking of petroleum.

(c) False: They are not constitutional isomers because they have different molecular formulas. Ethylene is C_2H_4, and acetylene is C_2H_2.

12.13 A saturated hydrocarbon contains only carbon-carbon single bonds. An unsaturated hydrocarbon contains one or more carbon-carbon double or triple bonds.

12.15 There are 6 alkenes with the molecular formula $C_5H_{10}O$.

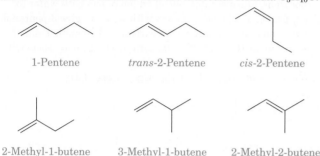

1-Pentene   trans-2-Pentene   cis-2-Pentene

2-Methyl-1-butene   3-Methyl-1-butene   2-Methyl-2-butene

12.17 There are several possibilities for each part. Here is one for each.

(a)   trans-4-Bromo-2-pentene   (b)   4-Bromo-1-pentene

12.19 (a) 109.5° 120°   (b) 120° —CH$_2$OH

(c) HC≡C—CH=CH$_2$ 180° 120° 120°   (d) 120°

12.21 (a) T (b) F (c) F (d) F (e) T (f) F
(b) False: Its name is 2-butene.
(c) False: To show *cis-trans* isomerism, each carbon of a double bond must have two different groups bonded to it.
(d) False: The four atoms bonded to the carbons atoms of the double bond all lie in the same plane.
(f) False: It has no stereocenters and no possibility for *cis-trans* isomerism.

12.23 (a) ⟋⟍⟋—Cl   (b) ⬡—CH$_3$

(c)   (d)

(e) ▷   (f)

12.25 (a) 2,5-Dimethyl-1-hexene
(b) 1,3-Dimethylcyclopentene

(c) 2-Methyl-1-butene (d) 2-Propyl-1-pentene

12.27 (a) The longest chain is four carbon atoms. The correct name is 2-methyl-1-butene.
(b) The ring is numbered incorrectly. The correct name is 4-isopropylcyclohexene.
(c) The ring is not numbered correctly. The correct name is 3-methyl-2-hexene.
(d) The chain is numbered incorrectly. The correct name is 2-ethyl-3-methyl-1-pentene
(e) The ring is numbered incorrectly. The correct name is 3,3-dimethylcyclohexene.
(f) The longest chain is seven carbon atoms. The correct name is 3-methyl-3-heptene.

12.29 Only (b) 2-hexene, (c) 3-hexene, and (e) 3-methyl-2-hexene will show *cis-trans* isomerism.

(a)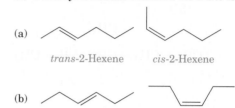

trans-2-Hexene   cis-2-Hexene

(b)

trans-3-Hexene   cis-3-Hexene

(c)

trans-3-Methyl-2-hexene   cis-3-Methyl-2-hexene

12.31 Here are drawings of the *cis* and *trans* isomers of cyclodecene.

cis-Cyclodecene   trans-Cyclodecene

12.33 Here are the *cis* and *trans* isomers:

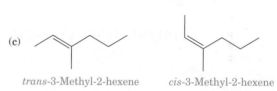
          H   H

Oleic acid

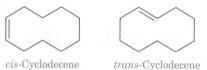   H
          H

Elaidic acid

12.35 Only parts (b) and (d) show *cis-trans* isomerism.

(b) CH$_3$ CH$_3$ (d) CH$_3$ CH$_3$

cis-1,2-dimethylcyclohexane   cis-4,5-dimethylcyclohexene

12.37 β-Ocimene has the molecular formula $C_{10}H_{16}$.

β-Ocimene

12.39 (a) T (b) T (c) T (d) T (e) F (f) T (g) T
(h) T (i) T (j) T (k) F (l) F (m) F (n) F
(e) False: A carbocation carbon has three bonds and a positive charge.
(k) False: Catalytic reduction of cyclohexene gives cyclohexane.
(l) False: H^+ comes from the acid catalyst, and —OH comes from H_2O.
(m) False: It is neither oxidation nor reduction. It is a hydration (addition).
(n) False: Both hydrations give 2-butanol.

12.41 (a) HBr (b) H_2O/H_2SO_4 (c) HI (d) Br_2

12.43 (a)

$$CH_3—CH_2—\overset{CH_3}{\underset{+}{C}}—CH_2—CH_3 \quad \text{and}$$
$$(3°)$$

$$CH_3—CH_2—\overset{CH_3}{\underset{+}{CH}}—CH—CH_3$$
$$(2°)$$

(b) $CH_3—CH_2—CH_2—\overset{+}{CH}—CH_3$ and
$(2°)$

$$CH_3CH_2—\overset{+}{CH}—CH_2—CH_3$$
$$(2°)$$

(c)
and
$(3°)$ $(2°)$

(d)
—CH_3 and
—CH_2^+
$(3°)$ $(1°)$

12.45 (a)
(b)

12.47 (a) $CH_3—\overset{CH_3}{\underset{|}{C}}=CH—CH_3$

(b) $CH_3—\overset{CH_3}{\underset{|}{C}}—CH=CH_2$

(c) $CH_2=CH—CH_2—CH_2—CH_3$

12.49 The only possibility for the formation of a 1° alcohol is if the molecule contains two =CH_2 groups (CH_2=CH_2). In ethylene, both carbons are =CH_2 and protonation of either gives a $CH_3—CH_2^+$ cation intermediate, which converts to ethanol, a 1° alcohol. For any other terminal alkene ($RCCH=CH_2$), H^+ will always add to the terminal carbon because it is the carbon of the double bond with the greater number of H atoms bonded to it. The result will be either a 2° or a 3° carbocation intermediate, which will lead to a 2° alcohol or a 3° alcohol.

12.51 Acid-catalyzed hydration of 2-pentene gives a mixture of 2-pentanol and 3-pentanol.

$$CH_2—CH=CH—CH_2—CH_3 + H_2O \xrightarrow{H_2SO_4}$$

$$CH_3—\overset{OH}{\underset{|}{CH}}—CH_2—CH_2—CH_3 +$$
2-Pentanol

$$CH_3—CH_2—\overset{OH}{\underset{|}{CH}}—CH_2—CH_3$$
3-Pentanol

Addition of a proton to 2-pentene gives two different 2° carbocation intermediates, one of which reacts with water to give 2-pentanol. The other reacts with water to give 3-pentanol.

Addition of a proton to either carbon of the double bond of 3-hexene gives the same 2° carbocation, which then reacts with a water molecule to give 3-hexanol.

12.53 (a) and (b) $CH_3—CH_2—CH_2—CH_2—CH_3$

(c)
(d)
—CH_3

12.55 (a) $CH_2=CH_2 + H_2 \xrightarrow{Pd} CH_3—CH_3$

(b) $CH_2=CH_2 + H_2O \xrightarrow{H_2SO_4} CH_3—CH_2OH$

(c) $CH_2=CH_2 + HBr \longrightarrow CH_3CH_2Br$

(d) $CH_2=CH_2 + Br_2 \longrightarrow Br—CH_2CH_2—Br$

(e) $CH_2=CH_2 + HCl \longrightarrow CH_3CH_2—Cl$

12.57 (a) F (b) F (c) T (d) T (e) T (f) F
(a) False: Polyethylene contains no carbon-carbon double bonds.
(b) False: There are no carbon-carbon double bonds in polyethylene. All C—C—C bond angles are 109.5°.
(f) False: Currently, neither polystyrene (PS) nor polyvinyl chloride (PVC) are recycled.

12.59 A *pheromone* is a chemical produced by an animal and serves as a stimulus to another individual of the same species for one or more behavioral responses.

12.61 The molecular weight of 11-tetradecenyl acetate is 254.4 amu, and its molar mass is 254.4 g/mol. 1.0×10^{-12} g of 11-tetradecenyl acetate is equal to 2.0×10^9 molecules.

$$1.0 \times 10^{-12}g \times \frac{mol}{254. \text{ g}} \times \frac{6.02 \times 10^{23} \text{ molecules}}{mol}$$
$$= 2 \times 10^9 \text{ molecules}$$

12.63 It is longer for the all-*trans* isomer.

12.65 V is a code letter for poly(vinyl chloride), also known as PVC. PP is the code for polypropylene, and PS is the code for polystyrene.

12.67 There are 13 carbon-carbon double bonds in lycopene, only 11 of which have the possibility for *cis-trans* isomerism. For the double bonds on either end of the long hydrocarbon chain, no *cis-trans* isomerism is possible.

12.69 (a)
(b)
—CH_3

(c)
(d)
=CH_2

12.71 Following is the equation for the acid-catalyzed hydration of 3-hexene. Because of the symmetrical structure of 3-hexene, the *cis* and *trans* isomers will both give 3-hexanol.

$$CH_3CH_2CH=CHCH_2CH_3 + H_2O \xrightarrow{H_2SO_4}$$

3-Hexene

$$CH_3CH_2\overset{\overset{\textstyle OH}{|}}{C}HCH_2CH_2CH_3$$

3-Hexanol

12.73 Reagents are shown over the reaction arrows:

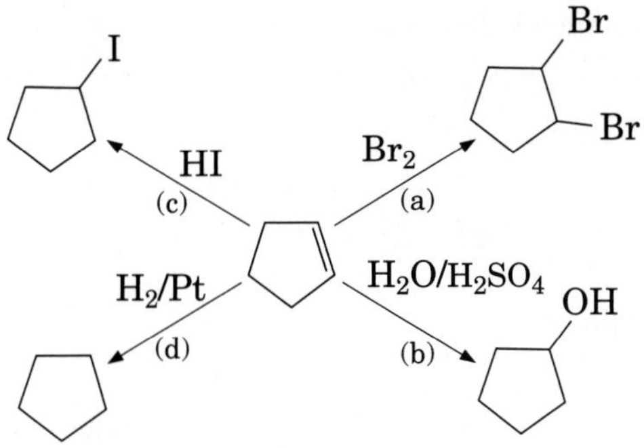

12.75 Two *cis-trans* isomers are possible for oleic acid, four for linoleic acid, and eight for linolenic acid. Note that each is a C_{18} fatty acid. The all-*cis* isomer of each is drawn below:

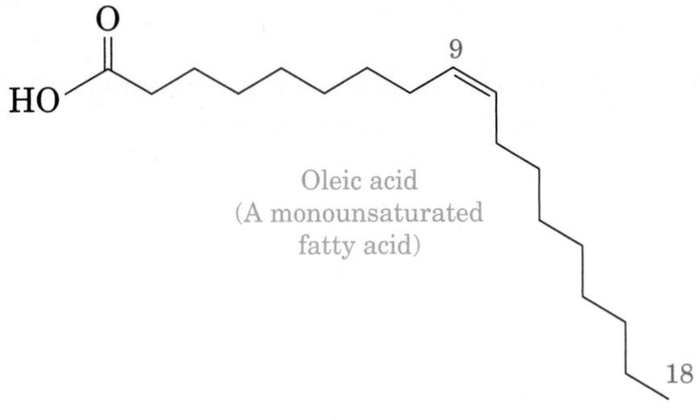

Oleic acid
(A monounsaturated
fatty acid)

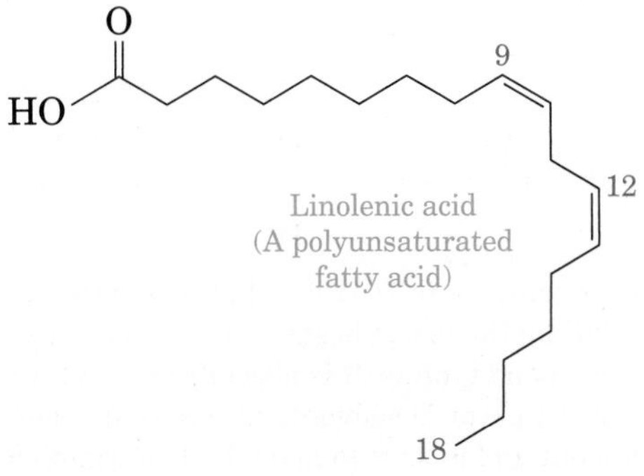

Linolenic acid
(A polyunsaturated
fatty acid)

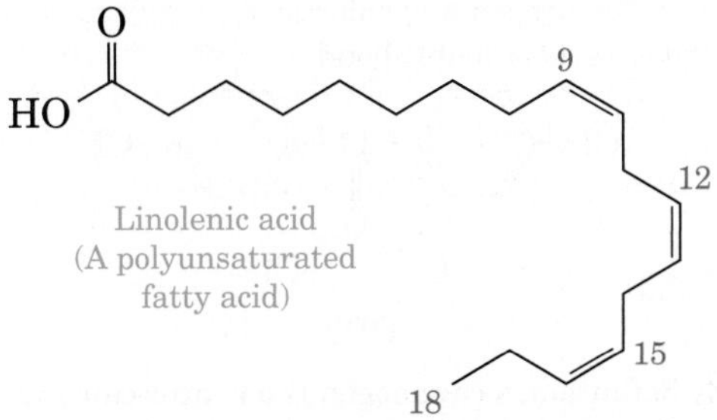

Linolenic acid
(A polyunsaturated
fatty acid)

12.77 Step 1: An oxidation that converts a carbon-carbon single bond of an alkane to a carbon-carbon single bond of an alkene.

Step 2: A hydration of a carbon-carbon double bond to give a 2° alcohol.

Step 3: An oxidation of a 2° alcohol to a ketone.

It is common in organic and biochemistry to call a carbon adjacent to a functional group an α-carbon and the next one a β-carbon. In this biochemical sequence, the β-carbon from the carbonyl group of the ester is converted from a —CH₂— group to a ketone group by an oxidation reaction, hence the term β-oxidation.

Chapter 13 Benzene and Its Derivatives

13.1 (a) 2,4,6-Tri-*tert*-butylphenol (b) 2,4-Dichloroaniline
(c) 3-Nitrobenzoic acid

13.3 A saturated compound contains only single covalent bonds. An unsaturated compound contains one or more double or triple bonds. The most common types of double and triple bonds are:

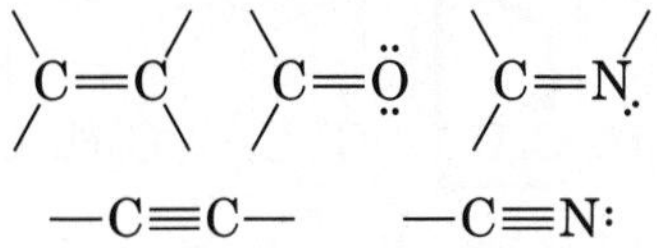

13.5 The members of each class of hydrocarbons contain fewer hydrogens than an alkane or cycloalkane with the same number of carbon atoms. Alternatively, each class of hydrocarbons contains one or more carbon-carbon double or triple bonds.

13.7 NO

13.9 (a) CH_4 (b) $CH_2=CH_2$

Methane   Ethene
(Ethylene)

(c) $HC\equiv CH$ (d)

Ethyne
(Acetylene)

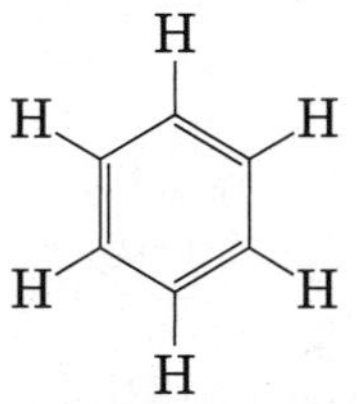

Benzene

13.11 Benzene consists of six carbons, each surrounded by three regions of electron density, which gives 120° for all bond angles. Bond angles of 120° in benzene can be maintained only if the molecule is planar. All C, H, and Cl atoms in 1,4-dichlorobenzene lie in the same plane.

13.13 (a) T (b) F (c) F (d) T (e) F (f) F
(b) False: *para* refers to a 1,4-disubstituted benzene ring.
(c) False: Benzene rings are planar, and therefore, *cis-trans* isomers are not possible.
(e) False: Benzene contains only one ring and is not a PAH (polynuclear aromatic hydrocarbon). A PAH must have two or more adjacent aromatic rings.
(f) False: Benzo[*a*]pyrene does not directly bind with DNA, but instead, its metabolic product, a diol epoxide, binds with DNA and causes mutations.

13.15 (a)

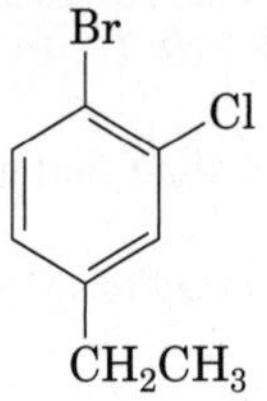

(b)

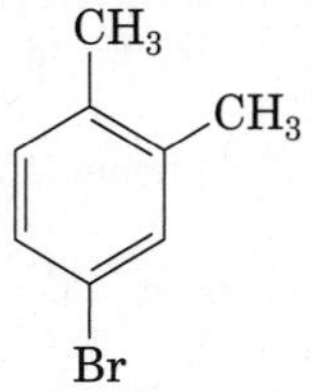

(c) 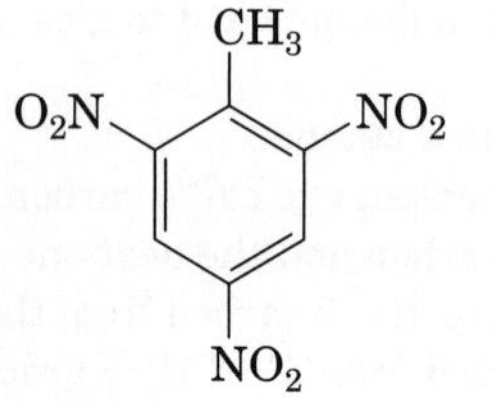 [structure: benzene ring with CH₃, two NO₂ groups ortho, and NO₂ para]

(d) 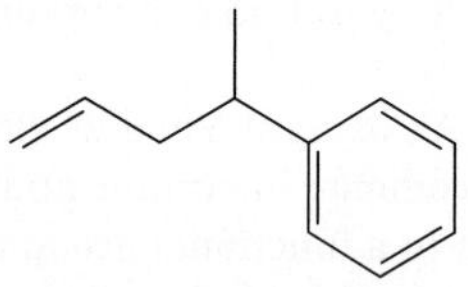 [structure: 4-phenyl-1-pentene type]

(e) 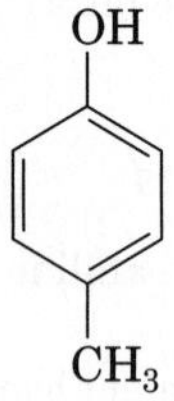 [structure: phenol with CH₃ para — OH]

(f) 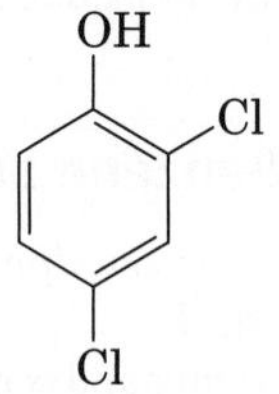 [structure: phenol with Cl ortho and Cl para]

13.17 The repeating unit is circled and shown in bold.

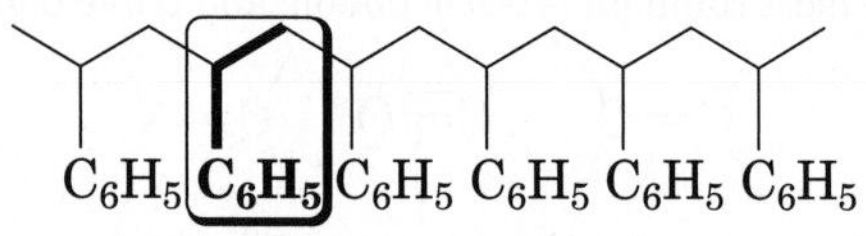

$$C_6H_5 \quad \boxed{C_6H_5} \quad C_6H_5 \quad C_6H_5 \quad C_6H_5 \quad C_6H_5$$

13.19 Only cyclohexene will react with a solution of bromine in dichloromethane. A solution of Br_2/CH_2Cl_2 is reddish-purple, whereas benzene and 1,2-dibromocyclohexane are both colorless. To tell which bottle contains which compound, place a small quantity of each compound in a test tube and to each, add a few drops of Br_2/CH_2Cl_2 solution. If the red color disappears, the compound is cyclohexene, which is converted to 1,2-dibromocyclohexane. If the reddish-purple color remains, the compound is benzene.

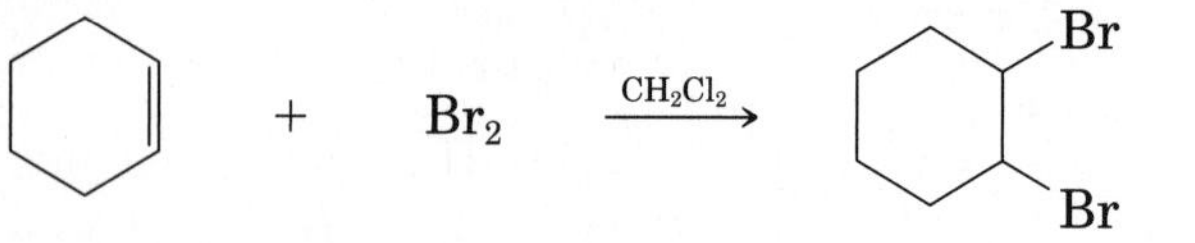

[reaction: Cyclohexene + Br₂ → (CH₂Cl₂) → 1,2-dibromocyclohexane]

Cyclohexene (colorless)   Bromine (red)   1,2-dibromocyclohexane (colorless)

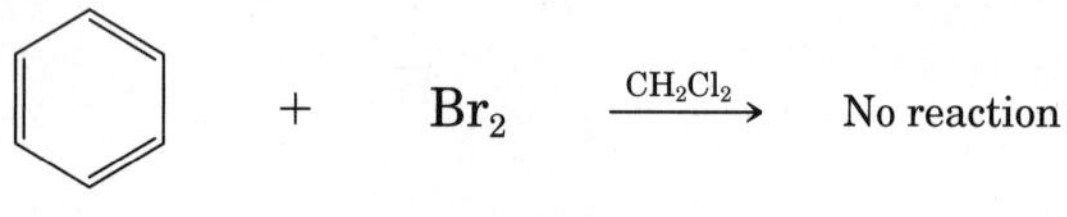

[reaction: Benzene + Br₂ → (CH₂Cl₂) → No reaction]

Benzene

13.21

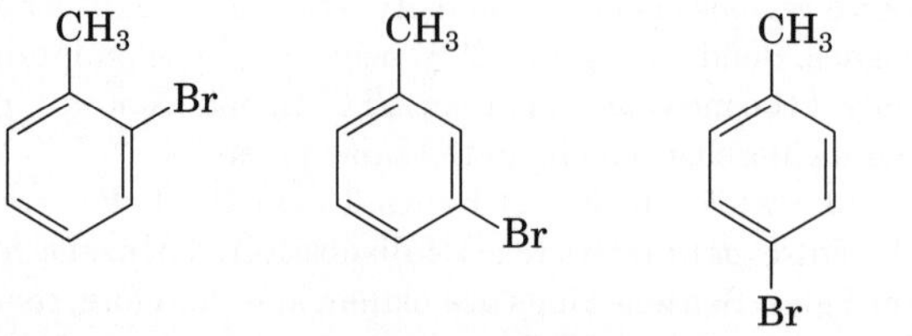

2-Bromotoluene (*o*-Bromotoluene)   3-Bromotoluene (*m*-Bromotoluene)   4-Bromotoluene (*p*-Bromotoluene)

13.23 (a) Nitration using HNO_3/H_2SO_4 followed by sulfonation using H_2SO_4. The order of the steps may also be reversed.
(b) Bromination using $Br_2/FeCl_3$ followed by chlorination using $Cl_2/FeCl_3$. The order of the steps may also be reversed.
13.25 (a) T (b) T (c) F (d) F (e) F (f) T (g) T (h) T (i) T
(c) False: The pK_a of phenol (10) is larger than that of acetic acid (4.7).
(d) False: In autoxidation, a C—H is converted to a C—O—O—H (hydroperoxide) group.
(e) False: A carbon radical has no charge.

13.27 Autoxidation is the reaction of a C—H group with oxygen, O_2, to form a hydroperoxide group, C—O—O—H.
13.29 Adding the equations for Step 2a and 2b and then canceling species that appear on both sides of the reaction arrow does, in fact, give the net equation for hydroperoxide formation.
13.31 The common group in vitamin E, BHT, and BHA is a substituted phenol.
13.33 (a) The functional groups in albuterol are a primary (1°) alcohol, a secondary (2°) alcohol, a secondary (2°) amine, and a phenolic —OH group.
(b) When albuterol is treated with one equivalent of NaOH, the phenolic —OH group is the only group acidic enough to react and form a sodium salt.

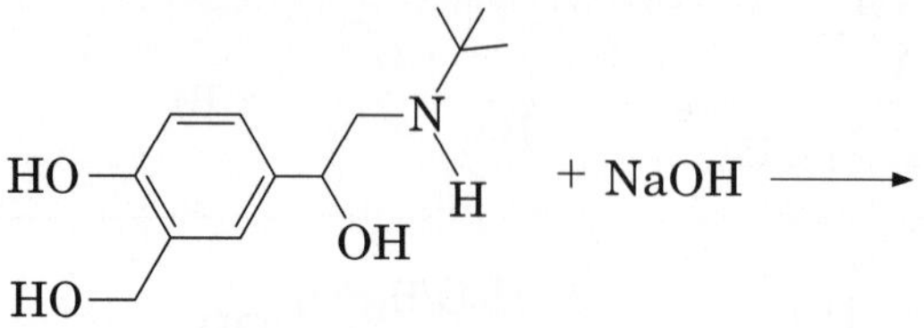

[structure: Albuterol + NaOH →]

Albuterol

[structure: sodium phenoxide salt + H₂O]

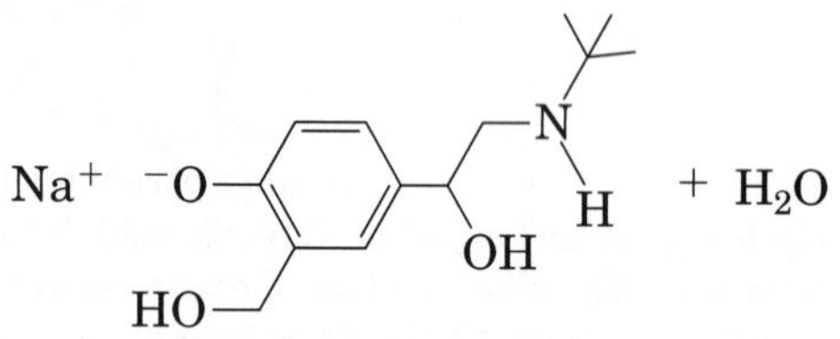

(c) When albuterol is treated with one equivalent of HCl, the 2° amine is converted to an ammonium salt.

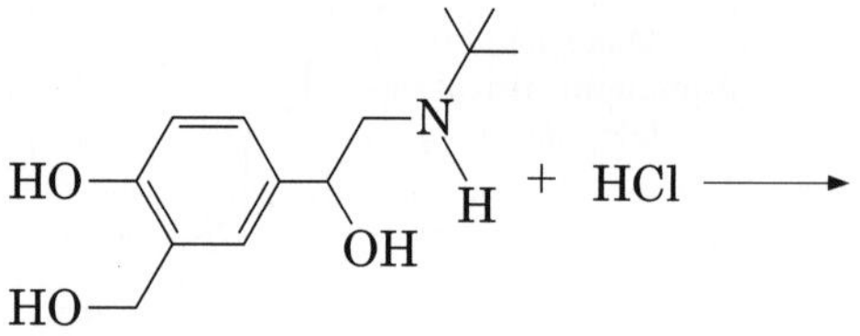

[structure: Albuterol + HCl →]

Albuterol

[structure: ammonium salt product]

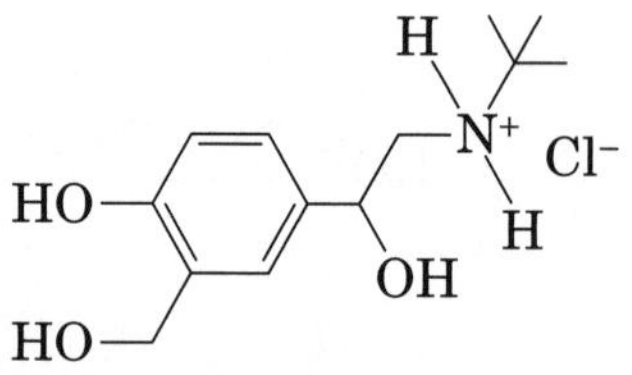

13.35 The advantages of DDT are that it is extremely effective in killing the insect hosts that transmit diseases such as malaria and typhus. It is also effective in killing crop-destroying insect pests. Disadvantages are that it persists in the environment and is toxic to many higher organisms.
13.37 Loss of hydrogen and chlorine from adjacent carbons gives a carbon-carbon double bond.

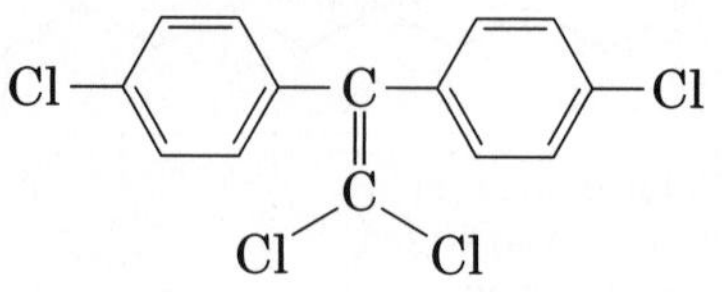

[structure: DDE]

DDE

13.39 By definition, a carcinogen is a cancer-causing substance. The carcinogens present in cigarette smoke belong to the class of compounds called polynuclear aromatic hydrocarbons (PAHs).

13.41 Cyclonite (RDX) has the largest percentage of nitrogen.

Explosive	Mol Wt	% N
TNT	227.1	18.50
Nitroglycerin	227.1	18.50
Cyclonite	222.1	37.84
PETN	316.1	17.71

13.43 The functional groups most responsible for the water solubility of these dyes are the ionic —SO_3^-—Na^+ groups in each.

13.45 (a) Both lycopene and β-carotene have an extended system of single-double-single bonds.
(b) The extended system of single-double-single bonds.
(c) Both lycopene and β-carotene are nonpolar covalent compounds and, therefore, are insoluble in water. The food dyes have some polar covalent bonds, but their water solubility is due primarily to the presence of ionic groups and the fact that they are ionic compounds.

13.47 The phenolic group is the —OH group bonded to the benzene ring, here shown in boldface.

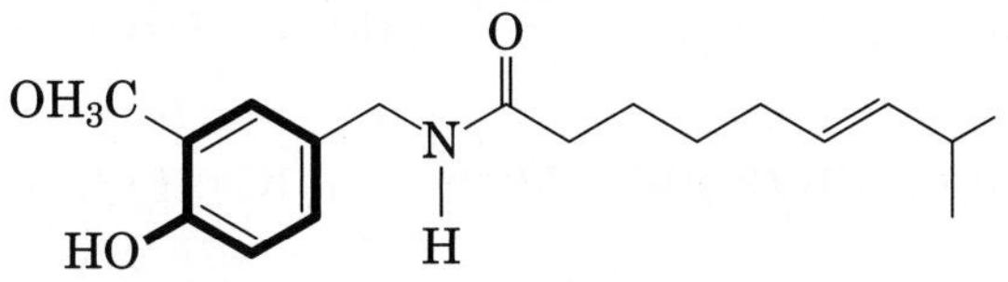

13.49 Capsaicin is used in medicine as a topical anesthetic.

13.51 (a)

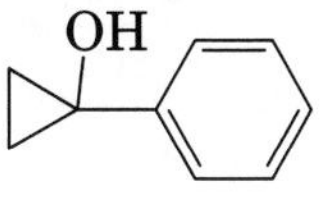

Phenylcyclopropanol

(b)

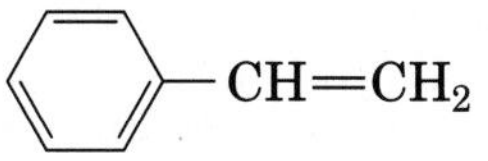

Styrene

(c)

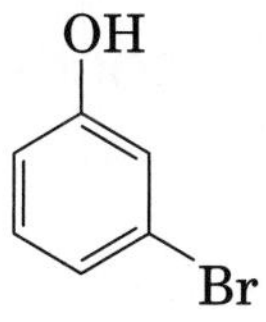

m-Bromophenol

(d)

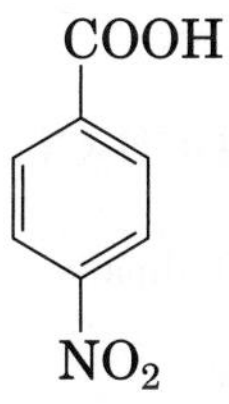

4-Nitrobenzoic acid

(e)

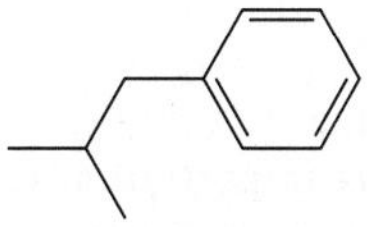

Isobutylbenzene

(f)

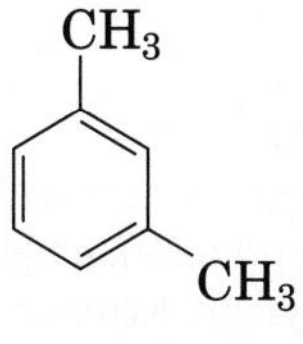

m-Xylene

13.53 (a)

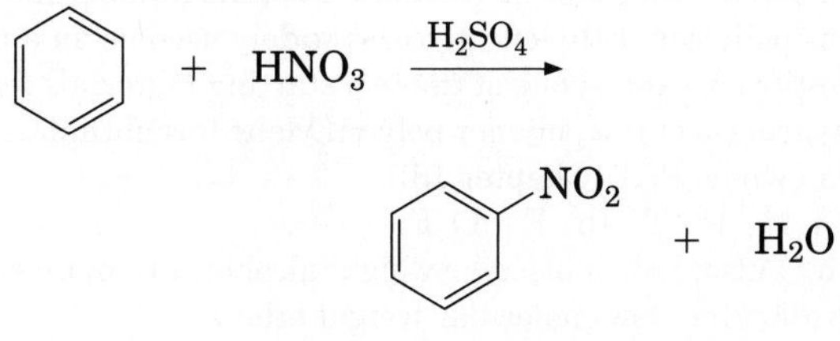

(b)

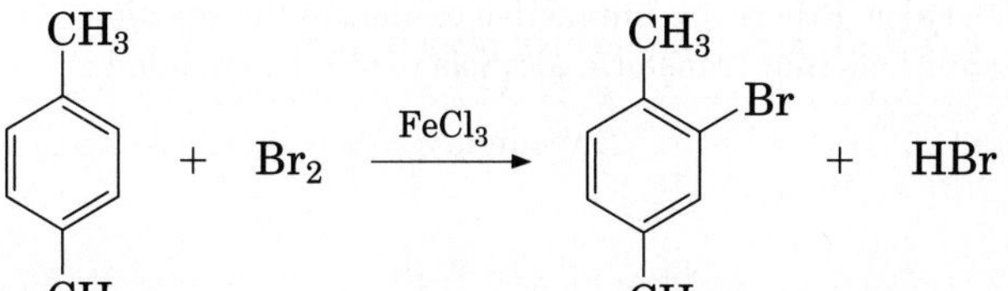

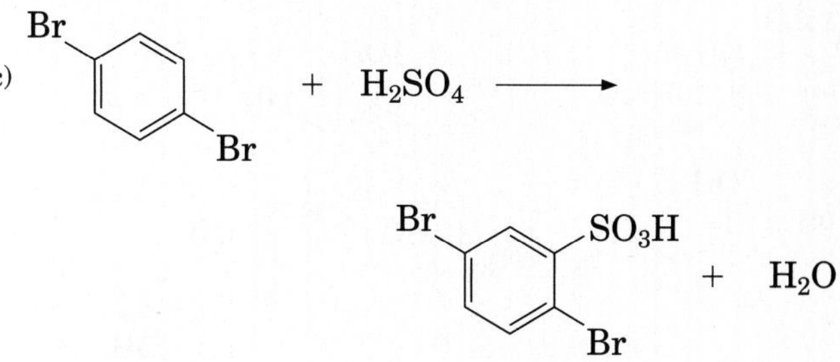

13.55

o-Bromotoluene

m-Bromotoluene   p-Bromotoluene

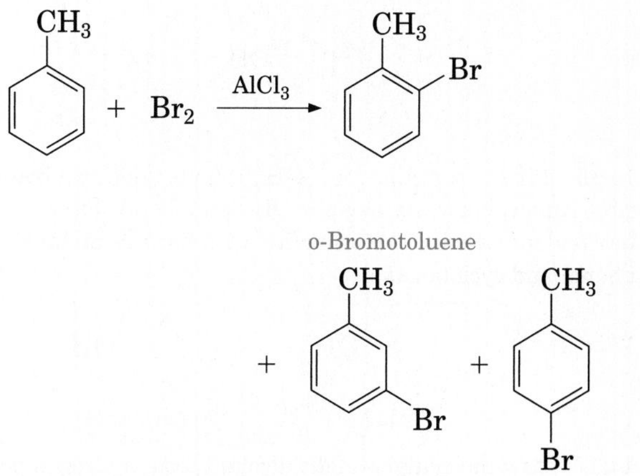

Chapter 14 Alcohols, Ethers, and Thiols

14.1 (a) 2-Heptanol (b) 2,2-Dimethyl-1-propanol
(c) *cis*-3-Isopropylcyclohexanol
14.2 (a) Primary (1°) (b) Secondary (2°) (c) Primary (1°)
(d) Tertiary (3°)
14.3 The structure of the major alkene product from each reaction is enclosed in a box. In each case, the major product contains the more substituted double bond.

(a)

(b)

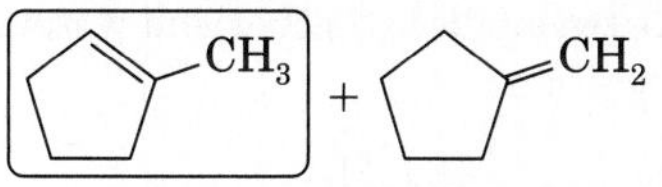

14.4

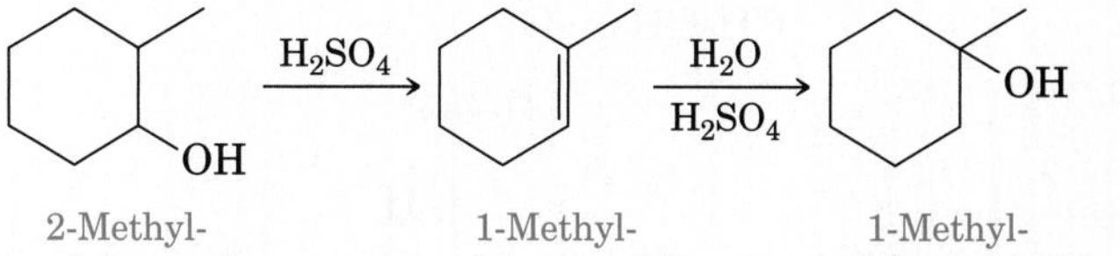

2-Methyl-cyclohexanol   1-Methyl-cyclohexene (C)   1-Methyl-cyclohexanol (D)

14.5 Each secondary alcohol is oxidized to a ketone.

(a) 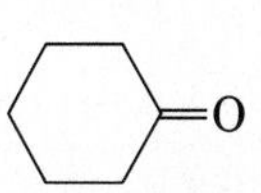   (b) $CH_3CCH_2CH_2CH_3$ (with O double bonded above C)

14.6 (a) Ethyl isobutyl ether (b) Cyclopentyl methyl ether
14.7 (a) 3-Methyl-1-butanethiol (b) 3-Methyl-2-butanethiol
14.9 The difference is in the number of carbon atoms bonded directly to the carbon bearing the –OH group. For primary alcohols, there is zero or one carbon group; for a secondary alcohol, there are two, and for a tertiary alcohol, there are three.
14.11 None of the alcohols in problem 14.10 is primary. The alcohols in parts (a) and (b) are tertiary. The alcohols in parts (c) and (d) are secondary.

14.13

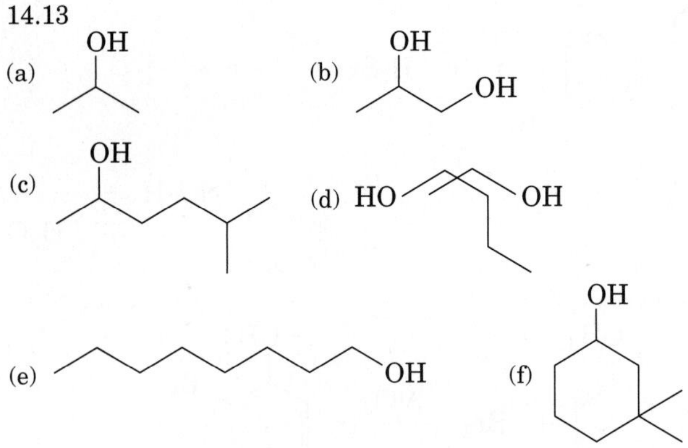

(a)

(b)

(c)

(d) HO⎯⎯OH (alkene diol)

(e)

(f)

14.15 Phenol contains an —OH group bonded to a benzene ring. An alcohol contains an —OH group bonded to a tetravalent carbon. Following are structural formulas for phenol and cyclohexanol.

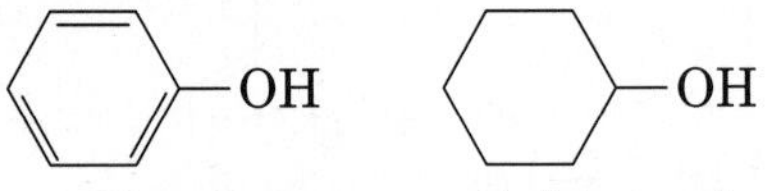

Phenol   Cyclohexanol

14.17 Low-molecular-weight alcohols form hydrogen bonds with water molecules through both the oxygen and hydrogen atoms of their —OH groups, and because of this association, they dissolve in water. Low-molecular-weight alkanes, alkenes, and alkynes are nonpolar compounds and do not associate in any way with water molecules. The large amount of hydrogen bonding between an alcohol molecule and water molecules makes the low-molecular-weight alcohols soluble in water. Low-molecular-weight alkanes, alkenes, and alkynes are insoluble in water.

14.19 As the molecular weight of an alcohol increases, the hydrocarbon part of the molecule becomes a larger and larger part of its surface area. The physical properties of alcohols, including their solubility in water, then become more like those of hydrocarbons of similar carbon skeletons.

14.21 Hydrogen bonding between diethyl ether and water is shown in the following illustration.

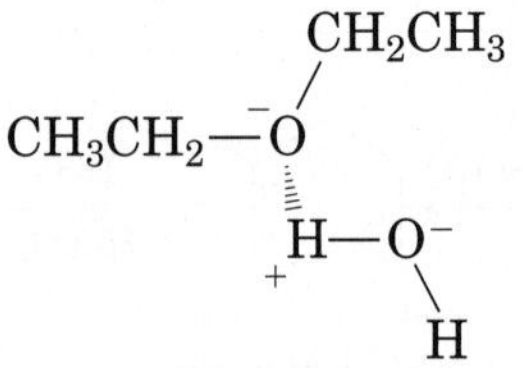

14.23 In order of increasing boiling point, they are:

$CH_3CH_2CH_2CH_3$ $CH_3CH_2OCH_2CH_3$

0°C   35°C

$CH_3CH_2CH_2OH$

97°C

14.25 The thickness (viscosity) of these three liquids is related to the degree of hydrogen bonding between their molecules in the liquid state. Hydrogen bonding is strongest between molecules of glycerol, weaker between molecules of ethylene glycol, and weakest between molecules of ethanol.

14.27 In order of decreasing solubility in water, they are:
(a) Ethanol > Diethyl ether > Butane
(b) 1,2-Hexanediol > 1-Hexanol > Hexane
14.29 (a) T (b) T (c) F (d) F (e) F (f) F (g) T (h) F (i) T (j) F
(c) False: Only phenols will react with strong bases to form water-soluble salts.
(d) False: Acid-catalyzed dehydration of cyclohexanol gives cyclohexene.
(e) False: The carbon-carbon double bond with the greater number of substituents predominates in the acid-catalyzed dehydration of alcohols.
(f) False: Acid-catalyzed dehydration of 2-butanol gives *trans*-2-butene as the major product.
(h) False: The oxidation of a 2° alcohol gives a ketone.
(j) False: The product will be 2-propanone (acetone).
14.31 Phenols (pK_a 10) are weak acids. Alcohols (pK_e 16-18) are considerably weaker acids than phenols, and are even weaker acids than water.

14.33

(a) $CH_3CH_2CH_2CH_2OH \xrightarrow[\text{heat}]{H_2SO_4} CH_3CH_2CH=CH_2 + H_2O$

(b) $CH_3CH_2CH_2CH_2OH \xrightarrow[H_2SO_4]{K_2Cr_2O_7} CH_3CH_2CH_2CO_2H$

14.35

(a)

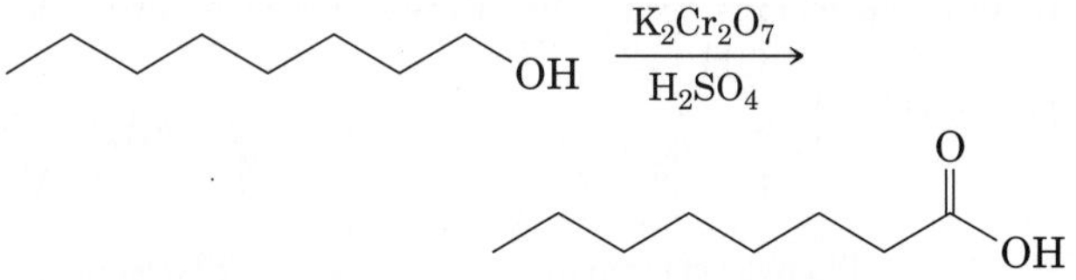

(b) $HOCH_2CH_2CH_2CH_2OH \xrightarrow[H_2SO_4]{K_2Cr_2O_7}$

14.37 (a) H_2SO_4, heat
(b) H_2O/H_2SO_4
(c) $K_2Cr_2O_7/H_2SO_4$
(d) HBr
(e) Br_2
(f) H_2/Pd
(g) $K_2Cr_2O_7/H_2SO_4$
(h) $K_2Cr_2O_7/H_2SO_4$
(i) $K_2Cr_2O_7/H_2SO_4$
14.39 Ethanol and ethylene glycol are important alcohols derived from ethylene. Ethanol is used as a solvent; is the alcohol in alcoholic beverages; is the starting material for the synthesis of diethyl ether, also an important solvent; and is blended with gasoline to improve octane number and reduce air pollution. Ethylene glycol is widely used as an automotive antifreeze and is one of the two starting materials for the synthesis of the polymer poly(ethylene terephthalate), better known as PET (Chapter 18).
14.41 (a) T (b) F (c) F
(b) False: Low-molecular-weight alcohols are more soluble in water than low-molecular-weight ethers.
(c) False: Ethers are unreactive to most of the reaction conditions that transform alcohols to other compounds.

14.43 (a) Ethyl isopropyl ether   (b) Dibutyl ether
(c) Diphenyl ether
14.45 (a) 2-Butanethiol   (b) 1-Butanethiol
(c) Cyclohexanethiol
14.47 1-Butanol has the higher boiling point because its molecules can associate in the liquid state by hydrogen bonding. The polarity of an —SH group is so low that there is little association among thiol molecules in the liquid state.

$$CH_3CH_2CH_2CH_2OH \qquad CH_3CH_2CH_2CH_2SH$$

1-Butanol   1-Butanethiol
(bp 117°C)   (bp 98°C)

14.49 (a) T (b) T (c) T (d) T (e) T (f) T
All statements are true. There are no false statements.
14.51 Nobel discovered that diatomaceous earth absorbs nitroglycerin so that it will not explode without a fuse.
14.53 Dichromate ion is reddish-orange; chromium(III) ion is green. When breath containing ethanol passes through a solution containing dichromate ion, ethanol is oxidized and dichromate ion is reduced to green chromium(III) ion.
14.55 Optimal bond angles about a tetravalent carbon atom and a divalent oxygen atom are 109.5°. In ethylene oxide, the C—C—O and C—O—C bond angles are compressed to approximately 60°, which results in high angle strain within the molecule.
14.57 The molecular formula for each is $C_3H_2ClF_5O$. They have the same molecular formula but a different connectivity of their atoms.

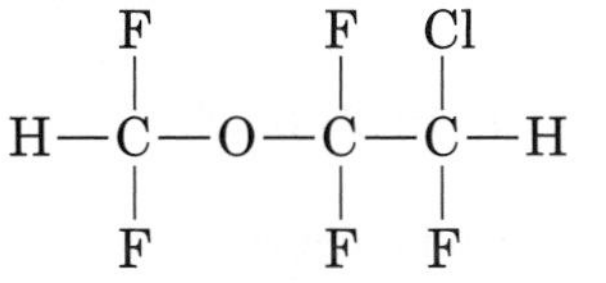

Enflurane 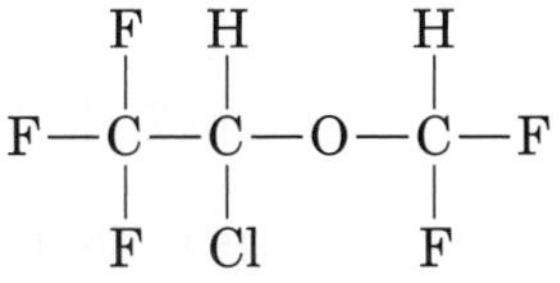   Isoflurane

14.59 $CH_3CH_2OH + 3O_2 \longrightarrow 2CO_2 + 3H_2O$
14.61 The eight isomeric alcohols with the molecular formula $C_5H_{12}O$ are:

$$CH_3CH_2CH_2CH_2CH_2OH$$

1-Pentanol

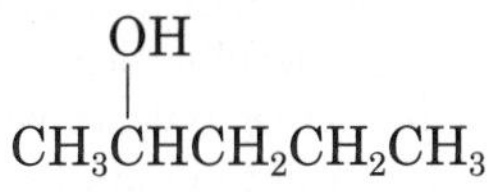

2-Pentanol

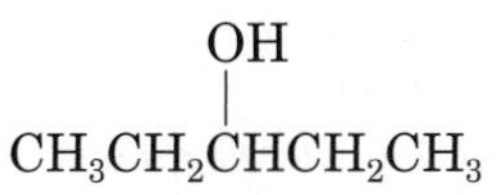

$$CH_3CH_2CHCH_2CH_3$$
|
OH

3-Pentanol

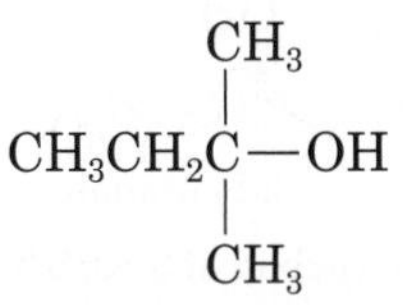

2-Methyl-2-butanol

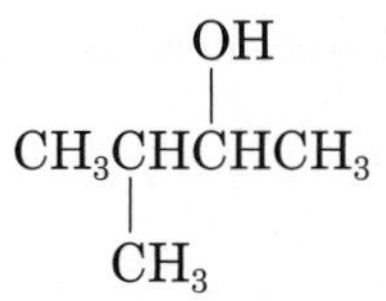

3-Methyl-2-butanol

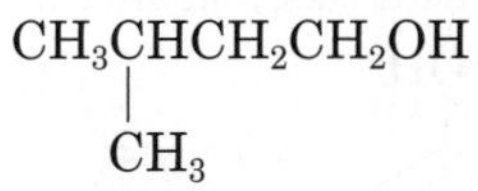

3-Methyl-1-butanol

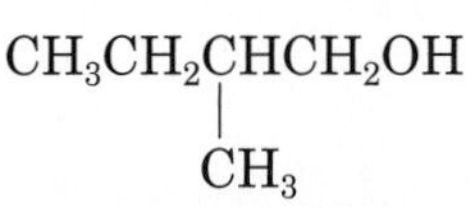

2-Methyl-1-butanol

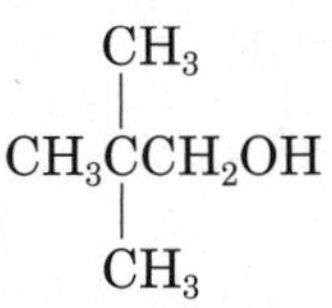

2,2-Dimethyl-1-propanol

14.63 Ethylene glycol has two —OH groups that allow each molecule to participate in hydrogen bonding, whereas 1-propanol has only one. The stronger intermolecular forces of attraction between molecules of ethylene glycol give it the higher boiling point.

$$CH_3CH_2CH_2OH \qquad HOCH_2CH_2OH$$

1-Propanol   Ethylene glycol
(bp 97°C)   (bp 198°C)

14.65 Arranged in order of increasing boiling point, they are:

$$CH_3CH_2CH_2CH_2CH_2CH_3 \qquad CH_3CH_2CH_2CH_2CH_2OH$$

Hexane   1-Pentanol
(bp 69°C)   (bp 138°C)

$$HOCH_2CH_2CH_2CH_2OH$$

1,4-Butanediol
(bp 230°C)

14.67 The solvent with the greater water solubility is circled.

(a) CH_2Cl_2 or $\boxed{CH_3CH_2OH}$

(b) $CH_3CH_2OCH_2CH_3$ or $\boxed{CH_3CH_2OH}$

14.69 Each can be prepared from 2-methylcyclohexanol (circled) as shown in the following flowchart.

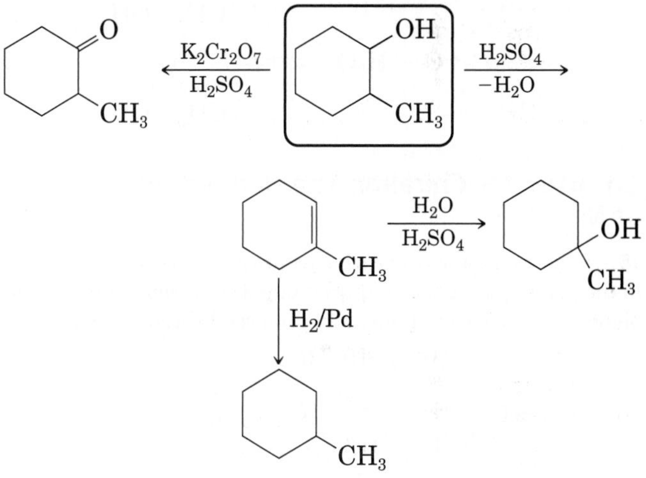

14.71 Each transformation can be done in two or three steps using reactions studied in the chapter.

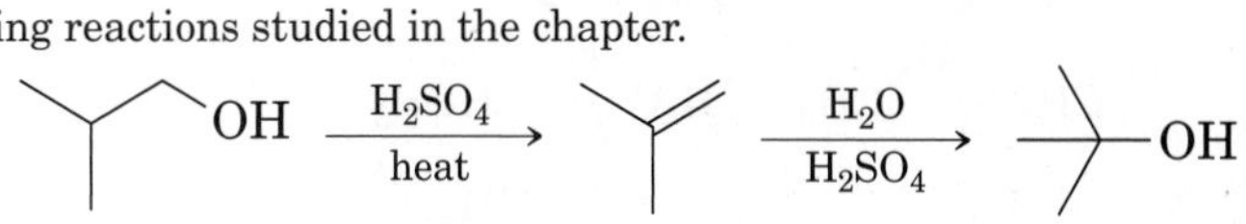

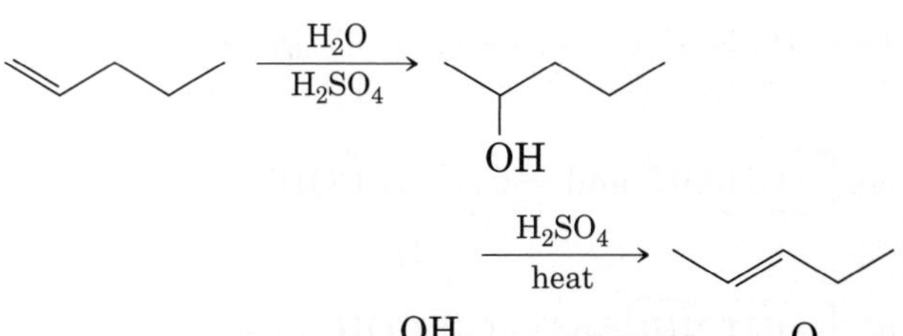

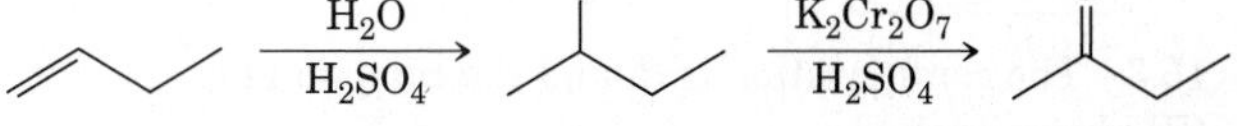

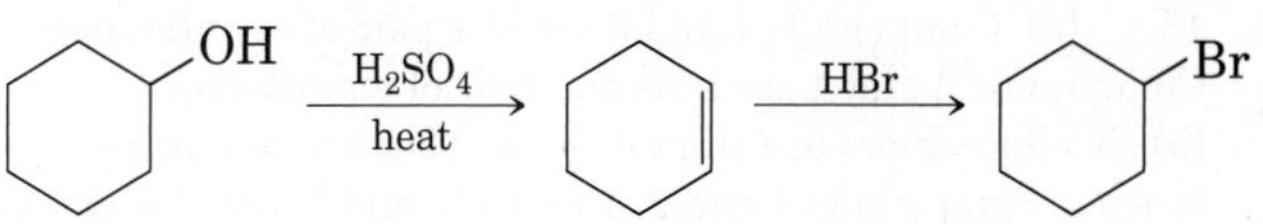

(e) (cyclopentane)–CH$_2$OH $\xrightarrow[\text{H}_2\text{SO}_4]{\text{K}_2\text{Cr}_2\text{O}_7}$ (cyclopentane)–C(=O)–OH

(f) (cyclopentane with CH$_3$ and OH) $\xrightarrow[\text{heat}]{\text{H}_2\text{SO}_4}$ (cyclopentene with CH$_3$) $\xrightarrow{\text{H}_2/\text{Pt}}$ (cyclopentane with CH$_3$)

14.73 (a) The three functional groups in cysteine are a 1° thiol, a 1° amine, and a carboxyl group.
(b) Below is the structural formula for cystine. Oxidation forms a disulfide bond.

Disulfide bond

HOCCHCH$_2$S—S—CH$_2$CHCOH

with O double bonds and NH$_2$ groups

Cystine

14.75 The hydrate results from the addition of H$_2$O to the carbonyl group:

$$\text{CH}_3\text{-C(=O)-CH}_3 + \text{H}_2\text{O} \rightleftharpoons \text{CH}_3\text{-C(OH)(OH)-CH}_3$$

with CH$_3$ OH / CH$_3$ OH

Chapter 15 Chirality: The Handedness of Molecules

15.1 The enantiomers of each part are drawn with two groups in the plane of the paper, a third group toward you in front of the plane, and the fourth group away from you behind the plane.

(a) COOH / C with H and CH$_3$ on cyclopentane; mirror image HOOC / C with H and H$_3$C on cyclopentane

(b) HO, H / C / CH$_3$CH, CH$_3$ with CH$_3$; mirror image H, OH / C / H$_3$C, CHCH$_3$ with CH$_3$

15.2 The group of higher priority in each set is circled.

(a) $\boxed{-\text{CH}_2\text{OH}}$ and $-\text{CH}_2\text{CH}_2\overset{\text{O}}{\overset{\|}{\text{C}}}\text{OH}$

(b) $\boxed{-\text{CH}_2\text{NH}_2}$ and $-\text{CH}_2\overset{\text{O}}{\overset{\|}{\text{C}}}\text{OH}$

15.3 The configuration is R, and the compound is (*R*)-glyceraldehyde.
15.4 (a) Compounds 1 and 3 are one pair of enantiomers. Compounds 2 and 4 are a second pair of enantiomers.
(b) Diastereomers are stereoisomers that are not mirror images. Compounds 1 and 2, 1 and 4, 2 and 3, and 3 and 4 are diastereomers.

15.5 Carbons 1 and 3 of 3-methylcyclohexanol are stereocenters. Therefore, $2^2 = 4$ stereoisomers are possible for this molecule. The *cis* isomer exists as one pair of enantiomers and the *trans* isomer exists as a second pair of enantiomers.

OH (on cyclohexane with two stereocenters marked *) CH$_3$

3-Methylcyclohexanol

15.6 Each stereocenter is marked by an asterisk, and the number of stereoisomers possible is shown under the structural formula.

(a) HO / HO (benzene ring)—CH$_2$CHCOOH with NH$_2$ and *

$2^1 = 2$

(b) CH$_2$=CHCHCH$_2$CH$_3$ with OH and *

$2^1 = 2$

(c) (cyclohexane)—OH and —NH$_2$ with two * marks

$2^2 = 4$

15.7 (a) T (b) T (c) T (d) F (e) T (f) T (g) T (h) T (i) T
(d) False: Constitutional isomers have a different connectivity of their atoms.
15.9 An achiral object has no handedness; it is an object whose mirror image is superposable on the original. Examples are methane, CH$_4$, and benzene, C$_6$H$_6$.
15.11 Both constitutional isomers and stereoisomers have the same molecular formula. Whereas stereoisomers have the same connectivity, constitutional isomers have a different connectivity of their atoms.
15.13 2-Pentanol has one stereocenter (carbon 2). 3-Pentanol has no stereocenter.

OH (2-Pentanol)   OH (3-Pentanol)

2-Pentanol   3-Pentanol

15.15 The carbon of a carbonyl group has only three groups bonded to it. To be a stereocenter, a carbon must have four different groups bonded to it.
15.17 Compounds (b), (c), and (d) contain stereocenters, here marked by asterisks, and are chiral.

(b) Cl—CH(OH)— with *
(c) (cyclopentane with CH$_3$ and OH and two *)
(d) (benzene)—CH(OH)—CH$_2$CH$_3$ with *

15.19 Following are the mirror images of each molecule.

(a)

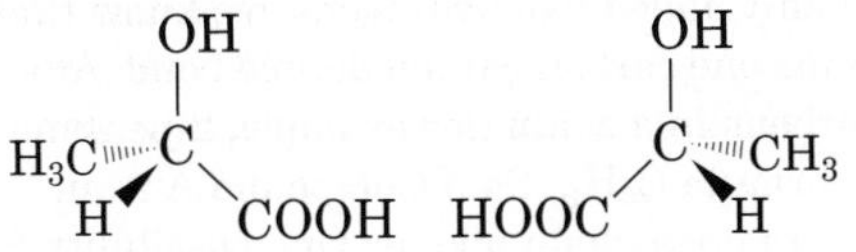

(b)

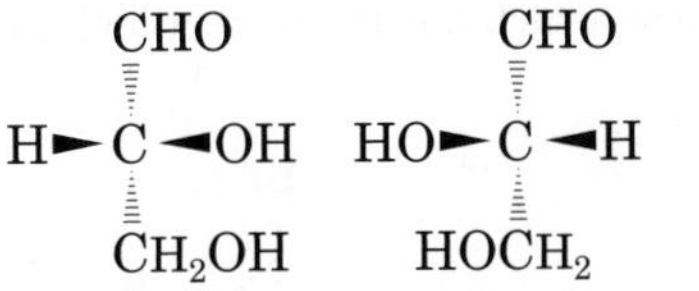

(c)

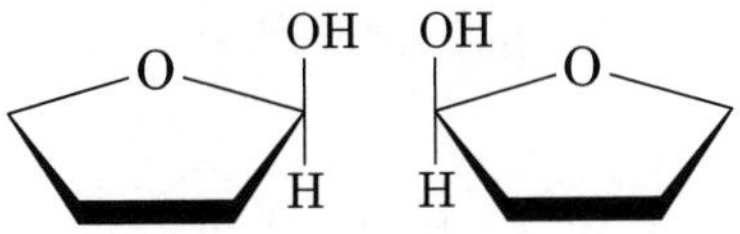

(d)

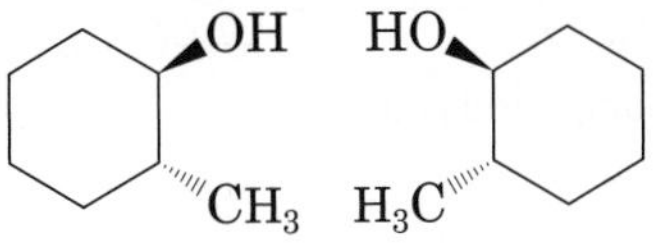

15.21 (a) T (b) F (c) T (d) F (e) T (f) T
(b) False: For a molecule with three stereocenters, $2^3 = 8$ stereoisomers are possible.
(d) False: 2-Pentanol is chiral, but 3-pentanol is achiral. Only 2-pentanol shows enantiomerism.
15.23 Parts (b) and (c) contain stereocenters.

(b) 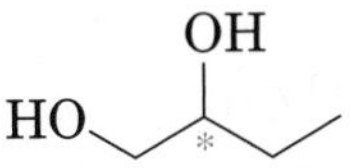   (c)

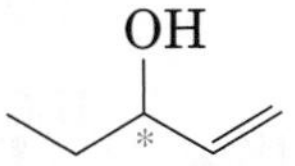

15.25 Stereocenters are marked with an asterisk. Under each is the number of stereoisomers possible.

(a)

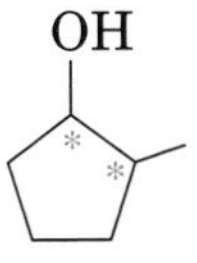

$2^2 = 4$

(Two pair of enantiomers)

(b)

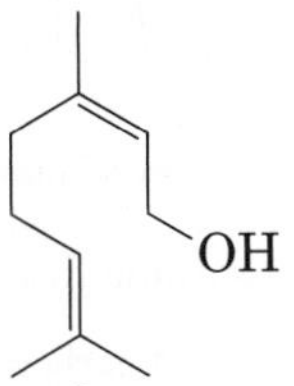

2

(One pair of cis + trans isomers)

(c)

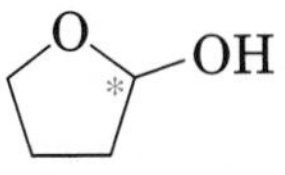

$2^1 = 2$

(One pair of enantiomers)

(d)

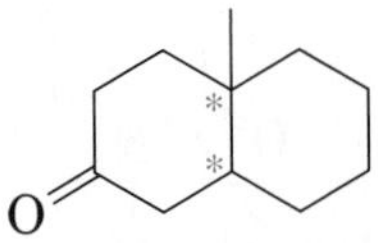

$2^2 = 4$

(Two pair of enantiomers)

15.27 The specific rotation of its enantiomer is +41°.
15.29 (a) T (b) T (c) F (d) T
(c) False: *cis-trans* stereoisomers do not need to be chiral; therefore, there is no requirement that they need to be optically active.
15.31 All three structures are chiral. The stereocenters in each are marked with an asterisk. Under the name of

each is the number of stereoisomers that are possible for it.

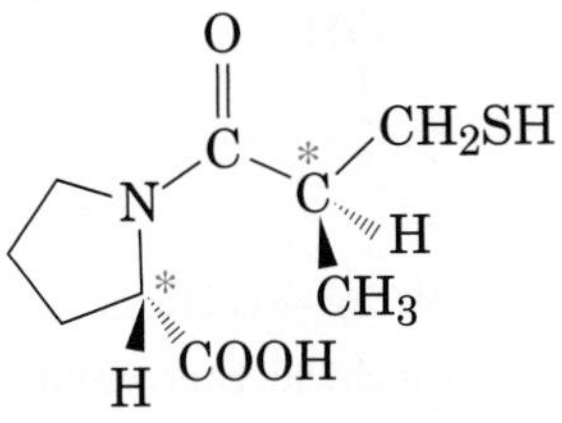

Captopril
($2^2 = 4$ stereoisomers)

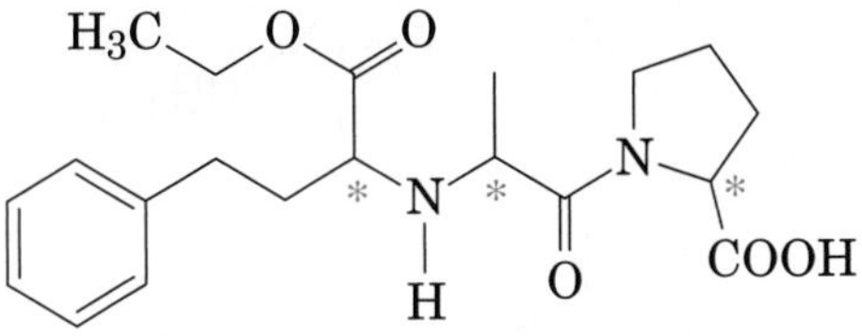

Enalopril (Altace)
($2^3 = 8$ stereoisomers)

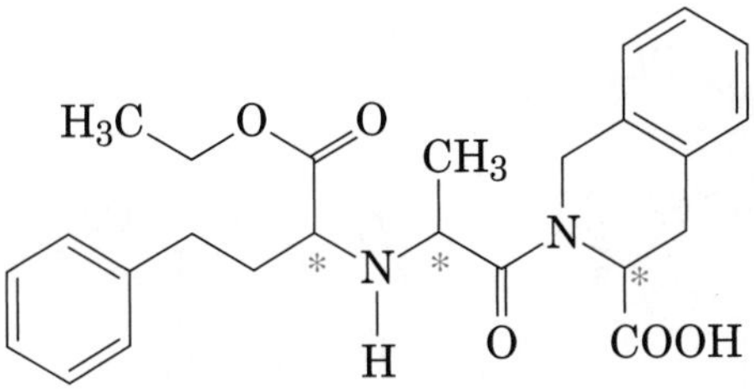

Quinapril (Accupril)
($2^3 = 8$ stereoisomers)

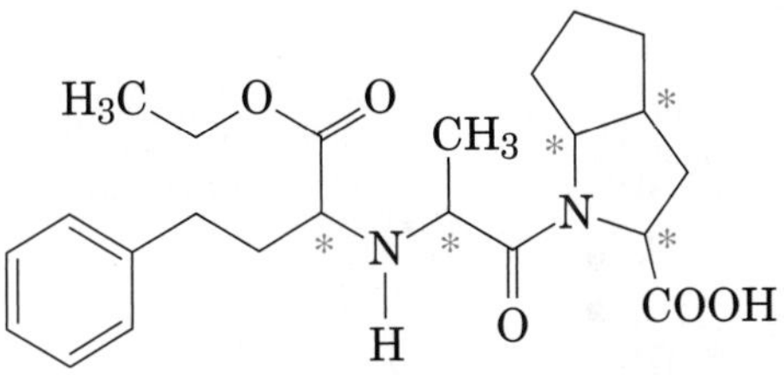

Ramipril (Vasotec)
($2^5 = 32$ stereoisomers)

The partial skeletons of the three (left) compared to that of Captopril (right).

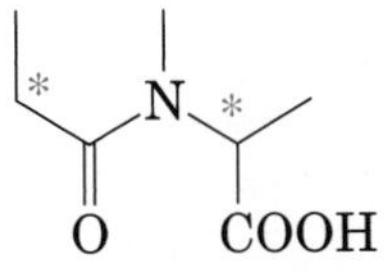

all of the molecules share structural features similar to captopril

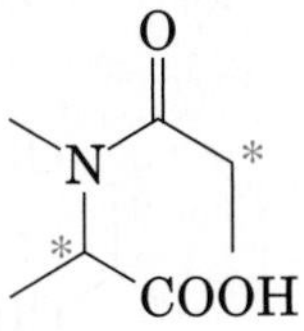

In addition, the second, third, and fourth molecules have a larger skeletal unit in common:

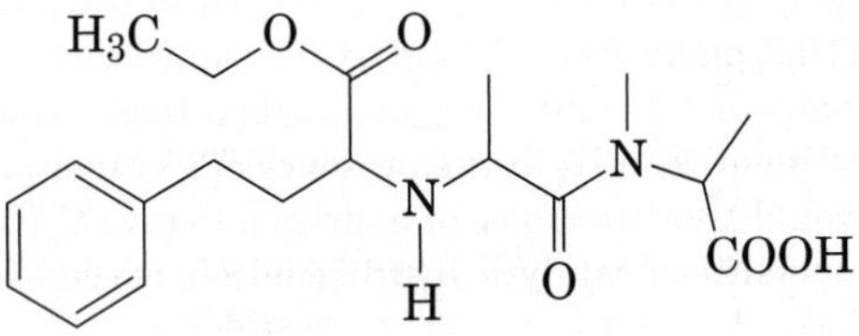

15.33 Only one alcohol with the molecular formula $C_6H_{14}O$ contains two stereocenters.

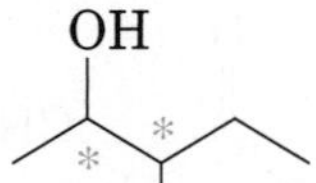

OH

3-Methyl-2-pentanol

15.35 All three drugs are chiral. Stereocenters are marked with asterisks.

(a)

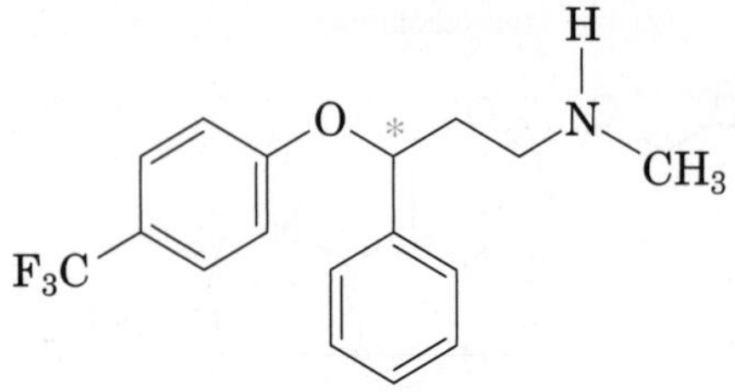

Fluoxetine
(Prozac)
($2^1 = 2$)

(b)

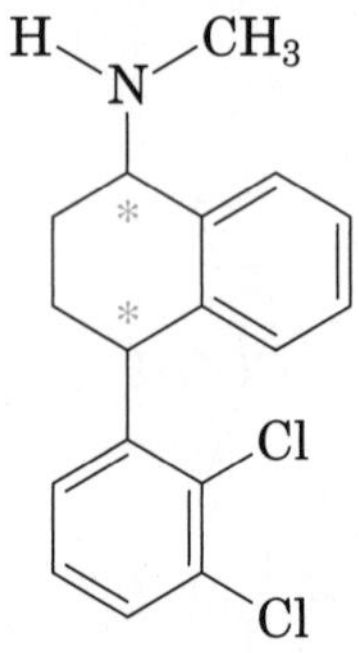

Sertraline
(Zoloft)
($2^2 = 4$)

(c)

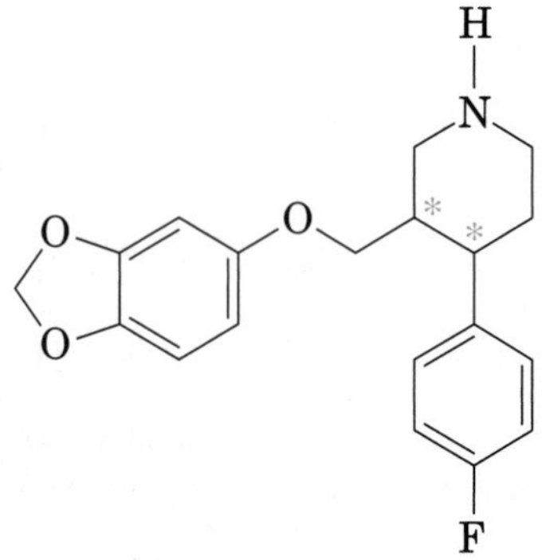

Paroxetine
(Paxil)
($2^2 = 4$)

15.37

(a)

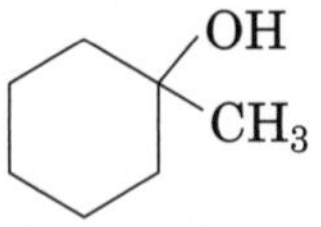

OH
CH₃

(b)

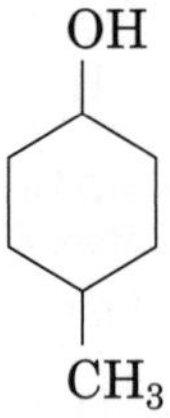

OH

CH₃

(c)

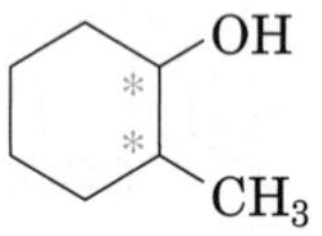

OH
CH₃

or

OH

CH₃

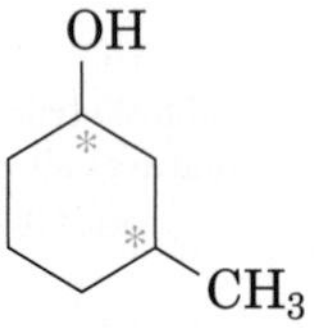

15.39 The majority have a right-handed twist because the machines that make them all impart the same twist.

15.41 Compound A could contain a carbon triple bond, but if it did, treatment with H_2 in the presence of a transition metal catalyst would add two moles of hydrogen to give C_5H_{12}. So from the evidence of catalytic hydrogenation, we can eliminate

the presence of a carbon-carbon triple bond, and because catalytic reduction only added two hydrogens, we know that Compound A contains one carbon-carbon double bond. Any alkene with five carbons in a chain (for example, 2-pentene) has the molecular formula C_5H_{10}. So if Compound A is an alkene, it cannot be an open-chain alkene. One possibility for Compound A is cyclopentene, which has the correct molecular formula.

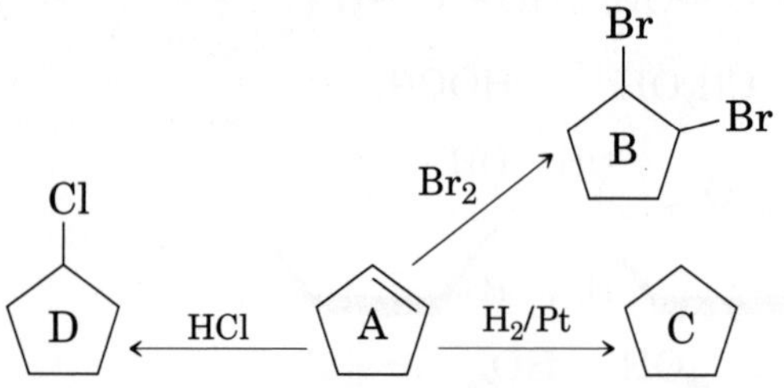

Cl
D ← HCl ← A → H₂/Pt → C
Br₂ → B (Br, Br)

Two other possibilities for Compound A are:

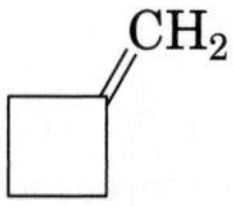

CH₂

and

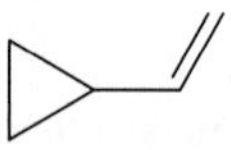

15.43 This molecule has seven stereocenters, here marked with asterisks. There are $2^7 = 256$ possible stereoisomers (128 pairs of enantiomers).

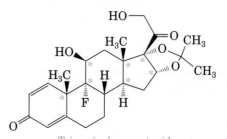

Triamcinolone acetonide

15.45 (a) Tamiflu contains five functional groups.

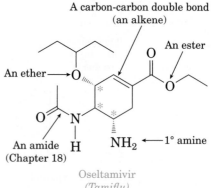

A carbon-carbon double bond
(an alkene)

An ether

An ester

An amide
(Chapter 18)

1° amine

Oseltamivir
(Tamiflu)

(b) Its molecular formula is $C_{16}H_{28}N_2O_4$.
(c) It is chiral and contains three stereocenters marked with asterisks and has $2^3 = 8$ possible stereoisomers. If it were isolated from a natural source, it would be optically active and would rotate the plane of polarized light. If, however, it were synthesized in the laboratory, it would be a mixture of stereoisomers—racemic and optically inactive.

(d) There are three stereocenters in Tamiflu. There is not enough information given in the original structure to determine the configuration of one of the stereocenters. To do so, you also need to know the orientation of the H atom at that stereocenter.

not enough information given to determine the configuration here

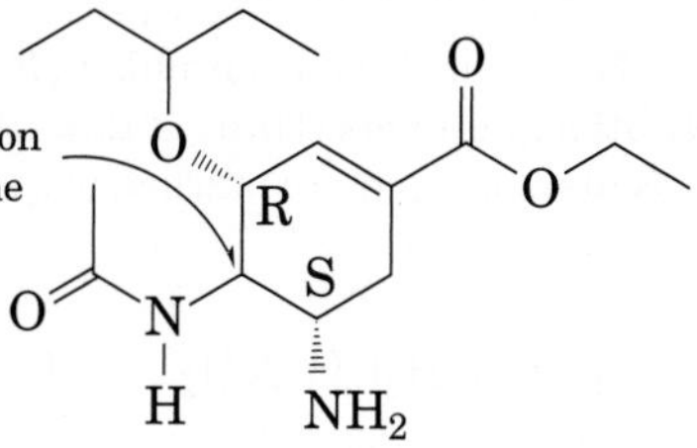

Oseltamivir
(Tamiflu)

(e) The enantiomer of Tamiflu is its nonsuperimposable mirror image.

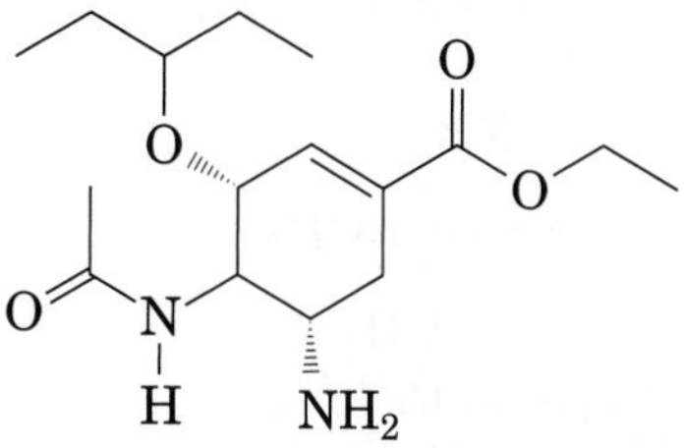

Oseltamivir
(Tamiflu)

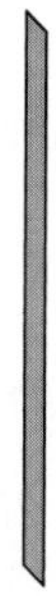

Mirror

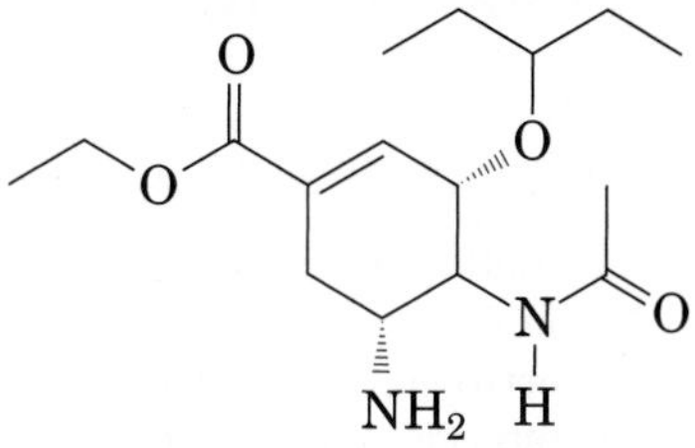

Mirror Image of Tamiflu

(f) A diastereomer is a stereoisomer that is not a mirror image. For a diastereomer of Tamiflu, invert the configuration at any one of the stereocenters. The diastereomers shown here differ in the configuration at the carbon atom bearing the 1° amine group.

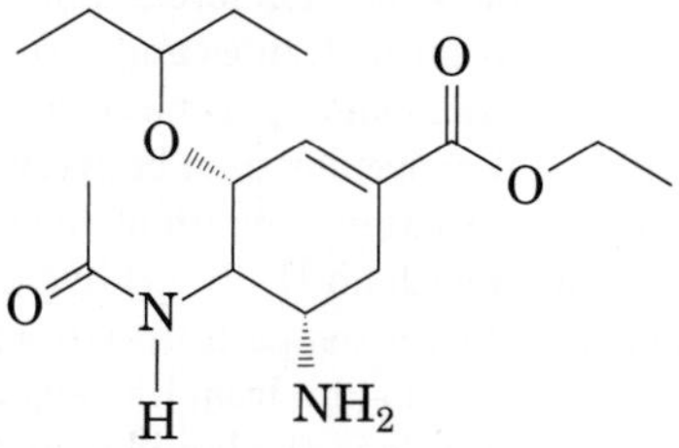

Oseltamivir
(Tamiflu)

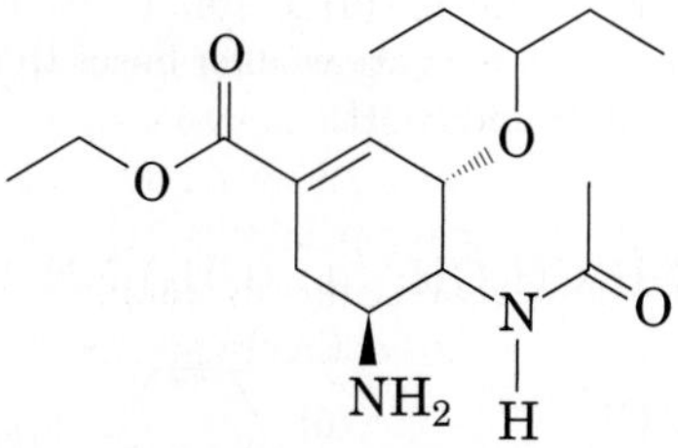

Diastereomer of Tamiflu
(a stereoisomer that is not
the mirror image)

(g) The orientation of each substituent on the six-membered ring is marked (e) for equatorial or (a) for axial. The more stable chair conformation is drawn on the left. In it, all groups on the ring are equatorial.

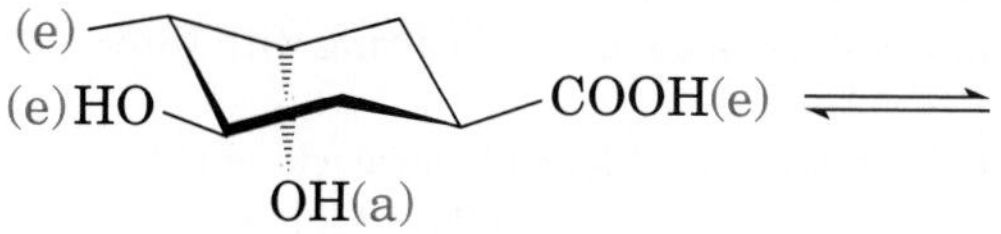

(e)HO
(e)HO
OH(a)

More stable chair
(e, e, e, a)

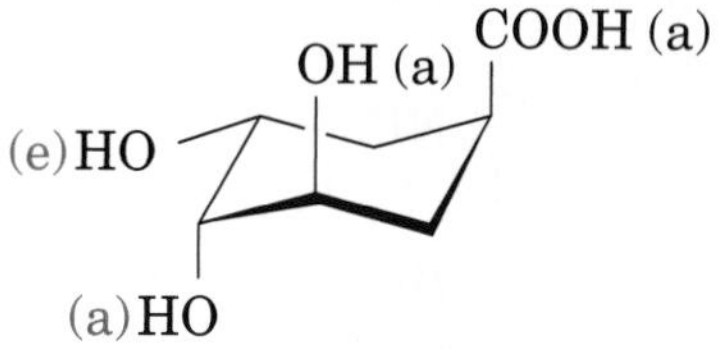

OH (a) COOH (a)
(e)HO
(a)HO

Less stable chair
(a, a, a, e)

Chapter 16 Amines

16.1 Pyrrolidine has nine hydrogens; its molecular formula is C_4H_9N. Purine has four hydrogens; its molecular formula is $C_5H_4N_4$.

16.2

(a) 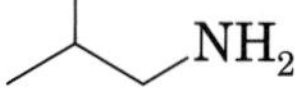 NH$_2$

(b) 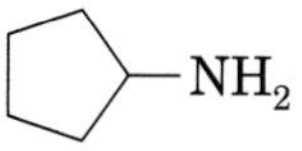 NH$_2$

(c) 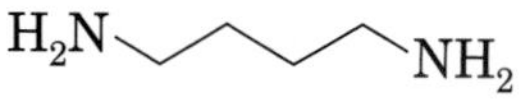 H$_2$N NH$_2$

16.3

(a) 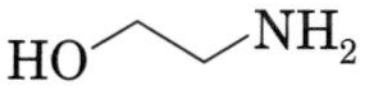 HO NH$_2$

(b)

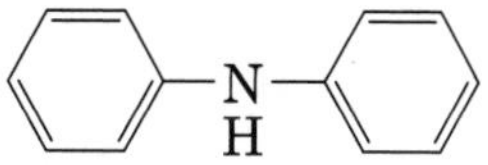

(c)

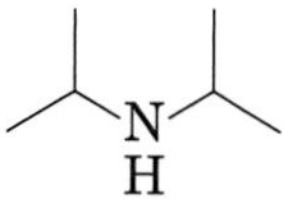

16.4 The stronger base is circled.

(a)  or

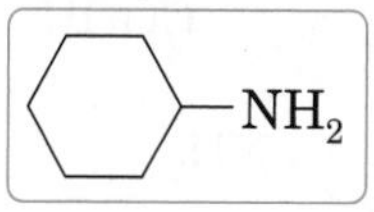

(b) CH$_3$NH$_2$ or NH$_2$

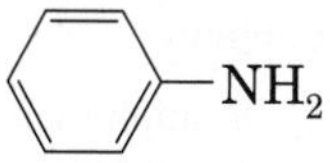

16.5 The product of each reaction is an ammonium salt.

(a) $(CH_3CH_2)_3\overset{+}{N}HCl^-$
Triethylammonium chloride

(b) 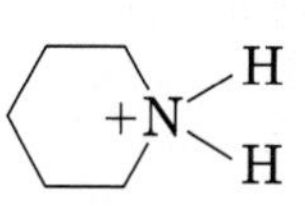
Piperidinium acetate

16.7 In the classification of alcohols, the terms 1°, 2°, and 3° refer to the number of carbons bonded to the carbon bearing the —OH group. In the classification of amines, the terms 1°, 2°, and 3° refer to the number of carbons bonded to the nitrogen atom of the amine.

16.9 Each compound has a six-membered ring with three double bonds.

16.11 Following is a structural formula for each amine:

(a) [structure: sec-butylamine with NH₂]

(b) $CH_3(CH_2)_6CH_2NH_2$

(c) [structure with NH₂]

(d) $H_2N(CH_2)_5NH_2$

(e) [benzene ring with NH₂ and Br]

(f) $(CH_3CH_2CH_2CH_2)_3N$

16.13 Each amine is classified by type.

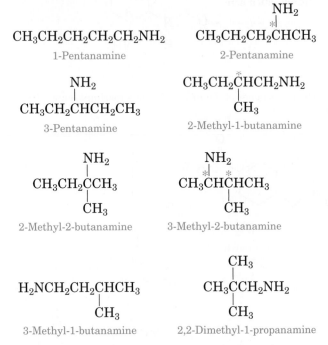

(a) [structure] 1° aliphatic amine, HO—, NH₂, heterocyclic aromatic amine

(b) [structure with ester O, O, and H₂N] 1° aromatic amine

(c) [structure] 3° aliphatic amine, CH₃, N, CH₃, O

(d) H₂N— ...COOH, NH₂ — 1° aliphatic amines

(e) [structure] 2° aromatic amine, 3° aliphatic amine, H, N, N, Cl, N — heterocyclic aromatic amine

(f) H_2N ...COOH — 1° aliphatic amine

16.15 For the molecular formula $C_5H_{13}N$, there are eight constitutional isomers. Three of these amines are chiral. The stereocenters in each are marked with asterisks.

$CH_3CH_2CH_2CH_2CH_2NH_2$
1-Pentanamine

$CH_3CH_2CH_2\overset{*}{\underset{|}{C}}HCH_3$ with NH₂
2-Pentanamine

$CH_3CH_2CHCH_2CH_3$ with NH₂
3-Pentanamine

$CH_3CH_2\overset{*}{C}HCH_2NH_2$ with CH₃
2-Methyl-1-butanamine

$CH_3CH_2CCH_3$ with NH₂ and CH₃
2-Methyl-2-butanamine

$CH_3\overset{*}{C}H\overset{*}{C}HCH_3$ with NH₂ and CH₃
3-Methyl-2-butanamine

$H_2NCH_2CH_2CHCH_3$ with CH₃
3-Methyl-1-butanamine

$CH_3CCH_2NH_2$ with CH₃ and CH₃
2,2-Dimethyl-1-propanamine

16.17 Both propylamine (a 1° amine) and ethylmethylamine (a 2° amine) have an N—H group, and hydrogen bonding occurs between their molecules in the liquid state. Because of this intermolecular force of attraction, these two amines have higher boiling points than trimethylamine (a 3° amine), which has no N—H bond and, therefore, cannot participate in intermolecular hydrogen bonding.

16.19 2-Methylpropane is a nonpolar hydrocarbon and the only attractive forces between its molecules in the liquid state are the very weak London dispersion forces. Both 2-propanol and 2-propanamine are polar molecules and associate in the liquid state by hydrogen bonding. Hydrogen bonding is stronger between alcohol molecules than between amine molecules because of the greater strength of an O—H----O hydrogen bond compared with an H—N----H hydrogen bond. It takes more energy (a higher temperature) to separate an alcohol molecule in the liquid state from its neighbors than to separate an amine molecule in the liquid state from its neighbors, and therefore, the alcohol has the higher boiling point.

16.21 (a) T (b) T (c) F (d) T (e) T (f) T (g) T
(c) False: Aliphatic amines are weaker bases than inorganic bases such as NaOH and KOH.

16.23

(a) $(CH_3)_3\overset{+}{N}CH_2CH_3OH^-$

(b) $(CH_3)_2NH_2^+I^-$

(c) $(CH_3)_4N^+Cl^-$

(d) [benzene ring]—$NH_3^+Br^-$

16.25 The stronger base is denoted with a rectangle. The determining factor is that aliphatic amines are stronger bases than aromatic amines.

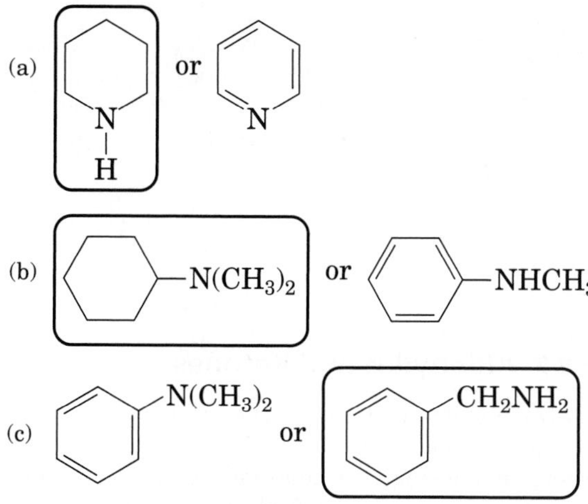

(a) [piperidine structure] or [pyridine structure]

(b) [cyclohexyl-N(CH₃)₂ structure] or [phenyl-NHCH₃ structure]

(c) [phenyl-N(CH₃)₂ structure] or [phenyl-CH₂NH₂ structure]

16.27 (a) Here is the completed Lewis structure for guanidine:

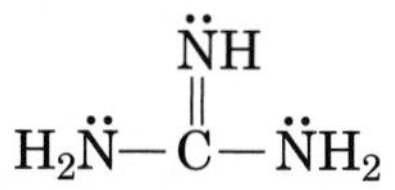

$$H_2\ddot{N}-\overset{\overset{\ddot{N}H}{\|}}{C}-\ddot{N}H_2$$
Guanidine

(b) Here are the three equivalent contributing structures. Each has an identical pattern of bonding:

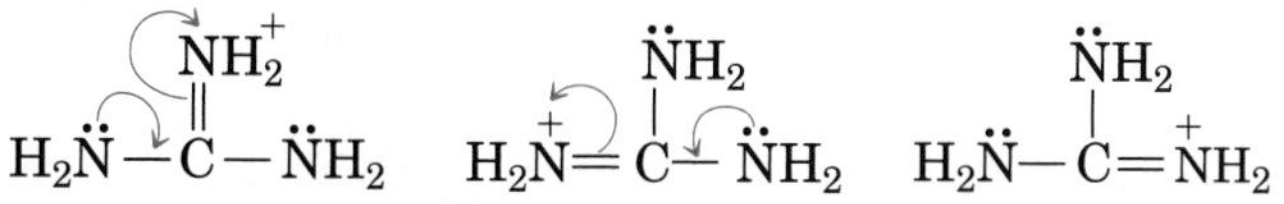

$$H_2\ddot{N}-\overset{\overset{+}{\overset{\ddot{N}H_2}{\|}}}{C}-\ddot{N}H_2 \quad H_2\overset{+}{N}=\overset{\overset{\ddot{N}H_2}{|}}{C}-\ddot{N}H_2 \quad H_2\ddot{N}-\overset{\overset{\ddot{N}H_2}{|}}{C}=\overset{+}{N}H_2$$

(c) If protonation occurs on the =NH nitrogen, then all three nitrogens become equivalent; that is, they have identical patterns of chemical bonding.
(d) Predict the N—C—N bond angle in the hybrid to be 120°.
(e) Guanidine is a considerably stronger base than ammonia.
16.29 Both compounds are insoluble in water. Treat each with aqueous HCl. The amine will react with HCl to form a water-soluble salt. The alcohol does not react with aqueous HCl and is insoluble in it. Thus, the amine dissolves in aqueous HCl, whereas the alcohol does not.
16.31 The primary aliphatic amine is the stronger base and forms a water-soluble salt in HCl. The salt is named pyridoxamine hydrochloride, which is an amine of pyridoxine, and a form of vitamin B6.

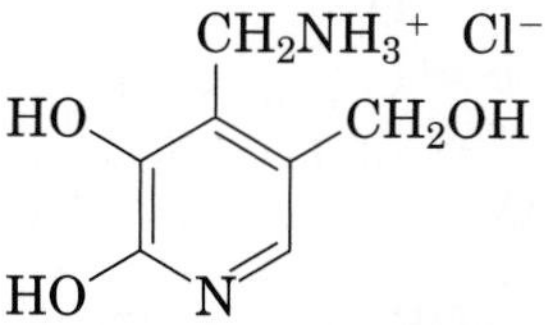

[pyridoxamine hydrochloride structure with CH₂NH₃⁺ Cl⁻, CH₂OH, two HO groups on pyridine ring]

Pyridoxamine hydrochloride

16.33 (a) Epinephrine has two phenolic —OH groups, whereas amphetamine has none. Epinephrine has a 2° alcohol group on its carbon side chain, whereas amphetamine has none. Epinephrine has a 2° amine, whereas amphetamine has a 1° amine. Finally, both are chiral, but their stereocenters are at different locations within the molecules.
16.35 Alkaloids are basic nitrogen-containing compounds found in the roots, bark, leaves, berries, and fruits of plants. In

almost all alkaloids, the nitrogen atom is present as a member of a ring; that is, it is present as a heterocyclic amine. By definition, alkaloids are nitrogen-containing bases, and therefore, they turn red litmus blue.
16.37 The tertiary aliphatic amine in the five-membered ring is the stronger base and is protonated by aqueous HCl.

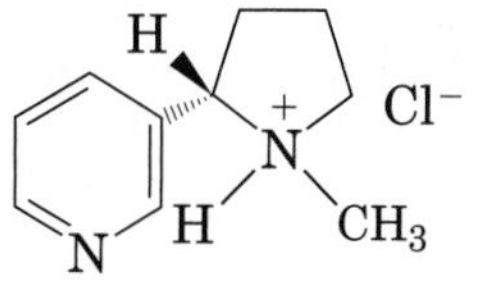

[(S)-Nicotine hydrochloride structure with Cl⁻, N⁺, CH₃]

(*S*)-Nicotine hydrochloride

16.39 The common structural feature is a benzene ring fused to a seven-membered ring containing two nitrogen atoms. This parental compound, from which a number of psychoactive compounds are synthesized, is named benzodiazepine.
16.41 The structural formula of 4-aminobutanoic acid is drawn on the left, showing the 1° amino group (a base) and the carboxyl group (an acid). It is drawn on the right as an internal salt called a zwitterion.

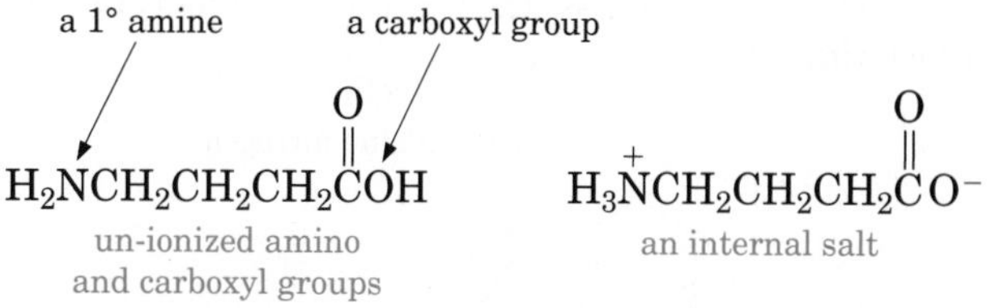

a 1° amine   a carboxyl group

$$H_2NCH_2CH_2CH_2\overset{\overset{O}{\|}}{C}OH \qquad \overset{+}{H_3}NCH_2CH_2CH_2\overset{\overset{O}{\|}}{C}O^-$$

un-ionized amino   an internal salt
and carboxyl groups

16.43 Amine salts are more soluble in water and body fluids and are less reactive to oxidation by atmospheric oxygen.
16.45 Following is one possible structural formula for each part:

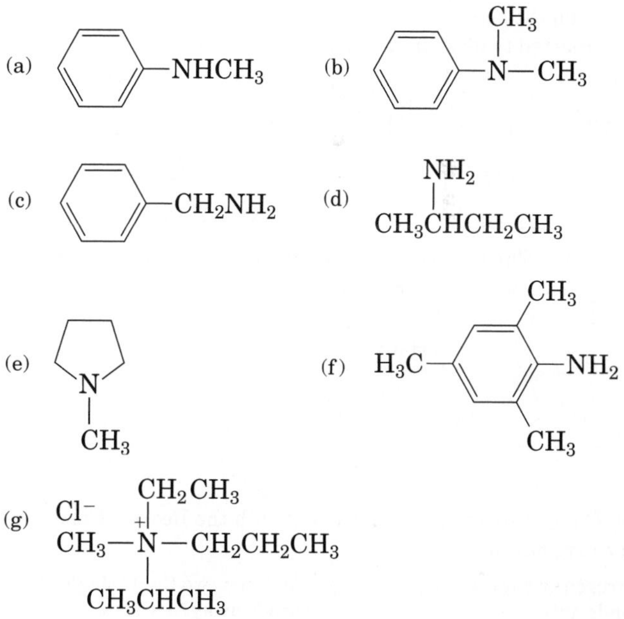

(a) [phenyl-NHCH₃ structure]

(b) [phenyl-N(CH₃)-CH₃ structure]

(c) [phenyl-CH₂NH₂ structure]

(d) $$\overset{\overset{NH_2}{|}}{CH_3CHCH_2CH_3}$$

(e) [N-methylpyrrolidine structure with N-CH₃]

(f) $$H_3C-\text{[benzene ring with CH}_3\text{ groups]}-NH_2$$

(g) $$\underset{CH_3CHCH_3}{\overset{CH_2CH_3}{\underset{|}{CH_3-\overset{+}{N}-CH_2CH_2CH_3}}} \quad Cl^-$$

16.47 (a) CH₃SH is the strongest acid.
(b) (CH₃)₂NH is the strongest base.
(c) CH₃OH has the highest boiling point.
(d) Molecules of CH₃OH form the strongest hydrogen bonds.
16.49 Both alcohols and amines can interact with water by hydrogen bonding. Because amines and alcohols with the same molecular weight have about the same solubility in water, the strength of hydrogen bonding between their molecules and water must be comparable.

16.51 (a) Following is a structural formula for 1-phenyl-2-amino-1-propanol.

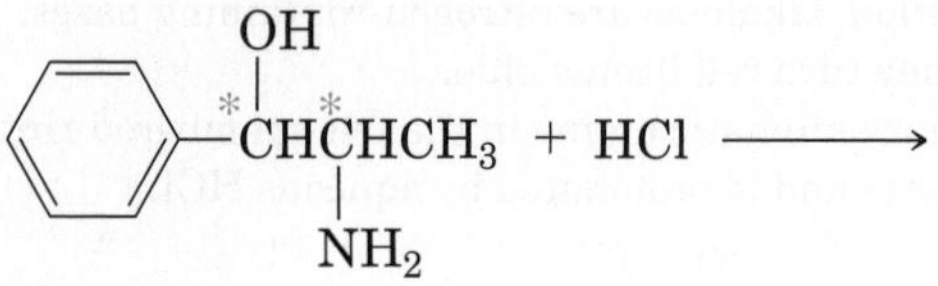

OH
|
(phenyl)—*CHCHCH₃ + HCl ⟶
         |
         NH₂*

1-Phenyl-2-amino-1-propanol

OH
|
(phenyl)—*CHCHCH₃*
         |
         NH₃⁺ Cl⁻

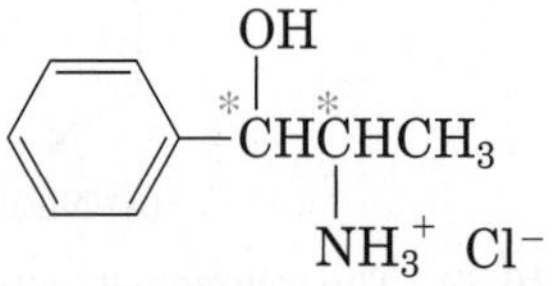

(b) The molecule has two stereocenters, and $2^2 = 4$ stereoisomers are possible; two pair of enantiomers.

16.53 Structural formula A is the better representation. The molecule has within its structure an —NH₂ group (a base) and a —COOH group (an acid). They react and form the internal salt seen in A.

16.55 The 2° aliphatic amine is more basic than the heterocyclic aromatic amine. The three stereocenters are marked with asterisks.

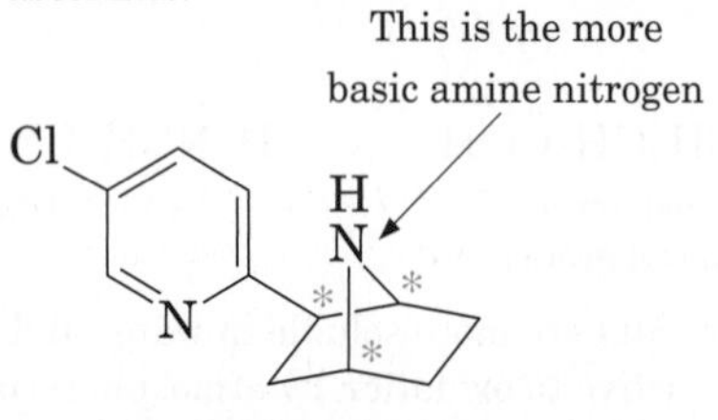

This is the more basic amine nitrogen

Epibatadine

16.57 Alanine is classified as an amino acid. Note that as drawn on the right, the 1° amino group and the carboxyl group have reacted to form an internal salt.

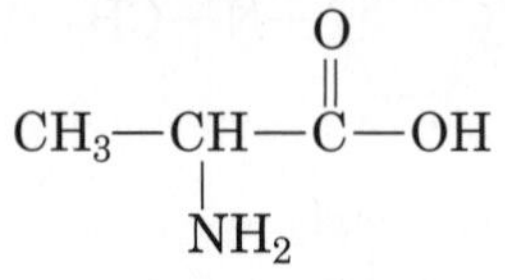

         O   O
         ‖   ‖
CH₃—CH—C—OH   CH₃—CH—C—O⁻
      | |
      NH₂   NH₃⁺

      Alanine 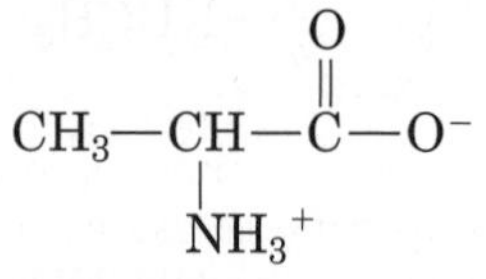   Alanine (internal salt)

16.59 (a) The following skeleton is the structural feature common to meperidine, methadone, and propoxyphene:

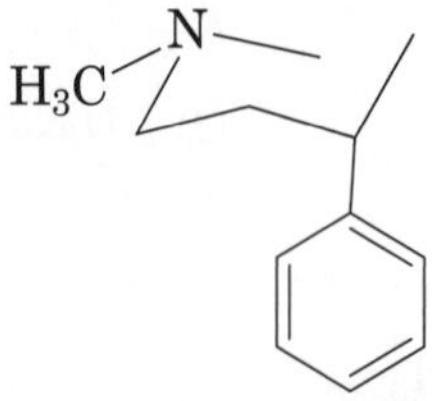

H₃C—N

(b) The features that are consistent with the Beckett-Casey rules are outlined:

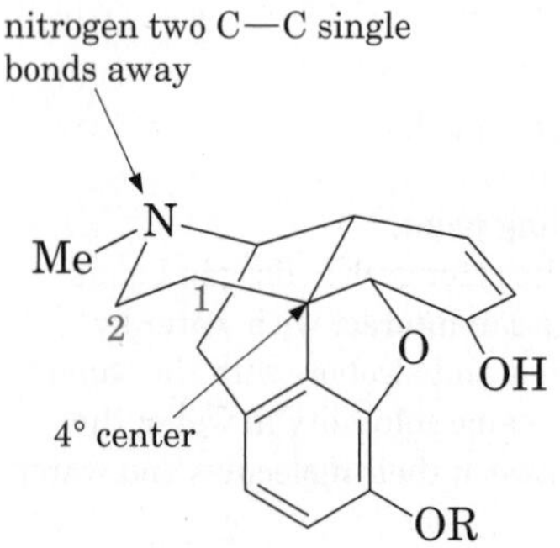

nitrogen two C—C single bonds away

Me—N ... OH ... OR
4° center

Morphine

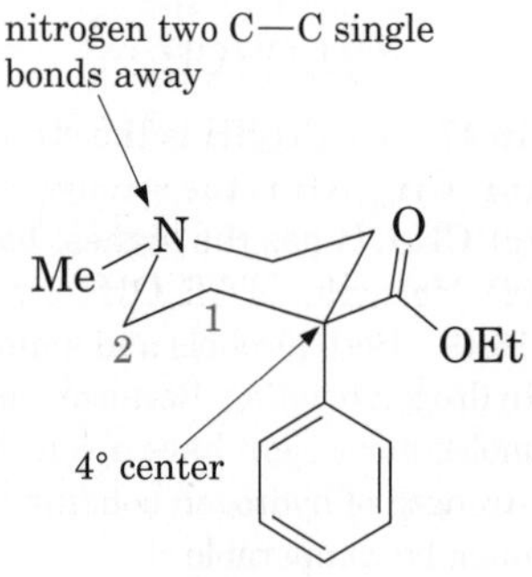

nitrogen two C—C single bonds away

Me—N ... O ... OEt
4° center

Meperidine (Demerol)

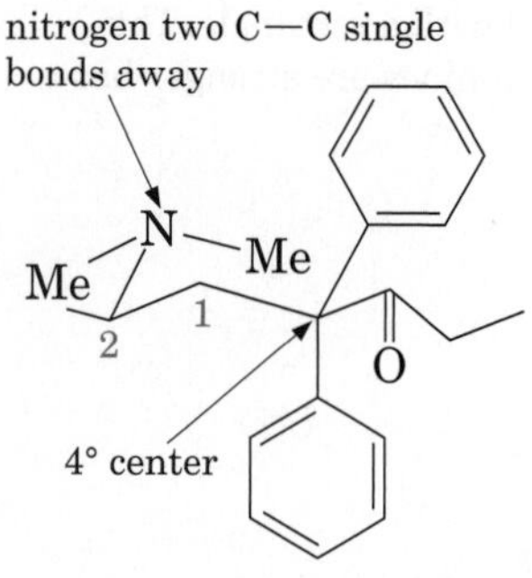

nitrogen two C—C single bonds away

Me—N—Me
1
2
O

4° center

Methadone

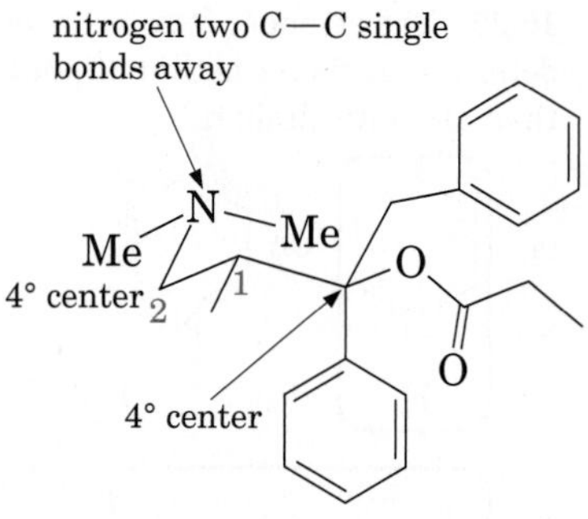

nitrogen two C—C single bonds away

Me—N—Me
4° center 2 1 O
         O

4° center

Propoxyphene (Darvon)

Chapter 17 Aldehydes and Ketones

17.1 (a) 3,3-Dimethylbutanal (b) Cyclopentanone
(c) 1-Phenyl-1-propanone

17.2 Following are line-angle formulas for the eight aldehydes with the molecular formula C₆H₁₂O. In the three that are chiral, the stereocenter is marked with an asterisk.

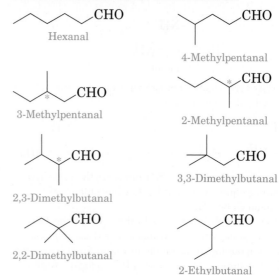

CHO
Hexanal

CHO
4-Methylpentanal

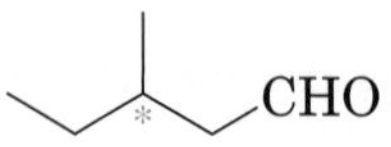

*CHO
3-Methylpentanal

*CHO
2-Methylpentanal

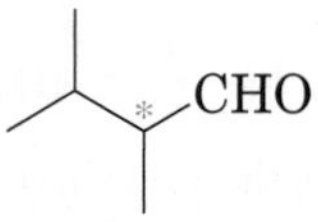

*CHO
2,3-Dimethylbutanal

CHO
3,3-Dimethylbutanal

CHO
2,2-Dimethylbutanal

CHO
2-Ethylbutanal

17.3 (a) 2,3-Dihydroxypropanal (b) 2-Aminobenzaldehyde
(c) 5-Amino-2-pentanone

17.4 Each aldehyde is oxidized to a carboxylic acid.

(a)
        O   O
        ‖   ‖
HO—C—   —C—OH

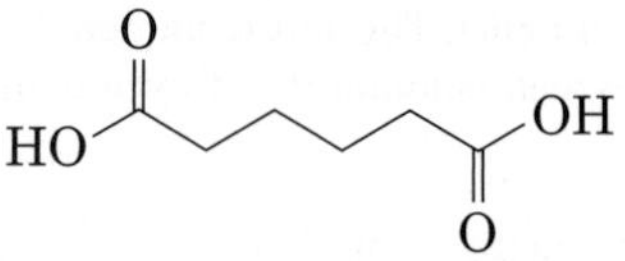

Hexanedioic acid
(Adipic acid)

(b)
            O
            ‖
(phenyl)—   —OH

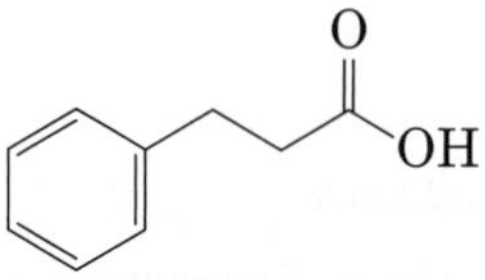

3-Phenylpropanoic acid

17.5 Each primary alcohol comes from the reduction of an aldehyde. Each secondary alcohol comes from the reduction of a ketone.

(a) (cyclohexanone)=O

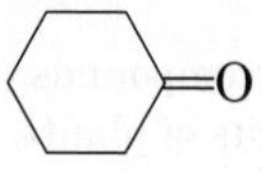

(b) CH₃O—⟨benzene⟩—CH₂CHO (with C=O)

(c) ⟨structure: O=C–CH₂CH₂CH₂–C=O⟩

17.6 Shown first is the hemiacetal and then the acetal.

Benzaldehyde   *A hemiacetal*

⟨acetal structure⟩ + H₂O

An acetal

17.7 (a) A hemiacetal formed from 3-pentanone (a ketone) and ethanol.
(b) Neither a hemiacetal nor an acetal. This compound is the dimethyl ether of ethylene glycol.
(c) A cyclic acetal derived from 5-hydroxypentanal and methanol.

17.8 Following is the keto form of each enol.

17.9 (a) T (b) T (c) T (d) F
(d) False: A carbonyl carbon of an aldehyde or a ketone has only three groups bonded to it. Therefore, it cannot be a stereocenter.

17.11 In an aromatic aldehyde, the —CHO group is bonded to an aromatic ring. In an aliphatic aldehyde, it is bonded to a tetrahedral carbon atom.

17.13 Compounds (b), (c), (d), and (f) contain carbonyl groups.

17.15 Of the four aldehydes with the molecular formula, only 2-methylbutanal is chiral. Its stereocenter is marked with an asterisk.

Pentanal   3-Methylbutanal
2-Methylbutanal   2,2-Dimethylpropanal

17.17

(a) H—C(=O)—H   (b) CH₃CH₂CHO

(c) ⟨structure⟩   (d) CH₃(CH₂)₈CHO

(e) ⟨4-hydroxybenzaldehyde⟩   (f) HOCH₂CHCHO with OH

17.19 (a) 4-Heptanone (b) 2-Methylcyclopentanone
(c) *cis*-2-Methyl-2-butenal (d) 2-Hydoxypropanal
(e) 1-Phenyl-2-propanone (f) Hexanedial

17.21 (a) The carbon chain is numbered incorrectly. The correct name is 2-butanone.
(b) The carbonyl group is on carbon 1, and therefore, the compound is an aldehyde. Its name is butanal.
(c) The longest (parent) carbon chain is five carbons long, and the correct name is pentanal.
(d) The carbon chain must be numbered to give the carbonyl group the lowest possible number. The correct name is 3,3-dimethyl-2-butanone.

17.23 (a) T (b) T (c) T (d) T
All statements are true.

17.25 The carbonyl group of acetone forms hydrogen bonds with water. These hydrogen bonds are sufficient to make acetone in any proportion soluble in water. 4-Heptanone contains a carbonyl, which, through its hydrogen bonding with water molecules, promotes water solubility. It also contains two 3-carbon hydrocarbon groups bonded to the carbonyl carbon, which inhibit water solubility. In 4-heptanone, the combined hydrophobic effect of the two hydrocarbon groups is greater than the hydrophilic effect of the single carbonyl group, making 4-heptanone insoluble in water.

17.27 Pentane is a nonpolar hydrocarbon, and the only attractive forces between its molecules in the liquid state are the very weak London dispersion forces. Pentane, therefore, has the lowest boiling point. Butanal and 1-butanol are both polar molecules. Because 1-butanol has a polar OH group, its molecules can associate by hydrogen bonding. The intermolecular attraction between molecules of 1-butanol are greater than those between molecules of butanal. 1-Butanol, therefore, has a higher boiling point than butanal.

Pentane (bp 36°C)   Butanal (bp 76°C)   1-Butanol (bp 117°C)

17.29 Acetone has no —OH group through which to form intramolecular hydrogen bonds.

17.31 Aldehydes are oxidized by this reagent to a carboxylic acid, ketones are not affected, and 2° alcohols are oxidized to ketones.

(a) CH₃CH₂CH₂COOH   (b) ⟨benzene⟩—C(=O)—OH

(c) NR   (d) ⟨cyclohexanone⟩

17.33 (a) Treat each with Tollens' reagent. Only pentanal gives a silver mirror.
(b) Treat each with $K_2Cr_2O_7/H_2SO_4$. Only 2-pentanol is oxidized (to 2-pentanone) by this reagent, which will cause the red color of $Cr_2O_7^{2-}$ ion to disappear and be replaced with the green color of Cr^{3+} ion.

17.35 The white solid is benzoic acid, C_6H_5COOH, formed by the oxidation of benzaldehyde in the air.

17.37 These experimental conditions reduce an aldehyde to a primary alcohol and a ketone to a secondary alcohol. Products (a) and (c) are chiral and will be formed as racemic mixtures.

(a) $CH_3\overset{\overset{\displaystyle OH}{|}}{C}HCH_2CH_3$

(b) $CH_3(CH_2)_4CH_2OH$

(c)

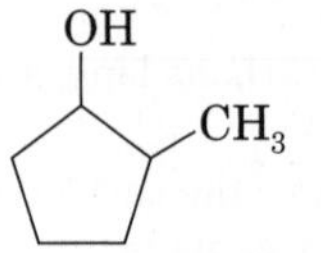

(d)

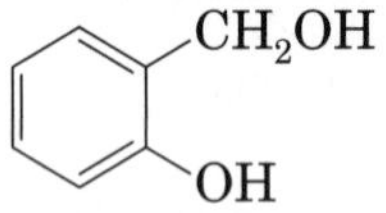

17.39 (a) Following is a structural formula for dihydroxyacetone:

$$HOCH_2-\overset{\overset{\displaystyle O}{||}}{C}-CH_2OH$$

1,3-Dihydroxy-2-propanone
(Dihydroxyacetone)

(b) Because dihydroxyacetone has two hydroxyl groups and one carbonyl group, all of which can interact with water molecules by hydrogen bonding, predict that it is soluble in water.
(c) Treatment of dihydroxyacetone with $NaBH_4$ followed by the addition of H_2O reduces the carbonyl group to a hydroxyl group.

$$HOCH_2-\overset{\overset{\displaystyle OH}{|}}{C}H-CH_2OH$$

1,2,3-Propanetriol
(Glycerol, glycerin)

17.41

(a) 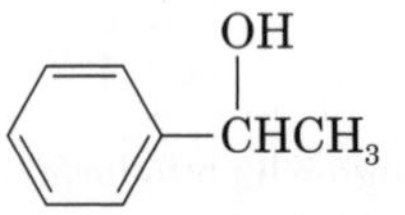 (b)

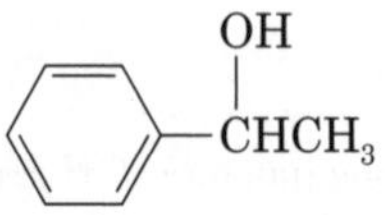

(c) No reaction (d) No reaction

17.43 Only parts (a), (b), (d), and (f) have an alpha hydrogen, and only these four can undergo keto-enol tautomerism.

17.45 Following are the keto forms of each enol:

(a)

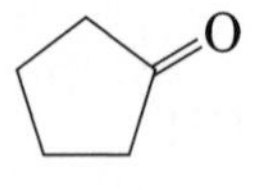

(b) $CH_3\overset{\overset{\displaystyle O}{||}}{C}CH_2CH_2CH_2CH_3$

(c)

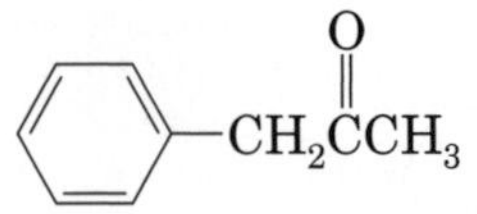

17.47 The characteristic structural feature of a hemiacetal is a carbon atom bonded to an —OH group and either an —OR or an —OAr group. The characteristic structural feature of an acetal is a carbon atom bonded to two —OR or —OAr groups.

17.49 (a) Hemiacetal (b) Acetal (c) Acetal
(d) Hemiacetal (e) Cyclic acetal (f) Acetal

17.51 Following are structural formulas for the products of each hydrolysis.

(a) $CH_3CH_2\overset{\overset{\displaystyle O}{||}}{C}CH_2CH_3 + HOCH_2CH_2OH$

(b)

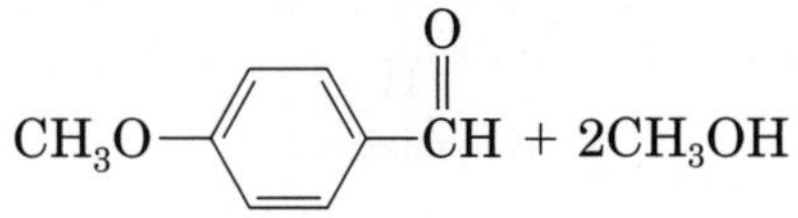

(c)

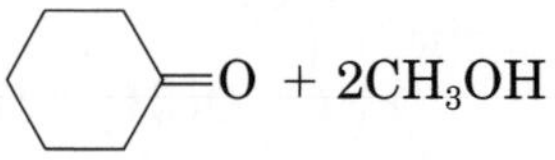

(d)

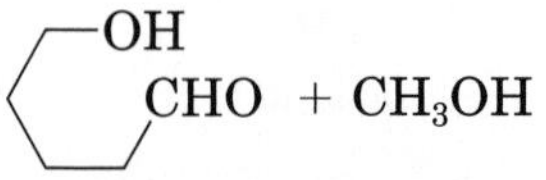

17.53 (a) Of the three carbon-carbon double bonds, only two show *cis/trans* isomerism, and both have the *trans* configuration.
(b) There are 14 stereocenters present in this molecule and 2^{14} stereoisomers arising from the 14 stereocenters plus 2 pairs of *cis-trans* isomers.
(c) No, there are no aromatic components in this molecule.
(d) Both of the marked stereocenters have the R configuration.
(e) The hemiacetal and ester functions are circled.

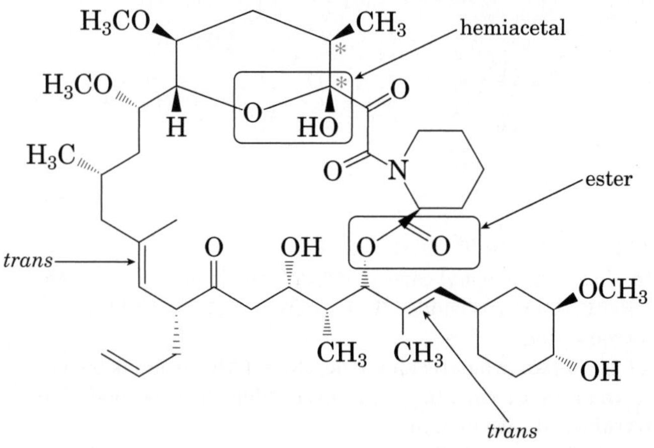

(f) Cleavage of the hemiacetal liberates an alcohol and a ketone, here shown in red.

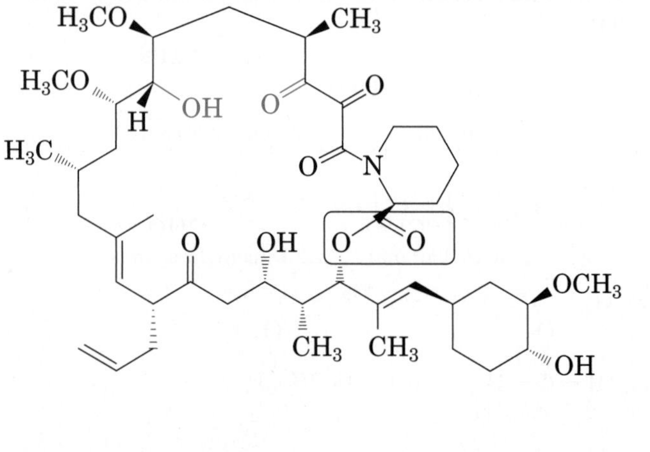

17.55 *Hydration* refers to the addition of one or more molecules of water to a substance. An example of hydration is the acid-catalyzed addition of water to propene to give 2-propanol. *Hydrolysis* refers to the reaction of a substance with water, generally with the breaking (lysis) of one or more bonds in the substance. As example of hydrolysis is the acid-catalyzed hydrolysis of an acetal with a molecule of water to give an aldehyde or a ketone and two molecules of alcohol. Another example of hydrolysis is the reverse of Fischer esterification (Chapter 18), namely the hydrolysis of an ester to give a carboxylic acid and an alcohol.

17.57 Reduction of a ketone gives a secondary (2°) alcohol. Reduction of an aldehyde gives a primary (1°) alcohol. It is not possible to obtain a tertiary (3°) alcohol by reduction of a carbonyl-containing compound.

(a) $CH_3-\overset{\overset{\displaystyle O}{\|}}{C}-CH_3$ (b) 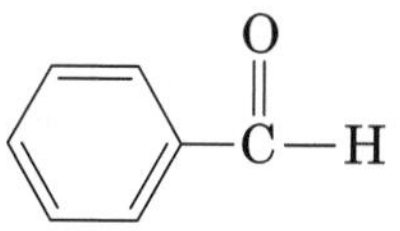 benzene ring $-\overset{\overset{\displaystyle O}{\|}}{C}-H$

(c) $H-\overset{\overset{\displaystyle O}{\|}}{C}-H$ (d) None exist

17.59 Acid-catalyzed hydration of propene involves addition of H^+ to propene to form a 2° carbocation intermediate. According to Markovnikov's rule, H^+ adds to the carbon of the double bond that already has the greater number or hydrogens, which in the case of propene, is carbon-1. This gives a 2° carbocation to which a molecule of water adds. The only product possible from the acid-catalyzed hydration of propene is 2-propanol.

17.61 Each conversion can be brought about in two steps: (a) acid-catalyzed hydration of the alkene to give an alcohol followed by (b) oxidation of the alcohol to a ketone.

(a) $CH_3CH_2CH_2CH=CH_2 \xrightarrow[\text{H}_2\text{SO}_4]{\text{H}_2\text{O}}$
1-Pentene

$CH_3CH_2CH_2\overset{\overset{\displaystyle OH}{|}}{C}HCH_3 \xrightarrow[\text{H}_2\text{SO}_4]{\text{K}_2\text{Cr}_2\text{O}_7} CH_3CH_2CH_2\overset{\overset{\displaystyle O}{\|}}{C}CH_3$
2-Pentanone

(b) 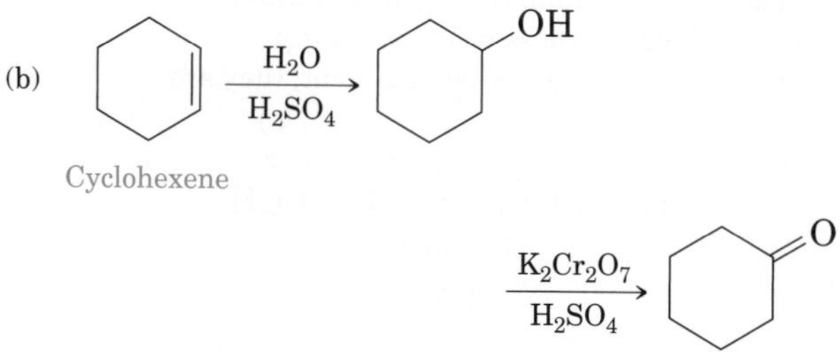 cyclohexene $\xrightarrow[\text{H}_2\text{SO}_4]{\text{H}_2\text{O}}$ cyclohexanol (OH)
Cyclohexene

$\xrightarrow[\text{H}_2\text{SO}_4]{\text{K}_2\text{Cr}_2\text{O}_7}$ cyclohexanone (O)
Cyclohexanone

17.63 The carbonyl group of each aldehyde or ketone is shown in color.

(a) $H\overset{\overset{\displaystyle O}{\|}}{C}CH_2CH_2CH_2\overset{\overset{\displaystyle O}{\|}}{C}CH_3$ (b) 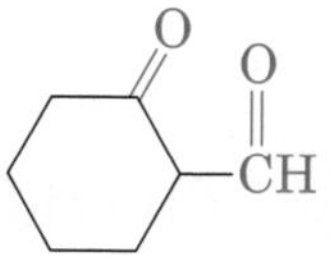 cyclohexanone with $-CH$

(c) $HOCH_2\overset{\overset{\displaystyle HO \ O}{| \ \|}}{C}H\overset{}{C}H$ (d) 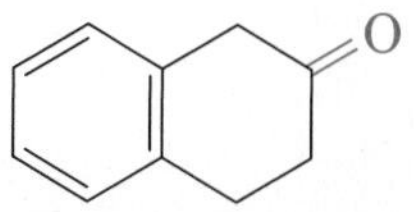 bicyclic structure with O

(e) 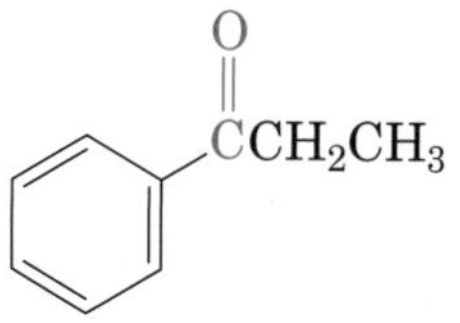 $\text{benzene ring}-\overset{\overset{\displaystyle O}{\|}}{C}CH_2CH_3$

(f) 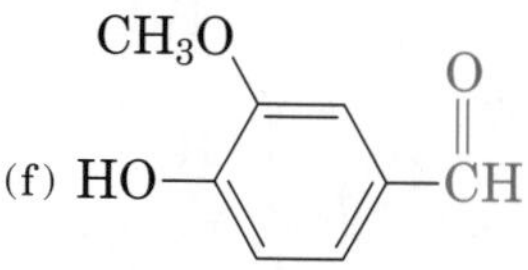 $HO-\text{benzene ring}(CH_3O)-\overset{\overset{\displaystyle O}{\|}}{C}H$

17.65 The one ketone and two aldehydes with the molecular formula C_4H_8O are:

$CH_3\overset{\overset{\displaystyle O}{\|}}{C}CH_2CH_3$ $CH_3CH_2CH_2\overset{\overset{\displaystyle O}{\|}}{C}H$ $CH_3\overset{\overset{\displaystyle O}{\|}}{C}H\overset{}{C}H$
$\qquad\qquad\qquad\qquad\qquad\qquad\qquad\qquad |$
$\qquad\qquad\qquad\qquad\qquad\qquad\qquad\quad CH_3$

17.67 2-Propanol has a higher boiling point than acetone because of the greater attraction between its molecules due to hydrogen bonding through its —OH group.

17.69 (a) Treat each compound with Tollens' regent. Only benzaldehyde will reduce Ag^+ to give a precipitate of silver metal that appears as a mirror.
(b) Treat each compound with Tollens' reagent. Only the aldehyde will reduce Ag^+ to give a precipitate of silver metal that appears as a mirror.

17.71

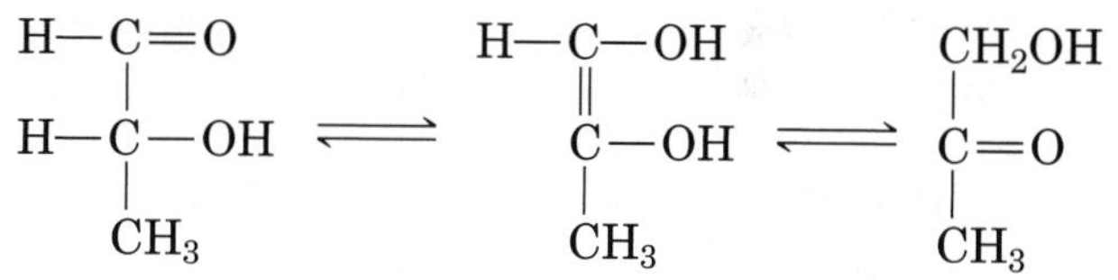

$\overset{\displaystyle H-C=O}{\underset{\displaystyle CH_3}{\overset{}{\underset{}{H-C-OH}}}} \rightleftharpoons \overset{\displaystyle H-C-OH}{\underset{\displaystyle CH_3}{\overset{\|}{\underset{}{C-OH}}}} \rightleftharpoons \overset{\displaystyle CH_2OH}{\underset{\displaystyle CH_3}{\overset{}{\underset{}{C=O}}}}$

α-Hydroxyaldehyde   An enediol   α-Hydroxyketone

17.73 (a) Carbon 5 provides the —OH group, and carbon 1 provides the —CHO group.
(b) Following is a structural formula for the free aldehyde form of glucose:

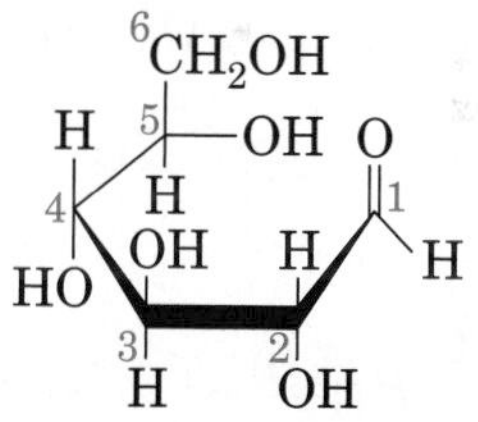

$\overset{6}{C}H_2OH$ ... (ring structure with positions labeled)

17.75 Testosterone contains one ketone, one secondary (2°) alcohol, and one carbon-carbon double bond. Methandrostenolone contains one ketone, one tertiary (3°) alcohol, and two carbon-carbon double bonds.
(a) The two differences in structural formula are that methandrostenolone has one additional carbon-carbon double bond and its alcohol is tertiary (3°) rather than secondary.
(b) Each stereocenter is marked with an asterisk. Each has six stereocenters and $2^6 = 64$ possible stereoisomers.

(a)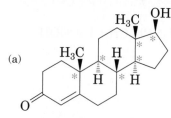

Testosterone

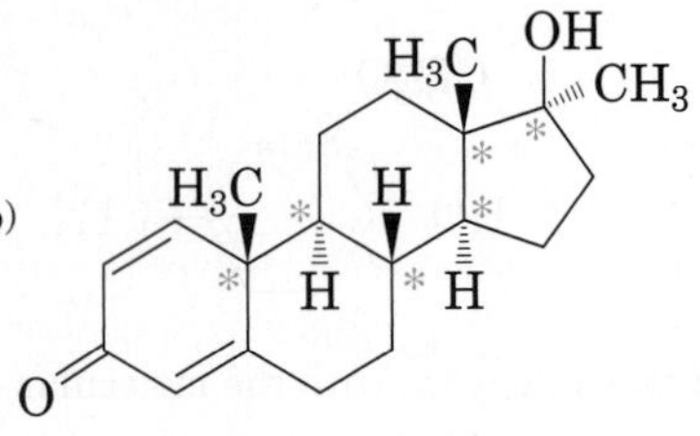

Methandrostenolone

17.77 Secondary alcohols are oxidized to ketones, and primary alcohols are oxidized to carboxylic acids.

(a) 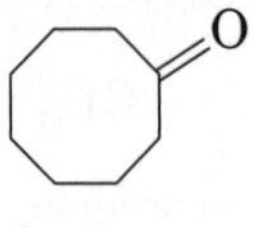   (b)

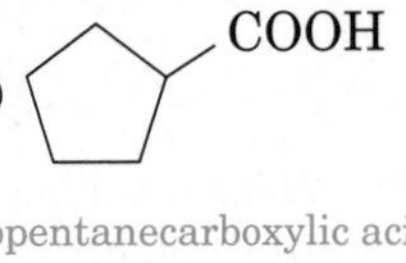

Cyclooctanone   Cyclopentanecarboxylic acid

Chapter 18 Carboxylic Acids

18.1 (a) 2,3-Dihydroxypropanoic acid
(b) 3-Aminopropanoic acid
(c) 3.5-Dihydroxy-3-methylpentanoic acid

18.2

(a) 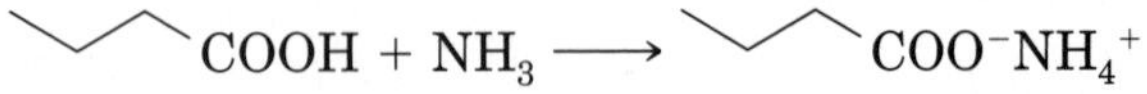 COOH + NH_3 ⟶ COO⁻NH_4^+

Butanoic acid   Ammonium butanoate

(b) 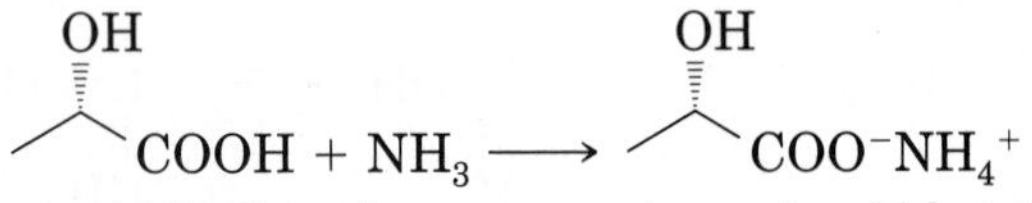 (S)-Lactic acid + NH_3 ⟶ Ammonium (S)-lactate

18.3

(a)

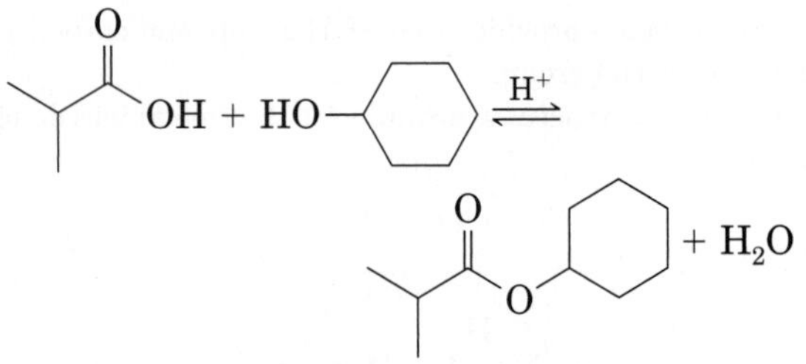

(b)

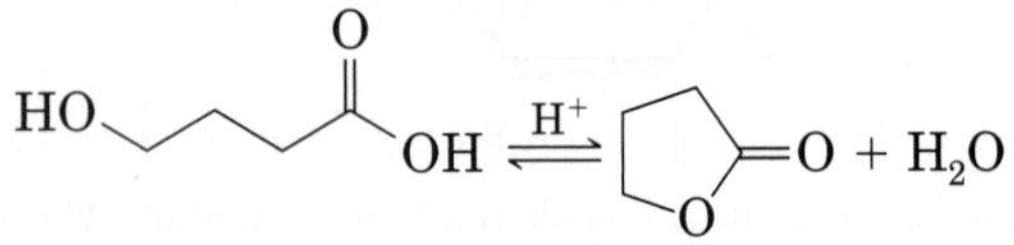

18.5 (a) T (b) T (c) T (d) F (e) T (f) T (g) T (h) T
(d) False: 3-methylbutanoic acid has no stereocenter and is therefore achiral.
18.7 (a) 3,4-Dimethylpentanoic acid
(b) 2-Aminobutanoic acid (c) Hexanoic acid
18.9

(a) 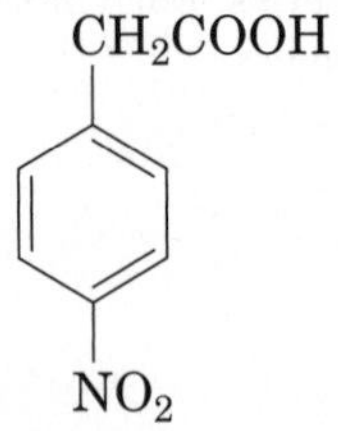   (b) $H_2NCH_2CH_2CH_2COOH$

(c) 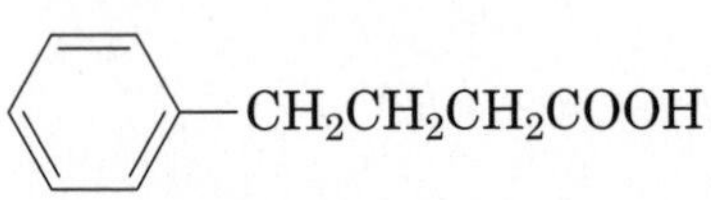 —$CH_2CH_2CH_2COOH$

(d)

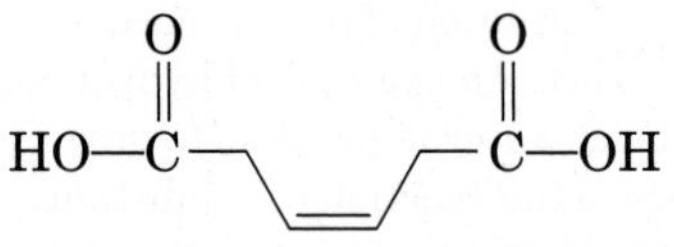

18.11

(a) 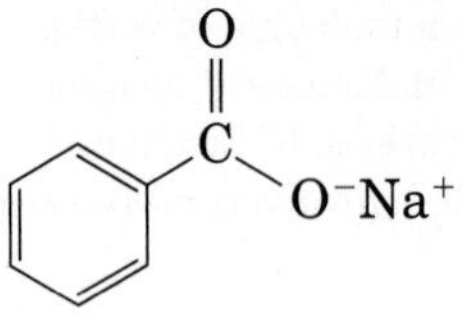   (b) 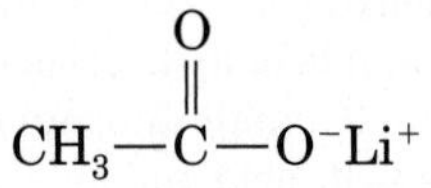 CH_3—C(=O)—O⁻Li^+

(c) CH_3—C(=O)—O⁻NH_4^+

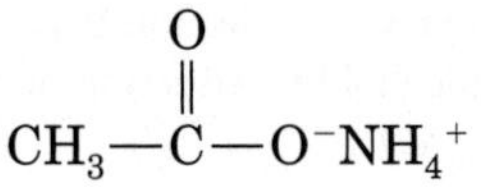

(d)

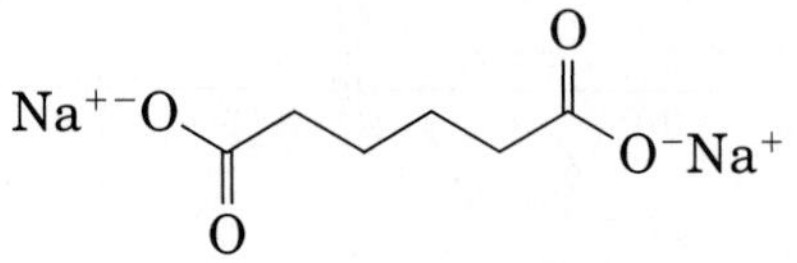

(e)

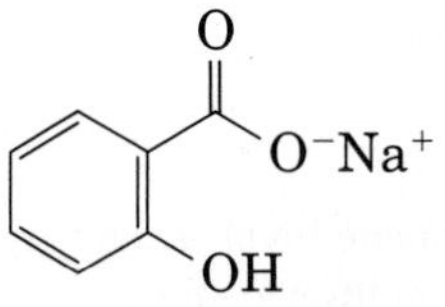

(f) $(CH_3CH_2CH_2COO^-)_2Ca^{2+}$

18.13

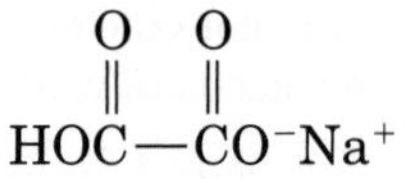

HOC(=O)—C(=O)O⁻Na^+

Monopotassium oxalate

18.15 The dimer drawn here shows two hydrogen bonds:

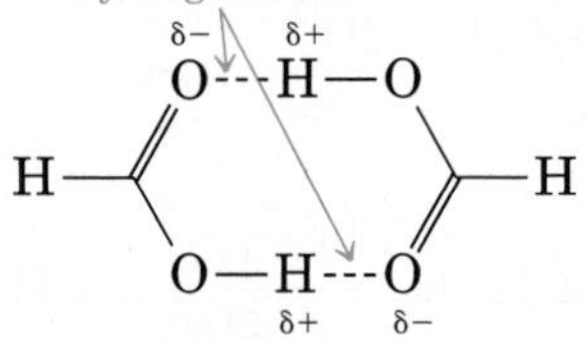

18.17 The carboxyl group contributes to water solubility; the hydrocarbon chain prevents water solubility.
18.19 In order of increasing boiling point, they are:

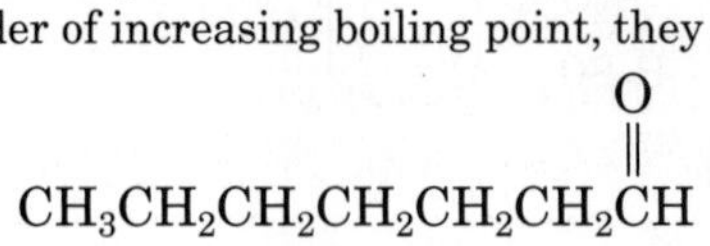

$CH_3CH_2CH_2CH_2CH_2CH_2CH$(=O)
Heptanal
(bp 153°C)

$CH_3CH_2CH_2CH_2CH_2CH_2CH_2OH$
1-Heptanol
(bp 176°C)

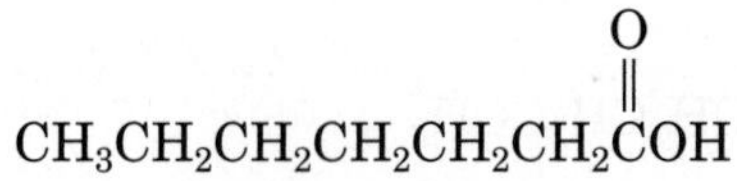

$CH_3CH_2CH_2CH_2CH_2CH_2COH$(=O)
Heptanoic acid
(bp 223°C)

18.21 In order of increasing solubility in water, they are decanoic acid, pentanoic acid, and acetic acid.

18.23 The structural features necessary to make a good detergent are (1) a hydrocarbon chain of 12–20 carbon atoms and (2) a polar end group that will not form insoluble salts with Ca^{2+}, Mg^{2+}, or Fe^{3+} ions.

18.25 Each has a long, hydrophobic, 16-carbon hydrocarbon chain and a polar end group that will not form insoluble salts in hard water.

18.27 Carboxylic acids (pK_a 4.0–5.0) are the strongest acids, and alcohols (pK_a 16–18) are the weakest acids. Phenols (pK_a 10) are intermediate in acidity.

18.29 Following are completed equations for these acid-base reactions:

(a)
$$\text{C}_6\text{H}_5\!-\!CH_2COOH + NaOH \longrightarrow$$
$$\text{C}_6\text{H}_5\!-\!CH_2COO^-Na^+ + H_2O$$

(b)
$$\text{CH}_2\!=\!\text{CHCH}_2\!-\!COOH + NaHCO_3 \longrightarrow$$
$$\text{COO}^-Na^+ + CO_2 + H_2O$$

(c)
COOH / OCH₃ (ortho)
$$+ NaHCO_3 \longrightarrow$$
COO⁻ + Na⁺ / OCH₃ (ortho)
$$+ CO_2 + H_2O$$

(d)
OH
$$CH_3\overset{|}{\text{C}}HCOOH + H_2NCH_2CH_2OH \longrightarrow$$
OH
$$CH_3\overset{|}{\text{C}}HCOO^- + H_3\overset{+}{\text{N}}CH_2CH_2OH$$

(e)
$$\text{COO}^-Na^+ + HCl \longrightarrow$$
$$\text{COOH} + NaCl$$

18.31 Formic acid reacts with $NaHCO_3$ in an acid-base reaction to convert formic acid to nonirritating sodium formate.

$$\underset{\text{Formic acid}}{H\!-\!\overset{\overset{\displaystyle O}{\|}}{C}\!-\!OH} + NaHCO_3 \longrightarrow$$

$$\underset{\text{Sodium formate}}{H\!-\!\overset{\overset{\displaystyle O}{\|}}{C}\!-\!O^-Na^+} + CO_2 + H_2O$$

18.33 (a) 1×10^{-3}
(b) 1
(c) 1×10^2
(d) 1×10^4
(e) 1×10^6

18.35 The pK_a of lactic acid is 4.07. At this pH, lactic acid would be present as 50% lactic acid, $CH_3CH(OH)COOH$, and 50% as lactate anion, $CH_3CH(OH)COO^-$. At pH 7.35 to 7.45, which is more basic than pH 4.07, lactic acid would be present as lactate anion, $CH_3CH(OH)COO^-$.

18.37 Recall from Chapter 8 that carboxylic acids are stronger acids than ammonium ion. Therefore, in part (a), the most acidic hydrogen is that of the COOH group.

(a)
$$\underset{\underset{NH_3^+}{|}}{CH_3\text{CHCOOH}} + NaOH \longrightarrow$$
$$\underset{\underset{NH_3^+}{|}}{CH_3\text{CHCOO}^-Na^+} + H_2O$$

(b)
$$\underset{\underset{NH_3^+}{|}}{CH_3\text{CHCOO}^-Na^+} + NaOH \longrightarrow$$
$$\underset{\underset{NH_2}{|}}{CH_3\text{CHCOO}^-Na^+} + H_2O$$

18.39 In part (a), the amino ($-NH_2$) group is the stronger base than the carboxylate ($-COO^-$) group.

(a)
$$\underset{\underset{NH_2}{|}}{CH_3\text{CHCOO}^-Na^+} + HCl \longrightarrow$$
$$\underset{\underset{NH_3^+Cl^-}{|}}{CH_3\text{CHCOO}^-Na^+}$$

(b)
$$\underset{\underset{NH_3^+}{|}}{CH_3\text{CHCOO}^-Na^+} + HCl \longrightarrow \underset{\underset{NH_3^+}{|}}{CH_3\text{CHCOOH}} + NaCl$$

18.41 Following are the structural formulas for the ester formed in each reaction:

(a) CH₃C(=O)O—CH₂CH(CH₃)₂ (isobutyl acetate structure)

(b) $CH_3\overset{\overset{\displaystyle O}{\|}}{C}O-$ cyclohexyl

(c)
COOCH₂CH₃ / COOCH₂CH₃ (ortho diester on benzene ring)

18.43 Following is a structural formula for methyl 2-hydroxybenzoate:

COOCH₃ / OH (ortho-substituted benzene ring)

Methyl 2-hydroxybenzoate
(Methyl salicylate)

18.45 Following are the expected organic products for each reaction:

(a) [benzyl group]—$CO_2^-Na^+$

(b) [benzyl group]—$CO_2^-Na^+$

(c) [benzyl group]—$CO_2^-NH_4^+$

(d) [benzyl group]—CH_2CH_2OH

(e) No reaction

(f) [benzyl group]—CO_2CH_3

(g) No reaction

18.47 Each product can be formed by Fischer esterification with the appropriate alcohol. Note that 4-aminobenzoic acid contains an aromatic amine group, which will react with the acid catalyst to form a salt. Therefore, after Fischer esterification is completed, sufficient base (NaOH) must be added to convert the amine salt to the free amine. Thus, formation of the ester and isolation of the final product requires two steps: (1) formation of the ester and (2) neutralization of the amine salt/acid catalyst.

[reaction scheme]

4-Aminobenzoic acid (p-Aminobenzoic acid)

1) CH_3OH/H_2SO_4
2) $NaOH / H_2O$
→ Methyl 4-aminobenzoate, Methylparaben

1) $CH_3CH_2CH_2OH/H_2SO_4$
2) $NaOH/H_2O$
→ Propyl 4-aminobenzoate, Propylparaben

18.49 The product is a 6-membered lactone.

$$CH_2CH_2CH_2CH_2C\text{—}OH \xrightarrow[\text{catalyst}]{\text{acid}} \text{a lactone} + H_2O$$
|
OH

5-Hydroxypentanoic acid   a lactone

18.51 Carboxylic acids can be formed by the oxidation of primary alcohols (—CH_2OH) or an aldehyde (CHO).

(a) [structure]—OH

(b) [structure]—H

(c) HO—[structure]—OH

18.53 (a) Only three of these intermediates are chiral: isocitric acid, oxalosuccinic acid, and malic acid. Isocitric acid has two stereocenters and has the possibility of $2^2 = 4$ stereoisomers (two pairs of enantiomers). Two of these intermediates are capable of *cis-trans* isomerism: aconitic acid and fumaric acid.

(b) **Step 1**: Dehydration of a 3° alcohol to give an alkene.
  Step 2: Hydration of an alkene to give an alcohol.
  Step 3: Oxidation of a 2° alcohol to a ketone.
(c) If the hydration were to follow Markovnikov's rule, citric acid would be formed. The fact that the hydration proceeds as it does is due to the presence of the enzyme. This enzyme positions aconitic acid at the reactive site on the enzyme surface in such a way that only one carbocation intermediate can be formed, and water is then added to it in a regiospecific way.
(d) **Step 4**: Decarboxylation of a β-ketoacid. Only one of the three carboxyl groups in oxalosuccinic acid has a carbonyl group beta to it.
  Step 5: Oxidative decarboxylation.
  Step 6: Oxidation of an alkane to an alkene.
  Step 7: Hydration of a carbon-carbon double bond to give a 2° alcohol.
  Step 8: Oxidation of a 2° alcohol to a ketone.
(e) This is a decarboxylation because CO_2 is lost, and it is an oxidation because an oxygen is added to the ketone carbonyl carbon to make it a new carboxylic acid.
(f) Yes, it is a decarboxylation, but it is not also an oxidation because no new oxygen is added.

Chapter 19 Anhydrides, Esters and Amides

19.1

(a) CH_3CNH—[cyclohexyl]

(b) [phenyl]—CNH_2

19.2 Under basic conditions, as in part (a), each carboxyl group is present as a carboxylate anion. Under acidic conditions, as in part (b), each carboxyl group is present in its un-ionized form.

(a) [benzene ring with $COCH_3$ and $COCH_3$ groups, both as esters] + 2NaOH $\xrightarrow{H_2O}$

[benzene ring with CO^-Na^+ and CO^-Na^+ groups] + 2CH_3OH

(b) [diketone ester structure] + $H_2O \xrightarrow{HCl}$

[diketo acid structure]—OH + CH_3CH_2OH

19.3 In aqueous NaOH, each carboxyl group is present as a carboxylic anion and each amine is present in its unprotonated form.

(a) $CH_3\overset{O}{\overset{\|}{C}}N(CH_3)_2$ + NaOH $\xrightarrow[\text{heat}]{H_2O}$

$CH_3\overset{O}{\overset{\|}{C}}O^-Na^+ + (CH_3)_2NH$

(b) 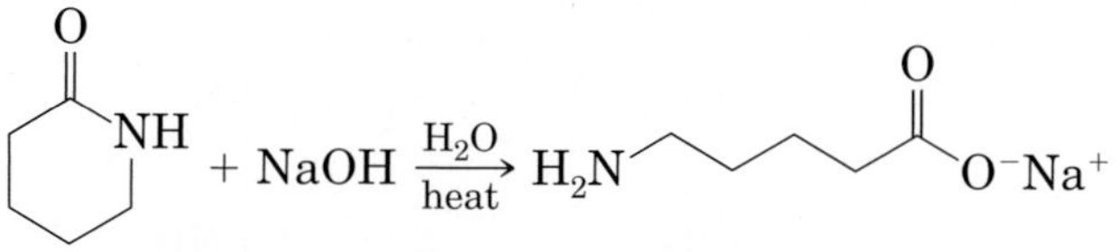 [lactam structure] + NaOH $\xrightarrow[\text{heat}]{H_2O}$ $H_2N \diagdown\diagup\diagdown\diagup\overset{O}{\overset{\|}{C}}O^-Na^+$

19.5 (a) Benzoic anhydride
  (b) Methyl decanoate
  (c) *N*-Methylhexanamide
  (d) 4-Aminobenzamide or *p*-aminobenzamide
  (e) Cyclopentyl ethanoate or cyclopentyl acetate
  (f) Ethyl 3-hydroxybutanoate

19.7 Each reaction brings about hydrolysis of the ester. Each product is shown as it would exist under the specified reaction conditions.

(a) 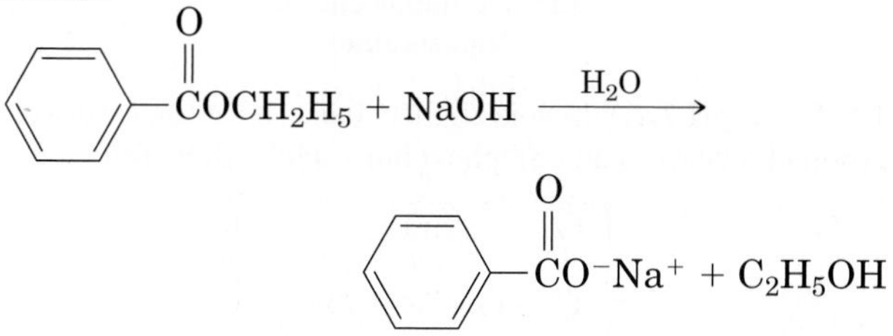 [phenyl]$-\overset{O}{\overset{\|}{C}}OCH_2H_5$ + NaOH $\xrightarrow{H_2O}$

[phenyl]$-\overset{O}{\overset{\|}{C}}O^-Na^+ + C_2H_5OH$

(b) 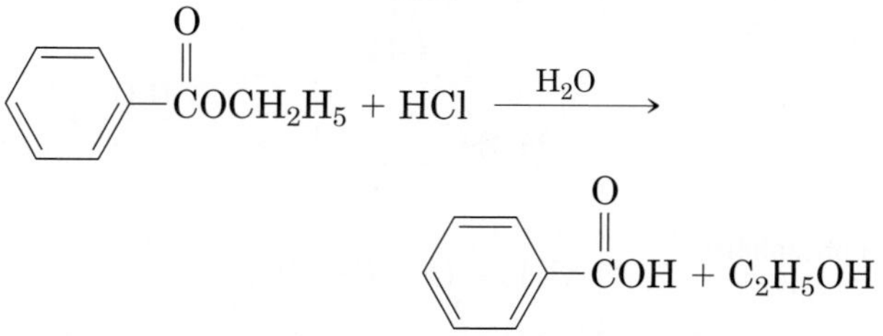 [phenyl]$-\overset{O}{\overset{\|}{C}}OCH_2H_5$ + HCl $\xrightarrow{H_2O}$

[phenyl]$-\overset{O}{\overset{\|}{C}}OH + C_2H_5OH$

19.9 Only a carboxylic acid (a) will react with sodium bicarbonate and produce bubbles of CO_2.

19.11 Phenacetin contains an amide.

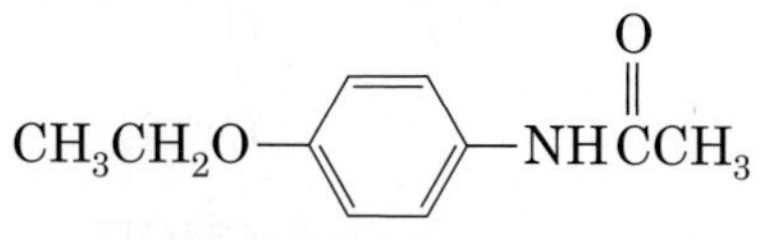

CH_3CH_2O-[phenyl]$-NH\overset{O}{\overset{\|}{C}}CH_3$

Phenacetin

19.13 (a) Aspartame is chiral. It has two stereocenters. Four stereoisomers (two pairs of enantiomers) are possible.

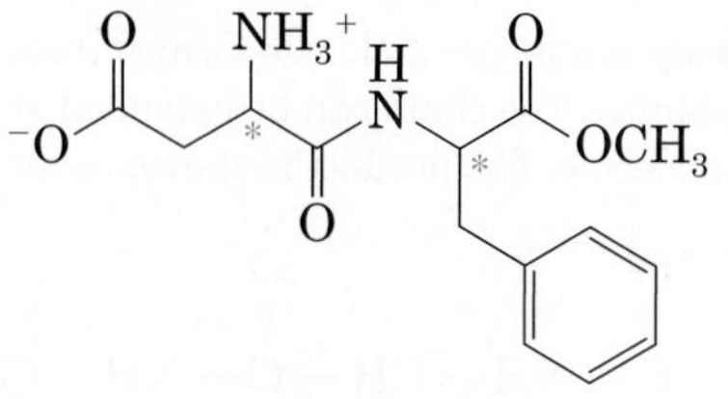

[Aspartame structure]

Aspartame

(b) Aspartame contains one carboxylate group, one 1° ammonium ion, one amide group, and one ester group.
(c) The net charge is zero.
(d) It is an internal salt. Therefore, expect it to be soluble in water.

(e) Hydrolysis in HCl

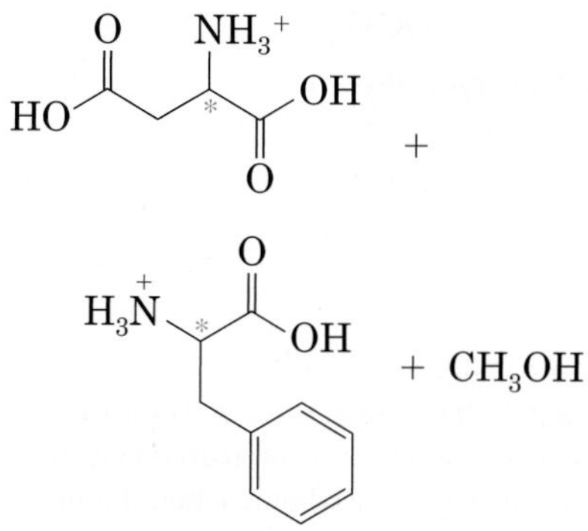

[structure] +

[structure] + CH_3OH

(f) Hydrolysis in NaOH

[structure] +

[structure] + CH_3OH

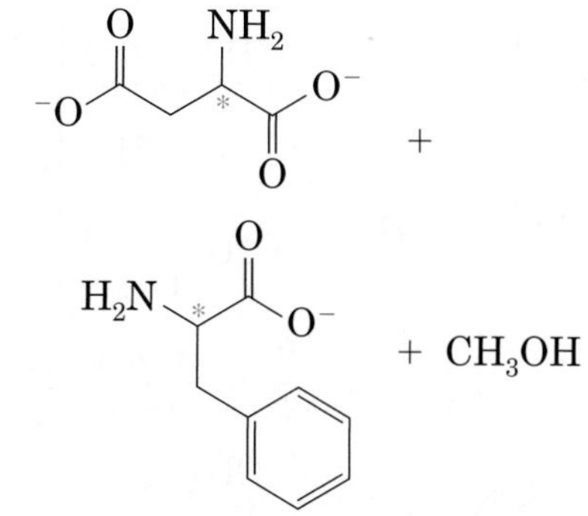

19.15 Following are sections of two parallel nylon-66 chains, with hydrogen bonds between N—H and C=O groups indicted by dashed lines.

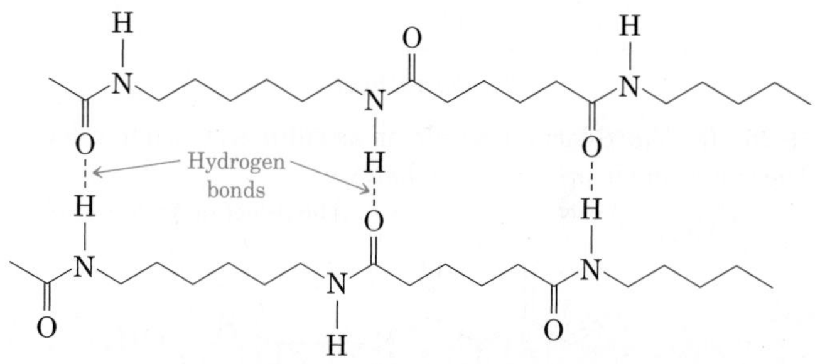

[nylon-66 structure with Hydrogen bonds labeled]

19.17 In the anhydrides of carboxylic acids, the functional group is two carbonyl (C=O) groups bonded to an oxygen atom. In an anhydride of phosphoric acid, the functional group is two phosphoryl (P=O) groups bonded to an oxygen atom.

19.19 Following is the structural formula of dihydroxyacetone phosphate as it would be ionized at pH 7.40.

$HOCH_2\overset{O}{\overset{\|}{C}}CH_2O\overset{O}{\overset{\|}{P}}O^-$
$\qquad\qquad\qquad\quad O^-$

Dihydroxyacetone phosphate

19.21 The products of hydrolysis in aqueous base (NaOH) are sodium phosphate and methanol.

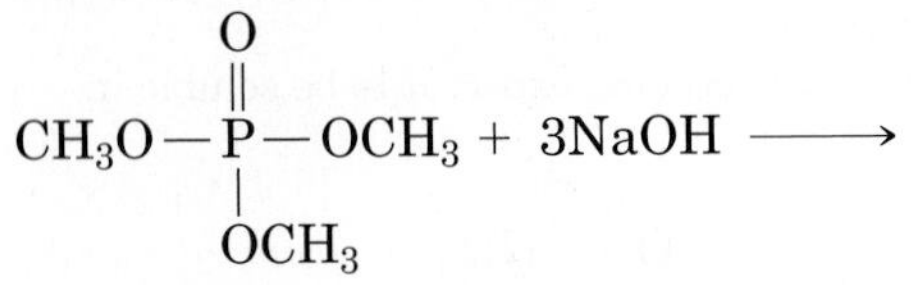

$$CH_3O-\overset{\overset{\displaystyle O}{\|}}{\underset{\underset{\displaystyle OCH_3}{|}}{P}}-OCH_3 + 3NaOH \longrightarrow$$

Trimethyl
phosphate

$$Na^+{}^-O-\overset{\overset{\displaystyle O}{\|}}{\underset{\underset{\displaystyle O^-Na^+}{|}}{P}}-O^-Na^+ + 3CH_3OH$$

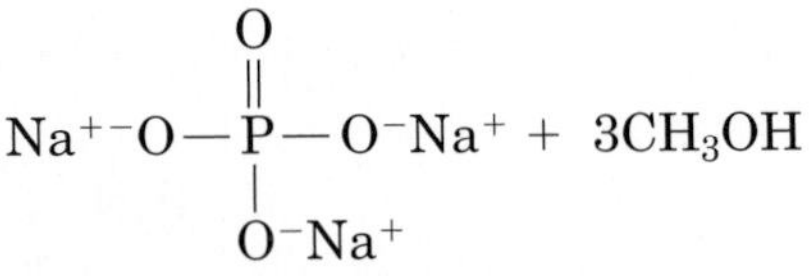

19.23 The boxes enclose their common structural features, which are a tetrasubstituted cyclopropane ring, an ester, and a trisubstituted carbon-carbon double bond.

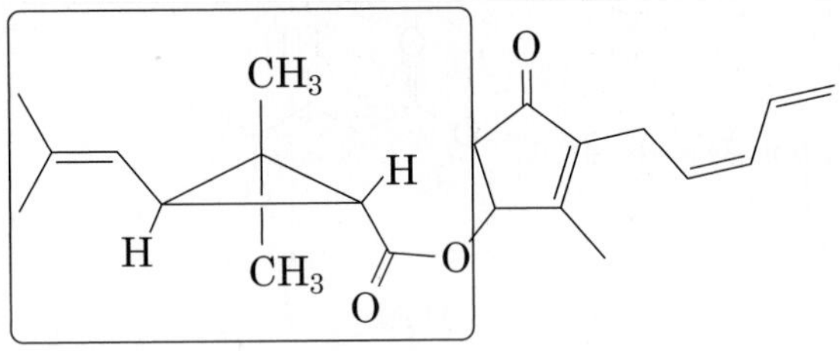

Pyrethrin I

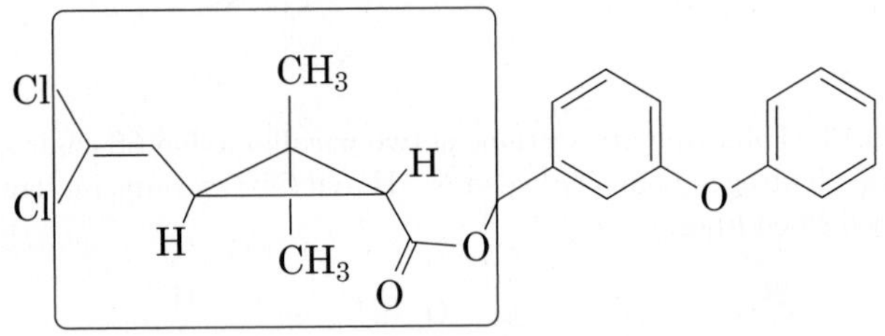

Permethrin

19.25 (a) The β-lactam ring in amoxicillin is shown in color. The same ring is present in cephalexin.

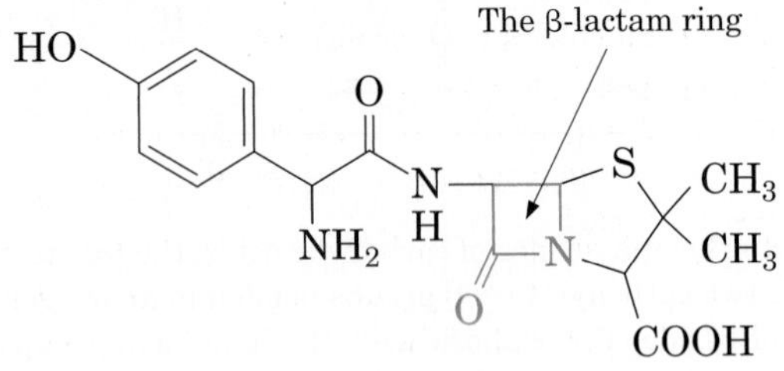

The β-lactam ring

Amoxicillin

(b) Hydrolysis of the β-lactam ring will generate a carboxyl group and a 2° heterocyclic amine group.

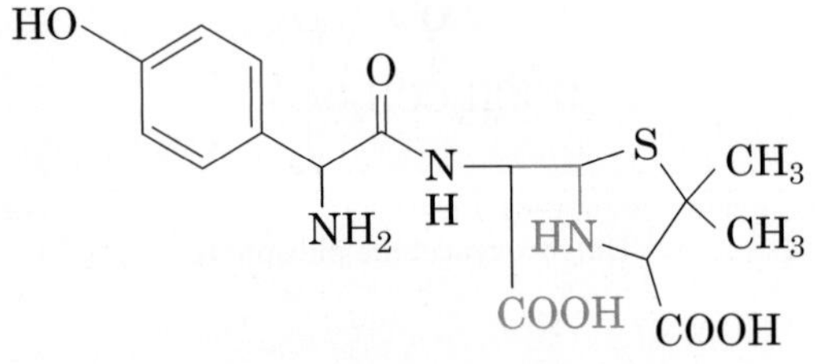

19.27 The two functional groups in aspirin are a carboxyl group and a phenolic ester.

19.29 Both aspirin and ibuprofen contain a carboxyl group and an aromatic ring. In ibuprofen, the aromatic ring is benzene. In naproxen, the aromatic component is naphthalene.

19.31 A *sunblock* prevents ultraviolet radiation from reaching protected skin by reflecting it away from the skin. A *sunscreen* prevents a portion of the sun's ultraviolet radiation from reaching protected skin by absorbing it and re-radiating it as heat.

19.33 The portions of each molecule that are derived from urea are circled.

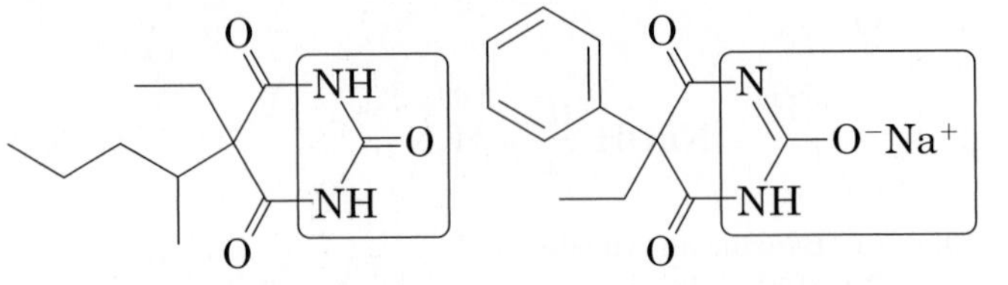

Pentobarbital   Phenobarbital

19.35 Benzocaine is an ethyl ester of 4-aminobenzoic acid.

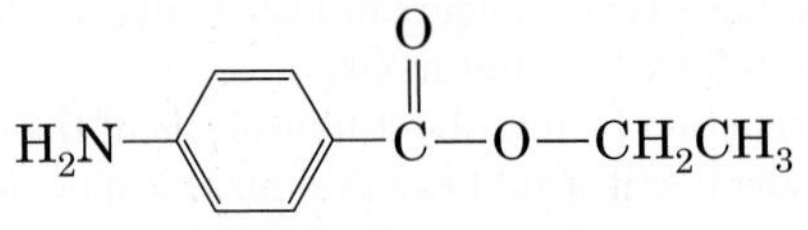

$$H_2N-\!\!\!\!\bigcirc\!\!\!\!-\overset{\overset{\displaystyle O}{\|}}{C}-O-CH_2CH_3$$

Ethyl 4-aminobenzoate
(Benzocaine)

19.37 At pH 7.4, which is slightly basic, both the carboxyl group of glyceric acid and phosphoric acid will be fully ionized.

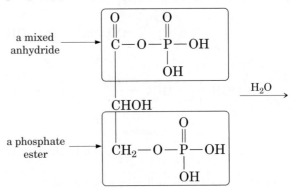

a mixed anhydride

a phosphate ester

1,2-Diphosphoglycerate

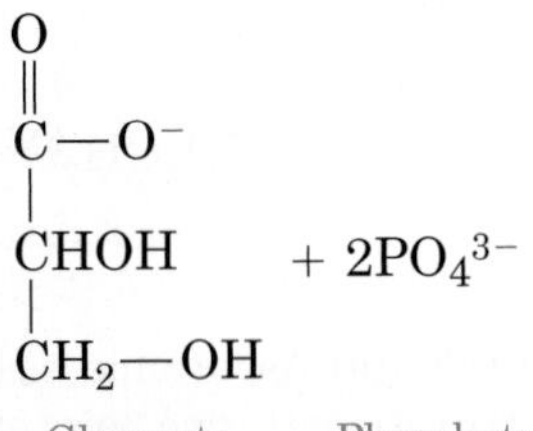

$$\overset{\overset{\displaystyle O}{\|}}{C}-O^-$$
$$CHOH \qquad + 2PO_4{}^{3-}$$
$$CH_2-OH$$

Glycerate   Phosphate

19.39 Following is a larger molecule formed from three molecules of alanine. The chain can be extended at either end to form a macromolecule. The product is shown as an internal salt.

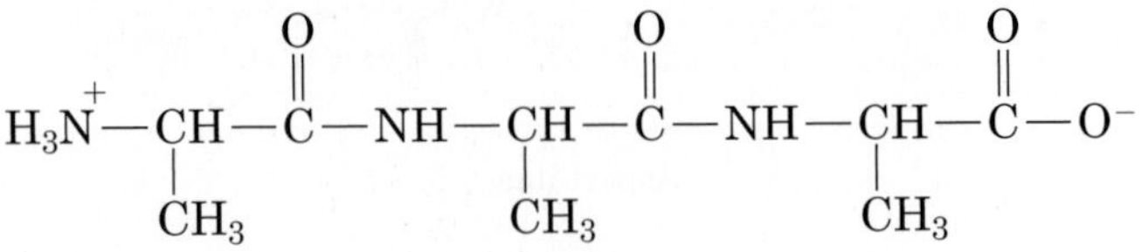

$$H_3\overset{+}{N}-\underset{\underset{\displaystyle CH_3}{|}}{CH}-\overset{\overset{\displaystyle O}{\|}}{C}-NH-\underset{\underset{\displaystyle CH_3}{|}}{CH}-\overset{\overset{\displaystyle O}{\|}}{C}-NH-\underset{\underset{\displaystyle CH_3}{|}}{CH}-\overset{\overset{\displaystyle O}{\|}}{C}-O^-$$

19.41 Key steps are acid-catalyzed hydration of ethylene to give ethanol and then oxidation of ethanol to acetic acid. Fischer esterification of acetic acid with ethanol gives ethyl acetate.

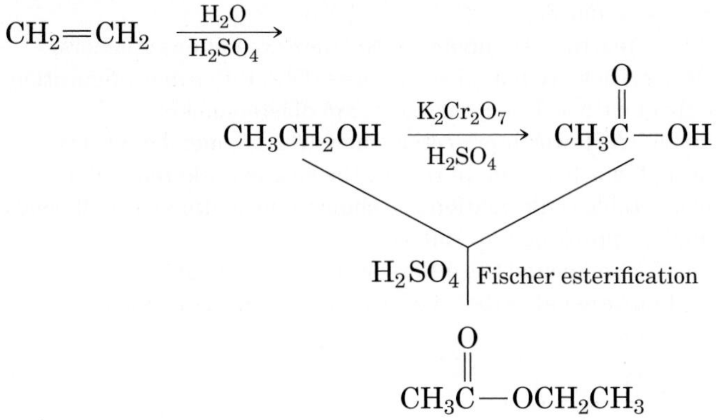

$CH_2{=}CH_2$
H_2O
H_2SO_4
CH_3CH_2OH
$K_2Cr_2O_7$
H_2SO_4
$CH_3C(=O){-}OH$
H_2SO_4
Fischer esterification
$CH_3C(=O){-}OCH_2CH_3$

Ethyl acetate

19.43 DEET can be synthesized from 3-methylbenzoic acid and diethylamine.

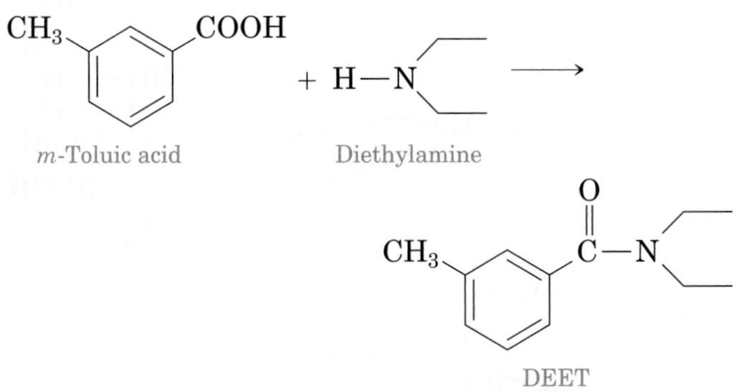

CH_3
COOH
+ H—N
m-Toluic acid
Diethylamine
CH_3
O
C—N
DEET

19.45 Following is the structural formula of barbital:

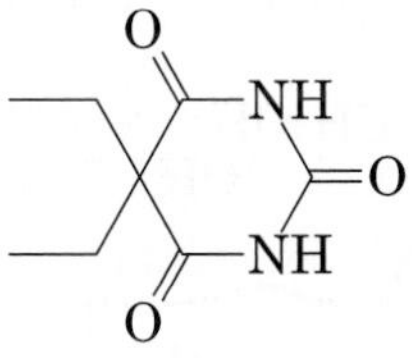

O
NH
O
NH
O

Barbital

19.47

(a) H_2N
(e)
C
NH
H
N
(a)
NH_2
6
O
1
OH (e)
COOH (e)
more stable chair
(e,e,e,a)
(a)
NH
C
H_2N
NH
(e)
NH_2
COOH (a)
O
6
1
OH (a)
less stable chair
(a,a,a,e)

(b) The ring of drug **A** is created by reaction of the 2° alcohol group with the aldehyde group of compound **B**. The new functional group created is a cyclic hemiacetal.

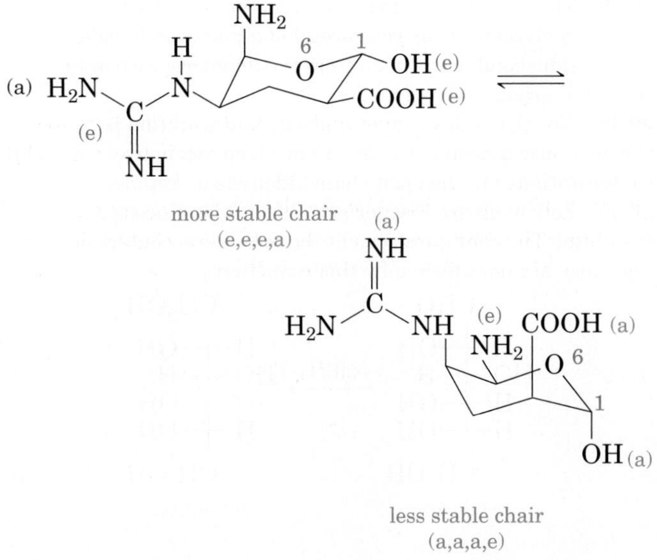

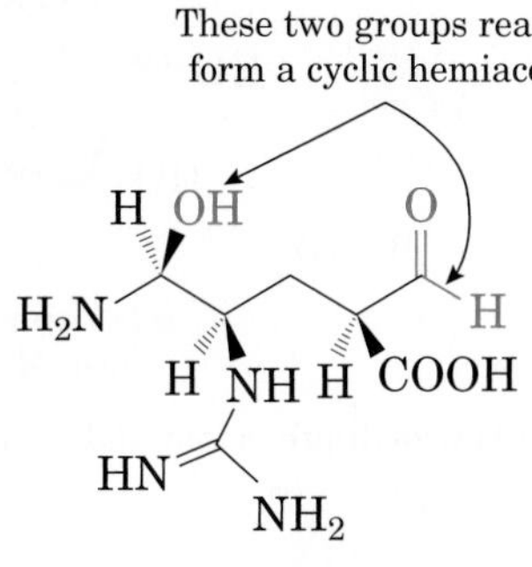

These two groups react to
form a cyclic hemiacetal
H OH
O
H_2N
H
H NH H COOH
HN
NH_2

Molecule B

Chapter 20 Carbohydrates

20.1 Following are Fischer projections for the four 2-ketopentoses. They consist of two pairs of enantiomers.

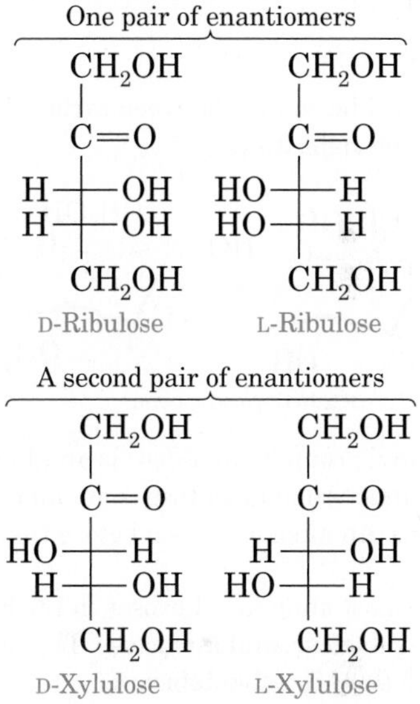

One pair of enantiomers
CH_2OH
C=O
H—OH
H—OH
CH_2OH
D-Ribulose
CH_2OH
C=O
HO—H
HO—H
CH_2OH
L-Ribulose
A second pair of enantiomers
CH_2OH
C=O
HO—H
H—OH
CH_2OH
D-Xylulose
CH_2OH
C=O
H—OH
HO—H
CH_2OH
L-Xylulose

20.2 D-Mannose differs in configuration from D-glucose only at carbon 2. One way to arrive at the structures of the α and β forms of D-mannopyranose is to draw the corresponding α and β forms of D-glucopyranose and then invert the configuration in each at carbon 2.

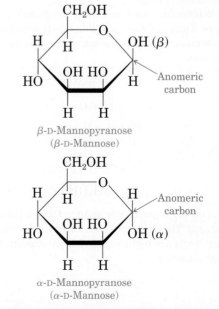

CH_2OH
O
H
H
OH (β)
OH HO
HO
H
Anomeric
carbon
H
H

β-D-Mannopyranose
(β-D-Mannose)

CH_2OH
O
H
H
H
Anomeric
carbon
OH HO
HO
OH (α)
H
H

α-D-Mannopyranose
(α-D-Mannose)

20.3 D-Mannose differs in configuration from D-glucose only at carbon 2.

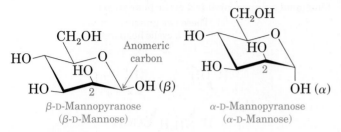

β-D-Mannopyranose (β-D-Mannose)

α-D-Mannopyranose (α-D-Mannose)

20.4 Following is a Haworth projection and a chair conformation for this glycoside.

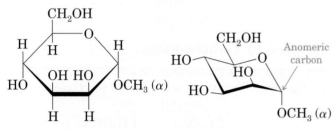

20.5 The β-glycosidic bond is between carbon 1 of the left unit and carbon 3 of the right unit.

Unit of β-D-glucopyranose   β-1,3-Glycosidic bond   Unit of α-D-glucopyranose

20.7 The carbonyl group in an aldose is an aldehyde. In a ketose, it is a ketone. An aldopentose is an aldose that contains five carbon atoms. An aldoketose is a ketose that contains five carbon atoms.

20.9 The three most abundant hexoses in the biological world are D-glucose, D-galactose, and D-fructose. The first two are aldohexoses. The third is a 2-ketohexose.

20.11 To say that they are enantiomers means that they are nonsuperposable mirror images.

20.13 The D or L configuration in an aldopentose is determined by its configuration at carbon 4.

20.15 Compounds (a) and (c) are D-monosaccharides. Compound (b) is an L-monosaccharide.

20.17 A 2-ketoheptose has four stereocenters and 16 possible stereoisomers. Eight of these are D-2-ketoheptoses, and eight are L-2-ketoheptoses. Following is one of the eight possible D-2-ketoheptoses.

$$
\begin{array}{c}
CH_2OH \\
| \\
C=O \\
HO \overset{*}{\longmapsto} H \\
HO \overset{*}{\longmapsto} H \\
H \overset{*}{\longmapsto} OH \\
H \overset{*}{\longmapsto} OH \\
CH_2OH
\end{array}
$$

20.19 In an amino sugar, one or more —OH groups are replaced by —NH_2 groups. The three most abundant amino sugars in the biological world are D-glucosamine, D-galactosamine, and N-acetyl-D-glucosamine.

20.21 (a) A pyranose is a six-membered cyclic hemiacetal form of a monosaccharide.
(b) A furanose is a five-membered cyclic hemiacetal form of a monosaccharide.

20.23 Yes, they are anomers. No, they are not enantiomers; that is, they are not mirror images. They differ in configuration only at carbon 1 and, therefore, are diastereomers.

20.25 A Haworth projection shows the six-membered ring as a planar hexagon. In reality, the ring is puckered and its most stable conformation is a chair conformation with all bond angles approximately 109.5°.

20.27 Compound (a) differs from D-glucose only in the configuration at carbon 4. Compound (b) differs only at carbon 3.

(a)

$$
\begin{array}{c}
CHO \\
H \longmapsto OH \\
HO \longmapsto H \\
HO \longmapsto H \\
H \longmapsto OH \\
CH_2OH
\end{array}
$$

D-Galactose

(b)

$$
\begin{array}{c}
CHO \\
H \longmapsto OH \\
H \longmapsto OH \\
H \longmapsto OH \\
H \longmapsto OH \\
CH_2OH
\end{array}
$$

D-Allose

20.29 The specific rotation of α-L-glucose is −112.2°.

20.31 A glycoside is a cyclic acetal of a monosaccharide. A glycosidic bond is the bond from the anomeric carbon to the —OR group of the glycoside.

20.33 No, glycosides cannot undergo mutarotation because the anomeric carbon is not free to interconvert between α and β configurations via the open-chain aldehyde or ketone.

20.35 Following are Fischer projections of D-glucose and D-sorbitol. The configurations at the four stereocenters of D-glucose are not affected by this reduction.

$$
\begin{array}{c}
CHO \\
H \longmapsto OH \\
HO \longmapsto H \\
H \longmapsto OH \\
H \longmapsto OH \\
CH_2OH
\end{array}
\xrightarrow{NaBH_4}
\begin{array}{c}
CH_2OH \\
H \longmapsto OH \\
HO \longmapsto H \\
H \longmapsto OH \\
H \longmapsto OH \\
CH_2OH
\end{array}
$$

D-Glucose   D-Sorbitol

20.37   Ribitol is the reduction product of D-ribose. β-D-ribose 1-phosphate is the phosphoric ester of the OH group on the anomeric carbon of β-D-ribofuranose.

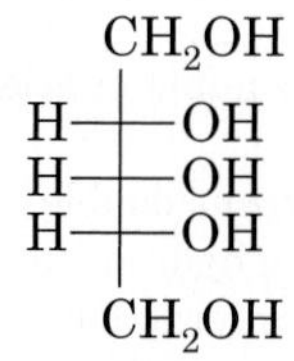

Ribitol

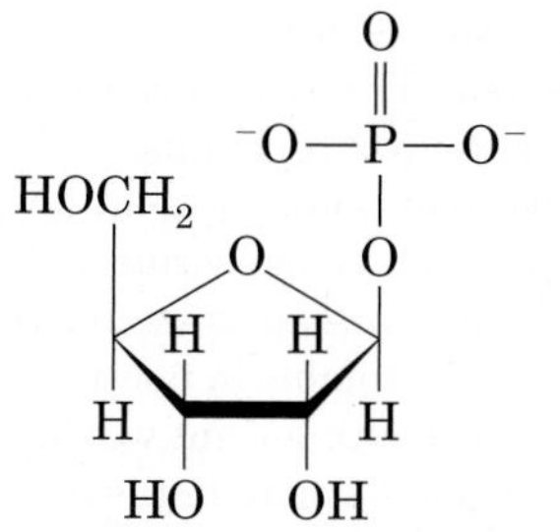

β-D-Ribofuranose 1-phosphate
(β-D-Ribose 1-phosphate)

20.39   To say that it is a β-1,4-glycosidic bond means that the configuration at the anomeric carbon (carbon 1 in this problem) of the monosaccharide unit forming the glycosidic bond is beta and that it is bonded to carbon 4 of the second monosaccharide unit. To say that it is an α-1,6-glycosidic bond means that the configuration at the anomeric carbon (carbon 1 in this problem) of the monosaccharide unit forming the glycosidic bond is alpha and that it is bonded to carbon 6 of the second monosaccharide unit.

20.41   (a) Both monosaccharide units are D-glucose.
(b) They are joined by a β-1,4-glycosidic bond.
(c) It is a reducing sugar and
(d) it undergoes mutarotation.

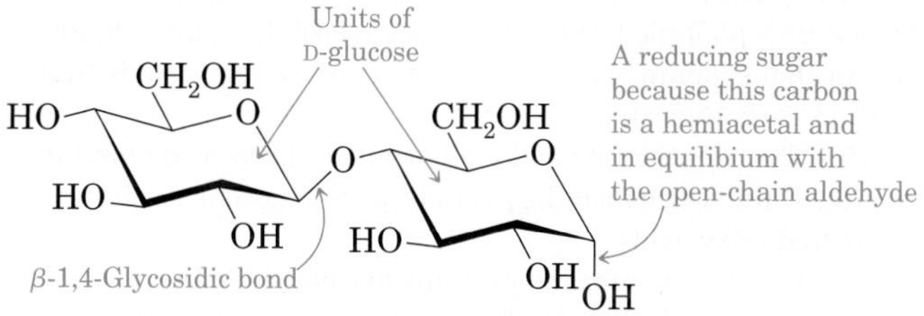

20.43   An oligosaccharide contains approximately six to ten monosaccharide units. A polysaccharide contains more—generally many more—than ten monosaccharide units.

20.45   The difference lies in the degree of chain branching. Amylose is composed of unbranched chains, whereas amylopectin is a branched network with the branches started by α-1, 6-glycosidic bonds.

20.47   Cellulose fibers are insoluble in water because the strength of hydrogen bonding of a cellulose molecule in the fiber with surface water molecules is not sufficient to overcome the intermolecular forces that hold it in the fiber.

20.49   (a) In these structural formulas, the CH₃CO (the acetyl group) is abbreviated Ac.

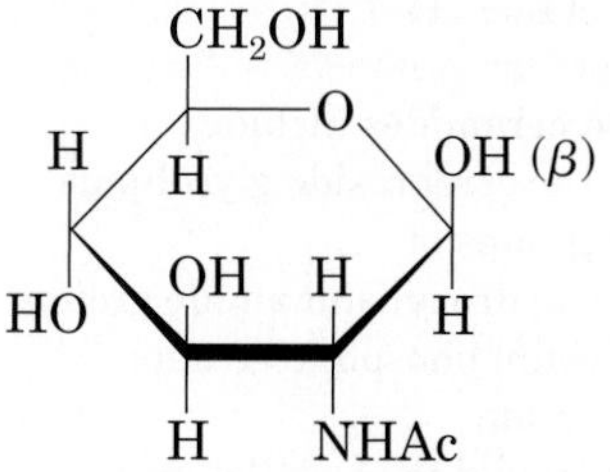

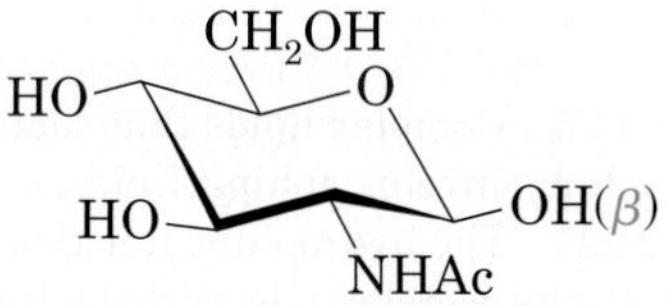

Following are Haworth and chair structures for this repeating disaccharide.

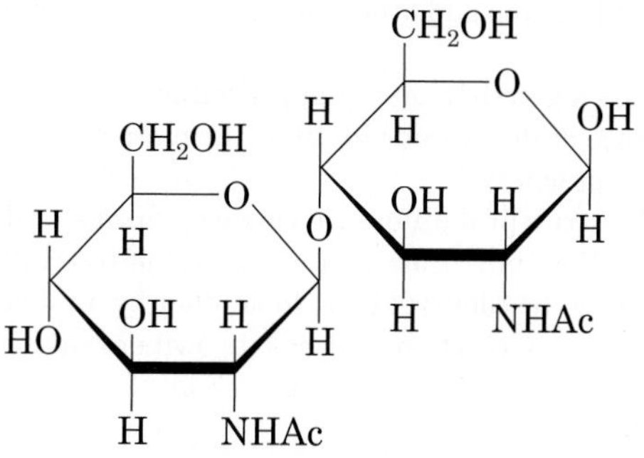

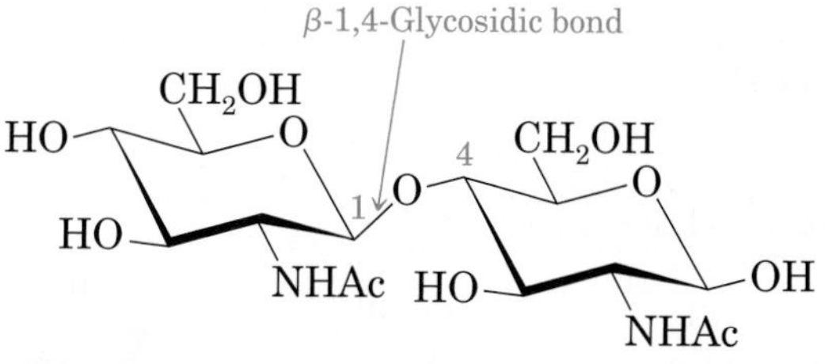

β-1,4-Glycosidic bond

20.51   Its lubricating power decreases.
20.53   With maturation, children develop an enzyme capable of metabolizing galactose. Thus, they are able to tolerate galactose as they mature. Until these children develop the ability to metabolize galactose, substituting sucrose for lactose replaces the galactose in lactose with fructose in sucrose.
20.55   L-Ascorbic acid is oxidized (there is loss of two hydrogen atoms) when it is converted to L-dehydroascorbic acid. L-Ascorbic acid is a biological reducing agent.
20.57   Types A, B, and O have in common D-galactose and L-fucose. Only type A has N-acetyl-D-glucosamine.
20.59   Mixing types A and B blood will result in coagulation.
20.61   Consult Table 20.1 for the structural formula of D-altrose and draw it. Then replace the —OH groups on carbons 2 and 6 with hydrogens.

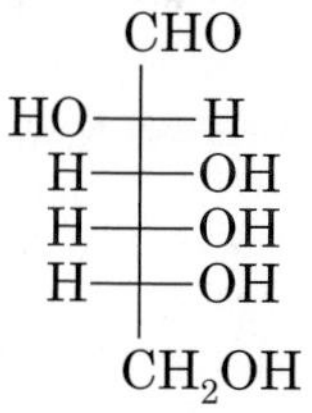

D-Altrose

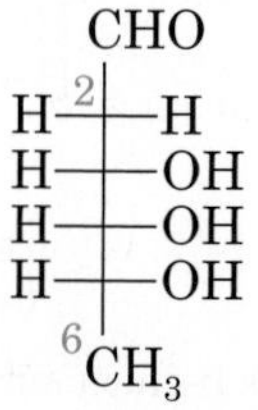

2,6-Dideoxy-D-Altrose
(D-Digitoxose)

20.63   The monosaccharide unit of salicin is D-glucose.
20.65   False. The molecular weight of carbohydrates in foods can vary widely. Sugars have relatively low molecular weights, but starches are polymers with high molecular weights.
20.67   The five-membered ring of fructose is nearly planar, so a Haworth projection is a good representation of its structure.

20.69 In starch, α-glycosidic bonds join one glucose moiety to another. Cellulose has β-glycosidic bonds. This difference means that humans and other animals can digest starch but not cellulose.

20.71 The ring system on the upper left is a sugar (glucose). The presence of the cyanide group is the main cause for concern about safety.

20.73 The structural difference between glucose and galactose is the configuration at carbon C4 of the ring. These two sugars can be converted one to another by an inversion of configuration in a reaction that has its own enzyme.

20.75

N-Acetyl-β-D-glucosamine

Chitin

20.77 Review Section 17.5 on keto-enol tautomerism and your answer to Problem 17.71. The intermediate in this conversion is an enediol; that is, it contains a carbon–carbon double bond with two —OH groups on it.

D-Glucose
6-phosphate

An enediol
intermediate

D-Fructose
6-phosphate

20.79 (a) The left unit is D-glucuonic acid, which is derived from D-glucose. The right unit is a sulfate ester derived from *N*-acetyl-D-galatosamine, which is in turn derived from D-galactosamine.
(b) The two units are joined by a β-1,3-glycosidic bond.

20.81 Cellulose molecules are held together by intermolecular hydrogen bonding. The addition of water disrupts this hydrogen bonding and compromises the strength of paper. Oil is nonpolar and does not interact with the cellulose hydroxyls in paper, therefore, does not affect the structural integrity of paper.

Chapter 21 Lipids

21.1 (a) It is an ester of glycerol and contains a phosphate group; therefore, it is a glycerophospholipid. Besides glycerol and phosphate, it has a myristic acid and a linoleic acid component. The other alcohol is serine. Therefore, it belongs to the subgroup of cephalins.
(b) The components present are glycerol, myristic acid, linoleic acid, phosphate, and serine.

21.3 Lipids are all characterized by being insoluble in polar solvents but soluble in non-polar solvents.

21.5 The melting point would increase. The *trans* double bonds would fit more in the packing of the long hydrophobic tails, creating more order and therefore more interaction between chains. This would require more energy to disrupt, and hence a higher melting point.

21.7 The diglycerides with the highest melting points will be the ones with two stearic acids (a saturated fatty acid). The lowest melting points will be the ones with two oleic acids (a monounsaturated fatty acid).

21.9 (b), because its molecular weight is higher.

21.11 lowest (c); then (b); highest (a)

21.13 The more long-chain groups, the lower the solubility; lowest (a); then (b); highest (c).

21.15 glycerol, sodium palmitate, sodium stearate, sodium linolenate

21.17 Complex lipids can be classified into two groups: phospholipids and glycolipids. Phospholipids contain an alcohol, two fatty acids, and a phosphate group. There are two types: glycerophospholipids and sphingolipids. In glycerophospholipids, the alcohol is glycerol. In sphingolipids, the alcohol is sphingosine. Glycolipids are complex lipids that contain carbohydrates.

21.19 The presence of *cis* double bonds in fatty acids produces greater fluidity because they cannot pack together as closely as saturated fatty acids.

21.21 Integral membrane proteins are embedded in the membrane. Peripheral membrane proteins are found on membrane surfaces.

21.23 A phosphatidyl inositol containing oleic acid and arachidonic acid:

21.25 Complex lipids that contain ceramides include sphingomyelin, sphingolipids, and the cerebroside glycolipids.

21.27 The hydrophilic functional groups of
(a) glucocerebroside: carbohydrate; hydroxyl and amide groups of the cerebroside. (b) Sphingomyelin: phosphate group; choline; hydroxyl and amide of ceramide.

21.29 Cholesterol crystals may be found in (1) gallstones, which are sometimes pure cholesterol, and (2) joints of people suffering from bursitis.

21.31 The carbon of the steroid D ring to which the acetyl group is bonded in progesterone undergoes the most substitution.

21.33 LDL from the bloodstream enters the cells by binding to LDL receptor proteins on the surface. After binding, the LDL is transported inside the cells, where cholesterol is released by enzymatic degradation of the LDL.

21.35 Removing lipids from the triglyceride cores of VLDL particles increases the density of the particles and converts them from VLDL to LDL particles.

21.37 When serum cholesterol concentration is high, the synthesis of cholesterol in the liver is inhibited and the synthesis of LDL receptors in the cell is increased. Serum cholesterol levels control the formation of cholesterol in the liver by regulating enzymes that synthesize cholesterol.

21.39 Estradiol (E) is synthesized from progesterone (P) through the intermediate testosterone (T). First, the D-ring acetyl group of P is converted to a hydroxyl group and T is produced. The methyl group in T, at the junction of the rings A and B, is removed and ring A becomes aromatic. The keto group in P and T is converted to a hydroxyl group in E.

21.41 Steroid structures are shown in Section 21.10. The major structural differences are at carbon 11. Progesterone has no substituents except hydrogen, cortisol has a hydroxyl group, cortisone has a keto group, and RU-486 has a large *p*-aminophenyl group. The functional group at carbon 11 apparently has little importance in receptor binding.

21.43 They have a steroid ring structure; they have a methyl group at carbon 13; they have a triply bonded group at carbon 17; and all have some unsaturation in the A ring, the B ring, or both.

21.45 Bile salts help solubilize fats. They are oxidation products of cholesterol themselves, and they bind to cholesterol, forming complexes that are eliminated in the feces.

21.47 (a) Glycocholate:

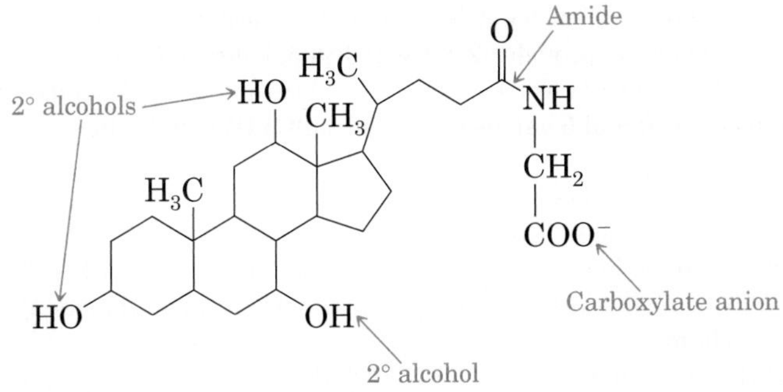

(b) Cortisone:

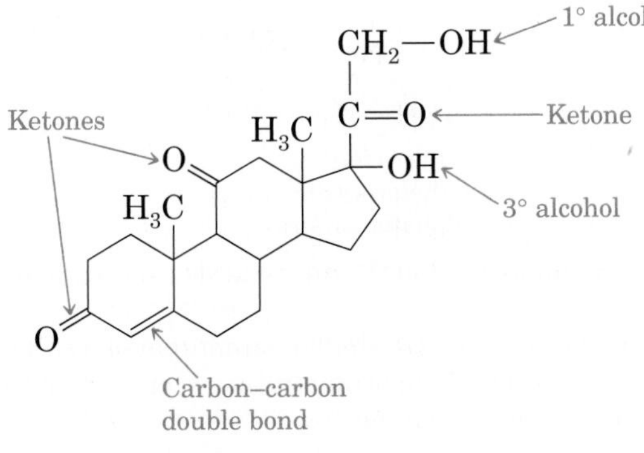

(c) PGE$_2$:

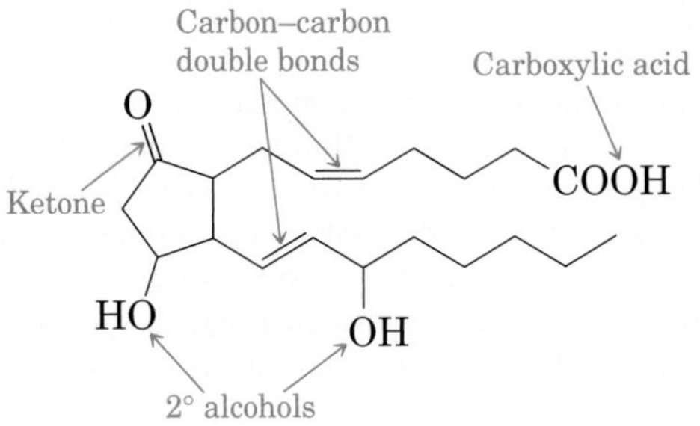

(d) Leukotriene B4:

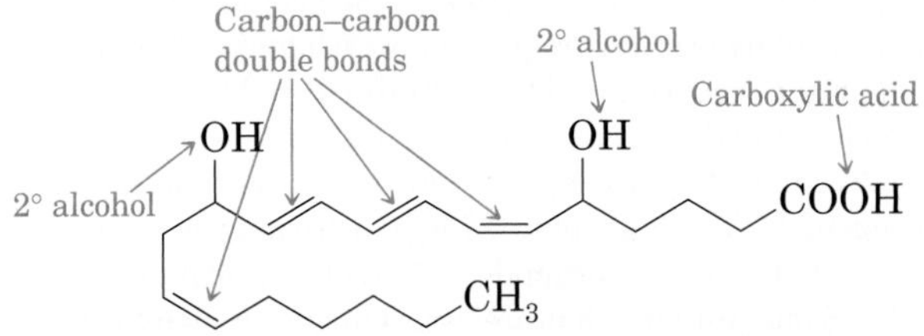

21.49 Aspirin slows the synthesis of thromboxanes by inhibiting the COX enzyme. Because thromboxanes enhance the blood clotting process, the result is that strokes caused by blood clots in the brain will occur less often.

21.51 Waxes consist primarily of esters of long-chain saturated acids and alcohols. Because of the saturated components, wax molecules pack more tightly than those of triglycerides, which frequently have unsaturated components.

21.53 The transporter is a helical transmembrane protein. The hydrophobic groups on the helices are turned outward and interact with the membrane. The hydrophilic groups of the helices are on the inside and interact with the hydrated chloride ions.

21.55 Ceramides with shorter chain lengths lead to less sensitivty to hypoxia in roundworms than ceramides with longer chain lengths. This is significant because of the link between cancer cell's survival under hypoxic conditions as well as the effect of strokes or heart attacks on cell survival.

21.57 α-D-galactose, β-D-glucose, β-D-glucose

21.59 Athletes use steroids in two principal ways. In power sports, such as sprinting a weight lifting, the use of steroids builds up muscle bulk, which leads to increased strength and speed. In endurance sports, the use of steroids allows for quicker recovery from exercise so that the athlete can decrease the time until the next training or racing session.

21.61 They prevent ovulation.

21.63 It inhibits prostaglandin formation by preventing ring closure.

21.65 NSAIDs inhibit cyclooxygenases (COX enzymes) that are needed for ring closure. Leukotrienes have no ring in their structure; therefore, they are not affected by COX inhibitors.

21.67 Fish contains high levels of omega-3 fatty acids, which inhibit the formation of certain prostaglandins and thromboxanes that cause blood clots in the heart.

21.69 (See Figure 21.2.) Polar molecules cannot penetrate the bilayer. They are insoluble in lipids. Nonpolar molecules can interact with the interior of the bilayer ("like dissolves like").

21.71 Both groups are derived from a common precursor, PGH$_2$, in a process catalyzed by the COX enzymes.

21.73 Coated pits are concentrations of LDL receptors on the surface of cells. They bind LDL and by endocytosis transfer it inside the cell.

21.75 In facilitated transport, a membrane protein assists in the movement of a molecule through the membrane with no requirement for energy. In active transport, a membrane protein assists in the process, but energy is required. ATP hydrolysis usually supplies the needed energy.

21.77 Aldosterone has an aldehyde group at the junction of the C and D rings. The other steroids have methyl groups.

21.79 The formula weight of the triglyceride is about 800 g/mol. This is 0.125 mol (100 g ÷ 800 g/mol = 0.125 mol). One mole of hydrogen is required for each mole of double bonds in the triglyceride. There are three double bonds, so the moles of hydrogen required for each 100 g = 0.125 mol × 3 = 0.375 mol of hydrogen gas. Converting to grams of hydrogen, 0.375 × 2 g/mol = 0.750 g hydrogen gas.

21.81 This lipid is a ceramide, a kind of sphingolipid.

21.83 Some proteins that are associated with membranes associate exclusively with one side of the membrane rather than the other.

21.85 Statements (c) and (d) are consistent with what is known about membranes. Covalent bonding between lipids and proteins [statement (e)] is not widespread. Proteins "float" in the lipid bilayers rather than being sandwiched between them [statement (a)]. Bulkier molecules tend to be found in the outer lipid layer [statement (b)].

21.87 Statement (c) is correct. Transverse diffusion is only rarely observed [statement (b)]. Proteins are bound to the inside and outside of the membrane [statement (a)].

21.89 Both lipids and carbohydrates contain carbon, hydrogen, and oxygen. Carbohydrates have aldehyde and ketone groups, as do some steroids. Carbohydrates have a number of hydroxyl groups, which lipids do not have to a great extent. Lipids have major components that are hydrocarbon in nature. These structural features imply that carbohydrates tend to be significantly more polar than lipids.

21.91 primarily lipid: olive oil and butter; primarily carbohydrate: cotton and cotton candy

21.93 The amounts are the key point here. Large amounts of sugar can provide energy. Fat burning due to the presence of taurine plays a relatively minor role because of the small amount.

21.95 The other ends of the molecules involved in the ester linkages in lipids, such as fatty acids, tend not to form long chains of bonds with other molecules.

21.97 Stem cells are undifferentiated proginator cells that can form many cell types. If they become uncontrolled, they can lead to diseases such as leukemia

21.99 The bulkier molecules tend to be found on the exterior of the cell because the curvature of the cell membrane provides more room for them.

21.101 The charges tend to cluster on membrane surfaces. Positive and negative charges attract each other. Two positive or two negative charges repel each other, so unlike charges do not have this repulsion.

Chapter 22 Proteins

22.1

Valine (Val)   Phenylalanine (Phe)

Valylphenylalanine (Val-Phe)

22.2 The overall charge on $[Pt(NH_3)_2Cl_2]$ is zero. The chloride ligands each provide a 1− charge and the ammonia ligands are uncharged. Thus, the charge on platinum must be 2+ for the total charge of the compound to equal zero.

22.3 Cobalt is in group 9 of the periodic table so it has 9 valence electrons when the charge is zero. However, $[Co(NH_3)_5Cl]^{2+}$ has a chloride ligand that has a 1− charge and 5 ammonia ligands providing zero charge. Thus, the charge on cobalt must be 3+ for the total charge on the compound to be 2+. That means that cobalt (III) only has 6 valence electrons. It needs 12 more electrons to achieve the 18 electron rule. Each ammonia ligand provides a pair of electrons through a coordinate covalent bond for a total of 10 electrons and the chloride ligand provides 2 more additional electrons for a grand total of 12 electrons. These 12 electrons on the ligands plus the original 6 valence electrons on Co III satisfy the 18 electron rule.

22.4 a salt bridge

22.5 (a) storage (b) movement

22.7 protection

22.9 Tyrosine has an additional hydroxyl group on the phenyl side chain.

22.11 arginine

22.13

Pyrrolidines (heterocyclic aliphatic amines)

22.15 They supply most of the amino acids we need in our bodies.

22.17 These structures are similar except that one of the hydrogens in the side chain of alanine has been replaced with a phenyl group in phenylalanine.

22.19 Amino acids are zwitterions; therefore, they all have positive and negative charges. These molecules are very strongly attracted to each other, so they are solids at low temperatures.

22.21 All amino acids have a carboxyl group with a pK_a around 2 and an amino group with a pK_a between 8 and 10. One group is significantly more acidic, and one is more basic. To have an un-ionized amino acid, the hydrogen would have to be on the carboxyl group and have vacated the amino group. Given that the carboxyl group is the stronger acid, this would never happen.

22.23

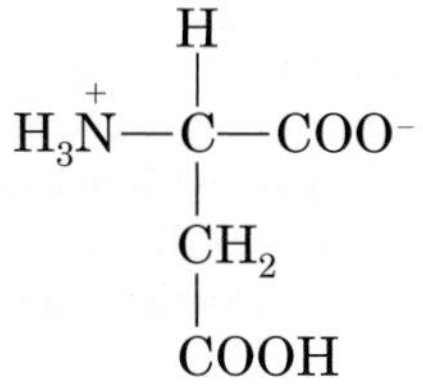

$$H_3\overset{+}{N}-\underset{\underset{COOH}{\underset{|}{CH_2}}}{\overset{\overset{H}{|}}{C}}-COO^-$$

22.25

$$H_2N-\underset{\underset{\overset{+}{NH_3}}{\underset{|}{(CH_2)_4}}}{\overset{\overset{H}{|}}{C}}-COO^-$$

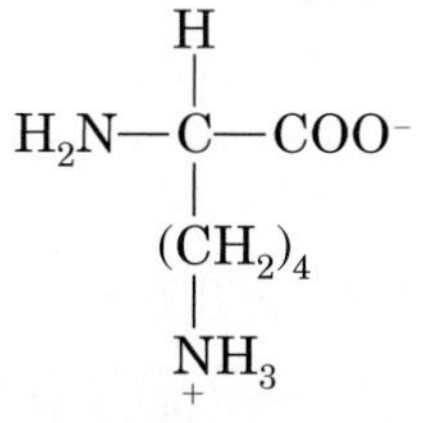

22.27 the side-chain imidazole

22.29 The side chain of histidine is an imidazole with a nitrogen that reversibly binds to a hydrogen. When dissociated, it is neutral; when associated, it is positive. Therefore, chemically it is a base, even though it does have a pK_a in the acidic range.

22.31 histidine, arginine, and lysine

22.33 Serine may be obtained by the hydroxylation of alanine. Tyrosine is obtained by the hydroxylation of phenyl-alanine.

22.35 Thyroxine is a hormone that controls overall metabolic rate. Both humans and animals sometimes suffer from low levels of thyroxine, causing lack of energy and tiredness.

22.37

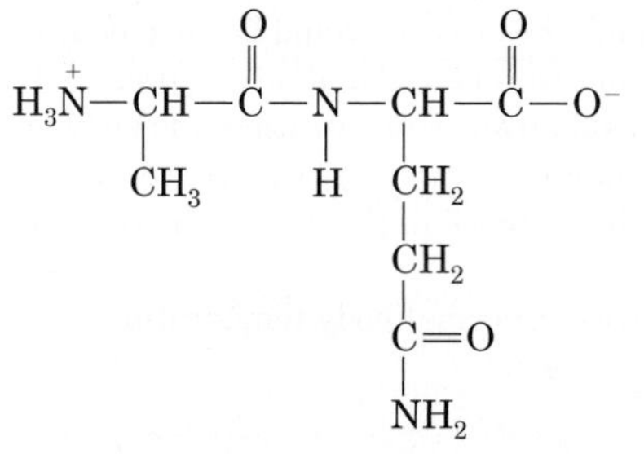

Alanylglutamine
(Ala-Gln)

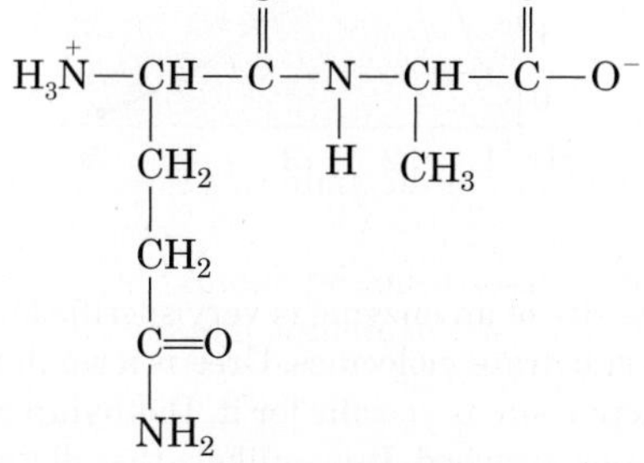

Glutaminylalanine
(Gln-Ala)

22.39

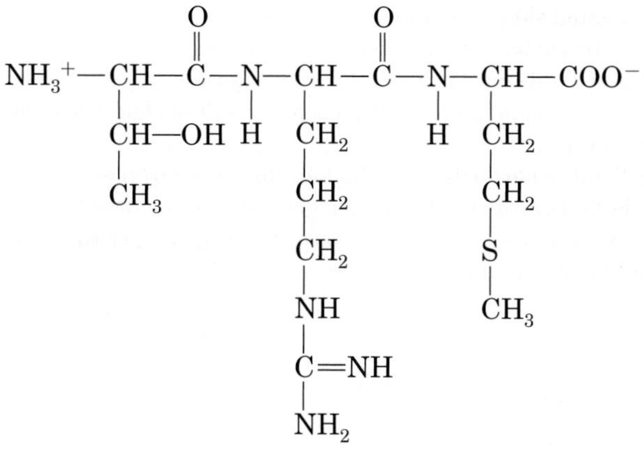

22.41 Only the peptide backbone contains polar units.

22.43

(a)

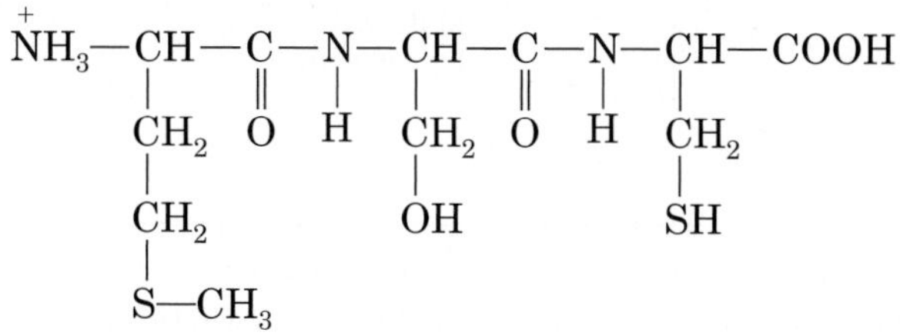

(b) pH 2 is shown above. At pH 7.0 would look like:

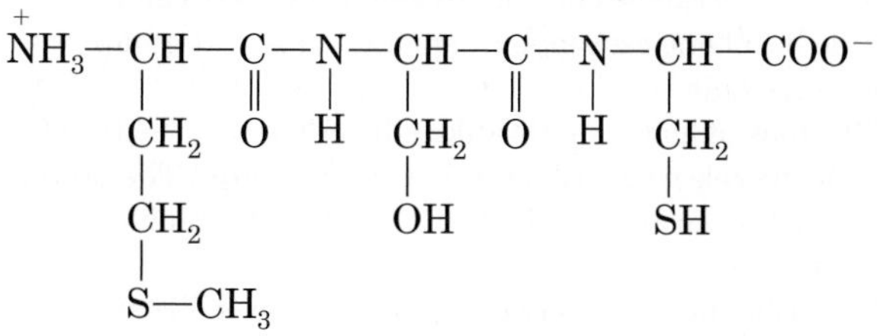

(c) At pH 10:

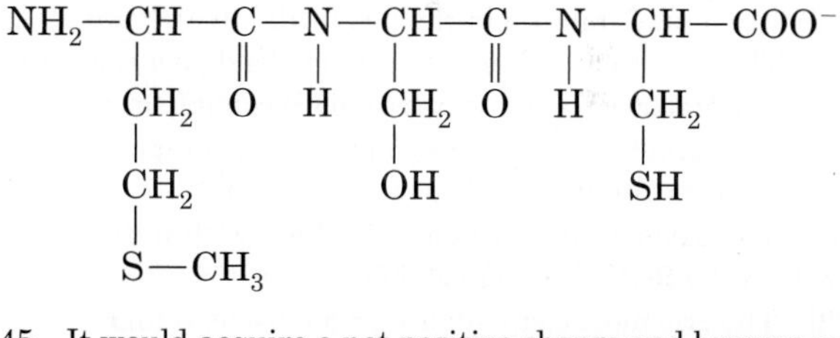

22.45 It would acquire a net positive charge and become more water-soluble.

22.47 (a) 256 (b) 160,000

22.49 valine or isoleucine

22.51 (a) secondary (b) quaternary (c) quaternary (d) primary

22.53 In $Fe(CO)_5$, both iron and carbon monoxide are electrically neutral. The iron atom has eight valence electrons. Each CO molecule contributes two electrons to a coordinate covalent bond. Since there are five CO molecules, they contribute a total of ten electrons. Thus, the 18-electron rule is satisfied.

22.55 Zinc is present as Zn^{2+}, with 10 electrons. Each NH_3 contributes two electrons, giving four more. The two chloride ions add two more electrons each, for another four. The total is 18.

22.57 Above pH 6.0, the COOH groups are converted to COO^- groups. The negative charges repel each other, disrupting the compact α-helix and converting it to a random coil.

22.59 (1) C-terminal end (2) N-terminal end (3) pleated sheet (4) random coil (5) hydrophobic interaction (6) disulfide bridge (7) α-helix (8) salt bridge (9) hydrogen bonds

22.61 (a) Fetal hemoglobin has fewer salt bridges between the chains.
(b) Fetal hemoglobin has a higher affinity for oxygen.
(c) Fetal hemoglobin has an oxygen saturation curve that is in between myoglobin and maternal hemoglobin, so the graph would look like the figure below:

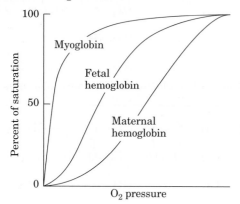

22.63 The heme and the polypeptide chain form the quaternary structure of cytochrome c. This is a conjugated protein.

22.65 the intramolecular hydrogen bonds between the peptide backbone carbonyl group and the N—H group

22.67 cysteine

22.69 Ions of heavy metals like silver denature bacterial proteins by reacting with cysteine —SH groups. The proteins, denatured by formation of silver salts, form insoluble precipitates.

22.71 (Chemical Connections 22A) Nutrasweet contains phenylalanine. People suffering from the genetic disease phenylketonuria must avoid phenylalanine as they cannot metabolize it, and buildup in the body will have severe effects.

22.72 The symptoms of hunger, sweating, and poor coordination accompany diabetes when hypoglycemia is coming on.

22.75 The abnormal form has a higher percentage of β-pleated sheet compared to the normal form.

22.77 The oxygen-binding behavior of myoglobin is hyperbolic, while that of hemoglobin is sigmoidal.

22.79 The two most common are prion diseases and Alzheimer's disease.

22.81 Even if it is feasible, it is not completely correct to call the imaginary process that converts α-keratin to β-keratin "denaturation." Any process that changes a protein from α to β requires at least two steps: (1) conversion from the α form to a random coil and (2) conversion from the random coil to the β form. The term "denaturation" describes only the first half of the process (Step 1). The second step would be called "renaturation." The overall process is called denaturation followed by renaturation. If we assume that the imaginary process actually occurs without passing through a random coil, then the term "denaturation" does not apply.

22.83 a quaternary structure because subunits are cross-linked

22.85 (a) hydrophobic (b) salt bridge (c) hydrogen bond (d) hydrophobic

22.87 glycine

22.89 one positive charge on the amino group

22.91 These amino acids have side chains that can catalyze organic reactions. They are polar or sometimes charged, and the ability to make hydrogen bonds or salt bridges can help catalyze the reaction.

22.93 Proteins can be denatured when the temperature is only slightly higher than a particular optimum. For this reason, the health of a warm-blooded animal is dependent on the body temperature. If the temperature is too high, proteins could denature and lose function.

22.95 Even if you know all of the genes in an organism, not all the genes code for proteins, nor are all genes expressed all the time.

22.97 A diet supplement full of collagen may help a person lose weight, but it would be of little use for repairing muscle tissue as collagen is not a good protein source. One-third of its amino acids are glycine, and another third are proline. Muscle repair requires high-quality protein to be effective.

Chapter 23 Enzymes

23.1 A catalyst is any substance that speeds up the rate of a reaction and is not itself changed by the reaction. An enzyme is a biological catalyst, which is either a protein or an RNA molecule.

23.3 Yes, lipases are not very specific.

23.5 because enzymes are very specific and thousands of reactions must be catalyzed in an organism

23.7 Lyases add water across a double bond or removes water from a molecule, thereby generating a double bond. Hydrolases use water to break an ester or amide bond, thereby generating two molecules.

23.9 (a) isomerase (b) hydrolase (c) oxidoreductase (d) lyase

23.11 *Cofactor* is more generic; it means a nonprotein part of an enzyme. A *coenzyme* is an organic cofactor.

23.13 In reversible inhibition, the inhibitor can bind and then be released. With noncompetitive inhibition, once the inhibitor is bound, no catalysis can occur. With irreversible inhibition, once the inhibitor is bound, the enzyme would be effectively dead, as the inhibitor could not be removed and no catalysis could occur.

23.15 No, at high substrate concentration, the enzyme surface is saturated and doubling of the substrate concentration will produce only a slight increase in the rate of the reaction or no increase at all.

23.17 (a) less active at normal body temperature (b) The activity decreases.

23.19 (a)

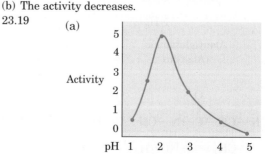

(b) 2
(c) zero activity

23.21 The active site of an enzyme is very specific for the size and shape of the substrate molecules. Urea is a small molecule, and the urease active site is specific for it. Diethylurea has the two ethyl groups attached. It is unlikely that diethylurea would fit into an active site specific for urea.

23.23 The amino acid residues most often found at enzyme active sites are His, Cys, Asp, Arg, and Glu.

23.25 The correct answer is (c). Initially, the enzyme does not have exactly the right shape for strongly binding a substrate, but the shape of the active site changes to better accommodate the substrate molecule.

23.27 Amino acid residues in addition to those at an enzyme active site are present to help form a three-dimensional pocket where the substrate binds. These amino acids act to make the size, shape, and environment (polar or nonpolar) of the active site just right for the substrate.

23.29 Caffeine is an allosteric regulator.

23.31 There is no difference. They are the same.

23.33

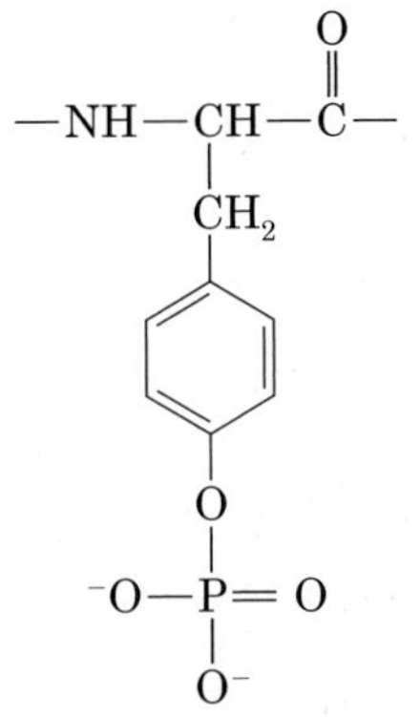

23.35

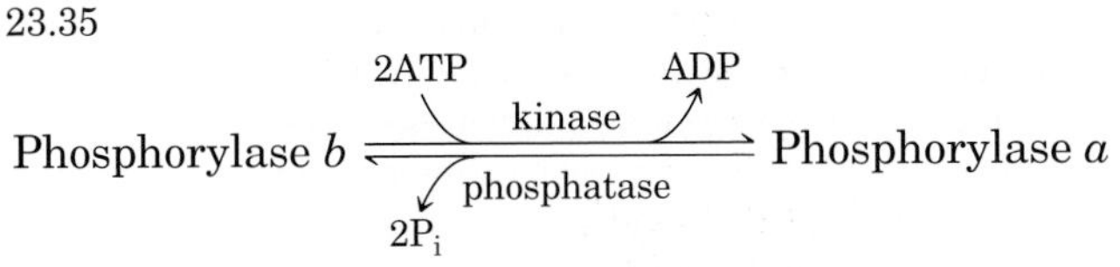

23.37 Glycogen phosphorylase is controlled by allosteric regulation and by phosphorylation. The allosteric controls are very fast, so that when the level of ATP drops, for example, there is an immediate response to the enzyme, allowing more energy to be produced. The covalent modification by phosphorylation is triggered by hormone responses. They are a bit slower, but more long-lasting and ultimately more effective.

23.39 Just as with lactate dehydrogenase, there are five isozymes of PFK: M_4, M_3L, M_2L_2, ML_3, and L_4.

23.41 Two enzymes that increase in serum concentration following a heart attack are creatine phosphokinase and aspartate aminotransferase. Creatine phosphokinase peaks earlier than aspartate aminotransferase and would be the best choice in the first 24 hours.

23.43 Serum levels of the enzymes AST and ALT are monitored for diagnosis of hepatitis and heart attack. Serum levels of AST are increased after a heart attack, but ALT levels are normal. In hepatitis, both enzymes are elevated. The diagnosis, until further testing, would indicate that the patient may have had a heart attack.

23.45 Chemicals present in organic vapors are detoxified in the liver. The enzyme alkaline phosphatase is monitored to diagnose liver problems.

23.47 It is not possible to administer chymotrypsin orally. The stomach would treat it just as it does all dietary proteins: degrade it by hydrolysis to free amino acids. Even if whole, intact molecules of the enzyme were present in the stomach, the low pH in the region would not allow activity for the enzyme, which prefers an optimal pH of 7.8.

23.49 A transition-state analog is built to mimic the transition state of the reaction. It is not the same shape as the substrate or the product, but rather something in between. The potency with which such analogs can act as inhibitors lends credence to the theory of induced fit.

23.51 Succinylcholine has a chemical structure similar to that of acetylcholine, so both can bind to the acetylcholine receptor of the muscle end plate. The binding of either choline causes a muscle contraction. However, the enzyme acetylcholinesterase hydrolyzes succinylcholine only very slowly compared to acetylcholine. Muscle contraction does not occur as long as succinylcholine is still present and acts as a relaxant.

23.53 The most common reactions of kinases that we study in this book are the ones that involve using ATP to phosphorylate another molecule, be it an enzyme or a metabolite of a pathway. One example would be glycogen phosphorylase kinase. This enzyme catalyzes the following reaction as described in Chemical Connections 23E:

$$\text{Phosphorylase} + \text{ATP} \longrightarrow \text{Phosphorylase-P} + \text{ADP}$$

Another example is hexokinase from the pathway called glycolysis (Chapter 28). Hexokinase catalyzes the following reaction:

$$\text{Glucose} + \text{ATP} \longrightarrow \text{glucose 6-P} + \text{ADP}$$

23.55 Many people have suffered psychological traumas that haunt them for many years or even their entire lives. If long-term memories could be selectively blocked, it would offer relief to patients suffering from something in their past.

23.57 In the enzyme pyruvate kinase, the $=CH_2$ of the substrate phosphoenolpyruvate sits in a hydrophobic pocket formed by the amino acids Ala, Gly, and Thr. The methyl group on the side chain of Thr, rather than the hydroxyl group, is in the pocket. Hydrophobic interactions are at work here to hold the substrate into the active site.

23.59 Researchers were trying to inhibit phosphodiesterases because cGMP acts to relax constricted blood vessels. This approach was hoped to help treat angina and high blood pressure.

23.61 Intestinal bacteria have many useful roles in the body, so it would not be a good idea to reduce their population. The side effects can be alleviated by inhibiting the enzyme in the bacteria that is responsible for the side effects.

23.63 Phosphorylase exists in a phosphorylated form and an unphosphorylated form, with the former being more active. Phosphorylase is also controlled allosterically by several compounds, including AMP and glucose. While the two act semi-independently, they are related to some degree. The phosphorylated form has a higher tendency to assume the R state, which is more active, and the unphosphorylated form has a higher tendency to be in the less active T state.

23.65 The iron in heme groups can change from one oxidation state to another and do so reversibly. This property allows it to play a role in a number of biochemical processes.

23.67 In the processing of cocaine by specific esterase enzymes, the cocaine molecule passes through an intermediate state. A molecule was designed that mimics this transition state. This transition-state analog can be given to a host animal, which then produces antibodies to the analog. When these antibodies are given to a person, they act like an enzyme and degrade cocaine.

23.69 Cocaine blocks the reuptake of the neurotransmitter dopamine, leading to overstimulation of the nervous system.

23.71 (a) Vegetables such as green beans, corn, and tomatoes are heated to kill microorganisms before they are preserved by canning. Milk is preserved by the heating process, pasteurization. (b) Pickles and sauerkraut are preserved by storage in vinegar (acetic acid).

23.73 High temperatures denature enzymes, and in many cases, the denaturation is irreversible. Any enzymes in a cooked hotdog are denatured and inactive.

23.75 The amino acid residues (Lys and Arg) that are cleaved by trypsin have basic side chains; thus, they are positively charged at physiological pH.

23.77 This enzyme works best at a pH of about 7.

23.79 a hydrolase

23.81 (a) The enzyme is called ethanol dehydrogenase or, more generally, alcohol dehydrogenase. It could also be called ethanol oxidoreductase.
(b) ethyl acetate esterase or ethyl acetate hydrolase

23.83 isozymes or isoenzymes

23.85 No, the direction a reaction goes is determined by the thermodynamics of the reaction, including the concentration of substrates and products. A reaction may only go in the forward direction in a metabolic pathway due to an overwhelming concentration of the substrates along with immediate removal of the products. However, the enzyme that catalyzes the reaction would catalyze the reaction in either direction if it is thermodynamically possible.

23.87 These supplements are taken by mouth. When the enzymes reach the stomach, they will be degraded by stomach acid, like all proteins. Then they will not have any activity.

23.89 The athlete may benefit from the stimulatory effect of caffeine, but in a long race, the athlete would also become dehydrated from the diuretic effect on the kidneys. One of the most important factors to endurance performance is hydration, so any substance that causes dehydration would be detrimental to performance in a long-distance event.

23.91 The structure of RNA makes it more likely to be able to adopt a wider range of tertiary structures, so it can fold up to form globular molecules similar to protein-based enzymes. It also has an extra oxygen, which gives it an additional reactive group to use in catalysis or an electronegative group, useful in hydrogen bonding.

Chapter 24 **Chemical Communications: Neurotransmitters and Hormones**

24.1 G-protein is an enzyme; it catalyzes the hydrolysis of GTP to GDP. GTP, therefore, is a substrate.

24.3 A chemical messenger operates between cells; secondary messengers signal inside a cell in the cytoplasm.

24.5 The concentration of Ca^{2+} in neurons controls the process. When it reaches $10^{-4}\ M$, the vesicles release the neurotransmitters into the synapse.

24.7 anterior pituitary gland

24.9 Upon binding of acetylcholine, the conformation of the proteins in the receptor changes and the central core of the ion channel opens.

24.11 The cobra toxin causes paralysis by acting as a nerve system antagonist. It blocks the receptor and interrupts the communication between neuron and muscle cell. The botulin toxin prevents the release of acetylcholine from presynaptic vesicles.

24.13 Calcium sparks (also called puffs) have to do with the actual concentration increase caused by allowing calcium into the cell. Calcium waves have to do with how long a calcium signal continues.

24.15 The calcium concentration increases 5 to 25 times.

24.17 Taurine is a β-amino acid; its acidic group is $-SO_2OH$ instead of $-COOH$.

24.19 The amino group in GABA is in the gamma position; proteins contain only alpha amino acids.

24.21 (a) norepinephrine and histamine
(b) They activate a secondary messenger, cAMP, inside the cell.
(c) amphetamines and histidine

24.23 It is phosphorylated by an ATP molecule.

24.25 Product from the MAO-catalyzed oxidation of dopamine:

24.27 (a) Amphetamines increase and (b) reserpine decreases the concentration of the adrenergic neurotransmitter.

24.29 the corresponding aldehyde

24.31 (a) the ion-translocating protein itself
(b) It gets phosphorylated and changes its shape.
(c) It activates the protein kinase that does the phosphorylation of the ion-translocating protein.

24.33 They are pentapeptides.

24.35 The enzyme is a kinase. The reaction is the phosphorylation of inositol-1,4-diphosphate to inositol-1,4,5-triphosphate:

$$P = -PO_3^{2-}$$

24.37 Cyclic AMP

24.39 Protein Kinase

24.41 Glucagon initiates a series of reactions that eventually activates protein kinase. The protein kinase phosphorylates two key enzymes in the liver, activating one and inhibiting the other. The combination of these effects lowers the level of fructose 2,6-bisphosphate, a key regulator of carbohydrate metabolism. Fructose 2,6-bisphosphate stimulates glycolysis and inhibits gluconeogenesis. Therefore, when fructose 2,6-bisphosphate is decreased, gluconeogenesis is stimulated and glycolysis is inhibited.

24.43 Insulin binds to insulin receptors on liver and muscle cells. The receptor is an example of a protein called a tyrosine kinase. A specific tyrosine residue becomes phosphorylated on the receptor, activating its kinase activity. The target protein called IRS is then phosphorylated by the active tyrosine kinase. The phosphorylated IRS acts as the second messenger. It causes the phosphorylation of many target enzymes in the cell. The effect is to reduce the level of glucose in the blood by increasing the rate of pathways that use glucose and slowing the rate of pathways that make glucose.

24.45 Most receptors for steroid hormones are located in the cell nucleus.

24.47 In the brain, steroid hormones can act as neurotransmitters.

24.49 Local injections of the toxin prevent release of acetylcholine in that area.

24.51 The neurofibrillar tangles found in the brains of Alzheimer's patients are composed of tau proteins. Mutated tau proteins, which normally interact with the cytoskeleton, grow into these tangles instead, thus altering normal cell structure.

24.53 Drugs that increase the concentration of the neurotransmitter acetylcholine may be effective in the treatment of Alzheimer's disease. Acetylcholinesterase inhibitors such as Aricept inhibit the enzyme that decomposes the neurotransmitter.

24.55 MRIs can detect the brain shrinkage that often accompanies the later stages of Alzheimer's. Spinal taps can be used to test spinal fluid for increased levels of tau proteins.

24.57 There are many drugs being tested. Some attempt to stop the production of the amyloid beta proteins or the tau proteins. Others try to eliminate the aggregates that form from these proteins. Another strategy is for a drug to elicit antibodies to the dangerous forms of the proteins.

24.59 The neurotransmitter dopamine is deficient in Parkinson's disease, but a dopamine pill would not be an effective treatment. Dopamine cannot cross the blood–brain barrier.

24.61 Drugs like Cogentin that block cholinergic receptors are often used to treat the symptoms of Parkinson's disease. These drugs lessen spastic motions and tremors.

24.63 Nitric oxide relaxes the smooth muscle that surrounds blood vessels. This relaxation causes increased blood flow in the brain, which in turn causes headaches.

24.65 Neurons adjacent to those damaged by the stroke begin to release glutamate and NO, which kills other cells in the region.

24.67 Insulin-dependent (Type I) diabetes is caused by insufficient production of insulin by the pancreas. Administration of insulin relieves symptoms of this type of diabetes. Non-insulin-dependent (Type II) diabetes is caused by a deficiency of insulin receptors or by the presence of inactive insulin receptors. Other drugs are used to relieve symptoms.

24.69 Monitoring glucose in the tears relieves the patient of taking many blood samples every day.

24.71 Moderate training increases the level and activity of the GLUT4 transporter.

24.73 The effect of exercise intensity is not very significant when compared to the effect of not doing any exercise at all.

24.75 Some possible dangers include enlargement of the prostate, increases in chromosomal abnormalities, breast cancer, and early onset of puberty.

24.77 Aldosterone binds to a specific receptor in the nucleus. The aldosterone–receptor complex serves as a transcription factor that regulates gene expression. Proteins for mineral metabolism are produced as a result.

24.79 Large doses of acetylcholine will help. Decamethonium bromide is a competitive inhibitor of acetylcholine esterase. The inhibitor can be removed by increasing substrate concentration.

24.81 Alanine is an α-amino acid in which the amino group is bonded to the same carbon as the carboxyl group. In β-alanine, the amino group is bonded to the carbon adjacent to the one to which the carboxyl group is bonded.

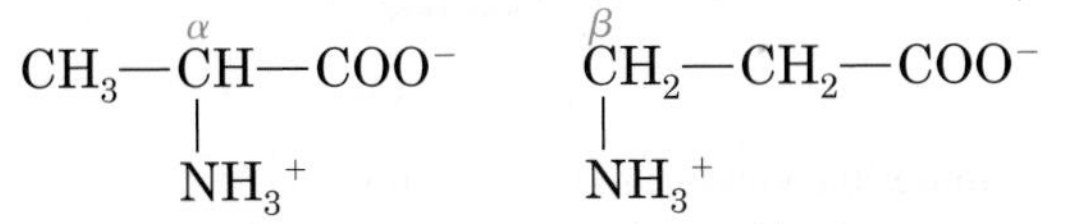

$$CH_3 \overset{\alpha}{-} CH - COO^- \qquad \overset{\beta}{CH_2} - CH_2 - COO^-$$
$$\quad\quad\quad | \qquad\qquad\qquad\quad |$$
$$\quad\quad NH_3^+ \qquad\qquad\qquad NH_3^+$$

Alanine β-Alanine
(an α-amino acid) (a β-amino acid)

24.83 Effects of NO on smooth muscle are as follows: dilation of blood vessels and increased blood flow; headaches caused by dilation of blood vessels in brain; increased blood flow in the penis, leading to erections.

24.85 Acetylcholine esterase catalyzes the hydrolysis of the neurotransmitter acetylcholine to produce acetate and choline. Acetylcholine transferase catalyzes the synthesis of acetylcholine from acetyl—CoA and choline.

24.87 The reaction shown below is the hydrolysis of GTP:

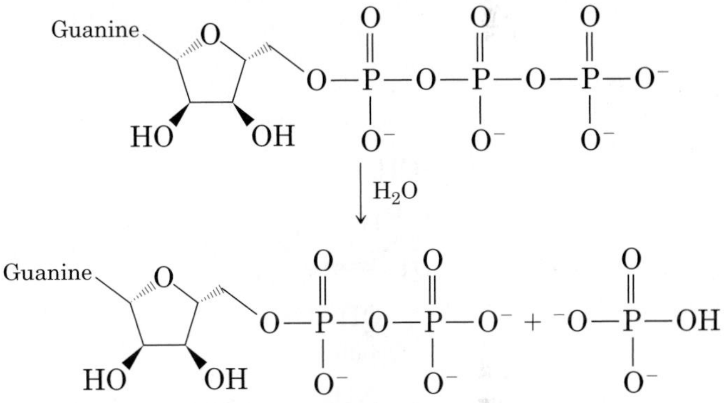

24.89 Ritalin increases serotonin levels. Serotonin has a calming effect on the brain. One advantage of this drug is that it does not increase levels of the stimulant dopamine.

24.91 Proteins are capable of specific interactions at recognition sites. This ability makes for useful selectivity in receptors.

24.93 Adrenergic messengers such as dopamine are derivatives of amino acids. For example, a biochemical pathway exists that produces dopamine from the amino acid tyrosine.

24.95 Insulin is a small protein. It would go through protein digestion if taken orally and would not be taken up as the whole protein.

24.97 Steroid hormones directly affect nucleic acid synthesis.

24.99 Chemical messengers vary in their response times. Those that operate over short distances, such as neurotransmitters, have short response times. Their mode of action frequently consists of opening or closing channels in a membrane or binding to a membrane-bound receptor. Hormones must be transmitted in the bloodstream, which requires a longer time for them to take effect. Some hormones can and do affect protein synthesis, which makes the response time even longer.

24.101 Having two different enzymes for the synthesis and breakdown of acetylcholine means that the rates of formation and breakdown can be controlled independently.

Chapter 25 Nucleotides, Nucleic Acids, and Heredity

25.1

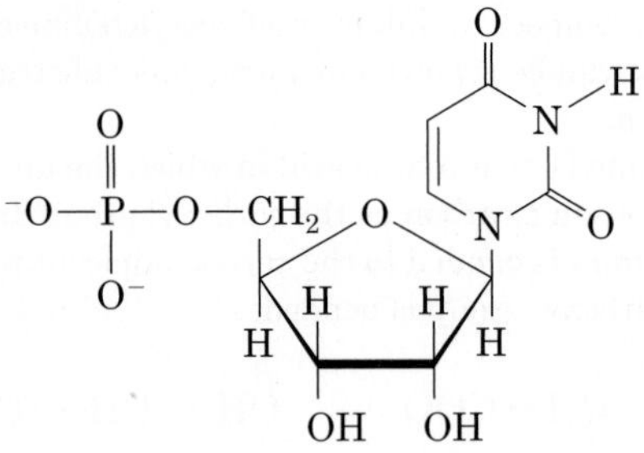

25.3 hemophilia, sickle cell anemia, etc.

25.5 (a) In eukaryotic cells, DNA is located in the cell nucleus and in mitochondria.
(b) RNA is synthesized from DNA in the nucleus, but further use of RNA (protein synthesis) occurs on ribosomes in the cytoplasm.

25.7 DNA has the sugar deoxyribose, while RNA has the sugar ribose. Also, RNA has uracil, while DNA has thymine.

25.9 Thymine and uracil are both based on the pyrimidine ring. However, thymine has a methyl substituent at carbon 5, whereas uracil has a hydrogen. All of the other ring substituents are the same.

25.11

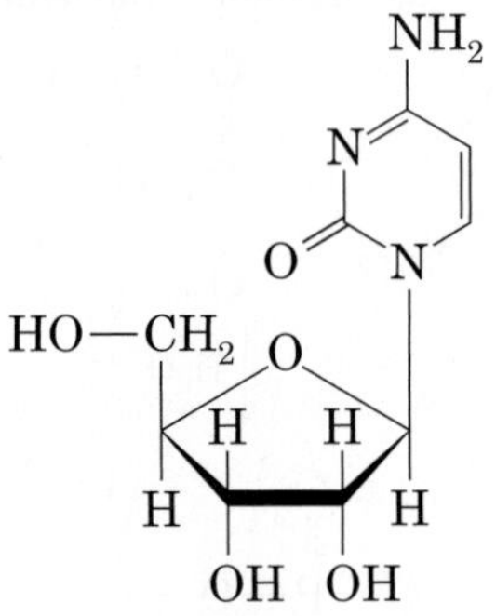

Cytidine

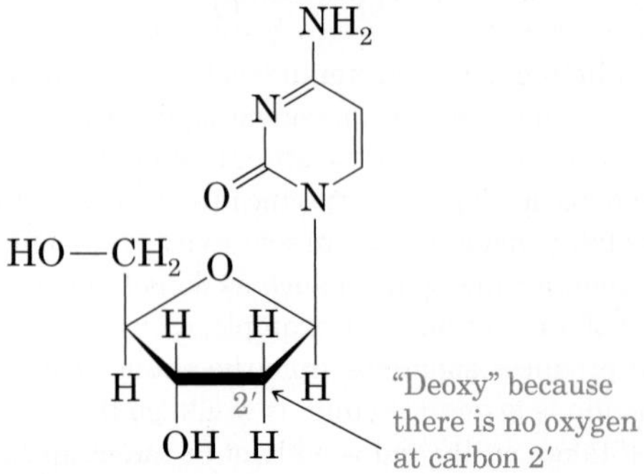

Deoxycytidine

"Deoxy" because there is no oxygen at carbon 2'

25.13 D-Ribose and 2-deoxy-D-ribose have the same structure except at carbon 2. D-Ribose has a hydroxyl group and hydrogen on carbon 2, whereas deoxyribose has two hydrogens.

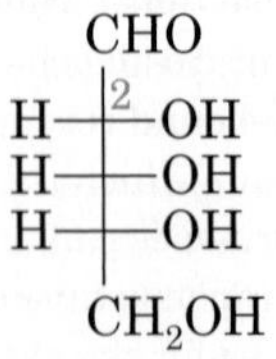

D-Ribose

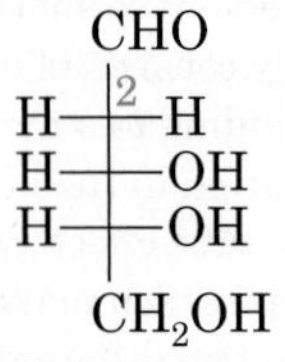

2-Deoxy-D-ribose

25.15 The name "nucleic acid" derives from the fact that the nucleosides are linked by phosphate groups, which are the dissociated form of phosphoric acid.

25.17 anhydride bonds

25.19 In RNA, carbons 3′ and 5′ of the ribose are linked by ester bonds to phosphates. Carbon 1 is linked to the nitrogen base with an N-glycosidic bond.

25.21 (a)

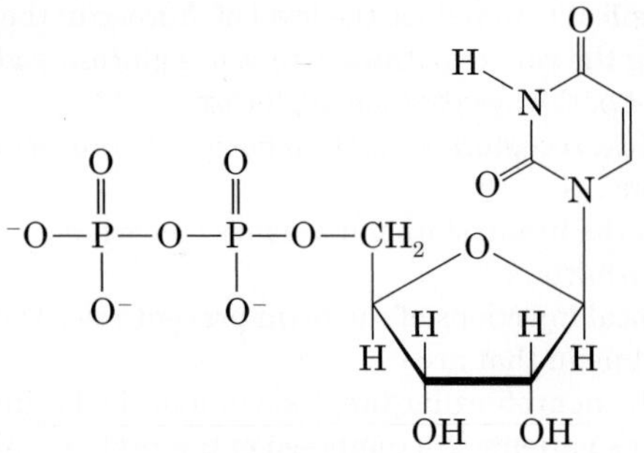

(b)

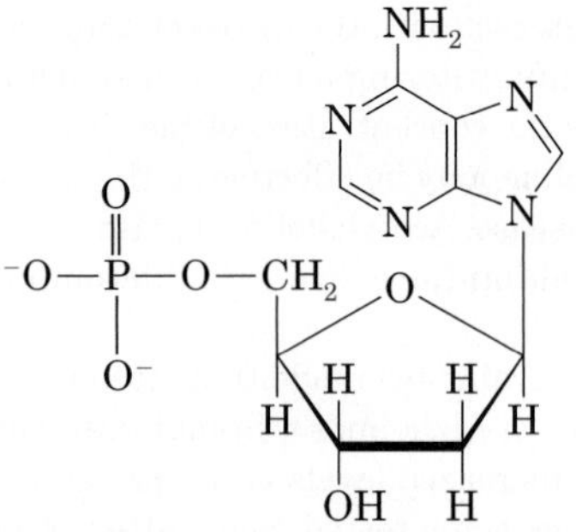

25.23 (a) One end will have a free 5′ phosphate or hydroxyl group that is not in phosphodiester linkage. That end is called the 5′ end. The other end, the 3′ end, will have a 3′ free phosphate or hydroxyl group.
(b) By convention, the end drawn to the left is the 5′ end. A is the 5′ end, and C is the 3′ end.
(c) The complementary strand would be GTATTGCCAT written from 5′ to 3′.

25.25 two

25.27 electrostatic interactions

25.29 The superstructure of chromosomes consists of many elements. DNA and histones combine to form nucleosomes that are wound into chromatin fibers. These fibers are further twisted into loops and minibands to form the chromosome superstructure.

25.31 the double helix

25.33 DNA is wound around histones, collectively forming nucleosomes that are further wound into solenoids, loops, and bands.

25.35 rRNA

25.37 mRNA

25.39 Ribozymes, or catalytic forms of RNA, are involved in post-transcriptional splicing reactions that cleave larger RNA molecules into smaller, more active forms. For example, tRNA molecules are formed in this way.

25.41 Small nuclear RNA is involved in splicing reactions of other RNA molecules.

25.43 Micro RNAs are 22 bases long and prevent transcription of certain genes. Small interfering RNAs vary from 22 to 30 bases and are involved in the degradation of specific mRNA molecules.

25.45 Immediately after transcription, messenger RNA contains both introns and exons. The introns are cleaved out by the action of ribozymes that catalyze splicing reactions on the mRNA.

25.47 no

25.49 the specificity between the base pairs, A-T and G-C

25.51

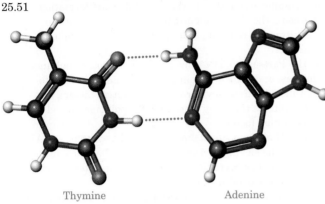

Thymine   Adenine

AT pair

25.53 four

25.55 In semiconservative DNA replication, the new daughter DNA helix is composed of one strand from the original (or parent) molecule and one new strand.

25.57

$$\text{Histone}-(CH_2)_4-NH_3^+ + CH_3-COO^- \xrightleftharpoons[\text{deacetylation}]{\text{acetylation}}$$

$$\text{Histone}-(CH_2)_4-NH-\overset{\overset{\displaystyle O}{\|}}{C}-CH_3$$

25.59 Helicases are enzymes that break the hydrogen bonds between the base pairs in double-helix DNA and thus help the helix to unwind. This prepares the DNA for the replication process.

25.61 pyrophosphate

25.63 The leading strand or continuous strand is synthesized in the 5′ to 3′ direction.

25.65 DNA ligase

25.67 from the 5′ to the 3′ direction

25.69 One of the enzymes involved in the DNA base excision repair (BER) pathway is an endonuclease that catalyzes the hydrolytic cleavage of the phosphodiester backbone. The enzyme hydrolyzes on the 5′ side of the AP site.

25.71 a β-N-glycosidic bond between the damaged base and the deoxyribose

25.73 Individuals with the inherited disease XP lack an enzyme involved in the NER pathway. They are not able to make repairs in DNA damaged by UV light.

25.75 5′ATGGCAGTAGGC3′

25.77 Ananda M. Chakrabarty, a General Electric engineer, filed for a patent on a strain of Pseudomonas bacteria that could break down oil slicks more efficiently.

25.79 No, biotech firms cannot actually own "your" genes, but they can currently own the rights to purified DNA samples containing the sequences of your genes.

25.81 DNA polymerase, the enzyme that makes the phosphodiester bonds in DNA, is not able to work at the end of linear DNA. This results in the shortening of the telomeres at each replication. The telomere shortening acts as a timer for the cell, allowing it to keep track of the number of divisions.

25.83 Because the genome is circular, even if the 5′ primers are removed, there will always be DNA upstream that can act as a primer for DNA polymerase to use as it synthesizes DNA.

25.85 A DNA fingerprint is made from the DNA of the child, the mother, and any prospective fathers and used to eliminate possible fathers.

25.87 Once a DNA fingerprint is made, each band in the child's DNA must come from one of the parents. Therefore, if the child has a band and the mother does not, then the father must have that band. In this way, possible fathers are eliminated.

25.89 One example might be that a life insurance company could raise your rates or refuse to give you insurance if your genetic profile had negative indicators. The same could happen with health insurance. Companies could begin to select for people with certain positive traits, thereby discriminating against everyone else. Having that information could lead to a new form of discrimination.

25.91 It began when Ventner and colleagues demonstrated they could transplant the DNA from one bacterial species to another. Then they created an artificial chromosome of the bacterium *Mycoplasma genitalium*, but unfortunately, the *M. genitalium* bacteria reproduced too slowly to be studied efficiently. So they switched species to the faster-growing *Mycoplasma mycoides*. Later, they demonstrated that they could transplant natural *M. mycoides* DNA into a close cousin, *M. capricolum*. They finished in 2010 by taking the synthetic version of the *M. mycoides* DNA and transplanting it into *M. capricolum* cells that had had their DNA removed.

25.93 Recent studies have shown that the species did not "die out"; instead, they interbred with modern humans. Thus, *H. neandertalis* DNA lives on.

25.95 Scientists are studying several genes that are related to cognitive issues. DYRK1A in modern humans is associated with Down syndrome. NRG3 is associated with schizophrenia. AUTS2 is associated with autism.

25.97 A three-dimensional pocket of ribonucleotides where substrate molecules are bound for catalytic reaction. Functional groups for catalysis include the phosphate backbone, ribose hydroxyl groups, and the nitrogen bases.

25.99 (a) The structure of the nitrogen base uracil is shown in Figure 25.1. It is a component of RNA. (b) Uracil with a ribose attached by an N-glycosidic bond is called uridine.

25.101 native DNA

25.103 mol % A = 29.3; mol % T = 29.3; mol % G = 20.7; mol % C = 20.7.

25.105 RNA synthesis is 5′ to 3′.

25.107 DNA replication requires a primer, which is RNA. Because RNA synthesis does not require a primer, it makes sense that RNA must have preceded DNA as a genetic material. This, added to the fact that RNA has been shown to be able to catalyze reactions, means that RNA can be both an enzyme and a heredity molecule.

25.109 The guanine-cytosine base pair has three hydrogen bonds, while the adenine-thymine base pair has only two. Therefore, it takes more energy to separate DNA strands with more G—C base pairs as it takes more energy to break their three hydrogen bonds.

25.111 DNA is the blueprint for all of the components of an organism. It is important that it have repair mechanisms because if it is wrong, all of its products will always be wrong. If correct DNA leads to incorrect RNA by some mutation, then the products of the RNA may be wrong. But RNA is short-lived, and the next time the RNA is produced, it will be correct. A good analogy is that of a cookbook. The words on the page are the DNA. How you read them is the RNA. If you misread the words, you may make the recipe wrong once. If the book is printed wrong, however, you will always make the recipe wrong.

Chapter 26 Gene Expression and Protein Synthesis

26.1 First, binding proteins must make the portion of the chromosome where the gene is less condensed and more accessible. Second, the helicase enzyme must unwind the double helix near the gene. Third, the polymerase must recognize the initiation signal on the gene.

26.2 (a) CAU and CAC (b) GUA and GUG

26.3 valine + ATP + tRNA$_{val}$

26.4 — CCT CGATTG —
— GGAGC TAAC —

26.5 (c); gene expression refers to both processes— transcription and translation.

26.7 Protein translation occurs on the ribosomes.

26.9 Helicases are enzymes that catalyze the unwinding of the DNA double helix prior to transcription. The helicases break the hydrogen bonds between base pairs.

26.11 The termination signal is at the 5′ end of the template strand that is being transcribed. It can also be said to be at the 3′ end of the coding strand.

26.13 The "guanine cap" methyl group is located on nitrogen number 7 of guanine.

26.15 No primer is necessary for transcription, but it is necessary for replication.

26.17 A sequence of conserved bases that helps the RNA polymerase identify where transcription should begin.

26.19 A DNA sequence that is eventually spliced out of the RNA and not part of the final RNA product of transcription.

26.21 on the messenger RNA

26.23 The main subunits are the 60S and the 40S ribosomal subunits, although these can be dissociated into even smaller subunits.

26.25 326

26.27 Leucine, arginine, and serine have the most, with six codons. Methionine and tryptophan have the fewest, with one apiece.

26.29 With only four available bases in DNA, if the genetic code had only two bases code for amino acids, there would be only 2^4, or 16, unique codons. This could account for only 16 of the 20 known standard amino acids.

26.31 The amino acid for protein translation is linked via an ester bond to the 3′ end of the tRNA. The energy for producing the ester bond comes from breaking two energy-rich phosphate anhydride bonds in ATP (producing AMP and two phosphates).

26.33 (a) The 40S subunit in eukaryotes forms the pre-initiation complex with the mRNA and the Met-tRNA that will become the first amino acid in the protein. (b) The 60S subunit binds to the pre-initiation complex and brings in the next aminoacyl-tRNA. The 60S subunit contains the peptidyl transferase enzyme.

26.35 Elongation factors are proteins that participate in the process of tRNA binding and movement of the ribosome on the mRNA during the elongation process in translation.

26.37 A special tRNA molecule is used for initiating protein synthesis. In prokaryotes, it is tRNA^{fmet}, which will carry a

formyl-methionine. In eukaryotes, there is a similar molecule, but it carries methionine. However, this tRNA carrying methionine for the initiation of synthesis is different from the tRNA carrying methionine for internal positions.

26.39 There are no amino acids in the vicinity of the nucleophilic attack that leads to peptide bond formation. Therefore, the ribosome must be using its RNA portion to catalyze the reaction, so it is a type of enzyme called a ribozyme.

26.41 Parts of the DNA involved are promoters, enhancers, silencers, and response elements. Molecules that bind to DNA include RNA polymerase, transcription factors, and other proteins that may bind the RNA polymerase and a transcription factor.

26.43 The active site of aminoacyl-tRNA synthases (AARS) contains the sieving portions that ensure that each amino acid is linked to its correct tRNA. The two sieving steps work on the basis of the size of the amino acid.

26.45 Both are DNA sequences that bind to transcription factors. The difference is largely due to our own understanding of the big picture. A response element controls a set of responses in a particular metabolic context. For example, a response element may activate several genes when the organism is challenged metabolically by heavy metals, by heat, or by a reduction in oxygen pressure.

26.47 Proteosomes play a role in post-translational degradation of damaged proteins. Proteins that are damaged by age or proteins that have misfolded are degraded by the proteosomes.

26.49 (a) Silent mutation: Assume the DNA sequence is TAT on the coding strand, which will lead to UAU on the mRNA. Tyrosine is incorporated into the protein. Now assume a mutation in the DNA to TAC. This will lead to UAC in mRNA. Again, the amino acid will be tyrosine. (b) Lethal mutation: the original DNA sequence is GAA on the coding strand, which transcribes into GAA on mRNA. This codes for the amino acid glutamic acid. The DNA mutation TAA will lead to UAA, a stop signal that incorporates no amino acid.

26.51 Yes, a harmful mutation may be carried as a recessive gene from generation to generation, with no individual demonstrating symptoms of the disease. Only when both parents carry recessive genes does an offspring have a 25% chance of inheriting the disease.

26.53 Restriction endonucleases are enzymes that recognize specific sequences on DNA and catalyze the hydrolysis of phosphodiester bonds in that region, thereby cleaving both strands of the DNA.

26.55 Mutation by natural selection is an exceedingly long, slow process that has occurred for centuries. Each natural change in the gene has been ecologically tested and found usually to have a positive effect or the organism is not viable. Genetic engineering, where a DNA mutation is done very fast, does not provide sufficient time to observe all of the possible biological and ecological consequences of the change.

26.57 The discovery of restriction enzymes allowed scientists to cut DNA at specific locations and link different pieces of DNA together. This led to the ability to clone foreign DNA into a host, leading to the ability to both amplify DNA of interest and also have it expressed. Without restriction enzymes scientists would not be able to express a human protein in a bacterial cell, for example, or to create the therapeutic gene used in gene therapy.

26.59 The viral coat is a protective protein covering around a virus particle. All of the components necessary to make the coat—for example, amino acids and lipids—come from the host.

26.61 An invariant site is a location in a protein that has the same amino acid in all species that have been studied. Studies of invariant sites help establish genetic links and evolutionary relationships.

26.63 A silent mutation is a change in the DNA that does not lead to a change in the DNA product. This can happen when there is a base change in the DNA but due to the redundancy of the genetic code, the change does not change the amino acid coded for.

26.65 A silent mutation may require a different tRNA molecule even though the same amino acid will be incorporated. The pace of the ribosome movement during translation may be different depending on the tRNA used, leading to the potential for different folding patterns in the protein produced.

26.67 The protein p53 is a tumor suppressor. When its gene is mutated, the protein no longer controls replication and the cell begins to grow at an increased rate.

26.69 The Duffy protein is found on the surface of human red blood cells. It acts as a docking protein for malaria, so mutations that lead to loss of the Duffy protein make the person resistant to malaria.

26.71 Several mutation types could affect production of protein Y. A mutation of gene X might change the protein sequence, such as we see in Duffy and sickle cell anemia. These changes can be minor or could lead to complete loss of function of the protein. Another mutation in gene X might be a silent mutation, but as we saw in Chemical Connections 26D, even a silent mutation could lead to a changed protein. Another possibility is that a mutation affects not the gene X directly, but rather the promoter for gene X. If the promoter region is mutated, it might lead to fewer RNA polymerase molecules binding and reduced expression of the protein. Mutations could also affect enhancer or silencer regions, altering the level of expression of protein Y.

26.73 It was a lung disease and it was felt that using the common cold virus, adenovirus, would be a safe and easy means of delivering the therapeutic gene to lung tissue.

26.75 By isolating the gene, scientists knew what corrections in the gene needed to be made so that a therapeutic gene could be developed for gene therapy trials and other forms of treatment.

26.77 VX-809 helps the mutated proteins make it to the cell membrane where they belong. VX-770 helps improve the function of the mutated CFTR protein.

26.79 (a) Transcription: The units include the DNA being transcribed, the RNA polymerases, and a variety of transcription factors.
(b) Translation: mRNA, ribosomal subunits, aminoacyl-tRNA, initiation factors, elongation factors

26.81 Hereditary diseases cannot be prevented, but genetic counseling can help people understand the risks involved in passing a mutated gene to their offspring.

26.83 (a) Plasmid: a small, closed circular piece of DNA found in bacteria. It is replicated in a process independent of the bacterial chromosome. (b) Gene: a section of chromosomal DNA that codes for a particular protein molecule or RNA.

26.85 Each of the amino acids has four codons. All of the codons start with G. The second base is different for each amino acid. The third base may be any of the four possible bases. The distinguishing feature for each amino acid is the second base.

26.87 The hexapeptide is Ala-Glu-Val-Glu-Val-Trp.

Chapter 27 Bioenergetics: How the Body Converts Food to Energy

27.1 ATP

27.3 (a) 2 (b) the outer membrane

27.5 Cristae are folded membranes originating from the inner membrane. They are connected to the inner membrane by tubular channels.

27.7 There are two phosphate anhydride bonds:

27.9 Neither; they yield the same energy.

27.11 It is a phosphate ester bond.

27.13 The two nitrogen atoms that are part of $C=N$ bonds are reduced to form $FADH_2$.

27.15 (a) ATP (b) NAD^+ and FAD (c) acetyl groups

27.17 An amide bond is formed between the amine portion of mercaptoethanolamine and the carboxyl group of pantothenic acid (see Figure 27.7).

27.19 No, the pantothenic acid portion is not the active part. The active part is the $-SH$ group at the end of the molecule.

27.21 Both fats and carbohydrates are degraded to acetyl coenzyme A.

27.23 α-ketoglutarate

27.25 Succinate is oxidized by FAD, and the oxidation product is fumarate.

27.27 Fumarase is a lyase (it adds water across a double bond).

27.29 No, but GTP is produced in Step ⑤.

27.31 It allows the energy to be released in small packets.

27.33 carbon–carbon double bonds occur in *cis*-aconitate and fumarate.

27.35 α-Ketoglutarate transfers its electrons to NAD^+, which becomes $NADH + H^+$.

27.37 mobile electron carriers of the electron transport chain: cytochrome c and CoQ

27.39 When H^+ passes through the ion channel, the proteins of the channel rotate. The kinetic energy of this rotatory motion is converted to and stored as the chemical energy in ATP.

27.41 This process takes place in the inner membranes of the mitochondria.

27.43 (a) 0.5 (b) 12

27.45 Ions reenter the mitochondrial matrix through the proton-translocating ATPase.

27.47 The F_1 portion of ATPase catalyzes the conversion of ADP to ATP.

27.49 The molecular weight of acetate = 59 g/mol, so 1 g acetate = $1 \div 59$ = 0.017 mol acetate. Each mole of acetate produces 12 moles of ATP [see Problem 27.43(b)], so 0.017 mol $\times$ 12 = 0.204 mol ATP. This gives 0.204 mol ATP $\times$ 7.3 kcal/mol = 1.5 kcal.

27.51 (a) Muscles contract by sliding the thick filaments (myosin) and the thin filaments (actin) past each other. (b) The energy comes from the hydrolysis of ATP.

27.53 ATP transfers a phosphate group to the serine residue at the active site of glycogen phosphorylase, thereby activating the enzyme.

27.55 No, it would harm humans because they would not synthesize enough ATP molecules.

27.57 This amount of energy (87.6 kcal) is obtained from 12 mol of ATP (87.6 kcal ÷ 7.3 kcal/mol ATP − 12 mol ATP). Oxidation of 1 mol of acetate yields 12 mol ATP. The molecular weight of CH_3COOH is 60 g/mol, so the answer is 60 g or 1 mol CH_3COOH.

27.59 The energy of motion appears first in the ion channel, where the passage of H^+ causes the proteins lining the channel to rotate.

27.61 They are both hydroxy acids.

27.63 Myosin, the thick filament in muscle, is an enzyme that acts as an ATPase.

27.65 Isocitrate has two stereocenters.

27.67 The ion channel is the F_0 portion of the ATPase; it is made of 12 subunits.

27.69 No, it largely comes from the chemical energy as a result of the breaking of bonds in the O_2 molecule.

27.71 It removes two hydrogens from succinate to produce fumarate.

27.73 The carbon dioxide that we exhale is released by the two oxidative decarboxylation steps in the citric acid cycle.

27.75 Because of the central role of citric acid in metabolism, it can be considered a good nutrient.

27.77 Complex II does not generate enough energy to produce ATP. The rest do.

27.79 Cancer cells process sugars at the point where they enter the central metabolic pathway, as do normal cells. In cancer cells, the breakdown products fuel the uncontrolled growth of the cancer, whereas they fuel controlled growth in normal cells.

27.81 When an animal is awake, cells are in a more active metabolic state than during periods of sleep. Circadian rhythms control the shift from one metabolic state to another.

27.83 Biosynthetic processes reverse the reaction of the oxidative breakdown of nutrients to provide energy. It is reasonable to expect that biosynthesis will require energy and that the overall reactions will be ones of reduction.

27.85 The reactions of the central metabolic pathway take place in the mitochondria. The reactions that break down the molecules, especially sugars, to the point at which they enter the central metabolic pathway take place outside the mitochondria. These reactions outside the mitochondria can provide energy for short times. Sustained effort requires energy from the reactions of the central metabolic pathway that take place within the mitochondria.

27.87 Citrate isomerizes to isocitrate to convert a tertiary alcohol to a secondary alcohol. Tertiary alcohols cannot be oxidized, but secondary alcohols can be oxidized to produce a keto group.

27.89 Iron is found in iron-sulfur clusters in proteins and is also part of the heme group of cytochromes.

27.91 Mobile electron carriers transfer electrons on their path from one large, less mobile protein complex to another.

27.93 ATP and reducing agents such as NADH and $FADH_2$, which are generated by the citric acid cycle, are needed for biosynthetic pathways.

27.95 Biosynthetic pathways are likely to feature reduction reactions because their net effect is to reverse catabolism, which is oxidative.

27.97 ATP is not stored in the body. It is hydrolyzed to provide energy for many different kinds of processes and thus turns over rapidly.

27.99 The citric acid cycle generates NADH and $FADH_2$, which are linked to oxygen by the electron transport chain.

Chapter 28 Specific Catabolic Pathways: Carbohydrate, Lipid, and Protein Metabolism

28.1 According to Table 28.2, the ATP yield from stearic acid is 146 ATP. This makes 146/18 = 8.1 ATP/carbon atom. For lauric acid (C_{12}):

Step ① Activation	−2 ATP
Step ② Dehydrogenation five times	10 ATP
Step ③ Dehydrogenation five times	15 ATP
Six C_2 fragments in common pathway	72 ATP
Total	95 ATP

95/12 = 7.9 ATP per carbon atom for lauric acid. Thus, stearic acid yields more ATP/C atom.

28.3 They serve as building blocks for the synthesis of proteins.

28.5 The two C_3 fragments are in equilibrium. As the glyceraldehyde phosphate is used up, the equilibrium shifts and converts the other C_3 fragment (dihydroxyacetone phosphate) to glyceraldehyde phosphate.

28.7 (a) Steps ① and ③ (b) Steps ⑥ and ⑨

28.9 ATP inhibition takes place at Step ⑨. It inhibits the pyruvate kinase by feedback regulation.

28.11 NADPH is the compound in question.

28.13 Each mole of glucose produces two moles of lactate, so three moles of glucose give rise to six moles of lactate.

28.15 According to Table 28.1, two moles of ATP are produced directly in the cytoplasm.

28.17 Two net ATP molecules are produced in both cases.

28.19 Enzymes that catalyze the phosphorylation of substrates using ATP are called kinases. Therefore, the enzyme that transforms glycerol to glycerol 1-phosphate is called glycerol kinase.

28.21 (a) The two enzymes are thiokinase and thiolase. (b) "Thio" refers to the presence of a sulfur-containing group, such as —SH. (c) Both enzymes insert a CoA—SH into a compound.

28.23 Each turn of fatty acid β-oxidation yields one acetyl CoA, one $FADH_2$, and one NADH. After three turns, $CH_3(CH_2)_4CO$—CoA remains from the original lauric acid; three acetyl CoA, three $FADH_2$, and three NADH + H^+ are produced.

28.25 Using data from Table 28.2, we obtain a figure of 112 moles of ATP for each mole of myristic acid.

28.27 The body preferentially uses carbohydrates as an energy source.

28.29 (a) The transformation of acetoacetate to β-hydroxybutyrate is a reduction reaction. (b) Acetone is produced by decarboxylation of acetoacetate.

28.31   It enters the citric acid cycle.

28.33   Oxidative deamination of alanine to pyruvate:

$$CH_3-CH-COO^- + NAD^+ + H_2O \longrightarrow$$
$$\overset{|}{NH_3^+}$$

$$CH_3-\underset{\underset{O}{\|}}{C}-COO^- + NADH + H^+ + NH_4^+$$

28.35   One of the nitrogens comes from ammonium ion through the intermediate carbamoyl phosphate. The other nitrogen comes from aspartate.

28.37   (a) The toxic product is ammonium ion.   (b) The body gets rid of it by converting it to urea.

28.39   Tyrosine is considered a glucogenic amino acid because pyruvate can be converted to glucose when the body needs it.

28.41   It is stored in ferritin and reused.

28.43   Muscle cramps come from lactic acid accumulation.

28.45   The bicarbonate/carbonic acid buffer counteracts the acidic effects of ketone bodies.

28.47   Protein turnover is important because there is no storage form of amino acids or proteins. Proteins are either used in the body or are degraded to recycle the amino acids they contain.

28.49   Ubiquitins are polypeptides that are found in many species. They target proteins for degradation.

28.51   The reaction is a transamination:

Phenylalanine   α-Ketoglutarate

Phenylpyruvate   Glutamate

28.53   Black and blue are due to the hemoglobin in congealed blood, green to biliverdin, and yellow to bilirubin.

28.55   Production of ethanol in yeast takes place as a result of glycolysis, giving a net yield of two ATP molecules for each mole of glucose metabolized.

28.57   Glucose can be converted to ribose by the pentose phosphate pathway.

28.59   The step in glycolysis in which a phosphate group is transferred from phosphoenolpyruvate (PEP) to ADP to produce ATP indicates that the energy of the phosphate group in PEP is higher than that in ATP.

28.61   Carbamoyl phosphate has an amide group and a phosphate group.

28.63   Pyruvate can be converted to oxaloacetate.

28.65   All the reactions of glycolysis are added algebraically to obtain the equation for the pathway. The presence of an oxidation reaction without an accompanying reduction reaction makes the equation for the whole pathway one of oxidation.

28.67   The catabolism of glycerol (part of fat catabolism) produces a glycerol phosphate, which also appears in glycolysis.

28.69   Table 28.1 takes into account the fact that glucose can be metabolized further by the citric acid cycle, which produces NADH and FADH$_2$. These coenzymes pass electrons to oxygen, giving rise to ATP in the process.

28.71   Lactate plays a key role in regenerating NAD$^+$.

28.73   Amino acids can be catabolized to yield energy, but usually only under starvation conditions.

28.75   catabolism, oxidative, energy-yielding; anabolism, reductive, energy-requiring

28.77   If you look at the balanced chemical equations for the two processes, they are the exact opposite of each other. They differ in that photosynthesis requires energy from the sun and occurs only in some organisms such as plants, whereas aerobic catabolism of glucose releases energy and occurs in organisms of all sorts.

28.79   Sugars are already partially oxidized, so their pathway of complete oxidation is further advanced, producing less energy.

28.81   The reactions of glycolysis take place in the cytosol. Because of their charge, the compounds that form a part of the pathway are not as prone to crossing the cell membrane to the exterior as they would be if they were uncharged. The reactions of the citric acid cycle take place in mitochondria, which have a double membrane. The intermediates of the citric acid cycle tend to stay within mitochondria even without a charge.

28.83   ATP production takes place in connection with the re-oxidation of the NADH and FADH$_2$ produced in the citric acid cycle.

Chapter 29   Biosynthetic Pathways

29.1   Different pathways allow for flexibility and overcome unfavorable equilibria. Separate control of anabolism and catabolism becomes possible.

29.3   The main biosynthesis of glycogen does not use inorganic phosphate because the presence of a large inorganic phosphate pool would shift the reaction to the degradation process such that no substantial amount of glycogen would be synthesized.

29.5   Photosynthesis is the reverse of respiration:

$$6CO_2 + 6H_2O \longrightarrow C_6H_{12}O_6 + 6O_2 \quad \text{Photosynthesis}$$
$$C_6H_{12}O_6 + 6O_2 \longrightarrow 6CO_2 + 6H_2O \quad \text{Respiration}$$

29.7   A compound that can be used for gluconeogenesis:
(a) from glycolysis: pyruvate
(b) from the citric acid cycle: oxaloacetate
(c) from amino acid oxidation: alanine

29.9   Glucose needs for the brain are met by gluconeogenesis, because the other pathways metabolize glucose, and only gluconeogenesis manufactures it.

29.11   Maltose is a disaccharide that is composed of two glucose units linked by an α-1,4-glycosidic bond.

$$\text{UDP-glucose} + \text{glucose} \longrightarrow \text{maltose} + \text{UDP}$$

29.13   UTP consists of uracil, ribose, and three phosphates.

29.15   (a) Fatty acid biosynthesis occurs primarily in the cytoplasm.   (b) No, fatty acid degradation occurs in the mitochondrial matrix.

29.17 In fatty acid biosynthesis, a three-carbon compound, malonyl ACP, is repeatedly added to the synthase.

29.19 Carbon dioxide is released from malonyl ACP, leading to the addition of two carbons to the growing fatty acid chain.

29.21 It is an oxidation step because the substrate is oxidized with concomitant removal of hydrogen. The oxidizing agent is O_2. NADPH is also oxidized during this step.

29.23 NADPH is bulkier than NADH because of its extra phosphate group; it also has two more negative charges.

29.25 No, the body makes other unsaturated fatty acids, such as oleic acid and arachidonic acid.

29.27 The activated components needed are sphingosine, acyl CoA, and UDP-glucose.

29.29 All of the carbons in cholesterol originate in acetyl CoA. A C_5 fragment called isopentenyl pyrophosphate is an important intermediate in steroid biosynthesis.

$$3 \text{ Acetyl-CoA} \longrightarrow \text{mevalonate}$$
$$C_2 \qquad\qquad\qquad C_6$$
$$\longrightarrow \text{isopentenyl pyrophosphate} + CO_2$$
$$C_5$$

29.31 The amino acid product is aspartic acid.

29.33 The products of the transamination reaction shown are valine and α-ketoglutarate.

$$(CH_3)_2CH-\overset{\overset{O}{\|}}{C}-COO^- + {}^-OOC-CH_2-CH_2-\underset{\underset{NH_3^+}{|}}{CH}-COO^- \longrightarrow$$

The keto form of valine 　　Glutamate

$$(CH_3)_2CH-\underset{\underset{NH_3^+}{|}}{CH}-COO^- + {}^-OOC-CH_2-CH_2-\overset{\overset{O}{\|}}{C}-COO^-$$

Valine 　　　α-Ketoglutarate

29.35 NADPH is the reducing agent in the process of carbon dioxide being incorporated into carbohydrates.

29.37 Acetyl-CoA carboxylase (ACC) is a key enzyme in fatty acid biosynthesis. It exists in two forms, one found in liver and one in muscle tissue. The one found in muscle affects weight loss and may become a target for anti-obesity drugs.

29.39 Higher glucose consumption can lead to higher levels of acetyl-CoA in the nucleus. This increase can, in turn, lead to increased acetylation of histones, having a marked effect on gene expression.

29.41 Behavior modification is one of the most useful ways to ensure long-term success of weight loss programs. Both immediate weight loss and long-term maintenance of lower weight are affected.

29.43 The bonds that connect the nitrogen bases to the ribose units are β-N-glycosidic bonds just like those found in nucleotides.

29.45 The amino acid produced by this transamination is phenylalanine.

29.47 The structure of a lecithin (phosphatidyl choline) is shown in Section 21.6. Synthesis of a molecule of this sort requires activated glycerol, two activated fatty acids, and activated choline. Each activation requires one ATP molecule, for a total number of four ATP molecules.

29.49 The compound that reacts with glutamate in a transamination reaction to form serine is 3-hydroxypyruvate. The reverse of the reaction is shown below:

$$\underset{\underset{CH_2OH}{|}}{\underset{\overset{|}{CH-NH_3^+}}{COO^-}} + \underset{\underset{\underset{COO^-}{|}}{\underset{\overset{|}{CH_2}}{CH_2}}}{\underset{\overset{|}{C=O}}{COO^-}} \longrightarrow \underset{\underset{CH_2OH}{|}}{\underset{\overset{|}{C=O}}{COO^-}} + \underset{\underset{\underset{COO^-}{|}}{\underset{\overset{|}{CH_2}}{CH_2}}}{\underset{\overset{|}{CH-NH_3^+}}{COO^-}}$$

Serine 　α-Keto-glutarate 　3-Hydroxy-pyruvate 　Glutamate

29.51 HMG-CoA is hydroxymethylglutaryl CoA. Its structure is shown in Section 29.4. Carbon 1 is the carbonyl group linked to the thio group of CoA.

29.53 Heme is a porphyrin ring with iron at the center. Chlorophyll is a porphyrin ring with magnesium at the center.

29.55 In the process of photosynthesis, the net result of carbohydrate synthesis is that six molecules of carbon dioxide are used to create a six-carbon sugar. In gluconeogenesis, the net result is reversal of glycolysis without involvement of carbon dioxide.

29.57 Fatty acid biosynthesis takes place in the cytoplasm, requires NADPH, and uses malonyl CoA. Fatty acid catabolism takes place in the mitochondrial matrix, produces NADH and $FADH_2$, and has no requirement for malonyl CoA.

29.59 Photosynthesis has high requirements for light energy from the sun.

29.61 Lack of essential amino acids would hinder the synthesis of the protein part. Gluconeogenesis can produce sugars even under starvation conditions.

29.63 Separation of catabolic and anabolic pathways allows for greater efficiency, especially in control of the pathways.

29.65 If laboratory rats are fed all the amino acids but minus one of the essential ones, they will be unable to synthesize protein. Administering the essential amino acid later will not be useful because the other amino acids have already been metabolized.

Chapter 30 Nutrition

30.1 No, nutrient requirements vary from person to person.

30.3 Sodium benzoate is not catabolized by the body; therefore, it does not comply with the definition of a nutrient—components of food that provide growth, replacement, and energy. Calcium propionate enters mainstream metabolism by conversion to succinyl-CoA and catabolism by the citric acid cycle and thus is a nutrient.

30.5 The Nutrition Facts label found on all foods must list the percentage of Daily Values for four important nutrients: vitamins A and C, calcium, and iron.

30.7 Chemically, fiber is cellulose, a polysaccharide that cannot be degraded by humans. It is important for proper operation of dietary processes, especially in the colon.

30.9 The basal caloric requirement is calculated assuming the body is completely at rest. Because most of us perform some activity, we need more calories than this basic minimum.

30.11   1833 Cal

30.13   No, using diuretics would be a temporary fix at best.

30.15   The product would be different-sized oligosaccharide fragments much smaller than the original amylose molecules.

30.17   No, dietary maltose, the disaccharide composed of glucose units linked by an α-1,4-glycosidic bond, is rapidly hydrolyzed in the stomach and small intestines. By the time it reaches the blood, it is the monosaccharide glucose.

30.19   linoleic acid

30.21   No, lipases degrade neither; they degrade triacylglycerols.

30.23   Yes, it is possible for a vegetarian to obtain a sufficient supply of adequate proteins; however, the person must be very knowledgeable about the amino acid content of vegetables, so as to allow for protein complementation.

30.25   Dietary proteins begin degradation in the stomach, which contains HCl in a concentration of about 0.5%. Trypsin is a protease present in the small intestines that continues protein digestion after the stomach. Stomach HCl denatures dietary protein and causes somewhat random hydrolysis of the amide bonds in the protein. Fragments of the protein are produced. Trypsin catalyzes hydrolysis of peptide bonds only on the carboxyl side of the amino acids Arg and Lys.

30.27   Carbohydrates and fats have forms that are stored for later use, such as glycogen and triacylglycerols. These can build up in times of plenty and are then used when caloric need exceeds caloric intake. Proteins have no such storage form. If you eat more amino acids than you can use to repair muscle tissue or replenish other proteins, they are not stored. They are instead converted to carbohydrates or fats.

30.29   It is expected that many of the prisoners will develop deficiency diseases in the near future.

30.31   Limes provided sailors with a supply of vitamin C to prevent scurvy.

30.33   Vitamin K is essential for proper blood clotting.

30.35   The only disease that has been proven scientifically to be prevented by vitamin C is scurvy.

30.37   Vitamins E and C and the carotenoids may have significant effects on respiratory health. This may be due to their activity as antioxidants.

30.39   There is a sulfur atom in biotin and in vitamin B_1 (also called thiamine).

30.41   Vitamin A is toxic in high doses. Some have suggested that limiting the concentration of the vitamins would protect people from their own tendency to think that more is always better.

30.43   The original Food Guide Pyramid did not consider the differences between types of nutrients. It assumed that all fats were to be limited and that all carbohydrates were healthy. The new guidelines recognizes that polyunsaturated fats are necessary and that carbohydrates from whole grains are better for you than those from refined sources. The new pyramid also recognizes the importance of exercise, which the original did not.

30.45   All proteins, carbohydrates, and fats in excess have metabolic pathways that lead to increased levels of fatty acids. However, there is no pathway that allows fats to generate a net surplus of carbohydrates. Thus, fat stores cannot be used to make carbohydrates when a person's blood glucose is low.

30.47   All effective weight loss is based on increasing activity while limiting caloric intake. However, it is more effective to concentrate on increasing activity than on limiting intake.

30.49   Theoretically speaking, if humans had the glyoxylate pathway, dieting would be easier. By eliminating the two decarboxylation steps of the citric acid cycle, there is no loss of carbon from the acetyl-CoA. Therefore, carbon compounds could be removed from the pathway to form glucose. A person could diet and use fat stores to power the body's systems and maintain blood glucose levels.

30.51   Iron is an important cofactor in many biological compounds. The most obvious is the part iron plays in hemoglobin. It is the iron that directly binds the oxygen that is the source of respiration for our metabolism. Iron must be consumed in the diet to maintain iron levels for hemoglobin and many other compounds.

30.53   Factors that affect absorption include the solubility of the compound of iron, the presence of antacids in the digestive tract, and the source of the iron.

30.55   arginine

30.57   carbohydrate loading before the event and consuming carbohydrates during the event

30.59   Caffeine acts as a central nervous system stimulant, which provides a feeling of energy that athletes often enjoy. In addition, caffeine reduces insulin levels and stimulates oxidation of fatty acids, which would be beneficial to endurance athletes. However, caffeine is also a diuretic and can lead to dehydration in long-distance events.

30.61   Cost is the most significant downside to organic food, as organic food can be up to 100% more expensive than non-organic. The type of food is also an important consideration as pesticides or other chemicals are transferred from the food to the consumer while others are not. For example, if a pesticide is concentrated in a banana peel, that is not as big a problem as if it were concentrated in the banana itself. Pesticides or other chemicals are more hazardous to children and pregnant women than to others.

30.63   Cannabinoid receptors are activated by three major groups of ligands—endocannabinoids (produced by the mammalian body), plant cannabinoids (such as THC, produced by the cannabis plant), and synthetic cannabinoids (such as HU-210).

30.65   While the mechanism is not known, studies have shown that patients who had diets higher in trans-fatty acids were almost 50% more likely to be depressed.

30.67   Studies have found that up to 30% of patients hospitalized for depression are deficient in vitamin B_{12}. A study of several hundred physically disabled women over the age of 65 found that women deficient in vitamin B_{12} were more likely to be severely depressed when compared with non-deficient women. A study of over three thousand elderly men and women with depression showed that those with vitamin B_{12} deficiency were almost 70% more likely to experience depression than those with normal vitamin B_{12} levels.

30.69   The vitamin pantothenic acid is part of CoA.

(a) Glycolysis: Pyruvate dehydrogenase uses CoA as a coenzyme.

(b) Fatty acid synthesis: The first step involves the enzyme fatty acid synthase.

30.71   Proteins that are ingested in the diet are degraded to free amino acids, which are then used to build proteins that carry out specific functions. Two very important functions include structural integrity and biological catalysis. Our proteins are constantly being turned over—that is, continuously being degraded and rebuilt using free amino acids.

30.73 The very tip of the Food Guide Pyramid displays fats, oils, and sweets, with the cautionary statement, "Use sparingly." We can omit sweets completely from the diet; however, complete omission of fats and oils is dangerous. We must have dietary fats and oils that contain the two essential fatty acids. The essential fatty acids may be present as components in other food groups—that is, the meat, poultry, and fish group.

30.75 Walnuts are not just a tasty snack—they are a healthy one. Walnuts have protein. In fact, nuts are included in a group of the U.S. Department of Agriculture's Food Guide Pyramid. Walnuts are also a good source of vitamins and minerals, including vitamins E and B, biotin, potassium, magnesium, phosphorus, zinc, and manganese.

30.77 No, the lecithin is degraded in the stomach and intestines long before it could get into the blood. The phosphoglyceride is degraded to fatty acids, glycerol, and choline, which are absorbed through the intestinal walls.

30.79 Patients who have undergone ulcer surgery are administered digestive enzymes that may have been lost during the procedure. The enzyme supplement should contain proteases to help break down proteins as well as lipases to assist in fat digestion.

Chapter 31 Immunochemistry

31.1 Examples of external innate immunity include action by the skin, tears, and mucus.

31.3 The skin fights infection by providing a barrier against penetration of pathogens. The skin also secretes lactic acid and fatty acids, both of which create a low pH, thereby inhibiting bacterial growth.

31.5 Innate immunity processes have little ability to change in response to immune dangers. The key features of adaptive (acquired) immunity are specificity and memory. The acquired immune system uses antibody molecules designed for each type of invader. In a second encounter with the same danger, the response is more rapid and more prolonged than the first response.

31.7 T cells originate in the bone marrow, but grow and develop in the thymus gland. B cells originate and grow in the bone marrow.

31.9 Macrophages are the first cells in the blood that encounter potential threats to the system. They attack virtually anything that is not recognized as part of the body, including pathogens, cancer cells, and damaged tissue. Macrophages engulf an invading bacterium or virus and kill it with NO, nitric oxide, and then digest it.

31.11 protein-based antigens

31.13 Class II MHC molecules pick up damaged antigens. A targeted antigen is first processed in lysosomes, where it is degraded by proteolytic enzymes. An enzyme, GILT, reduces the disulfide bridges of the antigen. The reduced peptide antigens unfold and are further degraded by proteases. The peptide fragments remaining serve as epitopes that are recognized by class II MHC molecules.

31.15 MHC molecules are transmembrane proteins that belong to the immunoglobulin superfamily. They are originally present inside cells until they become associated with antigens and move to the surface membrane.

31.17 If we assume that the rabbit has never been exposed to the antigen, the response will occur 1–2 weeks after the injection of antigen.

31.19 (a) IgE molecules have a carbohydrate content of 10–12%, which is equal to that of IgM molecules. IgE molecules have the lowest concentration in the blood. The blood concentration of IgE is about 0.01–0.1 mg/100 mL of blood. (b) IgE molecules are involved in the effects of hay fever and other allergies. They also offer protection against parasites.

31.21 The two Fab fragments would be able to bind to an antigen. These fragments contain the variable protein sequence regions and hence are able to change during synthesis against a specific antigen.

31.23 *Immunoglobulin superfamily* refers to all of the proteins that have the standard structure of a heavy chain and a light chain.

31.25 Antibodies and antigens are held together by weak noncovalent interactions: hydrogen bonds, electrostatic interactions (dipole–dipole), and hydrophobic interactions.

31.27 The DNA for the immunoglobulin superfamily has multiple ways of recombining during cell development. The diversity is a reflection of the number of permutations and ways of combining various constant regions, variable regions, joining regions, and diversity regions.

31.29 T cells carry on their surfaces unique receptor proteins that are specific for antigens. These receptors (TcR), which are members of the immunoglobulin superfamily, have constant and variable regions. They are anchored in the T-cell membrane by hydrophobic interactions. They are not able to bind antigens alone, but rather need additional protein molecules called cluster determinants that act as coreceptors. When TcR molecules combine with cluster determinant proteins, they form T-cell-receptor complexes (TcR complexes).

31.31 The components of the TcR complexes are (1) accessory protein molecules called cluster determinants and (2) the T-cell receptor.

31.33 CD4

31.35 They are adhesion molecules that help dock antigen-presenting cells to T cells. They also act as signal transducers.

31.37 Cytokines are glycoproteins that interact with cytokine receptors on macrophages, B cells, and T cells. They do not recognize and bind antigens.

31.39 Chemokines are a class of cytokines that send messages between cells. They attract leukocytes to the site of injury and bind to specific receptors on the leukocytes.

31.41 All chemokines are low-molecular-weight proteins that have four cysteine residues that are linked in very specific disulfide bonds: Cys1—Cys3 and Cys2—Cys4.

31.43 Edward Jenner discovered the concept of immunization by injecting serum from people with cowpox into people without either cowpox or smallpox and showing how it conferred protection against smallpox. His methods would be considered criminal today since he knowingly injected a boy with smallpox to show how the vaccine worked.

31.45 It was a derogatory term coined by some French scientists and literally meant "encowment" after Jenner's work with cowpox.

31.47 A subunit vaccine would be considered the safest, as it uses only pieces of the pathogen.

31.49 For a vaccine to work best, it must mimic an infection. The dendritic cells recognize the vaccine as foreign and release cytokines that will engage T cells and B cells in response.

31.51 Researchers have yet to perfect techniques for using vaccines to fight HIV, hepatitis C, tuberculosis, and malaria.

31.53 The LPS molecules were found to bind to and simulate the toll-like receptors on dendritic cells.

31.55   The T cells mature in the thymus gland. During maturation, those cells that fail to interact with MHC and thus cannot respond to foreign antigens are eliminated by a special selection process. T cells that express receptors that may interact with normal self antigens are eliminated by the same selection process.

31.57   A signaling pathway that controls the maturation of B cells is the phosphorylation pathway activated by tyrosine kinase and deactivated by phosphatase.

31.59   the cytokines and chemokines

31.61   helper T cells

31.63   It is hard to find because the virus mutates quickly. Also, one of its docking proteins changes conformation when it docks, so that antibodies elicited against undocked proteins are ineffective. It binds to several proteins that inhibit antiviral factors and cloaks its outer membrane with sugars that are very similar to the natural sugars found on host cells.

31.65   Vaccines rely on the immune system's ability to recognize a foreign molecule and make specific antibodies to it. HIV hides from the immune system in a variety of ways, and it changes often. The body makes antibodies, but they are not very effective at finding or neutralizing the virus.

31.67   HIV mutates quickly, which is one of the reasons it has been hard to make an effective antibody against it. If an antibody can be made that recognizes multiple antigens, then it would be more effective against a quickly mutating antigen.

31.69   Most cancer cells have specific proteins on their surface that help allow their identification as cancerous. Monoclonal antibodies are very specific for the molecules they will bind to, making them an excellent choice for a weapon against cancer. The antibodies will attack the cancer cell and only the cancer cell if the monoclonal antibody is specific enough.

31.71   Fluorescence labeling studies show that breast cancer cells have elevated levels of the HER2 protein. In addition, drugs designed to attack HER2 are very successful at targeting breast cancer cells.

31.73   Many cancers are linked to dimerization of specific cell receptors. Tyrosine kinase is a type of cell receptor that functions via dimerization. Specific monoclonal antibodies are being designed to block the dimerization of these tyrosine kinases.

31.75   Allergies to antibiotics can be very potent. A person may show no symptoms with the first exposure, but a second or third may produce severe reactions or even be fatal.

31.77   Sex workers in some countries use constant low doses of antibiotics in an attempt to avoid sexually transmitted diseases. The unfortunate side effect of this practice has been to allow the evolution of strains of gonorrhea that are antibiotic-resistant.

31.79   One of the molecules on the streptococcus bacteria resembles a protein found in the valves of the heart. The body's attempt to fight strep throat can lead to antibodies that attack not only the bacteria but also the person's own heart valves. This is the danger in rheumatic fever.

31.81   Stem cells can be transformed into other cell types. Scientists are working to find ways to use stem cells to repair damaged nerve tissue or brain tissue. In some animal models, brain cell function has been restored after a stroke by adding stem cells to the brain in the area of the damage.

31.83   The flu has been with us for hundreds of years and has caused multiple pandemics. Far more people have died from flu epidemics than have died from AIDS.

31.85   It depends on how "dangerous" is defined. The H5N1 flu is less transmissible, so far fewer humans have ever gotten it. However, it is more potent, and more than half of the ones who have gotten it have died.

31.87   The newest vaccines target a portion of the flu virus that is relatively stable, so the hope is that the new vaccines will be more effective against a wider variety of flu strains that all possess the viral target.

31.89   IgA molecules are the first line of defense since they are found in tears and mucous secretions. They can intercept invaders before they get into the bloodstream.

31.91   Chemokines (or, more generally, cytokines) help leukocytes migrate out of a blood vessel to the site of injury. Cytokines help the proliferation of leukocytes.

31.93   A compound called 12:13 dEpoB, a derivative of epothilon B, is being studied as an anticancer vaccine.

31.95   Tumor necrosis factor receptors are located on the surfaces of several cell types, but especially on tumor cells.

Glossary

A site (*Section 26.3*) The site on the large ribosomal subunit where the incoming tRNA molecule binds.

Absolute zero (*Section 1.4*) The lowest possible temperature; the zero point of the Kelvin temperature scale.

Abzyme (*Section 23.8*) An immunoglobin generated by using a transition state analog as an antigen.

Acetal (*Section 17.4C*) A molecule containing two —OR groups bonded to the same carbon.

Acetyl group (*Section 27.3*) The group CH_3CO—.

Achiral (*Section 15.1*) An object that lacks chirality; an object that is superposable on its mirror image.

Acid–base reaction (*Section 8.3*) A proton-transfer reaction.

Acid ionization constant (K_a) (*Section 8.5*) An equilibrium constant for the ionization of an acid in aqueous solution to H_3O^+ and its conjugate base. K_a is also called an **acid dissociation constant**.

Acid rain (*Chemical Connections 6A*) Rain with acids other than carbonic acid dissolved in it.

Acidosis (*Chemical Connections 8D*) A condition in which the pH of blood is lower than 7.35.

Acquired immunity (*Section 31.1*) The second line of defense that vertebrates have against invading organisms.

Actinide series (*Section 22.10*) The 14 elements (90–103) immediately following actinium in period 7 in which the $5f$ shell is being filled.

Activating receptor (*Section 31.8*) A receptor on a cell of the innate immune system that triggers activation of the immune cell in response to a foreign antigen.

Activation energy (*Section 7.2*) The minimum energy necessary to cause a chemical reaction.

Activation of an amino acid (*Section 26.5*) The process by which an amino acid is bonded to an AMP molecule and then to the 3'—OH of a tRNA molecule.

Activation of an enzyme (*Section 23.2*) Any process by which an inactive enzyme is transformed into an active enzyme.

Active site (*Section 23.3, Chemical Connections 23B*) A three-dimensional cavity of an enzyme with specific chemical properties to accommodate the substrate.

Activity series (*Section 8.6B*) The ranking of elements in order of their reducing abilities in aqueous solution.

Adaptive immunity (*Section 31.1*) Acquired immunity with specificity and memory.

Adenovirus (*Section 26.9*) A common vector used in gene therapy.

Adhesion molecule (*Section 31.5*) A protein that helps to bind an antigen to the T cell receptor.

Adrenergic neurotransmitter (*Section 24.4*) A monoamine neurotransmitter or hormone, the most common of which are epinephrine (adrenaline), serotonin, histamine, and dopamine.

Advanced glycation end products (*Section 22.7*) A chemical product of sugars and proteins linking together to produce an imine.

Affinity maturation (*Section 31.4*) The process of mutation of T cells and B cells in response to an antigen.

Agonist (*Section 24.1*) A molecule that mimics the structure of a natural neurotransmitter or hormone, binds to the same receptor, and elicits the same response.

AIDS (*Section 31.9*) *A*cquired *i*mmune *d*eficiency *s*yndrome. The disease caused by the human immunodeficiency virus, which attacks and depletes T cells.

Alcohol (*Section 10.4A*) A compound containing an —OH (hydroxyl) group bonded to a tetrahedral carbon atom.

Aldehyde (*Sections 10.4C and 17.1*) A compound containing a carbonyl group bonded to a hydrogen; a —CHO group.

Alditol (*Section 20.3*) The product formed when the CHO group of a monosaccharide is reduced to CH_2OH.

Aldose (*Section 20.1*) A monosaccharide containing an aldehyde group.

Aliphatic amine (*Section 16.1*) An amine in which nitrogen is bonded only to alkyl groups.

Aliphatic hydrocarbon (*Section 11.1*) An alkane.

Alkali metal (*Section 2.5C*) An element, except hydrogen, in Group 1A of the Periodic Table.

Alkaloid (*Chemical Connections 16B*) A basic nitrogen-containing compound of plant origin, many of which have physiological activity when administered to humans.

Alkalosis (*Chemical Connections 9E*) A condition in which the pH of blood is greater than 7.45.

Alkane (*Section 11.1*) A saturated hydrocarbon whose carbon atoms are arranged in an open chain—that is, not arranged in a ring.

Alkene (*Section 12.1*) An unsaturated hydrocarbon that contains a carbon–carbon double bond.

Alkyl group (*Section 11.4A*) A group derived by removing a hydrogen atom from an alkane; is given the symbol —R.

Alkyne (*Section 12.1*) An unsaturated hydrocarbon that contains a carbon–carbon triple bond.

Allosteric protein (*Chemical Connections 22G*) A protein that exhibits a behavior where binding of one molecule at one site changes the ability of the protein to bind another molecule at a different site.

Allosterism (Allosteric enzyme) (*Section 23.6*) An enzyme regulation in which the binding of a regulator on one site of the enzyme modifies the ability of the enzyme to bind the substrate at the active site. Allosteric enzymes often have multiple polypeptide chains with the possibility of chemical communication between the chains.

Allotrope (*Section 5.9*) An element that exists in different forms in the same physical state with different physical and chemical properties.

Alloy (*Section 6.2*) A homogeneous mixture of metals.

Alpha (α-) amino acid (*Section 22.2*) An amino acid in which the amino group is bonded to the carbon atom next to the —COOH carbon.

Alpha helix (*Section 22.9*) A type of repeating secondary structure of a protein in which the chain adopts a helical conformation stabilized by hydrogen bonding between a peptide backbone N—H and the backbone C=O four amino acids farther up the chain.

Alpha particle (α) (*Section 9.2*) A helium nucleus, He^{2+}, $^{4}_{2}He$.

Amine (*Section 11.4B*) An organic compound in which one, two , or three hydrogens of ammonia are replaced by carbon groups; RNH_2, R_2NH, or R_3N.

Amino acid (*Section 22.1*) An organic compound containing an amino group and a carboxyl group.

Amino acid neurotransmitter (*Section 24.5*) A neurotransmitter or hormone that is an amino acid.

Amino acid pool (*Section 28.1*) The free amino acids found both inside and outside cells throughout the body.

Aminoacyl tRNA synthase (*Section 26.3*) An enzyme that links the correct amino acid to a tRNA molecule.

Amino group (*Section 10.4B*) An —NH_2 group.

Amino sugar (*Section 20.1*) A monosaccharide in which an —OH group is replaced by an —NH_2 group.

Amorphous solid (*Section 5.9*) A solid whose atoms, molecules, or ions are not in an orderly arrangement.

Amphiprotic (*Section 8.3*) A substance that can act as either an acid or a base.

Amylase (*Section 30.3*) An enzyme that catalyzes the hydrolysis of α-1,4-glycosidic bonds in dietary starches.

Anabolism (*Section 27.1*) The biochemical process of building up larger molecules from smaller ones.

Aneuploid cell (*Chemical Connections 31C*) A cell with the wrong number of chromosomes.

Anion (*Section 3.2*) An ion with a negative electric charge.

Anode (*Section 6.6C*) The negatively charged electrode.

Anomeric carbon (*Section 20.2*) The hemiacetal carbon of the cyclic form of a monosaccharide.

Anomers (*Section 20.2*) Monosaccharides that differ in configuration only at their anomeric carbons.

Antagonist (*Section 24.1*) A molecule that binds to a neurotransmitter receptor but does not elicit the natural response.

Antibody (*Section 31.1*) A defense glycoprotein synthesized by the immune system of vertebrates that interacts with an antigen; also called an immunoglobulin.

Anticodon (*Section 26.3*) A sequence of three nucleotides on tRNA, also called a codon recognition site, complementary to the codon in mRNA.

Antigen (*Sections 31.1 and 31.3*) A substance foreign to the body that triggers an immune response.

Antigen-presenting cells (APCs) (*Section 31.2*) Cells that cleave foreign molecules and present them on their surfaces for binding to T cells or B cells.

Antisense strand (*Section 26.2*) The strand of DNA that acts as the template for transcription. Also called the template strand and the (-) strand.

AP site (*Section 25.7*) The ribose and phosphate that are left after a glycosylase removes a purine or pyrimidine base during DNA repair.

Apoenzyme (*Section 23.2*) The protein portion of an enzyme that has cofactors or prosthetic groups.

Aqueous solution (*Section 4.3*) A solution in which the solvent is water.

Ar- (*Section 13.1*) The symbol used for an aryl group.

Arene (*Section 13.1*) A compound containing one or more benzene rings.

Aromatic amine (*Section 16.1*) An amine in which nitrogen is bonded to one or more aromatic rings.

Aromatic compound (*Section 13.1*) A term used to classify benzene and its derivatives.

Aromatic sextet (*Section 13.1B*) The closed loop of six electrons (two from the second bond of each double bond) characteristic of a benzene ring.

Aryl group (*Section 13.1*) A group derived from an arene by removal of a hydrogen atom. Given the symbol Ar-.

Atom (*Section 2.3*) The smallest particle of an element that retains the chemical properties of the element.

Atomic mass unit (*Section 2.4A*) 1 amu = 1.6605×10^{-24} g. By definition, 1 amu is 1/12 the mass of a carbon atom containing 6 protons and 6 neutrons.

Atomic number (*Section 2.4C*) The number of protons in the nucleus of at atom.

Atomic weight (*Section 2.4E*) The weighted average of the masses, in atomic mss units, of the naturally occurring isotopes of the element.

Autoxidation (*Section 13.4C*) The reaction of a C—H group with oxygen, O_2, to form a hydroperoxide, R—OOH.

Avogadro's law (*Section 5.4*) Equal volumes of gases at the same temperature and pressure contain the same number of molecules.

Avogadro's number (*Section 4.6*) 6.02×10^{23} formula units per mole; the amount of any substance that contains the same number of formula units as the number of atoms in 12 g of carbon-12.

Axial position (*Section 11.7B*) A position on a chair conformation of a cyclohexane ring that extends from the ring parallel to the imaginary axis of the ring.

Axon (*Section 24.2*) The long part of a nerve cell that comes out of the main cell body and eventually connects with another nerve cell or tissue cell.

B cell (*Section 31.1*) A type of lymphocyte that is produced in and matures in the bone marrow. B cells produce antibody molecules.

Basal caloric requirement (*Section 30.2*) The caloric requirement for an individual at rest, usually given in Cal/day.

Base (*Section 8.1*) An Arrhenius base is a substance that ionizes in aqueous solution to give hydroxide (OH^-) ions.

Bases (*Section 25.2*) Purines and pyrimidines, which are components of nucleosides in DNA and RNA.

Becquerel (Bq) (*Section 9.5A*) A measure of radioactive decay equal to one disintegration per second.

Beta (β) oxidation (*Section 28.5*) The biochemical pathway that degrades fatty acids to acetyl CoA by removing two carbons at a time and yielding energy.

Beta particle (β) (*Section 9.2*) An electron, $^{0}_{-1}β$.

Beta (β-) pleated sheet (*Section 22.9*) A type of secondary protein structure in which the backbone of two protein chains in the same or different molecules is held together by hydrogen bonds.

Binary covalent compound (*Section 3.8A*) A compound containing two elements.

Binding protein (*Section 26.2*) A protein that binds to nucleosomes, making DNA more accessible for transcription.

Boiling point (*Section 6.8B*) The temperature at which the vapor pressure of a liquid is equal to the atmospheric pressure.

Boiling-point elevation (*Section 6.8B*) The increase in the boiling point of a liquid caused by adding a solute.

Bond angle (*Section 3.10*) The angle between two atoms bonded to a central atom.

Bonding electrons (*Section 3.7C*) Valence electrons involved in forming a covalent bond—that is, shared electrons.

Boyle's law (*Section 5.3A*) The volume of a gas at constant temperature is inversely proportional to the pressure applied to the gas.

Brønsted-Lowry acid (*Section 8.3*) A proton donor.

Brønsted-Lowry base (*Section 8.3*) A proton acceptor.

Brownian motion (*Section 6.8*) The random motion of colloidal-size particles.

Buffer (*Section 8.10*) A solution that resists change in pH when limited amounts of an acid or a base are added to it; an aqueous solution containing a weak acid and its conjugate base.

Buffer capacity (*Section 8.10*) The extent to which a buffer solution can prevent a significant change in the pH of a solution upon addition of an acid or a base.

Calorie (*Section 1.9*) The amount of heat necessary to raise the temperature of 1 g of liquid water by 1°C.

Carbocation (*Section 12.6A*) A species containing a carbon atom with only three bonds to it and bearing a positive charge.

Carbohydrate (*Section 20.1*) A polyhydroxyaldehyde or polyhydroxyketone or a substance that gives these compounds on hydrolysis.

Carbonyl group (*Section 10.4C*) A C=O group.

Carboxyl group (*Sections 10.4D and 18.1*) A —COOH group.

Carboxylic acid (*Section 10.4D*) A compound containing a —COOH group.

Carboxylic ester (*Section 10.4*) A derivative of a carboxylic acid in which the H of the carboxyl group is replaced by a carbon atom.

Carcinogen (*Section 26.7*) A chemical mutagen that can cause cancer.

Catabolism (*Section 7.1*) The biochemical process of breaking down molecules to supply energy.

Catalyst (*Section 7.4D*) A substance that increases the rate of a chemical reaction by providing an alternative pathway with a lower activation energy.

Cathode (*Section 6.6C*) The positively charged electrode.

Cation (*Section 3.2*) An ion with a positive electric charge.

Cell reprogramming (*Chemical Connections 31C*) A technique used in whole-mammal cloning in which a somatic cell is reprogrammed to behave like a fertilized egg.

Celsius scale (°C) (*Section 1.4*) A temperature scale based on 0° as the freezing point of water and 100° as the normal boiling point of water.

Central dogma (*Section 26.1*) A doctrine stating the basic directionality of heredity when DNA leads to RNA, which leads to protein. This doctrine is true in almost all life forms except certain viruses.

Chain-growth polymer (*Section 12.6*) A polymer formed by the stepwise addition of monomers to a growing polymer chain.

Chain reaction, nuclear (*Section 9.9*) A nuclear reaction that results from fusion of a nucleus with another particle (most commonly a neutron) followed by decay of the fused nucleus to smaller nuclei and more neutrons. The newly formed neutrons continue the process, which results in a chain reaction.

Chair conformation (*Section 11.6*) The most stable conformation of a cyclohexane ring; all bond angles are approximately 109.5°.

Chaperone (*Section 22.11*) A protein molecule that helps other proteins to fold into the biologically active conformation and enables partially denatured proteins to regain their biologically active conformation.

Charles's law (*Section 5.3B*) The volume of a gas at constant pressure is inversely proportional to the temperature in Kelvin.

Chemical equation (*Section 4.2*) A representation using chemical formulas of the process that occurs when reactants are converted to products.

Chemical equilibrium (*Section 7.5*) A state in which the rate of the forward reaction equals the rate of the reverse reaction.

Chemical kinetics (*Section 7.1*) The study of the rates of chemical reactions.

Chemical messenger (*Section 24.1*) Any chemical that is released from one location and travels to another location before acting. It may be a hormone, a neurotransmitter, or an ion.

Chemical property (*Section 1.1*) A chemical reaction that a substance undergoes.

Chemiosmotic theory (*Section 27.5*) Mitchell's proposal that electron transport is accompanied by an accumulation of protons in the intermembrane space of the mitochondrion, which in turn creates osmotic pressure; as protons flow from an area of high concentration to an area of low concentration, they are driven back to the mitochondrion under this pressure and generate ATP.

Chemistry (*Section 1.1*) The science that deals with matter.

Chemokine (*Section 31.6*) A chemotactic cytokine that facilitates the migration of leukocytes from the blood vessels to the site of injury or inflammation.

Chiral (*Section 15.1*) From the Greek *cheir*, meaning "hand"; an object that is not superposable on its mirror image.

Cholinergic neurotransmitter (*Section 24.1*) A neurotransmitter or hormone based on acetylcholine.

Chromatin (*Section 25.6*) A complex of DNA with histones and nonhistone proteins that exists in eukaryotic cells between cell divisions.

Chromosomes (*Section 25.1*) Structures within the nucleus of eukaryotes that contain DNA and protein and that are replicated as units during mitosis. Each chromosome is made up of one long DNA molecule that contains many heritable genes.

Cis (*Section 11.8*) A prefix meaning "on the same side."

Cis-trans isomers (*Sections 11.8 and 12.7*) Isomers that have the same (1) molecular formula, (2) connectivity of their atoms, but (3) a different arrangement of their atoms in space due to the presence of either a ring or a carbon–carbon double bond.

Cloning (*Section 25.8*) A process whereby DNA is amplified by inserting it into a host and having the host replicate it along with the host's own DNA.

Cluster determinant (*Section 31.5*) A set of membrane proteins on T cells that helps the binding of antigens to the T-cell receptors.

Coding strand (*Section 26.2*) The DNA strand that is not used as a template for transcription, but which has a sequence that is the same as the RNA produced. Also called the (+) strand and the sense strand.

Codon (*Section 26.3*) A three-nucleotide sequence on mRNA that specifies a particular amino acid.

Coenzyme (*Section 23.3*) An organic molecule, frequently a B vitamin, that acts as a cofactor.

Cofactor (*Section 23.3*) The nonprotein part of an enzyme necessary for its catalytic function.

Colligative property (*Section 6.8*) A property of a solution that depends only on the number of solute particles and not on the chemical identity of the solute particles.

Colloid (*Section 6.7*) A two-part mixture in which suspended solute particles range from 1 to 1000 nm in size.

Combined gas law (*Section 5.3C*) The pressure, volume, and temperature in Kelvin of two samples of the same gas are related by the equation $P_1V_1/T_1 = P_2V_2/T_2$.

Combustion (*Section 4.4*) Burning in air.

Competitive inhibition (*Section 23.3*) A mechanism of enzyme regulation in which an inhibitor competes with the substrate for the active site.

Complementary base pairs (*Section 25.3*) The combination of a purine and a pyrimidine base that hydrogen bond together in DNA.

Complete protein (*Section 30.5*) A protein source that contains sufficient quantities of all amino acids required for normal growth and development.

Condensation (*Section 5.7*) The change of a substance from the vapor or gaseous state to the liquid state.

Condensed structural formula (*Section 11.2*) A structural formula that shows all carbon and hydrogen atoms, as for example $CH_3CH_2CH_2CH_2CH_3$.

Configuration (*Section 11.8*) The arrangement of atoms about a stereocenter—that is, the relative arrangements of the parts of a molecule in space.

Conformation (*Section 11.7*) Any three-dimensional arrangement of atoms in a molecule that results from rotation about a single bond.

Conjugate acid (*Section 8.3*) According to the Brønsted-Lowry theory, a substance formed when a base accepts a proton.

Conjugate acid–base pair (*Section 8.3*) A pair of molecules or ions that are related to one another by the gain or loss of a proton.

Conjugate base (*Section 8.3*) According to the Brønsted-Lowry theory, a substance formed when an acid donates a proton to another molecule or ion.

Conjugated protein (*Section 22.12*) A protein that contains a nonprotein part, such as the heme part of hemoglobin.

Consensus sequence (*Section 26.2*) A sequence of DNA in the promoter region that is relatively conserved from species to species.

Constitutional isomers (*Section 11.3*) Compounds with the same molecular formula but a different order of attachment (connectivity) of their atoms.

Contributing structure (*Section 3.9B*) Representations of a molecule or ion that differ only in the distribution of valence electrons.

Control site (*Section 26.6*) A DNA sequence that is part of a prokaryotic operon. This sequence is upstream of the structural gene DNA and plays a role in controlling whether the structural gene is transcribed.

Conversion factor (*Section 1.5*) A ratio of two different units.

Coordinate covalent bond (*Section 22.10*) One in which both shared electrons come from one atom rather than one electron from each of two atoms.

Coordination compounds (*Section 22.10*) A kind of transition metal compound in which a central transition metal is bonded to a group of other ions or neutral molecules and may be electrically neutral or charged.

Cosmic rays (*Section 9.6*) High-energy particles, mainly protons, from outer space bombarding the Earth.

Covalent bond (*Section 3.4A*) A bond resulting from the sharing of electrons between two atoms.

Crenation (*Section 6.8C*) An osmotic process in which water flows out of red blood cells and into a solution through a semipermeable membrane, causing the cells to shrivel.

Crystalline solid (*Section 5.9*) A solid whose atoms, molecules, or ions are in an orderly arrangement.

Crystallization (*Section 5.7*) The formation of a solid from a liquid.

C-terminus (*Section 22.6*) The amino acid at the end of a peptide chain that has a free carboxyl group.

Curie (Ci) (*Section 9.5A*) A measure of radioactive decay equal to 3.7×10^{10} disintegrations per second.

Cyclic ether (*Section 14.3B*) An ether in which oxygen is one of the atoms of a ring.

Cycloalkane (*Section 11.6*) A saturated hydrocarbon that contains carbon atoms bonded to form a ring.

Cystine (*Section 22.4*) A dimer of cysteine in which the two amino acids are covalently bonded by a disulfide bond between their side chain —SH groups.

Cytokine (*Section 31.6*) A glycoprotein that traffics between cells and alters the function of a target cell.

Dalton's law (*Section 5.5*) The pressure of a mixture of gases is equal to the sum of the partial pressure of each gas in the mixture.

Debranching enzyme (*Section 30.3*) The enzyme that catalyzes the hydrolysis of the 1,6-glycosidic bonds in starch and glycogen.

Decarboxylation (*Section 18.5E*) The loss of CO_2 from a carboxyl (—COOH) group.

Decay, nuclear (*Section 9.3B*) The change of a radioactive nucleus of one element into the nucleus of another element.

Dehydration (*Section 14.2B*) The elimination of a molecule of water from an alcohol. An OH is removed from one carbon, and an H is removed from an adjacent carbon.

Dehydrogenase (*Section 23.2*) A class of enzymes that catalyze oxidation–reduction reactions, often using NAD^+ as the oxidizing agent.

Denaturation (*Section 22.13*) The loss of the secondary, tertiary, and quaternary structure of a protein by a chemical or physical agent that leaves the primary structure intact.

Dendrite (*Section 24.2*) A hair-like projection that extends from the cell body of a nerve cell on the opposite side from the axon.

Dendritic cells (*Sections 31.1, 31.2*) Important cells in the innate immune system that are often the first cells to defend against invaders.

Density (*Section 1.7*) The ratio of mass to volume for a substance.

Deoxyribonucleic acid (*Section 25.2*) The macromolecule of heredity in eukaryotes and prokaryotes. It is composed of chains of nucleotide monomers of a nitrogenous base, 2-deoxy-D-ribose, and phosphate.

Detergent (*Section 18.4D*) A synthetic soap. The most common are the linear alkylbenzene sulfonic acids (LAS).

Dextrorotatory (*Section 15.4B*) The clockwise (to the right) rotation of the plane of polarized light in a polarimeter.

Dialysis (*Section 6.8*) A process in which a solution containing particles of different sizes is placed in a bag made of a semipermeable membrane. The bag is placed in a solvent or solution containing only small molecules. The solution in the bag reaches equilibrium with the solvent outside, allowing the small molecules to diffuse across the membrane but retaining the large molecules.

Diastereomers (*Section 15.3A*) Stereoisomers that are not mirror images of each other.

Dietary Reference Intake (DRI) (*Section 30.1*) The current numerical system for reporting nutrient requirements; an average daily requirement for nutrients published by the U.S. Food and Drug Administration.

Diet faddism (*Section 30.1*) An exaggerated belief in the effects of nutrition upon health and disease.

Digestion (*Section 30.1*) The process in which the body breaks down large molecules into smaller ones that can then be absorbed and metabolized.

Diol (*Section 14.1B*) A compound containing two —OH (hydroxyl) groups.

Dipeptide (*Section 22.6*) A peptide made up of two amino acids.

Dipole (*Section 3.7B*) A chemical species in which there is a separation of charge; there is a positive pole in one part of the species and a negative pole in another part.

Dipole–dipole interaction (*Section 5.7B*) The attraction between the positive end of one dipole and the negative end of another dipole in the same or different molecule.

Diprotic acid (*Section 8.3*) An acid that can give up two protons.

Disaccharide (*Section 20.4*) A carbohydrate containing two monosaccharide units joined by a glycosidic bond.

Discriminatory curtailment diet (*Section 30.1*) A diet that avoids certain food ingredients that are considered harmful to the health of an individual—for example, low-sodium diets for people with high blood pressure.

Disulfide (*Section 14.4D*) A compound containing an —S—S— group.

D-monosaccharide (*Section 20.1*) A monosaccharide that, when written as a Fischer projection, has the —OH group on its penultimate carbon to the right.

DNA (*Section 25.2*) Deoxyribonucleic acid.

DNA fingerprint (*Chemical Connections 25C*) A pattern of DNA fragments generated by electrophoresis that is used in forensic science.

Double bond (*Section 3.7C*) A bond formed by sharing two pairs of electrons; represented by two lines between the two bonded atoms.

Double-headed arrow (*Section 3.9A*) A symbol used to show that the structures on either side of it are resonance-contributing structures.

Double helix (*Section 25.3*) The arrangement in which two strands of DNA are coiled around each other in a screw-like fashion.

Dynamic equilibrium (*Section 7.5*) A state in which the rate of the forward reaction equals the rate of the reverse reaction.

Effective collision (*Section 7.2*) A collision between two molecules or ions that results in a chemical reaction.

EGF (*Section 31.6*) Epidermal growth factor; a cytokine that stimulates epidermal cells during healing of wounds.

Electrolyte (*Section 6.4C*) A substance that, when dissolved in water, produces a solution that conducts electricity.

Electromagnetic spectrum (*Section 9.2*) The array of electromagnetic phenomena by wavelength.

Electron (*Section 2.4*) A subatomic particle with a mass of approximately 1/1837 amu and a charge of −1; it is found outside the nucleus.

Electron capture (*Section 9.3F*) A reaction in which a nucleus captures an extranuclear electron and then undergoes a nuclear decay.

Electron configuration (*Section 2.6C*) A description of the orbitals that the electrons of an atom occupy.

18-Electron rule (*Section 22.10*) When transition metals form compounds, maximum stability is achieved with 18 electrons in bonds and unshared pairs, analogous to the octet rule with second row elements.

Electronegativity (*Section 3.4B*) A measure of an atom's attraction for the electrons it shares in a chemical bond with another atom.

Electrophile (*Section 12.5A*) An electron-poor species that can accept a pair of electrons to form a new covalent bond.

Electrophoresis (*Chemical Connections 25C*) A laboratory technique involving the separation of molecules in an electric field.

Element (*Section 2.4A*) A substance that consists of identical atoms.

Elongation (*Section 26.2*) The phase of protein synthesis during which activated tRNA molecules deliver new amino acids to ribosomes where they are joined by peptide bonds to form a polypeptide.

Elongation factor (*Section 26.5*) A small protein molecule that is involved in the process of tRNA binding and movement of the ribosome on the mRNA during elongation.

Embryonal carcinoma cell (*Chemical Connections 31C*) A cell that is multipotent and is derived from carcinomas.

Embryonic stem cell (*Chemical Connections 31C*) Stem cells derived from embryonic tissue. Embryonic tissue is the richest source of stem cells.

Emulsion (*Section 6.7*) A system, such as fat in milk, consisting of a liquid with or without an emulsifying agent in an immiscible liquid, usually as droplets larger than colloidal size.

Enantiomers (*Section 15.1*) Stereoisomers that are nonsuperposable mirror images; refers to a relationship between pairs of objects.

End point (*Section 8.9*) The point in a titration where a visible change occurs.

Endocrine gland (*Section 24.2*) A gland such as the pancreas, pituitary, and hypothalamus that produces hormones involved in the control of chemical reactions and metabolism.

Endothermic reaction (*Section 4.8*) A chemical reaction that absorbs heat.

Energy (*Section 1.8*) The capacity to do work. The SI base unit is the joule (J).

Enhancer (*Section 26.6*) A DNA sequence that is not part of the promoter region that binds a transcription factor, enhancing transcription and speeding up protein production.

Enkephalin (*Section 24.6*) A pentapeptide found in nerve cells of the brain that acts to control the perception of pain.

Enol (*Section 17.5*) A molecule containing an —OH group bonded to a carbon of a carbon–carbon double bond.

Enzyme (*Section 23.1*) A biological catalyst that increases the rate of a chemical reaction by providing an alternative pathway with a lower activation energy.

Enzyme activity (*Section 23.4*) The rate at which an enzyme-catalyzed reaction proceeds, commonly measured as the amount of product produced per minute.

Enzyme specificity (*Section 23.1*) The limitation of an enzyme to catalyze one specific reaction with one specific substrate.

Enzyme-substrate complex (*Section 23.5*) A part of an enzyme reaction mechanism where the enzyme is bound to the substrate.

Epigenetics (*Chemical Connections 31C*) The study of heritable processes that alter gene expression without altering the actual DNA.

Epitope (*Section 31.3*) The smallest number of amino acids on an antigen that elicits an immune response.

Equatorial position (*Section 11.7B*) A position on a chair conformation of a cyclohexane ring that extends from the ring roughly perpendicular to the imaginary axis of the ring.

Equilibrium (*Section 5.8B*) A condition in which two opposing physical forces are equal.

Equilibrium constant (*Section 7.6*) A value calculated from the equilibrium expression for a given reaction indicating in which direction the reaction goes.

Equivalence point (*Section 8.9*) The point in an acid–base titration at which there is a stoichiometric amount of acid and base.

Ergogenic aid (*Chemical Connections 30D*) A substance that can be consumed to enhance athletic performance.

Essential amino acid (*Section 30.5*) An amino acid that the body cannot synthesize in the required amounts and so must be obtained in the diet.

Essential fatty acid (*Section 30.4*) A fatty acid required in the diet.

Ester (*Sections 10.4E and 18.5D*) A compound in which the —OH of a carboxyl group, RCOOH, is replaced by an —OR′ (alkoxy) group or —OAr (aryloxy) group.

Ether (*Section 14.3A*) A compound containing an oxygen atom bonded to two carbon atoms.

Excitatory neurotransmitter (*Section 24.4*) A neurotransmitter that increases the transmission of nerve impulses.

Exon (*Section 25.5*) A nucleotide sequence in mRNA that codes for a protein.

Exothermic reaction (*Section 4.8*) A chemical reaction that gives off heat.

Expression cassette (*Section 26.9*) A gene sequence containing a gene that was incorporated into a vector and introduced via gene therapy, replacing some of the vector's own DNA.

Extended helix (*Section 22.9*) A type of helix found in collagen, caused by a repeating sequence.

External innate immunity (*Section 31.1*) The innate protection against foreign invaders characteristic of the skin barrier, tears, and mucus.

Fact (*Section 1.2*) A statement based on experience.

Factor-Label method (*Section 1.5*) A way of doing conversions in which units are multiplied and divided.

Family in the Periodic Table (*Section 2.5*) The elements in a vertical column in the Periodic Table.

Fat (*Section 21.3*) A mixture of triglycerides containing a high proportion of long-chain, saturated fatty acids.

Fatty acid (*Section 18.4A*) A long, unbranched chain carboxylic acid, most commonly with 10–20 carbon atoms, derived from animal fats, vegetable oils, or the phospholipids of biological membranes. The hydrocarbon chain may be saturated or unsaturated. In most unsaturated fatty acids, the *cis* isomer predominates. *Trans* isomers are rare.

Feedback control (*Section 23.6*) A type of enzyme regulation where the product of a series of reactions inhibits the enzyme that catalyzes the first reaction in the series.

Fiber (*Section 30.1*) The cellulosic, non-nutrient component in our food.

Fibrous protein (*Section 22.1*) A protein used for structural purposes. Fibrous proteins are insoluble in water and have a high percentage of secondary structures, such as alpha helices and/or beta-pleated sheets.

Fischer esterification (*Section 18.5D*) The process of forming an ester by refluxing a carboxylic acid and an alcohol in the presence of an acid catalyst, commonly sulfuric acid.

Fischer projection (*Section 20.1B*) A two-dimensional representation of a sugar structure.

Fission, nuclear (*Section 9.9*) The fragmentation of a heavier nucleus into two or more smaller nuclei.

Formula weight (FW) (*Section 4.5*) The sum of the atomic weights of all atoms is a compound's formula expressed in atomic mass units (amu). Formula weight can be used for both ionic and molecular compounds.

Freezing-point depression (*Section 6.8A*) The decrease in the freezing point of a liquid caused by adding a solute.

Frequency (ν) (*Section 9.2*) The number of wave crests that pass a given point per unit of time.

Functional group (*Section 10.4*) An atom or group of atoms within a molecule that shows a characteristic set of physical and chemical properties.

Furanose (*Section 20.2*) A five-membered cyclic hemiacetal form of a monosaccharide.

Fusion, nuclear (*Section 9.8*) The combining of two or more nuclei to form a heavier nucleus.

Gamma ray (γ) (*Section 9.2*) A form of electromagnetic radiation characterized by very short wavelength and very high energy.

Gay-Lussac's law (*Section 5.3C*) The pressure of a gas at constant volume is directly proportional to its temperature in Kelvin.

Geiger-Müller counter (*Section 9.5*) An instrument for measuring ionizing radiation.

Gene (*Section 25.1*) The unit of heredity; a DNA segment that codes for a protein.

Gene expression (*Section 26.1*) The activation of a gene to produce a specific protein. It involves both transcription and translation.

Gene regulation (*Section 26.6*) The various methods used by organisms to control which genes will be expressed and when.

Gene therapy (*Section 26.9*) The process of treating a disease by introducing a functional copy of a gene to an organism that was lacking it.

General Transcription Factor (GTF) (*Section 26.6*) Proteins that make a complex with the DNA being transcribed and the RNA polymerase.

Genetic code (*Section 26.4*) The sequence of triplets of nucleotides (codons) that determines the sequence of amino acids in a protein.

Genetic engineering (*Section 26.8*) The process by which genes are inserted into cells.

Genome (*Chemical Connections 25D*) The complete DNA sequence of an organism.

Globular protein (*Section 22.1*) Protein that is used mainly for nonstructural purposes and is largely soluble in water.

Gluconeogenesis (*Section 29.2*) The process by which glucose is synthesized in the body.

Glycogenesis (*Section 29.2*) The conversion of glucose to glycogen.

Glycol (*Section 14.1B*) A compound with hydroxyl (—OH) groups on adjacent carbons.

Glycolysis (*Section 28.2*) The catabolic pathway in which glucose is broken down to pyruvate.

Glycoside (*Section 20.3*) A carbohydrate in which the —OH group on its anomeric carbon is replaced by an —OR group.

Glycosidic bond (*Section 20.3*) The bond from the anomeric carbon of a glycoside to an —OR group.

Gp120 (*Section 31.5*) A 120,000-molecular-weight glycoprotein on the surface of the human immunodeficiency virus that binds strongly to the CD4 molecules on T cells.

G-Protein (*Section 24.5*) A protein that is either stimulated or inhibited when a hormone binds to a receptor and subsequently alters the activity of another protein such as adenyl cyclase.

Gray (Gy) (*Section 9.6*) The SI unit of the amount of radiation absorbed from a source; 1 Gy = 100 rad.

Ground state electron configuration (*Section 2.6*) The electron configuration of the lowest energy of an atom.

Guanosine (*Section 25.2*) A nucleoside made of D-ribose and guanine.

Half-life, of a radioisotope (*Section 9.4*) The time it takes for one half of a sample of radioactive material to decay.

Halogen (*Section 2.5*) An element in Group 7A of the Periodic Table.

Haworth projection (*Section 20.2*) A representation of the cyclic structure of a sugar.

HDPE (*Section 12.7C*) High-density polyethylene.

Heat of combustion (*Section 4.8*) The heat given off in a combustion reaction.

Heat of reaction (*Section 4.8*) The heat given off or absorbed in a chemical reaction.

Helicase (*Section 25.6*) An unwinding protein that acts at a replication fork to unwind DNA so that DNA polymerase can synthesize a new DNA strand.

Helix-Turn-Helix (*Section 26.6*) A common motif for a transcription factor.

Helper T cells (*Section 31.2*) A type of T cell that helps in the response of the acquired immune system against invaders but does not kill infected cells directly.

Hemiacetal (*Section 17.4C*) A molecule containing a carbon bonded to one —OH and one —OR group; the product of adding one molecule of alcohol to the carbonyl group of an aldehyde or ketone.

Hemolysis (*Section 6.8C*) An osmotic process in which water flows into red blood cells through the cell's semipermeable membrane, causing the cells to burst.

Henderson-Hasselbalch equation (*Section 8.11*) A mathematical relationship between pH, the pK_a of a weak acid (represented by the general formula HA), and the concentrations of the weak acid and its conjugate base.

Henry's law (*Section 6.4C*) The solubility of a gas in a liquid is directly proportional to the pressure of the gas above the liquid.

Heterocyclic aliphatic amine (*Section 16.1*) A heterocyclic amine in which nitrogen is bonded only to alkyl groups.

Heterocyclic amine (*Section 16.1*) An amine in which nitrogen is one of the atoms of a ring.

Heterocyclic aromatic amine (*Section 16.1*) An amine in which nitrogen is one of the atoms of an aromatic ring.

Heterogeneous catalyst (*Section 8.4D*) A catalyst in a separate phase from the reactants—for example, the solid platinum, Pt(*s*), in the reaction between CO(*g*) and H_2(*g*) to produce CH_3OH(*l*).

Highly active antiretroviral therapy (HAART) (*Section 31.9*) An aggressive treatment against AIDS involving the use of several different drugs.

Histone (*Section 25.6*) A basic (pH > 7) protein that is found in complexes with DNA in eukaryotes.

HIV (*Sections 31.4 and 31.9*) Human immunodeficiency virus.

Homogeneous catalyst (*Section 8.4D*) A catalyst in the same phase as the reactants—for example, enzymes in body tissues.

Hormone (*Section 24.2*) A chemical messenger released by an endocrine gland into the bloodstream and transported there to reach its target cell.

Hybridization (*Section 25.8*) A process whereby two strands of nucleic acids or segments thereof form a double-stranded structure through hydrogen bonding of complementary base pairs.

Hybridoma (*Section 31.4*) A combination of a myeloma cell with a B cell to produce monoclonal antibodies.

Hydration (*Section 12.6B*) The addition of water.

Hydrocarbon (*Section 11.1*) A compound that contains only carbon and hydrogen atoms.

Hydrogen bond (*Section 5.7C*) A noncovalent force of attraction between the partial positive charge on a hydrogen atom bonded to an atom of high electronegativity, most commonly oxygen or nitrogen, and the partial negative charge on a nearby oxygen or nitrogen.

Hydrogenation (*Section 12.6D*) Addition of hydrogen atoms to a double or triple bond using H_2 in the presence of a transition metal catalyst, most commonly Ni, Pd, or Pt. Also called catalytic reduction or catalytic hydrogenation.

Hydrolase (*Section 23.2*) An enzyme that catalyzes a hydrolysis reaction.

Hydronium ion (*Section 8.1*) The H_3O^+ ion.

Hydrophobic interaction (*Section 22.11*) Interaction by London dispersion forces between hydrophobic groups.

Hydroxyl group (*Section 10.4A*) An —OH group bonded to a tetrahedral carbon atom.

Hygroscopic substance (*Section 6.6B*) A compound able to absorb water vapor from the air.

Hyperthermia (*Chemical Connections 1B*) Having a body temperature higher than normal.

Hyperthermophile (*Section 23.4*) An organism that lives at extremely high temperatures.

Hypothermia (*Chemical Connections 1B*) Having a body temperature lower than normal.

Hypothesis (*Section 1.2*) A statement that is proposed, without actual proof, to explain certain facts and their relationship.

Ideal gas (*Section 5.4*) A gas whose physical properties are described accurately by the ideal gas law.

Ideal gas constant (R) (*Section 5.4*) 0.0821·L·atm·mol^{-1}·K^{-1}.

Ideal gas law (*Section 5.4*) $PV = nRT$.

Immunogen (*Section 31.3*) Another term for antigen.

Immunoglobulin (*Section 31.4*) An antibody protein generated against and capable of binding specifically to an antigen.

Immunoglobulin superfamily (*Section 31.1*) A family of molecules based on a similar structure that includes the immunoglobulins, T cell receptors, and other membrane proteins that are involved in cell communications. All molecules in this class have a certain portion that can react with antigens.

Indicator, acid–base (*Section 8.8*) A substance that changes color within a given pH range.

Induced-fit model (*Section 23.5*) A model explaining the specificity of enzyme action by comparing the active site to a glove and the substrate to a hand.

Inhibition of enzymatic activity (*Section 23.3*) Any reversible or irreversible process that makes an enzyme less active.

Inhibitor (*Section 23.3*) A compound that binds to an enzyme and lowers its activity.

Inhibitory neurotransmitter (*Section 24.4*) A neurotransmitter that decreases the transmission of nerve impulses.

Inhibitory receptor (*Section 31.8*) A receptor on the surface of a cell of the innate immune system that recognizes antigens on healthy cells and prevents activation of the immune system.

Initiation of protein synthesis (*Section 26.5*) The first step in the process whereby the base sequence of a mRNA is translated into the primary structure of a polypeptide.

Initiation signal (*Section 26.2*) A sequence on DNA that identifies the location where transcription is to begin.

Innate immunity (*Section 31.1*) The first line of defense against foreign invaders, which includes skin resistance to penetration, tears, mucus, and nonspecific macrophages that engulf bacteria.

Interleukin (*Section 31.6*) A cytokine that controls and coordinates the action of leukocytes.

Internal innate immunity (*Section 31.1*) The type of innate immunity that is used once a pathogen has already penetrated a tissue.

International System of Units (SI) (*Section 1.4*) A system of units of measurement based in part on the metric system.

Intron (*Section 25.5*) A nucleotide sequence in mRNA that does not code for a protein.

Ion (*Section 2.8B*) An atom with an unequal number of protons and electrons.

Ion product of water, K_w (*Section 9.7*) The concentration of H_3O^+ multiplied by the concentration of OH^-; $[H_3O^+][OH^-] = 1 \times 10^{-14}$.

Ionic bond (*Section 3.4A*) A chemical bond resulting from the attraction between a positive ion and a negative ion.

Ionic compound (*Section 3.5A*) A compound formed by the combination of positive and negative ions.

Ionization energy (*Section 2.8B*) The energy required to remove the most loosely held electron from an atom in the gas phase.

Ionizing radiation (*Section 9.5*) Radiation that causes one or more electrons to be ejected from an atom or a molecule, thereby producing positive ions.

Isoelectric point (pI) (*Section 22.3*) The pH at which a molecule has no net charge.

Isoenzyme (*Section 23.6*) An enzyme that can be found in multiple forms, each of which catalyzes the same reaction. Also called an **isozyme**.

Isomerase (*Section 23.2*) An enzyme that catalyzes an isomerization reaction.

Isotonic (*Section 6.8C*) Solutions that have the same osmolarity.

Isozymes (*Section 23.6*) Two or more enzymes that perform the same functions but have different combinations of subunits—that is, different quaternary structures.

Joule (J) (*Section 1.9*) The SI base unit for heat; 4.184 J is 1 cal.

Ketone (*Sections 10.4C and 17.1*) A compound containing a carbonyl group bonded to two carbons.

Ketone body (*Chemical Connections 18C*) One of several ketone-based molecules—for example, acetone, 3-hydroxybutanoic acid (β-hydroxybutyric acid), and acetoacetic acid (3-oxobutanoic acid)—produced in the liver during overutilization of fatty acids when the supply of carbohydrates is limited.

Ketose (*Section 20.1*) A monosaccharide containing a ketone group.

Killer T cell (*Section 31.2*) A T cell that kills invading foreign cells by cell-to-cell contact. Also called cytotoxic T cell.

Kinase (*Section 23.6*) A class of enzymes that covalently modifies a protein with a phosphate group, usually through the —OH group on the side chain of a serine, threonine, or tyrosine.

Kinetic energy (*Section 1.8*) The energy of motion; energy that is in the process of doing work.

Kwashiorkor (*Section 30.5*) A disease caused by insufficient protein intake and characterized by a swollen stomach, skin discoloration, and retarded growth.

Lactam (*Section 19.1B*) A cyclic amide.

Lactone (*Section 19.1C*) A cyclic ester.

Lagging strand (*Section 25.6*) A discontinuously synthesized DNA that elongates in a direction away from the replication fork.

Lanthanide series (*Section 22.10*) The 14 elements (58–71) immediately following lanthanum in period 6 in which the 4*f* shell is being filled.

Law of conservation of energy (*Section 1.8*) Energy can be neither created nor destroyed.

LDPE (*Section 12.7B*) Low-density polyethylene.

Leading strand (*Section 25.6*) The continuously synthesized DNA strand that elongates toward the replication fork.

Le Chatelier's principle (*Section 7.7*) When a stress is applied to a system in chemical equilibrium, the position of the equilibrium shifts in the direction that will relieve the applied stress.

Leucine zipper (*Section 26.6*) A common motif for a transcription factor.

Leukocytes (*Section 31.2*) White blood cells, which are the principal parts of the acquired immunity system and act via phagocytosis or antibody production.

Levorotatory (*Section 15.4B*) The counterclockwise rotation of the plane of polarized light in a polarimeter.

Lewis dot structure (*Section 2.6F*) The symbol of the element surrounded by a number of dots equal to the number of electrons in the valence shell of an atom of that element.

Lewis structure (*Section 3.7C*) A formula for a molecule or an ion showing all pairs of bonding electrons as single, double, or triple lines and all nonbonding (unshared) electrons as pairs of Lewis dots.

Ligands (*Section 22.10*) Ions or neutral molecules bonded to a central transition metal in a coordination compound.

Ligase (*Section 23.2*) A class of enzymes that catalyzes a reaction joining two molecules. They are often called synthetases or synthases.

Limiting reagent (*Section 4.7*) The reactant that is consumed, leaving an excess of another reagent or reagents unreacted.

Line-angle formula (*Section 11.1*) An abbreviated way to draw structural formulas in which each vertex and line terminus represents a carbon atom and each line represents a bond.

Lipase (*Section 30.4*) An enzyme that catalyzes the hydrolysis of an ester bond between a fatty acid and glycerol.

Lipoprotein (*Section 20.9*) Spherically shaped clusters containing both lipid molecules and protein molecules.

L-monosaccharide (*Section 20.1*) A monosaccharide that, when written as a Fischer projection, has the —OH group on its penultimate carbon to the left.

Lock-and-key model (*Section 23.5*) A model for enzyme-substrate interaction based on the postulate that the active site of an enzyme is a perfect fit for the substrate.

London dispersion forces (*Section 5.7A*) Extremely weak attractive forces between atoms or molecules caused by the electrostatic attraction between temporary induced dipoles.

Lyase (*Section 23.2*) A class of enzymes that catalyzes the addition of two atoms or groups of atoms to a double bond or their removal to form a double bond.

Lymphocyte (*Sections 31.1 and 31.2*) A white blood cell that spends most of its time in the lymphatic tissues. Those that mature in the bone marrow are B cells. Those that mature in the thymus are T cells.

Lymphoid organs (*Section 31.2*) The main organs of the immune system, such as the lymph nodes, spleen, and thymus, that are connected by lymphatic capillary vessels.

Macrophage (*Sections 31.1 and 31.2*) An ameboid white blood cell that moves through tissue fibers, engulfing dead cells and bacteria by phagocytosis, and then displays some of the engulfed antigens on its surface.

Main group element (*Section 2.5A*) An element in the A groups (Groups 1A, 2A, and 3A–8A) of the Periodic Table.

Major groove (*Section 25.3*) The side of a DNA double helix that is narrower.

Major histocompatibility complex (MHC) (*Sections 31.2 and 31.3*) A transmembrane protein complex that brings the epitope of an antigen to the surface of the infected cell to be presented to the T cells.

Maloney murine leukemia virus (MMLV) (*Section 26.9*) A common vector used for gene therapy.

Marasmus (*Section 30.2*) Another term for chronic starvation, whereby the individual does not have adequate caloric intake. It is characterized by arrested growth, muscle wasting, anemia, and general weakness.

Markovnikov's rule (*Section 12.6A*) In the addition of HX or H_2O to an alkene, hydrogen adds to the carbon of the double bond having the greater number of hydrogens.

Mass (*Section 1.4*) The quantity of matter in an object; the SI base unit is the kilogram; often referred to as weight.

Matter (*Section 1.1*) Anything that has mass and takes up space.

Memory cell (*Section 31.2*) A type of T cell that stays in the blood after an infection is over and acts as a quick line of defense if the same antigen is encountered again.

Mercaptan (*Section 14.4B*) A common name for any molecule containing an —SH group.

Messenger RNA (mRNA) (*Section 25.4*) The RNA that carries genetic information from DNA to the ribosome and acts as a template for protein synthesis.

Meta (m) (*Section 13.2B*) Refers to groups occupying the 1 and 3 positions on a benzene ring.

Metabolic acidosis (*Chemical Connections 8D*) The lowering of the blood pH due to metabolic effects such as starvation or intense exercise.

Metabolism (*Section 27.1*) The sum of all chemical reactions in a cell.

Metal (*Section 2.5B*) An element that is a solid at room temperature (except for mercury which is a liquid), is shiny, conducts electricity, is ductile (they can be drawn into wires), is malleable, and forms alloys. In their reactions, metals tend to give up electrons.

Metal binding finger (*Section 26.6*) A type of transcription factor containing heavy metal ions, such as Zn^{2+}, that is involved in helping RNA polymerase bind to the DNA being transcribed.

Metalloid (*Section 2.5B*) An element that displays some of the properties of metals and some of the properties of nonmetals. Six elements are classified as metalloids.

Meter (*Section 1.4*) The SI base unit of length.

Methylene group (*Section 11.1*) A —CH_2— group.

Metric system (*Section 1.4*) A system in which measurements of parameter are related by powers of 10.

Micelle (*Section 18.4*) A spherical arrangement of molecules in aqueous solution such that their hydrophobic (water-hating) parts are shielded from the aqueous environment in the interior and their water-loving parts are on the surface of the sphere and in contact with the aqueous environment.

Micro RNA (miRNA) (*Section 25.4*) A small RNA of 22 nucleosides that is involved in the regulation of genes and the development of an organism.

Minisatellite (*Section 25.5*) A small repetitive DNA sequence that is sometimes associated with cancer when it mutates.

Minor groove (*Section 25.3*) The side of a DNA double helix that is wider.

Mirror image (*Section 15.1*) The reflection of an object in a mirror.

Molar mass (*Section 4.6*) The mass of one mole of a substance expressed in grams; the formula weight of a compound expressed in grams.

Molarity (*Section 6.5*) The number of moles of solute dissolved in 1 L of solution.

Mole (mol) (*Section 4.6*) The formula weight of a substance expressed in grams.

Molecular weight (MW) (*Section 4.5*) The sum of the atomic weights of all atoms in a molecular compound expressed in atomic mass units (amu).

Monoclonal antibody (*Section 31.4*) An antibody produced by clones of a single B cell specific to a single epitope.

Monomer (*Section 12.6A*) From the Greek *mono*, "single," and *meros*, "part"; the simplest nonredundant unit from which a polymer is synthesized.

Monoprotic acid (*Section 8.3*) An acid that can give up only one proton.

Monosaccharide (*Section 20.1*) A carbohydrate that cannot be hydrolyzed to a simpler compound.

Multiclonal antibodies (*Chemical Connections 31B*) The type of antibodies found in its serum after a vertebrate is exposed to an antigen.

Multipotent stem cell (*Chemical Connections 31C*) A stem cell capable of differentiating into many, but not all, cell types.

Mutagen (*Section 26.7*) A chemical substance that induces a base change or mutation in DNA.

Mutarotation (*Section 20.2*) The change in a specific rotation at a given wavelength that occurs when an α or β form of a carbohydrate is converted to an equilibrium mixture of the two forms.

Natural killer cell (*Sections 31.1 and 31.2*) A cell of the innate immune system that attacks infected or cancerous cells.

Negative modulation (*Section 23.6*) The process whereby an allosteric regulator inhibits enzymatic action.

Net ionic equation (*Section 4.3*) A chemical equation that does not contain spectator ions.

Neuron (*Section 24.1*) Another name for a nerve cell.

Neuropeptide Y (*Section 24.6*) A brain peptide that affects the hypothalamus and is an appetite-stimulating agent.

Neurotransmitter (*Section 24.2*) A chemical messenger between a neuron and another cell, which may be a neuron, a muscle cell, or the cell of a gland.

Neutralizing antibody (*Section 31.9*) A type of antibody that completely destroys its target antigen.

Neutron (*Section 2.4*) A subatomic particle with a mass of approximately 1 amu and a charge of zero; it is found in the nucleus.

Nonbonding electrons (*Section 3.7C*) Valence electrons not involved in forming covalent bonds—that is, unshared electrons.

Noncompetitive inhibition (*Section 23.3*) An enzyme regulation in which an inhibitor binds to the active site, thereby changing the shape of the active site and reducing its catalytic activity.

Nonmetal (*Section 12.5B*) An element that does not have the characteristic properties of a metal and, in its reactions, tends to accept electrons. Eighteen elements are classified as nonmetals.

Nonpolar covalent bond (*Section 3.7B*) A covalent bond between two atoms whose difference in electronegativity is less than 0.5.

Normal boiling point (*Section 5.8C*) The temperature at which a liquid boils under a pressure of 1 atm.

N-terminus (*Section 22.6*) The amino acid at the end of a peptide chain that has a free amino group.

Nuclear fission (*Section 9.9*) The process of splitting a nucleus into smaller nuclei.

Nuclear fusion (*Section 9.8*) Joining together atomic nuclei to form a heavier nucleus than the starting nuclei.

Nuclear reaction (*Section 9.3A*) A reaction that changes an atomic nucleus (usually to the nucleus of another element).

Nucleic acids (*Section 25.3*) A polymer composed of nucleotides.

Nucleophile (*Section 12.5A*) An electron-rich species that can donate a pair of electrons to form a new covalent bond.

Nucleophilic attack (*Section 23.5*) A chemical reaction where an electron-rich atom such as oxygen or sulfur bonds to an electron-deficient atom such as a carbonyl carbon.

Nucleoside (*Section 25.2*) The combination of a heterocyclic aromatic amine bonded by a glycosidic bond to either D-ribose or 2-deoxy-D-ribose.

Nucleosome (*Section 25.3*) Combinations of DNA and proteins.

Nucleotide (*Section 25.2*) A phosphoric ester of a nucleoside.

Nutrient (*Section 30.1*) Components of food and drink that provide energy, replacement, and growth.

Octet rule (*Section 3.2*) When undergoing chemical reactions, atoms of Group 1A–7A elements tend to gain, lose, or share electrons to achieve an election configuration having eight valence electrons.

Oil (*Section 21.2*) A mixture of triglycerides containing a high proportion of long-chain, unsaturated fatty acids.

Okazaki fragment (*Section 25.6*) A short segment of DNA made up of about 200 nucleotides in higher organisms and 2000 nucleosides in prokaryotes.

Oligosaccharide (*Section 20.4*) A carbohydrate containing from six to ten monosaccharide units, each joined to the next by a glycosidic bond.

Open complex (*Section 26.6*) The complex of DNA, RNA polymerase, and general transcription factors that must be formed before transcription can take place. In this complex, the DNA is being separated so that it can be transcribed.

Optically active (*Section 15.4B*) A compound that rotates the plane of polarized light.

Orbital (*Section 2.6A*) A region of space around a nucleus that can hold a maximum of two electrons.

Organic chemistry (*Section 10.1*) The study of the compounds of carbon.

Origin of replication (*Section 25.6*) The point in a DNA molecule where replication starts.

Ortho (o) (*Section 13.2B*) Refers to groups occupying the 1 and 2 positions on a benzene ring.

Osmolarity (*Section 6.8C*) Molarity multiplied by the number of particles in solution in each formula unit of solute.

Osmosis (*Section 6.8*) The passage of solvent molecules from a less concentrated solution across a semipermeable membrane into a more concentrated solution.

Osmotic pressure (Π) (*Section 6.8C*) The amount of external pressure that must be applied to a more concentrated solution to stop the passage of solvent molecules into it from across a semipermeable membrane.

Oxidation (*Section 4.4*) The loss of electrons; the gain of oxygen atoms or the loss of hydrogen atoms.

Oxidative deamination (*Section 28.8*) The reaction in which the amino group of an amino acid is removed and an α-ketoacid is formed.

Oxidizing agent (*Section 4.4*) An entity that accepts electrons in an oxidation–reduction reaction.

Oxidoreductase (*Section 23.2*) A class of enzymes that catalyzes an oxidation–reduction reaction.

Oxonium ion (*Section 12.6B*) An ion in which oxygen is bonded to three other atoms and bears a positive charge.

p53 (*Chemical Connections 26E*) A common and important tumor suppressor protein with a molecular weight of 53,000 that is found to be mutated in a large number of cancer types.

P site (*Section 26.5*) The site on the large ribosomal subunit where the current peptide is bound before peptidyl transferase links it to the amino acid attached to the A site during elongation.

Para (p) (*Section 13.2B*) Refers to groups occupying the 1 and 4 positions on a benzene ring.

Parenteral nutrition (*Chemical Connections 29A*) The technical term for intravenous feeding.

Partial pressure (*Section 5.5*) The pressure that a gas in a mixture of gases would exert if it were alone in a container.

Pentose phosphate pathway (*Section 28.2*) The biochemical pathway that produces ribose and NADPH from glucose-6-phosphate or, alternatively, releases energy.

Peptide (*Section 22.6*) A short chain of amino acids linked via peptide bonds.

Peptide backbone (*Section 22.7*) The repeating pattern of peptide bonds in a polypeptide or protein.

Peptide bond (*Section 22.6*) An amide bond that links two amino acids.

Peptidergic neurotransmitter (*Section 24.6*) A type of neurotransmitter or hormone that is based on a peptide, such as glucagon, insulin, and the enkephalins.

Peptidyl transferase (*Section 26.5*) The enzymatic activity of the ribosomal complex that is responsible for the formation of peptide bonds between the amino acids of the growing peptide.

Percent concentration (% w/v) (*Section 6.5A*) The number of grams of solute in 100 mL of solution.

Perforin (*Section 31.2*) A protein produced by killer T cells that punches holes in the membrane of target cells.

Period of the Periodic Table (*Section 2.5*) A horizontal row of the Periodic Table.

Peroxide (*Section 12.7B*) A compound that contains an —O—O— bond; for example, hydrogen peroxide, H—O—O—H.

pH (*Section 8.8*) The negative logarithm of the hydronium ion concentration; $pH = -\log[H_3O^+]$.

Phagocytosis (*Section 31.4*) The process by which large particulates, including bacteria, are pulled inside a white cell called a phagocyte.

Phase change (*Section 5.10*) A change from one physical state (gas, liquid, or solid) to another.

Phenol (*Section 13.4*) A compound that contains an —OH group bonded to a benzene ring.

Phenyl group (*Section 13.2A*) C_6H_5—, the aryl group derived by removing a hydrogen atom from benzene.

Pheromone (*Chemical Connections 12B*) A chemical secreted by an organism to influence the behavior of another member of the same species.

Photon (*Section 9.2*) The smallest unit of electromagnetic radiation.

Physical change (*Section 1.1*) A change in matter in which it does not lose its identity.

Physical property (*Section 1.1*) Characteristics of a substance that are not chemical properties; those properties that are not a result of a chemical change.

Plane-polarized light (*Section 15.4A*) Light vibrating in only parallel planes.

Plasma cell (*Section 31.2*) A cell derived from a B cell that has been exposed to an antigen.

Plasmids (*Section 26.8*) Small circular DNAs of bacterial origin often used to construct recombinant DNA.

Pluripotent stem cell (*Chemical Connections 31F*) A stem cell that is capable of developing into every cell type.

pOH (*Section 8.8*) The negative logarithm of the hydroxide ion concentration; $pOH = -\log[OH^-]$.

Polar covalent bond (*Section 3.7C*) A covalent bond between two atoms whose difference in electronegativity is between 0.5 and 1.9.

Polarimeter (*Section 15.4B*) An instrument for measuring the ability of a compound to rotate the plane of polarized light.

Polyatomic ion (*Section 3.3C*) An ion that contains more than one atom.

Polymer (*Section 12.7A*) From the Greek *poly*, "many", and *meros*, "parts"; any long-chain molecule synthesized by bonding together many single parts called monomers.

Polymerase chain reaction (PCR) (*Section 25.8*) An automated technique for amplifying DNA using a heat-stable DNA polymerase from thermophilic bacteria.

Polynuclear aromatic hydrocarbon (*Section 13.2D*) A hydrocarbon containing two or more benzene rings, each of which shares two carbon atoms with another benzene ring.

Polypeptide (*Section 22.6*) A long chain of amino acids bonded via peptide bonds.

Polysaccharide (*Section 20.4*) A carbohydrate containing a large number of monosaccharide units, each joined to the next by one or more glycosidic bonds.

Positive cooperativity (*Chemical Connections 22G*) A type of allosterism where the binding of one molecule of a protein makes it easier to bind another of the same molecule.

Positive modulation (*Section 23.6*) The process whereby an allosteric regulator increases enzymatic action.

Positron ($\beta+$) (*Section 9.3D*) A particle with the mass of an electron but a charge of +1, $_{+1}^{0}\beta$.

Positron emission tomography (PET) (*Section 9.7A*) The detection of positron-emitting isotopes in different tissues and organs; a medical imaging technique.

Postsynaptic membrane (*Section 24.2*) The membrane on the side of the synapse nearest the dendrite of the neuron receiving the transmission.

Post-transcription process (*Section 26.2*) A process such as splicing or capping that alters RNA after it is initially made during transcription.

Potential energy (*Section 1.8*) Energy that is being stored; energy that is available for later use.

Pressure (*Section 5.2*) The force per unit area exerted against a surface.

Presynaptic membrane (*Section 24.2*) The membrane on the side of the synapse nearest the dendrite of the axon of the neuron transmitting the signal.

Primary (1°) alcohol (*Section 10.4A*) An alcohol in which the carbon atom bearing the —OH group is bonded to only one other carbon group, a —CH_2OH group.

Primary (1°) amine (*Section 10.4B*) An amine in which nitrogen is bonded to one carbon group and two hydrogens.

Primary structure, of DNA (*Section 25.3*) The order of the bases in DNA.

Primary structure, of proteins (*Section 22.8*) The order of amino acids in a peptide, polypeptide, or protein.

Primer (*Section 25.6*) Short pieces of DNA or RNA that initiate DNA replication.

Principal energy level (*Section 2.6A*) An energy level containing orbitals of the same number (1, 2, 3, 4, and so forth).

Proenzyme (*Section 23.6*) An inactive form of an enzyme that must have part of its polypeptide chain cleaved before it becomes active.

Progenitor cells (*Chemical Connections 31D*) Another term for stem cells.

Prokaryote (*Section 25.6*) An organism that has no true nucleus or organelles.

Promoter (*Section 26.2*) An upstream DNA sequence that is used for RNA polymerase recognition and binding to DNA.

Prosthetic group (*Section 22.12*) The non-amino-acid part of a conjugated protein.

Proteasome (*Section 26.6*) A large protein complex that is involved in the degradation of other proteins.

Protein (*Section 22.1*) A long chain of amino acids linked via peptide bonds. There must usually be a minimum of 30 to 50 amino acids in a chain before it is considered a protein (instead of a peptide).

Protein complementation (*Section 30.5*) A diet that combines proteins of varied sources to arrive at a complete protein.

Protein microarray (*Chemical Connections 22F*) An automated technique used to study proteomics that is based on having thousands of protein samples imprinted on a chip.

Protein modification (*Section 23.6*) The process of affecting enzymatic activity by covalently modifying the enzyme, such as phosphorylating a particular amino acid.

Proteomics (*Chemical Connections 22F*) The collective knowledge of all the proteins and peptides of a cell or a tissue and their functions.

Proton (*Section 2.4A*) A subatomic particle with a charge of +1 and a mass of approximately 1 amu; found in a nucleus.

Pyranose (*Section 20.2*) A six-membered cyclic hemiacetal form of a monosaccharide.

Quaternary structure (*Section 22.12*) The organization of a protein that has multiple polypeptide chains, or subunits; refers principally to the way the multiple chains interact.

R– (*Section 11.4A*) A symbol used to represent an alkyl group.

R– (*Section 15.2*) From the Latin *rectus*, meaning "straight, correct"; used in the *R,S* system to show that when the lowest-priority group is away from you, the order of priority of groups on a stereocenter is clockwise.

R form (*Section 23.6*) The more active form of an allosteric enzyme.

R,S system (*Section 15.2*) A set of rules for specifying configuration about a stereocenter.

Racemic mixture (*Section 15.1*) A mixture of equal amounts of two enantiomers.

Rad (*Section 9.5*) *R*adiation *a*bsorbed *d*ose. The SI unit is the gray (Gy).

Radiation, nuclear (*Section 9.3*) Radiation emitted from a nucleus during nuclear decay. Includes alpha particles, beta particles, gamma rays, and positrons.

Radical (*Section 13.4*) An atom or a molecule with one or more unpaired electrons.

Radioactive (*Section 9.2*) Refers to a substance that emits radiation during nuclear decay.

Radioactive dating (*Chemical Connections 9A*) The process of establishing the age of a substance by analyzing radioisotope abundance as compared with a current relative abundance.

Radioactive isotope (*Section 9.3*) A radiation-emitting isotope of an element.

Radioactivity (*Section 9.2*) Another name for nuclear radiation. Includes alpha particles, beta particles, gamma rays, and positrons.

Random coils (*Section 22.9*) Proteins that do not exhibit any repeated pattern.

Rate constant (*Section 7.4B*) A proportionality constant, k, between the molar concentrations of reactants and the rate of reaction; rate = k[compound].

Reaction mechanism (*Section 12.6A*) A step-by-step description of how a chemical reaction occurs.

Receptor (*Section 24.1*) A membrane protein that can bind a chemical messenger and then perform a function such as synthesizing a second messenger or opening an ion channel.

Recognition site (*Section 26.3*) The area of the tRNA molecule that recognizes the mRNA codon.

Recombinant DNA (*Section 26.8*) DNAs from two sources that have been combined into one molecule.

Recommended Daily Allowance (RDA) (*Section 30.1*) *also* **Recommended Dietary Allowance**; an average daily requirement for nutrients published by the U.S. Food and Drug Administration.

Redox reaction (*Section 4.4*) An oxidation–reduction reaction.

Reducing agent (*Section 4.4*) An entity that donates electrons in an oxidation–reduction reaction.

Reducing sugar (*Section 20.3*) A carbohydrate that reacts with a mild oxidizing agent under basic conditions to give an aldonic acid; the carbohydrate reduces the oxidizing agent.

Reduction (*Section 4.4*) The gain of electrons; the loss of oxygen atoms or the gain of hydrogen atoms.

Regioselective reaction (*Section 12.6A*) A reaction in which one direction of bond forming or bond breaking occurs in preference to all other directions.

Regulator (*Section 23.6*) A molecule that binds to an allosteric enzyme and changes its activity. This change could be positive or negative.

Regulatory site (*Section 23.6*) A site other than the active site where a regulator binds to an allosteric site and affects the rate of reaction.

Rem (*Section 9.6*) *R*oentgen *e*quivalent for *m*an; a biological measure of radiation.

Replication (*Section 25.6*) The process whereby DNA is duplicated to form two exact replicas of an original DNA molecule.

Replication fork (*Section 25.6*) The point on a DNA molecule where replication is proceeding.

Residue (*Section 22.6*) Another term for an amino acid in a peptide chain.

Resonance (*Section 3.9*) A theory that many molecules and ions are best represented as hybrids of two or more Lewis contributing structures.

Resonance hybrid (*Section 3.9A*) A molecule best described as a composite of two or more Lewis structures.

Respiratory acidosis (*Chemical Connections 8D*) The lowering of the blood pH due to difficulty breathing.

Response element (*Section 26.6*) A sequence of DNA upstream from a promoter that interacts with a transcription factor to stimulate transcription in eukaryotes. Response elements may control several similar genes based on a single stimulus.

Restriction endonuclease (*Section 26.8*) An enzyme, usually purified from bacteria, that cuts DNA at a specific base sequence.

Retrovaccination (*Section 31.8*) A process whereby scientists have an antibody they want to use and try to develop molecules to elicit it.

Retrovirus (*Section 26.1*) A virus such as HIV that has an RNA genome.

Reuptake (*Section 24.4*) The transport of a neurotransmitter from its receptor back through the presynaptic membrane into the neuron.

Ribonucleic acid (RNA) (*Section 25.5*) A type of nucleic acid consisting of nucleotide monomers a nitrogenous base, D-ribose, and phosphate.

Ribosomal RNA (rRNA) (*Section 25.5*) The type of RNA that is complexed with proteins and makes up the ribosomes used in the translation of mRNA into protein.

Ribosome (*Section 25.4*) Small spherical bodies in the cell made of protein and RNA, the site of protein synthesis.

Ribozyme (*Section 23.1*) An enzyme that is made up of ribonucleic acid. The currently recognized ribozymes catalyze cleavage of part of their own sequences in mRNA and tRNA.

RNA (*Section 25.2*) Ribonucleic acid.

Roentgen (R) (*Section 9.6*) The amount of radiation that produces ions having 2.58×10^{-4} coulomb per kilogram.

S (*Section 15.2*) From the Latin *sinister*, meaning "left"; used in the *R,S* system to show that when the lowest-priority group is away from you, the order of priority of groups on a stereocenter is counterclockwise.

Saponification (*Section 18.4B*) The hydrolysis of an ester in aqueous NaOH or KOH to give an alcohol and the sodium or potassium salt of a carboxylic acid.

Satellites (*Section 25.5*) Short sequences of DNA that are repeated hundreds of thousands of times but do not code for any protein in RNA.

Saturated hydrocarbon (*Section 11.1*) A hydrocarbon that contains only carbon–carbon single bonds.

Saturation curve (*Section 23.4*) A graph of enzyme rate versus substrate concentration. At high levels of substrate, the enzyme becomes saturated and the velocity does not increase linearly with increasing substrate.

Scientific method (*Section 1.2*) A method of acquiring knowledge by testing theories.

Scintillation counter (*Section 9.5A*) An instrument containing a phosphor that emits light on exposure to ionizing radiation.

Secondary (2°) alcohol (*Section 10.4A*) An alcohol in which the carbon atom bearing the —OH group is bonded to two other carbon groups.

Secondary (2°) amine (*Section 10.4B*) An amine in which nitrogen is bonded to two carbon groups and one hydrogen.

Secondary messenger (*Section 24.1*) A molecule that is created or released due to the binding of a hormone or neurotransmitter, which then proceeds to carry and amplify the signal inside the cell.

Secondary structure of DNA (*Section 25.3*) Specific forms of DNA due to pairing of complementary bases.

Secondary structure of proteins (*Section 22.9*) Repeating structures within polypeptides that are based solely on interactions of the peptide backbone. Examples are the alpha helix and the beta-pleated sheet.

Semiconservative replication (*Section 25.6*) Replication of DNA strands whereby each daughter molecule has one parental strand and one newly synthesized strand.

Sense strand (*Section 26.2*) The DNA strand that is not used as a template for transcription but has a sequence that is the same as the RNA produced. Also called the coding strand and the (+) strand.

Severe Combined Immuno Deficiency (SCID) (*Section 26.9*) A disease caused by several possible missing enzymes that leads to the organism having no immune system.

Shell (*Section 2.6A*) All orbitals of a principal energy level of an atom.

Shine-Dalgarno Sequence (*Section 26.5*) A sequence on the mRNA that attracts the ribosome for translation.

SI (*Section 1.4*) International System of Units.

Side chains (*Section 22.7*) The unique part of an amino acid; the side chain is attached to the alpha carbon, and the nature of the side chain determines the characteristics of the amino acid.

Sievert (Sv) (*Section 9.6*) A biological measure of radiation. One sievert is the value of 100 rem.

Sieving portion (*Section 26.6*) A part of a ribosome that allows only certain tRNA molecules to enter.

Signal transduction (*Section 24.5*) A cascade of events through which the signal of a neurotransmitter or hormone delivered to its receptor is carried inside the target cell and amplified into many signals that can cause protein modification, enzyme activation, or the opening of membrane channels.

Significant figures (*Section 1.3*) Numbers that are known with certainty.

Silencer (*Section 26.6*) A DNA sequence that is not part of the promoter that binds a transcription factor suppressing transcription.

Single bond (*Section 3.7C*) A bond formed by sharing one pair of electrons; represented by a single line between two bonded atoms.

Small interfering RNA (siRNA) (*Section 25.4*) Small RNA molecules that are involved in the degradation of specific mRNA molecules.

Small nuclear RNA (snRNA) (*Section 25.4*) Small RNA molecules (100–200 nucleotides) located in the nucleus that are distinct from tRNA and rRNA.

Small nuclear ribonucleoprotein particles (snRNPs) (*Section 25.4*) Combinations of RNA and protein that are used in RNA splicing reactions.

Soap (*Section 18.4B*) A sodium or potassium salt of a fatty acid.

Solenoid (*Section 25.3*) A coil wound in the form of a helix.

Solubility (*Section 6.4*) The maximum amount of solute that can be dissolved in a solvent at a specific temperature and pressure.

Solute (*Section 6.2*) The substance or substances that are dissolved in a solvent to produce a solution.

Solvent (*Section 6.2*) A liquid in which a solute is dissolved to form a solution.

Specific gravity (*Section 1.7*) The density of a substance compared to water as a standard.

Specific heat (*Section 1.9*) The amount of heat (calories) necessary to raise the temperature of 1 g of a substance by 1°C.

Specificity (*Section 31.1*) A characteristic of acquired immunity based on the fact that cells make specific antibodies to a wide range of specific pathogens.

Spectator ion (*Section 4.3*) An ion that appears unchanged on both sides of a chemical equation.

Splicing (*Section 25.4*) The removal of an internal RNA segment and the joining of the remaining ends of the RNA molecule.

Standard temperature and pressure (STP) (*Section 5.4*) The pressure of one atmosphere and 0°C (273 K).

Step-growth polymerization (*Section 19.6*) A polymerization in which chain growth occurs in a stepwise manner between difunctional monomers—as, for example, between adipic acid and hexamethylenediamine to form nylon-66.

Stereocenter (*Section 15.1*) An atom, most commonly a tetrahedral carbon atom, at which exchange of two groups produces a stereoisomer.

Stereoisomers (*Section 11.7*) Isomers that have the same connectivity (the same order of attachment of their atoms) but different orientations of their atoms in space.

Stoichiometry (*Section 4.7*) The mass relationships in a chemical reaction.

(−) Strand (*Section 26.2*) The strand of DNA used as a template for transcription. Also called the template strand and the antisense strand.

(+) Strand (*Section 26.2*) The DNA strand that is not used as a template for transcription but has a sequence that is the same as the RNA produced. Also called the coding and the sense strand.

Strong acid (*Section 8.2*) An acid that ionizes completely in aqueous solution.

Strong base (*Section 8.2*) A base that ionizes completely in aqueous solution.

Structural formula (*Section 3.7C*) A formula showing how atoms in a molecule or ion are bonded to each other. Similar to a Lewis structure except that a structural formula shows only bonding pairs of electrons.

Structural genes (*Section 26.2*) Genes that code for the product proteins.

Sublimation (*Section 5.10*) A phase change from the solid state directly to the vapor state.

Subshell (*Section 2.6*) All the orbitals of an atom having the same principal energy level and the same letter designation (s, p, d, or f).

Substance P (*Section 24.6*) An 11-amino acid peptidergic neurotransmitter involved in the transmission of pain signals.

Substrate (*Section 23.3*) The compound or compounds whose reactions an enzyme catalyzes.

Substrate specificity (*Section 23.1*) The limitation of an enzyme to catalyze specific reactions with specific substrates.

Subunit (*Section 23.6*) An individual polypeptide chain of an enzyme that has multiple chains.

Sulfhydryl group (*Section 14.4A*) An —SH group.

Supersaturated solution (*Section 6.4*) A solution in which the solvent has dissolved an amount of solute beyond the maximum amount at a specific temperature and pressure.

Surface presentation (*Section 31.1*) The process whereby a portion of an antigen from a foreign pathogen that infected a cell is brought to the surface of the cell.

Surface tension (*Section 5.8A*) The layer on the surface of a liquid produced by the strength of the intermolecular attractions between the molecules of liquid at the surface layer.

Synapse (*Section 24.2*) A small aqueous space between the tip of a neuron and its target cell.

T cell (*Section 31.1*) A type of lymphoid cell that matures in the thymus and that reacts with antigens via bound receptors on its cell surface. T cells can differentiate into memory T cells or killer T cells.

T cell receptor (*Section 31.1*) A glycoprotein of the immunoglobulin superfamily on the surface of T cells that interacts with the epitope presented by the MHC (major histocompatibility complex).

T cell receptor complex (*Section 31.5*) The combination of T cell receptors, antigen, and cluster determinants (CD) that are all involved in the T cell's ability to bind antigen.

T form (*Section 23.6*) The form of an allosteric enzyme that is less active.

Tautomers (*Section 17.5*) Constitutional isomers that differ in the location of an H atom.

Template strand (*Section 26.2*) The strand of DNA used as a template for transcription. Also called the (−) strand and the antisense strand.

Termination (*Sections 26.2 and 26.5*) The final stage of translation during which a termination sequence on mRNA tells the ribosomes to dissociate and release the newly synthesized peptide.

Termination sequence (*Section 26.2*) A sequence of DNA that tells RNA polymerase to terminate synthesis.

Tertiary (3°) alcohol (*Section 10.4A*) An alcohol in which the carbon atom bearing the —OH group is bonded to three other carbon groups.

Tertiary (3°) amine (*Section 10.4B*) An amine in which nitrogen is bonded to three carbon groups.

Tertiary structure (*Section 22.11*) The overall 3-D conformation of a polypeptide chain, including the interactions of the side chains and the position of every atom in the polypeptide.

Theoretical yield (*Section 4.7*) The mass of product that should be formed in a chemical reaction according to the stoichiometry of the balanced equation.

Theory (*Section 1.2*) A hypothesis that is supported by evidence; a hypothesis that has passed tests.

Thiol (*Section 14.4A*) A compound containing an —SH (sulfhydryl) group bonded to a tetrahedral carbon atom.

Tissue necrosis factor (TNF) (*Section 31.6*) A type of cytokine produced by T cells and macrophages that has the ability to lyse susceptible tumor cells.

Titration (*Section 8.9*) An analytical procedure whereby we react a known volume of a solution of known concentration with a known volume of a solution of unknown concentration.

TNF (*Section 31.6*) Tumor necrosis factor; a type of cytokine produced by T cells and macrophages that has the ability to lyse tumor cells.

Trans (*Section 11.8*) A prefix meaning "across from."

Transamination (*Section 28.8*) The exchange of the amino group of an amino acid and a keto group of an α-ketoacid.

Transcription (*Section 25.4*) The process whereby DNA is used as a template for the synthesis of RNA.

Transcription factor (*Section 26.2*) Binding proteins that facilitate the binding of RNA polymerase to the DNA to be transcribed or that bind to a remote location and stimulate transcription.

Transfer RNA (tRNA) (*Section 25.4*) The RNA that transports amino acids to the site of protein synthesis on ribosomes.

Transferase (*Section 23.2*) A class of enzymes that catalyzes a reaction where a group of atoms such as an acyl group or amino group is transferred from one molecule to another.

Transition element (*Section 2.5A*) The elements in the B columns (Groups 3 to 12 in the new numbering system).

Transition state (*Sections 7.3 and 23.5*) An unstable species formed during the highest energy of a chemical reaction; a maximum on an energy diagram.

Transition-state analog (*Section 23.8*) A molecule constructed to mimic that which would form during the transition state of an enzyme-catalyzed reaction.

Translation (*Section 26.1*) The process in which information encoded in a mRNA is used to assemble a specific protein.

Translocation (*Section 26.5*) The part of translation where the ribosome moves down the mRNA a distance of three bases so that the new codon is on the A site.

Transmutation (*Section 9.3B*) Changing one element into another element.

Transporter (*Section 24.5*) A protein molecule carrying small molecules such as glucose or glutamic acid across a membrane.

Triglyceride (*Section 21.7*) A kind of lipid formed by bonding glycerol to three fatty acids with ester bonds.

Triple bond (*Section 3.7C*) A bond formed by sharing three pairs of electrons; represented by three lines between the two bonded atoms.

Triple helix (*Section 22.12*) The collagen triple helix is composed of three peptide chains. Each chain is itself a left-handed helix. These chains are twisted around each other in a right-handed helix.

Triprotic acid (*Section 8.3*) An acid that can give up three protons.

tRNA^{fmet} (*Section 26.5*) The special tRNA molecule that initiates translation.

Tumor suppression factor (*Chemical Connections 26F*) A protein that controls replication of DNA so that cells do not divide constantly. Many cancers are caused by mutated tumor suppression factors.

Tyndall effect (*Section 6. 7*) Light passing through and scattered by a colloid viewed at a right angle.

Unwinding proteins (*Section 25.6*) Special proteins that help unwind DNA so that it can be replicated.

Urea cycle (*Section 28.8*) A cyclic pathway that produces urea from ammonia and carbon dioxide.

Valence electron (*Section 2.6F*) An electron in the outermost occupied (valence) shell of an atom.

Valence shell (*Section 2.6F*) The outermost occupied shell of an atom.

Vapor pressure (*Section 5.8B*) The pressure of gas in equilibrium with its liquid form in a closed container.

Vesicle, synaptic (*Section 24.2*) A compartment containing a neurotransmitter that fuses with a presynaptic membrane and releases its contents when a nerve impulse arrives.

Vitamin (*Section 30.6*) An organic substance required in small quantities in the diet of most species, which generally functions as a cofactor in important metabolic reactions.

Volume (*Section 1.4*) The space that a substance occupies; the base SI unit is the cubic meter (m^3).

VSEPR model (*Section 3.10*) Valence-shell electron-pair repulsion model.

Wavelength (λ) (*Section 9.2*) The distance from the crest of one wave to the crest of the next.

Weak acid (*Section 8.2*) An acid that is only partially ionized in aqueous solution.

Weak base (*Section 8.2*) A base that is only partially ionized in aqueous solution.

Weight (*Section 1.4*) The result of a mass acted upon by gravity; the base unit of measure is a gram (g).

X-ray (*Section 9.2*) A type of electromagnetic radiation with a wavelength shorter than ultraviolet light but longer than gamma rays.

Zwitterion (*Section 22.3*) A molecule that has equal numbers of positive and negative charges, giving it a net charge of zero.

Zymogen (*Section 23.6*) An inactive form of an enzyme that must have part of its polypeptide chain cleaved before it becomes active; a proenzyme.

Index

Index page numbers in **boldface** refer to boldface terms in the text. Page numbers in *italics* refer to figures. Tables are indicated by a *t* following the page number. Boxed material is indicated by a *b* following the page number.